Emergency

AAOS

40th Anniversary
ORANGE BOOK SERIES

Tenth Edition

P9-DEU-470

Care and Transportation of the Sick and Injured

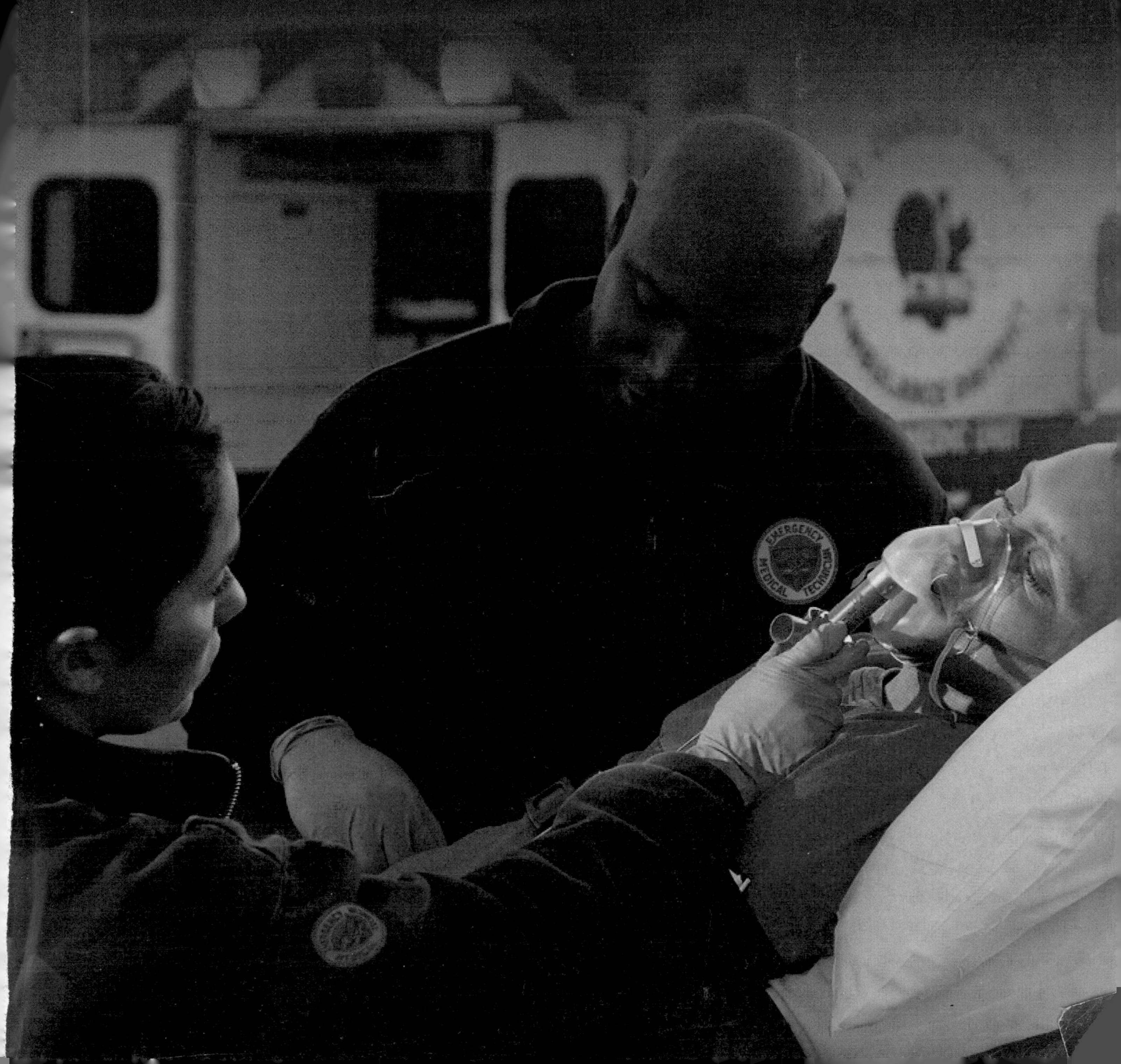

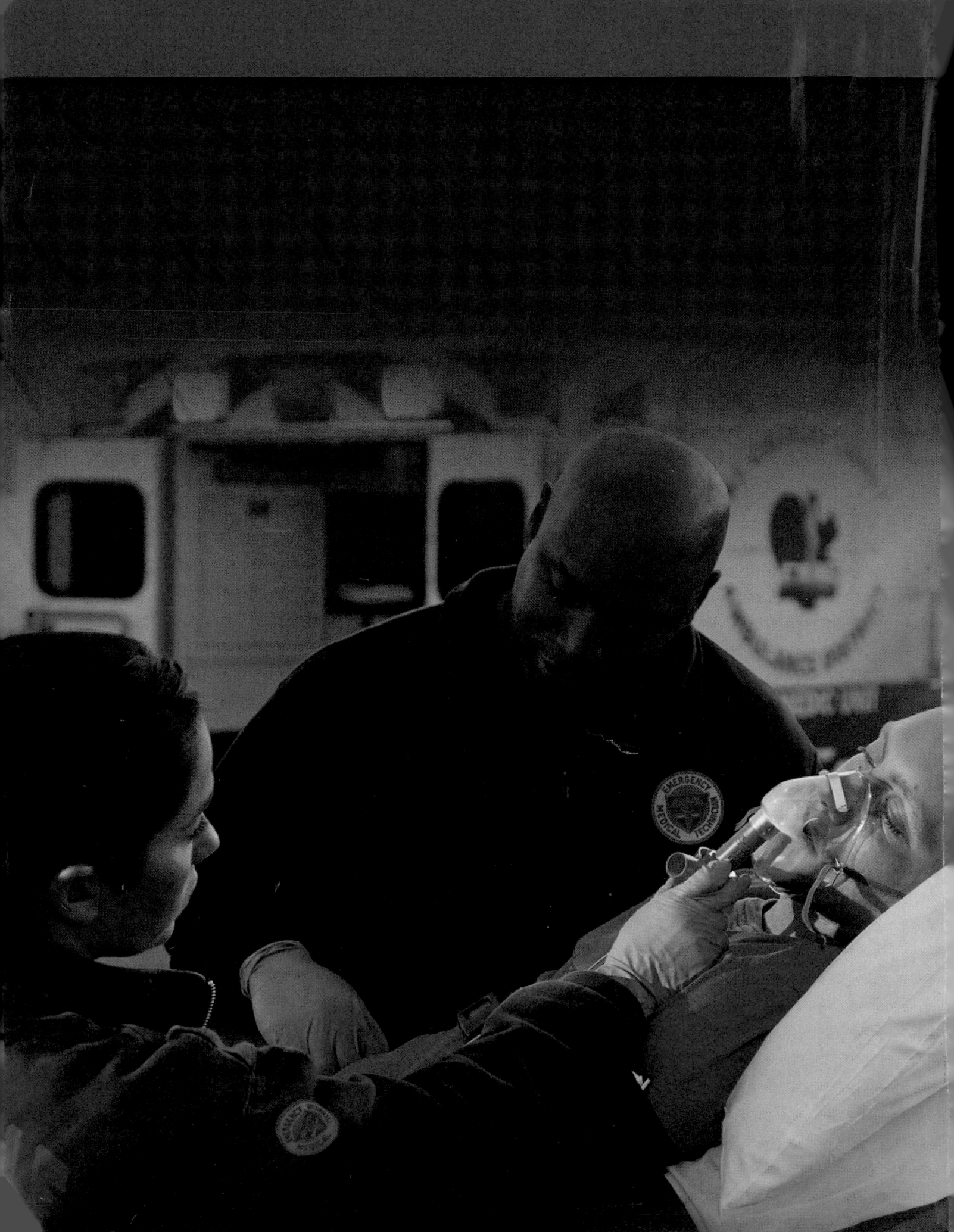

Emergency

Care and Transportation of the Sick and Injured

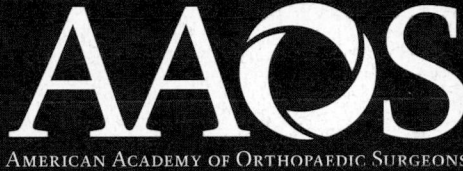

AAOS

AMERICAN ACADEMY OF ORTHOPAEDIC SURGEONS

Series Editor:

Andrew N. Pollak, MD, FAAOS

Editors:

Leaugeay Barnes, MS, CCEMT-P, NREMT-P
Joseph A. Ciotola, MD, FAAOS
Benjamin Gulli, MD, FAAOS

JONES AND BARTLETT PUBLISHERS

Sudbury, Massachusetts

BOSTON   TORONTO   LONDON   SINGAPORE

World Headquarters
Jones and Bartlett Publishers
40 Tall Pine Drive, Sudbury, MA 01776
978-443-5000
info@jbpub.com
www.jbpub.com

Jones and Bartlett Publishers Canada
6339 Ormindale Way
Mississauga, Ontario L5V 1J2
Canada

Jones and Bartlett Publishers International
Barb House, Barb Mews
London W6 7PA
United Kingdom

Substantial discounts on bulk quantities of Jones and Bartlett's publications are available to corporations, professional associations, and other qualified organizations. For details and specific discount information, contact the special sales department at Jones and Bartlett via the above contact information or send an e-mail to specialsales@jbpub.com.

AMERICAN ACADEMY OF ORTHOPAEDIC SURGEONS

Editorial Credits
Chief Education Officer: Mark W. Wieting
Director, Department of Publications: Marilyn L. Fox, PhD
Managing Editor: Barbara A. Scotese
Associate Senior Editor: Gayle Murray

AAOS Board of Directors, 2009-2010
President: Joseph D. Zuckerman, MD
First Vice President: John J. Callaghan, MD
Second Vice President: Daniel J. Berry, MD
Treasurer: Frederick M. Azar, MD
Thomas C. Barber, MD
Richard J. Barry, MD
Leesa M. Galatz, MD
M. Bradford Henley, MD, MBA
Michael L. Parks, MD
E. Anthony Rankin, MD
William J. Robb, III, MD
Michael F. Schafer, MD
David D. Teuscher, MD
Paul Tornetta III, MD
G. Zachary Wilhoit, MS, MBA
Karen L. Hackett, FACHE, CAE (*Ex-officio*)

Jones and Bartlett's books and products are available through most bookstores and online booksellers. To contact Jones and Bartlett Publishers directly, call 800-832-0034, fax 978-443-8000, or visit our website, www.jbpub.com.

Production Credits
Chief Executive Officer: Clayton E. Jones
Chief Operating Officer: Donald W. Jones, Jr.
President, Higher Education and Professional
  Publishing: Robert W. Holland, Jr.
Senior V.P., Sales and Marketing: James Homer
V.P., Design and Production: Anne Spencer
V.P., Manufacturing and Inventory Control: Therese Connell
Publisher: Kimberly Brophy
Acquisitions Editor—EMS: Christine Emerton
Managing Editor: Carol E. Guerrero
Associate Managing Editor: Amanda J. Green
Editor: Jennifer Deforge-Kling
Editorial Assistant: Kara Ebrahim
Senior Production Editor: Susan Schultz

Associate Production Editor: Sarah Bayle
Production Assistant: Tina Chen
Associate Photo Researcher: Jessica Elias
Director of Marketing: Alisha Weisman
Director of Sales, Public Safety Group: Matthew Maniscalco
Text Design: Anne Spencer
Cover Design: Kristin E. Parker
Cover Image: Cover photographed by Ray Kemp/911 imaging
  <www.911imaging.com>. Special thanks to Susan Hertzler, Terrance
  Jackson, Cass Wilson, AllMed, St. Charles County, MO Ambulance
  District; Photo of EMT patch: Stephen Coburn/ShutterStock, Inc.
Composition: diacriTech
Cover Printing: Courier Corporation
Text Printing and Binding: Courier Corporation

The procedures and protocols in this book are based on the most current recommendations of responsible medical sources. The American Academy of Orthopaedic Surgeons and the publisher, however, make no guarantee as to, and assume no responsibility for, the correctness, sufficiency, or completeness of such information or recommendations. Other or additional safety measures may be required under particular circumstances.

This textbook is intended solely as a guide to the appropriate procedures to be employed when rendering emergency care to the sick and injured. It is not intended as a statement of the standards of care required in any particular situation, because circumstances and the patient's physical condition can vary widely from one emergency to another. Nor is it intended that this textbook shall in any way advise emergency personnel concerning legal authority to perform the activities or procedures discussed. Such local determinations should be made only with the aid of legal council.

Notice: The patients described in "You are the Provider" and "Assessment in Action," throughout this text are fictitious.

Additional illustrations and photographic credits appear on pages 1565–1566, which constitute a continuation of the copyright page.

To order this product, use ISBN: 978-1-4496-3054-6 (paperback) or 978-1-4496-3056-0 (hardcover)

Library of Congress Cataloging-in-Publication Data
Emergency care and transportation of the sick and injured. —10th ed. / American Academy of Orthopaedic Surgeons; editors, Leaugeay Barnes,
Joseph A. Ciotola, Benjamin Gulli.
  p. ; cm.
Includes index.
ISBN 978-0-7637-7828-6 (pbk.)—ISBN 978-0-7637-7849-1 (hardcover)
1. Medical emergencies. 2. Transport of sick and wounded. I. Gulli, Benjamin. II. Ciotola, Joseph A. III. Barnes, Leaugeay. IV. American
Academy of Orthopaedic Surgeons.
[DNLM: 1. Emergency Medical Services. 2. Emergency Treatment. 3. Transportation of Patients. WX 215 E487 2011]
RC86.7.A43 2011
616.02'5—dc22
                                    2010002592

6048
Printed in the United States of America
 15 14 13 12 11   10 9 8 7 6 5 4 3

Brief Contents

Contents

Section 7: **Trauma**   744

Section 8: **Special Patient Populations**   1104

Section 9: EMS Operations   1282

Section 10: ALS Techniques   1460

Skill Drills

Tenth Edition Resources

Instructor Resources

Instructor's ToolKit CD-ROM

ISBN: 978-0-7637-9255-8

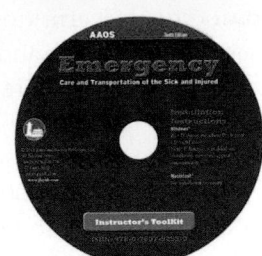

The CD includes:

- PowerPoint presentations with embedded video, animations, and check point questions
- Lecture outlines
- Teaching tips, enhancements, support materials, and preparation guidance
- Student activities and assignments
- Image and table bank
- Skill sheets
- Answers to end-of-chapter student questions

Instructor's TestBank CD-ROM

ISBN: 978-0-7637-9254-1

This powerful evaluation tool allows educators to gauge student competency through both general knowledge and critical-thinking questions. Each scenario-based, multiple-choice question is page-referenced to the *Tenth Edition*. Educators can originate tailor-made tests quickly and easily by selecting, editing, and printing a test along with an answer key.

Student Resources

Student Workbook

ISBN: 978-0-7637-9256-5

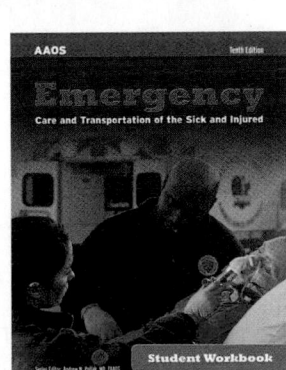

This resource is designed to encourage critical thinking and aid comprehension of the course material through a variety of activities:

- Matching, fill-in-the-blank, short-answer, and multiple-choice questions
- Skill Drill activities
- Complete the Patient Care Report
- Crossword puzzles
- And much more

EMT Field Guide, Third Edition

ISBN: 978-0-7637-5877-6

This handy reference covers basic information from patient management tips to guidelines on helping patients use medication. Special features include a prescription medication reference and documentation tips. The *EMT Field Guide* is pocket-sized, spiral-bound, and water-resistant for ready-reference in the field.

Technology Resources

Jones & Bartlett's Navigate

A New Paradigm for a New Generation of Students.

More than ever before, technology is impacting almost every facet of education, including how instructors teach and how students learn. Computers, mobile devices, and the ease of accessing the Internet at high speeds are enabling new learning paradigms and new instructional models that help instructors teach more efficiently and students learn more effectively. What's more, today's diverse students are more connected, more networked, and more "tech savvy" than any previous generation.

To help meet these changing needs of today's faculty and students, Jones & Bartlett introduces *Navigate*™, a groundbreaking new solution designed to help students and instructors enrich, enliven, and improve the entire learning experience. *Navigate* seamlessly combines authoritative content with curriculum, tutorial, assessment, and reporting tools in a web platform that helps students identify individual learning pathways, track learning progress, and improve measurable outcomes. *Navigate* aims to advance learning in traditional and nontraditional classrooms, while also preparing students and users for professional success beyond the classroom.

Navigate Learn

Navigate Learn lets instructors manage student learning through a complete online classroom environment. *Learn* includes a fully scoped and sequenced curriculum, developed to pedagogical course goals and framed by the successful "know-see-do-prove-report" learning paradigm. In addition, *Learn* offers a wide variety of additional content options such as topical indexes to search for new instructor resources, assignments, activities, lecture outlines, images, and discussion and assessment questions. *Learn* streamlines instruction with eBook access, assignment management, online submissions, and automatic grading. *Navigate Learn* also includes state-of-the-art reporting features, giving instructors a quick snap-shot of overall and individual student progress throughout the course.

With *Navigate Learn* Jones & Bartlett is making education more personal, flexible, accessible, and more efficient than ever before.

Navigate TestPrep

Navigate TestPrep is a dynamic online resource designed to help prepare students and professionals for state or national certification examinations by providing a series of self-study modules, practice examinations, and simulated certification examinations using case-based questions and detailed rationales. With *Navigate TestPrep*, students can create custom practice tests, review key topics covered in certification examinations, and receive instant feedback for each question, as well as an overview of results at the end of the examination.

■ Tenth Edition eBook/eWorkbook

ISBN: 978-0-7637-9405-7

Technology and content combine to create an online digital learning solution.

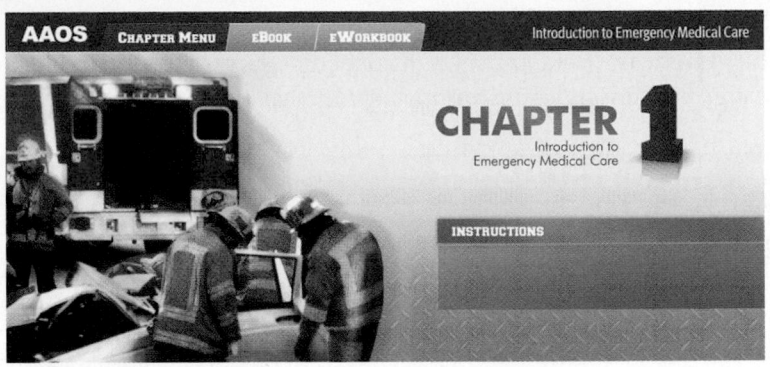

World-class content joins instructionally sound design in a user-friendly online interface to give students a truly interactive and engaging learning experience. Combining the complete content of the textbook and workbook into one, complete online solution, the *Tenth Edition eBook/eWorkbook* will transform EMT training.

The *Tenth Edition eBook/eWorkbook* is built upon the adult learning principles of Know, See, Do, and Prove.

Know: The content of the *Tenth Edition* is at the students' fingertips. Students can search for terms, highlight text, and take notes while reading chapters. Students can test their knowledge through check point questions, flashcards, and skill evaluation sheets.

See: In each chapter, students can see knowledge in action through 9-1-1 calls, anatomy review, animations, and skill videos.

Do: The *eWorkbook* offers a broad spectrum of activities to engage every learner's mind. Students can reinforce their general knowledge, hone their critical thinking skills, and perfect their psychomotor skills. The results are recorded in the integrated learning management system, *Navigate Learn.*

Prove: Educators can assess students' comprehension by originating a tailor-made examination from the *Instructor's Test Bank* in *Navigate Learn.*

■ EMT Interactive

ISBN: 978-0-7637-2242-5

Make better use of valuable classroom time with *EMT Interactive*. This tool allows educators to focus on students' psychomotor skills in the classroom, while students master the cognitive and didactic portions of the course online.

EMT Interactive is a series of online modules that bring EMT lectures to life through Flash animations and narration. The modules cover the entire EMT Education Standards.

■ www.EMT.EMSzone.com

www.EMT.EMSzone.com is specifically designed to complement *Emergency Care and Transportation of the Sick and Injured, Tenth Edition*. It provides a wealth of free resources, including Anatomy Review and Vital Vocabulary.

The code printed on the front inside cover of the textbook provides students with special user privileges. Educators can assign and track grades, if appropriate, for the following resources:

- Tenth Edition Audio Book in mp3 format
- Ambulance Calls
- Assessment in Action
- Chapter Pretests
- Interactive Skill Drills
- Skill Sheets
- Skill Video Clips

Acknowledgments

The American Academy of Orthopaedic Surgeons would like to acknowledge the contributors and reviewers of *Emergency Care and Transportation of the Sick and Injured, Tenth Edition.*

Editorial Board

Leaugeay Barnes, MS, CCEMT-P, NREMT-P
Director of Emergency Medical Services
Oklahoma City Community College
Oklahoma City, Oklahoma

Joseph A. Ciotola, MD, FAAOS
Department of Emergency Services
Queen Anne's County, Maryland
Benjamin Gulli, MD, FAAOS
Northwest Orthopaedic Surgeons
Robbinsdale, Minnesota

Andrew N. Pollak, MD, FAAOS
Medical Director, Baltimore County
   Fire Department
Associate Professor, University of Maryland
   School of Medicine
Baltimore, Maryland

Section Editors

David S. Becker, MA, EMT-P, EFO
Sanford-Brown College
St. Louis, Missouri

Al Benney, NREMT-P, STP
Hennepin Technical Community College
North Memorial Ambulance Service
Brooklyn Park, Minnesota

Sam Bradley, MICP
Sam Bradley and Associates
EMS QI and Training
Oakley, California

Timothy Brisbin, BSN, RN, NREMT-P
Director, Center for Prehospital Medicine
Department of Emergency Medicine
Carolinas Medical Center
Charlotte, North Carolina

Bruce Butterfras, MS Ed, LP
Assistant Professor, Director of Bachelor
   Degree Program
Department of Emergency Health Sciences
The University of Texas Health Science
   Center at San Antonio
San Antonio, Texas

Darby L. Copeland, EdD, RN, NREMT-P
Assistant Director
Parkway West CTC
Oakdale, Pennsylvania

Samuel A. Getz, Jr, BS, NREMT-P, EMSI
President and CEO, EMS/QS Systems
   Consulting Services
EMS Coordinator, Western Reserve
   Joint Fire District
Paramedic/Fire Fighter, Newton Falls
   Joint Fire District
Paramedic, Noga Ambulance Service
Austintown, Ohio

Carol Gupton, BS, NREMT-P
EMS Program Director
Emergency Provider Instruction
Omaha, Nebraska

Alan Heckman, BS, NREMT-P, NCEE
Program Coordinator
GEM-Emergency Medicine Institute
Lehigh Valley Hospital
Allentown, Pennsylvania

Bryan Hess, NREMT-P, CO-EMT-P
EMS Department & Education
   Program Director
Gunnison Valley Health
Adjunct Faculty
Western State College
Gunnison, Colorado

Catherine A. Parvensky Barwell, RN, PHRN, MEd
Integrated Learning Technology
West Chester, Pennsylvania

Chuck Sowerbrower, MEd, NREMT-P, NCEE
Sinclair Community College
Dayton, Ohio

Brian Williams, BS, NREMT-P, CCEMT-P
EMS Chief, Pembina Ambulance
Pembina, North Dakota

Contributors

Alan J. Azzara, JD, EMT-P
Westport Island, Maine

Art Breault
EMS Coordinator
Albany Medical Center Hospital
Albany, New York

Steven K. Frye
University of Maryland
Maryland Fire & Rescue Institute
College Park, Maryland

Tony Garcia, RN, CCEMT-P
Texas Engineering Extension Service
College Station, Texas

Richard K. Hilinski, BA, EMT-P
Community College of Allegheny County
Public Safety Institute
Pittsburgh, Pennsylvania

Don Kimlicka, NREMT-P, CCEMT-P
EMS Coordinator
St. Clare's Hospital
Weston, Wisconsin

Guy Peifer, BS, EMT-P
Paramedic Program Coordinator
Borough of Manhattan Community College
New York, New York

Sheri Polley
Conneault Lake, Pennsylvania

Stephen J. Rahm, NREMT-P
Bulverde-Spring Branch EMS
Boerne, Texas

Jose V. Salazar, MPH, NREMT-P
Loudoun County Department of Fire, Rescue
   and Emergency Management
Leesburg, Virginia

Annmary Thomas, MEd, NREMT-P
Jeff STAT EMS Training Center
Philadelphia, Pennsylvania

Brittany Ann Williams, MHSc, BSRT-NPS, NREMT-P
Associate Professor EMS
Santa Fe College
Gainesville, Florida

Reviewers

Jon Abrams, NREMT-P, BA, AAT
NAEMT Affiliate Faculty
Scottsdale, Arizona

Bradshaw Anderson, MA, EMT-B
EMS Educator
North Memorial Hospital
Robbinsdale, Minnesota

Kimberly Ayers, EMT-P
Moraine Valley Community College
Palos Hills, Illinois

Richard J. Bauknecht, AAS, Paramedic
Malcolm X College
Chicago, Illinois

Brandon R. Beck, BS, CCEMT-P
Gadsden State Community College
Gadsden, Alabama

Rob Bernini, EMT-P
Harrisburg Area Community College
Shumaker Public Safety Center
Harrisburg, Pennsylvania

Aimee Binning, EMT-I Advanced
Sublette County EMS
Pinedale, Wyoming

Paul A. Bishop, MPA, NREMT-P
Monroe Community College
Public Safety Training Facility
Rochester, New York

James M. Bobbitt, LP-CCP, MS
National College of Technical Instruction
Mesquite, Texas

Michael Bohanske, EMT-I/C
Boston University Emergency
    Medical Services
Boston, Massachusetts

Michael Bolin, NREMT-P
Great Plains Technology Center
Lawton, Oklahoma

James Brasiel, MD, MHA, MICP, CCEMT-P
Fire/EMS Program Instructor
Los Medanos College
Pittsburg, California

Melissa K. Brock, NREMT-P, BS, MS
Roanoke Valley Regional Fire-EMS
    Training Center
Roanoke, Virginia

Eileen Byrne, RNC, BC, WCC
Saint Barnabas Burn Foundation
Livingston, New Jersey

Brian S. Chamberlin, BS, FF, EMT-P/IC
Kennebec Valley EMS Council
Winslow, Maine

Joshua Chan, BA, NREMT-P
EMS Educator
Cuyuna Regional Medical Center
Crosby, Minnesota

Russ Christiansen, NREMT-P, CCEMT-P
Casper College
Casper, Wyoming

Harvey Conner, AS, NREMT-P
Oklahoma City Community College
Oklahoma City, Oklahoma

Peter Connick, REMT-P, EMT,I/C
Captain, Chatham Fire Rescue
Adjunct faculty, Cape Cod Community
College EMS Program
Chatham, Massachusetts

Heidi P. Cordi, MD, MPH, MS, EMT-P, FACE-P
Associate Medical Director
Emergency Medical Services
The New York Presbyterian Hospital
Elmsford, New York

Chris Coughlin, MEd, NREMT-P, FP-C
Glendale Community College
Glendale, Arizona

Patricia Courson, EMT-I
Director, Walla Walla County EMS
Walla Walla Community College
Walla Walla, Washington

John Creech, MEd, LP
Brazosport College EMS Program
Lake Jackson, Texas

W. Scott Crowley, BA, EMT-P
EMT Program Director, Phoenix College
Phoenix, Arizona

Jackilyn E. Cypher, RN, MSN, NREMT-P
Portland Community College Emergency
    Medical Services
Portland, Oregon

Jesse N. Davis, NREMT-P, I/C
ENMU-Roswell EMS Education Program
Roswell, New Mexico

Kathleen M. Dayton, MA, NREMT-I
Montgomery College
Rockville, Maryland

Bradley Dean, BBA, NREMT-P
Davidson County Emergency Services
Alamance Community College
Wake Forest University Baptist Medical
    Center, Trauma Department
Thomasville, North Carolina

Michael A. DeMello, BS, NREMT-P, I/C
Bristol Fire Department
North Providence Fire Department
New England Ambulance
Bristol, Rhode Island

Clyde Deschamp, PhD, NREMT-P
University of Mississippi Medical Center
Jackson, Mississippi

Eric Thomas Dotten, REMT-P
Simulations & Clinical Programs Coordinator
Emergency Medicine Learning &
    Resource Center
Orlando, Florida

Paul Duckworth, AAS, EMT-P
Emergency Training Academy
Augusta Fire Rescue
Augusta, Georgia

Michael Duell, EMS-I
Stark State College of Technology
North Canton, Ohio

Elizabeth R. Ehmling, BS, EMT-P, P-IC
Associate Professor, EMS Program
Chattanooga State Community College
Chattanooga, Tennessee

Sandra K. Eustice, NREMT-I, EMS-IC
Emergency Medical Services
    Department Chair
Chippewa Valley Technical College
Eau Claire, Wisconsin

Alexander Fein, BA, EMT-P
Penn Medicine Clinical Simulation Center
University of Pennsylvania Health System
Philadelphia, Pennsylvania

David Filipp, BS, NREMT-P, CCEMT-P
Mercy College of Health Sciences
Des Moines, Iowa

V. Josh Fremberg, BS, NREMT-P
The Pennsylvania State University
University Park, Pennsylvania

Michael J. Gannon, EMT-P, NCEE, Fire Fighter
Allegheny General Hospital
EMS Training and Coordination
Pittsburgh, Pennsylvania

Adiel Garcia, NREMT-P, CCEMT-P, FP-C
University of Texas at Brownsville
Brownsville, Texas

Alejandro Garcia, Licensed Paramedic
Wichita Falls Fire Department
    Hazmat Lieutenant
Vernon College FIRE/EMS Training
    Program Coordinator
Wichita Falls, Texas

Scott Garrett, REMT-P, Fire Chief
Westview-Fairforest Fire Department
Spartanburg, South Carolina

Ellen Garvey, MD, EMT-B
Pre-Hospital Emergency Care Educators
Hinsdale, New Hampshire

Kevin M. Garvey, BA, RN, EMT-I, I/C
PECE
Hinsdale, New Hampshire

Michelle Golba-Norek, BSN, RN, MICN, CEN
Raritan Bay Medical Center EMS
Perth Amboy, New Jersey

Matthew Goodman, CIEMT, BA, EMT-P
Program Director
California Institute of Emergency
    Medical Training
Signal Hill, California

Linda J. Gosselin, MS, REMT IC, Ed
MECTA Academy
Millbury, Massachusetts

Michael T. Greer, AAS, NREMT-P, Fire Fighter
Seeley Lake Rural Fire District
Big Sky Emergency Services Education
Seeley Lake, Montana

David Bradley Hall, MA, NREMT-P
Hinds Community College
Jackson, Mississippi

Mitchell Harrington, Captain, EMT-I
Bow Fire Department
Bow, New Hampshire

Robert Hawkes, MS, PA, NREMT-P
Southern Maine Community College
South Portland, Maine

Timothy M. Hellyer, MAT, EMT-P
Ivy Tech Community College–North Central
South Bend, Indiana

Rick Hilinski, BA, EMT-P
Community College of Allegheny County
    Public Safety Institute
Pittsburgh, Pennsylvania

Jonathan R. Hockman, EMS-IC, EMT-P, NREMT-P
Goldrush Consulting Services
Livonia, Michigan

Rob Holborn, EdD, NCEE
Reedy Creek Emergency Services
Lake Mary, Florida

Betty L. Holmes, BSE, EMT-P, I/C
Rich Mountain Community College
Mena, Arkansas

James B. Huettenmueller, BS, NREMT-P
Tulsa Technology Center
Tulsa, Oklahoma

Barbara L. Ireland, BA, AS-P
New Orleans EMS
New Orleans, Louisiana

Janelle Johnson, Paramedic/Instructor
MEMS Ambulance Service
Little Rock, Arkansas

Tom Jones, LP, BS
Lamar Institute of Technology
EMS Program Director
Beaumont, Texas

Edward J. Kalinowski, MEd, Dr PH
University of Hawaii
Kapiolani Community College
Department of Emergency Medical Services
Honolulu, Hawaii

Kevin Keen, AEMCA, ICP
Hamilton Emergency Services Fire Division
Ontario, Canada

Timothy M. Kimble, NREMT-P
Fauquier County Department of Fire,
    Rescue & Emergency Management
Warrenton, Virginia

David Jay Kleiman, NREMT-P, CCEMT-P
Paramedic Instructor
Cherokee County Fire-ES
Acworth, Georgia

**Shane Knox, MSc, HDip-EMT, Advanced
    Paramedic, Training Officer**
HSE-National Ambulance Service College
Galway, Ireland

John A. Kubincanek, EMT-P, AA, EMS-I
Cuyahoga Community College
Western Campus
Parma, Ohio

Stephen Lance, Training Officer, Medical Officer
Ouray County EMS, Ouray Mountain
    Rescue Team
Delta-Montrose Technical College, Ouray
    County EMS Training Group
Ouray, Colorado

David J. Leven, NYS EMT-B
University of Rochester Medical Center
Rochester, New York

Jon F. Levine, BA, EMT-P, I/C
EMS Training LTD
South Weymouth, Massachusetts

Ray Levy, MPH, EMT-I/C
Boston University Emergency Medical
    Services
Boston, Massachusetts

Paul J. Lolli, EMT-P, Deputy Fire Chief
City of Middletown Ohio, Division of Fire
Paramedic Education Coordinator, Butler
    Technology and Career Development
    Schools
State of Ohio Certified EMS and Fire
    Instructor and Instructor Trainer
Hamilton, Ohio

Steve Mackiewicz, NREMT-P, BS
Associate Dean EMS/Fire
Wisconsin Indianhead Technical College
Rice Lake, Wisconsin

Richard Main, BS, NREMT-P
National Center for Technical Instruction
Las Vegas, Nevada

Louis B. Mallory, MBA, REMT-P
Sante Fe College
Gainesville, Florida

Michael A. Marano, MD
Clinical Assistant Professor, New Jersey
    Medical School
Attending Surgeon, Burn Center
Saint Barnabas Medical Center
Livingston, New Jersey

Mark L. Marchetta, RN, BS, NREMT-P
Director, EMS Education and Trauma Services
Aultman Hospital
Canton, Ohio

Joao F. Mateus, EMT-B(I), RN, ASN, MICN
Less Stress Instructional Services
Roselle, New Jersey

Scott A. Matin, MBA, NREMT-P
Vice President
MONOC Mobile Health Services
Wall, New Jersey

Joe McConomy, MICP, EMT-B[I]
Senior EMT Instructor
Burlington County Emergency Services
    Training Center
Westampton, New Jersey

Craig McElhaney, NREMT-P
Miramar Fire-Rescue
Lead Paramedic Instructor, Broward College
Fort Lauderdale, Florida

John "Chip" McFadden, MEd, EMT
EMS Program Administrator
Warren County Community College
Washington, New Jersey

Bill McGrath, MPS, NREMT-P
EMS Department Chair
City College
Fort Lauderdale, Florida

William R. Montrie, AAS, EMT-P
Captain, Springfield Township Fire
    Department
EMM Training Program Coordinator
Owens Community College
Perrysburg, Ohio

Stephen J. Nardozzi, BA, EMT-P
Westchester Community College
Assistant Professor and Chairperson
Department of Prehospital EMS
Valhalla, New York

Angel J. Nater, MS, EMT-P
Seminole State College
Sanford, Florida

Michael Nemeth, AEMCA(f), ICP, CQIA
Seneca College of Applied Arts
    and Technology
Ontario, Canada

Chris Nollette, EdD, NREMT-P, L-P
Riverside Community College
Riverside, California

Justin Oakerson, AAS, NREMT-P
University of Texas Health Science Center at
    San Antonio
Department of Surgery/Division of
    Emergency Medicine
International Academy of Emergency
    Preparedness
San Antonio, Texas

C. Jill Oblak, MA, MBA, EMT-P
Community College of Allegheny County
Pittsburgh, Pennsylvania

Jim O'Connor, EMT-P
Columbus Ohio Division of Fire
Instructor, Eastland Fairfield Career and
    Technical Schools, AWDC
Groveport, Ohio

Douglas A. Paris, MEd, NREMT-P
Program Director, Emergency Medical
    Technology
Greenville Technical College
Greenville, South Carolina

Guy Peifer, BS, EMT-P
Paramedic Program Coordinator
Borough of Manhattan Community College
New York, New York

Stephen J. Phillipe Sr., BS, LEM, EMT-P
State of Louisiana Bureau of EMS
St. Rose, Louisiana

Mary L. Pilling, NREMT-I
EMS Instructor
Southwest Wisconsin Technical College
Fennimore, Wisconsin

David S. Pomerantz, NREMT-B, EMT-P
Education & Training Manager
Medics Ambulance Service
Deerfield Beach, Florida

Denise L. Quintrall, MEd
Mesa Community College
Mesa, Arizona

Wright N. Randolph Jr., CEP
Pima Community College
Tucson, Arizona

Barry Reed, MSA, RN, EMT-P, CEN, CCRN, CCEMT-P
Northwest Florida State College
Niceville, Florida

Christy Ridgill, Paramedic, AAS
EMS Instructor
Guilford Technical Community College
Jamestown, North Carolina

Don Royder, EMT-P, BS
EMS Training Program Training Manager
Texas Engineering Extension Service
College Station, Texas

Butch Russell, CCEMT-P, NREMT-P, IC
North East Mobile Health Services
Scarborough, Maine

Curt Schmittling, BS, EMT-P
EMT-P Program Coordinator
Southwestern Illinois College
Belleville, Illinois

Bernard J. Schweter, PhD, EMT-P, EMS-I
Cuyahoga Community College
Cleveland, Ohio

William D. Sheahan, MPA, EMT-P
EMS Coordinator
Livingston County EMS
Geneseo, New York

Jim Shedd, BA
RESA 3, Public Service Training
Dunbar, West Virginia

Richard Shok, BS, RN, EMS-I
Code One Training Solutions
Willimantic, Connecticut

Rintha Simpson, BS, NREMT-P
East Baton Rouge Parish Department of
Emergency Medical Services
Baton Rouge, Louisiana

Gursarn Singh, BS, EMS Director
University of Arkansas at Monticello College
of Technology
McGehee, Arkansas

Marc J. Sirkus, BBA, NREMT-P
Northern Virginia Community College
Medical Education Campus
Springfield, Virginia

Al M. Slarve, EMT-B, NREMT, AAS, AS, AA
Cochise College, Allied Health &
Fire Science Departments
Fry Fire District, Volunteer
Sierra Vista Fire Department, Volunteer
Sierra Vista, Arizona

John M. Slider, EMT-I, I/C
CEO, Matrix Training Center LLC
Carson City, Nevada

William Sugiyama, MA, RN, NREMT-P
EMS Fire Division Manager
City of Oakland Fire Department
Oakland, California

James K. Thompson, EMT-P
Emergency Medical Services
Upstate Carolina Medical Center
Gaffney, South Carolina

Alton L. Thygerson, EdD, FAWM
Professor of Health Science
Brigham Young University
Provo, Utah

John Todaro, BA, REMT-P, RN, TNS, NCEE
Director/Chief Operating Officer
Emergency Medicine Learning &
Resource Center
Orlando, Florida

Jeff Travers, EMT-P
Director of Public Safety
Great Oaks Institute
Cincinnati, Ohio

Ailsa R. Vogelsang, BTAS, NREMT-P
EMS/Fire Coordinator
Belmont Technical College
St. Clairsville, Ohio

Robert Wales, BS, CCEMT-P
Upstate EMS Council, Inc.
Easley, South Carolina

Bjorn Watsjold, MPH, EMT-I/C
Boston University Emergency
Medical Services
Boston, Massachusetts

Brad D. Weilbrenner, BS, NREMT-P
NH Division of Fire Standards & Training
and EMS
Concord, New Hampshire

Carl Weinstein, EMT-P, I/C
Bunker Hill Community College
Charlestown, Massachusetts

Joe Welsh, BS, CUSA, NREMT
President, CEO
Life-Savers, Inc.
Louisville, Kentucky

R. Greg West, MEd, EMT-I, EMS-I/C
North Shore Community College
Danvers, Massachusetts

Keith Widmeier, NREMT-P, CCEMT-P
Training Officer
Wayne County EMS
Monticello, Kentucky

Charlie Williams, EdS, EMT-P
Assistant Professor Emergency Medical
Technology
EMT Program Director, Walters State
Community College
Morristown, Tennessee

Mel Worth, EMT Course Coordinator
Atlantic County EMT Instruction, Inc.
Absecon, New Jersey

Captain Raymond L. Wright, BSN
U.S. Army Nurse Corps
Department of Combat Medic Training
Fort Sam Houston, Texas

Jeffrey Zuckernick, BS, MBA, NREMT
Professor, Program Director
University of Hawaii, KCC EMS
Honolulu, Hawaii

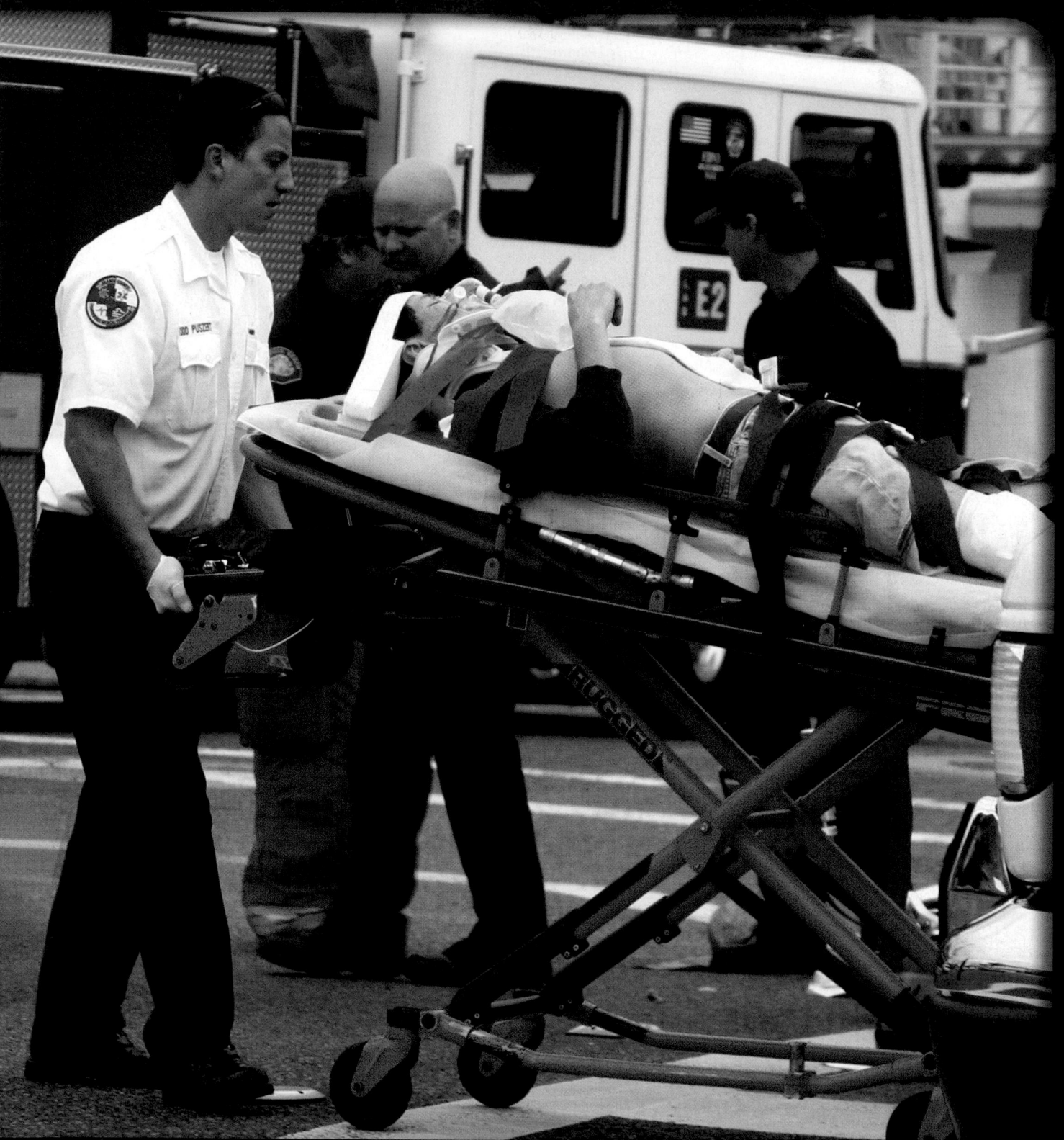

National EMS Education Standard Competencies

Preparatory

Applies fundamental knowledge of the emergency medical services (EMS) system, safety/well-being of the emergency medical technician (EMT), medical/legal, and ethical issues to the provision of emergency care.

Emergency Medical Services (EMS) Systems

- EMS systems (pp 14-23)
- History of EMS (pp 8-9)
- Roles/responsibilities/professionalism of EMS personnel (pp 23-24)
- Quality improvement (pp 17-19)
- Patient safety (pp 7-8)

Research

- Impact of research on emergency medical responder (EMR) care (pp 21-23)
- Data collection (pp 21-23)
- Evidence-based decision making (pp 21-23)

Public Health

Uses simple knowledge of the principles of illness and injury prevention in emergency care.

Knowledge Objectives

1. Define emergency medical services (EMS) systems. (p 5)
2. Discuss the four levels of EMT training and licensure. (pp 5-7)
3. Describe EMT licensure criteria, and understand that the Americans with Disabilities Act (ADA) applies to employment as an EMT. (p 8)
4. Discuss the historical background of the development of the EMS system. (pp 8-9)
5. Describe the levels of EMT training in terms of skill sets needed for each of the following: EMR, EMT, AEMT, and paramedic. (pp 9-13)
6. Understand the possible presence of other first responders at a scene with EMR training, some knowledge of first aid, or merely good intentions, and their need for direction. (p 12)
7. Name the 14 components of the EMS system. (pp 14-23)
8. Understand how medical direction of an EMS system works, and the EMT's role in the process. (pp 16-17)
9. Discuss the purpose of the EMS continuous quality improvement (CQI) process. (pp 17-19)
10. Characterize the EMS system's role in prevention and public education in the community. (pp 20-21)
11. Describe the roles and responsibilities of the EMT. (p 23)
12. Describe the attributes that an EMT is expected to possess. (pp 23-24)
13. Understand the impact of the Health Insurance Portability and Accountability Act (HIPAA) on patient privacy. (p 24)

Skills Objectives

There are no skills objectives for this chapter.

Introduction

This textbook serves as the primary resource for the emergency medical technician (EMT) course. It also discusses what will be expected of you during the course and what other requirements you will have to meet to be licensed or certified as an EMT in most states. You will also learn about the differences between first aid training, a Department of Transportation (DOT) emergency medical responder (EMR) training course, and the training courses for an EMT, advanced emergency medical technician (AEMT), and paramedic.

Emergency medical services (EMS) is a *system*. This system's key components and how they influence and affect the EMT and his or her delivery of emergency care are carefully discussed in this chapter. The administration, medical direction, quality control, and regulation of EMS services are also presented. The chapter ends with a detailed discussion of the roles and responsibilities of the EMT as a health care professional.

Course Description

You are about to enter an exciting field. The **emergency medical services (EMS)** system consists of a team of health care professionals who, in each area or jurisdiction, are responsible for and provide emergency care and transportation to the sick and injured **Figure 1-1**. Each emergency medical service is part of a local or regional EMS system that provides the many prehospital and hospital components required for the delivery of proper emergency medical care. The standards for prehospital emergency care and the individuals who provide it

Figure 1-1 As an EMT, you will be part of a larger team that responds to a variety of calls and provides a wide range of prehospital emergency care.

are governed by the laws in each state and are typically regulated by a state office of EMS.

When you successfully finish this course, you should be eligible to sit for your state's certification exam. A **certification** exam is used to ensure that all health care providers have at least the same basic level of knowledge and skill. Once you have passed this exam, you will be eligible to apply for state licensure. **Licensure** is how states control who is allowed to perform as a health care provider. It is the same principle as taking a test to certify you know how to drive in order to get a driver's license. Different states will refer to the authority granted to you to function as an EMT as licensure, certification, or credentialing. For the purposes of this text, the term *licensure* will be used.

In most states, individuals who work on an ambulance are categorized into four training and licensure levels: **emergency medical responder (EMR)**, **emergency medical technician (EMT)**, **advanced EMT (AEMT)**, and **paramedic**. An EMR has very basic

You are working your first shift as an EMT and are on duty with an experienced EMT and her paramedic partner. The crew is familiarizing you with the ambulance when the tone alert sounds, "EMS 4, respond to 325 Blossom Avenue for a woman with back pain." You and your crew proceed to the scene, which is located 4 miles from your station.

1. Do the roles and responsibilities of an EMT differ from those of a paramedic? If so, how?
2. Is there a difference between what you learned in your EMT class and the care you provide in the field?

training and provides care before the ambulance arrives. EMRs may also perform in an assistant role within the ambulance. An EMT has training in basic life support, including automated external defibrillation, use of airway adjuncts, and assisting patients with certain medications. An AEMT has training in specific aspects of **advanced life support (ALS)**, such as **intravenous (IV) therapy** and the administration of certain emergency medications. A paramedic has extensive training in ALS, including endotracheal intubation, emergency pharmacology, cardiac monitoring, and other advanced assessment and treatment skills.

Although the specific training and licensure requirements vary from one state to another, almost every state's requirements follow or exceed the guidelines recommended in the current National Highway Traffic Safety Administration (NHTSA) EMS Education Standards.

This textbook covers the material and skills identified in the 2009 *National EMS Education Standards*. This text also covers information needed for EMTs to perform the skills outlined in the 2005 National EMS Scope of Practice Model. In the United States, NHTSA is the federal administrative source for curriculum and related documents.

After you have successfully completed any prerequisites of your training institution, you are ready to take the EMT course. Like any introductory course, the EMT course covers a great deal of information and introduces many skills **Figure 1-2**. EMT courses include didactic instruction (knowledge), psychomotor instruction (skills laboratories), and clinical behavior/judgment (professionalism). Everything you learn in the course will be important to your ability to provide high-quality emergency care once you are licensed and ready to practice. In addition, the knowledge, understanding, and skills that you acquire in the EMT course will serve as a foundation for the additional knowledge and training that you will receive in future years.

In addition to the required core content, this text includes additional information that will help you understand and apply the material and skills included in the EMT course. Your instructor will furnish you with reading assignments. It is essential that you complete the assigned reading before each class. Your success in this course will depend on it.

In class, the instructor will review the key parts of the reading assignment and clarify and expand on them. He or she will also answer any questions you have and will clarify any points you or others find confusing. Unless you have carefully read the assignment and made notes before coming to class, you will not fully understand or benefit from the classroom presentation and discussions. Creating your own tools such as flashcards, study questions, and outlines will help you to retain important

information. It will also help to take better notes during class **Table 1-1**.

The EMT course will include four types of learning activities:

1. Reading assignments from the textbook, lecture presentations, and classroom discussions will provide you with the necessary knowledge base.
2. Step-by-step demonstrations will teach you hands-on skills that you then need to practice repeatedly in supervised small group workshops.
3. Summary skills sheets will help you memorize the sequence of steps in complex skills that contain a large number of steps or variations so you can perform the skill with no errors or omissions.
4. Case presentations and scenarios used in class will help you learn how to apply the knowledge and skills acquired to situations you will find in the field.

Figure 1-2 In the classroom, you will learn both didactic and practical skills to prepare you for various types of calls.

Table 1-1 **Study Tips for Using This Textbook**

- Complete each assignment diligently and carefully.
- Read the textbook like a textbook, not like a newspaper, magazine, or novel.
- Read each chapter several times and underline key points. Take notes!
- Ask your instructor to clarify any questions you note in your reading or in class.
- Take additional notes when the assigned material is expanded upon in class.
- Remember: The only absurd question is the one that a student has and fails to ask.

Words of Wisdom

The Star of Life

The National Highway Transportation Safety Administration (NHTSA) recognized the need for a symbol that would represent EMS as a critical public service and created the "Star of Life." The NHTSA holds priority rights to the use of this registered certification mark.

Adapted from the personal Medical Identification Symbol of the American Medical Association, each bar on the "Star of Life" represents an EMS function. The functions include:

1. Detection
2. Reporting
3. Response
4. On-scene care
5. Care in transit
6. Transfer to definitive care

The serpent and staff in the symbol portray the staff of Asclepius, an ancient Greek physician deified as the god of medicine. Overall, the staff represents medicine and healing, with the skin-shedding serpent being indicative of renewal.

The "Star of Life" has become synonymous with emergency medical care around the globe. This symbol can be seen as a means of identification on ambulances, emergency medical equipment, patches or apparel worn by EMS providers, and materials such as books, pamphlets, manuals, reports, and publications that either have a direct application to EMS or were generated by an EMS organization. It can also be found on road maps and highway signs indicating the location of or access to qualified emergency medical care.

Source: Adapted from US National Highway Traffic Safety Administration. www.ems.gov.

■ EMT Training: Focus and Requirements

What is an EMT? EMTs are the backbone of the EMS system in the United States. These men and women provide emergency care to the sick and injured. Some of the patients you will eventually treat are in life-threatening situations, whereas others require only supportive care. The skills needed to safely deliver this care are found within this text. Some of the subjects that will be discussed include:

- **Scene size-up:** EMS operates in a wide variety of environments that can create situations where EMS personnel can be injured—outside in the rain, inside a cluttered house, or anywhere in between. The EMT's primary job is to ensure that he or she is as safe as possible.
- **Patient assessment:** Patient assessment is the foundation of any EMS call. You must determine what is wrong with the patient. Patients can have many complaints and you will learn to determine which complaints are life threatening.
- **Treatment:** As an EMT, you will be providing oxygenation and medication therapies. You will control bleeding and assist patients in the delivery of their babies. In addition to hands-on skills, you will also learn how to manage people who are in emotional crisis, and to calm patients and to relieve some of their anxiety.
- **Packaging:** Most patients will need to be transported to a facility. This could mean a hospital, clinic, or other medical care facility. You will learn how to

You are the Provider: PART 2

You arrive at the scene, ensure that it is safe for you to enter, and make contact with the patient, a 59-year-old woman. She is sitting on her couch, is in obvious pain, and states that it has been ongoing for the past month. You assess the patient as your partner prepares to take her vital signs.

Recording Time: 0 Minutes	
Appearance	Grimaced appearance; obvious pain
Level of consciousness	Conscious and alert
Airway	Open; clear of secretions or foreign bodies
Breathing	Adequate rate and depth
Circulation	Radial pulse, normal rate and rhythm; skin is pink, warm, and dry

3. Is this patient experiencing a "true emergency?"

transport patients with a wide variety of illnesses and injuries.

- **EMS as a career:** Many of you are taking this course because you want to help people. To ensure all EMS providers have a long, healthy career, it is important to teach EMS providers how to take care of themselves. We will discuss job stresses and successful ways to cope with them.

Licensure Requirements

To be recognized and perform as an EMT, you must meet certain requirements. The specific requirements differ from state to state. You should ask your instructor, learning institute, or your state EMS official about the requirements in your state. Generally, the criteria to be licensed and employed as an EMT will include the following:

- High school diploma or equivalent
- Proof of immunization against certain communicable diseases
- Valid driver's license
- Successful completion of a recognized health care provider basic life support (BLS)/cardiopulmonary resuscitation (CPR) course
- Successful completion of a state-approved EMT course
- Successful completion of a state-recognized written certification examination
- Successful completion of a state-recognized practical certification examination
- Demonstrating that you can meet the mental and physical criteria necessary to be able to safely and properly perform all the tasks and functions described in the defined role of an EMT
- Compliance with other state, local, and employer provisions

The **Americans With Disabilities Act (ADA)** of 1990 protects individuals who have a disability from being denied access to programs and services that are provided by state or local governments and prohibits employers from failing to provide full and equal employment to the disabled. To obtain further information about the ADA and employment as an EMT, you should contact your state EMS office.

One of the primary responsibilities of each state is to ensure the safety of its residents. As such, states have requirements prohibiting individuals with certain types of legal histories from becoming EMS providers. The specific legal requirements, either misdemeanors and/or felonies, are created on a state-by-state basis. Contact your state EMS office for more information.

Special Populations

EMS systems must be capable of handling many different situations, including obstetric, pediatric, and geriatric emergencies. Proper procedures, drug dosages, and even assessment techniques are often different in children, adults, and older people.

Overview of the EMS System

History of EMS

As an EMT, you will be joining a long tradition of people who have provided emergency medical care to their fellow human beings. With the early use of motor vehicles in warfare, volunteer ambulances were organized and personnel went overseas to provide care for the wounded in World War I. In World War II, the military trained special corpsmen to provide care in the field and bring the casualties to aid stations staffed by nurses and physicians. In the Korean conflict, this evolved into the field medic and rapid helicopter evacuation to nearby Mobile Army Surgical Hospital units, where immediate surgical intervention was provided. Many advances in the immediate care of trauma patients resulted from the casualty experiences in the Korean and Vietnam conflicts.

Unfortunately, emergency care of the injured and ill at home had not progressed to a similar level. As late as the early 1960s, emergency ambulance service and care across the United States varied widely. In some places, it was provided by well-trained advanced first aid personnel who had well-equipped, modern ambulances. In a few urban areas, it was provided by hospital-based ambulance services that were staffed with interns and early forms of prehospital care providers. In many areas, the only emergency care and ambulance service was provided by the local funeral home using a hearse that could be converted to carry a cot and serve as an ambulance. In other places, the police or fire department used a station wagon that carried a cot and a first aid kit. In most cases, these vehicles were staffed by a driver and an attendant who had some basic first aid training. In the few areas where a commercial ambulance was available to transport the ill, it was usually similarly staffed and served primarily as a means to transport the patient to the hospital.

Many communities had no formal provision for prehospital emergency care or transportation. Injured persons were given basic first aid by police or fire personnel at the scene and were transported to the hospital in a police or fire officer's car. Customarily, patients with an acute illness were transported to the hospital by a relative or neighbor and were met by their family physician or an on-call hospital physician, who assessed them and then summoned any specialists and operating room staff that were needed. Except in large urban centers, most hospitals did not have the emergency department staff available today.

EMS as we know it today had its origins in 1966 with the publication of *Accidental Death and Disability: The Neglected Disease of Modern Society*, known more commonly as "The White Paper." This report, prepared jointly by the Committees on Trauma and Shock of the National Academy of Sciences/National Research Council, revealed to the public and Congress the serious inadequacy of prehospital emergency care and transportation in many areas. As a result, Congress mandated that two federal agencies address these issues. The NHTSA of the DOT, through the Highway Safety Act of 1966, and the Department of Health and Human Services, through the Emergency Medical Act of 1973, created funding sources and programs to develop improved systems of prehospital emergency care. This explains why EMS is administrated at the federal level through the DOT and not the Department of Health.

In the early 1970s, the DOT developed and published the first curriculum to serve as the guideline for EMT training. To support the EMT course, the American Academy of Orthopaedic Surgeons prepared and published the first EMT textbook—*Emergency Care and Transportation of the Sick and Injured*—in 1971, often called the Orange Book. The textbook you are reading is the tenth edition of that publication. Through the 1970s, following the recommended guidelines, each state developed the necessary legislation, and the EMS system expanded throughout the United States. During the same period, emergency medicine became a recognized medical specialty, and the fully staffed emergency departments that we know today became the accepted standard of care.

In the late 1970s, the DOT developed a recommended National Standard Curriculum for the training of paramedics and identified a part of the course to serve as training for EMTs.

During the 1980s, many areas enhanced the EMT National Standard Curriculum by adding EMTs with higher levels of training who could provide key components of ALS care and advanced lifesaving procedures. The availability of paramedics and ALS-level care on calls that require or benefit from advanced care has grown steadily in recent years. In addition, with the evolution in training and technology, the EMT and AEMT can now perform a number of important advanced skills in the field that were formerly reserved for only the paramedic.

This growth and sophistication of the EMS system did not come without its drawbacks. As each state sought to create a system that would meet the needs of its citizens, the definitions of EMS providers began to vary from state to state. For example, in some states EMTs were allowed to administer medications, while in other states they were not.

In the 1990s NHTSA again began an examination of EMS from a national perspective. With the counsel of EMS providers, physicians, fire chiefs, nurses, state administrators, educators, and other interested parties, NHTSA created the *EMS Agenda for the Future*. This important document creates a plan to standardize the levels of EMS education and EMS providers in an effort to ensure a more seamless delivery of EMS care across the country.

The skills you will be learning and the scope of practice that EMTs now enjoy are part of this national movement toward an EMS system that meets the needs of an ever-changing health care industry, and meets those needs through a safe and efficient method.

Levels of Training

As discussed earlier, licensure of EMTs is a state function subject to the laws and regulations of the state in which the EMT practices. Each state is granted the ability to control the functions of its licensed individuals. For this reason there remains some variation from state to state on the scope of EMT practice, as well as training and recertification requirements. Here is how the system is supposed to work from the federal level down to the local level.

At the federal level, NHTSA brought in experts from around the country to create the **National EMS Scope of Practice Model**. This document provides overarching guidelines as to what skills each level of EMS provider should be able to accomplish. Table 1-2 shows the guidelines from that model. The next step is at the state level. Because licensure is a state function, laws are enacted to regulate how EMS providers will operate and are then executed by the state level EMS administrative offices, which control licensure. Finally, the local medical director decides the day-to-day limits of EMS personnel Figure 1-3 . For example, the medications that will be carried on an ambulance or where patients are transported are the day-to-day operational concerns in which the medical director will have direct input.

The national guidelines are intended to create a more consistent delivery of EMS across the country. The only way a medical director can allow an EMT to perform a skill is if the state has already cleared that

Table 1-2 The Interpretive Guidelines: National EMS Scope of Practice Model

Airway and Breathing Minimum Psychomotor Skill Set			
EMR	**EMT**	**AEMT**	**Paramedic**
Oral airway	Humidifiers	Esophageal-tracheal intubation	BiPAP/CPAP
Bag-mask device	Partial rebreathing mask	Multilumen airways	Needle chest decompression
Sellick maneuver	Venturi mask		Chest tube monitoring
Head tilt–chin lift	Manually triggered ventilators		Percutaneous cricothyrotomy
Jaw-thrust	Automatic transport ventilators		$ETCO_2$/capnography
Modified chin lift	Oral and nasal airways		NG/OG tube
Obstruction, manual			Nasal and oral endotracheal intubation
Oxygen therapy			Airway obstruction removal by direct laryngoscopy
Nasal cannula			Positive end-expiratory pressure
Nonrebreathing mask			
Upper airway suctioning			
Assessment Minimum Psychomotor Skill Set			
Manual BP	Pulse oximetry	Blood glucose monitoring	ECG interpretation
	Manual and auto BP		Interpretive 12-lead
			Blood chemistry analysis
Pharmacologic Intervention Minimum Psychomotor Skill Set			
Medication Administration Routes ■ Unit dose auto-injector for self or peer care (MARK 1)	*Assisted Medications* ■ Assisting a patient in administering his/her own prescribed medications, including auto-injector	■ Peripheral IV insertion ■ IV fluid infusion ■ Pediatric IO insertion	■ Central line monitoring ■ IO insertion ■ Venous blood sampling
	Medication Administration Routes ■ Buccal ■ Oral	*Medication Administration Routes* ■ Aerosolized ■ SC ■ IM ■ Nebulized ■ SL ■ Intranasal ■ IV push or D_{50} and narcotic antagonist only	*Medication Administration Routes* ■ Endotracheal ■ IV (push and infusion) ■ Nasogastric (NG) ■ Rectal ■ IO ■ Topical ■ Accessing implanted central IV port
	Medications To Be Administered ■ Physician-approved over-the-counter medications (oral glucose, aspirin for chest pain or suspected ischemic origin)	*Medications To Be Administered* ■ SL nitroglycerin for chest pain of suspected ischemic origin ■ SQ and IM epinephrine for anaphylaxis	*Medications To Be Administered* ■ Physician-approved medications ■ Maintenance of blood administration

Continues

Table 1-2	The Interpretive Guidelines: National EMS Scope of Practice Model, continued		
Pharmacologic Intervention Minimum Psychomotor Skill Set			
EMR	**EMT**	**AEMT**	**Paramedic**
		▪ Glucagon and IV D_{50} for hypoglycemia ▪ Inhaled beta-agonist for dyspnea and wheezing ▪ Narcotic antagonist ▪ Nitrous oxide for pain relief	▪ Initiation of thrombolytics
Emergency Trauma Care Minimum Psychomotor Skill Set			
Manual cervical stabilization	Spinal immobilization		Morgan lens
Manual extremity stabilization	Seated spinal immobilization		
Eye irrigation	Long board		
Direct pressure	Extremity splinting		
Hemorrhage control	Traction splinting		
Emergency moves for endangered patients	Mechanical patient restraint		
	Tourniquet		
	MAST/PASG		
	Cervical collar		
	Rapid extrication		
Medical/Cardiac Care Minimum Psychomotor Skill Set			
CPR	Mechanical CPR		Cardioversion
AED	Assisted complicated delivery of an infant		Carotid massage
Assisted normal delivery of an infant			Manual defibrillation
			TC pacing

Abbreviations: AED, automated external defibrillator; BiPAP/CPAP, bilevel positive airway pressure/continuous positive airway pressure; BP, blood pressure; CPR, cardiopulmonary resuscitation; D_{50}, 50% dextrose in water; ECG, electrocardiogram; IM, intramuscular; IO, intraosseous; IV, intravenous; MAST/PASG, military antishock trousers/pneumatic antishock garments; NG, nasogastric; OG, orogastric; SL, sublingual; SQ, subcutaneous; TC, transcutaneous.

Note: The 2005 National EMS Scope of Practice Model serves as a foundation for states to build their own model. It is intended to illustrate the operation of each level of EMS provider and the progression from one level to another. It is not inclusive of every skill a state may allow.

skill. The medical director can limit scope of practice but cannot expand it beyond state law. Expanding the scope of practice requires state approval.

The EMR, EMT, AEMT, and paramedic curricula can be downloaded from NHTSA's web site at www.ems.gov. In addition, the National Registry of Emergency Medical Technicians (NREMT) is a nongovernmental agency that provides national standardized EMS testing and certification in much of the United States. Many states use the National Registry standards in certifying their EMTs and grant licensing reciprocity to NREMT-certified EMTs. It is important to remember, however, that EMS is regulated entirely by the state in which you are licensed.

■ Public Basic Life Support and Immediate Aid

With the development of EMS and increased awareness of the need for immediate emergency care, millions of laypeople have been trained in BLS/CPR. In addition to

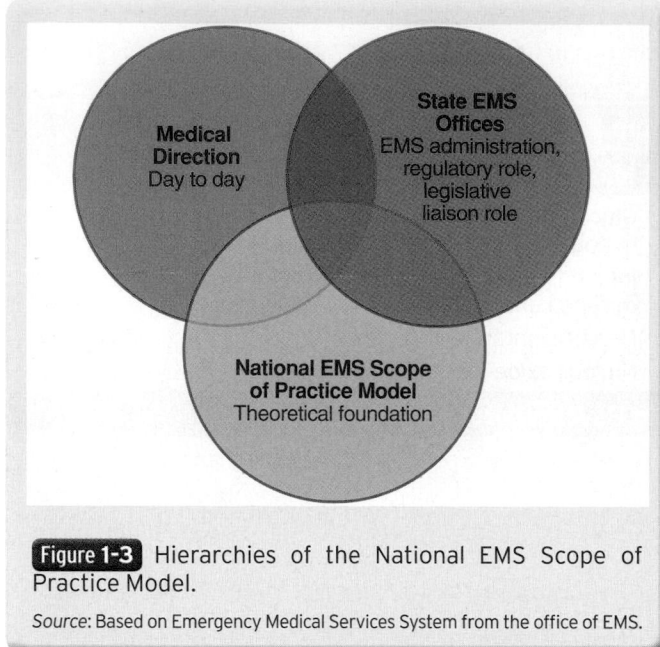

Figure 1-3 Hierarchies of the National EMS Scope of Practice Model.

Source: Based on Emergency Medical Services System from the office of EMS.

CPR, many individuals have taken first aid courses that include bleeding control and other simple skills that may be required to provide immediate essential care. These courses are designed to train individuals so those in the workplace—teachers, coaches, child care providers, and others—can provide the necessary critical care in the minutes before EMTs or other responders arrive at the scene.

In addition, many individuals, such as those who regularly accompany groups on camping trips or are in other situations where the arrival of EMS may be delayed because of remote location, are trained in advanced first aid. This course includes BLS and the essential additional care and packaging that may be necessary until the help of rescuers and EMTs can be obtained at a remote location.

One of the most dramatic recent developments in prehospital emergency care is the use of an **automated external defibrillator (AED)**. These remarkable devices, some no larger than a cellular phone, detect treatable life-threatening cardiac arrhythmias (ventricular fibrillation and ventricular tachycardia) and deliver the appropriate electrical shock to the patient. Designed to be used by the untrained layperson, these devices are now included at every level of prehospital emergency training.

■ Emergency Medical Responders

Because the presence of a person who is trained to initiate BLS and other urgent care cannot be ensured, the EMS system includes immediate care by EMRs, such as law enforcement officers, fire fighters, park rangers, ski patrollers, or other organized rescuers who often arrive at the scene before the ambulance, and EMTs **Figure 1-4**. EMR training provides these individuals with the skills necessary to initiate immediate care and assist the EMTs on their arrival. The course focuses on providing immediate BLS and urgent care with limited equipment. It also familiarizes the student with the additional procedures, equipment, and packaging techniques that EMTs may use and the EMR may be called on to assist.

In addition to professional EMRs, EMTs often encounter a variety of people on the scene eager to help. You will encounter Good Samaritans trained in first aid and CPR, physicians and nurses, and other well-meaning individuals with or without prior training and experience. Identified and used properly, these individuals can provide valuable assistance when you are short-handed. At other times, they can interfere with operations and even create problems or danger for themselves or others. It will be your task in your initial scene size-up to identify the various persons on the scene and orchestrate well-meaning attempts to assist.

■ Emergency Medical Technician

The EMT course requires approximately 150 hours (more in some states) and includes the essential knowledge and skills required to provide basic emergency care in the field. The course serves as the foundation on which additional knowledge and skills are built in AEMT training. On arrival at the scene, you and any other EMTs who have responded should assume responsibility for the

Figure 1-4 Emergency medical responders, such as law enforcement officers, are trained to provide immediate basic life support until EMTs arrive on the scene.

assessment and care of the patient, followed by proper packaging and transport of the patient to the emergency department if appropriate.

Advanced Emergency Medical Technician

The AEMT course and training are designed to add knowledge and skills in specific aspects of ALS to individuals who have been trained and have experience in providing emergency care as EMTs. These additional skills include IV therapy, use of advanced airway adjuncts, and, in many states, the knowledge and skills necessary to administer certain medications. The purpose of this level of EMS provider is to deliver an expanded range of skills over those of the EMT. In some parts of the United States, the availability of paramedics is limited. AEMTs help to fill the gap by providing limited ALS care to regions where paramedics are not available.

Paramedic

The paramedic has completed an extensive course of training that significantly increases knowledge and mastery of basic skills and covers a wide range of ALS skills **Figure 1-5**. This course ranges from 800 to more than 1,500 hours, usually equally divided between classroom and internship training. Increasingly, this training is offered within the context of an associate's degree or bachelor's degree college program.

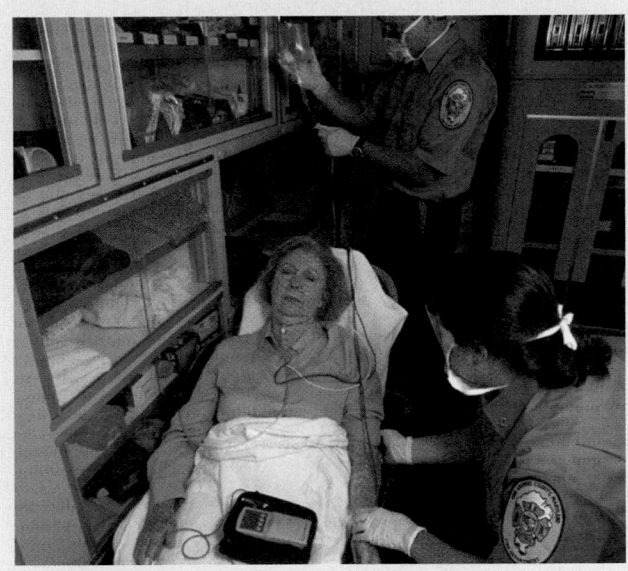

Figure 1-5 Paramedic training covers a wide range of ALS skills.

You are the Provider: PART 3

Your partner records the patient's vital signs on the patient care report as you ask the patient additional questions regarding her back pain. She tells you that her lower back began hurting about a month ago; however, she has never been evaluated by a physician. She denies injuring her back. She further denies any other symptoms or past medical history.

Recording Time: 4 Minutes	
Respirations	16 breaths/min; regular and unlabored
Pulse	88 beats/min; strong and regular
Skin	Pink, warm, and dry
Blood pressure	126/66 mm Hg
Oxygen saturation (Sao_2)	99% (on room air)

Your assessment of the patient's back does not reveal any obvious deformities, swelling, or bruising, and her vital signs are stable. The patient requests that you take her to the hospital.

4. The patient does not appear to be experiencing any life-threatening conditions. Should you transport her to the hospital?

Components of the EMS System

The *EMS Agenda for the Future* is a multidisciplinary, national review of all aspects of EMS delivery. The goal is to develop a more cohesive and consistent system across the country. In the document, there were 14 components of an EMS system outlined **Table 1-3**. NHTSA has taken these components and organized them in such a way to understand some of the interrelationships between the components.

Figure 1-6 demonstrates how the components interact. The tabs on the right side show the primary 9-1-1 components regarding EMS. Someone recognizes an emergency, 9-1-1 is activated, the ambulance is dispatched, and emergency care and transportation are administered. In the center are the essential aspects needed to allow the primary 9-1-1 components to function—finances, radios, computers, and people. The left side demonstrates the continuum of care from the prehospital environment to the emergency department and beyond. Finally, as the patient leaves the health care system, there are strategies of prevention and education to help ensure people live long and healthy lives. Understanding the 14 components will help you better understand how the EMS system works.

Table 1-3 **EMS Agenda for the Future Components of an EMS System**

EMS System	
1. Public Access	**8.** Communication Systems
2. Clinical Care	**9.** Human Resources
3. Medical Direction	**10.** Legislation and Regulation
4. Integration of Health Services	**11.** Evaluation
5. Information Systems	**12.** System Finance
6. Prevention	**13.** Public Education
7. EMS Research	**14.** Education Systems

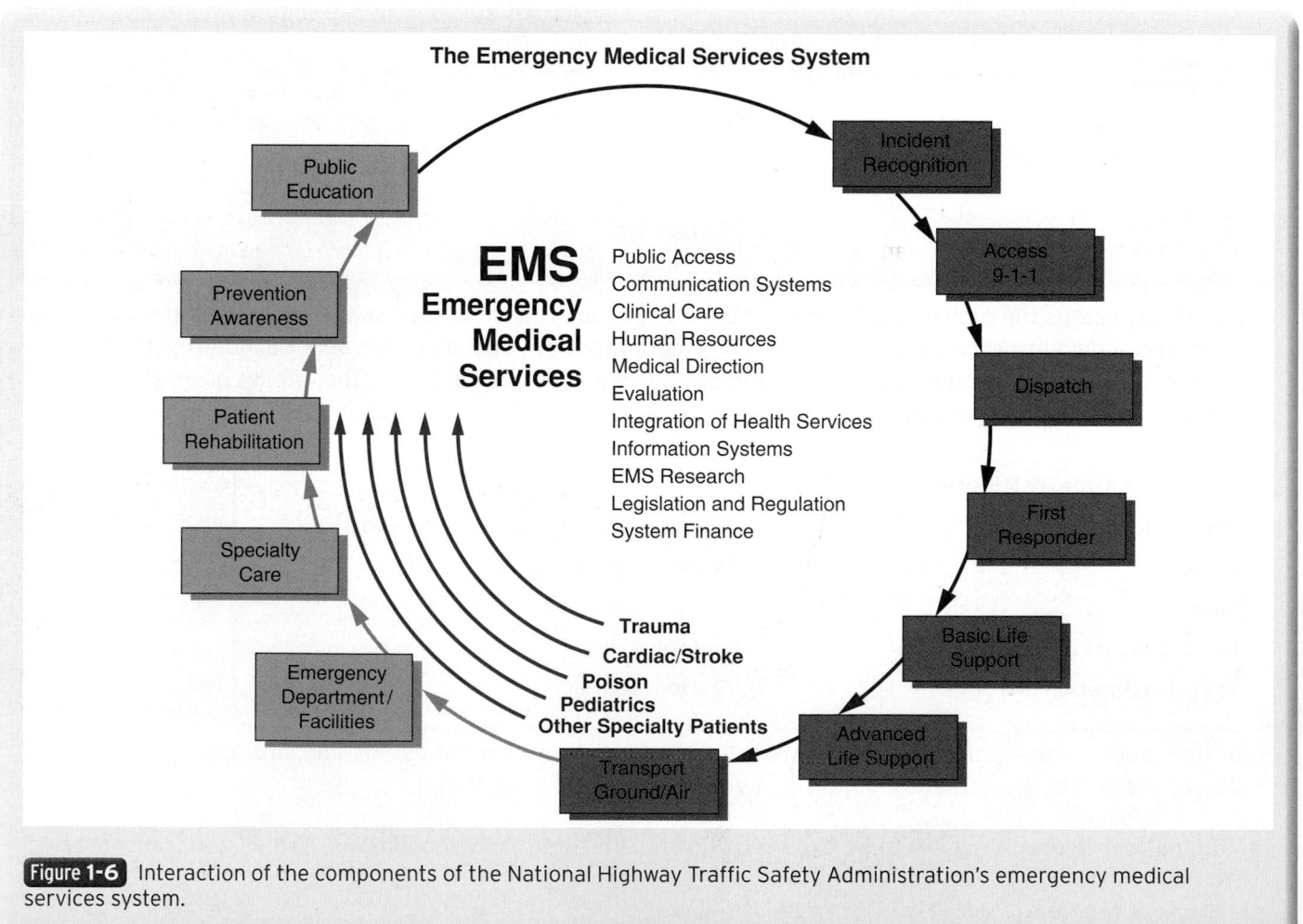

Figure 1-6 Interaction of the components of the National Highway Traffic Safety Administration's emergency medical services system.

Source: www.ems.gov

Public Access

Easy access to help in an emergency is essential. In most of the country, an emergency communication center that dispatches fire, police, rescue, and EMS units can be reached by dialing 9-1-1. At the communication center, trained dispatchers obtain the necessary information from the caller and, following dispatch protocols, dispatch the ambulance crew and other equipment and responders that may be needed **Figure 1-7**. This communication center is called a **public safety access point**.

In an enhanced 9-1-1 system, the address of the caller is displayed on a screen. The address remains on the screen until the dispatcher removes it. Therefore, if the caller is unable to speak or hangs up, the location remains displayed. However, many cellular phones do not yet have this capability. Most emergency communication centers also include special equipment so that individuals with speech or hearing disabilities can communicate with the dispatcher via a keyboard and printed messages. In some areas, rather than 9-1-1, a different special published emergency number may be used to call for EMS. Training the public in how to summon an EMS unit is an important part of the public education responsibility of each EMS service. Enhanced 9-1-1 systems are now becoming available that can identify not only the cellular phone number from which an emergency call is being placed, but also the exact geographic coordinates of the phone at the time the call is made. Such systems use global positioning system (GPS) technology. Because cellular phones capable of transmitting a GPS signal and a system capable of receiving that signal are both required, the technology will require additional time and resources to implement.

A system called **emergency medical dispatch (EMD)** has been developed to assist dispatchers in providing callers with vital instructions to help them deal with a medical emergency until the arrival of EMS crews. Dispatchers are provided with training and scripts to help them relay relevant instructions to the callers. The system also helps dispatchers select appropriately resourced units to respond to a request for assistance. It is the dispatcher's duty to relay all relevant and available information to the responding crews in a timely manner. Keep in mind, however, current technology does not allow the dispatcher to "see" what is actually going on at the scene, and it is not uncommon for you to find the reality of the call quite different from the dispatch information.

Communication Systems

With the information provided by the caller, the dispatcher will select the appropriate parts of the emergency system that need to be activated. In most municipalities, EMS is a part of the fire department. In others, it is a part of the police department or is an independent public or private service. In some areas, a contractor may provide either BLS or ALS service, while in other areas, a hospital-based program possibly covering several towns may provide the ambulance services.

New technologies are constantly being developed that can assist responders in locating their patients. As previously described, cellular telephones can be linked to GPS units to display their location. Responding units can transmit their position to a dispatcher and he or she can transmit the location of a call to a moving digital map in the unit, complete with turn-by-turn directions. Medical databases can be queried and patient information can be directly downloaded to the EMT's computer, or uploaded from the EMT's laptop to the database. Constant training and education are required to keep the EMT's knowledge of technological developments up to date.

Figure 1-7 Trained dispatchers obtain information about the call and then send responders to the scene as needed.

Safety

Practice makes perfect. Many seasoned EMS providers become a bit bored with checking the ambulance day after day. But take heart, by accomplishing your daily chores, you will learn where everything is and how it all works.

Clinical Care

The clinical care component describes the various pieces of equipment and scope of practice for using that equipment. As an EMT, you will use a wide range of different emergency equipment. During the EMT course, you w introduced to, and learn how to use, a variety of appl and devices that you may need on a call. You will also learn when the use of each is indicated and when it is of no benefit or may cause harm. Although the use of different models and brands of a given device will follow the same basic principles and methods, some variation and peculiarities may exist from one model to another. When you join a service, you should check each key piece of equipment before going on duty to ensure that it is in its assigned place, that it is working properly, and that you are familiar with the specific model carried on your ambulance.

Each EMT may be called on to drive the ambulance. Therefore, you must familiarize yourself with the roads in your **primary service area (PSA)** or sector. The PSA is the main area in which an EMS agency operates. Before going on duty, you should check all the equipment and supplies and communication equipment that the ambulance carries. It is also the EMT's responsibility to ensure that the ambulance is fully fueled, has sufficient oil and other key fluids, and that the tires are in good condition and properly inflated **Figure 1-8**. You should also test each of the driver's controls and each built-in unit and control in the patient compartment. If you have not driven the specific ambulance before, it is a good idea to take it out and become familiar with it before you respond to a call. Maintenance and safe driving of the ambulance are discussed in detail in Chapter 36.

■ Human Resources

The human resources component deals with people. Who delivers the care? How are these people compensated for their time and energy? How do other members of the medical community interact and participate within the EMS world? These are some of the questions that are discussed within the component of human resources.

Figure 1-8 Making sure the ambulance is fueled is part of an EMT's responsibility.

EMS in this component is examined as a profession. The overarching concept is to encourage the creation of EMS systems that provide an environment where talented people want to work and can turn their passion into a rewarding career.

Several objectives need to be accomplished to help make a career in EMS a lasting one. Efforts are being made to ensure that EMS providers can move from one state to another more seamlessly. From a global point of view, one of the core functions of a state is to provide and protect its citizens. This obligation has led to the creation of EMS levels that are unique to a particular state. Though effective for any one state, these idiosyncratic EMS levels make movement from one state to another complicated. One of the functions of the National Scope of Practice Model is to create stable foundations on which each level of EMS provider is grounded. The net effect is to encourage a more consistent definition of "what is an EMT" so providers can move more freely about the country.

The *EMS Agenda for the Future* encourages the creation of systems that help to protect the well-being of EMS providers. It also encourages systems to develop career ladders, allowing talented EMS providers ways to use that talent for many years.

■ Medical Direction

Each EMS system has a physician **medical director** who authorizes the EMTs in the service to provide medical care in the field. The appropriate care for each injury, condition, or illness that you will encounter in the field is determined by the medical director and is described in a set of written standing orders and protocols. Protocols are described in a comprehensive guide delineating the EMT's scope of practice. Standing orders are part of protocols and designate what the EMT is required to do for a specific complaint or condition.

The medical director provides the ongoing working liaison between the medical community, hospitals, and the EMTs in the service. If treatment problems arise or different procedures should be considered, these are referred to the medical director for his or her decision and action. To ensure the proper training standards are met, the medical director determines and approves the continuing education and training that are required of each EMT in the service and approves any that individuals obtain elsewhere.

Medical control is either off-line (indirect) or online (direct), as authorized by the medical director. Online medical control consists of direction given over the phone or radio directly from the medical director or designated physician. The medical direction can be transferred by the physician's designee; it does not have to be transferred by the physician himself or herself. Off-line medical control consists of standing orders, training, and supervision authorized by the medical director. Each EMT must know and follow the protocols developed by his or her medical director.

The service's protocols will also identify an EMS physician, usually at a local hospital, who can be reached by radio or telephone for medical control during a call. This is a type of direct online medical control. On some calls, once the ambulance has initiated any immediate urgent care and gives its radio report, the online medical control physician may either confirm or modify the proposed treatment plan or may prescribe any additional special orders that the EMTs are to follow for that patient. The point at which the EMTs should give their radio report or obtain online medical direction will vary.

■ Legislation and Regulation

Although each EMS system, medical director, and training program has latitude, their training, protocols, and practices must conform to the EMS legislation, rules, regulations, and guidelines adopted by each state. The state EMS office is responsible for authorizing, auditing, and regulating all EMS services, training institutions, courses, instructors, and providers within the state. In most states, the state EMS office obtains input from an advisory committee made up of representatives of the services, service medical directors, medical associations, hospitals, training programs, instructors' associations, EMT associations, and the public in that state.

At the local level, each EMS service operates in a designated PSA in which it is responsible for the provision of prehospital emergency care and the transportation of the sick and injured to the hospital.

EMS services are usually administered by a senior EMS official. Daily operations and overall direction of the service are provided by an appointed chief executive officer and several other officers who serve under him or her. When the EMS service is a part of a fire or police department, the department chief will usually delegate the responsibility for directing EMS to an assistant chief or other officer whose sole responsibility is to manage the EMS activities of the department. To provide clear guidelines, most services have written operating procedures and policies. When you join a service, you will be expected to learn and follow them.

The chief executive of the service is in charge of both the necessary administrative tasks (eg, scheduling, personnel, budgets, purchasing, vehicle maintenance) and the daily operations of the ambulances and crews. Except for medical matters, he or she operates as the chief (similar to a fire chief or police chief) of EMS for the service and the PSA that it covers.

■ Integration of Health Services

EMS does not work in a vacuum. EMS personnel travel to people's homes and to vehicle collisions. Once on scene, they deliver care and transport the patient to a care facility. Integration of health services means that the prehospital care you administer is coordinated with the care administered at the hospital. When you deliver a patient to the emergency department, you are simply transferring that patient to another care provider. The excellent care that you began is simply continued in the emergency department. This component helps to decrease errors, to increase efficiencies, and, most of all, to ensure the patient receives comprehensive continuity of care.

Words of Wisdom

A patient may only experience once what you may experience hundreds of times. Understand and be empathetic to the patient's anxiety.

■ Evaluation

The medical director is responsible for maintaining **quality control**, ensuring that all staff members who are involved in caring for patients meet appropriate medical care standards on each call. To provide the necessary quality control, the medical director and other involved staff review patient care reports, audit administrative records, and survey patients.

Continuous quality improvement (CQI) is a circular system of continuous internal and external reviews and audits of all aspects of an EMS system. To provide CQI, periodic run review meetings are held in which all those who are involved in patient care review the run reports and then discuss any areas of care that appear to need change or improvement. Positive feedback is also discussed. If a problem appears to be repeated by a single EMT or crew, the medical director will discuss the details with the individuals involved. The CQI process is designed to identify areas of improvement and, if necessary, assign remedial training or develop some other educational activity. The medical director is also responsible for ensuring that appropriate continuing education and training are available.

Information and skills in emergency medical care change constantly. You need refresher training or continuing education as new modalities of care, equipment, and understanding of critical illnesses and trauma develop. Equally, when you have not used a particular procedure for some time, skill decay may occur. Therefore, your medical director may establish a training program to correct the deficit. For example, an emergency department physician noted that despite their assessments, many EMTs were missing a high number of closed long bone fractures, resulting in poor prehospital care. A subsequent audit of calls led to a review and retraining session for assessment and care of fractures. This same process can apply to CPR or any other type of skill that you do not use often. Ensuring your skills and knowledge are current is one of the ongoing commitments of being an EMT.

Another function of the evaluation process is to determine ways to limit or eliminate human error. During the delivery of EMS, as with any occupation, there are times when errors can happen. Driving to the scene can be hazardous. As you are lifting and moving a patient, the patient can be dropped. Communicating with other EMTs or transferring the patient to the emergency department presents circumstances where errors can happen. Remember that errors can occur at any point during the call that can result in harm to the patient, public, and you.

Errors are not inevitable. If the circumstances of the errors are understood, it may be possible to eliminate or at least minimize them. There are many ways to examine medical errors. For the purpose of this text, errors have three possible sources. They can occur as a result of a rules-based failure, a knowledge-based failure, or a skills-based failure (or any combination of these). For example, does the EMT have the legal right to administer the particular medication needed by the patient? Has the medical director given permission to administer the drug? If not, a rules-based failure has occurred if the EMT assists with the administration. Does the EMT know all of the pertinent information about the medication being delivered? If not, a breakdown at this point, such as the administration of the wrong medication, would be referred to as a knowledge-based failure. Finally, is the equipment operating and being used properly? If not, a skills-based error has occurred. Any error can come from multiple sources.

How to limit errors requires work from both the EMS agency and EMS personnel. Agencies need to have clear protocols, which are detailed plans that describe how certain patient issues, such as chest pain or shortness of breath, are to be managed. These protocols need to be understood by all EMTs within the service.

The environment can be part of the reason for errors. Are there ways to limit distractions? How do we improve lighting so EMTs can see well? How organized is the equipment? Can the EMT find what he or she needs in a timely manner? Environmental considerations can be managed

You are the Provider: PART 4

The patient is placed onto the stretcher, placed in a position of comfort, and loaded into the ambulance. You and the paramedic are in the back with the patient as your EMT partner drives to the hospital. En route, the paramedic starts an intravenous (IV) line and administers pain medication to the patient. Shortly after the medication has been administered, you reassess the patient.

Recording Time: 12 Minutes

Level of consciousness	Conscious and alert
Respirations	16 breaths/min; regular and unlabored
Pulse	90 beats/min; strong and regular
Skin	Pink, warm, and dry
Blood pressure	120/62 mm Hg
Sao$_2$	97% (on room air)

Within a few minutes, the patient tells you that her back pain has subsided. She asks you if you think that her back pain could be a sign of a serious problem.

5. Is the paramedic required to contact medical control prior to administering any medications?

6. How should you answer the patient's question regarding her concern that she may have a serious condition?

using many approaches. Sometimes the solution is as easy as ensuring flashlights are available on all ambulances. Maybe this means having police assistance on certain types of EMS calls. Maybe an EMS supervisor can provide assistance. Perhaps a new type of equipment bag will provide better organization. Typically, when trying to reduce environmental factors regarding errors, this means having the right people with the right equipment in place.

EMTs themselves can also help to reduce errors. Your job is to protect the patient from harm and to deliver high-quality medical care. This is one of your most important responsibilities. You are a patient care advocate—you speak for patients on their behalf. Keeping this responsibility in mind will help you to limit errors.

There are other ways errors can be reduced. When you are about to perform a skill, ask yourself, "Why am I doing this?" Knowing the reason for your actions allows you time to reflect and make a more informed decision. Even within EMS, rarely do you have to act so quickly that you do not have a moment to consider what it is you are doing and why. If you have considered what to do and cannot come up with a solution, ask for help. Talk with your partner, contact medical control, or call your EMS supervisor.

Another way to help limit medical errors is to use "cheat sheets." Have a copy of your protocol book with you. Emergency physicians have many reference materials available to them. Physicians recognize they cannot memorize everything, so referencing a book or an Internet resource helps ensure the accuracy of their memory.

Finally, after a troublesome call, sit down and talk. Talk with your partner and/or your supervisor. Discussing the events that just happened provides an excellent avenue for learning. Your discussions can help lead to changes in protocol, how equipment is stocked, or even the purchase of new equipment.

■ Information Systems

EMS is not unlike any other profession in today's world. Without computers, the job would be much more difficult. An information system allows EMS providers to document the care that has been done. Once that information is stored electronically, it can be used to improve care. How many times has a department seen patients with chest pain? What is the average on-scene time for major trauma patients? How many AED runs has the department had? These questions and many more can be answered with the information from computerized medical records.

This information is now used for a variety of purposes. It is used to construct educational sessions for the department. Data from ambulance activity logs can be used to justify hiring more personnel. Examining the types of patients can provide the foundation for the purchase of new equipment and guide continuing education sessions. This information can also be combined with other computer resources, such as from a hospital, to determine patient outcome. Departments from around the country are sending information to Washington, DC, so a national snapshot of EMS can be obtained. The National EMS Information System (NEMSIS) information can be found at www.ems.gov. This information can be used to better plan for the needs of EMS systems today and in the future.

■ System Finance

All EMS departments need a funding system that allows them to continue to provide care; however, the type of system needed depends on many variables. There are several types of EMS departments around the country. *The Journal of Emergency Medical Services* annually reports on how EMS is delivered in the 200 largest cities within the United States. See **Table 1-4** for the breakdown of types of EMS services within the United States for the year 2008.

Table 1-4 Types of EMS Services that Transport Patients in the 200 Largest Cities Within the United States	
Type of Organization Providing EMS Transport Services	
Private organization	34.2%
Fire department	34.2%
City or county third service	18.0%
Hospital	5.4%
Public utility	2.7%
Public safety (police/fire/EMS)	1.8%
Law enforcement	0%
Source: Williams DM. 2008 JEMS 200-City Survey. *JEMS*. 2009, 32:36–51.	

These departments may have paid or volunteer personnel, or a mix of both. Financial resources are available for EMS departments through taxation, fee for service, paid subscription, donations, federal/state/local grants, fund-raisers, or combinations of same. Which financial system is used depends on the needs and makeup of each EMS department.

How are EMTs involved with the financial side of EMS? You may think the financial aspect belongs to those who work in the office. However, you may be asked to gather insurance information from patients or to get written permission from patients to bill their health insurance company. EMTs are involved in helping with fund-raisers, stuffing envelopes, or just making calls to potential subscribers to the service. Regardless of what type of system you will be working in, you will be helping the department secure its financial resources.

■ Education Systems

Your training will be conducted by many knowledgeable EMS educators. In most states, the instructors who are responsible for coordinating and teaching the EMT course and continuing education courses are approved and licensed by the state EMS office or agency. To be licensed in some states, an instructor must have extensive medical and educational training and teach for a designated period while being observed and supervised by an experienced instructor.

Generally, ALS training is provided in either a college, adult career center, or hospital setting. In most states, educational programs that provide ALS training must be approved by the state and have their own medical director. In these courses, many of the lectures and small group sessions are presented by the medical director or other physicians, nurses, and EMT instructors. In clinical sessions in which supervised practice is obtained in the emergency department or other in-hospital settings, students are also supervised directly by physicians and nurses.

The quality of care that you will provide depends on your ability and the quality of your training. Therefore, your instructor and the many others who developed and participated in your training program are key members of the emergency care team.

Once you no longer have the structured learning environment that is provided in your initial training course, you must assume responsibility for directing your own study and learning. As an EMT, you will be required to attend a certain number of hours of continuing education approved for EMTs each year to maintain, update, and expand your knowledge and skills. In many services, the required hours are provided by the training officer and medical director. In addition,

most EMS education programs and hospitals offer a number of regular continuing education opportunities in each region. You may also attend state and national EMS conferences to help keep you up-to-date about local, state, and national issues affecting EMS. Because there are many levels of licensing, you should ensure that the continuing education you receive is approved for the EMT. Whether you take advantage of these opportunities depends on you. You may decide to remain an EMT or you may want to achieve a higher level of training and certification, but whatever your choice is, the key to being a good EMT and providing high-quality care is your commitment to continual learning and increasing your knowledge and skills.

EMTs possess special knowledge and skills that are directed to the care of patients in emergency situations. The authority that is delegated to you to care for patients is a very special one. Maintaining your knowledge and skills is a substantial responsibility. Knowledge and skills that are learned in any profession weaken when they are not used on a continual basis. As an example, consider the steps involved in CPR. If you have not used these skills since your original training, it is unlikely you will perform CPR proficiently. Continuing education and refresher courses are measures you can take to maintain your skills and knowledge.

■ Prevention and Public Education

The next two components of the EMS system are often closely associated with each other. Prevention and public education are aspects of EMS where the focus is on public health. **Public health** examines the health needs of entire populations with the goal of preventing health problems. Although there are many definitions possible for public health, the prevention of health problems seems to provide a good overarching framework.

Health care in the United States is currently in a state of flux. The high-tech, on-demand style of care that is prevalent has two major drawbacks. One, it is very expensive. In the United States, more than 15% of the gross domestic product is accounted for by health care. Two, it may not deliver a better product. The US government reports that people born in the United States have an average life expectancy of 78 years. There are 49 other countries where people are living longer. If we are spending such large sums on health care, shouldn't we be living longer?

What needs to be addressed is the concept of prevention. Is it more expensive to treat a patient with a heart attack or to work with communities to help prevent the heart attack from happening? Or consider the scenario of an EMS provider working with a community to help get new traffic lights installed, thereby decreasing the incidence of vehicle collisions and subsequent injuries.

The concept of prevention can also apply to both the patient and the EMS provider. Eating right, exercising, and using other stress management techniques can help prevent medical emergencies. It may seem strange, but the goal of education should be to create an environment where the need for EMS is decreased.

The focus of the public health arm of health care is prevention. Public health works to prevent illness and injury, meaning being proactive. A good example of public health at work is the common product, salt. The next time you buy salt, look at the contents. In the United States, salt is sold with the additive iodine. It was discovered years ago that certain thyroid diseases, such as gout, are caused by a decrease in iodine levels within people's diets. The solution was to add this important element into a commonly used food source. Today, gout is rare within the United States. **Table 1-5** demonstrates other significant accomplishments of the public health system.

EMS is able to work with public health agencies on both primary and secondary prevention strategies. **Primary prevention** focuses on strategies that will prevent the event from ever happening. Polio was a devastating disease causing death and disability for thousands of Americans. It was discovered that a vaccine could be developed to prevent the disease. In the span of one generation, the disease was virtually eliminated. Vaccinations are a good example of primary prevention within public health.

In 2009, the World Health Organization declared the swine flu (H1N1) virus to be at pandemic levels, which meant that the virus had spread throughout the world. At the writing of this text, the Centers for Disease Control and Prevention has determined that the outbreak of this virus within the United States is limited. If a major outbreak of this virus were to occur in the United States, EMTs may be called on to assist in the administration of vaccinations. Other examples of primary prevention include ensuring people know the dangers of drinking and driving, and the harmful effects of using tobacco and other drugs.

In a **secondary prevention** strategy, the event has already happened. The question is how can we decrease the effects of the event? Helmets and seatbelts do not prevent the accident from happening, yet they do prevent serious injuries from occurring due to the accident. The next time you drive down a major roadway, take note of the construction of the guardrails. There have been significant changes in their construction over the years as more information has become available on what happens during a vehicle collision.

EMTs may also be involved in the surveillance of illnesses and injuries. The patient care reports that are generated by EMS personnel can be used to determine if a serious, widespread condition exists. For example, EMS is in a perfect position to provide statistical information to the local government about collisions. Injury surveillance data can be used to determine ways to improve a dangerous intersection, to prevent accidents from ever happening, or to limit the severity of injuries to drivers.

As discussed earlier, EMTs can help educate the public. People may not understand why an accident has happened. A parent allows her 15-month-old child to play outside with other children unsupervised. The child falls and cuts her hand. EMS arrives and the cause of the injury is obvious. EMTs can work with the parents professionally, respectfully, and kindly to help educate them on how to prevent this injury from occurring in the future.

The public may not understand the education that EMS providers have, and what services they can provide. EMTs can go to local schools and teach children to call 9-1-1 when there is a medical emergency. EMS personnel can work with local health care institutions to inform local residents when to call for an ambulance and when other transportation methods are more appropriate.

Teaching people how to perform CPR, how to help a choking victim, or even how to assist in the delivery of a baby are all aspects of public education. One of the important effects of public education is an increase in public respect for EMS. When people understand what it means to work on an ambulance and provide care to the sick and injured, they are more likely to consider EMS a vital part of the public health care system. This change in attitude can be powerful and lead to increased EMS funding and greater respect for EMS as a profession.

■ EMS Research

Why do EMTs perform the skills they do? How many ambulances does a city need? Should we remain on the scene and stabilize the patient or should we transport the patient rapidly? These questions and thousands more like them help determine the shape and impact of EMS on the community. The answers to these questions need to be derived from research. Unfortunately, many

Table 1-5 Examples of Public Health Accomplishments	
Vaccination programs	Clean drinking water
Fluoridation of water supplies	Seatbelt laws
Helmet laws	Tobacco use laws
Sewage systems	Restaurant inspections
Formation of the Food and Drug Administration	Prenatal screenings

of the tools and techniques that EMS providers use are borrowed from other health care settings without any research proving their effectiveness.

In the early days of EMS, it was believed that major trauma patients needed to be stabilized on the scene before they were transported. Paramedics would start IVs and use advanced airways. There was no foundation to support this behavior; it was assumed that this care needed to be done. After compiling significant amounts of prehospital EMS research, it was determined that major trauma patients needed to be transported to an operating room more than they needed an IV. Now EMS providers provide rapid transport of major trauma patients to trauma centers where they can get the surgical care they need. This is the power of EMS research. Evidence-based medical decision-making is the direction in which medical care is moving. It asks, can you prove that your therapy works?

EMTs will be involved in research typically through gathering data. You may be part of a study to determine how much oxygen should be given to patients with shortness of breath. You may be involved in a study to track the time it takes to get serious trauma patients to the emergency department. Your job will be to ensure you record all of the information about these patients carefully. The information gathered will then be analyzed by others to answer the question(s). The results could then be shared with the rest of the EMS community to change patient care practices. Traditional medical practice is based on such research.

Research can also be done at each EMS facility. EMS personnel can examine patient care records to determine where the department can improve. This information is then used to generate educational sessions for the EMTs or it can be used to plan public education/public prevention strategies. High-quality patient care should focus on procedures useful in improving patient outcomes through sound research.

It is important for EMS providers to stay up-to-date on the latest advances in medicine. Every 3 to 5 years, the American Heart Association unveils a revised set of guidelines based on large amounts of evidence. The American Heart Association is an excellent example of evidence-based medical decision-making in progress. These changes occur because more information is known. One word of caution: When reading new research, make sure you understand what the results mean.

Research information can be powerful, but it is often powerful within a very limited setting. A manufacturer of a defibrillator boasts that their new machine will terminate ventricular fibrillation on the first shock 95% of the time. On the basis of this information, you may want to immediately buy this new product. Terminating ventricular fibrillation is certainly a positive result, but does

You are the Provider: PART 5

The patient's condition has remained stable throughout transport, and she is now pain-free. After reassessing her, the paramedic asks you to call in your patient report to the receiving facility. Your estimated time of arrival is 8 minutes.

Recording Time: 19 Minutes	
Level of consciousness	Conscious and alert
Respirations	14 breaths/min; regular and unlabored
Pulse	70 beats/min; strong and regular
Skin	Pink, warm, and dry
Blood pressure	118/60 mm Hg
Sao$_2$	98% (on room air)

You deliver the patient to the emergency department in stable condition and give your verbal report to a staff nurse. The patient thanks you and your crew for taking such good care of her. You depart the hospital and return to service. On the way back to the station, the paramedic critiques your performance.

7. What is the purpose of an EMS call critique?

this defibrillator save more lives than other defibrillators? In this example, the manufacturer is reporting that the defibrillator is able to terminate ventricular fibrillation, not that the defibrillator is able to save more lives. People who do not examine the research will often make that hasty conclusion.

Be skeptical when reading research. Ask questions and conduct some of your own research. Conclusions that seem too good to be true are usually not true.

Roles and Responsibilities of the EMT

As an EMT, you will often be the first health care professional to assess and treat the patient; as such, you have certain roles and responsibilities Table 1-6 and are expected to possess certain attributes Table 1-7 . The guiding principle for EMS personnel is "everything you do needs to be done with the patient in mind." What is in the best interest of the patient? This is referred to as being a patient advocate.

Often, patient outcomes are determined by the care that you provide in the field and your identification of patients who need prompt transport. You are responsible for all aspects of EMS, from the preparation of the equipment to the delivery of care to providing a good example for others within the community.

Table 1-6 Roles and Responsibilities of the EMT

- Keep vehicles and equipment ready for an emergency.
- Ensure the safety of yourself, your partner, the patient, and bystanders.
- Emergency vehicle operation.
- Be an on-scene leader.
- Perform an evaluation of the scene.
- Call for additional resources as needed.
- Gain patient access.
- Perform a patient assessment.
- Give emergency medical care to the patient while awaiting the arrival of additional medical resources.
- Only move patients when absolutely necessary to preserve life.
- Give emotional support to the patient, the patient's family, and other responders.
- Maintain continuity of care by working with other medical professionals.
- Resolve emergency incidents.
- Uphold medical and legal standards.
- Ensure and protect patient privacy.
- Give administrative support.
- Constantly continue your professional development.
- Cultivate and sustain community relations.
- Give back to the profession.

Table 1-7 Professional Attributes of EMTs

Attribute	Description
Integrity	Consistency of action, a firm adherence to a code of honest behavior
Empathy	Being aware of and thoughtful toward the needs of others
Self-motivation	Being able to discover problems and solve them without someone directing you
Appearance and hygiene	Using your persona to project a sense of trust, professionalism, knowledge, and compassion
Self-confidence	The state of being where an EMT knows what he or she knows AND knows what he or she does not know; is able to ask for help
Time management	The ability to perform or delegate multiple tasks ensuring efficiency and safety
Communications	The ability to understand others and have them understand you
Teamwork and diplomacy	Being able to work with others; to know one's place within a team; to be able to communicate while giving respect to the listener
Respect	Placing others in high regard or importance; understanding that others are more important than you
Patient advocacy	Constantly keeping the needs of the patient at the center of care
Careful delivery of care	Paying attention to detail; making sure that what is being done for the patient is done as safely as possible

■ Professional Attributes

As an EMT, whether you are paid or a volunteer, you are a health care professional. Part of your responsibility is to make sure patient care is given a high priority without endangering your own safety or the safety of others. Another part of the responsibility to yourself, other EMTs, the patient, and other health care professionals is to maintain a professional appearance and manner at all times. Appearance, including uniforms, hair length, and tattoos, are usually regulated by the policies of your department **Figure 1-9**. Your attitude and behavior must reflect that you are knowledgeable and sincerely dedicated to serving anyone who is injured or in an acute medical emergency. A professional appearance and manner help to build confidence and ease the patient's anxiety. You will be expected to perform under pressure with composure and self-confidence. Patients and families who are under stress need to be treated with understanding, respect, and compassion.

Figure 1-9 **A.** A professional appearance and demeanor help build confidence and ease patient anxiety. **B.** An unprofessional appearance may promote distrust and incompetence.

Words of Wisdom

Professionalism extends beyond appearance and the activities you perform on a daily basis. As a professional, you have a responsibility to your partner, colleagues, patients, and profession to maintain a current level of knowledge.

Most patients will treat you with respect and appreciation, but some will not. Some patients are uncooperative, demanding, unpleasant, ungrateful, and verbally abusive. You must be nonjudgmental and overcome your instincts to react poorly to such behavior. Remember—when individuals are hurt, ill, under stress, frightened, despondent, under the influence of alcohol or drugs, or feel threatened, they will often react with inappropriate behavior, even toward those who are trying to help and care for them. Every patient, regardless of his or her attitude, is entitled to compassion, respect, and the best care that you can provide.

Most individuals in this country can obtain proper routine medical care when they are ill and are surrounded by relatives and friends who will help to take care of them. However, when you are called to a home for a medical problem that is clearly not an emergency, remember that for some individuals, calling an ambulance and being transported to the emergency department is the only way to obtain medical care.

As a new EMT, you will be given a lot of advice and training from the more experienced EMTs with whom you serve. Some may voice a callous disregard for some types of patients. You should not be influenced by the unprofessional attitude of these individuals, regardless of how experienced or skilled they appear.

As a health care professional and an extension of physician care, you are bound by patient confidentiality. You should not discuss your findings or any disclosures made by the patient with anyone but those who are treating the patient or, as required by law, the police or other social agencies. When discussing a call with others, you should be careful to avoid revealing any information that might disclose the name or identity of patients you have treated. Be careful not to gossip about calls and patients with others, even in your own home. The protection of patient privacy has drawn national attention with the passage of the **Health Insurance Portability and Accountability Act (HIPAA)**. You should be familiar with the requirements of this legislation, especially as it applies to your particular practice.

You are the Provider: SUMMARY

1. Do the roles and responsibilities of an EMT differ from those of a paramedic? If so, how?

The fundamental roles and responsibilities—providing *safe and effective* emergency medical care to the sick and injured, and transporting patients to an appropriate medical facility—are the same for all levels of EMS provider. Safe emergency medical care means that the treatment you provide will not cause further harm to the patient; it also means that you will not harm yourself, your partner, or bystanders in the process. Effective emergency medical care means that the treatment you provide, which is based on an accurate assessment, will have the desired effects—to stabilize the patient's condition and prevent it from getting worse.

The only difference between the EMT and paramedic is the level of care that is provided to the patient. EMTs have a fundamental knowledge of emergency care and provide basic life support (BLS), such as cardiopulmonary resuscitation, bleeding control, bandaging and splinting, and basic airway management. Paramedics have a comprehensive knowledge of emergency medical care, which is built on a solid knowledge of BLS, and provide advanced life support interventions, such as advanced airway management, cardiac monitoring, and medication administration. It is important to note that the quality of patient care the EMT provides is no different from that of the paramedic—just the level.

2. Is there a difference between what you learned in your EMT class and the care you provide in the field?

Your education and training is intended to prepare you to function as an entry-level competent EMT; therefore, your education should reflect the current practice of prehospital emergency medical care. The cognitive knowledge and psychomotor skills learned in the classroom are concepts that you will apply when caring for patients in the field. With experience and contact with many patients experiencing a variety of injuries and illnesses, your ability to apply the concepts learned in the classroom will be enhanced. You will develop your own "routine" regarding your general approach to patient care; however, you must be able to alter your routine based on the situation and the needs of the patient. Experience also enhances your critical thinking abilities. Critical thinking is a complex combination of skills; it includes the following characteristics:

- **Rationality:** Relying on reason rather than emotion; requiring evidence; ignoring no evidence; following evidence where it leads; and being more concerned about finding the best explanation than about being right; analyzing apparent confusion; and asking questions
- **Self-awareness:** Weighing the influences of your motives and biases and recognizing your own assumptions, prejudices, biases, or point of view
- **Open-mindedness:** Evaluating all reasonable inferences; considering a variety of possible viewpoints or perspectives; remaining open to alternative interpretations; accepting new explanations because they explain the evidence better, are simpler, or have fewer inconsistencies; accepting new priorities in

response to reevaluation of the evidence; and not rejecting unpopular views as out of hand

- **Judgment:** Recognizing the relevance and/or merit of alternative assumptions and perspectives and recognizing the extent and weight of evidence
- **Discipline:** Precise, meticulous, comprehensive, and exhaustive; resisting manipulation and irrational appeals; and avoiding snap judgments

3. Is this patient experiencing a "true emergency?"

An emergency can be defined as any event or situation that requires immediate intervention to minimize or prevent serious injury or death. An emergency to one person may not be an emergency to another. You must recognize that people call EMS when they perceive their situation as an emergency. In the interest of the patient, you should assume that an emergency exists unless a thorough and accurate assessment yields otherwise.

Although your primary assessment has not revealed any gross (obvious) life-threatening conditions, this does not mean that she is not sick or injured. Do not let the fact that she has been experiencing pain for the past month distract you from her present condition or situation; remember that she called 9-1-1 for a reason. Further assessment will be required to determine if she is experiencing an occult (hidden; not obvious) condition that may require immediate intervention. The EMT's job is to take care of patients whether a "true emergency" exists or not.

4. The patient does not appear to be experiencing any life-threatening conditions. Should you transport her to the hospital?

The patient has requested that you transport her to the hospital; therefore, you are legally obligated to do so. If you refuse to transport her, you could be held liable for abandonment. The absence of any obvious life-threatening conditions does not mean that she does not require further medical evaluation and treatment. She has been experiencing back pain for a month; this could indicate a serious underlying problem (ie, kidney disease or cancer) that can only be diagnosed in a hospital. You will often encounter patients who are not experiencing any life-threatening conditions but still require EMS treatment—even if it is just supportive—and transport to the hospital.

5. Is the paramedic required to contact medical control prior to administering any medications?

It depends on the EMS system's protocols. Some EMS protocols require prior contact before administering certain medications; others do not. Medical control is either off-line (indirect) or online (direct), as authorized by the EMS medical director. Online medical control consists of direction given over the phone or radio directly from the medical director or designated physician. Off-line medical control consists of standing orders, interventions that do not require prior contact with medical control, as authorized by the medical director. In this case, the paramedic started an IV line and

You are the Provider: SUMMARY, continued

administered medication without contacting medical control first. This indicates that he or she had standing orders to do so. The paramedic recognized the need for pain medication, has been appropriately educated and trained on the medication and how to administer it, and has been authorized by the medical director to administer it at his or her discretion. It is important to note that just because your EMS system has standing orders for certain interventions, you should always contact medical control if you have any questions or concerns or need advice. Be familiar with your EMS system's protocols.

6. How should you answer the patient's question regarding her concern that she may have a serious condition?

Honesty is a critical attribute of any EMS provider. Lying to a patient and providing false hope and reassurance is unethical and inhumane. In many cases, the most honest answer to a question is "I don't know." Do not speculate—based solely on your assessment—and tell her that she does or does not have a serious condition; you do not have the diagnostic equipment and resources needed to come to *any* conclusions. If you do not know the answer to a patient's question, do not be afraid to say "I don't know." Follow this up by reassuring her that you will give her the best medical care possible and that the physician at the hospital will do the same. Patients deserve to hear the truth; never tell them otherwise.

7. What is the purpose of an EMS call critique?

The purpose of an EMS call critique is to provide feedback regarding how you cared for the patient and met his or her physical and emotional needs. It should not be punitive or demeaning; it is an educational tool that will enable you to enhance your patient care skills. EMTs must be open to constructive criticism; this is how they learn and become more proficient emergency care providers. Informal, one-on-one critiques, such as what the paramedic is conducting with you after the call, are ideal learning opportunities because information about the call is still fresh in your mind. Formal critiques, such as those that are conducted as part of the EMS continuous quality improvement (CQI) process, are designed to ensure that safe and effective patient care is consistently provided by all EMS providers in the system. To provide CQI, periodic run review meetings are held in which all those who are involved in patient care review patient care reports and discuss any areas of care that appear to need change or improvement. Positive feedback should also be provided. If a problem appears to be repeated by a single EMT or crew, the medical director will discuss the details with the individual(s) involved. If deemed necessary by the medical director, he or she may assign remedial training or develop some other educational activity. Many EMS systems have a designated person, who is assigned by the medical director, to carry out these tasks.

EMS Patient Care Report (PCR)

Date: 3-23-09	**Incident No.:** 010109	**Nature of Call:** Back pain		**Location:** 325 Blossom Ave.
Dispatched: 0720	**En Route:** 0720	**At Scene:** 0723	**Transport:** 0735	**At Hospital:** 0750	**In Service:** 0801

Patient Information

Age: 59 **Sex:** F **Weight (in kg [lb]):** 64 kg (141 lb)	**Allergies:** None **Medications:** Ibuprofen **Past Medical History:** None **Chief Complaint:** Back pain

Vital Signs

Time: 0727	**BP:** 126/66	**Pulse:** 88	**Respirations:** 16	**Sao$_2$:** 99%
Time: 0735	**BP:** 120/62	**Pulse:** 90	**Respirations:** 16	**Sao$_2$:** 97%
Time: 0742	**BP:** 118/60	**Pulse:** 70	**Respirations:** 14	**Sao$_2$:** 98%

EMS Treatment (circle all that apply)

Oxygen @ __ L/min via (circle one): NC NRM Bag-Mask Device	**Assisted Ventilation**	**Airway Adjunct**	**CPR**
Defibrillation	**Bleeding Control**	**Bandaging**	**Splinting**	**Other** (IV line, pain medication by paramedic)

You are the Provider: SUMMARY, continued

Narrative
Dispatched for a 59-year-old female with back pain. On arrival at the scene, found the patient sitting on the couch in her living room. She was conscious and alert; her airway was patent; and her breathing was adequate. Patient complained of lower back pain, which has been present for the past month. She denied injuring her back; she further denied any other symptoms or past medical history. Medications include ibuprofen for pain. Assessment of patient's back revealed no gross evidence of deformity, swelling, or bruising. Pulse, sensory, and motor functions were grossly intact in all extremities. The patient stated that she has not been evaluated by a physician for her back pain; however, because it has progressively worsened, she called 9-1-1. Obtained vital signs, placed patient onto stretcher and placed her in position of comfort, loaded her into the ambulance, and began transport to the hospital. En route, paramedic started IV line and administered analgesia. Shortly after analgesia was administered, patient expressed relief of her pain. Reassessment revealed that she remained conscious and alert with stable vital signs. Provided reassurance and reassessment throughout remainder of transport. Delivered patient to emergency department without incident and gave verbal report to staff nurse. **End of report**

Prep Kit

- The standards for prehospital emergency care and the individuals who provide it are governed by the laws in each state and are typically regulated by a state office of EMS.

- The EMS ambulance is staffed by EMTs who have been trained to the emergency medical technician, advanced EMT (AEMT), or paramedic level according to recommended national standards and have been licensed by the state.

- An EMT has training in basic emergency care skills, including automated external defibrillation, use of airway adjuncts, and assisting patients with certain medications.

- An AEMT has training in specific aspects of advanced life support (ALS), such as intravenous therapy and the administration of certain emergency medications.

- A paramedic has extensive training in ALS, including endotracheal intubation, emergency pharmacology, cardiac monitoring, and other advanced assessment and treatment skills.

- Emergency medical responders, such as law enforcement officers, fire fighters, park rangers, ski patrollers, or other organized rescuers often arrive at the scene before the ambulance and EMTs.

- After the EMTs size up the scene and assess the patient, they provide the emergency care and transport that is indicated based on their findings and ordered by their medical director in the service's standing orders and protocols or by the physician who is providing online medical direction.

- The National EMS Scope of Practice Model, developed by the NHTSA, provides overarching guidelines as to what skills each level of EMS provider should be able to accomplish.

- The *EMS Agenda for the Future* is a multidisciplinary, national review of all aspects of EMS delivery that encourages the creation of systems that help to protect the well-being of EMS providers. It includes 14 components that make up an EMS system.

- You will often be the first health care professional to assess and treat the patient; as such, you have certain roles and are expected to possess certain attributes.

- EMT attributes include compassion and motivation to reduce suffering, pain, and mortality in those who are injured or acutely ill; a desire to provide each patient with the best possible care; commitment to obtain the knowledge and skills that this position requires; and the drive to continually increase your knowledge, skills, and ability.

- The EMT course that you are now taking will present the information and skills that you will need to pass the required certification examination needed to become a licensed EMT.

- Once you have completed the course, you must assume responsibility for directing your own study through continuing education provided by your service's training officer and medical director or through other opportunities available to you. Your commitment to continued learning is the key to being a good EMT.

- As a health care professional and an extension of physician care, you are bound by patient confidentiality.

Prep Kit, continued

Vital Vocabulary

advanced EMT (AEMT) An individual who has training in specific aspects of advanced life support, such as intravenous therapy, and the administration of certain emergency medications.

advanced life support (ALS) Advanced lifesaving procedures, some of which are now being provided by the EMT.

Americans With Disabilities Act (ADA) Comprehensive legislation that is designed to protect individuals with disabilities against discrimination.

automated external defibrillator (AED) A device that detects treatable life-threatening cardiac arrhythmias (ventricular fibrillation and ventricular tachycardia) and delivers the appropriate electrical shock to the patient.

certification A process in which a person, an institution, or a program is evaluated and recognized as meeting certain predetermined standards to provide safe and ethical care.

continuous quality improvement (CQI) A system of internal and external reviews and audits of all aspects of an EMS system.

emergency medical dispatch (EMD) A system that assists dispatchers in selecting appropriate units to respond to a particular call for assistance and in providing callers with vital instructions until the arrival of EMS crews.

emergency medical responder (EMR) The first trained individual, such as a police officer, fire fighter, lifeguard, or other rescuer, to arrive at the scene of an emergency to provide initial medical assistance.

emergency medical services (EMS) A multidisciplinary system that represents the combined efforts of several professionals and agencies to provide prehospital emergency care to the sick and injured.

emergency medical technician (EMT) An individual who has training in basic life support, including automated external defibrillation, use of a definitive airway adjunct, and assisting patients with certain medications.

Health Insurance Portability and Accountability Act (HIPAA) Federal legislation passed in 1996. Its main effect in EMS is in limiting availability of patients' health care information and penalizing violations of patient privacy.

intravenous (IV) therapy The delivery of medication directly into a vein.

licensure The process whereby a state allows individuals to perform a regulated act.

medical control Physician instructions that are given directly by radio or cell phone (online/direct) or indirectly by protocol/guidelines (off-line/indirect), as authorized by the medical director of the service program.

medical director The physician who authorizes or delegates to the EMT the authority to provide medical care in the field.

National EMS Scope of Practice Model A document created by the National Highway Traffic Safety Administration (NHTSA) that outlines the skills performed by various EMS providers.

paramedic An individual who has extensive training in advanced life support, including endotracheal intubation, emergency pharmacology, cardiac monitoring, and other advanced assessment and treatment skills.

primary prevention Efforts to prevent an injury or illness from ever occurring.

primary service area (PSA) The designated area in which the EMS service is responsible for the provision of prehospital emergency care and transportation to the hospital.

public health Focused on examining the health needs of entire populations with the goal of preventing health problems.

public safety access point A call center, staffed by trained personnel who are responsible for managing requests for police, fire fighting, and ambulance services.

quality control The responsibility of the medical director to ensure that the appropriate medical care standards are met by EMTs on each call.

secondary prevention Efforts to limit the effects of an injury or illness that you cannot completely prevent.

Assessment in Action

You are a new EMT who is employed by a major metropolitan fire-rescue department. It is your first day on the job and you are on your way to the station. You prepare by reflecting on some of the information that you learned in the classroom.

1. Which of the following agencies is the federal source for the EMT curriculum?
   A. Department of Health and Human Services
   B. National Highway Traffic Safety Administration
   C. Federal Emergency Management Agency
   D. Department of Transportation

2. The EMT's primary job is to:
   A. provide appropriate medical care.
   B. diagnose the patient's condition.
   C. ensure personal safety.
   D. provide transport to the closest hospital.

3. Explain the role of the "White Paper" in the development of EMS.

4. According to the National EMS Scope of Practice, EMTs are able to use which of the following airway adjuncts?
   A. Nasal cannula
   B. BIPAP\CPAP
   C. Oral endotracheal tube
   D. Positive end-expiratory pressure

5. Prior to going on duty the EMT should:
   A. check all of the equipment and supplies.
   B. ensure that the ambulance is fully fueled.
   C. test each of the driver's controls.
   D. all of the above.

6. Which of the following is an example of online medical control?
   A. Written protocols
   B. Standing orders
   C. Radio communication with the hospital
   D. Training exercises

7. To provide quality control, the medical director and other assigned staff need to do which of the following?
   A. Review patient care reports
   B. Audit administrative records
   C. Survey patients
   D. All of the above

8. What is continuous quality improvement (CQI) and how is it used to help ensure the safety of patients?

9. Why is EMS research a vital part in the evolution of management of trauma patients?

Workforce Safety and Wellness

National EMS Education Standard Competencies

Medicine

Applies fundamental knowledge to provide basic emergency care and transportation based on assessment findings for an acutely ill patient.

Infectious Diseases

Awareness of
- How to decontaminate equipment after treating a patient (pp 42-43)

Assessment and management of
- How to decontaminate the ambulance and equipment after treating a patient (pp 42-43)

Preparatory

Applies fundamental knowledge of the emergency medical services (EMS) system, safety/well-being of the emergency medical technician (EMT), medical/legal, and ethical issues to the provision of emergency care.

Workforce Safety and Wellness

- Standard safety precautions (pp 35-40)
- Personal protective equipment (pp 36-40)
- Stress management (pp 45-52)
  - Dealing with death and dying (pp 55-60)
- Prevention of response-related injuries (pp 62-67)
- Prevention of work-related injuries (pp 62-67)
- Lifting and moving patients (p 50)
- Disease transmission (pp 33-35)
- Wellness principles (pp 48-52)

Knowledge Objectives

1. Define "infectious disease" and "communicable disease." (p 33)
2. Describe the routes of disease transmission. (pp 33-35)
3. Understand the standard precautions that are used in treating patients to prevent infection. (pp 35-40)
4. Describe the steps to take for personal protection from airborne and bloodborne pathogens. (pp 35-40)

5. Understand the mode of transmission and the steps to prevent and/or deal with an exposure to hepatitis, tuberculosis, and HIV/AIDS. (pp 33-45)
6. Understand how immunity to infectious diseases is acquired. (pp 43-45)
7. Explain postexposure management of exposure to patient blood or body fluids, including completing a postexposure report. (p 45)
8. Understand the physiologic, physical, and psychological responses to stress. (pp 45-47)
9. Describe posttraumatic stress disorder (PTSD) and steps that can be taken, including critical incident stress management, to decrease the likelihood that PTSD will develop. (pp 46-47)
10. State the steps that contribute to wellness and their importance in managing stress. (pp 48-52)
11. Discuss workplace issues such as cultural diversity, sexual harassment, and substance abuse. (pp 52-54)
12. Understand the emotional aspects of emergency care. (pp 54-55)
13. Describe issues concerning care of the dying patient, death, and the grieving process of family members. (pp 55-58)
14. Understand the care of critically ill and injured patients. (pp 58-60)
15. Recognize the stress inherent in many situations, such as mass-casualty scenes. (pp 61-62)
16. Describe the steps necessary to determine scene safety and to prevent work-related injuries at the scene. (pp 62-67)
17. Discuss the different types of protective clothing worn to prevent injury. (pp 67-70)
18. Recognize the possibility of violent situations and the steps to take to deal with them. (pp 70-71)
19. Describe how to handle behavioral emergencies. (p 71)

Skills Objectives

1. Demonstrate proper handwashing techniques. (pp 35-37, Skill Drill 2-1)
2. Demonstrate how to properly remove gloves. (pp 36-38, Skill Drill 2-2)
3. Demonstrate the necessary steps to take to manage a potential exposure situation. (pp 40-42, Skill Drill 2-3)

Introduction

There is an ancient proverb, "Physician, heal thyself." As providers of health care, doctors need to look after themselves—in all respects—so that they can minister to others. An ill physician is in no position to render care as he or she was trained to do. That dictum applies to all health care providers and goes well beyond just physical issues. In caring for the critically ill and injured, there are many factors and situations that can interfere with an EMT's ability to treat the patient.

The personal health, safety, and well-being of all EMTs are vital to an EMS operation. As part of your training, you will learn how to recognize possible hazards and protect yourself from them. These hazards vary greatly, ranging from personal neglect to environmental and human-made threats to your health and safety. You will also learn how to cope with the mental and physical stress that result from caring for the sick and injured. Death and dying issues challenge you to deal with the realities of human weaknesses and the emotions of the survivors.

The emotional well-being of the EMT and the patient are intertwined, especially in high-stress rescues. This chapter covers both caring for the well-being of the patient and caring for yourself.

Infectious Diseases

As an EMT, you will be called on to treat and transport patients with a variety of infectious or communicable diseases. An **infectious disease** is a medical condition caused by the growth and spread of small, harmful organisms within the body. A **communicable disease** is a disease that can be spread from one person or species to another. Immunizations, protective techniques, and simple handwashing can dramatically minimize the health care provider's risk of **infection**. When these protective measures are used, the risk of the health care provider contracting a serious disease is negligible.

Words of Wisdom

All bloodborne diseases are communicable diseases but not all communicable diseases are bloodborne. For example, the **human immunodeficiency virus (HIV)** is both communicable and bloodborne, but chickenpox is communicable but not bloodborne.

Routes of Transmission

Whereas all infections result from an abnormal invasion of body spaces and tissues by germs, different germs use different means of attack, or mechanisms of transmission. **Transmission** is the way an infectious disease is spread. There are several ways infectious diseases can be transmitted, consisting of contact (direct or indirect), airborne, foodborne, and vector-borne (transmitted through insects or parasitic worms).

Contact transmission is the movement of an organism from one person to another through physical touch. There are two types of contact transmission: direct and indirect. **Direct contact** occurs when an organism is moved from one person to another through touching without any intermediary.

The scenario of a vehicle collision can help you understand how transmission occurs through direct contact. The driver of the vehicle has **hepatitis** B and is bleeding from an arm injury. The EMT caring for the patient is not wearing gloves and has a small unnoticed cut on his hand. As he handles a bloody dressing, the hepatitis

You are the Provider: PART 1

You have been working a regular EMS shift—24 hours on and 48 hours off—since becoming a certified EMT less than 6 months ago. You receive a call at 7:20 AM to 788 East Radcliffe for an unconscious child who is not breathing. You and your paramedic partner respond to the scene; an emergency medical responder (EMR) unit is dispatched at the same time. This is your first call involving a critically ill child.

1. How can you psychologically prepare yourself for this call?

virus can move from the victim's blood on the dressing into the EMT's body through the cut on his hand, thus infecting him Figure 2-1 . This is an example of direct contact where blood is the vehicle. **Bloodborne pathogens** are microorganisms that are present in human blood and can cause disease in humans. Another example of direct contact is sexual transmission. Patients who are infected with the human immunodeficiency virus (HIV) can transfer the virus to their partners during sex.

Words of Wisdom

Fungi are small, plant-like organisms, such as yeast. Fungi cause many common conditions such as athlete's foot and jock itch. Protozoa are single-cell animal-like microorganism. Protozoa cause malaria. Helminths are worms such as roundworms, pinworms, and hookworms. These worms are parasites that can infect people and cause serious health problems.

Indirect contact involves the spread of infection between the patient with an infection to another person through an inanimate object. The object that transmits the infection is called a fomite. Using the same patient from the example above, the EMT wore gloves. As the EMT was caring for the patient, blood got onto the ambulance stretcher. If the stretcher is not correctly cleaned afterwards, the virus remains on the stretcher and can be transmitted to someone else days later.

Needlesticks are another example of the spread of infection through indirect contact. In this case, the virus moves from the patient to the needle to the health care provider. This route of transmission was common many years ago before the advent of safety equipment such as needleless IV systems.

Airborne transmission involves spreading an infectious agent through mechanisms such as droplets or dust. The common cold is moved from person to person by coughing and sneezing. Interestingly, when a person sneezes, the moisture from the airway moves forcefully and quickly through a narrow opening. If the moisture droplets are large, they travel short distances and can be involved in direct contact transmission. If the moisture droplets are very small, they are turned into an aerosol and can now float in the air for long distances. Sneezing actually can transmit disease through direct contact and airborne routes Figure 2-2 .

Because of airborne transmission, it is unsanitary to use your hands to cover a cough or sneeze because the organism travels onto your hands. If you then touch a telephone, doorknob, or a patient, the organisms will travel. Using a tissue when coughing or sneezing is better for controlling the spread of organisms, but you then have a piece of paper full of organisms. One of the best techniques to avoid contaminating your hands is to cough or sneeze into your arm/sleeve. Since you do not touch objects with your inner arms, the risk of moving the organism to an object or person is reduced Figure 2-3 . The organisms are trapped in the fabric and will eventually die.

Foodborne transmission involves the contamination of food or water with an organism that can cause disease. When food is prepared, it is important to ensure that raw meats do not come into contact with other foods to prevent the spread of bacteria. It is also important that food is prepared and stored properly at all times to minimize the possibility of illness. Proper cleaning of food preparation surfaces before and after use also helps to decrease the likelihood of transmitting foodborne bacteria.

Vector-borne transmission involves the spread of infection by animals or insects that carry an organism from one person or place to another. The Black Death in Europe and Asia in the Middle Ages killed more than

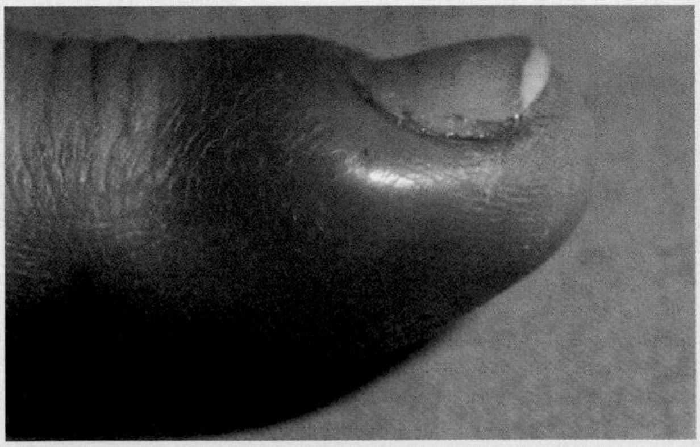

Figure 2-1 Finger infection resulting from not wearing gloves during patient contact.

Figure 2-2 Coughing and sneezing create droplets and aerosols.

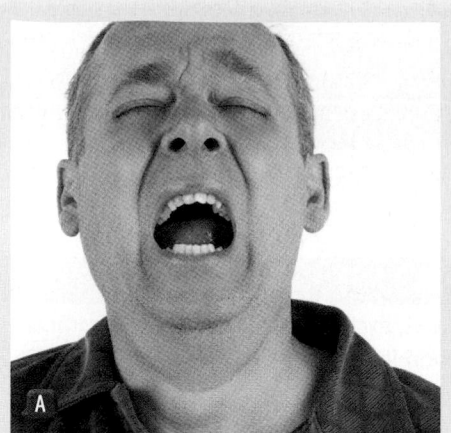

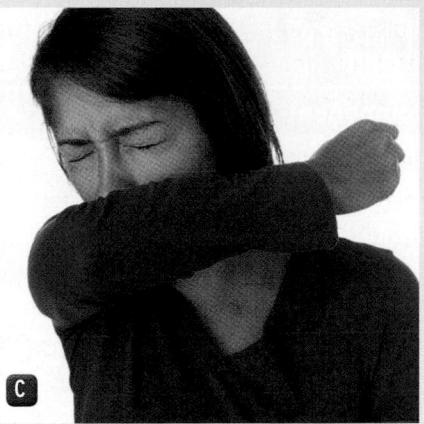

Figure 2-3 Coughing/sneezing techniques. **A.** Poor coughing/sneezing technique. **B.** Acceptable coughing/sneezing technique. **C.** Best coughing/sneezing technique.

25 million people. This disease is thought to have been caused by a flea that lived on rats. As the rats moved, so did their fleas, carrying the bubonic plague. Other vector-borne diseases include rabies and Lyme disease.

Risk Reduction and Prevention for Infectious and Communicable Diseases

Standard Precautions

The **Occupational Safety and Health Administration (OSHA)** develops and publishes guidelines concerning reducing risk in the workplace. It is also responsible for enforcing these guidelines. OSHA requires all EMTs to be trained in handling bloodborne **pathogens** and in approaching the patient who may have an infectious or communicable disease. Training must also be provided for issues including blood and body fluid precautions, airborne precautions, and **contamination** precautions.

Because health care workers are exposed to so many different kinds of infections, the **Centers for Disease Control and Prevention (CDC)** developed a set of **standard precautions** for health care workers to use in treating patients. These protective measures are designed to prevent health care workers from coming into contact with germs carried by patients. The CDC recommendation from 2007 is to assume that every person is potentially infected or can spread an organism that could be transmitted in the health care setting; therefore you must apply **infection control** procedures to reduce infection in patients and health care personnel. **Table 2-1** summarizes the CDC recommendations. You must also notify your **designated officer** if you are exposed.

Proper Hand Hygiene

Proper handwashing is the simplest yet most effective way to control disease transmission. You should always wash your hands before and after contact with a patient, even if you wear gloves. The longer the germs remain with you, the greater their chance of getting through your barriers. Although soap and water are not protective in all cases, in certain cases their use provides excellent protection against further transmission from your skin to others.

If no running water is available, you may use waterless handwashing substitutes **Figure 2-4**. If you use a waterless substitute in the field, make sure you wash your hands using soap and water at the hospital.

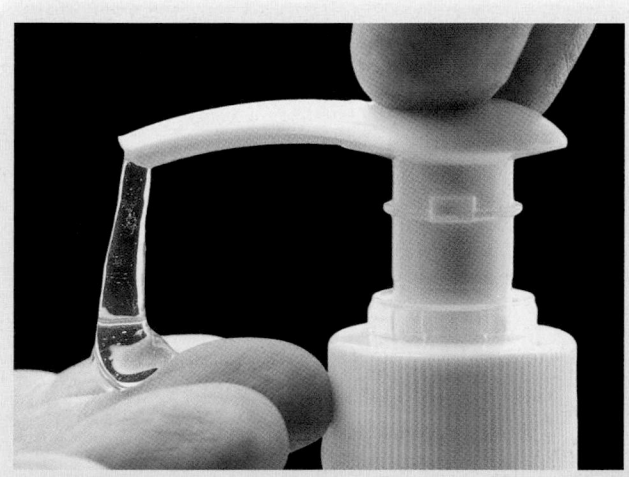

Figure 2-4 Use a waterless handwashing solution if there is no running water available. Be sure to wash your hands with soap and water once you arrive at the hospital.

Table 2-1 Standard Precautions for the Care of All Patients in All Health Care Settings, Centers for Disease Control and Prevention 2007

Component	Recommendation
Hand hygiene	▪ After touching blood, body fluids, secretions, excretions, or contaminated items ▪ Immediately after removing gloves ▪ Between patient contacts
Personal Protective Equipment (PPE)	
Gloves	▪ For touching blood, body fluids, secretions, excretions, or contaminated items ▪ For touching mucous membranes and non-intact skin
Gown	▪ During procedures and patient care activities when contact of the EMT's clothing/exposed skin to blood, body fluids, secretions, excretions, or contaminated items is anticipated
Mask, eye protection, face shield	▪ During procedures and patient care activities likely to generate splashes or sprays of blood, body fluids, secretions, or excretions. Examples include suctioning or endotracheal intubation
Patient Care Environment	
Soiled patient care equipment	▪ Handle in a manner that prevents transfer of microorganisms to others and to the environment ▪ Wear gloves if visibly contaminated ▪ Hand hygiene
Environmental controls	▪ Have procedures for the routine care, cleaning, and disinfection of environmental surfaces ▪ Special attention to frequently touched surfaces within the ambulance (handrails, seats, cabinets, doors)
Textiles and laundry	▪ Handle in a manner that prevents transfer of microorganisms to others and to the environment
Needles and other sharp objects	▪ Do not recap, bend, break, or hand-manipulate used needles ▪ Use safety features when available (needleless IV systems) ▪ Place sharps in puncture-resistant containers
Special Circumstances	
Patient resuscitation	▪ Use mouthpiece, resuscitation bag, or other ventilation devices to prevent contact with mouth and oral secretions
Respiratory hygiene/cough etiquette	▪ Instruct symptomatic patients to cover mouth/nose when sneezing or coughing ▪ Use tissues and dispose in no-touch receptacle ▪ Perform hand hygiene after touching tissues ▪ Place surgical mask on patient/provider ▪ If mask cannot be used, maintain special separation (> 3′) if possible

Follow the steps in **Skill Drill 2-1** for proper hand-washing:

1. Apply soap to hands.
2. Rub your hands together for at least 20 seconds to work up a lather. Pay particular attention to your fingernails.
3. Rinse your hands using warm water **Step 1**.
4. Dry your hands with a paper towel **Step 2**, and use the paper towel to turn off the faucet.

▪ Gloves

Gloves and eye protection are the minimum standard for all patient care if there is any possibility for **exposure** to blood or body fluids. Both vinyl and latex gloves provide adequate protection. Your department may prefer one

Skill Drill 2-1

Handwashing

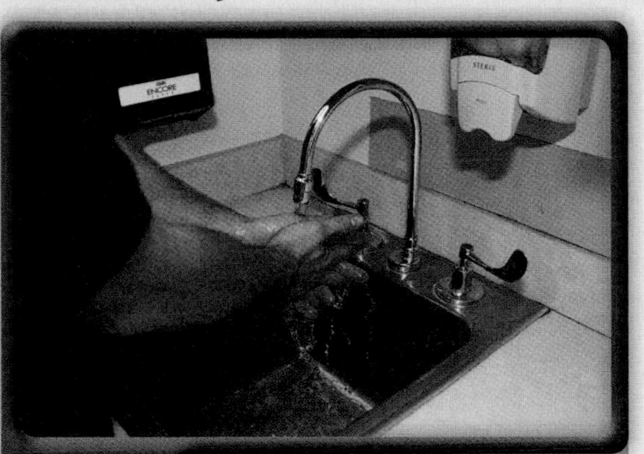

Step 1 Apply soap to hands. Rub hands together to work up a lather. Rinse both hands using warm water.

Step 2 Dry with a paper towel.

type of glove over the other, or you may choose yourself. You should evaluate each situation and choose the glove that works best. (Some individuals are allergic to latex. If you suspect that you are allergic, consult your supervisor for options.) Vinyl gloves may be best for routine procedures, and latex gloves may be best for invasive procedures. Change latex gloves if they have been exposed to motor oil, gasoline, or any petroleum-based product. Do not use petroleum jelly with latex gloves. Wear double gloves if there is substantial bleeding. You may also wear double gloves if you will be exposed to large volumes of other body fluids. Be sure to change gloves as you move from patient to patient. For cleaning and disinfecting the unit, you should use heavy-duty utility gloves Figure 2-5 . You should never use lightweight latex or vinyl gloves for cleaning.

Removing used latex or vinyl gloves requires a methodical technique to avoid contaminating yourself with the materials from which the gloves have protected you Skill Drill 2-2 .

Skill Drill 2-2

1. Begin by partially removing one glove. With your other gloved hand, pinch the first glove at the

wrist—being certain to touch only the outside of the first glove—and start to roll it back off your hand, inside out. Leave the exterior of the fingers on that first glove exposed Step 1 .

2. Use the still-gloved fingers of the first hand to pinch the wrist of the second glove and begin to

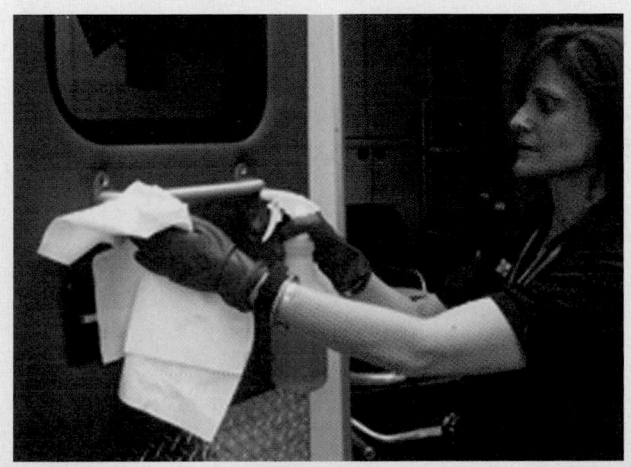

Figure 2-5 Use heavy-duty utility gloves to clean the unit. You should never use lightweight latex or vinyl gloves for cleaning.

pull it off, rolling it inside-out toward the fingertips as you did with the first glove (Step 2).

3. Continue pulling the second glove off until you can pull the second hand free (Step 3).

4. With your now-ungloved second hand, grasp the exposed inside of the first glove and pull it free of your first hand and over the now-loose second glove. Be sure that you touch only clean, interior surfaces with your ungloved hand (Step 4).

Gloves are the most common type of **personal protective equipment (PPE)**. In many EMS rescue operations, you must also protect your hands and wrists from injury. You may wear puncture-proof leather gloves, with latex gloves underneath. This combination will allow you free use of your hands with added protection from blood and body fluids. Remember that latex or vinyl gloves are considered medical waste and must be disposed of properly. Also remember that many patients

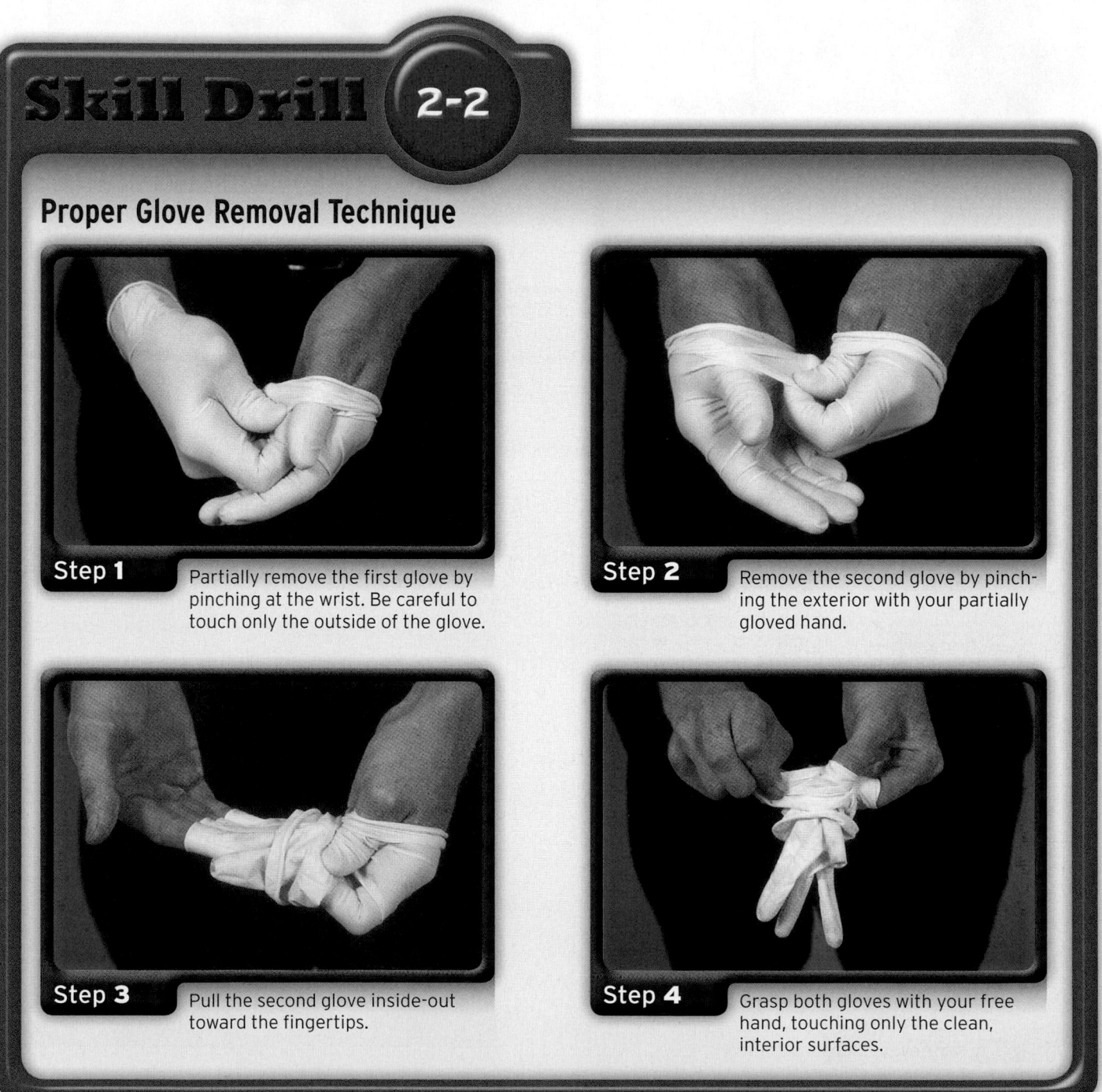

Skill Drill 2-2

Proper Glove Removal Technique

Step 1 Partially remove the first glove by pinching at the wrist. Be careful to touch only the outside of the glove.

Step 2 Remove the second glove by pinching the exterior with your partially gloved hand.

Step 3 Pull the second glove inside-out toward the fingertips.

Step 4 Grasp both gloves with your free hand, touching only the clean, interior surfaces.

have life-threatening allergies to latex gloves. Leather gloves must be treated as contaminated material until they can be properly decontaminated.

■ Gowns

Occasionally, you may need to wear a gown. A gown provides protection from extensive blood splatter. Gowns may be worn in situations such as field delivery of a baby or major trauma. However, wearing a gown may not be practical in many situations. In fact, in some instances, a gown may pose a risk for injury. Your department will likely have a policy regarding gowns. Be sure you know your local policy. There are times when a change of uniform is preferred because trying to clean off contaminants is difficult and sometimes impossible without professional cleaning and disinfection or disposing of the uniform entirely.

■ Eye Protection and Face Shields

Eye protection is important in case blood splatters toward your eyes Figure 2-6 . If this is a possibility, wearing goggles is your best protection. Individuals who wear prescription glasses will also need additional protection for their eyes. Prescription glasses offer little side protection. Obviously, contact lenses offer no added protection from splashing. Face shields will also provide good eye protection Figure 2-7 .

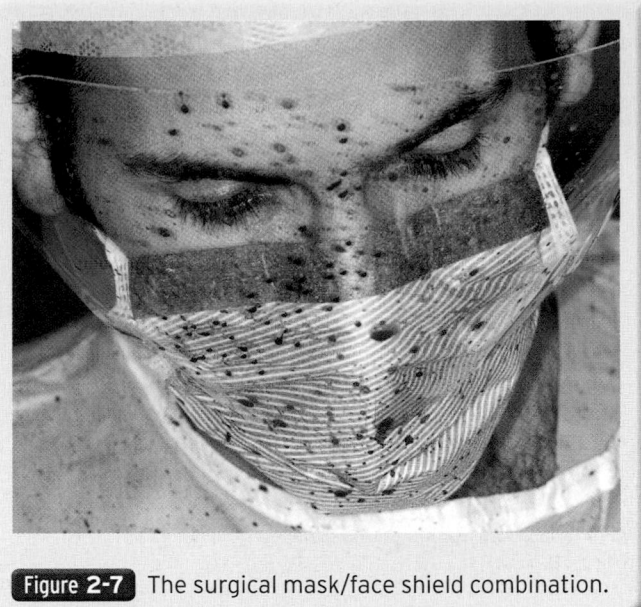

Figure 2-7 The surgical mask/face shield combination.

■ Masks, Respirators, and Barrier Devices

The use of masks is a complex issue, especially in light of OSHA and CDC requirements regarding protection from tuberculosis. You should wear a standard surgical mask if blood or body fluid spatter is a possibility. If you suspect that a patient has an airborne disease, you should place a surgical mask on the patient. However, if you suspect the patient has tuberculosis, place a surgical mask on the patient and a high-efficiency particulate air (HEPA) respirator on yourself Figure 2-8 . If the patient needs oxygen, place a nonrebreathing mask instead of a surgical mask on the patient and set the oxygen flow rate at 10 to 15 L/min. Do not place a HEPA respirator on the patient; it is unnecessary and uncomfortable. A simple surgical mask will reduce the risk of transmission of germs from the patient into the air. Use of a HEPA respirator should comply with OSHA standards, which state that facial hair, such as long sideburns or a mustache, will prevent a proper fit.

Although there are no documented cases of disease transmission

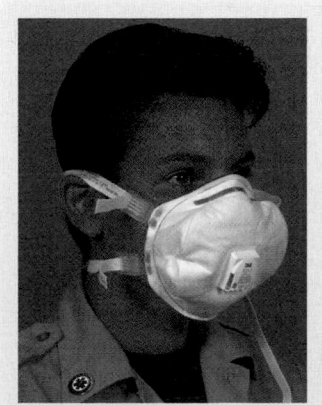

Figure 2-8 Wear a HEPA respirator if you treat a patient whom you suspect has tuberculosis.

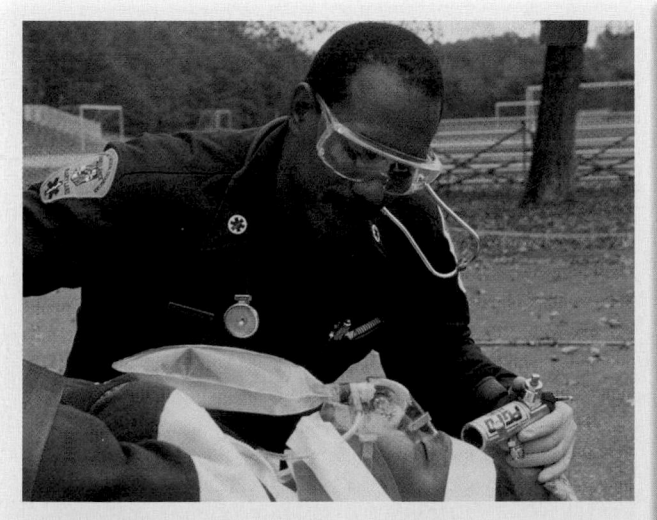

Figure 2-6 Wear eye protection to prevent blood splatter into your eyes.

to rescuers as a result of performing unprotected mouth-to-mouth resuscitation on a patient with an infection, you should use a pocket mask or bag-mask device Figure 2-9 . Mouth-to-mouth resuscitation is rarely necessary in a work situation.

Remember that outside surfaces of these items are considered contaminated after they have been exposed to the patient. You must make sure gloves, masks, gowns, and all other items that have been exposed to infectious processes or blood are properly disposed of according to local guidelines. If you are stuck by a needle, get blood or body fluids in your eye, or have significant body fluid contact with the patient, report the incident to your supervisor immediately.

■ Proper Disposal of Sharps

Be careful when handling needles, scalpels, and other sharp items. The spread of HIV and hepatitis in the health care setting can usually be traced to careless handling of sharps.

- Do not recap, break, or bend needles. Even the most careful individuals may expose themselves through a needlestick accidentally.
- Dispose of all sharp items that have been in contact with human secretions in approved, closed, rigid containers Figure 2-10 .

■ Employer Responsibilities

Your employer cannot guarantee a 100% risk-free environment. The risk of being exposed to or contracting a communicable disease is a part of your job. You have a right to know about diseases that may pose a risk to you. Remember, though, that your risk for infection is not high; however, OSHA regulations, especially for private and federal agencies, require that all employees be offered a workplace environment that reduces the risk

for exposure. Note that in some states that have their own OSHA plans, state and municipal employees must also be covered.

In addition to OSHA guidelines, other national guidelines and standards, including those from the CDC and National Fire Protection Agency (NFPA) Infection Control Standard 1581, address reducing the risk exposure to bloodborne pathogens (disease-causing organisms) and airborne diseases. These agencies set a standard of care for all fire and EMS personnel and apply whether you are a full-time paid employee or a volunteer. It is your responsibility to know your department's infection control plan and to use it Table 2-2 .

Establishing an Infection Control Routine

Infection control should be an important part of your daily routine. Follow the steps in Skill Drill 2-3 to manage potential exposure situations:

1. En route to the scene, make sure that PPE is out and available (Step 1).
2. On arrival, make sure the scene is safe to enter, then perform a rapid scan of the patient, noting whether any blood or body fluids are present.
3. Select the proper PPE according to the tasks you are likely to perform. Typically gloves will be used for all patient contacts (Step 2).

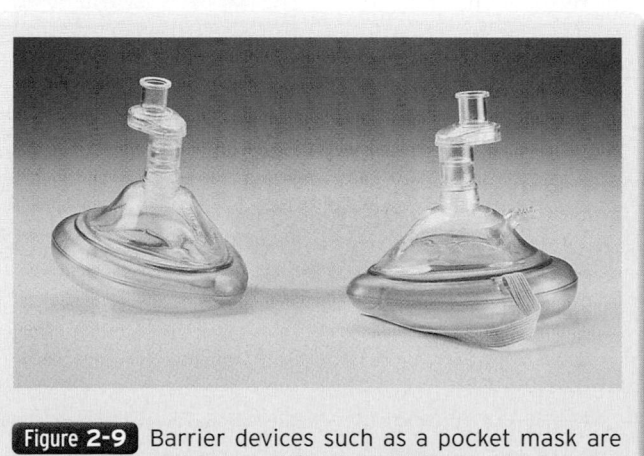

Figure 2-9 Barrier devices such as a pocket mask are necessary in providing artificial ventilations.

Figure 2-10 Properly dispose of sharps in a closed, rigid, marked container.

Table 2-2 Components of an Infection Control Plan

Determination of Exposure

- Determines who is at risk for ongoing contact with blood and other body fluids
- Creates a list of tasks that pose a risk for contact with blood or other body fluids
- Includes personal protective equipment (PPE) required by OSHA

Education and Training

- Explains why a qualified individual is required to answer questions about communicable diseases and infection control, rather than relying on packaged training materials
- Includes availability of an instructor able to train EMTs regarding bloodborne and airborne pathogens, such as hepatitis B and C, human immunodeficiency virus, syphilis, and tuberculosis
- Ensures that the instructor provides appropriate education, which is the best means for correcting many myths surrounding these issues

Hepatitis B Vaccine Program

- Spells out the vaccine offered, its safety and efficacy, record keeping, and tracking
- Addresses the need for postvaccine antibody titers to identify individuals who do not respond to the initial three-dose vaccination series

Personal Protective Equipment (PPE)

- Lists the PPE offered and why it was selected
- Lists how much equipment is available and where to obtain additional PPE
- States when each type of PPE is to be used for each risk procedure

Cleaning and Disinfection Practices

- Describes how to care for and maintain vehicles and equipment
- Identifies where and when cleaning should be performed, how it is to be done, what PPE is to be used, and what cleaning solution is to be used
- Addresses medical waste collection, storage, and disposal

Tuberculin Skin Testing/Fit Testing

- Addresses how often employees should undergo skin testing
- Addresses how often fit testing should be done to determine the proper size mask to protect the EMT from tuberculosis
- Addresses all issues dealing with HEPA respirator masks

Postexposure Management

- Identifies whom to notify when an exposure may have occurred, forms to be filled out, where to go for treatment, and what treatment is to be given

Compliance Monitoring

- Addresses how the service or department evaluates employee compliance with each aspect of the plan
- Ensures that employees understand what they are to do and why it is important
- States that noncompliance should be documented
- Indicates what disciplinary action should be taken in the face of continued noncompliance

Record Keeping

- Outlines all records that will be kept, how confidentiality will be maintained, and how, when, and by whom records can be accessed

4. Change gloves and wash hands between patients; do not unnecessarily delay treatment for use of PPE, thereby potentially putting patients at risk. Remove gloves and other gear after contact with the patient, unless you are in the patient compartment. Remember that good hand hygiene is always necessary.

5. Limit the number of people who are involved in patient care if there are multiple injuries and a substantial amount of blood at the scene.

6. If you or your partner is exposed while providing care, try to relieve one another as soon as possible so that you can seek care. Notify the designated officer and report the incident. This will also help to maintain confidentiality for both the patient and for you.

Be sure to routinely clean the ambulance after each run and on a daily basis. Cleaning is an essential part of the prevention and control of communicable diseases, ensuring removal of surface organisms that may remain in the unit. You should clean your unit as quickly as possible so it can be returned to service. Address the high-contact areas, including surfaces that were in direct contact with the patient's blood or body fluids or surfaces that you touched while caring for the patient after having contact with the patient's blood or body fluids. More information about decontaminating the ambulance can be found in Chapter 36, *Transport Operations*.

Whenever possible, cleaning should be done at the hospital. If you clean the unit back at the station, make sure you have a designated area with good ventilation. Any medical waste should be put in a red bag and disposed of at the hospital whenever possible. Any contaminated equipment that is left with the patient at the hospital should be cleaned by hospital staff or put in a red bag for transport and cleaning at the station.

You can use a bleach and water solution at a 1:10 dilution to clean the unit. The solution you mix should not have a strong odor of bleach if mixed correctly. A hospital-approved disinfectant that is effective against *Mycobacterium tuberculosis* can also be used. Use the cleaning solution in a bucket or use a pistol-handled spray container. Do not use alcohol or aerosol spray products to clean the unit. Pay attention to disinfectant directions.

Bleach solutions and most disinfectant agents will require air drying to be effective. Do not routinely go back over sprayed surfaces and dry them. Allow the sprayed surfaces to air dry unless otherwise indicated in the product directions.

Remove contaminated linen and place it into an appropriate bag for handling. Each hospital may have a different system for handling contaminated linen; you should learn hospital or department protocols **Figure 2-11** .

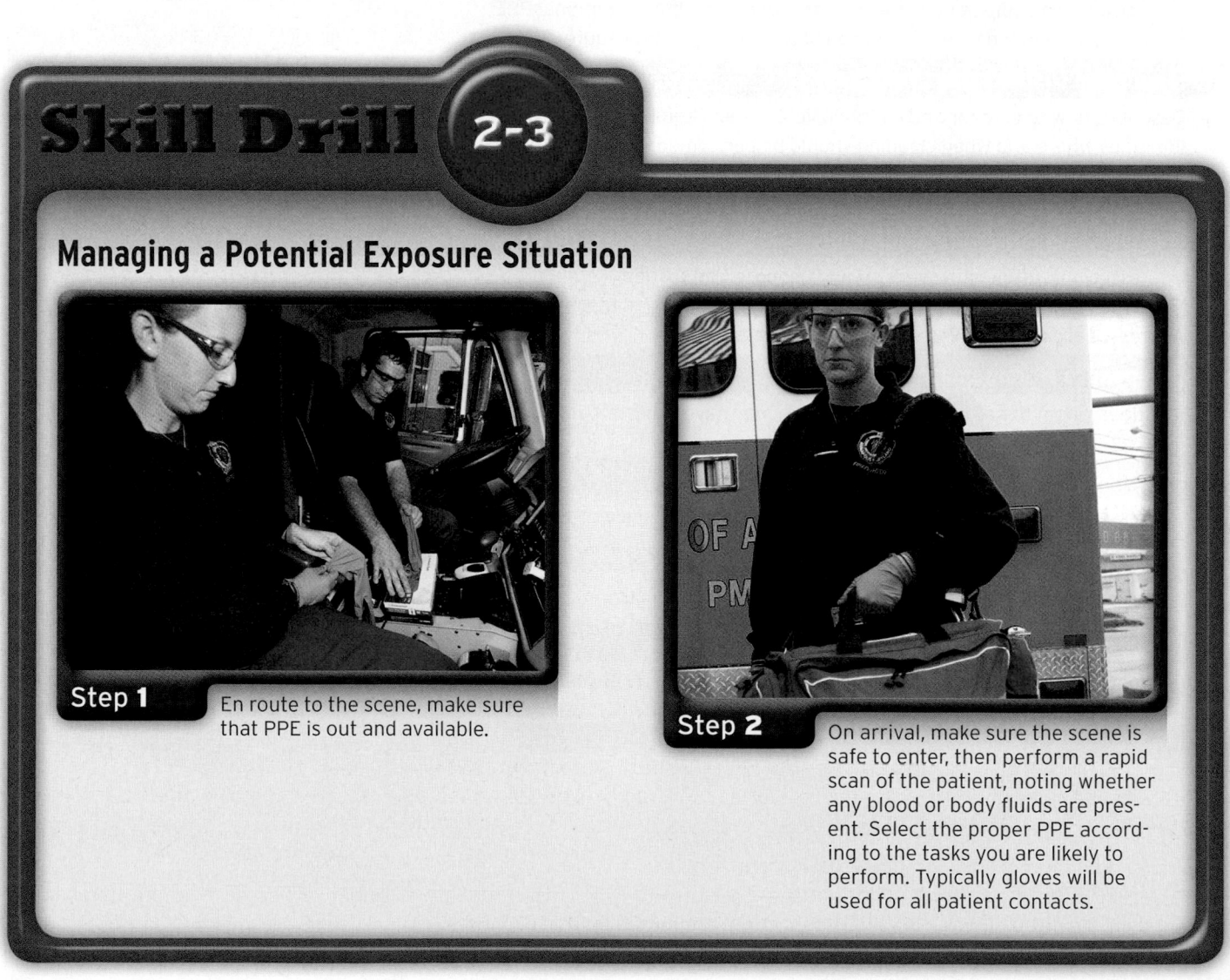

Skill Drill 2-3

Managing a Potential Exposure Situation

Step 1 En route to the scene, make sure that PPE is out and available.

Step 2 On arrival, make sure the scene is safe to enter, then perform a rapid scan of the patient, noting whether any blood or body fluids are present. Select the proper PPE according to the tasks you are likely to perform. Typically gloves will be used for all patient contacts.

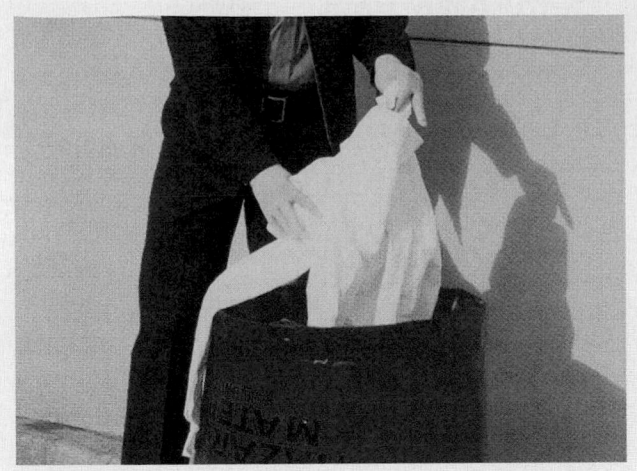

Figure 2-11 Contaminated linen and other wastes should be bagged appropriately and disposed of according to your local protocols.

Any reusable medical equipment should be properly cleaned and sterilized per your department's standard operating procedures. Keep in mind that in hospitals entire departments are devoted to sterilizing medical instruments. Proper sterilization requires the right tools and the right skills, so always carefully follow your department's procedures.

Learn the regulations defining medical waste in your area. The disposal of infectious waste, such as needles, sharps, and heavily soiled dressings, may vary from hospital to hospital and from state to state.

Immunity

Even if germs do reach you, you are not necessarily at risk for infection. For example, you may be **immune**, or resistant, to those particular germs. Immunity is a major factor in determining which **hosts** become ill from which germs Table 2-3. One way to gain immunity from many diseases today is to be immunized, or vaccinated, against them. Vaccinations have almost eliminated some childhood diseases, such as measles and polio.

Another way in which the body becomes immune to a disease is to recover from an infection from that germ. Afterward, the body's immune system recognizes and repels that germ when it shows up again. Once exposed, healthy individuals will develop lifelong immunity to many common pathogens. For example, a person who contracts and becomes infected with the hepatitis A virus may be ill for several weeks, but because an immunity will develop, the person will not get the illness again.

You are the Provider: PART 2

You arrive at the scene, enter the residence, and find two EMRs performing cardiopulmonary resuscitation (CPR) on the child, a 4-year-old girl. The child's mother tells you that when she went to wake her daughter up, she was unconscious and not breathing. She called 9-1-1 and started CPR. Your partner quickly assesses the child and asks you to open the jump kit.

Recording Time: O Minutes	
Appearance	Cyanotic; motionless
Level of consciousness	Unconscious and unresponsive
Airway	Small amount of vomitus in her mouth
Breathing	Absent
Circulation	Carotid pulse, absent; skin, cool and cyanotic

With CPR continuing, your partner prepares the cardiac monitor and asks you to suction the child's mouth and manage her airway. You quickly and effectively accomplish your assigned task, but notice that you are sweating profusely and can feel your heart racing.

2. What is stress? How does it manifest?

3. What phase of the stress response are you experiencing right now?

Sometimes, however, the immunity is only partial. Partial immunity protects against new infections. But germs that remain in the body from the first illness may still be able to cause the same disease again when the body is stressed or has some impairment in its immune system. For example, tuberculosis can cause a mild, unnoticeable infection before the body builds up a partial immunity. If the infection is never treated, the infection may be reactivated when immunity is weakened; however, such individuals are protected against a new infection from another person.

Humans seem unable to mount an effective immune response to some infections, such as HIV infection, which is infection with the human immunodeficiency virus that can progress to acquired immunodeficiency syndrome (AIDS).

Although hepatitis A immunization is not required by OSHA, you may wish to be vaccinated as a preventive measure. Hepatitis A vaccination is not necessary if you have had hepatitis A in the past. All these vaccines are effective and rarely cause side effects. Many communities require you to show proof that you are up to date with their immunizations.

Remember, germs that cause no symptoms in one person may cause serious illness in another.

■ Immunizations

As an EMT, you are at risk for acquiring an infectious or communicable disease. Using basic protective measures can minimize the risk. You are responsible for protecting yourself, so take an active role in achieving that goal.

Prevention begins by maintaining your personal health. Annual health examinations should be required for all EMS personnel. A history of all your childhood infectious diseases should be recorded and kept on file. Childhood infectious diseases include chickenpox, mumps, measles, rubella, and whooping cough. If you have not had one of these diseases, you must be immunized.

The CDC and OSHA have developed requirements for protection from bloodborne pathogens such as hepatitis B and human immunodeficiency viruses. You are required to get the hepatitis B vaccine or sign a waiver. Your employer is required by OSHA to provide the hepatitis B vaccine free of cost if you wish to receive it. An immunization program should be in place in your EMS system. Immunizations should be kept up to date and recorded in your file. Recommended immunizations include the following:

- Tetanus-diphtheria boosters (every 10 years)
- Measles, mumps, rubella (MMR) vaccine

Table 2-3 Immunity to Infectious Diseases

Type of Immunity	Characteristics	Examples	Comments
Lifelong	The illness will not recur.	Measles Mumps Polio Rubella Hepatitis A Hepatitis B	Infection or vaccination provides long-term immunity to new infection. A live vaccine is required for measles only.
Partial	The person who has recovered from a first infection is unlikely to get a new infection from another person but may develop illness from germs that lie dormant from the initial infection.	Chickenpox Tuberculosis	Infection provides lifelong immunity to the patient from acquiring a new infection, but the original illness may recur, or it may recur in a different way. In the case of chickenpox, which is caused by the herpes zoster virus, an infection may recur years later in the form of shingles.
None	Exposure confers no protection from reinfection. The infection may wear down the patient's resistance.	Gonorrhea Syphilis Human immunodeficiency virus (HIV) infection	No vaccine is available. Repeated infections are common. For example, there is effective immediate treatment for gonorrhea, and the germs may be eradicated; however, reinfection is likely if the high-risk practices (eg, unprotected sex) continue. For syphilis and HIV infection, the lack of immunity allows the germs to continue to cause damage within the host.

- Influenza vaccine (yearly)
- Hepatitis B vaccine
- Varicella (chickenpox) vaccine or having chickenpox

Other vaccines being investigated include pertussis (whooping cough) and *Staphylococcus aureus*. These vaccines are not currently recommended, but may be soon. You should also have a skin test for tuberculosis before you begin working as an EMT. The purpose of the test is to identify anyone who has been exposed to tuberculosis in the past. Testing should be repeated every year.

If you know that you will be transporting a patient who has a communicable disease, you have a definite advantage. This is when information in your health record will be valuable. If you have already had the disease or been vaccinated, you are not at risk. However, you will not always know whether a patient has a communicable disease. Therefore, you should always follow standard precautions if there is the possibility of exposure to blood or other body fluids.

General Postexposure Management

The likelihood of becoming infected during your performance of routine patient care is low. In the event that you are exposed to blood or other body substances despite all of your precautions, there are still preventative measures that you can take to protect your health. If you are exposed to a patient's blood or bodily fluids, first turn over patient care to another EMS provider. When it is safe to do so, clean the exposed area with soap and water. If your eyes were exposed, rinse them with water for at least 20 minutes as soon as possible.

Next, activate your department's infection control plan. This usually involves contacting a supervisor or your department's infection control officer to assist you. This person will help you to navigate the infection control process.

You will need to be screened to determine if there was a significant exposure to possible bloodborne pathogens. Just because you were exposed to a patient's blood or body fluids does not mean that there is a risk of infection. Typically, you will need a follow-up evaluation by a physician to determine if a significant exposure occurred. If the exposure was significant, blood may need to be drawn from both you and the patient to determine if any infectious agents were present.

You will have to complete an exposure report. Questions in the report may include: When did the event happen? What were you doing when you were exposed? What did you do after you were exposed? Completion of this paperwork will help relay critical information to the right people, resulting in help for you and possibly new protocols in the future to help prevent another incident.

Time is important! If you are exposed, let your supervisor or infection control officer know immediately. Some diseases will act quickly whereas others may lay dormant for a long time. The best way to reduce your risk of contracting a work-related disease is through early activation of your department's infection control plan.

Stress Management on the Job

EMS is a high-stress job. Understanding the causes of stress and knowing how to deal with stress is critical to your job performance, health, and interpersonal relationships. To prevent stress from affecting your life negatively, you need to understand what stress is, its physiologic effects, what you can do to minimize these effects, and how to deal with stress on an emotional level.

Stress is the impact of stressors on your physical and mental well-being. Stressors include emotional, physical, and environmental situations or conditions that may cause a variety of physiologic, physical, and psychological responses. The body's response to stress begins with an alarm response, followed by a stage of reaction and resistance, and then recovery or, if the stress is prolonged, exhaustion. This three-stage response is referred to as the **general adaptation syndrome**.

The physiologic responses involve the interaction of the endocrine and nervous systems, resulting in chemical and physical responses. This is commonly known as the fight-or-flight response. Positive stress, such as exercise, as well as negative forms of stress, such as shift work, long hours, or the frustration of losing a patient, all have the same physiologic manifestations. These include the following:

- Increased respirations and heart rate
- Increased blood pressure
- Dilated venous vessels near the skin surface (causes cool, clammy skin)
- Dilated pupils
- Tensed muscles
- Increased blood glucose levels
- Perspiration
- Decreased blood flow to the gastrointestinal tract

Situations that are stressful for EMS providers include the following:

- Dangerous situations
- Physical and psychological demands
- Critically ill or injured patients
- Dead and dying patients
- Overpowering sights, smells, and sounds
- Multiple patient situations
- Angry or upset patients, family, bystanders
- Unpredictability and demands of EMS
- Noncritical/non-9-1-1 patients

As you examine this list, you will see that some situations are clearly stressful: a car crash where a child is killed or a terrorist attack. Other situations may seem confusing. You may ask yourself why caring for noncritical patients is considered stressful. EMS providers today need to manage a large array of patients. One person's definition of an emergency may be quite different from another's. As EMTs begin their career in EMS, they oftentimes picture that all of their calls will be exciting life-and-death calls where they are able to save lives. The reality is that most patients are not critical and the care they need becomes rather routine. This can create stress in people who are unable to make the transition from the TV image of emergency medicine to its reality.

A new stressor for those who work in EMS is hospital waiting time. Emergency departments (EDs) around the country are dealing with greater and greater numbers of patients. Coupled with shortages of personnel like nurses, this increased patient load can lead to delays in care. This is where EMS comes into potential conflict. When the ambulance arrives at the hospital with a noncritical patient, the ED may not be able to accept the patient right away. As a result, EMTs may need to sit for hours in the ED hallway, waiting for a bed to open up so the patient can be transferred. This situation can certainly generate stress.

Reactions to stress can be categorized as acute, delayed, or cumulative. **Acute stress reactions** occur during a stressful situation. The EMT feels nervous, excited, and his or her ability to focus increases. This focus can be very helpful in managing a crisis situation. But if the stress of the situation becomes too great, the EMT is at risk of being caught up in emotional and physical reactions to stress. Picture stress as a wave in the ocean. If the crest of the wave is too high, the EMT can potentially drown if the stress goes unrecognized and is not relieved.

Delayed stress reactions manifest after the stressful event. During the crisis, the EMT is able to focus and function, but after things have calmed down, the EMT may be left with nervous, excited energy that continues to build and becomes a distraction. With both acute and delayed reactions, the important question to ask is how did the EMT manage these feelings during the stressful event? Was the EMT able to continue, managing the stress well and taking it in stride? Or, did the EMT not know how to manage the stress well, resulting in delayed stress reactions?

Cumulative stress reactions are the most important to understand. After the stressful event is over, is the EMT able to shake off the effects? Is the EMT still tired? Cumulative stress occurs when the EMT is exposed to prolonged or excessive stress. The EMT fights to remain in control and is successful, but is now starting to grow tired. Now the next stressful situation occurs. Each time,

the EMT finds it harder and harder to recover because the effects of the previous stress are tiring.

Cumulative stress can have physical symptoms such as fatigue, changes in appetite, gastrointestinal problems, or headaches. It may cause insomnia or hypersomnia, irritability, inability to concentrate, and hyperactivity or underactivity. Additionally, it may present with psychological reactions such as fear, dull or nonresponsive behavior, depression, guilt, oversensitivity, anger, irritability, and frustration. Often, today's fast-paced lifestyles compound these effects by not allowing a person to rest and recover after periods of stress. Prolonged or excessive stress has been proven to be a strong contributor to heart disease, hypertension, cancer, alcoholism, and depression.

Many people are subject to cumulative stress, whereby insignificant stressors accumulate to a larger stress-related problem. In the emergency services environment (EMS, police, fire fighters), stressors may also be sudden and more severe. Some events are unusually stressful or emotional, even by emergency services standards. These acute severe stressors result in what is referred to as critical incident stress. Events that can trigger critical incident stress include the following:

- Mass-casualty incidents
- Serious injury or traumatic death of a child
- Crash with injuries, caused by an emergency services provider while responding to or from a call
- Death or serious injury of a coworker in the line of duty

Posttraumatic stress disorder (PTSD) may develop after a person has experienced a psychologically distressing event. It is characterized by reexperiencing the event and overresponding to stimuli that recall the event. Stressful events in EMS are sometimes psychologically overwhelming. Some of the symptoms include depression, startle reactions, flashback phenomena, and dissociative episodes (eg, amnesia of the event).

A process called **critical incident stress management (CISM)** was developed to address acute stress situations and potentially decrease the likelihood that PTSD will develop after such an incident **Figure 2-12**. The process theoretically is used to confront the responses to critical incidents and defuse them, directing the emergency services personnel toward physical and emotional equilibrium. CISM can occur formally, as a debriefing for those who were on scene. In such situations, trained CISM teams of peers and mental health professionals may facilitate this. Additionally, CISM can occur at an ongoing scene in the following circumstances:

- When personnel are assessed for signs and symptoms of distress while resting
- Before reentering the scene

Figure 2-12 Critical incident stress management is sometimes employed to help providers relieve stress.

- During a scene demobilization in which personnel are educated about the signs of critical incident stress and given a buffer period to collect themselves before leaving

Safety

Coworkers often notice a change in behavior or attitude before a supervisor does. This is especially true in EMS, where close relationships develop between people who work together and share rooms, meals, and social interaction. Being a friend means helping a friend. Talk to your partner about changes in their behavior you may notice. If you are the EMT having trouble dealing with a crisis, remember, you are not alone. Talk with your partner.

Defusing sessions are the first to occur. These sessions are held during the event or immediately afterwards. A group informally discusses events that they experienced together. Defusing sessions are designed to educate the participants as to the expectations over the next few days, and give guidance on proper techniques to manage the feelings they may be experiencing. One example is to discourage drinking alcohol during this stressful time.

Debriefing sessions are held within 24 to 72 hours of a major incident. These meetings are held by a CISM team consisting of peers and mental health professionals. At the debriefing session, pent up emotions can be properly expressed. It is more likely that providers will be ready to express their emotions more freely a few days following the event.

One of the important rules associated with the debriefing session is to not turn it into an operational critique. No one is right. No one is wrong. No one is to blame. Only emotions about the specific event are to be relayed. These debriefing sessions may also have to be repeated at a later time.

CISM programs are located throughout the United States. CISM teams usually can be located by calling telephone directory assistance in your area and asking for CISM, or they can be requested through your employer. The International Critical Incident Stress Foundation, Inc is a company dedicated to limiting the effects of stress on EMS providers through education and support services. For more information, go to the Foundation's website at www.icisf.org.

Supporting patients in emergency situations is difficult. It is stressful for them but also for you. You are vulnerable to all the stresses that go with your profession. It is critical that you recognize the signs of cumulative stress so that it does not interfere with your work or life away from work, including your family life. The signs and symptoms of cumulative stress may not be obvious at first. Rather, they may be subtle and not present all the time Table 2-4.

■ Acute Management of Stress

When you are in a stressful situation, it is important to understand your focus. Your focus is patient care. Always ensure that excellent patient care is being maintained. This may mean recognizing that you are not alone. Your partner, police, a supervisor, or other additional personnel are often available to help you manage crisis situations. Make sure that you use these people so patients receive the best care and you will find that you will not be overwhelmed by the situation.

Your job is to remain professional at all times. Try and stay calm. Allow patients to express their feelings, including anger, without becoming angry yourself.

There are many methods of handling stress. Some are positive and healthy; others are harmful and

Table 2-4 Warning Signs of Stress

- Irritability toward coworkers, family, and friends
- Inability to concentrate
- Difficulty sleeping, increased sleeping, or nightmares
- Feelings of sadness, anxiety, or guilt
- Indecisiveness
- Loss of appetite (gastrointestinal disturbances)
- Loss of interest in sexual activities
- Isolation
- Loss of interest in work
- Increased use of alcohol
- Recreational drug use
- Physical symptoms such as chronic pain (headache, backache)
- Feelings of hopelessness

destructive. Americans consume more than 20 tons of aspirin per day, and doctors prescribe muscle relaxants, tranquilizers, and sedatives more than 90 million times per year to patients in the United States. Although these medications have legitimate uses, they do nothing to combat stress that may cause the medical problems described previously.

The term "stress management" refers to the tactics that have been shown to alleviate or eliminate stress reactions. These strategies may involve changing a few habits, changing your attitude, and perseverance **Table 2-5**.

A clue to the management of stress comes from the fact that it is not the event itself but the individual's reaction to it that determines how much it will strain the body's resources. Remember that stress is defined as anything you perceive as a threat to your equilibrium. Stress is an undeniable and unavoidable part of our everyday life. By understanding how it affects you physiologically, physically, and psychologically, you can manage it more successfully.

The following sections provide some suggestions for how to prevent the effects of stress from affecting you.

Table 2-5 Strategies to Manage Stress

- Minimize or eliminate stressors.
- Change partners to avoid a negative or hostile personality.
- Change work hours.
- Change the work environment.
- Cut back on overtime.
- Change your attitude about the stressor.
- Talk about your feelings with people you trust.
- Seek professional counseling if needed.
- Do not obsess over frustrating situations such as relapsing alcoholics and nursing home transfers; focus on delivering high-quality care.
- Try to adopt a more relaxed, philosophical outlook.
- Expand your social support system apart from your coworkers.
- Sustain friends and interests outside emergency services.
- Minimize the physical response to stress by employing various techniques, including:
  - A deep breath to settle an anger response
  - Periodic stretching
  - Slow, deep breathing
  - Regular physical exercise
  - Progressive muscle relaxation
  - Meditation
  - Limit intake of caffeine, alcohol, and tobacco use

Some of them may be useful in helping you prevent problems from developing. Others may help you solve problems should they develop.

■ Wellness and Stress Management

Anyone can respond to sudden physical stress for a short time. However, if stress is prolonged, and especially if physical action is not a permitted response, the body can quickly be drained of its reserves. This can leave it depleted of key nutrients, weakened, and more susceptible to illness.

Nutrition

Your body's three sources of fuel—carbohydrates, fat, and protein—are consumed in increased quantities during stress, particularly if physical activity is involved. The quickest source of energy is glucose, taken from stored glycogen in the liver. However, this supply will last less than a day. Protein, drawn primarily from muscle, is a long-term source of fuel. Tissues can use fat for energy. The body also conserves water during periods of stress. To do so, it retains sodium by exchanging and losing potassium from the kidneys. Other nutrients that are susceptible to depletion are the vitamins and minerals that are not stored by the body in substantial quantities. These include water-soluble B and C vitamins and most minerals.

As an EMT, you have little control of what stressors you will face on any given day. Consequently, stress in one form or another is an unavoidable part of your life. As you would study for a test, dress properly for a day of snow skiing, or train for a sporting event, you should physically prepare your body for stress. Physical conditioning and proper nutrition are the two variables over which you have absolute control. Muscles will grow and retain protein only with sufficient activity. Bones will not passively accumulate calcium. In response to the physical stress of exercise, bones store calcium and become denser and stronger. Regular, well-balanced meals are essential to provide the nutrients that are necessary to keep your body fueled **Figure 2-13**. Vitamin-mineral preparations that provide a balanced mix of all the nutrients may be necessary to supplement a less than perfectly balanced diet.

To perform efficiently, you must eat nutritious food. Food is the fuel that makes the body run. The physical exertion and stress that are a part of your job require a high energy output. If you do not have a ready source of fuel, your performance may be less than satisfactory. This can be dangerous for you, your partner, and your patient. Therefore, it is important for you to learn about and follow the rules of good nutrition.

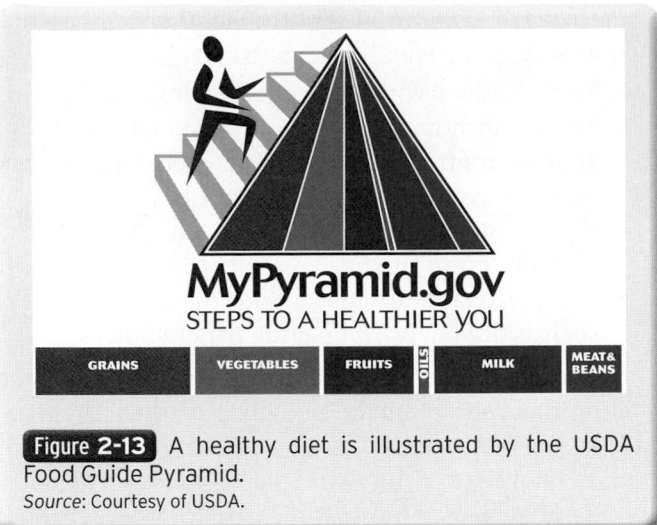

Figure 2-13 A healthy diet is illustrated by the USDA Food Guide Pyramid.
Source: Courtesy of USDA.

Words of Wisdom

As with most things in life, EMS comes down to balance. You need to understand that not all of the patients you care for will be critically ill or injured. This does not mean that they do not need care, only that they need a different kind of care. A thoughtful word or a hand on a shoulder can be powerful medicine. Care for each person, regardless of his or her complaint, as a person. Be satisfied with the rewards of simple compassion and you will find a home in EMS for many years to come.

Candy and soft drinks contain sugar. These foods are quickly absorbed and converted to fuel by the body. But simple sugars also stimulate the body's production of insulin, which reduces blood glucose levels. For some people, eating a lot of sugar can actually result in lower energy levels.

Complex carbohydrates rank next to simple sugars in their ability to produce energy. Complex carbohydrates such as pasta, rice, and vegetables are among the safest, most reliable sources for long-term energy production. However, some carbohydrates take hours to be converted into usable body fuel.

Fats are also easily converted to energy, but eating too much fat can lead to obesity, cardiac disease, and other long-term health problems. The proteins in meat, fish, chicken, beans, and cheese take several hours to convert to energy.

Carry an individual supply of high-energy food to help you maintain your energy levels **Figure 2-14**. Try eating several small meals throughout the day to keep your energy resources at constant high levels. Remember, however, that overeating may reduce your physical and mental performance. After a large meal, the blood that is needed for the digestive process is not available for other activities.

You must also make sure you maintain an adequate fluid intake **Figure 2-15**. Hydration is important for proper functioning. Fluids can be easily replenished by drinking any non-alcoholic, noncaffeinated fluid. Water is generally the best fluid available. The body absorbs it faster than any other fluid. Avoid fluids that contain high levels of sugar. These can actually slow the rate of fluid absorption by the body. They can also cause abdominal discomfort. One indication of adequate hydration is frequent urination. Infrequent urination or urine that has a deep yellow color indicates dehydration.

Figure 2-14 Carry a supply of high-energy food with you so that you can maintain your energy levels.

Exercise and Relaxation

A regular program of exercise will enhance the benefits of maintaining good nutrition and adequate hydration. When you are in good physical condition, you can handle job stress more easily. A regular program of exercise will increase your strength and endurance. To be healthy, you should engage in at least 30 minutes of physical activity most days of the week. Exercise should be moderate or vigorous to have a good health benefit. In other words, you should break a sweat **Figure 2-16**. You may wish to practice relaxation techniques, meditation, and visual imagery.

Your exercise routine should involve aspects of cardiovascular endurance, muscular strength building, and muscle flexibility. Endurance will ensure that your

Figure 2-15 Maintain an adequate fluid intake by drinking plenty of water or other nonalcoholic, caffeine-free fluids.

Figure 2-16 A regular program of exercise will increase strength and endurance.

cardiovascular system is able to provide your muscles and brain with needed oxygen. Strength and flexibility building ensures that the body is able to handle the requirements that you will place on it by lifting patients, performing CPR, and moving heavy equipment. Remember, if you do not use it, you will lose it. Exercise is critical to maintenance of a healthy body.

Safe Lifting Practices We have already discussed the physical requirements of the EMT. Lifting 125 lb can be difficult if you do not participate in a consistent exercise program. Lifting is one of the things that EMTs do often, so safe lifting techniques are critical to your health and well-being. Back injuries are a common reason for on-the-job injuries within EMS. Chapter 35, *Lifting and Moving Patients*, discusses lifting and moving in depth. For your health and well-being, remember these tips **Figure 2-17**:

- Preplan the move
- Bend your legs, *not* your waist
- Keep the weight close to your body
- Lift straight up using your legs, *not* your back

Sleep

Good productive sleep is as important as eating well and exercise in the maintenance of good heath. Sleep should be regular and uninterrupted. The number of hours is not nearly as important as the quality of sleep. Unfortunately, you may not have the luxury of sleeping throughout the night.

The signs that your sleep pattern is ineffective include:

- You fall asleep within seconds of lying down.
- Within an hour or so after an EMS call, you find yourself routinely fatigued. The excitement is over and now your adrenaline rush crashes.
- You are unable to make it through an entire day without severe fatigue.
- You are unable to concentrate on repetitive tasks such as driving or completing paperwork.

Actions you can take to improve your sleep include limiting your caffeine intake and tobacco use. Both agents have stimulating effects that can interrupt sleep. Limit your alcohol use. Alcohol is a depressant and encourages sleep. However, routine or excessive use of alcohol can change your sleep pattern, preventing deep sleep from occurring. Try to create as consistent a sleep cycle as possible. This may require naps. Many EMS providers are able to change their sleep pattern into several sleep episodes throughout the day.

Do not worry if you are unable to get 8 straight hours of sleep. Three sleep episodes of 2 to 3 hours each will provide similar effects. Each sleep episode needs to be more than 1 hour in length to encourage deep sleep. Finally, do not forget the effects of exercise and sleep. Routine exercise will promote the needed fatigue to slip into a restful sleep.

Disease Prevention

Besides sleep, diet, exercise, hydration, and all the other things that make up a healthy lifestyle, you need to be aware of your hereditary factors. Consider what you might know about your immediate family's and your

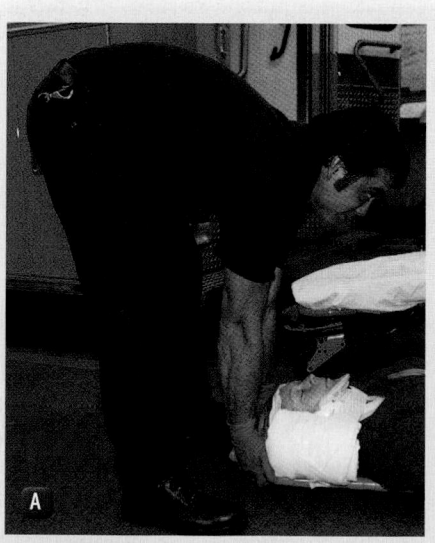
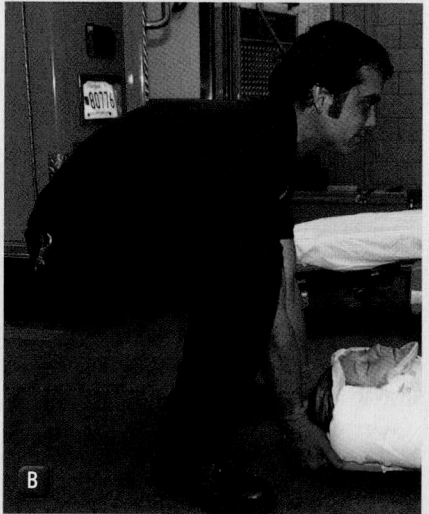

Figure 2-17 Lifting techniques. **A.** Poor lifting technique: the back is bent. **B.** Good lifting technique: the knee is bent.

ancestor's health. Alzheimer disease, chemical addiction, cancer, cardiac illness, hypertension, migraine, mental illness, and stroke all feature prominent hereditary factors. The most common of all heredity health factors are heart disease and cancer.

Share this information with your personal physician. Your physician is bound by the same oath of confidentiality that you are. Work with him or her to set up a schedule for health assessments, building them into your routine physical check-ups. Your physician should be your ally in screening for these diseases and in assessing your lifestyle as well as your heredity factors.

Knowing your hereditary factors will help you adjust your lifestyle to help prevent disease. For example, if diabetes runs in your family, exercise and diet are critical to your well-being. Maintaining a healthy weight and sustaining a consistent exercise routine will help minimize your risk of developing this disease.

Smoking If you don't already smoke, please don't start! If you do, please stop! Not only does this habit fly in the face of everything that EMS stands for, it also produces many of the most horrible cardiovascular and lung disasters that you will confront during your career. In addition, it sets an awful example for the public—especially to people who have breathing disorders such as asthma. And it makes you look and smell like anything but a professional caregiver.

Are you a smoker who is trying to quit? Several strategies can help you. First, try to cultivate a relationship with a mentor who was once truly addicted to smoking but who has successfully quit. Use that person as a support,

and draw on his or her advice and encouragement. There are also programs that attack a smoker's psychological dependency. These programs may include instructions and audio that provide ongoing support. Other options include therapy, hypnotism, and acupuncture.

Talk to your primary care doctor. Your doctor should be familiar with more techniques. All of these solutions are cheaper than cigarettes and their associated health risks.

Balancing Work, Family, and Health

As an EMT, you will often be called to assist the sick and injured any time of the day or night. Unfortunately, there is no rhyme or reason to the timing of illness, injury, or interfacility transfer. Volunteer EMTs may often be called away from family or friends during social activities. Shift workers may be required to be apart from loved ones for long periods of time. You should never let the job interfere excessively with your own needs. Find a balance between work and family; you owe it to yourself and to them. It is important to make sure that you have the time that you need to relax with family and friends.

It is also important to realize that coworkers, family, and friends often may not understand the stress caused by responding to EMS calls. As a result of a "bad call," you might not feel like going out to a movie or attending a family event that has been planned for some time. In these situations, help from a critical incident stress management team or information sessions conducted by the EMS unit's employee assistance program may assist you in resolving these problems.

You are the Provider: PART 3

The cardiac monitor reveals that the child has no cardiac activity (asystole). With CPR ongoing, your partner intubates the child and an intravenous (IV) line is inserted. The child's mother, who is standing back watching your efforts, is crying and keeps yelling at you, "Why isn't my daughter waking up? Why aren't you saving her?"

Recording Time: 5 Minutes

Respirations	Absent
Pulse	Absent
Skin	Cool and cyanotic
Blood pressure	Not obtainable
Oxygen saturation (Sao$_2$)	Not obtainable

4. How should you respond to the mother's question?

5. What stage of the grieving process is the mother experiencing?

When possible, rotate your schedule to give yourself time off. If your EMS system allows you to move from station to station, rotate to reduce or vary your call volume. Take vacations to provide for your good health so you will be able to respond the next time you are needed. If at any point you feel the stress of work is more than you can handle, seek help. You may want to discuss your stress informally with your family or coworkers. Help from more experienced team members can be invaluable. You may also wish to get help from peer counselors or other professionals. Seeking this help does not make you weak in the eyes of others. Rather, it shows that you are in control of your life.

Figure 2-18 Communicate with coworkers in a way that is sensitive and respectful of individual differences.

Workplace Issues

As our society continues to grow more and more culturally diverse, some groups that may have been satisfied in the past to accept and participate in mainstream American cultural traditions may seek instead to assert, preserve, and nurture their differences. As our society grows more culturally diverse, so do EMS workplaces. There will be challenges as these changes continue to occur. If you have any problem working with any particular group of people, you need to address this before finishing your EMT training. You are required to provide an equal standard of care to all patients and also need to be able to work efficiently and effectively with other health care professionals from a variety of different backgrounds.

■ Cultural Diversity on the Job

Each individual is different, and you should communicate with coworkers and patients in a way that is sensitive to everyone's needs **Figure 2-18**. Look at cultural diversity as a resource, and make the most of the differences among people in EMS, thus allowing them to provide optimum patient care. As the public safety workplace becomes more culturally diverse, changes may occur that could be considered disruptive. It is possible to build the strength of your workgroup through the use of diversity.

For many years, EMS and public safety have been dominated by Caucasian men. This trend continues to decline; more women and minorities are working in public safety. The proactive EMT understands the benefits of using cultural diversity to improve patient care and expects to work alongside workers with different backgrounds and to accept their differences.

Depending on your work experience, you may or may not have worked with people of varying backgrounds, attitudes, beliefs, and values.

Compared with traditional workplaces, EMS might seem like chaos. People who work in an office or manufacturing facility can reasonably expect to go to work every day, see the same people, and perform basically the same tasks. In EMS and public safety work, you are exposed to people in crisis. This exposure brings out traits and qualities your partners and coworkers use to manage their stress. Coworkers in traditional workplaces may not be willing to show this side of themselves to others. Debriefing after the call will help in adjusting to this environment.

Cultural diversity in EMS allows you to enjoy the benefits of accentuating the skills of a broad range of people. When you accept coworkers as individuals, the need to fit them into rigid roles is eliminated. To be more sensitive to cultural diversity issues, you must first be aware of your own cultural background. Ask yourself, "What are my own issues relative to race, color, religion, and ethnicity?" Because culture is not restricted to different nationalities, you should also consider age, disability, gender, sexual orientation, marital status, work experience, and education.

In sports, you play to your team's strengths. For example, in football, offensive lines have a fast side and a strong side, and they run plays toward either side depending on the situation. As part of an effective EMS team, you can make it part of your team culture to play to your group's strengths. This may be difficult to do; but once you begin the process, the benefits in terms of improved patient care are immeasurable.

■ Your Effectiveness as an EMT

To be an effective EMT, you need to discover the diverse cultural needs of your coworkers, as well as your patients and their families. Although it is unrealistic to expect you to become a cross-cultural expert with knowledge about all ethnicities, you should learn how to relate effectively.

Teamwork is essential in public safety and EMS. To work effectively as a team, you need to communicate to deal with cultural diversity issues.

As a health care professional, you should try to be a role model for new EMTs by showing them the value of diversity. If you are working with a coworker or patient from a particular cultural group, be careful about any opinion you may have formed about that group. Do not assume that there is a language barrier, and do not appear patronizing by saying, "Some of my best friends are… ." There are legitimate differences in how various cultures respond to stress. For example, you should be prepared to accept that people of different cultures might respond differently to the death of a loved one.

When you are working with patients or calling the hospital on the radio, other EMTs may be sensitive to how you treat patients from their cultural group. Therefore, when referring to patients, you should use the appropriate terminology. Avoid using terms such as "cripple," "deformed," "deaf," "dumb," "crazy," and "retard" when referring to patients. The word "handicapped" even has a negative connotation. Instead, use the term "disabled," and describe the specific disability.

You might want to consider taking multilingual training classes. This will not only be useful in communicating with your coworkers; it will also help to improve communication with your patients and sensitize you to the cultural richness of the people who are using the language.

Even the perception of discrimination can weaken morale and motivation and negatively affect the goal of EMS. Therefore, to achieve the benefits of cultural diversity in the EMS workplace, EMTs must understand how to communicate effectively with coworkers from various backgrounds.

■ Avoiding Sexual Harassment

The number of sexual harassment lawsuits skyrocketed in the 1990s because of increased media attention to the problem. Furthermore, guilty verdicts encouraged others to bring suits concerning conduct that once would have gone unchallenged.

Sexual harassment is any unwelcome sexual advance, request for sexual favors, or other verbal or physical conduct of a sexual nature when submitting is a condition of employment, submitting or rejecting is a basis for an employment decision, or such conduct substantially interferes with performance and/or creates a hostile or offensive work environment. Remember that even an overheard conversation can be construed as sexual harassment.

There are two types of sexual harassment: quid pro quo (the harasser requests sexual favors in exchange for something else, such as a promotion) and hostile work environment (jokes, touching, leering, requests for a date, talking about body parts). Seventy percent of sexual harassment incidents today are hostile work environment complaints. Remember, it does not matter the intent or the harasser. What matters is the perception of the act and the impact that behavior had on someone else. For many years, it was not uncommon to walk into a fire station and see sexually suggestive posters, calendars, or cartoons and to hear sexual jokes or comments. This situation is changing because it is not acceptable professional practice.

Because EMTs and other public safety professionals depend on each other for their safety, it is especially important for you to try to develop nonadversarial relationships with coworkers. Most EMS facilities and fire stations make arrangements for different bunkrooms for men and women. If this is not the case at your facility, you should discuss this with your supervisor and talk openly with coworkers of the opposite gender to allow for their privacy.

If you are concerned about a particular behavior, it may be helpful to ask yourself these questions: "Would I do or say this in front of my spouse, significant other, or parents?" "Would I want my family members to be exposed to this behavior?" "Would I want my behavior videotaped and shown on the evening news?"

If you have been harassed, you should report it to your supervisor immediately and keep notes of what happened and what was said. You should confront the harasser if you feel comfortable doing so; however, this may not be for everyone. If you are asked for a date, say, "I'm not interested." If remarks or touching offend you, say, "Please don't say/do that to me; it offends me."

■ Substance Abuse

In the past, part of the fire service ritual was to go back to the fire station after the fire, clean and maintain the equipment, and discuss the call. At some locations, having a few beers was not uncommon. EMS today is very different from the ambulance service of 20 years ago.

Drug and alcohol use in the workplace causes an increase in accidents and tension among workers, but most important, it can lead to poor treatment decisions. EMS personnel who use or abuse substances such as alcohol or marijuana are more likely to have problems with their work habits and their drivers' licenses may be revoked as a result. They may be absent from work more often than other workers. If the use or abuse has occurred within hours before the start of their shift, their ability to render emergency medical care may be lessened because of mental or physical impairment. Because of the seriousness of substance abuse or misuse, many EMS systems now require their personnel to undergo periodic random tests for illegal drug use. Since public safety workers depend so much on coworkers for their own safety, it is even more important that ways be found to manage this problem.

As an EMT, you will witness firsthand the tremendous effects of violence, trauma, and disease. Beyond CISM, members of the public safety community have a way of covering for each other. It is important to understand that the problem behavior will usually get worse before it gets better. Unfortunately, the stereotypical image of the alcoholic or addict lying in the gutter in an urban part of town often blinds EMS personnel to the existence of a coworker's drug or alcohol problem. Not all people with a substance abuse problem fit the stereotype.

As a member of the EMS team, you are responsible for responding to the community's emergency medical needs. Hazards in the EMS workplace are many. If you or one of the members of your team has an alcohol or other drug problem, these risks are increased. Furthermore, drug use that occurs off the job does not necessarily decrease the risk if a team member is showing up at work still under the effects of substance abuse. While varying state to state, a drug-related or alcohol-related arrest may result in the revocation of some or all driving privileges and even loss of EMT licensure. Because of the tremendous risk potential, it is critical that EMTs seek help or find a way to confront their partner or coworker even though there will be great pressure to allow the behavior to continue. Addicts and alcoholics develop great skill at covering their behavior; you might even decide not to bother your coworker because you feel that he or she has caught too many tough calls lately and needs to blow off some steam. Do not let this happen. You have to find a way to confront someone who has a substance abuse problem. Because of the tremendous hazard to patients, the public, and other team members, you have a legitimate right to confront coworkers with drug and alcohol problems.

When confronting a coworker with a potential drug or alcohol problem, make it clear to the coworker that if the problem is personal, it is the coworker's responsibility to take care of it. But you also have the power to assist this person. In many workplaces, coworkers are often in a position to notice a change in a coworker's behavior or attitude before a supervisor does. This is even more so the case in EMS because of the close relationship that develops between people who work together for many hours and share rooms and meals. This may allow you to help someone before his or her job performance is negatively affected.

Safety

Trust is your business. You will be given the privilege, and it is a privilege, to work with patients in their time of highest need. You must demonstrate that trust through consistent professionalism. Remember that you have support to help you make the right choice: your partner, your supervisor, your family.

To help reduce the potential for drug and alcohol use in the EMS workplace, EMTs can learn about alcohol and other drugs. Beyond following company policy, you and your coworkers can agree among yourselves what constitutes unacceptable behavior. The best time to confront these issues is usually after a call. Management sets the tone on these issues, but senior EMTs can also emphasize to new EMTs that drug and alcohol abuse will not be tolerated.

Employee assistance programs (EAPs) are often available for EMS personnel. These agencies are contracted with the EMS department to provide a wide array of mental health, substance abuse, crisis management, and counseling services. Talk with your supervisor to see what resources are available at your EMS department. Early intervention is the best bet to ensure a safe, alcohol- and drug-free workplace.

Emotional Aspects of Emergency Care

At times, even the most experienced health care providers have difficulty overcoming personal reactions and proceeding without hesitation. You may have patients that need to be removed from life-threatening situations, or you need to provide life support measures to patients who are severely injured. You may also be called on to recover human remains from highway accidents, aircraft disasters, or explosions Figure 2-19. In all of these situations, you must be calm and act responsibly as a member of the emergency medical care team. You must also realize that even though your personal emotions must be kept under control, these are normal feelings. Every EMT must deal with these feelings. The struggle to remain calm in the face of horrible circumstances contributes to the emotional stress of the job.

Death and Dying

Today, life expectancy has dramatically increased. In fact over two thirds of all deaths occur among those aged 65 years and older. The number one cause of all deaths today is attributed to heart disease. From the age of 1 year to 40 years, trauma is the leading cause of death. Death today is likely to occur either quite suddenly or after a prolonged terminal illness. The environment of death has changed since our nation's earlier days; it occurs in the home setting less frequently. The setting of death is somewhere else—in the hospital, a hospice, or a convalescent home, at the workplace, or on the highway. For this reason, we are less familiar with death than our ancestors were. While we all know we are going to die some day, some time, we tend to deny death in America. Illness can be much more drawn-out and much more removed from daily life. Life support systems and impersonal care remove the whole experience of death from most people's awareness. The mobility of families also makes it less likely there will be extended family support when death does occur.

You may have significant painful personal experience with death. No matter what the frequency of response

Figure 2-19 As an EMT, you will have to deal with the removal of bodies.

to emergency calls, death is something that every EMT will sometimes face. For some of you, it may be infrequent. Others, in urban settings, may see death many times in responding to motor vehicle crashes, drug overdoses, suicides, or homicides. Some EMTs may have to deal with the mass-casualty incident of an airplane crash or a hazardous materials incident. In all these situations, coming to grips with your thoughts, understandings, and adjustment to death is not only important personally, but also a function of delivering emergency medical care.

The Grieving Process

Everyone working as an EMT will experience grief at one time or another. This section discusses how to handle patient grief, as well as how to cope with your own grief that may result from a difficult call.

The death of a human being is one of the most difficult events for another human being to accept. If the survivor is a relative or close friend of the deceased, it is even more difficult. Emotional responses to the loss of a loved one or friend are appropriate and should be expected. In fact, it is expected that you will feel emotional about the death of a patient. Feelings and emotions are part of the grieving process. All of us experience these feelings after a stressful situation that causes us personal pain.

In 1969, Dr Elisabeth Kubler-Ross published research revealing people go through several stages of grieving. They are as follows:

1. **Denial.** Refusal to accept diagnosis or care, unrealistic demands for miracles, or persistent failure to understand why there is no improvement.
2. **Anger, hostility.** Projection of bad news onto the environment and commonly in all directions, at times almost at random. The person lashes out. Someone must be blamed, and those who are responsible must be punished. This is usually an ugly phase, and may even be inappropriately directed toward the EMT.
3. **Bargaining.** An attempt to secure a prize for good behavior or promise to change lifestyle. "I promise to be a 'perfect patient' if only I can live until 'x' event."
4. **Depression.** Internalized anger, hopelessness, and the desire to die. It rarely involves suicidal threats, complete withdrawal, or giving up long before the illness seems terminal. The patient is usually silent.
5. **Acceptance.** Acceptance of impending death by the patient, or the acceptance of the death of a loved one.

The stages may follow one another, occur simultaneously, or a person may jump back and forth between stages. The stages may last different amounts of time.

Even though the event (death) has not yet happened, the patient knows that it will happen. The patient has no control over this process. The patient will die whether or not he or she is ready to die. As an EMT, you may encounter situations in which the patient is close to death, and you may need to provide reassurance and emotional care.

■ What Can the EMT Do?

As patients and bystanders are grieving, you can do helpful things, and make simple suggestions. Ask whether there is anything that you can do that will be of help, such as calling a relative or religious advisor. Provide gentle and caring support. Reinforcing the reality of the situation is important. This can be accomplished by merely saying to a grieving person, "I am so sorry for your loss." It is not important that you have a well-rehearsed script, for it is not likely that your exact words or consolations will be remembered. Being honest and sincere are important.

Some statements of consolation tend to be trite, and some suggest a kind of silver lining behind the clouds. Although they may be said with the intention of making the person feel better about a situation, they also can be viewed as an attempt to diminish the person's grief. The grieving person needs to be validated. Statements like these can also indicate our inability to comprehend the profound sadness of grief because we have not experienced that kind of loss. If you have not experienced a death, it is okay to say so; do not pretend that you have.

Attempts to take grief away too quickly are not good. If you do not know how the person really feels, you should not say so. People may be offended by responses that give advice or explanations about the death **Table 2-6**. Statements such as "Oh, you shouldn't feel that way" are judgmental. If you judge what the grieving person is feeling, it is likely that he or she will stop talking with you. People feel what people feel. It is as simple and clean as that. Remember that anger is a stage of grieving. The anger may be directed at you. The anger seems irrational to everyone but the person grieving; therefore, it is necessary that you maintain a professional attitude and let the person grieve in his or her own way.

Statements and comments that suggest action on your part are generally helpful. These statements imply a sense of understanding; they focus on the grieving person's feelings. It is not necessary to go into an extensive discussion. All you need to do is be sincere and say, "I am so sorry for your loss. I just want you to know that I am

Table 2-6	Responding to Grief

Don't say...

- Give it time. Things will get better.
- You should not question God's will.
- You have to get on with your life.
- You have to keep on going.
- You can always have another child.
- You're not the only one who suffers.
- The living must go on.
- I know how you feel.

Try instead...

- I'm sorry for your loss.
- It is okay to be angry.
- It must be hard to accept.
- That must be painful for you.
- Tell me how you are feeling.
- If you want to cry, it's okay.
- People really cared for...

thinking about you." What people really appreciate is somebody who will listen to them. Simply ask, "Would you like to talk about how or what you are now feeling?" Then accept the response.

■ Dealing With the Patient and Family Members

There is no right or wrong way to grieve. Each person will experience grief and respond to it in his or her own way. Family members may express rage, anger, and despair. Many people will be rational and cooperative. Their concerns will usually be relieved by your calm, efficient manner. Your actions and words, even a simple touch, can communicate caring. While you must treat all patients with respect and dignity, use special care with dying patients and their families. Be concerned about their privacy and their wishes, and let them know that you take their concerns seriously. However, it is best to be honest with patients and their families; do not give them false hope.

■ Initial Care of the Dying, Critically Ill, or Injured Patient

Individuals who are in the process of dying as a result of trauma, an acute medical emergency, or a terminal disease will feel threatened. That threat may be related to their concern about survival. These concerns may involve feelings of helplessness, disability, pain, and separation **Table 2-7**.

Table 2-7	Concerns of the Dying, Critically Ill, or Injured Patient

- Anxiety
- Pain and fear
- Anger and hostility
- Depression
- Dependency
- Guilt
- Mental health problems
- Receiving unrelated bad news

Anxiety

Anxiety is a response to the anticipation of danger. The source of the anxiety is often unknown; but in the case of seriously injured or ill patients, the source is usually recognizable. What may increase the anxiety are the unknowns of the current situation. Patients may ask the following questions:

- What will happen to me?
- What are you doing?
- Will I make it?
- What will my disabilities be?

Patients who are anxious may have the following signs and symptoms:

- Emotional upset
- Sweaty and cool skin (diaphoretic)
- Rapid breathing (hyperventilating)
- Fast pulse (tachycardic)
- Restlessness
- Tension
- Fear
- Shakiness (tremulous)

For the anxious patient, time seems to be extended; seconds seem like minutes, and minutes seem like hours. Anxiety is never helpful to a patient, and can cause real physiological harm. It is your job to do everything you can to reduce your patient's anxiety and help your patient cope with what may be the most terrifying experience in his or her lifetime.

Pain and Fear

Pain and fear are closely interrelated. Pain often is associated with illness or trauma. Fear is generally thought of in relation to the oncoming pain and the outcome of the illness or trauma. It is often helpful to encourage patients to express their pains and fears, because expression begins the process of adjustment to the pain and acceptance of the emergency medical care that may be necessary. Some individuals have difficulty in openly admitting their fear. The fear may be expressed as bad dreams, withdrawal, tension, restlessness, "butterflies" in the stomach, or nervousness. In some cases, it may be expressed as anger.

Often you may be tempted to make light of a patient's pain and fear. It is easier to say to the stroke patient, "Oh, you'll be OK," than, "I'm sure you are really scared right now because you are not able to talk, but you should know that I am doing everything I can to help you." Making a connection with your patient through eye contact and the squeeze of a hand can often do more to allay fear than the most eloquent words.

Anger and Hostility

You may find that your patient is expressing anger with very demanding and complaining behavior. Often, this may be related to the fear and anxiety of the emergency medical care that is being given. In other situations, the fear is so acute that the patient may want to express anger toward you or others but is unable to do so because of the dependency factor. If you find that you are the target of the patient's anger, make sure you are safe; do not take the anger or insults personally. Be tolerant, and do not become defensive.

Anger may also be expressed physically, and you may be the target of the displaced aggression. If the patient or a relative becomes so emotionally upset that you are physically assaulted or you believe this could happen, back out of the situation. Such hostility must be contained. If emergency medical care is not possible under these circumstances, law enforcement intervention is required.

Depression

Depression is a natural physiological and psychological response to illness, especially if the illness is prolonged, debilitating, or terminal. Whether the depression is a temporary sadness or clinical depression that is long-term, there is, of course, little you can do to alleviate the pain of depression during the brief time the patient is being treated and transported. The best you can do in treating and transporting a patient experiencing depression is to be compassionate, supportive, and nonjudgmental.

Dependency

Dependency usually takes longer to develop than during the very brief relationships developed in EMS. When medical care is given to any individual, a sense of dependency may develop. Individuals who are placed in this position may feel helpless and become resentful. The resentfulness may arouse feelings of inferiority, shame, or weakness. Make every attempt to remain supportive and compassionate.

Guilt

Many patients who are dying, their families, or the caregivers of those patients may feel guilty over what has happened to them. Occasionally family members and/or long-term caregivers may feel a degree of relief when an extended illness is finally over. That relief may later turn into guilt. Most of the time, however, no one can explain these feelings. The magnitude of the guilt may be very great. Sometimes, feelings of guilt can result in a delay in seeking emergency medical care. Again, understanding the complex emotions that often come to the surface during times of emergency and stress may help you cope with some of the intense and often seemingly bizarre behavior you will encounter in your role as an EMT.

Mental Health Problems

As an EMT, you will be called on to treat and transport patients with mental health problems. These problems may be the cause of the patient's distress or may be caused by the stresses of physical illness or injury. Mental health problems such as disorientation, confusion, or delusions may develop in the dying patient. In these instances, the patient may display behavior inconsistent with normal patterns of thinking, feeling, or acting. Common characteristics of such behavior may include the following:

- Loss of contact with reality
- Distortion of perception—patients may have difficulty judging such common factors as time, distance, and relationships

- Regression—patients may regress to an earlier stage in their development, often infancy or childhood
- Diminished control of basic impulses and desires— patients may act out on their urges without being able to exercise the normal judgment expected as adults. For example, patients may become violent or inappropriately affectionate.
- Abnormal mental content, including delusions and hallucinations

The normal course of dying can cause a patient to seem disoriented. In some long-term situations, generalized personality deterioration may occur (see Chapter 20, *Psychiatric Emergencies,* for a discussion on mental health).

Receiving Unrelated Bad News

A patient who is in critical condition or is dying may not want to hear unrelated bad news, such as the death of someone close to him or her. Such news may depress the patient or cause the patient to give up hope.

Caring for Critically Ill and Injured Patients

When you are caring for a critically ill or injured patient, the patient needs to know who you are and what you are doing. Let the patient know you are attending to his or her immediate needs and these are your primary concerns at this moment Figure 2-20 . As soon as possible, explain

You are the Provider: PART 4

The child is placed onto the stretcher and loaded into the ambulance. Her mother is secured in the front seat of the ambulance. With resuscitative efforts continuing, you depart the scene and proceed to the hospital. The child's condition is reassessed en route.

Recording Time: 11 Minutes	
Level of consciousness	Unconscious and unresponsive
Respirations	Absent
Pulse	Absent
Skin	Cool and cyanotic
Blood pressure	Not obtainable
Sao$_2$	Not obtainable

6. How can poorly managed stress affect your physical well-being?

7. How can the EMT mitigate the stress associated with the job?

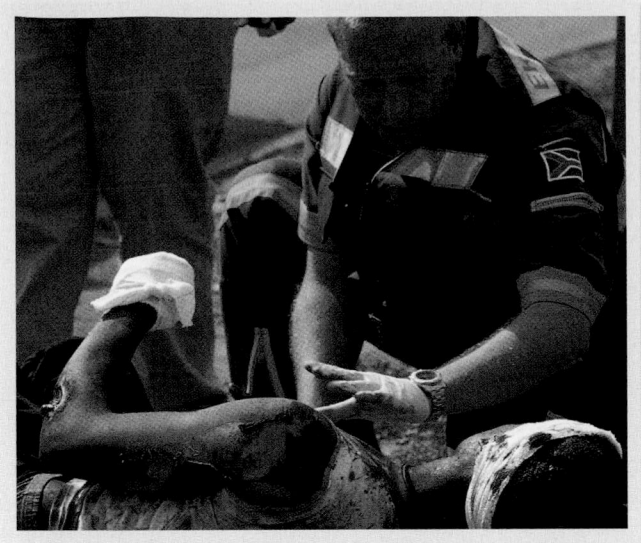

Figure 2-20 Let the patient know immediately that you are there to help.

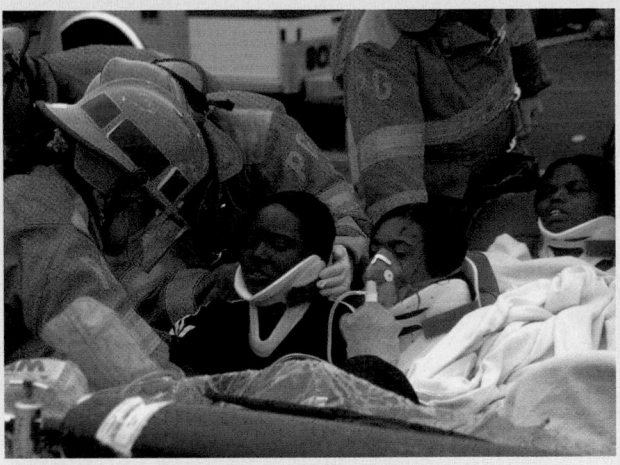

Figure 2-21 The aura of an emergency situation can be confusing and frightening to the patient. Make sure you explain to the patient what has happened.

to the patient what is going on. Confusion, anxiety, and other feelings of helplessness will be decreased if you keep the patient informed from the start.

Avoid Sad and Grim Comments

EMTs, other safety personnel, family members, and bystanders must avoid making grim comments about a patient's condition. Remarks such as "This is a bad one" or "The leg is badly damaged, and I think he will lose it" are inappropriate. These remarks may upset or increase anxiety in the patient and compromise possible recovery outcomes. This is especially true for the patient who may be able to hear but not to respond.

Orient the Patient

You should expect a patient to be disoriented in an emergency situation. The aura of the emergency situation—lights, sirens, smells, and strangers—is intense. The impact and effect of injuries or acute illness may cause the patient to be confused or unsettled. It is important for you to orient the patient to his or her surroundings Figure 2-21. Use brief, concise statements such as "Mr. Smith, you have had an accident, and I am now splinting your arm. I am John Foxworth of the New Britain EMS; I will be caring for you."

Be Honest

When approaching any patient, you must decide how much information each patient is able to understand and accept. You should be honest without additionally shocking the patient or giving information that is

unnecessary or that may not be understood. Simply explain what you are doing, and allow the patient to be part of the care being given; this can relieve feelings of helplessness as well as some of the fear.

Initial Refusal of Care

On occasion you may encounter a patient who refuses emergency medical care and insists you do nothing or leave him or her alone. In these cases, it is important to impress upon the patient the seriousness of his or her condition without causing undue alarm. If you say, "Everything will be okay," when it is obvious that it is not okay, you are not being truthful. Generally, seriously ill or injured patients know that they are in trouble; however, many people refuse care because of their inability to pay the medical expenses. Depending on department policy, the patient may be able to make payment arrangements.

Allow for Hope

In trauma and acute medical conditions, patients may ask you whether they are going to die. At these times, you may feel at a loss for words. You may also know, on the basis of past experience or in view of the seriousness of the present situation, that the prognosis is poor. But it is not your decision to tell the patient that he or she is dying. Statements such as "I don't know if you are going to die; let's fight this one out together" or "I am not going to give up on you, so do not give up on yourself" can be helpful to the patient. These statements transmit a sense of trust and hope, and they let the patient know that you are doing everything possible to save his or her life. If there is the slightest chance of hope remaining, you want that message transmitted in your attitude and in the statements you make to the patient.

Locate and Notify Family Members

Many patients will be concerned and ask you to notify their family or others close to them. The patient may or may not be able to assist you in doing this. You should make sure that an appropriate and responsible person makes an effort to locate the desired persons. Assuring the patient that someone is going to make these notifications may be a significant part of the patient's care because it will help to calm the patient.

Injured and Critically Ill Children

Injured and critically ill children who have life-threatening conditions should be cared for as any adult patient would be, insofar as assessing airway, breathing, and circulation (ABCs) and addressing immediate life threats. Due regard should be given to variations in height, weight, and size when providing emergency medical care to pediatric patients. Because of the increased commotion and the extraordinary nature of the emergency scene for a child, it is important that a relative or responsible adult accompany the child at all times to relieve anxiety and assist in care as appropriate.

Dealing With the Death of a Child

The death of a child is a tragic and dreaded event. It will not be unusual for you to think about the fact that the dead or dying child still has a lot more to do in life and should have many more years to live. In our society, we assume that only old people are supposed to die. In today's world, children die less frequently than they did in earlier times, so many people are unprepared for what they will feel when a child dies. You may think about your own children and those whom you know: nephews, nieces, grandchildren, and children of close friends. And you may think, "Why should this child, who is only 5 years old, die?"

Answering the difficult questions of your own mortality will help when dealing with the death of a child. Still, the death of a child will never be an easy subject to talk about. This will be especially so for the child's family, and as an EMT involved in a call that involves the death of a child, you will also likely experience stress.

One of your responsibilities may be to help the family through the initial period after the death. As an EMT, until more definitive and professional help can be arranged, you may be in the best position to help the family begin to cope with the loss. How a family initially deals with the death of a child will affect its stability and endurance. You can help a family through their initial period of grief and provide information about follow-up counseling and support services that are available.

Helping the Family

Whether the child has just died in your presence or was dead when you arrived, acknowledging the death is important. This should be done in a private place, even if that is inside an ambulance. Often, the parents cannot believe that the death is real, even if they have been preparing for it, as in the case of a terminal illness such as leukemia. Reactions vary, but shock, disbelief, and denial are common emotions and reactions. Some parents show little emotion at the initial news.

If it is possible and appropriate, find a place where the mother and father can hold the child. This is important in the parents' grieving process; it helps to lessen the sense of disbelief and makes the death real. Even if the parents do not ask to see the child, you should tell them that they may do so. Your decision in permitting the parents to see the child may need some discretion on your part. For example, in the case of a traumatic death in which there is significant disfigurement of the body, that decision might have to be delayed. The delay may involve having support services available or contacting the family physician or others who can help the parents through this difficult situation. This situation may also involve preparing the parents for what they will see and the changes brought on by rigor mortis or asphyxiation, for example.

Sometimes, you do not need to say much. In fact, silence can sometimes be more comforting than words. You can express your own sorrow. Do not overload grieving parents with a lot of information; at this point, they cannot handle it. Nonverbal communication, such as holding a hand or touching a shoulder, may also be valuable. Let the family's actions be your guide about what is appropriate. If you sense that the parents want to talk, it is important for you to encourage them to talk about their feelings.

Words of Wisdom

Patients don't care what you know until you show you care. Most patients are not technical experts. They will judge your treatment based on how you behave toward them.

Stressful Situations

Many situations, such as mass-casualty scenes, serious automobile crashes, excavation cave-ins, house fires, infant and child trauma, amputations, abuse of an infant/child/spouse/older person, and death of a coworker or other public safety personnel, will be stressful for everyone involved. During these situations, you must exercise extreme care in both your words and your actions. Be careful to present a professional demeanor in words and actions at the scene. Words that do not seem important, or that are said jokingly, may hurt someone. Conversations at the scene must be professional. You should not say, "Everything will be all right," or "There is nothing to worry about." A person who is trapped in a wrecked car, hurting from head to foot and worrying about a loved one, knows that all is not well. What will reassure the patient is your calm and caring approach to the emergency situation. Whether you are a brand new EMT or a seasoned veteran, patients expect you to bring some sense of order and stability to the terrifying chaos that has suddenly engulfed them. Briefly explain your plan of action to assist the patient in the crisis. Inform the patient that you need his or her help and the assistance of family members or bystanders to carry out your plan of action.

How a patient reacts to injury or illness may be influenced by certain personality traits. Some patients may become highly emotional over what may seem to be a minor problem. Others may show little or no emotion, even after serious injury or illness. Many other factors influence how a patient reacts to the stress of an EMS incident. Among these factors are the following:

- Socioeconomic background
- Fear of medical personnel
- Alcohol or substance abuse
- History of chronic disease
- Mental disorders
- Reaction to medication
- Age
- Nutritional status
- Feelings of guilt
- Past experience with illness or injury

You are not expected to always know why a patient is having an unusual emotional response. However, you can quickly and calmly assess the actions of the patient, family members, and bystanders. This assessment will help you to gain the confidence and cooperation of everyone at the scene. In addition, you should use a professional tone of voice and show courtesy, along with sincere concern and efficient action. These simple considerations will go far to relieve worry, fear, and insecurity on the part of everyone involved. Your calm reassurance will inspire confidence and cooperation. Compassion is also important, but you must be careful. Your professional judgment takes priority over compassion. For example, suppose a screaming child with no obvious life-threatening injuries is covered with another patient's blood. This frightened child appeals to your sense of compassion and thus gets your attention. In the meantime, an unconscious, nonbreathing adult nearby could die from lack of care.

Special Populations

When children are seriously ill or injured, family members and other people at the scene may be frantic. You need to remain calm and confident in your skills because this may be all that is needed to provide reassurance to those at the scene.

Patients must be given the opportunity to express their fears and concerns. You can easily relieve many of these concerns at the scene. Usually, patients are concerned about the safety or well-being of others who are involved in the accident and about the damage or loss of personal property. Your responses must be discreet and diplomatic, giving reassurance when appropriate. If a loved one has been killed or critically injured, you should wait, if possible, until clergy or the emergency department staff can give the patient the news. They can provide the necessary psychological support the patient needs after receiving this type of news.

Words of Wisdom

Calm reassurance on your part will inspire confidence and cooperation. Compassion can also be an important component of your care, but you must be careful that your compassion does not misdirect you to provide inappropriate care. Your professional judgment needs to take priority over compassion.

Some patients, especially children and the elderly, may be terrified or feel rejected when separated from family members by the uniformed EMS provider team. Other patients may not want family members to share their stress, see their injury, or witness their pain. It is usually best if parents are transported with their children and relatives accompany elderly patients. Medical attention for a child often requires adult consent. Treatment may be delayed if a caregiver is not transported with the child.

Religious customs or needs of the patient must also be respected. Some people will cling to religious medals or charms, especially if you make any attempt to remove them. Other people will express a strong desire for religious counsel, baptism, or last rites if death is near. You must try to accommodate these requests. Some people have religious convictions that strongly oppose the use of medications, blood, and blood products. If you obtain such information about your patient, it is imperative that you report it to the next level of care.

In the event of a death, you must handle the body with respect and dignity. It must be exposed as little as possible. Learn your local regulations and protocols about moving the body or changing its position, especially if you are at a possible crime scene. Even in these situations, cardiopulmonary resuscitation (CPR) and appropriate treatment must be given unless there are obvious signs of death.

■ Uncertain Situations

There will be times when you are unsure whether a true medical emergency exists. In these cases, contact medical control about the need to transport. If you cannot reach medical control, it is always best to transport the patient. For both ethical and medicolegal reasons, a physician must examine all patients who are transported and judge the degree of medical need.

Words of Wisdom

It is always best to ensure that the patient is an active participant in the medical care he or she receives. Give your patient the information he or she needs to make an informed decision. You will find that patients who are participating in their own care are more likely to be satisfied with the care they receive.

You must also realize that even the most minor symptoms may be early signs of severe illness or injury in your patient. Symptoms of many illnesses can be similar to those of substance abuse, hysteria, or other conditions. You must accept the patient's complaints and provide appropriate care until you are able to transfer care of the patient to a higher level (eg, paramedic, nurse, or physician). Your local protocols will direct your actions in these uncertain situations. When in doubt, err on the side of caution, acquire the patient's consent, and transport the patient to a medical facility.

Scene Safety

The personal safety of all those involved in an emergency situation is very important. In fact, it is so important that it is best you internalize the steps necessary to preserve personal safety so that your actions become automatic. A second accident at the scene or an injury to you or your partner creates more problems, delays emergency medical care for patients, increases the burden on the other EMTs, and may result in unnecessary injury or death.

You should begin protecting yourself as soon as you are dispatched. Before you leave the scene, begin preparing yourself mentally and physically. Make sure you wear seatbelts and shoulder harnesses en route to the scene. Also make sure to wear seatbelts and shoulder harnesses at all times during transport unless patient care makes it impossible **Figure 2-22**. Don the appropriate PPE prior to departing the ambulance. Many EMS units have mandatory seatbelt policies for the driver at all times, for all EMTs during transit to the scene, and for anyone who is riding with a patient.

Protecting yourself at the scene is also very important. A second accident may damage the ambulance and may result in injury to you, your partner, or additional injury to the patient. The scene must be well marked **Figure 2-23**. If law enforcement has not already done so, you should make sure the proper warning devices are placed at a sufficient distance from the scene. This will alert motorists coming from both directions that a crash has occurred. You should park the ambulance at a safe but convenient distance from the scene. Before attempting to access patients who are trapped in a vehicle, check the vehicle's stability. Then take any necessary measures to secure it. Do not rock or push on a vehicle to find out whether it will move. This can overturn the vehicle or send it crashing into a ditch. If you are uncertain about the safety of a crash scene, wait for appropriately trained individuals to arrive before approaching.

When working at night, you must have plenty of light. Poor lighting increases the risk of injury to both you and the patient. It also results in poor emergency

Figure 2-22 Wear seatbelts and shoulder harnesses en route to the scene.

Figure 2-23 Make sure the crash scene is well marked to prevent a second crash that may damage the ambulance or result in injury to you, your partner, or the patient.

on transportation vehicles and buildings, and labels are used on individual packages containing hazardous materials. The placards or labels are colored and diamond-shaped Figure 2-25. You should never approach any object marked with a placard or label. Remember, some hazardous materials may not be marked properly.

A specially trained and equipped hazardous materials team will be called to the scene to handle disposal of materials and removal of patients. You should not begin caring for patients until they have been moved away from the scene and are decontaminated or the scene is safe for you to enter.

medical care. Wearing reflective emblems or clothing will help to make you more visible at night and decrease your risk of injury Figure 2-24.

■ Scene Hazards

In the course of your career as an EMT, you will be exposed to many hazards. Some situations will be life threatening. In these cases, you must be properly protected, or you must take steps to avoid the hazard completely.

Hazardous Materials

Your safety is the most important consideration at a hazardous materials incident. On your arrival, you should look at the scene and try to read any labels, placards, and identification numbers from a distance, perhaps using binoculars. Placards are used

Safety

There are all kinds of things that can injure you when you are caring for patients. Your best protection against being injured is to carefully size up the scene and constantly check for potential hazards. Don't be foolish and blindly rush in before conducting a proper assessment.

The Department of Transportation (DOT) *Emergency Response Guidebook* is an important resource when dealing with a hazardous materials incident Figure 2-26. It lists common hazardous materials and the proper procedures for scene control and the emergency care of patients. Some state and local government agencies may also have information about hazardous materials commonly present in their areas. A copy of the guidebook and other information relevant to your area should be available in your unit or at the dispatch center. With these references, you should be able to begin proper emergency management as soon as the hazardous material is identified. Do not go into an area and risk exposure to yourself or your partner.

The following are general guidelines you should follow when dealing with scenes involving hazardous materials:

- Do not enter the scene if there is evidence of hazardous materials.
- Remain upwind and uphill of the scene.

Figure 2-24 Wear reflective emblems or clothing to help make you more visible at night and improve your safety in the dark.

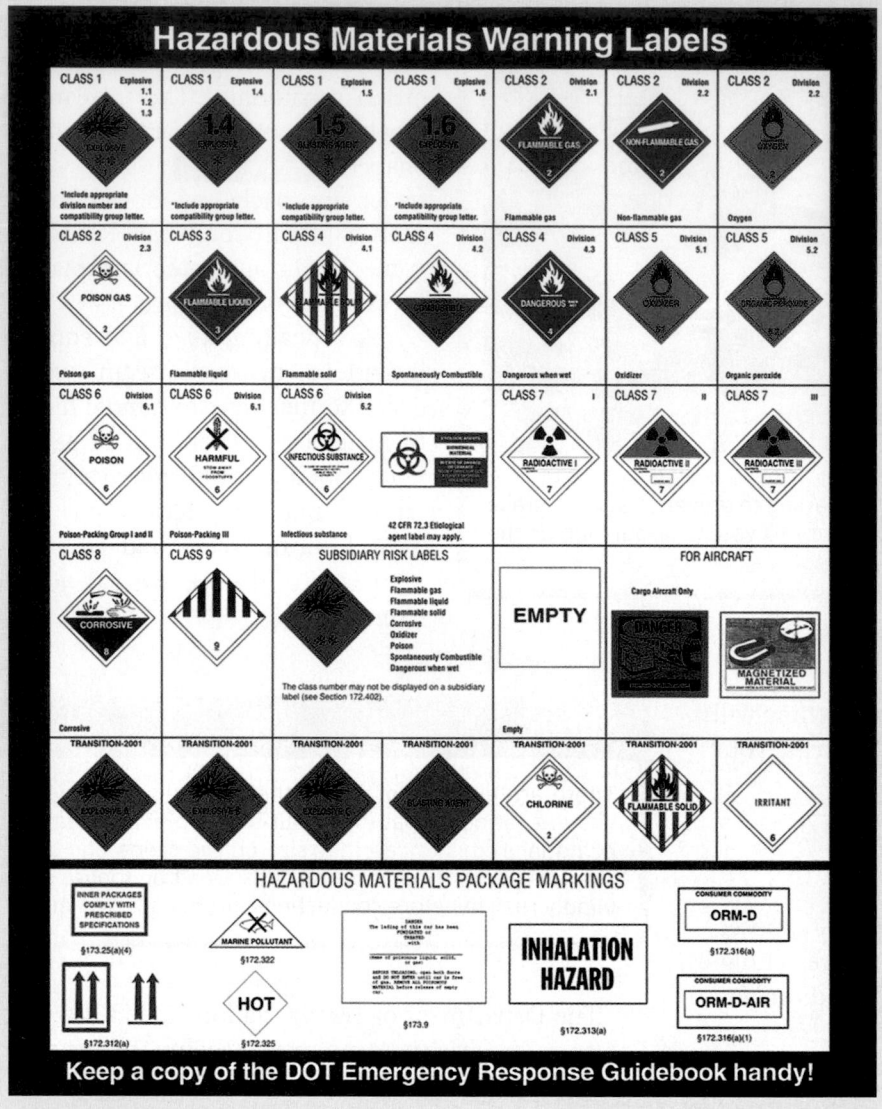

Figure 2-25 Hazardous materials safety placards and labels are colored and diamond-shaped.

Source: Courtesy of the US Department of Transportation.

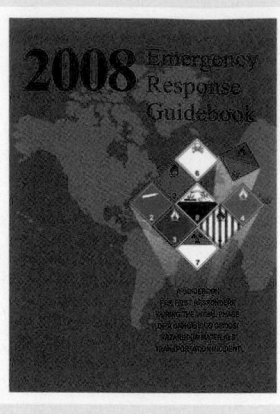

Figure 2-26 The DOT *Emergency Response Guidebook* lists many hazardous materials and the proper procedures for scene control and emergency care of patients.

Source: Courtesy of the US Department of Transportation.

- Keep your distance. This may mean retreating if you become aware of the true nature of the situation.
- Contact dispatch quickly.
- Request additional resources.
- Do not enter the scene until instructed to by trained hazardous materials responders.

Electricity

Electrical shock can be produced by human-made sources (power lines) or natural sources (lightning). No matter what the source, you must evaluate the risk to you and to the patient before you begin patient care.

Power Lines You should not touch downed power lines. Dealing with power lines is beyond the scope of EMT training. However, you should mark off a danger zone around the downed lines. Energized, or "live" power lines, especially high-voltage lines, behave in unpredictable ways. You need in-depth training to be able to handle the equipment that is used in an electrical emergency. The equipment also has specific storage needs and requires careful cleaning. Dirt or other contaminants can make this equipment useless or dangerous.

At the scene of a motor vehicle crash, above-ground and below-grade power lines may become hazards. Disrupted overhead wires are usually a visible hazard. You must be very careful even if you do not see sparks coming from the lines. Visible sparks are not always present in charged wires. The area around downed power lines is always a danger zone. This danger zone extends well beyond the immediate accident scene.

Use the utility poles as landmarks for establishing the perimeter of the danger zone. The danger zone must be a restricted area. Remember, the safety zone is one span of the power pole's distance. Only emergency personnel, equipment, and vehicles are allowed inside this area. Do not approach downed wires or touch anything that downed wires have come in contact with until qualified personnel have concluded that no risk of electrical injury exists. This may mean that you are unable to access a severely injured victim of a motor vehicle crash even though you can see and talk to him or her.

If you must enter this type of situation, be sure to wear the proper protective equipment according to the type of incident. A helmet and turnout gear Figure 2-27 are typically called for, though you cannot count on turnout gear for protection from electrical hazards. Other protective equipment may be needed.

Lightning Lightning is a complex natural phenomenon. You are unwise to think that, "lightning never strikes in the same place twice." If the right conditions remain, a repeat strike in the same area can occur.

Lightning is a threat in two ways: through a direct hit and through ground current. After the lightning bolt strikes, the current drains along the earth, following the most conductive pathway. Although you should avoid high ground to avoid a direct strike, to avoid being injured by a ground current, stay away from drainage ditches, moist areas, small depressions, and wet ropes. If you are involved in a rescue operation, you may need to delay it until the storm has passed. Recognize the warning signs just before a lightning strike. As your surroundings become charged, you may feel a slight tingling sensation on your skin, or your hair may even stand on end. In this situation, a strike may be imminent. Move immediately to the lowest possible area.

If you are caught in an open area, try to make yourself the smallest possible target for a direct hit or for ground current. To keep from being hit by the initial strike, stay away from projections from the ground, such as a single tree. Drop all equipment, particularly metal objects that project above your body. Avoid fences and other metal objects. These structures can transmit current from the initial strike over a long distance. Position yourself in a low crouch.

This position exposes only your feet to the ground current. If you sit, both your feet and your buttocks are exposed. Place an object made of nonconductive material, such as a blanket, under your feet. Get inside a car or your unit, if possible, as vehicles will protect you from lightning.

Fire

You will often be called to the scene of a fire. Therefore, you should understand some basic information about fire, if you do not know it already. There are seven common hazards in a fire:

- Smoke
- Oxygen deficiency
- High ambient temperatures
- Toxic gases
- Building collapse
- Equipment
- Explosions

Smoke is made up of particles of tar and carbon. These particles irritate the respiratory system on contact. Most smoke particles are trapped in the upper respiratory system, but many smaller particles enter the lungs. Some smoke particles not only irritate the airway, but also may be deadly. You must be trained in the use of appropriate airway protection, such as a self-contained breathing apparatus or a disposable short-term device, and have it available at all fire scenes Figure 2-28 .

Fire consumes oxygen. Particularly in a closed space, such as a room, fire may consume most of the available oxygen. This will make breathing difficult for anyone in that space. The high ambient temperatures in a fire can result in thermal burns and damage to the respiratory

Figure 2-27 Wear a helmet made of a certified electrical nonconductor material, making sure that the chin strap is fastened securely.

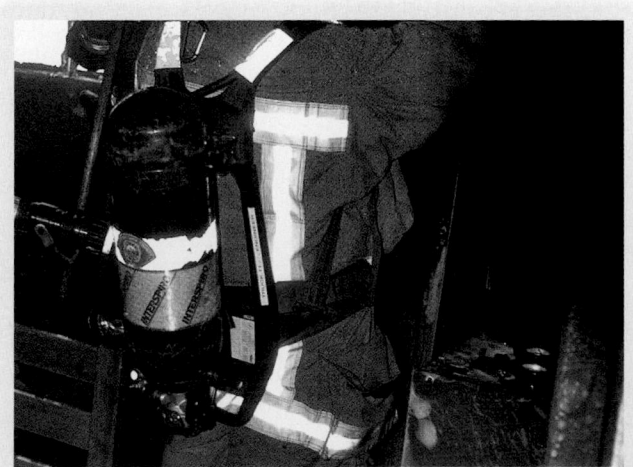

Figure 2-28 You should be trained in the use of self-contained breathing apparatus and have it available if you may be working near fire scenes.

system. Breathing air that is heated above 120°F (49°C) can damage the respiratory system.

A typical building fire emits a number of toxic gases, including carbon monoxide and carbon dioxide. Carbon monoxide is a colorless, odorless gas that is responsible for more fire deaths each year than any other by-product of combustion. Carbon monoxide combines with the hemoglobin in your red blood cells about 200 times more readily than oxygen does. It blocks the ability of the hemoglobin to transport oxygen to your body tissues. Carbon dioxide is also a colorless, odorless gas. Exposure causes increased respirations, dizziness, and sweating. Breathing concentrations of carbon dioxide of greater than 10% to 12% will result in death within a few minutes.

During and after a fire, there is always a possibility that all or part of the burned structure will collapse. Often, there are no warning signs. As an EMS provider, you should never enter a burning building without proper breathing apparatus and the approval of the incident commander or safety officer at the scene. Hasty entry into a burning structure may result in serious injury and possibly death. Once inside a burning building, you are subject to an uncontrolled, hostile environment. Fires are not selective about their victims. You must be extremely cautious whenever you are near a burning structure or

one in which a fire has just been placed under control. At any fire scene, follow the instructions of the incident commander and safety officer and never undertake any task (ie, enter a burning structure or initiate search and rescue) unless you have been properly trained to do so.

Fuel and fuel systems of vehicles that have been involved in crashes are also a hazard. Although this rarely happens, any leaking car fuel may ignite under the right conditions. If you see or smell a fuel leak, or people are trapped in the vehicle, you must coordinate appropriate fire protection equipment. Gasoline and other auto fluids are considered hazardous materials.

Make sure that you are properly protected if there is or has been a fire in the vehicle. Wear appropriate respiratory protection and thermal protection because the smoke from a vehicle fire contains many toxic by-products. The use of appropriate protective gear at a crash scene can reduce your risk of injury. Avoid using oxygen in or near a vehicle that is smoking, smoldering, or leaking fuel.

Vehicle Collisions

Vehicle collisions are common events for EMS providers. These environments provide some of the most unstable and potentially lethal situations an EMS provider will face. Traffic hazards are the first risk to consider. As you

You are the Provider: PART 5

Full resuscitative efforts are continued en route; however, the child has failed to respond to appropriate advanced life support (ALS) and basic life support (BLS) treatment. The child is reassessed and a radio report is called in to the receiving facility.

Recording Time: 18 Minutes

Level of consciousness	Unconscious and unresponsive
Respirations	Absent
Pulse	Absent
Skin	Cool and cyanotic
Blood pressure	Not obtainable
Sao$_2$	Not obtainable

The child is delivered to the emergency department and care is transferred to the attending physician. After an additional 15-minute period of resuscitative efforts in the emergency department, the child is pronounced dead. Later that evening, you find your paramedic partner in his dorm; he is crying and tells you that he does not want to talk right now.

8. Does the death of a child affect the EMT or paramedic differently than the death of an adult? If so, how?

9. How can you help your partner?

drive your ambulance to the scene of the accident, it is important to keep several things in mind. What is the flow of traffic near and around the collision? How will you be able to safely leave and move about the scene? Ideally, you should park your ambulance in a manner where you can easily leave the scene. Keep in mind that additional fire, rescue, and police vehicles also may be parked in the same area or they may be blocking your exit. Hydraulic and hose lines are just two examples of common blockages you may encounter.

If you are the first to arrive at the scene, use the ambulance itself as a shield to protect the scene. The ambulance can be relocated for easier exit once additional help arrives. Park at least 100′ away from all collision sites.

As you approach the scene, be very conscious about the flow of traffic. If needed, request police assistance to shut down the roadway. This will ensure a safe scene as you work with patients. Take notice if there are fluids leaking from the vehicles because they can be flammable. A more common problem with leaking fluids is slipping and sliding on the roadway.

How is the vehicle positioned? Is it stable? Cars and trucks can come to rest in a wide array of positions. As the center of gravity of the vehicle is raised, its ability to fall onto you increases. Standard approach for all vehicle collisions should be for fire fighters to first stabilize the car or truck to ensure safety for the passengers and any EMS providers.

Are there other hazards such as power lines? Downed lines can generate lethal electrical charges many feet away from vehicle collusions. If there are lines down, you should assume they are power lines and not approach. Call for additional resources to manage this hazard. Be aware that most electrical companies will not shut power down to the grid. Though this seems like a logical solution, how many injuries are caused by the unscheduled power outage? If people in their homes are on ventilators, this could create another emergency situation when the power is shut off.

Look closer at the scene. Where are the occupants? Does it appear that violence is present? Is there a good risk of violence? As you look at the vehicle, are there weapons inside? Do the passengers look suspicious? If you feel that there may be violence or if violence is obvious, have the police dispatched to assist you.

With proper equipment and training, you may enter the vehicle itself. Air bags can be another hazard. If the air bag has not deployed, there is a risk that it may accidently activate while you are in the vehicle. Air bags are typically rendered inoperable by the fire department when the power from the car battery is cut.

Your protective clothing will help you to remain safe while working in and around the vehicle collision. The risk of injuries from glass and sharp metal objects cannot be underestimated. Make sure if you are working inside the vehicle you have sufficient protective gear.

Protective Clothing: Preventing Injury

Wearing protective clothing and other appropriate gear is critical to your personal safety. Become familiar with the protective equipment that is available to you. Then you will know what clothing and gear are needed for the job. You will also be able to adapt or change items as the situation and environment change. Remember that protective clothing and gear provide protection only when they are in good condition. It is your responsibility to inspect your clothing and gear. Learn to recognize how wear and tear can make your equipment unsafe. Be sure to inspect equipment before you use it, even if you must do so at the scene.

Clothing that is worn for rescue must be appropriate for the activity and the environmental conditions where the activity will take place. For example, turnout gear worn for firefighting may be too restrictive for working in a confined space. In every situation involving blood and/or other body fluids, be sure to follow standard precautions. You must protect yourself and the patient by wearing gloves and eye protection, as well as any additional protective clothing that may be needed. EMS coats should provide a body fluids barrier if they were purchased after 1998.

Safety

American National Standards Institute (ANSI) requires that all EMS providers utilize a high-visibility public safety vest while on or near the road way.

Cold Weather Clothing

When dressing for cold weather, you should wear several layers of clothing. Multiple layers provide much better protection than a single thick cover. You have more flexibility to control your body temperature by adding or removing a layer. Cold weather protection should consist of at least the following three layers:

1. A thin inner layer (sometimes called the transport layer) next to your skin. This layer pulls moisture away from your skin, keeping you dry and warm. Underwear made of polypropylene or polyester material works well. Wool is the best fiber. The goal is to wick moisture away from the skin.

2. A thermal middle layer of bulkier material for insulation. Wool has been the material of choice for warmth, but newer materials, such as polyester pile, are also commonly used.

3. An outer layer that resists chilling winds and wet conditions, such as rain, sleet, or snow. The two top layers should have zippers to allow you to vent some body heat if you become too warm.

When choosing clothing to protect yourself from various weather conditions, pay attention to the type of material used. Cotton should be avoided in cold, wet environments. Cotton tends to absorb moisture, causing chilling from wetness. For example, if you wear cotton trousers and walk through wet grass, the cotton soaks up the moisture from the grass. This will chill you in cold weather. However, cotton is appropriate in warm, dry weather because it absorbs moisture and pulls heat away from the body.

As an outer layer in cold weather, you might consider plastic-coated nylon, as it provides good waterproof protection. However, it can also hold in body heat and perspiration, which makes you wet both inside and out. Newer, less airtight materials allow perspiration and some heat to escape while the material retains its water resistance. Avoid flammable or meltable synthetic material anytime there is any possibility of fire.

Turnout Gear

Turnout or bunker gear is a fire service term for protective clothing designed for use in structural firefighting environments Figure 2-29 . Turnout gear provides some protection by using different layers of fabric or other material to provide protection from the heat of fire. It also helps to reduce trauma from impact or cuts and keeps water away from the body. Like most protective clothing, turnout gear adds weight and reduces range of motion to some degree.

The exterior fabrics provide increased protection from cuts and abrasions. They also act as a barrier to high external temperatures. In cold weather, an insulated thermal inner layer of material that helps to retain body heat is recommended.

Figure 2-29 Turnout or bunker gear is protective clothing designed for use in firefighting.

Turnout gear or a bunker jacket provides minimal protection from electrical shock. However, it does protect you from heat, fire, possible flashover, and flying sparks. The front opening of the jacket should be fastened, and the jacket should be worn with the collar up and closed in front to protect your neck and upper chest. Proper fit is important so that you can move freely.

Gloves

Firefighting gloves will provide the best protection from heat, cold, and cuts Figure 2-30 . Yet these gloves reduce manual dexterity. In addition, firefighting gloves will not protect you from electrical hazards. In rescue situations, you must be able to use your hands freely to operate rescue tools, provide patient care, and perform other duties. Puncture-proof leather gloves, with latex gloves underneath, will permit free use of your hands with added protection from both injury and body fluids.

Helmets

You should wear a helmet any time you are working in a fall zone. A fall zone is an area where you are likely to encounter falling objects. The helmet should provide top and side impact protection. It should also have a secure chin strap Figure 2-31 . Objects will often fall one after another. If the strap is not secure, the first falling object may knock off your helmet. This leaves your head unprotected as the remaining objects fall.

Construction-type helmets are not well suited for rescue situations. They offer minimal impact protection and have inadequate chin straps. Modern fire helmets

Figure 2-30 Firefighting gloves protect your hands and wrists from heat, cold, and injury.

Figure **2-31** A helmet with top and side impact protection.

Figure **2-32** Boots should cover and protect your ankles, keeping out stones, debris, and snow. Steel-toed boots are preferred.

offer impact protection. However, the projecting brim at the back of the neck may get in your way in a rescue situation. In cold weather, you can lose a great deal of body heat if you are not wearing a hat or helmet. An insulated hat made from wool or a synthetic material can be pulled down over the face and the base of the skull to reduce heat loss in extremely cold weather.

In situations that may involve electrical hazard, you should always wear a helmet with a chin strap and face shield. The shell of the helmet should be made of a certified electrical nonconductor. The chin strap should not stretch. In fact, it should fasten securely so that the helmet stays in place if you are knocked down or a power line hits your head. You should also be able to lock the face shield on the helmet. This will protect your face and eyes from power lines and flying sparks. A standard fire turnout helmet should meet all of these needs.

■ Boots

Boots should protect your feet. They should be water resistant, fit well, and be flexible so that you can walk long distances comfortably. If you will be working outdoors, you should choose boots that cover and protect your ankles, keeping out stones, debris, and snow. Steel-toed boots are preferred Figure **2-32** . In cold weather, your boots must also protect you from the cold. Leather is one of the best materials for boots. However, other materials, such as Gore-Tex water-repellent fabric, are also very good. The soles of your boots must provide traction. Lug-type soles may grip well in snow, but they become very slippery when caked with mud.

Properly fitted boots and shoes are extremely important, because a minor annoyance can develop into a disabling injury. You may develop painful blisters if your feet slip around inside your boots. However, make sure you have enough room to wiggle your toes.

Boots should be puncture-resistant, protect the toes, and provide foot support. It may be difficult to obtain a good fit with firefighting boots; shoe inserts or sock layering may be needed to ensure a comfortable fit. Make sure the tops of your boots are sealed off to keep rain, snow, glass, or other materials from getting into your boots. Moisture increases blistering—wool or wicking socks help prevent feet from becoming wet.

Socks will keep your feet warm and provide some cushioning for you as you walk. In cold weather, two pairs of socks are generally preferable to one thick pair. A thin sock next to the foot helps to wick perspiration away to a thicker, outer sock. This tends to keep your feet warmer, drier, and generally more comfortable. When you purchase new shoes or boots, you may want to try them on while wearing the two pairs of socks to ensure a proper fit.

■ Eye Protection

The human eye is very fragile, and permanent loss of sight can occur from very minor injuries. You need to protect your eyes from blood and other body fluids, foreign objects, plants, insects, and debris from extrication. You may wear eyeglasses with side shields during routine patient care.

However, when tools are being used during extrication, you should wear a face shield or goggles. In these instances, prescription eyeglasses do not provide adequate protection. In snow or white sand, particularly at higher altitudes, you must protect your eyes from ultraviolet exposure. Specially designed glasses or goggles can provide this. In addition, your eye protection must be adaptable to the weather and the physical demands of the task. It is critical that you have clear vision at all times.

Ear Protection

Exposure to loud noises for long periods of time can cause permanent hearing loss. Certain equipment, such as helicopters, some extrication tools, and sirens, produces high levels of noise. Wearing soft foam industrial-type earplugs usually provides adequate protection.

Skin Protection

Your skin needs protection against sunburn while you are working outdoors. Long-term exposure to the sun increases the possibility of skin cancer. It may be considered simply an annoyance, but sunburn is a type of burn. In reflective areas such as sand, water, and snow, your risk of sunburn increases. Protect your skin by applying a sunscreen with a minimum rating of SPF 15.

Body Armor

The policy for most departments directs EMTs to avoid situations that may involve gun violence. EMS responders in some areas wear body armor (bulletproof vests) for personal protection. Several types of body armor are available. They range from extremely lightweight and flexible to heavy and bulky. The lighter vests do not stop large-caliber bullets. However, they offer more flexibility and are preferred by most law enforcement personnel. Lighter vests are commonly worn under a uniform shirt or jacket. The larger, heavier vests are worn on the outside of your uniform.

Violent Situations

The safety of you and your team is of primary concern. Civil disturbances, domestic disputes, and crime scenes, especially those involving gangs, can create many hazards for EMS personnel. Large gatherings of hostile or potentially hostile people are also dangerous. Several agencies will respond to large civil disturbances. In these instances, it is important for you to know who is in command and will be issuing orders **Figure 2-33**. However, you and your partner may be on your own when a group of people seems to grow larger and become increasingly hostile. In these cases, you should call law enforcement immediately if they are not already present. You may need to wait for law enforcement to arrive before you can begin treatment or safely approach a patient.

Remember that you and your partner must be protected from the dangers at the scene before you can

Figure 2-33 Several agencies may respond to large disturbances. It is important for you to know who is in command and will be issuing orders.

provide patient care. Law enforcement must make sure the scene is safe before you and your partner enter. A crime scene often poses potential problems for EMS personnel. If the perpetrator is still somewhere on the scene, this person could reappear and threaten you and your partner or attempt to further injure the patient you are treating. Bystanders who are trying to be helpful may interfere with your emergency medical care. Family members may be very distraught and not understand what you are doing when you attempt to splint an injured extremity and the patient cries out that what you are doing hurts. Be sure that you have adequate assistance from the appropriate public safety agency in these situations.

Sometimes EMTs will be at a scene where a dangerous situation is underway, such as a hostage situation or riot. In these instances, it may be necessary for EMS personnel to be protected from projectiles such as bullets, bottles, and rocks. Law enforcement personnel will ordinarily provide for concealment or cover of personnel who are involved in the response to the incident. **Cover and concealment** involve the tactical use of an impenetrable barrier for protection. EMTs should not be placed in a position that will endanger their lives or safety during such incidents. Do not depend on someone else for your safety.

Remember that your personal safety is of the utmost importance. You must thoroughly understand the risks of each environment you enter. Whenever you are in doubt about your safety, do not put yourself at risk. Never enter an unstable environment, such as a shooting, a brawl, a hostage situation, or a riot. Therefore, as part of your scene size-up, evaluate the scene for the potential for violence. If further violence is a possibility, call for additional resources. Failure to do so may put you and your partner at serious risk. When appropriate, allow law enforcement personnel to secure the scene before you approach; they have the necessary experience and expertise in handling these situations.

It is also important for you to remember that if you believe an event is a crime scene, you must attempt to maintain the chain of evidence. Make sure that you do not disturb the scene unless it is absolutely necessary in caring for the patient.

Behavioral Emergencies

The category of "behavioral emergencies" covers a wide range of situations. This catchall phrase includes emergencies that do not have a clear physical cause and that result in aberrant behavior. Often, the cause turns out to be physical; hypoglycemia, head trauma, hypoxia, and toxic ingestion can all cause altered mental status. Patients with psychiatric diseases, such as certain bipolar disorders or schizophrenia, may have altered sensorium or exhibit abnormal behavior.

Although most behavioral emergencies do not pose a threat to you, the potential of threat to either the patient or yourself still exists and you should use caution.

Although Chapter 20, *Psychiatric Emergencies*, goes into greater depth about psychiatric emergencies, consider these questions as you evaluate the patient in terms of a behavioral or psychiatric emergency that may lead to a violent patient reaction:

- How does this patient respond to you? Are your questions answered appropriately? Are the patient's vocabulary and expressions what you would expect under the circumstances?
- Is the patient withdrawn or detached? Is the patient hostile or friendly? Overly friendly?
- Does the patient understand why you are there?
- How is the patient dressed? Is the dress appropriate for the time of the year and occasion? Are the clothes clean? Dirty?
- Does the patient appear relaxed, stiff, or guarded? Are the patient's movements coordinated or jerky and awkward? Is there hyperactivity? Are the patient's movements purposeful, for example, in putting his or her clothes on? Are the actions aimless, such as sitting and rocking back and forth in a chair?
- Has the patient harmed herself or himself? Is there damage to the surroundings?
- What are the patient's facial expressions? Are they bland or flat, or are they expressive? Does the patient show joy, fear, or anger to appropriate stimuli? If so, to what degree?

It might not be possible for you to gather all of the information that these questions suggest. Sometimes, a patient who is experiencing a behavioral emergency will not respond at all. In those cases, the patient's facial expressions, pulse and respirations, tears, sweating, and blushing may be significant indicators of his or her emotional state.

The following principal determinants of violence, though not intended to be all-inclusive, are of value for the EMT:

- **Past history.** Has this patient previously exhibited hostile, overly aggressive, or violent behavior? This information should be solicited by EMS personnel at the scene or requested from law enforcement personnel, family, previous EMS records, or hospital information.
- **Posture.** How is this person sitting or standing? Does the patient appear to be tense, rigid, or sitting on the edge of the bed, chair, or wherever he or she is positioned? The observation of increased tension by physical posture is often a warning signal for hostility.
- **Vocal activity.** What is the nature of the speech the patient is using? Loud, obscene, erratic, and bizarre speech patterns usually indicate emotional distress. The patient who is conversing in quiet, ordered speech is not as likely to strike out against others as is the patient who is yelling and screaming.
- **Physical activity.** Perhaps one of the most demonstrative factors to look for is the motor activity of a person who is undergoing a behavioral crisis. The patient who is pacing, cannot sit still, or is displaying protection of his or her boundaries of personal space needs careful watching. Agitation is a prognostic sign to be observed with great care and scrutiny.

Other factors to take into consideration for potential violence include the following:

- Poor impulse control
- The behavior triad of truancy, fighting, and uncontrollable temper
- Instability of family structure, inability to keep a steady job
- Tattoos, such as those with gang identification or statements like "born to kill" or "born to lose"
- Substance abuse
- Functional disorder (If the patient says that he or she is hearing voices that say to kill, believe it!)
- Depression, which accounts for 20% of violent attacks
- Diagnosed illness such as bipolar disease

You are the Provider: SUMMARY

1. How can you psychologically prepare yourself for this call?

Regardless of your years of experience in EMS, you must prepare yourself psychologically and logistically when responding to *every* call. Even if the call is dispatched as something that appears to be minor, preparedness is critical; after all, you truly do not know what you are dealing with until you arrive at the scene and assess the patient. Responding to a call for a child—regardless of how it is dispatched—can make psychological preparedness especially difficult. Even the most experienced EMS providers have difficulty overcoming personal reactions and proceeding without hesitation when caring for sick or injured children.

You will experience anxiety during your response to the scene; this is a normal human reaction to a stressful event. The key is to recognize this, and to remain focused on the critical tasks that lie ahead. Instead of reacting negatively, channel your anxiety into a "positive psychological drive" that will make you even more determined to provide the best emergency medical care possible.

You and your partner should have a plan; clearly delineate each of your roles when you arrive at the scene. Discuss the skills and interventions that may need to be performed, the equipment that will be required, and whether or not additional resources will be needed. Doing so will help minimize confusion at the scene and the psychological stress that it can cause.

With experience, your ability to psychologically prepare for an EMS call should progressively get easier. *This does not mean that you are a cold person or any less human*; it means that you are proficient, can think and work under pressure, and are an effective emergency medical care provider.

2. What is stress? How does it manifest?

To prevent stress from negatively impacting your physical and psychological well-being, you must understand what stress is. Stress is the body's physiologic response to any kind of demand—good or bad—and is triggered by one or more stressors. A stressor is any emotional, physical, or environmental situation that causes a variety of physiologic, physical, and psychological responses.

The body's response to stress begins with an alarm response. When stress is placed on the body—in this case, attempting to resuscitate a child in cardiac arrest—the nervous system releases adrenaline into the bloodstream, causing the fight-or-flight response. Short-term stress—for example, treating a critically ill or injured patient—may give you the extra energy you need for optimum performance. Long-term, ineffectively managed stress, however, can negatively impact your emotional and physical well-being. The alarm response is followed by a phase of reaction and resistance, and then recovery or, if the stress is prolonged and ineffectively managed, exhaustion. This three-stage response to stress is called the general adaptation syndrome.

Positive stress (eustress), such as exercise, as well as negative stress (distress), such as shift work, long hours, or the frustration of losing a patient, all have the same physiologic manifestations; these include increased respirations and heart rate,

increased blood pressure, tensed muscles, increased blood glucose levels, and dilated blood vessels near the skin surface (causes cool, clammy skin), among others.

3. What phase of the stress response are you experiencing right now?

You are in an acutely stressful situation—attempting to resuscitate a child in cardiac arrest—and are experiencing the alarm response. Your nervous system is releasing adrenaline into the bloodstream, which is triggering the fight-or-flight response and causing your symptoms (eg, sweating, heart racing). As a result, your body has responded with a burst of energy, which is allowing you to carry out your assigned task of suctioning the child's mouth and managing her airway (the "fight" response). If you were experiencing the "flight" response, you would either "freeze" or try to escape the situation altogether. Being chased by a bear is an example of when your body would react with the flight response. The ability to effectively do your job—despite experiencing the symptoms of stress—indicates that you are able to work under pressure.

4. How should you respond to the mother's question?

Anger is often expressed by very demanding behavior and/or yelling. In this case, anger is a predictable response given the seriousness of the situation. EMS providers are often the target of the family member's anger. Do not take this personally, be tolerant, and do not become defensive.

Clearly, the situation looks grim. In a calm, professional, and caring manner, this information must be communicated to the child's mother. Reassure her that, although the situation is serious, you and your team are doing everything possible to save her child's life. Be honest, do not give her false hope, and do not make promises that you cannot deliver—for example, "Everything will be alright," or "There's nothing to worry about." Your actions and words, even a simple touch, can communicate caring.

5. What stage of the grieving process is the mother experiencing?

The child's mother is actually experiencing two stages of the grieving process simultaneously—denial and anger. There are five stages of the grieving process: denial, anger, bargaining, depression, and acceptance. Not all people grieve in this order and not all people experience all stages of the grieving process. Although unpleasant for both the family member and EMS provider, grieving is a healthy response to a bad situation.

A person in denial refuses to accept the seriousness of the situation, makes unrealistic demands for miracles, or persistently fails to understand why there is no improvement in his or her loved one's condition.

Anger is usually the ugliest stage of the grieving process. During this phase, the person lashes out—usually at the EMS provider. Someone must be blamed, and those who are responsible must be punished. Anger often manifests as hostility toward the provider. Some people may become physically

You are the Provider: SUMMARY, continued

abusive, in which case law enforcement should be summoned to the scene.

During the bargaining stage, the person often makes a promise to be a better person or change their lifestyle in exchange for a positive outcome—for example, "I promise to be a better parent if my child survives."

Depression is actually internalized anger, although the person usually remains silent. Denial, anger, and bargaining have brought no reprieve, so the person feels that the situation is hopeless.

The final stage of the grieving process is acceptance. Most family members do not experience this stage in the acute phase of the crisis, unless their loved one has a terminal illness and their death was expected. Family members generally will not readily accept that their child has unexpectedly died; some people never accept it.

6. How can poorly managed stress affect your physical well-being?

Most people can respond to sudden stress for a short time. With prolonged or poorly managed stress, however, the body can quickly be drained of its reserves. This leaves it depleted of key nutrients, weakened, and more susceptible to disease.

In addition to the emotional damage that poorly managed stress can cause (ie, depression, guilt, persistent anxiety), it has been proven to be a strong contributor to heart disease, hypertension, cancer, alcoholism, and drug abuse, among others.

7. How can the EMT mitigate the stress associated with the job?

Before you can manage stress, you must first recognize its signs and symptoms and identify the stressor(s) involved. Some stressors can be changed or eliminated altogether; others cannot. Caring for critically sick or injured patients is difficult. It is stressful for them, but also for you. Because you are vulnerable to all the stressors that accompany EMS, it is critical to recognize the manifestations of stress so that it does not interfere with your job or personal life.

The signs of chronic stress are not always obvious at first; they may be subtle and not present all the time. Warning signs include irritability toward coworkers, family, and friends; difficulty concentrating; insomnia, hypersomnia, or nightmares; anxiety; indecisiveness; loss of appetite; decreased sex drive; and loss of interest in work, among others.

A clue to the management of stress comes from the fact that it is not the event itself—that is, the stressor—but your reaction to it that determines how much it taxes your body's resources. Stress is an unavoidable part of everyday life; however, understanding how it affects you physiologically, physically, and psychologically can facilitate your ability to successfully manage it.

There are many useful and healthy strategies for managing stress; they may involve changing a few habits or your attitude. Behavioral tactics that have been shown to alleviate or eliminate the body's stress response include changing or eliminating the stressors (this is not always possible, especially in EMS), changing work hours, cutting back on overtime, changing your attitude about the stressor, developing a social network that does not involve your coworkers, and spending more time with your family.

There are also a number of exercises you can employ to minimize the physical response to stress, such as periodic stretching; slow, deep breathing; regular physical exercise; and progressive muscle relaxation. If you are experiencing difficulty managing the stress associated with your job, you should consider seeing a professional counselor.

The key to successful stress management is to find a strategy that works for *you* and to use that strategy frequently and consistently. Remember, the signs of stress are not always present; you may not feel stressed, despite the fact that you are.

8. Does the death of a child affect the EMT or paramedic differently than the death of an adult? If so, how?

The death of any patient is a tragic event. In our society, however, we assume that only older people are supposed to die; the death of a child is totally unacceptable, although an unfortunate reality.

Compared to earlier times, today's children die less frequently, so most people are unprepared for what they will feel when a child death does occur—including EMS personnel. Providers often think about their own children and/or other significant children in their lives (eg, niece, nephew, grandchild). It is common for EMS providers to feel that they did not do everything possible for the child, despite the fact that they indeed provided their best resuscitative efforts.

It is normal to feel sadness and depression following the death of a child; however, unlike the death of an older person, these feelings are often more profound. Children only account for about 10% of all EMS calls; therefore, the death of a child—expected or not—often catches the EMT off guard, resulting in a greater degree of stress and anxiety compared to what is experienced following the death of an adult.

9. How can you help your partner?

Your partner's behavior is consistent with a critical incident stress reaction. Many people are prone to cumulative stress. In the emergency services field, stressors are often sudden and more severe; therefore, many events are unusually stressful or emotional, even by emergency services standards. These acute severe stressors result in what is called critical incident stress. Events that can cause a critical incident stress reaction include mass-casualty incidents, serious injury or death of a child, and serious injury or death of a coworker in the line of duty, among others. No provider is immune to the sadness, despair, and depression that are caused by the death of a child—regardless of his or her years of experience in EMS.

Your partner could also be experiencing posttraumatic stress disorder (PTSD). PTSD occurs after a person experiences a

You are the Provider: SUMMARY, continued

psychologically distressing event that reminds him or her of a previous similar experience. Your partner is an experienced paramedic; therefore, it is likely that he has participated in the resuscitation of a child in the past; this event may have brought back the feelings of anxiety, sadness, and depression that he has previously experienced.

So, how do you help your partner? If he does not wish to talk, do not force the issue. He needs time to collect his thoughts and to grieve—just like the parents. However, you should reassure him that you are willing to listen; some people experience relief just by talking to a coworker, family member, or friend. In other cases, he or she may need to speak to a counselor.

You should alert your supervisor to your partner's crisis. If he is not emotionally fit to provide safe and effective emergency care, he should be replaced for the rest of the shift. In some cases, a grieving EMT or paramedic will become angry if his or her crisis is reported to the supervisor. However, you should reassure him or her that you reported the incident out of concern for his or her physical and emotional well-being. EMS personnel do not just look out for each other during an EMS call; they should also look out for each other after the call—even if it is just as a "sounding board."

EMS Patient Care Report (PCR)

Date: 4-3-09	**Incident #:** 020109	**Nature of Call:** Child not breathing		**Location:** 788 E. Radcliffe	
Dispatched: 0720	**En Route:** 0720	**At Scene:** 0725	**Transport:** 0736	**At Hospital:** 0752	**In Service:** 0812

Patient Information

Age: 4 **Sex:** F **Weight (in kg [lb]):** 19 kg (42 lb)	**Allergies:** None **Medications:** None **Past Medical History:** None **Chief Complaint:** Cardiopulmonary arrest

Vital Signs

Time: 0726	**BP:** Unobtainable	**Pulse:** Absent	**Respirations:** Absent	**Sao$_2$:** Unobtainable
Time: 0731	**BP:** Unobtainable	**Pulse:** Absent	**Respirations:** Absent	**Sao$_2$:** Unobtainable
Time: 0736	**BP:** Unobtainable	**Pulse:** Absent	**Respirations:** Absent	**Sao$_2$:** Unobtainable
Time: 0741	**BP:** Unobtainable	**Pulse:** Absent	**Respirations:** Absent	**Sao$_2$:** Unobtainable
Time: 0747	**BP:** Unobtainable	**Pulse:** Absent	**Respirations:** Absent	**Sao$_2$:** Unobtainable

EMS Treatment
(circle all that apply)

Oxygen @ 15 L/min via (circle one): NC NRM ~~Bag-Mask Device~~	(Assisted Ventilation)	(Airway Adjunct)		(CPR)
Defibrillation	**Bleeding Control**	**Bandaging**	**Splinting**	**Other:** Cardiac monitoring, IV, medication therapy, intubation

Narrative

9-1-1 dispatch for an unconscious child not breathing. On arrival at the scene, found 2 EMRs performing CPR on the child, a 4-year-old female. The child's mother stated that when she went to wake her up, she was unconscious, unresponsive, and not breathing; she called 9-1-1 and began CPR. The mother denies that her child has any significant past medical history or drug allergies; she further denies any recent trauma or potentially toxic ingestion. After 2 minutes of CPR, reassessment revealed that the child remained apneic and pulseless. Continued two-rescuer CPR and applied the cardiac monitor, which revealed asystole. Paramedic on scene successfully performed endotracheal intubation. An IV line was established and medications were administered per protocol. Performed resuscitative efforts at the scene for approximately 10 minutes, and then loaded the child into the ambulance and began transport. The child's mother accompanied her to the hospital, and was secured in the passenger's seat of the ambulance. Continued CPR and appropriate medication therapy en route. The child's condition remained unchanged; she remained apneic and pulseless and the electrocardiogram continued to show asystole. Delivered the child, whose condition remained unchanged, to the emergency department staff and gave verbal report to the attending physician. Provided emotional support to the child's mother and then returned to service. *End of report**

Prep Kit

Ready for Review

- A communicable disease is any disease that can be spread from person to person or animal to person.

- Infectious diseases can be transmitted by contact (direct or indirect), or they are airborne, foodborne, or vector-borne.

- Even if you are exposed to an infectious disease, your risk of becoming ill is small.

- Whether or not an acute infection occurs depends on several factors, including the amount and type of infectious organism and your resistance to that infection.

- You can take several steps to protect yourself against exposure to infectious diseases, including:
  - keeping up to date with recommended vaccinations
  - following standard precautions at all times
  - handling all needles and other sharp objects with great care

- Because it is often impossible to tell which patients have infectious diseases, you should avoid direct contact with the blood and body fluids of all patients.

- You should know what to do if you are exposed to an airborne or bloodborne disease. Your department's designated officer will be able to help you follow the protocol set up in your area.

- Infection control should be an important part of your daily routine. Be sure to follow the proper steps when dealing with potential exposure situations.

- If you think you may have been exposed to an infectious disease, see your physician (or your employer's designated physician) immediately.

- Recognizing the signs of stress is important for all EMTs.

- Common workplace issues include cultural diversity, sexual harassment, and substance abuse. You should know what to do to avoid or address these situations.

- EMTs will encounter death, dying patients, and the families and friends of those who have died.

- Scene hazards include potential exposure to the following:
  - Hazardous materials
  - Electricity
  - Fire

- At a hazardous materials incident, your safety is the most important consideration. Never approach an object labeled with a hazardous materials placard or label. Use binoculars to read the placards or labels from a safe distance.

- Do not begin caring for patients until they have been moved away from the scene and decontaminated by the hazardous materials team or the scene has been made safe for you to enter.

- There are seven common hazards in a fire:
  - Smoke
  - Oxygen deficiency
  - High ambient temperatures
  - Toxic gases
  - Building collapse
  - Equipment
  - Explosions

- Every patient encounter should be considered to be potentially dangerous. It is essential that you take all available precautions to minimize exposure and risk to scene hazards and infectious and communicable diseases.

- When signs of stress such as fatigue, anxiety, anger, feelings of hopelessness, worthlessness, or guilt, and other such indicators manifest themselves, behavioral problems can develop.

- Violent situations such as civil disturbances, domestic disputes, and crime scenes can create many hazards for EMS personnel.

- If you see the potential for violence during a scene size-up, call for additional resources.

Vital Vocabulary

acute stress reactions Reaction to stress that occurs during a stressful situation.

airborne transmission The spread of an organism in aerosol form.

bloodborne pathogens Pathogenic microorganisms that are present in human blood and can cause disease in humans. These pathogens include, but are not limited to, hepatitis B virus and human immunodeficiency virus (HIV).

Centers for Disease Control and Prevention (CDC) The primary federal agency that conducts and supports public health activities in the United States. The CDC is part of the US Department of Health and Human Services.

communicable disease A disease that can be spread from one person or species to another.

contamination The presence of infectious organisms on or in objects such as dressings, water, food, needles, wounds, or a patient's body.

cover and concealment The tactical use of an impenetrable barrier for protection.

critical incident stress management (CISM) A process that confronts the responses to critical incidents and defuses them, directing the emergency services personnel toward physical and emotional equilibrium.

cumulative stress reactions Prolonged or excessive stress.

delayed stress reactions Reaction to stress that occurs after a stressful situation.

designated officer The individual in the department who is charged with the responsibility of managing exposures and infection control issues.

direct contact Exposure or transmission of a communicable disease from one person to another by physical contact.

exposure A situation in which a person has had contact with blood, body fluids, tissues, or airborne particles in a manner that suggests disease transmission may occur.

foodborne transmission The contamination of food or water with an organism than can cause disease.

general adaptation syndrome The body's response to stress that begins with an alarm response, followed by a stage of reaction and resistance, and then recovery or, if the stress is prolonged, exhaustion.

hepatitis Inflammation of the liver, usually caused by a viral infection, that causes fever, loss of appetite, jaundice, fatigue, and altered liver function.

human immunodeficiency virus (HIV) Acquired immunodeficiency syndrome (AIDS) is caused by HIV, which damages the cells in the body's immune system so that the body is unable to fight infection or certain cancers.

host The organism or individual that is attacked by the infecting agent.

immune The body's ability to protect itself from acquiring a disease.

indirect contact Exposure or transmission of disease from one person to another by contact with a contaminated object.

infection The abnormal invasion of a host or host tissues by organisms such as bacteria, viruses, or parasites, with or without signs or symptoms of disease.

infection control Procedures to reduce transmission of infection among patients and health care personnel.

infectious disease A medical condition caused by the growth and spread of small, harmful organisms within the body.

Occupational Safety and Health Administration (OSHA) The federal regulatory compliance agency that develops, publishes, and enforces guidelines concerning safety in the workplace.

pathogen A microorganism that is capable of causing disease in a susceptible host.

personal protective equipment (PPE) Protective equipment that OSHA requires to be made available to the EMT. In the case of infection risk, PPE blocks entry of an organism into the body.

posttraumatic stress disorder (PTSD) A delayed stress reaction to a prior incident. This delayed reaction is often the result of one or more unresolved issues concerning the incident.

transmission The way in which an infectious disease is spread: contact, airborne, by vehicles, or by vectors.

standard precautions Protective measures that have traditionally been developed by the Centers for Disease Control and Prevention (CDC) for use in dealing with objects, blood, body fluids, or other potential exposure risks of communicable disease.

vector-borne transmission The use of an animal to spread an organism from one person or place to another.

Assessment
in Action

You and your partner are dispatched to a sick person. On arrival you find a conscious patient who is complaining of a fever, night sweats, and a cough. The patient also reports a history of tuberculosis. Your partner is assessing the patient and you notice that he has not taken standard precautions by donning personal protective equipment. Lately you have noticed that your partner is disinterested in his work and coming in late, taking unnecessary risks, and has taken to sitting alone at the station and not socializing with other members of the team. The job has been stressful lately; call volume has increased and you rarely have any downtime between calls.

1. What standard precautions should you take with a patient who has signs and symptoms of fever, night sweats, and a cough?
   A. Gloves only
   B. Mask only
   C. Gloves and mask
   D. Gloves, mask, and gown

2. Infectious diseases passed on by insects or parasitic worms are known as what type of transmission?
   A. Vector-borne
   B. Direct contact
   C. Indirect contact
   D. Foodborne

3. What is the name of the federal agency that conducts and supports public health?
   A. Occupational Safety and Health Administration (OSHA)
   B. Centers for Disease Control and Prevention (CDC)
   C. Drug Enforcement Administration (DEA)
   D. Federal Emergency Management Agency (FEMA)

4. Because your partner did not use standard precautions, he has been exposed to an infectious disease. What should he do?
   A. Ignore it because the risk of contamination is small.
   B. Report it to the hospital staff so they can isolate the patient.
   C. Report it to the infection control officer.
   D. Sanitize any equipment that was in contact with the patient.

5. What is contamination?

6. On the basis of your partner's actions, what is he most likely experiencing?
   A. Acute stress reaction
   B. Cumulative stress reaction
   C. Posttraumatic stress disorder
   D. Delayed stress reaction

7. What signs and symptoms is your partner displaying to show that he is under stress?

8. What are the long-term physical effects of stress?

9. What are the long-term psychological effects of stress?

10. What can be done to reduce stress?

Medical, Legal, and Ethical Issues

National EMS Education Standard Competencies

Preparatory

Applies fundamental knowledge of the emergency medical services (EMS) system, safety/well-being of the emergency medical technician (EMT), medical/legal, and ethical issues to the provision of emergency care.

Medical/Legal and Ethics

- Consent/refusal of care (pp 79-83)
- Confidentiality (pp 83-84)
- Advanced directives (pp 84-85)
- Tort and criminal actions (pp 90-93)
- Evidence preservation (p 94)
- Statutory responsibilities (pp 87-90)
- Mandatory reporting (pp 93-95)
- Ethical principles/moral obligations (pp 95-96)
- End-of-life issues (pp 84-87)

Knowledge Objectives

1. Define consent, and describe how it relates to decision making. (pp 79-80)
2. Differentiate expressed consent, implied consent, and involuntary consent. (pp 80-81)
3. Discuss the giving of consent by minors for treatment or transport. (p 81)
4. Describe local EMS system protocols for using forcible restraint. (pp 81-82)
5. Discuss the EMT's role and obligations if a patient refuses treatment or transport. (pp 82-83)

6. Understand that communication with patients is confidential, protected by the Health Insurance Portability and Accountability Act (HIPAA). (pp 83-84)
7. Discuss the importance of do not resuscitate (DNR) orders (advance directives) and provisions in the locality regarding EMS application. (pp 84-85)
8. Describe the physical, presumptive, and definitive signs of death. (pp 85-87)
9. Understand that organ donors are treated the same way as any other patients needing treatment, and the need to follow local protocols with such patients. (p 87)
10. Recognize the importance of medical identification insignia in treating the patient. (p 87)
11. Understand the scope of practice and standards of care. (pp 87-90)
12. Describe the EMT's legal duty to act. (p 90)
13. Discuss the issues of negligence, abandonment, assault and battery, and kidnapping and their implications for the EMT. (pp 90-92)
14. Explain the reporting requirements for special situations, including abuse, drug- or felony-related injuries, childbirth, and crime scenes. (pp 93-95)
15. Define ethics and morality, and discuss their implications for the EMT. (pp 95-96)
16. Understand the role and comportment of the EMT in court. (pp 96-97)

Skills Objectives

There are no skills objectives for this chapter.

Introduction

A basic principle of emergency care is to do no further harm. Any health care provider who acts in good faith and according to an appropriate standard of care usually avoids legal exposure. Providing emergency medical care in an organized system is a recent phenomenon.

__Emergency medical care__, or immediate care or treatment, is often provided by an EMT, who may be the first link in the chain of prehospital care. As the scope and nature of emergency medical care become more complex and widely available, litigation involving participants in EMS systems will no doubt increase. Providing competent emergency medical care that conforms with the standard of care taught to you will help you to avoid both civil and criminal actions. Consider the following situations:

- You are transporting a patient, and while the stretcher is being loaded into the ambulance, your partner slips, the stretcher crashes to the ground, and the patient is injured.
- You are about to begin treating a child, and the father commands you to stop.

What should you do? Even when emergency medical care is properly rendered, there are times when you may be sued by a patient who seeks to obtain relief, often in the form of a monetary award, for pain and suffering. Administrative action, such as suspension of your state EMT certificate, may be brought against you for failure to abide by the regulations of your state EMS agency. For these reasons, you must understand the various legal aspects of emergency medical care.

You must also consider ethical issues. As an EMT, should you stop and treat patients who were involved in an automobile crash while you are en route to another emergency call? Should you begin CPR on a patient who, according to the family, has terminal cancer? Should patient information be released to a patient's attorney on the telephone?

Consent

In the course of your career as an EMT, you will find in most circumstances that consent is required from every conscious adult before care can be started. A person receiving care must give permission, or __consent__, for treatment. If a person is conscious, rational, and capable of making informed decisions, he or she has a legal right to refuse care, even though ill or injured. A patient may also consent to some aspects of care and deny consent for others. If the patient refuses care, you may not care for the patient. In fact, doing so may be grounds for both criminal and civil action. Consent can be expressed (actual) or implied and can also apply to the care of a minor or a mentally incompetent patient.

The foundation of consent is decision making capacity. __Decision making capacity__ is the ability of a patient to understand the information you are providing to him or her, coupled with the ability to process that information and make an informed choice regarding medical care that is appropriate for him or her. It is important to keep in mind that the law allows the patient to make choices that may seem foolhardy and that might endanger the patient's life. The right of a patient to make decisions concerning his or her health is known as __patient autonomy__. The terms "decision making capacity" and "competence" are often used interchangably but there is a distinction: "competence" is generally regarded as a legal term and determinations regarding competence are typically made by a court of law, whereas "decision making capacity" is the term more commonly used in health care to determine

You are the Provider: PART 1

At 5:20 PM, you are dispatched to a grocery store at 1175 N. Main St. for a man with a severe headache. You respond to the scene, which is located only a few miles away. The weather is clear, the temperature is 90°F, and the traffic is heavy.

1. What is the difference between scope of practice and standard of care?

2. When does the EMT have a legal duty to act?

whether or not a patient is capable of making health care decisions.

The following factors should be considered when determining a patient's decision making capacity:

- Is the patient's intellectual capacity impaired by mental limitation or any type of dementia?
- Is the patient of legal age (18 years of age in most states)?
- Is the patient impaired by alcohol or drug intoxication or serious injury or illness?
- Does the patient appear to be experiencing significant pain?
- Are there any apparent hearing or visual problems?
- Is a language barrier present? Do you and your patient speak the same language?
- Does the patient appear to understand what you are saying? Does he or she ask rational questions that demonstrate an understanding of the information you are trying to share?

You should be familiar with various types of consent. These include expressed consent, implied consent, and involuntary consent.

■ Expressed Consent

Expressed consent (or actual consent) is the type of consent given when the patient verbally or otherwise acknowledges that he or she wants you to provide care or transport. Expressed consent may be nonverbal. For example, if you ask a patient if you can check his or her blood pressure, and the patient extends an arm to you, the patient is expressing consent nonverbally.

To be valid, the consent the patient provides must be **informed consent**, which means that you explained the nature of the treatment being offered, along with the potential risks, benefits, and alternatives to treatment, as well as potential consequences of refusing treatment, and the patient has given consent. The very nature of emergency care dictates that the information you provide to the patient in the field must be done quickly and will lack the detail and formality of consent that might occur in a hospital setting. Paramedics and others providing advanced life support must provide more detailed information to the patient because of the nature of the treatment they are offering. In such cases, there is a greater potential for side effects and other adverse responses associated with drug administration and other forms of advanced care.

Informed consent is valid if given orally, but it may be difficult to prove at a later point in time. Rarely do EMS providers have patients sign a consent form, so it is always advisable to document consent in your run report. Having someone witness the patient's consent may be helpful if the issue of consent is later challenged in court.

Remember that a patient may agree to certain types of emergency medical care but not to others. For example, a patient might agree to be removed from a car but refuse further care. An injured person might agree to emergency care at home but refuse to be transported to a medical facility. A patient may also change his or her mind at any time and withdraw consent to treatment and/or transport. In such cases, you must respect the patient's wishes and discontinue any further care.

■ Implied Consent

When a person is unconscious or otherwise incapable of making a rational, informed decision about care, and unable to give consent, the law assumes that the patient would consent to care and transport to a medical facility if he or she were able to do so **Figure 3-1**. Patients who are intoxicated by drugs or alcohol, mentally impaired, or suffering from certain conditions such as head injury might be included in this category. The legal principle that allows treatment under such circumstances is called **implied consent**. Implied consent applies only when a serious medical condition exists and should never be used unless there is a threat to life or limb. For this reason, the principle of implied consent is known as the **emergency doctrine**. However, sometimes what represents a "serious threat" may be unclear and it may become a legal question. This may result in legal proceedings and a **medicolegal** judgment, which should be supported by your best efforts to obtain consent and a thoroughly documented run report. In most instances, the law allows a spouse, a close relative, or next of kin to give consent for an injured person who is unable to do so and you should make every effort to obtain consent from an available relative before treating based on implied consent. It is also important to understand that if a patient being treated based on implied consent were to regain consciousness and appear capable of making an informed decision, the doctrine of implied consent would no longer apply.

■ Involuntary Consent

Assisting patients who are mentally ill or in behavioral (psychological) crisis or developmentally delayed is complicated. An adult patient who is mentally incompetent is not able to give informed consent. From a legal perspective, this situation is similar to those involving minors. Consent for emergency care should be obtained from someone who is legally responsible for the patient, such as a guardian or conservator. In many cases, however, such permission will not be readily obtainable. Many states

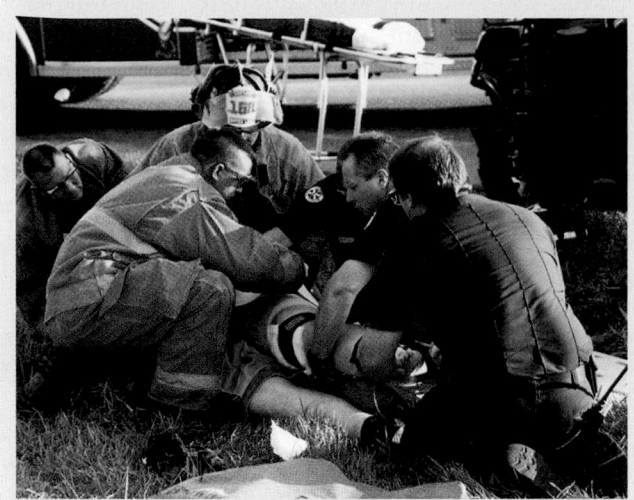

Figure 3-1 When a serious threat to life exists and the patient is unconscious or otherwise unable to give consent, the law assumes that the patient would give consent to care and transport to the hospital.

Figure 3-2 The law requires that a parent or a legal guardian give consent for treatment or transport of a minor. However, you must never withhold lifesaving care.

have protective custody statutes allowing such a person to be taken, under law enforcement authority, to a medical facility. Under certain conditions, law enforcement and prison officials are legally permitted to give consent for any individual who is incarcerated or has been placed under arrest. However, a prisoner who is conscious and capable of making decisions does not surrender the right to make medical decisions and may refuse care. Know the provisions in your area and involve online medical control in the process. Remember that when a serious medical emergency exists, you can assume that implied consent exists.

■ Minors and Consent

Because a minor might not have the wisdom, maturity, or judgment to give valid consent, the law requires that a parent or legal guardian give consent for treatment or transport **Figure 3-2**. However, in some states, a minor can give valid consent to receive medical care, depending on the minor's age and maturity. A great deal of confusion surrounds the issue of emancipated minors. **Emancipated minors** are individuals who, despite being under the legal age in a given state (in most cases the age is 18 years), can be legally treated as adults based on certain circumstances. For example, many states consider minors to be emancipated if they are married, if they are members of the armed services, or if they are parents. A minor who is a parent may also give consent for his or her own child. In addition, a minor is usually considered emancipated if living away from and no longer relying on his or her parents for support. A court may issue an order declaring a minor to be emancipated but this is

not commonly seen. You should know your state's laws concerning the issues surrounding emancipation.

If a minor is injured and requires medical treatment in a school or camp setting, teachers and school officials may act **in loco parentis**, which means in the position or place of a parent, and can legally give consent for treatment of the minor if a parent or guardian is not available. You should still make an effort to obtain consent from a parent or legal guardian whenever possible; however, if a true emergency exists and the parent or legal guardian is not available, the consent to treat the minor is implied, just as with an adult. You must never withhold lifesaving care for a minor because a person authorized to provide consent is not available.

■ Forcible Restraint

Forcible restraint is sometimes necessary when you are confronted with a patient who is in need of medical treatment and transportation but is combative and presents a significant risk of danger to himself, herself, or others **Figure 3-3**. Such behavior may result from an underlying psychiatric or behavioral condition, the effects of drugs, or a medical condition such as a head injury or hypoxia. Forcible restraint of such individuals is legally permissible and may be required before emergency care can be rendered. However, you must generally consult medical control for authorization to restrain or contact law enforcement personnel who have authority to restrain people. In some states, only a law enforcement officer may forcibly restrain an individual. You should be clearly informed about local laws. Restraint without legal authority exposes you to potential civil and criminal

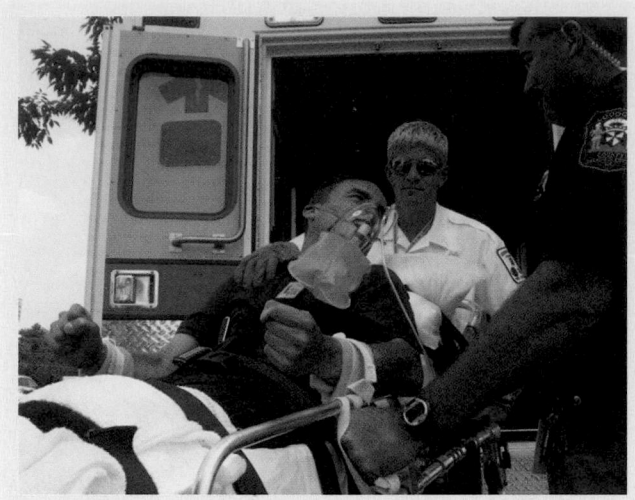

Figure 3-3 Be sure that you know the local laws about forcible restraint of a patient. In some states, only a law enforcement officer has the authority to restrain a patient.

applying restraints, it is important to remember, in terms of treatment, that if the patient is conscious and the situation is not urgent, consent is required. Restraints should only be considered if the patient has a medical condition that appears serious or suffers from an apparent behavioral disorder that poses a risk to the patient or others. After restraints are applied, they should not be removed en route unless they pose a risk to the patient, even if the patient promises to behave.

The Right to Refuse Treatment

Adults who are conscious, alert, and appear to have decision making capacity have the right to refuse treatment or withdraw from treatment at any time even if doing so may result in death or serious injury. Such patients present you with a dilemma. Should you provide care against their will? Should you leave them alone? Calls involving refusal of treatment are commonly litigated in EMS and require you to proceed very cautiously. You must be familiar with local policies regarding refusal of care. In all such cases, you should involve online medical control. A patient's decision to either accept or refuse treatment should be based on information that you provide. This information should include your assessment of what might be wrong with the patient, a description of the treatment that you feel is necessary, any possible risks of treatment, the availability of alternative treatments,

penalties. Restraint may be used only in circumstances of risk to the patient or others. When a patient is combative and poses a risk to the rescuer, it is advisable to wait for law enforcement to arrive on scene before attempting to treat the patient.

Your service should have clearly defined protocols to deal with situations involving restraint. Also, no matter what the situation is in terms of applying or not

You are the Provider: PART 2

On arriving at the scene, you find the patient, a 32-year-old man, sitting on the sidewalk outside the grocery store. He is grabbing both sides of his head, but looks up and acknowledges your presence. You begin to assess the patient as your partner opens the jump kit and prepares to take his vital signs.

Recording Time: 0 Minutes	
Appearance	Grabbing both sides of his head; in obvious pain
Level of consciousness	Conscious and alert
Airway	Open; clear of secretions or foreign bodies
Breathing	Increased respiratory rate; adequate depth
Circulation	Radial pulses, bilaterally strong and regular; skin is pink, warm, and dry

Without talking to the patient, your partner begins to take his blood pressure and applies the pulse oximetry probe to his finger.

3. Are you legally authorized to treat this patient? Why or why not?

4. How does informed consent differ from implied consent?

and the possible consequences of refusing treatment. Be sure that the patient understands everything that you say and encourage the patient to ask questions.

When treatment is refused, you must assess the patient's ability to make an informed decision. Ask and repeat questions, assess the patient's answers, and observe the patient's behavior. If the patient appears confused or delusional, you cannot assume that the decision to refuse is an informed refusal. Remember that no single assessment finding will enable you to determine that the patient is capable of making an informed decision about health care. As with most medical conditions, it is the constellation of findings that will support your conclusion. When in doubt, providing treatment is a much more defensible position than failing to treat a patient. However, do not endanger yourself to provide care and use the assistance of law enforcement to ensure your own safety.

Before leaving the scene where a patient has refused care, you should again encourage the patient to permit treatment and to call for the ambulance if he or she has a change of mind. Advise the patient to contact his or her personal physician as soon as possible. It is essential for you to ask the patient to sign a refusal of treatment form and to thoroughly document all refusals. Your documentation should include any assessment findings that you were able to make and all efforts that you made to obtain consent. Your documentation should also include a description of possible consequences of refusing treatment and transport. The patient's signature should be witnessed by a family member or law enforcement officer to protect you from a later claim for negligence or abandonment. Both of these terms are discussed later in this chapter. Notify medical control of your actions.

You may also be faced with a situation in which a parent refuses to permit treatment of an ill or injured child. In this situation, you must consider the emotional impact of the emergency on the parent's judgment. In this and virtually all cases of refusal, you can usually resolve the situation with patience and calm persuasion. You may also need the help of others, such as your supervisor, medical control, or law enforcement officials.

When you are not able to persuade the patient, guardian, conservator, or parent of a minor or mentally incompetent patient to proceed with treatment, you must obtain the signature of the individual who is refusing treatment on an official release form that acknowledges refusal. Document any assessment findings, the emergency care that you provided, your efforts to obtain consent, and the responses to your efforts. You must also obtain a signature from a witness to the refusal. Make every effort to have a responsible person, such as a law enforcement officer, serve as a witness to these events. Retain the documents with your records—they will be

Words of Wisdom

When a patient, parent, or guardian refuses treatment or transport, protect yourself with both a thorough patient care report (PCR) and an official refusal form. Have the patient or other refusing party sign the form, document in the PCR what you have done to ensure an informed refusal, and note the involvement of medical control in the situation. Be sure to submit the refusal form with your PCR.

important in the event a legal claim is filed later. If the person in authority refuses to sign a release form, inform medical control and thoroughly document the situation and the refusal. In some cases, parents who have refused to provide medical care for a child have been charged with child neglect. You might be called as a witness in such cases and you must be sure that all documentation is thorough and accurate.

Confidentiality

Communication between you and the patient is considered confidential and generally cannot be disclosed without permission from the patient or a court order. Confidential information includes the patient history, assessment findings, and treatment provided. Disclosure of such information without proper authorization may result in liability for **breach of confidentiality**. In most states, records may be released when a legal subpoena is presented or the patient signs a written release. Patient information may also be shared with billing personnel.

HIPAA

HIPAA is the acronym for the Health Insurance Portability and Accountability Act of 1996. Although this act had many aims, including improving the portability and continuity of health insurance coverage and combating waste and fraud in health insurance and the provision of health care, the section of the act that most affects EMS relates to patient privacy. The aim of this section of the act is to strengthen laws for the protection of the privacy of health care information and to safeguard patient confidentiality. As such, it provides guidance on what types of information is protected, the responsibility of health care providers regarding that protection, and penalties for breaching that protection.

HIPAA considers all patient information that you obtain in the course of providing medical treatment to a patient to be **protected health information (PHI)**. This includes not only medical information, but also any information that can be used to identify the patient. As an

EMT, you have an obligation to guard all protected health information from unlawful disclosure, either written or verbal.

PHI may be disclosed for purposes of treatment, payment, or operations. This means that you are permitted to report your assessment findings and treatment to other health care providers directly involved in the care of the patient. You may also release PHI required for third-party billing. Information may be used for internal quality improvement and training programs, but all identifying information must first be removed. There are also certain situations when you may be legally mandated to report your findings, such as in the case of child abuse or when you receive a subpoena. In most situations, except for treatment purposes, only the minimum amount of information necessary should be released. Failure to abide by the provisions of the HIPAA laws can result in civil and/or criminal action against your response agency and against you personally.

For specific policies, each EMS service is required to have a policy and procedure manual and a privacy officer who can answer questions. You can expect to receive further training on how this act impacts your specific response agency and resource hospital.

Advance Directives

Occasionally, you and your partner may respond to a call in which a patient is dying from an illness. When you arrive at the scene, you may find that family members do not want you to try to resuscitate the patient. Without valid written documentation from a physician, such as an advance directive or a **do not resuscitate (DNR) order** (also known as a "do not attempt resuscitation" order), this type of request places you in a very difficult position. A **competent** patient is able to make rational decisions about his or her well-being. An **advance directive** is a written document that specifies medical treatment for a competent patient, should he or she become unable to make decisions. Advanced directives are most commonly used when a patient becomes comatose. An advance directive is often referred to as a living will but may also be referred to as a **health care directive**. Not all advance directives are directions to withhold care. For example, a comfort care order is an advance directive that specifies care a person should receive in the event that they become incompetent. Such care may include nutrition and medication for pain.

DNR orders give you permission not to attempt resuscitation. Although laws can differ from state to state, generally speaking, to be valid, DNR orders must meet the following requirements:

- Clear statement of the patient's medical problem(s)
- Signature of the patient or legal guardian
- Signature of one or more physicians
- In some states, DNR orders contain an expiration date, whereas in others, no expiration date is included. DNR orders with expiration dates must be dated in the preceding 12 months to be valid.

Some patients may have named surrogates to make decisions for them regarding their health care in the event that they are incapacitated and unable to make such decisions for themselves. Such designations

You are the Provider: PART 3

Your partner reports that the patient's blood pressure is very high. The patient tells you that he has "blood pressure problems" and experiences a bad headache whenever he does not take his Prinivil–the medication he takes for his blood pressure. He does not want to go to the hospital and tells you that the clerk called 9-1-1, not him.

Recording Time: 4 Minutes	
Respirations	24 breaths/min; regular and unlabored
Pulse	110 beats/min; strong and regular
Skin	Pink, warm, and dry
Blood pressure	200/110 mm Hg
Oxygen saturation (Sao$_2$)	98% (on room air)

5. What should you do when a patient refuses treatment and/or transport?

6. Can you legally transport this patient against his will?

may be referred to as **durable powers of attorney for health care** or **health care proxies**. There are many different types of powers of attorney and not all authorize the exercise of medical decision making. Some powers of attorney simply authorize someone to handle the financial affairs of the person executing the power and others will apply only if the person executing the power is still competent. When presented with a power of attorney at the scene of a medical emergency, you must read it carefully to ascertain its meaning. If there is any question, you should contact online medical control for assistance. Do not delay emergency care while efforts to interpret the power of attorney are made. Keep in mind that a patient who remains conscious and competent does not surrender the right to make medical decisions. The person named in the power of attorney or health care proxy is only authorized to make decisions when the patient is no longer capable of doing so.

DNR does not mean "do not treat." Even in the presence of a DNR order, you are still obligated to provide supportive measures (oxygen, pain relief, and comfort) to a patient who is not in cardiac arrest, whenever possible. Each ambulance service, in consultation with its medical director and legal counsel, must develop a protocol to follow in these circumstances.

Because of terminal nursing home placement and hospice and home health programs, you may be faced with this situation often. Specific guidelines vary from state to state, but the following four statements may be considered general guidelines:

1. Patients have the right to refuse treatment, including resuscitative efforts, provided that they are able to communicate their wishes.
2. A written order from a physician is required for DNR orders to be valid in a health care facility.
3. You should periodically review state and local protocols and legislation regarding advance directives.
4. When you are in doubt or the written orders are not present, you have an obligation to resuscitate.

When presented with an advance directive, you should never become annoyed with family members and allow yourself to wonder, "Why did they bother to call 9-1-1 if they don't want us to do anything?" The patients, and their families, should be treated with the utmost respect and empathy. If information and support is what they called you for, be sure to provide it—it is part of your job.

Physical Signs of Death

Determination of the cause of death is the medical responsibility of a physician. There are both definitive

Words of Wisdom

EMT Oath

"Be it pledged as an Emergency Medical Technician, I will honor the physical and judicial laws of God and man. I will follow that regimen which, according to my ability and judgment, I consider for the benefit of patients and abstain from whatever is deleterious and mischievous, nor shall I suggest any such counsel. Into whatever homes I enter, I will go into them for the benefit of only the sick and injured, never revealing what I see or hear in the lives of men unless required by law.

I shall also share my medical knowledge with those who may benefit from what I have learned. I will serve unselfishly and continuously in order to help make a better world for all mankind.

While I continue to keep this oath unviolated, may it be granted to me to enjoy life, and the practice of the art, respected by all men, in all times. Should I trespass or violate this oath, may the reverse be my lot.
So help me God."

Written by Charles B. Gillespie, MD
Adopted by the National Association of Emergency Medical Technicians, 1978.

and presumptive signs of death. In many states, death is defined as the absence of circulatory and respiratory function. Many states have also adopted "brain death" provisions; these provisions refer to irreversible cessation of all functions of the brain and brain stem. Questions often arise as to whether to begin basic life support. In the absence of physician orders such as DNR orders, the general rule is: If the body is still warm and intact, initiate emergency medical care. An exception to this rule is cold temperature (hypothermia) emergencies. Hypothermia is a general cooling of the body in which the internal body temperature becomes abnormally low: below 95°F (35°C). It is considered a serious condition and is often fatal. At 86°F (30°C), the brain can survive without perfusion for about 10 minutes. When the core temperature drops to 82.4°F (28°C), the patient is in grave danger; however, individuals have survived hypothermic incidents with temperatures as low as 64°F (18°C). In cases of hypothermia, the patient should not be considered dead until he or she is warm and dead. Other conditions that may necessitate prolonged resuscitation include a healthy young person who has overdosed or a lightning strike victim.

Presumptive Signs of Death

Most medicolegal authorities will consider the presumptive signs of death that are listed in **Table 3-1** adequate, particularly when they follow a severe trauma or occur at

Table 3-1	**Presumptive Signs of Death**

- Unresponsiveness to painful stimuli
- Lack of a carotid pulse or heartbeat
- Absence of breath sounds
- No deep tendon or corneal reflexes
- Absence of eye movement
- No systolic blood pressure
- Profound cyanosis
- Lowered or decreased body temperature

the end stages of long-term illness such as cancer or other prolonged diseases. These signs would not be adequate in cases of sudden death due to hypothermia, acute poisoning, or cardiac arrest. Usually, in these cases, some combination of the signs is needed to declare death, not just one of them alone.

Definitive Signs of Death

Definitive or conclusive signs of death that are obvious and clear to even nonmedical persons include the following:

- Obvious mortal damage, such as a body in parts (decapitation)
- **Dependent lividity**: blood settling to the lowest point of the body, causing discoloration of the skin **Figure 3-4**
- **Rigor mortis**, the stiffening of body muscles caused by chemical changes within muscle tissue. It develops first in the face and jaw, gradually extending downward until the body is in full rigor. The rate of onset is affected by the body's ability to lose heat to its surroundings. A thin body loses heat

faster than a fat body. A body on a tile floor loses heat faster than a body wrapped up in a blanket in a bed. Rigor mortis occurs sometime between 2 and 12 hours after death

- **Putrefaction** (decomposition of body tissues). Depending on temperature conditions, this occurs sometime between 40 and 96 hours after death

Medical Examiner Cases

Involvement of the medical examiner, or the coroner in some states, depends on the nature and scene of the death. In most states, when trauma is a factor or the death involves suspected criminal or unusual situations such as hanging or poisoning, the medical examiner must be notified **Figure 3-5**. When the medical examiner or coroner assumes responsibility of the scene, that responsibility supersedes all others at the scene, including the family's. The following may be considered medical examiner's cases:

- When the person is dead on arrival (DOA) (sometimes referred to as dead on scene [DOS])
- Death without previous medical care or when the physician is unable to state the cause of death
- Suicide (self-destruction)
- Violent death
- Poisoning, known or suspected
- Death resulting from accidents
- Suspicion of a criminal act

If emergency medical care has been initiated, be sure to keep thorough notes of what was done or found. These records may be important during a subsequent investigation.

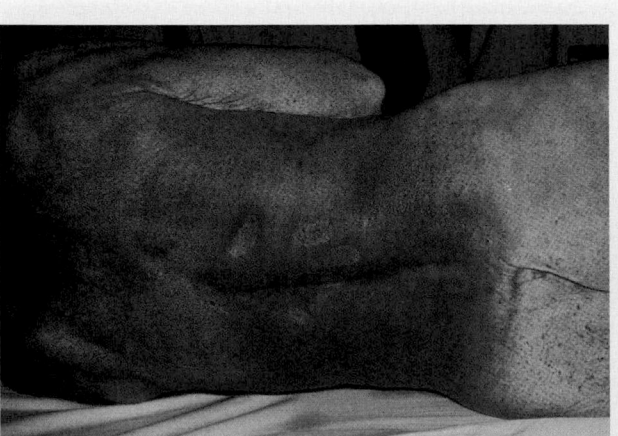

Figure 3-4 Dependent lividity is an obvious sign of death caused by discoloration of the body from pooling of the blood to the lower parts of the body.

Figure 3-5 When trauma is a factor or the death involves an unusual or a suspected criminal situation, the medical examiner is required.

In such instances, there is no urgent reason to move the body. The only immediate action that is required of you is to cover the body and prevent its disturbance. Local protocol will determine your ultimate action in these instances.

Special Situations

Organ Donors

You may be called to a scene involving a potential organ donor. An individual who has expressed a wish to donate organs is a potential organ donor. Consent to organ donation is voluntary and knowing. Consent is evidenced by either a donor card or a driver's license indicating that the individual wishes to be a donor **Figure 3-6**. You may need to consult with medical control when faced with this situation.

Organ/Tissue Donor Card

I wish to donate my organs and tissues. I wish to give:

☐ any needed organs and tissues ☐ only the following organs and tissues: ____________

Donor Signature ____________ Date ____________

Witness ____________

Witness ____________

Figure 3-6 The patient may be carrying a donor card or driver's license indicating that he or she wishes to be an organ donor.

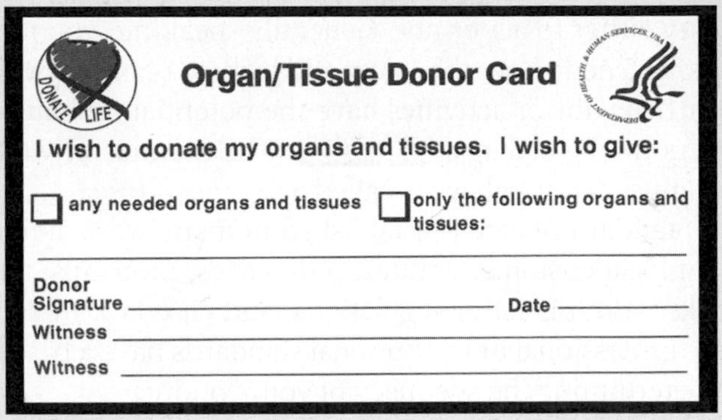

You should treat a potential organ donor in the same way that you would any other patient needing treatment. The fact that a patient is a possible donor does not mean that you should not use all means necessary to keep that patient alive. Organs that are often donated, such as a kidney, heart, or liver, need oxygen at all times; you must give oxygen to the possible donor or the organs will be damaged and become useless.

Remember that your priority is to save the patient's life. You may encounter potential organ donor situations at a mass-casualty incident. The potential organ donor should be triaged with other patients and assigned a category; the potential organ donor may have to have a lower priority than other less severely injured patients.

Be sure to learn what the specific protocols are in your area regarding these situations. Organ donors may not be able to be maintained properly in a mass-casualty incident. While unfortunate, the compromise of the organs cannot be avoided in this case.

Medical Identification Insignia

Many patients will carry important medical identification and information, often in the form of a bracelet, necklace, or card that identifies patient history information, such as whether the patient has a DNR order, allergies, diabetes, epilepsy, or some other serious condition **Figure 3-7**. This information is helpful to you in assessing and treating the patient.

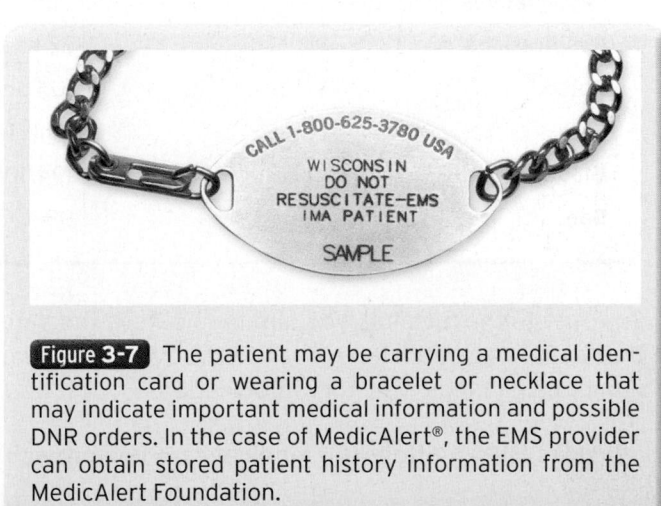

Figure 3-7 The patient may be carrying a medical identification card or wearing a bracelet or necklace that may indicate important medical information and possible DNR orders. In the case of MedicAlert®, the EMS provider can obtain stored patient history information from the MedicAlert Foundation.

Scope of Practice

The <u>scope of practice</u>, which is most commonly defined by state law, outlines the care you are able to provide for the patient. Your medical director further defines the scope of practice by developing protocols and standing orders. The medical director gives you the legal authorization to provide patient care through telephone or radio communication (online) or standing orders and protocols (off-line).

An EMT carrying out procedures for which he or she is not authorized under the enabling legislation is practicing outside his or her scope of practice, which may be considered negligence or, in some states, even a criminal offense. The scope of practice should not be confused with the standard of care, which is what a reasonable EMT in a similar situation would do.

You and other EMS personnel have a responsibility to provide proper, consistent patient care and to report problems, such as possible liability or exposure to infectious disease, to your medical director immediately.

Standards of Care

The law requires you to act or behave toward other individuals in a definite, definable way, regardless of the

activity involved. Under given circumstances, you have a duty either to act or not. Generally speaking, you must be concerned about the safety and welfare of others when your behavior or activities have the potential for causing others injury or harm Figure 3-8 . The manner in which you must act or behave is called a **standard of care**.

Standard of care is established in many ways, among them local customs, statutes, ordinances, protocols, textbooks, administrative regulations, and case law. In addition, professional or institutional standards have a bearing on determining the adequacy of your conduct.

■ Standards Imposed by Local Custom

The standard of care is how a reasonably prudent person with similar training and experience would act under similar circumstances, with similar equipment, and in the same or similar place. For example, the conduct of an EMT who is employed by an ambulance service is to be judged in comparison with the expected conduct of other EMTs from comparable ambulance services in the same geographic area. These standards are often based on locally accepted protocols.

As an EMT, you will not be held to the same standard of care as physicians or other more highly trained individuals. In addition, your conduct must be judged

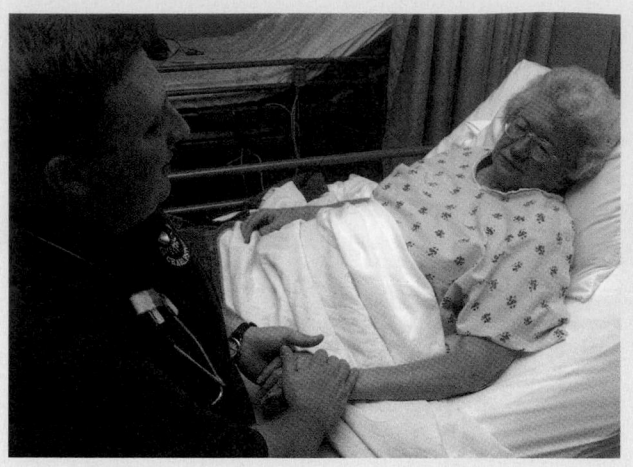

Figure 3-8 Act or behave toward others in a way that shows your concern about their safety and welfare.

in the light of the given emergency situation, taking into consideration the following factors:

- General confusion at the scene of the emergency
- The needs of other patients
- The type of equipment available

In this context, an **emergency** is a serious situation, such as an injury or an illness that arises suddenly,

You are the Provider: PART 4

After further discussion, the patient consents to EMS treatment and transport. After placing the patient on supplemental oxygen via nasal cannula, you place him onto the stretcher, and load him into the ambulance. After reassessing him, you begin transport to the hospital.

Recording Time: 10 Minutes	
Level of consciousness	Conscious and alert
Respirations	22 breaths/min; regular and unlabored
Pulse	104 beats/min; strong and regular
Skin	Pink, warm, and dry
Blood pressure	194/108 mm Hg
Sao$_2$	99% (on oxygen)

En route to the hospital, you dim the lights in the ambulance, apply a cool compress to his forehead, and ensure that he is in a comfortable position. You then obtain his address, social security number, date of birth, phone number, and medical history information.

7. Does HIPAA affect the medical care you provide to your patients?

threatens the life or welfare of a person or group of people, and requires immediate intervention **Figure 3-9**.

The prevailing custom of the community is an important element in determining the standard of emergency care required.

■ Standards Imposed by Law

In addition to local customs, standards of emergency medical care may be imposed by statutes, ordinances, administrative regulation, or case law. In many jurisdictions, violating one of these standards is said to create presumptive negligence. Therefore, you must become familiar with the particular legal standards that may exist in your state. In many states, this may take the form of treatment protocols published by a state agency.

■ Professional or Institutional Standards

In addition to standards imposed by law, professional or institutional standards may be admitted as evidence in determining the adequacy of an EMT's conduct. Professional standards include recommendations published by organizations and societies that are involved in emergency medical care. Institutional standards include specific rules and procedures of the EMS service, ambulance service, or organization to which you are attached.

Two notes of caution: First, you must be familiar with the standards of your organization. Second, if you are involved in formulating standards for a particular agency, they should be reasonable and realistic so that they do not impose an unreasonable burden on EMTs. Providing

the best emergency medical care should be every EMT's goal, but it is not realistic to have institutional standards that demand the best care.

Many standards of care may be imposed on you. State health department regulations usually govern the scope and level of training. Court decisions have resulted in case law defining standards of care. Professional standards are also imposed, such as the American Heart Association's standard for basic life support (BLS) and cardiopulmonary resuscitation (CPR) **Figure 3-10**.

Ordinary care is a minimum standard of care. In general, it is expected that anyone who offers assistance will exercise reasonable care and act prudently. If you act reasonably, according to the accepted standard, the risk of civil suit is small. If you apply the standard practices you have been trained to use, you can likely avoid liability. For example, various organizations have defined standards for performing CPR. If you deviate from these standards, you may be liable for civil and possibly criminal prosecution. In addition, state regulatory agencies that oversee EMS operations can sanction EMS personnel for deviating from the standard of care.

■ Standards Imposed by Textbooks

In the course of a lawsuit, an attorney will often ask an EMT if he or she recognizes various textbooks as being authoritative works in the field of EMS. Since virtually all EMS textbooks follow standards established by the National Highway Transportation Safety Administration (NHTSA), these textbooks are often recognized as contributing to the standard of care that is followed by EMTs. Local protocols, however, may differ from material

Figure 3-9 An emergency is a serious situation that arises suddenly, threatens the life or welfare of one or more individuals, and requires immediate intervention.

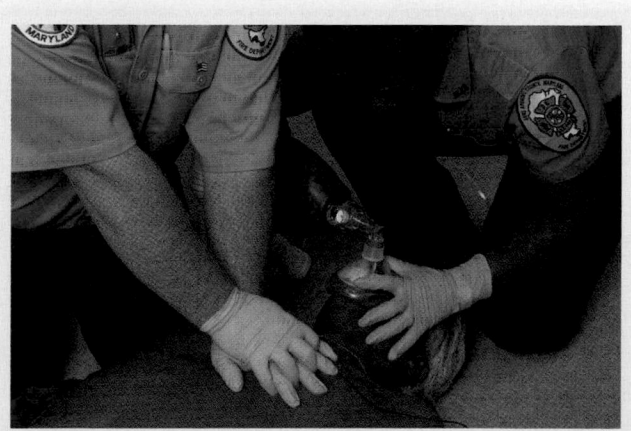

Figure 3-10 Many standards of care are imposed on you, such as those for performing basic life support and cardiopulmonary resuscitation.

presented in textbooks. When such differences occur, the EMT is bound to follow local protocols.

■ Standards Imposed by States

Medical Practices Act

In some states, EMS personnel are exempt from the licensure requirements of the Medical Practices Act because an EMT is regarded as a nonmedical professional. The practice of medicine is defined as the diagnosis and treatment of disease or illness. EMTs and others in the prehospital care chain assess the need for life support and begin care. Some states, however, have adopted legislation that establish the scope of practice for EMS providers. Therefore, as an EMT you must be aware of the standards established by legislation in your state so that you can be sure to provide care that is consistent with those standards.

Certification and Licensure

Some states provide certification, licensure, or credentialing of individuals who perform emergency medical care. **Certification** is the process by which an individual, institution, or program is evaluated and recognized as meeting certain predetermined standards to ensure safe and ethical patient care. Once certified, you are obliged to conform to the standards that are generally recognized nationally by various registry groups and provide an important link in nationwide EMS. **Licensure** is the process by which a competent authority, usually the state, grants permission to practice a job, trade, or profession. You must ensure that your certification or licensure remains current; skill levels must be kept up to date.

■ Duty to Act

Duty to act is an individual's responsibility to provide patient care. Responsibility comes from either statute or function. A bystander is under no obligation to assist a stranger in distress; there is no duty to act. There may be a duty to act in certain instances, including the following:

- You are charged with emergency medical response.
- Your service or department's policy states that you must assist in any emergency.

Once your ambulance responds to a call or treatment is begun, you have a legal duty to act. In most cases, if you are off duty and come upon a crash, you are not legally obligated to stop and assist patients. There may be some circumstances where this is not true and you

should be familiar with the laws and policies that apply in your service area.

■ Negligence

Negligence is the failure to provide the same care that a person with similar training would provide in the same or similar situation. It is deviation from the accepted standard of care that may result in further injury to the patient. Determination of negligence is based on the following four factors:

1. **Duty.** The EMT has an obligation to provide care and to do so in a manner that is consistent with the standard of care established by training and local protocols.
2. **Breach of duty.** There is a breach of duty when the EMT does not act within an expected and reasonable standard of care.
3. **Damages.** There are damages when a patient is physically or psychologically harmed in some noticeable way.
4. **Causation.** There must be a reasonable cause-and-effect relationship between the breach of duty and the damages suffered by the patient. An example is dropping the patient during lifting, causing a fracture of the patient's leg. If a person has a duty and abuses it, causing harm to another individual, the EMT, the agency, and/or the medical director may be sued for negligence. This is often referred to as **proximate causation**.

All four elements must be present for the legal doctrine of negligence to apply and for a plaintiff to prevail in a lawsuit against an EMS service or provider. It is also possible for an EMT or an EMS service to be held liable even when the plaintiff is unable to clearly demonstrate how an injury occurred under the theory of **res ipsa loquitor**. An EMT could be held liable under this theory if it can be shown that an injury occurred, that the instrumentality causing the injury was in the control of the EMT, and that such injuries generally do not occur unless there is negligence. For example, you and your partner are called to the home of a man with diabetes who has lapsed into unconsciousness. You find the patient lying on a couch with no visible signs of trauma. While loading the patient into the ambulance, your partner slips and the stretcher drops, causing the patient to sustain a facial laceration. The patient later files suit for negligence. Because the patient was unconscious, he is unable to describe exactly how he sustained a facial laceration. Under the theory of res ipsa loquitor, the patient may prevail in his lawsuit by showing that he was under your care, that he suffered an injury, and that his injury would not have occurred unless there was negligence.

In rare cases, the plaintiff may be able to establish liability by using the theory of **negligence per se**. This is a theory that may be used when the conduct of the person being sued is alleged to have occurred in clear violation of a statute. For example, if an EMT were to perform an advanced life support skill, such as the intravenous administration of a cardiac medication, the plaintiff might allege that this was negligence per se. In that case, the plaintiff would not have to establish the circumstances surrounding the EMT's conduct. There would be no need to show that the medication was inappropriate for the patient or that the EMT administered it in an improper manner.

All forms of negligence come under the general category known as **torts**. Torts are simply defined as civil wrongs. They are not within the jurisdiction of our criminal courts. Examples of tort actions other than negligence are suits for defamation of character and invasion of privacy.

Abandonment

Abandonment is the unilateral termination of care by the EMT without the patient's consent and without making any provisions for continuing care by a medical professional who is competent to provide care for the patient. Once care is started, you have assumed a duty that must not stop until an equally competent EMS provider assumes responsibility. Failure to peform that duty is a serious legal and ethical matter that exposes the patient to harm and can result in civil action against you.

For example, suppose you arrive at the scene of a single-car accident and begin care of two injured patients. A passerby tells you of a two-car accident farther down the road in which five people are injured. You turn care of the two injured patients from the first accident over to the passerby who is not a trained emergency care provider and leave to go to the other accident. Abandonment may have occurred because you did not turn care of the patients over to a person who is trained and competent to provide emergency care that meets the needs of the two patients. Consider the following general questions when you are faced with making a decision such as this one:

- What problems may develop from your actions?
- How might the patient's condition worsen if you leave?
- Does the patient need care?
- Are you neglecting your duty to your patient?
- Is the person assuming care capable of providing the level of care needed by the patient?

- Are you abandoning the patient if you leave the scene?
- Are you violating a standard of care?
- Are you acting prudently?

Surprisingly, abandonment may also take place in the emergency department where you are dropping off your patient. A part of your obligation as an EMT is to provide hospital personnel with a report of your assessment findings, the care you provided, and any changes in patient status that occurred during transport to the hospital. The failure to do so could result in a delay in treatment or a misdiagnosis. In such a case, a claim for abandonment might be filed against the EMT who failed to provide the report.

Assault and Battery and Kidnapping

Assault is defined as unlawfully placing a person in fear of immediate bodily harm. Threatening to restrain a patient who does not want to be transported could be considered assault. **Battery** is defined as unlawfully touching a person; this includes providing emergency care without consent. Assault and battery can be either civil or criminal in nature. Civil lawsuits for battery are common in health care. To sustain a criminal case of assault or battery, it is generally necessary to prove an intent to cause harm. The element of intent is rarely present in the case of an EMS provider; therefore, criminal cases of assault and/or battery are rare. **Kidnapping** is the seizing, confining, abducting, or carrying away of a person by force. In theory, this might include a situation where a patient is transported against his or her will. In reality, criminal charges of kidnapping are almost unheard of in EMS because the EMT is almost always acting in a good faith effort to provide care to the patient. It is far more likely that an EMT could be the target of a civil suit for **false imprisonment**. This is defined as the unauthorized confinement of a person that lasts for an appreciable period of time.

Serious legal problems may arise in situations in which a patient has not given consent for treatment. Battery could be considered if you apply a splint to a suspected fracture of the lower leg or use an EpiPen on a patient without the patient's consent. Under such circumstances, a patient might file suit for assault, battery, false

Words of Wisdom

If you place the welfare of the patient ahead of all other considerations, you will rarely if ever commit an unethical act in medical care.

imprisonment, or all three. Criminal charges are possible but far less likely. To protect yourself from these charges, make sure that you obtain expressed consent or that the situation allows for implied consent. Consult your medical director or service attorney if you have questions or doubt about a specific situation.

Defamation

As an EMT, you should also be aware of the laws involving defamation. **Defamation** is the communication of false information that damages the reputation of a person. Defamation that is in writing is referred to as **libel** and defamation that is spoken is known as **slander**. A legal claim for defamation could arise out of a false statement on a run report, inappropriate comments made during "station house" conversation, or sharing "war stories" with friends, relatives, or neighbors. To avoid liability for such a claim and to protect the confidentiality of patients, you must only communicate information about your patients to authorized persons and you should be sure that the information contained in your run reports and other documentation is accurate and relevant. Information should always be factual and you should never comment on your patient's personal information when it is not relevant to your assessment or treatment of the patient.

Good Samaritan Laws and Immunity

Most states have adopted **Good Samaritan laws**, which are based on the common law principle that when you reasonably help another person, you should not be liable for errors and omissions that are made in giving good faith emergency care. However, Good Samaritan laws do not necessarily protect you from a lawsuit. Good Samaritan provisions vary significantly from state to state and whereas some laws provide Good Samaritan protection for anyone who stops to render aid, other laws only provide protection for those with medical training. Good Samaritan statutes in some jurisdictions provide immunity from a lawsuit while others provide an affirmative defense if you are sued for rendering care. In most cases, they do not prohibit the filing of a lawsuit nor do they pertain to acts that could be considered wanton, gross, or willful negligence. To be protected by the provisions of a Good Samaritan law, several conditions must generally be met:

1. You acted in good faith in rendering care.
2. You rendered care without expectation of compensation.
3. You acted within the scope of your training.
4. You did not act in a grossly negligent manner.

You are the Provider: PART 5

While reassessing the patient, he admits to using cocaine and smoking marijuana a few weeks ago. You complete your reassessment and then call in your radio report to the receiving facility. The patient's blood pressure has improved and he tells you that his headache is not as bad as it was before.

Recording Time: 16 Minutes	
Level of consciousness	Conscious and alert
Respirations	18 breaths/min; regular and unlabored
Pulse	90 beats/min; strong and regular
Skin	Pink, warm, and dry
Blood pressure	166/94 mm Hg
Sao$_2$	99% (on oxygen)

You deliver the patient to the emergency department and give your oral report to the receiving nurse. After completing your patient care report, you and your partner return to service.

8. Are you required by law to report the patient's use of illegal substances to law enforcement personnel?

<u>Gross negligence</u> is defined as conduct that constitutes a willful or reckless disregard for a duty or standard of care.

Another group of laws grants immunity from liability to official EMS providers, such as EMTs. These laws, which vary from state to state, do not provide immunity when injury or damage is caused by gross negligence or willful misconduct. In most cases, immunity statutes apply to EMS services that are considered governmental agencies.

Most states have also adopted specific laws granting special privileges to EMS personnel, authorizing them to perform certain medical procedures. Many states also grant partial immunity to EMTs and physicians and nurses who give emergency instructions to EMS personnel via radio or other forms of communication. Consult your medical director for more information about the laws in your area.

Records and Reports

The government has formulated a policy to protect individuals with health regulations and statutes. Because certain individuals are in a position to observe and gather information about diseases, injuries, and emergency events, an obligation to compile such information and report it to certain agencies may be imposed. Even if there is no such requirement, you should compile a complete and accurate record of all incidents in which you come into contact with sick or injured patients **Figure 3-11**. Most medical and legal experts believe that a complete

and accurate record of an emergency medical incident is an important safeguard against legal complications. The absence of a record or a substantially incomplete record may mean that you have to testify about the events, your findings, and your actions relying on memory alone, which can prove to be wholly inadequate and embarrassing in the face of aggressive cross-examination.

The courts consider the following two rules of thumb regarding reports and records:

- If an action or procedure is not recorded on the written report, it was not performed.
- An incomplete or untidy report is evidence of incomplete or inexpert emergency medical care.

You can avoid both of these potentially dangerous presumptions by compiling and maintaining accurate reports and records of all events and patients. PCRs also help the EMS system evaluate individual and service provider performance. These reports are an integral part of most quality assurance programs.

Special Mandatory Reporting Requirements

Abuse of Children, Older Persons, and Others

All states and the District of Columbia have enacted laws to protect abused children, and some have added other protected groups such as the older population and

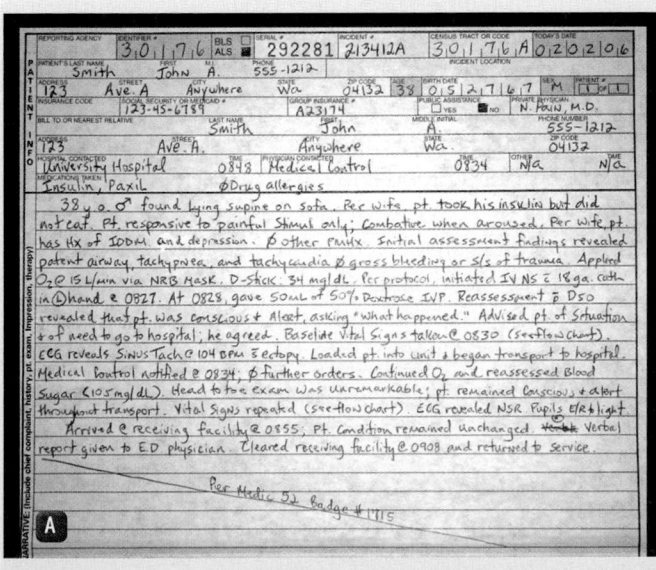

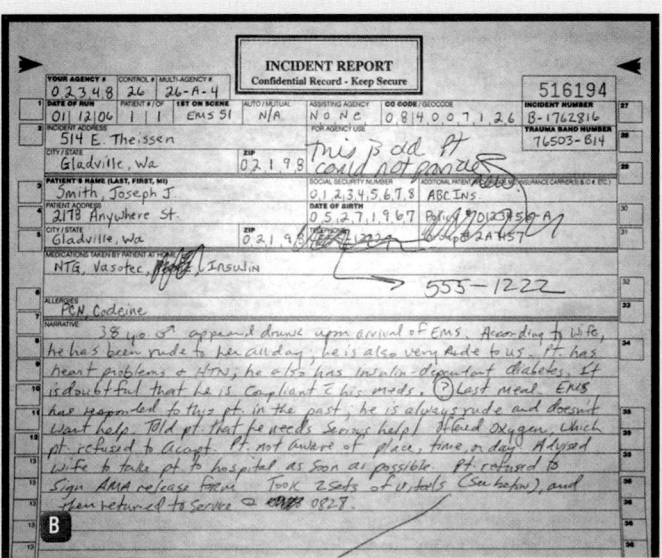

Figure 3-11 Filing a patient care report is a critical part of your responsibilities as a provider. **A.** Proper documentation. **B.** Improper documentation.

"at-risk" adults. Most states have a reporting obligation for certain individuals, ranging from physicians to any person. You must be aware of the requirements of the law in your state. Such statutes frequently grant immunity from liability for libel, slander, or defamation of character to the individual who is obligated to report, even if the reports are subsequently shown to be unfounded, as long as the reports are made in good faith.

■ Injury During the Commission of a Felony

Many states have laws requiring the reporting of any injury that is likely to have occurred during the commission of a crime, such as gunshot wounds, knife wounds, or poisonings. Again, you must be familiar with the legal requirements of your state.

■ Drug-Related Injuries

In some instances, drug-related injuries must be reported. These requirements may affect the EMT. However, it should be stressed that the US Supreme Court has held that drug addiction, in contrast to drug possession or sale, is an illness and not a crime. An injury as a result of a drug overdose, therefore, may not be within the definition of an injury resulting from a crime.

Some states, by statute, specifically establish confidentiality and excuse certain specified individuals from reporting drug cases, either to a government agency or to a minor's parents, if, in the opinion of those individuals, withholding reporting is necessary for the proper treatment of the patient. Once again, you must be familiar with the legal requirements of your state.

■ Childbirth

Many states require that anyone who attends at a live birth in any place other than a licensed medical facility report the birth. As before, you must be familiar with state requirements.

■ Other Reporting Requirements

Other reporting requirements may include attempted suicides, dog bites, certain communicable diseases, assaults, domestic violence, and sexual assault, such as rapes.

Special Populations

Elder abuse is as prominent as child abuse in our society. Do not forget to be observant and report any suspicious signs or symptoms to the proper authorities.

Most EMS agencies require that all exposures to infectious diseases be reported. You may be asked to transport certain patients in restraints, which may also need to be reported. Each of these situations can present significant legal problems. You should learn your local protocols regarding these situations.

Not only do the events that need to be reported vary significantly from state to state but so do the methods and procedures by which such reporting must take place. For example, although all states require that suspected child abuse be reported, some states require that the report be filed with law enforcement, others with a designated child protection agency, and yet others with the emergency department. There are often time provisions associated with reporting statutes. As has been noted earlier, it is important that you become familiar with reporting requirements of your state. Failure to report may result in disciplinary action, suspension of your privileges to practice as an EMT, a fine, or even criminal prosecution.

■ Scene of a Crime

If there is evidence at an emergency scene that a crime may have been committed, you must notify the dispatcher immediately so that law enforcement authorities can respond. Such circumstances should not stop you from providing lifesaving emergency medical care to the patient; however, your safety is a priority, so you must ensure that the scene is safe to enter. At times, you may have to transport the patient to the hospital before law enforcement arrives. While emergency medical care is being provided, you must be careful not to disturb the scene of the crime any more than absolutely necessary. Notes and drawings should be made of the position of the patient and of the presence and position of any weapon or other objects that may be valuable to the investigating officers. If possible, do not cut through holes in clothing that were caused by weapons or gunshot wounds. Avoid walking through blood and try to avoid leaving footprints in the dirt or grass at or near a crime scene. When a sexual assault is suspected, try to persuade the victim not to shower or clean himself or herself. You should confer periodically with local authorities and be aware of their wishes regarding actions you should take at the scene of the crime. It is best if these guidelines can be established by protocol.

■ The Deceased

In most states, EMTs do not have the authority to pronounce a patient dead. If there is any chance that life exists or that the patient can be resuscitated, you must make every effort to save the patient at the scene and during transport. However, at times death is obvious.

If a victim is clearly dead and the scene of the emergency may be where the crime was committed, do not move the body or disturb the scene.

Ethical Responsibilities

In addition to legal duties, EMTs have certain ethical responsibilities as health care providers. These responsibilities are to themselves, their coworkers, the public, and the patient. **Ethics** is the philosophy of right and wrong, of moral duties, and of ideal professional behavior. It is often referred to as the study of morality. **Morality** is a code of conduct that can be defined by society, religion, or a person, affecting character, conduct, and conscience. An entire field of ethics known as **bioethics** has evolved over the past several decades that addresses issues that arise in the practice of health care. Many such issues have drawn national attention, such as those dealing with termination of life support, rationing of medical resources, and physician-assisted suicide. From an EMS standpoint, ethics are not quite so complicated but nonetheless must be considered by all EMTs to some degree. As an EMT, you will be expected to conduct yourself in a manner that is consistent with the standards of your profession and to keep the best interests of your patients at the forefront of your conduct and decision making. The manner in which principles of ethics are incorporated into professional conduct is known as **applied ethics**.

You will encounter ethical dilemmas in the course of your employment that can be challenging to resolve. Examples might include the following:

- While returning from a call, you observe your partner drinking alcohol from a bottle that he has hidden in his jump kit.
- On arriving at the scene of an accident you find a patient who is seriously injured. You recognize him as the drunk driver who killed your sister several months earlier.
- You are dispatched to the home of a patient complaining of chest pain. Your partner recognizes the address and tells you not to hurry since this is one of your "frequent flyers" who constantly calls for no reason other than to get attention.
- You respond to the home of an elderly female patient in cardiac arrest. Despite the fact that the patient is 93 years old and was recently diagnosed with lung cancer, the family wants everything possible done to save her. Once the patient is loaded into the ambulance your partner tells you not to waste your time with this patient, considering her age and medical condition.

The manner in which you respond to each of these circumstances requires you to evaluate and apply your own ethical standards as well as those of your profession. Obviously, these choices can be difficult at times, particularly in those cases where your own personal standards of right and wrong do not necessarily agree with the standards of your profession. You might agree with your partner that it is futile to attempt to resuscitate the 93-year-old cancer patient but you also know that in the absence of a valid DNR order or direction from medical control, you are obliged to do so. You also know that you should report your partner who is drinking on the job, even if he is not just a partner but a good friend who has done you many personal favors and you understand that your report might cause him to lose his job.

Your behavior both on and off the job will be a reflection of your personal ethical standards. News stories that depict EMS personnel engaging in any immature or illegal activities serve to lessen the public's confidence in the services EMTs provide. Illegal drug use or selling drugs, inappropriate use of emergency vehicles, inappropriate visitors entertained at the station, and use of alcohol on duty can impact negatively on the EMT and EMS in general and should be strictly forbidden.

EMTs also may not stand by silently and watch as other EMS providers engage in misbehavior if they care about their patients, their coworkers, and the EMS system as a whole. Misconduct should be promptly reported to the appropriate chain of command. Similarly, EMTs are obligated to report medical errors they make or witness to the medical director or another appropriate person as soon as possible.

How can you make sure that you are acting ethically, especially with all the decisions you have to make in the field **Table 3-2**?

You must meet your legal and ethical responsibilities while caring for your patients' physical and emotional needs. Patient needs will vary depending on the situation, and you must be prepared to offer whatever physical and emotional support is necessary. In most cases that you will encounter as an EMT, there will be a rule, a law, or a policy that will guide your decision making and your actions. As a professional, you are bound to follow all such policies, rules, and laws even in those rare circumstances where your own personal sense of ethics might lead you to a different result.

One unquestionable responsibility you have is honest reporting. Absolute honesty in reporting is essential. You must provide a complete account of the events and the details of all patient care and professional duties. Accurate records are also important for quality improvement activities.

Table 3-2 Ethical Decision Making

1. Consider all options available to you and the consequence of each option.

2. What decisions have been made in the past regarding a similar situation? Is this a type of problem that reflects a rule, a law, or policy? Can an existing policy or rule be applied? This uses the concept of **precedence**, defined as basing current action on lessons, rules, or guidelines derived from previous similar experiences.

3. How would this action affect you if you were in your patient's or patient's family's place? This is a form of the Golden Rule. Was your decision one that you can honestly say was made in the best interest of the patient?

4. Would you feel comfortable having all prehospital care providers apply this action in all similar circumstances?

5. Can you supply a good justification for your action to:
   - Your peers?
   - The public?
   - Your supervisor?

6. How will the consequences of your decision provide the greatest benefit in view of all the alternatives?

7. Involve online medical control in your decision making.

Figure 3-12 Court discussions will be based on your documentations. Make sure your documentation is neat, thorough, and accurate.

The EMT in Court

As an EMT, there are a number of different circumstances that might cause you to end up in court, either as a witness or a defendant in a civil lawsuit or as a witness or defendant in a criminal case. Regardless of the circumstances, being in court is often stressful. As a witness in a civil case, you may be called on to testify about the condition of the plaintiff when you arrived at the scene of an accident and about the treatment that you provided. In a criminal case you may be asked to describe a crime scene, the injuries that you found when you examined a crime victim, or to testify concerning any admissions or statements made to you by a criminal defendant.

Whenever you are subpoenaed to testify in any court proceeding you should immediately notify the director of your service and legal counsel. As a witness you should remain neutral during your testimony. You are simply there to provide the facts as you observed them and not to take sides. In all likelihood, many of the questions that you will be asked will be based on the documentation you wrote at the time of the rescue call. Be sure to review your run report prior to your court appearance **Figure 3-12**.

As a defendant in either a civil or criminal proceeding, your involvement will obviously be far more significant and the outcome will have far greater consequence. In either case, you will definitely require the assistance of an attorney. In a civil suit, where you are being sued in your capacity as an employee or volunteer of an EMS service, your service or its insurance company generally will provide you with legal counsel.

A civil suit begins with the service of a summons and complaint. The complaint will set forth the details of the plaintiff's case and will provide the theory on which the plaintiff is relying to recover a judgment against you and your service. If served with a summons, you must bring this to the attention of the head of your service immediately, because the complaint must be responded to within a set period of time that is usually within 20 to 30 days. The response to the complaint is called an answer and it will generally deny the claims set forth in the complaint and set forth one or more defenses on behalf of you and your service. A defense is essentially a reason why the plaintiff should not recover a judgment against you. Depending on the nature of the case filed against you, the type of EMS service that you work for, and the state where you work, there may be different possible defenses available to you. These may include the defenses of statute of limitations, immunity, or contributory negligence.

The **statute of limitations** is the time within which a case must be commenced. For example, in many states, a claim for negligence must be commenced within 3 years. A case commenced beyond the 3-year period would be barred by the statute of limitations. In such a case, your attorney would include the defense of statute of limitations in the answer that is filed in response to the complaint.

Another possible defense is that of governmental immunity. **Governmental immunity** generally applies only to EMS services that are operated by municipalities or other governmental entities. If your service is covered by immunity, it may mean that you cannot be sued at all or it may limit the amount of the monetary judgment that the plaintiff may recover. State laws vary significantly on both the statute of limitations and immunity, and you should understand the laws that apply in your state.

Contributory negligence is a legal defense that may be raised when the defendant feels that the conduct of the plaintiff somehow contributed to any injuries or damages that were sustained by the plaintiff. For example, you are treating a patient with chest pain and you feel that the administration of aspirin is indicated. You ask the patient if he is allergic to aspirin and he says no. Shortly after you administer the aspirin the patient develops the signs and symptoms of a severe allergic reaction. Later in the hospital, the doctor advises you that the patient's medical chart history indicates that the patient has an allergy to aspirin. The patient states that he forgot he was allergic to aspirin. In this case, the defense of contributory negligence might be raised since it was the patient's forgetfulness and his denial of an aspirin allergy that contributed to his allergic reaction.

The next phase of the case is known as **discovery** and it is an opportunity for both sides to obtain information that will enable the attorneys to have a better understanding of the case and assist in negotiating a possible settlement or in preparing for trial. Discovery may include interrogatories, depositions, requests for production of documents, and physical examinations. **Interrogatories** are written questions that each side sends to the other and **depositions** are oral questions asked of parties and witnesses under oath. On completion of the discovery phase, the parties may try to negotiate a possible settlement. Most cases are settled and do not go to trial. If a settlement is not able to be negotiated, the case will be set down for trial. It is not uncommon for a case to take several years to get to trial.

At trial, each side will have an opportunity to present evidence that includes testimony of witnesses and documents such as medical reports and your run report. Witnesses may include experts such as physicians. Once both sides have concluded presenting evidence, a judge or jury will render a decision or verdict. If a judgment is rendered against you or your service, the plaintiff may be awarded several types of damages. These include the following:

1. **Compensatory damages**. These damages are intended to compensate the plaintiff for the injuries he or she sustained such as medical bills, damages to personal property, lost earnings, and pain and suffering.
2. **Punitive damages**. Punitive damages are not commonly awarded in negligence cases and are reserved for those cases where the defendant has acted intentionally or with a reckless disregard for the safety of the public.

In most cases, if a judgment is rendered against you, your service or its insurance carrier will pay the judgment.

There is also the possibility that an EMT could be arrested and charged with a criminal offense arising out of his or her employment as an EMS provider. Although these are rare occurrences, there have been EMTs who have been charged with crimes including theft of patient property, assault or sexual assault on a patient, operating a vehicle while under the influence of drugs or alcohol, or various drug-related offenses. Obviously, any arrest is considered very serious because a conviction could lead to imprisonment, the imposition of fines, and possible loss of the ability to practice as an EMT. Any EMT charged with a criminal offense should secure the services of a highly experienced criminal attorney immediately.

You are the Provider: SUMMARY

1. What is the difference between scope of practice and standard of care?

Scope of practice outlines the care that you are able to provide to a patient; it is most commonly defined by state law. Your EMS system medical director further defines your scope of practice through written protocols and standing orders. By doing so, he or she grants you legal authorization to perform certain patient care interventions under his or her medical license. The medical director, at his or her discretion, may expand or restrict your scope of practice. For example, he or she may train you to start intravenous (IV) lines—not a typical EMT skill—and grant you permission to start IVs in the EMS system he or she oversees.

Standard of care defines the accepted level of emergency care expected by reason of training and profession; it is written by legal or professional organizations so that patients are not exposed to unreasonable risk or harm. The law requires you to act or behave toward patients in a definite, definable way—regardless of the activity involved; the manner in which you do this is called the standard of care. An EMT's actions are compared to those of other EMTs with the same or similar training, under similar circumstances, and with the same or similar equipment. If your actions are grossly different than those expected from other EMTs, you may be held liable for not following the standard of care, even if no harm came to the patient.

You are the Provider: SUMMARY, continued

2. When does the EMT have a legal duty to act?

Duty to act is a medicolegal term relating to certain personnel who either by statute or by function have a responsibility to provide care.

Once your ambulance is dispatched to a call or treatment is initiated, you have a legal duty to act; this applies to *both* paid and volunteer EMTs. If you are charged with emergency medical response, such as when you are on duty, you have a legal duty to act if a call is received.

In some states, you may be legally obligated to stop and render aid, even if you are off duty. Other states do not require an off-duty EMT to act, although a moral obligation may exist. You must be familiar with the laws of the state in which you function regarding your off-duty actions. If in doubt, you should stop and provide care; few would argue that it is preferable to defend why you provided care as opposed to why you did not.

3. Are you legally authorized to treat this patient? Why or why not?

At this point, you have not obtained consent from the patient to begin treating him; in fact, you haven't even introduced yourself. Under most circumstances, you may not begin treatment of a mentally competent adult until he or she has given you permission, or consent, to do so. If the patient has decision making capacity—that is, he or she is conscious, alert, not under the influence of drugs or alcohol, and of legal age (18 years in most states)—you cannot legally provide care, even if the patient is obviously sick or injured. Providing care without the patient's consent may be grounds for both criminal and civil action, such as assault and battery.

If a patient is conscious, alert, and has decision making capacity, he or she must give you expressed (or actual) consent before you can legally begin treatment. Expressed consent is the type of consent in which the patient acknowledges that he or she wants you to provide treatment and transport. The patient may outright ask you to help him or her, or provide nonverbal cues, such as extending an arm to allow you to take a blood pressure.

4. How does informed consent differ from implied consent?

As previously discussed, you must obtain consent to treat any adult patient who has decision making capacity. Furthermore, the patient's consent must be informed, which means that he or she has been advised, in language he or she can understand, of the potential risks, benefits, and alternatives to treatment. The legal basis for the doctrine of informed consent assumes that the patient has a right to determine what will and what will not to be done to his or her body. Remember, however, that the patient must be of legal age and capable of making a rational decision.

Implied consent is based on the legal assumption that a critically ill or injured patient, who is physically unable to give consent (ie, unconscious, under the influence of drugs or alcohol), would consent to EMS treatment and transport if he or she were physically able to do so. Consent to treat is also implied when caring for a minor whose parents or caregivers are unable to be located; a minor cannot legally consent to or refuse medical care.

5. What should you do when a patient refuses treatment and/or transport?

When a patient refuses treatment and/or transport, it is not unreasonable to ask why he or she does not wish to be treated. Many people refuse treatment because of financial concerns or the fact that they are scared. Although these concerns are very real and should not be downplayed, it is important to inform the patient, in language he or she can understand, that the situation is potentially serious.

Consent must be informed—that is, the patient must be made aware of the potential benefits and risks of accepting treatment, as well as alternatives to treatment. A patient's refusal must also be informed; he or she must be made aware of the potential consequences of refusing treatment. Informing a patient of the potential consequences of accepting *and* refusing treatment will enable him or her to make an informed decision regarding his or her own health care.

In this case, you should explain that his high blood pressure and severe headache could indicate bleeding in the brain or some other potentially life-threatening condition, and that only a physician can diagnose his problem. Do not be afraid to advise the patient that his refusal could ultimately result in death; this is not a scare tactic, it is the truth and the patient has a right to hear it.

If, despite your best efforts to obtain consent to treat, a mentally competent adult still refuses, there is little else you can legally do. You should, however, inform medical control of the situation. In some cases, the physician may wish to speak directly to the patient.

Some patients will adamantly refuse treatment, despite all efforts to obtain consent. If this is the case, obtain the patient's signature on an official release form that acknowledges refusal; you should also obtain a signature from a witness to the refusal, such as a police officer. If the patient refuses to sign the release form, inform medical control and thoroughly document the situation.

Carefully document all patient refusals. Include your assessment findings, any treatment that you did provide, why the patient refused treatment, and what you did and/or said to attempt to obtain consent. Make sure the patient understands that he can always call 9-1-1 if he changes his mind.

6. Can you legally transport this patient against his will?

When a patient refuses treatment, you must assess his or her decision making capacity. Is his or her mental condition impaired? Is he or she under the influence of drugs or alcohol? Is he or she of legal age? Is he or she a danger to himself/herself or others? These are but a few of the questions that must be answered.

In this case, there is no evidence that the patient's decision making capacity is impaired, and although he needs medical

You are the Provider: SUMMARY, continued

attention for his headache and blood pressure, you cannot legally force him to accept it, nor can you transport him against his will.

Although mentally competent adults have the legal right to refuse treatment or withdraw from treatment at any time, this presents you with a dilemma. If you provide care against his will, you risk being accused of assault, battery, and false imprisonment. Conversely, if you leave the patient alone, you risk being accused of negligence and abandonment if his condition becomes worse.

The best course of action is to ensure that the patient is aware of the potential consequences of his refusal—namely, death—and contact medical control to apprise him or her of the situation.

7. Does HIPAA affect the medical care you provide to your patients?

The Health Insurance Portability and Accountability Act of 1996 (HIPAA) has many aims; however, the section of the act that most directly affects EMS relates to patient privacy. This section was designed to strengthen laws that protect a patient's personal health care information and to safeguard patient confidentiality.

HIPAA provides guidance on why types of information are protected (ie, name, address, social security number, medical history), the responsibility of health care providers regarding that protection, and penalties for breaching that protection.

HIPAA does not affect the medical care that you provide to a patient. It does, however, dramatically limit the ability of EMS providers to obtain follow-up information about patients they treat, including information that would serve to improve their knowledge of various medical conditions.

8. Are you required by law to report the patient's use of illegal substances to law enforcement personnel?

The patient's use of cocaine and marijuana is pertinent medical information that may have an impact on the care he receives at the hospital; therefore, it should be included in your patient care report and your oral report to the receiving facility. However, the US Supreme Court has held that drug use or addiction (in contrast to possession or sale) is an illness and not a crime. Therefore, you are not legally required to report the patient's admitted use of these substances to law enforcement personnel. If you are in doubt, consult with your EMS medical director. More important, you must be familiar with the reporting requirements of the state in which you function as an EMT.

EMS Patient Care Report (PCR)

Date: 4-19-09	Incident No.: 040109	Nature of Call: Headache		Location: 1175 N. Main St.	
Dispatched: 1720	En Route: 1721	At Scene: 1725	Transport: 1739	At Hospital: 1746	In Service: 1801

Patient Information

Age: 32 Sex: M Weight (in kg [lb]): 91 kg (200 lb)	Allergies: None Medications: Prinivil Past Medical History: Hypertension Chief Complaint: Severe headache

Vital Signs

Time: 1724	BP: 200/110	Pulse: 110	Respirations: 24	Sao$_2$: 98%
Time: 1734	BP: 194/108	Pulse: 104	Respirations: 22	Sao$_2$: 99%
Time: 1740	BP: 166/94	Pulse: 90	Respirations: 18	Sao$_2$: 99%

EMS Treatment (circle all that apply)

Oxygen @ 4 L/min via (circle one): (NC) NRM Bag-Mask Device		Assisted Ventilation	Airway Adjunct	CPR
Defibrillation	Bleeding Control	Bandaging	Splinting	Other: Dimmed lights, position of comfort

You are the Provider: SUMMARY, continued

Narrative
9-1-1 dispatch for a male patient with a severe headache. On arrival at the scene, found the patient, a 32-year-old male, sitting on the sidewalk outside convenience store, grabbing his head in pain. He was conscious and alert; his airway was patent, and his breathing was adequate. Patient states that his headache began a few hours earlier, and that he has not taken his prescribed antihypertensive medication. No trauma was involved in this incident. Past medical history significant for hypertension. He denies loss of consciousness, nausea, or any other symptoms. Patient was initially hesitant to consent to EMS treatment and transport. However, after the potential complications of his refusal were explained to him, he agreed to EMS treatment and transport. Obtained vital signs, applied oxygen at 4 L/min via nasal cannula, and performed further assessment, which was unremarkable. Placed patient onto stretcher, loaded him into the ambulance, dimmed the lights, and placed him in a position of comfort. Began transport to the hospital and monitored his condition en route. Patient admits to using cocaine and marijuana a few weeks ago, but he does not feel this is contributing to his condition. Patient remained conscious and alert during transport, and stated that his headache was improving. Reassessment of his vital signs revealed that his blood pressure had improved. Delivered patient to emergency department staff and gave oral report to charge nurse. **End of report**

Prep Kit

Ready for Review

- Under most circumstances, consent is required from every conscious adult before care can be started. The foundation of consent is decision making capacity.

- You should never withhold lifesaving care unless a valid do not resuscitate order is present.

- Because a minor might not have the wisdom, maturity, or judgment to give valid consent, the law requires that a parent or legal guardian give consent for treatment or transport.

- Adults who are conscious and alert and who appear to have decision making capacity have the right to refuse treatment or withdraw from treatment at any time, even if doing so may result in death or serious injury.

- Communication between you and the patient is considered confidential and generally cannot be disclosed without permission from the patient or a court order.

- Advanced directives, living wills, or health care directives are most commonly used when a patient becomes comatose.

- There are both definitive and presumptive signs of death. In many states, death is defined as the absence of circulatory and respiratory function.

- Consent to organ donation is evidenced by either a donor card or a driver's license indicating that the individual wishes to be a donor.

- Standard of care is established in many ways, among them local customs, statutes, ordinances, protocols, textbooks, administrative regulations, and case law. The scope of practice outlines the care you are able to provide for the patient.

- Once your ambulance responds to a call or treatment is begun, you have a legal duty to act. In most cases, if you are off duty and come upon a crash, you are not legally obligated to stop and assist patients.

- Determination of negligence is based on the following four factors: duty, breach of duty, damages, and causation. All four elements must be present for the legal doctrine of negligence to apply and for a plaintiff to prevail in a lawsuit against an EMS service or provider.

- Abandonment is the termination of care without the patient's consent and without making provisions for the transfer of care to a medical professional with skills at the same level or at a higher level than your own skills. Abandonment is legally and ethically a very serious act.

- Assault is defined as unlawfully placing a person in fear of immediate bodily harm. Battery is unlawfully touching a person; this includes providing emergency care without consent. To protect yourself from these charges, be sure to obtain expressed consent whenever possible.

- To avoid liability for defamation, you must only communicate information about your patients to authorized persons and you should be sure that the information contained in your run reports and other documentation is accurate and relevant.

- Good Samaritan laws are based on the common law principle that when you reasonably help another person, you should not be liable for errors and omissions that are made in giving good faith emergency care. Whereas some laws provide Good Samaritan protection for anyone who stops to render aid, others only provide protection for those with medical training.

- Records and reports are important; make sure that you compile a complete and accurate record of each incident. The courts consider an action or procedure that was not recorded on the written report as not having been performed, and an incomplete or untidy report is considered evidence of incomplete or inexpert medical care.

- You should know what the special reporting requirements are involving abuse of children, the elderly, and others; injuries related to crimes; drug-related injuries; and childbirth.

- You must meet your legal and ethical responsibilities while caring for your patients' physical and emotional needs.

- As an EMT, there are a number of different circumstances that might cause you to end up in court, either as a witness or a defendant in a civil lawsuit or as a witness or defendant in a criminal case.

Vital Vocabulary

abandonment Unilateral termination of care by the EMT without the patient's consent and without making provisions for transferring care to another medical professional with the skills and training necessary to meet the needs of the patient.

advance directive Written documentation that specifies medical treatment for a competent patient should the patient become unable to make decisions; also called a living will or health care directive.

applied ethics The manner in which principles of ethics are incorporated into professional conduct.

assault Unlawfully placing a patient in fear of bodily harm.

battery Touching a patient or providing emergency care without consent.

bioethics The study of ethics related to issues that arise in health care.

breach of confidentiality Disclosure of information without proper authorization.

certification A process in which a person, an institution, or a program is evaluated and recognized as meeting certain predetermined standards to provide safe and ethical care.

compensatory damages Damages awarded in a civil suit that are intended to restore the plaintiff to the same condition that he or she was in prior to the incident complained about in the lawsuit.

competent Able to make rational decisions about personal well-being.

consent Permission to render care.

contributary negligence A legal defense that may be raised when the defendant feels that the conduct of the plaintiff somehow contributed to any injuries or damages that were sustained by the plaintiff.

decision making capacity Ability to understand and process information and make a choice regarding appropriate medical care.

defamation The communication of false information about a person that is damaging to that person's reputation or standing in the community.

dependent lividity Blood settling to the lowest point of the body, causing discoloration of the skin.

depositions Oral questions asked of parties and witnesses under oath.

discovery The phase of a civil suit where the plaintiff and defense obtain information from each other that will enable the attorneys to have a better understanding of the case and which will assist in negotiating a possible settlement or in preparing for trial. Discovery includes depositions, interrogatories, and demands for production of records.

do not resuscitate (DNR) orders Written documentation by a physician giving permission to medical personnel to not attempt resuscitation in the event of cardiac arrest.

durable power of attorney for health care A type of advance directive executed by a competent adult that appoints another individual to make medical treatment decisions on his or her behalf in the event that the person making the appointment loses decision making capacity.

duty to act A medicolegal term relating to certain personnel who either by statute or by function have a responsibility to provide care.

emancipated minors A person who is under the legal age in a given state but, because of other circumstances, is legally considered an adult.

emergency A serious situation, such as injury or illness, that threatens the life or welfare of a person or group of people and requires immediate intervention.

emergency doctrine The principle of law that permits a health care provider to treat a patient in an emergency situation when the patient is incapable of granting consent because of an altered level of consciousness, disability, the effects of drugs or alcohol, or the patient's age.

emergency medical care Immediate care or treatment.

ethics The philosophy of right and wrong, of moral duties, and of ideal professional behavior.

expressed consent A type of consent in which a patient gives express authorization for provision of care or transport.

false imprisonment The confinement of a person without legal authority or the person's consent.

forcible restraint The act of physically preventing an individual from initiating any physical action.

Good Samaritan laws Statutory provisions enacted by many states to protect citizens from liability for errors and omissions in giving good faith emergency medical care, unless there is wanton, gross, or willful negligence.

governmental immunity If your service is covered by immunity, it may mean that you cannot be sued or it may limit the amount of the monetary judgment that the plaintiff may recover; generally applies only to EMS services that are operated by municipalities or other governmental entities.

gross negligence Conduct that constitutes a willful or reckless disregard for a duty or standard of care.

health care directive A written document that specifies medical treatment for a competent patient, should he or she become unable to make decisions. Also known as an advance directive or a living will.

health care proxies A type of advance directive executed by a competent adult that appoints another individual to make medical treatment decisions on his or her behalf in the event that the person making the appointment loses decision making capacity. Also known as a durable power of attorney for health care.

implied consent Type of consent in which a patient who is unable to give consent is given treatment under the legal assumption that he or she would want treatment.

informed consent Permission for treatment given by a competent patient after the potential risks, benefits, and alternatives to treatment have been explained.

in loco parentis Refers to the legal responsibility of a person or organization to take on some of the functions and responsibilities of a parent.

interrogatories Written questions that the defense and plaintiff send to one other.

kidnapping The seizing, confining, abducting, or carrying away of a person by force, including transporting a competent adult for medical treatment without his or her consent.

libel False and damaging information about a person that is communicated in writing.

licensure The process whereby a competent authority, usually the state, allows individuals to perform a regulated act.

medicolegal A term relating to medical jurisprudence (law) or forensic medicine.

morality A code of conduct that can be defined by society, religion, or a person, affecting character, conduct, and conscience.

negligence Failure to provide the same care that a person with similar training would provide.

negligence per se A theory that may be used when the conduct of the person being sued is alleged to have occurred in clear violation of a statute.

patient autonomy The right of a patient to make informed choices regading his or her health care.

precedence Basing current action on lessons, rules, or guidelines derived from previous similar experiences.

protected health information (PHI) Any information about health status, provision of health care, or payment for health care that can be linked to an individual. This is interpreted rather broadly and includes any part of a patient's medical record or payment history.

proximate causation When a person who has a duty abuses it, and causes harm to another individual, the EMT, the agency, and/or the medical director may be sued for negligence.

punitive damages Damages that are sometimes awarded in a civil suit when the conduct of the defendant was intentional or constituted a reckless disregard for the safety of the public.

putrefaction Decomposition of body tissues.

res ipsa loquitor When the EMT or an EMS service is held liable even when the plaintiff is unable to clearly demonstrate how an injury occurred.

rigor mortis Stiffening of the body; a definitive sign of death.

scope of practice Most commonly defined by state law; outlines the care you are able to provide for the patient.

slander False and damaging information about a person that is communicated by the spoken word.

standard of care Written, accepted levels of emergency care expected by reason of training and profession; written by legal or professional organizations so that patients are not exposed to unreasonable risk or harm.

statute of limitations The time within which a case must be commenced.

tort A wrongful act that gives rise to a civil suit.

Assessment in Action

You are dispatched to a home where you find a 65-year-old man complaining of chest pain. The patient is alert and oriented. Assessment of his vital signs shows a pulse rate of 110 beats/min and irregular, a blood pressure of 140/90 mm Hg, and a respiratory rate of 22 breaths/min. He describes the pain as crushing and tells you that it radiates to his left arm. He has an extensive cardiac history and is taking numerous medications. The patient states he does not want to go to the hospital and that his friend was overreacting by calling EMS. Your partner tells the patient he has no choice about transport and then forcibly places the patient on the stretcher against the patient's will. The patient becomes extremely agitated and your partner tells him to be quiet or he will be arrested and then taken to the hospital. When you arrive at the emergency department, the patient is emotionally and physically upset. The emergency department physician confirms that although the patient was noncritical, he now has a blood pressure that has spiked to 190/100 mm Hg. While you are in the emergency department, you overhear your partner talking to another EMS crew about this "crazy old man with chest pain."

1. The right of a patient to make decisions concerning his or her health is known as:
   A. decision-making capacity.
   B. patient autonomy.
   C. informed consent.
   D. expressed consent.

2. What factors must be present for a negligence case to be successful?

3. Was the Health Insurance Portability and Accountability Act (HIPAA) violated? If so, how?

4. With respect to negligence, the elevation in the patient's blood pressure is considered:
   A. causation.
   B. damage.
   C. breach of duty.
   D. in loco parentis.

5. Which of the following patients has the right to refuse treatment?
   A. Minor
   B. Unconscious patient
   C. Mentally incompetent adult
   D. Mentally competent adult

6. Which of the following types of consent allows treatment when the patient is unconscious?
   A. Mature consent
   B. Implied consent

   C. Informed consent
   D. Expressed consent

7. What is your best protection against lawsuits in refusal of care cases?
   A. Notify medical control.
   B. Write a well-documented prehospital care report.
   C. Make a verbal confirmation to the patient.
   D. Notify law enforcement.

8. Define breach of confidentiality.

9. When a person who has a duty abuses it and causes harm to another individual, this is referred to as:
   A. proximate causation.
   B. res ipsa loquitor.
   C. negligence per se.
   D. in loco parentis.

10. The manner in which you must act or behave is called:
    A. a duty to act.
    B. applied ethics.
    C. the standard of care.
    D. the Good Samaritan Act.

Communications and Documentation

National EMS Education Standard Competencies

Preparatory

Applies fundamental knowledge of the emergency medical services (EMS) system, safety/well-being of the emergency medical technician (EMT), medical/legal, and ethical issues to the provision of emergency care.

Therapeutic Communication

Principles of communicating with patients in a manner that achieves a positive relationship

- Interviewing techniques (pp 110-118)
- Adjusting communication strategies for age, stage of development, patients with special needs, and differing cultures (pp 108-109, 113-117)
- Verbal defusing strategies (p 111)
- Family presence issues (p 112)

EMS System Communication

Communication needed to

- Call for resources (pp 131-132)
- Transfer care of the patient (pp 117-118, 134)
- Interact within the team structure (pp 131-132)
- EMS communication system (pp 127-130)
- Communication with other health care professionals (pp 117-118, 132-136)
- Team communication and dynamics (pp 132-136)

Documentation

- Recording patient findings (pp 119-126)
- Principles of medical documentation and report writing (pp 119-126)

Medical Terminology

Uses foundational anatomical and medical terms and abbreviations in written and oral communication with colleagues and other health care professionals.

....

Knowledge Objectives

1. Describe factors and strategies to consider for therapeutic communication with patients. (pp 107-118)

2. Discuss the techniques of effective verbal communication. (pp 110-118)

3. Explain the skills that should be used to communicate with family members, bystanders, people from other agencies, and hospital personnel. (pp 110, 112, 113, 115-118)

4. Understand special considerations in communicating with older people, children, hearing-impaired patients, visually impaired patients, and non-English-speaking patients. (pp 113-117)

5. Describe the use of written communication and documentation. (pp 119-126)

6. Identify the information required in a patient care report (PCR). (pp 119-123)

7. Explain the legal implications of the patient care report. (pp 119, 123-125)

8. Understand how to document refusal of care, including the legal implications. (pp 124-125)

9. Discuss state and/or local special reporting requirements, such as for gunshot wounds, dog bites, and abuse. (p 125)

10. Understand the basic principles of the various types of communications equipment used in EMS. (pp 127-130)

11. Describe the use of radio communications, including the proper methods of initiating and terminating a radio call. (pp 130-136)

12. List the correct radio procedures in the following phases of a typical call: initial receipt of call, en route to call, on scene, arrival at hospital (or point of transfer), and return to service. (pp 130-136)

13. Give the proper sequence of information to communicate in radio delivery of a patient report. (p 134)

Skills Objectives

1. Demonstrate the techniques of successful cross-cultural communication. (pp 108-109)

2. Demonstrate completion of a patient care report. (pp 119-124)

3. Demonstrate how to make a simulated, concise radio transmission with dispatch. (pp 131-136)

Introduction

Communication is the transmission of information to another person—whether it is verbal or through body language. Effective communication is an essential component of prehospital care and is necessary to achieve a positive relationship with patients and coworkers.

Verbal communication skills are vitally important for EMTs. Your verbal skills will enable you to gather information from the patient and bystanders. They will also make it possible for you to effectively coordinate the variety of responders who are often present at the scene. Excellent verbal communication is also an integral part of transferring the patient's care to the nurses and physicians at the hospital.

Documentation is the written portion of the EMT's patient care interaction that becomes part of the patient's permanent medical record. It serves many purposes, including demonstrating that the care delivered was appropriate and within the scope and practice of the providers involved. Documentation also provides an opportunity to communicate the patient's story to others who may participate in the patient's care in the future. Adequate reporting and accurate records ensure the continuity of patient care. Complete patient records also guarantee proper transfer of responsibility, comply with the requirements of health departments and law enforcement agencies, and fulfill your organization's administrative needs. Reporting and record keeping duties are an essential aspect of patient care, although they are performed only after the patient's condition has been stabilized. Documentation in the field drives both funding and research for EMS. Seatbelts are a prime example. Studies gathered from record keeping in the early 1970s showed that patients have a significantly higher survival rate if seatbelts are used during motor vehicle accidents. Armed with this information, laws were passed to enforce seatbelt usage, and huge amounts of money were spent on educating the public.

Radio and telephone communications link you and your team with other members of the EMS, fire, and law enforcement communities. This link helps the entire team to work together more effectively and provides an important layer of safety and protection for each member of the team. You must know what your system can and cannot do, and you must be able to use your system efficiently and effectively.

This chapter describes the factors and strategies that you need to be an effective communicator, discusses a variety of effective methods of verbal communication, and provides guidelines for appropriate written documentation of patient care. The chapter concludes by identifying the kinds of communication equipment that are used, along with standard radio operating procedures and protocols. The roles of the Federal Communications Commission in EMS are also described.

Therapeutic Communication

How do we communicate? This simple question can be surprisingly complex as there are a number of things to consider during communication **Table 4-1**. **Therapeutic communication** uses various communication techniques and strategies, both verbal and nonverbal, to encourage patients to express how they are feeling and to achieve a positive relationship with the patient. This section will discuss the factors and strategies that are necessary for therapeutic communication.

People communicate in a variety of ways, such as through eye contact, body position, and facial expressions. Factors such as culture and age need to be taken

You are the Provider: PART 1

At 6:10 AM, you and your BLS unit are dispatched to 514 E. Bandera Street for a "sick person." You and your partner proceed to the scene, which is approximately 10 minutes away. En route, the dispatcher contacts you and states that she still has the caller on the phone.

1. What information should you ask the dispatcher to obtain from the caller?
2. Why is effective communication between the responding EMS unit and the dispatcher so important?

Words of Wisdom

The Shannon-Weaver communication model was developed to assist in the mathematical theory of communication for Bell Telephone Labs in the late 1940s **Figure 4-1**. Shannon and Weaver were trying to figure out the math involved in sending information through telephone lines. After its creation, it quickly became apparent that this model had application in areas other than math. Social scientists picked up this model and it remains a valuable tool in understanding the variables involved in human communications. In the communication model, the sender must take a thought, encode it into a message, and send the message to the receiver. The receiver then decodes the message and sends feedback to the sender.

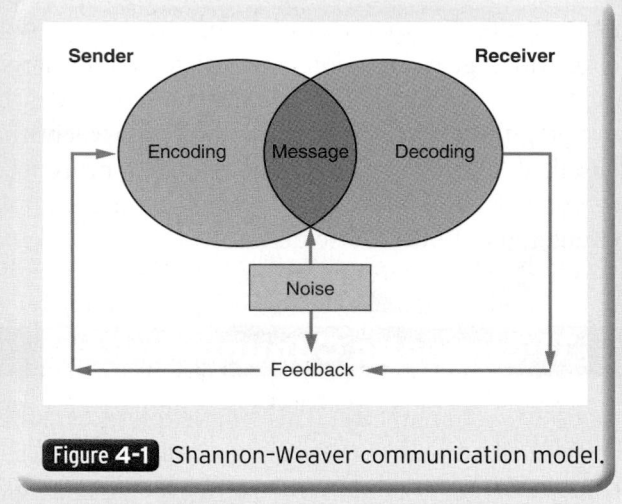

Figure 4-1 Shannon-Weaver communication model.

into consideration during communication. Patients with special needs may require you to consider alternative forms of communication. For example, if your patient is deaf and you cannot communicate using sign language, you may need to communicate by having the patient write down his feelings.

Table 4-1	Factors and Strategies to Consider During Communication
Age	Eye contact
Body language	Facial expression
Clothing	Gender
Culture	Posture
Educational background	Voice tempo
Environment	Volume

◼ Age, Culture, and Personal Experience

The thoughts of people are greatly influenced by their personal experiences. For example, an elderly person who often experiences great pain may view pain as more of an inconvenience than a problem. A child who has limited experience with pain would likely react much differently. People are taught to handle pain differently. How do people in certain cultures view illness and injury? Some cultures encourage people to express their emotions; others see it as a sign of weakness. These social and personal pressures will shape a person's thoughts.

People may talk, make gestures, or write a note to express how they are feeling. Again, culture and personal experience will shape how people communicate. "I am so sorry to bother you, but my chest hurts a little." "Hey! What took you so long? My chest is killing me! Are you going to help me or what?" Both of these messages talk about pain, but they also have much more information within them.

The tone, pace, and volume of the language certainly tell us about the mood of the person communicating. They also provide some insight into the perceived importance of the message. For example, the patient who is yelling at you may be angry, scared, or both. Take note not only of the words being spoken but how they are said.

The EMT needs to recognize that these concepts of body language and eye contact are often greatly affected by culture. In some cultures, direct eye contact is viewed as impolite, while in other cultures, it is impolite to look away while speaking.

People tend to translate the messages they receive using their own world view. **Ethnocentrism** occurs when you consider your own cultural values as more important when you are interacting with people of a different culture. If you are American, for example, you expect that a patient is hiding something, afraid, or untrustworthy if the patient looks away from you while you are talking. These conclusions may be true if the two people communicating are from the same culture. All aspects of communication—eye contact, social distances, body language, and even touching—have a cultural foundation. In Thailand, for example, the touching of the head is reserved for those who are very intimate. This cultural belief can present a problem for an EMT if the patient's head is bleeding.

Cultural imposition takes this idea to an extreme. Some health care providers may consciously or subconsciously force their cultural values onto their patient because they believe their values are better. For example, consider a child who is brought to the emergency room with red marks on his back from a traditional Asian healing practice called coining—rubbing hot coins on the child's back as a treatment for medical illness. The parents explain to the physician that the coining helped for a short time, but now the child seems to be getting sicker. The physician responds angrily to the parents, accusing them of poor parenting and insisting that their practices are harmful (although they aren't). This accusation reflects cultural imposition.

■ Nonverbal Communication

Facial Expressions, Body Language, and Eye Contact

Eye contact and body language are powerful. Consider how dogs interact. When two dogs meet for the first time, they look at each other. The position of the head, shoulders, tail, and back all help to communicate to the other dog. Before they get any closer, the dogs need to understand their new relationship. Who is dominating? Will you hurt me? These questions must be answered quickly.

People communicate using a similar technique. The body language we consciously or subconsciously choose provides more information than words alone. Consider the images in Figure 4-2. Without any words, it should be clear what the mood is of each of these people.

When you are treating a potentially hostile patient, it is important that you understand and be aware of your own body language. People tend to react to anger with anger. If you are dealing with an angry patient, the last thing you want to do is become angry. Be aware of your body language. Do not assume an aggressive posture. Make good eye contact, but do not stare. Speak calmly, confidently, and slowly. Give the patient choices, but limit those choices to ones you can live with. Do not be drawn into the verbal violence your patient may be projecting. Remember, your patient cannot make you angry if you do not allow it to happen.

It is important for you to be attentive to facial expressions, body language, and eye contact—your own and your patient's. These physical cues will help you and your patient to truly understand the message being sent.

Words of Wisdom

EMTs need to be dressed professionally. You are sending the message that you care without speaking a word. For example, your uniform should be pressed and your shirt tucked in.

Physical Factors

Various physical factors affect communication, which are referred to as noise. **Noise** is anything that dampens or obscures the true meaning of the message. Literal noise, or sounds in the environment, can make it difficult to understand the patient or for the patient to understand you. Lighting, distance, or obstacles are other factors that may affect your communication.

Proxemics is the study of space and how the distance between people affect communication Table 4-2. The degree to which people feel comfortable depends on with whom they are communicating. As a person gets closer, a greater and greater sense of trust must be established. When you finally enter someone's intimate space, there must be a high sense of trust.

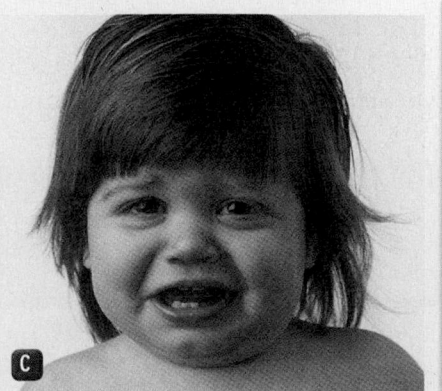

Figure 4-2 The effectiveness of body language. **A.** Happy. **B.** Angry. **C.** Sad.

Understanding how communication works and the importance of effective communication is important when gathering information from the patient. Your communication skills will be put to the test when you communicate with patients and/or families in emergency situations. Remember that someone who is sick or injured is scared and might not understand what you are doing or saying. Therefore, your gestures, body movements, and attitude toward the patient are critically important in gaining the trust of both patient and family.

■ Verbal Communication

As an EMT, you must master many communication skills, including those associated with radio operations and written communications. Skilled verbal communication with the patient and family, bystanders, and the rest of the health care team are an essential part of high-quality patient care. It will make it possible for you to effectively coordinate the variety of responders who are often present at the scene, transfer the patient's care to nurses and physicians at the hospital, and allow you to listen to fully understand the nature of the scene and the patient's problem. You must also be able to organize your thoughts to quickly and accurately verbalize instructions to the patient, bystanders, and other health care professionals.

One of the most fundamental aspects of what EMTs do is to ask patients questions. There are two types of questions: **open-ended questions** are ones in which a patient needs to provide some level of detail to give an answer, whereas **closed-ended questions** can be answered in very short or single-word responses. When first approaching your patient, you should use open-ended questions. "Good day. My name is Chuck and I am an EMT. What seems to be bothering you today?" Open-ended questions allow a free flow of conversation. They let the patient direct you to what is bothering him or her.

Closed-ended questions are important to use when patients are unable to provide long or complete answers to questions. Perhaps the patient is having severe breathing problems, or maybe the patient is a child who is scared and doesn't know what to say. In situations for which thoughtful answers are not possible, closed-ended questions are appropriate and are particularly useful when assessing a patient's condition. "Are you having trouble breathing? Do you take medications for your heart?"

With closed-ended questions, however, it is possible for the EMT to miss important issues if pertinent questions are not asked. Imagine how many ways a person can be sick or injured. Now imagine trying to come up with a single yes/no question for each sickness or injury. Closed-ended questions typically provide limited information, and you should consider the answers to these questions as only a starting point toward understanding the patient's condition.

When you are asking questions of the patient, be conscious of how many questions you are asking. "How are you doing today? Have you been feeling ill?" This common approach actually asks the patient two kinds of questions, one open-ended and one closed-ended. Often the patient will respond with a simple, "yes." To avoid this situation, it is best to ask a single question, wait for answer, and then proceed to another question.

There are many powerful communication tools you can use when trying to obtain information from patients. Sometimes patients will hide information, either consciously or unconsciously. Patients may be afraid, or they may be confused. The techniques in **Table 4-3** provide you with tools that will assist you in gathering patient information. They can be helpful to use not only in patients who are willing to share but in those who are resistant to sharing information.

When you are interviewing the patient, you can consider using touch as a means to communicate caring and compassion. Touch is a powerful tool; therefore, keep in mind that it should be used consciously and sparingly **Figure 4-3**. Many people will be uncomfortable with a stranger touching them suddenly. If you are going to touch the patient, approach slowly and touch the

Table 4-2   **Proxemics for the American Culture**

Space	Distance	Description
Intimate	Less than 18"	Whispering, touching; must be invited
Personal	18" to 4'	Conversations with close friends or family
Social	4' to 10'	Conversations with acquaintances
Public	10' to 25'	Interacting with strangers

Table 4-3 Communication Tools

Communication Tools	Definition	Example
Facilitation	Encouraging the patient to talk more or provide more information	EMT: "Can you tell me more about that? I am listening to you."
Silence	Not speaking	Giving your patient space and time to think and respond.
Reflection	Restating a patient's statement made to you to confirm your understanding	Patient: "I am so depressed that I could die." EMT: "I understand that you are feeling sad."
Empathy	Being sensitive to the patient's feelings and thoughts	Using eye contact and touching to reinforce your communication; adjusting tone and pace to allow for open communication
Clarification	Asking the patient to explain what he or she meant by an answer	Patient: "I just feel sick." EMT: "Can you please tell me what is feeling sick? Can you help me to understand what is going on?"
Confrontation	Making the patient who is in denial or in a mental state of shock focus on urgent and life-critical issues	Patient: "I am having pain in my chest, my back has been hurting me, I feel nauseated, and I ran out of my blood pressure medication." EMT: "Please tell me about your chest pain. We will talk about your other concerns in a moment."
Interpretation	Summing up your patient's complaint	EMT: "If I understand correctly, you have been feeling pain for the past 3 days, and it has gotten worse today." Patient: "That's right."
Explanation	Providing factual information to support a conversation	Patient: "I do not understand what is happening." EMT: "We have checked your blood sugar and blood pressure and both appear to be normal."
Summary	Providing the patient with an overview of the conversation and the steps you will be taking	EMT: "We will be taking you to the emergency department to care for your chest pain. I will be giving you some medication that should make you feel better."

patient's shoulder or arm. You may hold the patient's hand. This allows you to touch the patient, showing you care about what they are telling you, and also allows you to remain at a slight distance.

Avoid touching the patient's torso, chest, or face simply as a means of communication, because these areas are often viewed as intimate. Also, to touch these areas, you will need to get closer to the patient, potentially invading the patient's intimate space. Table 4-4 provides other tips on what to avoid when communicating with patients.

Patients can become hostile toward EMS providers. To help defuse these potentially escalating circumstances, stay calm. Talk to the patient openly and honestly. You will find that meeting hostility with calmness and confidence defuses a situation. Consider the safety of the scene. Decide whether you need police backup. Make sure that you have sufficient backup to provide safety for the patient and the crew. Then, with your backup clearly visible, calmly advise the patient what needs to be done. "Sir, I need you to sit on the ambulance cot now. Either you will sit on the cot or we will help you to the cot."

No one threatens the patient. No one moves toward the patient. In this rare circumstance, you are providing the patient with choices, while at the same time limiting those choices to ones you can accept.

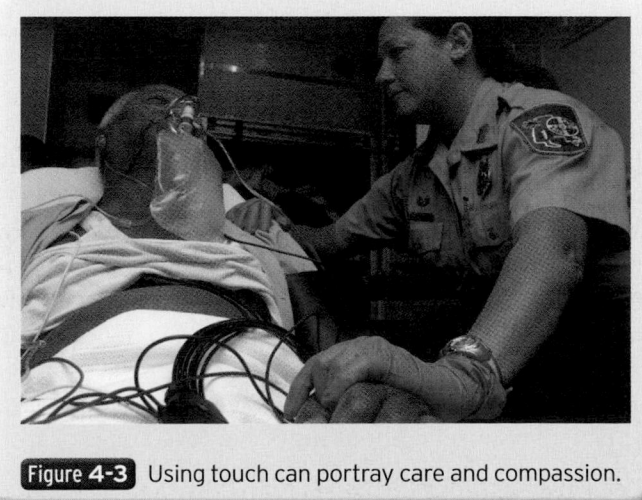

Figure 4-3 Using touch can portray care and compassion.

Table 4-4   Interview Techniques to Avoid

Improper Technique	Example	Comment
Providing false assurance or reassurance	EMT: "It will be okay." "This is nothing to worry about."	You really do not know that everything will be okay.
Giving unsolicited advice	EMT: "Well, if I were you, I wouldn't have called the ambulance at all."	This demeans the patient.
Asking leading or biased questions	EMT: "Are you telling me that this cut is the only reason you called the ambulance?"	Your patient deserves respectful communication. It is inappropriate for you to communicate in a way that suggests to the patient that an ambulance was not needed, even if that is what you believe.
Talking too much	The EMT talks to the patient without really listening to the patient, simply going through the motions.	You should guide the patient through the conversation. When the patient provides you information, you need to consider the information and move the conversation toward a goal.
Interrupting	Patient: "Well, I was having trouble breathing last month and . . ." EMT: "Can we move on to how you are feeling now?"	You may seem bored or annoyed that the patient is taking up your time.
Using "why" questions	EMT: "Why did you call the ambulance today?"	You may seem annoyed that the patient called you.
Using authoritative language	EMT: "Tell me what is wrong with you." "Just give me the details."	This language does not encourage open communication.
Speaking in professional jargon	EMT: "I think we will need to take you to the ED stat. We will give you ASA and NTG en route. Any questions?"	This type of communication confuses the patient. Most patients do not understand medical jargon.

Safety

We are constantly communicating, both consciously and unconsciously. As health care professionals, we all need to ensure that our body language reflects our words. Remember, people call us when they are in crisis. Part of being a good EMT is being able to project calm and control into a chaotic environment. You may not feel calm, but a good EMT will "never let them see you sweat."

The presence of family, friends, and bystanders during your interview of the patient can be valuable. Sometimes, however, well-meaning family members will speak for the patient, and, at times, you may need to ask the family member to allow the patient to answer. Ultimately, you will need to assess the situation and determine whether the additional people are helping you care for the patient or hindering your efforts. Do not be afraid to ask others to step outside or step aside for a moment while you talk with the patient. Take into account how the patient will feel without his or her loved ones nearby. Removing them may make the patient more anxious.

These ten Golden Rules will help you to calm and reassure your patient and provide a therapeutic rapport:

1. **Make and keep eye contact with your patient at all times.** Give the patient your undivided attention. This will let the patient know that he or she is your top priority. Look the patient straight in the eye to establish a **rapport**. Establishing a rapport is building a trusting relationship with your patient. This will make caring for the patient much easier.

2. **Provide your name and use the patient's proper name.** Introduce yourself and your partner. If your department provides you with a name tag, wear it. Ask the patient what he or she wishes to be called. Avoid using terms such as "honey" or "dear." Use a patient's first name only if the patient is a child or the patient asks you to use his or her first name. Rather, use a courtesy title, such as "Mr Peters," "Mrs Smith," or "Ms Butler." If you do not know the patient's name, refer to him or her as "sir" or "ma'am."

3. **Tell the patient the truth.** Even if you have to say something very unpleasant, telling the truth is better than lying. Lying will destroy the patient's trust in you and decrease your own confidence. You might not always tell the patient everything, but if the patient or a family member asks a specific question, you should answer truthfully. A direct question

deserves a direct answer. If you do not know the answer to a patient's question, say so. For example, a patient may ask, "Am I having a heart attack?" To which you would answer, "I don't know, but we will certainly get more information at the hospital. Right now, I am providing you with all the care needed for a person who may be having a heart attack."

4. **Use language that the patient can understand.** Do not talk up or down to the patient in any way. Avoid technical medical terms that the patient might not understand. For example, ask the patient whether he or she has a history of "heart problems." This will usually result in more accurate information than if you ask about "previous episodes of myocardial infarction" or a "history of cardiomyopathy."

5. **Be careful what you say about the patient to others.** You need to understand the relationship between the person you are talking with and the patient. Does the patient want you talking with this person? Ask the patient if it is okay to talk with this person. While speaking to others, ensure that you leave the general area of the patient if you must have a confidential conversation. Be mindful that sharing patient information may be a HIPAA violation.

6. **Be aware of your body language** **Figure 4-4**. Nonverbal communication is extremely important in dealing with patients. In stressful situations, patients may misinterpret your gestures and movements. Be particularly careful not to appear threatening. Instead, position yourself at the same level or at a lower level than the patient when practical. Remember that you should always conduct yourself in a calm, professional manner.

7. **Always speak slowly, clearly, and distinctly.** Pay close attention to your tone of voice.

8. **If the patient is hearing-impaired, face the person so that he or she can read your lips.** Do not shout at a person who is hearing-impaired. Shouting will not make it any easier for the patient to hear you. Shouting may frighten the patient and can end up making it even more difficult for the patient. Never assume that an older patient is hearing-impaired or otherwise unable to understand you. Also, never use "baby talk" with older patients or with anyone other than infants.

9. **Allow time for the patient to answer or respond to your questions.** Do not rush a patient unless there is immediate danger. Sick and injured people may not be thinking clearly and may need time to answer even simple questions. This is especially true when treating older patients.

10. **Act and speak in a calm, confident manner while caring for the patient.** Make sure you attend to the patient's pains and needs. Try to make the patient physically

Figure 4-4 Watch your body language because patients may misinterpret your gestures, movements, and stance.

comfortable and relaxed. Find out whether the patient is more comfortable sitting or lying down. Is the patient cold or hot? Does the patient want a friend or relative nearby?

Patients literally place their lives in your hands. They deserve to know that you can provide medical care and that you are concerned about their well-being. These ten Golden Rules will help provide a good foundation and will make it easier to gather information when the patient wants to talk.

Sometimes, you need to gather information from a reluctant audience. Patients may be defensive about their problems and may not want to talk about them because they are embarrassed. They may direct the conversation away from the true problem. With these patients, start the conversation as usual. Introduce yourself. Be open and compassionate. If you find yourself not getting any real answers, then consider one of the techniques in Table 4-3.

Communicating With Older Patients

According to the US Census Bureau, nearly 35 million individuals are older than 65 years. It is projected that by the year 2030, the geriatric population will be greater than 70 million, resulting in ever-increasing numbers of encounters with persons in this category. However, a person's actual age might not be the most important factor in making him or her geriatric. It is more important to determine a person's functional age. The functional age relates to the person's ability to function in daily activities, the person's mental state, health status, and activity pattern.

As an EMS provider, when you enter a scene to care for an older patient, you are being asked to take control. You have been called because a person needs help.

What you say and how you say it has an impact on the patient's perception of the call. You should present yourself as competent, confident, and concerned. You must take charge of the situation, but do so with compassion. You are there to listen and act on what you learn. Don't limit your assessment to the obvious problem. Oftentimes, older patients who express that they are not well or who are overly concerned about their health or general condition are at risk for a serious decline in their physical, emotional, or psychological state. Table 4-5 provides guidelines for interviewing an older patient.

Table 4-5　Interviewing an Older Patient

In general, when interviewing an older patient, the following techniques should be used:

- **Identify yourself.** Do not assume an older patient knows who you are.
- **Be aware of how you present yourself.** Frustration and impatience can be conveyed through body language.
- **Look directly at the patient.**
- **Speak slowly and distinctly.**
- **Explain what you are going to do before you do it.** Use simple terms to explain the use of medical equipment and procedures, avoiding medical jargon or slang.
- **Listen to the answer the patient gives you.**
- **Show the patient respect.** Refer to the patient as Mr, Mrs, or Miss.
- **Do not talk about the patient in front of him or her**; to do so gives the impression that the patient has no choice in his or her medical care. This is easy to forget when the patient has impaired cognitive (thought) processes or has difficulty communicating.
- **Be patient!**

You are the Provider: PART 2

You arrive at the scene and find the patient, an 83-year-old woman, sitting on the couch in her living room. She is conscious and alert and tells you that she started having light-headedness and nausea about an hour ago. As you begin your assessment, you note that she has hearing aids in both ears.

Recording Time: 0 Minutes	
Appearance	Calm; no obvious distress
Level of consciousness	Conscious and alert
Airway	Open; clear of secretions or foreign bodies
Breathing	Normal rate and depth; regular
Circulation	Radial pulses, strong and regular; skin is pink, warm, and dry

3. How can you maximize successful communication with a hearing-impaired patient?

4. Should your general approach to the assessment process be any different for this patient versus a younger patient? Why or why not?

Special Populations

A man has fallen and hurt his leg. The injury is not severe, so a vacuum splint is placed on his leg and he is moved to the ambulance. All this time, his 4-year-old daughter is watching. Suddenly she runs to the other side of the room and begins to cry. She watched the EMTs take her daddy away. What should you do?

Talk to the child. Have her come over to the ambulance cot and see that daddy is okay. Tell her about the splint and how it will make daddy's leg better. Let her touch her daddy and say goodbye. Let daddy tell her that he will be home soon. Sometimes the obvious patient is not always the *only* patient.

Generally speaking, older people think clearly, can give you a clear medical history, and are able to answer your questions appropriately Figure 4-5 . Do not assume that an older patient is senile or confused. Conversely, communicating with some older patients is extremely difficult, and you may encounter hostility, irritability, and, in fact, some confusion. Do not assume this to be normal behavior for an older patient. These signs may be caused by a simple lack of oxygen (hypoxia), brain injury including a cerebrovascular accident, unintentional drug overdose, or even hypovolemia. Never attribute altered mental status to old age. In addition, your older patients may have difficulty hearing or seeing you. Therefore, you need great patience and compassion when you are called on to care for such a patient. Think of the patient as someone's grandmother or grandfather—or even as yourself when you reach that age.

Approach an older patient slowly and calmly. Allow plenty of time for the patient to respond to your questions.

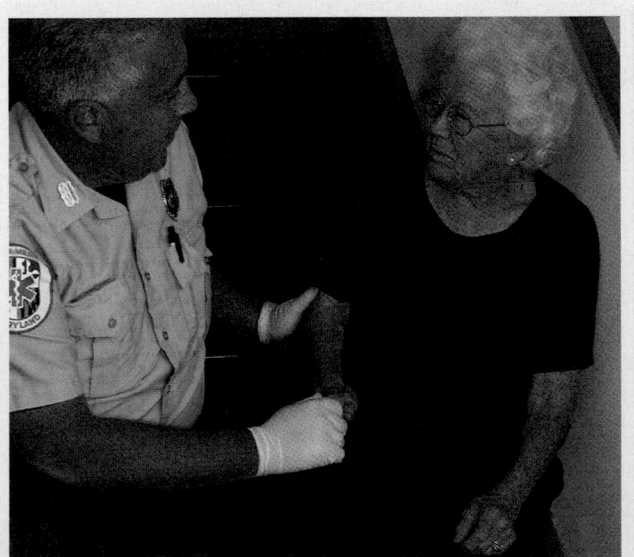

Figure 4-5 You need a great deal of compassion and patience when caring for older patients. Do not assume that the patient is senile or confused.

Watch for signs of confusion, anxiety, or impaired hearing or vision. The patient should feel confident that you are in charge and that everything possible is being done for him or her.

Older patients often do not feel much pain. An older person who has fallen or been injured may report no pain. In addition, older patients might not be fully aware of important changes in their body systems. Therefore, be especially vigilant for objective changes—no matter how subtle—in their condition. Objective changes are those that any observer would be able to witness. Respiratory rate, heart rate, sweating, or vomiting are all objective. Subjective finds are those that only the patient can experience, such as pain or nausea. Even minor changes in breathing or mental state may signal major problems.

When possible, give the patient time to pack a few personal items before leaving for the hospital. Be sure to locate hearing aids, glasses, or dentures before departure; it will make the patient's hospital stay far more pleasant. You should document on the patient care report that these items accompanied the patient to the hospital and the person to whom they were given in the emergency department.

Communicating With Children

Everyone who is thrust into an emergency situation becomes frightened to some degree. However, fear is probably most obvious and severe in children. Children may be frightened by your uniform, the ambulance, and the number of people who have suddenly gathered around. Even a child who says little may be very much aware of all that is going on.

Familiar faces and objects will help to reduce this fright. Let a child keep a favorite toy, doll, or security blanket to give the child some sense of control and comfort. Having a family member or friend nearby is also helpful. When not impractical due to the child's condition, it is often helpful to let the parent or a guardian hold the child during your evaluation and treatment. However, you will have to make sure this person will not upset the child or prevent the child from telling you important information. Sometimes, adult family members are not helpful because they become too upset by what has happened or the child will not share important information in front of them. An overly anxious parent or relative can make things worse. Be careful about selecting the proper adult for this role.

Children can easily see through lies or deceptions, so you must always be honest with them. Make sure that you explain to the child over and over again what and why certain things are happening. If treatment is going to hurt, such as applying a splint, tell the child ahead of time.

Respect a child's modesty. Children are often embarrassed if they have to undress or be undressed in front of strangers. This anxiety often intensifies during adolescence. When a wound or site of injury has to be exposed, try to do so out of the sight of strangers, and when appropriate be sure to have a parent or guardian present. Again, it is extremely important to tell the child what you are doing and why you are doing it.

You should speak to a child in a professional, yet friendly way. A child should feel reassured that you are there to help in every way possible. Maintain eye contact with a child, as you would with an adult, to let the child know that you are there to help and that you can be trusted **Figure 4-6** . It is helpful to position yourself at the child's level so you do not appear to tower above the child.

Communicating With Hearing-Impaired Patients

Patients who are hearing-impaired are usually not ashamed or embarrassed by their disability. Often, it is the people around a hearing-impaired person who have problems coping. Remember that you must be able to communicate with hearing-impaired patients so you can provide necessary or even lifesaving care.

Most hearing-impaired patients have normal intelligence and can usually understand what is going on around them, provided you can successfully communicate with them. Most patients who are hearing-impaired can read lips to some extent. Therefore, you should place yourself in a position so the patient can see your lips. Many hearing-impaired patients have hearing aids to help them communicate. Be careful that hearing aids are not lost during an accident or fall. Hearing aids may also be forgotten if the patient is confused or ill. Look around for one in the immediate area, or ask the patient or the family about use of a hearing aid.

Remember the following five steps to efficiently communicate with patients who are hearing-impaired:

1. Have paper and a pen available. This way, you can write down questions and the patient can write down answers, if necessary. Be sure to print so that your handwriting is not a communication barrier.
2. If the patient can read lips, you should face the patient and speak slowly and distinctly. Do not cover your mouth or mumble. If it is dark, consider shining a light on your face.
3. Never shout.
4. Be sure to listen carefully, ask short questions, and give short answers. Remember that although many hearing-impaired patients can speak distinctly, some cannot.
5. Learn some simple phrases in sign language. For example, knowing the signs for "sick," "hurt," and "help" may be useful if you cannot communicate in any other way **Figure 4-7** .

Communicating With Visually Impaired Patients

Like hearing-impaired patients, visually impaired and blind patients have usually accepted and learned to deal with their disability. Of course, visually impaired patients are not necessarily completely blind. Many can perceive light and dark or can see shadows or movement. Ask the patient whether he or she can see at all. Also remember, as with other patients who have disabilities, you should expect visually impaired patients to have normal intelligence.

As you begin caring for a visually impaired patient, explain everything you are doing in detail as you are doing it. Be sure to stay in physical contact with the patient as you begin your care. Hold your hand lightly on the patient's shoulder or arm, and try to avoid sudden movements. If the patient can walk to the ambulance, begin by placing his or her hand on your arm, taking care not to rush. Transport with the patient any mobility aids, such as a cane, to the hospital. A visually impaired person may have a guide dog. Guide dogs are easily identified by their special harnesses **Figure 4-8** . They are trained to not leave their masters and to not respond to strangers. A visually impaired patient who is conscious can tell you about the dog and give instructions for its care.

Figure 4-6 Maintain eye contact with a child to let the child know that you are there to help and that you can be trusted.

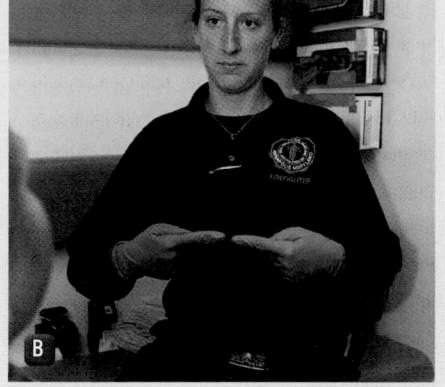

Figure 4-7 Learn simple phrases in sign language. **A.** Sick. **B.** Hurt. **C.** Help.

Figure 4-8 A guide dog is easily identified by its special harness.

The following rare situations are difficult to manage. If the patient is very stable, you should consider bringing the guide dog to the hospital in the back of the ambulance with the patient because it will help to alleviate some of the stress for both the patient and the dog. If the patient is unstable, the dog is unruly, or for other safety/patient care reasons it is inappropriate to transport the dog, then you should make arrangements for the care of the dog. Contact your supervisor for assistance. The exact method for managing a patient with a guide dog (or other medical care animal) will be outlined in your local department's policies and procedures. Follow your local protocols.

Communicating With Non-English-Speaking Patients

Part of patient care includes obtaining a medical history from the patient. You cannot skip this step simply because the patient does not speak English. Most patients who do not speak English fluently will still know certain important words or phrases.

Your first step is to find out how much English the patient can speak. Use short, simple questions and simple words whenever possible and avoid difficult medical terms. You can help patients better understand if you point to specific parts of the body as you ask questions. Speaking louder will not increase a patient's ability to understand you.

In many areas, particularly large urban centers, major segments of the population do not speak English. Your job will be much easier if you learn some common words and phrases in their language, especially common medical terms. Pocket cards that show the pronunciation of these terms are available. If the patient does not speak any English, find a family member or friend to act as an interpreter.

Communicating With Other Health Care Professionals

Effective communication between the EMT and health care professionals in the receiving facility is an essential cornerstone of efficient, effective, and appropriate patient care.

Your reporting responsibilities do not end when you arrive at the hospital. In fact, they have just begun. The transfer of care officially occurs during your oral report at the hospital. Once you arrive at the hospital, a hospital staff member will take responsibility for the patient from you **Figure 4-9**. However, you may only transfer the care of your patient to someone with at least your level of training. Once a hospital staff member is ready to take responsibility for the patient, you must provide that person with a formal oral report of the patient's condition.

Giving a report is a longstanding and well-documented end point in transferring the patient's care from one provider to another. Your oral report is usually given at the same time that the staff member is providing care for the patient. For example, a nurse or physician may start assessing the patient or help you to move the patient from the stretcher to an examination table. Therefore, you must report important information in a complete and precise

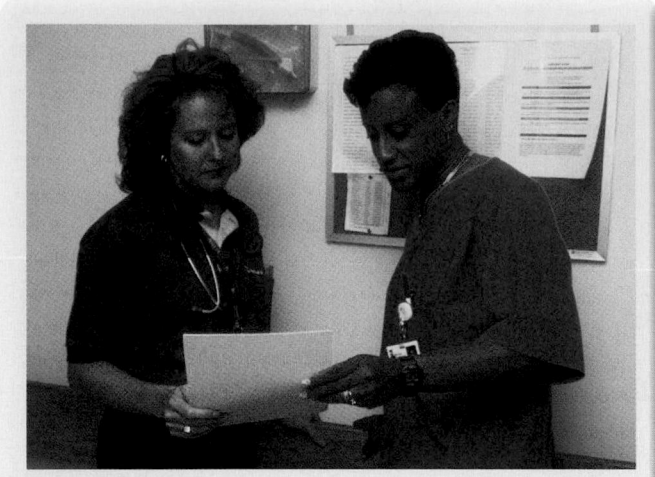

Figure 4-9 Once you arrive at the hospital, a staff member will take responsibility for the patient from you.

way. The following six components must be included in the oral report:

1. **Opening information** that includes the patient's name (if you know it), chief complaint, nature of the illness, or mechanism of injury. For example: "Good morning. This is Mrs McCarty. She is 65 years old and is complaining of back pain. She woke up around 3 AM, tripped, and fell into the bathtub after using the restroom."

2. **Detailed information** that was not provided during the radio report. For example: "She denies losing consciousness, states that she has no history of stroke, TIAs, or cardiac compromise, but has been feeling a little light-headed when she stands."

3. **Any important history** that was not already provided. For example: "Mrs McCarty lives by herself. She was unable to get out of the tub and was found by a hospice worker at 10 o'clock this morning. We suspect hypothermia because she had a core temperature of 94 degrees."

4. **The patient's response to treatment** given en route. It is especially important to report any changes in the patient or the treatment provided since your initial radio report. For example: "Oxygen was initiated by nonrebreathing face mask at 15 L/min. Although we suspected that her mid-back pain was a result of her leaning against the faucet of the bathtub for 7 hours, we put her in the Kendrick extrication device for both precautionary and extrication reasons. Hot packs wrapped in hand towels were used to help warm her up."

5. **Vital signs** assessed during transport and after the radio report. For example: "Her vitals include a blood pressure of 112/84 mm Hg, a pulse of 72 beats/min, respirations of 14 breaths/min, and core body temperature of 94 degrees at the time of transport. They are generally unchanged since then, except that her last temperature was 96 degrees."

6. **Other information** you may have gathered that was not important enough to report sooner. Information that was gathered during transport, patient medications you have brought with you, and any other details about the patient that were provided by family members or friends may be included. For example: "Mrs Woods, the home hospice worker, has contacted Mrs McCarty's family and followed us here to answer any questions."

You can use the same process described for giving an oral report if you need to transfer care during an EMS event. For example, if there are many patients, you may need to remain on the scene while someone else continues the assessment you began. Begin the oral report with a quick introduction, letting the other EMS provider know who you are and your level of licensure. You should then continue to transfer care just as you would inside the hospital.

You will also need to communicate routinely with many other professionals—police, social service personnel, fire personnel, and other EMS providers. Make sure your language and general demeanor are professional. Remember that federal laws protect a patient's right to privacy and that you should not give any health information about your patient to anyone other than those directly involved in the care of the patient.

As an EMT, you must be able to quickly and accurately find out what the patient needs and be able to tell others. Never forget that you are the vital link between the patient and the health care team.

Words of Wisdom

The Health Insurance Portability and Accountability Act (HIPAA) of 1996 established mandatory patient privacy rules and regulations to safeguard patient confidentiality. The act provides guidance on the types of information that are protected, the responsibility of health care providers regarding that protection, and penalties for breaching that protection.

Most personal health information is protected and should not be released without the patient's permission. These regulations apply to all forms of communication, written and verbal. To ensure that you are protecting your patient's right to confidentiality, do not give any information to anyone other than those directly involved in the care of the patient. Make sure you are aware of all policies and procedures governing your particular agency.

Written Communications and Documentation

The **patient care report (PCR)**, also known as a pre-hospital care report, is the legal document used to record all aspects of the care your patient received, from initial dispatch to arrival at the hospital. Either term can be used and both are acceptable. You may be able to complete the written report en route to the hospital if the trip is long enough and the patient needs minimal care. Usually, you will finish the written report after you have transferred care of the patient to an emergency department staff member. Be sure to leave the report at the emergency department before you depart to ensure continuity of care.

The information you collect during a call becomes part of the PCR, and that information is ultimately entered into a data pool. The National Emergency Medical Services Information System (NEMSIS) has been collecting prehospital care information for research purposes since the early 1970s. NEMSIS has identified specific data points needed to enable communication and comparison of EMS runs between agencies, regions, and states. The minimum data set includes both narrative components and check-off boxes Figure 4-10 .

If you go to www.ems.gov, you can see the national data set and discover interesting facts about delivery of EMS within the United States.

Because EMS systems track their own time, make sure that your watch is set with dispatch time at the beginning of the shift if that is procedure. Another way systems can manage this information is to contact the dispatcher and have him or her provide you with the time. Either way, it is important to be able to keep close track of time. Accurate documentation will depend on it.

You will begin gathering patient information as soon as you reach the patient. Continue collecting information as you provide care until you arrive at the hospital.

■ Patient Care Report

As discussed, a PCR helps ensure efficient continuity of patient care. This report describes the nature of the patient's injuries or illness at the scene and the initial treatment you provide. Although this report might not be read immediately at the hospital, it may be referred to later for important information. The report serves the following six functions:

1. Continuity of care
2. Legal documentation
3. Education
4. Administrative information
5. Essential research record
6. Evaluation and continuous quality improvement

Besides reporting the patient's condition on arrival at the scene and the care that was provided, a good PCR documents any changes in the patient's condition on arrival at the hospital. It is critical that you document everything in the clearest manner possible because the report serves multiple purposes. The information in the report will help to prove that you have provided a standard of care, and, in some instances, shows you have properly handled unusual or uncommon situations. Both objective and subjective information is included in this report.

The following are examples of information collected on a PCR:

- Chief complaint
- Level of consciousness (according to the AVPU scale) or mental status
- Vital signs
- Initial assessment
- Patient demographics (age, gender, ethnic background)

Should you ever be called to provide testimony concerning patient care, you and your PCR will be used to present evidence. As with your personal appearance, your PCR will reflect a professional or a nonprofessional image. A neat, concise, well-written document—including correct spelling and grammar—will reflect good patient care. Consider the adages "If you didn't write it, it didn't happen," or "If the report looks sloppy, the patient care was also sloppy."

These reports also provide valuable administrative information, such as that used for patient billing. Information included in PCRs can be used to evaluate response times, equipment usage, and other areas of administrative responsibility. The following are examples of administrative information gathered from a PCR:

- Time the incident was reported
- Time the EMS unit was notified
- Time the EMS unit arrived at the scene
- Time the EMS unit left the scene
- Time the EMS unit arrived at the receiving facility
- Time the patient care was transferred

Data may be obtained from the PCR to analyze causes, severity, and types of illness or injury requiring emergency medical care. These reports may also be used in an ongoing program for evaluation of the quality of patient care. All reports are periodically reviewed by your system. The purpose of these reviews is to make sure trauma triage and/or other prehospital care criteria have been met.

There are many requirements of a PCR Table 4-6 . Often, these requirements vary from jurisdiction to jurisdiction, mainly because different agencies obtain information

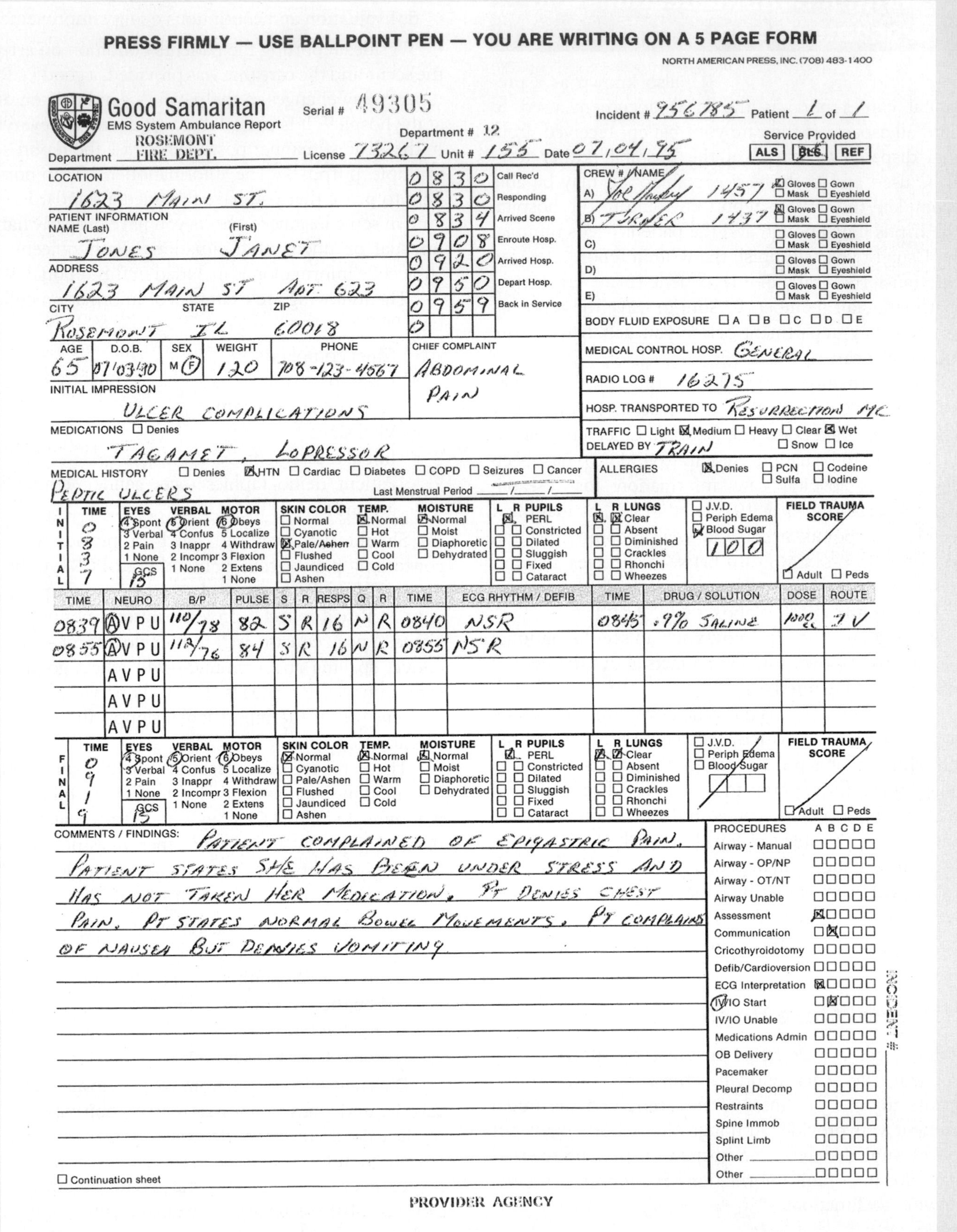

PRESS FIRMLY — USE BALLPOINT PEN — YOU ARE WRITING ON A 5 PAGE FORM

NORTH AMERICAN PRESS, INC. (708) 483-1400

Good Samaritan
EMS System Ambulance Report
ROSEMONT FIRE DEPT.

Serial # 49305
Department # 12
Incident # 956785 Patient 1 of 1

Department ROSEMONT FIRE DEPT. License 73267 Unit # 155 Date 07,04,95

Service Provided: ALS [BLS] REF

LOCATION				CREW # / NAME	
1623 Main St.	0830	Call Rec'd	A) Joe Shirley 1457	☒ Gloves ☐ Gown ☒ Mask ☐ Eyeshield
PATIENT INFORMATION	0830	Responding	B) Turner 1437	☐ Gloves ☐ Gown ☐ Mask ☐ Eyeshield
NAME (Last) (First) JONES JANET	0834	Arrived Scene	C)	☐ Gloves ☐ Gown ☐ Mask ☐ Eyeshield
ADDRESS	0908	Enroute Hosp.	D)	☐ Gloves ☐ Gown ☐ Mask ☐ Eyeshield
1623 Main St Apt 623	0920	Arrived Hosp.	E)	☐ Gloves ☐ Gown ☐ Mask ☐ Eyeshield
CITY STATE ZIP	0950	Depart Hosp.		
Rosemont IL 60018	0959	Back in Service	BODY FLUID EXPOSURE ☐ A ☐ B ☐ C ☐ D ☐ E	

AGE	D.O.B.	SEX	WEIGHT	PHONE	CHIEF COMPLAINT
65	07'03'30	M Ⓕ	120	708-123-4567	ABDOMINAL PAIN

MEDICAL CONTROL HOSP. GENERAL

INITIAL IMPRESSION ULCER COMPLICATIONS

RADIO LOG # 16275

HOSP. TRANSPORTED TO Resurrection MC

MEDICATIONS ☐ Denies TAGAMET, LOPRESSOR

TRAFFIC ☐ Light ☒ Medium ☐ Heavy ☐ Clear ☒ Wet
DELAYED BY TRAIN ☐ Snow ☐ Ice

MEDICAL HISTORY ☐ Denies ☒HTN ☐ Cardiac ☐ Diabetes ☐ COPD ☐ Seizures ☐ Cancer
PEPTIC ULCERS Last Menstrual Period __/__/__

ALLERGIES ☒Denies ☐ PCN ☐ Codeine ☐ Sulfa ☐ Iodine

INITIAL

TIME	EYES	VERBAL	MOTOR	SKIN COLOR	TEMP.	MOISTURE	L R PUPILS	L R LUNGS	
0837	4 Spont 3 Verbal 2 Pain 1 None	⑤Orient 4 Confus 3 Inappr 2 Incompr 1 None	⑥Obeys 5 Localize 4 Withdraw 3 Flexion 2 Extens 1 None	☐ Normal ☐ Cyanotic ☒ Pale/Ashen ☐ Flushed ☐ Jaundiced ☐ Ashen	☒ Normal ☐ Hot ☐ Warm ☐ Cool ☐ Cold	☒ Normal ☐ Moist ☐ Diaphoretic ☐ Dehydrated	☒ PERL ☐ Constricted ☐ Dilated ☐ Sluggish ☐ Fixed ☐ Cataract	☐ ☒ Clear ☐ Absent ☐ Diminished ☐ Crackles ☐ Rhonchi ☐ Wheezes	☐ J.V.D. ☐ Periph Edema ☒ Blood Sugar 100 FIELD TRAUMA SCORE ☐ Adult ☐ Peds
	GCS 15								

TIME	NEURO	B/P	PULSE	S	R	RESPS	Q	R	TIME	ECG RHYTHM / DEFIB	TIME	DRUG / SOLUTION	DOSE	ROUTE
0839	Ⓐ V P U	110/78	82	S	R	16	N	R	0840	NSR	0843	.9% SALINE	1000 cc	IV
0855	Ⓐ V P U	118/76	84	S	R	16	N	R	0855	NSR				
	A V P U													
	A V P U													
	A V P U													

FINAL

TIME	EYES	VERBAL	MOTOR	SKIN COLOR	TEMP.	MOISTURE	L R PUPILS	L R LUNGS	
0919	4 Spont 3 Verbal 2 Pain 1 None	⑤Orient 4 Confus 3 Inappr 2 Incompr 1 None	⑥Obeys 5 Localize 4 Withdraw 3 Flexion 2 Extens 1 None	☒ Normal ☐ Cyanotic ☐ Pale/Ashen ☐ Flushed ☐ Jaundiced ☐ Ashen	☒ Normal ☐ Hot ☐ Warm ☐ Cool ☐ Cold	☒ Normal ☐ Moist ☐ Diaphoretic ☐ Dehydrated	☒ PERL ☐ Constricted ☐ Dilated ☐ Sluggish ☐ Fixed ☐ Cataract	☒ ☒ Clear ☐ Absent ☐ Diminished ☐ Crackles ☐ Rhonchi ☐ Wheezes	☐ J.V.D. ☐ Periph Edema ☐ Blood Sugar FIELD TRAUMA SCORE ☐ Adult ☐ Peds
	GCS 15								

COMMENTS / FINDINGS: PATIENT COMPLAINED OF EPIGASTRIC PAIN. PATIENT STATES SHE HAS BEEN UNDER STRESS AND HAS NOT TAKEN HER MEDICATION. PT DENIES CHEST PAIN. PT STATES NORMAL BOWEL MOVEMENTS. PT COMPLAINS OF NAUSEA BUT DENIES VOMITING

☐ Continuation sheet

PROCEDURES	A B C D E
Airway - Manual	☐☐☐☐☐
Airway - OP/NP	☐☐☐☐☐
Airway - OT/NT	☐☐☐☐☐
Airway Unable	☐☐☐☐☐
Assessment	☒☐☐☐☐
Communication	☐☒☐☐☐
Cricothyroidotomy	☐☐☐☐☐
Defib/Cardioversion	☐☐☐☐☐
ECG Interpretation	☒☐☐☐☐
IV/IO Start	☐☒☐☐☐
IV/IO Unable	☐☐☐☐☐
Medications Admin	☐☐☐☐☐
OB Delivery	☐☐☐☐☐
Pacemaker	☐☐☐☐☐
Pleural Decomp	☐☐☐☐☐
Restraints	☐☐☐☐☐
Spine Immob	☐☐☐☐☐
Splint Limb	☐☐☐☐☐
Other ___	☐☐☐☐☐
Other ___	☐☐☐☐☐

PROVIDER AGENCY

Figure 4-10 The minimum data set includes patient and administrative information, including narrative components and check-off boxes.

from them. Although no universally accepted form exists, certain data points (uniform components of a PCR) are common in all areas. The benefits of collecting such information are significant, one being that national trends can be detected. For example, roughly 8% of the nation's EMS calls involve pediatric patients (ages 0 to 9 years). Of those patients, 11% will have a respiratory complaint. Such information is invaluable, and, when collected, form uniform data points.

Finally, PCRs are used by individual agencies to determine patterns of EMS responses. Busy times and high call volume areas can be predictive, and a thorough review of PCRs can set the stage for scheduling shifts and for system status management, including where units are placed.

■ Types of Forms

You will most likely use one of two types of forms. The first type is the traditional written form with check boxes

Table 4-6 Sample Uniform Components of a Patient Care Report

- Patient's name, gender, date of birth, and address
- Dispatched as (When was the ambulance called? What was the nature of the call as reported by the dispatcher?)
- Chief complaint
- Location of the patient when first seen (including specific details, especially if the incident is a car crash or when criminal activity is suspected)
- Rescue and treatment given before your arrival
- Signs and symptoms found during your patient assessment
- Care and treatment given by you at the site and during transport
- Vital signs
- SAMPLE history
- Changes in vital signs and condition
- Date of the call
- Time of the call
- Location of the call
- Time of dispatch
- Time of arrival at the scene
- Time of leaving the scene
- Time of arrival at the hospital
- Patient's insurance information
- Names and/or certification numbers of the EMTs who responded to the call
- Name of the base hospital involved in the run
- Type of run to the scene: emergency or routine

and a narrative section, as shown in Figure 4-10. The second type is a computerized version in which you fill in information electronically, such as on a computer or over the Internet Figure 4-11.

If your service uses written forms, be sure to fill in the boxes completely and avoid making stray marks on the sheet. Make sure that you are familiar with the specific procedures for collecting, recording, and reporting the information in your area. Regardless of which format is used to collect information, you must complete a narrative section. For this information to be valuable, it must be correct. Therefore, you should make every effort to ensure the information is accurate.

The narrative section of the PCR is arguably the most important portion. Here you will describe all of the facts related to the EMS call. Be sure to include significant negative findings and important observations about the scene. Do not record your conclusions about the incident. For example, you may write, "The patient admits to drinking today." This is a clear description that does not make any judgments about the patient's condition. You may also write that the patient "smelled of alcohol" but saying "the patient was drunk" is a judgment that may not be able to be supported. Choose your words carefully and thoughtfully. Make sure that what you write is not an opinion, but fact based on findings. Your job is to reproduce the important facts of the EMS call in writing.

The narrative section of the PCR needs to include the following information:

- Time of events
- Assessment findings
- Emergency medical care provided
- Changes in the patient after treatment
- Observations at the scene
- Final patient disposition
- Refusal of care
- Staff person who continued care

In written documentation, avoid radio codes and use only standard abbreviations, a list of which should be provided by your department. Remember, EMS personnel are not the only people who will be reading this document. Other hospital and billing personnel will need to read and understand what has been written. When information is of a sensitive nature, note the source of the information. Be sure to spell words correctly, especially medical terms. If you do not know the correct spelling of a particular word, find out how to spell it, or use another word. Also be sure to record the time with all assessment findings. Table 4-7 provides guidelines on how to write the narrative portion of your report. Whether you completed a medical or trauma assessment, the assessment-based approach follows each step of the assessment(s) as a guideline to narrative writing.

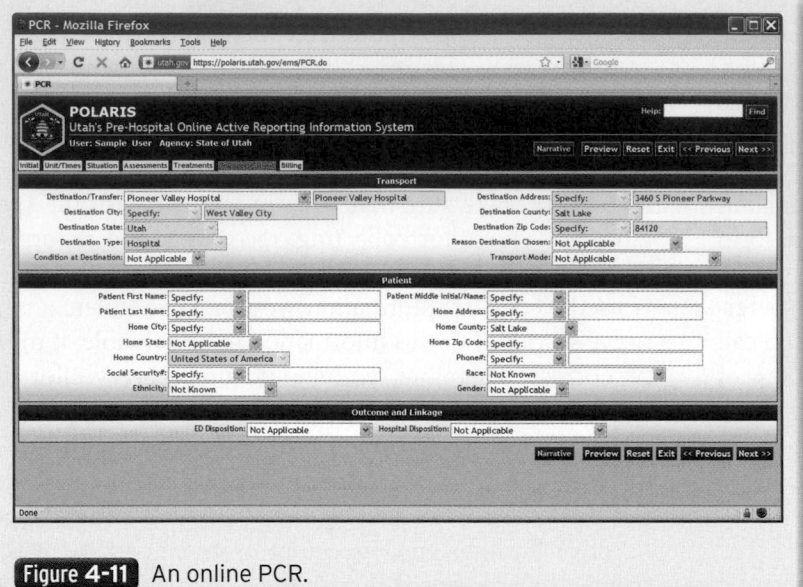

Figure 4-11 An online PCR.

Remember that the report form itself—and all the information in it—is considered a confidential document. Be sure you are familiar with state and local laws concerning confidentiality. All prehospital forms must be handled with care and stored in an appropriate manner once you have completed them. After you have completed a report, distribute copies to the appropriate locations, according to state and local protocol. In most instances, a copy of the report will remain at the hospital and will become part of the patient's medical record.

Depending on the requirements of the EMS system in which you work, you may not have the time to complete the full PCR while at the hospital. Even in these circumstances, however, a written record should be left with the patient. In these cases, most systems will

Table 4-7   How to Write a Narrative Report

Standard precautions	Were standard precautions initiated? If so, state which precautions were used and why.
Scene safety	Did you have to make your scene safe? If so, what did you do and why did you do it? Did this create a delay in patient care?
NOI/MOI	Simply state.
Number of patients	Record only when more than one patient is present; "This is patient 2 of 3."
Additional help	Did you call for help? If so, state why, at what time, and what time the help arrived. Was transport delayed?
Cervical spine	State what cervical spine precautions were initiated. You may want to include why; "Due to the significant MOI..."
Initial general impression	Simply record, if not already documented on the PCR.
Level of consciousness	Be sure to report LOC, any changes in LOC, and at what time changes occurred.
Chief complaint	Note and quote pertinent statements made by the patient and/or bystanders. This includes any pertinent denials; "Patient denies chest pain...."
Life threats	List all interventions and how the patient responded; "Assisted ventilations with O_2 (15 L/min) at 20 BPM with no change in LOC."
ABCs	Document what you found, and again, any interventions performed.
Oxygen	Record if O_2 was used, how it was applied, and how much was administered.
Primary, secondary, patient history, or reassessment	State the type of assessment used and any pertinent findings; "Secondary assessment revealed unequal pupils, crepitus to right ribs, and an apparent closed fracture of left tibia."
SAMPLE/OPQRST	Note and quote any pertinent answers.
Vital signs	Your service may want you to record vital signs in the narrative portion, as well as other places in the PCR.
Medical direction	Quote any orders given to you by medical control and who gave them.
Management of secondary injuries/treat for shock	Report all patient interventions, at what time they were completed, and how the patient responded.

Abbreviations: ABCs, airway, breathing, and circulation; LOC, level of consciousness; MOI, mechanism of injury; NOI, nature of illness; OPQRST, mnemonic used to facilitate taking a patient's symptoms; PCR, patient care report; SAMPLE, mnemonic used to facilitate taking a patient's symptoms.

Source: Reprinted with permission. Courtesy of Jay C. Keefauver.

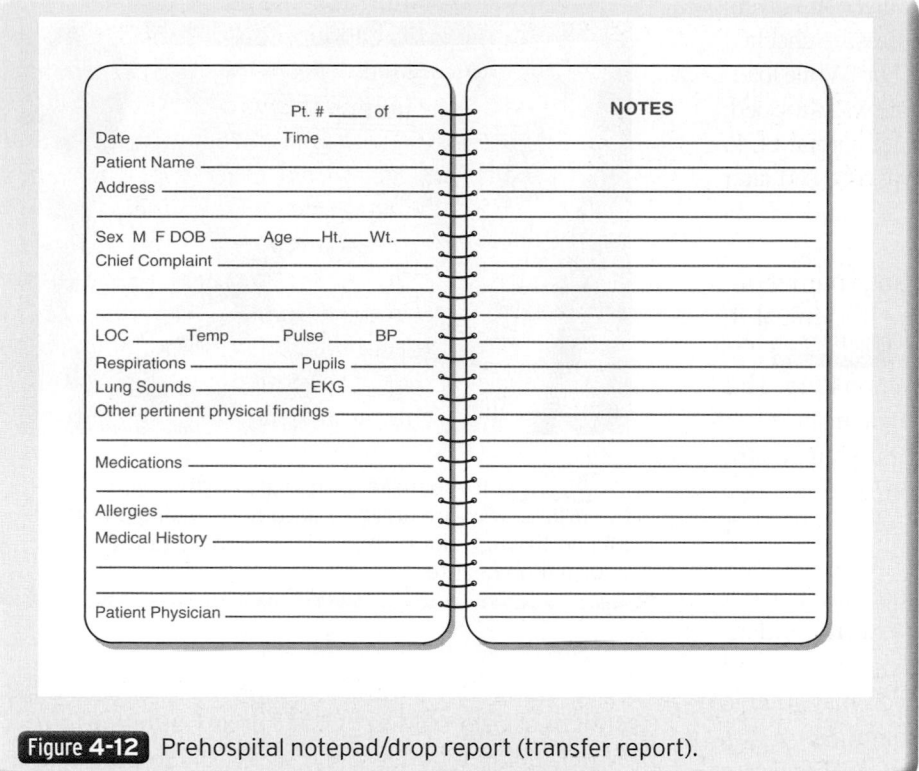

Figure 4-12 Prehospital notepad/drop report (transfer report).

have a "drop report or transfer report" **Figure 4-12** . These single-page, abbreviated forms are used as a memory aid during an EMS call. If you are unable to remain at the hospital to complete the PCR, copy these documents and leave them with the nurse or doctor.

■ Reporting Errors

Everyone makes mistakes. If you leave something out of a report or record information incorrectly, do not try to cover it up. Rather, write down what did or did not happen and the steps that were taken to correct the situation. Falsifying information on the PCR may result in suspension and/ or revocation of your certification/ license. Falsification may also have legal implications for the author. More important, falsifying information results in poor patient care, because other health care providers have a false impression of assessment findings or the treatment given. If you did not give the patient oxygen, do not chart that the patient was given oxygen.

Document only the vital signs that were actually taken. A classic case of improper documentation occurs with cardiac arrest patients. Under "Vital Signs," EMS providers will sometimes document: pulse 0 beats/min, respirations 0 breaths/min, and blood pressure 0/0 mm Hg. What the EMS provider actually documented was the application of a blood pressure cuff, inflating it, and deflating it while listening for a pulse. Someone may ask why the EMS provider took the time to check the blood pressure on a deceased person instead of performing CPR.

What if the wrong drug or the wrong dose was given to a patient? What if the patient is accidentally dropped? Unfortunately these things can and do happen. It is important that you document the event. Do not lie or cover it up. In your narrative, provide a factual account of what happened.

You are the Provider: PART 3

As your partner takes the patient's vital signs, you ask the patient further questions regarding her chief complaint. She denies any other complaints or past medical history and tells you that she only takes vitamins. You offer her supplemental oxygen via nasal cannula; however, she tells you that she does not feel that she needs it. Her blood glucose level is assessed and found to be 112 mg/dL.

Recording Time: 5 Minutes	
Respirations	20 breaths/min; regular and unlabored
Pulse	68 beats/min; strong and regular
Skin	Pink, warm, and dry
Blood pressure	122/62 mm Hg
Oxygen saturation (Sao_2)	98% (on room air)

5. What techniques can facilitate the interviewing process of an older patient?

For example: "Ordered: one sublingual nitroglycerin. Given: two sublingual nitroglycerin. Patient blood pressure checked following administration. No changes noted." or "While loading the patient into the ambulance, the patient was dropped. Patient was on the ambulance cot when it fell a total of 4′. Patient was not thrown off of cot. Patient was assessed after being dropped and complained of being scared and having neck pain. Hospital advised."

If you discover an error as you are writing your report, draw a single horizontal line through the error, initial it, and write the correct information next to it **Figure 4-13**. Do not try to erase or cover the error with correction fluid. This may be interpreted as an attempt to cover up a mistake.

If an error is discovered after you submit your report, follow the same process of error correction by drawing a single line through the error—preferably in a different color ink—initialing it, and dating it. Make sure to add a note with the correct information. If you accidentally left out information, begin the new section with the word "addendum," add the new information, and then add the date and your initials. When using a paper system, you may be able to add addendums using specific addendum forms.

When you do not have enough time to complete your report before the next call, you will need to fill it out later. If you are using a computerized documentation system, refer to the system's direction as to how to make an amendment to the original document. Most computerized systems will allow for amendments but will prevent erasure in a completed document.

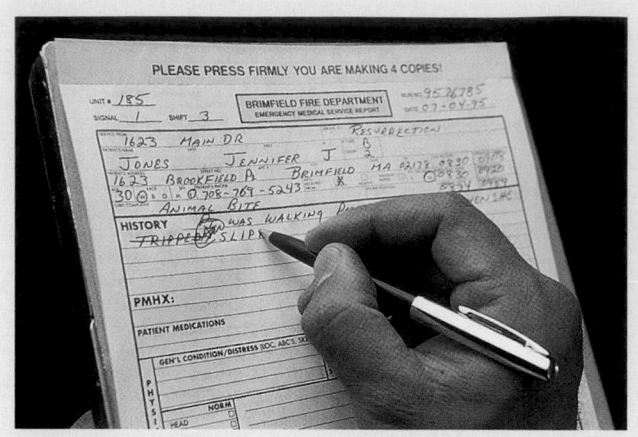

Figure 4-13 If you make a mistake in writing your report, the proper way to correct it is to draw a single horizontal line through the error, initial it, and write the correct information next to it.

■ Documenting Refusal of Care

Refusal of care is a common source of litigation in EMS; therefore, thorough documentation is crucial. Competent adult patients have the right to refuse treatment and, in fact, must specifically provide permission for treatment to be provided by EMS or any other health care provider. Before you leave the scene, try to persuade the patient to go to the hospital, and consult medical control as directed by local protocol. Also make sure the patient is an adult or an emancipated minor, able to make an informed decision, and is not under the influence of alcohol or other drugs or the effects of an illness or injury. Explain to the patient why it is important to be examined by a physician at the hospital. Also explain what may happen if the patient is not examined by a physician. If the patient still refuses, suggest other means for the patient to obtain proper care. Explain that you are willing to return. If the patient still refuses, document any assessment findings and emergency medical care given, and have the patient sign a refusal form **Figure 4-14**. You should also have a family member, police officer, or bystander sign the form as a witness. If the patient refuses to sign the refusal form, have a family member, police officer, or bystander sign the form verifying that the patient refused to sign.

Be sure to complete the PCR, including the patient assessment findings. You will need to document the advice you gave as to the risks associated with refusal of care. Report clinical information, such as the LOC, showing the competency of the person refusing care. Note pertinent patient comments and any medical advice given to the patient by the physician or medical control through phone or radio. Also include a description of the care that you wished to provide for the patient. There are many local variations of requirements for patient refusals. **Table 4-8** provides a reasonable list of items that should be included within the PCR of a patient refusal.

Refusal of care not only includes patients who do not wish to be transported to the hospital but also those

RELEASE FROM RESPONSIBILITY WHEN PATIENT REFUSES IV THERAPY

This is to certify that I, _____ , am refusing IV treatment. I acknowledge
patient's name
that I have been informed of the risk involved and hereby release the emergency medical services
provider(s), the physician consultant, and the consulting hospital from all responsibility for any
ill effects which may result from this action.

Witness _____   Signed _____
                                                patient name or nearest relative
Witness _____   _____
                                                relationship

RELEASE FROM RESPONSIBILITY WHEN PATIENT REFUSES SERVICE

This is to certify that I, _____ , am refusing the services offered by the
patient's name
emergency medical services provider(s). I acknowledge that I have been informed of the risk
involved and hereby release the emergency medical services provider(s), the physician consultant,
and the consulting hospital from all responsibility for any ill effects which may result from this action.

Witness _____   Signed _____
                                                patient name or nearest relative
Witness _____   _____
                                                relationship

**RELEASE FROM RESPONSIBILITY WHEN PATIENT REFUSES SERVICES
BUT ACCEPTS TRANSPORT**

This is to certify that I, _____ , am refusing _____
                        patient's name

_____ .

I acknowledge that I have been informed of the risk involved and hereby release the emergency
medical services provider(s), the physician consultant, and the consulting hospital from all
responsibility   for   any   ill   effects   which   may   result   from   this   action.

Witness _____   Signed _____
                                                patient name or nearest relative
Witness _____   _____
                                                relationship

Figure 4-14 A competent adult patient has the right to refuse medical treatment and must sign a refusal form.

who refuse a certain aspect of care. For example, a victim of a car crash may wish to be treated and transported but refuses to be fully immobilized. It is appropriate to carry out all other medical care and document the patient's refusal of spinal stabilization. Just because the patient refuses a cervical collar is no reason to deny oxygen. The same is true for the patient who wishes to use a local hospital when the injuries dictate transport to a trauma facility. Any time a patient refuses any part of the standard treatment, it needs to documented in the PCR.

■ Special Reporting Situations

In some situations, you may be required to file special reports with appropriate authorities. These situations may involved gunshot wounds, dog bites, certain infectious diseases, or suspected physical or sexual abuse. Learn your local requirements for reporting these incidents. Failure to report them may have legal consequences. It is important that the report be accurate, objective, and submitted in a timely manner. Also remember to keep a copy for your own records.

Another special reporting situation is a mass-casualty incident (MCI). The local MCI plan should have some means of temporarily recording important medical information (such as a triage tag that can be used later to complete the form). The standard for completing the form in an MCI is not the same as for a typical call. Your local plan should have specific guidelines.

A written report is the portion of the EMT's patient care interaction that becomes part of the patient's permanent medical record. It serves many purposes, including demonstrating that the care delivered was appropriate and within the scope of practice of the caregivers involved. Documentation also provides an opportunity to communicate the patient's story to others who may participate in the patient's care in the future. Adequate reporting and accurate records ensure the continuity of patient care. Complete patient records also guarantee proper transfer of responsibility, comply with the requirements of health departments and law enforcement agencies, and fulfill your organization's administrative needs. Reporting and recordkeeping duties are an essential aspect of patient care, although they are performed only after the patient's condition has been stabilized. Documentation in the field drives both funding and research for EMS. Seatbelts are a prime example. Studies gathered from recordkeeping in the early 1970s showed that patients had a significantly higher survival rate when seatbelts were used during motor vehicle accidents. Armed with this information, officials passed laws to enforce seatbelt usage, and large sums of money were spent on educating the public.

■ Medical Terminology

Medical terms are mainly derived from Latin. Many new EMS providers are confused as to why they need to learn this language, asking questions such as, "If a person has a sour stomach, why can we not simply write, 'patient complains of a sour stomach' and be done with it?"

Table 4-8 Components of a Thorough Patient Refusal Document

Complete assessment

Evidence the patient is able to make a rational, informed decision

Discussion with the patient as to what care/transportation the EMT would like to do

Discussion with the patient as to what may happen if he or she does not allow care or transportation. Typically these consequences should be listed and clear to include the possibility of severe illness/injury or death if care or transportation is refused.

Discussion with family/friend/bystanders to try to encourage the patient to allow care

Discussion with medical direction according to local protocol

Providing the patient with other alternatives: Going to see his or her family doctor, having a family member drive him or her to the hospital

Willingness of EMS to return

Signatures: Have a family member, police officer, or bystander sign the form as a witness. If the patient refuses to sign the refusal form, have a family member, police officer, or bystander sign the form verifying that the patient refused to sign.

If the patient refused care or did not allow a complete assessment, document that the patient did not allow for proper assessment and document whatever assessments were completed.

The reason to learn medical terms is to ensure accurate understanding by all people involved in the patient's care. What does sour stomach mean? Is there pain? Does the patient feel like he or she will vomit? Is the patient feeling like he or she does not want to eat? There are many possible options. Through medical terminology, all readers, no matter where they are located in the chain of care, will understand the patient's complaint. Gastritis, an inflammation of the stomach, is understood equally in Ohio, Florida, England, and Thailand. Medical personnel around the globe speak the same language—Latin.

Medical terminology uses a system that combines prefixes, root words, and suffixes to describe complaints or diseases. If a patient has a headache, this can be described in medical terms as cephalgia. *Ceph* (head) and *algia* (pain) are combined. When talking about medical issues with other health care providers, using such appropriate medical language will help ensure accurate understanding.

In the chapter on anatomy and physiology, it is clear that Latin prevails. The brain and spinal cord are part of the nervous system because neuro refers to nerve. There may be slight spelling changes, but the meaning remains. Common directional terms such as superior (toward the head), lateral (toward the side), and distal (away from the midline) are also found in the anatomy and physiology chapter. Taking a medical terminology course can certainly be helpful when working in medicine. Table 4-9 lists some of the terms that are commonly used.

Table 4-9 Common Medical Terms

Term	Meaning
A-, An-	Without
-algia	Pain
Arteri(o)-	Artery
Brady-	Slow
Cardi(o)-	Heart
Hem(o)-	Blood
Hyper-	Above
Hypo-	Under, deficient
-itis	Inflammation
Nas(o)-	Nose
Neur(o)-	Nerve
Poly-	Many
Retr(o)-	Backward or behind
Sub-	Under, Moderately
Supra-	Above
Tachy-	Fast
Trans-	Across
Vas(o)-	Vessel

Communications Systems and Equipment

Radio and telephone communications link you and your team with other members of the EMS, fire, and law enforcement communities. This link helps the entire team to work together more effectively and provides an important layer of safety and protection for each member of the team. You must know what your system can and cannot do, and you must be able to use your system efficiently and effectively. You must be able to send precise, accurate reports about the scene, the patient's condition, and the treatment that you provide.

As an EMT, you must be familiar with two-way radio communications and have a working knowledge of the mobile and hand-held portable radios that are used in your unit. You must also know when to use them and what to say when you are transmitting.

Base Station Radios

The dispatcher usually communicates with field units by transmitting through a fixed radio base station that is controlled from the dispatch center. A **base station** is any radio hardware containing a transmitter and receiver that is located in a fixed place. The base station may be used in a single place by an operator speaking into a microphone that is connected directly to the equipment. It also works remotely through telephone lines or by radio from a communications center. Base stations may include dispatch centers, fire stations, ambulance bases, or hospitals.

A two-way radio consists of two units: a transmitter and a receiver. Some base stations may have more than one transmitter and/or more than one receiver. They may also be equipped with one multichannel transmitter and several single-channel receivers. A **channel** is an assigned frequency or frequencies used to carry voice and/or data communications. Regardless of the number of transmitters and receivers, they are commonly called *base radios* or *stations*. Base stations usually have more power (often 100 watts or more) and higher, more efficient antenna systems than mobile or portable radios. This increased broadcasting range allows the base station operator to communicate with field units and other stations at much greater distances.

The base radio must be physically close to its antenna. Therefore, the actual base station cabinet and hardware are commonly found on the roof of a tall building or at the bottom of an antenna tower. The base station operator may be miles away in a dispatch center or hospital, communicating with the base station radio by dedicated lines or special radio links. A **dedicated line**, also known as a *hot line*, is used for specific point-to-point contact. This type of phone, typically located within an emergency department, is not on the main switchboard. EMS personnel are able to call the number directly without being placed on hold or transferred. This type of line makes recording medical command conversations much easier.

Mobile and Portable Radios

In the ambulance, you will use both mobile and portable radios to communicate with the dispatcher and/or medical control. An ambulance will often have more than one mobile radio, each on a different frequency **Figure 4-15**. One radio may be used to communicate with the dispatcher or other public safety agencies. A second radio is often used for communicating patient information to medical control.

A mobile radio is installed in a vehicle and usually operates at lower power than a base station. Most **VHF (very high frequency)** mobile radios operate between 30 and 300 MHz. **UHF (ultra-high frequency)** mobile radios operate between 300 MHz and 3,000 MHz. Radios that operate at 800 MHz are increasingly common in EMS systems. These systems provide a great amount of system flexibility without the need for vast numbers of frequencies. What was once accomplished with 30 separate frequencies can be done with less than 10. Mobile antennas are much closer to the ground than base station antennas, so communications from the unit are typically limited to 10 to 15 miles over average terrain.

Portable radios are hand-held devices that operate at 1 to 5 watts of power. Because the entire radio can be held in your hand, when in use, the antenna is often no higher than the EMT who is using the radio. The transmission range of a portable radio is more limited than that of mobile or base station radios. Portable radios are

Figure 4-15 Some ambulances have more than one mobile radio to allow communications with hospitals, mutual aid jurisdictions, and other agencies.

essential in helping to coordinate EMS activities at the scene of an MCI. They are also helpful when you are away from the ambulance and need to communicate with dispatch, another unit, or medical control Figure 4-16 .

■ Repeater-Based Systems

A **repeater** is a special base station radio that receives messages and signals on one frequency and then automatically retransmits them on a second frequency. Because a repeater is a base station (with a large antenna), it is able to receive lower power signals, such as those from a portable radio, from a long distance away. The signal is then rebroadcast with all the power of the base station Figure 4-17 . EMS systems that use repeaters usually have outstanding systemwide communications and are able to get the best signal from portable radios. There are also mobile repeaters that may be found in ambulances or placed in various areas around an EMS system area.

Figure 4-16 A portable radio is essential if you need to communicate with the dispatcher or medical control when you are away from the ambulance.

Figure 4-17 A message is sent from the control center to the transmitter by a land line. The radio carrier wave is picked up by the repeater for rebroadcast to outlying units. Return radio traffic is picked up by the repeater and rebroadcast to the control center.

At times, you may be able to communicate with a base station radio, but you will not be able to hear or transmit to another mobile unit that is also communicating with that base. Repeater base stations eliminate such problems. They allow two mobile or portable units that cannot reach each other directly to communicate through the repeater, using its greater power and antenna.

■ Digital Equipment

Although most people think of voice communications when they think of two-way radios, digital signals are also a part of EMS communications. Some EMS systems use telemetry to send an electrocardiogram from the unit to the hospital. With **telemetry**, electronic signals are converted into coded, audible signals. These signals can then be transmitted by radio or telephone to a receiver with a decoder at the hospital. The decoder converts the signals back into electronic impulses that can be displayed on a screen or printed. Another example of telemetry is a fax message.

Digital signals are also used in some kinds of paging and tone alerting systems because they transmit faster than spoken words and allow more choices and flexibility.

■ Cellular/Satellite Telephones

Whereas dispatchers communicate with field units by transmitting through a fixed radio base station, it is common for EMTs to communicate with receiving facilities by **cellular telephone**. These telephones are simply low-power portable radios that communicate through a series of interconnected repeater stations called *cells* (hence the name *cellular*). Cells are linked by a sophisticated computer system and connected to the telephone network. Another option is a satphone or satellite phone. These phones use a satellite, which receives and relays the signals, instead of a cell.

Many cellular systems make equipment and air time available to EMS services at little or no cost as a public service. The public is often able to call 9-1-1 or other emergency numbers on a cellular telephone free of charge. However, this easy access may result in overloading and jamming of cellular systems in mass-casualty and disaster situations.

When using these systems, ensure that a reference of commonly called numbers is available. Local hospitals, poison control, police services, and the number to the dispatcher should be readily available. Cellular and satellite systems also have areas of bad reception. As an EMT, it is important to be aware of any areas in which your equipment will not work.

As with all repeater-based systems, a cellular/satellite telephone is useless if the equipment fails, loses power, or is damaged by severe weather or other circumstances. Cellular and satellite phones use digital signals. This makes eavesdropping difficult but not impossible.

A **scanner** is a radio receiver that searches or "scans" across several frequencies until the message is completed. Although cellular/satellite telephones are more private than most other forms of radio communications, they can still be overheard. Therefore, you must always be careful to appropriately respect patient privacy and to speak in a professional manner every time you use any form of an EMS communications system.

■ Other Communications Equipment

Ambulances and other field units are usually equipped with an external public address system. This system may be a part of the siren or the mobile radio. The intercom between the cab and the patient compartment may also be a part of the mobile radio. These components do not involve radio wave transmission, but you must understand how they work and practice using them before you really need them.

EMS systems may use a variety of two-way radio hardware. Some systems operate VHF equipment in the **simplex** (push to talk, release to listen) mode. In this mode, radio transmissions can occur in either direction but not simultaneously in both. When one party transmits, the other can only receive. Once one party finishes transmitting, the other party can then reply. Other systems conduct **duplex** (simultaneous talk–listen) communications on UHF frequencies and cellular telephones. In the full duplex mode, radios can simultaneously transmit and receive communications on one channel. This is sometimes called "a pair of frequencies." A number of VHF and UHF channels, commonly called **MED channels**, are reserved exclusively for EMS use. However, hundreds of other commercial, local government, and fire services frequencies are also used for EMS communications.

Trunking, or 800-MHz, systems take advantage of the latest technologies in communications. Instead of being assigned to one or two frequencies, in a trunking system, many frequencies are assigned to a group. As the radio conversation begins, a computer selects the next open frequency and the EMT begins talking. When the EMT speaks a second time, he or she will likely be speaking on a different frequency because the computer is constantly monitoring for frequency load and reassigning transmissions to unused frequencies. These systems allow for greater traffic without greater numbers of frequencies. Therefore, you do not have to be worried about being able to transmit or receive. In a trunking system, the computer will switch you to another channel without your knowledge and you will operate the radio as you normally do.

You are the Provider: PART 4

The patient agrees to EMS transport, is placed on the stretcher, and is loaded into the ambulance. You cover her with a blanket to keep her warm and proceed to a hospital located 15 miles away. En route, you reassess her condition.

Recording Time: 11 Minutes	
Level of consciousness	Conscious and alert
Respirations	20 breaths/min; regular and unlabored
Pulse	74 beats/min; strong and regular
Skin	Pink, warm, and dry
Blood pressure	118/60 mm Hg
Sao$_2$	98% (on room air)

The patient's condition remains unchanged since your initial encounter. You contact the receiving facility and provide them with a radio report.

6. What are the components of a radio report to the hospital?

7. How does your oral report differ from your radio report?

Another type of communication system becoming popular is **mobile data terminals (MDT)** Figure 4-18. MDTs are small computer terminals inside ambulances that directly receive data from the dispatch center. These allow for greatly expanded communication capabilities. Instead of having to listen to the dispatcher and determine whether he said 11345 Main Street or 11354 Main Street, you look at the terminal where the address is displayed. Satellite communications can track your progress to the scene and can provide important scene information, such as known violent calls to this address, the nature of those calls, and the number of times the ambulance has been called.

Your ability to effectively communicate with other units or medical control depends on how well the weaker radio can "talk back." Base and repeater station radios often have higher antennas and much greater power than mobile or portable units do. This increased power ensures that signals are generally heard and understood from a far greater distance than the signal produced from a mobile unit. Remember, when you are at the scene, you may be able to clearly hear the dispatcher or hospital on your radio, but you may not be heard or understood when you transmit.

Even small changes in your location can significantly affect the quality of your transmission. Also remember that the location of the antenna is critically important for clear transmission. Commercial aircraft flying at 37,000′ can transmit and receive signals over hundreds of miles, yet their radios have only a few watts of power. The "power" comes from their antenna positioned at 37,000′.

The success of communications depends on the efficiency of your equipment. A damaged antenna or microphone often prevents high-quality communications. Check the condition and status of your equipment at the start of each shift, and then correct or report any problems.

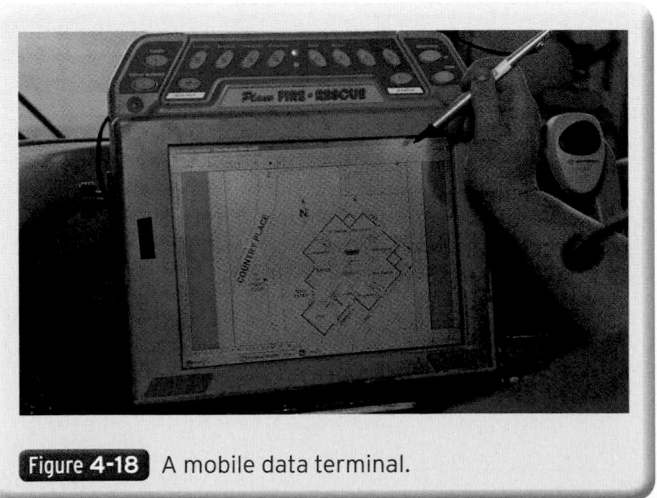

Figure 4-18 A mobile data terminal.

Radio Communications

All radio operations in the United States, including those used in EMS systems, are regulated by the **Federal Communications Commission (FCC)**. The FCC has jurisdiction over interstate and international telephone and telegraph services and satellite communications—all of which may involve EMS activity.

The FCC has five principal EMS-related responsibilities:

1. **Allocating specific radio frequencies for use by EMS providers.** Modern EMS communications began in 1974. At that time, the FCC assigned 10 MED channels in the 460- to 470-MHz (UHF) band to be used by EMS providers. These UHF channels were added to the several VHF frequencies that were already available for EMS systems. However, these VHF frequencies had to be shared with other "special emergencies" uses, including school buses and veterinarians. In 1993, the FCC created an EMS-only block of frequencies in the 220-MHz portion of the radio spectrum.

2. **Licensing base stations and assigning appropriate radio call signs for those stations.** An FCC license is usually issued for 5 years, after which time it must be renewed. Each FCC license is granted only for a specific operating group. Often, the longitude and latitude (locations) of the antenna and the address of the base station determine the call signs.

3. **Establishing licensing standards and operating specifications for radio equipment used by EMS providers.** Before it can be licensed, each piece of radio equipment must be submitted to the FCC by its manufacturer for type acceptance, based on established operating specifications and regulations.

4. **Establishing limitations for transmitter power output.** The FCC regulates broadcasting power to reduce radio interference between neighboring communications systems.

5. **Monitoring radio operations.** This includes making spot field checks to help ensure compliance with FCC rules and regulations.

The FCC's rules and regulations are written in technical and legal language and fill many volumes. Only a very small section (part 90, subpart C) deals with EMS communication issues. You are not responsible for reading these detailed and often confusing documents. For appropriate guidance on technical issues, contact your EMS system supervisor. In fact, many EMS systems look to radio and telephone communications experts for advice on technical issues.

■ Responding to the Scene

EMS communication systems may operate on several frequencies and use different frequency bands. Some EMS systems may even use different radios for various purposes. However, all EMS systems depend on the skill of the dispatcher. The dispatcher receives the first call to 9-1-1. You are part of the team that responds to calls once the dispatcher notifies your unit of an emergency.

The dispatcher has several important responsibilities during the alert and dispatch phase of EMS communications. The dispatcher must do all of the following:

- Properly screen and assign priority to each call (according to predetermined protocols).
- Select and alert the appropriate EMS response unit(s).
- Dispatch and direct EMS response unit(s) to the correct location.
- Coordinate EMS response unit(s) with other public safety services until the incident is over.
- Provide emergency medical instructions to the telephone caller (according to predetermined protocols) so that essential care (eg, CPR) may begin before the EMTs arrive.

When the first call to 9-1-1 comes in, the dispatcher must judge its relative importance to begin the appropriate EMS response using emergency medical dispatch protocols. First, the dispatcher must find out the exact location of the patient and the nature and severity of the problem. The dispatcher asks for the caller's telephone number, the patient's name and age, and other information, as directed by local protocol. Next, the dispatcher asks for some description of the scene, such as the number of patients or special environmental hazards.

From this information, the dispatcher will assign the appropriate EMS response unit(s) on the basis of local protocols to determine the level and type of response and the following factors:

- Dispatcher's determination of the nature and severity of the problem (Many emergency medical dispatch systems will determine this automatically based on a caller's answers to a defined series of questions.)
- Anticipated response time to the scene
- Level of training (EMR, EMT, AEMT, Paramedic) of available EMS response unit(s)
- Need for additional EMS units, fire suppression, rescue, a hazardous materials team, air medical support, or law enforcement

The dispatcher's next step is to alert the appropriate EMS response unit(s) **Figure 4-19**. Alerting these units

Figure 4-19 You will be assigned to a call by the dispatcher.

may be done in a variety of ways. The dispatcher may use the dispatch radio system to contact units that are already in service and monitoring the channel. Dedicated lines (hot lines) between the control center and the EMS station may also be used.

The dispatcher may also page EMS personnel. Pagers are commonly used in EMS operations to alert on-duty and off-duty personnel. **Paging** involves the use of a coded tone or digital radio signal and a voice or display message that is transmitted to pagers (beepers) or desktop monitor radios. Paging signals may be sent to alert only certain personnel or may be blanket signals that will activate all the pagers in the EMS service. Pagers and monitor radios are convenient because they are usually silent until their specific paging code is received. Alerted personnel contact the dispatcher to confirm the message and receive details of their assignments.

Once EMS personnel have been alerted, they must be properly dispatched and sent to the incident. Every EMS system should use a standard dispatching procedure. The dispatcher should give the responding unit(s) the following information:

- Nature and severity of the injury, illness, or incident
- Exact location of the incident
- Number of patients
- Responses by other public safety agencies
- Special directions or advisories, such as adverse road or traffic conditions, severe weather reports, or potential scene hazards
- Time at which the unit or units are dispatched

Your unit must confirm with the dispatcher that you have received the information and are en route to the scene. Local protocol will dictate whether it is the job of the dispatcher or your unit to notify other public safety agencies that you are responding to an emergency. In some areas, the emergency department is also notified when an ambulance responds to an emergency.

You should report to the dispatcher any problems during your response. You should also inform the dispatcher when you have arrived at the scene. The arrival report to the dispatcher should include any obvious details that you see during scene size-up. For example, you might say, "Dispatcher, BLS Unit Two is on scene at 3010 Mitchell Street. It is a blue house with long driveway." This information is particularly useful if additional units are responding to the same scene.

All radio communications during dispatch, as well as during other phases of operations, must be brief and easily understood. Speak in plain English. Your tone and pace should be relaxed and clear. Try to speak slowly so your voice can be easily understood. You do not need to use excessively polite language. Also, avoid wordiness. An example of an excessively wordy communication is: "Good morning dispatch, this is Ambulance 6-3-1. We are responding to 381 South Main Street. Have a good day." Although this sounds pleasant (and you should try to foster a good working environment with the dispatcher), this wastes radio time. Remember, the dispatcher's job is to field hundreds of calls an hour; therefore, you need to only report important information so that he or she can focus on what to do next. **Table 4-10** lists common instances for which EMS providers will need to use the radio to communicate with dispatch.

Table 4-11 lists tips to using the radio. Although these may change slightly from department to department, they provide a good foundation from which to begin.

■ Communicating With Medical Control and Hospitals

The principal reason for radio communication is to facilitate communication between you and medical control (and the hospital). Medical control may be located at the receiving hospital, another facility, or sometimes even in another city or state. You must, however, consult with medical control to notify the hospital of an incoming patient, request advice or receive orders from medical control, or to advise the hospital of special situations.

It is important to plan and organize your radio communication before you push the transmit button. Remember, a concise, well-organized report is the best method of accurately and thoroughly describing the patient and his or her medical condition to the providers who will be receiving the patient. It also demonstrates your competence and professionalism to all who hear your report. Well-organized radio communications with the hospital will engender confidence in the receiving facility's physicians and nurses, as well as others who are listening. In addition, the patient and family will be comforted by your organization and ability to communicate clearly. A well-delivered radio report puts you in control of the information, which is correct procedure.

You are the Provider: PART 5

With an estimated time of arrival at the hospital of 6 minutes, you reassess the patient and note that her condition has remained unchanged.

Recording Time: 16 Minutes	
Level of consciousness	Conscious and alert
Respirations	18 breaths/min; regular and unlabored
Pulse	72 beats/min; strong and regular
Skin	Pink, warm, and dry
Blood pressure	120/60 mm Hg
Sao$_2$	99% (on room air)

You arrive at the hospital and give your oral report to the charge nurse. After answering the nurse's questions, you complete your patient care report and return to service.

8. What functions does the patient care report serve?

Table 4-10 Typical EMS Communications With Dispatch

Phase of EMS Call	EMS Unit Communication
Initial receipt of call	Acknowledges call Responds to the call
En route to call	Requests assistance with directions, when needed Requests additional resources, when needed
On scene	Reports arrival at scene Check-ins; often a system will require EMS units to transmit every 20 minutes as a safety measure Requests additional resources, when needed Reports leaving scene
Arrival at hospital (or point of transfer)	Notifies dispatch of arrival at point of transfer
Return to service	Notifies dispatch when the unit is available for another call
Others	Some systems require EMS units to notify dispatch anytime they are not in station

Table 4-11 Tips When Using EMS Radio Communications

Turn radio on and adjust volume.
Ensure a clear frequency before speaking.
To speak, use the "press-to-talk" button, and wait 1 second before speaking.
Hold the microphone 2" to 3" from your mouth.
Address the unit you are calling, and provide the name of your unit.
The unit you call will signal that you can begin your transmission.
Use a clear, calm, and monotone voice and speak at a reasonable pace.
Keep the transmission brief.
Use clear text.
Avoid the use of codes or agency-specific terms.
Do not use useless or meaningless phrases, such as "be advised."
Limit saying "please," "thank you," and "you're welcome."
When transmitting numbers, such as an address, provide both the number and the individual digits ie, "Respond to 1381, 1-3-8-1, Main Street."
Remember that the airwaves are public and the use of scanners is popular.
Remain objective and impartial in describing patients.
Never use profanity; always be professional.
Use the words "affirmative" and "negative" instead of "yes" or "no."
Use the standard format for transmission of information.
When you are finished transmitting, indicate this by saying "over."
Do not provide a diagnosis of the patient's problem.
Use EMS frequencies only for EMS communications.
Monitor background noise.
Do not use names; protect the privacy of patients.

Hospital notification is the most common type of communication between you and the hospital. The purpose of these calls is to notify the receiving facility of the patient's complaint and condition **Figure 4-20** . On the basis of this information, the emergency department is able to appropriately prepare staff and equipment to receive the patient. This is primarily a one-way form of communication. You are telling the emergency department what to expect. You are not asking for advice or orders; you are simply notifying them.

Giving the Patient Report

The patient report should follow a standard format established by your EMS system. The report commonly includes the following seven elements:

1. **Your unit identification and level of services.** Example: "Columbus Fire 2-BLS."
2. **The receiving hospital and your estimated time of arrival.** Example: "Columbus Community Hospital, ETA 10 minutes," or "patient transport code" according to local protocols.
3. **The patient's age and gender.** Example: "An 86-year-old woman." The patient's name should not be given over the radio because it may be overheard. This would be a violation of the patient's privacy.
4. **The patient's chief complaint or your perception of the problem and its severity.** Example: "Patient complains of severe pelvic and less severe back pain."
5. **A brief history of the patient's current problem.** Example: "Patient fell into bathtub at 3 o'clock this morning and wasn't able to get out." Other important history information that may pertain to

the current problem should also be included, such as "The patient has diabetes and takes insulin."

6. **A brief report of physical findings.** This report should include level of consciousness, the patient's general appearance, pertinent abnormalities noted, and vital signs. Example: "The patient is alert and oriented, has pale skin color, and is cold to the touch. We noted crepitus in the pelvic girdle. Her blood pressure is 112/84 mm Hg, pulse is 72 beats/min, and respirations 14 breaths/min."
7. **A brief summary of the care given and any patient response.** Example: "We have immobilized her onto a backboard. She still has pulse, motor, and sensory function distally in all four extremities."

Be sure you report all patient information in an objective, accurate, and professional manner. People with scanners are listening. You could be successfully sued for slander if you describe a patient in a way that injures his or her reputation.

The Role of Medical Control

The delivery of EMS involves an impressive array of assessments, stabilization, and treatments. In some cases, you may assist patients in taking medications. AEMTs and paramedics go beyond this level by initiating medication therapy based on the patient's presenting signs. For logical, ethical, and legal reasons, the delivery of such sophisticated care must be done in association with physicians. For this reason, every EMS system needs input and involvement from physicians, including your system or department medical director, providing medical direction (medical control) for your EMS system. Medical control is either off-line (indirect) or online (direct), as authorized by the medical director. Medical control guides the treatment of patients in the system through protocols, direct orders, and advice, and postcall review.

Depending on how the protocols are written, you may need to call medical control for direct orders (permission) to administer certain treatments, to determine the transport destination of patients, or to be allowed to stop treatment and/or not transport a patient. In these cases, the radio or cellular phone provides a vital link between you and the expertise available through the base physician.

To maintain this link 24 hours a day, 7 days a week, medical control must be readily available on the radio at the hospital or on a mobile or portable unit when you call **Figure 4-21** . In most areas, medical control is provided by the physicians who work at the receiving hospital. However, many variations have developed across the country. For example, some EMS units receive medical direction from one hospital even though they are taking the patient to another hospital. In other

Figure 4-20 The patient report should be given in an objective, accurate, and professional manner.

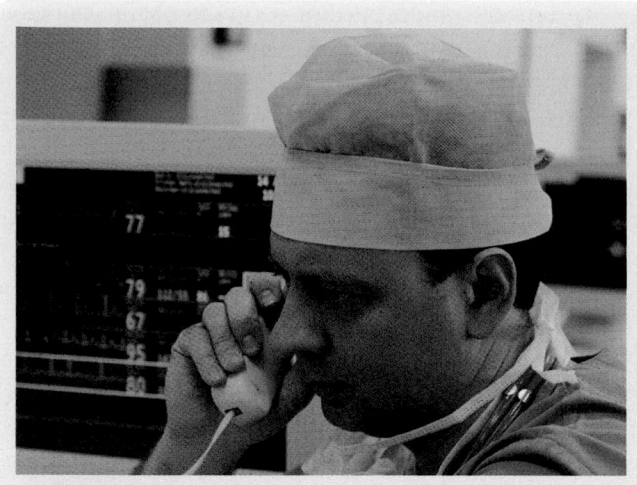

Figure 4-21 Medical control must be readily available on the radio at the hospital.

areas, medical direction may come from a freestanding center or even from an individual physician. Regardless of your system's design, your link to medical control is vital to maintain the high quality of care your patient requires and deserves.

Words of Wisdom

Some EMS systems will assign roles in their units: a primary person who is to speak on the radio and a primary person to do patient care. In these systems, all members of the crew must communicate very closely to make this work. In reality, EMTs are involved in every role, but the partial division of responsibilities can be efficient and effective. This approach is most common in systems that use extensive online medical control.

Calling Medical Control

You can use the radio in your unit or a portable radio to call medical control. A cellular telephone can also be used. Regardless of the type of communication, you should use a channel that is relatively free of other radio traffic and interference and one that will be recorded. Medical command communications create medical legal requirements that such conversations should be recorded. There are a number of ways to control access on ambulance-to-hospital channels. In some EMS systems, the dispatcher monitors and assigns appropriate, clear medical control channels. Other EMS systems rely on special communications operations, such as centralized medical emergency dispatch or resource coordination centers, to monitor and allocate the medical control channels.

Because of the large number of EMS calls to medical control, your radio report must be precise and well organized and must only contain important information. In addition, because you need specific directions on patient care, the information that you provide to medical control must be accurate. Remember, the physician on the other end bases his or her instructions on the information you provide.

You should never use codes when communicating with medical control unless you are directed by local protocol to do so. You should use proper medical terminology when giving your report. Most medical control systems handle many EMS agencies and will most likely not know your unit's special codes or signals.

To ensure complete understanding, once you receive an order from medical control, you must repeat the order back, word for word, and then receive confirmation. Whether the physician gives an order for medication or a specific treatment or denies a request for a particular treatment, you must repeat the order back, word for word. This "echo" exchange helps to eliminate confusion and the possibility of poor patient care. Orders that are unclear or seem inappropriate or incorrect should be questioned. Do not blindly follow an order that does not make sense to you. The physician may have misunderstood or may have missed part of your report. In that case, he or she may not be able to respond appropriately to the patient's needs.

Words of Wisdom

Orders that are unclear or seem inappropriate or incorrect should be questioned. Do not blindly follow an order that does not make sense to you.

Information About Special Situations

Depending on your system's procedures, you may initiate communication with one or more hospitals to advise them of an extraordinary call or situation. For instance, a small rural hospital may be better able to respond to multiple victims of a highway crash if it is notified when the ambulance is first responding. At the other extreme, an entire hospital system must be notified of any disaster, such as a plane or train crash, as early as possible to enable activation of its staff call-in system. These special situations may also include hazardous materials situations, rescues in progress, MCIs, or any other situation that could require special preparation on the part of the hospital. In some areas, mutual aid frequencies may be designated in MCIs so that responding agencies can communicate with one another on a common frequency.

When notifying the hospital(s) of any special situations, keep the following in mind: The earlier the notification, the better. You should ask to speak to the charge nurse or physician in charge, as he or she is best able to mobilize the resources necessary to respond. Also, whenever possible, provide an estimate of the number of individuals who may be transported to the facility. Be sure to identify any conditions the patient(s) may have that require special needs, such as burns or hazardous materials exposure, to assist the hospital in preparation. In many cases, hospital notification is part of a larger disaster or hazardous materials plan. Follow the plan for your system.

■ Maintenance of Radio Equipment

Like all other EMS equipment, radio equipment must be serviced by properly trained and equipped personnel. Remember that the radio is your lifeline to other public safety agencies (who function to protect you), as well as to medical control, and it must perform under emergency conditions. Radio equipment that is operating properly should be serviced at least once a year. Any equipment that is not working properly should be immediately removed from service and sent for repair. Outdated equipment should be removed from service as new equipment becomes available.

When you are beginning your shift, it is typical to check the ambulance to ensure that it is ready to go. You cannot assume that the crew before you left the ambulance well stocked and in operational readiness.

The radio is also an important component that needs to be checked to ensure that it is operating correctly and using the correct frequency.

Sometimes, radio equipment will stop working during a run. Your EMS system must have several backup plans and options. The goal of a backup plan is to make sure you can maintain contact when the usual procedures do not work. There are quite a few options.

The simplest backup plan relies on written standing orders. **Standing orders** are written documents that have been signed by the EMS system's medical director. These orders outline specific directions, permissions, and sometimes prohibitions regarding patient care. By their very nature, standing orders do not require direct communication with medical control. When properly followed, standing orders or formal protocols have the same authority and legal status as orders given over the radio. They exist to one extent or another in every EMS system and can be applied to all levels of EMS providers. Other backup plans can involve using a cell phone and calling the emergency department directly. The problem with this approach is that the conversation will probably not be recorded. Medical command conversations are often recorded for the purpose of quality improvement.

You are the Provider: SUMMARY

1. What information should you ask the dispatcher to obtain from the caller?

A "sick person" could be anything from a patient with the flu to a patient in cardiac arrest. For all you know, the patient could be experiencing a psychiatric crisis, in which case law enforcement should be dispatched to secure the scene before your arrival. After determining the nature of the patient's illness and gathering information that will maximize your own safety, your next priority is to determine if the patient is conscious and breathing. Try to ascertain the patient's age and gender, if possible. Although you will truly not know what you are dealing with until you arrive at the scene and assess the patient, you should capitalize on the fact that the dispatcher still has the caller on the phone. The more information you obtain prior to arrival at the scene, the better prepared you will be to care for the patient.

2. Why is effective communication between the responding EMS unit and the dispatcher so important?

After communication between the patient or caller and the dispatcher, the next link in the communication chain is between the responding EMS unit(s) and the dispatcher. The dispatcher should provide you with the information needed to mentally and logistically prepare for the call. In many cases, you will receive a great deal of information; in other cases, the information you receive will be minimal.

Communication between the responding EMS unit and the dispatcher is a two-way street. The dispatcher should advise you of the nature and severity of the injury, illness, or incident;

the exact location of the patient; the number of patients; whether other public safety agencies are responding; and special advisories, such as adverse road or traffic conditions, severe weather reports, or potential scene hazards. The responding EMS unit should also report any information to the dispatcher; this includes arrival at the scene and additional information that can only be obtained by viewing the scene (ie, scene size-up information, need for additional resources). The responding unit should survey the scene and report back to the dispatcher *before* exiting the ambulance, when possible; once you get out of the ambulance, communications may be hampered because of the transmitting limitations of your hand-held radio versus the mobile radio in the ambulance.

3. How can you maximize successful communication with a hearing-impaired patient?

First, you must determine the degree of the patient's hearing impairment; her hearing aids may allow her to hear normally. Do not assume that she is totally deaf! After determining her degree of hearing impairment, you must remember that most hearing-impaired patients have a normal intelligence level; provided that you successfully communicate with them, they usually understand what is going on around them.

Many hearing-impaired patients can read lips to some extent; therefore, you should position yourself where the patient can see your lips. The patient's ability to read lips and functional hearing aids can maximize your ability to successfully communicate with the hearing-impaired patient.

You are the Provider: SUMMARY, continued

Never shout in a hearing-impaired patient's ear. Patients with hearing impairment are usually unable to hear high-pitched sounds; yelling in his or her ear will only distort what you are trying to say.

Listen carefully, ask short questions, and give short answers whenever possible. Although many hearing-impaired patients can speak distinctly, some cannot.

If your efforts to verbally communicate with the patient prove unsuccessful, write down your questions on a piece of paper and ask the patient to write down his or her response. Be sure to print legibly, so that your handwriting is not a communications barrier.

Learn some simple phrases in sign language, such as "hurt," "sick," and "help" in case you cannot communicate in any other way.

4. Should your general approach to the assessment process be any different for this patient versus a younger patient? Why or why not?

Most older people think clearly, can give you a clear medical history, and can answer your questions; therefore, the patient should be your primary source of information unless it is determined that he or she cannot effectively communicate. A major fear that an older person experiences is losing his or her independence; if you ignore the patient and turn to a family member for information based on the preconceived notion that the patient will not understand you, you are simply reinforcing the patient's fear. Senility or confusion does not come automatically with age, and although some older patients may be hostile, irritable, or confused, this must not be assumed to be normal behavior caused by the aging process. These signs may be caused by cerebral hypoxia, stroke, drug overdose, or shock, among numerous others.

It is important to remember that, as a result of the natural process of aging, older patients may not experience pain as their younger counterparts. An older person who has fallen, for example, may not report any pain, despite the presence of an obvious injury.

Assess the older patient just as you would a younger patient; however, you may need to allow extra time for him or her to answer your questions. As with any patient, the older patient should feel confident that everything possible is being done for him or her.

5. What techniques can facilitate the interviewing process of an older patient?

Many of the same techniques used to interview younger patients can be used effectively to interview older patients. When interviewing an older patient, however, patience is even more critical. Identify yourself; do not assume that an older person—or any person for that matter—knows who you are. Remain aware of how you present yourself; frustration and impatience can be conveyed through body language.

When communicating with an older patient, look directly at him or her and speak slowly and distinctly. As previously discussed, do not increase the volume of your voice based on the assumption that the patient is hearing-impaired. After asking the patient a question, allow him or her ample time to answer it and then *actively listen* to his or her response.

As with any patient, show the older patient respect. Refer to him or her as Mr, Mrs, or Miss, unless he or she asks to be addressed otherwise.

Do not talk about the patient in front of him or her; doing so gives the patient the impression that he or she has no choice in his or her medical care. Again, this may only escalate the patient's fear of losing his or her independence.

6. What are the components of a radio report to the hospital?

The purpose of the radio report is to inform the receiving facility that you are transporting to their location and to provide an overview of the patient's condition so they can adequately prepare to receive the patient. Your radio report to the receiving hospital should be concise—brief in length, yet comprehensive in scope.

Identify your EMS system and unit number and then advise the nurse or physician that you are prepared to give a radio report. After they confirm that they can hear you, begin your radio report with the patient's age, gender, chief complaint, and level of consciousness. Next, provide a brief elaboration of the patient's chief complaint (eg, the history of present illness), your assessment findings, SAMPLE history, initial vital signs, and the most recent set of vital signs. Summarize any treatment that you provided and the patient's response, if any, to your treatment. Finally, give the hospital your estimated time of arrival and transport mode.

7. How does your oral report differ from your radio report?

Your reporting responsibility does not end when you arrive at the hospital; it just begins. Patient care transfer officially occurs during your oral report, not your radio report. Depending on the hospital and the patient's condition, the training of the person who assumes care of the patient will vary. However, you must transfer the care of your patient to someone with training that is equal to or higher than yours; in the hospital, this is a nurse or physician. Once a hospital staff member is ready to take responsibility for the patient, you must provide that person with a formal oral report of the patient's condition.

As previously discussed, your radio report should be brief, yet concise. Your oral report, however, should be more comprehensive. In many cases, your oral report is given at the same time the nurse or physician is providing care for the patient, such as assessing the patient or helping you move the patient from the stretcher to the hospital bed. Therefore, you must report important information in a complete, precise manner. The following components should be included in your oral report:

- The patient's name and the chief complaint, nature of illness, or mechanism of injury
- More detailed information than what you provided during your radio report, such as pertinent negatives and findings of a more detailed physical exam
- Any important medical history that was not already given
- The patient's response to treatment given en route. It is especially important to report any changes in the patient's condition or the treatment provided after your radio report
- The vital signs assessed during transport and after your radio report

You are the Provider: SUMMARY, continued

- Any other information that you obtained en route and after your radio report; for example, a list of medications that the patient is currently taking

---

8. What functions does the patient care report serve?

In addition to your radio and oral reports, you must also complete a formal written patient care report (PCR)before you leave the hospital. In some cases, you may be able to complete the PCR en route, if your transport time is long enough and the patient requires minimal care. Usually, you will finish the PCR after you have transferred patient care to an appropriate hospital staff member. A copy of the report must be left at the hospital.

The PCR helps to ensure efficient continuity of patient care. It describes the nature of the patient's injuries or illness at the scene, the treatment you provided initially and en route, vital signs, and the patient's condition on arrival at the hospital. Although the PCR may not be read immediately at the hospital, it may be referred to for important information. The PCR serves the following functions:

- Continuity of care
- Legal documentation
- Education
- Administrative information (ie, billing)
- Essential research information
- Evaluation and continuous quality improvement

The information in the PCR proves that you provided proper patient care. In some cases, it also shows that you properly handled unusual or uncommon situations. You should include both objective (what you find) and subjective (what the patient tells you) information in the PCR. A well-written, neat, and concise PCR—including correct spelling and grammar—reflects good patient care. Remember, if you didn't document it, you didn't do it. If the report looks sloppy, the care you provided will be assumed to have been the same.

EMS Patient Care Report (PCR)

Date: 4-16-09	**Incident No.:** 030109	**Nature of Call:** Sick person		**Location:** 514 E. Bandera	
Dispatched: 0610	**En Route:** 0160	**At Scene:** 0616	**Transport:** 0627	**At Hospital:** 0642	**In Service:** 0655

Patient Information

Age: 83 **Sex:** F **Weight (in kg [lb]):** 50 kg (110 lb)	**Allergies:** None **Medications:** Vitamins **Past Medical History:** None **Chief Complaint:** Light-headedness and nausea

Vital Signs

Time: 0621	**BP:** 122/62	**Pulse:** 68	**Respirations:** 20	**Sao₂:** 98%
Time: 0627	**BP:** 118/60	**Pulse:** 74	**Respirations:** 20	**Sao₂:** 98%
Time: 0632	**BP:** 120/60	**Pulse:** 72	**Respirations:** 18	**Sao₂:** 99%

EMS Treatment (circle all that apply)

Oxygen @ __ L/min via (circle one): NC NRM Bag-Mask Device		**Assisted Ventilation**	**Airway Adjunct**	**CPR**
Defibrillation	**Bleeding Control**	**Bandaging**	**Splinting**	**Other:** Blood glucose assessment, blanket for warmth

Narrative

Dispatched for a "sick person." Arrived on scene to find the patient, an 83-year-old female, sitting on the couch in her living room. She was conscious and alert; her airway was patent and her breathing was adequate. The patient complains of light-headedness and nausea that began approximately an hour ago. She denies vomiting or any other symptoms and further denies any significant medical problems. Medications include vitamins only; no prescribed medications or known drug allergies. Assessment did not reveal any gross abnormalities. Her blood glucose level was assessed and noted to be 112 mg/dL. Patient denies chest pain, shortness of breath, abdominal pain, or headache. Offered patient oxygen via nasal cannula; however, she wished not to receive it. Obtained vital signs and prepared patient for transport. Applied blanket because the patient stated that she was cold. Began transport and monitored patient's mental status and vital signs en route. Her condition remained unchanged. The patient was noted to be wearing hearing aids in both ears but was easy to communicate with. Arrived at the hospital and transferred patient care without incident. Oral report was given to staff nurse. Returned to service at 0655. **End of report**

Prep Kit

Ready for Review

- The Shannon–Weaver model of communication is a valuable tool in understanding the variables involved in human communications.

- There are many verbal and nonverbal factors and strategies that are necessary for therapeutic communication.

- Excellent communication skills are crucial in relaying pertinent information to the hospital before arrival.

- It is important to remember that people who are sick or injured may not understand what you are doing or saying. Therefore, your body language and attitude are very important in gaining the trust of both the patient and family. You must also take special care of individuals such as children; geriatric patients; and hearing-impaired, visually impaired, and non-English-speaking patients.

- EMTs must have excellent person-to-person communication skills. You should be able to interact with the patient and any family members, friends, or bystanders.

- You must complete a patient care report about the patient before you leave the hospital. This is a vital part of providing emergency medical care and ensuring the continuity of patient care. This information guarantees the proper transfer of responsibility, complies with the requirements of health departments and law enforcement agencies, and fulfills your administrative needs.

- Radio and telephone communication links you and your team to other members of the EMS, fire, and law enforcement communities. This enables your entire team to work together more effectively.

- An EMT must understand and be able to use many forms of communication. You must be familiar with two-way radio communications and have a working knowledge of mobile and hand-held portable radios. You must know when to use them and what type of information you can transmit.

- It is your job to know what your communication system can and cannot handle. You must be able to communicate effectively by sending precise, accurate reports about the scene, the patient's condition, and the treatment that you provide.

- Remember, the lines of communication are not always exclusive; therefore, you should speak in a professional manner at all times.

- Your reporting and recordkeeping duties are essential, but they should never come before the care of a patient.

Vital Vocabulary

base station Any radio hardware containing a transmitter and receiver that is located in a fixed place.

cellular telephone A low-power portable radio that communicates through an interconnected series of repeater stations called "cells."

channel An assigned frequency or frequencies that are used to carry voice and/or data communications.

close-ended questions Questions that can be answered in short or single word responses.

communication The transmission of information to another person—verbally or through body language.

cultural imposition When one person imposes his or her beliefs, values, and practices on another because he or she believe his or her ideals are superior.

dedicated line A special telephone line that is used for specific point-to-point communications; also known as a "hot line."

documentation The written portion of the EMT's patient interaction. This becomes part of the patient's permanent medical record.

duplex The ability to transmit and receive simultaneously.

ethnocentrism When a person considers his or her own cultural values as more important when interacting with people of a different culture.

Federal Communications Commission (FCC) The federal agency that has jurisdiction over interstate and international telephone and telegraph services and satellite communications, all of which may involve EMS activity.

MED channels VHF and UHF channels that the Federal Communications Commission has designated exclusively for EMS use.

mobile data terminals (MDT) Small computer terminals inside ambulances that directly receive data from the dispatch center.

noise Anything that dampens or obscures the true meaning of a message.

open-ended questions Questions for which the patient must provide detail to give an answer.

paging The use of a radio signal and a voice or digital message that is transmitted to pagers ("beepers") or desktop monitor radios.

patient care report (PCR) The legal document used to record all patient care activities. This report has direct patient care functions but also administrative and quality control functions. PCRs are also known as prehospital care reports.

proxemics The study of space between people and its effects on communication.

rapport A trusting relationship that you build with your patient.

repeater A special base station radio that receives messages and signals on one frequency and then automatically retransmits them on a second frequency.

scanner A radio receiver that searches or "scans" across several frequencies until the message is completed; the process is then repeated.

simplex Single-frequency radio; transmissions can occur in either direction but not simultaneously in both; when one party transmits, the other can only receive, and the party that is transmitting is unable to receive.

standing orders Written documents, signed by the EMS system's medical director, that outline specific directions, permissions, and sometimes prohibitions regarding patient care; also called protocols.

telemetry A process in which electronic signals are converted into coded, audible signals; these signals can then be transmitted by radio or telephone to a receiver with a decoder at the hospital.

therapeutic communication Verbal and nonverbal communication techniques that encourage patients to express their feelings and to achieve a positive relationship.

trunking Telecommunication systems that allow a computer to maximize utilization of a group of frequencies.

UHF (ultra-high frequency) Radio frequencies between 300 and 3,000 MHz.

VHF (very high frequency) Radio frequencies between 30 and 300 MHz; the VHF spectrum is further divided into "high" and "low" bands.

Assessment in Action

Rescue 31 responds to a home on a noisy, busy street for a dog bite injury. You arrive at a private residence and observe a teenager sitting on the front porch who is rocking a crying young boy in her arms. Law enforcement personnel are on-scene and tell you that it is safe to approach.

Your patient is a 4-year-old boy with a puncture wound to his right forearm secondary to a dog bite. His sister tells you that he was outside playing with a dog that suddenly bit him and ran off. Their parents are currently at work; however, law enforcement personnel have contacted the mother, who consented to treatment and is on her way home. The patient has an open, patent airway and a respiratory rate of 22 breaths/min. His pulse is 92 beats/min. While you await the arrival of the child's mother, you ask the boy a few questions as you clean and bandage the wound. Your partner gets a stuffed bear from the truck.

1. Why is documentation of the situation necessary?
   A. Demonstrates that the care delivered was appropriate
   B. Provides an opportunity to communicate the patient's story to others
   C. Ensures continuity of care
   D. All of the above

2. What is the most obvious physical factor that could affect communication in this situation?
   A. The dog ran off.
   B. The teenager is in charge.
   C. The parents are not present.
   D. There is a noisy, busy road.

3. Since the patient is a child, you will ask him close-ended questions. Which of the following is an example of a closed-ended question?
   A. How long ago were you playing with the dog?
   B. Does your arm hurt now?
   C. What does the pain feel like?
   D. What were you doing when you were bit by the dog?

4. You repeat the child's answer to your question to confirm your understanding of the situation. This is referred to as:
   A. explanation.
   B. facilitation.
   C. clarification.
   D. reflection.

5. Which of the following communication techniques will best facilitate care of the child?
   A. Avoid looking directly at him.
   B. Position yourself directly above him.
   C. Explain what you are going to do before you do it.
   D. Speak loudly and quickly.

6. Which of the following situations must be reported to the appropriate authorities?
   A. Dog bites
   B. Gunshot wounds
   C. Physical abuse
   D. All of the above

7. While you wait for the parent, you jot down some information on the PCR. What are the six functions that the PCR serves?

8. Describe some of the administrative information that can be found on a PCR.

9. Describe the importance of the administrative information.

10. When contacting the hospital you should:
    A. provide as much information as possible.
    B. begin speaking immediately after depressing the "press-to-talk" button.
    C. use only agency-specific codes.
    D. hold the microphone 8″ to 10″ from your mouth.

National EMS Education Standard Competencies

Preparatory

Applies fundamental knowledge of the emergency medical services (EMS) system, safety/well-being of the emergency medical technician (EMT), medical/legal and ethical issues to the provision of emergency care.

Anatomy and Physiology

Applies fundamental knowledge of the anatomy and function of all human systems to the practice of EMS.

Pathophysiology

Applies fundamental knowledge of the pathophysiology of respiration and perfusion to patient assessment and management.

Knowledge Objectives

1. Understand the body's topographic anatomy, including the anatomic position and the planes of the body. (pp 143-144)

2. Explain the following directional terms: anterior (ventral), posterior (dorsal), right, left, superior, inferior, proximal, distal, medial, lateral, superficial, and deep. (pp 144-145)

3. Describe the prone, supine, Fowler's, Trendelenburg's, and shock positions of the body. (p 147)

4. Identify the anatomy and physiology of the skeletal system. (pp 147-153)

5. Describe the physiology of the musculoskeletal system. (pp 154-155)

6. Discuss the anatomy and physiology of the respiratory system. (pp 155-163)

7. Discuss the anatomy and physiology of the circulatory system. (pp 163-173)

8. Discuss the anatomy and physiology of the nervous system. (pp 173-177)

9. Describe the anatomy and the physiology of the integumentary system. (pp 177-179)

10. Explain the anatomy and physiology of the digestive system. (pp 179-183)

11. Discuss the anatomy and physiology of the endocrine system. (pp 183-184)

12. Describe the anatomy and physiology of the urinary system. (pp 184-185)

13. Discuss the anatomy and physiology of the genital system. (pp 185-186)

14. Describe the life support chain, aerobic metabolism, and anaerobic metabolism. (pp 186-188)

15. Define pathophysiology. (pp 188-190)

Skills Objectives

There are no skills objectives for this chapter.

Introduction

A working knowledge of human anatomy is important for you as an EMT. By using the proper medical terms, you will be able to communicate correct information to medical professionals with the least possible confusion. At the same time, you need to be able to communicate with others who may or may not understand medical terms. Balancing these two facets is one of the most challenging aspects of your job. A basic understanding of human anatomy, physiology, and pathophysiology is essential so that you can meet these challenges.

This chapter begins with a discussion of topographic anatomy, or the landmarks on the surface of the body. The various parts of the body, or its anatomy, are then described. This information will provide you with the correct medical terms you will use in the field. Physiology, or the functions of the body or any of its parts, is also covered. Finally, pathophysiology is discussed, which describes how normal physiologic processes are affected by disease.

Topographic Anatomy

The surface of the body has many definite visible features that serve as guides or landmarks to the structures that lie beneath them. You must be able to identify the superficial landmarks of the body—its **topographic anatomy**—to perform an accurate assessment. But how do we know that everyone is looking at the body in the same orientation?

To accomplish this, the terms that are used to describe the topographic anatomy are applied to the body when it is in the **anatomic position**. This is a position of reference in which the patient stands facing you, arms at the side, with the palms of the hands forward. The anatomic position is used as a common starting point so that everyone is referring to the body in the same way. For example, you are looking at a person who is complaining of pain in his arm. Which left or right do you use? Your left or the patient's left? To be consistent, health care providers use the patient's left and right as the reference point.

The Planes of the Body

The anatomic planes of the body are imaginary straight lines that divide the body **Figure 5-1**. There are three main axes of the body depending on how it is sliced. Slicing

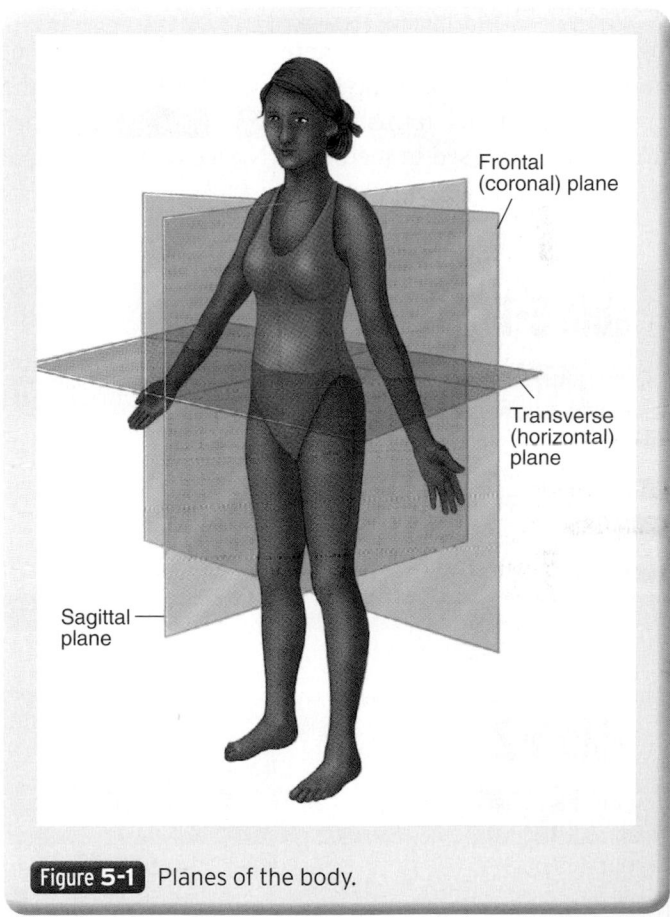

Figure 5-1 Planes of the body.

You are the Provider: PART 1

At 5:40 PM, you are dispatched to 322 Azalea Trail for a 60-year-old man with severe abdominal pain. The weather is overcast, the traffic is heavy, and your response time to the scene is approximately 6 minutes.

1. How will knowledge of anatomy and physiology help you provide appropriate patient care?

the body so that you have a front and back portion creates the frontal or **coronal plane**. If the body is sliced so the result is a top and bottom portion, this is referred to as the **transverse (axial) plane**. If the body is sliced so that you have a left and right portion, a **sagittal (lateral) plane** is formed.

The **midsagittal plane (midline)** is a special type of sagittal plane where the body is cut in half leaving equal left and right halves. Your nose and navel are found along this imaginary line. These planes help you to identify the location of internal structures and understand the relationships between and among the organs **Table 5-1**.

■ Directional Terms

When you are discussing where an injury is located or how a pain radiates in the body, you need to know the correct directional terms **Figure 5-2**. **Table 5-2** provides the basic terms used in medicine. Notice how directional terms are paired as "opposites."

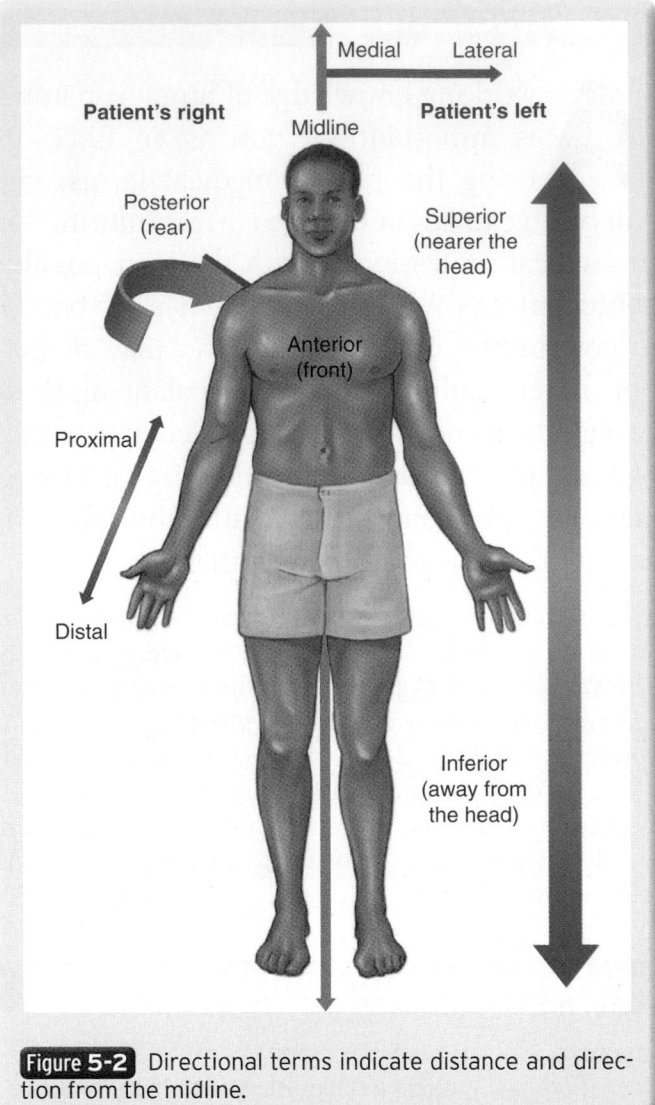

Figure 5-2 Directional terms indicate distance and direction from the midline.

Table 5-1 Planes of the Body

Plane of the Body	Description
Coronal	Front and back
Transverse	Top and bottom
Sagittal	Left and right
Midsagittal (midline)	Left and right—equal halves

Table 5-2 Directional Terms

Common Term	Directional Term	Definition
Front and back	Anterior (ventral) Posterior (dorsal)	The front surface of the body The back surface of the patient
Right and left	Right Left	The patient's right The patient's left
Top and bottom	Superior Inferior	Closest to the head Closest to the feet
Closest and farthest	Proximal Distal	Closest to the point of attachment Farthest from the point of attachment
Middle and side	Medial Lateral	Closest to the midline Farthest from the midline
In and out	Superficial Deep	Closest to the surface of the skin Farthest from the surface of the skin

Right and Left

The terms "right" and "left" refer to the patient's right and left sides, not to your right and left sides.

Superior and Inferior

The **superior** part of the body, or any body part, is the portion nearer to the head from a specific reference point. The part nearer to the feet is the **inferior** portion. These terms are also used to describe the relationship of one structure to another. For example, the knee is superior to the foot and inferior to the pelvis.

Lateral and Medial

Parts of the body that lie farther from the midline are called **lateral** (outer) structures. The parts that lie closer to the midline are called **medial** (inner) structures. For example, the knee has medial (inner) and lateral (outer) aspects (surfaces).

Proximal and Distal

The terms "proximal" and "distal" are used to describe the relationship of any two structures on an extremity. **Proximal** describes structures that are closer to the trunk. **Distal** describes structures that are farther from the trunk or nearer to the free end of the extremity. For example, the elbow is distal to the shoulder and proximal to the wrist and hand.

Superficial and Deep

Superficial means closer to or on the skin. **Deep** means farther inside the body and away from the skin.

Ventral and Dorsal

Ventral refers to the belly side of the body, or the anterior surface of the body. **Dorsal** refers to the spinal side of the body, or the posterior surface of the body, including the back of the hand. These terms are used less frequently than the terms **anterior** and **posterior**.

Palmar and Plantar

The front region of the hand is referred to as the palm or **palmar** surface. The bottom of the foot is referred to as the **plantar** surface.

Apex

The **apex (plural apices)** is the tip of a structure. For example, the apex of the heart is the bottom (inferior portion) of the ventricles in the left side of the chest.

■ Movement Terms

The following terms relate to movement Figure 5-3 .

- **Flexion** is the bending of a joint.
- **Extension** is the straightening of a joint.
- **Adduction** is motion toward the midline.
- **Abduction** is motion away from the midline.

Words of Wisdom

Using the correct anatomic terminology in your patient care report improves patient care by making the report more useful to hospital personnel and enhances your professional image as an EMT.

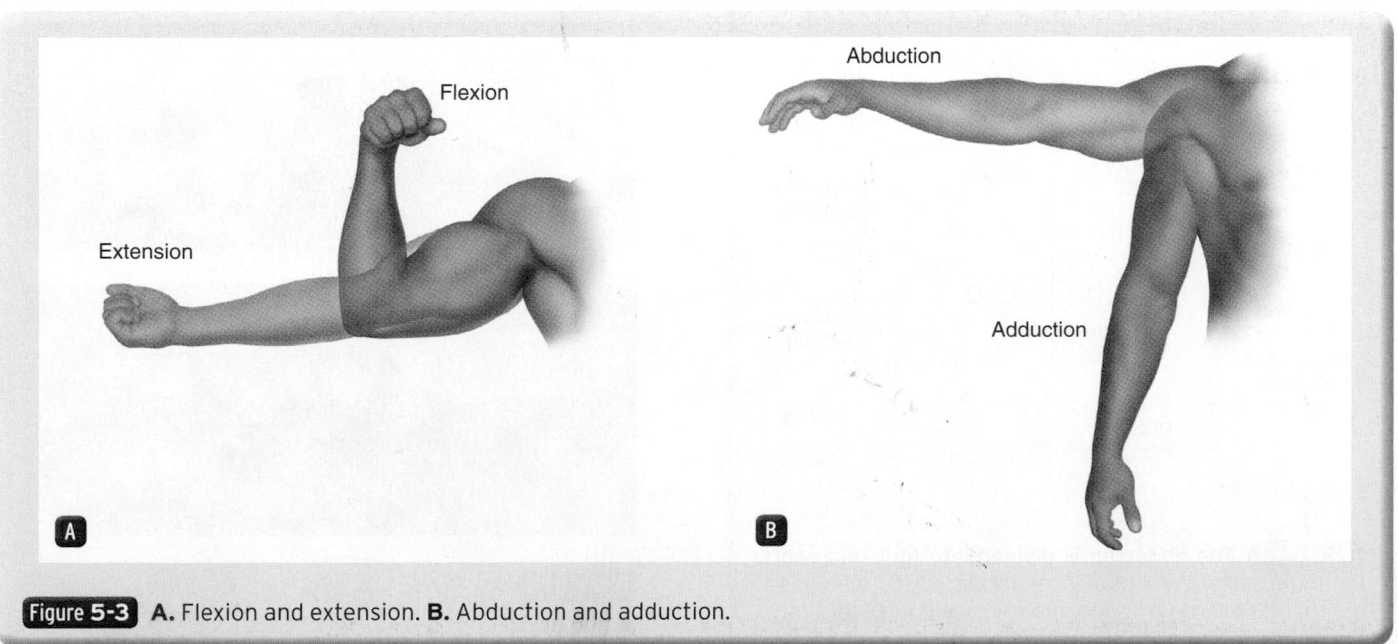

Figure 5-3 **A.** Flexion and extension. **B.** Abduction and adduction.

■ Other Directional Terms

Many structures of the body occur bilaterally. A body part that appears on both sides of the midline is **bilateral**. For example, the eyes, ears, hands, and feet are bilateral structures. This is also true for structures inside the body, such as the lungs and kidneys. Structures that appear on only one side of the body are said to occur unilaterally. For example, the spleen is on the left side of the body only, and the liver is on the right side. The terms unilateral and bilateral can also refer to something occurring on one side; for example, pain that is occurring on only one side of the body could be called unilateral pain.

As part of the assessment process, you will palpate the abdomen and report your findings. Therefore, it is important that you are able to describe the exact location of areas of the abdomen. The way to describe the sections of the abdominal cavity is by **quadrants**. Imagine two lines intersecting at the umbilicus, dividing the abdomen into four equal areas Figure 5-4. These are referred to as the right upper quadrant, left upper quadrant, right lower quadrant, and left lower quadrant. Remember that here, too, right and left refer to the patient's right and left, not yours.

It is important to learn all of these terms and concepts so that you can describe the location of any injury or assessment findings. When you use these terms properly, any other medical personnel who care for the patient will know immediately where to look and what to expect.

■ Anatomic Positions

You will use these terms to describe the position of the patient as you find him or her or when you are ready to transport the patient to the emergency department Figure 5-5.

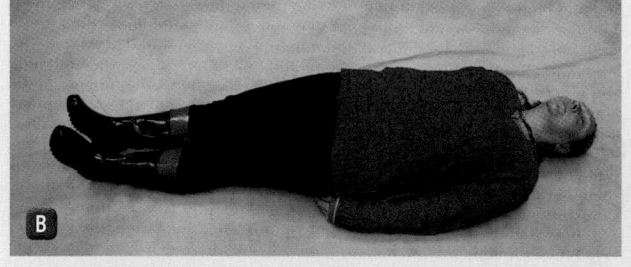

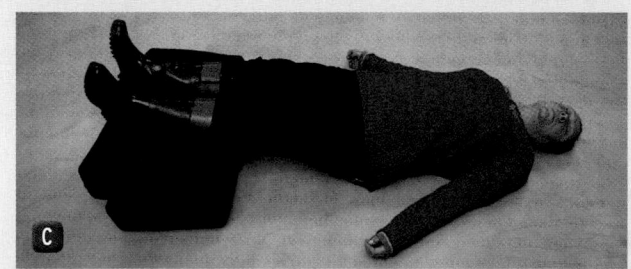

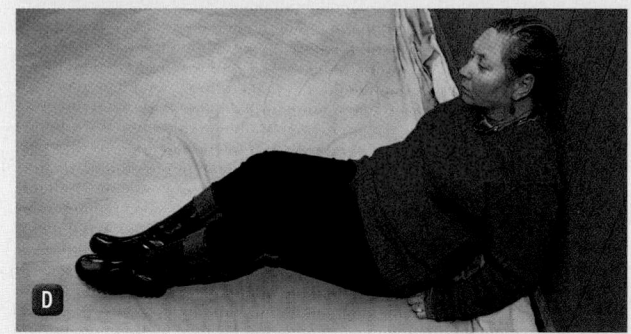

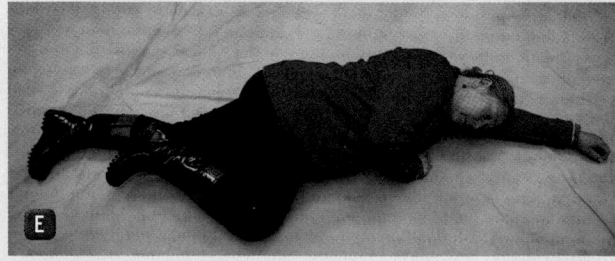

Figure 5-5 **A.** Prone. **B.** Supine. **C.** Shock position (modified Trendelenburg's position). **D.** Fowler's position. **E.** Recovery position.

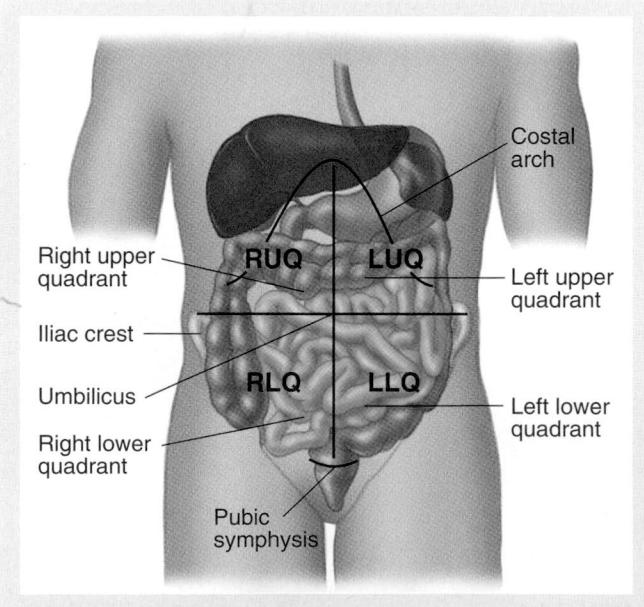

Figure 5-4 The abdomen is divided into four quadrants. RUQ indicates right upper quadrant; LUQ, left upper quadrant; RLQ, right lower quadrant; and LLQ, left lower quadrant.

Prone and Supine

These terms describe the position of the body. The body is in the prone position when lying face down; the body is in the supine position when lying face up.

Fowler's Position

The Fowler's position was named after a US surgeon, George R. Fowler, MD, at the end of the 19th century. Dr Fowler placed his patients in a semireclining position with the head elevated to help them breathe easier and to control the airway. A patient who is sitting up with the knees bent is therefore said to be in the Fowler's position.

Trendelenburg's Position

Trendelenburg's position was named after a German surgeon, Friedrich Trendelenburg, at the turn of the 20th century. Dr Trendelenburg frequently placed his patients in a supine position on an incline with their feet higher than their head to keep blood in the core of the body. Trendelenburg's position is a position in which the patient is on a backboard or stretcher with the feet 6″ to 12″ higher than the head.

Shock Position

In the shock position, or modified Trendelenburg's position, the head and **torso** (the trunk without the head and limbs) are supine, and the lower extremities are elevated 6″ to 12″ to help increase blood flow to the brain.

■ The Skeletal System: Anatomy

The **skeleton** gives us our recognizable human form and protects our vital internal organs. Bones constitute the major structure of the skeletal system. Connecting bones to each other is a structure called a **ligament**. **Tendons** connect muscles to bones. Finally, **cartilage** is the soft, semiflexible material that is found within the joints Table 5-3.

The skeletal system is divided into two main portions: the **axial skeleton** and the **appendicular skeleton**. The axial skeleton forms the foundation on which the arms and legs are hung. The axial skeleton is composed of

the skull, face, **thoracic cage**, and vertebral column. The arms and legs, their connection points, and the pelvis make up the appendicular skeleton Figure 5-6. The brain lies within the skull. The heart, lungs, and great vessels are enclosed in the **thorax**, which is part of the torso. Much of the liver and spleen are protected by the lower ribs. The spinal cord is contained within and protected by a bony spinal canal formed by the vertebrae.

The 206 bones of the skeleton provide a framework for the attachment of muscles. The skeleton is also designed to allow motion of the body. Bones come into contact with one another at joints where, with the help of muscles, the body is able to bend and move.

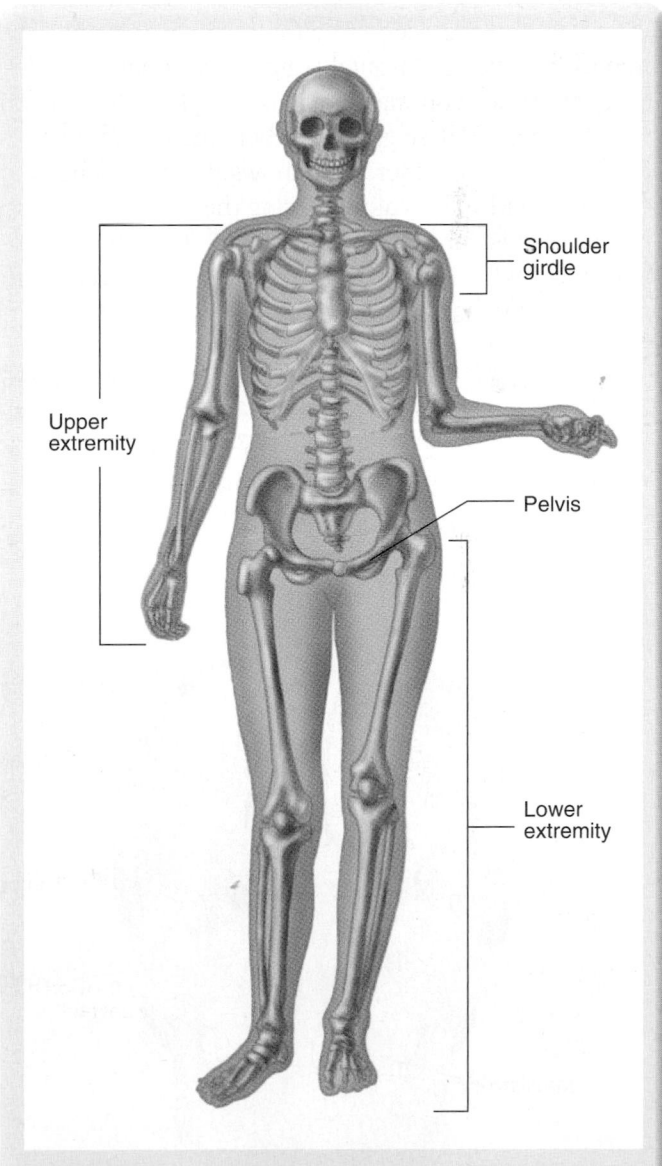

Figure 5-6 The 206 bones of the skeleton give us our form, protect our vital organs, and allow us to move. The axial skeleton runs in a straight line from the head to the pelvis. The appendicular skeleton is made up of the arms and legs and the pelvis.

Table 5-3	Support Structures Within the Skeletal System
Name	**Function**
Ligament	Connects bone to bone
Tendon	Connects muscle to bone
Cartilage	Cushion between bones

The Axial Skeleton

The Skull

The head is a great place to start since it is at the top of the body. The skull is composed of two groups of bones: the cranium—which protects the brain—and the facial bones. The **cranium** is composed of a number of thick bones that fuse together to form a shell above the eyes and ears that holds and protects the brain **Figure 5-7**. The brain connects to the spinal cord through a large opening at the base of the skull called the **foramen magnum** (Latin for "hole that is big").

Four major bones make up the cranium. The most posterior portion of the cranium is called the **occiput**. On each side of the cranium, the lateral portions are called the temples or **temporal regions**. Between the temporal regions and the occiput lie the **parietal regions**. The forehead is called the frontal region. If you have ever felt an infant's head, you may have noticed the soft spots on top of the head. These gaps are where the separate bones have yet to fuse together. This allows the infant's head to be molded without breaking during the birthing process.

The face is composed of 14 bones. The upper, non-moveable jawbones are called the **maxillae**, the cheek bones are called the **zygomas**, and the **mandible** is the lower, moveable portion of the jaw.

The **orbit** (eye socket) is made up of two facial bones: the maxilla and the zygoma. The orbit also includes the frontal bone of the cranium. Together, these bones form a solid bony rim that protrudes around the eye to protect it. If you look at the face from the side, you can see that the eyeball sits back in the orbit. The nose is made up of very short bones that form the bridge of the nose. Most of the nose is made of flexible cartilage. In fact, only the proximal one third of the nose is formed by bone.

The Spinal Column

The spinal column is the central supporting structure of the body and is composed of 33 bones, each called a vertebra. The **vertebrae** are named according to the section of the spine in which they lie and are numbered from top to bottom **Figure 5-8**. From the top down, the spine is divided into five sections:

- **Cervical spine**. The first seven vertebrae (C1 through C7) in the neck form the cervical spine. The skull rests on the first cervical vertebra (the atlas) and articulates with it.
- **Thoracic spine**. The next 12 vertebrae make up the thoracic spine. One pair of ribs is attached to each of the thoracic vertebrae.
- **Lumbar spine**. The next five vertebrae form the lumbar spine.
- **Sacrum**. The five sacral vertebrae are fused together to form one bone called the sacrum. The sacrum is joined to the iliac bones of the pelvis with strong ligaments at the sacroiliac joints to form the pelvis.
- **Coccyx**. The last four vertebrae, also fused together, form the coccyx, or tailbone.

The vertebrae are connected by ligaments, and between each vertebra is a cushion called the intervertebral disk. These ligaments and disks allow some

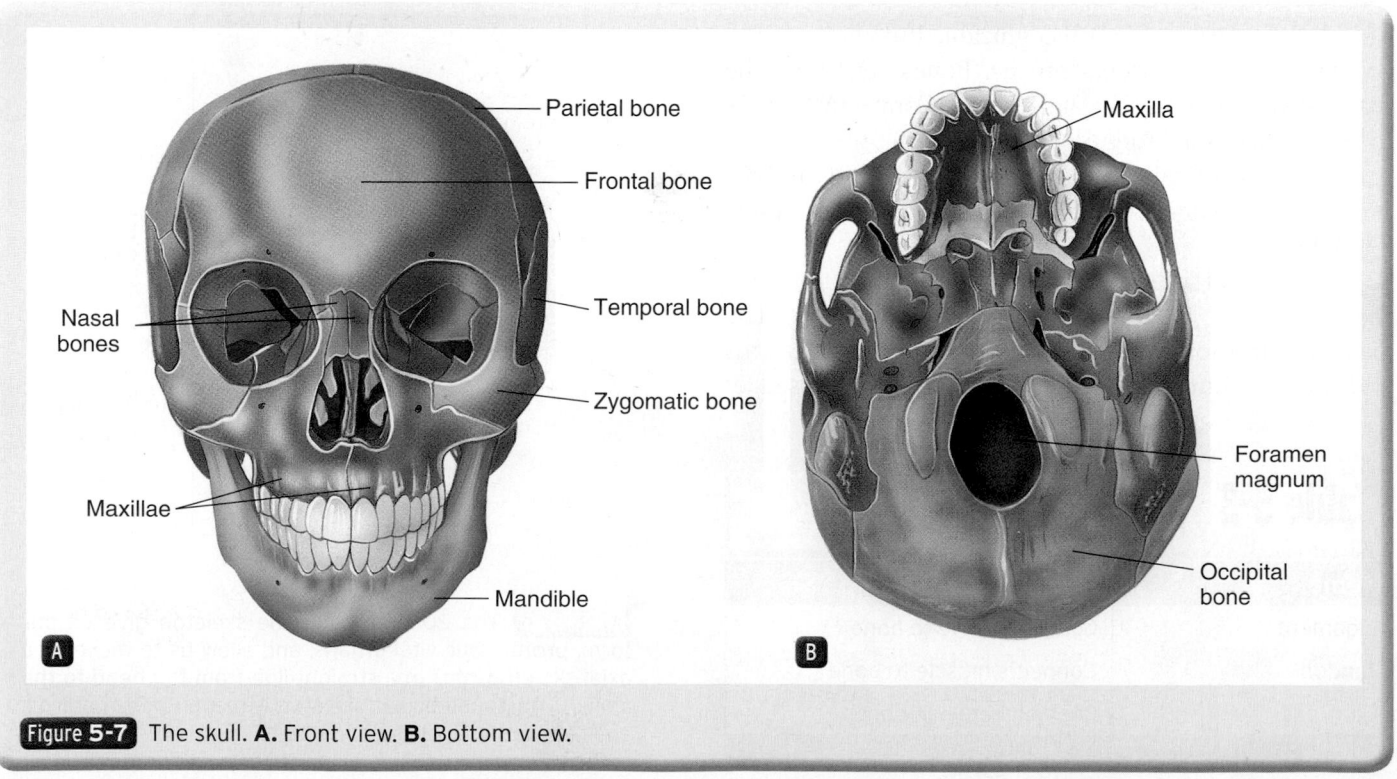

Parietal bone

Frontal bone

Nasal bones

Temporal bone

Zygomatic bone

Maxillae

Mandible

Maxilla

Foramen magnum

Occipital bone

A

B

Figure 5-7 The skull. **A.** Front view. **B.** Bottom view.

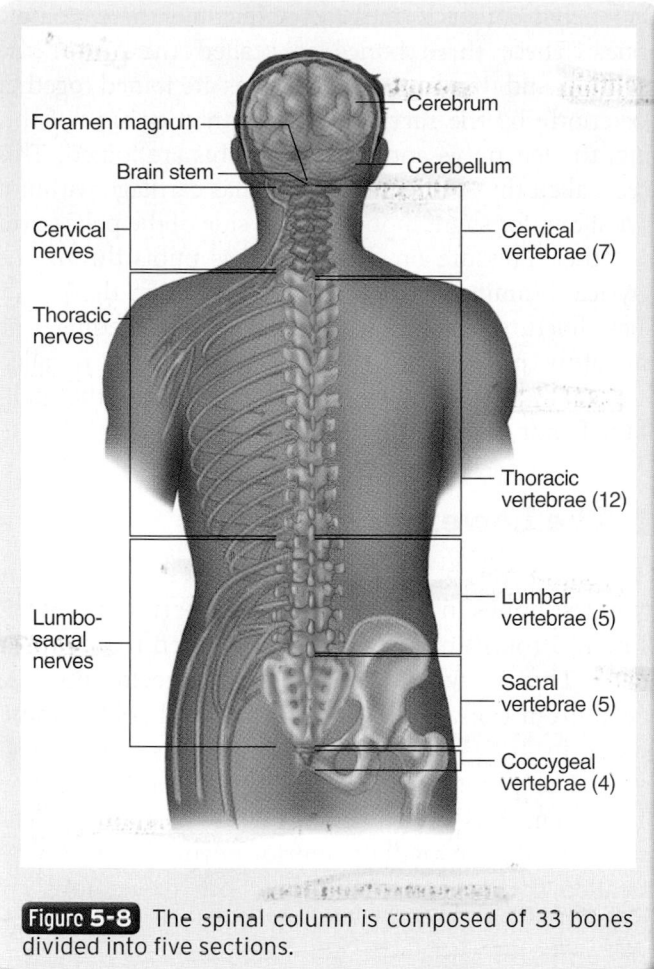

motion so the trunk can bend forward (flex) and back (extend), and they allow for rotation and lateral movement. However, they also limit motion of the vertebrae so that the spinal cord will not be injured. An injury to the spine may damage part of the spinal cord and its nerves that may not be protected by the vertebrae. Therefore, until the injury is stabilized, you must use extreme caution in caring for the patient to prevent injury to the spinal cord.

■ The Thorax

The **thoracic cavity** (chest) contains the heart, lungs, esophagus, and great vessels (the aorta and two venae cavae). It is formed by the 12 thoracic vertebrae (T1 through T12) and their 12 pairs of ribs.

Anteriorly, in the midline of the chest is the **sternum**. The superior border of the sternum forms the easily palpable jugular notch. This is the location where the trachea is entering the chest. The sternum has three components: the manubrium, the body, and the xiphoid process. The upper section of the sternum is called the **manubrium**. The body comprises the rest of the sternum except for a narrow, cartilaginous tip inferiorly, which is called the **xiphoid process** **Figure 5-9** .

■ **The Appendicular Skeleton**

The Upper Extremity

The upper extremity extends from the shoulder girdle to the fingertips and is composed of the arm, forearm, hand, and fingers. The joints are the elbow, wrist, and finger joints. The arm extends from the shoulder to the elbow, the forearm from the elbow to the wrist, and the hand from the wrist to the fingertips.

Shoulder Girdle

The **shoulder girdle** is where three bones come together, allowing the arm to be moved. These three bones are the **clavicle**, the scapula, and the humerus **Figure 5-10** . The clavicle (collarbone) overlies the superior boundaries of

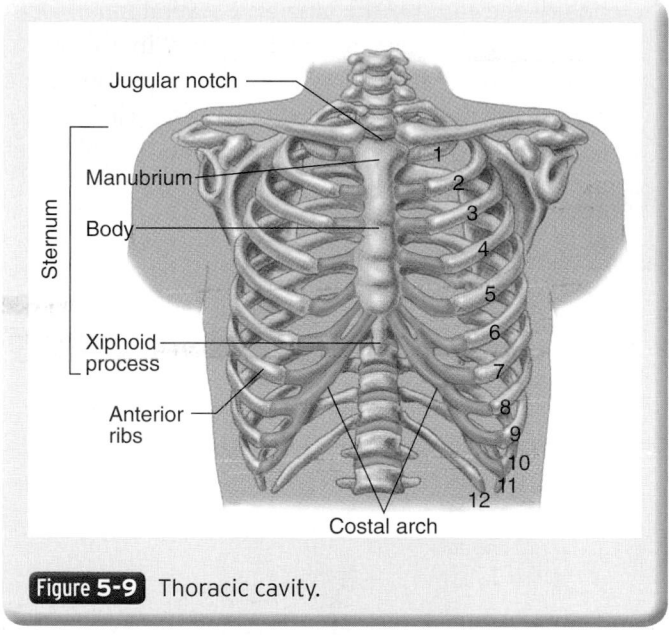

Figure 5-9 Thoracic cavity.

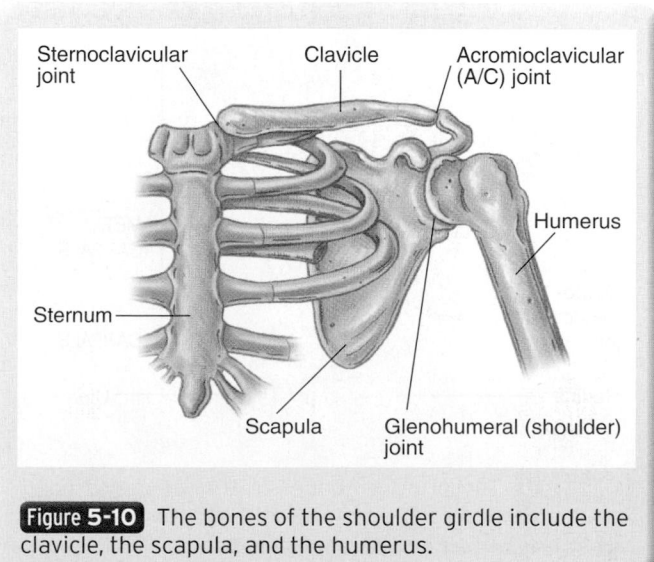

Figure 5-10 The bones of the shoulder girdle include the clavicle, the scapula, and the humerus.

the thorax in front and articulates (joins) posteriorly with the **scapula** (shoulder blade), which lies in the muscular tissue of the posterior thoracic wall. The inferior boundary of the thorax is the diaphragm, which separates the thorax from the abdomen.

Arm

The supporting bone of the arm is the **humerus**. Its long, straight shaft serves as an effective lever for heavy lifting. The forearm is composed of the radius and the ulna. The **ulna** is larger in the proximal forearm, and the **radius** is larger in the distal forearm. The radius lies on the lateral, or thumb, side of the forearm, and the ulna is on the medial or little finger side.

Wrist and Hand

The wrist is a modified ball-and-socket joint formed by the ends of the radius and ulna and several small wrist bones **Figure 5-11**. There are eight bones in the wrist, called carpal bones. Extending from the carpal bones are five metacarpals, which serve as a base for each of the five fingers, or digits. The fingers are composed of bones called the phalanges.

■ The Pelvis

The pelvis is a closed bony ring that consists of three bones: the sacrum and the two pelvic bones **Figure 5-12**.

Each pelvic bone is formed by the fusion of three separate bones. These three bones are called the **ilium**, the **ischium**, and the **pubis**. These bones are joined together posteriorly by the sacrum. On the anterior side of this ring, the left pubis and the right pubis are joined. This area, called the **pubic symphysis**, has cartilage within it that allows for slight motion of one side of the pelvis over the other. Pressure on the symphysis pubis during the physical examination can reveal fractures in the pelvis. These fractures can lead to life-threatening bleeding. The part of the pelvis where the lower leg connects is called the **acetabulum**. This is the "socket" in which the "ball" of the femur fits.

■ The Lower Extremity

The **femur** (thigh bone) is the longest and one of the strongest bones in the body. At the superior end of the bone is a round ball-like structure called the **femoral head**. This is where the femur connects into the acetabulum (pelvic girdle) by a ball-and-socket joint. Notice in **Figure 5-13** that the femur has two projections. The projection on the lateral/superior portion of the femur is called the **greater trochanter**. The projection on the medial/superior portion of the femur is called the **lesser trochanter**. Both projections are anchor points where the major muscles of the thigh connect to the femur.

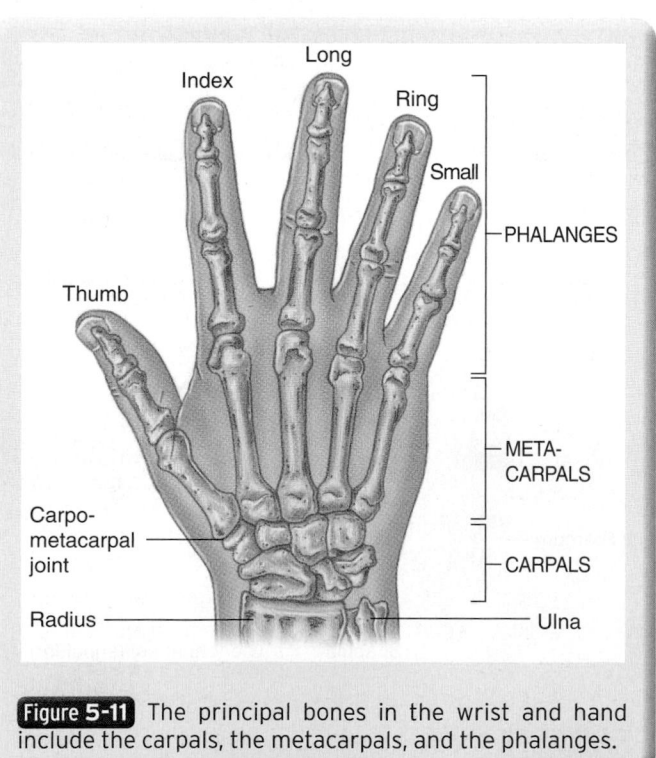

Figure 5-11 The principal bones in the wrist and hand include the carpals, the metacarpals, and the phalanges.

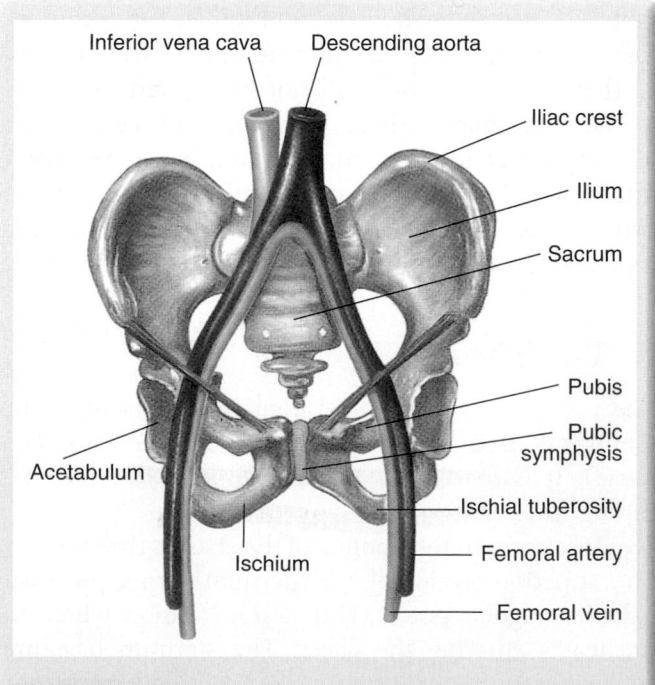

Figure 5-12 The pelvis is a closed bony ring that consists of the sacrum, ilium, ischium, pubis, acetabulum, and pubic symphysis.

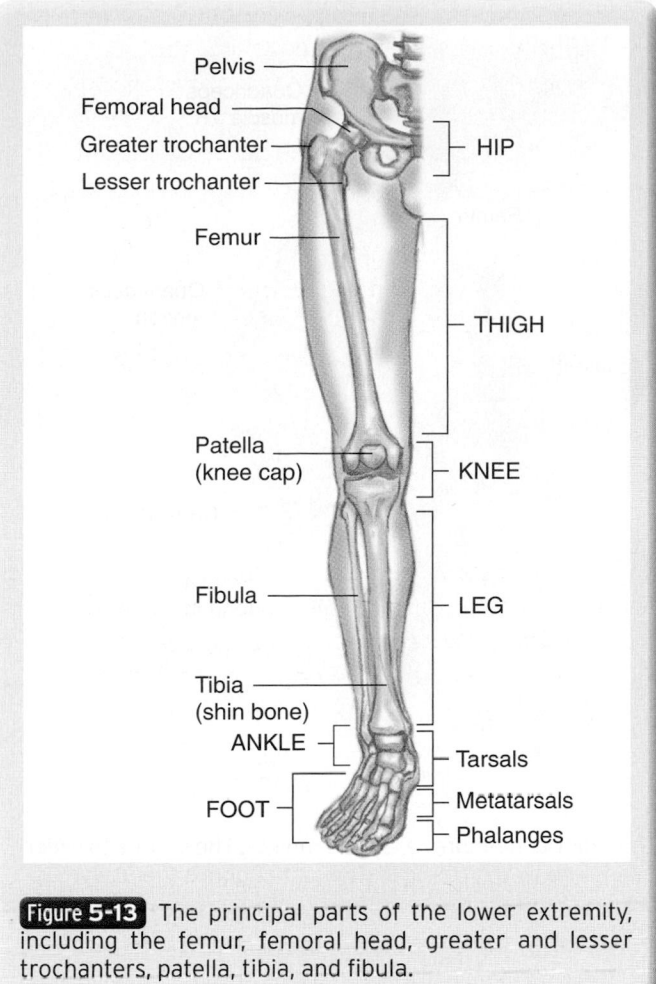

Figure 5-13 The principal parts of the lower extremity, including the femur, femoral head, greater and lesser trochanters, patella, tibia, and fibula.

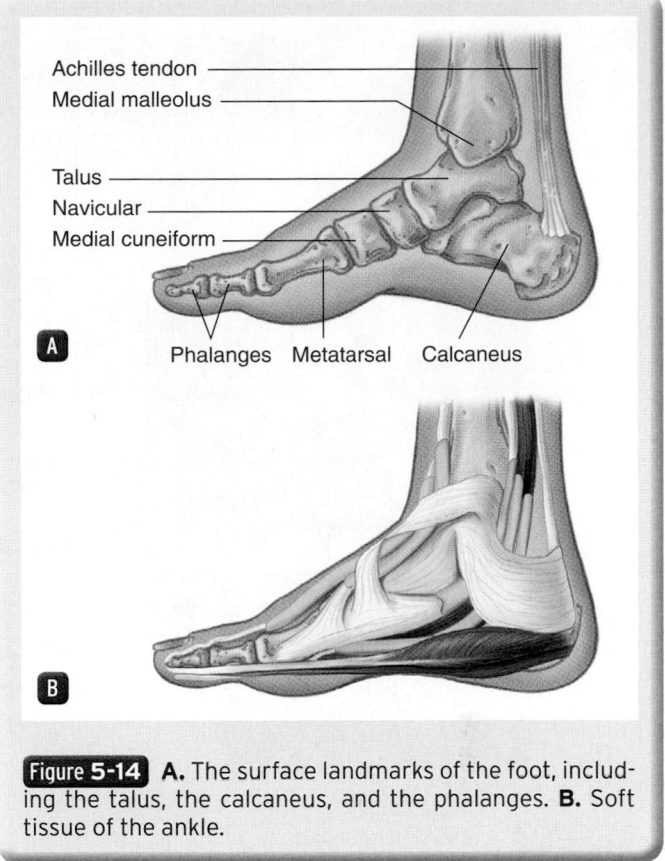

Figure 5-14 **A.** The surface landmarks of the foot, including the talus, the calcaneus, and the phalanges. **B.** Soft tissue of the ankle.

The joint that connects the upper leg to the lower leg is the knee. The knee is essentially a hinge joint, allowing only flexion and extension between the distal femur and the proximal tibia. Anterior to the knee is a specialized bone called the patella (kneecap). The lower leg lies between the knee and the ankle joint and is composed of the tibia and the fibula. The tibia (shin bone) is the larger bone and lies in the anterior of the leg. You can palpate the entire length of the tibia on the anterior surface of the leg just under the skin. The fibula lies on the lateral side of the leg. You can palpate the head of the fibula on the lateral aspect of the knee joint.

Ankle and Foot

The ankle is a hinge joint that allows flexion and extension of the foot on the leg **Figure 5-14**. The foot contains seven tarsal bones. The talus is one of the largest; the calcaneus, which forms the prominence of the heel, is the other large tarsal bone. Five metatarsal bones form the substance of the foot. The five toes are formed by 14 phalanges—two in the great toe and three in each of the smaller toes.

Joints

Wherever two long bones come in contact, a **joint (articulation)** is formed. A joint consists of the ends of the bones that make up the joint and the surrounding connecting and supporting tissue **Figure 5-15**. Most joints in the body are named by combining the names of the two bones that form that joint. For example, the sterno-clavicular joint is the articulation between the sternum and the clavicle. Most joints allow motion—for example, the knee, hip, and elbow—whereas some bones fuse with one another at joints to form a solid, immobile, bony structure. For example, the skull is composed of several bones that fuse as a child grows. An infant, whose skull bones are not yet fused, has fontanels (soft spots) between the bones. The fontanels close as the bones of the infant's skull fuse together. Some joints have slight, limited motion in which the bone ends are held together by fibrous tissue. Such a joint is called a **symphysis**.

The bone ends of a joint are held together by a fibrous sac called the **joint capsule**. This sac is composed of tissue called ligaments (bone to bone). At certain points around the circumference of the joint, the capsule is lax and thin so motion can occur. In other areas, it is quite thick and resists stretching or bending. A joint such as the **sacroiliac joint** that is virtually

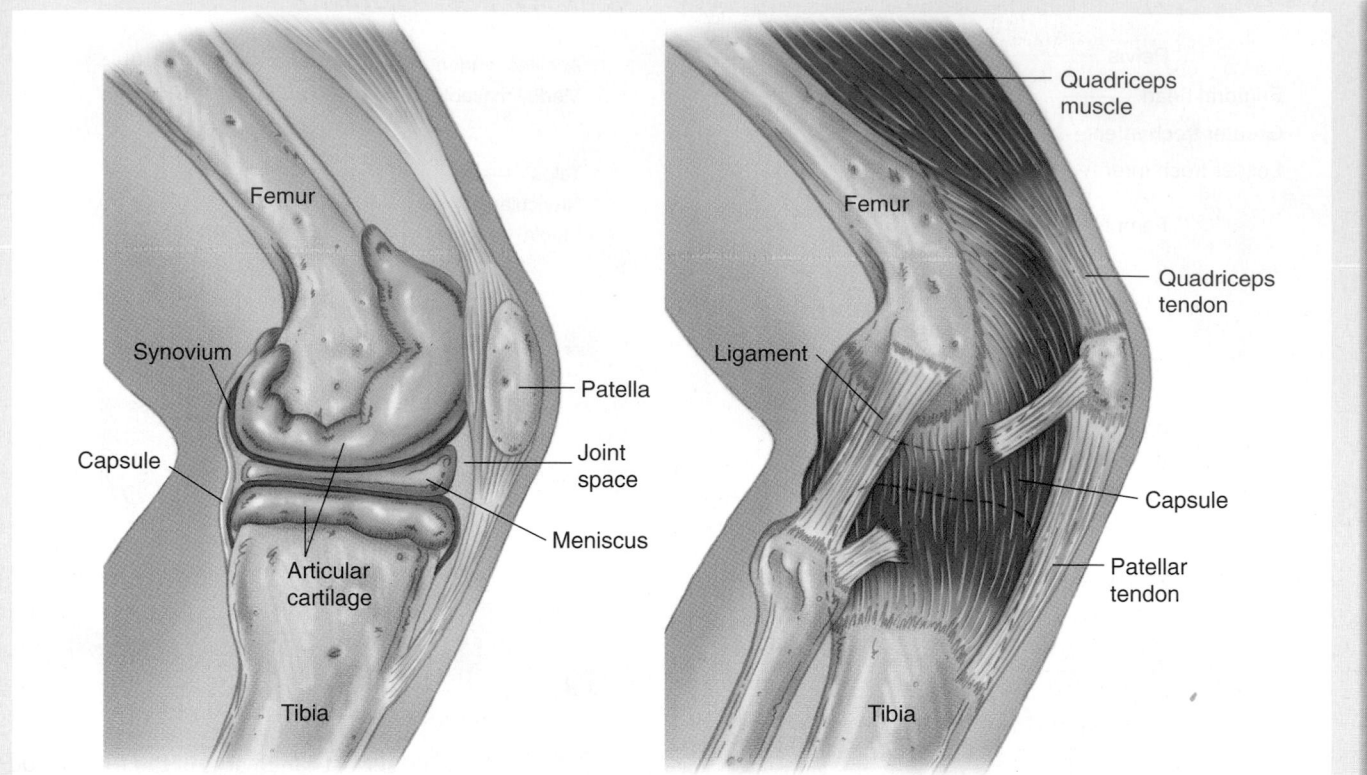

Figure 5-15 A joint consists of bone ends, the fibrous joint capsule, the synovial membrane, and ligaments. The degree to which a joint can move is determined by how the ligaments hold the bone ends and by the configuration of the bones themselves.

surrounded by tough, thick ligaments will have little motion, whereas a joint such as the shoulder, with few ligaments, will be free to move in almost any direction (and will, as a result, be more prone to dislocation). On the inner lining of the joint capsule is the **synovial membrane**. This special tissue is responsible for making a thick lubricant called **synovial fluid**. This "oil" allows the ends of the bones to glide over each other as opposed to rubbing and grating over each other.

The degree to which a joint can move is determined by the extent to which the ligaments hold the bone ends together and also by the configuration of the bone ends themselves. The shoulder joint is a **ball-and-socket joint**, which allows rotation and bending **Figure 5-16**. The finger joints, elbow, and knee are **hinge joints**, with motion restricted to one plane **Figure 5-17**. They can only **flex** (bend) and **extend** (straighten). Rotation is not possible because of the shape of the joint surfaces and the strong restraining ligaments on both sides of the joint. Although the amount of motion varies from joint to joint, all joints have a definite limit beyond which motion cannot occur. When a joint is forced beyond this limit, damage to some structure must occur. Either the bones that form the joint will break, or the supporting capsule and ligaments will be disrupted.

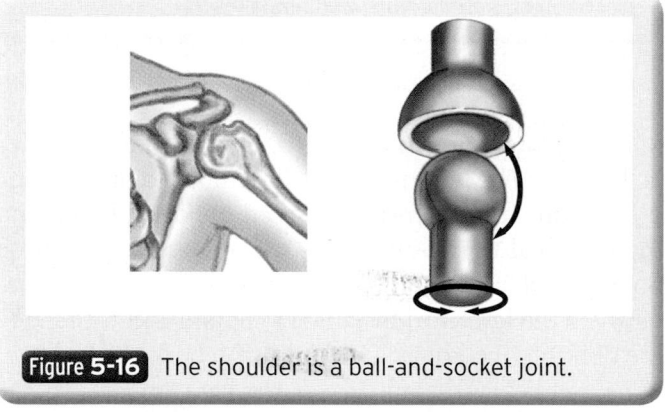

Figure 5-16 The shoulder is a ball-and-socket joint.

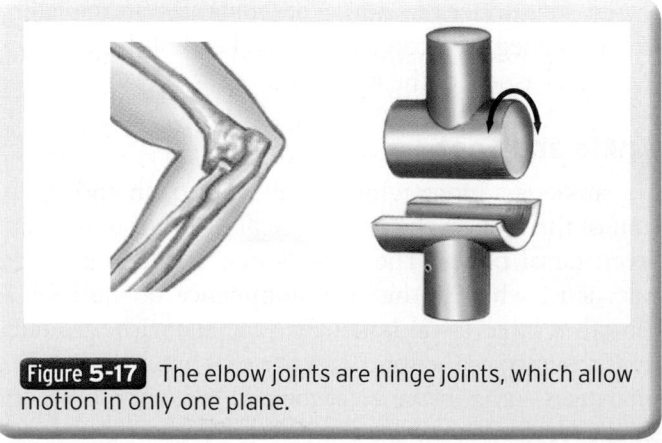

Figure 5-17 The elbow joints are hinge joints, which allow motion in only one plane.

The Skeletal System: Physiology

The skeletal system is responsible for several functions. The most notable functions are its ability to give the body shape, provide protection of fragile organs, and allow for movement. Another function of the skeletal system is the storage of calcium. Calcium is the main element the various bones cells use to create a structure that is hard and resilient. The bones are constantly being made and destroyed as stress is applied to them. Bones store and release calcium, and this calcium is important for other body systems.

The skeletal system also helps with the creation of various types of blood cells. In the marrow of certain types of bones, special cells are present that can transform themselves into red blood cells, white blood cells, and platelets. The cells, when stimulated, help to replace worn out cells in the blood.

The Musculoskeletal System: Anatomy

The human body is a well-designed system whose form, upright posture, and movement are provided by the **musculoskeletal system**. The term musculoskeletal refers to the bones and voluntary muscles of the body. The musculoskeletal system also protects the vital internal organs of the body. Muscles are a form of tissue that allows body movement. There are more than 600 muscles in the musculoskeletal system. The type of muscle found here is called skeletal muscle. Other types of muscle outside of the musculoskeletal system include **smooth muscle** and **cardiac muscle**. Smooth muscles are found within blood vessels and intestines. For example, when you hear your stomach growling, you really are hearing the rhythmic contractions of the smooth muscles of your intestines. Cardiac muscles are found only within the heart Figure 5-18 .

■ Skeletal Muscle

Skeletal muscle, so named because it attaches to the bones of the skeleton, forms the major muscle mass of the body. It is also called **voluntary muscle**, because all skeletal muscle is under direct voluntary control of the brain and can be stimulated to contract or relax at will. Movement of the body, like waving or walking, results from skeletal muscle contraction or relaxation. Usually, a specific motion is the result of several muscles contracting and relaxing simultaneously.

Most muscles within the body operate on the principle of antagonistic pairs. The muscles of the upper arm include the **biceps** muscle, which is located on the anterior aspect of the humerus. This muscle moves the lower part of the arm toward the head. If the muscle were working alone, the person would have little control over the speed of that movement. The way the body achieves control and fine movement is to have the biceps compete against another muscle group. The biceps competes with the **triceps** muscle, which is called the three-headed muscle of the arm because there are three bundles of muscle that join together at

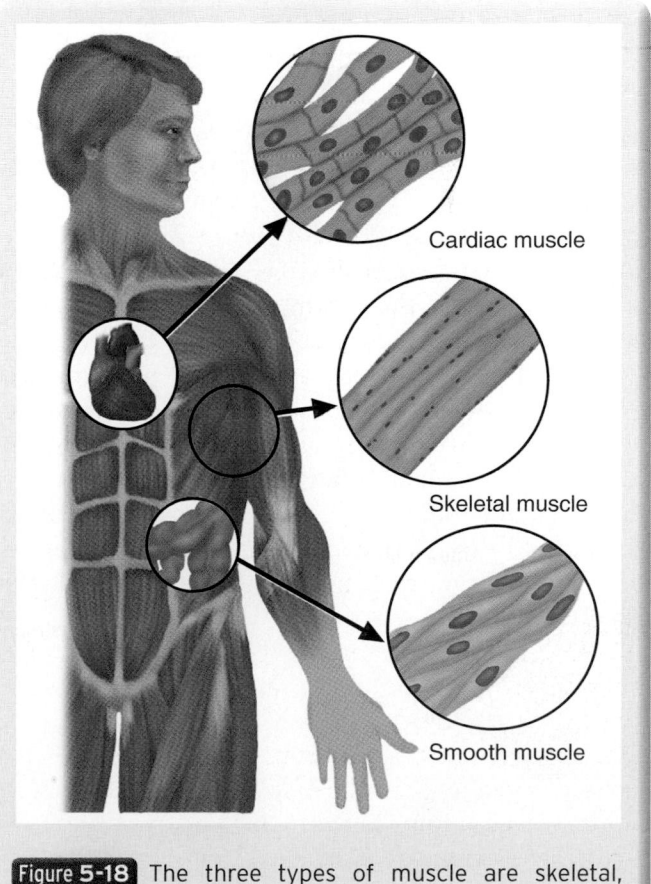

Cardiac muscle

Skeletal muscle

Smooth muscle

Figure 5-18 The three types of muscle are skeletal, smooth, and cardiac.

the elbow. Without the triceps, you would slap yourself in the face every time you bend your arm. The biceps works to slow the movement of the triceps as the arm is extended. To complete the discussion of the muscles, you need to understand some motion terms. Flexion is the bending of a joint, and extension is the straightening of a joint.

There are some important muscle groups to know. **Figure 5-19** and **Table 5-4** show the major muscles, their locations, and their functions.

The Musculoskeletal System: Physiology

The musculoskeletal system has several functions. A person's ability to move and be able to manipulate his or her environment is made possible by the contraction and relaxation of this system. A by-product of this movement is heat. When you get cold, you involuntarily shake your muscles, or shiver, to produce heat.

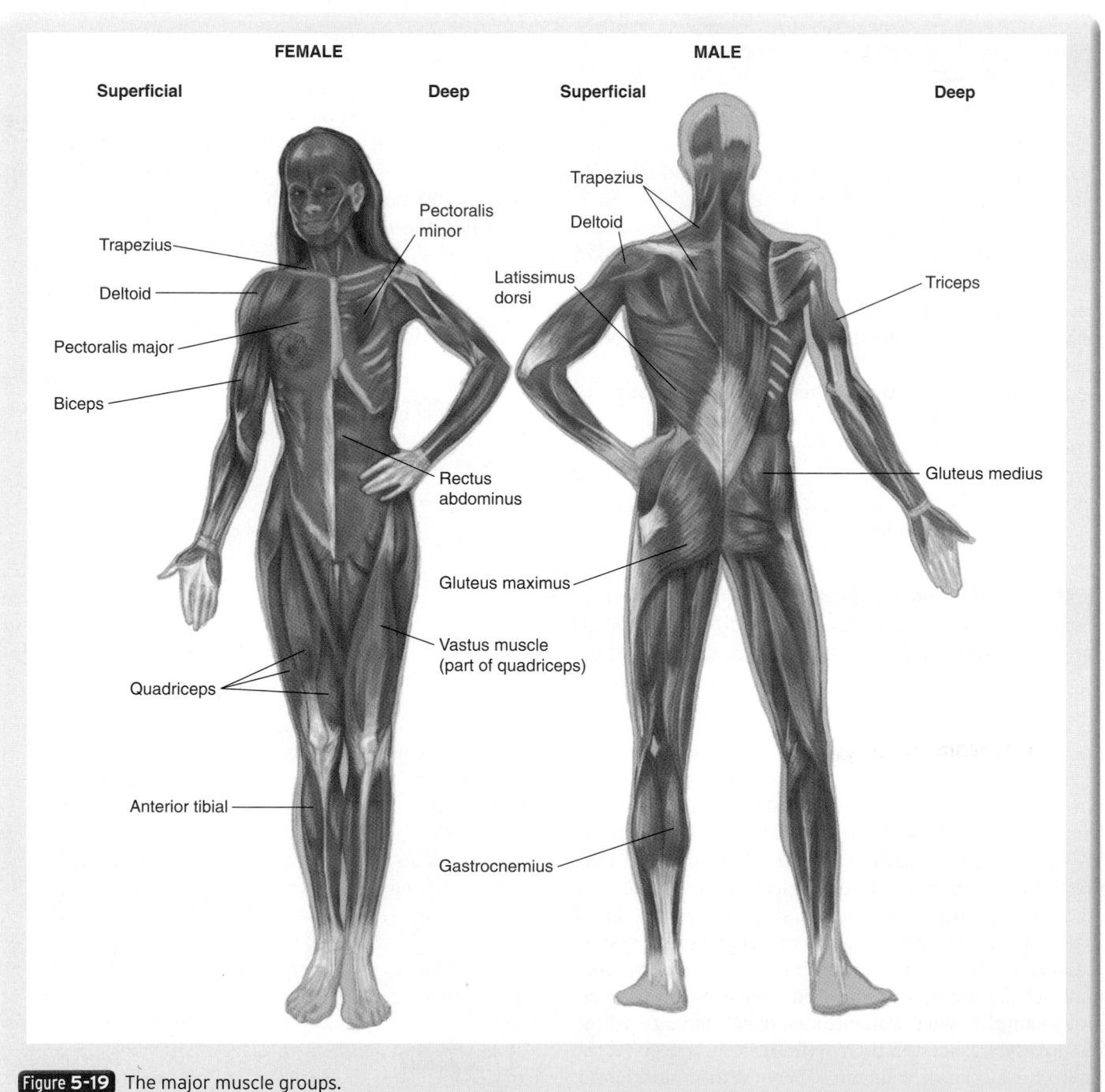

Figure 5-19 The major muscle groups.

Table 5-4 Muscles: Locations and Functions

Muscle Name	Location	Function
Biceps	Anterior, humerus	Flexes lower arm
Triceps	Posterior, humerus	Extends lower arm
Pectoralis	Anterior, thorax	Flexes and rotates arm
Latissimus dorsi	Posterior, thorax	Extends and rotates arm
Rectus abdominis	Anterior, abdomen	Flexes and rotates spine
Tibialis anterior	Anterior, tibia	Points toes toward head
Gastrocnemius	Posterior, tibia	Points toes away from head
Quadriceps (four separate muscles)	Anterior, femur	Extends lower leg
Biceps femoris	Posterior, femur	Flexes lower leg
Gluteus (three separate muscles)	Posterior, pelvis	Extends and rotates leg

Shivering is an essential function. Another function of muscles is to protect the structures under them, such as the intestines, which are protected by the rectus abdominus muscles.

The Respiratory System: Anatomy

The **respiratory system** consists of all the structures of the body that contribute to respiration, or the process of breathing Figure 5-20 . It includes the nose, mouth, throat, larynx, trachea, bronchi, and bronchioles, which are all air passages or airways. The system also includes the lungs, where oxygen is passed into the blood and carbon dioxide removed. Finally, the respiratory system includes the diaphragm, the muscles of the chest wall, and accessory muscles of breathing, which permit normal respiratory movement. In this text, the term "airway" usually refers to the upper airway or the passage above the larynx (voice box).

The Upper Airway

The structures of the upper airway are located anteriorly and at the midline. The upper airway includes the nose, mouth, tongue, jaw, oral cavity, larynx, and pharynx. The larynx is typically considered the dividing line between the upper and lower airway. The larynx is a rather complex arrangement of tiny bones, cartilage, muscles, and two vocal cords. The larynx does not tolerate any foreign solid or liquid material. A violent episode of coughing and spasm of the vocal cords will result from contact with solids or liquids. The nose and mouth lead to the oropharynx (throat). The pharynx is composed of the nasopharynx, oropharynx, and the laryngopharynx. The nostrils lead to the **nasopharynx** (above the roof of the mouth, or soft palate), and the mouth leads to the oropharynx. The nasal passages and nasopharynx warm, filter, and humidify air as a person breathes. Air enters through the mouth more rapidly and directly. As a result, it is less moist than air that enters through the nose.

Two passageways are located at the bottom of the pharynx: the esophagus behind and the **trachea** (windpipe) in front. Food and liquids enter the pharynx and pass into the esophagus, which carries them to the stomach. Air and other gases enter the trachea and go to the lungs.

Protecting the opening of the trachea is a thin, leaf-shaped valve called the **epiglottis**. This valve allows air to pass into the trachea but prevents food and liquid from entering the airway under normal circumstances. Air moves past the epiglottis into the larynx and the trachea.

The Lower Airway

The **Adam's apple**, or **thyroid cartilage**, is easily seen in the middle of the front of the neck. The thyroid cartilage is actually the anterior part of the larynx. Tiny muscles open and close the vocal cords and control tension on them. Sounds are created as air is forced past the vocal cords, making them vibrate. These vibrations make the sound. The pitch of the sound changes as the cords open and close. You can feel the vibrations if you place your fingers lightly on the larynx as you speak or sing. The vibrations of air are shaped by the tongue and muscles of the mouth to form understandable sounds. Immediately below the thyroid cartilage is the palpable **cricoid cartilage**. This is the location for using the Sellick maneuver to help in maintaining a proper airway.

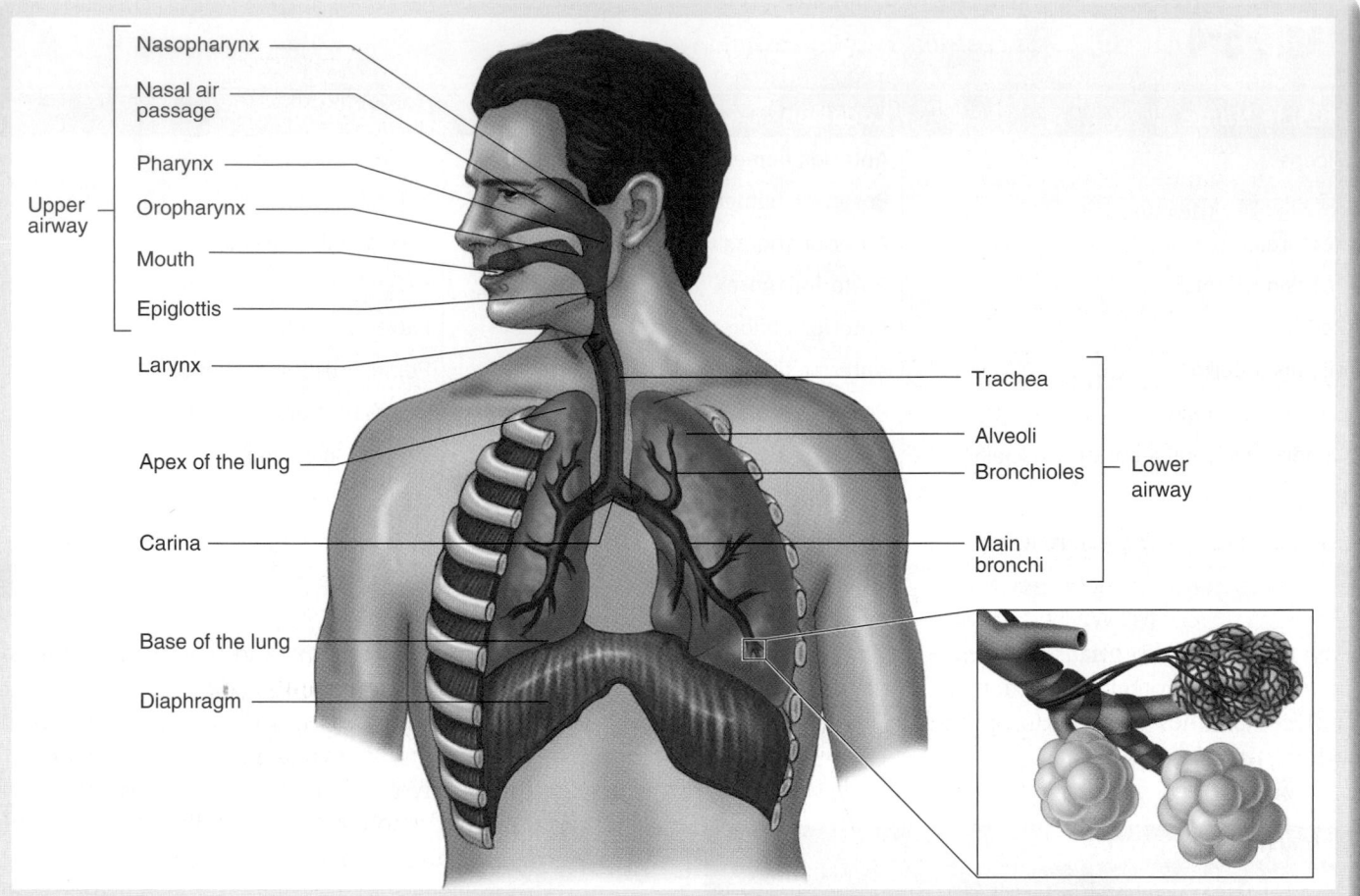

Figure 5-20 The respiratory system consists of all structures of the body that contribute to the process of breathing.

You are the Provider: PART 2

When you arrive at the scene, you find the patient lying on his side on the floor of his bedroom. His knees are drawn up to his abdomen, and he is in severe pain. As you assess the patient, your partner opens the jump kit and prepares to begin treatment.

Recording Time: 0 Minutes	
Appearance	Restless; diaphoretic; in severe pain
Level of consciousness	Conscious and alert
Airway	Open; clear of secretions and foreign bodies
Breathing	Increased rate; adequate depth
Circulation	Radial pulses, present and strong; skin, cool and clammy

The patient tells you that the pain is in the right upper quadrant of his abdomen and that it began suddenly about 20 minutes ago.

2. On the sole basis of the patient's chief complaint, which organ or organs should you suspect is/are the cause of his problem?

3. What additional questions should you ask to gather more information about his chief complaint?

Between the thyroid and cricoid cartilage lies the **cricothyroid membrane**, which can be felt as a depression in the midline of the neck just inferior to the thyroid cartilage. This is an important landmark when a needle airway device is being inserted. Needle airways may be inserted by paramedics to provide ventilation to patients who would otherwise die.

Below the cricoid cartilage is the trachea. The trachea is approximately 5″ long and is a semirigid, enclosed air tube made up of rings of cartilage that are open in the back. This enables food to pass through the esophagus, which lies right behind the trachea. The rings of cartilage keep the trachea from collapsing when air moves into and out of the lungs.

The trachea ends at the carina and divides into two smaller tubes. These tubes are the right and left mainstem bronchi, which enter the lungs. Each main bronchus immediately branches within the lung into smaller and smaller airways. Within the right lung, three major bronchi are formed. Within the left lung, there are only two bronchi. Each bronchus supplies air to one lobe of the lung. The final divisions of the bronchi are called bronchioles.

■ Lungs

The two lungs are held in place by the trachea, the arteries and veins, and the pulmonary ligaments. Each lung is divided into lobes. The right lung has three lobes: the upper, middle, and lower lobes. The left lung has an upper lobe and a lower lobe. Each lobe is divided further into segments. The bronchioles end in about 700 million tiny grapelike sacs called **alveoli** Figure 5-21 .

All of the respiratory structures that have been discussed so far have as their primary goal the movement of air to the alveoli. The exchange of oxygen and carbon dioxide occurs within these alveoli. They do all the work and are referred to as the functional unit of the respiratory system. The walls of the alveoli contain a network of tiny blood vessels (pulmonary capillaries) that carry the carbon dioxide from the body to the lungs and the oxygen from the lungs to the body.

The lungs cannot expand and contract themselves because they have no muscle. There is, however, a very definite mechanism in place to ensure that the lungs follow the motion of the chest wall and expand or contract with it. Covering each lung is a layer of very smooth, glistening tissue called **pleura** Figure 5-22 . Another layer of pleura lines the inside of the chest cavity. The two layers are called parietal pleura (lining the chest wall) and visceral pleura (covering the lungs). Between these two layers is a small amount of fluid that permits smooth gliding of the tissues. This is very similar in concept to how joints work.

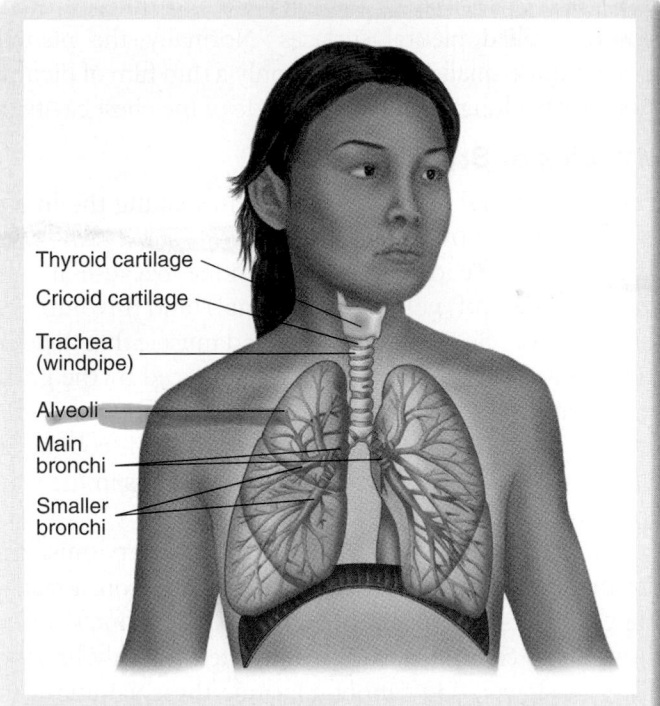

Figure 5-21 The lungs contain millions of air sacs (alveoli), which lie at the ends of air passages. Small blood vessels surround the alveoli, allowing for gas exchange.

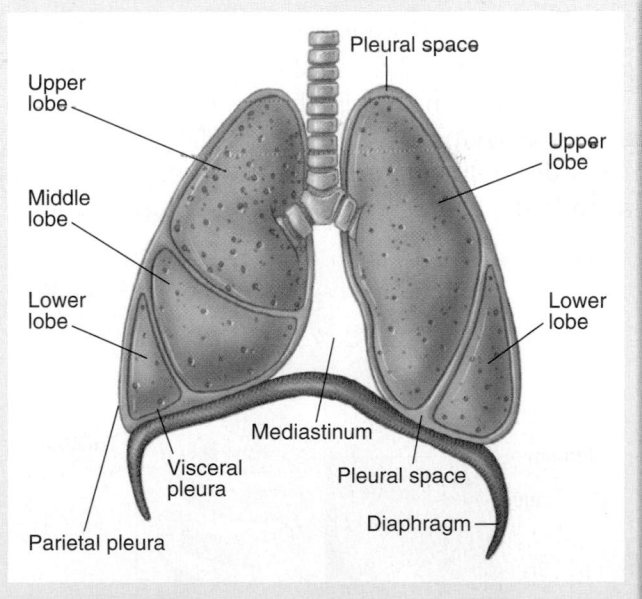

Figure 5-22 The pleura lining the chest wall and covering the lungs is an essential part of the breathing mechanism. The pleural space is not an actual space until blood or air leaks into it, causing the pleural surfaces to separate.

Between the parietal pleura and the visceral pleura is the **pleural space**, called a "potential" space rather than an actual space in the usual sense because normally these layers are in close contact everywhere. In fact, the layers are sealed tightly to one another by a thin film of fluid. When the chest wall expands, the lung is pulled with it and made to expand by the force exerted through these closely applied pleural surfaces. Normally, the pleural space is quite small and contains only a thin film of pleural fluid as each lung entirely fills its side of the chest cavity.

Muscles of Breathing

There are several muscles involved in making the lungs expand and contract. The primary muscle is called the **diaphragm**. The diaphragm is unique because it has characteristics of voluntary (skeletal) and involuntary (smooth) muscle. It is a dome-shaped muscle that divides the thorax from the abdomen and is pierced by the great vessels and the esophagus Figure 5-23 . It acts like a voluntary muscle when you take a deep breath, cough, or hold your breath. You control these variations in the way you breathe.

However, unlike other skeletal or voluntary muscles, the diaphragm performs an automatic function. Breathing continues during sleep and at all other times. Even though you can hold your breath or temporarily breathe faster or slower, you cannot continue these variations in breathing pattern indefinitely. When the concentration of carbon dioxide becomes too high, automatic regulation of breathing resumes. Therefore, although the diaphragm looks like voluntary skeletal muscle and is attached to the skeleton, it behaves, for the most part, like an **involuntary muscle**.

The other muscles involved in breathing are the intercostal muscles, the abdominal muscles, and the pectoral muscles. During inhalation, the diaphragm and intercostal muscles contract. When the diaphragm contracts, it moves down slightly, enlarging the thoracic cage from top to bottom. When the intercostal muscles contract,

they move the ribs up and out. These actions combine to enlarge the chest cavity in all dimensions. Pressure in the cavity then falls, and air rushes into the lungs. This is referred to as negative pressure breathing because air is essentially sucked into the lungs. This part of the cycle is active, requiring the muscles to contract.

During exhalation, the diaphragm and the intercostal muscles relax. Unlike inhalation, exhalation does not normally require muscular effort. As these muscles relax, all dimensions of the thorax decrease, and the ribs and muscles assume a normal resting position. When the volume of the chest cavity decreases, air in the lungs is compressed into a smaller space. Pressure is increased, and air is pushed out through the trachea. This phase of the cycle is passive.

The process of breathing is typically easy and requires little muscular effort. But, now imagine breathing through a straw and suddenly the diameter of the straw decreases. The smaller the diameter of the straw, the more effort you will now have to exert to move air. As the resistance in the airway increases, you will begin to use more muscle groups, namely your abdominal and pectoral muscles, to assist the diaphragm in moving that air.

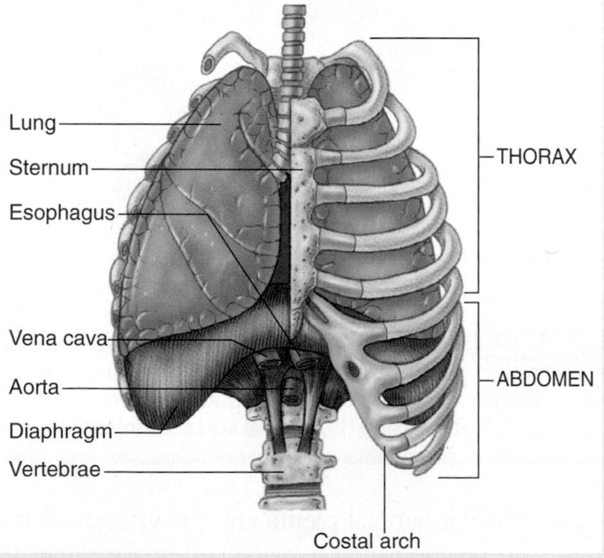

Words of Wisdom

When you are assessing a patient, make sure you assess both sides of the patient. It may seem like a waste of time to assess the left arm when the right arm is the one that is injured. However, the reason for assessing both sides is simple. You need to compare the sides to see whether there are differences. A deformity on one arm may be "normal" if the same deformity is found on the other arm. This idea of comparing sides applies to the respiratory system as well. You need to listen to both sides of the chest to determine the patient's lung sounds. Lungs sounds can change on only one side of the chest, or they can change on both sides of the chest. Use all of the information you obtain from both sides of the body to help you make your patient care decisions.

The Respiratory System: Physiology

The function of the respiratory system is to provide the body with oxygen and eliminate carbon dioxide. The exchange of oxygen and carbon dioxide takes place in the lungs and in the tissues. It is a complicated process that occurs automatically unless the airways or the lungs become diseased or damaged. There are two separate, yet interdependent, overall functions of the respiratory system: ventilation and respiration.

Ventilation is the simple movement of air between the lungs and the environment. You are providing ventilation when you administer oxygen to a patient who is not breathing. Ventilation is provided in the hope

Figure 5-23 The dome-shaped diaphragm divides the thorax from the abdomen. It is pierced by the great vessels and the esophagus.

Labels: Lung, Sternum, Esophagus, Vena cava, Aorta, Diaphragm, Vertebrae, THORAX, ABDOMEN, Costal arch that your patient will resume respiration. **Respiration** is the process of gas exchange. Respiration provides the much-needed oxygen to cells and removes the waste product carbon dioxide. This exchange of gases also helps to control the pH of the blood.

■ Respiration

As blood travels through the body, it gives its oxygen and nutrients to various tissues and cells. Oxygen passes from the blood through the capillaries to tissue cells. In the reverse process, carbon dioxide and cell waste pass from tissue cells through capillaries to the blood **Figure 5-24**.

Each time we take a breath, the alveoli receive a supply of oxygen-rich air. The oxygen then passes into a fine network of pulmonary capillaries, which are in close contact with the alveoli. In fact, the capillaries in the lungs are located in the walls of the alveoli. The walls of the capillaries and the alveoli are extremely thin. Thus, air in the alveoli and blood in the capillaries are separated by two very thin layers of tissue.

Oxygen and carbon dioxide pass rapidly across these thin tissue layers by diffusion. **Diffusion** is a passive process in which molecules move from an area with a higher concentration of molecules to an area of lower concentration. There are more oxygen molecules in the alveoli than in the blood. Therefore, the oxygen molecules move from the alveoli into the blood. Because there are more carbon dioxide molecules in the blood than in the inhaled air, carbon dioxide moves from the blood into the alveoli. This process is completely passive—nature does all the work by moving gas from areas of high concentration to areas of low concentration.

The blood does not use all the inhaled oxygen as it passes through the body. Exhaled air contains 16% oxygen and 3% to 5% carbon dioxide; the rest is nitrogen **Figure 5-26**. This 16% concentration of oxygen is adequate to support artificial ventilation. So as you provide artificial ventilations to a patient who is not breathing, the patient is receiving a 16% concentration of oxygen with each ventilation.

The Chemical Control of Breathing

The brain—or more specifically, the brain stem—controls breathing. This area is in one of the best-protected parts of the nervous system—deep within the skull. The nerves in this area act as sensors for the level of carbon dioxide in the blood and subsequently the spinal fluid. The brain automatically controls breathing if the level of carbon dioxide or oxygen in the arterial blood is too high or too low. In fact, adjustments can be made in just one breath. For these reasons, you cannot hold your breath indefinitely or breathe rapidly and deeply indefinitely.

Breathing occurs as the result of a buildup of carbon dioxide, which causes the pH to decrease in the cerebrospinal fluid. The cells are constantly working to eliminate carbon dioxide to regulate the acid–alkaline balance of the body. When the level of carbon dioxide becomes too high, a slight change occurs in the pH (the measure of acidity) of the cerebrospinal fluid. The medulla oblongata (a portion of the brain stem), which is sensitive to pH changes, stimulates the phrenic nerve, sending a signal to the diaphragm, thus causing you to breathe. You then exhale to reduce the level of carbon dioxide in your body. The primary reason you breathe is to lower

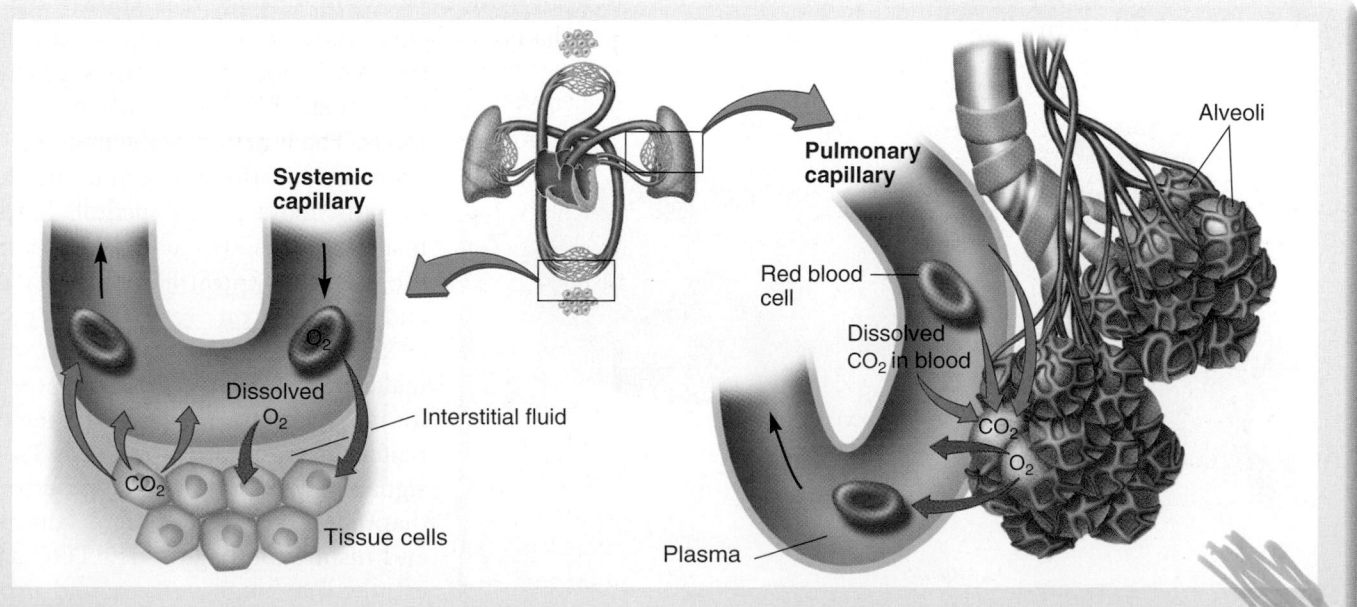

Figure 5-24 In the capillaries of the lungs, oxygen (O_2) passes from the blood to the tissue cells, and carbon dioxide (CO_2) and waste pass from the tissue cells to the blood. Diffusion occurs when molecules move from an area of higher concentration to an area of lower concentration.

Special Populations

The anatomy of the respiratory system in children is proportionally smaller and less rigid than in an adult Figure 5-25 . A child's nose and mouth are much smaller than those of an adult. The larynx, cricoid cartilage, and trachea are smaller, softer, and more flexible as well. This makes the mechanics of breathing much more delicate. A child's pharynx is also smaller and less deeply curved. The tongue takes up proportionally more space in a child's mouth than in an adult's mouth.

These anatomic differences are important for your assessment and treatment. For example, the smaller larynx of a child becomes obstructed more easily. The chest wall in children is softer. Therefore, children depend more heavily on the diaphragm for breathing. You will notice that a child's abdomen moves in and out considerably with each breath, especially in an infant. Infants younger than 1 month do not know how to breathe through the mouth. Smaller children also have proportionally larger heads compared with the rest of the body. This will affect the way you treat a suspected spinal injury. Therefore, as you assess and treat an infant or a child, you must carefully consider these differences.

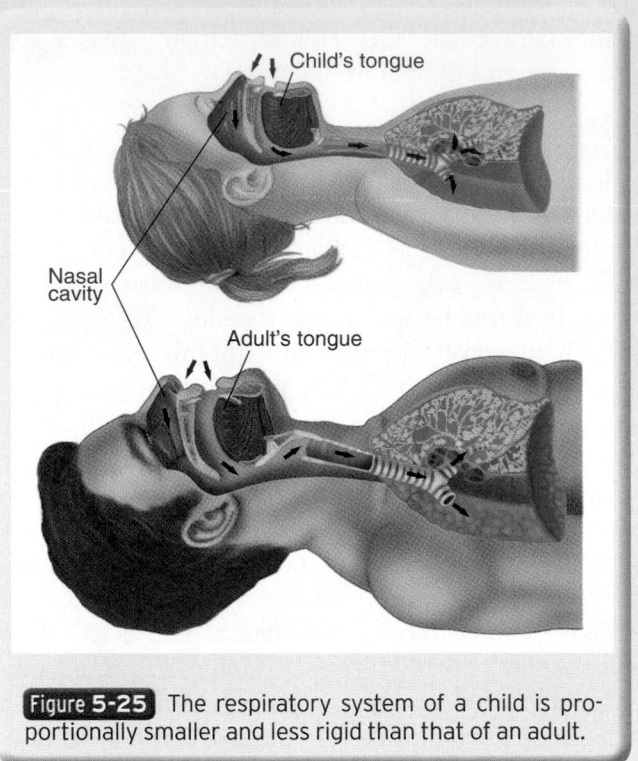

Figure 5-25 The respiratory system of a child is proportionally smaller and less rigid than that of an adult.

your level of carbon dioxide, not to increase your level of oxygen.

The body also has a "backup system" to control respiration called the **hypoxic drive**. When the oxygen level falls, this system will also stimulate breathing. There are areas in the brain, the walls of the aorta, and the carotid arteries that act as oxygen sensors. These sensors are easily satisfied by minimal levels of oxygen in the arterial blood. Therefore, the backup system, the hypoxic drive, is much less sensitive and less powerful than the carbon dioxide sensors in the brain stem.

The Nervous System Control of Breathing

The exact way breathing occurs is complicated and also poorly understood by science. It is known that the medulla oblongata is primarily responsible for initiating the ventilation cycle and is primarily stimulated by high carbon dioxide levels. The function of the medulla is to keep you breathing so you do not have to think about it. The medulla has two main portions that control breathing: the **dorsal respiratory group (DRG)** and the **ventral respiratory group (VRG)**. The DRG is the main pacemaker for breathing and is responsible for initiating inspiration. It sets the base pattern for respirations. The DRG sends signals down the phrenic nerve to the diaphragm. The diaphragm contracts, and inspiration begins. The DRG shuts off, the diaphragm relaxes, and expiration begins. The VRG helps to provide for forced inspiration or expiration as needed.

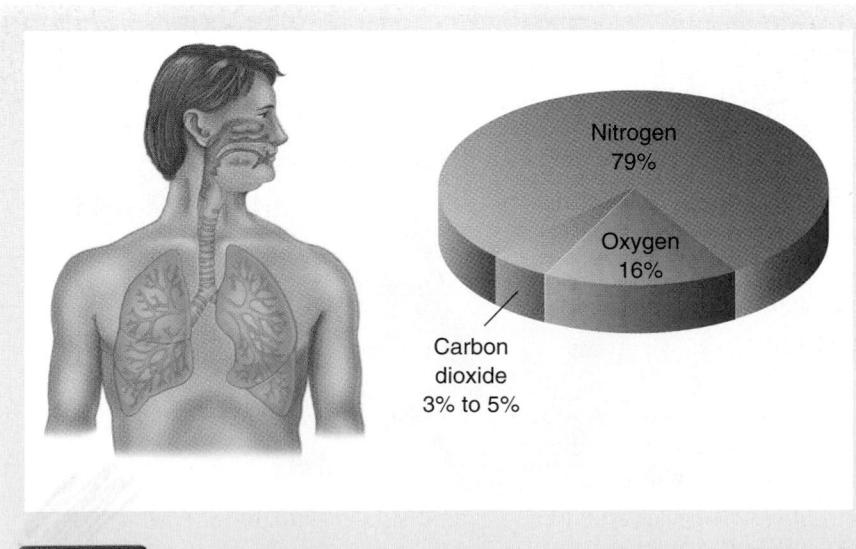

Figure 5-26 The components of exhaled air include oxygen, carbon dioxide, and nitrogen.

Special Populations

Normal breathing patterns in infants and children are essentially the same as those in adults. However, infants and children breathe faster than adults. An infant who is breathing normally will have respirations of 25 to 50 breaths/min. A child will have respirations of 15 to 30 breaths/min. Like adults, infants and children who are breathing normally will have smooth, regular inhalation and exhalation, equal breath sounds, and regular rise and fall movements on both sides of the chest.

Breathing problems in infants and children often appear the same as breathing problems in adults. Signs such as increased respirations, an irregular breathing pattern, unequal breath sounds, and unequal chest expansion indicate breathing problems in adults and children. Other signs that an infant or child is not breathing normally include the following:

- Muscle retractions, in which the muscles of the chest and neck are working extra hard in breathing
- Nasal flaring, in which the nostrils flare out as the child breathes
- Seesaw respirations in infants, in which the chest and abdominal muscles alternately contract to look like a seesaw

Exhalation becomes active when infants and children have trouble breathing. Normally, inhalation alone is the active, muscular part of breathing, as described earlier. However, with labored breathing, both inhalation and exhalation are hard work. With labored breathing, exhalation is not passive. Instead, air is forced out of the lungs during exhalation, and the child will often begin to wheeze. This type of labored breathing involves the use of the accessory muscles of breathing.

The pons, another area within the brain stem, helps regulate the DRG activities. The pons has two areas. The **pneumotaxic (pontine) center**, located in the superior portion of the pons, helps shut off the DRG, resulting in shorter, faster respirations. The **apneustic center**, located in the inferior portion of the pons, stimulates the DRG, resulting in longer, slower respirations. Both areas of the pons are used to help augment respirations during emotional or physical stress. The two areas of the medulla and the two areas of the pons work together to help you get the right amount of air when you need it.

The VRG, pneumotaxic center, and apneustic center are involved in changing the depth of inspiration, expiration, or both. How does the body know when to stop breathing in or out? When the VRG is causing you to take a forced inspiration, what prevents you from taking in so much air that you pop your lungs like a balloon? The answer is the **Hering-Breuer reflex**. Special stretch receptors in the chest wall are able to detect if the lungs are too full or too empty. The Hering-Breuer reflex stops the VRG, penumotaxic center, and apneustic centers from accidentally causing lung trauma.

Table 5-5 summarizes the nervous system functions regarding respirations.

■ Ventilation

A substantial amount of air can be moved within the respiratory system. **Figure 5-27** shows the typical volumes. An adult male has a total lung capacity of 6,000 mL (equivalent to three 2-liter bottles of soda). An adult female has about one third less total capacity because the lung size is smaller.

As you are reading this book, unless you just finished exercising, the amount of your air movement is approximately 500 mL. This is called **tidal volume**. Tidal

Table 5-5 Nervous System Control of Breathing

Name	Location	Function	Timing
Dorsal respiratory group (DRG)	Medulla	Causes inspiration when stimulated	Normal, resting respirations. Rhythmic, mechanical pattern
Ventral respiratory group (VRG)	Medulla	Causes forced expiration or inspiration	Speech, increased emotional or physical stress
Pneumotaxic (pontine) center	Pons	Inhibits the DRG; increases speed and depth of respirations	Increased emotional or physical stress
Apneustic center	Pons	Excites the DRG; prolongs inspiration, decreases rate	Increased emotional or physical stress
Hering-Breuer inflation reflex (stretch reflex)	Chest	Detects lung expansion to a point and then tells VRG and pneumotaxic and apneustic centers to stop	Increased emotional or physical stress
Hering-Breuer deflation reflex	Chest	Detects potential lung collapse and then tells VRG and pneumotaxic and apneustic centers to stop	Increased emotional or physical stress

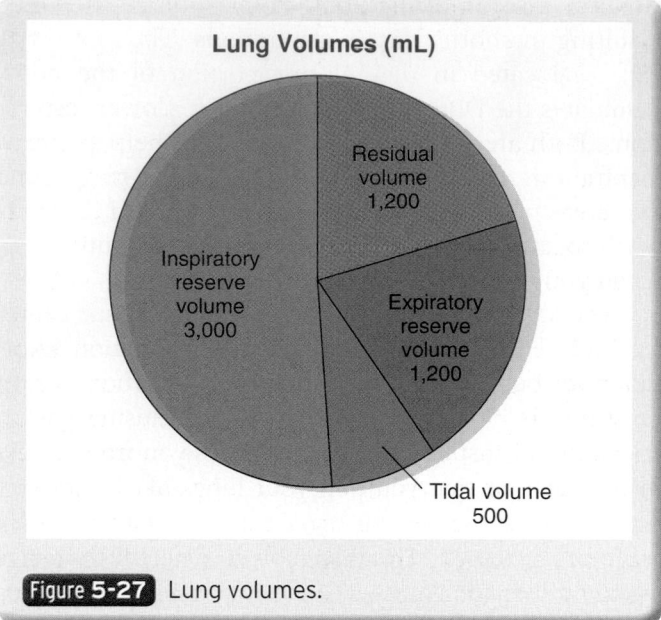

Figure 5-27 Lung volumes.

volume is the amount of air that is moved into or out of the lungs during a single breath. <u>Inspiratory reserve volume</u> is the deepest breath you can take after a normal breath. Conversely, <u>expiratory reserve volume</u> is the maximum amount of air that you can forcibly breathe out after a normal breath. Gas remains in the lungs simply to keep the lungs open. This is called the <u>residual volume</u>. This gas does not move during ventilation. Some residual volume is lost when a person is hit in the chest and has the "wind knocked out of him."

When you assist a patient's breathing, you move air in and out of the lungs. You will use a bag-mask device—a large bag filled with air that, when squeezed, pushes air out one end. The typical bag-mask device holds approximately 1,000 to 1,200 mL of air. Note that although a person's resting tidal volume is 500 mL, you need to use a bag-mask device that provides more than twice that volume. This is because of dead space.

<u>Dead space</u> is the portion of the respiratory system that has no alveoli, and, therefore, little or no exchange of gas between air and blood occurs. The mouth, trachea, bronchi, and bronchioles are all considered dead space. When you ventilate a patient with any device, you create more dead space. Gas must first fill the device before it can be moved into the patient.

When you are assessing your patient, you need to accurately determine whether he or she is having trouble breathing. Oftentimes, EMTs will look at the patient's respiratory rate; this rate, however, provides only part of the information that is needed. The depth of each breath is critical information to know when assessing ventilation. There is another measurement called minute volume that provides you with a more accurate determination of

effective ventilation. <u>**Minute volume**</u>, also referred to as minute ventilation, is easy to understand; it is the amount of air that moves in and out of the lungs in 1 minute minus the dead space.

Minute Volume = Respiratory Rate × Tidal Volume

This calculation helps you to determine how deeply a patient is breathing. While riding in the ambulance it will be difficult to determine the patient's exact tidal volume, but you will be able to estimate it. Consider the scenario of a patient who is breathing at a normal rate of 20 breaths/min. Yet, when you look at the patient's chest, it is barely moving. When you feel for air movement out of the mouth, you find very little movement. The patient is in trouble and needs your assistance now! Even though the patient's respiratory rate is normal, the amount of air being moved is inadequate. The minute volume is too low, and the patient needs ventilatory assistance. Always evaluate the amount of air being moved with each breath when assessing a patient's respirations.

■ Characteristics of Normal Breathing

You can think of a normal breathing pattern as a bellows system. Normal breathing should appear easy, not labored. As with a bellows that is used to move air to start a fire, breathing should be a smooth flow of air moving into and out of the lungs.

Normal breathing has the following characteristics:

- A normal rate and depth (tidal volume)
- A regular rhythm or pattern of inhalation and exhalation
- Good audible breath sounds on both sides of the chest
- Regular rise and fall movement on both sides of the chest
- Movement of the abdomen

■ Inadequate Breathing Patterns in Adults

An adult who is awake, alert, and talking to you has no immediate airway or breathing problems. However, you should keep supplemental oxygen on hand to assist with breathing should it become necessary. An adult who is not breathing well will appear to be working hard to breathe. This type of breathing pattern is called <u>**labored breathing**</u>. Labored breathing requires effort and may involve the accessory muscles. The person may also be breathing much slower (fewer than 12 breaths/min) or much faster (more than 20 breaths/min) than normal. An adult who is breathing normally will have respirations of 12 to 20 breaths/min Table 5-6.

With a normal breathing pattern, the accessory muscles are not being used. With inadequate breathing, a person, especially a child, may use the accessory muscles of the chest, neck, and abdomen. Other signs that a person is not breathing normally include the following:

- Muscle retractions above the clavicles, between the ribs, and below the rib cage, especially in children
- Pale or cyanotic (blue) skin
- Cool, damp (clammy) skin
- Tripod position Figure 5-28 (a position in which the patient is leaning forward onto two arms stretched forward)

A patient may also appear to be breathing after the heart has stopped. These occasional, gasping breaths are called **agonal gasps**. Agonal gasps occur when the respiratory center in the brain continues to send signals to the breathing muscles. These gasps are not adequate because they are slow and generally shallow. You should assist ventilations of patients with agonal gasps.

Table 5-6	Normal Respiratory Rate Ranges
Adults	12 to 20 breaths/min
Children	15 to 30 breaths/min
Infants	25 to 50 breaths/min

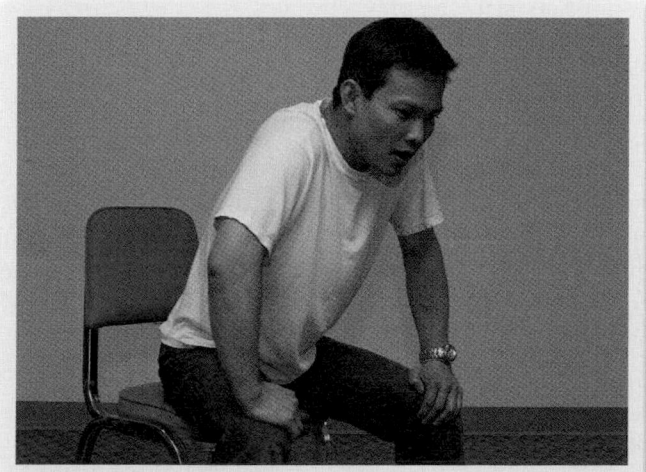

Figure 5-28 A patient in the tripod position will sit leaning forward on outstretched arms with the head and chin thrust slightly forward.

The Circulatory System: Anatomy

The **circulatory system** is a complex arrangement of connected tubes, including the arteries, arterioles, capillaries, venules, and veins Figure 5-29. Another name for this system is the cardiovascular (heart/blood vessels) system. The circulatory system is entirely closed, with capillaries connecting arterioles and venules. There are two circuits in the body: the **systemic circulation** in the body and the **pulmonary circulation** in the lungs. The systemic circulation, the circuit in the body, carries oxygen-rich blood from the left ventricle through the body and back to the right atrium. In the systemic circulation, as blood passes through the tissues and organs, it gives up oxygen and nutrients and absorbs cellular wastes and carbon dioxide. The cellular wastes are eliminated in passages through the liver and kidneys. The pulmonary circulation, the circuit in the lungs, carries oxygen-poor blood from the right ventricle through the lungs and back to the left atrium. In the pulmonary circulation, as blood passes through the lungs, it is refreshed with oxygen and gives up carbon dioxide.

■ The Heart

The **heart** is a hollow muscular organ approximately the size of the patient's clenched fist. It is made of a specialized muscle tissue called cardiac muscle or **myocardium** and actually works as two paired pumps, the left side being more muscular. A wall called the septum divides the heart down the middle into right and left sides. Each side of the heart is divided again into an upper chamber (**atrium**) and a lower chamber (**ventricle**). The left side of the heart, which pumps blood to the body, is a high-pressure pump; the right side supplies blood to the lungs and is a low-pressure pump.

The heart is an involuntary muscle. As such, it is under the control of the autonomic nervous system. However, it has its own electrical system and continues to function even without its central nervous system control. It is distinct from skeletal or smooth muscle in its requirement for a continuous supply of oxygen and nutrients.

The heart must function continuously from birth to death and has developed special adaptations to meet the needs of this continuous function. It can tolerate a serious interruption of its own blood supply for only a very few seconds before the signs of a heart attack develop. Thus, its blood supply is as rich and well distributed as possible.

Circulation

The heart receives the first blood distribution from the aorta. The two main coronary arteries have their openings

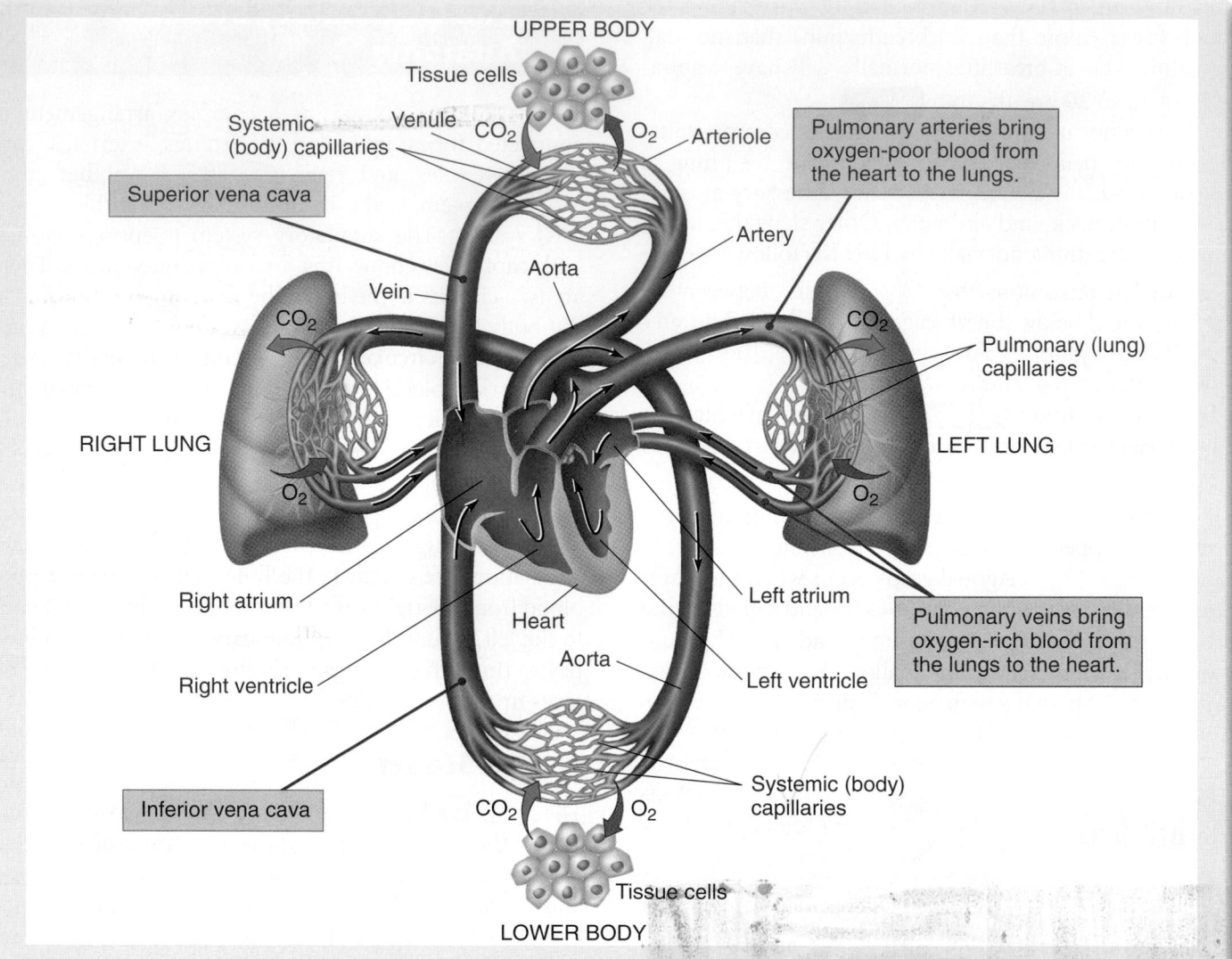

UPPER BODY

Tissue cells

Systemic (body) capillaries · Venule · CO_2 · O_2 · Arteriole

Superior vena cava

Pulmonary arteries bring oxygen-poor blood from the heart to the lungs.

Artery

Aorta

Vein

CO_2 · CO_2

RIGHT LUNG

Pulmonary (lung) capillaries

LEFT LUNG

O_2 · O_2

Right atrium

Heart

Left atrium

Pulmonary veins bring oxygen-rich blood from the lungs to the heart.

Right ventricle

Aorta

Left ventricle

Inferior vena cava

Systemic (body) capillaries

CO_2 · O_2

Tissue cells

LOWER BODY

Figure 5-29 The circulatory system includes the heart, arteries, veins, and interconnecting capillaries. The capillaries are the smallest vessels and connect venules and arterioles. At the center of the system, and providing its driving force, is the heart. Blood circulates through the body under pressure generated by the two sides of the heart.

immediately above the aortic valve at the beginning of the aorta where the pressures are highest **Figure 5-30**.

The right side of the heart receives blood from the veins of the body **Figure 5-31A**. The blood enters from the superior and inferior venae cavae into the right atrium and then passes through the tricuspid valve to fill the right ventricle. After the right ventricle is filled, the tricuspid valve closes to prevent backflow after the right ventricular muscle contracts. Contraction of the right ventricle causes blood to flow through the pulmonic valve into the pulmonary artery and the pulmonary circulation.

The left side receives oxygenated blood from the lungs through the **pulmonary veins** into the left atrium, where it passes through the mitral valve into the left ventricle **Figure 5-31B**. Contraction of this most muscular of the pumping chambers pumps the blood

through the aortic valve into the aorta and then to the arteries of the body.

The exit of each of the four heart chambers is governed by a one-way valve. The valves prevent the backflow of blood and keep it moving through the circulatory system in the proper direction. The **chordae tendineae** are thin bands of fibrous tissue that attach to the valves in the heart and prevent them from inverting. When a valve controlling the filling of a heart chamber is open, the other valve allowing it to empty is shut and vice versa. Normally, blood moves in only one direction through the entire system.

When a ventricle contracts (systole), the valve to the artery opens, and the valve between the ventricle and atrium closes. Blood is forced from the ventricle out into the pulmonary artery or aorta. At the end of contraction, the ventricle relaxes (diastole). Back pressure causes the

This is called the **stroke volume (SV)**. This is the amount of blood moved in one beat. In 1 minute, the entire blood volume of 5 to 6 L is circulated through all the vessels. This is the cardiac output (CO) or the amount of blood moved in 1 minute. Cardiac output is equal to heart rate times stroke volume. Mathematically, cardiac output can be expressed as follows:

$$CO = HR \times SV$$

Electrical Conduction System

A network of specialized tissue that is capable of conducting electrical current runs throughout the heart. The flow of electrical current through this network causes smooth, coordinated contractions of the heart. These contractions produce the pumping action of the heart. Each mechanical contraction of the heart is associated with two electrical processes. The first is depolarization, during which the electrical charges on the surface of the muscle cell change from positive to negative. The second is repolarization, during which the heart returns to its resting state and the positive charge is restored to the surface.

When the heart is working normally, the electrical impulse begins high in the atria at the sinoatrial node, then travels to the atrioventricular node and bundle of His, and moves through the Purkinje fibers to the ventricles. This movement produces a smooth flow of electricity through the heart, which depolarizes the muscle and produces a coordinated pumping contraction. The heart's electrical system becomes disturbed if part of the heart is oxygen deficient, is injured, or dies. As a result, the heart may not continue to beat properly. Blood pressure decreases, and a patient may lose consciousness.

■ Arteries

The arteries carry blood from the heart to all body tissues **Figure 5-32**. They branch into smaller arteries and then

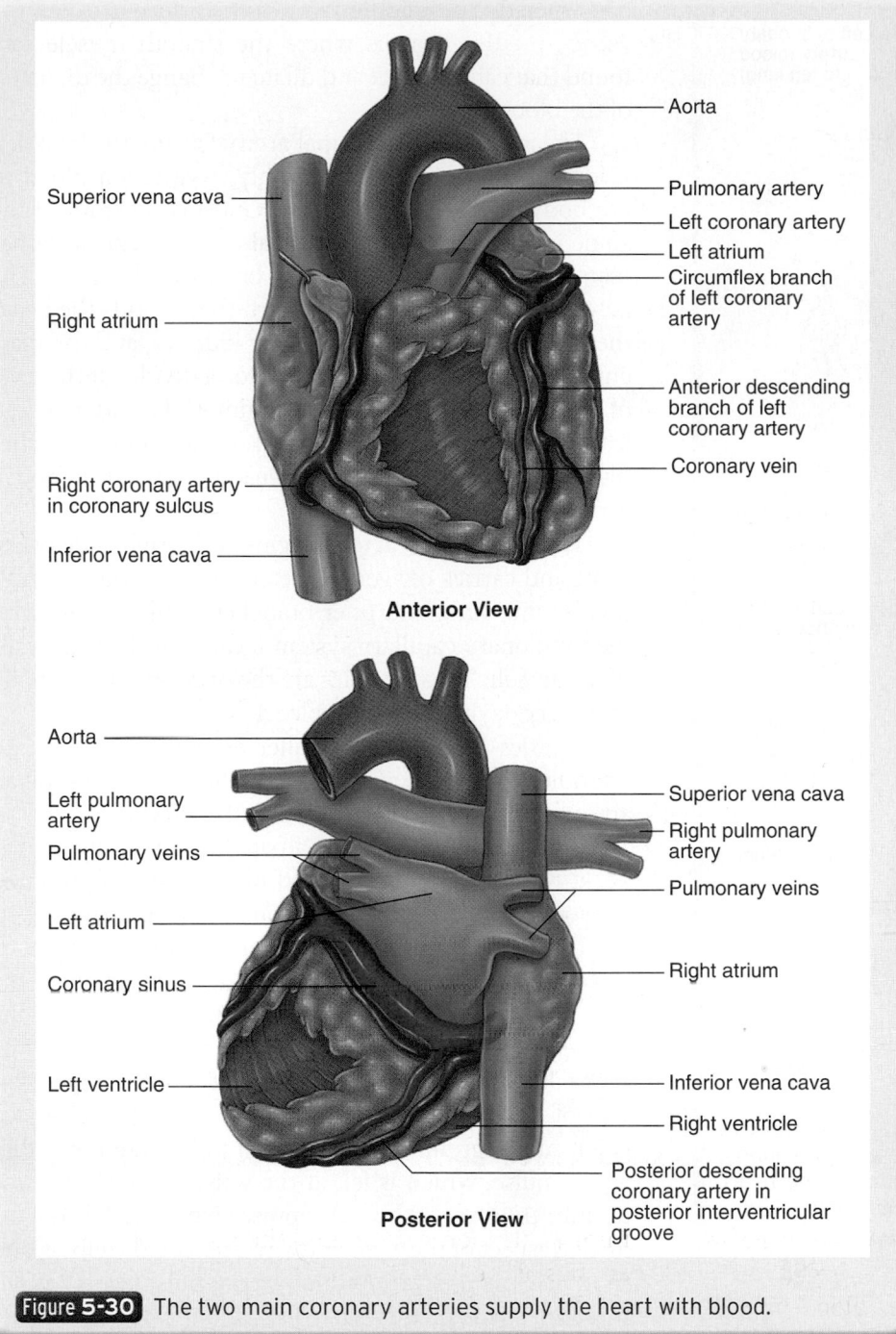

Anterior View

Aorta
Pulmonary artery
Left coronary artery
Left atrium
Circumflex branch of left coronary artery
Anterior descending branch of left coronary artery
Coronary vein
Superior vena cava
Right atrium
Right coronary artery in coronary sulcus
Inferior vena cava

Posterior View

Aorta
Left pulmonary artery
Pulmonary veins
Left atrium
Coronary sinus
Left ventricle
Superior vena cava
Right pulmonary artery
Pulmonary veins
Right atrium
Inferior vena cava
Right ventricle
Posterior descending coronary artery in posterior interventricular groove

Figure 5-30 The two main coronary arteries supply the heart with blood.

valve to the artery to close, and the entry valve to the ventricle opens as the ventricle relaxes. Blood then flows from the atrium into the ventricle. When the ventricle is stimulated to contract, the cycle is repeated.

Normal Heartbeat

In the normal adult, the resting heartbeat may range from 60 to 100 beats/min. A very well-conditioned athlete may have a normal resting **heart rate** of 50 to 60 beats/min. During vigorous physical activity, the heart rate may rise to as fast as 180 beats/min. At each beat, 70 to 80 mL of blood is ejected from the adult heart.

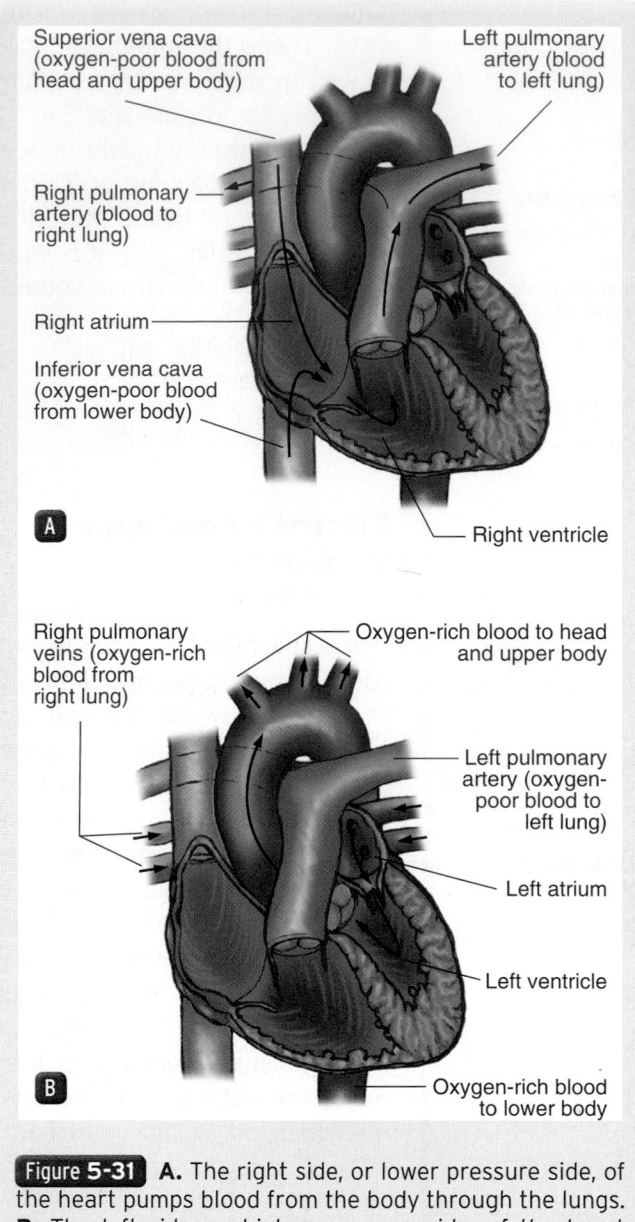

Superior vena cava (oxygen-poor blood from head and upper body)

Left pulmonary artery (blood to left lung)

Right pulmonary artery (blood to right lung)

Right atrium

Inferior vena cava (oxygen-poor blood from lower body)

A

Right ventricle

Right pulmonary veins (oxygen-rich blood from right lung)

Oxygen-rich blood to head and upper body

Left pulmonary artery (oxygen-poor blood to left lung)

Left atrium

Left ventricle

B

Oxygen-rich blood to lower body

Figure 5-31 **A.** The right side, or lower pressure side, of the heart pumps blood from the body through the lungs. **B.** The left side, or higher pressure side, of the heart pumps oxygen-rich blood to the rest of the body.

constructed. The middle layer of the artery is called the **tunica media**. Here is where the smooth muscles are found that can contract and dilate to change the diameter of the blood vessel.

The **aorta** is the principal artery leaving the back left side of the heart; it carries freshly oxygenated blood to the body. This blood vessel is found just in front of the spine in the chest and abdominal cavities. The aorta has many branches that supply the body's vital organs. The coronary arteries supply the heart, the carotids the head, the hepatic the liver, the renal the kidneys, and the mesenteric the digestive system. The aorta divides at the level of the umbilicus into the two common iliac arteries that lead to the lower extremities. All of the aorta's branches ultimately become arterioles leading into the body's capillary network.

The **pulmonary artery** begins at the right side of the heart and carries oxygen-depleted blood to the lungs. It divides into finer and finer branches until it meets with the pulmonary capillary system located in the thin walls of the alveoli. These arteries are the only ones in the body that carry oxygen-depleted blood.

Arteries branch into smaller arteries and then into arterioles. **Arterioles** are the smallest branches of an artery leading to the vast network of capillaries.

The **pulse**, which is palpated most easily at the neck, wrist, or groin, is created by the forceful pumping of blood out the left ventricle and into the major arteries. It is present throughout the entire arterial system. It can be felt most easily where the larger arteries near the skin can be pushed against a solid structure, like a bone or large muscle **Figure 5-33**. The central pulses are the **carotid artery** pulse, which can be felt at the upper portion of the neck, and the **femoral artery** pulse, which is felt in the groin. The peripheral pulses are the **radial artery** pulse, which is felt at the wrist at the base of the thumb; the **brachial artery** pulse, which is felt on the medial aspect of the arm, midway between the elbow and shoulder; the **posterior tibial artery** pulse, which is felt posterior to the medial malleolus; and the **dorsalis pedis artery** pulse, which is felt on the top of the foot.

■ Capillaries

In the body, there are billions of cells and billions of capillaries. **Capillary vessels** are fine end divisions of the arterial system that allow contact between the blood and the cells of the tissues. Oxygen and other nutrients pass from blood cells and plasma in the capillaries to the individual tissue cells through the very thin wall of the capillary. Carbon dioxide and other metabolic waste products pass in a reverse direction from the tissue cells to the blood to be carried away. Blood in arteries

into arterioles. The arterioles, in turn, branch into the vast network of capillaries. The walls of an artery are made of fine, circular muscle tissue. Some arteries are made of fine circular muscle and elastic tissue.

Arteries contract to accommodate loss of blood volume and increase blood pressure. Blood is supplied to tissues as they need it. For example, the digestive system is supplied with more blood after you eat a meal. The leg muscles are more heavily supplied when jogging. Some tissues need a constant blood supply, especially the heart, kidneys, and brain. Other tissues, such as the muscles in the extremities, the skin, and intestines, can function with less blood when at rest. The ability to respond to the needs of the body owes itself to the way arteries are

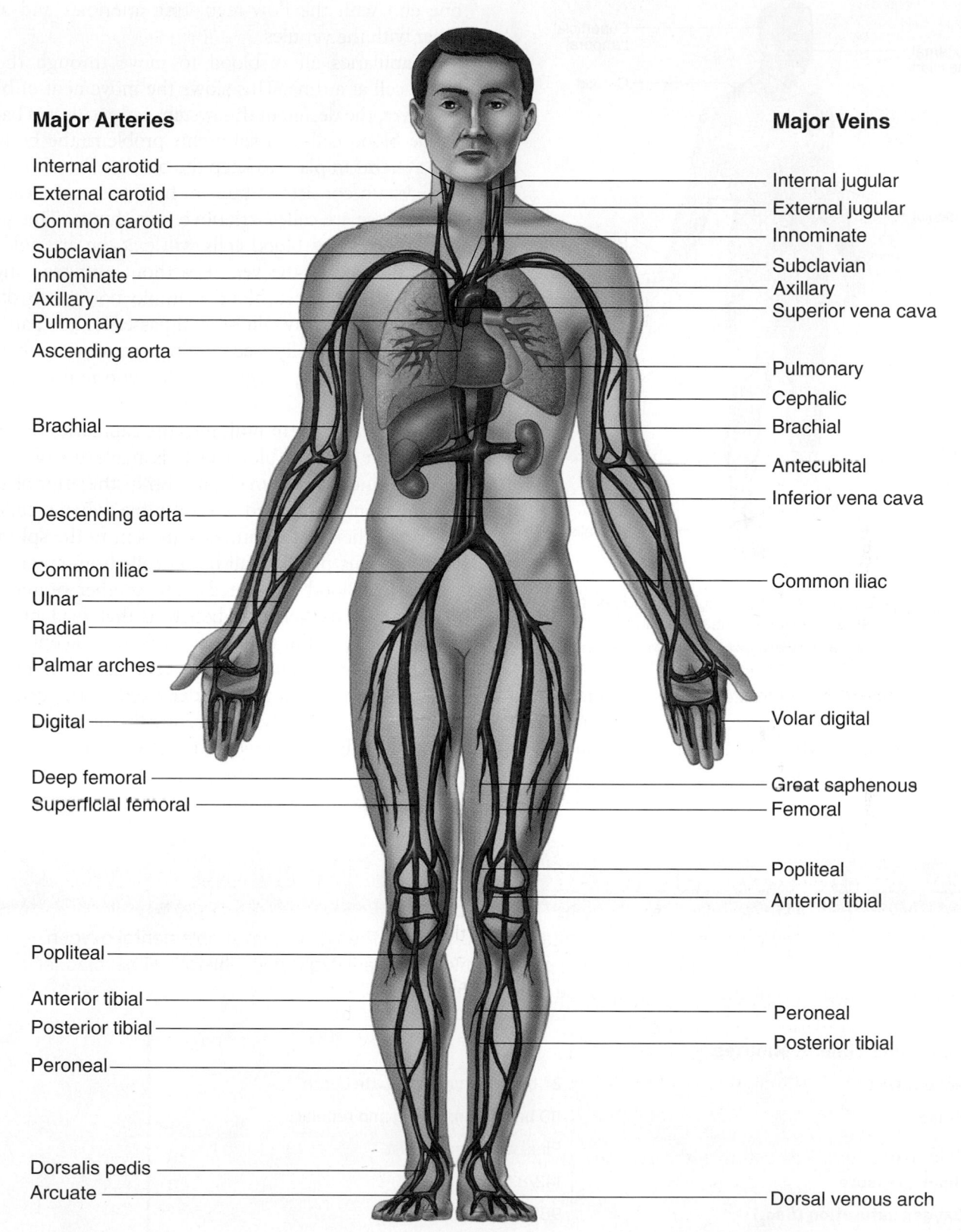

Major Arteries

Internal carotid
External carotid
Common carotid
Subclavian
Innominate
Axillary
Pulmonary
Ascending aorta

Brachial

Descending aorta

Common iliac
Ulnar
Radial

Palmar arches

Digital

Deep femoral
Superficial femoral

Popliteal

Anterior tibial
Posterior tibial

Peroneal

Dorsalis pedis
Arcuate

Major Veins

Internal jugular
External jugular
Innominate
Subclavian
Axillary
Superior vena cava

Pulmonary
Cephalic
Brachial

Antecubital

Inferior vena cava

Common iliac

Volar digital

Great saphenous
Femoral

Popliteal
Anterior tibial

Peroneal
Posterior tibial

Dorsal venous arch

Figure 5-32 The principal arteries supply blood to a vast network of smaller arteries and arterioles. Venules deliver oxygen-poor blood to the veins that return blood to the heart.

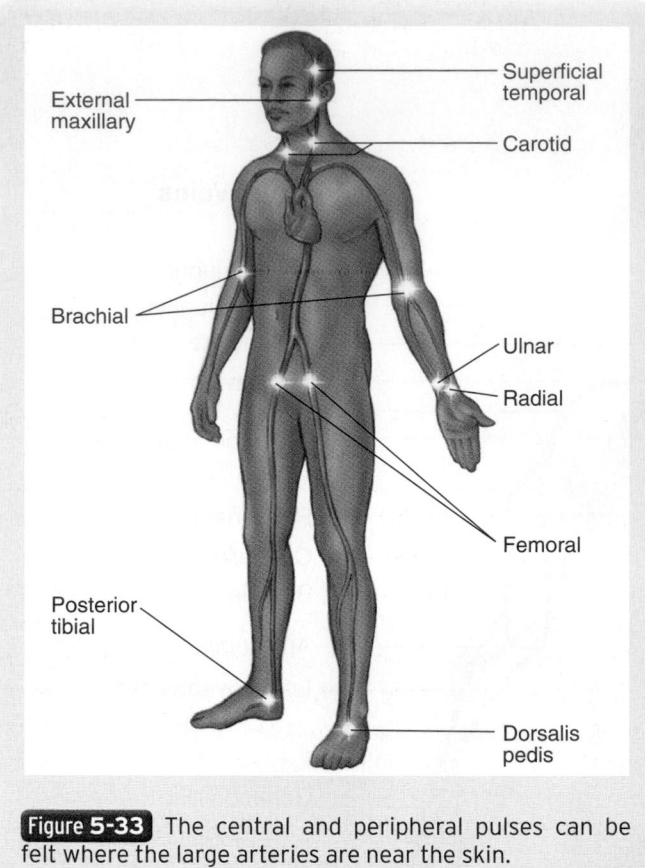

External maxillary

Superficial temporal

Carotid

Brachial

Ulnar

Radial

Femoral

Posterior tibial

Dorsalis pedis

Figure 5-33 The central and peripheral pulses can be felt where the large arteries are near the skin.

is characteristically bright red, because its hemoglobin is rich in oxygen. Blood in the veins is dark bluish red, because it has passed through a capillary bed and given up its oxygen to the cells. Capillaries connect directly at one end with the flow-regulating arterioles and at the other with the venules.

Capillaries allow blood to move through them a single cell at a time. This slows the movement of blood. However, the design of this system can result in a backup of the blood cells. To solve this problem, the body has two systems in place to keep the blood cells moving. The first system consists of built-in bypasses at the capillary level. These are called arteriovenous shunts. This means that some of the blood cells will exit the arteriole only to be shunted to the venule without ever reaching the true capillary. These blood cells do not get to offload their oxygen. It may take several passes for any particular blood cell to actually reach the true capillaries where the work of delivering oxygen supplies and removing waste is done.

The other system built into the capillaries that helps control the flow of blood cells is made up of **sphincters**. Sphincters are small muscles in the arterioles that can be opened or closed. For example, when a patient is bleeding, chemical commands are sent to the sphincters to close. This prevents all blood cells from entering the capillaries. Blood is shunted off to another venule while still carrying oxygen. The benefit is that available blood is shunted into the major blood vessels, keeping them filled. The detriment is that waste is not being removed and nutrients are not being delivered to the cells. This can go on temporarily, but eventually the tissues will be damaged if the cellular waste is not removed.

You are the Provider: PART 3

Your partner obtains and reports the patient's vital signs to you and then gives him supplemental oxygen. Following your assessment and with information provided by the patient regarding a history of gallbladder problems, you suspect that the gallbladder is the origin of his pain.

Recording Time: 2 Minutes	
Respirations	24 breaths/min; adequate depth
Pulse	110 beats/min; strong and regular
Skin	Pink, warm, and moist
Blood pressure	142/82 mm Hg
Oxygen saturation (Sao$_2$)	98% (on oxygen)

4. What is the function of the gallbladder?

5. What additional symptoms would you expect the patient to experience based on the function of the gallbladder?

Veins

Once oxygen-depleted blood passes through the network of capillaries, it moves to the venules, which are the smallest branches of the veins. The blood returns to the heart via a network of larger and larger veins. Veins have much thinner walls than arteries and are generally larger in diameter. The veins become larger and larger and ultimately form two major vessels, called the superior and inferior venae cavae. These two veins lie just to the right of the spine and collect blood just before it enters the heart. Because pressure generated by the heart dissipates as blood passes through the capillaries, venous blood flow is assisted by gravity, skeletal muscle contraction, and intrathoracic pressure changes from breathing. Unidirectional flow in the veins is governed by valves within the veins.

The **superior vena cava** carries blood returning from the head, neck, shoulders, and upper extremities. Blood from the abdomen, pelvis, and lower extremities passes through the **inferior vena cava**. The superior and inferior venae cavae join at the right atrium of the heart. The right ventricle receives blood from the right atrium and pumps it through the pulmonary arteries into the lungs. The venae cavae, aorta, and pulmonary arteries and veins are collectively known as the great vessels.

The body's ability to contract blood vessels is critical to survival. When you eat, you need more blood in your stomach and intestines. When you are running, you need more blood in your muscles. This ability to change blood flow is fundamental. What the body is actually doing is changing the size of the total blood volume container. A smaller container that has the same amount of liquid as the original container means a higher liquid pressure.

The state of the blood vessels, how dilated or constricted they are, is referred to as the **systemic vascular resistance (SVR)**. SVR is the resistance to blood flow within all of the blood vessels except the pulmonary vessels. The pathophysiology section of this chapter will discuss how various types of shock impact container size. In some types of shock, blood vessels dilate, the container becomes too large, and the patient's blood pressure falls dramatically **Table 5-7**.

Words of Wisdom

The terms *shock* and *hypoperfusion* are usually synonymous, at least when they are applied to multiple body systems. Localized hypoperfusion, such as from arterial occlusion (blockage), is *not* shock.

The Spleen

The spleen is a solid organ located under the rib cage in the left upper quadrant of the abdomen. The spleen is actually part of the lymphatic system, but because it

Table 5-7 Effects of Blood Vessel Diameter on Blood

State	Effects
Constricted blood vessel	Decreased size of container Increased pressure within container
Normal diameter	Balance of size and pressure
Dilated blood vessel	Increased size of container Decreased pressure within container

processes blood, it will be discussed here. At any one time there is about 450 mL of blood in the spleen. In the event of sudden blood loss, the body is able to squeeze the spleen and move this blood reservoir into the general circulation. Filtering is the basic duty of the spleen. Virtually all of the blood in the body passes though the spleen where it is filtered. Worn out blood cells, foreign substances, and bacteria are removed from the blood, and hemoglobin is recycled. Another function of this organ is to assist in the immune response.

The spleen is particularly susceptible to injury from blunt trauma because it is made of tissue that is delicate and because it is located directly under the flexible lower ribs, with very little soft tissue to cushion it. Therefore, it is one of the most frequently injured abdominal organs in patients with blunt trauma. Because the spleen is highly vascular, injury can lead to severe internal bleeding.

Blood Composition

Blood is a complex, thick, red fluid composed of plasma, red blood cells called erythrocytes, white blood cells called leukocytes, and platelets. The work of the circulatory system is to accomplish movement of blood, or perfusion.

Plasma is a sticky, yellow fluid that carries the blood cells and nutrients. This is the liquid portion of the blood. The primary components are water and proteins. All of the other components together make up 1% of the plasma:

- **Water:** Constitutes 92% of plasma
- **Proteins:** Constitute 7% of the plasma. The majority of this protein is albumin, which functions mainly to regulate oncotic pressure, and thereby controls the movement of water into and out of the circulation. Also includes clotting factors, enzymes, and some hormones
- **Oxygen:** Very little oxygen is dissolved in the plasma, almost all oxygen is bound to hemoglobin
- **Carbon dioxide:** Transported as bicarbonate in the plasma
- **Nitrogen:** The air that we breathe is mostly nitrogen; therefore, this gas is dissolved within the plasma

- **Nutrients:** Fuel for the cells
- **Cellular wastes:** Lactic acid, carbon dioxide, etc
- **Others:** Hormones, other cellular products

Red blood cells (erythrocytes) contain hemoglobin, which gives blood its red color. Hemoglobin is responsible for carrying oxygen. The majority of carbon dioxide is carried by conversion to carbonic acid, which dissolves in the plasma. A small amount of carbon dioxide is carried by hemoglobin. **White blood cells** (leukocytes) play a role in the body's immune defense mechanisms against infection. **Platelets** are tiny, disk-shaped elements that are much smaller than the cells. They are essential in the initial formation of a blood clot, the mechanism that stops bleeding.

The Circulatory System: Physiology

Blood pressure (BP) is the pressure the blood exerts against the walls of the arteries as it passes through them. When the cardiac muscle of the left ventricle contracts, it pumps blood from the ventricle into the aorta. This muscular contraction phase is called **systole**. When the muscle of the ventricle relaxes, the ventricle fills with blood. This phase is called **diastole**. The pulsed, forceful ejection of blood from the left ventricle of the heart into the aorta is transmitted through the arteries as a pulsatile pressure wave. This pressure wave keeps the blood moving through the body. The high and low points of the wave can be measured with a **sphygmomanometer** (blood pressure cuff) and are expressed numerically in millimeters of mercury (mm Hg). The high point is called the systolic blood pressure (measured as the heart muscle is contracting). The low point is called the diastolic blood pressure (measured when the heart muscle is in its relaxation phase). There are several pressures within the circulatory system that are essential to understanding how this system works. **Table 5-8** shows the various pressures from within the circulatory system and their meaning.

The average adult has approximately 6 L of blood in the vascular system. Children have less, 2 to 3 L, depending on their age and size. Infants have only about 300 mL. The loss of an amount of blood that may be insignificant for an adult could be fatal for an infant.

■ Normal Circulation in Adults

In all healthy people, the circulatory system is automatically adjusted and readjusted constantly so 100% of the capacity of the arteries, veins, and capillaries holds 100% of the blood at that moment. All of the vessels are never

Table 5-8 Cardiovascular Pressures

Name	Description	Clinical Significance
Systolic blood pressure	Pressure within the arteries when the heart is contracting; left ventricular force	Indicates heart pumping effectiveness Indicates blood available to the heart
Diastolic blood pressure	Pressure within the arteries when the heart is at rest	Indicates adequacy of the amount of blood vessel contraction (arterial) Indicates amount of blood within blood vessels
Pulse pressure	Difference between systolic blood pressure and diastolic blood pressure	Relationship between systolic and diastolic pressures; provides information about the body's response to stress
Preload	Amount of blood returning to the heart	Too little preload, and blood pressure falls Too high preload, and the heart cannot move blood effectively
Afterload	Pressure to be overcome when left ventricle contracts (pressure within the aorta)	Diastolic pressure is the same as afterload
Cardiac output (CO)	Amount of blood moved in 1 minute	$CO = SV \times HR$
Stroke volume (SV)	Amount of blood moved in one beat of the heart (left ventricle)	Weak left ventricle moves less blood per beat than a strong left ventricle
Systemic vascular resistance (SVR)	Resistance to blood flow within all of the blood vessels (except the pulmonary vessels)	The higher the SVR, the smaller the container, therefore, the higher the pressure of blood within the vessel

fully dilated or constricted. The size of arteries and veins is controlled by the nervous system, according to the amount of blood that is available and many other factors, to keep blood pressure normal at all times. Under the condition of normal pressure, with a system that can hold just 100% of the blood available, all parts of the system will have adequate blood supply all the time.

Perfusion is the circulation of blood in an organ or tissue in adequate amounts to meet the cells' current needs. Blood enters an organ or tissue through the arteries and leaves it through the veins **Figure 5-34**. Loss of normal blood pressure is an indication that blood is no longer circulating efficiently to every organ in the body. (However, a "good blood pressure" does not indicate that it is reaching all parts of the body.) There are many reasons for loss of blood pressure. The result in each case is the same: Organs, tissues, and cells are no longer adequately perfused or supplied with oxygen and food, and wastes can accumulate. Under these conditions, cells, tissues, and whole organs may die. The state of inadequate circulation, when it involves the entire body, is called **shock**, or hypoperfusion.

■ Inadequate Circulation in Adults

When a patient loses a small amount of blood, the arteries, veins, and heart automatically adjust to the smaller new

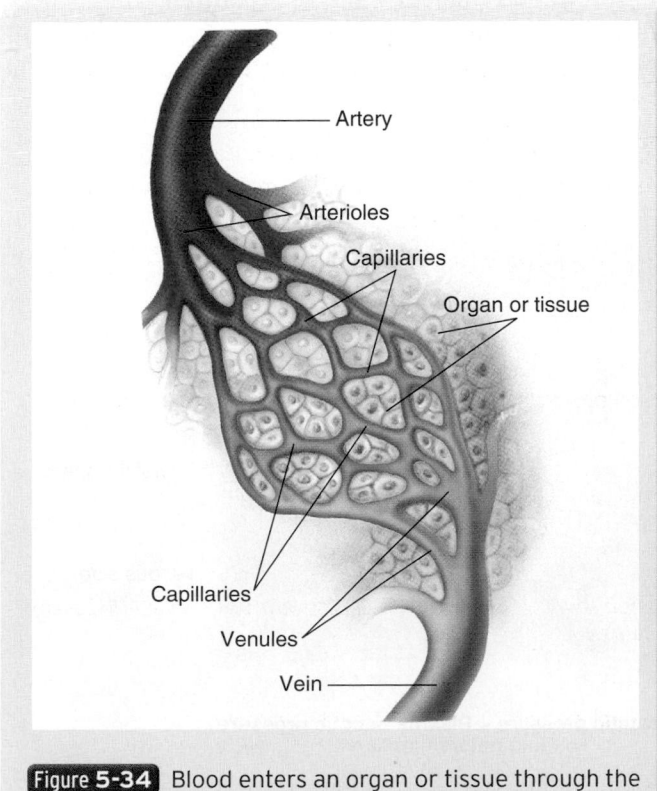

Figure 5-34 Blood enters an organ or tissue through the arteries and leaves through the veins. This process, called perfusion, provides adequate blood flow to the tissue to meet the cells' needs.

volume. The adjustment occurs in an effort to maintain adequate pressure throughout the circulatory system and maintain circulation for every organ. The adjustment occurs very rapidly after the loss, usually within minutes. Specifically, the vessels constrict to provide a smaller bed for the reduced volume of blood to fill. And the heart pumps more rapidly to circulate the remaining blood more efficiently. As the blood pressure falls, the pulse increases in an attempt to keep the cardiac output constant at 5 to 6 L per minute. If the loss of blood is too great, the adjustment fails, and the patient goes into shock.

■ The Function of Blood

Blood under pressure will gush or spurt intermittently from an artery and is bright red. When blood comes from a vein, it flows in a steady stream and is dark bluish red. From capillaries, blood will ooze at many tiny individual points. Clotting normally takes from 6 to 10 minutes.

Where is all of this blood distributed? This concept involves blood reservoirs. Most blood is unevenly distributed throughout the body. Approximately 30% of blood is found within the heart, arteries, and capillaries. Seventy percent of blood is found within the veins and venules. This may seem confusing, but if you remember that the heart and arteries are high-pressure systems and veins are low-pressure systems, it becomes clearer. As blood pressure falls, blood flow slows down and there is more blood in the veins. The blood flows away from the left ventricle and moves back to the right atria.

Consider the movement of blood and its ultimate function of perfusion. You know that capillaries are the smallest portions of the circulatory system where materials are able to exit and enter the bloodstream. Nutrients move from the capillaries into the interstitial space and into the cells. Wastes move from the cells through the interstitial space and into the capillaries. The **interstitial space** is the space between the cells.

Here is a simplified version of what is happening inside the capillary. The two main forces at work inside the capillary are **hydrostatic pressure** and **oncotic pressure**. Hydrostatic pressure is pressure exerted by a liquid and occurs when blood is moved through the artery at relatively high pressures. When that blood meets the capillary walls, the pressure of the fluid pushes against the walls to force fluid out of the capillary. The opposing force is oncotic pressure. Oncotic pressure is a form of osmotic pressure exerted by proteins in the blood plasma that usually tends to pull water into the circulatory system. These proteins tend to make the blood thicker. This thickness means that relative to the interstitial space, there is more water outside the capillary than inside. Diffusion occurs, and water seeks to move into the capillary.

Here is the entire process. Blood flows into the arterial side of the capillary. Water is forced out because the pressure is high. At the same time, water is trying to enter the capillary. Pressure on the arterial side is higher, so the hydrostatic pressure is also higher and water, carrying nutrients, leaves the capillary and enters the interstitial space. The hydrostatic pressure is greatly diminished, however, by the time the fluid reaches the venous side because the effort of pushing the fluid out of the capillary decreased its force. This decrease in pressure is beneficial because oncotic pressure is still pushing fluid into the capillary and the pressure is higher. Water, with all of the wastes from the cells, enters the venous side of the capillary. These wastes are then carried away **Figure 5-35**.

Another function of blood is the ability to clot. Coagulation, or clotting, occurs as the result of a very complex chemical process that creates small fibers near the injured blood vessel, trapping red blood cells. This chemical process involves platelets and clotting factors that are in the bloodstream. **Table 5-9** outlines the major functions of the blood.

■ Nervous System Control of the Cardiovascular System

The sympathetic nervous system is responsible for the fight-or-flight response. This powerful response has direct effects on the cardiovascular system. The sympathetic nervous system sends commands to the adrenal glands. There, two hormones, **epinephrine** and **norepinephrine**, are secreted to stimulate the heart and

Table 5-9	Functions of the Blood and the Components of Blood in Use
Function	**Component of the Blood in Use**
Fighting infection	White blood cells
Transporting oxygen	Red blood cells (hemoglobin)
Transporting carbon dioxide	Plasma
Controlling (buffering) pH	Chemicals within the plasma
Transporting wastes and nutrients	Plasma (water)
Clotting (coagulation)	Platelets and clotting factors in the plasma

blood vessels. The popular terms for these hormones are adrenaline and noradrenaline, respectively. Epinephrine and norepinephrine affect receptors within the heart and blood vessels. Their release allows better coping with dangerous or unexpected situations. Two types of receptors within the heart and blood vessels will be discussed here so you can understand how the nervous system controls the circulatory system.

The heart and blood vessels have **alpha-adrenergic receptors** and **beta-adrenergic receptors** within them. **Adrenergic** simply means related to the adrenal gland, where epinephrine and norepinephrine are made. The

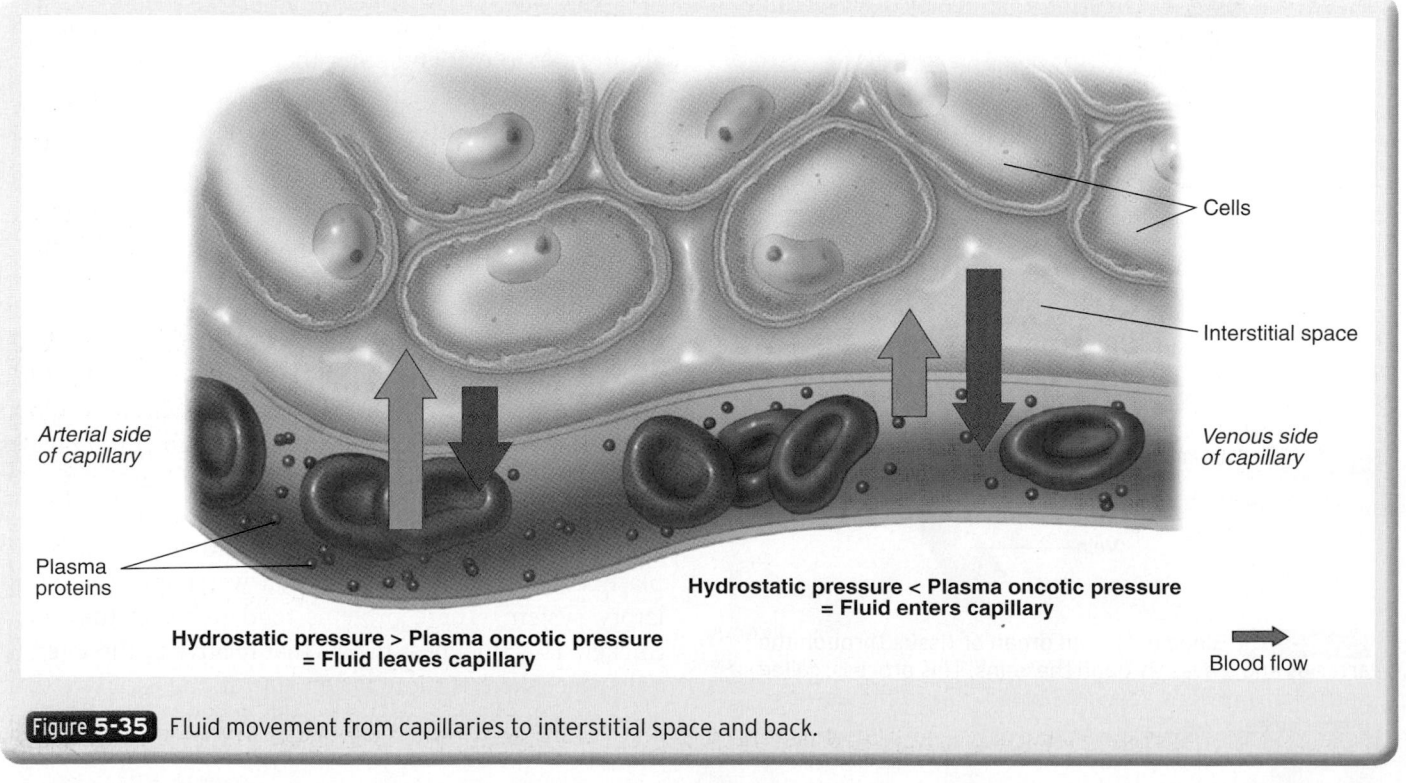

Figure 5-35   Fluid movement from capillaries to interstitial space and back.

alpha-adrenergic receptors are found in the blood vessels. When stimulated, the blood vessels constrict, thereby increasing blood pressure. The beta-adrenergic receptors are found in the heart and lungs. When beta-1 receptors are stimulated, they cause the heart to increase its rate and also squeeze harder with each contraction. This increases cardiac output. When beta-2 receptors are stimulated, the bronchi in the lungs dilate. This allows more air to be inhaled and exhaled; therefore, more oxygen is available to the cells of the body. Together, the alpha- and beta-adrenergic receptors prepare the body for fight or flight.

The parasympathetic nervous system also has effects on the cardiovascular system. When stimulated, this system causes the heart to slow and beat more weakly. Although the sympathetic and parasympathetic divisions function in opposition to each other, this opposition is considered complementary rather than antagonistic. The net effect is a dynamic body able to respond quickly Table 5-10 .

The brain needs to be able to know how the body is performing so that adjustments in the pressure exerted by circulating blood can be made. How is the brain alerted to what is happening at the feet, the liver, or the heart? Signals are sent through the nervous system from special pressure sensors (baroreceptors) spread throughout the body. These baroreceptors allow the brain to receive information about blood pressure. Remember, the main function of the cardiovascular system is to perfuse blood throughout the body. The main locations for these pressure receptors are found in the arch of the aorta and the carotid arteries. By measuring the pressure in these two locations, the body is ensuring that the most vulnerable and most important cells get oxygen.

With this information, the brain is able to act to maintain perfusion. To see the system in action, you may want to try this test. Kneel down to the floor. Now, as fast as you can, jump up. Did you pass out? Most likely you did not pass out because these systems with their pressure sensors are designed to maintain perfusion.

When you jumped up quickly, gravity was relocating the blood out of your brain. The baroreceptors detected the decrease in blood pressure at the carotid arteries. A signal was sent to the brain. The brain understood the implication of low blood pressure and immediately turned on the sympathetic nervous system. The blood vessels contracted, and the heart rate increased. The heart pumped harder. Your blood pressure returned to normal and may even have gone slightly high. Again, the baroreceptors detected this change. Signals were sent to the brain. The sympathetic nervous system was turned off. The parasympathetic nervous system was turned on. The heart rate slowed, and the force of the heart's contractions weakened. All of this happened in a fraction of a second. That is how responsive the cardiovascular system can be.

The Nervous System: Anatomy and Physiology

The **nervous system** is perhaps the most complex organ system within the human body. It is composed of two major structures, the brain and the spinal cord, and thousands of nerves that allow every part of the body to communicate. This system is responsible for fundamental functions such as controlling breathing, heart rate, and blood pressure. However, what makes the nervous system so special is that it allows the performance of higher level activity. Reading a good book, enjoying music, having a discussion with a friend, and even watching television require the brain to engage memory, understanding, and thought. Here is where the true complexity of the nervous system can be seen.

The nervous system is divided into two main portions: the **central nervous system (CNS)** and the peripheral nervous system. The **somatic nervous system** is the part of the peripheral nervous system that regulates activities over which there is voluntary control,

Table 5-10 Nervous System Effects on the Cardiovascular System			
Portion of Nervous System	**Receptor**	**Stimulation Area**	**Effect When Stimulated**
Sympathetic nervous system	Alpha-1	Blood vessels	Constricts blood vessels; skin becomes pale, cool, and clammy
	Beta-1	Heart	Increased heart rate
			Increased force of heart contraction
	Beta-2	Lungs	Bronchodilation
Parasympathetic nervous system	Heart	Heart	Decreased heart rate
			Decreased force of contraction

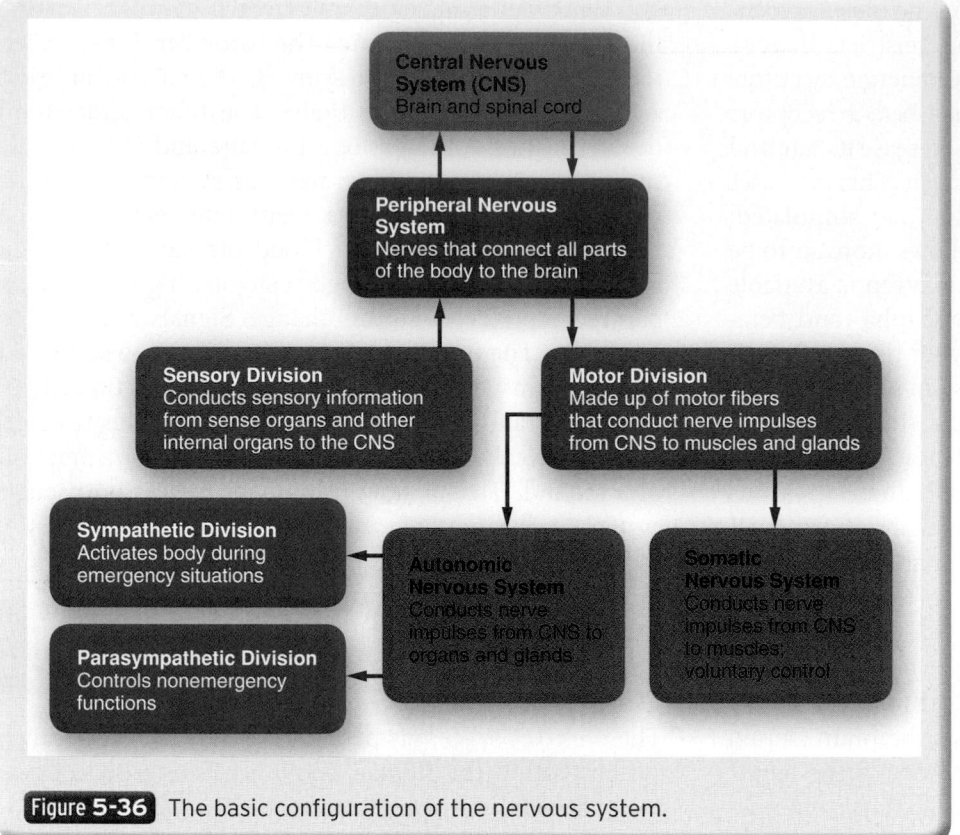

Figure 5-36 The basic configuration of the nervous system.

environment. In addition, the brain enables you to experience all the fine shadings of thought and feeling that make each of us an individual. The brain is subdivided into several areas, all of which have specific functions. Three major subdivisions of the brain are the cerebrum, the cerebellum, and the brain stem **Figure 5-37**.

The **cerebrum**, which is the largest part of the brain and is sometimes called the "gray matter," makes up about three fourths of the volume of the brain and is composed of four lobes: frontal, parietal, temporal, and occipital. The cerebrum on one side of the brain controls activities on the opposite side of the body. Each lobe of the cerebrum is responsible for a specific function. For example, one group of brain cells in the frontal lobe is responsible for the activity of all the voluntary muscles of the body. Brain cells in this area generate impulses that are sent along nerve fibers that extend from each cell into the spinal cord. An area in the parietal lobe has cells that receive sensory impulses from the peripheral nerves of the body. Other parts of the cerebrum are responsible for other body functions. For example, the occipital region, in the back of the cerebrum, receives visual impulses from the eyes; other areas control hearing, balance, and speech. Still other parts of the cerebrum are responsible for emotions and other characteristics of an individual's personality.

The **cerebellum**, which is located underneath the great mass of cerebral tissue, is sometimes called the "little brain." The major function of this area is to coordinate the various activities of the body, particularly body movements. Without the cerebellum, very specialized muscular activities such as writing would be impossible.

The **brain stem** is so called because the brain appears to be sitting on this portion of the CNS as a plant sits on its stem. The brain stem is the most primitive part of the CNS. It lies deep within the cranium and is the best-protected part of the CNS. The brain stem is the controlling center for virtually all body functions that are absolutely necessary for life. Cells in this part of the brain control cardiac, respiratory, and other basic body functions. One of the interesting operations of the brain stem is the regulation of consciousness. As you are reading this book, your **reticular activating system**

such as walking, talking, and writing. The **autonomic nervous system** controls the many body functions that occur without voluntary control. These activities include body functions such as digestion, dilation and constriction of blood vessels, sweating, and all other involuntary actions that are necessary for basic body functions. Thus, the nervous system as a whole can be divided anatomically into the central and peripheral nervous systems and functionally into somatic (voluntary) and autonomic (involuntary) components **Figure 5-36**.

Special Populations

Never assume that new or worsening confusion in a geriatric patient is purely the result of Alzheimer disease; you should always first consider potentially correctable causes such as new medications, infections, or myocardial infarction. An "apparent" emotional, psychological, or behavioral problem may have an organic cause, especially in the geriatric population.

■ The Central Nervous System

Brain

The **brain** is the controlling organ of the body. It is the center of consciousness. It is responsible for all of your voluntary body activities, your perception of your surroundings, and the control of your reactions to the

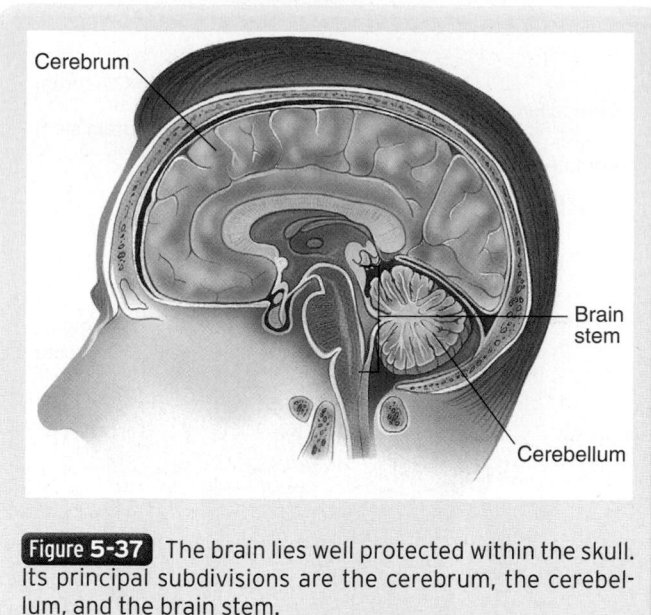

Cerebrum

Brain stem

Cerebellum

Figure 5-37 The brain lies well protected within the skull. Its principal subdivisions are the cerebrum, the cerebellum, and the brain stem.

in the midbrain is keeping you awake. The brain stem comprises three areas: the **midbrain**, the **pons**, and the **medulla oblongata**.

The brain has many other anatomic areas, all of which have specific and important functions. The brain receives a vast amount of information from the environment, sorts it all out, and directs the body to respond appropriately.

Many of the responses involve voluntary muscle action; others are automatic and involuntary. **Table 5-11** summarizes the major portions of the nervous system and their functions.

Cerebrospinal Fluid <u>Cerebrospinal fluid (CSF)</u> is a colorless fluid in and around the brain and spinal cord. CSF cushions these structures and filters out impurities and toxins; it also absorbs shocks. When forces are applied to the head, CSF allows the brain to shift in the skull without tearing. If a trauma patient has CSF leaking from the ears or nose, this is considered a significant finding, indicating a skull fracture.

Circulation in the Head The brain requires a constant flow of oxygenated blood to support brain function. Blood is supplied to the head through the carotid arteries, which can be palpated on either side of the neck. Deoxygenated blood drains from the head via the internal and external jugular veins.

Spinal Cord

The **spinal cord** is an extension of the brain stem **Figure 5-38**. Like the brain, the spinal cord contains nerve cell bodies, but the major portion of the spinal cord is made up of nerve fibers that extend from the cells of the brain. These nerve fibers transmit information to

Table 5-11 Structures of the Nervous System and General Functions

System	Major Structure	Subdivision	General Function
Central nervous system	Brain	Occipital lobe	Vision and storage of visual memories
		Parietal lobe	Sense of touch and texture; storage of those memories
		Temporal lobe	Hearing, smell, and language; storage of sound and odor memories
		Frontal lobe	Voluntary muscle control and storage of those memories
		Prefrontal area	Judgment and predicting consequences of actions, abstract intellectual functions
		Limbic system	Basic emotions, basic reflexes (chewing, swallowing, etc)
		Diencephalon (thalamus)	Relay center; filters important signals from routine signals
		Diencephalon (hypothalamus)	Emotions, temperature control, interface with endocrine system (hormone control)
	Brain stem	Midbrain	Level of consciousness, reticular activating system, muscle tone, and posture
		Pons	Respiratory patterning and depth
		Medulla oblongata	Heart rate, blood pressure, respiratory rate
	Spinal cord		Reflexes, relays information to and from body
Peripheral nervous system	Cranial nerves		Brain to body part; special peripheral nerves that connect directly to body parts
	Peripheral nerves		Brain to spinal cord to body part; receive stimulus from body, send commands to body

and from the brain. All fibers join together just below the brain stem to form the spinal cord. The spinal cord exits through a large opening at the base of the skull called the foramen magnum. It is encased within the spinal canal down to the level of the second lumbar vertebra. The spinal canal is created by an opening through the vertebrae, stacked one on another. Each vertebra surrounds the cord, and together the vertebrae form the bony spinal canal.

The principal function of the spinal cord is to transmit messages between the brain and the body. These messages are passed along the nerve fibers as electrical impulses, just as messages are passed along a telephone cable. The nerve fibers are arranged in specific bundles within the spinal cord to carry the messages from one specific area of the body to the brain and back.

Spinal Reflexes Within the spinal cord are cells with short fibers that connect the sensory nerves with the motor nerves. These are direct connections bypassing the brain. These cells are where spinal reflexes occur. In addition, these cells allow sensory and motor impulses to transmit from one nerve to another within the CNS.

An irritating stimulus to the sensory nerve, such as heat, will be transmitted from the sensory nerve along the connecting nerve directly to the motor nerve. This will stimulate the motor nerve. The muscle responds

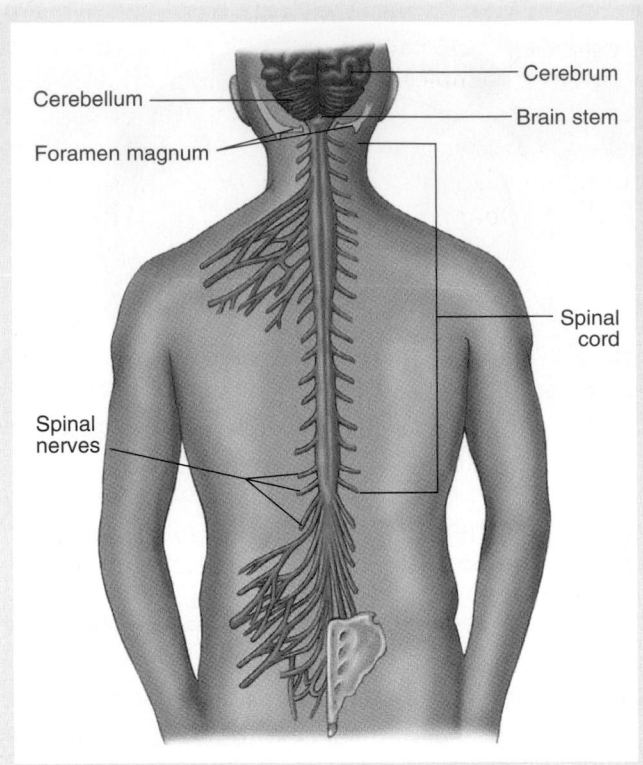

Figure 5-38 The spinal cord is a continuation of the brain stem. It exits the skull at the foramen magnum and extends down to the level of the second lumbar vertebra.

You are the Provider: PART 4

You complete your assessment of the patient, obtain his medical history, and prepare for transport. He remains conscious and alert, but is still experiencing severe pain. Shortly after loading him into the ambulance and departing the scene, you perform a reassessment.

Recording Time: 12 Minutes	
Level of consciousness	Conscious and alert
Respirations	24 breaths/min; adequate depth
Pulse	112 beats/min; strong and regular
Skin	Pink, warm, and moist
Blood pressure	138/88 mm Hg
Sao$_2$	97% (on oxygen)

You allow the patient to assume a position of comfort, which seems to help with his pain. With an estimated time of arrival at the hospital of 10 minutes, you call in your radio report.

6. How does knowledge of anatomy, physiology, and medical terminology facilitate communication with other health care professionals?

promptly, withdrawing the limb from the irritating stimulus even before this information can be transmitted to the brain. Technically, you do not "feel" the heat of the fire before you move your hand away. This process is in place to limit damage to the body. When a physician taps your knee with a rubber hammer, he or she is testing to see whether your reflex arc is intact.

■ The Peripheral Nervous System

Many of the cells in the CNS have long fibers that extend from the cell body out through openings in the bony covering of the spinal canal to form a cable of nerve fibers that link the CNS to the various organs of the body. These cables of nerve fibers make up the **peripheral nervous system (PNS)**. The PNS is divided into two portions. The first is the somatic nervous system that transmits signals from the brain to the voluntary muscles. As you turn the page of this text, you are accessing your somatic nervous system.

The other portion of the PNS is the autonomic nervous system, which, in turn, is split into two areas. The sympathetic nervous system is responsible for the "fight-or-flight" response, enabling you to fight if you find yourself in a dangerous situation or to run away. This fight-or-flight response generally increases the activity within your body so that your muscles are able to perform more effectively. An increased heart and respiratory rate, pupil dilation, and increased use of glucose all provide you with the resources to defend yourself or flee the scene.

The **parasympathetic nervous system**, the other half of the autonomic nervous system, generally slows down the body. When you are eating, your blood supply needs to move to your stomach and intestines so the food you eat can be processed. The parasympathetic nervous system slows your body's heart rate and respirations and allows your food to be properly digested.

There are two types of nerves within the peripheral nervous system. Sensory nerves carry information from the body to the CNS. **Motor nerves** carry information from the CNS to the muscles of the body.

Sensory Nerves

Sensory nerves of the body are quite complex. There are many types of sensory cells in the nervous system. One type forms the retina of the eye; others are responsible for the hearing and balancing mechanisms in the ear. Other sensory cells are located in the skin, muscles, joints, lungs, and other organs of the body. When a sensory cell is stimulated, it transmits its own special message to the brain. There are special sensory nerves to detect heat, cold, position, motion, pressure, pain, balance, light, taste, and smell, as well as other sensations. Specialized nerve endings are adapted for each cell so it perceives only one type of sensation and transmits only that message.

The sensory impulses constantly provide information to the brain about what the different parts of our body are doing in relation to our surroundings. Thus, the brain is continuously made aware of its surroundings. The cranial nerves supply sensations directly to the brain. Visual sensations (what we see) reach the brain directly by way of the optic nerve (the second cranial nerve) in each eye. The nerve endings for the optic nerve lie in the retina of the eye. The nerve endings are stimulated by light, and the impulses are carried along the nerve that passes through a hole in the back of the eye socket and carries impulses to the occipital portion of the brain.

When sensory nerve endings in the extremities are stimulated, the impulses are transmitted along a peripheral nerve to the spinal cord. The cell body of the peripheral nerve lies in the spinal cord. The impulse is then transmitted from that cell body to another nerve ending in the spinal cord and from there up the spinal cord to the sensory area in the parietal lobe of the brain, where the sensory information can be interpreted and acted on by the brain.

Motor Nerves

Each muscle in the body has its own motor nerve. The cell body for each motor nerve lies in the spinal cord, and a fiber from the cell body extends as part of the peripheral nerve to its specific muscle. Electrical impulses that are produced by the cell body in the spinal cord are transmitted along the motor nerve to the muscle and cause it to contract. The cell body in the spinal cord is stimulated by an impulse produced in the motor strip of the cerebral cortex. This impulse is transmitted along the spinal cord to the cell body of the motor nerve.

■ The Integumentary System (Skin): Anatomy

The skin is divided into two parts: the superficial epidermis, which is composed of several layers of cells, and the deeper dermis, which contains the specialized skin structures. Below the skin lies the **subcutaneous tissue** layer **Figure 5-39**. The cells of the epidermis are sealed to form a watertight protective covering for the body.

The **epidermis** is the most superficial layer of the skin and varies in thickness in different areas of the body. On the soles of the feet, the back, and the **scalp**, it is quite thick, but in some areas of the body, the epidermis is only two or three cell layers thick. The epidermis is

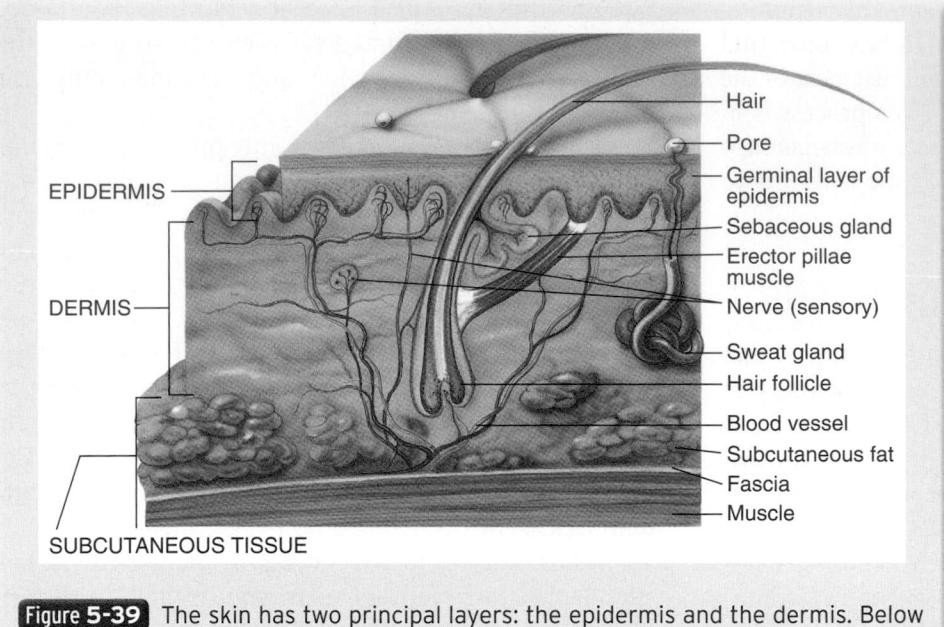

EPIDERMIS

DERMIS

SUBCUTANEOUS TISSUE

Hair
Pore
Germinal layer of epidermis
Sebaceous gland
Erector pillae muscle
Nerve (sensory)
Sweat gland
Hair follicle
Blood vessel
Subcutaneous fat
Fascia
Muscle

Figure 5-39 The skin has two principal layers: the epidermis and the dermis. Below the skin is a layer of subcutaneous tissue.

actually composed of several layers of cells. These layers can be separated into two regions. At the base of the epidermis is the **germinal layer**, which continuously produces new cells that gradually rise to the surface. On the way to the surface, these cells die and enter the **stratum corneal layer**. This is the dead layer of skin. Whereas the germinal layer has a blood supply, the stratum corneal layer does not. The journey from the germinal layer to the surface takes about 4 weeks. The outermost cells of the epidermis are constantly rubbed away and replaced by new cells produced by the germinal layer. The germinal layer also contains cells that produce pigment granules. These granules help to produce skin color.

Below the germinal layer is the **dermis**. Within the dermis lie many of the special structures of the skin: sweat glands, sebaceous (oil) glands, hair follicles, blood vessels, and specialized nerve endings.

Sweat glands produce sweat for cooling the body. The sweat is discharged onto the surface of the skin through small pores, or ducts, that pass through the epidermis. The sebaceous glands produce sebum, the oily material that seals the surface epidermal cells. The **sebaceous glands** lie next to hair follicles and secrete sebum along the hair follicle to the skin surface. In addition to providing waterproofing for the skin, sebum keeps the skin soft so it does not crack.

Hair follicles are the small organs that produce hair. The hair grows from the follicle along a shaft until it reaches the epidermal surface. A sebaceous gland is located along the hair shaft. Connected to the hair is a small muscle. The muscle pulls the hair into an erect

position when the individual is cold or frightened. Hair goes through stages of growth and rest. Each hair follicle on the scalp grows for about 3 years and then rests for about 1 to 2 years.

Blood vessels provide nutrients and oxygen to the skin. The blood vessels lie in the dermis. Small branches extend up to the germinal layer. A complex array of nerve endings also lies in the dermis. These specialized nerve endings are sensitive to environmental stimuli; they respond to these stimuli and send impulses along the nerves to the brain.

Beneath the skin, immediately under the dermis and attached to it, lies the subcutaneous tissue. The subcutaneous tissue is composed largely of fat. The fat serves as an insulator for the body and as a reservoir to store energy. The amount of subcutaneous tissue varies greatly from individual to individual. Beneath the subcutaneous tissue lie the muscles and the skeleton. The subcutaneous layer helps to anchor the skin to the structures below. As we age, the loss of the subcutaneous layer causes the skin to have limited support. This is why wrinkles form in the skin.

The skin covers the entire external surface of the body. The various orifices (openings to the body)—including the mouth, nose, anus, and vagina—are not covered by skin. Orifices are lined with mucous membranes. **Mucous membranes** are quite similar to skin in that they provide a protective barrier against bacterial invasion. Mucous membranes differ from skin in that they secrete **mucus**, a watery substance that lubricates the openings. Thus, mucous membranes are moist, whereas the skin is dry. A mucous membrane lines the entire gastrointestinal tract from the mouth to the anus.

The Integumentary System (Skin): Physiology

The skin, the largest single organ in the body, serves three major functions: to protect the body in the environment, to regulate the temperature of the body, and to transmit information from the environment to the brain.

The protective functions of the skin are numerous. Water makes up a large portion of the body. This water contains a delicate balance of chemical substances in solution. The skin is watertight and serves to keep this balanced internal solution intact. The skin also protects the body from the invasion of infectious organisms: bacteria, viruses, and fungi. These organisms are everywhere and are routinely found lying on the skin surface. However, they never penetrate the skin unless it is broken by injury; thus, the skin provides a constant protection against outside invaders.

The major organ for regulation of body temperature is the skin. Blood vessels in the skin constrict when the body is in a cold environment and dilate when the body is in a warm environment. In a cold environment, constriction of the blood vessels shunts the blood away from the skin to decrease the amount of heat radiated from the body surface. When the outside environment is hot, the vessels in the skin dilate, the skin becomes flushed or red, and heat radiates from the body surface.

Also, in a hot environment, sweat is secreted to the skin surface from the sweat glands. Evaporation of the sweat requires energy. This energy, as body heat, is taken from the body during the evaporation process, which causes the body temperature to fall. Sweating alone will not reduce body temperature; evaporation of the sweat must also occur.

Information from the environment is carried to the brain through a rich supply of sensory nerves that originate in the skin. Nerve endings that lie in the skin are adapted to perceive and transmit information about heat, cold, external pressure, pain, and the position of the body in space. The skin thus recognizes any changes in the environment. The skin also reacts to pressure, pain, and pleasurable stimuli.

The Digestive System: Anatomy

The digestive system, also called the gastrointestinal system, is composed of the gastrointestinal tract (stomach and intestines), mouth, salivary glands, pharynx, esophagus, liver, gallbladder, pancreas, rectum, and anus. The function of this system is **digestion**: the processing of food that nourishes the individual cells of the body. The organs of this system are found within the abdomen.

■ The Abdomen

The **abdomen** is the second major body cavity; it contains the major organs of digestion and excretion. The diaphragm separates the thoracic cavity from the abdominal cavity. Anteriorly and posteriorly, thick muscular abdominal walls create the boundaries of this space. Inferiorly, the abdomen is separated from the pelvis by an imaginary plane that extends from the pubic symphysis through the sacrum **Figure 5-40**. Some organs lie in the abdomen and the pelvis, depending on the posture of the patient.

The simplest and most common method of describing the portions of the abdomen is by quadrants, the four equal areas formed by two imaginary lines that intersect at right angles at the umbilicus. On the anterior abdominal wall, the quadrants thus formed are the right upper, right lower, left upper, and left lower **Figure 5-41**. The terms "right quadrant" and "left quadrant" refer to the patient's right and left. Pain or injury in a given quadrant usually arises from or involves the organs that lie in that quadrant. This simple means of designation will allow

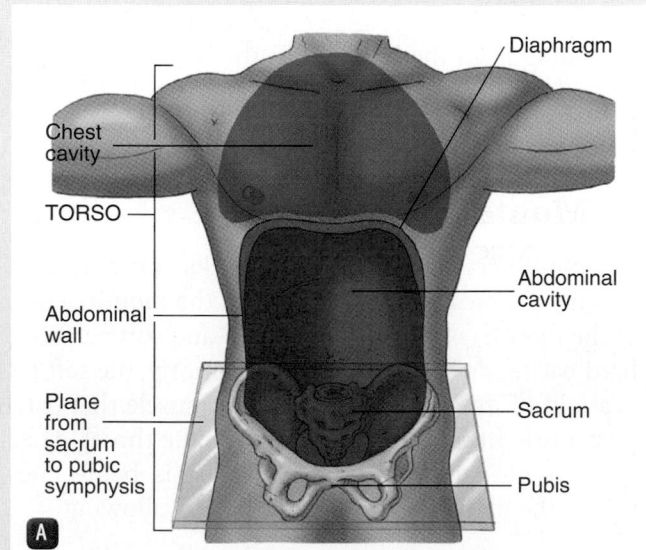

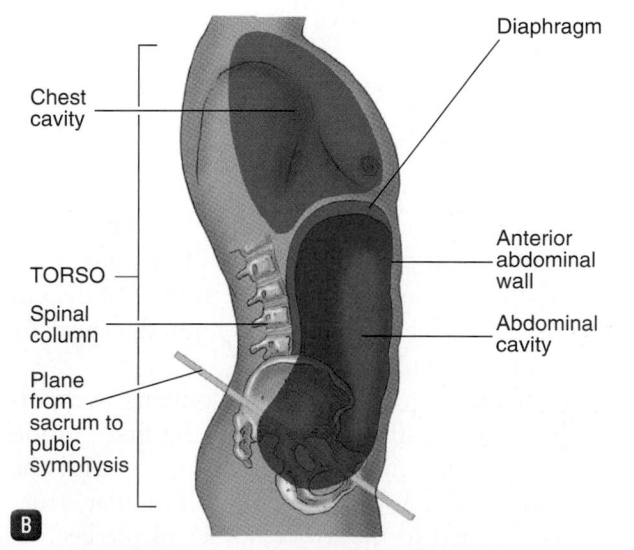

Figure 5-40 The boundaries of the abdomen are the anterior and posterior abdominal cavity walls, the diaphragm, and an imaginary plane from the pubic symphysis to the sacrum. The region below the plane is called the pelvic cavity. **A.** Anterior view. **B.** Lateral view.

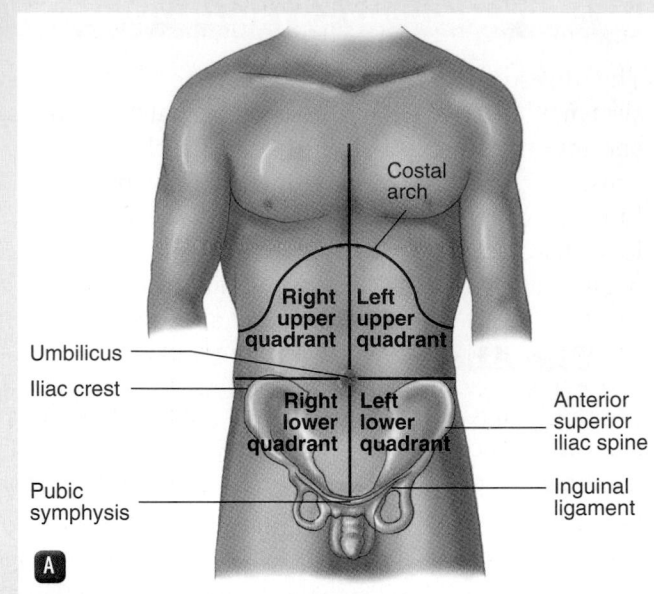

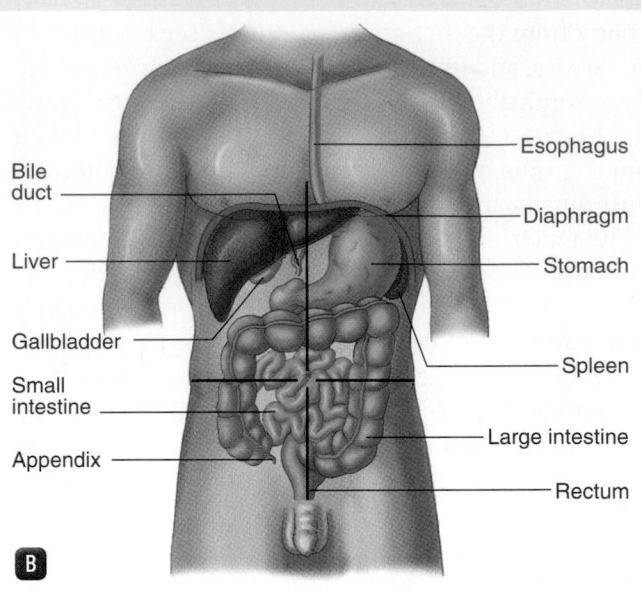

Figure 5-41 **A.** In the abdomen, quadrants are the easiest system for identifying areas. Major bony landmarks are also shown. **B.** Many of the organs in the abdomen lie in more than one quadrant.

you to identify injured or diseased organs that require emergency attention.

Organs and Vascular Structures

In the right upper quadrant (RUQ), the major organs are the liver, the gallbladder, and a portion of the colon. Most of the liver lies in this quadrant, almost entirely under the protection of the 8th to 12th ribs. The liver fills the entire anteroposterior depth of the abdomen in this quadrant. Therefore, injuries in this area are frequently associated with injuries of the liver.

In the left upper quadrant (LUQ), the principal organs are the stomach, the spleen, and a portion of the colon. The spleen is almost entirely under the protection of the left rib cage, whereas the stomach may sag well down into the left lower quadrant when full. The spleen lies in the lateral and posterior portion of this quadrant, under the diaphragm and immediately in front of the 9th to 11th ribs. The spleen is frequently injured, especially when these ribs are fractured.

The right lower quadrant (RLQ) contains two portions of the large intestine: the **cecum**, the first portion into which the small intestine (ileum) opens, and the ascending colon. The **appendix** is a small tubular structure that is attached to the lower border of the cecum. Appendicitis is the most frequent cause of tenderness and pain in this region. In the left lower quadrant (LLQ) lie the descending and the sigmoid portions of the colon.

Several organs lie in more than one quadrant. The small intestine, for instance, occupies the central part of the abdomen around the umbilicus, and parts of it lie in all four quadrants. The pancreas lies just behind the abdominal cavity on the posterior abdominal wall in both upper quadrants. The large intestine also traverses the abdomen, beginning in the RLQ and ending in the LLQ as it passes through all four quadrants. The urinary bladder lies just behind the pubic symphysis in the middle of the abdomen and, therefore, lies in both lower quadrants and also in the pelvis.

The kidneys and pancreas are called **retroperitoneal** organs because they lie behind the abdominal cavity **Figure 5-42**. They are above the level of the umbilicus, extending from the 11th rib to the 3rd lumbar vertebra on each side. The kidneys are approximately 5″ long and lie just anterior to the costovertebral angle.

■ Mouth

The mouth consists of the lips, cheeks, gums, teeth, and tongue. A mucous membrane lines the mouth. The roof of the mouth is formed by the hard and soft palates. The hard palate is a bony plate lying anteriorly; the soft palate is a fold of mucous membrane and muscle that extends posteriorly from the hard palate into the throat. The soft palate is designed to hold food that is being chewed within the mouth and to help initiate swallowing.

Salivary Glands

There are two **salivary glands** located under the tongue, one on each side of the lower jaw and one inside each cheek. They produce nearly 1.5 L of saliva daily. Saliva is approximately 98% water. The remaining 2% is composed of mucus, salts, and organic compounds. Saliva serves as

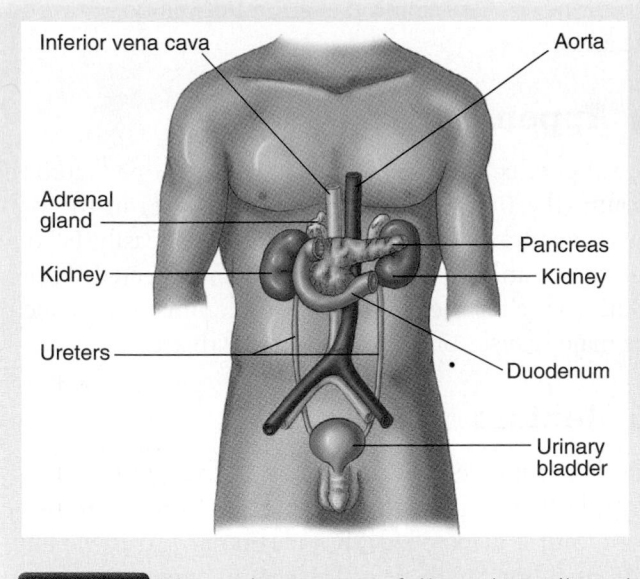

Inferior vena cava
Aorta
Adrenal gland
Kidney
Ureters
Pancreas
Kidney
Duodenum
Urinary bladder

Figure 5-42 The major organs of the retroperitoneal space lie behind the abdominal cavity, above the level of the umbilicus, and extend from the 11th rib to the 3rd lumbar vertebra. Note the bladder, inferior vena cava, and aorta also lie in this space.

a binder for the chewed food that is being swallowed and as a lubricant within the mouth. Saliva also contains certain digestive enzymes.

Oropharynx

The **oropharynx** is a tubular structure that extends vertically from the back of the mouth to the esophagus and trachea. An automatic movement of the pharynx during swallowing lifts the larynx to permit the epiglottis to close over it so that liquids and solids are moved into the esophagus and away from the trachea.

Esophagus

The **esophagus** is a collapsible tube about 10″ long that extends from the end of the pharynx to the stomach and lies just anterior to the spinal column in the chest. Contractions of the muscle in the wall of the esophagus propel food through it to the stomach. Liquids pass with very little assistance.

Stomach

The stomach is a hollow organ located in the left upper quadrant of the abdominal cavity, largely protected by the lower left ribs. Muscular contractions in the wall of the stomach and gastric juice, which contains a lot of mucus, convert ingested food to a thoroughly mixed semisolid mass, called **chyme**. The stomach produces approximately 1.5 L of gastric juice daily for this process. The principal function of the stomach is to receive food in large quantities intermittently, store it, and provide

for its movement into the small bowel in regular, small amounts. In 1 to 3 hours, the semisolid food mass derived from one meal is propelled by muscular contraction into the duodenum, the first part of the small intestine.

Pancreas

The **pancreas**, a flat, solid organ, lies below and behind the liver and stomach and behind the peritoneum. It is firmly fixed in position, deep within the abdomen, and is not easily damaged. It contains two kinds of glands, and the two portions of the pancreas are intertwined. One portion is exocrine, and it secretes nearly 2 L of pancreatic juice daily. This juice contains many enzymes that aid in the digestion of fat, starch, and protein. Pancreatic juice flows directly into the duodenum through the pancreatic ducts. The other portion of the gland is endocrine. It is called the islets of Langerhans, and this is where insulin is produced. Insulin regulates the amount of glucose in the blood.

Liver

The **liver** is a large, solid organ that takes up most of the area immediately beneath the diaphragm in the right upper quadrant and also extends into the left upper quadrant. It is the largest solid organ in the abdomen and has several functions. Poisonous substances produced by digestion are brought to the liver and rendered harmless. Factors that are necessary for blood clotting and for the production of normal plasma are formed here. Between 0.5 and 1 L of bile is made by the liver daily to assist in the normal digestion of fat. The liver is the principal organ for the storage of sugar or starch for immediate use by the body for energy. It also produces many of the factors that aid in the proper regulation of immune responses. Anatomically, the liver is a large mass of blood vessels and cells, packed tightly together. It is fragile and, because of its size, relatively easily injured. Blood flow in the liver is high, because all of the blood that is pumped to the gastrointestinal tract passes into the liver, through the portal vein, before it returns to the heart. In addition, the liver has a generous arterial blood supply of its own. Ordinarily, approximately 25% of the cardiac output of blood (1.5 L) passes through the liver each minute.

Bile Ducts

The liver is connected to the intestine by the **bile ducts**. The **gallbladder** is an outpouching from the bile ducts that serves as a reservoir and concentrating organ for bile produced in the liver. Together, the bile ducts and the gallbladder form the biliary system. The gallbladder discharges stored and concentrated bile into the duodenum through the common bile duct. The presence of food in the duodenum triggers a contraction of the gallbladder to empty it. The gallbladder usually contains about 60 to 90 mL of bile.

Small Intestine

The **small intestine** is the major hollow organ of the abdomen. The cells lining the small intestine produce enzymes and mucus to aid in digestion. Enzymes from the pancreas and the small intestine carry out the final processes of digestion. More than 90% of the products of digestion (amino acids, fatty acids, and simple sugars), together with water, ingested vitamins, and minerals are absorbed across the wall of the lower end of the small intestine into veins to be transported to the liver. The small intestine is composed of the duodenum, the jejunum, and the ileum. The duodenum, which is about 12″ long, is the part of the small intestine that receives food from the stomach. Here, food is mixed with secretions from the pancreas and liver for further digestion. Bile, produced by the liver and stored in the gallbladder, is emptied as needed into the duodenum. It is greenish black, but through changes during digestion, it gives feces its typical brown color. Its major function is in the digestion of fat. The jejunum and ileum together measure more than 20′ on average to make up the rest of the small intestine.

Large Intestine

The **large intestine**, another major hollow organ, consists of the cecum, the colon, and the rectum. About 5′ long, it encircles the outer border of the abdomen around the small bowel. The major function of the colon, the portion of the large intestine that extends from the cecum to the rectum, is to absorb the final 5% to 10% of digested food and water from the intestine to form solid stool, which is stored in the rectum and passed out of the body through the anus.

Appendix

The appendix is a tube 3″ to 4″ long that opens into the cecum (the first part of the large intestine) in the right lower quadrant of the abdomen. It may easily become obstructed and, as a result, inflamed and infected. Appendicitis, which is the term for this inflammation, is one of the major causes of severe abdominal distress.

Rectum

The lowermost end of the colon is the **rectum**. It is a large, hollow organ that is adapted to store quantities of feces until it is expelled. At its terminal end is the anus, a 2″ canal lined with skin. The rectum and anus are supplied with a complex series of circular muscles called sphincters that control, voluntarily and automatically, the escape of liquids, gases, and solids from the digestive tract. Table 5-12 provides a summary of the organs and functions of the digestive system.

The Digestive System: Physiology

Digestion of food, from the time it is taken into the mouth until essential compounds are extracted and delivered by the circulatory system to nourish all of the cells in the body, is a complicated chemical process. In succession,

Table 5-12   Digestive Organs and Functions

Organ/Structure	Function
Mouth	Mechanically breaks down food; begins chemical breakdown with saliva
Esophagus	Moves food from the mouth to the stomach; muscular and vascular structure
Stomach	Performs mechanical and chemical breakdown of food: food in, chyme out
Small intestine: duodenum, jejunum, and ileum	Major site for chemical breakdown of food; major absorption of water, fats, proteins, carbohydrates, and vitamins
Large intestine	Water absorption; formation of feces; bacterial digestion of food
Anus/rectum	Last portion of large intestine; sphincter to control release of feces
Liver	Production of bile; assists with carbohydrate, protein, and fat metabolism of nutrients within the bloodstream; vitamin storage and manufacture; detoxification of blood; elimination of waste
Pancreas	Exocrine: enzymes for protein, carbohydrate, and fat breakdown within the duodenum Endocrine: insulin and glucagon
Gallbladder	Storage of bile

different secretions, primarily **enzymes**, are added to the food by the salivary glands, the stomach, the liver, the pancreas, and the small intestine to convert the food into basic sugars, fatty acids, and amino acids. These basic products of digestion are carried across the wall of the intestine and transported through the portal vein to the liver. In the liver, the products are processed further and stored or transported to the heart through veins draining the liver. The heart then pumps the blood with these nutrients throughout the arteries to the capillaries, where the nutrients pass through the capillary walls to nourish the body's individual cells.

In normal routine activity, without any food or fluid ingestion at all, between 8 and 10 L of fluid is secreted daily into the gastrointestinal tract. This fluid comes from the salivary glands, stomach, liver, pancreas, and small intestine. In a healthy adult, about 7% of the body weight is delivered as fluid daily to the gastrointestinal tract. If significant vomiting or diarrhea occurs for more than 2 or 3 days, the person will lose a substantial portion of body composition and become severely ill.

The Endocrine System: Anatomy and Physiology

The brain controls the body through the nervous system, using electrical impulses, and the endocrine system, using hormones. The **endocrine system** is a complex message and control system that integrates many body functions. Endocrine glands release their hormones directly into the bloodstream Figure 5-43. Epinephrine, norepinephrine, and insulin are examples of hormones. Each endocrine gland produces one or

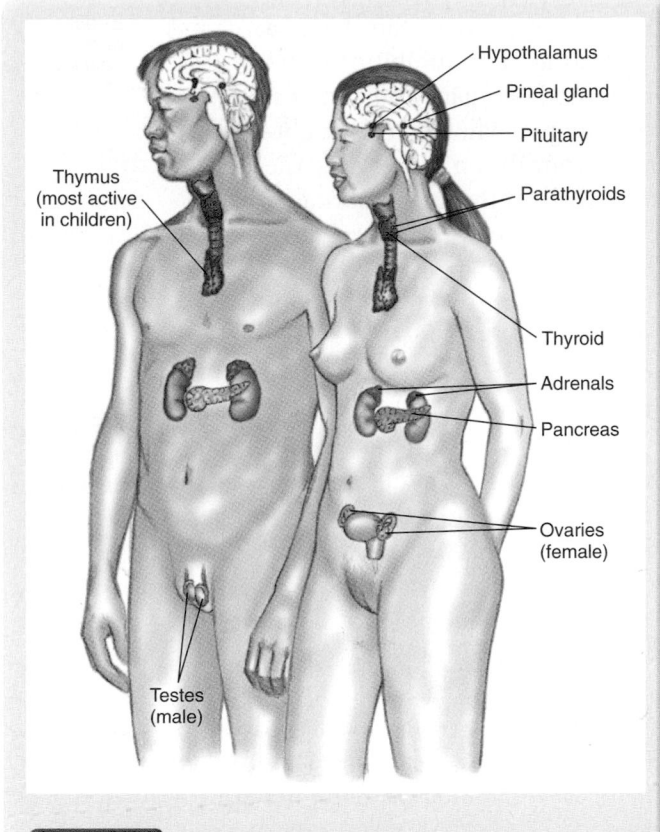

Figure 5-43 The endocrine system controls the release of hormones in the body.

You are the Provider: PART 5

Shortly before arriving at the hospital, you reassess the patient. He remains conscious and alert, and tells you that his pain is less severe than before. After transferring patient care to the attending physician, you later learn that the patient had an inflamed gallbladder, which was surgically removed.

Recording Time: 20 Minutes	
Level of consciousness	Conscious and alert
Respirations	20 breaths/min; adequate depth
Pulse	90 beats/min; strong and regular
Skin	Pink, warm, and moist
Blood pressure	132/80 mm Hg
Sao$_2$	99% (on oxygen)

7. Should your documentation of an EMS call differ from your verbal communication with other health care professionals? Why or why not?

more hormones. Each hormone has a specific effect on some organ, tissue, or process Table 5-13. The brain controls the release of hormones by the endocrine glands. **Hormones** can have a stimulating or an inhibiting effect on the body's organs and systems. For example, when you are frightened, your brain stimulates the adrenal gland through a hormone to release epinephrine and norepinphrine. This increases your blood pressure and heart rate. The resulting increase in blood pressure and heart rate decreases the amount of hormone released by the adrenal gland. The brain then reduces the amount of stimulation to the **adrenal glands**. Thus, a new steady state is achieved at heightened levels of alertness. This cycle is known as a feedback loop, and it helps keep the body's systems and functions in balance Figure 5-44.

Excesses or deficiencies in hormone levels cause various diseases. With endocrine diseases, specific body functions are increased, decreased, or absent. Diabetes mellitus is a common problem. Because production of the hormone insulin is deficient, the body is unable to use glucose normally. Insulin is responsible for rapidly moving glucose into cells. Without insulin, glucose moves slowly. This creates a series of complications as the body struggles to find a more readily available fuel for its cells.

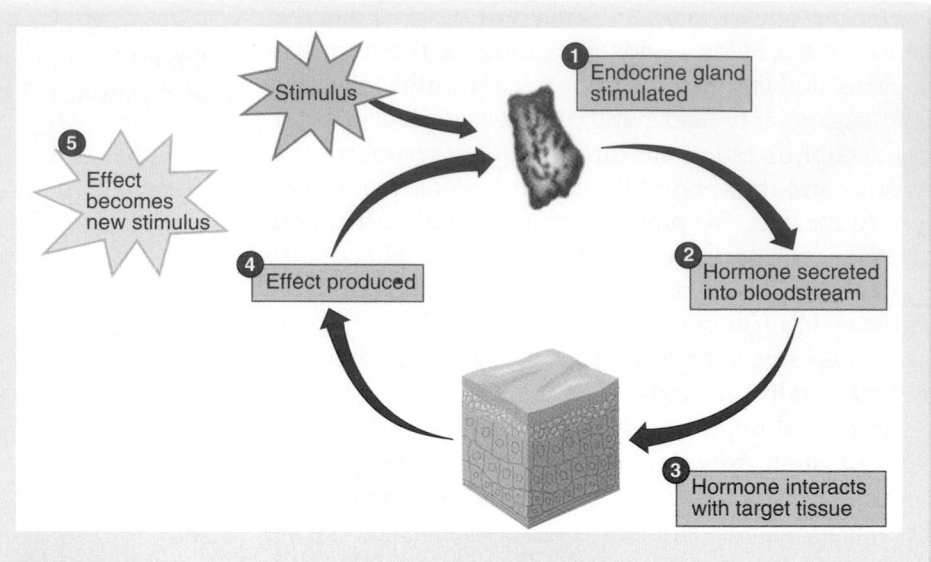

Figure 5-44 The endocrine system is tightly controlled with primary and secondary feedback loops to keep body systems in balance.

People with diabetes begin to burn fats and proteins to create the glucose that cells are craving. Interestingly, the end result is higher and higher blood glucose levels as glucose accumulates, unable to be moved efficiently into the cells. Chapter 17, *Endocrine and Hematologic Emergencies*, discusses how high blood glucose levels affect the body.

The Urinary System: Anatomy and Physiology

The **urinary system** controls the discharge of certain waste materials filtered from the blood by the kidneys.

Table 5-13 Endocrine Glands

Gland	Location	Function	Hormones Produced
Adrenal	Above the kidneys	Stress response, fight or flight	Epinephrine, norepinephrine, and others
Ovary	Female pelvis (two glands)	Regulates sexual function, characteristics, and reproduction	Estrogen and others
Pancreas	Retroperitoneal space	Regulates glucose metabolism and other functions	Insulin and others
Parathyroid	Neck (behind and beside the thyroid) (three to five glands)	Regulates serum calcium	Parathyroid hormone
Pituitary	Base of skull	Regulates all other endocrine glands	Multiple, controls other endocrine glands
Testes	Male scrotum (two glands)	Regulate sexual function, characteristics, and reproduction	Testosterone and others
Thyroid	Neck (over the larynx)	Regulates metabolism	Thyroxine and others

In the urinary system, the kidneys are solid organs; the ureters, bladder, and urethra are hollow organs **Figure 5-45**. The main functions of the urinary system are: (1) to control fluid balance in the body, (2) to filter and eliminate wastes, and (3) to control pH balance.

The body has two **kidneys** that lie on the posterior muscular wall of the abdomen behind the peritoneum in the retroperitoneal space. These organs rid the blood of toxic waste products and control its balance of water and salt. Blood flow in the kidneys is high. Nearly 20% of the output of blood from the heart passes through the kidneys each minute. Large vessels attach the kidneys directly to the aorta and the inferior vena cava. Waste products and water are constantly filtered from the blood to form urine. The kidneys continuously concentrate this filtered urine by reabsorbing the water as it passes through a system of specialized tubes within them. The tubes finally unite to form the **renal pelvis**, a cone-shaped collecting area that connects the ureter and the kidney. Normally, each kidney drains its urine into one ureter through which the urine passes to the bladder.

A **ureter** passes from the renal pelvis of each kidney along the surface of the posterior abdominal wall behind the peritoneum to drain into the urinary bladder. The ureters are small (0.2″ in diameter), hollow, muscular tubes. **Peristalsis**, a wavelike contraction of smooth muscle, occurs in these tubes to move the urine to the bladder.

The **urinary bladder** is located immediately behind the pubic symphysis in the pelvic cavity and is composed of smooth muscle with a specialized lining membrane. The two ureters enter posteriorly at its base on either side. The bladder empties to the outside of the body through the **urethra**. In the male, the urethra passes from the anterior base of the bladder through the penis. In the female, the urethra opens in front of the vagina. A healthy adult forms 1.5 to 2 L of urine every day. This waste is extracted and concentrated from the 1,500 L of blood that circulates through the kidneys daily.

The Genital System: Anatomy and Physiology

The **genital system** controls the reproductive processes by which life is created. The male genitalia, except for the **prostate gland** and the **seminal vesicles**, lie outside the pelvic cavity. The female genitalia, with the exception of the clitoris and labia, are contained entirely within the pelvis. The male and female reproductive organs have certain similarities and, of course, basic differences. They produce sperm and egg cells and reproductive hormones and play a significant role in sexual intercourse and reproduction.

The Male Reproductive System and Organs

The male reproductive system consists of the testicles, epididymis, vasa deferentia, prostate gland, seminal vesicles, and penis **Figure 5-46**. Each **testicle** contains specialized cells and ducts; some of these produce male hormones, and others develop sperm. The hormones are absorbed directly into the bloodstream from the testicles. The sperm are immature and are moved from the testicles to the epididymis so they can develop. During ejaculation, the sperm are carried through **vasa deferentia** (or vas deferens) to the urethra. Finally, the sperm are deposited by the penis.

The function of the reproductive system is to reproduce. Sperm are able to join with an egg to begin the process of life. In addition to reproduction, this system is

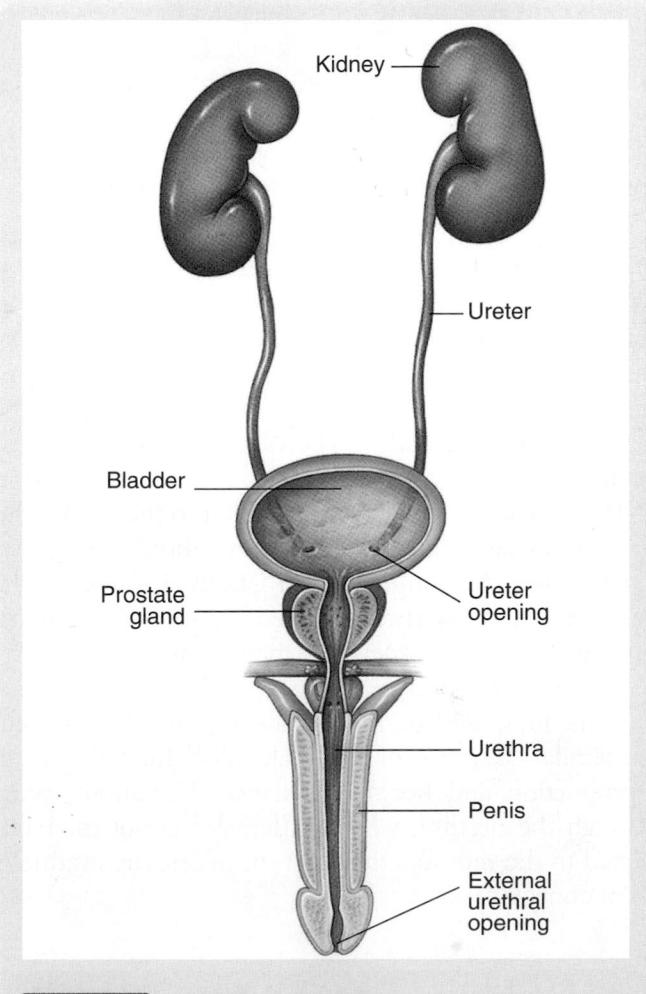

Figure 5-45 The urinary system lies in the retroperitoneal (behind the peritoneum) space behind the organs of the digestive system. The urinary system in males and females includes the kidneys, ureters, bladder, and urethra. This diagram shows the male urinary system.

Labels on figure: Kidney, Ureter, Bladder, Prostate gland, Ureter opening, Urethra, Penis, External urethral opening

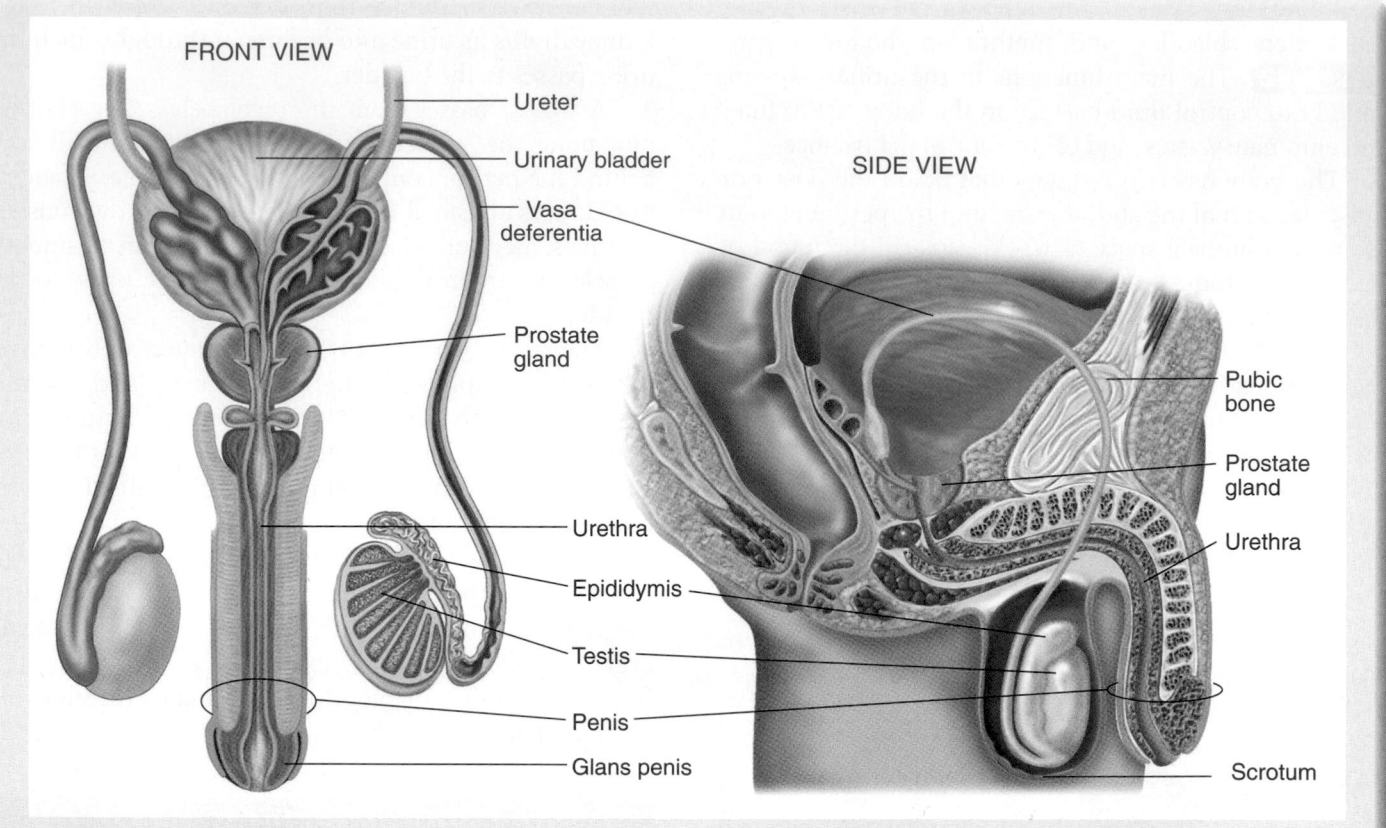

FRONT VIEW

Ureter

Urinary bladder

Vasa deferentia

SIDE VIEW

Prostate gland

Pubic bone

Prostate gland

Urethra

Urethra

Epididymis

Testis

Penis

Glans penis

Scrotum

Figure 5-46 The male reproductive system consists of the testicles, epidiymis, vasa deferentia, prostate gland, seminal vesicles, and penis.

also responsible for the production of sex hormones. Many of the physical characteristics of men, such as increased muscle mass, body hair, and deep voice, are attributed to the powerful effects of the hormones released by the testes. Finally, the penis, though part of the reproductive system, is also part of the urinary system. Any damage or infection to the penis can cause problems within the urinary bladder and/or the kidneys.

The Female Reproductive System and Organs

The female reproductive organs include the ovaries, fallopian tubes, uterus, cervix, and vagina **Figure 5-47**. The **ovaries**, like the testicles, produce sex hormones and specialized cells for reproduction. The female sex hormones are absorbed directly into the bloodstream. A specialized ovum, or egg cell, matures and is released regularly during the adult female's reproductive years. The ovaries release a mature egg approximately every 28 days. This egg travels through the fallopian tubes where fertilization normally occurs. The fallopian tubes exit into the uterus.

The **fallopian tubes** connect with the uterus and carry the ovum into the cavity of this organ. The

uterus is pear-shaped and hollow, with muscular walls. The narrow opening from the uterus to the vagina is the cervix. The **vagina** (birth canal) is a muscular distensible tube that connects the uterus with the vulva (the external female genitalia). The vagina receives the penis during sexual intercourse, when **semen** is deposited in it. The sperm in the semen may pass into the uterus and fertilize an egg, causing pregnancy. Should the pregnancy come to completion at about 40 weeks, the neonate will pass through the vagina and be born. The vagina also channels the menstrual flow from the uterus out of the body.

The functions of the female reproductive system are similar as those of the male reproductive system: reproduction and hormone balance. Urination occurs through the urethra, which in females is not interconnected to the reproductive tract. In males, the urethra is interconnected.

Life Support Chain

Cells are the foundation of the human body. Billions of cells compose the human body. Some cells make hair, other cells are involved in storing memory, and others

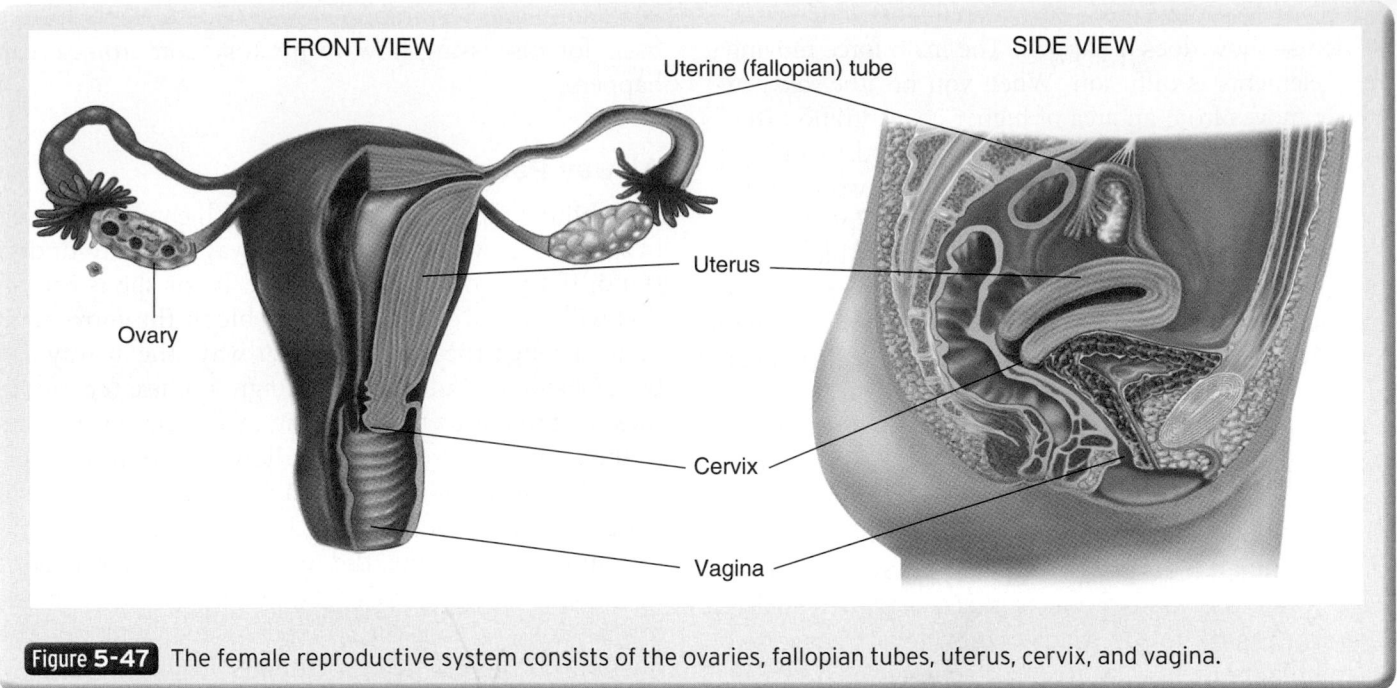

FRONT VIEW

SIDE VIEW

Uterine (fallopian) tube

Uterus

Ovary

Cervix

Vagina

Figure 5-47 The female reproductive system consists of the ovaries, fallopian tubes, uterus, cervix, and vagina.

help to move your eyes as you read this page. Cells with a common job grow close to each other and are called tissues. Groups of tissues that all perform similar or inter-related jobs form organs. A series of organs with similar jobs make up the body systems that we have been discussing in this chapter.

The body's cells, tissues, and organs, regardless of their function, all require oxygen, nutrients, and the removal of wastes to perform their job. Oxygen is brought to the cells through the respiratory and circulatory systems. Nutrients are made available to the body after we eat. The digestive system takes the food we eat and breaks it down into, among other things, glucose. Glucose is the primary fuel of the body. The circulatory system is the carrier of these supplies and wastes through the process of perfusion. If interference occurs in this delivery system, cells will become damaged or die.

The human body is designed to be able to handle a wide array of metabolic disturbances. Oxygen is a critical component for cells. Cells use oxygen to take the available nutrients and turn them into chemical energy. **Adenosine triphosphate (ATP)** is involved in energy metabolism and is used to store energy. Cells prefer to operate using oxygen because it provides the cells with 15 times as much ATP as when they operate without oxygen. The process that uses oxygen is called aerobic (meaning with air) metabolism. The waste products of **aerobic metabolism** are carbon dioxide and water. Some cells have become so specialized that they are unable to survive without constant supplies of oxygen. The brain and heart are just two examples. Without oxygen, brain cells will begin to die within 4 to 6 minutes.

Most cells in the body are able to continue to function, even without oxygen. This anaerobic (without air) state allows cells to operate despite no available oxygen. When illness or injury occurs, the body needs to shift available resources to areas in need while ensuring that critical areas, such as the brain and heart, have an uninterrupted supply of resources. Any time that available oxygen is limited to portions of the body, cells will switch to **anaerobic metabolism**. This occurs anytime you exercise vigorously and begin to feel a burning sensation. In this state, very limited amounts of energy are able to be released so the body must quickly correct the oxygen deficiency or risk cellular death. The most well known byproduct of anaerobic metabolism is lactic acid, which is the material that causes muscle burning during anaerobic exercise. **Lactic acid** is converted back to a useful energy source once oxygen is available. Anaerobic metabolism can be supported in most cells for only 1 to 3 minutes.

Words of Wisdom

Remember:
- When cells function with oxygen, they use aerobic metabolism.
  - They generate large amounts of ATP (cellular energy) and produce wastes of carbon dioxide and water.
- When cells function without oxygen, they use anaerobic metabolism.
  - They generate small amounts of ATP (cellular energy) and produce waste of lactic acid.

All of this movement of material—oxygen, waste, nutrients—how does it happen? The main force moving these elements is diffusion. When you breathe, oxygen simply moves from an area of higher concentration (the air) to one of lower concentration (your bloodstream).

Cells are surrounded by fluid that allows for easy movement of nutrients and wastes. A physical property of this fluid that is critical to cell survival is pH, which is the measure of acidity or alkalinity in a solution. Solutions that are high in pH (> 7.0) are considered alkaline. A common example is soap. Solutions that are low in pH (< 7.0) are considered acidic. Sulfuric acid in car batteries is one example. A solution that is neither acidic nor alkaline is considered neutral (pH 7.0). The body's cells want to exist in a near-neutral environment.

Your body spends a large amount of energy working to maintain a normal pH (normal physiologic pH is 7.35–7.45). The waste products of cells are often acidic, such as carbon dioxide. Carbon dioxide is transported by combining with water to create carbonic acid, which is more soluble in the plasma. The plasma also contains sodium bicarbonate, which is alkaline and helps to buffer or neutralize the acidic waste products of cells.

The lungs help to maintain the pH level in the body by controlling the level of carbon dioxide, and therefore the level of carbonic acid in the blood. If the blood becomes acidic, the respiratory centers in the brain stem will increase breathing to blow off more carbon dioxide. If too much carbon dioxide is blown off, then the body can become too alkaline, which is what happens during hyperventilation. The blood and lungs interact continuously to maintain pH balance in the body.

Pathophysiology

Pathophysiology is the study of the functional changes that occur when the body reacts to a particular disease. Many diseases can occur in patients. Diabetes is a disease of the pancreas. A stroke is a disease of the brain. Pneumonia is a disease of the lungs. An overview of the way the body responds to disease is discussed in this section. Specific pathophysiologic changes that occur with specific diseases are discussed in the chapters of Section 6: *Medical*.

■ Respiratory Compromise

Respiratory compromise is the inability of the body to move gas effectively, which can result in the body not getting enough oxygen (hypoxia), the body having a higher than normal level of carbon dioxide (hypercarbia), or both. Breathing deals with two main concepts:

ventilation and respiration. These two concepts form a basis for discussing how respiratory compromise can happen.

Airway Patency

The ability to move gas back and forth can be impaired in a variety of ways. A blocked airway is easy to understand. If a person chokes on what he or she is eating, this will partially or completely block the movement of air through the trachea. Other ways the airway can be blocked include other foreign bodies (eg, food, toys, or broken teeth), swelling in the airway, trauma to the mouth or neck, and swallowing blood or vomitus. The most common airway obstruction you will need to manage is blockage by the tongue. When a person is unconscious, the tongue relaxes and droops posteriorly in the mouth. The patient does not "swallow his tongue," but the relaxed tongue can block the opening to the trachea. Fortunately, by performing a head tilt–chin lift maneuver, the airway can now be opened. Opening a patient's airway will be discussed in later chapters.

Impairment of the muscles of breathing will impair the movement of gas. Neuromuscular diseases such as Guillain-Barré syndrome and myasthenia gravis can interfere with the ability of the brain to send signals to the diaphragm. Trauma can sever the phrenic nerve or damage the brain stem. If a patient's level of consciousness is too low, ventilation problems can occur. This means that any situation that decreases a patient's level of consciousness can have a direct effect on ventilation.

Ventilation can also be affected when only one part of the process is impaired. In asthma, patients tend to have a problem with exhalation, not inhalation. So, early in an asthma attack, the amount of carbon dioxide in the blood will rise. Typically, patients with asthma do not have great difficulty with oxygenation; their problem is more related to severe contraction of the muscles surrounding the lower airways.

Respiratory Compromise

Impairment in the movement of gas at the cellular level also forms the basis for disease. A change in the atmosphere can interfere with a person's ability to breathe. The air you breathe is 21% oxygen, and the air you exhale is 16% oxygen. This means there is only a 5% margin of safety for oxygen concentration in the air you breathe. If the oxygen concentration of the ambient air were 10%, your blood would have more oxygen in it than the gas you just inhaled. You would diffuse oxygen from your body to the lungs as opposed to the other way around. This gas composition change can occur at high altitudes where the pressure of oxygen in the air is low. It can also occur in closed environments where oxygen is displaced by another gas.

If the air you breathe is the correct composition, the only other way respiration can become compromised is with impaired movement of the gas across the cell membrane. If the patient has fluid in the alveoli, this fluid will prevent gas exchange. In pneumonia, mucus and infectious wastes form another type of barrier, preventing gas from accessing the alveoli. If the interstitial space is filled with fluid, this edema will actually increase the distance from the capillary to the alveoli. Because of the increased distance, it will take longer for the gas to move from inside the alveoli to inside the capillary. If one of the blood vessels bringing blood to the lungs is clogged, this will also affect the amount of gas diffused into and out of the blood.

A convenient way to understand respiratory compromise is to measure the ventilation/perfusion ratio to determine the severity of your patient's condition. This measurement, also called the **V̇/Q̇ ratio**, examines how much gas is being moved effectively (ventilation) and how much blood is gaining access to the alveoli (perfusion). A mismatch is said to occur when one of the two variables is not normal. For example, in a patient with pulmonary embolism, a sudden clog in the pulmonary artery prevents blood from gaining access to the alveoli. Part of the circulating blood does not receive air, and, therefore, gas is not exchanged. In pulmonary edema, gas is not able to move effectively through the alveoli into the blood. Some of the air that is breathed therefore does not reach the blood.

Effects of Respiratory Compromise on the Body

The net effect of respiratory compromise is the same. The brain will detect any increase in carbon dioxide levels. If increased respiration does not return the blood pH to normal, the patient will work harder to breathe. He or she will begin to use the accessory muscles of the chest, shoulders, back, and abdomen to move more air and obtain more oxygen. If the problem continues to worsen, oxygen levels will begin to fall. This will cause the brain to issue further commands to breathe. Once oxygen levels fall, the brain understands the situation is dire.

Decreased oxygen levels will force cells to move from aerobic metabolism to anaerobic metabolism. Remember, the heart and brain cells are unable to do this. Without a constant supply of oxygen, they will die in minutes. Anaerobic metabolism generates a fraction of the needed energy, and cellular functions will be slowed. Also, a by-product of anaerobic metabolism is lactic acid. If too much of this acid is created, the pH of the blood will drop. If the pH becomes too low, cells can no longer survive.

If the compromise is mild and gradual, the body will adapt. A compromise that is severe or lasts a long time can overwhelm the body's ability to adapt, and a compromise that is not corrected can exhaust the body's energy supplies and the patient may die **Table 5-14** .

Shock

Shock is a condition in which organs and tissue are not receiving an adequate flow of blood and oxygen, or perfusion. A patient in shock has difficulty transporting

Table 5-14 Summary of Respiratory Compromises

Category	Problem	Effect
Ventilation	Damage to the regulatory centers of the brain	Breathing pattern and rate become erratic
	Inability to exhale effectively	Carbon dioxide builds up in blood
	Inability to inhale effectively	Oxygen levels in the blood fall
	Injury to chest	Breathing depth decreases
	Obstruction of the airway	Decreased or absent movement of air
	Overdose/toxic exposure	Decreased level of consciousness leading to decreased breathing depth
	Unconsciousness	Breathing depth decreases
	Weakened respiratory muscles	Breathing depth decreases
Respiration	Fluid within the alveoli/pulmonary edema	Prevents gas from entering the alveoli
	Mucus or infectious wastes	Prevents gas from entering the alveoli
	Impaired blood flow to the lungs	Affects blood gaining access to lung tissue
Oxygenation	Decreased oxygen in the air breathed	Affects diffusion of gas
	Increased carbon dioxide in the air breathed	Affects diffusion of gas

oxygen in the blood, which allows the buildup of wastes. There are several ways a disease can have an impact on tissue perfusion. Essentially, the patient can have insufficient blood volume or a heart that does not pump effectively, or the patient's body can no longer control the blood vessels.

The most easily understood type of shock is the kind that results from the lack of blood volume, called hypovolemic shock. In trauma, patients can lose blood. This loss results in an inability to transport oxygen and nutrients. Loss of volume can also occur when patients have severe vomiting and diarrhea; the amount of water lost can be substantial, eventually leading to decreased amounts of circulating blood volume. In both cases, the effects of blood loss are similar. The ability to transport oxygen and nutrients is impaired.

The second type of shock is associated with heart function, called cardiogenic or obstructive shock. If the heart is not functioning correctly, wastes and nutrients can be prevented from being moved effectively. The heart can become weakened as the result of a myocardial infarction (heart attack). If the heart rate is very fast or very slow, this can also cause the blood to not move effectively. Blood pressure can drop, resulting in diminished perfusion. Even if the heart is working properly and the amount of blood volume is normal, perfusion problems in the blood vessels can still exist.

Issues regarding the dilation and constriction of blood vessels lead to the third type of shock, distributive shock. In severe allergic reactions, severe infections, or injuries to the nervous system, the patient can lose blood vessel control. If the blood vessels become more dilated, they hold more fluid and blood does not return to the heart as quickly as it should. Eventually the patient's blood pressure will drop and perfusion will decrease. A severe infection may also cause massive systemic vasodilation and increased capillary permeability. A spinal cord injury may result in vasodilation below the level of the injury.

If the blood vessels constrict too much, as in patients with very severe hypertension, the amount of work needed to overcome the very high blood pressure within the arteries is too much for the heart. This will overwhelm the heart and prevent it from moving effective amounts of blood. Perfusion will decrease.

Effects of Shock on the Body

The effects of inadequate perfusion on the body are very similar to those of respiratory compromise. The blood oxygen level falls, and the blood carbon dioxide level rises. A state of anaerobic metabolism will occur. The body will detect the decreased blood pressure by the baroreceptors, which will initiate the release of epinephrine and norepipehrine. The heart rate will increase,

the heart will beat more forcefully, and the blood vessels will constrict. The goal is to maintain blood pressure to the areas of the body that are unable to survive without oxygen: the brain and the heart.

As a person's blood pressure falls, the pressure in the capillaries (hydrostatic pressure) also falls. This provides a beneficial effect. The oncotic pressure remains unchanged, but with the decrease in hydrostatic pressure, interstitial fluid moves more forcefully into the capillaries. This movement helps to refill the blood vessels with the fluid sitting in the interstitial space. Volume therefore is restored so that the heart has enough liquid (blood and the interstitial fluid) to pump. The downside of this effect is that oxygen and nutrient supplies to the cells are diminished. Provided this state of decreased blood pressure does not continue for a long time, or is not profound, the body will be able to tolerate this situation and return itself to a state of health. Eventually the fluid taken from the interstitial space will be returned.

Alteration of Cellular Metabolism

The availability of fuel for the cells to use is interconnected with shock and respiratory compromise. As discussed previously, when there is inadequate oxygen, cells will create energy through anaerobic metabolism. This backup, temporary system allows cells to function at low energy levels for a short time. When perfusion is impaired, the availability of glucose for cells to turn into ATP is decreased. Most cells are able to use alternative fuel supplies to help bridge the gap until perfusion is restored.

When a person is engaged in strenuous exercise, the demand for glucose by the muscles exceeds the available supply. The body begins to burn fats and turns them into glucose to meet this need. As with anaerobic metabolism, this backup system has some important drawbacks. The major by-products of burning fats are acids, and this process requires the use of more energy than when using glucose for fuel. Therefore, there are more wastes to be removed, and it takes more out of the body to use this alternative fuel supply. Provided the other systems of the body are working correctly (respiratory and cardiovascular), the body should be able to maintain pressure within the system for a while. If the person has trouble breathing or perfusion problems, the amount of damage to the body can be extreme—even to the point of death.

Most cells are able to use alternative fuels. However, brain cells are not. They rely on a constant supply of glucose to function. If the supply of available glucose is dramatically decreased, brain cells will become damaged or die.

You are the Provider: SUMMARY

1. How will knowledge of anatomy and physiology help you provide appropriate patient care?

Knowledge of anatomy (the structure of the body and relation of body parts) and physiology (the function of the body and its parts) is critical for anyone who provides patient care—emergency or otherwise. When a patient complains of pain to any part of the body, knowledge of human anatomy will help you form a logical field impression regarding which organ or organs may be affected; knowledge of physiology will help you predict the negative effects the patient may experience based on the organ or organs affected. From this information, an appropriate treatment plan can be formulated and implemented.

Although you are not expected to diagnose a patient's problem, a strong fundamental knowledge of anatomy, physiology, and medical terminology will help you communicate the correct information to the emergency department physician or nurse.

2. On the sole basis of the patient's chief complaint, which organ or organs should you suspect is/are the cause of his problem?

The major organs in the right upper quadrant (RUQ) are the liver, gallbladder, and a portion of the large intestine (colon). Although your initial thoughts should focus on dysfunction of one or more of these organs, the patient's true problem may exist elsewhere in his abdomen and the pain just happens to be manifesting in his RUQ. Although your objectives are to recognize that the patient has an acute abdominal problem and to find and treat life-threatening conditions, you will need to ask additional questions to clarify his complaint. His answers to your questions will help you formulate a field impression—that is, what you believe to be his primary problem.

3. What additional questions should you ask to gather more information about his chief complaint?

After determining the patient's chief complaint (eg, the reason he or she called 9-1-1), you should ask the patient to elaborate on the complaint; this is called the history of present illness. The OPQRST (Onset, Palliation/provocation, Quality, Region/radiation, Severity, Timing of pain) mnemonic is a useful tool for this purpose.

The patient has already told you that his pain began suddenly, so the "O" in the OPQRST has already been established. Ask him if anything makes his pain better (palliates) or worse (provokes); patients with abdominal pain often draw their knees into their abdomen, which takes pressure off the abdominal muscles and may provide them slight relief of their pain. To determine if the pain radiates, ask the patient if the pain stays in the RUQ of his abdomen or moves/travels somewhere else; determine if he has referred pain by asking if he hurts anywhere else in addition to his RUQ. For example, gallbladder problems often present with pain to the RUQ and pain to the right shoulder; the pain does not radiate—it is in two separate locations. Assess the severity of his pain by using the 0 to 10 scale, with 0 being no pain and 10 being the worst pain ever experienced; pain severity should be assessed frequently,

especially after any interventions have been performed. The patient told you the pain began about 20 minutes ago; thus, the time of onset has already been established. Although chronic pain can indicate a serious underlying problem, you should be especially concerned that his pain began acutely.

Other questions to ask the patient should focus on common symptoms associated with abdominal pain, such as nausea and/or vomiting, diarrhea, and urinary difficulty, among others. When possible, try not to ask leading questions (ie, "Are you nauseated?"); instead, simply ask him if he has any other symptoms.

4. What is the function of the gallbladder?

The liver is connected to the intestine by the bile ducts. The gallbladder is an outpouching from the bile ducts that concentrates and stores bile, a digestive enzyme that is produced in the liver. The presence of food in the duodenum, the first part of the small intestine, triggers the gallbladder to contract and discharge bile into the duodenum to facilitate digestion.

5. What additional symptoms would you expect the patient to experience based on the function of the gallbladder?

The gallbladder contracts only when food enters the duodenum; patients with inflammation of the gallbladder (cholecystitis) typically have right upper quadrant pain within an hour or so after eating a meal. In many cases, the patient also complains of referred pain to the right shoulder. Other symptoms of gallbladder disease include nausea and vomiting and heartburn.

Most cases of cholecystitis occur when gallstones form and block the outlet of the gallbladder. In some cases, the gallstones spontaneously pass; however, if they do not, the patient experiences pain of varying intensity.

6. How does knowledge of anatomy, physiology, and medical terminology facilitate communication with other health care professionals?

As part of the health care team, everything you do should benefit the patient whose care you are charged with. An integral part of patient care is effective communication with other health care professionals. A strong knowledge of anatomy and physiology will directly benefit your patient. Your ability to speak the language of medicine will minimize communication barriers between you and other members of the health care team; again, the biggest benefit is to the patient.

Whether you are calling in your radio report from the ambulance or giving a verbal report at the hospital, the use of appropriate medical terminology not only ensures that the information you pass along is relevant and accurate—again, benefiting the patient—it also displays knowledge on your part, which earns you the respect of truly being a professional.

You should review human anatomy and physiology and medical terminology on a regular basis. Although the structure and function of the body and the terms used to describe it do not change, your ability to recall the information can deteriorate over time.

You are the Provider: SUMMARY, continued

7. Should your documentation of an EMS call differ from your verbal communication with other health care professionals? Why or why not?

You must extend the same relevant and accurate information verbally relayed to other health care professionals to your patient care report (PCR). When possible, use proper medical terminology when documenting the patient's complaint, history of present illness, medical history, and any treatment provided in the prehospital setting. Of course, if you are unsure of the correct medical term to accurately describe a particular aspect of the patient's complaint or physical examination, use plain English.

The PCR is read by the personnel who assume patient care from you and may have a direct impact on future care that the patient receives. The use of proper medical terminology, coupled with an accurate depiction of the care you provided, will facilitate continuity of the patient's care.

EMS Patient Care Report (PCR)

Date: 6-10-09	**Incident No.:** 050109	**Nature of Call:** Abdominal pain		**Location:** 322 Azalea Trail	
Dispatched: 1740	**En Route:** 1741	**At Scene:** 1747	**Transport:** 1759	**At Hospital:** 1809	**In Service:** 1817

Patient Information

Age: 60 **Sex:** M **Weight (in kg [lb]):** 84 kg (185 lb)	**Allergies:** Sulfa, Codeine, Contrast dye **Medications:** Zyrtec, Pepcid **Past Medical History:** Gallbladder problems **Chief Complaint:** Abdominal pain

Vital Signs

Time: 1749	**BP:** 142/82	**Pulse:** 110	**Respirations:** 24	**Sao$_2$:** 98%
Time: 1758	**BP:** 138/88	**Pulse:** 112	**Respirations:** 24	**Sao$_2$:** 97%
Time: 1808	**BP:** 132/80	**Pulse:** 90	**Respirations:** 20	**Sao$_2$:** 99%

EMS Treatment
(circle all that apply)

Oxygen @ 15 **L/min via (circle one):** NC **NRM** Bag-Mask Device		**Assisted Ventilation**	**Airway Adjunct**	**CPR**
Defibrillation	**Bleeding Control**	**Bandaging**	**Splinting**	**Other:** Position of comfort

Narrative

9-1-1 dispatch for a male patient with abdominal pain. On arrival at the scene, found the patient, a 60-year-old male, lying on his side on the floor of his bedroom with his knees drawn up into his abdomen. He was conscious and alert; his airway was patent, and his breathing was adequate. Patient states that the pain (8 on a 0 to 10 scale) began suddenly approximately 20 minutes ago. His past medical history is significant for gallbladder problems; no other medical history reported. Applied oxygen at 15 L/min via nonrebreathing mask and obtained vital signs. Further assessment of patient's abdomen revealed that it was soft; however, it was point tender to palpation of the RUQ. Patient denies chest pain, shortness of breath, nausea or vomiting, and any other symptoms. He further denies radiating and referred pain. Placed patient onto the stretcher, allowed him to assume a position of comfort, loaded him into the ambulance, and began transport to the hospital. En route, continued to monitor patient's condition, which remained unchanged. Vital signs reassessed and noted above. Shortly before arrival at the hospital, reassessment of the patient revealed that his vital signs remained stable and that his pain had decreased in severity. Transferred care of patient to receiving hospital without incident and gave verbal report to charge nurse. Departed the hospital and returned to service. **End of report**

Prep Kit

Ready for Review

- To properly care for your patients, you must have a through understanding of human anatomy and physiology so you can assess the patient's condition and communicate with hospital personnel and other health care providers.

- You must be able to identify superficial landmarks of the body and know what lies underneath the skin so that you can perform an accurate patient assessment.

- The skeleton gives the body its recognizable human form through a collection of bones, ligaments, tendons, and cartilage.

- The skeletal system provides protection for fragile organs, allows for movement, and gives the body its shape.

- The contraction and relaxation of the musculoskeletal system gives the body its ability to move.

- The respiratory system consists of all the structures of the body that contribute to the process of breathing. It includes the nose, mouth, throat, larynx, trachea, bronchi, and bronchioles.

- The function of the respiratory system is to provide the body with oxygen and eliminate carbon dioxide.

- The circulatory system is a complex arrangement of connected tubes, including the arteries, arterioles, capillaries, venules, and veins.

- The nervous system is perhaps the most complex organ system within the human body. It consists of the brain, spinal cord, and nerves.

- The skin is divided into two parts: the superficial epidermis, which is composed of several layers of cells, and the deeper dermis, which contains the specialized skin structures.

- The skin, the largest single organ in the body, serves three major functions: to protect the body in the environment, to regulate the temperature of the body, and to transmit information from the environment to the brain.

- The digestive system is composed of the gastrointestinal tract (stomach and intestines), mouth, salivary glands, pharynx, esophagus, liver, gallbladder, pancreas, rectum, and anus.

- Digestion of food, from the time it is taken into the mouth until essential compounds are extracted and delivered by the circulatory system to nourish all of the cells in the body, is a complicated chemical process.

- The endocrine system is a complex message and control system that integrates many body functions.

- The urinary system controls the discharge of certain waste materials filtered from the blood by the kidneys.

- The genital system controls the reproductive processes by which life is created.

- Pathophysiology is the study of how the body reacts to diseases.

Vital Vocabulary

abdomen The body cavity that contains the major organs of digestion and excretion. It is located below the diaphragm and above the pelvis.

abduction Motion of a limb away from the midline.

acetabulum The depression on the lateral pelvis where its three component bones join, in which the femoral head fits snugly.

Adam's apple The firm prominence in the upper part of the larynx formed by the thyroid cartilage. It is more prominent in men than in women.

adduction Motion of a limb toward the midline.

adenosine triphosphate (ATP) The nucleotide involved in energy metabolism; used to store energy.

adrenal glands Endocrine glands located on top of the kidneys that release adrenaline when stimulated by the sympathetic nervous system.

adrenergic Pertaining to nerves that release the neurotransmitter norepinephrine, or noradrenaline (such as adrenergic nerves, adrenergic response). The term also pertains to the receptors acted on by norepinephrine, that is, the adrenergic receptors.

aerobic metabolism Metabolism that can proceed only in the presence of oxygen.

agonal gasps Slow, gasping breaths, sometimes seen in dying patients.

alpha-adrenergic receptors Portions of the nervous system that when stimulated can cause constriction of blood vessels.

alveoli The air sacs of the lungs in which the exchange of oxygen and carbon dioxide takes place.

anaerobic metabolism The metabolism that takes place in the absence of oxygen; the principal product is lactic acid.

anatomic position The position of reference in which the patient stands facing you, arms at the side, with the palms of the hands forward.

anterior The front surface of the body; the side facing you in the standard anatomic position.

aorta The principal artery leaving the left side of the heart and carrying freshly oxygenated blood to the body.

apex (plural apices) The pointed extremity of a conical structure.

apneustic center Portion of the pons that increases the length of inspiration and decreases the respiratory rate.

appendicular skeleton The portion of the skeletal system that comprises the arms, legs, pelvis, and shoulder girdle.

appendix A small tubular structure that is attached to the lower border of the cecum in the lower right quadrant of the abdomen.

arterioles The smallest branches of arteries leading to the vast network of capillaries.

atrium One of the two upper chambers of the heart.

autonomic nervous system The part of the nervous system that regulates functions, such as digestion and sweating, that are not controlled voluntarily.

axial skeleton The part of the skeleton comprising the skull, spinal column, and rib cage.

ball-and-socket joint A joint that allows internal and external rotation, as well as bending.

beta-adrenergic receptors Portions of the nervous system that when stimulated can cause an increase in the force of contraction of the heart, an increased heart rate, and bronchial dilation.

biceps The large muscle that covers the front of the humerus.

bilateral In anatomy, a body part that appears on both sides of the midline.

bile ducts The ducts that convey bile between the liver and the intestine.

blood pressure (BP) The pressure that the blood exerts against the walls of the arteries as it passes through them.

brachial artery The major vessel in the upper extremity that supplies blood to the arm.

brain The controlling organ of the body and center of consciousness; functions include perception, control of reactions to the environment, emotional responses, and judgment.

brain stem The area of the brain between the spinal cord and cerebrum, surrounded by the cerebellum; controls functions that are necessary for life, such as respiration.

capillary vessels The tiny blood vessels between the arterioles and venules that permit transfer of oxygen, carbon dioxide, nutrients, and waste between body tissues and the blood.

cardiac muscle The heart muscle.

carotid artery The major artery that supplies blood to the head and brain.

cartilage The support structure of the skeletal system that provides cushioning between bones; also forms the nasal septum and portions of the outer ear.

cecum The first part of the large intestine, into which the ileum opens.

central nervous system (CNS) The brain and spinal cord.

cerebellum One of the three major subdivisions of the brain, sometimes called the "little brain"; coordinates the various activities of the brain, particularly fine body movements.

cerebrospinal fluid (CSF) Fluid produced in the ventricles of the brain that flows in the subarachnoid space and bathes the meninges.

cerebrum The largest part of the three subdivisions of the brain, sometimes called the "gray matter"; made up of several lobes that control movement, hearing, balance, speech, visual perception, emotions, and personality.

cervical spine The portion of the spinal column consisting of the first seven vertebrae that lie in the neck.

chordae tendineae Thin bands of fibrous tissue that attach to the valves in the heart and prevent them from inverting.

chyme The name of the substance that leaves the stomach. It is a combination of all of the eaten foods with added stomach acids.

circulatory system The complex arrangement of connected tubes, including the arteries, arterioles, capillaries, venules, and veins, that moves blood, oxygen, nutrients, carbon dioxide, and cellular waste throughout the body.

clavicle The collarbone; it is lateral to the sternum and anterior to the scapula.

coccyx The last three or four vertebrae of the spine; the tailbone.

coronal plane An imaginary plane where the body is cut into front and back parts.

cranium The area of the head above the ears and eyes; the skull. The cranium contains the brain.

cricoid cartilage A firm ridge of cartilage that forms the lower part of the larynx.

cricothyroid membrane A thin sheet of fascia that connects the thyroid and cricoid cartilages that make up the larynx.

dead space Any portion of the airway that does contain air and cannot participate in gas exchange, such as the trachea and bronchi.

deep Further inside the body and away from the skin.

dermis The inner layer of the skin, containing hair follicles, sweat glands, nerve endings, and blood vessels.

diaphragm A muscular dome that forms the undersurface of the thorax, separating the chest from the abdominal cavity. Contraction of the diaphragm (and the chest wall muscles) brings air into the lungs. Relaxation allows air to be expelled from the lungs.

diastole The relaxation, or period of relaxation, of the heart, especially of the ventricles.

diffusion Movement of a gas from an area of higher concentration to an area of lower concentration.

digestion The processing of food that nourishes the individual cells of the body.

distal Farther from the trunk or nearer to the free end of the extremity.

dorsal The posterior surface of the body, including the back of the hand.

dorsalis pedis artery The artery on the anterior surface of the foot between the first and second metatarsals.

dorsal respiratory group (DRG) A portion of the medulla oblongata where the primary respiratory pacemaker is found.

endocrine system The complex message and control system that integrates many body functions, including the release of hormones.

enzymes Substances catalysts designed to speed up the rate of specific biochemical reactions.

epidermis The outer layer of skin, which is made up of cells that are sealed together to form a watertight protective covering for the body.

epiglottis A thin, leaf-shaped valve that allows air to pass into the trachea but prevents food and liquid from entering.

epinephrine A hormone produced by the adrenal medulla that has a vital role in the function of the sympathetic nervous system.

esophagus A collapsible tube that extends from the pharynx to the stomach; contractions of the muscle in the wall of the esophagus propel food and liquids through it to the stomach.

expiratory reserve volume The amount of air that can be exhaled following a normal exhalation; average volume is about 1,200 mL.

extend To straighten.

extension The straightening of a joint.

fallopian tubes Long, slender tubes that extend from the uterus to the region of the ovary on the same side and through which the ovum passes from the ovary to the uterus.

femoral artery The principal artery of the thigh, a continuation of the external iliac artery. It supplies blood to the lower abdominal wall, external genitalia, and legs. It can be palpated in the groin area.

femoral head The proximal end of the femur, articulating with the acetabulum to form the hip joint.

femur The thighbone; the longest and one of the strongest bones in the body.

flex To bend.

flexion The bending of a joint.

foramen magnum A large opening at the base of the skull through which the brain connects to the spinal cord.

gallbladder A sac on the undersurface of the liver that collects bile from the liver and discharges it into the duodenum through the common bile duct.

genital system The reproductive system in males and females.

germinal layer The deepest layer of the epidermis where new skin cells are formed.

greater trochanter A bony prominence on the proximal lateral side of the thigh, just below the hip joint.

hair follicles The small organs that produce hair.

heart A hollow muscular organ that pumps blood through out the body.

heart rate The number of heartbeats during a specific time.

Hering-Breuer reflex A protective mechanism that terminates inhalation, thus preventing overexpansion of the lungs.

hinge joints Joints that can bend and straighten but cannot rotate; they restrict motion to one plane.

hormones Substances formed in specialized organs or glands and carried to another organ or group of cells in the same organism. Hormones regulate many body functions, including metabolism, growth, and body temperature.

humerus The supporting bone of the upper arm.

hydrostatic pressure The pressure of water against the walls of its container.

hypoxic drive A "backup system" to control respiration; senses drops in the oxygen level in the blood.

ilium One of three bones that fuse to form the pelvic ring.

inferior Below a body part or nearer to the feet.

inferior vena cava One of the two largest veins in the body; carries blood from the lower extremities and the pelvic and the abdominal organs to the heart.

inspiratory reserve volume The amount of air that can be inhaled after a normal inhalation; the amount of air that can be inhaled in addition to the normal tidal volume.

interstitial space The space in between the cells.

involuntary muscle The muscle over which a person has no conscious control. It is found in many automatic regulating systems of the body.

ischium One of three bones that fuse to form the pelvic ring.

joint (articulation) The place where two bones come into contact.

joint capsule The fibrous sac that encloses a joint.

kidneys Two retroperitoneal organs that excrete the end products of metabolism as urine and regulate the body's salt and water content.

labored breathing The use of muscles of the chest, back, and abdomen to assist in expanding the chest; occurs when air movement is impaired.

lactic acid A metabolic end product of the breakdown of glucose that accumulates when metabolism proceeds in the absence of oxygen.

large intestine The portion of the digestive tube that encircles the abdomen around the small bowel, consisting of the cecum, the colon, and the rectum. It helps regulate water balance and eliminate solid waste.

lateral In anatomy, parts of the body that lie farther from the midline. Also called outer structures.

lesser trochanter The projection on the medial/superior portion of the femur.

ligament A band of fibrous tissue that connects bones to bones. It supports and strengthens a joint.

liver A large solid organ that lies in the right upper quadrant immediately below the diaphragm; it produces bile, stores glucose for immediate use by the body, and produces many substances that help regulate immune responses.

lumbar spine The lower part of the back, formed by the lowest five nonfused vertebrae; also called the dorsal spine.

mandible The bone of the lower jaw.

manubrium The upper quarter of the sternum.

maxillae The upper jawbones that assist in the formation of the orbit, the nasal cavity, and the palate and hold the upper teeth.

medial Parts of the body that lie closer to the midline; also called inner structures.

medulla oblongata Nerve tissue that is continuous inferiorly with the spinal cord; serves as a conduction pathway for ascending and descending nerve tracts; coordinates heart rate, blood vessel diameter, breathing, swallowing, vomiting, coughing, and sneezing.

midbrain The part of the brain that is responsible for helping to regulate the level of consciousness.

midsagittal plane (midline) An imaginary vertical line drawn from the middle of the forehead through the nose and the umbilicus (navel) to the floor.

minute volume The amount of air that moves in and out of the lungs per minute minus the dead space. Also called minute ventilation.

motor nerves Nerves that carry information from the central nervous system to the muscles of the body.

mucous membranes The lining of body cavities and passages that communicate directly or indirectly with the environment outside the body.

mucus The opaque, sticky secretion of the mucous membranes that lubricates the body openings.

musculoskeletal system The bones and voluntary muscles of the body.

myocardium The heart muscle.

nasopharynx The part of the pharynx that lies above the level of the roof of the mouth, or palate.

nervous system The system that controls virtually all activities of the body, both voluntary and involuntary.

norepinephrine A neurotransmitter and drug sometimes used in the treatment of shock; produces vasoconstriction through its alpha-stimulator properties.

occiput The most posterior portion of the cranium.

oncotic pressure The pressure of water to move, typically into the capillary, as the result of the presence of plasma proteins.

orbit The eye socket, made up of the maxilla and zygoma.

oropharynx A tubular structure that extends vertically from the back of the mouth to the esophagus and trachea.

ovaries Female glands that produces sex hormones and ova (eggs).

palmar The forward facing part of the hand in the anatomic position.

pancreas A flat, solid organ that lies below the liver and the stomach; it is a major source of digestive enzymes and produces the hormone insulin.

parasympathetic nervous system A subdivision of the autonomic nervous system, involved in control of involuntary, vegetative functions, mediated largely by the vagus nerve through the chemical acetylcholine.

parietal regions The areas between the temporal and occipital regions of the cranium.

patella The kneecap; a specialized bone that lies within the tendon of the quadriceps muscle.

pathophysiology The study of how normal physiologic processes are affected by disease.

perfusion The circulation of oxygenated blood within an organ or tissue in adequate amounts to meet the cells' current needs.

peripheral nervous system The part of the nervous system that consists of 31 pairs of spinal nerves and 12 pairs of cranial nerves. These peripheral nerves may be sensory nerves, motor nerves, or connecting nerves.

peristalsis The wavelike contraction of smooth muscle by which the ureters or other tubular organs propel their contents.

plantar The bottom surface of the foot.

plasma A sticky, yellow fluid that carries the blood cells and nutrients and transports cellular waste material to the organs of excretion.

platelets Tiny, disk-shaped elements that are much smaller than the cells; they are essential in the initial formation of a blood clot, the mechanism that stops bleeding.

pleura The serous membranes covering the lungs and lining the thoracic cavity, completely enclosing a potential space known as the pleural space.

pleural space The potential space between the parietal pleura and the visceral pleura. It is described as "potential" because under normal conditions, the space does not exist.

pneumotaxic (pontine) center A portion of the pons that assists in creating shorter, faster respirations.

pons An organ that lies below the midbrain and above the medulla and contains numerous important nerve fibers, including those for sleep, respiration, and the medullary respiratory center.

posterior In anatomy, the back surface of the body; the side away from you in the standard anatomic position.

posterior tibial artery The artery just behind the medial malleolus; supplies blood to the foot.

prostate gland A small gland that surrounds the male urethra where it emerges from the urinary bladder; it secretes a fluid that is part of the ejaculatory fluid.

proximal Closer to the trunk.

pubic symphysis A hard bony and cartilaginous prominence found at the midline in the lowermost portion of the abdomen where the two halves of the pelvic ring are joined by cartilage at a joint with minimal motion.

pubis One of three bones that fuse to form the pelvic ring.

pulmonary artery The major artery leading from the right ventricle of the heart to the lungs; it carries oxygen-poor blood.

pulmonary circulation The flow of blood from the right ventricle through the pulmonary arteries and all of their branches and capillaries in the lungs and back to the left atrium through the venules and pulmonary veins; also called the lesser circulation.

pulmonary veins The four veins that return oxygenated blood from the lungs to the left atrium of the heart.

pulse The wave of pressure created as the heart contracts and forces blood out the left ventricle and into the major arteries.

quadrants The way to describe the sections of the abdominal cavity. Imagine two lines intersecting at the umbilicus dividing the abdomen into four equal areas.

radial artery The major artery in the forearm; it is palpable at the wrist on the thumb side.

radius The bone on the thumb side of the forearm.

rectum The lowermost end of the colon.

red blood cells Cells that carry oxygen to the body's tissues; also called erythrocytes.

renal pelvis A cone-shaped collecting area that connects the ureter and the kidney.

residual volume The air that remains in the lungs after maximal expiration.

respiration The inhaling and exhaling of air; the physiologic process that exchanges carbon dioxide from fresh air.

respiratory system All the structures of the body that contribute to the process of breathing, consisting of the upper and lower airways and their component parts.

reticular activating system Located in the upper brain stem; responsible for maintenance of consciousness, specifically one's level of arousal.

retroperitoneal Behind the abdominal cavity.

sacroiliac joint The connection point between the pelvis and the vertebral column.

sacrum One of three bones (sacrum and two pelvic bones) that make up the pelvic ring; consists of five fused sacral vertebrae.

sagittal (lateral) plane An imaginary line where the body is cut into left and right parts.

salivary glands The glands that produce saliva to keep the mouth and pharynx moist.

scalp The thick skin covering the cranium, which usually bears hair.

scapula The shoulder blade.

sebaceous glands Glands that produce an oily substance called sebum, which discharges along the shafts of the hairs.

semen Seminal fluid ejaculated from the penis and containing sperm.

seminal vesicles Storage sacs for sperm and seminal fluid, which empty into the urethra at the prostate.

sensory nerves The nerves that carry sensations of touch, taste, heat, cold, pain, and other modalities from the body to the central nervous system.

shock An abnormal state associated with inadequate oxygen and nutrient delivery to the metabolic apparatus of the cell.

shoulder girdle The proximal portion of the upper extremity, made up of the clavicle, the scapula, and the humerus.

skeletal muscle Muscle that is attached to bones and usually crosses at least one joint; striated, or voluntary, muscle.

skeleton The framework that gives the body its recognizable form; also designed to allow motion of the body and protection of vital organs.

small intestine The portion of the digestive tube between the stomach and the cecum, consisting of the duodenum, jejunum, and ileum.

smooth muscle Involuntary muscle; it constitutes the bulk of the gastrointestinal tract and is present in nearly every organ to regulate automatic activity.

somatic nervous system The part of the nervous system that regulates activities over which there is voluntary control.

sphincters Muscles arranged in circles that are able to decrease the diameter of tubes. Examples are found within the rectum, bladder, and blood vessels.

sphygmomanometer A device used to measure blood pressure.

spinal cord An extension of the brain, composed of virtually all the nerves carrying messages between the brain and the rest of the body. It lies inside of and is protected by the spinal canal.

sternum The breastbone.

stratum corneal layer The outermost or dead layer of the skin.

stroke volume (SV) The volume of blood pumped forward with each ventricular contraction.

subcutaneous tissue Tissue, largely fat, that lies directly under the dermis and serves as an insulator of the body.

superficial Closer to or on the skin.

superior Above a body part or nearer to the head.

superior vena cava One of the two largest veins in the body; carries blood from the upper extremities, head, neck, and chest into the heart.

sweat glands The glands that secrete sweat, located in the dermal layer of the skin.

symphysis A type of joint that has grown together forming a very stable connection.

synovial fluid The small amount of liquid within a joint used as lubrication.

synovial membrane The lining of a joint that secretes synovial fluid into the joint space.

systemic circulation The portion of the circulatory system outside of the heart and lungs.

systemic vascular resistance (SVR) The resistance that blood must overcome to be able to move within the blood vessels. SVR is related to the amount of dilation or constriction in the blood vessel.

systole The contraction, or period of contraction, of the heart, especially that of the ventricles.

temporal regions The lateral portions on each side of the cranium.

tendons The fibrous connective tissue that attaches muscle to bone.

testicle A male genital gland that contains specialized cells that produce hormones and sperm.

thoracic cage The chest or rib cage.

thoracic cavity The chest cavity that contains the heart, lungs, esophagus, and great vessels.

thoracic spine The 12 vertebrae that lie between the cervical vertebrae and the lumbar vertebrae. One pair of ribs is attached to each of the thoracic vertebrae.

thorax The chest cavity that contains the heart, lungs, esophagus, and great vessels.

thyroid cartilage A firm prominence of cartilage that forms the upper part of the larynx; the Adam's apple.

tibia The shin bone, the larger of the two bones of the lower leg.

tidal volume The amount of air moved in and out of the lungs in one relaxed breath; about 500 mL for an adult.

topographic anatomy The superficial landmarks of the body that serve as guides to the structures that lie beneath them.

torso The trunk without the head and limbs.

trachea The windpipe; the main trunk for air passing to and from the lungs.

transverse (axial) plane An imaginary line where the body is cut into top and bottom parts.

triceps The muscle in the back of the upper arm.

tunica media The middle and thickest layer of tissue of a blood vessel wall, composed of elastic tissue and smooth muscle cells that allow the vessel to expand or contract in response to changes in blood pressure and tissue demand.

ulna The inner bone of the forearm, on the side opposite the thumb.

ureter A small, hollow tube that carries urine from the kidneys to the bladder.

urethra The canal that conveys urine from the bladder to outside the body.

urinary bladder A sac behind the pubic symphysis made of smooth muscle that collects and stores urine.

urinary system The organs that control the discharge of certain waste materials filtered from the blood and excreted as urine.

vagina A muscular distensible tube that connects the uterus with the vulva (the external female genitalia); also called the birth canal.

vasa deferentia The spermatic duct of the testicles; also called vas deferens.

ventilation The movement of air between the lungs and the environment.

ventral The anterior surface of the body.

ventral respiratory group (VRG) A portion of the medulla oblongata that is responsible for modulating breathing during speech.

ventricle One of two lower chambers of the heart.

vertebrae The 33 bones that make up the spinal column.

voluntary muscle Muscle that is under direct voluntary control of the brain and can be contracted or relaxed at will; skeletal, or striated, muscle.

V̇/Q̇ ratio A measurement that examines how much gas is being moved effectively and how much blood is gaining access to the alveoli.

white blood cells Blood cells that have a role in the body's immune defense mechanisms against infection; also called leukocytes.

xiphoid process The narrow, cartilaginous lower tip of the sternum.

zygomas The quadrangular bones of the cheek, articulating with the frontal bone, the maxillae, the zygomatic processes of the temporal bone, and the great wings of the sphenoid bone.

Assessment in Action

You are dispatched to an unresponsive person at an automotive garage. On arrival, you find a middle-aged man lying on the ground by a tire rack.

1. The patient is lying on his back. What is the proper term for this position?
   - A. Supine
   - B. Prone
   - C. Recovery
   - D. Trendelenburg's

2. To determine the patient's level of consciousness, your partner rubs his knuckles on the patient's chest. The patient brings his arms up to his chest in response. This type of motion is called:
   - A. flexion.
   - B. abduction.
   - C. extension.
   - D. adduction.

3. After opening the patient's airway and confirming spontaneous respirations, you check for a pulse on his neck just lateral to the trachea. What artery are you using?
   - A. Radial
   - B. Carotid
   - C. Femoral
   - D. Brachial

4. You determine the patient has a strong pulse and perform a rapid assessment. While assessing the lower extremities, you observe marked swelling in the right thigh. What bone is in this location?
   - A. Fibula
   - B. Tibia
   - C. Femur
   - D. Patella

5. Which of the following substances is the main element that the bone cells use to create a hard and resilient structure?
   - A. Sodium
   - B. Potassium
   - C. Magnesium
   - D. Calcium

6. You perform an assessment while en route to the hospital. You note bruising and instability of the right cheekbone. What is the proper name for this bone?
   - A. Zygoma
   - B. Maxilla
   - C. Sphenoid
   - D. Mandible

7. The larger bone of the forearm at the wrist is called the:
   - A. humerus.
   - B. ulna.
   - C. radius.
   - D. clavicle.

8. The forearm makes up part of the elbow. The elbow is an example of what type of joint?
   - A. Hinge
   - B. Ball-and-socket
   - C. Saddle
   - D. Immovable

9. Based on your assessment findings, you suspect that your patient may have sustained a closed head injury. What structure of the brain is responsible for the level of consciousness and maintenance of vital signs?

10. Just before arriving at the hospital, you take your final set of vital signs. The patient's initial pulse rate was 89 beats/min and now it is 116 beats/min. What part of the nervous system is responsible for this increase?

Life Span Development

National EMS Education Standard Competencies

Preparatory

Applies fundamental knowledge of the emergency medical services (EMS) system, safety/well-being of the emergency medical technician (EMT), medical/legal, and ethical issues to the provision of emergency care.

Life Span Development

Applies fundamental knowledge of life span development to patient assessment and management.

• •

Knowledge Objectives

1. Understand the terms used to designate the following stages of life: infants, toddlers, preschoolers, school-age children, adolescents (teenagers), early adults, middle adults, and late adults. (pp 201-213)

2. Describe the major physiologic and psychosocial characteristics of an infant's life. (pp 201-204)

3. Describe the major physiologic and psychosocial characteristics of a toddler and preschooler's life. (pp 205-206)

4. Describe the major physiologic and psychosocial characteristics of a school-age child's life. (p 206)

5. Describe the major physiologic and psychosocial characteristics of an adolescent's life. (pp 206-207)

6. Describe the major physiologic and psychosocial characteristics of an early adult's life. (p 208)

7. Describe the major physiologic and psychosocial characteristics of a middle adult's life. (pp 208-209)

8. Describe the major physiologic and psychosocial characteristics of a late adult's life. (pp 209-213)

Skills Objectives

There are no skills objectives for this chapter.

Introduction

One of the most interesting things about humans is that we evolve—not just as a species, but as people over our life span. EMTs must be aware of both the obvious and subtle changes a person undergoes physically and mentally at various stages of life and understand how these changes may alter the approach to patient care.

Infants

As any parent can attest, **infants** (ages 1 month to 1 year) develop at a startling rate **Figure 6-1**. **Neonates** (from birth to 1 month) are covered in detail in Chapter 31, *Obstetrics and Neonatal Care*.

Physical Changes

Vital Signs

Table 6-1 lists the normal ranges of vital signs for various age groups. The general rule is the younger the person, the faster the pulse rate and respirations. At birth, a pulse rate of 90 to 180 beats/min and a respiratory rate of 30 to 60 breaths/min are considered normal. Within the first half hour after birth, a neonate's pulse rate often drops to 120 beats/min and the respiratory rate falls between 30 to 40 breaths/min. By age 1 year, the respiratory rate slows to 20 to

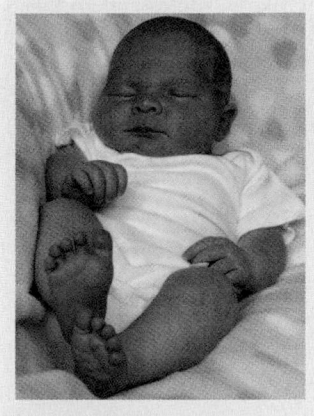

Figure 6-1 An infant.

Table 6-1 Vital Signs at Various Ages

Age	Pulse Rate (beats/min)	Respirations (breaths/min)	Systolic Blood Pressure (mm Hg)	Temperature (°F)
Neonate (0 to 1 month)	90 to 180	30 to 60	50 to 70	98 to 100
Infant (1 month to 1 year)	100 to 160	25 to 50	70 to 95	96.8 to 99.6
Toddler (1 to 3 years)	90 to 150	20 to 30	80 to 100	96.8 to 99.6
Preschool age (3 to 6 years)	80 to 140	20 to 25	80 to 100	98.6
School age (6 to 12 years)	70 to 120	15 to 20	80 to 110	98.6
Adolescent (12 to 18 years)	60 to 100	12 to 20	90 to 110	98.6
Early adult (19 to 40 years)	60 to 100	12 to 20	90 to 140	98.6
Middle adult (41 to 60 years)	60 to 100	12 to 20	90 to 140	98.6
Late adult (61 and older)	Depends on health	Depends on health	Depends on health	98.6

You are the Provider: PART 1

You and your partner are outside washing the ambulance when a man in his 50s pulls up in front of the ambulance bay door. He requests that you check his vital signs, which your EMS system offers as part of its community outreach program. As your partner is retrieving the blood pressure cuff and stethoscope from the ambulance, the man tells you that he is light-headed and needs to sit down. The time is 1:10 PM according to the dispatch operator, who acknowledges that you have a walk-in patient.

1. How does a patient's age affect your assessment?
2. What are some physical differences between middle adults and late adults?

30 breaths/min. Tidal volume in neonates starts at 6 to 8 mL/kg. By the end of the first year, the volume increases to 10 to 15 mL/kg.

Blood pressure directly corresponds to the patient's weight, so it typically increases with age. At birth, the average systolic blood pressure of a neonate is 50 to 70 mm Hg. By 1 year of age, it ranges between 70 and 95 mm Hg.

Weight

A neonate usually weighs 6 to 8 lb (3 to 3.5 kg) at birth. Remarkably, the head accounts for 25% of its body weight. In the first week after birth, neonates usually lose 5% to 10% of their birth weight due to fluid loss. By week 2, the neonate begins to gain weight. From here on, infants grow at a rate of about 30 g per day, doubling their weight by 4 to 6 months and tripling it by the end of the first year.

Special Populations

Infants often land head first when they fall because their heads account for 25% of their total body weight. Also, most infants cannot stretch out their arms in time to cushion or slow their fall. Keep this point in mind when considering spinal immobilization on an infant.

Cardiovascular System

Prior to birth, fetal circulation occurs through the placenta. During the birthing process, hormones and pressure changes help the neonate make the transition from fetal circulation to independent circulation. See Chapter 31, *Obstetrics and Neonatal Care*, for more information about fetal circulation.

Pulmonary System

Prior to a neonate's first breath, the lungs have never been inflated. A neonate's first breath is therefore forceful—it has to be!

Neonates are primarily "nose breathers." Infants younger than 6 months are particularly prone to nasal congestion, which can cause viral upper respiratory infections. If you receive a call for a baby choking, make sure the nasal passages are clear and unobstructed by mucus.

The rib cage of an infant is less rigid and the ribs sit horizontally. This explains the diaphragmatic breathing ("belly breathing") in infants.

Two other important anatomic points related to an infant's airway, when compared with an adult's, are the proportionally large size of the tongue and the

proportionally shorter and narrower airway. As a result of these factors, infants can much more easily occlude their airway than older children or adults can.

When providing bag-mask ventilations to an infant, you need to be aware that an infant's lungs are fragile. Ventilations that are too forceful can result in trauma from pressure, or **barotrauma**. Due to the large size of the infant's occiput, and the increased flexibility of the trachea, the airway can easily be inadvertently occluded by incorrect positioning, either over-extension or over-flexion. Infants also have very little reserves available to assist with breathing. The muscles they use to breathe are immature. They can manage normal requirements easily but can become fatigued when stressed. The number of alveoli in the infant's lungs is relatively low. Fortunately, the amount of oxygen that the infant needs is also relatively low. As the infant grows and moves more, the need for greater amounts of oxygen triggers a growth in the number of alveoli. However, in very small infants, respiratory problems can quickly turn life threatening. Infants who are struggling to breathe can quickly tire, become overheated, and even dehydrated.

Special Populations

When you are counting respirations in an infant, count the number of times the abdomen rises instead of concentrating solely on the chest rise.

Nervous System

Although the infant's nervous system is developed at birth, its evolution continues after birth. For example, the neonate lacks the ability to localize and isolate a particular response to sensation. When neonates are born, they tend to move their extremities together. They do not have independent arm or leg movements until many weeks later.

A neonate is born with certain reflexes. The **moro reflex** (startle reflex) happens when a neonate is caught off guard by something or someone; the neonate opens his or her arms wide, spreads the fingers, and seems to grab at things. A **palmar grasp** occurs when an object is placed into the neonate's palm. The **rooting reflex** takes place when something touches a neonate's cheek; the neonate will instinctively turn his or her head toward the touch. In conjunction with the **sucking reflex**, which occurs when a neonate's lips are stroked, these reflexes are often tested when feeding.

A neonate's **fontanelles** allow the head to be molded—for example, when the neonate passes through the birth canal Figure 6-2 . These three or four bones of the skull eventually bind together and form suture joints. The posterior fontanelle normally fuses by the

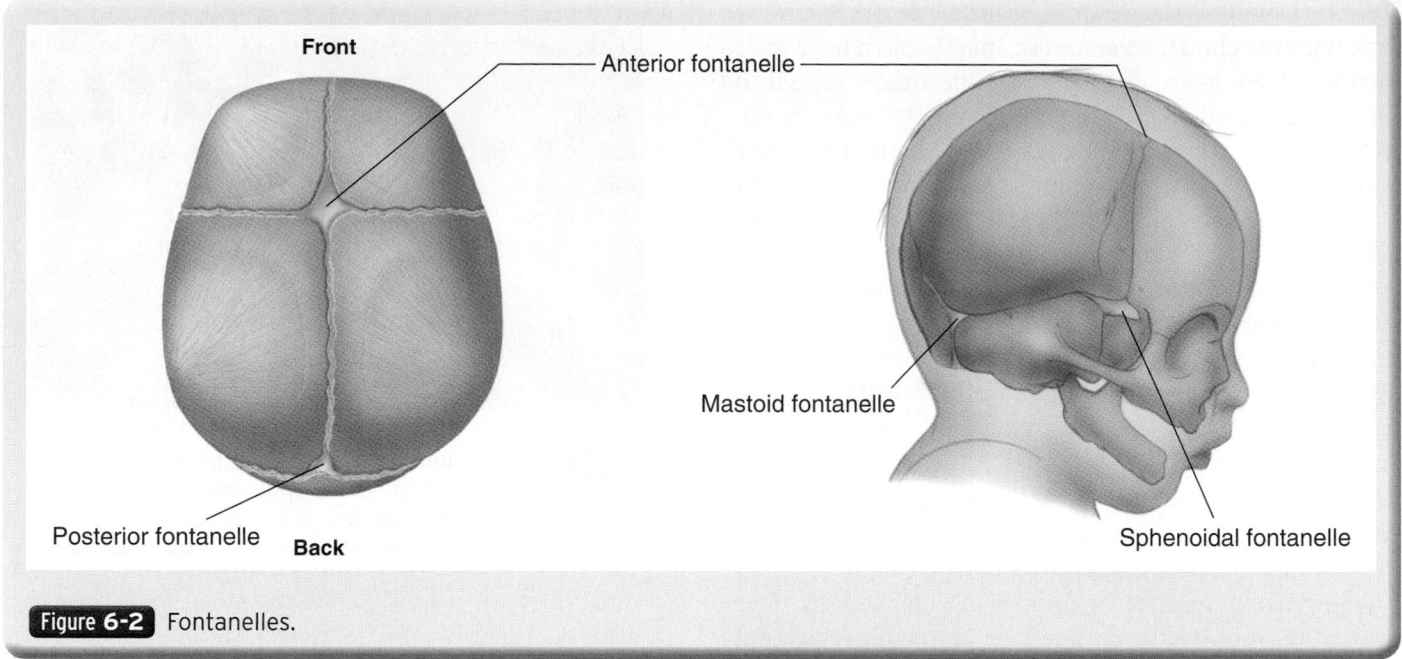

Front

Anterior fontanelle

Mastoid fontanelle

Posterior fontanelle **Back**

Sphenoidal fontanelle

Figure 6-2 Fontanelles.

third month of life. The anterior fontanelle fuses between 9 and 18 months of age. If either of the fontanelles is depressed, the infant is most likely dehydrated. A bulging fontanelle is indicative of increased intracranial pressure.

It is wonderful to watch an infant grow. At birth, the neonate is not able to do much without assistance. He or she cannot turn over or even focus his or her eyes beyond a very short distance. Sleep patterns begin to develop. But by 2 months of age, the infant is able to track objects with his or her eyes and should recognize familiar faces. At 6 months of age, the infant is able to sit upright and begins to make cooing and babbling sounds. By 12 months of age, the infant can walk with assistance and even knows his or her name.

Immune System

While in the womb, fetuses collect antibodies from the maternal blood. For the first year of life, the infant maintains some of the mother's immunities, so he or she has naturally acquired passive immunities. Infants can also receive antibodies via breastfeeding, further bolstering their immune system.

■ Psychosocial Changes

An infant's psychosocial development begins at birth and continues to evolve as the infant interacts with, and reacts to, the environment. Parents often obsess about whether their child is developing within the socially accepted norms. **Table 6-2** outlines typical ages at which major psychosocial changes are noticed.

In most infants, the primary method of communicating distress is through crying. Parents can often tell what is upsetting their child simply by listening to the

Table 6-2	Noticeable Characteristics at Various Ages
Age	**Characteristic**
2 months	Can recognize familiar faces; able to track objects with the eyes
3 months	Can bring objects to the mouth; can smile and frown
4 months	Reaches out to people; drools
5 months	Sleeps through the night; can tell family from strangers
6 months	Teething begins; sits upright in a chair; one-syllable words spoken
7 months	Afraid of strangers; mood swings
8 months	Responds to "no"; can sit alone; plays peek-a-boo
9 months	Pulls himself or herself up; places objects in mouth to explore them
10 months	Responds to his or her name; crawls efficiently
11 months	Starts to walk without help; frustrated with restrictions
12 months	Knows his or her name; can walk

tone of the child's crying—that is, they know the difference between tears for anger, frustration, pain, fear, hunger, discomfort, and sleepiness. Infants occasionally make another distinct cry—an alarming distressed cry. This cry may be heard when an unexpected event occurs, causing a situational crisis for the infant.

The key to having a happy, healthy infant is spending time with the child. Nevertheless, infants often have their own timetable as to when they will become attached to their parents and other family members. **Bonding**, or the formation of a close, personal relationship, is usually based on a **secure attachment**. A secure attachment occurs when an infant understands that parents or caregivers will be responsive to his or her needs. This realization encourages a child to reach out and explore, knowing that the parents will provide a "safety net."

Another type of attachment, referred to as **anxious-avoidant attachment**, is observed in infants who are repeatedly rejected. In this attachment style, children show little emotional response to their parents or caregivers and treat them as they would strangers. These children develop an isolated lifestyle where they do not have to depend on the support and care of others.

Separation anxiety is common in older infants. This normal reaction peaks between 10 and 18 months and involves clingy behavior and fear of unfamiliar places and people. Protesting by crying is another normal reaction in older infants. As infants become accustomed to their homes and families, they begin to need the security of a predictable environment. If the infant's environment is too unpredictable, the infant may despair and become withdrawn, which leads to trust issues.

Trust and mistrust refers to a stage of development from birth to about 18 months of age that involves an infant's needs being met by his or her parents or caregivers. When caregivers and parents provide an

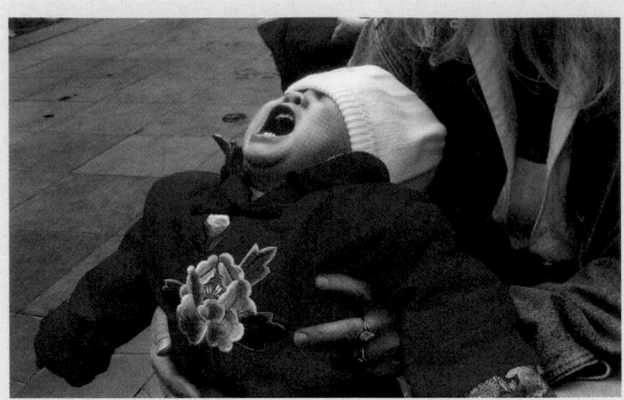

Figure 6-3 If an infant perceives that his or her parents or caregivers will not provide an organized, routine environment, behavior problems can develop.

organized, routine environment, the infant gains trust in those individuals. If the environment is not perceived as secure by the infant, a sense of mistrust will develop **Figure 6-3**.

Special Populations

When dealing with patients who are very young, try to keep their routine the same by keeping family and familiar items nearby.

You are the Provider: PART 2

After sitting the man down in a chair inside your station, he tells you that he has been stressed lately, although he does not know why, and that he has had several episodes of light-headedness over the past few days. He denies having chest pain, shortness of breath, or any other symptoms. As you are assessing the patient, your partner prepares to take his vital signs.

Recording Time: 0 Minutes	
Appearance	Calm
Level of consciousness	Conscious and alert
Airway	Open; clear of secretions or foreign bodies
Breathing	Normal rate; adequate depth
Circulation	Radial pulses, strong and regular; skin is pink, warm, and dry

3. What are some common psychosocial concerns experienced by middle adults?

Toddlers and Preschoolers

Physical Changes

In <u>toddlers</u> (ages 1 to 3 years), the pulse rate is 90 to 150 beats/min and the respiratory rate is 20 to 30 breaths/min, slower than the corresponding vital signs in infants, whereas the systolic blood pressure is higher (80 to 100 mm Hg). The average temperature of children this age is 96.8°F to 99.6°F, usually leveling off at 98.6°F by school age Figure 6-4 .

In <u>preschoolers</u> (ages 3 to 6 years), the pulse rate is 80 to 140 beats/min and the respiratory rate is 20 to 25 breaths/min. The systolic blood pressure is 80 to 100 mm Hg. At the same time, weight gain should level off Figure 6-5 .

A toddler's cardiovascular system is not dramatically different from an adult's. A toddler's lungs continue to develop more terminal bronchioles and alveoli. Although toddlers and preschoolers have more lung tissue, they do not have well-developed lung musculature. This anomaly prevents them from sustaining deep or rapid respirations for an extended period of time.

The loss of passive immunity is possibly the most obvious development at this stage of human life. "Colds" often develop that may manifest as gastrointestinal distress or upper respiratory tract infections. As toddlers spend more time around playmates and classmates, they acquire their own immunity as the body is exposed to various viruses and germs.

Neuromuscular growth also makes considerable progress at this age. Toddlers and preschoolers spend a great deal of time finding out exactly how to use their expansive nervous system and the muscles it controls by walking, running, jumping, and playing catch Figure 6-6 .

Figure 6-4 A toddler.

Figure 6-5 A preschooler.

Watching children play as they age from 1 to 6 years demonstrates how they move from gross motor activities (grabbing an object with the full palm) to fine motor activities (picking up a crayon). By the end of this stage, preschoolers will have a brain that weighs 90% of its final adult weight. In addition, all of this playing places stress on the muscles and bones. Consequently, muscle mass increases as does bone density.

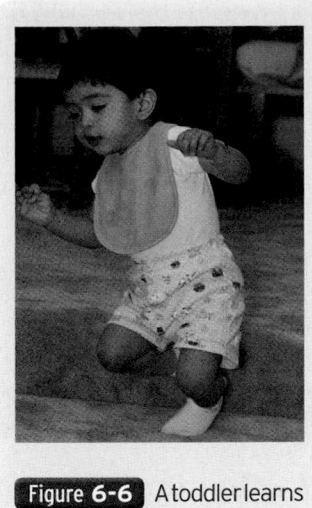

Figure 6-6 A toddler learns to walk, one of the major milestones in life.

This stage also includes the continued development of the renal system and of elimination patterns (ie, toilet training). Physiologically, toddlers have the neuromuscular control capable for bladder control by 12 to 15 months of age. However, the child may not be psychologically ready until 18 to 30 months of age. The average age for completion of toilet training is 28 months of age.

Other developments that occur during this time frame include the emergence of "baby" teeth. Teething (ie, teeth breaking through the gums) can be painful and accompanied by fever. In addition, parents and toddlers are enthralled with sensory development—for example, tickling.

Psychosocial Changes

This period of development is often exciting for parents. Toddlers or preschoolers are learning to speak and express themselves, thereby taking a major step toward independence. At the same time, toddlers are very attached to their parents and feel safe with them. Separation anxiety peaks between 10 and 18 months of age. It is fascinating to watch a child struggle through the conflict of wanting to play, yet wanting to be protected.

At 36 months of age, in most toddlers, basic language is mastered. By the age of 3 or 4 years, most children can use and understand full sentences. As they progress through this stage, they will go from using language to communicate what they want, to using language creatively and playfully.

This is also the time when toddlers begin to interact with other children and start to play games. Playing games teaches control, following of rules, and even competitiveness. Significant learning and development takes place by the child watching his or her peers during group outings, such as "play dates" with other children. By 18 to 24 months, toddlers begin to understand cause and effect. Of course, behavior observed on television and computers can also be learned, which is why some parents limit their children's viewing choices or the amount of time they devote to these activities. During this phase of development, children also learn to recognize sexual differences by observing their role models.

School-Age Children

Physical Changes

From ages 6 to 12 years, a **school-age** child's vital signs and body gradually approach those observed in adulthood **Figure 6-7**. The pulse rate is approximately 70 to 120 beats/min, the respiratory rate 15 to 20 breaths/min, and blood pressure is 80 to 110 mm Hg. Obvious physical traits and body function changes become apparent as most children grow about 4 lb (2 kg) and 2½″ (6 cm) each year. Permanent teeth also come in during this period and brain activity increases in both hemispheres.

Psychosocial Changes

School-age children are engaged in a lot of psychosocial growing up. Parents as a whole do not devote as much time to their children during this phase. Nevertheless, it

is at this critical time in human development that children learn various types of reasoning. In **preconventional reasoning**, children act almost purely to avoid punishment and to get what they want. In **conventional reasoning**, they look for approval from their peers and society. In **postconventional reasoning**, children make decisions guided by their conscience.

During this stage, children begin to develop their self-concept and self-esteem. Self-concept is our perception of ourselves; self-esteem is how we feel about ourselves and how we "fit in" with our peers.

Figure 6-7 A school-age child.

Adolescents (Teenagers)

Physical Changes

In **adolescents** (ages 12 to 18 years), vital signs begin to level off within the adult ranges, with a systolic blood pressure generally between 90 and 110 mm Hg,

You are the Provider: PART 3

As your partner takes the patient's vital signs, he tells you that he and his wife are taking care of his father, who is 82 years old and has Alzheimer disease. He further tells you that, although this situation is very stressful for him, he does not want to put his father in a nursing home. He is still light-headed and now complains of a headache. After applying oxygen via nasal cannula, you advise him that he should be transported to the hospital via EMS, but he tells you that he would rather drive himself.

Recording Time: 5 Minutes	
Respirations	14 breaths/min; regular and adequate
Pulse	76 beats/min; strong and regular
Skin	Pink, warm, and dry
Blood pressure	174/98 mm Hg
Oxygen saturation (Sao$_2$)	98% (on oxygen)

4. Are the patient's vital signs consistent with his age?

5. Why should you transport this patient to the hospital?

Figure 6-8 An adolescent.

a pulse rate between 60 and 100 beats/min, and respirations in the range of 12 to 20 breaths/min Figure 6-8 .

Adolescence is also the time of life when humans experience a 2- to 3-year growth spurt (ie, an increase in muscle and bone growth) and body changes. Growth begins with hands and feet, then moves to the long bones of the extremities, and finishes with growth of the torso. As a whole, boys experience this growth spurt later in life than girls. Girls finish their growth spurt by 16 years of age and boys by 18 years of age. When this period of growth has finished, however, boys are generally taller and stronger than girls. Muscle mass and bone density are nearly at adult levels.

One of the more subtle changes during adolescence is the maturation of the human reproductive system. Secondary sexual development begins, along with enlargement of the external sex organs. Pubic hair and axillary hair begin to appear. Voices start to change in range and depth. In females, the breasts and thighs increase in size as adipose (fat) tissue is deposited there. Menstruation begins during this time. Menarche, the first menstrual bleeding, occurs during this time; however, it is not uncommon to begin menstruation prior to becoming a teenager.

These changes in the endocrine and reproductive systems provide the platform for reproduction. By the middle of adolescence, boys are able to produce sufficient sperm and girls are able to develop eggs. Acne can also occur due to hormonal changes.

■ Psychosocial Changes

Adolescents and their families often deal with conflict as adolescents try to gain control of their lives from their parents. Privacy becomes an issue among adolescents, their siblings, and their parents. Self-consciousness also increases. Adolescents may struggle to create their own identity—to define themselves Figure 6-9 , for example, by dressing in a certain style of clothing to fit their personality. Adolescents use the feedback from their family and peers to help create their adult image. Adolescents are often caught between two worlds. They want to be treated like adults yet want to be cared for like younger children.

Figure 6-9 Adolescents want to fit in and may struggle to create identity.

Rebellious behavior can be part of an adolescent trying to find his or her own identity. Typically antisocial behavior and peer pressure tend to peak at around age 14 to 16 years. Smoking, illicit drug use, unprotected sex, and other high-risk behaviors also peak during this period. Children can try to exhibit self-control through what they eat, which can lead to eating disorders. Although these behaviors can be very troubling to parents, the adolescent is trying to determine if he or she is ready to take control of his or her own life. An adolescent's struggle toward independence may have setbacks that may be devastating. Patience and support from family and friends are essential in assisting an adolescent's transition into adulthood.

Adolescents may also show greater interest in sexual relations. Many adolescents are fixated on their public image and are terrified of being embarrassed. At this age, a code of personal ethics is developed, based partly on parents' ethics and values and partly on the influence of the adolescent's environment. At this tumultuous time, adolescents are at a higher risk than other populations for suicide and depression.

Special Populations

When you interview adolescents in the presence of their family, they may not tell you the complete truth in an attempt to protect their privacy or image. It is best to ask these patients certain questions in total privacy, where they feel they can answer without constraint.

Early Adults

Physical Changes

<u>Early adults</u> range in age from 19 to 40 years Figure 6-10 . Their vital signs do not vary greatly from those seen throughout adulthood. Ideally, the human pulse rate will average around 70 beats/min, the respiratory rate will stay in the range of 12 to 20 breaths/min, and the systolic blood pressure will be approximately between 90 and 140 mm Hg.

Figure 6-10 An early adult.

From age 19 years to shortly after 25 years, the human body should be functioning at its optimal level. Lifelong habits such as eating preferences, exercise, and tobacco use are solidified. At the beginning of this period, the body is working at peak efficiency, but as early adulthood continues, subtle erosion begins.

The disks in the spine begin to settle, and height can sometimes be affected, causing a "shrinking." Being able to eat anything without gaining weight becomes a thing of the past. Fatty tissue increases, which leads to weight gain. Muscle strength decreases, and reflexes slow.

Psychosocial Changes

Three words best describe a human's world during this stage of life: work, family, stress. During this period, adults strive to create a place for themselves in the world, and many do everything they can to "settle down." Along with this natural tendency to settle comes love and childbirth. Despite all of this stress and change, this age group enjoys one of the more stable periods of life.

Middle Adults

Physical Changes

<u>Middle adults</u> are ages 41 to 60 years Figure 6-11 . The average pulse rate for this age remains at 70 beats/min, the respiratory rate continues at 12 to 20 breaths/min, and the blood pressure also remains between 90 and 140 mm Hg. This group is vulnerable to vision and hearing loss. Cardiovascular health also becomes an issue in many of these persons, as does the greater incidence of cancer. In women, menopause—the cessation of menstruation—begins in the late 40s or early 50s. Middle adults

Figure 6-11 A middle adult.

You are the Provider: PART 4

After expressing your concern about the patient's health and advising him that driving himself to the hospital would not be safe, he agrees to be transported via EMS. You place the patient onto the stretcher, load him into the ambulance, and begin transport to a hospital located a short distance away. En route, you reassess his vital signs; assess his blood glucose level, which reads 100 mg/dL; and then call your radio report to the hospital.

Recording Time: 11 Minutes	
Level of consciousness	Conscious and alert
Respirations	14 breaths/min; regular and adequate
Pulse	80 beats/min; strong and regular
Skin	Pink, warm, and dry
Blood pressure	180/102 mm Hg
Sao$_2$	99% (on oxygen)

6. What additional treatment, if any, is required for this patient?

may begin having medical problems or be unaware of problems such as diabetes and hypertension. Medications or underlying conditions may affect patient response to treatments. Other concerns include an increase in cholesterol levels, a decrease in the efficiency of the heart, and problems with weight control. Many of the effects of aging can be diminished, however, with exercise and a healthy diet.

■ Psychosocial Changes

Middle adults tend to focus on achieving their life's goals, as they approach the halfway point in human life expectancy. After years of nurturing and living with children, parents must readjust their lifestyle as children leave home, commonly called the "empty nest" syndrome. Finances may become a worrisome issue, as people prepare for retirement while still managing everyday financial demands. During this time people often view crisis as a challenge to be overcome and not a threat to be avoided. Generally, their health is stable and they have the physical, emotional, and spiritual reserves to handle life's issues.

The parents of adults in this age group are getting older and now need care. Most of the elderly in the United States are cared for by family members inside the home. Therefore, a person in middle adulthood may need to manage children who are leaving for college while at the same time caring for parents who require greater assistance.

Late Adults

■ Physical Changes

Late adults include those ages 61 and older Figure 6-12. Life expectancy is constantly changing. In the early 1900s, life expectancy was 47 years. It is now approximately 78 years, with maximum life expectancy estimated at 120 years. The age to which a person will live is based on many factors. Perhaps surprisingly, the year you were born and the country you live in can have an effect on your life expectancy. These two facts are based on public health advances, changes within diets, attitudes regarding exercise, advances in medical care, access to that medical care, and personal behaviors.

Later in life, the vital signs depend on the patient's overall health, medical conditions, and medications taken. Today's late adults are staying active longer than their ancestors. Thanks to medical advances, they are often able to overcome numerous medical problems, but may need multiple medications to do so Figure 6-13.

Special Populations

Be patient when interviewing older patients. Some older patients may have physical, intellectual, and psychological barriers that may slow or interfere with effective communication.

Cardiovascular System

Cardiac function declines with age consequent to anatomic and physiologic changes that are largely related to **atherosclerosis**. In this disorder, which most commonly affects coronary vessels, cholesterol and calcium build-up inside the walls of blood vessels, forming plaque. The accumulation of plaque eventually leads to partial or complete blockage of blood flow. More than 60% of people older than 65 years have atherosclerotic disease. This can lead to decreased blood supply to the organs of the body.

Figure 6-12 A late adult.

Figure 6-13 Older people are often prescribed multiple medications to help them stay active.

Other age-related changes typically include a decrease in heart rate, a decline in cardiac output (the amount of blood circulated each minute), and the inability to elevate cardiac output to match the demands of the body. This translates into a heart that is less able to respond to exercise or disease. In the event of a life-threatening illness, the body typically needs to increase the heart rate to ensure adequate blood pressure. Because heart muscle may be weakened with age, the increase in heart rate can actually cause damage to the heart itself.

The vascular system also becomes stiff. For example, the diastolic blood pressure increases with age. Compensation for blood pressure changes is hampered because these vessels are less able to dilate and contract. As the blood vessels become stiffer, the heart must work harder to be able to move the blood effectively. These stiff blood vessels increase the workload of the heart.

Blood cells are also affected by aging. The body's blood cells originate from within the bone marrow. As a person ages, more of the bone marrow is replaced with fatty tissue. This replacement decreases the ability of the bones to manufacture more blood cells when needed. Although typically by itself this issue does not pose a problem, if an elderly person sustains trauma, the ability of the body to produce blood cells to replace those lost is diminished. Finally, functional blood volume gradually declines over time.

Respiratory System

In late adults, the size of the airway increases and the surface area of the alveoli decreases. The natural elasticity of the lungs also decreases, forcing individuals to use the muscles between their ribs, called intercostal muscles, more to breathe. As the elasticity of the lungs decreases, the overall strength of the intercostal muscles and diaphragm also decreases. These factors together make breathing more labor intensive for the elderly. One would think that a rigid chest would be more protecting, but this rigidity actually makes the chest more fragile. Instead of the chest being able to bend and give if struck, the calcified chest can fracture. As with all of the physical changes related to aging, however, the changes in the respiratory system are often gradual and go unnoticed until a severe, life-threatening condition occurs. The older person will then have less respiratory reserve on which to use in order to maintain adequate breathing.

Within the mouth and nose there is a gradual loss of the mechanisms that protect the upper airway. This leads to a decreased ability to clear secretions as well as decreased cough and gag reflexes. The cilia that line the airways diminish with age, while the innervation of the structures in the airway provides increasingly less sensation. Without the ability to maintain the upper airway, aspiration and obstruction become more likely.

When a younger patient inhales, the airway maintains its shape, allowing air to enter. As the smooth muscles of the lower airway weaken with age, strong inhalation can make the walls of the airway collapse inward and cause inspiratory wheezing. The collapsing airways result in low flow rates, because less air can move through the smaller airways, and air trapping, because air does not completely exit the alveoli (incomplete expiration). Also within the airways, the cells of the immune system are less functional. As a result of overall decreases in the metabolic activity of the elderly body, the white blood cells found within the airways are less aggressive at fighting invading organisms. This leads to an increased risk of lung infections.

By age 75 years, the vital capacity (the volume of air moved during the deepest inspiration and expiration) may amount to only 50% of the vital capacity of a young adult. Factors contributing to this decline include loss of respiratory muscle mass, increased stiffness of the thoracic cage, and decreased surface area available for the exchange of air.

Physiologically, vital capacity decreases and residual volume (the amount of air left in the lungs after expiration of the maximum possible amount of air) increases with age. A lifetime of breathing, especially breathing air with high levels of pollution, causes the accumulation of pollutants in the lungs. As a consequence, stagnant air remains in the alveoli and hampers diffusion of gases. The net effect is that the respiratory system is increasingly less able to handle the stresses of disease. This is why a simple cold, which for a 30-year-old would mean a runny nose and body aches, for an 80-year-old could mean pneumonia and possible death.

Endocrine System

As with the other systems of the body, the function of the endocrine system gradually declines. Insulin production begins to drop off and metabolism decreases. As people get older, they tend to slow down their physical activity. Unfortunately, they do not decrease their food intake. When a person gains weight, more insulin is needed to control the body's metabolism and blood glucose (sugar) level. The pancreas may not be able to produce enough insulin for the person's body size, which can lead to diabetes mellitus.

The reproductive systems of both men and women change with age. Men are able to produce sperm long into their 80s but the rigidity of the penis tends to decrease over time. It is unclear whether this decrease is due to aging itself or other diseases such as cardiovascular disease. Women have a decrease in the size of the uterus and vagina. Hormone production for both sexes gradually decreases as they age. Sexual desire may diminish with age but certainly does not cease.

Digestive System

Changes in gastric and intestinal function may inhibit nutritional intake and utilization in older adults. For example, taste bud sensitivity to salty and sweet sensations decreases. The sense of smell can also be diminished. In concert with a decreased taste response, elderly people may find food bland and flavorless.

Saliva secretion decreases, which reduces the body's ability to process complex carbohydrates. Older people may have loss of teeth that impacts their ability to chew. The ability of the intestines to contract and move food along diminishes with age. This can lead older adults to feel constipated or not hungry. Likewise, gastric acid secretion diminishes. Blood flow may drop by as much as 50%, decreasing the ability of the intestines to extract vitamins and minerals from digested food. Gallstones become increasingly common with age, and anal sphincter changes reduce elasticity and can produce fecal incontinence.

Renal Systems

In the kidneys, both structural and functional changes occur in the late adult. The filtration function of these organs, for example, declines by 50% from age 20 to 90 years. Kidney mass decreases by 20% over the same span. This is due in part to the decreased effectiveness of the blood vessels that supply blood to the nephrons. Nephrons are sophisticated capillaries that perform filtering in the kidney. One of the portions of the nephron is called the glomeruli. The decreased blood supply causes more abnormal glomeruli to be present as the person ages. The number of nephrons also declines between the ages of 30 to 80 years. This loss of renal function means a decrease in the ability to clear wastes from the body. It also means a decreased ability to conserve fluids when needed.

Nervous System

Nervous system changes can result in the most debilitating of age-related ailments. In the central nervous system, the brain weight may shrink 10% to 20% by age 80. Motor and sensory neural networks become slower and less responsive. The metabolic rate in the older brain does not change, however, and oxygen consumption remains constant throughout life. Generally, you have fewer brain cells (neurons) today than you did yesterday. If measured strictly by numbers of brain cells, infants are far more intelligent than any of us.

However, this is not how the brain works. Although it is true the elderly have a diminished number of brain cells, there is great flexibility in the operation of the brain. Interconnections between brain cells continue as people age. These new connections provide redundancy within the brain, allowing for loss of neurons without loss of knowledge or skill.

One of the consequences of the loss of neurons is a change in the sleep patterns of the elderly. Instead of sleeping through the night, the elderly may take a nap during the day and be up late at night. Their sleep cycle may move into a biphasic (two-phased) sleep cycle—sleep from 1:00 AM to 6:00 AM and then a nap from 12:00 PM to 3:00 PM.

The brain, which is surrounded by the meninges, takes up almost all of the space in the skull. Cerebrospinal fluid protects the brain inside these membranes. Unfortunately, age-related shrinkage creates a void between the brain and the outermost layer of the meninges, which provides room for the brain to move when stressed **Figure 6-14**. If trauma moves the brain forcefully, the bridging veins can tear and bleed. Bleeding can empty into this void and may go unnoticed for some time.

Functioning of the peripheral nervous system also slows with age. Sensation becomes diminished and misinterpreted. The ability to know where the body is in space, the kinesthetic sense, can be diminished. Increased reaction times cause longer delays between stimulation and motion. The resulting slowdown in reflexes and decreased kinesthetic sense may contribute to the incidence of falls and trauma. Nerve endings deteriorate, and the ability of the skin to sense the surroundings becomes hindered. Hot, cold, sharp, and wet items can all create dangerous situations because the body cannot sense them quickly enough.

Sensory Changes

Often it is assumed that the elderly are hard of hearing and have difficulty seeing. There are changes that diminish the effectiveness of the eyes and ears; however, most elderly individuals can hear well and are able to see clearly. They may need glasses or hearing aids, but it is wrong to assume that your older patient is deaf and nearly blind. Pupillary reaction and ocular movements become more restricted with age. The pupils are generally smaller in older patients, and the opacity of the eye's lens diminishes visual acuity and makes the pupils sluggish when responding to light. Visual distortions are also common in older people. Thickening of the lens makes it harder for the eye to focus, especially at close range. Peripheral fields of vision become narrower, and a greater sensitivity to glare constricts the visual field.

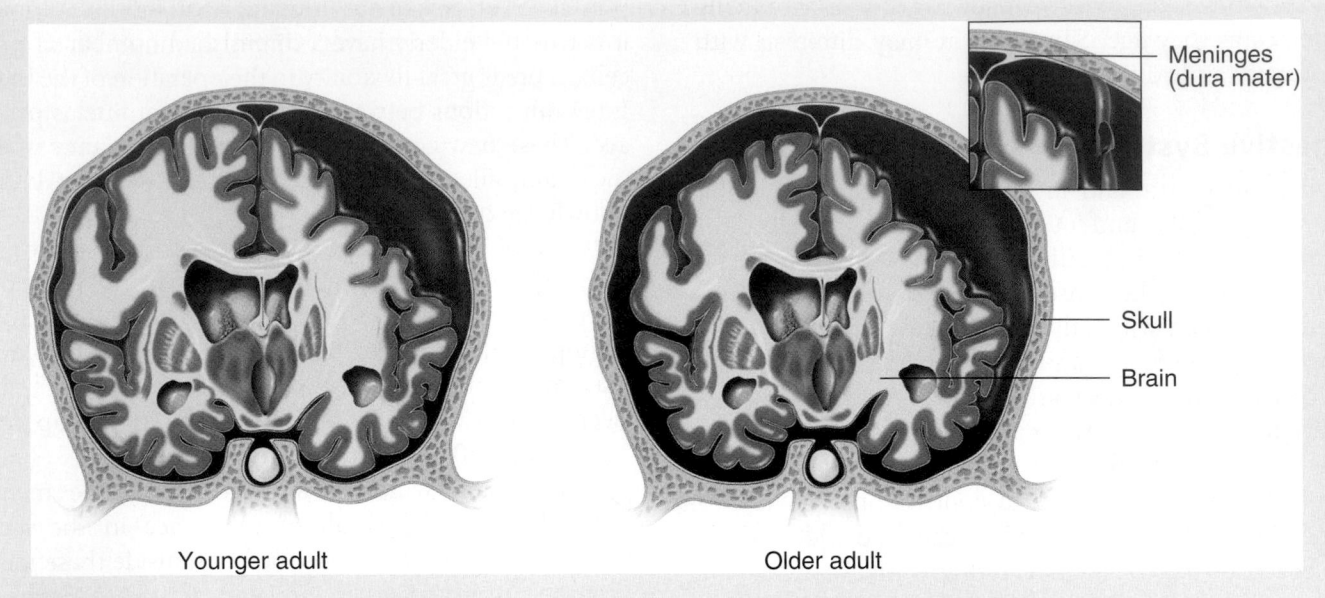

Younger adult   Older adult

Figure 6-14 Age-related atrophy or shrinkage of the brain results in a space between the brain and its cover, the dura mater. Bleeding into this area can occur because veins are stretched.

Hearing loss is about four times more common than loss of vision in late adults. Changes in several hearing-related structures may lead to a loss of high-frequency hearing, or even deafness.

Psychosocial Changes

EMTs should treasure their opportunities to spend time with and communicate with the elderly. Many of them have amazing stories and experiences to share with us, yet we often take them for granted. They share with us a great amount of wisdom, and we need to remind them of their self-worth. Until about 5 years before death, most late-stage adults retain high brain function. In the 5 years preceding death, however, mental function is presumed to decline, a theory referred to as the **terminal drop hypothesis**.

As the elderly population continues to grow, we have the responsibility to seek out unique ways to accommodate their needs during their last 20 to 40 years of life. Statistics indicate that 95% of the elderly live at home. They certainly may have the assistance of family, friends, or home health care, but they are relatively healthy, active, and independent. The increasing number of elderly in the United States as a result of the baby boom of the 1940s and 1950s has produced a need for additional assisted-living facilities. These facilities allow older adults to live in campus-based communities with people in their own age group, while enjoying the privacy of their own apartment and the security of nursing care, maintenance, and

food preparation, if desired **Figure 6-15**. Unfortunately, these facilities can be expensive.

Most people need to deal with financial issues throughout their lives. Few things in life produce more worry and stress than money problems. Late adults, in particular, may constantly worry about rising costs of health care and are often forced to make decisions such as whether to pay for groceries or their medication. Modern families often take less responsibility for their elderly family members than earlier generations did. Today, more than 50% of all single women in the

Figure 6-15 A small percentage of older adults live in assisted-living facilities.

United States who are 60 years of age or older are living at or below the poverty level. This problem remains to be resolved.

One of the important issues that the elderly need to face is their own mortality. The fact is everyone dies. Yet for most of us, this concept is an intellectual exercise with a distant connection to reality. Elderly persons witness their friends and loved ones, some of whom they have shared this life journey for half a century or more, die. Isolation and depression are challenges for the elderly.

Many elderly persons are happy and actively participating in life. With good financial resources and a good support system of family and friends, elderly individuals in their 80s can enjoy life and continue to feel productive.

You are the Provider: SUMMARY

1. How does a patient's age affect your assessment?

The fundamental concepts of patient assessment are the same for all age groups. However, factors such as physical development, communication skills, behavior, and vital sign values vary with age. An important part of "knowing your patient" involves understanding these age-related differences; they will affect how you direct your assessment and interpret the findings.

Communication with the patient is an integral part of the patient assessment process—especially the history-taking phase. Depending on the patient's age, communication can be relatively easy or extremely difficult. Adults are generally good historians and are able to provide reliable information, whereas infants and small children are unable to communicate their medical history, in which case you must rely on a parent or caregiver to obtain this information. If the parent of a sick or injured child is frantic, however, your ability to obtain the child's medical history may be hampered.

Older children, adolescents, and adults are usually able to tell you where it hurts or where their discomfort is. Although infants and small children cannot verbally communicate their pain or discomfort—other than crying—they often indirectly communicate by way of visual clues. For example, small children with a headache often grab the sides of their head; if they have ear pain caused by an infection, they often tug at the affected ear.

Certain medical conditions that are common in one age group are uncommon in others. Determining a patient's risk factors for disease is an important part of the history-taking process and affects your index of suspicion—the degree of concern you have that a patient may be experiencing a specific condition. For example, it is rare—although not impossible—for an otherwise healthy 8-year-old child with chest pain to be experiencing a cardiac problem because children generally have healthy hearts. However, you should be suspicious of a cardiac problem if a 55-year-old patient with a history of high blood pressure—a major risk factor for cardiovascular disease—presents with the same symptoms.

Understanding which illnesses are common among the various age groups will help you formulate a plausible field impression—that is, what you believe is wrong with the patient based on your assessment findings.

2. What are the physical differences between middle adults and late adults?

Middle adults (41 to 60 years of age) have reached the halfway point in human life expectancy. However, provided they are otherwise healthy, their vital signs and physical abilities usually remain consistent with those of early adults. Their average pulse rate is 70 beats/min, their respiratory rate is between 12 and 20 breaths/min, and their systolic blood pressure is between 90 and 140 mm Hg. In middle adults, factors such as increased cholesterol levels and issues with weight control commonly lead to concerns regarding cardiovascular health and an increased risk of developing diabetes. Many of the effects of aging during middle adulthood, however, can be mitigated with exercise, a proper diet, and other lifestyle changes (ie, cessation of smoking, weight control).

The age-related physical changes that occur in late adults (61 years of age and older) are more pronounced than they are in middle adults and affect nearly every organ and organ system in the body. Cardiac function declines due to changes that are often related to atherosclerotic heart disease. Atherosclerosis, a condition that most commonly affects the coronary arteries, is caused by an accumulation of cholesterol and calcium that forms plaque that eventually obstructs blood flow through the coronary arteries. More than 60% of people older than 65 years have atherosclerotic disease.

The vital signs of late adults are largely dependent on their underlying health and are often affected by medications they take for various conditions. In general, however, age-related vital sign changes include a decrease in heart rate and an increase in diastolic blood pressure that is the result of stiffening of the blood vessels, which makes them less able to dilate and contract. In late adults, elasticity of the lungs decreases, forcing the person to rely more on their intercostal muscles to breathe. In addition, the ribs become more rigid due to calcification, which adds to their breathing difficulty. In an attempt to compensate for these respiratory changes, an increase in the late adult's respiratory rate is commonly observed.

Other physical changes that occur during late adulthood include decreases in metabolism and insulin production, which can lead to diabetes, decreased gastrointestinal function, decreased taste bud sensation, decreased kidney mass and filtration (kidney function declines by 50% from age 20

You are the Provider: SUMMARY, continued

to age 90), nervous system changes (which includes a 10% to 20% decrease in brain weight by 80 years of age) and sensory and motor nerve deterioration, visual impairment, and hearing impairment, among others.

The anatomic and physiologic changes that occur between middle and late adulthood must be taken into consideration during your assessment. Disease does not automatically come with age; however, age-related changes in anatomy and physiology increase the risk for disease.

Knowledge of age-related anatomic and physiologic changes enhances your ability to discern typical findings from the atypical. Keep in mind, however, that compared to middle adults, late adults often present with fewer classic signs and symptoms of a wide variety of medical conditions.

3. What are some common psychosocial concerns experienced by middle adults?

During middle adulthood, people are approaching the halfway point in human life expectancy and they tend to focus on achieving their life goals. Many of their concerns center around finances—especially as they prepare to retire, but must still manage everyday finances; this often causes stress and uncertainty. However, most middle adults are in good health and have the physical and emotional reserves to meet these demands.

A unique psychosocial concern in the middle adult relates to their children and their parents. As their children move away from home—which forces them to readjust their lifestyle ("empty nest" syndrome)—their parents are getting older and now need care. Most middle adults prefer to care for their parents in their own home or in their parents' home; however, this often increases the stress and anxiety that they are already experiencing from other factors such as finances or retirement.

As previously discussed, you must know your patient. Each patient has his or her own psychosocial challenges and concerns, and you must factor these in with your overall patient care plan. The provision of psychological support is a critical part of the job you do as an EMT; it often reassures the patient that you are truly concerned about his or her well-being.

4. Are the patient's vital signs consistent with his age?

The patient's heart rate and respiratory rate are consistent with his age. However, his blood pressure is not. A typical middle adult's systolic blood pressure ranges between 90 and 140 mm Hg; the diastolic blood pressure usually ranges between 70 and 80 mm Hg.

Without any prior blood pressure measurements that have been recorded over time (ie, kept in a journal), you have no way of knowing if the patient's blood pressure is transiently elevated due to some acute stressor or if he has chronic hypertension. It is generally agreed on by most clinicians that hypertension exists when the blood pressure is persistently greater than 140/90 mm Hg.

Ask the patient if he keeps a journal of his vital signs. If he does, ask him what his blood pressure typically reads; clearly, hypertension cannot be diagnosed by a single blood pressure reading. If he keeps a journal and tells you that his current blood pressure is consistent with what it normally reads, ask him if he is under a physician's care or being treated with any medication. If he is not, you should advise him to be evaluated by a physician; he may have hypertension and not be aware of it. Hypertension is often referred to as the "silent killer," and a blood pressure of 174/98 mm Hg is abnormal in anyone, regardless of age.

5. Why should you transport this patient to the hospital?

This patient should *not* drive himself to the hospital. He experienced a syncopal episode, which could indicate a variety of underlying medical conditions—some of them potentially life-threatening. Syncope (fainting) could be caused by a cardiac arrhythmia, hypoglycemia, dehydration, a transient ischemic attack (TIA), or a hypertensive crisis, to name a few.

At the present time, the patient is light-headed, is complaining of a headache, and is hypertensive. He should be informed that if he drives himself to the hospital, he could experience another syncopal episode while driving; this would not only jeopardize his own safety, but the safety of other motorists on the road. As an EMT, your job is to look out for the safety of the patient and the public.

Do not attribute the patient's signs and symptoms to stress. Instead, assume that they signal a potentially life-threatening condition until proven otherwise. Inform the patient that transport via EMS is the wisest choice because you will be there to take care of him if his condition gets worse.

Although the patient is of legal age and has the decision-making capacity to legally refuse EMS transport, you should make a *concerted and sincere effort* to convince him to agree to EMS transport, and advise him that his refusal could potentially result in death.

6. What additional treatment, if any, is required for this patient?

Further treatment for this patient should be supportive. Continue to monitor his mental status and ABCs and make him comfortable. Dimming the lights in the back of the ambulance may provide him with some relief from his headache.

Remain alert for any changes in his neurologic status, such as slurred speech, unilateral weakness (weakness to one side of the body), or confusion, and contact the receiving facility if any changes are noted.

Although this patient does not require aggressive treatment, you are providing him with what he needs the most—transport to the hospital for evaluation by a physician.

EMS Patient Care Report (PCR)

Date: 7-17-09	Incident No.: 060109	Nature of Call: Vital sign check		Location: EMS Station 2	
Dispatched: 1310	En Route: 1310	At Scene: 1310	Transport: 1324	At Hospital: 1330	In Service: 1339

Patient Information

Age: 50 Sex: M Weight (in kg [lb]): 190 lb (86 kg)	Allergies: No known drug allergies Medications: Vitamins Past Medical History: None Chief Complaint: Light-headedness, fainting, headache

Vital Signs

Time: 1315	BP: 174/98	Pulse: 76	Respirations: 14	Sao$_2$: 98%
Time: 1325	BP: 180/102	Pulse: 80	Respirations: 14	Sao$_2$: 99%

EMS Treatment
(circle all that apply)

Oxygen @ 4 L/min via (circle one): (NC) NRM Bag-Mask Device	Assisted Ventilation	Airway Adjunct	CPR
Defibrillation	Bleeding Control	Bandaging	Splinting	Other: (Blood glucose level assessment)

Narrative

50-year-old male presented to EMS station 2 for routine vital sign check. Shortly after arrival, the patient stated that he felt light-headed and needed to sit down. He stated that he has been "stressed" about caring for his ill father, and has experienced several episodes of light-headedness over the past few days. He further stated that he experienced a syncopal episode earlier today, but did not know how long he was unconscious. Upon presentation, he was conscious and alert; his airway was patent and his breathing was adequate. He denies any past medical history and stated that he only wanted his vital signs checked. He further denies any medication allergies and states that he only takes vitamins. Initial vital signs revealed an elevated blood pressure. After vital sign assessment, the patient began complaining of a headache. Applied oxygen at 4 L/min via nasal cannula and reassessed his mental status; he remained conscious and alert. Advised patient that because of his elevated blood pressure, syncopal episode, light-headedness, and headache, EMS transport to the hospital for evaluation by a physician was prudent. He stated that he preferred to drive himself because he did not feel that transport via ambulance was necessary. Advised patient that driving himself was unsafe because he could experience another syncopal episode while driving. He was further advised that his signs and symptoms could signal a potentially life-threatening condition that only a physician could diagnose. After being informed of these potential consequences, the patient agreed to EMS transport. Placed patient onto the stretcher, loaded him into the ambulance, and began transport. The patient's condition remained unchanged en route. After dimming the lights in the back of the ambulance, he stated that his headache improved slightly, but he was still light-headed. Reassessed his vital signs and assessed his blood glucose level, which read 100 mg/dL. Duration of transport was uneventful, and the patient was delivered to the emergency department without incident. After giving verbal report to the charge nurse, Medic 2 returned to service. **End of report**

Prep Kit

- Whereas each developmental stage is marked by different physical and psychosocial changes and characteristics, infants (1 month to 1 year) develop at a startling rate.
- The vital signs of toddlers (ages 1 to 3 years) and preschoolers (ages 3 to 6 years) differ somewhat from those of an infant. During this stage, children learn to speak and express themselves.
- From ages 6 to 12 years, the school-age child's vital signs and body gradually approach those observed in adulthood. During this stage, children develop self-esteem.
- The vital signs of adolescents (ages 12 to 18 years) begin to level off within the adult ranges. Adolescents focus on creating their self-image.
- Early adults are those who are age 19 to 40 years. Early adults focus on work and family.
- Middle adults are those who are age 41 to 60 years. Middle adults focus on achieving life goals.
- Late adults are those who are age 61 years and older. Late adults focus on their mortality and the mortality of friends and loved ones.
- Vital signs do not vary greatly through adulthood.

Vital Vocabulary

adolescents Persons who are 12 to 18 years of age.

anxious-avoidant attachment A bond between an infant and his or her parent or caregiver in which the infant is repeatedly rejected and develops an isolated lifestyle that does not depend on the support and care of others.

atherosclerosis A disorder in which cholesterol and calcium build up inside the walls of the blood vessels, forming plaque, which eventually leads to partial or complete blockage of blood flow.

barotrauma Injury resulting from pressure disequilibrium across body surfaces, for example from too much pressure in the lungs.

bonding The formation of a close, personal relationship.

conventional reasoning A type of reasoning in which a child looks for approval from peers and society.

early adults Persons who are 19 to 40 years of age.

fontanelles Areas where the infant's skull has not fused together; usually disappear at approximately 18 months of age.

infants Persons who are from 1 month to 1 year of age.

late adults Persons who are 61 years old or older.

life expectancy The average amount of years a person can be expected to live.

middle adults Persons who are 41 to 60 years of age.

moro reflex An infant reflex in which, when an infant is caught off guard, the infant opens his or her arms wide, spreads the fingers, and seems to grab at things.

neonate Persons who are birth to 1 month of age.

nephrons The basic filtering units in the kidneys.

palmar grasp An infant reflex that occurs when something is placed in the infant's palm; the infant grasps the object.

postconventional reasoning A type of reasoning in which a child bases decisions on his or her conscience.

preconventional reasoning A type of reasoning in which a child acts almost purely to avoid punishment to get what he or she wants.

preschoolers Persons who are 3 to 6 years of age.

rooting reflex An infant reflex that occurs when something touches an infant's cheek, and the infant instinctively turns his or her head toward the touch.

school age A person who is 6 to 12 years of age.

secure attachment A bond between an infant and his or her parent or caregiver, in which the infant understands that his or her parents or caregivers will be responsive to his or her needs and take care of him or her when he or she needs help.

sucking reflex An infant reflex in which the infant starts sucking when his or her lips are stroked.

terminal drop hypothesis The theory that a person's mental function declines in the last 5 years of life.

toddlers Persons who are 1 to 3 years of age.

trust and mistrust A phrase that refers to a stage of development from birth to approximately 18 months of age, during which infants gain trust of their parents or caregivers if their world is planned, organized, and routine.

Assessment in Action

Rescue 29 responds to 5120 SE 20th Avenue for a report of altered mental status. You arrive at a private assisted-living facility where you are greeted in the lobby by one of the nurses. He tells you that the 65-year-old patient has been refusing to take his medication and became combative when the nurse tried to take his vital signs. The nurse asks you to wait in the lobby because he would like to give the patient notice that you will be coming to see him. While you wait, you think about some of the things you should consider in treating this patient.

1. Your patient is considered to be a late adult because he is:
   A. 55 and older.
   B. 57 and older.
   C. 60 and older.
   D. 61 and older.

2. You know the current life expectancy is approximately:
   A. 50 years.
   B. 65 years.
   C. 78 years.
   D. 92 years.

3. Describe some of the contributors that have an effect on your life expectancy.

4. Describe terminal drop hypothesis.

5. Hearing loss is about _____ times more common than loss of vision in late adults.
   A. Three
   B. Four
   C. Six
   D. Seven

6. _____ system changes can result in the most debilitating of age-related ailments.
   A. Nervous
   B. Renal
   C. Sensory
   D. Endocrine

7. What are some of the psychosocial issues people experience during late adulthood?

8. Vital signs of late adults typically depend on the:
   A. patient's immune system.
   B. overall health of the patient.
   C. patient's cardiovascular system.
   D. patient's diet.

9. Most of the elderly in the United States are cared for by:
   A. home health nurses.
   B. nursing homes.
   C. family members.
   D. an assisted-living facility.

10. Why is there an increased need for additional extended care facilities?

National EMS Education Standard Competencies

Pharmacology

Applies fundamental knowledge of the medications that the EMT may assist/administer to a patient during an emergency.

Principles of Pharmacology

- Medication safety (pp 227-228)
- Kinds of medications used during an emergency (pp 229-239)

Medication Administration

- Self-administer medication (pp 228-229)
- Peer-administer medication (pp 228-229)
- Assist/administer medications to a patient (pp 228-229)

Emergency Medications

- Names (p 222)
- Effects (pp 221-222)
- Actions (p 222)
- Indications (p 222)
- Contraindications (p 222)
- Complications (p 222)
- Routes of administration (pp 222-224)
- Side effects (p 222)
- Interactions (pp 230-231)
- Dosages for the medications administered (p 221)

Knowledge Objectives

1. Explain the actions of medications on the body and define the terms pharmacodynamics, intended effects, and indications. (pp 221-222)
2. Explain and give examples of medication contraindications and define the terms side effects, unintended effects, and untoward effects. (p 222)
3. Discuss the differences between a generic medication name and a trade medication name, and provide an example of each. (p 222)
4. Describe the enteral and parenteral routes of medication administration and explain how they differ. (pp 222-223)

5. Describe the following routes of medication administration and discuss their individual rates of absorption: rectal, oral, intravenous, intraosseous, subcutaneous, intramuscular, inhalation, sublingual, and transcutaneous. (pp 222-224)
6. Explain the solid, liquid, and gas forms of medication, provide examples of each, and discuss how the form of a medication dictates its route of administration. (pp 224-227)
7. Explain the "six rights" of medication administration and describe how each one relates to EMS. (pp 227-228)
8. Describe the role of medical direction in medication administration and explain the difference between direct orders (online) and standing orders (off-line). (pp 228-229)
9. Discuss the circumstances surrounding the administration of medication, including peer-assisted medication, patient-assisted medication, and EMT-administered medication. (p 229)
10. Give the generic and trade names, actions, indications, contraindications, routes of administration, side effects, interactions, and doses of ten medications that may be administered by an EMT in an emergency as dictated by state protocols and local medical direction. (pp 229-239)
11. Describe the medication administration considerations that must be applied to special populations, including pediatric, geriatric, and pregnant patients. (pp 229, 236, 238)
12. Describe the steps an EMT should follow when dispensing epinephrine to a patient using an auto-injector. (p 236)
13. Explain why determining what prescription and OTC medications a patient is taking is a critical aspect of patient assessment during an emergency. (p 239)

Skills Objectives

1. Demonstrate the process an EMT should follow when following the six rights of medication administration. (p 228)
2. Demonstrate how to administer oral medication to a patient. (pp 232-234, Skill Drill 7-1)
3. Demonstrate the administration of aspirin to a patient with chest pain. (pp 232-234, Skill Drill 7-1)
4. Demonstrate the administration of oral glucose to a patient with hypoglycemia. (pp 232-234, Skill Drill 7-1)
5. Demonstrate how to assist a patient with the sublingual administration of a medication. (pp 234-235)
6. Demonstrate how to administer epinephrine by injection. (pp 235-237)

Introduction

Administering medications is a serious business. Used appropriately, a medication may alleviate pain and improve a patient's well-being. However, used inappropriately, medication may cause harm and even death. As an EMT, you will be responsible for administering certain medications to patients and helping them to self-administer others. You will ask patients about their medications and allergies, and you will report this information to hospital personnel. This information will provide insight into a patient's medical history and ensure that patients do not receive medications for which they are allergic. To act without understanding how medications work is to place patients in danger.

This chapter describes the various forms of medications, the different ways in which they can be administered, and how they work. It then takes a close look at each of the seven forms of medications you may be asked to administer or help patients to self-administer. It will also explain when it is dangerous to administer these medications.

How Medications Work

Pharmacology is the science of drugs, including their ingredients, preparation, uses, and actions on the body. Although the terms "drugs" and "medications" are often used interchangeably, the term drugs may make some people think of narcotics or illegal substances. For this reason, you should use the word *medications*, especially

Words of Wisdom

It is important for you to become familiar with the "street" names of commonly used and abused drugs. Most users will not tell you they took methylendeioxymethamphetamine; most likely you will hear terms like ecstasy, XTC, rolling, or popping. Research or look up these common street names.

when interviewing patients and families. In general terms, a **medication** is a chemical substance that is used to treat or prevent disease or relieve pain.

Pharmacodynamics is the process by which a medication works on the body. Receptors are located throughout the body. These are sites on cells where chemicals can bind and cause reactions. When medications are given, they bind to these sites and either stimulate the receptor sites and cause a reaction or block the sites and prevent other chemicals from attaching. Thus, a medication can either increase or decrease a normal function of the body. A medication that causes stimulation of receptors is called an **agonist**. Medications that bind to a receptor and block other medications are called **antagonists**, or blockers.

The **dose** is the amount of the medication that is given. The dose depends on the patient's weight or age; adults and children will receive different amounts of the same medication. The dose also depends on the desired action of the medication. The **action** is the therapeutic effect or **intended effect** that a medication is expected to have on the body. These factors, among others, can help to explain why one dose of medication works quickly and efficiently on one patient and the same dose has little effect on another patient. Doses of medications may need to be decreased for infants because they have small bodies. Doses may also need to be decreased for the elderly because they cannot process medications as efficiently as younger people.

You are the Provider: PART 1

You and your partner are dispatched to a residence at 202 Cloudcroft Avenue for a woman with "diabetic complications." The time is 7:40 AM, the weather is clear, the traffic is light, and your response time to the scene is approximately 8 minutes.

1. What is pharmacology?
2. Why is knowledge of pharmacology important to patient care?

Indications are the reasons or conditions for which a particular medication is given. For example, nitroglycerin relaxes the walls of the blood vessels and may dilate the arteries. This increases the blood flow and the supply of oxygen to the heart muscle. In this way, nitroglycerin relieves the discomfort that can occur with the cardiac condition called angina. Therefore, nitroglycerin is indicated for chest pain associated with angina.

There are times when you should not give a medication, even if it usually is indicated for that person's condition. Such situations are called **contraindications**. A medication is contraindicated when it would harm the patient or have no positive effect on the patient's condition. For example, the administration of activated charcoal is indicated when a patient has swallowed a poison. Generally, activated charcoal, premixed with water, is used to prevent the body from absorbing a poison. However, activated charcoal would be contraindicated if the patient were unconscious and could not swallow.

Side effects are any actions of a medication other than the desired ones. There are two types of side effects: unintended effects and untoward effects. The reason that side effects are separated into two groups is because one group of side effects is simply bothersome and the other can be harmful. **Unintended effects** are the effects that are undesirable but pose little risk to the patient. **Untoward effects** are the effects that can be harmful to the patient.

Consider diphenhydramine (Benadryl). People take this medication for allergic reactions (indication). The medication is supposed to block the effects of histamine (intended effect). It has side effects of drying the mouth and sleepiness (unintended) and can increase the pressure of the fluid within the eye (untoward). Patients are told to expect drowsiness after taking this medication. They are also warned that if they have glaucoma, an eye condition causing increased fluid pressures, they should avoid this medication. Diphenhydramine is contraindicated for patients who have glaucoma because of the increased fluid pressures in their eyes.

■ Medication Names

Medications usually have two types of names. The **generic name** (such as ibuprofen) is a simple, clear, nonproprietary name. The generic name is not capitalized. Sometimes a medication is called by its generic name more often than by any of its trade names. For example, you may hear the term "nitroglycerin" used more often than the trade names Isordil and Nitrostat. All medications that are licensed for use in the United States are listed by their generic names in the *United States Pharmacopoeia*.

A **trade name** is the brand name that a manufacturer gives to a medication, such as Tylenol or Lasix. As a proper noun, a trade name begins with a capital letter. Trade names are used in every aspect of our daily lives, not just in medications. Well-known examples include Jell-O gelatin, Band-Aid adhesive bandages, and Hershey's chocolate candy. A medication may have many different trade names, depending on how many companies manufacture it. Advil, Nuprin, and Motrin all are trade names for the generic medication ibuprofen. A trade name sometimes is also designated by a raised registered symbol, that is, Advil®.

Medications may be **prescription medications** or **over-the-counter (OTC) medications**. Prescription medications are distributed to patients only by pharmacists according to a physician's order. Medications that are OTC may be purchased directly, such as from a discount store or supermarket, without a prescription. In recent years, the number of prescription medications that have become available OTC has increased dramatically.

You may come into contact with patients who have taken "street" drugs such as heroin or cocaine. Other patients may take herbal medications, enhancement drugs, or vitamin supplements. As we have discussed, the body's cells are configured to operate using chemical reactions; they do not discern between safe and unsafe pharmacologic agents. Any medication that a patient takes can be pharmacologically active and can cause an effect. As an EMT, you need to ask patients about any medications they are taking.

■ Routes of Administration

Medications can be given by a wide array of routes. To simplify this topic, the routes of medication administration are divided into two categories: enteral and parenteral. **Enteral medications** enter the body through the digestive system. Typically, the form of the medication will be a pill or a liquid such as cough medicine. Medications administered via this route tend to absorb slowly and are therefore not commonly used in an emergency setting. **Parenteral medications** enter the body by a route other than the digestive tract, the skin, or the mucous membranes. Parenteral medications are often in a liquid form and are generally administered using syringes and needles. These medications are absorbed much more quickly and offer a more predictable and measurable response.

Regardless of the route of administration of a medication, the end goal is to get that medication into the bloodstream. **Absorption** is the process by which medications travel through body tissues until they reach the bloodstream. Often the rate at which a medication is absorbed into the bloodstream depends on its route of administration. **Table 7-1** lists common routes of medication administration and rates of absorption.

- **Per rectum (PR)**. Per rectum means by rectum. This route of delivery is most commonly used with children because of easier administration and more reliable absorption. (Children often regurgitate some or all of a medication.) For similar reasons, many medications that are used for nausea and vomiting come in a rectal suppository form. Some medications to control seizures are administered PR when it is impossible to administer them intravenously. The PR route also is used to give some medications when the patient cannot swallow or is unconscious.

- **Oral**. Many medications are taken by mouth, or **per os (PO)**, and enter the bloodstream through the digestive system. This process often takes as long as 1 hour. One of the advantages of using this route is it is noninvasive. Patients are often much happier to take a pill than to have a needle stuck in them. It is also less expensive to use enteral medications than to use parenteral. However, the main disadvantage of this administration route is the unpredictability of medication absorption. If the patient has an upset stomach or diarrhea, the amount of medication that is absorbed will be altered.

- **Intravenous (IV) injection**. Intravenous means into the vein. Medications that need to enter the bloodstream immediately may be injected directly into a vein. This is the fastest way to deliver a chemical substance, but the IV route cannot be used for all chemicals. For example, aspirin, oxygen, and charcoal cannot be given by the IV route.

Table 7-1 Routes of Administration and Rates of Absorption

Route	Rate
Enteral	
Per rectum (PR)	Rapid
Ingestion (oral)	Slow
Parenteral	
Intravenous (IV)	Immediate
Intraosseous (IO)	Immediate
Subcutaneous (SC)	Slow
Intramuscular (IM)	Moderate
Inhalation	Rapid
Sublingual (SL)	Rapid
Transcutaneous	Slow

- **Intraosseous (IO) injection**. Intraosseous means into the bone. Medications that are given by this route reach the bloodstream through the bone marrow. Giving a medication by the IO route, into the marrow, requires drilling a needle into the outer layer of the bone. Because this is painful, the IO route is used most often in patients who are unconscious as a result of cardiac arrest or extreme shock. Most commonly, the IO route is used for children who have fewer available (or difficult to access) IV sites.

- **Subcutaneous (SC) injection**. Subcutaneous means beneath the skin. An SC injection is given into the tissue between the skin and the muscle. Because there is less blood here than in the muscles, medications that are given by this route are generally absorbed more slowly, and their effects last longer. An SC injection is a useful way to give medications that cannot be taken by mouth, as long as they do not irritate or damage the tissue. Daily insulin injections for patients with diabetes are given by the SC route. Some forms of epinephrine can be given by the SC route. (Subcutaneous sometimes is abbreviated as SQ or sub-Q.)

- **Intramuscular (IM) injection**. Intramuscular means into the muscle. Usually, medications that are administered by IM injection are absorbed quickly because muscles have a lot of blood vessels. However, not all medications can be administered by the IM route. Possible problems with IM injections are damage to muscle tissue and uneven, unreliable absorption, especially in people with decreased tissue perfusion or who are in shock.

  EMTs will typically use the IM route of medication administration with an auto-injector. These devices deliver a predetermined amount of medication into the patient when pressed firmly into the thigh. Examples of this delivery method would be the EpiPen auto-injector, which is used for allergic reactions, and the Mark-1 auto-injector, which is used for managing an exposure to certain chemicals. For more information on the EpiPen auto-injector, see Chapter 18, *Immunologic Emergencies*. For more information on the Mark-1 auto-injector, see Chapter 39, *Terrorism Response and Disaster Management*.

- **Inhalation**. Some medications are inhaled into the lungs so that they can be absorbed into the bloodstream more quickly. Others are inhaled because they work in the lungs. Generally, inhalation helps minimize the effects of the medication in other body tissues. Such medications come in the form of aerosols, fine powders, and sprays.

- **Sublingual (SL)**. Sublingual means under the tongue. Medications given by the SL route, such as nitroglycerin tablets, enter through the oral mucosa under the tongue and are absorbed into the bloodstream within minutes. This route is faster than the oral route, and it protects medications from chemicals in the digestive system, such as acids that can weaken or inactivate them.

- **Transcutaneous (transdermal)**. Transcutaneous means through the skin. Some medications can be absorbed transcutaneously, such as the nicotine in patches used by people who are trying to quit smoking. On occasion, a medication that also comes in another form is administered transcutaneously to achieve a longer-lasting effect. An example is an adhesive patch containing nitroglycerin.

- **Intranasal (IN)**. Intranasal is a relatively new format for the delivery of medication. In this route a medication is pushed through a specialized atomizer device called a **mucosal atomizer device (MAD)**. The liquid medication is turned into a spray and is administered into a nostril. Blood flow to the head and face is very high; therefore, absorption is rather quick with this route. Naloxone can be administered to some overdose patients via this route.

Table 7-2 lists the words that are used for routes of medication delivery, along with their meanings.

Safety

Make absolutely certain you follow standard precautions when administering any medication, particularly topical drugs. If the medication can be absorbed into the patient's skin, it can be absorbed into yours as well.

Medication Forms

The form of a medication usually dictates the route of administration. For example, a tablet or a spray cannot be given through a needle. The manufacturer chooses the form to ensure the proper route of administration, the timing of its release into the bloodstream, and its effects on the target organs or body systems. As an EMT, you should be familiar with the following seven medication forms.

Tablets and Capsules

Most medications that are given by mouth to adult patients are in tablet or capsule form Figure 7-1. Capsules are gelatin shells filled with powdered or liquid medication. If the capsule contains liquid, the shell is sealed and usually soft. If the capsule contains powder, the shell can usually

Figure 7-1 Tablets and capsules are typically taken by mouth and enter the bloodstream through the digestive system.

Table 7-2 Routes of Administration: Words and Their Meanings

This Word...	From These Latin Words...	Means
Inhalation	*inhalatio* (drawing air into the lungs)	inhaling or breathing in
Intramuscular (IM)	*intra* (into) and *muscularis* (of the muscles)	into muscle
Intraosseous (IO)	*intra* (into) and *osse* (bone)	into bone
Intravenous (IV)	*intra* (into) and *venosus* (of the veins)	into vein
Per os (PO)	*per* (by) and *os* (mouth)	by mouth
Per rectum (PR)	*per* (by) and *rectum* (rectum)	by rectum
Subcutaneous (SC)	*sub* (under) and *cutis* (skin)	under the skin
Sublingual (SL)	*sub* (under) and *lingua* (relating to the tongue)	under the tongue
Transcutaneous (transdermal)	*trans* (through) and *cutis* (skin)	through the skin
Intranasal	*intra* (into) and *nasal* (nose)	into the nose

be pulled apart. In tablets, the medication is compressed under high pressure. Tablets often contain other materials that are mixed with the medication.

Some tablets are designed to dissolve very quickly in small amounts of liquid so that they can be given sublingually and absorbed rapidly. An example is the sublingual nitroglycerin tablet used to treat chest pain in patients with cardiac conditions. These medications are especially useful in emergency situations. Sublingual medications are generally placed under the tongue. Generally, a medication that must be swallowed is less useful in an emergency because the digestive tract provides a slower route of delivery. For example, an oral pain medication is less useful than an IV pain medication when pain relief is needed within minutes.

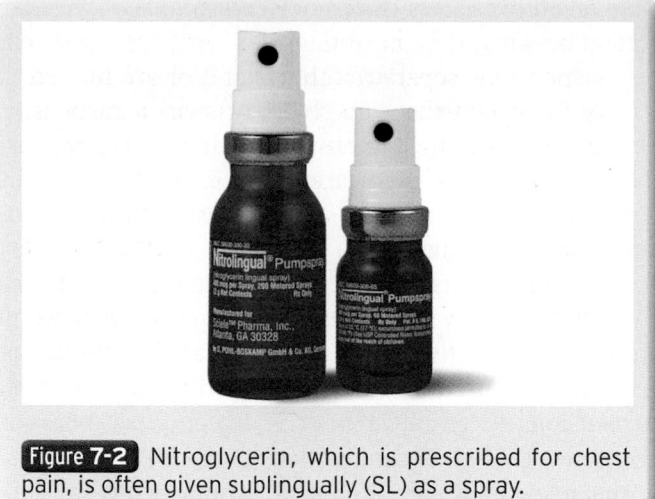

Figure 7-2 Nitroglycerin, which is prescribed for chest pain, is often given sublingually (SL) as a spray.

■ Solutions and Suspensions

A **solution** is a liquid mixture of one or more substances that cannot be separated by filtering or allowing the mixture to stand. Solutions can be given by almost any route. When given by mouth, solutions may be absorbed from the stomach fairly quickly because the medication is already dissolved. For example, you may need to help in the sublingual delivery of a nitroglycerin spray **Figure 7-2**. Many solutions can be given as an IV, IM, or SC injection.

If a patient has a severe allergic reaction, you may help to administer a solution of epinephrine using an auto-injector.

Many substances do not dissolve well in liquids. Some of these can be ground into fine particles and evenly distributed throughout a liquid by shaking or stirring. This type of mixture is called a **suspension**. An example is activated charcoal, which you may give to patients who

You are the Provider: PART 2

You arrive at the scene and find the patient, a 60-year-old woman, sitting in a recliner in her living room. Her son, who called 9-1-1, tells you that she is not acting right. He further tells you that he is not sure if she has eaten today. You assess the patient as your partner opens the jump kit and prepares to begin treatment.

Recording Time: 0 Minutes	
Appearance	Confused; diaphoretic; pale
Level of consciousness	Conscious but confused
Airway	Open; clear of secretions or foreign bodies
Breathing	Increased rate; shallow depth
Circulation	Radial pulses, rapid and weak; skin, pale and diaphoretic

As your partner gives the patient high-flow oxygen via nonrebreathing mask, her son tells you that she has type 2 diabetes, heart disease, and depression. Her prescribed medications include metformin (Glucophage), nitroglycerin (Nitrostat), and sertraline (Zoloft). You assess her blood glucose level, which reads 36 mg/dL.

3. Other than oxygen, does this patient require any other medications?

4. What are the "six rights" of medication administration?

have taken overdoses of certain medications or ingested certain poisons.

Suspensions separate if they stand or are filtered. It is very important that you shake or swirl a suspension before administering it to ensure that the patient receives the right amount of medication. For example, if you are a parent, you may have had to shake a suspension of oral antibiotic before giving it to your child.

Suspensions usually are administered by mouth but sometimes are given rectally. Occasionally, suspensions are applied directly to the skin to treat skin problems. You may have used calamine lotion in this way. Injectable suspensions are given via IM or SC injection only. Certain hormone shots or vaccinations are given this way because of the suspended particles. They cannot be given by IV injection because the suspended particles do not remain dissolved.

■ Metered-Dose Inhalers

If liquids or solids are broken into small enough droplets or particles, they can be inhaled. A **metered-dose inhaler (MDI)** is a miniature spray canister used to direct such substances through the mouth and into the lungs **Figure 7-3** . An MDI delivers the same amount of medication each time it is used. Because an inhaled medication usually is suspended in a propellant, the MDI must be shaken vigorously before the medication is administered. An MDI is often used by a patient with respiratory illnesses such as asthma or emphysema. See Chapter 13, *Respiratory Emergencies*, for more information on the MDI.

■ Topical Medications

Lotions, creams, and ointments are **topical medications**, that is, they are applied to the surface of the skin and affect only that area. Lotions contain the most water, and ointments contain the least. Lotions are absorbed the most rapidly and ointments the most slowly. Calamine lotion is an example of a medical lotion. Hydrocortisone cream, to diminish skin itching, is an example of a medical cream that can also be given in ointment form. Neosporin is an example of a first-aid ointment.

■ Transcutaneous Medications

Transdermal medications are designed to be absorbed through the skin, or transcutaneously. Medications such as nitroglycerin paste usually have properties or delivery systems that help to dilate the blood vessels in the skin and, thus, speed absorption into the bloodstream. In contrast with most topical medicines, which work directly on the application site, transdermal medications are usually intended for systemic (whole-body) effects. A note of caution: If you touch such a medication with your bare skin while administering it, you will absorb it just as readily as the patient will.

One of the newer delivery systems for transcutaneous medications is the adhesive patch. Patches attach to the skin and allow even absorption of a medication for many hours **Figure 7-4** . Prescription and OTC medications come in this form. Common examples are nitroglycerin, nicotine, some pain medications, and some oral contraceptives.

■ Gels

A **gel** is a semiliquid substance that is administered orally in capsule form or through plastic tubes. Gels usually have the consistency of pastes or creams but are transparent (clear). "Gelatinous" means thick and sticky, like gelatin. Depending on your local medical directives, as an EMT, you may give oral glucose in gel form to a patient with diabetes **Figure 7-5** .

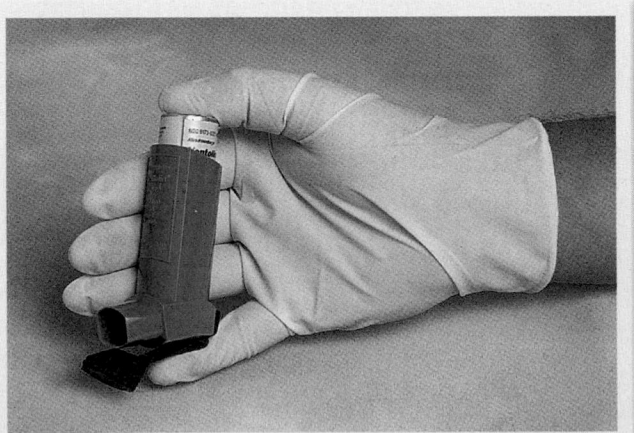

Figure 7-3 Some medications are inhaled into the lungs with a metered-dose inhaler so that they can be absorbed into the bloodstream more quickly.

Figure 7-4 Some medications are transcutaneous, or administered through the skin, such as the nitroglycerin patch shown.

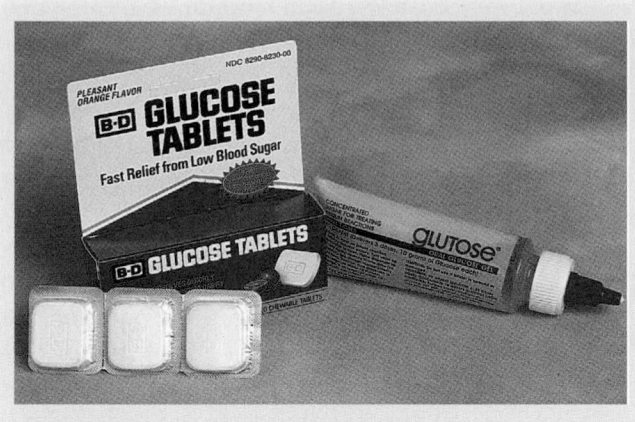

Figure 7-5 Oral glucose, used in diabetic emergencies, is available in gel and tablet forms.

Gases for Inhalation

Gaseous medications are neither solid nor liquid and most often are given in an operating room. The medication most commonly used in gas form outside the operating room is oxygen. You might not think of oxygen as a medication because it is all around us and we all use it. However, in its concentrated form, it is a potent medication that has systemic effects, that is, effects throughout the body **Figure 7-6**. You will deliver this gas usually through a nonrebreathing mask or a nasal cannula.

Words of Wisdom

When you document the use of oxygen, include the liter flow rate, the time oxygen (usually recorded in military time) was delivered, and the type of device used. For example, "09:15 Nonrebreathing mask at 15 L/min. Patient states shortness of breath is better." Also document the patient's response to oxygen administration.

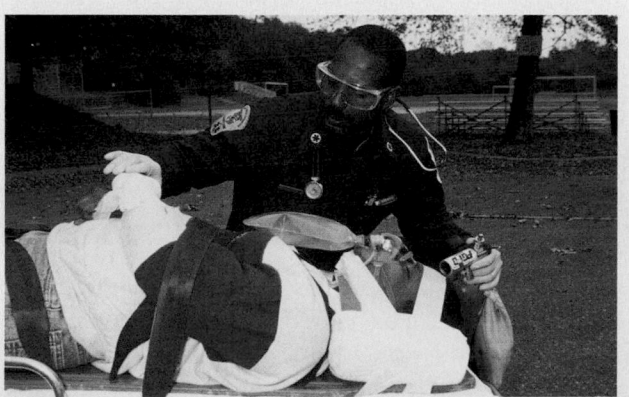

Figure 7-6 Oxygen is a potent medication that you will typically give through a nonrebreathing mask.

General Steps in Administering Medication

As an EMT, you may only administer medications for which you have an order from medical control. After the order has been confirmed, you must be familiar with the six general steps of administering any medication to a patient. These steps are the six rights of medication administration **Table 7-3**. After the medication has been administered, you will need to reassess the patient to see if it worked. You should look for side effects and then be prepared to document your actions and your findings.

Table 7-3 The "Six Rights" of Medication Administration
Right patient
Right medication
Right dose
Right route
Right time
Right documentation

When administering or assisting with the administration of patient medications, the EMT must have an order from medical control to do so. This order may be given to you directly, through online medical control via telephone or radio. If you are receiving orders for the administration of a medication, it is important that you repeat the order back to the physician. This is referred to as the echo technique and it is done to ensure that you heard the order correctly.

The following example illustrates the correct way to acknowledge orders for medications. EMT Johnson is talking on the radio to medical control. She has already given the physician all of the patient assessment information and vital signs. She now asks, "Do you have any orders?" Dr Ortez says, "Yes, please assist the patient with one nitroglycerin tab. Make sure that tablet goes under the patient's tongue. After 3 minutes, contact me again for more orders." EMT Johnson should reply, "Dr Ortez, I copy one nitroglycerin tablet administered under the tongue. I will contact you in 3 minutes for additional orders." Dr Ortez says, "That is correct, ABC Hospital clear."

If at any point while you are receiving an order for a medication you are confused or unclear about what to do, you should tell medical control. It is very important that you understand what the physician wants you to do. With all of the noise in the back of ambulance, the possible radio/cell phone disruptions, or simply the stress of caring for a very ill patient, it is important to err on the cautious side. If you are not sure what to do, tell the doctor you were unclear on the order and ask for the order to be repeated.

The other way to receive orders to administer or assist with medications is through indirect or off-line medical control. Protocols are documents that contain standing orders for the administration of certain medications. For example, your system may use a protocol that describes how the medical director wants you to deal with a patient who is having respiratory difficulties. Part of this protocol may direct you to use a nonrebreathing mask to deliver oxygen at 15 L/min. You may do this without calling online medical control if the patient meets the criteria of the protocol.

1. **Right patient.** For EMS, this safety check may seem unneeded if there is only one patient in the back of an ambulance. In that case, it will be rather obvious who is to be administered the medication. However, there will be times when you will be working with more than one patient and you will need to ensure that the patient who needs the medication is the person who receives the medication.

2. **Right medication.** Verify the proper medication and prescription. Once received, confirm the medication order and determine that the patient is still a candidate for the medication. Confirm again that the medication you are about to give is the correct medication. Carefully read the label. If it is the patient's own prescription, the bottle may show the trade name or the generic name. If you have any questions, contact online medical control. Make sure that the medication is the patient's own and does not belong to a friend or relative. You should never give a medication to a patient that has been prescribed for someone else.

3. **Right dose.** Verify the form and dose of the medication. You have confirmed your order and verified that the medication is the correct one to give. Now you must make sure that the form of the medication and the dose are correct. This is where it is important to pay very close attention to detail. If you are ordered to give 1,000 mg of acetaminophen, you will need to read the bottle to determine how many milligrams are in each tablet. If acetaminophen is available in 500-mg tablets, how many will you need to give the patient?

4. **Right route.** Verify the route of the medication. Now you must make sure the route matches the order you received. For example, suppose you are told to give the patient a sublingual nitroglycerin tablet. The patient's nitroglycerin tablet bottle is empty, but he has another bottle of nitroglycerin capsules. These are to be swallowed four times a day. The medication is the same, even the dose may be the same, but the route of delivery is different from the order given. You may not substitute the capsules for the tablets without specific orders from medical control.

5. **Right time.** Check the expiration date and condition of the medication. The last step before administering a medication is to make sure the expiration date has not passed. Prescription and OTC medications alike should have an expiration date on their labels. Check the date. If no date can be found, you should examine the medication with suspicion. In addition, if you find discoloration, cloudiness, or particles in a liquid medication, you should not use it. If a patient with asthma gives you an MDI and the expiration date on it is smudged, you should not use it. After the medication is administered, you will need to reassess the patient to see if it has worked. Does the patient still have the same complaint as before you administered the medication? Has it changed? Has the medication seemed to have made a difference? Depending on the medication and the route of administration, it may take seconds or hours for the medication to take effect. You should reassess the vital signs, especially heart rate and blood pressure, at least every 5 minutes or sooner if the patient's condition changes.

6. **Right documentation.** Remember the EMS rule: The work is not done until the paperwork is done. Once the medication has been given, you must document your actions and the patient's response. This includes the time you gave the medication and the name, dose, and route of administration. Did the patient's condition improve, worsen, or not change? Were there any side effects? A second EMS rule says, "If you did not write it down, it did not happen." If your performance should ever be questioned, documentation is your best defense.

Medication Administration and the EMT

There are several medications that may be carried on the EMS unit, including oxygen, oral glucose, activated charcoal, aspirin, and epinephrine. When used wisely, each can be a powerful tool. Keep in mind, however, that you may give these medications only according to standing orders in a protocol (off-line medical control) or a direct order (online medical control). Along with the several different medications that can be given by EMTs, there are several different routes of administration that will be used to deliver these medications to the patient.

Before specific medications are discussed, the circumstances surrounding the administration of the medications need to be discussed. Over the years, EMTs have been allowed increasing responsibility to work with medications, but this growth has come with some degree of confusion and worry. Many departments throughout the United States have strict controls on when an EMT is allowed to administer a medication. The circumstances are:

- **peer-assisted medication**
- **patient-assisted medication**
- **EMT-administered medication**

In peer-assisted medication administration, you are administering medication to yourself or your partner. It may be necessary for the EMS crew to receive medications because they were exposed to a toxic agent. In this case, you would first treat yourself and then your partner. Typically, the types of medications would be found in an auto-injector format.

In patient-assisted medication administration, you are assisting the patient with the administration of his or her own medication. In the case of nitroglycerin, EMTs will often work with the patient to help them take their own medication. Medications where the EMT commonly assists the patient are the EpiPen auto-injector, which is used for allergic reactions, an MDI bronchodilator that

Words of Wisdom

The following is a list of medications that, depending on local protocol, may be administered by EMTs.

Keep in mind that this list is ultimately set by the state in which you will be delivering patient care and the medical director of your ambulance department.

- Oxygen
- Activated charcoal
- Oral glucose
- Aspirin
- Epinephrine auto-injector
- Metered-dose inhaler medications
- Nitroglycerin

You may administer or help to administer medications only under the following conditions:

- Medical control gives you a direct order to administer a medication and/or the local medical protocols under which you are working permit you to administer that medication.
- The local medical protocols, developed by a medical physician under which you are working, include standing orders for the use of a medication in defined situations. It is imperative that you do not give or help patients take any other medications under any other circumstances.

Special Populations

Pediatric and geriatric patients often have slower absorption and elimination times, necessitating modification of the doses administered. Pregnant patients are limited in the medications they can take because of the risk to the fetus.

is often used for asthma, or nitroglycerin to relieve the pain of angina. Perhaps the patient cannot find their medication. Maybe the patient is so upset that he or she cannot open the pill bottle or hold the MDI steady. In this circumstance, the patient is trying to administer the medication, but the EMT needs to offer some help so the task can be completed.

The last circumstance is EMT-administered medications. Here the EMT is directly administering the medication to the patient. It can certainly be difficult to find the exact point where "assisting" a patient ends and actually administering a medication begins. The patient may be severely confused or unable to understand the need for the medication. Common medications that the EMT administers in this circumstance are oxygen, oral glucose, activated charcoal, and aspirin.

It is important for you to understand that the medication itself does not necessarily dictate whether or not you will be assisting or actually administering. Medical control, state guidelines, and local protocols will be the determining factors that define the role of the EMT. The EpiPen has often been an example of a medication that is both patient-assisted and EMT-administered. Refer to your local standards to obtain a listing of how and when EMTs can administer medications.

Medications Used by EMTs

The following is a discussion of medications that may be administered by EMTs. Again, your state, department, and medical director will ultimately define what medications are carried on your ambulance. **Table 7-4** provides an excellent overview of these medications and their actions, indications, contraindications, routes of administration, side effects, interactions, and doses.

The 2009 National EMS Education Standards recognize that some regions of the country may need their EMTs involved in the administration of additional medications aside from oxygen, oral glucose, activated charcoal, aspirin, and epinephrine. The exact list of medications that each EMT will be allowed to manage is controlled ultimately by the state and medical director of the ambulance. Acetaminophen, ibuprofen, and diphenhydramine are three common medications that are used

Table 7-4 EMT Medications

Generic/Trade	Action	Indications	Contraindications	Routes	Side Effects	Interactions	Adult Dose	Administration Concerns
Acetaminophen/ Tylenol	Analgesic and antifever	Relief of mild pain or fever, headache, muscle aches	Hypersensitivity	PO	Allergic reaction	Caution must be taken when EMTs are administering acetaminophen to avoid potential overdosing. Many over-the-counter medications contain acetaminophen	500 to 1,000 mg every 4 hours as needed; dose is weight based for children	Weight of child is more important than age
Activated charcoal/Actidose with Sorbitol	Adsorbs toxic substances in the digestive tract	Most oral poisonings; overdose	Decreased level of consciousness; overdose of corrosives, caustics, or petroleum substances.	PO	Nausea, vomiting, constipation, black stools	Bonds with and inactivates most medications/ substances in the digestive tract	1 to 2 g/kg	Stains; protect patient and provider clothing; do not give when giving other PO medications
Aspirin/Bayer	Anti- inflammatory agent and anti- fever agent; prevents plate- lets from clump- ing, thereby decreasing formation of new clots	Relief of mild pain, headache, muscle aches; chest pain when consider- ing myocardial infarction	Hypersensitivity, recent bleeding	PO	Nausea, vomiting, stomach pain, bleeding, allergic reactions	Caution should be used in patient who are taking anticoagulants	160 to 325 mg; 160 to 325 mg chewable tablets for chest pain	Do not administer for pain caused by trauma or for fevers in children; patients with chest pain must be able to chew tablets
Albuterol/ Proventil, Ventolin	Stimulates nervous sys- tem, causing bronchodilation	Asthma/ difficulty breathing with wheezing	Hypersensitivity, tachycardia, myocar- dial infarction	MDI/ inhala- tion	Hypertension, tachycardia, anxiety, restlessness	Increases effects of other nervous system stimulants	1 to 2 inhalations; wait 5 minutes before repeating dose	Patient must inhale all medication in 1 breath; coach patient to hold breath for 5 seconds after inhalation
Diphenhydramine/ Benadryl	Antihistamine (blocks hista- mine)	Mild allergic reactions	Asthma, glau- coma, pregnancy, hypertension, infants	PO	Sleepiness (although can stimulate chil- dren), dry mouth and throat	Do not take with alco- hol or MAO inhibitors (a type of psychiatric medication)	25 to 50 mg	Can use in severe allergic reaction; however, epine- phrine is adminis- tered first

Medication	Action	Indications	Contraindications	Route	Side Effects	Interactions	Dose	Special Considerations
Epinephrine/EpiPen	Stimulates nervous system, causing bronchodilation	Severe allergic reaction	Myocardial infarction, hypothermia, hypertension	IM (auto-injector)	Hypertension, tachycardia, anxiety, restlessness	Increases effects of other nervous system stimulants	1 auto-injector	Medication will last approximately 5 minutes; do not repeat dose; ensure ALS is en route for continuing treatment
Ibuprofen/Advil, Motrin, Nuprin	Nonsteroidal anti-inflammatory drug that reduces inflammation and fever, analgesic	Mild pain or fever, headache, muscle aches	Hypersensitivity	PO	Nausea, vomiting, stomach pain, bleeding, allergic reactions	Do not take with aspirin	200 to 400 mg every 4 to 6 hours; dose is weight based in children	Do not take for pain caused by trauma; weight of child is more important than age
Nitroglycerin/Nitrostat	Dilates blood vessels	Chest pain due to myocardial infarction or angina	Hypotension, having taken sildenafil (Viagra) or another treatment for erectile dysfunction within the past 24 hours; head injury	SL/spray	Headache, burning under tongue, hypotension, nausea	Increases dilating effects of other blood vessel-dilating medications	0.3 to 0.4 mg SL; 0.4 mg spray	Ensure ALS is en route
Oral glucose/Glutose	When absorbed, provides glucose for cell use	Low blood glucose (hypoglycemia)	Decreased level of consciousness, nausea, vomiting	PO	Nausea, vomiting	None	1/2 to 1 tube	Patient must have control of airway and be awake and able to follow commands
Oxygen (no trade name)	Reverses hypoxia, provides oxygen to be absorbed by lungs	Hypoxia or suspected hypoxia	Very rarely in patients with COPD. Do not use near open flames as oxygen will support combustion	Gas/inhalation	Decreased respiratory effort in rare cases in patients with COPD	Can support combustion	Use oxygen delivery devices to administer 28% to 100% oxygen	No open flames nearby; do not withhold oxygen from patients in respiratory distress

daily by millions of Americans. These medications have been deemed by the Food and Drug Administration to be safe for consumer use without medical supervision. When patients are suffering from minor pain, fevers, or allergic reactions, these medications can make the patient feel better.

■ Oral Medications

There are several medications EMTs may be asked to administer or to assist with administration. Activated charcoal, oral glucose, aspirin, and several over-the-counter medications can be administered by this route. As discussed, the advantages of this route are its ease of access and comfort level for the patient. One of the disadvantages of administering medications orally is that the digestive track can be easily affected by foods, stress, and illness. The speed of movement of food through the track dramatically changes the speed of absorption; this is considered a serious limitation. As with all medications, the EMT needs to start with the six rights. Follow these steps to perform oral medication administration Skill Drill 7-1 :

1. Take standard precautions.
2. If using a liquid, pour the desired amount into a calibrated cup.

3. Instruct the patient to swallow the medication (if a liquid) or chew (if a solid and it is appropriate). Provide water if administering a solid Step 1 .
4. Monitor the patient's condition and document.

Activated Charcoal

Many poisoning emergencies involve overdoses of medications taken by mouth. Many medications bind with activated charcoal, keeping the medications from being absorbed by the body. **Adsorption** means to bind to or stick to a surface, while absorption is the process by which medications travel through body tissues until they reach the bloodstream. **Activated charcoal** is ground into a very fine powder to provide the greatest possible surface area for binding. You will probably carry a container with a premixed suspension of activated charcoal powder and water in the EMS unit, if allowed by local protocol Figure 7-7 . The usual dose is 1 to 2 g/kg of body weight.

The bond between medication and charcoal is not permanent. Because the medication may break free and be absorbed into the bloodstream if activated charcoal remains in the digestive system throughout a normal day, charcoal is frequently suspended with another medication called sorbitol (a complex sugar). This suspension has a laxative effect that causes the entire mixture, including the medication, to move quickly through the digestive system.

Activated charcoal is given by mouth. Although sorbitol sweetens the suspension, the black charcoal makes

You are the Provider: PART 3

After administering the proper medication to the patient, you reassess her and note that her condition has improved. She is conscious and alert and asks you what happened. As you are explaining what happened to her, your partner takes her vital signs.

Recording Time: 5 Minutes	
Respirations	22 breaths/min; regular and adequate
Pulse	112 beats/min; strong and regular
Skin	Pink; slightly moist
Blood pressure	122/72 mm Hg
Oxygen saturation (Sao$_2$)	98% (on oxygen)

5. What medications are typically carried on an EMT ambulance?
6. What medications can the EMT help a patient self-administer?

Skill Drill 7-1

Oral Medication Administration

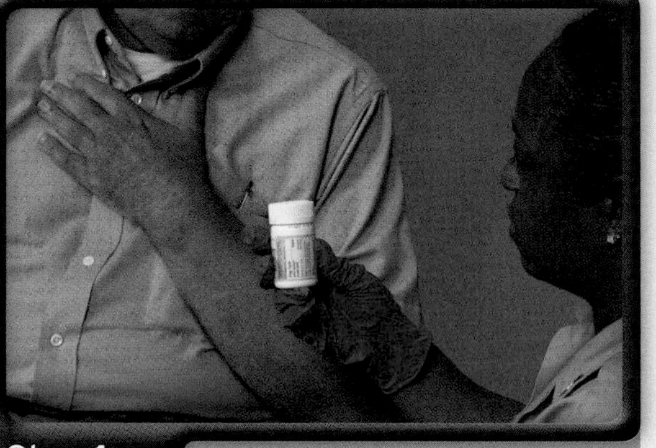

Step 1 Take standard precautions. Prepare the appropriate amount of medication. Instruct the patient to chew (if appropriate) or swallow the medication with water, if administering a pill or tablet.

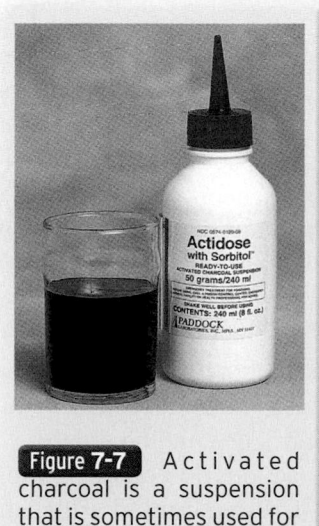

Figure 7-7 Activated charcoal is a suspension that is sometimes used for patients who have taken a medication overdose or swallowed a poison.

it look unappealing. For this reason, you should use a covered container and ask the patient to drink the fluid through a straw. Also, it is important to have a layer of protective clothing, such as a gown, over your EMT uniform. Activated charcoal is very staining to clothing, so protect yourself and the patient's clothes when administering it. This medication should not be given to anyone with an altered level of consciousness because of the risk of aspiration. Activated charcoal is not indicated for patients who have ingested an acid, an alkali, or a petroleum product.

Oral Glucose

Glucose is a sugar that our cells use as fuel. Although some cells can use other sugars, brain cells must have glucose.

If the level of glucose in the blood gets too low, a person can lose consciousness, have seizures, and ultimately die.

The medical term for an extremely low blood glucose level is **hypoglycemia**. Hypoglycemia can be caused by an excess of insulin, which is taken to control blood glucose levels. Patients with diabetes who use insulin regularly understand the effects of this medication on the body. The **oral glucose** that is carried in the EMS unit can counteract the effects of hypoglycemia in the same way as a candy bar or sweet drink, but faster. This is because common table sugar (sucrose) and fruit sugars (fructose) are complex sugars and must be broken down before they can be absorbed. Glucose is a simple sugar that is readily absorbed by the bloodstream.

As an EMT, you can give glucose only by mouth. Hospital personnel and advanced providers (AEMTs and paramedics) can give glucose through an IV line. Glucose is available as a gel designed to be spread on the mucous membranes between the cheek and gum; however, absorption through this route is not as quick as with injection. Because the patient may be conscious one moment and unconscious the next, you must be very careful when administering oral glucose. Never administer oral medications to an unconscious patient or to one who is unable to swallow or protect the airway.

See Chapter 17, *Endocrine and Hematologic Emergencies*, for more information on the administration of oral glucose.

Aspirin

Aspirin (acetylsalicylic acid or ASA) is an antipyretic (reduces fever), analgesic (reduces pain), and anti-inflammatory (reduces inflammation) and inhibits platelet aggregation (clumping). This last property makes it one of the most used medications today. Because research has shown that platelets aggregating under certain conditions in the coronary arteries is one of the direct causes of heart attack, patients at risk for coronary artery disease are often prescribed one or two "baby" (or children's) aspirins a day. During a potential heart attack, aspirin may be lifesaving.

Contraindications for aspirin include documented hypersensitivity to aspirin, pre-existent liver damage, bleeding disorders, and asthma. Because of the association of aspirin with Reye syndrome (a rare but serious condition that causes swelling in the brain and liver), it should not be given to children during episodes of fever-causing illnesses.

◼ Sublingual Medications

The sublingual route of administration has many advantages. Assuming the patient is awake and alert and able to follow commands, it is easy to talk with the patient and advise them to place a pill under his or her tongue. The head and face receive large amounts of blood flow, so absorption rates are relatively quick. Be aware, however, that any medication placed in the mouth requires constant evaluation of the airway. You must also be alert to any signs of choking on the pill. If the patient is uncooperative or unconscious, this route of medication administration should not be used.

Nitroglycerin

Many patients with cardiac conditions carry some form of fast-acting nitroglycerin to relieve the pain of angina. Nitroglycerin has been used medically since the 1800s. Nitroglycerin is typically the only medication that EMTs will help to administer sublingually. See Chapter 14, *Cardiovascular Emergencies*, for more information on how to administer nitroglycerin.

If you have ever run for a prolonged period, you probably remember your muscles developed a painful, heavy, burning sensation. This is because the demand for oxygen by the muscles exceeded the supply. When heart muscle develops a similar pain, it is called angina pectoris. The cause is the same: not enough oxygen; in this case because of a blockage or narrowing in the blood vessels that supply the heart. Occasionally, the cause is a spasm in these blood vessels. Unlike the runner with sore legs, of course, the heart muscle cannot stop and rest until the pain goes away.

The purpose of nitroglycerin is to increase blood flow by relieving the spasms or causing the arteries to dilate. It does this by relaxing the muscular walls of the coronary arteries and veins. Nitroglycerin also relaxes veins throughout the body, so less blood is returned to the heart and the heart does not have to work as hard each time it contracts. In short, blood pressure is decreased. Because of this, however, it is important that you always take the patient's blood pressure before administering nitroglycerin. If the systolic blood pressure is less than 100 mm Hg, the nitroglycerin may have the harmful effect of lowering the blood flow to the heart's own blood vessels. Even a patient who has adequate blood pressure should sit or lie down with the head elevated before taking this medication. If the patient is standing, he or she may faint when blood flow to the brain is reduced as the nitroglycerin starts to work. If a significant drop in the patient's blood pressure (15 to 20 mm Hg) occurs and the patient suddenly feels dizzy or sick, lay the patient down and raise the legs.

During a heart attack (myocardial infarction, or MI), a blood clot forms in a narrowed coronary artery. This blocks the blood flow to a section of the heart muscle. If the blockage is not cleared in time, that section of the heart muscle beyond the clot will die. If nitroglycerin no longer brings relief to a person in whom it has previously worked, the person may be experiencing an MI instead of an angina attack. Therefore, it is important to know how much nitroglycerin a patient has needed in the past to relieve chest pain and how much has been taken during the current emergency, including the use of nitroglycerin patches. Always report this information to medical control. Remember, you cannot administer this medication without clearance from medical control or standing orders.

There are important interactions to consider when administering nitroglycerin. The medication sildenafil (Viagra) can have potentially fatal interactions with nitroglycerin. When taken together, nitroglycerin and sildenafil can cause dramatic drops in blood pressure. Any medication that is similar to sildenafil can also have the same effect. Ask a patient who has been prescribed nitroglycerin if they have used sildenafil (Viagra), tadalafil (Cialis), vardenafil (Levitra), or any other medication that is used for the treatment of erectile dysfunction within the previous 24 hours. If he or she has, do not administer the nitroglycerin and report this to medical control.

Nitroglycerin has the following effects:

- Relaxes the muscular walls of coronary arteries and veins
- Results in less blood returning to the heart
- Decreases blood pressure
- Relaxes arteries throughout the body
- Often causes a mild headache after administration

Administering Nitroglycerin by Tablet Nitroglycerin is usually taken sublingually. The patient places a tiny tablet under the tongue, where it dissolves. The tablet should create a slight tingling or burning sensation. If the nitroglycerin has lost its usual "bite," it may have lost potency because of aging or improper storage. Be sure to check the expiration date on the bottle.

Sublingual nitroglycerin tablets should be stored in their original glass container with the cap screwed on tightly. Note that what looks like cotton in the container is actually rayon. If real cotton is placed in the container, it can absorb nitroglycerin, thus reducing the potency of the tablets. Other medications placed in the container can likewise rob nitroglycerin of its power. Exposure to light, heat, or air may degrade the strength of the medication as well. If you notice any signs of improper storage, be sure to include that information in the patient's medical history.

Administering Nitroglycerin by Metered-Dose Spray Some patients who take nitroglycerin use a metered-dose spray, which deposits medication on or under the tongue. Each spray is equivalent to one tablet. To ensure direct, proper dosing on the bottom of the tongue, do not use a spacer with the metered-dose canister when giving nitroglycerin by this method.

Whether using the tablets or the metered-dose spray, you should wait 5 minutes for a response before repeating the dose. Closely monitor the patient's vital signs, particularly the blood pressure. Give repeated doses per medical control and/or local protocol. Remember always to wear gloves when handling nitroglycerin tablets or spray; this medication can be absorbed by your skin.

Next, you must reconfirm that the patient can tolerate the medication. For example, suppose you have received and verified the order to give one sublingual nitroglycerin tablet to a patient with a cardiac condition. While you were getting the order, however, the patient begins to sweat more and becomes less responsive. Reassessment of the blood pressure reveals a pressure of 80/60 mm Hg. Using your knowledge about nitroglycerin, you decide not to give the medication. Instead, you notify medical control of the changes in the patient's condition and seek further orders.

Lastly, if the patient is having chest pain "just like my other heart attack," you may be able to apply oxygen, but you may have to call medical control or follow standing orders before helping to administer the patient's nitroglycerin spray. Knowing and understanding the local

> **Words of Wisdom**
>
> **General Steps in Administering Medication**
> 1. Obtain an order from medical control.
> 2. Verify the proper medication and prescription.
> 3. Verify the form, dose, and route of the medication.
> 4. Check the expiration date and condition of the medication.
> 5. Reassess the vital signs, especially heart rate and blood pressure, at least every 5 minutes or as the patient's condition changes.
> 6. Document.

protocols under which you will be working are absolutely essential. Refer back to Table 7-4 for a review of all of the medications and the important information needed for their administration.

■ Intramuscular Medications

The intramuscular (IM) route of administration provides quick and easy access to the circulatory system without the need for placing a needle within a vein. Blood flow to the muscles is relatively stable even during circumstances of severe illness or injury. This advantage makes the IM route an efficient means to deliver some medications. A major disadvantage for this route is the use of a needle and the subsequent pain that it can cause. Patients may be reluctant to use the needle for fear of pain or injury. With proper technique, the EMT can administer medications via the IM route and limit the amount of pain delivered to the patient. See Chapter 18, *Immunologic Emergencies*, for more information on the administration of epinephrine using an auto-injector.

Epinephrine

Epinephrine is the main hormone that controls the body's fight-or-flight response and is the primary medication that EMTs will be administering IM. Epinephrine is a sympathomimetic. A sympathomimetic mimics the effect of the sympathetic nervous system. The body releases epinephrine when there is sudden stress, such as during exercise or when the patient is suddenly scared. Because epinephrine is secreted by the adrenal glands, it is also known as adrenaline. Epinephrine has different effects on different body tissues and is used as a medication in several forms. Generally, epinephrine will increase the heart rate and blood pressure and dilate passages in the lungs. Therefore, it can ease breathing problems caused by the bronchial spasms that are common in asthma and allergic reactions. In a person who is close to anaphylactic shock as a result of an allergic reaction, epinephrine may also help to maintain the patient's blood pressure. This medication should not be given to patients with hypertension, hypothermia, or if you believe the patient may be having a myocardial infarction. Patients who are not wheezing or who have no signs of respiratory compromise or hypotension should not be given epinephrine.

Epinephrine has the following characteristics:

- Secreted naturally by the adrenal glands
- Dilates passages in the lungs
- Constricts blood vessels, causing increased blood pressure
- Increases heart rate and blood pressure

Refer back to Chapter 5, *The Human Body* for more information on epinephrine.

Administering Epinephrine by Injection Some states and EMS services now authorize the use of epinephrine by EMTs for the treatment of life-threatening anaphylaxis. In certain individuals, insect venom or other allergens cause the body to release histamine, which lowers blood pressure by relaxing the small blood vessels and allowing them to leak. The release of histamine may also cause wheezing from bronchial spasms and swelling of the airway tissues (edema), which make it difficult for the patient to breathe. Epinephrine acts as a specific antidote to histamine, countering both of these harmful effects. It constricts the blood vessels, allowing blood pressure to rise and reducing the swelling. In the lungs, it has the opposite effect; it dilates the air passages, so the flow of air is less restricted. You can also expect the patient's heart rate to increase after administration of epinephrine.

Epinephrine may be dispensed from an auto-injector, which automatically delivers a preset amount of the medication **Figure 7-8**. This is usually 0.3 mg of epinephrine. This is the method that you will most likely use. Be sure to familiarize yourself with the procedures for using the auto-injector on your unit. The general procedure is as follows:

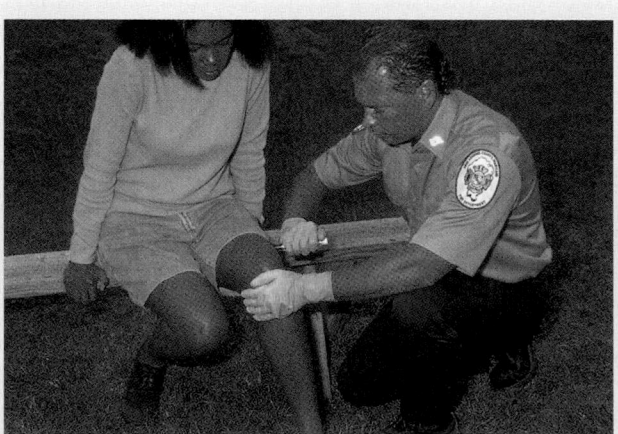

Figure 7-8 An EpiPen auto-injector may be used to administer a preset dose of epinephrine.

1. Grasp unit with the tip pointing downward.
2. Form a fist around the unit.
3. With the other hand, pull off the activation cap.
4. Hold the tip near the outer part of the patient's thigh.
5. Swing and jab firmly into the outer thigh so that the unit is perpendicular (at a 90° angle) to the thigh. Do not allow it to bounce.
6. Hold firmly in the thigh for several seconds.

Epinephrine causes a burning sensation where it is injected, and the patient's heart rate will increase after the injection, so be prepared for these side effects. As discussed, it is not the medication, but the other factors that will determine if a medication can be administered or patient-assisted. You have already read the general

Words of Wisdom

When documenting a medication, include the name of the medication, dose and route, and vital signs before and after administration. For example: 10:30 AM–vital signs: pulse, 88 beats/min; respirations, 18 breaths/min; blood pressure, 128/68 mm Hg; nitroglycerin, 0.4 mg SL. 10:35 AM–vital signs: pulse, 80 beats/min; respirations, 18 breaths/min; blood pressure, 124/60 mm Hg.

Special Populations

Geriatric patients often take many medications. They might also save medications left over from previous medical conditions. Make every effort to identify which medications are current and the conditions they are being used to treat. Ask family members to help distinguish current from outdated medications, or look at the expiration dates on the medication labels. Create a list of all of the patient's current medications or bring the medications with you to the emergency department.

Geriatric patients can become confused about their medication regimen. Uncertainty about whether they missed a dose may cause the patients to repeat the medication, possibly leading to an overdose. If you think an overdose has occurred, contact medical control.

Remember, medications can interact with each other, creating potentially harmful conditions. Even though a medication may be indicated for a special condition, it might be contraindicated in the presence of another medication. For example, if the patient is taking the heart medication propranolol (Inderal) and has an acute episode of shortness of breath, any asthma treatment might be made ineffective by the heart medication.

Although medications help people to recover from acute conditions and adjust to chronic diseases, they can pose serious problems for geriatric patients. You should distinguish current from previous medications, suspect accidental or intentional overdoses, and be prepared for potentially lethal medication interactions. Document all findings, and inform medical control.

information on epinephrine and its use by EMTs for anaphylaxis. Some services do not permit EMTs to carry epinephrine but do allow them to assist patients in administering their own epinephrine in life-threatening anaphylactic reactions. In addition, EMTs frequently may assist patients in administering epinephrine through their own MDIs for bronchospasm.

■ Inhalation Medications

Oxygen

All cells need <u>oxygen</u> to function properly. The heart and brain, especially, cannot function for long if oxygen levels decrease, which is why oxygen is an on-board medication for EMS units. If a patient is not breathing or is having trouble getting air into the lungs, you should administer supplemental oxygen. In general, you will be giving oxygen via a nonrebreathing mask at 10 to 15 L/min (or via nasal cannula at 2 to 6 L/min if the patient cannot tolerate a nonrebreathing mask). However, if the patient is not breathing, you must also provide artificial ventilations, so you will need to use a bag-mask device. Oxygen is usually delivered at 15 L/min with this technique.

Outside a hospital, the nonrebreathing mask is the preferred method of giving oxygen to patients who are experiencing significant respiratory difficulties or shock. With a good mask-to-mouth seal, this mask can provide up to 90% inspired oxygen Figure 7-9 . With a nasal cannula, oxygen flows through two small, tubelike prongs that fit into the patient's nostrils. This device can provide up to 44% inspired oxygen if the flowmeter is set at 6 L/min.

Remember that, although oxygen itself does not burn, it allows other things to burn. If there is extra oxygen in the air, objects will burn more easily. So make sure there are no open flames, lit cigarettes, or sparks in the area in which you are administering oxygen.

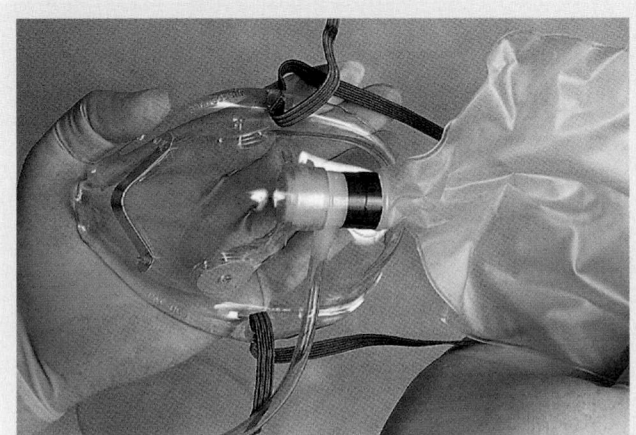

Figure 7-9 A nonrebreathing mask is the preferred method of giving oxygen because it provides up to 90% inspired oxygen.

You are the Provider: PART 4

The patient is placed onto the stretcher and loaded into the ambulance. She remains conscious and alert. Her son tells you that he has to retrieve some items from her house and will follow the ambulance in his car. Shortly before departing the scene, you reassess the patient's vital signs.

Recording Time: 13 Minutes	
Level of consciousness	Conscious and alert
Respirations	18 breaths/min; regular and adequate
Pulse	84 beats/min; strong and regular
Skin	Pink, warm, and slightly moist
Blood pressure	128/74 mm Hg
Sao$_2$	99% (on oxygen)

The patient tells you that she thinks she may have accidentally taken too much of her Glucophage. You reassess her blood glucose level and note that it is 94 mg/dL.

7. How can you determine the use for Glucophage?

8. Does the patient require further medication therapy at this point?

Metered-Dose Inhalers and Nebulizers

MDIs and nebulizers are used to administer liquid medications that have been turned into a fine mist by a flow of air or oxygen. With the medication atomized, it is breathed into the lungs and delivered to the alveoli. Blood flow to the alveoli is very high and absorption rates are very close to those found with IV medications. This route is fast and relatively easy to access. MDIs are commonly used because of their convenience and portability. The major disadvantage of an MDI is that the patient needs to be cooperative and control his or her breathing. If the patient is unconscious, an MDI cannot be used, although an EMT could use a nebulizer. Nebulizers are often used for more severe problems.

Medications Administered Using a Metered-Dose Inhaler

Sometimes, a respiratory condition such as asthma is not severe enough to require the use of epinephrine. In such cases, patients may use one of the epinephrine "cousins" that are more narrowly focused on the lungs. These medications are delivered with an MDI. An MDI requires a great deal of coordination, something that may be difficult to achieve when a person is having trouble breathing. Patients must aim properly and spray just as they start to inhale; however, most of the medication tends to end up on the roof of the patient's mouth. An

adapter, called a "spacer," which fits over the inhaler like a sleeve, can be used to avoid misdirecting the spray **Figure 7-10** . The inhaler fits into an opening on one end of the spacer's chamber, and the mouthpiece fits on the other end. The patient sprays the prescribed dose into the chamber and then breathes in and out of the mouthpiece until the mist is completely inhaled.

You can activate the spray by pressing the canister into the adapter just as the patient starts to inhale. If relief is not achieved, wait about 3 to 5 minutes and repeat this sequence according to the patient's prescription. Above all, it is important to ensure that the patient inhales all the medication in a single-sprayed dose. See Chapter 13, *Respiratory Emergencies*, for more information on using an MDI.

Asthma, also known as "reactive airway disease," can be a life-threatening condition. Therefore, some patients use epinephrine inhalers to relieve bronchial spasms quickly. The trade names of some of these inhalers are Primatene Mist, Bronitin Mist, Bronkaid Mist, and Medihaler-Epi. Because epinephrine tends to increase the heart rate and blood pressure, most patients with asthma use certain chemical cousins of epinephrine that produce fewer side effects. Metaproterenol (Alupent or Metaprel) and albuterol (Proventil or Ventolin) are beta-2 agonists. A beta-2 agonist acts more specifically than epinephrine on the bronchi of the lungs causing dilation with a lesser effect on the heart (beta-1 agonist). Give repeated doses per medical control and/or local protocol.

One of the most important factors you should be aware of when using the patient's MDI is that many patients are prescribed many types of MDIs. Patients with asthma or COPD may be using MDIs that are corticosteroids, others that are mucolytics, and still others that are cell stabilizers. There are at least 20 different MDIs on the market, and often patients may be prescribed three to six different individual medications. The only medication that will be effective during an acute attack of shortness of breath will be short-action bronchodilators. Albuterol

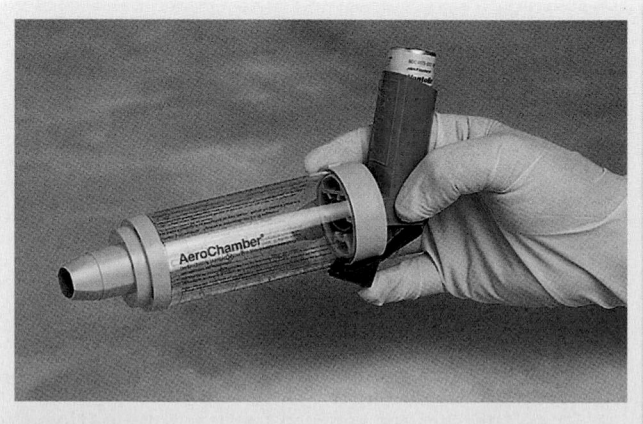

Figure 7-10 Some inhalers have spacer devices to better direct the medication spray.

(Proventil or Ventolin) remains a popular, effective, short-action bronchodilator. Be sure that you know the right drug—read the label of the MDI before you use or assist the patient in using it. If you are administering a medication using a nebulizer, you will know which bronchodilator is being administered because you will select it out of the medication bag. The most common bronchodilators are either albuterol or metaproterenol. The process involves placing the medication into a nebulizer and then running a flow of oxygen through the device, which will atomize the liquid and allow the patient to breathe in the medication.

Patient Medications

Part of your patient assessment includes finding out what medications your patient is currently taking. This information may provide vital clues to your patient's condition that may help guide your treatment or be extremely useful to the emergency department physician. Often, knowing what medications a patient takes may be the only way you can determine what chronic or underlying conditions your patient may have, such as when a patient is unable to relate his medical history to you. The patient may be unresponsive, confused, not knowledgeable of his or her medical history, uncooperative, or unable to communicate. Discovering what the patient takes and transporting the medications or a list of them with you

Special Populations

Polypharmacy is a term referring to the use of multiple medications by one person. It is not uncommon today to find patients, especially elderly patients, taking many medications on a regular basis. Often, the prescription regimens can be complex and confusing. The medications may be prescribed by multiple physicians. The person may also be taking nonprescription and herbal medicines. Add to this the possibility of failing memory and confusion, and the potential for overdosing, underdosing, and harmful interactions increases exponentially.

to the emergency department can be crucial in assessing your patient's needs.

In addition to prescription medications, patients often take nonprescription OTC medications and herbal medications. Many times, they do not consider these substances "medications" and will not report them to you unless you ask about them specifically. Yet, they may be as potent as prescription medications and can have interactions and effects on the patient's health and condition that are just as important. Be sure to ask specifically about these also. **Table 7-5** lists the top 20 prescribed medications and their uses.

Patients are naturally reluctant to tell you about any illegal drugs or medications they may have taken. It is important to ask, and you can assure them that your only interest in asking is to be able to treat them appropriately.

You are the Provider: PART 5

The patient's condition remains stable during transport. You reassess her vital signs and then call in your radio report to the hospital, which you will arrive at in approximately 8 minutes.

Recording Time: 19 Minutes	
Level of consciousness	Conscious and alert
Respirations	18 breaths/min; regular and adequate
Pulse	74 beats/min; strong and regular
Skin	Pink, warm, and dry
Blood pressure	126/72 mm Hg
Sao$_2$	98% (on oxygen)

You arrive at the hospital and give your verbal report to the charge nurse. The patient's son arrives shortly thereafter and presents the nurse with a plastic bag containing seven medications, including those that you have already noted. After further assessment and treatment in the emergency department, the patient is admitted for observation.

9. What does the term "polypharmacy" mean?

Table 7-5   Most Prescribed Medications in the United States in 2007

Trade Name/Generic Name	Use
Lipitor/atorvastatin	Lowers cholesterol
Singulair/montelukast	Helps prevent asthma attacks
Lexapro/escitalopram	Depression
Nexium/esomeprazole	Gastric reflux
Synthroid/levothyroxine	Decreased thyroid functioning
Plavix/clopidogrel	Prevents stroke and heart attack
Toprol/metoprolol	Lowers blood pressure
Prevacid/lansoprazole	Stomach ulcers
Vytorin/ezetimibe and simvastatin	Lowers cholesterol
Advair Diskus/fluticasone and salmeterol	Asthma
Zyrtec/cetirizine	Allergies
Effexor/venlafaxine	Depression
Protonix/pantoprazole	Gastric reflux
Diovan/valsartan	High blood pressure
Fosamax/alendronate	Osteoporosis
Zetia/ezetimibe	Lowers cholesterol
Crestor/rosuvastatin	Lowers cholesterol
Levaquin/levofloxacin	Antibiotic
Diovan HCT/valsartan and hydrochlorothiazide	Hypertension
Klor-Con/potassium chloride	Low potassium levels

You are the Provider: SUMMARY

1. **What is pharmacology?**

   Pharmacology is the scientific study of medications, including their ingredients, preparations (ie, how they are stored), therapeutic uses, and actions on the body. Several terms are used when discussing pharmacology. The *dose* is the amount of medication that is given to the patient; it is based on the patient's weight or age. The *action* is the therapeutic effect that the medication is expected to have on the body. *Indications* are the reasons or conditions for which a particular medication is given. *Contraindications* are the reasons or conditions for which a particular medication should not be given because it may cause further harm; a particular medication may be contraindicated for a given patient—even if it is generally indicated for the patient's condition. *Side effects* are any actions of a medication other than the desired effects; they are usually predictable and expected—although unpleasant for the patient—and may occur even when the medication is administered properly.

2. **Why is knowledge of pharmacology important to patient care?**

   Giving a medication to a patient without understanding how it will affect him or her is dangerous. Prior to administering *any* medication—including oxygen—you must understand what effect(s) it will have on the patient. In addition, you must perform a careful and accurate assessment to determine if medication therapy is even indicated.

   The patient may have a condition for which a particular drug is indicated; however, various factors that are unique to the patient (ie, known allergy to the drug, unstable vital signs, etc) may otherwise contraindicate it. The only way you will be able to determine this is through a careful assessment.

   It is easy enough to memorize the indications, contraindications, doses, and side effects of the drugs that you may administer as an EMT, but if you do not know how the drug will affect the patient's body, you should not be giving it. *Once you give it, you cannot take it back!*

You are the Provider: SUMMARY, continued

As an EMT, you must have a solid understanding of each drug that you may be called on to administer, as well as the medications that you may be asked to assist a patient with.

3. Other than oxygen, does this patient require any other medications?

Considering the fact that a normal blood glucose level is 80 to 120 mg/dL, a level of 36 mg/dL is significantly low (hypoglycemia) and would explain the patient's present mental status. Glucose is a sugar that the body's cells use as fuel. Some cells can use other sugars; however, the brain requires glucose. If the blood glucose level gets too low, a person can lose consciousness, experience seizures, and ultimately die.

One of the drugs that EMTs are trained to administer is oral glucose. Oral glucose (eg, Glutose, Insta-Glucose) counteracts the effects of hypoglycemia in the same way as a candy bar or sweet drink, only faster. Unlike complex sugars (eg, common table sugar [sucrose], fruit sugars [fructose]), which must be broken down by the body before they can be absorbed, glucose is a simple sugar that is readily absorbed in the bloodstream.

Oral glucose is available as a gel or as tablets. If authorized by medical control, you should administer oral glucose to any patient with a decreased level of consciousness who has a history of diabetes. The only contraindications to oral glucose are an inability to swallow and unconsciousness, because of the risk of aspiration.

4. What are the "six rights" of medication administration?

Prior to assisting a patient with his or her prescribed medication, as well as administering a drug off of your ambulance, you should review the six rights of medication administration, a tool used to promote safe and accurate medication administration.

- **Right patient:** Look at the medication label to ensure that it reads the same name as your patient. *Do not* give the medication to anyone other than the person whose name is on the medication.

- **Right drug:** While you are looking at the medication label, check to make sure it is the right medication for the patient's condition. For example, if a patient has asthma and is prescribed an MDI of albuterol (Proventil), ensure that the label reads albuterol.

- **Right dose:** Again, look at the medication label and take note of the dose. The dosing information should be on the medication container. If it is not, contact medical control.

- **Right route:** A medication given by the wrong route, even if it is the correct medication, may be ineffective or may even cause harm to the patient. Aspirin, for example, is administered orally and the patient is instructed to chew the tablets prior to swallowing them; whereas nitroglycerin is administered sublingually (under the tongue), and the patient is instructed to keep the tablet under the tongue until it completely dissolves.

- **Right time:** Medications that can be repeated must be given at the correct intervals. For example, nitroglycerin can be repeated up to three times, in 5-minute intervals, provided the patient's blood pressure is adequate and he or she still has chest pain or discomfort. After administering the medication, document the time. After the proper time has passed, contact medical control again if the drug needs to be readministered.

- **Right documentation:** After administering any medication to any patient, you must document the drug, dose, route, time(s) of administration, and reassessment findings after the medication has been given. Proper documentation will ensure that the receiving facility is aware of the medications the patient received in the field.

5. What medications are typically carried on an EMT ambulance?

There are five medications typically carried on an ambulance that is staffed by EMTs: oxygen, aspirin, oral glucose, activated charcoal, and epinephrine. Other medications may be carried on the ambulance, depending on local protocol.

When used properly, the medications you carry on the ambulance can be of tremendous benefit to the patient. When used improperly, however, they can cause further harm.

It is important to note that, just because these medications are carried on the ambulance, the EMT cannot administer them at will. They may only be given on the direct order of a physician (online medical control) or according to standing orders in your local protocol (off-line medical control).

6. What medications can the EMT help a patient self-administer?

There are certain medications that a patient may have prescribed to them that the EMT may be asked to help the patient self-administer. These include epinephrine auto-injectors, metered-dose inhaler medications (ie, albuterol [Proventil, Ventolin], metaproterenol [Alupent]), and nitroglycerin (Nitrostat).

As previously discussed, you must perform a careful assessment of your patient to determine if medication therapy is indicated. Just because the patient is prescribed a particular medication does not mean that it is indicated. For example, nitroglycerin—a vasodilator drug—is contraindicated if the patient's systolic blood pressure is less than 100 mm Hg. By dilating the patient's blood vessels, nitroglycerin may cause a dangerous drop in his or her blood pressure.

7. How can you determine the use for Glucophage?

The simplest and most obvious way of determining what her medication is used for is to simply ask her; she is conscious and alert and will likely be able to answer your question. If she is unsure what it is used for, refer to an EMT field guide or a drug reference text, if you have one on the ambulance. In some cases, you may have to contact medical control to determine what the medication is used for.

You are the Provider: SUMMARY, continued

As an EMT, you will often encounter patients who take numerous medications. Just because it is not one that you carry on the ambulance or are authorized to assist the patient in taking does not mean that you shouldn't try to determine what it is used for. Much information about a patient's medical history can be obtained by looking at the medications they are taking.

Metformin (Glucophage) is an oral hypoglycemic; it is commonly used by type 2 diabetics to help lower their blood glucose level. Although this is not a medication carried on an EMT ambulance, you should try to ascertain its use; doing so may help you determine what caused your patient's condition.

8. Does the patient require further medication therapy at this point?

The patient's level of consciousness has improved and her vital signs are stable compared with her initial presentation. Therefore, she does not require further medication therapy at this time. However, you must reassess her and look for signs that may require further medication therapy, in this case, oral glucose. The patient's original problem may recur, depending on your transport time to the hospital; if you do not reassess your patient, you may not detect a change in her condition.

9. What does the term "polypharmacy" mean?

Polypharmacy is a term referring to the use of multiple medications by the same patient. It is not uncommon to find patients, especially older patients, taking multiple medications on a regular basis; in many cases, the medications are prescribed by more than one physician.

In addition to prescription medications, patients often take over-the-counter (nonprescription) medications and herbal remedies; this often makes a patient's medication regimen complex and confusing.

The potential for inadvertent underdosing and overdosing and harmful drug interactions increases in patients who take multiple medications. Furthermore, the patient's primary problem may be the result of one or more of the medications they are taking.

The EMT should carry a field guide or similar reference that lists common prescription and nonprescription medications. In cases where the patient is unable to communicate with you and a reliable source (eg, family member, caregiver) is not available to answer your questions, the patient's medications can give you important clues as to their medical history.

EMS Patient Care Report (PCR)

Date: 7-5-09	Incident No.: 220109	Nature of Call: Diabetic complications		Location: 202 Cloudcroft Ave.	
Dispatched: 0740	En Route: 0741	At Scene: 0749	Transport: 0805	At Hospital: 0816	In Service: 0827

Patient Information

Age: 60 Sex: F Weight (in kg [lb]): 64 kg (140 lb)	Allergies: No known drug allergies Medications: Glucophage, Nitrostat, Zoloft Past Medical History: Diabetes, heart disease, depression Chief Complaint: Confused

Vital Signs

Time: 0754	BP: 122/72	Pulse: 112	Respirations: 22	Sao$_2$: 98%
Time: 0802	BP: 128/74	Pulse: 84	Respirations: 18	Sao$_2$: 99%
Time: 0808	BP: 126/72	Pulse: 74	Respirations: 18	Sao$_2$: 98%

EMS Treatment
(circle all that apply)

Oxygen @ 15 L/min via (circle one): NC (NRM) Bag-Mask Device		Assisted Ventilation	Airway Adjunct	CPR
Defibrillation	Bleeding Control	Bandaging	Splinting	Other (Oral glucose)

Narrative
Dispatched to a residence for a 60-year-old woman with "diabetic complications." Arrived on scene to find the patient sitting in a recliner in her living room. She was conscious, but confused. Her airway was patent and her breathing, although increased in rate, was producing adequate tidal volume. The patient's son, who called 9-1-1, advised that his mother has type 2 diabetes, and he is not sure when she last ate. Further past medical history includes heart disease and depression. Further assessment of patient revealed that her skin was cool, clammy, and pale. She was in no obvious respiratory distress and did not appear to be experiencing any pain. Applied oxygen at 15 L/min via nonrebreathing mask and obtained initial vital signs. Blood glucose level was assessed and read 36 mg/dL. After ensuring that the patient was able to swallow, administered one tube (15 g) of oral glucose. Placed patient onto stretcher, loaded her into the ambulance, and reassessed her status. Her level of consciousness had improved and her vital signs were stable. Began transport to the hospital and closely monitored the patient en route. Her airway and breathing remained adequate and her skin color and condition improved. The patient stated that she ate, but could not remember when. She further stated that she thinks that she may have accidentally taken too much of her Glucophage. Reassessed her blood glucose level, which read 94 mg/dL. Delivered patient to the emergency department without incident, gave verbal report to the charge nurse, and returned to service. **End of report**

Prep Kit

- Pharmacology is the science of drugs, including their ingredients, preparation, uses, and actions on the body.

- Medications may be administered through the following routes: intravenous, intramuscular, or subcutaneous injection; intranasal; orally; sublingually; intraosseously; transcutaneously; by inhalation; and by rectum.

- These routes of administration often determine the speed with which the medication takes effect.

- Medications come in seven forms: tablets and capsules, solutions and suspensions, metered-dose inhalers, topical medications, transdermal medications, gels, and gases.

- The administration of any medication requires approval by medical control, through direct orders given online or standing orders that are part of the local protocols.

- Once an order from medical control has been obtained, follow the steps in administering medications: Verify the patient, verify the proper medication, verify the dose, verify the route, and verify the time. Once the medication has been administered, reassess vital signs and document the patient's history, assessment, treatment, and response findings.

- Three medications are typically carried on the EMS unit: oxygen, oral glucose, and activated charcoal. Two medicines have recently been added to the list by some states and services: aspirin and epinephrine.

- There are three additional medications that you may help the patient self-administer: metered-dose inhaler medications, nitroglycerin, and epinephrine. Remember, though, that the medications may differ depending on local protocol.

- Knowing what medications a patient takes may be the only way you can determine what chronic or underlying conditions your patient may have.

Vital Vocabulary

absorption The process by which medications travel through body tissues until they reach the bloodstream.

action The therapeutic effect of a medication on the body.

activated charcoal An oral medication that binds and adsorbs ingested toxins in the gastrointestinal tract for treatment of some poisonings and medication overdoses. Charcoal is ground into a very fine powder that provides the greatest possible surface area for binding medications that have been taken by mouth; it is carried on the EMS unit.

adsorption The process of binding or sticking to a surface.

agonist A medication that causes stimulation of receptors.

antagonist A medication that binds to a receptor and blocks other medications.

aspirin (acetylsalicylic acid or ASA) A medication that is an antipyretic (reduces fever), analgesic (reduces pain), anti-inflammatory (reduces inflammation), and potent inhibitor of platelet aggregation (clumping).

contraindications Conditions that make a particular medication or treatment inappropriate, for example, a condition in which a medication should not be given because it would not help or may actually harm a patient.

dose The amount of medication given on the basis of the patient's size and age.

EMT-administered medication When the EMT directly administers the medication to the patient.

enteral medications Medications that enter the body through the digestive system.

epinephrine A medication that increases heart rate and blood pressure but also eases breathing problems by decreasing muscle tone of the bronchiole tree; you may be allowed to help the patient self-administer the medication.

gel A semiliquid substance that is administered orally in capsule form or through plastic tubes.

generic name The original chemical name of a medication (in contrast with one of its "trade names"); the name is not capitalized.

hypoglycemia An abnormally low blood glucose level.

indications The therapeutic uses for a specific medication.

inhalation Breathing into the lungs; a medication delivery route.

intended effect The effect that a medication is expected to have on the body.

intramuscular (IM) injection An injection into a muscle; a medication delivery route.

intranasal (IN) A delivery route in which a medication is pushed through a specialized atomizer device called a mucosal atomizer device (MAD) into the nare.

intraosseous (IO) Into the bone; a medication delivery route.

intravenous (IV) injection An injection directly into a vein; a medication delivery route.

medication A chemical substance that is used to treat or prevent disease or relieve pain.

metered-dose inhaler (MDI) A miniature spray canister through which droplets or particles of medication may be inhaled.

mucosal atomizer device (MAD) A device that is used to change a liquid medication into a spray and pushes it into a nostril.

nitroglycerin A medication that increases cardiac perfusion by causing arteries to dilate; you may be allowed to help the patient self-administer the medication.

oral By mouth; a medication delivery route.

oral glucose A simple sugar that is readily absorbed by the bloodstream; it is carried on the EMS unit.

over-the-counter (OTC) medications Medications that may be purchased directly by a patient without a prescription.

oxygen A gas that all cells need for metabolism; the heart and brain, especially, cannot function without oxygen.

parenteral medications Medications that enter the body by a route other than the digestive tract, skin, or mucous membranes.

patient-assisted medication When the EMT assists the patient with the administration of his or her own medication.

peer-assisted medication When the EMT adminsters medication to him or herself or to a partner.

per os (PO) Through the mouth; a medication delivery route; same as oral.

per rectum (PR) Through the rectum; a medication delivery route.

pharmacodynamics The process by which a medication works on the body.

pharmacology The study of the properties and effects of medications.

polypharmacy The use of multiple medications on a regular basis.

prescription medications Medications that are distributed to patients only by pharmacists according to a physician's order.

side effects Any effects of a medication other than the desired ones.

solution A liquid mixture that cannot be separated by filtering or allowing the mixture to stand.

subcutaneous (SC) injection Injection into the tissue between the skin and muscle; a medication delivery route.

sublingual (SL) Under the tongue; a medication delivery route.

suspension A mixture of ground particles that are distributed evenly throughout a liquid but do not dissolve.

topical medications Lotions, creams, and ointments that are applied to the surface of the skin and affect only that area; a medication delivery route.

trade name The brand name that a manufacturer gives a medication; the name is capitalized.

transcutaneous (transdermal) Through the skin; a medication delivery route.

unintended effect Actions that are undesirable but pose little risk to the patient.

untoward effects Actions that can be harmful to the patient.

Assessment in Action

Rescue 5 responds to 157 Lafayette Street for a 68-year-old woman who is complaining of feeling weak and dizzy. You are greeted at the front door of a private residence by the patient's neighbor. You are brought to the living room where you see the patient seated on the couch. Her skin looks pale, but she does not appear to be in any distress. She tells you that her physician prescribed a new high blood pressure pill and she took her first dose approximately 1 hour prior to the onset of symptoms.

1. The patient tells you that the name of her new medication is Tenormin. This name is an example of a:
   A. trade name.
   B. generic name.
   C. chemical name.
   D. official name.

2. The patient was prescribed Tenormin to lower her blood pressure. This is referred to as the medication's:
   A. indication.
   B. contraindication.
   C. side effect.
   D. intended effect.

3. The symptoms that occurred following the patient's dose of Tenormin are considered:
   A. indications.
   B. contraindications.
   C. side effects.
   D. intended effects.

4. While you are obtaining the SAMPLE history from the patient, she tells you that she has a history of diabetes mellitus and she takes insulin. Through what medication route is insulin administered?
   A. Intravenous
   B. Rectal
   C. Subcutaneous
   D. Intramuscular

5. Your partner is preparing to administer oxygen to the patient. What is the preferred method for delivering oxygen to a breathing patient?
   A. Nasal cannula
   B. Nonrebreathing mask
   C. Bag-mask device
   D. Venturi mask

6. Placing the patient on oxygen is considered what type of medication delivery?
   A. Peer-assisted
   B. EMT-administered
   C. Patient-assisted
   D. Supervisor-assisted

7. As you are preparing to place the patient on the stretcher she suddenly remembers that she uses an inhaler for "breathing problems." Which of the following is a common medication that is delivered via an inhaler?
   A. Nitroglycerin
   B. Clopidogrel
   C. Atorvastatin
   D. Albuterol

8. Which of the following is considered a major disadvantage of the use of a metered-dose inhaler?
   A. The inhaler needs to be kept at a cold temperature.
   B. Medications are not absorbed as well from the lung tissue.
   C. The patient must be cooperative to assist in its use.
   D. Use of a metered-dose inhaler is associated with many side effects.

9. What route of medication administration has the fastest onset of action? Why?

10. Why is it important to ask patients if they take any over-the-counter medications, vitamins, or herbal remedies in addition to prescription medications?

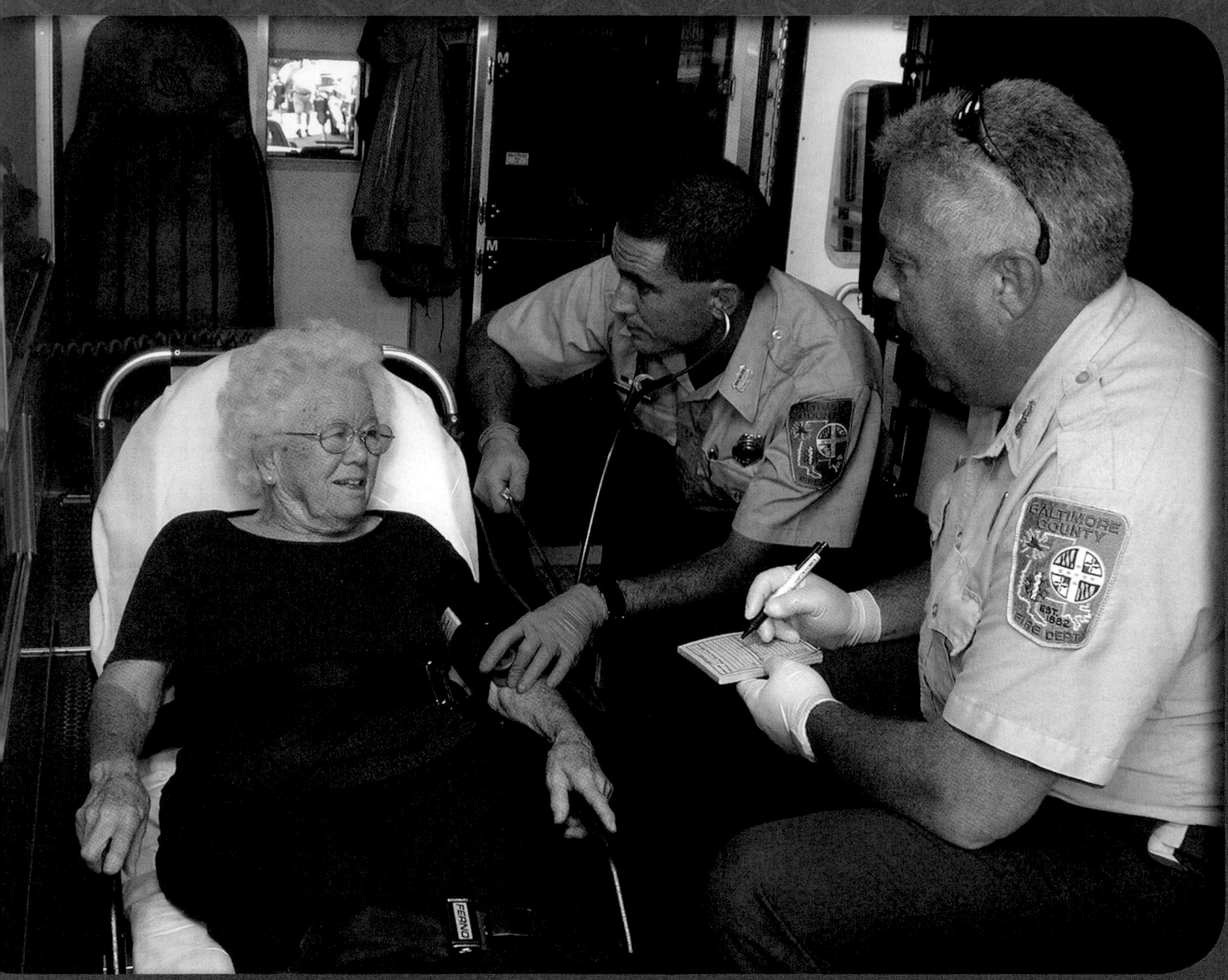

8 Patient Assessment

National EMS Education Standard Competencies

Assessment

Applies scene information and patient assessment findings (scene size-up, primary and secondary assessment, patient history, and reassessment) to guide emergency management.

Scene Size-up

- Scene safety (pp 255-256)
- Scene management
  - Impact of the environment on patient care (p 255)
  - Addressing hazards (p 256)
  - Violence (p 256)
  - Need for additional or specialized resources (pp 258-259)
  - Standard precautions (pp 257-258)
  - Multiple patient situations (p 257)

Primary Assessment

- Primary assessment for all patient situations (pp 261-278)
  - Level of consciousness (pp 262-265)
  - ABCs (pp 265-274)
  - Identifying life threats (pp 274-275)
  - Assessment of vital functions (pp 262-265)
  - Initial general impression (p 261)
- Begin interventions needed to preserve life (pp 274-275)
- Integration of treatment/procedures needed to preserve life (pp 275-278)

History Taking

- Determining the chief complaint (pp 280-282)
- Mechanism of injury/nature of illness (pp 256-257)
- Associated signs and symptoms (pp 280-282)
- Investigation of the chief complaint (pp 280-282)
- Past medical history (pp 280-282)
- Pertinent negatives (p 282)

Secondary Assessment

- Performing a rapid full-body scan (pp 291-296)
- Focused assessment of pain (pp 291, 296-304)
- Assessment of vital signs (pp 289-291)

Techniques of physical examination
- Respiratory system (p 296)
  - Presence of breath sounds (p 266)
- Cardiovascular system (pp 296-301)
- Neurologic system (p 301)
- Musculoskeletal system (pp 301-302)
- All anatomic regions (pp 302-304)

Monitoring Devices

- Obtaining and using information from patient monitoring devices including (but not limited to)
  - Pulse oximetry (pp 289-290)
  - Noninvasive blood pressure (p 290)

Reassessment

- How and when to reassess patients (p 306)
- How and when to perform a reassessment for all patient situations (p 306)

Knowledge Objectives

1. Identify the components of the patient assessment process and explain how the different causes and presentations of emergencies will affect how each step is performed by the EMT. (p 253)
2. Discuss some of the possible environmental, chemical, and biological hazards that may be present at an emergency scene, ways to recognize them, and precautions to protect personal safety. (pp 255-256)
3. Discuss the steps EMTs should take to survey a scene for signs of violence and protect themselves and bystanders from real or potential danger. (p 256)
4. Describe how to determine the mechanism of injury (MOI) or nature of illness (NOI) at an emergency and the importance of differentiating trauma patients from medical patients. (pp 256-257)
5. List the minimum standard precautions that should be followed and personal protective equipment (PPE) that should be worn at an emergency scene, including examples of when additional precautions would be appropriate. (pp 257-258)
6. Explain why it is important for EMTs to identify the total number of patients at an emergency scene and how this evaluation relates to determining the need for additional or specialized resources, implementation of the incident command system (ICS), and triage. (pp 258-259)
7. Describe the principal goals of the primary assessment process: to identify and treat life-threats and to determine if immediate transport is required. (pp 261-278)

8. Explain the process of forming a general impression of a patient as part of primary assessment and the reasons why this step is critical to patient management. (p 261)

9. Explain the importance of assessing a patient's level of consciousness (LOC) to determine altered mental status and give examples of different methods used to assess alertness, responsiveness, and orientation. (pp 262-265)

10. Describe the assessment of airway status in patients who are both responsive and unresponsive and give examples of possible signs and causes of airway obstruction in each case as well as the appropriate EMT response. (pp 265-266)

11. Describe the assessment of a patient's breathing status, including the key information the EMT must obtain during this process and the care required for patients who have both adequate and inadequate breathing. (pp 266-269)

12. List the signs of respiratory distress and respiratory failure. (p 269)

13. Describe the assessment of a patient's circulatory status, including the different methods for obtaining a pulse and appropriate management depending on the patient's status. (pp 270-272)

14. Explain the variations required to obtain a pulse in infant and child patients as compared with adult patients. (pp 270-272)

15. Describe the assessment of a patient's skin color, temperature, and condition, providing examples of both normal and abnormal findings and the information this provides related to the patient's status. (pp 272-274)

16. Discuss the process of assessing for and methods for controlling external bleeding. (p 274)

17. Discuss the steps used to identify and subsequently treat life-threatening conditions that endanger a patient during an emergency. (pp 274-275)

18. List the steps the EMT should follow during the rapid scan of a trauma patient, including examples of abnormal signs and appropriate related actions. (pp 275-277)

19. Explain the process for determining the priority of patient care and transport at an emergency scene and give examples of conditions that necessitate immediate transport. (pp 275, 277-278)

20. Discuss the importance of protecting a trauma patient's spine and identifying fractured extremities during patient packaging for transport. (pp 277-278)

21. Discuss the process of taking a focused history, its key components, and its relationship to the primary assessment process. (pp 280-281)

22. Describe examples of different techniques an EMT may use to obtain information from patients during the history taking process. (pp 280-287)

23. Discuss different challenges an EMT may face when taking a patient history on sensitive topics and strategies they may use to facilitate each situation. (pp 282-287)

24. Explain the purpose of performing a physical exam during secondary assessment, its components, special patient considerations, and methods for determining which aspects of the physical examination will be used. (pp 289-304)

25. Describe the purpose of a full-body scan and list the steps used during this process. (pp 291-295)

26. Explain situations in which patients may receive a focused assessment and then give examples by body system of what each focused assessment should include based on a patient's chief complaint. (pp 291, 296-304)

27. List normal blood pressure ranges for adults, children, and infants. (pp 296-301)

28. Explain the importance of performing a reassessment of the patient and the steps in this process. (p 306)

Skills Objectives

1. Demonstrate how to use the AVPU scale to test for patient responsiveness. (p 262)

2. Demonstrate how to evaluate a patient's orientation and document his or her status correctly. (pp 262-264)

3. Demonstrate how to test pupil reaction in response to light in a patient and document his or her status correctly. (pp 264-265)

4. Demonstrate the techniques for assessing a patient's airway and correctly obtain information related to respiratory rate, rhythm, quality/character of breathing, and depth of breathing. (pp 265-269)

5. Demonstrate how to assess a radial pulse in a responsive patient and an unresponsive patient. (pp 270-272)

6. Demonstrate how to assess a carotid pulse in an unresponsive patient. (pp 270-272)

7. Demonstrate how to palpate a brachial pulse in a child who is younger than 1 year (or a manikin). (pp 270-272)

8. Demonstrate how to obtain a pulse rate in a patient. (pp 270-272)

9. Demonstrate how to assess capillary refill in an adult or child older than 6 years. (pp 273-274)

10. Demonstrate how to assess capillary refill in an infant or child younger than 6 years; explain variations that would be required when assessing a newborn. (pp 272-274)

11. Demonstrate how to perform a rapid scan of a patient. (pp 275-277, Skill Drill 8-1)

12. Demonstrate the use of a pulse oximetry device to evaluate the effectiveness of oxygenation in the patient. (pp 289-290)

13. Demonstrate the use of electronic devices to assist in determining the patient's blood pressure in the field. (pp 290-291)

14. Demonstrate how to perform a full-body scan. (pp 291-295, Skill Drill 8-2)

15. Demonstrate how to measure blood pressure by auscultation. (pp 296-300, Skill Drill 8-3)

16. Demonstrate how to measure blood pressure by palpation. (p 300)

Patient Assessment

Scene Size-up

Ensure scene safety

Determine mechanism of injury/nature of illness

Take standard precautions

Determine number of patients

Consider additional/specialized resources

Primary Assessment

Form a general impression

Assess level of consciousness

Assess the airway: identify and treat life threats

Assess breathing: identify and treat life threats

Assess circulation: identify and treat life threats

Perform rapid scan

Determine priority of patient care and transport

History Taking

Investigate the chief complaint (history of present illness)

Obtain SAMPLE history

Secondary Assessment: Medical

Assess vital signs using the appropriate monitoring device

Systematically assess the patient

   Full-body scan and/or focused assessment

Secondary Assessment: Trauma

Assess vital signs using the appropriate monitoring device

Systematically assess the patient

   Full-body scan and/or focused assessment

Reassessment

Repeat the primary assessment

Reassess vital signs

Reassess the chief complaint

Recheck interventions

Identify and treat changes in the patient's condition

Reassess patient

- Unstable patients: every 5 minutes
- Stable patients: every 15 minutes

Introduction

The importance of patient assessment cannot be overemphasized. EMTs must master and be comfortable with the patient assessment process. Patient assessment is used, to some degree, in every patient encounter. EMTs develop and perfect their own assessment techniques as they complete their education and gain experience in the field. This chapter provides the framework and information necessary for you to be able to understand and conduct the patient assessment process. The assessment process is divided into five main parts:

1. Scene size-up
2. Primary assessment
3. History taking
4. Secondary assessment
5. Reassessment

Although the steps of the assessment process represent a logical approach to evaluation of a patient, the order in which they are performed is dictated by the patient's condition. For example, the same components of patient assessment used to evaluate a medical patient are used to assess a trauma patient. The differences lie in your findings and how you effectively care for your patient. Whether you are assessing a medical patient or a trauma patient, the key in both situations is to remain organized.

Rarely does one sign or symptom reveal to you the patient's status or underlying problem. Rather, it is the combination of many signs and symptoms that reveal the underlying problem or condition of your patient. A **symptom** is a subjective condition that the patient feels and tells you about. A **sign** is an objective condition that you can observe about the patient. Therefore, it is essential to have a basic understanding of the causes and presentations of emergencies so that you know what to look for.

For example, a man with chest pain may be having a heart attack. Or, he may have a lung infection, a pulmonary embolism, or a simple strained muscle in the chest. He may also have sustained chest trauma. He describes the pain as crushing, radiating down the left arm and up into the jaw, he is pale and soaked in sweat, the episode began while he was shoveling snow, and he has a history of coronary bypass surgery and has nitroglycerin in his pocket. On the basis of this information, your assessment should lean toward a diagnosis of myocardial infarction. In this example, you can see how it is essential to collect all pertinent information and be able to interpret how it fits together.

The patient assessment process is the ground on which all levels of EMT education are built and is the foundation of all patient care. EMS providers cannot effectively treat their patients if they cannot assess them. A strong foundation will assist you in the process of saving lives.

Words of Wisdom

Although the steps of patient assessment represent a logical approach to evaluation of a patient, the order in which they are performed is dictated by the patient's condition.

You are the Provider: PART 1

At 2:10 PM, you are dispatched to an apartment complex at 1004 West Almond for a "woman down." The dispatcher has no additional information to provide, and law enforcement personnel are en route to the scene. The weather is clear, the temperature is 90°F, and the traffic is heavy. Your response time to the scene is approximately 6 minutes.

1. **What are the components of patient assessment?**
2. **Will your assessment of the patient differ if she is injured versus ill? If so, how?**

Patient Assessment

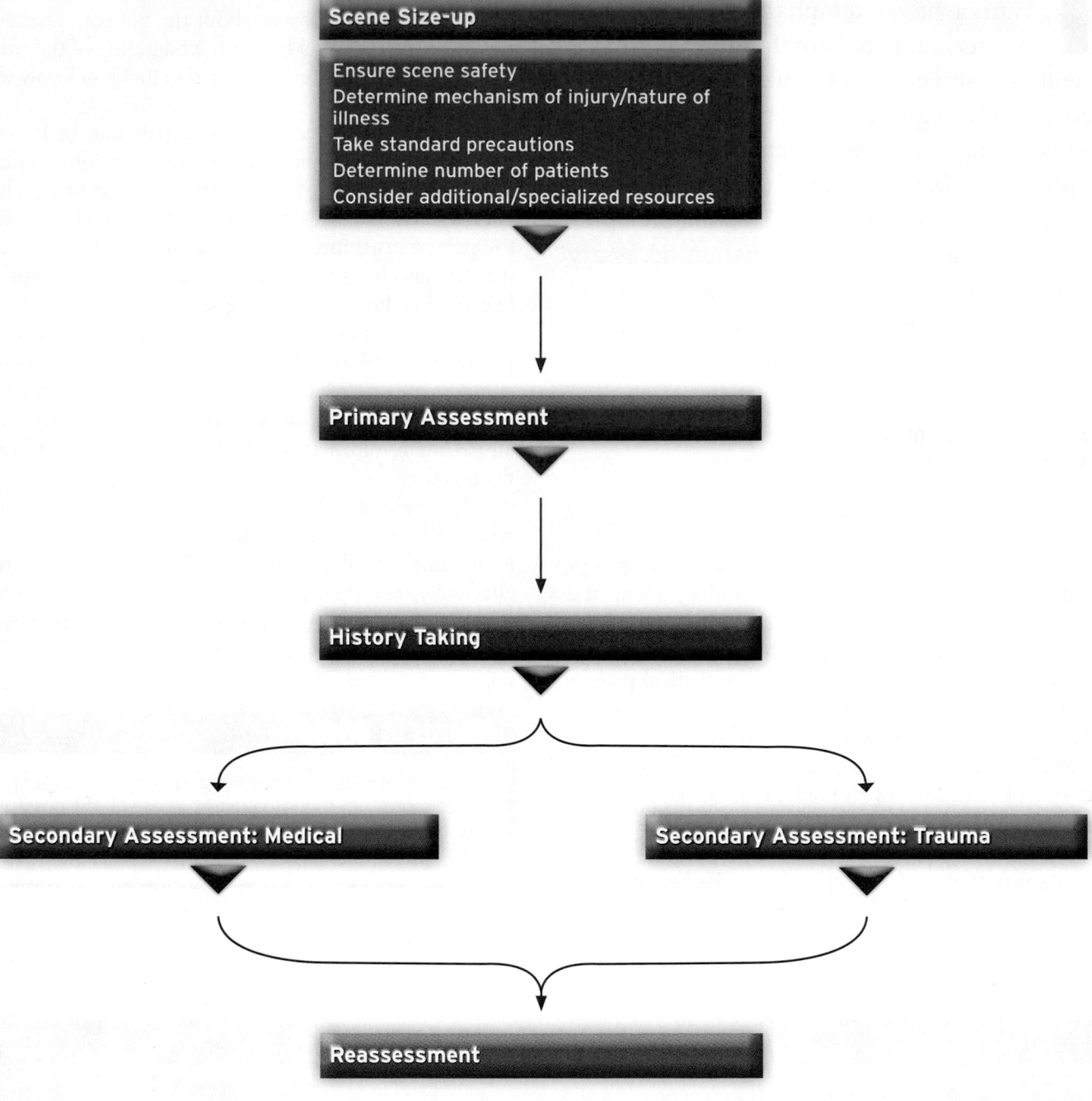

Scene Size-up

Ensure scene safety
Determine mechanism of injury/nature of illness
Take standard precautions
Determine number of patients
Consider additional/specialized resources

Primary Assessment

History Taking

Secondary Assessment: Medical   **Secondary Assessment: Trauma**

Reassessment

Scene Size-up

When you are alerted for an emergency call, your dispatcher will provide you with some basic information about the situation that requires your assistance. Your **scene size-up** begins here. The scene size-up is how you prepare for a specific situation. From the moment you are called into action until you finally reach your patient, you must consider a variety of things that will have an impact on how you begin to care for your patient. The scene size-up includes dispatch information and must be combined with an inspection of the scene to help you ensure safety and identify hazards, safety concerns, and the number of patients you may have, as well as additional resources you might need to safely and effectively care for the patient.

Ensure Scene Safety

The prehospital setting is not a controlled and isolated scene; it is unpredictable, dangerous, and unforgiving, and the outcome could be less than favorable. What seems relatively safe and secure can turn bad without notice. Every prehospital scene has a potential for injury and, if you or your partner becomes injured or incapacitated in such a way you cannot function at the emergency, you have taken away the initial resource the patient needs.

You need to ensure your safety first and that of your patient second. This begins by wearing, at a minimum, an American National Standards Institute 207 certified high-visibility public safety vest. The vest provides you appropriate visibility while minimizing interference with your other clothing and equipment. Look for possible dangers as you approach the scene and before you step out of the vehicle. Observe unstable surfaces, slopes, ice, water, and wet grass. Remember, you have to gain access to a patient, but typically the way you enter an area is also the same way you will leave. When you leave, you will be moving up to a 100-lb stretcher, if it is motorized, and possibly a man weighing 220 lb or more. If your footing was compromised going in, you are going to have an even more difficult time coming out.

Consider traffic safety issues and issues related to scene safety in vehicle rescue. Do not enter potentially dangerous scenes until a professional rescuer (eg, fire fighters, utility workers, hazardous materials technicians, or law enforcement personnel) has made the scene safe. Consider your ambulance a safe haven when caring for patients. Park your vehicle in an area far enough away but in a place that allows you rapid access to equipment and your patient.

Consider environmental conditions at the scene—in other words, take into account that the location of your patient may be outdoors, indoors, or in a public place and, thus, subject to a wide variety of conditions. You are obliged to provide protection for the patient. Is it cold, snowing, raining, or humid? Are there numerous onlookers at the scene? Consider these factors. Some areas of the United States can be covered with snow and ice for 4 months during the year, and other areas may have persistent periods of rain for 6 months. You may be aware of the weather and prepared for it, but also consider the physical terrain, such as mountains, wooded areas, gorges, bodies of water, rivers, lakes, streams, and islands. A significant portion of the United States is covered by hills, mountains, and sand; therefore, a good possibility exists that you will be faced with challenging conditions and decisions Figure 8-1 .

Working in unfavorable conditions and on unstable surfaces is a large part of prehospital care. Without knowing the infinite number of situations you may become involved in, a good rule to use when faced with a wide variety of possibilities is that any actions you may take to protect yourself (eg, heavy coats, rain gear, life jackets, air conditioned or heated vehicles) should also be considered for the patient. If you are putting on equipment to address environmental hazards, provide the patient with the same or similar equipment. If you move away from the scene to take cover from an environmental hazard, move the patient with you. Taking your time to stay focused on what you are doing will go a long way in preventing injuries to yourself and patients.

If appropriate, help protect bystanders from becoming patients as well. Many bystanders attempt to help during an emergency; always remember that they are not trained to handle complicated EMS equipment, illnesses, or injuries.

Figure 8-1 At times you may need to carry patients out of areas with unstable terrain.

Hazards come in many different forms, shapes, and sizes. Chemical and biologic hazards, electricity from downed electric lines or lighting, water hazards, fires, explosions, and potentially toxic environments containing carbon monoxide are just a few hazards you will encounter. Information that may help to determine potential hazards can be provided by dispatch. For example, you may be sent to a motor vehicle collision involving a semi truck carrying hazardous materials or to the scene of an industrial site that produces chemicals or even to a private residence that has an animal that poses a threat **Figure 8-2**. There are also hazards found at every motor vehicle collision scene. Consider the traffic that is moving around the crash scene. It is also likely that fluids are leaking from the vehicle. Antifreeze, oil, and gasoline are common chemicals found at the scene. When they come in contact with the roadway, it can make for a very slippery surface. As you walk on this surface with leather-soled shoes, the oil and asphalt make for an extremely hazardous surface. All of these examples may be relatively obvious, but be aware of the dangers that may not be so apparent.

Occasionally you and your partner will not be able to enter a scene safely. If the scene is unsafe, make it safe. If this is not possible, do not enter. If the hazard presents a substantial risk to your health and safety, you should request the appropriate assistance. If the scene is a potential crime scene, follow local protocols before entering. Ask for law enforcement personnel to accompany you when needed **Figure 8-3**.

On occasion, you may enter a scene that appears safe but quickly becomes unsafe. If this situation presents itself and you have the appropriate education and

Figure 8-3 If the scene is unsafe, request law enforcement support.

training to make the scene safe, do so. If not, be prepared to extricate yourself and your patient as quickly as possible, protecting the patient from further injury as much as possible.

Be aware of scenes that have the potential for violence. You will encounter violent patients, distraught family members, angry bystanders, gangs, and unruly crowds. When you enter the home of a patient, look around the immediate area. Are there weapons visible in the area to which a patient or others may have access? Weapons need not be typical like a knife or gun, they can also be items such as a screwdriver, hammer, or simple things sitting on the kitchen table or nightstand by the bed. Always be observant for such objects, and if they are not secured, make sure that you place yourself between the patient and the potential danger, thus preventing possible access to the object. Request the assistance of law enforcement personnel if the scene is unsafe with the potential for violence.

Determine Mechanism of Injury/Nature of Illness

As an EMT, you will be called to motor vehicle collisions or other situations in which patients may have sustained life-threatening traumatic injuries. Traumatic injuries are the result of physical forces applied to the body. To care for trauma patients properly, you must understand how traumatic injuries occur, or the **mechanism of injury (MOI)**. With a traumatic injury, the body has been exposed to some force or energy that has resulted in a temporary injury, permanent damage, or even death.

Figure 8-2 Evaluate the scene for hazards as soon as you arrive.

As you might expect, certain parts of the body are more easily injured than others. The brain and the spinal cord are very fragile and easy to injure. Fortunately, they are protected by the skull, the vertebrae, and several layers of soft tissues. The eyes are also easily injured. Even small forces on the eye may result in serious injury. The bones and certain organs are stronger and can absorb small forces without resulting injury. The net result of this information is that you can use the MOI as a kind of guide to predict the potential for a serious injury by evaluating three factors: the amount of force applied to the body, the length of time the force was applied, and the areas of the body that are involved.

You will commonly hear the terms "blunt trauma" and "penetrating trauma." With blunt trauma, the force of the injury occurs over a broad area, and the skin is usually not broken. However, the tissues and organs below the area of impact may be damaged. With penetrating trauma, the force of the injury occurs at a small point of contact between the skin and the object. The object pierces the skin and creates an open wound that carries a high potential for infection. The severity of injury depends on the characteristics of the penetrating object, the amount of force or energy, and the part of the body affected.

As an EMT, you will also care for many medical patients. Medical patients are patients who require EMS attention because of illnesses or conditions not caused by an outside force. For medical patients, you must examine the general type of illness the patient is experiencing, or the **nature of illness (NOI)**. There are similarities between the MOI and the NOI. Both require you to search for clues regarding how the incident occurred. You must make an effort to determine the general type of illness, which is often best described by the patient's chief complaint: the reason EMS was called. To quickly determine the NOI, talk with the patient, family, or bystanders about the problem. But at the same time, use your senses to check the scene for clues as to the possible problem. You may see open or spilled medication containers, poisonous substances, or unsanitary living conditions. You may smell an unusual or strong odor, such as the odor of fresh paint in a closed room. You may hear a hissing sound, such as a leak from a home oxygen system. Keep these observations of the scene in mind as you begin to assess a medical patient.

Be aware of scenes with more than one patient who are exhibiting similar signs or symptoms. An example would be an elderly couple experiencing flulike symptoms, headache, nausea, and vomiting. These symptoms may be indicative of carbon monoxide poisoning, which would be an unhealthy situation for you and your partner as well.

The Importance of the MOI and NOI

Considering the MOI or NOI early can be of value in preparing to care for your patient. For example, when you begin to gather equipment from the unit to treat your patient, what would you take for an older patient complaining of chest pain? How would that equipment differ from the equipment used for a pedestrian struck by a vehicle? The appearance of the scene may also guide you in your preparation. Other MOIs may include motor vehicle crashes, assaults, and stabbings or gunshot wounds. Examples of NOIs include seizures, heart attacks, diabetic problems, and poisonings. Family members, bystanders, or even law enforcement personnel may also provide important trauma or medical information to help you assist the patient.

During your prehospital assessment you may be tempted to categorize your patient immediately as a trauma or medical patient. Remember, the fundamentals of a good patient assessment are the same despite the unique aspects of trauma and medical care. If an unconscious patient is found at the bottom of a ladder, did he fall off the ladder, strike his head, and become unconscious or did he experience a medical problem that caused him to fall off the ladder and then lose consciousness? Early in the assessment, it can be difficult to identify with absolute certainty whether the problem is of a traumatic or medical origin. Although further assessment is needed to come to a conclusion, considering the MOI or NOI early will help you prepare for the rest of your assessment.

Take Standard Precautions

Standard precautions and **personal protective equipment (PPE)** need to be considered and adapted to the prehospital task at hand. Personal protective equipment includes clothing or specialized equipment that provides protection to the wearer. The type of PPE used will depend on the specific job duties required during a patient care interaction. For example, fire fighters may wear PPE such as steel boots, helmets, turnout gear, gloves, heat-resistant outerwear, and self-contained breathing apparatus designed to protect them from injury when performing a forced entry. Hazardous materials technicians may don a protective, encapsulated suit designed to prevent contamination by potentially lethal hazardous materials.

Standard precautions are protective measures that have traditionally been developed by the Centers for Disease Control and Prevention for use in dealing with objects,

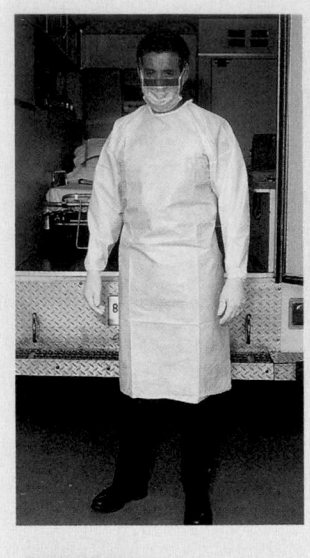

Figure 8-4 Proper protective equipment is vital when you are called to a scene in which you may be exposed to blood or other body fluids.

blood, body fluids, and other potential exposure risks of communicable disease. If you have a primary responsibility for patient care, you will need to follow standard precautions when receiving the call for help. It is a standard required in each and every patient encounter. These measures may not provide absolute protection from exposure to infectious diseases or bloodborne pathogens, but they are the most effective way to reduce a provider's risk of exposure. The concept of standard precautions assumes that all blood, body fluids (except sweat), nonintact skin, and mucous membranes may pose a substantial risk of infection. Remember the saying when dealing with body fluids and patients, "If it's wet, it's infectious."

When you step out of the EMS vehicle and before actual patient contact, standard precautions must have been taken or initiated **Figure 8-4**. After you have contact with a patient it will be too late to think about what precautions should have been considered. The use of standard precautions in EMS, including but not limited to consistent handwashing (with soap and water or with disinfectants), gloves, eye protection, a mask, and a gown, may be dictated by local standards or protocols. At a minimum, gloves must be in place before any patient contact. Remember that after contact with a patient, gloves may be contaminated by infectious materials, so consider this when handling EMS equipment with the same gloves used during patient contact. The use of eye protection may be necessary during patient interactions. Standard eye glasses may not offer enough protection because most are not designed with side splash guards; for that reason, eye wear should protect you from potential exposures from many different directions. Blood and body fluids that contain potentially infectious materials may become airborne; therefore, you should consider wearing a mask. A mask will provide protection from some airborne diseases, but its level of protection will depend on the type of mask, a proper fit, and your ability to apply and wear it properly. You must be appropriately educated in the use of standard precautions, which should include training in the many types of PPE used in different situations. If you are not trained in the application of PPE, you should not approach a scene or make patient contact but should call for additional help.

Determine Number of Patients

As part of the scene size-up, it is essential that you accurately identify the total number of patients. This evaluation is critical in determining your need for additional resources, such as fire fighters, specialized rescue group, a hazardous materials team, or additional ambulances. When there are multiple patients you should use the **incident command system**, call for additional units, and then begin triage **Figure 8-5**. The incident command system is a system implemented to manage disasters and mass-casualty incidents in which section chiefs, including finance, logistics, operations, and planning, report to the incident commander. **Triage** is the process of sorting patients based on the severity of each patient's condition. Once all patients have been triaged, you can begin to establish treatment and transport priorities. Usually the most experienced EMT is assigned to perform triage. This process will help you allocate your personnel, equipment, and resources to provide the most effective care to everyone. When a large number of patients are present or there are more patients than the responding unit can effectively handle, you should put your mass-casualty plan into action based on your local protocols. You should be familiar with the incident command system and understand your local protocols.

Consider Additional/Specialized Resources

Some trauma or medical situations may require more ambulances, whereas others may have a need for specialized resources. Basic life support units may be all that are needed for some patients; however, advanced life support (ALS) should be requested for patients with severe injuries or complex medical problems depending on available resources and local protocols. The ALS may

Figure 8-5 With multiple patients, you should use the incident command system, call for additional resources, and then begin triage.

also be needed to control traffic or intervene in domestic violence situations. Many police departments throughout the country have their officers trained as first responders; they know cardio-pulmonary resuscitation (CPR) and may even carry an automated external defibrillator (AED). In the perfect situation, law enforcement personnel should be the first to enter crime scenes and hostile environments. You should stage yourself and your vehicle at a safe distance until the scene has been secured.

Figure 8-6 Scenes involving toxic substances may require specially trained rescuers with extra protective equipment.

If any situation presents itself as a danger to you, your partner, or patient, you must retreat to the safety of the EMS vehicle. Be aware of the potential for danger at all times and understand when additional or specialized resources are required.

To determine if you require additional resources, you should ask yourself the following questions:

- How many patients are there?
- What is the nature of their conditions?
- Who contacted EMS?
- Does the scene pose a threat to you, your patient, or others?

Knowing how your EMS system is organized will help you determine the additional resources that may be required. The sooner these resources are identified, the sooner they can be requested.

be provided by AEMTs or paramedics, depending on how your EMS system is set up. Air medical support is another good resource for ALS. Follow your local protocols in requesting ALS resources.

Many resources in addition to fire suppression are often available through the fire department, including high-angle rescue, hazardous materials management, complex extrication from motor vehicle crashes, water rescue, and other specific types of rescue, such as swift water rescue **Figure 8-6**. Specialized equipment and gear will also be needed for each of these situations.

Search and rescue teams can be helpful in finding, packaging, and transporting patients over long distances or across unusual terrain. Law enforcement personnel may

Words of Wisdom

Universal precautions and *standard precautions* are two terms that use the same practice. They are protective measures that help protect EMTs and patients against possible exposure to communicable diseases.

Primary Assessment

Patient Assessment

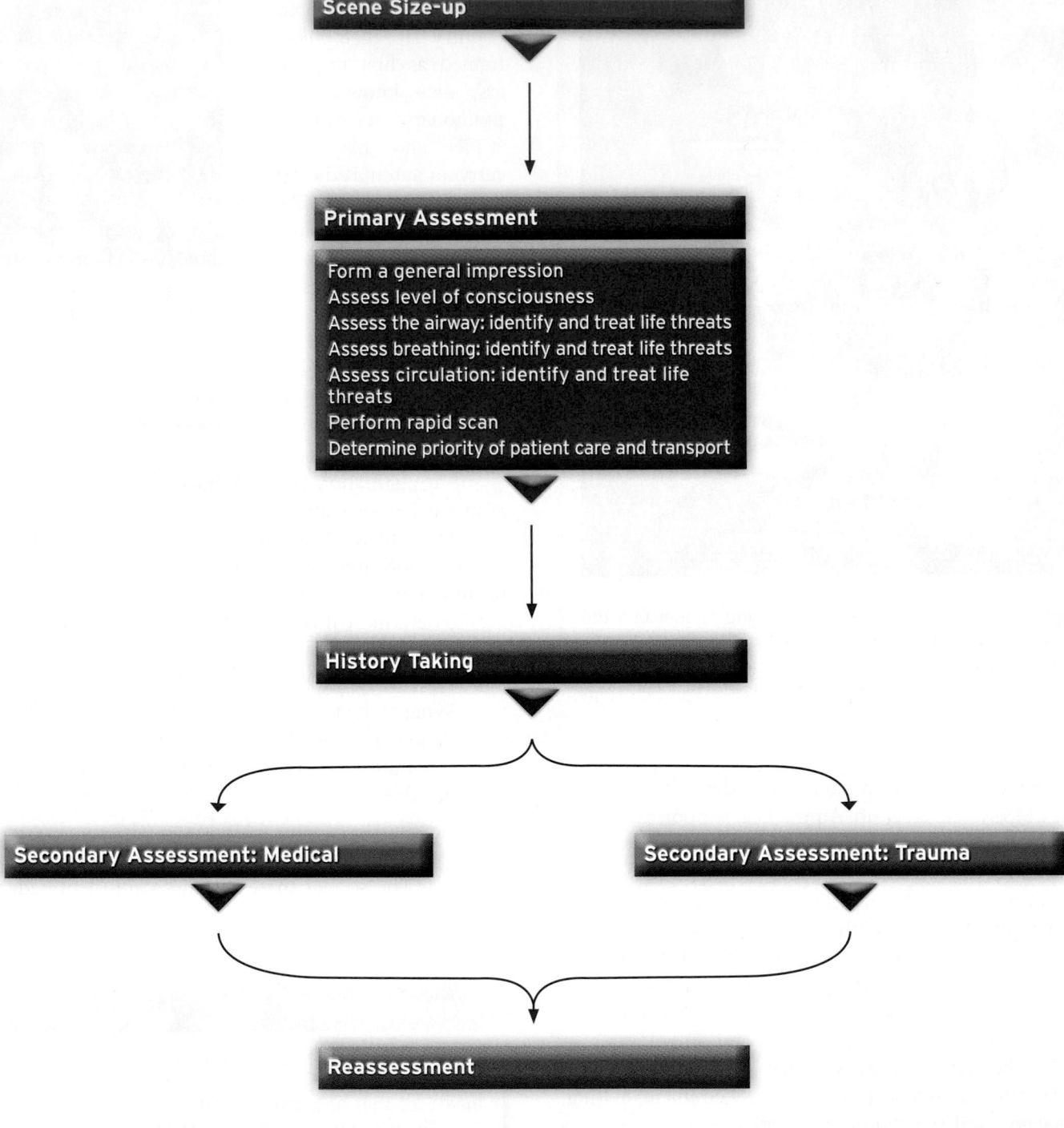

Scene Size-up

Primary Assessment

Form a general impression
Assess level of consciousness
Assess the airway: identify and treat life threats
Assess breathing: identify and treat life threats
Assess circulation: identify and treat life threats
Perform rapid scan
Determine priority of patient care and transport

History Taking

Secondary Assessment: Medical

Secondary Assessment: Trauma

Reassessment

Primary Assessment

During the scene size-up, you used dispatch information and your own evaluation of the scene to begin determining what happened. You also evaluated potential or actual scene hazards and threats, protected yourself and your team, and decided whether you needed additional resources. These steps are critical in the initiation of patient care. Patient assessment, however, begins when you greet your patient. The **primary assessment** has a single, critical, all-important goal: to identify and initiate treatment of immediate or potential life threats. The patient's vitals signs (level of consciousness; airway, breathing, and circulation [ABCs]) will determine the extent of your treatment at the scene. **Vital signs** are the key signs that are used to evaluate the patient's condition. Always give priority to the patient's level of consciousness and ABCs to ensure lifesaving treatment Figure 8-7 . From here you will be able to determine the priority of patient care and transport.

Form a General Impression

Anytime you meet someone new, you form an initial general impression about that person. Forming an initial **general impression** of your patient is a similar process, but it helps to focus your attention toward life-threatening problems. The initial general impression is formed to determine the priority of care and is based on your immediate assessment of the patient. This includes noting things such as the person's age, sex, race, level of distress, and overall appearance. You may anticipate different problems depending on the patient's age, sex, and race. A woman reporting abdominal pain, for example, may have more serious implications than a man with the same complaint because of the complexity of the female reproductive system. Write any important information down because it will be difficult to remember minor details later.

You should think of your initial general impression as a visual assessment, gathering information as you approach the patient. As you approach, make sure that the patient sees you coming to avoid surprising the patient or causing the patient to turn to see you, possibly making any injuries worse. Note the patient's position and whether the patient is moving or still.

When you reach your patient, place yourself at a lower position, if possible, to show respect for the patient and help the patient feel comfortable and less threatened as you begin your assessment Figure 8-8 . Refer to the patient by name. The initial general impression continues during your introduction. Introduce yourself to the patient by stating "Hi, I am Sam, an EMT with the fire department and I am here to help you." Is the patient able to respond to your greeting easily and appropriately? The patient's response can give you insight into the level of consciousness, airway patency, and respiratory status before you begin your examination.

After you introduce yourself, you should ask the patient about the chief complaint. Assess the patient's skin color and condition as you begin. For example, is the patient's skin pink, pale, gray, or cyanotic? Is it dry, clammy, or diaphoretic? The patient may direct you to a wound on his or her leg or demonstrate an airway problem by creating abnormal sounds when breathing. If a life-threatening problem is found, it should be treated immediately. Determine whether your patient's condition is stable, stable but potentially unstable, or unstable.

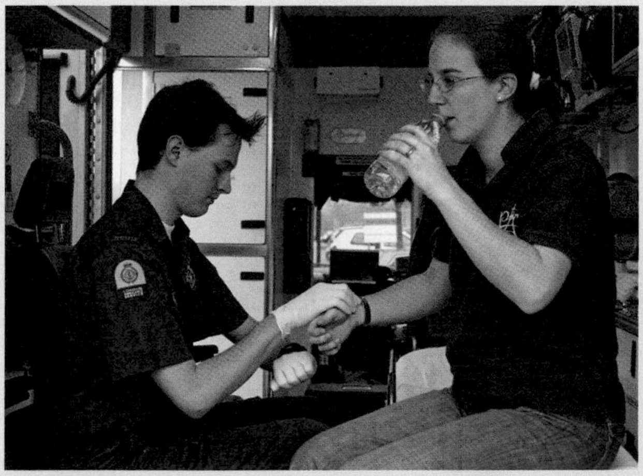

Figure 8-7 An assessment of vital functions is used to establish the patient's condition. This patient is being assessed for heat exhaustion.

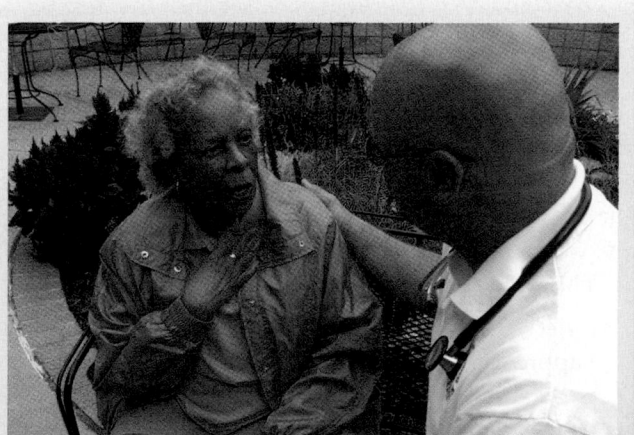

Figure 8-8 As you approach the patient, form an initial general impression of his or her overall condition.

Assess Level of Consciousness

The patient's level of consciousness (LOC) is considered a vital sign because it can tell a great deal about the patient's neurologic and physiologic status. The brain requires a constant supply of oxygen and glucose to function properly. In the primary assessment, you need to ascertain only the gross LOC by determining which of the following categories best fits your patient:

- Conscious with an unaltered LOC
- Conscious with an altered LOC
- Unconscious

When a patient is conscious with an altered LOC, it may indicate that inadequate perfusion and oxygenation or a chemical or neurologic problem is adversely affecting the brain and its ability to function. **Perfusion** is the circulation of blood within an organ or tissue. An altered LOC in a conscious patient can also be caused by medications, drugs, alcohol, or poisoning.

Your assessment of a patient who is unconscious should focus initially on problems with airway, breathing, and circulation, which are critical life threats, and then identify other emergency care that the patient may need. Sustained unconsciousness should warn you that a critical respiratory, circulatory, or central nervous system problem or deficit may exist, and you must assume that the patient has a potentially critical injury or potentially life-threatening condition. Therefore, after rapidly assessing the patient and providing emergency treatment, you should package the patient and provide rapid transport to the hospital.

When you assess a patient, you must determine the appropriateness of the patient's response by how well it demonstrates the patient's understanding and mental activity, not by how well it reflects your definition of socially acceptable behavior. Mental status and LOC can be evaluated in just a few seconds by testing for responsiveness and orientation.

A test for **responsiveness** assesses whether the patient is alert and how well a patient responds to external stimuli, including verbal stimuli (sound) and painful stimuli (such as pinching the patient's earlobe). One test for responsiveness uses the AVPU scale to assess how well a patient responds to stimuli.

The AVPU scale is based on the following criteria:

- **Alert.** The patient's eyes open spontaneously as you approach, and the patient appears to be aware of you and responsive to the environment. The patient appears to follow commands, and the eyes visually track people and objects.
- **Responsive to Verbal stimuli.** The patient's eyes do not open spontaneously. However, the patient's eyes do open to verbal stimuli, and the patient is able to respond in some meaningful way when spoken to.
- **Responsive to Pain.** The patient does not respond to your questions but moves or cries out in response to painful stimulus. There are appropriate and inappropriate methods of applying a painful stimulus based a great deal on personal preference Figure 8-9. Be aware that some methods may not give an accurate result if a spinal cord injury is present.
- **Unresponsive.** The patient does not respond spontaneously or to a verbal or painful stimulus. The patients usually have no cough or gag reflex and lack the ability to protect their airway. If you are in doubt about whether a patient is truly unresponsive, assume the worst and treat appropriately.

Words of Wisdom

The **AVPU scale** is a rapid method of assessing the patient's level of consciousness using one of the following four terms:

A Awake and Alert
V Responsive to Verbal stimuli
P Responsive to Pain
U Unresponsive

You should determine whether a patient is awake and alert. A patient who is not awake and alert but who is aroused and responds to your voice by opening his or her eyes, moaning, speaking, or moving is responding to verbal stimuli. A patient who does not respond to your normal speaking voice but who responds when you speak loudly is responding to loud verbal stimuli. Be sure to note how the patient responded. Tap a patient who is hearing impaired with your fingers repeatedly. If the patient responds, note that the patient is hearing impaired but responds to being tapped.

To determine whether a patient who does not respond to verbal stimuli will respond to a painful stimulus, you should gently but firmly pinch the patient's skin. An area where this technique works best is on the patient's ear, back of the upper arm (triceps), or the trapezius area (the muscle above the collar bone). A patient who moans or withdraws is responding to the painful stimulus. Be sure to note the type and location of the stimulus and how the patient responded.

If the patient does not respond to a painful stimulus on one side, try to elicit a response on the other side. Note that a patient who remains flaccid without moving or making a sound with no indication of hearing you is considered unresponsive.

For a patient who is alert and responsive to verbal stimuli, you should next evaluate orientation. **Orientation** tests mental status by checking a patient's memory and thinking ability. The most common test evaluates a patient's ability to remember four things:

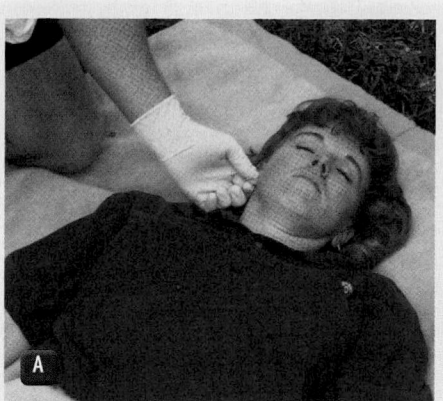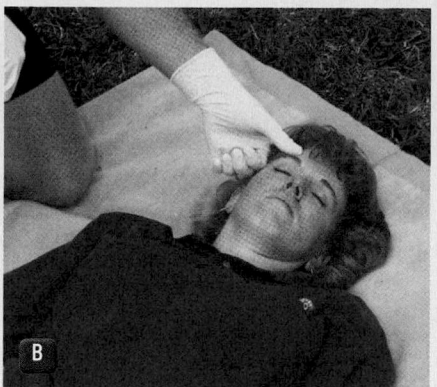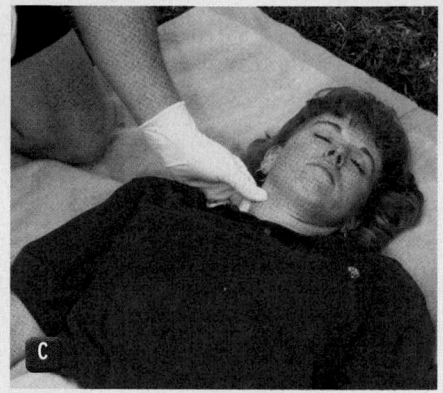

Figure 8-9 Methods of gauging a patient's responsiveness to painful stimuli. **A.** Gently but firmly pinch the patient's earlobe. **B.** Press down on the bone above the eye. **C.** Pinch the muscles of the neck.

- **Person.** The patient is able to remember his or her name.
- **Place.** The patient is able to identify his or her current location.
- **Time.** The patient is able to tell you the current year, month, and approximate date.
- **Event.** The patient is able to describe what happened (the MOI or NOI).

These questions were not selected at random. They evaluate long-term memory (person and place if the patient is at home), intermediate memory (place and time when asking year or month), and short-term memory (time when asking approximate date and event). If the patient knows these facts, the patient is said to be "alert and fully oriented," "alert and oriented to person, place,

time, and event," or "alert and oriented × 4." If a patient does not know these facts, he or she is considered less than fully oriented.

Use of the Glasgow Coma Scale (GCS) score can be helpful in providing additional information on patients with changes in mental status. The GCS (discussed further in Chapter 15, *Neurologic Emergencies*) uses parameters that test a patient's eye opening, best verbal response, and best motor response, which provide a numeric score that defines the severity of a patient's brain dysfunction **Table 8-1**. This information provides baseline data on the patient's overall neurologic status and can be a reliable predictor of the outcome of a patient with a brain injury. A modified GCS is used for children and infants, who

Table 8-1 Glasgow Coma Scale

Eye Opening		Best Verbal Response		Best Motor Response	
Spontaneous	4	Oriented conversation	5	Obeys commands	6
In response to speech	3	Confused conversation	4	Localizes pain	5
In response to pain	2	Inappropriate words	3	Withdraws to pain	4
None	1	Incomprehensible sounds	2	Abnormal flexion	3
		None	1	Abnormal extension	2
				None	1

Score: 13–15 may indicate mild dysfunction, although 15 is the score a person with no neurologic disabilities would receive.

Score: 9–12 may indicate moderate dysfunction.

Score: 8 or less is indicative of severe dysfunction.

Primary Assessment

respond differently from adults. When you are reporting the GCS score, you should document or report each section (eg, Eye opening: 3, Verbal response: 4, Motor response: 5 = GCS score of 12) to document baseline function in each area.

Pupils

The diameter and reactivity to light of the patient's pupils reflect the status of the brain's perfusion, oxygenation, and condition. The pupil is a circular opening in the center of the pigmented iris of the eye. The pupils are normally round and of approximately equal size and serve as optical diaphragms, adjusting their size depending on the available light. In normal room light, the pupil appears to be midsize. With less light, the pupils dilate, allowing more light to enter the eye, making it possible to see even in dim light. With high light levels or when a bright light is suddenly introduced, the pupils instantly constrict, allowing less light to enter, protecting the sensitive receptors in the inner eye from damage **Figure 8-10A**. When a brighter light is introduced into one eye (or higher levels of light enter one eye only), both pupils should constrict equally to the appropriate size for the pupil receiving the most light.

In the absence of any light, the pupils will become fully relaxed and dilated **Figure 8-10B**. When light is introduced, each eye sends sensory signals to the brain indicating the level of light it is receiving. Pupil size is regulated by a series of continuous motor commands that the brain automatically sends through the oculomotor nerves to each eye, causing both pupils to constrict to the same appropriate size. Normally, pupil size changes instantly to any change in light level.

A small number of the population exhibit unequal pupils (anisocoria). If the patient or family member cannot confirm the presence of this condition, you must assume the patient has depressed brain function as a result of central nervous system depression or injury if the pupils react in any of the following ways:

- Become fixed with no reaction to light
- Dilate with introduction of a bright light and constrict when the light is removed
- React sluggishly instead of briskly
- Become unequal in size **Figure 8-10C**
- Become unequal in size when a bright light is introduced into or removed from one eye

Depressed brain function can be caused by the following situations:

- Injury of the brain or brain stem
- Trauma or stroke
- Brain tumor

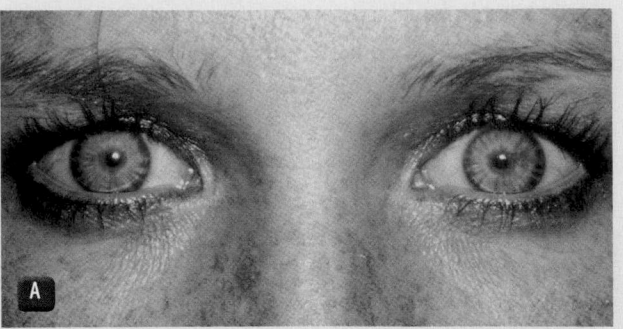
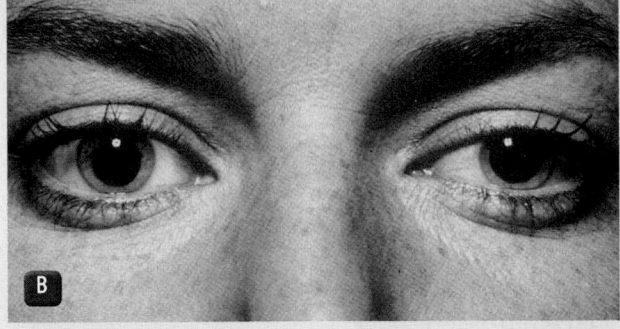
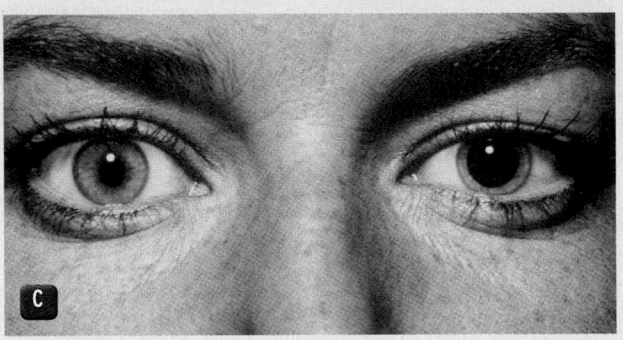

Figure 8-10 **A.** Constricted pupils. **B.** Dilated pupils. **C.** Unequal pupils.

- Inadequate oxygenation or perfusion
- Drugs or toxins (central nervous system depressants)

Opiates, which are one category of central nervous system depressants, cause the pupils to constrict so significantly, regardless of light, that they become so small as to be described as pinpoint. Intracranial pressure from intracranial bleeding may cause sufficient pressure against the oculomotor nerve on one side that the motor commands can no longer pass from the brain to that eye. When this occurs, the eye no longer receives commands to constrict, and its pupil becomes fully dilated and fixed. This is described as a blown pupil.

The letters PEARRL serve as a useful guide in assessing the pupils. They stand for the following:

P Pupils
E Equal

A And
R Round
R Regular in size
L React to Light

For patients with normal pupils, you can report "Pupils are Equal And Round, Regular in size, and react properly to Light" or "pupils = PEARRL." Describe any abnormal findings using the longer form, such as "Pupils are equal and round, the left pupil is fixed and dilated, the right pupil is regular in size and reacts to light."

Assess the Airway

As you move through the steps of the primary assessment, you must always be alert for signs of airway obstruction. Regardless of the cause, a mild or severe airway obstruction will result in inadequate or absent air flow into and out of the lungs. To prevent permanent damage to the brain, heart, and lungs, or even death, you must determine if the airway is open (patent) and adequate.

Responsive Patients

Patients of any age who are talking or crying have an open airway. However, watching and listening to how patients speak, particularly patients with respiratory problems, may provide important clues about the adequacy of their airway and the status of their breathing. For example, sounds of **stridor**, a brassy crowing sound prominent on inspiration, suggest a mildly occluded airway caused by swelling. High-pitched crowing sounds may indicate a mild airway obstruction from a foreign body. A conscious patient who cannot speak or cry most likely has a severe airway obstruction.

If you identify an airway problem, stop the assessment process and obtain a patent airway. This may be as simple as positioning the patient so the air moves in and out, suctioning liquids from the airway, or removing an obvious foreign body from the patient's mouth; it may be as complex as abdominal thrusts to remove a foreign body from the airway. Although airway and breathing problems are not the same, their signs and symptoms often overlap. If your patient has signs of difficulty breathing or is not breathing, you should immediately take corrective actions using appropriate airway management techniques.

Words of Wisdom

If there is a history of trauma or a potential history of trauma, you should open the airway using the modified jaw-thrust technique. If you can confirm that there was no traumatic event, then open the airway using the head tilt-chin lift technique.

Unresponsive Patients

With an unresponsive patient or a patient with a decreased LOC, you should immediately assess the patency of the airway. Unresponsive patients should be considered to have experienced a traumatic event. If there is a potential for trauma, use the modified jaw-thrust technique to open the airway. If it can be confirmed that

You are the Provider: PART 2

When you arrive on scene, a police officer directs you to a poorly kept apartment on the second floor. The scene is safe. You find the patient, a young woman, lying in a prone position on the floor in her kitchen. She was found by her neighbor, who became concerned when she did not answer the door. You carefully log roll the patient to a supine position and begin your assessment. An engine company arrives to provide assistance.

Recording Time: 0 Minutes	
Appearance	Pale; blood draining from the side of the mouth
Level of consciousness	Minimally responsive
Airway	Bloody secretions and vomitus in the mouth
Breathing	Slow, shallow, and gurgling
Circulation	Radial pulse, slow and weak; skin, cool and pale

3. Are spinal precautions indicated? Why or why not?

4. Which of these assessment findings requires your *most* immediate attention?

the patient did not experience a traumatic event, use the head tilt–chin lift technique to open and maintain a patent airway. Both of these techniques are described in Chapter 9, *Airway Management*. Another cause of airway obstruction in an unconscious patient could be relaxation of the tongue muscles, allowing the tongue to fall to the back of the throat. Dentures, blood clots, vomitus, mucus, food, and other foreign objects may also create an obstruction. If the airway is clear, you can continue your assessment.

Signs of airway obstruction in an unconscious patient include the following:

- Obvious trauma, blood, or other obstruction
- Noisy breathing, such as snoring, bubbling, gurgling, crowing, or other abnormal sounds (Normal breathing is quiet.)
- Extremely shallow or absent breathing (Airway obstructions may impair breathing.)

If any of the aforementioned conditions exist, the airway is considered not patent and you should open the airway using the appropriate method, the head tilt–chin lift or jaw-thrust maneuver, suction as necessary, and use an airway adjunct as necessary. The body will not be supplied the necessary oxygen needed to survive if the airway is not managed quickly and efficiently. Remember that airway positioning depends on the age and size of your patient. For trauma patients or patients with an unknown illness, you must manually stabilize the cervical spine while using the jaw-thrust maneuver to open the airway.

Assess Breathing

A patient's breathing status is directly related to the adequacy of his or her airway. Once you have made sure the patient's airway is open, make sure the patient's breathing is present and adequate. A patient who is breathing without assistance is said to have **spontaneous respirations** or spontaneous breathing. Each complete breath includes two distinct phases: inspiration and expiration. During inspiration (inhalation), the diaphragm and intercostal muscles contract and the chest rises up and out, drawing oxygenated air into the lungs. During expiration (exhalation), the muscles relax and the chest returns to its original position, releasing air with an increased carbon dioxide level out of the lungs. Inhalation and exhalation times occur in a 1:3 ratio; the active inhalation phase lasts one third the amount of time of the passive exhalation phase.

Breathing is a continuous process in which each breath regularly follows the last with no notable interruption. Breathing is normally a spontaneous, automatic process that occurs without conscious thought, visible effort, marked sounds, or pain. You will assess breathing by *watching* the patient's chest rise and fall, *feeling* for air through the mouth and nose during exhalation, and *listening* to **breath sounds** with a stethoscope over each lung. Chest rise and breath sounds should be equal on both sides of the chest.

When assessing breathing, you must obtain the following information:

- Respiratory rate
- Rhythm, regular or irregular
- Quality/character of breathing
- Depth of breathing

As you assess the patient's breathing, you should ask yourself the following questions:

- Does the patient appear to be choking?
- Is the respiratory rate too fast or too slow?
- Are the patient's respirations shallow or deep?
- Is the patient cyanotic (blue)?
- Do you hear abnormal sounds when listening to the lungs?
- Is the patient moving air into and out of the lungs on both sides?

Oxygen should always be administered to patients who are having difficulty breathing, but it may also be provided to patients who are breathing adequately. Positive-pressure ventilations should be performed for patients who are apneic or whose breathing is too slow or too shallow.

If a patient seems to develop difficulty breathing after your primary assessment, you should immediately reevaluate the airway. If the airway is open and breathing is present and adequate, you should consider administering supplemental oxygen. If breathing is present and inadequate (the normal rate is 12 to 20 breaths/min in adults) because respirations are too fast (generally more than 20 breaths/min), too shallow, or too slow (generally fewer than 12 breaths/min), you should administer supplemental oxygen. When respirations exceed 24 breaths/min or are fewer than 8 breaths/min, you should consider providing positive-pressure ventilations with an airway adjunct. Remember that air exchange is the critical issue, not the number of breaths.

Respiratory Rate

A normal respiratory rate varies widely in adults, ranging from 12 to 20 breaths/min. Children breathe at even faster rates. With practice, you will be able to estimate the rate and note whether it is too fast or too slow. At times it may be important to actually count the number of respirations during your primary assessment.

Respirations are determined by counting the number of breaths in a 30-second period and multiplying by two. The result equals the number of breaths per minute. For accuracy, you should count each breath at the same point in its cycle. This is most easily done by counting each peak chest rise. Although you can see peak chest rise, it is easier to place your hand on the patient's chest and feel it. However, be aware that a conscious patient who knows that you are evaluating his or her breathing will often override the automatic rate and depth by breathing more slowly and deeply. To prevent this from happening, you should check respirations in a conscious, alert patient without making the patient aware of what you are evaluating. This can be easily done by first taking a radial pulse and then, without releasing the wrist or otherwise suggesting a change, counting the chest rise that you see or feel as the patient's forearm rises and falls with the movement of the chest **Figure 8-11**. If the patient coughs, yawns, sighs, or talks during the 30-second period, you should wait a few seconds and start again. **Table 8-2** shows the normal range of respiratory rates of patients who are at rest.

Respiratory Rhythm

While counting the patient's respirations, also note the rhythm. If the time from one peak chest rise to the next is fairly consistent, respirations are considered regular. If respirations vary or change frequently, they are considered irregular. When you document the vital signs, be sure to note whether the patient's respirations were regular or irregular.

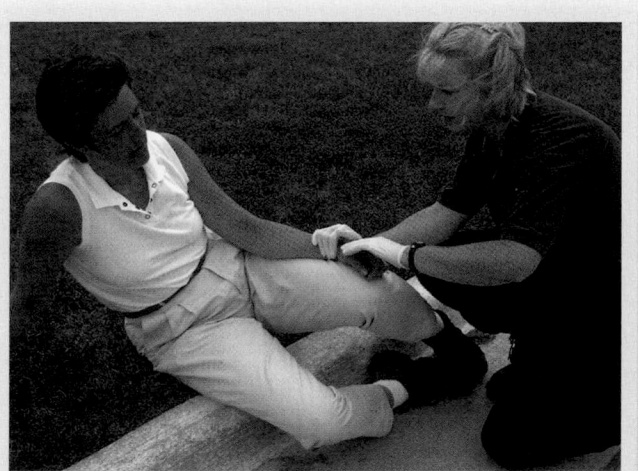

Figure 8-11 Assess respirations in a conscious patient by first taking a radial pulse and then, without releasing the patient's wrist, counting the chest rise and fall for 30 seconds.

Quality of Breathing

It may be helpful to listen to breath sounds on each side of your patient's chest early in the primary assessment. This can help identify the quality of air movement in both lungs. Decreased or absent breath sounds on one side of the chest and decreased movement in the rise and fall on one side indicate inadequate breathing.

Normal breathing is silent or, in a very quiet environment, accompanied only by the sounds of air movement at the mouth and nose. Through a stethoscope, normal breath sounds include only the sound of air movement through the bronchi accompanied by a soft, low-pitched murmur. Breathing accompanied by other sounds indicates a significant respiratory problem. When the upper airway has a mild obstruction by a foreign body or swelling, you may hear stridor, a harsh, high-pitched, crowing sound. If you can hear bubbling or gurgling, the patient probably has fluid in the airway. You should immediately suction the patient to avoid aspiration of fluid into the lungs. You may hear other sounds, like wheezes or snoring, or a musical sound indicative of a mild lower airway obstruction. A mild upper airway obstruction is usually a result of the tongue blocking the airway. The presence of any of these indicates that a serious respiratory problem exists. With a severe airway obstruction, the patient will not be able to move any air and will no longer be able to cough or talk. Sounds are caused by air moving through small spaces or fluid. If you hear no sounds, the patient may be moving no air.

A patient who coughs up thick, yellowish or greenish sputum (matter from the lungs) most likely has an advanced respiratory infection. A patient with a chest injury may cough up blood or frothy whitish or pinkish foamlike sputum. A patient with congestive heart failure may also cough up frothy sputum. The presence of either substance, regardless of its cause, indicates that an urgent, potentially critical cardiovascular and respiratory problem exists. The patient's condition may deteriorate rapidly to a point at which the patient can no longer breathe.

Table 8-2 Normal Ranges for Respirations	
Age	**Range (breaths/min)**
Adults and adolescents	12 to 20
Children (1 to 12 years)	15 to 30
Infants	25 to 50
Note: Ranges presented in other courses may vary.	

The following describes how and where to listen to assess breathing:

- First, remember that you can almost always hear a patient's breath sounds better from the patient's back; therefore, if the patient's back is accessible, listen (**auscultate**) there. If you have immobilized the patient or if the patient is in a supine position, listen from the front and sides **Figure 8-12**.
- Auscultate over the upper lungs (apices), the lower lungs (bases), and over the major airways (midclavicular and midaxillary lines).
- Lift the clothing or slide the stethoscope under the clothing. When you listen over clothing, you will primarily hear the sound of the stethoscope sliding over the fabric because breath sounds are muted by clothing.
- Place the diaphragm of the stethoscope firmly against the skin to hear the breath sounds.

What are you listening for? You may be able to identify one of the following sounds:

- **Normal breath sounds.** These are clear and quiet during inspiration and expiration.
- **Wheezing breath sounds.** These suggest an obstruction of the lower airways. Wheezing is a high-pitched whistling sound that is most prominent on expiration.
- **Rales**. Wet breath sounds may indicate cardiac failure. A moist crackling, usually on both inspiration and expiration, also called crackles.
- **Rhonchi**. Congested breath sounds may suggest the presence of mucus in the lungs. Expect to hear low-pitched, noisy sounds that are most prominent on expiration. The patient often reports a productive cough associated with these sounds.
- **Stridor.** This is often heard without a stethoscope and may indicate that the patient has an airway obstruction in the neck or upper part of the chest. Expect to hear a brassy, crowing sound that is most prominent on inspiration.

You can determine the quality or character of respirations as you are counting the number of respirations. **Table 8-3** shows four ways in which the quality or character can be described. Use your sense of hearing to listen for breath sounds or use the preferred method of auscultation, listening with a stethoscope. Note any adventitious lung sounds and treat the patient accordingly.

Depth of Breathing

The amount of air that the patient is exchanging depends on the rate and the tidal volume. **Tidal volume** is a measure of the depth of breathing and is the amount of air in milliliters that is moved into or out of the lungs during one breath. The depth of the breath determines whether the tidal volume is normal, less than normal, or more than normal.

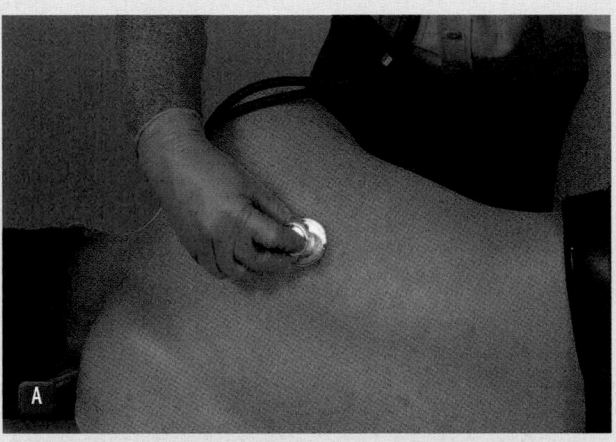

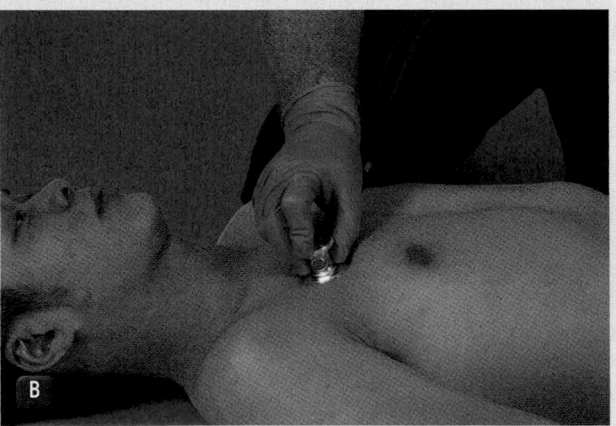

Figure 8-12 **A.** Listen to breath sounds from the patient's back if possible, over the apices, the bases, and the major airways. **B.** If the patient is immobilized or in a supine position, listen from the front and sides.

Table 8-3	Characteristics of Respirations
Normal	Breathing is neither shallow nor deep Equal chest rise and fall No use of accessory muscles
Shallow	Decreased chest or abdominal wall motion
Labored	Increased breathing effort Use of accessory muscles Possible gasping Nasal flaring, supraclavicular and intercostal retractions in infants and children
Noisy	Increase in sound of breathing, including snoring, wheezing, gurgling, crowing, grunting, and stridor

Observe how much effort is required for the patient to breathe. Normal respirations are not usually shallow or excessively deep. <u>Shallow respirations</u> can be identified by little movement of the chest wall (reduced tidal volume) or poor chest excursion. Deep respirations cause a significant rise and fall of the chest. You should document when the patient's respirations are shallow or deep; however, you do not have to record a normal depth of breathing.

The presence of <u>retractions</u> (indentation above the clavicles and in the spaces between the ribs) or the use of <u>accessory muscles</u> of respiration is a sign of inadequate breathing. Accessory muscles include the neck muscles (sternocleidomastoid), the chest pectoralis major muscles, and the abdominal muscles. <u>Nasal flaring</u> and seesaw breathing in pediatric patients indicate inadequate breathing. A patient who can speak only two or three words without pausing to take a breath, a condition known as <u>two- to three-word dyspnea</u>, has a serious breathing problem.

Normal breathing is an effortless process that does not affect a patient's speech, posture, or positioning. Speech is a good indicator of whether a conscious patient is having difficulty breathing. A patient who can speak smoothly without unusual extra pauses is breathing normally. However, a patient who can speak only one word at a time or must stop every two to three words to catch his or her breath is having significant difficulty breathing. Patients who are having marked difficulty breathing will instinctively assume a posture in which it is easier for them to breathe. There are two common postures that indicate the patient is trying to increase air flow. The first position is called the <u>tripod position</u>. In this position, a patient is sitting and leaning forward on outstretched arms with the head and chin thrust slightly forward; significant conscious effort is required for breathing. The second position is most commonly seen in children—the <u>sniffing position</u>. The patient sits upright with the head and chin thrust slightly forward, and the patient appears to be sniffing **Figure 8-13**.

Breathing that becomes progressively more difficult requires progressively more effort. When you can see that effort, the patient's breathing is described as <u>labored breathing</u>. Initially, labored breathing is characterized by the patient's position, concentration on breathing, and the increased effort and depth of each breath. As breathing becomes more labored, accessory muscles in the chest and neck are used, and the patient may make grunting sounds with each breath. In infants and small children, nasal flaring and supraclavicular and intercostal retractions are commonly associated with labored breathing. Sometimes the patient may be gasping.

Infants and small children may have labored breathing for a sustained period, will then often become

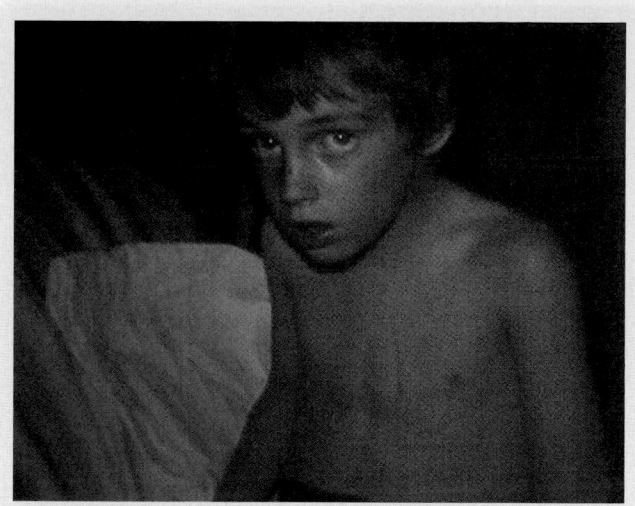

Figure 8-13 A patient in the sniffing position sits upright with the head and chin thrust slightly forward.

exhausted, and finally will no longer have the strength to maintain the necessary energy to breathe. In infants and small children, cardiac arrest is generally caused by respiratory arrest.

Respiratory distress occurs when a person, particularly a child, has difficulty breathing; therefore, the work of breathing is increased. Typically a person in respiratory distress presents with an increase in respiratory effort and rate. Respiratory failure occurs when the blood is inadequately oxygenated or ventilation is inadequate to meet the oxygen demands of the body. Respiratory arrest is the ultimate result of respiratory failure if it is not corrected **Table 8-4**.

Table 8-4 Signs of Respiratory Distress and Failure

Respiratory Distress	Respiratory Failure
Agitation, anxiety, restlessness	Lethargy, difficult to rouse
Stridor, wheezing	Tachypnea with periods of bradypnea or agonal respirations
Accessory muscle use; intercostal retractions, neck muscle use (sternomastoid)	Inadequate chest rise/poor excursion
Tachypnea	Inadequate respiratory rate or effort
Mild tachycardia	Bradycardia
Nasal flaring, seesaw breathing, head bobbing	Diminished muscle tone

Assess Circulation

Assessing circulation helps you to evaluate how well blood is circulating to the major organs, including the brain, lungs, heart, kidneys, and the rest of the body. A variety of problems can impair circulation, including blood loss, shock, and conditions that affect the heart and major blood vessels. Circulation is evaluated by assessing the pulse rate, pulse quality, and pulse rhythm. You will also need to identify external bleeding and evaluate the skin.

Assess Pulse

With each heartbeat, the ventricles contract, forcefully ejecting blood from the heart and propelling it into the arteries. Often referred to as a heartbeat, the **pulse** is the pressure wave that occurs as each heartbeat causes a surge in the blood circulating through the arteries. The pulse is most easily felt at a pulse point where a major artery lies near the surface and can be pressed gently against a bone or solid organ.

Your first consideration when taking a pulse is to determine whether the patient has a pulse or is pulseless. To determine if a pulse is present, you will need to **palpate** (feel) the pulse. Hold together your index and middle fingers and place their tips over a pulse point, pressing gently against the artery until you feel intermittent pulsations. In responsive patients who are older than 1 year, you should palpate the radial pulse at the wrist **Figure 8-14A**. In unresponsive patients older than 1 year, you should palpate the carotid pulse in the neck **Figure 8-14B**. When palpating the carotid pulse, you should place the fingertips of your index and middle fingers along the carotid artery in the groove between the trachea and the neck muscle. Palpate the carotid pulse on the same side of the patient that you are on; palpating on the opposite side of the patient may lead to undue pressure on the neck that may compromise respirations. Use caution when palpating the carotid

pulse in a responsive patient, especially an older patient. Only gentle pressure on one side of the neck should be used. Never press on the carotid arteries on both sides of the neck at the same time. Doing so can reduce circulation to the brain.

Sometimes, you may have to slide your fingertips a little to each side and press again until you feel a pulse. When palpating a pulse, do not allow your thumb to touch the patient. If you do so, you may mistake the strong pulsing circulation in your thumb for the patient's pulse.

Palpate the brachial pulse, located at the medial area (inside) of the upper arm, in children younger than 1 year **Figure 8-15**. With the infant lying supine, you can access the brachial pulse by elevating the arm over the infant's head. Because most infants have chubby arms, you need to press your adjacent fingertips firmly along the brachial artery, which lies parallel to the long axis of the upper arm, to be able to palpate the pulse.

If you cannot palpate a pulse in an unresponsive patient, begin CPR. If an AED is available, attach it and follow the voice prompts, following your local protocol. An AED is indicated for use on patients who have been assessed to be unresponsive, not breathing, and pulseless. An AED with special pediatric pads and a dose-attenuating system should be used on pediatric patients older than 1 year; if these are not available, and adult AED should be used. In infants (1 month to 1 year), it is preferable to perform manual debrillation or use a dose-attenuating system; if these are not available, an adult AED should be used. More information about this is available in Chapter 11, *BLS Resuscitation*.

If the patient has a pulse but is not breathing, provide ventilations at a rate of at least 10 breaths/min for adults and at least 12 breaths/min for an infant or a child. Continue to monitor the pulse to evaluate the effectiveness of your ventilations. If at any time the pulse is lost, start CPR and apply the AED if indicated. The apparent absence of a palpable pulse in a responsive patient is not caused by cardiac arrest. Therefore, never begin CPR or use an AED on a responsive patient.

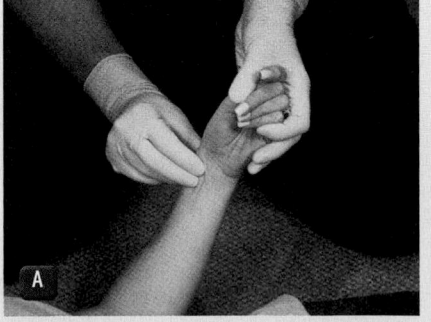
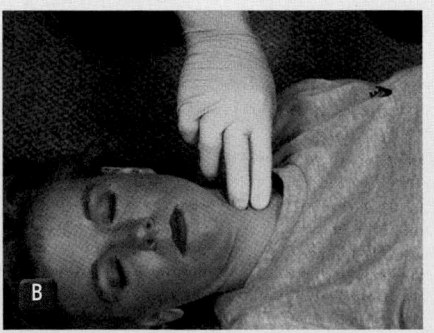

Figure 8-14 **A.** To palpate the radial pulse, place the tips of your first two fingers over the radial artery, pressing gently until you feel intermittent pulsations. **B.** To palpate the carotid pulse, place the tips of your first two fingers over the carotid artery, pressing gently until you feel intermittent pulsations.

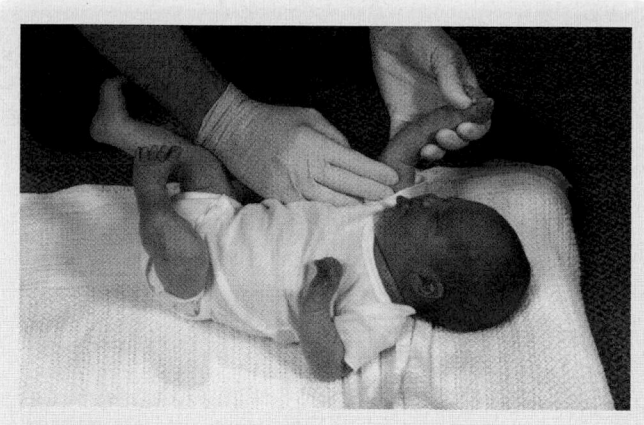

Figure 8-15 To palpate the brachial pulse in an infant, press firmly along the brachial artery on the inside of the upper arm.

generally the younger the patient, the faster the pulse rate. In well-conditioned athletes or in people taking heart medications such as beta-blockers, the pulse rate may be considerably lower. **Table 8-5** shows the normal ranges of pulse rates for adults and children.

To obtain the pulse rate in most patients, you should count the number of pulses felt in a 30-second period and then multiply by two. A pulse that is weak and difficult to palpate, irregular, or extremely slow should be palpated and counted for a full minute. A pulse rate is

After you have determined that a pulse is present, next determine its adequacy. This is done by assessing the pulse rate, pulse quality, and pulse rhythm.

Pulse Rate

For an adult, the normal resting pulse rate should be between 60 and 100 beats/min and could be as much as 100 beats/min in geriatric patients. In pediatric patients,

Table 8-5 Normal Ranges for Pulse Rate

Age	Range (beats/min)
Infant: 1 month to 1 year	100 to 160
Toddler: 1 to 3 years	90 to 150
Preschool age: 3 to 6 years	80 to 140
School age: 6 to 12 years	70 to 120
Adolescent: 12 to 18 years	60 to 100
Adult	60 to 100

Primary Assessment

You are the Provider: PART 3

Your partner begins assisting the patient's ventilations with high-flow oxygen while an EMT from the engine company obtains her vital signs. You ask the police officer to inspect the patient's apartment for anything suspicious. The neighbor has no knowledge of the patient's medical history. The patient's blood glucose level is assessed and reads 112 mg/dL.

Recording Time: 5 Minutes	
Respirations	6 breaths/min and shallow (baseline); ventilations are being assisted
Pulse	40 beats/min; weak and regular
Skin	Cool and pale
Blood pressure	80/60 mm Hg
Oxygen saturation (Sao$_2$)	95% (with assisted ventilation)

Your secondary assessment of the patient reveals no obvious signs of trauma, medical alert tags, or anything else that might explain her condition. The police officer did not find any pill bottles, drug paraphernalia, or anything else suspicious. Her driver's license shows that she is 22 years old. You hear on the radio that a paramedic unit has just cleared a scene and is approximately 17 minutes away.

5. Does the patient require further treatment at the scene? If so, what?

6. Should you remain at the scene and wait for the paramedic unit? Why or why not?

counted as beats per minute; however, in reporting the pulse rate, it is not necessary to state or write "beats per minute" after the number.

In assessing the pulse rate in an adult patient, a rate that is greater than 100 beats/min is described as **tachy-cardia**, and a rate of less than 60 beats/min is described as **bradycardia**.

Pulse Quality

You should always report the pulse's quality whenever reporting or recording the pulse. The pulse is generally palpated at the radial or carotid arteries in adults and at the brachial artery in infants, because it is normally strong and easily palpable at these locations. Therefore, if the pulse feels of normal strength, you should describe it as being strong. You should describe a stronger than normal pulse as "bounding" and a pulse that is weak and difficult to feel as "weak" or "thready." With a little experience, you will be able to easily make the necessary distinctions.

Pulse Rhythm

When you are assessing the pulse, you must also determine whether the rhythm is regular or irregular. When the interval between each ventricular contraction of the heart is short, the pulse is rapid. When the interval is longer, the pulse is slower. Regardless of the rate, the interval between each contraction should be the same, and the pulse should occur at a constant, regular rhythm. You should document this rhythm as regular.

The rhythm is considered irregular if the heart periodically has a premature or late beat or if a pulse beat is missed. Some people have a chronically irregular pulse; however, if an irregular pulse is found in a patient with signs and symptoms that suggest a cardiovascular problem, the patient likely needs advanced cardiac assessment and life support. Therefore, depending on your protocols, you should call for ALS backup, arrange for an intercept by paramedics, or initiate prompt transport to definitive care.

With practice, you will be able to assess whether the pulse is too slow, too fast, or irregular without actually counting the pulsations. This will help to speed up your assessment of the ABCs and allow you to focus on finding other potentially life-threatening problems. A pulse rate that is too slow or too fast may change decisions related to transporting your patient. The pulse should be easily felt at the radial or carotid artery and have a regular rhythm. If it is difficult to feel or irregular, the patient may have problems with his or her circulatory system that may need further evaluation later in your assessment.

The Skin

The skin has many functions. It helps maintain the water content of the body, acts as insulation and protection from infection, and has a role in regulating body temperature by changing the amount of blood circulating through the surface of skin.

Assessing the skin is one of the most important and most readily accessible ways of evaluating circulation and perfusion, blood oxygen level and body temperature. A normally functioning circulatory system perfuses the skin with oxygenated blood. A lack of perfusion or hypoperfusion will result in hypoxia of the brain, lungs, heart, and kidneys. In most situations, hypoperfusion is caused by shock. The degree of hypoperfusion and how long it lasts will determine if a patient will sustain permanent damage related to the hypoxia. Perfusion is assessed by evaluating a patient's skin color, temperature, moisture, and capillary refill.

Skin Color

Assessing the skin helps you to determine the adequacy of perfusion. Adequate perfusion meets the current needs of the cells; inadequate perfusion causes cells and tissues to die.

Many blood vessels lie near the surface of the skin. The skin's color is determined by the blood circulating through these vessels and the amount and type of pigment that is present in the skin. Blood is red when it is adequately saturated with oxygen. As a result, skin in lightly pigmented people is pinkish. The pigmentation in most people will not hide changes in the skin's underlying color, regardless of the person's race. In patients with deeply pigmented skin, changes in color may be apparent only in certain areas, such as the fingernail beds, the mucous membranes in the mouth, the lips, the underside of the arm and palm (which are usually less pigmented), and the conjunctiva of the eyes. The **conjunctiva** is the delicate membrane lining the eyelids, and it covers the exposed surface of the eye. In addition, the palms of the hands and soles of the feet should be assessed in infants and children.

Poor peripheral circulation will cause the skin to appear pale, white, ashen, or gray, possibly with a waxy translucent appearance like a white candle. Abnormally cold or frozen skin may also appear this way. When the blood is not properly saturated with oxygen, it appears bluish. Therefore, in a patient with insufficient air exchange and low levels of oxygen in the blood, the blood and vessels become bluish, and the lips, mucous membranes, nail beds, and skin over the blood vessels appear blue or gray. This condition is called **cyanosis**

Figure 8-16.

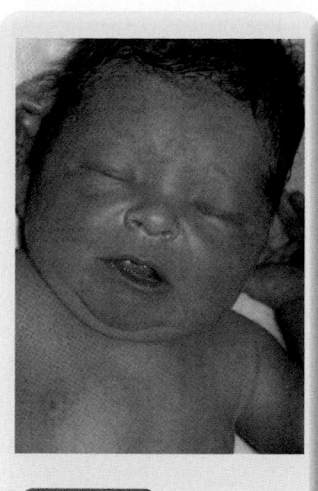

Figure 8-16 Cyanosis occurs when the patient has low levels of oxygen in the blood.

High blood pressure may cause the skin to be abnormally flushed and red. In some patients with extremely high blood pressure, all of the visible blood vessels will be so full that the skin will appear to be a dark reddish purple. A patient with carbon monoxide poisoning or a significant fever, heatstroke, sunburn, mild thermal burns, or other conditions in which the body is unable to properly dissipate heat will also appear to have red skin.

Changes in skin color may also result from chronic illness. Liver disease or dysfunction may cause __jaundice__, resulting in the patient's skin and sclera turning yellow. The __sclera__ is the normally white portion of the eye and may show color changes even before skin color change is visible.

Skin Temperature

Normal skin temperature will be warm to the touch (normal body temperature is 98.6°F). Abnormal skin temperatures are hot, cool, cold, and clammy. Clammy is considered cool and moist. When the patient has a significant fever, sunburn, or hyperthermia, the skin feels hot to the touch. The skin will feel cool when the patient is in early shock, has mild hypothermia, or has inadequate perfusion. With poor perfusion, the body pulls blood away from the surface of the skin and diverts it to the core of the body. The result is cool, pale, clammy skin; in your primary assessment, this is a good indication of hypoperfusion and inadequacy of circulatory system function (shock). The skin will feel cold when the patient is in profound shock, has hypothermia, or has frostbite.

Body temperature is normally measured with a thermometer in the hospital. However, in the field, feeling the patient's forehead with the back of your hand is usually adequate to determine whether the patient's temperature is elevated or decreased.

Skin Moisture

Dry skin is normal. Skin that is wet, moist (often called diaphoretic), or excessively dry and hot suggests a problem. In the early stages of shock, the skin will become slightly moist. Skin that is only slightly moist but not covered excessively with sweat is described as clammy, damp, or moist. When the skin is bathed in sweat, such as after strenuous exercise or when the patient is in shock, the skin is described as wet or __diaphoretic__.

Because the skin's color, temperature, and moisture are often related signs, you should consider them together. When recording or reporting your assessment of the skin, you should first describe the color, then the temperature, and last, whether the skin is dry, moist, or wet. For example, you could say or write, "Skin: pale, cool, and clammy."

Again, these characteristics are important findings in your primary assessment because hypoperfusion can lead to serious consequences if treatment is delayed or ignored.

Words of Wisdom

Remember to assess:
- Skin color
- Skin temperature
- Skin moisture

Capillary Refill

__Capillary refill__ is evaluated to assess the ability of the circulatory system to restore blood to the capillary system. When evaluated in an uninjured limb, capillary refill time (CRT) may provide an indication of the patient's level of perfusion. It should be kept in mind, however, that capillary refill can be affected by the patient's body temperature, position, preexisting medical conditions, and medications. Remember that other conditions, not related to the body's circulation, may also slow capillary refill. These conditions include, but are not limited to, the patient's age, exposure to a cold environment (__hypothermia__), frozen tissue (__frostbite__), and __vasoconstriction__ (narrowing of a blood vessel, such as with hypoperfusion or cold extremities). Injuries to bones and muscles of the extremities may cause local circulatory compromise, resulting in hypoperfusion of an extremity rather than hypoperfusion of the body in general.

To test capillary refill, place your thumb on the patient's fingernail with your fingers on the underside of the patient's finger and gently compress **Figure 8-17A**. The blood will be forced from the capillaries in the nail bed. Remove the pressure applied against the tip of the patient's finger. The nail bed will remain blanched and white for a brief period. As the underlying capillaries refill with blood, the nail bed will be restored to its normal pink color.

Capillary refill should be prompt, and the nail bed color should be pink. With adequate perfusion, the color in the nail bed should be restored to its normal pink color within 2 seconds, or about the time it takes to say "capillary refill" at a normal rate of speech Figure 8-17B . You should report and document the CRT as normal (2 seconds or less) You should suspect poor peripheral circulation when capillary refill takes more than 2 seconds or the nail bed remains blanched. In this case, you should report and document the CRT as delayed or CRT > 2. Capillary refill is not considered an accurate indication of perfusion in adult patients.

A bluish color may indicate that the capillaries are refilling with blood drawn from the veins rather than with oxygenated blood from the arteries, making the test invalid. You should also consider the capillary refill test invalid if the patient is in or has been exposed to a cold environment or if the patient is older. In both situations, delayed capillary refill may be normal.

To assess capillary refill in older infants and children younger than 6 years, press on the skin or nail bed and determine how long it takes for the pink color to return. In newborns and young infants, press on the forehead, chin, or sternum to determine capillary refill time. As with adults, normal capillary refill takes 2 seconds or less. However, it is a much more reliable indicator of cardiovascular status in children than it is in adults and should be recorded for all of your pediatric patients.

Assess and Control External Bleeding

Perform a rapid scan of the patient to identify any major external bleeding. In some cases, blood loss can be very rapid and can quickly result in shock or even death. Therefore, this step demands your immediate attention. Signs of blood loss include active bleeding from wounds and/or evidence of bleeding such as blood on the clothes or near the patient. Serious bleeding from a large vein may be characterized by steady blood flow. Bleeding from an artery is characterized by a spurting flow of blood. When you evaluate an unconscious patient, do

a sweep for blood quickly and lightly by running your gloved hands from head to toe, pausing periodically to see if your gloves are bloody.

Controlling external bleeding is often very simple. Initially, direct pressure with your gloved hand and soon thereafter a sterile bandage over the wound will control bleeding in most cases. This direct pressure stops the bleeding and helps the blood to **coagulate**, or clot naturally. Most often, bleeding can be adequately controlled by using direct pressure, along with elevating the extremity if bleeding is from the arms or legs. When direct pressure and elevation are not successful, you should apply a tourniquet. More information about applying a tourniquet is found in Chapter 23, *Bleeding*.

Identify and Treat Life Threats

Many conditions present an immediate threat to life, and key to your role as an EMT is to determine if a life threat is present and, if so, to quickly address it. In many situations, there is a process the body takes when reacting to a life threat.

The first observation that you will most likely make is that there will be a loss of meaningful communication between you and the patient. A dying person becomes less aware of his or her surroundings and stops making attempts to communicate. After a variable period, loss of consciousness occurs. The patient becomes totally unresponsive to external stimuli. The muscles become slack, among them the muscles of the jaw, thus permitting the tongue to sag against the posterior part of the throat. This in turn leads very quickly to airway obstruction. Air can no longer enter the lungs, and within a few minutes the patient stops breathing. The heart cannot continue to function without oxygen, and it stops beating. Within a few minutes, a number of brain cells are damaged, leading to irreversible brain damage. There are only a few conditions that cause sudden death: airway obstruction, respiratory arrest, cardiac arrest, and severe bleeding.

Often these conditions are reversible, but to reverse them, you have to be able to recognize them quickly and take immediate steps to correct them. This is the purpose of the primary assessment.

Lifesaving interventions include opening the airway. Airway patency is your number one priority. Assess the patient's breathing, and initiate ventilations in patients who have inadequate respirations or a respiratory rate greater than 24 or less than 8 breaths/min. Although there is a range of "normal" respiratory rates,

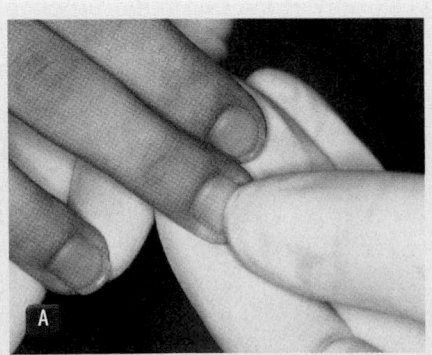
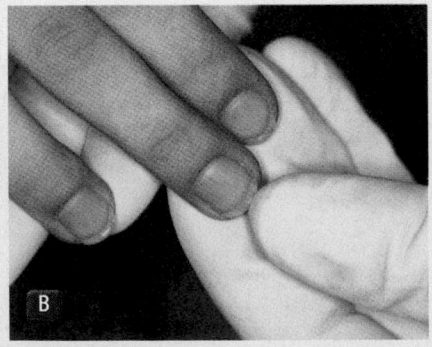

Figure 8-17 **A.** To test capillary refill, gently compress the fingertip until it blanches. **B.** Release the fingertip, and count until it returns to its normal pink color.

you should always assess your patient's overall condition to decide on appropriate treatment. It is important to assess the quality of respirations, the patient's mental status, skin color, appearance, and chest rise. Respirations that are shallow may not be ventilating and oxygenating the patient adequately. Next assess the patient's pulse. If you determine that a patient is unresponsive, not breathing, and does not have a pulse, you must begin CPR, starting with chest compressions. (Chapter 11, *BLS Resuscitation*, covers CPR in more detail.) The last lifesaving intervention is the detection of severe bleeding. Severe external bleeding must be controlled using the techniques of direct pressure, elevation, and a tourniquet if allowed by local protocol.

Perform a Rapid Scan

You will need to take 60 to 90 seconds and perform a rapid scan of the patient's body to identify injuries that must be managed and/or protected immediately. This is *not* a systematic or focused physical examination that will be performed during the secondary assessment. This is a rapid scan to identify injuries that must be managed and/or protected before and during packaging and loading the patient for transport.

To perform a rapid scan of the patient, follow the steps in **Skill Drill 8-1**. Remember, this should take no longer than 60 to 90 seconds!

Skill Drill 8-1

1. Assess the head, looking and feeling for DCAP-BTLS (Deformities, Contusions, Abrasions, Punctures/penetrations, Burns, Tenderness, Lacerations, and Swelling) and crepitus (Step 1).
2. Assess the neck, looking and feeling for DCAP-BTLS, jugular venous distention, tracheal deviation, and crepitus (Step 2). In trauma patients, you should now apply a cervical spinal immobilization device (Step 3). It is particularly important to assess the neck before covering it with a cervical collar.
3. Assess the chest, looking and feeling for DCAP-BTLS, paradoxical motion, and crepitus. You should also listen to breath sounds on both sides of the patient's chest (Step 4).
4. Assess the abdomen, looking and feeling for DCAP-BTLS, rigidity (firm or soft), and distention (Step 5).
5. Assess the pelvis, looking for DCAP-BTLS. If there is no pain, gently compress the pelvis downward and inward to look for tenderness and instability (Step 6).
6. Assess all four extremities, looking and feeling for DCAP-BTLS. Also assess bilaterally for distal pulses and the motor and sensory function (Step 7).

7. Assess the back and buttocks, looking and feeling for DCAP-BTLS. In all trauma patients you should maintain in-line stabilization of the spine while rolling the patient on his or her side in one motion (Step 8). If you are placing the patient on a backboard, it is particularly important that you check the back before you log roll the patient and before you place him or her onto a backboard.

Determine Priority of Patient Care and Transport

As you complete your primary assessment, you have to make some decisions about patient care and transport. The rapid scan will assist you in determining transport priority **Figure 8-18**. If you do not identify any injuries that require treatment or rapid transport when completing your assessment of the ABCs, you may find indications for rapid transport during your rapid scan of the patient's body. For example, you may identify an internal hemorrhage by the presence of a distended or firm abdomen or bilateral femoral fractures. These types of conditions would be indications for rapid transport.

Would you consider your patient a high, medium, or low priority for transport? Priority designation is used to determine if your patient needs immediate transport or will tolerate a few more minutes on scene. Patients with any of the following conditions are examples of high-priority patients and should be transported immediately:

- Difficulty breathing
- Poor general impression
- Unresponsive with no gag or cough reflex
- Severe chest pain
- Pale skin or other signs of poor perfusion
- Complicated childbirth

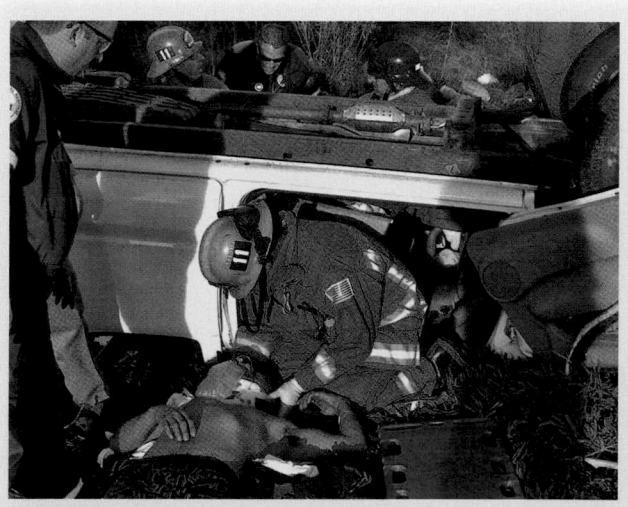

Figure 8-18 Identifying priority patients.

Skill Drill 8-1

Rapid Scan

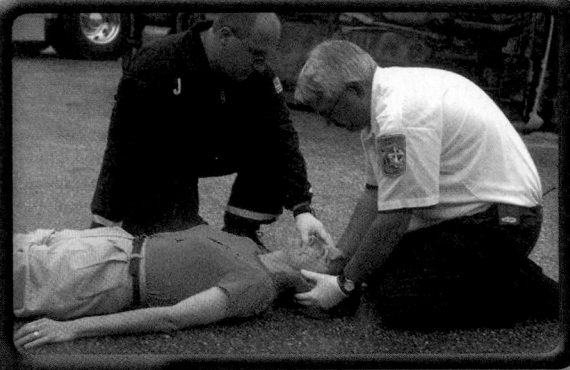

Step 1 Assess the head. Have your partner maintain in-line stabilization if trauma is suspected.

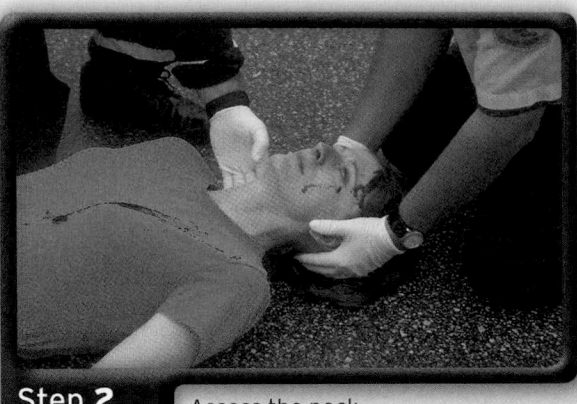

Step 2 Assess the neck.

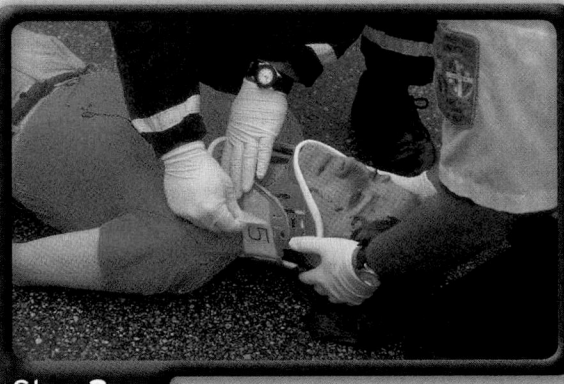

Step 3 Apply a cervical spinal immobilization device on trauma patients.

Step 4 Assess the chest. Listen to breath sounds on both sides of the chest.

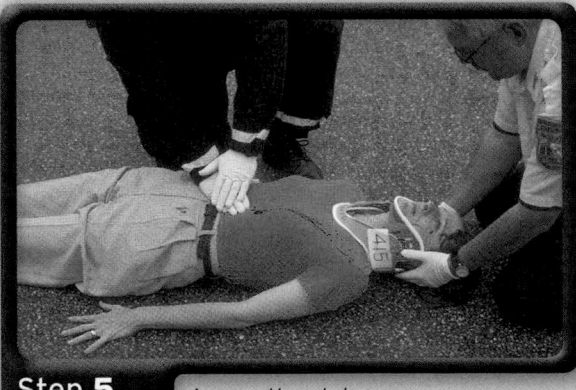

Step 5 Assess the abdomen.

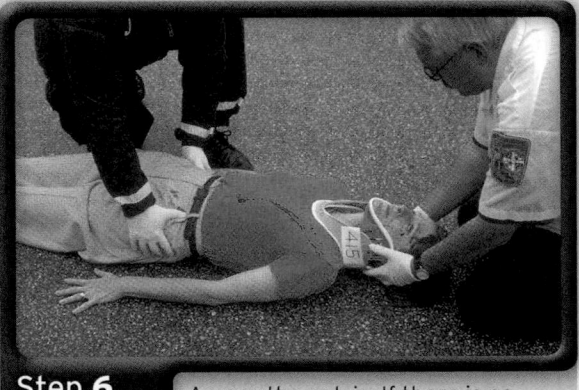

Step 6 Assess the pelvis. If there is no pain, gently compress the pelvis downward and inward to look for tenderness and instability.

Skill Drill 8-1

Rapid Scan, continued

Step 7 Assess all four extremities. Assess pulse and the motor and sensory function.

Step 8 Assess the back. In trauma patients, roll the patient in one motion.

- Uncontrolled bleeding
- Responsive but unable to follow commands
- Severe pain in any area of the body
- Inability to move any part of the body

Protecting the patient's spine and identifying fractured extremities are integral parts of packaging for transport. If a spinal injury is suspected or the MOI is significant enough to cause a possible injury, consider spinal immobilization early. If you are unsure if spinal immobilization is necessary, always err on the side of caution and immobilize the patient. These injuries can be made worse if you neglect to assess and treat them before moving the patient. Recognizing the need to transport serious trauma patients is of such importance that you may hear colleagues refer to the **Golden Period**. This refers to the time from injury to definitive care, during which treatment of shock and traumatic injuries should occur because survival potential is best Figure 8-19 . After the first 60 minutes, the body has increasing difficulty in compensating for shock and traumatic injuries. For

this reason, you should spend as little time as possible on scene with patients who have sustained significant or severe trauma. Aim to assess, stabilize, package, and begin transport to the appropriate facility within 10 minutes (often referred to as the "Platinum 10") after

Words of Wisdom

The "Golden Period" was previously referred to as the "Golden Hour." However, because many injured patients require definitive care in less than an hour, this is now referred to as the "Golden Period."

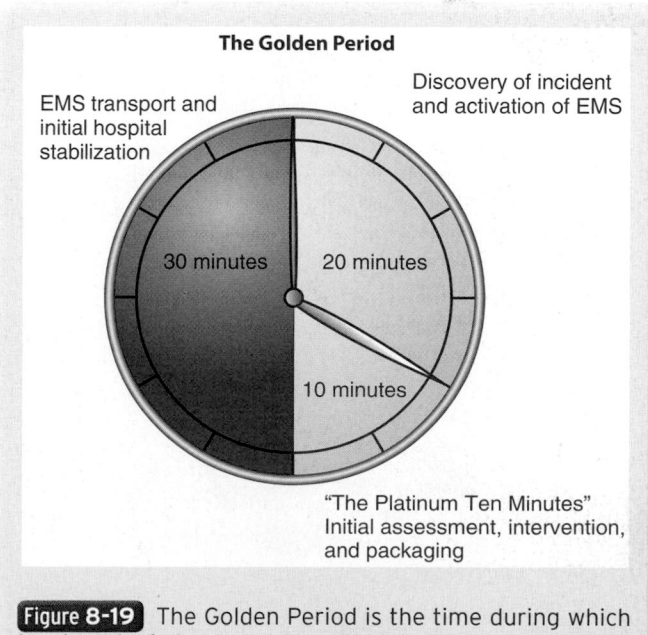

The Golden Period

EMS transport and initial hospital stabilization

Discovery of incident and activation of EMS

30 minutes

20 minutes

10 minutes

"The Platinum Ten Minutes" Initial assessment, intervention, and packaging

Figure 8-19 The Golden Period is the time during which treatment of shock or traumatic injuries is most critical and the potential for survival is best.

arrival on scene whenever possible (a difficult or complex extrication may obviously limit possibilities).

Some patients may benefit from remaining on scene and receiving continuing care. For example, an older patient with chest pain may be better served on scene by being administered nitroglycerin and waiting for an ALS vehicle than by immediate transport. Support from ALS should be called for if a unit is not already en route to the scene. Depending on the travel distance, an ALS unit can be met while transporting the critical patient. If ALS is delayed or farther away, coordinating a rendezvous may be a better decision for a high-priority patient. Your decision to stay on scene or transport immediately will be based on your patient's condition, the availability of more advanced help, the distance you must transport, and your local protocols.

Correct identification of high-priority patients is an essential aspect of the primary assessment and helps to improve patient outcome. While initial treatment is important, it is essential to remember that immediate transport is one of the keys to the survival of any high-priority patient. Transport should be initiated as soon as practical and possible.

Remember, the goal of your primary assessment is to identify and treat life threats, including management of airway, breathing, and circulation problems, as quickly as possible. Measuring vital signs more exactly is accomplished during the secondary assessment (discussed later), once time and life threats are less of an issue.

If the patient's condition is stable, you should reassess vital signs every 15 minutes until you reach the emergency department. If the patient's condition is unstable you should reassess vital signs every 5 minutes, or as often as the situation permits, looking for trends in the patient's condition, and treat for shock.

Do not be falsely reassured by apparently normal vital signs. The body has amazing abilities to compensate for severe injury or illness, especially in children and young adults. Even patients who have experienced severe medical or traumatic conditions may initially present with fairly normal vital signs. However, the body eventually loses its ability to compensate (decompensatory shock), and the vital signs may deteriorate rapidly, especially in children. In fact, this tendency for the vital signs to fall rapidly as the patient decompensates is the reason that it is important to frequently recheck and record vital signs. Treating a patient for shock before obvious signs of shock appear helps to reduce the overall effects of decompensatory shock and, therefore, to potentially increase your patient's survival.

Words of Wisdom

Reassess vital signs often, watching for trends that may indicate a patient is unable to compensate for his or her illness or injury. You should suspect shock in any patient exhibiting tachycardia and pale, cool, clammy skin and transport immediately.

Patient Assessment

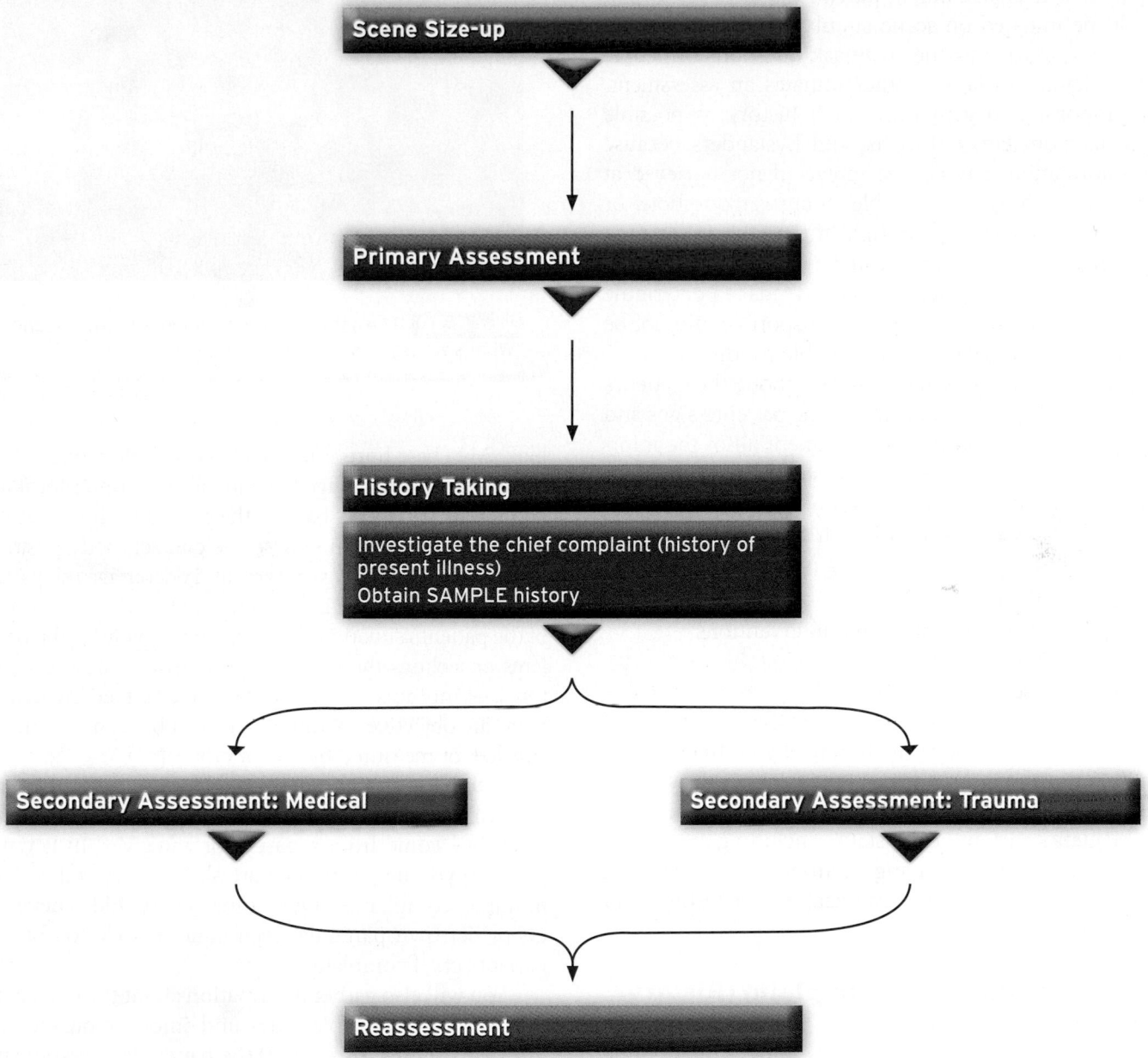

Scene Size-up

Primary Assessment

History Taking

Investigate the chief complaint (history of present illness)
Obtain SAMPLE history

Secondary Assessment: Medical

Secondary Assessment: Trauma

Reassessment

History Taking

History Taking

Although history taking is listed after the primary assessment, it is an integral part of the assessment and should be initiated on scene simultaneously with other tasks. You or your partner may ask questions of people in the vicinity while the other initiates an assessment. It is important to gather as much history as possible on scene from family, friends, and bystanders because this information may be lost forever if not retrieved at this time. If the patient is able to answer questions or a family member is transported in the ambulance with the patient, history taking can be expanded while en route. Sometimes history may be essential to determine the underlying illness or injury. Transport should not be delayed in patients who are in unstable condition.

History taking provides detail about the patient's chief complaint and an account of the patient's signs and symptoms. It is important to document all of the information gathered during this phase of the assessment process. This includes demographic information, past medical history, and current health status of the patient. Be sure to document the following information:

- Date of the incident
- All times of assessments and interventions
- Patient's age
- Patient's sex
- Patient's race
- Past medical history, including any pertinent information about the patient's condition, such as medical problems, traumatic injuries, and surgeries
- Patient's current health status, including diet, medications, drug use, living environment and hazards, physician visits for immunizations or testing, and family history

Investigate the Chief Complaint (History of Present Illness)

The patient's chief complaint is the most serious thing that the patient is concerned about. This is the reason why the patient or someone else called 9-1-1 Figure 8-20. To investigate the chief complaint, first begin by making introductions, make the patient feel comfortable, and obtain permission to treat; you can then start your investigation by asking a few simple, open-ended questions. Refer to the patient as Mr., Ms., or Mrs., using the patient's last name. Questions such as, "What seems to be the matter?" or "What's wrong today?" should produce a response that will help determine a chief complaint. These questions and others can help to elicit a response that may determine the patient's highest concern. The response is usually expressed in the patient's own words with simple answers

Figure 8-20 The patient's initial response to the question "What's wrong?" is the chief complaint.

like, "My chest hurts" or "I have been feeling weak." Use eye contact to encourage the patient to continue speaking, and repeat statements back to the patient to show that you understand the situation. Use eye contact, body position, and language to show you care, and encourage the patient to continue speaking. Do not interrupt, and be empathetic of the patient's situation. As discussed previously, the problems or feelings the patient reports to you are the symptoms. Symptoms cannot be felt or observed by others. Signs are objective conditions that can be seen, heard, felt, smelled, or measured by you or others Figure 8-21.

You must consider the wide range of age groups that you will interact with. Information from infants and children may come from a parent or caregiver. In geriatric patients, you may find they are slow to respond or have multiple complaints. Over time, every EMT develops his or her own particular technique or style to obtain a patient's chief complaint.

You will also gather information about the chief complaint from observable clues and information received from the original dispatch. If the patient is unresponsive, information about the patient, pertinent past medical history, and clues about the immediate incident may be obtained from family members present, a person who may have witnessed the situation, or medical alert jewelry Figure 8-22. Observable clues may include things such as the patient not being able to respond using full sentences and appearing to have some respiratory distress. These clues may indicate the patient's chief complaint is "difficulty breathing," or the clues may be part of a bigger problem that has to do with a lengthy history of cardiac problems.

For example, you are called to the home of an elderly man who fell. This information was provided by

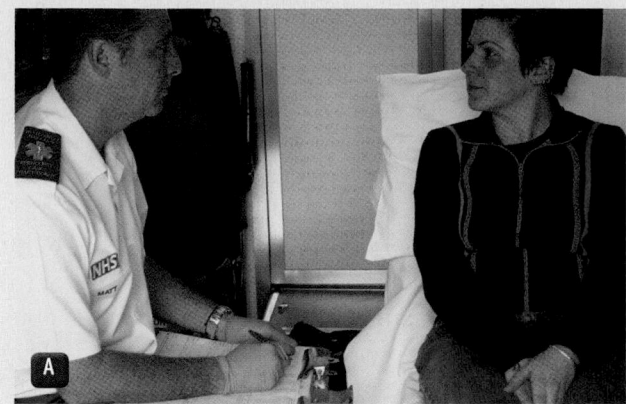

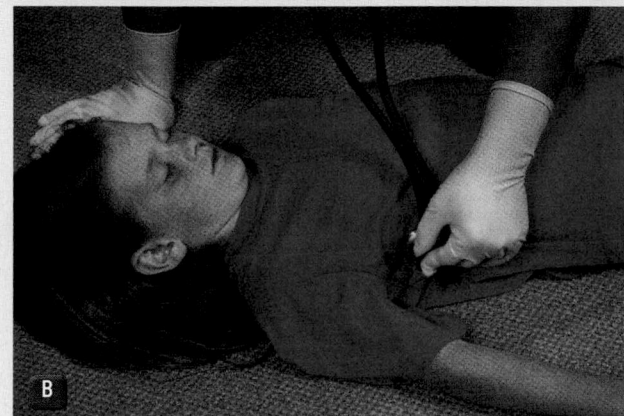

Figure 8-21 **A.** A symptom is a subjective condition that the patient feels and tells you about. **B.** A sign is an objective condition that you can observe about the patient.

Figure 8-22 If the patient is unresponsive, try to obtain a pertinent history or patient information from family or bystanders.

their lives. Sorting through the clues from the emergency scene itself, from the patient's complaints, and from the patient's signs and symptoms and past medical history will assist you in understanding the cause of your patient's problem and enable you to make appropriate, timely decisions about your patient's care. Remember to use family members, friends, bystanders, and medical identification tags to gain essential information concerning events leading up to the incident. The patient's history will help to tie together your findings from the primary assessment.

Obtain SAMPLE History

As discussed previously, the problems or feelings the patient reports to you are the symptoms. Symptoms cannot be felt or observed by others. Signs are objective conditions that can be seen, heard, felt, smelled, or measured by you or others. As you obtain a patient history from medical and trauma patients, you will need to know some of the standard techniques for questioning patients. By obtaining a <u>SAMPLE history</u>, a mnemonic used to gather a general past medical or trauma history, you will be able to gather important information from the patient. Use the mnemonic SAMPLE to obtain the following information:

S **Signs and symptoms.** What signs and symptoms occurred at the onset of the incident? Does the patient report pain?

A **Allergies.** Is the patient allergic to any medication, food, or other substance? What reactions did the patient have to any of them? If the patient has no known allergies, you should note this on the run report as "no known allergies" or "NKA."

the dispatcher, and you can use it to help process all of the clues that may be presented in what appears to be a simple fall. You find the patient lying at the bottom of the stairs. How many stairs are there, are they carpeted, and is the floor concrete, wood, or tile? Additional observable clues are used to determine a chief complaint. You note that the patient has an obvious deformity to his right arm, and your initial general impression is a possible fracture. Is this the patient's chief complaint, or is this the result of another problem? The patient states he did fall, which is how the injury occurred, and he is complaining of pain in the right arm. However, was the fall the result of tripping on a step, or was it associated with a medical problem such as dizziness, vertigo, or a syncopal episode, all of which may have caused the fall? It is your responsibility to look at all the possibilities and ask the appropriate questions to determine the patient's real chief complaint.

Patient interaction typically occurs during the worst possible time in the patient's life. These are emergency situations in which patients are afraid and confused. Oftentimes, patients assume this could be the end of

M Medications. What medication is the patient prescribed? What dosage is prescribed? How often does the patient take the medication? What prescription, over-the-counter, and herbal medications has the patient taken in the last 12 hours? This includes medications taken for birth control or erectile dysfunction. How much was taken and when? Does the patient take recreational drugs or drink alcohol?

P Pertinent past medical history. Does the patient have any history of medical, surgical, or trauma occurrences? Has the patient had a recent illness or injury, fall, or blow to the head? Is there important family history that should be known?

L Last oral intake. When did the patient last eat or drink? What did the patient eat or drink, and how much was consumed? Did the patient take any drugs or drink alcohol? Has there been any other oral intake in the last 4 hours?

E Events leading up to the injury or illness. What are the key events that led up to this incident? What occurred between the onset of the incident and your arrival? What was the patient doing when this illness started? What was the patient doing when this injury happened?

Another device that can be very helpful in the assessment of pain is the mnemonic **OPQRST**:

O Onset. When did the problem begin and what caused it?

P Provocation or palliation. Does anything make it feel better or worse? How are you most comfortable?

Q Quality. What is the pain like? Is it sharp, dull, crushing, tearing? Ask the patient to describe the pain.

R Region/radiation. Where does it hurt? Does the pain move anywhere?

S Severity. On a scale of 1 to 10, how would you rate your pain?

T Timing. Has the pain been constant or does it come and go? How long have you had the pain (often answered under "O," onset)? When did the pain start?

Identify Pertinent Negatives

While obtaining information in the process of achieving a thorough history, you must also document pertinent negatives. **Pertinent negatives** are negative findings that warrant no care or intervention. They also indicate that a thorough and complete examination and history were performed. Pertinent negatives will vary with each patient interaction. For example, you are assessing a patient with a history of asthma. The patient's chief complaint is dizziness or vertigo but in asking the patient about any breathing problems, the patient denies any shortness of breath or respiratory symptoms. Another example would

be during the assessment of a patient who is complaining of chest pain. The patient states the pain is stabbing or crushing, a 9 on a scale of 1 to 10, but denies radiation of the pain to the arm and jaw.

Taking History on Sensitive Topics
Alcohol and Drugs

The signs and symptoms a patient may present with while under the influence of alcohol or drugs may be confusing, hidden, or disguised. Many patients who abuse alcohol and/or drugs may deny having any problems. Families, friends, and coworkers may be unaware that a patient has any drug or alcohol troubles because patients often hide their dependency from these same people. The reasons patients deny using alcohol or drugs can vary greatly. It may be out of fear of losing their employment or driver's license, worry about what friends may think about them, and embarrassment or insecurity about their dependency.

The history that you gather from a chemically dependent patient may be unreliable **Figure 8-23**. If patients are not telling the people closest to them that they have a problem, you, as an outsider, may have even less success in obtaining information about a patient's current dependency. The signs and symptoms of alcohol or drug use may be masked by the patient's presentation. Use all of your senses when dealing with patient care.

Establish a strong rapport with your patients. Do not judge a patient who may have a chemical dependency, and be professional in your approach. Be honest, open, and, foremost, impress on the patient that information received will be kept in confidence. Then and only then, a patient may open up to you and provide information that can be valuable in the assessment and their treatment.

Figure 8-23 Many vehicle crashes involve alcohol. In these cases, the patient history may not be reliable.

Physical Abuse or Violence

All cases of suspected physical abuse or domestic violence must be reported to the appropriate authorities. Follow your local protocols when dealing with such cases. If you suspect a patient is a victim of physical abuse or domestic violence, do not accuse any person of being responsible for the situation. Instead, immediately involve law enforcement.

Because abuse and physical violence are very sensitive situations, look for hidden clues that such a situation exists. Information gathered at the scene, during the assessment process, and while transporting a patient may indicate violence or abuse.

What should you look for? When gathering a history, determine if the information provided by the patient and others present at the scene is inconsistent. Do you observe multiple injuries in various stages of healing? Are some bruises red, black, brown, or even green? In some cases, a victim of abuse or violence will not tell you what happened because of fear of further violence when EMS is not present. Victims may not answer your questions because the physical aggressor is still present and is answering questions for the patient. In these cases, separate the people present and interview both parties about the situation.

In cases of domestic violence, involvement can be extremely dangerous. If you determine that the emergency response is part of a domestic abuse situation, call law enforcement personnel immediately **Figure 8-24**.

When involved with cases of physical abuse, be very observant and open-minded, have a high index of suspicion, and be nonjudgmental. Documentation will be very important in cases of abuse and domestic violence. Your documentation should be an objective report of the facts. Avoid subjective, judgmental statements and

Figure 8-24 Do not handle potentially violent calls alone. Summon law enforcement personnel.

You are the Provider: PART 4

As you are packaging the patient and preparing to move her from her apartment, the paramedic unit is dispatched to another call. There are no other paramedic units in your district. You move the patient from her apartment and load her into the ambulance. With an EMT from the engine company assisting you in the back with the patient, you depart the scene and reassess the patient. The closest appropriate hospital is 13 minutes away.

Recording Time: 12 Minutes	
Level of consciousness	Unconscious and unresponsive
Respirations	6 breaths/min (baseline); ventilations are being assisted
Pulse	44 beats/min; weak and regular
Skin	Pale and cool; cyanosis noted around mouth
Blood pressure	78/58 mm Hg
Sao$_2$	88% (with assisted ventilation)

7. How has your patient's condition changed from the previous assessments?

8. What should you do in response to the patient's condition change?

include any pertinent statements made by the patient or others present using quotation marks. Remember, these prehospital situations will most likely involve some type of legal process later on. You may be summoned at a later date to provide testimony regarding what may have happened, which makes accurate and thorough documentation very important.

Sexual History

Obtaining information about a patient's sexual history may be limited because a number of factors may influence the details a patient may reveal. Religious beliefs, cultural stereotypes, and society's expectations may have a major role in patients not revealing a very personal side of their life, including practices considered by some people to be bizarre or exotic. In additional, some patients find sharing information regarding their sexual history with others very uncomfortable.

When would information about a patient's sexual history become important? As an EMT, you will be involved in the care of female patients reporting lower abdominal pain. You should consider all females of childbearing years who are reporting lower abdominal pain to be pregnant unless ruled out by history or other information. There are a number of questions to ask when faced with this prehospital scenario:

- When was your last menstrual period?
- Are your periods normal (Is there any vaginal discharge or bleeding not associated with a menstrual period)?
- Do you have urinary frequency or burning?
- What is the severity of cramping, and are there any foul odors?
- Is there a possibility you may be pregnant?
- Are you taking birth control pills?
- How many sexual partners do you have?

When dealing with a male patient, you must inquire about urinary symptoms:

- Is there pain associated with urination?
- Do you have any discharge, sores, or an increase in urination?
- Do you have burning or difficulty voiding?
- Has there been any trauma?
- Have you had recent sexual encounters?

In obtaining a history in all patients, ask about the potential for sexually transmitted diseases. The gathering of this information may be difficult and uncomfortable for the patient. Never be judgmental once this information is gathered. All patients should be and expect to be treated with compassion and respect. All information gathered from a patient for the purpose of determining a treatment plan is strictly confidential and should not be shared with others unless necessary in the process of treating a patient's medical or traumatic condition.

Special Challenges in Obtaining Patient History

Dealing with patient care, you will be faced with a number of challenges, many of them new and difficult. Each and every patient interaction should be viewed as a new experience and handled as an educational opportunity as well. As you participate in each patient encounter, remember that every patient interaction is a new experience.

Silence

Dealing with patients who say very little or say nothing at all can be difficult and frustrating. Patience is extremely important when dealing with patients and their emergency crises. Patients may be thinking about how to answer you, getting the facts straight, or assessing your crew to determine if they feel comfortable answering you. Using a close-ended question that requires a simple yes or no answer may work best. Consider whether the silence is a clue to the patient's chief complaint.

Always look for visual signs in the patient's environment that may indicate why a patient is not communicating. In addition, look for nonverbal clues, including facial expressions that may show pain or fear. Is the patient distressed or intimidated by your presence? How is the patient sitting or standing? Is there a communication problem? Is there a language problem? The number of reasons a patient may be silent during the prehospital encounter is endless. A good EMT will continue to assess the situation and determine a way to communicate with the patient.

Overly Talkative

On the other end of the spectrum is the patient who is extremely talkative. Some people just talk a lot, and gathering details about their medical condition may be difficult if they talk around your question or you have a difficult time refocusing the patient's conversation. Some possible causes as to why a patient may be overly talkative could include excessive caffeine consumption, nervousness, and ingestion of cocaine, crack, or methamphetamines.

Once you have allowed a talkative patient a chance to express himself or herself, you must keep the patient focused on the questions presented. Have the patient stick to the facts, and clarify statements for the purpose of making sure the information you are gathering is correct. Remember that there is no such thing as too much information.

Multiple Symptoms

The geriatric age group is a part of the population in which you can expect your patient to have multiple symptoms during a single patient encounter. Prioritize the patient's complaints as you would in triage; start with the most serious and end with the least serious. Always ask for additional information to determine why EMS was called.

Keep an open mind, and do not focus on one complaint or detail to determine a treatment plan. Always remember there may be a number of possible medical or traumatic causes for a patient's chief complaint.

Anxiety

When a person is involved in an emergency situation, it is natural for that person to appear excited or anxious. Many people have not been faced with a true emergency during their lifetime, and their reactions may be unpredictable. EMTs are trained to handle stressful situations. Your patient or bystander may be nervous, pacing, vocal, panicked, or, in some extreme cases, experiencing complete hysteria. It is your responsibility to deal not only with the emergency crisis at hand, but also with the people present who are having difficulties coping with the situation. Frequently, anxious patients can be observed in emergency scenes that involve a large number of patients, such as during a disaster. Anxiety also can be observed or encountered during a routine EMS call when family members or patients cannot cope.

EMTs can expect anxious patients to exhibit signs of psychological shock, such as pallor, diaphoresis, shortness of breath, numbness in the hands and feet, dizziness or light-headedness, and even loss of consciousness. Some anxious patients may have no real medical complaint but may be hiding or concealing something, such as trying to keep a family member, friend, or employer from discovering their dependency on alcohol or drugs. Or, the patient may have been involved in a physical abuse or domestic situation that he or she wants to keep quiet. In any situation involving an anxious patient, you must be aware of verbal and nonverbal clues. Is the patient making sense during a verbal conversation? Can the patient be calmed down, or is there a possibility the patient may need to be restrained?

During a crisis situation, reassure the patient that any nervous or anxious response is normal and can be overcome. It may be possible for you to control an anxious patient by simply smiling or using a delicate touch. Be confident in your approach, and have a positive demeanor. In many patient care interactions, your presence may be all that is required to calm the patient.

As in every response, safety is a paramount concern. Be aware that emergency responses involving anxious and possibly hysterical patients can turn violent. A confident but cautious EMT can prevent a bad situation from getting worse and professionally calm and control anxious patients, friends, and family members.

Anger and Hostility

Every patient encounter has the potential for violence and hostility, from a situation involving a 9-year-old boy who was hit by a vehicle to a 90-year-old grandmother experiencing chest pain. These emergency calls have a high potential for unexpected violence because friends, family, or bystanders may direct their anger and rage toward you. Do not to take this anger and frustration personally. More important, do not become angry yourself because "anger feeds anger."

In handling potentially violent situations, remain calm, reassuring, and gentle. Always be observant. Be aware of nonverbal clues, such as posture, position, and facial expressions. Look at the patient, and be aware of how the patient is positioned. Is the patient stiff, with hands clenched and feet wide apart? Is his or her body weight all on one leg? This may indicate the patient has assumed a position to allow him or her to kick.

It is not unusual for a patient, family member, or friend to vent hostility toward EMS responders. If the scene is not safe or secured, get it secured. Never let a potentially violent or hostile patient leave the room alone. Understand that everything in reach of a patient has the potential to be used as a weapon.

Intoxication

The number of EMS calls dealing with an intoxicated patient has increased over the years. When you attempt to gather a history for an intoxicated patient, be aware the information may not only be difficult to get, but could also be unreliable. An intoxicated patient may become very impatient with you when he or she is trying to provide you with information. As the patient's impatience increases, so does his or her anger level. Do not put an intoxicated patient in a position where he or she feels threatened and has no way out. As in other emergency cases, the potential for violence and a physical confrontation is high when a patient is intoxicated.

During the assessment and treatment of a patient who has consumed alcohol, be accepting, diplomatic, objective, and nonjudgmental. Because of the intoxication, the patient may not be telling you everything about how he or she feels. Alcohol dulls a patient's senses, which will make it difficult for an intoxicated patient to

History Taking

inform you that something feels painful. Treat the patient with dignity and respect despite the intoxication. You must never assume the patient's condition is the result of alcohol consumption when there may be an underlying medical or traumatic cause for the patient's presentation.

Crying

A crying patient is a breathing patient. But why do some patients cry? It could be a joyous expression or one of depression. A patient who cries may be sad, in pain, or emotionally overwhelmed. No matter the reason for crying, you need to remain calm and be patient, reassuring, and confident, and maintain a soft voice.

Your presence may make a crying patient feel more secure. In some extreme cases, additional diplomacy and verbal intervention will help the patient. No matter how you address a crying patient, as with all patients, be sympathetic and treat them with respect and dignity.

Depression

Depression is a common reason patients call EMS. In fact, according to the World Health Organization, depression is among the leading causes of disability worldwide. Some of the symptoms a patient will present with when depressed include sadness, a feeling of hopelessness, restlessness, and irritability. The patient may also have sleeping and eating disorders and a decreased energy level. Depression is a normal human response, but it can lead to harmful behavior. In the treatment of depression, be nonjudgmental and compassionate toward the patient's feelings. The most effective treatment in handling a patient's depression is being a good listener. Oftentimes, the patient just needs someone to talk to and someone to listen.

Confusing Behavior or History

Patients sometimes provide more or additional history information to hospital personnel because they are embarrassed or frightened about telling the EMTs and they may feel more comfortable talking with hospital staff. Whatever the situation may be, there are medical causes that you must be aware of that can cause a patient to report a confusing history. Conditions such as hypoxia, stroke, diabetes, trauma, medications, and other drugs could alter a patient's explanation of events. One of the most common causes of confusion is hypoxia. In geriatric patients, it is not uncommon to encounter a patient who has dementia, delirium, or Alzheimer disease. It is important to verify the normal mental status of each patient. Do not assume that because a patient is elderly that he or she has one of these conditions.

Confusing behavior is not a normal response. After you have properly assessed and treated any life threats, attempt to ask the patient again about the chief complaint or ask someone close to the patient, such as family members or friends, to provide additional details.

Limited Cognitive Abilities

Patients who have limited cognitive abilities are considered developmentally handicapped. These handicaps can range from those that are barely recognizable to those that are very severe. You should develop a habitual method for dealing with a patient who has limited cognitive abilities. First, assume you can get an adequate history. Keep your questions simple, and limit the use of medical terms. Be alert for partial answers, and keep asking questions. In cases of patients with severely limited cognitive function, rely on the presence of family, caregivers, and friends to supply answers to your questions.

Language Barriers

We live in a country that is a melting pot of people with diverse nationalities. Not everyone speaks the native language. You will encounter patients who do not speak English. For example, you respond to a call for an elderly woman who fell at a nursing home. The emergency response seems pretty straightforward until you discover that the patient is in the Alzheimer unit of the nursing home. When you ask the patient what happened, she answers in French. How will you ask the patient to describe what happened and what hurts?

The best answer is to find an interpreter, but it is not always that simple. First, determine whether the patient speaks or understands any English by asking the patient or others who may be present. Start by introducing yourself by using your name. Determine whether the patient understands who you are. If the patient is able to respond by giving you his or her name, the patient has the ability to understand some English, but more important, the patient has cognitive ability, which is the ability to understand. Remember that increasing the volume of your questions will not increase the patient's understanding of what you are asking him or her. Keep questions straightforward and brief. Simple is best in these patient situations. Use of hand gestures may be helpful.

Be aware of the language diversity in your community. Most hospitals have set up programs within the institution that identify various employees who can speak different languages. Provide the hospital with advanced notice that a non–English-speaking patient will be arriving. This will allow the hospital the opportunity to make arrangements for an interpreter.

Hearing Problems

Hearing disabilities in patients range from very slight to total deafness. Hearing problems can make the process of obtaining an in-depth history difficult. When you are treating a patient who has a hearing loss, ask questions slowly and clearly. You may want to use a stethoscope to function like a hearing aid; have the patient place the stethoscope in his or her ears and speak into the stethoscope bell, which will amplify the sound. Changing the pitch of your voice may also help the patient to hear you.

Oftentimes, a patient who has had a hearing disability for some time will have mastered the technique of reading lips. If the patient has a hearing aid, ask the patient to use it. Speak slowly and face-to-face with the patient. Some deaf patients will attempt to use sign language for communication, which can be difficult for others to understand; therefore, attempting to learn simple sign language during your career will help in the communication process. Probably the simplest way to communicate with a patient who has a hearing deficit is to use a pencil and paper. Write uncomplicated questions that require simple yes or no answers. If the patient cannot see clearly and has glasses, ask the patient to put them on.

Visual Impairments

When you enter the home of a visually impaired patient who has called for help, identify yourself verbally when entering. By announcing yourself when entering a residence, you are letting a patient know that help has arrived and any response from the patient may help you locate the patient's whereabouts. It is also a safe thing to do so patients are not surprised when a stranger appears in their home.

During the assessment and subsequent treatment of a patient who is visually impaired, it is important that you put any items that have been moved back into their previous position. Many visually impaired patients can move freely about their homes because they know exactly where everything is placed. If you move something, put it back.

During the assessment and history taking process, explain to the patient what is happening. Explain that you will be checking vital signs by feeling for the pulse, listening to breath sounds, and applying a blood pressure cuff to the patient's arm. Remember, you are a stranger to the patient, and an EMS vehicle is a foreign environment. A little communication can go a long way in easing the uncertainty in a visually impaired patient. If the patient is not able to provide you with all of the necessary information, try to find someone else who can.

History Taking

Patient Assessment

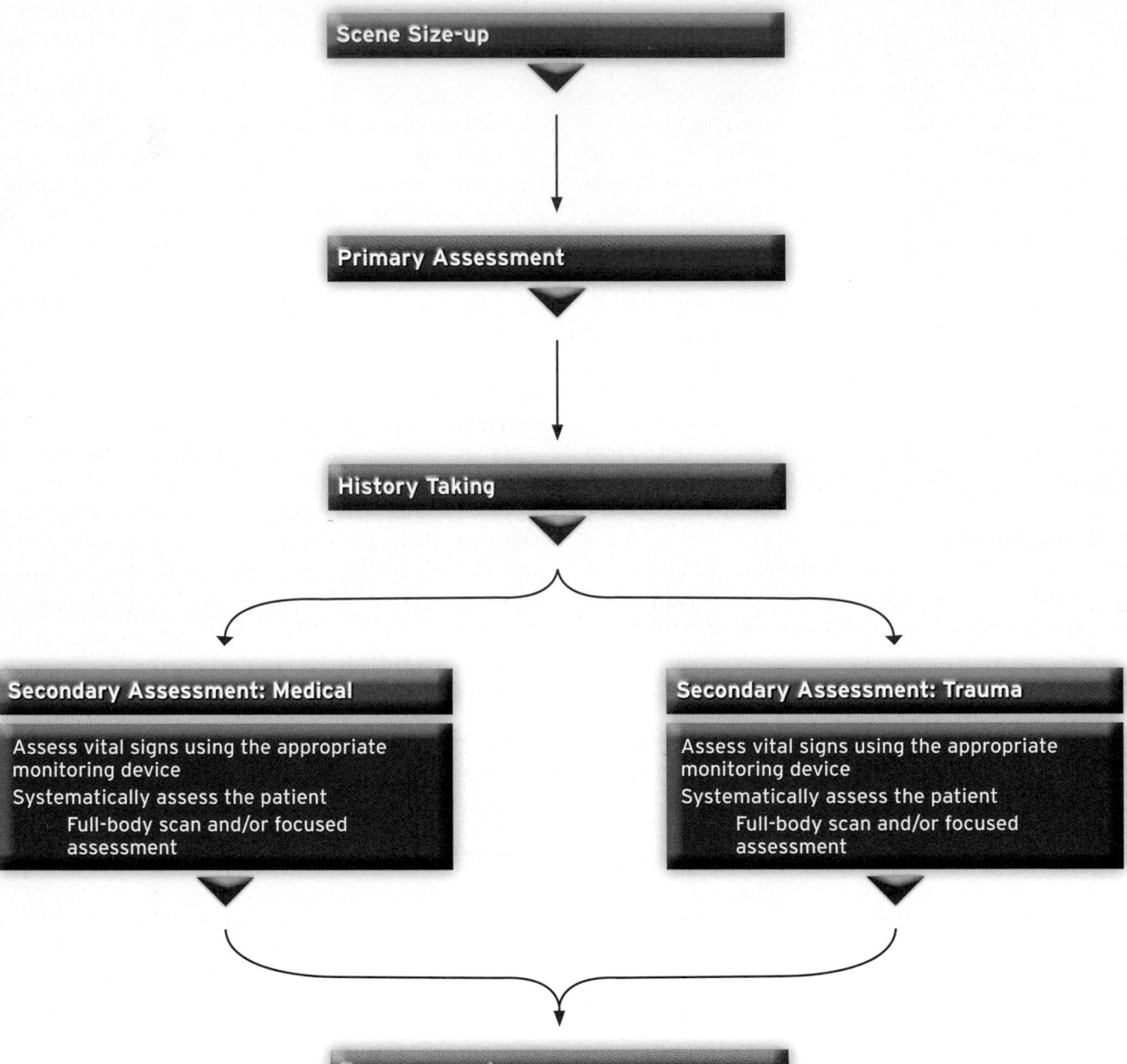

Scene Size-up

Primary Assessment

History Taking

Secondary Assessment: Medical

Assess vital signs using the appropriate monitoring device
Systematically assess the patient
    Full-body scan and/or focused assessment

Secondary Assessment: Trauma

Assess vital signs using the appropriate monitoring device
Systematically assess the patient
    Full-body scan and/or focused assessment

Reassessment

Secondary Assessment

Secondary Assessment

If the secondary assessment is not performed at the scene, it is performed in the back of the ambulance en route to the hospital. However, there will be situations where you may not have time to perform a secondary assessment at all if you have to continually manage life threats that were identified during the primary assessment. If the patient is in stable condition and has an isolated complaint, the secondary assessment may occur at the scene.

The purpose of the **secondary assessment** is to perform a systematic physical examination of the patient. The physical examination may be a systematic head-to-toe, full-body scan or a systematic assessment that focuses on a certain area or region of the body, often determined through the chief complaint. Circumstances will dictate which aspects of the physical examination will be used.

Words of Wisdom

Patients may feel vulnerable and exposed during a physical examination. Display compassion during this difficult time. It is also important to maintain the patient's modesty and body temperature.

As discussed previously, the physical examination you perform is based on the needs of your patient. The following are guidelines on how and what to assess during a physical examination:

- **Inspection.** Inspection is simply looking at your patient for abnormalities. This is done by looking for anything that may indicate a problem. For example, swelling in a lower extremity may indicate an acute injury or a chronic illness.
- **Palpation.** Palpation describes the process of touching or feeling the patient for abnormalities. At times palpation is gentle, and at other times it is firmer and will help you to identify where the patient has pain. Your fingertips are best suited for detecting texture and consistency, while the back of your hand is best suited for noting temperature.
- **Auscultation.** Auscultation is the process of listening to sounds the body makes by using a stethoscope. For example, when measuring a patient's blood pressure, you listen to the flow of blood against the brachial artery with the head of the stethoscope. This is auscultation of a blood pressure.

The mnemonic **DCAP-BTLS** will help remind you what to look for when inspecting and palpating various body regions. Each area of the body is evaluated for the following:

- Deformities
- Contusions
- Abrasions
- Punctures/penetrations
- Burns
- Tenderness
- Lacerations
- Swelling

An integral part of your physical examination is to compare findings on one side of the body with the other side when possible. For example, if your patient complains of a grating or grinding sensation in his or her arm or you note air bubbles under the skin that produce a crackling sound, check the other arm before determining that the sensation or noise is caused by fractured bone ends or joints rubbing together (**crepitus**). If one ankle appears swollen, look at the other. If one shoulder feels "out of joint," feel the other one to compare. When listening to breath sounds, listen to both sides of the chest. On some occasions, it may be helpful to use your nose during an examination. Odors can indicate anything from infections, to certain medical conditions, to scene safety threats.

Assess Vital Signs Using the Appropriate Monitoring Device

The use of monitoring equipment in the prehospital setting has continued to expand. EMTs at all levels use a wide variety of devices in the continuous monitoring of patients. It is important to remember that these devices are manufactured and subject to limitations and failures. These devices should never be used to replace your comprehensive assessment of your patient; think of these devices as simply adjuncts to the assessment and treatment of your patient. Obtaining and using information from patient monitoring devices includes, but is not limited, to data from pulse oximetry and noninvasive blood pressure monitoring.

Pulse Oximetry

Pulse oximetry is a newer assessment tool that is used to evaluate the effectiveness of oxygenation. The pulse oximeter is a photoelectric device that monitors the oxygen saturation of hemoglobin (the iron-containing portion of the red blood cell to which oxygen attaches) in the capillary beds **Figure 8-25**. The parts that make up the pulse oximeter include a monitor and a sensing probe. The sensing probe clips onto a finger or ear lobe. The light source must have unobstructed access to a capillary bed, so fingernail polish should be removed. Results appear as a percentage on the display screen. Normally, pulse oximetry values in ambient air will vary depending on the altitude, with the majority of values falling between 95% and 100%.

The goal of any oxygen therapy is to increase oxygen saturation to a normal level. This device is a useful assessment tool to determine the effectiveness of oxygen therapy, bronchodilator therapy, and use of the bag-mask device in certain conditions. However, the pulse oximeter does not take the place of good assessment skills and should not prevent the application of oxygen to any patient who reports difficulty breathing regardless of the pulse oximetry value seen on the monitor.

Because the device presumes adequate perfusion and numbers of red blood cells, any situation that causes vasoconstriction or loss of red blood cells (such as bleeding or anemia) will result in inaccurate or misleading values. The device also presumes that oxygen is saturating the hemoglobin. Therefore, any chemical that displaces oxygen (such as carbon monoxide) will also cause misleading values.

The pulse oximeter is a useful tool as long as you remember that the device is only a tool, not a substitute for a good assessment. The device should not be used when hypoperfusion or known anemia is present, carbon monoxide or exposure to other toxic inhalants has occurred, or the patient's extremities are cold.

Noninvasive Blood Pressure Measurement

The sphygmomanometer, or blood pressure cuff, is used in the measurement of the patient's blood pressure. This device consists of an inflatable cuff that occludes blood flow and a manometer (pressure meter) that is used to determine the pressure in the artery at various points in the physical examination. These two components are connected via tubing. In manual cuffs, a separate tube connects to an inflation bulb. The auscultatory method (listening) is the most common means of measuring a patient's blood pressure using a sphygmomanometer.

Oscillometric measurement, or electronic measurement, is another method of obtaining blood pressure readings on patients. An electronic device measures changes in pressure oscillations that occur during cuff deflation and are related to systolic, mean, and diastolic pressures. Two different types of electronic devices are used in the prehospital setting; the blood pressure cuff deflates differently in each device. The first device measures readings using linear deflation,

and the second type measures readings by using stepped deflation. An electronic blood pressure cuff that uses linear deflation allows a uniform decline in pressure in the cuff during deflation. Conversely, stepped deflation allows the cuff to deflate in small steps or intervals. Although both devices are accurate in the prehospital setting, stepped deflation tends to be more accurate in patients who are moving and in patients who may be hypotensive. Stepped deflation can release the pressure in the cuff in intervals at variable lengths, allowing the system to better detect oscillations.

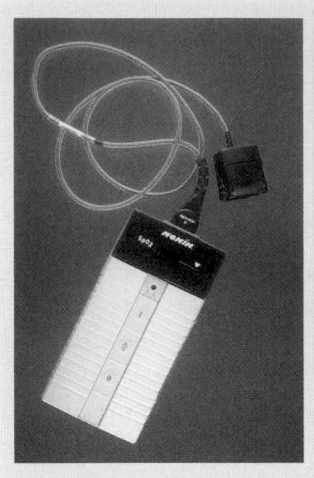

Figure 8-25 The pulse oximeter is a device that measures the saturation of oxygen in the blood as a percentage.

End-Tidal Carbon Dioxide

Pulse oximetry cannot measure the amount of oxygen being consumed by a patient's cells during cellular metabolism. Metabolism refers to the chemical reactions that occur in the body or cells to maintain life. To determine oxygen consumption, you will need to measure carbon dioxide (CO_2) levels. **Carbon dioxide** is the by-product of aerobic cellular metabolism and reflects the amount of oxygen being consumed during the process. **Capnography** is a noninvasive method that can quickly and efficiently provide information on a patient's ventilatory status, circulation, and metabolism. It can be used as a secondary tool during the confirmation of endotracheal intubation or the effectiveness of ongoing CPR.

End-tidal CO_2 is the partial pressure or maximal concentration of CO_2 at the end of an exhaled breath, which is expressed as a percentage of CO_2, or millimeters of mercury. The normal range is 35 to 45 mm Hg, or 5% to 6% CO_2. When CO_2 is absent when measured by capnography, it may indicate the endotracheal tube is in the wrong position or there is an absence or decrease in the level of CO_2 in the lungs, possibly from cardiac arrest, ineffective CPR, or shock. When cardiac output increases, the end-tidal CO_2 measurement will provide information on the adequacy of ventilations and circulation.

End-tidal CO_2 is measured or detected by colorimetry, capnometry, and capnography devices. **Colorimetric devices** come in different shapes and sizes but provide continuous end-tidal monitoring by displaying one of three colors. Purple indicates a CO_2 level of less than 0.5%, tan indicates a range of 0.5% to 2%, and yellow indicates a level of greater than 2%. A reading of yellow indicates adequate circulation. Remember to check these devices regularly for damage, cracks, and blockages caused by gastric secretions because this will affect the accuracy of the readings.

Capnometry and capnography provide a digital reading and waveform of end-tidal CO_2. The digital display of end-tidal CO_2 is expressed in millimeters of mercury or as a percentage of exhaled gas. Normal values should be in the range of 35 to 45 mm Hg. These devices are typically used in the prehospital setting as a secondary means to determine endotracheal placement, to maximize a patient's ventilatory status, and to avoid inadvertent hyperventilation of head-injured patients, which has been linked to poor outcomes.

Systematically Assess the Patient— Full-Body Scan

The full-body scan is a systematic head-to-toe examination. The goal of this process is to identify hidden injuries or identify causes that may not have been found during the 60- to 90-second rapid scan that took place during the primary assessment. Any patient who has sustained a significant MOI, is unconscious, or is in critical condition should receive this type of examination. An unconscious patient is unable to tell you what is wrong; therefore, this type of examination may give you clues to identify the problem.

To perform a full-body scan of a patient with no suspected spinal injuries, follow the steps in Skill Drill 8-2. To perform a full-body scan in which the patient has sustained significant trauma, be sure that spinal immobilization is still in place and follow the steps in Skill Drill 8-2.

Skill Drill 8-2

1. Look at the face for obvious lacerations, bruises, and deformities (Step 1).
2. Inspect the area around the eyes and eyelids (Step 2).
3. Examine the eyes for redness and for contact lenses. Assess the pupils using a penlight (Step 3).
4. Look behind the patient's ears to assess for bruising (Battle's sign) (Step 4).
5. Use the penlight to look for drainage of spinal fluid or blood in the ears (Step 5).
6. Look for bruising and lacerations about the head. Palpate for tenderness, depressions of the skull, and deformities (Step 6).
7. Palpate the zygomas for tenderness or instability (Step 7).
8. Palpate the maxillae (Step 8).
9. Check the nose for blood and drainage (Step 9).
10. Palpate the mandible (Step 10).
11. Assess the mouth and nose for cyanosis, foreign bodies (including loose teeth or dentures), bleeding, lacerations, and deformities (Step 11).
12. Check for unusual odors on the patient's breath (Step 12).
13. Look at the neck for obvious lacerations, bruises, and deformities. Observe for jugular vein distention (Step 13).
14. Palpate the front and the back of the neck for tenderness and deformity (Step 14).
15. Look at the chest for obvious signs of injury before you begin palpation. Be sure to watch for movement of the chest with respirations (Step 15).
16. Gently palpate over the ribs to elicit tenderness. Avoid pressing over obvious bruises and fractures (Step 16).
17. Listen for breath sounds over the midaxillary and midclavicular lines (Step 17).
18. Listen also at the bases and apices of the lungs (Step 18).
19. Look at the abdomen and pelvis for obvious lacerations, bruises, and deformities. Gently palpate the abdomen for tenderness. If the abdomen is unusually tense, you should describe the abdomen as rigid (Step 19).
20. Gently compress the pelvis from the sides to assess for tenderness (Step 20).
21. Gently press the iliac crests to elicit instability, tenderness, and/or crepitus (Step 21).
22. Inspect all four extremities for lacerations, bruises, swelling, deformities, and medical alert anklets or bracelets. Also assess distal pulses and motor and sensory function in all extremities (Step 22).
23. Assess the back for tenderness and deformities. Remember, if you suspect a spinal cord injury, use spinal precautions as you log roll the patient (Step 23).

Systematically Assess the Patient— Focused Assessment

A focused assessment is generally performed on patients who have sustained nonsignificant MOIs or on responsive medical patients. This type of examination is based on the chief complaint. For example, in a person reporting a headache, you should carefully and systematically assess the head and/or the neurologic system. A person with a laceration to the arm may need to only have that arm evaluated. The goal of a focussed assessment is to focus your attention on the immediate problem.

Skill Drill   8-2

Performing the Full-Body Scan

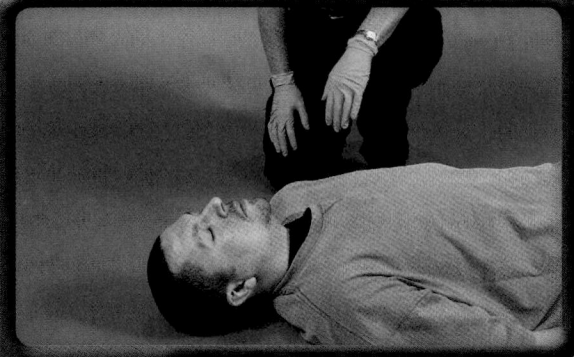

Step 1   Observe the face.

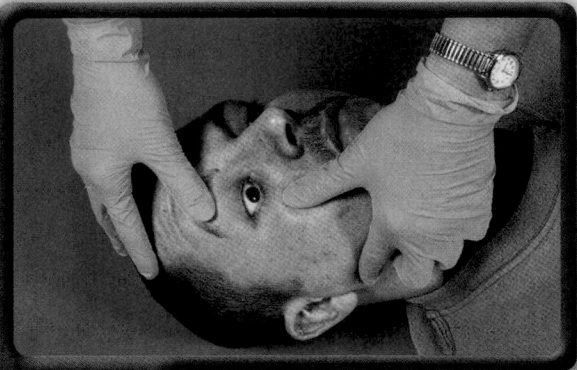

Step 2   Inspect the area around the eyes and eyelids.

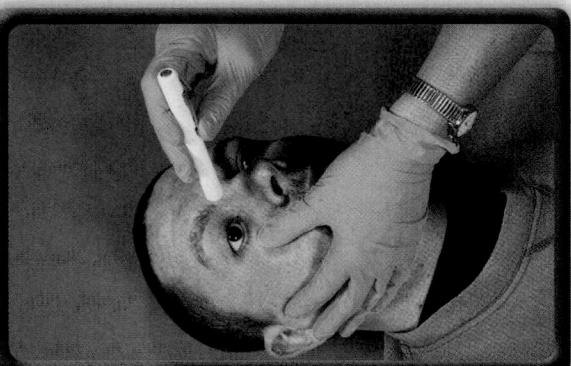

Step 3   Examine the eyes for redness and contact lenses. Check pupil function.

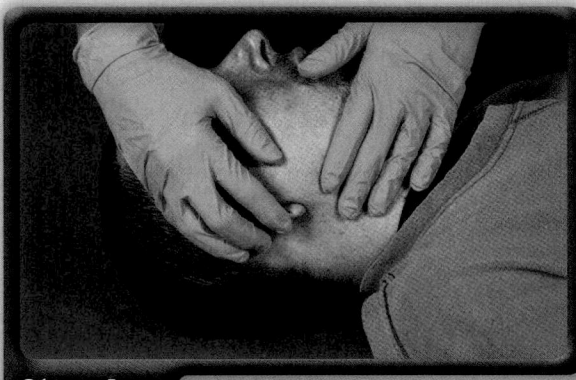

Step 4   Look behind the ears for Battle's sign.

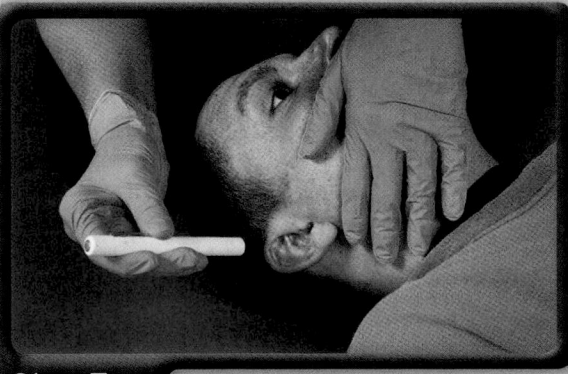

Step 5   Check the ears for drainage or blood.

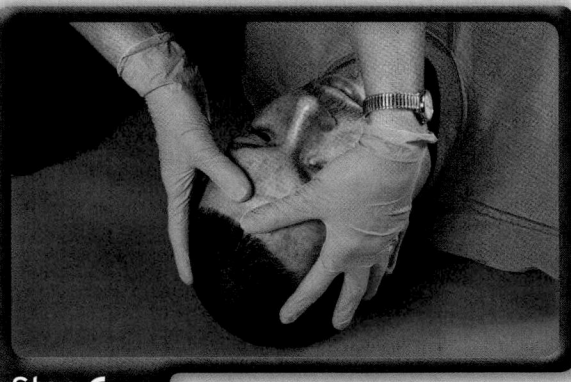

Step 6   Observe and palpate the head.

Skill Drill 8-2

Performing the Full-Body Scan, continued

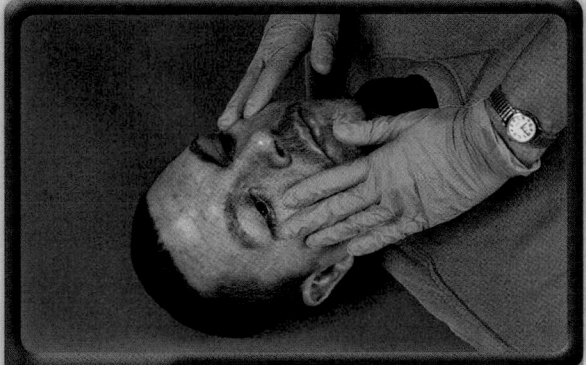

Step 7 Palpate the zygomas.

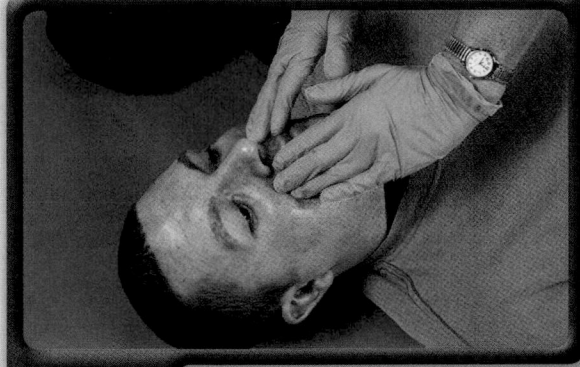

Step 8 Palpate the maxillae.

Step 9 Check the nose for blood and drainage.

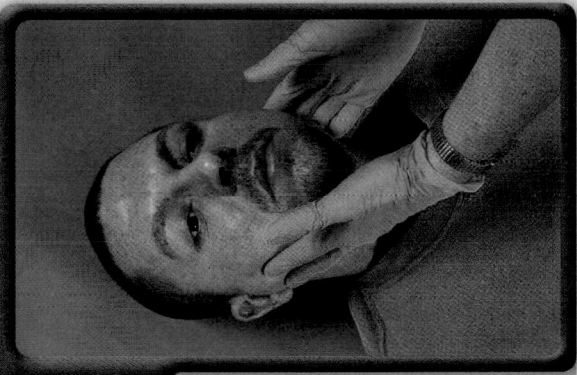

Step 10 Palpate the mandible.

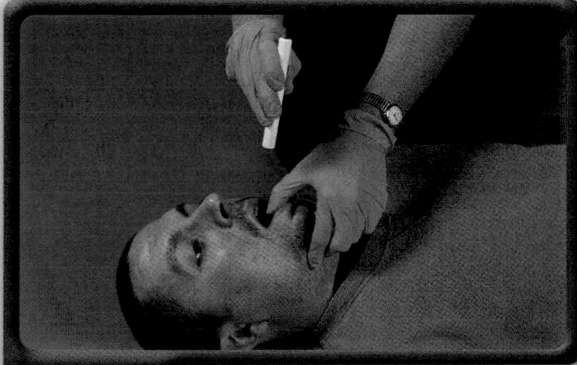

Step 11 Assess the mouth and nose.

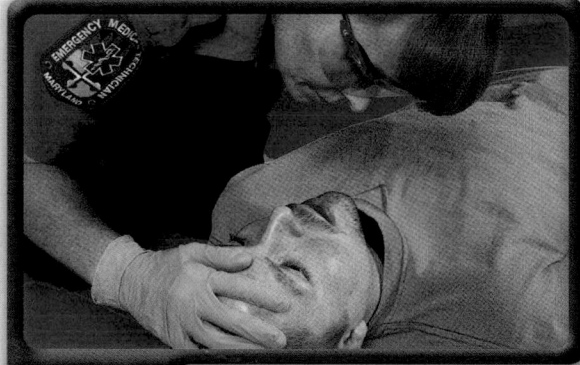

Step 12 Check for unusual breath odors.

Secondary Assessment

Skill Drill 8-2

Performing the Full-Body Scan, continued

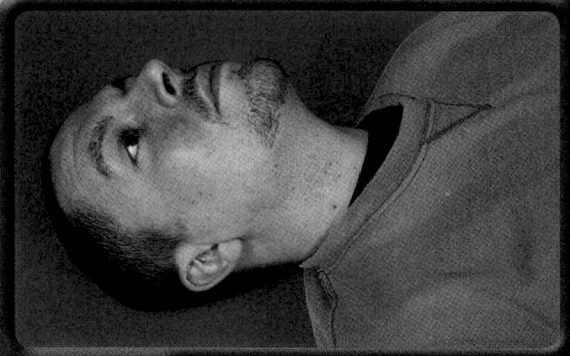

Step 13 Inspect the neck. Observe for jugular vein distention.

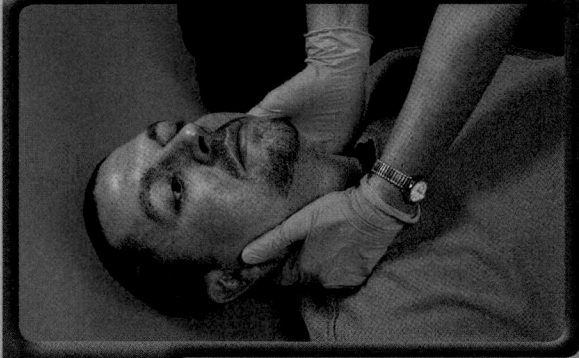

Step 14 Palpate the front and back of the neck.

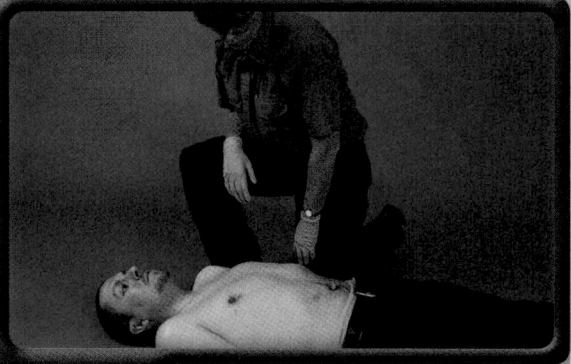

Step 15 Inspect the chest, and observe breathing motion.

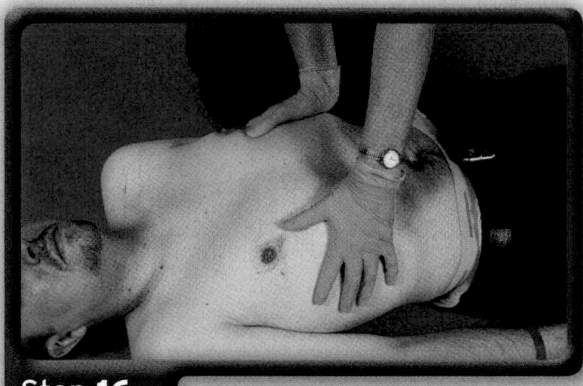

Step 16 Gently palpate over the ribs.

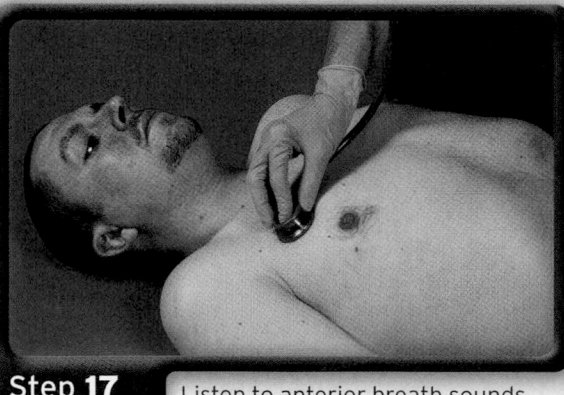

Step 17 Listen to anterior breath sounds (midaxillary, midclavicular).

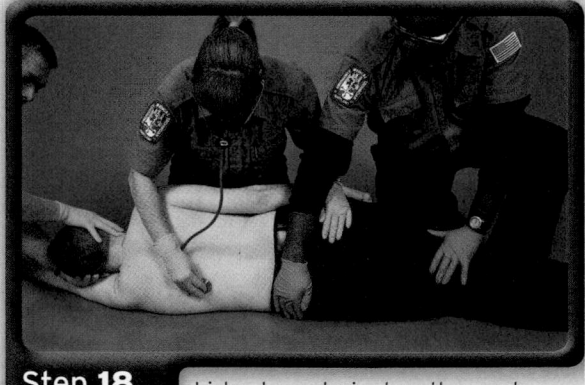

Step 18 Listen to posterior breath sounds (bases, apices).

Secondary Assessment

Skill Drill 8-2

Performing the Full-Body Scan, continued

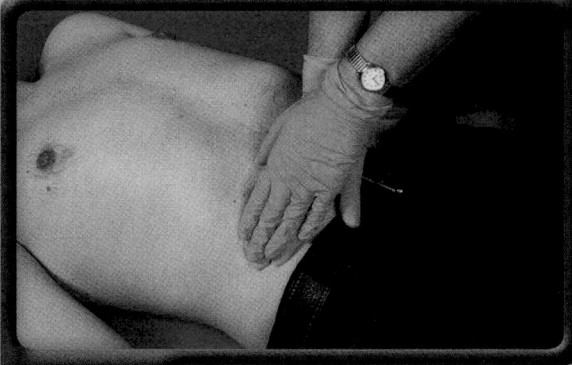

Step 19 Observe and then palpate the abdomen and pelvis.

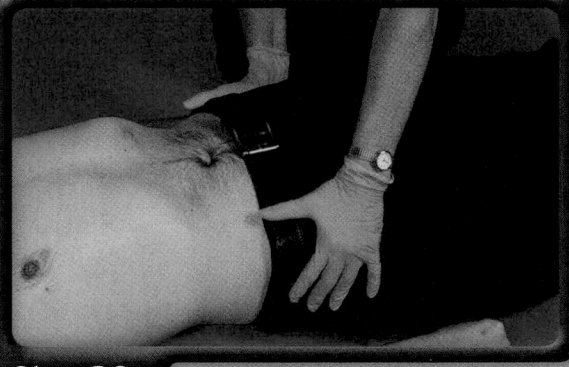

Step 20 Gently compress the pelvis from the sides.

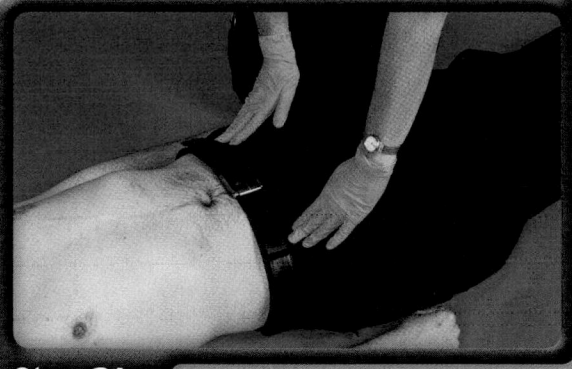

Step 21 Gently press the iliac crests.

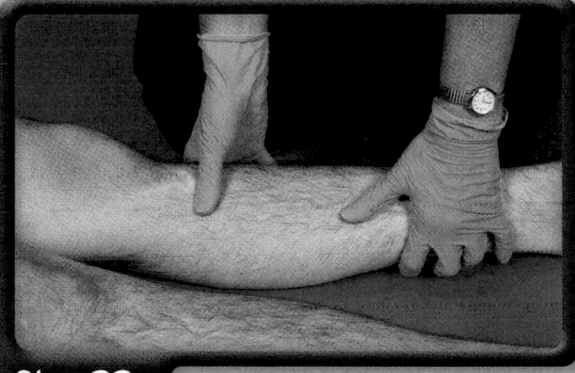

Step 22 Inspect the extremities; assess distal circulation and motor and sensory function.

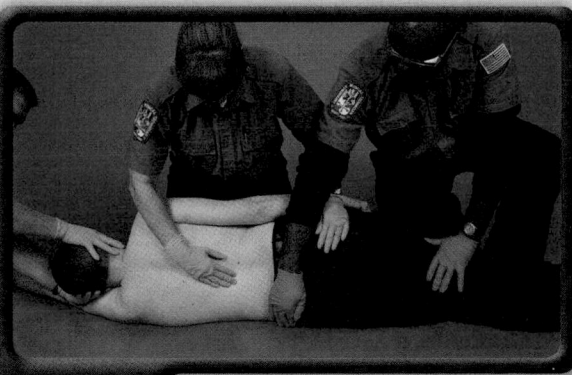

Step 23 Log roll the patient, and inspect the back.

Secondary Assessment

Secondary Assessment

Respiratory System

When the patient's chief complaint is focused on the respiratory system, you should have identified and managed life threats during the primary assessment. During the secondary assessment, you will perform an examination directed at obtaining clues that may indicate the cause of the respiratory symptoms. When you use examination techniques that focus specifically on the respiratory system, you may be able to determine which treatment to perform and protocols to follow.

Expose the patient's chest. Look again for signs of airway obstruction, as well as trauma to the neck and/or chest. Inspect the chest for overall symmetry. Does the right side of the chest look like the left side? Listen carefully to breath sounds, noting abnormalities. Measure the respiratory rate, chest rise and fall (for tidal volume), and effort. Look for retractions. Is the patient using accessory muscles to help with breathing, and is there increased work of breathing?

Because the location of this complaint is the chest, carefully reevaluate the pulse rate and skin and blood pressure (described in the next section). Inspect and palpate from the clavicles to the shoulder to the abdomen, and reassess breath sounds. Note any abnormalities found, and document those findings on the patient care report. With this information, you can develop a treatment plan and prioritize transport procedures.

Cardiovascular System

When the patient's chief complaint is associated with chest pain, a physical examination should include looking, listening, and feeling for abnormalities in the patient's thoracic region. Look for trauma to the chest, and listen for breath sounds. Reevaluate the pulse and respiratory rate and the blood pressure. Pay particular attention to rate, quality, and rhythm. Reevaluate the skin. Assessment of these functions will allow you to determine how well the cardiovascular and respiratory systems are functioning. Check and compare distal pulses to determine any right and left side differences. Consider auscultation for abnormal heart sounds; however, keep in mind that obtaining these sounds may be difficult in a noisy prehospital setting. Always remember that a patient's chief complaint may have a medical cause or could be the result of trauma.

Blood Pressure

Adequate blood pressure is necessary to maintain proper circulation and perfusion of the vital organ cells. **Blood pressure** is the pressure of circulating blood against the walls of the arteries. A decrease in the blood pressure may indicate one of the following:

- Loss of blood or its fluid components
- Loss of vascular tone and sufficient arterial constriction to maintain the necessary pressure even without any actual fluid or blood loss
- A cardiac pumping problem

When any of these conditions occurs and results in a drop in circulation, the body's compensatory mechanisms are activated, resulting in an increased heart rate and constriction of the arteries. Normal blood pressure is maintained, and by decreasing the blood flow to the skin and extremities, available blood volume is temporarily redirected to the vital organs so that they remain adequately perfused. However, as shock progresses, and the body's defense mechanisms can no longer keep up, the blood pressure will fall. Decreased blood pressure is a late sign of shock and indicates that the critical decompensated phase has begun. Any patient with a markedly low blood pressure has inadequate pressure to maintain proper perfusion of all of the vital organs and needs to have his or her blood pressure and perfusion restored immediately to a normal level.

When the blood pressure becomes elevated, the body's defenses act to reduce it. Some people have chronically high blood pressure from progressive narrowing of the arteries that occurs with age, and during an acute episode, their blood pressure may increase to even higher levels. Head injury or a number of other conditions may also cause blood pressure to rise to very high levels. Abnormally high blood pressure may result in a rupture or other critical damage in the arterial system.

Blood pressure contains two key separate components: systolic pressure and diastolic pressure. **Systolic pressure** is the increased pressure that is caused along the artery with each contraction (systole) of the ventricles and the pulse wave that it produces. **Diastolic pressure** is the residual pressure that remains in the arteries during the relaxing phase of the heart's cycle (diastole), when the left ventricle is at rest. Systolic pressure represents the maximum pressure to which the arteries are subjected, and the diastolic pressure represents the minimum amount of pressure that is always present in the arteries.

Early blood pressure gauges contained a column of mercury and a linear scale that was graduated in millimeters. Even though different gauges are used today, the blood pressure is still measured in millimeters of mercury (mm Hg). Blood pressure is reported as a fraction in the form of systolic pressure over diastolic pressure. Therefore, if the patient's systolic pressure is 120 and the diastolic pressure is 78, you would record it as "BP 120/78 mm Hg." You would report the patient's blood pressure verbally as "BP is 120 over 78."

You should avoid taking a blood pressure on an arm if the patient has an intravenous site, a central line catheter, or a port; has had a mastectomy on that side; or has an injury to that arm. You can ask the patient if any of these exist if they are not visible, for example, a mastectomy. If a patient has chronic renal failure and is undergoing dialysis, you may ask if the patient has a port.

A blood pressure cuff contains the following components **Figure 8-26** :

- A wide outer cuff designed to be fastened snugly around the entire arm or leg
- An inflatable wide bladder sewn into a portion of the cuff
- A ball-pump with a one-way valve that allows air to enter and a turn-valve that can be closed or, when opened, will allow air to be released at a controlled speed from the cuff
- A pressure gauge calibrated in millimeters of mercury, which indicates the pressure that exists in the cuff that is being applied against the underlying artery

Most agencies carry at least three sizes of blood pressure cuffs: adult, thigh, and pediatric **Figure 8-27** . You must be sure to select the appropriately sized cuff. A cuff that is too small may result in falsely high readings;

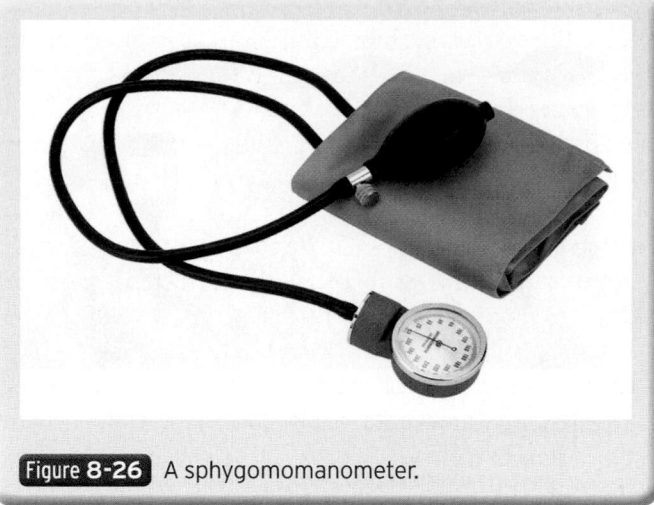

Figure 8-26 A sphygomomanometer.

a cuff that is too large may result in falsely low readings. The normal size cuff is designed to wrap around the arm 1 to 1.5 times and take up two thirds the length from the armpit to the crease in the elbow of most adults. Use a thigh cuff with patients who are obese or have exceptionally well-developed arm muscles or to take the

You are the Provider: PART 5

Following the appropriate interventions, your patient's oxygenation and ventilation status have improved; however, she is still bradycardic, hypotensive, and unresponsive. Because there was no one at the scene to provide information regarding her medical history, you continue to treat her based on her signs and symptoms, perform a reassessment, and call in your radio report to the receiving facility.

Recording Time: 17 Minutes	
Level of consciousness	Unconscious and unresponsive
Respirations	6 breaths/min (baseline); ventilations are being assisted
Pulse	40 beats/min; weak and regular
Skin	Pale and cool; cyanosis has resolved
Blood pressure	82/62 mm Hg
Sao$_2$	97% (with assisted ventilation)

You reassess the patient again just before arriving at the emergency department and note that her condition is unchanged. She is immediately evaluated by the staff physician, who determines that she has overdosed on numerous drugs—including narcotics. After further treatment in the emergency department, she was admitted to the intensive care unit for close observation.

9. What components of the SAMPLE history, if any, can you obtain when your patient is unresponsive? How would you obtain the information?

10. Why is reassessing your interventions so important?

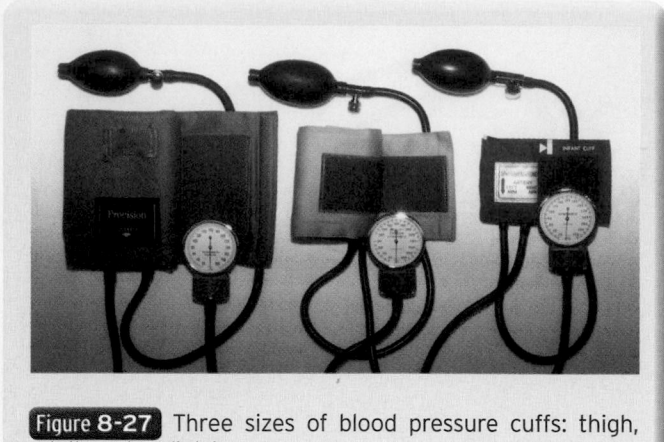

Figure 8-27 Three sizes of blood pressure cuffs: thigh, adult, and pediatric.

blood pressure of the thigh in patients who have injuries in both arms. Use a small pediatric cuff with children and exceptionally small adults. You should measure blood pressure in all patients older than 3 years.

The auscultatory method (listening) is the most common means of measuring a patient's blood pressure. A blood pressure cuff (sphygmomanometer) is applied to a patient's upper arm, allowing for the compression of the brachial artery when inflated. This compression creates turbulence and arterial vibrations that make sounds that can be heard using a stethoscope. These sounds are known as Korotkoff sounds. As the cuff is released, the blood flow returns to the artery, and Korotkoff sounds will be heard, denoting the systolic pressure. The disappearance of Korotkoff sounds indicates the diastolic pressure reading.

Follow the steps in **Skill Drill 8-3** to measure blood pressure by auscultation:

1. Follow standard precautions. Explain the procedure to the patient. Examine for ports, central lines, mastectomy, or injury to the arm. If any are present, use the other arm.

2. With the patient's arm exposed, extended, and with the palm up, place the appropriately sized cuff so that it lies across the upper arm and is located with its distal edge about 1″ above the antecubital space, or the crease at the inside of the patient's elbow. Make sure the center of the inflatable bladder, which is usually marked by an arrow on the cuff, lies over the brachial artery. Next, wrap the ends so that the cuff surrounds the upper arm snugly but not tightly. Secure the cuff with the Velcro fastener attached to it, making sure to rub your hand over the entire area where the two sides of the Velcro fastener are in contact (Step 1).

3. Once the cuff has been properly secured around the upper arm, the arm should be held at about the same level as the heart. With your nondominant hand, palpate the brachial artery (in the antecubital fossa, the anterior aspect of the elbow) to determine where to place the stethoscope (Step 2).

4. Place the bell (if one is present) of the stethoscope over the artery, and hold it firmly against the artery with the fingers of your nondominant hand. Hold the rubber ball-pump in the palm of your other hand and the turn-valve between your thumb and first finger (Step 3).

5. Close the valve tightly, and pump the ball-pump until you no longer hear pulse sounds. Continue pumping to increase the cuff's pressure by an additional 30 mm Hg. Next, slowly turn the valve, opening it until air is steadily escaping from the cuff and you see the needle of the gauge slowly drop. Watch the gauge, and listen carefully. Note the patient's systolic pressure as the reading on the gauge at which the "taps" or "thumps" of the pulse waves can first be heard clearly. As the pressure in the cuff is progressively reduced, pulse sounds will continue for a time, then suddenly disappear. Note the patient's diastolic pressure as the reading on the gauge at which the sounds stopped (Step 4).

6. As soon as the pulse sounds stop, open the valve, and release the remaining air quickly. Once you have finished measuring the blood pressure, you should document your findings and the time at which the blood pressure was taken. Blood pressure is most often measured by auscultation with the patient in a sitting or semisitting position. Be sure to note whether a different method or position was used. Occasionally when a patient's blood pressure is very low, you will continue to hear pulse sounds from the reading at which they started all the way until the gauge has reached 0. When this occurs, you should record the diastolic pressure as "0" or "all the way down" to indicate that it was heard until the gauge read 0 (Step 5).

Obtaining a patient's blood pressure accurately by auscultation may be difficult at times. Noisy environments, patient movement from tremors or seizures,

Skill Drill 8-3

Obtaining Blood Pressure by Auscultation

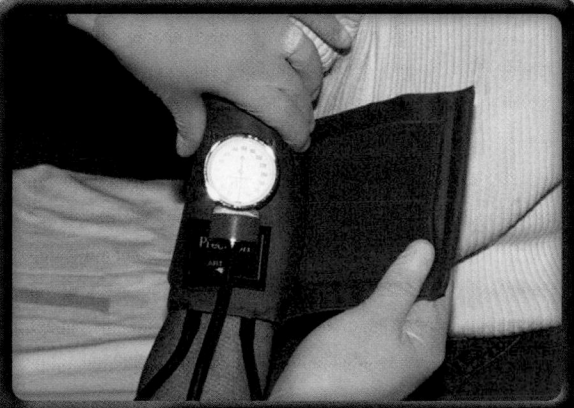

Step 1
Follow standard precautions. Check for ports, central lines, mastectomy, and injury to the arm. If any are present, use the other arm. Apply the cuff snugly. The lower border of the cuff should be about 1″ above the antecubital space.

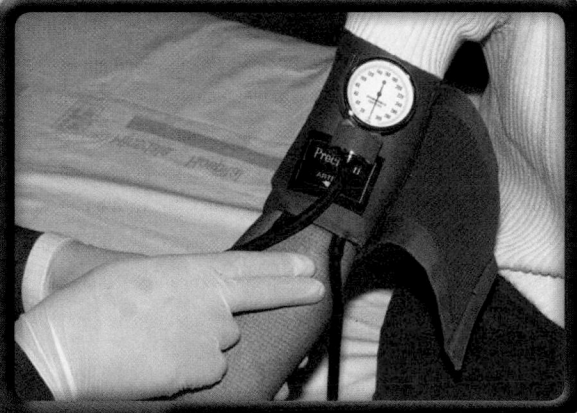

Step 2
Support the exposed arm at the level of the heart. Palpate the brachial artery.

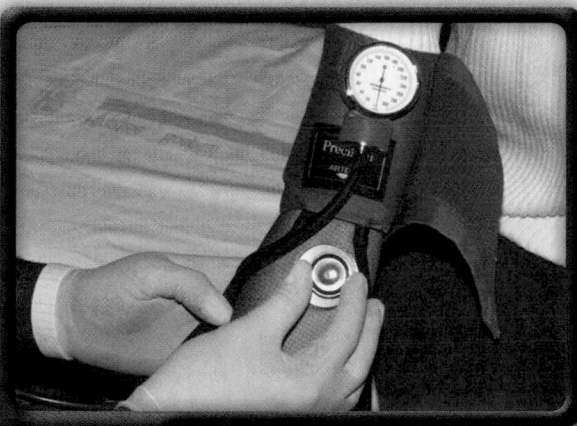

Step 3
Place the stethoscope over the brachial artery, and grasp the ball-pump and turn-valve.

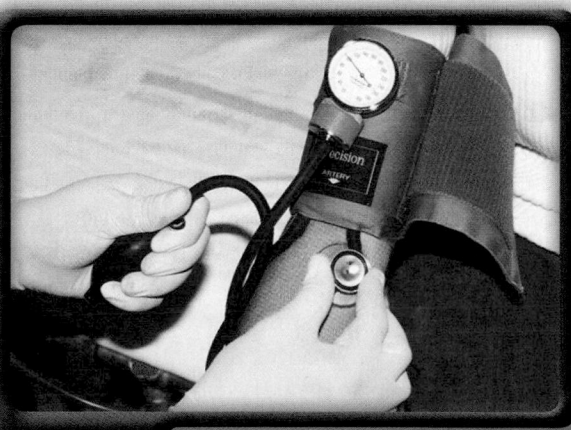

Step 4
Close the valve, and pump to 30 mm Hg above the point at which you stop hearing pulse sounds. Note the systolic and diastolic pressures as you let air escape slowly.

Skill Drill 8-3

Obtaining Blood Pressure by Auscultation, continued

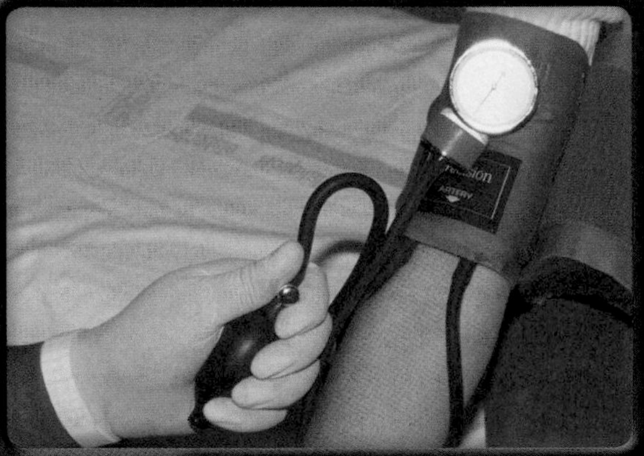

Step 5 Open the valve, and quickly release remaining air.

external vibrations from the EMS vehicle, and excessive noises may produce sounds that mimic Korotkoff sounds and provide inaccurate readings. Other variables that may make obtaining an accurate blood pressure reading nearly impossible are uncooperative adults, infants and children, and patients who are hypotensive with poor perfusion. In these cases, measure blood pressure by palpation.

The palpation (feeling) method does not depend on your ability to hear sounds and should be used in these cases to obtain a patient's blood pressure **Figure 8-28** . If possible, it is preferable that you first obtain a baseline auscultated blood pressure.

To measure blood pressure by palpation, secure the appropriately sized cuff around the patient's upper arm in the manner previously described. With your nondominant hand, palpate the patient's radial pulse on the same arm as the cuff, without moving your fingertips once you have located it, until you have completed taking the blood pressure.

While holding the ball-pump in your other hand, close the turn-valve and slowly inflate the cuff until the pulse disappears and then continue to inflate another 30 mm Hg. As the cuff inflates, you will no longer feel the pulse under your fingertips. Open the turn-valve so that air slowly escapes from the cuff, and carefully observe the gauge. When you can again feel the radial pulse under your fingertips, you should note the reading on the gauge as the patient's systolic blood pressure. You will not be able to determine the diastolic pressure with this method. Next, open the turn-valve further, and completely deflate the cuff. Document your findings, including the time, and note that the pressure was taken by palpation. On your patient care report, you can record the blood pressure as "120/P" and verbalize it as "120 palpated."

Normal Blood Pressure

Blood pressure levels vary with age and sex.

Table 8-6 serves as a guideline for normal blood pressure ranges.

A patient has **hypotension** when the blood pressure is lower than the normal range and

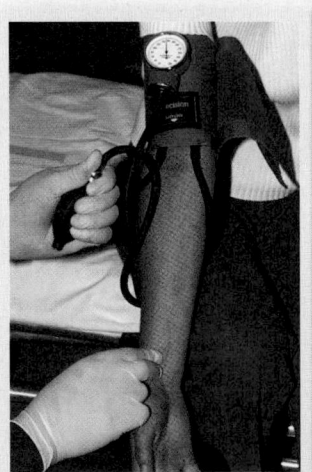

Figure 8-28 When using the palpation method, you should place your fingertips on the radial artery so that you feel the radial pulse.

Table 8-6 Normal Range for Blood Pressure

Age	Range, mm Hg
Adults	90 to 140 (systolic)
Children (ages 1 to 8 years)	80 to 110 (systolic)
Infants (newborn to age 1 year)	50 to 95 (systolic)

hypertension when the blood pressure is higher than the normal range.

Typically, you will see children less frequently than adults; therefore, you might not remember the normal ranges for the various age groups. It is a good idea to carry a chart with you that lists normal blood pressure ranges and other vital signs.

When assessing the patient's general circulation, the blood pressure, pulse, skin temperature, and capillary refill should not be assessed in an injured limb. However, once you have obtained these vital signs from an uninjured limb, you might want to compare the distal skin temperature, quality of the distal pulse, and/or capillary refill time in the injured limb with those found on the uninjured side. This information is useful in evaluating whether the injury may have compromised the circulation in the injured limb.

Neurologic System

Assessment of a patient's neurologic system can be very time consuming and detailed. A neurologic assessment should be performed any time you are confronted with a patient who has changes in mental status, a possible head injury, stupor, dizziness, drowsiness, or syncope. A neurologic assessment begins without even touching the patient. It can be as simple as talking with the patient, asking questions, and receiving an appropriate reply from the patient. This may be performed during the primary assessment.

Evaluate the level of consciousness and orientation to determine the patient's ability to think. Use the AVPU scale if appropriate. Determine the patient's mental status. Is the patient alert, oriented to person, place, time, and events? Is the patient responsive or unresponsive? Does the patient respond to verbal and painful stimuli? If the patient is responsive, evaluate speech for clarity, speed, organization, and logic. When evaluating speech, assess the patient's thought process and determine if he or she may be delusional or has unusual reasoning. Inspect the head for trauma. Pulse, blood pressure, and skin changes may indicate hypoperfusion of the brain. What is the

patient's activity level? What are the patient's mood and thought content? What do the patient's facial expressions tell you? Is the patient angry, fearful, depressed, anxious, or restless? Does the patient appear uncomfortable? Does the patient make incomprehensible or understandable statements? Is the patient's memory affected? Does the patient remember who family members are? What is the patient's perception or view on what is happening? These are all important considerations when assessing the neurologic system.

Now perform a hands-on assessment to determine motor response. How does the patient move? Check for bilateral muscle strength and weaknesses. Complete a thorough sensory assessment. Test for pain, sensations, and position, and compare distal and proximal motor and sensory responses and one side with the other. Remember that a physical examination that deals with a specific chief complaint can be streamlined to assess a specific area of concern.

Musculoskeletal System

An assessment of the patient's musculoskeletal system typically is done because of a chief complaint associated with some type of trauma. Do all extremities appear to be properly positioned, and do all extremities appear to be functioning normally? Assess for posture if standing, and look at joints, checking for range of motion. This should be done by asking the patient how much he or she can move the extremity or joint. Never force a painful joint to move. Always compare the right side with the left side, looking for weakness or atrophy, and assess equality of grip strength. Look for trauma to the abdomen and for distention. Palpate the abdomen for tenderness, rigidity, and patient **guarding**. Expose the site, and evaluate the pulse and motor and sensory function adjacent to and below the affected area.

Pelvis

Inspect the pelvis for symmetry and any obvious signs of injury, bleeding, and deformity. If the patient reports no pain, gently press downward and inward on the pelvic bones Figure 8-29 . Do not rock the pelvis; this action may result in motion of an unstable spine. If you feel any movement or crepitus or the patient reports pain or tenderness, severe injury may be present. Injuries to the pelvis and surrounding abdomen may bleed profusely, so continue to monitor the patient's skin color and vital signs, and be sure to give supplemental oxygen to minimize the effects of shock.

Extremities

Inspect each extremity for symmetry, cuts, bruises, swelling, obvious injuries, and bleeding Figure 8-30 . Also palpate

Secondary Assessment

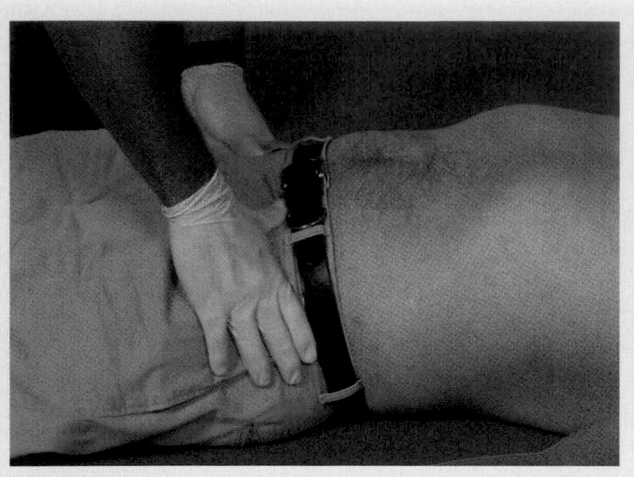

Figure 8-29 Inspect the pelvis for any obvious signs of injury, bleeding, and deformity.

along each extremity for deformities. Ask the patient about any tenderness or pain. As you evaluate the extremities, check for pulses, motor function, and sensory function:

- **Pulse.** Check the distal pulses on the foot (dorsalis pedis or posterior tibial) **Figure 8-31** and **Figure 8-32** and wrist. Assess the pulses in the lower extremities for rate, quality, and rhythm. Is the pulse fast, slow, or irregular? Is the pulse weak, thready, or bounding? Also check circulation. Evaluate the skin color and temperature in the hands and feet. Is it normal? How does it compare with the skin color and temperature of the other extremities? Pale or cyanotic skin may indicate poor circulation in that extremity.
- **Motor function.** Ask the patient to wiggle his or her fingers and toes. An inability to move a single extremity can be the result of a bone, muscle, or nerve injury. An inability to move several extremities

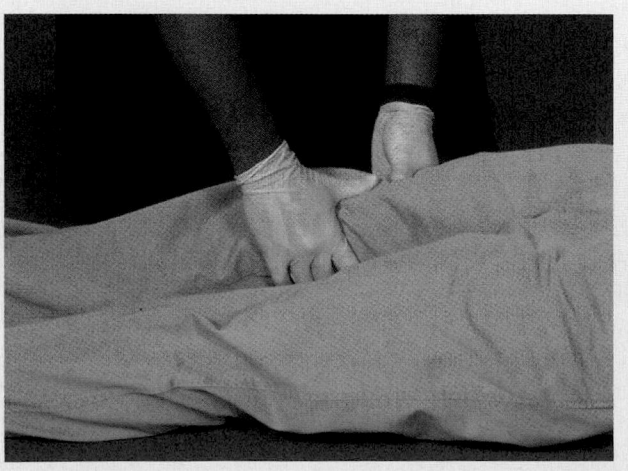

Figure 8-30 Inspect each extremity for cuts, bruises, swelling, obvious injuries, and bleeding.

may be a sign of a brain abnormality or spinal cord injury. Verify that spinal precautions are still in place.
- **Sensory function.** Evaluate sensory function in the extremity by asking the patient to close his or her eyes. Gently squeeze or pinch a finger or toe, and ask the patient to identify what you are doing. The inability to feel sensation in the extremity may indicate a local nerve injury. The inability to feel in several extremities may be a sign of a spinal cord injury. Ensure that you are maintaining spinal immobilization.

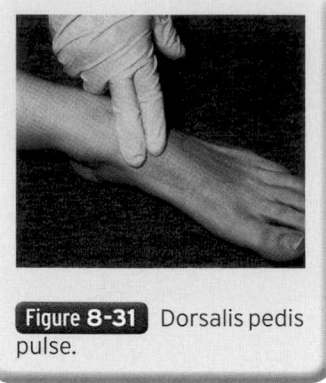

Figure 8-31 Dorsalis pedis pulse.

Posterior Body
Inspect the back for tenderness, deformity, symmetry, and open wounds **Figure 8-33**. Carefully palpate the spine from the neck to the pelvis for tenderness and deformity.

Anatomic Regions
Head, Neck, and Cervical Spine
Inspect for abnormalities of the head, neck, and cervical spine. Gently palpate the scalp and skull for any pain, deformity, tenderness, crepitus, and bleeding **Figure 8-34**. Ask a responsive patient if he or she feels any pain or tenderness. Look at the patient's face. Is it symmetrical, is there evidence of trauma, such as ecchymosis or hemotomas? Does the patient have any facial expressions such as a smile or grimace? Check the patient's eyes, and assess pupillary function, shape, and response. Are the pupils equal in size and reactive to light, or are they constricted, dilated, or unequal? Check the color of the sclera. Assess the patient's cheek bones (zygomas) for possible injury. Check the patient's ears and nose for fluid.

Next, before opening the patient's mouth, check the upper (maxillae) and lower (mandible) jaw. Once the patient's jaws have been assessed and it has been determined that movement will not create any additional pain or injury, open the patient's mouth, looking for any broken or missing

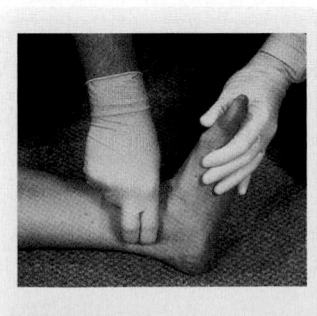

Figure 8-32 Posterior tibial pulse.

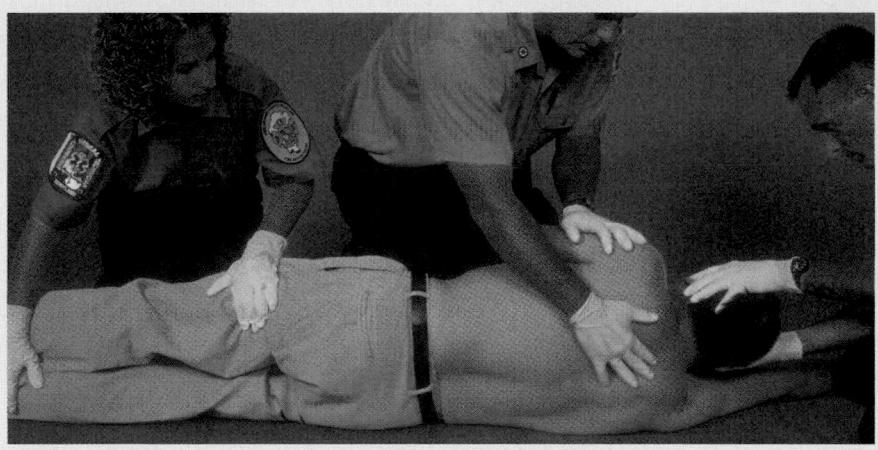

Figure 8-33 Feel the back for tenderness, deformity, and open wounds. Carefully palpate the spine from the neck to the pelvis for tenderness and deformity. Look under the clothing for obvious injuries, including bruising and bleeding.

a problem with blood returning to the heart. Report and record your findings carefully.

Chest

When assessing the chest, inspect, visualize, and palpate over the chest area for injury and signs of trauma, including bruising, tenderness, and swelling **Figure 8-36**.

When assessing breathing, watch for both sides of the chest to rise and fall together with normal breathing. Observe for abnormal breathing signs, including retractions or **paradoxical motion** (when only one section of the chest rises on inspiration while another area of the chest falls).

teeth. If blood and secretions have impaired the airway, this should have been corrected during the primary assessment. Before moving on to the neck, note any unusual odors that may be present in the patient's mouth. This may give an indication of what type of emergency you may be dealing with.

Next, check the neck for signs of swelling or bleeding. Palpate the neck for signs of trauma, such as deformities, bumps, swelling, bruising, and bleeding, as well as a crackling sound produced by air bubbles under the skin, also known as **subcutaneous emphysema** **Figure 8-35**. Also, in patients in whom spinal injury is not suspected, inspect for pronounced or distended jugular veins with the patient sitting at a 45° angle. This is a normal finding in a person who is lying down; however, jugular venous distention in a patient who is sitting up suggests

Retractions indicate the patient has some condition, usually medical, that is impairing the flow of air into and out of the lungs. Paradoxical motion is associated with a fracture of several ribs (flail), causing a section of the chest to move independently from the rest of the chest wall. Feel for grating of the bones as the patient breathes. Crepitus is often associated with rib fractures. Palpate the chest for subcutaneous emphysema, especially in cases of severe blunt chest trauma.

If the patient reports difficulty breathing or has evidence of trauma to the chest, auscultate breath and lung sounds. This helps you to evaluate air movement in and out of the lungs. To auscultate, you need a stethoscope. Make sure you place the ear pieces facing forward in your ears. The position of the patient will determine the way you proceed to check for breathing.

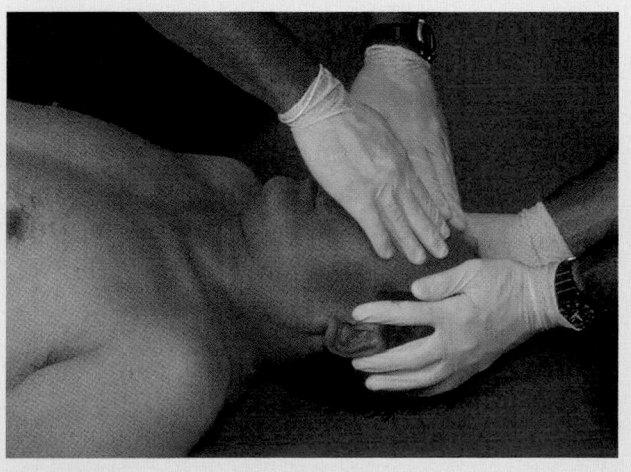

Figure 8-34 Gently palpate the head for any pain, deformity, tenderness, crepitus, and bleeding.

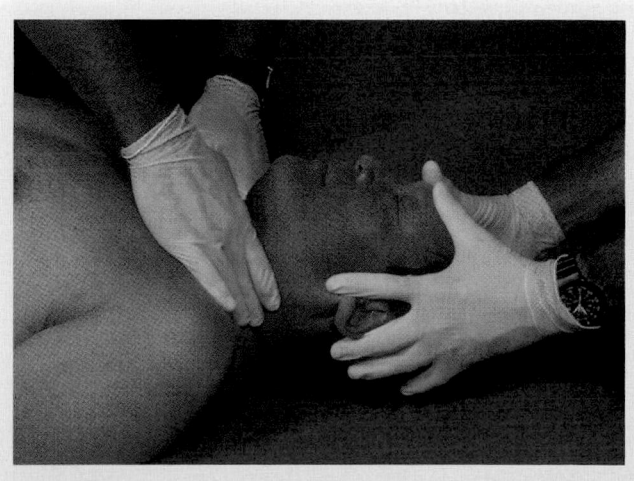

Figure 8-35 Gently palpate the neck.

Secondary Assessment

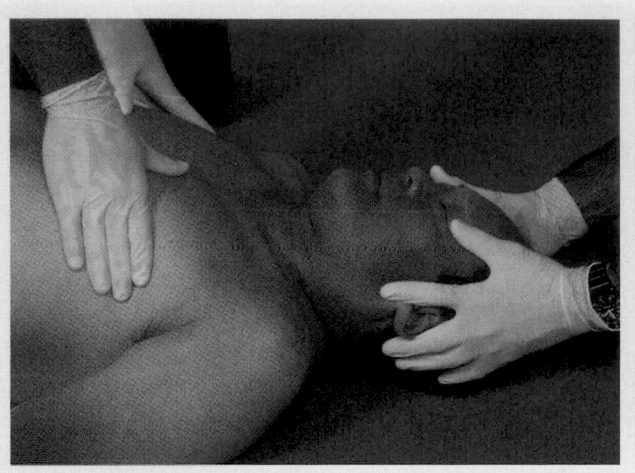

Figure 8-36 Inspect, visualize, and palpate over the chest area for injury and signs of trauma.

for symmetry, masses, tenderness, and bleeding. As you palpate the abdomen, use the terms "firm," "soft," "tender," or "distended" (swollen) to report your findings. If the patient is awake and alert, ask about pain as you perform the examination. The abdomen is broken into four quadrants, left upper quadrant (LUQ), left lower quadrant (LLQ), right upper quadrant (RUQ), and right lower quadrant (RLQ). Always start the palpation of the abdomen in the quadrant that is farthest from the patient's pain. Do not palpate obvious soft-tissue injuries, and be careful not to palpate too firmly. Assess for the presence of rebound tenderness.

The goal is to hear and document the presence or absence of breath sounds. It is important to compare one side with the other. If you believe the patient's breathing is abnormal, reassess breathing, and then ensure that the patient is receiving oxygen and, if appropriate, assist with ventilations.

Abdomen

Inspect and palpate the abdomen for any obvious injuries, bruising, and bleeding **Figure 8-37** . Be sure to palpate the front and back of the abdomen, evaluating

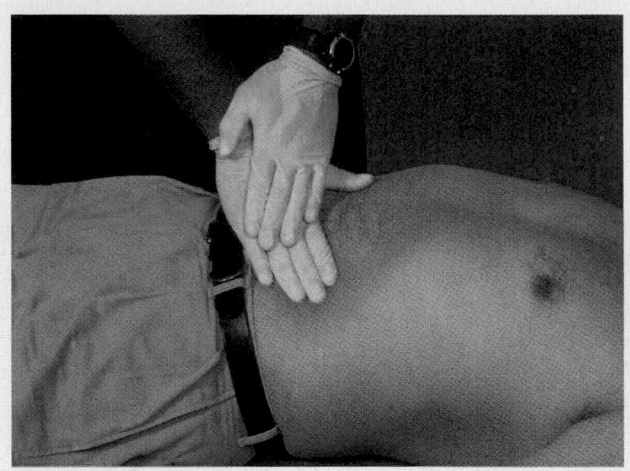

Figure 8-37 Palpate the abdomen, evaluating for tenderness and bleeding.

Patient Assessment

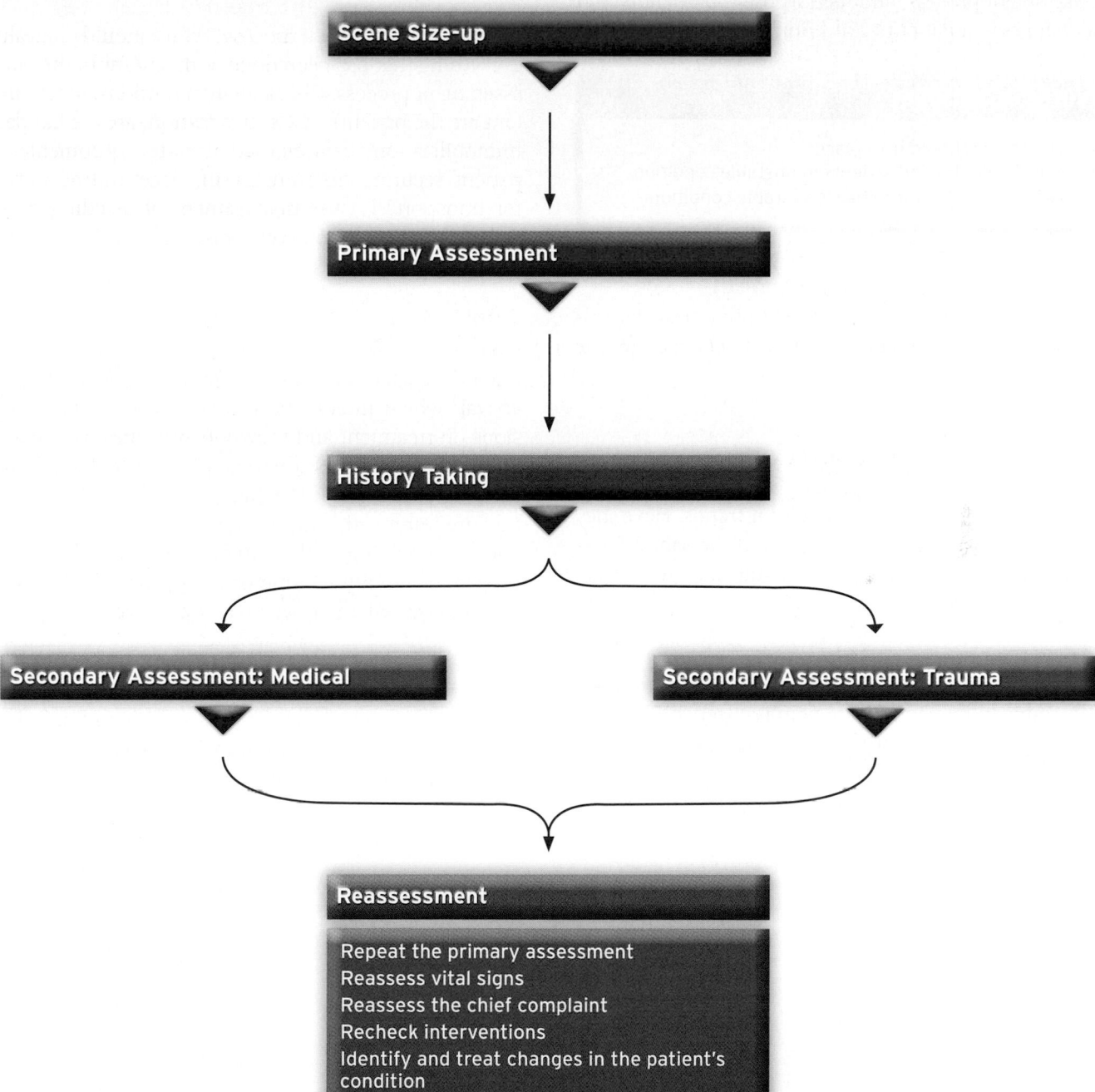

Scene Size-up

Primary Assessment

History Taking

Secondary Assessment: Medical

Secondary Assessment: Trauma

Reassessment

Repeat the primary assessment
Reassess vital signs
Reassess the chief complaint
Recheck interventions
Identify and treat changes in the patient's condition
Reassess patient
- Unstable patients: every 5 minutes
- Stable patients: every 15 minutes

Reassessment

A **reassessment** is performed at regular intervals during the assessment process, and its purpose is to identify and treat changes in a patient's condition.

Words of Wisdom

Reassessment should take place:
- every 5 minutes for patients in unstable condition
- every 15 minutes for patients in stable condition

Repeat the Primary Assessment

The reassessment procedure is to simply repeat the primary assessment to identify and treat changes in the patient's condition.

Reassess Vital Signs

Reassess and record vital signs. Compare the baseline vitals obtained during the primary assessment with any and all subsequent vital signs. Look for trends. Have they changed, improved, declined, or stayed the same? Reassess the mental status, airway, breathing, and circulation. Monitor skin color and temperature.

Reassess the Chief Complaint

With all EMT services, a reassessment must be completed to determine if the patient care plan is effective. The purpose is to ask and answer the following questions about the patient's chief complaint:

- Is the current treatment improving the patient's condition?
- Has an already identified problem gotten better?
- Has an already identified problem gotten worse?
- What is the nature of any newly identified problems?

Recheck Interventions

In the reassessment process, you should reevaluate everything that has been done to this point in the patient assessment process. Check all interventions. Most important are the patient's ABCs. In addition, are the bandages, immobilization devices, extrication equipment, and patient securing instruments in place and appropriate for transport? Ensure management of bleeding. Ensure adequacy of other interventions, and consider the need for new interventions.

Identify and Treat Changes in the Patient's Condition

No matter what the patient's condition was prior to your arrival, which interventions were used, or which decisions on treatment and transport priorities were made, a reassessment is necessary to help monitor changes in the patient's condition. If the changes in the patient's condition are improved, simply continue whatever treatments you are providing. If the patient's condition deteriorates, prepare to modify treatments as appropriate. Document any changes, whether negative or positive.

Reassess Patient

How and when to perform a reassessment depends on the patient's condition. A patient in unstable condition should be reassessed every 5 minutes, whereas a patient in stable condition should be reassessed every 15 minutes.

You are the Provider: SUMMARY

1. What are the components of patient assessment?

The treatment you provide is only as good as the assessment you perform. There are five components in the assessment process: scene size-up, primary assessment, history taking, secondary assessment, and reassessment. Each component has an integral role in your overall treatment of the patient. Although the steps of assessment represent a logical approach to evaluation of a patient, the order in which they are performed is dictated by the patient's condition.

You should never enter a scene without carefully evaluating its complexity and looking for possible hazards first; this is called the scene size-up. The scene size-up actually begins when you are dispatched. For example, if you are dispatched to a shooting, motor vehicle crash, "unknown illness or injury," or a hazardous materials incident, you know that the potential for danger is high. When you arrive at the scene, implement standard precautions (eg, gloves) and other personal protective equipment as dictated by the situation, note the mechanism of injury (MOI) or nature of illness (NOI), determine the number of patients, and request additional resources as needed. It is important to note that the scene size-up is *not* a one-time evaluation; you must constantly remain aware of your surroundings. Every scene has the potential to escalate to an extremely dangerous situation, regardless of its initial appearance.

After the scene size-up, the next component of the assessment process is the primary assessment, an integration of assessment and treatment aimed at identifying and correcting immediate life threats. The primary assessment begins by forming an initial general impression of the patient prior to making physical contact. This assessment of the patient, by visual examination only, will enable a general impression to be formed that allows classification of the patient's condition as one of the following: appears stable, appears stable but is potentially unstable, or appears unstable. After making physical contact with the patient, take appropriate spinal precautions (if trauma is suspected), and determine his or her level of consciousness using the AVPU scale. The next steps are critical: determining the adequacy of the patient's airway and breathing, assessing his or her circulatory status, and looking for and controlling major external bleeding. Remember, you must integrate your treatment with your assessment; if you find an airway problem, manage it immediately; the same applies if you find a circulation problem. In many cases, you and your partner will be working together so that the ABCs can be assessed and treated at the same time. The findings of your primary assessment will help you determine the appropriate transport priority; is transport required now, or is additional assessment and treatment at the scene appropriate?

If the patient is conscious, alert, and in stable condition, determine his or her chief complaint, ask further questions, and perform an assessment that is directly related to the chief complaint (history of present illness); the OPQRST mnemonic is a useful tool for gathering additional information regarding the patient's chief complaint. Other components of the history-taking process include the SAMPLE history and assessing the patient's vital signs. As with the primary assessment, history taking often occurs while you are treating the patient.

The secondary assessment is a systematic examination of the patient from head to toe and involves examining the major body systems in an orderly manner. It is designed to further assess the patient to find and treat injuries or conditions that were not grossly obvious during the primary assessment. Although emphasis is placed on areas suggested by the history of present illness and chief complaint or presenting injury, the secondary assessment may yield further conditions that require prompt attention. In most cases, the secondary assessment is performed in the back of the ambulance, while you are en route to the hospital.

During the assessment process, you will use a variety of monitoring devices to assist you in gathering additional information regarding the patient's condition. These include pulse oximetry, noninvasive blood pressure monitoring, and other devices that are specific to your EMS system. It is important to note that these devices, although useful in the overall assessment process, are adjuncts and must not replace your eyes and ears; they have limitations and should not be relied on solely.

The last component of the assessment process is reassessment. The purpose of reassessment is to identify and treat changes in the patient's condition in a timely manner. Components of reassessment include repeating the primary assessment, reassessing the patient's vital signs, reevaluating the patient's chief complaint, and reassessing your interventions to determine their effectiveness. The frequency with which you reassess a patient is determined by his or her condition. A patient in unstable condition should be reassessed *at least* every 5 minutes. A patient in stable condition should be reassessed *at least* every 15 minutes. Any change in the patient's condition should prompt you to reassess his or her condition.

2. Will your assessment of the patient differ if she is injured versus ill? If so, how?

The same components of patient assessment used to evaluate a medical patient are used to assess a trauma patient. The differences lie in what you find and how you treat it. For example, during the scene size-up of a trauma patient, you will evaluate the MOI to help predict the type and severity of injuries your patient may have; you will also consider taking appropriate spinal precautions. In medical patients, the NOI is assessed to help determine what category of medical condition you are dealing with (eg, cardiac, respiratory, endocrine), which will help guide your assessment in the appropriate direction.

Whether you are assessing a medical patient or a trauma patient, one key is to remain organized. Furthermore, you should assess the patient for medical conditions *and* injuries, especially if you suspect that the patient's injury was preceded by a medical condition. For example, a man whose car swerves off the road and strikes a tree could merely have been an accident, or it could have been preceded by a heart attack, stroke, or hypoglycemia, among other things. Likewise, a woman complaining of chest pain—leading you to suspect a cardiac condition—could have actually been struck in the chest.

Keep an open mind when you are assessing a patient, and direct your assessment, at least initially, to information gathered in the scene size-up, the patient's chief complaint, and his or her clinical presentation.

3. Are spinal precautions indicated? Why or why not?

Yes, spinal precautions should be implemented. The patient was found in a prone position on the floor, and there are no witnesses who can confirm how she got there. She could have a medical condition, a traumatic injury, or both. Spinal precautions should be considered any time the events that preceded a patient's collapse are unclear and/or there are no witnesses who can definitively rule out trauma—for example, a person who states that he or she caught the patient and placed him or her on the ground.

The patient could have experienced a seizure, fallen to the ground, and struck her head. By contrast, she could have felt weak, lain down on the floor, and then lost consciousness. Avoid the temptation to rule out trauma just because you do not see any obvious injuries, especially when there were no witnesses to the event.

Err on the side of caution, and protect the patient's spine. Manually stabilize her head in a neutral position, log roll her to a supine position as a unit, and open her airway with the jaw-thrust maneuver.

4. Which of these assessment findings requires your *most* immediate attention?

The patient's airway, which contains bloody secretions and vomitus, is in *immediate* jeopardy! Log roll her back onto her side, maintaining a neutral head position, and use suction to clear her airway. Suction for up to 10 seconds, and then reassess her airway status. If you hear gurgling, think suction!

Do not suction the patient's airway while he or she is in a supine position; doing so significantly increases the risk of aspiration by pushing secretions and vomitus farther into the airway and potentially into the trachea.

After ensuring that her airway is clear of secretions and vomitus, insert a simple airway adjunct—while keeping the airway open manually—to further help maintain airway patency. The airway adjunct you use depends on the patient's level of consciousness and the presence or absence of a gag reflex. The patient is minimally responsive; however, this does not always indicate an absent gag reflex. The eyelash reflex test can be used to help determine the presence or absence of a gag reflex. Gently stroke the eyelash and observe the lower eyelid; if it quivers, you should assume the patient has an intact gag reflex and insert a nasopharyngeal airway. If there is no response to the eyelash reflex test, *carefully* insert an oropharyngeal airway, but be ready to *quickly remove it if she begins to gag*.

5. Does the patient require further treatment at the scene? If so, what?

You have identified and corrected all immediately life-threatening conditions found during the primary assessment.

The patient's airway has been suctioned, an airway adjunct has been inserted, and her breathing is being assisted.

A secondary assessment has not revealed any injuries or conditions that warrant immediate attention, she is not wearing any medical alert jewelry, an inspection of her apartment has not revealed anything suspicious, and her blood glucose level has ruled out hyperglycemia and hypoglycemia.

At this point, you should continue to support her airway and breathing, monitor her circulatory status (ie, her pulse, which is slow), keep her warm, and be prepared to begin cardiopulmonary resuscitation and apply the automated external defibrillator. Apply spinal precautions (ie, cervical collar, backboard, straps, head immobilizer), and prepare for immediate transport. Remember, there were no witnesses to this event; err on the side of caution and protect the patient's spine.

You will often encounter patients who are minimally responsive or completely unresponsive; however, your assessment of the patient and his or her surroundings will not yield anything that caused his or her condition, and there will be no witnesses or family members who can explain what happened. In cases such as these, treat the patient symptomatically—that is, treat what you have found to be abnormal—and provide transport to an appropriate medical facility. En route, reassess the patient—in this case, *at least* every 5 minutes, and adjust your treatment as needed. If the patient's condition deteriorates, immediately repeat the primary assessment.

6. Should you remain at the scene and wait for the paramedic unit? Why or why not?

No, do not remain at the scene! Seventeen minutes is too long to remain at the scene with a patient in unstable condition, not to mention one who still needs to be moved from a second-floor apartment. You should advise the paramedic ambulance of the situation, but do so while you are preparing for immediate transport. The longer you remain at the scene, the greater the chance that the patient's condition will further deteriorate.

Correctly moving a patient down a flight of stairs, especially one who is secured to a backboard, can be time consuming. If the paramedic ambulance arrives before you depart the scene, it would clearly be prudent to transfer care to them. Paramedics have other assessment tools (eg, cardiac monitoring) and can perform other interventions (eg, advanced airway management, medication therapy) that may benefit the patient.

If you are ready to transport a patient in unstable condition and a paramedic unit's arrival will be delayed, transport the patient at once. You should, however, consider an intercept with them at a designated location. Your EMS system protocols should have a plan for coordinating a paramedic intercept when transporting a patient in unstable condition or a patient who requires care that is beyond your level of training.

7. How has your patient's condition changed from the previous assessments?

Your patient's condition has obviously deteriorated. Compared with earlier assessments, which revealed that she was minimally responsive, she is now unconscious and unresponsive. Furthermore, her oxygen saturation is declining despite assisted ventilation with high-flow oxygen, and cyanosis is developing around her mouth (perioral/circumoral cyanosis).

Her heart rate and blood pressure—although still unstable—are essentially unchanged compared with previous assessments. However, they must still be closely monitored for further deterioration.

The findings of your reassessment (eg, unresponsive, low Sao_2, cyanosis) point to a problem with her oxygenation and ventilation status that may be caused by more than one factor. Her head may not be properly positioned, the simple airway adjunct may need to be repositioned, her airway may be filling with blood or vomitus, or she is not being adequately ventilated.

8. What should you do in response to the patient's condition change?

Failure to reassess a patient at appropriate intervals—every 5 minutes if in unstable condition, and every 15 minutes if in stable condition—may cause you to miss even the most subtle signs of deterioration; by the time you intervene, it may be too late! Because of this, the criticality of reassessment cannot be overemphasized.

Deterioration of a patient's condition should immediately prompt you to repeat the primary assessment, which begins by reassessing airway, breathing, and circulation.

Make sure her head is correctly positioned. Look in her mouth for secretions, and remove them with suction if present. Reassess the position of the simple airway adjunct; is the nasal airway protruding from her nose? If an oral airway is in place, is it protruding from her mouth? Reassess the mask-to-face seal of the bag-mask device; is it adequate, or is there air leaking in all directions? Are you ventilating at the appropriate rate (10 to 12 breaths/min in an adult) with the appropriate volume (each breath delivered over 1 second—just enough to cause visible chest rise)?

A rapid, yet careful reassessment of the ABCs will often reveal the cause of the patient's status change, thus allowing you to rapidly correct it. However, do not wait for the patient's condition to deteriorate before reassessing him or her.

9. What components of the SAMPLE history, if any, can you obtain when your patient is unresponsive? How would you obtain the information?

The SAMPLE history is an integral part of the patient assessment process; it can provide information about the patient that may lead to a field impression—that is, what you think is wrong with the patient.

In the absence of any family members or bystanders who know the patient, it is usually not possible to obtain a complete and accurate SAMPLE history. However, there are certain components of the SAMPLE history that can be obtained from an unresponsive patient; obtaining this information relies on your good assessment skills and you taking a "thinking outside the box" approach.

Signs and symptoms can be established by simply assessing the patient; although signs and symptoms alone will not tell you what the patient's underlying problem is, they will enable you to direct your treatment accordingly.

Some patients wear a medical alert bracelet embossed with their drug allergies, if any, and/or any medications they are taking. Other patients may carry a medical alert card in their wallet or purse that details their entire medical history, including drug allergies and current medications. If you do not look for medical alert jewelry or cards, you are missing out on finding potential sources of information. The patient's environment can often give you clues about the medical history. Are there any prescription medication bottles present? Is there medical equipment present (eg, home oxygen, nebulizer) that indicates an underlying condition? It is wise to carry a field guide that lists some of the more common prescription medications; although an unresponsive patient cannot speak, you can learn something about his or her conditions based on the medications found at the scene.

Unresponsive patients cannot tell you when they last ate; therefore, you must assume that the stomach is full and that they are at risk for regurgitation and aspiration.

It is arguably easier to obtain the patient's SAMPLE history if he or she is conscious, alert, and able to provide you with the information. However, when your patient is not able to provide you with the information that you seek, it is not impossible to obtain some of the information if your assessment is thorough and you are observant of the patient's surroundings.

10. Why is reassessing your interventions so important?

The main purpose of reassessing the interventions you have performed on a patient is to determine their effectiveness. If your intervention has been effective, you should see an improvement in the patient's condition. If your intervention has not been effective, the patient's condition will have remained unchanged or will have deteriorated.

Intervention reassessment also enables you to determine if you need to make modifications to existing interventions, cease an intervention, or perform another intervention.

For example, if a conscious patient is receiving oxygen via a nonrebreathing mask, and your reassessment reveals that he is now unresponsive, you must perform several new interventions—in this case, manually opening his airway, ensuring that the airway is clear of secretions and foreign bodies, and inserting a simple airway adjunct. Further assessment will determine the need for additional interventions; if the patient is still breathing adequately, the nonrebreathing mask would likely continue to suffice, although the patient should be positioned on his or her side—a new intervention. However, if the patient's breathing status has deteriorated, you would have to modify the way you are oxygenating—in this case, switching from the nonrebreathing mask to assisted ventilation with a bag-mask device.

The mere performance of an intervention does not mean that it will cause your patient's condition to improve, nor does it mean that it will not cause his or her condition to deteriorate. Reassess, reassess, reassess!

You are the Provider: SUMMARY, continued

EMS Patient Care Report (PCR)

Date: 7-20-09	Incident No.: 010809	Nature of Call: Woman down		Location: 1004 W. Almond
Dispatched: 1410	En Route: 1410	At Scene: 1416	Transport: 1428	At Hospital: 1441	In Service: 1455

Patient Information

Age: 22 **Sex:** F **Weight (in kg [lb]):** 50 kg estimated (110 lb)	**Allergies:** Unknown **Medications:** Unknown **Past Medical History:** Unknown **Chief Complaint:** Unresponsive; unknown circumstances

Vital Signs

Time: 1421	BP: 80/60	Pulse: 40	Respirations: 6	Sao$_2$: 95%
Time: 1428	BP: 78/58	Pulse: 44	Respirations: 6	Sao$_2$: 88%
Time: 1433	BP: 82/62	Pulse: 40	Respirations: 6	Sao$_2$: 97%
Time: 1438	BP: 82/62	Pulse: 40	Respirations: 6	Sao$_2$: 97%

EMS Treatment
(circle all that apply)

Oxygen @ 15 L/min via (circle one): NC NRM (Bag-Mask Device)	(Assisted Ventilation)	(Airway Adjunct)	CPR
Defibrillation	Bleeding Control	Bandaging	Splinting	Other: (Suctioning, spinal precautions, blanket)

Narrative

9-1-1 dispatch for a "woman down." Law enforcement personnel responded and ensured the scene was secure prior to EMS arrival. On arrival at the scene, found the patient, a 22-year-old woman, lying in a prone position on the kitchen floor of her second-floor apartment. A neighbor was present but did not know what happened and has no knowledge of the patient's medical history. The patient was minimally responsive. Manually stabilized the patient's head, and log rolled her to a supine position. Opened her airway with the jaw-thrust maneuver, and noted bloody secretions and vomitus coming from her mouth. Immediately suctioned her oropharynx until clear of debris, and inserted a nasal airway. Assessment of her breathing revealed it to be slow and shallow; began assisting her ventilations with a bag-mask device attached to high-flow oxygen. Secondary assessment revealed no obvious signs of trauma, and law enforcement's search of the patient's apartment revealed no medication bottles, drug paraphernalia, or anything else suspicious. Unable to obtain SAMPLE history. Blood glucose level was assessed and read 112 mg/dL. A paramedic unit was available but was 17 minutes away from our location, so decision was made to continue treatment and begin immediate transport. Applied full spinal precautions, applied a blanket for warmth, moved the patient down a flight of stairs with the assistance of Engine Company 13, and loaded her into the ambulance. Began transport to hospital; an EMT from Engine Company 13 accompanied us with the patient to provide assistance. Reassessment revealed that the patient was now unresponsive, her oxygen saturation had declined, and cyanosis was developing around her mouth. Resuctioned her oropharynx, inserted an oral airway, and continued to assist her ventilations. Intervention reassessment revealed improvement in the patient's oxygenation status; however, she remained unresponsive, bradycardic, and hypotensive. Continued to assist patient's ventilations, and reassessed her condition every 3 to 5 minutes. Patient's condition remained unchanged throughout transport. Delivered her to the emergency department staff without incident, and gave verbal report to attending physician. Medic 80 returned to service at 1455. **End of report**

Prep Kit

Ready for Review

- The assessment process begins with the scene size-up, which identifies real or potential hazards. The patient should not be approached until these hazards have been dealt with in a way that eliminates or minimizes risk to the EMTs and the patient(s).
- The primary assessment is performed on all patients. It includes forming an initial general impression of the patient, including the level of consciousness, and identifies any life-threatening conditions to the ABCs. A rapid scan is performed to assist in prioritizing time and mode of transport. Any life threats identified must be treated before moving on to the next step of the assessment.
- Airway, breathing, and circulation are assessed to evaluate the patient's general condition.
- History taking includes an investigation of the patient's chief complaint or history of present illness. A SAMPLE history is generally taken during this step of the assessment process. This information may be obtained from the patient, family, friends, or bystanders.
- By asking several important questions, you will be able to determine the patient's signs and symptoms, allergies, medications, pertinent past history, last oral intake, and events leading up to the incident.

- The secondary assessment is a systematic physical examination of the patient. The physical examination may be a systematic head-to-toe, full-body scan or a systematic assessment that focuses on a certain area or region of the body, often determined through the chief complaint. Circumstances will dictate which aspects of the physical examination will be used. The secondary assessment is performed on scene or in the back of the ambulance en route to the hospital; there are times when you may not have time to perform a secondary assessment at all if the patient has serious life threats.
- The reassessment is performed on all patients. It gives you an opportunity to reevaluate the chief complaint and to reassess interventions to ensure that they are still being delivered correctly. Information from the reassessment may be used to identify and treat changes in the patient's condition.
- A patient in stable condition should be reassessed every 15 minutes, whereas a patient in unstable condition should be reassessed every 5 minutes.
- The assessment process is systematic and dynamic. Each assessment you perform will be slightly different, depending on the needs of the patient. The result will be a process that will enable you to quickly identify and treat the needs of all patients, both medical and trauma related, in a way that meets their unique needs.

Vital Vocabulary

accessory muscles The secondary muscles of respiration. They include the neck muscles (sternocleidomastoids), the chest pectoralis major muscles, and the abdominal muscles.

auscultate To listen to sounds within an organ with a stethoscope.

AVPU scale A method of assessing the level of consciousness by determining whether the patient is awake and alert, responsive to verbal stimuli or pain, or unresponsive; used principally early in the assessment process.

blood pressure The pressure of circulating blood against the walls of the arteries.

bradycardia A slow heart rate, less than 60 beats/min.

breath sounds An indication of air movement in the lungs, usually assessed with a stethoscope.

capillary refill A test that evaluates distal circulatory system function by squeezing (blanching) blood from an area such as a nail bed and watching the speed of its return after releasing the pressure.

capnography A noninvasive method that can quickly and efficiently provide information on a patient's ventilatory status, circulation, and metabolism.

capnometry The use of a capnometer, a device that measures the amount of expired carbon dioxide.

carbon dioxide Carbon dioxide is a component of air and typically makes up 0.3% of air at sea level. It is also a waste product exhaled during expiration by the respiratory system.

chief complaint The reason a patient called for help; also, the patient's response to questions such as "What's wrong?" or "What happened?"

coagulate To form a clot to plug an opening in an injured blood vessel and stop bleeding.

colorimetric devices Capnometer or end-tidal carbon dioxide detectors are devices that use a chemical reaction to detect the amount of carbon dioxide present in expired gases by changing colors (qualitative measurement rather than quantitative).

conjunctiva The delicate membrane that lines the eyelids and covers the exposed surface of the eye.

crepitus A grating or grinding sensation caused by fractured bone ends or joints rubbing together; also air bubbles under the skin that produce a crackling sound or crinkly feeling.

cyanosis A bluish gray skin color that is caused by a reduced level of oxygen in the blood.

DCAP-BTLS A mnemonic for assessment in which each area of the body is evaluated for Deformities, Contusions, Abrasions, Punctures/penetrations, Burns, Tenderness, Lacerations, and Swelling.

diaphoretic Characterized by profuse sweating.

diastolic pressure The pressure that remains in the arteries during the relaxing phase of the heart's cycle (diastole) when the left ventricle is at rest.

end-tidal CO_2 The amount of carbon dioxide present in exhaled breath.

focused assessment A type of physical assessment that is typically performed on patients who have sustained nonsignificant mechanisms of injury or on responsive medical patients. This type of examination is based on the chief complaint and focuses on one body system or part.

frostbite Damage to tissues as the result of exposure to cold; frozen or partially frozen body parts are frostbitten.

full-body scan A systematic head-to-toe examination that is performed during the secondary assessment on a patient who has sustained a significant mechanism of injury, is unconscious, or is in critical condition.

general impression The overall initial impression that determines the priority for patient care; based on the patient's surroundings, the mechanism of injury, signs and symptoms, and the chief complaint.

Golden Period The time from injury to definitive care, during which treatment of shock and traumatic injuries should occur because survival potential is best.

guarding Involuntary muscle contractions (spasms) of the abdominal wall in an effort to protect an inflamed abdomen; a sign of peritonitis.

history taking A step within the patient assessment process that provides detail about the patient's chief complaint and an account of the patient's signs and symptoms.

hypertension Blood pressure that is higher than the normal range.

hypotension Blood pressure that is lower than the normal range.

hypothermia A condition in which the internal body temperature falls below 95°F (35°C) after exposure to a cold environment.

incident command system A system implemented to manage disasters and mass- and multiple-casualty incidents in which section chiefs, including finance, logistics, operations, and planning, report to the incident commander. Also referred to as the incident management system.

jaundice Yellow skin or sclera that is caused by liver disease or dysfunction.

labored breathing Breathing that requires visibly increased effort; characterized by grunting, stridor, and use of accessory muscles.

mechanism of injury (MOI) The way in which traumatic injuries occur; the forces that act on the body to cause damage.

nasal flaring Flaring out of the nostrils, indicating that there is an airway obstruction.

nature of illness (NOI) The general type of illness a patient is experiencing.

OPQRST An abbreviation for key terms used in evaluating a patient's pain: Onset, Provocation or Palliation, Quality, Region/radiation, Severity, and Timing of pain.

orientation The mental status of a patient as measured by memory of person (name), place (current location), time (current year, month, and approximate date), and event (what happened).

palpate To examine by touch.

paradoxical motion The motion of the chest wall section that is detached in a flail chest; the motion is exactly the opposite of normal motion during breathing (ie, in during inhalation, out during exhalation).

perfusion Circulation of blood within an organ or tissue.

personal protective equipment (PPE) Clothing or specialized equipment that provides protection to the wearer.

pertinent negatives Negative findings that warrant no care or intervention.

primary assessment A step within the patient assessment process that identifies and initiates treatment of immediate and potential life threats.

pulse The pressure wave that occurs as each heartbeat causes a surge in the blood circulating through the arteries.

pulse oximetry An assessment tool that measures oxygen saturation of hemoglobin in the capillary beds.

rales A crackling, rattling breath sound that signals fluid in the air spaces of the lungs; also called crackles.

reassessment A step within the patient assessment process that is performed at regular intervals during the assessment process. Its purpose is to identify and treat changes in a patient's condition. A patient in unstable condition should be reassessed every 5 minutes, whereas a patient in stable condition should be reassessed every 15 minutes.

responsiveness The way in which a patient responds to external stimuli, including verbal stimuli (sound), tactile stimuli (touch), and painful stimuli.

retractions Movements in which the skin pulls in around the ribs during inspiration.

rhonchi Coarse, low-pitched breath sounds heard in patients with chronic mucus in the upper airways.

SAMPLE history A brief history of a patient's condition to determine signs and symptoms, allergies, medications, pertinent past history, last oral intake, and events leading to the injury or illness.

scene size-up A step within the patient assessment process that involves a quick assessment of the scene and the surroundings to provide information about scene safety and the mechanism of injury or nature of illness before you enter and begin patient care.

sclera The white portion of the eye; the tough outer coat that gives protection to the delicate, light-sensitive inner layer.

secondary assessment A step within the patient assessment process in which a systematic physical examination of the patient is performed. The examination may be a systematic full-body scan or a systematic assessment that focuses on a certain area or region of the body, often determined through the chief complaint.

shallow respirations Respirations that are charcterized by little movement of the chest wall (reduced tidal volume) or poor chest excursion.

sign Objective findings that can be seen, heard, felt, smelled, or measured.

sniffing position An upright position in which the patient's head and chin are thrust slightly forward to keep the airway open.

spontaneous respirations Breathing that occurs with no assistance.

standard precautions Protective measures that have traditionally been developed by the Centers for Disease Control and Prevention for use in dealing with objects, blood, body fluids, and other potential exposure risks of communicable disease.

stridor A harsh, high-pitched, crowing inspiratory sound, such as the sound often heard in acute laryngeal (upper airway) obstruction; may sound like crowing and be audible without a stethoscope.

subcutaneous emphysema The presence of air in soft tissues, causing a characteristic crackling sensation on palpation.

symptom Subjective findings that the patient feels but that can be identified only by the patient.

systolic pressure The increased pressure in an artery with each contraction of the ventricles (systole).

tachycardia A rapid heart rate, more than 100 beats/min.

tidal volume The amount of air (in milliliters) that is moved in or out of the lungs during one breath.

triage The process of establishing treatment and transportation priorities according to severity of injury and medical need.

tripod position An upright position in which the patient leans forward onto two arms stretched forward and thrusts the head and chin forward.

two- to three-word dyspnea A severe breathing problem in which a patient can speak only two to three words at a time without pausing to take a breath.

vasoconstriction Narrowing of a blood vessel.

vital signs The key signs that are used to evaluate the patient's overall condition, including respirations, pulse, blood pressure, level of consciousness, and skin characteristics.

Assessment in Action

You are dispatched to a motel for an unknown medical emergency. The room is small with two beds and two occupants. The patient is an approximately 33-year-old woman lying supine on the bed. She opens her eyes when you speak to her but is lethargic and complaining of fever, chills, vomiting, and diarrhea. The man in the room tells you she has been sick for several days and has not eaten in the past 2 days. Your assessment reveals pain and tenderness in the right lower abdominal quadrant. Her vital signs are a blood pressure of 100/60 mm Hg; a pulse rate of 140 beats/min, strong and bounding; and a respiratory rate of 24 breaths/min and shallow. Her skin is flushed and hot to the touch. The patient reports no significant medical history and does not take any medications.

1. What is the first concern when entering this room?
   - **A.** ABCs
   - **B.** Physical assessment
   - **C.** Scene safety
   - **D.** Vital signs

2. What is the difference between a sign and a symptom?

3. What is this patient responsive to?
   - **A.** Verbal stimuli
   - **B.** Painful stimuli
   - **C.** Noxious stimuli
   - **D.** Unresponsive

4. What is this patient's chief complaint?
   - **A.** Lower right quadrant pain
   - **B.** Fever and chills
   - **C.** Dyspnea
   - **D.** Palpitations

5. What does the mnemonic SAMPLE stand for?

6. What is a pertinent negative?
   - **A.** A negative finding that requires further care and/or intervention
   - **B.** A negative finding that requires advanced life support
   - **C.** A negative finding that implies another condition may be present
   - **D.** A negative finding that requires no further care or intervention

7. Which of the following terms would be used to describe a pulse rate of 140 beats/min?
   - **A.** Bradycardia
   - **B.** Dyspnea
   - **C.** Tachycardia
   - **D.** Tachypnea

8. Describe an objective finding for the patient.

9. Describe a subjective finding for the patient.

10. What is a primary assessment?
   - **A.** A systematic physical examination
   - **B.** Physical examination performed at regular intervals
   - **C.** A process that identifies life threats
   - **D.** A process that identifies the nature of illness

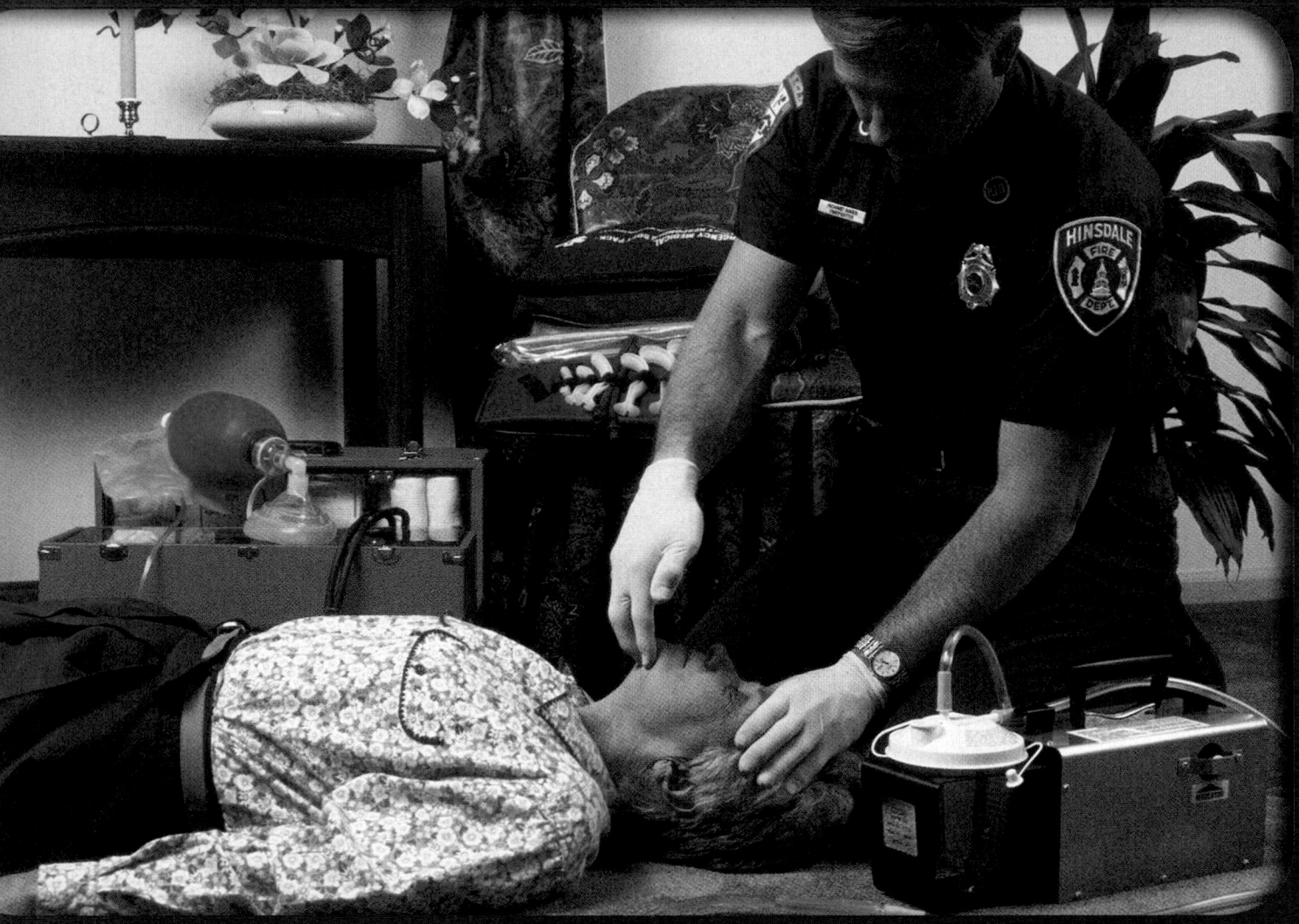

9 Airway Management

Airway Management

National EMS Education Standard Competencies

Airway Management, Respiration, and Artificial Ventilation

Applies knowledge of general anatomy and physiology to patient assessment and management in order to assure a patent airway, adequate mechanical ventilation, and respiration for patients of all ages.

Airway Management

- Airway anatomy (pp 319–323)
- Airway assessment (pp 331–335)
- Techniques of assuring a patent airway (pp 335–336)

Respiration

- Anatomy of the respiratory system (pp 331–335)
- Physiology and pathophysiology of respiration
  - Pulmonary ventilation (pp 324–327)
  - Oxygenation (p 327)
  - Respiration (pp 327–328)
    - External (pp 327–328)
    - Internal (p 328)
    - Cellular (p 328)
- Assessment and management of adequate and inadequate respiration (pp 333–335)
- Supplemental oxygen therapy (pp 346–353)

Artificial Ventilation

- Assessment and management of adequate and inadequate ventilation (pp 353–361)
- Artificial ventilation (pp 354–361)
- Minute ventilation (pp 325–326)
- Alveolar ventilation (p 325)
- Effect of artificial ventilation on cardiac output (p 355)

Pathophysiology

Applies fundamental knowledge of the pathophysiology of respiration and perfusion to patient assessment and management.

Knowledge Objectives

1. Describe the major structures of the respiratory system. (pp 319–323)
2. Discuss the physiology of breathing. (pp 323–328)
3. Give the signs of adequate breathing (p 331)
4. Give the signs of inadequate breathing. (pp 331–333)
5. Describe the assessment and care of a patient with apnea. (pp 333, 361–364)
6. Understand how to assess for adequate and inadequate respiration, including the use of pulse oximetry. (pp 333–335)
7. Understand how to assess for a patent airway. (pp 335–336)
8. Describe how to perform the head tilt–chin lift maneuver. (p 337)
9. Describe how to perform the jaw-thrust maneuver. (pp 337–338)
10. Explain how to measure and insert an oropharyngeal (oral) airway. (pp 338–340)
11. Describe how to measure and insert a nasopharyngeal (nasal) airway. (pp 340–342)
12. Understand the importance and techniques of suctioning. (pp 343–345)
13. Explain the use of the recovery position to maintain a clear airway. (p 346)
14. Describe the importance of giving supplemental oxygen to patients who are hypoxic. (p 346)
15. Understand the basics of how oxygen is stored and the various hazards associated with its use. (pp 346–351)
16. Describe the use of a nonrebreathing mask and state the oxygen flow requirements for its use. (pp 351–352)
17. Understand the indications for using a nasal cannula rather than a nonrebreathing face mask. (p 352)
18. Describe the indications for use of a humidifier during supplemental oxygen therapy. (p 353)
19. Explain the steps to take to perform mouth-to-mouth or mouth-to-mask ventilation. (pp 355–357)
20. Describe the use of a one-, two-, or three-person bag-mask device, and a manually triggered ventilation (MTV) device. (pp 356–361)
21. Describe the signs associated with adequate and inadequate artificial ventilation. (p 360)
22. Describe the use of continuous positive airway pressure (CPAP). (pp 361–364)
23. Understand how to recognize and care for a foreign body airway obstruction. (pp 366–368)

Skills Objectives

1. Demonstrate use of pulse oximetry. (pp 334–335, Skill Drill 9-1)
2. Demonstrate how to position the unconscious patient. (pp 335–336, Skill Drill 9-2)
3. Demonstrate the steps in performing the head tilt–chin lift maneuver. (p 337)
4. Demonstrate the steps in performing the jaw-thrust maneuver. (pp 337–338)
5. Demonstrate the insertion of an oral airway. (pp 339–340, Skill Drill 9-3)
6. Demonstrate the insertion of an oral airway with a 90° rotation. (pp 340–341, Skill Drill 9-4)
7. Demonstrate the insertion of a nasal airway. (pp 341–342, Skill Drill 9-5)
8. Demonstrate how to operate a suction unit. (p 344)
9. Demonstrate how to suction a patient's airway. (pp 344–345, Skill Drill 9-6)
10. Demonstrate how to place a patient in the recovery position. (p 346)
11. Demonstrate how to place an oxygen cylinder into service. (pp 349–350, Skill Drill 9-7)
12. Demonstrate the use of a partial rebreathing mask in providing supplemental oxygen therapy to patients. (p 352)
13. Demonstrate the use of a Venturi mask in providing supplemental oxygen therapy to patients. (pp 352–353)
14. Demonstrate the use of a humidifier in providing supplemental oxygen therapy to patients. (p 353)
15. Demonstrate how to assist a patient with ventilations using the bag-mask device for one and two rescuers. (p 354)
16. Demonstrate mouth-to-mask ventilation. (pp 356–357, Skill Drill 9-8)
17. Demonstrate the use of a manually triggered ventilation device to assist in delivering artificial ventilation to the patient (pp 360–361)
18. Demonstrate the use of an automatic transport ventilator to assist in delivering artificial ventilation to the patient. (p 361)
19. Demonstrate the use of CPAP. (pp 363–364, Skill Drill 9-9)

Introduction

Part of our existence is primarily based on the notion that we can adequately breathe oxygen. When the ability to breathe is disrupted, oxygen delivery to the body tissues and cells is compromised. Cells require a constant supply of oxygen to survive. Within seconds of being deprived of oxygen, vital organs such as the heart and brain may not function normally. Therefore, it is imperative that EMTs recognize airway and breathing inadequacies and correct them in a timely manner. Brain tissue will begin to die within 4 to 6 minutes.

Oxygen reaches body tissues and cells through two separate but related processes: breathing and circulation. During inhalation, oxygen moves from the atmosphere into the lungs, then passes from the air sacs in the lungs into the capillaries to oxygenate the blood. At the same time, carbon dioxide, produced by cells in the tissues of the body, moves from the blood into the air sacs through a process called **diffusion**. The blood, enriched with oxygen, travels through the body by the pumping action of the heart. The carbon dioxide then leaves our bodies during exhalation.

As an EMT, you must be able to locate the parts of the respiratory system, understand how the system works, and be able to recognize which patients are breathing adequately and which patients are breathing inadequately. This will enable you to determine how best to treat your patients.

This chapter reviews the anatomy, physiology, and pathophysiology of the respiratory system. It describes how to assess patients quickly and to carefully determine their airway and ventilation status. The equipment, procedures, and guidelines that you will need to manage a patient's airway and breathing are described in detail. You will learn several ways to open a patient's airway and specific techniques for removing foreign objects or fluids that may be blocking the airway. Because airway management equipment can be dangerous if used improperly, the chapter thoroughly discusses airway adjuncts, oxygen therapy devices, and artificial ventilation methods.

Anatomy of the Respiratory System

The respiratory system consists of all the structures in the body that make up the **airway** and help us breathe, or ventilate Figure 9-1. The airway is divided into the upper and lower airway. Structures that help us breathe include the diaphragm, the muscles of the chest wall, accessory muscles of breathing, and the nerves from the brain and spinal cord to those muscles. Ventilation is the exchange of air between the lungs and environment. The diaphragm and muscles of the chest wall are responsible for the regular rise and fall of the chest that accompany normal breathing.

Anatomy of the Upper Airway

The upper airway consists of all anatomic airway structures above the level of the vocal cords. These include the nose, mouth, jaw, oral cavity, pharynx, and larynx. Its major functions are to warm, filter, and humidify air as it enters the body through the nose and mouth. The pharynx (throat) is a muscular tube that extends from the nose and mouth to the level of the esophagus and trachea. The pharynx is composed of the nasopharynx, oropharynx, and the laryngopharynx (also called the hypopharynx) Figure 9-2. The laryngopharynx is the lowest portion of

You are the Provider: PART 1

You and your partner are dispatched to a residence at 145 Landa Street for a man with trouble breathing. The patient's wife, who called 9-1-1, told the dispatcher that her husband is "breathing funny" and is not responding to her appropriately. The time is 3:10 PM, the temperature outside is 39°F, and a fine mist is falling.

1. What is the function of the respiratory system?
2. What is the difference between ventilation and respiration?
3. How often should you assess a patient's airway and breathing status?

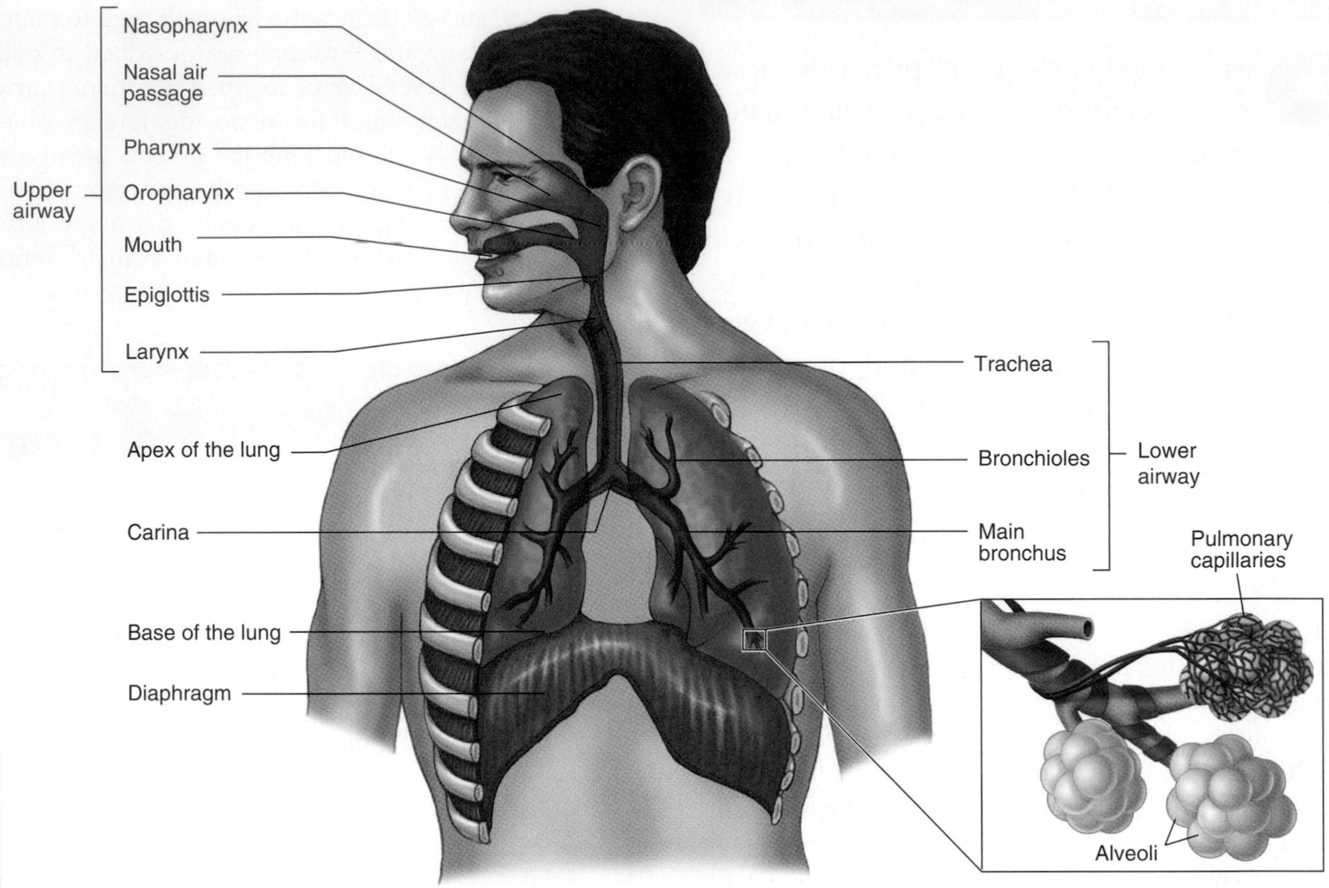

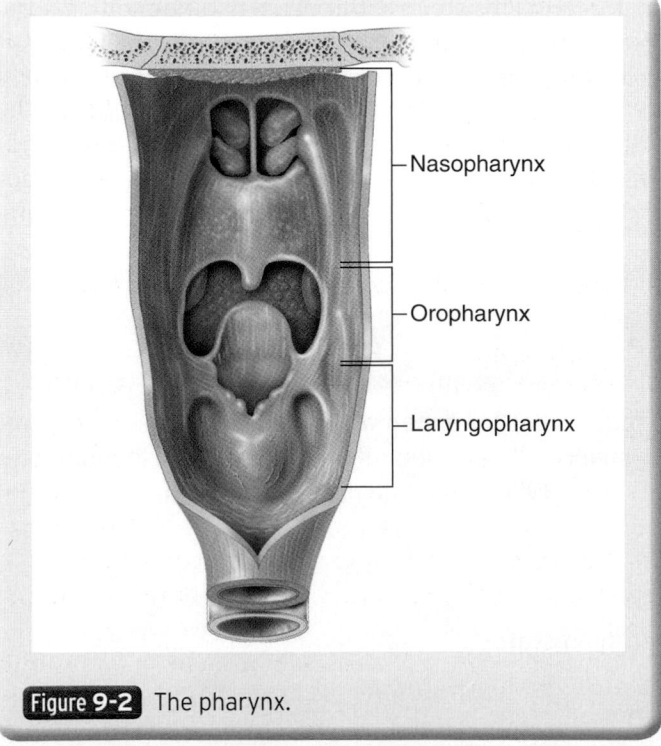

Nasopharynx

Nasal air passage

Pharynx

Oropharynx

Mouth

Epiglottis

Larynx

Upper airway

Apex of the lung

Carina

Base of the lung

Diaphragm

Trachea

Bronchioles

Main bronchus

Lower airway

Pulmonary capillaries

Alveoli

Figure 9-1 The upper and lower airways contain the structures in the body that help us breathe.

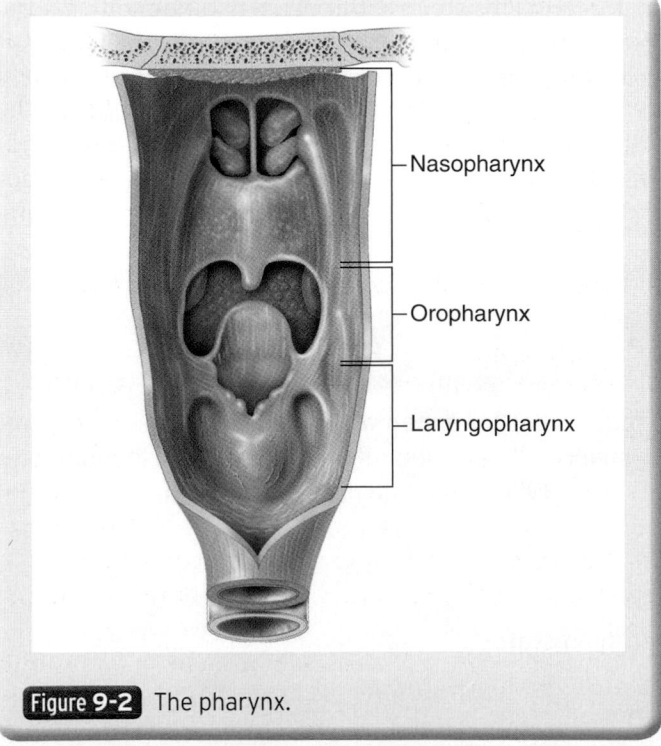

Nasopharynx

Oropharynx

Laryngopharynx

Figure 9-2 The pharynx.

the pharynx. At the base, it splits into two lumens, the larynx anteriorly and the esophagus posteriorly.

Nasopharynx

During inhalation, air typically enters the body through the nose and passes into the **nasopharynx**. The nasopharynx, which is formed by the union of the facial bones, is lined with a ciliated mucous membrane that keeps contaminants such as dust and other small particles out of the respiratory tract. In addition, the mucous membrane warms and humidifies air as it enters the body.

Oropharynx

The **oropharynx** forms the posterior portion of the oral cavity, which is bordered superiorly by the hard and soft palates, laterally by the cheeks, and inferiorly by the tongue **Figure 9-3**. The lips and mouth form the entrance to the oral cavity, which serves not only as an entrance for the respiratory system, but for the digestive system as well. Superior to the larynx, the epiglottis is

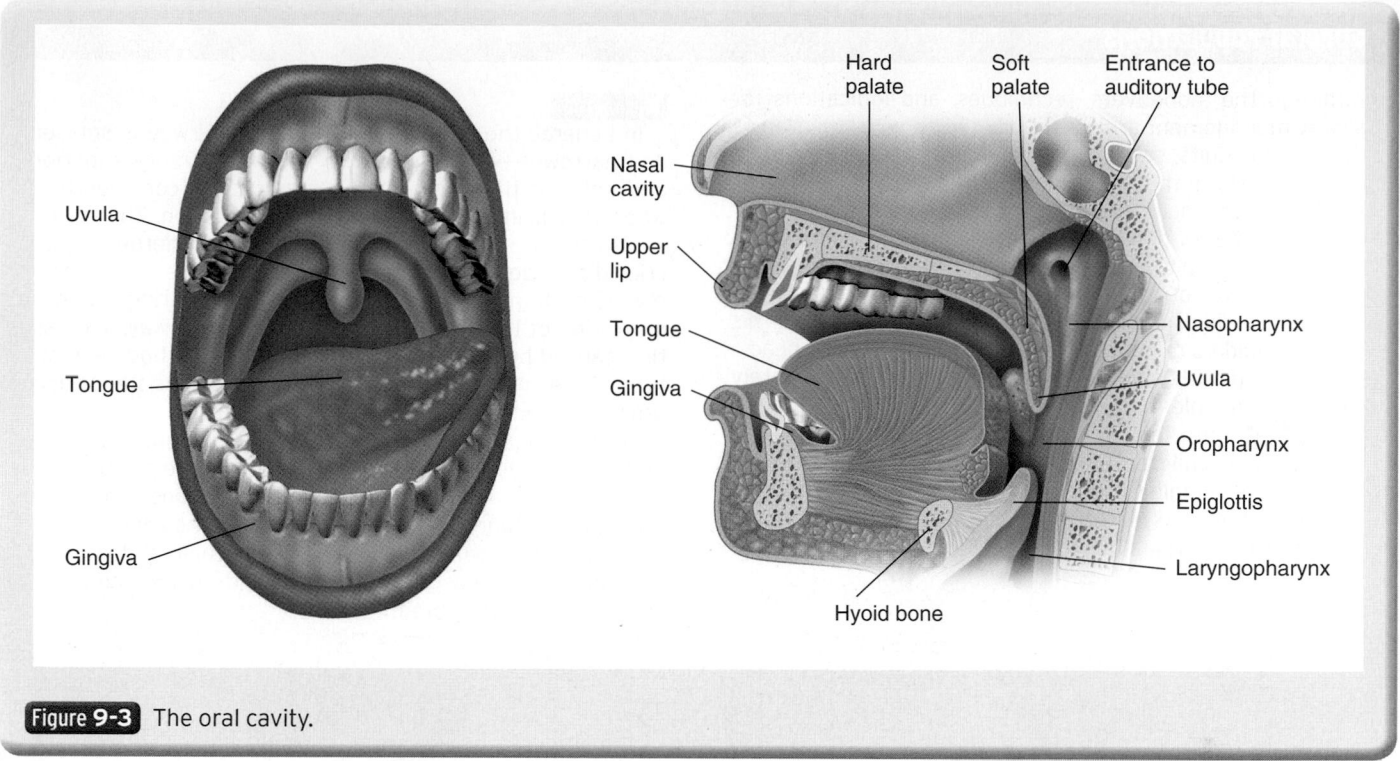

Figure 9-3 The oral cavity.

a leaf-shaped cartilaginous structure that helps separate the digestive system from the respiratory system. Its function is to prevent food and liquid from entering the larynx during swallowing. When swallowing occurs, the larynx is elevated and the epiglottis folds over the glottis to prevent **aspiration** of contents into the trachea.

Larynx

The **larynx** is a complex structure formed by many independent cartilaginous structures **Figure 9-6**. It marks where the upper airway ends and the lower airway begins.

The thyroid cartilage is a shield-shaped structure formed by two plates that join in a "V" shape anteriorly to form the laryngeal prominence known as the Adam's apple.

The cricoid cartilage, or cricoid ring, lies inferiorly to the thyroid cartilage; it forms the lowest portion of the larynx. The cricoid cartilage is the first ring of the trachea and the only lower airway structure that forms a complete ring.

The **glottis**, also called the glottic opening, is the space in between the vocal cords and the narrowest portion of the adult's airway. The lateral borders of the glottis are the **vocal cords**. These white bands of thin muscle tissue are partially separated at rest and serve as the primary center for speech production. In addition, the vocal cords contain defense reflexes that protect the lower airway, causing a spasmodic closure to the lower airway to prevent substances from entering the trachea (eg, water, vomitus).

■ Anatomy of the Lower Airway

The function of the lower airway is to exchange oxygen and carbon dioxide. Its external boundaries are the fourth cervical vertebra and the xiphoid process, which is the narrow cartilaginous lower tip of the sternum. Internally, it spans the glottis to the pulmonary capillary membrane.

The trachea, or windpipe, is the conduit for air entry into the lungs. This tubular structure is approximately 10 to 12 cm in length and consists of C-shaped cartilaginous rings. The trachea begins directly below the cricoid cartilage and descends anteriorly down the midline of the neck into the thoracic cavity. Once in the thoracic cavity, the trachea divides at the level of the **carina** into the two mainstem bronchi (right and left). The hollow bronchi are supported by cartilage and distribute air into the right and left lungs.

The lungs consist of the entire mass of tissue that includes the smaller bronchi, bronchioles, and alveoli **Figure 9-7**. The lungs are surrounded by a serous membrane called the pleura. All lung tissue is covered with a thin, slippery outer membrane called the **visceral pleura**. The **parietal pleura** lines the inside of the thoracic cavity. A small amount of fluid is found between these two layers and serves as a lubricant to prevent friction during breathing.

On entering the lungs, each bronchus divides into increasingly smaller bronchi, which in turn subdivide into bronchioles. The **bronchioles** are thin, hollow

Special Populations

Although the maneuvers, techniques, and indications for airway management are essentially the same in children as they are in adults, several anatomic differences in the child make mastery of these techniques difficult.

Infants and small children have a proportionately larger occiput (posterior portion of the cranium), which causes the head to flex when the child lies supine; this position itself can cause an airway obstruction. When positioning the airway of an infant or child, you should place a folded towel under his or her shoulders to maintain a neutral position of the head.

Compared with adults, children have a proportionately smaller mandible and a proportionately larger tongue Figure 9-4 . Both factors increase the incidence of airway obstruction in children.

The child's epiglottis is more floppy and omega-shaped than an adult's. As a consequence, it must be lifted out of the way to visualize the vocal cords for intubation Figure 9-5 .

In general, the infant's and the child's airway is smaller and narrower at all levels. The larynx lies more superior and anterior than an adult's—an important consideration when visualizing the vocal cords for intubation. The larynx is also funnel-shaped due to the narrow, underdeveloped cricoid cartilage. In children younger than 8 years, the narrowest portion of the airway is at the cricoid ring. Further narrowing of the child's inherently narrow airway, such as that caused by soft-tissue swelling or foreign body aspiration, can result in a major decrease in airway resistance and breathing inadequacy.

Children do not have well-developed chest musculature, and their ribs and cartilage are softer and more pliable than an adult's. As a result, the thoracic cavity cannot optimally contribute to lung expansion. Children rely heavily on their diaphragm for breathing, which moves their abdomen in and out. For this reason, infants and children are commonly referred to as "belly breathers."

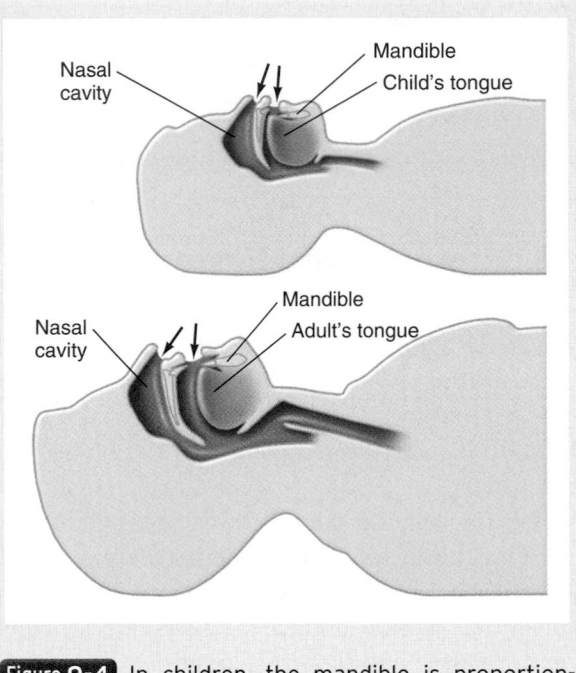

Figure 9-4 In children, the mandible is proportionately smaller and the tongue is proportionately larger than in an adult.

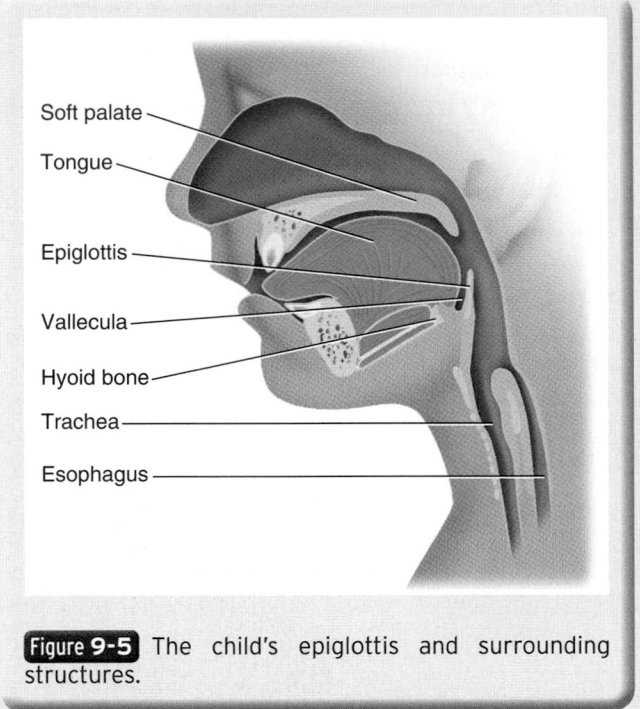

Figure 9-5 The child's epiglottis and surrounding structures.

tubes made of smooth muscle. The tone of these smooth muscles allows the bronchioles to dilate or constrict in response to various stimuli. The smaller bronchioles branch into alveolar ducts that end at the alveolar sacs.

The alveoli, located at the end of the airway, are millions of thin-walled, balloon-like sacs that serve as the functional site for the exchange of oxygen and carbon dioxide. Surrounding each of these sacs is an intricate bed of blood vessels, known as pulmonary capillaries. Oxygen diffuses through the lining of the alveoli into the pulmonary capillaries where, depending on adequate

blood volume and pressure, it is carried back to the heart for distribution to the rest of the body. At the same time, carbon dioxide (waste) diffuses from the pulmonary capillaries into the alveoli where it is exhaled and removed from the body.

The chest cage (thoracic cavity) contains the lungs, one on each side Figure 9-8 . The boundaries of the thorax are the rib cage anteriorly, superiorly, and posteriorly and the diaphragm inferiorly. Each individual rib plays a part in the overall protection of the thorax. In between each rib are intercostal muscles that can assist with breathing; however, they generally are not used unless the patient

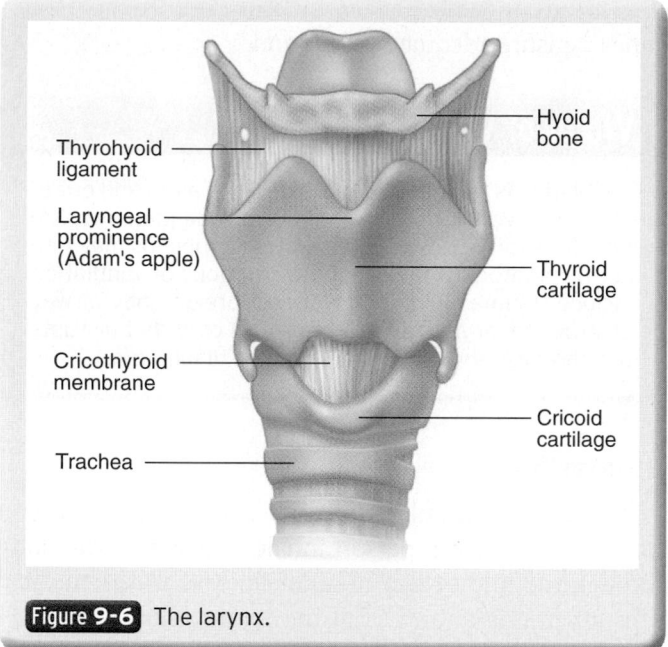

Figure 9-6 The larynx.

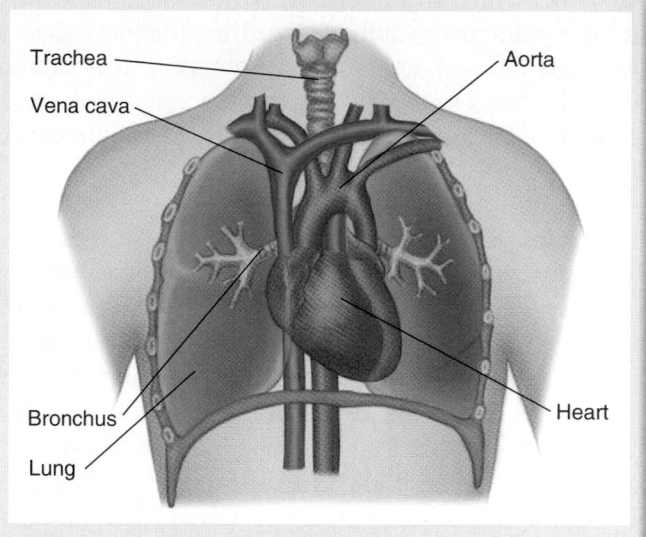

Figure 9-8 The thoracic cavity contains important anatomic structures for respiration, including the lungs and bronchi, heart, great vessels (the vena cava and aorta), and trachea.

is in respiratory distress. Within the chest cage, you will find the lungs, which hang freely within the chest cavity. Between the layers of the visceral and parietal pleura is a small amount of lubricating fluid that protects the lungs from friction during breathing. Between the lungs is a space called the **mediastinum**, which is surrounded by tough connective tissue. This space contains the heart, the great vessels, the esophagus, the trachea, the major bronchi, and many nerves. The mediastinum effectively separates the right lung space from the left lung space. In addition to the respiratory and circulatory structures found in the chest cage, an important structure of the nervous system is also found in the thorax—the **phrenic nerve**.

Originating from the cervical plexus of nerves in the neck, the phrenic nerve is one of the most important nervous structures in the body. The phrenic nerve innervates the diaphragm muscle, allowing it to contract. Contraction of the diaphragm occurs in a downward direction and is necessary for adequate breathing to occur.

Physiology of Breathing

The respiratory and cardiovascular systems work together to ensure that a constant supply of oxygen and nutrients are delivered to every cell in the body and carbon dioxide and waste products are removed from every cell. The following sections will describe the process of ventilation, oxygenation, and respiration; however, you first need to understand how the processes of breathing and circulation are connected.

As described earlier, air enters the body through the oral and nasal cavities and travels into the laryngopharynx. It passes through the vocal cords, into the glottis, and down the trachea, where it is distributed through the mainstem bronchi into the bronchioles of the lungs. This occurs because a negative pressure is created in the chest. Eventually the air reaches the alveolar sacs

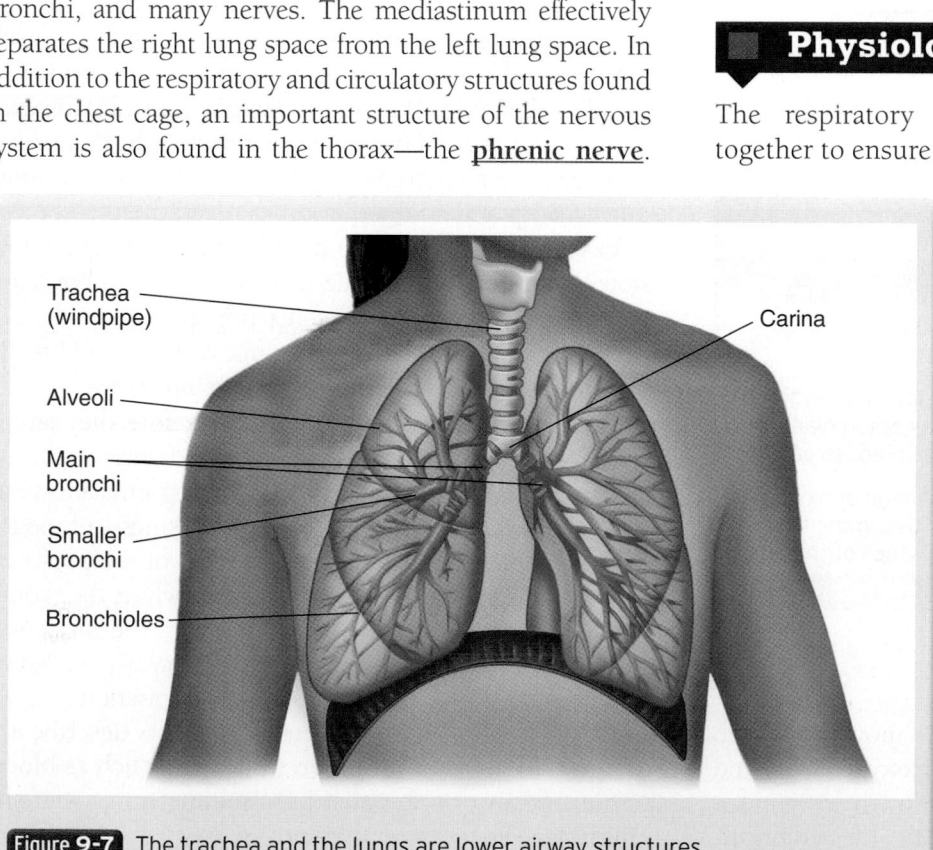

Figure 9-7 The trachea and the lungs are lower airway structures.

where the oxygen is diffused across the alveolar membrane into the pulmonary capillaries. At the same time, carbon dioxide is diffused across this membrane and is exhaled from the body. The oxygen in the pulmonary capillaries is transported back to the heart, where it is distributed to the rest of the body.

The heart pumps blood to the tissues of the body through a series of arteries and veins. Arteries carry oxygenated blood away from the heart and branch into arterioles and capillaries. Once in the capillaries, the exchange of nutrients and waste products takes place. Oxygen and nutrients leave the capillaries and enter the cells. At the same time, waste products, such as carbon dioxide, diffuse from the cells back into the blood of the capillaries. From here, the blood travels through a series of venules that connect to larger veins. All veins carry deoxygenated blood to the heart. The deoxygenated blood enters the right side of the heart through the right atrium, where it is pumped through the tricuspid valve, right ventricle, and pulmonary artery before being pumped to the lungs for oxygenation and removal of carbon dioxide. The oxygenated blood then travels through the pulmonary vein to the left atrium through the bicuspid valve and into the left ventricle, where it is again pumped to the rest of the body. Refer back to Chapter 5, *The Human Body*, for an illustration of this process.

It is important to understand that the respiratory and circulatory systems work together to facilitate oxygen delivery to the tissues of the body **Table 9-1**. When one of these systems is compromised, oxygen delivery is not effective and cellular death could result.

Table 9-1	Ventilation, Oxygenation, and Respiration
Function	**Definition**
Ventilation	The physical act of moving air into and out of the lungs
Oxygenation	The process of loading oxygen molecules onto hemoglobin molecules in the bloodstream
Respiration	The actual exchange of oxygen and carbon dioxide in the alveoli as well as the tissues of the body

■ Ventilation

Pulmonary ventilation, the process of moving air into and out of the lungs, is necessary for oxygenation and respiration to occur. Adequate, continuous ventilation is essential for life and therefore is one of the highest priorities in treating any patient. If a patient is not breathing,

or is breathing inadequately, you must immediately intervene to ensure adequate ventilation.

Words of Wisdom

Ventilation is the physical act of moving air in and out of the lungs. Ventilation is required for adequate respiration. If ventilation is adequate, other problems may hinder respiration. Examples of interruptions of ventilation include trauma such as flail chest, foreign body airway obstruction, or an injury to the spinal cord that disrupts the phrenic nerve that innervates the diaphragm.

Inhalation

The active, muscular part of breathing is called **inhalation**. When a person inhales, air enters the body through the mouth and nose and moves to the trachea. This air travels to and from the lungs, filling and emptying the alveoli. During inhalation, the diaphragm and intercostal muscles contract. When the diaphragm contracts, it moves down slightly and enlarges the thoracic cage from top to bottom, and when the intercostal muscles contract, they lift the ribs up and out. The combined actions of these structures enlarge the thorax in all directions. Take a deep breath to see how your chest expands.

The diaphragm is a specialized skeletal muscle. Innervated by the phrenic nerve, the diaphragm is attached to the costal arch and the vertebrae and functions as a voluntary and involuntary muscle. It acts as a voluntary muscle when taking a deep breath, coughing, or holding your breath—all actions that are under our control. However, unlike other skeletal or voluntary muscles, the diaphragm also performs an automatic function. Breathing continues during sleep and at all other times. Even though you can hold your breath or temporarily breathe more quickly or slowly, you cannot continue these variations in breathing indefinitely. When the concentration of carbon dioxide rises within the blood, the autonomic regulation of breathing resumes under the control of the brain stem.

The lungs have no muscle tissue; therefore, they cannot move on their own. They need the help of other structures to be able to expand and contract during inhalation and exhalation. Therefore, the ability of the lungs to function properly is dependent on the movement of the chest and supporting structures. These structures include the thorax, the thoracic cage (chest), the diaphragm, the intercostal muscles, and the accessory muscles of breathing. Accessory muscles are secondary muscles of respiration.

Partial pressure is the term used to describe the amount of gas in air or dissolved in fluid, such as blood. Partial pressure is measured in millimeters of mercury (mm Hg). The partial pressure of oxygen in air residing in the alveoli (Pa_{O_2}) is 104 mm Hg. Carbon dioxide

(CO_2) enters the alveoli from the blood and causes a CO_2 partial pressure of 40 mm Hg.

Oxygenated arterial blood from the heart has a partial pressure of oxygen (Pao_2) that is lower than the partial pressure of carbon dioxide ($Paco_2$) in the pulmonary capillaries. The body attempts to equalize the partial pressure, which results in oxygen diffusion across the membrane into the blood; carbon dioxide diffuses into the alveoli and is eliminated as waste during exhalation. Oxygen and carbon dioxide both diffuse until partial pressure in the air and blood is equal. This process occurs in reverse when the arterial blood reaches the tissues. Oxygen diffuses into the tissue fluid and then into the cells, and carbon dioxide diffuses out of the cells into the tissue fluid and blood.

The air pressure outside the body, called the atmospheric pressure, is normally higher than the air pressure within the thorax. During inhalation, the thoracic cage expands and the air pressure within the thorax decreases, creating a slight vacuum. This pulls air in through the trachea, causing the lungs to fill. When the air pressure outside equals the air pressure inside, air stops moving. Gases, such as oxygen, will move from an area of higher pressure to an area of lower pressure until the pressures are equal. At this point, the air stops moving, and inhalation stops.

It may help you to understand this if you think of the thoracic cage as a bell jar in which balloons are suspended. In this example, the balloons are the lungs. The base of the jar is the diaphragm, which moves up and down slightly with each breath. The ribs, which are the sides of the jar, maintain the shape of the chest. The only opening into the jar is a small tube at the top, similar to the trachea. During inhalation, the bottom of the jar moves down slightly, causing a decrease in pressure in the jar and creating a slight vacuum. As a result, the balloons fill with air **Figure 9-9**.

The entire process of inspiration is focused on delivering oxygen to the alveoli. However, not all of the air you breathe actually reaches the alveoli. The volume of air that reaches the alveoli is referred to as alveolar ventilation. **Alveolar ventilation** is determined by subtracting the amount of *dead space* air from the *tidal volume*. **Tidal volume**, a measure of depth of breathing, is the amount of air in milliliters (mL) that is moved into or out of the lungs during a single breath. The average tidal volume for a man is approximately 500 mL. Breathing becomes deeper as the tidal volume responds to the increased metabolic demand for oxygen. However, as noted previously, not all inspired air reaches the alveoli for gas exchange. **Dead space** is the portion of the tidal volume that does not reach the alveoli and thus does not participate in gas exchange. Anatomic dead space contains the air that remains in the mouth, nose, trachea, bronchi, and larger bronchioles. This can add up to approximately 150 mL in an adult.

Minute ventilation, also referred to as minute volume, is the amount of air moved through the lungs in 1 minute minus the dead space. Minute volume can be calculated by subtracting the dead space from the tidal volume, then multiplying that number by the respiratory rate. Therefore, if a patient has a respiratory rate of 12 breaths/min, a tidal volume of 500 mL per breath, and a dead space of 150 mL, the minute volume would be 4,200 mL (4.2 L). It is important to note that variations in tidal volume, respiratory rate, or both will affect minute volume. For example, if a patient is breathing at a rate of 12 breaths/min, but the tidal volume is reduced (shallow breathing), minute volume will decrease. Likewise, if a

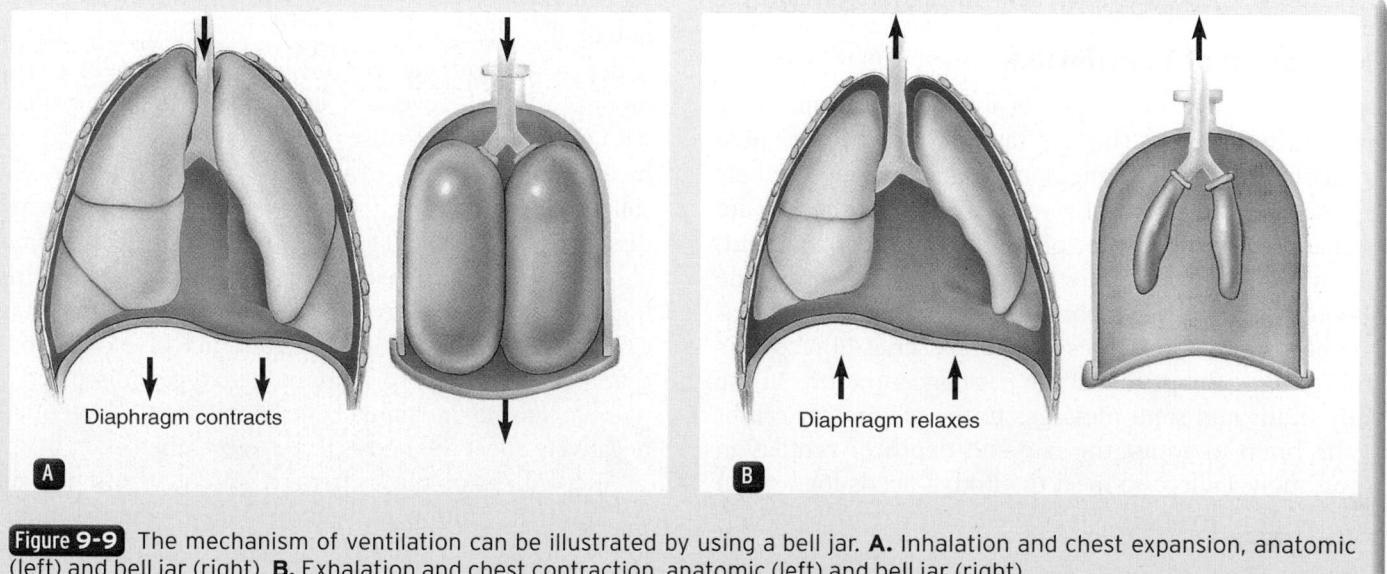

Figure 9-9 The mechanism of ventilation can be illustrated by using a bell jar. **A.** Inhalation and chest expansion, anatomic (left) and bell jar (right). **B.** Exhalation and chest contraction, anatomic (left) and bell jar (right).

patient is breathing at a rate of 12 breaths/min and the tidal volume increases (deep breathing), minute volume will increase.

<u>Vital capacity</u> refers to the amount of air that can be forcibly expelled from the lungs after breathing deeply. However, even if you exhale forcefully, you cannot completely empty your lungs of air. The air that remains after maximal expiration is known as the **residual volume**, which is approximately 1,200 mL in the average adult male. This is one of the reasons why lay rescuers are able to perform cardiopulmonary resuscitation (CPR) without providing ventilations and still manage to circulate oxygen.

Exhalation

Unlike inhalation, **exhalation** does not normally require muscular effort; therefore, it is a passive process. During exhalation, the diaphragm and the intercostal muscles relax. In response, the thorax decreases in size, and the ribs and muscles assume a normal resting position. When the size of the thoracic cage decreases, air in the lungs is compressed into a smaller space. The air pressure within the thorax then becomes higher than the outside pressure, and the air is pushed out through the trachea.

Remember that air will reach the lungs only if it travels through the trachea. This is why clearing and maintaining an open airway is so important. Clearing the airway means removing obstructing material, tissue, or fluids from the nose, mouth, and throat. Maintaining the airway means keeping the airway **patent** so that air can enter and leave the lungs freely **Figure 9-10**.

Air may also pass into the chest cavity through an abnormal opening in the throat or chest wall as a result of trauma, remaining outside the bronchi and never reaching the alveoli. In later chapters, you will learn how to recognize and manage these potentially life-threatening conditions.

Regulation of Ventilation

The body's need for oxygen is dynamic, meaning it is constantly changing. The respiratory system must be able to accommodate the changes in oxygen demand by altering the rate and depth of ventilation. These changes are regulated primarily by the pH of the cerebrospinal fluid, which is directly related to the amount of carbon dioxide dissolved in the plasma portion of the blood. The regulation of ventilation involves a complex series of receptors and feedback loops that sense gas concentrations in the body fluids and send messages to the respiratory center in the brain to adjust the rate and depth of ventilation accordingly. Failure to meet the body's needs for oxygen may result in **hypoxia**. Hypoxia is an extremely dangerous condition in which the tissues and cells of the body do not get enough oxygen. If this process is not corrected, patients may die quite quickly.

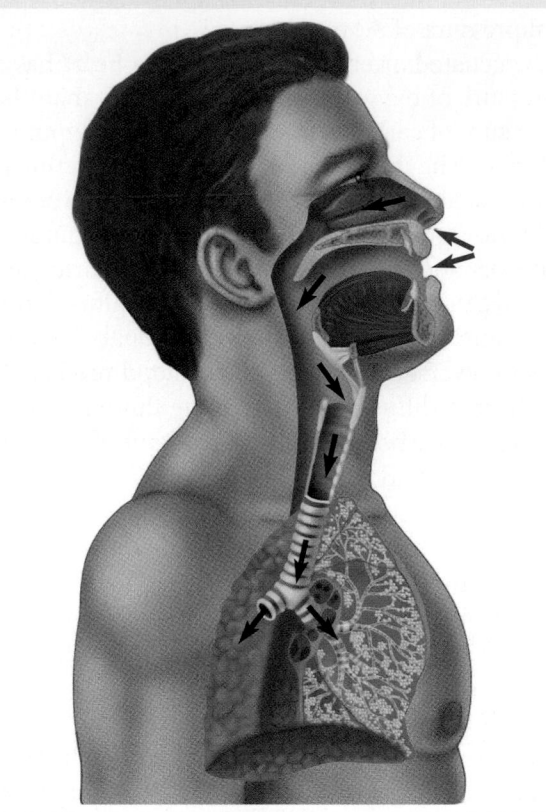

Figure 9-10 Air reaches the lungs only if it travels through the trachea. Maintaining the airway means keeping the airway patent so that air can enter and leave the lungs freely.

For most people, the drive to breathe is based on pH changes (related to carbon dioxide levels) in the blood and cerebrospinal fluid. However, patients with chronic obstructive pulmonary diseases (COPD) have difficulty eliminating carbon dioxide through exhalation; thus, they always have higher levels of carbon dioxide. This potentially alters their drive for breathing. The theory is that respiratory centers in the brain gradually accommodate to high levels of carbon dioxide. In patients with COPD, the body uses a "backup system" to control breathing. This theory of secondary control of breathing, called **hypoxic drive**, is based on levels of oxygen dissolved in plasma. This is different from the primary control of breathing that uses carbon dioxide as the driving force. Hypoxic drive is typically found in end-stage COPD. Providing high concentrations of oxygen over time will increase the amount of oxygen dissolved in plasma. However, many believe this could potentially negatively affect the body's drive to breathe.

Regardless of the current research, it still remains certain that caution should be taken when administering high concentrations of oxygen to patients with obstructive pulmonary disease. However, it is important to remember that high concentrations of oxygen should never be withheld from any patient who needs it. Patients with severe respiratory and/or circulatory compromise should receive high concentrations of oxygen regardless of their underlying medical conditions.

Patients who are breathing inadequately will show varying signs and symptoms of hypoxia. The onset and degree of tissue damage caused by hypoxia often depend on the quality of ventilations. Early signs of hypoxia include restlessness, irritability, apprehension, fast heart rate (tachycardia), and anxiety. Late signs of hypoxia include mental status changes, a weak (thready) pulse, and cyanosis. Conscious patients will complain of shortness of breath (**dyspnea**) and may not be able to talk in complete sentences. The best time to give a patient oxygen is before signs and symptoms of hypoxia appear.

■ Oxygenation

Oxygenation is the process of loading oxygen molecules onto hemoglobin molecules in the bloodstream. Adequate oxygenation is required for internal respiration to take place; however, it does not guarantee internal respiration is taking place. Oxygenation requires that the air used for ventilation contains an adequate percentage of oxygen. While you generally cannot oxygenate without ventilation, it is possible to ventilate without oxygenation. This situation occurs in places where oxygen levels in the breathing air have been depleted, such as in mines and confined spaces. Ventilation without adequate oxygenation also occurs in climbers who ascend too quickly to an altitude of lower atmospheric pressure. At high altitudes, the percentage of oxygen remains the same, but the atmospheric pressure makes it difficult to adequately bring sufficient amounts of oxygen into the body.

Words of Wisdom

Oxygenation can be disrupted through carbon monoxide poisoning. Carbon monoxide has a much greater affinity for hemoglobin than oxygen (250 times more), thus not allowing for proper transport of oxygen to tissues.

■ Respiration

All living cells perform a specific function and need energy to survive. Cells take energy from nutrients through a series of chemical processes. The name given to these processes as a whole is **metabolism (cellular respiration)**. During metabolism, each cell combines nutrients (such as sugar) and oxygen and produces energy and waste products, primarily water and carbon dioxide. Each cell in the body requires a continuous supply of oxygen and a regular means of disposing of waste (carbon dioxide). The body provides for these requirements through respiration.

Respiration is the process of exchanging oxygen and carbon dioxide. This exchange occurs by diffusion, a process in which a gas moves from an area of greater concentration to an area of lower concentration. In the body, gases diffuse rapidly across a short distance of only micrometers, and the diffusion occurs rapidly.

External Respiration

External respiration (pulmonary respiration) is the process of breathing fresh air into the respiratory system and exchanging oxygen and carbon dioxide between the alveoli and the blood in the pulmonary capillaries **Figure 9-11**.

Fresh air that is inspired into the lungs contains about 21% oxygen, 78% nitrogen, and 0.3% carbon dioxide. As this air reaches the alveoli, it comes into contact with a combination of phospholipids called **surfactant**. Surfactant reduces surface tension within the alveoli and keeps them expanded, thus making it easier for the gas exchange between oxygen and carbon dioxide to take place. It is important to remember that while adequate ventilation is necessary for external respiration to take place, it does not guarantee that external respiration is being achieved.

Once the oxygen crosses the alveolar membrane, it is bound to hemoglobin, an iron-containing molecule that has a great affinity for oxygen molecules. Found in red blood cells, hemoglobin molecules low in oxygen concentration are pumped from the right side of the heart into the capillaries of the pulmonary circulation. The capillaries surround alveoli containing high concentrations

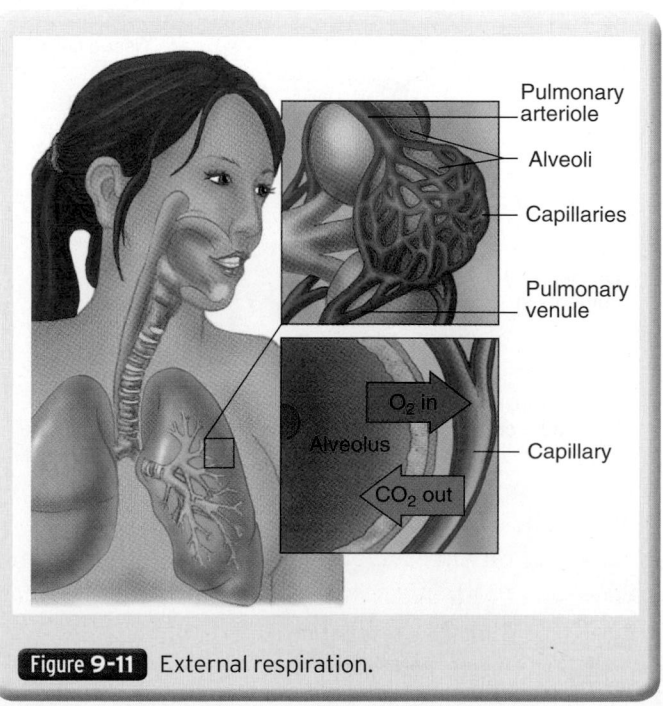

Figure 9-11 External respiration.

of oxygen (from inspired air). The hemoglobin molecules pick up fresh oxygen as it crosses the alveolar membrane and transport it back to the left side of the heart, where it is pumped out to the rest of the body. Under normal conditions, 96% to 100% of the hemoglobin receptor sites contain oxygen.

Internal Respiration

The exchange of oxygen and carbon dioxide between the systemic circulatory system and the cells of the body is called __internal respiration__. As blood travels through the body, it supplies oxygen and nutrients to various tissues and cells. As the oxygenated blood travels through the arteries and capillaries, the oxygen passes from the blood in the capillaries to tissue cells, while carbon dioxide and cell waste pass in the opposite direction: from tissue cells through capillaries and into the veins **Figure 9-12**.

Every cell in the body needs a constant supply of oxygen to survive. Whereas some tissues are more resilient than others, eventually all cells will die if deprived of oxygen **Figure 9-13**. To deliver adequate amounts of oxygen to the tissues of the body, sufficient levels of external ventilation and perfusion must take place.

In the presence of oxygen, the mitochondria of the cells convert glucose into energy through a process known as __aerobic metabolism__. Energy in the form of adenosine triphosphate (ATP) is produced through a series of processes known as the Krebs cycle and oxidative phosphorylation. Together, these chemical processes yield nearly 40 molecules of energy-rich ATP for each molecule of glucose metabolized. Without adequate oxygen, the cells do not completely convert glucose

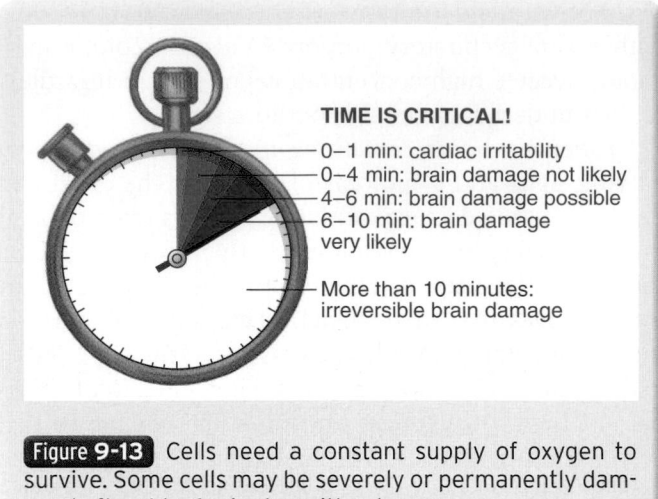

TIME IS CRITICAL!

- 0–1 min: cardiac irritability
- 0–4 min: brain damage not likely
- 4–6 min: brain damage possible
- 6–10 min: brain damage very likely
- More than 10 minutes: irreversible brain damage

Figure 9-13 Cells need a constant supply of oxygen to survive. Some cells may be severely or permanently damaged after 4 to 6 minutes without oxygen.

into energy, and lactic acid and other toxins accumulate in the cell. This process, __anaerobic metabolism__, cannot meet the metabolic demands of the cell. Although another intracellular process, glycolysis, also contributes to ATP production and does not require oxygen, this process results in less ATP production, and lactic acid waste products and toxins are produced. If this process is not corrected, the cells will eventually die. This is why adequate levels of perfusion (circulation of blood within an organ or tissue) and external ventilation must be present for aerobic internal respiration to take place. However, while these elements are necessary for internal respiration, they do not guarantee that aerobic internal respiration will take place.

Words of Wisdom

Oxygen is one of the most powerful medications for saving lives that you will carry in your ambulance. Don't be afraid to use it!

When the mitochondria within each cell use oxygen to convert glucose to energy, carbon dioxide, the main waste product, accumulates in the cell. Carbon dioxide is then transported through the circulatory system and back to the lungs for exhalation.

Understanding the process of ventilation, oxygenation, and respiration is an important concept for the EMT. The overall goal of these mechanisms is to deliver an adequate supply of oxygen to the cells of the body. When one of these processes fails or becomes disrupted, cells are going to die. By recognizing the signs and symptoms of inadequate tissue perfusion and oxygenation, the EMT can immediately intervene and correct a potentially life-threatening condition.

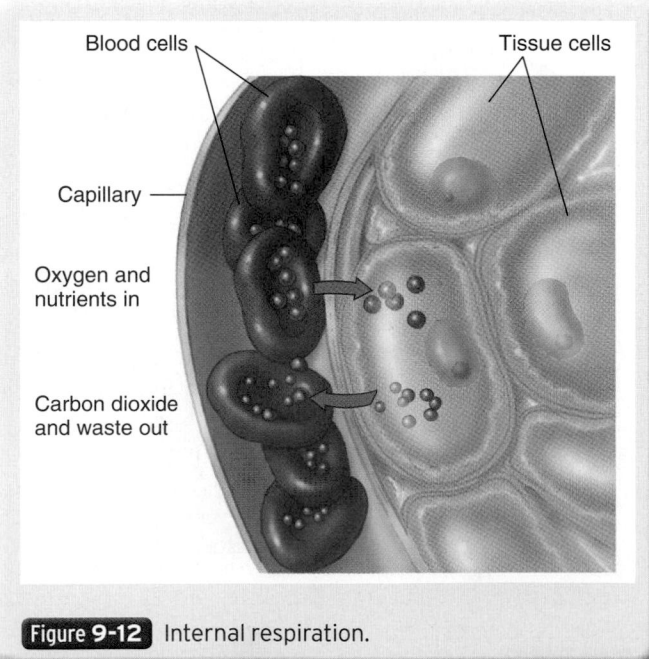

Blood cells

Tissue cells

Capillary

Oxygen and nutrients in

Carbon dioxide and waste out

Figure 9-12 Internal respiration.

Pathophysiology of Respiration

Multiple conditions inhibit the body's ability to effectively deliver oxygen to the cells. Disruption of pulmonary ventilation, oxygenation, and respiration will cause immediate effects on the body. As an EMT, you need to recognize these conditions and correct them in a timely manner.

Factors in the Nervous System

Chemical factors are commonly involved in respiratory control issues because of the level of complexity of the human body. A complex series of chemical reactions are constantly taking place. For example, <u>chemoreceptors</u> monitor the levels of oxygen, carbon dioxide, hydrogen ions, and the pH of the cerebrospinal fluid and then provide feedback to the respiratory centers to modify the rate and depth of breathing based on the body's needs at any given time. Central chemoreceptors in the medulla respond quickly to slight elevations in carbon dioxide, or a decrease in the pH of the cerebrospinal fluid. The peripheral chemoreceptors, located in the carotid arteries and the aortic arch, are sensitive to decreased levels of oxygen in arterial blood as well as to low pH levels.

When serum carbon dioxide or hydrogen ion levels increase because of medical or traumatic conditions involving the respiratory system, chemoreceptors stimulate the dorsal and ventral respiratory groups in the medulla to increase the respiratory rate, thus removing more carbon dioxide or acid from the body. The dorsal respiratory group is responsible for initiating inspiration based on the information received from the chemoreceptors. The ventral respiratory group is primarily responsible for motor control of the inspiratory and expiratory muscles.

In addition, the dorsal respiratory group and the ventral respiratory group are affected by the apneustic center and the pneumotaxic (pontine) center of the pons. The apneustic center stimulates the dorsal respiratory group, resulting in longer, slower respirations. The pneumotaxic center helps in shutting off the dorsal respiratory group, resulting in shorter, faster respirations. If one item in this process is disrupted, then the respiratory process will be affected.

Ventilation/Perfusion Ratio and Mismatch

The lung has a functional role of placing ambient air in proximity to circulating blood to permit gas exchange by simple diffusion. To accomplish this action, air and blood flow must be directed to the same place at the same time. In other words, ventilation and perfusion must be matched. A failure to match ventilation and perfusion, or $\dot{V}/\dot{Q}$ ratio mismatch, lies behind most abnormalities in oxygen and carbon dioxide exchange.

In most patients, the normal resting minute ventilation is approximately 6 L/min. About one third of this volume fills dead space; therefore, resting alveolar ventilation is approximately 4 L/min. However, pulmonary artery blood flow is approximately 5 L/min. This yields an overall ratio of ventilation to perfusion of 4/5 L/min or 0.8 L/min. Since neither ventilation nor perfusion is distributed equally, both are distributed to dependent regions at rest. However, the increase in gravity-dependent flow is more marked with perfusion (blood) than with ventilation (air). Hence, the ratio of ventilation to perfusion is highest at the apex of the lung and lowest at the base.

When ventilation is compromised but perfusion continues, blood passes over some alveolar membranes without gas exchange taking place; therefore, not all alveoli are enriched with oxygen. This, in turn, results in a lack of oxygen diffusing across the membrane and into blood circulation. Along the same lines, carbon dioxide is also not able to diffuse across the membrane and is recirculated in the bloodstream. This condition results in a $\dot{V}/\dot{Q}$ ratio mismatch and could lead to severe hypoxemia if this problem is not recognized and treated.

Similar problems can occur when perfusion across the alveolar membrane is disrupted. Even though the alveoli are filled with fresh oxygen, disruption in blood flow does not allow for optimal exchange in gases across the membrane. This results in less oxygen absorption in the bloodstream and less carbon dioxide removal. This $\dot{V}/\dot{Q}$ ratio mismatch can also lead to hypoxemia and you need to provide immediate intervention to prevent further damage or death.

Factors Affecting Pulmonary Ventilation

Maintaining a patent airway is critical to the delivery of oxygen to the tissues of the body. There are many intrinsic and extrinsic factors that cause airway obstructions. Intrinsic conditions such as infections, allergic reactions, and unresponsiveness (tongue obstruction) can cause significant restrictions on the ability to maintain an open airway. Swelling from infections and allergic reactions can be fatal if not aggressively managed with medications and possibly advanced airway maneuvers. The tongue is the most common airway obstruction in the unresponsive patient. This airway obstruction, while easily corrected, can result in hypoxia and hinder adequate tissue perfusion. Snoring respirations and the position of the head and/or neck are good indicators that the tongue may be obstructing the airway. Prompt correction of this obstruction is necessary for adequate oxygenation.

Some factors affecting pulmonary ventilation are not necessarily directly part of the respiratory system. The central and peripheral nervous systems play key roles in the regulation of breathing. Interruptions to these systems can have a drastic effect on the ability to breathe efficiently. Medications that depress the central nervous system lower the respiratory rate and tidal volume. This lower rate and volume will decrease the overall minute volume as well as alveolar ventilation. As a result, this increases the amount of carbon dioxide in the respiratory and circulatory systems, resulting in an overall increase of carbon dioxide levels in the bloodstream, known as **hypercarbia**. Trauma to the head and spinal cord can also interrupt nervous control of ventilation, resulting in decreased respiratory function and even failure. In addition to medications and trauma, conditions such as muscular dystrophy can also affect nervous control. This disease causes degeneration of muscle fibers resulting in a gradual weakening of muscles, slowing motor development, and loss of muscle contractility. Curvature of the spine is also likely in patients with muscular dystrophy and can impair pulmonary function.

Patients with allergic reactions not only suffer from a potential airway obstruction from swelling, but may also have a decrease in pulmonary ventilation from bronchoconstriction. As the bronchioles constrict, air is forced through smaller lumens resulting in decreased ventilation. This condition is also found in patients suffering from COPD, such as asthma and emphysema.

Extrinsic factors affecting pulmonary ventilation can include trauma or foreign body airway obstruction. Trauma to the airway or chest requires immediate evaluation and intervention. Blunt or penetrating trauma and burns can disrupt airflow through the trachea and into the lungs, quickly resulting in oxygenation deficiencies. In addition, trauma to the chest wall can result in structural damage to the thorax, leading to inadequate pulmonary ventilation. Swelling, punctures, and bruising have a tremendous effect on the ability to deliver oxygen to the alveoli and into the bloodstream. Proper airway management and high concentrations of oxygen are crucial to the outcome in these situations.

Factors Affecting Respiration

External elements in the environment can affect the overall process of respiration. For proper respiration to take place at the cellular level, both oxygenation and perfusion need to function efficiently.

External Factors Adequate respiration requires proper ventilation and oxygenation. Here, external factors such as atmospheric pressure and the partial pressure of oxygen in the ambient air play a key role in the overall process of respiration. At high altitudes, the percentage of oxygen remains the same, but the partial pressure decreases because the total atmospheric pressure decreases. The low partial pressure of oxygen can make it difficult (or impossible) to adequately oxygenate tissue, thus interrupting internal respiration. In addition, closed environments, such as mines and trenches, may also have decreases in ambient oxygen, resulting in poor oxygenation and respiration.

Carbon monoxide, along with other toxic and poisonous gases, displaces oxygen in the environment and makes proper oxygenation and respiration difficult. Carbon monoxide, in particular, has a much greater affinity for hemoglobin than oxygen (250 times more), thus not allowing for proper transport of oxygen to tissues.

Internal Factors Conditions that reduce the surface area for gas exchange also decrease the body's oxygen supply, leading to inadequate tissue perfusion. Medical conditions such as pneumonia, pulmonary edema, and COPD/emphysema may also result in a disturbance of cellular metabolism. These conditions decrease the surface area of the alveoli either by damaging the alveoli or by leading to an accumulation of fluid in the lungs.

Nonfunctional alveoli inhibit the diffusion of oxygen and carbon dioxide. As a result, blood entering the lungs from the right side of the heart bypasses the alveoli and returns to the left side of the heart in an unoxygenated state, a condition called **intrapulmonary shunting**.

Drowning victims and/or patients with pulmonary edema have fluid in the alveoli. This accumulation of fluid inhibits adequate gas exchange at the alveolar membrane and results in decreased oxygenation and respiration. In addition, exposure to certain environmental conditions, like high altitudes, or occupational hazards, such as epoxy resins, over time can result in fluid accumulation or other abnormal conditions, resulting in an overall decrease in respiration. These conditions can interrupt the process of aerobic respiration at the cellular level, resulting in anaerobic respiration and an increase in lactic acid accumulation.

Other conditions affecting cells of the body include hypoxia, hypoglycemia (low blood glucose), and infection. As oxygen and glucose levels decrease, the body is unable to maintain a homeostatic balance with regard to energy production. At this point, the energy production cannot meet the needs of the body, and cellular death is likely if the condition is not corrected. Infection also increases the metabolic needs of the body and disrupts homeostasis. If not corrected, the cells will die as well.

Circulatory Compromise

For respiration to take place, the circulatory system must function efficiently to deliver oxygen to the tissues of the body. When this system becomes compromised, the perfusion of oxygen is not enough to meet the oxygen demands of the tissues.

Obstruction of blood flow to individual cells and tissue is typically related to trauma emergencies an EMT may encounter. These conditions include pulmonary embolism, a simple or **tension pneumothorax**, open pneumothorax (sucking chest wound), hemothorax, and hemopneumothorax. All of these conditions limit the ability for gas exchange at the tissue level as a result of their effects on the respiratory and circulatory systems. In addition, conditions such as heart failure and cardiac tamponade inhibit the ability of the heart to effectively pump oxygenated blood to the tissues.

Blood loss and anemia, a deficiency of red blood cells, result in a decreased ability of blood to carry oxygen. Without sufficient circulating red blood cells, the hemoglobin molecules do not have enough sites for binding.

When the body is in a state of shock, oxygen is not being delivered to the cells efficiently. Hypovolemic shock is an abnormal decrease in blood volume that causes inadequate oxygen delivery to the body. In contrast, vasodilatory shock is not determined by the amount of circulating blood, but by the size of the blood vessels. As the diameter of the blood vessels increases, the blood pressure in the circulatory system decreases. As the systemic blood pressure falls, oxygen is not delivered to the tissues in an effective manner. Both forms of shock result in poor tissue perfusion that leads to anaerobic metabolism. Any patient suspected of being in shock should be treated aggressively to prevent further interruptions to tissue perfusion.

Patient Assessment

Recognizing Adequate Breathing

Breathing is something that all individuals do every day; yet, most of the time, you are not aware of your own breathing or the breathing of others around you. Breathing should be a smooth flow of air moving into and out of the lungs. As a general rule, unless you are directly assessing the patient's airway, you should not be able to see or hear a patient breathe. Signs of normal (adequate) breathing for adult patients are as follows:

- A normal rate (between 12 and 20 breaths/min)
- A regular pattern of inhalation and exhalation
- Clear and equal lung sounds on both sides of the chest (**bilateral**)
- Regular and equal chest rise and fall (chest expansion)
- Adequate depth (tidal volume)

Recognizing Abnormal Breathing

An adult who is awake, alert, and talking to you generally has no immediate airway or breathing problems. However, you should always have supplemental oxygen and a **bag-mask device** or pocket mask close at hand to assist with breathing if this becomes necessary. An adult who is breathing normally will have respirations of 12 to 20 breaths/min Table 9-2 .

You are the Provider: PART 2

You arrive at the patient's residence, enter his home, and find him sitting on the couch. He is a 55-year-old man with a history of congestive heart failure and high blood pressure. He is conscious, but appears sleepy, and can only speak in two-word sentences before stopping to catch his breath. As your partner opens the jump kit, you assess the patient.

Recording Time: 0 Minutes	
Appearance	Obvious breathing difficulty; skin is moist
Level of consciousness	Conscious; appears sleepy
Airway	Open; no secretions or foreign bodies
Breathing	Labored breathing; rapid respiratory rate
Circulation	Skin, cool and moist; radial pulse, rapid and weak

4. Is this patient's airway patent?

5. Is he breathing adequately? Why or why not?

6. How should you manage his present airway and breathing status?

Table 9-2	Normal Respiratory Rate Ranges
Adults	12 to 20 breaths/min
Children	15 to 30 breaths/min
Infants	25 to 50 breaths/min

Note: These ranges are per the NHTSA 2009 EMT National EMS Education Standards. Ranges presented in other courses may vary.

The adult patient who is breathing slower (fewer than 12 breaths/min) than normal should be evaluated for inadequate breathing by assessing the depth of his or her respirations. A patient with shallow depth of breathing (reduced tidal volume) may require assisted ventilations, even if his or her respiratory rate is within normal limits.

A patient with inadequate breathing may appear to be working hard to breathe. This type of breathing pattern is called **labored breathing**. It requires effort and, especially among children, may involve the use of accessory muscles. Accessory muscles include the neck muscles (sternocleidomastoid), the chest pectoralis major muscles, and the abdominal muscles Figure 9-14 .

Words of Wisdom

The respiratory status of a patient is so important that it should be noted at the beginning of your radio report, after mental status. Any changes during treatment or transport should be immediately reported to the receiving hospital. Respiratory status along with any changes should also be clearly documented in your patient care report.

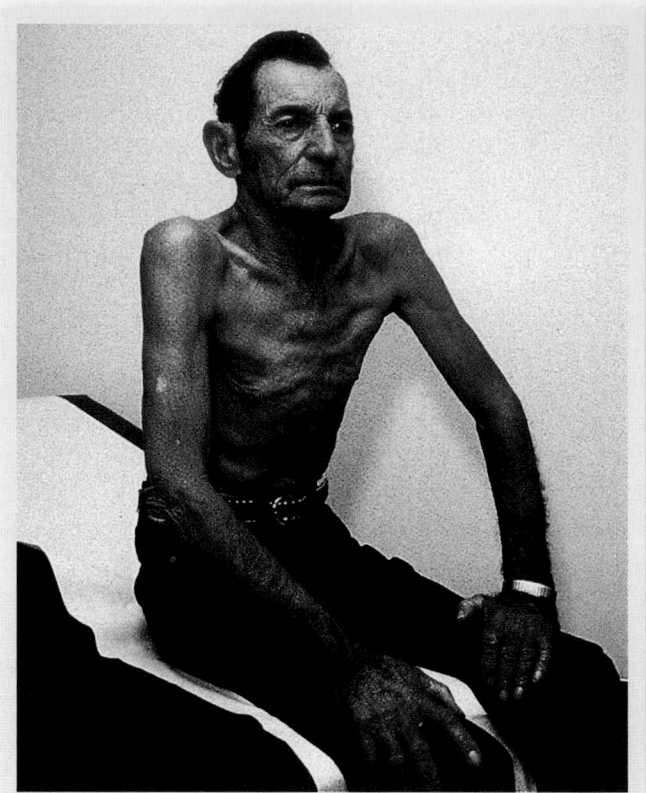

Figure 9-14 The accessory muscles of breathing are used when a patient is having difficulty breathing, but not during normal breathing. The accessory muscles include the sternocleidomastoid, pectoralis major, and abdominal muscles.

These muscles are not used during normal breathing. More information about recognizing labored breathing and respiratory distress in children is found in later chapters. Signs of inadequate breathing in adult patients are as follows:

- Respiratory rate of fewer than 12 breaths/min or more than 20 breaths/min in the presence of shortness of breath (dyspnea)
- Irregular rhythm, such as a patient taking a series of deep breaths followed by periods of apnea
- Diminished, absent, or noisy auscultated breath sounds
- Reduced flow of expired air at the nose and mouth
- Unequal or inadequate chest expansion, resulting in reduced tidal volume
- Increased effort of breathing—use of accessory muscles
- Shallow depth (reduced tidal volume)

- Skin that is pale, cyanotic (blue), cool, or moist (clammy)
- Skin pulling in around the ribs or above the clavicles during inspiration (**retractions**)

When you are assessing a patient with a potential airway compromise, pay particular attention to the external environment and take that into consideration when examining your patient. Conditions such as high altitude and enclosed spaces alter the partial pressure of oxygen in the environment, thus making the process of oxygenation difficult for the patient. In addition, poisonous gases, such as carbon monoxide, displace oxygen in the environment and alter the overall metabolism of the patient. It is important to recognize these potential situations and take them into consideration when deciding on appropriate treatment for the patient.

You should be aware that a patient may appear to be breathing after his or her heart has stopped. These occasional, gasping breaths are called **agonal gasps**. They occur when the respiratory center in the brain continues to send signals to the respiratory muscles. These gasps do not provide adequate oxygen because they are infrequent, gasping respiratory efforts. In patients with agonal gasps, you will need to provide artificial ventilations and, most likely, chest compressions.

Some patients may have irregular respiratory breathing patterns that are related to a specific condition. For example, Cheyne-Stokes respirations are often seen in patients with stroke and patients with serious head injuries **Figure 9-15** .

Cheyne-Stokes respirations are an irregular respiratory pattern in which the patient breathes with an increasing rate and depth of respirations that is followed by a period of **apnea**, or lack of spontaneous breathing, followed again by a pattern of increasing rate and depth of respiration. Serious head injuries may also cause changes in the normal respiratory rate and pattern of breathing. The result may be irregular, ineffective respirations that may or may not have an identifiable pattern (**ataxic respirations**). Patients experiencing a metabolic or toxic disorder may display other irregular respiratory patterns such as Kussmaul respirations. Kussmaul respirations are characterized as deep, gasping respirations commonly seen in patients with metabolic acidosis.

Whereas rapid breathing is a compensatory mechanism to help patients in respiratory distress, some patients are so ill that their body is not able to compensate for their respiratory distress. The patients may look like they are compensating; however, no clinical improvement will be noticeable. EMTs need to be vigilant when monitoring patients in respiratory distress because their condition may decline rapidly.

Patients with inadequate breathing have inadequate minute volume and need to be treated immediately. This condition is most easily recognized in patients who are unable to speak in complete sentences when at rest or who have a fast or slow respiratory rate, both of which may result in a reduction in tidal volume. Emergency medical care includes airway management, supplemental oxygen, and ventilatory support.

■ Assessment of Respiration

Respiration is the actual exchange of oxygen and carbon dioxide at the tissue level. Even though a patient may be

ventilating appropriately, the process of respiration may be compromised. Therefore, EMTs must assess for signs of adequate and inadequate respiration in their patients. These signs and symptoms will guide you in the overall assessment of these patients.

As stated earlier, there are external factors that may disrupt the process of respiration. Areas involving poor oxygenation, such as enclosed spaces, do not provide adequate oxygen levels for patients and, as a result, hinder respiration. EMTs should be aware of the patient's environment and assess the quality of ambient air when approaching the patient. High altitudes and poisonous gases should always be considered when assessing respiration. These factors can dramatically affect respiration and alter metabolism in your patient. Some EMS services carry hand-held carbon monoxide detectors to aid in the process of assessing ambient air. However, if you feel the quality of the ambient air is not safe, remove yourself and the patient (if possible) from the scene immediately and contact the appropriate resource. If there is more than one patient with similar symptoms, consider the presence of poisonous or toxic gases. Remove yourself, your partner, and any patients immediately. For example, carbon monoxide poisoning may be present in a residence with gas heating. If more than one family member complains of a headache, nausea, vomiting, and fatigue, consider an environmental hazard first rather than multiple concurrent cases of the flu.

A patient's level of consciousness and skin color are excellent indicators of respiration. During normal respiration, oxygen and carbon dioxide diffuse in and out of tissues and allow aerobic metabolism to take place. When you are assessing the brain and skin tissues, it will be apparent if the patient has adequate oxygen levels reaching these areas. A patient presenting with an altered level of consciousness may not have adequate oxygen levels reaching the brain. This lack of oxygen can cause rapid changes in the patient's mental status. Therefore, when treating patients with an altered mental status, always consider the possibility that these patients may not be getting adequate oxygen levels to their brain and that you need to consider the possible underlying causes. However, keep in mind to determine a baseline mental status on the patient. Some patients naturally have an abnormal mental status because of a previous medical condition. Ask family members what the patient's normal mental status is.

Just as an altered level of consciousness is indicative of inadequate respiration, the same is true for patients with poor skin color. As oxygen fails to reach the skin tissue of the body, either from a lack of perfusion or poor oxygenation, the color of the skin changes to reflect the poor level of oxygenation. Pale skin and mucous membranes, commonly referred to as pallor, are typically associated

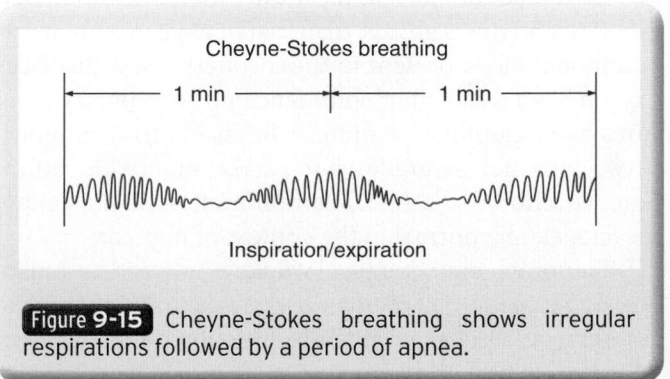

Figure 9-15 Cheyne-Stokes breathing shows irregular respirations followed by a period of apnea.

with poor perfusion caused by illness or shock. As this condition worsens, cyanosis becomes noticeable first peripherally, in the fingertips, and then centrally, in the mucous membranes and around the lips. Eventually, if the poor perfusion or oxygenation is not corrected, anaerobic metabolism will take place. This could cause the skin to become marked with blotches of different colors, commonly referred to as mottling.

Whereas assessment of a patient's baseline mental status and color of the skin and mucous membranes provide good indicators of respiration, EMTs should also consider proper oxygenation when assessing patients. Oxygenation is the process of loading oxygen molecules onto hemoglobin molecules in the bloodstream. Several methods can be used to assess proper oxygenation, including assessing skin color and mental status and the more recent use of **pulse oximetry**.

Oxygen saturation (Sao_2) is the measure of the percentage of hemoglobin molecules that are bound in arterial blood. Because hemoglobin delivers 97% of the oxygen delivered to the body's tissues, oxygen saturation is an excellent indication of the amount of oxygen available to the end organs.

In the past few years the pulse oximeter has become standard equipment in the treatment of emergency patients **Figure 9-16**. The pulse oximeter provides a rapid, reliable, noninvasive, real-time indication of respiratory efficiency. Although its results must not be used without conducting an overall clinical assessment of the patient, careful use of the pulse oximeter provides valuable information about a patient's oxygenation status. This device can be used to assess the adequacy of oxygenation during positive-pressure ventilation and assess the overall impact of interventions on your patient.

A pulse oximeter measures the percentage of hemoglobin saturation. Under normal conditions, the Sao_2 should be 98% to 100% while breathing room air. Although no definitive threshold for normal values exists, an Sao_2 of less than 96% in a nonsmoker may indicate

hypoxemia. An Sao_2 of 90% generally requires treatment unless the patient has a chronic condition causing perpetually low oxygen saturations. Pulse oximeters are highly reliable in Sao_2 readings above 85%; however, readings below that are less reliable but certainly indicate profound hypoxemia.

Pulse oximetry is considered a routine vital sign and can be used as part of any patient assessment. Whereas there are no true contraindications to using pulse oximetry, EMTs must be aware of the limitations associated with this device. To function properly, the pulse oximeter must find a pulsation in the selected tissue. The most commonly used site is a finger. Follow the steps in **Skill Drill 9-1** to measure pulse oximetry:

Skill Drill 9-1

1. Clean the patient's finger, and remove nail polish as needed. Place the index or middle finger into the pulse oximeter probe. Turn on the pulse oximeter, and note the LED reading of the Sao_2 (**Step 1**).
2. Palpate the radial pulse to ensure that it correlates with the LED display on the pulse oximeter (**Step 2**).

In patients with significant vasoconstriction or very low perfusion states (including cardiac arrest), there may not be enough peripheral perfusion to be detected by the sensor. In these cases, move the sensor to a more central location (bridge of the nose or ear lobe). Always consult the manufacturer's guidelines for proper placement and troubleshooting of these devices. An inaccurate pulse oximetry reading may be caused by the following:

- Hypovolemia
- Anemia
- Severe peripheral vasoconstriction (chronic hypoxia, smoking, or hypothermia)
- Time delay in detecting respiratory insufficiency
- Dark or metallic nail polish
- Dirty fingers
- Carbon monoxide poisoning

Carbon monoxide has an affinity for hemoglobin that is 200 to 250 times greater than that of oxygen. When carbon monoxide is present in the inspired gas, it displaces oxygen from the hemoglobin. Since pulse oximetry measures hemoglobin saturation, it is unable to distinguish between oxygen saturation and carbon monoxide saturation. Therefore, in cases of carbon monoxide poisoning, the Sao_2 can be normal in the context of hypoxia.

The pulse oximeter is a valuable adjunct to aid in decision making, but is not a replacement for a complete assessment. Because of many factors, the pulse oximeter may give falsely high or low readings. When you are

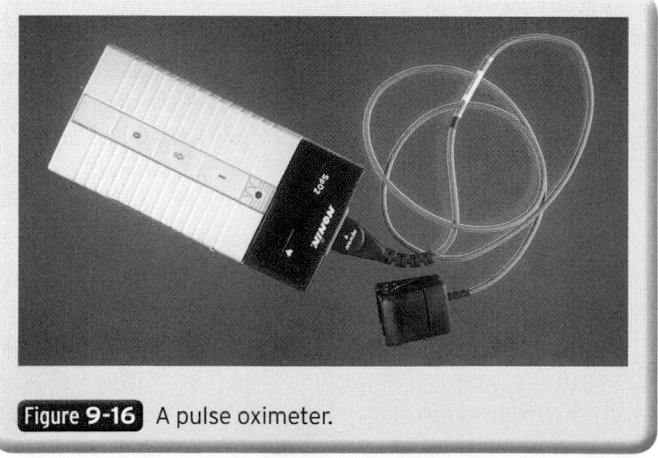

Figure 9-16 A pulse oximeter.

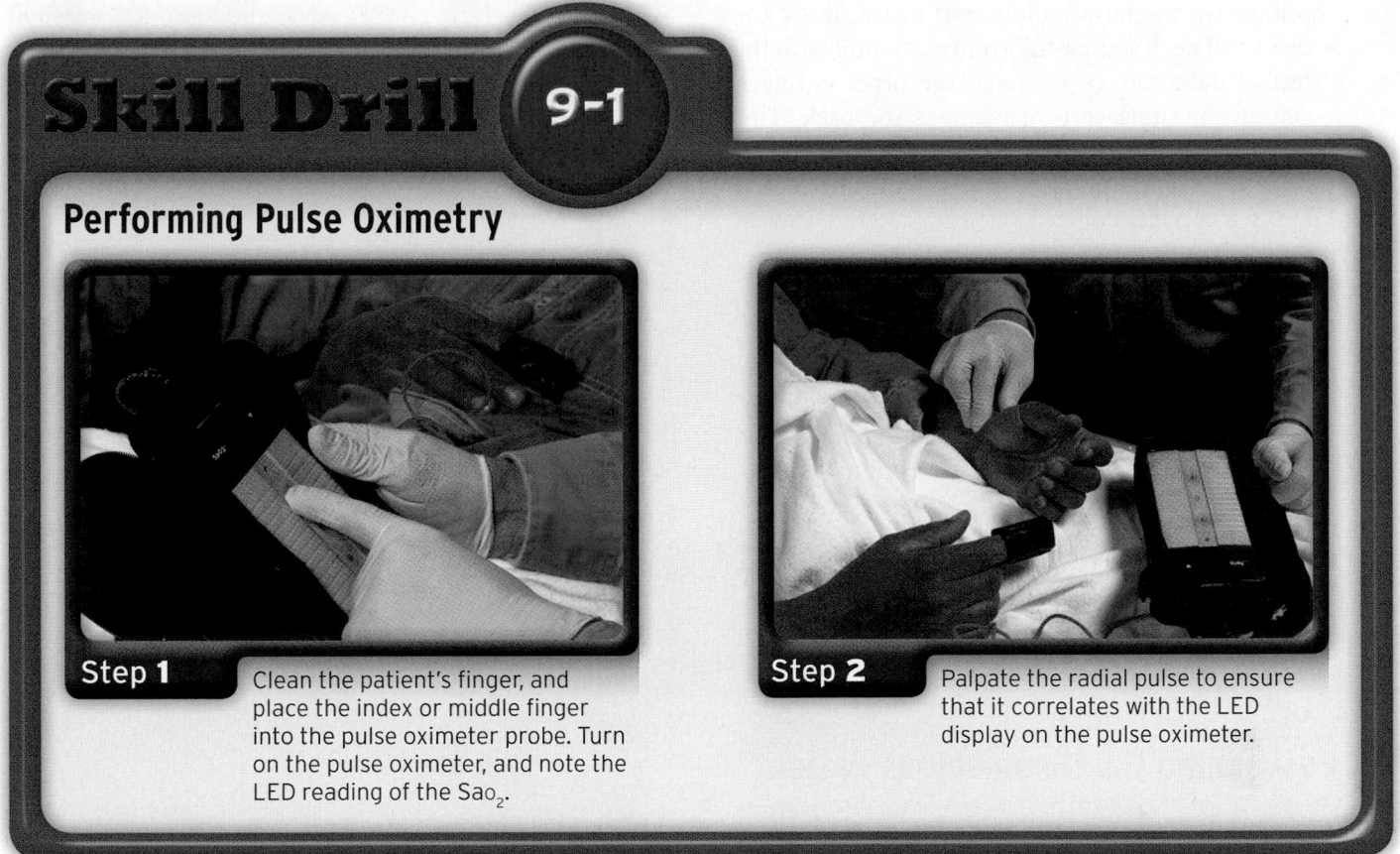

Skill Drill 9-1

Performing Pulse Oximetry

Step 1 Clean the patient's finger, and place the index or middle finger into the pulse oximeter probe. Turn on the pulse oximeter, and note the LED reading of the Sa_{O_2}.

Step 2 Palpate the radial pulse to ensure that it correlates with the LED display on the pulse oximeter.

conducting a complete patient assessment, consider using pulse oximetry readings as one additional measure to use while obtaining all of the other comprehensive information you need. Assess the patient for signs and symptoms of adequate oxygenation. If a patient has signs, such as cyanosis, pale or clammy skin, or symptoms, such as shortness of breath and normal Sa_{O_2}, treat the patient's condition, not the environment.

Opening the Airway

Emergency medical care begins with ensuring an open airway. If you cannot immediately open and maintain a patent airway, you cannot provide effective patient care. Regardless of the patient's condition, the airway must remain patent at all times.

When you respond to a call and find an unconscious patient, you need to quickly assess for a pulse and breathing; if the patient has a pulse, you need to determine whether breathing is adequate. Remember that airway and breathing are two separate components that are closely related to each other; however, you must understand that adequate breathing does not always equate to an adequate airway. To most effectively open the airway and assess breathing, the patient should be in the supine position. However, if your patient is in a position that delays placement in a supine position (for example, entrapped in a vehicle), the patient's airway must be opened and

assessed in the position in which you find the patient. If your patient is found in the prone position (lying face down), he or she must be repositioned to allow for assessment of airway and breathing and to begin CPR should it become necessary. Today health care workers are taught to begin CPR with compressions if cardiac arrest is suspected. The patient should be log rolled as a unit so the head, neck, and spine all move together without twisting. Unconscious patients, especially when there are no witnesses who can rule out trauma, should be moved as a unit because of the potential for spinal injury. To position the unconscious patient in order to open the airway, follow the steps in **Skill Drill 9-2**.

Skill Drill 9-2

1. Kneel beside the patient. Make sure you kneel far enough away so that the patient, when rolled toward you, does not come to rest in your lap. Place your hands behind the patient's head and neck to provide in-line stabilization of the cervical spine as your partner straightens the patient's legs **Step 1**.

2. Have your partner place his or her hands on the patient's far shoulder and hip **Step 2**.

3. As you call the count to control movement, have your partner turn the patient toward you by

pulling on the far shoulder and hip. Control the head and neck so that they move as a unit with the rest of the torso. In this way, the head and neck stay in the same vertical plane as the back. This single motion will minimize aggravation of any potential spine injury. At this point, you should apply a cervical collar. Place the patient's arms at his or her side (Step 3).

4. Once the patient is positioned, maintain an open airway and check for breathing (Step 4).

In an unconscious patient, the most common airway obstruction is the patient's tongue, which falls back into the throat when the muscles of the throat and tongue relax Figure 9-17 . Dentures (false teeth), blood, vomitus, mucus,

Words of Wisdom

Causes of airway obstruction include:
- Relaxation of the tongue in an unresponsive patient
- Foreign objects—food, small toys, dentures
- Blood clots, broken teeth, or damaged oral tissue following trauma
- Airway tissue swelling—infection, allergic reaction
- Aspirated vomitus (stomach contents)

food, and other foreign objects may also create an airway obstruction. Therefore, you should always be prepared to help clear and maintain a patent (open) airway.

Skill Drill   9-2

Positioning the Unconscious Patient

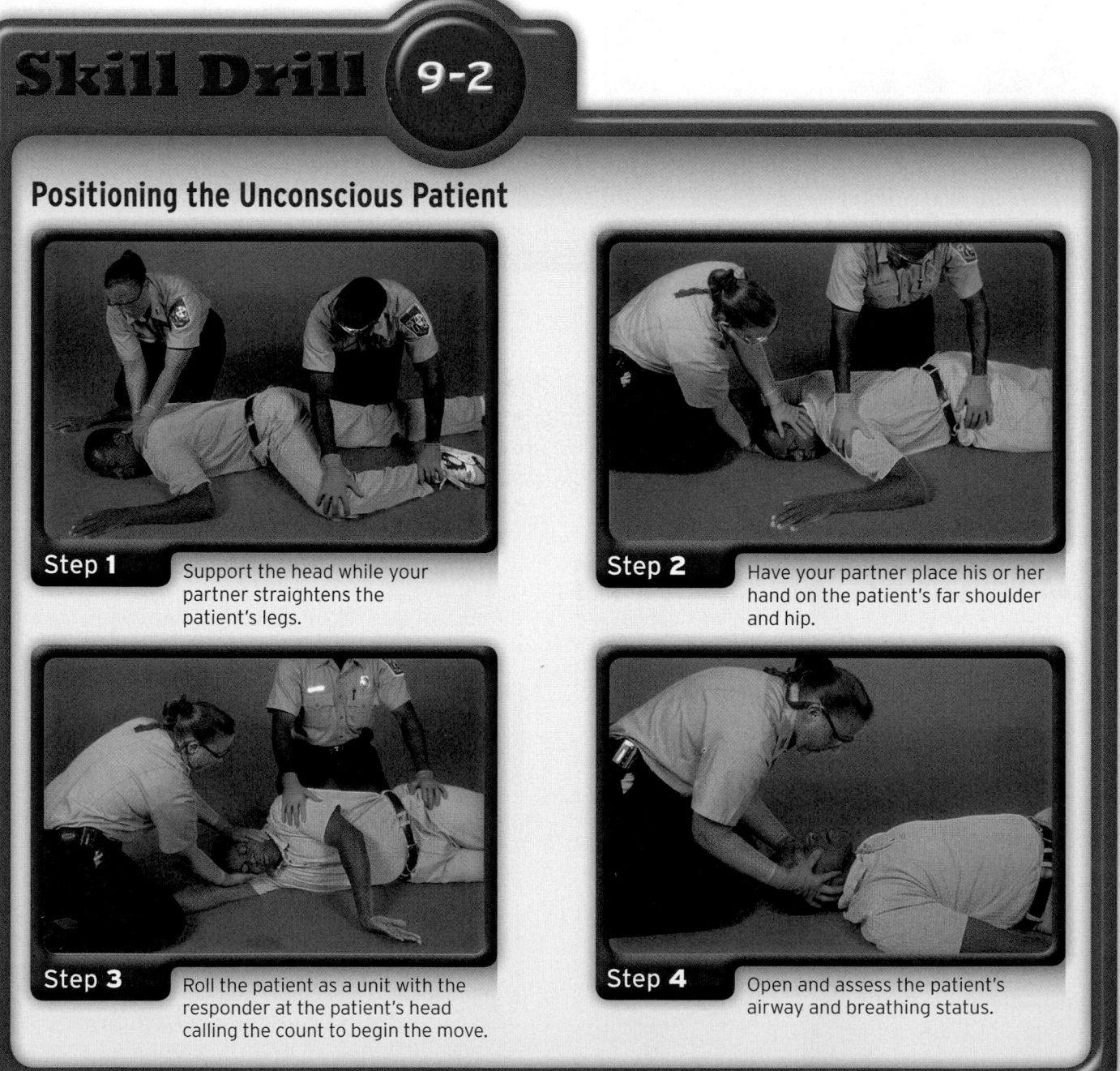

Step 1 Support the head while your partner straightens the patient's legs.

Step 2 Have your partner place his or her hand on the patient's far shoulder and hip.

Step 3 Roll the patient as a unit with the responder at the patient's head calling the count to begin the move.

Step 4 Open and assess the patient's airway and breathing status.

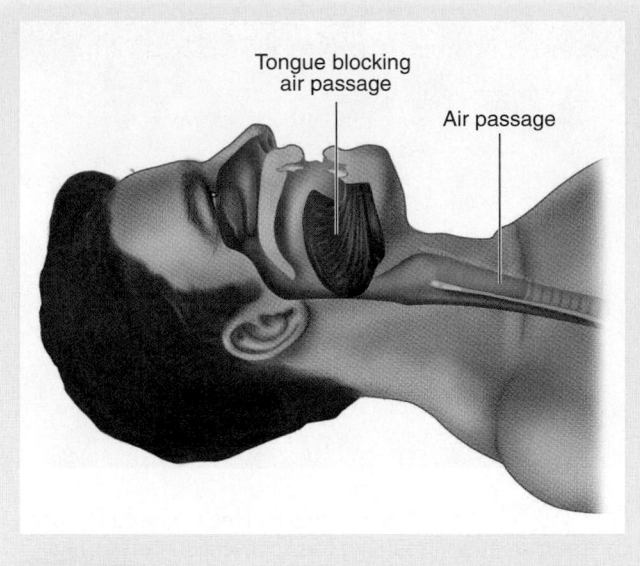

Figure 9-17 The most common airway obstruction is the patient's own tongue, which falls back into the throat when the muscles of the throat and tongue relax.

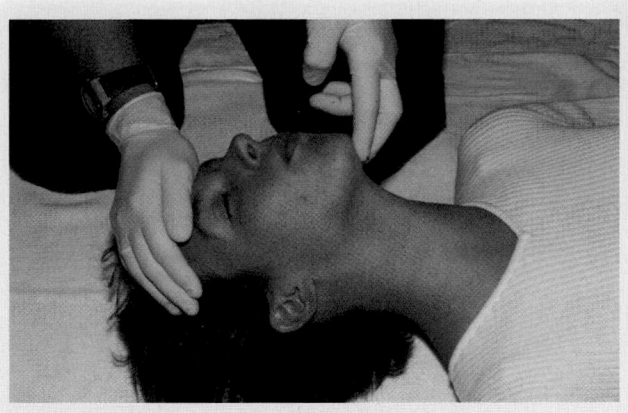

Figure 9-18 The head tilt–chin lift maneuver is a simple technique for opening the airway in a patient without a suspected cervical spine injury.

Head Tilt–Chin Lift Maneuver

Opening the airway to relieve an obstruction can often be done quickly and easily by simply tilting the patient's head back and lifting the chin in what is known as the **head tilt–chin lift maneuver**. For patients who have not sustained or are not suspected of having sustained trauma, this simple maneuver is sometimes all that is needed for the patient to resume breathing.

To perform the head tilt–chin life maneuver, follow these steps:

1. With the patient in a supine position, position yourself beside the patient's head.
2. Place the heel of one hand on the patient's forehead, and apply firm backward pressure with your palm to tilt the patient's head back. This extension of the neck will move the tongue forward, away from the back of the throat, and clear the airway if the tongue is blocking it.
3. Place the fingertips of your other hand under the lower jaw near the bony part of the chin. Do not compress the soft tissue under the chin, as this may block the airway.
4. Lift the chin upward, bringing the entire lower jaw with it, helping to tilt the head back. Do not use your thumb to lift the chin. Lift so that the teeth are nearly brought together, but avoid closing the mouth completely. Continue to hold the forehead to maintain the backward tilt of the head **Figure 9-18**.

Jaw-Thrust Maneuver

The head tilt–chin lift maneuver will open the airway in most patients. However, if you suspect a cervical spine injury, use the jaw-thrust maneuver. The **jaw-thrust maneuver** is a technique to open the airway by placing the fingers behind the angle of the jaw and lifting the jaw upward. You can easily seal a mask around the mouth while doing the jaw-thrust maneuver. This is the method of choice for patients with suspected cervical spine injuries. Refer to Chapter 26, *Head and Spine Injuries,* for a more detailed discussion of these types of injuries.

Perform the jaw-thrust maneuver in an adult using the following steps **Figure 9-19**:

1. Kneel above the patient's head. Place your fingers behind the angles of the lower jaw, and move the jaw upward. Use your thumbs to help position the lower jaw to allow breathing through the mouth and nose.
2. The completed maneuver should open the airway with the mouth slightly open and the jaw jutting forward.

Patients who have a pulse may start to breathe on their own once the airway has been opened. Assess whether breathing has returned by using the look, listen, and feel technique **Figure 9-20**.

With complete airway obstruction, there will be no movement of air. However, you may see the chest and abdomen rise and fall considerably with the patient's frantic attempts to breathe. This is why the presence of chest wall movement alone does not indicate if adequate breathing is present. Regular chest wall movement indicates a respiratory effort is present. Observing chest and abdominal movement is often difficult with a fully clothed patient. You may see little, if any, chest movement, even with normal breathing. This is particularly true in some patients with chronic lung disease if you discover that there is no movement of air.

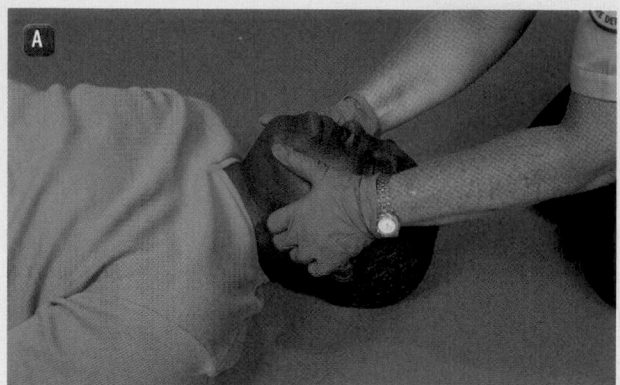

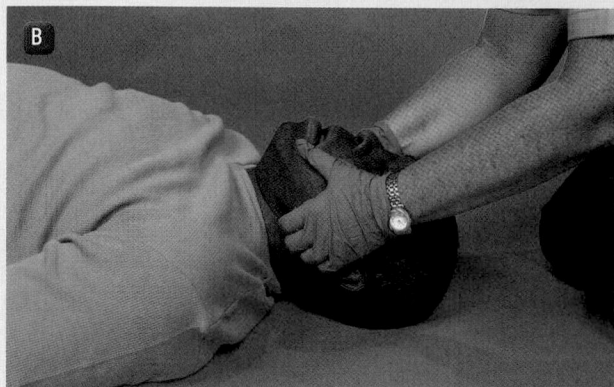

Figure 9-19 Performing the jaw-thrust maneuver. **A.** Kneeling above the patient's head, place your fingers behind the angles of the lower jaw, and move the jaw upward. Use your thumbs to help position the lower jaw. **B.** The completed maneuver should look like this.

Basic Airway Adjuncts

The primary function of an airway adjunct is to prevent obstruction of the upper airway by the tongue and allow the passage of air and oxygen to the lungs.

■ Oropharyngeal Airways

An <u>oropharyngeal (oral) airway</u> has two principal purposes. The first is to keep the tongue from blocking the upper airway. The second is to make it easier to suction the oropharynx if necessary. Suctioning is possible through an opening down the center or along either side of the oropharyngeal airway **Figure 9-21**.

Indications for the oral airway include the following:

- Unresponsive patients without a gag reflex (breathing or apneic)
- Any apneic patient being ventilated with a bag-mask device

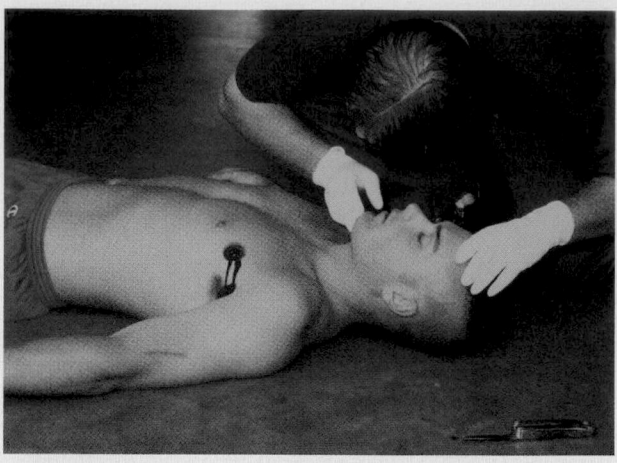

Figure 9-20 The look, listen, and feel technique can be used to assess whether breathing has spontaneously returned.

Contraindications for the oral airway include the following:

- Conscious patients
- Any patient (conscious or unconscious) who has an intact gag reflex

The <u>gag reflex</u> is a protective reflex mechanism that prevents food and other particles from entering the airway. If you try to insert an oral airway in a patient with an intact gag reflex, the result may be vomiting or a spasm of the vocal cords. If the patient gags while you are attempting to insert an oral airway, immediately remove the adjunct and be prepared to log roll the patient and suction the oropharynx, should vomiting occur. An oral airway is also a safe, effective way to help maintain the airway of a patient with a possible spinal injury. The use of an oral airway may make manual airway maneuvers such as the head tilt–chin lift and the jaw-thrust easier

Figure 9-21 An oral airway is used for unconscious patients who have no gag reflex. It keeps the tongue from blocking the airway and makes suctioning the airway easier.

to maintain; however, manual maneuvers are often still needed to ensure that the airway remains open.

You must clearly understand when and how this device is used. If the oropharyngeal airway is too large, it could actually push the tongue back into the pharynx, blocking the airway. Conversely, an oral airway that is too small could block the airway directly, just like any foreign body obstruction. The following steps should be used when inserting an oropharyngeal airway Skill Drill 9-3 .

Skill Drill 9-3

1. To select the proper size, measure from the patient's earlobe or angle of the jaw to the corner of the mouth on the side of the face Step 1 .
2. Open the patient's mouth with the cross-finger technique. Hold the airway upside down with your other hand. Insert the airway with the tip facing the roof of the mouth Step 2 .

Skill Drill 9-3

Inserting an Oral Airway

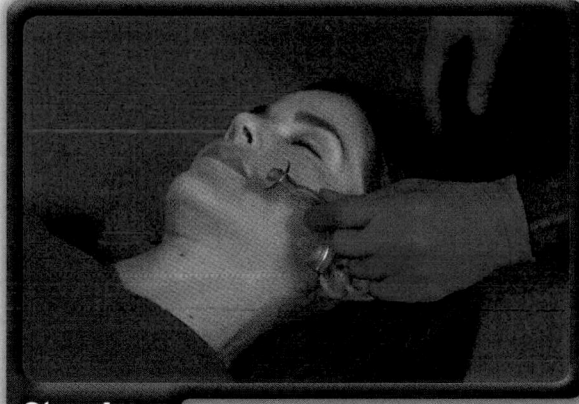

Step 1 Size the airway by measuring from the patient's earlobe to the corner of the mouth.

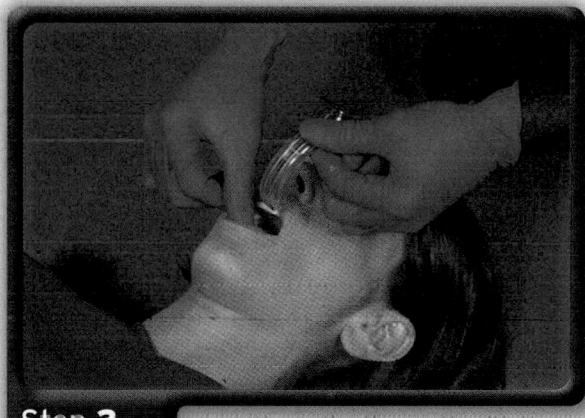

Step 2 Open the patient's mouth with the cross-finger technique. Hold the airway upside down with your other hand. Insert the airway with the tip facing the roof of the mouth.

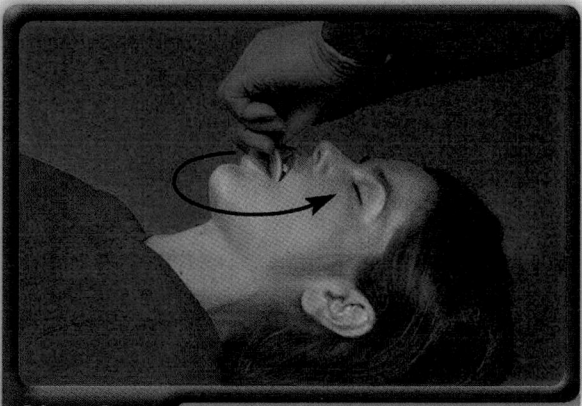

Step 3 Rotate the airway 180°. Insert the airway until the flange rests on the patient's lips and teeth. In this position, the airway will hold the tongue forward.

3. Rotate the airway 180°. When inserted properly, the airway will rest in the mouth with the curvature of the airway following the contour of the anatomy. The flange should rest against the lips or teeth, with the other end opening into the pharynx (Step 3).

Take care to avoid injuring the hard palate (roof of the mouth) as you insert the airway. Roughness can cause bleeding that may aggravate airway problems or even cause vomiting.

Special Populations

In children, the only acceptable method of inserting an oral airway is to use a tongue blade to hold the tongue down while inserting the airway. Because the airways of children are undeveloped, rotating an oropharyngeal airway in the posterior pharynx may cause damage. For more discussion on pediatric airways, see Chapter 32, *Pediatric Emergencies*.

If you encounter difficulty while inserting the oral airway, insert the airway with a 90° rotation **Skill Drill 9-4** :

1. Use a tongue depressor or blade, bite stick, or a wooden tongue blade to depress the tongue, ensuring the tongue remains forward (Step 1).

2. Insert the oral airway sideways from the corner of the mouth, until the flange reaches the teeth (Step 2).

3. Rotate the oral airway 90°, removing the depressor, blade, or bite stick as you exert gentle backward pressure on the oral airway until it rests securely in place against the lips and teeth (Step 3).

In some cases, a patient may become responsive and regain the gag reflex after you have inserted an oral airway. If this occurs, gently remove the airway by pulling it out, following the normal curvature of the mouth and throat. Be prepared for the patient to vomit. Have suction available, and log roll the patient onto his or her side and allow any fluids to drain out.

■ Nasopharyngeal Airways

A **nasopharyngeal (nasal) airway** is usually used with an unresponsive patient or a patient with an altered level of consciousness who has an intact gag reflex and is not able to maintain his or her airway spontaneously **Figure 9-22** .

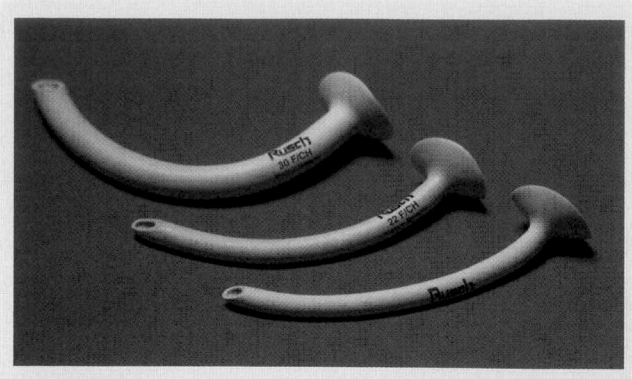

Figure 9-22 A nasal airway is better tolerated than is an oral airway by patients who have an intact gag reflex.

Special Populations

When managing the airway of an older patient, you must be aware of the presence of dentures or other dental appliances. If dentures are tight-fitting and allow for effective airway management, they should be left in place. However, if the dentures are loose, they must be removed to avoid potential airway obstruction.

Patients with an altered mental status or who have just had a seizure may also benefit from this type of airway. If a patient has sustained severe trauma to the head or face, you should consult medical control before inserting a nasopharyngeal airway. Extreme care must be used with such trauma patients. If the nasal airway is accidentally pushed through a hole caused by a fracture of the base of the skull, called the cribiform plate, it may penetrate into the brain.

This type of airway is usually better tolerated by patients who have an intact gag reflex. It is not as likely as the oropharyngeal airway to cause vomiting. You should coat the airway well with a water-soluble lubricant before it is inserted. Be aware that slight bleeding may occur even when the airway is inserted properly. However, you should never force the airway into place.

Indications for the nasopharyngeal airway include the following:

- Semiconscious or unconscious patients with an intact gag reflex
- Patients who otherwise will not tolerate an oropharyngeal airway

Contraindications for the nasopharyngeal airway include the following:

- Severe head injury with blood draining from the nose
- History of fractured nasal bone

Skill Drill 9-4

Inserting an Oral Airway With a 90° Rotation

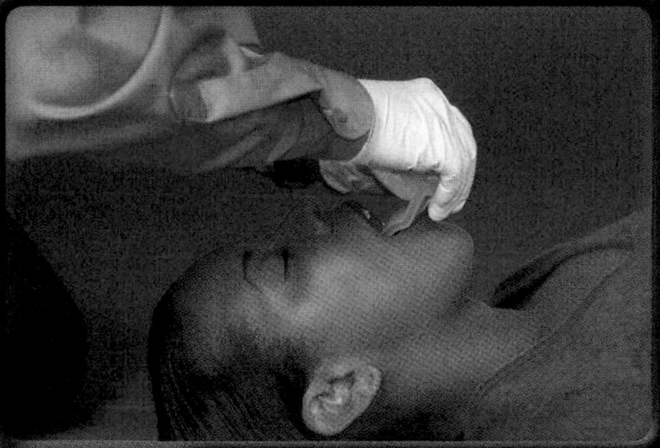

Step 1 Depress the tongue so that it remains forward.

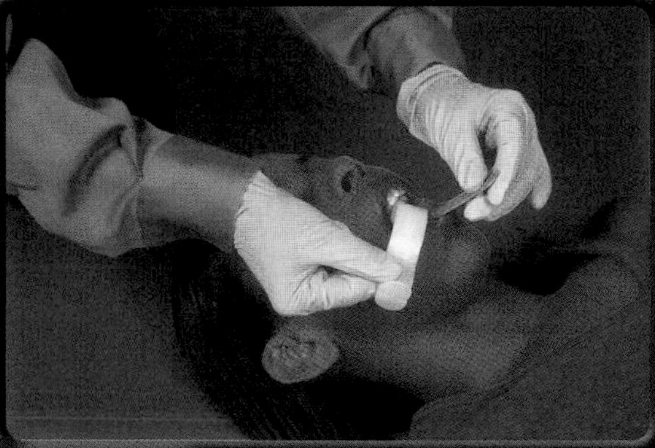

Step 2 Insert the oral airway sideways from the corner of the mouth, until the flange reaches the teeth.

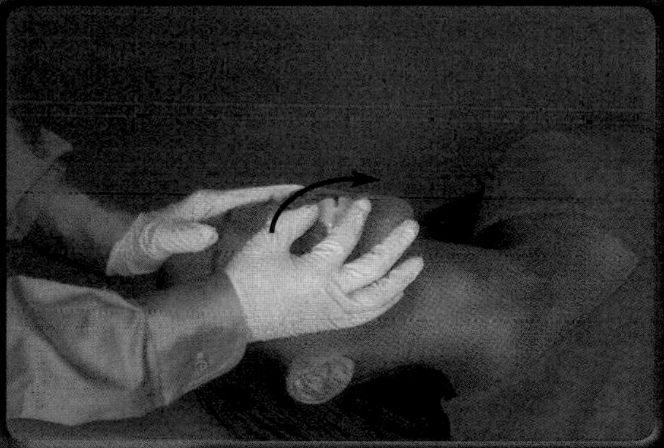

Step 3 Rotate the oral airway at a 90° angle. Remove the bite stick as you exert gentle backward pressure on the oral airway until it rests securely in place against the lips and teeth.

Follow these steps to ensure correct placement of the nasopharyngeal airway Skill Drill 9-5 :

Skill Drill 9-5

1. Before inserting the airway, be sure you have selected the proper size. Measure from the tip of the patient's nose to the earlobe. In almost all individuals, one nostril is larger than the other Step 1 .

2. The airway should be placed in the larger nostril, with the curvature of the device following the curve of the floor of the nose. If using the right nare, the bevel should face the septum Step 2 . If using the left nare, insert the airway with the tip of the airway pointing upward, which will allow the bevel to face the septum.

3. Advance the airway gently Step 3 . If using the left nare, insert the nasal airway until resistance is met. Then rotate the nasal airway 180° into position. This rotation is not required if using the right nostril.

4. When completely inserted, the flange rests against the nostril. The other end of the airway opens into the posterior pharynx (Step 4). If the patient becomes intolerant of the nasal airway, you may have to remove it. Gently withdraw the airway from the nasal passage. Precautions similar to those used when removing an oral airway should be followed.

Skill Drill　9-5

Inserting a Nasal Airway

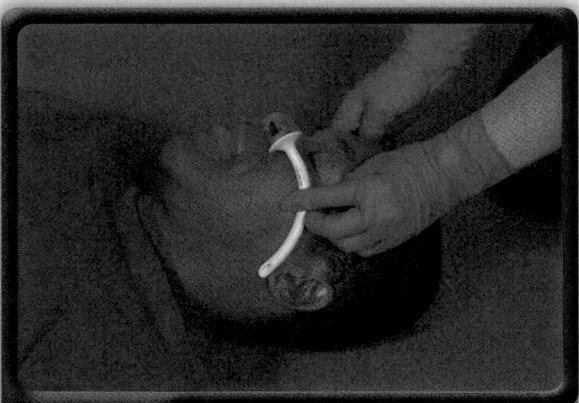

Step 1　Size the airway by measuring from the tip of the nose to the patient's earlobe. Coat the tip with a water-soluble lubricant.

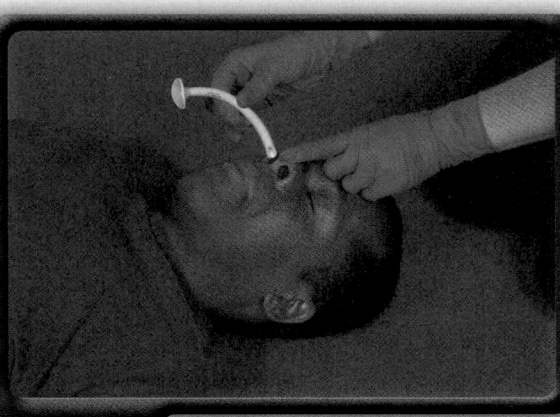

Step 2　Insert the lubricated airway into the larger nostril with the curvature following the floor of the nose. If using the right nare, the bevel should face the septum. If using the left nare, insert the airway with the tip of the airway pointing upward, which will allow the bevel to face the septum.

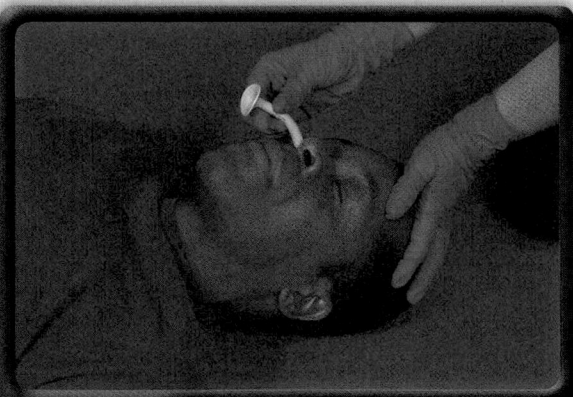

Step 3　Gently advance the airway. If using the left nare, insert the nasopharyngeal airway until resistance is met. Then rotate the nasopharyngeal airway 180° into position. This rotation is not required if using the right nostril.

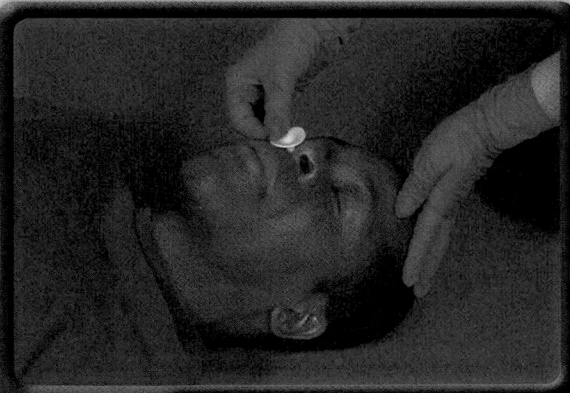

Step 4　Continue until the flange rests against the nostril. If you feel any resistance or obstruction, remove the airway and insert it into the other nostril.

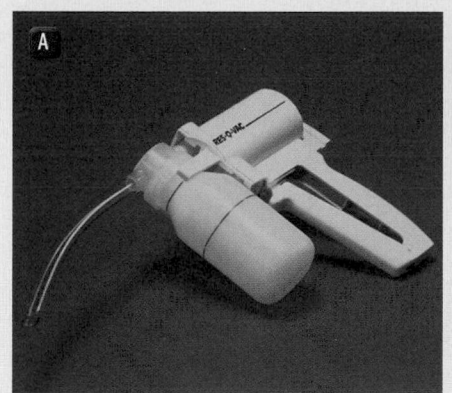
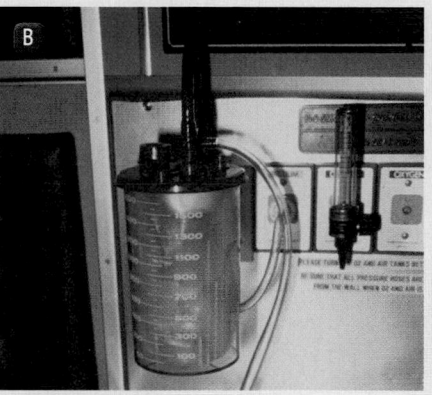
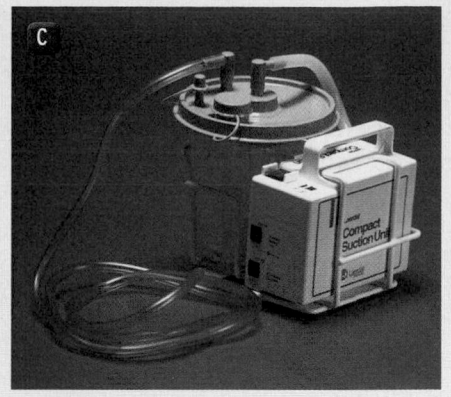

Figure 9-23 Suctioning equipment is essential for resuscitation. **A.** Hand-operated unit. **B.** Fixed unit. **C.** Portable unit.

Suctioning

You must keep the airway clear so that you can ventilate the patient properly. If the airway is not clear, you will force the fluids and secretions into the lungs and possibly cause a complete airway obstruction. Therefore, suctioning is your next priority. If you have any doubt about the situation, remember this rule: If you hear gurgling, the patient needs suctioning!

Suctioning Equipment

Portable, hand-operated, and fixed (mounted) suctioning equipment is essential for resuscitation **Figure 9-23**.

A portable suctioning unit must provide enough vacuum pressure and flow to allow you to suction the mouth and nose effectively. Hand-operated suctioning units with disposable chambers are reliable, effective, and relatively inexpensive. A fixed suctioning unit should generate airflow of more than 40 L/min and a vacuum of more than 300 mm Hg when the tubing is clamped.

A portable or fixed suctioning unit should be fitted with the following:

- Wide-bore, thick-walled, nonkinking tubing
- Plastic, rigid pharyngeal suction tips, called **tonsil tips** or Yankauer tips
- Nonrigid plastic catheters, called French or whistle-tip catheters
- A nonbreakable, disposable collection bottle
- A supply of water for rinsing the tips

A **suction catheter** is a hollow, cylindrical device that is used to remove fluids from the patient's airway. A tonsil-tip catheter is the best kind of catheter for suctioning the oropharynx in adults and is preferred for

You are the Provider: PART 3

Your partner begins treating the patient while you obtain his vital signs and ask his wife what he was doing when his respiratory distress began. She tells you that he began complaining of slight shortness of breath the day before, but it suddenly worsened today when he was sitting on the couch reading the newspaper.

Recording Time: 2 Minutes	
Respirations	30 breaths/min, labored
Pulse	120 beats/min, weak
Skin	Cool and moist; cyanosis around the mouth
Blood pressure	126/60 mm Hg
Oxygen saturation (Sao$_2$)	85% (on oxygen)

7. What is cyanosis? What does it indicate?

8. What does the patient's oxygen saturation indicate?

infants and children. The plastic tips have a large diameter and are rigid, so they do not collapse Figure 9-24.

Tips with a curved contour allow for easy, rapid placement in the oropharynx. Soft plastic, nonrigid catheters, sometimes called French or whistle-tip catheters, are used to suction the nose and liquid secretions in the back of the mouth and in situations in which you cannot use a rigid catheter, such as for a patient with a **stoma** Figure 9-25. A stoma is an opening through the skin that goes into an organ or other structure.

For example, a rigid catheter could break off a patient's tooth, whereas a flexible catheter may be inserted along the cheeks without injury. Before you insert any catheter, make sure to measure for the proper size. Use the same technique as you would use when measuring for an oropharyngeal airway. Be careful not to touch the back of the airway with a suction catheter. This can activate the gag reflex, causing vomiting, and increase the possibility of aspiration.

Words of Wisdom

Anytime there are fluids in the airway, there are risks for aspiration. Aspiration may increase the potential of mortality by 30% to 70%.

■ Techniques of Suctioning

You should inspect your suctioning equipment regularly to make sure it is in proper working condition. Turn on the suction, clamp the tubing, and make sure that the unit generates a vacuum of more than 300 mm Hg. Check that a battery-charged unit has charged batteries. Ensure that your suctioning equipment is at the patient's head and is easily accessible. Follow these general steps to operate the suction unit:

1. Check the unit for proper assembly of all its parts.
2. Turn on the suctioning unit and test it to ensure a vacuum pressure of more than 300 mm Hg.

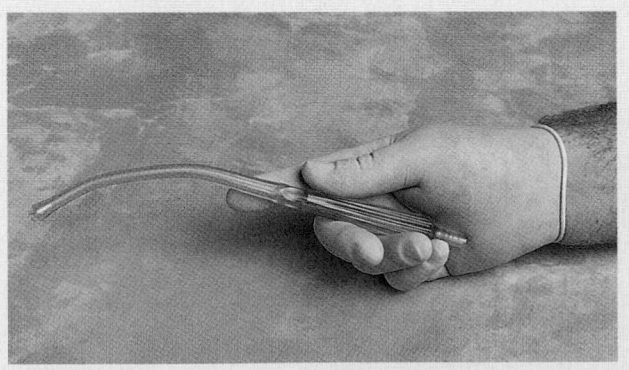

Figure 9-24 Tonsil-tip catheters are the best for suctioning because they have wide-diameter tips and are rigid.

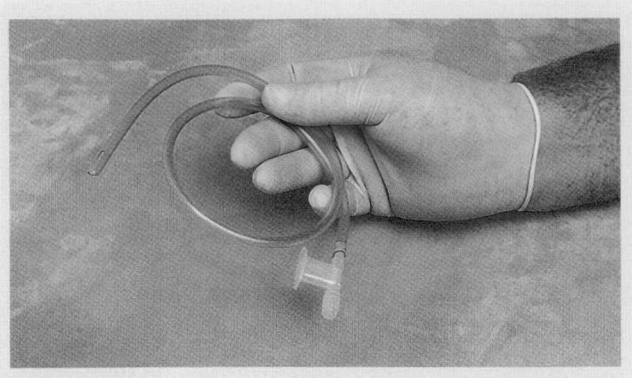

Figure 9-25 French, or whistle-tip, catheters are used in situations in which rigid catheters cannot be used, such as with a patient who has a stoma, patients whose teeth are clenched, or if suctioning the nose is necessary.

3. Select and attach the appropriate suction catheter to the tubing.

Never suction the mouth or nose for more than 15 seconds at one time for adult patients, 10 seconds for children, and 5 seconds for infants. Suctioning removes oxygen from the airway along with the obstructive material and can result in hypoxia. Rinse the catheter and tubing with water to prevent clogging of the tube with dried vomitus or other secretions. Repeat suctioning only after the patient has been adequately ventilated and reoxygenated.

You should use extreme caution when suctioning a conscious or semiconscious patient. Put the tip of the suction catheter in only as far as you can visualize. Be aware that suctioning may induce vomiting.

To properly suction a patient, follow the steps in Skill Drill 9-6:

Skill Drill 9-6

1. Turn on the assembled suction unit Step 1.
2. Measure the catheter to the correct depth by measuring the catheter from the corner of the patient's mouth to the edge of the earlobe or angle of the jaw Step 2.
3. Open the patient's mouth using the cross-finger technique or tongue-jaw lift, and insert the tip of the catheter to the depth measured Step 3.
4. Insert the catheter to the premeasured depth and apply suction in a circular motion as you withdraw the catheter. Do not suction an adult for more than 15 seconds Step 4.

At times, a patient may have secretions or vomitus that cannot be suctioned quickly and easily, and some

Skill Drill 9-6

Suctioning a Patient's Airway

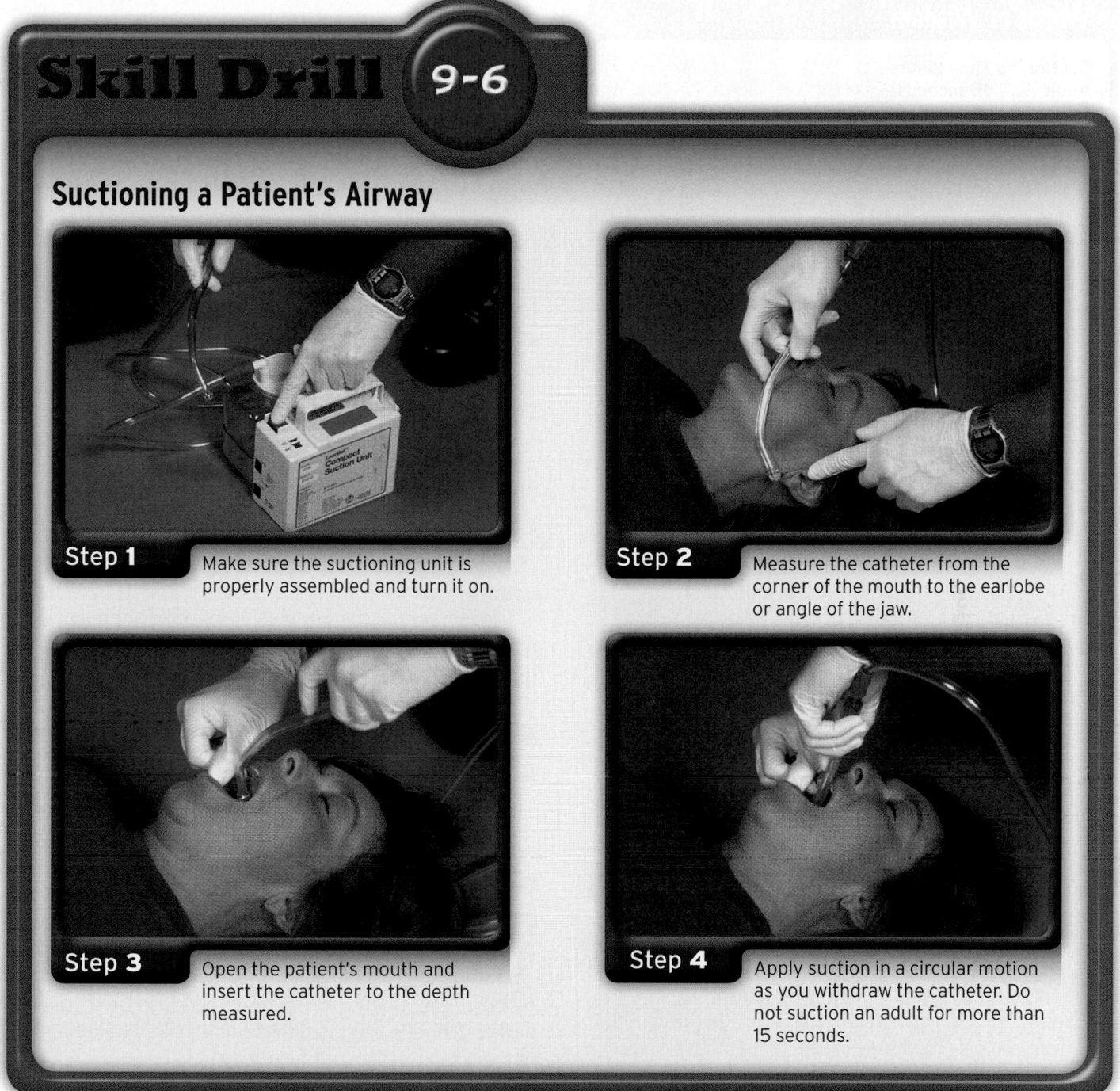

Step 1 Make sure the suctioning unit is properly assembled and turn it on.

Step 2 Measure the catheter from the corner of the mouth to the earlobe or angle of the jaw.

Step 3 Open the patient's mouth and insert the catheter to the depth measured.

Step 4 Apply suction in a circular motion as you withdraw the catheter. Do not suction an adult for more than 15 seconds.

suction units cannot effectively remove solid objects such as teeth, foreign bodies, and food. In these cases, you should remove the catheter from the patient's mouth, log roll the patient to the side, and then clear the mouth carefully with your gloved finger. Only attempt to remove an object if it is visible during examination of the open mouth; blind sweeps of the back of the oropharynx may push an object further down in the airway, making the obstruction worse. A patient who requires assisted ventilations may also produce frothy secretions as quickly as you can suction them from the airway. In this situation,

you should suction the patient's airway for 15 seconds (less time in infants and children), and then ventilate the patient for 2 minutes. This alternating pattern of suctioning and ventilating should continue until all secretions have been cleared from the patient's airway. Continuous ventilation is not appropriate if vomitus or other particles are present in the airway.

You should clean and decontaminate your suctioning equipment after each use according to the manufacturer's guidelines. Place all disposable suctioning equipment (such as catheter, suction tubing) in a biohazard bag.

Words of Wisdom

Suctioning Time Limits
Adult   15 seconds
Child   10 seconds
Infant   5 seconds

Maintaining the Airway

The **recovery position** is used to help maintain a clear airway in an unconscious patient who is not injured and is breathing on his or her own with a normal respiratory rate and adequate tidal volume (depth of breathing) **Figure 9-26**.

Take the following steps to put the patient in the recovery position:

1. Roll the patient onto the left side so that the head, shoulders, and torso move at the same time without twisting.
2. Place the patient's extended left arm and right hand under his or her cheek.

For patients who have resumed spontaneous breathing after being resuscitated, the recovery position will prevent the aspiration of vomitus. However, this position is not appropriate for patients with suspected spinal injuries, nor is it adequate for patients who are unconscious and require ventilatory assistance. You must reposition such patients to provide adequate access to the airway while maintaining appropriate spinal stabilization.

Supplemental Oxygen

You should always give supplemental oxygen to patients who are hypoxic because they are not getting enough oxygen to the tissues and cells of the body.

Some tissues and organs, such as the heart, central nervous system, lungs, kidneys, and liver, need a constant supply of oxygen to function normally. *Never withhold oxygen from any patient who might benefit from it, especially if you must assist ventilations.*

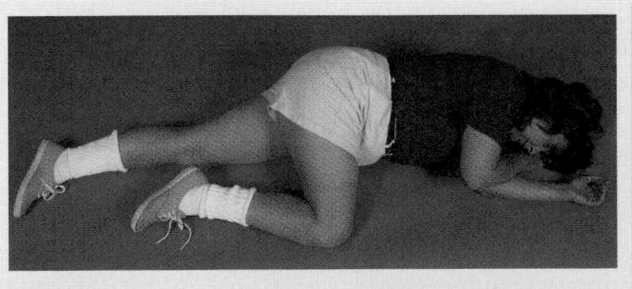

Figure 9-26 In the recovery position, the patient is rolled onto his or her left side.

Figure 9-27 Oxygen tanks for medical use will have a series of letters and numbers stamped into the metal on the collar of the cylinder.

When ventilating any patient in cardiac or respiratory arrest, you must always use high-concentration supplemental oxygen.

Supplemental Oxygen Equipment

In addition to knowing when and how to give supplemental oxygen, you must understand how oxygen is stored and the various hazards associated with its use.

Oxygen Cylinders

The oxygen that you will give to patients is usually supplied as a compressed gas in green, seamless, steel or aluminum cylinders. Some cylinders may be silver or chrome with a green area around the valve stem on top. Newer cylinders are often made of lightweight aluminum or spun steel; older cylinders are much heavier.

Check to make sure that the cylinder is labeled for medical oxygen. You should look for letters and numbers stamped into the metal on the collar of the cylinder **Figure 9-27**. Of particular importance are the month and year stamps, which indicate when the cylinder was last tested. Generally, aluminum cylinders are tested every 5 years; composite cylinders are tested every 3 years.

Oxygen cylinders are available in several sizes. The two sizes that you will most often use are the D (or super D) and M cylinders **Figure 9-28**. The D (or super D) cylinder can be carried from your unit to the patient. The M tank remains on board your unit as a main supply tank. Other sizes that you will see are A, E, G, H, and K **Table 9-3**. The length of time you can use an

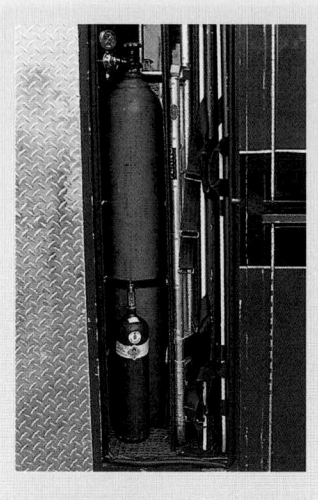

Figure 9-28 The cylinders that are most commonly found on an ambulance are the D (or super D) and M size cylinders.

oxygen cylinder depends on the pressure in the cylinder and the flow rate. A method of calculating cylinder duration is shown in **Table 9-4**.

Liquid Oxygen

Like all gases, oxygen changes from a gas to a liquid when cooled. Liquid oxygen is becoming more commonly used as an alternative to compressed gas oxygen. Liquid oxygen containers tend to be more expensive than compressed oxygen tanks; however, the containers hold a large volume of oxygen and do not need to be filled as often. Liquid oxygen units also weigh less than aluminum or steel tanks. For these reasons, many people who receive long-term oxygen therapy use liquid oxygen units. Unfortunately, liquid oxygen tanks generally need to be kept upright and have special requirements for filling, large-volume storage, and cylinder transfer.

Safety Considerations

Compressed gas cylinders must be handled carefully because their contents are under pressure. Cylinders are fitted with pressure regulators to make sure that patients receive the right amount and type of gas. Make sure that

Table 9-3 Oxygen Cylinder Sizes Carried on the Ambulance

Size	Volume, Liters
D	350
Super D	500
E	625
M	3,000
G	5,300
H, A, K	6,900

the correct pressure regulator is firmly attached before you transport the cylinders. A puncture or hole in the tank can cause the cylinder to become a deadly missile. Do not handle a cylinder by the neck assembly alone. Cylinders should be secured with mounting brackets when they are stored on the ambulance. Oxygen cylinders that are in use during transport should be positioned and secured to prevent the tank from falling and to prevent damage to the valve-gauge assembly.

Pin-Indexing System

The compressed gas industry has established a **pin-indexing system** for portable cylinders to prevent an oxygen regulator from being connected to a carbon dioxide cylinder, a carbon dioxide regulator from being connected to an oxygen cylinder, and so on. In preparing to administer oxygen, always check to be sure that the pin-holes on the cylinder exactly match the corresponding pins on the regulator.

Table 9-4 Oxygen Cylinders: Duration of Flow

Formula

$$\frac{(\text{Gauge pressure in psi} - \text{Safe residual pressure}) \times \text{Cylinder constant}}{\text{Flow rate in L/min}} = \text{Duration of flow in minutes}$$

Safe residual pressure = 200 psi

Cylinder constant for a given cylinder size:

A = 3.14	G = 2.41
D = 0.16	H = 3.14
E = 0.28	K = 3.14
M = 1.56	

Determine the life of an M cylinder that has a pressure of 2,000 psi and a flow rate of 10 L/min.

$$\frac{(2{,}000 - 200) \times 1.56}{10} = \frac{2{,}808}{10} = 281 \text{ min, or } 4 \text{ h } 41 \text{ min}$$

psi = pounds per square inch.

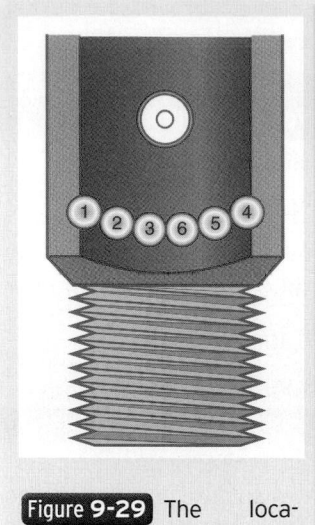

Figure 9-29 The locations of the pin-indexing safety system holes in a cylinder valve face. Each cylinder of a specific gas has a given pattern and a given number of pins.

The pin-indexing system features a series of pins on a yoke that must be matched with the holes on the valve stem of the gas cylinder. The arrangement of the pins and holes varies for different gases according to accepted national standards **Figure 9-29**. Other gases that are supplied in portable cylinders, such as acetylene, carbon dioxide, and nitrogen, use regulators and flowmeters that are similar to those used with oxygen. Each cylinder of a specific gas type has a given pattern and a given number of pins. These safety measures make it impossible for you to attach a cylinder of nitrous oxide to an oxygen regulator. The oxygen regulator will not fit.

The outlet valves on portable oxygen cylinders are designed to accept yoke-type pressure-reducing gauges, which conform to the pin-indexing system **Figure 9-30**.

The safety system for the large cylinders is known as the **American Standard System**. In this system, oxygen cylinders are equipped with threaded gas outlet valves. The inside and outside thread sizes of these outlets vary depending on the gas in the cylinder. The cylinder will not accept a regulator valve unless it is properly threaded to fit that regulator. The purpose of these safety devices is the same as in the pin-indexing system: to prevent the accidental attachment of a regulator to a wrong cylinder.

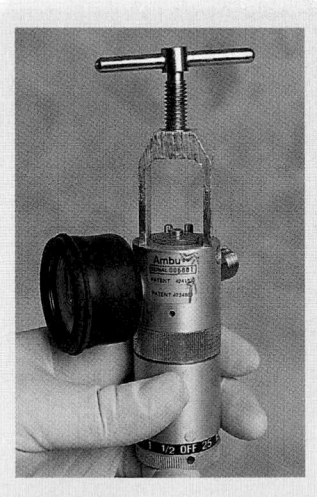

Figure 9-30 A yoke-type pressure-reducing gauge is used with a portable oxygen cylinder.

Pressure Regulators

The pressure of the gas in a full oxygen cylinder is approximately 2,000 psi. This is far too much pressure to be safe or useful for your purposes. Pressure regulators reduce the pressure to a more useful range, usually 40 to 70 psi. Most pressure regulators in use today reduce the pressure in a single stage, although multistage regulators exist. A two-stage regulator will reduce the pressure first to 700 psi and then to 40 to 70 psi.

After the pressure is reduced to a workable level, the final attachment for delivering the gas to the patient is usually one of the following:

- A quick-connect female fitting that will accept a quick-connect male plug from a pressure hose or ventilator/resuscitator
- A flowmeter that will permit the regulated release of gas measured in liters per minute

Flowmeters

Flowmeters are usually permanently attached to pressure regulators on emergency medical equipment. The two types of flowmeters that are commonly used are pressure-compensated flowmeters and Bourdon-gauge flowmeters.

A pressure-compensated flowmeter incorporates a float ball within the tapered calibrated tube. The flow of gas is controlled by a needle valve located downstream from the float ball. This type of flowmeter is affected by gravity and must always be maintained in an upright position for an accurate flow reading **Figure 9-31**.

The Bourdon-gauge flowmeter is not affected by gravity and can be used in any position **Figure 9-32**. It is actually a pressure gauge that is calibrated to record flow rate. The major disadvantage of the flowmeter is that it does not compensate for backpressure. Therefore, it will usually record a higher flow rate when there is any obstruction to gas flow downstream. This type of flowmeter is generally now considered outdated. New flowmeters incorporate a fixable setting with either a dial or a knob that sets the flow. In these regulators, a Bourdon-gauge is not necessary.

■ Operating Procedures

To place an oxygen cylinder into service, follow the steps in **Skill Drill 9-7**:

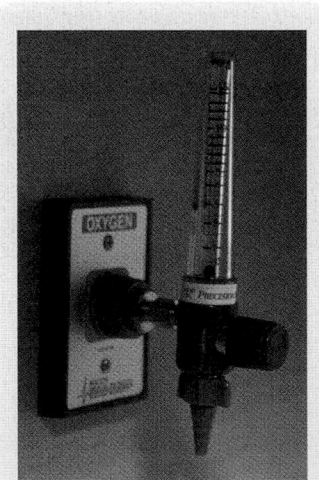

Figure 9-31 A pressure-compensated flowmeter contains a float ball that rises or falls according to the gas flow within the tube. It must be maintained in an upright position for an accurate reading.

Figure 9-32 The Bourdon-gauge flowmeter is not affected by gravity and can be used in any position.

1. Inspect the cylinder and its markings. If the cylinder was commercially filled, it will have a plastic seal around the valve stem covering the opening in the stem. Remove the seal, and inspect the opening to make sure that it is free of dirt and other debris. The valve stem should not be sealed or covered with adhesive tape or any petroleum-based substances. These can contaminate the oxygen and can contribute to combustion when mixed with pressurized oxygen.

   "Crack" the cylinder by slowly opening and then reclosing the valve to help make sure that the dirt particles and other possible contaminants do not enter the oxygen flow. Never face the tank toward yourself or others when cracking the cylinder. Open the tank by attaching a tank key to the valve and rotating the valve counterclockwise. You should be able to hear clearly the rush of oxygen coming from the tank. Close the tank by rotating the valve clockwise (Step 1).

2. Attach the regulator/flowmeter to the valve stem after clearing the opening. On one side of the valve stem, you will find three holes. The larger one, on top, is a true opening through which the oxygen flows. The two smaller holes below it do not extend to the inside of the tank. They provide stability to the regulator. Following the design of the pin-indexing system, these two holes are very precisely located in positions that are unique to the oxygen cylinders.

   Above the pins on the inside of the collar is the actual port through which oxygen flows from the cylinder to the regulator. A metal-bound elastomeric sealing washer (also called a gasket) is placed around the oxygen port to optimize the airtight seal between the collar of the regulator and the valve stem (Step 2). In the past, crush gaskets made of plastic and nylon were used, but are no longer recommended. If used, crush gaskets can be used only once and then they must be replaced.

3. Place the regulator collar over the cylinder valve, with the oxygen port and pin-indexing pins on the side of the valve stem that has the three holes. Open the screw bolt just enough to allow the collar to fit freely over the valve stem. Move the regulator so that the oxygen port and the pins fit into the correct holes on the valve stem. The screw bolt on the opposite side should be aligned with the dimple depression. As you hold the regulator securely against the valve stem, hand tighten the screw bolt until the regulator is firmly attached to the cylinder. At this point, you should not see any open spaces between the sides of the valve stem and the interior walls of the collar (Step 3).

4. With the regulator firmly attached, open the cylinder completely, check for air leaking from the regulator–oxygen cylinder connection, and read the pressure level on the regulator gauge. Most portable cylinders have a maximum pressure of approximately 2,000 psi. Most EMS services consider a cylinder with less than 500 to 1,000 psi to be too low to keep in service. Learn your department's policies in this regard and follow them.

   The flowmeter will have a second gauge or a selector dial that indicates the oxygen flow rate. Several popular types of devices are widely used. Attach the selected oxygen device to the flowmeter by connecting the universal oxygen connecting tubing to the "Christmas tree" nipple on the flowmeter. Most oxygen-delivery devices come with this tubing permanently attached; however, some oxygen masks do not. You must attach this tubing to the oxygen-delivery device if it is not already attached (Step 4).

Safety

Slowly open the oxygen tank after attaching the regulator and check for leaks. Remember that although oxygen itself is not combustible, it supports combustion, and any ignition source may cause fire or an explosion in an oxygen-rich environment—especially if oxygen is being released too quickly from the cylinder at the time or if the seal between the regulator and oxygen cylinder is not secure.

Skill Drill 9-7

Placing an Oxygen Cylinder Into Service

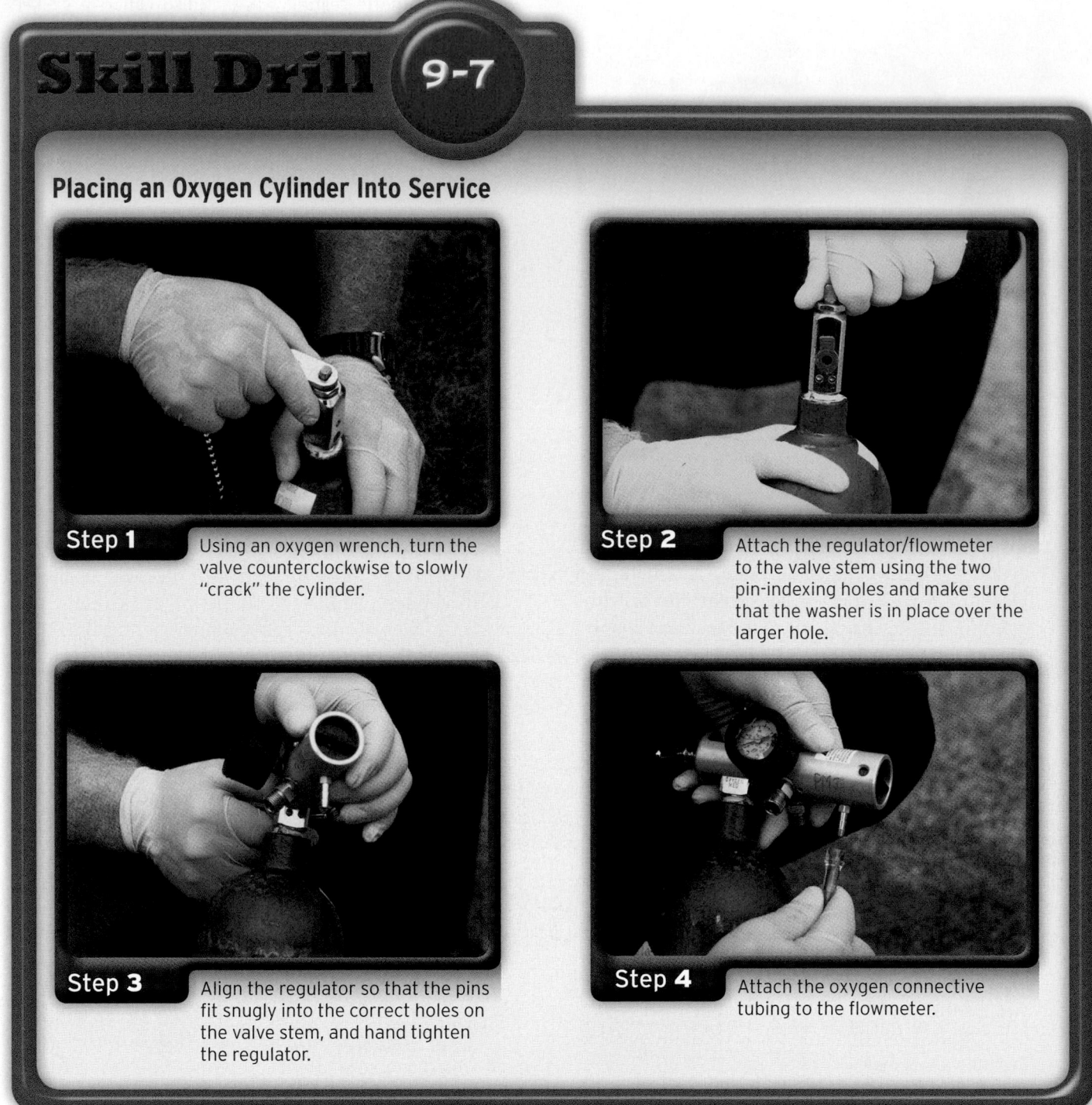

Step 1 Using an oxygen wrench, turn the valve counterclockwise to slowly "crack" the cylinder.

Step 2 Attach the regulator/flowmeter to the valve stem using the two pin-indexing holes and make sure that the washer is in place over the larger hole.

Step 3 Align the regulator so that the pins fit snugly into the correct holes on the valve stem, and hand tighten the regulator.

Step 4 Attach the oxygen connective tubing to the flowmeter.

Open the flowmeter to the desired flow rate. Flow rates will vary based on the oxygen-delivery device being used. Remember that you must be completely familiar with the equipment before attempting to use it on a patient. Once the oxygen is flowing at the desired rate, apply the oxygen device to the patient and make any necessary adjustments. Monitor the patient's response to the oxygen and to the oxygen device, and periodically recheck the regulator gauge to make sure there is sufficient oxygen in the cylinder. Disconnect the tubing from the flowmeter nipple and turn off the cylinder valve when oxygen therapy is complete or when the patient has been transferred to the hospital and is using the hospital's oxygen system. In a few seconds, the sound of oxygen flowing from the nipple will cease. This indicates that all the pressurized oxygen has been removed from the flowmeter. Turn off the flowmeter. The gauge on the regulator should read zero with the tank valve closed. This confirms that there is no pressure left above the valve stem. As long as there is a pressure reading on the regulator gauge, it is not safe to remove the regulator from the valve stem.

Hazards of Supplemental Oxygen

Oxygen does not burn or explode. However, it does support combustion. The more oxygen is around, the faster the combustion process. A small spark, even a glowing cigarette, can become a flame in an oxygen-rich atmosphere. Therefore, you must keep any possible source of fire away from the area while oxygen is in use. Make sure the area is adequately ventilated, especially in industrial settings where hazardous materials may be present and where sparks are easily generated. Be extremely cautious in any enclosed environment in which oxygen is being administered, as an oxygen-rich environment increases the chance of fire if a spark or flame is introduced. A bystander who is smoking or sparks generated during vehicle extrication are possible sources of ignition. Never leave an oxygen cylinder standing unattended. The cylinder can be knocked over, injuring the patient or damaging the equipment.

Oxygen-Delivery Equipment

In general, the oxygen-delivery equipment that is used in the field should be limited to nonrebreathing masks, bag-mask devices, and nasal cannulas, depending on local protocol. However, you may encounter other devices during transports between medical facilities.

Nonrebreathing Masks

The **nonrebreathing mask** is the preferred way of giving oxygen in the prehospital setting to patients who

Words of Wisdom

Oxygen-Delivery Devices

Device	Flow Rate	Oxygen Delivered
Nasal cannula	1 to 6 L/min	24% to 44%
Nonrebreathing mask	10 to 15 L/min	Up to 90%
Bag-mask device with reservoir	15 L/min	Nearly 100%
Mouth-to-mask device	15 L/min	Nearly 55%

are breathing adequately but are suspected of having or showing signs of hypoxia. With a good mask-to-face seal, it is capable of providing up to 90% inspired oxygen.

The nonrebreathing mask is a combination mask and reservoir bag system. Oxygen fills a reservoir bag that is attached to the mask by a one-way valve. The system is called a nonrebreathing mask because the exhaled gas escapes through flapper valve ports at the cheek areas of the mask **Figure 9-33**. These valves prevent the patient from rebreathing exhaled gases.

In this system, you must be sure that the reservoir bag is full before the mask is placed on the patient. Adjust the flow rate so that the bag does not fully collapse when the patient inhales, to about two thirds of the bag volume, or 10 to 15 L/min. Make sure the bag stays inflated. Should the bag collapse when the patient inhales, increase the flow rate of oxygen. In addition, if oxygen therapy is

You are the Provider: PART 4

A rescue unit arrives at the scene to assist you and your partner. As you are preparing to load the patient onto the stretcher, you note that his level of consciousness has markedly diminished and he is making a snoring sound when he breathes. You immediately reassess him.

Recording Time: 6 Minutes	
Level of consciousness	Responsive only to pain
Respirations	10 breaths/min, shallow, snoring
Pulse	130 beats/min, weak
Skin	Cool and moist, cyanotic
Blood pressure	118/54 mm Hg
Sao_2	76% (on oxygen)

9. What should be your *most* immediate action?

10. How should you adjust your treatment of the patient?

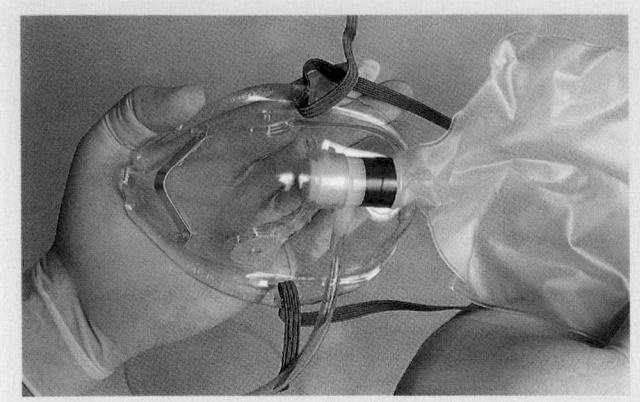

Figure **9-33** The nonrebreathing mask contains flapper valve ports at the cheek areas of the mask to prevent the patient from rebreathing exhaled gases.

discontinued, remove the mask from the patient's face. Leaving the mask in place, while oxygen is not flowing, allows the patient to rebreathe exhaled carbon dioxide. Use a pediatric nonrebreathing mask, which has a smaller reservoir bag, with infants and children, as they will inhale a smaller volume.

■ Nasal Cannulas

A **nasal cannula** delivers oxygen through two small, tube-like prongs that fit into the patient's nostrils Figure **9-34**. This device can provide 24% to 44% inspired oxygen when the flowmeter is set at 1 to 6 L/min. For the comfort of your patient, flow rates above 6 L/min are not recommended with the nasal cannula.

The nasal cannula delivers dry oxygen directly into the nostrils, which, over prolonged periods, can cause dryness or irritate the mucous membrane lining of the nose. Therefore, when you anticipate a long transport time, you should consider the use of humidification.

A nasal cannula has limited use in the prehospital setting. For example, a patient who breathes through the mouth or who has a nasal obstruction will likely get little or no benefit from a nasal cannula. Always try to give high-flow oxygen through a nonrebreathing mask if you suspect that a patient may have hypoxia, coaching him or her if necessary. If the patient will not tolerate a nonrebreathing mask, you will have to use a nasal cannula, which some patients find more comfortable. As always, a good assessment of your patient will guide your decision.

■ Partial Rebreathing Masks

The partial rebreathing mask is similar to a nonrebreathing mask except that there is no one-way valve between the mask and the reservoir. Consequently, patients rebreathe a small amount of their exhaled air. This has some benefit when you want to increase the patient's partial pressure of carbon dioxide, which makes this the ideal mask for patients who you think are suffering from hyperventilation syndrome. The oxygen enriches the air mixture and delivers a gas mix of approximately 80% to 90% oxygen and 2% to 3% carbon dioxide. You can easily convert a nonrebreathing mask to a partial rebreathing mask by removing the one-way valve between the mask and the reservoir bag.

■ Venturi Masks

A Venturi mask has a number of attachments that enable you to vary the percentage of oxygen delivered to the patient while a constant flow is maintained from the regulator Figure **9-35**. This is accomplished by the Venturi principle, which causes air to be drawn into the flow of oxygen as it passes a hole in the line. The Venturi mask is a medium-flow device that delivers 24% to 40% oxygen, depending on the manufacturer.

The main advantage of the Venturi mask is the use of its fine adjustment capabilities in the long-term management of physiologically stable patients. However, in the emergency setting, such fine adjustments

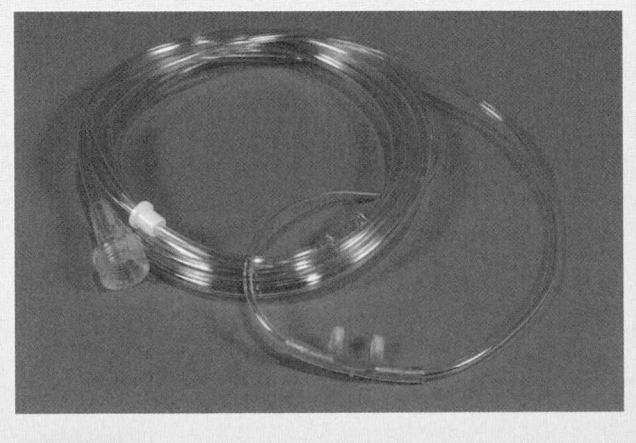

Figure **9-34** The nasal cannula delivers oxygen directly through the nostrils.

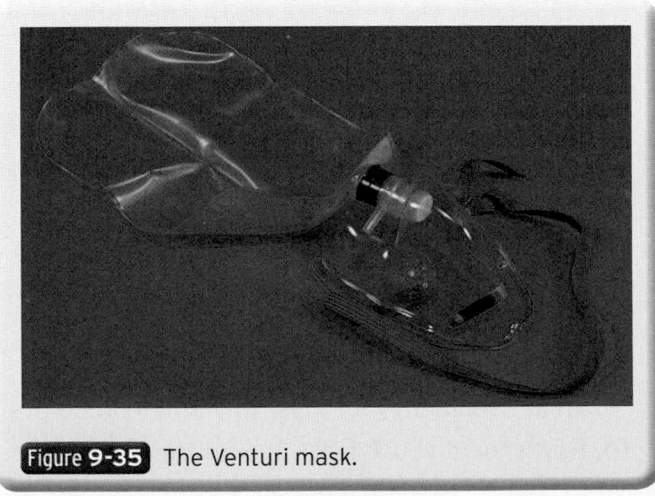

Figure **9-35** The Venturi mask.

are not necessary. When you need to adjust the oxygen concentration in an emergency, it is typically done by adjusting the flow rate or changing the delivery device.

■ Tracheostomy Masks

Patients with tracheostomies do not breathe through their mouth and nose. A face mask or nasal cannula therefore cannot be used to treat them. Masks designed specifically for these patients cover the tracheostomy hole and have a strap that goes around the neck. These masks are usually available in intensive care units, where many patients have tracheostomies, and may not be available in an emergency setting. If you do not have a tracheostomy mask, you can improvise by placing a face mask over the stoma. Even though the mask is shaped to fit the face, you can usually get an adequate fit over the patient's neck by adjusting the strap **Figure 9-36** .

■ Humidification

Some EMS systems provide humidified oxygen to patients during extended transport or for certain conditions such as croup **Figure 9-37** . However, humidified oxygen is usually indicated only for long-term oxygen therapy. Dry oxygen is not considered harmful for short-term use. An oxygen humidifier consists of a small bottle of water through which the oxygen leaving the cylinder becomes moisturized before it reaches the patient. Because the humidifier must be kept in an upright position, however, it is practical only for the fixed oxygen unit in the ambulance. Therefore,

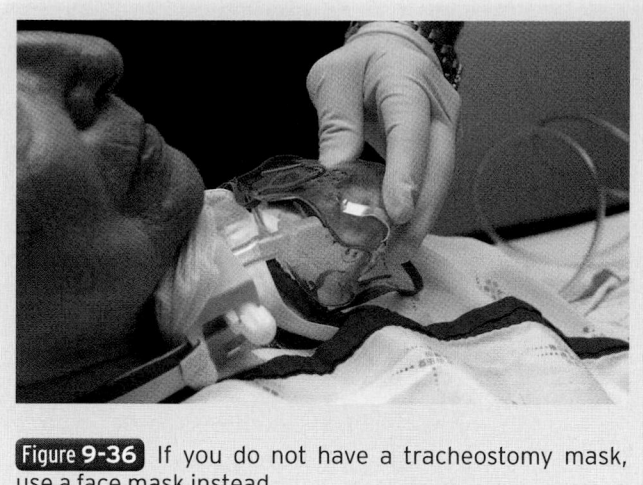

Figure 9-36 If you do not have a tracheostomy mask, use a face mask instead.

many EMS systems do not use humidified oxygen in the prehospital setting. Always refer to medical control or local protocols for guidance involving patient treatment issues.

■ Assisted and Artificial Ventilation

Obviously, a patient who is not breathing needs artificial ventilation and 100% supplemental oxygen. Assisted and artificial ventilation are probably the most important skills in EMS—at any level. Too often emphasis is placed on advanced airway techniques, making the basic

You are the Provider: PART 5

You continue your treatment of the patient, load him into the ambulance, and proceed to the hospital. A member of the rescue team, who is also an EMT, accompanies you to assist with patient care. After reassessing the patient, you call in your report to the emergency department.

Recording Time: 10 Minutes	
Level of consciousness	Responsive only to pain
Respirations	14 breaths/min, shallow
Pulse	100 beats/min, stronger
Skin	Warm and moist, less cyanotic
Blood pressure	124/62 mm Hg
Sao$_2$	92% (on oxygen)

11. Has your treatment improved the patient's condition? If so, how do you know?

12. What is the appropriate technique for ventilating an apneic adult?

13. What are the dangers of hyperventilating a patient?

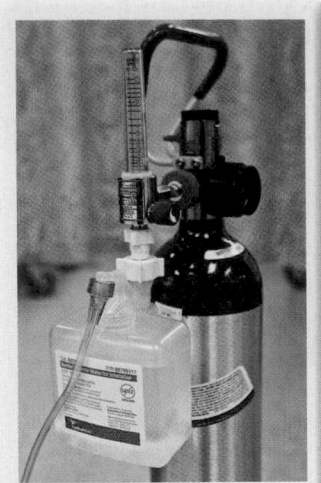

Figure 9-37 Giving humidified oxygen may be preferred with long transport times. However, the use of this type of oxygen-delivery system is not universal in all EMS systems.

airway maneuvers seem ineffective. This cannot be further from the truth. Basic airway and ventilation techniques are extremely effective when administered appropriately. Mastery of these techniques at the EMT level is imperative.

Patients who are breathing inadequately, such as those who are breathing too fast or too slowly with reduced tidal volume, are typically unable to speak in complete sentences. An irregular breathing pattern will also require artificial ventilation to assist patients in maintaining adequate minute volume. Keep in mind that fast, shallow breathing can be just as dangerous as very slow breathing. Fast, shallow breathing moves air primarily in the larger airway passages (dead air space) and does not allow for adequate exchange of air and carbon dioxide in the alveoli. Patients with inadequate breathing require assisted ventilations with some form of positive-pressure ventilation. Remember to follow standard precautions as needed when managing the patient's airway.

Words of Wisdom

Methods of Ventilation (listed in order of preference)
- Mouth-to-mask with one-way valve
- Two-person bag-mask device with reservoir and supplemental oxygen
- Manually triggered ventilation device (flow-restricted, oxygen-powered ventilation device)
- One-person bag-mask device with oxygen reservoir and supplemental oxygen

Note: This order of preference has been stated because research has shown that personnel who infrequently ventilate patients have great difficulty maintaining an adequate seal between the mask and the patient's face.

■ Assisting Ventilation in Respiratory Distress/Failure

When a patient is in severe respiratory distress or respiratory failure and not breathing adequately, you must intervene quickly to prevent further deterioration of the patient. Two treatment options are available

in these situations: assisted ventilation and continuous positive airway pressure (CPAP). CPAP is discussed later in this chapter; the focus of this section will be on assisted ventilation.

The purpose of assisted ventilations is to improve the overall oxygenation and ventilatory status of the patient. These patients are no longer able to maintain adequate oxygen levels for the body and need assistance to prevent further hypoxia.

You need to be familiar with the signs and symptoms associated with inadequate ventilation. Signs of altered mental status and inadequate minute volume are indications for assisted ventilation. In addition, excessive accessory muscle use and fatigue from labored breathing are signs of potential respiratory failure. Patients exhibiting these signs need immediate treatment.

Follow these steps to assist a patient with ventilations using a bag-mask device:

1. Explain the procedure to the patient.
2. Place the mask over the patient's nose and mouth.
3. Squeeze the bag each time the patient breathes, maintaining the same rate as the patient.
4. After the initial 5 to 10 breaths, slowly adjust the rate and deliver an appropriate tidal volume.
5. Adjust the rate and tidal volume to maintain an adequate minute volume.

■ Artificial Ventilation

Patients who are in respiratory arrest need immediate treatment. Without it, patients will die. However, the act of breathing for a patient, or artificial ventilation, is not a skill you should take lightly. Once you determine that a patient is not breathing, you should begin artificial ventilation immediately. The methods that you may use to provide artificial ventilation include the mouth-to-mask technique; a one-, two-, or three-person bag-mask device; and the manually-triggered ventilation device.

Normal Ventilation Versus Positive-Pressure Ventilation

The EMT must understand that while artificial ventilations are necessary to sustain life, they are not the same as normal breathing. As discussed earlier, the act of air moving in and out the lungs is based on pressure changes within the thoracic cavity. During normal ventilation, the diaphragm contracts and negative pressure is generated in the chest cavity. This essentially sucks air into the chest from the trachea in an attempt to equalize the pressure in the chest with the atmospheric pressure. However, positive-pressure ventilation generated by a device, such as a bag-mask device, forces air into the chest cavity from the external environment, rather than based on pressure

Table 9-5 Normal Ventilation Versus Positive-Pressure Ventilation

	Normal Ventilation	Positive-Pressure Ventilation
Air movement	Air is sucked into the lungs due to the negative intrathoracic pressure created when the diaphragm contracts.	Air is forced into the lungs through a means of mechanical ventilation.
Blood movement	Normal breathing allows blood to naturally be pulled back to the heart.	Intrathoracic pressure is increased, not allowing blood to be adequately pulled back to the heart. This causes the amount of blood pumped by the heart to be reduced.
Airway wall pressure	Not affected during normal breathing.	More volume is required to have the same effects as normal breathing. As a result, the walls are pushed out of their normal anatomic shape.
Esophageal opening pressure	Not affected during normal breathing.	Air is forced into the stomach, causing gastric distention that could result in vomiting and aspiration.
Overventilation	Overventilation is not typical of normal breathing.	Forcing volume and rate results in increased intrathoracic pressure, gastric distention, and decrease in cardiac output (hypotension).

changes. This difference between normal ventilation and positive-pressure ventilation can create some challenges for the EMT **Table 9-5**.

The physical act of the chest wall expanding and retracting during breathing serves to aid the circulatory system in returning blood back to the heart. During normal ventilation, the chest wall movement works similar to a pump. The pressure changes in the thoracic cavity help draw venous return back to the heart. However, when positive-pressure ventilation is initiated, more air is needed to achieve the same oxygenation and ventilatory effects of normal breathing. This increase in airway wall pressure causes the walls of the chest cavity to push out of their normal anatomic shape. As a result, there is an increase in the overall intrathoracic pressure within the chest cavity. This pressure increase affects the venous return of blood back to the heart. The blood flow is decreased due to the increased pressure in the chest. This causes poor venous return to the heart, and the amount of blood pumped out of the heart is reduced. Therefore, it is imperative that the EMT regulate the rate and volume of artificial ventilations to help prevent this drop in cardiac output. Cardiac output is a function of stroke volume and heart rate, such that cardiac output = stroke volume × heart rate. Stroke volume is the amount of blood ejected by the ventricle in one cardiac cycle. The heart rate is assessed by taking the pulse for 1 minute. The cardiac output is the amount of blood ejected by the left ventricle in 1 minute.

Another difference between normal ventilation and positive-pressure ventilation is the control of airflow. When a person breathes, air enters the trachea, and generally, not the esophagus. However, the force generated from positive-pressure ventilation allows air to enter not only the trachea, but the esophagus as well. Ventilations that are too forceful can lead to excessive air in the stomach. This potential complication will be discussed shortly.

Words of Wisdom

Ventilation Rates*
Adult	1 breath per 5 to 6 seconds
Child	1 breath per 3 to 5 seconds
Infant	1 breath per 3 to 5 seconds

*For apneic patients with a pulse.

Mouth-to-Mouth and Mouth-to-Mask Ventilation

As you learned in your CPR course, mouth-to-mouth ventilations are now routinely done with a barrier device, such as a mask or face shield. A **barrier device** is a protective item that features a plastic barrier placed on a patient's face with a one-way valve to prevent the backflow of secretions, vomitus, and gases. Barrier devices provide adequate protection **Figure 9-38**. Mouth-to-mouth ventilations without a barrier device should be provided only in extreme situations. Performing mouth-to-mask

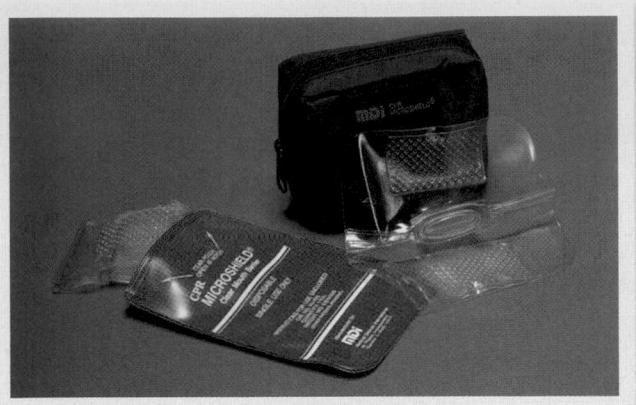

Figure 9-38 Barrier devices such as a plastic shield or a pocket mask with a one-way valve provide adequate protection.

ventilations with a pocket mask containing a one-way valve is a safer method to prevent possible disease transmission.

A mask with an oxygen inlet provides oxygen during mouth-to-mask ventilation to supplement the air from your lungs. Remember that the gas you exhale contains 16% oxygen. With the mouth-to-mask system, however, the patient gets the additional benefit of significant oxygen enrichment with inspired air. This system also frees both your hands to help keep the airway open and helps you to provide a better seal between the mask and face, thus delivering adequate tidal volume.

The mask may be shaped like a triangle or a doughnut, with the apex (top) placed across the bridge of the nose. The base (bottom) of the mask is placed in the groove between the lower lip and the chin. In the center of the mask is a chimney with a 15/22-mm connector.

Follow these steps to use mouth-to-mask ventilation **Skill Drill 9-8**:

1. Kneel at the patient's head. Open the airway using the head tilt–chin lift maneuver or the jaw-thrust maneuver if trauma is suspected. Insert an oral or nasal airway to help maintain airway patency. Connect the one-way valve to the face mask and place the mask on the patient's face. Make sure the top is over the bridge of the nose and the bottom is in the groove between the lower lip and the chin. Hold the mask in position by placing your thumbs over the top part of the mask and your index fingers over the bottom half. Grasp the lower jaw with the remaining three fingers on each hand, making an airtight seal by pulling the lower jaw into the mask. Maintain an upward and forward pull on the lower

jaw with your fingers to keep the airway open. This method of securing the mask to the patient's face is known as the EC clamp method **Step 1**.
2. Take a deep breath and exhale through the open port of the one-way valve. Breathe slowly into the patient's mask until you observe adequate chest rise **Step 2**.
3. Remove your mouth, and watch for the patient's chest to fall during passive exhalation **Step 3**.

You know that you are providing adequate ventilations if you see the patient's color improving, the chest rise adequately, and do not meet resistance when ventilating. You should also hear and feel air escape as the patient exhales. Make sure that you are providing the correct number of breaths per minute for the patient's age.

To increase the oxygen concentration, administer high-flow oxygen at 15 L/min through the oxygen inlet valve of the mask. This, when combined with your exhaled breath, will deliver approximately 55% oxygen to the patient. If supplemental oxygen is available, deliver a tidal volume of approximately 500 to 600 mL (6 to 7 mL/kg) over 1 to 2 seconds. If supplemental oxygen is not available, tidal volumes of approximately 700 to 1,000 mL (10 mL/kg) should be delivered over 2 seconds.

The Bag-Mask Device

With an oxygen flow rate of 15 L/min and an adequate mask-to-face seal, a bag-mask device with an oxygen reservoir can deliver nearly 100% oxygen **Figure 9-39**. Most bag-mask devices on the market today include modifications or accessories (reservoirs) that permit the delivery of oxygen concentrations approaching 100%; however, the device can deliver only as much volume as you can squeeze out of the bag by hand. The bag-mask device

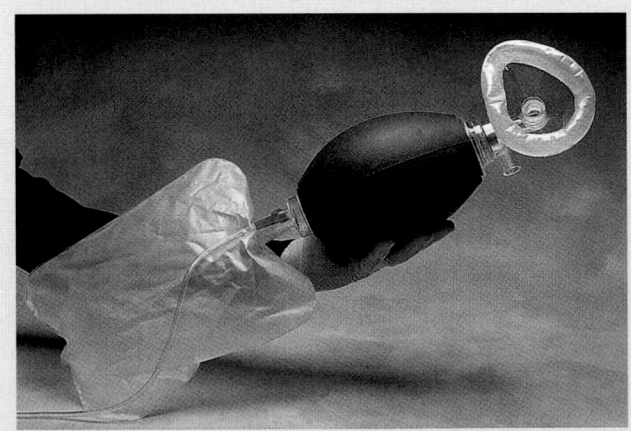

Figure 9-39 A bag-mask device with an oxygen reservoir can deliver nearly 100% oxygen if a good seal between the mouth and mask is achieved and if supplemental oxygen is used.

Skill Drill 9-8

Performing Mouth-to-Mask Ventilation

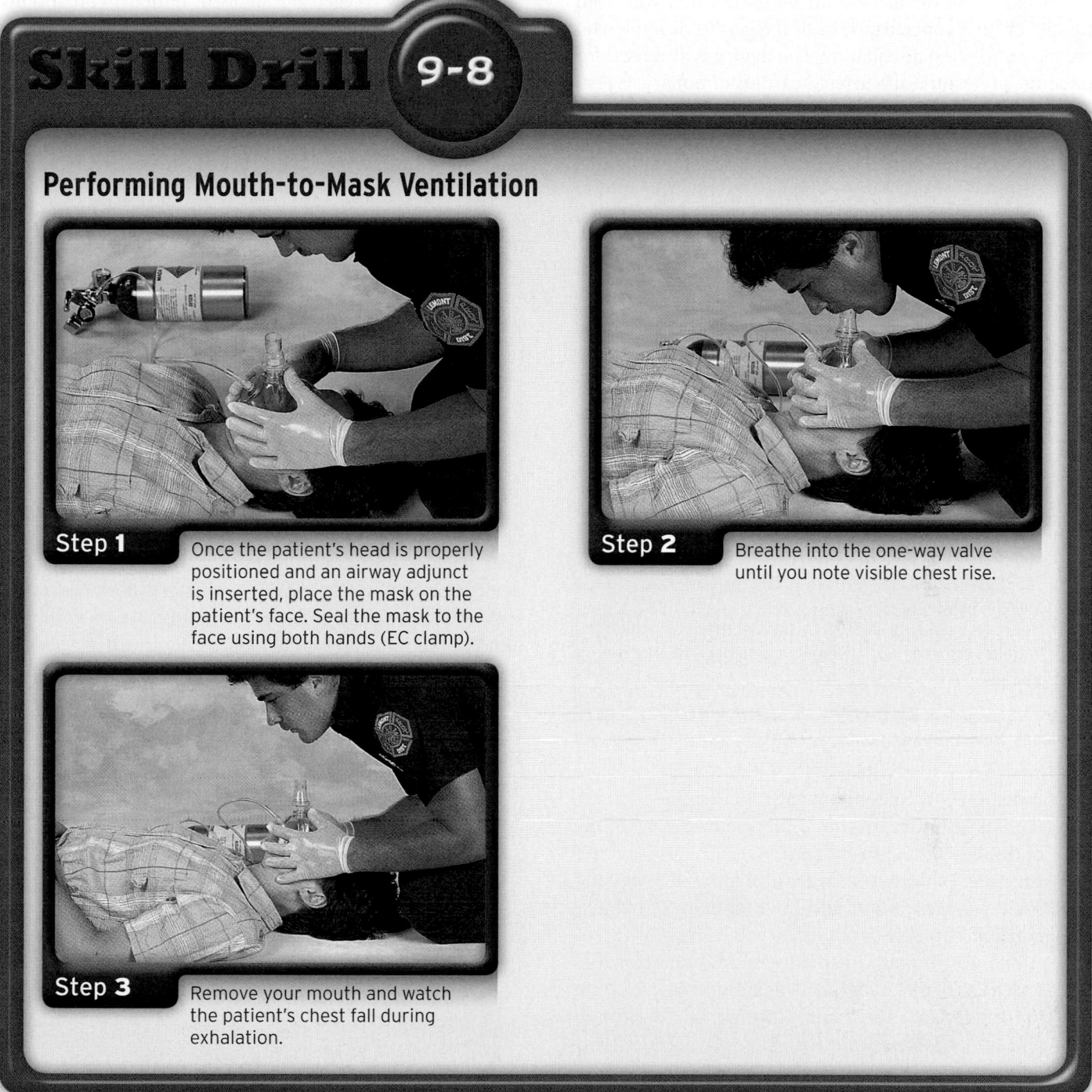

Step 1 Once the patient's head is properly positioned and an airway adjunct is inserted, place the mask on the patient's face. Seal the mask to the face using both hands (EC clamp).

Step 2 Breathe into the one-way valve until you note visible chest rise.

Step 3 Remove your mouth and watch the patient's chest fall during exhalation.

provides less tidal volume than mouth-to-mask ventilation; however, it delivers a much higher concentration of oxygen. The bag-mask device is the most common method used to ventilate patients in the field. While an experienced EMT will be able to supply adequate tidal volume with a bag-mask device, as a new EMT you should develop proficiency at ventilating airway-training manikins before using a bag-mask device on a patient. If you have difficulty adequately ventilating a patient with a bag-mask device, you should immediately switch to an alternative method of ventilation, such as the mouth-to-mask technique.

Words of Wisdom

Volume Capabilities of the Bag-Mask Device

Size	Amount, mL
Adult	1,200 to 1,600
Pediatric	500 to 700
Infant	150 to 240

A bag-mask device should be used when you need to deliver high concentrations of oxygen to patients who are not ventilating adequately. The device is also used for patients in respiratory arrest, cardiopulmonary arrest, and respiratory failure. The bag-mask device may be used with or without oxygen. However, to ensure the highest concentration of delivered oxygen, you must attach supplemental oxygen and a reservoir. You should use an oral or nasal airway adjunct in conjunction with the bag-mask device.

Words of Wisdom

Indications That Artificial Ventilation Is Adequate
- Visible and equal chest rise and fall with ventilation
- Ventilations delivered at the appropriate rate
  - 10 to 12 breaths/min for adults*
  - 12 to 20 breaths/min for infants and children*
- Heart rate returns to normal range
- Patient's color is improving (pink)

Indications That Artificial Ventilation Is Inadequate
- Minimal or no chest rise and fall
- Ventilations are delivered too fast or too slowly for patient's age
- Heart rate does not return to normal range
- Patient's color remains cyanotic, mottled, or deteriorates

*In apneic patients with a pulse.

Bag-Mask Device Components All adult bag-mask devices should have the following components:

- A disposable self-refilling bag
- No pop-off valve, or if one is present, the capability of disabling the pop-off valve
- An outlet valve that is a true valve for nonrebreathing
- An oxygen reservoir that allows for delivery of high-concentration oxygen
- A one-way, no-jam inlet valve system that provides an oxygen inlet flow at a maximum of 15 L/min with standard 15/22-mm fittings for face mask and endotracheal tube (or other advanced airway adjunct) connection
- A transparent face mask
- Ability to perform under extreme environmental conditions, including extreme heat or cold

The total volume in the bag of an adult bag-mask device is usually 1,200 to 1,600 mL. The pediatric bag contains 500 to 700 mL, and the infant bag holds 150 to 240 mL.

The volume of air (oxygen) delivered to the patient is based on one key observation—chest rise and fall. This, in essence, is generally the only means of assessing tidal volume in the field. In most situations, you will be using the bag-mask device attached to high-flow oxygen (15 L/min). When using the bag-mask device

with high-flow oxygen on an adult patient, you should squeeze the bag enough to cause a noticeable rise of the patient's chest—a volume of 400 to 600 mL (approximately 6 to 7 mL/kg) over 1 to 2 seconds. When oxygen is not available, higher tidal volume amounts are required to cause good rise of the patient's chest—700 to 1,000 mL (approximately 10 mL/kg) over 2 seconds. By delivering smaller tidal volumes when the bag-mask device is used with oxygen, the risk of gastric distention (and associated complications of vomiting and aspiration) is reduced.

There are two issues to consider with this approach. It is not practical for the EMT to accurately measure tidal volume in milliliters per kilogram for each patient ventilated in the field. There is also a significant risk of hypoxia when ventilating with smaller volumes. For these reasons, the key is to watch for good chest rise and fall—let these observations determine the appropriate amount of volume to deliver.

Bag-Mask Device Technique Whenever possible, you and your partner should work together to provide bag-mask device ventilation. One EMT can maintain a good mask seal by securing the mask to the patient's face with two hands while the other EMT squeezes the bag. Ventilation using a bag-mask device is a challenging skill: It may be very difficult for one EMT to maintain a proper seal between the mask and the face with one hand while squeezing the bag well enough to deliver an adequate volume to the patient. This skill can be difficult to maintain if you do not have many opportunities to practice. Effective one-person bag-mask device ventilation requires considerable experience. Also, performance of this skill depends on having enough personnel to carry out other actions that need to be done at the same time, such as chest compressions, putting the stretcher in place, or helping to lift the patient onto the stretcher.

Follow these steps to use the two-person bag-mask device technique:

1. Kneel above the patient's head. If possible, your partner should be at the side of the head to squeeze the bag while you hold a seal between the mask and the patient's face with two hands.
2. Maintain the patient's neck in an extended position unless you suspect a cervical spine injury. In that case, you should immobilize the patient's head and neck and use the jaw-thrust maneuver. Have your partner hold the head, or, if you are alone, use your knees to immobilize the head.
3. Open the patient's mouth, and suction as needed. Insert an oral or nasal airway to maintain airway patency.
4. Select the proper mask size.

5. Place the mask on the patient's face. Make sure the top is over the bridge of the nose and the bottom is in the groove between the lower lip and the chin. If the mask has a large, round cuff around the ventilation port, center the port over the patient's mouth. Inflate the collar to obtain a better fit and seal to the face if necessary.

6. Hold the mask in position by placing the thumbs over the top part of the mask and the index fingers over the bottom half.

7. Bring the lower jaw up to the mask with the last three fingers of your hand. This will help to maintain an open airway. Make sure you do not grab the fleshy part of the neck, as you may compress structures and create an airway obstruction. If you think the patient may have a spinal injury, make sure your partner immobilizes the cervical spine as you move the lower jaw.

8. Connect the bag to the mask if you have not already done so.

9. Hold the mask in place while your partner squeezes the bag with two hands until the patient's chest rises ▌Figure **9-40**▐. If a spinal injury is suspected, immobilize the patient's head and neck with your forearms while maintaining an adequate mask-to-face seal with your hands. Continue squeezing the bag once every 5 seconds in an adult and once every 3 seconds for infants and children.

10. If you are alone, hold your index finger over the lower part of the mask, your thumb over the upper part of the mask, and then use your remaining fingers to pull the lower jaw into the mask. This is known as the EC-clamp method and will maintain

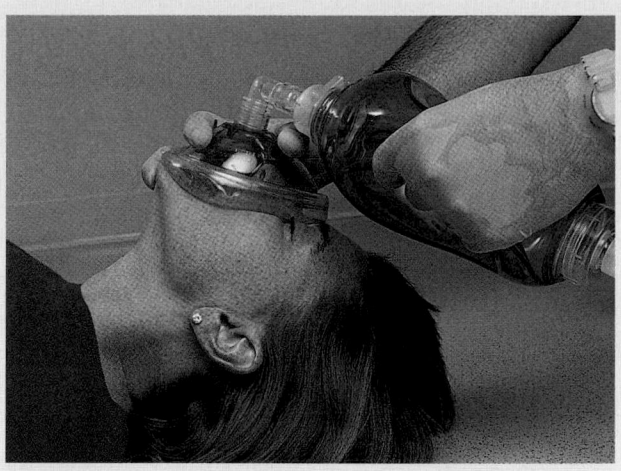

▌Figure **9-41**▐ Maintain the seal of the mask to the face using the EC-clamp method if you must ventilate alone.

an effective face-to-mask seal ▌Figure **9-41**▐. Use the head tilt–chin lift maneuver to make sure the neck is extended. If spinal injury is suspected, stabilize the patient's head in a neutral in-line position with your knees as you pull the patient's lower jaw into the mask. Squeeze the bag in a rhythmic manner once every 5 seconds with your other hand. Continue squeezing the bag once every 5 seconds for an adult and once every 3 seconds for infants and children.

When using the device to assist ventilations of a patient who is breathing too slowly (hypoventilation) with reduced tidal volume, you should squeeze the bag as the patient tries to breathe in. Then, for the next 5 to 10 breaths, slowly adjust the rate and delivered tidal volume until an adequate minute volume is achieved.

To assist respirations of a patient who is breathing too fast (hyperventilation) with reduced tidal volume, you must first explain the procedure to the patient if the patient is coherent. Initially assist respirations at the rate at which the patient has been breathing, squeezing the bag each time the patient inhales. Then, for the next 5 to

▌Figure **9-40**▐ With two-person bag-mask device ventilation, you should hold the mask in place while your partner squeezes the bag with two hands until the patient's chest rises.

Words of Wisdom

Ventilation is the physical act of moving air in and out of the lungs. Ventilation is required for adequate respiration. If ventilation is adequate, other problems may hinder respiration. Examples of interruptions of ventilation include trauma such as a flail chest, foreign body airway obstruction, and an injury to the spinal cord that disrupts the phrenic nerve that innervates the diaphragm.

Words of Wisdom

The Sellick maneuver, also known as cricoid pressure, has been used to inhibit the flow of air into the stomach (and thus reduce gastric distention) and reduce the chance of aspiration by helping block the regurgitation of gastric contents from the esophagus. In this maneuver, a rescuer applies cricoid pressure on the patient by placing the thumb and index finger on either side of the cricoid cartilage (at the inferior border of the larynx) and pressing down.

According to several studies cited in the 2010 American Heart Association Guidelines, cricoid pressure may actually *impede* ventilation and not completely prevent aspiration. For this reason, the procedure is generally not recommended. Be sure to follow your local protocol regarding the use of the Sellick maneuver.

10 breaths, slowly adjust the rate and the delivered tidal volume until an adequate minute volume is achieved.

As you are assisting ventilations with a bag-mask device, you should evaluate the effectiveness of your delivered ventilations. You will know that artificial ventilations are not adequate if the patient's chest does not rise and fall with each ventilation, the rate at which you are ventilating is too slow or too fast, or the heart rate does not return to normal. If the patient's chest does not rise and fall, you may need to reposition the head or use an airway adjunct.

When using a bag-mask device or any other ventilation device, be alert for **gastric distention**, inflation of the stomach with air. To prevent or alleviate distention, you should do the following: (1) ensure that the patient's airway is appropriately positioned, (2) ventilate the patient at the appropriate rate, and (3) ventilate the patient with the appropriate volume.

If the patient's stomach appears to be distending, you should reposition the head. In a patient with a possible spinal injury, you should reposition the jaw rather than the head (that is, use the jaw-thrust). If too much air is escaping from under the mask, reposition the mask for a better seal. If the patient's chest still does not rise and fall after you have made these corrections, check for an airway obstruction. If an obstruction is not present, you should attempt ventilations using an alternative method, such as the mouth-to-mask technique.

The bag-mask device may also be used in conjunction with an endotracheal tube or with other advanced airway devices. Advanced airway techniques are beneficial when a good seal is difficult to maintain, the patient has a cervical spine injury, or the patient's condition warrants. These techniques are discussed later in the text.

Manually Triggered Ventilation Devices

Another method of providing artificial ventilation is with a **manually triggered ventilation device** Figure 9-42 . These devices, also known as flow-restricted, oxygen-powered ventilation devices, are widely available and have been used in EMS for several years. The major advantage to this device is that it allows a single rescuer to use both hands to maintain a mask-to-face seal while providing positive-pressure ventilation. It also reduces rescuer fatigue associated with using a bag-mask device on extended transports. However, recent findings suggest that manually triggered ventilation devices are associated with difficulty in maintaining adequate ventilation without assistance and should not be used routinely because of the high incidence of gastric distention and possible damage to structures within the chest cavity. Another disadvantage is that a special unit and additional training are required when using the manually triggered ventilation device on infants and children. In addition, this device *should not* be used on patients with COPD or suspected cervical spine or chest injuries. Because the rescuer is not actively squeezing a bag, it is difficult to assess for lung compliance. As a result, the rescuer should take extra care when ventilating; the high ventilatory pressures generated by the device may damage lung tissue if not carefully monitored. Typical

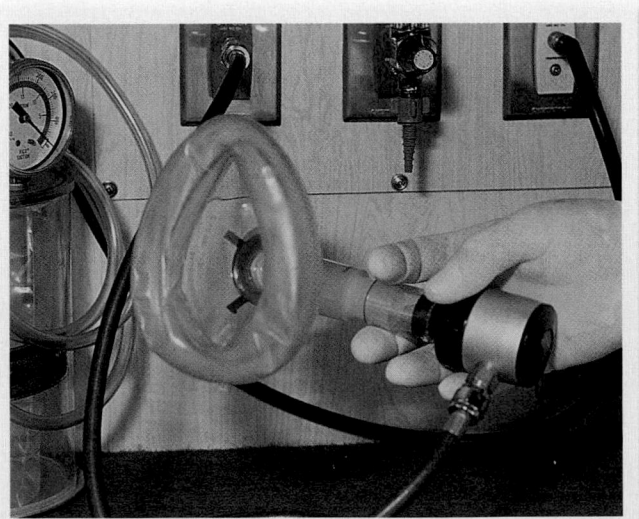

Figure 9-42 A manually triggered ventilation device can provide up to 100% oxygen.

adult ventilation consumes 5 L/min of oxygen versus the manually triggered device at 15 to 25 L/min.

Manually Triggered Ventilation Device Components Manually triggered ventilation devices should have the following components:

- A peak flow rate of 100% oxygen at up to 40 L/min
- An inspiratory pressure safety release valve that opens at approximately 60 cm of water and vents any remaining volume to the atmosphere or stops the flow of oxygen
- An audible alarm that sounds whenever you exceed the relief valve pressure
- The ability to operate satisfactorily under normal and varying environmental conditions
- A trigger (or lever) positioned so that both your hands can remain on the mask to provide an airtight seal while supporting and tilting the patient's head and keeping the jaw elevated

Learning how to use these devices correctly requires proper training and considerable practice. As with bag-mask devices, you must make sure there is an effective seal between the patient's face and mask. The amount of pressure that is necessary to ventilate a patient adequately will vary according to the size of the patient, the patient's lung volume, and the condition of the lungs. A patient with COPD will need greater pressure to receive adequate volume than would be necessary for a patient with normal lungs. Pressures that are too great can cause a **pneumothorax**. Always follow local medical protocols carefully when you use these devices.

Automatic Transport Ventilator/ Resuscitator

The **automatic transport ventilator (ATV)** is essentially a manually triggered ventilation device attached to a control box that allows the variables of ventilation to be set. Although the ATV lacks the sophisticated control of a hospital ventilator, it frees the EMT to perform other tasks, such as maintaining a mask seal or ensuring continued patency of the airway. You can even perform non–airway-related tasks if the patient has an advanced airway in place and is being ventilated with the ATV. However, even though an ATV is helpful to an EMT, a bag-mask device and mask should always be prepared and ready for use should a malfunction occur with the ATV.

Most models have adjustments for respiratory rate and tidal volume. In most cases, the respiratory rate is set at the midpoint or average for the patient's age. Tidal volume is usually estimated using the formula of 6 to 7 mL/kg because ATVs are oxygen-powered and provide oxygen-enriched breathing gas. The tidal volume can be adjusted based on the patient's chest

rise and physiologic response. ATVs are considered volume-cycled/rate-controlled ventilators. This means that they deliver a preset volume at a preset respiratory rate, although this does not guarantee that all of the volume is being delivered to the lungs.

Like the manually triggered ventilation device, the ATV is generally oxygen-powered, although some models may require an external power source. Whereas this device does require oxygen, it generally consumes 5 L/min of oxygen, unlike a bag-mask device that uses 15 to 25 L/min. In addition, just like the manually triggered ventilation device, the ATV has a pressure relief valve, which can lead to hypoventilation in patients with poor lung compliance, increased airway resistance, or airway obstruction. **Compliance** is the ability of the alveoli to expand when air is drawn in during inhalation; poor lung compliance is the inability of the alveoli to fully expand during inhalation.

Whereas ATVs potentially free the EMT to perform other tasks, constant reassessment of the patient is necessary. Barotrauma is a common complication associated with manually triggered ventilation devices and the ATV. In addition, the EMT needs to assess for full chest recoil when using an ATV. This step is not only essential with patients in respiratory arrest, but with patients in cardiac arrest receiving chest compressions as well.

Continuous Positive Airway Pressure

Continuous positive airway pressure (CPAP) is a noninvasive means of providing ventilatory support for patients experiencing respiratory distress. Many people who have been diagnosed with obstructive sleep apnea wear a CPAP unit at night to maintain their airway while they sleep Figure 9-43 . Over the past several years, the use

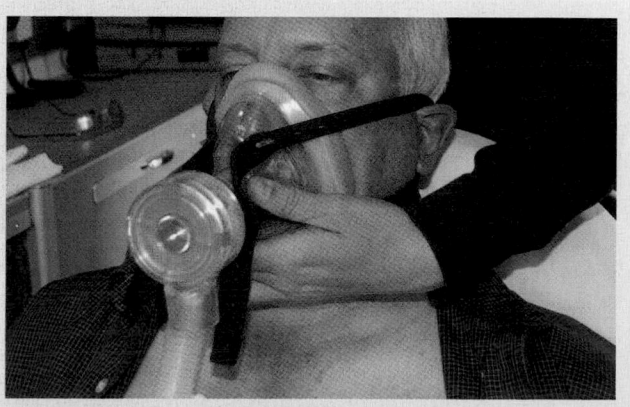

Figure 9-43 Many people who have been diagnosed with obstructive sleep apnea wear a CPAP unit at night to maintain their airway while they sleep.

of CPAP in the prehospital environment has proven to be an excellent adjunct in the treatment of respiratory distress associated with obstructive pulmonary disease and acute pulmonary edema. Typically, many of the patients would be managed with advanced airway devices, such as endotracheal intubation. Research has shown that there is a significant increase in morbidity and mortality when patients with these conditions receive intubation for their condition in the field. CPAP offers an alternative means for providing ventilatory assistance to patients and helps to decrease the overall morbidity and mortality. Because of the simplicity of the device and its great benefit to the patient, CPAP is becoming widely used at the EMT level. However, not all EMTs are trained in the use of CPAP. Follow your local protocols.

Mechanism

CPAP increases pressure in the lungs, opens collapsed alveoli, pushes more oxygen across the alveolar membrane, and forces interstitial fluid back into the pulmonary circulation. Studies of this treatment have shown positive results in patients with obstructive pulmonary diseases and those with acute pulmonary edema. The therapy is typically delivered through a face mask that is held to the head with a strapping system. A good seal with minimal leakage between the face and mask is essential.

Many CPAP systems use oxygen as the driving force to deliver the positive ventilatory pressure to the patient. Frequently check the oxygen regulator when administering CPAP; depending on the flow and the patient's respiratory rate, some CPAP units will empty a D cylinder in as little as 5 to 10 minutes.

The face mask is fitted with a pressure-relief valve that determines the amount of pressure delivered to the patient (such as 5 cm H_2O). The result is similar to hanging your head out the window while driving on the highway. This results in a high inspiratory flow and the need to push a pressure valve open with exhalation. While this may appear to require a great deal of effort on the part of a patient who is already in distress, many patients make a dramatic turnaround when CPAP is applied.

Because CPAP increases pressure inside the chest, it reduces the amount of blood flow returning to the heart. As the pressure in the thorax increases, the venous flow of blood returning to the heart meets the resistance of the increased pressure in the chest. The result is a decrease in the workload of the heart and a drop in cardiac output. This is not common with lower levels of CPAP; however, caution should be used when considering CPAP in patients with potentially low blood pressure. Continually reassess patients for sudden drops in blood pressure.

Indications

CPAP is indicated for patients experiencing respiratory distress in which their own compensatory mechanisms are not enough to keep up with their oxygen demand. Whereas most patients improve after the application of CPAP, it is important to remember that CPAP is merely treating the symptoms and not necessarily the underlying pathology.

The following are some general guidelines for CPAP candidates:

- Patient must be alert and able to follow commands
- Patient is displaying obvious signs of moderate to severe respiratory distress (eg, accessory muscle use, tripod position) from an underlying pathology, such as pulmonary edema or obstructive pulmonary disease (ie, COPD)
- Patient is breathing rapidly, such that it affects overall minute volume (greater than 26 breaths/min)
- Pulse oximetry reading is less than 90%

Whereas these guidelines should be considered when assessing the need for CPAP, it is important that you follow your local guidelines and protocols.

Contraindications

CPAP has proven to be immensely beneficial to patients experiencing respiratory distress from acute pulmonary edema or obstructive pulmonary disease; however, there are times when CPAP is not appropriate for the patient.

The following are general contraindications for CPAP use:

- A patient who is in respiratory arrest
- Signs and symptoms of a pneumothorax or chest trauma
- A patient who has a tracheostomy
- Active gastrointestinal bleeding or vomiting
- Patient is unable to follow verbal commands

In addition to these contraindications, the EMT should always reassess the patient for signs of deterioration and/or respiratory failure. CPAP is an excellent tool to assist with ventilation; however, not all patients will improve with this device. Once signs of respiratory failure become apparent or the patient is no longer able to follow commands, CPAP should be removed from the patient, and positive-pressure ventilation with a bag-mask device attached to high-flow oxygen should be initiated.

Application

Several varieties of CPAP units are available to EMS services; however, most follow the same general guidelines for use and set up. CPAP units are generally composed of a generator, a mask, a circuit that contains corrugated tubing, a bacteria filter, and a one-way valve. During the expiratory phase, the patient exhales against a resistance called **positive end-expiratory pressure (PEEP)**. Within the CPAP generator is a valve that determines the amount of PEEP, however, some CPAP models have PEEP valves that connect separately. Depending on the device, the PEEP is controlled by the EMT manually adjusting the PEEP using a manometer or predetermined by a fixed setting on the PEEP valve. A PEEP of 7.0 to 10.0 cm H_2O is generally an acceptable therapeutic range for a patient on CPAP. Always consult the operations manual of a particular CPAP device for proper assembly instructions.

Since most CPAP units are powered by oxygen, it is important to have a full cylinder of oxygen when using CPAP. Some CPAP units use a continuous flow of oxygen, while others use oxygen on more of a demand basis. In either situation, continuously monitor the amount of available oxygen in your cylinder. A typical CPAP unit will deplete a full D cylinder of oxygen in 15 to 30 minutes, depending on the fraction of inspired oxygen (FIO_2) setting. Therefore, proper planning for oxygen consumption is necessary when considering applying CPAP. In addition, some of the newer CPAP devices allow the provider to adjust the FIO_2. Most CPAP devices are set to deliver a fixed FIO_2 of 30% to 35%; however, some can deliver as high as 80%.

Follow the steps in Skill Drill 9-9 to use a CPAP:

1. Connect the circuit to the CPAP generator. Make sure your generator is connected to an oxygen source and/or a power source if required (Step 1).
2. Connect the face mask to the circuit tubing (Step 2). Once the system is connected, check to see if there is an on/off button. Some of the newer models have this feature. Make sure the device is set in the "on" position before you apply CPAP to the patient.
3. Confirm the device is working, and place the mask over the patient's mouth and nose, creating as much of an airtight seal as possible. This can be a rather difficult task depending on your patient. Many patients will resist the application of a mask to their face while in severe respiratory distress. Explain the application to the patient and coach him or her through the initial application of the mask. In fact, allowing the patient to hold the

mask to his or her face initially may actually be beneficial in alleviating some of the stress associated with CPAP application (Step 3).
4. Once the mask is on the face, use the strapping mechanism to secure it to the patient's head, making sure the seal between the mask and face remains. Consult the manufacturer's guidelines for specific strapping instructions (Step 4).
5. Adjust the PEEP valve and the FIO_2 accordingly to maintain adequate oxygenation and ventilation. With CPAP in place, the patient's oxygenation should improve and the work of breathing should decrease. Constant reassessment of patients for signs of deterioration is essential (Step 5).

■ Complications

The application and administration of CPAP is a relatively easy process. However, some patients may find CPAP claustrophobic and will resist the application. As patients become more hypoxic, the application of a mask to their face is sometimes perceived as suffocation, rather than helping them breathe. In any event, it is important to explain the application to patients and coach them through the process. Do not force the mask on patients. This will create a higher level of anxiety and increase their oxygen demand. Coach patients through the application of CPAP, allowing them to adjust to the situation. Coaching patients is not always an easy task; it takes practice and a willingness to work closely with your patient during a rather difficult time.

Due to the high volume of pressure generated by CPAP, there is the possibility of causing a pneumothorax. Whereas some literature suggests this is not likely, EMTs should be aware of this risk and continually assess their patients for signs and symptoms of a pneumothorax.

In addition to a pneumothorax, high pressure in the chest can lower a patient's blood pressure. As the intrathoracic pressure increases, venous blood returning to the heart meets resistance from the increased pressure in the chest. This can result in a sudden drop in blood pressure. While this is not common with lower levels of CPAP, continuous monitoring of blood pressure is necessary.

CPAP has shown positive results consistently with patients experiencing moderate and severe respiratory distress; however, there are still cases in which patients deteriorate. It is important that you reassess the patient for signs of deterioration. If the patient is no longer able to follow verbal commands and/or goes into respiratory failure/arrest, you must act quickly to remove CPAP and begin positive-pressure ventilation using a bag-mask device attached to high-flow oxygen.

Skill Drill 9-9

Using the CPAP

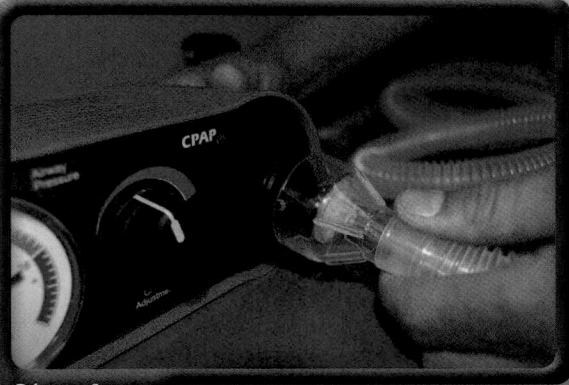

Step 1
Connect the circuit to the CPAP generator.

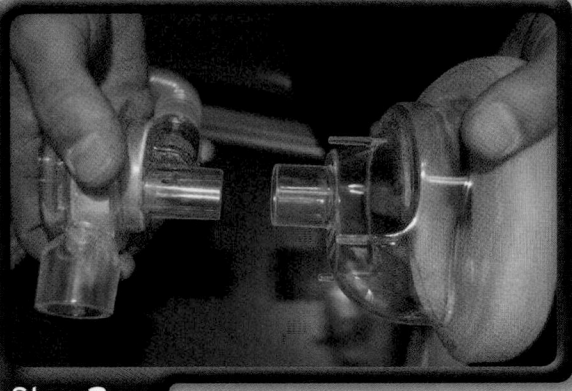

Step 2
Connect the face mask to the circuit tubing.

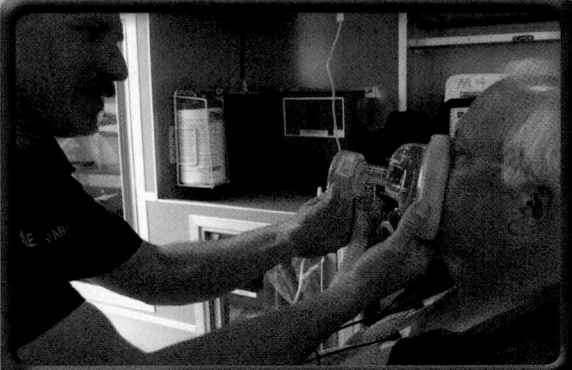

Step 3
Confirm that the device is on before you apply it to the patient's face. Place the mask over the patient's mouth and nose or allow the patient to hold it to his or her mouth and nose.

Step 4
Use the strapping mechanism to secure the CPAP to the patient's head. Make sure there is a tight seal.

Step 5
Adjust the PEEP valve and the F_{IO_2} accordingly to maintain adequate oxygenation and ventilation. Reassess the patient.

Special Considerations

Gastric Distention

Gastric distention occurs when artificial ventilation fills the stomach with air. Although it most commonly affects children, it also affects adults. Gastric distention is most likely to occur when you ventilate the patient too forcefully or too rapidly with a bag-mask or pocket mask device. It may also occur when the airway is obstructed as a result of a foreign body or improper head position. For this reason, you should give slow, gentle breaths during artificial ventilation over 1 second (enough to see the chest rise) in the adult patient. As compliance decreases, you will notice it becoming increasingly difficult to squeeze the bag-mask device to get air into the lungs. Slight gastric distention is not of concern; however, severe inflation of the stomach is dangerous because it may cause vomiting and increase the risk of aspiration during CPR. Gastric distention can also significantly reduce the lung volume by elevating the diaphragm, especially in infants and children. Gastric distention is a common complication associated with the use of manually triggered ventilation devices, a key reason why this device is not highly recommended.

If the patient's stomach becomes distended as the result of rescue breathing, you should recheck and reposition the airway, and watch for rise and fall of the chest wall as you perform rescue breathing. Continue slow rescue breathing without attempting to expel the stomach contents. If gastric distention makes it impossible to ventilate the patient and an ALS provider is not available to perform orogastric tube or nasogastric tube decompression, consider applying pressure over the upper abdomen. Manual decompression should only be used as a last resort. Applying manual pressure over the patient's upper abdomen will likely result in vomiting; therefore, if vomiting occurs, turn the patient's entire body to the side, suction and/or wipe out the mouth with your gloved hand, and return the patient back to a supine position so that you can continue rescue breathing.

Stomas and Tracheostomy Tubes

Bag-mask device ventilation may also need to be used for patients who have had a laryngectomy (surgical removal of the larynx). These patients have a permanent tracheal stoma (an opening in the neck that connects the trachea directly to the skin) Figure 9-44. This type of stoma, known as the **tracheostomy**, is an opening at the center front and base of the neck. Many patients who have had a laryngectomy will have other openings in the neck, according to the type of operation performed. You should ignore any opening other than the midline tracheal stoma. The midline opening is the only one that can be used to put air into the patient's lungs.

You are the Provider: PART 6

On arrival at the emergency department, your reassessment reveals that the patient's condition has improved significantly. His eyes are open, he responds to verbal stimuli, and his breathing, although still somewhat labored, has improved.

Recording Time: 17 Minutes	
Level of consciousness	Eyes open; responsive to verbal stimuli
Respirations	16 breaths/min, slightly labored, adequate depth
Pulse	88 beats/min, strong and regular
Skin	Pink, warm, and moist
Blood pressure	132/72 mm Hg
Sao$_2$	97% (on oxygen)

You transfer care of the patient to the emergency department physician, who tells you that he believes the patient is experiencing acute exacerbation of his congestive heart failure. After further treatment in the emergency department, the patient is admitted to the intensive care unit.

14. How does respiratory failure differ from respiratory arrest?

15. Was your patient experiencing respiratory failure or respiratory arrest?

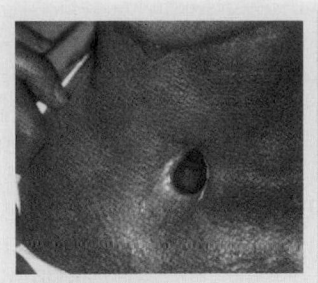

Figure 9-44 A tracheal stoma typically lies in the midline of the neck. The midline opening is the only one that can be used to deliver oxygen to the patient's lungs.

Neither the head tilt–chin lift nor the jaw-thrust maneuver is required for ventilating a patient with a stoma. If the patient has a tracheostomy tube, you should ventilate through the tube with a bag-mask device (the standard 22/15 adapter on the bag-mask device will fit onto the tube in the tracheal stoma) and 100% oxygen attached directly to the bag-mask device. If the patient has a stoma and no tube is in place, use an infant or child mask with your bag-mask device to make a seal over the stoma. Seal the patient's mouth and nose with one hand to prevent a leak of air through the upper airway when you ventilate through a stoma. Release the seal of the patient's mouth and nose for exhalation. This allows the air to exhale through the upper airway.

If you are unable to ventilate a patient who has a stoma, try suctioning the stoma and the mouth with a French or soft-tip catheter before giving the patient artificial ventilation through the mouth and nose. If you seal the stoma during mouth-to-mouth ventilation, the ability to ventilate the patient may be improved, or it may help to clear any obstructions.

Foreign Body Airway Obstruction

A foreign body that *completely* blocks the airway in a patient is a true emergency that will result in death if not treated immediately. In an adult, sudden foreign body airway obstruction usually occurs during a meal. In a child, it occurs while eating, playing with small toys, or crawling around the house. An otherwise healthy child who has sudden difficulty breathing has probably aspirated a foreign object.

By far, the most common airway obstruction in an unconscious patient is the tongue, which relaxes and falls back into the throat. There are other causes of airway obstruction that do not involve foreign bodies in the airway. These include swelling (from infection or acute allergic reaction) and trauma (tissue damage from injury). With airway obstruction from medical conditions such as infection and acute allergic reactions, repeated attempts to clear the airway as if there were a foreign body will be

unsuccessful and potentially dangerous. These patients require specific emergency medical care for their condition; therefore, rapid transport to the hospital is critical.

■ Recognition

Early recognition of airway obstruction is crucial for the EMT to be able to provide emergency medical care effectively. Obstruction from a foreign body can result in a **mild airway obstruction** or a **severe airway obstruction**.

Patients with a mild airway obstruction are still able to exchange air but will have varying degrees of respiratory distress. Great care must be taken to prevent a mild airway obstruction from becoming a severe airway obstruction. The patient will usually have noisy breathing and may be coughing. You should assess the patient and determine whether the patient has **good air exchange** or **poor air exchange**.

With good air exchange, the patient can cough forcefully, although you may hear **wheezing** (the production of whistling sounds during respiration) between coughs. Wheezing is usually indicative of a mild lower airway obstruction. As long as the patient can breathe, cough forcefully, or talk, you should not interfere with the patient's efforts to expel the foreign object on his or her own. Continue to monitor the patient closely and encourage the patient to continue coughing. Abdominal thrusts are usually not effective for dislodging a partial obstruction. Attempts to remove the object manually could force the object farther down into the airway and cause a severe airway obstruction. Continually reassess the patient's condition and be prepared to provide treatment if the air exchange becomes poor or a mild obstruction becomes a severe obstruction.

With poor air exchange, the patient has a weak, ineffective (not forceful) cough and may have increased difficulty breathing, **stridor** (a high-pitched noise heard primarily on inspiration), and cyanosis. Stridor is an indication of a mild upper airway obstruction. You must quickly recognize this situation and provide immediate care.

For patients with mild airway obstruction with poor air exchange, treat immediately as if there is a severe airway obstruction.

Patients with a severe airway obstruction cannot breathe, talk, or cough. One sure sign of a severe obstruction is the sudden inability to speak or cough during or immediately after eating. The person may clutch or grasp his or her throat (universal distress signal), begin to turn cyanotic, and have extreme difficulty breathing **Figure 9-45** . There is little or no air movement. Ask the conscious patient, "Are you choking?" If the patient nods "yes," provide immediate treatment. If the obstruction is

Figure 9-45 The universal sign of choking is a person who grasps his or her throat and has difficulty breathing.

enough to clear the airway; however, if you are unable to ventilate the patient after several attempts (no chest rise and fall) or you feel resistance while ventilating, consider the possibility of an airway obstruction. Resistance to ventilation can also be due to poor lung compliance. As discussed, compliance is the ability of the alveoli to expand when air is drawn in during inhalation; poor lung compliance is the inability of the alveoli to fully expand during inhalation.

> ## Words of Wisdom
>
> **Possible Causes of Airway Obstruction**
> - Relaxation of the tongue in an unconscious patient
> - Aspirated vomitus (stomach contents)
> - Foreign objects—food, small toys, dentures
> - Blood clots, bone fragments, or damaged tissue after an injury
> - Airway tissue swelling—infection, allergic reaction

not cleared quickly, the amount of oxygen in the patient's blood will decrease dramatically. If not treated, the patient will become unconscious and die.

Some patients with a severe airway obstruction will be unconscious as you form your general impression. You may not know that an airway obstruction is the cause of their condition. There are many other causes of unconsciousness and respiratory failure, including stroke, heart attack, trauma, seizures, and drug overdoses. A complete and thorough patient assessment by you, therefore, is essential to providing appropriate emergency medical care.

If the patient is found unresponsive, does not appear to be breathing, and does not have a pulse, you should begin CPR with chest compressions. When you open the airway and attempt two ventilations following chest compressions, it will be obvious to you that the airway is blocked **Figure 9-46**. The compressions may have been

■ Emergency Medical Care for Foreign Body Airway Obstruction

Perform the head tilt–chin lift maneuver to clear an obstruction that has been caused by the tongue and throat muscles relaxing back into the airway in any person who is found unconscious. This should be performed on unresponsive patients with adequate or inadequate breathing, who are not suspected of having spinal trauma. If spinal trauma is suspected, you should open the airway with a jaw-thrust maneuver. Large pieces of vomited food, mucus, loose dentures, or blood clots in the mouth should be swept forward and out of the mouth with your gloved index finger. When available, suctioning should be used to maintain a clear airway.

Abdominal thrusts are the most effective method of dislodging and forcing an object out of the airway of a conscious adult or child. Residual air, which is always present in the lungs, is compressed upward and used to expel the object. You should use abdominal thrusts until the object dislodges or the patient becomes unconscious.

For the unresponsive patient with a severe foreign body airway obstruction, reassess to confirm apnea and inability to ventilate. Begin chest compression just as you would for CPR, following the 30 compressions to 2 breaths ratio. At the completion of the 30 compressions, perform a tongue-jaw lift by grasping the jaw with your thumb and index finger. Place your thumb onto the tip of the patient's lower teeth and tongue while placing your index finger under the bony portion of the chin. Be careful not to compress the soft tissues under the chin. Pull the jaw/mouth open and look at the back of the oropharynx for any foreign objects. If an object is observed, remove it with a gloved index finger or suction. Only attempt to

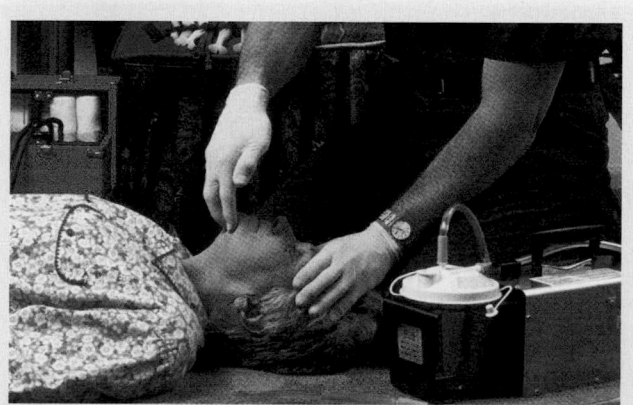

Figure 9-46 When you open the airway and attempt ventilations, it will be obvious to you if the airway is still blocked.

remove an object if it is visible during examination of the open mouth; blind sweeps of the back of the oropharynx may push an object further down in the airway, making the obstruction worse. Once the object(s) is removed or if no object was seen during the tongue-jaw lift, attempt to ventilate. If you are still unable to ventilate, repeat the process.

If you are unable to clear a severe airway obstruction with your initial attempts, begin rapid transport and continue your efforts to relieve the obstruction with abdominal thrusts (chest compressions in the unresponsive patient) on the way to the hospital.

Remember to treat patients with a mild airway obstruction with poor air exchange as if they have a severe airway obstruction.

Patients with a mild airway obstruction and good air exchange should be monitored closely for deterioration of their condition. If the patient is unable to clear the obstruction and remains conscious, support (or let the patient control) the airway position that is most efficient and comfortable. Provide supplemental oxygen, and transport to the hospital.

Words of Wisdom

If spinal trauma is suspected, open the airway using the jaw-thrust maneuver.

■ Dental Appliances

Many dental appliances can cause an airway obstruction. If a dental appliance, such as a crown or bridge, dentures, or even a piece or sections of braces, has become loose, you should manually remove it before providing ventilations. Simple manual removal may relieve the obstruction and allow the patient to breathe on his or her own.

Providing bag-mask device or mouth-to-mask ventilation is usually much easier when dentures can be left in place. Leaving the dentures in place provides more "structure" to the face and will generally assist you in being able to provide a good face-to-mask seal, thus delivering adequate tidal volume. However, loose dentures make it difficult to perform artificial ventilation by any method and can easily obstruct the airway. Therefore, dentures and dental appliances that do not stay in place should be removed. Dentures and appliances may become loose or be completely out of place following an accident or as you are providing care. Periodically reassess the patient's airway to make sure these devices are firmly in place. If dentures become dislodged, if possible, place in a container and transport with the patient.

■ Facial Bleeding

Airway problems can be especially challenging in patients with serious facial injuries Figure 9-47. Because the blood supply to the face is so rich, injuries to the face can result in severe tissue swelling and bleeding into the airway. Control bleeding with direct pressure and suction as necessary.

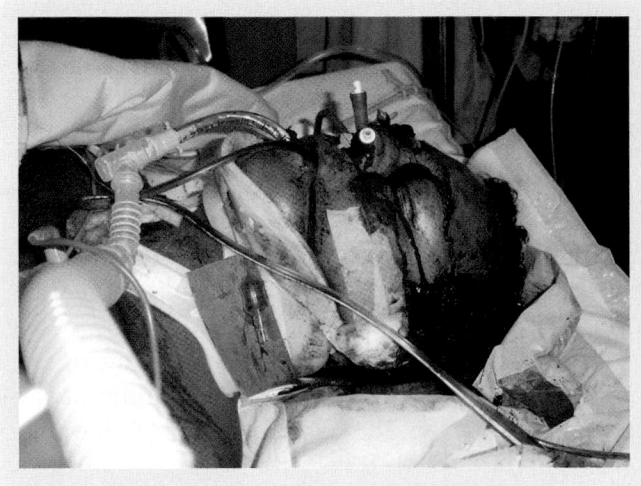

Figure 9-47 Airway problems can be especially challenging in patients with serious facial injuries.

You are the Provider: SUMMARY

1. What is the function of the respiratory system?

The function (physiology) of the respiratory system, as critical as it is, is quite simplistic: to bring oxygen into the lungs and remove carbon dioxide from the lungs. Failure of the respiratory system will compromise oxygenation—the loading of oxygen onto the hemoglobin molecule—and cause carbon dioxide to accumulate in the blood. If the body's cells do not receive oxygen, they will metabolize carbon dioxide and produce dangerous acids.

2. What is the difference between ventilation and respiration?

Ventilation is the movement of air into and out of the lungs. Normal, unassisted breathing occurs through a process called negative-pressure ventilation. When the diaphragm and intercostal muscles contract, the thorax enlarges and pressure within it falls; this creates a slight vacuum, which pulls air into the lungs. By contrast, positive-pressure ventilation occurs when air is pushed into the lungs, such as when performing artificial ventilation with a bag-mask device or pocket face mask.

Respiration is the process of exchanging oxygen and carbon dioxide. Pulmonary (external) respiration is the exchange of oxygen and carbon dioxide between the blood and alveoli in the lungs. Cellular (internal) respiration is the exchange of oxygen and carbon dioxide between the blood and the body's cells. Adequate cellular respiration relies on adequate ventilation, oxygenation, and pulmonary respiration.

3. How often should you assess a patient's airway and breathing status?

Assessment of a patient's airway and breathing status should be an ongoing process, from the time you initially encounter the patient until he or she is delivered to the emergency department. A patient's airway may be patent and his or her breathing may be adequate initially; however, this can change acutely. Frequent assessments allow you to detect airway or breathing problems and correct them immediately.

4. Is this patient's airway patent?

A patent airway is one that is open and free of secretions or foreign bodies. At the present time, the patient's airway is patent. However, a patent airway does not mean that the patient is breathing adequately. Carefully assess your patient!

5. Is he breathing adequately? Why or why not?

A patient who is breathing adequately is able to move enough air into and out of the lungs to adequately oxygenate the blood. Adequate breathing in the adult is characterized by a respiratory rate of between 12 and 20 breaths/min; adequate depth (tidal volume); the ability to speak in complete sentences; a regular pattern of inhalation and exhalation; skin that is pink, warm, and dry; pink mucous membranes; and a normal mental status.

The patient in this scenario is *not* breathing adequately. His respirations are severely labored, he can only speak in two-word sentences, and his mental status is decreased (ie, he is "sleepy"). An altered mental status in a patient with respiratory distress indicates that the brain is hypoxic—that is, it is not receiving enough oxygen.

6. How should you manage his present airway and breathing status?

If a patient is unable to draw enough air into the lungs on his own through the process of negative-pressure ventilation, you must provide some form of positive-pressure ventilation. In a conscious or semiconscious patient, this involves applying a bag-mask device to the patient's face, and advising him that each time he takes a breath, you will squeeze the bag.

Continuous positive airway pressure (CPAP) may also be beneficial to this patient. CPAP delivers positive pressure to the spontaneously breathing patient during the respiratory cycle; it is especially useful in treating patients with breathing difficulty secondary to congestive heart failure and pulmonary edema. With CPAP, the patient exhales against positive pressure that is delivered through a tight-fitting mask; this helps prevent atelectasis (alveolar collapse), forces fluid from the alveoli, and improves gas exchange in the lungs.

Left untreated, inadequate breathing will cause the patient to continue to deteriorate—potentially to the point where he stops breathing completely. If the patient's level of consciousness decreases further, manually keep his airway open and consider inserting an airway adjunct (eg, oral or nasal airway).

7. What is cyanosis? What does it indicate?

Cyanosis is a dark blue or purple color of the skin; it indicates hypoxemia—a deficiency of oxygen in the blood. Highly oxygenated blood is bright red, which gives the patient's skin a pink appearance. If the blood is poorly oxygenated, as with inadequate breathing, it assumes a dark red or purple color. Cyanosis is commonly seen on the face, nail beds, and mucous membranes. It is important to note that cyanosis is a later sign of hypoxemia, so its absence does not rule out an airway or breathing problem.

8. What does the patient's oxygen saturation indicate?

Oxygen saturation (Sao_2) refers to the amount of oxygen that is saturating the hemoglobin (the iron-containing portion of the red blood cell to which oxygen attaches) in the capillary beds. Oxygen saturation is measured with a pulse oximeter—a device that sends a beam of light through the capillary bed and measures the density of the blood. Pulse oximetry values normally range between 95% and 100%. An Sao_2 of between 90% and 94% indicates mild to moderate hypoxemia, while an Sao_2 of less than 90% indicates significant hypoxemia. Your patient's Sao_2 of 85% indicates that a significant portion of his blood is not being oxygenated; you must correct this immediately by ensuring that he receives high-flow oxygen. Pulse oximetry is a useful tool for assessing oxygenation provided you remember that it is only a tool, not a substitute for a good assessment.

9. What should be your *most* immediate action?

Snoring respirations indicate that the patient's tongue has fallen back into his throat and is partially blocking his airway. In fact, the tongue is the most common cause of upper airway obstruction in semi-conscious and unconscious patients. Your patient's airway is no longer patent, and you must take immediate corrective action. Opening the patient's airway with the head tilt-chin lift maneuver is the quickest, most effective way of accomplishing this. Since his level of consciousness has decreased, an airway adjunct should also be inserted to help maintain his airway. In this case, a nasopharyngeal airway would be the best choice; although his level of consciousness has significantly decreased, he is not completely unconscious and likely has a gag reflex. Inserting an oropharyngeal airway may result in vomiting and aspiration.

10. How should you adjust your treatment of the patient?

If you weren't already assisting the patient's ventilations, you certainly need to now! Ventilate the patient with 100% oxygen, ensure a good mask-to-face seal, and observe for visible chest rise during each ventilation. His decreased level of consciousness; slow, shallow breathing; and profoundly low oxygen saturation make it clear that he is rapidly progressing toward respiratory arrest. This is likely because he is fatigued from laboring to breathe for a prolonged period. If a patient cannot bring enough oxygen into the lungs, hypoxia will develop. Hypoxia—a dangerous condition in which the tissues and cells of the body are deprived of oxygen—can rapidly cause death unless promptly treated.

11. Has your treatment improved the patient's condition? If so, how do you know?

Compared with earlier assessments, the patient's condition has improved. His skin color and condition are improving, his oxygen saturation is increasing, his heart rate and quality are improving, and his respiratory rate (unassisted), although still shallow, has increased. He still requires assisted ventilation; however, if you continue to ventilate him adequately with high-flow oxygen, you can keep him

stable until his underlying problem can be evaluated and definitively treated by a physician. Do not let your guard down, however, because his condition could just as easily deteriorate again.

12. What is the appropriate technique for ventilating an apneic adult?

When ventilating an apneic adult, you should deliver each breath over a period of 1 second—just enough to produce visible chest rise—at a rate of 10 to 12 breaths/min (one breath every 5 to 6 seconds).

13. What are the dangers of hyperventilating a patient?

Hyperventilation (ventilating too fast or with too much force) can have several negative effects on the patient and should be avoided. Hyperventilation increases the incidence of gastric distention, thus increasing the threat of aspiration if regurgitation occurs. Hyperventilation also hyperinflates the lungs; this effect may reduce the amount of blood that returns to the heart and cause cardiac output to fall. Proper ventilation involves delivering each breath over a period of 1 second—just enough to cause visible chest rise—at the appropriate rate (10 to 12 breaths/min for adults; 12 to 20 breaths/min for infants and children).

14. How does respiratory failure differ from respiratory arrest?

Respiratory failure is just that—failure of the respiratory system to bring oxygen into the lungs and remove carbon dioxide. The patient is breathing; however, it is inadequate and requires ventilation assistance. If not promptly treated, respiratory failure can rapidly deteriorate to respiratory arrest. In respiratory arrest, the patient is no longer breathing; he or she is apneic (cessation of breathing). Patients with respiratory arrest need immediate artificial ventilation; otherwise, cellular death will occur and cause cardiopulmonary arrest.

15. Was your patient experiencing respiratory failure or respiratory arrest?

Initially, the patient was experiencing respiratory distress; however, his condition deteriorated to respiratory failure. This was evidenced by a decrease in his level of consciousness, slowing of his respirations, and falling oxygen saturation. Because he was treated promptly and appropriately, he never experienced respiratory arrest (eg, he never stopped spontaneously breathing).

EMS Patient Care Report (PCR)

Date: 3-4-09	Incident No.: 090109	Nature of Call: Respiratory		Location: 145 Landa St.	
Dispatched: 1510	En Route: 1510	At Scene: 1515	Transport: 1525	At Hospital: 1532	In Service: 1541

Patient Information

Age: 55 Sex: M Weight (in kg [lb]): 75 kg (165 lb)	Allergies: None Medications: Digoxin, Vasotec, Lasix Past Medical History: Congestive heart failure, HTN Chief Complaint: Respiratory distress

Vital Signs

Time: 1517	BP: 126/60	Pulse: 120	Respirations: 30	Sao$_2$: 85%
Time: 1523	BP: 118/54	Pulse: 130	Respirations: 10	Sao$_2$: 76%
Time: 1527	BP: 124/62	Pulse: 100	Respirations: 14	Sao$_2$: 92%
Time: 1534	BP: 132/72	Pulse: 88	Respirations: 16	Sao$_2$: 97%

EMS Treatment
(circle all that apply)

Oxygen @ 15 L/min via (circle one): NC NRM Bag-Mask Device	Assisted Ventilation	Airway Adjunct	CPR
Defibrillation	Bleeding Control	Bandaging	Splinting	Other

Narrative

911 dispatch for a male patient with "breathing problems." Arrived on scene to find the patient, a 55-year-old male, in obvious respiratory distress. He was conscious, but appeared sleepy. Skin was cool, pale, and moist. Began assisting patient's breathing with a bag-mask device and high-flow oxygen secondary to poor respiratory effort and signs of hypoxemia. Vital signs obtained and noted above. According to the patient's wife, he began complaining of shortness of breath the day before, but it worsened today. His past medical history is significant for CHF and HTN. As preparations were being made to transport, reassessment revealed that patient's LOC had markedly decreased (responsive only to pain), his respiratory rate (unassisted) had decreased, his oxygen saturation decreased, and he developed cyanosis to the facial area. Manually opened airway, inserted nasal airway device, and continued to assist breathing. Began transport to emergency department and continued treatment en route. Assessment of assisted ventilation revealed that breath sounds were audible bilaterally with a stethoscope and chest rise was visible with each ventilation. On arrival at the emergency department, the patient's LOC had improved; he was conscious and responded to verbal stimuli, and his oxygen saturation and skin condition had markedly improved. Patient was resistant to assisted ventilation; therefore, high-flow oxygen was continued via nonrebreathing mask. Delivered patient to hospital staff and gave verbal report to attending physician.
End of report

Prep Kit

- The upper airway includes the nose, mouth, jaw, oral cavity, pharynx, and larynx. Its function is to warm, filter, and humidify air as it enters the nose and mouth.

- The lower airway includes the trachea and lungs and its function is to exchange oxygen and carbon dioxide.

- Adequate breathing for an adult features a normal rate of 12 to 20 breaths/min, a regular pattern of inhalation and exhalation, adequate depth, bilaterally clear and equal lung sounds, and regular and equal chest rise and fall.

- Inadequate breathing for an adult features a respiratory rate of fewer than 12 breaths/min or more than 20 breaths/min, shallow depth (reduced tidal volume), an irregular pattern of inhalation and exhalation, and breath sounds that are diminished, absent, or noisy.

- Patients who are breathing inadequately show signs of hypoxia, a dangerous condition in which the body's tissues and cells do not have enough oxygen.

- Patients with inadequate breathing need to be treated immediately. Emergency medical care includes airway management, supplemental oxygen, and ventilatory support.

- Basic techniques for opening the airway include the head tilt–chin lift maneuver or, if trauma is suspected, the jaw-thrust maneuver.

- One basic airway adjunct is the oropharyngeal or oral airway, which keeps the tongue from blocking the airway in unconscious patients with no gag reflex. If the oral airway is not the proper size or is inserted incorrectly, it can actually cause an obstruction.

- Another basic airway adjunct is the nasopharyngeal or nasal airway, which is usually used with patients who have a gag reflex and is better tolerated than the oral airway.

- Suctioning is the next priority after opening the airway. Rigid tonsil-tip catheters are the best catheters to use when suctioning the pharynx; soft plastic catheters are used to suction the nose and liquid secretions in the back of the mouth.

- The recovery position is used to help maintain the airway in patients without traumatic injuries who are unconscious and breathing adequately.

- You must provide immediate artificial ventilations with supplemental oxygen to patients who are not breathing on their own. Patients with inadequate breathing may also require artificial ventilations to maintain effective tidal volume.

- Handle compressed gas cylinders carefully; their contents are under pressure. Always make sure the correct pressure regulator is firmly attached before transporting a cylinder. The pin-indexing safety system features a series of pins on a yoke that must be matched with the holes on the valve stem of the gas cylinder. Pressure regulators reduce the pressure of gas in an oxygen cylinder to between 40 and 70 psi. Pressure-compensated flowmeters and Bourdon-gauge flowmeters permit the regulated release of gas measured in liters per minute.

- When oxygen therapy is complete, disconnect the tubing from the flowmeter nipple and turn off the cylinder valve, then turn off the flowmeter. As long as there is a pressure reading on the regulator gauge, it is not safe to remove the regulator from the valve stem. Keep any possible source of fire away from the area while oxygen is in use.

- Nasal cannulas and nonrebreathing masks are used most often to deliver oxygen in the field. The nonrebreathing mask is the delivery device of choice for providing supplemental oxygen to patients who are breathing adequately but are suspected of having or are showing signs of hypoxia. With a flow rate set at 15 L/min and the reservoir bag preinflated, the nonrebreathing mask can provide more than 90% inspired oxygen. If the patient will not tolerate a nonrebreathing mask, apply a nasal cannula.

- The methods of providing artificial ventilation include mouth-to-mask ventilation, two-person bag-mask device ventilation, manually triggered ventilation device, and one-person bag-mask device ventilation. The manually triggered ventilation device is not a recommended ventilation device by most standards. Combined with your own exhaled breath, mouth-to-mask ventilation will give your patient up to 55% oxygen; a bag-mask device with an oxygen reservoir and supplemental oxygen can deliver nearly 100% oxygen.

- CPAP is a noninvasive method of providing ventilatory support for patients in respiratory distress or suffering from sleep apnea.

- When you are providing artificial ventilation, remember that ventilating too forcefully can cause gastric distention. Slow, gentle breaths during artificial ventilation can help to prevent gastric distention. Patients who have a tracheal stoma or a tracheostomy tube need to be ventilated through the tube or the stoma.

- Foreign body airway obstruction usually occurs during a meal in an adult or while a child is eating, playing with small objects, or crawling about the house. The earlier you recognize an airway obstruction, the better. You must learn to recognize the difference between airway obstruction caused by a foreign object and that caused by a medical condition.

- Foreign body airway obstructions are classified as being mild or severe. Patients with a mild airway obstruction are able to move adequate amounts of air and should be left alone. Patients with a severe airway obstruction cannot move any air at all and require immediate treatment. Perform abdominal thrusts on conscious adults and children with a severe airway obstruction. If the patient becomes unconscious, open the airway and look in the mouth (do not perform blind finger sweeps), attempt to ventilate the patient, and perform chest compressions if ventilations are unsuccessful.

- Check for loose dental appliances in a patient before assisting ventilations. Loose appliances should be removed to prevent them from obstructing the airway. Tight-fitting appliances should be left in place.

Vital Vocabulary

aerobic metabolism Metabolism that can proceed only in the presence of oxygen.

agonal gasps Occasional, gasping breaths that occur after the heart has stopped.

airway The upper airway tract or the passage above the larynx, which includes the nose, mouth, and throat.

alveolar ventilation The volume of air that reaches the alveoli. It is determined by subtracting the amount of *dead space* air from the *tidal volume*.

American Standard System A safety system for large oxygen cylinders, designed to prevent the accidental attachment of a regulator to a cylinder containing the wrong type of gas.

anaerobic metabolism The metabolism that take place in the absence of oxygen; the principle product is lactic acid.

apnea Absence of spontaneous breathing.

aspiration In the context of airway, the introduction of vomitus or other foreign material into the lungs.

ataxic respirations Irregular, ineffective respirations that may or may not have an identifiable pattern.

automatic transport ventilator (ATV) A ventilation device attached to a control box that allows the variables of ventilation to be set. It frees the EMT to perform other tasks while the patient is being ventilated.

bag-mask device A device with a one-way valve and a face mask attached to a ventilation bag; when attached to a reservoir and connected to oxygen, delivers more than 90% supplemental oxygen.

barrier device A protective item, such as a pocket mask with a valve, that limits exposure to a patient's body fluids.

bilateral A body part or condition that appears on both sides of the midline.

bronchioles Subdivision of the smaller bronchi in the lungs; made of smooth muscle and dilate or constrict in response to various stimuli.

carina Point at which the trachea bifurcates (divides) into the left and right mainstem bronchi.

chemoreceptors Monitor the levels of O_2, Co_2, and the pH of the cerebrospinal fluid and then provide feedback to the respiratory centers to modify the rate and depth of breathing based on the body's needs at any given time.

compliance The ability of the alveoli to expand when air is drawn in during inhalation.

continuous positive airway pressure (CPAP) A method of ventilation used primarily in the treatment of critically ill patients with respiratory distress; can prevent the need for endotracheal intubation.

dead space The portion of the tidal volume that does not reach the alveoli and thus does not participate in gas exchange.

diffusion A process in which molecules move from an area of higher concentration to an area of lower concentration.

dyspnea Shortness of breath.

exhalation The passive part of the breathing process in which the diaphragm and the intercostal muscles relax, forcing air out of the lungs.

external respiration The exchange of gases between the lungs and the blood cells in the pulmonary capillaries; also called pulmonary respiration.

gag reflex A normal reflex mechanism that causes retching; activated by touching the soft palate or the back of the throat.

gastric distention A condition in which air fills the stomach, often as a result of high volume and pressure during artificial ventilation.

glottis The space in between the voal cords that is the narrowest portion of the adult's airway; also called the glottic opening.

good air exchange A term used to distinguish the degree of distress in a patient with a mild airway obstruction. With good air exchange, the patient is still conscious and able to cough forcefully, although wheezing may be heard.

head tilt–chin lift maneuver A combination of two movements to open the airway by tilting the forehead back and lifting the chin; not used for trauma patients.

hypercarbia Increased carbon dioxide level in the bloodstream.

hypoxia A dangerous condition in which the body tissues and cells do not have enough oxygen.

hypoxic drive A condition in which chronically low levels of oxygen in the blood stimulate the respiratory drive; seen in patients with chronic lung diseases.

inhalation The active, muscular part of breathing that draws air into the airway and lungs.

internal respiration The exchange of gases between the blood cells and the tissues.

intrapulmonary shunting Bypassing of oxygen-poor blood past nonfunctional alveoli to the left side of the heart.

jaw-thrust maneuver Technique to open the airway by placing the fingers behind the angle of the jaw and bringing the jaw forward; used for patients who may have a cervical spine injury.

labored breathing Breathing that requires greater than normal effort; may be slower or faster than normal and usually requires the use of accessory muscles.

larynx A complex structure formed by many independent cartilaginous structures that all work together; where the upper airway ends and the lower airway begins; also called the voice box.

manually triggered ventilation device A fixed flow/rate ventilation device that delivers a breath every time its button is pushed; also referred to as a flow-restricted, oxygen-powered ventilation device.

mediastinum Space within the chest that contains the heart, major blood vessels, vagus nerve, trachea, major bronchi, and esophagus; located between the two lungs.

metabolism (cellular respiration) The biochemical processes that result in production of energy from nutrients within the cells.

mild airway obstruction Occurs when a foreign body partially obstructs the patient's airway. The patient is able to move adequate amounts of air, but also experiences some degree of respiratory distress.

minute ventilation The volume of air moved through the lungs in 1 minute minus the dead space; calculated by multiplying tidal volume (minus dead space) and respiratory rate; also referred to as minute volume.

nasal cannula An oxygen-delivery device in which oxygen flows through two small, tubelike prongs that fit into the patient's nostrils; delivers 24% to 44% supplemental oxygen, depending on the flow rate.

nasopharyngeal (nasal) airway Airway adjunct inserted into the nostril of an unresponsive patient, or a patient with an altered level of consciousness who is unable to maintain airway patency independently.

nasopharynx The nasal cavity; formed by the union of facial bones and protects the respiratory tract from contaminants.

nonrebreathing mask A combination mask and reservoir bag system that is the preferred way to give oxygen in the prehospital setting; delivers up to 90% inspired oxygen and prevents inhaling the exhaled gases (carbon dioxide).

oropharyngeal (oral) airway Airway adjunct inserted into the mouth of an unresponsive patient to keep the tongue from blocking the upper airway and to facilitate suctioning the airway, if necessary.

oropharynx Forms the posterior portion of the oral cavity, which is bordered superiorly by the hard and soft palates, laterally by the cheeks, and inferiorly by the tongue.

oxygenation The process of delivering oxygen to the blood by diffusion from the alveoli following inhalation into the lungs.

parietal pleura Thin membrane that lines the chest cavity.

partial pressure The term used to describe the amount of gas in air or dissolved in fluid, such as blood.

patent Open, clear of obstruction.

phrenic nerve Nerve that innervates the diaphragm; necessary for adequate breathing to occur.

pin-indexing system A system established for portable cylinders to ensure that a regulator is not connected to a cylinder containing the wrong type of gas.

pneumothorax A partial or complete accumulation of air in the pleural space.

poor air exchange A term used to describe the degree of distress in a patient with a mild airway obstruction. With poor air exchange, the patient often has a weak, ineffective cough, increased difficulty breathing, or possible cyanosis and may produce a high-pitched noise during inhalation (stridor).

positive end-expiratory pressure (PEEP) Mechanical maintenance of pressure in the airway at the end of expiration to increase the volume of gas remaining in the lungs.

pulse oximetry An assessment tool that measures oxygen saturation of hemoglobin in the capillary beds.

recovery position A side-lying position used to maintain a clear airway in unconscious patients without injuries who are breathing adequately.

residual volume The air that remains in the lungs after maximal expiration.

respiration The process of exchanging oxygen and carbon dioxide.

retractions Movements in which the skin pulls in around the ribs during inspiration.

severe airway obstruction Occurs when a foreign body completely obstructs the patient's airway. Patients cannot breathe, talk, or cough.

stoma An opening through the skin and into an organ or other structure; a stoma in the neck connects the trachea directly to the skin.

stridor A high-pitched noise heard primarily on inspiration.

suction catheter A hollow, cylindrical device used to remove fluid from the patient's airway.

surfactant A liquid protein substance that coats the alveoli in the lungs, decreases alveolar surface tension, and keeps the alveoli expanded; a low level in a premature infant contributes to respiratory distress syndrome.

tension pneumothorax A life-threatening collection of air within the pleural space; the volume and pressure have both collasped the involved lung and caused a shift of the mediastinal structures to the opposite side.

tidal volume The amount of air (in mL) that is moved in or out of the lungs during one breath.

tonsil tips Large, semirigid suction tips recommended for suctioning the pharynx; also called Yankauer tips.

tracheostomy Surgical opening into the trachea.

ventilation Exchange of air between the lungs and the environment, spontaneously by the patient or with assistance from another person, such as an EMT.

visceral pleura Thin membrane that covers the lungs.

vital capacity The amount of air that can be forcibly expelled from the lungs after breathing in as deeply as possible.

vocal cords Thin white bands of tough muscular tissue that are lateral borders of the glottis and serve as the primary center for speech production.

wheezing The production of whistling sounds during expiration such as occurs in asthma and bronchiolitis.

Assessment in Action

Y ou and your partner have just arrived at a local golf course for a 56-year-old man with a chief complaint of shortness of breath. The patient is seated at a table in the clubhouse. He appears to be using his accessory muscles to breathe.

1. Which of the following are considered accessory muscles?
   A. Diaphragm
   B. Sternocleidomastoid
   C. Intercostals
   D. Deltoids

2. What part of the airway serves as the functional site for gas exchange?
   A. Trachea
   B. Bronchi
   C. Alveoli
   D. Carina

3. As you approach the patient, he begins to shout "Hurry up and take care of me, will you!" Early signs of hypoxia include:
   A. bradycardia.
   B. cyanosis.
   C. hypertension.
   D. irritability.

4. What supplemental oxygen device should you have your partner apply to the patient?
   A. Nasal cannula
   B. Nonrebreathing mask
   C. Venturi device
   D. Bag-mask device

5. Your patient tells you that he has a history of chronic obstructive pulmonary disease (COPD), diabetes mellitus, and high cholesterol. Patients with COPD are stimulated to breathe by the:
   A. hypoxic drive.
   B. carbon dioxide mechanism.
   C. cerebrospinal fluid levels.
   D. brain stem backup.

6. While en route to the emergency department, the patient's level of consciousness diminishes, his respiratory rate decreases to 4 breaths/min, his lips have a blue tint, and the pulse oximeter is now reading 65%. You should begin delivering artificial breaths with a bag-mask device at a rate of one breath every:
   A. 3 to 5 seconds.
   B. 5 to 6 seconds.
   C. 8 to 10 seconds.
   D. 12 to 15 seconds.

7. Which of the following may produce an inaccurate pulse oximeter reading?
   A. Volume overload
   B. Hyperventilation
   C. Chronic hypoxia
   D. False fingernails

8. When ventilating a patient, the volume of air delivered to the patient is based on:
   A. the heart rate.
   B. pulse oximetry.
   C. skin color.
   D. chest rise.

9. How can gastric distention be prevented when performing artificial ventilation?

10. What safety considerations must you be aware of when handling oxygen cylinders?

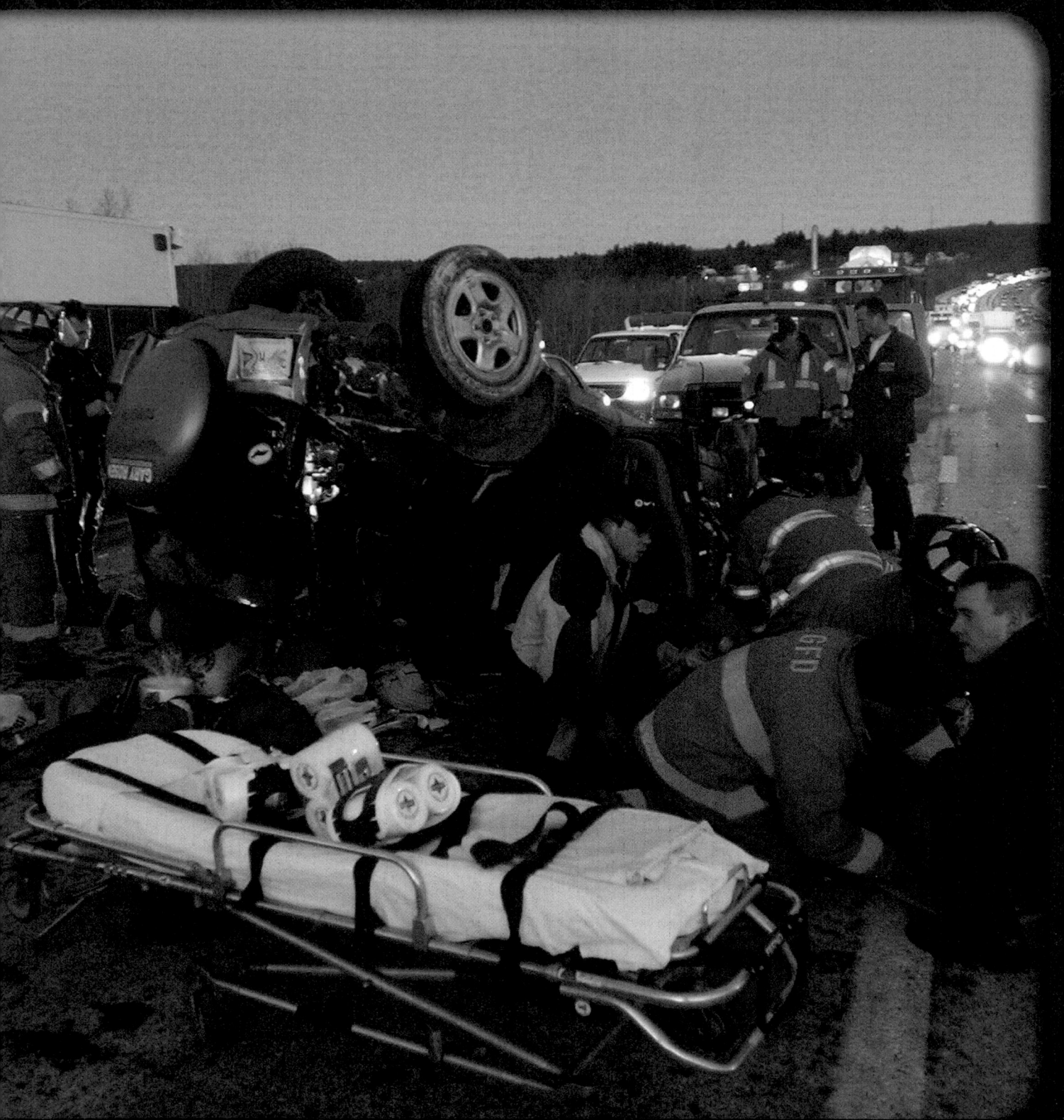

National EMS Education Standard Competencies

Shock and Resuscitation

Applies a fundamental knowledge of the causes, pathophysiology, and management of shock, respiratory failure or arrest, cardiac failure or arrest, and post resuscitation management.

Pathophysiology

Applies fundamental knowledge of the pathophysiology of respiration and perfusion to patient assessment and management.

Knowledge Objectives

1. Understand the pathophysiology of shock (hypoperfusion). (pp 381-384)
2. Recognize the causes of shock. (pp 384-388)
3. Describe the various types of shock. (pp 385-389)
4. Describe the signs and symptoms of shock. (pp 389-390)
5. Discuss patient assessment for shock. (pp 390-392)
6. Describe the steps to follow in the emergency care of the patient with signs and symptoms of shock. (pp 393-399)

Skills Objectives

1. Demonstrate how to control shock. (pp 393-389, Skill Drill 10-1)
2. Demonstrate how to complete an EMS patient care report for a patient with bleeding and/or shock. (pp 393, 403)

Introduction

Shock has a number of meanings. For example, it is often said that a person who has been frightened or received bad news is in shock. An electric current passing through the body delivers a shock. In this chapter, **shock** (hypoperfusion) describes a state of collapse and failure of the cardiovascular system. When the circulation of blood in the body becomes inadequate, the oxygen and nutrient needs of the cells cannot be met. In the early stages of shock, the body will attempt to maintain **homeostasis** (a balance of all systems of the body); however, as shock progresses, blood circulation slows and eventually ceases. This abnormal state of inadequate oxygen and nutrient delivery to the cells of the body causes organs and then organ systems to fail. If not treated promptly, shock can be fatal.

Shock can occur because of several medical or traumatic events such as a heart attack, severe allergic reaction, an automobile crash, or a gunshot wound. As an EMT, you will respond to these different types of emergencies to provide care and transportation for these patients. Therefore, you must be constantly alert to the signs and symptoms of shock. In general, you cannot go wrong assuming that every patient is in shock or may go into shock; treat every patient for shock.

This chapter begins with a close-up look at perfusion, the function that fails in shock. Next it looks at the physiologic causes of shock and describes each of its major forms. Finally, it discusses the emergency treatment of shock in general and of each kind of shock in particular. See Chapter 11, *BLS Resuscitation,* for resuscitation techniques.

Pathophysiology

Perfusion

Perfusion is the circulation of blood within an organ or tissue in adequate amounts to meet the cells' current needs for oxygen, nutrients, and waste removal. The body is perfused via the circulatory system. The circulatory system is a complex arrangement of connected tubes, including the arteries, arterioles, capillaries, venules, and veins. There are two circuits in the body: the systemic circulation in the body and the pulmonary circulation in the lungs. The systemic circulation, the circuit in the body, carries oxygen-rich blood from the left ventricle through the body and back to the right atrium. In the systemic circulation, as blood passes through the tissues and organs, it gives up oxygen and nutrients and absorbs cellular wastes and carbon dioxide. Perfusion is an important part of the process by which waste products such as carbon dioxide made by the cells are removed.

It is also important to understand the physiology of respiration because organs, tissues, and cells must have adequate oxygenation or they may die. Each time you take a breath, the alveoli, which are microscopic, thin-walled air sacs, receive a supply of oxygen-rich air. The oxygen then dissolves in the blood plasma and attaches to the blood's hemoglobin. The oxygenated blood passes through the alveolar wall into the walls of a fine network of pulmonary capillaries that are in close contact with the alveoli. If the oxygenated blood is not properly circulated, some of the cells and organs will not receive proper nutrients, possibly resulting in cell death.

Oxygen and carbon dioxide pass rapidly across these thin tissue layers through diffusion. Diffusion is a

You are the Provider: PART 1

At 8:22 PM, your alert tones sound, "Medic 4, respond to the Cedar Hills Urgent Care Clinic at 1111 Cedar Hills Dr. for a 39-year-old female who is going into shock." You and your partner proceed to the clinic, which is approximately 9 minutes from your station. It is cloudy outside, the temperature is 66°F, and the traffic is light.

1. **What additional information should you attempt to gather about the patient while en route to the clinic?**

2. **What is shock and how does it relate to perfusion?**

passive process in which molecules move from an area with a higher concentration of molecules to an area of lower concentration. There are more oxygen molecules in the alveoli than in the blood. Therefore, the oxygen molecules move from the alveoli into the blood. Because there are more carbon dioxide molecules in the blood than in the inhaled air, carbon dioxide moves from the blood into the alveoli.

Just like oxygen, carbon dioxide is dissolved in the plasma and attaches to the blood's hemoglobin. The body takes the carbon dioxide, combines it with water, and creates sodium bicarbonate. Sodium bicarbonate concentrations become very high just as the blood is moving toward the lungs. Once it reaches the lungs, the sodium bicarbonate breaks down and the carbon dioxide is exhaled. All of this action takes place to maintain the delicate balance between the gases. If there is a disturbance in the transportation of carbon dioxide, dangerous waste products will build up in the cells and organs leading to cell or organ death.

Shock, or hypoperfusion, refers to a state of collapse and failure of the cardiovascular system that leads to inadequate circulation. Like internal bleeding, shock is an unseen underlying life threat caused by a medical disorder or traumatic injury. To protect vital organs, the body attempts to compensate by directing blood flow from organs that are more tolerant of low flow (such as the skin and intestines) to organs that cannot tolerate low blood flow (such as the heart, brain, and lungs). If the conditions causing shock are not promptly addressed, the patient will soon die. However, you can recognize the signs and symptoms of shock early and initiate treatment for this life threat soon after the onset of shock.

The cardiovascular system consists of three parts: a pump (the heart), a set of pipes (the blood vessels or arteries that act as the container), and the contents of the container (the blood) **Figure 10-1**. These three parts can be referred to as the "perfusion triangle" **Figure 10-2**. When a patient is in shock, one or more of the three parts is not working properly.

Blood is the vehicle for carrying oxygen and nutrients through the vessels to the capillary beds to tissue cells, where these supplies are exchanged for waste products. For this process to happen, the vessels (container) must be intact. Blood contains red blood cells, white blood cells, platelets, and a liquid called plasma. As discussed in Chapter 5, *The Human Body*, red blood cells are responsible for the transportation of oxygen to the cells and for transporting carbon dioxide (a waste product of cellular metabolism) away from the cells to the lungs where it is exhaled and removed from the body. White blood cells help the body to fight infection. Platelets are

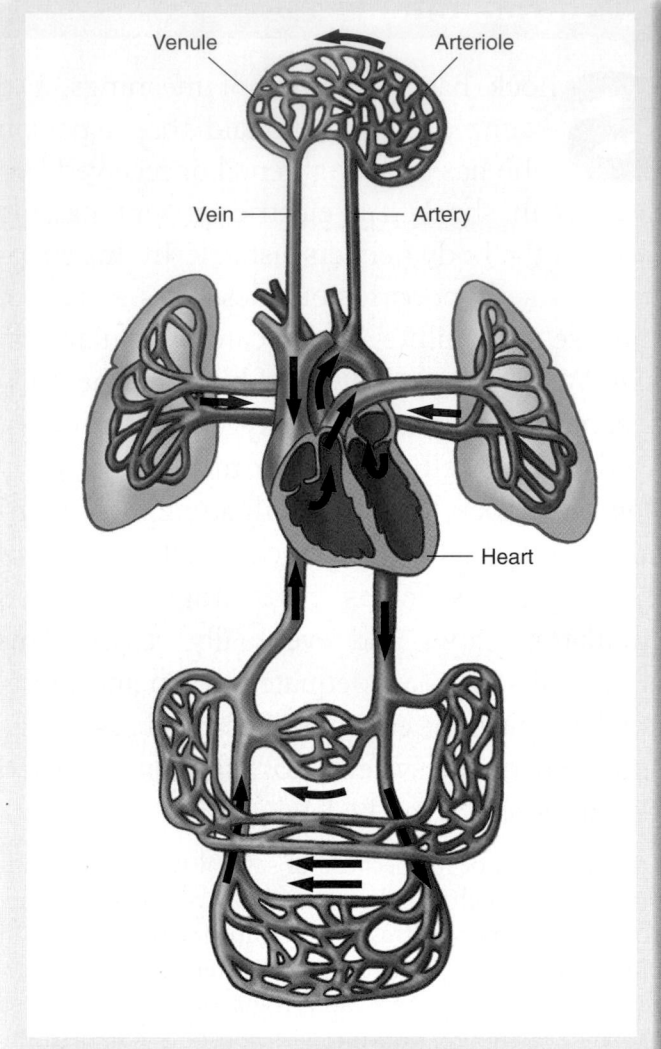

Figure 10-1 The cardiovascular system consists of three parts: the pump (heart), the container (vessels), and the contents (blood). The blood carries oxygen and nutrients through the vessels to the capillary beds, where they are exchanged for waste products.

responsible for forming blood clots. In the body, a blood clot forms depending on one of the following principles: retention of blood due to blockage in blood circulation (blood stasis), changes in the vessel wall (such as a wound), and the blood's ability to clot (as the result of a disease process or medication). When injury occurs to tissues in the body, platelets begin to collect at the site of injury; this causes the red blood cells to become sticky and clump together. As the red blood cells begin to clump, another substance in the body called fibrinogen reinforces the red blood cells. This is the final step in the formation of a blood clot. Blood clots are an important response from the body to control blood loss; however, be aware that clots are unstable and prone to rupture because blood keeps moving as a result of pressure that

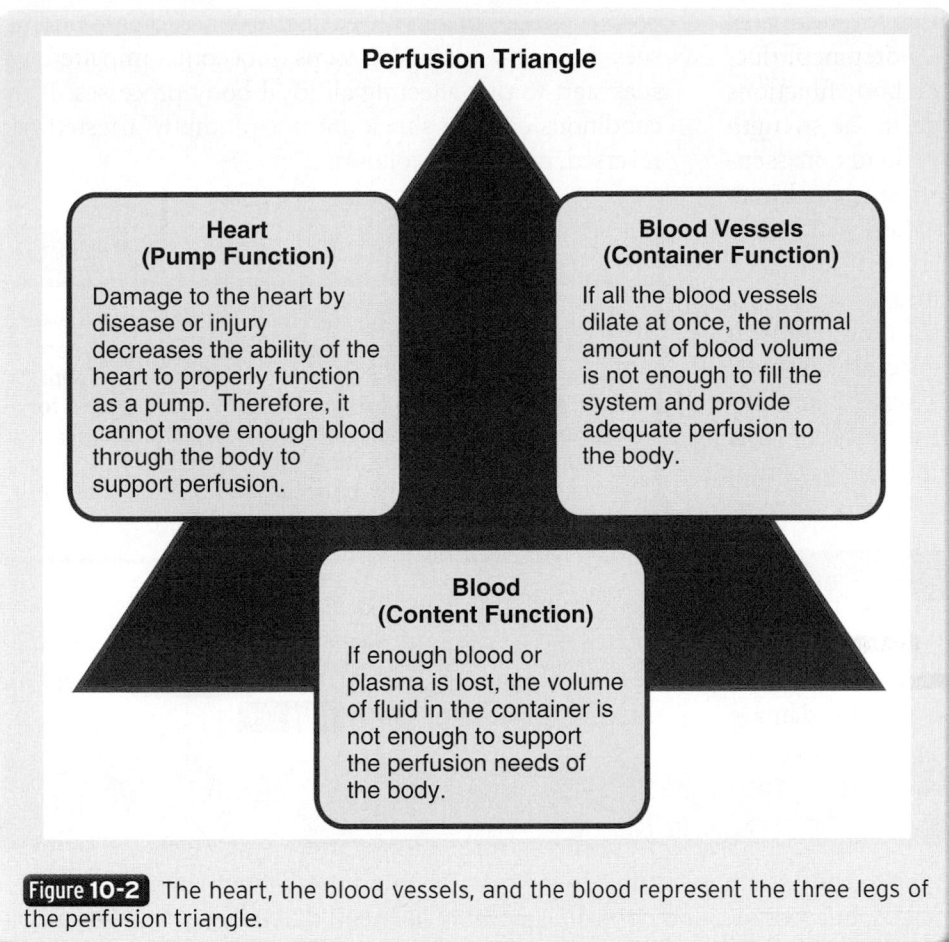

Perfusion Triangle

Heart (Pump Function)

Damage to the heart by disease or injury decreases the ability of the heart to properly function as a pump. Therefore, it cannot move enough blood through the body to support perfusion.

Blood Vessels (Container Function)

If all the blood vessels dilate at once, the normal amount of blood volume is not enough to fill the system and provide adequate perfusion to the body.

Blood (Content Function)

If enough blood or plasma is lost, the volume of fluid in the container is not enough to support the perfusion needs of the body.

Figure 10-2 The heart, the blood vessels, and the blood represent the three legs of the perfusion triangle.

is generated by the contractions of the heart and the actions of the blood vessels as they dilate and constrict. This pressure, which we call blood pressure, is usually carefully controlled by the body so that there is always sufficient circulation, or perfusion, in the various tissues and organs. Blood pressure is, in fact, a rough measure of perfusion.

Words of Wisdom

Capillary hydrostatic pressure tends to force fluids through capillary walls, whereas interstitial fluid hydrostatic pressure pushes fluid back into the cells.

Remember that blood pressure is really the pressure of blood within the vessels at any one time. The *systolic* pressure is the peak arterial pressure, or pressure generated every time the heart contracts; the *diastolic* pressure is the pressure maintained within the arteries while the heart rests between heartbeats.

Blood flow through the capillary beds is regulated by the capillary sphincters, circular muscular walls

that constrict and dilate. These **sphincters** are under the control of the **autonomic nervous system**, which regulates involuntary functions such as sweating and digestion. Capillary sphincters also respond to other stimuli such as heat, cold, the need for oxygen, and the need for waste removal. Keep in mind that, under normal circumstances, not all cells have the same needs at the same time. For example, the stomach and intestines have a high need for blood flow during and shortly after eating, when digestion is at a peak. Between meals, blood flow is lessened, and blood is diverted to other areas. The brain, by contrast, needs a constant and consistent supply of blood to function.

Thus, regulation of blood flow is determined by cellular need and is accomplished by vessel constriction or dilation, together with sphincter constriction or dilation. Maintenance of blood flow, or perfusion, is accomplished by the heart, blood vessels, and blood working together.

Perfusion requires more than just having a working cardiovascular system, however. It also requires adequate oxygen exchange in the lungs, adequate nutrients in the form of glucose in the blood, and adequate waste removal, primarily through the lungs. Carbon dioxide is one of the primary waste products of cellular work (metabolism) in the body and is removed from the body by the lungs. This is the reason adequate ventilation and oxygenation is one of your primary concerns. The body has neural and hormonal mechanisms in place to help support the respiratory and cardiovascular systems when the need for perfusion of vital organs is increased. These mechanisms, including the autonomic nervous system and certain chemicals called hormones, are triggered when the body senses that the pressure in the system is falling. The sympathetic side of the autonomic nervous system, which is responsible for the fight-or-flight response, will assume more control of the body's functions during a state of shock. The parasympathetic nervous system is a division of the autonomic nervous system that controls involuntary functions by sending signals to the cardiac, smooth, and glandular muscles. This response by the autonomic nervous system causes the release of hormones such as epinephrine and norepinephrine. These hormones cause changes in certain body functions such as an increase in the heart rate and in the strength of cardiac contractions and vasoconstriction in nonessential areas, primarily in the skin and gastrointestinal tract (peripheral vasoconstriction). Together, these actions are designed to maintain pressure in the system and, as a result, sustain perfusion of all vital organs.

Eventually, there is also a shifting of body fluids to help maintain pressure within the system. However, the response of the autonomic nervous system and hormones comes within seconds. It is this response that causes all the signs and symptoms of shock in a patient.

Causes of Shock

Shock can result from many conditions, including bleeding respiratory failure, acute allergic reactions, and overwhelming infection. In all cases, however, the damage

occurs because of insufficient perfusion of organs and tissues. As soon as perfusion stops or becomes impaired, tissues start to die, affecting all local body processes. If the conditions causing shock are not promptly arrested and reversed, death soon follows.

Words of Wisdom

Shock is a complex physiologic process that gives subtle signs to its presence before it becomes severe. These early signs relate very closely to the events that lead to more severe shock, so it is important for you to know the underlying processes thoroughly. If you understand what causes shock, you will be able to recognize it in many patients before it gets out of control.

Understanding the basic physiologic causes of shock will better prepare you to treat it **Figure 10-3**. There are three basic causes of shock **Table 10-1**.

You are the Provider: PART 2

You arrive at the clinic and are escorted to the patient by a clinic technician. You find the patient lying supine on an examination table. She is conscious, but restless, and her skin is notably pale and diaphoretic. She has a blanket covering her, her legs are elevated, and she is receiving oxygen via a nasal cannula at 4 L/min. Several attempts at establishing intravenous (IV) access were unsuccessful. Your assessment of the patient reveals the following:

Recording Time: 0 Minutes	
Appearance	Restless, pale, and diaphoretic
Level of consciousness	Conscious and alert, but restless
Airway	Open; clear of secretions or foreign bodies
Breathing	Increased rate; adequate depth
Circulation	Radial pulses, weak and rapid; skin is cool, pale, and diaphoretic

The clinic physician tells you that the patient presented approximately 15 minutes ago complaining of abdominal pain and rectal bleeding, which apparently started about 24 hours ago. There is no history of trauma, she has a history of irritable bowel syndrome, she takes lubiprostone (Amitiza) and dicyclomine hydrochloride (Bentyl), and she is allergic to codeine.

3. **On the basis of your assessment, does this patient require any changes in the treatment she is currently receiving?**

4. **How do the patient's signs and symptoms correlate with the body's response to inadequate perfusion?**

A **Pump failure**
Causes: Heart attack, trauma to heart, obstructive causes

B **Low fluid volume**
Causes: Trauma to vessels or tissues, fluid loss from GI tract (vomiting/diarrhea can also lower the fluid component of blood)

C **Poor vessel function**
Causes: Infection, drug overdose (narcotic), spinal cord injury, anaphylaxis

Figure 10-3 There are three basic causes of shock and impaired tissue perfusion. **A.** Pump failure occurs when the heart is damaged by disease, injury, or obstructive causes. The heart may not generate enough energy to move the blood through the system. **B.** Low fluid volume, often a result of bleeding, leads to inadequate perfusion. **C.** The blood vessels can dilate excessively so that the blood within them, even though it is of normal volume, is inadequate to fill the system and provide efficient perfusion.

Table 10-1 Causes of Shock

Pump Failure
- Cardiogenic shock
- Obstructive shock

Poor Vessel Function
- Distributive shock
  - Septic shock
  - Neurogenic shock
  - Anaphylactic shock
  - Psychogenic shock

Low Fluid Volume
- Hypovolemic shock
  - Hemorrhagic shock
  - Nonhemorrhagic shock

Types of Shock

Cardiogenic Shock

Cardiogenic shock is caused by inadequate function of the heart, or pump failure. Circulation of blood throughout the vascular system requires the constant pumping action of a normal and vigorous heart muscle. Many diseases or injury can cause destruction or inflammation of this muscle. Within certain limits, the heart can adapt to these problems. If too much muscular damage occurs, however, as sometimes happens after a heart attack, the heart no longer functions well. A major effect is the backup of blood into the lungs. The resulting buildup of fluid within the pulmonary tissue is called pulmonary edema. **Edema** is the presence of abnormally large amounts of fluid between cells in body tissues, causing swelling of the affected area **Figure 10-4**. Pulmonary edema leads to impaired ventilation, which may be manifested by an increased respiratory rate and abnormal lung sounds.

The muscular contraction of the heart moves blood through the vessels at distinct pressures. For blood to circulate efficiently throughout the entire system, there

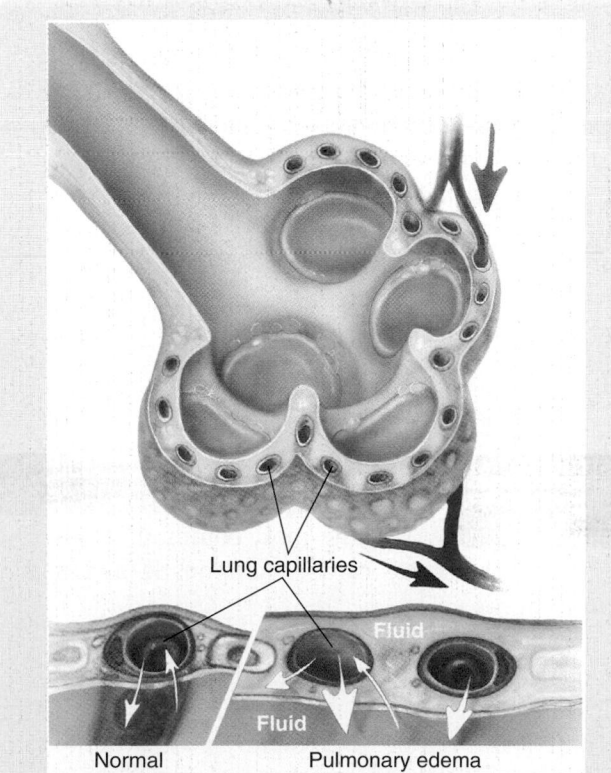

Lung capillaries

Fluid

Fluid

Normal   Pulmonary edema

Figure 10-4 Pulmonary edema develops as a result of fluid buildup within the pulmonary tissue. The edema causes swelling and leads to impaired ventilation.

must be the right amount of pressure and an adequate number of heartbeats. For this reason, the heart has its own electrical system that initiates and regulates its beating. Disease or injury can damage or destroy this system, causing irregular and uncoordinated beats, beats that are too slow (fewer than 60 beats/min), or beats that are too fast (more than 100 beats/min).

Cardiogenic shock develops when the heart cannot maintain sufficient output (cardiac output) to meet the demands of the body. Cardiac output is the volume of blood that the heart can pump per minute, and it is dependent upon several factors. First, the heart must have adequate strength, which is largely determined by the ability of the heart muscle to contract. This ability to contract is referred to as **myocardial contractility**. Second, the heart must receive adequate blood to pump. As the volume of blood coming to the heart increases, the precontraction pressure in the heart builds up. This precontraction pressure is known as **preload**. As preload increases, the volume of blood within the ventricles increases, which causes the heart muscle to stretch. When the muscle is stretched, myocardial contractility increases, leading to greater force of contraction and increased cardiac output. Lastly, the resistance to flow in the peripheral circulation must be appropriate. The force or resistance against which the heart pumps is known as **afterload**. In general, as afterload increases, cardiac output decreases. Increased afterload may also cause the heart to overwork while trying to maintain adequate cardiac output. High afterload is often the reason that heart failure develops in patients with hypertension. Cardiogenic shock may result from low cardiac output due to high afterload, low preload, poor contractility, or any combination of the three.

■ Obstructive Shock

Obstructive shock results when conditions that cause mechanical obstruction of the cardiac muscle also impact pump function. Two of the most common examples of obstructive shock are cardiac tamponade and tension pneumothorax.

A collection of fluid between the pericardial sac and the myocardium is a cardiac tamponade, also called a pericardial tamponade. It is caused by blunt or penetrating trauma and can progress rapidly. Cardiac tamponade occurs when blood leaks into the tough fibrous membrane known as the pericardium, causing an accumulation of blood within the pericardial sac. This accumulation leads to compression of the heart. Because the pericardium has a limited ability to stretch, each contraction of the heart allows more blood accumulation between the heart and the sac. The accumulated blood prevents the heart from opening up to allow complete refilling. Continued pressure within the pericardial sac obstructs the flow of

blood into the heart, resulting in decreased outflow from the heart. Signs and symptoms of cardiac tamponade are referred to as Beck's Triad, the presence of jugular vein distention, muffled heart sounds, and systolic and diastolic blood pressure starting to merge (systolic pressure drops and the diastolic pressure rises).

Another obstructive condition occurs with a tension pneumothorax. A tension pneumothorax is caused by damage to the lung tissue. This damage allows air normally held within the lung to escape into the chest cavity. If a pneumothorax is allowed to continue untreated, a sufficient amount of air will accumulate within the chest cavity and begin applying pressure to the structures in the mediastinum. The primary organs in this area are the heart and great vessels (aorta and vena cava). When the trapped air begins to shift the chest organs toward the uninjured side, a pneumothorax becomes known as a tension pneumothorax, which is a very serious and life-threatening condition. As pressure from one side of the chest begins to push the mediastinum toward the other side, the vena cava loses its ability to stay fully expanded. This mechanical compression of the vessel leads to reduced return of blood to the heart. The patient becomes anxious and short of breath. The heart and respiratory rates increase and become shallower. Blood pressure drops. You may notice difficulty when attempting to ventilate the patient with a bag-mask device as well. The affected side will have decreased or absent lung sounds and the patient will become cyanotic. Tracheal deviation may be a late sign of tension pneumothorax.

■ Distributive Shock

Distributive shock results when there is widespread dilation of the small arterioles, small venules, or both. As a result, the circulating blood volume pools in the expanded vascular beds and tissue perfusion decreases. The four most common types of distributive shock are septic shock, neurogenic shock, anaphylactic shock, and psychogenic shock.

Septic Shock

Septic shock occurs as a result of severe infections, usually bacterial, in which toxins (poisons) are generated by the bacteria or by infected body tissues. In this condition, the toxins damage the vessel walls, causing increased cellular permeability. The vessel walls leak and are unable to contract well. Widespread dilation of vessels, in combination with plasma loss through the injured vessel walls, results in shock.

Septic shock is a complex problem. First, there is an insufficient volume of fluid in the container, because much of the plasma has leaked out of the vascular system (hypovolemia). Second, the fluid that has leaked out often collects in the respiratory system, interfering with

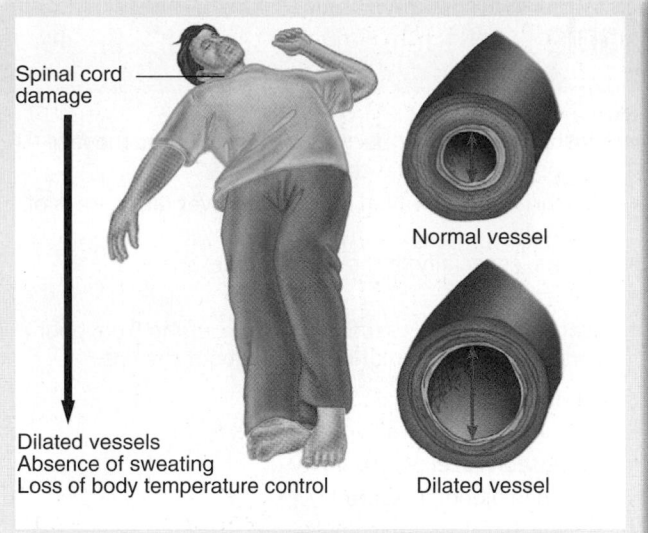

ventilation. Third, the vasodilation leads to a larger-than-normal vascular bed to contain the smaller-than-normal volume of intravascular fluid.

Septic shock is almost always a complication of a very serious illness, injury, or surgery.

Neurogenic Shock

Damage to the spinal cord, particularly at the upper cervical levels, may cause significant injury to the part of the nervous system that controls the size and muscular tone of the blood vessels. Neurogenic shock is usually the result. Although not as common, there are medical causes as well. These include brain conditions, tumors, pressure on the spinal cord, and spina bifida. In neurogenic shock, the muscles in the walls of the blood vessels are cut off from the sympathetic nervous system and nerve impulses that cause them to contract. Therefore, all vessels below the level of the spinal injury dilate widely, increasing the size and capacity of the vascular system Figure 10-5 and causing blood to pool. The available 6 L of blood in the body can no longer fill the enlarged vascular system. Even though no blood or fluid has been lost, perfusion of organs and tissues becomes inadequate, and shock occurs. In this condition, a radical change in the size

Spinal cord damage

Normal vessel

Dilated vessels
Absence of sweating
Loss of body temperature control

Dilated vessel

Figure 10-5 Damage to the spinal cord can cause significant injury to the part of the nervous system that controls the size and muscle tone of blood vessels. If the muscles in the blood vessels are cut off from their impulses to contract, the vessels dilate widely, increasing the size and capacity of the vascular system. The blood in the body can no longer fill the enlarged vessels; inadequate perfusion results.

of the vascular system has caused shock. Characteristic signs of this type of shock are the absence of sweating below the level of injury and normal warm skin.

Anaphylactic Shock

Anaphylaxis, or anaphylactic shock, occurs when a person reacts violently to a substance to which he or she has been sensitized. Sensitization means becoming sensitive to a substance that did not initially cause a reaction. Do not be misled by a patient who reports no history of allergic reaction to a substance on first or second exposure. Each subsequent exposure after sensitization tends to produce a more severe reaction.

Instances that cause severe allergic reactions commonly fall into the following four categories of exposure:

- Injections (tetanus antitoxin, penicillin)
- Stings (honeybee, wasp, yellow jacket, hornet)
- Ingestion (shellfish, fruit, medication)
- Inhalation (dust, pollen)

Anaphylactic reactions can develop within minutes or even seconds after contact with the substance to which the patient is allergic. The signs of such allergic reactions are very distinct and not seen with other forms of shock. Table 10-2 lists the signs of anaphylactic shock in the order in which they typically occur. Note that cyanosis (bluish color of the skin) is a late sign of anaphylactic shock.

Words of Wisdom

With neurogenic shock, many other functions that are under the control of the same part of the nervous system are also lost. The most important of them, in an acute injury setting, is the ability to control body temperature. Body temperature in a patient with neurogenic shock can rapidly fall to match that of the environment. In many situations, significant hypothermia occurs, severely complicating the situation. Hypothermia is a condition in which the internal body temperature falls below 95°F (35°C), usually after prolonged exposure to cool or freezing temperatures. Maintenance of body temperature is always an important element of treatment for a patient in shock.

Words of Wisdom

Microcirculation is a term used to describe the small vessels in the vasculature that are embedded within organs and responsible for the distribution of blood within tissues. True capillaries are part of microcirculation. They branch off the arterioles and allow for exchange between cells and circulation. The arteriole-venule shunt are short vessels that connect the arteriole and venule at opposite sides. The main functions of microcirculation include the regulation of blood flow and tissue perfusion, blood pressure, tissue fluid, delivery of oxygen, removal of carbon dioxide, and the regulation of body temperature and inflammation. Variations of microcirculation can influence the capillaries on the local, neural, and hormonal levels.

Table 10-2 Signs of Anaphylactic Shock

Skin
- Flushing, itching, or burning, especially over the face and upper part of the chest
- Urticaria (hives), which may spread over large areas of the body
- Edema, especially of the face, tongue, and lips
- Pallor
- Cyanosis (a bluish cast to the skin resulting from poor oxygenation of circulating blood) about the lips

Circulatory System
- Dilation of peripheral blood vessels
- Increased vessel permeability
- A drop in blood pressure
- A weak, barely palpable pulse
- Dizziness
- Fainting and coma

Respiratory System
- Sneezing or itching in the nasal passages
- Tightness in the chest, with a persistent dry cough
- Wheezing and dyspnea (difficulty breathing)
- Secretions of fluid and mucus into the bronchial passages, alveoli, and lung tissue, causing coughing
- Constriction of the bronchi; difficulty drawing air into the lungs
- Forced expiration, requiring exertion and accompanied by wheezing
- Cessation of breathing

In anaphylactic shock, there is no loss of blood, no mechanical vascular damage, and only a slight possibility of direct cardiac muscular injury. Instead, there is widespread vascular dilation, increased permeability, and bronchoconstriction. The combination of poor oxygenation and poor perfusion in anaphylactic shock may easily prove fatal.

Psychogenic Shock

A patient in **psychogenic shock** has had a sudden reaction of the nervous system that produces a temporary, generalized vascular dilation, resulting in fainting, or **syncope**. Blood pools in the dilated vessels, reducing the blood supply to the brain; as a result, the brain ceases to function normally, and the patient faints. While there are many causes of syncope, it is important to realize that some are of a serious nature but others are not. Causes of syncope that are potentially life threatening result from events such as an irregular heartbeat or a brain **aneurysm**. Other non–life-threatening events that cause syncope may be the receipt of bad news or experiencing fear or unpleasant sights (like the sight of blood).

Hypovolemic Shock

Hypovolemic shock is the result of an inadequate amount of fluid or volume in the system. There are hemorrhagic and non-hemorrhagic causes of hypovolemic shock. Injuries may result in hemorrhagic shock, while vomiting and diarrhea may result in non-hemorrhagic hypovolemic shock.

Hypovolemic shock also occurs with severe thermal burns. In this case, it is intravascular plasma (the colorless part of the blood) that is lost, leaking from the circulatory system into the burned tissues that lie adjacent to the injury. Likewise, crushing injuries may result in the loss of blood and plasma from damaged vessels into injured tissues.

Dehydration, the loss of water or fluid from body tissues, can cause or aggravate shock. Fluid loss may be a result of severe vomiting and/or diarrhea. Patients who are very young or elderly are particularly susceptible to fluid loss and therefore at risk for developing shock through dehydration. People who exercise in hot weather and are not accustomed to it may experience dehydration if they do not drink enough fluids. In these circumstances, the common factor is an insufficient volume of blood within the vascular system to provide adequate circulation to the organs of the body.

Respiratory Insufficiency

A patient with a severe chest injury, such as flail chest, or obstruction of the airway, may be unable to breathe in an adequate amount of oxygen. This affects the ventilation process of respiration; enough oxygen cannot be inspired to meet the metabolic demand.

An insufficient concentration of oxygen in the blood can produce shock as rapidly as vascular causes, even if the volume of blood, the volume of the vessels, and the action of the heart are all normal. Without oxygen, the organs in the body cannot survive, and their cells promptly start to deteriorate.

Certain types of poisoning may affect the ability of cells to metabolize or carry oxygen. Carbon monoxide has a 250 times greater affinity for hemoglobin than oxygen. If a patient is in an environment where they inhale carbon monoxide, it will bind to the hemoglobin, forming carboxyhemoglobin, rather than allowing oxygen to bind. This results in a hypoxic state if not corrected. Cyanide impairs the ability of cells to metabolize oxygen within the cell and cellular asphixia may occur.

Anemia occurs when there is an abnormally low number of red blood cells. Red blood cells contain hemoglobin, an iron-containing pigment. Hemoglobin transports oxygen from the lungs to the tissues. Each hemoglobin molecule is able to carry four molecules of oxygen. Anemia may be the result of either chronic or acute bleeding, a deficiency in certain vitamins or minerals, or an underlying disease process. If anemia is present, tissues may be hypoxic because the blood may not be able to carry adequate oxygen, even though the hemoglobin is fully saturated. In this situation, a pulse oximeter may indicate that there is adequate saturation, even though the tissues are hypoxic. This type of hypoxia is known as hypoxemic hypoxia.

The Progression of Shock

Although you cannot see shock, you can see its signs and symptoms Table 10-3 . The early stage of shock, while the body can still compensate for blood loss, is called **compensated shock**. The late stage, when blood pressure is falling, is called **decompensated shock**. The last stage, when shock has progressed to a terminal stage, is called **irreversible shock**. A transfusion during irreversible shock will not save the patient's life.

Remember that blood pressure may be the last measurable factor to change in shock. As we have seen, the body has several automatic mechanisms to compensate for initial blood loss and to help maintain blood pressure. Thus, by the time you detect a drop in blood pressure, shock is well developed. This is particularly true in infants and children, who can maintain their blood pressure until they have lost more than half their blood volume. By the time blood pressure drops in infants and children who are in shock, they are close to death.

The EMT must also use caution when caring for elderly patients. As a result of the aging process, elderly patients generally have more serious complications than younger patients. Although illness is a common complaint among the elderly, you must understand that it

Table 10-3 Progression of Shock

Compensated Shock
- Agitation
- Anxiety
- Restlessness
- Feeling of impending doom
- Altered mental status
- Weak, rapid (thready), or absent pulse
- Clammy (pale, cool, moist) skin
- Pallor, with cyanosis about the lips
- Shallow, rapid breathing
- Air hunger (shortness of breath), especially if there is a chest injury
- Nausea or vomiting
- Capillary refill of longer than 2 seconds in infants and children
- Marked thirst

Decompensated Shock
- Falling blood pressure (systolic blood pressure of 90 mm Hg or lower in an adult)
- Labored or irregular breathing
- Ashen, mottled, or cyanotic skin
- Thready or absent peripheral pulses
- Dull eyes, dilated pupils
- Poor urinary output

You are the Provider: PART 3

You continue with the treatment initiated by the clinic; however, you remove the nasal cannula and apply high-flow oxygen via a nonrebreathing mask. Your partner takes the patient's vital signs and reports them to you and the clinic physician.

Recording Time: 4 Minutes	
Respirations	24 breaths/min; shallow
Pulse	120 beats/min; weak
Skin	Pale, cool, and diaphoretic
Blood pressure	108/58 mm Hg
Oxygen saturation (Sao$_2$)	95% (on oxygen)

You bring the stretcher into the room and prepare the patient for immediate transport. The patient remains conscious and alert, but is becoming increasingly restless and tells you that she is extremely thirsty.

5. Is the patient in compensated or decompensated shock? How can you tell?

is not just part of aging. Keep in mind the following signs of the normal aging process when managing geriatric patients:

- The central nervous system often has a delayed response.
- The cardiovascular system has a variety of changes that result in a decrease in the efficiency of the system. On assessment, be alert for higher resting heart rates and irregular pulse rates.
- The respiratory system has significant changes as the elasticity of the lungs and their size and strength decrease. On assessment, be alert for higher respiratory rates, lower tidal volume, and a decreased gag reflex. In addition, you must remember that cervical arthritis may be present and that dentures may cause an airway obstruction.
- The skin becomes thinner, drier, less elastic, and more fragile, thus providing less protection and thermal regulation (cold and hot).
- The renal system decreases in function and may not respond well to unusual demands such as illness.
- The gastrointestinal system sustains changes in gas-tric motility that may lead to slower gastric emptying.

Treating a pediatric or geriatric patient in shock is no different than treating any other shock patient:

1. Provide in-line spinal stabilization if indicated. If spinal immoblization is not indicated, maintain the patient in a position of comfort.
2. Suction as necessary and provide high-flow oxygen via a nonrebreathing mask.
3. Control bleeding.
4. Maintain body temperature.
5. Provide rapid transportation.

You should expect shock in many emergency medical situations. For example, you would expect shock to accompany massive external or internal bleeding. You should also expect shock if a patient has any one of the following conditions:

- Multiple severe fractures
- Abdominal or chest injury
- Spinal injury
- A severe infection
- A major heart attack
- Anaphylaxis

Words of Wisdom

Taking and recording frequent vital signs—and observing perfusion indicators such as skin condition and mental status—will give you a window into the progression of shock. Use your documentation to remind you to suspect shock early and treat it aggressively.

Patient Assessment for Shock

Scene Size-up

Scene Safety

As you approach the scene, be alert to potential hazards to your safety such as downed power lines, fast moving traffic, or anything else that threatens your safety. Once on scene, look for and address hazards and threats to the safety of the crew, bystanders, and the patient. If this is a trauma scene or bleeding is suspected, put on gloves and eye protection, at a minimum. Put several pairs of gloves in your pocket for easy access in case your gloves tear or there are multiple patients with bleeding.

At vehicle crashes, ensure that there is no leaking fuel in the area where you will be working and that energized electrical lines are not close to where you will be working. If you are called to a two-car collision, how many patients are possible? Two or eight? Do you have the necessary resources available? Consider early what you may need, and verify as you begin to form your initial general impression. The sooner you call for help, the sooner it will arrive.

In incidents involving violence, such as assaults or gunshot wounds, make sure that police are on scene. At times you may need to stage several blocks away until law enforcement personnel have secured the area.

Mechanism of Injury/Nature of Illness

When you first see the patient, observe the scene and patient for clues to determine the nature of the illness or the mechanism of injury. Medical complaints typically involve only one patient, but always ensure that you only have one patient to care for. It is not uncommon that trauma incidents involve more than one patient; obtain an accurate account of all patients.

Primary Assessment

The primary assessment for a patient with suspected shock should include a rapid scan of the patient to determine level of consciousness, identify and manage life-threatening concerns, and determine priority of the patient and transport. Treating according to the ABCs is always a good choice. Threats to airway, breathing, or circulation are considered life threatening and must be treated immediately to prevent mortality. In some situations, significant bleeding may require management before applying oxygen for a person with adequate breathing. Significant bleeding, internal or external, is an immediate life threat. If the patient has obvious life-threatening bleeding, it must be controlled quickly and treatment of shock begun as quickly as possible. The decision on what to treat first will come with experience.

Provide high-flow oxygen to assist in perfusion of damaged tissues. If the patient has signs of hypoperfusion, treat aggressively and provide rapid transport to the hospital. Request advanced life support (ALS) as necessary to assist with more aggressive shock management. Do not delay transport of the seriously injured trauma patient to complete non-lifesaving treatments in the field, such as splinting extremity fractures; instead, complete these types of treatments en route to the hospital.

Form a General Impression

When you first visualize your patient, quickly form an initial general impression. This includes age, sex, signs of distress, obvious life-threatening injuries, abnormal positioning, and skin color. These observations will help you develop an early sense of urgency for care of a patient who appears "sick."

Once you are close to the patient, determine the need for manual spinal immobilization and assess the patient's level of consciousness using the AVPU (*Alert* to person, place, and day; responsive to *Verbal* stimuli; responsive to *Pain*; *Unresponsive*) scale. A patient who has an altered level of consciousness (LOC) may need emergency airway management. If the patient is awake and alert, determine a chief complaint.

Airway and Breathing

Next, quickly assess the airway to ensure it is patent. If the patient is awake and answering questions, the airway is patent. Be alert to abnormal airway sounds such as gurgling (suction the airway) or stridor, indicating partial airway obstruction. If the patient is awake and answering questions, an airway adjunct is not needed; consider an adjunct such as an oropharyngeal or nasopharyngeal airway for a patient with an altered LOC.

Next, you must quickly assess breathing in the patient. You must inspect and palpate the chest wall to assess for DCAP-BTLS (Deformities, Contusions, Abrasions, Punctures/penetrations, Burns, Tenderness, Lacerations, and Swelling). Observe the patient for signs of accessory muscle use such as the muscles of the neck, intercostal retractions, or abnormal use of the abdominal muscles. An increased respiratory rate is often an early sign of impending shock. You must assess the patient's breath sounds with a stethoscope, listening for wheezes or other abnormal breath sounds. Once you have quickly completed this assessment of breathing, give the patient high-flow oxygen, or, if needed, assist respirations with a bag-mask device.

Circulation

Next you must quickly assess the patient's circulatory status. Check for the presence of a distal pulse. If you cannot obtain a distal pulse, assess for a central pulse.

Make a rapid determination if the pulse is fast, slow, weak, strong, or altogether absent. A rapid pulse suggests compensated shock. In shock or compensated shock, the skin may be cool, clammy, or ashen. If the patient has no pulse and is not breathing, immediately begin cardiopulmonary resuscitation (CPR). In trauma patients, ensure you have assessed for and identified any life-threatening bleeding; if serious bleeding is discovered, treat it at once. You must also quickly assess skin temperature, condition, and color; also check for capillary refill time.

Transport Decision

Once you have assessed perfusion, you can determine whether the patient should be treated as high priority, whether ALS is needed, and which facility to transport to. Trauma patients with shock, or a suspicious mechanism of injury (MOI), generally should go to a trauma center. Sometimes, local protocols dictate that a patient should be transported to the nearest hospital for stabilization prior to transfer to a definitive treatment center.

History Taking

Investigate Chief Complaint

After the life threats have been managed during the primary assessment, determine the chief complaint. The EMT should obtain a medical history and be alert for injury-specific signs and symptoms as well as any pertinent negatives such as loss of sensation.

SAMPLE History

You must now quickly obtain a SAMPLE history from the patient. Remember, if the patient has a significant change in LOC before arrival at the hospital, you should provide the hospital personnel with this important information.

Secondary Assessment

The secondary assessment is a more detailed, comprehensive examination of the patient that is used to uncover injuries that may have been missed during the primary assessment. The secondary assessment begins by repeating the primary assessment followed by a focused assessment. In some instances, such as a critically injured patient or short transport time, you may not have time to conduct a secondary assessment.

Physical Examinations

If significant trauma has likely affected multiple systems, start with a full-body scan to be sure that you have identified all injuries. Next assess the respiratory system. This should involve looking, listening, and feeling. When assessing the respiratory system, look at the patient and ask yourself the following questions:

1. Is the patient's respiratory rate and quality within normal limits?
2. What is the patient's skin color and condition?
3. Are there any signs of increased respiratory efforts such as retractions, nasal flaring, stridor, or use of accessory muscles?

Next, listen for air movement at the patient's mouth and nose. Then listen to breath sounds with a stethoscope. Breath sounds should be clear and equal bilaterally, anteriorly, and posteriorly. Determine the patient's rate and quality of respiration. Finally assess asymmetrical chest wall movement.

You must be able to quickly assess the pulse rate and quality; determine the skin condition, color, and temperature; and check the capillary refill time.

Assess the neurologic system to gather baseline data on your patient. This examination should include:

- level of consciousness—use AVPU
- pupil size and reactivity
- motor response
- sensory response

Assess the musculoskeletal system by doing a detailed full-body scan. Look for DCAP-BTLS. Assess the chest, abdomen, and extremities for hidden bleeding and injuries. Log roll the patient and assess the posterior torso for injuries as well. Once the back has been assessed, the patient can be log rolled back down onto a backboard, followed by complete spinal stabilization. Log rolling and securing the patient to a backboard or other full-body stabilization device should take into consideration injuries found during the primary assessment.

Assess all anatomic regions looking for the following signs/symptoms:

- Be alert for raccoon eyes, Battle's sign, and/or drainage of blood or fluid from the ears or nose.
- Check the neck for jugular vein distention and tracheal deviation. Be alert for patients with a stoma or tracheostomy.
- Check the pelvis for stability.
- Check the abdomen, feel all four quadrants for tenderness or rigidity. If the abdomen is tender, expect internal bleeding.
- Check the extremities and record pulse, motor, and sensory function.

If your patient is a trauma patient with a significant mechanism of injury or multiple injuries, one who gives you a poor initial general impression, or you found problems in the primary assessment, perform a rapid full-body scan. If your patient has a medical problem but is not responsive or problems were noted in the primary assessment, perform a rapid full-body scan. These scans should be performed quickly but thoroughly to ensure that you do not miss any significant or life-threatening problems or delay needed care.

If your patient has only a simple mechanism of injury, such as a twisted ankle, focus your examination on the area affected. Whether your examination is rapid or focused, if a life-threatening problem is found, treat it immediately.

When time permits and the patient's condition is stable, perform a thorough examination of the patient. This includes a complete neurologic assessment.

Vital Signs

Obtain a complete set of baseline vital signs. If the patient's condition is unstable or could become unstable, reassess vital signs every 5 minutes. If the patient is in stable condition, reassess vital signs every 10 to 15 minutes. Baseline vital signs will help you trend changes in your patient.

Monitoring Devices

In addition to hands-on assessment, you should use monitoring devices to quantify the patient's oxygenation and circulatory status. Use a noninvasive technique to monitor blood pressure and a pulse oximeter to evaluate the effectiveness of oxygenation. It is recommended to assess the patient's blood pressure with a sphygmomanometer (blood pressure cuff) and stethoscope (manually), before using a noninvasive blood pressure monitor, to establish a baseline blood pressure and to determine the accuracy of the noninvasive blood pressure monitor.

Reassessment

This portion of patient assessment is very important in patient care. The rule of thumb is assess—intervene—reassess. This portion of the assessment revisits the primary assessment, the vital signs, the chief complaint, and any treatment performed on the patient, including oxygen administration. You must assess the patient to determine whether the interventions you performed are having any effect on the patient. This step prepares you to present the patient at the hospital with a complete, concise account of the patient encounter and care.

Interventions

You must determine what interventions are needed for your patient at this point based on the findings of your assessment. You should focus on supporting the cardiovascular system. Treating for shock early and aggressively by providing oxygen may limit the harm to your patient that can result from inadequate perfusion.

Communication and Documentation

Patients who are in decompensated shock will need rapid interventions to restore adequate perfusion. The hospital may or may not have suggestions on how best to support a patient's failing cardiovascular system. Most of the interventions used to treat shock do not require a specific physician's order; however, some do. Determine, based on the signs and symptoms found in your assessment, whether your patient is in compensated or decompensated shock. Document these findings after you have treated for shock.

Words of Wisdom

The Trendelenburg's position may help temporarily increase perfusion in some shock patients, but it has never been shown to improve survival. In patients with chest injuries or difficultly breathing, this position may worsen symptoms. In patients with head injury, it may cause a detrimental increase in intracranial pressure. In patients with lower extremity or hip fractures, elevation of the legs may cause increasing pain. As always, follow your local protocols or consult medical control.

Emergency Medical Care for Shock

You must begin immediate treatment for shock as soon as you realize that the condition may exist. Follow the steps in **Skill Drill 10-1**:

Skill Drill 10-1

1. As with any type of patient care, you should begin by following standard precautions and by making sure the patient has an open airway. Maintain manual in-line stabilization if necessary, and check breathing and pulse. Comfort, calm, and reassure the patient, while maintaining the patient in the supine position. Never allow patients to eat or drink anything prior to being evaluated by a physician. Patients who have had a severe heart attack or who have lung disease may find it easier to breathe in a sitting or semisitting position (Step 1).

2. Next, control all obvious external bleeding. Place dry, sterile dressings over the bleeding sites, and secure with bandages. If direct pressure is not rapidly successful in the control of bleeding from an extremity, apply a tourniquet proximal to the bleeding site according to local protocol (Step 2).

3. Splint the patient on a backboard. Do not delay transport by applying individual splints in the field. If possible, splint individual extremity fractures during transport. This minimizes pain, bleeding, and discomfort, all of which can aggravate shock. It also prevents the broken bone ends

You are the Provider: PART 4

The patient is placed onto the stretcher and loaded into the ambulance. You quickly gather the patient records from the clinic physician and begin transport to a hospital that is only 10 minutes away. En route, you continue with your treatment and reassess her condition.

Recording Time: 11 Minutes	
Level of consciousness	Responsive to pain only
Respirations	30 breaths/min; shallower
Pulse	130 beats/min; absent radial pulses (carotid pulse present)
Skin	Pale, cool, and diaphoretic
Blood pressure	84/44 mm Hg
Sao$_2$	89% (on oxygen)

6. How has your patient's condition changed?

7. Are adjustments in your current interventions required? If so, what?

Skill Drill 10-1

Treating Shock

Step 1 Keep the patient supine, open the airway, and check breathing and pulse.

Step 2 Control obvious external bleeding. Apply a tourniquet, if necessary, to achieve rapid control of blood loss from extremities.

Step 3 Splint the patient on a backboard. Splint any broken bones or joint injuries during transport.

Step 4 Give high-flow oxygen if you have not already done so, and place blankets under and over the patient.

from further damaging adjacent soft tissue. In general, splinting will make it easier to move the patient. Handle the patient gently and no more than is necessary ⟨Step 3⟩.

4. Remember that inadequate ventilation may be the primary cause of shock or a major factor in

its development. Always provide oxygen, assist with ventilations, and use airway control adjuncts as needed, and continue to monitor the patient's breathing. To prevent the loss of body heat, place blankets under and over the patient. Be careful not to overload the patient with covers or attempt to warm the body too much; it is best for the patient to maintain a normal body temperature. Do not use external heat sources, such as hot water bottles or heating pads. They may harm a patient in shock by causing vasodilation and decreasing blood pressure even more (Step 4).

5. Once you have positioned the patient on a backboard or stretcher, consider placing the patient in the Trendelenburg's position. This technique is easily accomplished by raising the foot of the backboard or stretcher about 6" to 12". If the patient is not on a backboard and no lower extremity fractures or pelvic fractures are suspected, consider placing the patient in a shock position. This is accomplished by elevating the patient's legs 6" to 12" by propping them up on several blankets or other stable objects. These positions may help to return blood from the extremities back to the core of the body where it is most needed. Raising the lower extremities any higher may aggravate the patient's breathing because the abdominal organs push against the diaphragm. Take care not to use the Trendelenburg position or shock position for patients who have associated chest injury or intra-abdominal injury, which may be aggravated by causing the abdominal contents to push against the diaphragm and further impair breathing.

6. Transport the patient and treat additional injuries en route.

Words of Wisdom

You are never wrong to treat for shock, and many patients will experience some degree of shock. Consider whether or not you need to treat for shock for each patient you encounter.

Do not give the patient anything by mouth, no matter how urgently you are asked. To relieve the intense thirst that often accompanies shock, give the patient a moistened piece of gauze to chew or suck. Never give a patient in shock an alcoholic drink or other depressant. A stimulant, such as coffee, also has little value in treating shock.

Accurately record the patient's vital signs approximately every 5 minutes throughout treatment and transport. It is essential to transport trauma patients to the ED as rapidly as possible for definitive treatment. The Golden Period refers to the first 60 minutes after injury, which is thought to be a critically important period for the early resuscitation and treatment of severely injured trauma patients. This concept underscores the importance of rapid evaluation, stabilization, and transport.

The goal of EMS is to limit on-scene time (time on-scene until transport to hospital is started) to 10 minutes or less. Remember to speak calmly and reassuringly to a conscious patient throughout assessment, care, and transport.

Table 10-4 lists the general supportive measures for the major types of shock. Not every measure is used for every type of shock.

Words of Wisdom

There is little evidence to support the use of pneumatic antishock garments. If used at all, they should be used cautiously and limited to pelvic fractures or as a lower-body air splint.

Treating Cardiogenic Shock

The patient who is in shock as a result of a heart attack does not require a transfusion of blood, intravenous fluids, or elevation of the legs. There is already a greater volume of blood in circulation than the heart can handle. The damaged heart muscle simply cannot generate the necessary power to pump blood throughout the circulatory system.

Keep in mind that chronic lung disease will aggravate cardiogenic shock. If the patient has chronic obstructive pulmonary disease and heart disease, oxygenation of the blood passing through the lungs is impaired. Because fluid is collecting in the lungs, this patient is often able to breathe better in a sitting or semisitting position and may tell you so.

Usually, patients with cardiogenic shock do not have any injury, but they may be having chest pain. Such a patient may have taken nitroglycerin before your arrival and may want to take more. Before helping the patient self-administer nitroglycerin, be sure to consult with medical control for instructions. You will also need to perform an accurate assessment to ensure that the patient's blood pressure meets the criteria for this medication. If the blood pressure is too low, nitroglycerin may increase the problem. Remember that patients in cardiogenic shock usually have a low blood pressure. Other signs include a weak, irregular pulse; cyanosis about the lips and underneath the fingernails; anxiety; and nausea.

Treatment of cardiogenic shock should begin by placing the patient in the position in which breathing is easiest as you give high-flow oxygen. Be ready to assist ventilations as necessary, and have suction nearby in case the patient vomits. Provide prompt transport to the emergency department. Remember also to approach a patient who has had a suspected heart attack with

Table 10-4 Types of Shock

Type of Shock	Examples of Potential Causes	Signs and Symptoms	Treatment
Cardiogenic	Inadequate heart function Disease of muscle tissue Impaired electrical system Disease or injury	Chest pain Irregular pulse Weak pulse Low blood pressure Cyanosis (lips, under nails) Cool, clammy skin Anxiety Rales Pulmonary edema	Position comfortably Administer oxygen Assist ventilations Transport promptly

Obstructive	Mechanical obstruction of the cardiac muscle causing a decrease in cardiac output
	1. Tension pneumothorax
	2. Cardiac tamponade
		Dependent on cause:
		■ Dyspnea
		■ Rapid, weak pulse
		■ Rapid, shallow breaths
		■ Decreased lung compliance
		■ Unilateral, decreased, or absent breath sounds
		■ Decreased blood pressure
		■ Jugular vein distention
		■ Subcutaneous emphysema
		■ Cyanosis
		■ Tracheal deviation toward affected side
		■ Beck's triad (cardiac tamponade):
		  – Jugular vein distention
		  – Narrowing pulse pressure
		  – Muffled heart tones
			Dependent on cause:
			■ ALS assist and/or rapid transport

Septic	Severe bacterial infection	Warm skin Tachycardia Low blood pressure	Transport promptly Administer oxygen en route Provide full ventilatory support Consider elevating legs Keep patient warm
Neurogenic	Damaged cervical spine, which causes widespread blood vessel dilation	Bradycardia (slow pulse) Low blood pressure Signs of neck injury	Secure airway Spinal stabilization Assist ventilations Administer high-flow oxygen Preserve body heat Transport promptly
Anaphylactic	Extreme life-threatening allergic reaction	Can develop within seconds Mild itching or rash Burning skin Vascular dilation Generalized edema Coma Rapid death	Manage the airway Assist ventilations Administer high-flow oxygen Determine cause Assist with administration of epinephrine Transport promptly

Continues

Table 10-4 Types of Shock, continued

Type of Shock	Examples of Potential Causes	Signs and Symptoms	Treatment
Psychogenic (fainting)	Temporary, generalized vascular dilation Anxiety, bad news, sight of injury or blood, prospect of medical treatment, severe pain, illness, tiredness	Rapid pulse Normal or low blood pressure	Determine duration of unconsciousness Record initial vital signs and mental status Suspect head injury if patient is confused or slow to regain consciousness Transport promptly
Hypovolemic	Loss of blood or fluid	Rapid, weak pulse Low blood pressure Change in mental status Cyanosis (lips, under nails) Cool, clammy skin Increased respiratory rate	Secure airway Assist ventilations Administer high-flow oxygen Control external bleeding Consider elevating legs Keep warm Transport promptly
Respiratory insufficiency	Severe chest injury, airway obstruction	Rapid, weak pulse Low blood pressure Change in mental status Cyanosis (lips, under nails) Cool, clammy skin Increased respiratory rate	Secure airway Clear air passages Assist ventilations Administer high-flow oxygen Transport promptly

calm reassurance. Frequently checking for a pulse in an unresponsive patient is important to identify early whether an automated external defibrillator is needed.

Treating Obstructive Shock

As discussed previously, two of the most common examples of obstructive shock are cardiac tamponade and tension pneumothorax.

Increasing cardiac output should be the priority in treating cardiac tamponade. As the heart is being squeezed by the increasing pressure in the pericardium, the preload must be increased. Apply high-flow oxygen. Oxygen should never be withheld from a patient who needs it; however, you must weigh the need for positive-pressure ventilations against the possibility of hypoventilation. The only definitive treatment for cardiac tamponade is surgery; however, pericardiocentesis, which involves penetrating the pericardium with a needle and withdrawing the accumulated blood from the pericardial sac, is the only practical advanced life support (ALS) prehospital approach. This procedure is rarely performed in the field, so early recognition along with rapid transport or ALS management, if available, is the key treatment available to EMT providers.

In treating tension pneumothorax, high-flow oxygen via nonrebreathing mask should be applied early to prevent hypoxia. Be cautious about providing positive-pressure ventilation to a patient with a tension pneumothorax, as the increase of air will increase the pressure in the chest. Usually the only action that can prevent eventual death from a tension pneumothorax is decompression of the injured side of the chest, relieving the pressure in the chest and allowing the heart to expand fully again. Needle chest decompression is a skill many ALS providers are allowed to perform. Rapid transport or ALS management, if available, is the key treatment available to EMT providers.

Treating Septic Shock

The proper treatment of septic shock requires complex hospital management, including antibiotics. If you suspect that a patient has septic shock, you must use appropriate standard precautions and transport as promptly as possible. Use high-flow oxygen during transport. Ventilatory support may be necessary to maintain adequate tidal volume. Use blankets to conserve body heat.

Treating Neurogenic Shock

Shock that accompanies spinal cord injury is best treated by a combination of all known supportive measures. The patient who has sustained this kind of injury usually will require hospitalization for a long time.

Emergency treatment must be directed at obtaining and maintaining a proper airway, providing spinal immobilization, assisting inadequate breathing as needed, conserving body heat, and providing the most effective circulation possible.

This patient usually is not losing blood. However, the capacity of his or her blood vessels has become significantly larger than the volume of blood they contain. Slight elevation of the foot end of the spine board may help bring the blood that is pooling in the vessels of the legs to the vital organs. Placing the patient's arms across his or her chest without moving the spine will also return some pooled blood. Be sure to monitor the patient for breathing problems, and, if they appear, lower the spine board. Supplemental oxygen will boost the concentration of oxygen in the blood. If respirations are weak or inadequate, provide assisted ventilations. Keep the patient as warm as possible with blankets, because the injury may have disabled the body's normal temperature controls. Transport promptly.

■ Treating Anaphylactic Shock

The only really effective treatment for a severe, acute allergic reaction is to administer epinephrine by way of subcutaneous or intramuscular injection. For more information on the emergency care for allergic reactions, see Chapter 18, *Immunologic Emergencies*. A patient who is aware of having a specific sensitivity may carry a bee-sting kit containing epinephrine Figure 10-6 . If he or she is unable to inject the medication, you may have to do so if you are allowed by local protocol. If the patient's signs and symptoms recur or the patient's condition deteriorates, you should repeat the injection after consulting with medical control.

Promptly transport the patient to the emergency department while providing all possible support, primarily supplemental oxygen and ventilatory assistance. You should also try to find out what agent caused the reaction (for example, a drug, an insect bite or sting, a food item) and how it was received (for example, by mouth, by inhalation, or by injection). The severity of allergic reactions can vary greatly, with symptoms ranging from mild itching to profound coma and rapid death. Keep in mind that a mild reaction may worsen suddenly or over time. Consider requesting ALS backup, if available.

■ Treating Psychogenic Shock

In an uncomplicated case of fainting, once the patient collapses and becomes supine, circulation to the brain is usually restored and with it, a normal state of functioning. Remember that psychogenic shock can significantly

You are the Provider: PART 5

You ask your partner to call ahead to the hospital because you are busy caring for the patient and cannot free up your hands. The noninvasive automatic vital sign machine records another set of vital signs. With an estimated time of arrival at the hospital of 5 minutes, you reassess the patient.

Recording Time: 16 Minutes	
Level of consciousness	Responsive to pain only
Respirations	30 breaths/min and shallow (baseline); ventilations are being assisted
Pulse	128 beats/min; absent radial pulses (carotid pulse present)
Skin	Pale, cool, and diaphoretic
Blood pressure	80/40 mm Hg
Sao$_2$	96% (with assisted ventilation; on oxygen)

You arrive at the hospital and give your report to the charge nurse. Intravenous access is rapidly obtained, the patient is quickly assessed by the attending physician, and additional treatment is given.

8. What part of the patient's perfusion triangle has failed?

9. How does shock caused by content failure differ from shock caused by container failure?

Figure 10-6 Patients who are allergic to bee stings often carry commercial bee-sting kits, such as an intramuscular injector or auto-injector, containing epinephrine.

worsen other types of shock. If the attack has caused the patient to fall, you must check for injuries, especially in older patients. However, you should also assess the patient thoroughly for any other abnormality. If, after regaining consciousness, the patient is unable to walk without weakness, dizziness, or pain, you should suspect another problem, such as head injury. You should transport this patient promptly.

Be sure to record your initial observations of vital signs and level of consciousness. In addition, try to learn from bystanders whether the patient complained of anything before fainting and how long he or she was unconscious.

Treating Hypovolemic Shock

The emergency treatment of hypovolemic or hemorrhagic shock includes the control of all obvious external bleeding. To prevent continued bleeding, you must apply sufficient pressure to control obvious external bleeding, splint any bone and joint injuries, and ensure that you use great care to handle the patient gently. If there are no fractured extremities, head injuries, or chest injuries, consider raising the legs 6″ to 12″, keeping the torso in a horizontal position. This may increase blood flow to the heart from the lower body and will keep unwanted pressure off the diaphragm.

Although you cannot control internal bleeding in the field, you must recognize its existence and provide aggressive general support. Secure and maintain an airway, and provide respiratory support, including supplemental oxygen and, if needed, assisted ventilations. Start the oxygen as soon as you suspect shock, and continue it during transport; with too little circulating blood, additional oxygen may be lifesaving. Be sure the patient does not aspirate blood or vomitus. Most important, you must transport the patient as rapidly as possible to the emergency department.

Treating Respiratory Insufficiency

In treating the patient who is in shock as a result of inadequate respiration, you must immediately secure and maintain the airway. Clear the mouth and throat of anything obstructing the air passages, including mucus, vomitus, and foreign material. If necessary, provide ventilations with a bag-mask device. Give supplemental oxygen, and transport the patient promptly.

You are the Provider: SUMMARY

1. What additional information should you attempt to gather about the patient while en route to the clinic?

Information of "a woman going into shock" tells you very little. The patient could have a severe injury and is bleeding internally, she could be experiencing a severe allergic reaction, or she may have simply fainted.

When you are dispatched to *any* call and the information provided is minimal, attempt to gather additional information while you are en route. In many cases, the dispatch operator will provide additional patient information—without you asking for it—as it becomes available to him or her. In other cases, you will need to ask the dispatch operator to try to make contact with the caller to obtain a patient update. Many EMS systems use an emergency medical dispatcher (EMD); if this is the case, he or she should be able to provide you with more detailed information, as well as give pre-arrival instructions to the caller.

First and foremost, you should try to ascertain if the patient is conscious. If she is conscious, inquire about breathing and a pulse. If she is unconscious, not breathing, and pulseless, consider requesting additional assistance depending on your local protocol. Ask if the patient is sick, injured, or both. Do not assume that "a patient going into shock" means that the patient is injured; there are many nontraumatic causes of shock. Any additional information you attempt to obtain beyond the patient's ABCs will depend on your response time to the scene.

Information is knowledge; the more knowledge you have about your patient before you arrive at the scene, the better the position you will be in to "map out" a patient care plan with your partner. Obviously, however, you will not *ultimately* know what situation you have until you see the patient.

You are the Provider: SUMMARY, continued

2. What is shock and how does it relate to perfusion?

Before discussing shock, a review of perfusion is in order. Perfusion is the delivery of blood and oxygen and other essential nutrients to the body's cells to keep them alive. While delivering these essential components to the body's cells, waste products such as carbon dioxide are removed from the cell and eliminated from the body.

Adequate perfusion is the responsibility of the "perfusion triangle," which consists of three essential components—a functioning pump (the heart), adequate volume (the blood and water), and an intact container (the blood vessels). The respiratory system is also a critical component for adequate perfusion; if oxygen cannot get into the lungs, the heart cannot pump it through the blood vessels and to the cells.

Shock is a state of inadequate perfusion (hypoperfusion) of blood through the body's tissues and to the cells, and is the result of failure of one or more components of the perfusion triangle or the respiratory system. The "type" of shock that a patient is experiencing indicates the component of the perfusion triangle that has failed. For example, cardiogenic shock occurs when the heart is unable to effectively pump blood throughout the body, hypovolemic shock occurs when there is inadequate volume (blood or water) to deliver oxygen to the cells, and neurogenic shock occurs when the nerves that regulate the diameter of the blood vessels fail, resulting in widespread vasodilation (container failure).

Other types of shock are the result of failure of more than one component of the perfusion triangle or respiratory system. For example, anaphylactic shock—the result of a severe allergic reaction—is caused by failure of the respiratory system and blood vessels; massive amounts of histamines and leukotrienes that are released during a severe allergic reaction constrict the bronchioles, which impairs breathing and dilates the blood vessels, thus decreasing the pressure with which blood is delivered to the cells (container failure).

Regardless of the "type" of shock a patient is experiencing, the end result is the same—inadequate perfusion of the body's tissues and cells, which will lead to death if untreated.

3. Based on your assessment, does this patient require any changes in the treatment she is currently receiving?

The patient clearly has signs of shock—restlessness; tachypnea; tachycardia; weak radial pulses; and cool, pale, diaphoretic skin—and the treatment that has been provided thus far is essentially appropriate. Her legs are elevated, she is being kept warm, and she is receiving oxygen. However, patients with signs of shock need *high-flow* oxygen. She is presently receiving oxygen via a nasal cannula; this should be changed to a nonrebreathing mask with the flow rate set at 15 L/min.

Remember what shock is: inadequate perfusion to the body's cells. Based on her presentation—abdominal pain and rectal bleeding—you should suspect blood loss as the cause of her shock. Blood carries oxygen; if the blood volume decreases, so does the ability of oxygen to get to the cells. Providing a high concentration of oxygen will oxygenate the red blood cells that remain in the circulatory system.

Oxygenating the blood and circulating oxygenated blood are two related but very different processes. In addition to receiving high-flow oxygen—which will help oxygenate the patient's blood—she needs IV access and volume replacement to help circulate the oxygenated blood. The physician has been unable to obtain IV access, and since IV therapy is beyond the EMT's level of training, you must prepare the patient for immediate transport; she may require a blood transfusion at the hospital. If your transport time to the hospital will be prolonged, you should consider an intercept with a unit staffed by an AEMT or paramedic; they are trained to establish IV access.

4. How do the patient's signs and symptoms correlate with the body's response to inadequate perfusion?

During shock, the body mounts a physiologic response aimed at maintaining adequate perfusion; most of the physiologic compensation that occurs during shock is the result of increased activity of the sympathetic nervous system—the part of the nervous system that prepares the body to handle stress.

Receptors located throughout the body constantly monitor blood pressure as well as the levels of oxygen and carbon dioxide in the blood. Under normal conditions, these receptors are constantly receiving input and sending messages to the brain. The brain then sends signals to the sympathetic nervous system and respiratory muscles, where fine-tune adjustments are made to the respiratory rate, heart rate, and blood pressure.

If perfusion to the body decreases for any reason, the signals sent from the receptors to the brain and the signals sent from the brain to the nervous system and respiratory muscles are more intense, resulting in a predictable series of signs and symptoms that appear when a patient is in shock.

Restlessness, perhaps one of the earliest signs of shock, is caused by a decrease in oxygen to the brain. The brain does not know where the underlying problem is; it only knows that it is not receiving enough oxygen. As a result, the number of signals it sends to the respiratory muscles increases, which causes an increase in the patient's respiratory rate (tachypnea).

In shock, the sympathetic nervous system releases two hormones—epinephrine and norepinephrine—in greater quantities than normal. These hormones are responsible for the signs and symptoms of shock that are classically seen.

Epinephrine, also called adrenalin, causes an increase in heart rate and cardiac contractility; as a result, blood is pumped faster and with greater force throughout the body to compensate for decreased perfusion. Norepinephrine causes the blood vessels to constrict (vasoconstriction), thus maintaining the patient's blood pressure. Early in shock, blood is shunted from areas of lesser need (ie, the skin and muscles) to areas of greater need (ie, heart, lungs, liver, kidneys). When blood is shunted away from the skin by vasoconstriction, it turns pale (pallor) and becomes cool to the touch. When sympathetic nervous system activity increases, sweat gland activity increases as well, resulting in diaphoresis (profuse sweating).

5. Is the patient in compensated or decompensated shock? How can you tell?

In the early stages of shock, the body can still compensate for inadequate perfusion through the mechanisms previously discussed; therefore he or she is said to be in compensated shock. Signs of compensated shock, which your patient is currently experiencing, include restlessness, anxiety, or agitation; tachycardia; rapid, weak (thready) peripheral pulses; tachypnea; and marked thirst. In compensated shock, however, the patient's systolic blood pressure is maintained—usually above 90 mm Hg in adults. Your patient's current blood pressure is 108/58 mm Hg.

The next stage of shock—decompensated shock—is marked by a falling blood pressure. In decompensated shock, the body's compensatory mechanisms are no longer able to maintain adequate perfusion. It is important to note that the patient's blood pressure is often the last measurable factor to change in shock; by the time a low blood pressure (hypotension) is detected, shock is well developed. Other signs of decompensated shock include absent peripheral pulses; dilated pupils; ashen, mottled, or cyanotic skin; and a decreasing level of consciousness.

The last stage of shock—irreversible shock—is a terminal event. Widespread cellular death has occurred, which causes multiple organs to fail. Survival is unlikely—even with rapid transport and aggressive treatment at the hospital.

The key to treating shock and reducing mortality from it is recognizing it in its earliest possible stage, initiating immediate prehospital treatment, and rapidly transporting the patient to the hospital for definitive care.

6. How has your patient's condition changed?

Your patient's condition has changed for the worse. Her level of consciousness has decreased (responsive to pain only), her respirations have increased in rate and decreased in depth (reduced tidal volume), her heart rate has increased and her radial pulses are no longer palpable, her blood pressure is below 90 mm Hg (84/44 mm Hg), and her oxygen saturation has fallen—despite the administration of high-flow oxygen.

Based on these reassessment findings, your patient is now in decompensated shock with inadequate breathing. As previously discussed, decompensated shock occurs when the body's compensatory mechanisms begin to fail and are no longer able to maintain adequate perfusion. At the clinic, the patient had signs of shock; however, she was still conscious and alert, although restless, and her systolic blood pressure was maintained. Patients can decompensate within a matter of minutes; this fact underscores the criticality of frequent reassessments.

7. Are adjustments in your current interventions required? If so, what?

In terms of shock treatment, you are doing everything that you can. The patient is being kept warm with a blanket, her legs are elevated 6″ to 12″, and she is receiving high-flow oxygen.

An intercept with an ALS unit is clearly not practical; you are too close to the hospital. However, the patient's breathing is no longer adequate (30 breaths/min and shallow) and requires assistance.

Because of her decreased level of consciousness, you should insert a simple airway adjunct. In this case, a nasopharyngeal airway would be appropriate; although she is only responsive to pain, she likely has an intact gag reflex. Begin assisting her ventilations with a bag-mask device attached to high-flow oxygen and monitor her closely for signs of improvement or further deterioration.

Many EMS systems carry noninvasive blood pressure monitoring devices that automatically take the patient's blood pressure and other vital signs. If you have this capability, you should set the device to reassess the patient's vital signs *at least* every 5 minutes, or as deemed appropriate by the patient's condition. You are the only EMT in the back with the patient; managing her airway and assisting her ventilations clearly has priority over obtaining a manual blood pressure.

8. What part of the patient's perfusion triangle has failed?

There are three components to the perfusion triangle, each of which must function adequately at all times to maintain adequate perfusion: the heart (pump), the blood vessels (container), and the blood (content).

Recalling the patient's chief complaint—abdominal pain and rectal bleeding—she is in shock secondary to blood loss (hemorrhagic shock). Therefore, the content function of the perfusion triangle has failed. If enough blood or plasma is lost, internally or externally, the volume of fluid that remains in the container (blood vessels) will not be able to carry sufficient amounts of oxygen to the cells to adequately perfuse them.

9. How does shock caused by content failure differ from shock caused by container failure?

Shock caused by content failure refers to insufficient oxygen delivery to the cells because of inadequate volume, and is therefore called hypovolemic shock. Hypovolemia is a generic term that simply means low volume; it could be blood, plasma, water, or a combination. Common causes of hypovolemic shock include blunt or penetrating trauma, burns, and dehydration. Shock that is caused by blood loss specifically is called hemorrhagic shock.

Blood carries oxygen and oxygen is required to keep the body's tissues and cells alive. If a significant amount of blood is lost—internally or externally—the cells will not receive the amount of oxygen they need to survive. Furthermore, the waste products of oxygen metabolism (eg, carbon dioxide) that return to the heart and lungs via the blood for elimination from the body will build up in the cells and organs, leading to cell or organ death.

The signs of hypovolemic shock represent the sympathetic nervous system's compensatory response previously discussed, and include tachycardia; pale, cool, clammy skin; tachypnea; restlessness, agitation, or anxiety; and as a late sign, hypotension.

You are the Provider: SUMMARY, continued

Shock caused by container failure refers to inadequate perfusion because of excessive dilation of the blood vessels, resulting in a decrease in pressure within the circulatory system. Although the volume of blood has not changed, the container that it circulates within has increased; this is referred to as relative hypovolemia. In other words, relative to the now larger container, the normal volume of blood is insufficient to fill the system and provide adequate perfusion. Common causes of shock caused by container failure include anaphylaxis, overdose with drugs that suppress the nervous system (ie, narcotics), and spinal cord injury.

A unique type of shock caused by container failure occurs when the spinal cord is injured (spinal shock). The nerves that innervate the sympathetic nervous system arise from the spinal cord in the thoracic region. If the spinal cord is injured at or above this level, the body has lost a crucial compensatory mechanism—the sympathetic nervous system. Recall that it is the sympathetic nervous system that releases epinephrine and norepinephrine, which increases the heart rate, causes the heart to beat with greater force, and constricts the blood vessels—effects that are aimed at maintaining adequate perfusion. The classic signs of shock seen in patients with hypovolemia—specifically, tachycardia, pallor, and diaphoresis—are caused by sympathetic nervous system's effect on the heart and blood vessels. However, if the nervous system has no way of communicating with the body, these classic signs will be absent. Instead, the patient's skin is usually warm and dry and the heart rate is normal or low.

It is important to remember that regardless of the underlying cause of shock—pump failure, content failure, container failure, or a combination of these—inadequate tissue perfusion, and death if left untreated, is the end result.

EMS Patient Care Report (PCR)

Date: 7-21-09	**Incident No.:** 011009	**Nature of Call:** Woman in shock	**Location:** 1111 Cedar Hills Dr.	
Dispatched: 2022	**En Route:** 2023	**At Scene:** 2032	**Transport:** 2043	**At Hospital:** 2053	**In Service:** 2105

Patient Information

Age: 39 **Sex:** F **Weight (in kg [lb]):** 73 kg (160 lb)	**Allergies:** Codeine **Medications:** Amitiza, Bentyl **Past Medical History:** Irritable bowel syndrome **Chief Complaint:** Abdominal pain and rectal bleeding

Vital Signs

Time: 2036	**BP:** 108/58	**Pulse:** 120	**Respirations:** 24	**Sao$_2$:** 95%
Time: 2043	**BP:** 84/44	**Pulse:** 130	**Respirations:** 30	**Sao$_2$:** 89%
Time: 2048	**BP:** 80/40	**Pulse:** 128	**Respirations:** 30	**Sao$_2$:** 96%

EMS Treatment
(circle all that apply)

Oxygen @ 15 L/min via (circle one): NC ~~NRM~~ (switched to Bag-Mask Device)	(Assisted Ventilation)	(Airway Adjunct)	**CPR**
Defibrillation	**Bleeding Control**	**Bandaging**	**Splinting**	**Other:** (Blanket for warmth, elevation of the legs)

Narrative
9-1-1 dispatch to Cedar Hills Urgent Care Clinic for a "woman going into shock." Arrived on scene and found the patient, a 39-year-old female, lying supine on an exam table. She was conscious and alert, but restless. Her airway was patent and her breathing, although increased in rate, was producing adequate tidal volume. Her skin was cool, pale, and diaphoretic. According to the clinic physician, the patient presented 15 minutes prior to complaining of abdominal pain and rectal bleeding, which began approximately 24 hours ago. Prior to EMS arrival, clinic personnel applied oxygen at 4 L/min via nasal cannula, elevated the patient's legs, and applied a blanket. The physician advised that she attempted to establish IV access several times, but was unsuccessful. Further assessment of the patient revealed that her abdomen was diffusely tender to palpation, and she was actively bleeding from the rectum. Remainder of assessment was unremarkable. The patient denies abdominal trauma, and states that she has a history of irritable bowel syndrome. Removed nasal cannula, applied high-flow oxygen via nonrebreathing mask, obtained vital signs, and moved patient to the ambulance. Departed the scene and continued treatment en route. Reassessment revealed that the patient's mental status had decreased; she was now only responsive to painful stimuli. Her blood pressure had decreased, her radial pulses were absent (carotid pulse was present), her respirations were more rapid and shallow, and her Sao_2 had fallen to 89%. Inserted a nasopharyngeal airway and began assisting her ventilations with a bag-mask device attached to high-flow oxygen. Partner notified the receiving facility and gave radio report. Continued to assist patient's ventilations and monitor her vital signs. She remained hypotensive and tachycardic; however, her Sao_2 increased to 96% with assisted ventilation. Delivered patient to the emergency department staff without incident and gave verbal report to charge nurse. Medic 4 returned to service at 2105. **End of report**

Assessment and Emergency Care of Shock

Scene Size-up

Scene Safety	Ensure scene safety and address hazards. Standard precautions should include a minimum of gloves and eye protection. Consider the number of patients, the need for additional help/ALS, and cervical spine stabilization.
Mechanism of Injury (MOI)/ Nature of Illness (NOI)	Determine the MOI/NOI. Observe the scene and look for indicators of the MOI such as falls, motor vehicle crashes, gunshot wounds, or stabbings. Observe the patient for signs of NOI such as urticaria, chest pain, or fever.

▼ ▼

Primary Assessment

Form a General Impression	Determine level of consciousness and find and treat any immediate life threats. Observe overall appearance of patient and body position. Determine priority of care based on the MOI/NOI. If the patient has a poor general impression, call for ALS assistance. A rapid scan will help you identify and manage life threats.
Airway and Breathing	If a cervical spine injury is suspected, open the airway using a modified jaw-thrust and ensure the airway is patent. A patient with an altered level of consciousness may need emergency airway management; consider inserting a properly sized oropharyngeal or nasopharyngeal airway. Assess for gurgling, stridor, or rales. Quickly assess the chest for DCAP-BTLS, accessory muscle use, intercostal and abdominal muscle use, and treat any threats to life. Provide high-flow oxygen at 15 L/min and evaluate the depth and rate of the respiratory cycle, providing ventilatory support as needed.
Circulation	Evaluate distal pulse rate and quality; observe skin color, temperature, and condition; look for life-threatening bleeding and treat accordingly. If distal pulses are not palpable, assess for a central pulse. Place the patient in a supine position. Consider the Trendelenburg's or shock position if no contraindications exist. Prevent heat loss by using blankets. Serious bleeding must be treated at once.
Transport Decision	If the patient has an airway or breathing problem, significant external bleeding, or signs and symptoms of internal bleeding, consider rapid transport. Suspected shock patients or those with a suspicious MOI should go to a trauma center. ALS providers can treat these patients with intravenous fluids to support circulation (shock) problems. If anaphylactic shock is suspected, determine if the patient has a prescribed EpiPen auto-injector before leaving the scene. Do not delay transport to manage non-life-threatening injuries; instead treat en route to the hospital.

▼ ▼

NOTE: The order of the steps in this section differs depending on whether the patient is conscious or unconscious. The following order is for a conscious patient. For an unconscious patient, perform a primary assessment, perform a rapid full-body scan, obtain vital signs, and if possible, obtain the past medical history from a family member, bystander, or emergency medical identification device.

History Taking

Investigate Chief Complaint	Investigate the chief complaint. Identify signs and symptoms and pertinent negatives. Be alert for injuries and life threats. Observe for signs and symptoms of shock. Monitor patient for change in mental status. If possible, ask OPQRST and SAMPLE questions.

▼ ▼

Secondary Assessment

Physical Examinations	Perform a systematic full-body scan beginning with the head, looking for DCAP-BTLS. Assessment should be rapid if the patient has a poor general impression. Inspect, palpate, and auscultate the chest, focusing on the respiratory effort and adequacy of ventilation. Perform a thorough neurologic examination assessing the pupils, motor response, and sensory response in all extremities. Assess the musculoskeletal system for DCAP-BTLS. Assess the abdomen for signs of internal bleeding. Log roll and inspect the posterior torso for injuries.
Vital Signs	Obtain baseline vital signs, monitoring trends. Repeat every 5 to 15 minutes depending on patient impression. Vitals signs should include blood pressure, pulse rate and quality, respiration rate and quality, and skin assessment for perfusion. Note patient's level of consciousness. Use pulse oximetry, if available, to assess the patient's perfusion status.

Reassessment

Interventions	Consider cervical spine precautions. Repeat the primary assessment and reassess vital signs and chief complaint. Check interventions and treatment rendered. Airway control using adjuncts may be necessary. Assist breathing as required, administering high-flow oxygen. Control all bleeding and circulation problems. Support the cardiovascular system and treat for shock early. Consider placing patients who are on a backboard in the Trendelenburg's position by elevating the board 6″ to 12″ if no contraindications, such as head injury or chest injury, are present. Prevent body heat loss by placing blankets under and over the patient. Splint bone or joint injuries. Do not give the patient anything by mouth. Do not delay transport.
Communication and Documentation	Contact medical control with a radio report. Include a thorough description of the MOI/NOI and the position the patient was found in. Include treatments performed and patient response. Follow local protocols. Be sure to document any changes in patient status and the time. Document the reasoning for your treatment and the patient's response.

NOTE: Although the steps below are widely accepted, be sure to consult and follow your local protocols.

Assessment and Emergency Care of Shock, continued

Shock

Treating Cardiogenic Shock

This type of shock is a failure of the pump (heart) and is often the result of a myocardial infarction. Since the heart is no longer an effective pump, fluid backs up in the body and the lungs.

1. Assess ABCs. Patient will often have rales (fluid in the lungs).
2. Patient is often complaining of chest pain.
3. Administer high-flow oxygen via a nonrebreathing mask.
4. Place the patient in a sitting or semi-sitting position to assist breathing.
5. Do not administer nitroglycerin if blood pressure is low; contact medical control.
6. Keep the patient calm, request ALS if available, and transport promptly.
7. Keep alert for the need to assist ventilation, perform cardiopulmonary resuscitation, or defibrillate.

Treating Obstructive Shock

This type of shock is usually caused by cardiac tamponade or tension pneumothorax. Patient requires management by ALS providers or more complex management at the hospital.

1. Request ALS.
2. In treating cardiac tamponade, weigh the need for positive-pressure ventilations against the possibility of hypoventilation. In treating tension pneumothorax, high-flow oxygen should be applied early to prevent hypoxia.
3. Prompt transport to the closest emergency department is essential.

Treating Septic Shock

A systemic infection causes the blood vessels to become leaky and dilate, causing the container to enlarge. Patient requires complex management in the hospital.

1. Assess and manage life threats to the ABCs.
2. Administer high-flow oxygen.
3. Prevent heat loss.
4. Transport as promptly as possible.

Assessment and Emergency Care of Shock, continued

Shock

Treating Neurogenic Shock

This type of shock involves an injury to the central nervous system, causing the patient's blood vessels to dilate (container gets bigger). Even though the blood pressure drops, there is no blood loss.

1. Suspect neurogenic shock if the MOI is suspicious.
2. Maintain cervical spine stablization and airway control with a modified jaw-thrust.
3. Provide oxygen and assist breathing as necessary.
4. Provide spinal immobilization.
5. Consider elevating the foot end of the backboard slightly to help move blood into the vital organs if no contraindications, such as head injury or chest injury, exist.
6. Prevent body heat loss.
7. Transport promptly to a trauma center.

Treating Anaphylactic Shock

Severe allergic reactions can rapidly progress to anaphylactic shock. The body's response to the allergen causes widespread vasodilation.

1. Request ALS.
2. Be prepared to assist the patient with their prescribed epinephrine auto-injector.
3. Oxygenate and ventilate the patient as necessary.
4. Prompt transport to the closest emergency department is essential.

Treating Psychogenic Shock

A sudden reaction of the nervous system causes a temporary vasodilation, resulting in syncope (fainting). It is important to investigate other possible causes of the syncopal episode. Once supine, the patient regains consciousness.

1. Perform a thorough primary assessment to identify any possible life-threatening causes of the syncope.
2. Administer oxygen, elevate the legs, and prevent heat loss.
3. Perform a secondary assessment to identify any injuries that may have occurred when the patient collapsed.
4. Patient should be transported for evaluation.

Assessment and Emergency Care of Shock, continued

Shock

Treating Hypovolemic Shock

This type of shock is caused by a loss of blood or body fluids. It should be suspected first whenever a patient presents with signs and symptoms of shock. Blood loss may be external or internal secondary to a traumatic injury. Body fluids can be lost due to burns, excessive vomiting, or diarrhea.

1. Management of a patient with hypovolemic shock focuses on preventing further blood or fluid loss.
2. Manage threats to the ABCs.
3. Control external bleeding with direct pressure, pressure dressings, and tourniquets.
4. Internal bleeding is difficult to manage. Splinting injured extremities may slow blood loss.
5. The Trendelenburg's or shock position may be used to assist with perfusion if no contraindications exist.
6. High-flow oxygen should be administered all hypovolemic patients.
7. Prompt transport to a trauma center is required. Do not delay transport.

Treating Respiratory Insufficiency

Patients in shock as a result of respiratory insufficiency require immediate airway maintenance and oxygen.

1. Clear obstructions and suction airway as required.
2. Give supplemental oxygen and assist ventilations if necessary.
3. Transport promptly.

Prep Kit

Ready to Review

- Perfusion requires an intact cardiovascular system and a functioning respiratory system.

- Remember, most types of shock (hypoperfusion) are caused by dysfunction in one or more parts of the perfusion triangle:
  - The pump (the heart)
  - The pipes, or container (blood vessels)
  - The content, or volume (blood)

- Shock (hypoperfusion) is the collapse and failure of the cardiovascular system, when blood circulation slows and eventually stops.

- Blood is the vehicle for carrying oxygen and nutrients through the vessels to the capillary beds to tissue cells, where these supplies are exchanged for waste products.

- Blood contains red blood cells, white blood cells, platelets, and a liquid called plasma.

- The *systolic* pressure is the peak arterial pressure, or pressure generated every time the heart contracts; the *diastolic* pressure is the pressure maintained within the arteries while the heart rests between heartbeats.

- The various types of shock are cardiogenic, obstructive, septic, neurogenic, anaphylactic, psychogenic, and hypovolemic.

- Signs of compensated shock include anxiety or agitation; tachycardia; pale, cool, moist skin; increased respiratory rate; nausea and vomiting; and increased thirst. If there is any question on your part, treat for shock. It is never wrong to treat for shock.

- Signs of decompensated shock include labored or irregular respirations, ashen gray or cyanotic skin color, weak or absent distal pulses, dilated pupils, and profound hypotension.

- Remember, by the time a drop in blood pressure is detected, shock is usually in an advanced stage.

- Anticipate shock in patients who may have the following conditions:
  - Severe infection
  - Significant blunt force trauma or penetrating trauma
  - Massive external bleeding or index of suspicion for major internal bleeding
  - Spinal injury
  - Chest or abdominal injury
  - Major heart attack
  - Anaphylaxis

- Treating a pediatric or geriatric patient in shock is no different than treating any other shock patient.

- Treat all patients suspected to be in shock from any cause as follows and in this order:
  - Open and maintain the airway.
  - Provide high-flow oxygen and as needed, provide bag-mask assisted ventilations.
  - Control all obvious external bleeding.
  - Consider placing the patient in the shock position or, if on a backboard or stretcher, in the Trendelenburg's position as long as no contraindications exist.
  - Maintain normal body temperature with blankets.
  - Provide prompt transport to the appropriate hospital.

Vital Vocabulary

afterload The force or resistance against which the heart pumps.

anaphylactic shock Severe shock caused by an allergic reaction.

anaphylaxis An unusual or exaggerated allergic reaction to foreign protein or other substances.

aneurysm A swelling or enlargement of a part of an artery, resulting from weakening of the arterial wall.

autonomic nervous system The part of the nervous system that regulates involuntary functions, such as heart rate, blood pressure, digestion, and sweating.

cardiogenic shock Shock caused by inadequate function of the heart, or pump failure.

compensated shock The early stage of shock, in which the body can still compensate for blood loss.

cyanosis Bluish color of the skin resulting from poor oxygenation of the circulating blood.

decompensated shock The late stage of shock when blood pressure is falling.

dehydration Loss of water from the tissues of the body.

distributive shock A condition that occurs when there is widespread dilation of the small arterioles, small venules, or both.

edema The presence of abnormally large amounts of fluid between cells in body tissues, causing swelling of the affected area.

homeostasis A balance of all systems of the body.

hypothermia A condition in which the internal body temperature falls below 95°F (35°C), usually as a result of prolonged exposure to cool or freezing temperatures.

hypovolemic shock Shock caused by fluid or blood loss.

irreversible shock The final stage of shock, resulting in death.

myocardial contractility The ability of the heart muscle to contract.

neurogenic shock Circulatory failure caused by paralysis of the nerves that control the size of the blood vessels, leading to widespread dilation; seen in patients with spinal cord injuries.

obstructive shock Shock that occurs when there is a block to blood flow in the heart or great vessels, causing an insufficient blood supply to the body's tissues.

perfusion Circulation of blood within an organ or tissue in adequate amounts to meet the cells' current needs.

preload The precontraction pressure in the heart as the volume of blood builds up.

psychogenic shock Shock caused by a sudden, temporary reduction in blood supply to the brain that causes fainting (syncope).

sensitization Developing a sensitivity to a substance that initially caused no allergic reaction.

septic shock Shock caused by severe infection, usually a bacterial infection.

shock A condition in which the circulatory system fails to provide sufficient circulation to enable every body part to perform its function; also called hypoperfusion.

sphincters Circular muscles that encircle and, by contracting, constrict a duct, tube, or opening.

syncope Fainting.

Assessment in Action

You are dispatched to a person who is complaining of chest pain and shortness of breath. You arrive to find a conscious 58-year-old female who is sitting up and complaining of severe chest pain and shortness of breath. Physical examination shows that her skin is cool and clammy and her pulse is rapid, weak, and irregular. Her breathing is labored, with a respiratory rate of 24 breaths/min. Lung sounds show rales (crackles) in all fields. Blood pressure is 90/60 mm Hg.

1. What type of shock is this person experiencing?
   - A. Hypovolemic
   - B. Anaphylactic
   - C. Cardiogenic
   - D. Neurogenic

2. Define shock.

3. What stage of shock is this patient in?
   - A. Compensated
   - B. Decompensated
   - C. Irreversible
   - D. Emotional

4. What treatment should be initiated first for this patent?
   - A. High-flow oxygen
   - B. Administration of nitroglycerin
   - C. Rapid transport
   - D. Maintain warmth

5. The patient is having trouble breathing. What position should the patient be placed in?
   - A. Supine
   - B. Fowler's position
   - C. Trendelenburg's position
   - D. Left lateral recumbent

6. Define irreversible shock.

7. What part of the patient's cardiovascular system is failing?
   - A. The fluid (blood)
   - B. The container (vessels)
   - C. The pump (heart)
   - D. The control (nervous system)

8. Define cardiogenic shock.

9. Which of the following terms is used to describe a balance of all body systems?
   - A. Hypothermia
   - B. Autonomic nervous system
   - C. Perfusion
   - D. Homeostasis

10. Define perfusion.

CHAPTER

11

BLS Resuscitation

National EMS Education Standard Competencies

Shock and Resuscitation

Applies a fundamental knowledge of the causes, pathophysiology, and management of shock, respiratory failure or arrest, cardiac failure or arrest, and post resuscitation management.

Knowledge Objectives

1. Explain the elements of basic life support (BLS), how it differs from advanced life support (ALS), and the urgency surrounding its rapid application. (pp 413-414, 416-417)
2. Explain the goals of cardiopulmonary resuscitation (CPR) and when it should be performed on a patient. (p 414)
3. Explain the system components of CPR, the four links in the American Heart Association chain of survival, and how each one relates to maximizing the survival of a patient. (pp 414-415)
4. Discuss guidelines for circumstances that require the use of and automated external defibrillator (AED) on both adult and pediatric patients experiencing cardiac arrest. (p 415)
5. Explain three special situations related to the use of automated external defibrillation. (p 416)
6. Describe the proper way to position an adult patient to receive basic life support. (pp 417-418)
7. Describe the two techniques an EMT may use to open an adult patient's airway and the circumstances that would determine when each technique would be used. (pp 419-421)
8. Describe the purpose of external chest compressions. (pp 418-419)
9. Describe the recovery position and circumstances that would warrant its use as well as situations in which it would be contraindicated. (p 421-422)
10. Describe the process of providing artificial ventilations to an adult patient using a barrier device, ways to avoid gastric distention, and modifications required for a patient with a stoma. (pp 422-424)
11. Explain the steps in providing one-rescuer adult CPR. (pp 424-426)
12. Explain the steps in providing two-rescuer adult CPR, including the method for switching positions during the process. (pp 424-428)
13. Describe the different mechanical devices that are available to assist emergency responders in delivering improved circulatory efforts during CPR. (pp 428-429)
14. Describe the different possible causes of cardiopulmonary arrest in children. (pp 429-430)
15. Explain the four steps of pediatric basic life support (BLS) procedures and how they differ from procedures used in an adult patient. (pp 429-435)
16. Describe the ethical issues related to patient resuscitation, providing examples of when not to start CPR on a patient. (pp 436-437)
17. Explain the various factors involved in the decision to stop CPR once it has been started on a patient. (p 436-437)

18. Explain common causes of foreign body airway obstruction in both children and adults and how to distinguish mild or partial airway obstruction from complete airway obstruction. (pp 437-441)
19. Describe the different methods for removing a foreign body airway obstruction in an infant, child, and adult, including the procedure for a patient with an obstruction who becomes unconscious. (pp 438-443)

Skills Objectives

1. Demonstrate how to reposition an unconscious adult for airway management. (pp 417-418, Skill Drill 11-1)
2. Demonstrate how to check for a pulse at the carotid artery in an unresponsive patient. (p 417)
3. Demonstrate how to perform external chest compressions in an adult. (pp 419-420, Skill Drill 11-2)
4. Demonstrate how to perform a head tilt-chin lift maneuver on an adult patient. (pp 419-421)
5. Demonstrate how to perform a jaw-thrust maneuver on an adult patient. (p 420-421)
6. Demonstrate how to place a patient in the recovery position. (pp 421-422)
7. Demonstrate how to perform rescue breathing in an adult with a simple barrier device. (pp 422-423)
8. Demonstrate how to perform one-rescuer adult CPR. (pp 424-426, Skill Drill 11-3)
9. Demonstrate how to perform two-rescuer adult CPR. (pp 424-428, Skill Drill 11-4)
10. Demonstrate how to perform a head tilt-chin lift maneuver on a pediatric patient. (p 434)
11. Demonstrate how to perform a jaw-thrust maneuver on a pediatric patient. (p 434)
12. Demonstrate the use of mechanical devices that assist emergency responders in delivering improved circulatory efforts during CPR. (pp 428-429)
13. Demonstrate how to perform rescue breathing on a child. (pp 434-435)
14. Demonstrate how to perform rescue breathing on an infant. (pp 432-433)
15. Demonstrate how to perform external chest compressions on an infant. (pp 431-433, Skill Drill 11-5)
16. Demonstrate how to perform CPR in a child who is between 1 year of age and the onset of puberty. (pp 432-433, Skill Drill 11-6)
17. Demonstrate how to remove a foreign body airway obstruction in a conscious adult patient using abdominal thrusts (Heimlich maneuver). (p 439)
18. Demonstrate how to remove a foreign body airway obstruction in a conscious pregnant or obese patient using chest thrusts. (pp 439-440)
19. Demonstrate how to remove a foreign body airway obstruction in a conscious child older than 1 year using abdominal thrusts (Heimlich maneuver). (pp 440-441)
20. Demonstrate how to remove a foreign body airway obstruction in an unconscious child. (pp 441-443, Skill Drill 11-7)
21. Demonstrate how to remove a foreign body airway obstruction in an infant. (pp 441-442)

Introduction

The principles of basic life support (BLS) were introduced in 1960. Since then, the specific techniques have been reviewed and revised every 5 to 6 years. The updated guidelines are published in peer-reviewed journals: *Circulation* in the United States and *Resuscitation* in Europe. The most recent revision occurred as a result of the 2010 Conference on Cardiopulmonary Resuscitation and Emergency Cardiac Care. The information in this chapter follows the 2010 guidelines and presents a review of BLS.

This chapter begins with a definition and general discussion of BLS. The chapter then discusses methods for opening and maintaining an airway, providing artificial ventilation to a person who is not breathing, providing artificial circulation to a person with no pulse, and removing a foreign body airway obstruction. Each of these topics is followed by a review of the changes in technique that are necessary to treat infants and children. A discussion of the methods of preventing the transmission of infectious diseases during cardiopulmonary resuscitation (CPR) is provided in Chapter 2, *Workforce Safety and Wellness*. A discussion of the anatomy and physiology of the respiratory and cardiovascular systems can be found in Chapter 5, *The Human Body*.

Words of Wisdom

Although your chances of contracting a disease during CPR training or actual CPR on a patient are very low, common sense and Occupational Safety and Health Administration (OSHA) guidelines both demand that you take reasonable precautions to prevent unnecessary exposure to infectious disease. Using standard precautions makes the risk of contracting disease from CPR extremely low.

Elements of BLS

Basic life support (BLS) is noninvasive emergency lifesaving care that is used to treat medical conditions, including airway obstruction, respiratory arrest, and cardiac arrest. This care focuses on what is often termed the ABCs: airway (obstruction), breathing (respiratory arrest), and circulation (cardiac arrest or severe bleeding). If cardiac arrest is suspected, the order becomes CAB because chest compressions are essential and must be started as quickly as possible **Figure 11-1**. BLS follows a specific sequence for adults and for infants and children. Ideally, only

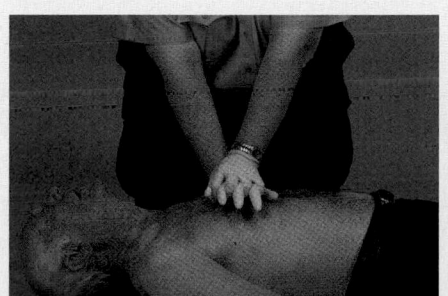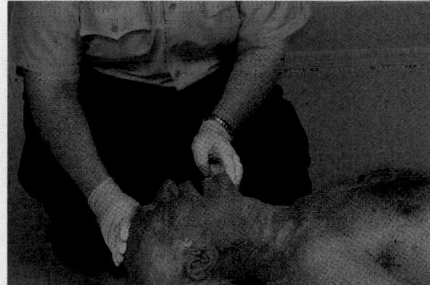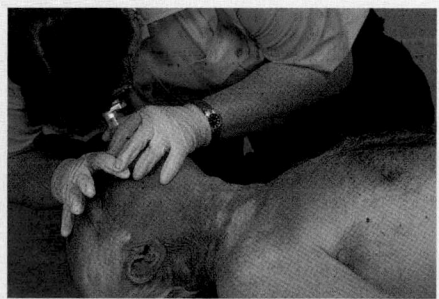

Figure 11-1 CAB: Chest compressions, airway, and breathing.

You are the Provider: PART 1

At 2:45 PM, you and your partner respond to a local supermarket where a middle-aged man reportedly collapsed in the parking lot. While you are en route to the scene, dispatch advises you that bystander CPR is in progress. Your response time is less than 5 minutes.

1. **What should you immediately do on receiving this update from dispatch?**
2. **What should be your initial actions on arriving at this scene?**

seconds should pass between the time you recognize that a patient needs BLS and the start of treatment. Remember, brain cells die every second that they are deprived of oxygen. Permanent brain damage is possible if the brain is without oxygen for 4 to 6 minutes. After 6 minutes without oxygen, brain damage is likely Figure 11-2 .

If a patient is not breathing well or at all, you may simply need to open the airway. Very often, this helps the patient to breathe normally again. However, if the patient has no pulse, you must combine artificial ventilation with artificial circulation beginning with compressions. If breathing stops before the heart stops, the patient will have enough oxygen in the lungs to stay alive for several minutes. But when cardiac arrest occurs first, the heart and brain stop receiving oxygen immediately.

Cardiopulmonary resuscitation (CPR) is used to establish artificial ventilation and circulation in a patient who is not breathing and has no pulse. The steps for CPR include the following:

1. First, restore circulation by means of chest compressions to circulate blood through the body.
2. After performing 30 high-quality compressions at least 2″ deep in an adult and at the rate of at least 100 per minute, open the airway with the jaw-thrust or head tilt–chin lift maneuver.
3. Last, restore breathing by means of rescue breathing (mouth-to-mouth ventilation, mouth-to-nose ventilation, or the use of mechanical ventilation devices).

The goal of CPR is to restore spontaneous breathing and circulation; however, advanced procedures such as medications and defibrillation are often necessary for this to occur. For CPR to be effective, you must be able to easily identify a patient who is in respiratory and/or cardiac arrest and immediately begin BLS measures Figure 11-3 .

BLS differs from **advanced life support (ALS)**, which involves advanced lifesaving procedures, such as

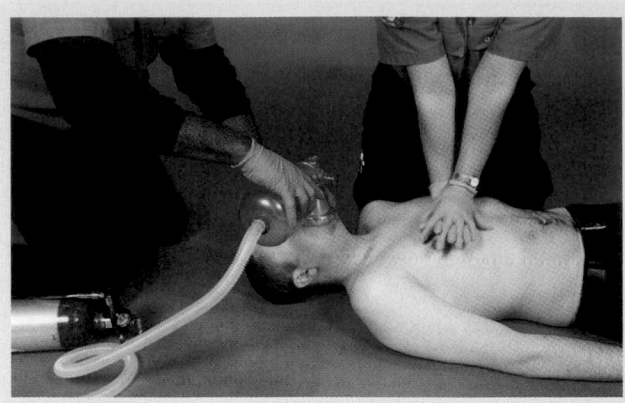

Figure 11-3 You must quickly identify patients in respiratory and/or cardiac arrest so that BLS measures can begin immediately.

cardiac monitoring, administration of intravenous fluids and medications, and use of advanced airway adjuncts. However, when done correctly, BLS can maintain life for a short time until ALS measures can be started. In some cases, such as choking, near drowning, or lightning injuries, early BLS measures may be all that is needed to restore a patient's pulse and breathing. Of course, these patients still require transport to the hospital for evaluation.

The BLS measures are only as effective as the person who is performing them. Whereas your skills may be very good immediately after training, as time goes on, skills will deteriorate unless you practice them regularly.

The System Components of CPR

Chain of Survival

Few people who experience cardiac arrest in the prehospital environment survive unless a rapid sequence of events takes place. The American Heart Association has determined an ideal sequence of events, termed the chain of survival, that if taken can improve the chance of successful resuscitation of a patient who has an occurrence of sudden cardiac arrest Figure 11-4 . The five links in the chain of survival are as follows:

- **Early access.** This link requires public education and awareness in the recognition of early warning signs of a cardiac emergency and immediate activation of EMS. "Access EMS by calling 9-1-1" is the first step in the chain so that emergency responders are dispatched to the scene quickly, thus allowing the other links of the chain to occur. In addition, 9-1-1 prearrival instructions can be given on notification and the emergency dispatcher can direct the caller to provide CPR compressions as needed.

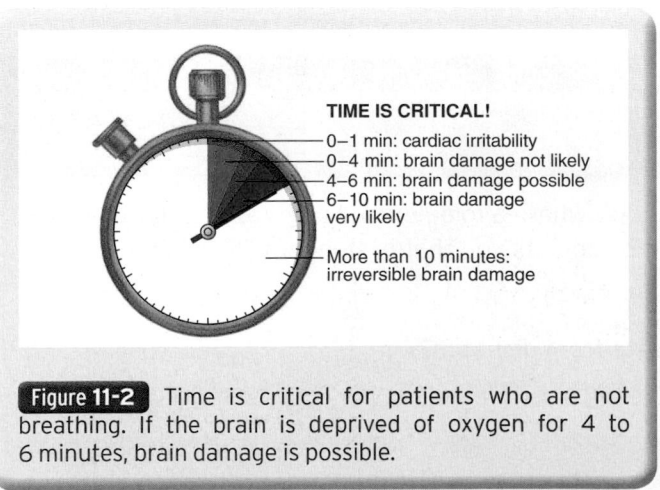

TIME IS CRITICAL!

0–1 min: cardiac irritability
0–4 min: brain damage not likely
4–6 min: brain damage possible
6–10 min: brain damage very likely

More than 10 minutes: irreversible brain damage

Figure 11-2 Time is critical for patients who are not breathing. If the brain is deprived of oxygen for 4 to 6 minutes, brain damage is possible.

- **Early CPR.** Immediate bystander CPR is essential for successful resuscitation of a person in cardiac arrest. CPR will keep blood, and therefore oxygen, flowing to the vital organs to keep them alive until the other components of the chain are available. The more people trained in CPR in the community, the better the chances of CPR being administered quickly to a person in cardiac arrest. Research shows that quick initiation of CPR can double to triple the chances of survival. The lay public as well as emergency responders should all be trained in CPR. In addition, all family members should be trained and ready to provide CPR if the need ever arises because 75% of sudden cardiac arrests occur in the home.

- **Early defibrillation.** Of all the links, early defibrillation offers the best opportunity to achieve a successful patient outcome. Automated external defibrillators (AEDs) have become readily available in many schools, fitness clubs, concert venues, sports arenas, and government buildings.

- **Early advanced care.** Advanced cardiac life support includes advanced airway placement, manual defibrillation, intravenous (IV) or intraosseous (IO) access, and administration of medications, which together increase the chance of a successful resuscitation.

- **Integrated post-arrest care.** The final step in the chain of survival is integrated post-arrest care. This refers to controlling temperature to optimize neurologic recovery in the field and maintaining glucose levels in the patient who is hypogylcemic. It also includes cardiopulmonary and neurologic support at the hospital, therapeutic hypothermia, percutaneous coronary interventions when indicated, and an electroencephalogram to detect seizure activity.

If any one of the links in the chain is absent, the patient is more likely to die. For example, few patients survive cardiac arrest if CPR is not administered within the first few minutes of the arrest. Likewise, if the time from cardiac arrest to defibrillation is more than 10 minutes, the chance of survival is minimal. The best chance of survival occurs when all links in the chain are strong.

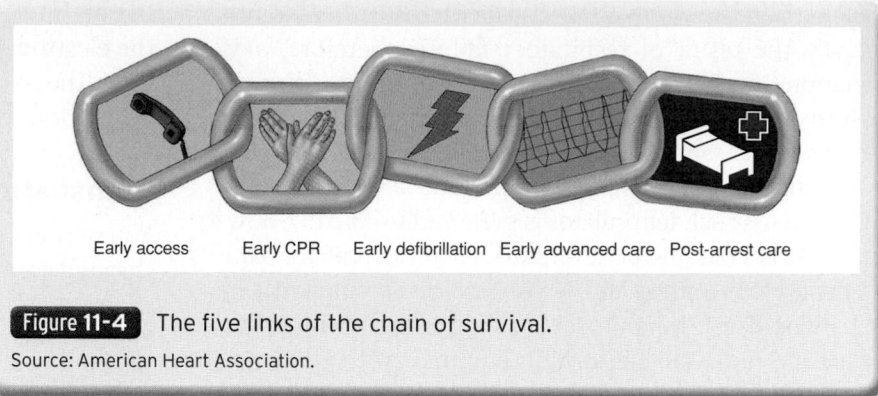

Figure 11-4 The five links of the chain of survival.
Source: American Heart Association.

Early access · Early CPR · Early defibrillation · Early advanced care · Post-arrest care

Automated External Defibrillation

Most prehospital cardiac arrests occur as the result of a sudden cardiac rhythm disturbance (arrhythmia), such as ventricular fibrillation (V-fib) or pulseless ventricular tachycardia (V-tach). The normal heart rhythm is known as normal sinus rhythm. Ventricular fibrillation is the disorganized twitching of the ventricles, resulting in no blood flow and a state of cardiac arrest. Ventricular tachycardia is a rapid contraction of the ventricles that does not allow for normal filling. As mentioned previously, according to the American Heart Association, early defibrillation is the link in the chain of survival that is most likely to improve survival rates. For each minute the patient remains in V-fib or pulseless V-tach, there is a 7% to 10% less chance of survival.

The automated external defibrillator (AED) should be applied to any cardiac arrest patient as soon as it is available. If indicated, defibrillation should be performed immediately. The simple design of the AED makes it easy for emergency medical responders and lay persons to use with very little training.

When a patient is in cardiac arrest, you should begin CPR starting with chest compressions and apply an AED as soon as it is available. Chapter 14, *Cardiovascular Emergencies*, covers AED use in detail.

Words of Wisdom

When operating an AED, make sure that no one is injured, including yourself. Be sure no one is touching the patient. Do not defibrillate a patient who is in pooled water. Do not defibrillate someone who is touching metal that others are touching. Finally, carefully remove any medication patches from a patient's chest with your gloved hands, and wipe the area with a dry towel before defibrillation to prevent ignition of the patch.

AED Usage in Children

AEDs can safely be used in children using the pediatric-sized pads and a dose-attenuating system (energy reducer). However, if these are unavailable, you should use an adult AED. During CPR, the AED should be applied to infants or children after the first five cycles of

CPR have been completed. Cardiac arrest in children is usually the result of respiratory failure; therefore, oxygenation and ventilation are vitally important. After the first five cycles of CPR, the AED should be used to deliver shocks in the same manner as with an adult patient.

If the child is between 1 month and 1 year of age (an infant), a manual defibrillator is preferred to an AED; however, this is a paramedic-level skill. Therefore, call for paramedic backup immediately if you suspect an infant may be in cardiac arrest. If paramedic backup with a manual defibrillator is not available, an AED equipped with a pediatric dose attenuator is preferred. If neither is available, an AED without a pediatric dose attenuator may be used.

Words of Wisdom

AEDs are becoming more and more accessible in the community. Be familiar with your local protocols on pediatric defibrillation. Your service may use a pediatric AED or an AED with a pediatric adapter.

Remember, if the child is past the onset of puberty (12 to 14 years of age signified by breast development on females and underarm, chest, and facial hair on males), use the adult CPR sequence, including the use of an adult AED.

Special AED Situations

Safety to you, others at the scene, and the patient should always be a priority. As such, it is important to keep some factors in mind when using an AED unit.

Pacemaker

You may encounter a patient who has an implanted defibrillator or pacemaker that delivers shocks directly to the heart if necessary. These patients usually have a high risk of sudden cardiac arrest. It is easy to recognize these devices because they create a hard lump beneath the skin in the chest, near the heart. If the electrical pads are placed directly over the device, it may block any shock delivered by the AED unit. Therefore, if you identify an implanted defibrillator or pacemaker, you should place the AED electrodes at least 1″ to the side of the device.

Occasionally, the implanted device will deliver shocks to the patient. If you observe the patient's muscles twitching as if just shocked, wait 30 to 60 seconds before delivering a shock from the AED.

Wet Patients

Water is a good conductor of electricity. Therefore, the AED unit should not be used in water. If the patient's chest is covered with water, the electrical current may move across the skin rather than between the pads to the patient's heart. If the patient is in water, pull him or her out of the water and quickly dry the skin before attaching the electrodes. If the patient is in a small puddle of water or in the snow, the AED can be used, but the patient's chest should be dry.

Transdermal Medication Patches

You may encounter a patient who is receiving medication through a transdermal medication patch. The medication is absorbed through the skin. The patch could block the electrical current to the heart and may cause a burn to the skin. To prevent this, remove the patch and wipe the skin to remove the medication residue prior to attaching the AED pads.

▌ Assessing the Need for BLS

As always, begin by surveying the scene. Is the scene safe? How many patients are there? What is your initial impression of the patients? Are there bystanders who may have additional information? What is the mechanism of injury or nature of illness? Do you suspect trauma? If you were dispatched to the scene, does the dispatch information match what you are seeing?

Because of the urgent need to start CPR in a pulseless, nonbreathing patient, you must complete a primary assessment as soon as possible and begin CPR with chest compressions. The first step is determining unresponsiveness and checking for breathing **Figure 11-5**. Clearly, a patient who is conscious does not need CPR. A person who is unresponsive may or may not need CPR; to determine unresponsiveness, gently tap the patient on the shoulder and shout "are you okay?" If the patient does not respond to verbal or physical stimulation, he or she is unconscious.

If you suspect the presence of a cervical spine injury, you must take steps to protect the spinal cord from further injury as you perform CPR. If there is even a remote

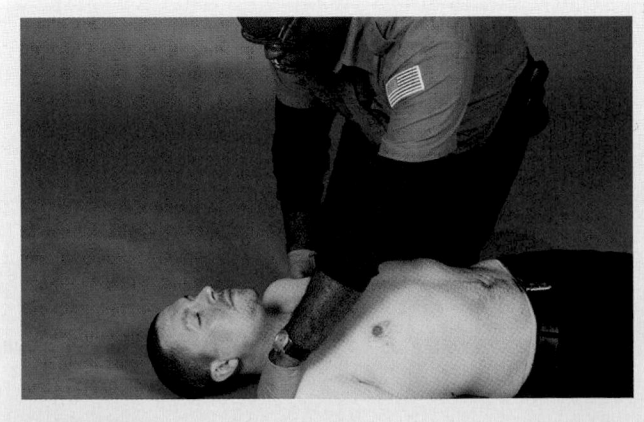

Figure 11-5 Assess an unresponsive patient by first attempting to rouse him or her by tapping on the shoulder.

possibility of this type of injury, you should begin taking appropriate precautions during the primary assessment.

The basic principles of BLS are the same for infants, children, and adults. For the purposes of BLS, anyone younger than 1 year is considered an infant. A child is between 1 year of age and the onset of puberty (12 to 14 years of age). Adulthood is from the onset of puberty and older. Children vary in size. Some small children may best be treated as infants, some larger children as adults. There are two basic differences in providing CPR for infants, children, and adults. The first is that the emergencies in which infants and children require CPR usually have different underlying causes. The second is that there are anatomic differences in adults, children, and infants, such as smaller airways in infants and children than in adults.

Although cardiac arrest in adults usually occurs before respiratory arrest, the reverse is true in infants and children. In most cases, cardiac arrest in children results from respiratory arrest. If untreated, respiratory arrest will quickly lead to cardiac arrest and death. Respiratory arrest in infants and children has a variety of causes, including aspiration of foreign bodies into the airway, such as parts of hot dogs, peanuts, candy, or small toys; airway infections, such as croup and epiglottitis; near-drowning incidents or electrocution; and sudden infant death syndrome (also known as SIDS).

Positioning the Patient

The next step in providing CPR is to position the patient to ensure that the airway is open. For CPR to be effective, the patient must be lying supine on a firm, flat surface, with enough clear space around the patient for two rescuers to perform CPR. If the patient is crumpled up or lying face down, you will need to reposition him or her. The few seconds that you spend to position the patient properly will greatly improve the delivery and effectiveness of CPR.

Follow the steps in Skill Drill 11-1 to reposition an unconscious adult for airway management:

Skill Drill 11-1

1. Kneel beside the patient. You and your partner must be far enough away so that the patient, when rolled toward you, does not come to rest in your lap Step 1.
2. Place your hands on either side of the patient's head and neck to protect the cervical spine if you suspect spinal injury. Your partner places his or her hands on the patient's distant shoulder and hip Step 2.

3. Your partner turns the patient toward him or her by pulling on the patient's distant shoulder and hip. Control the head and neck so that they move as a unit with the rest of the torso. This single motion will allow the head, neck, and back to stay in the same vertical plane and will minimize aggravation of any spinal injury Step 3.
4. Place the patient in a supine position, with the legs straight and both arms at the sides Step 4.

If possible, log roll the patient onto a long backboard as you are positioning him or her for CPR. This device will provide support during transport and emergency care. Once the patient is properly positioned, you can easily assess airway, breathing, circulation, and the need for defibrillation and start CPR if necessary.

Check for a Pulse

Once you have determined that the patient is unresponsive and not breathing or not breathing normally (ie, only gasping), and the patient has been properly positioned for management, you will need to quickly check the patient's pulse and begin chest compressions.

Cardiac arrest is determined by the absence of a palpable pulse at the carotid artery. Feel for the carotid artery by locating the larynx at the front of the neck and then sliding two fingers toward one side (the side closest to you). The pulse is felt in the groove between the larynx and the sternocleidomastoid muscle, with the pads of the index and middle fingers held side by side Figure 11-6. Light pressure is sufficient to palpate the pulse. Check the pulse for at least 5 seconds but no longer than 10 seconds; if a pulse cannot be felt, begin chest compressions.

If the patient has a pulse, but is not breathing, provide rescue breaths (described later in this chapter), at a rate of 10 to 12 breaths/min or one every 5–6 seconds for an adult and one every 3–5 seconds for an infant or child.

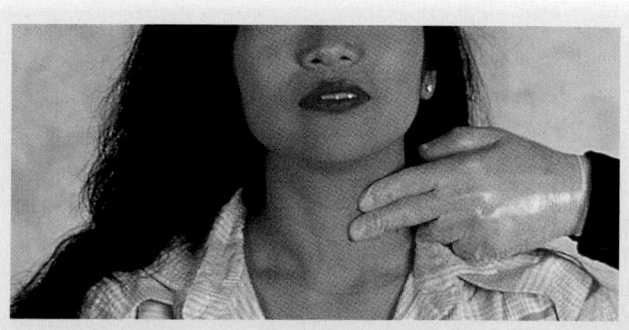

Figure 11-6 Feel for the carotid artery by locating the larynx, then slide your index and middle fingers toward one side. You can feel the pulse in the groove between the larynx and sternocleidomastoid muscle.

Skill Drill   11-1

Positioning the Patient

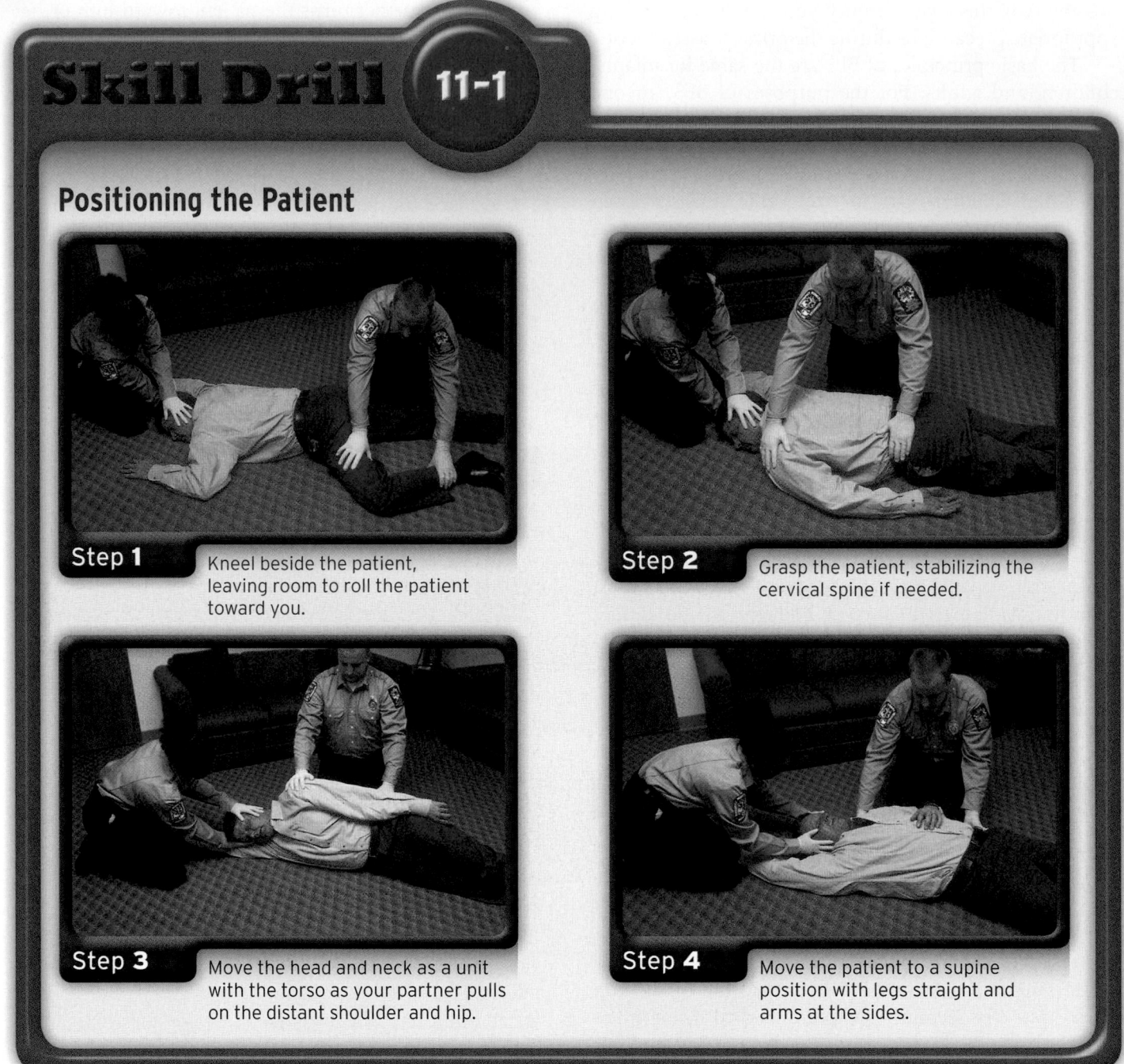

Step 1 Kneel beside the patient, leaving room to roll the patient toward you.

Step 2 Grasp the patient, stabilizing the cervical spine if needed.

Step 3 Move the head and neck as a unit with the torso as your partner pulls on the distant shoulder and hip.

Step 4 Move the patient to a supine position with legs straight and arms at the sides.

■ Provide External Chest Compressions

Chest compressions are administered by applying rhythmic pressure and relaxation to the lower half of the sternum. The heart is located slightly to the left of the middle of the chest between the sternum and the spine **Figure 11-7**. Compressions squeeze the heart, thereby acting as a pump to circulate blood. When artificial ventilations are provided, the blood that is circulated through the lungs by chest compressions is likely to receive adequate oxygen to maintain tissue perfusion. However, even when external chest compressions are performed as proficiently as possible, they circulate only one third of the blood that is normally pumped by the heart, so it is very important to perform compressions properly.

Prior to administering external chest compressions, place the patient on a firm, flat surface, in a supine position. The patient's head should not be elevated at a level above the heart because this will further reduce blood flow to the brain. The surface can be the ground, the floor, or a backboard on a stretcher. You cannot perform chest compressions adequately on a bed; therefore, a patient who is in bed should be moved to the floor or have a board placed under the back.

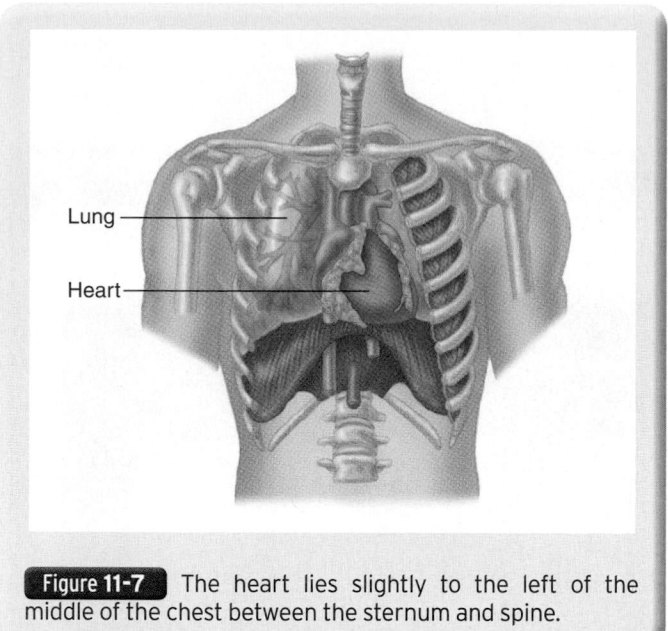

Figure 11-7 The heart lies slightly to the left of the middle of the chest between the sternum and spine.

Proper Hand Position

Correct hand position is established by placing the heel of one hand on the sternum in the center of the chest (lower half of the sternum). Follow the steps in **Skill Drill 11-2**:

Skill Drill 11-2

1. Place the heel of one hand on the sternum in the center of the chest **Step 1**.
2. Place the heel of your other hand over the first hand **Step 2**.
3. With your arms straight, lock your elbows, and position your shoulders directly over your hands. Your technique may be improved or made more comfortable if you interlock the fingers of your lower hand with the fingers of your upper hand; either way, your fingers should be kept off the patient's chest.
4. Depress the sternum at least 2″ (in adults), using direct downward movement and then rising gently upward **Step 3**. It is important that you allow the chest to return to its normal position. Compression and relaxation should be of equal duration.

Proper Compression Technique

Complications from chest compressions are rare but can include fractured ribs, a lacerated liver, and a fractured sternum. Although these injuries cannot be entirely avoided, you can minimize the chance that they will

occur if you use good, smooth technique and proper hand placement.

Proper compressions begin by locking your elbows, with your arms straight, and positioning your shoulders directly over your hand so that the thrust of each compression is straight down on the sternum. Depress the sternum at least 2″ in an adult, avoiding a rocking motion and rising gently upward. This motion allows pressure to be delivered vertically down from your shoulders. Vertical downward pressure produces a compression that must be followed immediately by an equal period of relaxation. The ratio of time devoted to compression versus relaxation should be 1:1.

The actual motions must be smooth, rhythmic, and uninterrupted **Figure 11-8A**. Short, jabbing compressions are not effective in producing artificial blood flow. Do not remove the heel of your hand from the patient's chest during relaxation, but make sure that you completely release pressure on the sternum so that it can return to its normal resting position between compressions **Figure 11-8B**.

Assessing Airway and Breathing

Opening the Airway in Adults

Without an open airway, rescue breathing will not be effective. As discussed in Chapter 9, *Airway Management*, there are two techniques for opening the airway in adults: the head tilt–chin lift maneuver and the jaw-thrust maneuver. Open the airway with the head tilt–chin lift maneuver if there is no indication of a spinal injury. If spinal injury is suspected, use the jaw-thrust maneuver.

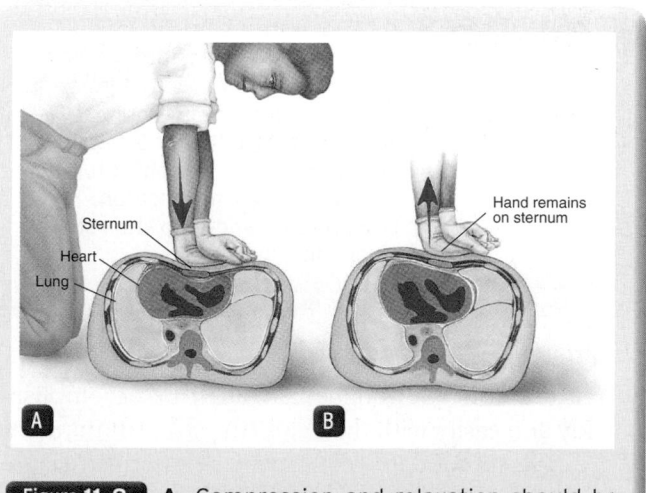

Figure 11-8 **A.** Compression and relaxation should be rhythmic and of equal duration. **B.** Pressure on the sternum must be released so that the sternum can return to its normal resting position between compressions. However, do not remove the heel of the hand from the sternum.

Skill Drill 11-2

Performing Chest Compressions

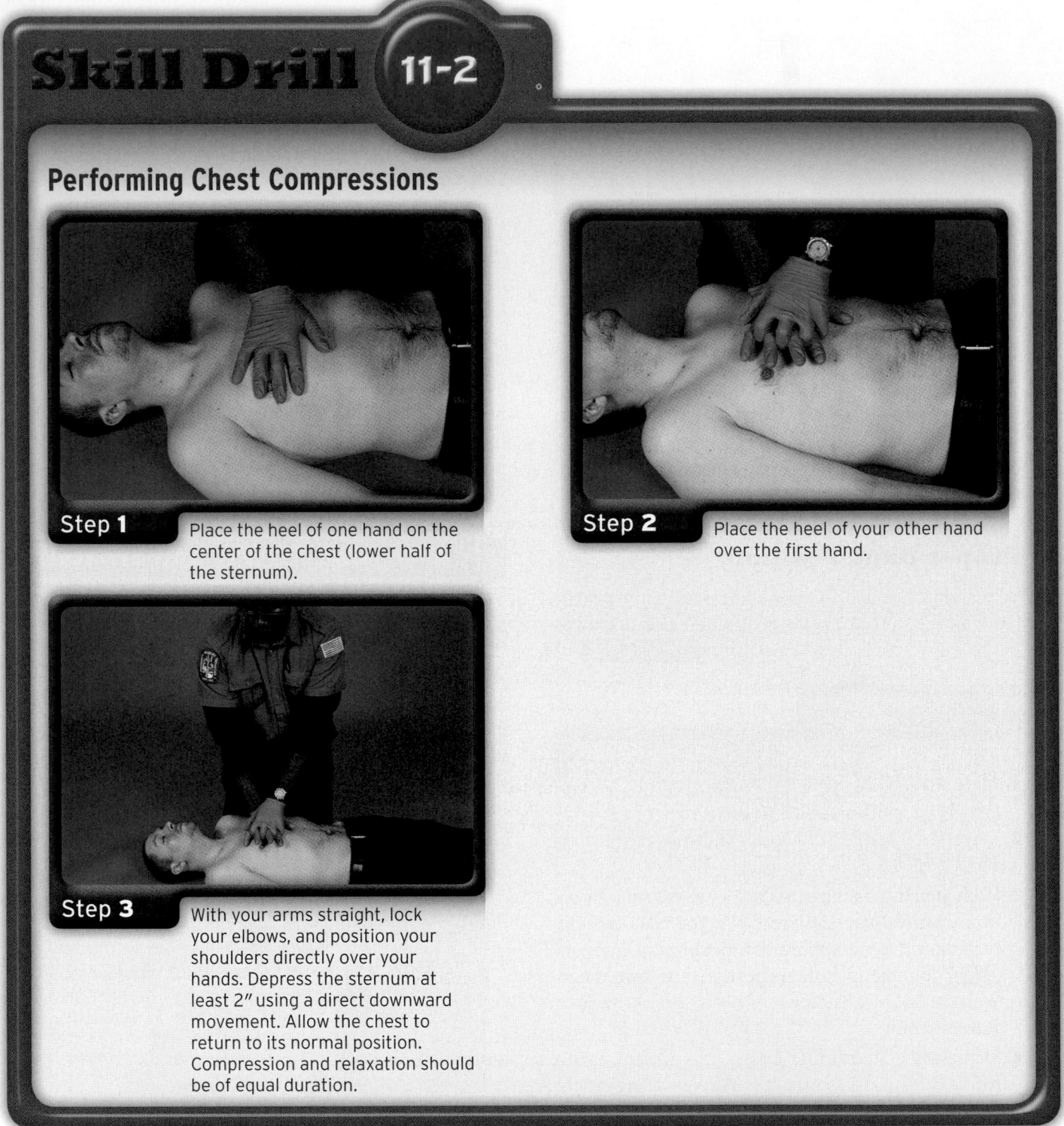

Step 1 Place the heel of one hand on the center of the chest (lower half of the sternum).

Step 2 Place the heel of your other hand over the first hand.

Step 3 With your arms straight, lock your elbows, and position your shoulders directly over your hands. Depress the sternum at least 2" using a direct downward movement. Allow the chest to return to its normal position. Compression and relaxation should be of equal duration.

Opening the airway to relieve an obstruction caused by relaxation of the tongue can often be accomplished quickly and easily with the **head tilt–chin lift maneuver** **Figure 11-9** . In patients who have not sustained trauma, this simple maneuver is sometimes all that is required for the patient to resume breathing. If the patient has any foreign material or vomitus in the mouth, you should quickly remove it. Wipe out any liquid materials from the mouth with a piece of cloth held by your index and

middle fingers; use your hooked index finger to remove any solid material. **Figure 11-10** reviews how to perform the head tilt–chin lift maneuver in an adult.

The head tilt–chin lift maneuver is effective for opening the airway in most patients. In patients with suspected spinal injury, you want to minimize movement of the patient's neck. In this case, perform a **jaw-thrust maneuver**. To perform a jaw-thrust maneuver, place your fingers behind the angles of the patient's lower jaw

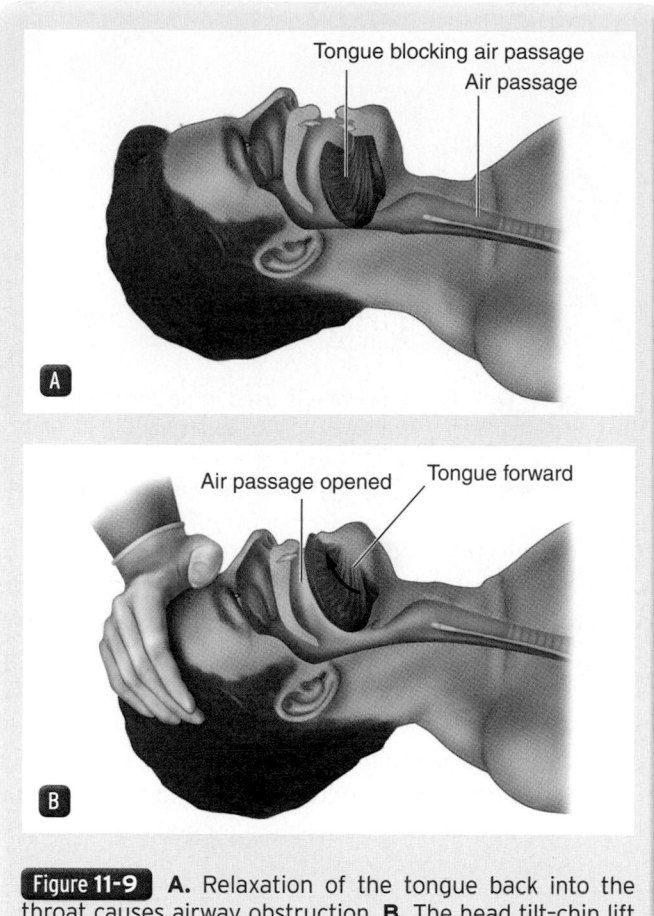

Figure 11-9 **A.** Relaxation of the tongue back into the throat causes airway obstruction. **B.** The head tilt–chin lift maneuver combines two movements of opening the airway.

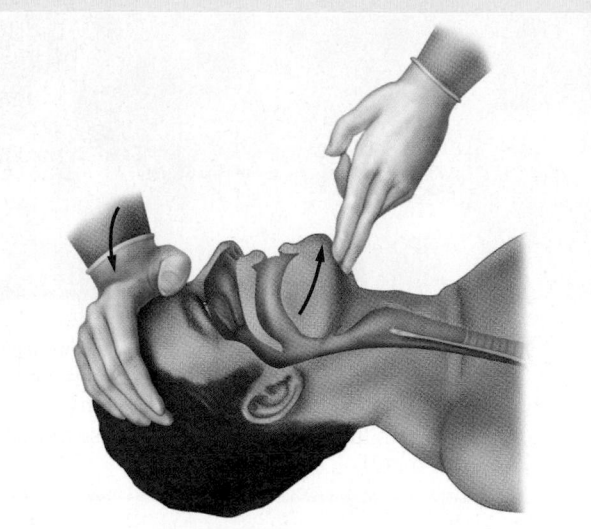

Figure 11-10 To perform the head tilt–chin lift maneuver, place one hand on the patient's forehead and apply firm backward pressure with your palm to tilt the head back. Next, place the tips of the index and middle fingers of your other hand under the lower jaw near the bony part of the chin. Lift the chin upward, bringing the entire lower jaw with it, helping to tilt the head back.

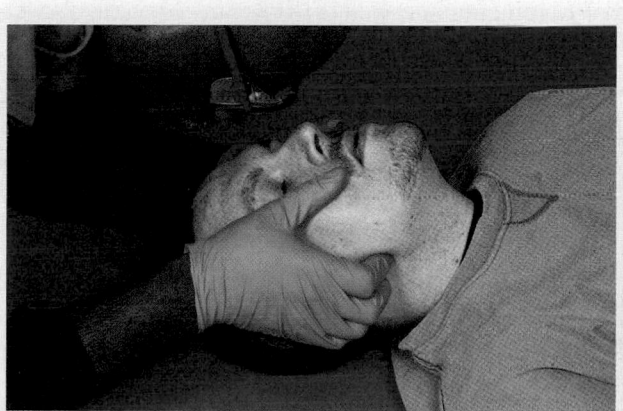

Figure 11-11 To perform the jaw-thrust maneuver, maintain the head in neutral alignment and place your fingers behind the angles of the lower jaw, and move the jaw upward. The completed maneuver should look like this.

and then move the jaw upward. Keep the head in a neutral position as you move the jaw upward and open the mouth. If the patient's mouth remains closed, you can use your thumbs to pull the patient's lower lip down, to allow breathing. If the jaw thrust fails to open the airway, the head tilt–chin lift should be used to open the airway. An open airway is a primary goal when caring for trauma patients and must be attained to improve survival. Figure 11-11 reviews how to perform the jaw-thrust maneuver.

Recovery Position

If the patient is breathing on his or her own, and has no signs of trauma, you should place him or her in the **recovery position**. This position helps to maintain a clear airway in a patient with a decreased level of consciousness who has not sustained traumatic injuries and is breathing adequately on his or her own Figure 11-12. It also allows vomitus to drain from the mouth. Roll the patient onto his or her side so that the head, shoulders, and torso move as a unit, without twisting. Then place the top hand under his or her cheek. Never place a patient who has a suspected head or spinal injury in the

recovery position because maintenance of spinal alignment in this position is not possible and further spinal cord injury could result.

Breathing

A lack of oxygen, combined with too much carbon dioxide in the blood, is lethal. To correct this condition, you must provide slow, deliberate ventilations that last 1 second. This gentle, slow method of ventilating the patient prevents air from being forced into the stomach.

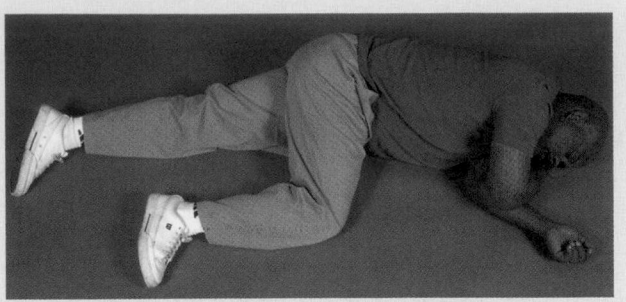

Figure 11-12 The recovery position is used to maintain an open airway in an adequately breathing patient with a decreased level of consciousness who has had no traumatic injuries. It allows vomitus, blood, and any other secretions to drain from the mouth.

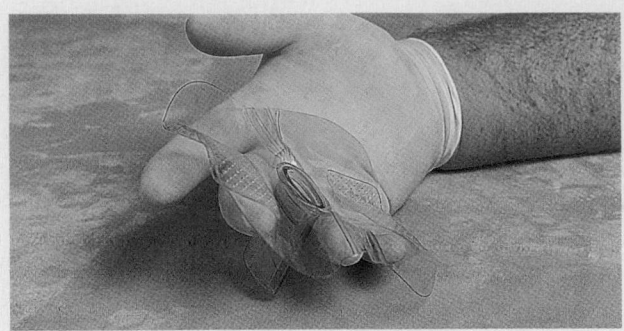

Figure 11-13 A barrier device is used in performing ventilation because it prevents exposure to saliva, blood, and vomitus.

■ Provide Artificial Ventilations

Ventilations can be given by one or two EMS providers. A barrier device should be used when you are administering ventilations. In the prehospital environment, ventilations are routinely provided using barrier devices, such as a pocket mask or a bag-mask device. These devices feature a plastic barrier that covers the patient's mouth and nose and a one-way valve to prevent exposure to secretions and exhaled contaminants **Figure 11-13**. Such devices also pro-

vide good infection control. If a mask is not available, a face shield or some other type of physical barrier should be used. Providing ventilations without a barrier device should be performed only in extreme conditions. You should use devices that supply supplemental oxygen when possible. Devices that have an oxygen reservoir will provide higher percentages of oxygen to the patient. Regardless of whether you are ventilating the patient with or without supplemental oxygen, you should observe the chest for good rise to assess the effectiveness of your ventilations.

You are the Provider: PART 2

You arrive at the scene and find two bystanders performing CPR on the patient, who appears to be in his late 40s. A second BLS ambulance is en route to the scene and will arrive in about 5 minutes. You perform a primary assessment as your partner opens the AED.

Recording Time: 0 Minutes	
Appearance	Motionless; cyanosis to face
Level of consciousness	Unconscious and unresponsive
Airway	Open; clear of secretions or foreign bodies
Breathing	Absent
Circulation	No carotid pulse; skin, cool and pale; no gross bleeding

Your partner takes over performing CPR. One of the bystanders tells you that the patient was about to get in his car when he suddenly grabbed his chest, slumped against the car, and eased himself to the ground. By the time the bystander got to him, he was unconscious and not breathing. The bystander further tells you that he immediately called 9-1-1 and then began CPR.

3. What links in the chain of survival have been maintained at this point?

4. Why is it so critical to minimize interruptions in CPR?

Ventilations need to be delivered at a rate and depth that is not excessive so as to not cause increased intrathoracic pressure. Increased intrathoracic pressure impedes venous return to the right side of the heart, thus decreasing the effectiveness of CPR because overall blood flow is reduced, resulting in the heart and brain receiving decreased amounts of oxygen.

You should perform rescue breathing in an adult with a simple barrier device in the following manner Figure 11-14 :

1. Open the airway with the head tilt–chin lift maneuver (nontrauma patient).
2. Press on the forehead to maintain the backward tilt of the head. Pinch the patient's nostrils together with your thumb and index finger.
3. Depress the lower lip with the thumb of the hand that is lifting the chin. This will help to keep the patient's mouth open.
4. Open the patient's mouth widely, and place the barrier device over the patient's mouth and nose.
5. Take a deep breath, and then make a tight seal with your mouth around the barrier device. Give two slow rescue breaths, each lasting 1 second.
6. Remove your mouth, and allow the patient to exhale passively. Turn your head slightly to watch for movement of the patient's chest.

When using the jaw-thrust maneuver to open the airway (in suspected neck or spine injury), positioning yourself at the patient's head will facilitate simultaneous cervical spine stabilization and adequate ventilation. Keep the patient's mouth open with both thumbs, and seal the nose by placing your cheek against the patient's nostrils Figure 11-15 . Note that this maneuver is somewhat difficult; practicing with a manikin will help you gain familiarity with this technique.

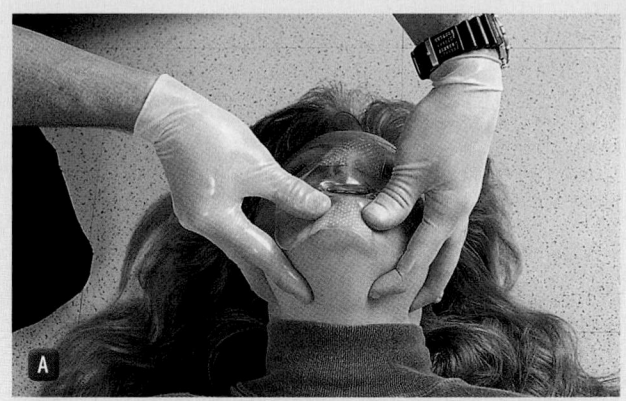

Figure 11-15 **A.** If you use the jaw-thrust maneuver to open the airway, keep the patient's mouth open with both thumbs as you move from above the patient's head to the side. **B.** Seal the nose by placing your cheek against the patient's nostrils.

Words of Wisdom

Ventilation is the physical act of moving air in and out of the lungs. Ventilation is required for adequate respiration. If ventilation is adequate, other problems may hinder respiration. Examples of interruptions of ventilation include trauma such as flail chest, foreign body airway obstruction, and an injury to the spinal cord that disrupts the phrenic nerve that innervates the diaphragm.

Stoma Ventilation

Patients who have undergone surgical removal of the larynx often have a permanent tracheal stoma at the midline in the neck. In this case, a stoma is an opening that connects the trachea directly to the skin Figure 11-16 . Because it is at the midline, the stoma is the only opening that will move air into the patient's lungs; you should ignore any other openings. Patients with a stoma should be ventilated with a bag-mask device or pocket mask device directly over the stoma.

Not all stomas are disconnected from the nose and mouth. If air leakage through the nose and mouth

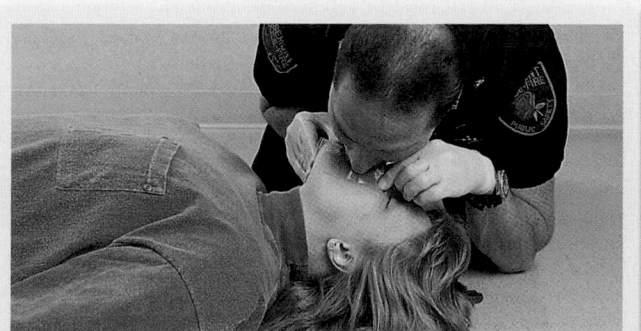

Figure 11-14 To perform ventilations, ensure that you make a tight seal with your mouth around the barrier device and then give two slow, gentle breaths, each lasting 1 second.

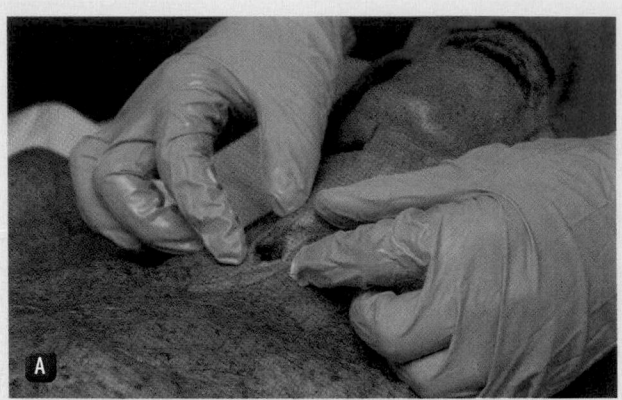

Figure 11-16 **A.** This stoma connects the trachea directly to the skin. **B.** Use a bag-mask device or pocket mask device to ventilate a patient with a stoma.

interferes with ventilation through the stoma, cover the nose and mouth with your hand. Use a pediatric or infant mask to ventilate through the stoma.

Gastric Distention

Artificial ventilation may result in the stomach becoming filled with air, a condition called **gastric distention**. Although it occurs more easily in children, it also happens frequently in adults. Gastric distention is likely to occur if you ventilate too fast or with too much pressure as you ventilate. If you give too much air, or if the patient's airway is not opened adequately, the excess of gas opens up the collapsible tube, the esophagus, allowing gas to enter the stomach. Therefore, it is important for you to give slow, gentle breaths. Such breaths are also more effective in ventilating the lungs. Serious inflation of the stomach is dangerous because it can cause the patient to vomit during CPR. It can also reduce lung volume by elevating the diaphragm.

If massive gastric distention interferes with adequate ventilation, you should contact medical control. Check

the airway again and reposition the patient, watch for rise and fall of the chest, and avoid giving forceful breaths. If gastric distention makes it impossible to ventilate the patient and paramedics are not available to perform orogastric or nasogastric tube decompression, medical control may order you to roll the patient on his or her side and provide gentle manual pressure to the abdomen to expel air from the stomach. Have suction readily available, and be prepared for copious amounts of vomitus.

One-Rescuer Adult CPR

When you are providing CPR alone, you must give both artificial ventilations and chest compressions in a ratio of compressions to ventilations of 30:2. To perform one-rescuer adult CPR, follow the steps in Skill Drill 11-3 :

Skill Drill 11-3

1. Determine unresponsiveness and breathlessness and call for additional help Step 1 .
2. Position the patient properly (supine) on a flat surface.
3. Determine pulselessness by checking the carotid pulse. Check the pulse for no more than 10 seconds Step 2 .
4. If there is no pulse, begin CPR until an AED is available. Place your hands in the proper position for delivering external chest compressions, as described previously Step 3 . Give 30 chest compressions at a rate of at least 100 per minute for an adult. Each set of 30 compressions should take about 17 seconds.
5. Open the airway according to your suspicion of spinal injury Step 4 .
6. Give two ventilations of 1 second each and observe for visible chest rise Step 5 .
7. Continue cycles of 30 chest compressions and two ventilations until additional personnel arrive or the patient starts to move.

Two-Rescuer Adult CPR

You and your team should be able to perform one-rescuer and two-rescuer CPR with ease. Two-rescuer CPR is always preferable because it is less tiring and facilitates effective chest compressions. In fact, a team approach to CPR and AED use is far superior to the one-rescuer approach. Once one-rescuer CPR is in progress, additional rescuers can be added very easily. Prior to assisting

Skill Drill 11-3

Performing One-Rescuer Adult CPR

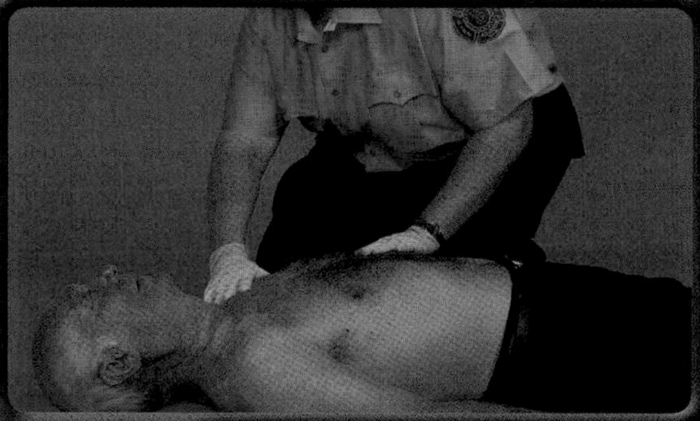

Step 1 Determine unresponsiveness and breathlessness and call for help.

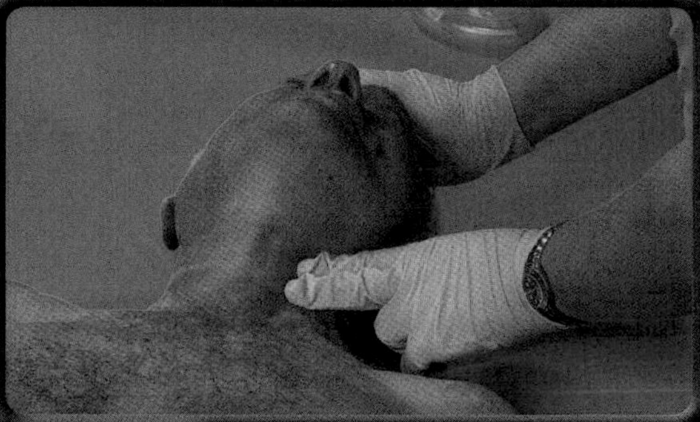

Step 2 Check for a carotid pulse for no more than 10 seconds.

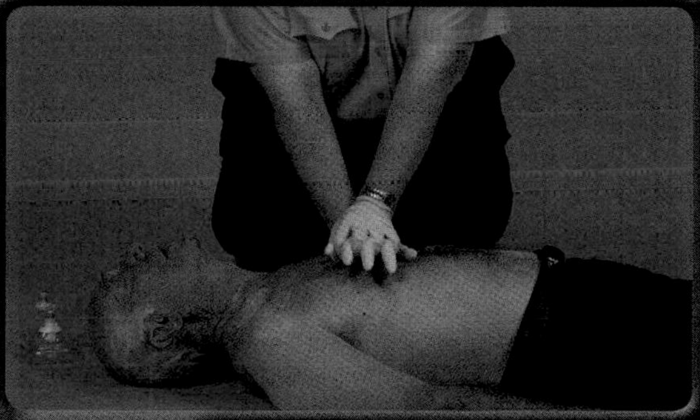

Step 3 If there is no pulse, begin CPR until an AED is available. Give 30 chest compressions at a rate of at least 100 per minute.

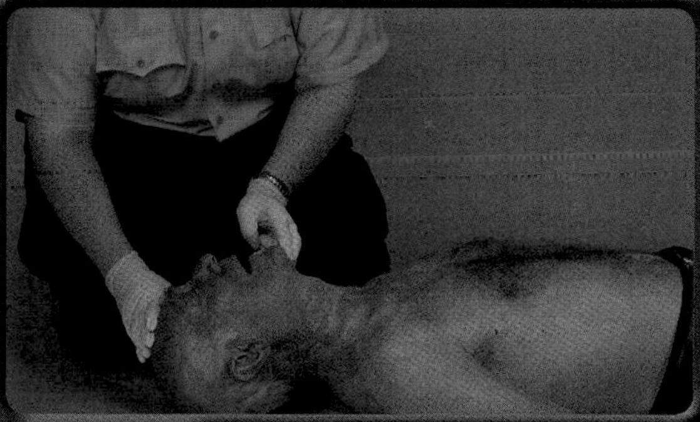

Step 4 Open the airway according to your suspicion of spinal injury.

with CPR, a second rescuer should apply the AED and then set up airway adjuncts including a bag-mask device and suction, and should insert an oral airway. If CPR is in progress, the second rescuer should enter the procedure after a cycle of 30 compressions and two ventilations. To perform two-rescuer adult CPR, follow the steps in **Skill Drill 11-4**:

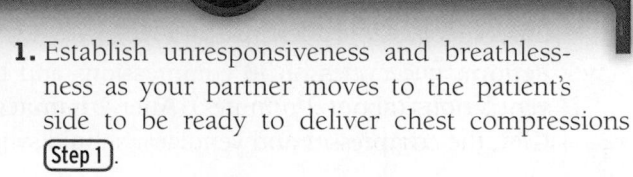

Skill Drill 11-4

1. Establish unresponsiveness and breathlessness as your partner moves to the patient's side to be ready to deliver chest compressions **Step 1**.

Skill Drill 11-3

Performing One-Rescuer Adult CPR, continued

Step 5 Give two ventilations of 1 second each and observe for visible chest rise. Continue cycles of 30 chest compressions and two ventilations until additional personnel arrive or the patient starts to move.

2. If the patient is unresponsive and not breathing, check the carotid pulse for no more than 10 seconds (Step 2). If the patient has no pulse and an AED is available, apply it now.

3. Begin CPR, starting with chest compressions. Give 30 chest compressions at a rate of at least 100 per minute (Step 3).

4. Open the airway according to your suspicion of spinal injury (Step 4).

5. Give two ventilations of 1 second each and observe for visible chest rise (Step 5).

6. Perform five cycles of 30 compressions and two ventilations (about 2 minutes). After 2 minutes of CPR, the compressor and ventilator should switch positions. The switch time should take no longer than 5 seconds.

7. Continue cycles of 30 chest compressions and two ventilations until ALS personnel take over or the patient starts to move.

■ Switching Positions

Switching rescuers during CPR is beneficial to the quality of compressions administered to the patient. After five cycles of CPR (about 2 minutes), the rescuer providing compressions will begin to tire and compression quality will begin to suffer. It is therefore recommended to switch the rescuer doing compressions every 2 minutes. If there are only two rescuers, the rescuers will switch positions. If additional

Skill Drill 11-4

Performing Two-Rescuer Adult CPR

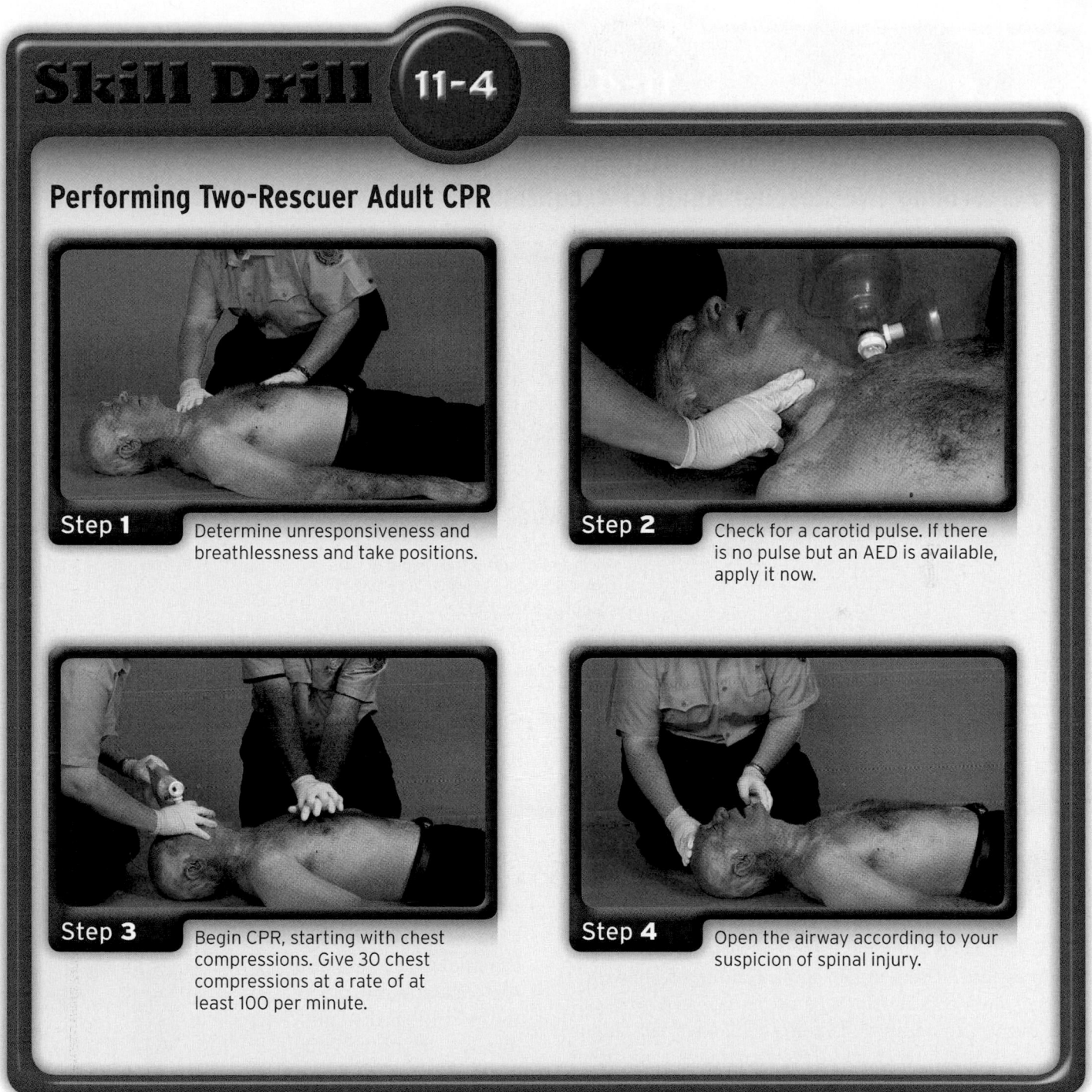

Step 1 Determine unresponsiveness and breathlessness and take positions.

Step 2 Check for a carotid pulse. If there is no pulse but an AED is available, apply it now.

Step 3 Begin CPR, starting with chest compressions. Give 30 chest compressions at a rate of at least 100 per minute.

Step 4 Open the airway according to your suspicion of spinal injury.

rescuers are available, rotating the rescuer providing compressions every five cycles (2 minutes) is required. During switches, every effort should be made to minimize the time that no compressions are being administered. This should be approximately 5 seconds but no more than 10 seconds of a break in between the compression cycle.

The switch between the two rescuers can be easily accomplished. Rescuer one should finish the cycle of 30 compressions while rescuer two moves to the opposite side of the chest and moves into position to begin compressions. Rescuer one should deliver two rescue breaths and then rescuer two should take over compressions by administering 30 chest compressions. Rescuer one will then deliver two ventilations and the CPR cycles will continue as needed until the next 2-minute mark (five cycles) is reached, at which time the process will be repeated.

Skill Drill 11-4

Performing Two-Rescuer Adult CPR, continued

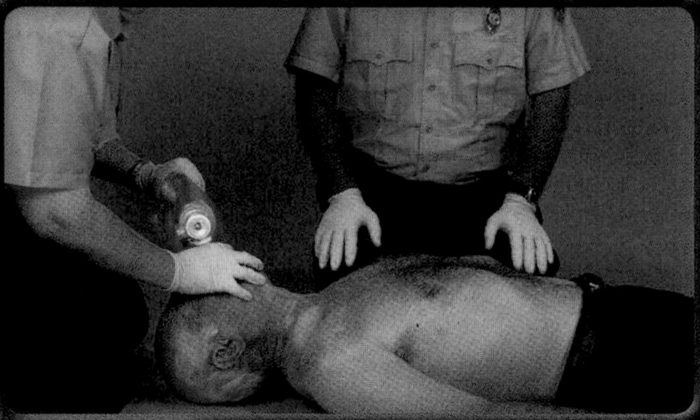

Step 5 Give two ventilations of 1 second each and observe for visible chest rise. Continue cycles of 30 chest compressions and two ventilations (switch roles every five cycles) until ALS personnel take over or the patient starts to move.

Devices to Assist Circulation

The effectiveness of CPR is dependent on the amount of blood circulated throughout the body as a result of chest compressions. Even under ideal conditions, however, manual chest compressions cannot equate to normal cardiac output. In addition, factors such as rescuer fatigue or inaccurate depth or rate of compressions can further impede the resuscitation process.

Several mechanical devices are now available to assist emergency responders in delivering improved compressions when providing CPR. Although improved patient outcomes have not yet been documented, these devices may be considered for use as an adjunct to CPR when used by properly trained personnel for patients in cardiac arrest in the prehospital or in-hospital setting.

Impedance Threshold Device

An **impedance threshold device (ITD)** is a valve device placed between the endotracheal tube and a bag-mask device. It is designed to limit the air entering the lungs during the recoil phase between chest compressions **Figure 11-17**. This results in negative intrathoracic pressure that draws more blood toward the heart, ultimately resulting in improved cardiac filling and circulation during each chest compression. It has been shown to improve short-term survival in adults when combined with other adjuncts to circulation in the management of a cardiac arrest. It has not been shown to improve long-term survival or the neurologic status of the patient.

For use in a non-intubated patient, studies suggest that the ITD could be used with a face mask; however, a tight seal is essential to achieve the desired effect.

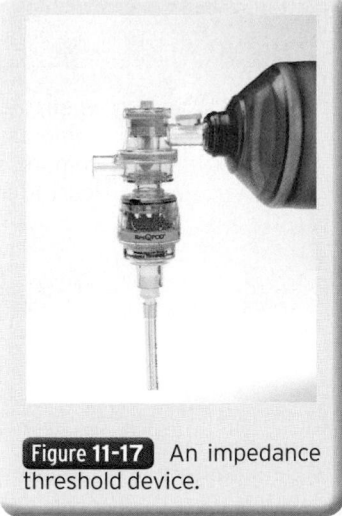

Figure 11-17 An impedance threshold device.

Although increased survival rates have not been proven, the use of ITDs may improve the effectiveness of CPR when used by trained rescuers.

Mechanical Piston Device

A **mechanical piston device** is a device that depresses the sternum via a compressed gas-powered plunger mounted on a backboard **Figure 11-18**. The patient is positioned supine on the backboard, with the piston positioned on top of the patient with the plunger centered over the patient's thorax in the same manner as with manual chest compressions. The device is then secured to the backboard.

The device allows rescuers to configure the depth and rate of compressions, resulting in uniform delivery. This frees the rescuer to complete other tasks and eliminates rescuer fatigue that results from continuous delivery of manual chest compressions. These devices have been around for many years. The latest versions of these devices offer the provider the option of just providing compressions using a battery instead of an oxygen or compressed air system, thus eliminating the tanks and hoses.

Load-Distributing Band CPR or Vest CPR

The **load-distributing band (LDB)** is a circumferential chest compression device composed of a constricting band and backboard **Figure 11-19**. The device is either electrically or pneumatically driven to compress the heart by putting inward pressure on the thorax.

As with the mechanical piston device, use of the device frees the rescuer to complete other tasks. The device is lighter than the early version mechanical piston devices and can be easier to apply. The end result is sup-

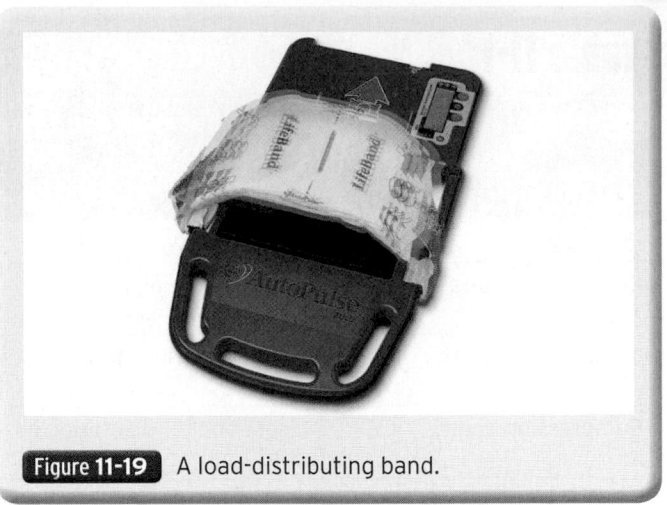

Figure 11-19   A load-distributing band.

posedly improved hemodynamics in the patient when used by properly trained emergency responders, though studies have demonstrated no improvement in short-term survival and worse neurologic outcome when the device was used. Further studies are needed.

Infant and Child CPR

In most cases, cardiac arrest in infants and children follows respiratory arrest, which triggers hypoxia and ischemia of the heart. Children consume oxygen two to three times as rapidly as adults. Therefore, you must first focus on opening an airway and providing artificial ventilation. Often, this will be enough to allow the child to resume spontaneous breathing and, thus, prevent cardiac arrest. Therefore, airway and breathing are the focus of pediatric basic life support (BLS) **Table 11-1**.

Respiratory problems leading to cardiopulmonary arrest in children can have a number of different causes, including:

- Injury, both blunt and penetrating
- Infections of the respiratory tract or another organ system
- A foreign body in the airway
- Submersion
- Electrocution
- Poisoning or drug overdose
- Sudden infant death syndrome

Pediatric BLS can be divided into four steps:

1. Determining responsiveness
2. Circulation
3. Airway
4. Breathing

Note that neonatal patients are defined as birth to age 1 month, and infants as age 1 month to 1 year. Neonatal resuscitation is covered in Chapter 31, *Obstetrics and Neonatal Care*.

Figure 11-18   A mechanical piston device.

Table 11-1   Review of Pediatric BLS Procedures

Procedure	Infants (between age 1 month and 1 year[a])	Children (1 year to onset of puberty[b])
Circulation		
Pulse check	Brachial artery	Carotid or femoral artery
Compression area	Just below the nipple line	In the center of the chest, in between the nipples
Compression width	Two fingers or two thumb encircling-hands technique	Heel of one or both hands
Compression depth	At least one third anterior-posterior diameter (about 1 1/2″)	At least one third anterior-posterior diameter (about 2″)
Compression rate	At least 100/min	At least 100/min
Compression-to-ventilation ratio (until advanced airway is inserted)	30:2 (one rescuer); 15:2 (two rescuers)[c]	30:2 (one rescuer); 15:2 (two rescuers)[c]
Foreign body obstruction	Responsive: Back slaps and chest thrusts Unresponsive: CPR	Responsive: Abdominal thrust Unresponsive: CPR
Airway		
	Head tilt-chin lift; jaw thrust if spinal injury is suspected	Head tilt-chin lift; jaw thrust if spinal injury is suspected
Breathing		
Ventilations	1 breath every 3 to 5 seconds (12 to 20 breaths/min). About 1 second per breath. Visible chest rise.	1 breath every 3 to 5 seconds (12 to 20 breaths/min.) About 1 second per breath. Visible chest rise.

[a]The AHA defines neonatal patients as birth to age 1 month, and infants as age 1 month to 1 year. Neonatal resuscitation is covered in Chapter 31, *Obstetrics and Neonatal Care.*
[b]Onset of puberty is approximately 12 to 14 years of age, as defined by secondary characteristics (eg, breast development in girls and armpit hair in boys).
[c]Pause compressions to deliver ventilations.

■ Determining Responsiveness and Breathing

Never shake a child to determine whether he or she is responsive, especially if there is the possibility of a neck or back injury. Instead, gently tap the child on the shoulder, and speak loudly **Figure 11-20**. If a child is responsive but struggling to breathe, allow him or her to remain in whatever position is most comfortable.

If you find an unresponsive, apneic, and pulseless child while you are alone and not on duty, perform CPR beginning with chest compressions for approximately five cycles (2 minutes), and then stop to call the EMS system. Why not call right away, as you would with an adult? Because, as mentioned previously, cardiopulmonary arrest in children is most often the result of respiratory failure, not a primary cardiac event. Therefore, they will require immediate restoration of oxygenation,

ventilation, and circulation, which can be accomplished by immediately performing five cycles (about 2 minutes) of CPR before activating the EMS system.

■ Circulation

After determining responsiveness and checking breathing, you need to assess circulation. As with an adult, you should first check for a palpable pulse in a large central artery. Absence of a palpable pulse in a major central artery means that you must begin external chest compressions. You can usually palpate the carotid or femoral pulse in children older than 1 year, but it is difficult in infants. Therefore, in infants, palpate the brachial artery, which is located on the inner side of the arm, midway between the elbow and shoulder. Place your thumb on the outer surface of the arm between the elbow and shoulder. Then place the tips of

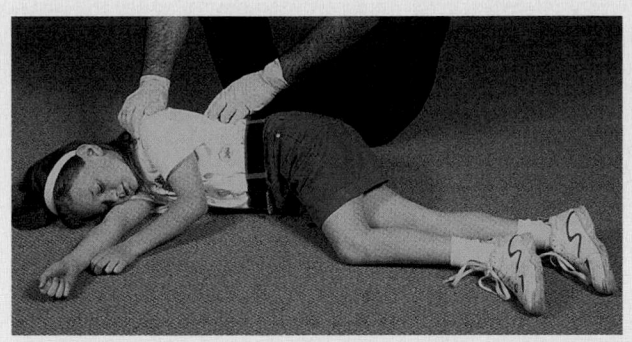

Figure 11-20 Never shake a child to determine responsiveness. Rather, gently tap on the shoulder (child) or tap the soles of the feet (infant), and speak loudly.

your index and middle fingers on the inside of the biceps, and press lightly toward the bone. Take at least 5 seconds but no more than 10 seconds to assess for a pulse. If the infant or child is not breathing, the pulse is often too slow (less than 60 beats/min) or absent altogether; therefore CPR will be required.

As with an adult, an infant or child must be lying on a hard surface for effective chest compressions. If you need to carry an infant while providing CPR, your forearm and hand can serve as the flat surface. Your palm should support the infant's head. In this way, the infant's shoulders are elevated, and the head is slightly tilted back in a position that will keep the airway open. However, you must ensure that the infant's head is not higher than the rest of the body.

The technique for chest compressions in infants and children differs because of a number of anatomic differences, including the position of the heart, the size of the chest, and the fragile organs of a child. The liver is relatively large, immediately under the right side of the diaphragm, and very fragile, especially in infants. The spleen, on the left, is much smaller and much more fragile in children than in adults. These organs are easily injured if you are not careful in performing chest compressions, so be sure that your hand position is correct before you begin. The chest of an infant is smaller and more pliable than that of an older child or adult; therefore, you should only use two fingers to compress the chest. In children, especially those older than 8 years of age, you can use the heel of one or both hands to compress the chest.

Follow these steps to perform infant chest compressions **Skill Drill 11-5** :

You are the Provider: PART 3

With CPR ongoing, you open the AED pads and prepare to apply them to the patient's chest. You note that the patient has a medication patch on the right upper part of his chest. You also see a bulge with a scar over it on the left upper part of his chest. You apply the AED pads, analyze the patient's cardiac rhythm, and receive a "shock advised" message. After delivering the shock, you and your partner resume CPR. The backup ambulance arrives and one of the EMTs assesses the quality of your CPR.

Recording Time: 4 Minutes	
Level of consciousness	Unconscious and unresponsive
Respirations	Absent (baseline); two breaths are being given after every 30 chest compressions; chest rise is visible with each breath
Pulse	Absent (baseline); femoral pulse is palpable with chest compressions
Skin	Pale
Blood pressure	Not measurable
Oxygen saturation (Sao$_2$)	Not measurable

5. Should you remove the medication patch or leave it in place? Why or why not?

6. What does the bulge and scar over the patient's chest indicate? How will this affect the way you treat the patient?

1. Place the infant on a firm surface, using one hand to keep the head in an open airway position. You can also use a pad or wedge under the shoulders and upper body to keep the head from tilting forward.

2. Imagine a line drawn between the nipples. Place two fingers in the middle of the sternum, just below the nipple line (Step 1).

3. Using two fingers, compress the sternum about one third the anterior-posterior diameter of the chest (approximately 1½″ in most infants). Compress the chest at a rate of at least 100 per minute.

4. After each compression, allow the sternum to return briefly to its normal position. Allow equal time for compression and relaxation of the chest. Do not remove your fingers from the sternum, and avoid jerky movements (Step 2).

Coordinate rapid compressions and ventilations in a 30:2 ratio if working alone, and 15:2 if working with another health care provider, making sure the infant's chest fully recoils in between compressions and that the chest visibly rises with each ventilation. You will find this easier to do if you use your free hand to keep the head in the open airway position. If the chest does not rise, or rises only a little, use a chin lift to open the airway. Reassess the infant for signs of spontaneous breathing or a pulse after five cycles (about 2 minutes) of CPR.

Skill Drill 11-6 shows the steps for performing CPR in children between 1 year of age and the onset of puberty:

1. Place the child on a firm surface. Place the heel of one or two hands in the center of the chest, in between the nipples. Avoid compression over the lower tip of the sternum, which is called the xiphoid process (Step 1).

2. Compress the chest about one third the anterior-posterior diameter of the chest (approximately 2″ in most children) at a rate of at least 100 per minute. With pauses for ventilation, the actual number of compressions delivered will be about 80 per minute. In between compressions, allow the chest to fully recoil. Compression and relaxation time should be the same duration. Use smooth movements. Hold your fingers off the child's ribs, and keep the heel of your hand(s) on the sternum.

Skill Drill 11-5

Performing Infant Chest Compressions

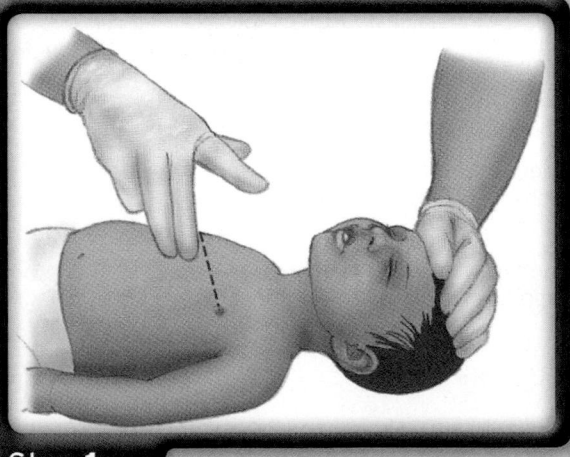

Step 1 Position the infant on a firm surface while maintaining the airway. Place two fingers in the middle of the sternum with one finger touching the nipple line.

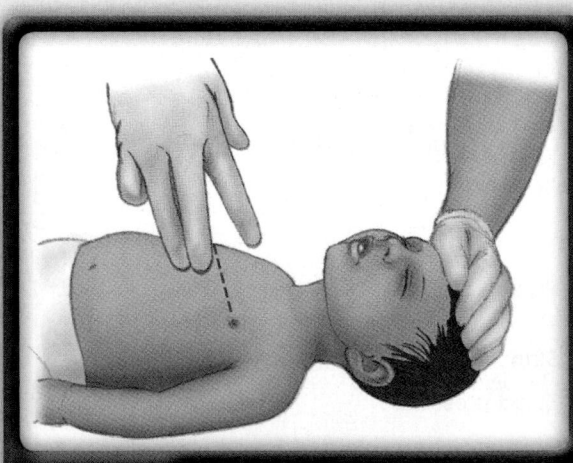

Step 2 Use two fingers to compress the chest one third to one half its depth at a rate of at least 100 per minute. Allow the sternum to return to its normal position between compressions.

3. Coordinate rapid compressions and ventilations in a 30:2 ratio for one rescuer and 15:2 for two rescuers, making sure the chest rises with each ventilation. At the end of each cycle, pause for two ventilations (Step 2).

4. After five cycles (about 2 minutes) assess for signs of breathing or a pulse. If there is no pulse and you have an AED, apply it now.

5. If the child regains a pulse of greater than 60 beats/min and resumes effective breathing, place him or her in a position that allows for frequent reassessment of the airway and vital signs during transport (Step 3).

Switching rescuer positions is the same for children as it is for adults, every five cycles (2 minutes) of CPR.

Remember, if the child is past the onset of puberty, use the adult CPR sequence, including the use of the AED.

Skill Drill 11-6

Performing CPR on a Child

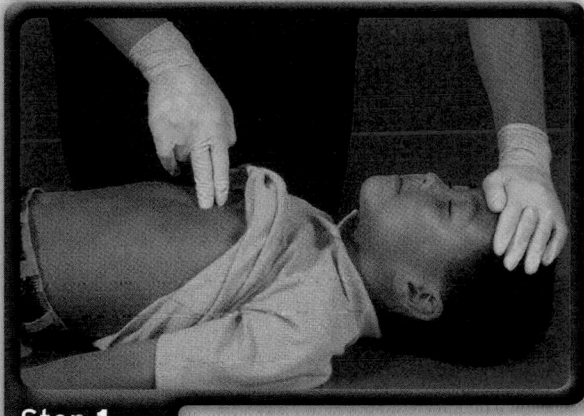

Step 1 Place the child on a firm surface. Place the heel of one or both hands in the center of the chest, in between the nipples, avoiding the xiphoid process.

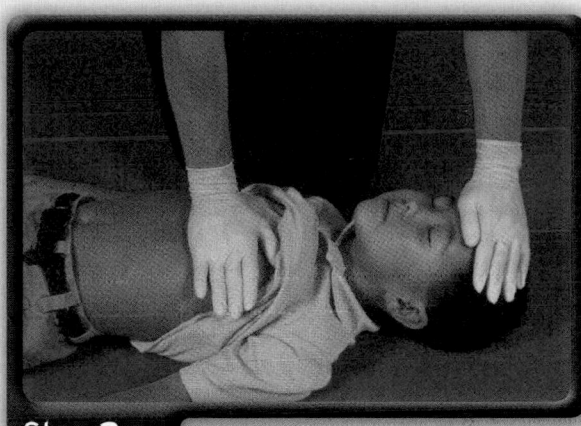

Step 2 Compress the chest about one third the anterior-posterior diameter of the chest at a rate of at least 100 times/min. Coordinate compressions with ventilations in a 30:2 ratio (one rescuer) or 15:2 (two rescuers), pausing for ventilations.

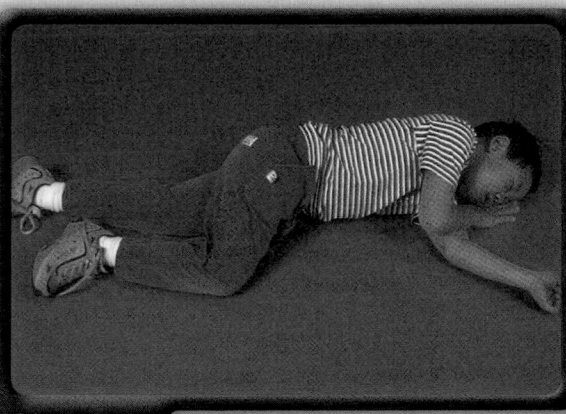

Step 3 If there is no pulse, apply your AED. If the child regains a pulse of greater than 60 beats/min and resumes effective breathing, place him or her in a position that allows for frequent reassessment of the airway and vital signs during transport.

■ Airway

Children (infants and toddlers) often put toys and other objects, as well as food, in their mouths; therefore, foreign body obstruction of the upper airway is common. You must make sure that the upper airway is open when managing pediatric respiratory emergencies or cardio-pulmonary arrest. If the child is unresponsive and lying in a supine position, the airway may become obstructed when the tongue and throat muscles relax and the tongue falls backward.

If the child is unresponsive but breathing adequately, place him or her in the recovery position to maintain an open airway and allow drainage of saliva, vomitus, or other secretions from the mouth Figure 11-21. Do not use this position if you suspect a spinal injury unless you can secure the child to a backboard that can be tilted to the side. Do not attempt to open the airway at all if the child is responsive and breathing, but in a labored fashion. Instead, provide immediate transport to the nearest hospital.

Opening the airway in an infant or child is done by using the same techniques as used for an adult. However, because a child's neck is so flexible, the techniques should be slightly modified. The jaw-thrust maneuver without a head tilt is the best method to use if you suspect a spinal injury in a child. If a second rescuer is present, he or she should immobilize the child's cervical spine. If spinal injury is not suspected, use the head tilt–chin lift maneuver but modified so that, as you tilt the head back, you are moving it only into the neutral position or a slightly extended position Figure 11-22.

Head Tilt-Chin Lift Maneuver

Perform the head tilt-chin lift technique in a child in the following manner:

1. Place one hand on the child's forehead, and tilt the head back gently, with the neck slightly extended.
2. Place two or three fingers (not the thumb) of your other hand under the child's chin, and lift the jaw

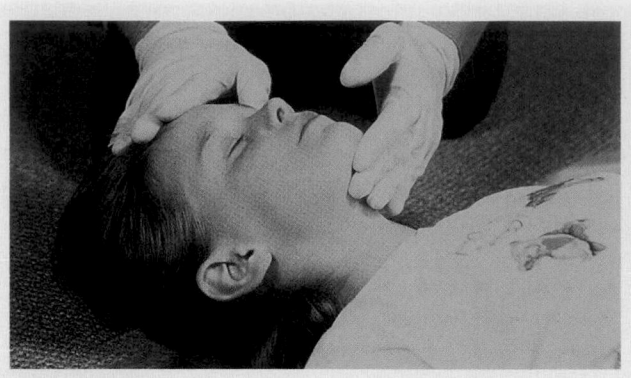

Figure 11-22 Use the head tilt-chin lift maneuver to open the airway in a child who has not sustained a traumatic injury. Do not overextend the neck.

upward and outward. Do not close the mouth or push under the chin; either move may obstruct rather than open the airway.
3. Remove any visible foreign body or vomitus.

Jaw-Thrust Maneuver

Perform the jaw-thrust maneuver in a child in the following manner:

1. Place two or three fingers under each side of the angle of the lower jaw; lift the jaw upward and outward.
2. If the jaw thrust alone does not open the airway and cervical spine injury is not a consideration, tilt the head slightly. If cervical spine injury is suspected, use a second rescuer to immobilize the cervical spine.

Remember that the head of an infant or young child is disproportionately large in comparison with the chest and shoulders. As a result, when a child is lying flat on his or her back, especially on a backboard, the head will bend forward (hyperflexion) onto the upper chest. This position can partially or completely obstruct the upper airway. To avoid this possibility, place a wedge of padding under the child's upper chest and shoulders (torso).

■ Provide Rescue Breathing

Once the airway is open, take at least 5 seconds but no more than 10 seconds to determine whether the child is breathing spontaneously Figure 11-23.

If an infant or small child is breathing, provide immediate transport. Again, a child who is in respiratory distress should be allowed to stay in whatever position is most comfortable. Larger children who are unresponsive and breathing with difficulty should be kept in the recovery position if possible.

If an infant or child is not breathing but has a pulse, provide rescue breathing while keeping the airway open. If you are using mouth-to-mouth resuscitation with an infant,

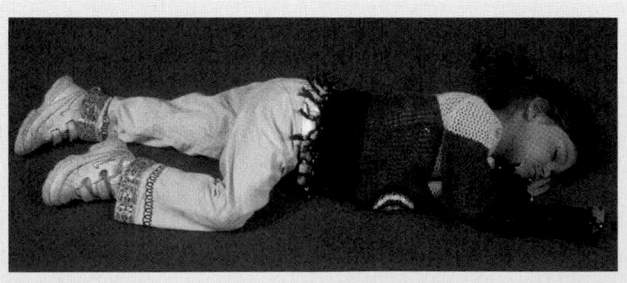

Figure 11-21 A child who is unconscious but breathing should be placed in the recovery position to allow saliva or vomitus to drain from the mouth.

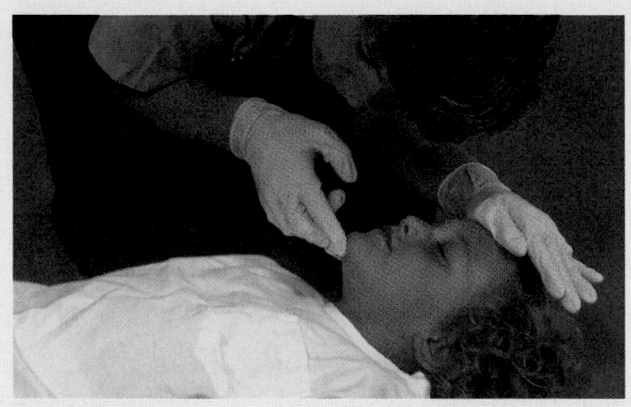

Figure 11-23 After you have opened the airway, determine whether the child is breathing spontaneously.

Children in respiratory distress are often struggling to breathe. As a result, they usually position themselves in a way that keeps the airway open enough for air to move. Let the child stay in that position as long as his or her breathing remains adequate. If you and your partner arrive at the scene and find that the infant or child is not breathing or has cyanosis, immediate management, including rescue breathing and supplemental oxygen, is essential. Consider requesting additional assistance, if available.

For infants, the preferred technique of rescue breathing is mouth-to-nose-and-mouth ventilation. With this technique, a seal must be made over the mouth and nose. Various masks and other barrier devices are recommended for this technique. If the patient is a large child (1 to 8 years old) for whom a tight seal cannot be made over both mouth and nose, you should provide mouth-to-mouth ventilation as you would for an adult.

Once you have made an airtight seal over the mouth, give two gentle breaths, each lasting 1 second. These initial breaths will help you assess for airway obstruction and expand the lungs. Because the lungs of infants and children are much smaller than those of adults, you do not need to blow in a large amount of air. Limit the amount of air to that needed to cause the chest to rise.

Remember, too, that a child's airway is smaller than that of an adult. Therefore, there is greater resistance to air flow. As a result, you will need to use slightly more ventilatory pressure to inflate the lungs. You will know you are giving the correct amount of air volume when you see the chest rise. Infants and children should be ventilated once every 3 to 5 seconds, or 12 to 20 breaths/min.

If air enters freely with your initial breaths and the chest rises, the airway is clear. You should then check the pulse. If air does not enter freely, you should check the airway for obstruction. Reposition the patient to open the airway, and attempt to give another breath. If air still does not enter freely, you must take steps to relieve the obstruction.

place your mouth over the infant's mouth and nose to create a seal. If you are using a bag-mask device to assist ventilations in an infant, use the proper sized mask and the technique described earlier.

In a child with tracheostomy (breathing) tubes in the neck, remove the mask from the bag-mask device and connect it directly to the tracheostomy tube to ventilate the child. If a bag-mask device is unavailable, a mask, barrier device, or your mouth over the tracheostomy site can be used. Place your hand firmly over the child's mouth and nose to prevent the artificial breaths from leaking out of the upper airway.

Words of Wisdom

An injured child with serious airway or breathing problems is likely to need full-time attention from two EMTs. The need for a driver, and often for added help with patient care, makes it important for you to start arranging early for backup from another unit—possibly even before you arrive at the scene.

Words of Wisdom

AEDs are becoming more and more accessible in the community. Be familiar with your local protocols on pediatric defibrillation. Your service may use a pediatric AED or an AED with a pediatric adapter.

Interrupting CPR

CPR is an important holding action that provides minimal circulation and ventilation until the patient can receive definitive care in the form of defibrillation or further care at the hospital. No matter how well CPR is performed, however, it is rarely enough to save a patient's life. If ALS is not available at the scene, you must provide transport based on your local protocols, continuing CPR on the way. En route to the hospital, you should consider requesting a rendezvous with ALS personnel, if available. This will provide ALS care to the patient earlier, improving his or her chance for survival. Note however, that not all EMS systems have ALS support available to them, especially in rural settings.

Try not to interrupt CPR for more than a few seconds, except when it is absolutely necessary. For example, if you have to move a patient up or down stairs, you should continue CPR until you arrive at the head or foot of the stairs, interrupt CPR at an agreed-on signal, and move quickly to the next level where you can resume CPR. Do not move the patient until all transport arrangements are made so that your interruptions of CPR can be kept to a minimum.

When Not to Start BLS

As an EMT, it is your responsibility to start CPR in virtually all patients who are in cardiac arrest. There are only two general exceptions to the rule.

First, you should not start CPR if the patient has obvious signs of death. Obvious signs of death include an absence of a pulse and breathing, along with any one of the following findings:

- Rigor mortis, or stiffening of the body after death
- Dependent lividity (livor mortis), a discoloration of the skin caused by pooling of blood **Figure 11-24**
- Putrefaction or decomposition of the body
- Evidence of nonsurvivable injury, such as decapitation, dismemberment, or burned beyond recognition.

Rigor mortis and dependent lividity develop after a patient has been dead for a long period.

Second, you should not start CPR if the patient and his or her physician have previously agreed on do not resuscitate (DNR) orders or no-CPR orders **Figure 11-25**. This may apply only to situations in which the patient is known to be in the terminal stage of an incurable disease. In this situation, CPR serves only to prolong the patient's death. However, this can be a complicated issue. Advance directives, such as living wills, may express the patient's wishes; however, these documents may not be readily producible by the patient's family or caregiver.

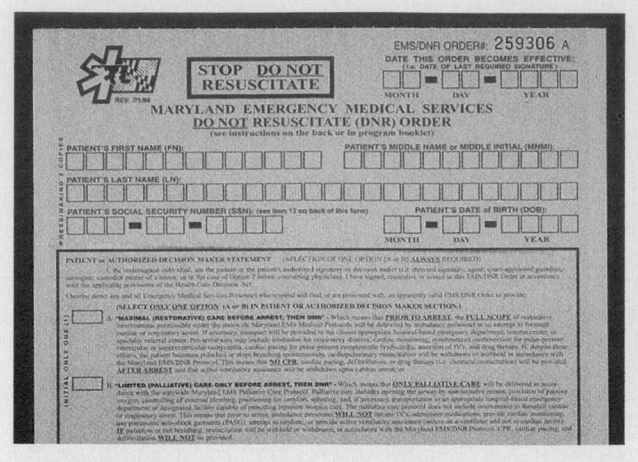

Figure 11-25 You should not start CPR if the patient and his or her physician have previously agreed on DNR or no-CPR orders. Learn your local protocols for treating terminally ill patients.

In such cases, the safest course is to assume that an emergency exists and begin CPR under the rule of implied consent and contact medical control for further guidance. Conversely, if a valid DNR document or living will is produced, resuscitative efforts may be withheld. Learn your local protocols and the standards in your system for treating terminally ill patients. Some EMS systems have computer notes on patients who are preregistered with the system. These notes usually specify the amount and extent of treatment that is desired. Other states have specific EMS DNR forms that allow EMS providers to withhold care when the patient, family, and physician have agreed in advance that such a course is most appropriate. It is essential that you understand your local protocols and are aware of the specific restrictions these advance directives imply.

In all other cases, you should begin CPR on anyone who is in cardiac arrest. It is usually impossible to know how long the patient has been without oxygen to the brain and vital organs. Factors such as air temperature and the basic health of the patient's tissues and organs can affect their ability to survive. Therefore, most legal advisers recommend that, when in doubt, always give too much care rather than too little care. You should always start CPR if any doubt exists.

When to Stop BLS

You are not responsible for making the decision to stop CPR. Once you begin CPR in the field, you must continue until one of the following events occurs:

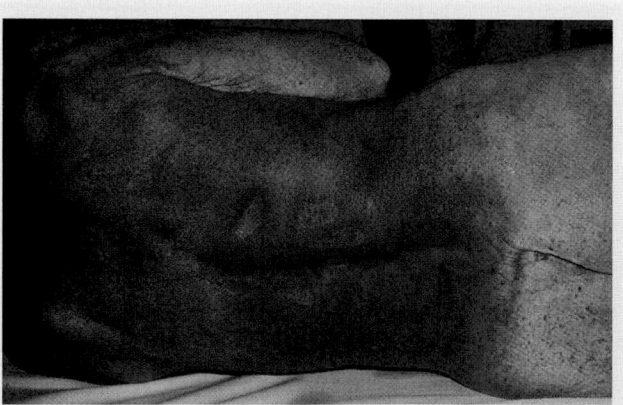

Figure 11-24 Dependent lividity is an obvious sign of death, caused by blood settling to the areas of the body not in firm contact with the ground. The lividity in this figure is seen as purple discoloration of the back, except in areas that are in firm contact with the ground (scapula and buttock).

S The patient *Starts* breathing and has a pulse.

T The patient is *Transferred* to another person who is trained in BLS, to ALS-trained personnel, or to another emergency medical responder.

O You are *Out* of strength or too tired to continue.

P A *Physician* who is present or providing online medical direction assumes responsibility for the patient and gives direction to discontinue CPR.

"Out of strength" does not mean merely weary; rather, it means that the person providing CPR is no longer physically able to perform CPR. In short, CPR should always be continued until the patient's care is transferred to a physician or higher medical authority in the field. In some cases, your medical director or a designated medical control physician may order you to stop CPR on the basis of the patient's condition.

Every EMS system should have clear standing orders or protocols that provide guidelines for starting and stopping CPR. Your medical director and your system's legal adviser should agree on these protocols, which should be closely administered and reviewed by your medical director.

Words of Wisdom

Correct handling of situations when you choose not to start CPR on a patient in cardiac arrest begins with compliance with your protocols and ends with you providing detailed documentation. In particular, record physical examination signs that led to your decision and make reference to the protocol that states these signs are a reason not to start. If extenuating circumstances such as entrapment physically prevent you from making resuscitation attempts, record the conditions thoroughly. These decisions occasionally give rise to questions that can often be put to rest immediately with reference to a well-written report.

Foreign Body Airway Obstruction in Adults

Occasionally, a large foreign body will be aspirated and block the upper airway. An airway obstruction may be caused by various things, including relaxation of the throat muscles in an unconscious patient, vomited or regurgitated stomach contents, blood, damaged tissue after an injury, dentures, or foreign bodies such as food or small objects.

Large objects that cannot be removed from the airway with suction, such as loose dentures, large pieces of food, or blood clots, should be swept forward and out with your gloved index finger. Suctioning can then be used as needed to keep the airway clear of thinner secretions such as blood, vomitus, and mucus.

Recognizing Foreign Body Airway Obstruction

An airway obstruction by a foreign body in an adult usually occurs during a meal. In children, it usually occurs during mealtime or at play. Children commonly choke on peanuts, large bits of a hot dog, or small toys. If the foreign body is not removed quickly, the lungs will use up their oxygen supply and unconsciousness and death will follow. Treatment is based on the severity of airway obstruction the patient is experiencing.

Mild Airway Obstruction

Patients with a mild (partial) airway obstruction are able to exchange adequate amounts of air, but still have signs of respiratory distress. Breathing may be noisy; however, the patient usually has a strong, effective cough. Leave these patients alone! Your main concern is to prevent a mild airway obstruction from becoming a severe airway obstruction. The abdominal thrust is not indicated in patients with a mild airway obstruction.

For the patient with a mild airway obstruction, you should first encourage him or her to cough or to continue coughing if they are already doing so. Do not interfere with the patient's own attempts to expel the foreign body. Instead, give 100% oxygen with a nonrebreathing mask and provide prompt transport to the hospital. Closely monitor the patient and observe for signs of a severe airway obstruction (weak or absent cough, decreasing level of consciousness, cyanosis).

Conscious Patients

A sudden, severe airway obstruction is usually easy to recognize in someone who is eating or has just finished eating. The person is suddenly unable to speak or cough, grasps his or her throat, turns cyanotic, and makes exaggerated efforts to breathe. Air is not moving into and out of the airway or the air movement is so slight that it is not detectable. At first, the patient will be conscious and able to clearly indicate the nature of the problem. Ask the patient, "Are you choking?" The patient will usually answer by nodding yes. Alternatively, he or she may use the universal sign to indicate airway blockage **Figure 11-26** .

If there is a minimal amount of air movement, you may hear a high-pitched sound called stridor. This occurs when the object is not fully occluding the airway, but the small amount of air entering the lungs is not enough to sustain life and the patient will eventually lose consciousness if the obstruction is not relieved.

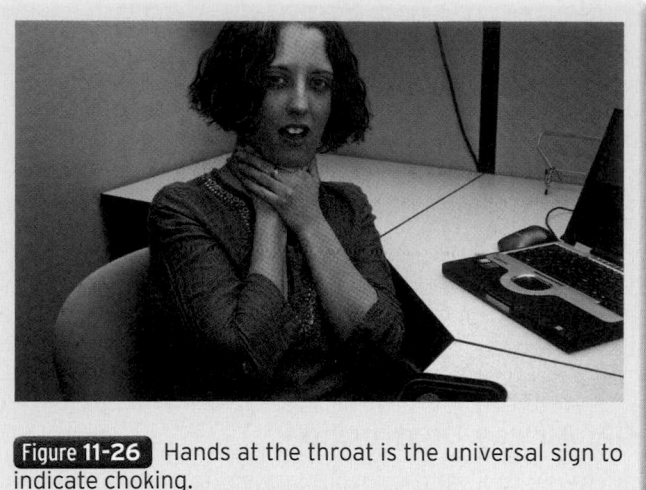

Figure 11-26 Hands at the throat is the universal sign to indicate choking.

are not effective. If you feel resistance to blowing into the patient's lungs or pressure builds up in your mouth, the patient probably has some type of obstruction.

■ Removing a Foreign Body Airway Obstruction in an Adult

The manual maneuver recommended for removing severe airway obstructions in conscious adults and children older than 1 year of age is the abdominal-thrust maneuver (the Heimlich maneuver). This technique creates an artificial cough by causing a sudden increase in intrathoracic pressure when thrusts are applied to the subdiaphragmatic region; it is a very effective method for removing a foreign body that is obstructing the airway.

Unconscious Patients

When you discover an unconscious patient, your first step is to determine whether he or she is breathing and has a pulse. The unconsciousness may be caused by airway obstruction, cardiac arrest, or a number of other problems. Remember that you must first clear the patient's airway, making sure it is open and unobstructed, before checking for a pulse and addressing other problems, such as cardiac arrest.

You should suspect an airway obstruction if the standard maneuvers to open the airway and ventilate the lungs

Conscious Patients

Abdominal-Thrust Maneuver The <u>abdominal-thrust maneuver</u>, also called the Heimlich maneuver, is the preferred way to dislodge a severe airway obstruction in conscious adults and children older than 1 year of age. The goal is to compress the lungs upward and force the residual air, which is always present in the lungs, to flow upwards and expel the object. In conscious patients with a severe airway obstruction, you should repeat abdominal thrusts until the foreign body is expelled or the patient becomes unconscious. Each thrust should be deliberate, with the intent of relieving the obstruction.

You are the Provider: PART 4

After 2 minutes of CPR, you reanalyze the patient's cardiac rhythm and receive a "no shock advised" message. You and your partner immediately resume CPR. During CPR, your partner ventilates the patient with a bag-mask device and high-flow oxygen. As she attempts to insert an oral airway, the patient starts to gag. You quickly reassess him.

Recording Time: 7 Minutes	
Level of consciousness	Unconscious and unresponsive
Respirations	Occasional agonal breaths; 4 breaths/min
Pulse	100 beats/min; strong carotid pulse; absent radial pulses
Skin	Skin color is improving
Blood pressure	70/40 mm Hg
Sao_2	82% (on oxygen)

7. How should you continue to treat this patient?

8. Because the patient is no longer in cardiac arrest, should you remove the AED pads? Why or why not?

To perform abdominal thrusts on a conscious adult **Figure 11-27**, use the following technique:

1. Stand behind the patient, and wrap your arms around his or her abdomen. Straddle your legs outside the patient's legs. This will allow you to easily slide the patient to the ground in the event he or she becomes unconscious.
2. Make a fist with one hand; grasp the fist with the other hand. Place the thumb side of the fist against the patient's abdomen just above the umbilicus.
3. Press your fist into the patient's abdomen with a quick inward and upward thrust.
4. Continue abdominal thrusts until the object is expelled from the airway or the patient becomes unconscious.

Chest Thrusts You can perform the abdominal-thrust maneuver safely on all adults and children. However, for women in advanced stages of pregnancy and patients who are very obese, you should use chest thrusts instead.

To perform chest thrusts on the conscious adult, use the following technique **Figure 11-28**:

1. Stand behind the patient with your arms directly under the patient's armpits, and wrap your arms around the patient's chest.
2. Make a fist with one hand; grasp the fist with the other hand. Place the thumb side of the fist against the patient's sternum, avoiding the xiphoid process and the edges of the rib cage.
3. Press your fist into the patient's chest with backward thrusts until the object is expelled or the patient becomes unconscious.

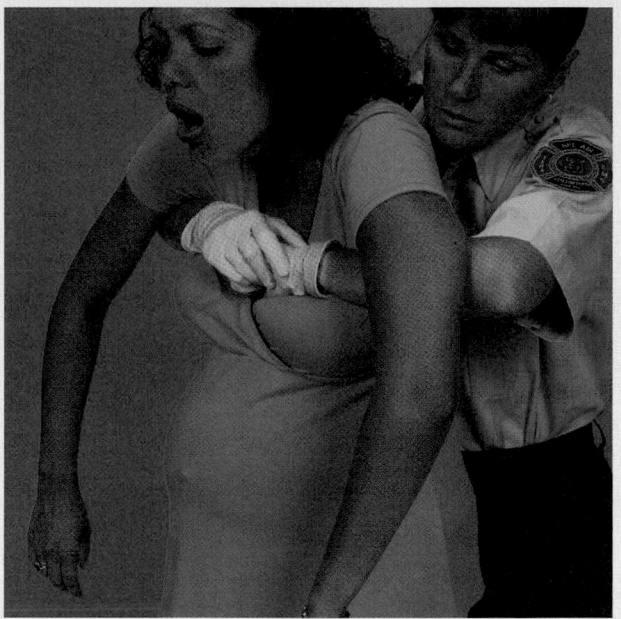

Figure 11-28 Removal of a foreign body obstruction in a conscious adult using chest thrusts. Stand behind the patient and wrap your arms around the patient's chest. Place the thumb side of one fist against the chest while holding your fist with your other hand. Press your fists into the patient's chest with backward thrusts.

4. If the patient becomes unconscious, you should begin CPR, starting with chest compressions **Figure 11-29**.

Words of Wisdom

If a conscious choking patient is found lying on the floor, abdominal thrusts can be administered by straddling the patient's legs, placing your hands just above the umbilicus, and giving rapid thrusts inward and upward under the rib cage, using the palm of your hand with your other hand on top of it.

Conscious Patients Who Become Unconscious

A patient with an airway obstruction may become unconscious and require additional care. Knowing that the patient had an obstruction should prompt you to open the airway and look for an obstruction before completing the additional steps of resuscitation. Use the following steps to manage the patient:

1. Open the patient's airway.
2. Look in the patient's mouth. If you see the foreign object, remove it. If an object is not seen, begin CPR compressions.
3. Assess for breathing.

Figure 11-27 The abdominal-thrust maneuver in a conscious adult. Stand behind the patient and wrap your arms around the patient's abdomen. Place the thumb side of one fist against the patient's abdomen while holding your fist with your other hand. Press your fists into the patient's abdomen, using inward and upward thrusts.

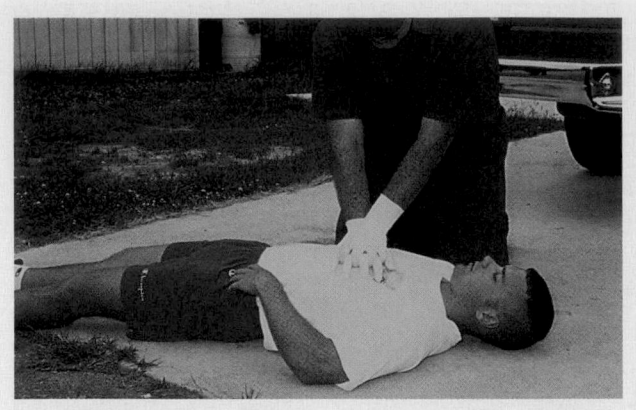

Figure 11-29 An unconscious patient with an airway obstruction requires CPR.

4. If the patient is not breathing, attempt to give one ventilation. If the air does not go in, reposition the patient's head and attempt one more ventilation.
5. If the air still does not go in, give 30 chest compressions.
6. Look in the patient's mouth to see if you can visualize the object. If you see the object, remove it. If not, attempt to ventilate.
7. Continue steps 4 and 5 until the object is removed and air flow is established. Once you can successfully ventilate the patient, check for a pulse. If there is no pulse, begin CPR with compressions.

Unconscious Patients

When a victim is found unconscious, it is unlikely that you will know what caused the problem. Begin the steps of CPR by determining unresponsiveness and beginning chest compressions. Perform 30 chest compressions and then open the airway and look in the mouth. If an object is visible, attempt to remove it. Never perform blind finger sweeps on any patient; doing so may push the obstruction further into the airway. After opening the airway and looking inside the mouth, reattempt to ventilate the patient. Continue the process of chest compressions, opening the airway, looking inside the mouth, and attempting to ventilate until the airway is clear or advanced life support help arrives.

Words of Wisdom

It is likely that if you find an unresponsive victim you will not know whether choking was the initial cause. Begin the steps of CPR and assess during the breathing phase to determine if there is good air flow by looking at chest rise. If the chest does not rise, reposition the airway, look inside the mouth, and ventilate again.

Foreign Body Airway Obstruction in Infants and Children

Airway obstruction is a common problem in infants and children. It is usually caused by a foreign body such as food or a toy, or by an infection, such as croup or epiglottitis, resulting in swelling and narrowing of the airway. You should try to identify the cause of the obstruction as soon as possible. In patients who have signs and symptoms of an airway infection, you should not waste time trying to dislodge a foreign body. The child needs 100% oxygen with a nonrebreathing mask and immediate transport to the emergency department.

A previously healthy child who is eating or playing with small toys or an infant who is crawling about the house and who suddenly has difficulty breathing has probably aspirated a foreign body. As in adults, foreign bodies may cause a mild or a severe airway obstruction.

With a mild airway obstruction, the patient can cough forcefully, although there may be wheezing between coughs. As long as the patient can breathe, cough, or talk, you should not interfere with his or her attempts to expel the foreign body. As with the adult, encourage the child to continue coughing. Administer 100% oxygen with a nonrebreathing mask (if tolerated) and provide transport to the emergency department.

You should intervene only if signs of a severe airway obstruction develop, such as a weak, ineffective cough, cyanosis, stridor, absent air movement, or a decreasing level of consciousness.

■ Removing a Foreign Body Airway Obstruction in a Child

Conscious Child

If you determine a child older than 1 year has an airway obstruction, stand or kneel behind the child and provide abdominal thrusts in the same manner as an adult, but use less force, until the object is expelled or the child becomes unconscious. If the child becomes unconscious, follow the same steps as for the unconscious adult.

To perform the abdominal-thrust maneuver in a conscious child who is in a standing or sitting position, follow these steps **Figure 11-30** :

1. Kneel on one knee behind the child, and circle both of your arms around the child's body. Prepare to give abdominal thrusts by placing your fist just above the patient's umbilicus and well below the lower tip of the sternum. Place your other hand over that fist.
2. Give the child abdominal thrusts in an upward direction. Be careful to avoid applying force to the lower rib cage or sternum.

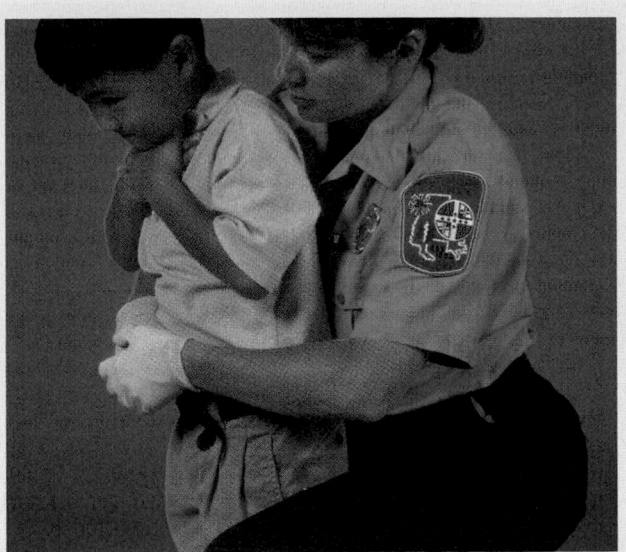

Figure 11-30 To perform the abdominal-thrust maneuver, kneel behind the child, wrap your arms around his or her body, and place your fist just above the umbilicus and well below the lower tip of the sternum.

5. Attempt rescue breathing. If the first attempt fails, reposition the head and try again.

6. If the airway remains obstructed, begin CPR starting with chest compressions.

If you manage to clear the airway obstruction in an unconscious child but he or she still has no spontaneous breathing or circulation, perform CPR.

Unconscious Child

An unconscious child older than 1 year who has an airway obstruction is managed in the same manner as an adult. **Skill Drill 11-7** demonstrates the steps for removing a foreign body airway obstruction in an unconscious child:

3. Repeat this standing technique until the child expels the foreign body or loses consciousness.

4. If the child becomes unconscious, position the child on a hard surface. Open the airway using the head tilt–chin lift maneuver and look inside the child's mouth. If you can see the foreign body, try to remove it.

Skill Drill 11-7

1. Place the child in a supine position on a firm, flat surface Step 1 .

2. Open the airway using the head tilt–chin lift maneuver and look inside the child's mouth for the obstruction Step 2 .

3. Attempt rescue breathing. If the first try is unsuccessful, reposition the child's head and try again Step 3 .

4. If ventilation is still unsuccessful, begin CPR Step 4 .

You are the Provider: PART 5

You package the patient, load him into the ambulance, and begin transport to a hospital located 5 miles away. An EMT from the backup ambulance accompanies you in the back and continues rescue breathing. En route, you reassess the patient and then call your radio report to the receiving hospital.

Recording Time: 12 Minutes	
Level of consciousness	Unconscious and unresponsive
Respirations	8 breaths/min; shallow depth
Pulse	94 beats/min; strong carotid pulse, weak radial pulses
Skin	Pink, cool, and dry
Blood pressure	86/66 mm Hg
Sao$_2$	95% (on oxygen)

9. Would an impedance threshold device benefit your patient at this point?

10. What further treatment is indicated for this patient?

5. Place the heel of one hand on the center of the chest (lower half of the sternum).

6. Administer 30 chest compressions. Compressions should be one third to one half the depth of the chest.

7. Open the airway using the head tilt–chin lift maneuver and look inside the child's mouth. If you see the object, remove it (Step 5).

8. Repeat the process starting at Step 3.

■ Removing a Foreign Body Airway Obstruction in Infants

Conscious Infants

Abdominal thrusts are not recommended for conscious infants with an airway obstruction because of the risk of injury to the immature organs of the abdomen. Instead, perform back slaps and chest thrusts to try to clear a severe airway obstruction in a conscious infant, as follows (Figure 11-31):

1. Hold the infant face down, with the body resting on your forearm. Support the infant's jaw and face with your hand, and keep the head lower than the rest of the body.

2. Deliver five back slaps between the shoulder blades, using the heel of your hand.

3. Place your free hand behind the infant's head and back, and turn the infant face up on your other forearm and thigh, sandwiching the infant's body between your two hands and arms. The infant's head should remain below the level of the body.

4. Give five quick chest thrusts in the same location and manner as chest compressions, using two fingers placed on the lower half of the sternum. For larger infants, or if you have small hands, you can perform this step by placing the infant in your lap and turning the infant's whole body as a unit between back slaps and chest thrusts.

5. Check the airway. If you can see the foreign body now, remove it. If not, repeat the cycle as often as necessary.

6. If the infant becomes unconscious, begin CPR, remembering to look in the airway before ventilations each time.

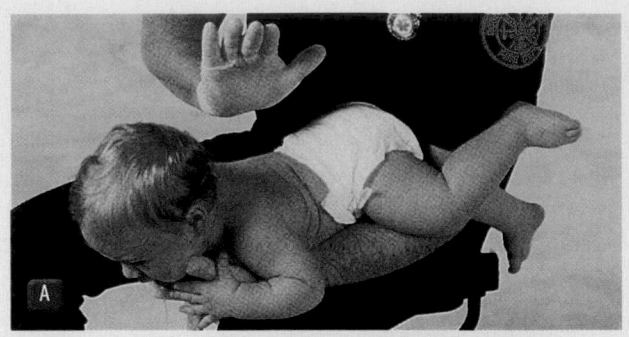
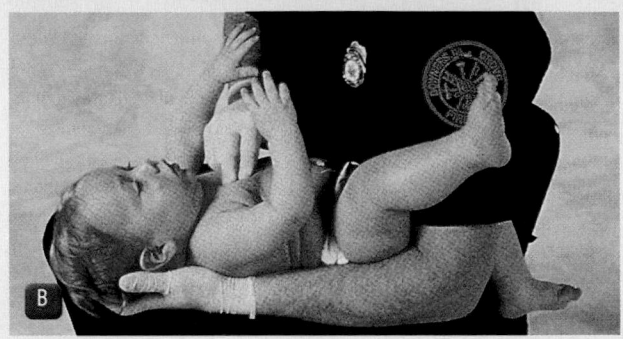

Figure 11-31 **A.** Hold the infant face down with the body resting on your forearm. Support the jaw and face with your hand, and keep the head lower than the rest of the body. Give the infant five back slaps between the shoulder blades, using the heel of your hand. **B.** Give the infant five quick chest thrusts, using two fingers placed on the lower half of the sternum.

As with the adult and child, if the infant loses consciousness, look inside the mouth. If you see the object, remove it. If not, begin CPR with 30 compressions. If there is no pulse, or the pulse is less than 60 beats per minute, continue the process of compressions, looking in the mouth, and attempting ventilations until the obstruction is relieved and then assess for a pulse.

Unconscious Infants

Begin CPR but include one extra step: Look inside the infant's airway each time before ventilating and remove the object if seen.

Skill Drill 11-7

Removing a Foreign Body Airway Obstruction in an Unconscious Child

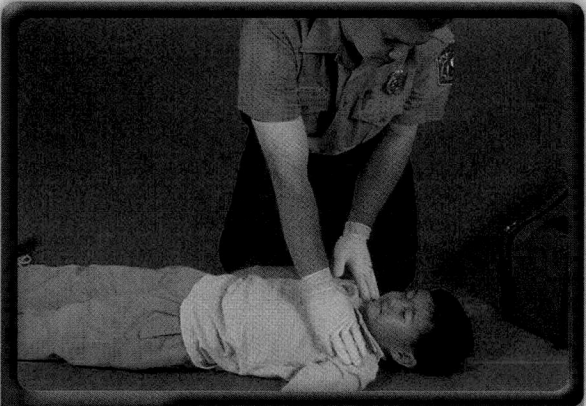

Step 1 Position the child on a firm, flat surface.

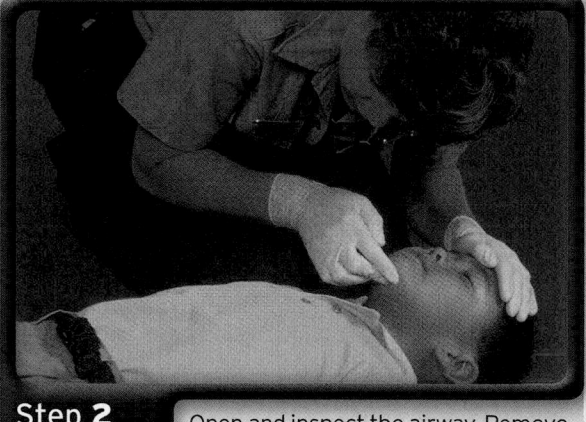

Step 2 Open and inspect the airway. Remove any foreign object that you can see.

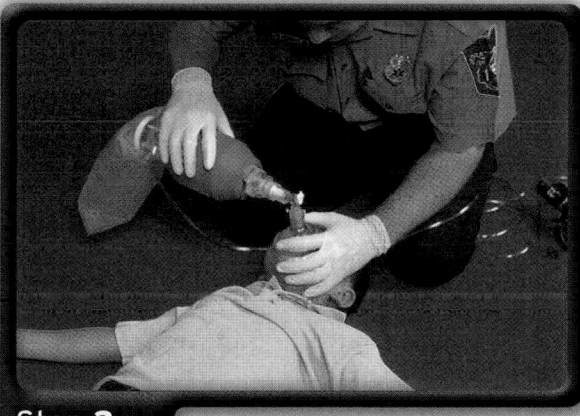

Step 3 Attempt rescue breathing. If unsuccessful, reposition the head and try again.

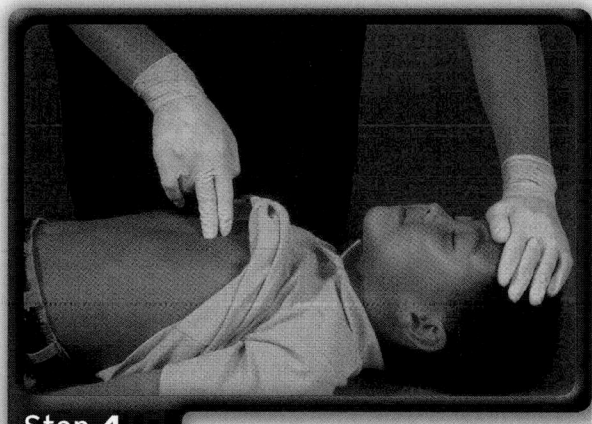

Step 4 If ventilation is still unsuccessful, begin CPR. Locate the proper hand position on the chest of the child.

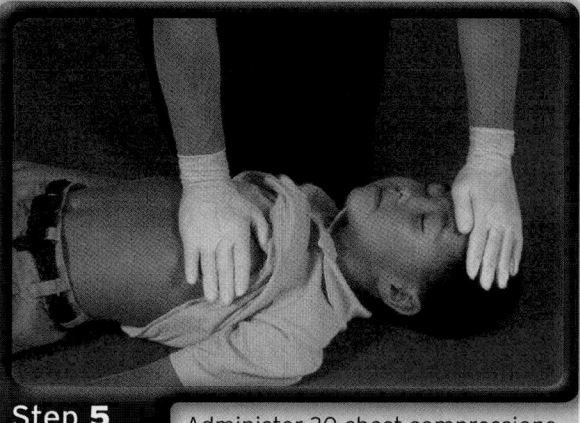

Step 5 Administer 30 chest compressions and look inside the child's mouth. If you see the object, remove it.

You are the Provider: SUMMARY

1. What should you immediately do on receiving this update from dispatch?

Once you are informed that CPR is in progress, you should immediately request additional assistance. Effective treatment of a patient in cardiac arrest requires adequate personnel at the scene and during transport.

The type of backup you receive (ie, EMT versus paramedic) will depend on your EMS system and the resources that are available to you. An advanced life support (ALS) ambulance staffed with paramedics would be optimum; paramedics are able to establish vascular access, administer various cardiac medications, and perform cardiac monitoring and advanced airway management. Combined with early, high-quality CPR and defibrillation, early advanced care increases the patient's chance for survival.

Some EMS systems provide basic life support (BLS) only; they are not staffed with AEMTs or paramedics. In very small EMS systems, only one EMT ambulance may cover a large area. If you do not have access to other EMTs or to paramedics, you should request assistance from the fire department. Fire departments are often staffed with at least one or two EMRs who are able to perform CPR and assist with certain BLS interventions.

Regardless of the resources available to you, you should request them as soon as possible—in this case, as soon as you are advised that CPR is in progress. One EMT cannot *effectively* treat a cardiac arrest patient during transport; he or she would have to perform continuous CPR, which would arguably result in rescuer fatigue and decreased chest compression effectiveness.

In some EMS systems, two ambulances are automatically dispatched to calls that may involve a patient in cardiac arrest (ie, an "unconscious" person). In other systems, EMRs are dispatched to assist the ambulance. As an EMT, you must be familiar with the resources that are available to you and know when it is appropriate to request them.

2. What should be your initial actions on arriving at this scene?

After ensuring your own safety, you should approach this patient as you would any other patient, by performing a primary assessment. Although the dispatcher has advised you that bystander CPR is in progress, you must still assess the patient to confirm that he is indeed apneic and pulseless and requires CPR.

Your primary assessment should take only a few seconds, just long enough to confirm that the patient is in cardiac arrest. If so, begin CPR immediately, apply the AED as soon as it is available, and analyze the patient's cardiac rhythm. To avoid interrupting CPR, you should apply the AED pads around your partner's hands as he or she is performing chest compressions.

If the AED advises you to shock, deliver the shock and immediately resume CPR, starting with chest compressions. If the AED does not advise you to shock, immediately resume CPR, starting with chest compressions. During CPR, ask the bystanders if they witnessed the event and determine if they know anything about the patient (ie, past medical history, events leading up to the arrest).

Regardless of how a call is dispatched and whether or not you are assuming patient care from bystanders or other health care providers, it is important for you to always perform a primary assessment of the patient.

3. What links in the chain of survival have been maintained at this point?

Few people who experience prehospital cardiac arrest survive unless a rapid sequence of events takes place within a very narrow time frame. This sequence of events—called the chain of survival—emphasizes the most critical elements for treating cardiac arrest patients. There are five links in the chain of survival: early access, early CPR, early defibrillation, early advanced care, and post-arrest care.

Early access requires public education and awareness in the recognition of the early warning signs of a cardiac emergency and immediate activation of EMS. This link has been maintained because the bystander quickly recognized that the patient was experiencing a cardiac emergency and immediately called 9-1-1.

Early CPR is an essential component of the chain of survival. High-quality CPR will help keep blood, and therefore oxygen, flowing to the vital organs (ie, heart, brain), thus helping to keep these organs viable until additional care can be provided. Although CPR alone rarely resuscitates a cardiac arrest patient, in its absence, the chance of survival is low. This link in the chain of survival has also been maintained because the bystander began immediate CPR after calling 9-1-1.

The third link in the chain of survival, early defibrillation, has also been maintained because an AED is present and the patient can be quickly defibrillated if needed. Of all the links in the chain of survival, early defibrillation has the most profound impact on patient survival because approximately 70% of patients who experience sudden cardiac arrest are in a shockable rhythm (eg, ventricular fibrillation [V-fib] or pulseless ventricular tachycardia [V-tach]) initially. With early access and early CPR, defibrillation may successfully terminate these lethal cardiac arrhythmias in a significant number of patients. It is important to note that ventricular fibrillation and pulseless ventricular tachycardia are transient arrhythmias; they quickly deteriorate to asystole, a nonshockable rhythm. For each minute that defibrillation is delayed, the patient's chance for survival decreases by 7% to 10%.

The fourth link in the chain of survival, early advanced care, can only be maintained if you have ALS personnel at the scene or are very close to a hospital. Advanced level care includes interventions such as intravenous therapy, cardiac drug administration, cardiac monitoring, manual defibrillation, and advanced airway management. In this case, you do not have ALS personnel available at the scene; therefore, it is critical that you transport the patient as soon as possible.

If *any one* of the links in the chain of survival is absent, the patient's chance for survival decreases. If CPR is not performed within the first few minutes of the arrest, for example, the patient's chance of survival decreases. Likewise, if the administration of defibrillation is delayed for more than 10 minutes, the patient's chance of survival decreases. The patient's chance for survival is greatest when *all the links* in the chain are strong.

4. Why is it so critical to minimize interruptions in CPR?

Even when CPR is performed optimally (eg, proper rate and depth), chest compressions only deliver about one third of a person's normal cardiac output. It is critical to perform CPR correctly; the depth of compressions should consistently be at least 2" in the adult and the rate of chest compressions should be at least 100 per minute.

If CPR is performed properly and with minimal interruption, it is often enough to keep the patient's vital organs viable until defibrillation and more advanced care can be provided. Of course, this assumes that defibrillation and advanced care are provided within a short period of time.

Within a few seconds of stopping chest compressions, the pressure generated in the arteries drops to near zero. With this fact in mind, frequent or prolonged interruptions in chest compressions will not even provide the minimum perfusion needed to keep the vital organs viable. This has clearly been linked to low survival rates from cardiac arrest.

As soon as cardiac arrest has been confirmed, it is crucial to begin CPR immediately and apply the AED as soon as it is available. Even when the AED pads are being applied, chest compressions should continue; the rescuer should apply the pads around his or her partner's hands.

5. Should you remove the medication patch or leave it in place? Why or why not?

It is not uncommon to encounter a patient who is wearing a medication patch. With the transdermal patch, medication is absorbed through the skin. A number of medications are delivered via the transdermal route, including fentanyl (a narcotic analgesic) for chronic pain, and nitroglycerin (a vasodilator) for angina, among others.

The patch in this patient is located on his right upper chest, which is where you will place one of the AED pads. Because of its location, the patch could interfere with the electrical current to the heart and may cause skin burns. To prevent this complication, you should remove the patch, wipe any residue from the skin, and then apply the AED pads. Do not forget to observe standard precautions!

6. What does the bulge and scar over the patient's left chest indicate? How will this affect the way you treat the patient?

Patients who are at high risk for certain cardiac arrhythmias and cardiac arrest may have an automatic implantable cardioverter/defibrillator (AICD) or pacemaker. The AICD will deliver shocks directly to the heart when it detects a lethal cardiac arrhythmia. Implanted pacemakers are used to increase the patient's heart rate if it falls below a given value.

It is easy to recognize these devices because they create a hard lump or bulge on the patient's chest and usually have a scar over them. Most AICDs and pacemakers are placed beneath the skin in the upper left aspect of the chest, just below the clavicle.

If the AED pads are placed directly over the device, shocks delivered by the AED may be less effective. Therefore, if you identify an AICD or pacemaker, you should place the AED pad at least 1" away from the device. Because most of these devices are implanted in the upper left chest, this should not be an issue. The pads are placed to the right of the upper sternum and to the lower left chest, just below the nipple, so they should be well beyond 1" from the device.

Occasionally, the AICD will deliver shocks to the patient; when it does, you will see the patient's muscles twitch. However, because the electricity delivered by the device is so low, it should not pose a threat to your safety. Follow your local protocols regarding patients with AICDs or implanted pacemakers.

7. How should you continue to treat this patient?

You have restored a pulse in your patient; however, his breathing is not adequate. Agonal breaths are slow, occasional breaths are ineffective and do not produce adequate minute volume.

Some patients may have an intact gag reflex, despite being unconscious and unresponsive; in these cases, an oropharyngeal airway is contraindicated. You should insert a nasopharyngeal airway and continue to provide rescue breathing. Deliver one breath every 5 to 6 seconds (10 to 12 breaths/min); each breath should be delivered over 1 second (just enough to produce visible chest rise). Closely and carefully monitor the patient's pulse and be prepared to resume CPR if necessary.

You should assume that the patient has a full stomach and have suction ready in case he regurgitates. Remember that mortality increases significantly if aspiration occurs.

It is important to avoid hyperventilating the patient. Hyperventilation, which is defined as ventilating too fast *or* with too much force, can cause several negative effects. It increases the incidence of gastric distention, which increases the risks of regurgitation and aspiration. It also hyperinflates the lungs, which reduces the amount of blood that returns to the heart (preload), and subsequently reduces the amount of blood that is ejected from the left ventricle (stroke volume).

8. Because the patient is no longer in cardiac arrest, should you remove the AED pads? Why or why not?

Although the patient is not in cardiac arrest, he is still at high risk for redeveloping cardiac arrest. Therefore, you should not remove the AED pads; simply turn the AED off, continue rescue breathing, and prepare the patient for immediate transport.

If cardiac arrest recurs en route to the hospital, you should first ask your partner to stop the ambulance while you begin CPR. Remember, the AED will not analyze the cardiac rhythm if it detects movement.

As soon as your partner is in the back of the ambulance to assist you, turn the AED back on, reanalyze the patient's cardiac rhythm, deliver a shock if indicated, and resume CPR. Follow your local protocols or contact medical control regarding how many shocks you should deliver prior to initiating transport from the scene or resuming transport if the patient redevelops cardiac arrest en route to the hospital.

You are the Provider: SUMMARY, continued

9. Would an impedance threshold device (ITD) benefit your patient at this point?

An impedance threshold device (ITD) is a valve device that is placed between the endotracheal (ET) tube and bag-mask device. It limits the amount of air that enters the thoracic cavity during the recoil phase between chest compressions. This results in negative intrathoracic pressure, which facilitates blood return to the heart (preload) and ultimately increases the amount of blood that is ejected from the heart during each chest compression. Despite the fact that the ITD attaches to the ET tube, it is a circulatory-assist device, not a ventilation device.

The ITD is only used for patients who are *apneic and pulseless*. At this point, your patient has a pulse and is breathing—albeit slowly and shallowly; therefore, the ITD is not indicated. However, if he redevelops cardiac arrest, the emergency physician may elect to use the ITD after he or she intubates the patient.

Although use of the ITD can help improve the effectiveness of CPR, it is important to note that the device is *not* a replacement for high-quality CPR, which includes pushing hard and fast, allowing the chest to fully recoil in between compressions, and limiting interruptions in CPR. If return of spontaneous circulation occurs, the ITD must be removed.

10. What further treatment is indicated for this patient?

Further treatment for your patient should consist of *careful* monitoring, keeping in mind that he remains at high risk for recurrence of cardiac arrest. In patients who are conscious and alert, the presence of a pulse is obvious; however, when a patient is unconscious and unresponsive, you must reassess for a pulse frequently.

Unconscious, unresponsive patients are at increased risk for regurgitation, which could lead to aspiration and increased mortality. Vigilantly monitor the patient's airway status and be prepared to turn his head to the side if he regurgitates. Maintain his airway with manual positioning and a basic airway adjunct, in this case, a nasal airway.

Although the patient is breathing, his breaths are slow and shallow. Slow, shallow (reduced tidal volume) respirations will not produce adequate minute volume and need to be assisted; therefore, continue to assist the patient's ventilations with a bag-mask device, but *do not hyperventilate him*. Deliver each breath over 1 second while observing for visible chest rise. Monitoring his oxygen saturation (Sao_2) level and heart rate will help you determine if your assisted ventilations are adequate.

You are the Provider: SUMMARY, continued

As mentioned earlier, do not remove the AED pads. Turn the AED off but be prepared to stop the ambulance if he redevelops cardiac arrest.

Depending on your local protocols, you may consider elevating the patient's legs 6″ to 12″ in an attempt to improve his blood pressure (currently 86/66 mm Hg) and to improve cerebral perfusion.

EMS Patient Care Report (PCR)				
Date: 12-29-09	**Incident No.:** 011109	**Nature of Call:** Cardiac arrest		**Location:** 123 Wilshire Ave.
Dispatched: 1445	**En Route:** 1447	**At Scene:** 1454	**Transport:** 1508	**At Hospital:** 1518	**In Service:** 1528

Patient Information

Age: 48 Sex: M Weight (in kg [lb]): 77 kg (170 lb)	Allergies: Unknown Medications: Unknown Past Medical History: Unknown Chief Complaint: Cardiac arrest

Vital Signs

Time: 1454	BP: N/A	Pulse: 0	Respirations: 0	Sao₂: N/A
Time: 1458	BP: N/A	Pulse: 0	Respirations: 0	Sao₂: N/A
Time: 1501	BP: 70/40	Pulse: 100	Respirations: 4	Sao₂: 82%
Time: 1508	BP: 86/66	Pulse: 94	Respirations: 8	Sao₂: 95%

EMS Treatment
(circle all that apply)

Oxygen @ 15 L/min via (circle one): NC NRM **Bag-Mask Device**	**Assisted Ventilation**	**Airway Adjunct**	**CPR**
Defibrillation	Bleeding Control	Bandaging	Splinting	Other:

Narrative

Medic 51 dispatched to grocery store parking lot for "CPR in progress." On arrival at the scene, found two bystanders performing CPR on the patient, a 48-year-old male. Medic 48 was dispatched to the scene to assist. Primary assessment revealed that the patient was apneic and pulseless. Continued one-rescuer CPR for 2 minutes while the AED was being prepared. Per one of the bystanders, the patient was about to get in his car when he suddenly grabbed his chest, slumped against the car, and eased himself to the ground. There was no trauma involved. The bystander further stated that by the time he got to the patient, he was unconscious and without pulse or breathing. After 2 minutes of CPR, analyzed patient's cardiac rhythm with the AED and received a shock advised message. Delivered single shock and immediately resumed CPR. Medic 48 arrived at scene and assisted with CPR and airway management. The patient's past medical history was unknown; although he had an AICD and was wearing a medication patch, which was removed. Continued CPR for 2 minutes, reanalyzed the patient's cardiac rhythm, and received a no shock advised message. Continued CPR and attempted to insert an oral airway; however, the patient began to gag. Immediate reassessment revealed that he had a strong carotid pulse, but was not breathing adequately. Inserted a nasal airway, continued ventilations at 12 breaths/min, packaged the patient, and loaded him into the ambulance. EMT from Medic 48 assisted with patient care en route to the hospital. En route, reassessed patient and found that he remained unconscious and unresponsive; his respiratory rate increased, but the depth of his breathing remained shallow. Continued assisted ventilation and called in radio report to the receiving facility. Monitored the patient's pulse, elevated his lower extremities in an attempt to improve his blood pressure, and delivered him to the emergency department without incident. Gave verbal report to attending physician. Medic 51 cleared the hospital and returned to service at 1528. *End of report*

- Basic life support (BLS) is noninvasive emergency lifesaving care that is used to treat medical conditions, including airway obstruction, respiratory arrest, and cardiac arrest.

- BLS care focuses on what is often termed the ABCs: airway (obstruction), breathing (respiratory arrest), and circulation (cardiac arrest or severe bleeding).

- Cardiopulmonary resuscitation (CPR) is used to establish artificial ventilation and circulation in a patient who is not breathing and has no pulse.

- The goal of CPR is to restore spontaneous breathing and circulation; however, advanced procedures such as medications and defibrillation are often necessary for this to occur.

- Advanced life support (ALS) involves advanced lifesaving procedures, such as cardiac monitoring, administration of intravenous fluids and medications, and use of advanced airway adjuncts.

- The five links in the chain of survival are early access, early CPR, early defibrillation, early advanced care, and post-arrest care.

- The automated external defibrillator (AED) should be applied to any nontrauma cardiac arrest patient as soon as it is available.

- When using an AED on a child between 1 and 8 years of age, you should use pediatric-sized pads and a dose-attenuating system (energy reducer). If these are not available, an adult AED should be used. In infants (1 month to 1 year), it is preferable to perform manual defibrillation or use a dose-attenuating system. If these are not available, an adult AED should be used.

- As an EMT, it is your responsibility to start CPR in virtually all patients who are in cardiac arrest. There are only two general exceptions to the rule: You should not start CPR if the patient has obvious signs of death and you should not start CPR if the patient and his or her physician have previously agreed on do not resuscitate (DNR) or no-CPR orders.

- You are not responsible for making the decision to stop CPR. Once you begin CPR in the field, you must continue until one of the following events occurs:
  - S, the patient Starts breathing and has a pulse.
  - T, the patient is Transferred to another person who is trained in BLS, to ALS-trained personnel, or to another emergency medical responder.
  - O, you are Out of strength or too tired to continue.
  - P, a Physician who is present or providing online medical direction assumes responsibility for the patient and gives direction to discontinue CPR.

- An airway obstruction may be caused by various things, including relaxation of the throat muscles in an unconscious patient, vomited or regurgitated stomach contents, blood, damaged tissue after an injury, dentures, or foreign bodies such as food or small objects.

- The manual maneuver recommended for removing severe airway obstructions in the conscious adult and child is the abdominal-thrust maneuver (the Heimlich maneuver).

abdominal-thrust maneuver The preferred method to dislodge a severe airway obstruction in adults and children; also called the Heimlich maneuver.

advanced life support (ALS) Advanced lifesaving procedures, some of which are now being provided by the EMT.

basic life support (BLS) Noninvasive emergency lifesaving care that is used to treat medical conditions, including airway obstruction, respiratory arrest, and cardiac arrest.

cardiopulmonary resuscitation (CPR) The combination of rescue breathing and chest compressions used to establish adequate ventilation and circulation in a patient who is not breathing and has no pulse.

gastric distention A condition in which air fills the stomach, often as a result of high volume and pressure during artificial ventilation.

head tilt–chin lift maneuver A combination of two movements to open the airway by tilting the forehead back and lifting the chin; not used for trauma patients.

impedance threshold device (ITD) A valve device placed between the endotracheal tube and a bag-mask device that limits the amount of air entering the lungs during the recoil phase between chest compressions.

jaw-thrust maneuver Technique to open the airway by placing the fingers behind the angle of the jaw and bringing the jaw forward; used for patients who may have a cervical spine injury.

load-distributing band (LDB) A circumferential chest compression device composed of a constricting band and backboard that is either electrically or pneumatically driven to compress the heart by putting inward pressure on the thorax.

mechanical piston device A device that depresses the sternum via a compressed gas-powered plunger mounted on a backboard.

recovery position A side-lying position used to maintain a clear airway in unconscious patients without injuries who are breathing adequately.

Assessment in Action

You and your partner will be teaching basic life support to a new class of EMT students in the morning. You use your downtime to prepare your lesson plan and develop review questions.

1. The links in the chain of survival include:
   A. early access.
   B. early defibrillation.
   C. early advanced care.
   D. all of the above.

2. Which of the following rhythms will the automated external defibrillator shock?
   A. Normal sinus rhythm
   B. Atrial fibrillation
   C. Asystole
   D. Ventricular fibrillation

3. What is the minimum amount of time that should be spent checking for spontaneous breathing in an unconscious child?
   A. 5 seconds
   B. 10 seconds
   C. 15 seconds
   D. 20 seconds

4. Each artificial breath should be delivered over a period of how many seconds?
   A. 4
   B. 3
   C. 2
   D. 1

5. What is the proper compression-to-ventilation ratio for adult one-rescuer cardiopulmonary resuscitation (CPR)?
   A. 15:2
   B. 30:2
   C. 50:2
   D. 100:2

6. When checking for a pulse in an infant, you should palpate which of the following arteries?
   A. Carotid
   B. Femoral
   C. Brachial
   D. Dorsalis pedis

7. When you are performing CPR on an adult or child, you should reassess the patient for return of respirations and/or circulation approximately every _____ minutes.
   A. 5
   B. 3
   C. 2
   D. 1

8. What is the preferred method of removing a foreign body in a conscious adult?
   A. Back slaps
   B. Abdominal thrusts
   C. Chest compressions
   D. Manual removal

9. After you have started CPR in the field, under what circumstances can you stop?

10. Explain why the presence of gastric distention is dangerous.

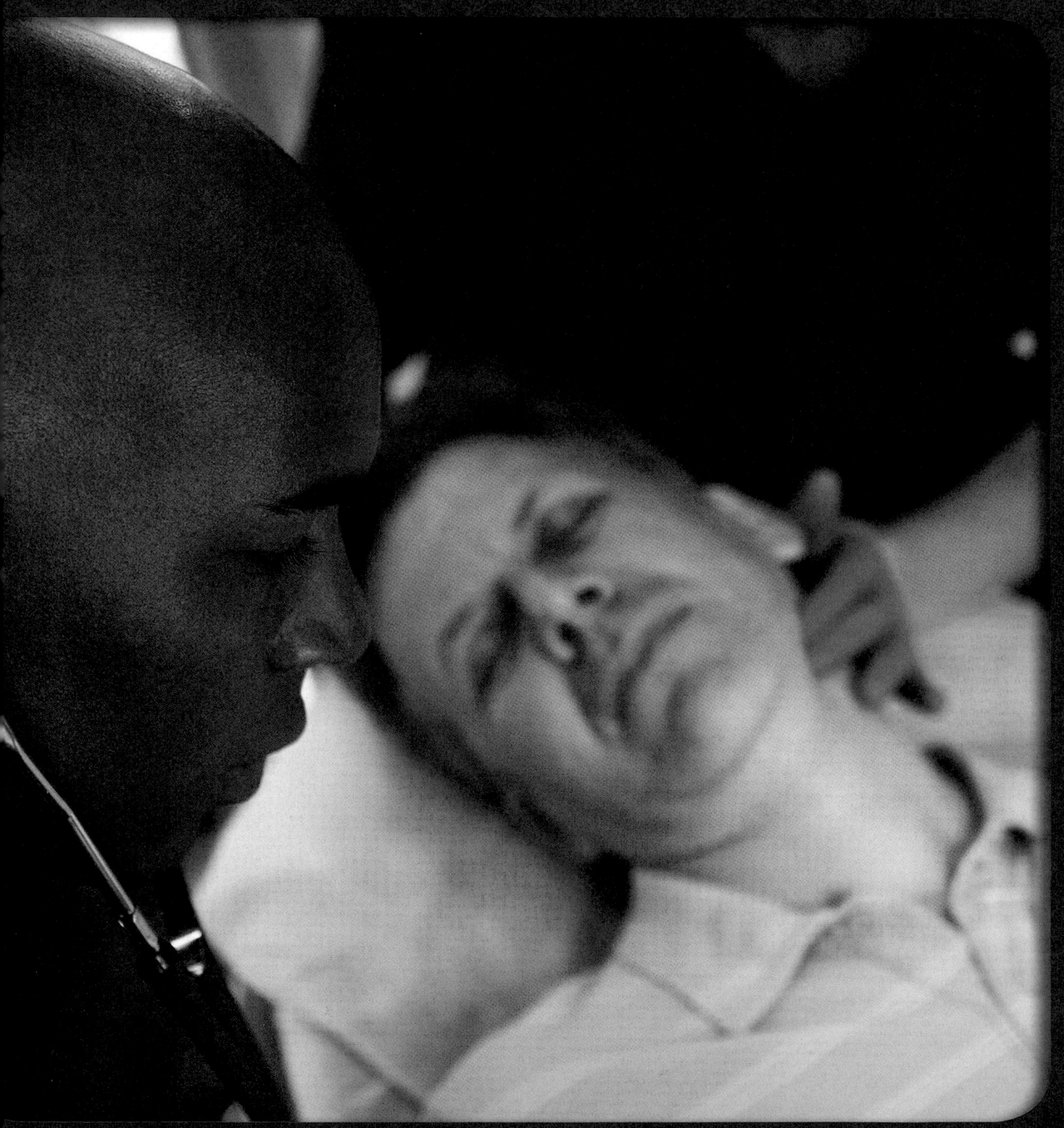

Medical Overview

National EMS Education Standard Competencies

Medicine
Applies fundamental knowledge to provide basic emergency care and transportation based on assessment findings for an acutely ill patient.

Medical Overview
Assessment and management of a
- Medical complaint (pp 453-460)

Pathophysiology, assessment, and management of medical complaints to include
- Transport mode (pp 459-460)
- Destination decisions (p 460)

Infectious Diseases
Awareness of
- A patient who may have an infectious disease (pp 461-466)

Assessment and management of
- A patient who may have an infectious disease (pp 460-466)

Knowledge Objectives
1. Differentiate between medical emergencies and trauma emergencies, remembering that some patients may have both. (p 453)
2. Name the various categories of common medical emergencies and give examples. (pp 453-454)
3. Describe the evaluation of the nature of illness (NOI). (pp 453-454)
4. Discuss the assessment of a patient with a medical emergency. (pp 454-458)
5. Explain the importance of transport time and destination selection for a medical patient. (pp 459-460)
6. Define "infectious disease" and "communicable disease." (p 461)
7. Describe the routes of transmission for an infectious disease. (p 460)
8. Discuss diseases of special concern and their routes of transmission, including herpes simplex, HIV/AIDS, syphilis, hepatitis, meningitis, tuberculosis, whooping cough, MRSA, hantavirus, West Nile virus, SARS, avian flu, and H1N1 (swine flu). (pp 461-466)

Skills Objectives
There are no skills objectives for this chapter.

Introduction

Patients who need EMS assistance generally have experienced either a medical emergency or trauma emergency; in some cases both have occurred. **Trauma emergencies** involve injuries resulting from physical forces applied to the body. **Medical emergencies** involve illnesses or conditions caused by disease. While it is important for you to be able to make the distinction between medical and trauma patients, it is equally important for you to remember that patients may have a combination of medical and trauma conditions affecting their health. For example, a person who has a heart attack while driving may have a collision, or a diabetic patient whose blood glucose level is too low may fall and be injured. This chapter discusses medical emergencies. Chapter 22, *Trauma Overview*, discusses trauma emergencies.

Types of Medical Emergencies

There are many types of medical emergencies Table 12-1. Respiratory emergencies occur when patients have trouble breathing or when the amount of oxygen supplied to the tissues is inadequate. Diseases that can lead to respiratory emergencies include asthma, emphysema, and chronic bronchitis. Cardiovascular emergencies are caused by conditions affecting the circulatory system. The most common examples that require EMS intervention include heart attacks and congestive heart failure. Neurologic emergencies involve the brain and may be caused by a seizure, stroke, or fainting (syncope). Many

gastrointestinal conditions can result in a call to EMS for help. The most well-known condition is appendicitis, although there are many others, including diverticulitis and pancreatitis. A urologic emergency can involve kidney stones. The most common endocrine emergencies are caused by complications of diabetes mellitus. Hematologic (blood) emergencies may be the result of sickle cell disease or various types of blood clotting disorders such as hemophilia. Immunologic emergencies involve the body's response to foreign substances. When the body overreacts to a foreign substance, it is commonly referred to as an allergic reaction. Allergic reactions are a type of immunologic medical emergency that can range from fairly minor to life threatening. Toxicologic emergencies, including poisoning and substance abuse, result in other types of medical emergencies. Some medical emergencies are caused by psychological or behavioral problems. Behavioral emergencies may be especially difficult to deal with because patients often do not present with typical signs and symptoms. Gynecologic conditions are a special category of medical emergencies that involve the female reproductive organs. These conditions will most likely be very challenging for you because there is little that you can do to treat patients with gynecologic conditions in the prehospital setting. The chapters in this section discuss each of these medical emergencies.

Patient Assessment

Assessment of a medical patient is similar to the assessment of a trauma patient but with a different focus. Whereas trauma assessments focus on the mechanism of injury or physical injuries, most of which are visible through a physical examination, medical patient assessment is focused on the **nature of illness (NOI)**, symptoms, and the patient's chief complaint. When you are assessing a patient, establish an accurate medical history.

You are the Provider: PART 1

Your unit is dispatched to 125 Green Hills Drive for a 36-year-old man with a fever and chills. The time is 1:25 PM, there is a fine mist falling, the temperature is 72°F, and the traffic is moderate. You and your partner respond; the scene is located approximately 10 minutes away.

1. **What observations should you make on arriving at the scene but before making physical contact with the patient?**

Table 12-1 Common Medical Emergencies

Type of Medical Emergency	Examples of Condition
Respiratory	Asthma, emphysema, chronic bronchitis
Cardiovascular	Heart attack, congestive heart failure
Neurologic	Seizure, stroke, syncope
Gastrointestinal	Appendicitis, diverticulitis, pancreatitis
Urologic	Kidney stones
Endocrine	Diabetes mellitus
Hematologic	Sickle cell disease, hemophilia
Immunologic	Anaphylactic reaction (severe allergy to bee stings or bites), food allergy
Toxicologic	Substance abuse, food or plant poisoning
Psychiatric	Alzheimer disease, schizophrenia, depression
Gynecologic	Vaginal bleeding, sexually transmitted disease, pelvic inflammatory disease

Information received from dispatch can be helpful in anticipating what you might find when you arrive on scene, but it is conceivable that what appears to be a traumatic emergency may in fact be a medical emergency or vice versa. Use the dispatch information to guide your initial response, but do not get locked into a preconceived idea of the patient's condition strictly from what the dispatcher tells you. During assessment, be aware of several challenges. It is possible that a patient has sustained an injury that distracts you from an underlying condition. For example, a patient may have a medical condition that resulted in a motor vehicle crash, or the patient may have sustained a large laceration and you fail to recognize that the patient has had a hypoglycemic event. Tunnel vision occurs when you become focused on one aspect of the patient's condition and exclude all others, which may cause you to miss an important injury or illness.

Patients may sometimes be uncooperative or even hostile toward those who respond to care for them. Patients may be fearful, angry, and confused and may take out their frustrations on you. It is important that you maintain a professional, calm, nonjudgmental demeanor at all times.

You are obligated as a medical professional to refrain from labeling patients and displaying personal biases. Never assume that you know what the problem is, even when you are treating patients who frequently call for EMS. This attitude could result in missing a serious condition. For example, an intoxicated patient may call 9-1-1 regularly and then call at another time after a fall, resulting in a serious head injury. The head injury may be overlooked if the EMT assumes the call is a response to intoxication only. Labeling a patient is demeaning and detrimental to the EMT and the patient. Personal biases should never affect your management of a patient. Any biases you may have need to be worked on and resolved. As discussed in Chapter 8, *Patient Assessment*, the major components of patient assessment include the following:

- Scene size-up
- Primary assessment
- History taking
- Secondary assessment
- Reassessment

Scene Size-up

Scene Safety

As you approach the scene, you must complete a scene size-up. The most important aspect of this step is to make sure the scene is safe. Although hazards are not as obvious with medical emergencies as with trauma situations, they still exist and must be considered. Patients who are substance abusers may need help in potentially dangerous locations; some patients may have guard dogs or weapons that could be a threat to you. Many other situations that may potentially affect your safety might arise. Therefore, remain conscious of your safety and the safety of your crew and patient before you enter a scene and throughout the call.

It is also important that you use standard precautions when you respond to an emergency, including wearing gloves and other protective equipment. As soon as possible after your arrival, determine the number of patients who need assistance. In most medical cases, there will only be one patient, but anticipate the possibility of more patients and be prepared. Finally, consider whether you need additional help. If you anticipate needing air transport, an advanced life support (ALS) unit, or police assistance, call for them immediately if you have not already done so, so that they will arrive as soon as possible.

Mechanism of Injury/Nature of Illness

Determine the NOI. What signs and symptoms is the patient experiencing? Evaluation of the NOI for a medical patient will provide you with an index of suspicion for different types of serious and/or life-threatening underlying illnesses. The **index of suspicion** is your awareness and concern for potentially serious underlying and unseen injuries or illness.

Primary Assessment

Form a General Impression

As you approach a medical patient, you should develop a general impression of his or her condition and identify life threats. With experience, you will be able to recognize patients with serious conditions from your first contact with them. Perform a rapid scan of the patient. Visual clues include apparent unconsciousness, obvious severe bleeding, or extreme difficulty breathing. Do not let a relatively normal impression lull you into complacency, however, because the conditions of many medical patients may not appear serious at first.

As you approach the patient, quickly determine his or her level of consciousness using the AVPU (*Alert* to person, place, and day; responsive to *Verbal* stimuli; responsive to *Pain*; *Unresponsive*) scale. If the patient is alert on your approach, you can infer several things about his or her condition, but you must always complete the remainder of the primary assessment. If the patient is unconscious as you approach, try to see if you can get a response to verbal stimuli by speaking to the patient while using a gentle touch. If the patient does not respond to your verbal stimulation, pinch the patient's ear or use the trapezius squeeze (a pinch on the muscle that runs along the side of the neck to the shoulders) to see whether the patient responds. If there is no response to verbal or painful stimuli, you should consider the patient unresponsive and quickly continue the assessment.

Airway and Breathing

In conscious patients, ensure the airway is open and they are breathing adequately. Check the respiratory rate, depth, and quality. It is a good idea to consider applying oxygen at this time if there is any indication that breathing has been affected. When in doubt, apply oxygen. For unconscious patients, make sure to open the airway using the proper technique for their condition and take several seconds to evaluate their breathing. Medical patients often require oxygen, so consider having your partner administer oxygen at this time. Unconscious patients may need airway adjuncts and ventilatory assistance with a bag-mask device. The appropriate airway adjunct should be utilized for an unconscious patient. Your partner can assist with treatment while you continue with the primary assessment.

Circulation

Quickly assess the circulation in a conscious patient by checking the radial pulse and observing the patient's

You are the Provider: PART 2

You arrive at the residence and knock on the patient's door. His wife answers and escorts you to the bedroom, where you find the patient in a semisitting position in his bed. He is conscious and alert, is covered with several blankets, and is shivering. Your primary assessment reveals the following information:

Recording Time: 0 Minutes	
Appearance	Flushed skin, shivering
Level of consciousness	Conscious and alert
Airway	Open; clear of secretions and foreign bodies
Breathing	Increased rate; adequate depth
Circulation	Radial pulses, rapid and strong; skin, flushed and hot to the touch

Your partner pulls a nonrebreathing mask from the jump kit; however, the patient states that he does want anything covering his face. He does, however, consent to oxygen via a nasal cannula. The patient tells you that he began feeling ill the day before but then started running a fever last night. Other than chills and generalized weakness, he denies any other symptoms. He took 800 mg of ibuprofen approximately 2 hours ago, and his wife took his temperature shortly before your arrival; it read 102.6°F.

2. On the basis of your general impression and primary assessment findings, does this patient require immediate transport?

3. How should you proceed with your care of the patient?

skin color, temperature, and condition 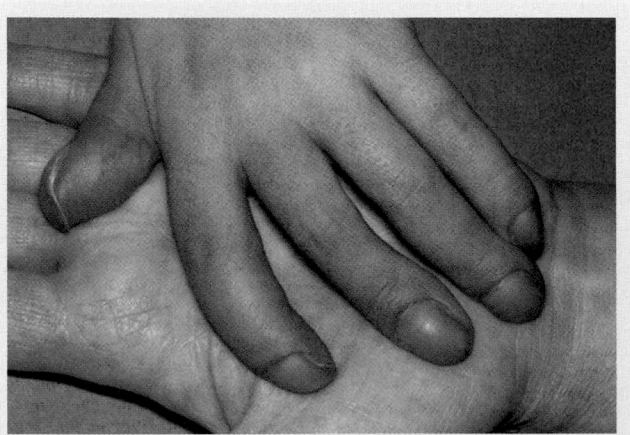. For unconscious patients, assess the circulation at the carotid artery because generally this is the site of the strongest pulse and it is relatively easy to palpate on a supine person. Also quickly glance around the patient to identify any life threats such as severe bleeding or injury to the chest that affects the breathing. If any life threats are found, address them immediately.

Transport Decision

Once you have completed the primary assessment, you should have enough information to make a preliminary transport descision. The following patients should be considered in serious condition and in need of rapid transport: patients who are unconscious or who have an altered mental status, patients with airway or breathing problems, and patients with obvious circulation problems such as severe bleeding or signs of shock. Patients identified as needing rapid transportation still require additional assessment and care.

If the patient does not meet the criteria for rapid transport at this time, you should continue your assessment on scene and prepare for transport when you have completed the assessment and treatment. If you find that your patient's condition deteriorates during the primary assessment, prepare the patient for rapid transport and complete the assessment en route to the emergency department.

History Taking

Investigate Chief Complaint

With a medical patient, history taking may be the only way to determine what the problem is or what may be causing the problem. It is imperative to gather a thorough history from the patient and any family, friends, or bystanders who may have pertinent information. Family members may be the only people aware that an elderly patient sustained a head injury the previous week or that a patient has a history of drug abuse. Bystanders may have seen clues prior to the 9-1-1 call that will lead you and the hospital staff to identify the cause of a patient's condition. First inquire about the patient's chief complaint by asking a question such as, "Why did you call for EMS today?" Listen as the patient responds to you and record as much of this information as possible for later reference, especially for your written report. Investigate the NOI by asking questions about the chief complaint. Identifying signs and symptoms associated with the chief complaint will often help you determine the nature of the condition. Ask about the history of the present illness and ask follow-up questions such as, "Has anything like this ever happened before?" and if the patient answers yes, then ask, "What was done at that time?" and "How does this episode compare with previous episodes?"

If your patient is unconscious, survey the scene for evidence of medication containers or medical devices that the patient may have been using. In addition, try to obtain as much of the patient's medical history as possible from family members, friends, bystanders, or from the scene itself. Family members or friends may know the patient's allergies, medications, or medical conditions. Ask whether the patient was complaining of any symptoms before he or she lost consciousness. If possible, have a family member accompany you to the hospital to answer questions there as well.

SAMPLE History

As you continue to gather information, remember to obtain a SAMPLE history and to ask questions about the patient's chief complaint using the OPQRST mnemonic.

- **O** Onset, that is, when did the problem begin and what caused it?
- **P** Provocation or palliation, that is, does anything make it feel better? Worse?
- **Q** Quality, that is, what is the pain like? Sharp, dull, crushing, tearing? Is it steady?
- **R** Region/radiation, that is, where does it hurt? Does the pain move anywhere?
- **S** Severity, that is, on a scale of 1 to 10, how would you rate your pain? Has the pain changed?
- **T** Timing of pain, that is, has the pain been constant, or does it come and go? How long have you had the pain (often answered under "O," onset).

Ask patients to identify all the symptoms they are experiencing. Make sure you record any allergies, medical conditions, and any medications they take **Figure 12-2**.

Figure 12-1 Skin color can provide an early and fast indication of several disease processes. Cyanosis presents as blue skin.

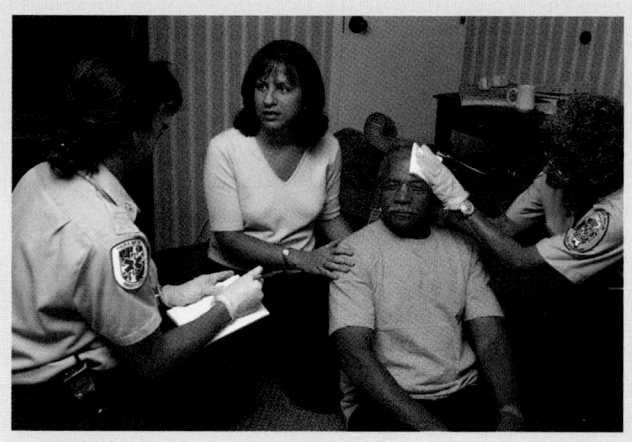

Figure 12-2 History taking is an important part of the assessment process.

Sometimes elderly patients will report taking numerous medications. In those situations, it is best to take the medications with you to the hospital, but remember that the medications need to be listed in your report. Ask patients about their medical history to help determine the current problem and to also help identify any other conditions that might cause complications. To obtain a complete history, you may need to ask about specific conditions such as heart problems, breathing problems, and blood sugar (glucose) problems and whether the patient is taking any medications for these conditions or any other conditions. Also, ask patients who are taking medications whether they are compliant with their drug regimen. The purpose of these questions is to obtain the most complete medical history possible. While you are obtaining the history from the patient, look around the scene for evidence that may also help you determine the history of the patient.

Secondary Assessment

In some cases in which the patient is critically ill or injured or the transport time is short, you may not have time to conduct a secondary assessment. In other cases, the secondary assessment may occur on scene or en route to the emergency department.

Physical Examinations

Conscious medical patients seldom need a full-body scan or head-to-toe examination, but all conscious patients should undergo a limited or focused assessment based on their chief complaint. For example, you should check for pulse, motion, and sensation in all of the patient's extremities and check the patient's pupillary reactions if you suspect a neurologic problem.

Unconscious patients are unable to tell you what is wrong, so you should always perform a full-body scan or head-to-toe examination to obtain clues to assess the problem. Medical alert jewelry will provide you with valuable information. Also look for any evidence of trauma.

Begin by carefully examining the head, scalp, and face. Look for evidence of possible trauma, and monitor the patient for any signs of pain with palpation throughout the assessment. Examine the head and face for symmetry, making sure to check the pupils for equality and reactivity to light. Pupils should be constricted in areas of bright light and dilated in lower light situations. Look at the conjunctiva of the eyes for moisture and the ears and nose for any drainage. Look for nasal flaring, and examine the mouth for foreign bodies (including loose teeth or dentures) and pink, moist mucosa.

Examine the neck closely for any evidence of accessory muscle use with respirations. Check for jugular vein distention and tracheal deviation, which can be indicators of respiratory or cardiac problems. While you are examining the neck, make sure to move any clothing so that you can check for a stoma. Also, look for medical alert jewelry.

Next, assess the chest and abdomen. At the chest, make sure to check breath sounds and heart sounds. Ensure that the patient is breathing adequately with equal chest rise and fall on each side. Palpate the chest and abdomen carefully to identify any areas of tenderness or swelling. Look for medication patches on the chest or abdomen and any implanted medical devices, which usually can be palpated just under the skin. Check for rigidity and distention in the abdomen, and look for scars on the chest or abdomen that might indicate previous surgeries. Finally, check the pelvis and genital area for signs of pain or incontinence.

Palpate the legs and arms for swelling and other abnormalities, making sure to check for distal motion, sensation, and circulation. Note any scars or "track marks" along the veins, which is an indicator of intravenous drug use. Look for medical alert jewelry at the wrists as well. Finally, examine the patient's back to note any irregularities, pain, or scars. At this point, your full assessment of the patient should be complete and treatment of non–life-threatening conditions should be instituted. Treatment will depend on the condition(s) and your local protocols.

Vital Signs

Obtaining a good set of vital signs is critical. Often your partner can begin this process while you are asking about the medical history. Assess the pulse for rate, quality, and regularity at the most appropriate site, either at the radial artery if the patient is conscious or at the carotid artery if the patient is unconscious. If no radial pulse is present, check the carotid artery for the presence of a pulse. Assess respirations as you assess the pulse to prevent the patient from modifying his or her respirations in response to your observation. Identify the rate, quality, and regularity of the respirations and any difficulties that may be apparent. Finally, obtain an initial blood pressure, measuring both systolic and diastolic pressures.

Monitoring Devices

Consider using the automatic blood pressure cuff for future assessments at regular intervals. Depending on your local protocol, other important information to consider obtaining includes a blood glucose level and a pulse oximetry reading. End-tidal CO_2 ($ETCO_2$) monitoring should be considered if the patient complains of respiratory problems or if you suspect a stroke.

Reassessment

Once the assessment and treatment have been completed, reassessment should begin and continue throughout transport. During the reassessment you should repeat the primary assessment and reassess the chief complaint. Look for any changes in the level of consciousness; reassess the airway, breathing, and circulation; and reexamine the transport decision. Consider the need for ALS backup. Obtain another full set of vital signs every 5 minutes for unstable patients or every 15 minutes for stable patients. Reassessment should also include repeating your physical examination to identify and treat changes in the patient's condition.

Interventions

Finally, the reassessment includes reviewing all treatments that have been performed. Reassess oxygen delivery, any bandages or splints applied, and any other treatment that has been performed.

You are the Provider: PART 3

Your partner obtains the patient's vital signs. The patient agrees to transport and asks you to take him to a hospital that he has been to before, which is located 25 miles away. There is another hospital located only 10 miles away.

Recording Time: 5 Minutes	
Respirations	22 breaths/min; regular and adequate
Pulse	110 beats/min; strong and regular
Skin	Flushed; hot to the touch
Blood pressure	124/70 mm Hg
Oxygen saturation (Sao_2)	99% (on oxygen)

4. Is it appropriate to transport the patient to the hospital he requested, or should you transport him to the closer facility?

Communication and Documentation

Document any changes that have developed as a result of the treatments and, if needed, adjust any of the treatments accordingly. Reassessment is an important step in patient assessment to modify care accordingly and so that you have the most current information on the patient's condition when you arrive at the hospital.

Management: Transport and Destination

Most medical emergencies require a level of treatment beyond that available in the prehospital setting. Also, the treatments depend on an accurate diagnosis of the exact medical condition, which may require advanced testing available in a hospital. The primary prehospital treatments for medical emergencies address the symptoms more than the actual disease process.

Depending on local protocol, it may be beyond the scope of an EMT to administer medications to a patient. In a few limited circumstances, such as the administration of nitroglycerin to a patient with chest pain or to a patient for whom it has been prescribed, an exception may be made. Another exception may be granted to allow an EMT to assist a patient with a prescribed metered-dose inhaler when it is required because of respiratory difficulty.

Administration of medications that are stored in the ambulance is also limited for EMTs and is dependent on the state and local protocols in the area. A few of these protocols include administering aspirin for patients having chest pain, administering oral glucose to a diabetic with a low blood glucose level, and possibly administering albuterol to a patient with respiratory difficulty. The administration of activated charcoal to a patient who has ingested a poison is also allowed when it may be beneficial. Each of these situations and any other administration of medication by an EMT requires that direct permission be obtained from medical control. Without an accurate diagnosis, it cannot be determined if the proper condition exists for the medication. The process of obtaining permission includes completing a thorough assessment of the patient before calling medical control. After you give a proper report to the physician and obtain permission, the medication may be administered. Never administer any medication without first obtaining permission from medical control, and always follow your state and local protocols.

You may also use an automated external defibrillator (AED) on a patient who is pulseless and apneic. In some cases of cardiac arrest, immediate treatment with an AED may provide the best option to resuscitate the patient. Using an AED, as well as administering any medications carried on the ambulance, requires advanced training. The AED will be discussed in more detail in Chapter 14,

Cardiovascular Emergencies. It is important to familiarize yourself with the equipment and medications carried on your ambulance in order to use them appropriately under a medical director's instruction.

■ Scene Time

In many cases, the time on scene may be longer for medical patients than for trauma patients. If the patient is not in critical condition, you should gather as much information as possible from the scene so that you can transmit that information to the physician at the emergency department. "Critical" patients include those with altered mental status, airway or breathing difficulties, or any sign of circulatory compromise. In addition, a patient who is very old or very young may be considered critical even if they appear to be fairly stable. Critical patients always need rapid transport. The time on scene should be limited to 10 minutes or less.

■ Type of Transport

Serious consideration should be given as to how to best transport a medical patient. If a life-threatening condition exists, the transportation should include lights and sirens, but if the patient is not critical, careful consideration should be given to nonemergency transport. Many patients experiencing a medical emergency do not have immediate life-threatening conditions; therefore, they can be transported without the use of lights and sirens. This is a much safer method of transport and will often result in arrival only a few minutes later than an emergency transport using lights and sirens.

Differentiating a high-priority transport from a low-priority transport is often a skill developed with experience, but it is also a skill that can be learned. A good rule of thumb for determining the priority of transport is to consider the results of the patient's primary assessment. Patients with an altered mental status, especially if it is still present at the completion of your assessment and treatment, should be considered a high-priority transport. Patients with circulatory compromise, including signs and symptoms of shock, should also be considered a high-priority transport. Most patients with circulatory problems cannot be stabilized in the prehospital setting and need to receive treatment at a hospital quickly but safely. Patients with respiratory difficulty generally require high-priority transport; however, if your patient has responded well to your initial treatment, such as oxygen and albuterol administration, sirens may not be required.

Modes of transport ultimately come in one of two categories: ground **Figure 12-3** or air **Figure 12-4** . Ground transportation EMS units are generally staffed by EMTs and paramedics. Air transportation EMS units or critical

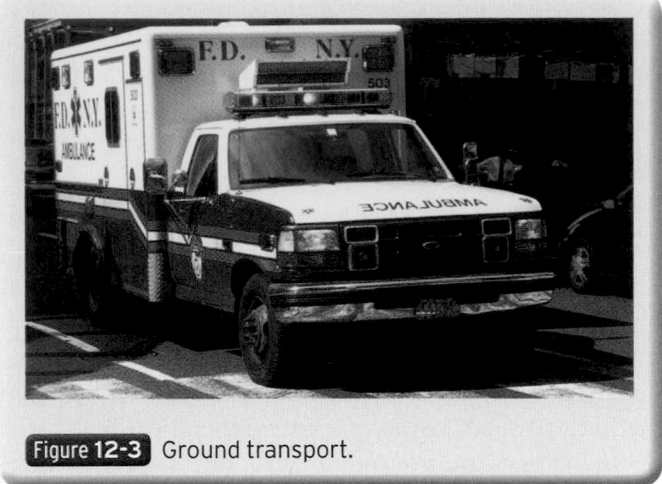

Figure 12-3 Ground transport.

Figure 12-4 Air transport.

care transport units are generally staffed by critical care transport professionals such as critical care nurses and paramedics. Whereas it is not as common to summon an air ambulance for a medical patient, there are many instances where it is advisable. In rural areas with long ground transport times, patients who have possibly experienced a heart attack, a stroke, or a complication of pregnancy could benefit from air transport. Children with serious medical conditions can also benefit from air transport. When you are considering ALS support for a patient, compare the total time for a ground ALS unit to respond and transport to the time required for an ALS helicopter to respond and transport. The quickest mode of transport should be chosen.

■ Destination Selection

It is generally appropriate to select the closest hospital with an emergency department as your destination.

However, there are times when the closest hospital is not necessarily the most appropriate choice. Patients with chest pain as the result of a heart attack may need a facility that is capable of performing heart catheterization, which may not necessarily be available at the closest hospital. If the patient is in cardiac arrest or experiences cardiac arrest during transport, you should immediately reroute to the closest hospital with emergency facilities. Stroke patients can also benefit from specialized hospital selection. Although most hospitals now have designated stroke teams, taking a possible stroke patient to a hospital without this arrangement might be unwise.

Many medical patients will benefit from being transported to a hospital capable of handling their particular condition. Also, some medical patients may benefit from on-scene treatment provided by advanced EMS personnel such as paramedics. It is important to recognize early on when paramedics are needed on scene so that they can be called to respond in a timely manner.

■ Infectious Diseases

As discussed in Chapter 2, *Workforce Safety and Wellness*, you will be called on to treat and transport patients with a variety of infectious or communicable diseases. Chapter 2 discussed the routes of transmission and standard precautions that responders need to take to reduce risk and increase prevention. This chapter discusses the management, awareness, assessment, and of a patient who may have a communicable or infectious disease. Chapter 36, *Transport Operations*, discusses decontamination techniques for transport.

■ General Assessment Principles

The assessment of a patient suspected to have an infectious disease should be approached much like any other medical patient. First, the scene must be sized up and standard precautions taken. Once you can be assured that the scene is safe, proceed with the primary assessment by assessing the patient's mental status and airway, breathing, and circulation and by prioritizing treatment of the patient. With most patients who have a potentially infectious disease in the prehospital setting, the next step is to gather patient history, using OPQRST to elaborate on the patient's chief complaint. Typical chief complaints include fever, nausea, rash, pleuritic chest pain, and difficulty breathing. Obtain a SAMPLE history and a set of baseline vital signs, paying particular attention to medications the patient is currently taking and the events leading up to today's problem. Also ask whether the patient has recently traveled. Always show respect for the feelings of the patient, family members, and others at the scene.

General Management Principles

The general management of the patient with a suspected infectious disease first focuses on any life-threatening conditions that were identified in the primary assessment (airway maintenance, oxygen and ventilatory assistance, bleeding control, and circulatory support). Remember to be empathetic. Because most of these patients will have a fever of unexplained origin or mild breathing problems, place the patient in the position of comfort on the stretcher to keep warm. Remember to use standard precautions for your own safety. Always follow your agency's exposure control plan in cleaning equipment and properly discard any disposable supplies as well as linens.

Words of Wisdom

An infectious disease is a medical condition caused by the growth and spread of small harmful organisms within the body. A communicable disease is a disease that can be spread from one person or species to another. Most of these diseases are much harder to be infected with than is commonly believed. In addition, there are many immunizations, protective techniques, and devices that can be used to minimize your risk of infection. When these protective measures are used, the risk of your contracting a serious infectious disease is negligible.

Words of Wisdom

Causes of Infectious Disease

Type of Organism	Description	Example
Bacteria	Grow and reproduce outside the human cell in the appropriate temperature and with the appropriate nutrients	*Salmonella*
Viruses	Smaller than bacteria; multiply only inside a host and die when exposed to the environment	Human immunodeficiency virus
Fungi	Similar to bacteria in that they require the appropriate nutrients and organic material to grow	Mold
Protozoa (parasites)	One-celled microscopic organisms, some of which cause disease	Amoebas
Helminths (parasites)	Invertebrates with long, flexible, rounded, or flattened bodies	Worms

Common or Serious Communicable Diseases

Herpes Simplex

Herpes simplex is a common virus strain carried by humans. Eighty percent of individuals carrying the virus are asymptomatic, but symptomatic infections can be serious and are on the rise, especially in immunocompromised patients. The primary mode of infection is through close personal contact, so standard precautions are generally sufficient to prevent spread to or from health care workers.

HIV Infection

Exposure to the virus that causes acquired immunodeficiency syndrome (AIDS) is the most feared infection risk for EMTs. It is this prospect that led to the development of standard precautions. There is no vaccine to protect against human immunodeficiency virus (HIV) infection, and despite great progress in drug treatments, AIDS is still fatal. Fortunately, it is not easily transmitted in your work setting. For example, it is far less contagious than hepatitis B. HIV infection is a potential hazard only when deposited on a mucous membrane or directly into the bloodstream.

This can occur via sexual contact or exposure to blood or body fluids, meaning your risk of infection is limited to exposure to an infected patient's blood and body fluids. Exposure can take place in the following ways:

- The patient's blood is splashed or sprayed into your eyes, nose, or mouth or into an open sore or cut, however tiny; even a microscopic opening in the skin is an invitation for infection with a virus.
- You have blood from the infected patient on your hands and then touch your own eyes, nose, mouth, or an open sore or cut.
- A needle used to inject the patient breaks your skin. The risk to you from a single injection, even with a hollow-bore needle, is small, probably less than 1 in 1,000. However, this is by far the most dangerous form of exposure.
- Broken glass at a motor vehicle collision or other incident may penetrate your glove (and skin), which may have already been covered with blood from an infected patient.

Many patients who are infected with HIV do not show any symptoms. This is why the government requires health care workers to wear certain types of gloves any time they are likely to come into contact with secretions or blood from any patient. You should always put on the proper type of gloves before leaving the ambulance to care for a patient. In addition, you must take great care in handling and disposing of needles and scalpels so that you and others are not inadvertently exposed to them. Finally, you should cover any open wounds that you have whenever you are on the job.

If you have any reason to think that a patient's blood or secretions may have entered your system, especially through inoculation with a patient's blood, you should seek medical advice as soon as possible. If you know that the patient is infected with HIV, your physician may suggest immediate treatment to try to prevent you from becoming infected. However, if the patient is an unlikely candidate for HIV infection, your physician may recommend that you and the patient be tested before you undergo therapy. As scientists learn more about HIV infection, testing and treatment recommendations change. It is important that you immediately see your doctor (or your program's designated doctor) any time you are potentially exposed to a communicable or infectious disease. Know the policy for your system, and take time now to consider what you would do in the event of exposure.

■ Syphilis

Although syphilis is commonly thought of as a sexually transmitted disease, it is also a bloodborne disease. There

is a small risk for transmission through a contaminated needlestick injury or direct blood-to-blood contact. If treated with penicillin, the individual is considered non-communicable within 24 to 48 hours.

The initial infection with syphilis produces a lesion called a chancre. Chancres are most commonly located in the genital region.

■ Hepatitis

The term hepatitis refers to an inflammation (and often infection) of the liver. Hepatitis can be caused by a number of different viruses and toxins. Early signs of viral hepatitis include loss of appetite, vomiting, fever, fatigue, sore throat, cough, and muscle and joint pain. Several weeks later, jaundice (yellow eyes and skin) and right upper quadrant abdominal pain develop Figure 12-5. The severity of toxin-induced hepatitis depends on the amount of agent absorbed and the duration of exposure. Toxin-induced hepatitis is not contagious. There is no sure way to tell which patients with hepatitis have a contagious form of the disease and which do not. Table 12-2 shows the characteristics of different types of hepatitis, from which you can assess your risk of exposure. Hepatitis A can be transmitted only from a patient who has an acute infection, whereas hepatitis B and hepatitis C can

You are the Provider: PART 4

You place the patient onto the stretcher, load him into the ambulance, and begin transport to the hospital. En route, you reassess his vital signs and temperature, which reads 100.2°F. He is still receiving oxygen at 4 L/min via nasal cannula.

Recording Time: 12 Minutes	
Level of consciousness	Conscious and alert
Respirations	22 breaths/min; regular and adequate
Pulse	104 beats/min; strong and regular
Skin	Flushed, warm, and moist
Blood pressure	122/68 mm Hg
Sao$_2$	98% (on oxygen)

You set the automatic vital sign device to reassess his vital signs every 10 minutes. The patient tells you that he thinks his fever is breaking because he is sweating. En route, his vital signs and overall condition are reassessed and remain unchanged.

5. Should you set the vital sign machine to reassess his vital signs at shorter intervals? Why or why not?

6. On the basis of the patient's chief complaint, what additional information can you obtain by using the OPQRST mnemonic?

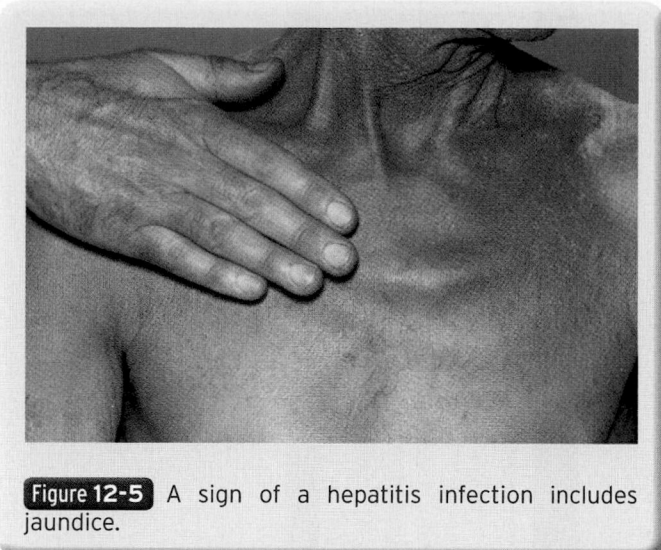

Figure 12-5 A sign of a hepatitis infection includes jaundice.

be transmitted from long-term carriers have no signs of illness. A carrier is a person (or animal) in whom an infectious organism has taken up permanent residence and may or may not cause any active disease. Carriers may never know that they harbor the organism; however, they can infect other individuals.

Hepatitis A is transmitted orally through oral or fecal contamination. This means that, generally, you must eat or drink something that is contaminated with the virus. Contamination is the presence of an infectious organism on or in an object. The organisms that cause hepatitis B and C are transmitted through vehicles other than food or water. For example, these organisms may enter the body through a transfusion or needlestick with infected blood, which puts health care workers at high risk for contracting hepatitis B, the more contagious and virulent form. **Virulence** is the strength or ability of a pathogen to produce disease. Hepatitis B is far more contagious than HIV. For this reason, vaccination with hepatitis B vaccine is highly recommended for EMTs. Unfortunately, not everyone who is vaccinated develops immediate immunity to the virus. Sometimes, but not always, an additional dose will provide immunity. You should be tested after vaccination to determine your immune status.

If you are stuck with a needle or injured in some other way while caring for a patient who might have hepatitis, see your doctor immediately.

■ Meningitis

Meningitis is an inflammation of the meningeal coverings of the brain and spinal cord. Patients with meningitis will have signs and symptoms such as fever, headache, stiff neck, and altered mental status. It is an uncommon but very frightening infectious disease. Meningitis can be caused by viruses or bacteria, most of which are not contagious. However, one form, meningococcal meningitis, is highly contagious. The meningococcus

bacterium colonizes the human nose and throat and only rarely causes an acute infection. When it does, it can be lethal. Patients with this kind of infection often have red blotches on their skin; however, many patients with forms of meningitis that are not contagious also have red blotches.

Only laboratory tests can sort out the different forms of meningitis; therefore, you should take standard precautions with any patient who is suspected of having meningitis. Gloves and a mask will go a long way to prevent the patient's secretions from getting into your nose and mouth. Again, the risk of infection is small, even if the organism is transmitted. For this reason, vaccines, which are available for most types of meningococcus, are rarely used. Meningitis can be treated at the emergency department with antibiotics.

After treating a patient with meningitis, you should contact your employer health representative. Many states consider meningitis "reportable" and will notify you that one of your patients was diagnosed with meningitis. Prophylactic treatment is then in order for you.

■ Tuberculosis

Most patients who are infected with *Mycobacterium tuberculosis* (the tubercle bacillus) are well most of the time. If the disease involves the brain or kidneys, the patient is only slightly contagious. In the United States, however, **tuberculosis** is a chronic mycobacterial disease that usually strikes the lungs. Disease that occurs shortly after infection is called primary tuberculosis. Except in infants, this infection is not usually serious. After the primary infection, the tubercle bacillus is rendered dormant by the patient's immune system. However, even after decades of lying dormant, this germ can reactivate. Reactive tuberculosis is common and can be much more difficult to treat, especially because an increasing number of tuberculosis strains have grown resistant to most antibiotics.

Although tuberculosis is often hard to distinguish from other diseases, patients who pose the highest risk almost invariably have a cough. Therefore, for your safety, you should consider respiratory tuberculosis to be the only contagious form because it is the only one that is spread by airborne transmission. The droplets that are produced by coughing are not the real problem. The real problem is the droplet nuclei, which are the remnants of the droplets after the excess water has evaporated. These particles are tiny enough to be totally invisible and can remain suspended in the air for a long time. In fact, as long as these particles are shielded from ultraviolet light, they can remain alive for decades. Particles that are the size of droplet nuclei are not stopped by routine surgical masks. Inhaled, they are carried directly to the alveoli of the lungs, where the bacteria may begin to grow. HEPA masks are required to stop droplet nuclei.

Table 12-2 Characteristics of Hepatitis

Type	Route of Infection	Incubation Period	Chronic Infection	Vaccine and Treatment	Comments
Viral Hepatitis					
Hepatitis A (infectious)	Fecal-oral, infected food or drink	2–6 wk	Chronic condition does not exist	Vaccine is available; no treatment is available	Mild illness, approximately 2% of patients die. After acute infection, the patient has life-long immunity
Hepatitis B	Blood, sexual contact, saliva, urine, breast milk	4–12 wk	Chronic infection affects up to 10% of patients and up to 90% of newborns who have the disease	Vaccine is available; treatment is minimally effective	Up to 30% of patients may become chronic carriers. Patients are asymptomatic and without signs of liver disease, but they may infect others. Approximately 1% to 2% of patients die
Hepatitis C	Blood, sexual contact	2–10 wk	Chronic infection affects 90% of patients	No vaccine is available; treatment is minimally effective	Cirrhosis of the liver develops in 50% of patients with chronic hepatitis C. Chronic infection increases the risk of cancer of the liver
Hepatitis D	Blood, sexual contact	4–12 wk	Chronic infection is common	No vaccine is available; no treatment is available	Occurs only in patients with active hepatitis B infection. Fulminant disease may develop in 20% of patients
Toxin-Induced Hepatitis					
Medications, drugs, and alcohol	Inhalation, skin or mucous membrane exposure, oral ingestion, or intravenous administration	Within hours to days following exposure	Some chemicals may initiate an inflammatory response that continues to cause liver damage long after the chemical is out of the body	No vaccine is available; treatment is to stop exposure. In patients with an overdose of acetaminophen, certain drugs may minimize liver injury if given early enough	This type of hepatitis is not contagious. Patients with toxin-induced hepatitis may have liver damage, such as jaundice. Not every exposure to a toxin will cause liver damage

Special Populations

Everyone has body defenses that help protect against getting sick, but the aging process can pose a threat to the body's natural defense mechanisms against invading microorganisms. As a person ages, his or her physical defenses weaken or are eliminated. The skin's thinning and loss of supportive collagen, along with a reduction in the number of blood vessels, allows bacteria or viruses to enter the body with less resistance. The respiratory system cannot trap and eliminate bacteria and viruses in the airways as efficiently as it once did. Finally, the gastrointestinal system allows easier entry for bacteria or viruses through the intestines. As the body ages, physical barriers to entry weaken, the immune system deteriorates, and invading organisms are not as easily identified as abnormal. Infectious agents can take hold in elderly patients much more easily because of reduced defenses.

When transporting an elderly patient, protect the patient from the environment because extremes in heat or cold can further reduce the body's defenses. If your patient has a cold or the flu, protect yourself. However, remember that your defense system is probably much stronger than that of the patient.

Why is tuberculosis not more common than it is? After all, absolute protection from infection with the tubercle bacillus does not exist. Everyone who breathes is at risk. According to the Centers for Disease Control and Prevention, one third of the world's population is infected with tuberculosis. The vaccine for tuberculosis, called BCG, is only rarely used in the United States. Under normal circumstances, however, the mechanism of transmission used by M tuberculosis is not very efficient. Infected air is easily diluted with uninfected air. M tuberculosis is one of those germs that typically causes no illness in a new host. In fact, many patients with tuberculosis do not even transmit the infection to family members. However, in crowded environments with poor ventilation, the disease spreads more easily.

If you are exposed to a patient who is found to have pulmonary tuberculosis, you will be given a tuberculin skin test. This simple skin test determines whether a person has been infected with M tuberculosis. A positive result means that exposure has occurred; it does not mean that the person has active tuberculosis. It takes at least 6 weeks for the bacteria to show up in the laboratory test. So if you are tested for the disease within a few weeks of the exposure and your results are positive, this means that you had already acquired the infection at an earlier time from somebody else. You will probably never identify the source. Most transmissions occur silently, so it is necessary that you have tuberculin skin tests regularly. If the infection is found before you become ill, preventive therapy is almost 100% effective. Usually, a daily dose of

the medication isoniazid will prevent the development of active infection.

■ Whooping Cough

Whooping cough, also called pertussis, is an airborne disease caused by bacteria that mostly affects children younger than 6 years. Signs and symptoms include fever and a "whoop" sound that occurs when the patient tries to inhale after a coughing attack.

The best way to prevent exposure to whooping cough is to place a mask on the patient and on yourself.

■ Methicillin-Resistant *Staphylococcus Aureus*

Methicillin-resistant *Staphylococcus aureus* (MRSA) is a bacterium that causes infections and is resistant to most antibiotics. In health care settings, MRSA is believed to be transmitted from patient to patient via unwashed hands of health care providers. Studies have shown that 5% to 15% of health care providers carry MRSA in their nares; the pathogen can subsequently be transferred to skin and other areas of the body through a break in the skin. Surfaces contaminated with MRSA do not seem to be important in transmission. Factors that increase the risk for developing MRSA include antibiotic therapy, prolonged hospital stays, a stay in intensive care or burn unit, and exposure to an infected patient.

The incubation period for MRSA appears to be between 5 and 45 days. The communicable period varies, as patients who have active infection may carry MRSA for months. MRSA results in soft-tissue infections. Its signs and symptoms may involve localized skin abscesses, and sepsis may be found in older patients with the infection.

■ New and Emerging Diseases

Hantavirus

Newly recognized diseases, such as those caused by hantavirus (a rare but deadly virus transmitted through rodent urine and droppings) and enteropathogenic *Escherichia coli* (a common cause of pediatric diarrhea in developing countries), are being reported. These diseases are not transmitted from person to person directly; rather, they are carried by a vehicle, such as food, or a vector, such as rodents.

West Nile Virus

Although not a newly discovered illness, West Nile virus has caused some concern in the past. This virus' vector is the mosquito, and it affects humans and birds. The virus is actually tracked by tests done on birds suspected of being killed by the virus. These diseases are not communicable and do not pose a risk to you during patient care.

SARS

A virus that has caused significant concern in the recent past is known as **SARS (severe acute respiratory syndrome)**. SARS is a serious, potentially life-threatening viral infection caused by a recently discovered family of viruses. SARS usually starts with flulike symptoms, which may progress to pneumonia, respiratory failure, and, in some cases, death. The SARS virus strain probably spread from Guangdong province in southern China to Hong Kong, Singapore, and Taiwan. Canada has experienced a significant outbreak in the Toronto area. SARS is thought to be primarily transmitted by close person-to-person contact. Most cases have involved persons who lived with or cared for a person with SARS or who had exposure to contaminated secretions from a SARS patient.

Avian Flu

Avian (bird) flu is caused by a virus that occurs naturally in the bird population. This virus is carried in the intestinal tract of wild birds and does not usually cause illness. However, in domestic bird populations (eg, chickens, ducks, and turkeys), it is very contagious. Birds acquire the illness from contact with contaminated excretions or surfaces that are contaminated with excretions. If an infected bird is used for food and is cooked, it does not pose a risk to those who eat it.

The first case of this flu was reported in Hong Kong in 1997; 18 people became infected and 6 died in the outbreak. In the cases that have occurred since then, the death rate is approximately 25 percent. No rapid human-to-human cases of this disease have been reported. Instead, the cases occuring involving humans have involved close contact with infected birds. The transmission risk for humans is quite low.

H1N1

Another virus of recent concern is the H1N1 virus that was initially identified as the "swine flu." This virus is not new, having been present for years in animals. H1N1 is contagious. Whereas this virus is new to humans, it is only one type of influenza among the hundreds of other strains of influenza that exist and infect humans regularly. Many deaths have been caused by the H1N1 virus, although deaths caused by other influenza viruses also have occurred. The most positive effect of the outbreak of H1N1 virus has been a greater awareness on the part of the general public of the routes of transmission of contagious diseases. This increased awareness could result in a reduction of all communicable diseases, not only H1N1.

You are the Provider: PART 5

With an estimated time of arrival at the hospital of 5 minutes, you reassess the patient and call in your radio report. The hospital acknowledges your report and is awaiting your arrival. The patient tells you that he feels better, but is very thirsty. You reassess his temperature and obtain a reading of 99.8°F.

Recording Time: 21 Minutes	
Level of consciousness	Conscious and alert
Respirations	18 breaths/min; regular and adequate
Pulse	90 beats/min; strong and regular
Skin	Warm and moist; less flushed
Blood pressure	128/72 mm Hg
Sao$_2$	98% (on oxygen)

The patient is delivered to the hospital without incident, and you give your verbal report to a staff nurse, who assumes care of the patient. After cleaning the ambulance, you return to service.

7. How and why do the patient's vital signs differ from your initial readings?

Words of Wisdom

The ability of your EMS service to support you in the event of exposure to a communicable disease depends on your understanding of how an exposure can occur and your immediate reporting of exposure to potentially infectious materials. Make notes right away to ensure that you remember all pertinent information, and report the possible exposure immediately after the response, following your service's guidelines.

Conclusion

Although trauma patients often present with dramatic signs and symptoms, the assessment and treatment you provide for them is fairly straightforward. The assessment and treatment of medical patients on the other hand can be very challenging and interesting because of the nature of medical conditions. The condition of a medical patient may not be as apparent as in a trauma patient and, therefore, treatment may not be as straightforward. You must remember that delays of any kind in an attempt to diagnose a condition can be harmful to the patient and thus are not recommended. Your best approach is to keep calm, use your patient assessment skills, treat the patient's symptoms, report to medical control, and transport the patient safely to the emergency department. Finally, keep in mind that patients sometimes have more than one isolated problem, so you must be prepared to handle any combination of conditions, including conditions of medical patients who have been involved in traumatic situations.

You are the Provider: SUMMARY

1. **What observations should you make on arriving at the scene but before making physical contact with the patient?**

   When arriving at any scene, your first priority is to assess for any actual or potential hazards that could pose a risk to the safety of you and your partner. These hazards can range from loose gravel in the driveway that could cause you to slip to a vicious dog or violent person. Maximizing your safety at the scene can range from simply watching where and how you walk to summoning additional personnel (eg, law enforcement, animal control). Remember to use standard precautions *before* making contact with the patient!

   After ensuring personal safety, assess the environment in which the patient is found. Is it well kept? Are there any indications that the patient has a medical condition, such as medication bottles, home oxygen, or nebulizers? Is there only one patient? Not only will assessment of the patient's environment potentially reveal any underlying medical problems, it will also give you an idea of how the patient takes care of himself or herself.

   As you approach the patient, form a general impression—a "quick glance" that will help you rapidly recognize life-threatening conditions before making physical contact with the patient. Apparent unconsciousness, obvious external bleeding, and severe difficulty breathing are but a few of the visual clues that you may recognize during the initial general impression. Do not become complacent if the initial general impression does not reveal any obvious life threats; many patients—medical and trauma—may not appear serious at first glance.

   After visually assessing the scene, the patient's environment, and the patient, you should then proceed with the primary assessment.

2. **On the basis of your general impression and primary assessment findings, does this patient require immediate transport?**

   Once the initial general impression and primary assessment are complete, you should have enough information to determine if immediate transport is required or if additional assessment and treatment at the scene are appropriate. If you decide to further assess and treat the patient at the scene, however, you must carefully monitor his or her condition and be prepared to transport immediately.

   Your patient is clearly sick; he has a high fever and reports weakness. However, he is conscious and alert and does not have any airway, breathing, or circulation problems. His heart rate and respiratory rate are both increased; however, his heart rate is strong and palpable at the radial artery, and his breathing is producing adequate tidal volume. Tachypnea and tachycardia are common physiologic responses to fever. At the present time, there are no signs indicating the need for immediate transport.

3. **How should you proceed with your care of the patient?**

   You have already determined the patient's chief complaint and have begun initial treatment. Because his condition is stable, immediate transport is not indicated; therefore, you should proceed by inquiring about the history of his present illness, taking his vital signs, obtaining a SAMPLE history, and performing a secondary assessment. The secondary assessment of a medical patient should primarily focus on his or her chief complaint and presenting signs and symptoms.

   A thorough patient history is one of the most beneficial aspects of the patient assessment process; it is the primary component in the overall assessment of medical patients because it allows you to identify other signs and symptoms associated with the patient's chief complaint (history of present illness). A good patient history will often help you determine the nature of the patient's condition, which helps ensure that the proper care will be provided. Obtaining a thorough and accurate medical history, does, however, require a balance of knowledge and skill.

   An important aspect of the patient history includes pertinent negatives and pertinent positives. A pertinent negative is any sign or symptom that commonly accompanies a particular condition, but is absent. For example, patients with chest pain commonly report shortness of breath as well. However, if a patient with chest pain denies shortness of breath when asked about it, you have found a pertinent negative.

A pertinent positive is any sign or symptom that commonly accompanies a particular condition and is present—for example, a patient with chest pain who also states that he or she is short of breath when asked about it.

The SAMPLE history is used in conjunction with the history of present illness to obtain further information about the patient's chief complaint. It also identifies any drug allergies, significant past medical history, last oral intake, and the events that preceded the patient's condition.

Vital signs are often obtained by your partner while you are assessing the patient. Obtain a baseline set of vital signs—including pulse oximetry, and if indicated, a blood glucose level—and compare future readings with the initial set (trending). Trending a patient's vital signs and reassessing his or her overall condition at regular intervals often helps you determine if the patient's condition is unchanged, has improved, or has worsened.

The secondary assessment is designed to identify any signs or symptoms of illness that you may not have detected in the primary assessment. It involves a systematic assessment of the patient from head to toe, or a more focused assessment with emphasis on his or her chief complaint. Depending on the patient's chief complaint and overall condition, a complete secondary assessment may not be necessary.

4. Is it appropriate to transport the patient to the hospital he requested, or should you transport him to the closer facility?

Generally speaking, patients should be transported to the hospital of their choice when at all possible. Ultimately, however, the destination facility should be dictated by the patient's condition. Patients in stable condition should be transported to the facility of their choice, and patients in unstable condition should be transported to the closest appropriate facility.

Your patient is in stable condition—that is, he has no airway, breathing, or circulation problems. Therefore, it is not unreasonable to comply with his request and transport him to the hospital of his choice. However, you should inform him that if his condition worsens, it may be necessary to divert to a closer facility. If his wife will be following you in her personal vehicle, ask for her mobile phone number, if she has one, so you can contact her should diversion to a closer facility become necessary.

In some EMS systems, a prolonged transport time may take their only ambulance out of service. If this is the case, explain to the patient that his or her medical records can be obtained by the closer facility and recommend transport there. If the patient insists on going to a hospital that will take your unit out of service for a prolonged period, contact your EMS supervisor and, if dictated by local protocol, your medical director, to inform them of the situation. In some areas, neighboring EMS systems with more than one ambulance may be able to

provide mutual aid and cover your jurisdiction while you are out of service.

5. Should you set the vital sign machine to reassess his vital signs at shorter intervals? Why or why not?

You should reassess *any* patient at regular intervals—whether he or she is in stable or unstable condition; doing so will allow you to identify and treat changes in his or her condition in a timely manner.

You should reassess a patient in stable condition—to include his or her vital signs—*at least* every 15 minutes or as often as practical depending on his or her condition. Patients in unstable condition should be reassessed every 5 minutes, or as often as practical depending on their condition.

On the basis of your patient's stable condition, reassessing his vital signs every 10 minutes is appropriate at this time. If his condition worsens, you can always set the machine to record his vital signs at shorter intervals.

Vital signs are only one component of reassessment. You should also monitor the patient's level of consciousness and other parameters (eg, skin condition and temperature, breathing status, pulse regularity and strength) en route; in many cases, these parameters change when a patient's condition is deteriorating before the vital signs change.

6. On the basis of the patient's chief complaint, what additional information can you obtain by using the OPQRST mnemonic?

The OPQRST mnemonic is a valuable tool to obtain further information about the patient's chief complaint. It is typically used early in the assessment process but can certainly be used—at least in part—during reassessment of the patient. Not every component of the OPQRST mnemonic will apply to every patient; however, there are some components that will.

Your patient's chief complaint was fever, chills, and weakness. The onset of his symptoms can easily be established by asking, "Did your symptoms occur suddenly, or did they develop over time?" An acute onset of fever versus fever that developed slowly is important to note and can aid the emergency department physician in his or her diagnosis.

The presence of provoking or palliating factors can also be established; ask the patient if there is anything that makes him feel better or worse. In a patient with a fever, ask him or her if any antipyretics (fever-reducing medications), such as ibuprofen or acetaminophen, were taken and if they seemed to help. If there is a particular position that improves or worsens his symptoms, what is it?

Because the patient is not complaining of pain or discomfort, there is no "quality" to determine. However, if the patient begins to report pain or discomfort, ask him to describe it. Use open-ended questions when assessing the quality of a patient's pain to obtain the most reliable description. Avoid asking leading questions, such as "Is your pain sharp?" Radiating or referred pain also cannot be established in this patient because he is not experiencing pain.

You are the Provider: SUMMARY, continued

The severity of the patient's condition—as perceived by him or her—is completely subjective because all patients perceive things differently. The 0 to 10 scale is an excellent tool to use for patients with or without pain. If pain is involved, ask the patient to assign it a number initially and then ask again at regular intervals. If the patient is not in pain and is "just not feeling good," the 0 to 10 scale can still be used to assess his or her perception of the severity of the situation.

Establishing the time the symptoms began is related to the onset, but is more specific; it can easily be established in this patient. When establishing the onset, you are determining if the symptoms began acutely or developed over time. When establishing the time of onset, you are asking for a specific time that the symptoms began (eg, yesterday around 3:00 PM).

7. How and why do the patient's vital signs differ from your initial readings?

Reassessing a patient's vital signs is obviously important; understanding how and why they have changed is of equal importance.

Vital signs are called "vital" signs for a reason. When assessed in conjunction with the chief complaint, a proper physical examination, and medical history, they often help reinforce what you believe the patient's primary problem is (eg, your field impression).

The patient's blood pressure has remained consistent throughout your encounter with him. However, his heart rate, respiratory rate, and skin condition are different from previous readings.

His chief complaint was fever, chills, and weakness. Fever usually causes chills, which cause the patient to expend a lot of energy and can make him or her feel weak. Fever also increases a person's metabolic rate, resulting in the production of more heat energy. Physiologically, the body responds to an increased metabolic rate by increasing its vital functions—namely, respirations and heart rate. Thus, when the patient's temperature was 102.6°F, his body was responding with tachypnea and tachycardia. However, as his temperature reduced to 99.8°F, so did his metabolic rate and, thus, his respiratory rate and heart rate.

When a person is actively "running a fever," the skin is typically flushed (red), abnormally warm or hot, and dry. However, as the fever begins to subside, sweating occurs, which is the body's way of removing heat through evaporation.

EMS Patient Care Report (PCR)

Date: 7-29-09	Incident No.: 011109	Nature of Call: Sick person		Location: 125 Green Hills Dr.
Dispatched: 1325	En Route: 1325	At Scene: 1335	Transport: 1348	At Hospital: 1406	In Service: 1417

Patient Information

Age: 36 Sex: M Weight (in kg [lb]): 80 kg (175 lb)	Allergies: No known drug allergies Medications: Ibuprofen Past Medical History: None Chief Complaint: Fever, chills, and weakness

Vital Signs

Time: 1340	BP: 124/70	Pulse: 110	Respirations: 22	Sao₂: 99%
Time: 1352	BP: 122/68	Pulse: 104	Respirations: 22	Sao₂: 98%
Time: 1402	BP: 128/72	Pulse: 90	Respirations: 18	Sao₂: 98%

EMS Treatment
(circle all that apply)

Oxygen @ 4 L/min via (circle one): (NC) NRM Bag-Mask Device	Assisted Ventilation	Airway Adjunct	CPR
Defibrillation	Bleeding Control	Bandaging	Splinting	Other: Position of comfort

Narrative

9-1-1 dispatch for a 36-year-old male with fever and chills. On arrival at the scene, found the patient in a semisitting position in his bed. He was conscious and alert, his airway was patent, and his breathing was adequate. Patient states that he began feeling bad the day before but then began running a fever last night. He also complains of chills and weakness, but denies any other symptoms. Approximately 2 hours prior to EMS arrival, patient took 800 mg of ibuprofen. His wife took his temperature just prior to EMS arrival and noted a reading of 102.6°F. Patient would not accept oxygen via nonrebreathing mask but consented to oxygen at 4 L/min via nasal cannula. Obtained vital signs and performed additional assessment. Patient's skin was noted to be flushed, hot, and dry. His breath sounds were clear to auscultation bilaterally and he denied a cough. He further denied any significant past medical history and stated no allergies to medications. Completed assessment and treatment at the scene and began transport to the hospital with patient in position of comfort. En route, continued to monitor patient's condition and vital signs as indicated. He remained conscious and alert, and his vital signs improved; he was less tachypneic and tachycardic. Patient stated that he felt better but was very thirsty. Reassessment of his temperature revealed a reading of 100.2°F. At this point, it was also noticed that patient began to sweat and his skin was less flushed. Remainder of transport was uneventful. Reassessed vital signs shortly before arrival at the hospital and also reassessed his temperature; a reading of 99.8°F was noted. Delivered patient to emergency department without incident and gave verbal report to staff nurse. Medic 14 returned to service at 1417. **End of report**

Prep Kit

Ready for Review

- Trauma emergencies are injuries that are the result of physical forces applied to the body. Medical emergencies require EMS attention because of illnesses or conditions not caused by an outside force.

- The assessment of a medical patient is similar to the assessment of a trauma patient but with a different focus. Whereas a trauma assessment focuses on physical injuries, most of which are visible through a physical examination, medical patient assessment is usually more focused on symptoms and depends more on establishing an accurate medical history.

- Many medical patients may not appear to be seriously ill at first glance.

- For conscious medical patients, obtaining a thorough patient history can be one of the most beneficial aspects of the patient assessment. Try to determine the nature of the illness by asking questions about the patient's chief complaint.

- Conscious medical patients seldom need a full-body scan, but all should get a focused examination based on their chief complaint. On the other hand, you should always perform a full-body scan on unconscious patients; this head-to-toe assessment may give you clues to help identify the problem.

- Most medical emergencies require a level of treatment beyond what is available in the prehospital setting. Also, the treatments depend on an accurate diagnosis of the exact medical condition; therefore, advanced testing in the hospital may be required.

- If the patient is not in critical condition, you should gather as much information as possible from the scene so that you can transmit that information to the physician at the emergency department.

- Many medical emergency patients do not have immediate life-threatening conditions. If a life-threatening condition exists, transportation should include the use of lights and sirens, but if that is not the case, careful consideration should be given to nonemergency transport.

- Modes of transport ultimately come in one of two categories: ground or air.

- Many medical patients will benefit from being transported to a specific hospital capable of handling their particular condition.

- Because it is often impossible to tell which patients have infectious diseases, you should avoid direct contact with the blood and body fluids of all patients.

- If you think you may have been exposed to an infectious disease, see your physician (or your employer's designated physician) immediately.

- Six infectious diseases of special concern are:
  - HIV infection
  - Hepatitis B
  - Meningitis
  - Tuberculosis
  - SARS
  - H1N1

- Infection control should be an important part of your daily routine. Be sure to follow the proper steps when dealing with potential exposure situations.

Vital Vocabulary

herpes simplex Virus caused by human herpesviruses 1 and 2, characterized by small blisters whose location depends on the type of virus. Type 2 results in blisters on the genital area, while type 1 results in blisters in nongenital areas.

index of suspicion Awareness that unseen life-threatening injuries or illness may exist.

medical emergencies Life threats that require EMS attention because of illnesses or conditions not caused by an outside force.

meningitis An inflammation of the meningeal coverings of the brain and spinal cord; it is usually caused by a virus or a bacterium.

nature of illness (NOI) The general type of illness a patient is experiencing.

SARS (severe acute respiratory syndrome) Potentially life-threatening viral infection that usually starts with flulike symptoms.

trauma emergencies Injuries that are the result of physical forces applied to the body.

tuberculosis A chronic bacterial disease, caused by *Mycobacterium tuberculosis*, that usually affects the lungs but can also affect other organs such as the brain and kidneys.

virulence The strength or ability of a pathogen to produce disease.

Assessment in Action

You are dispatched to an unknown medical emergency in an apartment complex. On arrival, you find a 65-year-old man sitting at the kitchen table. From the doorway, he appears to be pale. He is complaining of feeling weak, and he states that he has been sick for a week with flulike symptoms of fever, vomiting, and general body aches. Past medical history includes a myocardial infarction 2 years ago and diabetes.

1. What is the patient's chief complaint?
   A. Vomiting
   B. Flu
   C. Weakness
   D. Body aches

2. What is the nature of his illness?
   A. Unknown medical
   B. Flu
   C. Weakness
   D. Body aches

3. The visual impression of the patient from the doorway is known as:
   A. scene size-up.
   B. SAMPLE history.
   C. AVPU.
   D. general impression.

4. What mnemonic is used to gather a patient's information?
   A. AVPU
   B. SAMPLE
   C. OPQRST
   D. DUMBELS

5. What does OPQRST stand for?

6. What standard precautions should be taken with this patient?
   A. Gloves only
   B. Gloves and mask
   C. Gloves, mask, and goggles
   D. Mask only

7. List the six infectious diseases of concern for EMS personnel.

8. After you assess the ABCs, what should be done for this patient?
   A. Transport
   B. Full-body scan
   C. Vital signs
   D. Reassessment

9. Which of the following best describes a communicable disease?
   A. The growth and spread of small harmful organisms within the body
   B. A disease that can be spread from one person or species to another
   C. A disease that is capable of being transmitted from one person to another
   D. Presence of infectious organisms on or in objects

10. Your transport decision should be based on the:
    A. secondary assessment.
    B. past medical history.
    C. physical examination.
    D. primary assessment.

National EMS Education Standard Competencies

Medicine

Applies fundamental knowledge to provide basic emergency care and transportation based on assessment findings for an acutely ill patient.

Respiratory

Anatomy, signs, symptoms, and management of respiratory emergencies including those that affect the

- Upper airway (pp 475-482)
- Lower airway (pp 475-482)

Anatomy, physiology, pathophysiology, assessment, and management of

- Epiglottitis (pp 480-509)
- Spontaneous pneumothorax (pp 487-506)
- Pulmonary edema (pp 480-505)
- Asthma (pp 484-510)
- Chronic obstructive pulmonary disease (pp 483-506)
- Environmental/industrial exposure (pp 489-490)
- Toxic gas (p 490)
- Pertussis (pp 481, 511)
- Cystic fibrosis (p 511)
- Pulmonary embolism (pp 488-489, 507)
- Pneumonia (pp 480-511)
- Viral respiratory infections (pp 490, 510)

Knowledge Objectives

1. List the structures and functions of the upper and lower airways, lungs, and accessory structures of the respiratory system. (pp 475-476)
2. Explain the physiology of respiration and list the signs of normal breathing. (pp 476-478)
3. Discuss the pathophysiology of respiration and provide examples of the common signs and symptoms a patient with inadequate breathing may present with in an emergency situation. (pp 476-478)
4. Explain the special patient assessment and care considerations that are required for geriatric patients who are experiencing respiratory distress. (pp 479, 508-512)

5. Describe different respiratory conditions that cause dyspnea, including their causes, assessment findings and symptoms, complications, and specific prehospital management and transport decisions. (pp 479-511)
6. List and review the characteristics of infectious diseases that are frequently associated with dyspnea. (pp 480-481)
7. Describe the assessment of a patient who is in respiratory distress and the relationship of the assessment findings to patient management and transport decisions. (pp 491-500)
8. List and define five different types of adventitious breath sounds, their signs and symptoms, and the disease process associated with each one. (pp 492-493)
9. Describe the primary emergency medical care of a person who is in respiratory distress. (pp 500-502)
10. State the generic name, medication forms, dose, administration, indications, actions, and contraindications for medications that are administered via metered-dose inhalers (MDI) and small-volume nebulizers. (pp 500-505)
11. Discuss some epidemic and pandemic considerations related to the spread of influenza type A and strategies EMTs should employ to protect themselves from infection during a possible crisis situation. (pp 507-508)
12. Explain the special patient assessment and care considerations that are required for pediatric patients who are experiencing respiratory distress. (pp 508-512)

Skills Objectives

1. Demonstrate the process of history taking to obtain more information related to a patient's chief complaint based on a case scenario. (pp 495-496)
2. Demonstrate how to use the OPQRST assessment to obtain more specific information about a patient's breathing problem.(p 496)
3. Demonstrate how to use the PASTE assessment to obtain more specific information about a patient's breathing problem. (p 496)
4. Demonstrate how to assist a patient with the administration of a metered-dose inhaler. (pp 503-504, Skill Drill 13-1)
5. Demonstrate how to assist a patient with the administration of a small-volume nebulizer. (pp 504-505, Skill Drill 13-2)

Introduction

As an EMT, you will encounter often the patient complaint of **dyspnea**, when a patient reports feeling short of breath or has difficulty breathing. It is a symptom of many different conditions, from the common cold or asthma to heart failure and pulmonary embolism. You may or may not be able to determine what is causing dyspnea in a particular patient; this can be difficult even for physicians in a hospital setting. Also, several different problems may be contributing to a patient's dyspnea at the same time, including some that are serious or life threatening. Even without a definitive diagnosis, however, you may still be able to save the patient's life.

This chapter begins with a basic review of respiratory anatomy and physiology as defined more thoroughly in Chapter 9, *Airway Management*. It then looks at common medical problems that can impede normal respiratory functioning and cause dyspnea. This chapter then goes on to explain specific strategies for you to use to assess a patient who has difficulty breathing, using the patient assessment template and organized approach established in Chapter 8, *Patient Assessment*. You will learn the signs and symptoms of each condition, and topics such as foreign body and anatomic airway obstruction, lung infections, and chronic airway disease will be covered. You should keep all these medical possibilities in mind as you obtain the patient's history and perform a physical assessment; these processes will be described in detail in this chapter. The information that you collect will help you to decide on the proper treatment, which can differ according to the probable cause of the dyspnea.

Remember, the sensation of not getting enough air can be terrifying, regardless of its cause. As an EMT, you should be prepared to fully treat your patient, addressing not just the symptom and the underlying problem, but also the anxiety that it produces.

Anatomy of the Respiratory System

The respiratory system consists of the structures of the body that contribute to the breathing process Figure 13-1. Structures that help us breathe include the diaphragm, the muscles of the chest wall, the accessory muscles of breathing, and the nerves from the brain and spinal cord to those muscles.

The upper airway consists of all anatomic airway structures above the level of the vocal cords. These include the nose, mouth, jaw, oral cavity, pharynx, and larynx. Air enters the upper airway through the nose and mouth, and it is here that the air is filtered, warmed, and humidified. The upper airway ends at the larynx, where it is protected by the epiglottis. This leaf-shaped valve diverts food and fluid into the esophagus and air into the trachea. Air then moves through the trachea and into the lung.

The principal function of the lungs is **respiration**, which is the exchange of oxygen and carbon dioxide. When air reaches the lower airway, it travels through the trachea and into each lung, to the bronchus (larger airways), then on to the bronchioles (smaller airways), and finally into the alveoli. The alveoli are microscopic, thin-walled air sacs where the actual exchange of oxygen and carbon dioxide occurs.

You are the Provider: PART 1

It is 4:30 AM when your alert tones sound, "Medic 81, respond to 109 East Lawler for a 72-year-old woman with shortness of breath." You recognize the address as one to which you have responded on numerous occasions. The woman lives alone; has emphysema, hypertension, and gout; and always refuses EMS transport. You and your partner proceed to the scene. The weather is clear, and the temperature is 65°F.

1. What is emphysema? What is the typical cause?
2. Why is it especially significant that *this* patient called 9-1-1?

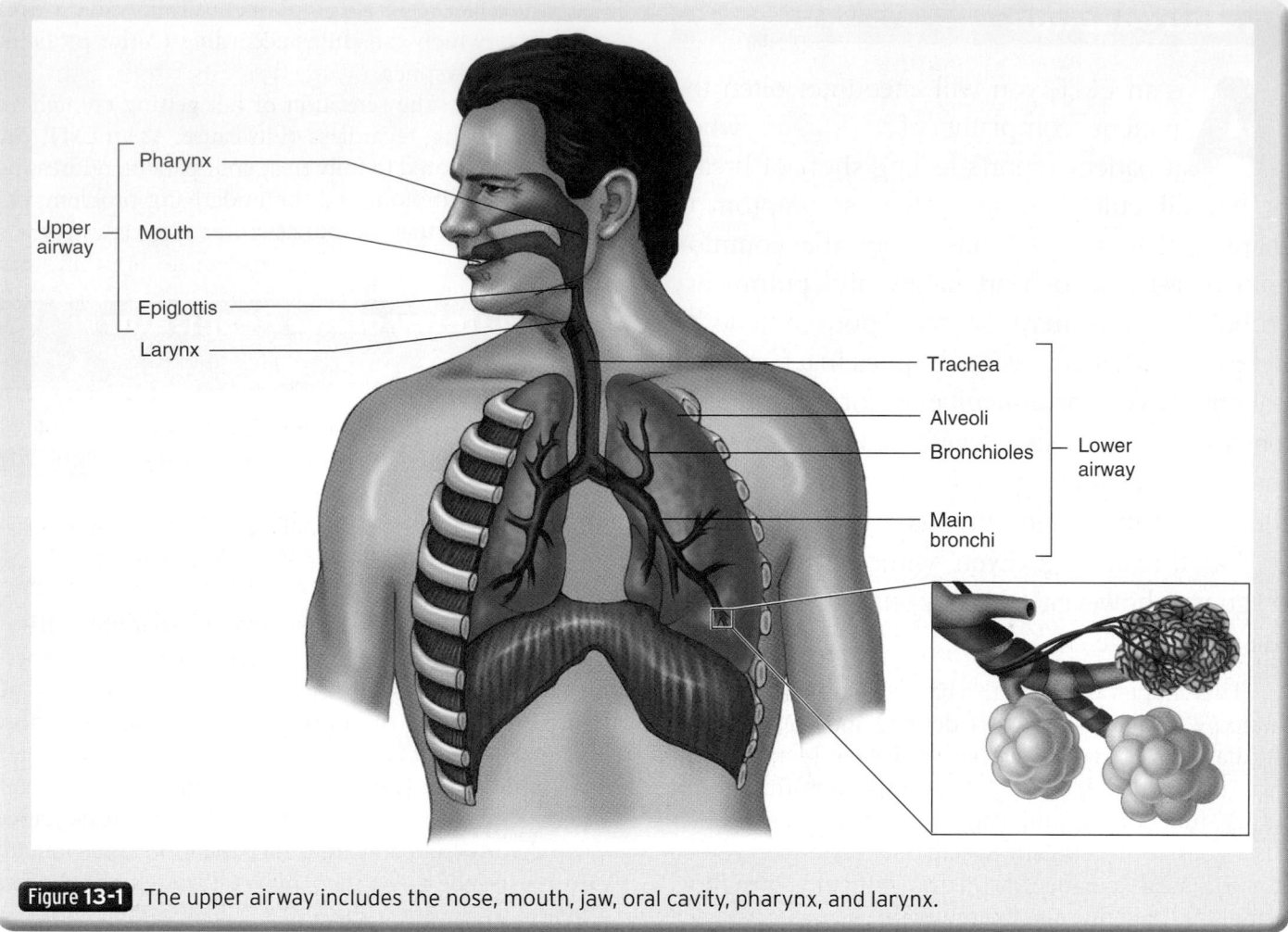

Figure 13-1 The upper airway includes the nose, mouth, jaw, oral cavity, pharynx, and larynx.

Labels on figure:
- Upper airway: Pharynx, Mouth, Epiglottis, Larynx
- Trachea, Alveoli, Bronchioles, Main bronchi (Lower airway)

Physiology of Respiration

As discussed in Chapter 9, *Airway Management*, the two processes that occur during respiration are inspiration, the act of breathing in or inhaling, and expiration, the act of breathing out, or exhaling. During respiration, oxygen is provided to the blood, and carbon dioxide is removed from it. In healthy lungs, this exchange of gases takes place rapidly at the level of the alveoli **Figure 13-2** . The alveoli lie against the pulmonary capillary vessels, and as oxygen enters the alveoli from inhalation, it passes freely through tiny passages in the alveolar wall into these capillaries. The oxygen is then carried to the heart, which then pumps the oxygen around the body. Carbon dioxide produced by the body's cells returns to the lungs in the blood that circulates through and around the alveolar air spaces. The carbon dioxide diffuses back into the alveoli and travels back up the bronchial tree and out through the upper airways during exhalation **Figure 13-3** . Again, carbon dioxide is "exchanged" for oxygen, which travels in exactly the opposite direction (during inhalation).

Through this whole process of respiration, the brainstem constantly senses the level of carbon dioxide in the arterial blood. The level of carbon dioxide bathing the brainstem stimulates a healthy person to breathe. If the level of carbon dioxide drops too low, the person automatically breathes at a slower rate and less deeply. As a result, less carbon dioxide is expired, allowing carbon dioxide levels in the blood to return to normal. If the level of carbon dioxide in the arterial blood rises above normal, the person breathes more rapidly and more deeply. When more fresh air is brought into the alveoli, more carbon dioxide diffuses out of the bloodstream, thereby lowering the level of carbon dioxide in the blood.

Pathophysiology

The pathophysiology of respiration refers to conditions under which body processes are not working as they should and, as a result, interfere with normal respiration. Abnormal or pathologic conditions in the

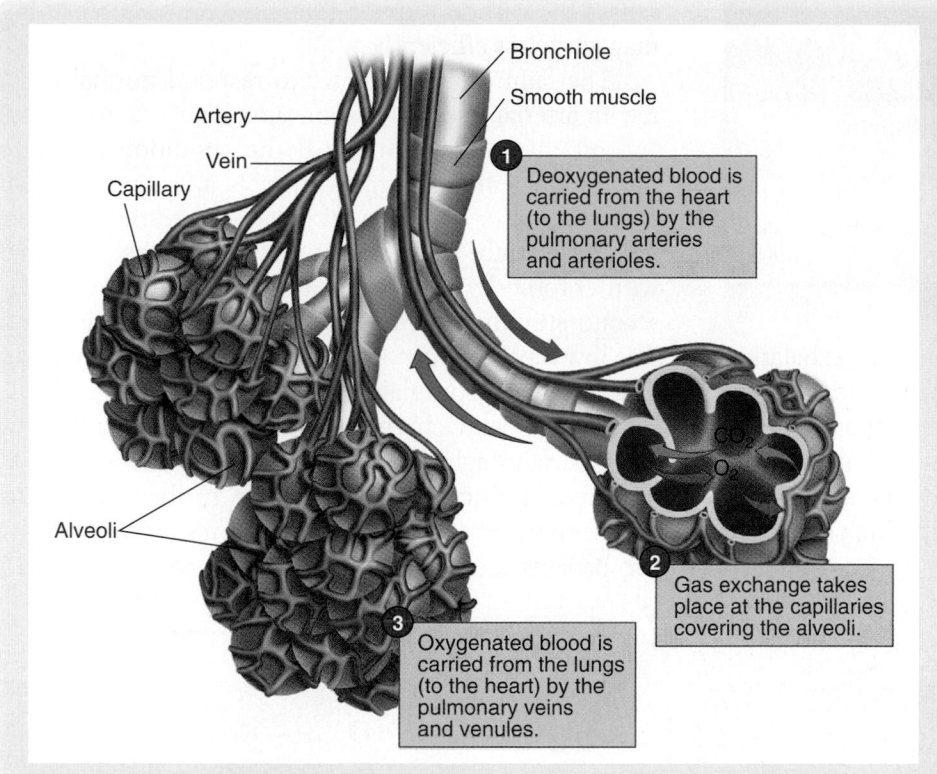

1 Deoxygenated blood is carried from the heart (to the lungs) by the pulmonary arteries and arterioles.

2 Gas exchange takes place at the capillaries covering the alveoli.

3 Oxygenated blood is carried from the lungs (to the heart) by the pulmonary veins and venules.

Figure 13-2 An enlarged view of a single alveolus (air sac) showing where the exchange of oxygen and carbon dioxide between air in the sac and blood in the pulmonary capillaries takes place.

anatomy of the airway, disease processes, and traumatic conditions can prevent the proper exchange of oxygen and carbon dioxide. In addition, the pulmonary blood

vessels themselves may have abnormalities that interfere with blood flow and thus with the transfer of gases.

Regardless of the reason for breathing difficulty, the critical issue is that you must be able to immediately recognize the signs and symptoms of inadequate breathing and know what to do about it. **Table 13-1** gives the signs of normal (adequate) breathing for adults.

Table 13-2 lists the clues that will help you determine whether your patient is having difficulty breathing.

Table 13-3 provides key signs and symptoms to help you recognize and differentiate between different respiratory-related complaints.

■ Carbon Dioxide Retention and Hypoxic Drive

When discussing the pathophysiology of respiration and the fact that body processes do not always work as they should, you will sometimes encounter patients who have an elevated level of carbon dioxide in their arterial blood.

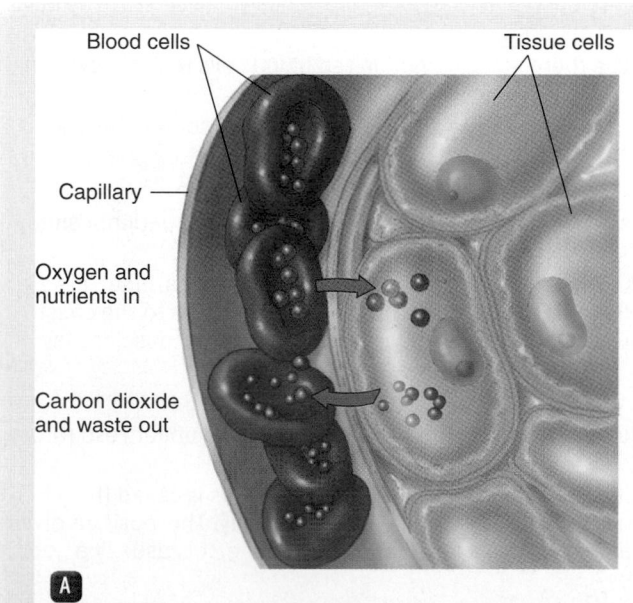

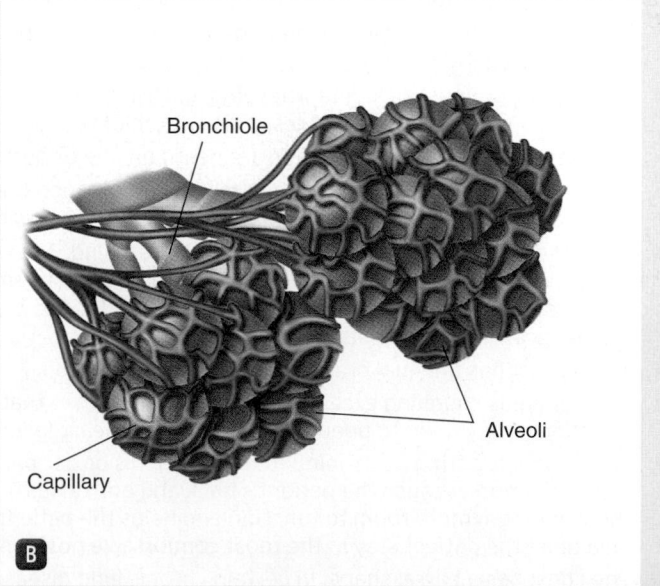

Figure 13-3 The exchange of oxygen and carbon dioxide in respiration. **A.** Oxygen passes from the blood through capillaries to tissue cells. Carbon dioxide passes from tissue cells through capillaries to the blood. **B.** In the lungs, oxygen is picked up by the blood, and carbon dioxide is given off.

Table 13-1 Signs of Normal Breathing

- A normal rate (between 12 and 20 breaths/min)
- A regular pattern of inhalation and exhalation
- Clear and equal lung sounds on both sides of the chest
- Regular and equal chest rise (chest expansion) and fall
- Adequate depth (tidal volume)

The level can rise for a number of reasons. The exhalation process may be impaired by various types of lung disease. The body may also produce too much carbon dioxide, either temporarily or chronically, depending on the disease or abnormality. If, for a period of years, arterial carbon dioxide levels rise to an abnormally high level and remain there, the respiratory center in the brain, which senses the carbon dioxide level and controls breathing, may work less efficiently.

The failure of this center to respond normally to a rise in arterial levels of carbon dioxide is due to chronic **carbon dioxide retention**. If the condition is severe, respiration will stop unless there is a secondary drive, called **hypoxic drive**, to stimulate the respiratory center. Fortunately, a second stimulus does help patients with chronically high blood carbon dioxide levels; this second stimulus is a low level of oxygen in the blood. The low blood oxygen level causes the respiratory center to respond and stimulate respiration.

Patients with chronic lung diseases frequently have a chronically high level of carbon dioxide in the blood. Therefore, giving too much oxygen to them may actually depress, or completely stop, the respirations. Unless the patients are unresponsive, use a more conservative

Table 13-2 Signs and Symptoms of Inadequate Breathing

- The patient complains of difficult breathing or shortness of breath.
- The patient has an altered mental status associated with shallow or slow breathing.
- The patient appears anxious or restless. This can happen if the brain is not getting enough oxygen for its needs.
- The respiratory rate is too fast (respirations of more than 20 breaths/min).
- The respiratory rate is too slow (respirations of less than 12 breaths/min); you may need to assist ventilations with a bag-mask device.
- The breathing rhythm is irregular. Because the brain controls breathing, an irregular breathing rhythm may indicate a head injury. In this situation, the patient will most likely be unresponsive.
- The skin is pale, cool, or clammy.
- The skin is blue (cyanotic). The tongue, nail beds, and inside the lips are good places to look for cyanosis. These areas all have a large collection of blood vessels and thin skin; thus, cyanosis is more apparent.
- The conjunctivae are pale. The patient may be short of breath because there are not enough red blood cells to carry oxygen to the tissues.
- Respiratory sounds, including wheezing, gurgling, snoring, stridor, or crowing, may be heard. Adventitious sounds can be associated with many types of respiratory problems.
- Decreased or noisy breath sounds are heard on one or both sides of the chest.
- The patient cannot speak more than few words between breaths. Ask the patient "How are you doing?" If the patient cannot speak at all, he or she most likely has a respiratory emergency that will need immediate attention.
- You observe muscle retractions or labored breathing. The patient is using the accessory muscles to assist breathing. If the patient is using only the diaphragm to breathe, suspect damage to the nerves that carry breathing commands to the chest muscles; the diaphragm may be getting the command to breathe, but because of spinal cord injury, the chest muscles may not be receiving the signals.
- The patient has unequal or inadequate chest expansion.
- The patient is coughing excessively, which can indicate that the patient has a condition ranging from a mild upper respiratory infection or hay fever to pneumonia, asthma, or heart failure.
- The patient is sitting up, leaning forward with his or her palms flat on the bed or the arms of the chair. This is called the tripod position because the patient's back and both arms are working together to support the upper body. This position gives the diaphragm more room to function and helps the patient to use accessory muscles to assist breathing. It is usually a good idea to let the patient stay in the most comfortable position.
- The chest has a barrel shape. In certain chronic lung diseases, air has been gradually and continuously trapped within the lung in increasing amounts; therefore, the distance from the front of the lung to the back of it gets longer, nearly equaling the side-to-side distance. A barrel chest may indicate a long history of breathing problems.
- The patient has pursed lips or nasal flaring.

Table 13-3	Signs and Symptoms Seen in Various Respiratory Conditions
Condition	**Signs and Symptoms**
Asthma	■ Wheezing on inspiration/expiration ■ Bronchospasm
Anaphylaxis	■ Flushed skin or hives ■ Generalized edema ■ Decreased blood pressure ■ Laryngeal edema with dyspnea
Bronchitis	■ Chronic cough ■ Wheezing ■ Cyanosis ■ Productive cough
Congestive heart failure	■ Dependent edema ■ Rales ■ Paroxysmal nocturnal dyspnea
Croup	■ Fever ■ Barking cough ■ Mostly seen in pediatric patients
Emphysema	■ Barrel chest ■ Pursed lip breathing ■ Dyspnea on exertion
Pneumonia	■ Dyspnea ■ Chills, fever ■ Cough ■ Dark sputum
Pneumothorax	■ Sudden chest pain with dyspnea ■ Decreased lung sounds/affected side
Pulmonary embolus	■ Sharp, pinpoint pain ■ Dyspnea ■ Sudden onset ■ After childbirth or surgery
Tension pneumothorax	■ Progressive shortness of breath ■ Increasing altered level of consciousness ■ Neck vein distention ■ Tracheal deviation
Pertussis (whooping cough)	■ Coughing spells ■ "Whooping" sound ■ Fever ■ Mostly seen in pediatric patients

Special Populations

As a result of the normal aging process and natural physiologic changes, geriatric patients have greater difficulties with the exchange of carbon dioxide and oxygen. In respiratory emergencies, it is necessary to initiate oxygen therapy early in the assessment and treatment process and continue its administration throughout the care of geriatric patients.

Causes of Dyspnea

Many different medical problems may cause dyspnea. You need to be aware that if the patient's problem is severe and the brain is deprived of oxygen, he or she may not be alert enough to complain about shortness of breath. More commonly, altered mental status is a sign of **hypoxia** of the brain.

Patients often have breathing difficulty and/or hypoxia with the following medical conditions:

- Upper or lower airway infection
- Acute pulmonary edema
- Chronic obstructive pulmonary disease (COPD)
- Asthma
- Hay fever
- Anaphylaxis
- Spontaneous pneumothorax
- Pleural effusion
- Prolonged seizures
- Obstruction of the airway
- Pulmonary embolism
- Hyperventilation syndrome
- Environmental/industrial exposure
- Carbon monoxide poisoning
- Infectious diseases

As you treat patients with disorders of the lung, you should be aware that one or more of the following situations most likely exists:

- Gas exchange between the alveoli and pulmonary circulation is obstructed by fluid in the lung, infection, or collapsed alveoli (**atelectasis**).
- The alveoli are damaged and cannot transport gases properly across their own walls.
- The air passages are obstructed by muscle spasm, mucus, or weakened floppy airway walls.
- Blood flow to the lungs is obstructed by blood clots.
- The pleural space is filled with air or excess fluid, so the lungs cannot properly expand.

All of these conditions prevent the proper exchange of oxygen and carbon dioxide. In addition, the pulmonary blood vessels themselves may have abnormalities that

approach by administering low-flow oxygen, adjusting it higher until symptoms have improved. Do *not* withhold oxygen for fear of depressing or stopping breathing in a patient with COPD. It may be dangerous to withhold oxygen from a patient complaining of dyspnea. Monitor the patient closely. If respirations slow, lower the liters per minute. You should be prepared to provide positive-pressure ventilation if the patient becomes apneic or unresponsive.

interfere with blood flow and thus with the transfer of gases.

Besides shortness of breath, a patient with dyspnea may also report the sensation of chest tightness and air hunger. Air hunger is when a person reports the feeling of "not getting enough air" and has a strong need to breathe. Chest tightness is described as an uncomfortable feeling in the chest, and it is commonly reported by patients with asthma.

Dyspnea is also a common complaint in patients with cardiopulmonary diseases. In some cases, it may be caused by physical exertion that has been made difficult because the patient's heart is damaged. Congestive heart failure is a troublesome cause of breathlessness because the heart is not pumping efficiently and, therefore, the body does not have adequate oxygen. Another condition commonly associated with congestive heart failure is pulmonary edema, in which the alveoli are filled with fluid.

Severe pain itself can cause a patient to experience rapid, shallow breathing without the presence of a primary pulmonary dysfunction. In some patients, breathing deeply causes pain because it causes expansion of the chest wall.

When you assess your patient for complaints of dyspnea, ask about chest pain; conversely, when you are evaluating your patient for chest pain, ask about dyspnea.

■ Upper or Lower Airway Infection

Infectious diseases causing dyspnea may affect all parts of the airway. Some cause mild discomfort; others obstruct the airway to the point that patients require a full range of respiratory support. Problems that impair the flow of air through the airways are problems of respiration. Difficulty providing adequate oxygen to the tissues due to a lack of oxygen in the air is a problem of **oxygenation**. If the air does not contain an adequate amount of oxygen, oxygenation cannot occur. In general, the problem causing the dyspnea is always some form of obstruction, either to the flow of air in the major passages (colds, diphtheria, epiglottitis, and croup) or to the exchange of gases between the alveoli and the capillaries (pneumonia). See **Table 13-4** for infectious diseases that may be associated with complaints of dyspnea.

In patients with infectious diseases, you will be in close contact, so be diligent about your personal use of

Table 13-4 Infectious Diseases Associated With Dyspnea

Disease	Characteristics
Bronchitis	■ An acute or chronic inflammation of the air passages (bronchi and bronchioles) often due to infection, usually associated with productive cough, and usually presents without fever. ■ Accumulation of fluid within the air passages, as well as swelling of the walls, restricts air flow and may lead to signs of asthma such as wheezing. It is often associated with rhonchi. Crackles are not usually present unless pneumonia has developed. ■ The breathing pattern in bronchitis does not indicate major airway obstruction, but the patient may experience tachypnea, an increase in the breathing rate, which is an attempt to compensate for the reduced amount of normal lung tissue and for the buildup of fluid.
Common cold	■ A viral infection usually associated with swollen nasal mucous membranes and the production of fluid from the sinuses and nose ■ Dyspnea is not severe; patients complain of "stuffiness" or difficulty breathing through the nose.
Tuberculosis (TB)	■ A disease that can lay dormant in a person's lungs for decades, then reactivate. ■ Dangerous because many TB strains are resistant to many antibiotics. ■ Spread by cough. Droplet nuclei can remain intact for decades. ■ Use a high-efficiency air particulate, or HEPA, respirator.
Diphtheria	■ Although the disease has been well controlled in the past decade, it is still highly contagious and serious when it occurs. ■ The disease causes the formation of a diphtheritic membrane lining the pharynx that is composed of debris, inflammatory cells, and mucus. This membrane can rapidly and severely obstruct the passage of air into the larynx.
Pneumonia	■ An acute bacterial or viral infection of the lung that damages lung tissue, usually associated with fever, cough, and production of sputum ■ Fluid also accumulates in the surrounding normal lung tissue, separating the alveoli from their capillaries. (Sometimes, fluid can also accumulate in the pleural space.)

Continues

Table 13-4 Infectious Diseases Associated With Dyspnea, continued

Disease	Characteristics
	■ The lung's ability to exchange oxygen and carbon dioxide is impaired. ■ The breathing pattern in pneumonia does not indicate major airway obstruction, but the patient may experience tachypnea, an increase in the breathing rate, which is an attempt to compensate for the reduced amount of normal lung tissue and for the buildup of fluid.
Epiglottitis Figure 13-4	■ An inflammation of the epiglottis due to bacterial infection that can produce severe swelling of the flap over the larynx ■ In preschool and school-aged children especially, the epiglottis can swell to two to three times its normal size. ■ The airway may become almost completely obstructed, sometimes quite suddenly. ■ Stridor (harsh, high-pitched, continued rough, barking inspiratory sounds) may be heard late in the development of airway obstruction. ■ Acute epiglottitis in adults is characterized by a severe sore throat. ■ The disease is now much less common than it was 20 years ago because of a vaccine that can help to prevent most cases. Occasionally, the constellation of symptoms of epiglottitis can appear in an adult or geriatric patient, especially with other issues that affect the patient's ability to fight off disease.
Croup Figure 13-5	■ An inflammation and swelling of the whole airway (pharynx, larynx, and trachea) typically seen in children between the ages of 6 months and 3 years ■ The common signs of croup are stridor and a seal-bark cough, which signal a significant narrowing of the air passage of the trachea that may progress to significant obstruction. ■ Croup often responds well to the administration of humidified oxygen. ■ Croup is rarely seen in adults because the airways are larger.
Respiratory syncytial virus	■ A major cause of illness in young children ■ Causes an infection of the lungs and breathing passages ■ Can lead to other serious illnesses that affect the lungs or heart, such as bronchiolitis and pneumonia ■ Highly contagious and spread through droplets ■ Survives on surfaces, including hands and clothing ■ Look for signs of dehydration. ■ Humidified oxygen is helpful if available.
Pertussis (whooping cough)	■ Pertussis is an airborne bacterial infection that affects mostly children younger than 6 years. ■ Patient will be feverish and exhibit a "whoop" sound on inspiration after a coughing attack. ■ Highly contagious through droplet infection ■ Coughing spells, which can last for more than a minute, in which the child may turn red or purple ■ Pertussis/whooping cough does not cause the typical whooping illness in adults. It causes a severe upper respiratory infection that could be an entry pathway to pneumonia in older people.
Severe acute respiratory syndrome (SARS)	■ A virus that has caused significant concern ■ SARS is a serious, potentially life-threatening viral infection caused by a recently discovered family of viruses best known as the second most common cause of the common cold. ■ SARS usually starts with flulike symptoms, and may progress to pneumonia, respiratory failure, and, in some cases, death. ■ SARS is thought to be transmitted primarily by close person-to-person contact.
Influenza type A	■ Virus that has crossed the animal/human barrier and has infected humans ■ Flu that has the potential to spread at a pandemic level
Meningococcal meningitis	■ An inflammation of the meningeal coverings of the brain and spinal cord that can be highly contagious ■ The bacteria can be spread through the exchange of respiratory and throat secretions through coughing and sneezing. ■ The effects are lethal in some cases. Victims who survive can be left with brain damage, hearing loss, or learning disabilities. ■ Patients may present with flulike symptoms, but unique to meningitis are high fever, severe headache, photophobia (light sensitivity), and a stiff neck in adults. Patients sometimes have an altered level of consciousness and can have red blotches on the skin. ■ Use respiratory protection, and report any potential cases.

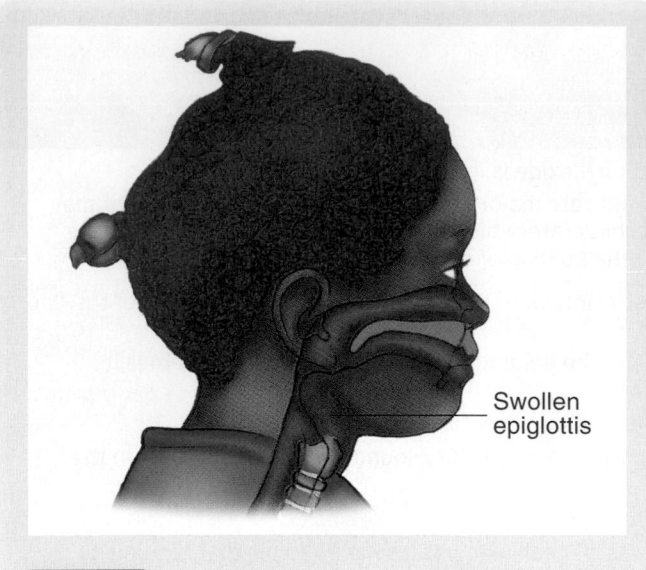

Figure 13-4 Acute epiglottitis is caused by a bacterial infection that results in severe swelling of the epiglottis. The epiglottis is massively swollen and almost fully obstructs the airway.

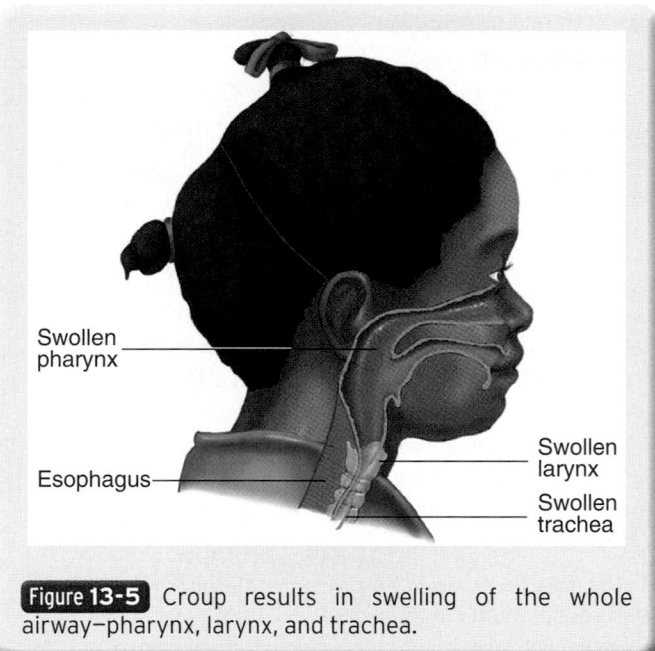

Figure 13-5 Croup results in swelling of the whole airway—pharynx, larynx, and trachea.

appropriate personal protective equipment (PPE). A minimum of gloves, eye protection, and a surgical mask or HEPA respirator should be mandatory. Gowns can be considered in some situations. Be aware of your local guidelines.

■ Acute Pulmonary Edema

Sometimes, the heart muscle is so injured after a heart attack or other illness that it cannot circulate blood properly. In these cases, the left side of the heart cannot remove blood from the lung as fast as the right side delivers it. As a result, fluid builds up within the alveoli and in the lung tissue between the alveoli and the pulmonary capillaries. This accumulation of fluid is referred to as **pulmonary edema**, and it is usually a result of congestive heart failure. By physically separating the alveoli from the pulmonary capillary vessels, the edema interferes with the exchange of carbon dioxide and oxygen **Figure 13-6**. Not enough space is left in the lung to allow for slow, deep breaths. The patient usually experiences dyspnea with rapid, shallow respirations. In the most severe cases, you will see frothy pink sputum at the nose and mouth.

Pulmonary edema is one of the most common causes of hospital admission in the United States. It is not uncommon for a patient to have repeated bouts.

Not all patients with pulmonary edema have heart disease. Poisonings from inhaling large amounts of smoke or toxic chemical fumes can produce pulmonary edema, as can traumatic injuries of the chest and exposure to high altitudes. In these cases, fluid collects in the alveoli and lung tissue in response to damage of the tissues of the lung or the bronchi.

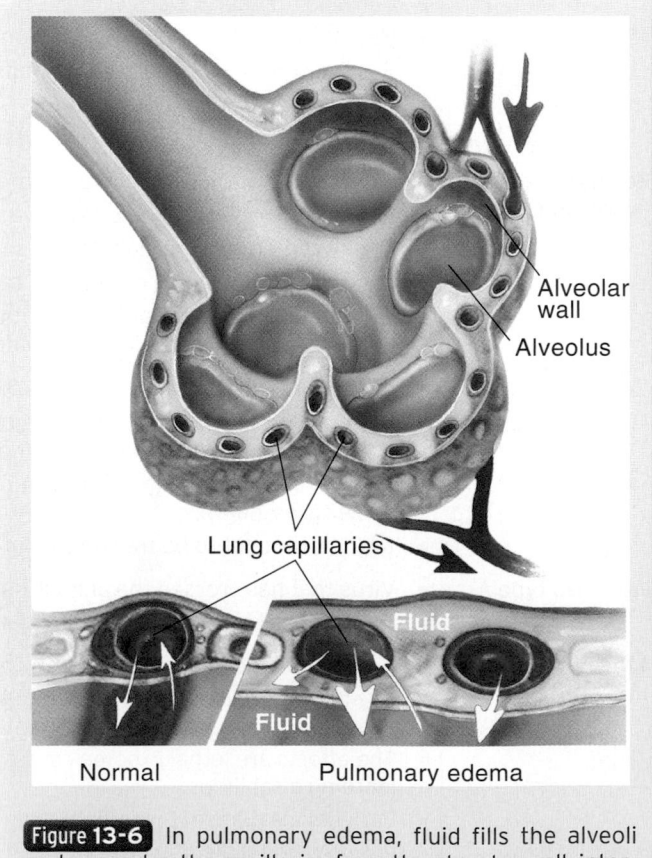

Figure 13-6 In pulmonary edema, fluid fills the alveoli and separates the capillaries from the alveolar wall, interfering with the exchange of oxygen and carbon dioxide.

■ Chronic Obstructive Pulmonary Disease

Chronic obstructive pulmonary disease (COPD) is a slow process of dilation and disruption of the airways and alveoli caused by chronic bronchial obstruction. According to the US Department of Health and Human Services, an estimated 12.1 million adults 25 years or older are reported to have the diagnosis of COPD. It is the fourth leading cause of death.

COPD may be a result of direct lung and airway damage from repeated infections or inhalation of toxic agents such as industrial gases and particles, but most often it results from cigarette smoking. Although it is well known that cigarettes are a direct cause of lung cancer, their role in the development of COPD is far more significant and less well publicized.

Tobacco smoke is a bronchial irritant and can create chronic bronchitis, an ongoing irritation of the trachea and bronchi. With bronchitis, excess mucus is constantly produced, obstructing small airways and alveoli. Protective cells and lung mechanisms that remove foreign particles are destroyed, further weakening the airways. Chronic oxygenation problems can also lead to right-sided heart failure and fluid retention, such as edema in the legs.

Pneumonia develops easily when the air passages are persistently obstructed. Ultimately, repeated episodes of irritation and pneumonia cause scarring in the lungs and some dilation of the obstructed alveoli, leading to COPD Figure 13-7 .

Another type of COPD is called emphysema. Emphysema is a loss of the elastic material around the air spaces as a result of chronic stretching of the alveoli when inflamed airways obstruct easy expulsion of gases. Smoking can also directly destroy the elasticity of the lung tissue. Normally, lungs act like a spongy balloon that is inflated; once they are inflated, they will naturally recoil because of their elastic nature, expelling gas rapidly. However, when they are constantly obstructed or when the "balloon's" elasticity is diminished, air is no longer expelled rapidly, and the walls of the alveoli eventually fall apart, leaving large "holes" in the lung that resemble a large air pocket or cavity. This condition is called emphysema.

Most patients with COPD have elements of both chronic bronchitis and emphysema. Some patients will have more elements of one condition than the other; few patients will have only emphysema or bronchitis. Therefore, most patients with COPD will chronically produce sputum, have a chronic cough, and have difficulty expelling air from their lungs, with long expiration

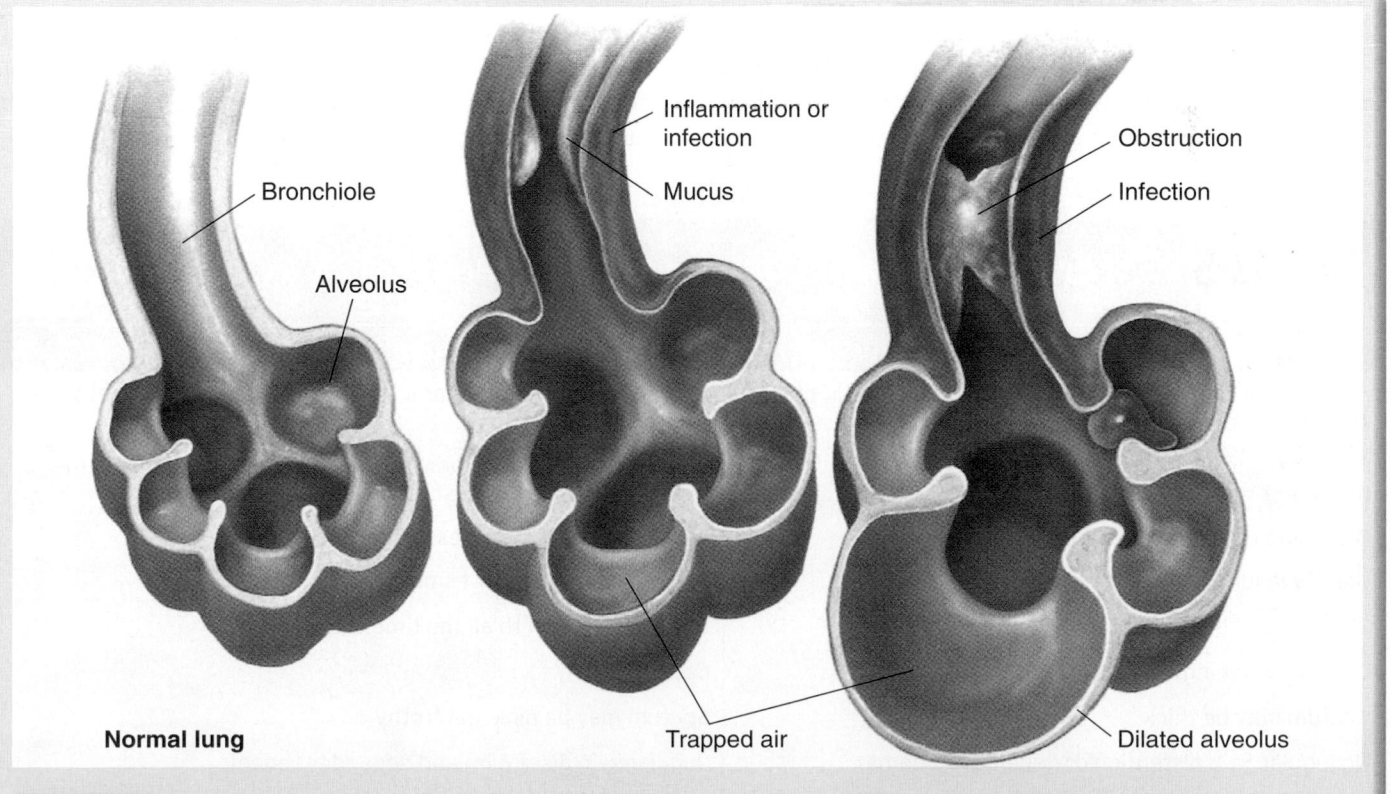

Bronchiole

Alveolus

Inflammation or infection

Mucus

Obstruction

Infection

Normal lung

Trapped air

Dilated alveolus

Figure 13-7 Repeated episodes of irritation and inflammation in the alveoli result in the obstruction, scarring, and some dilation of the alveolar sac characteristic of chronic obstructive pulmonary disease.

phases and wheezing. Patients present with abnormal breath sounds such as rales, crackles, rhonchi, and wheezes.

Wet Lungs Versus Dry Lungs and "Cardiac Asthma"

There is sometimes confusion in the differential diagnosis of the "wet lungs" of pulmonary edema caused most often by congestive heart failure and the "dry lungs" of COPD.

Suppose you are called to assist an 80-year-old man who has had shortness of breath for 45 minutes. Physical examination reveals that his pulse and respirations are elevated and you can see that he has pedal edema and jugular vein distention. His lung sound check reveals wheezing. He has a history of hypertension, congestive heart failure, and myocardial infarction; however, he has no history of smoking.

What supports an initial general impression of congestive heart failure are the pedal edema, jugular vein distention, and history of congestive heart failure. Are patients with congestive heart failure supposed to have rales rather than wheezing **Table 13-5**? Lung sounds are very helpful but can also be confusing. In a case in which the alveoli are so full of fluid, bubbles (the condition that gives the sound of rales) cannot form. The bronchi also become constricted, which produces wheezing. The wheezing this patient is experiencing is called "cardiac asthma," which is all the more confusing because the patient has no history of asthma.

Patients with COPD wheeze because of bronchial constriction and present with shortness of breath. Their breathing gets progressively worse over time, and they have the most trouble breathing on exertion. Patients with COPD have chronic coughing and thick sputum. They do not have jugular vein distention or dependent edema and are usually long-term smokers with a thin, barrel chest appearance. Their medications would include home oxygen, bronchodilators, and corticosteroids.

What should lead you in the direction of congestive heart failure in this patient are the elevated blood pressure, the pedal edema, and a history of congestive heart failure. Unlike a typical patient with COPD, he has no history of smoking and takes diuretics and preload reducers.

Patients with COPD have a slower onset of symptoms because their disease is worsened by infection and other stressors. Patients with congestive heart failure experience a fluid overload in the lung, which develops quickly from a failing pump.

As you try to discern between COPD and congestive heart failure, do not assume that all COPD patients have wheezing and all congestive heart failure patients have rales; keep an open mind so that you do not miss other important differences. The best advice is to treat the patient, not the lung sounds. In some cases, a patient with COPD may have air passages that are so constricted you do not hear anything.

■ Asthma, Hay Fever, and Anaphylaxis

Asthma, hay fever, or allergic rhinitis, and anaphylaxis (anaphylactic shock) are the result of an allergic reaction to an inhaled, ingested, or injected substance. The substance itself (allergen) is not the cause of the allergic

Table 13-5 Major Differences Between COPD and Congestive Heart Failure

COPD	Congestive Heart Failure
A disease of the lung characterized by shortness of breath and wheezing	A disease of the heart characterized by shortness of breath, edema, and weakness
Home oxygen, bronchodilators, and steroids used for treatment	Diuretics prescribed to help promote cardiac function and to reduce fluid loads on the heart
Breathing progressively worse over time	Sudden onset of shortness of breath
Usually in long-term smokers	Patient may or may not smoke
Shortness of breath mostly on exertion	Shortness of breath all the time
Chronic coughing	Coughing
Sputum may be thick	Sputum may be pink and frothy
No jugular vein distention or dependent edema	Jugular vein distention and dependent edema
Patient usually thin with a barrel chest	May have distended abdomen

reaction; rather, it is an exaggerated response of the body's immune system to that substance that causes it. In some cases, however, there is no identifiable allergen that triggers the body's immune system.

Asthma

Asthma is an acute spasm of the smaller air passages, called bronchioles, associated with excessive mucus production and with swelling of the mucous lining of the respiratory passages **Figure 13-8**. In 2006, 3,563 deaths were attributed to asthma. Approximately 22.9 million Americans (including 6.7 million children) had asthma in 2007. Asthma affects people of all ages, but the highest prevalence rate is seen in children 5 to 17 years of age.

Asthma produces a characteristic wheezing as patients attempt to exhale through partially obstructed air passages; wheezing is indicative of a partial lower airway obstruction. These same air passages open easily during inspiration. The wheezing may be so loud that you can hear it without a stethoscope. In other cases, the airways are so blocked that no air movement is heard. In severe cases, the actual work of exhaling is very tiring, and cyanosis and/or respiratory arrest may quickly develop.

An acute asthma attack may be caused by an allergic response to specific foods or some other allergen. Between attacks, patients may breathe normally. Asthma attacks may also be caused by severe emotional stress, exercise,

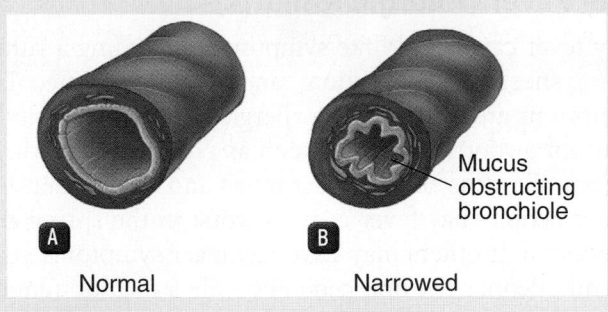

Figure 13-8 Asthma is an inflammation of the lungs associated with excessive mucus production and swelling of the bronchioles. **A.** Cross-section of a normal bronchiole. **B.** The bronchiole in spasm; a mucus plug has formed and partially obstructed the bronchiole.

and respiratory infections. In its most severe form, an allergic reaction can produce anaphylactic shock. This, in turn, may cause respiratory distress that is severe enough to result in coma and death.

Most patients with asthma are familiar with their symptoms and know when an attack is imminent. Typically, they will have appropriate medication with them. Depending on your local protocols, you may be allowed to assist an asthma patient with an inhaler or nebulizer. You should listen carefully to what a patient with asthma tells you; they often know exactly what they need.

You are the Provider: PART 2

After arriving at the scene and entering the patient's house, you immediately smell cigarette smoke. There are numerous full ashtrays in the living room. The patient is sitting on the edge of her couch; she is wearing a nasal cannula attached to home oxygen, is smoking a cigarette, and is experiencing obvious breathing difficulty. She tells you, in broken sentences, that her shortness of breath has worsened. You perform an assessment as your partner prepares to begin treatment.

Recording Time: 0 Minutes	
Appearance	Obvious breathing difficulty; breathing through pursed lips
Level of consciousness	Conscious and alert
Airway	Open; no secretions or foreign bodies
Breathing	Rapid and labored
Circulation	Radial pulse, rapid and weak; skin, pink, warm, and dry

3. What should be your *most* immediate action?
4. How does emphysema differ from chronic bronchitis?

Hay Fever (Allergic Rhinitis)

Hay fever causes coldlike symptoms, including a runny nose, sneezing, congestion, and sinus pressure. The symptoms are caused by an allergic response, usually to outdoor airborne allergens such as pollen or sometimes indoor allergens such as dust mites and pet dander. For many people, hay fever is at its worst in the spring and summer, but others may have hay fever symptoms year-round. People do not generally call 9-1-1 or request an ambulance for simple hay fever symptoms, but hay fever is included in this discussion of allergic conditions because it affects so many people. People with hay fever tend to be atopic, meaning that they are more likely to have other allergies, and they may also have a higher incidence of severe reactions, including anaphylaxis.

Anaphylactic Reactions

Patients with or without asthma may have severe allergic reactions. Anaphylaxis is a severe allergic reaction characterized by airway swelling and dilation of blood vessels all over the body, which may significantly lower blood pressure Figure 13-9 . This condition is sometimes referred to as anaphylactic shock. Anaphylaxis may be associated with widespread itching and signs and symptoms similar to asthma. The airway may swell so much that breathing problems can progress from extreme difficulty in breathing to total airway obstruction in a matter of a few minutes. Most anaphylactic reactions occur within 30 minutes of exposure to the allergen, which can be anything from eating nuts to receiving a penicillin injection. For some patients, the episode of anaphylaxis may represent the first time they were aware of sensitivity to the substance. Therefore, they may not know what caused the reaction. In other cases, the patient may be aware of what substance he or she is sensitive to but is not aware that an exposure has occurred, such as eating a food item that was not supposed to contain nuts. In severe cases, epinephrine is the treatment of choice. Patients may or may not have their own prescribed

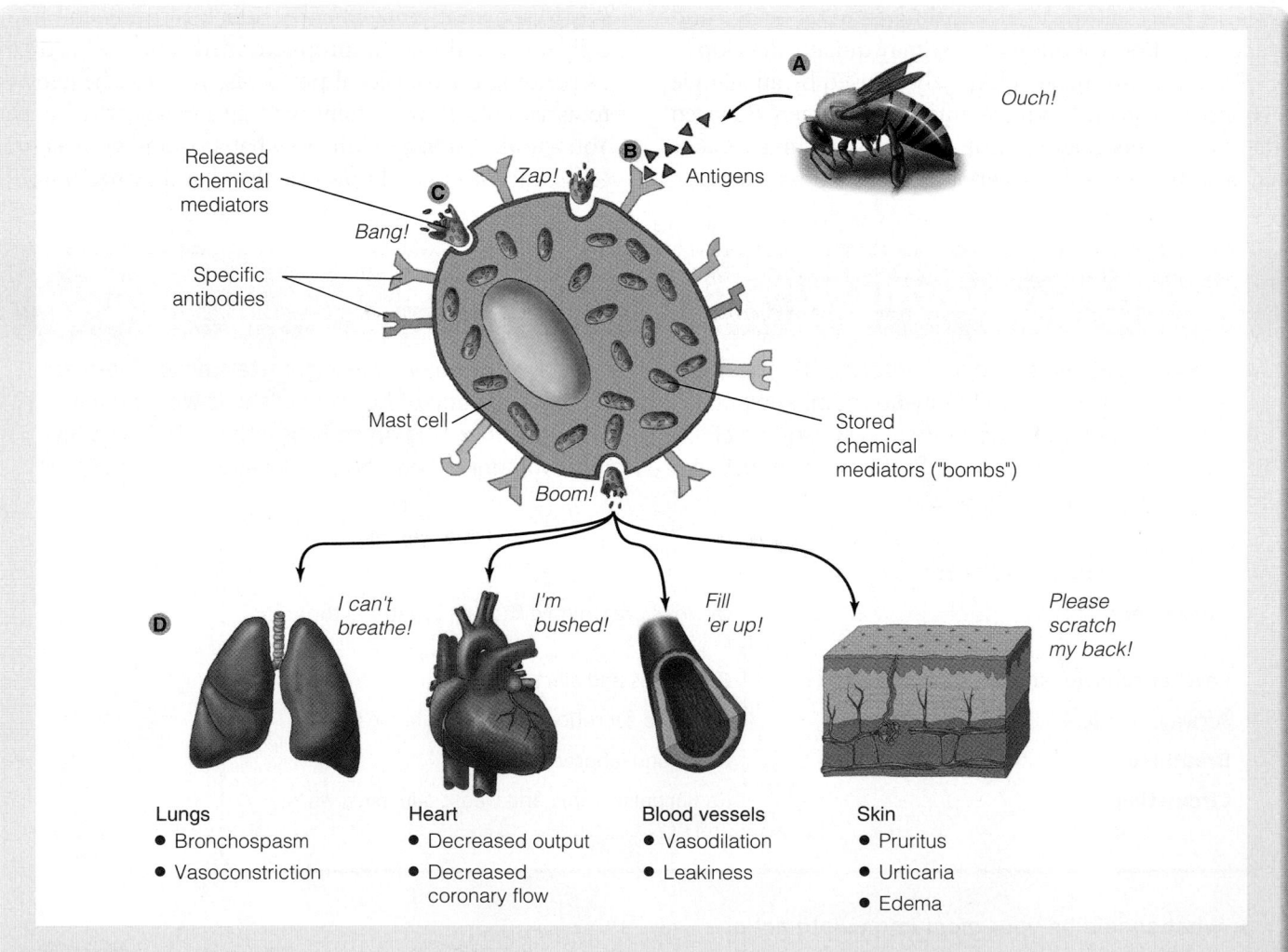

Figure 13-9 The sequence of events in anaphylaxis. **A.** The antigen is introduced into the body. **B.** The antigen-antibody reaction at the surface of a mast cell. **C.** Release of mast cell chemical mediators. **D.** Chemical mediators exert their effects on end organs.

EpiPen auto-injector. Oxygen and antihistamines are also useful. As always, medical direction should guide appropriate therapy. For more information about anaphylaxis and the EpiPen auto-injector, see Chapter 18, *Immunologic Emergencies*.

■ Spontaneous Pneumothorax

<u>Pneumothorax</u> is a partial or complete accumulation of air in the pleural space. Pneumothorax is most often caused by trauma, but it can also be caused by some medical conditions. In these cases, the condition is called a "spontaneous" pneumothorax.

Normally, the "vacuum" pressure in the pleural space keeps the lung inflated. When the surface of the lung is disrupted, however, air escapes into the pleural cavity, and the negative vacuum pressure is lost. The natural elasticity of the lung tissue causes the lung to collapse. The accumulation of air in the pleural space may be mild or severe **Figure 13-10**.

Spontaneous pneumothorax may occur in patients with certain chronic lung infections or in young people born with weak areas of the lung. Patients with emphysema and asthma are at high risk for spontaneous pneumothorax when a weakened portion of lung ruptures, often during severe coughing. Tall, thin males are also more susceptible than the rest of the population to developing spontaneous pneumothorax, particularly while performing strenuous activities, such as heavy lifting.

A patient with a spontaneous pneumothorax becomes dyspneic (short of breath) and might complain of <u>pleuritic chest pain</u>, a sharp, stabbing pain on one side that is worse during inspiration and expiration or with certain movement of the chest wall. By listening to the chest with a stethoscope, you can sometimes detect that breath sounds are absent or decreased on the affected side. However, altered breath sounds are very difficult to detect in a patient with severe emphysema. Spontaneous pneumothorax may be the cause of sudden dyspnea in a patient with underlying emphysema. A spontaneous pneumothorax has the potential to evolve into a life-threatening pneumothorax. You should continually reassess for anxiety, increased dyspnea, hypotension, absent or severely decreased breath sounds on one side, the presence of jugular vein distention, and cyanosis.

■ Pleural Effusion

A <u>pleural effusion</u> is a collection of fluid outside the lung on one or both sides of the chest. It compresses the lung or lungs and causes dyspnea **Figure 13-11**. This fluid may collect in large volumes in response to any irritation, infection, congestive heart failure, or cancer. Though it can build up gradually, over days or even weeks, patients often report that their dyspnea came on suddenly. Pleural effusions should be considered as a contributing diagnosis in any patient with lung cancer and shortness of breath.

When you listen with a stethoscope to the chest of a patient with dyspnea resulting from pleural effusion, you will hear decreased breath sounds over the region of the chest where fluid has moved the lung away from the chest wall. The patients frequently feel better if they are sitting upright. Nothing will really relieve their

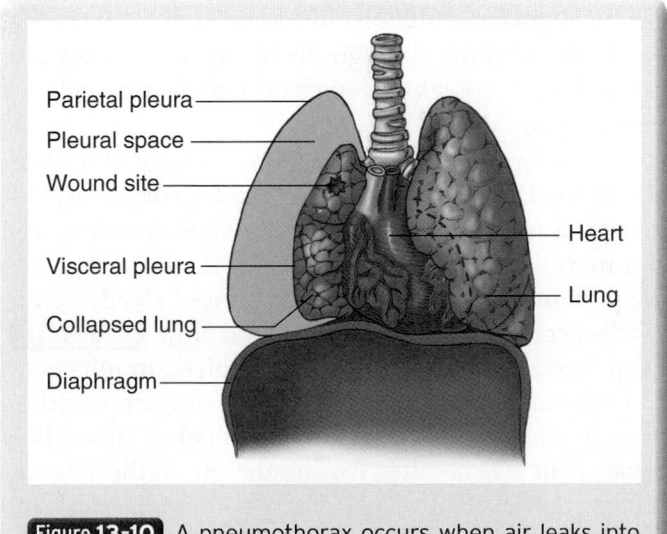

Figure 13-10 A pneumothorax occurs when air leaks into the pleural space from an opening in the chest wall or the surface of the lung. The lung collapses as air fills the pleural space and the two pleural surfaces are no longer in contact.

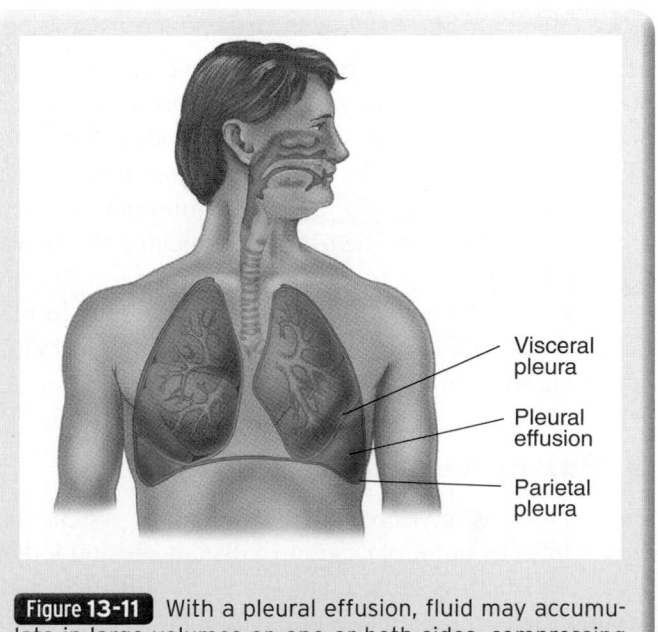

Figure 13-11 With a pleural effusion, fluid may accumulate in large volumes on one or both sides, compressing the lungs and causing dyspnea.

symptoms, however, except removal of the fluid, which must be done by a physician in the hospital.

■ Prolonged Seizures

A seizure is caused by a disruption in the electrical activity in the brain. A tonic-clonic seizure (formerly called a grand mal seizure) is one in which the patient has a sudden loss of consiousness, tonic-clonic movement of the body, and often incontinence. In typical cases, the seizure lasts only minutes and the patient regains consciousness. There are cases, however, in which the patient continues to have seizures every few minutes without regaining consciousness and/or the seizure lasts longer than 30 minutes. This situation is called status epilipticus and could be life threatening.

During a seizure that lasts 1 to 3 minutes, the patient will have some impairment in his or her ability to breathe and/or will bite the tongue. In a prolonged seizure, a number of other airway issues emerge. While unconscious, the patient has no control over the airway. Blood and saliva can be aspirated into the airway. If the patient is on his or her back, the tongue can block the airway. The constant muscle contractions will interfere with the ability of the chest to expand, and the patient will hypoventilate and become hypoxic. The patient could also become apneic. Other effects of a prolonged seizure could include hyperthermia, dehydration, and hypoglycemia. A prolonged seizure is a life-threatening situation that needs immediate intervention and aggressive airway management.

■ Obstruction of the Airway

As an EMT, you should always be aware of the possibility that a patient with dyspnea may have a mechanical obstruction of the airway and be prepared to treat it quickly. In semiconscious and unconscious individuals, the obstruction may be the result of aspiration of vomitus or a foreign object **Figure 13-12A** or improper positioning of the head so that the tongue is blocking the airway **Figure 13-12B**.

Always consider upper airway obstruction from a foreign body first in patients who were eating just before becoming short of breath.

■ Pulmonary Embolism

An **embolus** is anything in the circulatory system that moves from its point of origin to a distant site and lodges there, obstructing subsequent blood flow in that area. Beyond the point of obstruction, circulation can be completely cut off or at least markedly decreased, which can result in a serious, life-threatening condition. Emboli

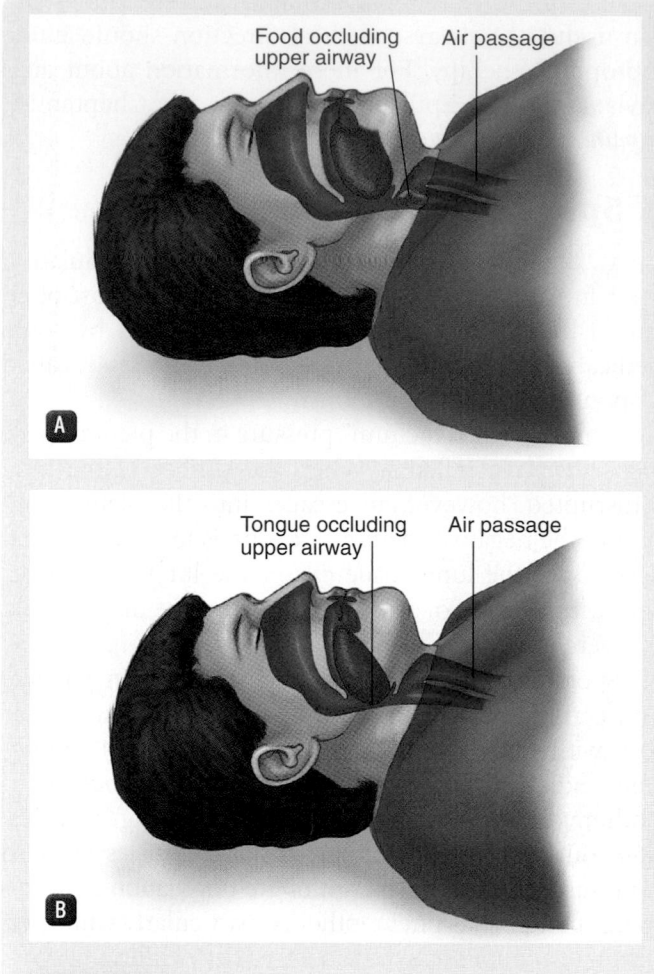

Figure 13-12 **A.** Foreign body obstruction occurs when an object, such as food, is lodged in the airway. **B.** Mechanical obstruction also occurs when the head is not properly positioned, causing the tongue to fall back into the throat.

can be fragments of blood clots in an artery or vein that break off and travel through the bloodstream. They also can be foreign bodies that enter the circulation, such as a bubble of air.

A **pulmonary embolism** is the passage of a blood clot formed in a vein, usually in the legs or pelvis, that breaks off and circulates through the venous system. The clot moves through the right side of the heart and into the pulmonary artery, where it becomes lodged, significantly decreasing or blocking blood flow **Figure 13-13**. Even though the lung is actively involved in inhalation and exhalation of air, no exchange of oxygen or carbon dioxide takes place in the areas of blocked blood flow because there is no effective circulation. In this circumstance, the level of arterial carbon dioxide rises, and the oxygen level may drop enough to cause cyanosis. The severity of cyanosis and dyspnea is directly related to the size of the embolism and the amount of tissue affected.

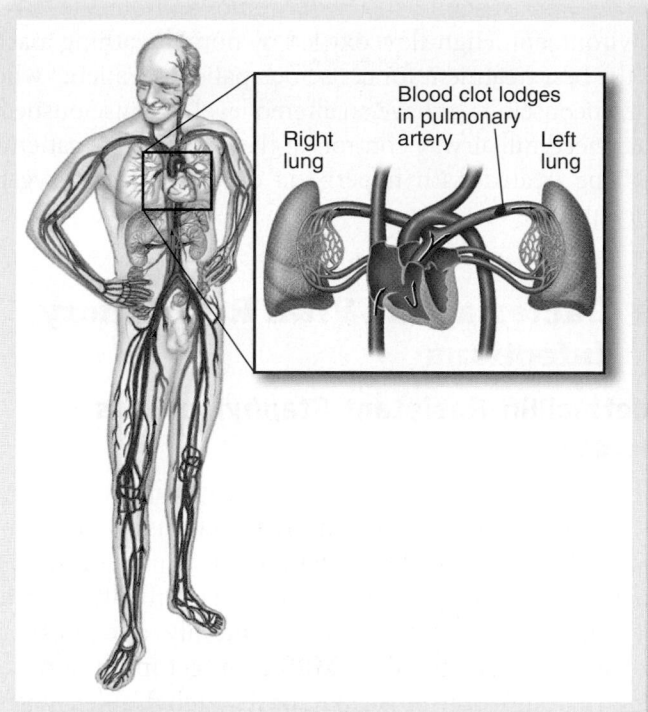

Figure 13-13 A pulmonary embolus is a blood clot from a vein that breaks off, circulates through the venous system, and moves through the right side of the heart into the pulmonary artery. Here, it can become lodged and significantly obstruct blood flow.

Pulmonary emboli may occur as a result of damage to the lining of vessels, a tendency for blood to clot unusually fast, or, most often, slow blood flow in a lower extremity. Slow blood flow in the legs is usually caused by long-term bed rest, which can lead to the collapse of veins. Patients whose legs are immobilized following a fracture or recent surgery are at risk for pulmonary emboli for days or weeks after the incident. Only rarely do pulmonary emboli occur in active, healthy individuals.

Although they are fairly common, pulmonary emboli are difficult to diagnose. According to the US Department of Health and Human Services, 100,000 cases of pulmonary embolism occur each year in the United States. Symptoms and signs of pulmonary emboli include the following:

- Dyspnea
- Acute chest pain
- Hemoptysis (coughing up blood)
- Cyanosis
- Tachypnea
- Varying degrees of hypoxia

With a large enough embolus, complete, sudden obstruction of the output of blood flow from the right side of the heart can result in sudden death.

Hyperventilation

Hyperventilation is defined as overbreathing to the point that the level of arterial carbon dioxide falls below normal. This may be an indicator of major, life-threatening illness. For example, a patient with diabetes who has a very high blood glucose level, a patient who has taken an overdose of aspirin, or a patient with a severe infection is likely to hyperventilate. In these cases, rapid, deep breathing is the body's attempt to stay alive. The body is trying to compensate for **acidosis**, the buildup of excess acid in the blood or body tissues that results from the primary illness. Because carbon dioxide, mixed with water in the bloodstream, can add to the blood's acidity, lowering the level of carbon dioxide helps to compensate for the other acids.

Similarly, in an otherwise healthy person, blood acidity can be diminished by excessive breathing because it "blows off" too much carbon dioxide. The result is a relative lack of acids. The resulting condition, **alkalosis**, is the buildup of excess base (lack of acids) in the body fluids.

Alkalosis is the cause of many of the symptoms associated with **hyperventilation syndrome (panic attack)**, including anxiety, dizziness, numbness, tingling of the hands and feet, and even a sense of dyspnea despite the rapid breathing. Although hyperventilation can be the response to illness and a buildup of acids, hyperventilation syndrome is not the same thing. Instead, this syndrome occurs in the absence of other physical problems. It commonly occurs when a person is experiencing psychological stress and affects some 10% of the population at one time or another. The respirations of an individual who is experiencing hyperventilation syndrome may be as high as 40 shallow breaths/min or as low as only 20 very deep breaths/min.

The decision whether hyperventilation is being caused by a life-threatening illness or a panic attack should not be made outside the hospital. Initially, you can verbally instruct the patient to slow his or her breathing; however, if that does not work, give supplemental oxygen and provide transport to the hospital where physicians will determine the cause of the hyperventilation.

Environmental/Industrial Exposure

Many accidental exposures that cause inhalation injury and dyspnea occur at industrial sites. Pesticides, cleaning solutions, chemicals, chlorine, and other gases can be accidentally released and inhaled by employees. **Carbon monoxide** is an odorless, highly poisonous gas that results from incomplete oxidation of carbon in combustion. It is produced in industrial settings by vehicles, gasoline-powered tools, and heaters. Sometimes chemicals like ammonia and chlorine bleach are mixed and create a hazardous by-product.

In many cases, industrial sites have their own medical/fire/hazardous materials teams that are very familiar with all the chemicals used at their site and know what to do in case of an exposure. They will begin immediate decontamination and medical care. In these cases, the patient needs to be decontaminated by trained responders prior to your taking responsibility.

Once the patient is decontaminated, gather information from the first responders about the substance and the cause of dyspnea. Assess the patient, paying special attention to lung sounds. Inhalation injuries can cause aspiration pneumonia that can result in eventual pulmonary edema. The inhaled substance can also cause lung damage. Blood coming from the airway is a particularly ominous sign.

Carbon Monoxide Poisoning

Toxic gases can also affect people outside the industrial setting. One common type of exposure is carbon monoxide. Natural gas has an odor, but carbon monoxide is odorless and tasteless, and you cannot see it. Carbon monoxide is the leading cause of accidental poisoning deaths in the United States. People who survive carbon monoxide poisoning can have permanent brain damage.

Carbon monoxide is produced by household appliances such as gas water heaters, space heaters, grills, and generators and is even present in cigarette smoke. A common cause of carbon monoxide poisoning occurs at the onset of cold weather when people turn heaters on for the first time. The combined effects of incomplete combustion and generally poor or no ventilation in a cold weather–sealed building result in the perfect setting for the production of carbon monoxide. Other common sources of carbon monoxide poisoning are smoke from fires and motor vehicle exhaust. Some people will attempt suicide by closing the car in the garage, turning the vehicle on, and inhaling the exhaust.

People who are exposed to carbon monoxide may think they have the flu. They initially complain of headache, dizziness, fatigue, and nausea and vomiting. They may complain of dyspnea on exertion and chest pain and display nervous system symptoms like impaired judgment, confusion, or even hallucinations. The worst exposures may result in syncope or seizure.

When assessing the scene, ensure that you do not become contaminated or put yourself at risk of exposure. Consider toxic gas exposure if more than one patient in the same environment is experiencing the same signs and symptoms. The symptoms of patients will start to be relieved as soon as they are removed from the toxic environment. High-flow oxygen by nonrebreathing mask is the best treatment for conscious patients. Patients who are unconscious or have an altered level of consciousness may need full airway control. In the worst cases, patients will be treated with hyperbaric or pressurized oxygen therapy.

■ Bacterial and Viral Respiratory Infections

Methicillin-Resistant *Staphylococcus aureus*

Methicillin-resistant *Staphylococcus aureus* (MRSA) is a bacterium that can cause infections in different parts of the body and is transmitted by different routes, including the respiratory route. It is difficult to treat because it is resistant to many commonly used antibiotics, especially methicillin. MRSA can get into the body, often through non-intact skin or through droplets when patients cough productively. Occasionally it can cause serious problems like infected wounds or pneumonia. MRSA infections are most common in people who have weak immune systems and are staying in hospitals or living in nursing homes. Infections can also appear around surgical wounds or invasive devices, like feeding tubes and catheters.

Tuberculosis

Tuberculosis (TB) is an infection caused by a bacterium called *Mycobacterium tuberculosis*. Tuberculosis most commonly affects the lungs but can also be found in almost any organ of the body, particularly the kidneys, bones, and lining of the brain and spinal cord (meninges). In some cases, TB can remain in an inactive (dormant) state for years without causing symptoms or being infectious to other people. However, when the person is in a state of weakened immunity, the TB can become active again. The patient may not even be aware he or she is harboring the disease.

Patients with active TB will report fever, coughing, fatigue, night sweats, and weight loss. If the lung infection becomes more severe, the patient will experience shortness of breath, coughing, productive sputum, bloody sputum, and chest pain.

In dense populations such as homeless people, prison inmates, and nursing home residents, TB has a higher prevalence. TB is also found in persons who abuse intravenous drugs or alcohol, and people whose immune systems are compromised by an infection such as HIV. People at risk of contracting the disease are those who are in close contact with individuals who have active TB or are in contact with people from countries that have a high prevalence of TB. Health care providers are also at risk.

If you suspect that your patient may have active TB, you need to wear (at a minimum) your gloves, eye protection, and a HEPA respirator. These respirators are fit-tested to the individual to ensure that no contaminated air can pass through and are designed to filter out TB bacteria.

Patient Assessment

Assessment of patients in respiratory distress should be conducted as a calm and systematic process. The patients are usually quite anxious, and they may be some of the most ill and challenging patients you will encounter.

Scene Size-up

Scene Safety

Your first thought as an EMT should always be the consideration of standard precautions and use of PPE. Proper protective equipment is vital any time there is a potential for exposure to blood, body fluids, or respiratory secretions. The patient may have a respiratory infection that could be passed to you through sputum and/or air droplets. The minimum PPE when treating patients with respiratory distress should be examination gloves, eye protection, and a HEPA respirator. If you suspect the patient has a respiratory disease, a face shield and/or gown may also be used.

Scene safety may be as simple as ensuring safe access to the patient and deciding on the best way to safely lift and move the patient. Or, you may need to consider that the respiratory emergency may have been caused by a toxic substance that was inhaled, absorbed, or ingested. Even in a medical situation, there is always a need to consider the potential for violence.

Once you have determined that the scene is safe, you need to determine how many patients there are and whether you need additional or specialized resources. Frequently, in situations where there are multiple people with dyspnea, you should consider the possibility of an airborne hazardous material release.

Mechanism of Injury/Nature of Illness

If the nature of illness (NOI) is in question, ask why 9-1-1 was activated. By questioning the patient, family and/or bystanders, you should be able to determine the NOI. The NOI is often based on a history of chronic medical problems.

Primary Assessment

The purpose of the primary assessment is to perform a rapid scan in order to identify immediate life threats, which would include problems with the patency of the airway, conditions that inhibit normal breathing, and issues that prevent adequate circulation. If any major problem is identified, treat it immediately. A decision for early and rapid transport must be made if the results of the primary assessment reveal life-threatening issues.

Pulse, respirations, blood pressure, skin color, capillary refill, level of consciousness, and pain measurement are important factors to note when evaluating a patient with respiratory problems. It is essential to look at the whole clinical picture when evaluating the patient in respiratory distress and not fixate on any one vital sign or symptom. For patients in respiratory distress, the body initially compensates by increasing the respiratory and heart rates. If the body is able to maintain adequate oxygenation, it will be able to maintain the patient's level of consciousness, skin color, and capillary refill time. Blood pressure will vary with the patient's baseline status and condition. It is often elevated in patients with pulmonary edema as the result of congestive heart failure.

Form a General Impression

As you approach and begin interacting with the patient, you need to gain a general impression of the patient. What is his or her age and position? The more distress a dyspneic patient is in, the more the patient will want to sit up. In a worst-case scenario, you will arrive to see the patient leaning forward with his or her hands on the knees. This is called the tripod position, and it is a good indicator that your patient is in significant distress.

Does the patient appear calm? Is he or she anxious or restless? Does the patient appear listless and tired? How severe is his or her breathing complaint? This initial impression will help you decide whether the patient's condition is stable or unstable. A stable condition generally will not deteriorate during treatment and transport, for example, a patient who has had pneumonia for 3 days and is being transported to the hospital to receive intravenous antibiotics. Conversely, an unstable condition may deteriorate during treatment and transport, for example, a patient who has been stung by a bee and is experiencing increasing difficulty in breathing.

At the same time, you will be determining the patient's level of consciousness. Use the AVPU (*Alert* to person, place, and day; responsive to *Verbal* stimuli; responsive to *Pain; Unresponsive*) scale to check for responsiveness. If the patient is alert or responding to verbal stimuli, you know that the brain is still receiving oxygen. Now is a good time to ask the patient about his or her chief complaint. If the patient is responsive only to painful stimuli or unresponsive, the brain may not be oxygenating well and the potential for an airway or breathing problem is more likely. If there is no gag or cough reflex, you need to immediately assess the patient's airway status. Within seconds you will be able to determine if there are any immediate threats to life.

Airway and Breathing

Assess the airway to make sure it is patent and adequate. Air must flow in and out of the chest easily for the airway to be considered patent or adequate. If there is any question about airway patency, the airway must be opened immediately using the head tilt–chin lift maneuver in nontrauma patients and the jaw-thrust method for patients who could possibly have spinal trauma.

If the airway is adequate or patent, next evaluate the patient's breathing. If the patient is breathing, ensure that the breathing is adequate. What are the rate, rhythm, and quality of the respirations? Is the rate within normal limits? Is the patient using accessory muscles to assist the respiratory effort, and can you see retractions? Is there abdominal breathing? What is the depth of breathing, and is the tidal volume adequate? Is there adequate rise and fall of the chest? What are the color, temperature, and condition of the patient's skin? Are the patient's respirations labored? If the patient can only speak one or two words at a time before gasping for a breath, ventilations are considered labored. If the respiratory effort is inadequate, you must provide the necessary intervention. If the patient is in respiratory distress, place him or her in a position that best facilitates breathing and begin administering oxygen at 15 L/min via nonrebreathing mask, unless contraindicated because of preexisting medical conditions. If the patient's breathing has inadequate depth or the rate is too slow, ventilations may need to be assisted with a bag-mask device.

If the patient is not breathing, give two ventilations immediately. Are there snoring sounds heard in an unresponsive patient? If so, reposition the airway and insert an oral or nasal airway if necessary to maintain a patent airway. If stridorous sounds are heard, position the patient so he or she can breathe easily. If gurgling sounds are heard, suction as necessary. As you ventilate a patient who is not breathing, you need to evaluate whether your ventilations are adequate enough to meet the oxygen needs of your patient.

Ask yourself the following questions:

1. Is the air going in?
2. Does the chest expand with each breath?
3. Does the chest fall after each breath?
4. Is the rate adequate for the age of your victim?

If the answer to any of these questions is "no," something is wrong. Try to reposition the patient and insert an oral airway to keep the tongue from blocking the airway. Reposition the patient's head. Reassess your hand position and face mask seal. Slow down or speed up your ventilation rate. Refer to Chapter 9, *Airway Management*, for a review of positive-pressure ventilation techniques. Remember that you will need to continue to monitor the airway for fluid, secretions, and other problems as you move on to assess the adequacy of your patient's breathing.

Lung and Breath Sounds

Obtaining lung sounds or breath sounds is one of the most important vital signs for your patient in respiratory distress. Listen over the bare chest. Trying to listen over clothing or chest hair may give you inaccurate information. The diaphragm of the stethoscope must be in firm contact with the skin. If your patient is lying down, bring him or her to a sitting position, which is a better position for assessing lung sounds.

You need to determine whether your patient's breath sounds are normal (**vesicular breath sounds**, **bronchial breath sounds**) or decreased, absent, or abnormal (**adventitious breath sounds**). With your stethoscope, check lung sounds on the right and left sides of the chest, and compare each apex (top) of the lung with the opposite apex and each base (bottom) of the lung with the opposite base [Figure 13-14]. When listening on the patient's back, place the stethoscope head between and below the scapulae, not over them, or you will not have an accurate assessment.

Make sure that you listen for a full respiratory cycle so you can detect the adventitious sounds that may be heard at the end of the inspiratory or expiratory phase. When assessing for fluid collection, pay special attention to the lower lung fields. Start from the bottom up and determine at which level you start hearing clear breath sounds.

You want to hear clear flow of air in both lungs. Not hearing the flow of air is considered an absent lung sound. The lack of air movement in the lung is a significant finding. Listen carefully and do not confuse absent lung sounds with clear lung sounds. See [Table 13-6] for examples of lung sounds, the diseases that may be associated with them, and important signs and symptoms.

Snoring sounds are indicative of a partial upper airway obstruction, usually in the oropharynx. **Wheezing** is a sound that indicates constriction and/or inflammation in

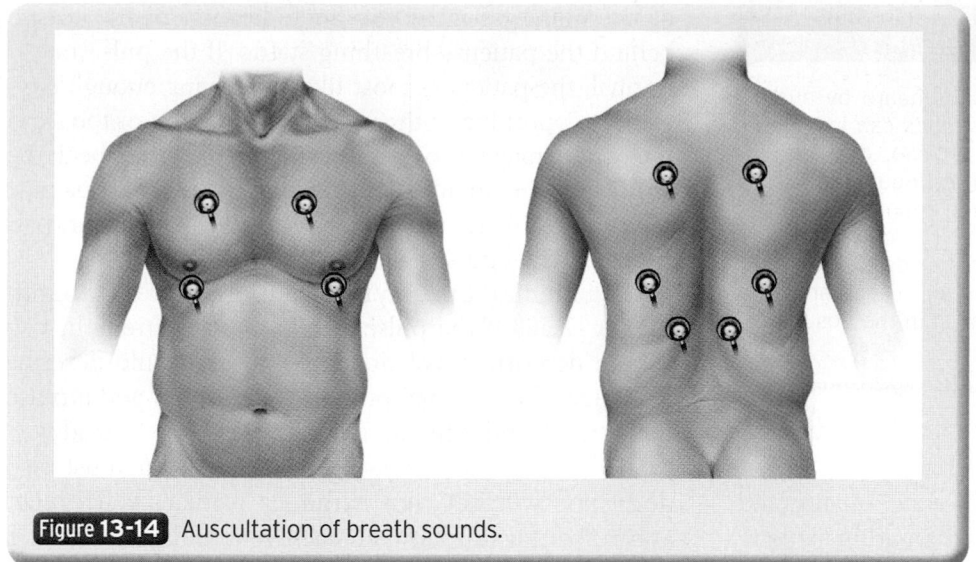

Figure 13-14 Auscultation of breath sounds.

bubbling sound typically heard on inspiration. There are high-pitched sounds called "fine" crackles and low-pitched sounds called "coarse" crackles. These sounds are often a result of congestive heart failure, pulmonary edema, or fluid in the lungs caused by congestive heart failure.

<u>Rhonchi</u> are lower pitched sounds caused by secretions or mucus in the larger airway. The sound resembles rattling or is sometimes referred to as "junky" lung sounds. This can be heard with infections such as pneumonia, bronchitis, or in cases of aspiration.

the bronchus. Wheezing is generally heard on exhalation as a high-pitched, almost musical or whistling sound. This sound is commonly heard in patients with asthma and sometimes in patients with COPD.

<u>Rales</u> or crackles are the sounds of air trying to pass through fluid in the alveoli. It is a crackling or

<u>Stridor</u> is the high-pitched sound heard on inspiration as air tries to pass through an obstruction in the upper airway. This sound indicates a partial obstruction of the trachea and is seen in patients with anatomic or foreign body airway obstruction. Strider is also an indication of a partial upper airway obstruction (in the trachea).

Table 13-6 Signs, Symptoms, and Adventitious Lung Sounds Associated With Specific Respiratory Diseases

Lung Sounds	Disease	Signs and Symptoms
Wheezes	Asthma Chronic obstructive pulmonary disease Congestive heart failure/pulmonary edema Pneumonia Bronchitis Anaphylaxis	Dyspnea Productive or nonproductive cough Dependent edema, pink frothy sputum Fever, pleuritic chest pain Clear or white sputum Hives and stridor, nonproductive cough
Rhonchi	Chronic obstructive pulmonary disease Pneumonia Bronchitis	Productive cough Fever, pleuritic chest pain Clear or white sputum
Rales (crackles)	Congestive heart failure/pulmonary edema Pneumonia	Dependent edema, pink frothy sputum Fever, pleuritic chest pain
Stridor	Croup Epiglottitis	Fever, barking cough Fever, sore throat, drooling
Decreased or absent breath sounds	Asthma Chronic obstructive pulmonary disease Pneumonia Hemothorax: Shock, respiratory distress Pneumothorax: Fever, pleuritic chest pain Atelectasis: Fever, decreased oxygen saturation	Nonproductive cough, dyspnea Productive cough Fever, pleuritic chest pain Blood in chest Air in chest Collapsed lung

Words of Wisdom

Adventitious breath sounds are sounds heard by auscultation of abnormal lungs. These sounds can include wheezing, rales, rhonchi, gurgling, snoring, crackling, and stridor. Being able to hear and distinguish different kinds of breath sounds can give you important clues as to what is wrong with your patient. The only way to develop your ability to identify breath sounds is through practice. Ask your instructor if you can accompany a physician, nurse, or respiratory therapist in the hospital to help you develop this ability.

Circulation

Next, you will need to assess the pulse rate, quality, and rhythm. In assessing the pulse rate in an adult patient, a rate that is greater than 100 beats/min is described as tachycardia. An increased pulse rate is the body's way of responding to respiratory distress and can be an indicator of shock. Tachycardia is also a normal response to pain, fear, excitement, and exertion. It is your responsibility to carefully assess the patient to determine if respiratory distress or shock exists. A rate of less than 60 beats/min is described as bradycardia. A slow pulse rate could mean problems with the cardiac conduction system, a medication reaction, organophosphate poisoning, or decompensation. If your patient is breathing, he or she will have a pulse; however, evaluating the adequacy

of the pulse can give you an indication of the reason behind the patient's breathing status. If the pulse rate is normal, the patient is most likely receiving enough oxygen to support life. If the pulse rate is too fast or too slow, the patient may not be getting enough oxygen. Check the radial pulse in an adult. If no radial pulse is felt, palpate the carotid pulse. In a child 1 year old or younger, palpate a brachial pulse.

Determine the quality of the pulse. Is it strong, bounding, or weak? If the pulse feels of normal strength, you should describe it as being strong. You should describe a stronger than normal pulse as "bounding" and a pulse that is weak and difficult to feel as "weak" or "thready."

When you are assessing the pulse, you must also determine whether the rhythm is regular or irregular. When the interval between each ventricular contraction of the heart is short, the pulse is rapid. When the interval is longer, the pulse is slower. No matter what the rate, the interval between each contraction should be the same, and the pulse that results should occur at a constant, regular rhythm. Irregular beats could indicate a cardiac problem.

Assessing a patient's circulation includes an evaluation for the presence of shock and bleeding. Respiratory distress in a patient could be caused by an insufficient number of red blood cells to transport the oxygen. Assess capillary refill in infants and children. Normal capillary refill is less than 2 seconds; abnormal capillary refill is

You are the Provider: PART 3

The patient's vital signs are obtained by your partner as you continue your assessment. You notice that she is breathing through pursed lips and has a prolonged exhalation phase, and cyanosis is present in her fingernail beds. You auscultate her lung sounds and hear scattered wheezing in all lung fields. When you talk to her, you note that she is now confused, is slow to answer your questions, and appears fatigued.

Recording Time: 3 Minutes	
Respirations	28 breaths/min, labored; prolonged exhalation phase
Pulse	120 beats/min; weak
Skin	Cyanotic, cool, clammy
Blood pressure	116/54 mm Hg
Oxygen saturation (Sao$_2$)	83% (on oxygen)

5. Why do patients with emphysema breathe through pursed lips?

6. What does a prolonged exhalation phase indicate in patients with obstructive lung disease?

7. What treatment is indicated for the patient at this point?

greater than 2 seconds. Capillary refill is not considered a reliable assessment tool in the adult patient.

Assess the patient's perfusion by evaluating skin color, temperature, and condition. The patient's skin color is assessed by looking at the nail beds, lips, and eyes. In a white person, normal skin is pink. Abnormal skin conditions are pale, cyanotic or blue-gray, flushed or red, and jaundiced or yellow.

Assess the patient's skin temperature by feeling the skin. Normal is warm. Abnormal skin temperatures are hot, cool, cold, and clammy. Assess the patient's skin condition. Normal is dry; abnormal is moist or wet.

A loss of perfusion may be caused by chronic anemia, a wound, internal bleeding, or simply shock overwhelming the body's ability to compensate for the illness. At this point, you need to check on your interventions. Is the oxygen bottle hooked up to the mask? Is the oxygen turned on? Is the flow rate adequate (10 to 15 L/min)? Is there a good face-mask seal? Is the chest rising and falling with each breath? Is the airway blocked with vomit or the tongue? Control any bleeding no matter how mild, and treat your patient for shock.

You know enough now to be able to identify any life threats in your patient. They would include any of the following signs or symptoms:

- Problems with the ABCs
- Poor initial general impression
- Unresponsiveness
- Potential hypoperfusion or shock
- Chest pain associated with a low blood pressure
- Severe pain anywhere
- Excessive bleeding

Transport Decision

If the patient's condition is unstable and there is a possible life threat, address the life threat and proceed with rapid transport. This means you will keep your scene time short, providing only lifesaving interventions. Perform a secondary assessment en route to the hospital. If the patient's condition is stable and there are no life threats, you may decide to perform a thorough assessment (a secondary assessment) on scene, after obtaining the patient history.

▶ History Taking

Investigate Chief Complaint

The next step of your assessment will provide more information specific to the patient's chief complaint (history of present illness) through history taking. The information you obtain during history taking will be subjective and objective. Subjective is what the patient expresses (symptoms), and objective is what you observe (signs). Both sets of information are important in building a general assessment. Rule out any findings that warrant no care or intervention. It is important to report pertinent negatives to health care providers or receiving facility staff members. A pertinent negative is any sign or symptom that commonly accompanies a particular condition, but is absent. Examples of pertinent negatives would be a patient in respiratory distress who denies chest pain, or one with severe chest pain who denies shortness of breath, or one with severe chest pain but who denies radiation of chest pain.

Find out what the patient has done for the breathing problem. Does the patient have home oxygen? Does the patient use a prescribed inhaler or a small-volume nebulizer? If so, when was it used last? How many doses have been taken? Does the patient use more than one inhaler or treatment? Be sure to record the name of each device and when it was used.

Different respiratory complaints offer different clues and different challenges. Patients with chronic conditions may have long periods in which they are able to live relatively normal lives but then sometimes experience acute worsening of their conditions. That is when you are called, and it is important for you to be able to determine your patient's baseline status, in other words, his or her normal condition, and what is different this time that made the patient call you. For example, patients with COPD (emphysema and chronic bronchitis) do not cope well with pulmonary infections because the existing airway damage makes them unable to cough up the mucus or sputum produced by the infection. The chronic lower airway obstruction makes it difficult for the patient to breathe deeply enough to clear the lungs. Gradually, the arterial oxygen level falls, and the carbon dioxide level rises. If a new infection of the lung occurs in a patient with COPD, the arterial oxygen level may fall rapidly. In a few patients, the carbon dioxide level may become high enough to cause sleepiness. In these cases, patients require respiratory support and careful administration of oxygen.

The patient with COPD usually has a long history of dyspnea with a sudden increase in shortness of breath. There is rarely a history of chest pain. More often, the patient will remember having had a recent "chest cold" with fever and either an inability to cough up mucus or a sudden increase in sputum. If the patient is able to cough up sputum, it will be thick and is often green or yellow. The blood pressure of patients with COPD is normal; however, the pulse is rapid and occasionally irregular. Pay particular attention to the respirations. They may be rapid, or they may be very slow.

Patients with asthma may have different "triggers," meaning different causes of acute attacks. These include allergens, cold, exercise, stress, infection, and noncompliance with medication prescriptions. It is important to try to determine what may have triggered the attack so that it can be treated appropriately. For example, an asthma attack that occurred while your patient was jogging in the cold will probably not respond to antihistamines, whereas one brought on by a reaction to pollen might.

Patients with congestive heart failure often walk a fine line between compensating for their diminished cardiac capacity and decompensating. Many take several medications, most often including diuretics ("water pills") and blood pressure medications. Your history taking should include obtaining a list of all their medications and paying special attention to the events leading up to the present problem.

SAMPLE History

With patients in respiratory distress, many of the SAMPLE questions can be answered by the family or bystanders if they are present. Limit the number of questions to pertinent ones—a patient who is in respiratory distress does not need to be using any additional air to answer questions.

Be sure to ask the following questions about a patient in respiratory distress:

- What is the patient's general state of health?
- Has the patient had any childhood or adult diseases?
- Have there been any surgeries or recent hospitalizations?
- Are there any psychiatric or mental health illnesses?
- Have there been any traumatic injuries?

To help determine the cause of your patient's problem, be a detective. Look for medications, medical alert bracelets, environmental conditions, and other clues to what may be causing the problem. Each part of the SAMPLE history may give you clues, so be sure to be thorough. For example, you forget to ask about allergies, only to find out later that your patient has a severe allergy to cat dander and that her 8-year-old son had been playing with a cat shortly before the onset of her problem. You would have missed important and possibly lifesaving information.

OPQRST

The OPQRST assessment, generally used for determining the specifics of pain, can also be modified to obtain more specific information about the breathing problem. Begin by asking the patient to describe the problem. Ask open-ended question such as, "What can you tell me about your breathing?" Pay close attention to OPQRST and include the following questions:

- When did the breathing problem begin (Onset)?
- What makes the breathing difficulty worse (Provocation or palliation)?

- How does the breathing feel (Quality)?
- Does the discomfort move (Radiation/region)?
- How much of a problem is the patient having (Severity)?
- Is the problem continuous or intermittent? If it is intermittent, how frequently does it occur and how long does it last (Timing)?

PASTE

A specific alternative assessment for a complaint of shortness of breath or difficulty breathing uses the mnemonic PASTE:

P *Progression*. Similar to the O in OPQRST, you want to know if the problem started suddenly or has worsened over time.

A *Associated chest pain*. Dyspnea can be a significant symptom of a cardiac problem.

S *Sputum*. Has the patient been coughing up sputum? Mucuslike sputum could indicate a respiratory infection, pink frothy sputum is indicative of fluid in the lungs, and a problem like a pulmonary embolus may not result in any sputum at all.

T *Talking tiredness*. This is an indicator of how much distress the patient is in. Ask the patient to repeat a sentence and see how many words he or she can speak without needing to take a breath. The assessment results would be reported that the patient "speaks in full sentences" or, perhaps, "speaks in two- to three-word sentences."

E *Exercise tolerance*. Ask the patient a question about what he or she was able to do before this problem started, like walk across the room, and then ask if the patient could do it now. If the answer is "no," then it is another indicator that your patient is in distress. Exercise tolerance will decrease as the breathing problem and hypoxia increase.

Secondary Assessment

The secondary assessment is a more in-depth assessment of body systems, and it addresses the specific chief complaint, for example, difficult breathing (dyspnea) or shortness of breath. The secondary assessment includes a physical examination and the taking of vital signs.

In respiratory emergencies as in all other emergencies, you should only proceed to history taking and the secondary assessment once all life threats have been identified and treated during the primary assessment. If you are busy treating airway or breathing problems, you may not have the opportunity to proceed to a physical examination prior to arriving at the emergency department. Never compromise the assessment and treatment of airway and breathing problems to conduct a physical examination.

Sometimes it is not possible to quickly and definitively determine what is causing your patient's respiratory distress. If your patient is a 20-year-old woman at a picnic who rapidly develops difficulty breathing and hives after being stung by a bee, you have a clear-cut diagnostic picture. Conversely, if your patient is an older woman in a nursing home who is receiving 12 medications and has a cough and increasing shortness of breath that developed during the past week, you have a more perplexing problem. Keep an open mind, gather as complete a history as possible, and perform a physical examination.

Physical Examinations

Additional pieces to the assessment and treatment puzzle may be revealed during the physical examination. For example, you are treating a patient in acute respiratory distress who is breathing at a rate of 40 breaths/min and has audible wheezing. On the basis of this information, you may be unsure as to whether the patient is in congestive heart failure or is having an asthma attack. The physical examination may provide you with some clues, such as a consistently elevated blood pressure and swollen legs and feet (pedal edema) that would lead you in the direction of congestive heart failure.

Blood pressure should be auscultated (by listening) when possible to obtain the systolic and diastolic numbers. If you are in an environment where you cannot hear well enough to auscultate the blood pressure, then palpation (by feeling) is an alternative. It is preferable to auscultate when you can because palpation does not provide a diastolic value that may be pertinent when a patient has a condition such as hypertension.

Signs of Chronic Obstructive Pulmonary Disease

Assume you are assessing a patient with COPD. What would you notice? Patients with COPD are usually older than 50 years. They will always have a history of recurring lung problems and are almost always long-term cigarette smokers. Patients may complain of tightness in the chest and constant fatigue. Because air has been gradually and continuously trapped in their lungs in increasing amounts, their chests often have a barrel-like appearance. Patients with COPD often use accessory muscles to breathe **Figure 13-15**. If you listen to the patient's chest with a stethoscope, you will hear abnormal breath sounds. Patients with COPD will often exhale through pursed lips as a strategy to keep airways open longer. Digital clubbing (abnormal enlargement of the ends of the fingers) is also a sign of COPD.

Table 13-7 outlines the differences in patients with COPD (emphysema and chronic bronchitis) and patients with congestive heart failure.

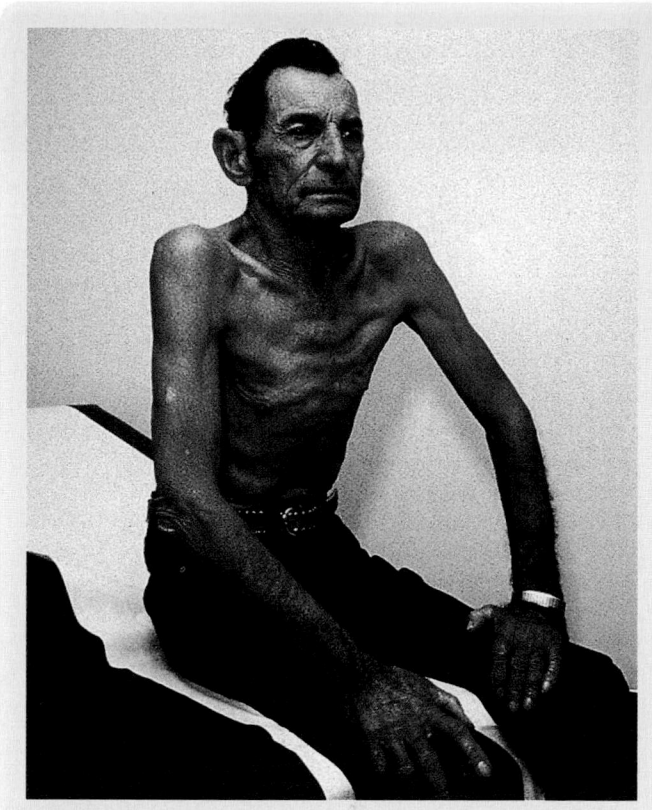

Figure 13-15 Patients with COPD often use accessory muscles and pursed lips for breathing. Notice, also, that this patient is sitting in the tripod position.

Vital Signs

You will want to conduct an in-depth assessment when a patient complains of shortness of breath. In addition to the signs of air hunger present in all patients with respiratory distress, such as tripod positioning, rapid breathing, and use of accessory muscles, restriction of the small lower airways in patients with asthma often causes wheezing. Patients may have a prolonged expiratory phase of breathing as they attempt to exhale trapped air from the lungs. In severe cases, you may actually not hear wheezing because of insufficient air flow. As your patient tires from the effort of breathing and oxygen levels drop, the respiratory and heart rates may actually drop, and your patient may seem to relax or go to sleep. These signs indicate impending respiratory arrest, and you must act immediately.

When patients with congestive heart failure decompensate, they will often experience pulmonary edema, as fluid backs up in their circulatory system and into the lungs. High blood pressure and low cardiac output often trigger this "flash" (sudden) pulmonary edema. The patients are among the most sick, frightened, and worrisome patients you will encounter. They are literally drowning in their own fluid. In addition to the classic

Table 13-7   **Comparison of Chronic Obstructive Pulmonary Disease (Emphysema and Chronic Bronchitis) and Congestive Heart Failure**

	Emphysema	Chronic Bronchitis	Congestive Heart Failure
Pathophysiology	■ Destruction of the airways distal to the bronchiole ■ Destruction of the pulmonary capillary bed ■ Decreased ability to oxygenate the blood ■ Lower cardiac output and hyperventilation ■ Development of muscle wasting and weight loss	■ Excessive mucus production with airway obstruction ■ Pulmonary capillary bed undamaged ■ Compensation by decreasing ventilation and increasing cardiac output ■ Poorly ventilated lungs, leading to hypoxemia ■ Increased carbon dioxide retention	■ Damaged ventricles and failure of heart as a pump ■ Attempt by heart to compensate with increased rate ■ Enlarged left ventricle ■ Backup of fluid into the body as the heart fails to pump adequately
Appearance	■ Thin appearance with barrel chest ■ Use of accessory muscles and tripod position ■ "Puffing" (pursed-lip) style of breathing ■ Usually pink skin	■ May be obese ■ Use of accessory muscles ■ Difficulty with expiration ■ Often cyanotic	■ Abdominal distention ■ Dependent edema (sacral or pedal) ■ Tachycardia ■ Increased respiratory rate ■ Anxiety ■ Inability to lie flat ■ Ashen or cyanotic
Level of consciousness	Normal or altered	Normal or altered	Confusion
Neck veins	Flat	Distended when heart failure also present	Distended
Skin color	Pink	Blue, often cyanotic	Blue
Lung condition	Dry	Wet when heart failure also present	Wet
Lung sounds	Wheezing	Rhonchi, wheezing, or rales	Rhonchi, wheezing, or rales
Blood pressure	Normal for age	Normal to high	Normal to high
Cough	Little or none	Frequent cough	With possible fluid expectoration
Sputum	No mucus	Excessive mucus	Pink, frothy sputum

signs of respiratory distress, they may have pink, frothy sputum coming from the mouth. They will have adventitious lung sound, most often wet (rales, rhonchi, crackles) but sometimes dry sounding (wheezes). Their legs and feet may be swollen (pedal edema) from the backup of fluid.

Remember that the brain needs a constant, adequate supply of oxygen to function normally. In patients with respiratory distress, you will notice an altered level of consciousness when the oxygen level drops. This may manifest itself as confusion, lack of coordination, bizarre behavior, or even combativeness. A change in affect or level of consciousness is one of the early warning signs of respiratory inadequacy.

When there is inadequate oxygen in the blood, the body will attempt to divert blood from the extremities

to the core to help keep the vital organs, including the brain, functioning. This action by the body will result in pale skin and delayed capillary refill in the hands and feet. Capillary refill that takes longer than 2 seconds is considered delayed. Feel for the skin temperature, and look for color changes in the extremities and in the core of the body. Cyanosis is a late sign and can be seen first in the lips and mucous membranes. Cyanosis is an ominous sign that requires immediate, aggressive intervention.

When you are performing a secondary assessment on the respiratory system, look for overall symmetry of the chest, adequate rise and fall of the chest, and evidence of retractions or accessory muscle use. Does the patient have increased work of breathing? Assess breath sounds, and do a physical assessment if warranted.

A secondary assessment of the cardiovascular system, especially when there is associated chest pain, should include checking and comparing distal pulses, reassessing the skin condition, and being alert for bradycardia and tachycardia.

It is important to assess the neurologic system because the level of consciousness can change. Check the patient's mental status, and determine if the patient's activity can be described as anxious or restless. If so, that would be an indicator of hypoxia. Does the patient have clear thought processes? Disorientation may be another indicator of hypoxia.

Words of Wisdom

Never compromise the assessment and treatment of airway and breathing problems to conduct a physical examination.

Monitoring Devices

Use monitoring devices if you have them available, including, but not limited to, a pulse oximeter. Pulse oximetry is an effective diagnostic tool when used in conjunction with experience, good assessment skills, and clinical judgment. Pulse oximeters measure the percentage of hemoglobin that is saturated by oxygen. In patients with normal levels of hemoglobin, pulse oximetry can be an important tool in evaluating oxygenation. To use pulse oximetry properly, it is important for you to be able to evaluate the quality of the reading and correlate it with the patient's condition. For example, it is doubtful a patient with congestive heart failure in severe respiratory distress will be able to maintain a pulse oximetry reading of 98% or that a conscious, alert, active patient with good skin color can be "maintained" by a pulse oximetry reading of 80%.

If you get a good reading consistent with your patient's condition, the pulse oximeter can help you

Words of Wisdom

It is important to be aware of conditions that can skew pulse oximeter results. Bright light, darkly pigmented skin, and nail polish can cause errors in the readings. Remember that it only measures the percentage of hemoglobin that is saturated with oxygen. Therefore, a patient with a low hemoglobin level, such as an anemic or hypovolemic patient, may have 100% oxygen saturation. This means that the hemoglobin is saturated, but the reading does not tell you that the hemoglobin level in the bloodstream is not sufficient to sustain organ function. Other conditions that may cause false readings are sickle cell disease and carbon monoxide poisoning.

determine the severity of the respiratory component of the patient's problem. Also, if the reading goes steadily up or down, it can give you an indication of improvement or deterioration of ventilatory status, often even prior to its manifestation in the patient's appearance or vital signs.

Reassessment

Once the assessment and treatment have been completed, you need to reassess the patient and closely watch patients with shortness of breath. Repeat the primary assessment, and maintain an open airway. Monitor the patient's breathing, and reassess circulation.

Determine if there have been changes in the patient's condition. Confirm the adequacy of interventions and patient status. Is the current treatment improving the patient's condition? Has an already identified problem improved? Has an already identified problem gotten worse? What is the nature of any newly identified problems?

If the changes you find are improvements, simply continue the treatments; however, if your patient's condition deteriorates, prepare to modify treatments. Be prepared to assist ventilations with a bag-mask device. Monitor the skin color and temperature. Reassess and record vital signs at least every 5 minutes for a patient in unstable condition and/or after the patient uses an inhaler. If the patient's condition is stable and no life threat exists, vital signs should be obtained at least every 15 minutes.

Interventions

Now that you have completed the physical examination and have gathered information about your patient with difficulty breathing, it is time to provide interventions for the problems that are not immediate life threats. Your interventions may be based on standing orders, or you should contact the hospital and ask for specific directions. Remember, interventions for immediate life threats should be been completed during the primary assessment and should not require contacting the hospital first. Interventions for respiratory problems may include the following:

- Providing oxygen via a nonrebreathing mask at 15 L/min
- Providing positive-pressure ventilations using a bag-mask device, pocket mask, or a flow-restricted oxygen-powered ventilation device
- Using airway management techniques such as an oropharyngeal airway, a nasopharyngeal airway, suctioning, or airway positioning
- Positioning the patient in a high Fowler's position or a position of choice to facilitate breathing

- Assisting with respiratory medications found in a patient-prescribed metered-dose inhaler or a small-volume nebulizer

Some of these interventions were performed in the primary assessment to address life threats. Others are used to support breathing problems until definitive care can be provided at the hospital. Some of your interventions may even correct the problem.

Communication and Documentation

Contact medical control with any change in level of consciousness or difficulty breathing. Depending on local protocol, contact medical control prior to assisting with any prescribed medications. Be sure to document any changes (and at what time) and any orders given by medical control.

Emergency Medical Care

Management of respiratory distress involves continuing awareness of scene safety and the use of standard precautions. Management of ABCs and positioning are primary treatments along with oxygen and suction.

You will usually administer oxygen. If a patient complains of breathing difficulty, you should administer supplemental oxygen immediately. All responsive patients breathing more than 30 breaths/min or fewer than 8 breaths/min should receive high-flow oxygen (defined as 15 L/min) via a bag-mask device.

Take great care in monitoring the patient's respirations as you provide oxygen. Reevaluate the respirations and the patient's response to oxygen repeatedly, at least every 5 minutes, until you reach the emergency department. In a person with a chronically high carbon dioxide level (eg, certain patients with COPD), this is critical, because the supplemental oxygen may cause a rapid rise in the arterial oxygen level. This, in turn, may abolish the secondary respiratory oxygen drive and cause respiratory arrest.

As stated previously, there is concern about suppression of the "hypoxic" drive to breathe in some patients with COPD. Unless the patients are unresponsive, a more conservative approach is suggested. In patients who have long-standing COPD and probable carbon dioxide retention, administration of low-flow oxygen (2 L/min) is a good place to start, with adjustments to 3 L/min, then 4 L/min, and so on, until symptoms have improved (for example, the patient has less dyspnea or a better mental status). You may want to use an end-tidal carbon dioxide ($ETCO_2$) detector. Some electronic $ETCO_2$ detectors can evaluate $ETCO_2$ in the spontaneously breathing patient via a specialized nasal cannula–type device. When in doubt, err on the side of more oxygen, and monitor the patient closely.

Do *not* withhold oxygen for fear of depressing or stopping breathing in a patient with COPD who needs oxygen. A decreased respiratory rate after administration of oxygen does not necessarily mean that the patient no longer needs the oxygen; he or she may need it even more. If respirations slow and the patient becomes unconscious, you should assist breathing with a bag-mask device.

Continue to use pulse oximetry, if available, to adjust oxygen therapy. Always provide emotional support to the patient who is anxious. Always speak with assurance and assume a concerned, professional approach to reassure the patient, who is probably very frightened.

Words of Wisdom

Some states allow EMTs to administer inhalers or assist patients in the administration of their own inhalers. With this increased scope of practice comes an increased responsibility to know the names, doses, indications, contraindications, side effects, and precautions of the numerous inhalers available for a variety of conditions. Patients sometimes do not know the difference between their "rescue" inhalers (immediately effective medication, such as albuterol) and their maintenance inhalers (such as corticosteroids, which have no immediate effect). It is essential, then, that you know the difference!

Metered-Dose Inhaler and Small-Volume Nebulizer

Patients who call for help because of breathing difficulty are likely to have had the same problem before. They probably have prescribed medications to use that are delivered by an inhaler or small-volume nebulizer. If so, you may be able to help them use these devices. Some of the most common medications used for shortness of breath are called inhaled beta-agonists, which dilate breathing passages. The following medications may be administered via a **metered-dose inhaler (MDI)** (a miniature spray canister used to direct such substances through the mouth and into the lungs). Typical trade names you will encounter are Proventil, Ventolin, Alupent, Metaprel, and Brethine. The generic name for Proventil and Ventolin is albuterol; for Alupent and Metaprel, it is metaproterenol; and for Brethine, it is terbutaline. Medications typically administered by small-volume nebulizer include, but are not limited to, albuterol, metaproterenol, and epinephrine.

The **small-volume nebulizer** works by providing a means for a fine mist of aerosolized medicine to get deep into the patient's lungs and start to work quickly. The patient inhales the mist through a mouthpiece. When the medicine is breathed in correctly, it goes directly into the lungs.

Medical Control

Consult medical control (online), or go by standing orders (off-line). Remember to report what the medication is, when the patient last gave himself or herself a treatment, how many puffs or how much medication was used at that time, and what the label states regarding dosage. If medical control or standing orders permit, you may assist the patient to self-administer the medication. Be certain that the inhaler belongs to the patient, it contains the correct medication, the expiration date has not passed, and the correct dose is being administered. Administer repeated doses of the medication if the maximum dose has not been exceeded and the patient is still experiencing shortness of breath.

More skill is required by EMTs to assist with a small-volume nebulizer than an MDI because the device must be assembled prior to use. An oxygen tank is also required to administer the aerosolized medication. The patient may have a tank available, or you will need to use your own tank. Keep in mind also that the equipment is sterile, and aseptic technique must be used when setting up the device.

Indications and Contraindications

Before helping a patient to self-administer any MDI or small-volume nebulizer medication, make sure that the medication is indicated, that is, the patient has signs and symptoms of shortness of breath. The most common use for an MDI is asthma, and a small-volume nebulizer is used in asthma, bronchiolitis, COPD, and anaphylaxis. Check that there are no contraindications for its use, such as the following:

- The patient is unable to help coordinate inhalation with depression of the trigger on an MDI or is too confused to effectively administer medication through a small-volume nebulizer. These devices will be only minimally effective when patients are in respiratory failure and have only minimal air movement.
- The MDI or small-volume nebulizer is not prescribed for this patient.
- You did not obtain permission from medical control and/or it is not permissible by local protocol.
- The patient has already met the maximum prescribed dose before your arrival.
- The medication is expired.
- There are other contraindications specific to the medication.

Actions

Most of the medications used relax the muscles that surround the air passages in the lungs, leading to enlargement (dilation) of the airways and easier movement of air. See **Table 13-8** for a list of medications used for acute symptoms and medications used for chronic symptoms. The medications used for acute symptoms are designed

You are the Provider: PART 4

After initiating the appropriate treatment, you place the patient onto the stretcher, load her into the ambulance, and begin transport to the hospital. You reassess her and note that her condition has acutely deteriorated. You insert a nasopharyngeal airway and begin assisting her ventilations with a bag-mask device and high-flow oxygen.

Recording Time: 9 Minutes	
Level of consciousness	Responsive only to pain
Respirations	12 breaths/min; shallow
Pulse	132 beats/min; weak
Skin	Cool and dry; cyanosis of the nail beds and around the lips
Blood pressure	108/50 mm Hg
Sao_2	77% (on oxygen)

8. Why is cyanosis a later sign of hypoxemia in patients with emphysema?

9. Why does tachycardia develop in hypoxemic patients?

Table 13-8 Respiratory Inhalation Medications

Medication			Indications			Use: Acute Versus Chronic Disease	
Generic Drug Name	Trade Names	Action	Asthma	Bronchitis	COPD	Acute	Chronic
Albuterol	Proventil, Ventolin, Volmax	Dilates bronchioles	Yes	Yes	Yes	Yes	No
Beclomethasone dipropionate	Beclovent, Beconase, Qvar, Vanceril	Anti-inflammatory, reduces swelling	Yes	No	No	No	Yes
Cromolyn sodium	Crolom, Gastrocom, Intal, Nasalcrom	Decreases release of histamines	Yes	No	No	No	Yes
Fluticasone propionate	Cutivate, Flonase, Flovent	Anti-inflammatory, reduces swelling	Yes	No	No	No	Yes
Fluticasone propionate, salmeterol xinafoate	Advair Diskus	Decreases secretions	Yes	No	No	No	Yes
Ipratropium bromide	Atrovent	Dilates bronchioles	Yes	Yes	Yes	Yes	No
Metaproterenol sulfate	Alupent, Metaprel	Dilates bronchioles	Yes	Yes	Yes	Yes	No
Salmeterol xinafoate	Serevent	Dilates bronchioles	Yes	Yes	Yes	No	Yes

to give the patient rapid relief from symptoms if the condition is reversible. Medications used for chronic symptoms are administered for preventive measures or as maintenance doses. The medications for long-term use will provide little relief of acute symptoms.

Side Effects

Common side effects of inhalers used for acute shortness of breath include increased pulse rate, nervousness, and muscle tremors.

If the patient has a prescribed MDI or small-volume nebulizer, read the label carefully to make sure that the medication is to be used for shortness of breath and that it has, in fact, been prescribed by a physician Figure 13-16. When in doubt, consult medical control.

Dose and Route

Medication from an inhaler is delivered through the respiratory tract to the lung. The dose is one puff for an

MDI and continuation of the small-volume nebulizer until all the medication has been administered or the patient no longer feels the need for the medication.

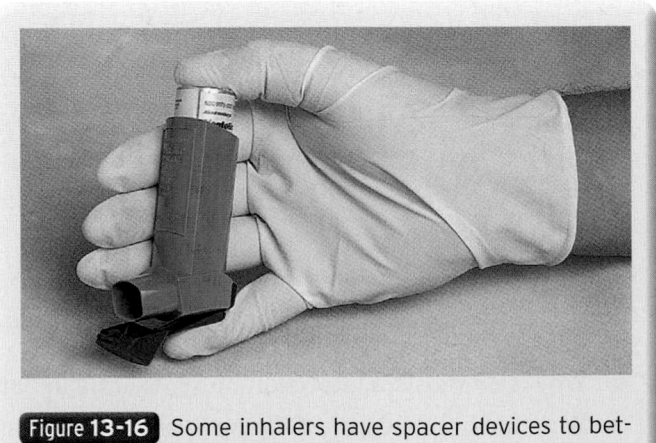

Figure 13-16 Some inhalers have spacer devices to better direct the medication spray.

Administration of a Metered-Dose Inhaler

To help a patient self-administer medication from an inhaler, follow the steps in **Skill Drill 13-1**:

1. Follow standard precautions.
2. Obtain an order from medical control or local protocol.

3. Check that you have the right medication, right patient, right dose, and right route and that the medication is not expired.
4. Make sure that the patient is alert enough to use the inhaler.
5. Check to see whether the patient has already taken any doses.
6. Make sure the inhaler is at room temperature or warmer (**Step 1**).
7. Shake the inhaler vigorously several times.
8. Stop administering supplemental oxygen, and remove any mask from the patient's face.
9. Ask the patient to exhale deeply and, before inhaling, to put his or her lips around the opening of the inhaler (**Step 2**).

Skill Drill 13-1

Assisting a Patient With a Metered-Dose Inhaler

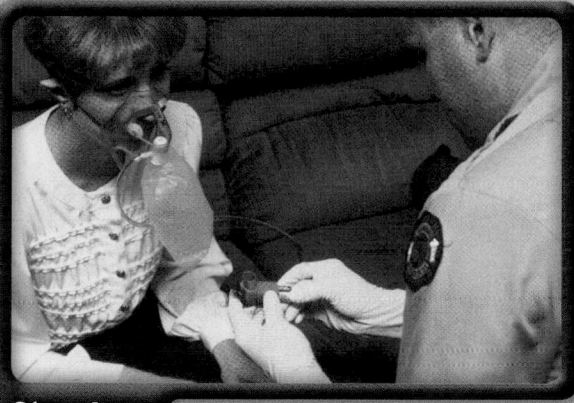

Step 1 Check to make sure you have the correct medication for the correct patient. Check the expiration date. Ensure inhaler is at room temperature or warmer.

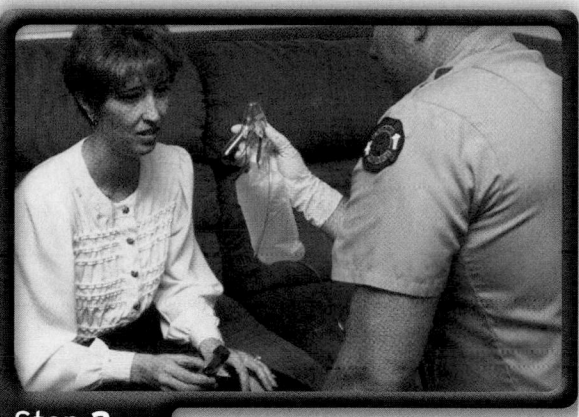

Step 2 Remove oxygen mask. Hand inhaler to patient. Instruct about breathing and lip seal.

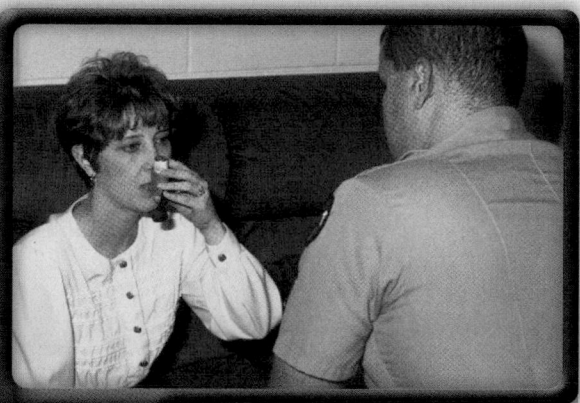

Step 3 Instruct patient to press inhaler and inhale one puff. Instruct about breath holding.

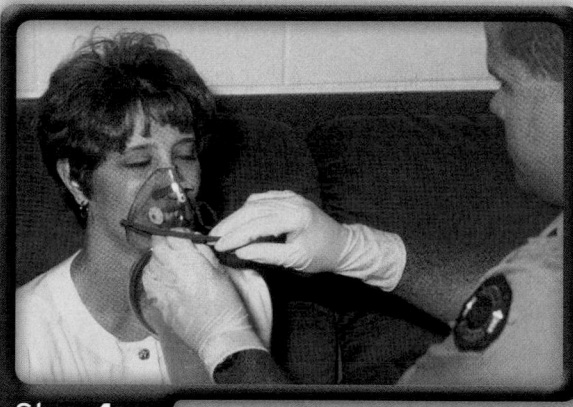

Step 4 Reapply oxygen. After a few breaths, have patient repeat dose if order or protocol allows.

10. Have the patient depress the hand-held inhaler as he or she begins to inhale deeply.

11. Instruct the patient to hold his or her breath for as long as is comfortable to help the body absorb the medication (Step 3).

12. Continue to administer supplemental oxygen.

13. Allow the patient to breathe a few times, then repeat a second dose per direction from medical control or local protocol (Step 4).

■ Administration of Small-Volume Nebulizer

To help a patient self-administer medication from a small-volume nebulizer, follow the steps in **Skill Drill 13-2**:

1. Follow standard precautions.
2. Obtain an order from medical control or local protocol.

Skill Drill 13-2

Assisting a Patient With a Small-Volume Nebulizer

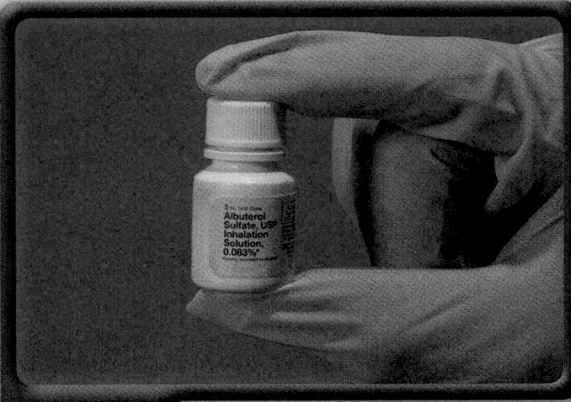

Step 1 Check to make sure you have the correct medication for the correct patient. Check the expiration date. Confirm you have the correct patient.

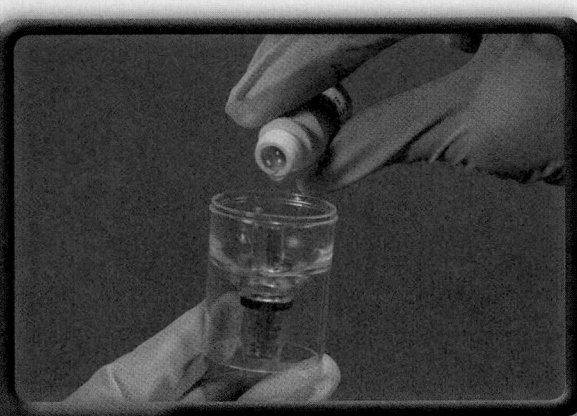

Step 2 Insert the medication into the container on the nebulizer. In some cases, sterile saline may be added (about 3 mL) to achieve the optimum volume of fluid for the nebulized application.

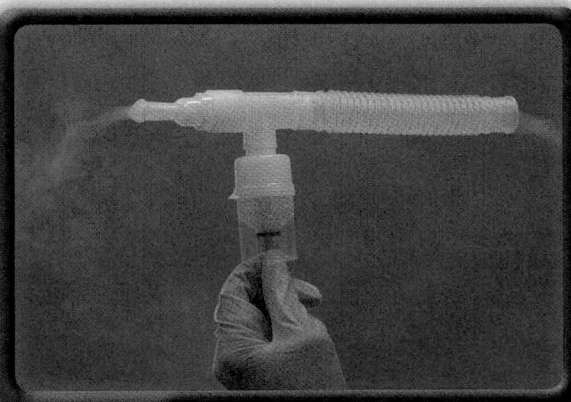

Step 3 Attach the medication container to the nebulizer, mouthpiece, and tubing. Attach oxygen tubing to the oxygen tank. Set the flowmeter at 6 L/min.

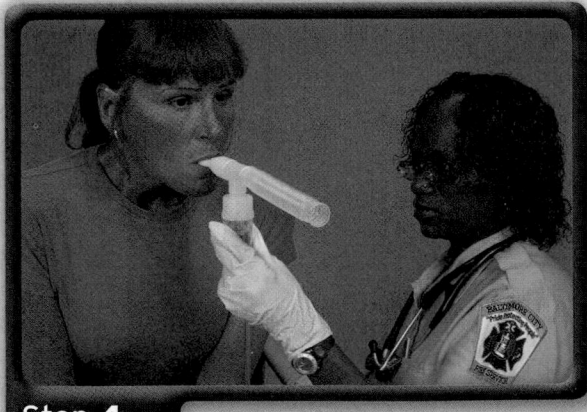

Step 4 Instruct the patient on how to breathe.

3. Check that you have the right medication, right patient, right dose, and right route and that the medication is not expired. Ensure there are no issues with contamination, discoloration, or clarity of the medication (Step 1).

4. Make sure that the patient is alert enough to use the device.

5. Check to see whether the patient has already taken any treatments.

6. If assisting to assemble the device, maintain aseptic technique.

7. Open the medication container on the nebulizer, and insert the medication (generally the whole volume of the medication). In some cases, sterile saline may be added (about 3 mL) to achieve the optimum volume of fluid for the nebulized application (Step 2).

8. Attach the medication container to the nebulizer, mouthpiece, and tubing. Attach oxygen tubing to the oxygen tank.

9. Adjust oxygen flow to 6 L/min to establish misting effect (Step 3).

10. Stop administering supplemental oxygen, and remove nonrebreathing mask from the patient's face.

11. Ask the patient to put his or her lips around the mouthpiece of the device, inhale the mist, and hold it for 3 to 5 seconds before exhaling (Step 4).

12. When the mist dissipates and the medication has been used or the patient is no longer experiencing shortness of breath, discontinue use of the device.

13. Place the nonrebreathing mask back on the patient if the patient continues to complain of shortness of breath.

14. Reassess vital signs, and document your actions and the patient's response.

15. Consult with medical control and/or follow local policy if repeated doses are necessary.

Treatment of Specific Conditions

Upper or Lower Airway Infection

Dyspnea associated with acute infections is quite common. Except in the patient with pneumonia, acute bronchitis, or epiglottitis, it is rarely serious. The acute congestion and stuffiness of a common cold hardly ever require emergency care. Indeed, most people with colds treat themselves with over-the-counter medications. However, individuals with a common cold who have underlying problems such as asthma or heart failure may experience a worsening of their condition as a result of the additional stress of the infection. In addition, medications for colds may also have stressful side effects, such as agitation, increased heart rate, and increased blood pressure.

For patients with upper airway infections and dyspnea, administer humidified oxygen (if available). Do not attempt to suction the airway or place an oropharyngeal airway in a patient with suspected epiglottitis. These maneuvers may cause a spasm and complete airway obstruction. Transport the patient promptly to the hospital. Allow the patient to sit in the position that is most comfortable. For someone with epiglottitis, this is usually sitting upright and leaning forward in the "sniffing position" **Figure 13-17**. To force a patient with epiglottitis to lie supine may cause upper airway obstruction that could result in death.

■ Acute Pulmonary Edema

Dyspnea caused by acute pulmonary edema may be associated with cardiac disease or direct lung damage. In either case, administer 100% oxygen, and, if necessary, carefully suction any secretions from the airway. The best position for a conscious patient who has a myocardial infarction or direct lung injury is the position in which it is easiest to breathe. Usually, this is sitting up. An unconscious patient with acute pulmonary edema may require full ventilatory support, including airway, positive-pressure ventilation with oxygen, and suctioning. CPAP has proven to be immensely beneficial to patients experiencing respiratory distress from acute pulmonary edema. Provide prompt transport to the emergency department.

■ Chronic Obstructive Pulmonary Disease

Patients with COPD may be semiconscious or unconscious from hypoxia, a condition in which the body's cells and tissues do not get enough oxygen, or from carbon dioxide retention. Patients with COPD often find

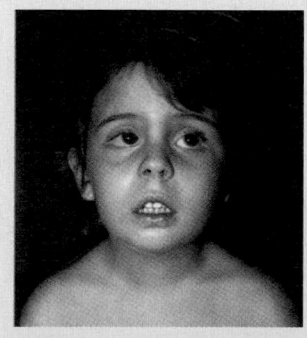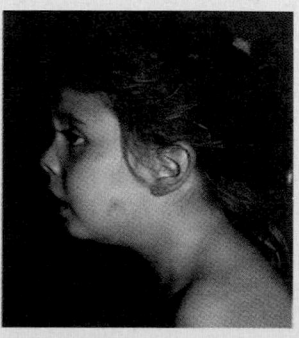

Figure 13-17 A child with epiglottitis may be more comfortable sitting up and leaning forward.

breathing difficult when lying down. Assist with the patient's prescribed inhaler if there is one. Oftentimes a patient with COPD will overuse an inhaler, so watch for side effects. Transport patients with COPD as promptly as possible to the emergency department, allowing them to sit upright if this is most comfortable.

Asthma, Hay Fever, and Anaphylaxis

Many lung problems are incorrectly labeled "asthma"; therefore, your assessment of the patient is critical. Asthma is often a recurring pathologic condition. Confirm whether the patient is able to breathe normally at other times. If possible, ask family members to describe the patient's asthma. Even if they only identify wheezing as a problem, be aware that some forms of heart failure, foreign body aspiration, toxic fumes inhalation, or allergic reactions may cause wheezing.

As you assess the patient's vital signs, note that the pulse rate will be normal or elevated, the blood pressure may be slightly elevated, and respirations will be increased. Ask questions about how and when the symptoms began.

As you care for the patient, be prepared to suction large amounts of mucus from the mouth and to administer oxygen. If you do suction, do not withhold oxygen for more than 15 seconds for adult patients, 10 seconds for a child, and 5 seconds for an infant. Allow some time for oxygenation between suction attempts. If the patient is unconscious, you may have to provide airway management.

If the patient has medication, such as an inhaler for an asthma attack, you may help with its administration, as directed by local protocol. Even patients who use their inhaler may continue to get worse. You need to reassess breathing frequently and be prepared to assist ventilations in severe cases. If you must assist ventilations in a patient who is having an asthma attack, use slow, gentle breaths. Remember, the problem in asthma is getting the air out of the lungs, not into them. Resist the temptation to squeeze the bag hard and fast. Always assist with ventilations as a last resort, and then provide only about 10 to 12 shallow breaths/min.

A prolonged asthma attack that is unrelieved may progress into a condition known as *status asthmaticus*. The patient is likely to be frightened, frantically trying to breathe, while using all the accessory muscles. Status asthmaticus is a true emergency, and the patient must be given oxygen and transported immediately to the emergency department.

The effort to breathe during an asthma attack is very tiring, and the patient may be exhausted by the time you arrive at the hospital. An exhausted patient may have stopped feeling anxious or even struggling to breathe. This patient is not recovering; he or she is at a very critical stage and is likely to stop breathing. Aggressive airway management, oxygen administration, and prompt transport are essential in this situation. Advanced life support (ALS) should be considered. Follow local protocol.

The patient with hay fever is not likely to need emergency treatment unless the condition has worsened from generalized cold symptoms. Manage the airway, and give oxygen according to the level of distress.

An anaphylactic reaction is a life-threatening emergency and must be treated as such. The first step should be to remove the offending agent. For example, if the patient is allergic to peanuts and is being exposed to peanuts through inspiration, you may need to remove the patient from the room because you may not be able to eliminate the peanut allergen from the air. If the patient has a stinger from a bee sting still in place, you may need to remove the stinger. Remember to scrape the stinger off because you can inject more venom into the patient if you pinch or squeeze the stinger.

Maintain the airway—the airway is always a priority regardless of the situation. If the patient is still awake, allow him or her to assume a position that does not compromise breathing. Use an appropriate oxygen device for supplemental oxygen administration, and consider early transport. Be prepared to assist breathing as needed. Rapid transport and the early administration of epinephrine, if allowed by protocol, should be a priority. Because epinephrine has immediate action, it can rapidly reverse the effects of anaphylaxis.

Spontaneous Pneumothorax

Patients with spontaneous pneumothorax may have severe respiratory distress, or they may have no distress at all and complain only of pleuritic chest pain. Provide supplemental oxygen, and provide prompt transport to the hospital. Like most dyspneic patients, those with spontaneous pneumothorax are usually more comfortable sitting up. Monitor the patient carefully, watching for any sudden deterioration in the respiratory status. Be ready to support the airway, assist respirations, and give full cardiopulmonary support if it becomes necessary.

Pleural Effusion

Treatment of pleural effusion consists of removal of fluid collected outside the lung, which must be done by a physician in a hospital setting. However, you should provide oxygen and other routine support measures to these patients.

Prolonged Seizures

Your ability to manage the patient's airway is compromised while he or she is having a seizure. The patient requires medication to stop the seizure activity before you can manage the airway; therefore, the patient needs to get to a hospital, or an ALS unit needs to get to you as quickly as possible. Aggressive airway management will be required for this patient as soon as you are able, followed by rapid transport to the nearest emergency department.

■ Obstruction of the Airway

If the patient is a small child or someone who was eating just before dyspnea developed, you may assume that the problem is an inhaled or aspirated foreign body. If the patient is old enough to talk but cannot make any noise, upper airway obstruction is the likely cause.

Upper airway obstruction may be either partial or complete. If your patient is able to talk and breathe, the wisest course may be to provide supplemental oxygen and transport carefully in a position of comfort to the hospital. As long as the patient is able to obtain sufficient oxygen, avoid doing anything that might turn a partial airway obstruction into a complete airway obstruction.

There is no condition more immediately life threatening than a complete airway obstruction. The obstructing body must be removed before any other actions will be effective. Clear the patient's upper airway according to basic life support guidelines. Opening the airway with the head tilt–chin lift maneuver may solve the problem. You should perform this maneuver only after you have ruled out a head or neck injury. If simply opening the airway does not correct the breathing problem, you will have to assess the upper airway for the obstruction. Then, whether or not you are successful in clearing the airway, administer supplemental oxygen and transport the patient promptly to the emergency department.

■ Pulmonary Embolism

Because a considerable amount of lung tissue may not be functioning, supplemental oxygen is mandatory in a patient with a pulmonary embolism. Place the patient in a comfortable position, usually sitting, and assist breathing as necessary. Hemoptysis, if present, is usually not severe, but any blood that has been coughed up should be cleared from the airway. The patient may have an unusually rapid and possibly irregular heartbeat. Transport the patient to the emergency department promptly. Be aware that pulmonary emboli may cause cardiac arrest.

■ Hyperventilation

When you respond to a patient who is hyperventilating, complete a primary assessment and gather a history of the event. Is the patient having chest pain? Is there a history of cardiac problems or diabetes? You must always assume a serious underlying problem even if you suspect that the underlying problem is stress. Do not have the patient breathe into a paper bag, even though it was thought to be the technique for managing hyperventilation syndrome. In theory, breathing into a paper bag causes the patient to rebreathe exhaled carbon dioxide, allowing the level of carbon dioxide in the blood to return to normal. In fact, if the patient is hyperventilating because of a serious medical problem, this maneuver could make things worse. A patient with underlying pulmonary disease who breathes into a bag may become severely hypoxic. Treatment should instead consist of reassuring the patient in a calm, professional manner; supplying supplemental oxygen; and providing prompt transport to the emergency department. Patients who hyperventilate need to be evaluated in the hospital setting.

■ Environmental/Industrial Exposure

The commonality in these kinds of respiratory problems is the inhalation of a toxic chemical. There are many different types of chemicals, different types of presentations, and certainly different levels of severity. Ensure that all patients are decontaminated prior to your providing treatment. Treat with oxygen, adjuncts, and suction on the basis of presentation, level of consciousness, and level of distress that is observed in your patient.

■ Epidemic and Pandemic Considerations

An **epidemic** occurs when new cases of a disease occur in a human population and substantially exceed what is "expected," based on recent experience. A **pandemic** is an outbreak that occurs on a global scale. A flu pandemic occurs when a new influenza virus emerges for which people have little or no immunity. The disease can spread easily from person to person, cause serious illness, and be found in multiple countries in a short time. Obviously, there would be no specific vaccine immediately available.

■ Influenza Type A

Influenza type A is primarily an animal respiratory disease that has mutated to infect humans. Human infections have occurred and are spreading. In 2009, the H1N1 strain of influenza type A became pandemic. Like seasonal flu, it may make chronic medical conditions worse. All strains of influenza type A are transmitted by direct contact with nasal secretions and aerosolized droplets from coughing and sneezing by infected people.

Many potentially serious diseases can be passed by the respiratory route; therefore, you need to be especially compliant with wearing of PPE (gloves, eye protection, and HEPA respirator at a minimum). Viruses can live for several days on surfaces, so frequent handwashing is also important. Maintain your vaccinations and stay up to date on the latest Centers for Disease Control and Prevention recommendations. Place a surgical mask on patients with suspected or confirmed respiratory disease. Wear a HEPA respirator during any aerosol-generating procedures such as suctioning of airway secretions, cardiopulmonary resuscitation, and endotracheal intubation.

Age-Related Assessment and Management

Foreign Body Aspiration

Upper airway obstruction is very common in young children who are putting everything in their mouths. If you have evidence of a partial or complete airway obstruction in a young child, especially a crawling baby, consider that the child may have swallowed and choked on a small object. Perform the appropriate airway clearing technique specific to the age of the child.

Another scenario to consider is that an object passed through the airway and has been aspirated (inhaled) into the lung. This problem will not be as obvious as an airway obstruction.

Most deaths from foreign body aspiration occur in patients who are younger than 5 years, and most of them are infants. Typical items aspirated include balloons, small balls, and small parts of toys. Toddlers may aspirate pieces of food like hot dogs or peanuts.

One sign of aspiration in a child may be an abnormality in the voice. The aspirated object will most likely go down the right mainstem bronchus. If the bronchus is fully obstructed, the lung could collapse. Aspiration pneumonia may also develop.

Provide oxygen, and transport any child with a suspected aspiration. An x-ray will be needed to confirm the aspiration, its location, and the treatment.

For an elderly person, the normal process of aging creates conditions that contribute to breathing problems. For example, weakening of the airway musculature can cause decreased breathing capacity. Decreased cough and gag reflexes cause a decreased ability to clear secretions. Difficulty in swallowing means the risk for aspiration is markedly increased. Elderly people can aspirate food or oral secretions that, in many cases, can develop into a potentially life-threatening aspiration pneumonia.

Tracheostomy Dysfunction

Children with chronic pulmonary medical conditions may use a home ventilator that is connected by a tracheostomy tube. This tube is placed in the neck and can sometimes become obstructed by secretions, mucus, or foreign bodies. Other tracheostomy tube complications include bleeding, leaking, dislodgement, and infection. Your main goal is to establish a patent airway. Place the patient in a position of comfort and provide suctioning to clear the obstruction. If you are unable to clear the airway, consider ALS intervention. Once the obstruction is clear, oxygenate the patient and treat based on the patient's presentation.

Geriatric patients may have a tracheostomy tube in place because of airway obstruction, laryngeal cancer, severe infection, trauma, or the inability to manage secretions. As with children, the tube can become obstructed by secretions, foreign bodies, or airway swelling. The stoma itself can become infected. Establishing airway patency is the immediate goal.

Croup

Croup is due to inflammation and swelling of the pharynx, larynx, and trachea. This disease is often secondary to an acute viral infection of the upper respiratory tract and is typically seen in children between ages 6 months and 3 years. It is easily passed between children.

The disease starts with a cold, cough, and a low-grade fever that develop over 2 days. The hallmark signs of croup are stridor and a seal-bark cough, which is a signal of significant narrowing of the air passage of the trachea that may progress to significant obstruction. Peak seasonal outbreaks of this desease occur in the late fall and during the winter.

Croup is rarely seen in adults because their breathing passages are larger and are able to accommodate the inflammation and mucus production without producing symptoms. The airways of adults are wider, and the supporting tissue is firmer than in children.

Croup often responds well to the administration of humidified oxygen.

Epiglottitis

Epiglottitis is a serious inflammation of the epiglottis, usually due to a bacterial infection that produces severe swelling of the flap over the larynx. Although it may be seen at any age, it is much more predominant in children. In preschool and school-aged children especially, the epiglottis can swell to two to three times its normal size. The airway is at risk of becoming completely obstructed. Patients with epiglottitis will look very sick. Epiglottitis usually has a sudden onset in an otherwise healthy child;

children with this infection look ill, complain of a very sore throat, and have a high fever. They will usually be in the tripod position and drooling. Patients will also have stridor, high-pitched inspiratory sounds indicating partial airway obstruction.

Treat the children gently and try not to do anything that will make them cry. Keep them in a position of comfort, and give them high-flow oxygen. *Do not* put anything in their mouths as this could create a complete airway obstruction.

Occasionally, the constellation of symptoms of epiglottitis can also appear in an adult or a geriatric patient, especially if the patient has other issues like diabetes, which affects his or her ability to fight off disease. In adults, epiglottitis, or supraglottitis, can be caused by different bacterial or viral organisms. Acute epiglottitis in an adult or geriatric patient can be potentially life threatening and misdiganosed because it is primarily recognized as a disease of pediatric patients. Deterioration can occur quickly in adults with acute epiglottitis. You should be very concerned if your patient is an adult presenting with stridor or any other sign of anatomic airway obstruction. Management should be focused on maintaining a patent airway.

Asthma

Asthma is a common childhood illness. When assessing a pediatric patient, look for retractions of the skin above the sternum and between the ribs. Retractions are typically easier to see in children than in adults. Cyanosis is a late finding in children.

Keep in mind that a cough is not always a symptom of a cold; it could signal pneumonia or asthma. Even if you do not hear much wheezing, the presence of a cough can indicate that some degree of reactive airway disease or an acute asthma attack may be taking place.

The emergency care of a child with shortness of breath is the same as it is for an adult, including the use of supplemental oxygen. However, many small children will not tolerate (or may refuse to wear) a face mask. Rather than fighting with the child, provide blow-by oxygen by holding the oxygen mask in front of the child's face or ask the parent to hold the mask **Figure 13-18**.

Many children with asthma also will have prescribed hand-held MDIs. Use these inhalers just as you would with an adult. Pediatric patients are more likely to use spacers to assist in inhaler use. Treat as in adult asthma.

You are the Provider: PART 5

With an estimated time of arrival at the hospital of 8 minutes, you ask your partner to radio in the patient report. The patient's level of consciousness has not changed; however, the cyanosis around her mouth and in her nail beds has resolved and her oxygen saturation has improved. You complete your reassessment and continue treatment.

Recording Time: 15 Minutes	
Level of consciousness	Responsive to pain only
Respirations	12 breaths/min; shallow
Pulse	118 beats/min; weak
Skin	Cool and dry; cyanosis has resolved
Blood pressure	112/70 mm Hg
Sao$_2$	90% (on oxygen)

You deliver the patient to the emergency department staff and give your verbal report to the nurse. Because of the patient's decreased level of consciousness and the need for ongoing ventilation assistance, the physician elects to intubate her. She is diagnosed with acute exacerbation of her emphysema and is admitted to the intensive care unit.

10. How can positive-pressure ventilation cause a decrease in a patient's blood pressure?

11. What does the term "exacerbation" mean?

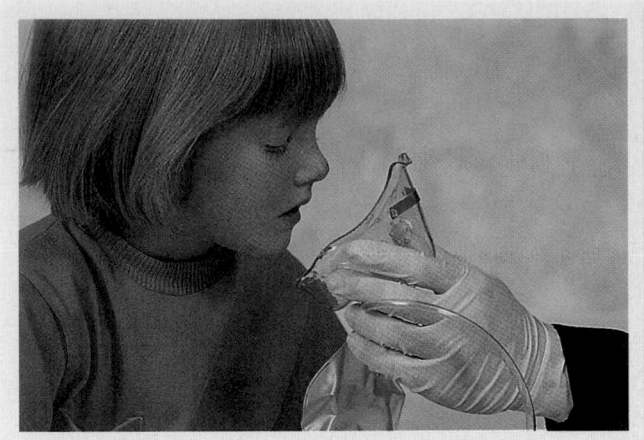

Figure 13-18 Because children may refuse to wear an oxygen mask, you may have to hold the mask in front of the child's face. If the child still refuses, enlist the parents' help.

Asthma in an older patient causes bronchospasm, swelling of the lining of the airways, and an accumulation of secretions. Attacks are easily triggered by air pollutants, viral infections, allergens, and sometimes something as simple as exposure to cold air. Asthma, as any chronic disease, can become life threatening in an older person, especially in patients who have problems with airway control. The condition is made worse by anxiety and dehydration, which is typical in older people. Geriatric patients with asthma tend to have both inspiratory and expiratory wheezes.

■ Bronchiolitis

Bronchiolitis is a specific viral illness that usually occurs in newborns and toddlers, often caused by respiratory syncytial virus (RSV) that causes inflammation of the bronchioles. Bronchiolitis occurs during the first 2 years of life and is more common in boys. The infections are most widespread in the winter and early spring. Bronchioles, the tiny airways that lead to the alveoli in the lungs, become inflamed, swell, and fill with mucus. The airways of infants and young children can become easily blocked. Bronchiolitis is often a precursor to asthma.

Although more commonly associated with children, certain types of bronchiolitis can affect adults. Bronchiolitis obliterans is a rare and life-threatening form of nonreversible obstructive lung disease in which the bronchioles are plugged with granulation tissue. Bronchiolitis obliterans with organizing pneumonia is an inflammation of the bronchioles that also affects surrounding tissue in the lung. A more accurate term would be **pneumonitis**, or inflammation of the lung. Predisposing causes of this disease may include radiation therapy, bone marrow transplantation, connective tissue disorders, and certain medications.

A patient with this disease may have no signs or symptoms but, in some cases, it may cause shortness of breath and fever that present gradually over a course of several weeks.

Respiratory Syncytial Virus

Respiratory syncytial virus (RSV) is a major cause of illness in young children, creating an infection in the lungs and breathing passages. The more serious infections found in premature infants and children with depressed immune systems can lead to other serious illnesses that affect the lungs or heart. An RSV infection can cause respiratory illnesses like bronchiolitis and pneumonia.

The RSV is highly contagious and spread through droplets when the patient coughs or sneezes. The virus can also survive on surfaces, including hands and clothing. The infection tends to spread rapidly through schools and in child care centers.

When assessing a child, look for signs of dehydrations. Infants with RSV often refuse liquids, so look for signs and symptoms of dehydration. Treat airway and breathing problems as appropriate. Humidified oxygen is helpful if available.

The RSV can also cause severe upper respiratory infections and typical asthma symptoms in adults and geriatric patients.

■ Pneumonia

According to the World Health Organization, pneumonia is the leading cause of death in children worldwide. **Pneumonia** is a general term that refers to an infection of the lungs. Pneumonia is often a secondary infection, meaning it begins after an upper respiratory tract infection like a cold or sore throat. It can also occur from chemical injury after an accidental ingestion or a direct lung injury from a near drowning. Therapies like intubation and tracheostomy can create conditions likely to cause pneumonia. Children with diseases causing immunodeficiency have an increased predisposition for pneumonia. Symptoms of pneumonia vary, depending on the age of the person and the cause of the illness. Children will present with unusually rapid breathing or will breathe with grunting or wheezing sounds. Other signs of labored breathing are tracheal indrawing and retractions, in which the chest moves in rather than expands during inhalation. In severe cases, the lips and fingernails may be bluish or gray. If the pneumonia is in the lower part of the lungs near the abdomen, there may be fever, abdominal pain, and vomiting rather than dyspnea.

Bacterial pneumonia will come on more quickly and results in higher fevers, putting the child at risk for febrile seizures. A viral pneumonia presents more gradually and is less severe. Wheezing may be present in viral pneumonia. Treatment includes airway support and providing supplemental oxygen.

Pneumonia especially affects people who are chronically and terminally ill and people who have a lowered resistance because the fluids in the lung interfere with the ability of the alveoli to oxygenate the blood. Some types of pneumonia are infectious, with symptoms such as a cough that produces greenish or yellow sputum, a high fever, and sometimes shaking chills. Remember to wear respiratory protection when you are assessing a patient with a potentially infectious respiratory disease.

You will need to evaluate the pathophysiology through history and possible risk factors. The factors that predispose patients to pneumonia include institutional residence or hospitalization, chronic disease processes, immune system compromise, and a history of COPD or cancer. Other factors include a history of inhaled toxins or aspiration of material into the lung.

Associated signs and symptoms include exertional dyspnea, a productive cough, chest discomfort and pain, wheezing, headache, nausea and vomiting, musculoskeletal pain, weight loss, and confusion.

On your physical assessment of the patient, you will most likely see cyanosis and pallor, dry skin, possible fever, decreased skin turgor, and pale, dry mucosa. The patient's heart rate will be tachycardic with possible hypotension, and an assessment of the lungs will reveal diminished breath sounds with the adventitious sounds of wheezing, rales, or rhonchi.

If possible, a core temperature should be obtained to determine the presence of fever. Pulse oximetry readings, if available, will most likely be low. The patient should be treated with airway, ventilatory, and circulatory support. Use oxygen with appropriate adjuncts for the patient, and provide supportive measures if needed. Evaluate patient treatment through reassessment and prepare for possible deterioration in the patient's condition.

■ Pertussis

Pertussis (whooping cough) is an airborne bacterial infection that mostly affects children younger than 6 years. It is highly contagious and is passed through droplet infection.

A patient with pertussis will be feverish and exhibit a "whoop" sound on inspiration after a coughing attack. Symptoms are generally similar to colds, but coughing spells can last for more than a minute in which the child may turn red or purple. This may frighten the parents into calling 9-1-1.

Some infants and younger children with pertussis should be treated in a hospital because they are at greater risk for complications like pneumonia, which occurs mostly in children younger than 1 year. In infants younger than 6 months, pertussis can sometimes be life threatening.

Children with pertussis may vomit or not want to eat or drink. Watch for signs of dehydration. You may have to suction thick secretions to clear the airway. Give oxygen by the most appropriate means.

Pertussis in adults or geriatric patients does not cause the typical whooping illness that it does in infants and toddlers. However, it can cause a very severe upper respiratory infection, which in older people can result in pneumonia. The infection can cause coughing spells that last for weeks and can be so severe that patients find it hard to breathe, eat, or sleep. In the worst cases of geriatric infection, coughing can lead to cracked ribs. For patients who are already weak from other chronic conditions, pertussis can lead to hospitalization. The disease has become a serious enough issue that physicians are becoming more aggressive about immunizing adults with the pertussis vaccine.

Pertussis is easily prevented with a vaccine. All EMTs should check their immunization status and/or get a booster.

■ Cystic Fibrosis

Cystic fibrosis (CF) is a genetic disorder that affects the lungs and digestive system. CF disrupts the normal function of cells that make up the sweat glands in the skin and that also line the lungs and the digestive and reproductive systems. The disease predisposes the child to repeated lung infections.

The disease process in CF disrupts the essential balance of salt and water necessary to maintain a normal coating of fluid and mucus inside the lungs and other organs. The end result is that the mucus becomes thick, sticky, and hard to move. The mucus holds germs, causing the lungs to become infected.

In CF, the child's symptoms range from sinus congestion to wheezing and asthma-like complaints. The child may develop a chronic cough that produces thick, heavy, discolored mucus. As lung function decreases, so does the ability to breathe effectively. The child often has dyspnea; this generally results in the parents calling EMS. Treat the child with suction and oxygen using age-appropriate adjuncts.

Cystic fibrosis often causes death in childhood because of chronic pneumonia secondary to the very thick, pathologic mucus in the airway. It also causes malabsorption of nutrients in the intestines. Because of advances in treatment, the life expectancy for CF patients becomes better each year. Adults with CF are predisposed to other medical conditions, including arthritis, osteoporosis, diabetes, and liver problems.

Special Populations

Most geriatric patients take medications, sometimes many, to treat various ailments that are part of the aging process. Some of these medications will blunt the body's normal reactions to stress and the mechanisms the body uses to compensate for respiratory compromise and hypoxia. For example, beta-blockers, used for a variety of conditions, prevent the heart from speeding up and the veins from constricting to compensate for a loss of blood pressure or oxygenation. Keep this in mind when evaluating vital signs in geriatric patients.

■ Congestive Heart Failure

After a heart attack or other illness, the heart muscle may be so injured that it cannot circulate blood properly. The heart is not able to maintain cardiac output that meets the needs of the body, thus the heart is failing as a pump.

In these cases, the left side of the heart cannot remove blood from the lungs as fast as the right side delivers it. As a result, fluid builds up within the alveoli and in the lung tissue between the alveoli and the pulmonary capillaries. This results in pulmonary edema.

Patient risk factors for congestive heart failure include hypertension and a history of coronary artery disease and/or atrial fibrillation, a condition in which the atria no longer contract, but instead quiver.

In most cases, patients have a long-standing history of chronic congestive heart failure that can be kept under control with medication. However, an acute onset may occur if the patient stops taking the medication, eats food that is too salty, or has a stressful illness, a new heart attack, or an abnormal heart rhythm.

Signs and symptoms of congestive heart failure include the patient reporting difficulty breathing with exertion because the heart cannot keep up with the body's need for oxygen. Patients may also report a sudden attack of respiratory distress that wakes them at night when they are in a reclining position. This is caused by fluid accumulation in the lungs. Patients also complain of coughing, feeling suffocated, cold sweats, and tachycardia.

In your assessment, you might find the patient has cool, diaphoretic, cyanotic skin and you will hear adventitious breath sounds like crackles, wheezing, or rales. The patient's pulse will be tachycardic. The patient may have hypertension early, followed by deterioration to hypotension as a late finding.

Treatment should consist of airway, ventilatory, and circulatory support. Provide oxygen with adjuncts appropriate to the patient's condition, and prepare for the next level of deterioration.

Continuous positive airway pressure (CPAP) is a noninvasive means of providing ventilatory support for patients experiencing respiratory distress associated with obstructive pulmonary disease and acute pulmonary edema. As discussed in Chapter 9, *Airway Management*, CPAP increases pressure in the lungs, opens collapsed alveoli, pushes more oxygen across the alveolar membrane, and forces interstitial fluid back into the pulmonary circulation. The CPAP systems use oxygen to deliver the positive ventilatory pressure to the patient. Many patients show dramatic improvement with the use of CPAP. CPAP can be used for patients who have moderate to severe respiratory distress from an underlying disease, such as pulmonary edema or obstructive pulmonary disease, are alert and able to follow commands, are breathing at a rate of more than 26 breaths/min, or have a pulse oximetry reading of less than 90%. One potential contraindication to the use of CPAP is low blood pressure. Because of the increased pressure inside the chest, blood flow returning to the heart is diminished. CPAP is also not used in patients in respiratory arrest or who have signs and symptoms of a pneumothorax or chest trauma, a tracheostomy, have a decreased level of consciousness inability to follow commands, or have active gastrointestinal bleeding. Continue to reassess patients using CPAP for signs of deterioration and/or respiratory failure.

You are the Provider: SUMMARY

1. What is emphysema? What is the typical cause?

Emphysema, a form of chronic obstructive pulmonary disease (COPD), is a disease of the respiratory system in which destructive changes occur in the alveoli. The inner walls of normal alveoli are lined with a protein lubricant called surfactant, which allows the alveoli to expand and recoil with little resistance. In emphysema, surfactant is progressively destroyed; this causes increased breathing difficulty and less efficient gas exchange in the lungs. Although emphysema is an irreversible condition, its symptoms can be reduced and the disease progression slowed with lifestyle changes (eg, quitting smoking) and certain medications.

The single most common cause of emphysema is heavy, long-term cigarette smoking. Other causes include frequent pulmonary infections and long-term exposure to toxic agents, such as by working in an industrial plant for a long period.

2. Why is it especially significant that *this* patient called 9-1-1?

Patients with chronic diseases do not call 9-1-1 when they feel better; they call when something has gotten worse. The patient could be experiencing an acute exacerbation of her emphysema, a secondary respiratory illness to which she is predisposed (ie, pneumonia), or complete respiratory failure. Just because she has refused EMS transport in the past does not mean that she will this time; she has a known respiratory illness, which you should assume has gotten worse, and she should be treated no differently from any other patient with a respiratory emergency.

3. What should be your *most* immediate action?

Oxygen and lit cigarettes—or any other source of fire—do not go together! Ask the patient to immediately extinguish her cigarette and then continue your assessment. Although oxygen is not flammable or explosive, it does support the process of combustion. A small spark or lit cigarette can become a flame in an oxygen-rich atmosphere. Oxygen will cause a fire to burn more vigorously, as well as hotter. The patient could literally light her face on fire; you and your partner could be injured as well.

4. How does emphysema differ from chronic bronchitis?

As previously discussed, emphysema is a disease in which the inner walls of the alveoli are progressively destroyed, resulting in a loss of their elasticity. Chronic bronchitis is caused by ongoing inflammation of the bronchi. With chronic bronchitis, excess mucus is constantly produced, which obstructs the bronchioles and alveoli. As a result, pulmonary gas exchange is less efficient. Many patients with chronic bronchitis have a chronic productive cough (a cough that produces phlegm). In some patients, however, the cough reflex is weakened; this causes sputum to settle in the lungs and become infected, resulting in pneumonia. Emphysema and chronic bronchitis are both forms of COPD, and are both usually caused by heavy, long-term cigarette smoking.

5. Why do patients with emphysema breathe through pursed lips?

With emphysema, the force of exhalation causes closure of the small airways, causing air to be trapped in the alveoli. The harder the patient tries to push air out, the more it gets trapped in the alveoli. Chronic air trapping in the lungs explains why many patients with long-term emphysema have a characteristic barrel-shape appearance to their chest. Over time, patients with emphysema learn that if they push air out slowly at a lower pressure, they can exhale more air than if they try to push it out faster and harder. One of the ways they do this is by breathing through pursed lips during exhalation. Breathing in this way allows the patient to push air out slowly under controlled pressure.

6. What does a prolonged exhalation phase indicate in patients with obstructive lung disease?

A prolonged exhalation phase indicates that the patient is experiencing difficulty exhaling air from the lungs, which, as a result, causes chronic air trapping in the lungs. The inhalation-to-exhalation ratio in healthy people during normal breathing is typically 1:2. In other words, it takes about twice as long to exhale as it does to inhale. Depending on the severity of their disease process, patients with obstructive lung disease may have an I:E ratio of 1:4, 1:5, or longer. Because the patient's bronchioles significantly narrow when he or she exhales, wheezing—a whistling sound—is often heard during the exhalation phase while auscultating the lungs.

7. What treatment is indicated for the patient at this point?

Compared with the patient's condition during your initial assessment, it has deteriorated. She is now confused and slow to answer your questions, which indicates decreased oxygen delivery to the brain. Her oxygen saturation level of 83% indicates significant hypoxemia, and the fatigue indicates that she is less able to compensate for her condition. She clearly needs a higher concentration of oxygen than what her nasal cannula is supplying. Apply a nonrebreathing mask, set the flow rate at 12 to 15 L/min, and reassess her condition. If it does not improve, assisted ventilation with a bag-mask device may be necessary. If you must assist the patient's breathing, however, ensure that you allow *complete* exhalation between positive-pressure breaths. Remember, patients with emphysema have a lot of air trapped in the alveoli.

8. Why is cyanosis a later sign of hypoxemia in patients with emphysema?

Patients with emphysema maintain chronically low blood oxygen levels and chronically elevated carbon dioxide levels. In many patients, this leads to excessive red blood cell production, a condition called polycythemia, which is why the patients tend to maintain a pink skin color, even in the presence of hypoxemia. This, and the fact that many patients with emphysema breathe through pursed lips, is why they are often referred to as "pink puffers." Unlike otherwise healthy people, patients with emphysema often do not develop cyanosis until

You are the Provider: SUMMARY, continued

significantly more hemoglobin is desaturated (not carrying oxygen). The absence of cyanosis does not rule out hypoxemia in any patient, especially a patient with emphysema.

9. Why does tachycardia develop in hypoxemic patients?

Whenever the body's demand for oxygen increases and its supply decreases (eg, hypoxemia), the nervous system increases the production of epinephrine from the adrenal glands. Epinephrine (adrenalin) is a hormone that causes tachycardia (rapid heart rate) and an increase in the strength of cardiac contraction. Tachycardia is a critical physiologic compensatory mechanism that circulates oxygenated blood faster, thus helping to maintain adequate perfusion of the body's vital organs. However, if the underlying cause of the patient's hypoxemia is not corrected, the nervous system, which also requires oxygen, will no longer be able to compensate and the patient's heart rate will begin to fall.

10. How can positive-pressure ventilation cause a decrease in a patient's blood pressure?

If you will recall from Chapter 9, *Airway Management*, negative-pressure ventilation—the process that occurs with normal breathing—involves contraction of the diaphragm and intercostal muscles and a decrease in intrathoracic pressure; as a result, air is pulled into the lungs. Positive-pressure ventilation involves pushing air into the lungs, as with artificial ventilation. It is critical to perform positive-pressure ventilation correctly. Deliver each breath over a period of 1 second, just enough to produce visible chest rise, at a rate that is appropriate for the patient (10 to 12 breaths/min for adults; 12 to 20 breaths/min

for infants and children). If positive-pressure ventilation is delivered too fast or with too much force (hyperventilation), the resultant hyperinflation of the lungs and increase in intrathoracic pressure may squeeze the heart and impair blood return to the right atrium. If blood return is impaired, the amount of blood that is pumped from the left ventricle per contraction (stroke volume) will be impaired as well. As a result, the patient's blood pressure (and perfusion status) will deteriorate.

11. What does the term "exacerbation" mean?

The term "exacerbation" means to intensify or worsen in severity. In acute exacerbation of COPD, no copathologic (secondary) condition exists that would clearly explain the patient's sudden deterioration (ie, congestive heart failure, pneumonia). Patients with COPD often experience acute exacerbation of their disease secondary to a change in environmental conditions, such as weather, humidity, or sudden activation of central heating or cooling in the home. As with diseases such as asthma, COPD can also be exacerbated by certain triggers, such as cat dander, dust, and seasonal allergens. In some cases, acute exacerbation is idiopathic (of unknown cause). As the patient's disease progresses, he or she will eventually reach a point at which the lungs simply cannot support oxygenation and ventilation (end-stage COPD). In end-stage COPD, it can be difficult to determine whether the patient is experiencing an exacerbation that can be resolved or if he or she has reached the end of the disease process. This will not affect your treatment, however, which involves airway management and ensuring adequate oxygenation and ventilation.

EMS Patient Care Report (PCR)

Date: 3-12-09	**Incident No.:** 130309	**Nature of Call:** Respiratory		**Location:** 109 East Lawler	
Dispatched: 0430	**En Route:** 0432	**At Scene:** 0440	**Transport:** 0449	**At Hospital:** 0510	**In Service:** 0519

Patient Information

Age: 72 **Sex:** F **Weight (in kg [lb]):** 50 kg (110 lb)	**Allergies:** Sulfa, ibuprofen, aspirin **Medications:** Oxygen, Combivent, albuterol, lisinopril **Past Medical History:** Emphysema, hypertension, gout **Chief Complaint:** Trouble breathing

Vital Signs

Time: 0443	**BP:** 116/54	**Pulse:** 120	**Respirations:** 28	**Sao$_2$:** 83%
Time: 0449	**BP:** 108/50	**Pulse:** 132	**Respirations:** 12	**Sao$_2$:** 77%
Time: 0455	**BP:** 112/70	**Pulse:** 118	**Respirations:** 12	**Sao$_2$:** 90%

EMS Treatment (circle all that apply)

Oxygen @ <u>15</u> L/min via (circle one): NC NRM (Bag-Mask Device)	(Assisted Ventilation)	(Airway Adjunct)	CPR
Defibrillation	**Bleeding Control**	**Bandaging**	**Splinting**	**Other:**

You are the Provider: SUMMARY, continued

Narrative

Dispatched to the residence of a 72-year-old female with difficulty breathing. On arrival at the scene, found the patient sitting on the edge of her couch; she was wearing home oxygen via nasal cannula and was smoking a cigarette. Immediately asked patient to extinguish cigarette and continued with assessment. Patient could only speak in two-word sentences and was experiencing marked respiratory distress. Past medical history significant for emphysema, hypertension, and gout. The patient states that she is normally short of breath; however, it suddenly worsened today. Further assessment revealed pursed-lip breathing, prolonged exhalation, and scattered wheezing on auscultation. Obtained vital signs and applied high-flow oxygen via nonrebreathing mask because of signs of worsened hypoxemia. Patient takes numerous medications and states that she has been compliant with all of them. Placed patient onto stretcher, loaded her into the ambulance, and reassessed her condition. Patient's level of consciousness had markedly decreased (responsive to pain only), her respirations became slow and shallow, her oxygen saturation decreased, and she developed cyanosis around her mouth and to her nail beds. Inserted nasal airway and began assisting her ventilations with a bag-mask device and high-flow oxygen. Began transport to hospital and continued treatment. En route, noted that patient's oxygen saturation had improved and her cyanosis had resolved; however, her LOC remained unchanged. Continued treatment and reassessment until arrival at the hospital. Transferred patient care to emergency department staff w/o incident. Verbal report given to staff nurse. **End of report**

Assessment and Emergency Care of Respiratory Emergencies

Scene Size-up

Scene Safety	Ensure scene safety and safe access to the patient. Consider that the patient may be in distress because of exposure to a toxic substance. Standard precautions should include a minimum of gloves and eye protection. A HEPA respirator should be considered if there is evidence of a communicable disease. Determine the number of patients; toxic environments may produce multiple patients. Assess the need for additional resources such as ALS, police, or specialized units.
Mechanism of Injury (MOI)/ Nature of Illness (NOI)	Determine the NOI: Observe the scene and look for indicators of the NOI. Ensure that the respiratory emergency is not the result of a traumatic event. Usually the NOI can be determined by the patient's chief complaint or by asking family members or bystanders. If the NOI is not evident, observe for signs of urticaria, chest pain, and fever. Look for a medical identification device.

Primary Assessment

Form a General Impression	Perform a rapid scan of the patient and observe overall appearance of the patient, age, and body position. Is the patient in the tripod position? Does the patient have a barrel chest? Observe work of breathing and circulation. Pale skin and cyanosis are indicators of poor perfusion. Determine level of consciousness using the AVPU scale. Is the patient calm or anxious? Is he or she able to speak in full sentences? Identify immediate threats to life. Determine priority of care based on the MOI/NOI. If the patient has a poor general impression, call for ALS assistance. A rapid visual scan will help you identify and manage life threats.
Airway and Breathing	Ensure the airway is open, clear, and self-maintained. If needed, open and maintain the airway using a modified jaw-thrust when a cervical spine injury is suspected and a head tilt–chin lift maneuver in nontraumatic situations. A patient with an altered level of consciousness may need emergency airway management. Consider inserting a properly sized oropharyngeal or nasopharyngeal airway. Assess for gurgling or stridor. Suction as needed. Evaluate the patient's ventilatory status for rate and depth of breathing, respiratory effort, and tidal volume. Quickly assess the chest for DCAP-BTLS, accessory muscle use, and intercostal and abdominal muscle use, and treat any threats to life. Inspect for urticaria (hives), which may be present during an anaphylactic reaction. Assess lung sounds, and determine whether they are normal, decreased, abnormal, or absent. Administer high-flow oxygen at 15 L/min, providing ventilatory support as needed.
Circulation	Evaluate distal pulse rate, quality (strength), and rhythm. Tachycardia may be an indicator of respiratory distress or shock. Bradycardia might occur with a cardiac emergency, medication reaction, or poisoning. Observe skin color, temperature, and condition; look for life-threatening bleeding, and treat accordingly. The transport of oxygen may be reduced from a lack of red blood cells. If distal pulses are not palpable, assess for a central pulse.
Transport Decision	If the patient has an airway or breathing problem, signs and symptoms of internal bleeding, or other life threats, manage him or her immediately and consider rapid transport, performing the secondary assessment en route to the hospital. For stable patients without life threats, perform a thorough assessment and history on scene. Do not delay transport to manage non-life-threatening conditions; instead treat en route to the hospital.

NOTE: The order of the steps in this section differs depending on whether the patient is conscious or unconscious. The following order is for a conscious patient. For an unconscious patient, perform a primary assessment, perform a full-body scan, obtain vital signs, and obtain the past medical history from a family member, bystander, or emergency medical identification device.

Assessment and Emergency Care of Respiratory Emergencies, continued

History Taking

Investigate Chief Complaint	Investigate the chief complaint. Monitor patient for changes in mental status. Ask OPQRST, SAMPLE, and PASTE questions. SAMPLE can also be obtained from family, bystanders, and medical identification devices. Identify pertinent negatives. Determine if patient has done anything for the breathing problem. If an inhaler was used, ask how many doses and when. Is the patient coughing? Is the patient able to sleep lying down? Is the patient a smoker?

Secondary Assessment

Physical Examinations	Perform a systematic examination beginning with the head, looking for DCAP-BTLS. Assessment should be rapid if the patient has a poor general impression. Inspect, palpate, and auscultate the chest, focusing on the respiratory effort and adequacy of ventilation. The sounds you hear when you auscultate the lungs will help you determine lung function. Accessory muscle use, nasal flaring, pursed lips, confusion, and tachypnea (rapid breathing) are signs of respiratory distress. Look for hives and rashes. Examine the skin for color; pallor and cyanosis are indicators of hypoxia (low oxygen level). Monitor patient's mental status for changes.
Vital Signs	Obtain baseline vital signs and repeat every 5 to 15 minutes depending on patient impression; monitor trends. Vital signs should include blood pressure by auscultation, pulse rate and quality, respiration rate and quality, and skin assessment for perfusion. Note patient's level of consciousness. Use pulse oximetry, if available, to assess the patient's perfusion status.

Reassessment

Interventions	Reassess the primary assessment, vital signs, and chief complaint. Airway control using adjuncts may be necessary. Assist breathing as required, administering high-flow oxygen. Assist patient with prescribed MDI or EpiPen auto-injector. If authorized and indicated, assist with small-volume nebulizer. Check interventions and treatment rendered; be prepared to modify treatments. Support the cardiovascular system. Do not delay transport.
Communication and Documentation	Contact medical control with a radio report when necessary. Include a thorough description of the NOI and position the patient was found in. Include treatments performed and patient response. Be sure to document any changes in patient status and the time. Follow local protocols. Document the reasoning for your treatment and the patient's response.

NOTE: Although the following steps are widely accepted, be sure to consult and follow your local protocols. Take appropriate standard precautions when treating all patients.

Respiratory Emergencies

General Management of Respiratory Emergencies

Managing life threats to the patient's ABCs and ensuring the delivery of high-flow oxygen are the primary concerns with any respiratory emergency. Patients breathing at a rate of less than 8 breaths/min or greater than 30 breaths/min should have ventilations assisted with a bag-mask device. Continually assess the patient's mental status, and provide emotional support as needed. Transport in the position of comfort. For all respiratory emergencies, make sure you have taken the appropriate standard precautions, including the use of a HEPA respirator.

Assessment and Emergency Care of Respiratory Emergencies, continued

Respiratory Emergencies

Upper or Lower Airway Infection

Dyspnea from an upper airway infection may be from croup or epiglottitis. Patients should receive humidified oxygen if available. Patients who are sitting forward, seem lethargic, or are drooling may have epiglottitis. Do not force the patient to lie down or attempt to suction or insert an oropharyngeal airway because this may cause a spasm and a complete airway obstruction. Transport should be rapid.

Lower airway infections may be from the common cold, bronchitis, or pneumonia. Patients need supplemental oxygen, monitoring of vital signs, and transport to the hospital.

Acute Pulmonary Edema

Congestive heart failure or a toxic inhalation may cause the patient to have pulmonary edema. Place the patient in a position of comfort, usually sitting up. Administer high-flow oxygen and provide ventilatory support and suctioning as needed. Continuous positive airway pressure can be initiated if you are authorized to do so, or request ALS personnel to perform the procedure. Provide prompt transport to the hospital.

Chronic Obstructive Pulmonary Disease

Patients with COPD may be semiconscious or unconscious from hypoxia, a condition in which the body's cells and tissues do not get enough oxygen, or from carbon dioxide retention. They may appear to be in respiratory distress and/or be cyanotic. They may have pursed lips and may be using accessory muscles to breathe, including those in the neck and shoulders. Assist with the patient's prescribed inhaler if there is one. Document time and effect on patient with each use. Oftentimes, a patient with COPD will overuse an inhaler; watch for side effects. Transport patients with COPD as promptly as possible to the emergency department, allowing them to sit upright if this is most comfortable; breathing may be difficult when lying down. Treat with full-flow oxygen via nonrebreathing mask at 15 L/min and be aware that some patients have a reverse breathing reflex; too much oxygen can eliminate their stimulus to breathe. CPAP can be initiated if you are authorized to do so, or request ALS to perform the procedure.

Asthma, Hay Fever, and Anaphylaxis

Not all wheezing is the result of asthma! Obtain a thorough history from the patient or family. If the patient is wheezing and has asthma, assist with the patient's prescribed inhaler or administer a small-volume nebulizer containing albuterol. Provide supplemental oxygen and provide ventilatory support as needed. Patients whose asthma progresses to status asthmaticus require immediate transportation. Be prepared to assist their ventilations because they may become too exhausted to breathe.

Hay fever usually requires only support and transport, but if the condition has worsened from generalized cold symptoms, the patient may require supplemental oxygen and airway support.

Anaphylaxis is a true emergency that requires rapid intervention and transport. Airway, oxygen, and ventilatory support are paramount. Determine if the patient has a prescribed EpiPen auto-injector and assist with administration by placing it in the patient's hand. Guide the EpiPen auto-injector to the patient's thigh (lateral aspect) at a 90° angle and administer the medication. Hold the auto-injector in place for 10 seconds. Transport promptly. Reassess the patient's condition en route to the hospital.

Assessment and Emergency Care of Respiratory Emergencies, continued

Respiratory Emergencies

Pneumothorax

A pneumothorax may occur spontaneously or may be the result of a traumatic event. Place the patient in a position of comfort, and support the ABCs. Provide prompt transport, monitor the patient carefully, and be prepared to assist ventilations and provide cardiopulmonary resuscitation if necessary.

Pleural Effusion

Treatment consists of removal of the fluid collected outside the lung. This must be performed by a physician in a hospital. Provide oxygen and support the ABCs, place the patient in a position of comfort, and transport promptly.

Obstruction of the Airway

Managing an airway obstruction is a priority. Use age-appropriate basic life support foreign body airway obstruction maneuvers to clear the airway. Administer supplemental oxygen, and transport the patient to the closest hospital. Some patients do not want to go to a hospital after the obstruction is cleared. Encourage them to be transported for evaluation of possible injury to the airway.

Pulmonary Embolism

Pulmonary embolism causes a ventilation-perfusion mismatch. The gas exchange is not able to take place, and the patient will become hypoxic. A sitting position is usually preferred by the patient. Ensure the airway is clear; hemoptysis should be cleared as it occurs. Provide supplemental oxygen and provide ventilatory support as needed. Cardiopulmonary arrest may occur with a pulmonary embolism.

Hyperventilation

Gather a thorough history, and attempt to determine the underlying cause because the hyperventilation may be the result of a serious problem. Do not have the patient breathe into a paper bag; this maneuver could make things worse. Instead, reassure the patient, administer supplemental oxygen, and provide prompt transport to the hospital.

Prep Kit

Ready for Review

- Dyspnea is a common complaint that may be caused by numerous medical problems, including infections of the upper or lower airways, acute pulmonary edema, chronic obstructive pulmonary disease, spontaneous pneumothorax, asthma, allergic reactions, pleural effusion, mechanical obstruction of the airway, pulmonary embolism, and hyperventilation.

- Each of these lung disorders has the ability to interfere with the exchange of oxygen and carbon dioxide that takes place during respiration. This interference may be in the form of damage to the alveoli, separation of the alveoli from the pulmonary vessels by fluid or infection, obstruction of the air passages, or air or excess fluid in the pleural space.

- Patients with long-standing lung diseases often have chronically high levels of blood carbon dioxide; in some cases, giving too much oxygen to them may depress or stop respirations. However, judicious use of oxygen is always an important priority in patients with dyspnea.

- Patients often develop breathing difficulty and/or hypoxia with the following medical conditions: upper or lower airway infection, acute pulmonary edema, chronic obstructive pulmonary disease, hay fever, asthma, anaphylaxis, spontaneous pneumothorax, and pleural effusion.

- Infectious diseases associated with dyspnea include epiglottitis, bronchitis, tuberculosis, pneumonia, and pertussis.

- Lung and breaths sounds are some of the most important vital signs you should assess when treating a patient in respiratory distress.

- Signs and symptoms of breathing difficulty include unusual breath sounds (wheezing, stridor, rales, and rhonchi); nasal flaring; pursed-lip breathing; cyanosis; inability to talk; use of accessory muscles to breathe; and sitting in the tripod position, which allows the diaphragm the most room to function.

- Interventions for respiratory problems may include the following:
    - Oxygen via a nonrebreathing mask at 15 L/min, positive-pressure ventilations using a bag-mask device, pocket mask, or a flow-restricted oxygen-powered ventilation device
    - Airway management techniques such as use of an oropharyngeal airway, a nasopharyngeal airway, suctioning, or airway positioning
    - Positioning the patient in a high Fowler's position or a position of comfort to facilitate breathing
    - Assistance with respiratory medications found in a prescribed MDI or a small-volume nebulizer (Consult medical control to assist with its use, or follow standing orders if the orders allow for this.)

- Remember, a patient who is breathing rapidly may not be getting enough oxygen as a result of respiratory distress from a variety of problems, including pneumonia or a pulmonary embolism; trying to "blow off" more carbon dioxide to compensate for acidosis caused by a poison, a severe infection, or a high level of blood glucose; or having a stress reaction.

- In every case, prompt recognition of the problem, administration of oxygen, and prompt transport are essential.

Vital Vocabulary

acidosis The buildup of excess acid in the blood or body tissues that results from a primary illness.

adventitious breath sounds Abnormal breath sounds such as wheezes, rhonchi, and rales.

alkalosis The buildup of excess base (lack of acids) in the body fluids.

allergen A substance that causes an allergic reaction.

anaphylaxis (anaphylactic shock) An extreme, life-threatening systemic allergic reaction that may include shock and respiratory failure.

asthma An acute spasm of the smaller air passages, called bronchioles, associated with excessive mucus production and with swelling of the mucous lining of the respiratory passages.

atelectasis Collapse of the alveolar air spaces of the lungs.

bronchial breath sounds Normal breath sounds made by air moving through the bronchi.

bronchiolitis Inflammation of the bronchioles that usually occurs in children younger than 2 years and is often caused by the respiratory syncytial virus.

bronchitis An acute or chronic inflammation of the lung that may damage lung tissue; usually associated with cough and production of sputum and, depending on its cause, sometimes fever.

carbon dioxide retention A condition characterized by a chronically high blood level of carbon dioxide in which the respiratory center no longer responds to high blood levels of carbon dioxide.

carbon monoxide An odorless, highly poisonous gas that results from incomplete oxidation of carbon in combustion.

chronic bronchitis Irritation of the major lung passageways from infectious disease or irritants such as smoke.

chronic obstructive pulmonary disease (COPD) A slow process of dilation and disruption of the airways and alveoli caused by chronic bronchial obstruction.

common cold A viral infection usually associated with swollen nasal mucous membranes and the production of fluid from the sinuses and nose.

croup An inflammatory disease of the upper respiratory system that may cause a partial airway obstruction and is characterized by a barking cough; usually seen in children.

diphtheria An infectious disease in which a membrane forms, lining the pharynx; this lining can severely obstruct the passage of air into the larynx.

dyspnea Shortness of breath or difficulty breathing.

embolus A blood clot or other substance in the circulatory system that travels to a blood vessel where it causes a blockage.

emphysema A disease of the lungs in which there is extreme dilation and eventual destruction of the pulmonary alveoli with poor exchange of oxygen and carbon dioxide; it is one form of chronic obstructive pulmonary disease.

epidemic Occurs when new cases of a disease occur in a human population and substantially exceed what is "expected," based on recent experience.

epiglottitis A disease in which the epiglottis becomes inflamed and enlarged and may cause an upper airway obstruction.

hay fever An allergic response usually to outdoor airborne allergens such as pollen or sometimes indoor allergens such as dust mites or pet dander; also called allergic rhinitis.

hyperventilation Rapid or deep breathing that lowers the blood carbon dioxide level below normal.

hyperventilation syndrome (panic attack) This syndrome occurs in the absence of other physical problems. The respirations of a person who is experiencing hyperventilation syndrome may be as high as 40 shallow breaths/min or as low as only 20 very deep breaths/min.

hypoxia A condition in which the body's cells and tissues do not have enough oxygen.

hypoxic drive Backup system to control respirations when oxygen levels fall.

influenza type A Virus that has crossed the animal/human barrier and has infected humans, recently reaching a pandemic level with the H1N1 strain.

meningococcal meningitis An inflammation of the meningeal coverings of the brain and spinal cord; can be highly contagious.

metered-dose inhaler (MDI) A miniature spray canister used to direct medications through the mouth and into the lungs.

methicillin-resistant *Staphylococcus aureus* **(MRSA)** A bacterium that can cause infections in different parts of the body; it is transmitted by different routes, including the respiratory route and is particularly dangerous because of its resistance to methicillin.

oxygenation The process of delivering oxygen to the blood.

pandemic An outbreak that occurs on a global scale.

pertussis (whooping cough) An airborne bacterial infection that affects mostly children younger than 6 years. Patients will be feverish and exhibit a "whoop" sound on inspiration after a coughing attack; highly contagious through droplet infection.

pleural effusion A collection of fluid between the lung and chest wall that may compress the lung.

pleuritic chest pain Sharp, stabbing pain in the chest that is worsened by a deep breath or other chest wall movement; often caused by inflammation or irritation of the pleura.

pneumonia An infectious disease of the lung that damages lung tissue.

pneumonitis Inflammation of the lung.

pneumothorax A partial or complete accumulation of air in the pleural space.

pulmonary edema A buildup of fluid in the lungs, usually as a result of congestive heart failure.

pulmonary embolism A blood clot that breaks off from a large vein and travels to the blood vessels of the lung, causing obstruction of blood flow.

rales Crackling, rattling breath sounds signaling fluid in the air spaces of the lungs.

respiration The exchange of oxygen and carbon dioxide.

respiratory syncytial virus (RSV) A virus that causes an infection of the lungs and breathing passages; can lead to other serious illnesses that affect the lungs or heart, such as bronchiolitis and pneumonia. RSV is highly contagious and spread through droplets.

rhonchi Coarse breath sounds heard in patients with chronic mucus in the airways.

severe acute respiratory syndrome (SARS) Potentially life-threatening viral infection that usually starts with flulike symptoms.

small-volume nebulizer A respiratory device that holds liquid medicine that is turned into a fine mist. The patient inhales the medication into the airways and lungs as a treatment for conditions like asthma.

stridor A harsh, high-pitched, barking inspiratory sound often heard in acute laryngeal (upper airway) obstruction.

tuberculosis (TB) A disease that can lay dormant in a person's lungs for decades, then reactivate; many strains are resistant to many antibiotics. TB is spread by cough.

vesicular breath sounds Normal breath sounds made by air moving in and out of the alveoli.

wheezing A high-pitched, whistling breath sound, characteristically heard on expiration in patients with asthma or chronic obstructive pulmonary disease.

Assessment in Action

You are dispatched to a church for a person in respiratory distress. On arrival you and your partner are led to an office where you observe a 67-year-old woman seated upright in a chair and leaning her arms against a desk for support. She is anxious, pale, and diaphoretic with audible gurgling on exhalation. Initial vital signs are a blood pressure of 176/90 mm Hg, a pulse rate of 117 beats/min, and a respiratory rate of 28 breaths/min and labored.

1. Given the patient's presentation, you suspect which of the following conditions?
   A. Acute asthma attack
   B. Pneumonia
   C. Pulmonary edema
   D. Chronic bronchitis

2. In addition to heart disease, pulmonary edema can be caused by:
   A. toxic fumes.
   B. chest trauma.
   C. high altitude exposure.
   D. all of the above.

3. A patient with a history of pulmonary edema may use which of the following medications on a daily basis?
   A. Metered-dose inhaler
   B. Diuretic
   C. Nitroglycerin
   D. Epinephrine

4. Which of the following breath sounds is commonly heard in a patient with pulmonary edema?
   A. Rales
   B. Pleural friction rub
   C. Stridor
   D. Grunting

5. Pulmonary edema is caused by which of the following actions?
   A. Early collapse of the alveoli
   B. Acute bronchoconstriction
   C. Fluid buildup within the alveoli
   D. Upper airway inflammation

6. You ask your partner to administer oxygen. What is the most appropriate means for oxygen delivery?
   A. Nasal cannula at 2 to 6 L/min
   B. Nonrebreathing mask at 15 L/min
   C. Venturi mask at 8 L/min
   D. Bag-valve mask at 15 L/min

7. What color is the sputum you may observe while treating a patient with pulmonary edema?
   A. Brown
   B. Yellow
   C. Green
   D. Pink

8. While you are on the way to the hospital, you note the patient is becoming sleepy and her respiratory rate is decreasing. What is the most appropriate action?
   A. Continue to monitor the patient and do nothing.
   B. Prepare to assist ventilations with a bag-mask device.
   C. Give the patient oxygen by nasal cannula.
   D. Prepare to insert a multi-lumen airway.

9. Explain how pulmonary edema interferes with the transfer of oxygen from the alveoli to the red blood cells.

10. Identify the signs and symptoms present in the above patient that are indicative of inadequate breathing.

National EMS Education Standard Competencies

Pathophysiology

Applies fundamental knowledge of the pathophysiology of respiration and perfusion to patient assessment and management.

Medicine

Applies fundamental knowledge to provide basic emergency care and transportation based on assessment findings for an acutely ill patient.

Cardiovascular

Anatomy, signs, symptoms, and management of

- Chest pain (pp 525-555)
- Cardiac arrest (pp 525-530, 545-555)

Anatomy, physiology, pathophysiology, assessment, and management of

- Acute coronary syndrome (pp 525-534)
  - Angina pectoris (pp 525-532)
  - Myocardial infarction (pp 525-534)
- Aortic aneurysm/dissection (pp 525-530, 537-538)
- Thromboembolism (pp 525-532)
- Heart failure (pp 525-530, 535-537)
- Hypertensive emergencies (pp 525-530, 537-538)

Knowledge Objectives

1. Understand the basic anatomy and physiology of the cardiovascular system. (pp 525-530)
2. Describe the anatomy, physiology, pathophysiology, assessment, and management of angina pectoris. (pp 525-532)
3. Describe the anatomy, physiology, pathophysiology, assessment, and management of thromboembolism. (pp 525-532)
4. Describe the anatomy, physiology, pathophysiology, assessment, and management of myocardial infarction. (pp 525-534)
5. Understand the anatomy, signs and symptoms, and management of hypertensive emergencies. (pp 525-530, 537-538)
6. Describe the anatomy, physiology, pathophysiology, assessment, and management of aortic aneurysm/dissection. (pp 525-530, 537-538)
7. Discuss the pathophysiology of the cardiovascular system. (pp 531-538)
8. Understand the relationship between airway management and the patient with cardiac compromise. (p 538)
9. Explain patient assessment procedures for cardiovascular problems. (pp 538-543)
10. Give the indications and contraindications for the use of nitroglycerin. (pp 542-544)

11. Recognize that many patients will have had cardiac surgery and may have implanted pacemakers. (pp 544-545)
12. Define "cardiac arrest." (p 545)
13. Give the indications and contraindications for use of an automated external defibrillator (AED). (pp 545-547)
14. Explain the relationship of age and weight to defibrillation. (p 546)
15. Discuss the different types of AEDs. (pp 546-547)
16. Give the advantages of using AEDs. (pp 546-547)
17. Describe the difference between the fully automated and the semiautomated defibrillator. (pp 546-547)
18. Explain the use of remote, adhesive defibrillator pads. (pp 546-547)
19. Recognize that not all patients in cardiac arrest require an electric shock. (pp 546-547)
20. Explain the circumstances that may result in inappropriate shocks from an AED. (pp 546-547)
21. Explain the reason not to touch the patient, such as by delivering CPR, while the AED is analyzing the heart rhythm and delivering shocks. (p 547)
22. Understand the reasons for early defibrillation. (p 548)
23. Describe AED maintenance procedures. (pp 548-550)
24. Explain the role played by medical direction in the use of AEDs. (p 550)
25. Understand the importance of practice and continuing education with the AED. (p 550)
26. Explain the need for a case review of each incident in which an AED is used. (p 550)
27. Understand quality improvement goals relating to AEDs. (p 550)
28. Discuss the procedures to follow for standard operation of the various types of AEDs. (pp 550-554)
29. Describe the emergency medical care for the patient with cardiac arrest. (pp 550-555)
30. Describe the components of care following AED shocks. (pp 553-554)
31. Explain criteria for transport of the patient for advanced life support (ALS) following CPR and defibrillation. (p 555)
32. Discuss the importance of coordinating with advanced life support (ALS) personnel. (p 555)

Skills Objectives

1. Demonstrate how to assess and provide emergency medical care for a patient with chest pain or discomfort. (pp 538-543)
2. Demonstrate the administration of nitroglycerin. (pp 542-544, Skill Drill 14-1)
3. Demonstrate how to perform maintenance of an AED. (pp 548-550)
4. Demonstrate how to perform AED and CPR. (pp 551-553, Skill Drill 14-2)

Introduction

The American Heart Association reports that cardiovascular disease claimed 864, 480 lives in the United States in 2005. This is 35.3% of all deaths, or approximately 1 of every 2.8 deaths. Heart disease has been the leading killer of Americans since 1900.

It is important for EMS providers to understand that many deaths caused by cardiovascular disease occur because of problems that may have been avoided by people living more healthful lifestyles and by access to improved medical technology. We can help to reduce these numbers of deaths with better public awareness, early access, increased numbers of laypeople trained in cardiopulmonary resuscitation (CPR), public access to defibrillation devices, and the recognition of the need for advanced life support (ALS) services.

This chapter begins with a brief description of the heart and how it works. It then discusses the relationship between chest pain and ischemic heart disease. It explains how to recognize and treat acute myocardial infarction (classic heart attack) and its complications—sudden death, cardiogenic shock, and congestive heart failure (CHF). The use of nitroglycerin and aspirin are described. The last part of the chapter is devoted to the use and maintenance of the automated external defibrillator (AED).

Anatomy and Physiology

The heart is a relatively simple organ with a simple job. It has to pump blood to supply oxygen-enriched red blood cells to the tissues of the body. The heart is divided down the middle into two sides (left and right) by a wall called the septum. Each side of the heart has an **atrium**, or upper chamber, to receive incoming blood, and a **ventricle**, or lower chamber, to pump outgoing blood Figure 14-1 . Blood leaves each of the four chambers of the heart through a one-way valve. These valves keep the blood moving through the circulatory system in the proper direction. The **aorta**, the body's main artery, receives the blood ejected from the left ventricle and delivers it to all the other arteries so that they can carry blood to the tissues of the body.

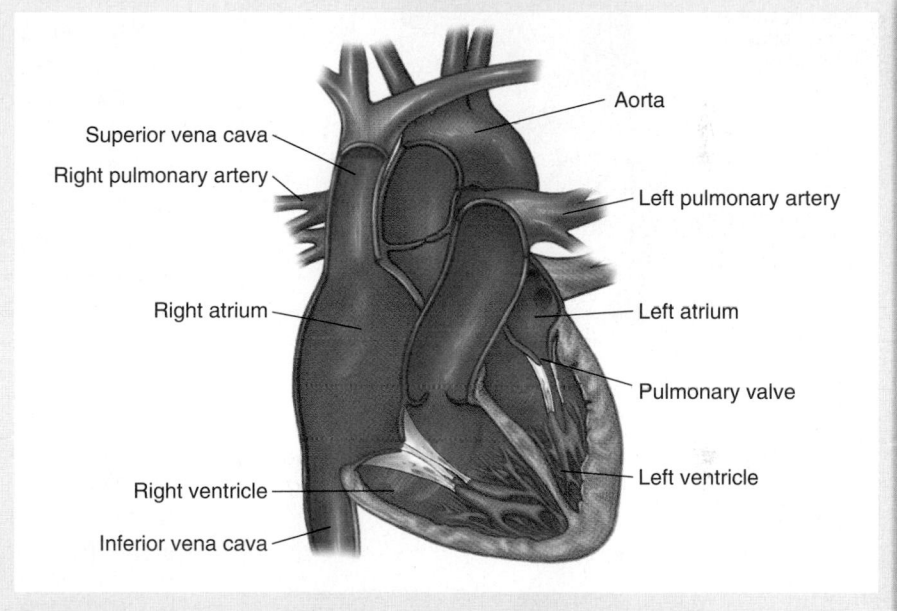

Figure 14-1 The heart is a four-chambered muscle that pumps blood to all parts of the body.

You are the Provider: PART 1

You and your partner are returning to your station after completing a call when you are dispatched to 1152 Blanco Road for a 60-year-old woman with chest pain. Dispatch advises you that the patient's son, who called 9-1-1, stated that she has a history of heart problems. You proceed to the scene, which is approximately 5 minutes away. The time is 9:42 AM, traffic is light, the weather is clear, and the temperature is 80°F.

1. What is the function of the heart?
2. What does the heart require to function effectively?
3. What should you include in your primary assessment of a patient with cardiac problems?

The right side of the heart receives oxygen-poor (deoxygenated) blood from the veins of the body **Figure 14-2A**. Blood from the vena cava enters the right atrium, which then fills the right ventricle. After contraction of the right ventricle, blood flows into the pulmonary artery and the pulmonary circulation in the lungs, where the blood is oxygenated. As the blood reaches the lungs, it receives fresh oxygen from the alveoli and carbon dioxide waste is removed from the blood and moved into the alveoli. The blood then returns to the heart through the pulmonary veins. The left side of the heart receives oxygen-rich (oxygenated) blood from the lungs through the pulmonary veins **Figure 14-2B**. Blood enters the left atrium and then passes into the left ventricle. This side of the heart is more muscular than the other side because it must pump blood into the aorta and all the other arteries of the body.

The heart contains more than muscle tissue. The heart's electrical system controls heart rate and enables the atria and ventricles to work together **Figure 14-3**. Normal electrical impulses begin in the sinus node, which is in the upper part of the right atrium and is also known as the sinoatrial node. The impulses travel across both atria, causing them to contract. Between the atria and the ventricles, the impulses cross a bridge of special electrical tissue called the atrioventricular node. Here, the signal is slowed for about one to two tenths of a second to allow blood time to pass from the atria to the ventricles. The impulses then exit the atrioventricular node and spread throughout both ventricles via Purkinje fibers, causing the ventricular muscle cells to contract.

Cardiac muscle cells have a special characteristic called **automaticity** that is not found in any other type of muscle cells. Automaticity allows a cardiac muscle cell to contract spontaneously without a stimulus from a nerve source. Normal impulses in the heart start at the sinoatrial node. As long as impulses come from the sinoatrial node, the other myocardial cells will contract when the impulse reaches them. If no impulse arrives, however, the other myocardial cells are capable of creating their own impulses and stimulating a contraction of the heart, although at a generally slower rate.

The stimulus, which originates in the sinoatrial node, is controlled by impulses from the brain, which arrive by way of the **autonomic nervous system**. The

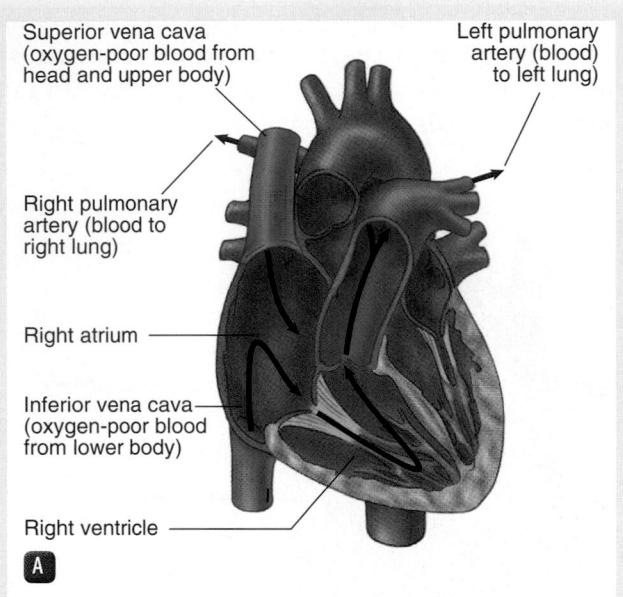

Superior vena cava (oxygen-poor blood from head and upper body)

Left pulmonary artery (blood) to left lung)

Right pulmonary artery (blood to right lung)

Right atrium

Inferior vena cava (oxygen-poor blood from lower body)

Right ventricle

A

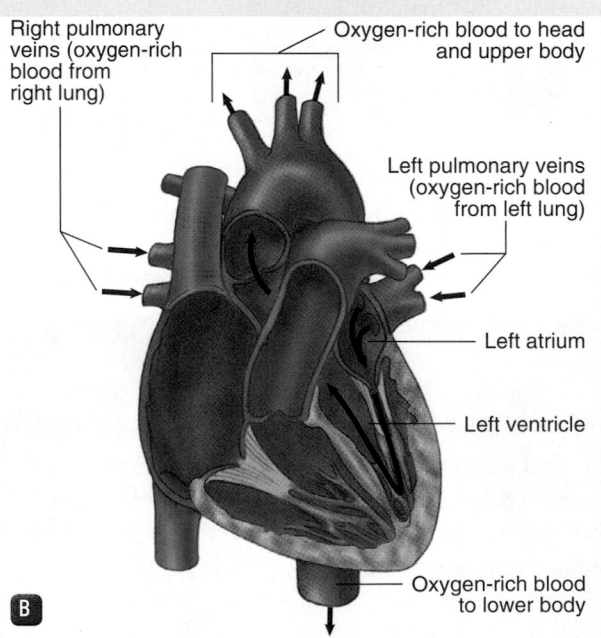

Right pulmonary veins (oxygen-rich blood from right lung)

Oxygen-rich blood to head and upper body

Left pulmonary veins (oxygen-rich blood from left lung)

Left atrium

Left ventricle

Oxygen-rich blood to lower body

B

Figure 14-2 **A.** The right side of the heart receives oxygen-poor blood from the venous circulation. **B.** The left side of the heart receives oxygen-rich blood from the lungs through the pulmonary veins.

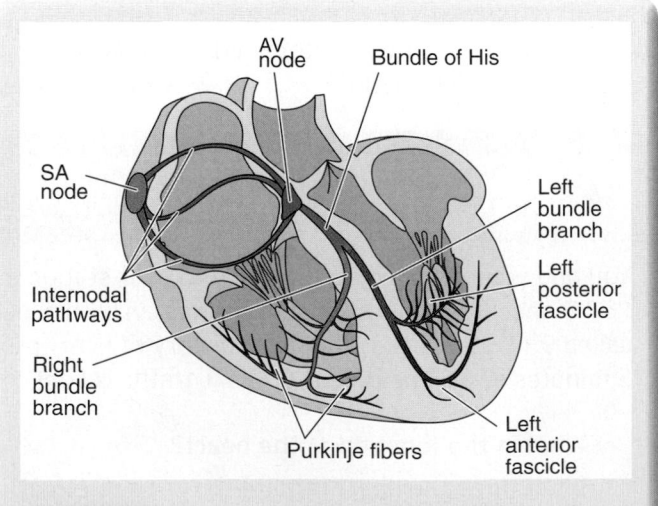

AV node

Bundle of His

SA node

Left bundle branch

Left posterior fascicle

Internodal pathways

Right bundle branch

Purkinje fibers

Left anterior fascicle

Figure 14-3 The electrical conduction system of the heart controls most aspects of heart rate and enables the four chambers to work together.

autonomic nervous system is the part of the brain that controls the functions of the body that do not require conscious thought, such as the heartbeat, respirations, dilation and constriction of blood vessels, and digestion of food. The autonomic nervous system has two parts, the **sympathetic nervous system** and the **parasympathetic nervous system**. The sympathetic nervous system is also known as the "fight-or-flight" system and makes adjustments to the body to allow for physical activity. The sympathetic nervous system speeds up the heart rate, increases respiratory rate and depth, dilates blood vessels in the muscles, and constricts blood vessels in the digestive system. The parasympathetic nervous system directly opposes the sympathetic nervous system. The parasympathetic nervous system slows the heart and respiratory rates, constricts blood vessels in the muscles, and dilates blood vessels in the digestive system. Normally, these two systems balance each other, but in times of stress, the sympathetic nervous system gains primary control, whereas in times of relaxation, the parasympathetic system takes control.

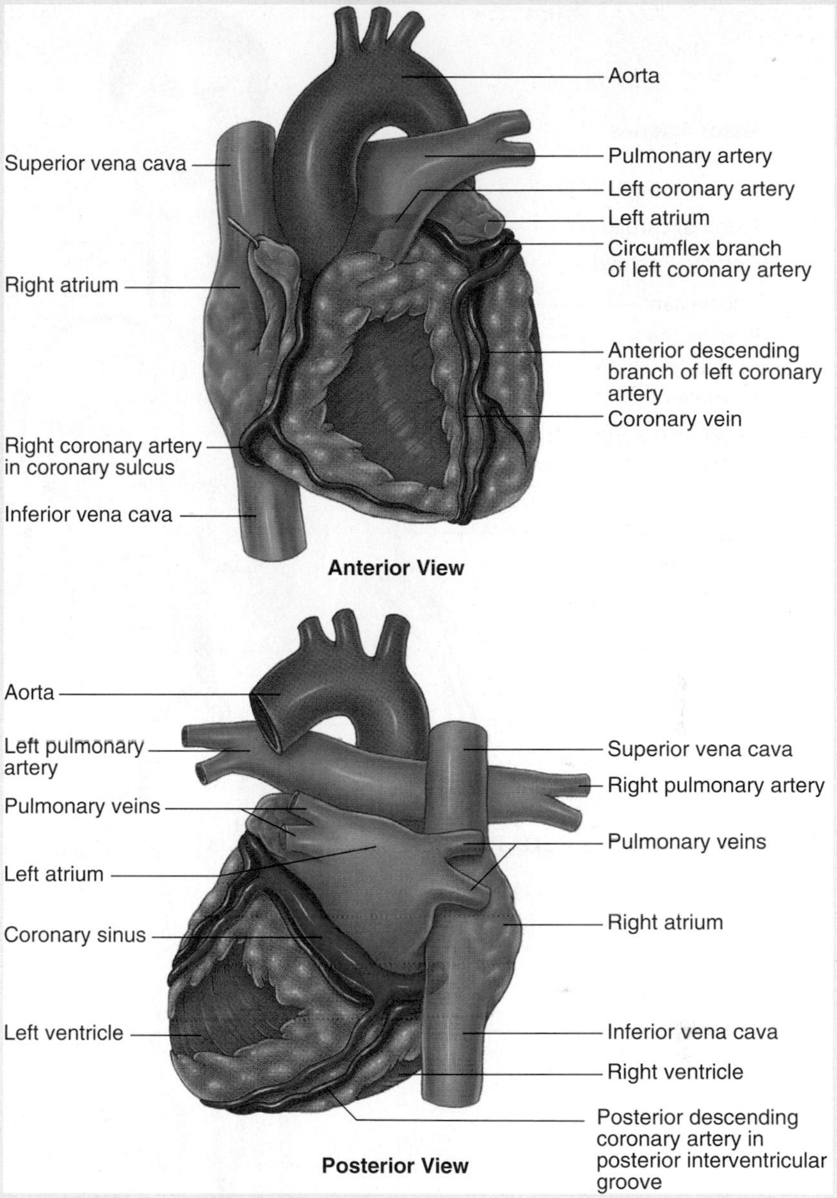

■ Circulation

To carry out its function of pumping blood, the **myocardium**, or heart muscle, must have a continuous supply of oxygen and nutrients. During periods of physical exertion or stress, the myocardium requires more oxygen. The heart must increase cardiac output to meet the increased metabolic requirements of the body. Cardiac output is increased by increasing the heart rate or stroke volume. The **stroke volume** is the volume of blood ejected with each ventricular contraction. In the normal heart, increased oxygen demand of the myocardium itself is supplied by **dilation**, or widening, of the coronary arteries, which increases blood flow. The **coronary arteries** are the blood vessels that supply blood to the heart muscle **Figure 14-4**. They start at the first part of the aorta, just above the **aortic valve**. The right coronary artery supplies blood to the right ventricle and, in most people, the bottom part, or inferior wall, of the left

Figure 14-4 The coronary arteries carry the blood supply to the heart.

ventricle. The left coronary artery divides into two major branches, both of which supply the left ventricle.

Two major arteries branching from the upper aorta supply blood to the head and arms **Figure 14-5**. The right and left carotid arteries supply the head and brain with blood. The subclavian arteries (under the clavicles) supply blood to the upper extremities. As the subclavian artery enters each arm, it becomes the brachial artery, the major vessel that supplies blood to each arm. Just below the elbow, the brachial artery divides into two major branches: the radial and ulnar arteries, supplying blood to the hands.

At the level of the umbilicus, the descending aorta divides into two main branches called the right and left

Major Arteries

Major Veins

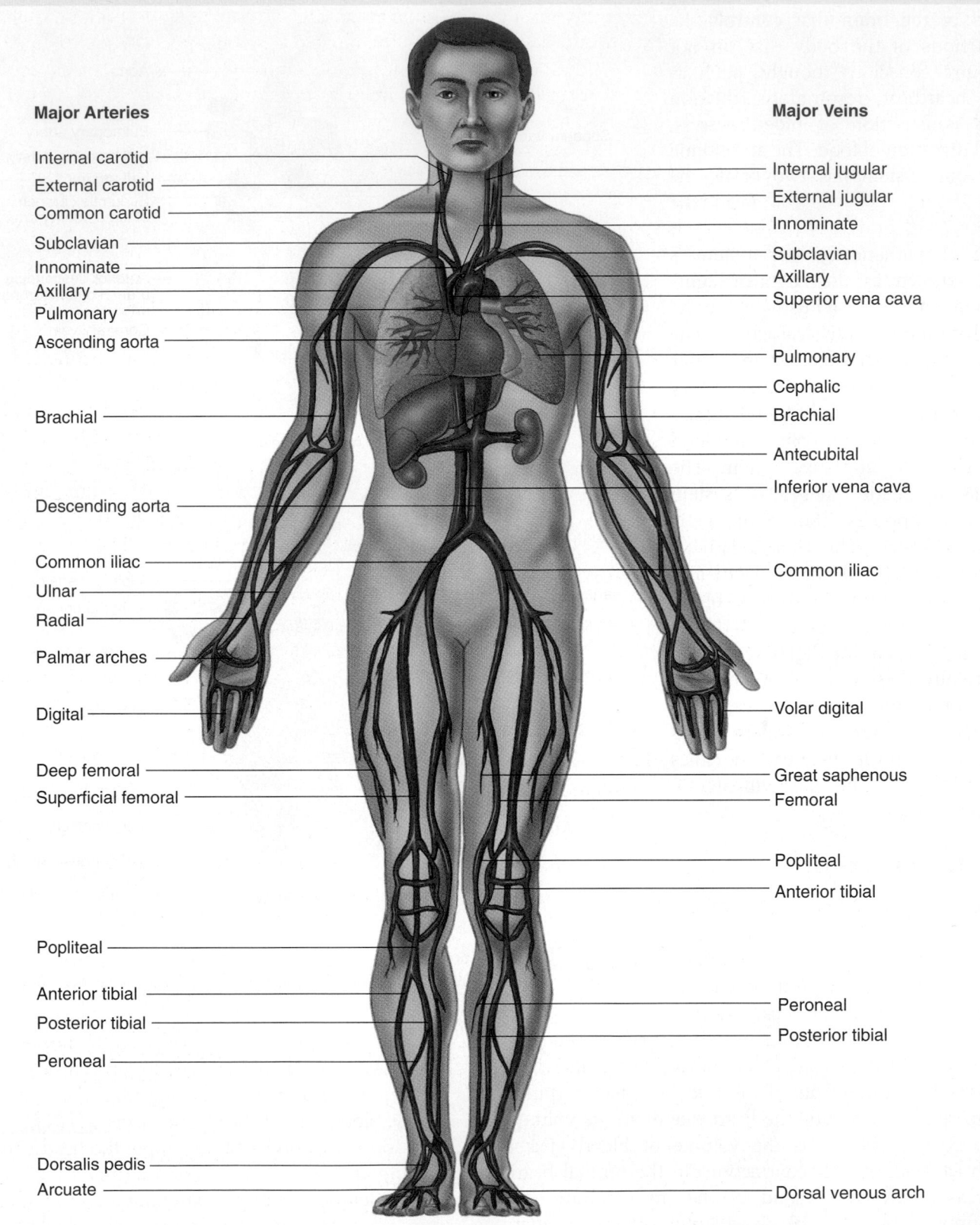

Internal carotid

External carotid

Common carotid

Subclavian

Innominate

Axillary

Pulmonary

Ascending aorta

Brachial

Descending aorta

Common iliac

Ulnar

Radial

Palmar arches

Digital

Deep femoral

Superficial femoral

Popliteal

Anterior tibial

Posterior tibial

Peroneal

Dorsalis pedis

Arcuate

Internal jugular

External jugular

Innominate

Subclavian

Axillary

Superior vena cava

Pulmonary

Cephalic

Brachial

Antecubital

Inferior vena cava

Common iliac

Volar digital

Great saphenous

Femoral

Popliteal

Anterior tibial

Peroneal

Posterior tibial

Dorsal venous arch

Figure 14-5 The major arteries of the body carry oxygen-rich blood to all parts of the body.

iliac arteries, which supply blood to the groin, pelvis, and legs. As the iliac arteries enter the legs through the groin, they become the right and left femoral arteries. At the level of the knee, the femoral artery divides into the **anterior** (front) and **posterior** (back) tibial arteries and the peroneal artery, supplying blood to the feet.

After blood travels through the arteries, it enters smaller and smaller vessels called arterioles and capillaries. The capillaries are tiny blood vessels about one cell thick that connect arterioles to venules. Capillaries, which are found in all parts of the body, allow the exchange of nutrients and waste at the cellular level. As the blood passes through the capillaries, it gives up oxygen to the tissues and picks up carbon dioxide and other waste products to be removed from the body.

Venules are the smallest branches of veins. After traveling through the capillaries, oxygen-poor blood enters the system of veins, starting with the venules, on its way back to the heart. The veins become larger and larger and eventually form the two large venae cavae: the superior vena cava and the inferior vena cava. The **superior** (upper) vena cava carries blood from the head and arms back to the right atrium. The **inferior** (lower) vena cava carries blood from the abdomen, kidneys, and legs back to the right atrium. The superior and inferior venae cavae join at the right atrium of the heart, where blood is eventually returned into the pulmonary circulation for oxygenation.

Blood consists of several types of cells and fluid **Figure 14-6**. Red blood cells are the most numerous and give the blood its color. Red blood cells carry oxygen to the body's tissues and remove carbon dioxide. Larger white blood cells help to fight infection. Platelets, which help the blood to clot, are much smaller than either red or white blood cells. Plasma is the fluid that the cells float in. It is a mixture of water, salts, nutrients, and proteins.

Blood pressure is the pressure of circulating blood against the walls of the arteries. Systolic blood pressure is the maximum pressure generated by the left ventricle as it contracts, which is known as systole. As the left ventricle relaxes in the stage known as diastole, the arterial pressure falls. When the aortic valve closes, blood flow between the left ventricle and the aorta stops. The diastolic blood pressure is the pressure exerted against the walls of the arteries while the left ventricle is at rest. Remember that the top number in a blood pressure reading is the systolic pressure, and the bottom number is the diastolic, or resting, pressure. The cardiac cycle consists of one systolic and one diastolic period.

As the blood passes through an artery during systole and if pressure is placed on the artery by a finger pressing it against a firm surface, a pulse will be felt. Pulses felt in the extremities, such as the radial and the posterior tibial, are called peripheral pulses, whereas pulses near the trunk of the body, such as the femoral and carotid pulses, are known as central pulses.

The rate of cardiac contractions can be increased or decreased by the autonomic nervous system. The heart also has the ability to increase or decrease the volume of blood it pumps with each contraction based on the autonomic nervous system response. To obtain an accurate measure of the efficiency of the heart, we have to measure the volume of blood pumped and the heart rate. This is determined by calculating the cardiac output. The **cardiac output** is calculated by multiplying the heart rate by the volume of blood ejected with each contraction, or the stroke volume. This is the volume of blood that passes through the heart in 1 minute and is the best measure of the output of the heart. In the field, we have no way of directly measuring the volume of blood being pumped; therefore, we must rely on the heart rate and the strength of the pulse to estimate the cardiac output.

The constant flow of oxygenated blood to the tissues is known as **perfusion**. Good perfusion requires three primary components. The first is a well-functioning heart, or "pump." The heart must operate at an appropriate rate because a rate that is too slow or too fast will reduce the volume of blood circulated and, thus, reduce the cardiac output. When the heart beats too rapidly, there is not enough time between contractions for the heart to

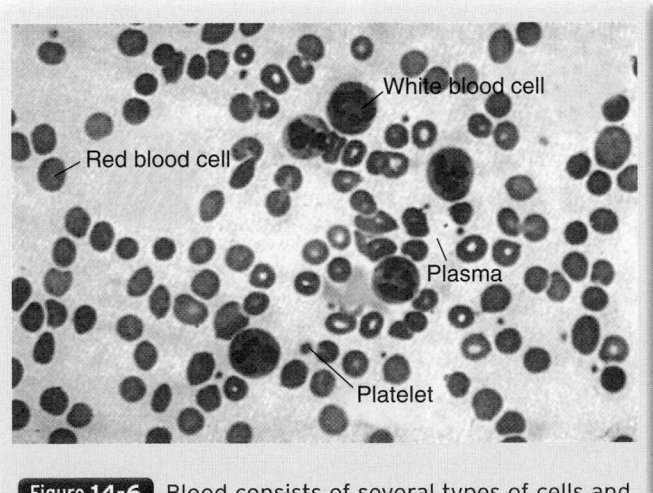

Figure 14-6 Blood consists of several types of cells and fluids, including red blood cells, white blood cells, and platelets.

refill completely, and when the heart beats too slowly, the volume of blood circulated per minute decreases due to the slow pulse rate. The second component of good perfusion is an adequate volume of "fluid," or blood. If blood is lost through hemorrhage, the reduced volume will limit the amount of tissue that can be perfused. Third, the blood must be carried in a properly sized "container." This means that the blood vessels must be appropriately constricted to match the volume of blood available so that circulation can occur without problems. If the blood vessels dilate, thereby increasing the size of the container, perfusion will be reduced. If there is a problem with the functioning of the heart, the functioning of the blood vessels, or the volume of blood, perfusion will fall, which will lead to cellular death and, eventually, death of the patient.

Words of Wisdom

Cardiac Output = Heart Rate × Stroke Volume
Cardiac output is the amount of blood pumped out of the left ventricle in 1 minute.

Heart rate is the number of times the heart contracts in 1 minute.

Stroke volume is the volume of blood pumped out by the left ventricle in one cardiac cycle.

Stroke volume is affected by preload, afterload, and contractility. Preload is related to the venous return to the right atrium. Afterload is associated with systemic vascular resistance, which is a function of the constriction of the systemic blood vessels. As the blood vessels constrict, it is harder for the ventricle to push the blood into them. Contractility refers to how forcefully the heart contracts.

Words of Wisdom

Pulsation

As the left ventricle contracts, it ejects a forceful wave of blood through the arteries. You can feel that wave in areas where the artery lies over a bone and is near the surface of the skin. This wave of blood is called the pulse. The evaluation of a patient's pulse is important in the assessment and treatment of cardiovascular emergencies. EMTs should be skilled at finding multiple pulse points and should compare proximal and distal pulses bilaterally, when applicable, to determine any differences in quality or strength that could indicate the patient's condition is progressing to decompensated shock.

Common places to feel for a pulse include the following **Figure 14-7**:

- The carotid pulse can be felt in the neck, two finger-breadths on either side of the Adam's apple (thyroid cartilage), and should be taken on the side closest to the EMT. Do not assess both carotid pulses at the same time.
- The femoral pulse can be felt in the groin at the crease dividing the lower abdomen from the leg.
- The brachial pulse can be felt on the medial aspect of the elbow at the level of the crease. This is the pulse that you listen for when you take blood pressure. Pulsations also can be palpated on the medial side of the arm between the elbow and armpit.
- The radial pulse can be felt on the thumb side of the wrist, about one finger width above the wrist crease.
- The posterior tibial pulse can be felt on the inside of the ankle, just behind the medial malleolus. The medial malleolus is the bony bump at the end of the tibia.
- The dorsalis pedis pulse can be felt at the top of the foot. This artery is not in the same place in all people. To find the pulse, place your hand across the top of the foot just below the ankle crease. Once you feel something that might be a pulse, use your fingertips to confirm that finding.

Practice feeling for these pulses on yourself and on friends and family members.

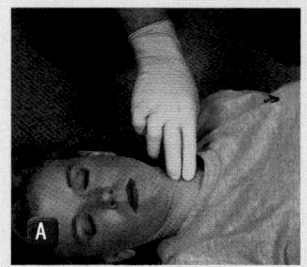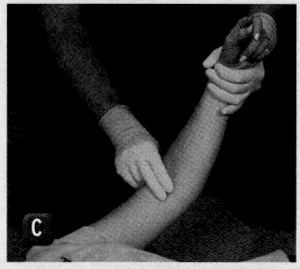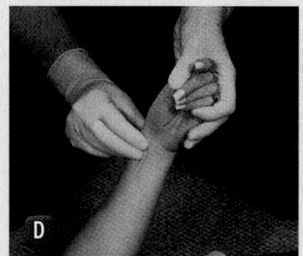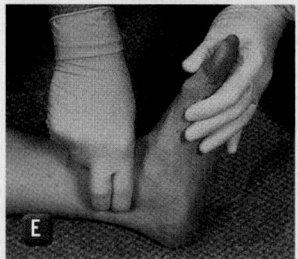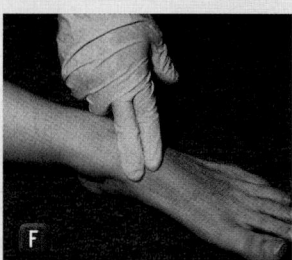

Figure 14-7 Common pulse points. **A.** The carotid pulse is felt in the neck. **B.** The femoral pulse is felt in the groin area. **C.** The brachial pulse can be felt on the inside of the upper arm. **D.** The radial pulse can be felt on the thumb side of the wrist. **E.** The posterior tibial pulse can be felt on the inside of the ankle. **F.** The dorsalis pedis pulse can be felt at the top of the foot.

Pathophysiology

Chest pain or discomfort that is related to the heart usually stems from a condition called <u>ischemia</u>, which is decreased blood flow, in this case, to the heart. Because of a partial or complete blockage of blood flow through the coronary arteries, heart tissue fails to get enough oxygen and nutrients. The tissue soon begins to starve and, if blood flow is not restored, eventually dies. Ischemic heart disease, then, is disease involving a decrease in blood flow to one or more portions of the heart muscle.

■ Atherosclerosis

Most often, the low blood flow to heart tissue is caused by coronary artery atherosclerosis. <u>Atherosclerosis</u> is a disorder in which calcium and a fatty material called cholesterol build up and form a plaque inside the walls of blood vessels, obstructing flow and interfering with their ability to dilate or contract `Figure 14-8`. Eventually, atherosclerosis can even cause complete <u>occlusion</u>, or blockage, of a coronary artery. Atherosclerosis usually involves other arteries of the body as well.

The problem begins when the first deposit of cholesterol is deposited on the inside of an artery. This may happen during the teenage years. As a person ages, more of this fatty material is deposited; the <u>lumen</u>, or the inside diameter of the artery, narrows. As the cholesterol deposits grow, calcium deposits can form as well. The inner wall of the artery, which is normally smooth and elastic, becomes rough and brittle with these atherosclerotic plaques. Damage to the coronary arteries may become so extensive that they cannot accommodate increased blood flow at times of maximum need.

For reasons that are still not completely understood, a brittle plaque will sometimes develop a crack, exposing the inside of the atherosclerotic wall. Acting like a torn blood vessel, the ragged edge of the crack activates the blood-clotting system, just as when an injury has caused bleeding. In this situation, however, the resulting blood clot will partially or completely block the lumen of the artery. If it does not occlude the artery at that location, the blood clot may break loose and begin floating in the blood, becoming what is known as a thromboembolism. A <u>thromboembolism</u> is a blood clot that is floating through blood vessels until it reaches an area too narrow for it to pass, causing it to stop and block the blood flow at that point. Tissues downstream from the blood clot will have a lack of oxygen (hypoxia). If blood flow is resumed in a short time, the hypoxic tissues will recover. However, if too much time goes by before blood flow is resumed, the tissues will die. If a blockage occurs in a coronary artery, the condition is known as an <u>acute myocardial infarction (AMI)</u>, a classic heart attack `Figure 14-9`. <u>Infarction</u> means the death of tissue. The same sequence may also cause the death of cells in other organs, such as the brain. The death of heart muscle can lead to severe diminishment of the heart's ability to pump or cause it to stop completely (<u>cardiac arrest</u>).

In the United States, coronary artery disease is the number one cause of death for men and women. The peak incidence of heart disease occurs between ages

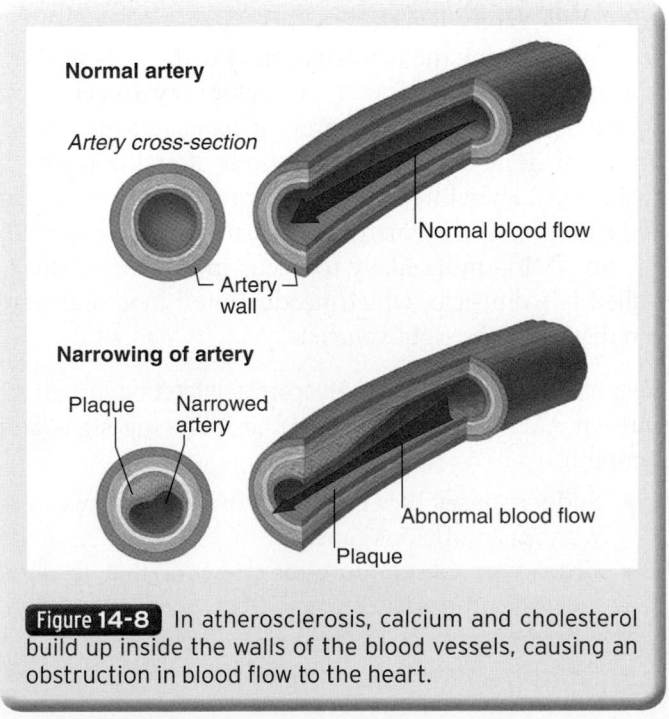

Figure 14-8 In atherosclerosis, calcium and cholesterol build up inside the walls of the blood vessels, causing an obstruction in blood flow to the heart.

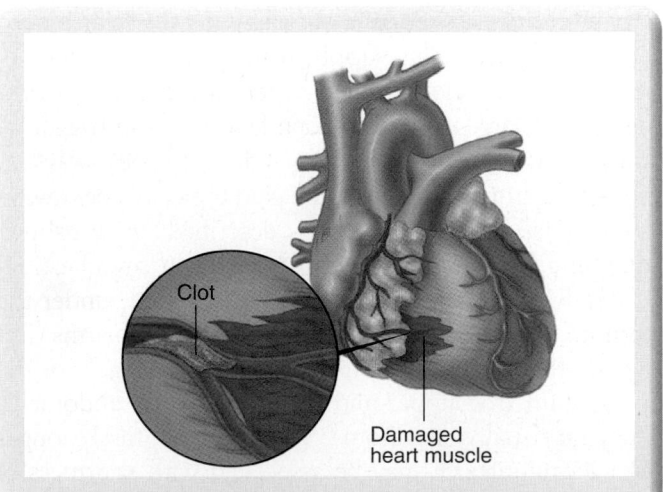

Figure 14-9 An acute myocardial infarction (heart attack) occurs when a blood clot prevents blood flow to an area of the heart muscle. If left untreated, death of heart tissue can result.

40 and 70 years, but it can also strike teens and people in their 90s. You must be alert to the possibility that, although less likely, a 26-year-old with chest pain could actually be having an AMI, especially if he or she has a higher than usual risk.

Factors that place a person at higher risk for an AMI are called *risk factors.* The major controllable factors are cigarette smoking, high blood pressure, elevated cholesterol level, elevated blood glucose level (diabetes), lack of exercise, and stress. The major risk factors that cannot be controlled are older age, family history of atherosclerotic coronary artery disease, and male sex.

■ Acute Coronary Syndrome

Many patients who call for EMS assistance because of chest pain have acute coronary syndrome. **Acute coronary syndrome**, also called ACS, is a term used to describe a group of symptoms caused by myocardial ischemia. As discussed earlier, myocardial ischemia is a decrease in blood flow to the heart, which leads to chest pain through reduction of oxygen and nutrients to the tissues of the heart. This can be a temporary situation known as *angina pectoris*, or a more serious condition, an AMI. Because the signs and symptoms of these two conditions are very similar, they are treated basically the same under the designation of acute coronary syndrome. To understand them better, we will look at each one separately.

Angina Pectoris

Chest pain does not always mean that a person is having an AMI. When, for a brief time, heart tissues are not getting enough oxygen, the pain is called **angina pectoris**, or angina. Although angina can result from a spasm of an artery, it is most often a symptom of atherosclerotic coronary artery disease. Angina occurs when the heart's need for oxygen exceeds its supply, usually during periods of physical or emotional stress when the heart is working hard. A large meal or sudden fear may also trigger an attack. When the increased oxygen demand goes away (eg, the person stops exercising), the pain typically goes away.

Anginal pain is typically described as crushing, squeezing, or "like somebody standing on my chest." It is usually felt in the mid portion of the chest, under the sternum. However, it can radiate to the jaw, the arms (frequently the left arm), the mid portion of the back, or the epigastrium (the upper-middle region of the abdomen). The pain usually lasts from 3 to 8 minutes, rarely longer than 15 minutes. It may be associated with shortness of breath, nausea, or sweating. It disappears promptly with rest, supplemental oxygen, or nitroglycerin, all of which decrease the need or increase the supply of oxygen to the heart. Although angina pectoris is frightening, it does not mean that heart cells are dying, nor does it usually lead to

death or permanent heart damage. It is, however, a warning that you and the patient should take seriously. Even with angina, because the oxygen supply to the heart is diminished, the electrical system can be compromised, and the person is at risk for significant cardiac rhythm problems.

Angina can be further differentiated into "stable" and "unstable" angina. Unstable angina is characterized by pain in the chest of coronary origin that occurs in response to progressively less exercise or fewer stimuli than ordinarily required to produce angina. If untreated, it can lead to AMI. Stable angina is characterized by pain in the chest of coronary origin that is relieved by the things that normally relieve it in a given patient, such as resting or taking nitroglycerin. EMS usually becomes involved when stable angina becomes unstable, such as when a patient whose pain is normally relieved by sitting down and taking one nitroglycerin tablet has taken three tablets with no relief. Keep in mind that it can be difficult, even for physicians in hospitals, to distinguish between the pain of angina and the pain of an AMI. Patients experiencing chest pain, therefore, should always be treated as if they are having an AMI.

Acute Myocardial Infarction

The pain of an AMI signals the actual death of cells in the area of the heart muscle where blood flow is obstructed. Once dead, the cells cannot be revived. Instead, they will eventually turn to scar tissue and become a burden to the beating heart. This is why fast action is so critical in treating a heart attack. The sooner the blockage can be cleared, the fewer the cells that may die. About 30 minutes after blood flow is cut off, some heart muscle cells begin to die. After about 2 hours, as many as half of the cells in the area can be dead; in most cases, after 4 to 6 hours, more than 90% will be dead. In many cases, however, opening the coronary artery with "clot-busting" (thrombolytic) medications or angioplasty (mechanical clearing of the artery) can prevent damage to the heart muscle if done within the first hour after the onset of symptoms. Therefore, immediate prehospital treatment and transport to the emergency department are essential.

An AMI is more likely to occur in the larger, thick-walled left ventricle, which needs more blood and oxygen than does the right ventricle.

Signs and Symptoms of Acute Myocardial Infarction A patient with an AMI may show any of the following signs and symptoms:

- Sudden onset of weakness, nausea, and sweating without an obvious cause
- Chest pain, discomfort, or pressure that is often crushing or squeezing and that does not change with each breath
- Pain, discomfort, or pressure in the lower jaw, arms, back, abdomen, or neck

- Irregular heartbeat and <u>syncope</u> (fainting)
- Shortness of breath, or dyspnea
- Pink, frothy sputum (indicating possible pulmonary edema)
- Sudden death

The Pain of Acute Myocardial Infarction The pain of an AMI differs from the pain of angina in three ways:

- It may or may not be caused by exertion but can occur at any time, sometimes when a person is sitting quietly or even sleeping.
- It does not resolve in a few minutes; rather, it can last between 30 minutes and several hours.
- It may or may not be relieved by rest or nitroglycerin.

Note that not all patients who are having an AMI experience pain or recognize it when it occurs. In fact, about a third of patients never seek medical attention. This can be attributed, in part, to the fact that people are afraid of dying and do not want to face the possibility that their symptoms may be serious (cardiac denial). Middle-aged men, in particular, are likely to minimize their symptoms. However, a few patients, particularly older people, women, and people with diabetes, do not experience any pain during an AMI but have other common complaints associated with ischemia discussed earlier. Others may feel only mild discomfort and call it "indigestion." It is not uncommon for the only complaint, especially in older women, to be fatigue. Heart disease is the number one killer of women in the United States, and EMTs should consider AMI even when the classic symptom of chest pain is not present. This is also true for elderly people and people with diabetes.

Therefore, when you are called to a scene where the chief complaint is chest pain, complete a thorough assessment, no matter what the patient says. Patients with cardiac risk factors should also be carefully assessed if they have any of the associated symptoms, even if no chest pain is present. Any complaint of chest discomfort is a serious matter. In fact, the best thing you can do is to assume the worst.

Physical Findings of Acute Myocardial Infarction and Cardiac Compromise The physical findings of AMI vary, depending on the extent and severity of heart muscle damage. The following are common:

- **General appearance.** The patient often appears frightened. There may be nausea, vomiting, and a cold sweat. The skin is often pale or ashen gray because of poor cardiac output and the loss of perfusion, or blood flow through the tissue. Occasionally, the skin will have a bluish tint, called cyanosis; this is the result of poor oxygenation of the circulating blood.

- **Pulse.** Generally, the pulse rate increases as a normal response to pain, stress, fear, or actual injury to the myocardium. Because arrhythmias are common in an AMI, you may feel an irregularity or even a slowing of the pulse. The pulse may also be dependent on the area of the heart that has been affected by the AMI. Damage to the inferior area of the heart often presents with bradycardia.

Words of Wisdom

Documenting exactly how a patient describes chest discomfort, in the patient's own words, is a valuable source of information for hospital staff. Remember OPQRST (Onset, Provocation/palliation, Quality, Radiation, Severity, Time of onset).

- **Blood pressure.** Blood pressure may fall as a result of diminished cardiac output and diminished capability of the left ventricle to pump. However, most patients with an AMI will have a normal or, most likely, elevated blood pressure.
- **Respiration.** Respirations are usually normal unless the patient has CHF. In that case, respirations may become rapid and labored. A complaint of difficulty breathing is common with cardiac compromise, so even if the rate seems normal, look at the work of breathing, and treat the patient as if respiratory compromise were present.
- **Mental status.** Patients with AMIs sometimes experience an almost overwhelming feeling of impending doom. If a patient tells you, "I think I am going to die," pay attention.

Consequences of Acute Myocardial Infarction An AMI can have three serious consequences:

- Sudden death
- Cardiogenic shock
- CHF

Sudden Death Approximately 40% of all patients with an AMI do not reach the hospital alive. Sudden death is usually the result of cardiac arrest, in which the heart fails to generate effective blood flow. Although you cannot feel a pulse in someone experiencing cardiac arrest, the heart may still be twitching, though erratically. The heart is using up energy without pumping any blood. Such an abnormality of heart rhythm is a ventricular <u>arrhythmia</u>, known as ventricular fibrillation.

A variety of other lethal and nonlethal arrhythmias may follow an AMI, usually within the first hour. In most cases, premature ventricular contractions, or extra beats in the damaged ventricle, occur. Premature ventricular contractions by themselves may be harmless and are common among healthy people, as well as sick people.

Other arrhythmias can be much more dangerous. These include the following Figure 14-10 :

- **Tachycardia**. Rapid beating of the heart, 100 beats/min or more.
- **Bradycardia**. Unusually slow beating of the heart, 60 beats/min or less.
- **Ventricular tachycardia**. Rapid heart rhythm, usually at a rate of 150 to 200 beats/min. The electrical activity starts in the ventricle instead of the atrium. This rhythm usually does not allow adequate time between beats for the left ventricle to fill with blood. Therefore, the patient's blood pressure may fall, and the pulse may be lost altogether. The patient may also feel weak or light-headed or may even become unresponsive. In some cases, existing chest pain may worsen or chest pain that was not there before onset of the arrhythmia may develop. Most cases of ventricular tachycardia will be sustained and may deteriorate into ventricular fibrillation.
- **Ventricular fibrillation**. Disorganized, ineffective quivering of the ventricles. No blood is pumped through the body, and the patient usually becomes unconscious within seconds. The only way to convert this arrhythmia is to defibrillate the heart. To **defibrillate** means to shock the heart with a specialized electrical current in an attempt to stop the chaotic, disorganized contraction of the myocardial cells and allow them to start again in a synchronized manner to restore a normal rhythmic beat. Defibrillation is highly successful in terms of saving a life if delivered within the first few minutes of sudden death. If a defibrillator is not immediately available, CPR must be initiated until the defibrillator arrives. Even if CPR is begun at the time of collapse, chances of survival diminish 10% each minute until defibrillation is accomplished.

If uncorrected, unstable ventricular tachycardia or ventricular fibrillation will eventually lead to **asystole**, the absence of all heart electrical activity. Without CPR, asystole may occur within minutes. Asystole usually reflects a long period of ischemia, and nearly all patients you find in asystole will die.

■ Cardiogenic Shock

Shock is a simple concept but one that few people without medical training really understand. For that reason,

You are the Provider: PART 2

You arrive at the scene and are escorted by the patient's son to her bedroom. She is sitting up in bed with her fist clutched against her chest. She is conscious and alert, but is notably anxious. Her skin is pale and diaphoretic. Your partner opens the jump kit as you assess the patient.

Recording Time: 0 Minutes	
Appearance	Anxious; notably diaphoretic
Level of consciousness	Conscious and alert
Airway	Open; clear of secretions and foreign bodies
Breathing	Increased respiratory rate; adequate depth
Circulation	Radial pulse, rapid and irregular; skin, pale and diaphoretic

After confirming that she has not taken any medication and that she is not allergic to any medications, you give the patient two low-dose aspirins to chew and swallow. As you continue your assessment and further inquire about her medical history, your partner applies high-flow oxygen via nonrebreathing mask and prepares to take her vital signs. She tells you that she had a heart attack 3 years ago; has high blood pressure; and takes enalapril, nitroglycerin, and one aspirin per day.

4. Why is aspirin given to patients with an acute cardiac event?

5. What type of medication is nitroglycerin? How may it help relieve chest pain, pressure, or discomfort?

6. When is nitroglycerin indicated for a patient? What is the typical dose?

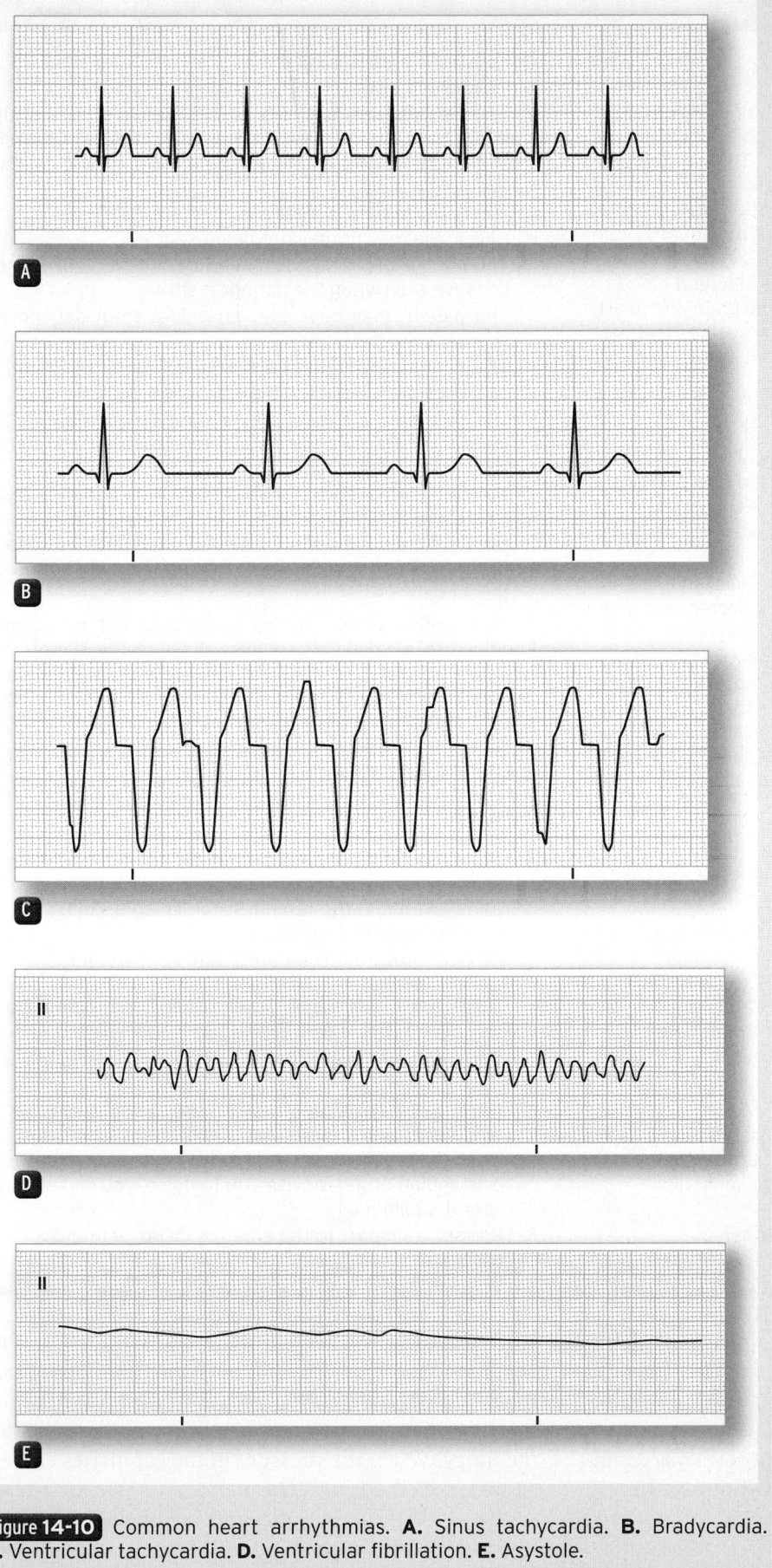

Figure 14-10 Common heart arrhythmias. **A.** Sinus tachycardia. **B.** Bradycardia. **C.** Ventricular tachycardia. **D.** Ventricular fibrillation. **E.** Asystole.

Chapter 10 is devoted to a discussion of shock. The discussion of shock in this chapter is limited to that associated with cardiac problems; however, many other medical problems may cause shock as well.

For EMTs, shock is a critical concept. Shock is present when body tissues do not get enough oxygen, causing body organs to malfunction. In **cardiogenic shock**, often caused by a heart attack, the problem is that the heart lacks enough power to force the proper volume of blood through the circulatory system. Cardiogenic shock is more commonly found in an AMI that affects the inferior and posterior regions of the heart of the left ventricle because this provides circulation to the majority of the body. Cardiogenic shock can occur immediately or as late as 24 hours after the onset of the AMI. The various signs and symptoms of cardiogenic shock are produced by the improper functioning of the body's organs. The challenge for you is to recognize shock in its early stages, when treatment is much more likely to be successful.

■ Congestive Heart Failure

Failure of the heart occurs when the ventricular heart muscle is so damaged that it can no longer keep up with the return flow of blood from the atria. **Congestive heart failure (CHF)** can occur any time after a myocardial infarction, heart valve damage, or long-standing high blood pressure, but it usually happens between the first few hours and the first few days after a heart attack.

Just as the pumping function of the left ventricle can be damaged by coronary artery disease, it can also be damaged by diseased

Words of Wisdom

Cardiogenic Shock
Signs and symptoms

- One of the first signs of shock is anxiety or restlessness as the brain becomes relatively starved for oxygen. The patient may complain of "air hunger." Think of the possibility of shock when the patient is saying that he or she cannot breathe. Obviously, the patient can breathe, because he or she can talk. However, the patient's brain is sensing that it is not getting enough oxygen.
- As the shock continues, the body tries to send blood to the most important organs, such as the brain and heart, and away from less important organs, such as the skin. Therefore, you may see pale, clammy skin in patients with shock.
- As the shock gets worse, the body will attempt to compensate by increasing the amount of blood pumped through the heart. Therefore, the pulse rate will be higher than normal. In severe shock, the heart rate usually, but not always, is greater than 120 beats/min. As the shock progresses, the pulses may become irregular and weak.
- Shock can also be characterized by rapid and shallow breathing, nausea and vomiting, and a decrease in body temperature.
- Finally, as the heart and other organs begin to malfunction, the blood pressure will fall below normal. A systolic blood pressure less than 90 mm Hg is easy to recognize, but it is a late finding that indicates decompensated shock. Do not assume that shock is not present just because the blood pressure is normal (compensated shock).

Treatment of Patients With Cardiogenic Shock
Take the following steps when treating patients with signs and symptoms of cardiogenic shock:

1. Position the patient comfortably. Most patients with heart failure will be more comfortable in a semi-Fowler's position (head and knees slightly elevated); however, patients with low blood pressure may not tolerate a semiupright position but may be more comfortable and more alert in a supine position.
2. Administer high-flow oxygen.
3. Assist ventilations as necessary.
4. Cover the patient with sheets or blankets as indicated to preserve body heat. Be sure to cover the patient's head in cold weather; this is where the most heat is lost.
5. Provide prompt transport to the emergency department.

Congestive Heart Failure
Signs and symptoms

- The patient finds it easier to breathe when sitting up. When the patient is lying down, more blood is returned to the right ventricle and lungs, causing further pulmonary congestion.
- Often, the patient is mildly or severely agitated.
- Chest pain may or may not be present.
- The patient often has distended neck veins that do not collapse even when the patient is sitting.
- The patient may have swollen ankles from dependent edema (backup of fluid).
- The patient generally will have high blood pressure, a rapid heart rate, and rapid respirations.
- The patient will usually be using accessory breathing muscles of the neck and ribs, reflecting the additional hard work of breathing.
- The fluid surrounding small airways may produce rales (crackles), best heard by listening to either side of the patient's chest, about midway down the back. In severe CHF, these soft sounds can be heard even at the top of the lung.

Once CHF develops, it can be treated but not cured. Regular use of medications may alleviate the symptoms. However, patients with CHF often become ill again and are frequently hospitalized. Approximately half will die within 5 years of the onset of symptoms.

Treatment of Congestive Heart Failure
Treat a patient with CHF the same way as a patient with chest pain:

1. Take the vital signs, and give oxygen by nonrebreathing mask with an oxygen flow of 10 to 15 L/min.
2. Allow the patient to remain sitting in an upright position with the legs down.
3. Be reassuring; many patients with CHF are quite anxious because they cannot breathe.
4. Patients who have had problems with CHF before will usually have specific medications for its treatment. Gather these medications, and take them along to the hospital.
5. Nitroglycerin may be of value if the patient's systolic blood pressure is more than 100 mm Hg. If the patient has been prescribed nitroglycerin, and medical control or standing orders advise you to do so, you can administer it sublingually.
6. Prompt transport to the emergency department is essential.

---

heart valves or chronic hypertension. In any of these cases, when the muscle can no longer contract effectively, the heart tries other ways to maintain an adequate cardiac output. Two specific changes in heart function occur: The heart rate increases, and the left ventricle enlarges in an effort to increase the amount of blood pumped each minute.

When these adaptations can no longer make up for the decreased heart function, CHF eventually develops.

It is called "congestive" heart failure because the lungs become congested with fluid once the heart fails to pump the blood effectively. Blood tends to back up in the pulmonary veins, increasing the pressure in the capillaries of the lungs. When the pressure in the capillaries exceeds a certain level, fluid (mostly water) passes through the walls of the capillary vessels and into the alveoli. This condition is called pulmonary edema. It may occur suddenly, as in an AMI, or slowly over months, as in chronic CHF.

Sometimes, in patients with an acute onset of CHF, severe pulmonary edema will develop, in which the patient has pink, frothy sputum and severe dyspnea.

If the right side of the heart is damaged, fluid collects in the body, often showing up as swelling in the feet and legs. The collection of fluid in the part of the body that is closest to the ground is called <u>dependent edema</u>. The swelling causes relatively few symptoms other than discomfort. However, chronic dependent edema may indicate underlying heart disease even in the absence of pain or other symptoms.

■ Hypertensive Emergencies

Hypertension is defined as any systolic blood pressure greater than 140 mm Hg or a diastolic blood pressure greater than 90 mm Hg. Another cardiac-related condition is a hypertensive emergency. A <u>hypertensive emergency</u> usually occurs only with a systolic pressure greater than 160 mm Hg or a rapid rise in the systolic pressure. Because blood pressure cannot be directly felt by the patient, the signs and symptoms of hypertensive emergency are related to the effects of the hypertension. Some patients with chronic hypertension may not experience signs or symptoms until their systolic pressure is significantly higher than this value. One of the most common signs is a sudden severe headache. Often described as "the worst headache I have ever felt," this may also be a sign of cerebral hemorrhage. Other signs and symptoms include strong bounding pulse, ringing in the ears, nausea and vomiting, dizziness, warm skin (dry or moist), nosebleed, altered mental status, and even the sudden development of pulmonary edema. Untreated hypertensive emergencies can lead to a stroke or a dissecting aortic aneurysm.

If you suspect your patient is experiencing a hypertensive emergency, attempt to make him or her comfortable and monitor the blood pressure regularly. Position the patient with the head elevated, and transport rapidly to the emergency department. Depending on the distance and time involved in transport, you should consider ALS assistance for the patient. Paramedics may be able to administer medications to lower the blood pressure to a safer level. If ALS personnel can be on the scene quickly, contact them early and allow them to transport the patient from the scene. If the transport distance is long, consider asking for an ALS unit to meet you along the way and take over patient care and transportation from that point. Remember that getting the patient with a hypertensive emergency to the hospital as quickly and safely as possible is the best prehospital treatment you can provide.

An <u>aortic aneurysm</u> is a weakness in the wall of the aorta. The aorta dilates at the weakened area, which makes it susceptible to rupture. A <u>dissecting aneurysm</u> occurs when the inner layers of the aorta become separated, allowing blood (at high pressures) to flow between the layers. Uncontrolled hypertension is the primary cause of dissecting aortic aneurysms. This separation of layers weakens the wall of the aorta significantly, making it more likely to be ruptured under conditions of continued high blood pressure. If the aorta ruptures, the amount of internal blood loss will be so large that the patient will die almost immediately. The signs and symptoms of a dissecting aortic aneurysm include very sudden chest pain located in the anterior part of the chest or in the back between the shoulder blades. It may be difficult to differentiate the chest pain of a dissecting aortic aneurysm from that of an AMI, but a number of distinctive features may help. The pain from an AMI is often preceded by other symptoms—nausea, indigestion, weakness, and sweating—and tends to come on gradually, getting more severe with time and often described as "pressure" rather than "stabbing." By contrast, the pain of a dissecting aortic aneurysm usually comes on full force from one minute to the next Table 14-1. A patient with a dissecting aortic aneurysm also may

Table 14-1	AMI Versus Dissecting Aortic Aneurysm	
	AMI	**Dissecting Aneurysm**
Onset of pain	Gradual, with additional symptoms	Abrupt, without additional symptoms
Quality of pain	Tightness or pressure	Sharp or tearing
Severity of pain	Increases with time	Maximal from the outset
Timing of pain	May wax and wane	Does not abate once it has started
Region/radiation	Substernal; back is rarely involved	Back possibly involved, between the shoulder blades
Clinical signs	Peripheral pulses equal	Blood pressure discrepancy between arms or decrease in a femoral or carotid pulse

exhibit a difference in blood pressure between arms or diminished pulses in the lower extremities. Aortic aneurysms are almost impossible to diagnose in the prehospital setting, but you must consider them a possibility in any patient with significant hypertension. Transport the patient without delay.

Patient Assessment

While en route to the scene, consider the minimum and maximum standard precautions that will be needed. The precautions can be as simple as gloves for a patient with chest pain or full precautions for a patient in cardiac arrest. Remember, the patient's condition can change rapidly between the time you are dispatched and your arrival.

Scene Size-up

Scene Safety

Do not let your guard down on medical calls. Always ensure that the scene is safe for you, your partner, your patient, and bystanders. As you approach the scene, look for and address any hazards, and assess the scene for the potential of violence. Determine how many patients there are. From the nature of the call and first glance at your patient, determine the necessary standard precautions and whether you will need additional resources to assist in moving the patient(s).

Mechanism of Injury/Nature of Illness

Identification of the nature of illness is important to get your patient assessment started in the right direction. Use the information you get from the dispatcher, clues at the scene, and comments of bystanders or family members to begin to develop an idea about the type of problem your patient might be experiencing. For patients with cardiac problems, the clues often include a report of chest pain, difficulty breathing, or sudden loss of consciousness. Once you establish a preliminary nature of illness, you will be able to guide your assessment to find the important information much more effectively. Just remember not to become fixated on a specific condition at this early point in the assessment; sometimes the situation turns out to be very different from how it initially appeared.

Primary Assessment

Form a General Impression

As you approach the patient, form a general impression of his or her condition to recognize and address life threats. You will likely begin by determining whether the patient is responsive. Perform a rapid scan of the patient. If the patient is unresponsive and is not breathing, begin CPR, starting with chest compressions, and call for an AED. Use of the AED is discussed in the section on cardiac arrest later in this chapter. Generally, an AED should be applied if the patient is pulseless, not breathing (apneic), and unresponsive. Consider calling for ALS backup if needed.

Airway and Breathing

Unless the patient is unresponsive, the airway will most likely be patent. Responsive patients should be able to maintain their own airway. Some episodes of cardiac compromise may produce dizziness or even fainting spells (syncope). If dizziness or fainting has occurred, consider the possibility of a spinal injury from a fall. Assess and treat the patient as appropriate.

Assess the patient's breathing to determine if it is adequate to provide enough oxygen to an ailing heart. If the rate is too fast or too slow, the depth of respiration seems to be too shallow, or the patient is struggling to breathe, respirations are inadequate. Listen for abnormal breath sounds at this time because these can also be important indicators of respiratory distress. Some patients feel short of breath even though there are no obvious signs of respiratory distress. In either situation, apply oxygen with a nonrebreathing mask at 10 to 15 L/min. If the patient is not breathing or has inadequate breathing, ensure adequate ventilations with a bag-mask device and 100% oxygen.

Patients experiencing pulmonary edema may require positive-pressure ventilation with a bag-mask device or continuous positive airway pressure (CPAP). CPAP is the most effective way to assist a person with CHF to breathe effectively and prevent an invasive airway management technique. You should be aware of the indications and contraindications of CPAP and be competent in utilizing this equipment. Once intubated, these patients may be difficult to wean from the ventilator.

Circulation

Assess the patient's circulation. Determine the rate and quality of the patient's pulse. Is the pulse rhythm regular or irregular? Is the pulse too fast or too slow? If you find abnormalities in the pulse, you should be more suspicious. Assess the patient's skin condition, color, moisture, and temperature, as well as the capillary refill time. Changes in perfusion may indicate more serious cardiac compromise. Begin treatment for cardiogenic shock early to reduce the workload of the heart. Place the patient in a comfortable position, usually sitting up and well supported. Provide reassurance that appropriate treatment is being given for the condition to reduce the patient's anxiety. Make a very quick scan of the body to see if there is any major bleeding that needs to be controlled or any significant edema in the extremities. If severe bleeding is identified, use direct pressure to control the bleeding, and bandage appropriately.

Transport Decision

Make a transport decision based on whether you were able to stabilize life threats during the primary assessment. The remainder of the assessment can be performed en route, if time allows. Generally speaking, most patients with chest pain should be transported immediately. Whether to transport using the lights and siren is determined for each patient individually and may be partially based on the estimated transport time. As a general rule, however, patients with cardiac problems should be transported in the most gentle, stress-relieving manner possible. Very little time is saved by the using the lights and siren, but you can do a lot to calm your patient and reduce the release of heart-damaging adrenaline through your reassurance and by creating a ride to the hospital that is as pleasant as possible. Try not to allow the patient to exert himself or herself, strain, or walk. If necessary, lift the patient, using care.

Your decision of where to transport the patient to will depend on your local protocol. Patients are generally transported to the closest appropriate facility. If your service is served by one hospital, the transport decision is easy. In larger urban areas, there may be several hospitals within the service areas. Some medical directors have written protocols requiring patients with suspected cardiac emergencies to be transported to medical centers with certain capabilities, such as emergency angioplasty. Others require the patient to be transported to the nearest facility for stabilization before transporting to a specialty hospital. Be sure you know your local protocol.

Words of Wisdom

Athletes may have a slower (bradycardic < 60 beats/min) heart rate as a normal physiologic change. Tachycardia (> 100 beats/min) is a normal physiologic response to exercise to ensure adequate tissue perfusion. Pain, fear, and excitement may also cause a person to be tachycardic. It is important that you assess the patient to determine whether a bradycardic or tachycardic heart rate is appropriate for the patient.

History Taking

Investigate Chief Complaint

Once you have stabilized life threats, you will want to determine and investigate the chief complaint and know more about the history of the present illness. For a conscious medical patient, begin with taking a brief past history, identifying associated signs and symptoms, and identifying pertinent negatives. Friends or family members who are present often have helpful information.

Remember that not all patients experiencing an AMI have the same signs and symptoms. A chief complaint

You are the Provider: PART 3

The patient took two of her prescribed nitroglycerin tablets before her son called 9-1-1; however, she is still experiencing chest pain, which she rates as a 7 on a scale of 0 to 10. Your partner takes her vital signs as you perform a more focused examination, inquire about her past medical history, and prepare for further treatment.

Recording Time: 2 Minutes	
Respirations	20 breaths/min; adequate depth
Pulse	118 beats/min; strong and irregular
Skin	Pale, cool, and diaphoretic
Blood pressure	150/90 mm Hg
Oxygen saturation (Sao$_2$)	98% (on oxygen)

7. What is significant about the patient's vital signs?
8. Should you give her additional nitroglycerin? Why or why not?

of chest pain or discomfort, shortness of breath, or dizziness should be taken seriously. Many patients who suspect that something is wrong experience restlessness, appear anxious, and perhaps have a sense of impending doom. Act professionally; be calm. Speak to the patient in a normal voice that is neither too loud nor too soft. Let the patient know that trained responders, including you, are present to provide care and that he or she will soon be taken to the hospital. Remember, some patients may act carefree, while others may be demanding. Most patients, however, are frightened. Your professional attitude may be the single most important factor in winning the patient's cooperation and helping the patient through this event. Patients often have a good idea about what is happening, so do not lie and offer false reassurance. If asked, "Am I having a heart attack?" you can say, "I do not know for sure, but in case you are, we are taking care of you. We are going to help you now by giving oxygen, and we will be taking you to the hospital. You are in good hands."

Begin by asking questions about the current situation. Determine whether the patient is experiencing chest pain or discomfort and whether there are any other signs and symptoms. Determine whether the patient is having respiratory difficulty because this is common among patients with chest pain. If the patient is experiencing dyspnea, find out whether it is related to exertion and whether it is related to the patient's position. Often patients with chest pain experience worse difficulty breathing when they are lying down. Also determine whether the dyspnea is continuous or if it changes, especially with deep breathing. Note whether the patient has a cough and whether the cough produces sputum. Ask about other signs and symptoms that are commonly found such as nausea and vomiting, fatigue, headache, and palpitations (a feeling of the heart "skipping a beat" or racing). Make sure to ask about any trauma the patient might have experienced during the last few days. Be sure to record your findings, including those that are negative (known as pertinent negatives).

SAMPLE History

If the patient is responsive, begin obtaining the SAMPLE history and asking the following questions specific to a cardiovascular emergency:

- Have you ever had a heart attack?
- Have you been told that you have heart problems?
  - Have you ever been diagnosed with angina, heart failure, or heart valve disease?
  - Have you ever had high blood pressure?
  - Have you ever been diagnosed with an aneurysm?
  - Do you have any respiratory diseases such as emphysema or chronic bronchitis?
  - Do you have diabetes or have you ever had any problems with your blood sugar?
  - Have you ever had kidney disease?
- Do you have any risk factors for coronary artery disease, such as smoking, high blood pressure, or high-stress lifestyle?
  - Is there a family history of heart disease?
  - Do you currently take any medications?

The SAMPLE history provides basic information on the patient's overall medical history. You will want to determine as many signs and symptoms as you can. For example, you may determine that the patient has chest pain at rest or absence of chest pain with respirations or movement. The more signs and symptoms a patient has, the easier it is to identify a particular problem. In addition, ask whether the patient has had the same pain before. If so, ask "Do you take any medications for the pain?" and "Do you have any of the medication with you?" If the patient has had a heart attack or angina before, ask whether the pain is similar.

Make sure to ask about allergies because the patient will very likely be given medication in the hospital. If the patient is taking medications, determine whether they are prescribed, over the counter, and/or recreational drugs. Even when a patient may not be able to articulate his or her exact medical condition, knowing the patient's medications may give you important clues. For example, a patient may say he has "heart problems." You see that he is taking furosemide (Lasix), digoxin, and amiodarone (Cordarone). Furosemide is a diuretic, digoxin increases the strength of heart contractions, and amiodarone controls certain types of arrhythmias. These drugs are most often prescribed together for patients with CHF and may alert you to carefully evaluate the lungs for the presence of rales or crackles, which indicate fluid in the lungs and a need to increase the amount of oxygen being delivered. When you ask about medical conditions next, be sure to ask whether the patient takes medications for any other condition he or she identifies. Also if the patient tells you that he or she takes prescription medications, always ask what condition these are taken for. Asking about the last oral intake may seem unnecessary, but this information can be very important; it is always better to have too much information rather than not enough. Also remember to ask about any "home remedies" the patient might have used because this information can be important too.

OPQRST

Be sure to include the OPQRST questions when you are obtaining the symptoms as part of the SAMPLE history. Using OPQRST helps you to understand the details of specific complaints, such as chest pain **Table 14-2**.

Table 14-2   OPQRST Mnemonic for Assessing Pain

Onset	When did the problem begin, and what caused it?
Provocation/palliation	Ask what makes the pain or discomfort worse. Is it positional? Does a deep breath or palpation of the chest make it worse? Did you take anything for it (including anything nonprescribed)?
Quality	Ask what type of pain it is. Let the patient use his or her own words to describe what is happening. Try to avoid supplying the patient with only one option. Do not ask, "Does it feel like an elephant is sitting on your chest?" Instead, say, "Tell me what the pain feels like." If the patient cannot answer an open-ended question, then provide a list of alternatives: "There are lots of different kinds of pain. Is your pain more like heaviness, pressure, burning, tearing, dull ache, stabbing, or needlelike?"
Region/radiation	Ask where the pain is and whether the pain travels to another part of the body.
Severity	Ask the patient to rate the pain on a simple scale. Often, a scale ranging from 0 to 10 is used; a 10 represents the worst pain imaginable. Do not use the patient's answer to determine whether the pain has a serious cause. Instead, use it to check whether the pain is getting better or worse. After a few minutes of oxygen or administration of nitroglycerin, ask the patient to rate the pain again.
Timing	Find out how long the pain lasts when it is present and whether it has been intermittent or continuous.

Secondary Assessment

The purpose of the secondary assessment is to perform a physical examination of the patient. The physical examination may be a systematic full-body scan or an assessment that focuses on a certain area or region of the body. For example, the assessment of a conscious patient with chest pain would likely focus on the patient's cardiac and respiratory systems, which are the most commonly affected in patients with chest pain. Circumstances will dictate which aspects of the physical examination will be used.

Physical Examinations

A physical examination of a patient with chest pain would focus primarily on the cardiovascular system. Evaluate the patient's circulation by assessing pulses at various locations, and assess skin color, temperature, and condition. Is the skin cool or moist? How do the mucous membranes look? Are they pink, ashen, or cyanotic? Are the pulses of equal strength bilaterally? Does the patient have any edema in the extremities, especially the lower extremities? All of these physical findings can help identify poor circulation, which may be caused by a failure of the cardiovascular system.

In addition to the cardiovascular system, examine the respiratory system for signs of inadequate ventilation. These two systems are closely related, and some problems with the respiratory system can be caused by cardiovascular issues. Are the lung sounds clear? Wet-sounding lungs indicate fluid is being moved into the lungs from the circulatory system, possibly because of a problem with the heart. Are the breath sounds equal? Are the neck veins distended? Is the trachea deviated, or is it midline? The answers to these questions can help determine whether a problem exists with the lungs or with the heart. While the physical examination is not usually as important as the history in a patient with a possible cardiac problem, it may produce important clues to the patient's condition.

Vital Signs

Measure and record the patient's vital signs, including airway, breathing, and circulation. You must obtain readings for systolic and diastolic blood pressures. If available, use pulse oximetry. Pulse oximetry may not give an accurate measurement if the patient has poor circulation, has been exposed to a toxic chemical, or is in cardiac arrest, but it should be used and the readings noted for all patients with possible cardiac problems.

If you have access to continuous blood pressure monitoring, be sure to use it as well, making sure you get an accurate manual blood pressure first. Repeat the vital signs at appropriate intervals, and use the settings on the automatic blood pressure monitoring machine to remind you when it is time to recheck and record the vital signs. Be sure to note the time that each set of vital signs is taken.

Reassessment

Repeat the primary assessment by checking to see whether the patient's chief complaint and condition have improved or are deteriorating. Vital signs should be reassessed at least every 5 minutes or any time significant changes in the patient's condition occur. It is essential to monitor the patient with a suspected AMI closely because sudden cardiac arrest is always a risk. If cardiac arrest occurs, you must be ready to begin automated defibrillation or chest compressions immediately. If an AED is immediately available, use it; if not, perform CPR until the AED is available, as discussed in the later section on cardiac arrest. Reassess your interventions to see whether they are helping and whether the patient's condition is improving. Reassessment will also determine whether further interventions are indicated or contraindicated.

Interventions

Your treatment of the patient begins with proper positioning. As mentioned before, some patients will not tolerate being positioned supine, so they should be allowed to sit up (leaning back on the stretcher). Also loosen tight clothing, trying to make the patient as comfortable as possible.

You should be giving the patient oxygen by this time, but if you are not, then you should do it now. For patients with mild dyspnea, a nasal cannula may be all that is needed, whereas patients with more serious respiratory difficulty will respond better to a nonrebreathing mask. A patient who is unconscious or in obvious respiratory distress may need assistance with breathing. Use a bag-mask device or a positive-pressure ventilation device such as positive end-expiratory pressure, CPAP, bilevel positive airway pressure, or a manual or automatic transport ventilator if available and you have been approved to use one of these methods in your service. Be aware that ALS may be required to support the use of positive end-expiratory pressure, CPAP, bilevel positive airway pressure, and transport ventilators.

Depending on local protocol, prepare to administer low-dose (sometimes called baby or children's) aspirin and assist with prescribed nitroglycerin. Aspirin (acetylsalicylic acid) prevents clots from forming or getting bigger. Administer low-dose aspirin according to local protocol. Low-dose aspirin comes in 81-mg chewable tablets. The recommended dose is 162 mg (two tablets) to 324 mg (four tablets). Be sure you have verified that the patient is not allergic to aspirin before you give it, because many people are. Also, ask the patient if he or she has any history of internal bleeding such as stomach ulcers, and, if so, contact medical control before giving the patient aspirin.

Nitroglycerin relives the pain of angina. Nitroglycerin comes in several forms—as a small white pill, placed sublingually (under the tongue); as a spray, also taken sublingually; and as a skin patch applied to the chest Figure 14-11 . In any form, the effect is the same. Nitroglycerin relaxes the muscle of blood vessel walls, dilates coronary arteries, increases blood flow and the supply of oxygen to the heart muscle, and decreases the workload of the heart. Nitroglycerin also dilates blood vessels in other parts of the body and can sometimes cause low blood pressure and/or a severe headache. Other side effects include changes in the patient's pulse rate, including tachycardia or bradycardia. You should therefore take the patient's blood pressure within 5 minutes after each dose. If the systolic blood pressure is less than 100 mm Hg, do not give more medication. Other contraindications include the presence of a head injury, use of erectile dysfunction drugs within the previous 24 hours, and the maximum prescribed dose has already been given (usually three doses).

Check the condition of the medication and its expiration date, and do not administer contaminated or expired medications. Also make sure the medication is prescribed for your patient. Occasionally, patients will try to take medications prescribed for their spouse or a friend if they think it will help them. Be sure to wear gloves when handling nitroglycerin tablets, or spray because it is easily absorbed through the skin. If you handle tablets with bare fingers or get the spray on your fingers, it may be absorbed into your body, causing you to experience a very painful headache. If the patient has a nitroglycerin patch on when you arrive, be sure to carefully remove it if the patient is hypotensive or in cardiac arrest (before use of AED).

After you obtain permission from medical control, help the patient administer prescribed nitroglycerin. Nitroglycerin works in most patients within 5 minutes.

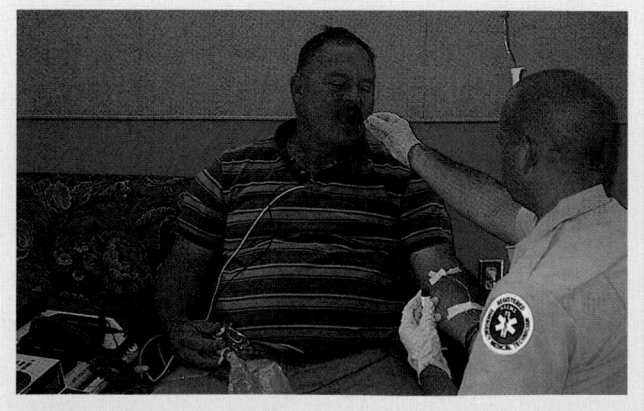

Figure 14-11 Nitroglycerin used to treat angina pectoris comes in many forms, including tablets and sprays.

Most patients who have been prescribed nitroglycerin carry a supply with them. Nitrostat is one trade name for nitroglycerin. Patients take one dose of nitroglycerin under the tongue whenever they have an episode of angina that does not immediately go away with rest. If the pain is still present after 5 minutes, patients are typically instructed by their physicians to take a second dose. If the second dose does not work, most patients are told to take a third dose and then call for EMS. If the patient has not taken all three doses, you can help to administer the medication, if you are allowed to do so by local protocol.

Be aware that nitroglycerin will lose its potency over time, especially if exposed to light. Patients who take it only rarely may keep a bottle in their pocket for months. It may lose its potency even before its expiration date. When the nitroglycerin tablet loses its potency, patients may not feel the fizzing sensation when the tablet is placed under their tongue, and they may not experience the normal burning sensation and headache that often accompany nitroglycerin administration. Note that the fizzing only occurs with a potent tablet, not with the spray form.

To safely assist the patient with nitroglycerin, follow the steps listed below Skill Drill 14-1:

1. Obtain an order from medical control—online or off-line protocol.
2. Take the patient's blood pressure. Administer nitroglycerin only if the systolic blood pressure is greater than 100 mm Hg (Step 1).
3. Check that you have the right medication, the right patient, and the right delivery route. Check the expiration date. Make sure the patient has no contraindications, such as having taken medication for erectile dysfunction in the past 24 hours.
4. Ask the patient about the last dose he or she took and its effects. Make sure that the patient understands the route of administration. Be prepared to have the patient lie down to prevent fainting if the nitroglycerin substantially lowers the patient's blood pressure (the patient gets dizzy or feels faint) (Step 2).
5. Ask the patient to lift his or her tongue. Place the tablet or spray the dose under the tongue (while wearing gloves), or have the patient do so. Have the patient keep his or her mouth closed with the tablet or spray under the tongue until it is dissolved and absorbed. Caution the patient against chewing or swallowing the tablet (Step 3).

6. Recheck the blood pressure within 5 minutes. Record the medication and the time of administration. Reevaluate the chest pain, and note the response to the medication. If the chest pain persists and the patient still has a systolic blood pressure greater than 100 mm Hg, repeat the dose every 5 minutes as authorized by medical control. In general, a maximum of three doses of nitroglycerin is given for any one episode of chest pain (Step 4).

Transport the patient. Early, prompt transport to the emergency department is critical so that treatments such as clot-busting medications or angioplasty can be initiated. To be most effective, these treatments must be started as soon as possible after the onset of the attack. If the patient does not have prescribed nitroglycerin, complete your patient assessment and prepare to transport. Be sure that this process does not consume too much time. Do not delay transport to assist with administration of nitroglycerin. The drug can be given en route.

Communication and Documentation

Alert the emergency department staff about the status of your patient's condition and your estimated time of arrival. Report to the hospital by radio or cellular telephone while en route. Include information about the patient's history, vital signs, the reassessment of vital signs, medications taken, and any treatment you are giving. Follow the instructions of medical control. Describe the patient's condition to the emergency department staff on arrival.

It is important to document your assessment of the patient. You must record the interventions performed. All interventions should be initiated according to protocol. If the intervention required an order from medical control, document the intervention and/or medication requested and whether approval was granted. It must be clear in your documentation that the patient was reassessed appropriately following any intervention. The patient's response to the intervention and the time of each intervention must also be recorded. On completing your documentation, obtain the medical control physician's signature (if required by local protocol) showing approval of medication administration.

Heart Surgeries and Pacemakers

During the last 20 years, hundreds of thousands of open-heart surgeries have been performed to bypass damaged segments of coronary arteries in the heart. In a coronary artery bypass graft, a blood vessel from the chest or leg is sewn directly from the aorta to a coronary artery beyond the point of the obstruction. Another procedure is the percutaneous transluminal coronary angioplasty, which aims to dilate, rather than bypass, the coronary artery. In this procedure, usually called an angioplasty or balloon angioplasty, a tiny balloon is attached to the end of a long, thin tube. The tube is introduced through the skin into a large artery, usually in the groin, and then threaded into

Skill Drill 14-1

Administration of Nitroglycerin

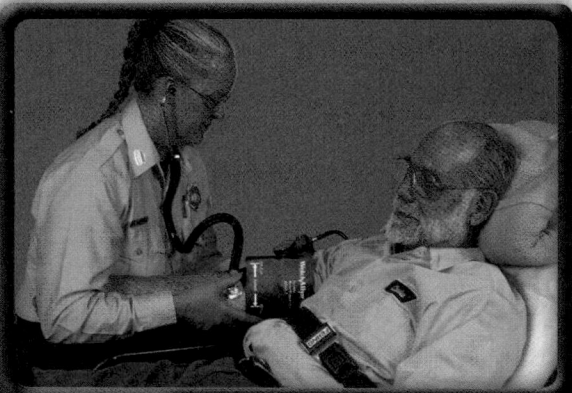

Step 1 Obtain an order from medical control. Take the patient's blood pressure. Administer nitroglycerin only if the systolic blood pressure is greater than 100 mm Hg.

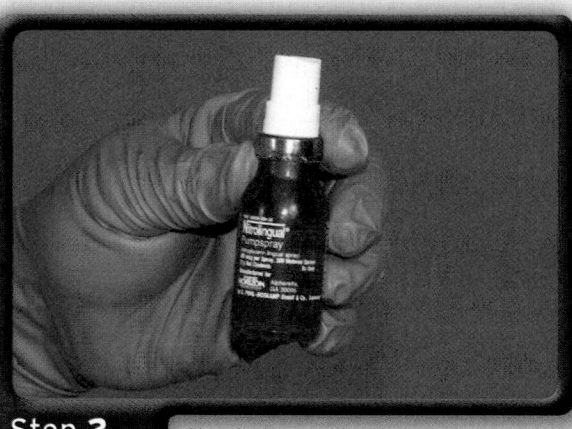

Step 2 Check the medication and expiration date. Ask the patient about the last dose he or she took and its effects. Make sure that the patient understands the route of administration. Prepare to have the patient lie down to prevent fainting.

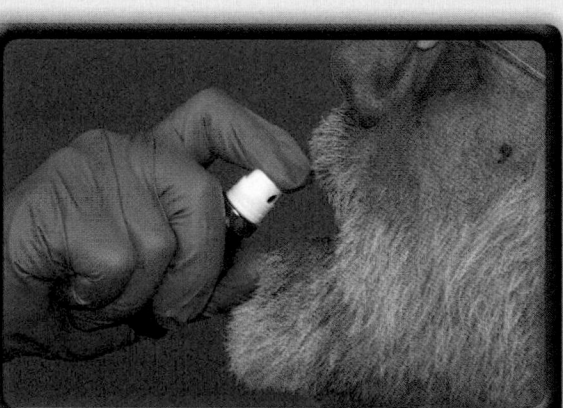

Step 3 Ask the patient to lift his or her tongue. Place the tablet or spray the dose under the tongue (while wearing gloves), or have the patient do so. Have the patient keep his or her mouth closed with the tablet or spray under the tongue until it is dissolved and absorbed. Caution the patient against chewing or swallowing the tablet.

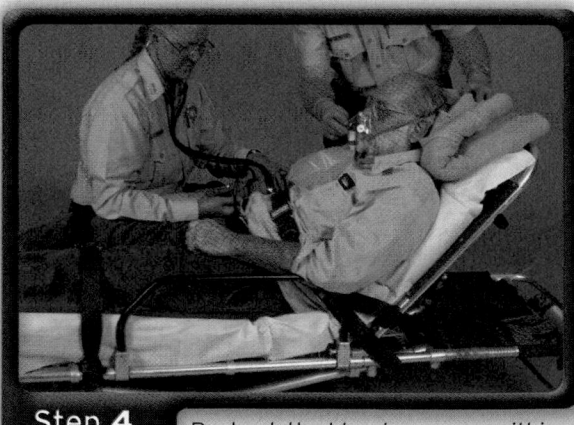

Step 4 Recheck the blood pressure within 5 minutes. Record each medication and the time of administration. Reevaluate the chest pain and blood pressure, and repeat treatment if necessary.

the narrowed coronary artery, with radiographs serving as a guide. Once the balloon is in position inside the coronary artery, it is inflated. The balloon is then deflated, and the tube is removed from the body. Sometimes, a metal mesh called a stent is placed inside the artery instead of or after the balloon. The stent is left in place permanently to help keep the artery from narrowing again.

A patient who has had an AMI or angina will almost certainly have had one of these procedures. Patients who have had a bypass graft will have a long surgical scar on the chest from the operation. Patients who have had an angioplasty or a coronary artery stent usually will not. However, newer "keyhole" surgical techniques may not produce a large scar. You should not assume that a patient who has a small scar has not had bypass surgery. Chest pain in a patient who has had any of these procedures should be treated in the same manner as chest pain in patients who have not had any heart surgery. Carry out all the described tasks, and transport the patient promptly to the emergency department of the hospital. If CPR is required, perform it in the usual way, regardless of the scar on the patient's chest. Likewise, if indicated, an AED should be used as well.

Many people with heart disease in the United States have cardiac pacemakers to maintain a regular cardiac rhythm and rate. Pacemakers are inserted when the electrical control system of the heart is so damaged that it cannot function properly. These battery-powered devices deliver an electrical impulse through wires that are in direct contact with the myocardium. The generating unit is generally placed under a heavy muscle or a fold of skin; it typically resembles a small silver dollar under the skin in the left upper portion of the chest **Figure 14-12**.

Normally, you do not need to be concerned about problems with pacemakers. Thanks to modern technology, an implanted unit will not require replacement or a battery charge for years. Wires are well protected and rarely broken. In the past, pacemakers sometimes malfunctioned when a patient got too close to an electrical radiation source, such as a microwave oven, but this is no longer the case. Every patient with a pacemaker should be aware of the precautions, if any, that must be taken to maintain its proper functioning.

If a pacemaker does not function properly, as when the battery wears out, the patient may experience syncope, dizziness, or weakness because of an excessively slow heart rate. The pulse ordinarily will be less than 60 beats/min because the heart is beating without the stimulus of the pacemaker and without the regulation of its own electrical system, which may be damaged. In these circumstances, the heart tends to assume a fixed slow rate that is not fast enough to allow the patient to function normally. A patient with a malfunctioning pacemaker should be promptly transported to the emergency department; repair of the problem may require surgery. When an AED is used, the patches should not be placed directly over the pacemaker. This will ensure a better flow of electricity through the patient's body.

■ Automatic Implantable Cardiac Defibrillators

More and more patients who survive cardiac arrest due to ventricular fibrillation have a small automatic implantable cardiac defibrillator implanted. Some patients who are at particularly high risk for a cardiac arrest have them as well. These devices are attached directly to the heart and can prolong the lives of certain patients. They continuously monitor the heart rhythm, delivering shocks as needed. Regardless of whether a patient having an AMI has an automatic implantable cardiac defibrillator, he or she should be treated like all other patients having an AMI. Treatment should include performing CPR and using an AED if the patient goes into cardiac arrest. Generally, the electricity from an automatic implantable cardiac defibrillator is so low that it will have no effect on rescuers and, therefore, should not be of concern to you **Figure 14-13**.

■ Cardiac Arrest

Cardiac arrest is the complete cessation of cardiac activity—electrical, mechanical, or both. It is indicated in the field by the absence of a carotid pulse. Until the advent of CPR and external defibrillation in the 1960s, cardiac arrest was virtually always a terminal event. Now, although it is still infrequent for a patient to survive a cardiac arrest without neurologic damage, great strides have been made in resuscitation science during the last 40 years.

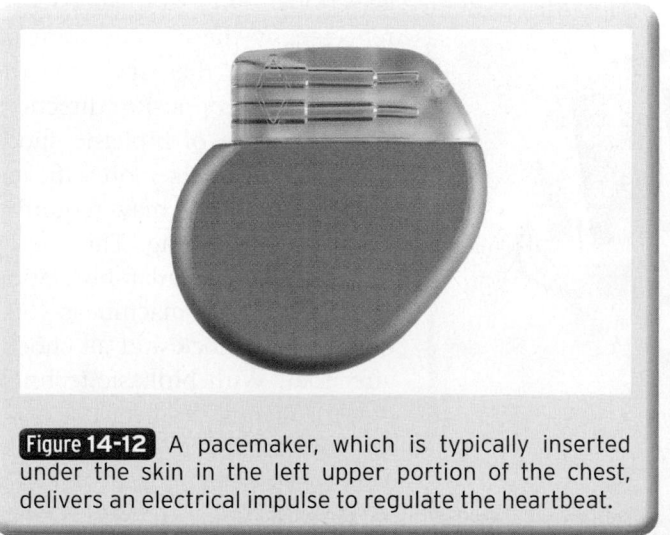

Figure 14-12 A pacemaker, which is typically inserted under the skin in the left upper portion of the chest, delivers an electrical impulse to regulate the heartbeat.

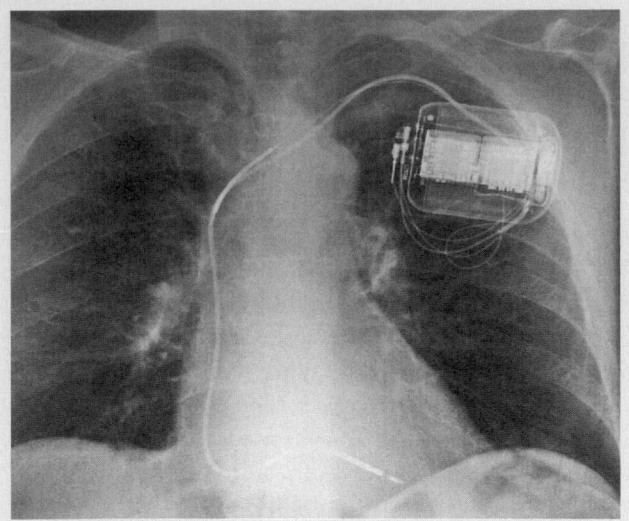

Figure 14-13 An automatic implantable cardiac defibrillator is attached directly to the heart and continuously monitors heart rhythm, delivering shocks as needed. The electricity from the defibrillator is so low that it has no effect on rescuers.

Special Populations

A pediatric patient with chest pain is not a common call. It is usually associated with a pediatric patient who has a preexisting condition, usually congenital. In pediatric situations, it is vital that EMTs see family members or caregivers as a valuable source of information.

Cardiac arrest in infants and children is usually the result of respiratory failure, not a primary cardiac event. However, the American Heart Association has determined that AEDs are safe to use in infants and children. If the patient is 8 years old or less, pediatric-sized pads and a dose-attenuating system (energy reducer) are preferred. However, if these are unavailable, a regular adult AED can be used. If the child is between 1 month and 1 year of age (an infant), a manual defibrillator is preferred to an AED. If a manual defibrillator is not available, an AED equipped with a pediatric dose attenuator is preferred. If neither is available, an AED without a pediatric dose attenuator may be used.

■ Automated External Defibrillation

In the late 1970s and early 1980s, scientists developed a small computer that could analyze electrical signals from the heart and determine when ventricular fibrillation was taking place. This development, along with improved battery technology, made the automated portable defibrillator possible—a device that can automatically administer an electrical shock to the heart when needed.

The AED machines come in different models with different features **Figure 14-14**. All of them require a certain degree of operator interaction, beginning with applying the pads and turning the machine on. The operator also has to push a button to deliver an electrical shock, regardless of the model. Many AEDs use a computer voice synthesizer to advise the EMT which steps to take on the basis of the AED's analysis. Some have a button that tells the computer to analyze the heart's electrical rhythm; other models start doing this as soon as they are turned on. In the United States, the majority of the AEDs are semiautomated. Even though most defibrillators are now semiautomated, we are using the term automated external defibrillator (AED) as the general term to describe all of these machines. There are few actual AEDs left; all manufacturers are producing only semiautomated external defibrillators.

AEDs also come equipped to give a monophasic shock or a biphasic shock. Monophasic means to send the energy in one direction, from negative to positive. Current in the biphasic waveform flows in positive and negative directions. This two-directional flow of current is reflected by the current going in one direction, then reversing the flow in the opposite direction. The advantage of biphasic shock is that it produces more efficient defibrillation and may require a lower energy setting. The energy setting for ventricular fibrillation on a monophasic machine is 360 J for the first shock and all shocks after that. With biphasic technology, the energy can be set at 120 J for the first shock and all shocks after that or can start at 120 J for the first shock and then escalate

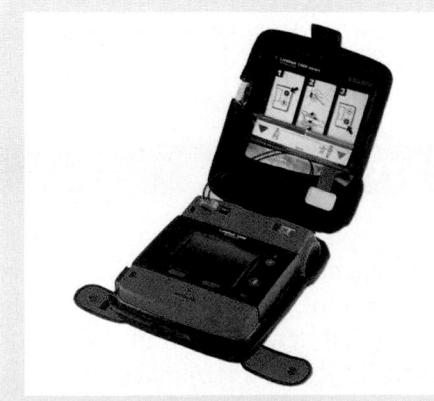
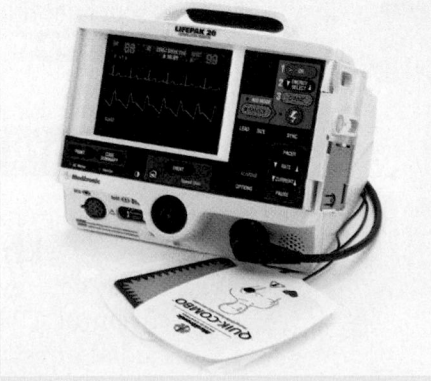

Figure 14-14 Automated external defibrillators vary in their design, features, and operation.

to 200 J for subsequent shocks. The optimum energy setting for the biphasic AED is still being studied, and no recommendation is currently supported in the literature. The computer inside the AED is specially programmed to recognize rhythms that require defibrillation to correct, most commonly ventricular fibrillation. The AEDs are extremely accurate. It would be extremely rare for the AED to recommend a shock when a shock is not required, and they rarely fail to recommend one when it would be helpful. Therefore, if the AED recommends a shock, you can believe that it is indicated.

When an error occurs, it is usually the operator's fault. The most common error is not having a charged battery. To avoid this problem, many defibrillator companies have built smarter machines that will warn the operator that the battery is unlikely to work. However, some of the older models do not have this feature. You should check the AED daily and exercise the battery as often as the manufacturer recommends.

Another error occurs when the AED is applied to a patient who is moving. The computer may be unable to tell the difference between electrical signals from the heart and electrical signals from the arms and chest muscles that are moving. The way to avoid this error is to apply the AED only to pulseless, unresponsive patients and to stay clear of the patient (do not touch the patient) during analysis and shocking.

A third error can occur when the AED is applied to a responsive patient with a rapid heart rate. Most computers identify a regular rhythm faster than 150 or 180 beats/min as ventricular tachycardia, which should be shocked. Sometimes, though, a patient has another heart rhythm that should not be shocked but that is fast enough to confuse the computer. Again, to avoid this problem, you should apply the AED only to unresponsive patients with no pulse.

Automated external defibrillation offers EMTs a number of advantages. First, of course, the machine is fast, and it delivers the most important treatment for a patient in ventricular fibrillation: an electrical shock. It can be delivered within 1 minute of your arrival at the patient's side. Second, you will find that an AED is easy to operate. ALS providers do not have to be on the scene to provide this definitive care.

Current AEDs offer two other advantages. The shock can be given through remote, adhesive defibrillator pads, which are safer for you than paddles. Also, the pad area is larger than paddles, which means that the transmission of electricity is more efficient. Usually, there are pictures on the pads to remind you where they go on the patient's chest. As a safety measure, make sure the patient is not lying on wet ground or touching metal objects when he or she is being shocked.

Not all patients in cardiac arrest require an electrical shock. Although the cardiac rhythm of all patients in cardiac arrest should be analyzed with an AED, some do not have shockable rhythms (eg, pulseless electrical activity and asystole). Asystole (flatline) indicates that no electrical activity remains. Pulseless electrical activity usually refers to a state of cardiac arrest despite an organized electrical complex. In both cases, CPR should be initiated as soon as possible beginning with chest compressions.

You are the Provider: PART 4

After completing the remainder of your assessment and initial treatment, you place the patient onto the stretcher, load her into the ambulance, and proceed to a hospital located 20 miles away. You ask your partner to notify the hospital to give the staff a "heads up" as you reassess the patient.

Recording Time: 10 Minutes	
Level of consciousness	Conscious and alert; still anxious
Respirations	18 breaths/min; adequate depth
Pulse	84 beats/min; strong and irregular
Skin	Pale and cool; less diaphoretic
Blood pressure	136/84 mm Hg
Sao$_2$	96% (on oxygen)

9. Why is early notification of the receiving facility so important for patients with an acute coronary event?

10. Should you apply the AED to determine if she is experiencing a cardiac arrhythmia? Why or why not?

Rationale for Early Defibrillation

Few patients who experience sudden cardiac arrest outside a hospital survive unless a rapid sequence of events takes place. The chain of survival is a way of describing the ideal sequence of events that can take place when such an arrest occurs.

The five links in the chain of survival are as follows **Figure 14-15**:

- Recognition of early warning signs and immediate activation of EMS
- Early CPR with emphasis on chest compressions
- Early defibrillation
- Early advanced cardiac life support
- Integrated post-arrest care

If any one of the links in the chain is absent, the patient is more likely to die. For example, few patients benefit from defibrillation when more than 10 minutes elapse before administration of the first shock or if CPR is not performed in the first 2 to 3 minutes. If all links in the chain are strong, the patient has the best possible chance of survival. The link that is the best determinant for survival is the third link—early defibrillation.

CPR helps patients in cardiac arrest because it prolongs the period during which defibrillation can be effective. Rapid defibrillation has successfully resuscitated many patients with cardiac arrest due to ventricular fibrillation. However, defibrillation works best if it takes place within 2 minutes of the onset of the cardiac arrest. To try to achieve better survival rates among cardiac arrest victims, many communities are exploring the idea that nontraditional first responders should be trained to administer early defibrillation. These responders would include police officers, security personnel, lifeguards, maintenance workers, and flight attendants. As an EMT, you should support these efforts to shorten the interval until defibrillation. Remember, seconds really matter when a patient is in cardiac arrest.

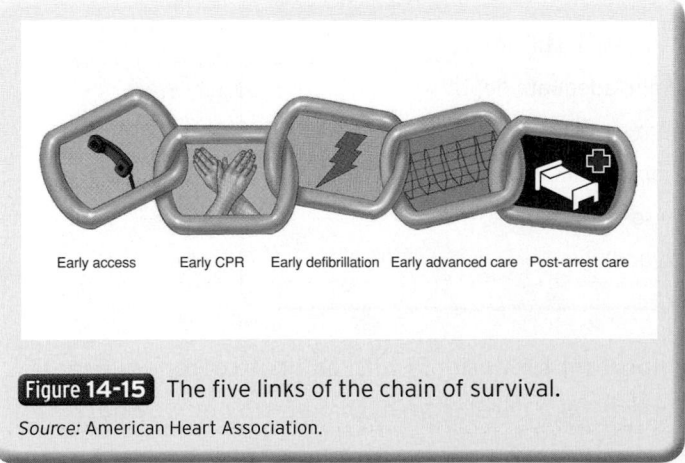

Early access   Early CPR   Early defibrillation   Early advanced care   Post-arrest care

Figure 14-15 The five links of the chain of survival.
Source: American Heart Association.

The final step in the chain of survival is integrated post-arrest care. This refers to controlling temperature to optimize neurologic recovery in the field and maintaining glucose levels in the patient who is hypoglycemic. It also includes cardiopulmonary and neurologic support at the hospital as well as other advanced assessment techniques and interventions when indicated.

Integrating the Automated External Defibrillator and Cardiopulmonary Resuscitation

Because most cardiac arrests occur in the home, a bystander at the scene may already have started CPR before you arrive. For this reason, you must know how to work the AED into the CPR sequence. Remember that the AED is not very complex; it may not be able to distinguish other movements from ventricular fibrillation. Therefore, do not touch the patient while the AED is analyzing the heart rhythm and delivering shocks. Stop CPR, and let the AED do its job.

Automated External Defibrillator Maintenance

One of your primary missions as an EMT is to deliver an electrical shock to a patient in ventricular fibrillation. To accomplish this mission, you need to have a functioning AED. You must become familiar with the maintenance procedures required for the brand of AED your service uses. Read the operator's manual. If your defibrillator does not work on the scene, someone will want to know what went wrong. That person may be your system's administrator, your medical director, the local newspaper reporter, or the family's attorney. You will be asked to show proof that you maintained the defibrillator properly and attended any mandatory in-service sessions.

The main legal risk in using the AED is failing to deliver a shock when one was needed. The most common reason for this failure is that the battery did not work, usually because it was not properly maintained. Another problem is operator error. This means not pushing the analyze or shock buttons when the machine advises you to do so or failing to apply the AED to a patient in cardiac arrest. Of course, the AED is like any other manufactured item. It can fail, although this is rare. Ideally, you will encounter any such failure while doing routine maintenance, not while caring for a patient in cardiac arrest. Check your equipment, including your AED, at the beginning of each shift. Ask the manufacturer for a checklist of items that should be checked daily, weekly, or less often **Figure 14-16**.

If the AED fails while you are caring for a patient, you must report the problem to the manufacturer and the US Food and Drug Administration. Be sure to follow

AUTOMATED EXTERNAL DEFIBRILLATOR
Inspection Checklist

Serial # _____   Date _____   Time _____

Model # _____   Inspected by _____

Item	Pass	Fail
Exterior/Cables		
Nothing stored on top of unit		
Carry case intact and clean		
Exterior/LCD screen clean and undamaged		
Cables/connectors clean and undamaged		
Cables securely attached to unit		
Batteries		
Unit charger is plugged in and operational (if applicable)		
Fully charged battery in unit		
Fully charged spare battery		
Spare battery charger plugged in and operational (if applicable)		
Valid expiration date on both batteries		
Supplies		
Two sets of electrodes		
Electrodes in sealed packages with valid expiration dates		
Razor		
Hand towel		
Alcohol wipes		
Memory/voice recording device—module, card, microcassette		
Manual override—module, key (if applicable)		
Printer paper (if applicable)		
Operation		
Unit self-test per manufacturer's recommendation/instructions		
Display (if applicable)		
Visual indicators		
Verbal prompts		
Printer (if applicable)		
Attach AED to simulator/tester		
Recognizes shockable rhythm		
Charges to correct energy level within manufacturer's specifications		
Delivers charge		
Recognizes nonshockable rhythm		
Manual override system in working order (if applicable)		

Signature:

Figure 14-16 A sample checklist for the automated external defibrillator (AED).

Special Populations

Like the other body systems, the cardiovascular system undergoes changes as we get older. The heart, like other major organs, will show the effects of aging. As the heart's muscle mass and tone decrease, the amount of blood pumped out of the heart per beat is decreased. The residual (reserve) capacity of the heart is also reduced; therefore, when the vital organs of the body need additional blood flow, the heart cannot meet the need. When blood flow to the tissues is decreased, the organs suffer. If blood flow to the brain is inadequate, the patient may complain of weakness, fatigue, or dizziness and may experience syncope (fainting).

The power to the heart muscle can fail. The heart runs on electricity and has its own electrical system. Under normal conditions, electrical impulses travel throughout the heart, resulting in the contraction of the heart muscle and the pumping of blood from the heart's chambers. With aging, the electrical system can deteriorate, causing the heart's contraction to weaken or, if blood flow to the heart muscle is affected, extra beats to form. With decreased strength of contraction, the heartbeat is weaker and blood flow to the tissues is reduced. If extra beats are produced, the patient's heart rhythm will be irregular. Although some irregular heart rhythms are not harmful, others can be potentially lethal.

The arteries are also affected by aging. Arteriosclerosis (hardening of the arteries) can develop, affecting perfusion of the tissues. There is an increased chance of heart attack or stroke due to decreased blood flow or plaque formation (atherosclerosis) in the narrowed arteries.

Patients with diabetes can experience reduced circulation to the hands and feet, which makes peripheral pulses harder to detect. It also puts the hands and feet at particular risk for infection and ulceration.

In some older patients with angina or AMI, particularly people with diabetes, chest pain is absent, and the clinical picture can be confused with other, noncardiac conditions.

The cardiovascular system is affected by aging. You should be aware of the changes, seeking to determine what is normal versus what is chronic versus what is an acute condition for the individual patient. Sometimes, the weakening of the heart muscle, the deterioration of its electrical system, and the hardening of the arteries make the task of assessing and caring for older patients more difficult.

the appropriate EMS procedures for notifying these organizations.

Medical Direction

Defibrillation of the heart is a medical procedure. Although AEDs have made the process of delivering electricity much simpler, there is still a benefit in having a physician's involvement. The medical director of your service or the instructor he or she designates should help to teach you how to use the AED. At the very least, he or she should approve the written protocol that you will follow in caring for patients in cardiac arrest. In most states, completion of AED training in an EMT course is not permitted without approval by state laws, rules, and local medical direction authority.

There should be a review of each incident in which the AED is used. After returning from the hospital or the scene, discuss with the rest of the team what happened. This discussion will help all members of the team learn from the incident. Review such events by using the written report, any voice-electrocardiographic tape recorder, and the device's solid-state memory modules and magnetic tape recordings, if applicable.

There should also be a review of the incident by your service's medical director or quality improvement officer. Quality improvement involves people using AEDs and the responsible EMS system managers. This review should focus on speed of defibrillation, that is, the time from the call to the shock. Few systems will achieve the ultimate goal: shocking 100% of patients within 1 minute of the call. However, all systems continuously work on improving patient care. Mandatory continuing education with skill competency review is generally required for EMS providers, with a continuing competency skill review every 3 to 6 months for EMTs.

Emergency Medical Care for Cardiac Arrest

Preparation

When dispatch reports an unresponsive patient with CPR being performed, the AED is probably one of the first pieces of equipment you will obtain from the ambulance. As the operator of the AED, you are responsible for making sure that the electricity injures no one, including yourself. Remote defibrillation using pads allows you to distance yourself safely from the patient. As long as you place the pads in the correct position and make sure no one is touching the patient, you should be safe. Do not defibrillate a patient who is in pooled water. Although there is some danger to you if you are also in the water, there is another problem. Electricity follows the path of least resistance; instead of traveling between the pads and through the patient's heart, it will diffuse into the water. Therefore, the heart will not receive enough electricity to cause defibrillation. You can defibrillate a soaking wet patient, but try first to dry the patient's chest. Do not defibrillate someone who is touching metal that others are touching, and carefully remove a nitroglycerin patch from a patient's chest and wipe the area with a dry towel before defibrillation to prevent ignition of the patch. It is often helpful to shave a hairy patient's chest before pad placement to increase conductivity. Be sure to consult local protocols for issues such as pad placement and preparation of the pad site.

Determine the nature of illness and/or mechanism of injury. If the incident involves trauma, perform spinal stabilization as you begin the primary assessment. Is there only one patient? If you are in a tiered system and the patient is in cardiac arrest, call for ALS assistance.

■ Performing Defibrillation

If you witness a patient's cardiac arrest, begin CPR starting with chest compressions and attach the AED as soon as it is available. However, if the patient's cardiac arrest was not witnessed, especially if the call-to-arrival interval is greater than 5 minutes, you should perform five cycles (about 2 minutes) of CPR before applying the AED. The rationale for this is that the heart is more likely to respond to defibrillation within the first few minutes of the onset of ventricular fibrillation. If the arrest interval is prolonged, however, metabolic waste products accumulate within the heart, energy stores are rapidly depleted, and the chance of successful defibrillation is reduced. Therefore, a 2-minute period of CPR before applying the AED to patients with prolonged (> 5 minutes) cardiac arrest can "prime the pump," thus restoring oxygen to the heart, removing metabolic waste products, and increasing the chance of successful defibrillation. The steps for using the AED are listed here and shown in **Skill Drill 14-2**:

Skill Drill 14-2

1. If CPR is in progress, assess the effectiveness of chest compressions by palpating for a carotid or femoral pulse. It is important to limit the amount of time compressions are interrupted. If the patient is responsive, do not apply the AED.

2. If the patient is unresponsive and CPR has not been started yet, begin providing chest compressions and rescue breaths at a ratio of 30 compressions to 2 breaths, continuing until an AED arrives and is ready for use **Step 1**. It is important to start chest compressions and use the AED as soon as possible. Compressions provide vital blood flow to the heart and brain, improving the patient's chance of survival.

3. Turn on the AED **Step 2**. Remove clothing from the patient's chest area. Apply the pads to the chest: one just to the right of the breastbone (sternum) just below the collarbone (clavicle), the other on the left lower chest area with the top of the pad 2″ to 3″ below the armpit. Do not place the pads on top of breast tissue. If necessary, lift the breast out of the way and place the pad underneath. Ensure that the pads are attached to the patient cables

You are the Provider: PART 5

The patient is still conscious and alert and appears less anxious. She tells you that her chest pain has decreased in severity and is now a 3 on a 0 to 10 scale. After reassessing her, you contact the receiving facility and give the staff a patient update.

Recording Time: 17 Minutes	
Level of consciousness	Conscious and alert; less anxious
Respirations	16 breaths/min; adequate depth
Pulse	80 beats/min; strong and irregular
Skin	Pink, cool, and dry
Blood pressure	128/78 mm Hg
Sao$_2$	98% (on oxygen)

You deliver the patient to the emergency department, where the cardiac team greets you and assumes care of the patient. The physician obtains a 12-lead electrocardiogram and determines that she is experiencing an acute myocardial infarction. Within 15 minutes, she is taken to the cardiac catheterization laboratory, where two coronary stents were successfully placed.

11. What is the difference between angina pectoris and an acute myocardial infarction?

12. Can an EMT distinguish angina pectoris from an acute myocardial infarction?

(and that they are attached to the AED in some models). Plug in the pads connector to the AED.

4. Stop CPR.

5. State aloud, "Clear the patient," and ensure that no one is touching the patient.

6. Push the Analyze button, if there is one, and wait for the AED to determine whether a shockable rhythm is present.

7. If a shock is not advised, perform five cycles (about 2 minutes) of CPR beginning with chest compressions and then reanalyze the cardiac rhythm. If a

Skill Drill 14-2

AED and CPR

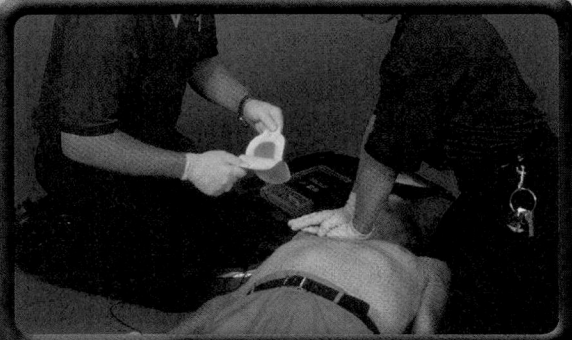

Step 1 Assess compression effectiveness if CPR is already in progress. If the patient is unresponsive and CPR has not been started yet, begin providing chest compressions and rescue breaths at a ratio of 30 compressions to two breaths, continuing until an AED arrives and is ready for use.

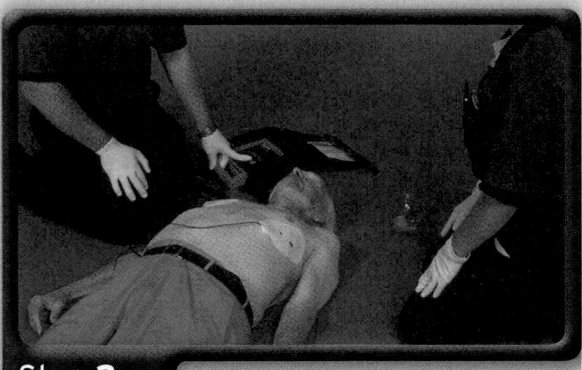

Step 2 Turn on the AED. Apply the AED pads to the chest and attach the pads to the AED. Stop CPR.

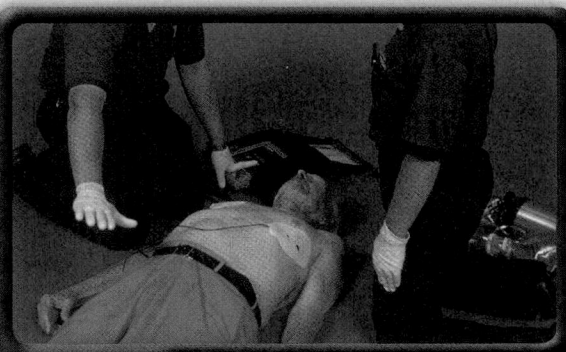

Step 3 Verbally and visually clear the patient. Push the Analyze button, if there is one. Wait for the AED to analyze the cardiac rhythm. If no shock is advised, perform five cycles (2 minutes) of CPR and then reanalyze the cardiac rhythm. If a shock is advised, recheck that all are clear, and push the Shock button. After the shock is delivered, immediately resume CPR beginning with chest compressions.

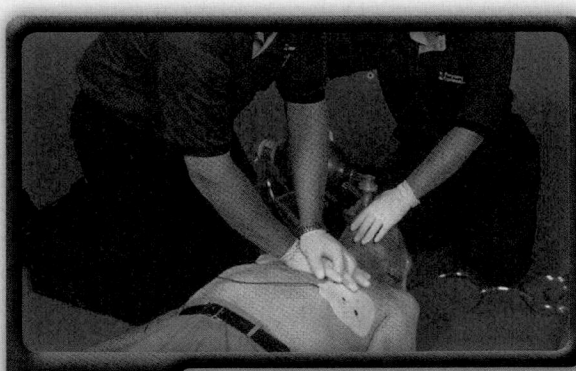

Step 4 After five cycles (2 minutes) of CPR, reanalyze the cardiac rhythm. Do not interrupt chest compressions for more than 10 seconds.

Skill Drill 14-2

AED and CPR, continued

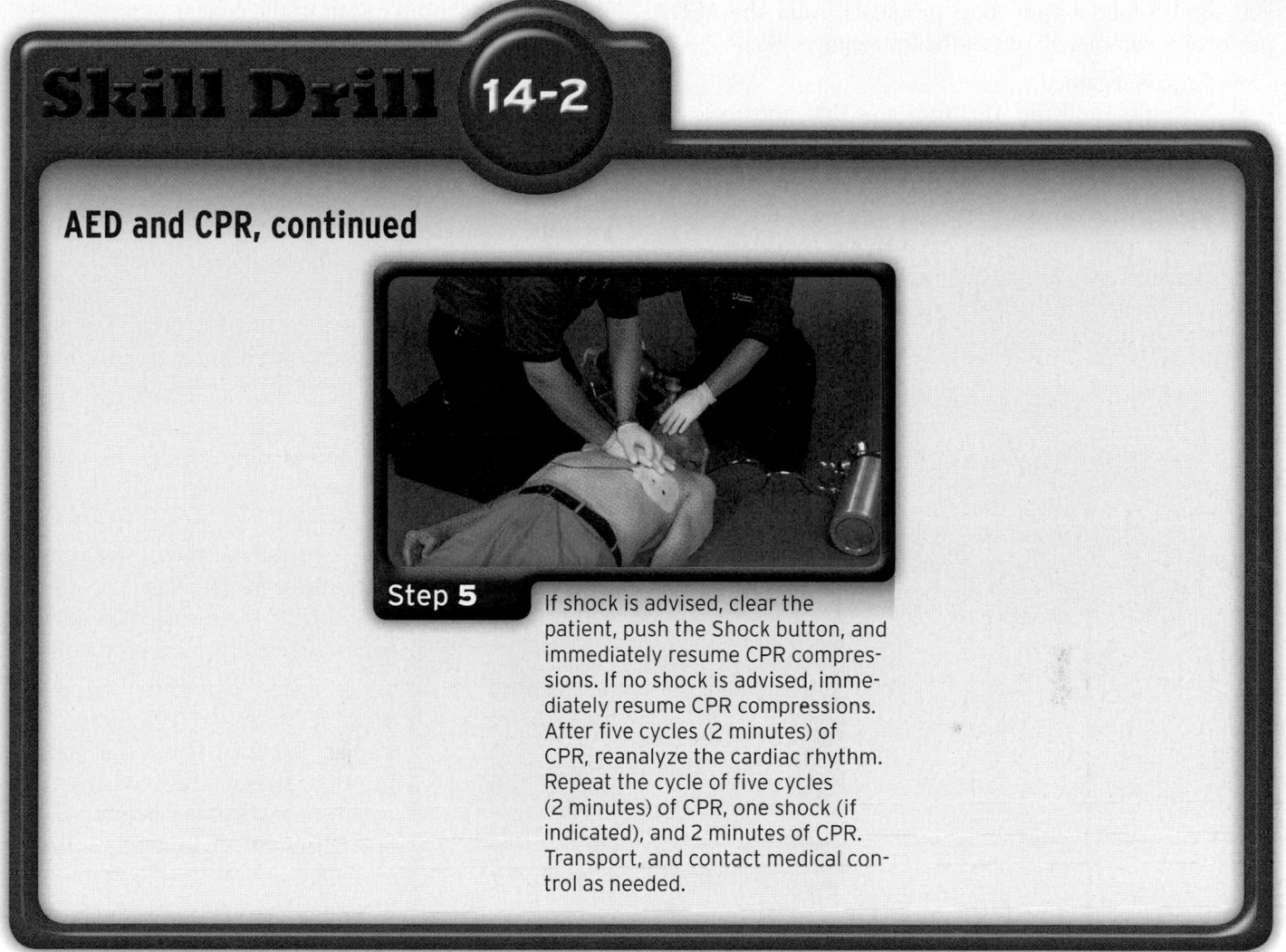

Step 5

If shock is advised, clear the patient, push the Shock button, and immediately resume CPR compressions. If no shock is advised, immediately resume CPR compressions. After five cycles (2 minutes) of CPR, reanalyze the cardiac rhythm. Repeat the cycle of five cycles (2 minutes) of CPR, one shock (if indicated), and 2 minutes of CPR. Transport, and contact medical control as needed.

shock is advised, reconfirm that no one is touching the patient and push the Shock button.

8. After the shock is delivered, immediately resume CPR, beginning with chest compressions (Step 3).

9. After five cycles (about 2 minutes) of CPR, reanalyze the patient's cardiac rhythm (Step 4). Do not interrupt chest compressions for more than 10 seconds.

10. If the AED advises a shock, clear the patient, push the Shock button, and immediately resume CPR compressions. If no shock is advised, immediately resume CPR, beginning with chest compressions.

11. Gather additional information about the arrest event.

12. After five cycles (2 minutes) of CPR, reassess the patient.

13. Repeat the cycle of 2 minutes of CPR, one shock (if indicated), and 2 minutes of CPR.

14. Transport, and contact medical control as needed (Step 5).

If the AED advises no shock and the patient has a pulse, check the patient's breathing. If the patient is breathing adequately, give 100% oxygen via nonrebreathing mask and transport. If the patient is not breathing adequately, provide artificial ventilation with a bag-mask device or pocket mask device attached to 100% oxygen and transport. Ensure that proper airway techniques are used at all times.

If the patient has no pulse, perform five cycles (approximately 2 minutes) of CPR beginning with chest compressions. After 2 minutes of CPR, reanalyze the patient's cardiac rhythm. If the AED advises to shock, deliver one shock followed immediately by CPR, beginning with chest compressions. Repeat these steps if needed.

If the patient has no pulse and the AED advises no shock, perform five cycles (approximately 2 minutes) of CPR beginning with chest compressions. After five cycles (2 minutes) of CPR, reanalyze the patient's cardiac rhythm. If no shock is advised, continue CPR. Transport the patient, and contact medical control as needed.

After Automated External Device Shocks

The care of the patient after the AED delivers a shock depends on your location and EMS system; therefore, you should follow your local protocols. After the AED protocol is completed, one of the following is likely:

- Pulse is regained.
- No pulse, and the AED indicates that no shock is advised.
- No pulse, and the AED indicates that a shock is advised.

Patients who do not regain a pulse on the scene of the cardiac arrest usually do not survive. What you do with these patients, again, depends on your EMS system.

Whether you should transport the patient or wait for ALS to arrive should be in the local protocols established by medical control. If paramedics or another ALS service is responding to the scene, the best option usually is to stay where you are and continue the sequence of shocks and CPR. Administering CPR while patients are being moved or transported is usually not effective. The best chance for patient survival occurs when the patient is resuscitated where found, unless the location is unsafe.

If an ALS service is not responding to the scene and your local protocols agree, you should begin transport when one of the following occurs:

- The patient regains a pulse.
- Six to nine shocks have been delivered (or as directed by local protocol).
- The machine gives three consecutive messages (separated by 2 minutes of CPR) that no shock is advised (or as directed by local protocol).

If you transport a patient while performing CPR, you need a plan for managing the patient in the ambulance. Ideally, you will have two EMTs in the patient compartment while a third drives. You may deliver additional shocks at the scene or en route with the approval of medical control. Keep in mind that AEDs cannot analyze the rhythm while the vehicle is in motion; nor is it as safe to defibrillate in a moving ambulance. Therefore, you should come to a complete stop if more shocks are needed. Be sure to memorize the protocol of your EMS service **Figure 14-17**.

Cardiac Arrest During Transport

If you are traveling to the hospital with an unconscious patient, check the pulse at least every 30 seconds. If a pulse is not present, take the following steps:

1. Stop the vehicle.
2. If the AED is not immediately ready, perform CPR, beginning with chest compressions, until it is available.
3. Analyze the rhythm.

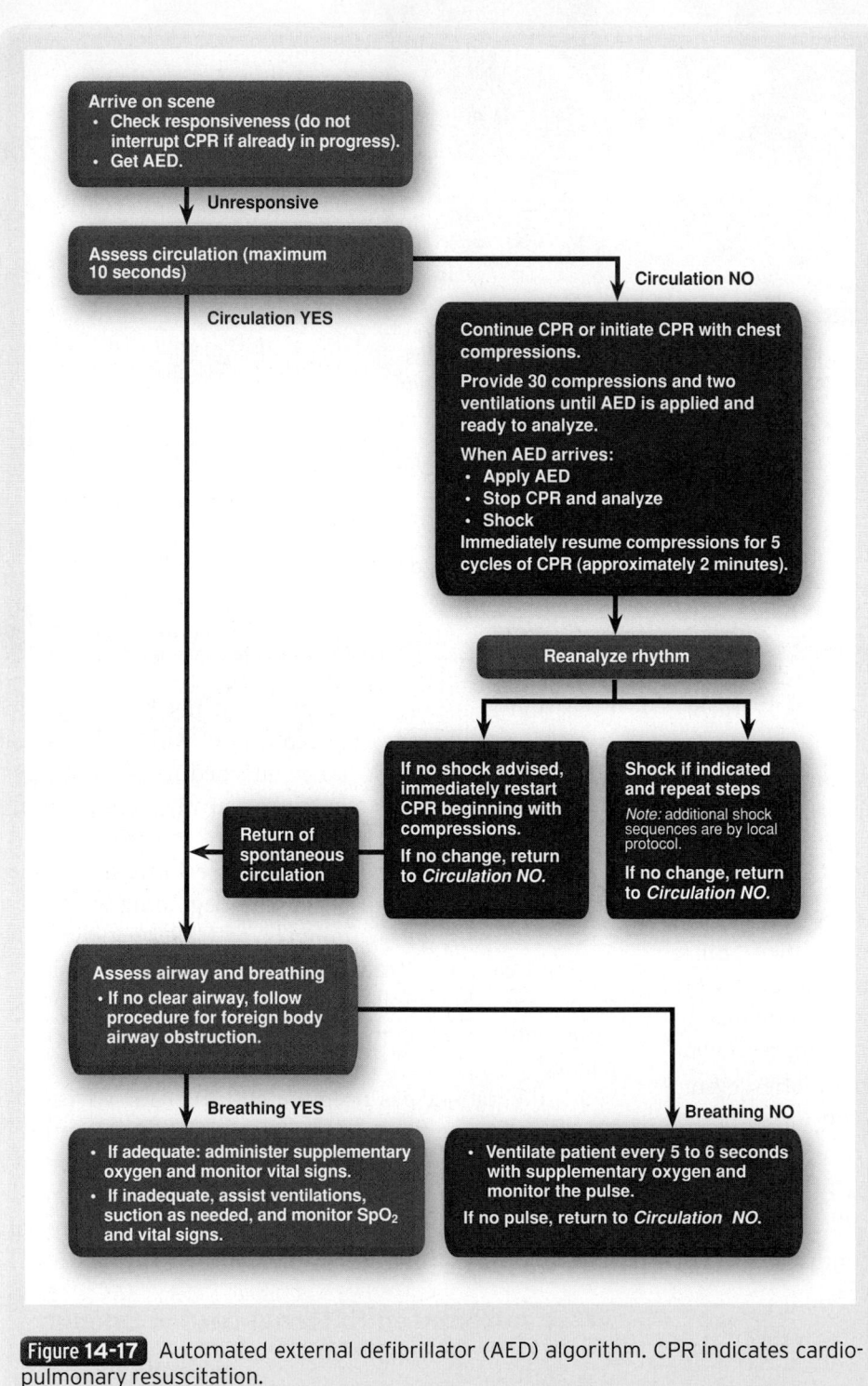

Figure 14-17 Automated external defibrillator (AED) algorithm. CPR indicates cardiopulmonary resuscitation.

4. Deliver one shock, if indicated, and immediately resume CPR.

5. Continue resuscitation according to your local protocol.

If you are en route with a conscious adult patient who is having chest pain and becomes unconscious, take the following steps:

1. Check for a pulse.

2. Stop the vehicle.

3. If the AED is not immediately ready, perform CPR, beginning with chest compressions, until it is ready.

4. Analyze the rhythm.

5. Deliver one shock, if indicated, and immediately resume CPR.

6. Begin compressions, and continue resuscitation according to your local protocol, including transporting the patient.

Coordination With Advanced Life Support Personnel

The time to defibrillation is critical to survival after cardiac arrest. As an EMT equipped with an AED, you have the one tool that a dying patient in ventricular fibrillation needs most. Furthermore, it is very difficult to hurt someone with an AED. Therefore, if you have an AED available, do not wait for the paramedics to arrive to administer a shock to a patient in ventricular fibrillation. Waiting might seem like a good idea. It is not. It is throwing away the patient's best chance for survival.

If the patient is unresponsive and does not have a pulse, apply the AED, and push the Analyze button (if there is one) as quickly as you can. Notify the ALS personnel as soon as possible after you recognize a cardiac arrest, but do not delay defibrillation. After the paramedics arrive at the scene, you should interact with them according to your local protocols.

Safety

Always remember when using an AED, there are several safety items to review.

1. Be aware of the surface the patient is lying on. Wet and metal surfaces may conduct electricity, making defibrillation of the patient dangerous to EMTs.

2. What is the age of the patient? Use pediatric AED pads when appropriate.

3. Does the patient have a medication patch in the area the AED pads will be placed? If so, remove the medication patch, wipe the area clean, and then attach the AED pad.

4. Does the patient have an implanted pacemaker or internal defibrillator in the same area the AED pads will be placed? If so, place the AED pad below the pacemaker or defibrillator, or place the pads in anterior and posterior positions.

You are the Provider: SUMMARY

1. What is the function of the heart?

The heart receives deoxygenated blood from the body, sends it to the lungs to be reoxygenated, and then pumps highly oxygenated blood throughout the entire body. The heart must pump effectively to ensure that the body's tissues and cells receive an uninterrupted supply of oxygen and that metabolic waste (eg, carbon dioxide) is removed from the tissues and cells and returned, through the heart, for elimination from the body by the respiratory system.

2. What does the heart require to function effectively?

Like any other critical organ or muscle, the heart requires a constant supply of oxygen, which it receives from the coronary arteries. It also relies on electricity to stimulate the contraction of the muscular layer of the heart (myocardium). Unlike any other muscle in the body, cardiac muscle generates its own electrical impulses—a process called automaticity—that it sends throughout the heart via the cardiac conduction system. Adequate blood volume is also required for effective cardiac function. As blood returns to the heart, it enters the chambers, stretches their walls, and causes them to contract with greater force. If blood volume is low, the heart will stretch less, and its contractile force will decrease.

3. What should you include in your primary assessment of a patient with cardiac problems?

Your primary assessment of a patient with cardiac problems should be no different from that of any other patient: to find and immediately correct problems with airway, breathing, and circulation. In addition, you should look for signs of impaired cardiac function, such as an irregular heartbeat, a fast or slow heart rate, a weak (thready) pulse, and poor skin condition (eg, pallor, diaphoresis).

4. Why is aspirin given to patients with an acute cardiac event?

Aspirin (acetylsalicylic acid [also known as ASA]) has clearly been shown to reduce mortality and morbidity from acute myocardial infarction (AMI); unless the patient is allergic to aspirin, it should be given as soon as possible if an acute cardiac event is suspected. An AMI occurs when an atherosclerotic plaque ruptures and occludes a coronary artery. When this occurs, platelets rush to the area and aggregate (clump together), which further occludes the coronary artery. Aspirin makes the platelets less "sticky," which makes them less likely to aggregate. Although aspirin will not dissolve the existing clot that is occluding the coronary artery, it may help prevent it from getting larger by reducing the amount of platelet aggregation.

5. What type of medication is nitroglycerin? How may it help relieve chest pain, pressure, or discomfort?

Nitroglycerin is a vasodilator; it works by relaxing the smooth muscle that regulates the diameter of the blood vessels, causing them to dilate (open). Nitroglycerin is used by patients with coronary artery disease who are experiencing chest pain, pressure, or discomfort; it dilates the coronary arteries and increases blood flow to the heart. As a result, the myocardial oxygen supply and demand are rebalanced, and the pain subsides or resolves completely. In some cases, however, nitroglycerin does not relieve the patient's chest pain; in a patient with a cardiac history, this should make you more suspicious that he or she is experiencing an AMI.

6. When is nitroglycerin indicated for a patient? What is the typical dose?

Nitroglycerin is indicated for patients with coronary artery disease who experience chest pain, pressure, or discomfort. Many patients with coronary artery disease have prescribed nitroglycerin, which they self-administer. If the patient has not taken any of his or her prescribed nitroglycerin, the EMT may assist the patient in doing so after ensuring that his or her systolic blood pressure is at least 100 mm Hg and that approval from medical control has been obtained. Nitroglycerin is supplied in many forms; however, tablets and spray are the most common. The dose is one 0.4-mg tablet or metered-dose spray every 5 minutes, up to three doses. Because nitroglycerin is a vasodilator, it can cause hypotension; therefore, it is important to reassess with patient's blood pressure within a few minutes after administering nitroglycerin to ensure that it is at least 100 mm Hg. Nitroglycerin is contraindicated in patients with a systolic blood pressure of less than 100 mm Hg and in patients who have taken drugs for erectile dysfunction (eg, sildenafil [Viagra], tadalafil [Cialis], and vardenafil [Levitra]) within the past 24 to 36 hours. Drugs for erectile dysfunction are also vasodilators; if given together with nitroglycerin, significant hypotension may occur.

7. What is significant about the patient's vital signs?

Pale, cool, clammy (diaphoretic) skin is not exclusive to a cardiac problem; however, in the context of the patient's chief complaint and history of heart problems, it is highly suggestive that her chest pain is of a cardiac origin. An irregular heartbeat indicates a disturbance in the cardiac electrical conduction system (dysrhythmia); again, in the context of her chief complaint and cardiac history, this should further increase your index of suspicion that she is experiencing a cardiac event. An irregular heartbeat in a patient with a cardiac problem could indicate an impending life-threatening dysrhythmia (ie, ventricular fibrillation or ventricular tachycardia). The patient's rapid heart rate (tachycardia) and relatively elevated blood pressure (150/90 mm Hg) are also clinically significant; they indicate that her heart is working harder than normal. As the heart rate and blood pressure increase, the heart consumes and requires more oxygen. If the heart is already deprived of oxygen, the patient's condition could worsen.

8. Should you give her additional nitroglycerin? Why or why not?

Despite taking two of her prescribed nitroglycerin doses before your arrival, the patient is still experiencing significant chest pain (7 on a 0 to 10 scale). Because her systolic blood pressure is well above 100 mm Hg, you should contact medical control and request permission to assist her with one more nitroglycerin dose. Remember to reassess her blood pressure within a few minutes after administering the medication.

9. Why is early notification of the receiving facility so important for patients with an acute coronary event?

The longer it takes to reestablish blood flow distal to an occluded artery, the greater the amount of cardiac muscle damage (hence the phrase, "time is muscle"). Early reperfusion—with fibrinolytic medications (clot busters) or cardiac catheterization and stent placement—has clearly been shown to minimize the amount of cardiac damage and improve the patient's outcome. The earlier you notify the receiving facility that you are transporting a patient with a possible AMI, the more time the staff will have to allocate the resources needed to facilitate rapid cardiac reperfusion. The reperfusion strategy used is determined by the physician; the EMT's job is to recognize that the patient may be experiencing an AMI, provide immediate lifesaving care, promptly notify the receiving facility, and transport without delay.

10. Should you apply the AED to determine if she is experiencing a cardiac arrhythmia? Why or why not?

No. The automated external defibrillator (AED) is applied *only* to patients who are apneic and pulseless (eg, in cardiac arrest). At present, your patient is breathing and has a pulse. Even if you did apply the AED, it would not analyze her cardiac rhythm; an AED will not analyze the cardiac rhythm if it detects patient movement. You should have the AED readily available in case she experiences cardiac arrest, but its application is not indicated at this point.

11. What is the difference between angina pectoris and an acute myocardial infarction?

Angina pectoris occurs when the heart's demand for oxygen exceeds its available supply (ischemia), resulting in chest pain or discomfort. It signals the presence of coronary artery disease due to atherosclerosis, a condition in which one or more coronary arteries is partially occluded. Angina is typically triggered by exertion, which increases myocardial oxygen consumption and demand. When the patient ceases exertion, oxygen supply and demand are rebalanced and the pain resolves, usually in less than 15 minutes. In more severe cases, a combination of rest and nitroglycerin are required for resolution of the patient's chest pain or discomfort.

An AMI occurs when a portion of the heart muscle is completely deprived of oxygen because of complete occlusion of one or more coronary arteries. Unlike angina, the chest pain, pressure, or discomfort associated with an AMI typically does not resolve with rest or nitroglycerin and persists for greater

You are the Provider: SUMMARY, continued

than 15 minutes. The patient experiencing an AMI needs prompt treatment in the hospital, which is aimed at removing the clot in the coronary artery and reestablishing distal blood flow. Fibrinolytic medications (clot busters) or cardiac catheterization and placement of a stent (a device that keeps the coronary artery open) are used to accomplish this. If not promptly treated, the portion of the myocardium distal to the occluded coronary artery will undergo necrosis (permanent tissue death).

12. Can an EMT distinguish angina pectoris from an acute myocardial infarction?

The signs and symptoms of angina and an AMI are essentially the same and usually cannot be distinguished without advanced diagnostic procedures. In both conditions, the chest pain or discomfort may be described as a feeling of pressure or heaviness. Other signs and symptoms that accompany an AMI, such as diaphoresis, nausea, and a feeling of impending doom, are also commonly present in patients with angina. Although the pain caused by angina commonly subsides in less than 15 minutes with rest and/or nitroglycerin, it can persist for more than 15 minutes in some cases. Paramedics can use 12-lead electrocardiography to help distinguish angina from an AMI, but even these findings are not totally conclusive. The patient requires physician evaluation, blood analysis, and other tests to diagnose an AMI. The EMT should assume that any patient with nontraumatic chest pain or discomfort is experiencing an AMI until ruled out by a physician.

EMS Patient Care Report (PCR)

Date: 3-10-09	Incident No.: 130209	Nature of Call: Cardiac		Location: 1152 Blanco Road	
Dispatched: 0942	En Route: 0942	At Scene: 0950	Transport: 1000	At Hospital: 1017	In Service: 1029

Patient Information

Age: 60	Allergies: No known drug allergies
Sex: F	Medications: Vasotec, aspirin (ASA), nitroglycerin (NTG)
Weight (in kg [lb]): 54 kg (121 lb)	Past Medical History: Heart attack, hypertension
	Chief Complaint: Chest pain

Vital Signs

Time: 0952	BP: 150/90	Pulse: 118	Respirations: 20	Sao$_2$: 98%
Time: 1002	BP: 136/84	Pulse: 84	Respirations: 18	Sao$_2$: 96%
Time: 1012	BP: 128/78	Pulse: 80	Respirations: 16	Sao$_2$: 98%

EMS Treatment
(circle all that apply)

Oxygen @ <u>15</u> L/min via (circle one): NC **(NRM)** Bag-Mask Device	Assisted Ventilation	Airway Adjunct	CPR
Defibrillation	Bleeding Control	Bandaging	Splinting	Other: (324 mg ASA)

Narrative

Dispatched for a 60-year-old female with chest pain. Upon arrival at the scene, found the patient sitting up in her bed with her fist clenched against her chest. She was conscious and alert, although anxious. Her airway was patent, and her breathing was adequate. She was markedly diaphoretic; had pale, cool skin; and had a rapid, irregular pulse. Patient states that she had a heart attack 3 years ago and has hypertension. She is presently taking Vasotec, NTG, and one (1) ASA per day and states that she has been compliant with her medications. Patient took two doses of her prescribed NTG before EMS arrival; however, she states that the medication had no effect; she presently describes her pain as a "7" on a scale of 0 to 10. Administered 324 mg ASA (no known drug allergies), applied oxygen at 15 L/min via nonrebreathing mask, and obtained vital signs. Contacted medical control, who authorized the administration of one more NTG dose. Placed patient onto stretcher, loaded her into the ambulance, and began transport to the hospital. Contacted hospital shortly after departing the scene and advised that we were transporting patient with possible AMI. En route to the hospital, allowed patient to assume position of comfort and reassessed her vital signs. She was still complaining of chest pain (3/10); however, her pulse rate, although still irregular, was notably slower. Reassessment of her skin revealed that it was pink, cool, and dry, and she was noted to be less anxious. Continued to monitor patient's condition throughout transport; there was no gross evidence of deterioration, and she remained conscious and alert. Delivered her to emergency department staff w/o incident. Upon arrival at the hospital, we were greeted by the cardiac team, who assumed patient care. Gave verbal report to charge nurse and returned to service. **End of report**

Assessment and Emergency Care of Cardiovascular Emergencies

Scene Size-up

Scene Safety	Ensure scene safety and safe access to the patient. Standard precautions should include a minimum of gloves. Determine the number of patients. ALS should be requested. Assess the need for additional resources.
Mechanism of Injury (MOI)/ Nature of Illness (NOI)	Determine the MOI/NOI. Ensure that the cardiac emergency is not the result of a traumatic event. Dispatch information, observations at the scene, and comments from family or bystanders will help you develop an idea of the NOI. Usually the NOI can be determined by the patient's chief complaint. Chest pain, difficulty breathing, and syncope are some indicators that the NOI may be a cardiac emergency.

▼   ▼

Primary Assessment

Form a General Impression	Observe overall appearance of the patient, age, body position, and responsiveness. Observe work of breathing and circulation. Pale skin and cyanosis are indicators of poor perfusion. Determine the level of consciousness using the AVPU scale. Is the patient calm or anxious? Is the patient able to speak in full sentences? Identify immediate threats to life. Determine priority of care based on the MOI/NOI. If the patient is unconscious, determine whether CPR is needed. If the patient has a poor general impression, call for ALS assistance. A rapid visual examination will help you identify and manage life threats.
Airway and Breathing	Ensure the airway is open, clear, and self-maintained. If needed, open and maintain the airway using a modified jaw-thrust if a cervical-spine injury is suspected or a head-tilt–chin-lift in nontrauma patients. A patient with an altered level of consciousness may need emergency airway management; consider inserting a properly sized oropharyngeal airway in an unconscious patient or use a nasopharyngeal airway if the patient has an intact gag reflex. Assess for gurgling and stridor. Suction as needed. Evaluate the patient's ventilatory status for rate and depth of breathing, respiratory effort, and tidal volume. Quickly assess the chest for DCAP-BTLS, accessory muscle use, and intercostal and abdominal muscle use, and treat any threats to life. Assess lung sounds and determine whether they are normal, decreased, abnormal, or absent. Patients with cardiac problems may present with crackles or rales, indicating fluid (edema) in the lungs. Administer high-flow oxygen at 15 L/min, providing ventilatory support as needed.
Circulation	Evaluate distal pulse rate, quality (strength), and rhythm. Tachycardia may be an indicator of shock or a myocardial infarction. Bradycardia might be due to cardiogenic shock as the heart attempts to reduce oxygen demand. Observe skin color, temperature, and condition and capillary refill time. Look for and treat any external bleeding. The transport of blood and oxygen may be reduced if cardiac output is low. If distal pulses are not palpable, assess for a central pulse. **Note:** If the patient is unresponsive, not breathing, and does not have a pulse (cardiac arrest), follow the CAB protocol and begin CPR, starting with chest compressions.
Transport Decision	If the patient has an airway or a breathing problem, severe chest discomfort, signs and symptoms of internal bleeding, or other life threats, manage the problems immediately, and consider rapid transport, performing the secondary assessment en route to the hospital. Delayed transport in a cardiac emergency will cause more cardiac muscle damage. For patients without life threats and in stable condition, perform a thorough assessment and history on scene. Do not allow the patient to exert himself or herself by walking to the ambulance; a carrying device should be used. Transportation should be to a cardiac care facility if one is close, otherwise take the patient to the nearest hospital for evaluation and stabilization. Do not delay transport to manage non–life-threatening conditions, instead treat en route to the hospital. Lights and sirens might cause anxiety and stress to a patient with a cardiac problem; patient condition and distance from the hospital are factors you can use to evaluate the need to use lights and siren.

▼   ▼

NOTE: The order of the steps in this section differs depending on whether the patient is conscious or unconscious. The following order is for a conscious patient. For an unconscious patient, perform a primary assessment, perform a full-body scan, obtain vital signs, and obtain the past medical history from a family member or medical ID.

Assessment and Emergency Care of Cardiovascular Emergencies, continued

History Taking

Investigate Chief Complaint	Investigate the chief complaint. Conscious patients can supply you with a brief history; ask OPQRST and SAMPLE questions. Find out when the problem started; if anything makes it feel better or worse; and if the patient has pain, what type of pain it is, and if it radiates. Ask the patient to rate the pain on a scale of 0 to 10, with 10 being the worst pain ever experienced, and how long the pain lasts. Ask the patient about previous heart attacks, heart problems, high blood pressure, aneurysms, lung disorders, diabetes, and kidney disease. Ascertain whether there are any risk factors for coronary disease, and obtain a family history. SAMPLE can also be obtained from family members and medical alert tags. The patient may have more than one complaint, such as chest pain and difficulty breathing; try to determine which presented first. Patients having a cardiac emergency may also complain of dizziness, appear anxious, or have a sense of impending doom. Some patients may deny that the symptoms are cardiac-related, and others may appear frightened. Nausea, vomiting, fatigue, headache, arm pain, jaw pain, and palpitations are other complaints the patient may have. Identify pertinent negatives. Determine if there was any trauma and if the pain increases on inhalation (pleuritic) or movement. Maintain a calm, professional attitude; be honest; and provide reassurance. Place the patient in a position of comfort.

Secondary Assessment

Physical Examinations	Focus primarily on the cardiovascular system. Assess central and peripheral pulse quality. Examine the skin color; pallor and cyanosis are indicators of hypoxia (low oxygen level). Also assess skin temperature and condition. Look for edema in the extremities, which may be an indicator of cardiovascular failure. Inspect, palpate, and auscultate the chest, focusing on the respiratory effort and adequacy of ventilation. Crackles (rales) heard on auscultation of lung sounds and jugular vein distention are indicators of possible heart failure.
Vital Signs	Obtain baseline vital signs, and repeat depending on patient impression, monitoring trends. Vital signs should include blood pressure by auscultation, pulse rate and quality, respiratory rate and quality, and skin assessment for perfusion. Use pulse oximetry, if available, to assess the patient's perfusion status, keeping in mind that readings for patients with poor circulation may not be accurate.

Reassessment

Interventions	Patients who are unresponsive, without breathing or a pulse, need CPR. An automated external defibrillator (AED) should be applied as soon as it becomes available. Airway control using adjuncts may be necessary. Patients who are conscious should be placed in a position of comfort, usually sitting up. Loosen any tight clothing. Reassess the primary assessment findings, vital signs, and chief complaint. Assist breathing as required, administering high-flow oxygen. CPAP might be indicated for patients with congestive heart failure (CHF). If permitted, administer 162 to 324 mg of chewable low-dose aspirin, and if the blood pressure is adequate, assist the patient in taking his or her prescribed nitroglycerin. Do not delay transport.
Communications and Documentation	Contact medical control and/or the receiving hospital with a radio report. Include a thorough description of the MOI/NOI, the position in which the patient was found, and vital signs. Include treatments performed and patient response. Document interventions performed and any changes in patient status and the time the interventions and changes occurred. Follow local protocols. Document the reasoning for your treatment and the patient's response. Obtain a signature from the hospital physician or nurse on patient transfer.

NOTE: Although the following steps are widely accepted, be sure to consult and follow your local protocols. Follow standard precautions when treating all patients.

Assessment and Emergency Care of Cardiovascular Emergencies, continued

Cardiovascular Emergencies

General Management of Cardiovascular Emergencies

Managing life threats to the patient's ABCs and ensuring the delivery of high-flow oxygen are primary concerns with any cardiovascular emergency. If the patient is unconscious, determine whether CPR is needed; if CPR is needed, provide 30 chest compressions, then open the airway and provide rescue breaths. In conscious patients, obtain a thorough history. "Time is muscle," so rapid transport will be needed to a cardiac care facility for patients presenting with signs and symptoms of a myocardial infarction. If local protocols allow, administer aspirin and assist the patient in taking his or her prescribed nitroglycerin. Be prepared to defibrillate if the patient becomes pulseless.

Cardiogenic Shock

Shock is a state of hypoperfusion. The hypoperfusion from cardiogenic shock is due to failure of the pump (heart). The container (blood vessels) is intact, and the fluid (blood) is still present within the container. EMTs need to be able to recognize cardiogenic shock over other types of shock because the management is different. The first clue is that there is no mechanism of injury. Chest pain is usually the chief complaint. The pulse may be irregular. The patient may have respiratory distress due to fluid buildup in the lungs (pulmonary edema) due to poor cardiac output. As with other shocks, the blood pressure is low. *Do not* place this patient in the shock or Trendelenburg's position because it will increase the workload of the heart and cause increased fluid collection in the lungs. Place the patient in a position of comfort. Administer high-concentration oxygen. Request ALS support if transport is delayed. *Do not* give nitroglycerin; the blood pressure is already low. If a specialty center is close by, transport there; if not, transport to the nearest hospital.

Congestive Heart Failure

Fluid in the lungs is called pulmonary edema. Pulmonary edema can be caused by cardiac failure after an AMI (CHF) or a noncardiogenic cause such as a toxic inhalation. In either case, the outcome is the same, fluid in the lungs prevents the efficient exchange of oxygen and carbon dioxide. The patient will present with respiratory distress, usually severe, and appear very anxious. The skin will be cool, pale, and moist. The patient's blood pressure is often high unless the AMI is so severe as to cause cardiogenic shock. Patients with a history of CHF will often sleep with multiple pillows or upright in a recliner. Jugular vein distention is common. The patient needs high-flow oxygen. Assisted ventilation or CPAP is often helpful. Assist the patient in taking his or her prescribed nitroglycerin if medical control or protocol allows, ensuring the systolic blood pressure is more than 100 mm Hg before giving the nitroglycerin. Patients experiencing pulmonary edema may require positive-pressure ventilation with a bag-mask device or CPAP. CPAP is the most effective way to assist a person with CHF to breathe effectively and prevent an invasive airway management technique. Transport promptly to the closest emergency department.

Assessment and Emergency Care of Cardiovascular Emergencies, continued

Cardiovascular Emergencies

Cardiac Arrest

When cardiac activity ceases completely, the condition is called cardiac arrest. The patient will not be breathing and will not have a palpable pulse. The most common cause of sudden cardiac death is an electrical disturbance (arrhythmia) called ventricular fibrillation. Treatment of ventricular fibrillation requires defibrillation with an AED; apply an AED as quickly as possible and follow the voice commands. If shock (defibrillation) is indicated, clear everyone from around the patient, including yourself, and deliver the shock. Immediately begin CPR for 2 minutes, performing 30 compressions and then two breaths, and repeat for five cycles. Support from ALS should be requested as soon as possible. While performing CPR, it is important that compressions are not interrupted, that they are delivered at a rate of at least 100 per minute, and that they are of adequate depth. Ventilations can be delivered using a pocket mask or other barrier device or a bag-mask device. If using a bag-mask device, ensure that you have an open airway and a good mask seal. It is best to have a second rescuer, if one is available, assist when using a bag-mask device.

Ready for Review

- The heart is divided down the middle into two sides, right and left, each with an upper chamber called the atrium and a lower chamber called the ventricle.

- The heart valve that keeps blood moving through the circulatory system in the proper direction is the aortic valve, which lies between the left ventricle and the aorta, the body's main artery.

- The heart's electrical system controls heart rate and helps the atria and ventricles work together to pump the blood.

- During periods of exertion or stress, the myocardium requires more oxygen. The oxygen is supplied by dilation of the coronary arteries, which increases blood flow.

- Common places to feel for a pulse include the carotid, femoral, brachial, radial, posterior tibial, and dorsalis pedis arteries.

- Low blood flow to the heart is usually caused by coronary artery atherosclerosis, a disease in which cholesterol plaques build up inside blood vessels, eventually occluding them.

- Occasionally, a brittle plaque in an artery will crack, causing a blood clot to form. Heart tissue downstream suffers from a lack of oxygen and, within 30 minutes, will begin to die. This condition is called an acute myocardial infarction (AMI), or heart attack.

- Heart tissues that are not getting enough oxygen but are not yet dying can cause pain called angina. The pain of an AMI is different from the pain of angina in that it can come at any time, not just with exertion; it lasts up to several hours, rather than just a few moments; and it is not relieved by rest or nitroglycerin.

- In addition to crushing chest pain, signs of AMI include sudden onset of weakness, nausea, and sweating; sudden arrhythmia; pulmonary edema; and even sudden death.

- Heart attacks can have three serious consequences. One is sudden death, usually the result of cardiac arrest caused by abnormal heart rhythms called arrhythmias. These include tachycardia, bradycardia, ventricular tachycardia, and, most commonly, ventricular fibrillation.

- The second consequence is cardiogenic shock. Symptoms include restlessness; anxiety; pale, clammy skin; pulse rate higher than normal; and blood pressure lower than normal. Patients with these symptoms should receive oxygen, assisted ventilations as needed, and immediate transport.

- The third consequence of AMI is congestive heart failure, in which damaged heart muscle can no longer contract effectively enough to pump blood through the system. The lungs become congested with fluid, breathing becomes difficult, the heart rate increases, and the left ventricle enlarges.

- Signs include swollen ankles from dependent edema, high blood pressure, rapid heart rate and respirations, rales (crackles), and, sometimes, the pink sputum and dyspnea of pulmonary edema.

- Treat a patient with congestive heart failure as you would a patient with chest pain. Monitor the patient's vital signs. Give the patient oxygen via nonrebreathing face mask. Allow the patient to remain sitting up.

- When treating patients with chest pain, obtain a SAMPLE history, following the OPQRST mnemonic to assess the pain; measure and record vital signs; ensure the patient is in a comfortable position, usually semireclining or half sitting up; administer prescribed nitroglycerin and oxygen; and transport the patient, reporting to medical control as you do.

- If a patient is not responsive, is not breathing, and does not have a pulse, you may perform the following, depending on the patient's age and your local protocol:
  - Unresponsive adult or child older than 8 years, perform automated external defibrillation.
  - Unresponsive child younger than 8 years, perform automated external defibrillation with pediatric pads and dose attenuator; if neither is available, an adult AED may be used.
  - Unresponsive infant between the ages of 1 month and 1 year, use a manual defibrillator. If a manual defibrillator is not available, an AED equipped with pediatric pads and a dose attenuator may be used. If neither is available, an adult AED may be used.

- The AED requires the operator to apply the pads, power on the unit, follow the AED prompts, and press the shock button as indicated. The computer inside the AED recognizes rhythms that require shocking and will not mislead you.

- The three most common errors in using certain AEDs are failure to keep a charged battery in the machine; applying the AED to a patient who is moving, squirming, or being transported; and applying the AED to a responsive patient with a rapid heart rate.

- Do not touch the patient while the AED is analyzing the heart rhythm or delivering shocks.

- Effective CPR and early defibrillation with an AED are critical interventions to the survival of a patient in cardiac arrest. Begin CPR starting with chest compressions and apply the AED as soon as it is available.

- If an advanced life support (ALS) service is responding to the scene, stay where you are and continue CPR and defibrillation as needed. If ALS is not responding, you should begin transport if the patient regains a pulse, if you have delivered six to nine shocks, or if the AED gives three consecutive messages (separated by 2 minutes of CPR) that no shock is advised. Follow your local protocols regarding when it is appropriate to transport the patient.

- If an unconscious patient has a pulse but the pulse is lost during transport, you must stop the vehicle, reanalyze the rhythm, and defibrillate again or begin CPR, as appropriate.

- The chain of survival, which is the sequence of events that must happen for a patient with cardiac arrest to have the best chance of survival, includes recognition of early warning signs and immediate activation of EMS, early CPR, early defibrillation, early advanced care, and integrated post-arrest care. Seconds count at every stage.

Vital Vocabulary

acute coronary syndrome A term used to describe a group of symptoms caused by myocardial ischemia; includes angina and myocardial infarction.

acute myocardial infarction (AMI) A heart attack; death of heart muscle following obstruction of blood flow to it. Acute in this context means "new" or "happening right now."

angina pectoris Transient (short-lived) chest discomfort caused by partial or temporary blockage of blood flow to the heart muscle.

anterior The front surface of the body; the side facing you in the standard anatomic position.

aorta The main artery, which receives blood from the left ventricle and delivers it to all the other arteries that carry blood to the tissues of the body.

aortic aneurysm A weakness in the wall of the aorta that makes it susceptible to rupture.

aortic valve The one-way valve that lies between the left ventricle and the aorta and keeps blood from flowing back into the left ventricle after the left ventricle ejects its blood into the aorta; one of four heart valves.

arrhythmia An irregular or abnormal heart rhythm.

asystole The complete absence of heart electrical activity.

atherosclerosis A disorder in which cholesterol and calcium build up inside the walls of blood vessels, eventually leading to partial or complete blockage of blood flow.

atrium One of two (right and left) upper chambers of the heart. The right atrium receives blood from the vena cava and delivers it to the right ventricle. The left atrium receives blood from pulmonary veins and delivers it to the left ventricle.

automaticity The ability of cardiac muscle cells to contract without stimulation from the nervous system.

autonomic nervous system The part of the nervous system that controls the involuntary activities of the body such as the heart rate, blood pressure, and digestion of food.

bradycardia A slow heart rate, less than 60 beats/min.

cardiac arrest A state in which the heart fails to generate effective and detectable blood flow; pulses are not palpable in cardiac arrest, even if muscular and electrical activity continues in the heart.

cardiogenic shock A state in which not enough oxygen is delivered to the tissues of the body, caused by low output of blood from the heart. It can be a severe complication of a large acute myocardial infarction, as well as other conditions.

cardiac output A measure of the volume of blood circulated by the heart in 1 minute, calculated by multiplying the stroke volume by the heart rate.

congestive heart failure (CHF) A disorder in which the heart loses part of its ability to effectively pump blood, usually as a result of damage to the heart muscle and usually resulting in a backup of fluid into the lungs.

coronary arteries The blood vessels that carry blood and nutrients to the heart muscle.

defibrillate To shock a fibrillating (chaotically beating) heart with specialized electrical current in an attempt to restore a normal, rhythmic beat.

dependent edema Swelling in the part of the body closest to the ground, caused by collection of fluid in the tissues; a possible sign of congestive heart failure.

dilation Widening of a tubular structure such as a coronary artery.

dissecting aneurysm A condition in which the inner layers of an artery, such as the aorta, become separated, allowing blood (at high pressures) to flow between the layers.

hypertensive emergency An emergency situation created by excessively high blood pressure, which can lead to serious complications such as stroke or aneurysm.

infarction Death of a body tissue, usually caused by interruption of its blood supply.

inferior The part of the body or any body part nearer to the feet.

ischemia A lack of oxygen that deprives tissues of necessary nutrients, resulting from partial or complete blockage of blood flow; potentially reversible because permanent injury has not yet occurred.

lumen The inside diameter of an artery or other hollow structure.

myocardium The heart muscle.

occlusion A blockage, usually of a tubular structure such as a blood vessel.

parasympathetic nervous system The part of the autonomic nervous system that controls vegetative functions such as digestion of food and relaxation.

perfusion The flow of blood through body tissues and vessels.

posterior The back surface of the body; the side away from you in the standard anatomic position.

stroke volume The volume of blood ejected with each ventricular contraction.

superior The part of the body or any body part nearer to the head.

sympathetic nervous system The part of the autonomic nervous system that controls active functions such as responding to fear (also known as the "fight-or-flight" system).

syncope A fainting spell or transient loss of consciousness.

tachycardia A rapid heart rate, more than 100 beats/min.

thromboembolism A blood clot that has formed within a blood vessel and is floating within the bloodstream.

ventricle One of two (right and left) lower chambers of the heart. The left ventricle receives blood from the left atrium (upper chamber) and delivers blood to the aorta. The right ventricle receives blood from the right atrium and pumps it into the pulmonary artery.

ventricular fibrillation Disorganized, ineffective twitching of the ventricles, resulting in no blood flow and a state of cardiac arrest.

ventricular tachycardia A rapid heart rhythm in which the electrical impulse begins in the ventricle (instead of the atrium), which may result in inadequate blood flow and eventually deteriorate into cardiac arrest.

Assessment in Action

You are dispatched to a person complaining of chest pain and shortness of breath. You arrive to the residence to find a 56-year-old man sitting up in the kitchen. His skin is ashen, and he is diaphoretic. The patient describes the pain as substernal and crushing. On a scale of 0 to 10, he states his pain is an 8. The patient had a myocardial infarction 2 years ago, and he has angina, hypertension, and a high cholesterol level. His medications include nitroglycerin, furosemide, and atorvastatin. His vital signs are as follows: pulse, 140 beats/min and irregular; respiratory rate, 28 breaths/min; and blood pressure, 90/50 mm Hg. His lungs have crackles (rales).

1. The patient's difficulty breathing and crackles are due to blood backing up in which part of the body?
   A. The heart
   B. The lungs
   C. The vessels
   D. The arteries

2. What is ischemia?
   A. The death of tissue in an organ or other body part
   B. The oozing of plasma out of the bloodstream and into the tissues
   C. A disorder of the hemoglobin that leads to oversaturation of oxygen in the tissues
   D. An inadequate blood supply to a body part

3. What is an infarct?
   A. Tissue death in an organ or other body part
   B. A thromboembolism
   C. A diseased area between the ventricles of the heart that allows blood to flow between the ventricles
   D. An obstruction of a blood vessel

4. On the basis of the information given, which of the following is the most likely type of shock the patient is experiencing?
   A. Neurogenic
   B. Vasogenic
   C. Cardiogenic
   D. Hypovolemic

5. Describe the path of blood through the heart and lungs.

6. Treatment for this patient includes which of the following?
   A. Oxygen, furosemide, and nitroglycerin
   B. Oxygen and transport in a position of comfort
   C. Oxygen and nitroglycerin
   D. Oxygen and transport in Trendelenburg's position

7. Nitroglycerin will cause which of the following?
   A. Vasoconstriction
   B. Bronchoconstriction
   C. Vasodilation
   D. Bronchodilation

8. What is cardiogenic shock?

9. The buildup of fats or calcium on vessel walls is known as which of the following?
   A. Atherosclerosis
   B. Thrombosis
   C. Aneurysm
   D. Embolus

10. The primary pacemaker of the heart is (are) the:
    A. atrioventricular node.
    B. bundle of His.
    C. Purkinje fibers.
    D. sinoatrial node.

National EMS Education Standard Competencies

Medicine

Applies fundamental knowledge to provide basic emergency care and transportation based on assessment findings for an acutely ill patient.

Neurology

Anatomy, presentations, and management of

- Decreased level of responsiveness (pp 567-568, 576-577, 586)
- Seizure (pp 567-568, 573-576, 586)
- Stroke (pp 567-568, 570-573, 585-586)

Anatomy, physiology, pathophysiology, assessment, and management of

- Stroke/transient ischemic attack (pp 567-568, 570-573, 578-586)
- Seizure (pp 567-568, 573-576, 578-586)
- Status epilepticus (pp 567-568, 573-576, 578-586)
- Headache (pp 567-570, 578-585)

Knowledge Objectives

1. Discuss the anatomy and physiology of the brain and spinal cord. (pp 567-568)
2. Discuss the different types of headaches, the possible causes of each, and how to distinguish a harmless headache from a potentially life-threatening condition. (pp 568-570)
3. List the various ways blood flow to the brain may be interrupted and cause a cerebrovascular accident (CVA). (p 570)
4. Discuss the causes of ischemic strokes, hemorrhagic strokes, and transient ischemic attacks (TIA) and their similarities and differences. (pp 570-572)
5. List the general signs and symptoms of stroke and identify those symptoms that manifest if the left hemisphere of the brain is affected, if the right hemisphere of the brain is affected, and if there is bleeding in the brain. (p 572)
6. Discuss three conditions with symptoms that mimic stroke and the assessment techniques the EMT may use to identify them. (pp 572-573)
7. Define a generalized seizure, partial seizure, and status epilepticus, including their effects on a patient and how they differ from each other. (pp 573-574)
8. Describe the different phases of a seizure. (p 574)

9. List the different types of seizures and their possible causes. (pp 573-575)
10. Explain why it is important for the EMT to recognize when a seizure is occurring or whether one has already occurred in a patient and to identify other problems that may be associated with the seizure. (p 575)
11. Describe the postictal state and the specific patient care interventions that may be necessary to assist the patient. (p 576)
12. Define altered mental status, its various possible causes, and the patient assessment considerations that apply to each. (pp 576-577)
13. Discuss the special considerations required for pediatric patients who exhibit altered mental status. (p 577)
14. Discuss scene safety considerations when responding to a patient with a neurologic emergency. (p 578)
15. Describe the steps involved in performing a primary assessment of a patient who is experiencing a neurologic emergency and the necessary interventions that may be required to address all life threats. (pp 578-579)
16. Describe the process of history taking for a patient who is experiencing a neurologic emergency and explain how this process varies depending on the nature of the patient's illness. (pp 579-581)
17. Describe the secondary assessment of patient who is experiencing a neurologic emergency. (pp 581-583)
18. Discuss how to use a stroke assessment tool to identify a stroke patient rapidly, giving examples of two commonly used tools. (pp 581-582)
19. List the key information an EMT must obtain and document for a stroke patient during assessment and reassessment. (pp 583-585)
20. Explain why a patient who is suspected of having a stroke is placed on stroke alert and requires treatment within the first 3 to 6 hours after the stroke begins. (pp 584-585)
21. Discuss special considerations for geriatric patients who are experiencing a neurologic emergency. (p 585)
22. Describe the patient management, treatment, and transport of patients who are experiencing headaches, stroke, seizure, and altered mental status. (pp 585-586)

Skills Objectives

1. Demonstrate how to use a stroke assessment tool such as the Cincinnati Prehospital Stroke Scale to test a patient for aphasia, facial weakness, and motor weakness. (pp 581-582)

Introduction

Stroke is the third leading cause of death in the United States, after heart disease and cancer. It is common in geriatric patients. Current statistics show that more men than women have strokes, but strokes are more likely to be fatal in women. Other contributing factors for stroke include family history and race—African Americans, Hispanics, and Asians have a higher risk for stroke. Fortunately, there are revolutionary treatments available for stroke, and many hospitals are certified stroke centers. Some patients can avoid the devastating consequences of an acute stroke if they can get to the hospital in time for treatment.

Seizures and altered mental status may also occur when there is a disorder in the brain. Seizures may occur as a result of a recent or an old head injury, a brain tumor, a metabolic problem, or simply a genetic disposition. Your ability to recognize when a seizure has occurred or is occurring is critical for the patient because you can then provide the appropriate management.

Altered mental status (AMS) is a common presentation in patients with a wide variety of medical problems. Although it is tempting, you should not make assumptions about the cause of AMS. There can be many possible causes; some are obvious, some are not: intoxication, head injury, hypoxia, stroke, metabolic disturbances, and many more. Obviously, treatment varies widely as well. Patients with AMS present a particular challenge in that they may be difficult to handle and frustrating to treat. Your professionalism is paramount in these situations.

This chapter describes the structure and function of the brain and the most common causes of brain disorders, including stroke, transient ischemic attack (TIA), seizure, headache, and AMS. It then discusses the signs and symptoms of each condition. You will also learn how to approach and assess a patient with a neurologic emergency and why prompt transport to an appropriate medical facility is so important.

Anatomy and Physiology

The brain is the body's computer. It controls breathing, speech, and all other body functions. All of your thoughts, memories, wants, needs, and desires reside in the brain. Different parts of the brain perform different functions. For example, some parts of the brain receive input from the senses, including sight, hearing, taste, smell, and touch; others control the muscles and movement; while others control the formation of speech.

The brain is divided into three major parts: the brain stem, the cerebellum, and the largest part, the cerebrum Figure 15-1 . The brain stem controls the most basic functions of the body, such as breathing, blood pressure, swallowing, and pupil constriction. Located just behind the brain stem, the cerebellum controls muscle and body coordination. It is responsible for coordinating complex tasks that involve many muscles, such as standing on one foot without falling, walking, writing, picking up a coin, and playing the piano.

The cerebrum, located above the cerebellum, is divided down the middle into the right and left cerebral hemispheres. Each hemisphere controls activities on the opposite side of the body. The front part of the cerebrum controls emotion and thought, and the middle part controls touch and movement. The back part of the cerebrum processes sight. In most people, speech is controlled on the left side of the brain near the middle of the cerebrum.

Messages sent to and from the brain travel through nerves. Twelve cranial nerves run directly from the brain to various parts of the head, such as the eyes,

You are the Provider: PART 1

At 6:23 PM, your unit is dispatched to 106 Scottie Drive for a 58-year-old woman who is having a seizure. You respond to the scene, which is located approximately 4 miles from your station. The weather is cloudy, the traffic is moderate, and the temperature is 87°F. While en route, you and your partner discuss the different types of seizures.

1. On the basis of the dispatch information, what type of seizure is the patient most likely experiencing?
2. What are some common causes of seizures in this patient's age group?

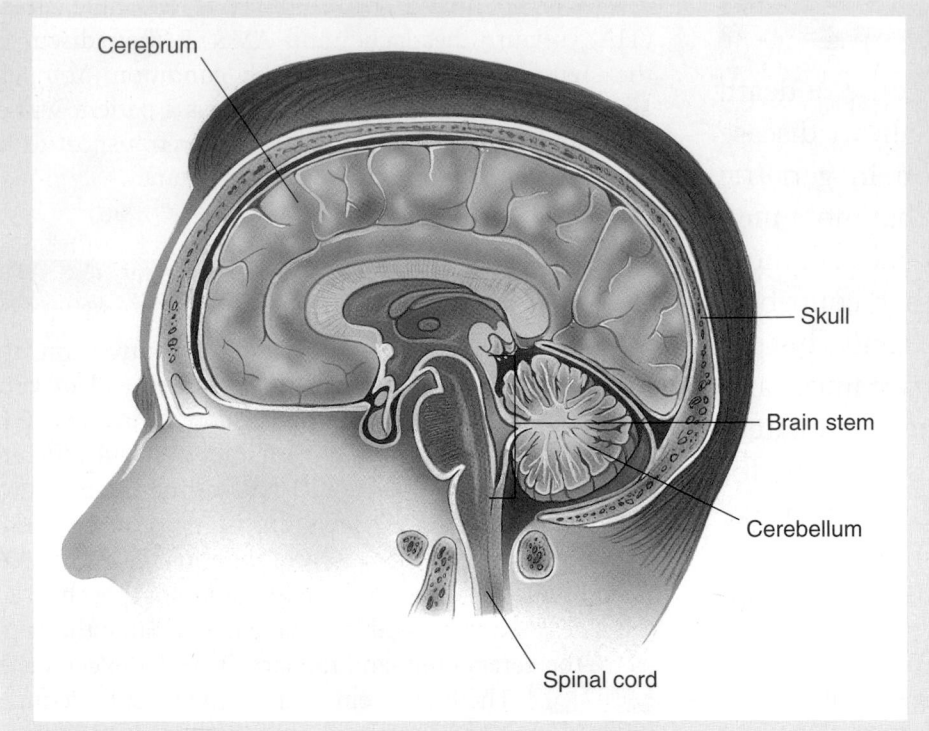

Figure 15-1 The brain lies well protected within the skull. Its major parts are the cerebrum, the cerebellum, and the brain stem.

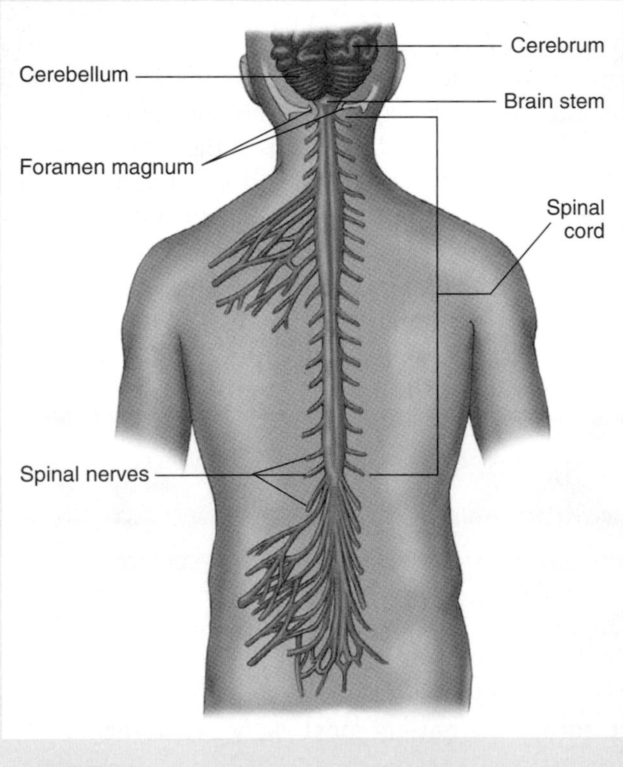

Figure 15-2 The spinal cord is the continuation of the brain stem. It exits the skull at the foramen magnum and extends down to the level of the second lumbar vertebra.

ears, nose, and face. All of the rest of the nerves join in the spinal cord and exit the brain through a large opening in the base of the skull called the foramen magnum **Figure 15-2**. At each vertebra in the neck and back, two nerves, called spinal nerves, branch out from the spinal cord and carry signals to and from the body.

Pathophysiology

Many different disorders can cause brain dysfunction or other neurologic symptoms and can affect the patient's level of consciousness, speech, and voluntary muscle control. The brain is most sensitive to changes in oxygen, glucose, and temperature levels. A significant change in any one of these three areas will result in a neurologic change. As a general rule, if the problem is caused primarily by disorders in the heart and lungs, the entire brain will be affected. For example, when the patient has no blood flow (cardiac arrest), he or she will go into a **coma** and can have permanent brain damage within minutes. However, if the primary problem is in the brain, such as a poor blood supply to one side of the brain, the patient may have signs and symptoms affecting only one side of the body. A low oxygen level in the bloodstream will affect the entire brain, often causing anxiety, restlessness, and confusion.

Stroke is a common neurologic condition that is potentially treatable. Other brain disorders include infection and tumor, which may themselves cause seizures, AMS, and headaches. The information in this chapter will help you better understand, communicate with, and care for patients who have experienced a neurologic emergency.

Headache

One of the most common complaints you will hear from your patients in terms of pain is headache. Because it is subjective, a headache can be a symptom of another condition or it can be considered a neurologic condition on its own. Millions of people experience a headache every year, but only a small percentage of these are caused by a serious medical condition. The brain and skull do not actually sense pain because neither contains pain receptors. The pain associated with a headache is felt from the surrounding areas of the face, scalp, meninges (membranes that cover the brain and spinal cord), larger blood vessels, and the muscles of the head, neck, and face.

Most headaches are harmless and do not require emergency medical care. However, headaches can be caused by a stroke, tumor, infection of the central nervous system, or hypertension, all of which are potentially life-threatening conditions. A patient who presents with a sudden, severe headache or a headache with other signs and symptoms should be considered in serious condition and requires a complete assessment and transport to the hospital. An incident in which you have more than one patient reporting a headache may indicate carbon monoxide poisoning.

Tension headaches, migraines, and sinus headaches are the most common types of headaches and are not considered life threatening, although they may be debilitating for the patient. Tension headaches are the most common type of headache. These headaches are caused by muscle contractions in the head and neck and are attributed to stress. Patients usually describe the pain as squeezing, dull, or as an ache. This type of headache does not have any associated symptoms and usually does not require medical attention.

Migraine headaches are the second most common type of headache and are thought to be caused by changes in the blood vessel size in the base of the brain. Adults and children have migraine headaches. Women experience migraines three times more often than do men. Frequently, the patient will have a history of migraines and will tell you that this episode is similar to one in the past. Pain from a migraine headache is usually described as pounding, throbbing, and pulsating. Migraines are often associated with visual changes such as flashing lights or partial vision loss. The patient will often have nausea and vomiting. Migraine headaches can last for several days.

Sinus headaches are caused by pressure that is the result of fluid accumulation in the sinus cavities. Patients may also have coldlike signs and symptoms of nasal congestion, cough, and fever if they also have a sinus infection. Patients may report increased pain when they bend over or when their head is moved forward. This type of headache is self-limiting, and prehospital emergency care is not required.

Although most headaches are not life-threatening situations, there are patients with a chief complaint of headache who will require medical attention. Serious conditions that include headache as a symptom are hemorrhagic stroke (bleeding in the brain), brain tumors, and meningitis. You should be concerned if the patient complains of a sudden onset, severe headache or a sudden headache that has associated symptoms. Headaches

You are the Provider: PART 2

You arrive at the scene and are greeted by the patient's sister. She tells you that they were having a conversation when the patient suddenly grabbed both sides of her head and then began "shaking all over." The patient is lying on the floor in her living room with a pillow behind her head. She is conscious, but confused, and is complaining of a severe headache. You perform an assessment as your partner opens the jump bag.

Recording Time: 0 Minutes	
Appearance	Slightly pale; skin is dry
Level of consciousness	Conscious, but confused
Airway	Open; clear of secretions and foreign bodies
Breathing	Rapid rate; adequate depth
Circulation	Radial pulse, rapid and bounding

Your partner applies high-flow oxygen via a nonrebreathing mask and prepares to take the patient's vital signs. The patient's sister tells you that she has never had seizures before. The patient is wearing a medical alert bracelet, which identifies her medical history of high blood pressure, heart disease, and type 2 diabetes.

3. What additional questions should you ask the patient's sister?

4. What prehospital assessments can you perform to try to determine the cause of the patient's seizure?

5. Other than oxygen, what additional treatment is indicated at this point?

with fever, seizures, or AMS or following head trauma are potentially life threatening.

A sudden, severe headache, often described as the worst pain the patient has ever had, could be a sign of a hemorrhagic stroke. The blood from a ruptured blood vessel irritates the tissues of the brain and can cause increased intracranial pressure, resulting in severe head pain. This type of pain may initially be localized and then become more diffuse as the irritation in the meninges spreads. You should suspect a stroke in patients with a severe headache, seizures, and AMS. Early signs of increased intracranial pressure include headache, vomiting, AMS, and seizures. Increasing intracranial pressure can be from a hemorrhagic stroke (a blood vessel swells and ruptures) or from head trauma that may have occurred hours or days before this event. Your patient assessment should include asking the patient if he or she has experienced any recent head trauma.

Bacterial meningitis, an inflammation of the meninges caused by a bacterial infection, is a central nervous system infection in which the patient may complain of a headache, stiff neck, fever, and sensitivity to light. This is a serious condition requiring prompt medical attention. Use standard precautions, and provide supportive care of the ABCs. Provide a quiet, darkened environment, when possible.

Stroke

A **cerebrovascular accident (CVA)**, or **stroke**, is an interruption of blood flow to the brain that results in the loss of brain function. Lacking oxygen, brain cells stop working and begin to die. Medical science currently has little to offer in the way of treatment once these cells are dead. However, it may take several hours or more for brain cells to die, even when it appears that severe disability will occur. In some cases, a trickle of blood may still be getting through to the affected area of the brain. This blood may supply enough oxygen to keep cells alive but not enough to allow them to work properly. For example, if cells that are responsible for controlling the left arm are starved for oxygen, the patient will not be able to move that arm. The brain cells will develop **ischemia**, a lack of oxygen that causes the cells not to function properly. If normal blood flow is restored to that area of the brain, in time, the patient may regain full use and control of the arm.

Unfortunately, many patients having a stroke deny or ignore their symptoms and delay seeking medical attention. The delay in seeking care can result in devastating consequences because "time is brain."

There are two main types of stroke: ischemic and hemorrhagic. An ischemic stroke is due to the direct blockage of blood flow through the cerebral arteries. This blockage can be due to **thrombosis**, where a clot forms at the site of blockage, or due to an **embolus**, where the clot forms in a remote area (such as a diseased heart) and then travels to the site of blockage. In hemorrhagic stroke, a blood vessel ruptures, causing increased pressure in the brain and subsequent brain damage.

■ Types of Stroke

Ischemic Stroke

Ischemic stroke is the most common type of stroke, accounting for more than 80% of all strokes combined. When blood flow to a particular part of the brain is cut off by a blockage (clot) inside a blood vessel, the result is an ischemic stroke. This can be from a thrombosis or an embolism that blocks blood flow. As with coronary artery disease, atherosclerosis in the blood vessels is often the cause. Atherosclerosis is a disorder in which calcium and cholesterol build up, forming a plaque inside the walls of the blood vessels. This plaque may obstruct blood flow and interfere with the vessel's ability to dilate. Eventually, atherosclerosis may cause complete occlusion (blockage) of an artery `Figure 15-3`. In other cases, an atherosclerotic plaque in the carotid artery in the neck will rupture. A blood clot will form over the crack in the plaque, sometimes growing large enough to completely block all blood flow through that artery. Deprived of oxygen, parts of the brain supplied by the artery will stop working. Patients with ischemic strokes may have dramatic

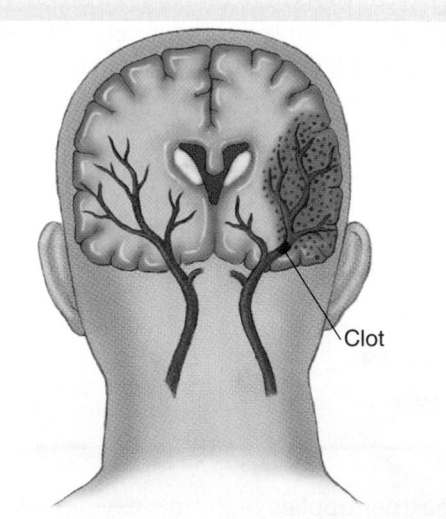

Figure 15-3 Atherosclerosis can damage the wall of a cerebral artery, producing narrowing and/or a clot. When the vessel is narrowed or completely blocked, blood flow to that part of the brain may be blocked, and the cells begin to die because of the lack of adequate oxygenation.

symptoms, including loss of movement on the side of the body opposite from the side where the blockage has occurred.

Even if the blockage in the carotid artery is not complete, smaller pieces of the clot may embolize (break off and be carried by the normal flow of blood) deep into the brain, heart, or lungs. If this piece of clot ends up in the brain, it may become lodged in a smaller branch of a blood vessel. This cerebral embolism then blocks blood flow Figure 15-4 . Depending on the location of the lodged clot, the patient may experience a range of symptoms from nothing at all to complete paralysis.

Hemorrhagic Stroke

Hemorrhagic stroke accounts for 10% to 20% of all strokes and occurs as a result of bleeding inside the brain. The free blood then forms a clot, which squeezes the brain tissue next to it. When that tissue is compressed, oxygenated blood cannot get into the area and the brain cells begin to die.

Certain people are at higher risk of hemorrhagic stroke, which commonly occurs in people experiencing stress or exertion. The people who are at highest risk are those who have very high blood pressure or long-term elevated blood pressure that is not treated. After many years of high pressure, the blood vessels in the brain weaken. If a vessel ruptures, the bleeding in the brain will increase the pressure inside the cranium.

Cerebral hemorrhages are often fatal, although proper treatment of high blood pressure can help to prevent this long-term damage to the blood vessels, reducing morbidity and mortality.

Some people are born with weaknesses in the walls of the arteries, called aneurysms. An **aneurysm** is a swelling or enlargement of part of an artery resulting from weakening of the arterial wall. Figure 15-5 is an angiogram of a cerebral aneurysm. Many people with an aneurysm have a sudden onset of a severe headache. The headache is from the irritation of blood on the brain tissue after the artery swells and ruptures. When a hemorrhagic stroke occurs in an otherwise healthy young person, the likely cause is often a weakness in a blood vessel called a berry aneurysm. This type of aneurysm resembles a tiny balloon (or berry) that juts out from the artery. When the aneurysm is overstretched and ruptures, blood spurts into an area between two of the coverings of the brain called the subarachnoid space. These types of strokes are called subarachnoid hemorrhages. Again, patients with this type of stroke experience a sudden severe headache, typically described as the worst headache they have ever had. If the patient gets to the hospital rapidly, surgeons may be able to repair the aneurysm. However, like other brain bleeding and cerebral hemorrhage, this condition is often fatal.

Transient Ischemic Attack

In some patients, normal processes in the body will break up a blood clot in the brain. When that happens quickly, blood flow is restored to the affected area, and the patient will regain use of the affected body part; however, this often indicates that the patient has a serious medical condition that may prove fatal. When stroke symptoms go away on their own in less than 24 hours, the event is called a **transient ischemic attack (TIA)**. Some patients call these ministrokes.

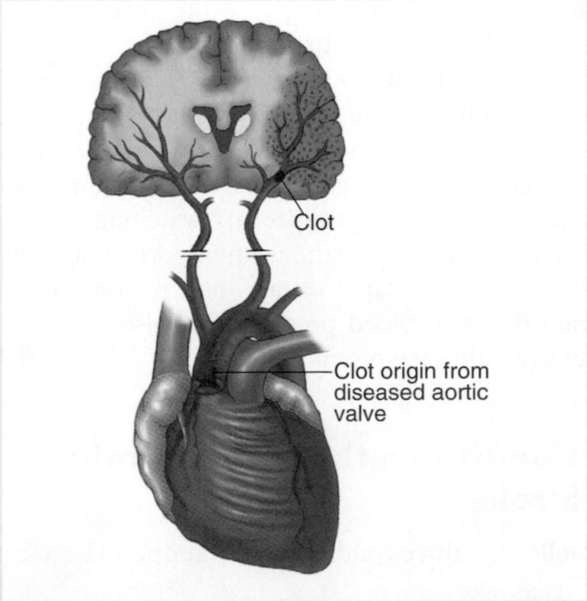

Clot

Clot origin from diseased aortic valve

Figure 15-4 An embolus, a blood clot often formed on a diseased heart valve, can travel through the body's vascular system, lodge in a cerebral artery, and cause a stroke.

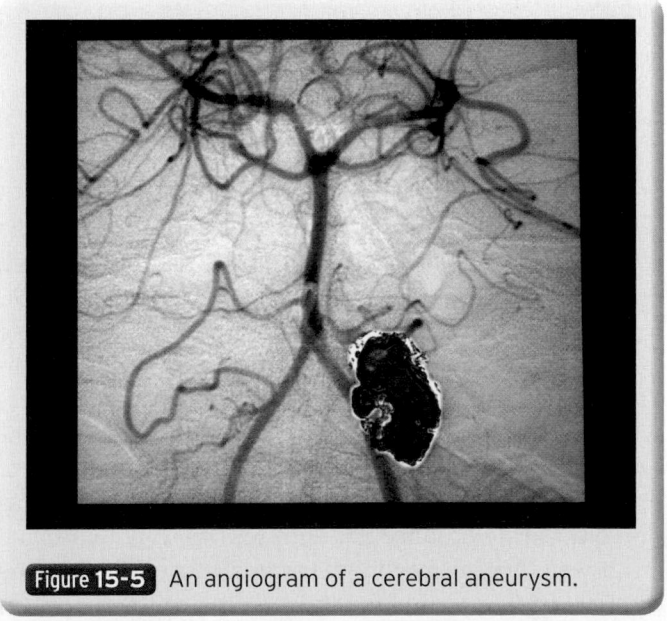

Figure 15-5 An angiogram of a cerebral aneurysm.

Although most patients with TIAs do well, every TIA is an emergency. It may be a warning sign that a larger, significant stroke may occur in the future. Approximately one third of patients who have a TIA will experience a stroke soon after the TIA. For this reason, all patients with a TIA should be evaluated by a physician to determine whether preventive action should be taken.

■ Signs and Symptoms of Stroke

The general signs and symptoms of stroke include the following:

- Facial drooping
- Sudden weakness or numbness in the face, arm, leg, or one side of the body
- Loss of movement and sensation on one side of the body
- Lack of muscle coordination (ataxia)
- Sudden vision loss in one eye, blurred and double vision
- Difficulty swallowing (a primary reason for good airway management in a patient with a stroke)
- Decreased or increased level of responsiveness
- Speech disorders (dysphasia)
- Difficulty expressing thoughts or inability to speak (expressive aphasia)
- Slurred speech (**dysarthria**)
- Difficulty understanding others (receptive aphasia)
- Decreased or absent movement in one or more extremities
- Sudden and severe headache
- Sudden loss of balance or trouble walking
- Confusion
- Dizziness
- Weakness
- Combativeness
- Restlessness
- Tongue deviation
- Coma

Left Hemisphere

If the left cerebral hemisphere has been affected by a stroke, the patient may have a speech disorder called **aphasia**, an inability to produce or understand speech. Speech problems can vary widely. Some patients will have trouble understanding speech but be able speak clearly. You can detect this problem by asking the patient a question such as "What day is today?" In response, the patient may say, "Green." The speech is clear, but it does not make sense. Other patients will be able to understand the question but cannot produce the right sounds so as to provide an answer. Only grunts or other incomprehensible sounds emerge. Strokes that affect the left side of the brain can also cause paralysis of the right side of the body.

Right Hemisphere

If the right cerebral hemisphere of the brain is not getting enough blood, patients will have trouble moving the muscles on the left side of the body. Usually, they will understand language and be able to speak, but their words may be slurred and hard to understand.

Interestingly, patients with right hemisphere strokes may be completely oblivious to their problem. If you ask the patients to lift their left arm and they cannot, they will lift their right arm instead. They seem to have forgotten that the left arm even exists. This symptom is called neglect. Patients with a problem affecting the back part of the cerebrum may neglect certain parts of their vision. Generally, this is hard to detect in the field because of the patient's ability to compensate without conscious effort. Nevertheless, you should be aware of the possibility. Try to sit or stand on the patient's "good" side because he or she may be unable to see things on the "bad" side.

The problem of neglect causes many patients who have had large strokes to delay seeking help. Strokes may not be painful; therefore, a patient may be unaware that he or she has a problem until a family member or friend points out that some part of the patient's body is not working correctly.

Bleeding in the Brain

Patients who have bleeding in their brain (cerebral hemorrhage) may have very high blood pressure. High blood pressure can cause the bleeding, but many times the high blood pressure is a compensatory response to the bleeding itself. The blood pressure increases as the body attempts to force more oxygen to the area of the brain where the damage is occurring. Remember, the brain is located inside a box (skull) with only a few openings. When bleeding is occurring inside the brain, the pressure inside the skull increases. The body must increase the blood pressure to get blood to the brain's tissues. Monitoring the blood pressure and watching for a trend of increasing blood pressure is an important sign. Blood pressure may then taper off and return to normal. Significant drops in blood pressure may also occur as the patient's condition worsens.

■ Conditions that May Mimic Stroke

The following three conditions may appear to be a stroke:

- Hypoglycemia
- A **postictal state** (a period following a seizure that lasts between 5 and 30 minutes, characterized by labored respirations and some degree of AMS)
- Subdural or epidural bleeding (a collection of blood near the skull that presses on the brain)

Because oxygen and glucose are needed for brain metabolism, a patient with hypoglycemia may present like a patient who is having a stroke. With good patient assessment, you should find out whether the patient's medical history includes diabetes. All patients with an AMS should have their blood glucose level checked if allowed by your local protocol.

A patient who has experienced a seizure may look like a patient who is having a stroke. This is often referred to as the postictal state. However, in most cases, a patient having a seizure will recover rapidly, within several minutes.

Subdural and epidural bleeding usually occur as a result of trauma. The dura is a leathery covering over the brain, next to the skull. A fracture near the temples may cause an artery to bleed on top of the dura, resulting in pressure on the brain **Figure 15-6A**. The onset of this epidural bleeding is usually very rapid after injury. In other cases, the veins just below the dura may be torn and bleed; this is referred to as subdural bleeding **Figure 15-6B**. The onset of subdural bleeding occurs more slowly, sometimes over a period of several days.

With subdural and epidural bleeding, the onset of strokelike signs and symptoms may be subtle; the patient or family may not even remember the original injury that is causing the bleeding.

Words of Wisdom

Epidural hemorrhage can be associated with a period of time (usually minutes to hours) after a head injury where the patient has normal mental status, only to deteriorate markedly. This period between injury and deterioration is called a lucid interval.

Seizures

The incidence of seizures in emergency medical care is high. EMS systems have reported that as many as 30% of their 9-1-1 calls involve a patient having a seizure. In the United States, it is estimated that 4 million people have epilepsy. A **seizure**, or convulsion, is a temporary alteration in consciousness. The different types of seizures can be classified as generalized, partial, and status epilepticus.

A **generalized seizure**, formerly called a grand mal seizure, is typically characterized by unconsciousness and a generalized severe twitching of all of the body's muscles that lasts several minutes or longer. This type of seizure results from abnormal discharges from large areas of the brain, usually involving both hemispheres.

In other cases, the seizure may simply be characterized by a brief lapse of consciousness in which the patient seems to stare and not to respond to anyone. This type of seizure does not involve any changes in motor activity and is called a petit mal or absence seizure.

A **partial seizure** begins in one part of the brain and is classified as simple or complex. In a simple partial seizure, there is no change in the patient's level of consciousness. Patients may complain of numbness, weakness, or dizziness. Their senses may also be involved, with patients reporting visual changes and unusual smells or tastes. A simple partial seizure may also cause twitching of the muscles and the extremities that may spread slowly from one part of the body to another, but it is not characterized by the dramatic severe twitching and muscle movements seen in a generalized seizure. Patients may also have brief paralysis.

In a complex partial seizure, the patient has an AMS and does not interact normally with his or her environment. This type of seizure results from abnormal discharges from the temporal lobe of the brain. Other characteristics may be lip smacking, eye blinking, and isolated convulsions or jerking of the body or one part of the body such as an arm. Patients also may experience unpleasant smells and visual hallucinations, exhibit uncontrollable fear, and perform repetitive physical behavior such as constant sitting and standing.

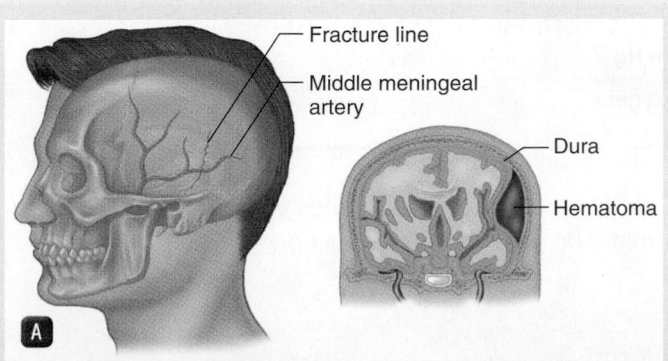
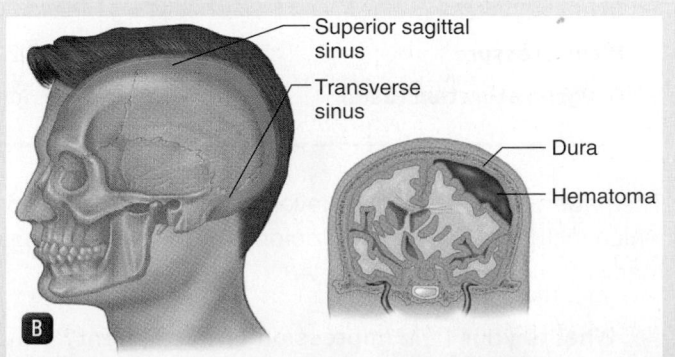

Figure 15-6 Trauma to the head can result in intracranial bleeding. **A.** Bleeding outside the dura and under the skull is epidural. **B.** Bleeding beneath the dura but outside the brain is subdural.

Some seizures occur on only one side of the body. Others begin on one side and gradually progress to a generalized seizure that affects the entire body. Depending on the type of seizure, the patient may have no loss of consciousness but still experience body shaking or muscle tremors. Most people with lifelong or chronic seizures tolerate these events reasonably well without complications, but in some situations, seizures may signal life-threatening conditions.

Often, a patient may experience a warning prior to the event, referred to as an **aura**. The seizure is characterized by sudden loss of consciousness, chaotic muscle movement and tone, and apnea. The patient may also experience a tonic phase, usually lasting only seconds, in which there will be a period of constant muscle contraction and trembling, tongue biting, bladder incontinence, or bowel incontinence. During a **tonic-clonic seizure**, the patient may exhibit bilateral movement characterized by muscle rigidity and relaxation usually lasting 1 to 3 minutes. Throughout a tonic-clonic seizure, the patient exhibits tachycardia, hyperventilation, sweating, and intense salivation. Most seizures last 3 to 5 minutes and are followed by a lengthy period (5 to 30 minutes or more) called a postictal state, in which the patient is unresponsive at first and gradually regains consciousness. The postictal state is over when the patient regains a complete return of his or her normal level of consciousness. In most cases, the patient will gradually begin to recover and awaken but appear dazed, confused, and fatigued. In contrast, an absence (formerly called petit mal) seizure can last for just seconds, after which the patient fully recovers with only a brief lapse of memory of the event. Seizures lasting more than 5 minutes are likely to progress to status epilepticus. Seizures that continue every few minutes without the person regaining consciousness or last longer than 30 minutes are referred to as **status epilepticus**. Recurring or prolonged seizures should be considered potentially life-threatening situations in which patients need emergency medical care. If the patient does not regain consciousness or the seizure continues, protect the patient from harming himself or herself, and call for advanced life support (ALS) backup. The patients need advanced airway management and medication to stop the seizure.

■ Causes of Seizures

Some seizure disorders, such as epilepsy, are congenital, which means that the patient was born with the condition. Other types of seizures may be caused by high fevers, structural problems in the brain, or metabolic or chemical problems in the body **Table 15-1**. There is also a percentage of the population who will experience a seizure in which the cause cannot be determined (idiopathic). Epileptic seizures usually can be controlled with medications such

---

You are the Provider: PART 3

Your partner reports the patient's vital signs. The patient tells you that she is nauseated and that her headache, which is still severe, is located on both sides of her head. She further tells you that the last thing she remembers was the sudden, severe headache. When she woke up, she was lying on the floor with a pillow under her head. Her sister tells you that she caught the patient before she struck the ground.

Recording Time: 4 Minutes	
Respirations	14 breaths/min; adequate depth
Pulse	100 beats/min; strong and regular
Skin	Pink, warm, and moist
Blood pressure	200/112 mm Hg
Oxygen saturation (Sao$_2$)	96% (on oxygen)

The patient's blood glucose level is assessed and noted to be 97 mg/dL. Her sister hands you her medication list, which includes benazepril, clopidogrel (Plavix), and metformin. The patient says she is noncompliant with her medication regimen.

6. What is your field impression of this patient? Why?

7. On the basis of your field impression, for which additional signs and symptoms should you monitor the patient?

Table 15-1 Common Causes of Seizures

Type	Cause
Epileptic	Congenital origin
Structural	Tumor (benign or cancerous)
	Infection (brain abscess)
	Scar tissue from injury
	Head trauma
	Stroke
Metabolic	Hypoxia
	Abnormal blood chemical values
	Hypoglycemia
	Poisoning
	Drug overdose
	Sudden withdrawal from alcohol, medications
Febrile	Sudden high fever

as phenytoin (Dilantin), phenobarbital, or carbamazepine (Tegretol). Patients with epilepsy will often have seizures if they stop taking their medications or if they do not take the prescribed dose on a regular basis.

Seizures may also be caused by an abnormal area in the brain, such as a benign or cancerous tumor, an infection (brain abscess), or scar tissue from some type of injury. These seizures are said to have a structural cause. Seizures from a metabolic cause can result from abnormal levels of certain blood chemicals (eg, extremely low sodium level), **hypoglycemia** (low blood glucose level), poisons, drug overdoses, or sudden withdrawal from routine heavy alcohol or sedative drug use or even from prescribed medications. Phenytoin, a drug that is used to control seizures, can cause seizures itself if the person takes too much.

Seizures can also result from sudden high fevers, particularly in children. **Febrile seizures** are usually very unnerving for parents to observe but are generally well tolerated by the child. Nevertheless, you must transport a child who has had a febrile seizure because this condition needs to be evaluated in the hospital. The fact that a second seizure may occur is worrisome, and if it occurs, the patient requires rapid transport to the hospital so possible causes can be identified.

■ The Importance of Recognizing Seizures

Regardless of the type or the cause of a seizure, it is important for you to recognize when a seizure is occurring or whether one has already occurred. You must also determine whether this episode differs from any previous ones. For example, if the previous seizure occurred on only one side of the body and this seizure occurs over the entire body, an additional or new problem may be involved. In addition to recognizing that seizure activity has occurred and/or that something different may now be occurring, you must also recognize the postictal state and the complications of seizures.

Because most seizures involve a vigorous twitching of the muscles, the muscles use a lot of oxygen. This excessive demand consumes oxygen that is needed for the vital functions of the body. As a result, there is a buildup of acids in the bloodstream, and the patient may turn cyanotic (bluish lips, tongue, and skin) from the lack of oxygen. Often, the seizures themselves prevent the patient from breathing normally, making the problem worse. In a patient with diabetes, the blood glucose value may drop because of the excessive muscular contraction of a seizure. If approved in your system, monitor the blood glucose level closely after a patient with diabetes has a seizure

Words of Wisdom

Be on the lookout for patients who may behave violently during the postictal phase. Although most patients who have had a seizure pose no threat to EMS personnel, signs of alcohol or drug abuse should heighten your awareness of the potential for dangerous behavior.

Recognizing seizure activity also means looking at other problems associated with the seizure. For example, the patient may have fallen during the seizure episode and have an injury; head injury is the most serious possibility. Patients having a generalized seizure also may experience **incontinence**, meaning that they may lose bowel and bladder control. Therefore, one clue that unresponsive or confused patients may have had a seizure is to find that they were incontinent. Although incontinence is possible with other medical conditions, sudden incontinence is very likely a sign that a seizure has occurred. When the patients regain consciousness, they are naturally embarrassed by this temporary loss of control. Do what you can to minimize this discomfort by covering the patient and assuring him or her that incontinence is part of the loss of control that accompanies a seizure.

Words of Wisdom

Physician examination of a patient who has had a seizure depends heavily on reports of the seizure pattern and changes in that pattern. Record all pertinent information about the seizure in terms of duration, areas of body movement, and possible triggering factors. This requires effective interviewing of available witnesses, family members, and/or caregivers.

The Postictal State

Once a seizure has stopped, the patient's muscles relax, becoming almost flaccid, or floppy, and the breathing becomes labored (fast and deep) in an attempt to compensate for the buildup of acids in the bloodstream. By breathing faster and more deeply, the body can balance the acidity in the bloodstream. With normal circulation and liver function, the acids clear away within minutes, and the patient will begin to breathe more normally. The longer and more intense the seizures are, the longer it will take for this imbalance to correct itself. Likewise, longer and more severe seizures will result in longer postictal unresponsiveness and confusion. Once the patient regains a normal level of consciousness, the postictal state is over.

Words of Wisdom

Interventions during the postictal phase are critical. Patients may be unable to maintain an open airway because of the relaxed and exhausted state they are in; therefore, patient positioning, clearing the airway of secretions, and being alert for the potential to aspirate are critical tasks for you to perform to best achieve positive patient outcomes.

In some situations, the postictal state may be characterized by **hemiparesis**, or weakness on one side of the body, resembling a stroke. Unlike the typical stroke, hypoxic hemiparesis soon resolves. Most commonly, the postictal state is characterized by lethargy and confusion to the point that the patient may be combative. You must be prepared for these circumstances, in your approach to scene control and in your treatment of the patient's symptoms. If the patient's condition does not improve, you should consider other possible underlying conditions such as hypoglycemia or infection.

Special Populations

Status epilepticus is harmful at any age, but because of the normal aging process, geriatric patients are at greater risk of hypoxia, hypotension, and/or cardiac dysrhythmias.

Altered Mental Status

Aside from stroke and seizures, the most common type of neurologic emergency that you will encounter is a patient with an AMS. Simply put, AMS means that the patient is not thinking clearly or is incapable of being aroused.

In some cases, the patient will be unconscious; in others, the patient may be alert but confused Figure 15-7. The range of problems is wide, and the causes are many, including hypoglycemia, hypoxemia, intoxication, drug overdose, unrecognized head injury, brain infection, body temperature abnormalities, brain tumors, and overdoses and/or poisonings.

Causes of Altered Mental Status

Hypoglycemia

The clinical picture of patients with AMS caused by hypoglycemia is very complex. Patients can have signs and symptoms that mimic stroke and seizures. Because oxygen and glucose are needed for brain function, hypoglycemia can mimic conditions in the brain such as those associated with stroke. In these cases, the patient may have hemiparesis, similar to what occurs as a result of a stroke. The principal difference, however, is that a patient who has had a stroke may be alert and attempting to communicate normally, whereas a patient with hypoglycemia almost always has an altered or decreased level of consciousness Figure 15-8.

Patients with hypoglycemia commonly, but not always, take medications that lower the blood glucose level. Thus, if the patient appears to have signs and symptoms of stroke and an AMS, you should report your findings to medical control and treat the patient accordingly. Remember that patients with a decreased level of consciousness should not be given anything by mouth.

Patients with hypoglycemia can also experience seizures, and you may arrive at the scene to find a patient in a postictal state: confused and disoriented or

Figure 15-7 A patient with an altered mental status can be unconscious in some cases; in others, the patient may be alert but confused.

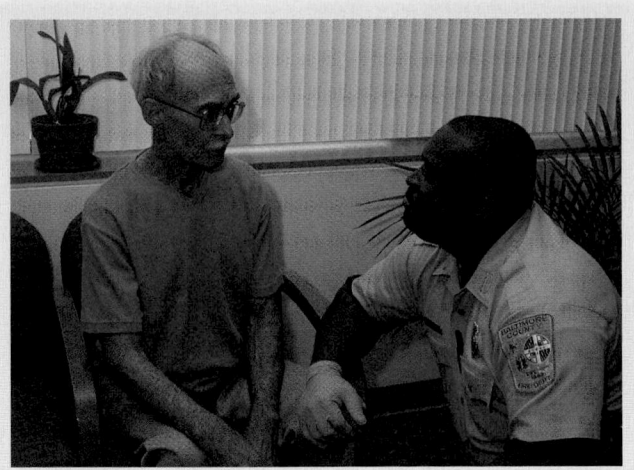

Figure 15-8 During your assessment of a patient with an altered or decreased level of consciousness, consider the possibility of hypoglycemia.

unresponsive. The mental status of a patient who has had a typical seizure is likely to improve. However, in a patient with hypoglycemia, the mental status is not likely to improve, even after several minutes. You should consider the possibility of hypoglycemia in a patient who has had a seizure, especially if the patient has a history of diabetes.

Likewise, you should consider hypoglycemia in a patient who has AMS after an injury such as a motor vehicle crash, even when there is the possibility of an accompanying head injury. As with any other patient,

Words of Wisdom

Always remember that altered mental status is a symptom, not a disease.

Words of Wisdom

A helpful mnemonic to use when reviewing the possible causes of altered mental status is "Tips on the Vowels:"
T Trauma
I Infection
P Psychogenic causes
S Seizure, syncope

A Alcohol
E Electrolytes
I Insulin
O Opiates
U Uremia (toxic condition resulting from kidney disease)

you should look for medical identification bracelets or medications that might confirm your suspicions.

Other Causes of Altered Mental Status

Other causes of AMS include unrecognized head injury and severe alcohol intoxication. Your consideration of other possibilities becomes important because a patient with AMS may be combative and refuse treatment and transport. You should be prepared for difficult patient encounters and follow local protocols for dealing with these situations.

In most cases, a patient who appears intoxicated most likely is just that; however, you must consider other problems as well. A person with chronic alcoholism can have liver function, blood clotting, and immune system abnormalities, causing a predisposition to intracranial bleeding, brain and bloodstream infections, and hypoglycemia.

Psychological problems and complications of medications are also possible causes of AMS. A person who appears to have a psychological problem may also have an underlying medical condition.

Infections are another possible cause of AMS, particularly infections involving the brain or bloodstream. Infections in these areas are life threatening and require immediate attention. Patients may not demonstrate the typical signs of infection, such as fever, particularly if they are very young or very old or have an impaired immune system. An AMS can also be caused by a drug overdose or poisoning.

Special Populations

Children can have altered mental status caused by strokes, seizures, high or low blood glucose levels, infection (eg, meningitis), poisoning, or tumors. Hemorrhagic strokes are usually caused by congenital defects in blood vessels; these defects are called berry aneurysms. Ischemic strokes can be caused by disorders such as sickle cell anemia. Children who have sickle cell anemia are at a particularly high risk for ischemic stroke. Treat stroke and AMS in children the same way that you do in adults.

As mentioned earlier in this chapter, seizures can result from sudden high fevers, particularly in children. Remember that although febrile seizures are generally well tolerated by children, you must transport the patient to the hospital. The possibility of a second seizure makes transport mandatory so that if other problems develop, the child is in the hospital and can receive immediate definitive care.

If you suspect that a patient with AMS has hypoglycemia and you are trained and approved to do so, you should use your glucometer to test the blood glucose level and treat the patient according to local protocols. Remember, the patient requires close monitoring, particularly of the airway, en route to the hospital.

Patient Assessment

Scene Size-up

Scene Safety

Your dispatcher may obtain a lot of information about your patient or very little. Dispatchers are frequently given information regarding a seizure by the caller. Even if the caller has never seen a seizure before, on the basis of the caller's description of the convulsions, the dispatcher will recognize that a seizure is taking place.

In some calls, the description of the patient's signs and symptoms will give you a fairly good idea of what the problem may be (the patient has slurred speech or one-sided paralysis), or the description may be vague (patient has a headache). There are many ways to describe a patient with an AMS. The most significant difference between an AMS and other emergencies is that your patient cannot tell you reliably what is wrong, and there may be more than one cause. Try to make an early determination if the call is medical or trauma related because this sets the tone and approach to the patient.

Do not be distracted by the seriousness of the situation or by frightened family members who want you to rush. Look first for threats to your safety, and follow standard precautions.

Consider the need for spinal precautions based on dispatch information and your assessment of the scene as you approach the patient. Many calls involving a neurologic emergency require ALS if it is available. Call for additional resources early.

Mechanism of Injury/Nature of Illness

You will need to look for clues to help you determine the nature of illness. There are special considerations for a patient with a suspected neurologic emergency that include an evaluation of the patient's environment, assessing for any signs of potential trauma (mechanism of injury), indications of a previous medical condition such as diabetic supplies or medical alert tags, and evidence of a seizure. Be aware of indications of the nature of illness—did anyone witness what happened? When was the last time anyone saw the patient appearing healthy? Is the patient's bed or furniture in disarray? Most patients with a neurologic emergency display a change in their level of consciousness and their ability to interact with their environment and others. You will need to be alert for other signs and symptoms related to the present illness to help determine the cause of the emergency.

Primary Assessment

As you begin to assess the patient, remember that your first priority is to look for and treat any life-threatening conditions. Perform a rapid scan. There are many reasons why patients become unresponsive or have an altered level of consciousness, especially from a neurologic cause. Your sound approach to assessing the airway, breathing, and circulation will have a significant impact on how well these systems respond to your care and treatment.

You are the Provider: PART 4

An engine crew arrives shortly before you load the patient onto the stretcher. Shortly after loading her into the ambulance, you reassess her and note that her condition has deteriorated. One of the engine crew EMTs accompanies you in the back of the ambulance, and you proceed to the hospital.

Recording Time: 10 Minutes	
Level of consciousness	Unconscious and unresponsive
Respirations	6 breaths/min; snoring, irregular, and shallow
Pulse	60 beats/min; bounding
Skin	Pink, warm, and dry
Blood pressure	198/110 mm Hg
Sao$_2$	94% (on oxygen)

8. What should be your most immediate action?

9. What additional treatment does this patient require?

Form a General Impression

As you approach the patient, gather information from the scene (is this medical or trauma?) and note the patient's body position and level of consciousness. This information will often help set the pace of your call as you are able to determine early on how critical the situation may be. A patient found lying on the ground in an unnatural position is more likely to be a patient with a potentially life-threatening condition than one found sitting up in bed. In a call that indicates that a seizure is taking place, you should be able to tell whether the patient is still having a seizure. Unless you are stationed extremely close to where the patient is located and have an arrival time of a minute or less, most seizures should be over by the time you arrive. If the seizure is still occurring, the potentially life-threatening condition of status epilepticus may be present. If the patient is in a postictal state, he or she may be unresponsive or starting to regain awareness of the surroundings. Determining the patient's level of consciousness should be first in the list of assessment actions for anyone with an AMS. To assess the patient's level of consciousness, use the AVPU (*Alert* to person, place, and day; responsive to *Verbal* stimuli; responsive to *Pain*; *Unresponsive*) scale.

Airway and Breathing

As with any other situation, you should focus on the patient's airway and breathing on arrival. Strokes affect how the body functions in many ways. Patients may have difficulty swallowing and are at risk for choking on their own saliva. Evaluate the airway of an unresponsive patient to make sure it is patent and will remain that way Figure 15-9 . If the patient requires assistance maintaining an airway, consider an oropharyngeal or

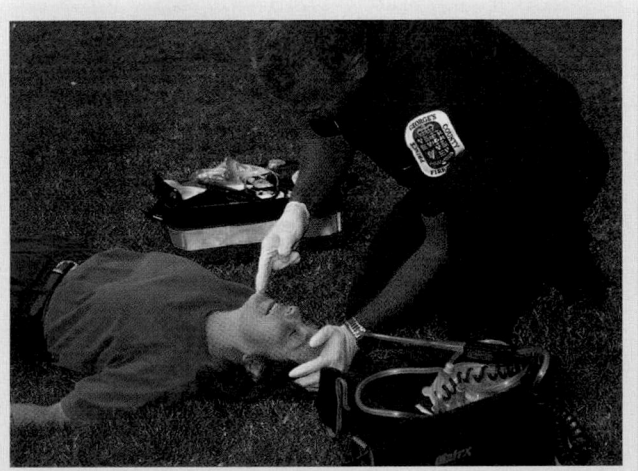

Figure 15-9 Securing and maintaining the airway in a patient who is unconscious is critical; also be sure to have suction readily available in case the patient vomits.

nasopharyngeal airway. Provide suction, and position the patient to prevent aspiration. If you determine that the patient cannot protect his or her airway, place the patient in the recovery position to help prevent secretions from entering the airway. Suction as necessary.

A patient who has had or is having a seizure may have been eating or chewing gum at the time of the seizure, and there may be a foreign-body obstruction. Bystanders may have tried to put objects in the patient's mouth to keep the person from "swallowing the tongue," even though this practice is not advised.

Assess the patient's breathing. All patients with an AMS, regardless of the cause, should receive high-flow oxygen. Seizures cause patients to use up oxygen quickly, resulting in hypoxia. Again, in the immediate postictal state following a seizure, you should anticipate rapid, deep respirations and an accompanying fast heart rate as a result of the stress of the severe convulsions. However, the respirations and the heart rate should begin to slow to normal rates after several minutes.

Circulation

Circulation should be confirmed as normal or treated as necessary. Your assessment of the patient's circulation should begin with checking the pulse if the patient is unresponsive. If no pulse is found, immediately begin cardiopulmonary resuscitation, and attach an automated external defibrillator. If the patient is responsive, determine whether the pulse is fast or slow, weak or strong. Is the patient in shock? Oxygen administration is helpful for limiting the effects of hypoperfusion to the brain. Evaluate the patient quickly for external bleeding. It is unlikely a patient with a stroke has sustained trauma, but it is possible with a patient who has had a seizure. You should consider this possibility and assess appropriately.

Transport Decision

You should establish your priorities of care based on your assessment of the patient's level of consciousness and ABCs. How the patient is presenting will guide you as to whether you stay at the scene for further assessment or proceed to immediate transport. If you suspect the patient is experiencing a stroke, you should rapidly transport the patient to an appropriate facility to ensure that every chance is available to reduce the disability caused by an ischemic stroke.

History Taking

Investigate Chief Complaint

If the patient is unresponsive, you will need to gather any history of the present illness from family or bystanders. If no one is around, quickly look for explanations for the AMS (eg, signs of trauma, medical alert tags, track marks, environmental clues such as empty alcohol or medication containers).

To determine the chief complaint in a responsive patient, begin by asking the patient what happened. Look for signs and symptoms that may indicate a cause for his or her AMS, such as a stroke, or if there is any evidence of a seizure (incontinence, bitten tongue). Evaluate the patient's speech. Is the patient making any sense? Is speech slurred?

If you know that the patient has had a seizure and is now in a postictal state, you will not be able to obtain a history from the patient. Look for any obvious trauma or explanations as to why the patient may have had a seizure.

SAMPLE History

If the patient is responsive and breathing, obtain a SAMPLE history. Also try to speak with family or friends who may be able to explain the events leading up to the AMS **Figure 15-10**, remembering that time can be critical in a neurologic emergency. Make a special effort to determine the exact time that the patient last appeared to be healthy. In the case of a patient having a stroke, this information will help physicians decide whether it is safe to begin certain treatments that must be given within the first hours after the onset of stroke symptoms. You may be the only person with the opportunity to speak with bystanders to obtain this critical information. Many times, you will be able to find out only that the patient was healthy when he or she went to sleep the night before. Note that in such cases, the time the patient was last seen to be healthy was at bedtime, not when the patient awoke with symptoms. Collect or list all medications the patient has taken. When possible, determine allergies and the patient's last oral intake.

Although a patient who has had a stroke may appear to be unconscious and unable to speak, the patient may still be able to hear and understand what is taking place. Therefore, you should avoid making unnecessary or inappropriate remarks. Try to communicate with the patient by looking for indications that the patient can understand you, such as a glance, gaze, motion or pressure of the hand, effort to speak, or head nod. Reassure your patient that you understand that communication between the two of you may be difficult at this point but that you will provide him or her with continuous information as to what you and the other team members are doing. Establishing effective communication can help you to calm the patient and lessen the fear that accompanies an inability to communicate **Figure 15-11**. Try to keep in mind that the patient has just experienced a potentially life-threatening event and that anxiety, frustration, and embarrassment may inhibit communication with you.

With patients who have had a seizure, your SAMPLE history should reveal if the patient has a history of seizures. If so, it is important to find out how the patient's seizures typically occur and whether this episode differs in some way from previous episodes. You should also ask what medications the patient has been taking. If the patient takes phenytoin (Dilantin) and phenobarbital, he or she most likely has a seizure disorder. You might find that the patient ran out of medication or stopped taking the medication for a time. Patients who have a history of seizures *and* diabetes may use up all the glucose in the body to fuel the seizure.

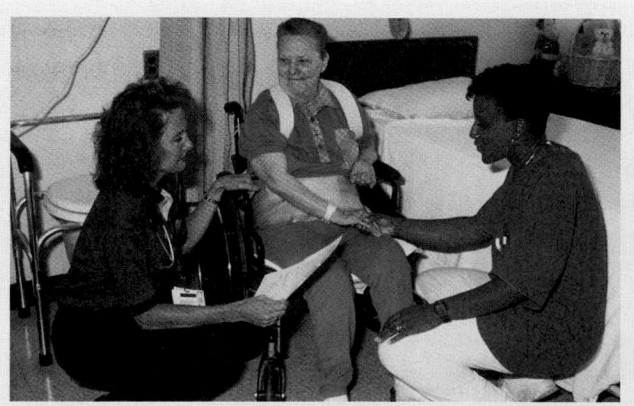

Figure 15-10 Try to speak with family members or bystanders who may know what happened. They may also be able to tell you when the patient last appeared normal.

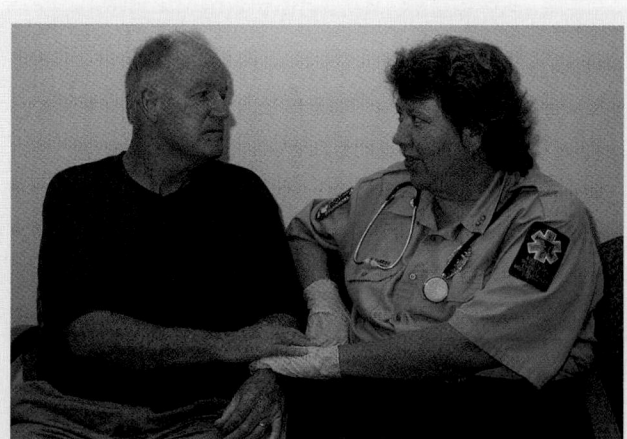

Figure 15-11 Make a special effort to establish communication with a patient who may have had a stroke. Look for indications that the patient understands you, such as a glance, gaze, squeeze of the hand, efforts to speak, or nodding of the head.

If the patient does not have a history of seizures and now suddenly has a seizure, a serious condition, such as a brain tumor, intracranial bleeding, or serious infection should be suspected. During this part of the patient assessment process, it is also the time to determine whether the patient takes medications that lower blood glucose, such as insulin or oral hypoglycemic agents. In other situations, you may want to inquire about drug use or exposure to poisons.

Words of Wisdom

When you are assessing a patient who might have had a stroke, it is critical to pinpoint when the symptoms first started. Typically there is a 3-hour treatment window in which outcomes are greatly improved if the patient receives treatment. You must alert the hospital prior to arrival when the symptoms first began.

Safety

Be aware of the potential for a patient to become violent during a neurologic emergency. When alcohol or drugs are part of a patient's SAMPLE history, the potential for this type of behavior increases.

Secondary Assessment

Physical Examinations

Your assessment of the patient should continue with a full-body scan, paying particular attention to the system involved. If you suspect your patient is having a stroke, then you should direct particular attention to your neurologic assessment. As always, your secondary assessment should include a complete set of vital signs using the monitoring devices you have available.

Stroke Assessment

You should use a stroke assessment tool as part of your secondary assessment. Many EMS services use the Cincinnati Prehospital Stroke Scale **Table 15-2** or the Los Angeles Prehospital Stroke Screen **Table 15-3** to rapidly identify stroke in the field. You should become familiar with what is used in your local protocol. Stroke scales evaluate the face, arms, and speech. If the patient does not have a normal response to these evaluations, you should strongly suspect a stroke. Rapid transport to a designated stroke center is indicated.

You are the Provider: PART 5

You continue treatment en route to the hospital. You reassess the patient and note that her condition has deteriorated further. Her pupils are unequal, and she is exhibiting decorticate posturing (flexion posturing to painful stimulus). You call in your report to the receiving facility with an estimated time of arrival of 6 minutes.

Recording Time: 17 Minutes	
Level of consciousness	Unconscious and unresponsive
Respirations	6 breaths/min; irregular and shallow
Pulse	64 beats/min; bounding
Skin	Pink, warm, and dry
Blood pressure	194/104 mm Hg
Sao₂	98% (on oxygen)

You arrive at the hospital and transfer patient care to the emergency department staff. After further treatment in the emergency department, a computed tomographic scan of her brain was performed and revealed a massive intracerebral hemorrhage. Despite aggressive treatment in the intensive care unit, the patient died the next day.

10. What do unequal pupils indicate?

11. On the basis of your last assessment, what is the patient's Glasgow Coma Scale score?

Table 15-2 Cincinnati Prehospital Stroke Scale

Test	Normal	Abnormal
Facial Droop (Ask patient to show teeth or smile.)	Both sides of face move equally well.	One side of face does not move as well as other.
Arm Drift (Ask patient to close eyes and hold both arms out with palms up.)	Both arms move the same, or both arms do not move.	One arm does not move, or one arm drifts down compared with the other side.
Speech (Ask patient to say, "The sky is blue in Cincinnati.")	Patient uses correct words with no slurring.	Patient slurs words, uses inappropriate words, or is unable to speak.

Table 15-3 Los Angeles Prehospital Stroke Screen

Criteria	Yes	Unknown	No
1. Age > 45 y	❏	❏	❏
2. History of seizures or epilepsy absent	❏	❏	❏
3. Symptoms < 24 h	❏	❏	❏
4. At baseline, patient is not wheelchair-bound or bedridden	❏	❏	❏
5. Blood glucose between 60 and 400 mg/dL	❏	❏	❏
6. Obvious asymmetry (right versus left) in any of the following three exam categories (must be unilateral):	❏	❏	❏
	Equal	**Right Weak**	**Left Weak**
Facial smile/grimace	❏	❏ Droop	❏ Droop
Grip	❏	❏ Weak grip ❏ No grip	❏ Weak grip ❏ No grip
Arm strength	❏	❏ Drifts down ❏ Falls rapidly	❏ Drifts down ❏ Falls rapidly

Interpretation: If criteria 1-6 are marked yes, the probability of a stroke is 97%.

To test speech, ask the patient to repeat a simple phrase such as "The sky is blue in Cincinnati." If the patient does this correctly, you know that he or she understands and can produce speech. If the patient cannot repeat the phrase, the problem may be with either function: understanding speech or producing it.

To test facial movement, ask the patient to show his or her teeth (or gums if there are no teeth). Watch to see that both sides of the face around the mouth move equally. If only one side is moving well, you know that something is wrong with the control of the muscles on the other side.

To test arm movement, ask the patient to hold both arms in front of his or her body, palms up toward the sky, with eyes closed and without moving. During the next 10 seconds, watch the patient's hands. If you see one side drift down toward the ground, you know that side is weak. If both arms stay up and do not move, you know that both sides of the brain are functioning.

If both arms fall to the ground, you have not really learned anything. Perhaps the patient did not understand your instructions. Try the arm test again, but this time move the patient's arms into position yourself.

All patients with an AMS (stroke, TIA, seizure, of unknown cause) should also have a Glasgow Coma Scale (GCS) score calculated **Table 15-4**.

Vital Signs

Patients with significant intracranial bleeding (hemorrhagic stroke) may have a great deal of pressure in the skull that is compressing the brain, thus slowing

Table 15-4 Glasgow Coma Scale

Eye Opening		Best Verbal Response		Best Motor Response	
Spontaneous	4	Oriented conversation	5	Obeys commands	6
In response to speech	3	Confused conversation	4	Localizes pain	5
In response to pain	2	Inappropriate words	3	Withdraws to pain	4
None	1	Incomprehensible sounds	2	Abnormal flexion	3
		None	1	Abnormal extension	2
				None	1

Score: 13-15 may indicate mild dysfunction, although 15 is the score a person with no neurologic disabilities would receive.

Score: 9-12 may indicate moderate dysfunction.

Score: 8 or less is indicative of severe dysfunction.

the pulse and causing respirations to be erratic. Blood pressure is usually high to compensate for poor perfusion in the brain. Changes in pupil size and reactivity indicate significant bleeding and pressure on the brain. If the patient has an AMS (regardless of the cause), you should check the blood glucose level if you have the equipment available and your local protocol allows.

During most active seizures, it is impossible to evaluate vital signs, nor is this the priority when a patient is having a seizure. Unless the situation is unusual, vital signs in a postictal state will approximate normal. Obtain pulse rate, rhythm, and quality; respiratory rate, rhythm, and quality; blood pressure; skin color, temperature, and condition; and pupil size and reactivity.

Monitoring Devices

As stated previously, if the patient has an AMS (regardless of the cause), you should check the patient's blood glucose level if you have the equipment available and your local protocol allows. Most commonly, this is accomplished using a portable blood glucose monitor (glucometer), similar to the one your patient may use at home. The portable blood glucose monitor measures the glucose level in whole blood, using capillary or venous samples. Chapter 17, *Endocrine and Hematologic Emergencies,* discusses the use of a glucometer in more detail.

You may also use noninvasive blood pressure methods to monitor blood pressure. It is recommended that the you always assess the patient's first blood pressure manually with a sphygmomanometer (blood pressure cuff) and a stethoscope.

Reassessment

The reassessment should focus on reassessing the ABCs, vital signs, and interventions provided so far.

Patients who have had a stroke can lose their airway or stop breathing without warning. Multiple interventions may be necessary. The effectiveness of airway adjuncts, positive-pressure ventilations, and other treatments can only be determined with immediate and continuous observation after providing the intervention. If an intervention is not working, try something else.

You have established baseline vital signs already in your assessment, as well as a GCS score. Now is the time to compare that baseline information with updated information. Any changes may indicate if treatments are effective. Watch carefully for changes in pulse, blood pressure, respirations, and GCS scores.

Words of Wisdom

The following is key information to document for a patient who may have had a stroke:
- Time of onset of the signs and symptoms
- Results of the Glasgow Coma Scale
- Results of a stroke assessment tool (Cincinnati or Los Angeles)
- Changes noted on reassessment

Establishing the time of onset is critical because it helps determine whether the patient is a candidate for treatment with clot-dissolving drugs.

Interventions

When your patient shows signs and/or symptoms of stroke, seizure, hypoglycemia, or hypoxia, these conditions typically can be more easily identified, and treatment options are readily available. With other neurologic emergencies, the cause of the patient's symptoms will not always be obvious to you and you may not be able to determine the cause. It is possible that the hospital will need more time and diagnostic testing to determine the cause. This may make it difficult for you to provide definitive treatment in the field. Most of your interventions will be based on your assessment findings. For example, if the blood glucose level is low, you may give oral glucose according to protocol, or if a patient is unresponsive, you may need to position him or her in the recovery position to protect the airway. Remember never to give anything orally to a patient with decreased mental status or a patient who is unable to swallow normally, as this may result in aspiration. Your best treatment in these situations is to perform a thorough assessment and maintain the ABCs.

In most patients with a suspected stroke, physicians in the emergency department need to determine whether there is bleeding in the brain. If there is no bleeding, the patient may be a candidate for clot-dissolving medication that may help brain cells survive. The only reliable way to tell whether there is bleeding in the brain is with a special type of imaging test called computed tomography (CT) of the head. Blood is usually easy to see on the CT scan.

Some EMS systems designate specific hospitals for patients who may be having a stroke. These institutions have CT scanner technicians, radiologists, and neurosurgeons on duty 24 hours a day.

Most hospitals have only one CT scanner and may not have CT technicians available 24 hours a day. It is important that you recognize the signs and symptoms of a stroke and notify the hospital staff as early as possible if you have a "stroke alert" patient. If the emergency department staff knows that you are transporting a possible stroke patient, they can call in the technician if needed or may be able to free up the CT scanner so it is immediately available. Keep in mind that most treatments for stroke must be started as soon as possible after the onset of the event **Table 15-5**. Few, if any, current treatments do any good if they are started more than 3 hours after the stroke begins. Even if 3 hours have passed, prompt action on your part is essential.

In most situations, patients who have had a seizure require definitive evaluation and treatment in the hospital. Unless the patient has a well-established history of seizures and is completely alert and oriented, supplemental oxygen is strongly advised, not only to provide extra oxygen, but also to prevent the possibility of a recurrent seizure.

Seizures are usually limited in how long they last. Most seizures will not require a significant amount of intervention on your part because the seizure will have ended by the time you arrive. For patients who are having a seizure, protect them from harm, maintain a clear airway by suctioning as necessary, and provide oxygen as quickly as possible. If trauma is suspected, provide spinal immobilization. With recurrent seizures, protect

Table 15-5   Tips on Patient Care

- Patients who experience a transient ischemic attack (TIA) may exhibit most of the same signs and symptoms as patients who are having a stroke. These signs and symptoms can last from minutes up to 24 hours. Therefore, the signs of stroke that you note on arrival may gradually disappear. Patients who appear to have had a TIA should be transported for further evaluation.
- Place the patient's affected or paralyzed extremity in a secure and safe position during patient movement and transport.
- Some patients who have had a stroke may be unable to communicate, but they can often understand what is being said around them. Be aware of this possibility.
- New therapies for stroke must be used as soon as possible after the start of symptoms. Minimize time on the scene, and notify the receiving hospital as soon as possible.

the patient from further injury, and manage the airway once the seizure ceases.

For patients who continue to have a seizure, as in status epilepticus, suction the airway, provide positive-pressure ventilations, and transport quickly to the hospital. If you have the option to rendezvous with ALS, you should do so. The ALS providers have medications that can stop a prolonged seizure.

In all cases, you should show patience and tolerance because many of the patients are likely to be confused and occasionally frightened. Many patients who experience seizures are frustrated with their condition and may refuse transport. Kindness and professional behavior are required to help convince the patient that transport is necessary for definitive care.

Communication and Documentation

Notify the receiving facility of your patient's chief complaint and your assessment findings. Most designated stroke centers will want you to call a "stroke alert" for patients you have assessed and found to be having a stroke (check local protocol). This will alert the stroke team members at the hospital and give them time to assemble their resources to treat the patient without delay. Be sure to communicate the time that the patient was last seen to be healthy, the findings of your neurologic examination, and the time you anticipate arriving at the hospital.

A key piece of information to document is the time of onset of the patient's signs and symptoms. If the diagnosis is an ischemic stroke, time of onset of the signs and symptoms is critical in determining whether the patient is a candidate for treatment with clot-dissolving drugs. It is also important to document your findings from your stroke scale and the score results of the GCS, along with any changes you found in your reassessment. Document airway management and interventions performed, including the position in which the patient was placed.

For patients who have had a seizure, give a description of the seizure activity if known. Include bystanders' comments if they witnessed the seizure. Document the onset and duration of the seizure. Did the patient notice or express noticing an aura? Record any evidence of trauma and interventions performed. Document whether this is the patient's first seizure or whether the patient has a history of seizures. If the patient has a history of seizure activity, how often does he or she have them, and is there any history of status epilepticus? When you are documenting your interventions, record the time the intervention was performed, the patient's response to the intervention, and the findings of continued reassessments.

Emergency Medical Care

■ Headache

As discussed earlier, most headaches are harmless and do not require emergency medical care. However, you should be concerned if the patient complains of a sudden-onset, severe headache or a sudden headache that has associated symptoms. Headaches with fever, seizures, AMS, or following head trauma are potentially life threatening. Complete a thorough patient assessment, and transport the patient to the hospital.

Migraine

Treatment of a migraine headache is supportive; however, you should always assess the patient for other signs and symptoms that might indicate a more serious condition. Applying high-flow oxygen, if tolerated, may help ease the patient's condition. When possible, provide a darkened and quiet environment because patients are sensitive to light and sound. Do not use lights and sirens during transport.

■ Stroke

Management of a patient having a stroke in the field is based on supporting the ABCs and providing rapid transport to a stroke center. Depending on the location of the stroke in the brain and the signs and symptoms, the patient may require manual airway positioning. Patients may have difficulty swallowing and controlling their own secretions; therefore, use suction as needed. Provide

Special Populations

Over time, the brain gradually deteriorates and shrinks as a part of the normal aging process. These processes increase the risk of head injury from minor forces because the brain can more readily impact the inside of the skull as a result of the increased space and because the veins that connect the brain to the dura are stretched. A reduced brain mass can also reduce the patient's mental status and capacity. A smaller brain can impair memory function, and with lapses in short-term memory, a geriatric patient will often ask the same or similar questions repeatedly.

When you are called to care for a geriatric patient with an AMS, consider the possibility of a stroke or transient ischemic attack (TIA). At the scene of a motor vehicle crash involving an older driver, consider a stroke or TIA as the precipitating factor in the crash. Be alert for altered mental status and unusual pupil responses (eg, constricted pupils in dim light, unequal pupils).

Take special note of complaints of a headache. Although geriatric patients get tension headaches, they are far less common in the older population. You should consider any headache as potentially serious.

As with the general population, older people can also experience seizures. Remember that seizures are not necessarily caused by epilepsy. You should consider and assess for the possibility of a drug overdose, stroke, head injury, or central nervous system infection. Status epilepticus in a geriatric patient can have harmful effects such as hypoxia, irregular heart rhythm, hypotension, elevated body temperature, low blood glucose level, and, if the patient vomits, aspiration.

Remember that the geriatric patient is at higher risk for central nervous system illnesses and injuries, including brain injury, TIA, stroke, and seizures. Do not be surprised to find a serious head injury from what you might consider a simple bump on the head.

high-flow oxygen, and monitor the patient's oxygen saturation with a pulse oximeter. A patient's paralyzed extremities will require protection from harm. The patient may not be able to feel his or her extremities or move them out of harms way as you package and move the patient for transport. You should continuously talk to the patient and inform him or her of what is going on. Most patients having or who have had a stroke understand what is going on but may not be able to communicate with you. The patient may not be able to physically speak, or when he or she does, inappropriate words may come out. Regardless, the patient will be scared. Reassure the patient, and provide emotional support throughout the call.

Thrombolytic therapy (clot dissolvers) may reverse stroke symptoms and even stop the stroke if given within 2 to 3 hours of the onset of symptoms. These therapies may not work for all patients, and they cannot be given to patients with bleeding-type (hemorrhagic) strokes.

Because hospital personnel will ultimately make these treatment decisions, you should proceed under the assumption that an area of the brain can still be saved. The sooner the treatment is begun, the better the prognosis for the patient.

Spend as little time at the scene as possible. Remember, stroke is an emergency and "time is brain." There may be treatment available for the patient at the hospital, and rapid transport is essential to maximize the possibility of recovery. If available in your area, you should consider transporting the patient to a designated stroke center.

■ Seizure

Most patients who have had a seizure will be in a postictal state on your arrival. For those patients who are still having a seizure, continue to assess and treat the ABCs. It may be necessary to maintain the patient's airway with manual airway positioning. Use suction to clear the airway of any excessive secretions or vomitus. You should consider using a nasopharyngeal airway in a patient having a seizure if you can insert it safely. Oxygen is rapidly consumed by the body during seizure activity. You should monitor the patient's oxygen saturation level with a pulse oximeter, and apply high-flow oxygen. Administer oxygen even if you are unable to get an accurate pulse oximetry reading because of the patient's seizure activity, shaking, or tremors. Provide patients with emotional support.

It is difficult to safely package a patient having a seizure for transport. Make sure that prior to packaging you have assessed the patient for trauma and have taken appropriate spinal precautions if indicated. Protecting the patient from his or her surroundings is essential. Never attempt to restrain a patient having a seizure. Injury could result from tonic-clonic movement. Use soft materials for padding, and move any objects out of the way that may harm your patient.

Not every patient who has had a seizure wants to be transported. It is usually in the best interest of the patient to be evaluated by a physician in the emergency department after a seizure. Your goal is to encourage the patient to be seen by a physician. Should the patient refuse transport, you should be prepared to discuss the situation with the hospital staff on the radio prior to releasing the patient. Ask yourself the following questions if a patient in a postictal state refuses transport:

- Is the patient awake and completely oriented after a seizure (GCS score of 15)?
- Does your assessment reveal no indication of trauma or complications from the seizure?
- Has the patient ever had a seizure before?
- Was this seizure the "usual" seizure in every way (length, activity, recovery)?
- Is the patient currently being treated with medications and receiving regular evaluations by a physician?

If the answer to all of these questions is "yes," you may consider agreeing to a patient's refusal for transport if the patient can be released to a responsible person and monitored. If any one of the questions has a "no" answer, strongly encourage the patient to be transported and evaluated. Follow your local protocols for patients who refuse care and transport.

■ Altered Mental Status

The signs and symptoms of AMS can vary widely from simple confusion to coma. No matter what the cause, you should consider AMS to be an emergency that requires immediate attention, even when it appears that it may be caused by alcohol intoxication or minor head trauma. Your care should include determining the cause (mechanism of injury versus nature of illness), spinal immobilization as indicated, airway and ventilation support, and transport to the appropriate facility.

You are the Provider: SUMMARY

1. On the basis of the dispatch information, what type of seizure is the patient most likely experiencing?

A seizure (convulsion) is a temporary alteration in consciousness. Seizures are caused by a massive discharge of neurons in the brain and can be the result of a variety of conditions. Seizures in adults are typically generalized seizures, formerly referred to as grand mal seizures. Generalized seizures are characterized by unconsciousness and generalized severe twitching of all of the body's muscles (tonic-clonic movement). A generalized seizure can last for several minutes or longer. Generalized seizures indicate a problem with the entire brain.

Focal motor seizures can also occur in adults, although they are not as common as generalized seizures. Focal motor seizures are characterized by spasm or jerking of a single muscle or muscle group. For example, the seizure may only affect a single extremity. The seizure may remain localized to a single body part, or it may spread to adjacent muscles (Jacksonian seizure). Focal motor seizures indicate a problem in an isolated part of the brain.

Another type of seizure is the absence seizure (formerly called petit mal seizure), which typically occurs in children between 4 and 12 years. Absence seizures are often characterized by a brief lapse of consciousness without generalized body twitching. In an absence seizure, the patient often has a blank stare and typically does not respond to voice command. Other characteristics of absence seizures include lip smacking and eye blinking.

2. What are some common causes of seizures in this patient's age group?

Seizures can occur from a variety of underlying problems. Seizures in adults are typically caused by one of three underlying problems: epilepsy, structural brain problems, or metabolic derangements. Febrile seizures are rare in adults. Epilepsy is generally caused by a congenital problem, although many epileptic seizures are of unknown cause (idiopathic). Structural causes of seizures include brain tumors and abscesses, head trauma, and stroke. Metabolic causes of seizures include cerebral hypoxia, hypoglycemia, drug overdose, poisoning, and alcohol withdrawal.

3. What additional questions should you ask the patient's sister?

It has already been established that the patient does not have a history of seizures. However, there are still some important questions that need to be answered. You should ask what the patient was doing and what position she was in when the seizure began. Was she sitting or standing up? Seizures are often mistaken for fainting; however, fainting typically occurs while the patient is standing, whereas seizures can occur in any position. Many patients experience an aura—a strange feeling, smell, or other sensation—that warns of the impending seizure. Although there was no mention of an aura by the patient or her sister, this does not rule out a seizure.

Determine how long the seizure lasted; seizures can last for several minutes, whereas patients who faint typically regain consciousness quickly after the episode. If the patient was unconscious following the seizure, determine for how long. Generalized seizures are typically followed by a postictal phase, in which the patient appears dazed, confused, sleepy, or fatigued; in some cases, the patient may be combative. Fainting is not associated with a postictal phase. Ask about any recent history of head trauma and whether the patient was injured during the seizure. Did she fall to the ground and strike her head? If she was lying down when the episode began, did she strike her head or any other body part on a solid object? If the patient struck her head, which could also have injured her spine, apply spinal immobilization precautions. Obtain as much information as possible regarding the events before, during, and after the seizure, and relay that information to the emergency department staff.

4. What prehospital assessments can you perform to try to determine the cause of the patient's seizure?

In most cases, you will not be able to determine the underlying cause of the patient's seizure in the prehospital setting. However, there are a few assessments you can perform and observations you can make that may increase your index of suspicion. Assess the patient's blood glucose level, if trained to do so, to rule out hypoglycemia as the cause of her seizure. Normal blood glucose levels range between 80 and 120 mg/dL (milligrams per deciliter [100 mL]). If the brain is deprived of glucose, its cells will be irritable, potentially resulting in a seizure.

If the patient is conscious and able to follow commands, test the patient by using the Cincinnati Prehospital Stroke Scale; this test consists of three assessments: facial droop, speech, and arm drift. The patient's face should be symmetrical (equal on both sides) when she smiles and shows her teeth. Ask the patient to repeat a simple phrase to determine if her speech is poorly articulated or slurred. If the patient is uninjured and is able to sit or stand, ask her to place her arms in front of her body, palms up, and close her eyes. Observe for one arm to slowly drift and pronate (turn palm down). If any one of these assessments is abnormal, you should suspect a stroke, specifically an acute ischemic stroke.

You should also assess and closely monitor the patient's vital signs and neurologic status. For example, if she experienced a hemorrhagic stroke (ruptured cerebral artery), her vital signs may be indicative of increased intracranial pressure (ICP), such as hypertension, bradycardia, and abnormal breathing. A decreasing level of consciousness, abnormal pupils, and posturing (decorticate [flexor] or decerebrate [extensor]) also indicate increased ICP. You should also frequently assess the patient's Glasgow Coma Scale (GCS) score; the GCS is an excellent tool to use when assessing a patient's neurologic status.

5. Other than oxygen, what additional treatment is indicated at this point?

Unless the patient was injured during the seizure or you have identified an underlying cause of her seizure that can

You are the Provider: SUMMARY, continued

be treated in the prehospital setting (eg, hypoglycemia), additional treatment is mainly supportive. Maintaining a patent airway and ensuring adequate oxygenation and ventilation are your highest priorities. Provide a calm, quiet environment; reassure and reorient the patient as needed; avoid any loud or bright stimulus (may cause another seizure); and safely transport her to the hospital. Continuously monitor the patient's ABCs, level of consciousness, and vital signs.

6. What is your field impression of this patient? Why?

From the patient's sister's description, it is likely that she experienced a seizure. Her present signs and symptoms (confusion; sudden, severe headache; nausea) and medical history (poorly controlled hypertension) should make you suspicious that she is experiencing a hemorrhagic stroke, which likely caused the seizure.

Hemorrhagic strokes account for approximately 10% of all strokes; they occur as a result of bleeding inside the brain secondary to a leaking or ruptured cerebral artery. As blood leaks from the artery, it collects in the brain and squeezes the brain tissue. When brain tissue is compressed, oxygenated blood cannot get to that area, and the surrounding cells become ischemic (deprived of oxygen).

Patients who are at highest risk for a hemorrhagic stroke are patients who have very high blood pressure (her initial blood pressure was 200/112 mm Hg) or long-term hypertension that is poorly controlled (the patient admitted to being noncompliant with her medication regimens).

Patients experiencing a hemorrhagic stroke usually present with a sudden, severe headache, which they typically describe as the "worst headache of their life." The headache is the result of irritation by blood of the brain tissue after the cerebral artery swells and leaks or ruptures. As blood collects in the cranium, the ICP increases. Signs of increased ICP include changes in level of consciousness, nausea and vomiting, seizures, and high blood pressure, among others; your patient is experiencing all of these signs and symptoms.

7. On the basis of your field impression, for which additional signs and symptoms should you monitor the patient?

Rupture of a large cerebral artery is often rapidly fatal; therefore, your patient's present condition suggests that she has a leaking cerebral artery and blood is slowly accumulating in her brain tissue. Over time, chronic, poorly controlled high blood pressure has caused a weakening in the arterial wall (aneurysm). Eventually, the arterial wall perforates (tears), and blood begins to leak.

As blood continues to accumulate in the brain and the ICP increases, the patient's level of consciousness will deteriorate; therefore, the level of consciousness is the single most important assessment parameter to monitor. Because of the cerebral ischemia caused by the increased ICP, the patient may experience another seizure. As the ICP increases further, the blood pressure often increases and the heart rate commonly decreases. Pressure on the brain stem may cause irregular and ineffective breathing; therefore, assisted

ventilation may be necessary. It is critical to continuously monitor your patient's condition and be prepared to intervene if her condition deteriorates.

8. What should be your most immediate action?

Airway, airway, airway! Snoring respirations indicate partial obstruction of the airway by the tongue. Performing the head tilt–chin lift maneuver is the quickest way to correct the problem. The patient is now unconscious and unresponsive; insert an airway adjunct (eg, oral or nasal airway) to help maintain airway patency. Patients with increased ICP often vomit; remain alert to this possibility and have suction readily available. Regardless of the situation, you must ensure that the patient's airway remains patent at all times. No airway, no patient—it's that simple!

9. What additional treatment does this patient require?

After establishing a patent airway, your next priority is to assist the patient's breathing. A slow (6 breaths/min), irregular breathing pattern will not support adequate minute volume; therefore, you must deliver some form of positive-pressure ventilation. Positive-pressure ventilation can be provided with a pocket face mask and one-way valve or a bag-mask device. Be sure to attach 100% oxygen to the ventilation device you will be using.

It is important to ventilate the patient at the appropriate rate with the proper volume. Deliver each breath during a period of 1 second (just enough to produce visible chest rise) at a rate of 10 to 12 breaths/min. Do not hyperventilate the patient; doing so can have several negative consequences. Hyperventilation hyperinflates the lungs, which can impair blood return to the right atrium and cause a decrease in cardiac output. Hyperventilation also increases the risks of regurgitation and aspiration. In addition to the risks already discussed, hyperventilation can be especially detrimental in patients with intracerebral bleeding and increased intracranial pressure; it causes cerebral vasoconstriction, which shunts blood (and oxygen) away from the brain. This decrease in cerebral perfusion may cause further injury to the brain.

10. What do unequal pupils indicate?

In the context of a traumatic brain injury or hemorrhagic stroke, unequal pupils are an ominous sign. It indicates significantly increased ICP and compression of one of the oculomotor nerves, the nerves that control the pupillary response. The affected pupil is often fully dilated (blown) and does not constrict when a light source is shone into it.

11. On the basis of your last assessment, what is the patient's Glasgow Coma Scale score?

The GCS assesses three parameters: eye opening, verbal response, and motor response. Your last assessment revealed that the patient was unconscious and unresponsive (she did not open her eyes and was unresponsive to all stimuli) and was exhibiting decorticate (abnormal flexion) posturing. Therefore, she would receive a GCS score of 5, based on the following values (the numeric value for each component is in bold):

You are the Provider: SUMMARY, continued

Eye opening:

Spontaneous: 4
Responsive to speech: 3
Responsive to pain: 2
None: 1

Best verbal response:

Oriented conversation: 5
Confused conversation: 4
Inappropriate words: 3

Incomprehensible sounds: 2
None: 1

Best motor response:

Obeys commands: 6
Localizes pain: 5
Withdraws from pain: 4
Abnormal flexion: 3
Abnormal extension: 2
None: 1

EMS Patient Care Report (PCR)

Date: 3-16-09	Incident No.: 140109	Nature of Call: Seizure		Location: 106 Scottie Drive	
Dispatched: 1823	En Route: 1823	At Scene: 1827	Transport: 1839	At Hospital: 1952	In Service: 2009

Patient Information

Age: 58 Sex: F Weight (in kg [lb]): 77 kg (170 lb)	Allergies: Penicillin, codeine Medications: Benazepril, Plavix, metformin Past Medical History: Hypertension, heart disease, type 2 diabetes mellitus Chief Complaint: Severe headache and nausea

Vital Signs

Time: 1831	BP: 200/112	Pulse: 100	Respirations: 14	Sao$_2$: 96%
Time: 1837	BP: 198/110	Pulse: 60	Respirations: 6	Sao$_2$: 94%
Time: 1844	BP: 194/104	Pulse: 64	Respirations: 6	Sao$_2$: 98%

EMS Treatment
(circle all that apply)

Oxygen @ 15 L/min via (circle one): NC (NRM) (Bag-Mask Device)	(Assisted Ventilation)	(Airway Adjunct)	CPR
Defibrillation	Bleeding Control	Bandaging	Splinting	Other

Narrative

Dispatched for a 58-year-old woman experiencing a seizure. On arrival at the scene, found the patient lying supine on her living room floor with a pillow under her head. She was conscious, but confused and complained of nausea and the "worst headache of her life." The patient's sister witnessed the episode and stated that she suddenly grabbed both sides of her head and then began "shaking all over." There was no trauma involved; the patient's sister states that she caught her before she struck the ground. Per the sister, the patient has never experienced a seizure before today. Past medical history significant for hypertension, heart disease, and type 2 diabetes mellitus. Medications listed above; patient admits to being noncompliant with her prescribed medications. Applied high-flow oxygen via nonrebreathing mask and obtained vital signs. Blood glucose level was assessed and noted to be 97 mg/dL. Further assessment did not reveal any gross evidence of a seizure (eg, tongue-biting, urinary incontinence). Engine 60 arrived at the scene to provide assistance. As the patient was being loaded into the ambulance, she became unconscious and unresponsive. Reassessment revealed that her respirations were slow, irregular, and shallow. Inserted an oral airway and began assisting the patient's ventilations with a bag-mask device and high-flow oxygen. Requested assistance from engine 60 EMT and began transport to the hospital. Continued to assist the patient's ventilations en route and reassessed her vital signs. Shortly before arriving at the hospital, reassessed the patient and noted that her pupils were unequal and she began exhibiting decorticate posturing. Assigned a Glasgow Coma Scale score of 5. Expeditiously transferred patient to the emergency department staff and gave verbal report to attending physician. **End of report**

Assessment and Emergency Care of Neurologic Emergencies

Scene Size-up

Scene Safety	Ensure scene safety and safe access to the patient. Take appropriate standard precautions. Review dispatch information; emergency calls dispatched as a medical condition may have an associated mechanism of injury. Establish early in the call whether the patient's condition is medical or trauma. Determine the number of patients and the need for additional resources. A neurologic emergency may be the result of exposure to hazardous materials or from another type of incident. Advanced life support (ALS) is often required at neurologic emergencies. Call for ALS early. Their support can be canceled if it is determined they are not needed. Consider cervical spine stabilization if the mechanism of injury is unknown.
Mechanism of Injury (MOI)/ Nature of Illness (NOI)	Determine the MOI/NOI. Observe the scene, and look for clues that will assist you in determining the MOI/NOI. The nature of the problem may not be apparent until more information is gathered. Ask family members or bystanders about the event. Assess the patient for signs of trauma; evaluate the patient's environment. Look for signs of medical conditions that may be indicated by the patient's medications, medical supplies, and medical alert bracelet or tag.

▼   ▼

Primary Assessment

Form a General Impression	Note the position of the patient, and identify any immediate life threats. Determine priority of care based on the MOI/NOI. If the patient has a poor general impression, call for ALS assistance. A rapid scan of the patient will help you identify and manage life threats. Assess the patient's level of consciousness using the AVPU scale. Obtain the chief complaint from the patient if possible.
Airway and Breathing	Ensure the airway is open, clear, and self-maintained. The airway of an unresponsive patient needs to be opened and maintained using a modified jaw-thrust if spinal injury is suspected. A patient with an altered level of consciousness may need emergency airway management; consider inserting a properly sized oropharyngeal or nasopharyngeal airway. Suction the airway as required. Place the patient in the recovery position to prevent aspiration. Observe for the possibility of a foreign body airway obstruction. Evaluate the patient's ventilatory status for rate and depth of breathing, respiratory effort, and tidal volume. Patient respirations of less than 12 breaths/min or more than 20 breaths/min are considered inadequate, and the patient may require assistance breathing. Continuously monitor oxygen saturation, and look for additional signs of hypoxia. Administer high-flow oxygen at 15 L/min, providing ventilatory support as needed. Do not place anything in the patient's mouth.
Circulation	Check the patient's pulse; if no pulse is present, begin cardiopulmonary resuscitation immediately. Evaluate distal pulse for rate, quality (strength), and rhythm. Observe skin color, temperature, and condition. Assess capillary refill time; if greater than 2 seconds, treat the patient aggressively for shock. Look for evidence of life-threatening bleeding, and treat the patient accordingly.
Transport Decision	If the patient has an airway or breathing problem, signs and symptoms of bleeding, or other life threats, manage them immediately and consider rapid transport, performing the secondary assessment en route to the hospital. If the patient has a paralyzed extremity, ensure it is placed in a safe, secure position.

▼   ▼

NOTE: The order of the steps in this section differs depending on whether the patient is conscious or unconscious. The following order is for a conscious patient. For an unconscious patient, perform a primary assessment, perform a rapid full-body scan, obtain vital signs, and obtain the past medical history from a family member, bystander, or an emergency medical identification device.

Assessment and Emergency Care of Neurologic Emergencies, continued

History Taking

Investigate Chief Complaint	Investigate the chief complaint and gather a history once you have identified and treated life threats. Look for signs and symptoms that may indicate a cause for the patient's condition. Evaluate the patient's facial expressions and speech. Identify associated symptoms and pertinent negatives. Ask OPQRST and SAMPLE questions, focusing on the events surrounding the incident. Use the TIPS-AEIOU mnemonic to help you identify possible causes of altered mental status. Determine when the patient last appeared healthy. Collect or list all medications the patient is taking or has stopped taking. Inquire about the use of drugs or alcohol. Information can also be obtained from family members, bystanders, and medical alert tags if the patient is not able to provide the information.

Secondary Assessment

Physical Examinations	Perform a systematic prehospital head-to-toe examination beginning with the head, looking for DCAP-BTLS. Assessment should be rapid if the patient has a poor general impression. Patients suspected of having a stroke should have a focused neurologic examination performed, such as the full-body scan or Cincinnati Stroke Scale.
Vital Signs	On completion of the primary assessment and treatment of immediate life threats, obtain the patient's baseline vital signs. Vital signs should include blood pressure by auscultation, pulse rate and quality, respiration rate and quality, pupil evaluation, and skin assessment for perfusion. Note the patient's level of consciousness. Determine the patient's Glasgow Coma Scale score. Use pulse oximetry, if available, to assess the patient's perfusion status. Tachycardia or hypotension may indicate hypoperfusion. Bradycardia and/or erratic breathing may indicate increasing intracranial pressure, as do unequal pupils. Bradypnea may indicate cerebral hypoxia and impending cardiopulmonary arrest. Check the patient's blood glucose level if local protocol allows. Reassess the patient's vital signs every 5 minutes to observe trends. A widening pulse pressure (the difference between the systolic and diastolic pressures) is a sign of increasing intracranial pressure.

Reassessment

Interventions	If a spine or head injury is suspected, immobilize the spine. To ensure an open airway, position an oropharyngeal or nasopharyngeal airway if necessary. Provide oxygen via nonrebreathing mask or bag-mask device as required. Control any external bleeding and treat for shock. Reassess the chief complaint, primary assessment, vital signs, and any interventions already performed. Is the oxygen delivery sufficient and is the patient breathing adequately? Is the pulse oximetry value improving? If hypoglycemia is suspected to be the cause of an altered mental status and the patient is awake and able to swallow, administer oral glucose. If stroke is suspected, time is crucial. Administer high-flow oxygen and provide rapid transport to the appropriate facility. Vital signs should be repeated every 5 minutes and the results compared with those obtained earlier. If the patient is awake and alert, place him or her in a position of comfort. Unresponsive patients should be placed in the recovery position, unless spinal injury is suspected, to prevent aspiration and facilitate drainage of secretions. Stroke patients should be positioned with the head slightly elevated. Continuously observe and reassess the patient's condition during transport to enable management of worsening conditions. Document changes noted during your assessment and reassessment of the patient.

Assessment and Emergency Care of Neurologic Emergencies, continued

Communication and Documentation	Contact medical control and the receiving hospital with a radio report that includes notification of a neurologic emergency. Some hospitals require a "stroke alert." Include a thorough description of the MOI/NOI and position in which the patient was found. Include treatments performed and patient response. Be sure to document the patient's chief complaint, physical findings, history, time of initial onset, and any changes in patient status and the time they occurred. Document the scene observations on arrival. Follow local treatment protocols.
	Communicating with a patient experiencing a neurologic emergency is sometimes difficult. Patients may appear unresponsive or have difficulty speaking, but might still be able to hear and understand you. Avoid unnecessary or inappropriate comments. If possible, establish an effective form of communication. Calm and reassure the patient.

NOTE: Although the steps below are widely accepted, be sure to consult and follow your local protocols. Follow standard precautions when treating all patients.

Neurologic Emergencies

General Management of Neurologic Emergencies

Managing life threats to the patient's ABCs is the primary concern during any neurologic emergency. The general impression of the patient is very important. Note the patient's overall appearance, work and pattern of breathing, and the position in which the patient was found. The management of specific neurologic emergencies begins with the following steps:

1. Ensure scene safety, taking the appropriate standard precautions.
2. Determine whether the call indicates a medical or trauma event.
3. Open, clear, and maintain the patient's airway.
4. Inspect, palpate, and auscultate the chest.
5. Administer high-concentration oxygen via a nonrebreathing mask or bag-mask device as appropriate.
6. Determine appropriate interventions based on the chief complaint and findings of the physical examination.
7. Transport to the appropriate treatment facility.

Headache

Headache may be a sign of a potentially life-threatening situation. Look for signs of stroke, meningitis, and possible carbon monoxide poisoning. Perform a complete patient assessment that includes obtaining vital signs. Administer high-flow oxygen. Place the patient in a position of comfort, usually in a darkened, quiet environment. Transport the patient to the hospital.

Stroke (Cerebrovascular Accident)/Transient Ischemic Attack

Determine the time of symptom onset. Observe the patient's position and protect any affected extremities. Adminster high-flow oxygen via a nonrebreathing mask or bag-mask device if the patient's breathing is inadequate. Obtain a history of illness using OPQRST. Obtain a SAMPLE history. A transient ischemic attack has occurred if stroke symptoms end on their own within 24 hours of onset. Perform a field stroke test such as the Cincinnati Prehospital Stroke Scale or Los Angeles Prehospital Stroke Screen. Monitor vital signs every 5 minutes; watch for trends in blood pressure measurements. Be alert for conditions that may mimic stroke. Transport the patient to a stroke specialty center. Do not delay transport.

Assessment and Emergency Care of Neurologic Emergencies, continued

Neurologic Emergencies

Seizures

Seizures may be the result of epilepsy, fever, metabolic disorders, or blood chemical problems or may be idiopathic. Recognize that a seizure is occuring and determine whether the seizure is new onset or similar to previous seizures. If possible, determine the part of the body where the seizure activity was first noticed. Ensure an open airway but do not place anything into the patient's mouth; a nasopharyngeal airway can be inserted. Administer high-flow oxygen via a nonrebreathing mask. If local protocol allows, obtain and record the patient's blood glucose level. Do not restrain a patient having a seizure. Provide emotional support, and transport to a hospital. Patients refusing transport after a seizure should be encouraged to seek medical follow-up.

Altered Mental Status

There are many causes of altered mental status. The TIPS-AEIOU mnemonic will assist you in determining the cause. Hypoglycemia can cause an altered mental status and may mimic stroke or seizure. Hypoglycemia should be ruled out in patients with altered mental status or a decreased level of consciousness. Seizures may occur in patients whose blood glucose level drops and becomes very low. Head injury may also be the cause of an altered mental status. Assess the patient for a possible MOI, and perform a full-body scan. Changes in the patient's mental status may also be the result of drug use or alcohol intoxication. The patient's history may provide information about psychological disorders that may be causing the altered mental status. Any patient with an altered mental status requires immediate attention. Support the ABCs, provide spinal immobilization if indicated, and transport the patient to the appropriate hospital.

Prep Kit

- The cerebrum, the largest part of the brain, is divided into right and left hemispheres, each controlling the opposite side of the body.

- Different parts of the brain control different functions. The front part of the cerebrum controls emotion and thought; the middle part controls touch and movement; and the back part of the cerebrum is involved with vision. In most people, speech is controlled on the left side of the brain, near the middle of the cerebrum.

- Many different disorders can cause brain or other neurologic symptoms. As a general rule, if the problem is primarily in the brain, only part of the brain will be affected. If the problem is in the heart or lungs, the whole brain will be affected.

- Stroke is a significant brain disorder because it is common and potentially treatable.

- Seizures and altered mental status are also common, and you must learn to recognize the signs and symptoms of each condition.

- Other causes of neurologic dysfunction include coma, infections, and tumors.

- Strokes occur when part of the blood flow to the brain is suddenly cut off; within minutes, brain cells begin to die.

- Signs and symptoms of stroke include receptive and/or expressive aphasia, slurred speech (dysarthria), muscle weakness or numbness on one side of the body, facial droop, and sometimes high blood pressure.

- You should always perform at least three neurologic tests on patients you suspect of having a stroke: testing speech, facial movement, and arm movement.

- In a transient ischemic attack (TIA), normal body processes break up the blood clot, restoring blood flow and ending symptoms in less than 24 hours. However, patients experiencing a TIA are at high risk for a completed stroke.

- Because current treatments for stroke must be administered within 1 to 3 hours (and preferably within 2 hours) of the onset of symptoms to be most effective, you should provide prompt transport.

- Always notify the hospital as soon as possible that you are bringing in a patient with a possible stroke, so that staff there can prepare to test and treat the patient without delay.

- Seizures are characterized by unconsciousness and generalized twitching of all or part of the body.

- There are types of seizures that you should learn to recognize: generalized, partial, and status epilepticus.

- Most seizures last between 3 and 5 minutes and are followed by a postictal state in which the patient may be unresponsive, have labored breathing, have hemiparesis, and may have been incontinent.

- It is important for you to recognize the signs and symptoms of seizures so that you can provide the emergency department staff with information as you transport the patient.

- Altered mental status is a common neurologic problem that you will encounter as an EMT. Signs and symptoms vary widely, as do the causes for this condition.

- Among the most common causes of altered mental status are hypoglycemia, intoxication, drug overdose, and poisoning.

- As you assess a patient with an altered mental status, do not always assume intoxication; hypoglycemia is just as likely a cause. Prompt transport with close monitoring of vital signs en route is indicated.

Vital Vocabulary

aneurysm A swelling or enlargement of part of a blood vessel, resulting from weakening of the vessel wall.

aphasia The inability to understand and/or produce speech.

aura A sensation experienced prior to a seizure; serves as a warning sign that a seizure is about to occur.

cerebrovascular accident (CVA) An interruption of blood flow to the brain that results in the loss of brain function. Also called a stroke.

coma A state of profound unconsciousness from which one cannot be roused.

dysarthria Slurred speech.

embolus Clotting that forms in a remote area and travels to the site of blockage.

febrile seizures Seizures that result from sudden high fevers, particularly in children.

generalized seizure A seizure characterized by severe twitching of all of the body's muscles that may last several minutes or more; formerly known as a grand mal seizure.

hemiparesis Weakness on one side of the body.

hemorrhagic stroke One of the two main types of stroke; occurs as a result of bleeding inside the brain.

hypoglycemia A condition characterized by a low blood glucose level.

incontinence Loss of bowel and/or bladder control; may be the result of a generalized seizure.

ischemia A lack of oxygen in the cells of the brain that causes them to not function properly.

ischemic stroke One of the two main types of stroke; occurs when blood flow to a particular part of the brain is cut off by a blockage (eg, a clot) inside a blood vessel.

partial seizure A seizure affecting a limited portion of the brain.

postictal state A period following a seizure that lasts between 5 and 30 minutes; characterized by labored respirations and some degree of altered mental status.

seizure Generalized, uncoordinated muscular activity associated with loss of consciousness; a convulsion.

status epilepticus A condition in which seizures recur every few minutes or last more than 30 minutes.

stroke An interruption of blood flow to the brain that results in the loss of brain function; also called a cerebrovascular accident (CVA).

thrombosis Clotting of the cerebral arteries that may result in the interruption of cerebral blood flow and subsequent stroke.

tonic-clonic seizure A type of seizure that features rhythmic back-and-forth motion of an extremity and body stiffness.

transient ischemic attack (TIA) A disorder of the brain in which brain cells temporarily stop working because of insufficient oxygen, causing strokelike symptoms that resolve completely within 24 hours of onset.

Assessment in Action

You are dispatched to a residence for an unknown medical emergency. You arrive to find a 68-year-old man sitting up at the kitchen table, but he is leaning slightly to the left. He reports that he has had a headache for the past 3 days. As you are assessing him, you notice that his speech is slurred and he is not moving his left side. You ask the patient to lift his left arm; he lifts his right instead. A check of his vital signs shows a pulse of 110 beats/min, a blood pressure of 170/110 mm Hg, and respirations of 22 breaths/min. His past medical history includes diabetes and hypertension. His blood glucose level is 70 mg/dL.

1. The patient is exhibiting what type of speech pattern?
   A. Ataxic
   B. Wernicke
   C. Broca
   D. Dysarthria

2. What type of neurologic examination should you perform on this patient?
   A. Chicago Stroke Assessment
   B. Philadelphia Stroke Assessment
   C. Cincinnati Prehospital Stroke Scale Assessment
   D. Camden Stroke Assessment

3. What is the difference between a cerebrovascular accident (stroke) and a transient ischemic attack?

4. Which of the following occurs when blood flow to a particular part of the brain is cut off by a blockage?
   A. Ischemic stroke
   B. Hemorrhagic stroke
   C. Seizure
   D. Status epilepticus

5. What is the name of the condition in which the patient forgets about the injured side?
   A. Hemiparesis
   B. Neglect
   C. Aphasia
   D. Ataxia

6. Describe how to perform a stroke assessment.

7. Aside from a possible cerebrovascular accident, what else could be causing this patient's problem?
   A. Postictal state
   B. Meningitis
   C. Hypoglycemia
   D. Brain tumor

8. On the basis of the patient's condition, what should you be most concerned with?
   A. Maintaining the ABCs
   B. Figuring out the type of stroke
   C. Administering oral glucose
   D. Performing further assessment

9. What is the window of time for a stroke patient to receive definitive treatment to improve the patient's outcome?
   A. 1 to 3 hours
   B. 3 to 6 hours
   C. 6 to 9 hours
   D. 9 to 12 hours

10. What does "Tips on the Vowels" stand for?

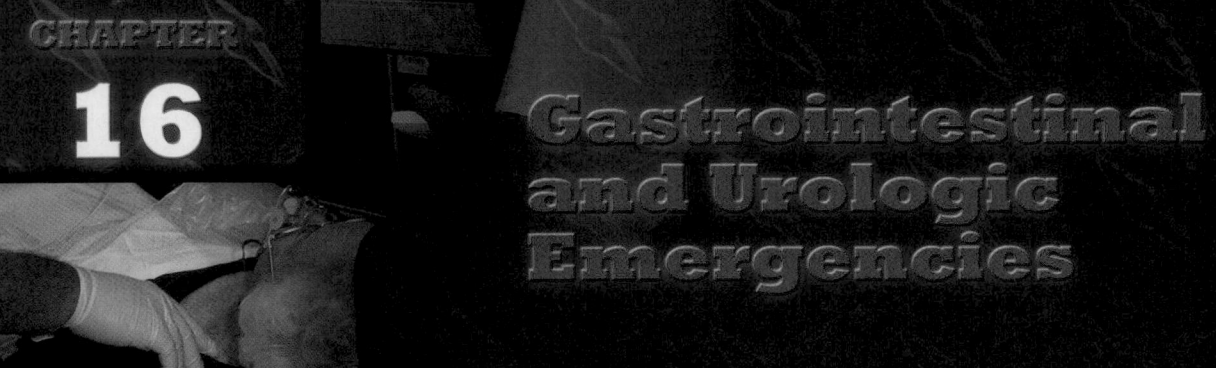

National EMS Education Standard Competencies

Medicine
Applies fundamental knowledge to provide basic emergency care and transportation based on assessment findings for an acutely ill patient.

Abdominal and Gastrointestinal Disorders
Anatomy, presentations, and management of shock associated with abdominal emergencies
- Gastrointestinal bleeding (pp 604, 612)

Anatomy, physiology, pathophysiology, assessment, and management of
- Acute and chronic gastrointestinal hemorrhage (pp 604, 608, 612)
- Peritonitis (pp 601-602, 610-612)
- Ulcerative diseases (pp 603, 612)

Genitourinary/Renal
- Blood pressure assessment in hemodialysis patients

Anatomy, physiology, pathophysiology, assessment, and management of
- Complications related to
  - Renal dialysis (pp 612-613)
  - Urinary catheter management (not insertion) (p 613)
- Kidney stones (p 606)

Knowledge Objectives
1. Understand the basic anatomy and physiology of the gastrointestinal, genital, and urinary systems. (pp 599-601)
2. Define the term "acute abdomen." (p 602)
3. Describe pathologic conditions of the gastrointestinal, genital, and urinary systems. (pp 601-607)
4. Explain the concept of referred pain. (pp 602-603)
5. Understand that abdominal pain can arise from other body systems. (pp 602-603)

6. List the most common abdominal emergencies, with the most common locations of direct and referred pain. (p 603)
7. Identify the signs and symptoms, and common causes, of an acute abdomen. (pp 601-607)
8. Explain the procedures to follow for patient assessment of gastrointestinal and urologic emergencies. (pp 607-612)
9. Describe the emergency medical care of the patient with gastrointestinal or urologic emergencies. (p 612)
10. Describe the procedures to follow in managing the patient with shock associated with abdominal emergencies. (p 612)
11. Explain the procedures to follow in the assessment and management of acute and chronic gastrointestinal hemorrhage, peritonitis, and ulcerative diseases. (p 612)
12. Understand the principles of kidney dialysis. (pp 612-613)

Skills Objectives
1. Demonstrate the assessment of a patient's abdomen. (pp 610-611)

Introduction

Abdominal pain is a common complaint, but the cause is often difficult to identify, even for a physician. As an EMT, you do not need to determine the exact cause of acute abdominal pain, but it is helpful for you to understand the pathophysiology and the signs and symptoms of common illnesses. You need to be able to recognize a life-threatening problem and act swiftly in response. Remember, the patient is in pain and is probably anxious, requiring your skills of rapid assessment and emotional support.

Anatomy and Physiology

The abdominal cavity contains solid and hollow organs that make up the gastrointestinal, genital, and urinary systems **Figure 16-1**. Solid organs include the liver, spleen, pancreas, kidneys, and ovaries (in women). Technically, organs such as the kidneys, ovaries, and the pancreas are retroperitoneal (behind the peritoneum). However, because they lie next to the peritoneum, they can cause abdominal pain. An injury to a solid organ can cause shock and bleeding because of the amount of blood vessels that the organ holds.

Hollow organs include the gallbladder, stomach, small intestine, large intestine, and urinary bladder. If there is a breach into one of these hollow organs, the contents of the organ will leak and contaminate the abdominal cavity.

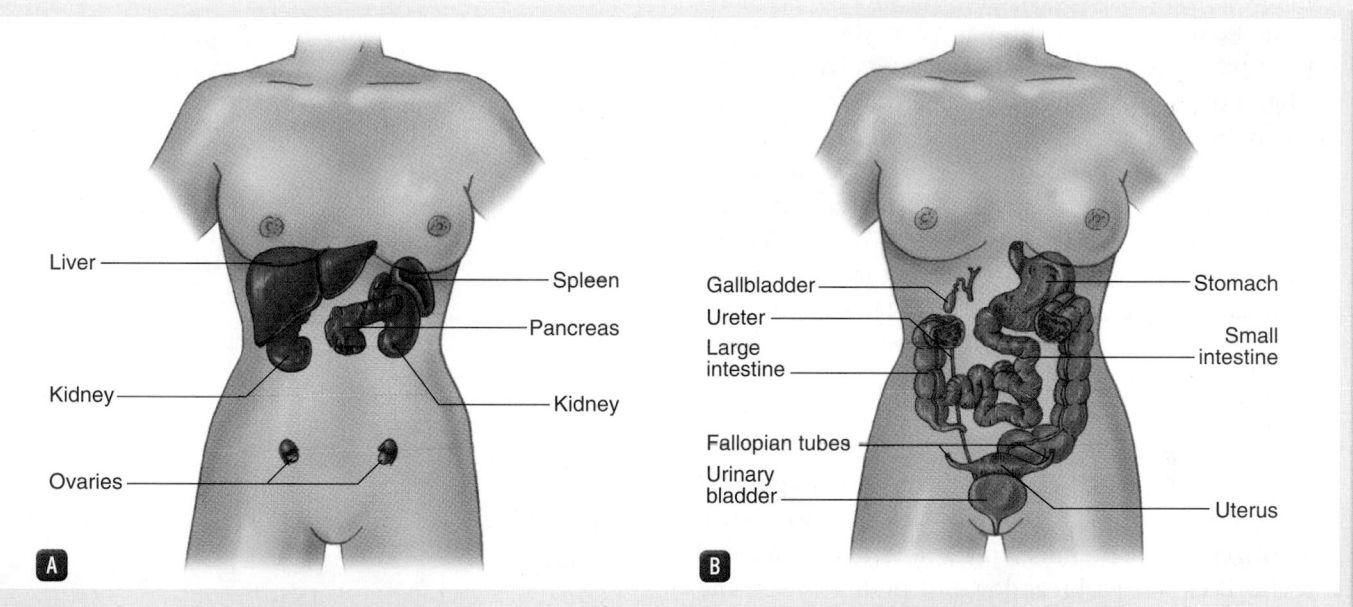

Figure 16-1 The solid and hollow organs of the abdomen. **A.** Solid organs include the liver, spleen, pancreas, kidneys, and ovaries (in women). **B.** Hollow organs include the gallbladder, stomach, small intestine, large intestine, and bladder.

You are the Provider: PART 1

At 3:20 AM you and your partner are dispatched to 1500 East River Road, Apartment 5, for a 79-year-old man with abdominal pain. You proceed to the scene, which is approximately 8 miles from your station. The weather is clear, the temperature is 67°F, and the traffic is light.

1. What is the definition of an acute abdomen?
2. What is the EMT's role in treating a patient with abdominal pain?

The Gastrointestinal System

The gastrointestinal system is responsible for the digestion process. Digestion begins when the food is put into the mouth and chewed; the salivary glands secrete saliva and begin to break the food down, then it is swallowed. The food travels down the esophagus to the stomach. The stomach is the main organ of the digestive system. Most digestion takes place in the stomach, where gastric juices break the food down to a form that can be used by the body.

The liver assists in digestion by secreting bile, which aids in the digestion of fats. The liver also filters toxic substances produced by digestion, creates glucose stores, and produces substances necessary for blood clotting and immune function. The gallbladder is a hollow pouch that acts as a reservoir for bile.

Food then travels down into the small intestine, which consists of three sections: the duodenum, jejunum, and ileum. The duodenum is where digestive juices from the pancreas and liver mix together. The pancreas aids digestion by secreting juice that contains enzymes that help break down starches, fats, and proteins. It also releases amylase. Amylase is responsible for breaking down starches into sugar and is present in saliva and the duodenum. In the duodenum, amylase comes from the pancreas, where it is manufactured. Bicarbonate, an alkali, is part of the buffering system that neutralizes stomach acid in the duodenum. Bicarbonate is also produced in the pancreas. Insulin too is produced in the pancreas, which regulates the amount of glucose in the bloodstream.

The jejunum plays a major role in the absorption of digestive products. In fact, the jejunum comprises much of the surface area of the small intestine and does much of the work. Because the breakdown of proteins, fats, and starches occurs earlier in the digestive system, the jejunum and ileum are a mostly pH-neutral environment for absorption. The jejunum connects the duodenum to the ileum, and the ileum is where soluble molecules are absorbed into the blood. It is here that proteins, fats, and starches are reduced to amino acids, fatty acids, and simple sugars.

The food that was not broken down and used as nutrients then moves into the colon, or large intestine, as waste products. A rhythmic movement called peristalsis moves the waste matter through the intestines. Water is absorbed and stool is formed that passes through the rectum to the anus, where it is defecated.

The spleen is also located in the abdomen but has no digestive system function. The spleen is part of the lymphatic system and plays a significant role in relation to red blood cells and the immune system. It assists in the filtration of blood, aids in the development of red blood cells, and serves as a blood reservoir. The spleen also produces antibodies.

The Genital System

The abdominal space also holds the male and female reproductive organs. The male reproductive system consists of the testicles, epididymis, vasa deferentia, seminal vesicles, prostate gland, and penis. The female reproductive system includes the ovaries, fallopian tubes, uterus, cervix, and the vagina.

The Urinary System

The urinary system controls the discharge of certain waste materials filtered from the blood by the kidneys. In the urinary system, the kidneys are solid organs; the ureters, bladder, and urethra are hollow organs **Figure 16-2**.

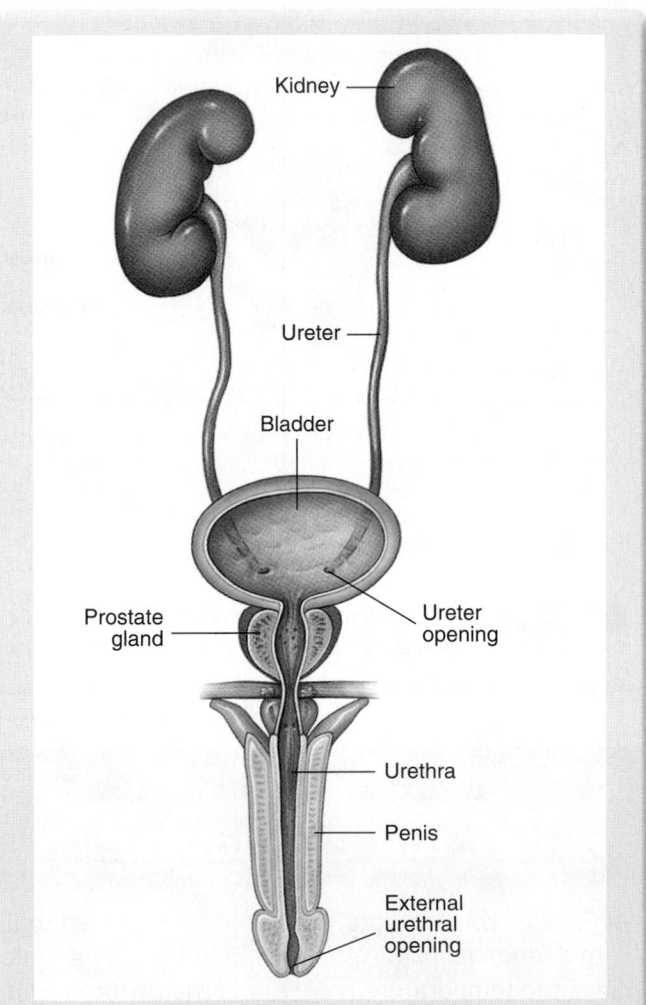

Figure 16-2 The urinary system lies in the retroperitoneal space behind the organs of the digestive system. The urinary system in males and females includes the kidneys, ureters, bladder, and urethra. This diagram shows the male urinary system.

Ordinarily, the urinary and genital systems are considered together because they share many organs. One system can directly affect the other. For example, if the prostate gland in the male genital system enlarges, then the urethra will narrow, weakening the bladder, and eventually leading to urinary retention.

The body contains two kidneys, one on each side, that lie on the posterior muscular wall of the abdomen behind the peritoneum in the retroperitoneal space. The kidneys play an important role in the regulation of acidity and blood pressure. Blood pressure regulation is associated with the kidney's ability to remove sodium chloride from the body and therefore excess fluid. The most common cause of secondary hypertension is kidney disease. The kidneys rid the body of toxic wastes and control the body's balance of fluid and electrolytes. Blood flow in the kidneys is high. Nearly 20% of the output of blood from the heart passes through the kidneys each minute. Large vessels attach the kidneys directly to the aorta and the inferior vena cava. Waste products and water are constantly filtered from the blood to form urine. The kidneys continuously concentrate this filtered urine by reabsorbing the water as it passes through a system of specialized tubes within them. The tubes finally unite to form the renal pelvis, a cone-shaped collecting area that connects the ureter and the kidney. Normally, each kidney drains its urine into one ureter through which the urine passes to the bladder.

A ureter passes from the renal pelvis of each kidney along the surface of the posterior abdominal wall behind the peritoneum to drain into the urinary bladder. The ureters are small (0.2″ in diameter), hollow, muscular tubes. Peristalsis, a wavelike contraction of smooth muscle, occurs in these tubes to move the urine to the bladder.

The urinary bladder is located immediately behind the pubic symphysis in the pelvic cavity and is composed of smooth muscle with a specialized lining membrane. The two ureters enter posteriorly at its base on either side. The bladder empties to the outside of the body through the urethra. In the male, the urethra passes from the anterior base of the bladder through the penis. In the female, the urethra opens at the front of the vagina. The normal adult forms 1.5 to 2 L of urine every day. This waste is extracted and concentrated from the 1,500 L of blood that circulate through the kidneys daily.

Pathophysiology

The abdominal cavity is lined by a membrane called the **peritoneum**. The peritoneum also covers the organs

You are the Provider: PART 2

When you arrive at the scene and enter the patient's residence, you find him lying on the couch on his side with his knees drawn up to his abdomen. He is markedly diaphoretic and pale and is in obvious severe pain. You introduce yourself and begin your assessment.

Recording Time: 0 Minutes	
Appearance	Lying on his side, diaphoretic, in obvious pain
Level of consciousness	Conscious and alert; restless
Airway	Open; clear of secretions or foreign bodies
Breathing	Rapid, shallow respirations
Circulation	Radial pulse, weak and rapid; skin is pale and diaphoretic

Your partner administers oxygen at 15 L/min via a nonrebreathing mask as you continue your assessment. The patient tells you that his abdominal pain began suddenly and has been severe from the onset. He describes the pain as a "tearing" sensation that radiates to his lower back. He denies nausea, vomiting, or any other symptoms. As you examine his abdomen, your partner prepares to take his vital signs.

3. What is the proper technique of assessing a patient's abdomen? What should you assess for?

4. What is the difference between radiating pain and referred pain?

of the abdomen. The parietal peritoneum lines the abdominal cavity, and the visceral peritoneum covers the organs themselves. The abdominal space normally contains a small amount of peritoneal fluid to bathe and lubricate the organs in the abdominal cavity. Any foreign material, such as blood, pus, bile, pancreatic juice, or amniotic fluid, can cause irritation of the peritoneum, called **peritonitis**.

Acute abdomen is a medical term referring to the sudden onset of abdominal pain, often associated with severe, progressive problems that require medical attention. Peritonitis will usually develop if the acute abdomen is not treated, and can be fatal.

Peritonitis typically causes **ileus**, or paralysis of the muscular contractions that normally propel material through the intestine. The retained gas and feces, in turn, cause abdominal distention. In the presence of such paralysis, nothing that is eaten can pass normally out of the stomach or through the bowel. In this situation, the only way the stomach can empty itself is by **emesis**, or vomiting. For this reason, peritonitis is almost always associated with nausea and vomiting. These symptoms do not point to a particular cause because they can accompany almost every type of gastrointestinal disease or injury.

To gauge the degree of distention, simply look at the patient's abdomen. Distention begins shortly after muscular contractions of the bowel have ceased. The patient's pulse and blood pressure may change significantly or not at all. These findings reflect the severity of the process, its duration, and the amount of fluid lost into the abdomen.

Peritonitis is also associated with a loss of body fluid into the abdominal cavity. The loss of fluid usually results from abnormal shifts of fluid from the bloodstream into body tissues. This fluid shift decreases the volume of circulating blood and may lead to decreased blood pressure or even shock. The patient may have normal vital signs or, if the peritonitis has progressed farther, the patient may present with tachycardia and hypotension. When peritonitis is accompanied by hemorrhage, the signs of shock are much more apparent.

Fever may or may not be present, depending on the cause of the peritonitis. Patients with **diverticulitis** or cholecystitis (inflammation of the gallbladder) may have a substantial elevation in temperature. However, patients with acute appendicitis may have a temperature within normal limits until the appendix ruptures and an abscess starts to form.

■ Abdominal Pain

Two different types of nerves supply the peritoneum, and therefore abdominal pain can have different qualities. The parietal peritoneum is supplied by the same nerves from the spinal cord that supply the skin of the abdomen; it can therefore perceive much the same sensations: pain, touch, pressure, heat, and cold. These sensory nerves can easily identify and localize a point of irritation. In contrast, the visceral peritoneum is supplied by the autonomic nervous system. These nerves are far less able to localize sensation. What this means for you is that your patient will not be able to localize and describe exactly where the pain is. The visceral peritoneum is stimulated when distention or contraction of the hollow abdominal organs activates the stretch receptors. Patients sometimes describe it as a "deep" pain. Other painful sensations that occur because of an irritated visceral peritoneum may be perceived at a distant point on the surface of the body, such as the back or shoulder. This phenomenon is called **referred pain**.

Referred pain is the result of connections between the body's two separate nervous systems. The nerves connecting the somatic nervous system and autonomic nervous system cause the stimulation of the autonomic nerves to be perceived as stimulation of the spinal sensory nerves. For example, acute cholecystitis may cause pain in the right shoulder because the autonomic nerves serving the gallbladder lie near the spinal cord at the same anatomic level as the spinal sensory nerves that supply the skin of the shoulder **Figure 16-3** .

The most common abdominal emergencies, with most common locations of direct and referred pain, are listed in **Table 16-1** .

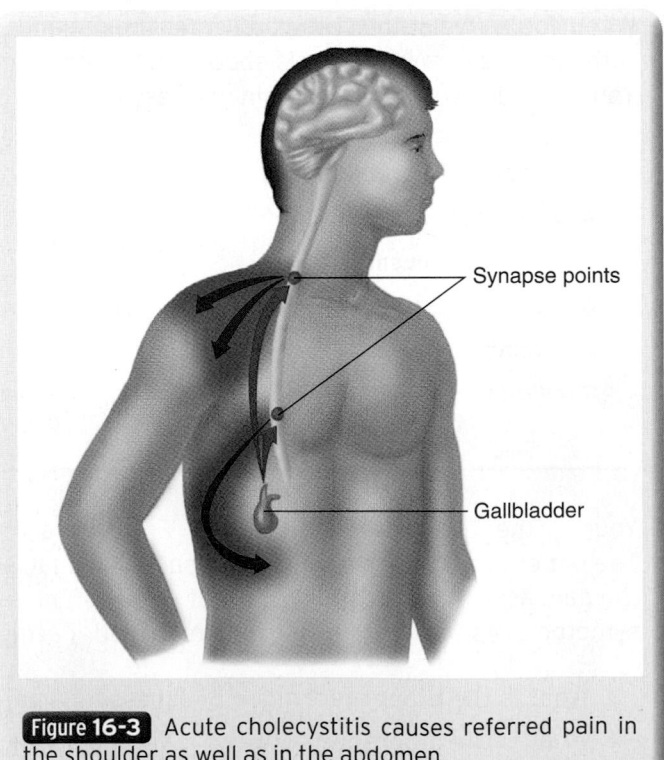

Figure 16-3 Acute cholecystitis causes referred pain in the shoulder as well as in the abdomen.

Table 16-1 Common Abdominal Conditions

Condition	Localization of Pain
Appendicitis	Right lower quadrant (direct); around navel (referred); rebounding pain (pain felt on the rebound after palpation)
Cholecystitis	Right upper quadrant (direct); right shoulder (referred)
Ulcer	Upper midabdomen or upper part of back
Diverticulitis	Left lower quadrant
Abdominal aortic aneurysm (ruptured or dissecting)	Low part of back and lower quadrants
Cystitis (inflammation of the bladder)	Lower midabdomen (retropubic)
Kidney infection	Costovertebral angle
Kidney stone	Right or left flank, radiating to genitalia
Pancreatitis	Upper abdomen (both quadrants); back
Pneumonia	Referred pain to the upper abdomen
Hernia	Anywhere in the abdominal area
Peritonitis	Anywhere in the abdominal area

Causes of Acute Abdomen

Almost any problem with an abdominal organ can cause an acute abdomen. Some of the more common causes are discussed here. Because the visceral peritoneum is usually irritated first, early abdominal pain tends to be vague and poorly localized. As the parietal peritoneum becomes irritated, pain becomes more severe and may be more specifically located.

Ulcers

The stomach and duodenum are subjected to high levels of acidity. To prevent damage to these organs, protective layers of mucus line both organs. In peptic ulcer disease (PUD), the protective layer is eroded, allowing the acid to eat into the organ itself over the course of weeks, months, or even years.

Most peptic ulcers are the result of infection of the stomach with *Helicobacter pylori*. Another major cause is chronic use of nonsteroidal anti-inflammatory drugs (NSAIDs). Alcohol and smoking can also affect the severity of PUD by increasing gastric acidity.

PUD affects both men and women equally, but tends to occur more often in the older population. As people age, the immune system's ability to fight infection decreases, making infection more likely. The geriatric population, in general, also uses NSAIDs frequently for arthritis and other musculoskeletal conditions.

Patients with peptic ulcers experience a classic sequence of burning or gnawing pain in the stomach that subsides or diminishes immediately after eating and then reemerges 2 to 3 hours later. The pain usually presents in the upper abdomen, but sometimes may be found below the sternum. With some patients, the pain occurs immediately after eating. Nausea, vomiting, belching, and heartburn are common symptoms. If the erosion is severe, gastric bleeding can occur, resulting in hematemesis and melena (black, tarry stools containing blood).

Some ulcers will heal without medical intervention, but often complications can occur from bleeding or perforation (a hole through the wall of the stomach). More serious ulcerative conditions can cause severe peritonitis and an acute abdomen.

Gallstones

The gallbladder is a storage pouch for digestive juices and waste from the liver. Gallstones can form and block the outlet from the gallbladder, causing pain. Sometimes the blockage will pass, but if not, it can lead to severe inflammation of the gallbladder, called **cholecystitis**. This is a condition in which the wall of the gallbladder becomes inflamed. In severe cases, the gallbladder may rupture, causing inflammation to spread and irritate surrounding structures such as the diaphragm and bowel. This condition presents as a constant, severe pain in the right upper or midabdominal region and may refer to the right upper back, shoulder area, or flank. The pain may steadily increase for hours or may come and go. Cholecystitis commonly produces symptoms about 30 minutes after a particularly fatty meal and usually at night. Other symptoms include general gastrointestinal distress such as nausea and vomiting, indigestion, bloating, gas, and belching.

Pancreatitis

The pancreas forms digestive juices and is also the source of insulin. Inflammation of the pancreas is called pancreatitis. **Pancreatitis** can be caused by an obstructing gallstone, alcohol abuse, and other diseases. Severe pain may present in the upper left and right quadrants and may often radiate to the back. Other signs and symptoms accompanying the pain are nausea and vomiting, abdominal distention, and tenderness. Complications like sepsis or hemorrhage can occur, in which case assessment may also reveal fever or tachycardia.

Appendicitis

The appendix is a small recess in the large intestine. Inflammation or infection in the appendix is called **appendicitis**, and is a frequent cause of acute abdomen. This inflammation can eventually cause the tissues to die and/or rupture, causing an abscess, peritonitis, or shock. Initially, the pain caused by appendicitis is more generalized, dull, and diffuse and may center in the umbilical area. The pain later localizes to the right lower quadrant of the abdomen. Appendicitis can also cause referred pain. The patient may also report nausea and vomiting, anorexia (lack of appetite for food), fever, and chills. A classic symptom of appendicitis is rebound tenderness. Rebound tenderness is a result of peritoneal irritation. This can be assessed by pressing down gently and firmly on the abdomen. The patient will feel pain when the pressure is released. Women who are pregnant may not exhibit this symptom.

Gastrointestinal Hemorrhage

Bleeding within the gastrointestinal tract is a symptom of another disease, not a disease itself. Gastrointestinal hemorrhage can be acute, which may be shorter term and more severe, or chronic which may be of longer duration and less severe. All complaints of bleeding should be considered serious.

A gastrointestinal hemorrhage can occur in the upper or lower gastrointestinal tract. Bleeding in the upper gastrointestinal tract occurs from the esophagus to the upper small intestine. In the esophagus, problems might include esophagitis, esophageal varices, or Mallory-Weiss syndrome.

Lower gastrointestinal bleeding occurs between the upper part of the small intestine and the anus. Bowel inflammation, diverticulitis, and hemorrhoids are common causes of bleeding in the lower gastrointestinal tract.

Esophagitis

Esophagitis occurs when the lining of the esophagus becomes inflamed by infection or from the acids in the stomach (gastroesophageal reflux disease). The patient may report pain with swallowing and complain of feeling like an object is stuck in his or her throat. Additional symptoms include heartburn, nausea, vomiting, and sores in the mouth. In the worst cases, bleeding can occur from the small capillary vessels within the esophageal lining or the main blood vessels.

Esophageal Varices

Esophageal varices occur when the amount of pressure within the blood vessels surrounding the esophagus increases. The esophageal blood vessels eventually deposit their blood into the portal system. If the liver becomes damaged and blood cannot flow through it easily, blood begins to back up into these portal vessels, dilating the vessels and causing the capillary network of the esophagus to begin leaking. If pressure continues to build, the vessel walls may fail, causing bleeding.

You are the Provider: PART 3

The patient's vital signs are obtained and recorded. Your assessment of his abdomen reveals that it is tender to palpation and guarding is present. As your partner retrieves the stretcher from the ambulance, the patient tells you that he has high blood pressure, depression, and had his appendix removed 30 years ago. You note that he is becoming more restless and is still experiencing intense pain, which he describes as a 10 on a scale of 0 to 10 (10/10).

Recording Time: 3 Minutes	
Respirations	28 breaths/min; shallow
Pulse	124 beats/min; regular
Skin	Pale, cool, and diaphoretic
Blood pressure	98/60 mm Hg
Oxygen saturation (Sao$_2$)	96% (on oxygen)

5. What do the patient's vital signs indicate?

6. What do you suspect is the cause of the patient's abdominal pain?

In industrialized countries, alcohol is the main cause of portal hypertension. Long-term alcohol consumption damages the interior of the liver (cirrhosis), leading to slower blood flow. In developing countries, viral hepatitis is the main cause of liver damage.

Presentation of esophageal varices takes two forms. Initially, the patient shows signs of liver disease—fatigue, weight loss, jaundice, anorexia, edema in the abdomen, abdominal pain, nausea, and vomiting. This very gradual disease process takes months to years before the patient reaches a state of extreme discomfort.

By contrast, the rupture of the varices is far more sudden. The patient will complain of sudden-onset discomfort in the throat. He or she may have severe difficulty swallowing, vomiting of bright red blood, hypotension, and signs of shock. If the bleeding is less dramatic, hematemesis (vomiting blood) and melena (black, tarry stools) are likely. Regardless of the speed of bleeding, damage to these vessels can be life threatening. Spontaneous rupture is often life threatening and significant blood loss at the scene may be evident. Major ruptures can lead to death in a matter of minutes.

Mallory-Weiss Syndrome

Mallory-Weiss syndrome may lead to severe hemorrhage. In this condition, the junction between the esophagus and the stomach tears, causing severe bleeding and potentially death. Primary risk factors include alcoholism and eating disorders. Mallory-Weiss syndrome affects both men and women equally, but is more prevalent in older adults and older children.

Vomiting is the principal symptom. In women, this syndrome may be associated with severe vomiting related to pregnancy. The extent of the bleeding can range from very minor bleeding, resulting in very little blood loss, to severe bleeding and extreme fluid loss. In extreme cases, patients may experience signs and symptoms of shock, upper abdominal pain, hematemesis, and melena.

Gastroenteritis

Acute gastroenteritis comprises a family of conditions revolving around a central theme of infection combined with diarrhea, nausea, and vomiting. Bacterial and viral organisms can cause this condition. These organisms typically enter the body through contaminated food or water. Patients may begin to experience an upset stomach and diarrhea as soon as several hours or several days after contact with the contaminated matter. The disease can then run its course in 2 to 3 days or continue for several weeks.

Gastroenteritis is not an infectious disease but has all of the hallmarks of its acute (infectious) cousin. Patients with this condition experience nausea, vomiting, and diarrhea from a noninfectious cause, such as medications, toxins from shellfish, or chemotherapy.

Diarrhea is the principal symptom in both types of gastroenteritis. Patients may experience large dumping-type diarrhea or frequent small liquid stools. The diarrhea may contain blood and/or pus, and it may have a foul odor or be odorless. Abdominal cramping is frequently reported. Nausea, vomiting, fever, and anorexia are also present. If the diarrhea continues, dehydration will result. As the volume of fluid loss increases, the likelihood of shock increases.

Diverticulitis

Diverticulitis was first recognized around 1900, when the types of foods people ate began to change dramatically. In particular, the amount of fiber within the US diet plummeted as the amount of processed foods eaten increased.

As the amount of fiber consumed as part of the diet decreases, the consistency of the normal stool becomes more solid. This hard stool requires more intestinal contractions, subsequently increasing pressure within the colon. In this environment, small defects within the colonic wall that would otherwise never pose a problem now fail, resulting in bulges in the wall. These small outcroppings eventually turn into pouches, called diverticula. As feces travel through the colon, some may become trapped within these pouches. When bacteria grow there, they cause localized inflammation and infection.

The main symptom of diverticulitis is abdominal pain, which tends to be localized to the left side of the lower abdomen. Classic signs of infection include fever, malaise, body aches, chills, nausea, and vomiting. Bleeding is rare with this condition. Because of the local infections of these pouches, adhesions may develop, narrowing the diameter of the colon and resulting in constipation and bowel obstruction.

Hemorrhoids

Hemorrhoids are created by swelling and inflammation of the blood vessels surrounding the rectum. They are a common problem, with almost half the population having at least one hemorrhoid by age 50 years. Hemorrhoids may result from conditions that increase pressure on the rectum or irritation of the rectum. Pregnancy, straining at stool, and chronic constipation cause increased pressure. Diarrhea can cause irritation.

Hemorrhoids present as bright red blood during defecation. This bleeding tends to be minimal and is easily controlled. Additionally, patients may experience itching and a small mass on the rectum. Typically, this mass is a clot formed in response to the mild bleeding.

Urinary System

Issues in the urinary system can cause acute abdominal pain. Bladder inflammation, called **cystitis**, is common, especially in women. This condition is generally caused by a bacterial infection, and can be referred to as a **urinary tract infection (UTI)**. A bladder infection can be painful. Patients with cystitis usually have lower quadrant abdominal pain. They also may report an urgency and frequency in urination, and pressure and pain around the bladder. If the infection is severe, the urethra can become inflamed, causing urinary retention. When you are assessing a patient with cystitis, the patient may report tenderness when you are palpating the abdomen over the bladder (just above the pubic bone). Cystitis can become a serious health problem if the infection spreads to the kidneys.

Kidneys

The kidneys play a major role in maintaining homeostasis, or keeping all body systems in balance. The kidneys preserve this balance by eliminating waste from the blood. When the kidneys fail, the patient loses the ability to excrete waste from the body, leading to a condition called **uremia**. This means that the waste product, urea, which is normally excreted into the urine, remains in the blood.

Urine drains from the kidney through a narrow tube called the ureter through to the bladder. Chemicals may crystallize in the urine and form **kidney stones** (renal calculi). Kidney stones can grow over time and if a stone passes into the ureter, it can cause a blockage. Pressure will build up behind the kidney stone and cause swelling in the kidney. Patients with a kidney stone blockage may initially report vague discomfort in the flank, but the pain can become quite intense within 60 minutes and typically will radiate to the groin. These patients are often agitated and restless as they try to get into a comfortable position to relieve the pain. They may also report nausea and vomiting. The pain from kidney stones is often caused when the stone moves within the ureter. In many cases, the stone will pass on its own, but in other cases it may have to be surgically removed (or broken up). A slight amount of blood in the urine (hematuria) after the stone passes may be present because of irritation of the ureter.

Kidney (renal) failure can be acute or chronic. Acute renal failure is a sudden (possibly over a period of days) decrease in function. It occurs from a variety of causes including hemorrhage, dehydration, trauma, shock, sepsis, heart failure, medications, drug abuse, and kidney stones. Acute renal failure can be reversed with prompt diagnosis and treatment.

Chronic renal failure is irreversible. It is progressive and develops over months and years. It is often caused by diabetes or hypertension. The kidney tissue shrinks and function diminishes. Eventually the patient requires dialysis or a kidney transplant to remove waste products from the bloodstream.

Patients with chronic renal failure exhibit several signs and symptoms, beginning with an altered level of consciousness. In later stages, seizures and coma are possible. Additional signs and symptoms include lethargy, nausea, headaches, cramps, and edema in the extremities and face because of fluid imbalances.

Female Reproductive Organs

Gynecologic problems are a common cause of acute abdominal pain. Always consider that a woman with lower quadrant abdominal pain and tenderness may have a problem related to her ovaries, fallopian tubes, or uterus. Chapter 21, *Gynecologic Emergencies*, covers gynecologic emergencies in depth.

Other Organ Systems

The aorta lies immediately behind the peritoneum. In older people, the wall of the aorta sometimes develops weak areas that swell to form an abdominal aortic

Special Populations

Geriatric patients are as susceptible to the acute abdomen as younger adults. However, the signs and symptoms in geriatric patients might be different. Because of altered pain sensation, geriatric patients with an acute abdomen may not feel any discomfort or may describe the discomfort as mild, even in severe conditions.

Because the older patient has decreased body temperature regulation and response, the patient with an acute abdomen, including peritonitis, may not have a fever. However, if a fever is present, it can be minimal.

Because of the older patient's response to the acute abdomen, a delay in identifying the condition and seeking medical attention is possible, putting the patient at risk for complications. You should ask about the patient's medical history, especially the history of recent illness, to identify a potential illness. Ask about abdominal discomfort, when the patient last had a bowel movement, whether she or he was constipated or had diarrhea. Inquire if the patient has had previous bowel obstructions. Inquire as to when the patient last ate, how much fluid he or she has consumed, and whether he or she has vomited. Many geriatric patients think that a few cups of coffee a day is adequate fluid intake, but coffee (especially caffeinated) causes vasoconstriction and dehydration within the digestive system. Quickly determining the severity of the patient's problem can hasten proper treatment and recovery.

aneurysm (AAA). A pulsating mass may be felt in the abdomen, although this is a rare sign and is often hard to detect. Use extreme caution when trying to assess or detect this condition. The development of an aneurysm is rarely associated with symptoms because it occurs slowly, but if the aneurysm tears and ruptures, massive hemorrhage may occur, and the patient will present with signs of acute peritoneal irritation. The patient may also report radiation of severe pain to the back because the peritoneum can be stripped away from the wall of the main abdominal cavity by the hemorrhage. Back pain is a common symptom when an aneurysm has started to expand and the aortic linings begin to tear. Back pain that cannot be easily explained should be investigated closely in patients who are suspected of having an AAA. The patient generally describes the pain as "tearing" which is different than most other complaints of abdominal pain. The association of acute abdominal signs and symptoms of shock requires prompt transportation. Because this is a fragile situation with a large, leaking artery, avoid unnecessary or vigorous palpation of the abdomen and do not aggressively treat the patient for shock because these actions can cause a small tear to expand. Remember to handle the patient gently during transport.

Pneumonia, especially in the lower parts of the lung, may cause both ileus and abdominal pain. In this situation, the problem lies in an adjacent body cavity, but the intense inflammatory response can reflect in the abdomen.

A **hernia** is a protrusion of an organ or tissue through a hole or opening into a body cavity where it does not belong. Hernias can occur as a result of the following:

- A congenital defect, as around the umbilicus
- A surgical wound that has failed to heal properly
- A natural weakness in an area such as in the groin

Hernias do not always produce a mass or lump that the patient will notice. At times, the mass will disappear back into the body cavity in which it belongs. In this case, the hernia is said to be reducible. If the mass cannot be pushed back within the body, it is said to be incarcerated.

Reducible hernias pose little risk to the patient; some people live with them for years. When a hernia is incarcerated, however, its contents may become seriously compressed by the surrounding tissue, eventually compromising the blood supply. This situation, called **strangulation**, is a serious medical emergency. Immediate surgery is required to remove any dead tissue and repair the hernia.

The following signs and symptoms indicate a serious hernia problem:

- A formerly reducible mass is no longer reducible
- Pain at the hernia site
- Tenderness when the hernia is palpated
- Red or blue skin discoloration over the hernia

Any of these signs and symptoms is cause for prompt transport to the emergency department.

Words of Wisdom

An acute abdomen usually indicates peritonitis, in which generalized signs can make it challenging to determine exactly where the problem lies, even for physicians. Knowing abdominal assessment steps well, and recording your findings in detail, are important early factors in the process that leads to diagnosis.

Special Populations

Causes of abdominal pain are difficult enough to determine in adults who can provide a good history, but for children who can only tell you they have a "stomachache" it is even more problematic. It is hard for a parent or caregiver to provide accurate information when pain is so subjective. Confirm with the parent or caregiver the details of the medical history and whether the current problem could be an exacerbation (return) of a chronic problem. Chapter 32, *Pediatric Emergencies*, covers the acute abdomen in pediatric patients in depth.

Abdominal pain could mean an infection, be related to something the child ate, or indicate a poisoning. Look for clues that may indicate if the child ingested something poisonous. Consider environmental causes like spider bites (black widow or brown recluse) or metabolic issues like diabetic complications. Confirm the duration and location of the pain and if there has been vomiting.

Assess the child's appearance. Ask if there has been diarrhea or any kind of rash. It is always wise to transport a child with abdominal pain for further assessment.

Patient Assessment

Scene Size-up

Scene Safety

Your first action should always be to ensure that the scene is safe. Follow standard precautions with a minimum of gloves and eye protection. Consider donning a gown and covering your shoes with disposable, protective covers because often there may be feces and urine on the floor and some patients may have active projectile vomiting.

As you proceed to the patient, observe the scene for safety threats to yourself and your partner and determine the number of patients at the scene. If your call involves going to the patient's home and he or she does not come to the door, the patient may have had a syncopal episode (fainted). Request police assistance to help you gain access to the patient. Consider the need for additional or specialized medical resources. Call for additional resources earlier rather than later.

Mechanism of Injury/Nature of Illness

You will need to look for clues to help you determine the nature of illness (NOI) or the mechanism of injury. Acute abdomen can be the result of violence, such as blunt or penetrating trauma, so always be vigilant. Chapter 28, *Abdominal and Genitourinary Injuries*, discusses traumatic injuries in detail. Clues will help you develop an early index of suspicion for life threats. For example, a pale and sweating patient who reports tearing pain may have an AAA. Observe the scene closely and interview bystanders or family members if the NOI is not obvious. In some cases, your senses can help give you a clue as to the NOI. For example, gastrointestinal bleeding often has a characteristic odor that you will learn to recognize.

▶ Primary Assessment

As you begin to assess the patient, remember that your first priority is to look for and treat any life-threatening conditions. Assess the patient's level of consciousness. Assess the patient's ABCs; threats to airway, breathing, or circulation are considered life threatening and must be treated immediately. Perform a rapid scan. Note the position of the patient. Commonly the patient will have his or her knees drawn up to help alleviate the pain associated with acute abdomen. Consider necessary treatment and transportation options and the need for early advanced life support (ALS) assistance.

Form a General Impression

Approach the patient and ask him or her about the chief complaint. A description of the current problem in the patient's own words should help you to identify where to begin your assessment. If the chief complaint indicates a life-threatening problem, assess and treat it immediately. If the chief complaint is a minor problem, it should wait until you have had a chance to assess for and treat any potential life threats. The patient's level of consciousness, using the AVPU (*Alert* to person, place, and day; responsive to *Verbal* stimuli; responsive to *Pain*; *Unresponsive*) scale, should be included in your general impression.

Airway and Breathing

Ensure that the patient's airway is clear and that the patient's respirations are adequate. Administer oxygen to the patient when needed. As a result of the abdominal pain, the patient may show shallow or inadequate respirations because deep breaths often intensify the pain.

Circulation

When you are assessing the patient's circulation, remember to assess for major bleeding. Ask the patient about blood in the vomit (hematemesis) or black, tarry stools (melena). The patient's pulse rate and quality, as well as skin condition, may indicate shock. Check the pulses in both arms because a difference in pulse strength may indicate an AAA.

You are the Provider: PART 4

After providing further treatment, you place the patient onto the stretcher, load him into the ambulance, and proceed to the closest appropriate hospital, which is located 20 miles away. En route, you reassess the patient.

Recording Time: 12 Minutes	
Level of consciousness	Conscious and alert; restless
Respirations	28 breaths/min; shallow
Pulse	130 beats/min; weak and regular
Skin	Cool, pale, and diaphoretic
Blood pressure	100/62 mm Hg
Sao$_2$	98% (on oxygen)

7. Are there any special considerations for this patient? If so, what are they?

Shock may be caused by hypovolemia or may be the result of a severe infection (septic). If evidence of shock (inadequate perfusion) is present, interventions should include high-flow oxygen, elevating the patient's legs 6″ to 12″ by placing a pillow or blanket underneath the knees and lower legs or to a position of comfort, and keeping the patient warm. Ensure that you provide prompt treatment for life threats and do not delay in providing transport.

Transport Decision

Certain patients should be transported quickly. These include patients who have airway, breathing, or circulation problems, including problems with pulse and perfusion, and patients with suspected internal bleeding. Included in the group to package quickly and transport rapidly are patients who have a poor general impression, especially pediatric and geriatric patients. Pale, cool, diaphoretic skin, tachycardia, hypotension, and altered level of consciousness are all signs of significant illness.

Ensure that the ride during transport is as gentle as possible for the patient. Drive smoothly and steadily. Rapid driving can result in increased vehicle movement, potentially aggravating and possibly worsening the patient's abdominal pain.

▶ History Taking

Investigate Chief Complaint

The chief complaint is often based on a patient's previous history of chronic medical problems. This history includes information expressed in the patient's own words (subjective) and what you observe through physical assessment (objective). The patient history should also include the patient's general state of health, surgeries or recent hospitalizations, and any traumatic injuries.

SAMPLE History

If the patient is responsive, begin with obtaining the SAMPLE history. Ask the following questions specific to the Signs and symptoms of a gastrointestinal or urologic emergency:

- **Nausea and vomiting.** Do you feel nauseous? Have you vomited? How many times? Over what period of time? Was there red blood? Did it look like coffee grounds?
- **Changes in bowel habits.** Has there been any change in your bowel habits? Have you been constipated? Did the stool look dark and tarry? Have you had diarrhea? Was there any red blood in it?
- **Urination.** Have you been urinating more or less often? Is there pain when you urinate? Is the color dark or unusual? Is there an unusual odor?

- **Weight loss.** Have you lost weight recently? How many pounds?
- **Belching or flatulence.** Have you experienced belching or flatulence? For how long?
- **Pain.** What does the pain feel like? How long have you had this pain? Is the pain constant or intermittent?
- **Other.** Ask about any other signs or symptoms related to this complaint, such as "Are there any changes you have noted recently that may be contributing to your pain?"
- **Concurrent chest pain.** If the patient reports chest pain, use OPQRST (Onset, Provocation/palliation, Quality, Region/radiation, Severity, and Timing of pain) to ask the patient what makes the pain better or worse.

Continue with the SAMPLE history. Does the patient have any Allergies? What are the patient's current Medications? Determine the patient's general state of health through the Pertinent past history. Has the patient experienced this kind of abdominal pain before? If the patient is female and of childbearing age, determine the date of her last menstrual period. This will determine if the patient could possibly be pregnant or raise the suspicion of an ectopic pregnancy. Has the patient had any surgery or recent hospitalizations?

Ask the patient about his or her Last oral intake. It is important to determine whether the patient has ingested any substance that could be causing the acute abdomen. If eating causes pain, discomfort, vomiting, or diarrhea, the patient will eat less often or stop eating. Do not give the patient anything by mouth. Food or fluid may only aggravate many of the symptoms. Also, the presence of food in the stomach increases the risk of aspiration.

Finally, determine the Events that led up to the patient's present illness. It is important to determine whether this is a medical emergency or related to trauma. Therefore, you need to question the patient about any recent trauma.

The SAMPLE history may not affect the interventions you perform, but it will help provide needed information for the physician in the emergency department to aid in determining the cause of the acute abdomen.

Words of Wisdom

Consider pertinent negatives, which are a record of normal findings that warrant no care or intervention. It is important to know and document that the patient denies shortness of breath symptoms or denies radiation of chest pain.

Secondary Assessment

If the secondary assessment is not performed at the scene, it is performed in the back of the ambulance en route to the hospital. However, there will be situations when you may not have time to perform a secondary assessment if you have to continually manage life threats that were identified during the primary assessment. If the patient is stable and has an isolated complaint, the secondary assessment may occur at the scene.

In some situations, patients are comfortable only when lying in one particular position, which tends to relax muscles adjacent to the inflamed organ and thus lessen the pain. Therefore, the position of the patient may provide you with an important clue. For example, a patient with appendicitis may draw up the right knee. A patient with pancreatitis may lie curled up on one side.

Physical Examinations

Information gathered in the history-taking portion of the patient assessment may be used to focus your physical examination of the abdomen. A normal abdomen is soft and not tender to the touch. Pain and tenderness are the most common symptoms of an acute abdomen. The pain may be sharply localized or diffuse and will vary in its severity. Localized pain gives a clue to the problem organ or area causing it. Tenderness may be minimal or so great that the patient will not allow you to touch the abdomen. In some instances, the muscles of the abdominal wall become rigid in an involuntary effort to protect the abdomen from further irritation. This boardlike muscle spasm, called **guarding**, can be seen with major problems such as a perforated ulcer or pancreatitis.

Remember, the patient with peritonitis usually has abdominal pain, even when lying quietly. The patient may have difficulty breathing and may take rapid, shallow breaths because of the pain. Usually, you will find tenderness on palpation of the abdomen or when the patient moves. The degree of pain and tenderness is usually related directly to the severity of peritoneal inflammation.

Use the following steps to assess the abdomen:

1. Explain to the patient what you are going to do in terms of assessing the abdomen.
2. Place the patient in a supine position with the legs drawn up and flexed at the knees to relax the abdominal muscles, unless there is any trauma, in which case the patient will remain supine and stabilized. Determine whether the patient is restless or quiet, and whether motion causes pain.
3. Expose the abdomen and visually assess it. Does the abdomen appear distended (enlarged)? Do you see any pulsating masses (indicates an aortic aneurysm)? Is there bruising to the abdominal wall?
4. Ask the patient where the pain is most intense. Palpate in a clockwise direction beginning with the quadrant *after* the one the patient indicates is tender or painful; end with the quadrant the patient indicates is tender or painful. If the most painful area is palpated first, the patient may guard against

You are the Provider: PART 5

With an estimated time of arrival at the hospital of 9 minutes, you reassess the patient and then call in your radio report. The patient remains conscious and alert, but restless, and is still experiencing 10/10 abdominal pain.

Recording Time: 17 Minutes	
Level of consciousness	Conscious and alert; restless
Respirations	28 breaths/min; shallow
Pulse	128 beats/min; weak and regular
Skin	Cool, clammy, and diaphoretic
Blood pressure	96/58 mm Hg
Sao$_2$	97% (on oxygen)

The patient's condition is unchanged on arrival at the hospital. You give your verbal report to the charge nurse. After further assessment and treatment in the emergency department, the patient is taken to surgery. You later learn that he had an expanding abdominal aortic aneurysm, which was successfully repaired.

8. Could you have done anything definitively for this patient in the field?

further examination, making your assessment more difficult and less reliable.

5. Remember to be very gentle when palpating the abdomen. Occasionally, an organ within the abdomen will be enlarged and very fragile and rough palpation could cause further damage. If you see a pulsating mass, do not touch it; doing so could cause the aorta to rupture.

6. Palpate the four quadrants of the abdomen gently to determine whether each quadrant is tense (guarded) or soft when palpated **Figure 16-4**.

7. Note whether the pain is localized to a particular quadrant or diffuse (widespread).

8. Palpate and wait for the patient to respond, looking for a facial grimace or a verbal "ouch." Do not ask the patient, "Does it hurt here?" as you palpate.

9. Determine whether the patient exhibits rebound tenderness (may be tender when direct pressure is applied, but very painful when pressure is released). This is an indicator of peritonitis. When you are palpating for rebound tenderness, you should use extreme caution.

10. Determine whether the patient can relax the abdominal wall on command.

11. Guarding and rigidity may be detected. Guarding is tensing of the abdominal wall muscles.

Vital Signs

Findings of a high respiratory rate with a normal pulse rate and blood pressure may indicate the patient is unable to ventilate properly because deep breathing causes pain. A high respiratory rate and pulse rate with signs of shock, such as pallor and diaphoresis (profuse sweating), may indicate septic or hypovolemic shock.

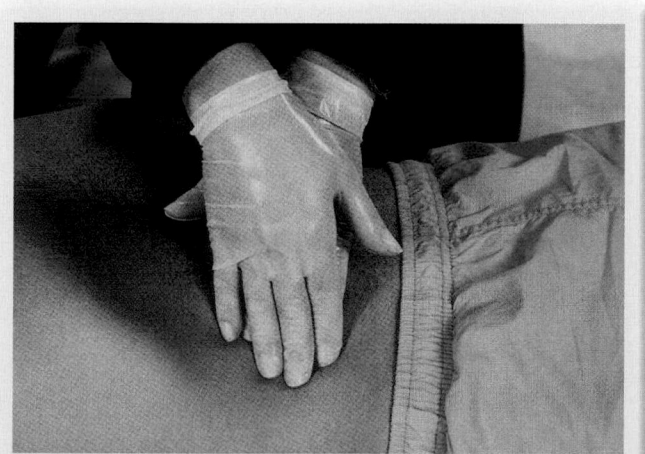

Figure 16-4 Check for tenderness or rigidity by gently palpating the abdomen.

Special Populations

Elderly patients may not exhibit rigidity or guarding like a younger adult. Abdominal pain can sometimes be related to cardiac conditions. Abdominal pain is frequently caused by bowel impaction or obstruction. Obstructions can be very serious and can lead to bowel ruptures that can be and often are life threatening. Elderly patients may also not exhibit the same pain response ability because of deterioration of their sensory systems. Provide transport to an appropriate facility that can meet the needs of a geriatric patient.

Words of Wisdom

When palpating the abdomen of a patient with abdominal complaints, it is important to palpate clockwise, beginning in the quadrant next to the area of the described pain.

Monitoring Devices

Use pulse oximetry and noninvasive blood pressure devices when these monitoring devices are available. It is recommended that you always assess the patient's first blood pressure manually with a sphygmomanometer (blood pressure cuff) and stethoscope.

Reassessment

Because it is often difficult to determine the cause of an acute abdominal emergency, it is extremely important to reassess your patient frequently to determine whether the patient's condition has changed. Remember, the condition of a patient with an acute abdomen can change rapidly from stable to unstable.

Vital signs must be reassessed and compared with the patient's baseline vital signs. If anything changes en route to the hospital, manage the problem and document any changes or additional treatment.

Reassess the patient and then ask and answer the following questions (where appropriate):

- Has the patient's level of consciousness changed?
- Has the patient become more anxious?
- Have the skin signs begun to change?
- Has the pain gotten better or worse?
- Has bleeding become worse or better?
- Is current treatment improving the patient's condition?
- Has an already identified problem gotten better?
- Has an already identified problem gotten worse?
- What is the nature of any newly identified problems?

Interventions

Interventions generally include treatment for shock and providing emotional support. Administer oxygen, cover the patient with a blanket for warmth, and provide gentle transport for the patient without delay. Place the patient in a position of comfort. You will find that patients want to be supine with their knees drawn up. If the patient wants to lay on his or her side, try to make that possible. Be sure that you can observe and maintain the patient's airway because vomiting is common. If the patient's pain is extreme or he or she is showing significant signs of shock, consider the use of ALS assistance (if available) for intravenous fluids and pain management. If transport time is extended and rapid transport is needed, consider air medical if available.

Communication and Documentation

Communicate with the receiving hospital early to allow hospital staff to recruit the resources necessary to treat your patient on arrival. Carefully document your findings in your patient care report and relay all relevant information to the receiving physician or nurse. This information should include updated vital signs, changes in the patient's level of consciousness, and any new or worsening complaints.

Emergency Medical Care

Although you cannot treat the causes of acute abdomen, you can take steps to provide comfort and lessen the effects of shock by reassuring the patient and making the patient feel at ease. Treat the patient for shock even when obvious signs of shock are not apparent. Position patients who are vomiting to maintain a patent airway. Contain the vomitus to prevent the spread of infections (by using a biohazard bag). Airborne bacteria and viruses produced from vomiting can be easily transmitted to others. Ensure you are wearing gloves, eye protection, and a gown to prevent contamination of yourself, and wear a mask to prevent breathing in any infectious organisms. When you have released your patient to the hospital staff, clean the ambulance and any equipment you have used, preferably with an antibacterial cleaner. Do not forget to wash your hands even though you were wearing gloves.

Providing the patient low-flow oxygen often decreases the nausea. If the patient is having problems breathing, high concentrations of oxygen are more appropriate. Elevate the patient's legs to facilitate blood flow to the core of the body and improve circulation. Loosen restrictive clothing and transport gently in a position of comfort. You should constantly reassess your patient's condition for signs of deterioration.

Kidney Dialysis

The only definitive treatment in chronic cases of kidney failure is peritoneal dialysis or hemodialysis. In these processes, the patient's blood is filtered and cleansed of the toxins and then returned to the body. The treatment eliminates waste, normalizes the blood chemistry, and reduces excess fluid. If a patient misses a dialysis treatment, weakness and pulmonary edema can be the first in a series of conditions that can become progressively more serious if normal balance is not returned to the patient's body.

The only time you will most likely see such a machine is if your service transports patients to and from dialysis centers. If there is a dialysis machine in a private residence, treatments will most likely be performed by a trained dialysis technician.

In hemodialysis, the patient's blood circulates through a dialysis machine that functions in much the same way as the normal kidneys. Most patients undergoing long-term hemodialysis have some sort of shunt, ie, a surgically created connection between a vein and an artery. The patient is connected to the dialysis machine through this shunt, which allows blood to flow from the body into the dialysis machine and back to the body. Some patients have an internal shunt (fistula), which is an artificial connection (graft) between a vein and an artery that is usually located in the forearm or upper arm Figure 16-5 .

In peritoneal dialysis, large amounts of specially formulated dialysis fluid are infused into (and back out of) the abdominal cavity. This fluid stays in the cavity for 1 to 2 hours, allowing equilibrium to occur. Peritoneal dialysis is very effective but carries a high risk of peritonitis.

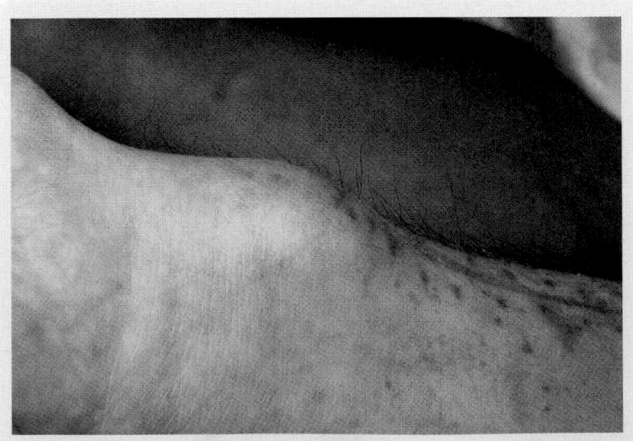

Figure 16-5 Expose and visualize fistulas or shunts to determine if there is an infection or you need to control bleeding.

With proper training, however, peritoneal dialysis can be performed in the home.

The adverse effects of dialysis include hypotension, muscle cramps, nausea and vomiting, hemorrhage from the access site, and infection at the access site. If your call involves a patient on dialysis, start with the ABCs: assess and manage the airway, breathing, and circulation. Provide high-flow oxygen and manage any bleeding from the access site. Position the patient sitting up in cases of pulmonary edema or supine if the patient is in shock and transport promptly.

Many dialysis patients also have urinary catheters. The catheter is placed in the bladder so the urine can run into a bag. These catheters can often be a source of infection. The patient may report fever and general malaise (illness) in addition to any symptoms specific to kidney failure. Leave the device in place. Treat any signs and symptoms and transport the patient for further evaluation.

During transport, unless there is a life-threatening event, make all attempts to deliver the patient to a hospital with dialysis capability.

You are the Provider: SUMMARY

1. What is the definition of an acute abdomen?

Acute abdomen is a term used to describe the sudden (acute) onset of abdominal pain that is not caused by a traumatic injury. It is generally associated with severe, progressive symptoms that require medical attention. Acute abdominal pain can be caused by dysfunction of one or more of the abdominal organs, such as the liver, spleen, gallbladder, stomach, pancreas, kidneys, large or small intestines, or appendix. Peritonitis, irritation and inflammation of the abdominal lining, can occur if the acute abdomen is not treated, and can be fatal.

2. What is the EMT's role in treating a patient with abdominal pain?

The underlying cause of a patient's abdominal pain, acute or chronic, is often difficult to identify—even for a physician. As an EMT, it is far more important for you to recognize life-threatening conditions and provide prompt emergency care than it is to attempt to identify the underlying cause of the patient's pain. Patients who are in pain, especially when the pain occurs suddenly, are often very anxious and scared; emotional support on your part is important.

3. What is the proper technique of assessing a patient's abdomen? What should you assess for?

Although assessment of the abdomen may help localize the source of a patient's pain, it should not be prolonged. Place the patient in a supine position with the legs drawn up and flexed at the knees; this position will relax the abdominal muscles and may alleviate some of his or her pain. Look at the abdomen first; does it appear distended (enlarged)? Do you see any pulsating masses (indicates an aortic aneurysm)? Is there bruising to the abdominal wall? Ask the patient where the pain is most intense and assess that area last. If the most painful area is palpated first, the patient may guard against further examination, making your assessment more difficult and less reliable. Gently palpate the four abdominal quadrants to determine whether each quadrant is rigid or soft, or tender, and note the presence of any masses. Pay particular attention to the patient's facial expressions when palpating each abdominal quadrant; they

may yield valuable information. Note whether the pain is localized to a particular quadrant or diffuse (widespread). Determine whether the patient can relax the abdominal wall on command; if he or she cannot, the abdomen is said to be guarded. Avoid vigorous palpation of the abdomen; doing so will only cause the patient more pain and can worsen his or her condition, especially if one of the abdominal organs is enlarged and fragile.

4. What is the difference between radiating pain and referred pain?

Radiating pain "moves" from its point of origin to other parts of the body, such as the pain from acute myocardial infarction, which often radiates to the neck, jaw, or down one of the arms. With radiating pain, there is pain at point A and point B, with a "trail" of pain in between the two points.

Referred pain originates in a particular organ but is described or perceived by the patient as pain in a different location. For example, the origin of pain associated with cholecystitis—inflammation of the gallbladder—is usually the right upper quadrant of the abdomen. However, the patient commonly reports pain in the right shoulder. In some cases, the patient reports pain in both the right upper quadrant and the right shoulder. With referred pain, the origin of pain is point A, but the patient feels or perceives the pain at point B. In other cases, there is pain at point A and point B; however, unlike radiating pain, there is no pain in between the two points.

It is important to distinguish between radiating and referred pain; some conditions are associated with radiating pain, whereas others are associated with referred pain. When determining if the patient's pain radiates, ask him or her "Does the pain stay in one place or does it move anywhere else?" When asking a patient about referred pain, ask him or her "Do you have pain anywhere else?"

5. What do the patient's vital signs indicate?

The patient's vital signs indicate shock. His respirations are rapid (tachypnea); he has a rapid heart rate (tachycardia);

You are the Provider: SUMMARY, continued

his skin is cool, pale, and diaphoretic (clammy); and his blood pressure—considering the fact the he has a history of hypertension—is low. Whether the patient has an intra-abdominal infection (peritonitis) or intra-abdominal bleeding, the end result, if untreated, will be the same—death!

As previously discussed, it is more important to recognize life-threatening conditions (ie, shock) than it is to try to determine the exact cause of a patient's abdominal pain, or any pain for that matter. You must begin immediate treatment aimed at maintaining adequate perfusion, such as applying high-flow oxygen (your partner has already done this), keeping him warm, and preparing for rapid transport. Depending on your local protocols, you may also consider elevating his legs 6" to 12".

6. What do you suspect is the cause of the patient's abdominal pain?

Although the exact cause of the patient's pain can only be determined by a physician, the pain that the patient is describing, a "tearing" sensation that radiates to the lower back, is characteristic of acute aortic dissection. There are other clues that reinforce this field impression. He has a history of hypertension, which is a major risk factor for aortic dissection, and the pain is of maximum intensity from the onset—also characteristic of aortic dissection.

The aorta, like all arteries, has three layers. Aortic dissection, or separation of the aortic wall, occurs when the inner layer of the aorta tears. With each pulsatile wave caused by left ventricular contraction, blood is forced into the torn aortic wall, putting pressure on the middle and outer layers of the aorta. Eventually, a weakening in the aortic wall, called an aneurysm, will develop.

The typical patient with an abdominal aortic aneurysm (AAA) is a man in his late 60s or older. As long as the aneurysm is not expanding, the patient usually will be asymptomatic. When the aneurysm starts to expand, however, the patient presents with a sudden onset of abdominal pain, which is classically described as a ripping, tearing, or searing sensation in the abdomen that radiates to the back. When an aortic aneurysm starts expanding and

producing symptoms, rupture may be imminent. If aortic rupture occurs, the patient often bleeds to death (exsanguination) very quickly. If the aneurysm is leaking, however, blood will accumulate in the abdominal cavity and cause signs of shock; this is what may be happening to your patient.

7. Are there any special considerations for this patient? If so, what are they?

As with any patient exhibiting signs of shock, your priority is to provide transport to an appropriate medical facility without delay. In addition, patients with a suspected aortic aneurysm must be handled carefully; avoid rough driving and unnecessary bumps in the road. Because this is a fragile situation, with a large, leaking artery, avoid further palpation of the abdomen. Some patients with an abdominal aortic aneurysm have a pulsating mass that can be palpated (and sometimes seen) near the umbilicus. If you see a pulsating mass, do *not* touch it; doing so could cause the aorta to rupture.

Avoid anything that will make the patient more anxious; anxiety causes increases in heart rate and blood pressure. An acute increase in blood pressure, even a slight one, may be all that is needed to cause an aortic rupture. In this particular patient, you should avoid elevating the patient's legs, unless otherwise instructed to do so by medical control. Elevating the lower extremities may cause a surge of blood back to the heart, resulting in an increase in blood pressure.

8. Could you have done anything definitively for this patient in the field?

No. Definitive care (eg, surgically repairing the aneurysm) can only be provided at the hospital. Although paramedics can start intravenous lines and give pain medications, these interventions are simply aimed at controlling pain and partially treating shock, not fixing the aneurysm. As previously discussed, your role as an EMT is to recognize that the patient's condition is serious, provide emergency medical treatment, and transport without delay.

EMS Patient Care Report (PCR)

Date: 3-20-09	Incident No.: 150109	Nature of Call: Abdominal pain		Location: 1500 E. River Rd., Apt. 5
Dispatched: 0320	En Route: 0321	At Scene: 0333	Transport: 0345	At Hospital: 0410	In Service: 0420

Patient Information

Age: 79 Sex: M Weight (in kg [lb]): 86 kg (190 lb)	Allergies: None Medications: Toprol, Paxil Past Medical History: Hypertension, depression, appendectomy Chief Complaint: Abdominal pain

Vital Signs

Time: 0335	BP: 98/60	Pulse: 124	Respirations: 28	Sao$_2$: 96%
Time: 0345	BP: 100/62	Pulse: 130	Respirations: 28	Sao$_2$: 98%
Time: 0350	BP: 96/58	Pulse: 128	Respirations: 28	Sao$_2$: 97%

You are the Provider: SUMMARY, continued

EMS Treatment (circle all that apply)				
Oxygen @ <u>15</u> L/min via (circle one): NC (NRM) Bag-Mask Device	**Assisted Ventilation**	**Airway Adjunct**		**CPR**
Defibrillation	**Bleeding Control**	**Bandaging**	**Splinting**	**Other:** (Thermal management)

Narrative
Dispatched for 59-year-old male with abdominal pain. On arrival at the scene, found the patient lying on his side on a couch in his living room; his knees were drawn up to his abdomen. He was conscious and alert, but very restless and in severe pain. His skin was cool, pale, and diaphoretic. The patient states that the pain, which began suddenly, is a "tearing" sensation in his abdomen. He further states that the pain radiates to his lower back. Pain severity, per patient, is a 10 on a scale of 0 to 10. Patient denies nausea, vomiting, or any other symptoms. History is significant for hypertension, depression, and an appendectomy 30 years ago. Applied high-flow oxygen via nonrebreathing mask, obtained vital signs, and performed further assessment. Vital signs revealed tachycardia, tachypnea, a relatively low blood pressure, and an oxygen saturation level of 96%. Assessment of abdomen revealed that it was diffusely tender to palpation and guarded. No abdominal distention was noted. Covered patient with blanket to keep him warm, placed him onto the stretcher, loaded him into the ambulance, and began transport to the hospital. Continued oxygen therapy en route and continuously monitored his condition. He remained conscious and alert, but restless, and stated that his pain was still a 10/10. Called radio report to the hospital; no further orders were given by the attending physician. Delivered patient to the emergency department without incident. Verbal report was given to charge nurse. **End of report**

Assessment and Emergency Care of Gastrointestinal and Urologic Emergencies

Scene Size-up

Scene Safety	Ensure scene safety and safe access to the patient. Standard precautions should include a minimum of gloves and eye protection if there is vomiting. Consider a gown and shoe covers if other bodily fluids are involved. Determine the number of patients. Assess the need for additional or specialized medical resources.
Mechanism of Injury (MOI)/ Nature of Illness (NOI)	Determine the NOI. Interview the patient, family, and/or bystanders to ensure there is not a traumatic cause (mechanism of injury) for the abdominal pain.

Primary Assessment

Form a General Impression	Inquire about the chief complaint and observe the patient's overall body position (are the patient's knees drawn up?). Observe the work of breathing and circulation. Determine the level of consciousness using the AVPU scale. Identify immediate threats to life. Determine priority of care based on the NOI. If the patient has a poor general impression, call for ALS assistance. A rapid SCAN will help you identify and manage life threats.
Airway and Breathing	Ensure the airway is open, clear, and self-maintained. Evaluate the patient's ventilatory status for rate and depth of breathing, respiratory effort, and tidal volume. Administer high-flow oxygen at 15 L/min, providing ventilatory support as needed. Hypoxia may cause changes in the patient's mental state. If vomiting is a possibility, place the patient in the recovery position if no spinal injury is suspected.
Circulation	Observe skin color, temperature, and condition; look for life-threatening bleeding and treat accordingly. Evaluate the distal pulse rate, quality (strength), and rhythm. Inquire about bloody vomitus or stool.
Transport Decision	If the patient has an airway or breathing problem, signs and symptoms of bleeding, or other life threats, treat the patient immediately and transport, performing the secondary assessment en route to the hospital.

NOTE: The order of the steps in this section differs depending on whether the patient is conscious or unconscious. The following order is for a conscious patient. For an unconscious patient, perform a primary assessment, perform a full-body scan, obtain vital signs, and obtain the past medical history from a family member, bystander, or emergency medical identification device.

History Taking

Investigate Chief Complaint	Investigate the chief complaint. Monitor the patient for changes in mental status. Ask OPQRST and SAMPLE questions. SAMPLE can also be obtained from family, bystanders, and medical alert tags. Inquire about associated symptoms of an acute abdomen or urologic emergency. Remember that a cardiac event can present as perceived abdominal pain by the patient. With female patients, inquire about the last menstrual period and the possibility of pregnancy.

Assessment and Emergency Care of Gastrointestinal and Urologic Emergencies, continued

Secondary Assessment

Physical Examinations	Perform a systematic physical examination or a focused examination. Advise the patient of your assessment actions prior to performing any examination. When assessing the abdomen, remember to perform the examination in this specific order: look (for abnormalities), listen (to bowel sounds), and feel (for pain with light palpation or rebound tenderness). Abdominal pain that is referred to the shoulder could be a sign of internal bleeding. Do not delay transport to perform the physical examination at the scene.
Vital Signs	Obtain baseline vital signs as soon as practical. Vital signs should include blood pressure by auscultation, pulse rate and quality, respiration rate and quality, pupils, and skin assessment for perfusion. Note the patient's level of consciousness. Use pulse oximetry, if available, to assess the patient's perfusion status.

Reassessment

Interventions	Repeat the primary assessment, vital signs, and confirm the chief complaint. Treat for shock and provide emotional support. Assist breathing as required, administering high-flow oxygen. Place the patient in a position of comfort.
Communication and Documentation	Contact medical control/receiving hospital with a radio report; many hospitals require additional personnel and a separate treatment area. Include a thorough description of the NOI and the position the patient was found in. Include treatments performed and patient response. Be sure to document the patient's distress, answers to your questions, and any changes in patient status and the time. Follow local protocols. Document the reasoning for your treatment and the patient's response.

NOTE: Although the following steps are widely accepted, be sure to consult and follow your local protocols. Take appropriate standard precautions when treating all patients.

Gastrointestinal and Urologic Emergencies

General Management of Gastrointestinal and Urologic Emergencies

1. Explain to the patient what you are going to do in terms of assessing the abdomen.
2. Establish and maintain a patent airway. Provide oxygen (low-flow reduces nausea). Monitor for vomiting and protect the airway against aspiration.
3. Allow the patient to assume a position of comfort. You will find that most patients want to be supine with their knees drawn up to relax the abdominal muscles, unless there is any trauma, in which case the patient will remain supine and stabilized.
4. Obtain SAMPLE history and vital signs.
5. Palpate the four quadrants of the abdomen gently to determine whether each quadrant is tense (guarded) or soft when palpated.
6. Determine whether the patient can relax the abdominal wall on command.
7. Request ALS support when intravenous fluids or pain management is necessary.

Prep Kit

Ready for Review

- The acute abdomen is a medical emergency, requiring prompt but gentle transport.

- The pain, tenderness, and abdominal distention associated with an acute abdomen may be signs of peritonitis, which may be caused by any condition that allows pus, blood, feces, urine, gastric juice, intestinal contents, bile, pancreatic juice, amniotic fluid, or other foreign material to lie within or adjacent to the peritoneum.

- In addition to abdominal disease or injury, problems in the gastrointestinal, genital, and urinary systems may also cause peritonitis.

- Signs and symptoms of acute abdomen include pain, nausea, vomiting, and a tense, distended abdomen.

- Pain is common directly over the inflamed area of the peritoneum, or it may be referred to another part of the body. Referred pain occurs because of the connections between the two different nervous systems supplying the parietal peritoneum and the visceral peritoneum.

- Do not give the patient with an acute abdomen anything by mouth.

- A patient in shock or with any life-threatening condition should be transported without delay. Call for advanced life support assistance if your patient's condition deteriorates during transport.

Vital Vocabulary

acute abdomen A condition of sudden onset of pain within the abdomen, usually indicating peritonitis; immediate medical or surgical treatment is necessary.

appendicitis Inflammation of the appendix.

cholecystitis Inflammation of the gallbladder.

cystitis Inflammation of the bladder.

diverticulitis Inflammation in small pockets at weak areas in the muscle walls.

emesis Vomiting.

guarding Involuntary muscle contractions (spasm) of the abdominal wall; an effort to protect the inflamed abdomen.

hernia The protrusion of a loop of an organ or tissue through an abnormal body opening.

ileus Paralysis of the bowel, arising from any one of several causes; stops contractions that move material through the intestine.

kidney stones Solid crystalline masses formed in the kidney, resulting from an excess of insoluble salts or uric acid crystallizing in the urine; may become trapped anywhere along the urinary tract.

pancreatitis Inflammation of the pancreas.

peritoneum The membrane lining the abdominal cavity (parietal peritoneum) and covering the abdominal organs (visceral peritoneum).

peritonitis Inflammation of the peritoneum.

referred pain Pain felt in an area of the body other than the area where the cause of pain is located.

strangulation Complete obstruction of blood circulation in a given organ as a result of compression or entrapment; an emergency situation causing death of tissue.

uremia Severe kidney failure resulting in the buildup of waste products within the blood. Eventually brain functions will be impaired.

urinary tract infection (UTI) An infection, usually of the lower urinary tract (urethra and bladder) that occurs when normal flora bacteria enter the urethra and grow.

Assessment in Action

You and your partner are called to a daycare center for a 6-year-old boy complaining of abdominal pain. On arrival, you find the boy lying on his left side, curled up in a fetal position, and crying.

1. Which of the following is an example of a solid organ found within the abdominal cavity?
   A. Gallbladder
   B. Spleen
   C. Lungs
   D. Small intestine

2. The abdominal cavity is lined with a membrane called the:
   A. pleura.
   B. meninges.
   C. peritoneum.
   D. pericardium.

3. Which of the following is a common cause of abdominal pain in children?
   A. Cholecystitis
   B. Kidney stones
   C. Appendicitis
   D. Gastrointestinal bleeding

4. The appendix is located in what quadrant of the abdomen?
   A. Upper right
   B. Lower right
   C. Upper left
   D. Lower left

5. As you palpate the child's right lower quadrant of the abdomen, he screams in pain when you bring your hand up away from his abdomen. This pain is known as:
   A. guarding.
   B. referred pain.
   C. phantom pain.
   D. rebound tenderness.

6. Peritonitis can be caused by the presence of what substance in the abdomen?
   A. Pus
   B. Blood
   C. Bile
   D. All of the above

7. Signs and symptoms of peritonitis include:
   A. swelling of the abdomen.
   B. increased appetite.
   C. bradycardia.
   D. high blood pressure.

8. What is the last vital sign to change in the setting of shock?
   A. Heart rate
   B. Blood pressure
   C. Respiratory rate
   D. Pulse oximetry reading

9. What information should you ask the patient about when assessing the nature of illness?

10. Discuss the appropriate management for this patient.

National EMS Education Standard Competencies

Medicine

Applies fundamental knowledge to provide basic emergency care and transportation based on assessment findings for an acutely ill patient.

Endocrine Disorders

Awareness that

- Diabetic emergencies cause altered mental status (p 632)

Anatomy, physiology, pathophysiology, assessment, and management of

- Acute diabetic emergencies (pp 622-632)

Hematology

Anatomy, physiology, pathophysiology, assessment, and management of

- Sickle cell crisis (pp 633-634)
- Clotting disorders (pp 634-635)

Knowledge Objectives

1. Describe the anatomy and physiology of the endocrine system and its main function in the body. (p 621)
2. Define and explain the terms diabetes, low blood glucose, and high blood glucose and distinguish between the two types of diabetes and how their onset patterns differ. (pp 621-622)
3. Discuss the role of glucose as a major source of energy for the body and its relationship to insulin. (pp 622-623)
4. Describe the differences and similarities between hyperglycemic and hypoglycemic diabetic emergencies, including their onset, signs and symptoms, and management considerations. (pp 624-626)
5. Explain some age-related considerations when managing a pediatric patient who is experiencing a hypoglycemic crisis. (p 626)
6. Discuss the steps the EMT should follow when conducting a primary and secondary assessment of a patient with an altered mental status who is a suspected diabetic patient. (pp 626-629)

7. Explain the process for assessing and managing the airway of a patient with an altered mental status, including ways to differentiate a hyperglycemic patient from a hypoglycemic patient. (pp 627, 629, 632-633)
8. Describe the interventions for providing emergency medical care to both a conscious and unconscious patient with an altered mental status and a history of diabetes who is having a hypoglycemic crisis. (p 629)
9. Describe the interventions for providing emergency medical care to both a conscious and unconscious patient with an altered mental status and a history of diabetes who is having a hyperglycemic crisis. (p 629)
10. Explain when it is appropriate to obtain medical direction when providing emergency medical care to a diabetic patient. (p 629)
11. Provide the generic and trade names, form, dose, administration, indications and contraindications for giving oral glucose to a patient with a decreased level of consciousness who has a history of diabetes. (pp 630-631)
12. Explain some age-related considerations when managing a geriatric patient who has undiagnosed diabetes. (p 630)
13. Discuss the composition and functions of blood. (p 633)
14. Describe the pathophysiology of sickle cell disease and the four main types of sickle cell crises. (pp 633-634)
15. Describe the assessment and management of a patient with suspected sickle cell disease. (pp 635-636)
16. Describe two types of blood clotting disorders, and the risk factors, characteristics, and management of each. (pp 634-636)

Skills Objectives

1. Demonstrate the assessment and care of a patient with hypoglycemia and a decreased level of consciousness. (pp 624-626, 630-632)
2. Demonstrate how to administer glucose to a patient with an altered mental status. (pp 631-632, Skill Drill 17-1)

Introduction

The endocrine system directly or indirectly influences almost every cell, organ, and function of the body. Consequently, patients with an endocrine disorder often are seen with a multitude of signs and symptoms that require a thorough assessment and immediate treatment. This chapter also discusses hematologic emergencies, which rarely occur in most EMS systems. Although hematologic disorders can be difficult to assess and treat in a prehospital setting, your actions may not only offer support, but you may save the patient's life.

Anatomy and Physiology

The **endocrine system** is a complex message and control system that includes a network of glands that produce and secrete messengers called hormones. **Endocrine glands** (endo means *inside*) secrete or release chemicals that are used inside the body. A **hormone** is a chemical substance produced by a gland that has special regulatory effects on other organs and tissues. The main function of the endocrine system and its hormone messengers is to maintain homeostasis, which is stability in the body's internal environment. For example, maintaining homeostasis requires a response to any change in the body, such as low glucose or calcium levels in the blood. Endocrine disorders can be caused by either hypersecretion (overproduction) or hyposecretion (underproduction) of a gland.

Pathophysiology

Diabetes is a common disease, affecting about 7% of the population. Diabetes is a metabolic disorder that involves abnormalities in the body's ability to use glucose (sugar). Consequently, the body is unable to control the level of glucose in the blood. Without treatment, blood glucose levels become too high, which can cause coma and death. If properly treated, most people with diabetes can live a relatively normal life. However, diabetes can have severe complications, including blindness, cardiovascular disease, and kidney failure, that affect the length and quality of life. Also, the treatment to lower high blood glucose levels can exceed the patient's need and cause a life-threatening state of hypoglycemia (low blood glucose). Therefore, as an EMT, you need to know the signs and symptoms of a blood glucose level that is either too high or too low, so you can administer the proper lifesaving treatment.

The two types of diabetes are discussed as well as how they are controlled, including the role of glucose and insulin. You will learn how to distinguish between hyperglycemia and hypoglycemia, conditions that have signs and symptoms that often resemble each other. Complications related to diabetes, such as seizures and altered mental status, are also briefly discussed.

■ Diabetes

The word "diabetes" literally means "a passer through; a siphon." Medically, the term refers to a metabolic disorder in which the body's ability to metabolize simple carbohydrates (glucose) is impaired. **Glucose**, or dextrose, is one of the basic sugars used in the body and, in conjunction with oxygen, is the primary fuel for cellular metabolism.

You are the Provider: PART 1

At 7:20 AM, you and your partner are dispatched to 445 Landon Way for a patient with diabetic complications. You proceed to the scene, with a response time of approximately 6 minutes. When you arrive at the residence, you find the patient, a 56-year-old man, lying on the couch. He is conscious, but appears confused, and his breathing appears deep and rapid. His wife tells you that he has type 2 diabetes and has been sick for the past few days.

1. What is diabetes mellitus (DM)? What is the central problem in patients with DM?

2. What is the difference between type 1 and type 2 diabetes?

In a person with diabetes, signs and symptoms include significant thirst, passage of large quantities of urine containing glucose, and deterioration of body functions. The central problem in diabetes is the lack or ineffective action of **insulin**, a hormone that is normally produced by the endocrine glands on the pancreas that enables glucose to enter the cells. Without insulin, cells begin to "starve" because insulin is needed, like a key, to allow glucose into the cells.

The full name of diabetes is **diabetes mellitus**, which means "sweet diabetes." This refers to the presence of glucose (sugar) in the urine. Diabetes mellitus is a metabolic disorder in which the body cannot metabolize glucose, usually because of the lack of insulin; the result is a wasting of glucose in the urine. *Diabetes insipidus*, a rare condition, also involves excessive urination, but here the missing hormone is one that regulates urinary fluid reabsorption. In this text, the term "diabetes" always refers to diabetes mellitus.

Left untreated, diabetes leads to a wasting of body tissues and death. Even with medical care, some patients with particularly aggressive forms of diabetes will die relatively young of one or more complications of the disease. Most patients with diabetes, however, live a normal life span, but they must be willing to adjust their lives, especially their eating habits and activities, to the demands of the disease.

Types of Diabetes

Diabetes is a disease with two distinct onset patterns. It may become evident when the patient is a child, or it may develop in later life, usually when the patient is middle-aged.

In **type 1 diabetes**, patients do not produce insulin. They need daily injections of supplemental, synthetic insulin to control their levels of blood glucose. This type of diabetes typically develops during childhood; because of this, in the past it was called "juvenile-onset diabetes" or insulin-dependent diabetes mellitus (IDDM). However, type 1 diabetes can, in many cases, develop in later life as well. Patients with type 1 diabetes are more likely to have metabolic problems and organ damage, such as blindness, heart disease, kidney failure, and nerve disorders.

In **type 2 diabetes**, which usually appears later in life, patients produce inadequate amounts of insulin or they may produce a normal amount, but the insulin does not function effectively. Although some patients with type 2 diabetes may require supplemental insulin, many patients can be treated with diet, exercise, and non–insulin-type oral medications (hypoglycemic agents), such as chlorpropamide (Diabinese), tolbutamide (Orinase), glyburide (Micronase), glipizide (Glucotrol),

metformin (Glucophage), and rosiglitazone (Avandia). These medications stimulate the pancreas to produce more insulin and thus lower blood glucose levels. In some patients, these medications can lead to hypoglycemia (an abnormally low level of blood glucose), particularly when patient activity and exercise levels are too vigorous or excessive. In the past, type 2 diabetes was called non–insulin-dependent diabetes mellitus (NIDDM), or adult (maturity)-onset diabetes. Again, some patients with type 2 diabetes may, in fact, require insulin.

The two types of diabetes are equally serious, although type 2 diabetes is often easier to regulate. Both can affect many tissues and functions other than those of the glucose-regulating mechanism. Both require life-long medical management. Type 1 diabetes is considered an autoimmune problem, in which the body becomes allergic to and, therefore, destroys the insulin-producing cells of the endocrine glands in the pancreas. The severity of diabetic complications is related to how high the average blood glucose level is and how early in life the diabetes begins.

Type 2 diabetes is much more common than type 1 diabetes and is becoming more prevalent in today's society. Obesity is one risk factor that has become more frequent in the population.

The Role of Glucose and Insulin

Glucose is the major source of energy for the body. All cells need glucose to function properly, and some cells will even cease to function without it. Having a constant supply of glucose is as important as having a constant supply of oxygen to the brain. Without glucose, or with very low levels, brain cells rapidly suffer permanent damage. As stated earlier, insulin is needed to allow glucose to enter individual cells, with the exception of brain cells, to fuel their functions. For this reason, insulin is said to be a "cellular key" **Figure 17-1** .

Without insulin, glucose from food remains in the blood and the levels gradually become extremely high. This condition is called hyperglycemia. Once the blood glucose level reaches 200 mg/dL or more, or twice the usual amount (normal is 80 to 120 mg/dL), excess glucose is excreted by the kidney. This process requires a large amount of water. The loss of water in such large amounts causes the classic symptoms of uncontrolled diabetes, the "3 Ps":

- **Polyuria**: frequent and plentiful urination
- **Polydipsia**: frequent drinking of liquid to satisfy continuous thirst (following the loss of excessive amounts of body water)
- **Polyphagia**: excessive eating as a result of cellular "hunger"; seen only occasionally

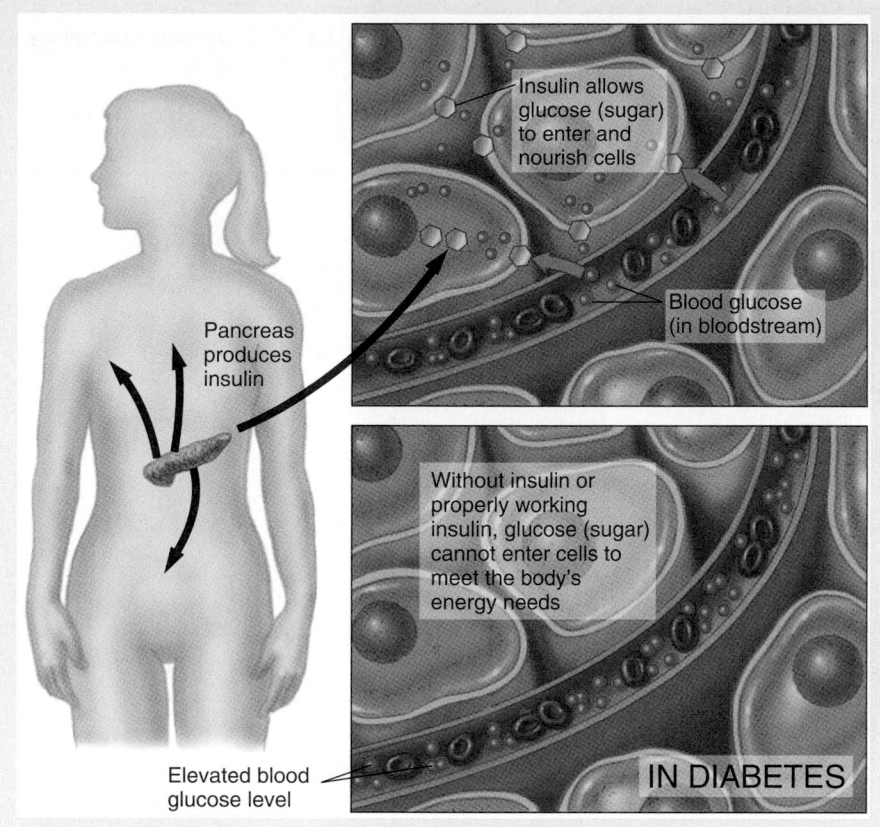

Figure 17-1 Diabetes is defined as the lack of or the ineffective action of insulin. Without insulin, cells begin to "starve" because insulin is needed to allow glucose to enter and nourish the cells.

Within the figure:
- Pancreas produces insulin
- Insulin allows glucose (sugar) to enter and nourish cells
- Blood glucose (in bloodstream)
- Without insulin or properly working insulin, glucose (sugar) cannot enter cells to meet the body's energy needs
- Elevated blood glucose level
- IN DIABETES

Without glucose to supply energy for cells, the body must turn to other fuel sources. The most abundant source is fat. Unfortunately, when fat is used as an immediate energy source, chemicals called *ketones* and *fatty acids* are formed as waste products and are hard for the body to excrete. As they accumulate in blood and tissue, certain ketones can produce a dangerous condition called **acidosis**. The form of acidosis seen in uncontrolled diabetes is called **diabetic ketoacidosis (DKA)**, in which an accumulation of certain acids occurs when insulin is not available in the body. DKA is considered to be a type of hyperglycemic crisis. DKA is more commonly found in type 1 diabetes because the body does not produce any insulin. Signs and symptoms of DKA include weakness, nausea, vomiting, abdominal pain, a weak and rapid pulse, and a type of deep, rapid breathing called **Kussmaul respirations**. These respirations help the body "blow off" excess acids. Diabetics suffering from DKA may have an altered mental status and may have a sweet smell to their breath caused by the ketones. When the acid levels in the body become too high, individual cells will cease to function. If the patient is not given proper fluid and insulin to reverse fat metabolism and restore the use of glucose as a source of energy, ketoacidosis will progress to unconsciousness, hyperglycemic crisis, and eventually, death.

Type 2 diabetes more often results in HHNC, hyperosmolar hyperglycemic nonketotic coma. The onset of HHNC is typically slower and occurs over a longer period than DKA. The sweet smell is not present on the breath. Because the body is producing some insulin, the body does not burn fat for energy; therefore, ketones are not produced. The body tries to get rid of the excess sugar in the urine. The fluid follows the sugar, causing dehydration.

As we have seen, diabetes mellitus is treatable; however, treatment must be tailored for the individual patient. The patient's need for glucose must be balanced with the available supply of insulin by testing either the blood or urine. Most type 1 diabetic patients monitor their blood glucose levels several times a day with a glucometer, a credit-card-sized device. A drop of blood, usually from the fingertip, is touched to a disposable sensor and read by the device. The readings are in milligrams per deciliter of blood; remember that the normal blood glucose level is between 80 and 120 mg/dL. Among the new measuring devices under development is one that is worn like a wristwatch or used like a pulse oximeter. Currently, EMTs are allowed to use glucometers in some systems across North America **Figure 17-2**; however, glucose test strips, in which a drop of blood is placed on a paper strip that changes color, may still be used in

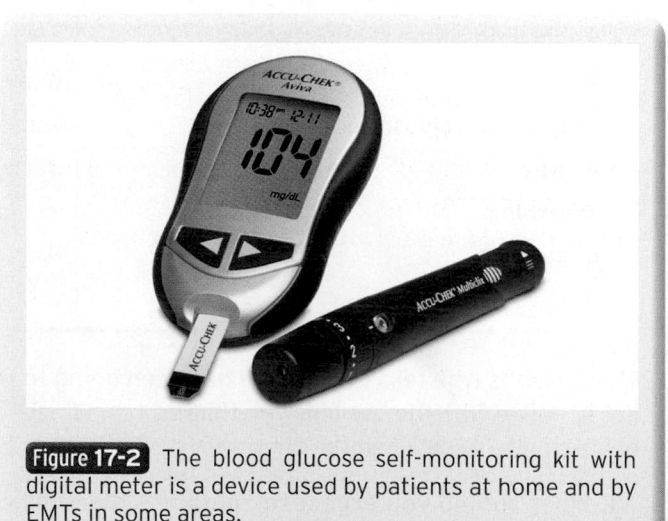

Figure 17-2 The blood glucose self-monitoring kit with digital meter is a device used by patients at home and by EMTs in some areas.

some systems. Test strips do not provide the accuracy of glucometers, and their readings should be used with caution.

■ Hyperglycemia and Hypoglycemia

Two conditions can lead to a diabetic emergency: hyperglycemia and hypoglycemia. **Hyperglycemia** is a state in which the blood glucose level is above normal. **Hypoglycemia** is a state in which the blood glucose level is below normal. Extremes of hyperglycemia and hypoglycemia can lead to diabetic emergencies Figure 17-3 . Prolonged hyperglycemia with exceptionally high glucose levels results in diabetic ketoacidosis, a type of hyperglycemic crisis. Hypoglycemia, on the other hand, will progress to unresponsiveness and eventually hypoglycemic crisis.

The signs and symptoms of hyperglycemia and hypoglycemia can be quite similar Table 17-1 . For example, staggering and an intoxicated appearance or complete unresponsiveness are signs and symptoms of both. Note that your assessment of these potential emergencies should not prevent you from providing prompt care and transport as detailed in this chapter. However, in such urgent emergencies, the earlier clues are gathered, the better for the patient. With specific information about the type of emergency that is presenting, you can help the hospital to prepare prompt, definitive care for the patient.

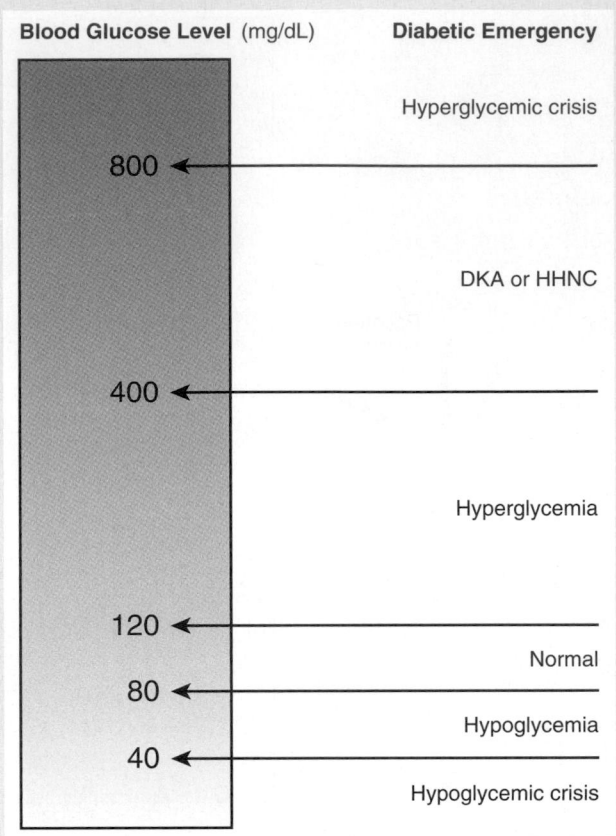

Figure 17-3 The two most common diabetic emergencies, hyperglycemic crisis and hypoglycemic crisis, develop when the patient has either too much or too little glucose in the blood, respectively.

You are the Provider: PART 2

You perform a primary assessment of the patient as your partner applies high-flow oxygen via a nonrebreathing mask. You attempt to obtain more information from the patient, but he is slow to answer your questions. His wife tells you that he takes rosiglitazone (Avandia) to control his blood glucose and lisinopril for high blood pressure.

Recording Time: 0 Minutes	
Appearance	Weak and confused
Level of consciousness	Conscious but confused
Airway	Open; clear of secretions or foreign bodies
Breathing	Increased rate and depth
Circulation	Radial pulses, rapid and weak; skin, warm and dry with poor turgor

The patient's wife tells you that he has been going to the bathroom frequently and has been drinking large quantities of water. She further tells you that his rosiglitazone (Avandia) refill is ready to be picked up at the pharmacy; however, he has not had the chance to go get it. Your partner assesses his blood glucose level with the glucometer.

3. What should you expect the patient's blood glucose level to read? Why?

4. What is causing the patient's frequent urination and deep, rapid breathing?

Table 17-1 Characteristics of Diabetic Emergencies

	Hyperglycemia	Hypoglycemia
History		
Food intake	Excessive	Insufficient
Insulin dosage	Insufficient	Excessive
Onset	Gradual (hours to days)	Rapid, within minutes
Skin	Warm and dry	Pale, cool, and moist
Infection	Common	Uncommon
Gastrointestinal tract		
Thirst	Intense	Absent
Hunger	Absent	Intense
Vomiting	Common	Uncommon
Respiratory system		
Breathing	Rapid, deep (Kussmaul respirations)	Normal or rapid
Odor of breath	Sweet, fruity	Normal
Cardiovascular system		
Blood pressure	Normal to low	Normal to low
Pulse	Rapid, weak, and thready	Rapid, weak
Nervous system		
Consciousness	Restlessness, possibly progressing to coma; abnormal or slurred speech; unsteady gait	Irritability, confusion, seizure, or coma; unsteady gait
Treatment		
Response	Gradual, within 6 to 12 hours following medical treatment	Immediately after administration of glucose

Hyperglycemic Crisis (Diabetic Coma)

<u>Hyperglycemic crisis</u> is a state of unconsciousness resulting from several problems, including ketoacidosis, hyperglycemia, and dehydration resulting from excessive urination. Too much blood glucose by itself does not always cause a hyperglycemic crisis, but on some occasions, excess blood glucose can lead to it.

A hyperglycemic crisis may occur in a patient who is not under medical treatment, who takes an insufficient amount of insulin, who markedly overeats, or who is undergoing some sort of stress that may involve an infection, illness, overexertion, fatigue, or drinking alcohol. Usually, ketoacidosis develops over a period of time, from hours to days. The patient may ultimately be found with the following physical signs:

- Kussmaul respirations (rapid, deep respirations)
- Dehydration, as indicated by dry, warm, "tenting" skin and sunken eyes
- A sweet or fruity (acetone) odor on the breath, caused by the unusual waste products in the blood (ketones)

- A rapid, weak ("thready") pulse
- A normal or slightly low blood pressure
- Varying degrees of unresponsiveness
- Weakness, nausea, and vomiting
- Polyuria (excessive urination), polydipsia (excessive thirst), polyphagia (excessive eating)

Hypoglycemic Crisis (Insulin Shock)

In a <u>hypoglycemic crisis</u>, also known as insulin shock when it occurs in patients taking supplemental insulin or medications that stimulate the pancreas to produce more insulin, the problem is hypoglycemia, or insufficient levels of glucose in the blood. When insulin levels remain high, glucose is rapidly taken out of the blood to fuel the cells. If glucose levels fall too low, there may be an insufficient amount to supply the brain. The mental status of the patient declines (hypoglycemic crisis) and he or she may become aggressive or display unusual behavior. If blood glucose remains low, unconsciousness and permanent brain damage can quickly follow.

A hypoglycemic crisis occurs when the patient has done one of the following:

- Taken too much insulin
- Taken a regular dose of insulin but has not eaten enough food
- Had an unusual amount of activity or vigorous exercise and used up all available glucose

A hypoglycemic crisis may also occur after the patient vomits a meal after he or she took a regular dose of insulin. At times, hypoglycemia may occur with no identifiable predisposing factor.

Children who have diabetes may pose a particular management problem. First, their high levels of activity mean that they can use up circulating glucose more quickly than adults do, even after a normal insulin injection. Second, they do not always eat correctly and on schedule. As a result, a hypoglycemic crisis can develop more often and more severely in children than in adults.

Words of Wisdom

> Hyperglycemia in pediatric patients can cause cerebral edema, which is the number one cause of diabetic-related death in pediatrics.

Hypoglycemia develops much more quickly than hyperglycemia. In some instances, it can occur in a matter of minutes. Hypoglycemia can be associated with the following signs and symptoms:

- Normal to shallow or rapid respirations
- Pale, moist (clammy) skin
- Diaphoresis (sweating)
- Dizziness, headache
- Rapid pulse
- Normal to low blood pressure
- Altered mental status (aggressive, confused, lethargic, or unusual behavior)
- Anxious or combative behavior
- Hunger
- Seizure, fainting, or coma
- Weakness on one side of the body (may mimic stroke)
- Rapid changes in mental status

Both extremes of hyperglycemia and hypoglycemia produce unconsciousness and, in some instances, death. But they call for very different treatment. Hyperglycemia is a complex metabolic condition that usually develops over time and involves all the tissues of the body. Correcting this condition may take many hours in a well-controlled hospital setting. Hypoglycemia, however, is an acute condition that can develop rapidly. A patient with diabetes who has taken his or her standard insulin

dose and missed lunch may have a hypoglycemic crisis before dinner. The condition is just as quickly reversed by giving the patient glucose. Without the glucose, however, the patient can suffer permanent brain damage. Minutes count.

Although most individuals with diabetes understand and manage their disease well, emergencies can occur. In addition to hyperglycemic and hypoglycemic crises, patients with diabetes may have "silent," or painless heart attacks, a possibility that you should always consider. Their only symptom may be "not feeling so well."

Patient Assessment of Diabetes

Scene Size-up

Scene Safety

Evaluate scene safety as you arrive on scene and as you approach the patient. Make sure that all hazards are addressed. Remember that diabetic patients often use syringes to administer insulin. It is possible you may be stuck by a used needle that was not disposed of properly. Insulin syringes on the bed stand, insulin bottles in the refrigerator, a plate of food, or glass of orange juice are important clues that may help you decide what is possibly wrong with your patient. Evaluate each situation quickly and make sure the necessary personal protective equipment is readily available. Standard precautions should consist of gloves and eye protection at a minimum. As you approach, question bystanders on events leading to your arrival. Determine whether this is your only patient and whether trauma was involved. Decide whether you will need any additional resources.

Mechanism of Injury/Nature of Illness

Although your report from dispatch may be for a patient with an altered mental status, keep open the possibility that trauma may have occurred because of a medical incident. Determine mechanism of injury and/or nature of illness. Do not let your guard down even on what appears to be a routine call.

Primary Assessment

Form a General Impression

Perform a rapid scan of the patient in order to form a general impression of the patient. How does the patient look? Does he or she appear anxious, restless, or listless? Is the patient apathetic or irritable? Is the patient interacting with his or her environment appropriately? These initial observations may lead you to suspect high or low blood glucose values. Identify life threats, and provide lifesaving interventions, particularly airway management. Determine the patient's level of consciousness using the AVPU (*Alert* to person, place, and day; responsive to *Verbal* stimuli; responsive to *Pain; Unresponsive*) scale. If a suspected diabetic patient is unresponsive, call for advanced life support (ALS) immediately. An unconscious patient may have undiagnosed diabetes. In patients with an altered mental status, you may be able to determine whether diabetes is present in the field by assessing the patient's blood glucose level if you have the proper equipment and training. Perform cervical spine stabilization, when necessary, and provide rapid transport. At the emergency department, diabetes and its complications can be quickly diagnosed.

Remember that even though a person has diabetes, the diabetes may not be causing the current problem; heart attack, stroke, or another medical emergency may be the cause. For this reason, you must always carry out a thorough, careful primary assessment, paying attention to the ABCs.

Airway and Breathing

As you are forming your general impression, assess the patient's airway and breathing. Patients showing signs of inadequate breathing or altered mental status should receive high-flow oxygen at 12 to 15 L/min via nonrebreathing mask. A patient who is hyperglycemic may have rapid, deep respirations (Kussmaul respirations) and sweet, fruity breath. A patient who is hypoglycemic will have normal or shallow to rapid respirations. If the patient is not breathing or is having difficulty breathing, open the airway and insert an airway adjunct, administer oxygen, and assist ventilations. Continue to monitor the airway as you provide care.

Circulation

Once you have assessed airway and breathing and have performed the necessary lifesaving interventions, check the patient's circulatory status. A patient with dry and warm skin indicates hyperglycemia, whereas a patient with moist and pale skin indicates hypoglycemia. The patient in hypoglycemic crisis will have a rapid, weak pulse.

Transport Decision

Whether you decide to transport at this stage of the assessment will depend on the patient's level of consciousness and the ability to swallow. Patients with an altered mental status and impaired ability to swallow should be transported promptly. Patients who have the ability to swallow and are conscious enough to maintain their own airway may be further evaluated on scene and interventions performed.

History Taking

Investigate Chief Complaint

Investigate the chief complaint or the history of the present illness. Responsive patients usually are able to provide their own medical history. If the patient has eaten but has not taken insulin, it is more likely that hyperglycemia is developing. If the patient has taken insulin but has not eaten, the problem is more likely to be hypoglycemia. A patient

You are the Provider: PART 3

Your partner reports that the patient's blood glucose level reads "high." You continue to assess the patient as your partner obtains his vital signs. The patient's wife calls his endocrinologist, who requests that you transport him to the closest hospital. A community hospital is located about 15 miles away.

Recording Time: 5 Minutes	
Respirations	30 breaths/min; deep
Pulse	120 beats/min; weak radial pulses
Skin	Pink, warm, and dry; poor turgor
Blood pressure	112/54 mm Hg
Oxygen saturation (Sao$_2$)	95% (on oxygen)

5. What other factors can cause hyperglycemia in patients with diabetes?

6. What is the difference between hyperglycemic crisis (DKA) and hypoglycemic crisis (insulin shock)? How can you distinguish one from the other?

with diabetes will often know what is wrong. If the patient is not thinking or speaking clearly (or is unconscious), ask a family member or bystander the same questions.

Physical signs such as tremors, abdominal cramps, vomiting, a fruity breath odor, or a dry mouth may guide you in determining whether the patient is hypoglycemic or hyperglycemic.

SAMPLE History

You will need to obtain a SAMPLE history from your patient. In addition, be sure to ask the following questions of a known diabetic patient:

- Do you take insulin or any pills that lower your blood sugar?
- Have you taken your usual dose of insulin (or pills) today?
- Have you eaten normally today?
- Have you had any illness, unusual amount of activity, or stress?

When you are assessing a patient who might have diabetes, check to see whether he or she has an emergency medical identification device—a wallet card, necklace, or bracelet—or ask the patient or a family member. Remember that the environment, bystanders, and medical identification devices may provide important clues about your patient's condition.

Secondary Assessment

In some instances where the patient is critically ill or injured or the transport time is short, you may not have time to conduct a secondary assessment. In other instances, the secondary assessment may occur on scene, or en route to the emergency department.

Physical Examinations

First, assess unresponsive patients from head to toe with a full-body scan, looking for clues to their condition. The patient may have experienced trauma resulting from dizziness or from changes in level of consciousness. Next, reassess the patient's vital signs.

As in every call, you should perform a secondary assessment when time permits. With unconscious patients or patients with an altered mental status, you must assume the role of detective and look for problems or injuries that are not obvious because the patient is unable to communicate these to you. Although an altered mental status may be caused by a blood glucose level that is too high or too low, the patient may have sustained trauma or have another metabolic problem. An altered mental status may also be caused by something else, such as intoxication, poisoning, or a head injury. A systematic examination of the patient may provide you with information essential to proper patient care.

When you suspect a diabetes-related problem, a secondary assessment should focus on the patient's mental status and ability to swallow and protect the airway. Obtain a Glasgow Coma Scale score to track the patient's neurologic status.

Vital Signs

Obtain a complete set of vital signs, including a measurement of the patient's blood glucose level using a glucometer, if available and local protocols allow. In hypoglycemia, respirations are normal to rapid, pulse is weak and rapid, and skin is typically pale and clammy with a low blood pressure. In hyperglycemia, respirations are deep and rapid; pulse is rapid, weak, and thready; and skin is warm and dry with a normal blood pressure. At times the blood pressure may be low. It should be easier for you to identify abnormal vital signs when you know the blood glucose level is too high or too low. Remember, the patient may have abnormal vital signs and a normal blood glucose value. When this is the case, something else may be causing the patient's altered mental status, vomiting, or other complaints.

Monitoring Devices

Because hyperglycemia or hypoglycemia may be the cause of your patient's decreased level of consciousness, it is important to obtain your patient's blood glucose level using a monitoring device, if protocols allow. Most commonly this is accomplished via a portable blood glucose monitor (glucometer), similar to the one your patient may use at home. The portable blood glucose monitor measures the glucose level in whole blood using either capillary or venous samples.

It is important to read and understand the operator's manual before use because the specifications of the device may vary depending on the manufacturer. Some glucometers indicate low (Lo) when they detect a glucose reading less than 20 mg/dL, whereas others display Lo when they detect a reading less than 30 mg/dL. Conversely, the same is true with a high (Hi) reading; some glucometers read Hi at 550 mg/dL and some at 600 mg/dL; therefore, it is important to know both the upper and lower ranges at which your glucometer functions.

The normal range for glucose levels in blood in nonfasting adults and children is 80 to 120 mg/dL; the blood glucose level in neonates should be above 70 mg/dL.

In a patient experiencing a diabetic emergency, a pulse oximeter is a useful device that will assist you in assessing the patient's perfusion status. By using pulse oximetry, you will be able to determine the percentage of oxygen saturation in the bloodstream, which will assist in identifying the patient's degree of respiratory distress. However, remember that pulse oximetry is just another tool in the EMT's toolbox. The decision to apply oxygen to a patient experiencing a diabetic emergency should be based on a careful assessment of the patient's airway and breathing, not solely on pulse oximetry readings.

Reassessment

It is important to reassess the diabetic patient frequently to assess changes. Is there an improvement in the patient's mental status? Are the ABCs still intact? How is the patient responding to the interventions performed? How must you adjust or change the interventions? In many patients with diabetes, you will note marked improvement with appropriate treatment. Document each assessment, your findings, the time of the interventions, and any changes in the patient's condition. Base your administration of glucose on serial readings if you have access to a glucometer. If a glucometer is unavailable, a deteriorating level of consciousness indicates that you need to provide more glucose. Again, the use of glucometers and the administration of glucose will be based on your service's protocols and standing orders.

Interventions

If your patient is hypoglycemic, conscious, and able to swallow without the risk of aspiration, you should encourage him or her to drink juice or other drinks that contain sugar. Do not be afraid to give too much sugar. Do not give sugar-free drinks that are sweetened with saccharin or other synthetic sweetening compounds, as they will have little or no effect. If you are permitted by local protocol, you may also administer a highly concentrated sugar gel, such as oral glucose or intramuscular (IM) glucagon, which is squirted between the patient's cheek and gums or placed between the cheek and gum on a tongue depressor. The patient will usually become more alert within minutes. Remember that even when the patient responds after receiving glucose, he or she may still need additional treatment. Therefore, you must transport the patient to the hospital as soon as possible.

If your hypoglycemic patient is unconscious, or if there is any risk of aspiration, the patient will need intravenous (IV) glucose or intramuscular (IM) glucagon, which you are not authorized to give. Your responsibility is to provide prompt transport to the hospital, where the proper care can be given. If you are working in a tiered system, AEMTs and paramedics are able to start an IV line and administer IV glucose.

If no one else is present and you know that the unconscious patient has diabetes, you must use your knowledge of the signs and symptoms to decide whether the problem is hypoglycemia or hyperglycemia. Remember, however, this assessment should not prevent you from providing prompt treatment and transport. The primary visible difference will be the patient's breathing—deep, sighing respirations in hyperglycemia and normal or rapid respirations in hypoglycemia. The patient with diabetes who is unconscious and having seizures is more likely to be in a hypoglycemic crisis.

Safety

Before you give a conscious patient anything to drink or administer instant glucose, you must ensure that there is no danger of aspiration. One rule of thumb: if patients can lift the cup or squirt the glucose into their own mouths, they are most likely not in danger of aspiration. Watch them carefully!

A patient in hypoglycemic crisis (rapid onset of altered mental status, hypoglycemia) needs sugar immediately. A patient in hyperglycemic crisis (acidosis, dehydration, hyperglycemia) needs insulin and IV fluid therapy. These patients need prompt transport to the hospital for appropriate medical care.

When there is any doubt about whether a conscious patient with diabetes is going into hypoglycemic or hyperglycemic crisis, most protocols will err on the side of giving glucose, even though the patient may have hyperglycemia or diabetic ketoacidosis. Untreated hypoglycemia will result in loss of consciousness and can quickly cause significant brain damage or death. The condition of a patient in hypoglycemic crisis is far more critical and far more likely to cause permanent problems compared with the condition of a patient with hyperglycemia or diabetic ketoacidosis. Furthermore, the amount of sugar that is typically given to a patient with hypoglycemic crisis is unlikely to make a patient in diabetic ketoacidosis significantly worse. When in doubt, consult medical control.

Determining whether the blood glucose level is too high or too low in a known diabetic patient can be difficult when signs and symptoms are confusing and you have no way to test for a blood glucose value. In these situations, perform a thorough assessment and contact the hospital to help sort out the signs and symptoms. The hospital should be a resource for you to help problem-solve situations and provide guidance on how to manage your patient.

Safety

Managing problems related to diabetes and altered mental status poses minimal risk to you because exposure to body fluids is generally very limited. However, some patients can become confused and even aggressive at times. Follow standard precautions, as you would with any other patient. Always use gloves and carefully wash your hands after obtaining and checking a blood sample or performing airway techniques.

Special Populations

You may encounter a geriatric patient who has undiagnosed diabetes. The patient is likely to report that he or she has not been feeling well for a while but has not seen a physician. A patient with undiagnosed diabetes or one who is in denial or ignores the advice of his or her physician may call 9-1-1 when the signs and symptoms become annoying. Nonhealing wounds (which can lead to infection), blindness, renal failure, atypical (silent) myocardial infarction presentation, and other complications are associated with poorly controlled or uncontrolled diabetes. As an EMT, you may be the first to recognize and suggest medical treatment to a geriatric patient who might otherwise ignore his or her condition. It is important that you recognize the signs and symptoms of diabetes.

Communication and Documentation

Communication with hospital staff is important for continuity of care. Hospital personnel need to be informed about the patient's history, the present situation, your assessment findings, and your interventions and their results.

Your run report is the only legal document you have to say that appropriate care was provided. Document clearly your assessment findings as the basis for your treatment. Patients who refuse transport because you "cured" them with oral glucose may require even more thorough documentation. Follow your local protocols for patients who refuse treatment or transport.

Emergency Medical Care for Diabetic Emergencies

Giving Oral Glucose

Oral glucose is a commercially available gel that dissolves when placed in the mouth **Figure 17-4**. One toothpaste-type tube of gel equals one 30-g dose. Trade names for the gel include Glutose and Insta-Glucose. Glucose gel acts to increase a patient's blood glucose levels. If authorized by your system, you should administer glucose gel to any patient with a decreased level of consciousness who has a history of diabetes. The only contraindications to oral glucose are an inability to swallow and unconsciousness, because aspiration (inhalation of the substance) can occur. Oral glucose itself has no side effects if it is

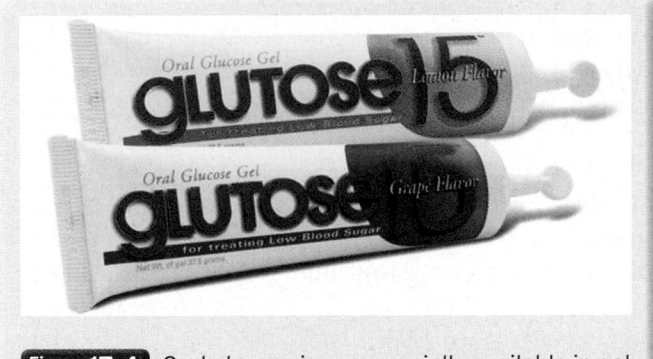

Figure 17-4   Oral glucose is commercially available in gel and tablet form. One tube of gel equals one 15-g dose.

You are the Provider: PART 4

Because of the patient's signs and symptoms, history, and a glucometer reading that indicates a level of "high," you determine that oral glucose is not indicated. The patient is placed onto the stretcher and loaded into the ambulance. Shortly after departing the scene, you reassess his mental status and vital signs.

Recording Time: 11 Minutes	
Level of consciousness	Conscious but confused
Respirations	30 breaths/min; deep
Pulse	124 beats/min; weak radial pulses
Skin	Pink, warm, and dry; poor turgor
Blood pressure	108/56 mm Hg
Sao$_2$	96% (on oxygen)

7. What additional treatment should you provide to this patient?

administered properly; however, the risk of aspiration in a patient who does not have a gag reflex is substantial. A conscious patient (even if confused) who does not really need glucose will not be harmed by it. Therefore, do not hesitate to give glucose under these circumstances.

As always, be sure to wear gloves before placing anything into a patient's mouth. After you have confirmed that the patient is conscious and able to swallow and have obtained an online or off-line order, follow these steps to administer oral glucose Skill Drill 17-1 :

1. Examine the tube to ensure that it is not open or broken. Check the expiration date Step 1.
2. Squeeze a generous amount onto the bottom third of a bite stick or tongue depressor Step 2.
3. Open the patient's mouth.
4. Place the tongue depressor on the mucous membranes between the cheek and gum, with the gel side next to the cheek Step 3. Once the gel is

Skill Drill 17-1

Administering Glucose

Step 1 Make sure that the tube of glucose is intact and has not expired.

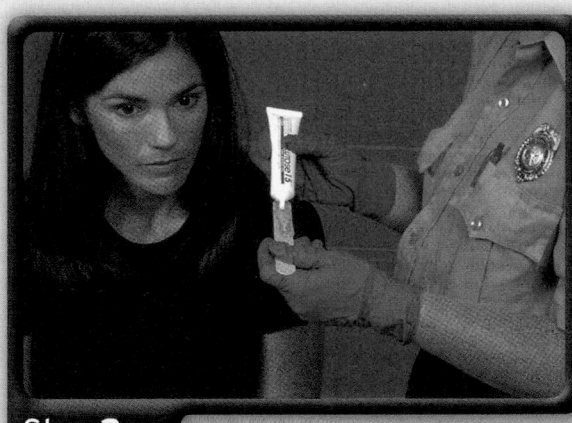

Step 2 Squeeze a generous amount of oral glucose onto the bottom third of a bite stick or tongue depressor.

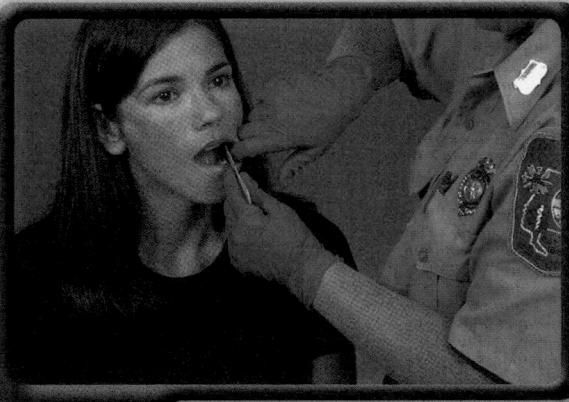

Step 3 Open the patient's mouth. Place the tongue depressor on the mucous membranes between the cheek and the gum with the gel side next to the cheek. Repeat until the entire tube has been used.

dissolved, or if the patient loses consciousness or has a seizure, remove the tongue depressor. Repeat until the entire tube has been used. Note that the patient should not swallow the glucose; it acts more quickly when dissolved in the mouth.

Reassess the patient's condition regularly after giving glucose, even if you see rapid improvement. Watch for airway problems, sudden loss of consciousness, or seizures. Provide prompt transport to the hospital; do not delay transport just to give additional oral glucose.

Words of Wisdom

Diabetes is a systemic disease affecting all tissues of the body, especially the kidneys, eyes, small arteries, and peripheral nerves. Therefore, you are likely to be called to treat patients with a variety of complications of diabetes, such as heart disease, visual disturbances, renal failure, stroke, and ulcers or infections of the feet or toes. With the exception of heart attack and stroke, most of these will not be acute emergencies. Considering that diabetes is a major risk factor for cardiovascular disease, individuals with diabetes should always be suspected of having a potential for heart attack, particularly older patients, even when they do not present with classic symptoms such as chest pain and shortness of breath.

Problems Associated With Diabetes

Problems associated with diabetes include seizures, altered mental status, and airway problems. Remember to consider diabetic emergencies in patients who present with these emergencies.

Seizures

Although seizures are rarely life threatening, you should consider them to be very serious, even in patients with a history of chronic seizures. Seizures, which may be brief or prolonged, are caused by infections, poisoning, hypoglycemia, trauma, or decreased levels of oxygen, or they may be idiopathic (of unknown cause). In children, they may be caused by fever or undiagnosed epilepsy. Although brief seizures are not harmful, they may indicate a more dangerous and potentially life-threatening underlying condition. Because seizures can be the result of a head injury, consider trauma as a cause. In the patient with diabetes, you should also consider hypoglycemia.

Emergency medical care of seizures includes ensuring that the airway is clear and placing the patient on his or her side, if there is no possibility of cervical spine trauma. Do not attempt to place anything in the patient's mouth (eg, a bite stick or an oral airway). Be sure to have suctioning equipment ready in case the patient vomits. Provide oxygen or artificial ventilation if the patient is cyanotic or appears to be breathing inadequately, and provide prompt transport.

Altered Mental Status

Although altered mental status is often caused by complications of diabetes, it may also be caused by a variety of other conditions, including poisoning, infection, head injury, part of the postictal state (period following a seizure), and decreased perfusion to the brain. In diabetes, altered mental status can be caused by hypoglycemia and by ketoacidosis.

Begin emergency medical care of altered mental status by ensuring that the airway is clear. Be prepared to provide artificial ventilation and suctioning in case the patient vomits, and provide prompt transport.

Alcoholism

Occasionally, patients with hypoglycemia or hyperglycemia are thought to be intoxicated, especially if their condition has caused a motor vehicle crash or other incident. Confined by police in a "drunk tank," a patient with diabetes is at risk. In such situations, an emergency medical identification bracelet, necklace, or card may help to save the patient's life. Often, only a blood glucose test performed at the scene or in the emergency department will identify the real problem. In some EMS systems, you will be trained and allowed to perform the blood glucose testing at the scene. Otherwise, you must always suspect hypoglycemia in any patient with an altered mental status.

Certainly, diabetes and alcoholism can coexist in a patient. But you must be alert to the similarity in symptoms of acute alcohol intoxication and diabetic emergencies. Likewise, hypoglycemia and a head injury can coexist, and you must appreciate the potential for hypoglycemia even when the head injury is obvious.

Relationship to Airway Management

Patients with an altered mental status, particularly those who are difficult to awaken, are at risk for losing their gag reflex. When the gag reflex is not working, patients cannot expel foreign materials in their mouth (including vomit), and their tongues will often relax and obstruct the airway. Therefore, you must carefully monitor the airway in patients with hyperglycemia, hypoglycemia, or in a diabetic complication such as stroke or seizure. Place the patient in a lateral recumbent position, and make sure suction is readily available.

Hematologic Emergencies

Hematology is the study and prevention of blood-related diseases, such as sickle cell disease or hemophilia. To understand how hematologic disorders affect the body, the EMT must be familiar with the composition of blood. Blood is "the fluid of life." Without it, people would not be able to live.

Anatomy and Physiology

Blood and Plasma

Blood is made up of two main components: cells and plasma. Red blood cells, also known as erythrocytes, make up 47% of the blood volume in males and 42% in females. Within the red blood cells, hemoglobin is responsible for carrying oxygen to the tissues. White blood cells, also called leukocytes, are also found in the bloodstream. These cells are the "cleaners" of the body, traveling throughout in response to infection and dead cells. Platelets are small cells in the blood that are essential for clot formation. When damage occurs to a blood vessel, platelets are sent to the site of injury to assist in creating a blood clot to stop the bleeding. All of these components—red blood cells, white blood cells, and platelets—are suspended in a straw-colored fluid called plasma, which helps to transport cells throughout the body.

Pathophysiology

Sickle Cell Disease

Sickle cell disease is an inherited blood disorder that affects the red blood cells. It is predominantly found in African Americans. The hemoglobin within the red blood cells allows the cells to carry oxygen from the lungs to the rest of the body. Normal red blood cells contain hemoglobin A.

People who have sickle cell disease have an abnormal type of red blood cells that contains hemoglobin S. These red blood cells, instead of being the normal, round shape, become oblong, or sickle shaped Figure 17-5 . This shape makes the red blood cells poor oxygen carriers, which means a patient with the disease may experience hypoxia, or a lack of oxygen in the body's cells and tissues. The odd shape may also cause red blood cells to lodge in small blood vessels or in the spleen, causing the organ to swell and rupture, which can lead to death. Normally,

You are the Provider: PART 5

You reassess the patient and then call your radio report to the receiving hospital. The patient is still conscious, but confused. Your estimated time of arrival at the hospital is 8 minutes.

Recording Time: 17 Minutes	
Level of consciousness	Conscious, but confused
Respirations	28 breaths/min; deep
Pulse	118 beats/min; weak radial pulses
Skin	Pink, warm, and dry; poor turgor
Blood pressure	110/58 mm Hg
Sao$_2$	97% (on oxygen)

You arrive at the hospital and transfer patient care to the attending physician. After further assessment and treatment in the emergency department, the patient is admitted to the medical intensive care unit.

8. What treatment is provided at the hospital for patients with hyperglycemic crisis that cannot be provided in the prehospital setting?

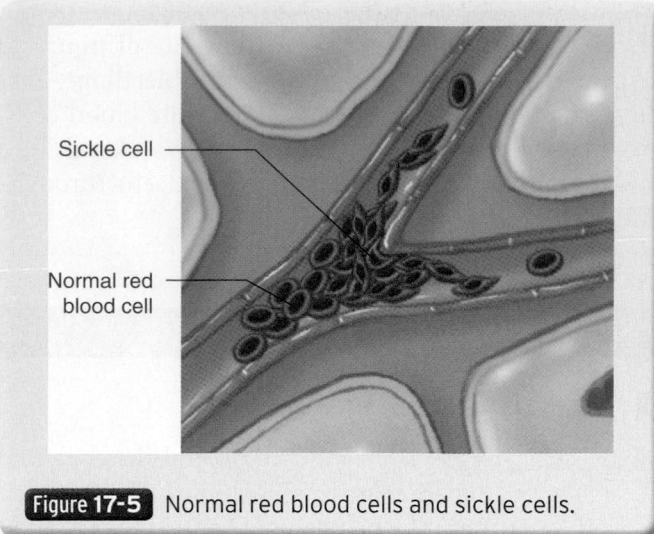

Figure 17-5 Normal red blood cells and sickle cells.

red blood cells live for 120 days before they are replaced by new cells; hemoglobin S cells live only 16 days.

There are four main types of sickle cell crises:

- A **vaso-occlusive crisis** results from blood flow to an organ becoming restricted, causing pain, ischemia, and often organ damage. Most vaso-occlusive crises last between 5 and 7 days. Frequently, circulation to the spleen becomes obstructed as a result of its narrow vessels and function of removing damaged red blood cells.
- An **aplastic crisis** is a worsening of the patient's baseline anemia (lack of circulating red blood cells in the body), which causes tachycardia, pallor, and fatigue. This may be caused by the parvovirus B19, which affects the production of red blood cells, nearly stopping new production for 2 to 3 days.
- A **hemolytic crisis** is an acute accelerated drop in the patient's hemoglobin level. Caused by red blood cells breaking down at a faster than normal rate, this type of crisis is common in patients with glucose-6-phosphate dehydrogenase deficiency (a common enzyme deficiency).
- A **splenic sequestration crisis** is caused by painful, acute enlargement of the spleen, causing the abdomen to become very hard and bloated.

Complications of Sickle Cell Disease

Patients with chronic sickle cell attacks are prone to severe, life-threatening complications of which the EMT must be aware. Although some of these complications take days to weeks to develop, some complications are acute and life threatening. Some of the potential complications of sickle cell disease are as follows:

- Cerebral vascular attack
- Gallstones
- Jaundice

- Avascular necrosis
- Splenic infections
- Osteomyelitis
- Opiate tolerance
- Leg ulcers
- Retinopathy
- Chronic pain
- Pulmonary hypertension
- Chronic renal failure

■ Clotting Disorders

A clotting disorder is a condition in which there is an abnormality in clotting of the blood. The development of a blood clot is called **thrombosis** and can occur in either arterial or venous blood vessels. The patient's symptoms are related to the part of the vascular system in which the clot occurs, the size of the clot, and whether the clot becomes dislodged and travels to another part of the body.

Thrombophilia

Thrombophilia, or the tendency to develop blood clots, affects a large number of people around the world and affects approximately 5% to 7% of the Caucasian population of European descent in the United States.

Thrombosis is a common medical problem. Currently, an estimated 2 million people experience a deep venous thrombosis, or the formation of a clot in a deep vein, each year in the United States. In addition, nearly 50% of the patients experience long-term health consequences that adversely affect their quality of life and require millions of dollars of treatment.

Thrombosis may manifest itself as the formation of a blood clot in a blood vessel or in one of the chambers of the heart. Deep venous thromboses are a leading cause of death in hospitalized patients. This form of clot develops for the first time in 200,000 to 300,000 patients annually during hospitalization because of their lack of mobility. Nearly 40% of the patients have a complication known as pulmonary embolism (a clot that travels to the lung and obstructs a significant amount of blood flow to the organ).

Many patients with thrombophilia receive medications called blood thinners that decrease the tendency to form a clot. Examples of these medications include aspirin, heparin, and warfarin (Coumadin). Typically, pediatric patients do not experience blood clots.

The following are some risk factors for increased clotting:

- Recent surgery
- Impaired mobility
- Congestive heart failure
- Cancer
- Respiratory failure

- Infectious diseases
- Age, older than 40 years
- Being overweight/obesity
- Smoking
- Oral contraceptive use

Hemophilia

Hemophilia is a genetic disorder that is usually inherited from the mother. In people with hemophilia, the body is not able to control bleeding by developing spontaneous clots as normal, resulting in an increased bleeding time. This condition occurs predominately in males and occurs in approximately 1 in every 5,000 to 10,000 births. The disease is classified into two primary types:

- **Hemophilia A.** The most common type, hemophilia A is due to low levels of factor VIII.
- **Hemophilia B.** This second most common type is associated with a deficiency of factor IX.

The levels of factors VIII and IX determine the severity of the disease.

Both type A and type B have the same signs and symptoms. Acute and chronic bleeding can occur at any time and may or may not be life threatening. Any injury or illness that can cause bleeding should not be taken lightly in a person with hemophilia. Spontaneous intracranial bleeding is common in hemophilia and is a major cause of death.

Patient Assessment of Hematologic Disorders

Scene Size-up

Scene Safety

Although your report from dispatch may be for a patient with an unknown medical problem, most patients presenting with a sickle cell crisis have had a crisis before and will relay that information to the dispatcher. As you approach the scene, ensure your safety by assessing for hazards. Standard precautions should consist of gloves and eye protection at a minimum. Remember to evaluate each situation quickly and make sure the necessary personal protective equipment is readily available.

Determine whether this is your only patient and whether trauma was involved. Decide whether you will need any additional resources. Patients experiencing a vaso-occlusive crisis are often in extreme pain and would benefit from ALS providers being able to administer analgesics.

Mechanism of Injury/Nature of Illness

Remember to keep open the possibility that trauma may have occurred because of a medical incident. Determine mechanism of injury and/or nature of illness.

Primary Assessment

An African American patient or any patient of Mediterranean descent who complains of severe pain may have undiagnosed sickle cell disease.

Perform cervical spine stabilization, if necessary. Remember that even though a person has a history of sickle cell disease, sickle cell disease may not be causing the current problem; trauma or another type of medical emergency may be the cause. For this reason, you must always perform a thorough, careful primary assessment, paying attention to the ABCs and immediately correcting any life-threatening issues.

Form a General Impression

Perform a rapid scan of the patient in order to form an initial general impression of the patient. How does the patient look? Does the patient appear anxious, restless, or listless? Is the patient apathetic or irritable? Determine the patient's level of consciousness.

Airway and Breathing

As you are forming your general impression, assess the patient's airway and breathing. Patients showing signs of inadequate breathing or altered mental status should receive high-flow oxygen at 12 to 15 L/min via nonrebreathing mask. A patient who is experiencing a sickle cell crisis may have increased respirations as a result of severe pain or exhibit signs of pneumonia. If the patient is having difficulty breathing, open the airway and insert an airway adjunct, administer oxygen, and assist ventilations. Continue to monitor the airway as you provide care.

Circulation

Once you have assessed the airway and breathing and have performed the necessary interventions, check the patient's circulatory status. An increased heart rate represents a compensatory mechanism, in an attempt to "force" the sickled cells through smaller vasculature.

In patients with suspected hemophilia, be alert for signs of acute blood loss such as pallor, weak pulse, and hypotension. Note any bleeding of unknown origin, such as nosebleeds, bloody sputum, and blood in the urine or stool. Owing to blood loss, patients with hemophilia may exhibit signs of hypoxia.

Transport Decision

Whether you decide to rapidly transport the patient will depend on the severity of the patient's pain and the patient's wishes. Patients with a history of sickle cell disease, but who have not had a crisis in some time, may require emotional support and refuse transport. However, transport to an emergency department should always be recommended to any patient who is experiencing a sickle cell crisis or hemophilia.

History Taking

Investigate Chief Complaint

If the patient is conscious, what is the chief complaint or history of present illness?

Responsive medical patients are able to provide their own medical history to help you identify a cause for their severe pain. Physical signs, such as swelling of the fingers and toes, priapism, and jaundice may guide you in determining whether the patient is experiencing a sickle cell crisis. Also important to ascertain is whether the pain is isolated to a single location or if pain is felt throughout the entire body. Is the patient having any visual disturbances? Is the patient experiencing any gastrointestinal problems, such as nausea, vomiting, or abdominal cramping? Is the patient reporting any chest pain or shortness of breath?

SAMPLE History

In a patient with known sickle cell disease, ask the following questions in addition to obtaining a SAMPLE history:

- Have you had a crisis before?
- When was the last time you had a crisis?
- How did your last crisis resolve?
- Have you had any illness, unusual amount of activity, or stress lately?

Secondary Assessment

The secondary assessment may be performed on scene, en route to the emergency department, or not at all. This will depend on transport time and the patient's condition.

Physical Examinations

Next, systematically examine the patient, focusing on major joints at which cells congregate, and obtain your patient's baseline vital signs. Evaluate and document mental status using the AVPU scale.

Vital Signs

Obtain a complete set of vital signs, including a measurement of the patient's oxygen saturation level. In patients experiencing a sickle cell crisis, respirations are normal to rapid, pulse is weak and rapid, and skin is typically pale and clammy with a low blood pressure.

Monitoring Devices

Use pulse oximetry, if available. However, keep in mind that the oxygen saturation reading you obtain may be inaccurate as a result of the patient's anemic state.

Reassessment

It is important to reassess the patient frequently to determine if there have been changes in his or her condition.

For example, are there changes in the patient's mental status? Are the ABCs still intact? How is the patient responding to the interventions performed? Should you adjust or change the interventions? In many patients, you will note marked improvement with appropriate treatment. Document each assessment, your findings, the time of the interventions, and any changes in the patient's condition.

Interventions

Supplemental oxygen should be administered via nonrebreathing mask at 12 to 15 L/min in an attempt to hypersaturate the remaining hemoglobin and increase the level of perfusion that has been decreased by the sickled cells or hemophilia.

Once arriving at the hospital, care for sickle cell patients can include analgesics for pain, penicillin to prevent infection, IV fluid for hydration, and, depending on the severity of the crisis, a blood transfusion.

Distinguishing a true sickle cell crisis from other nonspecific causes of pain can be difficult. In these situations, perform a thorough assessment and contact the hospital to help sort out the signs and symptoms. The hospital should be a resource for you to help problem-solve situations and provide guidance on how to manage your patient.

Hospital care for a patient with hemophilia may include IV therapy to treat hypotension, and a transfusion of plasma. Analgesics may also be appropriate.

Communication and Documentation

Communication with hospital staff is important for continuity of care. Hospital personnel need to be informed about the patient's history, the present situation, your assessment findings, and your interventions and their results.

Your run report is the only legal document you have to say that appropriate care was provided. Document clearly your assessment findings as the basis for your treatment. Follow your local protocols for patients who refuse treatment or transport.

Emergency Medical Care for Hematologic Disorders

Emergency care for patients with hematologic disorders is mainly supportive and symptomatic. Patients showing signs of inadequate breathing or altered mental status should receive high-flow oxygen at 12 to 15 L/min via nonrebreathing mask and should be placed in a position of comfort and transported rapidly to the hospital.

You are the Provider: SUMMARY

1. What is diabetes mellitus (DM)? What is the central problem in patients with DM?

Diabetes mellitus (DM), or "sweet diabetes," refers to the presence of glucose—one of the body's simple sugars—in the urine. It is a metabolic disorder in which the body's ability to metabolize simple carbohydrates (glucose) is impaired. Glucose, along with oxygen, is the major source of fuel in the production of adenosine triphosphate (ATP), or energy, for the body; all cells need it to function properly. The brain needs an equal and constant supply of oxygen and glucose to function.

The central problem in patients with diabetes is the lack or ineffective action of insulin. Insulin is a hormone that is produced by the islets of Langerhans—endocrine glands on the pancreas—that allows the cellular uptake of glucose from the bloodstream. Without insulin, the cells begin to starve because insulin is needed, like a key, to unlock the cells and let glucose in. For this reason, insulin is referred to as the "cellular key."

2. What is the difference between type 1 and type 2 diabetes?

Diabetes has two distinct onset patterns. It may develop during childhood, or later in life—usually during middle adulthood.

Type 1 diabetes (formerly known as type I diabetes, insulin-dependent diabetes, and juvenile-onset diabetes) develops when the pancreas produces no insulin at all. As a result, patients with type 1 diabetes require daily injections of supplemental, synthetic insulin to control their blood glucose levels. Type 1 diabetes typically develops in children; however, it can also occur later in life.

Type 2 diabetes (formerly called type II diabetes, non-insulin-dependent diabetes, and adult-onset diabetes), typically develops later in life. In type 2 diabetes, the pancreas produces inadequate amounts of insulin; in other cases, the pancreas produces a normal amount of insulin, but it does not function effectively. Although some patients with type 2 diabetes may require supplemental insulin injections, most patients can be treated with diet, exercise, and oral hypoglycemic medications. Oral hypoglycemic medications, such as metformin (Glucophage), glyburide (Micronase), tolbutamide (Orinase), and rosiglitazone (Avandia), help lower blood glucose levels by increasing insulin production in the pancreas or decreasing glucose production in the liver.

Although both types of diabetes are equally serious and require life-long management, type 2 diabetes is usually easier to regulate. In general, diabetes can lead to other metabolic problems and organ damage, such as blindness, heart disease, kidney failure, and nervous system disorders. The severity of diabetic complications is related to how the patient manages his or her disease, how high the average blood glucose level is, and how early in life the disease develops.

3. What should you expect the patient's blood glucose level to read? Why?

Rosiglitazone (Avandia) is an oral hypoglycemic medication that helps lower blood glucose levels by increasing the sensitivity of liver, fat, and muscle cells to insulin, thus enabling them to remove glucose from the body more effectively.

Normal blood glucose levels range between 80 and 120 mg/dL (milligrams per deciliter [100 milliliters] of blood). However, if the patient is noncompliant with a regimen of oral hypoglycemic medication, glucose will accumulate in the blood and the blood glucose level will become abnormally high (hyperglycemia). Therefore, your assessment of his blood glucose level via the glucometer yielded a high numeric value. If the level is severely elevated (> 300 mg/dL, but variable, depending on the glucometer), it may simply read "high" on the glucometer.

4. What is causing the patient's frequent urination and deep, rapid breathing?

Without insulin, glucose gradually accumulates in the blood—eventually leading to extremely high levels. Once the blood glucose levels reach 200 mg/dL or more, excess glucose is excreted from the body by the kidneys. This process requires large amounts of water; therefore, the patient experiences frequent and plentiful urination (polyuria), which leads to dehydration. Dehydration causes the patient to drink large amounts of water to satisfy their continuous thirst (polydipsia).

In the absence of glucose in the cells, the body must turn to other fuel sources, the most abundant of which is fat. However, when fat is metabolized by the cells, fatty acids and chemicals called ketones are produced. Ketones and fatty acids are difficult for the body to excrete; therefore, they accumulate in the blood, resulting in a dangerous condition called acidosis. The form of acidosis seen in uncontrolled diabetes is called diabetic ketoacidosis (DKA), also known as diabetic coma. This is a form of hyperglycemic crisis. Although seen only occasionally, patients with DKA may present with excessive hunger (polyphagia) because the body's cells are literally starving for sugar.

In an attempt to remove ketoacids from the blood, the respiratory system increases its function, which manifests as deep, rapid breathing (Kussmaul respirations). In addition, the patient often has a fruity (acetone) or sweet breath odor, which indicates the respiratory elimination of ketoacids.

5. What other factors can cause hyperglycemia in patients with diabetes?

There are many factors that can cause hyperglycemia in patients with diabetes—some of which are more common to one type of diabetes than the other.

In type 1 diabetes, hyperglycemia commonly occurs if the patient inadvertently takes too little of his or her prescribed insulin—even if he or she follows a diet that is appropriate for diabetics. Hyperglycemia is virtually guaranteed if the patient eats too much sugar-containing food *and* takes too little of his or her prescribed insulin.

In type 2 diabetes, hyperglycemia commonly occurs if the patient is noncompliant with his or her prescribed oral hypoglycemic medication, as is the case with the patient in this case study. Hyperglycemia can also occur if the patient eats too much sugar-containing food—especially if he or she is also noncompliant with his or her prescribed hypoglycemic medication.

You are the Provider: SUMMARY, continued

Hyperglycemia occurs naturally as the result of infection, illness, inflammation, and psychological stress. However, it is typically more severe in patients with type 1 or type 2 diabetes. Any type of stress—physiological or psychological—causes the body to release epinephrine (adrenalin), which, among other things, increases circulating blood glucose levels.

When assessing a patient with suspected or documented (ie, by glucometer) hyperglycemia, it is important to try to determine the underlying cause. If the patient has type 1 diabetes, did he or she take the appropriate dose of his or her insulin? If the patient has type 2 diabetes, has he or she been compliant with his or her prescribed hypoglycemic medication? Has the patient been ill, had an infection, or experienced psychological stress? Document your findings and pass them along to the receiving facility; doing so will help the physician direct his or her treatment accordingly.

6. What is the difference between hyperglycemic crisis (DKA) and hypoglycemic crisis (insulin shock)? How can you distinguish one from the other?

Hyperglycemic crisis is a state of unconsciousness that results from several problems, including hyperglycemia, ketoacidosis, and dehydration secondary to excessive urination; these processes were discussed earlier in this case study.

Not all patients experiencing a hyperglycemic crisis are truly comatose; however, their level of consciousness is typically altered. Furthermore, hyperglycemia by itself does not always cause a crisis; however, when it does occur, hyperglycemia is the key contributing factor. This is why DKA is always considered a type of hyperglycemic crisis.

As with all types of hypoglycemic crisis, the underlying problem is a low circulating blood glucose level (hypoglycemia). When insulin levels remain high, glucose is rapidly taken out of the bloodstream to fuel the cells. If blood glucose levels get too low, there may an insufficient amount available to supply the brain. Left untreated, a hypoglycemic crisis can rapidly cause unconsciousness, permanent brain damage, and death.

A hypoglycemic crisis typically occurs when the patient takes too much of his or her insulin, takes a regular dose of insulin but does not eat, or has exercised vigorously and has depleted his or her body of available glucose.

It is important to note that hypoglycemia is not exclusive to patients who have diabetes. Some patients are prone to a drop in blood glucose levels, such as those who exercise vigorously and/or go for prolonged periods of time without eating. In this case, the body is using more glucose than it can produce and the patient is simply said to be "hypoglycemic."

A key aspect in distinguishing a hyperglycemic crisis from a hypoglycemic crisis is the time of symptom onset. The processes that lead to a hyperglycemic crisis—hyperglycemia, ketoacidosis, and dehydration—typically progress over hours to days. By contrast, a hypoglycemic crisis has an acute onset—often over a period of a few minutes.

A hyperglycemic crisis and hypoglycemic crisis also present with uniquely different signs and symptoms. Signs and symptoms of a hyperglycemic crisis include tachycardia; signs of dehydration (ie, warm, dry skin, poor skin turgor, and sunken eyes); deep, rapid breathing (Kussmaul respirations), which indicates the respiratory system's attempt to eliminate ketones from the body; a sweet or fruity (acetone) breath odor; and mental status changes ranging from confusion to coma.

A hypoglycemic crisis presents with signs and symptoms similar to hypoxemia and shock, and include rapid, shallow respirations; pale, cool, clammy (diaphoretic) skin; tachycardia; weakness, which may be confined to one side of the body and mimic a stroke; and varying degrees of mental status change, including confusion, irritability, combativeness, seizures, and coma.

The primary visible difference that helps you differentiate a hyperglycemic crisis from a hypoglycemic crisis is the patient's breathing. In hyperglycemic crisis, the patient's respirations are deep and rapid; in hypoglycemic crisis, they are normal or rapid (but not deep). When you are treating a patient with a diabetic emergency, you must rely on your knowledge of the signs and symptoms previously discussed; however, this assessment should not delay prompt treatment and transport.

7. What additional treatment should you provide to this patient?

The patient's signs and symptoms clearly point to hyperglycemic crisis, specifically diabetic ketoacidosis (DKA). The glucometer reads "high," which indicates that his blood glucose level is most likely in excess of 300 mg/dL; he has deep, rapid breathing (Kussmaul respirations); and he is dehydrated, which is the result of excessive urination secondary to a high blood glucose level.

Hyperglycemic crisis requires definitive care that can only be provided at the hospital. Prehospital treatment at the EMT level is aimed at providing supportive care (ie, maintaining the ABCs) and promptly transporting the patient to the hospital. En route, closely monitor the patient's mental status and breathing adequacy; if his respirations become slow and/or shallow—especially if his mental status deteriorates further—you should assist his ventilations with a bag-mask device.

Some patients with hyperglycemic crisis become so dehydrated that they develop hypovolemic shock; therefore, it is important to closely monitor the patient's perfusion status (eg, heart rate, peripheral pulse quality, blood pressure, mental status). If signs of shock are observed, keep the patient warm and elevate his lower extremities 6″ to 12″. Although the patient is extremely thirsty, do not give him anything to drink; doing so increases the risk of aspiration if he vomits.

8. What treatment is provided at the hospital for patients with hyperglycemic crisis that cannot be provided in the prehospital setting?

Hyperglycemic crisis occurs when there is little or no insulin in the body to facilitate the cellular uptake of glucose. Therefore, patients with a hyperglycemic crisis desperately need insulin to move glucose from the bloodstream into the cells. However, insulin is rarely, if ever, administered in the prehospital

You are the Provider: SUMMARY, continued

setting—even by paramedics. This underscores the importance of performing a rapid assessment, initiating treatment without delay, and promptly transporting the patient to the hospital.

Significant hyperglycemia causes the patient to excrete large amounts of water from the body (diuresis), resulting in varying degrees of dehydration. Although this can be treated in the prehospital setting with intravenous (IV) therapy and fluid replacement, IV therapy is beyond the EMT's typical scope of practice. If your transport time will be prolonged, consider an intercept with an advanced life support unit, if available;

AEMTs and paramedics are trained to start IVs and administer fluids.

Hyperglycemic crisis is a complex medical problem that causes numerous complications; it cannot be treated in the prehospital setting and it certainly cannot be changed quickly. Insulin is needed to restore circulating blood glucose to a normal level and IV fluids are needed to correct dehydration. Electrolyte (ie, sodium, potassium, magnesium) abnormalities may also occur because of the excessive urination and must be promptly corrected.

EMS Patient Care Report (PCR)

Date: 09-19-09	**Incident No.:** 011609	**Nature of Call:** Diabetic complications		**Location:** 445 Landon Way	
Dispatched: 0720	**En Route:** 0720	**At Scene:** 0726	**Transport:** 0737	**At Hospital:** 0751	**In Service:** 0801

Patient Information

Age: 56 **Sex:** M **Weight (in kg [lb]):** 91 kg (200 lb)		**Allergies:** No known drug allergies **Medications:** Avandia, lisonopril **Past Medical History:** Type 2 diabetes, Hypertension **Chief Complaint:** Weakness and confusion

Vital Signs

Time: 0731	BP: 112/54	Pulse: 120	Respirations: 30	Sao$_2$: 95%
Time: 0737	BP: 108/56	Pulse: 124	Respirations: 30	Sao$_2$: 96%
Time: 0743	BP: 110/58	Pulse: 118	Respirations: 28	Sao$_2$: 97%

EMS Treatment
(circle all that apply)

Oxygen @ 15 L/min via (circle one): NC **NRM** Bag-Mask Device	Assisted Ventilation	Airway Adjunct	CPR
Defibrillation	Bleeding Control	Bandaging	Splinting	Other

Narrative

9-1-1 dispatch for a patient with "diabetic complications." Arrived on scene and found the patient, a 56-year-old male, lying on the couch in his living room. He was conscious, but confused. His airway was patent, and his breathing was deep and rapid. The patient's wife advised that he has type 1 diabetes and hypertension, and that he has been sick for the past few days. She further advised that he has not taken his prescribed oral hypoglycemic medication (Avandia) for the past 2 days; since then, he has been urinating excessively and has been drinking a lot of water. Further assessment of the patient revealed that his radial pulse was rapid and weak; his skin was pink, warm, and dry; and he had poor skin turgor. Applied high-flow oxygen via nonrebreathing mask at 15 L/min and performed further assessment. Breath sounds were clear to auscultation bilaterally, pupils were equal and reactive to light, there were no gross signs of trauma, and the patient had a sweet, fruity odor on his breath. Vital signs were obtained and blood glucose level was assessed and noted to read "high." The patient's wife spoke with his physician, who requested EMS transport to the closest appropriate facility. Patient was placed onto stretcher and loaded into the ambulance. Mental status and vital signs were reassessed and transport was begun. Continued to monitor patient en route; his mental status and vital signs remained unchanged. Continued oxygen therapy, which maintained his oxygen saturation above 95%. Notified receiving facility of our impending arrival; no further medical direction was given. Delivered patient to emergency department; his condition was unchanged. Gave verbal report to attending physician, transferred patient care, and returned to service. **End of report**

Assessment and Emergency Care of Diabetic Emergencies

Scene Size-up

Scene Safety	Ensure scene safety and address hazards. Standard precautions should include a minimum of gloves and eye protection. Consider the number of patients, the need for additional help/ALS, and cervical spine stabilization.
Mechanism of Injury/Nature of Illness	Determine the mechanism of injury (MOI)/nature of illness (NOI).

▼ ▼

Primary Assessment

Form a General Impression	Perform a rapid scan of the patient to determine level of consciousness and identify any life threats. Determine priority of care based on environment and MOI/NOI. Assess ABCs.
Airway and Breathing	Ensure patent airway. Insert airway adjunct, if indicated. Provide high-flow oxygen at 12 to 15 L/min. Evaluate rate, rhythm, and quality of the respirations and provide ventilations, as needed.
Circulation	Evaluate pulse rate, rhythm, and quality; observe skin color, temperature, and condition and treat accordingly.
Transport Decision	Rapid transport.

▼ ▼

NOTE: The order of the steps in this section differs depending on whether the patient is conscious or unconscious. The order below is for a conscious patient. For an unconscious patient, perform a primary assessment, obtain vital signs, and if possible, obtain the past medical history before transport.

History Taking

Investigate Chief Complaint	Investigate the chief complaint (history of present illness). Identify signs and symptoms and pertinent negatives. If the patient is not thinking or speaking clearly (or is unconscious), talk to a family member or bystander. Ask pertinent SAMPLE and OPQRST questions. Be sure to ask if and what interventions were taken before your arrival, how many interventions, and at what time. Ascertain if patient has taken insulin or pills for diabetes. Has the patient been compliant with his or her diet and medication regimens? Inquire about any recent illness, physical activity, and/or stress.

▼ ▼

Assessment and Emergency Care of Diabetic Emergencies, continued

Secondary Assessment

Physical Examinations	Perform a systematic examination assessing all regions. Assess respiratory, cardiovascular, neurologic, and musculoskeletal systems. Determine the blood glucose level and mental status using AVPU.
Vital Signs	Take vital signs, noting skin color and temperature, as well as patient's level of consciousness.

Reassessment

Interventions	A conscious patient who is able to swallow can be given fluids with a high sugar content or a highly concentrated glucose gel, as protocols allow.
Communication and Documentation	Contact medical control with a radio report, informing of the patient's condition and blood glucose level(s). Relay any change in level of consciousness or difficulty breathing. Be sure to document any changes, the time they occurred, and blood glucose readings.

NOTE: Although the steps below are widely accepted, be sure to consult and follow your local protocol.

Diabetic Emergencies

Administering Glucose

1. Examine the tube to ensure that it is not open or broken. Check the expiration date.
2. Squeeze a generous amount onto the bottom third of a bite stick or tongue depressor.
3. Open the patient's mouth. Place the tongue depressor on the mucous membranes between the cheek and gum, with the gel side next to the cheek.

Assessment and Emergency Care of Hematologic Disorders

Scene Size-up

Scene Safety	Ensure scene safety and address hazards. Standard precautions should include a minimum of gloves and eye protection. Consider the number of patients, the need for additional help/ALS, and cervical spine stabilization.
Mechanism of Injury/Nature of Illness	Determine the mechanism of injury (MOI)/nature of illness (NOI).

▼   ▼

Primary Assessment

Form a General Impression	Perform a rapid scan of the patient to determine level of consciousness and identify any life threats. Determine priority of care based on environment and patient's chief complaint. Assess ABCs.
Airway and Breathing	Ensure patent airway. Insert airway adjunct, if indicated. Provide high-flow oxygen at 12 to 15 L/min. Evaluate rate, rhythm, and quality of the respirations, and provide ventilations, as needed.
Circulation	Evaluate pulse rate, rhythm, and quality; observe skin color, temperature, and condition, and treat accordingly. Determine if bleeding is present and control if life threatening.
Transport decision	Rapid transport.

▼   ▼

History Taking

Investigate Chief Complaint	Investigate the chief complaint. Identify signs and symptoms and pertinent negatives. Ask pertinent SAMPLE and OPQRST questions. Be sure to ask when last sickle cell attack occurred. Has the patient been compliant with his or her medication regimen? Inquire about any recent illness, physical activity, and/or stress.

▼   ▼

Secondary Assessment

Physical Examinations	Perform a full-body scan, assessing all regions. Assess respiratory, cardiovascular, neurologic, and musculoskeletal systems. Determine the Glasgow Coma Scale score.
Vital Signs	Take vital signs, noting skin color and temperature, as well as patient's level of consciousness.

▼   ▼

Reassessment

Interventions	Apply supplemental oxygen at 12 to 15 L/min via nonrebreathing mask.
Communication and Documentation	Contact medical control with a radio report, informing of the patient's condition. Relay any change in level of consciousness or difficulty breathing. Be sure to document any changes and the time when they occurred.

Prep Kit

Ready for Review

- Diabetes is a disorder of glucose metabolism or difficulty metabolizing carbohydrates, fats, and proteins.

- There are two types of diabetes. Type 1 diabetes typically develops in childhood and requires daily insulin to control blood glucose. Type 2 diabetes typically develops in middle age and often can be controlled with diet, activity, and oral medications.

- Both types of diabetes are serious systemic diseases, especially affecting the kidneys, eyes, small arteries, and peripheral nerves.

- Patients with diabetes have chronic complications that place them at risk for other diseases, such as heart attack, stroke, and infections. Most often, however, you will be called on to treat the acute complications of blood glucose imbalance. These include hyperglycemia (excess blood glucose) and hypoglycemia (insufficient blood glucose).

- Hyperglycemia is typically characterized by excessive urination and resulting thirst, in conjunction with the deterioration of body tissues.

- Hyperglycemia is usually associated with dehydration and ketoacidosis and can result in marked rapid (often deep) respirations; warm, dry skin; a weak pulse; and a fruity breath odor. Hyperglycemia must be treated in the hospital with insulin and IV fluids.

- Symptoms of hypoglycemia classically include confusion; rapid respirations; pale, moist skin; diaphoresis; dizziness; fainting; and even coma and seizures. This condition is rapidly reversible with the administration of glucose or sugar. Without treatment, however, permanent brain damage and death can occur.

- Because a blood glucose level that is either too high or too low can result in altered mental status, you must perform a thorough history and patient assessment to determine the nature of the problem. When the problem cannot be determined, it is best to treat the patient for hypoglycemia.

- Be prepared to give oral glucose to a conscious patient who is confused or has a slightly decreased level of consciousness; however, do not give oral glucose to a patient who is unconscious or otherwise unable to swallow properly or protect his or her own airway.

- Remember, in all cases, providing emergency medical care and prompt transport is your primary responsibility.

- Sickle cell disease is a blood disorder that affects the shape of red blood cells.

- Symptoms of sickle cell disease are typically characterized by pain in the joints, fever, respiratory distress, and abdominal pain.

- Hemoglobin A is considered normal hemoglobin; hemoglobin S is considered an abnormal type of hemoglobin and is responsible for sickle cell crisis.

- Patients with sickle cell disease have chronic complications that place them at risk for other diseases, such as heart attack, stroke, and infection. Most often, however, you will be called on to treat the acute complications of severe pain.

- Patients with hemophilia are not able to control bleeding because clots do not develop as they should.

- Emergency care in the prehospital setting is supportive for patients with sickle cell disease or a clotting disorder such as hemophilia.

Vital Vocabulary

acidosis A pathologic condition that results from the accumulation of acids in the body.

aplastic crisis A condition in which the body stops producing red blood cells; typically caused by infection.

diabetes mellitus A metabolic disorder in which the ability to metabolize carbohydrates (sugars) is impaired, usually because of a lack of insulin.

diabetic ketoacidosis (DKA) A form of hyperglycemia in uncontrolled diabetes in which certain acids accumulate when insulin is not available.

endocrine glands Glands that secrete or release chemicals that are used inside the body.

endocrine system Regulates metabolism and maintains homeostasis.

glucose One of the basic sugars; it is the primary fuel, in conjunction with oxygen, for cellular metabolism.

hematology The study and prevention of blood-related disorders.

hemolytic crisis A rapid destruction of red blood cells that occurs faster than the body's ability to create new cells.

hemophilia A congenital abnormality in which the body is unable to produce clots, which results in uncontrollable bleeding.

hormone A chemical substance produced by a gland that regulates the activity of organs and tissues.

hyperglycemia An abnormally high glucose level in the blood.

hyperglycemic crisis A state of unconsciousness resulting from several problems, including ketoacidosis, dehydration because of excessive urination, and hyperglycemia.

hypoglycemia An abnormally low glucose level in the blood.

hypoglycemic crisis Severe hypoglycemia resulting in changes in mental status.

insulin A hormone produced by the islets of Langerhans (endocrine gland located throughout the pancreas) that enables glucose in the blood to enter cells; used in synthetic form to treat and control diabetes mellitus.

Kussmaul respirations Deep, rapid breathing; usually the result of an accumulation of certain acids when insulin is not available in the body.

polydipsia Excessive thirst that persists for long periods, despite reasonable fluid intake; often the result of excessive urination.

polyphagia Excessive eating; in diabetes, the inability to use glucose properly can cause a sense of hunger.

polyuria The passage of an unusually large volume of urine in a given period; in diabetes, this can result from the wasting of glucose in the urine.

sickle cell disease A hereditary disease that causes normal, round red blood cells to become oblong, or sickle shaped.

splenic sequestration crisis An acute, painful enlargement of the spleen caused by sickle cell disease.

thrombophilia A tendency toward the development of blood clots as a result of an abnormality of the system of coagulation.

thrombosis A blood clot, either in the arterial or venous system.

type 1 diabetes The type of diabetic disease that typically develops in childhood and requires synthetic insulin for proper treatment and control.

type 2 diabetes The type of diabetic disease that typically develops in later life and often can be controlled through diet and oral medications.

vaso-occlusive crisis Ischemia and pain caused by sickle-shaped red blood cells that obstruct blood flow to a portion of the body.

Assessment in Action

You are working a detail for the women's state volleyball championship at the local university gym when you are dispatched to the visiting team locker room for a player who is feeling weak and dizzy. You are met at the locker room door by the coach who points to a young woman lying supine on a bench. The coach informs you that the player has a history of diabetes and takes insulin on a regular basis. As you approach the patient, you run through some of the information that you know about diabetes.

1. Diabetes is a metabolic disorder in which the body's ability to metabolize _____ is impaired.
   A. Protein
   B. Fats
   C. Glucose
   D. Electrolytes

2. A chemical substance produced by a gland that has special regulatory effects on other body organs and tissues is called a(n):
   A. neurotransmitter.
   B. hormone.
   C. regulator.
   D. inhibitor.

3. Patients with diabetes who overexert themselves are prone to:
   A. rapid drops in their ability to sweat.
   B. rapid increases in their blood pressure.
   C. rapid drops in their blood glucose levels.
   D. rapid increases in their blood glucose levels.

4. Describe some of the problems that patients with type 1 diabetes are likely to have.

5. Which of the following organs can rapidly suffer permanent damage when the body's glucose level is too low?
   A. Brain
   B. Heart
   C. Kidney
   D. Liver

6. Based on the information described in the scenario, what question should you ask this patient first?
   A. Are you having any pain?
   B. When was the last time you ate?
   C. Do you feel short of breath?
   D. When was your last menstrual cycle?

7. What other signs and symptoms may your patient exhibit?
   A. A slow, bounding pulse
   B. Rapid, deep respirations
   C. Nausea and vomiting
   D. Pale, moist skin

8. Your partner prepares to perform a fingerstick to check the patient's glucose level. Normal glucose levels are between:
   A. 20 to 45 mg/dL
   B. 50 to 75 mg/dL
   C. 80 to 120 mg/dL
   D. 200 to 350 mg/dL

9. Appropriate interventions for your patient should include which of the following?
   A. Placement in the shock position
   B. Insertion of an oropharyngeal airway
   C. Have the patient drink a glass of orange juice
   D. Administration of her next dose of insulin

10. Is a patient who is experiencing a hypoglycemic crisis more critical than a patient with hyperglycemia? Explain your answer.

National EMS Education Standard Competencies

Medicine
Applies fundamental knowledge to provide basic emergency care and transportation based on assessment findings for an acutely ill patient.

Immunology
Recognition and management of shock and difficulty breathing related to
- Anaphylactic reactions (pp 649-659)

Anatomy, physiology, pathophysiology, assessment, and management of
- Hypersensitivity disorders and/or emergencies (pp 647-659)
- Anaphylactic reactions (pp 647-659)

Knowledge Objectives
1. Understand and define the terms allergic reaction and anaphylaxis. (p 647)
2. Explain the difference between a local and a systemic response to allergens. (pp 647-648)
3. Differentiate the primary assessment for a patient with a systemic allergic or anaphylactic reaction and a local reaction. (pp 649-654)
4. Describe the five categories of stimuli that could cause an allergic reaction or an extreme allergic reaction. (pp 647-648)
5. Discuss the steps in the primary assessment that are specific to a patient who is having an allergic reaction. (pp 650-652)
6. Explain the importance of managing the ABCs of a patient who is having an allergic reaction. (pp 650-652)
7. Explain the factors involved when making a transport decision for a patient having an allergic reaction. (pp 650-652)
8. Explain the rationale, including communication and documentation considerations, when determining whether to administer epinephrine to a patient who is having an allergic reaction. (pp 649-654)

9. Review the process for providing emergency medical care to a patient who is experiencing an allergic reaction. (pp 654-659)
10. Describe some age-related contraindications to using epinephrine to treat an allergic reaction in a geriatric patient. (p 650)

Skills Objectives
1. Demonstrate how to remove the stinger from a honey bee sting and proper patient management following its removal. (p 649)
2. Demonstrate how to use an EpiPen auto-injector. (pp 655-656, Skill Drill 18-1)
3. Demonstrate how to use a Twinject. (pp 657-659, Skill Drill 18-2)

Introduction

Every year, at least 1,000 Americans die of allergic reactions. When managing allergy-related emergencies, you must be aware of the possibility of acute airway obstruction and cardiovascular collapse and be prepared to treat these life-threatening complications. You must also be able to distinguish between the body's usual response to a sting or bite and an allergic reaction, which may require epinephrine. Your ability to recognize and manage the many signs and symptoms of allergic reactions may be the only thing standing between a patient's life and imminent death.

This chapter describes **immunology**, the study of the body's immune system, and the five categories of stimuli that may provoke allergic reactions. You will learn what to look for in assessing patients who may be having an allergic reaction and how to care for them, including administration of epinephrine. The chapter then describes insect bites and stings and their management.

Anatomy and Physiology

The **immune system** protects the human body from substances and organisms that are foreign to the body. Without the immune system for protection, life as you know it would not exist. You would be under constant attack from any type of invader, such as a bacterium or virus that wanted to make your body a home. Fortunately, most people have immune systems that are well equipped to detect unauthorized visits or invading attacks by foreign substances. Once a foreign substance invades the body, the body goes on alert and initiates a series of responses to inactivate the invader.

Pathophysiology

Contrary to what many people think, an **allergic reaction**, an exaggerated **immune response** to any substance, is not caused directly by an outside stimulus, such as a bite or sting. Rather, it is a reaction by the body's immune system, which releases chemicals to combat the stimulus. Among these chemicals are **histamines** and **leukotrienes**, both of which contribute to an allergic reaction. An allergic reaction may be mild and local, involving hives, itching, or tenderness, or it may be severe and systemic, resulting in shock and respiratory failure.

Anaphylaxis is an extreme allergic reaction that is life threatening and involves multiple organ systems. In severe cases, anaphylaxis can rapidly result in death. One of the most common signs of anaphylaxis is **wheezing**, a high-pitched, whistling breath sound that is typically heard on expiration, usually resulting from bronchospasm/bronchoconstriction and increased mucus production. Also present is widespread urticaria, or hives. **Urticaria** consists of small areas of generalized itching or burning that appear as multiple, small, raised areas on the skin Figure 18-1 . You may also note hypotension as a result of hypovolemic shock due to increased capillary permeability.

Given the right person and the right circumstances, almost any substance can trigger the body's immune system and cause an allergic reaction: animal bites, food, latex, and many other substances can be **allergens**. The most common allergens, however, fall into the following five general categories:

- **Insect bites and stings.** When an insect bites you and injects the bite with its venom, the act is called **envenomation** or, more commonly, a sting. The sting of a honeybee, wasp, ant, yellow jacket, or hornet may cause a severe reaction with the swiftness of an injected medication. The reaction may be local, causing swelling and itchiness in the surrounding tissue, or it may be

You are the Provider: PART 1

You and your partner are standing by at a corporate picnic when a 33-year-old man presents to you after being stung on the leg by a hornet. He tells you that he broke out in hives the last time he was stung by a hornet. Your general impression reveals that he is conscious, but anxious, and he is covered with hives.

1. What causes an allergic reaction?

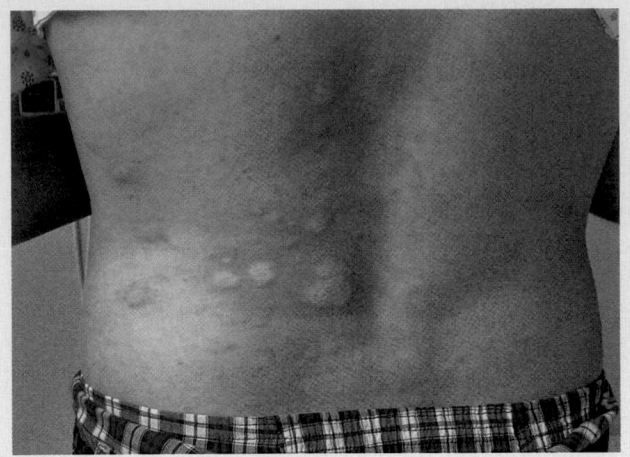

Figure 18-1 Urticaria, or hives, may appear following a sting and is characterized by multiple, small, raised areas on the skin. Urticaria may be one of the warning signs of impending anaphylactic reaction.

systemic, involving the entire body. Such a total body reaction would be considered an anaphylactic reaction.

- **Medications.** Injection of medications such as penicillin may cause an immediate (within 30 minutes) and severe allergic reaction **Figure 18-2**. However, reactions to oral medications, such as oral penicillin, may be slower in onset (more than 30 minutes) but equally severe. The fact that a person has taken a medication once without experiencing an allergic reaction is no guarantee that he or she will not have an allergic reaction to it the next time around. In fact, just the opposite is true. A person will typically experience an allergic reaction *after* becoming sensitized by the first exposure.
- **Plants.** Individuals who inhale dusts, pollens, or other plant materials to which they are sensitive may experience a rapid and severe allergic reaction.

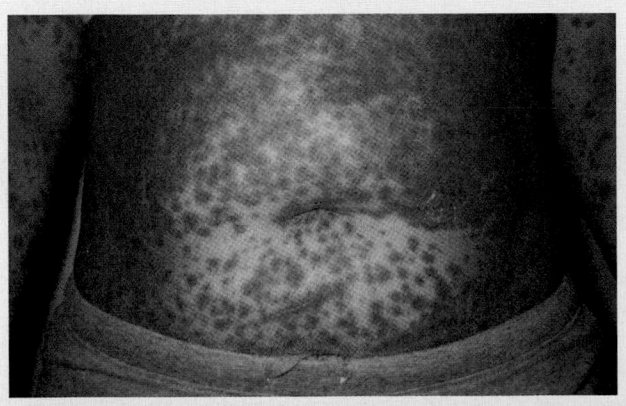

Figure 18-2 A severe allergic reaction to medication.

- **Food.** Eating certain foods, such as shellfish or nuts, may result in a relatively slow (more than 30 minutes) reaction that still can be quite severe. The person may be unaware of the exposure or inciting agent.
- **Chemicals.** Certain chemicals, makeup, soap, latex, and various other substances can cause severe allergic reactions.

Insect Stings

There are more than 100,000 species of bees, wasps, and hornets. Deaths from anaphylactic reactions to stinging insects far outnumber deaths from snakebites. The stinging organ of most bees, wasps, yellow jackets, and hornets is a small hollow spine projecting from the abdomen. Venom can be injected through this spine directly into the skin. The stinger of the honeybee is barbed, so the bee cannot withdraw it **Figure 18-3A**. Therefore, the bee leaves a part of its abdomen embedded with the stinger and dies shortly after flying away. Wasps and hornets have no such handicap; they can sting repeatedly **Figure 18-3B**. Because these insects usually fly away after stinging, it is often impossible to identify which species was responsible for the injury.

Some ants, especially the fire ant (*Formicoidea*) **Figure 18-4A**, also strike repeatedly, often injecting a particularly irritating **toxin**, or poison, at the bite sites. It is not uncommon for a patient to sustain multiple ant bites, usually on the feet and legs, within a very short time **Figure 18-4B**.

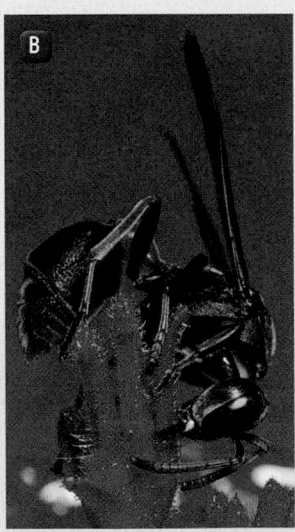

Figure 18-3 Most stinging insects inject venom through a small, hollow spine that projects from the abdomen. **A.** The stinger of the honeybee is barbed and cannot be withdrawn once the bee has stung someone. **B.** The wasp's stinger is unbarbed, meaning that it can inflict multiple stings.

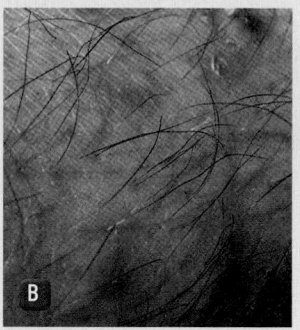

Figure 18-4 **A.** The fire ant. **B.** Fire ants inject an irritating toxin at multiple sites. Bites are generally found on the feet and the legs and appear as multiple small, raised pustules.

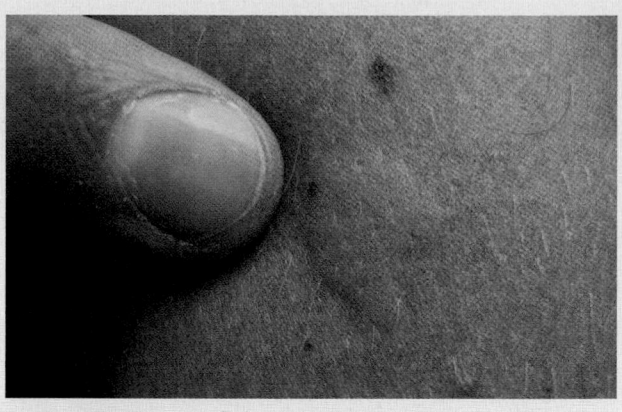

Figure 18-5 A wheal is a whitish, firm elevation of the skin that occurs after an insect sting or bite.

Signs and symptoms of insect stings and bites include sudden pain, swelling, localized heat, and redness in light-skinned individuals, usually at the site of injury. There may be itching and sometimes a **wheal**, which is a raised, swollen, well-defined area on the skin Figure 18-5 . There is no specific treatment for these injuries, although applying ice sometimes makes them less irritating. The swelling associated with an insect bite may be dramatic and sometimes frightening to patients. However, these local manifestations are usually not serious.

Because the stinger of the honeybee remains in the wound, it can continue to inject venom for up to 20 minutes after the bee has flown away. In caring for a patient who has been stung by a honeybee, you should gently attempt to remove the stinger and attached muscle by scraping the skin with the edge of a sharp, stiff object such as a credit card Figure 18-6 . Generally, you should not use tweezers or forceps because squeezing may cause the stinger to inject still more venom into the wound. Gently wash the area with soap and water or a mild antiseptic. Try to remove any jewelry from the area before swelling begins. Position the injection site slightly below the level of the heart, and apply ice or cold packs to the area, but not directly on the skin, to help relieve pain and slow the absorption of the toxin. Be alert for vomiting or any signs of shock or allergic reaction, and do not give the patient anything by mouth. Place the patient in the shock position, and give oxygen if needed. Monitor the patient's vital signs, and be prepared to provide further support as needed.

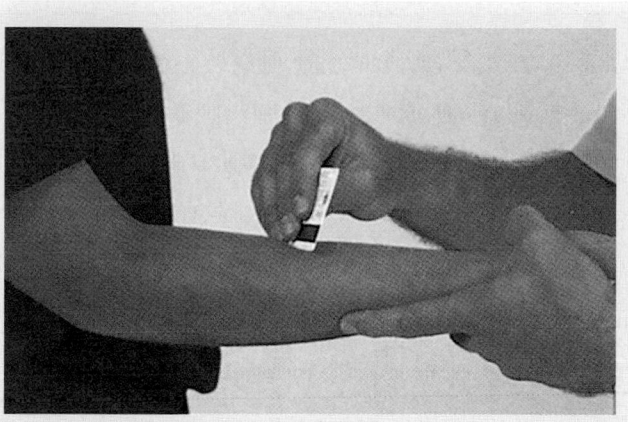

Figure 18-6 To remove the stinger of a honeybee, gently scrape the skin with the edge of a sharp, stiff object such as a credit card.

Anaphylactic Reaction to Stings

Approximately 5% of all people are allergic to the venom of the bee, hornet, yellow jacket, or wasp. This type of allergy, which accounts for about 200 deaths per year, can cause very severe reactions, including anaphylaxis. Patients may experience generalized itching and burning, widespread urticaria, wheals, swelling about the lips and tongue, bronchospasm and wheezing, chest tightness and coughing, dyspnea, anxiety, abdominal cramps, and hypotension Figure 18-7 . Occasionally, respiratory failure occurs.

If untreated, such an anaphylactic reaction can proceed rapidly to death. In fact, more than two thirds of patients who die of anaphylaxis do so within the first half hour, so speed on your part is essential.

■ **Patient Assessment of an Immunologic Emergency**

▶ **Scene Size-up**

Scene Safety

First and foremost, ensure the scene is safe. Assess the impact of hazards on patient care, including environmental hazards, and address them. Also assess the situation

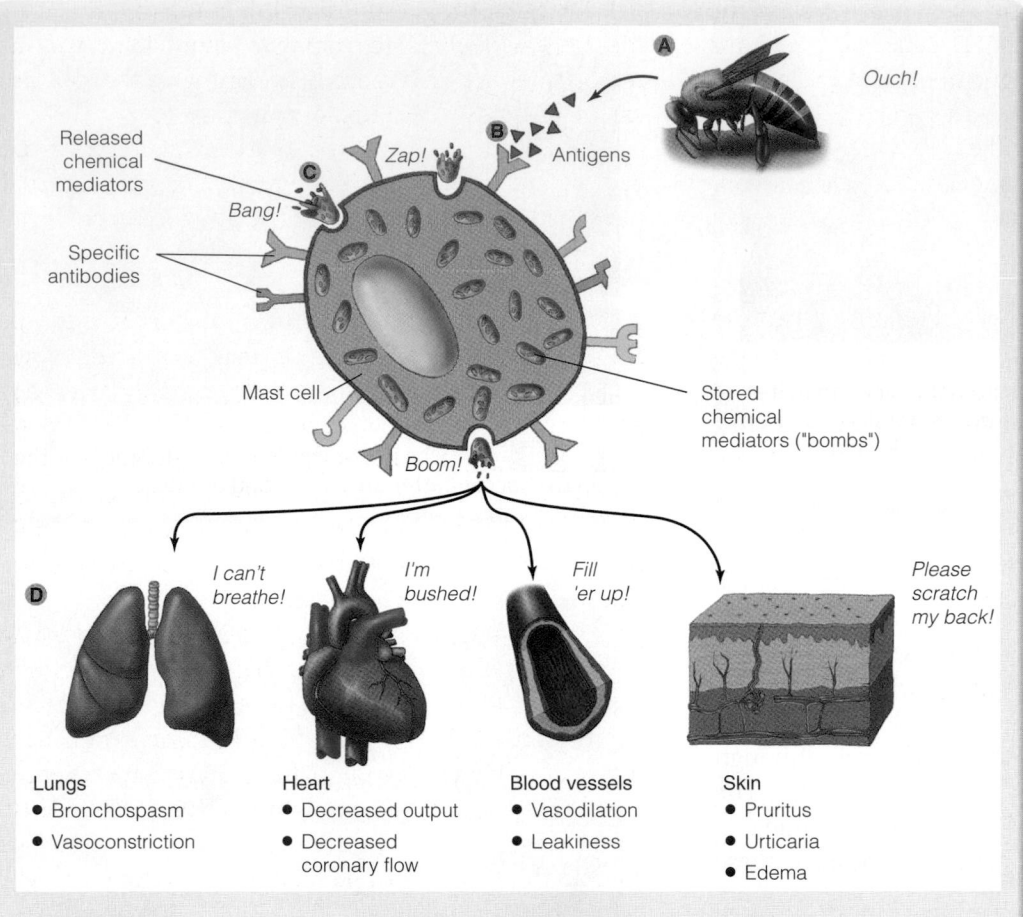

Figure 18-7 The sequence of events in anaphylaxis. **A.** The antigen is introduced into the body. **B.** The antigen-antibody reaction at the surface of a mast cell. **C.** Release of mast cell chemical mediators. **D.** Chemical mediators exert their effects on end organs.

Labels within figure:

- Released chemical mediators
- Specific antibodies
- Mast cell
- *Zap!*
- *Bang!*
- B *Antigens*
- *Ouch!*
- Stored chemical mediators ("bombs")
- *Boom!*
- *I can't breathe!*
- *I'm bushed!*
- *Fill 'er up!*
- *Please scratch my back!*

Lungs
- Bronchospasm
- Vasoconstriction

Heart
- Decreased output
- Decreased coronary flow

Blood vessels
- Vasodilation
- Leakiness

Skin
- Pruritus
- Urticaria
- Edema

Special Populations

When you encounter a geriatric patient experiencing anaphylaxis, obtaining a complete and accurate medical history is imperative. Because of the potential side effects of epinephrine, such as increased pulse rate, increased myocardial oxygen demand, and increased workload of the heart, you must weigh the risk versus benefit in epinephrine administration. If the patient has a history of cardiac problems, such as a previous heart attack, or coronary artery disease, the administration of epinephrine is relatively contraindicated. This means that you could potentially cause harm to the patient if he or she receives epinephrine. In situations such as these, if available, online medical control should be contacted for guidance.

for potential violence. The patient's environment or the activity he or she was performing may indicate the source of the reaction, such as a sting or bite from an insect, a food allergy at a restaurant, or a new medication regimen. A respiratory problem reported by dispatch may be an allergic reaction. If many people are affected, however, it could be an inhaled poison or terrorist event. Never enter a scene where more than one person is experiencing the same symptoms with similar onset. Follow standard precautions with a minimum of gloves and eye protection. As you proceed to the patient, observe for safety threats to yourself and your partner and determine the number of patients at the scene. Consider the need for additional or specialized resources. Call for additional resources earlier rather than later.

Mechanism of Injury/Nature of Illness

Although your report from dispatch may be for a patient with an allergic reaction, keep open the possibility that trauma may have occurred because of a medical incident. Determine mechanism of injury and/or nature of illness. Look for bee stingers or contact with chemicals and other indications of a reaction.

Do not let your guard down, even on what appears to be a routine call.

Primary Assessment

When a patient presents with an allergic reaction, you should perform a rapid scan of the patient to identify and treat any immediate or potential life threats. By paying careful attention to the patient's ABCs, you will be able to maintain breathing and circulation until you are able to transport the patient to the emergency department.

Form a General Impression

Allergic reactions may present as respiratory distress or as cardiovascular distress in the form of shock. Patients experiencing a severe allergic reaction will often be very anxious and feel like they are going to die. If your first impression finds the person anxious and in distress, call for advanced life support (ALS) back-up if available. Some patients who are known to be severely allergic to bee stings, certain medications, or other substances wear a medical identification tag. You will need to assess the patient's level of consciousness. If conscious, patients will provide this information and identification as you ask about their chief complaint. Some patients may have even begun self-treatment with their own medications. If they are unresponsive or have a decreased level of consciousness, immediately evaluate and treat their airway, breathing, and circulation.

Airway and Breathing

The most severe form of allergic reactions, anaphylaxis, can cause rapid swelling of the upper airway. You may have only a few minutes to assess the airway and provide lifesaving measures; however, not all allergic reactions are anaphylactic reactions. Work quickly to assess the patient to determine the severity of the symptoms. Position the conscious patient in a tripod position leaning forward. This will help to facilitate air entry into the lungs and may help the patient to relax. Quickly listen to the lungs on each side of the chest. If wheezing or a silent chest is heard, the lower airways are also closing, preventing oxygen from entering the circulatory system. Do not hesitate to initiate high-flow oxygen therapy. You may have to assist with ventilations for a patient in severe respiratory distress with a severe allergic reaction. This can be done in a semiresponsive or an unresponsive patient. The positive-pressure ventilations you provide will force air through the swelling in the throat and into the lungs while you are waiting for more definitive treatment. In severe situations such as these, the definitive care needed is an injection of epinephrine.

If necessary, be prepared to use standard airway procedures and positive-pressure ventilation according to the principles identified in Chapter 9, *Airway Management.*

Circulation

Although respiratory complaints are most common, some patients in anaphylaxis may not present with severe respiratory symptoms but primarily with signs and symptoms of circulatory distress, such as hypotension. Palpating a radial pulse will help you to identify how the circulatory system is responding to the reaction. If the patient is unresponsive and without a pulse, begin basic life support measures or use an automatic external defibrillator if necessary. Assess for a rapid pulse rate; pale, cool, cyanotic or red, moist skin; and delayed capillary refill times that indicate hypoperfusion. Your initial treatment for shock should include oxygen, positioning in a shock position, and maintaining normal body temperature. The definitive treatment for anaphylactic shock is epinephrine. Trauma is unlikely with allergic reactions, but if trauma has occurred, bandage all bleeding sites, and take spinal precautions when appropriate.

You are the Provider: PART 2

You perform a primary assessment of the patient. He tells you that he is allergic to hornets and has an epinephrine auto-injector (EpiPen); however, it is in his car on the other side of the park. He also tells you that his entire body itches.

Recording Time: 0 Minutes	
Appearance	Anxious; widespread hives
Level of consciousness	Conscious and alert, but anxious
Airway	Open, clear of obstructions or foreign bodies
Breathing	Normal rate; unlabored
Circulation	Radial pulse, normal rate and strong; skin, flushed and covered with hives

Your partner attempts to apply oxygen via a nonrebreathing mask; however, the patient will not accept it because he feels claustrophobic when using it. He will, however, tolerate administration of oxygen via a nasal cannula.

2. Is this patient experiencing a local or a systemic allergic reaction?

3. On what area(s) of the patient's body should you focus your secondary assessment?

Transport Decision

Always provide prompt transport for any patient who may be having an allergic reaction. Take with you all medications and auto-injectors the patient has at the time. Make your transport decision based on findings in the primary assessment. If the patient has signs of respiratory distress or shock, treat those conditions and transport. If the patient is calm and has no signs of respiratory distress or shock after contact with a substance that causes an allergic reaction, continue with the assessment.

History Taking

Investigate Chief Complaint

Investigate the patient's chief complaint or history of present illness. Identify any associated signs and symptoms such as wheezing or a rash **Table 18-1**. Identify the pertinent negatives, such as lack of nausea or vomiting, or no chest pain.

SAMPLE History

If the patient is responsive, begin with obtaining the SAMPLE history and asking the following questions specific to an allergic reaction:

- Have any interventions already been completed?
- Do you have any prescribed, preloaded medications for allergic reactions? Patients who have had severe allergic reactions in the past may carry an epinephrine auto-injector or antihistamines, such as chlorpheniramine or diphenhydramine (Benadryl). Susceptible patients may also carry bronchodilator inhalers, such as albuterol, or other allergy medications.
- Do you have any respiratory symptoms? These are the most troubling because a patient's condition can rapidly deteriorate from respiratory distress to arrest.
- Do you have other symptoms such as itching, rash, hives, pallor, or bite or sting marks, or have you experienced any confusion?
- Have you had previous allergic reactions, asthma, or hospitalizations?
- What did you last eat? This may help you determine the cause of the reaction. For example, peanuts, chocolate, and shellfish can be potent allergens.
- What were you doing or what were you exposed to before the onset of symptoms? This information may be key to effective treatment.

Secondary Assessment

Physical Examinations

The secondary assessment may help direct treatment. As in all emergencies, your assessment of the patient experiencing an allergic reaction should include a systematic head-to-toe or focused assessment to determine hidden trauma or other unrelated medical problems.

Perform evaluations of the respiratory system. Thoroughly assess breathing, including increased work of breathing, use of accessory muscles, head bobbing, tripod positioning, nostril flaring, and grunting. Carefully auscultate both the trachea and the chest.

Table 18-1 Common Signs and Symptoms of Allergic Reaction

Respiratory System	Cardiovascular System	Skin	Other Findings
■ **Shortness of breath (dyspnea)** ■ Sneezing or an itchy, runny nose (initially) ■ Tightness in the chest or throat ■ Irritating, persistent dry cough ■ Hoarseness ■ Respirations that become rapid, labored, or noisy ■ Wheezing and/or stridor (progressing to a silent chest with anaphylaxsis)	■ **Decrease in blood pressure as the blood vessels dilate (hypotension)** ■ **Increase in pulse rate (initially) (tachycardia)** ■ Pale skin as the vascular system fails ■ Loss of consciousness and coma	■ Flushing, itching, or burning skin, especially common over the face and upper part of the chest ■ Urticaria over large areas of the body, may be internal or external ■ Swelling, especially of the face, neck, hands, feet, and/or tongue, local or generalized ■ Swelling and cyanosis or pallor around the lips ■ Warm, tingling feeling in the face, mouth, chest, feet, and hands	■ Anxiety; a sense of impending doom ■ Abdominal cramps ■ Headache ■ Itchy, watery eyes ■ Dizziness ■ Decreasing mental status

* Key indicators of anaphylaxis are indicated by bold type.

Wheezing occurs because of narrowing of the air passages, which is mainly the result of contraction of muscles around the bronchioles in reaction to the allergen, and mobilization of mucus in an attempt to "push" out the allergen. Exhalation, normally the passive, relaxed part of breathing, becomes harder as the patient tries to cough up the secretions or move air past the constricted airways. The fluid in the air passages and the constricted bronchi together produce the wheezing sound. Breathing rapidly becomes more difficult, and the patient may even stop breathing. Prolonged respiratory difficulty can cause a rapid heartbeat (tachycardia), shock, and even death. **Stridor**, a harsh, high-pitched inspiratory sound, occurs when swelling in the upper airway (near the vocal cords and throat) closes off the airway and can eventually lead to total obstruction.

Assess the circulatory system. Remember, the presence of hypoperfusion (shock) or respiratory distress indicates that the patient is having a severe enough allergic reaction that it can lead to death.

Carefully assess the skin for swelling, rash, hives, or signs of the source of the reaction: bite, sting, or contact marks. A rapidly spreading rash can be concerning because it may indicate a systemic reaction. Red, hot skin may also indicate a systemic reaction as the blood vessels lose their ability to constrict and blood moves to the extremities. If this reaction continues, the body will have difficulty supplying blood and oxygen to the vital organs, and one of the first signs will be altered mental status as the organs are deprived of oxygen and glucose.

Vital Signs

Vital signs help determine whether the body is compensating for stress. Assess baseline vital signs, including pulse, respirations, blood pressure, skin, pupils, and oxygen saturation. Rapid, labored breathing indicates airway obstruction. Rapid respiratory and pulse rates may indicate respiratory distress or systemic shock. Fast pulses and hypotension are ominous signs, indicating systemic vascular collapse and shock. Skin signs may be an unreliable indicator of hypoperfusion because of rashes and swelling.

Monitoring Devices

In a patient experiencing an allergic reaction, pulse oximetry is a useful method that you can use to assess the patient's perfusion status. By using pulse oximetry, you can determine the percentage of oxygen saturation in the bloodstream, which will assist in identifying the degree of respiratory distress. However, it is important to remember that pulse oximetry is just another tool in your toolbox. The decision to apply oxygen to a patient experiencing an allergic reaction should be based on a careful assessment of the patient's airway and breathing, not solely on the pulse oximetry readings.

You are the Provider: PART 3

As your partner goes to retrieve the patient's EpiPen, you assess his vital signs. You note that his respirations have increased and are becoming labored, and he is tachycardic. He is still conscious, but is now confused, and he tells you that it feels like he has a lump in his throat.

Recording Time: 5 Minutes	
Respirations	28 breaths/min; labored
Pulse	120 beats/min; weak at the radial artery
Skin	Pale and cool; widespread hives
Blood pressure	88/60 mm Hg
Oxygen saturation (Sao$_2$)	88% (on oxygen)

Further assessment reveals expiratory wheezing during auscultation of his lungs and swelling to his face and lips. You change the nasal cannula to high-flow oxygen via a nonrebreathing mask as your partner returns with the patient's EpiPen.

4. What is the pathophysiology of the patient's acute deterioration?
5. What are the therapeutic effects of epinephrine when given for anaphylaxis?

Reassessment

Repeat the primary assessment. Reassess the patient's vital signs. The patient experiencing a suspected allergic reaction should be monitored with vigilance because deterioration of the patient's condition can be rapid and fatal. Special attention should be given to any signs of airway compromise, including increasing work of breathing, stridor, and wheezing. The patient's anxiety level should be monitored because increased anxiety is a good indication that the reaction may be progressing. Also, watch the skin for signs of shock, including pallor and diaphoresis, as well as for flushing because of vascular collapse. Serial vital signs are important indicators when evaluating your patient's status. Any increase in the respiratory or pulse rate or decrease in blood pressure should be noted. Reassess the chief complaint.

Interventions

To treat allergic reactions, you must first identify how much distress the patient is in. Some allergic reactions will produce severe signs and symptoms in a matter of minutes and threaten the patient's life. Other allergic reactions have a slower onset and cause less severe distress. Epinephrine and ventilatory support are required for severe reactions. Milder reactions, without respiratory or cardiovascular distress, may only require supportive care, such as oxygen. In either situation, the patient should be transported to a medical facility for further evaluation.

Recheck your interventions. If you administered epinephrine, what was the effect? Is the patient's condition improving? Do you need to consider a second dose? You may need to give more than one injection of epinephrine if you note that the patient has decreasing mental status, increased breathing difficulty, or a decreasing blood pressure. Be sure to consult medical control first. Identify and treat changes in the patient's condition.

Communication and Documentation

When to contact medical control depends on your assessment findings and the urgency of care required. In some allergic reactions, you may use standing orders to administer epinephrine before ever calling medical control. At other times, the reaction may be less severe and you may question whether the patient needs an injection of epinephrine. Medical control will be most helpful in the latter situation. Follow your local protocols, which may guide you in providing lifesaving care without needing to contact medical control.

Your documentation not only should include the signs and symptoms found during your assessment, but also should clearly show why you chose to provide the care you did. If anyone should question your care, your documentation will show the reasoning for what you did. Be complete in your documentation, including not only assessment findings and treatment, but also the patient's response to your treatment.

Words of Wisdom

While one EMT is getting oxygen ready, the other should be assisting the patient into a comfortable position, generally in a high-Fowler position, in an effort to maximize ventilations. These measures will help perfusion to the brain while easing respiratory effort.

Emergency Medical Care of Immunologic Emergencies

If the patient appears to be having a severe allergic (or anaphylactic) reaction, you should administer basic life support at once and provide prompt transport to the hospital, reassessing vital signs every 5 minutes if the patient's condition is unstable or every 15 minutes if the patient's condition is stable. Place the patient in a position of comfort. Consider the Trendelenburg's position for hypotensive patients and patients showing signs and symptoms of shock. You may want to request ALS back-up if you work in a tiered response system. In addition to providing oxygen, you should be prepared to maintain a patent airway or administer cardiopulmonary resuscitation. If the allergic reaction was caused by a sting and the stinger is still present, attempt to scrape it away using a hard object, such as a credit card. Do not use an object such as tweezers to remove the stinger because this will force more toxins into the wound. Placing ice over the injury site has been thought to slow absorption of the toxin and diminish swelling, but ice packs placed directly on the skin may freeze the skin and cause more damage. Like any other attempt to reduce swelling with ice, you should be careful not to overdo the icing. In some areas, you may be allowed to administer epinephrine or assist the patient with epinephrine administration.

Epinephrine

The body normally produces epinephrine. **Epinephrine** is a sympathomimetic. This means it mimics the sympathetic (flight or fight) response. Epinephrine has various properties that cause the blood vessels to constrict, which reverses vasodilation and hypotension; this, in turn, elevates the diastolic pressure and improves coronary blood flow. Other properties of epinephrine increase cardiac contractility and relieve brochospasm in the lungs. Because epinephrine has immediate action, it can rapidly reverse the effects of anaphylaxis. The indications for administering epinephrine include a severe allergic reaction and hypersensitivity to an exposed substance.

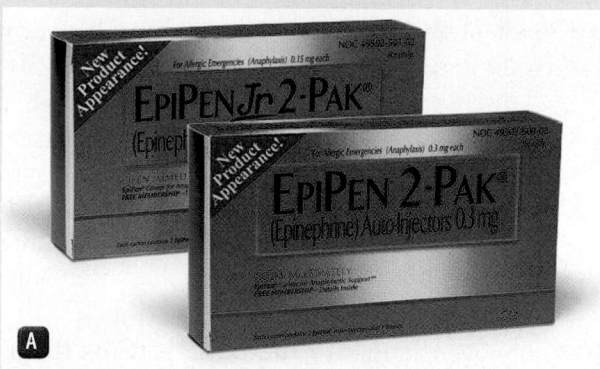
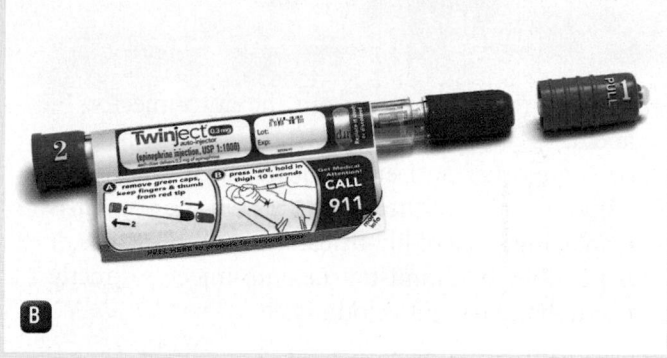

Figure 18-8 Patients who experience severe allergic reactions often carry their own epinephrine, which comes predosed in an auto-injector or a standard syringe. **A.** EpiPen auto-injectors. **B.** Twinject auto-injector.

All allergic emergency kits should contain a prepared, auto-injectable syringe of epinephrine, ready for intramuscular injection, along with instructions for its use **Figure 18-8**.

If the patient is known to have an allergy, he or she may carry their own kit. If the patient is able to use the auto-injector on his or her own, your role is limited to assisting. Your EMS service may or may not allow you to help patients self-administer epinephrine. In some places, the medical director may authorize you to carry an epinephrine auto-injector (EpiPen or Twinject).

The adult EpiPen system delivers 0.3 mg of epinephrine via an automatic needle and syringe system; the infant-child system delivers 0.15 mg.

The Twinject auto-injector contains two doses of epinephrine and it is also available in two strengths, 0.15 mg of epinephrine for those weighing 33 to 66 lb, and 0.3 mg of epinephrine for those who weigh 66 lb or more.

To use, or help the patient use, the auto-injector, you should first receive a direct order from medical control or follow local protocols or standing orders. Follow standard precautions, and make sure the medication has been prescribed specifically for that patient. If it has not or if it has expired or is discolored, do not give the medication, inform medical control, and provide immediate transport.

Once you have done these things, follow the steps in **Skill Drill 18-1** to use an EpiPen auto-injector.

You are the Provider: PART 4

Your protocols allow you to administer epinephrine without contacting medical control first. After confirming that the drug is prescribed to the patient and that it is not expired, you administer it in the lateral aspect of his thigh. After loading him into the ambulance, you reassess him and began transport to the hospital.

Recording Time: 10 Minutes	
Level of consciousness	Conscious and alert
Respirations	22 breaths/min; less labored
Pulse	124 beats/min; stronger at the radial artery
Skin	Pink, warm, and dry; hives are still present
Blood pressure	104/66 mm Hg
Sao$_2$	95% (on oxygen)

6. In addition to the patient's vital signs, what else should you reassess?

7. Is additional treatment indicated for the patient at this time? If so, what?

1. Remove the safety cap from the auto-injector, and, if possible, wipe the patient's thigh with alcohol or some other antiseptic. However, do not delay administration of the drug [Step 1]. If the patient is displaying signs of life-threatening anaphylaxsis, it is possible to administer the auto-injector directly through the patient's clothing.

2. Place the tip of the auto-injector against the lateral part of the patient's thigh, midway between the waist and the knee [Step 2].

3. Push the injector firmly against the thigh until the injector activates. Hold steady pressure to prevent kickback from the spring in the syringe, and prevent the needle from being pushed out of the injection site too soon. Hold the injector in place until the medication has been injected (10 seconds) [Step 3].

4. Remove the injector from the patient's thigh and dispose of it in the proper biohazard container.

5. Record the time and dose of the injection on your run sheet.

Skill Drill   18-1

Using an EpiPen Auto-injector

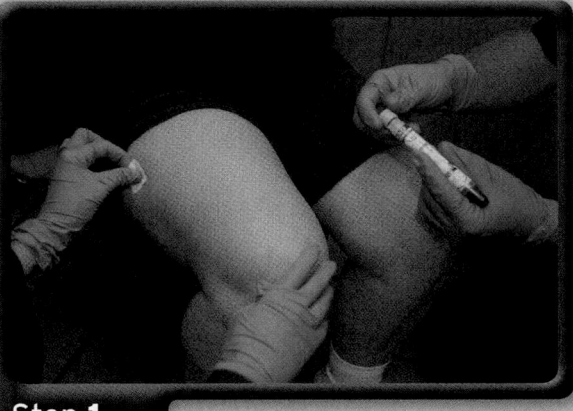

Step 1 Remove the auto-injector's safety cap, and quickly wipe the thigh with antiseptic.

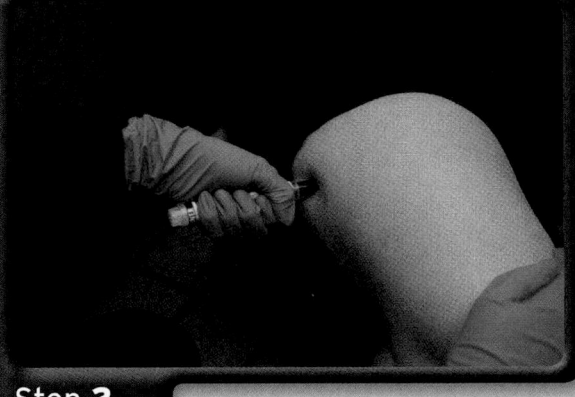

Step 2 Place the tip of the auto-injector against the lateral part of the thigh.

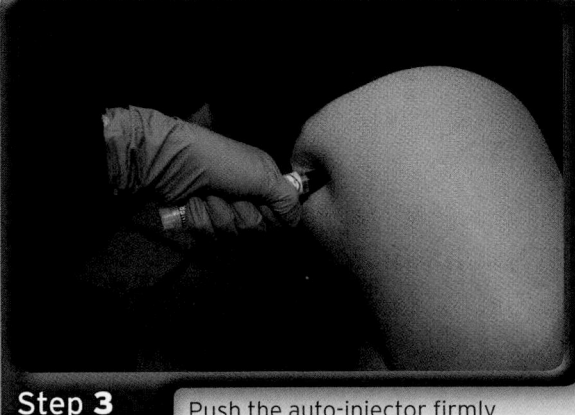

Step 3 Push the auto-injector firmly against the thigh, and hold it in place until all the medication has been injected.

Words of Wisdom

Allergic reactions and responses to bites and stings can progress quickly to life threats. With good care, severe signs and symptoms may subside just as quickly. Performing a multisystem examination and documenting your findings is important before and after treatment. Give particular attention to skin signs and respiratory, circulatory, and mental functioning.

6. Reassess and record the patient's vital signs after using the auto-injector.

7. If the patient's signs and symptoms do not improve after 5 minutes and the patient has another auto-injector, consider assisting the patient with the administration of a second (and final) dose of epinephrine.

   If you administer the Twinject, make the same general preparations you make for the EpiPen: Get an order from medical control, follow standard precautions,

Words of Wisdom

If your medical director and protocols allow it and your patient has an inhaler and an epinephrine auto-injector, one EMT can help administer the inhaler while the other administers the epinephrine.

and ensure that the medication belongs to this patient, is not discolored, and the expiration date has not passed. Follow the steps in **Skill Drill 18-2** to administer epinephrine from a Twinject.

Skill Drill 18-2

1. Remove the auto-injector from the container (Step 1).
2. Clean the administration site with an alcohol preparation. Pull off green cap "1" to expose a round red tip. Do not cover the rounded tip with your hand. Pull off green cap "2" (Step 2).
3. Place the round red tip against the lateral part of the thigh. The injection can be administered outside of clothing if necessary. Once the needle has entered the skin, press hard for 10 seconds (Step 3).
4. Remove the Twinject. Check to see whether the needle is visible. If the needle is *not* visible, the dose was not administered and all the steps should be repeated.
5. If symptoms recur or have not improved within 10 minutes, administer the second dose. Carefully unscrew and remove the red tip. Hold the blue plastic, pulling the syringe out of the barrel

You are the Provider: PART 5

The patient's condition continues to improve throughout transport. He still has hives, but they are scattered and seem to be resolving. You perform a reassessment and then call your radio report to the receiving hospital.

Recording Time: 18 Minutes	
Level of consciousness	Conscious and alert
Respirations	18 breaths/min; unlabored
Pulse	114 beats/min; strong and regular
Skin	Pink, warm, and dry; scattered hives
Blood pressure	128/72 mm Hg
Sao$_2$	97% (on oxygen)

The patient is delivered to the emergency department staff and you give your verbal report to the charge nurse. The attending physician asks you how much epinephrine the patient received in the field.

8. **What is the dose and concentration of epinephrine contained in an adult EpiPen?**
9. **How else might you administer epinephrine for patients experiencing anaphylaxis?**

without touching the needle. Slide the yellow collar off the plunger without pulling on the plunger (Step 4).

6. Insert the needle into the skin on the lateral part of the thigh and push the plunger down (Step 5).

Other kits may contain oral or intramuscular antihistamines, agents that block the effect of histamine. These work relatively slowly, within several minutes to 1 hour. Because epinephrine can have an effect within 1 minute, it is the primary way to save the life of someone having a severe anaphylactic reaction.

Skill Drill 18-2

Using a Twinject Auto-injector

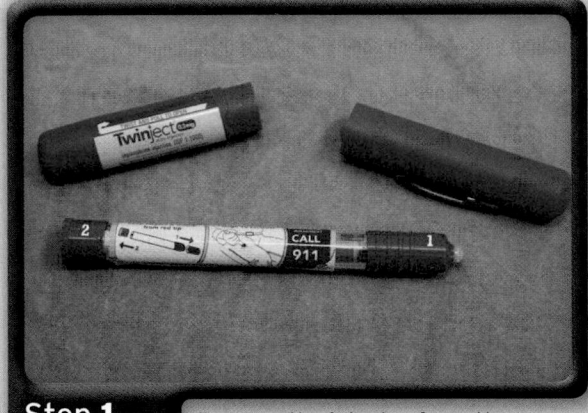

Step 1 Remove the injector from the container.

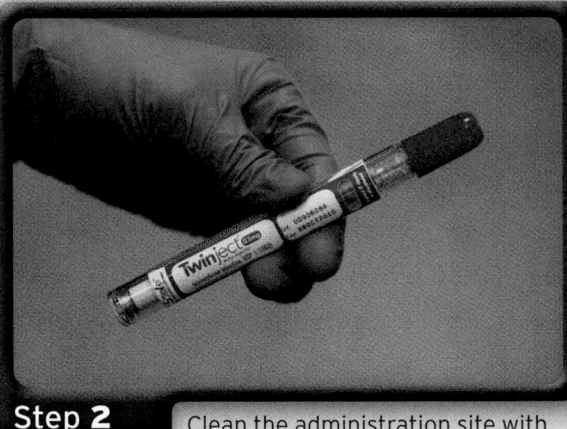

Step 2 Clean the administration site with an alcohol preparation. Pull off green cap "1" to expose a round red tip. Do not cover the rounded tip with your hand. Pull off green cap "2."

Step 3 Place the round red tip against the lateral part of the thigh. The injection can be administered outside of clothing if necessary. Once the needle has entered the skin, press hard for 10 seconds. Remove the Twinject. Check to make sure the needle is visible. If the needle is not visible, repeat the steps.

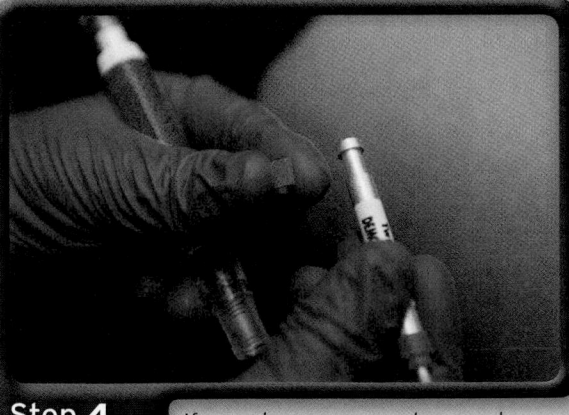

Step 4 If symptoms recur or have not improved within 10 minutes, repeat the dose. Carefully unscrew and remove the red tip. Hold the blue plastic, pulling the syringe out of the barrel without touching the needle. Slide the yellow collar off the plunger without pulling on the plunger.

Because epinephrine constricts blood vessels, it may cause the patient's blood pressure to rise significantly. Other side effects include increased pulse rate, anxiety, cardiac arrhythmias, pallor, dizziness, chest pain, headache, nausea, and vomiting. In a life-threatening situation, the administration of epinephrine outweighs the risk of side effects. Remember that patients who are not wheezing or who have no signs of respiratory compromise or hypotension should not be given epinephrine.

After you have provided emergency care, provide prompt transport while closely monitoring the patient's vital signs. Remember that all patients with suspected anaphylaxis should be given high-flow, high-concentration oxygen.

Skill Drill 18-2

Using a Twinject Auto-injector, continued

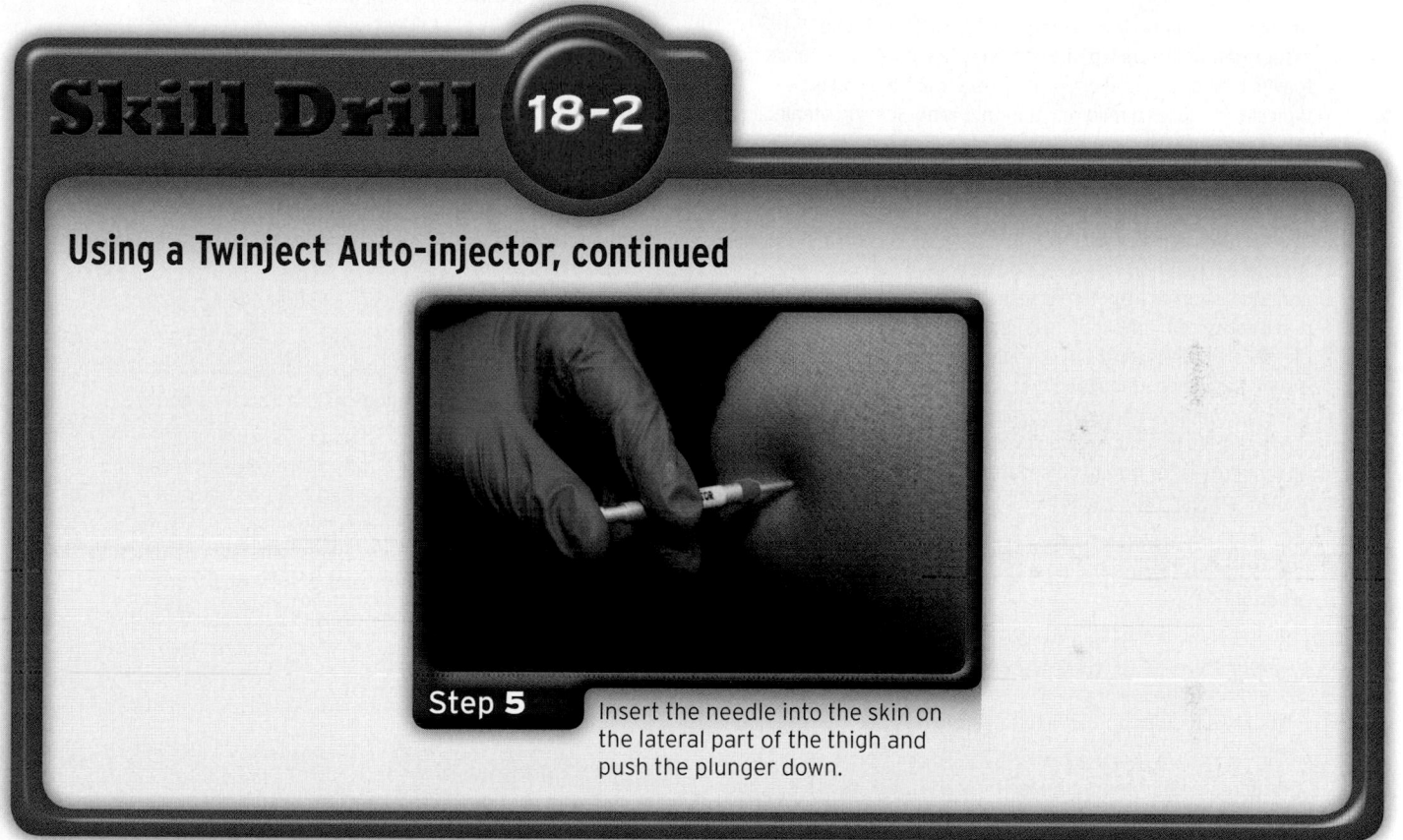

Step 5 Insert the needle into the skin on the lateral part of the thigh and push the plunger down.

You are the Provider: SUMMARY

1. What causes an allergic reaction?

Before an allergic reaction can occur, the body must be exposed to an allergen (or antigen), which triggers the immune system to produce an antibody. Most allergens are harmless substances and do not pose a threat to life. Given the right person and circumstances, almost any substance can cause an allergic reaction. Insect venom, food, latex gloves, medications, plants, and many other substances all can be allergens.

An allergic reaction is an exaggerated immune system response that occurs when a person is reexposed to a substance to which he or she is allergic. Initial exposure to an allergen typically produces a mild reaction that is not life-threatening; this is because the immune system must first recognize the allergen as being foreign and create an antibody against it (sensitization). However, on subsequent exposure to the same allergen, the allergic reactions are typically more severe.

An allergic reaction is not caused directly by an outside stimulus, such as insect venom, medication, or food. Rather, it is a reaction by the body's immune system, which releases chemicals that combat the stimulus. Among these chemicals are histamines and leukotrienes, which are responsible for the signs and symptoms of an allergic reaction. An allergic reaction may be mild and localized, involving itching, burning, or tenderness, or it may be severe and systemic, resulting in shock and respiratory failure.

2. Is this patient experiencing a local or a systemic allergic reaction?

The presence of widespread urticaria (hives) indicates that the patient is experiencing a systemic allergic reaction. Systemic reactions vary in severity, and can range from diffuse (widespread) hives and itching to cardiovascular collapse and death.

A local reaction is characterized by tenderness, redness, itching, and swelling at and immediately adjacent to the bite or sting. In many cases, the reaction is not "allergic" in nature—it is simply irritation and inflammation that is caused by the bite or sting itself.

It is important to perform a careful and thorough assessment of patients who are exposed to something to which they have a confirmed allergy. A seemingly local and mild reaction can become systemic and severe within a matter of minutes.

3. On what area(s) of the patient's body should you focus your secondary assessment?

Further assessment of the patient should focus on body systems that are commonly affected by an allergic reaction—the respiratory and circulatory systems and the skin. In most cases, a severe allergic reaction occurs within minutes of exposure; however, it may be delayed for up to an hour in some patients.

Your primary assessment has revealed no immediate threats to your patient's airway, breathing, or circulation; however, the presence of a widespread rash indicates a systemic reaction and warrants a more thorough assessment. As you continue to assess the patient, look for clinical signs that indicate a worsening reaction and be prepared to assist ventilations and treat for shock.

Signs of respiratory system involvement include respirations that become rapid, labored, or noisy; wheezing; stridor; an irritating, persistent dry cough; hoarseness; and tightness in the chest or throat.

Signs of circulatory system involvement include tachycardia (initially), followed by pallor, dizziness, and hypotension. A decreasing level of consciousness indicates a decrease in cerebral blood flow; this is usually secondary to vascular dilation and hypotension.

The patient already has widespread hives and his skin is flushed. However, you should further assess the skin by looking for swelling—especially of the face, tongue, neck, hands, and feet. If the patient reports a warm, tingling feeling in the face, mouth, chest, feet, and hands, this should also be cause for concern.

Other signs and symptoms of an allergic reaction include anxiety; abdominal cramps; headache; and itchy, watery eyes.

4. What is the pathophysiology of the patient's acute deterioration?

As previously discussed, a mild allergic reaction can deteriorate into a severe allergic reaction within a matter of minutes; this is exactly what is happening to your patient. He is experiencing anaphylaxis—an extreme allergic reaction that involves multiple organ systems and causes shock.

The term "anaphylaxis" means without protection; the immune system response is severely exaggerated to the point where it is causing the patient more harm than good. Anaphylactic shock can rapidly result in death if not immediately treated.

In contrast to a mild systemic allergic reaction, in which case the immune system releases limited amounts of histamines and leukotrienes into the bloodstream, anaphylaxis occurs when the immune system releases massive amounts of these chemicals. Histamines are primarily responsible for the negative effects on the respiratory and circulatory systems.

Swelling of the face, neck, hands, and feet that occurs during anaphylaxis is called angioedema, and is caused by vasodilation and increased capillary permeability. In other words, the capillaries lose their ability to retain fluid, so it leaks out into the subcutaneous (fatty) layer of the skin. Stridor—a harsh, high-pitched inspiratory sound—occurs when swelling near the vocal cords causes narrowing of the upper airway and can eventually lead to complete obstruction.

Wheezing occurs because of narrowing of the bronchioles in the lungs, which is mainly caused by contraction of the muscles around the bronchioles in response to the massive release of histamines. Exhalation, normally the passive, relaxed part of breathing, becomes more difficult as the patient attempts to move air through the constricted bronchioles. As the bronchoconstriction worsens, oxygenation and ventilation are impaired and hypoxemia develops. Clinical signs of hypoxemia, which your patient is now experiencing, include altered mental status, tachycardia, cyanosis, and a low oxygen saturation (Sao_2).

Tachycardia indicates that the body is attempting to compensate for decreased perfusion and hypoxemia by releasing more epinephrine (adrenaline) into the bloodstream to pump more blood to the body's organs, tissues, and cells.

Hypotension occurs because of widespread vasodilation and a decrease in arterial pressure—again, in response to the body's massive release of histamines. As the blood pressure falls, the brain and other vital organs are deprived of oxygen.

5. What are the therapeutic effects of epinephrine when given for anaphylaxis?

Epinephrine (adrenaline)—a hormone that is normally produced by the body—works to rapidly increase the heart rate, dilate the bronchioles in the lungs, and raise the blood pressure by constricting the blood vessels. During anaphylaxis, however, the body may not produce enough epinephrine to enable these actions; therefore, epinephrine is administered to compensate for the body's slow response.

Epinephrine does not stop the allergic reaction itself; it reverses the negative effects of bronchoconstriction and vasodilation, which are caused by the reaction. Therefore, when epinephrine is administered to the patient, it dilates the bronchioles, which improves breathing, and constricts the blood vessels, which increases the blood pressure and improves perfusion.

In patients experiencing a moderate allergic reaction (ie, wheezing, hives, adequate blood pressure and mental status), a single dose of epinephrine is usually sufficient in resolving their symptoms. However, in anaphylaxis (ie, wheezing, hives, angioedema, hypotension, altered mental status), the patient may require more than one dose.

It is also important to be aware of the side effects of epinephrine. These include tachycardia, pallor, dizziness, chest pain, headache, anxiety, and nervousness. The benefits of epinephrine, however, clearly outweigh its side effects when given to patients experiencing a life-threatening allergic reaction.

6. In addition to the patient's vital signs, what else should you reassess?

The patient's mental status and vital signs have clearly improved following the administration of his epinephrine. However, further reassessment is required to determine what, if any, additional treatment is required.

Ask him if he still feels like he has a lump in his throat; this was likely the result of mild upper airway swelling caused by angioedema and *must* be reassessed. Even though he did not present with obvious external angioedema, you should still

reassess his face, lips, tongue, neck, and other parts of his body for swelling.

Auscultate his breath sounds to determine if wheezing is still present. Scattered wheezing may still be heard, even though the patient is not exhibiting any outward signs of respiratory distress.

Reassess his skin to determine if his hives are resolving or if they are still present. In most cases, hives will persist, at least to some degree, following the administration of epinephrine. You will usually notice improvement in the patient's breathing and perfusion status (eg, mental status, blood pressure, peripheral pulse quality) before you see resolution of hives. Remember, epinephrine does not stop the body's release of histamines; it reverses the life-threatening effects that the histamines have on the respiratory and circulatory systems. Therefore, patients are usually given an antihistamine (ie, diphenhydramine [Benadryl]) after epinephrine.

Reassessment of any patient is an integral part of emergency medical care. You must assess the effectiveness of your intervention(s), which will allow you to make decisions regarding additional treatment that may be required.

7. Is additional treatment indicated for the patient at this time? If so, what?

Other than oxygen, further drug therapy is not indicated for the patient at this time. His mental status, blood pressure, breathing effort, and oxygen saturation have all improved. You should reassess his condition at the appropriate intervals, make him comfortable, and continue transport to the hospital.

Some EMTs carry small-volume nebulizers on the ambulance, and are allowed to administer bronchodilator medications, such as albuterol (Proventil), to patients with residual wheezing following treatment for an allergic reaction. Follow your local protocols or contact medical control as needed if you carry bronchodilators and your reassessment reveals that the patient has residual, scattered wheezing.

If the patient's condition deteriorates, consider an intercept with a paramedic unit, depending on your location and transport time to the hospital. Paramedics carry epinephrine and can administer it via the subcutaneous (SC), intramuscular (IM), or intravenous (IV) routes. They can also provide advanced airway management, such as intubation. If a paramedic intercept is not an option, support airway, breathing, and circulation, update the receiving facility, and get the patient to the hospital without delay.

8. What is the dose and concentration of epinephrine contained in an adult EpiPen?

The adult EpiPen contains 0.3 milligrams (mg) of a 1:1,000 concentration for intramuscular (IM) injection. A 1:1,000 concentration contains 1 mg of epinephrine per 1 milliliter (mL). Therefore, 0.3 mL contains 0.3 mg of epinephrine—all of which is injected into the patient's thigh.

You are the Provider: SUMMARY, continued

9. How else might you administer epinephrine for patients experiencing anaphylaxis?

Many patients with known allergies carry epinephrine. The Twinject contains a prefilled syringe of epinephrine; the dose and concentration is 0.3 mg of epinephrine for those weighing 66 lb or more and 0.15 mg for those weighing 33 to 66 lb. Because epinephrine can have an effect within 1 minute,

it is the single most important drug to give to a patient with anaphylaxis.

In some EMS systems across the country, EMTs carry epinephrine on the ambulance and can administer it to patients with allergic reactions and anaphylaxis. Follow your local protocols or contact medical control as needed.

EMS Patient Care Report (PCR)

Date: 10-3-09	**Incident No.:** 011709	**Nature of Call:** Allergic reaction		**Location:** 1444 City Park Drive	
Dispatched: 1310	**En Route:** 1310	**At Scene:** 1310	**Transport:** 1320	**At Hospital:** 1338	**In Service:** 1349

Patient Information

Age: 33 **Sex:** M **Weight (in kg [lb]):** 73 kg (160 lb)	**Allergies:** Hornet stings; no known drug allergies **Medications:** Prescribed EpiPen **Past Medical History:** Allergic reaction to hornet sting **Chief Complaint:** Hives and itching following hornet sting

Vital Signs

Time: 1315	**BP:** 88/60	**Pulse:** 120	**Respirations:** 28	**Sao$_2$:** 88%
Time: 1320	**BP:** 104/66	**Pulse:** 124	**Respirations:** 22	**Sao$_2$:** 95%
Time: 1328	**BP:** 128/72	**Pulse:** 114	**Respirations:** 18	**Sao$_2$:** 97%

EMS Treatment
(circle all that apply)

Oxygen @ 15 **L/min via (circle one):** NC (NRM) Bag-Mask Device	**Assisted Ventilation**	**Airway Adjunct**	**CPR**
Defibrillation	**Bleeding Control**	**Bandaging**	**Splinting**	**Other** Epinephrine 0.3 mg via EpiPen

Narrative

Medic 85 was standing by at a park event when a 33-year-old male presented with generalized hives and itching after being stung on the leg by a hornet. The patient was conscious and alert; his airway was patent and his breathing was adequate. Patient states that he is allergic to hornets and has an EpiPen; however, it was located in his car on the other side of the park. He further states that the same thing happened to him the last time he was stung by a hornet. Partner retrieved patient's EpiPen while further assessment was performed. Patient's breath sounds were clear to auscultation bilaterally; there was no gross evidence of swelling to the face, tongue, neck, hands, or feet. Patient denied chest tightness, difficulty breathing, or any other symptoms. Patient denied any other past medical history. Attempted to give oxygen via nonrebreathing mask; however, the patient stated that he was claustrophobic. He would, however, tolerate oxygen via nasal cannula. As assessment continued, patient's condition began to deteriorate. He remained conscious, but was now confused. He began experiencing respiratory distress and stated that it felt like he had a "lump" in his throat. Reassessment revealed obvious respiratory distress, widespread wheezing to auscultation of his lungs, hypotension, and a falling oxygen saturation. Changed nasal cannula to nonrebreathing mask and set the flow rate at 15 L/min. The patient was unable to self-administer his EpiPen; therefore, it was given by EMS, following standing orders, in the lateral aspect of his right thigh; dose given was 0.3 mg of 1:1,000 concentration. Placed patient onto the stretcher, loaded him into the ambulance, and began transport to the hospital. En route, reassessed the patient and noted that his symptoms began to resolve; his mental status improved and he stated that it was easier to breathe. Blood pressure and oxygen saturation also improved. Hives were still present, although they appeared to be resolving. Continued to monitor his condition throughout transport; he continued to improve and was delivered to the emergency department staff without incident. Gave verbal report to charge nurse and returned to service. *End of report*

Assessment and Emergency Care of Immunologic Emergencies

Scene Size-up

Scene Safety	Ensure scene safety and address hazards. Consider the number of patients, the need for additional help/ALS, and cervical spine stabilization. Standard precautions should include a minimum of gloves and eye protection.
Mechanism of Injury/Nature of Illness	Determine the mechanism of injury (MOI)/nature of illness (NOI).

▼

Primary Assessment

Form a General Impression	Determine level of consciousness, and perform a rapid scan to find and treat any immediate threats to life. Determine priority of care based on the ABCs. If the patient appears anxious or fears death, call for ALS assistance.
Airway and Breathing	Ensure patent airway. Provide high-flow oxygen at 15 L/min. If possible, place in a tripod position and evaluate depth and rate of the respiratory cycle and provide ventilatory support as needed.
Circulation	Evaluate pulse rate and quality; observe skin color, temperature, and condition, and treat accordingly.
Transport Decision	Rapid transport.

▼

History Taking

Investigate Chief Complaint	Investigate the chief complaint (history of present illness). Identify signs and symptoms and pertinent negatives. Ask SAMPLE questions and determine if patient has a prescribed auto-injector(s)/inhaler. Be sure to ask if and what interventions were taken before your arrival, how many interventions were performed, and at what time.

▼

NOTE: The order of the steps in this section differs depending on whether the patient is conscious or unconscious. The following order is for a conscious patient. For an unconscious patient, perform a primary assessment, perform a full-body scan, obtain vital signs, and, if possible, obtain the past medical history before transport.

Assessment and Emergency Care of Immunologic Emergencies, continued

Secondary Assessment

Physical Examinations	Perform a systematic assessment of the patient, focusing on the respiratory drive, adequate ventilation, the adequacy and effectiveness of the circulatory system, and the patient's mental status.
Vital Signs	Take vital signs, noting skin color and temperature as well as patient's level of consciousness. Use pulse oximetry, if available, to assess the patient's perfusion status.

Reassessment

Interventions	Repeat the primary assessment and reassess interventions performed. Reassess vital signs and the chief complaint. Support the patient as needed. Consider the use of oxygen, positive-pressure ventilations, adjuncts, and proper positioning of the patient. Assist with the use of auto-injector(s) or inhaler as defined by local protocols.
Communication and Documentation	Contact medical control with a radio report. In some cases, per standing orders, you may treat the patient with epinephrine. In other cases you will need to contact medical control for direction. Follow local protocols. Be sure to document any changes in patient status and the time. Document the reasoning for your treatment and the patient's response.

NOTE: Although the steps below are widely accepted, be sure to consult and follow your local protocols.

Allergic Reactions

Using an Auto-injector

1. Remove the auto-injector's safety cap, and quickly wipe the thigh with antiseptic.
2. Place the tip of the auto-injector against the lateral part of the thigh.
3. Push the auto-injector firmly against the thigh, and hold it in place until all the medication is injected (about 10 seconds).

Using Twinject

1. Remove the injector from the container.
2. Clean the administration site with an alcohol preparation. Pull off green cap "1" to expose a round red tip. Do not cover the rounded tip with your hand.
3. Pull off green cap "2."
4. Place the round red tip against the lateral part of the thigh. The injection can be administered outside of clothing if necessary. Once the needle has entered the skin, press hard for 10 seconds.
5. Remove the Twinject. Check to see whether the needle is visible. If the needle is *not* visible, the dose was not administered and all the steps should be repeated.
6. If symptoms recur or have not improved within 10 minutes, repeat the dose. Carefully unscrew and remove the red tip. Hold the blue plastic, pulling the syringe out of the barrel without touching the needle. Slide the yellow collar off the plunger without pulling on the plunger.
7. Insert the needle into the skin on the lateral part of the thigh and push the plunger down.

Prep Kit

Ready for Review

- An allergic reaction is a response to chemicals the body releases to combat certain stimuli, called allergens.
- Allergic reactions occur most often in response to five categories of stimuli: insect bites and stings, medications, food, plants, and chemicals.
- The reaction may be mild and local, involving itching, redness, and tenderness, or it may be severe and systemic, including shock and respiratory failure.
- Anaphylaxis is a life-threatening allergic reaction mounted by multiple organ systems, which must be treated with epinephrine.
- Wheezing and skin wheals can be signs of anaphylaxis.

- People who know that they are allergic to bee, hornet, yellow jacket, or wasp venom often carry a bee-sting kit that contains epinephrine in an auto-injector. You may help to administer this medication in this form with authorization from medical control.
- All patients with suspected anaphylaxis require oxygen.
- When assessing a person who may be having an allergic reaction, you should check for flushing, itching, and swelling skin, hives, wheezing and stridor, a persistent cough, a decrease in blood pressure, a weak pulse, dizziness, abdominal cramps, and headache.
- Always provide prompt transport to the hospital for any patient who is having an allergic reaction. Remember that signs and symptoms can rapidly become more severe. Carefully monitor the patient's vital signs en route; be especially alert for airway compromise.

Vital Vocabulary

allergens Substances that cause an allergic reaction.

allergic reaction The body's exaggerated immune response to an internal or surface agent.

anaphylaxis An extreme, life-threatening systemic allergic reaction that may include shock and respiratory failure.

envenomation The act of injecting venom.

epinephrine A substance produced by the body (commonly called adrenaline), and a drug produced by pharmaceutical companies that increases pulse rate and blood pressure; the drug of choice for an anaphylactic reaction.

histamines Substances released by the immune system in allergic reactions that are responsible for many of the symptoms of anaphylaxis, such as vasodilation.

immune response The body's response to a substance perceived by the body as foreign.

immune system The body system that includes all of the structures and processes designed to mount a defense against foreign substances and disease-causing agents.

immunology The study of the body's immune system.

leukotrienes Chemical substances that contribute to anaphylaxis; released by the immune system in allergic reactions.

stridor A harsh, high-pitched respiratory sound, generally heard during inspiration, that is caused by partial blockage or narrowing of the upper airway.

toxin A poison or harmful substance.

urticaria Small spots of generalized itching and/or burning that appear as multiple raised areas on the skin; hives.

wheal A raised, swollen, well-defined area on the skin resulting from an insect bite or allergic reaction.

wheezing A high-pitched, whistling breath sound, usually caused by a constriction of the smaller tubes of the lungs and typically heard on expiration.

Assessment in Action

Rescue 11 responds to 1284 NW 152 Avenue for a report of respiratory distress. On arrival, a visibly upset woman meets you at the door. She informs you that her 10-year-old son and his friends were playing football in the backyard when he felt a sudden pain on his arm and began having difficulty breathing. She brings you and your partner to the back porch where you see your patient leaning forward in a chair with his elbows on his knees; he is struggling to catch his breath. He is unable to answer your questions without stopping to catch his breath. Your partner immediately applies a nonrebreathing mask as you complete the primary assessment. The child's respiratory rate is 26 breaths/min and shallow, and his pulse is 130 beats/min and regular. While your partner conducts a systematic full-body scan, you obtain vital signs and obtain a SAMPLE history from the mother. He has had severe allergic reactions to bee stings in the past and is prescribed an EpiPen Jr. Your partner informs you that he heard wheezes in all lung fields and observed numerous raised red spots on the child's arms and chest.

1. The raised red spots are known as:
   A. blebs.
   B. pustules.
   C. urticaria.
   D. varicosities.

2. The stinger of a honeybee should be removed by:
   A. scraping the skin with the edge of a credit card.
   B. using a tweezer.
   C. applying rubbing alcohol.
   D. shaving the affected area.

3. What is one of the chemicals released into the body during an allergic reaction?
   A. Histamine
   B. Antihistamine
   C. Carbon dioxide
   D. Glucose

4. What is the most definitive treatment for this patient?
   A. Albuterol
   B. Epinephrine
   C. Aspirin
   D. Nitroglycerin

5. A high-pitched, whistling sound caused by broncho-constriction is called:
   A. wheezing.
   B. rhonchi.
   C. stridor.
   D. crackles.

6. Which of the following is a common potent allergen?
   A. Red meat
   B. Milk
   C. Shellfish
   D. Oranges

7. Why is epinephrine the drug of choice in the treatment of a severe allergic reaction?

8. When is the administration of epinephrine relatively contraindicated?

9. What side effects may be observed following the administration of epinephrine?

10. Should you assist the patient with the delivery of his EpiPen Jr.?

National EMS Education Standard Competencies

Medicine
Applies fundamental knowledge to provide basic emergency care and transportation based on assessment findings for an acutely ill patient.

Toxicology
- Recognition and management of
  - Carbon monoxide poisoning (pp 671-673)
  - Nerve agent poisoning (p 685)
- How and when to contact a poison control center (p 671)
- Anatomy, physiology, pathophysiology, assessment, and management of
  - Inhaled poisons (pp 671-673)
  - Ingested poisons (pp 674-675)
  - Injected poisons (pp 675-676)
  - Absorbed poisons (pp 673-674)
  - Alcohol intoxication and withdrawal (pp 680-681)

Knowledge Objectives
1. Define toxicology, poison, and overdose. (p 669)
2. Describe how poisons enter the body. (pp 670-676)
3. Identify the common signs and symptoms of poisoning. (pp 669-670)
4. Describe the assessment and treatment of the patient with suspected poisoning. (pp 676-689)
5. Describe the assessment and treatment of the patient with a possible overdose. (pp 669-670, 676-689)
6. Explain the use of activated charcoal, including indications, contraindications, and the need to obtain approval from medical control before administering it. (pp 674-675, 678-679)
7. Identify the main types of specific poisons and their effects, including alcohol, opioids, sedative-hypnotic drugs, inhalants, sympathomimetics, marijuana, hallucinogens, anticholinergic agents, and cholinergic agents. (pp 679-686)
8. Describe the assessment and treatment for the patient with suspected food poisoning. (pp 686-687)
9. Describe the assessment and treatment for the patient with suspected plant poisoning. (pp 687-689)
10. Understand the role of airway management in the patient suffering from poisoning or overdose. (p 677)
11. Discuss the use of activated charcoal. (pp 674-675, 678-679)

Skills Objectives
1. Demonstrate the steps in the assessment and treatment of the patient with suspected poisoning. (pp 676-679)
2. Demonstrate the steps in the assessment and treatment of the patient with suspected overdose. (pp 678-679)
3. Demonstrate the steps required to administer activated charcoal. (pp 674-675, 678-679)

Introduction

Every day, each of us comes into contact with things that are potentially poisonous. This is not surprising when you consider that almost any substance may be a poison in certain circumstances. Different doses can turn even a remedy into a poison. Consider a common substance such as aspirin. When taken in recommended doses, it is a safe and effective analgesic. Too much aspirin, however, can result in death.

Acute poisoning affects some 5 million children and adults each year. Chronic poisoning, often caused by abuse of medications and other substances, including tobacco and alcohol, is much more common. Fortunately, deaths caused by poisoning are fairly rare. Rates of death as the result of poisoning in children have decreased steadily since the 1960s, when safety caps were introduced for drug bottles and containers. Deaths caused by poisoning in adults, though, have been rising, the majority the result of drug abuse.

In this chapter, the term "poisoned" includes acute and chronic poisonings. As an EMT, you must recognize that patients with either type of problem may have a variety of injuries. Although you cannot stop a chronic substance abuse problem in a patient, you may be able to prevent death caused by the acute effects of a poison.

This chapter discusses how to identify a patient who has been poisoned and how to gather clues about the poison. Also described are the different ways in which a poison is introduced into the body. The chapter then discusses the signs, symptoms, and treatment of specific poisons, including sedatives and **opioids** (medicines with actions similar to morphine). Food poisoning and plant poisoning are also discussed.

Identifying the Patient and the Poison

Toxicology is the study of toxic or poisonous substances. A **poison** is any substance whose chemical action can damage body structures or impair body function. A poison can be introduced into the body through a variety of means. Poisons act by changing the normal metabolism of cells or by actually destroying them. Poisons may act acutely, as in an overdose of heroin, or chronically, as in years of alcohol or other substance abuse. **Substance abuse** is the misuse of any substance to produce a desired effect (for example, cocaine intoxication). A common complication of substance abuse is **overdose**, when a patient takes a toxic dose of a drug.

Your primary responsibility to the patient who has been poisoned is to recognize that a poisoning has occurred. Keep in mind that very small amounts of some poisons can cause considerable damage or death. If you have even the slightest suspicion that a patient has taken a poisonous substance, you should notify medical control and begin emergency treatment at once. Discussion of issues relating to suicide is covered in Chapter 20, *Psychiatric Emergencies*.

Symptoms and signs of poisoning or overdose vary according to the specific agent, as shown in **Table 19-1**. Some poisons cause the pulse to speed up, whereas others cause it to slow down; some poisons cause the pupils to dilate, while others cause the pupils to constrict. If respiration is depressed or difficult, cyanosis may occur. Some chemical compounds will irritate or burn the skin or mucous membranes, resulting in burning or blistering. The presence of such injuries at the patient's mouth strongly suggests the **ingestion** (swallowing) of a poison,

You are the Provider: PART 1

At 2:20 AM, you are dispatched to apartment 9B on 213 West Maple for an overdose. After briefly staging two blocks away from the scene, law enforcement arrives and advises you that the scene is secure and that the patient, a young man, is unconscious. A fine mist is falling, the temperature is 44°F, and the traffic is light.

1. In addition to providing immediate lifesaving treatment, what else should your actions consist of when you arrive at this scene?

2. What is a toxidrome? How can knowledge of various toxidromes improve the care you provide to a patient?

Table 19-1 Toxidromes: Typical Signs and Symptoms of Specific Overdoses

Agent	Signs and Symptoms
Opioid (Examples: heroin, oxycodone)	<ul><li>Hypoventilation or respiratory arrest</li><li>Pinpoint pupils</li><li>Sedation or coma</li><li>Hypotension</li></ul>
Sympathomimetics (Examples: epinephrine, albuterol, cocaine, methamphetamine)	<ul><li>Hypertension</li><li>Tachycardia</li><li>Dilated pupils</li><li>Agitation or seizures</li><li>Hyperthermia</li></ul>
Sedative-hypnotics (Examples: diazepam [Valium], secobarbital [Seconal], flunitrazepam [Rohypnol])	<ul><li>Slurred speech</li><li>Sedation or coma</li><li>Hypoventilation</li><li>Hypotension</li></ul>
Anticholinergics (Examples: atropine, Jimson weed)	<ul><li>Tachycardia</li><li>Hyperthermia</li><li>Hypertension</li><li>Dilated pupils</li><li>Dry skin and mucous membranes</li><li>Sedation, agitation, seizures, coma, or delirium</li><li>Decreased bowel sounds</li></ul>
Cholinergics (Examples: pilocarpine, nerve gas)	<ul><li>Excess defecation or urination</li><li>Muscle fasciculations</li><li>Pinpoint pupils</li><li>Excess lacrimation (tearing) or salivation</li><li>Airway compromise</li><li>Nausea or vomiting</li></ul>

such as lye. If possible, consider asking the patient the following questions:

- What substance did you take?
- When did you take it (or become exposed to it)?
- How much did you ingest?
- What actions have been taken?
- How much do you weigh?

Try to determine the nature of the poison. Look around the immediate area because objects at the scene may provide clues: an overturned bottle, a needle or syringe, scattered pills, chemicals, even an overturned or damaged plant. The remains of any nearby food or drink may also be important. Place any suspicious material in a plastic bag and take it with you to the hospital, along with any containers you find.

Containers at the scene can provide critical information. In addition to the name and concentration of the drug, a pill bottle label may list specific ingredients, the number of pills that were originally in the bottle, the name of the manufacturer, and the dose that was prescribed.

This information can help emergency department physicians to determine how much has been ingested and what specific treatment may be required. For certain food poisonings, a food container that lists the name and location of the maker or the vendor may be of equal importance in saving the life of the patient and possibly other people.

If the patient vomits, examine the contents for pill fragments. Ensure that you are wearing proper personal protective equipment for this activity. Note and document anything unusual that you see. You should try to collect the material, called **vomitus**, in a separate plastic bag so that it can be analyzed at the hospital.

How Poisons Get Into the Body

Emergency care for a patient who has been poisoned may include actions on your part that range from reassuring an anxious parent to instituting cardiopulmonary resuscitation (CPR). Most often, you will not be administering a specific

Words of Wisdom

Poison Centers

Several hundred poison centers are located in the United States. The phone number of your local poison center is typically found on the inside cover of your local phone book. The telephone number for the Poison Help hotline is 1 (800) 222-1222. Staff persons at every center have access to information about virtually all of the commonly used medications, chemicals, and substances that could possibly be poisonous. They know the appropriate emergency treatment for each, including the **antidote**, if there is one. An antidote is a substance that will counteract the effects of a particular poison.

If you believe that a patient has been poisoned, you should immediately provide the poison center with all relevant information: when the poisoning occurred; evidence found at the scene; a description of the suspected poison, including the amount involved; and the patient's size, weight, and age. If necessary, medical control can contact the regional poison center for you and relay specific instructions back to you.

A medical toxicologist is a physician who specializes in caring for patients who have been poisoned. About 100 of these specialists work in special hospitals called medical toxicology treatment centers, located throughout the United States. At times, your medical control may divert a patient who meets certain poisoning criteria to one of these centers instead of to the closest hospital.

You and your medical control center should know the telephone number of your regional poison center and have it available in the event you encounter an unexpected case of poisoning.

antidote because most poisons do not have one. Therefore, in general, the most important treatment you can perform for a poisoning is diluting and/or physically removing the poisonous agent. How you do this depends on how the poison gets into the patient's body in the first place. Essentially, the four avenues to consider are as follows:

- Inhalation Figure 19-1A
- Absorption (surface contact) Figure 19-1B
- Ingestion Figure 19-1C
- Injection Figure 19-1D

Of the four avenues of poisoning, injection often can be the most worrisome one in terms of treatment. In the case of a patient who has inhaled a poison, you can administer oxygen, and you can give activated charcoal to a patient who has ingested a poison. In the case of a patient who has contacted a poison, you can flood the skin with water and wash out the patient's eyes. However, when a poison has been injected, it is difficult to remove or dilute the poison, a fact that makes these cases especially urgent. Conversely, all routes of poisoning can be deadly, and each should be thought of as being equally serious.

Always consult medical control before you proceed with the treatment of any poisoning victim.

■ Inhaled Poisons

Patients who have inhaled poison, including natural gas, sewer gas, certain pesticides, carbon monoxide,

You are the Provider: PART 2

You enter the apartment and find the patient, a 22-year-old man, lying supine on his couch. He appears to be unconscious, has secretions draining from his mouth, and is making a snoring sound. A police officer advises you that she found several empty pill bottles on the kitchen table. As your partner opens the jump kit, you perform a primary assessment.

Recording Time: 0 Minutes	
Appearance	Supine on the couch; motionless
Level of consciousness	Unconscious and unresponsive
Airway	Oral secretions; snoring respirations
Breathing	Slow rate; shallow depth; snoring sound
Circulation	Radial pulses, slow and weak; skin is cool, pale, and dry; no gross bleeding

3. What should your immediate treatment be for this patient?

4. On the basis of the patient's initial presentation, what type of drug should you suspect that he overdosed on?

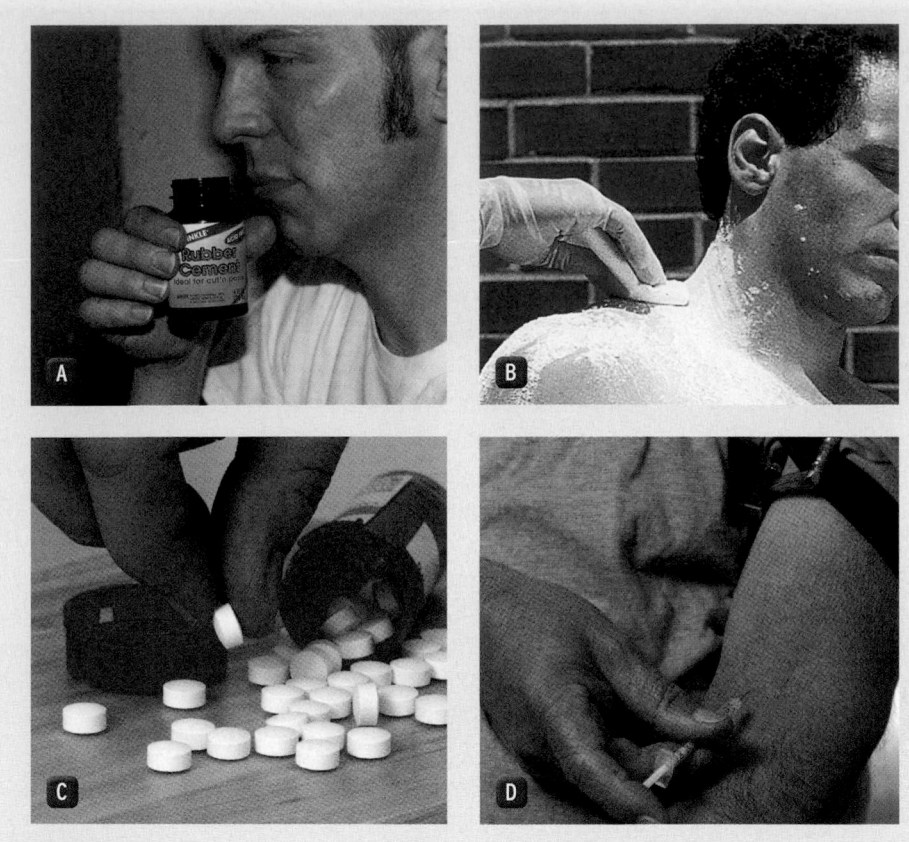

Figure 19-1 There are four routes by which a poison can enter the body. **A.** Inhalation. **B.** Absorption (surface contact). **C.** Ingestion. **D.** Injection.

following signs and symptoms: burning eyes, sore throat, cough, chest pain, hoarseness, wheezing, respiratory distress, dizziness, confusion, headache, or stridor in severe cases. The patient may also have seizures or an altered mental status. Some inhaled agents cause progressive lung damage, even after the patient has been removed from direct exposure; the damage may not be evident for several hours. Meanwhile, it may take 2 or 3 days or more of intensive care to reestablish normal lung function. For this reason, all patients who have inhaled poison require immediate transport to an emergency department. Be prepared to use supplemental oxygen via a nonrebreathing mask and/or ventilatory support with a bag-mask device, if necessary. Make sure a suctioning unit is available in case the patient vomits.

As with other poisonings, it is helpful to take the containers, bottles, and labels with you when you transport the patient to the hospital. Often patients use inhaled poisons to commit suicide. A common technique is for the patient to sit inside a vehicle with the engine running in an enclosed garage. The exhaust fumes from the vehicle contain high levels of carbon monoxide that will cause the patient to lose consciousness and eventually stop breathing. A recent variation on the use of automobiles for suicide involves people using a tightly sealed vehicle as a type of "gas chamber." These people mix fairly common household chemicals inside the

chlorine, or other gases, should be moved into fresh air immediately. Depending on how long the patient was exposed, he or she may require supplemental oxygen **Figure 19-2** . Always use self-contained breathing apparatus to protect yourself from poisonous fumes. If you are not specifically trained in the use of this apparatus or do not have appropriately fit-tested equipment available, defer exposure to hazardous environments where inhalational toxins are present to appropriately trained and equipped personnel. Patients may need to be decontaminated by specially trained personnel after they are removed from the toxic environment. The patient's clothing should be removed in this process because it may contain trapped gases that can be released, exposing you to the toxin. You cannot administer emergency care until this step has been completed and there is no danger of the poison contaminating you.

Some inhaled poisons, such as carbon monoxide, are odorless and produce severe hypoxia without damaging or even irritating the lungs. Others, such as chlorine, are very irritating to the tissues and cause airway obstruction and pulmonary edema. The patient may have the

Words of Wisdom

Any time there is more than one patient and no evidence of mechanism of injury, be suspicious. Toxic fumes may be odorless and colorless or may seem harmless, such as in the case of sewer gas. If the substance is in the atmosphere, it will affect the rescuers as well as the victims. An EMT who is incapacitated is no good to anyone. Be suspicious of toxic fumes when encountering patients with changes in level of consciousness, especially at an industrial site or enclosed space.

Figure 19-2 Patients who have inhaled poisons may need supplemental oxygen and prompt transport to the emergency department.

vehicle to produce hydrogen sulfide gas, which is quickly fatal. When you approach the vehicle and open the door, you may be overcome by the gas as well. If you suspect this type of scenario has taken place, contact hazardous materials responders and have them remove the victim.

■ Absorbed and Surface Contact Poisons

Poisons that come in contact with the surface of the body can affect the patient in many ways. Many corrosive substances will damage the skin, mucous membranes, or eyes, causing chemical burns, telltale rashes, or lesions. Acids, alkalis, and some petroleum (hydrocarbon) products are very destructive. Other substances are absorbed into the bloodstream through the skin and have systemic effects, just like medications or drugs taken via the oral or injectable routes. Other substances such as poison ivy or poison oak may just cause an itchy rash without being dangerous to the patient's health. It is important, therefore, to distinguish between contact burns and contact absorption.

Words of Wisdom

Absorption of toxic substances through the skin is a common problem in agriculture and manufacturing. Most solvents and "cides"—such as insecticides, herbicides, and pesticides—are toxic and can be readily absorbed through the skin.

Signs and symptoms of absorbed poisoning include a history of exposure, liquid or powder on a patient's skin, burns, itching, irritation, redness of the skin in light-skinned individuals, or typical odors of the substance.

Emergency treatment for a typical contact poisoning includes the following two steps:

1. Avoid contaminating yourself or others.
2. While protecting yourself from exposure, remove the irritating or corrosive substance from the patient as rapidly as possible.

Remove all clothing that has been contaminated with poisons or irritating substances, thoroughly brush off any dry chemicals, flush the skin with running water, and then wash the skin with soap and water. When a large amount of material has been spilled on a patient, flooding the affected part for at least 20 minutes may be the fastest and most effective treatment. If the patient has a chemical agent in the eyes, you should irrigate them quickly and thoroughly. To avoid contaminating the other eye as you irrigate the affected eye, make sure that the fluid runs from the bridge of the nose outward Figure 19-3 . This action should be started initially on the scene and continued during transport.

Many chemical burns occur in industrial settings, where showers and specific protocols for handling surface burns are available. If you are called to such a scene, trained people usually will be there to assist you. Do not spend time trying to neutralize substances on the skin with additional chemicals. This action may actually be more harmful. Instead, wash the substance off immediately with plenty of water. Obtain material safety data sheets from industrial sites and transport them with the patient, if available.

Figure 19-3 If chemical agents are in the patient's eyes, irrigate the eyes quickly and thoroughly, ensuring that the irrigation fluid runs from the bridge of the nose outward. (Use of a nasal cannula is pictured.)

Words of Wisdom

During the scene size-up, if you find that a suspected toxic or hazardous material is present, call for specialized resources such as the Hazardous Materials Team. Never approach a contaminated patient unless you have specialized hazardous materials training and are using the appropriate specialized personal protective equipment. Also, during a hazardous materials incident, it is imperative that all contaminated patients be thoroughly decontaminated prior to transport to the hospital. Failure to do so will result in the risk of contamination for the entire emergency department and staff. Remember that everyone at the scene who is exposed to the hazardous material must be thoroughly decontaminated before leaving the scene. Chapter 38, *Incident Management*, discusses hazardous materials and decontamination in detail.

The only time you should not irrigate the contact area with water is when a patient has been contaminated with a poison that reacts violently with water, such as phosphorus or elemental sodium. These substances ignite when they come into contact with water. Instead, brush the chemical off the patient, remove contaminated clothing, and apply a dry dressing to the burn area. Be sure to wear appropriate protective gloves and the proper protective clothing.

Provide prompt transport to the emergency department for definitive care. En route, continue irrigation and provide oxygen if possible.

■ Ingested Poisons

Approximately 80% of all poisoning is by mouth (ingestion). Ingested poisons include liquids, household cleaners, contaminated food, plants, and, in most cases, drugs. Ingested poisoning is usually accidental in children and, except for contaminated food, deliberate in adults. Plant poisonings are common among children, who like to explore and often bite the leaves of various bushes or shrubs.

The signs and symptoms of ingested poisons vary greatly with the type of poison, the age of the patient, and the time that has passed since the ingestion. Small children may respond by crying if the poison is an acid or alkaline, and these types of poisons often cause burns around the mouth. Gastrointestinal pain may be present in some cases, and patients may vomit before or after your arrival. If the patient has an altered mental status, it is urgent that you protect the patient from aspirating if he or she vomits. Other signs and symptoms depend on the substance involved; for example, some poisons may cause cardiac arrhythmias whereas others may cause seizures. It is important to treat these signs and symptoms and notify the poison center and medical control of the patient's condition.

Your goal as an EMT is to rapidly remove as much of the poison as possible from the gastrointestinal tract. For most poisoned patients, this emergency treatment is sufficient until further care can be provided at the emergency department.

Words of Wisdom

Be aware that some chemicals react with water. Although small amounts can usually be flushed safely with large quantities of water, larger amounts of such chemicals can give off toxic fumes or explode when wet. Be sure to check the relevant warnings and placards.

In the past, syrup of ipecac was used to induce vomiting, but today it is recommended in only a few situations in which the risk of losing consciousness is clearly low. Because syrup of ipecac induces vomiting, people who have ingested substances that may cause diminished alertness over time might vomit and inhale the vomitus into the lungs as they lose consciousness. As a result, syrup of ipecac is usually not carried on ambulances. Today, many EMS systems allow you to carry activated charcoal on the unit. Activated charcoal comes as a suspension that binds to the poison in the stomach and carries it out of the system. Therefore, it is more effective and safer than syrup of ipecac. Because activated charcoal is an inky, messy fluid, you may have to do some coaxing to get the patient to drink it; try to give it in a covered cup with a straw **Figure 19-4**. Remember, you should never force this (or any other) liquid into a patient's mouth.

Although every poison will result in a specific set of symptoms and signs, you should always immediately assess the airway, breathing, and circulation of every patient who has been poisoned. Many patients have died

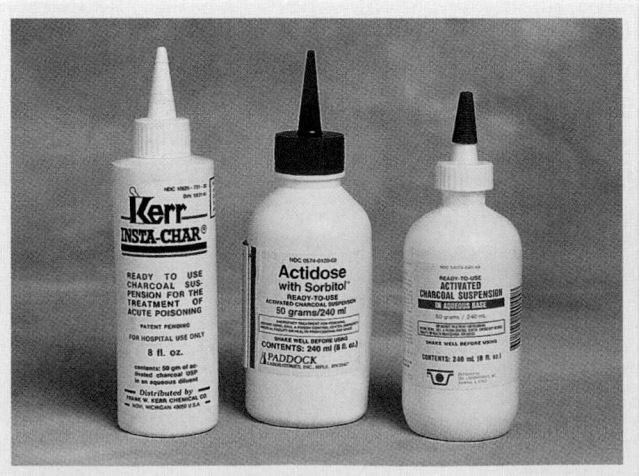

Figure 19-4 Activated charcoal comes as a premixed suspension that you should give, if local protocol allows, in a covered cup with a straw.

as a result of problems with the ABCs that might have been managed easily. Be prepared to provide aggressive ventilatory support and CPR to a patient who has ingested an opiate, a sedative, or a barbiturate, each of which can cause depression of the central nervous system (CNS) and slow breathing. Whenever poisoning is involved, you should provide prompt transport to the emergency department. The patient may need intravenous (IV) support and other treatments that can be given only in the hospital. If you work in a tiered system, ALS backup also may be appropriate because these providers often carry and can administer additional medications and therapies.

> ### Words of Wisdom
>
> While one EMT explains the activated charcoal treatment to the patient, the other EMT can prepare a large plastic garbage bag to hang on the patient as a bib. This will help contain the charcoal solution if the patient vomits.

■ Injected Poisons

Poisoning by injection is usually the result of drug abuse, such as heroin or cocaine **Figure 19-5** . Contrary to the thinking of television detectives, the only other parties who are likely to have injected a patient with poison are insects and animals.

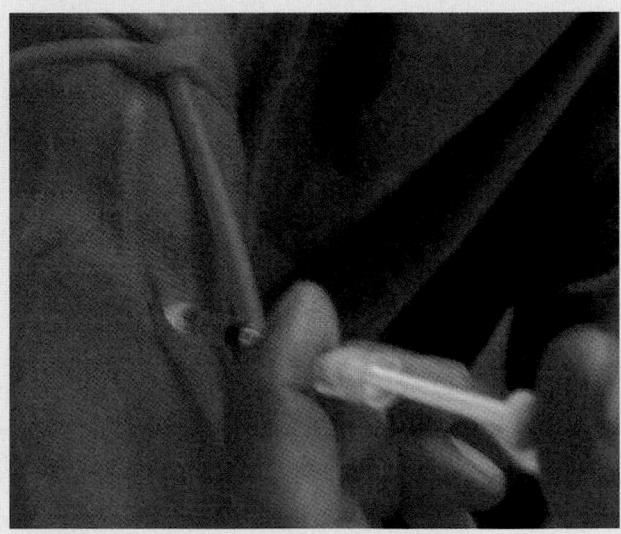

Figure 19-5 Injected poisons are impossible to dilute or remove from the body in the field; therefore, prompt transport to the emergency department is critical.

You are the Provider: PART 3

A paramedic unit was a short distance away and arrives to provide assistance. As your partner supports the patient's breathing, you perform a secondary assessment. The only abnormal finding is the patient's pupils, which are bilaterally constricted (pinpoint). While one of the paramedics establishes an intravenous (IV) line, you assess the patient's vital signs.

Recording Time: 6 Minutes	
Respirations	6 breaths/min (baseline); ventilations are being assisted
Pulse	40 beats/min; weak and regular
Skin	Cool, pale, and dry
Blood pressure	76/50 mm Hg
Oxygen saturation (Sao$_2$)	91% (on oxygen)

There are three empty medication bottles, which are labeled oxycodone, 30 mg; Tylenol, 500 mg; and ibuprofen, 200 mg. The oxycodone is prescribed to someone else, was filled the day before, and originally contained 60 tablets. After the IV line is established, the paramedic prepares to administer naloxone (Narcan) to the patient. The patient's blood glucose level is assessed, and is noted to read 112 mg/dL.

5. Would activated charcoal benefit this patient? Why or why not?
6. What is naloxone? Why is it being given to this patient?

Words of Wisdom

Take time at the scene to make thorough notes about the nature of the poisoning. Having compiled and documented this information, you can then quickly use it to state the type and amount of substance and the time and route of exposure in your radio, verbal, and written reports. Clear notes that can be handed over on arrival at the emergency department will also be appreciated by busy hospital staff.

Signs and symptoms of poisoning by injection can have a multitude of presentations, including weakness, dizziness, fever, chills, and unresponsiveness, or the patient may be easily excited.

In general, injected poisons are impossible to dilute or remove because they are usually absorbed quickly into the body or cause intense local tissue destruction. If you suspect that rapid absorption has occurred, monitor the patient's airway, provide high-flow oxygen, and be alert for nausea and vomiting. Remove rings, watches, and bracelets from areas around the injection site if swelling occurs. Prompt transport to the emergency department is essential. Take all containers, bottles, and labels with the patient to the hospital.

Patient Assessment

Scene Size-up

Scene Safety

When you have a situation that involves a toxicologic emergency, a well-trained dispatcher is of great value. Dispatchers with an appropriate set of protocols and excellent interrogation skills can obtain important information pertaining to a poisoning call that will help you anticipate the proper protection needed to ensure your safety. The dispatcher may be able to obtain information pertaining to the number of patients involved, whether additional resources are needed, and whether trauma is involved. If this information is not obtained before your arrival, you must take the time to assess the scene thoroughly to ensure your safety and to determine the nature of the illness and any mechanism of injury, the number of patients involved, the need for additional resources, and whether spine stabilization is required.

Because of the risk of possible cross-contamination by poisons that can be inhaled, absorbed, ingested, and injected, you must take appropriate standard precautions.

Use the appropriate personal protective equipment necessary to avoid being contaminated.

Mechanism of Injury/Nature of Illness

Most calls that involve poisoning will include information provided by the dispatcher to indicate the nature of the illness. Other calls may require some detective work on your part to determine if a poisoning has occurred.

As you approach the scene, you should look for clues that might indicate the substance and/or poison involved. Ask yourself the following questions:

- Are there medication bottles lying around the patient and the scene? If so, is there medication missing that might indicate an overdose?
- Are there alcoholic beverage containers present?
- Are there syringes or other drug paraphernalia on the scene?
- Is there an unpleasant or odd odor in the room? If so, is the scene safe? This could be a clue to an inhaled poison too.
- Is there a suspicious odor and/or drug paraphernalia present that may indicate the presence of a drug laboratory? Drug laboratories can be very volatile, so ensure scene safety **Figure 19-6**.

The location of the patient may help contribute to identifying a possible poisoning, and other clues such as empty pill bottles or open bottles of cleaners near the patient may provide further information to help you determine what happened. Keep a constant observant eye on the surroundings, and keep an open mind when questioning the patient or bystanders to avoid coming to mistaken conclusions.

Figure 19-6 A laboratory capable of producing large quantities of methamphetamine.

Primary Assessment

Form a General Impression

To best determine the severity of the patient's condition, first obtain a general impression of the patient, assess his or her level of consciousness, and determine any life threats. With substance abuse and poisonings, do not be fooled into thinking that a conscious, alert, and oriented patient is in stable condition and has no apparent life threats. The patient may have a harmful or even lethal amount of poison in his or her system that has not had time to produce systemic reactions. A primary assessment that reveals a patient with signs of distress and/or altered mental status gives you early confirmation that the poisonous substance is causing systemic reactions.

Airway and Breathing

Quickly ensure that the patient has an open airway and adequate ventilation. Do not hesitate to begin oxygen therapy for the patient. If the patient is unresponsive to painful stimuli, you need to consider inserting an airway adjunct to ensure an open airway. Have suction available; these patients are susceptible to vomiting. You may also have to assist a patient's ventilations with a bag-mask device because some substances act as depressants on the body's systems. As you assess and manage the patient's airway and breathing, you must consider the potential for spinal injury. Spinal precautions in an unresponsive patient must begin when the airway is first opened and be continued when positive-pressure ventilations are needed.

Circulation

Once the airway and breathing have been assessed and appropriate interventions performed, assess the patient's circulatory status. You will find variations in a patient's circulatory status depending on the substance involved. Assess the pulse and skin condition. Some poisons are stimulants, and others are depressants. Some poisons will cause vasoconstriction and others, vasodilation. Although bleeding may not be obvious, alterations in consciousness may have contributed to trauma and bleeding.

Transport Decision

Patients with obvious alterations in the ABCs or patients you have determined have a poor general impression should be considered for immediate transport. A delay on the scene to further assess and treat patients is rarely indicated. Some industrial settings may have specific decontamination stations and antidotes available at the site. The majority of the time, decontamination and antidote administration will have been initiated by the industrial response team before your arrival and should not delay rapid transport. Consider decontamination of the patient before transport depending on the poison your patient was exposed to. This action is necessary if a patient continues to off-gas or the treating crew has the potential to become exposed in the confined space of the ambulance during transit. Decontamination is especially important when transporting exposed patients in a helicopter.

History Taking

Investigate Chief Complaint

After the life threats have been managed during the primary assessment, investigate the chief complaint or history of present illness. You should obtain the patient's medical history. In many situations, this can be performed in the ambulance en route to the hospital. If your patient is responsive and can answer questions, begin with an evaluation of the exposure and the SAMPLE history. If the patient is not responsive, attempt to obtain the history from other sources, such as friends or family members. Medical identification jewelry and cards in wallets may also provide information about the patient's medical history.

SAMPLE History

In these situations, the SAMPLE history guides you in what to focus on as you continue to assess the patient's complaints, the physical examination helps to explain what is happening outside the patient's body, and the vital signs tell you what is happening inside the body. These three assessments are important in that they give you direction in the interventions your patient might need.

In addition to the SAMPLE history, you should ask the following questions:

- What is the substance involved? If you know the substance involved, you will be better able to access the appropriate resource, such as the poison center, to determine lethal doses, time before harmful effects begin, effects of the substance at toxic levels, and appropriate interventions.
- When did the patient ingest or become exposed to the substance? This will let you know if and when the harmful effects will begin. This will also let the emergency physician know what harmful effects can be reversed and which ones cannot because of the length of time the patient has been exposed to the substance.
- How much did the patient ingest or what was the level of exposure? With this information, the poison center will be able to inform you whether the patient has had a harmful or lethal dose.
- Over what period did the patient take the substance? Did the patient take the substance all at once or over minutes or hours?

- Has the patient or a bystander performed any intervention on the patient? Has the intervention helped? The patient's or bystander's intervention may cause more complications. The emergency physician will also need to know this information to be able to adjust interventions accordingly.
- How much does the patient weigh? If activated charcoal is indicated, you will need to determine the dose based on the patient's weight. The antidote or neutralizing agent given by the emergency physician may be based on the patient's weight as well. For the physical examination, assess the affected body systems, giving particular attention to the respiratory and cardiovascular systems.

Secondary Assessment

The secondary assessment is a more detailed, comprehensive examination of the patient that is used to uncover issues that may have been missed during the primary assessment. In some instances, such as a critically ill patient or a short transport time, you may not have time to conduct a secondary assessment.

Physical Examinations

Your physical examination should focus on the area of the body involved with the poisoning or the route of exposure. For example, if a person has ingested a poison, inspect the mouth for indications of poisoning. Are there burns from caustic chemicals? Are there plant or pill fragments? If the person's skin came in contact with a poison, is there a rash or burns? How large an area is involved? If a respiratory exposure occurred, auscultate the lungs. Is there good air movement in and out of the lungs? Do you hear any wheezing or crackles? Much of what you should focus on in your physical examination is based on the route of exposure and the particular drug or chemical the patient was exposed to. Take the time to become knowledgeable about the effects of general classes of drugs and chemicals so that you will become familiar with specific and common poisons.

Management of the ABCs during the primary assessment is the priority assessment and treatment goal. These interventions take precedence over a thorough physical examination. However, once the ABCs have been addressed and managed, conducting a thorough physical examination will often provide additional information on the exposure the patient experienced. A general review of all body systems may help to identify systemic problems. This review should be performed, at a minimum, on patients with extensive chemical burns or other significant trauma and on patients who are unresponsive.

Vital Signs

A complete set of baseline vital signs is an important tool for you to use to determine how your patient is

doing. Many poisons have no outward indications of the seriousness of the exposure. Alterations in the level of consciousness, pulse, respirations, blood pressure, and skin are more sensitive indicators that something serious is wrong. Be aware that exposure to carbon monoxide may produce false pulse oximetry readings.

Reassessment

The condition of patients exposed to poisons may change suddenly and without warning. You should continually reassess the adequacy of the patient's ABCs. Repeat the vital signs, and compare them with the baseline set obtained earlier in your assessment. Evaluate the effectiveness of interventions you have provided. If your assessment has provided necessary information about the poisonous substance, you may be able to anticipate changes in the patient's condition. If the patient has consumed a harmful or lethal dose of a poisonous substance, you must repeat the assessment of vital signs every 5 minutes, or constantly if needed. If the patient is in stable condition and there are no life threats, reassess every 15 minutes. If the poison or the level of exposure (eg, the number and type of pills taken) is unknown, careful and frequent reassessment is mandatory.

Interventions

The treatment you provide for poisoned patients depends a great deal on what they were exposed to, how they were exposed, and other signs and symptoms found in your assessment. Supporting the ABCs is your most important task. Some poisons can be easily diluted or decontaminated before transport or en route to the hospital. Dilute airborne exposures with oxygen, remove contact exposures with copious amounts of water unless contraindicated, and consider activated charcoal for ingested poisons. Contact your medical control or a poison center to discuss treatment options for particular poisonings.

Communication and Documentation

Once you have completed your primary assessment, history taking, and secondary assessment, contact medical control to request necessary interventions. Report to the hospital as much information as you have about the poison or chemical that the patient was exposed to. If a material safety data sheet is immediately available in a work setting, take it with you to the hospital. If it is not immediately available, ask the company to fax it to the receiving hospital while you are en route. This will help to identify and quickly make available specific interventions and potential antidotes.

Emergency Medical Care

First, ensure scene safety by following standard precautions and performing external decontamination. Remove tablets or fragments from the patient's mouth, and wash or brush poison from the patient's skin. Treatment focuses on support: assessing and maintaining the patient's ABCs and monitoring the patient's breathing. Provide oxygen to the patient, and perform assisted ventilations if necessary. Keep the patient warm, treat for shock as necessary, and transport promptly to the nearest appropriate hospital.

In some cases, you will give activated charcoal to patients who have ingested poison, if approved by medical control or local protocol. Activated charcoal is not indicated for patients who have ingested an acid, an alkali, or a petroleum product; who have a decreased level of consciousness and cannot protect their airway; or who are unable to swallow.

Remember that activated charcoal adsorbs, or sticks to, many commonly ingested poisons, preventing the **toxin** (poison) from being absorbed into the body by the stomach or intestines. If local protocol permits, you will likely carry plastic bottles of premixed suspension, each containing up to 50 g of activated charcoal. Some common trade names for the suspension form are InstaChar, Actidose, and LiquiChar. The usual dose for an adult or child is 1 g of activated charcoal per kilogram of body weight. The usual adult dose is 25 to 50 g, and the usual pediatric dose is 12.5 to 25 g.

Before you give a patient charcoal, obtain approval from medical control. Next, shake the bottle vigorously to mix the suspension. The medication looks like mud, so it is best to cover the outside of the container so that the fluid is not visible and ask the patient to drink it with a straw. You might need to persuade the patient to drink it, particularly if the patient is a child, but never force it. If the patient takes a long time to drink the mixture, you will have to shake the container frequently to keep the medication mixed. Once the patient has finished, discard the container from which the charcoal was administered. Be sure to record the time when you administered the activated charcoal.

The major side effect of ingesting activated charcoal is black stools. If the patient has ingested a poison that causes nausea, he or she may vomit after taking activated charcoal, and the dose will have to be repeated. As you reassess the patient, be prepared for vomiting, nausea, and possible airway problems.

Specific Poisons

Over time, a person who routinely misuses a substance may need increasing amounts of it to achieve the same result. This is called developing a **tolerance** to the substance. A person with an **addiction** has an overwhelming desire or need to continue using the substance, at whatever cost, with a tendency to increase the dose. This does not happen only with the classic drugs of abuse, such as cocaine. Almost any substance can be abused, including laxatives, nasal decongestants, vitamins, and food.

You are the Provider: PART 4

The paramedic administers 2 mg in 0.5 mg increments slowly of naloxone to the patient via the IV route. After about 30 seconds, the patient starts moving and begins pushing the bag-mask device away from his face. Your partner applies a nonrebreathing mask at 15 L/min, which the patient tolerates. His condition and vital signs are reassessed, and preparations for transport are made.

Recording Time: 11 Minutes	
Level of consciousness	Conscious, but sleepy
Respirations	12 breaths/min; adequate depth
Pulse	64 beats/min; regular and stronger
Skin	Cool; color is improving
Blood pressure	100/52 mm Hg
Sao$_2$	96% (on oxygen)

7. What other issues about this patient should concern you?

8. How would this patient's presentation have differed had he overdosed on a sympathomimetic?

The importance of safety awareness and standard precautions in caring for victims of drug abuse cannot be stressed enough. Known drug abusers have a fairly high incidence of serious and undiagnosed infections, including human immunodeficiency virus and hepatitis. These patients, when intoxicated, may bite, spit, hit, or otherwise injure you, causing you to come into contact with their blood and other body fluids. Always be sure to wear appropriate protective equipment. A calm, professional approach on your part can defuse frightening situations, but keep your safety and that of your team uppermost in mind. Expect the unexpected and remember: The drug user, not the drug, can pose the greatest threat.

■ Alcohol

The most commonly abused drug in the United States is alcohol **Figure 19-7**. It affects people from all walks of life and kills more than 200,000 people each year. More than 40% of all traffic fatalities or injuries, 67% of murders, and 33% of suicides are related to alcohol, which impairs the capacity to think and function rationally. Alcoholism is one of the greatest national health problems, along with heart disease, cancer, and stroke.

Alcohol abuse can result in many long-term effects. The most common effect is liver damage with estimates that up to 90% of heavy drinkers will develop some level of hepatitis and 10% to 20% of alcoholics will develop cirrhosis. Other long-term effects include an increased incidence of pancreatitis, development of erosive gastritis, and an increased risk for breast and colorectal cancer. The long-term abuse of alcohol leads to atrophy

of the cerebrum, possibly resulting in permanently reduced mental function. Although alcohol is often seen as a substance to promote sexual activity, alcohol actually decreases the ability to respond to sexual stimulation and long-term use can lead to impotence and sterility.

Alcohol is a powerful CNS depressant. It is a **sedative**, a substance that decreases activity and excitement, and a **hypnotic**, meaning that it induces sleep. In general, alcohol dulls the sense of awareness, slows reflexes, and reduces reaction time. It may also cause aggressive and inappropriate behavior and lack of coordination. However, a person who appears intoxicated may have other medical problems as well. Look for signs of head trauma, toxic reactions, or uncontrolled diabetes. Severe acute alcohol ingestion may cause hypoglycemia, which may contribute to the symptoms. At the very least, you should assume that all intoxicated patients are experiencing a drug overdose and require a thorough examination by a physician. In most states, such patients cannot legally refuse transport.

Alcohol potentiates many other drugs and is commonly not the only drug taken. Over-the-counter drugs, including antihistamines and diet medications, can cause serious problems when combined with alcohol.

If a patient exhibits signs of serious CNS depression, you must provide respiratory support. This may be difficult, however, because depression of the respiratory system can also cause **emesis**, or vomiting. The vomiting may be very forceful or even bloody (**hematemesis**) because large amounts of alcohol irritate the stomach. Internal bleeding should also be considered if the patient appears to be in shock (hypoperfusion) because blood

Figure 19-7 Alcohol intoxication causes altered mental status, slowed reflexes, and impaired reaction time.

Special Populations

Drug and alcohol abuse among teenagers is one of the most common problems in society today. Most teenagers are encouraged to experiment with drugs through the most powerful coercion tool of the teen years, peer pressure. Many times older teenagers will encourage younger teenagers, who are looking for acceptance, to try combinations of drugs. In other cases, an older teenager who has developed a tolerance, will give too large of a dose of a drug to a first-time user, resulting in overdose. Many teenagers are so motivated by peer pressure, they will do things they know are not safe just to gain acceptance from these peers. Often teenagers will lie about taking drugs or what drugs they have taken out of fear of being arrested. You must reassure them that your intent is only to give them the best treatment possible.

You should also be aware that some teenagers use drugs to attempt suicide. Unfortunately teen suicide attempts are far too common. Do not be judgmental with these teenagers and treat them as you would any other suicidal patient, with empathy and patience.

might not clot effectively in a patient who has a prolonged history of alcohol abuse.

A patient in alcohol withdrawal may experience frightening hallucinations, or **delirium tremens (DTs)**, a syndrome characterized by restlessness, fever, sweating, disorientation, agitation, and even seizures. These conditions may develop if patients no longer have their daily source of alcohol. Alcoholic hallucinations come and go. A patient with an otherwise fairly clear mental state may see fantastic shapes or figures or hear odd voices. Such auditory and visual hallucinations often precede DTs, which are a much more severe complication.

About 1 to 7 days after a person stops drinking or when alcohol consumption levels are decreased suddenly, DTs may develop. Patients may experience one or more of the following signs and symptoms:

- Agitation and restlessness
- Fever
- Sweating
- Tremors
- Confusion and/or disorientation
- Delusions and/or hallucinations
- Seizures

Provide prompt transport after you have completed your assessment and given necessary care. A person who is experiencing hallucinations or DTs is extremely ill. Should seizures develop, treat them as you would any other seizure. The patient should not be restrained, although you must protect him or her from self-injury. Give the patient oxygen, and watch carefully for vomiting; have suction ready. Hypovolemia may develop because of sweating, fluid loss, insufficient fluid intake, or vomiting associated with DTs. If you see signs of hypovolemic shock, elevate the patient's feet slightly, clear the airway, and turn the patient's head to one side to minimize the chance of aspiration during transport. These patients may not respond appropriately to suggestions or conversation; they are often confused and frightened. Therefore, your approach should be calm and relaxed. Reassure the patient, and provide emotional support.

Safety

In situations that involve toxic substances, your safety is paramount. Always be aware of the environment. When dealing with patients who have taken illegal drugs, be cautious and be prepared for unexpected violence.

■ Opioids

The pain relievers called opioid analgesics are named for the opium in poppy seeds, the origin of heroin, codeine, and morphine. On the list of frequently abused

Table 19-2 Common Opioid Drugs
Butorphanol (Stadol)
Codeine
Fentanyl derivatives ("China White")
Heroin
Hydrocodone (Vicodin)
Hydromorphone (Dilaudid)
Meperidine (Demerol)
Methadone (Dolophine)
Morphine
Oxycodone (Percocet)
Oxycodone hydrochloride (OxyContin)
Pentazocine (Talwin)
Propoxyphene (Darvon)

drugs, they have been joined by a number of synthetic opioids, with origins in the laboratory. These include meperidine (Demerol), hydromorphone (Dilaudid), propoxyphene (Darvon), oxycodone (Percocet), oxycodone hydrochloride (OxyContin), hydrocodone (Vicodin), and methadone (Dolophine) **Table 19-2**. Most of these drugs have legitimate medical uses. With the exception of heroin, which is illegal in the United States, many addicts may have started using many of the opioids with an appropriate medical prescription.

These agents are CNS depressants and can cause severe respiratory depression. When administered intravenously, however, they produce a characteristic "high" or "kick." Tolerance develops rapidly, so some users may require massive doses to experience the same high. In general, emergency medical problems related to opioids are caused by respiratory depression, including a decreased volume of inspired air and decreased respirations. These drugs often cause nausea and vomiting and may lead to the development of hypotension. Although seizures are uncommon, they can occur and an overdose can result in the patient entering a comatose condition. Patients typically appear sedated or unconscious and cyanotic with pinpoint pupils. Whereas all of these signs and symptoms may be present with other drugs, the pinpoint pupils are the most commonly accepted sign of opiate abuse.

Treatment includes supporting the airway and breathing. You may try to arouse patients by talking loudly to them or shaking them gently. Always open the airway, give supplemental oxygen, and be prepared for vomiting. You will not be able to do much to adequately address hypotension associated with poor circulation caused by these drugs; however, placing the patient supine with his or her feet elevated may help until advanced help arrives. Many home remedies are believed to reverse the respiratory depression associated with heroin overdose, including applying ice to the groin or forcing milk into the mouth. Do not attempt any of these actions because they are not effective, and they frequently complicate the clinical picture. Nevertheless, you should be aware that a patient's friends may have attempted inappropriate methods of resuscitation. The only effective antidote to reverse the symptoms and signs of opioid overdose are certain narcotic antagonists such as naloxone (Narcan). Patients will respond within 2 minutes to naloxone when it is given intravenously. Naloxone is usually administered by paramedics or by physicians at the emergency department.

■ Sedative-Hypnotic Drugs

Barbiturates and benzodiazepines have been a part of legitimate medicine for a long time. They are easy to obtain and relatively cheap. People sometimes solicit prescriptions from several physicians for the same hypnotics or a variety of sedative-hypnotics Table 19-3. These drugs are CNS depressants and alter the level of consciousness, with effects similar to those of alcohol so that the patient may appear drowsy, peaceful, or intoxicated. By themselves, these drugs do not relieve pain, nor do they produce a specific high, although users often take alcohol or an opioid at the same time to boost their effects.

In general, these agents are taken by mouth. Occasionally, however, contents of capsules are suspended or dissolved in water and injected to produce a rather sudden state of ease and contentment. Use of IV sedative-hypnotic drugs quickly induces tolerance, so the person requires increasingly larger doses. You are less likely to be called on to treat an acute overdose in someone who chronically abuses these drugs; however, you may be called to a scene of an attempted suicide in which the patient has taken large quantities of these drugs. In these situations, patients will have marked respiratory depression and may be in a coma.

Sedative-hypnotic drugs may also be given to unsuspecting people as a "knock-out" drink, or "Mickey Finn." More recently, drugs such as flunitrazepam (Rohypnol) have been abused as a "date rape drug," causing an unwary person to become sedated and even unconscious. The person later awakens, confused and unable to remember what happened.

In general, your treatment of patients who have overdosed with sedative-hypnotics and have respiratory depression is to provide airway clearance, ventilatory assistance, and prompt transport. Give supplemental oxygen, and be ready to assist ventilation. You may attempt to stimulate the person by speaking loudly or gently shaking him or her; remember to watch for vomiting.

A specific antidote is available for acute benzodiazepine overdose. It is called flumazenil and is given intravenously. Although it will reverse the sedation and respiratory depression of the benzodiazepine sedative-hypnotics, it will have no effect on the signs and symptoms of overdose from ethyl alcohol or barbiturates. Almost always, flumazenil is administered in the hospital after a physician's assessment. As multidrug use becomes more common, you may find it increasingly difficult to determine what agents patients have taken. Your best approach is to treat any obvious injuries or illnesses, keeping in mind that drug use may complicate the picture and make full life support necessary. Focus on the ABCs, especially the possibility of airway problems (relaxation of the tongue, causing obstruction), vomiting, respiratory depression, and, in severe cases, cardiac arrest.

Table 19-3 Examples of Sedative-Hypnotic Drugs

Barbiturates	Benzodiazepines	Others
Amobarbital (Amytal)	Alprazolam (Xanax)	Carisoprodol (Soma)
Butabarbital (Butisol)	Chlordiazepoxide (Librium)	Chloral hydrate ("Mickey Finn")
Pentobarbital (Nembutal)	Diazepam (Valium)	Cyclobenzaprine (Flexeril)
Phenobarbital (Luminal)	Flunitrazepam (Rohypnol)	Ethchlorvynol (Placidyl)
Secobarbital (Seconal)	Lorazepam (Ativan)	Ethyl alcohol (drinking alcohol)
	Oxazepam (Serax)	Glutethimide (Doriden)
	Temazepam (Restoril)	Hydrocarbon inhalants
		Isopropyl alcohol (rubbing alcohol)
		Meprobamate (Equagesic)

Abused Inhalants

Many abused inhalants produce several of the same CNS effects as do other sedative-hypnotics, but these agents are inhaled instead of ingested or injected. Some of the more common agents include acetone, toluene, xylene, and hexane, which are found in glues, cleaning compounds, paint thinners, and lacquers. Similarly, gasoline and various halogenated hydrocarbons, such as Freon, used as propellants in aerosol sprays, are also abused as inhalants. None of these inhalants are medications. Because these are products that can be bought in hardware stores, they are commonly abused by teenagers seeking an alcohol-like high. The effective dose and the lethal dose are very close, making these extremely dangerous drugs. The low cost and relative availability make them favorites of children and curious experimenters. Unfortunately, this is an often lethal combination.

Always use special care in dealing with a patient who may have used inhalants. Effects of inhalants range from mild drowsiness to coma, but unlike most other sedative-hypnotics, these agents may often cause seizures. Also, halogenated hydrocarbon solvents can make the heart hypersensitive to the patient's own adrenaline, putting the patient at high risk for sudden cardiac death because of ventricular fibrillation; even the action of walking may release enough adrenaline to cause a fatal ventricular arrhythmia. You must try to keep such patients from struggling with you or exerting themselves. Give supplemental oxygen, and use a stretcher to move the patient. Prompt transport to the hospital is essential; monitor vital signs en route.

Sympathomimetics

Sympathomimetics are CNS stimulants that mimic the effects of the sympathetic (fight-or-flight) nervous system. These stimulants frequently cause hypertension, tachycardia, and dilated pupils. A **stimulant** is an agent that produces an excited state. Amphetamine and methamphetamine ("ice") are commonly taken by mouth. They are also injected by drug abusers in many cases. They typically are taken to make the user "feel good," improve task performance, suppress appetite, or prevent sleepiness. They may just as easily produce irritability, anxiety, lack of concentration, or seizures. Other common examples include phentermine hydrochloride, an appetite suppressant, and amphetamine sulfate (Benzedrine), taken for weight control, narcolepsy, and chronic fatigue syndrome. Caffeine, theophylline, and phenylpropanolamine (a nasal decongestant) are all mild sympathomimetics. So-called designer drugs, such as ecstasy and Eve, are also frequently abused in certain areas of the United States.

Sympathomimetic drugs are frequently called "uppers" Table 19-4. A person using one of these agents

Table 19-4	Street Names for Sympathomimetics
Street Name	**Drug Name**
Adam	3,4-Methylenedioxymethamphetamine (MDMA)
Bennies	Amphetamines
Crank	Crack cocaine, heroin, amphetamine, methamphetamine, methcathinone
DOM	4-Methyl-2,5-dimethoxyamphetamine
Ecstasy	MDMA
Eve	MDMA
Fen-phen	Phentermine
Golden eagle	4-Methylthioamphetamine
Ice	Cocaine, crack cocaine, smokable methamphetamine, methamphetamine, MDMA, phencyclidine (PCP)
MDA	Methaqualone
Meth	Methamphetamine
Speed	Crack cocaine, amphetamine, methamphetamine
STP	PCP
Uppers	Amphetamines

may display disorganized behavior, restlessness, and sometimes anxiety or great fear. Paranoia and delusions are common with sympathomimetic abuse.

Cocaine, also called coke, crack, crystal, snow, freebase, rock, gold dust, blow, and lady, may be taken in a number of different ways. Classically, it is inhaled into the nose and absorbed through the nasal mucosa, damaging tissue, causing nosebleeds, and ultimately destroying the nasal septum. It can also be injected intravenously or subcutaneously (skin-popping). Cocaine can be absorbed through all mucous membranes and even across the skin. In any form, the immediate effects of a given dose last less than an hour.

Another method of abusing cocaine is by smoking it. Crack is pure cocaine. It melts at 93°F (34°C) and vaporizes at a slightly higher temperature. Therefore, crack is easily smoked. In this form, it reaches the capillary network of the lungs and can be absorbed into the body in seconds. The immediate outflow of blood from the heart speeds the drug to the brain, so its effect is felt at once. Smoked crack produces the most rapid means of absorption and, therefore, the most potent effect.

Cocaine is one of the most addicting substances known. Its immediate effects include excitement and euphoria. Acute cocaine overdose is a genuine emergency because patients are at high risk for seizures and cardiac arrhythmias. Chronic cocaine abuse may cause hallucinations; patients with "cocaine bugs" think that bugs are crawling out of their skin.

In caring for patients who have been poisoned with sympathomimetics, be aware that their severe agitation can lead to tachycardia and hypertension. Patients may also be paranoid, putting you and other health care providers in danger. Law enforcement officers should be at the scene to restrain the patient, if necessary. Do not leave the patient unattended and unmonitored during transport.

All of these patients need prompt transport to the emergency department because of the risk of seizures, cardiac arrhythmias, and stroke. You may see blood pressure measurements as high as 250/150 mm Hg. Give supplemental oxygen and be ready to provide suctioning. If the patient is already having a seizure, protect him or her against self-injury.

■ Marijuana

The flowering hemp plant, *Cannabis sativa*, called marijuana, is abused throughout the world. It has been estimated that as many as 20 million people use marijuana daily in the United States. Inhaling marijuana smoke from a cigarette or pipe produces euphoria, relaxation, and drowsiness. It also impairs short-term memory and the capacity to do complex thinking and work. In some people, the euphoria progresses to depression and confusion. An altered perception of time is common, and anxiety and panic can occur. With very high doses, patients experience hallucinations.

A person who has been using marijuana rarely needs transport to the hospital. Exceptions may include someone who is hallucinating, very anxious, or paranoid. In these cases, you must reassure the patient and transport the patient with a minimum amount of excitement. However, you should be aware that marijuana is often used as a vehicle to get other drugs into the body. For example, it may be covered with crack or PCP, also known as "angel dust."

■ Hallucinogens

<u>Hallucinogens</u> alter a person's sensory perceptions Table 19-5 . The classic hallucinogen is lysergic acid diethylamide (LSD). Abuse of another hallucinogen, PCP, or angel dust, is relatively uncommon among young adults. Phencyclidine is a dissociative anesthetic that is easily synthesized and highly potent. Its effectiveness by oral, nasal, pulmonary, and intravenous routes makes it easy to add to other street drugs. It is dangerous because it causes severe behavioral changes in which individuals often inflict injury on themselves.

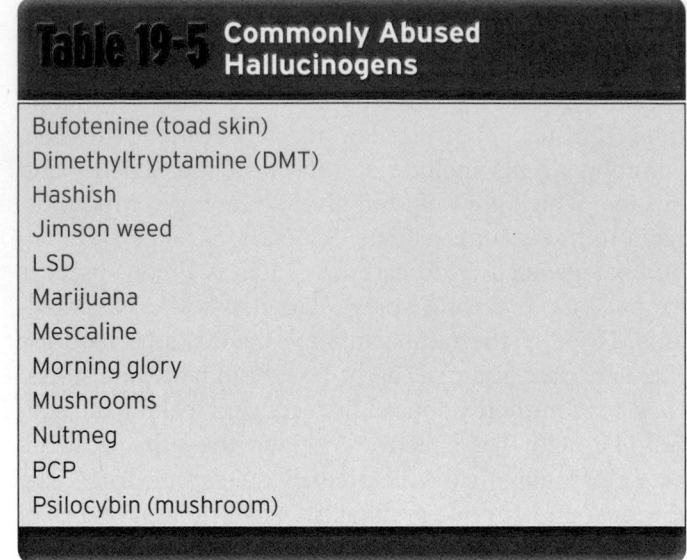

Table 19-5 | Commonly Abused Hallucinogens

Bufotenine (toad skin)
Dimethyltryptamine (DMT)
Hashish
Jimson weed
LSD
Marijuana
Mescaline
Morning glory
Mushrooms
Nutmeg
PCP
Psilocybin (mushroom)

All these agents cause visual hallucinations, intensify vision and hearing, and generally separate the user from reality. The user, of course, expects that the altered sensory state will be pleasurable. Often, however, it can be terrifying. At some point, you are bound to encounter patients who are having a "bad trip." They usually will be hypertensive, tachycardic, anxious, and probably paranoid.

Many hallucinogens have sympathomimetic properties. Indeed, your care for a patient who is having a bad reaction to a hallucinogenic agent is the same as that for a patient who has taken a sympathomimetic. Use a calm, professional manner, and provide emotional support. Do not use restraints unless you or the patient is in danger of injury. Follow the guidelines specified by local authorities. These patients may suddenly experience hallucinations or odd perceptions, so you must watch them carefully throughout transport. Never leave a patient who has taken a hallucinogen unattended and unmonitored. Provide a great deal of reassurance, and request ALS assistance.

■ Anticholinergic Agents

The classic picture of a person who has taken too much of an anticholinergic medication is "hot as a hare, blind as a bat, dry as a bone, red as a beet, and mad as a hatter." These are medications that have properties that, among other effects, block the parasympathetic nerves. Common drugs with a significant anticholinergic effect include atropine, diphenhydramine (Benadryl), Jimson weed, and certain tricyclic antidepressants. With the exception of Jimson weed, these medications usually are not abused drugs but may be taken as an intentional overdose. You will find that it is often difficult to distinguish between an anticholinergic overdose and a sympathomimetic overdose. Both groups of patients may be agitated and tachycardic and have dilated pupils. Once a pure anticholinergic poisoning has been diagnosed, the patient may be treated with physostigmine intravenously by staff in the emergency department, depending on the severity of the situation.

As newer, safer antidepressants such as fluoxetine (Prozac) and sertraline (Zoloft) are added to the market, you can expect to see fewer overdoses of tricyclic antidepressants such as amitriptyline (Elavil) and imipramine (Tofranil). In addition to its anticholinergic effects, a tricyclic antidepressant overdose may cause more serious and life-threatening effects because the medication may block the electrical conduction system in the heart, leading to lethal cardiac arrhythmias. Patients with acute tricyclic antidepressant overdose must be transported immediately to the emergency department; they may go from appearing "normal" to seizure and death within 30 minutes. The seizures and cardiac arrhythmias caused by a severe tricyclic antidepressant overdose are best treated in the hospital with IV sodium bicarbonate. If you work in a tiered system, you should consider calling for ALS backup when you are en route to the scene.

Cholinergic Agents

The "nerve gases" designed for chemical warfare are cholinergic agents. These agents overstimulate normal body functions that are controlled by the parasympathetic nerves, resulting in salivation, mucous secretion, urination, crying, and an abnormal heart rate. You are unlikely to encounter nerve gases. However, you may be called to care for patients who have been exposed to one of the organophosphate insecticides (pesticides) or certain wild mushrooms, which are also cholinergic agents. The signs and symptoms of cholinergic drug poisoning are easy to remember because of the mnemonic DUMBELS:

D Defecation
U Urination
M Miosis (constriction of the pupils)
B Bronchorrhea (discharge of mucus from the lungs)
E Emesis
L Lacrimation (tearing)
S Salivation

Alternatively, you can use the mnemonic SLUDGEM:

S Salivation
L Lacrimation
U Urination
D Defecation
G GI (gastrointestinal) irritation
E Eye constriction/emesis
M Muscle twitching/Miosis (pinpoint pupils)

In poisonings, patients will have excessive amounts of these normal functions and body secretions. In addition, patients may have either bradycardia or tachycardia.

The most important consideration in caring for a patient who has been exposed to an organophosphate insecticide or some other cholinergic agent is to avoid exposure yourself. Because these agents may cling to a patient's clothing and skin, decontamination may take priority over immediate transport to the emergency department. Hospital staff or paramedics can use the anticholinergic drug atropine to dry up the patient's secretions. In the meantime, your priorities after decontamination are to decrease the secretions in the mouth and trachea that threaten to suffocate the patient and provide airway support. Depending on your local EMS protocol, this can be treated as a hazardous materials (HazMat) situation.

The military has developed antidotes to nerve gas agents that can be administered if they are available and indicated. The most common of these antidotes are the Mark I kit and the DuoDote kit. The indications for these are a known exposure to nerve agents or organophosphates with manifestation of signs and symptoms. The kits consist of an auto-injector of atropine and one of 2-PAM chloride (pralidoxime chloride). The auto-injectors are activated in the outer thigh of the patient. Removal of the patient from the source of the exposure is also critical in these cases. If your service carries these antidote kits, you should receive training on their proper use prior to being cleared to administer them.

Miscellaneous Drugs

Accidental or intentional overdose with cardiac medications has become common only because there are so many patients who have these medications prescribed for them. Children may ingest these medications at their grandparent's house, thinking they are candy. Another common scenario is elderly patients who have forgotten they have already taken their medication and take a second dose. Occasionally people wanting to commit suicide will take cardiac medications if that is all they have available. The signs and symptoms of cardiac medication overdose depend on the medication ingested. These drugs may cause bleeding, cardiac arrhythmias, unconsciousness, and even cardiac arrest. Most of these medications are very powerful, so contacting the poison center as soon as possible is important. It is likely you will be given an order to administer activated charcoal, but make sure to check with the poison center first.

Although not as common as it was 30 years ago, aspirin poisoning remains a potentially lethal condition. Ingesting too many aspirin tablets, acutely or chronically, may result in nausea, vomiting, hyperventilation, and ringing in the ears. Patients with this problem are frequently anxious, confused, tachypneic, hyperthermic, and in danger of having seizures. They should be transported quickly to the hospital.

Overdosing with acetaminophen is also very common, probably because acetaminophen is available in so many different preparations, such as Tylenol. The good news is that acetaminophen is generally not very toxic. A healthy patient could ingest 140 mg of acetaminophen for every kilogram of body weight without serious adverse effects. The bad news is that the symptoms of an overdose generally do not appear until it is too late. For example, massive liver failure may not be apparent for a full week. In addition, patients may not provide the information necessary for a correct diagnosis. For this reason, gathering information at the scene is very important. By finding an empty acetaminophen bottle, you may save a patient's life. If given early enough (before liver failure occurs), a specific antidote may prevent liver damage.

Be extremely careful in dealing with a child who has ingested a poisonous substance. Although such incidents usually do not lead to death, family members may be distraught, and your professional attitude will help to ease the tension. Remember, however, that a single swallow of some substances can kill a child.

Some alcohols, including methyl alcohol and ethylene glycol, are even more toxic than ethyl alcohol (drinking alcohol). Methyl alcohol is found in dry gas products and Sterno; ethylene glycol is found in some antifreeze products. Both cause a "drunken" feeling. Left untreated, both will also cause severe tachypnea, blindness (methyl alcohol), renal failure (ethylene glycol), and eventually death. Even ethyl alcohol (typical drinking alcohol) can stop a patient's breathing if taken in too high a dose or too fast, particularly in children. Although they may be used as a substitute by a chronic alcoholic who is unable to obtain ethyl alcohol, they are more often taken by

someone attempting suicide. In either case, immediate transport to the emergency department is essential. **Table 19-6** lists the most common fatally ingested poisons.

Table 19-6   Fatal Ingested Poisons

Benzocaine
Calcium channel blockers (verapamil, nifedipine, diltiazem)
Camphor
Chloroquine
Hydrocarbon solvents
Diphenoxylate-atropine (Lomotil)
Methanol and ethylene glycol
Methylsalicylate (oil of wintergreen)
Phenothiazines (eg, Thorazine)
Quinine
Theophylline
Tricyclic antidepressants (amitriptyline [Elavil], imipramine [Tofranil], nortriptyline [Pamelor])
Tetrahydrozoline (Visine)

Food Poisoning

The term "ptomaine poisoning" was coined in 1870 to indicate poisoning by a class of chemicals found in rotting food. It is still used today in many news accounts of food poisoning. Food poisoning is almost always caused by eating food that is contaminated by bacteria. The food

You are the Provider: PART 5

You begin transport to the hospital, which is located a short distance away. A paramedic accompanies you in the back of the ambulance because the patient has an IV line and is on a cardiac monitor. You reassess the patient, including his vital signs, and then call in your radio report to the hospital. Your estimated time of arrival is 6 minutes.

Recording Time: 17 Minutes	
Level of consciousness	Conscious, but sleepy
Respirations	14 breaths/min; adequate depth
Pulse	72 beats/min; strong and regular
Skin	Pink, warm, and dry
Blood pressure	108/60 mm Hg
Sao$_2$	97% (on oxygen)

9. What additional treatment is required for this patient?

10. What information should you relay to the hospital staff during your verbal report?

Table 19-7	Common Sources of Food Poisoning

Bacillus cereus
Campylobacter
Clostridium botulinum toxin
Clostridium perfringens
Cryptosporidium
Enterococcus
Escherichia coli
Giardia lamblia
Rotavirus
Salmonella
Shigella
Staphylococcus toxin
Vibrio parahaemolyticus
Yersinia enterocolitica

may appear perfectly good, with little or no decay or odor to suggest danger.

There are two main types of food poisoning. In one, the organism itself causes disease; in the other, the organism produces toxins that cause disease **Table 19-7**. A toxin is a poison or harmful substance produced by bacteria, animals, or plants.

One organism that produces direct effects of food poisoning is the *Salmonella* bacterium. The condition called salmonellosis is characterized by severe gastrointestinal symptoms within 72 hours of ingestion, including nausea, vomiting, abdominal pain, and diarrhea. In addition, patients with salmonellosis may be systemically ill with fever and generalized weakness. Some people are carriers of certain bacteria; although they may not become ill themselves, they may transmit diseases, particularly if they work in the food services industry. Usually, proper cooking kills bacteria, and proper cleanliness in the kitchen prevents the contamination of uncooked foods.

The more common cause of food poisoning is the ingestion of powerful toxins produced by bacteria, often in leftovers. The bacterium *Staphylococcus*, a common culprit, is quick to grow and produce toxins in foods that have been prepared in advance and kept too long, even in the refrigerator. Foods prepared with mayonnaise, when left unrefrigerated, are a common vehicle for the development of staphylococcal toxins. Usually, staphylococcal food poisoning results in sudden gastrointestinal symptoms, including nausea, vomiting, and diarrhea. Although time frames may vary from person to person, these symptoms usually may start within 2 to 3 hours after ingestion or as long as 8 to 12 hours after ingestion.

The most severe form of toxin ingestion is botulism. This often-fatal disease usually results from eating improperly canned food, in which the spores of *Clostridium* bacteria have grown and produced a toxin. The symptoms of botulism are neurologic: blurring of vision, weakness, and difficulty in speaking and breathing. Botulism can also cause muscle paralysis and is typically fatal when it reaches the muscles of respiration. Symptoms of botulism may develop as long as 4 days after ingestion or as early as the first 24 hours.

In general, you should not try to determine the specific cause of acute gastrointestinal problems. After all, severe vomiting may be a sign of a self-limiting food poisoning, a bowel obstruction requiring surgery, or another poison, such as copper, arsenic, zinc, cadmium, scombrotoxin (fish poison), or *Clitocybe* or *Inocybe* mushrooms. Instead, you should gather as much history as possible from the patient and transport him or her promptly to the hospital. When two or more persons in one group have the same illness, you should take along some of the suspected food. In advanced cases of botulism, you may have to assist ventilation and give basic life support.

Plant Poisoning

Several thousand cases of poisoning from plants occur each year, some severe. Many household plants are poisonous if ingested; children have been known to nibble on the leaves **Table 19-8**. Some poisonous plants cause

Table 19-8	Common Toxic Plants
Scientific Name	**Common Name**
Abrus precatorius	Jequirity bean/rosary pea
Cicuta species	Water hemlock/wild carrot
Colchicum autumnalel	Autumn crocus
Conium maculatum	Poison hemlock
Convallaria majalis	Lily of the valley
Datura species	Jimson weed/stinkweed
Dieffenbachia	Dumbcane
Digitalis purpurea	Foxglove
Nerium oleander	Oleander or rose laurel
Nicotiana glauca	Tree tobacco
Phoradendron	Mistletoe
Phytolacca americana	Pokeweed
Rhododendron	Rhododendron or azalea
Ricinus communis	Castor bean
Solarium nigrum	Nightshade
Zygadenus species	Death camas

Figure 19-8 The toxins in these common poisonous plants are often ingested or absorbed through the skin. **A.** Dieffenbachia. **B.** Mistletoe. **C.** Castor bean. **D.** Nightshade. **E.** Foxglove. **F.** Rhododendron.

Special Populations

In an accidental overdose or poisoning, a geriatric patient may have become confused about his or her drug regimen. He or she may have forgotten that the medication had been taken, repeating the dose a number of times. Or the patient could have forgotten the physician's instructions to discard leftover medication and might have taken the current and the older drug, resulting in an increase in effects or an unwanted drug interaction.

A geriatric patient may also intentionally overdose in an attempt to commit suicide. Geriatric patients have been known to ingest common household chemicals such as insecticides, acetaminophen, aspirin, or caustic substances in an attempt to end their lives. Be alert for any indication of an intentional overdose or poisoning, even though the patient might deny an attempted suicide.

In considering any poisoning, remember the basics. Because of the aging process, the absorption of poisons may change. For example, decreased gastric mobility may delay absorption of ingested poisons, limiting systemic effects, but may result in increased damage to the stomach. Alcoholism is common in the elderly and alcohol can exacerbate the effects of other ingested substances. Also,

elderly patients with alcoholism are more susceptible to making mistakes when taking their medications and may have more trouble metabolizing medication because of liver damage.

If an elderly person inhales a poison, even in tiny quantities, lung damage can be severe. Consider the decreased lung capacity and ability to exchange oxygen and carbon dioxide in an older patient's lungs. Pulmonary function could be worsened to potentially fatal levels with the inhalation of minute amounts of poison.

For poisons that are absorbed by or injected into the skin, reduced circulation to the skin can decrease or delay absorption into the body. Watch for an increased reaction or irritation at the skin site.

In a geriatric patient, the liver may not be able to metabolize the poison as effectively or the kidneys may not be able to excrete the poison as quickly. In either case, the drug or poison remains in the body for a longer period, causing additional tissue damage. When a medication is not metabolized or excreted as quickly as before, the drug could accumulate to toxic levels and, ultimately, become fatal in lesser doses than in a younger person.

Figure 19-8 **G.** Jimson weed. **H.** Death camas. **I.** Poison ivy. **J.** Poison oak. **K.** Pokeweed. **L.** Rosary pea. **M.** Poison sumac.

local irritation of the skin; others can affect the circulatory system, the gastrointestinal tract, or the CNS. It is impossible for you to memorize every plant and poison, let alone their effects Figure 19-8 . You can and should do the following:

1. Assess the patient's airway and vital signs.
2. Notify the regional poison center for assistance in identifying the plant.
3. Take the plant to the emergency department.
4. Provide prompt transport.

Irritation of the skin and/or mucous membranes is a problem with the common houseplant called dieffenbachia, which resembles "elephant ears." When chewed, a single leaf may irritate the lining of the upper airway enough to cause difficulty swallowing, breathing, and speaking. For this reason, dieffenbachia has been called "dumbcane." In rare circumstances, the airway may be completely obstructed. Emergency medical treatment of dieffenbachia poisoning includes maintaining an open airway, giving oxygen, and transporting the patient promptly to the hospital for respiratory support.

You should continue to assess the patient for airway difficulties throughout transport. If necessary, provide positive-pressure ventilation.

Special Populations

Small children or toddlers are often the victims of accidental poisoning through their natural curiosity. This age of children commonly put objects in their mouth as a means to learn about them. When they gain access to toxic substances, they will almost always ingest them. Most parents are aware of these dangers and properly secure cabinets or store substances out of reach, but often grandparents or older adults without children will store household chemicals in easy-to-reach lower cabinets without locks. This creates a serious hazard when young children visit and begin to "explore." It is fortunate that most accidental poisonings of children are not fatal. Avoid blaming the adults and focus instead on treating the child in the most appropriate manner.

You are the Provider: SUMMARY

1. In addition to providing immediate lifesaving treatment, what else should your actions consist of when you arrive at this scene?

Although law enforcement personnel advised you that the scene is safe, you should always remain aware of your surroundings, especially when dealing with situations that may involve violence (ie, drug overdose). Although the patient is reportedly unconscious, the presence of weapons, which people often use to protect their drugs, cannot be ruled out, nor can the presence of other people, who may also be under the influence of drugs or alcohol. Be safe!

On arriving at the scene, you must rapidly assess the patient and begin immediate lifesaving treatment as needed. In some cases, it will be obvious that an overdose occurred (eg, empty medication bottles, the presence of drug paraphernalia [needle and syringe]); in other cases, it will not be so obvious.

Because the patient is unconscious, you will obviously get no information from him. However, if there is another person present, you must try to determine what was taken; how long ago it was taken; by which route it was taken (eg, ingestion, injection, etc); what actions, if any, were taken prior to your arrival; and an estimate of the patient's weight. This information is vital when determining the most appropriate treatment.

If you are unable to obtain specific information regarding the overdose, proceed with patient care, focusing your treatment on the ABCs. Bear in mind that when multiple drugs are taken, their combined effects often produce an unpredictable response; the patient may appear to be unconscious, only to wake up in fits of rage, and then lapse back into unconsciousness. Again, safety on your part is paramount. If you are able to determine what the patient took, begin treatment, consider requesting ALS support, and contact the poison center per your local protocol.

2. What is a toxidrome? How can knowledge of various toxidromes improve the care you provide to a patient?

A toxin (or poison) is any substance that can damage body structures or impair body function. Toxins change the normal metabolism of cells or destroy them outright, resulting in death. Toxins may act acutely, as in an overdose of heroin, or chronically, as with years of alcohol or other substance abuse.

A toxidrome is a series of signs and symptoms that indicate a particular type of toxic exposure. When assessing a patient who has overdosed on a medication or other substance, a careful assessment may help you identify a toxidrome, thereby allowing you to direct your treatment accordingly.

An overdose of a central nervous system (CNS) depressant, such as a narcotic or sedative-hypnotic, would be expected to cause hypoventilation, hypotension, bradycardia, and a decreased level of consciousness. By contrast, overdose on a CNS stimulant, such as methamphetamine or cocaine, would be expected to cause the opposite effects: hypertension, agitation or violent behavior, tachycardia, and hyperventilation. It is

the distinct difference in these two clinical presentations that will help you differentiate a CNS depressant toxidrome from a CNS stimulant toxidrome.

Many people overdose on a variety of medications or substances, each of which acts differently in the body, and in many cases, the drugs oppose each other. In these cases, a clear toxidrome may not be identifiable, and your treatment will consist of supportive care (eg, maintaining the ABCs) and rapid transport only.

If a toxidrome can be identified, your treatment may be more specific; in addition to supportive care and rapid transport, there may be a specific antidote or reversal agent that can be given to the patient. If this is the case, you should either summon an ALS unit to the scene or coordinate an ALS intercept en route to the hospital. Although few antidotes and reversal agents are available in the prehospital setting, those that are available are typically given by paramedics, unless otherwise directed by local protocol.

3. What should your immediate treatment be for this patient?

First, you must move the patient from the couch to the floor; he needs treatment that would be too difficult to perform in his present position. Observe proper lifting technique; if you and your partner cannot safely lift and move him, ask a police officer to assist you.

Your patient is unconscious and unresponsive; he has secretions in his mouth, and his respirations are slow, shallow, and snoring. His airway and ventilation status is inadequate and requires immediate attention. Place the patient on his left side and suction his oropharynx to prevent aspiration. After his airway is clear of secretions, return him to a supine position, manually open his airway, and insert an airway adjunct. If he does not have a gag reflex, insert an oral airway; otherwise, insert a nasal airway.

Slow, shallow breathing will not produce adequate minute volume; if not treated or inappropriately treated, he may become more hypoxemic than he already is and may stop breathing altogether. He needs some form of positive-pressure ventilation; assist his breathing with a bag-mask device (10 to 12 breaths/min) attached to high-flow oxygen. *Do not hyperventilate him*; doing so increases the risk of aspiration if he vomits and may compromise cardiac output because of compression of the heart from hyperinflation of the lungs.

Unconscious patients are at an increased risk for vomiting, which could lead to aspiration. When ventilating an unresponsive patient whose airway is not secured with an advanced device (ie, endotracheal tube), posterior pressure should be applied to the cricoid cartilage (Sellick maneuver); this will help compress the esophagus and minimize the amount of air that enters the stomach.

It is obvious that this patient requires aggressive treatment. You should request additional personnel to respond to the scene, ideally, an AEMT or paramedic. The patient may require advanced airway management; depending on further assessment findings, he may also require medication therapy.

4. On the basis of the patient's initial presentation, what type of drug should you suspect that he overdosed on?

On the basis of the patient's initial presentation—unconsciousness, hypoventilation, bradycardia—you should suspect that he has overdosed on a CNS depressant drug.

CNS depressant drugs generally fall into three basic categories: opiates (narcotics), such as oxycodone (OxyContin), heroin, morphine, and codeine; barbiturates, such as phenobarbital (Luminal), pentobarbital (Nembutal), secobarbital (Seconal), and amobarbital (Amytal); and benzodiazepines, such as diazepam (Valium), lorazepam (Ativan), clonazepam (Klonopin), and temazepam (Restoril).

One or more of these CNS depressant drugs, when taken in excess, can result in severe respiratory depression, cardiovascular collapse, and death. In some cases, patients may overdose on one particular CNS depressant; in other cases, they take numerous quantities of different types of CNS depressants.

Although the mechanism of action (eg, how they work) of opiates/narcotics, barbiturates, and benzodiazepines are somewhat different, they all suppress the function of the central nervous system and have very similar clinical manifestations.

As an EMT, your job is to recognize the signs and symptoms (eg, the toxidrome) that are associated with these types of drugs, begin immediate treatment to support the ABCs, and transport the patient to the hospital without delay.

5. Would activated charcoal benefit this patient? Why or why not?

There are several reasons why activated charcoal (InstaChar, Actidose, LiquiChar) is *not* indicated for this patient. First, the patient is unconscious, is unable to protect his own airway, and clearly cannot swallow. Pouring anything into his mouth would lead to aspiration, thus substantially increasing his chance of death!

On the basis of the patient's clinical presentation, it is clear that the medications he ingested are no longer in his stomach. They are now systemic and are causing compromise of his breathing and circulatory status. When a substance is ingested, it typically remains in the stomach for about an hour, after which time it passes into the small intestine and is digested.

Even if the patient was conscious, but was showing signs of CNS depression, activated charcoal would likely be ineffective. The presence of systemic signs and symptoms following overdose of a medication indicates that it has already been digested and is active in the circulatory system.

Other contraindications to administration of activated charcoal include ingestion of a corrosive (strong acid or alkali) substance and known esophageal disease (eg, varices, cancer).

In some cases, ALS providers *may* insert a nasogastric tube (a tube that is passed through the nose and into

the stomach) and instill activated charcoal. However, this procedure is governed by local protocol and the estimated time of ingestion.

6. What is naloxone? Why is it being given to this patient?

Naloxone (Narcan) is an opiate/narcotic antagonist—that is, it blocks opiate receptors in the body and prevents the uptake of narcotics into the central nervous system. One of the drugs the patient has ingested—oxycodone—is a powerful narcotic; therefore, naloxone is indicated.

Naloxone is given to patients who present with a decreased level of consciousness and are suspected of having ingested or injected a narcotic. Its fast-acting effects quickly reverse the CNS depression caused by narcotic overdose. However, compared to most narcotics, naloxone has a shorter duration of action and its effects may wear off before the narcotic does. Therefore, patients often need more than one dose, especially if they are under the influence of multiple narcotics.

It is important to note that naloxone is *only* effective in reversing the CNS depressant effects caused by excessive narcotic ingestion or injection; it will not reverse CNS depression caused by benzodiazepines or barbiturates. If a patient has ingested several types of CNS depressant drugs, the effects of naloxone may not be noticed. In such cases, continue to support oxygenation, ventilation, and circulation, and rapidly transport the patient to the hospital.

Naloxone can be administered via the intravenous (IV), intraosseous (IO), intramuscular (IM), and intranasal (IN) routes. Although EMTs typically do not administer naloxone, it is important to understand the drug and why it is being given. In some EMS systems, EMTs are authorized by local protocol to administer naloxone.

7. What other issues about this patient should concern you?

Although the patient's condition has improved following the administration of naloxone, the fact still remains that he also ingested large quantities of acetaminophen (Tylenol) and ibuprofen. By no means is he "out of the woods."

Acetaminophen (APAP)—the active ingredient in Tylenol—is extremely toxic to the liver when taken in large doses. Although it may not have an acute effect, if not promptly treated in the hospital, the patient may die of liver failure. Treatment for APAP toxicity in the field is supportive; the patient must be transported to the hospital, where an antidote can be given.

Ibuprofen—the active ingredient in Motrin and Advil—is irritating to the gastric lining. Even in therapeutic doses, prolonged ibuprofen use may cause conditions such as gastric ulcers and gastrointestinal bleeding. In toxic doses (> 400 mg/kg), the risk of this is clearly higher, as are coma and death. Like APAP, ibuprofen is also toxic to the liver.

When you encounter a patient who has ingested an over-the-counter medication, such as ibuprofen and acetaminophen

(Tylenol), you have no point of reference regarding how many tablets were in the bottle and when it was bought or filled. Therefore, you should assume that the patient ingested the entire amount, especially if he or she is unconscious and is unable to confirm otherwise.

Ibuprofen is typically supplied in 200-mg tablets and extra-strength Tylenol is supplied in 500-mg tablets. Whereas it may take as many as 140 ibuprofen tablets to cause toxicity in a 150-lb (68-kg) person, as few as 14 extra-strength Tylenol tablets can cause toxicity in a person of the same weight.

8. How would this patient's presentation have differed had he overdosed on a sympathomimetic?

Unlike CNS depressants, which are commonly called "downers," sympathomimetics are CNS stimulants; thus, they are commonly called "uppers." A sympathomimetic is any substance that mimics the effects of the sympathetic (fight or flight) nervous system. When the sympathetic nervous system is stimulated, it releases epinephrine and norepinephrine, resulting in hypertension, tachycardia, and restlessness or agitation.

Had the patient overdosed on a sympathomimetic, his clinical presentation would have been the exact opposite. Instead of depression of his vital functions, such as breathing, heart rate, and blood pressure, he would have experienced a significant increase in these functions. Furthermore, he would likely have experienced paranoia, delusions, and disorganized behavior.

Common sympathomimetics include phentermine hydrochloride, an appetite suppressant that is used for weight control; amphetamine sulfate (Benzedrine), which is also used for weight control, but is also used to treat narcolepsy and chronic fatigue syndrome; a drug combination of amphetamine and dextroamphetamine (Adderall) and methylphenidate hydrochloride (Ritalin), which are used in the treatment of attention-deficit hyperactivity disorder (ADHD); theophylline (a bronchodilator); caffeine; and phenylpropanolamine (nasal decongestant). These drugs contain various amounts of amphetamine, a CNS stimulant.

Methamphetamine (ice), cocaine (coke, blow, snow, rock, gold dust), and crank (crack cocaine) are examples of illegal sympathomimetics. Methamphetamine is commonly taken by mouth; however, many drug abusers inject it. Cocaine may be taken in a number of different ways. Classically, it is inhaled (snorted) into the nose and absorbed into the bloodstream via the nasal mucosa. It can also be injected into the subcutaneous layer of the skin (skin-popping) or melted down, vaporized, and smoked.

Sympathomimetics are commonly taken to make the user "feel good," improve task performance, prevent sleepiness, or suppress the appetite. However, these drugs may just as easily produce irritability, anxiety, lack of coordination, and seizures. Taken in excess, they can cause death secondary to massive sympathetic nervous system stimulation and cardiac dysrhythmias (ie, ventricular fibrillation).

9. What additional treatment is required for this patient?

Further treatment for this patient is mainly supportive; closely monitor his ABCs and be alert for the recurrence of CNS depression (eg, decreased level of consciousness, hypoventilation, bradycardia, hypotension). Monitor his vital signs at regular intervals.

As previously discussed, naloxone is a short-acting reversal agent when compared with most opiate/narcotics. Assisted ventilation and additional naloxone will be required if the patient lapses back into CNS depression. The paramedic, who is monitoring the patient's cardiac rhythm, may need to give additional medications if the patient develops a cardiac dysrhythmia.

Bear in mind that oxycodone, ibuprofen, and acetaminophen (Tylenol) are the suspected drugs that the patient overdosed on; this information is based on the empty medication bottles that were found at the scene. However, he may have ingested additional drugs that may have a more delayed onset of action.

If time and patient condition permits, you should inquire about any past medical history he may have. This information is relevant because the drugs he overdosed on could exacerbate, or be exacerbated by, certain underlying medical conditions. You should also provide emotional support to the patient; do not judge him because he attempted suicide. He is still a patient and deserves the same compassion, empathy, and professional care that you would provide to any other patient.

10. What information should you relay to the hospital staff during your verbal report?

Your verbal (hand-off) report at the hospital should be more in-depth than what was provided over the radio. You should advise the receiving nurse or physician of how you found the patient, what you initially did to treat him, and how he responded to your treatment.

Inform the hospital staff of the patient's condition en route to the hospital, being sure to advise them of *any* changes—good or bad—that may have occurred after you gave your radio report.

When you are treating a patient who has overdosed, you should take all medication bottles found at the scene—full or empty—to the hospital and turn them over to the receiving nurse or physician. Additional assessment at the hospital will include a toxicology screen, which involves drawing blood specimens to detect other substances, legal or illegal, that may be in the patient's body.

A clear and concise verbal report is an integral part of the patient care process; it helps ensure continuity of care and often helps the emergency department physician in his or her decision about further treatment.

You are the Provider: SUMMARY, continued

EMS Patient Care Report (PCR)

Date: 11-30-09	**Incident No.:** 011909	**Nature of Call:** Overdose			**Location:** 213 W. Maple, Apt. 9B	
Dispatched: 0220	**En Route:** 0222	**At Scene:** 0226	**Transport:** 0240	**At Hospital:** 0249		**In Service:** 0301

Patient Information

Age: 22 **Sex:** M **Weight (in kg [lb]):** 70 kg (155 lb)	**Allergies:** No known drug allergies **Medications:** None that were prescribed to the patient **Past Medical History:** None **Chief Complaint:** Drug ingestion

Vital Signs

Time: 0231	**BP:** 76/50	**Pulse:** 40	**Respirations:** 6	**Sao2:** 91%
Time: 0237	**BP:** 100/52	**Pulse:** 64	**Respirations:** 12	**Sao2:** 96%
Time: 0243	**BP:** 108/60	**Pulse:** 72	**Respirations:** 14	**Sao2:** 97%

EMS Treatment
(circle all that apply)

Oxygen @ <u>15</u> L/min via (circle one): NC (NRM) (Bag-Mask Device)	(Assisted Ventilation)	(Airway Adjunct)	CPR
Defibrillation	**Bleeding Control**	**Bandaging**	**Splinting**	**Other:** IV therapy, 2 mg Narcan, cardiac monitoring, blood glucose assessment

Narrative

Medic 90 dispatched to an apartment complex for an overdose. Law enforcement secured the scene prior to EMS arrival. Found the patient, a 22-year-old male, lying supine on his couch. He was unconscious and unresponsive; there were secretions in his airway and his breathing was slow, shallow, and was making a snoring sound. Moved patient to floor, suctioned his oropharynx, manually maintained an open airway, and inserted an airway adjunct. Began assisted ventilation via bag-mask device and high-flow oxygen. Patient's pulse was weak and slow. ALS unit 44 arrived at the scene to provide assistance. Three empty medication bottles were found adjacent to the patient; they contained oxycodone 30 mg, ibuprofen 200 mg, and Tylenol 500 mg. Obtained vital signs as paramedic from ALS 44 established IV access. Secondary assessment revealed pinpoint pupils, but was otherwise unremarkable. No medical alert bracelets were found on the patient. Blood glucose level was assessed and was noted to read 112 mg/dL. After IV was established, paramedic administered 2 mg Narcan. Reassessment of patient revealed that his level of consciousness was improving and his heart rate, blood pressure, and oxygen saturation improved. He would no longer tolerate assisted ventilation, but would tolerate high-flow oxygen via nonrebreathing mask. Cardiac rhythm was monitored by ALS 44 paramedic. Placed patient onto the stretcher, loaded him into the ambulance, and began transport. Paramedic from ALS 44 accompanied patient and EMT in the back of the ambulance. En route, continued to monitor patient's condition and vital signs. He remained conscious, but sleepy. His airway was patent and his breathing and pulse were adequate. Patient stated that he intended to kill himself, but refused to state why. Provided emotional support, continued oxygen therapy, and monitored patient's vital signs for the duration of the transport. Patient was delivered to the emergency department staff without incident; his condition was markedly improved. Gave verbal report to charge nurse and provided him with the empty medication bottles found at the scene. Medic 90 cleared the hospital and returned to service at 0301. *End of report*

Assessment and Emergency Care of Toxicologic Emergencies

Scene Size-up

Scene Safety	Ensure scene safety and safe access to the patient. Be aware of potential violence or the possibility of a crime scene and call for law enforcement. Standard precautions should include a minimum of gloves. Determine the number of patients and assess the need for additional resources. Pay close attention to the patient and the surroundings; patients with an altered mental status may unexpectedly become violent. Look for evidence of environmental toxins. Be observant for possible weapons. Identify yourself clearly, and remain calm and confident.
Mechanism of Injury (MOI)/ Nature of Illness (NOI)	Determine the NOI. Observe the scene and look for indicators of a toxic exposure (empty containers, multiple patients, strange odors, etc). Consider that a toxin may be the reason for the person's behavior because certain toxins can produce an altered mental status.

▼   ▼

Primary Assessment

Form a General Impression	Your assessment of a toxicologic emergency patient should begin at the door. Are there empty medication, alcohol, or chemical containers around the patient? Observe the work of breathing and circulation. Determine the level of consciousness using the AVPU scale. Identify immediate threats to life. Determine priority of care based on the NOI. If the patient has a poor general impression, call for ALS assistance. A rapid scan will help you identify and manage life threats.
Airway and Breathing	Ensure the airway is open, clear, and self-maintained. Unresponsive patients will need the airway opened and maintained using a modified jaw-thrust maneuver if cervical spine injury is suspected; use the head tilt–chin lift maneuver in nontrauma patients. A patient with an altered level of consciousness may need emergency airway management; consider inserting a properly sized oropharyngeal or nasopharyngeal airway. Evaluate the patient's ventilatory status for rate and depth of breathing, respiratory effort, and tidal volume. Administer high-flow oxygen at 15 L/min, providing ventilatory support as needed. Hypoxia may cause changes in the patient's mental state. If vomiting is a possibility, place the patient in the recovery position if no spinal injury is suspected.
Circulation	Observe skin color, temperature, and condition; look for life-threatening bleeding and treat accordingly. Evaluate the distal pulse rate, quality (strength), and rhythm. Tachycardia may be an indicator of a toxic exposure, but it may also indicate respiratory distress or shock. Bradycardia may also occur as a result of a toxic exposure.
Transport Decision	If the patient has an airway or breathing problem, signs and symptoms of bleeding, or other life threats, manage them immediately and consider rapid transport, performing the secondary assessment en route to the hospital. Depending on the NOI, patients should be decontaminated prior to transport to decrease the dose the patient received and to limit the spread of contamination.

▼   ▼

NOTE: The order of the steps in this section differs depending on whether the patient is conscious or unconscious. The following order is for a conscious patient. For an unconscious patient, perform a primary assessment, a full-body scan, obtain vital signs, and obtain the past medical history from a family member, bystander, or emergency medical identification device.

Assessment and Emergency Care of Toxicologic Emergencies, continued

History Taking

Investigate Chief Complaint	Investigate the chief complaint. Monitor the patient for changes in mental status. Ask OPQRST and SAMPLE questions. SAMPLE can also be obtained from family, bystanders, and medical alert tags. If the toxicologic event was intentional, treat the patient as a psychiatric emergency as well. When the toxic substance is not immediately evident, determine if the patient is exhibiting signs and/or symptoms consistent with a toxidrome; the patient's signs and symptoms often can provide clues to the nature of the toxic exposure.

Secondary Assessment

Physical Examinations	The physical examination should focus on the area of the body involved with the poisoning or the route of exposure and the particular drug or chemical the patient was exposed to. An examination of all body systems may help to identify systemic problems.
Vital Signs	Obtain baseline vital signs as soon as practical. Vitals signs should include blood pressure by auscultation, pulse rate and quality, respiration rate and quality, pupils, and skin assessment for perfusion. Note the patient's level of consciousness. Use pulse oximetry, if available, to assess the patient's perfusion status. Vitals signs are important; preexisting medical conditions can be worsened by a toxic exposure.

Reassessment

Interventions	Repeat the primary assessment, vital signs, and assessment of the chief complaint. Assist breathing as required, administering high-flow oxygen. Be prepared for sudden deterioriation as the toxin spreads within the body. If the patient has consumed a harmful or lethal dose of a poisonous substance, you must repeat the assessment of vital signs every 5 minutes, or constantly if needed. If the patient is in stable condition and there are no life threats, reassess every 15 minutes.
Communication and Documentation	Contact medical assessment of the or a poison center for information as permitted by local protocal. Contact medical control/receiving hospital with a radio report; many hospitals require additional personnel and a separate treatment area. Include a thorough description of the NOI and the position the patient was found in. Include treatments performed and patient response. Be sure to document the patient's distress, answers to your questions, attitude toward emergency care providers, and any changes in patient status and the time. Follow local protocols. Document the reasoning for your treatment and the patient's response. If restraints were necessary, thoroughly document your rationale and the patient condition/status while being restrained.

Assessment and Emergency Care of Toxicologic Emergencies, continued

NOTE: Although the following steps are widely accepted, be sure to consult and follow your local protocols. Take appropriate standard precautions when treating all patients.

Toxicologic Emergencies

General Management of Toxicologic Emergencies

1. Rescues from a toxic environment should only be performed by trained rescuers wearing appropriate personal protective equipment.
2. Establish and maintain a patent airway. Provide high-flow oxygen. Monitor for vomiting and protect against aspiration.
3. Obtain SAMPLE history and vital signs. Ascertain what toxin may be involved.
4. Request ALS when necessary.
5. Take all containers, bottles, and labels of poisons to the receiving hospital.
6. For patients who have taken alcohol, opiods, sedative-hypnotics, or abused inhalants, monitor the level of consciousness and airway patency because these drugs produce central nervous system depression and respiratory depression.
7. For abused inhalants, patients are prone to seizures and ventricular fibrillation.
8. For stimulants or anticholinergics, it is critical to monitor patients for hypertension and/or other cardiovascular effects.
9. For cholinergic agents, decontamination is a necessity. Monitor the patient for excessive respiratory secretions and seizures.
10. For patients who have plant poisoning, contact the regional poison center for assistance in identifying the plant.
11. For patients who have food poisoning, transport the food suspected to be responsible for the poisoning.

Administer activated charcoal for poisonous ingestions according to local protocol. Follow these steps:

1. Do not give activated charcoal if the patient exhibits an altered mental status, has ingested a substance for which charcoal is contraindicated, or is unable to swallow.
2. Obtain an order from medical direction or follow protocol.
3. Shake the activated charcoal container well.
4. Place the activated charcoal suspension in a covered cup with a straw and ask the patient to drink. The dose for infants and children is 12.5 to 25 g and the dose for adults is 25 to 50 g.

Prep Kit

Ready for Review

- Poisons act acutely or chronically to destroy or impair body cells.

- If you believe a patient may have taken a poisonous substance, you should support the ABCs and notify medical control.

- Management of the patient also entails collecting any evidence of the type of poison that was used and taking it to the hospital; diluting and physically removing the poisonous agent; providing respiratory support; and transporting the patient promptly to the hospital.

- Emergency treatment may include administration of an antidote, usually at the hospital, if an antidote exists.

- A poison can be introduced into the body in one of four ways:
  - Inhalation
  - Absorption (surface contact)
  - Ingestion
  - Injection

- It is difficult to remove or dilute injected poisons, a fact that makes these cases especially urgent.

- Always consult medical control before you proceed with the treatment of any poisoning victim.

- Move patients who have inhaled poison into the fresh air; be prepared to use supplemental oxygen via a non-rebreathing mask and/or ventilatory support via a bag-mask device.

- With absorbed or surface contact poisons, be sure to avoid contaminating yourself. You should remove all contaminated substances and clothing from the patient, and flood the affected part.

- Approximately 80% of all poisonings are by ingestion, including plants, contaminated food, and most drugs. In general, activated charcoal should be used in these patients.

- People who abuse a substance can develop a tolerance to it or can develop an addiction.

- The most commonly abused drug in the United States is alcohol. It can depress the central nervous system and can cause respiratory depression. You must support the airway in such cases, and be prepared for the patient to vomit.

- Opioids, sedative-hypnotic drugs, and abused inhalants can also depress the central nervous system and can cause respiratory depression.

- Take special care with patients who have used inhalants because the drugs may cause seizures or sudden death.

- Sympathomimetics, including cocaine, stimulate the central nervous system, causing hypertension, tachycardia, seizures, and dilated pupils. Patients who have taken these drugs may be paranoid, as may patients who have taken hallucinogens.

- Anticholinergic medications, often taken in suicide attempts, can cause a person to become hot, dry, blind, red-faced, and mentally unbalanced. An overdose of tricyclic antidepressants can lead to cardiac arrhythmias.

- The symptoms of cholinergic medications, which include organophosphate insecticides, can be remembered by the mnemonic DUMBELS, for excessive Defecation, Urination, Miosis, Bronchorrhea, Emesis, Lacrimation, and Salivation; or SLUDGE, for Salivation, Lacrimation, Urination, Defecation, Gastrointestinal irritation, and Eye constriction/emesis.

- Two main types of food poisoning cause gastrointestinal symptoms.
  - In one type, bacteria in the food directly cause disease, such as salmonellosis; in the other, bacteria such as *Staphylococcus* produce powerful toxins, often in leftover food.
  - The most severe form of toxin ingestion is botulism; the first neurologic symptoms may appear as late as 4 days after ingestion.

- Plant poisoning can affect the circulatory system, the gastrointestinal system, and the central nervous system. Some plants, such as the dieffenbachia, irritate the skin or mucous membranes and may cause obstruction of the airway.

Vital Vocabulary

addiction A state of overwhelming obsession or physical need to continue the use of a drug or agent.

antidote A substance that is used to neutralize or counteract a poison.

delirium tremens (DTs) A severe withdrawal syndrome seen in alcoholics who are deprived of ethyl alcohol; characterized by restlessness, fever, sweating, disorientation, agitation, and seizures; can be fatal if untreated.

emesis Vomiting.

hallucinogens Agents that produce false perceptions in any one of the five senses.

hematemesis Vomiting blood.

hypnotic A sleep-inducing effect or agent.

ingestion Swallowing; taking a substance by mouth.

opioids Any drug or agent with actions similar to morphine.

overdose An excessive quantity of a drug which, when taken or administered, can have toxic or lethal consequences.

poison A substance whose chemical action could damage structures or impair function when introduced into the body.

sedative A substance that decreases activity and excitement.

stimulant An agent that produces an excited state.

substance abuse The misuse of any substance to produce some desired effect.

tolerance The need for increasing amounts of a drug to obtain the same effect.

toxicology The study of toxic or poisonous substances.

toxin A poison or harmful substance produced by bacteria, animals, or plants.

vomitus Vomited material.

Assessment in Action

You and your partner are dispatched in the early morning hours to an apartment for a possible overdose. The door is answered by an anxious woman who takes you to a bedroom where you find middle-aged man lying supine across the bed. He has snoring respirations at a rate of 6 breaths/min. The woman tells you that she came home approximately 20 minutes before you arrived and found her husband as described.

1. What is your first priority?
   - A. Confirm the woman's account of events.
   - B. Open the patient's airway.
   - C. Begin a detailed assessment.
   - D. Take vital signs.

2. As you approach the patient's bed, you observe numerous beer cans, an empty whiskey bottle, and two empty prescription containers scattered by the bedside. What is the most commonly abused drug in the United States?
   - A. Cocaine
   - B. Methamphetamine
   - C. Heroin
   - D. Alcohol

3. What is the most frequently seen long-term effect of alcohol abuse?
   - A. Kidney cancer
   - B. Liver damage
   - C. Peptic ulcers
   - D. Pancreatitis

4. The woman states that her husband has been extremely depressed lately because of losing his job following a car accident that left him with a back injury. He has been taking pain killers regularly for the past month and had his prescriptions refilled yesterday. Which of the following is an example of an opioid drug?
   - A. Diazepam (Valium)
   - B. Phenobarbital (Luminal)
   - C. Oxycodone hydrochloride (OxyContin)
   - D. Cyclobenzaprine (Flexeril)

5. What is the most commonly accepted sign of opioid abuse?
   - A. Respiratory depression
   - B. Seizures
   - C. Pinpoint pupils
   - D. Hypotension

6. As you open the patient's airway you detect a strong alcohol-like odor. Severe acute alcohol ingestion may cause:
   - A. a decreased blood glucose level.
   - B. an increased reaction time.
   - C. heightened awareness.
   - D. high blood pressure.

7. You are able to maintain a patent airway with an oropharyngeal airway. The patient now has a respiratory rate of 10 breaths/min, a pulse rate of 65 beats/min, and a blood pressure of 80/40 mm Hg. Treatment for this patient should now include:
   - A. airway support and supplemental oxygen.
   - B. observing the patient for vomiting.
   - C. placing the patient in a position of comfort.
   - D. all of the above.

8. On the basis of the patient's condition, vital signs should be assessed every _____ minutes.
   - A. 5
   - B. 10
   - C. 15
   - D. 20

9. What clues may be at the scene that could help you identify the substance(s) ingested by the patient?

10. In addition to obtaining the SAMPLE history, what information might be helpful in determining what is going on with your patient?

Psychiatric Emergencies

National EMS Education Standard Competencies

Medicine

Applies fundamental knowledge to provide basic emergency care and transportation based on assessment findings for an acutely ill patient.

Psychiatric

Recognition of

- Behaviors that pose a risk to the EMT, patient, or others (pp 704, 706-710)
- Basic principles of the mental health system (pp 702-703)
- Assessment and management of
  - Acute psychosis (p 708)
  - Suicidal/risk (pp 708-709)
  - Agitated delirium (pp 709-710)

Knowledge Objectives

1. Discuss the myths and realities concerning psychiatric emergencies. (p 701)
2. Discuss general factors that can cause alteration in a patient's behavior. (p 701)
3. Define a behavioral crisis. (p 702)
4. Understand the magnitude of mental health problems in society. (pp 702-703)
5. Understand the main principles of how the mental health system functions. (pp 702-703)
6. Describe the two basic categories of diagnosis that a physician will use. (p 703)
7. Discuss special considerations for assessing and managing a behavioral crisis or psychiatric emergency. (pp 703-708)
8. Define acute psychosis. (p 708)
9. Define schizophrenia. (p 708)
10. Describe the care for a psychotic patient. (p 708)
11. Explain how to recognize the behavior of a patient at risk of suicide, and discuss the management of such a patient. (pp 708-709)
12. Define agitated delirium and describe the care for a patient with agitated delirium. (pp 709-710)
13. Discuss the medical and legal aspects of managing a psychiatric emergency. (pp 710-712)
14. Describe methods used to restrain patients. (pp 712-713)
15. Explain the safe management of a potentially violent patient. (pp 713-714)

Skills Objectives

1. Demonstrate the techniques used to mechanically restrain a patient. (pp 712-713)

Introduction

As an EMT, you can expect to deal often with patients undergoing a psychological or behavioral crisis. The crisis may be the result of the emergency situation, mental illness, mind-altering substances, stress, or many other causes. This chapter discusses various kinds of psychiatric emergencies and behavioral crises, including those involving overdoses, violent behavior, and mental illness. You will learn how to assess a person who exhibits signs and symptoms of a psychiatric emergency or a behavioral crisis and what kind of emergency care may be required in these situations. The chapter also covers legal concerns in dealing with disturbed patients. Finally, it describes how to identify and manage a potentially violent patient, including the use of restraints.

Myth and Reality

Everyone develops some symptoms of mental illness at some point in life, but that does not mean everyone develops mental illness. Perfectly healthy people may have some of the symptoms and signs of mental illness from time to time. Therefore, you should not jump to the conclusion that you are mentally disturbed when you behave in certain ways that are discussed in this chapter. With that said, you also should not jump to this same type of conclusion about a patient in any given situation.

The most common misconception about mental illness is that if you are feeling "bad" or "depressed,"

you must be "sick." That is simply untrue. There are many perfectly justifiable reasons for feeling depressed, including divorce, loss of a job, or the death of a relative or friend. For a teenager who just broke up with his girlfriend of 12 months, it is altogether normal to withdraw from ordinary activities and to feel "blue." This is a normal reaction to a crisis situation. However, when a person finds that Monday morning blues last until Friday, week after week, he or she may indeed have a mental health problem.

Many people believe that all individuals with mental health disorders are dangerous, violent, or otherwise unmanageable. This is also untrue. Only a small percentage of people with mental health problems fall into these categories. As an EMT, however, you may be exposed to a higher proportion of violent patients because you are seeing people who are, by definition, considered to be having an emergency; otherwise, you probably would not be seeing them. You have been called because family members or friends felt unable to manage the patient by themselves. The problem may be a result of the use or abuse of drugs or alcohol, or the situation could be that the patient has a long history of mental illness and is reacting to a particularly stressful event.

Although you cannot determine what has caused a person's crisis, you may be able to predict whether the person will become violent. The ability to predict violence is one of your more important assessment tools.

Words of Wisdom

Many people have mental disorders, and many people who do not currently have a mental disorder may have one at some point in their lives. This is no cause for shame. With so many stressors in today's society, it is quite understandable.

You are the Provider: PART 1

At 7:20 PM, you are dispatched to a residence at 517 East Bandera for a man who, according to a neighbor, is sitting in his yard "acting bizarre." Law enforcement is en route to the scene but has not yet arrived. You and your partner acknowledge the call, get into the ambulance, and proceed to the scene. The weather is clear, the temperature is 70°F, and the traffic is moderate.

1. How should you and your partner proceed to this call?
2. Are behavioral emergencies always caused by an underlying psychiatric condition?

Defining a Behavioral Crisis

<u>Behavior</u> is what you can see of a person's response to the environment: his or her actions. Sometimes, it is obvious what a person is responding to: A person is punched, and he or she runs away or bursts into tears or hits back. Sometimes, it is less clear, such as when someone is depressed for very complex emotional and biologic reasons.

Most of the time, people respond to the environment in reasonable ways. Over the years, they have learned to adapt to a variety of situations in daily life, including stresses and strains. This is called adjustment. There are times, however, when the stress is so great that the normal ways of adjusting do not work. When this happens, a person's behavior is likely to change, even if only temporarily. The new behavior may not be appropriate, or "normal."

The definition of a **behavioral crisis** or emergency is any reaction to events that interferes with the **activities of daily living** or has become unacceptable to the patient, family, or community. For example, when someone experiences an interruption of the daily routine, such as bathing, dressing, and eating, chances are his or her behavior has become a problem. For that person, at that time, a behavioral crisis may exist. If the interruption of daily routine tends to recur on a regular basis, the behavior is also considered a mental health problem. It is then a pattern, rather than an isolated incident.

For example, a person who experiences a panic attack after having a heart attack is not necessarily mentally ill. Likewise, you would expect a person who is fired from a job to have some sort of reaction, often sadness and depression. These problems are short-term and isolated events. However, when a person reacts with a fit of rage, attacking people and property or going on a "bender" for a week, this behavior has gone beyond what society considers appropriate or normal. That person is clearly undergoing a behavioral emergency. Usually, if an abnormal or disturbing pattern of behavior lasts for at least a month, it is regarded as a matter of concern from a mental health standpoint. For example, chronic **depression**, a persistent feeling of sadness and despair, may be a symptom of a mental or physical disorder. This type of long-term problem would be labeled a mental health disorder.

When a **psychiatric emergency** arises, the patient may show agitation or violence or become a threat to himself, herself, or others. This is more serious than a more typical behavioral crisis that causes inappropriate behavior such as interference with activities of daily living or bizarre behavior. When an immediate threat to the person involved or to others in the immediate area, including family, friends, bystanders, and EMTs, is present, this situation should be considered a psychiatric emergency. For example, a person might respond to the death of a spouse by attempting suicide. On the other hand, although this is a major life disruption, it does not have to involve violence or harm to a person. Disruption can take many forms (eg, divorce or loss of a job); not all involve violence, nor are they all psychiatric emergencies.

Words of Wisdom

The medicolegal issues associated with responses to a behavioral crisis put added emphasis on your providing thorough and specific documentation. Record detailed, objective findings that support the conclusion of abnormal behavior (eg, withdrawn, will not talk, crying uncontrollably) and quote the patient's own words when appropriate, for example, "Life isn't worth it any more," or "The voices are telling me to kill people." Avoid judgmental statements; these create the impression that you based your care on personal bias rather than the patient's needs.

The Magnitude of Mental Health Problems

According to the National Institute of Mental Health, at one time or another, one in five Americans has some type of **psychiatric disorder**, an illness with psychological or behavioral symptoms that may result in impaired functioning. The mental health system in the United States provides many levels of assistance to people with psychological problems. Common emotional issues such as marital conflict and parenting issues can often be resolved with the assistance of a professional counselor. More serious issues such as clinical depression are often handled by a psychologist who has special training and preparation for dealing with more complex psychological conditions. Some psychological conditions require the services of a psychiatrist who can prescribe medication for the treatment of the most severe psychological conditions, like schizophrenia and bipolar disorder. Most psychological problems can be handled through outpatient visits; however, some people require hospitalization in specialized psychiatric units within the hospital.

Psychiatric disorders have many possible underlying causes. These include social and situational stress such as divorce or death of a loved one, psychiatric diseases such as schizophrenia, physical illnesses such as diabetic emergencies, chemical problems such as alcohol or drug use, or biologic disturbances such as electrolyte imbalances.

Sometimes these conditions can be compounded by noncompliance with prescribed medication regimens.

Pathology

As an EMT, you are not responsible for diagnosing the underlying cause of a behavioral crisis or psychiatric emergency. However, you should know the two basic categories of diagnosis that a physician will use: organic (physical) and functional (psychological).

■ Organic

Organic brain syndrome is a temporary or permanent dysfunction of the brain caused by a disturbance in the physical or physiologic functioning of the brain tissue. Causes of organic brain syndrome include sudden illness; recent trauma to the head; seizure disorders; drug and alcohol intoxication, overdose, or withdrawal; and diseases of the brain, such as Alzheimer disease and meningitis.

Altered mental status can arise from a low level of blood glucose, lack of oxygen, inadequate blood flow to the brain, and excessive heat or cold. An altered mental status, or a change in the way a person thinks or behaves, may be one indicator of a psychiatric disease such as bipolar disorder.

■ Functional

A **functional disorder** is one in which the abnormal operation of an organ cannot be traced to an obvious change in the actual structure or physiology of the organ or organ system. Something has gone wrong, but the root cause cannot be identified as the working of the organ itself. Schizophrenia, anxiety conditions, and depression are good examples of psychiatric disorders. There may be a chemical or physical cause for these disorders, but it is not obvious or well understood.

Words of Wisdom

A patient displaying bizarre behavior may actually have an acute medical illness that is the cause, or a partial cause, of the behavior. Recognizing this possibility may allow you to save a life.

Safe Approach to a Behavioral Crisis

All regular EMT skills—assessment, providing care, patient approach, obtaining the history, and patient communication—are used in a behavioral crisis. However, other management techniques also come into play.

You are the Provider: PART 2

Law enforcement personnel arrive at the scene and radio you that it is safe for you to enter. You find the patient, a 44-year-old man, sitting on the lawn in front of his house; one of the police officers is trying to talk to him. He appears sad and withdrawn and is rocking back and forth. You introduce yourself to the patient and perform a primary assessment.

Recording Time: 0 Minutes	
Appearance	Sad, withdrawn appearance
Level of consciousness	Conscious and alert
Airway	Open; clear of secretions and foreign bodies
Breathing	Normal rate; adequate depth
Circulation	Normal pulse rate; skin is pink and moist; no obvious bleeding

One of the police officers advises you that the patient allowed them to check his person for any weapons, and that he does not have any. The patient tells you that he "has a lot of problems," but nobody will listen to him.

3. What should be your most immediate concern with this patient?

4. How should you proceed with your assessment of this patient?

There is not room in this chapter for a full discussion of these techniques, but you should follow the general guidelines listed in **Table 20-1** to ensure your safety at the scene of a behavioral crisis or psychiatric emergency.

Patient Assessment

Scene Size-up

Scene Safety

The first things for you to consider at the scene of a psychiatric emergency or behavioral crisis are your safety and the patient's response to the environment. Some situations may be more serious than others and, therefore, more threatening to your safety. Is the situation unduly dangerous to you and your partner? Do you need immediate law enforcement backup? Should you stage until law enforcement personnel have secured the scene? Does the patient's behavior seem typical or normal for the circumstances (such as hysterical behavior on receiving notice of the sudden death of a loved one)? Are there legal issues involved (crime scene, consent, refusal)? For example, a patient who has just been assaulted has good reason to be fearful of other people, including you. Make sure to take appropriate standard precautions. Request any additional resources you may need (law enforcement, additional personnel) early. You can always send them away if they are not needed. Be vigilant, and avoid tunnel vision.

Mechanism of Injury/Nature of Illness

Determine the mechanism of injury and/or nature of illness. For example, a patient with diabetes may have an altered mental status because of a low glucose level.

Words of Wisdom

Never leave a patient alone who may be experiencing a behavioral emergency.

Table 20-1 Safety Guidelines for a Behavioral Crisis or a Psychiatric Emergency

- **Be prepared to spend extra time.** It may take longer to assess, listen to, and prepare the patient for transport. Remember to always treat the patient with respect.
- **Have a definite plan of action.** Decide who will do what. If restraint is needed, how will it be accomplished? Avoid restraint unless it is absolutely necessary.
- **Identify yourself calmly.** Try to gain the patient's confidence. Ask questions in a low, calm voice, and be patient in your attitude. Reassure the patient that you are there to help.
- **Be direct.** State your intentions and what you expect of the patient. Let the patient know what you are doing and maintain good eye contact.
- **Assess the scene.** If the patient is armed or has potentially harmful objects in his or her possession, have these removed by law enforcement personnel before you provide care.
- **Stay with the patient.** Do not let the patient leave the area, and do not leave the area yourself unless law enforcement personnel can stay with the patient. Try to remove from the area any stimulus that is distressing to the patient.
- **Encourage purposeful movement.** Help the patient to get dressed and gather appropriate belongings to take to the hospital.
- **Express interest in the patient's story.** Let the patient tell you what happened or what is going on now in his or her own words. However, do not play along with auditory or visual disturbances or hallucinations.
- **Do not get too close to the patient.** Everyone needs personal space, so avoid unnecessary physical contact. You want to be able to move quickly if the patient becomes violent or tries to run away, but otherwise do not make quick moves. Do not physically talk down to or directly confront the patient. A squatting, 45° angle approach is usually not confrontational but the position may hinder your movements. Do not allow the patient to get between you and the exit.
- **Avoid fighting with the patient.** Do not threaten or belittle the patient. Remember, the patient is not responding to you in a normal manner; he or she may be wrestling with internal forces over which neither of you has control. You and others may be stimulating these inner forces without knowing it. If you can respond with understanding to the feeling that the patient is expressing, whether the feeling is anger or fear or desperation, you may be able to gain his or her cooperation. If possible, try to involve a friend or family member whom the patient trusts. Always try to talk the patient into cooperation.
- **Be honest and reassuring.** If the patient asks whether he or she has to go to the hospital, the answer should be, "Yes, that is where you can receive medical help."
- **Do not judge.** You may see patient behavior that you dislike. Set those feelings aside and concentrate on providing emergency medical care to your patient.

Primary Assessment

Form a General Impression

Begin your assessment from the doorway or from a distance. How does the patient appear? Calm? Agitated? Awake or sleepy? Begin with an introduction of who you are, and let the patient know that you are there to help. Allow the patient to tell you what happened or how he or she feels. Perform a rapid scan.

Observe the patient closely. Does the patient answer slowly with single word answers or rapidly in long, rambling sentences? Is the patient sitting, slumped in a chair; hunched and shuffling around the room; or rigid and standing perfectly still? Is the patient alert and oriented? Use the AVPU (*Alert* to person, place, and day; responsive to *Verbal* stimuli; responsive to *Pain*; *Unresponsive*) scale to check for alertness. To determine orientation, ask the patient, "Where are you?" and "What is the reason you are here?" The answers to these questions will allow you to begin to establish a rapport with the patient; this rapport is critical to the success of your further actions.

Almost every situation, medical or trauma, will have some behavioral component. For example, a depressed person may slit his or her wrists, causing traumatic bleeding. A patient experiencing a heart attack may have difficulty breathing and an increased pulse due to anxiety. It is just as important to treat mental illness as it is to treat the medical or traumatic problem; however, the focus of the primary assessment is assessing and treating life threats.

Airway and Breathing

If your patient is in physical distress, assess the airway to make sure it is patent and adequate. Next, evaluate the patient's breathing. Provide the appropriate interventions based on your assessment findings. Some behavioral situations will involve a compromised airway and inadequate breathing because of a suicide attempt from ingesting a handful of sleeping pills with alcohol.

Circulation

Next, you will need to assess the pulse rate, quality, and rhythm. Obtain the systolic and diastolic blood pressures when possible. Assessing a patient's circulation includes an evaluation for the presence of shock and bleeding. Assess the patient's perfusion by evaluating skin color, temperature, and condition.

Transport Decision

Unless your patient is unstable from a medical problem or trauma, prepare to spend time at the scene with your patient. Depending on your local protocol, there may be a specific facility to which patients with mental problems are transported.

History Taking

Investigate Chief Complaint

When a medical patient is conscious, the next step of your assessment is to investigate the chief complaint and then obtain a SAMPLE history. At this stage of your assessment, the majority of your time will be spent asking the patient about his or her medical history.

In trying to determine the reason for the patient's behavioral state, your assessment should consider three major areas as possible contributors:

- Is the patient's central nervous system functioning properly? For example, the patient may be experiencing diabetic problems, particularly hypoglycemia. This situation could cause the patient to behave in an unusual or irrational manner.
- Are hallucinogens or other drugs or alcohol a factor? Does the patient see strange things? Is everything distorted? Do you smell alcohol on the patient's breath?
- Are psychogenic circumstances, symptoms, or illness (caused by mental rather than physical factors) involved? These might include the death of a loved one, severe depression, history of mental illness, threats of suicide, or some other major interruption of activities of daily living.

SAMPLE History

A complete and careful SAMPLE history will be helpful in treating your patient and passing on information to personnel at the receiving facility. You may be able to elicit information not available to the hospital staff. Ask specifically about previous episodes, treatments, hospitalizations, and medications related to behavioral problems Table 20-2.

Is Alzheimer disease or another type of dementia a possible cause? In geriatric patients, consider Alzheimer disease and dementia as possible causes of abnormal behavior. In these cases, it is essential to obtain information from relatives, friends, or extended care facility staff. Determining the patient's baseline mental status will be essential in guiding your treatment and transport decisions and will also be extremely helpful to hospital personnel.

Family, friends, and observers may be of great help in answering these questions. Together with your observations and interaction with the patient, they should provide enough data for you to assess the situation. This assessment has two primary goals: recognizing major threats to life and reducing the stress of the situation as much as possible.

Reflective listening is a technique frequently used by mental health professionals to gain insight into a patient's

Table 20-2   Questions to Ask in Evaluating a Mental Health Disorder

- Does the patient answer your questions appropriately?
- Does the patient's behavior seem appropriate?
- Does the patient seem to understand you and the surroundings?
- Is the patient withdrawn or detached? Hostile or friendly? Elated or depressed?
- Are the patient's vocabulary and expressions what you would expect under the circumstances?
- Does the patient seem aggressive or dangerous to you or others?
- Is the patient's memory intact? Check orientation to time, place, person, and event: What day, month, and year is it? Who am I?
- Does the patient express disordered thoughts, delusions, or hallucinations?

thinking. It involves repeating, in question form, what the patient has said, encouraging the patient to expand on the thoughts. Although it often requires more time to be effective than is available in an EMS setting, it may be a helpful tool for you to use when other techniques are unsuccessful at gathering the patient's history.

Words of Wisdom

When assessing a patient in a psychiatric emergency or behavioral crisis, it can be very useful to gather information separately from a relative or caregiver. Splitting up the history-taking process in this way often yields valuable information and can help reduce the potential for violence when there is tension between the people involved. However, if the patient is threatening or uncontrolled, do not leave the room to obtain a more detailed history unless additional people such as law enforcement personnel are there to help.

Secondary Assessment

Physical Examinations

In an unconscious medical patient, begin with a full-body scan to look for a reason for the unresponsiveness. Follow this rapid check for hidden life threats with a detailed full-body scan and obtain a complete set of vital signs. Then gather what history you can from others. When physically examining a patient with a behavioral emergency, remember to check for track marks indicating drug abuse and for signs of self-mutilation.

A physical examination for a behavioral problem may be difficult to perform but may provide clues to the patient's state of mind and thinking. Some patients

welcome physical contact as reassuring, but others may feel acutely threatened. Avoid touching the patient without permission. In fact, this is a good practice for all patients.

Sometimes even a patient who is conscious in a behavioral crisis or psychiatric emergency will not respond at all to your questions. In those cases, you may be able to tell quite a lot about the patient's emotional state from facial expressions, pulse rate, and respirations. Tears, sweating, and blushing may be significant indicators of state of mind such as sadness, nervousness, or embarrassment. Also, make sure that you look at the patient's eyes; a patient who has a blank gaze or rapidly moving eyes may be experiencing central nervous system dysfunction **Figure 20-1**.

A behavioral crisis or psychiatric emergency puts tremendous stress on a person's coping mechanisms. The person is actually incapable of responding reasonably to the demands of the environment. This state may be temporary, as in an acute illness like drug-induced hallucinations, or longer lived, as in a complex, chronic mental illness such as schizophrenia. The patient's perception of reality may be compromised or distorted.

Vital Signs

Obtain vital signs when doing so will not exacerbate your patient's emotional distress. Make every effort to assess blood pressure, pulse, respirations, skin, and pupils. Remember that a behavioral crisis can be caused or precipitated by physiologic problems such as head injuries or diabetic disorders, and they can aggravate preexisting conditions. Do not forget that the physical person and the emotional person are one.

Monitoring Devices

When it will not exacerbate your patient's emotional distress, you may use monitoring devices to quantify

Figure 20-1 Making eye contact with a patient can provide useful clues about a patient's emotional state.

your patient's oxygenation and circulatory status. It is recommended that you always assess the patient's first blood pressure with a sphygmomanometer (blood pressure cuff) and a stethoscope. A pulse oximetry device, if available, can be used to assess the patient's perfusion status.

Reassessment

Never let your guard down. Most patients you are called to treat and transport with emotional complaints pose no danger to you or others on your crew, but it is impossible to determine this while on the scene. Remember that many patients experiencing a behavioral crisis will act spontaneously. Be prepared to intervene quickly. If restraints are necessary, reassess and document the patient's respirations, as well as pulse and motor and sensory function in all restrained extremities, every 5 minutes. Respiratory and circulatory problems have been known to occur in combative patients who are restrained. When available, have additional personnel such as law enforcement officers or fire fighters accompany you in the back of the ambulance during transport. This provides you with additional assistance should the patient's behavior change rapidly.

Words of Wisdom

Behavioral changes may be the result of or a symptom of a treatable medical condition, such as diabetes or a stroke.

Interventions

As much as your heart may go out to an emotionally distressed patient, there often is little you will be able to do for the patient during the short time you will be treating him or her. Your job is to diffuse and control the situation and safely transport your patient to the hospital. Intervene only as much as it takes to accomplish these tasks. Be caring and careful. If you have determined that it is necessary to restrain your patient, release the restraints only if necessary to provide patient care.

Communication and Documentation

Try to give the receiving hospital advance warning when a patient experiencing a psychiatric emergency is coming in. Many hospitals require extra preparation to ensure that appropriate staff and rooms are available. Report whether restraints will be required when the patient arrives at the hospital. Document thoroughly and

You are the Provider: PART 3

The patient tells you that his wife was killed in a car accident 1 year ago today and that he has been to numerous counseling sessions over the past year, but they have not seemed to help him. He further tells you that his employer does not seem to care about his problems and has threatened to fire him unless he "snaps out of it." He allows your partner to take his vital signs but refuses to allow you to perform any other assessment or treatment.

Recording Time: 9 Minutes	
Respirations	16 breaths/min; adequate depth
Pulse	88 beats/min; strong and regular
Skin	Pink, warm, and moist
Blood pressure	144/84 mm Hg
Oxygen saturation (Sao$_2$)	98% (on room air)

The patient denies chest pain, shortness of breath, or any other physical symptoms. He tells you that it is extremely difficult for him to even get out of bed and face each day and that he feels as though his life no longer serves any purpose.

5. What is your field impression of this patient?

6. What care can you provide to this patient in the field?

carefully. Think about what you are going to write before you write it, so that you can describe what are often confusing scenes as clearly as possible. Because psychiatric emergencies have few or no physical signs, yours may be the only documentation about the patient's distress. Because psychiatric emergencies are fraught also with legal dangers, document everything that occurred on the call, particularly situations that required restraint. When restraints are required to protect you or the patient from harm, include why and what type of restraints were used. This information is essential if the case is reviewed for medicolegal reasons.

Acute Psychosis

Psychosis is a state of delusion in which the person is out of touch with reality. Affected people live in their own reality of ideas and feelings. To the person experiencing a psychotic episode, the line between their reality and fantasy is blurred. That reality may make patients belligerent and angry toward others. Patients may become silent and withdrawn as they give all their attention to the voices and feelings within. Psychotic episodes occur for many reasons; the use of mind-altering substances is one of the most common causes, and that experience may be limited to the duration of the substance within the body. Other causes include intense stress, delusional disorders, and, more commonly, schizophrenia. Some psychotic episodes last for brief periods; others last a lifetime.

Schizophrenia

Schizophrenia is a complex disorder that is not easily defined or easily treated. The typical onset occurs during early adulthood, with symptoms becoming more prominent over time. Some people diagnosed with schizophrenia display signs during early childhood; their disease may be associated with brain damage or may have other causes. Other influences thought to contribute to this disorder include genetics and psychological and social influences. Persons with schizophrenia may experience symptoms including delusions, hallucinations, a lack of interest in pleasure, and erratic speech.

Dealing with a psychotic patient is difficult. The usual methods of reasoning with a patient are unlikely to be effective because the psychotic person has his or her own rules of logic that may be quite different from nonpsychotic thinking. Follow these guidelines in dealing with a psychotic patient:

- Determine if the situation is a danger to yourself or others.
- Identify yourself clearly. ("I'm Gloria. I'm an EMT with the ambulance service, and this is my partner, Stan. We've come to see if we can help. Can you tell us about your problem?")
- Be calm, direct, and straightforward. Your calmness and confidence can do a great deal toward calming the patient.
- Maintain an emotional distance. Do not touch the patient, and do not be overly friendly or effusively reassuring. Convey a calm attitude.
- Do not argue. Do not challenge patients regarding the reality of their beliefs or the validity of their perceptions. Do not go along with their delusions simply to humor them, but do not make an issue of the delusions. Talk about real things.
- Explain what you would like to do. ("Let's walk downstairs to the ambulance.")
- Involve people the patient trusts, such as family or friends, to gain patient cooperation.

Suicide

The single most significant factor that contributes to suicide is depression. Any time that you encounter an emotionally depressed patient, you must consider the possibility of suicide. The risk factors for suicide are listed in **Table 20-3**.

It is a common misconception that people who threaten suicide never commit it. This is not correct. Suicide is a cry for help. Threatening suicide is an indication that someone is in a crisis that he or she cannot handle alone. Immediate intervention is necessary.

Whether the patient has any of these risk factors, you must be alert to the following warning signs:

- Does the patient have an air of tearfulness, sadness, deep despair, or hopelessness that suggests depression?
- Does the patient avoid eye contact, speak slowly or haltingly, and project a sense of vacancy, as if he or she really is not there?
- Does the patient seem unable to talk about the future? Ask the patient whether he or she has any vacation plans. Suicidal people consider the future so uninteresting that they do not think about it; people who are seriously depressed consider the future so distant that they may not be able to think about it at all.
- Is there any suggestion of suicide? Even vague suggestions should not be taken lightly, even if presented as a joke. If you think that suicide is a possibility, do not hesitate to bring the subject up. You will not "give the patient ideas" if you ask directly, "Are you considering suicide?"
- Does the patient have any specific plans relating to death? Has the patient recently prepared a will? Given away significant possessions or advised close

Table 20-3 Risk Factors for Suicide

- Depression at any age, including feeling trapped, purposeless, or hopeless
- Previous suicide attempt (About 80% of successful suicides were preceded by at least one attempt.)
- Current expression of wanting to commit suicide or sense of hopelessness; specific plan for suicide
- Family history of suicide
- Older than 40 years, particularly for single, widowed, divorced, alcoholic, or depressed people (Men in this category who are older than 55 years have an especially high risk and are very often successful if they make an attempt.)
- Recent loss of spouse, significant other, family member, or support system
- Chronic debilitating illness or recent diagnosis of serious illness
- Feeling anxious, agitated, angry, reckless, or aggressive; also dramatic mood changes such as from depression to agitation
- Financial setback, loss of job, police arrest, imprisonment, or some sort of social embarrassment
- Substance abuse, particularly with increasing use
- Children of an alcoholic or abusive parent
- Withdrawal from family and friends or a lack of social support, resulting in isolation
- Anniversary of death of loved one, job loss, marriage after the death of a spouse, and so forth
- Unusual gathering or new acquisition of things that can cause death, such as purchase of a gun, a large volume of pills, or increased use of alcohol

friends what he or she would like done with them? Arranged for a funeral service? These are critical warning signs.

Consider also the following additional risk factors for suicide:

- Are there any unsafe objects in the patient's hands or nearby (for example, a sharp knife, glass, poisons, or gun)?
- Is the environment unsafe (for example, an open window in a high-rise building, a patient standing on a bridge or precipice)?
- Is there evidence of self-destructive behavior (for example, partially cut wrists, large alcohol or drug intake)?
- Is there an imminent threat to the patient or others?
- Is there an underlying medical problem?
- Are there cultural or religious beliefs promoting suicide?
- Has there been trauma?

On the basis of your observations and conversation with the patient, you may need to determine if interventions such as restraints are needed. Remember, a suicidal patient may be homicidal as well. Do not jeopardize your life or the lives of your fellow EMTs. If you have reason to believe that you are in danger, you must obtain police intervention. In the meantime, try not to frighten the patient or make him or her suspicious. Remember, the most important service you can provide for a suicidal patient is compassionate transportation to a medical facility where the patient can receive proper treatment.

Safety

Patients with suicidal thoughts, especially patients who have made a threat or unsuccessful attempt, may not be thinking clearly and may behave in very unpredictable ways. Some recognize that if they get into the ambulance or enter the hospital, they will not have the opportunity to complete their threat or gesture. Therefore, they may make a last effort to kill themselves. Suicidal/homicidal patients will not hesitate to hurt you or your partner. Be very careful how you assess the situation, making certain that you, your team, and the patient are safe.

Agitated Delirium

A problem sometimes encountered in an EMS response is **agitated delirium**. Delirium is a condition of impairment in cognitive function that can present with disorientation, hallucinations, or delusions. Agitation is a behavior that is characterized by restless and irregular physical activity. Although patients experiencing delirium are generally not dangerous, if they exhibit agitated behavior they may strike out irrationally. One of the most important factors to consider in these cases is your personal safety.

The symptoms of agitated delirium may include hyperactive irrational behavior with inattentiveness and possible vivid hallucinations. Common physical symptoms include hypertension, tachycardia, diaphoresis, and dilated pupils. Because hallucinations are erroneous perceptions of reality, the patient may perceive you as a threat. Agitation is recognized as a biologic attempt to release nervous tension and can produce sudden, unpredictable physical actions in your patient.

If you think that you can safely approach the patient, be very calm, supportive, and empathetic. Be an active listener by nodding, indicating understanding, and by limiting your interruptions of the patient's comments. It is extremely important to approach the patient slowly and purposefully and to respect the patient's territory. Limit physical contact with the patient as much as possible. It is also imperative that the patient not be left unattended, unless the situation becomes unsafe for you or your partner.

Use careful interviewing to assess the patient's cognitive functioning. Try to indirectly determine the patient's orientation, memory, concentration, and judgment by asking simple questions such as "When did you first begin to notice these feelings?" Through interviewing, try to determine what the patient is thinking. Are the patient's thoughts disorganized? For example, does the patient begin to answer your question and then drift off only to begin discussing a childhood friend? Is the patient experiencing delusions or hallucinations? Does the patient have any unusual worries or fears? For example, does the patient express anxiety if you go too close to a pile of old newspapers?

Pay particular attention to the patient's ability to communicate clearly, and make notes on the patient's apparent mood. Is the patient anxious, depressed, elated (extremely happy or joyful) under inappropriate circumstances, or agitated? Pay attention to the patient's appearance, dress, and personal hygiene.

If you determine that the patient requires restraint because he or she is a threat to himself or herself or others, make sure you have adequate, well-trained personnel available to help you before approaching the patient.

If the patient appears to be experiencing an overdose, take all medication bottles or illegal substances with you to the medical facility. The patient should be transported to a hospital with psychiatric facilities capable of handling the condition. Whenever possible, refrain from using lights and sirens because these sights and sounds may aggravate the patient's condition.

Medicolegal Considerations

The medical and legal aspects of emergency medical care become more complicated when the patient is undergoing a behavioral crisis or psychiatric emergency. Nevertheless, legal problems are greatly reduced with an emotionally disturbed patient who consents to care. Gaining the patient's confidence is, therefore, a critical task for you.

Mental incapacity can take many forms: unconsciousness (as a result of hypoxia, alcohol, or drugs), temporary but severe stress, and depression. Once you have determined that a patient has impaired mental capacity, you must decide whether he or she requires immediate emergency medical care. A patient in a mentally unstable condition may resist your attempts to provide care. Nevertheless, you must not leave this patient alone. Doing so may result in harm to the patient and expose you to civil

You are the Provider: PART 4

The patient is initially reluctant to allow you to transport him to the hospital. However, after you express your concern about his safety and well-being, he consents to transport. Other than allowing you to reassess his vital signs, he tells you that he would prefer to just be taken to the hospital; he does not want to be physically assessed.

Recording Time: 20 Minutes	
Level of consciousness	Conscious and alert
Respirations	16 breaths/min; adequate depth
Pulse	76 beats/min; strong and regular
Skin	Pink, warm, and moist
Blood pressure	138/86 mm Hg
Sao$_2$	98% (on room air)

7. Should you perform a physical assessment of this patient despite the fact that he requested that you not do so? Why or why not?

action for abandonment or negligence. In such situations, you should request that law enforcement personnel handle the patient. Another reason for seeking law enforcement support is for the patient who resists treatment; such a patient often threatens EMTs and others. Violent or dangerous people must be taken into custody by law enforcement before emergency care can be rendered.

■ Consent

When a patient is not mentally competent to grant consent for emergency medical care, the law assumes that there is implied consent. For example, the consent of an unconscious patient is implied if life or health is at risk. The law refers to this as the emergency doctrine: Consent is implied because of the necessity for immediate emergency treatment. In a situation that is not immediately life threatening, emergency medical care or transportation may be delayed until the proper consent is obtained.

In cases involving psychiatric emergencies, however, the matter is not always clear-cut. Does a life-threatening emergency exist or not? If you are not sure, you should request the assistance of law enforcement personnel.

■ Limited Legal Authority

As an EMT, you have limited legal authority to require or force a patient to undergo emergency medical care when no life-threatening emergency exists. Patients have the right to refuse care. However, most states have legal statutes regarding the emergency care of mentally ill and drug-impaired people. These statutory provisions permit law enforcement personnel to place the person in protective custody so that emergency care can be given. You should be familiar with your local and state laws regarding these situations.

The typical provision may state that:

> Any police officer who has reasonable cause to believe that a person is mentally ill and dangerous to himself, herself, or others or gravely disabled . . . may take such person into custody and take or cause such person to be taken to a general hospital for emergency examination . . .

Again, because these provisions vary, you should become familiar with the provisions in your state.

The general rule of law is that a competent adult has the right to refuse treatment, even if lifesaving care is involved. In psychiatric cases, however, a court of law would probably consider your actions in providing lifesaving care to be appropriate, particularly if you have a reasonable belief that the patient would harm himself, herself, or others without your intervention. In addition, a patient who is in any way impaired, whether by mental illness, medical condition, or intoxication,

Special Populations

As the population ages, you will begin to see more patients older than 65 years. In responding to an increasing number of geriatric patients, you will probably notice some behavioral or psychiatric problems, including depression, dementia, and delirium. These mental status changes can affect your ability to thoroughly assess and treat an ill or injured geriatric patient. Understanding the causes of altered behavior in geriatric patients will help you in patient care.

Depression is one of the more common mental status problems that you will see in older people. As an EMT, you can recognize a problem and perhaps prevent a suicide in a depressed older person.

Depression has a number of causes. A major illness such as cancer or dementia can lead to depression. Furthermore, medications can induce a feeling of depression, possibly because of an interaction with other drugs. In addition, changes in the endocrine system, such as menopause, can elicit depression. Depression might also be caused by an imbalance in brain chemicals.

With all the possible causes of depression, an older adult can feel helpless and hopeless. A depressed person can be argumentative or placid. He or she might trivialize complaints, not wanting to be a bother to anyone. Someone who sees no way out of his or her situation may turn to suicide. Be alert for a suicidal gesture or ideation, even though it may not be obvious.

Although depression can create behavioral problems in geriatric patients, dementia is another cause of abnormal behavior. The most common cause of dementia is primary progressive dementia, also known as Alzheimer dementia. It is estimated that 12.5% of the population older than 65 years and a much higher percentage of the population older than 85 years has Alzheimer dementia. Currently, there is no cure for Alzheimer dementia, but there are medications that can slow progress of the disease.

During the progression of the disease, the patient can develop openly hostile behavior, kicking, yelling, pinching, and hitting you, your partner, or the patient's caregiver. You might need to restrain a violent patient, but do so gently and only to the point at which the violent behavior stops.

Other causes of altered behavior include diabetic emergencies, heat- and cold-related illnesses, poisoning and overdose, strokes and transient ischemic attacks, and infection. Although the mechanism is not understood, urinary tract infection and constipation can each alter an older person's behavior.

As the EMT responding to a call for help, you should accept the possibility of depression in a geriatric patient. Do not discount the patient's feelings or devalue his or her emotions. Be alert for a suicidal gesture, and pay attention to any statements about death. To get the patient's cooperation, you can elicit his or her help in providing care for the acute illness or injury. A smile and a touch can go a long way in alleviating fear in all of your patients, especially older patients.

may not be considered competent to refuse treatment or transportation. These situations are among the most perilous you will encounter from a legal standpoint. When in doubt, consult your supervisor, police, or medical control. Always maintain a pessimistic attitude toward your patient's condition—assume the worst and hope for the best. Err on the side of treatment and transport. It is far easier to defend yourself against charges of battery than it is to justify abandonment.

Restraint

Ordinarily, restraint of a person must be ordered by a physician, a court, or a law enforcement officer. If you restrain a person without authority in a nonemergency situation, you expose yourself to a possible lawsuit and to personal danger. Legal actions against you can involve charges of assault, battery, false imprisonment, and violation of civil rights. You may use restraints only to protect yourself or others from bodily harm or to prevent the patient from causing injury to himself or herself **Figure 20-2**. In either case, you may use only reasonable force as necessary to control the patient, something that different courts may define differently. For this reason, you should always consult medical control and contact law enforcement personnel for help before restraining a patient.

In fact, you probably should always involve law enforcement personnel if you are called to assist a patient in a severe behavioral crisis or psychiatric emergency. They will provide physical backup in managing the patient and serve as the necessary witnesses and legal authority to restrain the patient. A patient who is restrained by law enforcement personnel is in their custody.

Always try to transport a disturbed patient without restraints if possible. Once the decision has been made to restrain a patient, however, you should carry it out quickly. Be aware of standard precautions. If the patient is spitting, place a surgical mask over his or her mouth.

Make sure you have adequate help to restrain a patient safely. At least four people should be present to carry out the restraint, each being responsible for one extremity. Before you begin, discuss the plan of action. As you prepare to restrain the patient, stay outside the patient's range of motion.

In subduing a disturbed patient, use the minimum force necessary. You should avoid acts of physical force that may cause injury to the patient. The level of force will vary, depending on the following factors:

- The degree of force that is necessary to keep the patient from injuring himself, herself, and others.
- A patient's sex, size, strength, and mental status including the possibility of drug-induced states. Phenylcyclohexylpiperidine (PCP) use may make the patient especially difficult to restrain.
- The type of abnormal behavior the patient is exhibiting. You should use only restraint devices that have been approved by your state's health department for this purpose; soft, wide, leather, or cloth restraints are preferred to police-type handcuffs.

Acting at the same time, the police officers should secure the patient's extremities with approved equipment. Somebody, preferably you or your partner, should continue to talk to the patient throughout the process. Remember to treat the patient with dignity and respect at all times. Also, monitor the patient for vomiting, airway obstruction, and cardiovascular stability because the patient cannot fend for himself or herself. Drug or alcohol intoxication may cause violent behavior but then lead to such physical problems as well. Never place your patient facedown because it is impossible to adequately monitor the patient and this position may inhibit the breathing of an impaired or exhausted patient. Be careful not to place restraints in such a way that respiration is compromised. Reassess airway and breathing continuously. You should make frequent checks of circulation on all restrained extremities, regardless of patient position **Figure 20-3**. Document the reason for the restraint and the technique that was used. Be especially careful if a combative patient suddenly becomes calm and cooperative. This is the time not to relax but to secure the situation. The patient may suddenly become combative again and injure

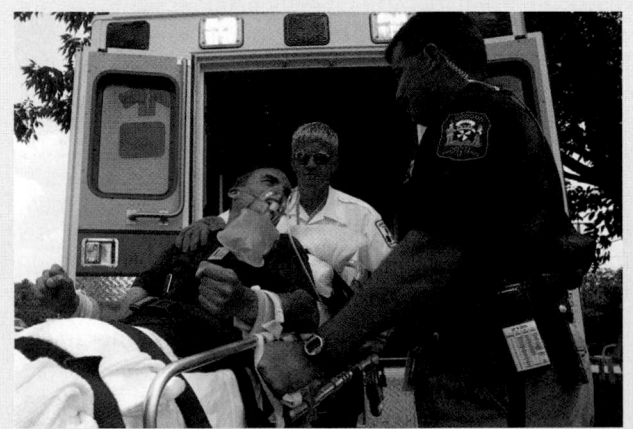

Figure 20-2 You may use restraints only to protect yourself or others or to prevent a patient from causing injury to himself or herself.

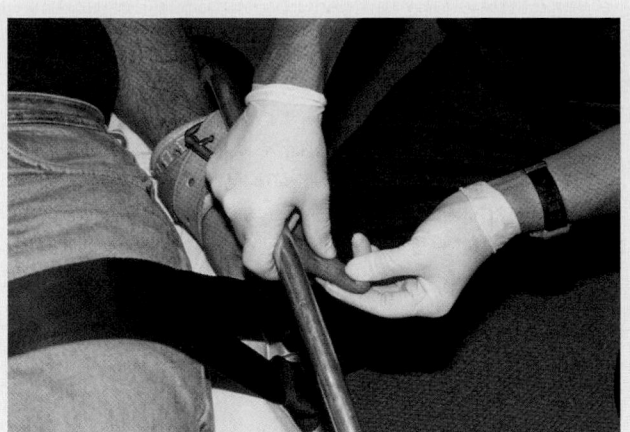

Figure 20-3 Assess circulation frequently while a patient is restrained.

someone. Keep in mind that you may use reasonable force to defend yourself against an attack by an emotionally disturbed patient. It is extremely helpful to have (and document) witnesses in attendance even during transport to protect against false accusations. EMTs have been accused of sexual misconduct and other physical abuse in such circumstances.

The Potentially Violent Patient

Violent patients make up only a small percentage of the patients undergoing a behavioral or psychiatric crisis. However, the potential for violence is always an important consideration for you Figure 20-4 .

Words of Wisdom

When working with a potentially hostile or violent patient, remove everyone who is not needed, such as family, friends, or bystanders, from the scene. This will prevent injury or involvement of others.

Use the following list of risk factors to assess the level of danger:

- **History.** Has the patient previously exhibited hostile, overly aggressive, or violent behavior? Ask people at the scene, or request this information from law enforcement personnel or family.
- **Posture.** How is the patient sitting or standing? Is the patient tense, rigid, or sitting on the edge of his or her seat? Such physical tension is often a warning signal of impending hostility.

You are the Provider: PART 5

You depart the scene and begin transport to a hospital located 8 miles away. The patient remains conscious and alert but is still obviously withdrawn and sad. You ask him additional questions about his present situation, but he does not answer you; he briefly looks up at you and then looks back down. You reassess his vital signs and then call your radio report in to the receiving facility.

Recording Time: 30 Minutes	
Level of consciousness	Conscious and alert
Respirations	16 breaths/min; adequate depth
Pulse	72 beats/min; strong and regular
Skin	Pink, warm, and dry
Blood pressure	130/80 mm Hg
Sao$_2$	99% (on room air)

You arrive at the destination hospital and give your verbal report to the charge nurse. After transferring patient care to the hospital staff, you return to service.

8. What factors should you consider before transporting a patient with an emotional crisis?

9. If the patient does not answer your questions, should you continue to encourage him to talk? Why or why not?

Figure 20-4 The potential for violence is an important consideration for EMTs.

- **The scene.** Is the patient holding or near potentially lethal objects such as a knife, gun, glass, poker, or bat (or near a window or glass door)?
- **Vocal activity.** What kind of speech is the patient using? Loud, obscene, erratic, and bizarre speech patterns usually indicate emotional distress. Someone using quiet, ordered speech is not as likely to strike out as someone who is yelling and screaming.
- **Physical activity.** The motor activity of a person undergoing a psychiatric crisis may be the most telling factor of all. A patient who has tense muscles, clenched fists, or glaring eyes; is pacing; cannot sit still; or is fiercely protecting personal space requires careful watching. Agitation may predict a quick escalation to violence.

Other factors to consider in assessing a patient's potential for violence include the following:

- Poor impulse control
- A history of truancy, fighting, and uncontrollable temper
- Tattoos, especially those with gang identification, prison tattoos, or statements such as "Born to Kill" or "Born to Lose"
- Substance abuse
- Depression, which accounts for 20% of violent attacks
- Functional disorder (If the patient says that voices are telling him or her to kill, believe it.)

Special Populations

In general, children may experience behavioral crisis as commonly as adults, but often the children's situations are managed by their parents or caregivers. If you are called to help with a child experiencing a behavioral crisis, it is imperative to listen to the caregiver and follow his or her lead on how to best approach the child. Aggressive behavior in children, especially when it seems to be a pattern, may be a symptom of an underlying medical or psychological condition. As a precaution against them hurting themselves or others, the children need a thorough evaluation from a mental health professional. Although some children with a behavioral crisis may be physically large, do not make the mistake of assuming you can treat them like adults.

One specific behavioral problem that is common among teenagers is suicide. Teenagers are one of the largest age groups to attempt suicide. Although we sometimes tend to view a teenager's problems as minor, the problems often appear insurmountable to them. It is important to never discount a teenager's comments about suicide as being "just an attempt to get attention."

Common factors that lead to suicide attempts in adults are often also found in teenagers. One of the common issues is dealing with the termination of a relationship. Teenagers are just beginning to relate to others in an intimate way, so when a relationship ends, they often do not know how to handle the apparent rejection. Adults who attempt suicide may have drug or alcohol problems, and these problems are common among teenagers as well. Teenage suicide victims may have a history of disciplinary problems or may have a very unstable home life. Another factor that sometimes comes into play with teenage suicide victims is social pressures. Peer approval is one of the most important aspects of a teenager's life, and teenagers who seem to have poor relationships with their peers may be at a higher risk for suicide. Another risk factor to consider is that children of parents who commit suicide are more likely to attempt it themselves.

Expression of thoughts of suicide and attempts at suicide by teenagers should always be taken seriously. Even if a teenager appears to be using suicide as an "attention getter," the teenager should be treated as if he or she is earnest. Never disregard suicide comments, even if a parent insists that the child "is faking." Take action to ensure that teenagers are treated for attempting or considering suicide because it is very important for their long-term emotional well-being.

You are the Provider: SUMMARY

1. How should you and your partner proceed to this call?

This is not a scene that you should enter without the protection of law enforcement! The patient's reported behavior, "acting bizarre," is a very broad description that could indicate any number of problems. However, in the interest of your own safety, you should assume that the patient is a danger to himself or others.

Never approach a scene of actual or potential violence until law enforcement personnel have arrived and deemed the scene safe for you to enter. In this case, you and your partner should stage two or three blocks away from the scene and wait for law enforcement personnel to arrive. Proceed to the scene *only* after they have notified you by radio and given you the "all clear."

2. Are behavioral emergencies always caused by an underlying psychiatric condition?

A multitude of factors can affect a person's behavior; the presence of an underlying psychiatric condition is but one. Conditions such as drug use, hypoxemia, hypoglycemia, head trauma, and brain tumors (among others) can profoundly affect a person's behavior; these conditions may even cause the patient to become violent.

Behavior is defined as a person's response to his or her environment; how he or she reacts. Sometimes, it is obvious what a person is responding to (eg, being punched and hitting back or running away in tears). In other cases, it is less obvious, such as when a person is depressed for very complex reasons.

In most cases, people respond to environmental changes reasonably and rationally. They have learned to adapt to a variety of situations in daily life, including stresses and strains; this process is called adjustment. There are times, however, when the stress is so great that the usual ways of adjusting simply do not work. As a result, the person's behavior is likely to change—even if it is only temporary—and this new behavior may not be appropriate or "normal."

A behavioral crisis or emergency is defined as any reaction to an event (or events) that interferes with the activities of daily living or has become unacceptable to the patient, his or her family and friends, or the community in general.

If a person is fired from his or her job, you would expect him or her to experience some kind of behavioral reaction, usually sadness and depression. Although these feelings would be distressing to the person, they are a predictable response given the situation. Furthermore, the person's family and friends would likely not find his or her behavior to be unacceptable. This type of behavioral reaction is an example of a short-term isolated event that stems from a sudden life change, not an underlying psychiatric condition. In contrast, if a person reacts to a situation by attacking people or property, he or she has gone beyond what society considers appropriate or normal behavior and is, therefore, experiencing a behavioral "emergency."

Typically, if an abnormal or disturbing pattern of behavior lasts for at least a month, it is regarded as a matter of concern from a mental health perspective. For example, a person who has *persistent* delusions, hallucinations, disorganized speech, and bizarre behavior that cannot be linked to an underlying medical condition or environmental cause likely has a psychiatric condition.

3. What should be your most immediate concern with this patient?

Personal safety should be your primary concern when caring for *any* patient; however, this is especially true when dealing with patients who are displaying abnormal or bizarre behavior.

When caring for a patient who is experiencing a behavioral crisis, you must always consider his or her potential for violence. There are certain behaviors and risk factors that you should look for when assessing a patient's potential for violence: patient history, posture, and verbal and physical activity.

Does the patient have a history of hostile, overly aggressive, or violent behavior? This information may be obtained from friends, family members, or law enforcement personnel. In this case, you should ask the patient's neighbor; since he lives next door to the patient, he may be able to tell you how the patient usually behaves. Ask law enforcement personnel if they have responded to this patient's residence in the past for similar behavior.

Note how the patient is sitting or standing. Is he or she tense, rigid, or sitting on the edge of his or her seat? Your patient is sitting on the ground, rocking back and forth; this could indicate a general state of nervousness or increasing agitation.

Observe the patient's speech. Loud, obscene, erratic, or bizarre speech patterns are clear indicators of emotional distress. Although a patient with quiet, organized speech is less likely to turn violent than a person who is yelling and screaming, an otherwise "calm" patient can become acutely agitated, potentially to the point of violence.

The motor activity of a patient with a behavioral crisis or psychiatric emergency is perhaps the most telling factor of all. A patient who has tense muscles, clenched fists, or glaring eyes; is pacing; or cannot sit still must be carefully observed.

Whereas a person's physical size can be intimidating, there is no correlation between a person's physical size or body build and his or her potential for violence.

4. How should you proceed with your assessment of this patient?

The patient has something to say; he states that he "has a lot of problems," but also notes that nobody will listen to him. Therefore, you should use the most important assessment tool that you have: listening. Actively listen to what the patient is saying; not only may this give you clues as to the underlying cause of his behavior, it also reassures him that you *are* listening.

Reflective listening, a technique commonly used by mental health professionals to gain insight into a patient's thinking, involves repeating, in question form, what the patient has said and encouraging the patient to expand on his or her thoughts.

Do not assume that the patient will not offer any information regarding his medical history. When the patient is talking, *carefully listen* to what he is saying. Are his responses to your questions appropriate? Does he seem to understand you and his surroundings? Are his vocabulary and expressions what you would expect under the circumstances? Does he express disordered thoughts, delusions, or hallucinations?

Express sincere interest in what the patient is saying. Let him tell you what happened or what is going on in his own words. Do not interrupt him if he is in the middle of a sentence; doing so may tell him that you are simply "going through the motions" of listening but that you really do not care about what he has to say. Allow him to finish what he is saying before you ask him any questions—just as you would with any other patient.

When you are caring for a patient with a behavioral crisis, you must be prepared to spend extra time with him or her. It often takes longer to assess, listen to, and prepare the patient for transport.

5. What is your field impression of this patient?

The field impression is what the EMT thinks is wrong with the patient and is based on a number of factors—mainly physical assessment of the patient and his or her medical history. On the basis of the EMT's field impression, he or she can begin treatment that is most appropriate for the situation.

Your patient is displaying clear signs of depression. Despite attending counseling sessions, he is unable to cope with the loss of his wife; this inability to cope has brought him to the point of lacking the desire to even get out of bed in the morning. Furthermore, he no longer feels as though he has a purpose in life. He has experienced emotional distress for a year, and today—the anniversary of his wife's death—has precipitated an acute emotional crisis.

Although the patient does not appear to be experiencing any physical problems, he is clearly emotionally overwhelmed. *Depression is the single most significant factor that contributes to suicide;* therefore, he requires immediate intervention.

6. What care can you provide to this patient in the field?

As an EMT, your job is to take care of patients, whether their problem is purely physical, purely emotional, or a combination of both. The fact that your patient is severely depressed places him in a high-risk category for suicide; he should *not* be left alone!

Patients who are experiencing a behavioral crisis require a great deal of emotional support. As previously discussed, you will often need to spend extra time with the patient; most of this time is spent listening to what he or she has to say.

A psychological crisis can be as debilitating as a severe medical condition or traumatic injury; however, the treatment for such a crisis is usually complex and beyond the EMS provider's training and experience. During the short period of time that you will be caring for the patient, there is little you will be able to do. Your job is to diffuse and control the situation and *safely* transport the patient to the hospital. Intervene only as much as it takes to accomplish these tasks. Treatment for the patient will consist mainly of emotional support and active listening.

7. Should you perform a physical assessment of this patient despite the fact that he requested that you not do so? Why or why not?

The patient—although severely depressed—still has decision-making capacity; he is conscious and alert, is able to answer your questions appropriately, and is not displaying any psychotic behavior (eg, hallucinations, delusions). Therefore, he maintains the legal right to refuse a physical examination; touching him without his consent could lead to allegations of assault and battery against you.

Unless there is an accompanying physical complaint, a detailed full-body scan is rarely indicated in a patient with an emotional crisis; in fact, it may be detrimental to your gaining the patient's trust.

A critical aspect in treating patients with a behavioral crisis is to keep them as calm as possible. To perform a physical examination on a patient who is refusing to allow you to do so will only make him or her become agitated, and it significantly increases the risk of violent behavior. Simply transport the patient to the hospital, observe his behavior en route, and provide emotional support.

8. What factors should you consider before transporting a patient with an emotional crisis?

On arriving at the scene of a patient with an emotional crisis, you must assess him or her for signs of potential violence; these were discussed earlier in this case study. These observations should continue throughout the *entire* patient encounter. Even though this patient is calm right now, this could easily change and he could become acutely violent. The worst time for this to happen is in the back of the ambulance, where you will be the only EMT! Do not let your guard down when caring for a patient with an emotional crisis; continuously monitor his or her behavior.

If you have reason to believe that the patient is at an increased risk for becoming violent—for example, if he or she starts to become agitated at the scene—you should have other authorized personnel (ie, police officer) ride in the back of the ambulance for your own safety. If the patient becomes violent, it may be necessary to restrain him; this would clearly be difficult for one person to do. Although a minimum of four personnel should be used to restrain a violent patient, two people are clearly better than one.

If it is necessary to restrain the patient, use just enough force to effectively accomplish the task; do not use excessive force. Unless it is absolutely necessary to provide patient care, do not release the restraints—regardless of any promises the patient makes to calm down.

9. If the patient does not answer your questions, should you continue to encourage him to talk? Why or why not?

Some patients who are experiencing a behavioral crisis talk excessively (eg, in bipolar disorder), whereas others talk very little or not at all (as in depression). If the patient wants to talk, you should encourage him or her to do so. As previously discussed, many patients find relief—even if it is only

You are the Provider: SUMMARY, continued

temporary—by simply having someone to talk to. However, if the patient prefers not to talk, *do not force the issue*.

Depressed patients are typically withdrawn and are not very talkative. Instead of persistently encouraging a patient to talk, a tactic that could create agitation and possibly escalate the situation to a violent one, you should continue to monitor his or her behavior.

Remember your ultimate goal: to *safely* transport the patient to a facility where he or she can receive the psychological care that is needed. Whether the patient is talking or not, the potential for the situation to turn violent is minimized as long as he or she remains calm.

EMS Patient Care Report (PCR)

Date: 11-23-09	Incident No.: 012009	Nature of Call: Behavioral crisis	Location: 517 E. Bandera

Dispatched: 1920	En Route: 1921	At Scene: 1937	Transport: 2000	At Hospital: 2011	In Service: 2021

Patient Information

Age: 44	Allergies: No known drug allergies
Sex: M	Medications: Paxil, Ambien
Weight (in kg [lb]): 80 kg (175 lb)	Past Medical History: None
	Chief Complaint: Apparent emotional distress

Vital Signs

Time: 1946	BP: 144/84	Pulse: 88	Respirations: 16	Sao$_2$: 98%
Time: 1957	BP: 138/86	Pulse: 76	Respirations: 16	Sao$_2$: 98%
Time: 2007	BP: 130/80	Pulse: 72	Respirations: 16	Sao$_2$: 99%

EMS Treatment
(circle all that apply)

Oxygen @ __ L/min via (circle one): NC NRM Bag-Mask Device	Assisted Ventilation	Airway Adjunct	CPR
Defibrillation	Bleeding Control	Bandaging	Splinting	Other: (Emotional support)

Narrative

Medic 1780 dispatched to a residence for a man who is "acting bizarre." Staged at 5th Street and Elm until law enforcement arrived on scene and advised that it was safe to enter. On arriving at the scene, found the patient, a 44-year-old man, sitting on the lawn in front of his house, rocking back and forth. He was conscious and alert; however, he appeared sad and was clearly withdrawn. His airway was patent, and his breathing was adequate. A law enforcement official advised EMS that the patient had been checked for any weapons on his person and that none were found. Introduced EMS crew to patient and performed primary assessment; no gross abnormalities noted during this examination. The patient advised that today is the 1-year anniversary of his wife's death and that he finds it extremely difficult to even get out of bed in the morning. He further advised that he feels as though his life has no purpose anymore. According to the patient, he has been to several counseling sessions over the past year, but they have not seemed to help him. He further advised that his employer does not seem to care about his problems and has threatened to fire him unless his depressed behavior improves. Patient would only consent to assessment of his vital signs; he would not consent to further assessment or treatment. Visual examination did not reveal any gross injuries or life-threatening conditions, and the patient continued to remain conscious, alert, calm, and oriented and was able to answer questions appropriately. Because of the patient's significantly depressed state, advised him that transport to the emergency department was advisable. Initially, he refused EMS transport; however, after explaining to him that there could be an underlying medical problem contributing to his depression, he consented to transport and vital sign monitoring only; he maintained his refusal of a physical examination. Careful assessment of the patient did not reveal any gross indicators of potential violence. Patient walked to the ambulance on his own accord and was safely secured to the stretcher. Began transport to the hospital and continued to monitor the patient's behavior and vital signs en route. No changes in the patient's status were noted en route, and he remained calm but did not want to talk. Remainder of transport was uneventful. Delivered patient to emergency department staff and gave verbal report to charge nurse. After transferring patient care to the hospital staff, Medic 1780 returned to service. *End of report*

Assessment and Emergency Care of Psychiatric Emergencies

Scene Size-up

Scene Safety	Ensure scene safety and safe access to the patient. Be aware of potential violence or the possibility of a crime scene and call for law enforcement. Standard precautions should include a minimum of gloves. Determine the number of patients, and assess the need for additional resources. Pay close attention to the patient, and the surroundings; suicidal/homicidal patients will not hesitate to hurt you or your partner. Look for medications and signs of drug or alcohol use. Be observant for possible weapons. Identify yourself clearly, and remain calm and confident.
Mechanism of Injury (MOI)/ Nature of Illness (NOI)	Determine the MOI/NOI. Observe the scene, and look for indicators of an NOI/MOI. Abnormal behavior may be caused by Alzheimer disease, hypoglycemia, cerebrovascular accident, or environmental concerns. Consider that an MOI may be the reason for the person's behavior. Behavioral emergencies may also be the result of substance abuse, delusional disorders, and schizophrenia.

▼   ▼

Primary Assessment

Form a General Impression	Your assessment of a behavioral emergency patient should begin at the door. Note the patient's behavior and attitude. Observe the overall appearance of the patient, age, and body position. Is the patient sitting, hunched over, shuffling around, or standing still? Is the patient in a defensive position, or does he or she seem agitated? Is the patient holding or near a weapon? Note the patient's facial expression. Observe for tears, sweating, nervousness, or embarrassment. Look for gang identification. Observe the work of breathing and circulation. Determine the level of consciousness using the AVPU scale. Identify immediate threats to life. Determine priority of care based on the MOI/NOI. If the patient has a poor general impression, call for advanced life support assistance. A rapid scan will help you identify and manage life threats.
Airway and Breathing	Ensure the airway is open, clear, and self-maintained. Unresponsive patients will need their airway opened and maintained using a modified jaw-thrust if cervical spine injury is suspected and a head tilt–chin lift in nontrauma patients. A patient with an altered level of consciousness may need emergency airway management; consider inserting a properly sized oropharyngeal or nasopharyngeal airway. Evaluate the patient's ventilatory status for rate and depth of breathing, respiratory effort, and tidal volume. Administer high-flow oxygen at 15 L/min, providing ventilatory support as needed. Hypoxia may cause changes in the patient's mental state.
Circulation	Observe skin color, temperature, and condition; look for life-threatening bleeding and treat accordingly. Suicidal patients might have injured themselves. Evaluate distal pulse rate, quality (strength), and rhythm. Tachycardia may be an indicator of a behavioral emergency, but it may also indicate respiratory distress or shock. Bradycardia might be from a medication reaction or poisoning.
Transport Decision	If the patient has an airway or breathing problem, signs and symptoms of bleeding, or other life threats, manage them immediately and consider rapid transport, performing the secondary assessment en route to the hospital. Be prepared to spend time on the scene for patients without life threats who are stable. Some patients may feel threatened during the physical examination; therefore, remain calm, and be supportive and empathetic. It is a good idea to have an additional patient care provider in the back of the ambulance with you.

▼   ▼

NOTE: The order of the steps in this section differs depending on whether the patient is conscious or unconscious. The following order is for a conscious patient. For an unconscious patient, perform a primary assessment, perform a rapid full-body scan, obtain vital signs, and obtain the past medical history from a family member, bystander, or emergency medical identification device.

Assessment and Emergency Care of Psychiatric Emergencies, continued

History Taking

Investigate Chief Complaint	Investigate the chief complaint. Monitor the patient for changes in mental status. Ask OPQRST and SAMPLE questions. SAMPLE can also be obtained from family, bystanders, and medical alert tags. Once a traumatic event has been ruled out, history taking will help you determine the cause of the patient's behavioral state. The patient may be hypoglycemic from diabetes; might have a history of Alzheimer disease or dementia; alcohol or drug use may be a factor; or the patient may be depressed because of the loss of a job or a loved one. Ask about previous psychiatric emergencies, hospitalizations, and prescribed medications. If it is safe to do so, as you elicit the history from the patient, you should have your partner gather information separately from a family member. Listen to your patient, noting what he or she is saying and how he or she is saying it. Take the patient seriously.

Secondary Assessment

Physical Examinations	If the patient is unconscious, perform a systematic full-body scan beginning with the head, looking for DCAP-BTLS. Assessment should be rapid if the patient has a poor general impression. It is often difficult to perform a physical examination during a behavioral crisis or psychological emergency. Patients may feel threatened by your presence and actions. Obtain consent before attempting any examination or procedure and explain what you are going to do. Observe facial expressions; look for tears, sweating, or blushing. Note the patient's eyes and pupils; a central nervous system dysfunction may be indicated if the patient has a blank stare or rapid eye movement. Monitor the patient's mental status for sudden changes. Hallucinations may be a sign of the patient's compromised perception of reality.
Vital Signs	Obtain baseline vital signs if it is safe, the patient allows it, and it will not exacerbate the patient's emotional state. Vital signs should include blood pressure by auscultation, pulse rate and quality, respiration rate and quality, and skin assessment for perfusion. Note the patient's level of consciousness. Use pulse oximetry, if available, to assess the patient's perfusion status. Vital signs are important; the behavioral emergency may be the result of an injury or preexisting medical condition.

Reassessment

Interventions	Be acutely aware of changes in the patient's mental state; patients experiencing a behavioral crisis may act spontaneously and could become a danger to you and your crew. Reassess the primary assessment, vital signs, and chief complaint. Assist breathing as required, administering high-flow oxygen.
Communication and Documentation	Contact medical control/receiving hospital with a radio report; many hospitals require additional personnel and a separate treatment area. Include a thorough description of the MOI/NOI and the position the patient was found in. Include treatments performed and patient response. If the patient is restrained, the hospital needs to know this information. Be sure to document the patient's distress, answers to your questions, attitude toward emergency care providers, and any changes in patient status and the time. Follow local protocols. Document the reasoning for your treatment and the patient's response. If restraints were used, include why and what type of restraint was used.

NOTE: Although the following steps are widely accepted, be sure to consult and follow your local protocols. Take appropriate standard precautions when treating all patients.

Assessment and Emergency Care of Psychiatric Emergencies, continued

Psychiatric Emergencies

General Management of Psychiatric Emergencies

Managing life threats to the patient's ABCs and ensuring the delivery of high-flow oxygen are primary concerns with any psychiatric emergency. Without an underlying medical or trauma cause, there is usually little hands-on patient care for the EMT to perform. Competent adults have the right to refuse treatment and transport, but in the case of psychiatric emergencies, you may have a reasonable belief that the patient may harm himself, herself, or others. If this is the case, contact law enforcement to take the patient into custody. If restraint is required, consult with medical control, ensure law enforcement personnel are at the scene, and make sure that there are at least four people present. Use only approved restraint devices. Do not transport the patient in a prone position because the patient may experience severe respiratory distress or cardiac arrest, often called positional asphyxia.

Prep Kit

Ready for Review

- A behavioral crisis is any reaction to events that interferes with the activities of daily living or has become unacceptable to the patient, family, or community.

- During a psychiatric emergency, a patient may show agitation or violence or become a threat to himself, herself, or others. This is more serious than the more typical behavioral crisis that causes inappropriate behavior such as interference with activities of daily living or bizarre behavior.

- According to the National Institute of Mental Health, at one time or another, one in five Americans has some type of psychiatric disorder, an illness with psychological or behavioral symptoms that may result in impaired functioning.

- Psychiatric disorders have many possible underlying causes including social or situational stress such as divorce or death of a loved one; psychiatric disorders such as schizophrenia; physical illnesses such as diabetic emergencies; chemical problems such as alcohol or drug use; or biologic disturbances such as electrolyte imbalances. Sometimes these conditions can be compounded by noncompliance with prescribed medication regimens.

- As an EMT, you are not responsible for diagnosing the underlying cause of a behavioral crisis or psychiatric emergency.

- Your job is to diffuse and control the situation and safely transport your patient to the hospital. Intervene only as much as it takes to accomplish these tasks. Be caring and careful.

- To the person experiencing a psychotic episode, the line between their reality and fantasy is blurred.

- The threat of suicide requires immediate intervention. Depression is the most significant risk factor for suicide.

- Patients experiencing delirium are generally not dangerous, but if they exhibit agitated behavior they may strike out irrationally. One of the most important factors to consider in these cases is your personal safety.

- A patient in mentally unstable condition may resist your attempts to provide care. In such situations, request that law enforcement personnel handle the patient. Another reason for seeking law enforcement support is for a patient who resists treatment; such a patient often threatens EMTs and others. Violent or dangerous people must be taken into custody by the police before emergency care can be rendered.

- Always consult medical control and contact law enforcement personnel for help before restraining a patient. If restraints are required, use the minimum force necessary. Assess the airway and circulation frequently while the patient is restrained.

Vital Vocabulary

activities of daily living The basic activities a person usually accomplishes during a normal day, such as eating, dressing, and bathing.

agitated delirium A condition of disorientation, confusion, and possible hallucinations coupled with purposeless, restless physical activity.

altered mental status A change in the way a person thinks and behaves that may signal disease in the central nervous system or elsewhere in the body.

behavior How a person functions or acts in response to his or her environment.

behavioral crisis The point at which a person's reactions to events interfere with activities of daily living; this becomes a psychiatric emergency when it causes a major life interruption, such as attempted suicide.

depression A persistent mood of sadness, despair, and discouragement; may be a symptom of many different mental and physical disorders, or it may be a disorder on its own.

functional disorder A disorder in which there is no known physiologic reason for the abnormal functioning of an organ or organ system.

organic brain syndrome Temporary or permanent dysfunction of the brain, caused by a disturbance in the physical or physiologic functioning of brain tissue.

psychiatric disorder An illness with psychological or behavioral symptoms and/or impairment in functioning caused by a social, psychological, genetic, physical, chemical, or biologic disturbance.

psychiatric emergency An emergency in which abnormal behavior threatens a person's own health and safety or the health and safety of another person, for example when a person becomes suicidal, homicidal, or has a psychotic episode.

psychosis A mental disorder characterized by the loss of contact with reality.

Assessment in Action

You are called to a convenience store for a "person acting crazy." Law enforcement personnel have arrived on scene before you arrive and have deemed the scene safe. You approach and observe a deputy sheriff standing beside a disheveled, agitated middle-aged man who is pacing back and forth, waving his arms around, and speaking to someone whom apparently only he can see. He is shouting "Stay back or the alien leaders will have you demolished."

1. Although it is not your responsibility to diagnose the patient, what condition do you think the patient may have?
   A. Bipolar disorder
   B. Alzheimer disease
   C. Schizophrenia
   D. Major depression

2. Schizophrenia is an example of what type of disorder?
   A. Organic brain
   B. Environmental
   C. Functional
   D. Situational

3. The onset of schizophrenia typically occurs during:
   A. early childhood.
   B. adolescence.
   C. early adulthood.
   D. late adulthood.

4. Which of the following is considered a potential cause of psychiatric disorders?
   A. Social stresses
   B. Biologic disturbances
   C. Chemical problems
   D. All of the above

5. You try to encourage the patient to go to the hospital and be evaluated. He refuses, stating that "it is against alien orders." Consent for medical treatment when the patient is not mentally competent is considered what type of consent?
   A. Expressed
   B. Implied
   C. Shared
   D. Informed

6. The patient finally consents to being evaluated at the hospital. As you and your partner guide the patient toward the ambulance, he becomes combative and shouts, "The master alien has ordered me to disintegrate you." On the basis of the change in your patient's demeanor, what should you do?
   A. Tackle the patient and have the deputy handcuff him.
   B. Ask law enforcement personnel for help in restraining the patient.
   C. Request law enforcement personnel to use pepper spray on the patient.
   D. Refuse to transport the patient, and leave the scene.

7. What is the minimum number of people required to appropriately restrain a patient?
   A. 2
   B. 4
   C. 6
   D. 8

8. A restrained patient should not be placed in a face-down position because:
   A. it will increase the level of the patient's agitation.
   B. the patient may chew through the restraints.
   C. breathing may become inhibited and lead to respiratory arrest.
   D. the patient can sue you for false imprisonment.

9. Discuss the information that you should document once the decision to restrain a patient has been made.

Gynecologic
Emergencies

National EMS Education Standard Competencies

Medicine

Applies fundamental knowledge to provide basic emergency care and transportation based on assessment findings for an acutely ill patient.

Gynecology

- Recognition and management of shock associated with
  - Vaginal bleeding (pp 728, 729, 731, 732)
- Anatomy, physiology, assessment findings, and management of
  - Vaginal bleeding (pp 725-726, 728, 729, 731, 732)
  - Sexual assault (to include appropriate emotional support) (pp 732-734)
  - Infections (pp 725-728)

Knowledge Objectives

1. Describe the anatomy and physiology of the female reproductive system, including the developmental changes that occur during puberty and menopause. (pp 725-726)
2. Discuss the special, age-related patient management considerations an EMT should provide for both younger and older female patients who are experiencing gynecologic emergencies. (p 726)

3. List three common examples of gynecologic emergencies, including their causes, risk factors, assessment findings, and patient management considerations. (pp 726-728, 732-734)
4. Discuss the assessment and management of a patient who is experiencing a gynecologic emergency, including a discussion of specific assessment findings. (pp 728-731)
5. Explain the general management of a gynecologic emergency in relation to patient privacy and communication. (pp 731-732)
6. Give examples of the different types of personal protective equipment EMTs should use when treating patients with gynecologic emergencies. (p 732)
7. Discuss the special considerations and precautions an EMT must observe when arriving at the scene of a suspected case of sexual assault or rape. (pp 732-734)
8. Discuss the assessment and management of a patient who has been sexually assaulted, including the additional steps an EMT must take on behalf of the patient. (pp 732-734)

Skills Objectives

There are no skills objectives for this chapter.

Introduction

The most obvious difference between men and women is that women are uniquely designed to conceive and give birth. This difference makes women susceptible to a number of problems that do not occur in men. This chapter examines a few of those problems. Female anatomy is discussed first, followed by problems that may be encountered in the prehospital setting. Vaginal bleeding is briefly discussed. The principles of managing a woman who has been the victim of sexual assault are also addressed.

Anatomy and Physiology

The **ovaries** are the primary female reproductive organ Figure 21-1. The ovaries are located on each side of the lower abdomen and produce an ovum, or egg, that, if fertilized, will develop into a fetus. Each ovary produces an ovum in alternating months and releases it into the fallopian tube. Some women experience a minor cramping pain when the ovum is released (**ovulation**). The **fallopian tubes** connect each ovary with the uterus and are the primary location for fertilization of the ovum. The **uterus** is the muscular organ where the fetus grows during pregnancy. As the ovum moves slowly down the fallopian tube, sperm moving up the tube can surround it, and one sperm fertilizes it. When an ovum is fertilized

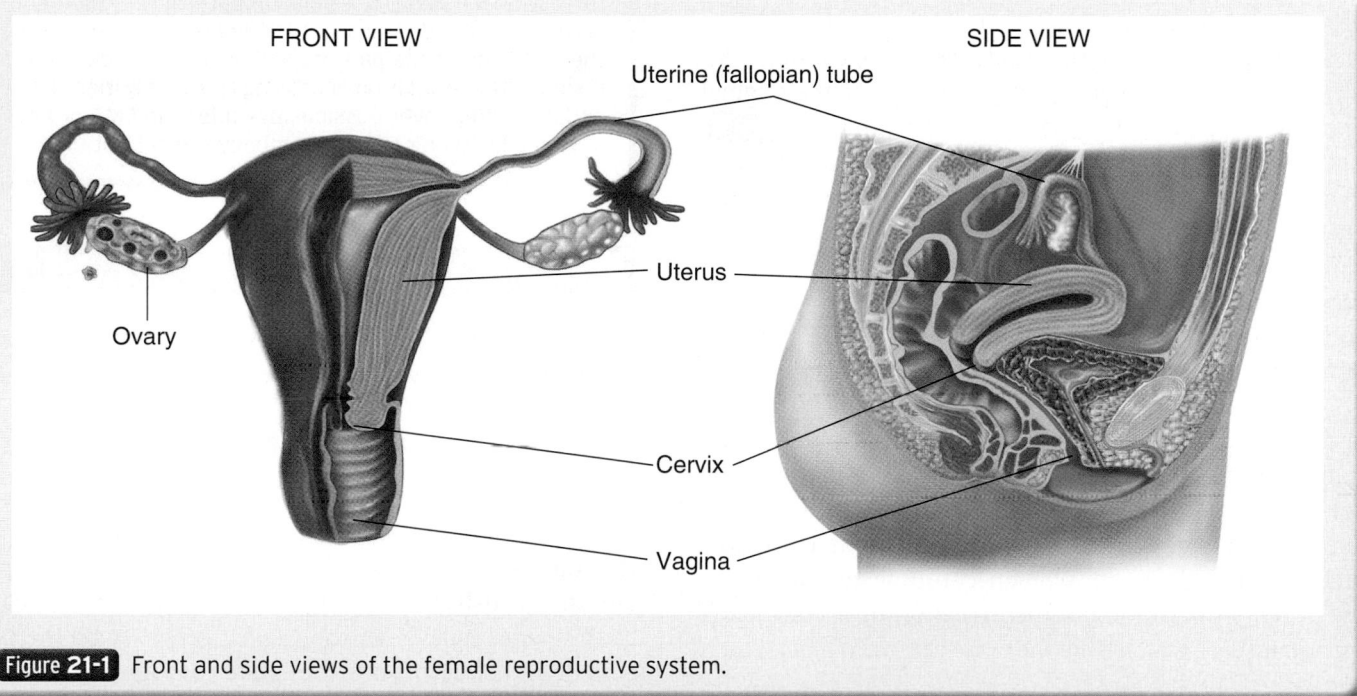

FRONT VIEW   SIDE VIEW

Uterine (fallopian) tube

Uterus

Ovary

Cervix

Vagina

Figure 21-1 Front and side views of the female reproductive system.

You are the Provider: PART 1

At 5:55 AM, law enforcement personnel request your assistance at 4300 West Avenue for a young woman who was assaulted. You and your partner respond to the scene, which is located about 4 miles away. While you are en route, an on-scene law enforcement officer radios you over a private channel and advises that the patient has been sexually assaulted and that the scene is safe. The weather is overcast, the temperature is 78°F, and the traffic is light.

1. What factors should you consider while responding to this call?
2. What are some unique aspects about assessing and treating a woman who has been sexually assaulted?

in the fallopian tube, the developing embryo travels into the uterus where the lining of the walls of the uterus have become engorged with blood in anticipation of receiving a fertilized ovum. Here, the embryo attaches to the uterine wall and continues to grow. The narrowest portion of the uterus, the **cervix**, opens into the vagina. The **vagina** is the outermost cavity of a woman's reproductive system and forms the lower part of the birth canal. In the vagina, sperm is deposited from the male penis. From here, the sperm pass through the cervix into the uterus and eventually up the fallopian tubes.

If the ovum is not fertilized in the fallopian tube, it continues to travel into the uterus. Because fertilization has not occurred within about 14 days of ovulation, the lining of the uterus begins to separate and menstruation occurs. The menstrual flow consists of blood from the separated lining of the uterus and lasts about 1 week. The process of ovulation and menstruation is controlled by female hormones primarily produced in the ovaries.

The external female genitalia consist of the vaginal opening just posterior to the urethral opening Figure 21-2. The **labia majora** and **labia minora** are folds of tissue that surround the urethral and vaginal openings. At the anterior end of the labia is the clitoris, and at the posterior end is the anus. The **perineum** is the area of skin between the vagina and the anus. The labia are extremely vascular and can be injured, but because of their location, they seldom are except in cases of sexual abuse.

When a female reaches puberty, she begins to ovulate and experience menstruation. The onset of menstruation is called menarche and usually occurs between the ages of 11 and 16 years, although it can occur earlier or later. Any female who has reached menarche is capable of becoming pregnant. Women continue to experience the cycle of ovulation and menstruation until they reach menopause, at which time the cycle ceases. Women reach menopause at widely varying ages, but it commonly occurs by around the age of 50 years.

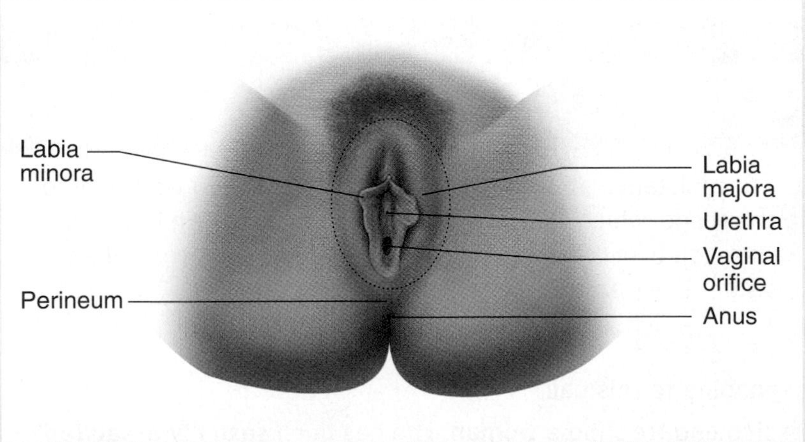

Special Populations

The onset of menarche in girls can be an emotionally and physically disturbing event. It is not uncommon for this event to be preceded by cramping pain that can be misinterpreted by the girl who has not experienced menstruation yet. Most girls have learned about the menstrual process from parents or from health classes at school, but it is possible that some are still unprepared when it finally occurs. Girls who have perhaps led a very sheltered life or who do not have a female parent in the home are more likely to be surprised by the onset of their menstruation. Parents may also be in denial, insisting that their little girl is too young to be experiencing menstruation.

Approach the patient (and her parents) in the most professional manner possible. Empathize with their concerns, and provide transport to the hospital to help allay the concerns of the parents and also to help determine if some other condition is causing or contributing to the situation. Whenever possible, use a female EMT or have a female family member accompany the patient.

Pathophysiology

The causes of gynecologic emergencies are varied and range from sexually transmitted diseases to trauma.

Pelvic Inflammatory Disease

Pelvic inflammatory disease (PID) is an infection of the female upper organs of reproduction—specifically, the uterus, ovaries, and fallopian tubes—that occurs almost exclusively in sexually active women. Disease-causing organisms enter the vagina, generally by the process of sexual activity, and migrate through the opening of the cervix and into the uterine cavity. The infection may then expand to the fallopian tubes (producing scarring that can lead to life-threatening ectopic pregnancy or sterility), eventually involving the ovaries (leading to the development of a life-threatening abscess). Ectopic pregnancy is a pregnancy that develops outside the uterus, most often in the fallopian tube. Although PID itself is seldom a threat to life, its ultimate consequences of an ectopic pregnancy or an abscess can be lethal. The most common presenting sign of PID is generalized lower abdominal pain. Other signs and symptoms include an abnormal and often foul-smelling vaginal discharge, increased pain with intercourse, fever, general malaise, and nausea and vomiting.

Labia minora
Labia majora
Urethra
Vaginal orifice
Perineum
Anus

Figure 21-2 The external genitalia of the female reproductive system.

Sexually Transmitted Diseases

Sexually transmitted diseases can lead to more serious conditions, such as PID. For example, untreated gonorrhea and chlamydia often progress to PID.

Chlamydia is caused by the bacterium *Chlamydia trachomatis*. It is a common sexually transmitted disease, affecting an estimated 2.8 million Americans each year. Although the symptoms of chlamydia are usually mild or absent, some women may have symptoms including lower abdominal pain, low back pain, nausea, fever, pain during intercourse, and/or bleeding between menstrual periods. Chlamydial infection of the cervix can spread to the rectum, leading to rectal pain, discharge, or bleeding. Left untreated, the disease can progress to PID. In rare cases, chlamydia causes arthritis that may be accompanied by skin lesions and inflammation of the eye and urethra.

Bacterial vaginosis is one of the most common conditions to afflict women. In this infection, normal bacteria in the vagina are replaced by an overgrowth of other bacterial forms. Symptoms may include itching, burning, or pain and may be accompanied by a "fishy," foul-smelling discharge. Left untreated, bacterial vaginosis can lead to premature birth or low birth weight in case of pregnancy, make the patient more susceptible to more serious infections, and result in PID.

You are the Provider: PART 2

You arrive at the scene and find the patient, a 25-year-old woman, sitting on the ground near her car. She is conscious and alert but is crying. Her shirt is torn, and she is nude from the waist down. A police officer covered her with a blanket. After introducing yourself and your partner, you perform a primary assessment.

Recording Time: 0 Minutes	
Appearance	Obvious emotional distress
Level of consciousness	Conscious and alert
Airway	Open; clear of secretions and foreign bodies
Breathing	Normal rate; adequate depth
Circulation	Radial pulses, normal rate and strong; skin is pink, warm, and moist; no obvious bleeding

A police officer tells you that the patient was apparently assaulted in her vehicle and that the last thing she remembered was talking to a young man she met while listening to music and drinking a margarita with her friends at a local nightclub at about 9:00 PM last night.

3. Does this patient require immediate medical treatment?

4. On the basis of the police officer's report, what should you suspect regarding the events that preceded the patient's assault?

Gonorrhea is caused by *Neisseria gonorrhoeae*, a bacterium that can grow and multiply rapidly in the warm, moist areas of the reproductive tract, including the cervix, uterus, and fallopian tubes in women and in the urethra in women and men. The bacterium can also grow in the mouth, throat, eyes, and anus. Symptoms, which are generally more severe in men than in women, appear approximately 2 to 10 days after exposure. Women may be infected with gonorrhea for months but experience virtually no symptoms until the infection has spread to other parts of the reproductive system. When symptoms do appear in women, they generally manifest as painful urination, with associated burning or itching; a yellowish or bloody vaginal discharge, usually with a foul odor; and blood associated with vaginal intercourse. More severe infections may present with cramping and abdominal pain, nausea and vomiting, and bleeding between periods; these symptoms indicate that the infection has progressed to PID. Rectal infections generally present with anal discharge and itching, plus occasional painful bowel movements with fecal blood spotting. Infection of the throat (for which oral sex is the introducing factor) usually results in mild symptoms consisting of painful or difficult swallowing, sore throat, swollen lymph glands, and fever. Headache and nasal congestion may also be present. If the infection is not treated, the bacterium may enter the bloodstream and spread to other parts of the body, including the brain.

Vaginal Bleeding

Because menstrual bleeding is a monthly occurrence in most females, vaginal bleeding that is the result of other causes may be initially overlooked. Some possible causes of vaginal bleeding include abnormal menstruation, vaginal trauma, ectopic pregnancy, spontaneous abortion (miscarriage), cervical polyps, and even cancer. Trauma to the internal female genitalia from any cause other than vaginal penetration is rare because of the location of these organs deep within the pelvis. Injuries to the vagina and external genitalia are very painful and serious because of the large quantity of nerves and blood vessels in this area. In contrast, internal bleeding from polyps or cancer, while also very serious, may be relatively painless.

Ectopic pregnancy and spontaneous abortion are two conditions that can cause vaginal bleeding in women who do not appear to be pregnant and who may not realize that they are pregnant. These potentially life-threatening conditions will be covered in Chapter 31, *Obstetrics and Neonatal Care*. All cases of vaginal bleeding should be taken seriously, and the patient should be transported to the hospital for a thorough gynecologic evaluation.

Patient Assessment

Obtaining an accurate and detailed patient assessment is critical when dealing with gynecologic issues. You may not be able to make a specific diagnosis in the field, but a thorough patient assessment will help determine just how sick the patient is and whether lifesaving measures should be initiated. This is especially true when dealing with abdominal pain.

Women have many of the same conditions that cause abdominal pain in men—for example, ulcers and appendicitis. In addition, there are numerous gynecologic causes of abdominal pain. An old medical axiom states, "Anyone who neglects to consider a gynecologic cause in a woman of childbearing age who complains of abdominal pain will miss the diagnosis at least 50% of the time." Missing the diagnosis may be fatal for the patient.

Scene Size-up

Scene Safety

Every emergency call—including calls involving gynecologic emergencies—begins with a thorough scene size-up. Is the scene safe? Will you need assistance? How many patients do you have? What is the nature of illness (NOI)? Have you taken standard precautions? Gynecologic emergencies can be very messy, sometimes involving large amounts of blood and body fluids contaminated with organisms that cause communicable diseases.

Where or in what position is the patient found **Figure 21-3** ? If she is at home, what is the condition of the residence? Is it clean, filthy, or wrecked? Do you see evidence of a fight? Are alcohol, tobacco products, or drug paraphernalia present? Are there pictures of loved

Figure 21-3 Note the position of the patient during your assessment.

ones or, conversely, a noticeable absence of pictures? Does the patient live alone or with other people? All information you obtain will contribute to your assessment of the patient's overall health and the safety of the scene. In the case of a crime scene, you also may be required to testify in court regarding the conditions on your arrival. Involve the police if any type of assault is suspected. In cases of sexual assault, it is important to have a female EMT to provide patient care, so consider calling for one early if you and your partner are both male.

Mechanism of Injury/Nature of Illness

Often the NOI or mechanism of injury in patients with gynecologic problems will be understood from the dispatch information, such as in cases of sexual assault. In other patients, the exact nature of the condition will not emerge until patient history information has been gathered. For example, your patient may present with vague symptoms such as abdominal pain, and you will not be able to determine the exact nature of the problem until you have gathered more information during the patient history.

Primary Assessment

Form a General Impression

The general impression is an important aspect of patient assessment. As you approach the patient, you should be able to tell if she is generally in stable or unstable condition by performing a rapid scan. You will use this information to help you as you proceed with your further assessment. Use the AVPU (*Alert* to person, place, and day; responsive to *Verbal* stimuli; responsive to *Pain*; *Unresponsive*) scale to determine the patient's level of consciousness.

Airway and Breathing

Always evaluate the airway and breathing immediately to ensure that they are adequate, and treat any airway or breathing problem that is identified according to established guidelines and local protocol. Identifying and treating life threats takes precedence over all other assessment and treatment.

Circulation

It is important to carefully assess the circulation in all patients. Palpating a pulse and evaluating skin color, temperature, and moisture can help identify the patient who might have blood loss. If the patient has experienced a significant blood loss because of vaginal bleeding, she may not be demonstrating obvious signs of shock but may still be hypovolemic. If the patient has a weak or rapid pulse or has pale, cool, or diaphoretic skin, place the patient in a supine position with her legs elevated. Cover the patient to keep her warm, and then provide transport to the emergency department for treatment.

You are the Provider: PART 3

The patient remains conscious and alert but is still in obvious emotional distress. She gives you consent to take her vital signs. A female police officer is present when you ask her if she is injured. She tells you that she is experiencing vaginal pain but denies vaginal bleeding. The only obvious injuries you can see are minor abrasions to her forearms.

Recording Time: 8 Minutes	
Respirations	14 breaths/min; adequate depth
Pulse	72 beats/min; strong and regular
Skin	Pink, warm, and moist
Blood pressure	110/70 mm Hg
Oxygen saturation (Sao$_2$)	98% (on room air)

The patient tells you that she wants to go take a shower and change her clothes. She also tells you that she needs to urinate. She asks a police officer to call a friend of hers and ask the friend to come to the scene.

5. What additional assessment should you perform on this patient?

6. How should you respond to the patient's request to take a shower, change her clothes, and use the bathroom?

Transport Decision

Most cases of gynecologic emergencies are not life threatening. However, if signs of shock exist because of bleeding, then rapid transport is warranted. The remainder of the assessment can be performed en route to the hospital.

History Taking

Investigate Chief Complaint

Begin by inquiring about the patient's chief complaint, realizing that some of the questions you must ask may be considered extremely personal. Be sensitive to the patient's feelings, and ensure that her privacy and dignity are protected. Gynecologic emergencies can be highly embarrassing for the patient, and many women may be extremely uncomfortable about discussing their sexual history in front of strangers or even close family members. An adolescent girl may want to keep her sexual history from her parents.

SAMPLE History

Obtain a SAMPLE history beginning with her current symptoms. Make note of any allergies she has or any medications she might be taking, such as birth control pills, and any birth control devices she is using. Ask the patient about medical conditions, and ask specifically about her last menstrual period. This will help determine if the patient is possibly pregnant. Make sure to inquire about the possibility of sexually transmitted diseases and the possibility of pregnancy. Find out when she last ate or drank and what events led up to her calling for EMS. Use her NOI, her chief complaint, and her answers to your other questions to lead your further questioning. For example, if she answers that she is sexually active, ask her about birth control and also about symptoms of pregnancy. If she has vaginal bleeding, determine how many pads she is using per hour. This information can help in creating an estimate of how much blood has been lost.

Secondary Assessment

The secondary assessment may be performed on scene, en route to the emergency department, or in some instances, not at all. If the patient is critically ill or injured or the transport time is short, you may not have time to conduct this part of the patient assessment process.

Physical Examinations

Your physical examination of a gynecologic patient should be limited and professional. Be sure to protect the woman's privacy during the physical examination.

Few women are comfortable with having their body exposed to a crowd of family, neighbors, EMTs, police officers, or fire fighters. Limit the personnel present to those required to perform the necessary tasks, and show the patient you respect her by being an advocate for her modesty. You also serve as a role model for other EMS providers when you act this way.

Focus your physical examination on the NOI and the patient's chief complaint. If vaginal bleeding is the NOI, you should visualize the bleeding and ask about its quality and quantity. Make sure to use external pads to control the bleeding, and keep the possibility of hypoperfusion or shock in mind. Always ask if there is pain associated with the vaginal bleeding or discharge. You should never insert anything into the vagina to control bleeding, including a tampon.

Vaginal discharge is another condition that should be observed if possible. Make observations about the discharge, and verify with the patient any qualities she noticed and the history of the discharge.

Fever, nausea, and vomiting are common with many medical conditions but should be considered especially significant with gynecologic emergencies. Any report of syncope on the part of the patient, especially if she is complaining of vaginal bleeding, is considered significant. Treat the patient reporting this symptom as being in shock until proven otherwise.

Words of Wisdom

During the patient assessment, explain what you are about to do and why to the patient. This will help gain the trust and confidence of the patient.

Vital Signs

Assess the patient's vital signs, including heart rate, rhythm, and quality; respiratory rate, rhythm, and quality; skin color, temperature, and condition; capillary refill time; and blood pressure. Pay special attention to the presence of tachycardia and hypotension, which could indicate hemorrhage.

Monitoring Devices

Use the appropriate monitoring devices to track the patient's condition. Use of pulse oximetry should be routine even if the patient is showing no signs of difficult breathing. Also consider using noninvasive blood pressure monitoring to continuously track the patient's blood pressure. It is recommended that you always assess the patient's first blood pressure manually with a sphygmomanometer (blood pressure cuff) and stethoscope.

Reassessment

Repeat the primary assessment. Reassess the patient's vital signs and the chief complaint. Reassessment of the patient's vital signs every 5 minutes may identify hypoperfusion from excessive blood loss. If the patient shows signs of shock, intervene with treatment.

How is the patient's condition improving with the interventions? Identify and treat any changes in the patient's condition. For example, if the patient appears to be losing consciousness, position her in the supine position, and perform a reassessment. Finally, pay specific attention to the needs of your patient, and accommodate her desire for conversation or silence. Provide her with calm reassurance. Explain to her that the hospital staff will be sympathetic to her condition and will be well qualified to treat her condition.

Interventions

There are very few interventions that can or should be done for a patient with a gynecologic emergency. If the patient has vaginal bleeding, she should be treated for hypoperfusion or shock. Keep her warm, place her in a supine position with her legs elevated, and provide her with supplemental oxygen even if she is not experiencing difficulty breathing. Then transport her promptly to the hospital.

Communication and Documentation

Be sure to notify staff at the receiving hospital of all relevant information, including the possibility of pregnancy, so that a proper response can be prepared. Carefully document the patient's condition, her chief complaint, the scene, and all interventions, especially in cases of sexual assault.

Words of Wisdom

Gynecologic emergencies can occur at any age during a woman's lifetime. As an EMT, there is little you can do in this type of emergency; therefore, you should focus on assessing and correcting the patient's ABCs and consider rapid transport as an important part of the call.

Emergency Medical Care

Whenever you deal with patients who have gynecologic problems, you must be sure to maintain the patient's privacy as much as possible. If the patient is in a public place, consider moving the patient to the ambulance. Make sure you communicate appropriately and gain the patient's confidence.

You are the Provider: PART 4

The patient refuses to go to the hospital and tells you that all she wants to do is take a shower and change her clothes. Her friend arrives at the scene and asks if she can talk to the patient alone. After a brief conversation, the patient tells you that she will go to the hospital, but only if her friend can accompany her. You advise her that this will be fine. You secure her to the stretcher, reassess her vital signs, and begin transport.

Recording Time: 28 Minutes	
Level of consciousness	Conscious and alert
Respirations	14 breaths/min; adequate depth
Pulse	80 beats/min; strong and regular
Skin	Pink, warm, and dry
Blood pressure	118/68 mm Hg
Sao$_2$	99% (on room air)

7. How should you respond to the patient's refusal to go to the hospital?

8. Is it the EMT's responsibility to determine why the patient was sexually assaulted?

Excessive internal vaginal bleeding can have many causes and can possibly lead to hypoperfusion or shock. Determining the cause of the bleeding should be of less importance than treating for shock and transporting the patient to an appropriate facility. Control any internal vaginal bleeding with the use of sanitary pads on the external genitalia. Most women will use sanitary pads to control the bleeding before you arrive, so you may continue that approach. However, if the woman has a tampon in place, it is not necessary to have her remove it. Vaginal bleeding is rarely significant enough to cause hemorrhagic shock, but the patient should be treated for shock nevertheless. Apply oxygen, keep the patient warm, and transport promptly to the hospital.

The external genitals have a rich nerve supply, making injuries very painful. Treat any external lacerations, abrasions, and tears with moist, sterile compresses, using local pressure to control bleeding and a diaper-type bandage to hold the dressings in place. Leave any foreign bodies in place after stabilizing them with bandages. Under no circumstances should you pack or place dressings in the vagina. Continue to assess the patients while transporting them to the emergency department. Contusions and other blunt trauma will require careful in-hospital evaluation.

Words of Wisdom

Gynecologic emergencies usually involve bleeding problems. Personal protective equipment (PPE), including gloves, eye protection, and a mask, must be considered.

Assessment and Management of Specific Conditions

Pelvic Inflammatory Disease

A patient with PID will complain of abdominal pain. The pain generally starts during or after normal menstruation, so inquiring about the date of the patient's last menstrual period is an important component of the patient's history. The pain may be described as "achy" and may be made worse by walking. Other symptoms may include vaginal discharge, fever and chills, and pain or burning on urination. Patients often present with a distinctive gait when they walk, which is sometimes called the "PID shuffle."

Prehospital treatment is limited, and nonemergency transport is usually recommended. As stated earlier, PID itself is seldom a threat to life, but it is serious enough to require transport and evaluation in the hospital.

Safety

Remember that many sexually transmitted diseases can also be transmitted by contact with blood. Some examples of these diseases include syphilis, many types of hepatitis, and acquired immunodeficiency syndrome.

Sexual Assault

Unfortunately, **sexual assault** and **rape** are all too common occurrences. In the United States, one of every three women will be raped in her lifetime, and one of every four will be sexually molested, often before the age of 12 years. EMTs called on to treat a victim of sexual assault, molestation, or actual or alleged rape face many complex issues, ranging from obvious medical ones to serious psychological and legal issues. You may be the first person the victim has contact with after the encounter, and how the situation is managed from first contact throughout treatment and transport may have lasting effects for the patient and you. Professionalism, tact, kindness, and sensitivity are of paramount importance.

Because rape is a crime, you can generally expect police involvement early in the situation. In many cases, EMS may be called by the police. Police officers generally have rudimentary medical training, with many states requiring at least basic training at the first responder (Emergency Medical Responder) level. Nevertheless, primary training for police officers focuses on investigation of the crime, not patient care.

A rape victim has just experienced a major trauma of her body and mind. The last thing a victim wants to do is give a concise, detailed report of what she has just experienced, and attempting to gather patient information in this manner most likely will cause the victim to "shut down." Whenever possible, a female rape victim should be given the option of being treated by a female EMT because the patient may be experiencing ambivalent feelings toward men; these feelings will hinder the patient assessment and the patient's well-being.

The job of the police is to solve the crime, arrest the perpetrator, and see justice served. Your job, as the EMT, is to deal with the medical aspects of the case and to act as the patient advocate. In this capacity, it is important for you to focus on several key issues.

The first issue is the medical treatment of the patient. Is she physically injured? Are any life-threatening injuries present? Does the patient complain of any pain?

The second issue is your psychological care of the patient. Do not cross-examine her or attempt to obtain information for the benefit of the police. These issues will be handled later by the hospital staff and police. Do not pass judgment on the patient, and protect her from the judgment of others on the scene. A crime has been committed, and you need to remain aware of that fact. Many women report feeling violated when subjected to interrogation, criticism, or disbelief.

Last, remember that you are at a crime scene. Although your job is to treat the medical aspects of the incident and not collect evidence, you still have a responsibility to preserve evidence. Do not cut through any clothing or throw away anything from the scene. Place bloodstained articles in separate paper (not plastic) bags. Obtain evidentiary bags from the police if necessary. Paper bags allow wet items to dry naturally, whereas plastic allows mold to grow and may destroy biologic evidence.

It may also be necessary to gently persuade the patient to not clean herself. Victims tend to want to "wash away" the humiliation and embarrassment of the assault. Valuable evidence can be destroyed in this process. The patient also needs to be discouraged from urinating, changing clothes, moving her bowels, or rinsing out her mouth. She will need to be photographed by law enforcement personnel as well, and the photographic record needs to be as accurate as possible. If the patient cannot be dissuaded from taking these actions, respect her feelings. Some patients may refuse transport altogether, and they have the right to do so. In such cases, you should follow your system's refusal of treatment policy or procedure for sexual assault victims without judging or being condescending to the patient. Your compassion is the best tool to gain the patient's confidence to get further help.

Offer to call the local rape crisis center for the patient. Many communities have rape crisis centers with victim advocates on-call. Getting a professional advocate to the scene may help the patient deal with the trauma, and the advocate can better explain the necessities of evidence preservation in more compassionate detail. Many victim advocates are rape-trauma survivors themselves. They can also provide support to the patient in the hospital during any additional physical examinations.

Take the patient's history, and limit any physical examination to a brief survey for life-threatening injuries. Treat all other injuries such as contusions or lacerations according to appropriate procedures and protocols for your EMS system. Follow standard precautions. Expose and examine the vaginal area only if there is evidence of bleeding that needs to be treated. Take care to shield the patient from any possible curious onlookers. Examine and interview the patient with a minimum of people present, moving her to the ambulance if necessary.

The patient report is a legal document and, should the case result in an arrest and subsequent trial, may be subpoenaed. Keep the report concise, and record only what the patient stated in her own words. Use quotation marks to indicate that you are reporting the patient's

You are the Provider: PART 5

You reassess the patient en route to the hospital; her mental status and vital signs indicate that she is stable. Her friend provides emotional support to her and assures her that what happened was not her fault. You call your radio report to the receiving facility and give them your estimated time of arrival.

Recording Time: 38 Minutes	
Level of consciousness	Conscious and alert
Respirations	16 breaths/min; adequate depth
Pulse	76 beats/min; strong and regular
Skin	Pink, warm, and dry
Blood pressure	122/72 mm Hg
Sao$_2$	98% (on room air)

9. Why is it important to transport a sexual assault victim to the hospital, even if she has no obvious injuries?

Table 21-1 Treatment Principles for Sexual Assault

In addition to the usual treatment principles that apply to all victims, you should follow these special steps with patients who have been sexually assaulted:

1. You must document the patient's history, assessment, treatment, and response to treatment in detail because you may have to appear in court as long as 2 or 3 years later. Do not speculate. Record only the facts.

2. Make airway maintenance a major priority.

3. Complete the SAMPLE history objectively.

4. Follow any crime scene policy established by your system to protect the scene and any potential evidence for police, particularly that for evidence collection. If the patient will tolerate being wrapped in a sterile burn sheet, this may help investigators to find any hair, fluid, or fiber from the alleged offender.

5. Do not examine the genitalia unless there is major bleeding. If an object has been inserted into the vagina or rectum, do not attempt to remove it.

6. To reduce the patient's anxiety, make sure the EMT is the same sex as the patient, whenever possible.

7. Discourage the patient from bathing, voiding, or cleaning any wounds until the hospital staff has completed an assessment. Handle the patient's clothes as little as possible, placing articles and any other evidence in paper bags. If the patient insists on urinating, ask the patient to do so in a sterile urine container (if available). Also, deposit the toilet paper in a paper bag. Seal and mark the bag for the police. This can be critical evidence.

version of events. Do not insert your own "opinion" as to whether the patient was raped or offer any conclusions that would validate or invalidate the patient's account of the event. Focus on the facts. Record all of your observations during the physical exam—the patient's emotional state, the condition of her clothing, obvious injuries, and so forth. Bear in mind that rape is a legal diagnosis, not a medical diagnosis. The medical team can establish only whether sexual intercourse occurred; a court must decide whether intercourse was inflicted forcibly on the victim, against her will. **Table 21-1** lists the treatment principles you should use when dealing with a victim of sexual assault.

Often the most important intervention for sexual assault patients is comforting reassurance and transport to a facility that has employees who are certified to perform the proper physical examination in this type of case. Reminding the patient that she is safe with you and that the hospital staff and the police will take good care of her may help to reassure her. Sometimes just the presence of a female EMT can be emotionally helpful. Do not insist that she talk to you, but listen carefully and nonjudgmentally if she does want to talk. Remember that victims of sexual assault also need medical assistance; therefore, treat the medical injuries but also remember to ensure the patient's privacy and provide her emotional support.

You are the Provider: SUMMARY

1. What factors should you consider while responding to this call?

As with any call, your first priority should be to ensure that the scene is safe for you to enter; this is especially true when you respond to an assault or any other call that has a higher than usual risk for violence. Although you have been advised by an on-scene law enforcement officer that the scene is safe, you must still remain aware of your surroundings when you arrive. Although the scene may be safe initially, it can quickly turn violent.

Your next consideration should be for the patient. Although you should avoid asking specific questions over the radio regarding sexual assault, you should attempt to determine if the patient is conscious and if she appears to have any life-threatening injuries or major bleeding. As with any patient, the more information you obtain while en route to the scene, the better prepared you will be to provide immediate care when you arrive.

Last, you should consider the fact that you are responding to a crime scene. While your job is to treat the *medical* aspects of the incident and not collect evidence, you still have a responsibility to preserve evidence, to the extent possible, without sacrificing patient care.

2. What are some unique aspects about assessing and treating a woman who has been sexually assaulted?

Whereas the medical treatment you provide to a sexual assault victim follows the same principles as for any other patient—identifying and treating life-threatening injuries—it is important to remember that the emotional effects on the patient are devastating.

EMTs called on to treat victims of sexual assault face many complex issues, ranging from the obvious medical ones to serious psychological and legal issues. How the situation is managed from first contact and then throughout treatment and transport may have lasting effects for the patient and you. You must remain professional, tactful, kind, and sensitive at all times.

A sexual assault victim has experienced a major trauma of her body *and* mind. You should expect that the last thing she wants to do is provide a concise, detailed report of what happened. Whereas some patients will provide more information than others, you should not force the issue if the victim does not want to talk about what she has experienced; doing so may cause her to completely "shut down."

Whenever possible, a female sexual assault victim should be assessed and treated by a female EMT. It is common for the patient to experience ambivalent feelings toward men in general; these feelings may hinder the patient assessment process and the patient's well-being.

If there are two male EMTs on your ambulance, and the patient is not critically injured, you should request a female EMT to respond to the scene. If this is not an option, ask the patient if there is a female friend that you can call; if there is, she should be present during your assessment and treatment of the patient. Another potential option is to request a female police officer, if available, to respond to the scene.

3. Does this patient require immediate medical treatment?

Your primary assessment has not revealed any obvious immediately life-threatening injuries or conditions; therefore, immediate emergency medical treatment is not indicated *at this point*.

The patient is conscious and alert, although she is emotionally distressed. Her airway is patent, and her breathing is adequate. Her radial pulses are strong and of a normal rate. There is no obvious bleeding that requires your attention, and her skin is pink, warm, and moist.

Although your primary assessment was unremarkable for gross (obvious) life-threatening injuries, this does *not* mean that the patient is not injured. She could have occult (hidden) injuries, such as internal bleeding, that have yet to manifest with signs and symptoms.

A secondary assessment may reveal less obvious injuries, some of which may be potentially life threatening. However, you should anticipate that the patient will be reluctant to allow you to perform a more detailed physical examination; in fact, she may absolutely refuse further assessment.

4. On the basis of the police officer's report, what should you suspect regarding the events that preceded the patient's assault?

Sexual assault is an acutely overwhelming emotional catastrophe, and some patients experience amnesia as an involuntary emotional protective reflex. However, this type of amnesia is typically limited to the assault itself, not several hours before the assault occurred. It is now 5:55 AM, and the patient has absolutely no recollection of the events that occurred after 9:00 PM the night before. You should be suspicious that the patient was unknowingly given a drug by the person who perpetrated this crime; this may have been the young man she met at the nightclub or a random person—male or female.

The use of drugs to facilitate sexual assault (eg, rape) is by no means a new tactic used by criminals. Alcohol was probably the first drug used for this purpose, and it remains a common element at many rape scenes. However, the drugs of choice for commission of a crime of rape are "club drugs," such as ketamine, 3,4-methylenedioxymethamphetamine (ecstasy), gamma-hydroxybutyric acid (GHB), and flunitrazepam (Rohypnol).

Although the exact pharmacology of these drugs is beyond the scope of this case study, a general knowledge of how these drugs act on a person's system is important for the EMT. All of the club drugs (also called "date rape drugs") previously mentioned are used to lower the victim's inhibitions, thus making the person more vulnerable and inducing amnesia. They are most commonly placed into the victim's drink, where they quickly dissolve and are generally tasteless. The onset of action varies with each drug, ranging anywhere from 30 to 60 minutes, and the effects can last for many hours. This effectively allows the perpetrator to commit the crime, leaving the victim with no recollection of the event (retrograde amnesia).

5. What additional assessment should you perform on this patient?

As with any patient, the extent of your secondary assessment is based on your suspicion of injuries or conditions that may not have been grossly apparent during the primary assessment. In sexual assault victims, however, it is unlikely that the patient will consent to a full-body scan. Therefore, you should limit any physical examination of the patient to a brief survey for life-threatening injuries.

Most of your additional assessment will be performed by asking the patient questions rather than performing a hands-on examination. In the unlikely event that the patient consents to a more in-depth assessment, you should do everything possible to protect her privacy. Conduct your interview and assessment with a minimum of people present, and move her to the ambulance if necessary. As previously discussed, a female EMT should conduct the assessment. If a female EMT is not available, a female police officer or friend should be present if possible.

The patient in this scenario complains of vaginal pain but denies vaginal bleeding. The external genitalia should not be exposed and examined—whether the EMT is a male or a female—unless there is evidence of severe bleeding that requires immediate treatment.

It is important to note that a patient with decision-making capacity has the legal right to refuse *any and all* assessment and treatment; sexual assault victims are no exception. You should expect that the last thing the patient will want is to be touched by *anyone*.

6. How should you respond to the patient's request to take a shower, change her clothes, and use the bathroom?

It is very common for victims of sexual assault to want to take a shower, change their clothes, rinse their mouth out, or douche. These actions stem from the desire to "wash away" the humiliation and embarrassment of the assault.

However, it is important to note that valuable evidence may be lost if the patient takes any measures to clean herself up; therefore, you should discourage—*not disallow*—her from doing so. The patient should also be discouraged from urinating or moving her bowels; doing so may destroy any DNA evidence that may have been left behind from vaginal or anal penetration. Law enforcement personnel will also need to photograph the victim, and the photographic record needs to be as accurate as possible. Make every attempt to explain to the patient that she potentially has evidence on or inside of her that may be used to identify the perpetrator. If, despite your best efforts, the patient cannot be dissuaded from cleaning herself up, you must respect her feelings and avoid forcing the issue.

7. How should you respond to the patient's refusal to go to the hospital?

In many cases, sexual assault victims will refuse EMS transport, and some of them will refuse *any and all* assessment and treatment. Provided the patient has decision-making capacity, she has the legal right to do so.

If the patient refuses EMS transport, do not simply accept the refusal and leave. You must still ensure that she is aware of the potential consequences of her refusal. She is complaining of vaginal pain; this could indicate a significant internal injury that potentially could be fatal; she *must* be made aware of this fact.

If she still refuses EMS transport, try to persuade her to allow you to call a friend or relative who can take her to the hospital. If she will not allow anyone to take her to the hospital, try to persuade her to allow you to call a friend or relative who can stay with her, or better yet, with whom she can stay. If a family member or trusted friend is present at the scene, advise the patient that the family member or friend can ride in the back of the ambulance with her to the hospital. It is important to avoid patient refusals whenever possible.

Although the patient in this scenario does not recall being assaulted, her signs and symptoms (eg, vaginal pain) and the way she presented—torn shirt and nude from the waist down—are clear indicators that she was sexually assaulted. Patients who recall being sexually assaulted often benefit from being removed from the scene in which the assault occurred because it keeps them from having to constantly relive the experience by being subjected to the environment where the assault occurred. The best way to do this is to convince her to go to the hospital; doing so will remove her from the scene *and* allow her to be evaluated by a physician.

8. Is it the EMT's responsibility to determine why the patient was sexually assaulted?

Absolutely not! It is law enforcement's responsibility to investigate the incident, arrest the perpetrator, and see justice served. Police officers will have questions pertaining to the incident; however, the prehospital setting is not the appropriate venue for this. The EMT's job is to address the physical and emotional needs of the patient and to serve as a patient advocate.

Victims of sexual assault require a great deal of psychological support; they do *not* need to be cross-examined or interrogated. Many women report feeling "reraped" after being subjected to interrogation, criticism, or any other questions relating to why the sexual assault occurred. Maintain a nonjudgmental attitude, and protect the patient from the judgment of others at the scene.

It does not matter how the patient was dressed, what her local reputation is, where she was, or what she was doing when the assault occurred. No one deserves to be assaulted, and the concept of a woman "deserving" to be raped because of her looks or actions is as ludicrous as a person "deserving" to be assaulted for wearing the shirt of a rival sports team. The fact remains that a crime has been committed; you must remain aware of this and focus your questions and assessment on the patient's physical well-being.

9. Why is it important to transport a sexual assault victim to the hospital, even if she has no obvious injuries?

Sexual assault victims should be transported to the hospital for several reasons. First and foremost, the patient should have a medical evaluation by a physician to rule out injuries or conditions that were not detected or not present in the field. Injuries such as internal abdominal bleeding can have a delayed onset of symptoms for up to several hours. Furthermore, the patient will need initial and follow-up screening for sexually transmitted diseases.

Some hospitals have access to certified personnel who are trained to perform the proper examination on patients who have been sexually assaulted. These personnel are often referred to as SANE (Sexual Assault Nurse Examiner) nurses. You should be aware of the hospitals in your area that are

staffed by SANE nurses. In some cases, a SANE nurse can meet you at the receiving facility if it does not have a SANE nurse on staff.

After the patient has undergone an appropriate medical evaluation and further assessment by a specially trained nurse, the hospital can refer the patient to a variety of patient advocacy groups, such as rape crisis centers, or can arrange for a counselor to come to the hospital. Many victim advocates are rape-trauma survivors themselves and are thus able to communicate with utmost compassion.

EMS Patient Care Report (PCR)

Date: 11-11-09	**Incident No.:** 211109	**Nature of Call:** Assault		**Location:** 4300 West Ave.	
Dispatched: 0555	**En Route:** 0557	**At Scene:** 0602	**Transport:** 0632	**At Hospital:** 0642	**In Service:** 0650

Patient Information

Age: 25 **Sex:** F **Weight (in kg [lb]):** 50 kg (110 lb)	**Allergies:** None **Medications:** No known drug allergies **Past Medical History:** None **Chief Complaint:** Vaginal pain; amnesia

Vital Signs

Time: 0610	**BP:** 110/70	**Pulse:** 72	**Respirations:** 14	**Sao$_2$:** 98%
Time: 0630	**BP:** 118/68	**Pulse:** 80	**Respirations:** 14	**Sao$_2$:** 99%
Time: 0640	**BP:** 122/72	**Pulse:** 76	**Respirations:** 16	**Sao$_2$:** 98%

EMS Treatment (circle all that apply)

Oxygen @ __ L/min via (circle one): NC NRM Bag-Mask Device	**Assisted Ventilation**	**Airway Adjunct**	**CPR**
Defibrillation	**Bleeding Control**	**Bandaging**	**Splinting**	**Other:** Limited assessment, emotional support

Narrative

Medic 86 was requested by law enforcement to respond to a residence for a young female who was assaulted. While en route to the scene, law enforcement advised that the patient had evidently been sexually assaulted. Arrived on scene and found the patient, a 25-year-old female, sitting on the ground next to her car with a blanket wrapped around her. She was conscious and alert, although clearly emotionally upset. Her airway was patent, her breathing was adequate, and no obvious bleeding was noted. Prior to EMS arrival, the patient advised law enforcement officials that she has no recollection of the events that occurred after 2100 the night before, when she was at a nightclub with her friends. She was found with her shirt torn and was nude from the waist down, so law enforcement personnel wrapped her with a blanket before EMS arrival. The patient complains of vaginal pain but denies vaginal bleeding. The only obvious injuries noted were several small abrasions to her forearms. The patient would not consent to a secondary assessment or treatment; she would only allow assessment of her vital signs. She denied significant medical history and medication allergies. The patient stated that she did not want to go to the hospital, via EMS or any other method of transportation. She further stated that she wanted to take a shower and change her clothes. Advised patient that this was not advisable because of the possibility of destroying potential evidence; however, she stated that she did not care and only wanted to clean herself up. Further advised the patient of the need for evaluation at the hospital because hidden injuries, some of which could be life threatening, could not be ruled out in the prehospital setting. The patient requested law enforcement to summon a friend of hers to the scene. After talking to her friend, she consented to EMS transport only; she further requested that her friend accompany her in the back of the ambulance. Began transport to the hospital and monitored the patient en route. She remained conscious and alert, and her vital signs remained stable. Provided emotional support, with the assistance of her friend, until delivery at the emergency department. Delivered patient to hospital and gave verbal report to staff nurse. Medic 86 cleared the hospital and returned to service at 0650. *End of report*

Assessment and Emergency Care of Gynecologic Emergencies

Scene Size-up

Scene Safety	Ensure scene safety and safe access to the patient. Be aware of potential violence or the possibility of a crime scene and call for law enforcement. If sexual assault is suspected, a same-sex EMT should be available to perform the assessment and provide care. Standard precautions should be taken because gynecologic emergencies often involve blood and body fluids. Determine the number of patients, and assess the need for additional resources. Pay close attention to the patient and the surroundings. Look for medications and signs of drug or alcohol use and note the condition of the living environment and whether the patient lives alone.
Mechanism of Injury (MOI)/ Nature of Illness (NOI)	Determine the MOI/NOI. Observe the scene and look for indicators that will assist you in determining the MOI/NOI. The nature of the problem may not be readily apparent until more information is gathered.

Primary Assessment

Form a General Impression	Your assessment of a patient with a gynecologic emergency should begin at the door. Perform a rapid scan to ascertain if the patient's condition is stable or unstable. Note the patient's behavior and the surroundings. Observe the patient's age and body position. Is the patient sitting, hunched over, or shuffling around (which may indicate pelvic inflammatory disease)? Determine the level of consciousness using the AVPU scale. Identify immediate threats to life. Determine the priority of care based on the MOI/NOI. If the patient has a poor general impression, call for advanced life support assistance. A rapid visual examination will help you identify and manage life threats; keep alert for signs of shock.
Airway and Breathing	Ensure the airway is open, clear, and self-maintained. Unresponsive patients will need the airway opened and maintained using a modified jaw-thrust manuever, especially if a spinal injury is suspected. For nontrauma patients, perform a head tilt–chin lift manuever. A patient with an altered level of consciousness may need emergency airway management. Consider inserting a properly sized oropharyngeal or nasopharyngeal airway. Evaluate the patient's ventilatory status for rate and depth of breathing, respiratory effort, and tidal volume. Administer high-flow oxygen at 15 L/min, providing ventilatory support as needed.
Circulation	Observe skin color, temperature, and condition; look for life-threatening vaginal bleeding, and treat accordingly. Evaluate distal pulse rate, quality (strength), and rhythm. Tachycardia may be an indicator of compensated shock. Treat for shock by placing the patient in a supine position, elevate the legs, maintain body temperature, and continue oxygen administration.
Transport Decision	If the patient has an airway or breathing problem, signs and symptoms of bleeding, or other life threats, manage them immediately and consider rapid transport, performing the secondary assessment en route to the hospital. Sexual assault patients may feel uneasy during the physical examination; remain calm, and be supportive and empathetic. It is a good idea to have a same-sex care provider in the back of the ambulance with you.

NOTE: The order of the steps in this section differs depending on whether the patient is conscious or unconscious. The following order is for a conscious patient. For an unconscious patient, perform a primary assessment, perform a rapid full-body scan, obtain vital signs, and obtain the past medical history from a family member, bystander, or emergency medical identification device.

Assessment and Emergency Care of Gynecologic Emergencies, continued

History Taking

Investigate Chief Complaint	Investigate the chief complaint. Necessary questions may be very personal; ensure patient privacy while remaining sensitive to the patient's feelings. Ask OPQRST and SAMPLE questions. SAMPLE can also be obtained from family, bystanders, and medical alert tags. Ask the patient about the use of birth control medications or birth control devices. Ask about the possiblity of the patient being pregnant, when her last menstrual period was, and the possibility of a sexually transmitted disease. Does the patient have any burning with urination? Is there an unusual discharge or foul odor? If vaginal bleeding was the chief complaint, ask about the number of pads she has already used.

Secondary Assessment

Physical Examinations	If the patient is unconscious, perform a systematic full-body examination beginning with the head, looking for DCAP-BTLS. Assessment should be rapid if the patient has a poor general impression. Patients may feel threatened or embarrassed by your presence and actions during a gynecologic emergency. Remain empathetic and professional at all times. Obtain consent before attempting any examination or procedure, and explain what you are going to do. Vaginal bleeding or vaginal discharge should be observed if possible.
Vital Signs	Obtain baseline vital signs. Vital signs should include blood pressure by auscultation, pulse rate and quality, respiration rate and quality, and skin assessment for perfusion. Note the patient's level of consciousness. Use pulse oximetry, if available, to assess the patient's perfusion status. Tachycardia or hypotension may indicate hypoperfusion.

Reassessment

Interventions	Repeat the primary assessment, checking the vital signs and chief complaint. Vital signs should be repeated every 5 minutes if blood loss is suspected to identify hypoperfusion. Place the patient in a position of comfort unless shock is suspected, then place the patient supine and treat accordingly. If the patient has vaginal bleeding, do not insert anything into the vagina; instead, use external pads and treat for shock. Keep the patient warm, and provide supplemental oxygen even if the patient is not experiencing difficulty breathing. If the patient was sexually assaulted, limit the physical examination and treat only life-threatening injuries.
Communication and Documentation	Contact medical control/receiving hospital with a radio report. Many hospitals require a sexual assault nurse examiner be notified and separate treatment areas provided for sexual assault cases. Include a thorough description of the MOI/NOI and the position the patient was found in. Include treatments performed and patient response. Be sure to document the patient's distress, answers to your questions, attitude toward emergency care providers, any changes in patient status, and the time. Document the scene observations on your arrival. Follow local protocols. Documentation of a sexual assault case needs to be concise, including statements made by the patient in her own words, without inserting your own opinion of what happened. Record injuries noted, scene findings, and the patient's emotional state.

NOTE: Although the steps that follow are widely accepted, be sure to consult and follow your local protocols. Take appropriate standard precautions when treating all patients.

Assessment and Emergency Care of Gynecologic Emergencies, continued

Gynecologic Emergencies

General Management of Gynecologic Emergencies

Managing life threats to the patient's ABCs is the primary concern with any gynecologic emergency. Gynecologic emergencies are rarely life threats, but always be alert for signs and symptoms of shock and manage as per local protocol. Women of childbearing age complaining of abdominal pain may have a life-threatening obstetric emergency. Vaginal bleeding should be controlled with pads placed on the external genitalia. Do not remove tampons the patient may have placed prior to EMS arrival. Other gynecologic emergencies EMS may be called for often involve a disease-causing organism. Signs and symptoms range from abdominal pain, lower back pain, pain with intercourse, burning during urination, fever, general malaise, "fishy" foul-smelling vaginal discharge, and bleeding. Remain compassionate to the patient's situation, maintain the patient's privacy, and transport to the appropriate medical facility.

Prep Kit

Ready for Review

- Women's bodies are uniquely designed to conceive and give birth. This difference makes women susceptible to a number of problems that do not occur in men.

- If fertilization of the ovum does not occur within about 14 days of ovulation, the lining of the uterus begins to separate, and menstruation occurs for about 1 week.

- When a girl reaches puberty, she begins to ovulate and experience menstruation.

- Women continue to experience the cycle of ovulation and menstruation until they reach menopause.

- The causes of gynecologic emergencies are varied and range from sexually transmitted diseases to trauma.

- Pelvic inflammatory disease is an infection of the female upper organs of reproduction: the uterus, ovaries, and fallopian tubes. It is the most common gynecologic reason why women access emergency medical services.

- Sexually transmitted diseases can lead to more serious conditions, such as pelvic inflammatory disease.

- Because menstrual bleeding is a monthly occurrence in most females, vaginal bleeding that is the result of other causes may be initially overlooked. Some possible causes of vaginal bleeding include abnormal menstruation, vaginal trauma, ectopic pregnancy, spontaneous abortion, cervical polyps, ectopic pregnancy, miscarriage, and even cancer.

- There are very few interventions that can or should be done in the prehospital setting with a gynecologic emergency.

- Whenever you deal with patients who have gynecologic problems, you must be sure to maintain the patient's privacy as much as possible.

- EMTs called on to treat a victim of sexual assault, molestation, or actual or alleged rape face many complex issues, ranging from obvious medical ones to serious psychological and legal issues. You may be the first person the victim has contact with after the encounter, and how the situation is managed from first contact throughout treatment and transport may have lasting effects for the patient and you. Professionalism, tact, kindness, and sensitivity are of paramount importance.

Vital Vocabulary

bacterial vaginosis An overgrowth of bacteria in the vagina; characterized by itching, burning, or pain, and possibly a "fishy" smelling discharge.

cervix The lower third, or neck, of the uterus.

chlamydia A sexually transmitted disease caused by the bacterium *Chlamydia trachomatis*.

fallopian tubes The tubes that connect each ovary with the uterus and are the primary location for fertilization of the ovum.

gonorrhea A sexually transmitted disease caused by *Neisseria gonorrhoeae*.

labia majora Outer fleshy "lips" covered with pubic hair that protect the vagina.

labia minora Inner fleshy "lips" devoid of pubic hair that protect the vagina.

ovaries The primary female reproductive organs that produce an ovum, or egg, that, if fertilized, will develop into a fetus.

ovulation The process in which an ovum is released from a follicle.

pelvic inflammatory disease (PID) An infection of the fallopian tubes and the surrounding tissues of the pelvis.

perineum The area of skin between the vagina and the anus.

rape Sexual intercourse inflicted forcibly on another person, against that person's will.

sexual assault An attack against a person that is sexual in nature, the most of common of which is rape.

uterus The muscular organ where the fetus grows, also called the womb; responsible for contractions during labor.

vagina The outermost cavity of a woman's reproductive system; the lower part of the birth canal.

Assessment in Action

It is 10:00 PM and you and your partner are dispatched to the parking lot of a high-rise office building for an assault patient. On arrival, you find a young female patient seated in the back seat of a police cruiser; she is wrapped in a blanket and crying. One of the officers on scene stops you before you can approach the car and tells you that the scene is safe. The officer informs you that the patient was abducted in the parking lot and was raped. You approach the police car to talk to the patient.

1. The primary assessment of the patient should include:
   A. a rapid scan to identify any life-threatening injuries.
   B. an extensive trauma examination.
   C. an extensive medical examination.
   D. both an extensive trauma and a medical examination.

2. The patient's clothes were ripped during the assault. What should you do with any of her clothing items that you remove during your examination and treatment?
   A. Leave them in the possession of a family member or friend.
   B. Transport them to the hospital in a red biohazard bag.
   C. Turn them over to the police as evidence.
   D. Throw them away to prevent crime scene contamination.

3. The patient tells you that she feels blood coming from the vaginal area. She gives you permission to look and you observe moderate bleeding from the vagina. What should you do to control the bleeding?
   A. Apply direct pressure using trauma dressings.
   B. Pack the vagina with trauma dressings.
   C. Ask the patient to insert a tampon.
   D. Place a sanitary pad over the vaginal opening.

4. Other than trauma, what are some of the causes of vaginal bleeding?

5. The patient repeatedly asks you if she can take a shower. Why is it important that a patient who has been assaulted not be allowed to shower or go to the bathroom?
   A. Evidence will be destroyed.
   B. On-scene time will be prolonged.
   C. The patient may leave the scene.
   D. A female escort may not be available.

6. In addition to providing comfort and reassurance to the patient, what should be your most important intervention?
   A. Encourage the patient to recount her experience.
   B. Transport the patient to a hospital certified to perform the proper examination.
   C. Provide your opinion of crimes of sexual assault.
   D. Encourage the patient to press charges against her assailants.

7. Careful documentation is always important. What information should be included in your patient care report?
   A. Patient condition
   B. Description of the scene
   C. Interventions performed
   D. All of the above

8. Why is it important not to be judgmental when caring for a patient who has been sexually assaulted?

9. How can you reduce the anxiety of a patient who has been raped?

10. Some sexually transmitted diseases can be transmitted by contact with blood. Name some of these diseases.

National EMS Education Standard Competencies

Trauma

Applies fundamental knowledge to provide basic emergency care and transportation based on assessment findings for an acutely injured patient.

Trauma Overview

Pathophysiology, assessment, and management of the trauma patient

- Trauma scoring (p 767)
- Rapid transport and destination issues (pp 764-768)
- Transport mode (p 765)

Multisystem Trauma

Recognition and management of
- Multisystem trauma (p 762)

Pathophysiology, assessment, and management of
- Multisystem trauma (p 762)
- Blast injuries (pp 760-762)

Knowledge Objectives

1. Define the term mechanism of injury (MOI) and explain its relationship to potential energy, kinetic energy, and work. (pp 747-749)
2. Define the terms blunt and penetrating trauma and provide examples of the mechanism of injury that would cause each one to occur. (pp 749-759)
3. Describe the five types of motor vehicle collisions, the injury patterns associated with each one, and how each relates to the index of suspicion of life-threatening injuries. (pp 750-757)
4. Discuss the three specific factors to consider during assessment of a patient who has been injured in a fall, plus additional considerations for pediatric and geriatric patients. (pp 757-758)
5. Discuss the affects of high-, medium-, and low-velocity penetrating trauma on the body and how an understanding of each type helps the EMT form an index of suspicion about unseen life-threatening injuries. (pp 758-759)
6. Discuss primary, secondary, tertiary, and miscellaneous blast injuries and describe the anticipated damage each one will cause to the body. (pp 760-762)
7. Describe multisystem trauma and the special considerations that are required for patients who fit this category, and provide a general overview of multisystem trauma patient management. (p 762)
8. Outline the major components of trauma patient assessment, including considerations related to whether the method of injury was significant or nonsignificant. (pp 762-764)
9. Discuss the special assessment considerations related to a trauma patient who has injuries in each of the following areas: head, neck and throat, chest, and abdomen. (pp 762-764)
10. Describe trauma patient management in relation to scene time and transport selection and list the Association of Air Medical Services criteria for the appropriate use of emergency air medical services. (pp 764-768)
11. Describe the American College of Surgeons' Committee on Trauma classification of trauma centers and how it relates to making an appropriate destination selection for a trauma patient. (pp 765-768)

Skills Objectives

There are no skills objectives for this chapter.

Introduction

According to the National Institutes of Health, traumatic injuries are the leading cause of death in the United States among people younger than age 40. Proper prehospital evaluation and care can do much to minimize suffering, long-term disability, and death from trauma. As discussed in Chapter 12, *Medical Overview*, patients who need EMS assistance generally fall in one of two categories, either medical or trauma, although occasionally one may result from the other or both may exist.

Trauma emergencies occur as a result of physical forces applied to the body. **Medical emergencies** occur when the patient has an illness or condition that is not caused by an outside force. This chapter introduces the basic physical concepts that dictate how traumatic injuries occur and how they affect the human body. When you understand these concepts, you will be better able to size up a crash scene and use that information as a vital part of patient assessment. This chapter begins with a basic discussion of energy and trauma. Next, different types of crashes and their impact on the body are explained. By assessing a vehicle that has crashed, you can often determine what happened to the passengers at the time of impact, which may allow you to predict what injuries the passengers sustained at the time of impact. Evaluation of the mechanism of injury (MOI) for the trauma patient will provide you with an index of suspicion for different types of serious and/or life-threatening underlying injuries. Certain injury patterns occur with certain types of injury events. The **index of suspicion** is your awareness and concern for potentially serious underlying and unseen injuries.

Energy and Trauma

Traumatic injury occurs when the body's tissues are exposed to energy levels beyond their tolerance **Figure 22-1**. The **mechanism of injury (MOI)** is the way in which traumatic injuries occur; it describes the forces (or energy transmission) acting on the body that cause injury. Three concepts of energy are typically associated with injury (not including thermal energy, which causes burns): potential energy, kinetic energy, and work. When considering the effects of energy on the human body, it is important to remember that energy can be neither created nor destroyed, but can only be converted or transformed. It is not the objective of this section to help you to reconstruct the scene of a motor vehicle crash. Rather, you should have a sense of the effects of work on the body and understand, in a broad sense, how that work is related to potential and kinetic energy. For example, when you are assessing a patient who fell, you need not calculate the speed at which the person hit the ground.

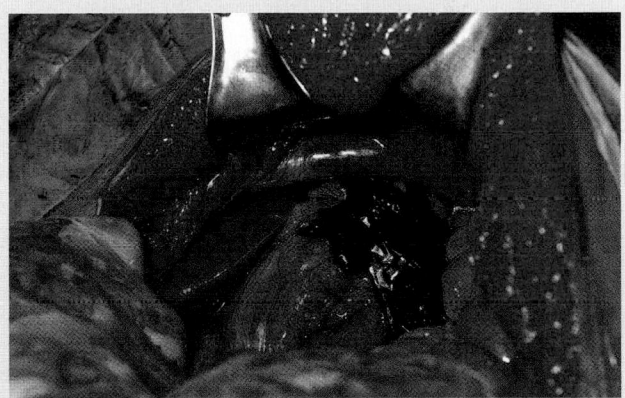

Figure 22-1 Traumatic injury occurs when the body's tissues are exposed to energy levels beyond their tolerance. This photo shows a ruptured spleen.

You are the Provider: PART 1

At 3:20 PM, you and your partner are dispatched to a motor vehicle collision, in which a passenger car reportedly struck a tree head-on at an unknown rate of speed. The dispatch operator reports that the patient is still in the vehicle, but it is unknown if the person is entrapped. Law enforcement personnel and two engine companies have also been dispatched to the scene. Your response time is 8 minutes, the weather is clear, and the traffic is heavy.

1. On the basis of the information from the dispatch operator, can you predict the type of injuries the patient may have? If so, how?
2. Why is it important to try to determine the speed a vehicle was traveling at the time of impact?

Words of Wisdom

Newton's Laws
Newton's First Law

Newton's first law states that objects at rest tend to stay at rest and objects in motion tend to stay in motion unless acted on by some force. The first part of the law is fairly clear. An object such as an empty soda can will not move spontaneously unless some force, such as a gust of wind, acts on it. An example will help to illustrate the second part. In a car going 30 mph, the passengers and the car are moving at 30 mph. The passengers do not feel as though they are moving because they are not moving relative to the car. However, when the car strikes a concrete barrier and comes to a sudden stop, the passengers continue to travel at 30 mph. They stay in motion until they are acted on by an external force—most likely the windshield, steering wheel, or dashboard. To appreciate the severity of the impact, think of the driver as sitting motionless while a steering wheel rams into his or her chest at 30 mph. Now consider that the same thing happens to the driver's internal organs. They also are in motion, traveling at 30 mph relative to the ground, until they are acted on by an external force, in this case the sternum, rib cage, or other body structure. This scenario illustrates the three collisions that are associated with blunt trauma.

Newton's Second Law

Newton's second law states that force (F) equals mass (M) times acceleration (A), that is, $F = M \times A$, in which acceleration is the change in velocity (speed) that occurs over time. Therefore, it is not so much that "speed kills" but that the change in velocity with respect to time generates the forces that cause injury. Simply put, it is not the fall, but the sudden stop at the bottom, that causes the injury.

In the example of the car traveling at 30 mph, it takes about 3 seconds for the car to decrease its speed from 30 mph to 0 mph when the driver applies the brakes smoothly. If he or she is properly restrained by well-adjusted seatbelts, the driver slows, or decelerates, at the same rate as the car. But if the car is stopped not by braking but by hitting a large tree and the driver is not restrained, his or her body will continue to stay in motion at 30 mph until it is stopped by an external force, in this case, the steering wheel. Although the change in the body's velocity is the same as when the car was braking smoothly in 3 seconds (30 to 0 mph), that change now takes place in about 0.01 second. Because the period of deceleration is 300 times less, the average force of impact is 300 times greater. This means that the force is approximately 150 times the force of gravity. Imagine a force 150 times your body weight slamming into your chest.

Now consider the same car striking the same tree, but this time, the driver is restrained with a shoulder and lap belt. The driver is essentially tied to the car and stops during the same period the car stops. It takes some time, although brief, to crush the front of the car and bring it to a halt. The car comes to a stop in approximately 0.05 second. The change in the driver's velocity is the same (30 to 0 mph), but the longer period of deceleration results in a g force of only 30 times that of gravity. This is still a substantial force, but it is much less than the force that is experienced by the unrestrained driver. More to the point, it is survivable.

In a final example, the car and driver, as before, are traveling at 30 mph, and the driver is properly restrained with a three-point seatbelt. In this case, however, the car is also equipped with an air bag. When the car hits the tree and suddenly stops, the driver's upper body initially continues forward at 30 mph. The body is partially slowed by the lap and shoulder belts but is finally brought to rest by the air bag. The upper body compresses the air bag, which stops the body's forward motion in about 0.1 second. Thus, the air bag stretches the duration of impact by 0.05 second, buying the body even more time, and the force on the upper body drops to approximately 15 times that of gravity.

The air bag has another advantage. The force of its impact is applied over a much larger area than the area that is affected by the steering wheel or the shoulder belt, shrinking the force per unit area. This point can be illustrated by an analogy. A person standing on one toe on a sheet of ice applies a concentrated load in a very small area, thus breaking the ice and falling through. If the person lies flat on the ice, he or she greatly expands the contact area and reduces the stress on the ice, which, depending on conditions, should not break. The dual action of the air bag (distributing the force of impact over a greater area and increasing the duration of impact) results in less severe injuries.

Newton's Third Law

Newton's third law states that for every action, there is an equal and opposite reaction. Therefore, if you push on a door, the door pushes back (reacts) with an equal force but in the opposite direction. In the case of a dented A-pillar, the force of the driver's head was sufficient to dent the strong metal. But in terms of patient assessment, the more important point is the reaction force of the pillar on the head. Newton's third law states that the two forces are equal but occur in opposite directions. In other words, the head was essentially hit by an A-pillar traveling at 30 mph. Similarly, it takes a substantial force to collapse a steering wheel. When you notice a collapsed steering wheel during scene size-up, you should suspect serious chest injuries even if the driver initially has no visible signs of chest injury. Often, reading the scene and understanding the basic principles of energy transfer will give you as clear a picture of the patient's potential injuries and injury severity as the actual physical patient assessment.

However, it is important to estimate the height from which he or she fell and to appreciate the injury potential of the fall.

Work is defined as force acting over a distance. For example, the force needed to bend metal multiplied by the distance over which the metal is bent is the work that crushes the front end of a vehicle that is involved in a frontal impact. Similarly, forces that bend, pull, or compress tissues beyond their inherent limits result in the work that causes injury.

The energy of a moving object is called **kinetic energy**. Kinetic energy reflects the relationship between the mass (weight) of the object and the velocity (speed) at which it is traveling. Kinetic energy is expressed as:

$$\text{Kinetic energy} = \frac{\text{mass}}{2} \times \text{velocity}^2$$

$$\text{or,} \quad KE = \frac{m}{2} \times v^2$$

Remember that energy cannot be created or destroyed, only converted. In the case of a motor vehicle crash, the kinetic energy of the speeding vehicle is converted into the work of stopping the car, usually by crushing the vehicle's exterior **Figure 22-2**. Similarly, the passengers of the vehicle have kinetic energy because they were traveling at the same speed as the vehicle. Their kinetic energy is converted to the work of bringing them to a stop. It is this work on the passengers that results in injury. Notice that, according to the equation for kinetic energy, the energy that is available to cause injury *doubles* when an object's weight doubles but *quadruples* when its speed doubles. When a car's speed increases from 50 mph to 70 mph, the energy that is available to cause injury doubles. This point is even clearer when considering gunshot wounds. The speed of the bullet (high-velocity compared with low-velocity) has a greater impact on producing injury than the mass (size) of the bullet. This is why it is so important to report to the hospital the type of firearm that was used in a shooting. The amount of kinetic energy that is converted to do work on the body dictates the severity of the injury. High-energy injuries often produce such severe damage that patients require immediate transport to an appropriate facility to have any hope of survival.

Figure 22-2 The kinetic energy of a speeding car is converted into the work of stopping the car, usually by crushing the car's exterior.

Potential energy is the product of mass (weight), force of gravity, and height and is mostly associated with the energy of falling objects. A worker on a scaffold has some potential energy because he or she is some height above the ground. If the worker falls, potential energy is converted into kinetic energy. As the worker hits the ground, the kinetic energy is converted into work, that is, the work of bringing the body to a stop and thereby fracturing bones and damaging tissues.

Mechanism of Injury Profiles

Different types of MOIs will produce many types of injuries. Examples of nonsignificant injuries include injury to an isolated body part or a fall without the loss of consciousness. Examples of significant MOIs include injury to more than one body system (**multisystem trauma**), falls from heights, motor vehicle and motorcycle crashes, car versus pedestrian (or bicycle or motorcycle), gunshot wounds, and stabbings. Whether one body system or more than one system is involved, you should maintain a high index of suspicion for serious unseen injuries.

Blunt and Penetrating Trauma

Traumatic injuries can be divided into two separate categories: blunt trauma and penetrating trauma. Either type of trauma may occur from a variety of MOIs. It is important for you to consider unseen as well as visible, obvious injuries with either type of trauma. **Blunt trauma** is the result of force (or energy transmission) to the body that causes injury without penetrating the soft tissues or internal organs and cavities. **Penetrating trauma** causes injury by objects that primarily pierce and penetrate the surface of the body and cause damage to soft tissues, internal organs, and body cavities.

Blunt Trauma

Blunt trauma results from an object making contact with the body. Any object, for example a baseball bat, can cause blunt trauma if it is moving fast enough. Motor vehicle crashes and falls are two of the most common MOIs for blunt trauma. When providing care for your patient, you should be alert to signs of skin discoloration or complaints of pain because these may be the only signs of blunt trauma. You also should maintain a high index of suspicion during patient assessment for hidden injuries in patients with blunt trauma.

Vehicular Collisions

Motor vehicle crashes are classified traditionally as frontal (head-on), rear-end, lateral (T-bone), rollovers, and rotational (spins). The principal difference among these collision types is the direction of the force of impact; also, with spins and rollovers, there is the possibility of multiple impacts. Motor vehicle crashes typically consist of a series of three collisions. Understanding the events that occur during each one of these three collisions will help you be alert for certain types of injury patterns. The three collisions in a typical impact are as follows:

1. **The collision of the car against another car, a tree, or some other object.** Damage to the car is perhaps the most dramatic part of the collision, but it does not directly affect patient care, except possibly to make extrication difficult **Figure 22-3**. However, it does provide information about the severity of the collision and, therefore, has an indirect effect on patient care. The greater the damage to the car, the greater the energy that was involved and, therefore, the greater the potential to cause injury to the patient. By assessing the vehicle that has crashed, you can often determine the MOI, which may allow you to predict what injuries may have happened to the passengers at the time of impact according to forces that acted on their bodies. When you arrive at the crash scene and perform your scene size-up, quickly inspect the severity of damage to the vehicle(s). If there is significant damage to a vehicle, your index of suspicion for the presence of life-threatening injuries should automatically increase. A great amount of force is required to

You are the Provider: PART 2

When you arrive at the scene, fire and law enforcement personnel are already present. An air transport helicopter has been placed on standby. The front of the vehicle is smashed all the way up to the windshield. The driver, a young male, was unrestrained and is still in the driver's seat; he appears to be unconscious and his face is covered with blood. As your partner accesses the patient from the back seat and manually stabilizes his head, you perform a primary assessment.

Recording Time: 0 Minutes	
Appearance	Bleeding from the head, face, and mouth; pale skin; labored breathing
Level of consciousness	Responsive only to pain
Airway	Blood in the oropharynx
Breathing	Rapid and labored
Circulation	Radial pulses, weak and rapid; skin is cool, pale, and clammy

Despite the severity of exterior damage to the vehicle, the patient is not entrapped and there is no interior intrusion into the passenger compartment. You suction the patient's mouth, apply a cervical collar, and prepare to rapidly extricate him from the vehicle. Responders from one of the engine companies have prepared a backboard, straps, and a lateral head immobilizer.

3. **Where will you most likely find damage to the vehicle's interior based on the patient's signs and symptoms?**

Figure 22-3 The first collision in a typical impact is that of the vehicle against another object (in this case, a utility pole). The appearance of the vehicle can provide you with critical information about the severity of the crash. The greater the damage to the vehicle, the greater the energy that was involved.

crush and deform a vehicle, cause intrusion into the passenger compartment, tear seats from their mountings, and collapse steering wheels. Such damage suggests the presence of high-energy trauma.

2. **The collision of the passenger against the interior of the car.** Just as the kinetic energy produced by the vehicle's mass and velocity is converted into the work of bringing the vehicle to a stop, the kinetic energy produced by the passenger's mass and velocity is converted into the work of stopping his or her body **Figure 22-4**. Just like the obvious damage to the exterior of the car, the injuries that result are often dramatic and usually immediately apparent during your scene size-up or primary assessment. Common passenger injuries include lower extremity fractures (knees into the dashboard), flail chest

Figure 22-4 The second collision in a typical impact is that of the passenger against the interior of the car. The appearance of the interior of the car can provide you with information about the severity of the patient's injuries.

(rib cage into the steering wheel), and head trauma (head into the windshield). Such injuries occur more frequently if the passenger is not restrained. But even when the passenger is restrained with a properly adjusted seatbelt, injuries can occur, especially in lateral and rollover impacts.

3. **The collision of the passenger's internal organs against the solid structures of the body.** The injuries that occur during the third collision may not be as obvious as external injuries, but they are often the most life threatening. For example, as the passenger's head hits the windshield, the brain continues to move forward until it comes to rest by striking the inside of the skull. This results in a compression injury (or bruising) to the anterior portion of the brain and stretching (or tearing) of the posterior portion of the brain **Figure 22-5**. This is an example of a **coup-contrecoup brain injury** **Figure 22-6**. Similarly, in the thoracic cage, the heart may slam into the sternum, which may rupture the aorta and cause fatal bleeding.

Understanding the relationship among the three collisions will help you make the connections between the amount of damage to the exterior of the vehicle and potential injury to the passenger. For example, in a high-speed collision that results in massive damage to the vehicle, you should suspect serious injuries to the passengers, even if the injuries are not readily apparent. A number of potential physical problems may develop as a result of traumatic injuries. Your initial general impression of the patient and the evaluation of the MOI can help direct lifesaving care and provide critical information to the hospital staff. Therefore, if you see a contusion on the patient's forehead and the windshield is starred and pushed out, you should strongly suspect an injury to the brain. After you inform medical control about the damage to the windshield, hospital staff can prepare for the patient by being ready to perform a computed

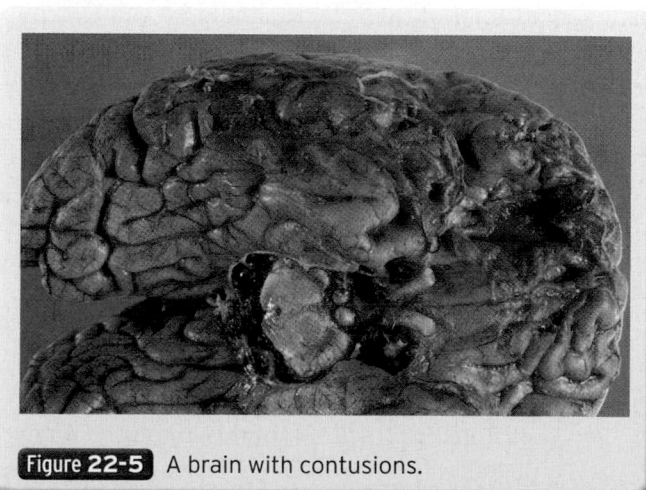

Figure 22-5 A brain with contusions.

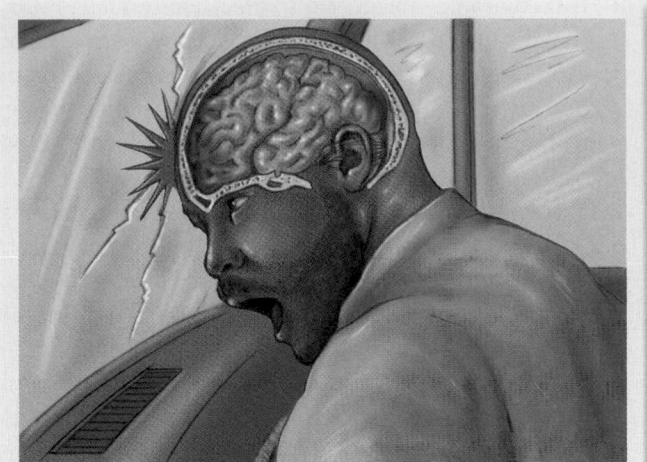

Figure 22-6 The third collision in a typical impact is that of the passenger's internal organs against the solid structures of the body. In this illustration, the brain continues its forward motion and strikes the inside of the skull, resulting in a compression injury to the anterior portion of the brain and stretching of the posterior portion.

tomography scan of the brain. Without your input, the physician might have found the brain injury anyway, but it might have not been detected until the brain had swollen sufficiently to cause clinical signs of the injury. Whenever there is a significant impact to the head, you should also suspect a spinal injury.

Words of Wisdom

When you are assessing trauma incidents, the MOI is a crucial element of history taking. Be alert to the extent of damage to the interior and exterior of the vehicles involved in crashes. Use this observation to paint a picture of the scene in written and verbal communication.

The amount of damage that is considered significant varies, depending on the type of collision, but any substantial deformity of the vehicle should be enough cause for you to consider transporting the patient to a trauma center. Significant mechanisms of injury include the following findings:

- Death of an occupant in the vehicle
- Severe deformity of vehicle or intrusion into vehicle
  - Severe deformities of the frontal part of a vehicle, with or without intrusion into the passenger compartment
  - Moderate intrusions from a lateral (T-bone) type of accident
  - Severe damage from the rear
  - Collisions in which rotation is involved (rollover and spins)

- Altered mental status
- Ejection from the vehicle

Damage to the vehicle that was involved and information obtained during the scene size-up are not the only clues you can use to determine crash severity. Clearly, if one or more of the passengers are dead, you should suspect that the other passengers have sustained serious injuries, even if the injuries are not obvious. Therefore, you should focus on treating life-threatening injuries and providing rapid transport to a trauma center, because these passengers have likely experienced the same amount of force that caused the death of the others. Polaroid or digital pictures of the crash scene may provide valuable information to the staff and treating physicians at the trauma center.

Frontal Collisions

Understanding the MOI after a frontal collision first involves evaluation of the supplemental restraint system, including seatbelts and air bags. You should determine whether the passenger was restrained by a full and properly applied three-point restraint. In addition, you should determine whether the air bag was deployed. Identifying the types of restraints used and whether air bags were deployed will help you identify injury patterns related to the supplemental restraint systems.

When properly applied, seatbelts are successful in restraining the passengers in a vehicle and preventing a second collision inside the motor vehicle. In addition, they may decrease the severity of the third collision, that of the passenger's organs with the chest or abdominal wall. The protective abilities of seatbelts are further enhanced by deployment of the air bags. Air bags provide the final capture point of the passengers and decrease the severity of **deceleration** injuries by allowing seatbelts to be more compliant and by gently cushioning the occupant and providing even more "ride down" as the body slows, or decelerates.

Remember that air bags decrease injury to the chest, face, and head very effectively. However, you should still suspect that other serious injuries to the extremities (resulting from the second collision) and to internal organs (resulting from the third collision) have occurred. Most new motor vehicles are manufactured with air bag safety systems. These safety devices enhance the safety and survival of forward-facing occupants inside the vehicle during a collision. In an emergency braking event, or collision, the air bag inflates very quickly. Because a rear-facing car seat is in proximity to the dashboard, rapid inflation of the air bag could cause serious injury or death to an infant. All children who are shorter than 4'9" should ride in the rear seat or, in the case of a pickup truck or other single-seated vehicle, the air bag should be turned off.

When you are providing care to an occupant inside a motor vehicle, it is important to remember that if the air bag did not inflate during the accident, it may deploy during extrication. If this occurs, you may be seriously injured or even killed. Extreme caution must be used when extricating a patient in a vehicle with an air bag that has not deployed.

You should also remember that supplemental restraint systems can cause harm whether they are used properly or improperly. For example, some older models of vehicles have seatbelts that buckle automatically at the shoulder but require the passengers to buckle the lap portion; these can result in the body "submarining" forward underneath the shoulder restraint when the lap portion is not attached. This movement of the body can cause the lower extremities and the pelvis to crash into the dashboard because that part of the body is unrestrained. Seatbelts may also cause unseen injuries, particularly in pediatric patients. Seatbelts are designed to be worn over the iliac crests of the pelvis in order to distribute the force over the bony surface. Hip dislocations may result if seatbelts are worn too low. Internal injuries can occur when the belt is worn too high, resulting in damage to abdominal organs Figure 22-7 . Lumbar spine fractures are also possible, particularly in elderly patients.

When passengers are riding in vehicles equipped with air bags but are not restrained by seatbelts, they are often thrown forward in the act of emergency braking. As a result, they come into contact with the air bag and/or the doors at the time of deployment. This MOI is also responsible for some severe injuries to children who are riding unrestrained in the front seats of vehicles.

In addition, some passengers may pass out before impact, and you may find them lying against the air bag when it deploys. When you encounter these types of situations, you should look for abrasions and/or traction-type injuries on the face, lower part of the neck, and chest Figure 22-8 .

Contact points are often obvious as you perform a simple quick evaluation of the interior of the vehicle. If there is no intrusion into the passenger compartment, you might see that an unrestrained front-seat passenger in a frontal collision has come into contact with the dashboard or the instrument panel at the knees, thus transferring loads from the knees through the femur to the pelvis and hip joint Figure 22-9A . The chest and/or abdomen may also hit the steering wheel Figure 22-9B . In addition, the passenger's face often hits the steering wheel or the passenger may launch forward and up, hitting the windshield and/or the roof header in the area of the visors Figure 22-9C . Signs of most of these injuries can be found by simply inspecting the interior of the vehicle during extrication of the patient.

Rear-End Collisions

Rear-end impacts are known to cause whiplash-type injuries, particularly when the passenger's head and/or neck is not restrained by an appropriately placed headrest Figure 22-10 . On impact, the passenger's body and torso move forward. As the body is propelled forward, the head and neck are left behind because the head is relatively heavy, and they appear to be whipped back relative to the torso. As the vehicle comes to rest, the unrestrained passenger moves forward, striking the dashboard. In this type of collision, the cervical spine and surrounding area may be injured. The cervical spine is less tolerant

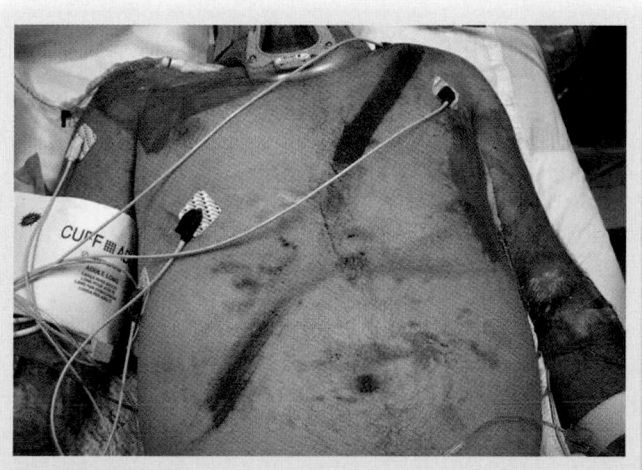

Figure 22-7 Injuries may result if the seatbelt is worn too high or too low around the waist.

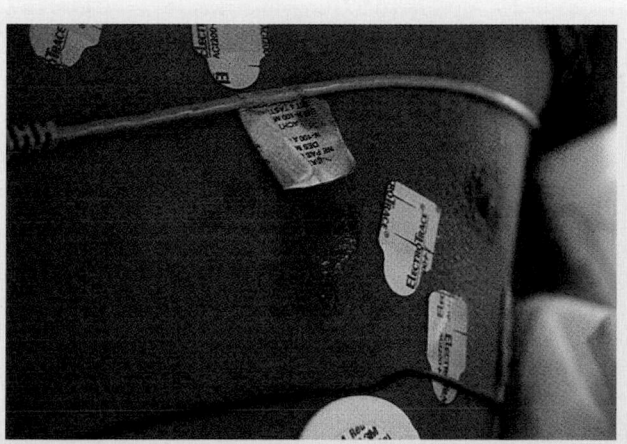

Figure 22-8 Air bags can cause injury in frontal collisions, specifically, abrasions and traction-type injuries to the face, neck, and chest.

Figure 22-9 Mechanism of injury and condition of the vehicle interior suggest likely areas of injury. **A.** The knees can strike the dashboard, resulting in a hip fracture or dislocation. **B.** Serious chest and abdominal injuries can result from striking the steering wheel. **C.** Head and spinal injuries can result when the face and head strike the windshield.

Figure 22-10 Rear-end impacts often cause whiplash-type injuries, particularly when the head and/or neck is not restrained by a headrest.

of damage when it is bent back. Headrests decrease extension of the head and neck during a collision and, therefore, help reduce injury. Other parts of the spine and the pelvis may also be at risk for injury. In addition, the patient may sustain an acceleration-type injury to the brain, that is, the third collision of the brain within the skull. Passengers in the back seat wearing only a lap belt might have a higher incidence of injuries to the thoracic and lumbar spine.

Lateral Collisions

Lateral or side impacts (commonly called T-bone collisions) are a very common cause of death associated with motor vehicle crashes. When a vehicle is struck from the side, it is typically struck above its center of gravity and begins to rock away from the side of the impact. This results in the passenger sustaining a lateral whiplash injury **Figure 22-11**. The movement is to the side, and the passenger's shoulders and head whip toward the intruding vehicle. This action may thrust the shoulder, thorax, and upper extremities, and, more important, the skull against the doorpost or the window. The cervical spine has little tolerance for lateral bending.

If there is substantial intrusion into the passenger compartment, you should suspect your patient to have lateral chest and abdomen injuries on the side of the impact, as well as possible fractures of the lower extremities, pelvis, and ribs. In addition, the organs within the abdomen are at risk because of a possible third collision. Approximately 25% of all severe injuries to the aorta that occur in motor vehicle crashes are a result of lateral collisions.

Figure 22-11 In a lateral collision, the car is typically struck above its center of gravity and begins to rock away from the side of impact. This causes a type of lateral whiplash in which the passenger's shoulders and head whip toward the intruding vehicle.

Figure 22-12 Passengers who have been ejected or partially ejected may have struck the interior of the car many times before ejection.

Rollover Crashes

Certain vehicles, such as large trucks and some sport utility vehicles, are more prone to rollover crashes because of their high center of gravity. Injury patterns that are commonly associated with rollover crashes differ, depending on whether the passenger was restrained or unrestrained. The most unpredictable types of injuries are caused by rollover crashes in which an unrestrained passenger may have sustained multiple strikes within the interior of the vehicle as it rolled one or more times. The most common life-threatening event in a rollover is ejection or partial ejection of the passenger from the vehicle Figure 22-12 . Passengers who have been ejected may have struck the interior of the vehicle many times before ejection. The passenger may also have struck several objects, such as trees, a guardrail, or the vehicle's exterior, before landing. Passengers who have been partially ejected may have struck both the interior and exterior of the vehicle and may have been sandwiched between the exterior of the vehicle and the environment as the vehicle rolled. Ejection and partial ejection are significant mechanisms of injury; in these cases, you should prepare to care for life-threatening injuries.

Even when restrained, passengers can sustain severe injuries during a rollover crash, although the patterns of injury tend to be more predictable, and when the restraint system is properly used, ejection from the vehicle is prevented. A passenger on the outboard side of a vehicle that rolls over is at high risk for injury because of the centrifugal force (the patient is pinned against

the door of the vehicle). Rollover crashes can also cause injury when the roof of the vehicle hits the ground during the rollover; a passenger who is restrained can still move far enough toward the roof to make contact and sustain a spinal cord injury. Therefore, rollover crashes are dangerous for both restrained and, to a greater degree, unrestrained passengers because these crashes provide multiple opportunities for second and third collisions.

Rotational Collisions

Rotational collisions (spins) are conceptually similar to rollovers. The rotation of the vehicle as it spins provides opportunities for the vehicle to strike objects such as utility poles. For example, as a vehicle spins and strikes a pole, the passengers experience not only the rotational motion, but also a lateral impact.

■ Car Versus Pedestrian

Car-versus-pedestrian collisions often result in patients who have graphic and apparent injuries, such as broken bones; however, this type of accident can cause serious unseen injuries to underlying body systems. Therefore, you must maintain a high index of suspicion for unseen injuries. A thorough evaluation of the MOI is critical. Your first step should be to estimate the speed of the vehicle that struck the patient; next determine whether the patient was thrown through the air and at what distance or whether the patient was struck and pulled under the vehicle. You should evaluate the vehicle that struck the patient for structural damage that might indicate contact points with the patient and alert you to potential injuries. Multisystem injuries are common after this type of event. ALS backup should be summoned for any patients who have or are thought to have sustained a significant MOI.

Car Versus Bicycle

In a car-versus-bicycle collision, you should evaluate the MOI in much the same manner as car-versus-pedestrian collisions. However, additional evaluation of damage to and the position of the bicycle is warranted. If the patient was wearing a helmet, you should inspect the helmet for damage and suspect potential injury to the head Figure 22-13 . Presume that the patient has sustained an injury to the spinal column, or spinal cord, until proven otherwise at the hospital. Spinal stabilization must be initiated and maintained during the encounter. When practical, the patient should be rolled on to his or her side to allow for an appropriate assessment to the posterior side of the body.

Car Versus Motorcycle

In a motorcycle crash, any structural protection afforded to the victims is not derived from a steel cage, as is the case in an automobile, but from protective devices worn by the rider, that is, helmet, leather or abrasion-resistant clothing, and boots. While helmets are designed to protect against impact forces to the head, they transmit any impact into the cervical spine, and as such, do not protect against severe cervical injury. Leather and synthetic

Figure 22-13 If the patient's bike helmet is damaged, suspect head and spine injuries.

gear worn over the body was initially designed to protect professional riders in competition, where falls tend to be controlled and result in long sliding mechanisms on hard surfaces rather than multiple collisions against road objects and other vehicles. Leather clothing will protect mostly against road abrasion but offers no protection against blunt trauma from secondary impacts. In a street crash, collisions occur usually against other larger vehicles or stationary objects.

When you are assessing the scene of a motorcycle crash, attention should be given to the deformity of the

You are the Provider: PART 3

The patient is removed from the vehicle onto the backboard, and you perform a quick secondary assessment. Your partner applies high-flow oxygen via a nonrebreathing mask while one of the EMTs from one of the engine companies obtains the patient's vital signs.

Recording Time: 5 Minutes	
Respirations	22 breaths/min; labored
Pulse	120 beats/min; weak radial pulses
Skin	Cool, clammy, and pale
Blood pressure	84/64 mm Hg .
Oxygen saturation (Sao$_2$)	97% (on oxygen)

The patient's level of consciousness is still markedly decreased; he only responds to pain. He has a large hematoma and laceration to his forehead, which is covered with a sterile dressing. He also has crepitus and bruising to his chest. A Level I trauma center is 30 miles away, but there is a Level III trauma center only 15 miles away.

4. Should the patient be transported to the Level I trauma center or the Level III trauma center? Why?

5. What other transport factors should you consider with this patient?

motorcycle, the side of most damage, the distance of skid in the road, the deformity of stationary objects or other vehicles, and the extent and location of deformity in the helmet. These findings can be helpful in estimating the extent of trauma in a patient.

There are four types of motorcycle impacts.

- **Head-on collision:** The motorcycle strikes another object and stops its forward motion while the rider and parts of the motorcycle that are broken off continue their forward motion until stopped by an outside force, such as drag from the road or another opposing force from a secondary collision.
- **Angular collision:** The motorcycle strikes an object or another vehicle at an angle so that the rider sustains direct crushing injuries to the lower extremity between the object and the motorcycle. This usually results in severe open and comminuted lower extremity injuries with severe neurovascular compromise, often requiring surgical amputation.
- **Ejection:** The rider will travel at high speed until stopped by a stationary object, another vehicle, or by road drag. Severe abrasion injuries (road rash) down to bone can occur with drag. An unpredictable combination of blunt injuries can occur from secondary collisions.
- **Controlled crash:** A technique used to separate the rider from the body of the motorcycle and the object to be hit is referred to as laying the bike down. It was developed by motorcycle racers and adapted by street bikers as a means of achieving a controlled crash. As a collision approaches, the motorcycle is turned flat and tipped sideways at 90° to the direction of travel so that one leg is dropped to the grass or asphalt. This slows the occupant faster than the motorcycle, allowing for the rider to become separated from the motorcycle. If properly protected with leather or synthetic abrasion-resistant gear, injuries should be limited to those sustained by rolling over the pavement and any secondary collision that may occur. When executed properly, this maneuver prevents the rider from being trapped between the bike and the object. However, a rider unable to clear the bike will continue into the vehicle, often with devastating results.

■ Falls

The injury potential of a fall is related to the height from which the patient fell. Falls are common MOIs for blunt trauma. The greater the height of the fall, the greater the potential for injury. A fall from more than 15′ or 3 times the patient's height is considered significant. The patient lands on the surface just as an unrestrained passenger smashes into the interior of a vehicle. The internal organs travel at the speed of the patient's body before it hits the ground and stop by smashing into the interior of the body. Again, as in a motor vehicle crash, it is these internal injuries that are the least obvious during assessment but pose the gravest threat to life. Therefore, you should suspect internal injuries in a patient who has fallen from a significant height, just as you would in a patient who has been in a high-speed motor vehicle crash. Always consider syncope or other underlying medical causes of the fall.

Special Populations

When your patient is a child, the following constitute a significant MOI:
- Falls of greater than 10′ without loss of consciousness
- Falls of less than 10′ with loss of consciousness
- Medium- to high-speed vehicle collision (> 25 mph)

Also note that small children are top-heavy, so they tend to land on their heads even from small falls.

Patients who fall and land on their feet may have less severe internal injuries because their legs may have absorbed much of the energy of the fall **Figure 22-14**.

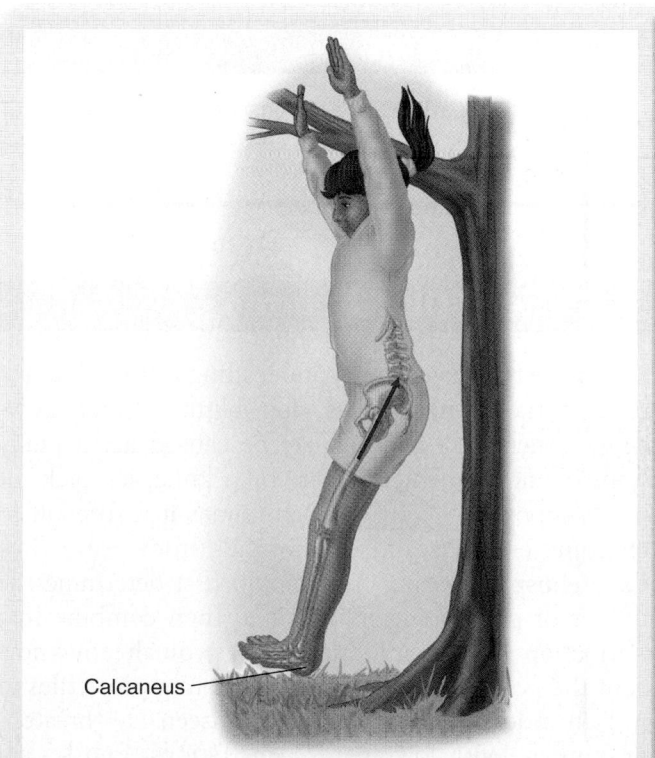

Calcaneus

Figure 22-14 When a patient falls and lands on his or her feet, the energy is transmitted to the spine, sometimes producing a spinal injury in addition to injuries to the legs and pelvis.

However, as a result, they may have very serious injuries to the lower extremities and pelvic and spinal injuries from energy that the legs do not absorb. Patients who fall onto their heads, as in diving accidents, will likely have serious head and/or spinal injuries. In either case, a fall from a significant height is a serious event with great injury potential, and the patient should be evaluated thoroughly. Take the following factors into account:

- The height of the fall
- The type of surface struck
- The part of the body that hit first, followed by the path of energy displacement

Some texts consider falls to be the most common form of trauma. Many falls, especially those by older persons, are not considered "true" trauma, even though bones may be broken. Often, these falls occur as a result of a fracture. Older patients often have osteoporosis, a condition in which the musculoskeletal system can fail under relatively low stress because the bones are structurally weakened. Because of this condition, an older patient can sustain a fracture while in a standing position and then fall as a result. Therefore, an older patient may have actually sustained a fracture before the fall. These cases do not constitute true high-energy trauma unless the patient fell from a significant height.

Special Populations

Many geriatric patients are seriously injured from falls. Completely assess older patients for all possible injuries, even from low-impact falls.

Penetrating Trauma

Penetrating trauma is the second leading cause of trauma death in the United States after blunt trauma. Low-energy penetrating trauma may be caused accidentally by impalement or intentionally by a knife, ice pick, or other weapon Figure 22-15 . Many times it is difficult to determine entrance and exit wounds from **projectiles** in a prehospital setting. You should first determine the number of penetrating injuries and then combine that information with the important things you already know about the potential pathway of penetrating projectiles to form an index of suspicion about unseen life-threatening injuries. With low-energy penetrations, injuries are caused by the sharp edges of the object moving through the body and are, therefore, close to the object's path. Weapons such as knives, however, may have been deliberately moved around internally, causing more damage than the external wound might suggest.

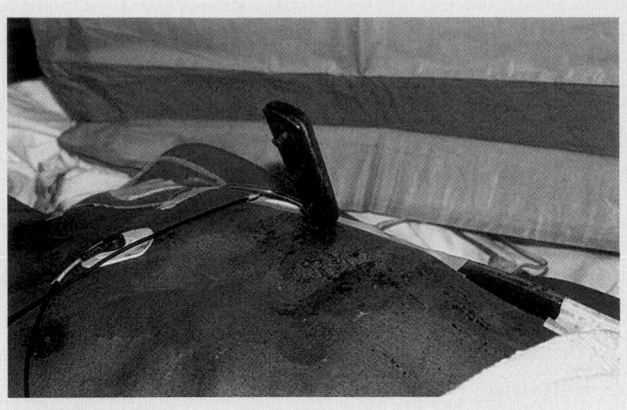

Figure 22-15 Injuries from low-energy penetrations, such as a stab wound, are caused by the sharp edges of the object moving through the body.

In medium- and high-velocity (speed) penetrating trauma, the path of the projectile (usually a bullet) may not be as easy to predict. This is because the bullet may flatten out, tumble, or even ricochet within the body before exiting. The path the projectile takes is referred to as a **trajectory**. Fragmentation, especially frangible bullets that are designed to disintegrate into tiny particles on impact, will increase damage as multiple fragments increase the likelihood of multiple organs/vessels sustaining injury. Full metal jacket bullets cause less damage than fragmented rounds because of their tendency to pass through the body's tissues. Also, because of a bullet's speed, pressure waves emanate from the bullet, causing damage remote from its path. There is often additional damage caused by the object moving inside the body and not along the suspected pathway. This phenomenon, called **cavitation**, can result in serious injury to internal organs distant to the actual path of the bullet Figure 22-16 . There are two types of cavitation, temporary and permanent. Temporary cavitation is caused by the acceleration of the bullet and causes a stretching of the tissues. Permanent cavitation is caused by the bullet path and remains once the projectile has passed through the tissue. You must remain alert during assessment because patients

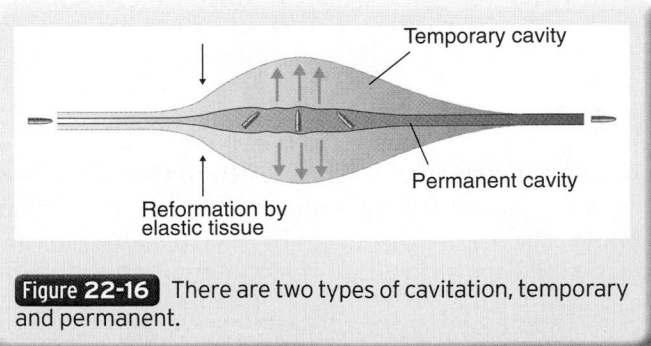

Figure 22-16 There are two types of cavitation, temporary and permanent.

will exhibit various signs and symptoms depending on the organ(s) struck.

The relationship between distance and the severity of injury varies depending on the type of weapon involved, such as a rifle, pistol, or shotgun. Air resistance, often referred to as **drag**, slows the projectile, decreasing the depth of penetration and thus reducing damage to the tissues. Much like a boat moving through water, the bullet disrupts not only the tissues that are directly in its path but also those in its wake. Therefore, the area that is damaged by medium- and high-velocity projectiles can be many times larger than the diameter of the projectile itself Figure 22-17 . This is one reason that exit wounds are often many times larger than entrance wounds. As with motor vehicle crashes, the energy available for a bullet to cause damage is more a function of its speed than its mass (weight). If the mass of the bullet is doubled, the energy that is available to cause injury is doubled. If the velocity of the bullet is doubled, the energy that is available to cause injury is quadrupled. For this reason, it is important for you to try to determine the type of weapon that was used. Although it is not necessary (or always possible) for you to distinguish between medium- and high-velocity injuries, any information regarding the type of weapon that was used should be relayed to medical control.

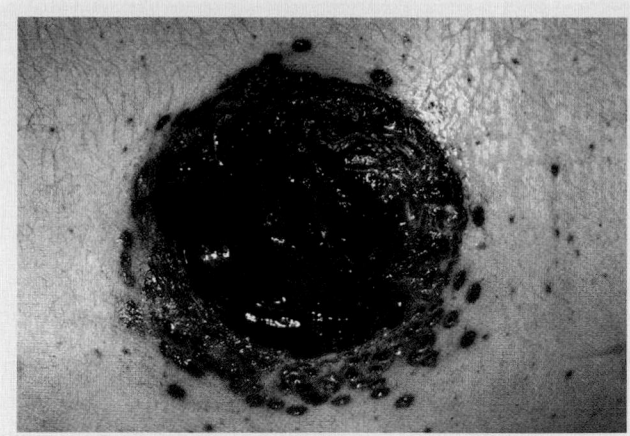

Figure 22-17 The area damaged by high-velocity projectiles, such as bullets, can be many times larger than the diameter of the projectile itself.

Medium-velocity injuries may be caused by handguns and some rifles, whereas high-energy injuries may be caused by a military weapon. Police at the scene may be a useful source of information regarding the caliber of weapon.

Table 22-1 summarizes how to recognize developing problems in trauma patients.

Table 22-1 Recognizing Developing Problems in Trauma Patients

Mechanism of Injury	Signs and Symptoms	Index of Suspicion
Blunt or penetrating trauma to the neck	<ul><li>Noisy or labored breathing</li><li>Swelling of the face or neck</li></ul>	<ul><li>Significant bleeding or foreign bodies in the upper or lower airway, causing obstruction</li><li>Be alert for airway compromise</li></ul>
Significant chest wall blunt trauma from motor vehicle crashes, car-versus-pedestrian, and other crashes; penetrating trauma to the chest wall	<ul><li>Significant chest pain</li><li>Shortness of breath</li><li>Asymmetrical chest wall movement</li></ul>	<ul><li>Cardiac or pulmonary contusion</li><li>Pneumothorax or hemothorax</li><li>Broken ribs, causing breathing compromise</li></ul>
Any significant blunt force trauma from motor vehicle crashes or penetrating injury	<ul><li>Blunt or penetrating trauma to the neck, chest, abdomen, or groin</li><li>Blows to the head sustained during motor vehicle crashes, falls, or other incidents, producing loss of consciousness, altered mental status, inability to recall events, combativeness, or changes in speech patterns</li><li>Difficulty moving extremities; headache, especially with nausea and vomiting</li></ul>	<ul><li>Injuries in these regions may tear and cause damage to the large blood vessels located in these body areas, resulting in significant internal and external bleeding</li><li>Be alert to the possibility of bruising to the brain and bleeding in and around the brain tissue, which may cause the development of excess pressure inside the skull around the brain</li></ul>
Any significant blunt force trauma, falls from a significant height, or penetrating trauma	<ul><li>Severe back and/or neck pain, history of difficulty moving extremities, loss of sensation or tingling in the extremities</li></ul>	<ul><li>Injury to the bones of the spinal column or to the spinal cord</li></ul>

Blast Injuries

Although most commonly associated with military conflict, blast injuries are also seen in civilian practice in mines, shipyards, chemical plants, and, increasingly, in association with terrorist activities. People who are injured in explosions may be injured by any of four different mechanisms **Figure 22-18**:

- **Primary blast injuries.** These injuries are due entirely to the blast itself, that is, damage to the body is caused by the pressure wave generated by the explosion. When the victim is close to the blast, the blast wave causes disruption of major blood vessels and rupture of major organs. Hollow organs are the most susceptible to the pressure wave.
- **Secondary blast injuries.** Damage to the body results from being struck by flying debris, such as shrapnel from the device or from glass or splinters, that has been set in motion by the explosion. Objects are propelled by the force of the blast and strike the victim, causing injury. These objects can travel great distances and be propelled at tremendous speeds, up to nearly 3,000 mph for conventional military explosives.

- **Tertiary blast injuries.** These injuries occur when the patient is hurled by the force of the explosion against a stationary object. A "blast wind" also causes the patient's body to be hurled or thrown, causing further injury. This physical displacement of the body is also referred to as ground shock when the body impacts the ground. In some cases, wind injuries can amputate limbs.
- **Miscellaneous blast injuries.** These injuries include burns from hot gases or fires started by the blast, respiratory injury from inhaling toxic gases, and crush injury from the collapse of buildings, among others.

Most patients who survive an explosion will have some combination of the four types of injury mentioned. The discussion here will be confined to primary blast injuries because these injuries are the ones that are most easily overlooked.

■ Tissues at Risk

Organs that contain air, such as the middle ear, lung, and gastrointestinal tract, are most susceptible to pressure changes. The junction between tissues of different densities and exposed areas such as head and neck tissues are prone to injury as well. The ear is the organ system that is most sensitive to blast injuries. The **tympanic membrane** evolved to detect minor changes in pressure and will rupture at pressures of 5 to 7 pounds per square inch above atmospheric pressure. Thus, the tympanic membranes are a sensitive indicator that you can use to help determine the possible presence of other blast injuries. The patient may complain of ringing in the ears, pain in the ears, or some loss of hearing, and blood may be visible in the ear canal. Dislocation of structural components of the ear, such as the ossicles conforming the inner ear, may occur. Permanent hearing loss is possible.

Pulmonary blast injuries are defined as pulmonary trauma (consisting of contusions and hemorrhages) that results from short-range exposure to the detonation of explosives. When the explosion occurs in an open space, the patient's side that was

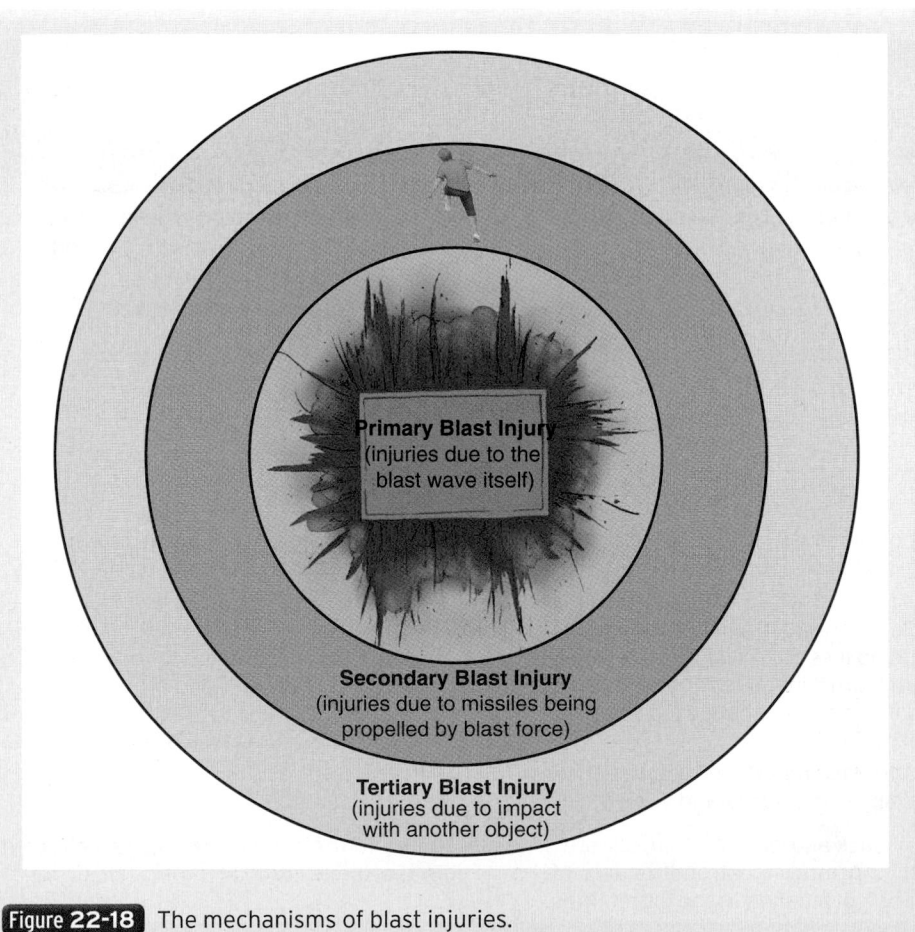

Figure 22-18 The mechanisms of blast injuries.

toward the explosion is usually injured, but the injury can be bilateral when the victim is located in a confined space. The patient may complain of tightness or pain in the chest and may cough up blood and have tachypnea or other signs of respiratory distress. Subcutaneous emphysema (crackling under the skin) can be detected over the chest through the use of palpation, indicating air in the thorax. Pneumothorax is a common injury and may require emergency decompression (which will be covered in Chapter 27, *Chest Injuries*) in the field for your patient to survive. Pulmonary edema may ensue rapidly. If there is any reason to suspect lung injury in a blast victim (even just the presence of a ruptured eardrum), administer oxygen. Avoid giving oxygen under positive pressure, however (that is, by demand valve) because that may simply increase the damage to the lung. Be cautious as well with intravenous fluids, which may be poorly tolerated in patients with this type of lung injury and result in pulmonary edema.

One of the most concerning pulmonary blast injuries is **arterial air embolism**, which occurs on alveolar disruption with subsequent air embolization into the pulmonary vasculature. Even small air bubbles can enter a coronary artery and cause myocardial injury. Air embolisms to the cerebrovascular system can produce disturbances in vision, changes in behavior, changes in state of consciousness, and a variety of other neurologic signs.

Solid organs are relatively protected from shock wave injury but may be injured by secondary missiles or a hurled body. Hollow organs, however, may be injured by the same mechanisms that damage lung tissue. Petechiae, or pinpoint hemorrhages that show up on the skin, to large hematomas are the dominant form of pathology. Perforation or rupture of the bowel and colon is a risk. Underwater explosions result in the most severe abdominal injuries.

Neurologic injuries and head trauma are the most common causes of death from blast injuries. Subarachnoid (beneath the arachnoid layer covering the brain) and subdural (beneath the outermost covering of the brain) hematomas are often seen. Permanent or transient neurologic deficits may be secondary to concussion, intracerebral bleeding, or air embolism. Instant but transient unconsciousness, with or without retrograde amnesia, may be initiated not only by head trauma, but also by cardiovascular problems. Bradycardia and hypotension are common after an intense pressure wave from an explosion. This is a vagal nerve–mediated form of cardiogenic shock without compensatory vasoconstriction (for example, vasovagal syncope).

You are the Provider: PART 4

The decision is made to transport the patient by air based on his injuries and the news of heavy traffic causing a lengthy delay in transporting him to the Level I trauma center. After full spinal precautions are applied, the patient is loaded into the ambulance and immediately reassessed.

Recording Time: 10 Minutes	
Level of consciousness	Responds only to pain
Respirations	30 breaths/min; severely labored
Pulse	130 beats/min; absent radial pulses
Skin	Cool, clammy, and pale
Blood pressure	80/50 mm Hg
Sao$_2$	89% (on oxygen)

Your partner begins assisting the patient's ventilations with a bag-mask device and high-flow oxygen and oral suctioning is performed as needed to keep the patient's airway clear of blood. The patient flexes his arms in response to pain, but does not open his eyes or respond verbally when you talk to him. An EMT from one of the engine companies drives the ambulance to the landing zone, which is about a mile away.

6. What trauma scoring systems are commonly used to assess the severity of a trauma patient's condition? How would you apply them to this patient?

Extremity injuries, including traumatic amputations, are common. Other injuries are often associated with tertiary blasts. Patients with traumatic amputation by postblast wind are likely to sustain fatal injuries secondary to the blast. In present day combat, improved body armor has increased the number of survivors of blast injuries from shrapnel wounds to the torso. The number of severe orthopaedic and extremity injuries, however, has increased. In addition, whereas body armor may limit or prevent shrapnel from entering the body, it also "catches" more energy from the blast wave, possibly resulting in the victim being thrown backward, thus increasing the potential for spine and spinal cord injury.

Although blast injuries have usually been the domain of military surgeons, they often occur in industrial settings and are, unfortunately, more common today owing to the increased use of explosives as a tool for urban terrorism and, in the United States, from methamphetamine laboratory explosions. Although civilian blast injuries in an industrial or mining setting used to be mostly characterized by blast injuries and burns, terrorist bombs often contain shrapnel. As an EMT, you and other EMS and trauma services personnel should be fully educated and aware of what to expect in these scenarios.

Multisystem Trauma

Multisystem trauma is a term that describes a person who has been subjected to multiple traumatic injuries involving more than one body system such as head and spinal trauma, chest and abdominal trauma, or chest and multiple extremity trauma. You must recognize patients who fit into this classification and provide rapid treatment and transportation, and alert medical control as to the nature of the patient's injuries so that the trauma center is prepared prior to your arrival. Multisystem trauma patients have a high level of morbidity and mortality; therefore, they require teams of physicians to treat their injuries. These teams may include specialists such as neurosurgeons, thoracic surgeons, and orthopaedic surgeons.

Golden Principles of Prehospital Trauma Care

As with any EMS call, your main priority in managing multisystem trauma is to ensure your safety, the safety of your crew, and the patient. Next, you must determine the need for additional personnel or equipment, evaluate the kinematics of the mechanism of injury, and identify and appropriately manage life threats. Once these steps have been completed, you can focus on patient care. Begin by assessing and managing the airway, including ventilatory support and high-flow oxygen while maintaining cervical spine stabilization. Ensure that basic shock therapy, such as controlling hemorrhages and stopping arterial bleeding, is completed. If bleeding cannot be controlled rapidly by direct pressure, consider the use of a tourniquet. If the patient is profusely bleeding, this must be controlled to ensure sufficient perfusion of organs and tissues.

Once threats to the ABCs are corrected, place the patient on a long backboard and transport immediately. If the patent is entrapped, consider the use of rapid extrication techniques. In most patients with multisystem trauma, definitive care requires surgical intervention; therefore, on-scene time should be limited to 10 minutes or less. This is referred to as the "platinum 10 minutes." During transport, obtain a SAMPLE history and complete a secondary assessment. Most care can be provided in transport. However, keep in mind that your patient has sustained multisystem trauma and the order in which you usually provide treatment and care may need to be adjusted depending on the needs of the patient. For critically injured patients, consider ALS intercept and/ or air medical transportation. Regardless of the mode of transport, ensure that the patient is transported to an appropriate facility and that the facility is notified as soon as possible. Specific standards of care in regard to multisystem trauma will be addressed in detail in respective chapters.

Words of Wisdom

Spend as little time as possible on scene with patients who have sustained significant trauma. After the first 60 minutes, the body has increasing difficulty in compensating for shock and traumatic injuries. This was previously referred to as the "Golden Hour." Because many injured patients require definitive care in less than an hour, this is now referred to as the "Golden Period."

Patient Assessment

Identifying life-threatening illnesses and injuries as soon as possible has proven to improve patient outcomes. As an EMT, you must apply this knowledge as well as the appropriate assessment skills to assess, triage, manage, and transport patients with traumatic injuries to the most appropriate facility. The major components of patient assessment include the following:

- Scene size-up
- Primary assessment
- History taking
- Secondary assessment
- Reassessment

When you are caring for a patient who has experienced a significant MOI and the patient is considered to be in serious or critical condition, you should perform a rapid full-body scan or rapid head-to-toe examination. Any patient who has suffered a nonsignificant MOI should receive an assessment focused on the chief complaint. The human body is divided into areas (or systems) based on body function, and its internal organs are subject to unseen injuries when force is applied to the body. For example, the brain may have bruising, the heart and lungs may have bruising or unseen bleeding, and the organs of the abdomen may have life-threatening bleeding. The following sections discuss the assessment of various body systems.

Injuries to the Head

The brain lies well protected within the skull. However, when the head is injured from trauma, disability and unseen injury to the brain may occur. The brain itself may tear or become bruised, causing bleeding. The blood vessels around the brain may also tear and produce bleeding. Bleeding or swelling inside the skull from brain injury is often life threatening; therefore, your assessment must include conducting frequent neurologic examinations. Some patients will not have obvious signs or symptoms, such as changes in pupillary size and reactivity, of unseen brain injury until minutes or hours after the injury has occurred.

Injuries to the Neck and Throat

The neck and throat contain many structures that are susceptible to injuries from trauma that could be serious or deadly to your patients. In this region of the human body, the trachea (or windpipe) may become torn or swell after an injury to the neck or deviate after an injury to the lungs. These types of injuries may result in an airway problem that could quickly become a serious life threat because it interferes with the patient's ability to breathe; therefore, your assessment must include frequent physical examination looking for DCAP-BTLS in the neck region. In addition, you should also assess for jugular venous (vein) distention and tracheal deviation.

The neck also contains large blood vessels that supply the brain with oxygen-rich blood. When a neck injury occurs, swelling may prevent blood flow to the brain and cause injury to the central nervous system, even though the brain may not have been directly affected by the initial force that caused the injury to the neck. If a penetrating injury to the neck results in an open wound, the patient may have significant bleeding, or air may be drawn into the circulatory system. If air enters the veins, this may result in air embolism, which may lead to cardiac arrest if the air enters the heart. A crushing injury to the upper part of the neck may cause the cartilages of the upper airway and larynx to fracture. This can lead to the leakage of air into the soft tissue of the neck. When air is

trapped in subcutaneous tissue, it produces a crackling sound called *subcutaneous crepitation*. Either air in the circulation or an airway cartilage fracture may cause rapid death.

Injuries to the Chest

The chest contains the heart, the lungs, and the large blood vessels of the body. When injury occurs to this area of the body, many life-threatening injuries may occur. For example, blunt trauma to the chest can fracture rib(s) or the sternum. When ribs are broken and the chest wall does not expand normally during breathing, this interferes with the body's ability to obtain sufficient amounts of oxygen for the cells. Bruising may occur to the heart and cause an irregular heartbeat. Depending on the severity of the trauma, the large vessels of the heart may be torn inside the chest; causing massive unseen bleeding that can quickly kill the trauma patient. In some chest injuries the lungs become bruised, thus interfering with normal oxygen exchange in the body.

Some chest injuries result in air collecting between the lung tissue and the chest wall. As air accumulates in this space, the lung tissue becomes compressed, again interfering with the body's ability to effectively exchange oxygen. This injury is called a pneumothorax. If left untreated or unrecognized, the lung tissue becomes squeezed under pressure until the heart is also squeezed and can no longer pump blood. This condition is called a tension pneumothorax and is a life-threatening emergency. Some patients develop bleeding in this portion of the chest. Instead of air collecting in this space, blood collects here and causes interference with breathing. This condition is called a hemothorax and it also poses a threat to the patient's life.

A penetration or perforation of the integrity of the chest is called an open chest wound. As air enters the chest cavity, the natural pressure balance within the chest cavity is no longer equal. If left untreated, shock and/or death will result. Regardless of the particular injury, it is imperative that you reassess a trauma patient's chest region every 5 minutes. The assessment should include DCAP-BTLS, lung sounds, and chest rise and fall. Some patients will not have obvious signs or symptoms such as absent breath sounds or respiratory difficulty immediately.

Injuries to the Abdomen

The abdomen is an area of the human body that contains many organs vital to body function. These organs also require a very high amount of blood flow so they can perform the functions necessary for life. The organs of the abdomen and retroperitoneum (the space immediately behind the true abdomen) can be classified into two simple categories: solid and hollow. The solid organs include the liver, spleen, pancreas, and kidneys. The hollow organs include the stomach, large and small intestines, and urinary bladder.

When injuries from trauma occur in this region of the body, serious and life-threatening problems may occur. The solid organs may tear, lacerate, or fracture. This causes serious bleeding into the abdomen that can quickly cause death. Be alert for a trauma patient who complains of abdominal pain—it may be a symptom of abdominal bleeding. Also be alert to vital signs that begin to worsen; this can be a sign of serious, unseen bleeding inside the abdominal region of the body.

When the hollow organs of the body have been injured, they may rupture and leak acidlike chemicals used for digestion into the abdomen. This not only causes pain, but the patient also may eventually develop a life-threatening infection.

The abdomen also contains large blood vessels that supply the organs of this region and the lower extremities with oxygen-rich blood. Occasionally these vessels rupture or tear and cause serious unseen bleeding that may cause death. Some patients, particularly healthy young adults, are able to compensate longer than others from blood loss; therefore, you should always maintain a high index of suspicion when the mechanism of injury suggests injury to the abdominal region. This is best accomplished by reassessing the abdominal region using DCAP-BTLS.

Management: Transport and Destination

Caring for victims of traumatic injuries requires you to have a solid understanding of the trauma system in the United States. You need to have a good working knowledge of the resources available to you, including the most optimal methods of rapid transport and trauma centers that can best provide definitive care.

Scene Time

Because survival of critically injured trauma patients is time dependent, you should limit on-scene time to the minimum amount necessary to correct life-threatening injuries and package the patient. Optimally, on-scene time for critically injured patients should be less than 10 minutes—the "platinum ten." The following criteria will help you identify a critically injured patient:

- Dangerous MOI
- Decreased level of consciousness
- Any threats to airway, breathing, or circulation

Patients who present with these criteria or who are very young or old or have chronic illnesses should also be considered to be high risk, thus requiring rapid treatment and transport.

You are the Provider: PART 5

The air transport helicopter arrives at the landing zone approximately 5 minutes after your unit. After reassessing the patient, you give your verbal report to the flight paramedic and transfer patient care.

Recording Time: 15 Minutes	
Level of consciousness	Responds only to pain
Respirations	30 breaths/min; severely labored
Pulse	140 beats/min; absent radial pulses
Skin	Cool, clammy, and pale
Blood pressure	74/50 mm Hg
Sao$_2$	95% (on oxygen)

After further assessment and treatment, the helicopter personnel load the patient into the aircraft and depart the scene. You later learn that the patient had intrathoracic and intracranial bleeding and multiple rib fractures. He was taken to surgery, and was in critical condition in the surgical intensive care unit.

7. How does the level of trauma care provided by the paramedic differ from that of the EMT?

Type of Transport

As discussed in Chapter 12, *Medical Overview*, modes of transport ultimately come in one of two categories: ground or air. Ground transportation EMS units are generally staffed by traditional EMTs and paramedics. Air transportation EMS units or critical care transport units are generally staffed by critical care transport professionals such as critical care nurses and paramedics **Figure 22-19** .

The Association of Air Medical Services (AAMS) and MedEvac Foundation International identifies the following criteria in the white paper, *Air Medicine: Accessing the Future of Healthcare*, for the appropriate use of emergency air medical services for trauma patients.

- There is an extended period required to access or extricate a remote (eg, injured hiker, snowmobiler, or boater) or trapped patient (eg, in a crashed car) which depletes the time window to get the patient to the trauma center by ground.
- Distance to the trauma center is greater than 20 to 25 miles.
- The patient needs medical care and stabilization at the ALS level, and there is no ALS-level ground ambulance service available within a reasonable time frame.
- Traffic conditions or hospital availability make it unlikely that the patient will get to a trauma center via ground ambulance within the ideal time frame for best clinical outcome.
- There are multiple patients who will overwhelm resources at the trauma center(s) reachable by ground within the time window.

Figure 22-19 A helicopter may be used to transport patients quickly to a trauma center.

- EMS systems require bringing a patient to the nearest hospital for initial evaluation and stabilization, rather than bypassing those facilities and going directly to a trauma center. This may add delay to definitive surgical care and necessitate air transport to mitigate the impact of that delay.
- There is a mass-casualty incident.

These recommendations are not to be understood as fully encompassing, but more so as to serve as a guideline for local decision makers to develop more comprehensive protocols for the use of air medical transport. You should always follow your local protocols when determining what type of patient transportation is appropriate.

Destination Selection

You will often be summoned to accident scenes to transport critically ill trauma patients to definitive care. For this reason, it is important for you to be familiar with how the American College of Surgeons' Committee on Trauma classifies trauma care. Trauma centers are classified into Levels I through IV with Level I having the most resources followed by Levels II, III and IV, respectively **Table 22-2** .

A Level I facility is a regional resource center and generally serves large cities or heavily populated areas. Level I facilities must be capable of providing every aspect of trauma care from prevention through rehabilitation; therefore, the facility must have adequate personnel and resources. Because of the extensive requirements, most Level I facilities are university-based teaching hospitals.

A Level II facility is typically located in less population-dense areas. Level II centers are expected to provide initial definitive care, regardless of injury severity. These facilities can be academic institutions or a public/private community facility. Because of its location and resources, a Level II trauma center may not be able to provide the same comprehensive care as a Level I trauma center.

Level III facilities serve communities that do not have access to Level I or II facilities. Level III facilities provide assessment, resuscitation, emergency care, and stabilization. A Level III facility must have transfer agreements with a Level I or II trauma center, and must have protocols in place to transfer patients whose needs exceed the resources of the facility.

Level IV facilities are typically found in remote outlying areas where no higher level of care is available. These facilities provide advanced trauma life support prior to transfer to a higher level trauma center. Such a facility may be a clinic urgent care facility, with or without a physician.

Table 22-2 Key Elements for Trauma Centers

Level	Definition	Key Elements
Level I	A comprehensive regional resource that is a tertiary care facility; capable of providing total care for every aspect of injury—from prevention through rehabilitation	1. 24-hour in-house coverage by general surgeons 2. Availability of care in specialties such as orthopaedic surgery, neurosurgery, anesthesiology, emergency medicine, radiology, internal medicine, and critical care 3. Should also include cardiac, hand, pediatric, and microvascular surgery and hemodialysis 4. Provides leadership in prevention, public education, and continuing education of trauma team members 5. Committed to continued improvement through a comprehensive quality assessment program and organized research to help direct new innovations in trauma care
Level II	Able to initiate definitive care for all injured patients	1. 24-hour immediate coverage by general surgeons 2. Availability of orthopaedic surgery, neurosurgery, anesthesiology, emergency medicine, radiology, and critical care 3. Tertiary care needs such as cardiac surgery, hemodialysis, and microvascular surgery may be referred to a Level I trauma center 4. Committed to trauma prevention and continuing education of trauma team members 5. Provides continued improvement in trauma care through a comprehensive quality assessment program
Level III	Ability to provide prompt assessment, resuscitation, and stabilization of injured patients and emergency operations	1. 24-hour immediate coverage by emergency medicine physicians and prompt availability of general surgeons and anesthesiologists 2. Program dedicated to continued improvement in trauma care through a comprehensive quality assessment program 3. Has developed transfer agreements for patients requiring more comprehensive care at a Level I or Level II trauma center 4. Committed to continuing education of nursing and allied health personnel or the trauma team 5. Must be involved with prevention and have an active outreach program for its referring communities
Level IV	Ability to provide advanced trauma life support (ATLS) before transfer of patients to a higher level trauma center	1. Include basic emergency department facilities to implement ATLS protocols and 24-hour laboratory coverage 2. Transfer to higher level trauma centers follows the guidelines outlined in formal transfer agreements 3. Committed to continued improvement of these trauma care activities through a formal quality assessment program 4. Involved in prevention, outreach, and education within its community

Although an inclusive trauma system should leave no facility without a direct link to a Level I or II facility, all facilities are expected to provide the same high quality of initial care regardless of the classification level.

Trauma centers are categorized as either adult trauma centers or pediatric trauma centers, but not necessarily both. Pediatric trauma centers are not nearly as common as adult trauma centers. When transporting a pediatric trauma patient, you must be certain to transport your patient to a pediatric trauma center if there is one in your area; do not make the mistake of transporting a pediatric patient to an adult trauma center when a pediatric trauma center is available.

The American College of Surgeons' Committee on Trauma provides criteria for Level I trauma patient classification Table 22-3. When one or more of the criteria listed below are present in the trauma patient, he or she is classified as a Level I trauma patient.

Although the American College of Surgeons' Committee on Trauma does not cite required criteria for a

Words of Wisdom

It is imperative for the EMT to have a strong understanding of trauma scoring systems to appropriately classify patients. The __trauma score__ calculates a number from 1 to 16, with 16 being the best possible score. It takes into account the __Glasgow Coma Scale (GCS) score__, respiratory rate, respiratory expansion, systolic blood pressure, and capillary refill (Figure 22-20). The GCS is an evaluation tool used to determine level of consciousness, which evaluates and assigns point values (scores) for eye opening, verbal response, and motor response; these scores are then totaled and help to effectively predict patient outcomes. The trauma score relates to the likelihood of patient survival. However, this scoring system does not accurately predict survivability in patients with severe head injuries because motor and verbal deficits make those criteria difficult to assess; in its place, the Revised Trauma Score (RTS), discussed next, is used.

Revised Trauma Score

The numeric scoring of trauma patients for determining the severity of their injury is common practice in the health care profession. When the various scoring systems were created, it was thought that the implementation of the scoring system would assist in rapidly identifying the severity of the patient's injuries. There are several different trauma scoring systems. The one that is the most commonly used for patients with head trauma is the __Revised Trauma Score (RTS)__ because it is heavily weighted to compensate for major head injury without multisystem injury or major physiologic changes.

The RTS is a physiological scoring system that is also used to assess the severity of a trauma patient's injuries. Objective data used to calculate the RTS includes the GCS score, systolic blood pressure (SBP), and respiratory rate (RR). In addition to assessing injury severity, the RTS has also demonstrated reliability in predicting survival in patients with severe injuries. The highest RTS a patient can receive is 12; the lowest is 0. The RTS is calculated as follows:

GLASGOW COMA SCALE

Eye Opening

Spontaneous	4
To Voice	3
To Pain	2
None	1

Verbal Response

Oriented	5
Confused	4
Inappropriate Words	3
Incomprehensible Words	2
None	1

Motor Response

Obeys Command	6
Localizes Pain	5
Withdraws (pain)	4
Flexion (pain)	3
Extension (pain)	2
None	1

Glasgow Coma Score Maximum Total 15 _____

Figure 22-20 The Glasgow Coma Scale is one method of evaluating level of consciousness. Note that the lower the score, the more severe the extent of brain injury.

GCS	SBP	RR	Value
13 to 15	> 89 mm Hg	10 to 29 breaths/min	4
9 to 12	76 to 89 mm Hg	> 29 breaths/min	3
6 to 8	50 to 75 mm Hg	6 to 9 breaths/min	2
4 to 5	1 to 49 mm Hg	1 to 5 breaths/min	1
3	0	0	0

Level II patient, they do provide recommendations, which are listed in Table 22-4 .

■ Special Considerations

Because traumatic injuries are as varied as the mechanisms that cause them, it is almost impossible for you to prepare for every possible situation that you may face during your career. In all situations, you must remain calm, complete an organized assessment, correct life-threatening injuries, and do no harm. You should never hesitate to contact ALS backup or medical control for guidance.

Table 22-3 American College of Surgeons Criteria for a Level I Patient

- Confirmed blood pressure of less than 90 mm Hg at any time in adults, and age-specific hypotension in children
- Respiratory compromise, obstruction, and/or intubation
- Receiving blood to maintain vital signs
- Emergency physician's discretion
- Glasgow Coma Scale (GCS) score of less than or equal to 8 with mechanism attributed to trauma
- Gunshot wound to the abdomen, neck, or chest

Table 22-4 American College of Surgeons Recommendations for a Level II Patient

Patient characteristic/condition indicators	1. Glasgow Coma Scale (GCS) score of less than 14 when associated with trauma 2. Respiratory rate of less than 10 or more than 29 breaths/min (less than 20 breaths/min in infant younger than 1 year of age) when associated with trauma 3. Penetrating wounds (other than gunshot wounds) to the head, neck, torso, and extremities proximal to the elbow and knee 4. Flail chest 5. Combination of trauma with burns 6. Two or more proximal long bone fractures 7. Pelvic fractures 8. Limb paralysis and/or spinal cord injury 9. Amputation proximal to the wrist and/or ankle

Mechanism of injury indicators	1. High-speed vehicle crash
	  ■ Initial speed of greater than 40 mph
	  ■ Major vehicle deformity
	  ■ Intrusion into the passenger compartment
	2. Ejection from the vehicle
	3. Death in same passenger compartment
	4. Extrication time of greater than 20 minutes
	5. Falls of greater than 20′ or significant falls in children or elderly
	6. Vehicle rollover
	7. Car-versus-pedestrian or car-versus-bicycle impact of greater than 5 mph
	8. All-terrain vehicle (ATV) or motorcycle crash of greater than 20 mph or separation of rider from ATV or motorcycle
	Pediatric indicators include:
	1. Falls of greater than 10′ without loss of consciousness
	2. Falls of less than 10′ with loss of consciousness
	3. Medium- to high-speed vehicle collision (> 25 mph)

Consider Level II classification with the following preexisting conditions	1. Age younger than 5 years or older than 55 years 2. Cardiac disease, respiratory disease 3. Type 1 diabetes mellitus, cirrhosis of the liver, morbid obesity 4. Pregnancy 5. Immunosuppressed patients 6. Patients with a bleeding disorder or on anticoagulants

You are the Provider: SUMMARY

1. **On the basis of the information from the dispatch operator, can you predict the type of injuries the patient may have? If so, how?**

Although you will not ultimately know the type and severity of injuries the patient has until you arrive at the scene, information provided by the dispatch operator can influence your index of suspicion. Index of suspicion is your concern for potentially serious injuries based on the mechanism of injury (MOI). The MOI is the way in which traumatic injuries occur; it describes the forces acting on the body that cause injury.

At this point, prior to your arrival at the scene, you know that the incident involves a head-on (frontal) collision with a tree, and the patient is still in the vehicle. What you do not know is the speed of the vehicle at the time of impact, whether or not the patient was restrained, or whether or not the air bags deployed. *Just because the patient is still in the vehicle does not indicate that he or she was restrained.* Until you arrive at the scene and gather more information, you should assume the worst—the patient was traveling at a high rate of speed and was not restrained.

To be able to predict the type of injuries a patient may have sustained following a frontal collision, it is essential to understand the forces that are involved in this type of collision. Three collisions occur during a frontal impact: collision of the vehicle against another object, collision of the occupant against the interior of the vehicle, and collision of the occupant's internal organs against the solid structures of the body. These three collisions occur whether the occupant was restrained or not, although a properly restrained occupant's injuries are usually less severe.

Damage to the vehicle (the first collision) is perhaps the most dramatic part of the collision. It provides information about the severity of the collision. The greater the damage is to the vehicle, the greater the energy that was involved and, therefore, the greater the potential to cause injury. When you arrive at the scene and conduct your scene size-up, quickly inspect the vehicle and judge the severity of damage; if it is significant, your index of suspicion for serious injury to the patient should increase.

Just as the kinetic energy produced by the vehicle's mass (weight) and velocity (speed) is converted into the work of bringing the vehicle to an abrupt stop (rapid deceleration), the kinetic energy produced by the occupant's mass and velocity is converted into the work of bringing his or her body to a stop (the second collision). Just like the obvious damage that occurs to the exterior of the vehicle during the first collision, the injuries that result from the second collision are often obvious during your primary assessment of the patient. Common injuries that occur during the second collision include lower extremity and pelvic fractures (when the knees impact the dashboard); flail chest, rib fractures, pneumothoraces (when the chest impacts the steering wheel), and head and neck trauma (when the head impacts the windshield).

The type of injury experienced by the unrestrained occupant depends on the path he or she took at the time of impact. The down and under path results in trauma to the abdomen, pelvis, and lower extremities when the abdomen impacts the lower part of the steering wheel and the knees impact the dashboard. The up-and-over path results in head and neck trauma (and possibly chest trauma) because the occupant is propelled over the steering wheel—sometimes impacting it with his or her chest—and then striking the windshield with his or her head. Direct impact with the steering wheel results in blunt chest and/or abdominal trauma.

The injuries that occur during the third collision may not be as obvious as external injuries, but these injuries are usually the most life threatening. For example, if the chest impacts the steering wheel, the thoracic organs continue their forward motion until they collide with the inside of the chest cavity. As a result of these forces, shearing injuries of the great vessels (eg, aorta, vena cava) or injury to the heart as it impacts with the sternum may occur. If the occupant's head strikes the windshield, the brain continues its forward motion until it strikes the inside of the skull; this results in compression injuries to the anterior part of the brain and stretching or tearing of the posterior part of the brain.

Understanding the relationship among the three collisions allows you to make a connection between the severity of damage to the exterior of the vehicle and the potential injury to the occupant(s). A high-speed frontal collision that causes massive damage to the vehicle should increase your index of suspicion for serious injuries to the occupant(s)—even if the injuries are not immediately apparent.

2. **Why is it important to try to determine the speed a vehicle was traveling at the time of impact?**

To appreciate how a vehicle's speed affects the potential for injury to its occupant(s), a review of kinetic energy and deceleration is in order.

Kinetic energy (the energy of motion) is calculated as follows: $KE = \frac{1}{2}mv^2$, where m = mass (weight) and v = velocity (speed). Energy cannot be created or destroyed, only converted. In a frontal collision, the kinetic energy of the moving vehicle is converted into the work of stopping it—usually by crushing its exterior. The occupant in the vehicle has kinetic energy because his or her body was traveling at the same speed as the vehicle. Therefore, the kinetic energy of the occupant is converted into the work of stopping him or her; it is this work on the occupant that causes injury. According to the equation for kinetic energy ($KE = \frac{1}{2}mv^2$), the energy available to cause injury doubles when the object's weight doubles, but *quadruples when the object's speed doubles*.

Now, let's discuss deceleration. If a car is traveling at 30 mph, it takes about 3 seconds for the car to decrease its speed from 30 mph to 0 mph when the driver applies the brakes smoothly. If the driver is properly restrained, he or she slows (decelerates) at the same rate as the vehicle. However, if the vehicle is suddenly stopped (rapid deceleration) by striking a fixed object (ie, tree, bridge pillar) instead of braking, and the driver is not restrained, his or her body will continue to travel at 30 mph until it is stopped by an external force, such as the steering wheel, dashboard, or windshield.

Although the change in the occupant's velocity is the same as when the vehicle was braking smoothly in 3 seconds (30 to 0 mph), that change now takes place in about 0.01 second. Because the period of deceleration is 300 times less, the average force of impact is 300 times greater.

The relationship between a vehicle's speed and its deceleration is best described in these terms: the faster the vehicle is traveling and the quicker it stops, the greater the potential for serious injury to the unrestrained occupant(s).

In most cases, the speed at which a vehicle was traveling at the time of impact can only be estimated—usually based on witness observation and the severity of vehicular damage. However, in some cases, the speedometer cable will get crimped or otherwise caught in between two pieces of metal, causing the speedometer to "freeze" at or near the speed the vehicle was traveling at the time of impact.

You are the Provider: SUMMARY, continued

3. Where will you most likely find damage to the vehicle's interior based on the patient's signs and symptoms?

The patient's signs and symptoms indicate, at a minimum, injury to his head and chest. He is bleeding from the head, face, and mouth; his level of consciousness is markedly decreased; and his respirations are rapid and labored. Assessment of the vehicle's interior will most likely reveal damage to more than one area; he likely took the up and over path, striking his chest on the steering wheel in the process, and impacting the windshield with his head and face.

Contact points—parts of the interior of the vehicle that the patient struck during impact—are often obvious and usually can be identified by a quick evaluation of the passenger compartment while the patient is being extricated. However, if interior intrusion is present (not the case in this incident), contact points can be difficult to distinguish from damage caused by impact with the object that the vehicle struck.

Inspection of the interior of the vehicle in this incident will most likely reveal a deformed steering wheel (top, bottom, or both) that occurred when the patient's chest—and maybe abdomen—impacted it. You will also likely find an outward bulge of the windshield with a typical "starburst fracture" where the patient's head and face made impact with it; you may also find some of the patient's hair caught in the windshield and this is a clear indicator of patient impact. Keep in mind that these may not be the only contact points; they are simply the most likely areas based on the patient's injuries. His knees may have also impacted the dashboard, resulting in injury to the lower extremities and pelvis.

4. Should the patient be transported to the Level I trauma center or the Level III trauma center? Why?

You are caring for a multisystem trauma patient; that is, he has experienced trauma that affects more than one body system. The obvious trauma to his head and his level of consciousness indicate a traumatic brain injury, the bruising and crepitus to his chest and his labored breathing indicate intrathoracic trauma, and his vital signs indicate a general state of shock. He requires the highest level of trauma care available—a Level I trauma center.

Trauma centers are classified as Levels I through IV, with Level I having the most resources, followed by Levels II, III, and IV, respectively.

A Level I facility is a regional resource that generally serves large cities or heavily populated areas. It must be capable of providing every aspect of trauma care from prevention through rehabilitation, and must be continuously staffed with emergency department physicians, trauma surgeons, operating room personnel, radiologists, orthopaedic surgeons, a comprehensive surgical intensive care unit, and 24-hour access to a computed tomographic scan, among others resources and personnel.

A Level III facility serves communities that do not have access to a Level I facility. A Level III facility provides assessment and resuscitation, emergency care, and stabilization; however, it does not provide definitive patient care that a Level I facility

offers (ie, surgery). Therefore, a Level III facility must have a transfer agreement with a Level I trauma center, and it must have protocols in place to transfer patients whose needs exceed its resources.

According to the Golden Period, a critical trauma patient should be in an appropriate definitive care facility (eg, a trauma center) *within 1 hour after the injury occurs*. For each minute that definitive care is delayed, the patient's chance of survival decreases. Since a Level III facility is not staffed by in-house trauma surgeons and other resources that the patient needs, the Level I facility is the most obvious transport destination. The method by which you transport the patient in a timely manner requires preplanning and specific protocols that address transport issues.

5. What other transport factors should you consider with this patient?

As previously discussed, the patient should be transported to a Level I trauma center; however, the closest one is 30 miles away. Ground transport in heavy traffic will cause a delay in patient care. The question is: Can the patient afford this delay based on the condition he is in?

Although transporting the patient by ground to the Level III facility that is only 15 miles away will get the patient to a hospital quicker, he will not be able to receive the definitive care that his injuries may require. This situation requires you to make a decision as to how to best transport the patient to a Level I facility in the most expedient manner.

In this scenario, you are faced with two transport options: Try to coordinate a paramedic intercept and transport by ground to the Level I facility, or request air medical transport. Coordinating a paramedic intercept and transporting the patient by ground will get advanced life support care to the patient quicker, but it will not get him to the Level I facility any quicker. Conversely, air medical transport via a helicopter will get advanced life support personnel (ie, paramedic, critical care nurse) to the patient *and* get him to the Level I facility in a shorter period of time.

6. What trauma scoring systems are commonly used to assess the severity of a trauma patient's condition? How would you apply them to this patient?

Numeric scoring of trauma patients for determining injury severity is a common practice in both the hospital and prehospital settings. It is important for you to have a strong understanding of these scoring systems to accurately classify the severity of a patient's injuries. The two most commonly used numeric trauma scoring systems are the Glasgow Coma Scale (GCS) and the Revised Trauma Score (RTS).

The GCS, a detailed assessment of a patient's level of consciousness, is based on three independent measurements: eye opening, verbal response, and motor response; each measurement is assigned a numeric value. The highest score a patient can receive on the GCS is 15; the lowest is 3. The lower the GCS score, the more severe the brain injury. The GCS is calculated as follows:

You are the Provider: SUMMARY, continued

Eye Opening	
Spontaneous	4
To Voice	3
To Pain	2
None	1
Verbal Response	
Oriented	5
Confused	4
Inappropriate Words	3
Incomprehensible Words	2
None	1
Motor Response	
Obeys Command	6
Localizes Pain	5
Withdraws (pain)	4
Flexion (pain)	3
Extension (pain)	2
None	1
Glasgow Coma Score Total	

A GCS score of 13 to 15 indicates a mild brain injury, a score of 9 to 12 indicates a moderate brain injury, and a score of 8 or less indicates a severe brain injury. It is important to note that *a single assessment of a trauma patient's GCS score cannot reliably capture his or her clinical progression*. Obtain a baseline GCS score and then frequently (at least every 5 minutes) reassess it. Document all GCS scores you obtained in the field, including the times they were obtained, and relay them to the receiving facility.

The Revised Trauma Score (RTS) is a physiologic scoring system that is also used to assess the severity of a trauma patient's injuries. Objective data used to calculate the RTS include the Glasgow Coma Scale (GCS) score, systolic blood pressure (SBP), and respiratory rate (RR). In addition to assessing injury severity, the RTS has also demonstrated reliability in predicting survival in patients with severe injuries. The highest RTS a patient can receive is 12; the lowest is 0. The RTS is calculated as follows:

GCS	SBP	RR	Value
13 to 15	> 89 mm Hg	10 to 29 breaths/min	4
9 to 12	76 to 89 mm Hg	> 29 breaths/min	3
6 to 8	50 to 75 mm Hg	6 to 9 breaths/min	2
4 to 5	1 to 49 mm Hg	1 to 5 breaths/min	1
3	0	0	0

You are the Provider: SUMMARY, continued

To assess the patient in this scenario, let's apply these numeric trauma scoring systems. First, you must calculate his GCS. The patient does not open his eyes, even when a painful stimulus is applied; therefore, he receives a 1 for eye opening. For verbal response, he also receives a 1 because he does not respond when you talk to him. For motor response, he receives a 3; he responds to pain by flexing his arms (decorticate posturing). Currently, the patient's GCS score is 5, which indicates a severe brain injury.

To calculate the patient's RTS, you will use his GCS, along with his systolic blood pressure and respiratory rate. He has already been assigned a GCS score of 5; therefore, he is assigned a numeric value of 1. His systolic blood pressure is 80 mm Hg; therefore, he is assigned a numeric value of 3. His respiratory rate is 34 breaths/min; therefore, he is assigned a numeric value of 3. On the basis of these parameters, the patient's RTS is 7.

Like the GCS, you should frequently reevaluate the RTS. The information obtained from numeric trauma scoring is valuable to the physicians at the receiving facility and is often used to direct further assessment and treatment of the patient after he or she is received in the emergency department. In the field, numeric trauma scoring is used to help determine the most appropriate transport destination, and, for advanced level providers (eg, AEMT, paramedic), as a tool to determine whether certain interventions (ie, advanced airway management) should be performed.

7. How does the level of trauma care provided by the paramedic differ from that of the EMT?

Regardless of your level of training and certification, the most important intervention you can provide to a critically injured patient is rapid transport to an appropriate medical facility. As mentioned several times in this case study, definitive care cannot be provided in the field. All levels of EMS providers must be able to recognize a critically injured patient, begin immediate treatment, and quickly select the most appropriate transport mode and destination.

The level of trauma care provided by the EMT versus personnel with a higher level of training (ie, AEMT, paramedic) differs mainly in the emergency treatment interventions that can be performed. For example, paramedics are trained to provide advanced airway management (ie, endotracheal intubation), intravenous therapy, and to administer certain emergency medications, among others. Although these additional skills can be of great benefit to the patient, they are not definitive care interventions. Paramedics cannot repair a lacerated liver or stop bleeding in the brain; thus, their focus on trauma care should be no different from the EMT—to recognize injuries, stabilize the patient, and provide rapid transport.

In many cases, the EMT will be called on to assist the paramedic in performing advanced level skills. Depending on local protocols, EMTs may even be able to perform additional skills as deemed necessary by the EMS system medical director.

EMS Patient Care Report (PCR)

Date: 9-1-09	Incident No.: 012109	Nature of Call: Motor vehicle crash		Location: 2100 Block Hwy 46	
Dispatched: 1520	En Route: 1520	At Scene: 1528	Transport: 1538	At Landing Zone: 1540	In Service: 1552

Patient Information

Age: 20 Sex: M Weight (in kg [lb]): estimated at 68 kg (150 lb)	Allergies: Unknown Medications: Unknown Past Medical History: Unknown Chief Complaint: Multiple traumatic injuries

Vital Signs

Time: 1533	BP: 84/64	Pulse: 120	Respirations: 22	Sao$_2$: 97%
Time: 1538	BP: 80/50	Pulse: 130	Respirations: 30	Sao$_2$: 89%
Time: 1543	BP: 74/50	Pulse: 140	Respirations: 30	Sao$_2$: 95%

EMS Treatment (circle all that apply)

Oxygen @ 15 L/min via (circle one): NC **(NRM)** Bag-Mask Device	**(Assisted Ventilation)**	Airway Adjunct	CPR
Defibrillation	**(Bleeding Control)**	**(Bandaging)**	Splinting	**Other:** (Thermal management, suction, full spinal precautions)

You are the Provider: SUMMARY, continued

Narrative
Dispatched for a motor vehicle versus tree head-on collision. Rescue assignment and law enforcement was dispatched as well. Arrived at the scene and noted that a small passenger vehicle made frontal impact with a large tree. Damage to the front of the vehicle was significant. The driver, a 20-year-old male, was still in the vehicle; however, he was unrestrained. Driver and passenger side air bags both deployed, and patient was not entrapped. Partner accessed patient through back seat and manually stabilized his head. Primary assessment revealed that the patient was responsive only to pain. He had blood in his oropharynx, a large hematoma and laceration with active bleeding to his forehead, and facial bleeding. His respirations were rapid and labored. Suctioned the patient's oropharynx, controlled the bleeding on his forehead, applied cervical collar, and rapidly extricated him from the vehicle. Due to the MOI and patient's clinical status, requested air transport. Applied oxygen @ 15 L/min via nonrebreathing mask and performed secondary assessment, which revealed diffuse bruising and crepitus to the chest. Breath sounds were diminished over the left side of the chest. Pelvis and upper and lower extremities were unremarkable for gross injury. Pupils were dilated and sluggish to react. Engine 3 firefighter reported interior damage to the steering wheel and a starburst fracture to the windshield with evidence of human hair. Applied full spinal precautions and a blanket for warmth, and loaded patient into the ambulance. Reassessment revealed that his respiratory rate had increased, his breathing effort was more labored, and his oxygen saturation had decreased. Began assisting his ventilations with a bag-mask device and high-flow oxygen. Engine 3 EMT drove ambulance to landing zone (LZ) to meet with air transport helicopter. Continued to reassess patient every 3 to 5 minutes and noted no change in his clinical status. Contacted air medical helicopter via radio and provided patient status update. Continued to assist patient's ventilations and suctioned his oropharynx as needed to maintain airway patency. Vital signs were also reassessed, as noted above. After a brief wait at the LZ, air transport helicopter arrived. Gave verbal report to flight paramedic, and transferred patient care to the flight crew. Helicopter departed the LZ at 1550, and EMS 3 returned to service at 1552. **End of report**

Prep Kit

- Determine the mechanism of injury (MOI) as quickly as possible; this will assist you in developing an index of suspicion for the seriousness of your patient's unseen injuries.
- Three concepts of energy are typically associated with injury: potential energy, kinetic energy, and work.
- Traumatic injuries can be described as blunt trauma or penetrating trauma.
- Motor vehicle crashes are classified traditionally as frontal (head-on), lateral (T-bone), rear-end, rotational (spins), and rollovers.
- In every crash there are three collisions that occur:
  - The collision of the vehicle against some type of object
  - The collision of the passenger against the interior of the vehicle
  - The collision of the passenger's internal organs against the solid structures of the body
- Maintain a high index of suspicion for serious injury in the patient who has been involved in a motor vehicle collision or a motor vehicle collision with significant damage to the vehicle, has fallen from a significant height, or has sustained penetrating trauma to the body.
- Communicate mechanism of injury (MOI) findings in the written patient care report and verbally to hospital staff; this will ensure that appropriate treatment for potential serious injuries continues for the patient at the hospital.
- People who are injured in explosions may have injuries that are classified as primary blast injuries, secondary blast injuries, tertiary blast injuries, and/or miscellaneous blast injuries.
- A patient who has sustained a significant mechanism of injury (MOI) and is considered to be in serious or critical condition should receive a rapid full-body scan or rapid head-to-toe examination. Any patient who has sustained a nonsignificant mechanism of injury (MOI) should receive an assessment focused on the chief complaint.
- Caring for victims of traumatic injuries requires the EMT to have a solid understanding of the trauma system in the United States. This includes transport time, transport destination, and selection of type of transport.

Vital Vocabulary

arterial air embolism Air bubbles in the arterial blood vessels.

blunt trauma An impact on the body by objects that cause injury without penetrating soft tissues or internal organs and cavities.

cavitation A phenomenon in which speed causes a bullet to generate pressure waves, which cause damage distant from the bullet's path.

coup-contrecoup brain injury A brain injury that occurs when force is applied to the head and energy transmission through brain tissue causes injury on the opposite side of original impact.

deceleration The slowing of an object.

drag Resistance that slows a projectile, such as air.

Glasgow Coma Scale (GCS) score An evaluation tool used to determine level of consciousness, which evaluates and assigns point values (scores) for eye opening, verbal response, and motor response, which are then totaled; effective in helping predict patient outcomes.

index of suspicion Awareness that unseen life-threatening injuries may exist when determining the mechanism of injury.

kinetic energy The energy of a moving object.

mechanism of injury (MOI) The forces or energy transmission applied to the body that cause injury.

medical emergencies Emergencies that require EMS attention because of illnesses or conditions not caused by an outside force.

multisystem trauma Trauma that affects more than one body system.

penetrating trauma Injury caused by objects, such as knives and bullets, that pierce the surface of the body and damage internal tissues and organs.

potential energy The product of mass, gravity, and height, which is converted into kinetic energy and results in injury, such as from a fall.

projectile Any object propelled by force, such as a bullet by a weapon.

pulmonary blast injuries Pulmonary trauma resulting from short-range exposure to the detonation of explosives.

Revised Trauma Score (RTS) A scoring system used for patients with head trauma.

trajectory The path a projectile takes once it is propelled.

trauma emergencies Emergencies that are the result of physical forces applied to a patient's body.

trauma score A score that relates to the likelihood of patient survival with the exception of a severe head injury. It calculates a number from 1 to 16, with 16 being the best possible score. It takes into account the Glasgow Coma Scale (GCS) score, respiratory rate, respiratory expansion, systolic blood pressure, and capillary refill.

tympanic membrane The eardrum; a thin, semitransparent membrane in the middle ear that transmits sound vibrations to the internal ear by means of auditory ossicles.

work The product of force times distance.

Assessment in Action

You are dispatched to a motor vehicle collision on a rural road. On arrival you see a single vehicle in an embankment. The scene appears to be safe. The car has front-end damage and a starred windshield. You find an unrestrained driver sitting in the driver's seat. He opens his eyes when you speak to him; however; he is confused and unable to tell you what happened, but he follows your commands. Examination reveals a laceration on the center of his forehead and a large bruise on his chest. Pulses and motor and sensory function is present in all extremities. His pulse is rapid, and breathing is slightly labored.

1. What is your first concern in this situation?
   A. History taking
   B. Physical examination
   C. Scene size-up
   D. Reassessment

2. What type of injury was sustained to the patient's chest?
   A. Blunt
   B. Penetrating
   C. Crushing
   D. Barotrauma

3. What other injuries might you suspect in this patient?

4. What is a coup-contrecoup injury?

5. Does this patient meet the criteria for rapid transport? Why or why not?

6. What is the Glasgow Coma Scale score for this patient?
   A. 15
   B. 14
   C. 13
   D. 12

7. What are the three collisions involved in a frontal motor vehicle accident?

8. The Revised Trauma Score is a scoring system that is used for patients with what type of trauma?
   A. Abdominal
   B. Head
   C. Chest
   D. Extremity

9. What is Newton's Third Law?
   A. Objects at rest stay at rest, an object in motion stays in motion.
   B. Force equals mass times acceleration.
   C. An object can only stretch so far before it fails.
   D. For every action there is an equal and opposite reaction.

10. What does index of suspicion mean?
    A. Life-threatening injuries are present with a significant mechanism of injury.
    B. The patient has an illness that is not caused by an outside force.
    C. Awareness that unseen life-threatening injuries may exist when determining the mechanism of injury.
    D. Awareness that the mechanism of injury will rule out any life-threatening injuries.

National EMS Education Standard Competencies

Trauma

Applies fundamental knowledge to provide basic emergency care and transportation based on assessment findings for an acutely injured patient.

Bleeding

Recognition and management of

- Bleeding (pp 781-796)

Pathophysiology, assessment, and management of

- Bleeding (pp 777-796)

Pathophysiology

Applies fundamental knowledge of the pathophysiology of respiration and perfusion to patient assessment and management.

Knowledge Objectives

1. Understand the general structure of the circulatory system and the function of its different parts, including the heart, arteries, veins, and interconnecting capillaries. (pp 777-780)
2. Explain the significant bleeding that may be caused by blunt force trauma, including the importance of perfusion. (pp 780-781)
3. Discuss bleeding and the possibility of hypovolemic shock, including the signs of shock. (pp 781-782)
4. Explain the importance of following standard precautions when treating patients with external bleeding. (p 781)

5. Describe the characteristics of external bleeding, including the identification of the following types of bleeding: arterial, venous, and capillary (pp 782-783).
6. Identify the signs and symptoms of internal bleeding. (pp 783-784)
7. Explain how to determine the nature of the illness (NOI) for internal bleeding, including identifying possible traumatic and nontraumatic sources. (pp 783-784)
8. Explain how to conduct a primary assessment, including identification of life threats beyond bleeding, ensuring a patent airway, and making the transport decision. (pp 784-785)
9. Discuss internal bleeding in terms of the different mechanisms of injury and their associated internal bleeding sources. (p 786)
10. Explain how to conduct a secondary assessment on a patient with external or internal bleeding, including physical examination, vital signs, and use of monitoring devices. (pp 786-787)
11. Explain the emergency medical care of the patient with external bleeding. (pp 788-794)
12. Explain the emergency medical care of the patient with internal bleeding. (pp 794-796)

Skills Objectives

1. Demonstrate emergency medical care of the patient with external bleeding using direct pressure. (pp 788-789, Skill Drill 23-1)
2. Demonstrate emergency medical care of the patient with external bleeding using a commercial tourniquet. (pp 790-792, Skill Drill 23-2)
3. Demonstrate emergency medical care of the patient with epistaxis, or nosebleed. (pp 793-794, Skill Drill 23-3)
4. Demonstrate emergency medical care of the patient who shows signs and symptoms of internal bleeding. (pp 794-796, Skill Drill 23-4)

Introduction

After managing the airway, recognizing bleeding and understanding how it affects the body are perhaps the most important skills you will learn as an EMT. Bleeding can be external and obvious or internal and hidden. Either way, it is potentially dangerous, first causing weakness and, if left uncontrolled, eventually shock and death. The most common cause of shock following trauma is bleeding. Generally the shock from trauma is caused at least in part from bleeding.

This chapter will help you understand how the cardiovascular system reacts to blood loss. The chapter begins with a brief review of the anatomy and function of the cardiovascular system. It then describes the signs, symptoms, and emergency medical care of both external and internal bleeding. The chapter concludes with a discussion on the relationship between bleeding and hypovolemic shock.

Anatomy and Physiology of the Cardiovascular System

The cardiovascular system circulates blood to all of the body's cells and tissues, delivering oxygen and nutrients and carrying away metabolic waste products **Figure 23-1**. Cells in the brain, spinal cord, and heart cannot tolerate a lack of blood for more than a few minutes. Cells in other organs, such as the lungs and kidneys, can survive for almost an hour while skeletal muscle cells may survive for 2 hours in a state of

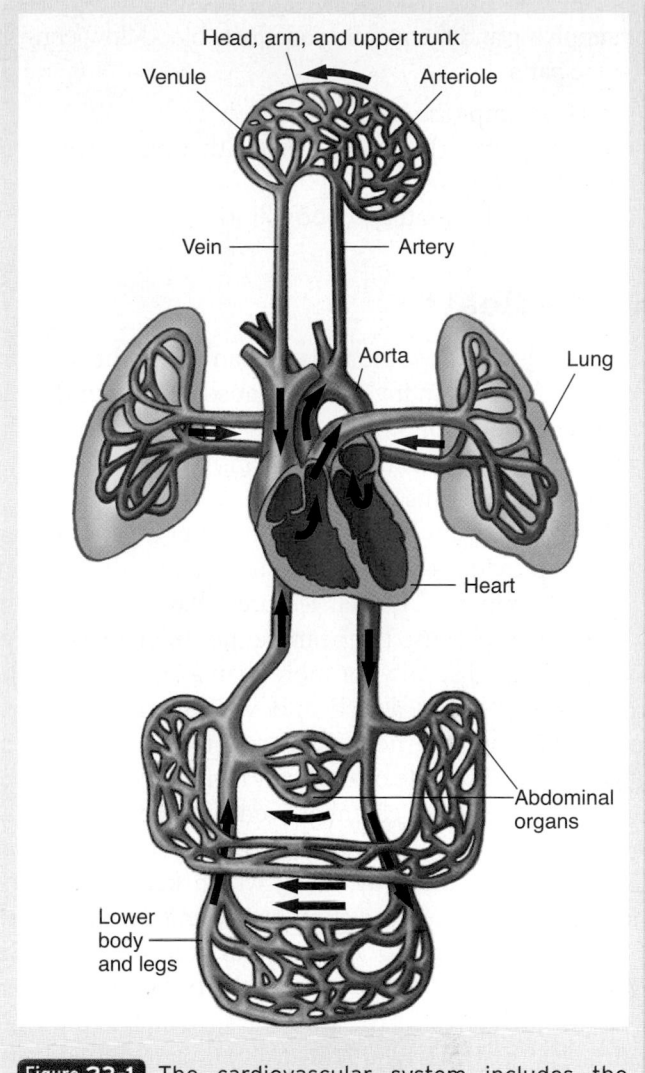

Figure 23-1 The cardiovascular system includes the heart, arteries, veins, and interconnecting capillaries.

inadequate perfusion. After that, their cells begin to die. This can lead to a permanent loss of function or, if enough cells die, death.

You are the Provider: PART 1

At 4:20 PM, you are dispatched to a woodworking shop at 517 East Graham for a 32-year-old man with severe bleeding from the arm. The exact mechanism of injury is unknown. You and your partner respond to the scene with a response time of approximately 6 minutes.

1. **What are the functions of arteries? What major arteries are located in the upper extremity?**

2. **Why is arterial bleeding more severe than venous bleeding?**

The cardiovascular system, the main system responsible for supplying and maintaining adequate blood flow, consists of three parts:

- The pump (the heart)
- A container (the blood vessels that reach every cell in the body)
- The fluid (blood and body fluids)

■ The Heart

The heart is a hollow muscular organ about the size of a clenched fist. It is an involuntary muscle that is under the control of the autonomic nervous system, but it has its own regulatory system. Thus, it can function even if the nervous system shuts down.

The heart is always working; all other organs depend on it to provide a rich blood supply. For this reason, it has a number of special features that other muscles do not. First, because the heart cannot tolerate a disruption of its blood supply for more than a few seconds, the heart muscle needs a rich and well-distributed blood supply. Second, the heart works as two paired pumps **Figure 23-2**. Each side of the heart has an upper chamber (atrium) and a lower chamber (ventricle), both of which pump blood. Blood leaves each chamber of a normal heart through a one-way valve, which keeps the blood moving in the proper direction by preventing backflow.

The right side of the heart receives oxygen-poor (deoxygenated) blood from the veins of the body. Blood enters the right atrium from the vena cava, then fills the right ventricle. After the right ventricle contracts, blood flows into the pulmonary artery and the pulmonary circulation. The now oxygen-rich (oxygenated) blood returns to the left side of the heart from the lungs through the pulmonary veins. Blood enters the left atrium, then passes into the left ventricle. This side of the heart is more muscular than the other because it must pump blood into the aorta and on to the arteries throughout the body. It is important to remember that the left ventricle is responsible for providing 100% of the body with oxygen-rich blood.

■ Blood Vessels and Blood

There are five types of blood vessels:

- Arteries
- Arterioles
- Capillaries
- Venules
- Veins

As blood flows out of the heart, it passes into the **aorta**, the largest **artery** in the body. The arteries become smaller as they move away from the heart. The smaller vessels that connect the arteries and capillaries are called

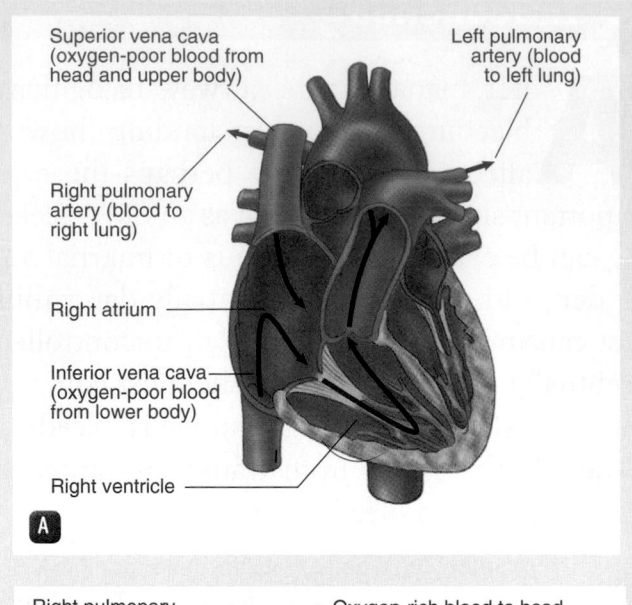

Superior vena cava (oxygen-poor blood from head and upper body)

Left pulmonary artery (blood to left lung)

Right pulmonary artery (blood to right lung)

Right atrium

Inferior vena cava (oxygen-poor blood from lower body)

Right ventricle

A

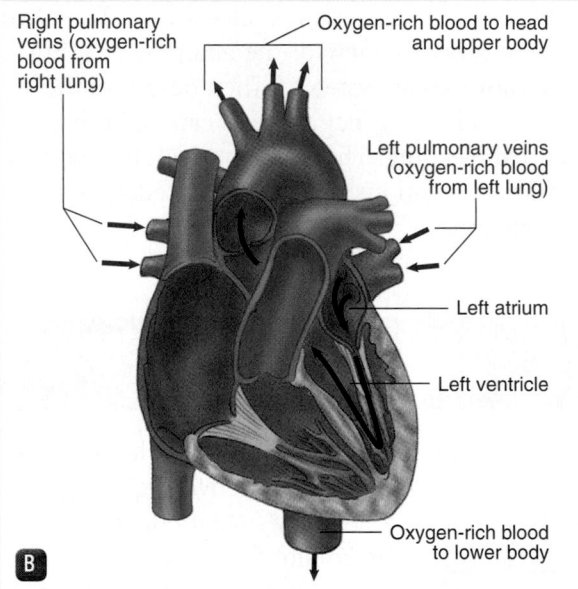

Right pulmonary veins (oxygen-rich blood from right lung)

Oxygen-rich blood to head and upper body

Left pulmonary veins (oxygen-rich blood from left lung)

Left atrium

Left ventricle

Oxygen-rich blood to lower body

B

Figure 23-2 **A.** The left side of the heart circulates oxygen-rich blood to all parts of the body. It is the more muscular of the two pumps because it must pump blood into the aorta and into the arteries. **B.** The right side of the heart circulates blood from the body to the lungs.

arterioles. **Capillaries** are small tubes, with the diameter of a single red blood cell, that pass among all the cells in the body, linking the arterioles and the **venules**. Blood leaving the distal side of the capillaries flows into the venules. These small, thin-walled vessels empty into the **veins**, and the veins then empty into the vena cava. This is the process that returns blood in the venous side of the circulatory system to the heart. Oxygen and nutrients easily pass from the capillaries into the cells, and waste and carbon dioxide diffuse from the cells into the capillaries **Figure 23-3**. This transportation system allows the body to rid itself of waste products.

each cell of the surrounding tissue; when the muscles are closed (constricted), there is no capillary blood flow. The muscles dilate and constrict in response to conditions such as fright, heat, cold, a specific need for oxygen, and the need to dispose of metabolic waste. In a healthy individual, all the vessels are never fully dilated or fully constricted at the same time.

The last part of the cardiovascular system is the contents of the container, or the blood. Blood contains red cells, white cells, platelets, and a liquid called plasma **Figure 23-4**. As discussed in Chapter 5, *The Human Body*, red blood cells are responsible for the transportation of oxygen to the cells and for transporting carbon dioxide (a waste product of cellular metabolism) away from the cells to the lungs, where it is exhaled and removed from the body. Platelets are responsible for forming blood clots. In the body, a blood clot forms depending on one of the following principles: blood stasis, changes in the vessel wall (such as a wound), or the blood's ability to clot (due to a disease process or medication). When injury occurs to tissues in the body, platelets will begin to collect at the site of injury; this causes red blood cells to become sticky and clump together. As the red blood cells begin to clump, another substance in the body called fibrinogen reinforces the red blood cells. This is the final step in formation of a blood clot. Blood clots are an important response from the body to control blood loss. Certain medical conditions that interfere with the normal clotting process will be discussed later in this chapter.

The autonomic nervous system monitors the body's needs from moment to moment and adjusts the blood flow by adjusting vascular tone as required. During emergencies, the autonomic nervous system automatically redirects blood away from other organs to the heart, brain, lungs, and kidneys. Thus, the cardiovascular system is dynamic and constantly adapting to changing

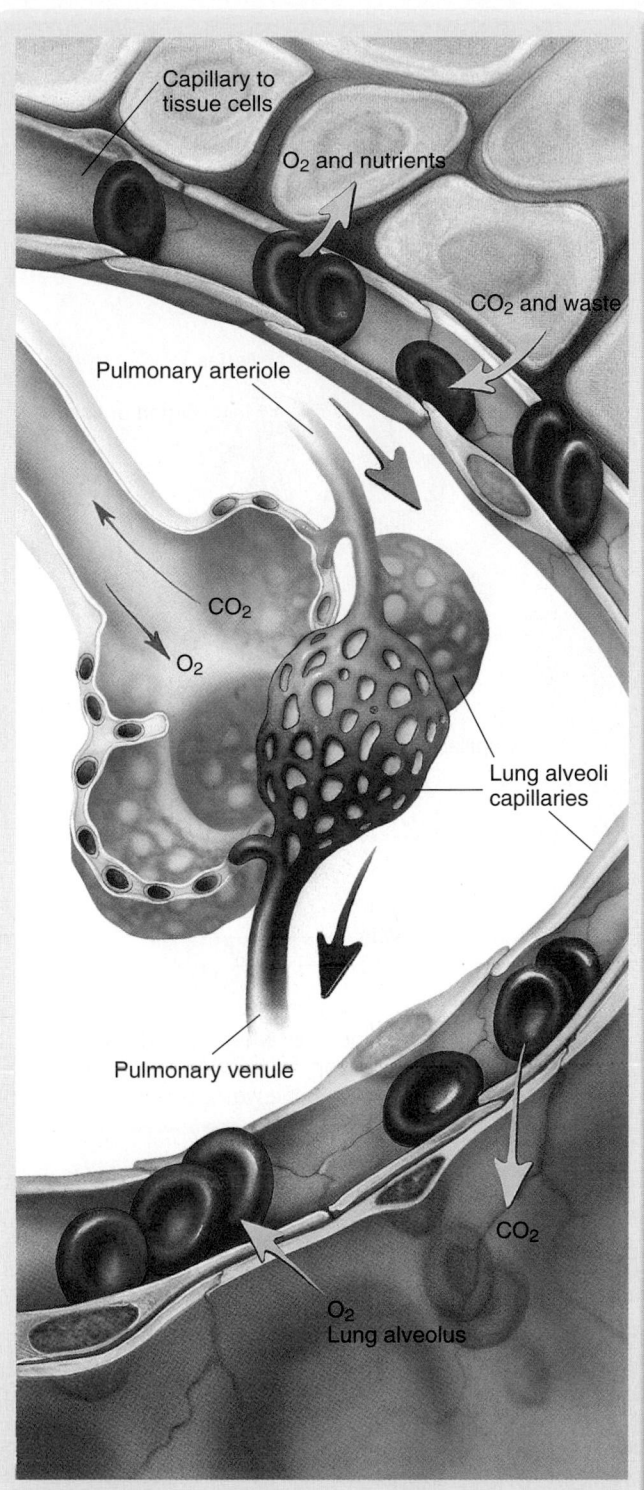

Figure 23-3 Oxygen and nutrients pass easily from the capillaries into the cells, and waste and carbon dioxide diffuse from the cells into the capillaries (top). Oxygen and carbon dioxide pass freely between the lungs and capillaries (bottom).

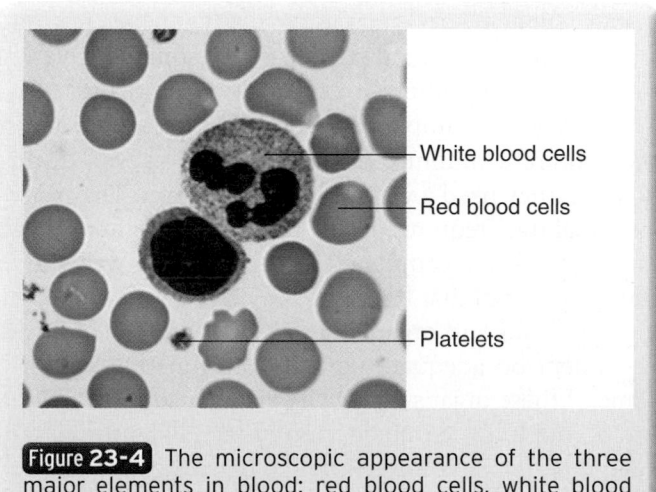

Figure 23-4 The microscopic appearance of the three major elements in blood: red blood cells, white blood cells, and platelets.

At the arterial ends of the capillaries and in the arteries themselves are circular muscular walls, which constrict and dilate automatically under the control of the autonomic nervous system. When these muscles open (dilate), blood passes into the capillaries in proximity to conditions in the body to maintain homeostasis and perfusion. At times, the system fails to provide sufficient circulation for every body part to perform its function. This condition is called **hypoperfusion**, or **shock**.

Pathophysiology and Perfusion

Blunt force trauma may cause injury and significant bleeding that is unseen inside a body cavity or region, such as when injury occurs to the liver or the spleen. These injuries cause the patient to lose significant amounts of blood, causing hypoperfusion without visible bleeding. In penetrating trauma, the patient may have only a small amount of bleeding that is visible; however, the patient may have sustained injury to internal organs that will produce significant bleeding that is unseen by you and may cause death quickly. Both of these situations are examples of serious internal bleeding, in which blood volume and supply have been interrupted to the cells of the body; this interruption is the cause of hypoperfusion (or shock) in the trauma patient.

Perfusion is the circulation of blood within an organ or tissue in adequate amounts to meet the cells' current needs for oxygen, nutrients, and waste removal. Blood enters an organ or tissue first through the arteries, then the arterioles, and finally the capillary beds **Figure 23-5**. While passing through the capillaries, the blood delivers nutrients and oxygen to the surrounding cells and picks up the wastes they have generated. Then the blood leaves the capillary beds through the venules and finally reaches the veins, which take the blood back to the heart. Oxygen and carbon dioxide exchange takes place in the lungs.

Blood must pass through the cardiovascular system at a speed that is fast enough to maintain adequate circulation throughout the body and slow enough to allow each cell time to exchange oxygen and nutrients for carbon dioxide and other waste products. Although some tissues, such as the lungs and kidneys, never rest and require a constant blood supply, most require circulating blood only intermittently, especially when active. Muscles are a good example. When you sleep, they are at rest and require a minimal blood supply. However, during exercise, they need a very large blood supply. The gastrointestinal tract requires a high flow of blood after a meal. After digestion is completed, it can do quite well with a small fraction of that flow.

All organs and organ systems of the human body are dependent on adequate perfusion to function properly. Some of these organs receive a very rich supply of blood and do not tolerate interruption of blood supply for very long. If perfusion is interrupted to these organs and damage occurs to the organ tissue, dysfunction and failure of that organ system will occur. Death of an organ system

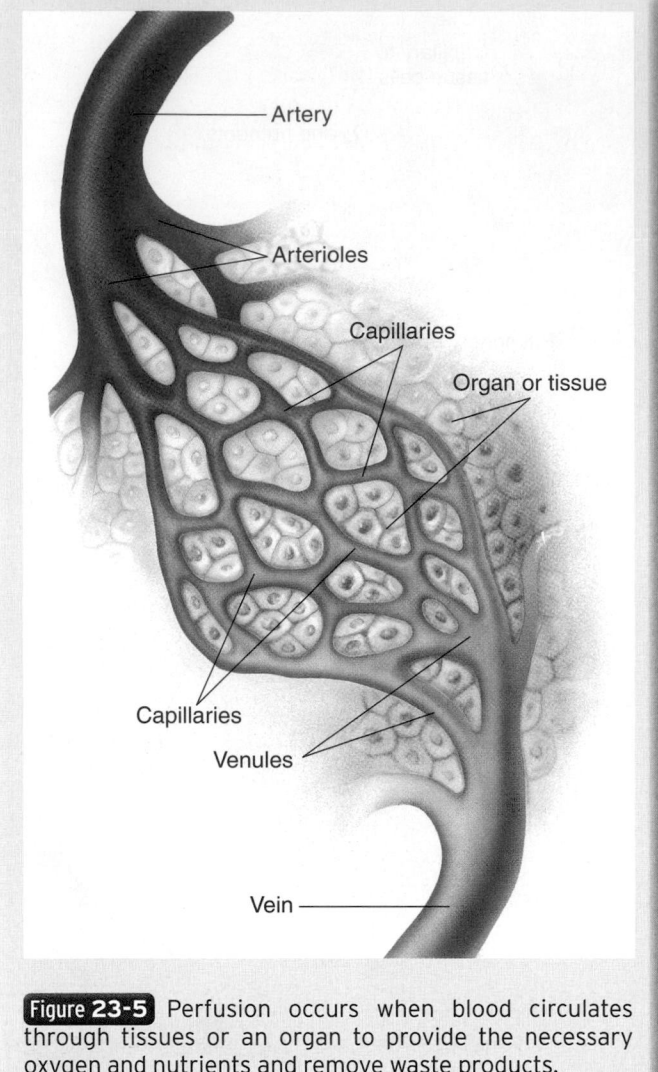

Figure 23-5 Perfusion occurs when blood circulates through tissues or an organ to provide the necessary oxygen and nutrients and remove waste products.

can quickly lead to death of the organism, the human. Emergency medical care is designed to support adequate perfusion to these organs and their systems, listed in **Table 23-1**, until the patient arrives at the hospital.

Table 23-1	Organs and Corresponding Organ Systems
Organ	**Organ System**
Heart	Cardiovascular system
Brain	Central nervous system
Lungs	Respiratory system
Kidneys	Renal system

The heart requires constant perfusion to function properly. The brain and spinal cord can be injured after 4 to 6 minutes without perfusion. It is important to remember that cells of the central nervous system do not have the capacity to regenerate. Kidneys can be damaged after 45 minutes of inadequate perfusion. Skeletal muscle demonstrates evidence of injury after 2 hours of inadequate perfusion. The gastrointestinal tract can tolerate slightly longer periods of inadequate perfusion. These times are based on a normal body temperature (98.6°F [37.0°C]). An organ or tissue that is considerably colder may be better able to resist damage from hypoperfusion.

External Bleeding

__Hemorrhage__ means bleeding. External bleeding is visible hemorrhage. Examples include nosebleeds and bleeding from open wounds. As an EMT, you must understand how to control external bleeding.

The Significance of External Bleeding

When patients have serious external blood loss, it is often difficult to determine the amount of blood that is present. This is a difficult task because blood will look different on different surfaces, such as when it is absorbed in clothing or when it has been diluted when mixed in water. Always attempt to determine the amount of external blood loss, but the presentation and assessment of the patient will direct the care and treatment the patient will receive from you as an EMT.

Words of Wisdom

Signs and Symptoms of Hypovolemic Shock
- Rapid, weak pulse
- Low blood pressure (late sign)
- Changes in mental status
- Cool, clammy skin
- Cyanosis (lips, oral membranes, nail beds)

The body will not tolerate an acute blood loss of greater than 20% of blood volume. The typical adult has approximately 70 mL of blood per kilogram of body weight, or 6 L (10 to 12 pints) in a body weighing 80 kg (175 lb). If the typical adult loses more than 1 L of blood (about 2 pints), significant changes in vital signs will occur, including increasing heart and respiratory rates and decreasing blood pressure. Because infants and children have less blood volume to begin with, the same effect is seen with smaller amounts of blood loss. For example, a 1-year-old has a total blood volume of about 800 mL. Significant symptoms of blood loss will occur after only 100 to 200 mL of blood loss. To

Safety

Remember that a bleeding patient may expose you to potentially infectious body fluids; therefore, you must always follow standard precautions when treating patients with external bleeding. Wear gloves and eye protection in all situations, and wear a gown and mask if there is a risk of blood splatter **Figure 23-6**. Avoid direct contact with body fluids if possible. Take special care if you have an open sore, cut, scratch, or ulcer. Also remember that frequent, thorough handwashing between patients and after every run is a simple yet important protective measure. You will be called to respond to emergencies involving more than one patient who needs emergency care. As you complete the assessment and care for each patient, remember to place clean gloves on your hands. Always keep spare gloves with you when responding to these incidents. This approach to patient care will greatly minimize the chance that you could cause cross-contamination of body fluids and blood between patients you may be caring for.

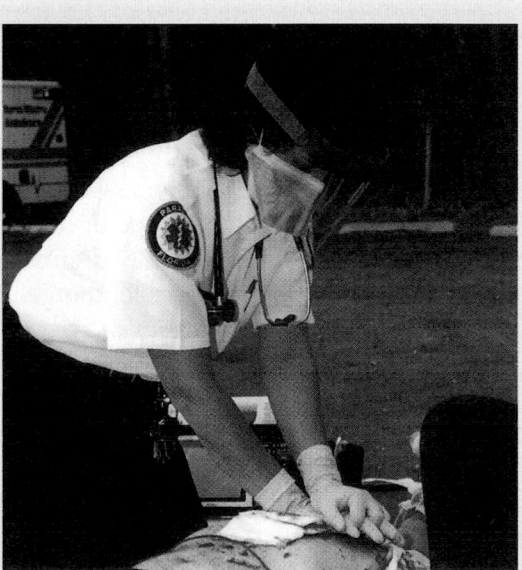

Figure 23-6 Your safety is paramount; therefore, you should always wear proper protective equipment when caring for a patient who is bleeding.

put this in perspective, a soft drink can holds roughly 355 mL of liquid.

How well people compensate for blood loss is related to how rapidly they bleed. A healthy adult can comfortably donate 1 unit (500 mL) of blood during a period of 15 to 20 minutes and adapts well to this decrease in blood volume. However, if a similar blood loss occurs in a much shorter period, the person may rapidly develop __hypovolemic shock__, a condition in which low blood volume results in inadequate perfusion and even death. The body simply cannot compensate for such a rapid blood loss. The age and preexisting health of the patient should also be considered.

You should consider bleeding to be serious if the following conditions are present:

- It is associated with a significant mechanism of injury (MOI).
- The patient has a poor general appearance and is calm.
- Assessment reveals signs and symptoms of shock (hypoperfusion).
- You note a significant amount of blood loss.
- The blood loss is rapid.
- You cannot control the bleeding.

In any situation, blood loss is an extremely serious problem. It demands your immediate attention as soon as you have cleared the airway and managed the patient's breathing.

Characteristics of External Bleeding

Injuries and some illnesses can disrupt blood vessels and cause bleeding. Typically, bleeding from an open artery (arterial bleeding) is brighter red (high in oxygen) and spurts in time with the pulse. The pressure that causes the blood to spurt also makes this type of bleeding difficult to control. As the amount of blood circulating in the body drops, so does the patient's blood pressure and, eventually, the arterial spurting.

Blood from an open vein (venous bleeding) is darker (low in oxygen) and flows slowly or severely, depending on the size of the vein. Because it is under less pressure, most venous blood does not spurt and is easier to manage; however, it can be profuse and life threatening. Capillary blood (bleeding from damaged capillary vessels) is dark red and oozes from a wound steadily but slowly. Venous and capillary blood is more likely to clot spontaneously than arterial blood **Figure 23-7**.

On its own, bleeding tends to stop rather quickly, within about 10 minutes, in response to internal mechanisms and exposure to air. When a person is cut, blood flows rapidly from the open vessel. Soon afterward, the cut ends of the vessel begin to narrow (**vasoconstriction**), reducing the amount of bleeding. Then a clot forms, plugging the hole and sealing the injured portions of the vessel. This process is called **coagulation**. Bleeding will never stop if a clot does not form, unless the injured vessel is completely cut off from the main blood supply.

Despite the efficiency of this system, it may fail in certain situations. Movement, medications, removal of bandages, and the external environment or body temperature commonly affect the blood's clotting factors. For example, a number of medications, including aspirin, interfere with normal clotting. With a severe injury, the damage to the vessel may be so large that a clot cannot completely block the hole. Sometimes only part of the vessel wall is cut, preventing it from constricting. In these cases, bleeding will continue unless it is stopped by external means. Occasionally, blood loss occurs very rapidly. In these cases, the patient might die before the body's defenses, such as clotting, could help.

A very small portion of the population lacks one or more of the blood's clotting factors. This condition is called **hemophilia**. There are several forms of hemophilia, most of which are hereditary and some of which are severe. Sometimes bleeding may occur spontaneously in hemophilia. Because the patient's blood does not clot, all injuries, no

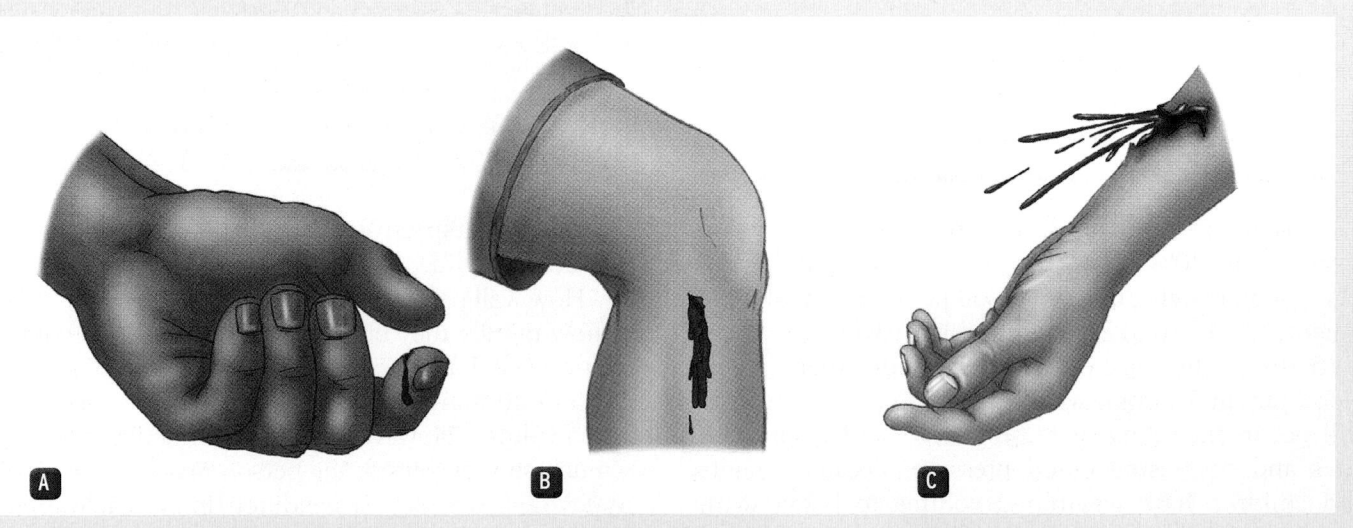

Figure 23-7 **A.** Bleeding from capillary vessels is dark red and oozes from the wound slowly but steadily. **B.** Venous bleeding is darker than arterial bleeding and flows steadily. **C.** Arterial bleeding is characteristically brighter red and spurts in time with the pulse.

Words of Wisdom

If a bandage has already been applied to control bleeding before you arrive on the scene, obtain a description of the wound and the amount of bleeding from the patient or bystanders.

matter how trivial, are potentially serious. A patient with hemophilia should be transported immediately.

Internal Bleeding

Internal bleeding is any bleeding in a cavity or space inside the body. It can be very serious, especially because you might not be aware that it is happening. Injury or damage to internal organs commonly results in extensive internal bleeding, which can cause hypovolemic shock before you realize the extent of blood loss. A person with a bleeding stomach ulcer may lose a large amount of blood very quickly. Similarly, a person who has a lacerated liver or a ruptured spleen may lose a considerable amount of blood within the abdomen. Yet the patient has no outward signs of bleeding.

Broken bones, especially broken ribs, also may cause serious internal blood loss. Sometimes this bleeding extends into the chest cavity and the soft tissues of the chest wall. A broken femur can easily result in the loss of 1 L or more of blood into the soft tissues of the thigh. Often the only signs of such bleeding are local swelling and bruising due to the accumulation of blood around the ends of the broken bone. Severe pelvic fractures may result in life-threatening hemorrhage.

You must always be alert to the possibility of internal bleeding and assess the patient for related signs and symptoms, particularly if the MOI is severe. If you suspect that a patient is bleeding internally, you should promptly transport him or her to the hospital.

Mechanism of Injury for Internal Bleeding

A high-energy MOI should increase your index of suspicion for the possibility of serious unseen injuries such as internal bleeding in the abdominal cavity. Internal bleeding is possible whenever the MOI suggests that severe forces affected the body. These forces include blunt and penetrating trauma. Internal bleeding commonly occurs as a result of falls, blast injuries, and automobile or motorcycle crashes. Remember that internal bleeding can result from penetrating trauma as well.

As you assess a patient, look for signs of injury using DCAP-BTLS (Deformities, Contusions, Abrasions,

Punctures/Penetrations, Burns, Tenderness, Lacerations, and Swelling) over the chest or abdomen, including contusions, abrasions, lacerations, and other signs of injury or deformity. You should always suspect internal bleeding in a patient who has penetrating injury or blunt trauma.

Nature of Illness for Internal Bleeding

Internal bleeding is not always caused by trauma. Many illnesses can cause internal bleeding. Some of the more common causes of nontraumatic internal bleeding include bleeding ulcers, bleeding from the colon, ruptured ectopic pregnancy, and aneurysms.

Abdominal tenderness, guarding, rigidity, pain, and distention are frequent in these situations but are not always present. In older patients, dizziness, faintness, or weakness may be the first sign of nontraumatic internal bleeding. Ulcers or other gastrointestinal problems may cause vomiting of blood or bloody diarrhea or urine.

It is not as important for you to know the specific organ involved as it is to recognize that the patient is in shock and respond appropriately.

Signs and Symptoms of Internal Bleeding

The most common symptom of internal bleeding is pain. Significant internal bleeding will generally cause swelling in the area of bleeding. Intra-abdominal bleeding will often cause pain and distention. Bruising is a sign of internal bleeding. It is most common in head, extremity, and pelvic injuries and can be a sign of significant abdominal trauma. Bleeding into the chest may cause dyspnea in addition to tachycardia and hypotension. A bruise is also called a **contusion**, or **ecchymosis**. A **hematoma**, a mass of blood in the soft tissues beneath the skin, indicates bleeding into soft tissues and may be the result of a minor or a severe injury. Bruising or ecchymosis may not be present initially, and the only sign of severe pelvic or abdominal trauma may be redness, skin abrasions, or pain.

Bleeding, however slight, from any body opening is serious. It usually indicates internal bleeding that is not easy to see or control. Bright red bleeding from the mouth or rectum or blood in the urine (hematuria) may suggest serious internal injury or disease. Nonmenstrual vaginal bleeding is always significant.

Other signs and symptoms of internal bleeding in both trauma and medical patients include the following:

- **Hematemesis**. This is vomited blood. It may be bright red or dark red, or, if the blood has been partially digested, it may look like coffee-grounds vomitus.
- **Melena**. This is a black, foul-smelling, tarry stool that contains digested blood.

- **Hemoptysis**. This is bright red blood that is coughed up by the patient.
- **Pain, tenderness, bruising, guarding, or swelling.** These signs and symptoms may mean that a closed fracture is bleeding.
- **Broken ribs, bruises over the lower part of the chest, or a rigid, distended abdomen.** These signs and symptoms may indicate a lacerated spleen or liver. Patients with an injury to either organ may have referred pain in the right shoulder (liver) or left shoulder (spleen). You should suspect internal abdominal bleeding in a patient with referred pain.

The first sign of hypovolemic shock (hypoperfusion) is a change in mental status, such as anxiety, restlessness, or combativeness. In nontrauma patients, weakness, faintness, or dizziness on standing is another early sign. Changes in skin color or pallor (pale skin) are seen often in both trauma and medical patients. Later signs of hypoperfusion suggesting internal bleeding include the following:

- Tachycardia
- Weakness, fainting, or dizziness at rest
- Thirst
- Nausea and vomiting
- Cold, moist (clammy) skin
- Shallow, rapid breathing
- Dull eyes
- Slightly dilated pupils that are slow to respond to light
- Capillary refill of more than 2 seconds in infants and children
- Weak, rapid (thready) pulse
- Decreasing blood pressure
- Altered level of consciousness

Patients with these signs and symptoms are at risk. Some may be in danger. Even if their bleeding stops, it could begin again at any moment. Therefore, prompt transport is necessary.

Patient Assessment for External and Internal Bleeding

Scene Size-up

Scene Safety

As you approach the patient, be alert to potential hazards to yourself and the crew, bystanders, and the patient(s). At vehicle crashes, ensure that there is no leaking fuel in the area where you will be working and that energized electrical lines are not close to where you will be working. In incidents involving violence, such as assaults or gunshot wounds, make sure that police are on scene. At times you may need to stage several blocks away until law enforcement personnel have secured the area.

Follow standard precautions. Place several pairs of gloves in your pocket for easy access in case your gloves tear or there are multiple patients with bleeding. If you are entering a residence, be alert for anxious bystanders and family members because they may become hostile. Ensure that you are only going to have to provide care for one patient. Consider early on what you may need, and verify as you begin your assessment.

Mechanism of Injury/Nature of Illness

Determine the nature of the illness (NOI) (such as bloody emesis or bloody stool), or the MOI (such as a turned-over step stool). Consider the need for manual spinal stabilization and the need for additional resources, such as an advanced life support unit. Be sure to also consider environmental factors in your decision making. For example, caring for a sick or injured victim of a car crash on a clear, sunny day is a bit different than treating the same victim during a snowstorm. Extreme hot or cold weather can worsen a patient's overall condition.

Special Populations

In older patients, dizziness, syncope, or weakness may be the first sign of nontraumatic internal hemorrhage.

Primary Assessment

In patients with suspected significant blood loss from a visible wound or from unseen internal bleeding, you must not be distracted from identifying life threats. The EMT should treat the patient according to the ABCs and provide treatment needed to preserve life. The management of life-threatening concerns during the primary assessment is determined by asking yourself, "What is going to kill my patient first?" For example, in some situations, significant bleeding may need management before applying oxygen for a person with adequate breathing. The decision on what to treat first will come with experience. Treating according to the ABCs is always a good choice.

Form a General Impression

As you approach a trauma patient, you must note important indicators that may alert you to the seriousness of the patient's condition. For example, patients with external bleeding may have blood stains on their clothing. Be aware of obvious signs of injury and distress (such as facial grimace), along with determining gender and age. Perform a rapid scan of the patient. Assess skin color. Pale or gray, cool, moist skin suggests a perfusion problem. Determine the patient's level of consciousness using the AVPU scale (Awake and alert; responsive to Verbal stimuli or Pain; Unresponsive). Is the patient able to speak? This will indicate whether or not the airway is patent. What is the mental status of the patient? These indicators will help you determine whether the patient is sick or not so sick; this assists you in developing an index of suspicion for serious illness or injuries related to internal bleeding.

Airway and Breathing

Consider the need for spinal stabilization. At the same time, ensure a patent airway, look for adequate breathing, and check for breath sounds. If necessary, provide the patient with high-flow oxygen or assist ventilation with a bag-mask device or nonrebreathing mask, depending on the patient's level of consciousness and rate and quality of breathing. If the patient is unconscious, the airway may be obstructed.

Circulation

You must be able to quickly assess pulse rate and quality; determine the skin condition, color, and temperature; and check the capillary refill time to help establish the potential for internal bleeding and shock. When life-threatening external bleeding is seen, you must begin the steps necessary to control the external bleeding and treatment of shock should begin as quickly as possible. Non–life-threatening bleeding, such as with abrasions, can be bandaged later in your assessment as necessary. Significant bleeding, internal or external, is an immediate life threat. Treat the patient for shock if needed by applying oxygen, improving circulation, and maintaining a normal body temperature.

Transport Decision

The results of your initial general impression and assessment of the ABCs will help you develop a sense of urgency for the patient and guide you in your transport decision to manage the patient on scene or manage the patient on the way to the hospital. For example, if the patient has signs and symptoms of internal bleeding or airway or breathing problems, you must transport quickly to the appropriate hospital for treatment by a physician. The condition of patients who may have significant bleeding will quickly become unstable. Signs such as tachycardia, tachypnea, low blood pressure, weak pulse, and clammy skin are signs of impending circulatory collapse and imply the need for rapid transport.

► History Taking

Investigate Chief Complaint

After the primary assessment is complete, investigate the chief complaint and be alert for signs or symptoms of other injuries due to the MOI and/or NOI. Internal bleeding can be found in both medical and trauma patients. If the bleeding is severe, you may have identified it in

You are the Provider: PART 2

You arrive at the scene and find the patient standing outside in front of the shop. He has a towel wrapped around his left wrist; however, it is soaked in blood and you can see a large amount of blood on the ground. He is conscious and alert, but anxious, and tells you that he cut his wrist on a table saw when his arm slipped and ran into the blade.

Recording Time: 0 Minutes	
Appearance	Anxious
Level of consciousness	Conscious and alert
Airway	Open; clear of secretions or foreign bodies
Breathing	Increased rate; adequate depth
Circulation	Bleeding from the left wrist; skin is cool, pale, and dry; pulse is rapid and strong

3. Is the patient effectively controlling the bleeding from his injury?

4. What should be your initial treatment priority?

the primary assessment and begun treatment and rapid transport to the hospital. If the signs and symptoms of internal bleeding are not as obvious as described previously, you will need to look more carefully in this step of the patient assessment process Table 23-2 . In a responsive trauma patient who has an isolated injury with a limited MOI, consider a focused assessment before assessing vital signs and obtaining a history.

When you encounter a patient who is bleeding, it is important to avoid focusing only on the bleeding. With significant trauma, you should assess the entire patient, looking for fractures and other problems. Determine if there are any preexisting illnesses.

SAMPLE History

Obtain a SAMPLE history from your patient. Be sure to ask the patient if he or she takes blood-thinning medications. If so, be aware that bleeding will generally be more profuse and more difficult to control. If the patient is unresponsive, obtain history information from medical alert tags or ask bystanders if they have any information.

Look for signs and symptoms of shock (hypoperfusion) and determine how much blood has been lost.

Secondary Assessment

As described earlier, the secondary assessment is a detailed, comprehensive examination of the patient to uncover injuries or illness that may have been missed during the primary assessment. The EMT should record vital signs, complete a focused assessment of pain, and attach appropriate monitoring devices. In some instances, such as a critically injured patient or a short transport time, there may not be time to conduct a secondary assessment.

Physical Examinations

When performing a secondary assessment, the examination should include a systematic full body scan. Assess the respiratory system. Specifically assess the airway for patency and determine the rate and quality of respirations. In the neck, look for distended neck veins and a deviated trachea. In the chest, check for paradoxical movement of the chest wall and bilateral breath sounds.

Table 23-2   The Mechanism of Injury: Indicators of Internal Bleeding

Mechanism of Injury	Potential Internal Bleeding Sources
Fall from a ladder striking the head	Head injury or hematoma
Fall from a ladder striking the extremities	Possible fractures; consider chest injury
Child struck by a car	Head trauma, chest and abdominal injuries, leg fractures
Fall on an outstretched arm	Possible broken bone or joint injury
Child thrown or falls from a height	Children usually have a head-first impact, causing head injury
Unrestrained driver in head-on collision	Head and neck, chest, abdomen injuries Knees, femur, hip, and pelvis injuries
Unrestrained front-seat passenger, side impact collision with intrusion into vehicle	Humerus broken exposing the chest wall (possible flail chest); pelvis and acetabulum injuries
Unrestrained driver crushed against steering column	Chest and abdomen injuries, ruptured spleen, neck trauma
Road bike or mountain bike (over the handlebars)	Fractured clavicle, road rash, head trauma if no helmet
Abrupt motorcycle stop, causing rider to catapult over the handlebars	Fractured femurs, head and neck injuries
Diving into the shallow end of a swimming pool	Head and neck injuries
Assault or fight	Punching or kicking injury to chest, abdomen, and the face
Blast or explosion	Injury from direct strike with debris; indirect and pressure wave in enclosed space. External injuries are dependent upon the anatomic area of the body injured. Internally, air-containing organs such as the middle of the ears and lungs are the most susceptible to injury.

Assess the cardiovascular system, specifically the rate and quality of pulses.

Assess the neurologic system to formulate baseline data to guide further decisions. This examination should include level of consciousness, pupil size and reactivity, motor response, and sensory response.

Assess the musculoskeletal system. Perform a detailed full body examination. Look for DCAP-BTLS to be sure that you have found all of the problems and injuries quickly.

Assess all anatomic regions. When you are examining the head, be alert for raccoon eyes, Battle's sign, and/or drainage of blood or fluid from the ears or nose. In the abdomen, feel all four quadrants for tenderness or rigidity. In the extremities, record pulse, motor, and sensory function.

Vital Signs

You must assess baseline vital signs to observe the changes that may occur during treatment. A systolic blood pressure of less than 100 mm Hg with a weak, rapid pulse should suggest to you the presence of hypoperfusion in a patient who may have significant bleeding. Cool, moist skin that is pale or gray is an important sign that the patient is experiencing a perfusion problem. Because infants and children have less blood volume to begin with, the same effect is seen with smaller amounts of blood loss.

In geriatric patients, the pulse rate may not increase with early shock; therefore, if possible, try to determine the patient's normal baseline blood pressure and circulatory status.

Monitoring Devices

In addition to hands-on assessment, the EMT should use monitoring devices to quantify oxygenation and circulatory status. The EMT may use a noninvasive technique to monitor blood pressure and a pulse oximeter to evaluate the effectiveness of oxygenation. It is recommended that the EMT always assess the patient's blood pressure with a sphygmomanometer and stethoscope (manually) before using a noninvasive blood pressure monitor to establish a baseline blood pressure and to determine the accuracy of the noninvasive blood pressure machine.

Reassessment

The reassessment is an important tool to see how your patient is doing over time. Reassess the patient, especially in the areas that showed abnormal findings during the primary assessment. The signs and symptoms of internal bleeding are often slow to present because of their covert nature. Children especially will compensate well for blood loss and then "crash" quickly. The reassessment is your best opportunity to determine whether your patient's condition is improving or getting worse. Assess the effectiveness of any interventions and treatments provided to the patient.

Vital signs show how well your patient is doing internally. In all cases of severe bleeding, obtain the patient's vital signs every 5 minutes. Is the patient's airway still patent and breathing still adequate? Is the oxygen helping the patient to breathe easier? Is your treatment for shock resulting in better perfusion of the vital organs? Is the bandage controlling the bleeding?

Interventions

Whenever you suspect significant bleeding, either external or internal, provide high-flow oxygen. If significant bleeding is visible, begin the steps to control external bleeding, as shown in Skill Drill 23-1. Using multiple methods to control external bleeding usually works best. If the patient has signs of hypoperfusion, provide aggressive treatment for shock and rapid transport to the appropriate hospital. If internal bleeding is suspected, apply high-flow oxygen via a nonrebreathing mask and provide rapid transport to the hospital. See Skill Drill 23-4 for additional steps to take.

You should not delay transport of a patient to complete an assessment, particularly when significant bleeding is present, even if the bleeding is controlled. The assessment can be started during transport.

Communication and Documentation

In patients with severe external bleeding, it is important to recognize, estimate, and report the amount of blood loss that has occurred and how rapidly or over what period of time it occurred. This can be a challenge to estimate, especially if the surface the patient is on is wet or absorbs fluids or if the environment is dark. For example, you may report that approximately one quart of blood was lost or that the bleeding soaked through three trauma dressings. Report this information to hospital personnel during transport to allow the hospital to evaluate needed resources, such as the availability of surgical suites, surgeons, and other specialty providers. Your transfer report at the hospital should update hospital personnel on how your patient has responded to your care. Be sure your paperwork reflects all of the patient's injuries and the care you have provided.

With internal bleeding, describe the MOI/NOI and the signs and symptoms that make you think internal bleeding is occurring. Report this information to the emergency department personnel to allow them to prepare to treat the patient on arrival. Communicate with the hospital on your findings and the interventions used to improve the patient's condition. Be sure to document all of the patient's injuries, the care provided, and the patient's response to the care. Give the information to emergency department personnel.

Emergency Medical Care for External Bleeding

As you begin to care for a patient with obvious external bleeding, remember to follow standard precautions. This includes, at the very least, gloves and eye protection and often a mask and possibly a gown. As with all patient care, make sure that the patient has an open airway and is breathing adequately. Provide high-flow oxygen to the patient. You may then concentrate on controlling the bleeding. In some cases, obvious life-threatening bleeding may be present and should be addressed as an immediate life threat and controlled as quickly as possible.

Several methods are available to control external bleeding. Start with the most commonly used; these include the following:

- Direct, even pressure and elevation
- Pressure dressings and/or splints
- Tourniquets

It will often be useful to combine these methods.

Most cases of external bleeding can be controlled simply by applying direct local pressure to the bleeding site. This method is by far the most effective way to control external bleeding. Pressure stops the flow of blood and permits normal coagulation to occur. You may apply pressure with your gloved fingertip or hand over the top of a sterile dressing if one is immediately available. If there is an object protruding from the wound, apply bulky dressings to stabilize the object in place, and apply pressure as best you can. Never remove an impaled object from a wound. Hold uninterrupted pressure for at least 5 minutes.

Elevate a bleeding extremity by as little as 6″ while applying direct pressure. In most cases, the combination of direct pressure and elevation will stop the bleeding. Remember to never elevate an open fracture to control bleeding. Fractures can be elevated after splinting, and splinting helps control bleeding.

Once you have applied a dressing to control bleeding, create a pressure dressing to maintain the pressure by firmly wrapping a sterile, self-adhering roller bandage around the entire wound. Use 4″ × 4″ sterile gauze pads for small wounds and sterile universal dressings for larger wounds.

Cover the entire dressing above and below the wound. Stretch the bandage tight enough to control bleeding. If you were able to palpate a distal pulse before applying the dressing, you should still be able to palpate a distal pulse on the injured extremity after applying the pressure dressing. If bleeding continues, the dressing is probably not tight enough. Do not remove a dressing until a physician has evaluated the patient. Instead, apply additional manual pressure through the dressing. Then add more gauze pads over the first dressing, and secure them both with a second, tighter roller bandage.

Bleeding will almost always stop when the pressure of the dressing exceeds arterial pressure. This will assist in controlling bleeding and helping blood to clot.

If direct pressure fails to immediately stop hemorrhage, apply a tourniquet above the level of the bleeding.

You are the Provider: PART 3

Bleeding from the patient's injury has been controlled. While you further assess the patient, your partner applies high-flow oxygen, obtains the patient's vital signs, and inquires about his past medical history. The patient denies having any medical problems and states that he does not take any medications.

Recording Time: 5 Minutes	
Respirations	24 breaths/min; regular and adequate
Pulse	120 beats/min; regular and strong
Skin	Cool, pale, and dry
Blood pressure	104/60 mm Hg
Oxygen saturation (Sao$_2$)	95% (on oxygen)

5. What are the components of the cardiovascular system?

6. What factors determine the severity of external bleeding?

If this is not possible because the bleeding is too far proximal, apply direct pressure and hold it until you arrive at the hospital.

Skill Drill 23-1 illustrates the basic techniques to control external bleeding:

Skill Drill 23-1

1. Follow standard precautions.
2. Maintain the airway with cervical spine immobilization if the mechanism of injury suggests the possibility of spinal injury.

3. Apply direct pressure over the wound with a dry, sterile dressing **Step 1**.
4. Apply a pressure dressing **Step 2**.
5. If direct pressure and a pressure dressing are not immediately effective, apply a tourniquet above the level of the bleeding **Step 3**.
6. Apply high-flow oxygen as necessary, once hemorrhage is controlled.

Much of the bleeding associated with broken bones occurs because the sharp ends of the bones cut muscles and other tissues. As long as a fracture remains unstable, the bone ends will move and continue to injure partially clotted vessels. Therefore, stabilizing a

Skill Drill 23-1

Controlling External Bleeding

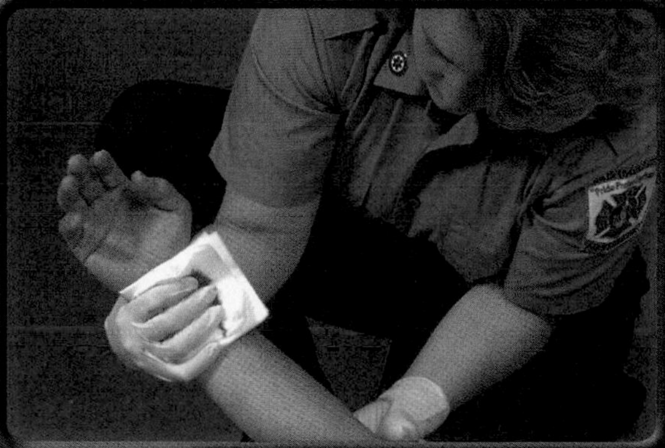

Step 1 Apply direct pressure over the wound with a dry, sterile dressing. Elevate the injury above the level of the heart if no fracture is suspected.

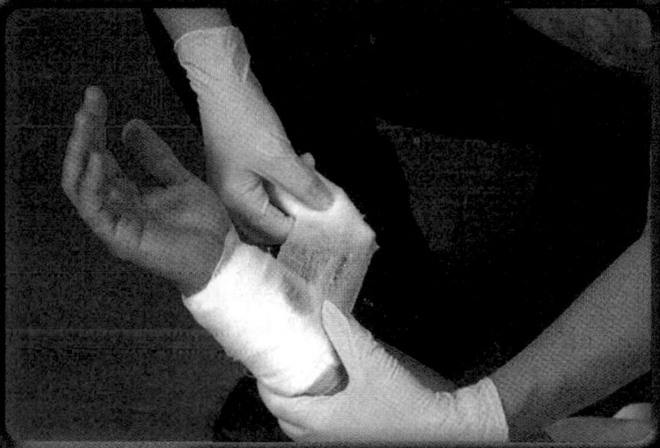

Step 2 Apply a pressure dressing.

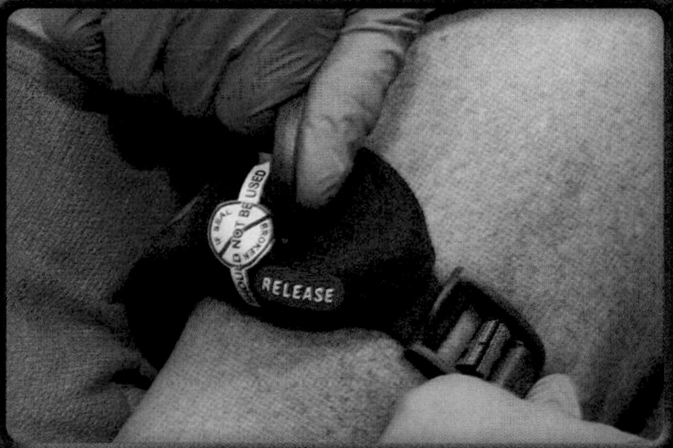

Step 3 If direct pressure with a pressure dressing does not control bleeding, apply a tourniquet above the level of the bleeding.

fracture and decreasing movement is a high priority in the prompt control of bleeding. Often, simple splints will quickly control bleeding associated with a fracture **Figure 23-8**. If not, you may need to use another splinting device, such as an air splint or a tourniquet, discussed next.

Recent studies have brought into question the effectiveness of using pressure points in severe external hemorrhage. If allowed by local protocol and policy, you should move to the use of a **tourniquet** without attempting pressure point control. If a tourniquet is deemed necessary, it should be applied quickly and not released until a physician is present.

■ Tourniquets

The tourniquet is especially useful if a patient has substantial bleeding from an extremity injury below the

Figure 23-8 Use of a simple splint will often control bleeding associated with a fracture. As long as a fracture is not immobilized, the bone ends are free to move and may continue to injure partially clotted vessels.

Words of Wisdom

Historically, if direct pressure and elevation proved ineffective, EMS providers were advised to apply pressure to a proximal arterial pressure point. A **pressure point** is a spot where a blood vessel lies near a bone. This technique should be considered interesting from a historic perspective only. Because a wound usually draws blood from more than one major artery, proximal compression of a major artery rarely stops bleeding completely. In rare cases, it may help to slow the loss of blood. You would need to be thoroughly familiar with the location of the pressure points for this to work **Figure 23-9**. Even if you are familiar, there is no real evidence that this is an effective or safe method to control potentially fatal hemorrhage. If the patient has an open fracture of an extremity, bleeding can be substantial. Consider a tourniquet early if bleeding is not easily controlled with direct pressure or if pressure results in excessive pain. The method used to control severe external bleeding may be governed by local protocol; regardless of the method, it must be quick and effective. Remember that uncontrolled bleeding may result in shock and death. Patients can and do bleed to death from extremity injuries. It is imperative that you use effective techniques to stop bleeding when you encounter it.

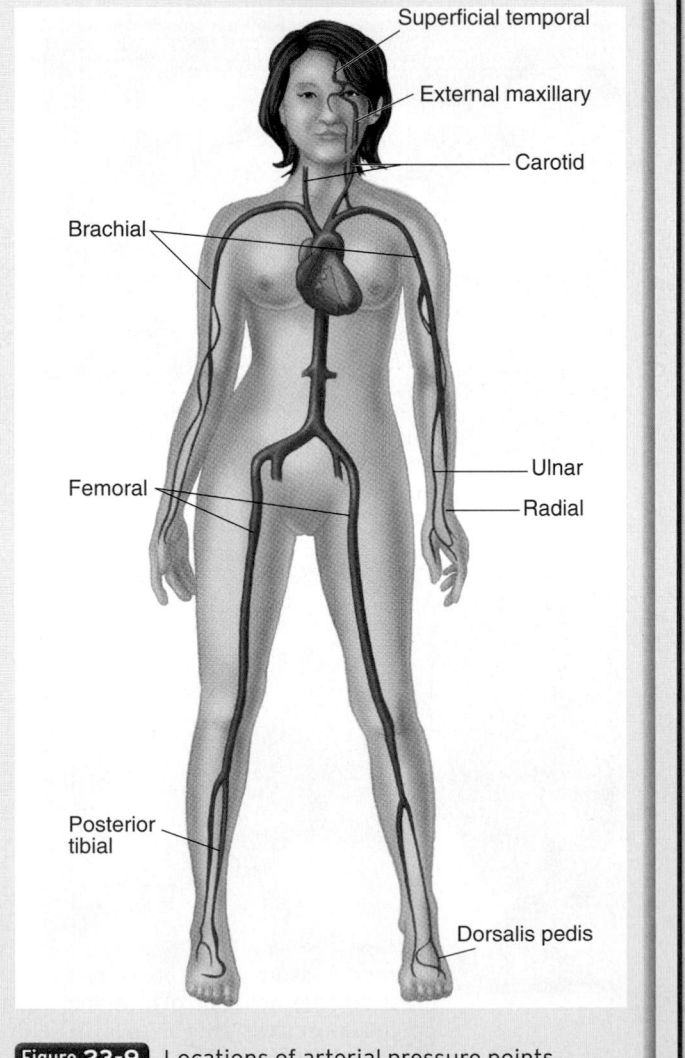

Figure 23-9 Locations of arterial pressure points.

Words of Wisdom

Hemostatic agents such as Celox, HemCon, and Quik-Clot are primarily utilized in the military to promote hemostasis or, in other words, to stop profuse bleeding. The agent may be granules poured into a wound or contained in a dressing. The agent absorbs the water component of blood thereby concentrating the clotting factors, activating platelets, and enhancing the coagulation cascade. Some of these agents have an exothermic affect that can damage the surrounding tissue.

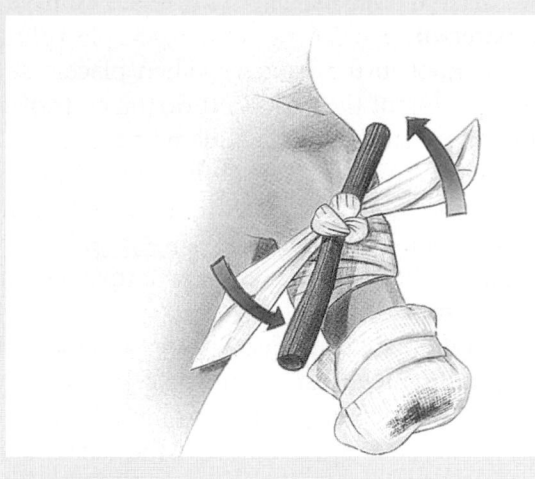

Figure 23-10 Twist the stick or rod to tighten the tourniquet until the bleeding has stopped; then stop twisting.

axilla or groin. Follow the steps in **Skill Drill 23-2** to apply a commercial tourniquet.

Skill Drill 23-2

1. Follow standard precautions.
2. Hold direct pressure over the bleeding site.
3. Place the tourniquet around the extremity just above the bleeding site **Step 1**.
4. Click the buckle into place and pull the strap tight.
5. Turn the tightening dial clockwise until pulses are no longer palpable distal to the tourniquet or until bleeding has been controlled **Step 2**.

6. To release the tourniquet at the hospital, or if otherwise instructed by medical control, push the release button and pull the strap back. Be aware that bleeding may rapidly return upon tourniquet release and that you should be prepared to reapply it immediately if necessary.

If a commercial tourniquet is not available, follow these steps to apply a tourniquet using a triangular bandage and a stick or rod:

1. Fold a triangular bandage until it is 4" wide and six to eight layers thick.
2. Wrap the bandage around the extremity twice. Choose an area only slightly proximal to the

Skill Drill 23-2

Applying a Commercial Tourniquet

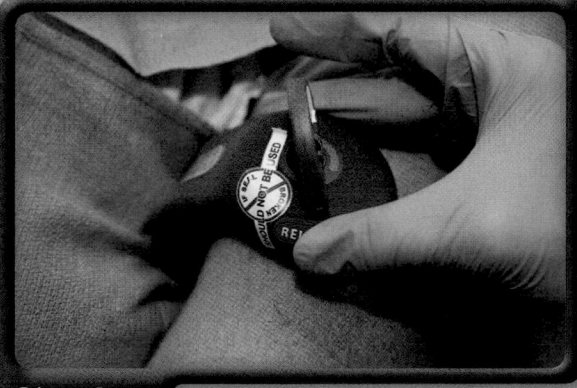

Step 1 Hold pressure over the bleeding site and place the tourniquet just above the injury.

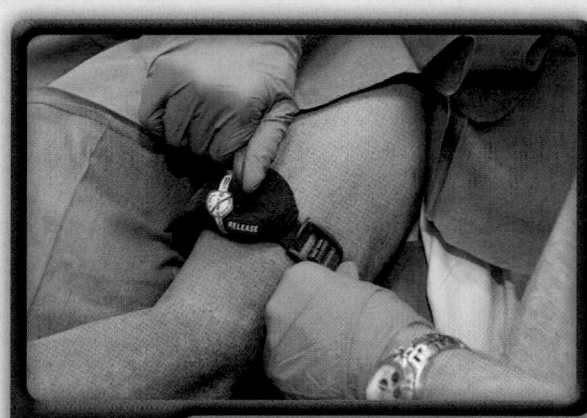

Step 2 Click the buckle into place, pull the strap tight, and turn the tightening dial clockwise until pulses are no longer palpable distal to the tourniquet or until bleeding has been controlled.

bleeding to reduce the amount of tissue damage to the extremity.

3. Tie one knot in the bandage. Then place a stick or rod on top of the knot, and tie the ends of the bandage over the stick in a square knot.

4. Use the stick or rod as a handle, and twist it to tighten the tourniquet until the bleeding has stopped; then stop twisting **Figure 23-10**.

5. Secure the stick in place, and make the wrapping neat and smooth.

6. Write "TK" (for "tourniquet") and the exact time (hour and minute) that you applied the tourniquet on a piece of adhesive tape. Use the phrase "time applied." Securely fasten the tape to the patient's forehead. Notify hospital personnel on your arrival that your patient has a tourniquet in place. Record this same information on the ambulance run report form.

7. As an alternative, you can use a blood pressure cuff as an effective tourniquet. Position the cuff proximal to the bleeding point, and inflate it just enough to stop the bleeding. Leave the cuff inflated. If you use a blood pressure cuff, monitor the gauge continuously to make sure that the pressure is not gradually dropping. You may have to clamp the tube with a hemostat leading from the cuff to the inflating bulb to prevent loss of pressure.

Whenever you apply a tourniquet, make sure you observe the following precautions:

- Do not apply a tourniquet directly over any joint. Keep it as close to the injury as possible.
- Make sure the tourniquet is tightened securely.

- Never use wire, rope, a belt, or any other narrow material. It could cut into the skin.
- Use wide padding under the tourniquet if possible. This will protect the tissues and help with arterial compression.
- Never cover a tourniquet with a bandage. Leave it open and in full view.
- Do not loosen the tourniquet after you have applied it. Hospital personnel will loosen it once they are prepared to manage the bleeding.

■ Splints

Air splints can control internal or external bleeding associated with severe injuries, such as fractures **Figure 23-11**. They also stabilize the fracture itself. An air splint acts like a pressure dressing applied to an entire extremity rather than to a small, local area. Air splints are also commonly referred to as soft splints or pressure splints. Once you have applied an air splint, be sure to monitor circulation in the distal extremity. Use only

Words of Wisdom

Research indicates that a pelvic compression device is an effective method to reduce the width of pelvic ring fractures. Overcompression has not been identified as an issue to date. The decrease in the width of the fracture will assist in the control of internal bleeding resulting from the fracture, specifically an open book fracture of the pelvis.

You are the Provider: PART 4

The patient is placed onto the stretcher and loaded into the ambulance. He remains conscious and alert, but is still anxious. You place him in a supine position, elevate his legs, and cover him with a blanket. Shortly before departing the scene, you reassess him and obtain another set of vital signs.

Recording Time: 10 Minutes	
Level of consciousness	Conscious and alert; anxious
Respirations	24 breaths/min; regular and adequate
Pulse	116 beats/min; strong and regular
Skin	Cool, pale, and dry
Blood pressure	112/70 mm Hg
Sao$_2$	98% (on oxygen)

7. How might a patient's outcome be affected if bleeding is internal rather than external?

8. What are the signs and symptoms of internal bleeding?

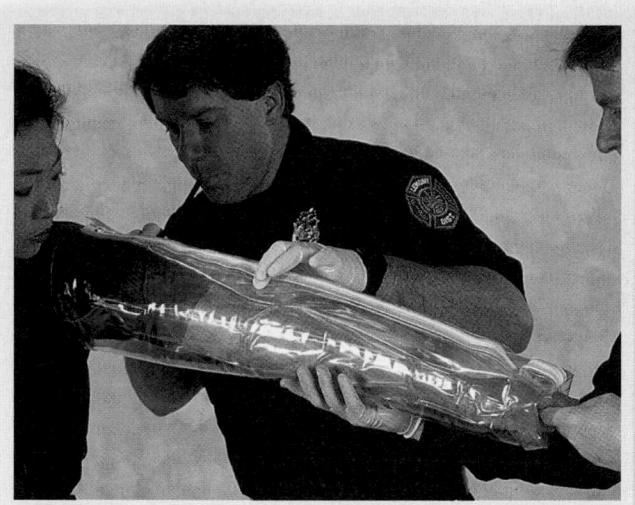

Figure 23-11 Air splints can also be used to control bleeding because they act as a pressure bandage for the entire extremity.

approved, clean, or disposable valve stems when orally inflating air splints.

Rigid splints can help stabilize fractures as well as reduce pain and prevent further damage to soft-tissue injuries. Once you have applied a rigid splint, be sure to monitor circulation in the distal extremity.

Traction splints are designed to stabilize femur fractures. When the EMT pulls traction to the ankle, counter-traction is applied to the ischium and groin. This reduces the thigh muscle spasms and prevents one end of the fracture from impacting or overriding the other. Be sure to pad these areas well to prevent applying excessive pressure to the soft tissue of the pelvis. Once you have applied a traction splint, be sure to monitor circulation in the distal extremity.

■ Bleeding From the Nose, Ears, and Mouth

Several conditions can result in bleeding from the nose, ears, and/or mouth, including the following:

- Skull fracture
- Facial injuries, including those caused by a direct blow to the nose
- Sinusitis, infections, nose drop use and abuse, dried or cracked nasal mucosa, or other abnormalities
- High blood pressure
- Coagulation disorders
- Digital trauma (nose picking)

Epistaxis, or nosebleed, is a common emergency. Occasionally, it can cause enough of a blood loss to send a patient into shock. Keep in mind that the blood you see may be only a small part of the total blood loss. Much

of the blood may pass down the throat into the stomach as the patient swallows. A person who swallows a large amount of blood may become nauseated and start vomiting the blood, which is sometimes confused with internal bleeding. Most nontraumatic nosebleeds occur from sites in the septum, the tissue dividing the nostrils. You can usually handle this type of bleeding effectively by pinching the nostrils together. **Skill Drill 23-3** illustrates the basic techniques to control epistaxis.

1. Follow standard precautions.
2. Help the patient to sit, leaning forward, with the head tilted forward. This position stops the blood from trickling down the throat or being aspirated into the lungs.
3. Apply direct pressure for at least 15 minutes by pinching the fleshy part of the nostrils together. This is the preferred method. This technique may also be self-administered by the patient **(Step 1)**.
4. Placing a rolled 4″ × 4″ gauze bandage between the upper lip and the gum is another option. Have the patient apply pressure by stretching the upper lip tightly against the rolled bandage and pushing it up into and against the nose. If the patient is unable to do this effectively, use your gloved fingers to press the gauze against the gum **(Step 2)**.
5. Keep the patient calm and quiet, especially if he or she has high blood pressure or is anxious. Anxiety tends to increase blood pressure, which could worsen the nosebleed.
6. Apply ice over the nose.
7. Maintain the pressure until the bleeding is completely controlled, usually no more than 15 minutes (assuming that this is the patient's only problem). Most often, failure to stop a nosebleed is the result of releasing the pressure too soon **(Step 3)**.
8. Provide prompt transport once the bleeding has stopped.
9. If you cannot control the bleeding, if the patient has a history of frequent nosebleeds, or if there is a significant amount of blood loss, transport the patient immediately. Assess the patient for signs and symptoms of shock. Treat appropriately for shock, and administer oxygen via mask, if necessary.

Bleeding from the nose or ears following a head injury may indicate a skull fracture. In these cases, you should not attempt to stop the blood flow. This bleeding may be difficult to control. Applying excessive pressure to the injury may force the blood leaking through the ear or nose to collect within the head. This could increase the pressure on the brain and possibly cause permanent damage. If you suspect a skull fracture, loosely cover the bleeding site with a sterile gauze pad to collect the blood and help keep contaminants away from the site. There is always a risk of infection to the brain. Apply light compression by wrapping the dressing loosely around the head **Figure 23-12**. If blood or drainage contains cerebrospinal fluid, a characteristic staining of the dressing, much like a target or halo, will occur **Figure 23-13**.

■ Emergency Medical Care for Internal Bleeding

Controlling internal bleeding or bleeding from major organs usually requires surgery or other procedures that must be done in the hospital. It is important for you to remain calm and reassure the patient. Keeping the patient as still and quiet as possible assists the body's clotting process. Next, if spinal injury is not suspected, place the patient in the shock position. Provide high-flow oxygen;

Skill Drill 23-3

Controlling Epistaxis

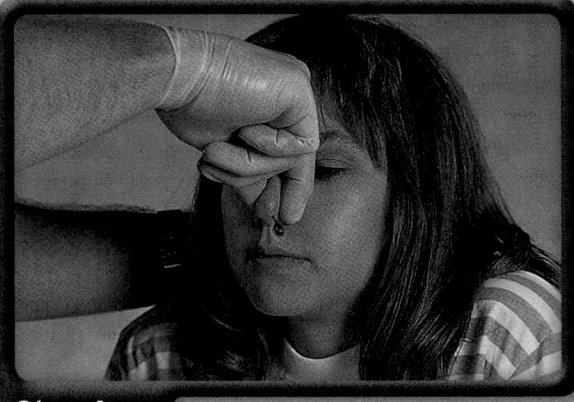

Step 1 Position the patient sitting, leaning forward. Apply direct pressure, pinching the fleshy part of the nostrils together.

Step 2 Alternative method: Use pressure with a rolled gauze bandage between the upper lip and gum. Calm the patient.

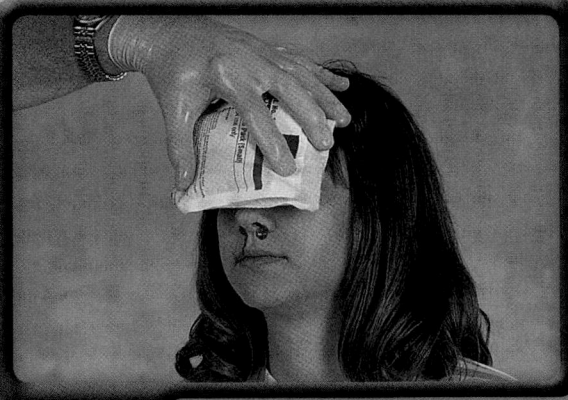

Step 3 Apply ice over the nose. Maintain pressure until bleeding is controlled. Provide prompt transport after bleeding stops. Transport immediately if indicated. Assess and treat for shock, including oxygen, as needed.

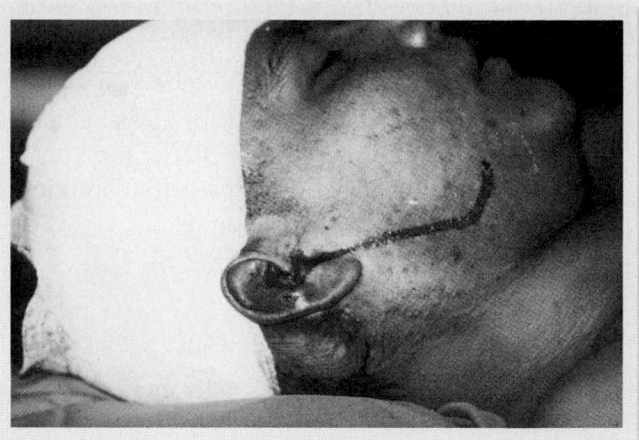

Figure 23-12 Bleeding from the ear after a head injury may indicate a skull fracture. Loosely cover the bleeding site with a sterile gauze pad, and apply light compression by wrapping the dressing loosely around the head.

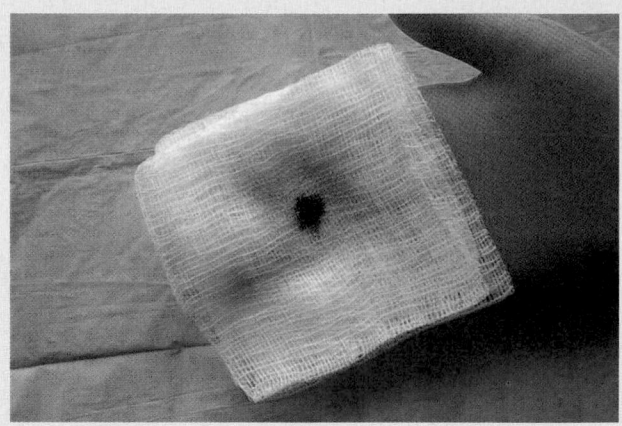

Figure 23-13 When cerebrospinal fluid is present in blood or drainage, a stain in the shape of a target or halo will appear.

also maintain body temperature. You can usually control internal bleeding into the extremities quite well in the field simply by splinting the extremity, usually most effectively with an air splint, and you should never use a tourniquet to control the bleeding from closed, internal, soft-tissue injuries. Follow the steps in **Skill Drill 23-4** to care for patients with possible internal bleeding.

1. Follow standard precautions.
2. Maintain the airway with cervical spine immobilization if a mechanism of injury suggests the possibility of spinal injury.
3. Administer high-flow oxygen and provide artificial ventilation as necessary. (Step 1)

You are the Provider: PART 5

You continue to monitor the patient en route to the hospital and reassess his condition as appropriate. After reassessing the patient and his vital signs, you call your radio report into the receiving facility.

Recording Time: 17 Minutes	
Level of consciousness	Conscious and alert; restless
Respirations	20 breaths/min; regular and adequate
Pulse	110 beats/min; strong and regular
Skin	Cool, pale, and dry
Blood pressure	114/68 mm Hg
Sao$_2$	97% (on oxygen)

The patient is delivered to the hospital and you give your report to the attending physician. An intravenous line is started, the patient is given normal saline to improve his perfusion status, and he is admitted for observation.

9. How does the body typically respond to blood loss?

4. Control all obvious external bleeding.

5. Treat suspected internal bleeding in an extremity by applying a splint. (Step 2)

6. Monitor and record the vital signs at least every 5 minutes. (Step 3)

7. Give the patient nothing (not even small sips of water) by mouth.

8. Elevate the legs 6″ to 12″ in nontrauma patients to help the blood return to the vital organs.

9. Keep the patient warm.

10. Provide immediate transport for all patients with signs and symptoms of shock (hypoperfusion). Report any changes in the patient's condition to emergency department personnel.

Skill Drill 23-4

Controlling Internal Bleeding

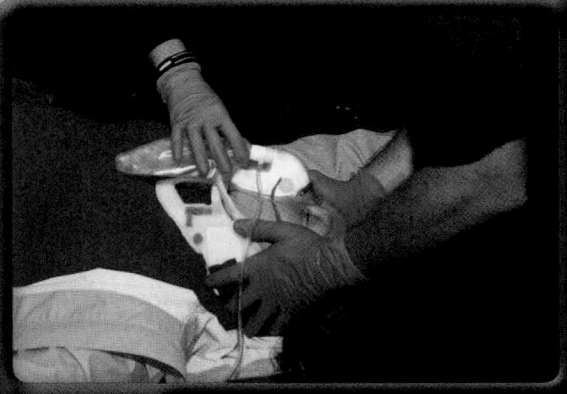

Step 1 Follow standard precautions. Maintain the airway and be alert for cervical spine injury. Administer oxygen and provide ventilation as necessary.

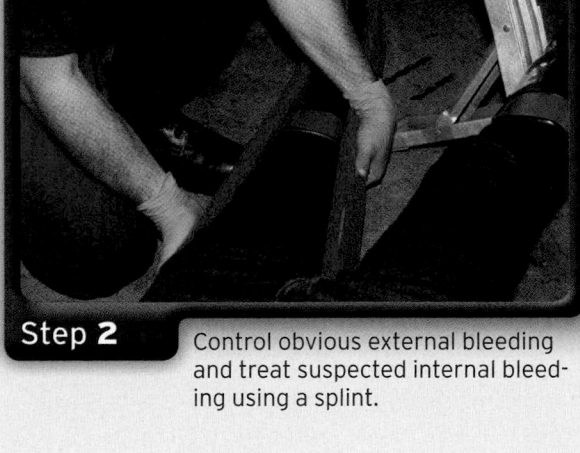

Step 2 Control obvious external bleeding and treat suspected internal bleeding using a splint.

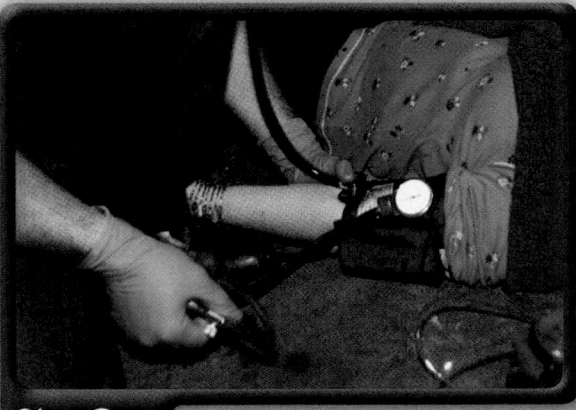

Step 3 Monitor vital signs and elevate the legs 6″ to 12″ in nontrauma patients to ensure blood is flowing to vital organs.

You are the Provider: SUMMARY

1. What are the functions of arteries? What major arteries are located in the upper extremity?

Arteries are high-pressure blood vessels that distribute oxygenated blood throughout the body. The largest artery in the body, the aorta, arises from the left ventricle and branches into smaller arteries and arterioles that deliver oxygen to the body's tissues and cells. In general, arteries carry highly oxygenated blood away from the heart; an exception to this is the pulmonary artery, which carries *deoxygenated* blood from the right ventricle to the lungs where it is reoxygenated.

Two major arteries are located in the upper extremity, the radial artery, which is located on the thumb-side (lateral) aspect of the wrist, proximal to the hand, and the brachial artery, which is located on the inner (medial) aspect of the arm, just proximal to the elbow.

2. Why is arterial bleeding more severe than venous bleeding?

Blood flow through the arteries is driven by contraction of the powerful left ventricle. Pressure in the arteries is much higher than pressure in the veins (high capacitance, low-pressure blood vessels that return deoxygenated blood to the heart).

Because blood flow through the arteries is much higher, blood loss is generally more rapid and severe. Arterial bleeding is also more difficult to control than venous bleeding. Oxygen loss is more severe from arterial bleeding than it is from venous bleeding; this is because arterial blood carries a higher concentration of oxygen than do the veins.

The color of blood and characteristic of the bleeding are often clues to the type of blood vessel that is injured. Venous blood is dark red and flows from the injury site, whereas arterial blood is bright red and spurts from the wound each time the left ventricle contracts.

3. Is the patient effectively controlling the bleeding from his injury?

As evidenced by the blood-soaked towel and large amount of blood on the ground, it is clear that the patient is *not* effectively controlling the bleeding from his injury. Furthermore, you do not know how much blood he has lost because he is standing outside—not in the area where the injury occurred. The fact that he is anxious and has cool, pale skin suggests significant external blood loss.

4. What should be your initial treatment priority?

You must take *immediate* action to control the patient's bleeding. His airway is patent, as evidenced by the fact that he is conscious, alert, and talking.

In most cases, direct pressure will control both venous and arterial bleeding. However, if direct pressure alone is ineffective, continued direct pressure and elevation of the extremity above the level of the heart typically controls the bleeding.

If direct pressure and elevation are ineffective in immediately controlling severe external bleeding, apply a tourniquet and transport. Only after the tourniquet is applied should the EMTs administer oxygen and consider splinting the arm.

5. What are the components of the cardiovascular system?

The cardiovascular system—the system responsible for supplying and maintaining adequate blood flow to the body's tissues and cells—consists of three components: the heart (pump), the container (the blood vessels), and the fluid (blood and body fluids). These components of the cardiovascular system are interdependent—that is, they rely on each other mutually to perform a common function.

6. What factors determine the severity of external bleeding?

Several factors determine the severity of external bleeding. The single most influential factor is the type and size of the blood vessel that is injured. A lacerated brachial artery, for example, will bleed more severely than a small vein in the leg. As previously discussed, arteries are under high pressure, while veins are under low pressure.

How the vessel is injured is also a determining factor in the severity of the bleeding and the difficulty in controlling it. A longitudinal laceration—one that extends in the direction of the length of the blood vessel—usually bleeds more profusely and is more difficult to control than a transverse laceration—one that is directly across the blood vessel.

The patient's blood pressure and heart rate can also affect the severity of external bleeding. For example, a patient with a blood pressure of 190/90 mm Hg and a heart rate of 120 beats/min would likely bleed more profusely than a patient with a blood pressure of 70/40 mm Hg and a heart rate of 50 beats/min. The greater the pressure on the arterial wall and the faster the heart rate, the more rapid the bleeding tends to be.

Certain aspects of a patient's medical history also can impact the severity of external bleeding. For example, patients who take blood-thinning medications (ie, warfarin [Coumadin]) or those with a bleeding disorder (ie, hemophilia) tend to bleed faster because it takes longer for their blood to clot. For this and other reasons, it is important to obtain an accurate medical history from the patient.

7. How might a patient's outcome be affected if bleeding is internal rather than external?

Internal bleeding is hidden and cannot be controlled in the prehospital setting. Many patients with internal bleeding do not present with signs or symptoms of shock until a significant amount of blood has been lost.

Overall, patients with internal bleeding have a higher mortality rate than those with external bleeding. Most of these deaths are the result of intrathoracic or intra-abdominal bleeding in which surgical intervention is delayed. Internal bleeding can also be caused by multiple long bone fractures and pelvic fractures.

You must always be alert to the possibility of internal bleeding and assess the patient for related signs and symptoms, particularly if the mechanism of injury is significant. Remember this: if a trauma patient is in shock but does not have any obvious external signs of injury, suspect internal bleeding!

You are the Provider: SUMMARY, continued

8. What are the signs and symptoms of internal bleeding?

Since internal bleeding cannot be seen outright, you must rely on your assessment skills and careful evaluation of the mechanism of injury. Signs and symptoms of internal bleeding are essentially those of shock: restlessness or anxiety; cool, pale, clammy skin; tachycardia; rapid, shallow breathing; thirst; and as a late sign, hypotension.

External indicators of internal bleeding in both medical and trauma patients include hematemesis (vomiting blood), melena (dark, tarry stools), and hemoptysis (coughing up blood). Other indicators of internal bleeding, which are more common in trauma patients, include redness or bruising, swelling, or tenderness over the injured area.

increasing heart and respiratory rates, and as a later sign, a decreasing blood pressure.

A loss of circulating blood volume is sensed by receptors in the body, which send messages to the nervous system. The nervous system, specifically, the sympathetic nervous system, releases epinephrine and norepinephrine. Norepinephrine constricts the peripheral blood vessels (vasoconstriction), thus shunting blood from areas of lesser need (ie, skin and muscles) to areas of greater need (ie, heart, brain, kidneys, liver). If blood loss continues, however, the body's compensatory mechanisms will eventually fail, the patient's blood pressure will fall, and he or she will die.

9. How does the body typically respond to blood loss?

If the typical adult loses more than 1 L of blood (about 2 pints), significant changes in vital signs will occur, including

---

EMS Patient Care Report (PCR)

Date: 6-30-09	Incident No.: 220109	Nature of Call: Laceration		Location: 517 E. Graham
Dispatched: 1620	En Route: 1621	At Scene: 1627	Transport: 1642	At Hospital: 1655	In Service: 1704

Patient Information

Age: 32 Sex: M Weight (in kg [lb]): 82 kg (180 lb)	Allergies: No known drug allergies Medications: None Past Medical History: None Chief Complaint: Laceration to left wrist

Vital Signs

Time: 1637	BP: 104/60	Pulse: 120	Respirations: 24	Sao$_2$: 95%
Time: 1642	BP: 112/70	Pulse: 116	Respirations: 24	Sao$_2$: 98%
Time: 1649	BP: 114/68	Pulse: 110	Respirations: 20	Sao$_2$: 97%

EMS Treatment
(circle all that apply)

Oxygen @ 15 L/min via (circle one): NC (NRM) Bag-Mask Device	Assisted Ventilation	Airway Adjunct	CPR
Defibrillation	(Bleeding Control)	(Bandaging)	Splinting	(Other) Shock treatment

Narrative

Dispatched for a patient with severe bleeding from the arm. Arrived on scene to find the patient, a 32-year-old male, standing in front of his place of employment, a woodworking shop. He was conscious and alert, but notably anxious. His airway was patent and his breathing, although increased, was producing adequate tidal volume. Patient had blood-soaked towel wrapped around his left wrist and an impressive amount of blood was on the ground where he was standing. Patient stated that his hand slipped while he was working with a table saw and his left wrist ran across the blade. Immediately applied direct pressure to patient's wrist with sterile dressing and elevated his left arm. This intervention successfully controlled the bleeding; a pressure dressing was then applied to maintain bleeding control. Applied oxygen at 15 L/min via nonrebreathing mask and obtained vital signs, as noted above. Further assessment revealed that patient's skin was cool, pale, and dry. Patient denied significant past medical history and further denied taking any medications. Placed patient onto stretcher, covered him with a blanket, elevated his lower extremities, and placed him into the ambulance. Reassessed patient's vital signs and began transport to the hospital. Continued to monitor patient's condition en route; he remained conscious and alert, although anxious, and his vital signs remained stable. Reassessed bandaged wound and noted that the bleeding remained controlled. Called report to receiving facility to inform them of our arrival. Delivered patient to hospital without incident. Verbal report given to charge nurse. **End of report**

Assessment and Emergency Care of Internal and External Bleeding

	External Bleeding	Internal Bleeding
Scene Size-up		
Scene Safety	Ensure scene safety. If incident involved violence, ensure that police are on scene. Consider if additional resources are needed. Wear at least gloves and eye protection to protect from bleeding.	Ensure scene safety. Consider if additional resources are needed. Follow standard precautions.
Mechanism of Injury/Nature of Illness	Determine the MOI/NOI.	High-energy MOI should increase your index of suspicion for possible internal bleeding.
Primary Assessment		
Form General Impression	Check for responsiveness and look for blood stains or other obvious signs of external bleeding. Assess skin color. Manage significant visible bleeding.	Suspect internal bleeding after blunt or penetrating trauma. Determine level of consciousness using AVPU and check the patient's mental status. Assess skin color. Consider the need for manual spinal immobilization.
Airway and Breathing	Ensure a patent airway, look for adequate breathing, and check for breath sounds. If necessary, provide high-flow oxygen or assist ventilation once significant bleeding is controlled.	Ensure a patent airway, look for adequate breathing, and check for breath sounds. If necessary, provide high-flow oxygen or assist ventilation.
Circulation	Assess pulse rate and quality, skin color and temperature, and check capillary refill time. Control external bleeding with direct pressure and elevation, or use of a tourniquet. Treat for shock if needed by applying oxygen, improving circulation, and maintaining normal temperature.	Assess pulse rate and quality, skin color and temperature, and check capillary refill time. Treat the patient for shock if needed by applying oxygen, improving circulation, and maintaining normal temperature.
Transport Decision	Transport quickly if breathing problem or significant bleeding exists.	If you suspect internal bleeding or signs of shock are present, promptly transport to the hospital.
History Taking		
Investigate Chief Complaint	Ask the patient about the chief complaint, if responsive. Attempt to determine the amount of blood loss.	Ask the patient what happened.

Assessment and Emergency Care of Internal and External Bleeding, continued

	External Bleeding	**Internal Bleeding**
Secondary Assessment		
Physical Examinations	Perform a systematic full-body scan. Assess respiratory, cardiovascular, neurologic, musculoskeletal (using DCAP-BTLS), and anatomic regions.	Perform a systematic full-body scan. Assess respiratory, cardiovascular, neurologic, musculoskeletal (using DCAP-BTLS), and anatomic regions. Look for bruising, pain, abdominal distention, and guarding.
Vital Signs	Assess vital signs. Look for signs of shock: systolic blood pressure less than 100 mm Hg with weak, rapid pulse. Pale or gray, cool, moist skin suggests a perfusion problem.	Assess vital signs. Look for signs of shock: systolic blood pressure less than 100 mm Hg with weak, rapid pulse. Pale or gray, cool, moist skin suggests a perfusion problem.
Reassessment		
Interventions	Repeat the primary assessment and reassess interventions performed. Reassess vital signs and the chief complaint. In cases of severe bleeding, obtain vital signs at least every 5 minutes while providing high-flow oxygen. Control significant bleeding and if signs of shock are present, treat aggressively. Determine whether patient's condition is improving or deteriorating.	Repeat the primary assessment and reassess interventions performed. Internal bleeding is often slow to present. Reassess vital signs and the chief complaint. Provide high-flow oxygen. Determine whether patient's condition is improving or deteriorating.
Communication and Documentation	Report approximate amount of blood lost, how rapidly, and over what period of time. Communicate interventions performed, and how patient has responded to care.	Describe the MOI/NOI and signs and symptoms that make you suspect internal bleeding is occurring. Communicate interventions performed, and how patient has responded to care.

NOTE: Although the steps below are widely accepted, be sure to consult and follow your local protocol.

Emergency Care

External Bleeding

Steps to Caring for Patient With External Bleeding

1. Follow standard precautions—at least gloves and eye protection.
2. Maintain cervical stabilization if MOI suggests possible spinal injury.
3. Administer high-flow oxygen as necessary, once significant bleeding is controlled.
4. Control external bleeding using as many of the following means as necessary:
   - Direct pressure, elevation, and pressure dressings
   - Tourniquets
   - Splints
5. Apply direct local pressure to bleeding site, elevate the bleeding extremity, and apply a pressure dressing.
6. If bleeding is not immediately controlled with the use of direct pressure, apply a tourniquet. Follow local protocol for approved methods of bleeding control.

Assessment and Emergency Care of Internal and External Bleeding, continued

Emergency Care

External Bleeding, continued

Applying a Commercial Tourniquet

1. Follow standard precautions.
2. Hold direct pressure over the bleeding site.
3. Place the tourniquet around the extremity just above the bleeding site.
4. Click the buckle into place and pull the strap tight.
5. Turn the tightening dial clockwise until pulses are no longer palpable distal to the tourniquet or until bleeding is controlled.

Treating Epistaxis

1. Follow standard precautions.
2. Help the patient to sit, leaning forward, with the head tilted forward.
3. Apply direct pressure for at least 15 minutes by pinching nostrils together.
4. Keep the patient calm and quiet.
5. Apply ice over the nose.
6. Maintain the pressure until bleeding is completely controlled.
7. Provide prompt transport.
8. If bleeding cannot be controlled, transport patient immediately. Treat for shock and administer oxygen via mask if necessary.

Internal Bleeding

Steps to Caring for Patient With Internal Bleeding

1. Follow standard precautions.
2. Maintain the airway with cervical immobilization if MOI suggests possible spinal injury.
3. Administer high-flow oxygen and provide artificial ventilation as necessary.
4. Control all obvious external bleeding.
5. Apply a splint to an extremity where internal bleeding is suspected.
6. Monitor and record vital signs at least every 5 minutes.
7. Give the patient nothing by mouth.
8. Elevate the legs 6" to 12" in nontrauma patients.
9. Keep the patient warm.
10. Provide immediate transport for patients with signs and symptoms of shock. Report changes in condition to hospital personnel.

Prep Kit

- Perfusion is the circulation of blood in adequate amounts to meet each cell's current needs for oxygen, nutrients, and waste removal.
- The three arms of the perfusion triad must be functioning to meet this demand: a working pump (heart), a set of intact pipes (blood vessels), and fluid volume (enough oxygen-carrying blood).
- Hypoperfusion, or shock, occurs when one or more of these three arms is not working properly and the cardiovascular system fails to provide adequate perfusion.
- Both internal and external bleeding can cause shock. You must know how to recognize and control both.
- The methods to control bleeding, in order, are:
  - Direct local pressure
  - Elevation
  - Pressure dressing
  - Tourniquet
  - Splinting device
- Bleeding from the nose, ears, and/or mouth may result from a skull fracture. Other causes include high blood pressure and sinus infection. Evaluate the MOI and consider the more serious problem of skull fracture.
- Bleeding around the face always presents a risk for airway obstruction or aspiration. Maintain a clear airway by positioning the patient appropriately and using suction when indicated.
- If bleeding is present at the nose and a skull fracture is suspected, place a gauze pad loosely under the nose.
- If bleeding from the nose is present and a skull fracture is not suspected, pinch both nostrils together for 15 minutes. If the patient is awake and has a patent airway, place a gauze pad inside the upper lip against the gum.
- Any patient you suspect of having internal bleeding or significant external bleeding should be transported promptly.
- If the mechanism of injury is significant, be alert to signs of unseen bleeding in the chest or abdomen—signs such as serious bruising or symptoms such as complaints of difficulty breathing or abdominal pain.
- Signs of serious internal bleeding include the following:
  - Vomiting blood (hematemesis)
  - Black tarry stools (melena)
  - Coughing up blood (hemoptysis)
  - Distended abdomen
  - Broken ribs

aorta The main artery that receives blood from the left ventricle and delivers it to all the other arteries that carry blood to the tissues of the body.

arterioles The smallest branches of arteries leading to the vast network of capillaries.

artery A blood vessel, consisting of three layers of tissue and smooth muscle that carries blood away from the heart.

capillaries The small blood vessels that connect arterioles and venules; various substances pass through capillary walls, into and out of the interstitial fluid, and then on to the cells.

coagulation The formation of clots to plug openings in injured blood vessels and stop blood flow.

contusion A bruise, or ecchymosis.

ecchymosis Discoloration of the skin associated with a closed wound; bruising.

epistaxis A nosebleed.

hematemesis Vomited blood.

hematoma A mass of blood in the soft tissues beneath the skin.

hemophilia A congenital condition in which the patient lacks one or more of the blood's normal clotting factors.

hemoptysis Coughing up blood.

hemorrhage Bleeding.

hypoperfusion A condition that occurs when the level of tissue perfusion decreases below that needed to maintain normal cellular functions; also called shock.

hypovolemic shock A condition in which low blood volume, due to massive internal or external bleeding or extensive loss of body water, results in inadequate perfusion.

melena Black, foul-smelling, tarry stool containing digested blood.

perfusion Circulation of blood within an organ or tissue in adequate amounts to meet the current needs of the cells.

pressure point A point where a blood vessel lies near a bone.

shock A condition in which the circulatory system fails to provide sufficient circulation so that every body part can perform its function; also called hypoperfusion.

tourniquet The bleeding control method used when a wound continues to bleed despite the use of direct pressure and elevation; useful if a patient is bleeding severely from a partial or complete amputation.

vasoconstriction Narrowing of a blood vessel, such as with hypoperfusion or cold extremeties.

veins The blood vessels that carry blood from the tissues to the heart.

venules Very small, thin-walled vessels.

Assessment in Action

Your unit is dispatched to a roadside construction site for a blast-related injury. The fire department arrives before you and radios to tell you that the scene is safe. On your arrival, you are informed that your patient is a 46-year-old man who had been blasting rock and had set the fuse too short. As he was leaving the area to seek cover from the explosion, he was blown forward onto a gravel area. He tells you that he remembers everything and he did not lose consciousness. He also indicates that the entire front of his body hurts and he can't hear very well. He denies having any past medical history or allergies and does not take any medications.

On examination, you find minor bleeding from his ears and some cuts and bruises to his arms. As you remove his clothing, you find that his chest and abdomen are bruised. He complains of increasing pain and experiences severe trouble breathing. As you begin your transport, you notice that he is now presenting with hematemesis, cool and clammy skin, tachycardia, and hypotension.

1. Does the mechanism of injury create the suspicion of serious injury prior to your arrival?

2. What is the first important factor to consider in this scenario?
   A. Scene safety
   B. Mechanism of injury
   C. Level of consciousness
   D. Apparent injuries

3. After considering this, what factor should you next consider?
   A. Scene safety
   B. Mechanism of injury
   C. Level of consciousness
   D. Apparent injuries

4. Is your patient's complaint of frontal body pain significant on your primary assessment?

5. The minor bleeding from his ears is most likely an indication of:
   A. a skull fracture.
   B. internal hemorrhaging.
   C. cardiac distress.
   D. an ocular cavity.

6. You determine that your patient is experiencing internal bleeding. What should you do first?
   A. Apply pressure dressings
   B. Immobilize the injury
   C. Apply oxygen
   D. Apply cold packs

7. Is your patient's pain likely to be a result of internal or external injuries? Explain your answer.

8. What condition is likely when signs of hypotension, tachycardia, and cool, clammy skin are found?
   A. Internal bleeding
   B. Shock
   C. Central nervous system depression
   D. Intracranial bleeding

9. Effective primary treatment of this patient should consist of:
   A. tourniquet use.
   B. direct pressure.
   C. rapid transport.
   D. Trendelenburg's positioning.

10. Trendelenburg's positioning:
   A. moves waste from the legs to the core.
   B. moves blood from the legs to the core.
   C. allows a more comfortable transport position.
   D. creates a platform for fluid diffusion.

National EMS Education Standard Competencies

Trauma
Applies fundamental knowledge to provide basic emergency care and transportation based on assessment findings for an acutely injured patient.

Soft-Tissue Trauma
Recognition and management of
- Wounds (pp 806-822)
- Burns
  - Electrical (pp 822-825, 826-828, 831-836)
  - Chemical (pp 822-826, 831-836)
  - Thermal (pp 822-825, 828-829, 831-836)
- Chemicals in the eye and on the skin (pp 822-826, 831-836)

Pathophysiology, assessment, and management
- Wounds
  - Avulsions (pp 806-807, 809-819)
  - Bite wounds (pp 806-807, 821-822)
  - Lacerations (pp 806-807, 809-819)
  - Puncture wounds (pp 806-807, 809-819)
  - Incisions (pp 806-807, 809-819)
- Burns
  - Electrical (pp 806-807, 822-825, 826-828, 831-836)
  - Chemical (pp 806-807, 822-826, 831-836)
  - Thermal (pp 806-807, 822-825, 828-829, 831-836)
  - Radiation (pp 806-807, 822-825, 830-836)
- Crush syndrome (pp 806-808, 811-817)

Knowledge Objectives
1. Discuss the anatomy of the skin, including the layers of the skin. (pp 805-806)
2. Understand the functions of the skin. (p 806)
3. Describe the three types of soft-tissue injuries. (pp 806-807)
4. Describe the types of closed soft-tissue injuries. (pp 808-809)
5. Describe the types of open soft-tissue injuries. (pp 809-811)
6. Discuss the assessment of both closed and open injuries. (pp 811-817)
7. Describe the relationship between airway management and the patient with closed and open injuries. (pp 813-814)
8. Describe the emergency medical care for closed and open injuries. (pp 817-822)
9. Explain the emergency medical care for a patient with an open wound to the abdomen. (p 819)
10. Discuss the emergency medical care for a patient with an impaled object. (pp 820-821)
11. Discuss the emergency medical care for neck injuries. (p 821)
12. Discuss the management of small animal bites, human bites, and rabies. (pp 821-822)
13. Explain how the seriousness of a burn is related to its depth and extent. (pp 822-825)
14. Define and give the characteristics of superficial, partial-thickness, and full-thickness burns. (p 824)
15. Describe and discuss the emergency management of chemical, electrical, thermal, inhalation, and radiation burns. (pp 825-831)
16. Explain the steps involved in the assessment of burns. (pp 831-834)
17. Describe the emergency medical care for burn injuries. (pp 825-831, 834-836)
18. Understand the functions of sterile dressings and bandages. (pp 836-837)

Skills Objectives
1. Demonstrate the emergency medical care of closed soft-tissue injuries. (p 817)
2. Demonstrate the emergency medical care of a patient with an open chest wound. (pp 817-819)
3. Demonstrate how to control bleeding from an open soft-tissue injury. (pp 817-819, Skill Drill 24-1)
4. Demonstrate the emergency medical care of a patient with an open abdominal wound. (p 819)
5. Demonstrate how to stabilize an impaled object. (pp 820-821, Skill Drill 24-2)
6. Demonstrate how to care for a burn. (pp 834-836, Skill Drill 24-3)
7. Demonstrate the emergency medical care of a patient with a chemical, electrical, thermal, inhalation, or radiation burn. (pp 825-831, 834-836)

Introduction

As an EMT, you will regularly be called to care for victims with soft-tissue injuries. These injuries can be as simple as a cut or scrape or as serious as a life-threatening internal injury. It is important for you to not allow yourself to be distracted by dramatic open wounds and make the critical mistake of neglecting other life-threatening conditions such as airway obstructions. It is your responsibility as an EMT to assess and treat each of these injuries within the current standard of care guidelines.

The soft tissues of the body can be injured through a variety of mechanisms. A blunt injury occurs when the energy exchange between the patient and an object is more than the tissues can tolerate, as can happen in an automobile collision that leads to the person striking the steering wheel. A blunt injury does not penetrate the skin. A penetrating injury occurs when an object, such as a bullet or knife, breaks through the skin and enters the body. Barotrauma injuries occur from sudden or extreme changes in air pressure, such as can occur during a scuba diving emergency. Burns may also result in soft-tissue injuries.

Soft-tissue trauma is the leading form of injury. Open wounds account for approximately 6.5 million emergency department (ED) visits, and nearly 5 million patients present with contusions (bruises). In fact, wound care is one of the most frequently performed procedures in EDs across the United States. Most of these injuries require basic interventions such as wound irrigation, dressing, bandaging, and limited suturing.

Death as the result of soft-tissue injury is often related to hemorrhage or infection. Uncontrolled hemorrhage can quickly lead to shock and death. When the skin barrier is breached, invading pathogens—bacteria, fungi, and viruses—can cause local or systemic infection. Infection can be life or limb threatening, especially in people with diabetes.

Soft-tissue injuries and their associated complications can often be prevented through the use of simple protective actions. For example, using gloves when working with abrasive materials helps to prevent skin injuries. To reduce injuries in the workplace, safety measures have been implemented that include the use of safety devices to prevent interaction between machine parts and body parts. Teaching children to avoid using sharp objects also helps prevent injury. Plastic scissors, plastic knives, and plastic drinking cups are all designed to reduce the risk of cuts and other skin injuries among children.

This chapter will discuss the various types of soft-tissue injuries and the appropriate assessment and treatment of this classification of injuries.

The Anatomy and Physiology of the Skin

The skin is our first line of defense against external forces and infection. It is also the largest organ in the body. Although it is relatively tough, skin is still quite susceptible to injury. Injuries to soft tissues range from simple bruises and abrasions to serious lacerations and amputations. Soft-tissue injury may result in loss of soft tissue, exposing deep structures such as blood vessels, nerves, and bones. In all instances, you must control bleeding, prevent further contamination to decrease the risk of infection, and protect the wound from further damage. Therefore, you must know how to apply dressings and bandages to various parts of the body.

Skin varies in thickness, depending on a person's age and the skin's location. The skin of the very young

You are the Provider: PART 1

You and your partner are standing by at the scene of a structural fire when fire fighters bring over a 45-year-old man who was rescued from the burning house. The patient is wrapped in a blanket. He is conscious and alert, is in severe pain, and has soot all over his face.

1. What should be your most immediate priority?
2. What is a thermal burn? Are thermal burns always caused by fire?

and very old is thinner than the skin of a young adult. The skin covering the scalp, the back, and the soles of the feet is quite thick, while the skin of the eyelids, lips, and ears is very thin. Thin skin is more easily damaged than thick skin.

■ Anatomy

The skin has two principal layers: the epidermis and the dermis **Figure 24-1**. The **epidermis** is the tough, external layer that forms a watertight covering for the body. The epidermis is itself composed of several layers. The cells on the surface layer of the epidermis are constantly worn away. They are replaced by cells that are pushed to the surface when new cells form in the germinal layer at the base of the epidermis. Deeper cells in the germinal layer contain pigment granules. Along with blood vessels in the dermis, these granules produce skin color.

The **dermis** is the inner layer of the skin. It lies below the germinal cells of the epidermis. The dermis contains the structures that give the skin its characteristic appearance: hair follicles, sweat glands, and sebaceous glands. The sweat glands act to cool the body. They discharge sweat onto the surface of the skin through small pores, or ducts, that pass through the epidermis. Sebaceous glands produce sebum, the oily material that waterproofs the skin and keeps it supple. Sebum travels to the skin's surface along the shaft of adjacent hair follicles. Hair follicles are small organs that produce hair. There is one follicle for each hair, each connected with a sebaceous gland and a tiny muscle. This muscle pulls the hair erect whenever a person is cold or frightened.

Blood vessels in the dermis provide the skin with nutrients and oxygen. Small branches reach up to the germinal cells, but no blood vessels penetrate farther into the epidermis. There are also specialized nerve endings within the dermis.

The skin covers all external surfaces of the body. The various openings in the body, including the mouth, nose, anus, and vagina, are not covered by skin. Instead, these openings are lined with **mucous membranes**. These membranes are similar to skin in that they, too, provide a protective barrier against bacterial invasion. But mucous membranes differ from skin in that they secrete a watery substance that lubricates the openings. Therefore, mucous membranes are moist, whereas skin is dry.

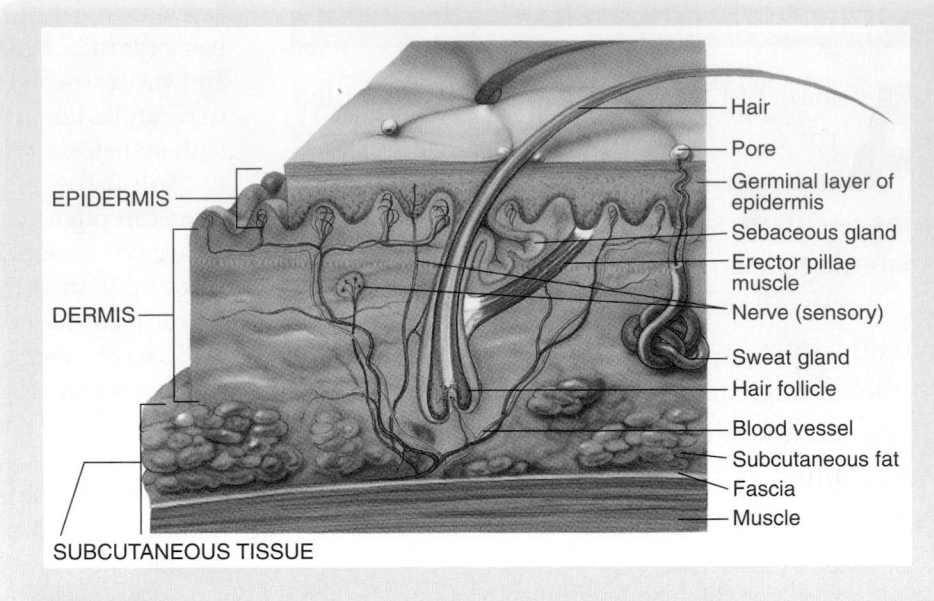

EPIDERMIS

DERMIS

SUBCUTANEOUS TISSUE

Hair
Pore
Germinal layer of epidermis
Sebaceous gland
Erector pillae muscle
Nerve (sensory)
Sweat gland
Hair follicle
Blood vessel
Subcutaneous fat
Fascia
Muscle

Figure 24-1 The skin is composed of a tough external layer called the epidermis and a vascular inner layer called the dermis.

■ Physiology

The skin serves many functions. It protects the body by keeping pathogens out and water in and assisting in body temperature regulation. The nerves in the skin report to the brain on the environment and on many sensations.

The skin is also the body's major organ for regulating temperature. In a cold environment, the blood vessels in the skin constrict, diverting blood away from the skin and decreasing the amount of heat that is radiated from the body's surface. In hot environments, the vessels in the skin dilate. The skin becomes flushed or red, and heat radiates from the body's surface. Also, sweat glands secrete sweat. As the sweat evaporates from the skin's surface, the body temperature drops, and the person begins to cool down.

Any break in the skin allows bacteria to enter and raises the possibilities of infection, fluid loss, and loss of temperature control. Any one of these problems can cause serious illness and even death.

Pathophysiology

Soft tissues are often injured because they are exposed to the environment. There are three types of soft-tissue injuries:

- **Closed injuries**, in which soft-tissue damage occurs beneath the skin or mucous membrane but the surface of the skin or mucous membrane remains intact.

- **Open injuries**, in which there is a break in the surface of the skin or the mucous membrane, exposing deeper tissues to potential contamination.
- **Burns**, in which the soft tissue damage occurs as a result from thermal heat, frictional heat, toxic chemicals, electricity, or nuclear radiation.

■ Pathophysiology of Closed and Open Injuries

Healing of wounds is a natural process that involves several overlapping stages, all directed toward the larger goal of maintaining homeostasis. Ultimately, the goal is for the body to return to a functional state, although the injured area may not always be restored to the preinjury condition.

Among the primary concerns in wound healing is the cessation of bleeding. Loss of blood, internal or external, hinders the provision of vital nutrients and oxygen to the affected area. It also impairs the tissue's ability to eliminate wastes. The end result is abnormal or absent function, which interferes with homeostasis. To stop the flow of blood, the vessels, platelets, and clotting cascade must work in unison.

In inflammation (the next stage of wound healing), additional cells move into the damaged area to begin repair. White blood cells migrate to the area to combat pathogens that have invaded exposed tissue. Foreign products and bacteria are also removed from the body. Similarly, lymphocytes (a type of white blood cell) destroy bacteria and other pathogens. Mast cells release histamine as part of the body's response in the early stages of inflammation. Histamine causes dilation of blood vessels, increasing blood flow to the injured area and resulting in a reddened, warm area immediately around the site. Histamine makes capillaries more permeable, and swelling may occur as fluid seeps out of these "leaky" capillaries. Inflammation ultimately leads to the removal of foreign material, damaged cellular parts, and invading microorganisms from the wound site.

In the outer layer of skin, cells are stacked in layers. To replace the area damaged in a soft-tissue injury, a new layer of cells must be moved into this region. This is the next stage of wound healing. Cells quickly multiply and redevelop across the edges of the wound. Except in cases of clean incisions, the appearance of the restructured area seldom returns to the preinjury state. For example, large wounds or injuries that result in significant disruption of the skin will often not complete this process. In persons with lightly pigmented skin, a pink line of scar tissue may signal the presence of collagen, a structural protein that has reinforced the damaged tissue. Despite the changed appearance, the function of the area may be restored to near normal.

During the next stage of wound healing, new blood vessels form as the body attempts to bring oxygen and nutrients to the injured tissue. New capillaries bud from intact capillaries that lie adjacent to the damaged skin. These vessels provide a conduit for oxygen and nutrients and serve as a pathway for waste removal. Because they are new and delicate, bleeding might result from a very minor injury. It may take weeks to months for the new capillaries to be as stable as preexisting vessels.

Collagen is a tough, fibrous protein found in scar tissue, hair, bones, and connective tissue. In the last stage of wound healing, collagen provides stability to the damaged tissue and joins wound borders, thereby closing the open tissue. Unfortunately, collagen cannot restore the damaged tissue to its original strength.

Words of Wisdom

Wound healing does not always follow the pattern described. Infection or an abnormal scar may develop, excessive bleeding may occur, or healing may be slow.

■ Pathophysiology of Burns

Burns are diffuse soft-tissue injuries created by destructive energy transfer via radiation, thermal, or electrical energy. Thermal burns can occur when skin is exposed to temperatures higher than 111°F (44°C). In general, the severity of a thermal injury correlates directly with temperature, concentration, or amount of heat energy possessed by the object or substance and the duration of exposure. For example, solids generally have higher heat content than gases, so exposure to a hot solid (such as the rack inside an oven) typically causes a more significant burn than exposure to hot gases (such as those coming out of an oven). Burns are a progressive process: The greater the heat energy, the deeper the wound.

Exposure time is another important factor. Thermal injury can occur to unresponsive or paralyzed patients from seemingly innocuous heat sources such as heating pads, transcutaneous oxygen sensors, and heat lamps left unattended for long periods.

It may be difficult to evaluate the amount of heat energy or the amount of exposure time in many cases. The temperature of a fire may vary tremendously from the floor to the ceiling. Although most people reflexively limit the amount of time exposed to such heat, if clothing is on fire or the person is trapped or unconscious, exposure time will be longer.

Closed and Open Injuries

Closed Injuries

Closed soft-tissue injuries are characterized by a history of blunt trauma, pain at the site of injury, swelling beneath the skin, and discoloration. Such injuries can vary from mild to quite severe.

A **contusion**, or bruise, is an injury that causes bleeding beneath the skin but does not break the skin. Contusions result from blunt forces striking the body. The epidermis remains intact, but cells within the dermis are damaged, and small blood vessels are usually torn. The depth of the injury varies, depending on the amount of energy absorbed. As fluid and blood leak into the damaged area, the patient may have swelling and pain. The buildup of blood produces a characteristic blue or black discoloration called **ecchymosis** Figure 24-2.

A **hematoma** is blood that has collected within damaged tissue or in a body cavity Figure 24-3. A hematoma occurs whenever a large blood vessel is damaged and bleeds rapidly. It is usually associated with extensive tissue damage. A hematoma can result from a soft-tissue injury, a fracture, or any injury to a large blood vessel. In severe cases, the hematoma may contain more than a liter of blood.

A **crushing injury** occurs when a great amount of force is applied to the body Figure 24-4. The extent of the damage depends on how much force is applied and the amount of time over which it is applied. In addition to causing some direct soft-tissue damage, continued compression of the soft tissues will cut off their circulation, producing further tissue destruction. For example, if a patient's legs are trapped under a collapsed pile of rocks, damage to the leg tissues will continue until the rocks are removed.

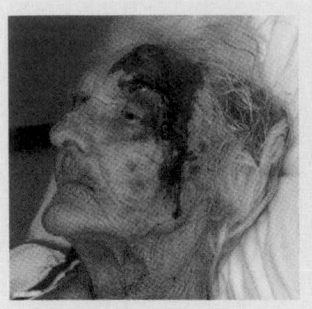

Figure 24-3 A hematoma develops whenever a large blood vessel is damaged and bleeds rapidly.

When an area of the body is trapped for longer than 4 hours and arterial blood flow is compromised, **crush syndrome** can develop. Crush syndrome is significant metabolic derangement that can lead to renal failure and death.

Another form of compression can result from the swelling that occurs whenever tissues are injured. The cells that are injured leak watery fluid into the spaces between the cells. The pressure of the fluid may become great enough to compress the tissue and cause further damage. This is especially true if the blood vessels become compressed, cutting off blood flow to the tissue. This condition is called **compartment syndrome**. Excessive swelling often follows significant injury to the extremities. The hallmark sign of compartment syndrome is pain out of proportion to the injury.

Severe closed injuries can also damage internal organs. The greater the amount of energy absorbed from the blunt force, the greater the risk of injury to deeper

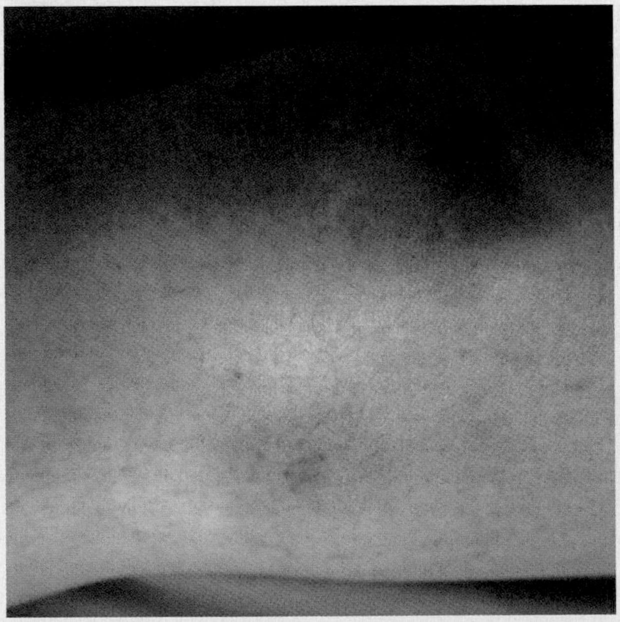

Figure 24-2 Contusions, more commonly known as bruises, occur as a result of a blunt force striking the body. The buildup of blood produces a characteristic blue or black discoloration (ecchymosis).

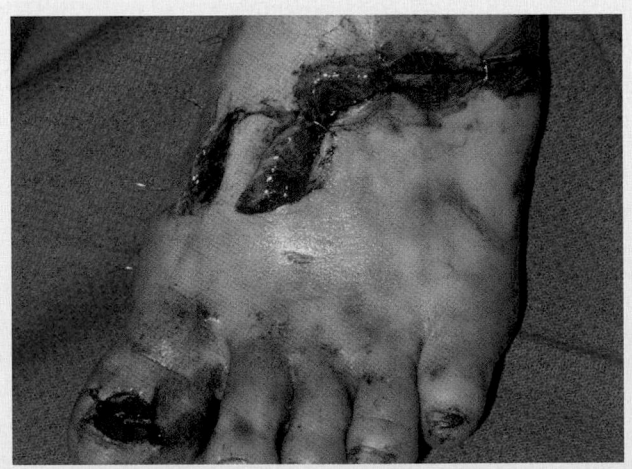

Figure 24-4 The damage associated with a crush or compression injury varies depending on the direct damage to the soft tissues and on how long the tissue was cut off from circulation.

structures. Therefore, you must assess all patients with closed injuries for more serious hidden injuries. Remain alert for signs of shock or internal bleeding, and begin treatment of these conditions if necessary.

■ Open Injuries

Open injuries differ from closed injuries in that the protective layer of skin is damaged. This can produce extensive bleeding. More important, however, a break in the protective skin layer or mucous membrane means that the wound is contaminated and may become infected. <u>Contamination</u> describes the presence of infectious organisms (pathogens) or foreign bodies, such as dirt, gravel, or metal, in the wound. You must address excessive bleeding and contamination in your treatment of open soft-tissue wounds. There are four types of open soft-tissue wounds that you must be prepared to manage:

- Abrasions
- Lacerations
- Avulsions
- Penetrating wounds

An <u>abrasion</u> is a wound of the superficial layer of the skin, caused by friction when a body part rubs or scrapes across a rough or hard surface. An abrasion usually does not penetrate completely through the dermis, but blood may ooze from the injured capillaries in the dermis. Known by a variety of names, including road rash, road burn, strawberry, and rug burn, abrasions can be extremely painful **Figure 24-5**.

A <u>laceration</u> is a jagged cut caused by a sharp object or a blunt force that tears the tissue, whereas an <u>incision</u> is a sharp, smooth cut. The depth of the injury can vary, extending through the skin and subcutaneous tissue, even into the underlying muscles and adjacent nerves and blood vessels **Figure 24-6**. Lacerations and incisions may appear linear (regular) or stellate (irregular) and may occur along with other types of soft-tissue injury. Lacerations or incisions that involve arteries or large veins may result in severe bleeding.

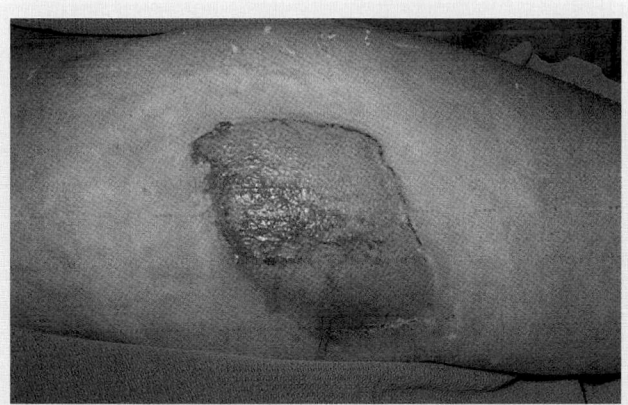

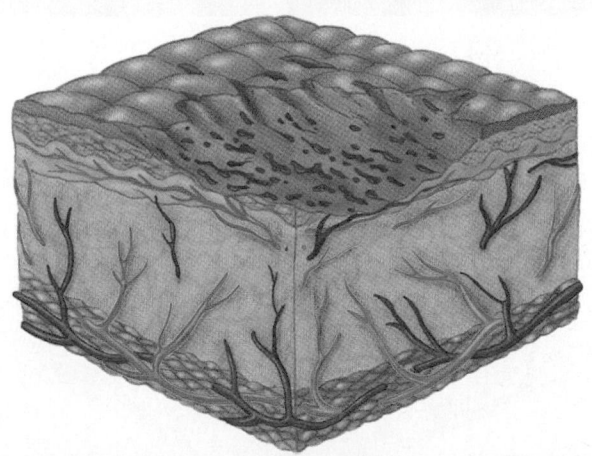

Figure 24-5 Abrasions usually do not penetrate completely through the dermis, but blood may ooze from the capillaries. These wounds are typically superficial and result from rubbing or scraping across a hard, rough surface.

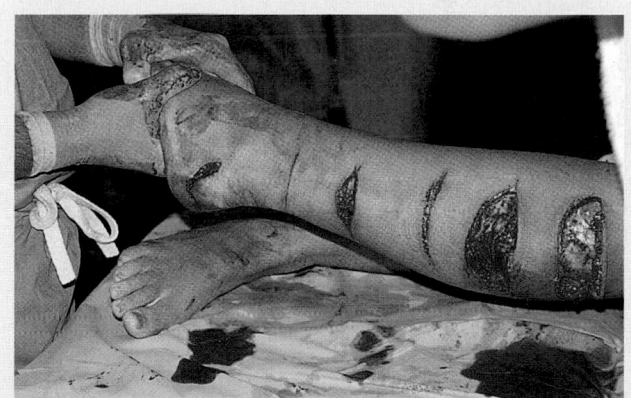

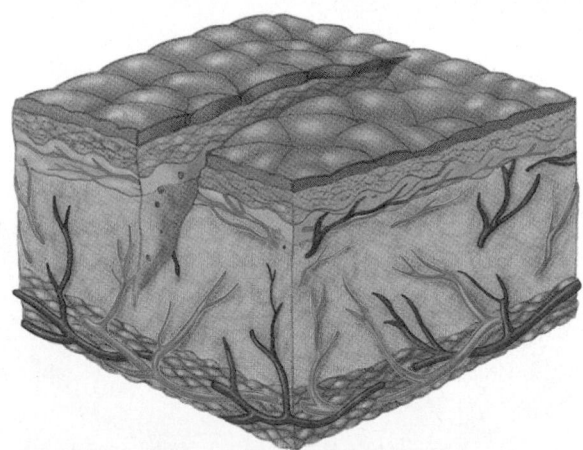

Figure 24-6 Lacerations vary in depth and can extend through the skin and subcutaneous tissue to the underlying muscles, nerves, and blood vessels. These wounds can be smooth or jagged as a result of a cut by a sharp object or a blunt force that tears the tissue.

An **avulsion** is an injury that separates various layers of soft tissue (usually between the subcutaneous layer and **fascia**) so that they become either completely detached or hang as a flap **Figure 24-7**. Often there is significant bleeding. If the avulsed tissue is hanging from a small piece of skin, the circulation through the flap may be at risk. If you can, replace the flat avulsed flap in its original position as long as it is not visibly contaminated with dirt and/or other foreign materials. If an avulsion is complete, you should wrap the separated tissue in sterile gauze and take it with you to the emergency department. This type of avulsion often poses serious infection concerns. Never remove an avulsion skin flap, regardless of its size.

An **amputation** is an injury in which part of the body is completely severed. Chapter 29, *Orthopaedic Injuries*, covers this topic in detail. We usually think of amputations as involving the upper and lower extremities. But other body parts, such as the scalp, ear, nose, penis, or lips, may also be totally avulsed, or amputated. You can easily control the bleeding from some amputations, such as the fingers, with direct pressure and pressure dressings. If an avulsion involves a large area of muscle mass, such as a thigh, there may be massive bleeding. In this situation, you need to treat the patient for hypovolemic shock. See Skill Drill 23-1: Controlling External Bleeding in Chapter 23, *Bleeding*.

A **penetrating wound** (or puncture wound) is an injury resulting from a sharp, pointed object, such as a knife, ice pick, splinter, or bullet. Such objects leave relatively small entrance wounds, so there may be little external bleeding **Figure 24-8**. However, these objects can damage structures deep within the body and cause unseen bleeding. If the wound is to the chest or abdomen, the injury can cause rapid, fatal bleeding. Assessing the amount of damage a puncture wound has created is very difficult and is reserved for the physician at the hospital.

Stabbings and shootings often result in multiple penetrating injuries. You must assess these patients carefully to identify all wounds. Since a penetrating object can pass completely through the body, always count the number of penetrating injuries (or holes),

Figure 24-7 Avulsions are injuries characterized by either complete separation of tissue or tissue hanging as a flap. Significant bleeding is common.

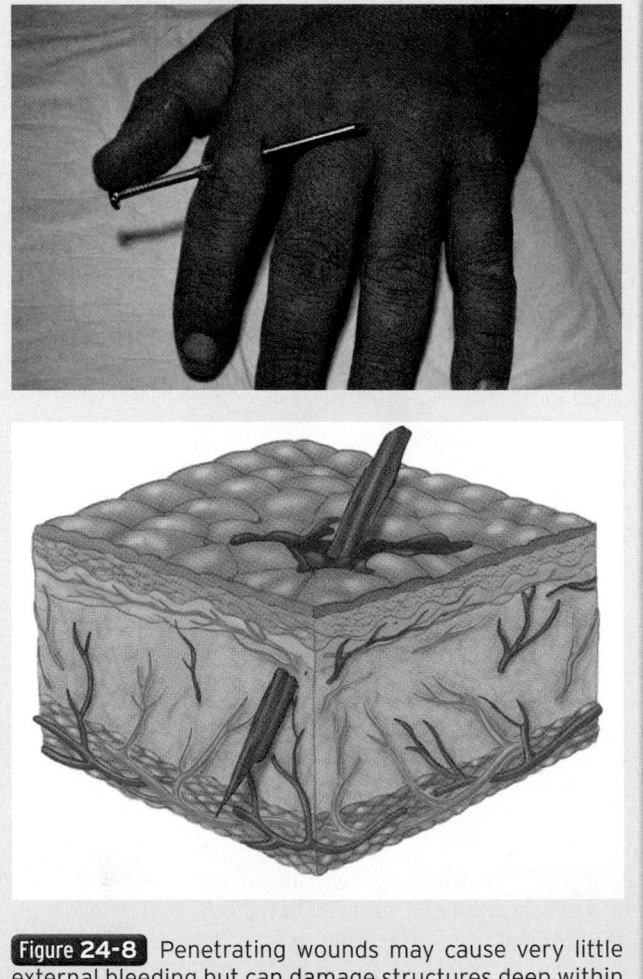

Figure 24-8 Penetrating wounds may cause very little external bleeding but can damage structures deep within the body.

especially with gunshot wounds. Entrance wounds and exit wounds may be difficult to tell apart in a prehospital setting, especially with the different types of ammunition available. While entrance wounds are often smaller than exit wounds **Figure 24-9**, it is better to simply count the number of penetrating injuries, and leave the distinction between entrance and exit to the physician who is working in a more controlled environment. Gunshot wounds have some unique characteristics that require special care. The amount of energy transmitted by a gunshot injury is directly related to the speed of the bullet. Thus, it is important to find out the type of gun that was used in the shooting. Sometimes, the patient or bystanders can tell you how many rounds were fired. This information can help hospital personnel to better care for the patient. Shotgun wounds create multiple paths of missiles (shot) and create a larger surface area and volume of tissue damage.

Many cases involving shootings end up in court at some point, and you may be called to testify. For this reason, you must carefully document the circumstances surrounding any gunshot injury, the patient's condition, and the treatment you give.

As with closed wounds caused by crushing, open wounds caused by crushing may involve damaged internal organs or broken bones, as well as extensive soft-tissue damage **Figure 24-10**. Whereas external bleeding may be minimal, internal bleeding may be severe, or even life threatening. The crushing force damages soft tissues as well as vessels and nerves. This frequently results in a painful, swollen, deformed area.

Blast injuries, as discussed in Chapter 22, *Trauma Overview*, may also often result in multiple penetrating injuries. The mechanism of injury (MOI) from a blast is generally due to three factors:

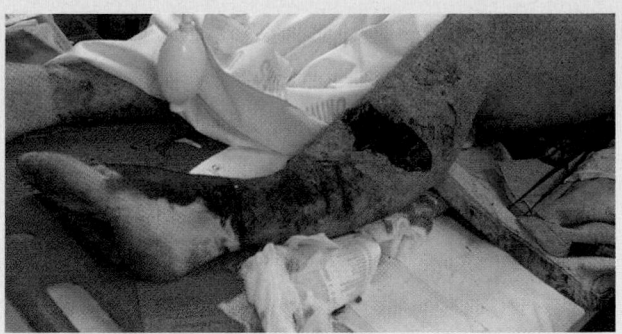

Figure 24-10 A crushing open wound is characterized by extensive tissue damage and deformity that is often accompanied by swelling and extreme pain.

- **Primary blast injury:** Due entirely to the blast itself; damage to the body is caused by the pressure generated by the explosion.
- **Secondary blast injury:** Damage results from the victim being struck by flying debris, propelled by the force of the blast.
- **Tertiary blast injury:** The victim is thrown or hurled by the force of the explosion into an object or onto the ground.

It is imperative for you to conduct a complete primary and secondary assessment to determine what type(s) of injuries are sustained from a blast injury and treat appropriately.

Patient Assessment of Closed and Open Injuries

Assessing closed injuries is much more difficult than assessing open injuries. Therefore, anytime you observe bruising, swelling, or deformity, or the patient is reporting pain, the possibility of a closed injury should be considered.

The assessment of an open injury is generally easier than the assessment of a closed injury because you can see the injury. Open wounds can be defined as an injury in which there is a break in the surface of the skin or the mucous membrane, exposing deeper tissues to potential contamination. You must use caution so as to not let a non–life-threatening gruesome injury distract you from recognizing that the patient has another injury that is considered life threatening.

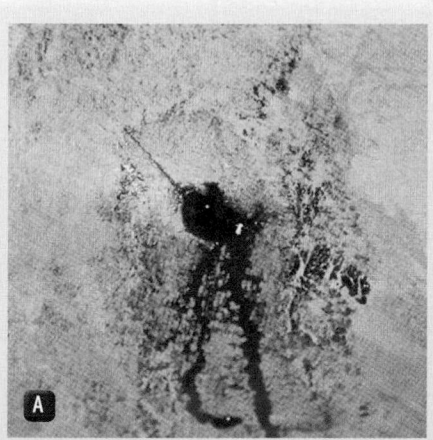
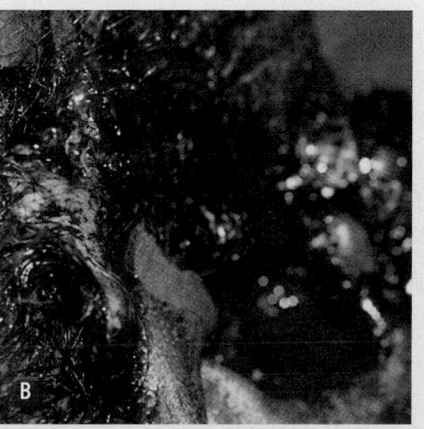

Figure 24-9 **A.** An entrance wound from a gunshot may have burns around the edges. **B.** An exit wound is often larger than an entrance wound and is associated with greater damage to soft tissues locally.

Words of Wisdom

Extremities that are painful, swollen, or deformed should be splinted. When you are splinting these types of injuries, remember to assess the patient's pulses, motor, and sensory function before and after applying the splint. You should also evaluate the mechanism of injury. Because diffuse or generalized soft-tissue injuries can be critical, all patients with a significant mechanism of injury should be considered to have internal bleeding and shock until proven otherwise by the emergency department staff.

Scene Size-up

Scene Safety

As you arrive on scene, observe the scene for hazards and threats to the safety of the crew, bystanders, and the patient. Assess the impact of hazards on patient care and address the hazards. Assess for the potential for violence, and assess for environmental hazards.

Ensure that you and your crew have taken standard precautions—a minimum of gloves and eye protection. Open soft-tissue injuries can be very messy. Control of the blood and bloody contaminants can be difficult unless you are careful about what and where you touch. Apply standard precautions before you approach the scene to minimize your direct exposure to body fluids. Because of the color of blood and how well it soaks through clothing, you can often identify patients with an open injury as you approach the scene. However, blood can be hidden under thick clothing such as denim and leather. Eye exposures may occur from splashes and droplets at a busy scene. Eye protection is required when managing open injuries. Place several pairs of gloves in your pocket for easy access in case your gloves tear or there are multiple patients with bleeding. Determine the number of patients, and consider if you need additional or specialized resources on the scene.

Mechanism of Injury/Nature of Illness

As you observe the scene, look for indicators of the MOI. This helps you develop an early index of suspicion for underlying injuries in a patient who has sustained a significant MOI. As you put together information from dispatch and your observations of the scene, consider how the MOI produced the injuries expected. For example, in a vehicle crash, a patient who has sustained abrasions and lacerations to the face from an impact with the steering wheel or windshield may have experienced enough force to injure the cervical spine as well. In this case and in many trauma situations, spinal precautions should be used very early on in your care of the patient. The MOI may also provide indications of safety threats. For example, gunshot wounds may indicate angry and violent individuals. Make sure the scene is safe, and consider requesting additional help early.

You are the Provider: PART 2

The patient is ambulatory, so you immediately move him to the ambulance that is parked a short distance away and complete your primary assessment. Your partner obtains additional information from the fire fighters who rescued him.

Recording Time: 2 Minutes	
Appearance	Shivering; in obvious pain
Level of consciousness	Conscious and alert
Airway	Open; clear of secretions or foreign bodies
Breathing	Normal rate; adequate depth; unlabored
Circulation	Increased pulse rate; strong at the radial site; no obvious bleeding

You apply high-flow oxygen via a nonrebreathing mask and then carefully remove clothing that is not adhered to his skin so you can assess the severity of his burns.

3. What additional information should be obtained from the fire fighters who rescued the patient?
4. How are thermal burns classified? What are the characteristics of each type of burn?

Primary Assessment

The primary assessment for a patient with a closed or open injury should focus on identifying and managing life-threatening concerns. Perform a rapid scan of the patient. Treating according to the ABCs is always a good choice. Threats to airway, breathing, or circulation are considered life threatening and must be treated immediately to prevent mortality. In some situations, significant bleeding may require management before applying oxygen for a person with adequate breathing. Significant bleeding, internal or external, is an immediate life threat. If the patient has obvious life-threatening bleeding, it must be controlled quickly and treatment of shock begun as quickly as possible. The decision on what to treat first will come with experience.

Providing high-flow oxygen to patients with closed soft-tissue injuries may help reduce the effects of shock and assist in perfusion of damaged tissues, particularly in crush injuries. If the patient has signs of hypoperfusion, treat aggressively for shock and provide rapid transport to the hospital. Request advanced life support (ALS) as necessary to assist with more aggressive shock management. Do not delay transport of a seriously injured trauma patient to complete nonlifesaving treatments in the field, such as splinting extremity fractures; instead, complete these types of treatments en route to the hospital.

If you discover an open injury with significant bleeding, cover the wound and control the bleeding as quickly as possible. If the bleeding is not significant, such as from an abrasion, it can be treated later during the secondary assessment. You may also need to provide spinal stabilization and assistance with breathing or perfusion problems.

Form a General Impression

As you approach the trauma patient with a suspected closed injury, important indicators will alert you to the seriousness of the patient's condition. Is the patient awake and interacting with his or her surroundings, or lying still, making no sounds? Does the patient have any apparent life threats? What color is the patient's skin? Is he or she responding to you appropriately or inappropriately? Your general impression will help you develop an index of suspicion for serious injuries and determine how urgently your patient needs care.

Trauma patients with closed soft-tissue injuries may have what appear to be minor injuries; however, you must not be distracted from looking for more serious hidden injuries. For example, if you observe that the patient has a hematoma on the head and a decreased level of consciousness, the patient may have a serious head injury.

Check for responsiveness using the AVPU scale (*Alert* to person, place, and day; responsive to *Verbal* stimuli;

responsive to *Pain*; *Unresponsive*). Assessing a patient's mental status is generally easy and can be done by asking the patient about his or her chief complaint. If the patient is alert, this should help direct you to any apparent life threats. If the patient is not alert, determine if he or she responds to verbal or painful stimuli or if he or she is unresponsive. An unresponsive patient may indicate a life-threatening condition. You should administer high-flow oxygen via a nonrebreathing mask to all patients whose level of consciousness is less than alert and oriented and provide immediate transport to the ED.

Trauma patients with open injuries may present with obvious significant injuries that indicate a serious condition. However, other injuries may not be as obvious but may still indicate a very serious condition. Your "impression" about how the patient is doing is based on information as simple as the patient's age, the MOI, and his or her level of consciousness. Observations such as bleeding from open injuries, skin color and condition, and gasping respirations also contribute to your general impression and help to determine your treatment priorities and the urgency of care needed. A good question to ask yourself is, "How sick is my patient based on what I know right now?"

Airway and Breathing

Next, ensure that the patient has a clear and patent airway. Because trauma was involved, protect the patient from further spinal injury as you manage the airway by preventing the head and torso from moving. If the patient is unresponsive or has a significantly altered level of consciousness, consider inserting an oropharyngeal airway or nasopharyngeal airway.

You must also quickly assess the patient for adequate breathing. Inspect and palpate the chest wall for DCAP-BTLS (Deformities, Contusions, Abrasions, Punctures/penetrations, Burns, Tenderness, Lacerations, and Swelling). If a soft-tissue injury is discovered on the chest or abdomen, check for clear and symmetric breath sounds and then provide high-flow oxygen, or provide assisted ventilations using a bag-mask device as needed, depending on the level of consciousness and if your patient is breathing inadequately.

Open soft-tissue injuries of the face and neck have a potential to interfere with the effectiveness of the airway and breathing. Evaluate the patient's voice and speaking ability to identify throat injuries. If an open injury is found on the chest, evaluate for air movement through the wound in the form of bubbling or sucking sounds that indicate a deep penetrating injury. Quickly place an occlusive dressing over the wound. Provide high-flow oxygen or assisted ventilations with a bag-mask device as needed, depending on the patient's level of consciousness and on the adequacy of the patient's breathing. Monitor the patient for signs of increasing respiratory distress that may require you to relieve pressure built up under the dressing (caused by a pneumothorax).

Although soft-tissue injuries often have a gruesome appearance, you must not be distracted by a non–life-threatening injury while overlooking a life-threatening injury. For example, a patient with a gruesome crush injury to the foot may have less obvious injuries that may interfere with airway and breathing, leading to death.

Circulation

You must quickly assess the patient's pulse rate and quality; determine the skin condition, color, and temperature; and check the capillary refill time. These assessments will help you determine the presence of circulatory problems or shock. Closed soft-tissue injuries may not always have visible signs of bleeding. Because most of the bleeding is occurring inside the body, shock may be present. Your assessment of the pulse and skin will give you an indication as to how aggressively you need to treat your patient for shock.

If visible significant bleeding is seen, you must begin the steps necessary to control bleeding. Significant bleeding is an immediate life threat and must be controlled quickly using appropriate methods. In dark environments, bleeding can be hard to see because of its color. Thick clothing may also hide bleeding. After you consider the MOI and form suspicions as to where bleeding may occur, expose that part of the body. Blood flowing freely from veins in a large gash can be as much of a threat as blood spurting from an artery.

You may ask yourself if bleeding should be controlled before you administer the patient oxygen. After discovering what injuries your patient has, you must then decide where the priorities lie. For example, control of oozing blood from damaged capillaries in an abrasion may be controlled later if more important problems are at hand.

If significant trauma has likely affected multiple systems, start with a rapid scan of the patient to be sure that you have found all of the problems and injuries. A rapid 60- to 90-second scan may identify factors that assist you in determining whether a patient requires rapid transport. Begin with the head and neck while manually holding the head in place. When you are done, apply a cervical spine immobilization device if you have not done so already. If you identify conditions that have the potential to become unstable, such as a distended abdomen or femur fractures, the patient requires rapid and immediate transport.

You are the Provider: PART 3

The patient tells you that he sustained the burns when he was trying to escape from the burning house. Both of the exits were blocked by fire and debris, so he ran into the bedroom. However, he was unable to exit through the window because of burglar bars he had installed; this was when the fire fighters found him. He denies losing consciousness. Your partner assesses his vital signs as you perform a secondary assessment.

Recording Time: 6 Minutes	
Respirations	14 breaths/min; adequate depth; unlabored
Pulse	108 beats/min; strong and regular
Skin	Red, warm, and dry; burns to the torso and arms
Blood pressure	166/86 mm Hg
Oxygen saturation (Sao$_2$)	98% (on oxygen)

Your secondary assessment reveals superficial and partial-thickness burns to his anterior chest and abdomen, and partial- and full-thickness burns to both of his arms, including his hands. His face is covered with soot, and his facial hair and the hair just above his hairline are singed. You do not see any obvious skin burns to his facial area, and the remainder of your assessment does not reveal any other injuries. The patient denies having difficulty breathing or any other symptoms other than pain.

5. **What percentage of the patient's body surface area has been burned?**
6. **What factors should you consider when determining the severity of a burn?**

Transport Decision

Determine whether your patient needs immediate transport or stabilization on scene. If the patient you are treating has an airway or breathing problem or signs and symptoms of shock or internal bleeding, you must consider rapid transport to the hospital for treatment or request ALS support. Whereas treatment performed during the primary assessment is directed at quickly addressing life threats, you should not delay transport of a trauma patient, particularly if the patient has a closed soft-tissue injury that may be a sign of a more serious deeper injury. Patients with a significant MOI may require a secondary assessment to identify these injuries.

Although most patients do not require immediate load-and-go transportation, there are certain conditions for which treatment is limited in the field, and, therefore, immediate transport is the better choice. The following list will help to guide you in determining types of patients that need immediate transportation:

- Poor initial general impression
- Altered level of consciousness
- Dyspnea
- Abnormal vital signs
- Shock
- Severe pain

It is easy for you to become distracted when a patient has significant soft-tissue injuries, there is a large amount of blood, and the patient is most likely frightened and may be screaming. However, at this point you need to focus on the problems at hand and follow the protocols you have learned.

Patients who have visible significant bleeding or signs of significant internal bleeding may quickly become unstable. Treatment must be directed at quickly addressing life threats and providing rapid transportation to the closest appropriate hospital. Signs such as tachycardia; tachypnea; weak pulse; and cool, moist, and pale skin are signs of hypoperfusion and imply the need for rapid transport. You should be alert to these signs and reassess your priority and transport decision if they develop.

History Taking

Investigate Chief Complaint

After the life threats have been managed during the primary assessment, investigate the chief complaint or history of present illness. Obtain a medical history and be alert for injury-specific signs and symptoms as well as any pertinent negatives such as no pain or loss of sensation.

Make every attempt to obtain a SAMPLE history from your patient. Using OPQRST (Onset, Provocation or palliation, Quality, Radiation, Severity, Time of onset) may

provide some background on isolated extremity injuries. You have the opportunity to interview the patient well before the ED physician's examination. Any information you receive will be very valuable if the patient loses consciousness.

If the patient is not responsive, attempt to obtain the history from other sources, such as friends or family members. Medical identification jewelry and cards in wallets may also provide information about the patient's medical history.

Typical signs of an open injury include bleeding, break(s) in the skin, shock, hemorrhage, and disfigurement or loss of a body part. Typically symptoms include pain and/or burning at the injury site. Conditions such as anemia (low quantity of hemoglobin in the blood) and hemophilia (a disorder in which blood has a diminished ability to clot) can complicate open soft-tissue injuries. Medications such as aspirin and other blood-thinning medications frequently taken by older patients may interfere with clotting and make bleeding control difficult. If the injury was self-inflicted, the patient may also have a mental health problem.

Secondary Assessment

The secondary assessment is a more systematic head-to-toe or focused examination of the patient that is used to reveal injuries that may have been missed during the primary assessment. In some instances such as a critically injured patient or a short transport time, you may not have time to conduct a secondary assessment. In other instances, the secondary assessment may occur en route to the ED.

Physical Examinations

Assessment of the respiratory system should involve looking, listening, and feeling for signs of airway problems. Look at the patient and ask yourself the following questions:

1. Is the patient in a tripod position?
2. What is the skin's color and condition?
3. Are there any signs of increased respiratory efforts such as retractions, nasal flaring, pursed lip breathing, or use of accessory muscles?

Next, listen for air movement at the patient's mouth and nose. Then listen to breath sounds with a stethoscope. Breath sounds should be clear and equal bilaterally, anteriorly, and posteriorly. Determine the patient's rate and quality of respiration. Finally, assess for asymmetric chest wall movement.

You must be able to quickly assess pulse rate and quality; determine the skin condition, color, and temperature; and check the capillary refill time.

Assess the neurologic system to gather baseline data on your patient. This examination should include:

- level of consciousness—use AVPU
- pupil size and reactivity
- motor response
- sensory response

Assess the musculoskeletal system by performing a detailed full-body scan. Look for DCAP-BTLS. Assess the chest, abdomen, and extremities for hidden bleeding and injuries. Log roll the patient, and assess the posterior torso for injuries. Once the back has been assessed, the patient can be log rolled back down onto a backboard, followed by complete spinal stabilization. Log rolling and securing the patient to a backboard or other full-body stabilization device should take into consideration injuries found during the primary assessment.

Assess all anatomic regions, looking for the following signs/symptoms:

- Be alert for raccoon eyes, Battle's sign, and/or drainage of blood or fluid from the ears or nose.
- Check the neck for jugular vein distention and tracheal deviation. Be alert for patients with a stoma or tracheostomy.
- Check the pelvis for stability.
- Check the abdomen; feel all four quadrants for tenderness or rigidity. If the abdomen is tender, expect internal bleeding.

- Check the extremities, and record pulse, motor, and sensory function.

Vital Signs

Patients who have hidden injuries under a closed soft-tissue injury may have internal bleeding and may rapidly become unstable. It is important to reassess the vital signs to identify how quickly the patient's condition is changing. Signs such as tachycardia, tachypnea, low blood pressure, weak pulse, and cool, moist, and pale skin indicate hypoperfusion and imply the need for rapid treatment at the hospital. Remember that soft-tissue injuries, even without a significant MOI, can cause shock. The reassessment of your patient's vital signs will give you a good understanding of how well or how poorly your patient is tolerating the injury.

Monitoring Devices

In addition to hands-on assessment, you should use monitoring devices to quantify your patient's oxygenation and circulatory status. You may also use noninvasive methods to monitor the blood pressure. It is recommended that you always assess the patient's first blood pressure manually with a sphygmomanometer (blood pressure cuff) and stethoscope.

Reassessment

Repeat the primary assessment. Reassess vital signs and the chief complaint. Are the airway, breathing,

You are the Provider: PART 4

After caring for the patient's burns, you cover him with a blanket and begin transport to the hospital. You contact medical control as soon as you leave the scene, and you are advised to provide transport to the emergency department because the closest burn center is 75 miles away. You reassess the patient, including his vital signs.

Recording Time: 12 Minutes	
Level of consciousness	Conscious and alert, but anxious
Respirations	22 breaths/min; becoming labored; voice is becoming hoarse
Pulse	120 beats/min; strong and regular
Skin	Red, warm, and dry
Blood pressure	158/84 mm Hg
Sao$_2$	95% (on oxygen)

7. What is the proper treatment for the patient's burns?

8. How has the patient's condition changed? What should you do now?

and circulation still adequate? Recheck patient interventions. Are the treatments you provided for problems with the ABCs still effective? Reassessing a patient with an open soft-tissue injury is extremely important, especially if you did not put the bandage on the patient's injury. Frequently, other emergency care personnel may have dressed and bandaged the wound before your arrival. You may need to add additional dressings over the original dressing or bandages. Assess all bandaging frequently. If blood continues to soak through bandages, use additional methods to control bleeding as discussed later in the chapter. How is the patient's condition improving with the interventions? Identify and treat changes in the patient's condition.

Interventions

Closed soft-tissues injuries can be life threatening if not appropriately treated. Assess and manage all threats to the patient's airway, breathing, and circulation. All patients with a closed injury should receive oxygen via a nonre-breathing mask.

Although most open soft-tissues injuries are not serious, if not appropriately treated, they can lead to substantial blood loss and even shock. By appropriately treating open soft-tissue injuries, you can minimize the common complications such as bleeding, shock, pain, and infection. You should expose all wounds, cleanse the wound surface, control bleeding, and be prepared to treat the patient for shock.

Extremities that are painful, swollen, or deformed should be splinted. When splinting these types of injuries, remember to assess the patient's pulses and motor and sensory function before and after applying the splint.

Communication and Documentation

Your communication and documentation must include a description of the MOI and the position in which you found the patient when you arrived on scene. In patients who have open injuries with severe external bleeding, it is important to recognize, estimate, and report the amount of blood loss that has occurred and how rapidly or how much time has passed since the bleeding started. This is a challenge, especially if the surface is wet, absorbs fluids, or is dark. You should attempt to report blood loss using terms that you are comfortable with and that will be easily understood by other personnel. For example, you may say "approximately a liter was lost," or "the bleeding has soaked through three trauma dressings." It is not as important how you describe it, but that you describe it accurately. You must include the location and description of any soft-tissue injuries or other wounds you have located and treated. Describe the size and depth of the injury. Provide an accurate account of how you treated these injuries. Your ability to communicate and document clearly and accurately enables the physicians and nurses at the hospital to continue quality care.

Emergency Medical Care for Closed Injuries

Small contusions require no special emergency medical care. More extensive closed injuries may involve significant swelling and bleeding beneath the skin, which could lead to hypovolemic shock. Before treating a closed injury, make sure to follow standard precautions.

Soft-tissue injuries may look rather dramatic. However, you must still focus on airway and breathing first. Always maintain the airway and provide oxygen to patients with potentially serious injuries. If the patient has inadequate breathing, you may have to assist ventilations with a bag-mask device. Treat a closed soft-tissue injury by applying the mnemonic RICES:

- **Rest.** Keep the patient as quiet and as comfortable as possible.
- **Ice.** Using ice or a cold pack slows bleeding by causing blood vessels to constrict and also reduces pain.
- **Compression.** Applying pressure over the injury site slows bleeding by compressing the blood vessels.
- **Elevation.** Raising the injured part just above the level of the patient's heart decreases swelling.
- **Splinting.** This decreases bleeding and also reduces pain by immobilizing a soft-tissue injury or an injured extremity.

In addition to using these measures to control bleeding and swelling, you should also be alert for signs of developing shock, including anxiety or agitation, changes in mental status, increased heart rate, increased respiratory rate, diaphoresis, cool or clammy skin, and decreased blood pressure. Any or all of these signs may indicate internal bleeding resulting from injuries to internal organs. If the patient exhibits signs and symptoms of shock, treat accordingly.

Emergency Medical Care for Open Injuries

Before you begin caring for a patient with an open wound, you should be sure to protect yourself by following standard precautions. Wear gloves, eye protection, and, if necessary, a gown and a mask. Remember that you must make sure the patient has an open airway and administer high-flow oxygen as necessary. If life-threatening bleeding is observed, assign a team member to apply direct pressure over the wound to control the bleeding. Then assess the severity of the wound. If the wound is in the chest or upper abdomen, place an occlusive dressing on the wound.

Your treatment priorities are the primary assessment and to begin lifesaving interventions. This includes controlling the bleeding, which can be extensive and severe. Several methods are available to control open injuries or external bleeding. Start with the most commonly used; these include the following:

- Direct, even pressure and elevation
- Pressure dressings and/or splints
- Tourniquets

It will often be useful to combine these methods. Follow the steps in **Skill Drill 24-1** to control bleeding from an extremity:

1. Apply direct pressure with a dry, sterile dressing (Step 1).

Skill Drill 24-1

Controlling Bleeding From an Open Soft-Tissue Injury

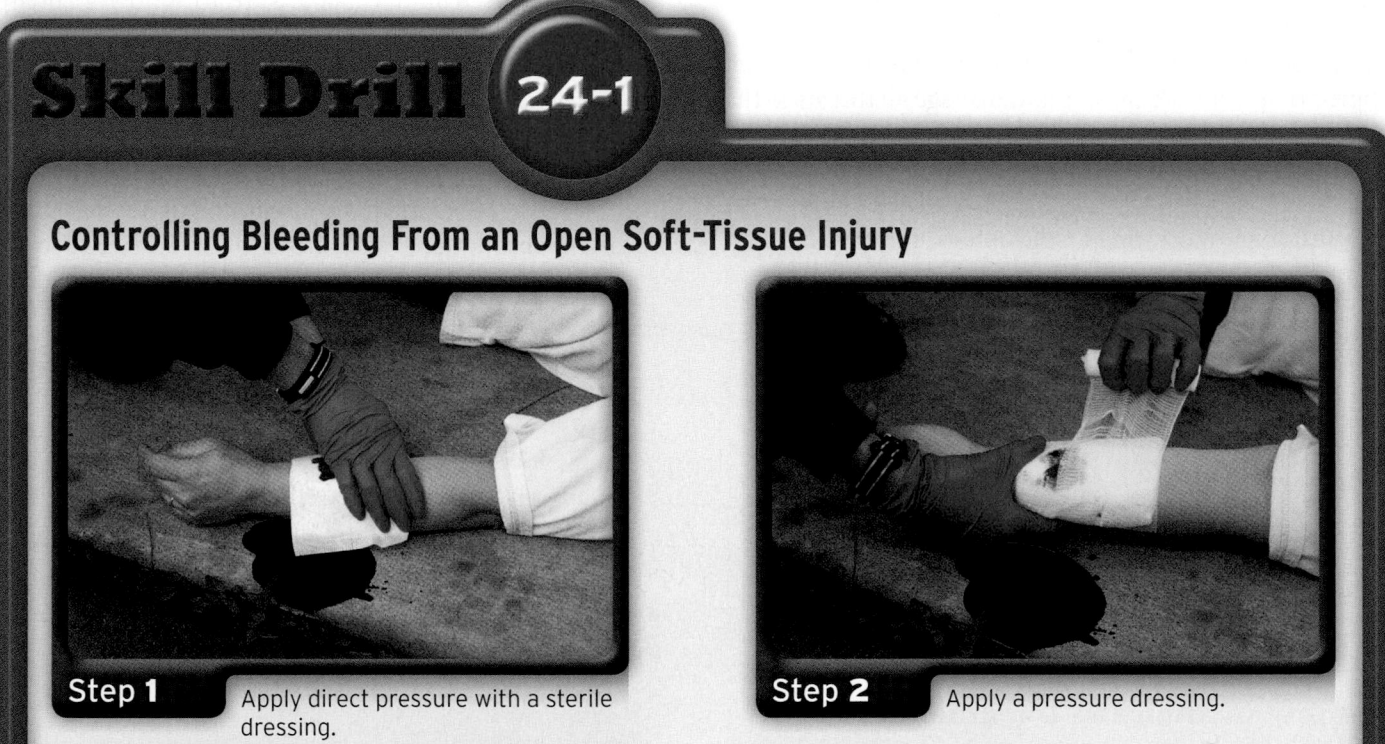

Step 1 Apply direct pressure with a sterile dressing.

Step 2 Apply a pressure dressing.

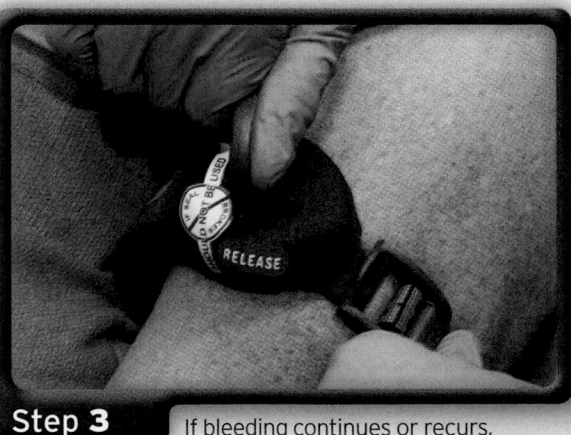

Step 3 If bleeding continues or recurs, apply a tourniquet above the level of bleeding.

2. Apply a pressure dressing (Step 2).

3. If bleeding continues or recurs, apply a tourniquet above the level of bleeding (Step 3).

All open wounds are assumed to be contaminated and present a risk of infection. By applying a sterile dressing, you are reducing the risk of further contamination. This keeps foreign material, such as hair, clothing, and dirt, out of the wound and decreases the risk of infection. In general, you should not try to remove material from an open wound, no matter how dirty the wound is. Rubbing, brushing, or washing an open wound can cause additional bleeding. Chemical burns and contamination should be flushed to remove remaining chemicals. Only hospital personnel should clean out an open wound. To prevent a wound from drying, you may apply sterile dressings moistened with sterile saline solution and then cover the moist dressing with a dry, sterile dressing.

Often, you can better control bleeding from open soft-tissue wounds by splinting the extremity, even if there is no fracture. Splinting can also help you to keep the patient calm and quiet, as it typically reduces pain. In addition, splinting keeps sterile dressings in place, minimizes damage to an already injured extremity, and makes moving the patient easier.

Keep in mind that a patient who is bleeding significantly from an open wound is at risk for hypovolemic shock. You must be alert for this possibility and provide treatment, as needed, in all cases of significant trauma and in patients with moderate to severe bleeding.

■ Abdominal Wounds

An open wound in the abdominal cavity may expose internal organs. In some cases, the organs may even protrude through the wound, an injury called an **evisceration** (Figure 24-11). Do not touch or move the exposed organs. Instead, cover the wound with sterile gauze moistened with sterile saline solution and secure with an occlusive dressing (Figure 24-12). Because the open abdomen radiates body heat very effectively, and because exposed organs lose fluid rapidly, you must keep the organs moist and warm. If you do not have gauze compresses, you may use moist sterile dressings, covered and secured in place with a bandage and tape. Do not use any material that is adherent or loses its substance when wet, such as toilet paper, facial tissue, paper towels, or absorbent cotton. If the patient's legs and knees are uninjured, and spinal injury is *not* suspected, flex them to relieve pressure on the abdomen. Most patients with abdominal wounds require immediate transport to a trauma center, depending on the local protocol.

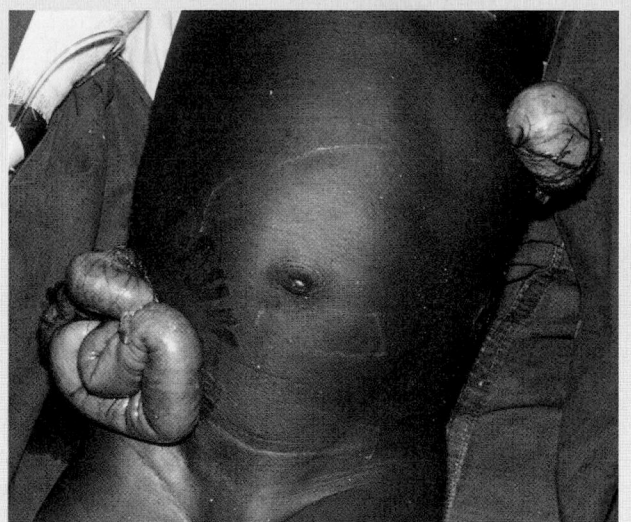

Figure 24-11 An abdominal evisceration is an open wound to the abdomen in which organs protrude through the wound.

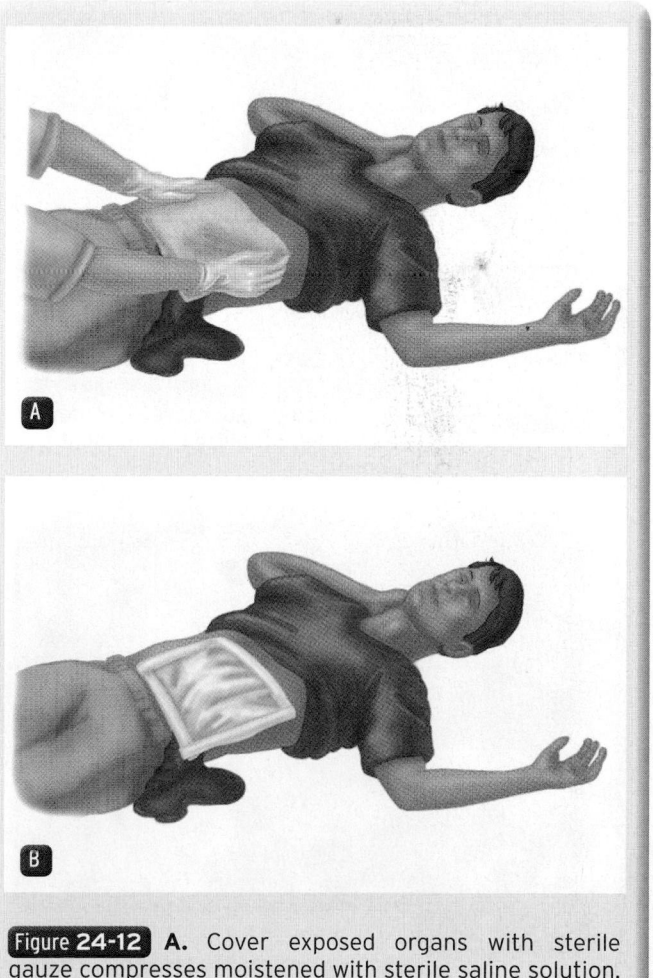

Figure 24-12 **A.** Cover exposed organs with sterile gauze compresses moistened with sterile saline solution. **B.** Place an occlusive dressing over the compresses, and secure it in place by taping all four sides.

■ Impaled Objects

Occasionally, a patient will have an object, such as a knife, fishhook, wood splinter, or piece of glass, impaled in his or her body. To treat this, follow the steps in Skill Drill 24-2:

1. Do not attempt to move or remove the object unless it is impaled through the cheek causing airway obstruc-

tion, or if the object is in the chest and interferes with cardiopulmonary resuscitation (CPR). In most cases, a surgeon will have to remove the object; removing it in the field may cause more bleeding or damage nerves, blood vessels, or muscles within the wound. Stabilize the impaled body part (Step 1).

2. Remove any clothing covering the injury. Control bleeding with direct pressure, and apply a bulky dressing to stabilize the object. Some combination of soft dressings, gauze, and tape may be effective, depending on the location and size of the object. To prevent further injury, manually secure the object by incorporating it into the dressing (Step 2).

Skill Drill 24-2

Stabilizing an Impaled Object

Step 1 Do not attempt to move or remove the object. Stabilize the impaled body part.

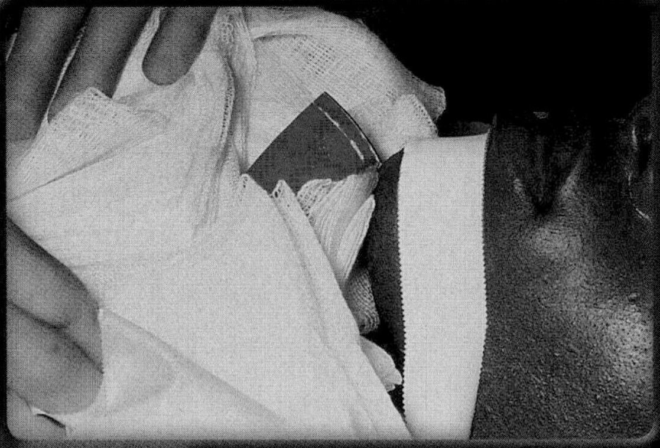

Step 2 Control bleeding, and stabilize the object in place using soft dressings, gauze, and/or tape.

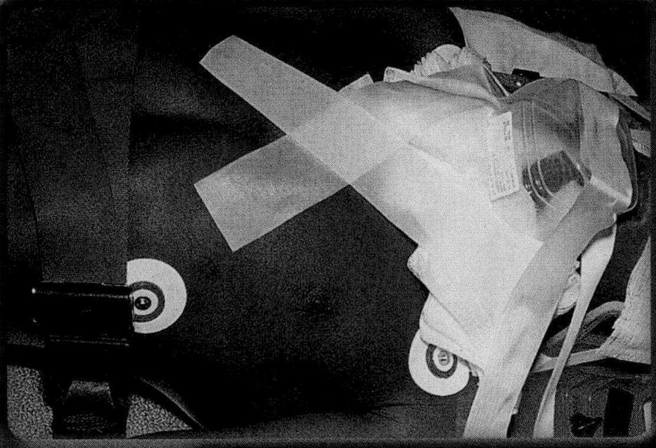

Step 3 Tape a rigid item over the stabilized object to prevent it from movement during transport.

3. Protect the impaled object from being bumped or moved during transport by taping a rigid item such as a plastic cup, a section of a plastic water bottle, or a supply container over the stabilized object and its bandaging (Step 3).

The only exceptions to the rule of not removing an impaled object are an object in the cheek that obstructs breathing and an object in the chest that interferes with CPR. If the object is very long, cut off (shorten) the exposed portion, first securing it to minimize motion and, thus, internal damage and pain. Once the object has been secured and the bleeding is under control, provide prompt transport.

Neck Injuries

An open neck injury can be life threatening. If the veins of the neck are open to the environment, they may suck in air (Figure 24-13). If enough air is sucked into a blood vessel, it can actually block the flow of blood in the lungs, sending the patient into cardiac arrest. This condition is called air embolism. To control bleeding and prevent the possibility of air embolism, cover the wound with an occlusive dressing. Apply manual pressure, but do not compress both carotid vessels at the same time; if you do, this may impair circulation to the brain and cause a stroke. Secure a pressure dressing over the wound by wrapping roller gauze loosely around the neck and then firmly through the opposite axilla (Figure 24-14).

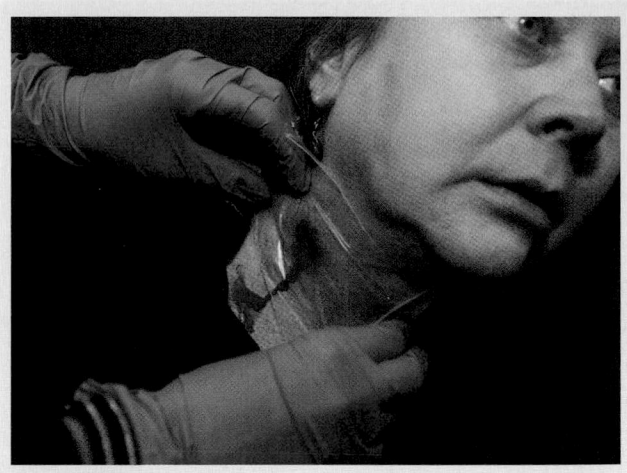

Figure 24-14 Cover neck wounds with an airtight dressing, and apply manual pressure. Be sure that you do not compress both carotid arteries at the same time, as this may impair circulation to the brain.

Bites

Small-Animal Bites and Rabies

At times you may be called to care for a person who has been bitten by a small animal such as a dog, cat, raccoon, squirrel, or other small nonlivestock animal.

Most people who are bitten by small animals do not report the incident to a physician, believing that these bites are not serious. They can be very serious, however. A small animal's mouth is heavily contaminated with virulent bacteria. You should consider all small animal bites as contaminated and potentially infected wounds that may require antibiotics, tetanus prophylaxis, and suturing (Figure 24-15). Occasionally, small animal bites result in mangled, complex wounds that require surgical repair. For these reasons, all small animal bites should be evaluated by a physician. Place a dry, sterile dressing over the wound, and promptly transport the patient to the ED. If an arm or leg was injured, splint that extremity. Often, the patient will be extremely upset and frightened, a situation that calls for reassurance on your part.

A major concern with small animal bites is the spread of rabies, an acute, potentially fatal viral infection of the central nervous system that can affect all warm-blooded animals. Although rabies is extremely rare today, particularly with widespread inoculation of pets, it still exists. Stray dogs that have not been inoculated can be carriers of the disease, as can squirrels, bats, foxes, skunks, and raccoons. The virus is in the saliva of a rabid, or infected, animal and is transmitted through

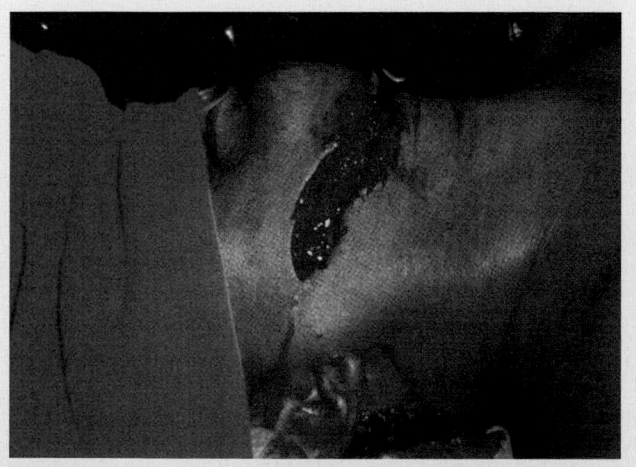

Figure 24-13 Open injuries to the neck can be very dangerous. If veins are open to the environment, they can suck in air, resulting in a potentially fatal condition called air embolism.

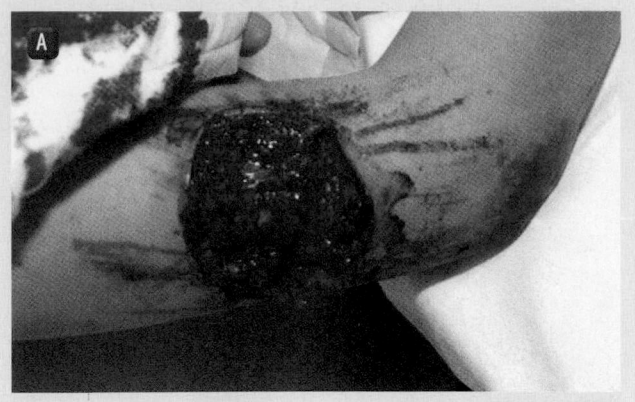

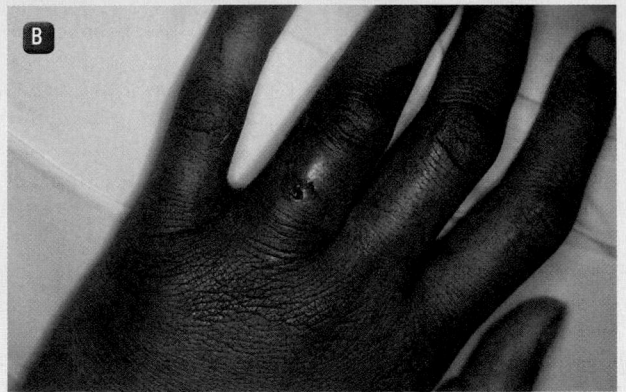

Figure 24-15 Small animal bite wounds should be examined at the hospital, as these wounds are heavily contaminated with virulent bacteria. **A.** Dog bite. **B.** Cat bite.

biting or licking an open wound. Infection can be prevented in a person who has been bitten by such an animal only by a series of special vaccine injections, a painful procedure that must be started soon after the bite. Since animals that have rabies do not always demonstrate symptoms immediately, a person's only chance to avoid the vaccine is to find the animal and turn it over to the health department for observation and/or testing. Refer to your local animal control procedures.

Children, particularly young ones, may be seriously injured or even killed by dogs. These dogs are not always vicious or **rabid**; sometimes a child unknowingly provokes the animal. However, you must assume that the animal may turn and attack you as well. Therefore, you generally should not enter the scene until the animal has been secured by the police or an animal control officer. Then you may carry out the necessary emergency care and transport the child to the ED.

Human Bites

The human mouth, more so than even the small animal's mouth, contains an exceptionally wide range

of virulent bacteria and viruses. For this reason, you should regard any human bite that has penetrated the skin as a very serious injury. Similarly, any laceration caused by a human tooth can result in a serious, spreading infection **Figure 24-16**. Remember this if you treat someone who has been punched in the mouth: the person who delivered the punch may also need treatment.

The emergency treatment of bites consists of the following steps:

1. Apply a dry, sterile dressing.
2. Promptly immobilize the area with a splint or bandage.
3. Provide transport to the ED for surgical cleansing of the wound and antibiotic therapy.

Burns

As an EMT, you will often provide care to patients who have been burned. Burns account for more than 10,000 deaths a year. Burns are also among the most serious and painful of all injuries. A burn occurs when the body, or a body part, receives more radiant energy than it can absorb, resulting in an injury. Potential sources of this energy include heat, toxic chemicals, and electricity. The proper emergency care of a burn may increase a patient's chances of survival and decrease the risk or duration of a long-term disability. Although a burn may be the patient's most obvious injury, you should always perform a complete assessment to determine whether there are other serious injuries. Finally, keep in mind that children, elderly patients, and patients with chronic illnesses are more likely to experience shock from burn

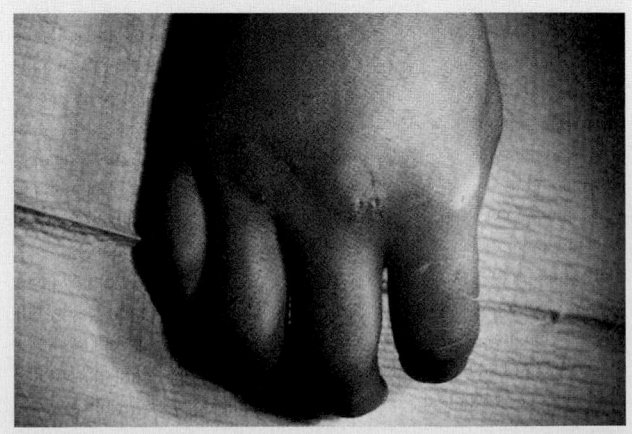

Figure 24-16 Human bites can result in serious, spreading infection. Thus, patients must be evaluated at the hospital.

injuries. It is important for you to be prepared to treat accordingly.

Complications of Burns

There are several complications that can result secondary to a burn injury, all of which can be life threatening. The skin serves as a barrier between the environment and the body. When a person is burned, this barrier is destroyed; the victim is now at a high risk for infection, hypothermia, hypovolemia, and shock. Burns to the airway are of significant importance because the loose mucosa in the hypopharynx can swell and lead to complete airway obstruction. Circumferential burns of the chest can compromise breathing. Circumferential burns of an extremity can lead to neurovascular compromise and irreversible damage if not appropriately treated. If you suspect any complications, ALS backup should be summoned.

Burn Severity

The seriousness of a burn may influence the choice of a treatment facility. Five factors will help you to determine the severity of a burn.

1. What is the depth of the burn?
2. What is the extent of the burn?

These first two factors are the most important. After gauging these, ask yourself the following remaining questions.

3. Are any critical areas (face, upper airway, hands, feet, genitalia) involved? Also included in critical areas are any circumferential burns, which are burns that go completely around a body part such as an arm, foot, or chest.
4. Does the patient have any preexisting medical conditions or other injuries?
5. Is the patient younger than 5 years or older than 55 years?

If the answer to any of these last three questions is yes, you should upgrade the classification Table 24-1 .

Keep in mind that burns to the face are of particular importance owing to the potential of airway involvement. In addition, burns to the hands or feet or over joints are also considered serious because of the potential of loss of function as the result of scarring.

Table 24-1 Classification of Burns in Adults

Severe Burns

- Full-thickness burns involving the hands, feet, face, upper airway, or genitalia or circumferential burns of other areas
- Full-thickness burns covering more than 10% of the body's total surface area
- Partial-thickness burns covering more than 30% of the body's total surface area
- Burns associated with respiratory injury (smoke inhalation or inhalation injury)
- Burns complicated by fractures
- Burns on patients younger than 5 years or older than 55 years that would be classified as "moderate" on young adults

Moderate Burns

- Full-thickness burns involving 2% to 10% of the body's total surface area (excluding hands, feet, face, genitalia, and upper airway)
- Partial-thickness burns covering 15% to 30% of the body's total surface area
- Superficial burns covering more than 50% of the body's total surface area

Minor Burns

- Full-thickness burns covering less than 2% of the body's total surface area
- Partial-thickness burns covering less than 15% of the body's total surface area
- Superficial burns covering less than 50% of the body's total surface area

Special Populations

Geriatric Needs

When you are treating geriatric patients with burns, it is important to be vigilant for the possibility of abuse. Geriatric patients who are institutionalized, disoriented, or incapable of clear communication are particularly susceptible to abuse.

Signs of abuse in a geriatric patient include evidence of multiple injuries in various stages of healing (eg, multiple bruises of different colors, new and old fractures involving more than one extremity), injuries that do not seem to correspond to the history provided by caregivers, and burns associated with a suspicious history.

Burns that appear in a "pattern" are suspicious for intentional injuries. Multiple, small circular burns may be indicative of cigarette or cigar injuries. Other patterns may indicate irons, stovetops, or other hot surfaces not easily encountered accidentally. Scalding injuries to the hands or feet may also be indicative of abuse. It is important to remember that these injuries are often inflicted in areas not readily seen. If the situation is suspicious for geriatric abuse, be sure to fully examine the patient under his or her clothing for signs of abuse. As always, appropriate support and transport of the patient in a timely manner remain a priority.

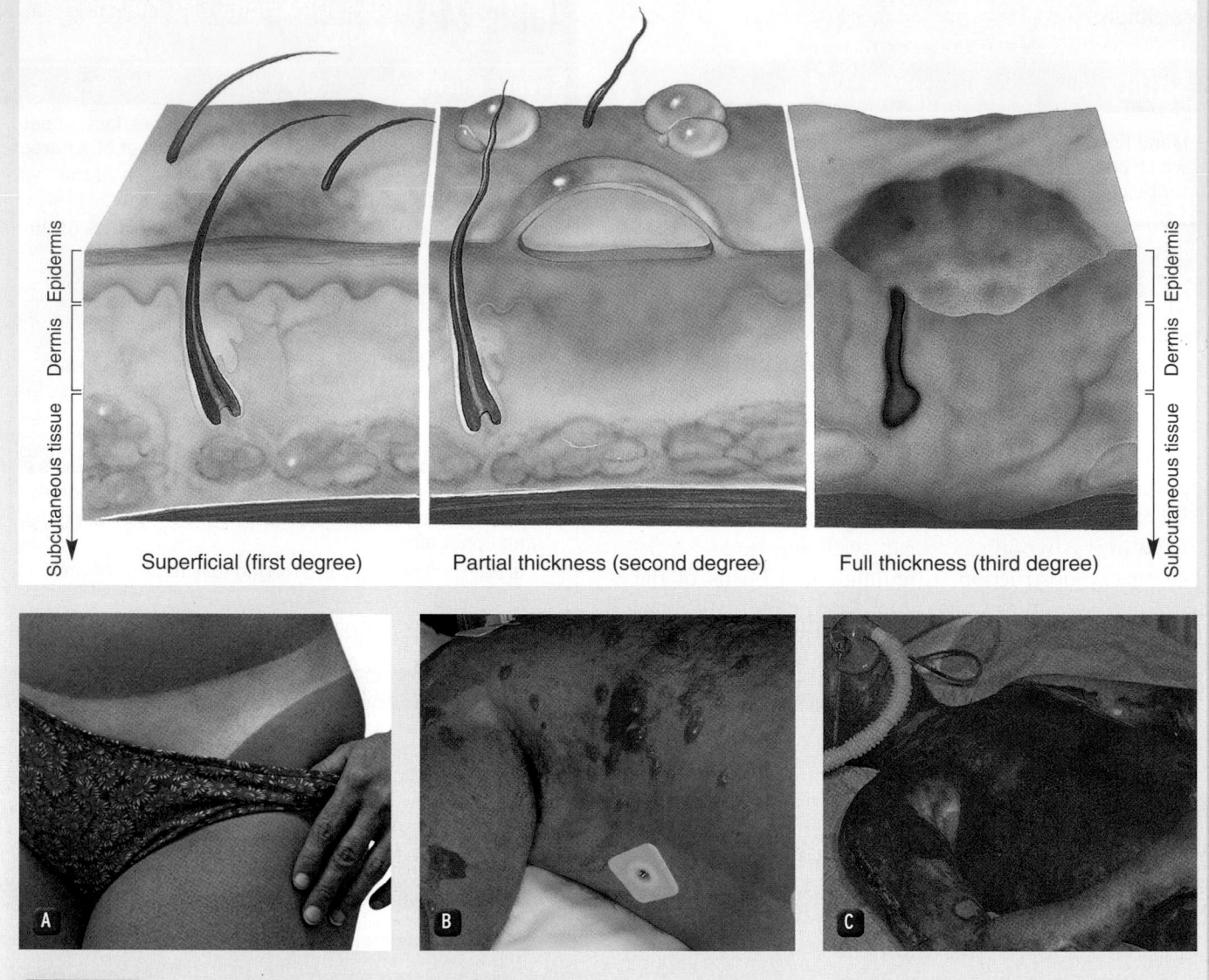

Figure 24-17 Classification of burns. **A.** Superficial or first-degree burns involve only the epidermis. The skin turns red but does not blister or actually burn through. **B.** Partial-thickness or second-degree burns involve some of the dermis, but they do not destroy the entire thickness of the skin. The skin is mottled, white to red, and is often blistered. **C.** Full-thickness or third-degree burns extend through all layers of the skin and may involve subcutaneous tissue and muscle. The skin is dry, leathery, and often either white or charred.

Depth

Burns are first classified according to their depth **Figure 24-17**. You must be able to identify the following three types of burns:

- **Superficial (first-degree) burns** involve only the top layer of skin, the epidermis. The skin turns red but does not blister or actually burn through. The burn site is painful. A sunburn is a good example of a superficial burn.
- **Partial-thickness (second-degree) burns** involve the epidermis and some portion of the dermis. These burns do not destroy the entire thickness of the skin, nor is the subcutaneous tissue injured. Typically, the skin is moist, mottled, and white to red. Blisters are present. Partial-thickness burns cause intense pain.

- **Full-thickness (third-degree) burns** extend through all skin layers and may involve subcutaneous layers, muscle, bone, or internal organs. The burned area is dry and leathery and may appear white, dark brown, or even charred. Some full-thickness burns feel hard to the touch. Clotted blood vessels or subcutaneous tissue may be visible under the burned skin. If the nerve endings have been destroyed, a severely burned area may have no feeling. However, the surrounding, less severely burned areas may be extremely painful.

A pure full-thickness burn is unusual. Severe burns are typically a combination of superficial, partial-thickness, and full-thickness burns. Superficial burns heal well without scarring. Small partial-thickness burns also heal without scarring. However, deep partial-thickness burns and all full-thickness burns are prone to scarring and may be best managed surgically.

Significant airway burns are also serious. They may be associated with singed hair within the nostrils, soot around the nose and mouth, hoarseness, and hypoxia.

It may be impossible to accurately estimate the depth of a particular burn shortly after injury. Even experienced burn surgeons sometimes underestimate or overestimate the extent of a particular burn.

Extent

One quick way to estimate the surface area that has been burned is to compare it to the size of the patient's palm, which is roughly equal to 1% of the patient's total body surface area. This technique is called the rule of palm. Another useful measurement system is the **rule of nines**, which divides the body into sections, each of which is approximately 9% of the total surface area Figure 24-18. Remember that the head of an infant or child is relatively larger than the head of an adult, and the legs are relatively smaller.

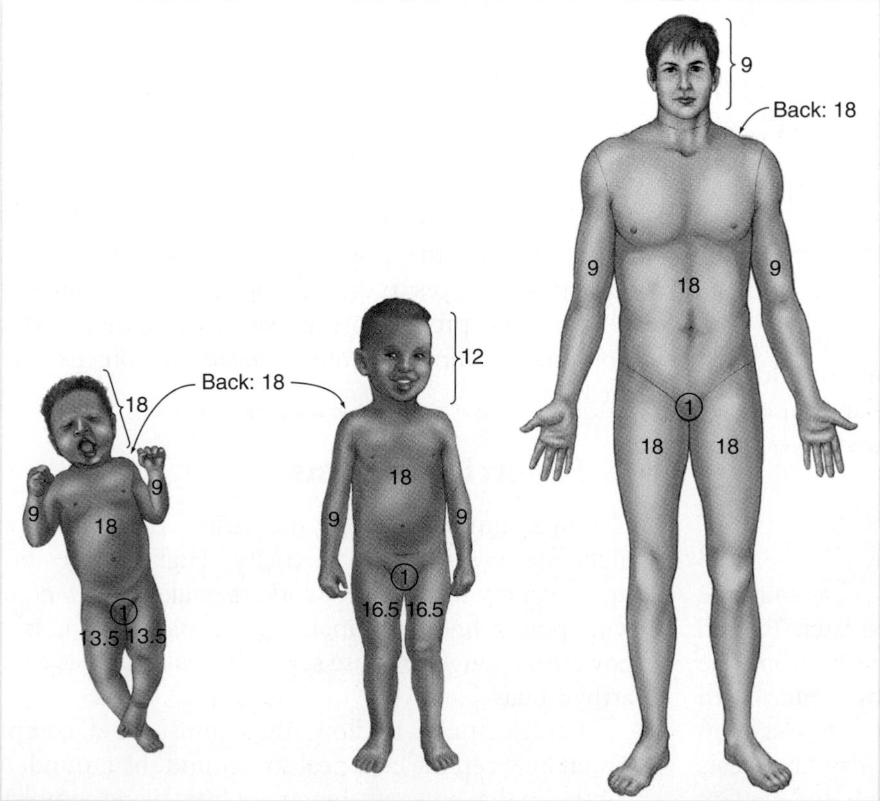

Figure 24-18 The rule of nines is a quick way to estimate the amount of surface area that has been burned. It divides the body into sections, each representing approximately 9% of the total body surface area. The proportions differ for infants, children, and adults.

Special Populations

Pediatric Needs
Burns to children are generally considered more serious than burns to adults Table 24-2. This is because infants and children have more surface area relative to total body mass, which means greater fluid and heat loss. In addition, children do not tolerate burns as well as adults do. Children are also more likely to go into shock, develop hypothermia, and experience airway problems because of the unique differences associated with their ages and anatomy.

Many burns in infants and children result from child abuse. The classic burn resulting from deliberate immersion involves the hands and wrists, as well as the feet, lower legs, and buttocks. Similarly, burns around the genitals and multiple cigarette burns should be viewed as possible abuse. You should report all suspected cases of abuse to the proper authorities, especially those where a significant delay in evaluation and treatment is evident (see Chapter 32, *Pediatric Emergencies*).

■ Chemical Burns

A chemical burn can occur whenever a toxic substance contacts the body. Most chemical burns are caused by strong acids or strong alkalis. The eyes are particularly vulnerable to chemical burns Figure 24-19. Sometimes the fumes alone from strong chemicals can cause burns, especially to the respiratory tract. The severity of the burn is directly related to the type of chemical, the concentration of the chemical, and the duration of the exposure.

To prevent exposure to hazardous materials, you must wear the appropriate chemical-resistant gloves and eye protection whenever you are caring for a patient with a chemical burn. Be particularly careful not to get any chemical, dry or liquid, on yourself or on your uniform; consider wearing a protective gown when this is a possibility. Remember that exposure risk is also present when you are cleaning up after a call. In cases of severe chemical burns or exposure, consider mobilization of the hazardous materials (HazMat) team, if appropriate.

Treatment for chemical burns can be specific to the chemical agent. If available, read all of the labels of the chemical agent. Do not

Table 24-2	Classification of Burns in Infants and Children
Severe Burns	
■ Any full-thickness burn ■ Partial-thickness burns covering more than 20% of the body's total surface area	
Moderate Burns	
■ Partial-thickness burns covering 10% to 20% of the body's total surface area	
Minor Burns	
■ Partial-thickness burns covering less than 10% of the body's total surface area	

Figure 24-20 Brush dry chemicals off the patient before you flush the burned area with water.

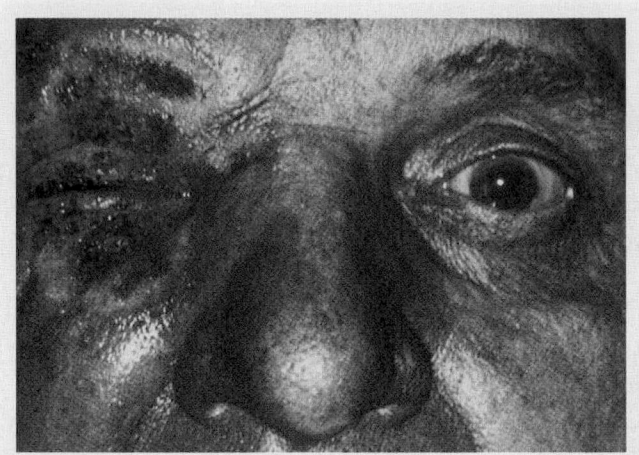

Figure 24-19 The eyes are particularly vulnerable to chemical burns.

risk exposure while attempting to gather information on the chemical. If the exposure occurs at an industrial site, such as a chemical manufacturing plant, an expert should be on-site and should be able to provide you with valuable information on the chemical.

Management of Chemical Burns

The emergency care of a chemical burn is basically the same as that for a thermal burn, discussed later. To stop the burning process, remove any chemical from the patient. A dry chemical that is activated by contact with water may damage the skin more when it is wet than when it is dry. Therefore, always brush dry chemicals off the skin and clothing before flushing the patient with water **Figure 24-20**. Remove the patient's clothing, including shoes, stockings, gloves, and any jewelry or glasses, because there may be small amounts of chemicals in the creases.

For liquid chemicals, immediately begin to flush the burned area with large amounts of water **Figure 24-21**, taking care not to contaminate uninjured areas or make the patient hypothermic. Never direct a forceful stream of water from a hose at the patient; the extreme water pressure may mechanically injure the burned skin. Continue flooding the area with gallons of water for 15 to 20 minutes after the patient says the burning pain has stopped. If the patient's eye has been burned, hold the eyelid open while flooding the eye with a gentle stream of water **Figure 24-22**. Flush the eyes from the inside corners to the outside to prevent cross contamination. If only one eye has been affected, turn the patient's head to that side and flush. If both eyes are affected, consider hooking up a nasal cannula to a bag of saline in order to flush both eyes simultaneously. The prongs can be placed on the bridge of the nose in order to flush from the inside corners of the eyes to the outside corners. Be careful not to touch the prongs to the eye or surrounding tissue. Continue flushing the contaminated area on the way to the hospital.

■ Electrical Burns

Electrical burns may be the result of contact with high- or low-voltage electricity. High-voltage burns may occur when utility workers make direct contact with power lines. Ordinary household current is still powerful enough to cause severe burns as well as cardiac arrhythmias.

For electricity to flow, there must be a complete circuit between the electrical source and the ground. Any substance that prevents this circuit from being completed, such as rubber, is called an insulator. Any substance that allows a current to flow through it is called a conductor. The human body, which is primarily water, is a good conductor. Thus, electrical burns occur when the body,

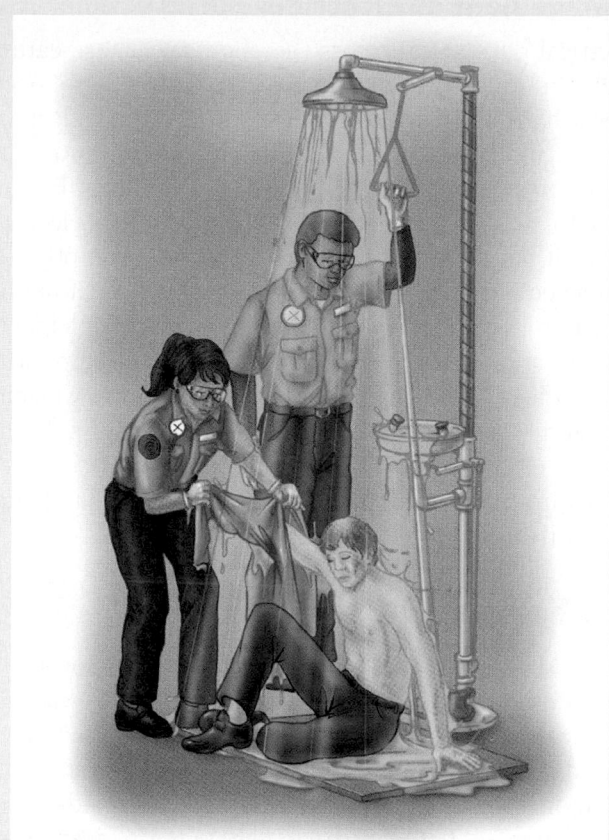

Figure 24-21 Flush the burned area with large amounts of water for 15 to 20 minutes after the patient says that the burning pain has stopped. Be careful to avoid contaminating uninjured areas.

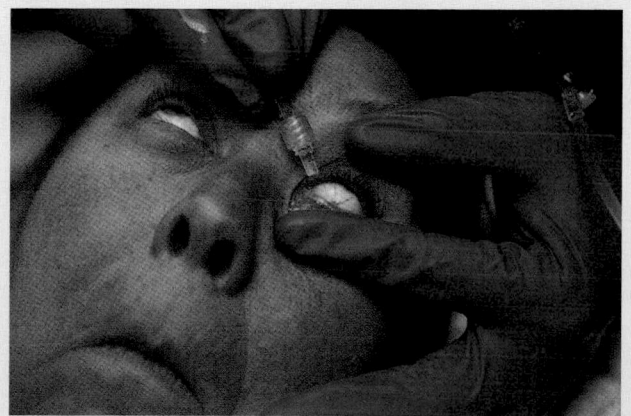

Figure 24-22 Flood the affected eye with a gentle stream of water. Hold the eyelids open, a challenging task because the patient's reflex is to keep the eye shut. Take care to prevent any of the chemical from getting into the other eye during flushing.

or a part of it, completes a circuit connecting a power source to the ground **Figure 24-23**.

The type of electric current, magnitude of current (amperage), and voltage have effects on the seriousness

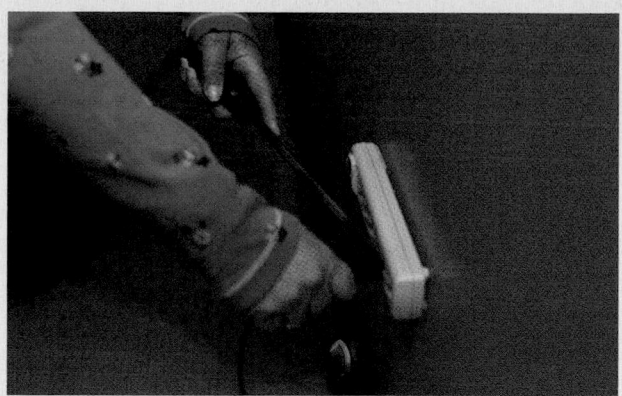

Figure 24-23 The human body is a good conductor of electricity. An electrical burn usually occurs when the body, acting as a conductor, completes a circuit.

of burns. When an electric current enters the body, the skin is burned at the entrance wound as well as everywhere along the path until the current grounds and exits the body. In addition to tissues damaged by the heat, significant chemical changes take place in the nervous, cardiovascular, and muscular systems of the body, causing disruption of the body's normal functions and/or even system failure.

Your safety is of particular importance when you are called to the scene of an emergency involving electricity. Obviously, you can be fatally injured by coming into contact with power lines. But you can also be fatally injured by touching a patient who is still in contact with a live power line or any other electrical source. For this reason, you must never attempt to remove someone from an electrical source unless you are specially trained to do so. Likewise, you should never move a downed power line unless you have the special training and equipment necessary for the job. Before even approaching someone who may still be in contact with a power line or an electrical appliance, make certain that the power is turned off. Always assume that any downed power line is live.

A burn injury appears where the electricity enters (an entrance wound) and exits (an exit wound) the body. The entrance wound may be quite small **Figure 24-24A**, but the exit wound can be extensive and deep **Figure 24-24B**. Always look for both entrance and exit wounds. There are two dangers specifically associated with electrical burns. First, there may be a large amount of deep tissue injury. Electrical burns are always more severe than the external signs indicate. The patient may have only a small burn to the skin but may have massive damage to the deeper tissues, organs, and the nervous system **Figure 24-25**. Second, the patient may go into cardiac or respiratory arrest from the electric shock.

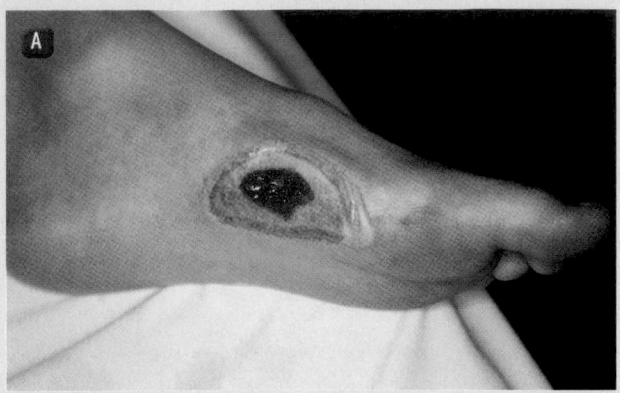

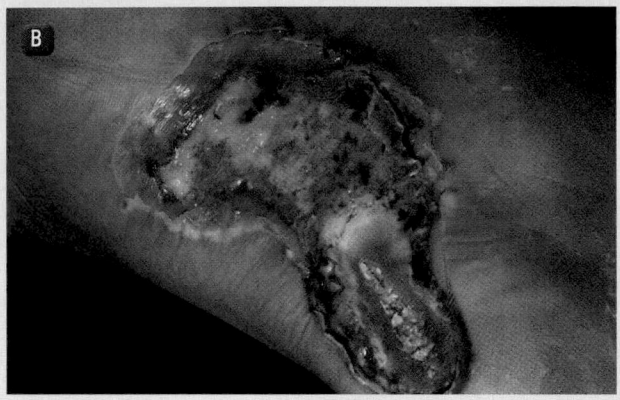

Figure 24-24 Electrical burns, like gunshot wounds, have entrance and exit wounds. **A.** An entrance wound is often quite small. **B.** The exit wound can be extensive and deep.

Management of Electrical Burns

Electrical current can cross the chest and cause cardiac arrest or arrhythmias. Cardiac arrest can also occur after a lightning strike, which is a form of an electrical burn. If indicated, begin CPR on the patient and apply the automated external defibrillator. Although CPR may need to be quite prolonged in patients with electrical burns, it has a high success rate if started promptly. You should be prepared to defibrillate if necessary. If neither CPR nor defibrillation is indicated, give supplemental oxygen, and monitor the patient closely for respiratory and cardiac arrest. Treat the soft-tissue injuries by placing dry, sterile dressings on all burn wounds and splinting suspected fractures. Provide prompt transport; all electrical burns are potentially severe injuries that require further treatment in the hospital.

■ Thermal Burns

<u>**Thermal burns**</u> are caused by heat (as opposed to electricity, chemicals, or radiation). Many different situations can cause thermal burns, and all pose a safety hazard to responding emergency care providers. Most commonly, thermal burns are caused by scalds or an open flame. A <u>**flame burn**</u> is very often a deep burn, especially if a person's clothing catches fire. Hot liquids produce scald injuries. A <u>**scald burn**</u> is most commonly seen in children and handicapped adults but can happen to anyone, particularly while cooking. Scald burns often cover large surface areas of the body because liquids can spread quickly. Coming in contact with hot objects produces a <u>**contact burn**</u>. Ordinarily, reflexes protect a person from prolonged exposure to a very hot object, so contact burns are rarely deep unless the patient was prevented from drawing away from the hot object (for example, unconscious, intoxicated, restrained, or impaired).

A <u>**steam burn**</u> can produce a topical (scald) burn. Minor steam burns are common when microwaving food covered with plastic wrap. When the plastic is peeled away, hot steam escapes directly onto the person's hand. Steam (that is, gaseous water) is also responsible for causing airway burns.

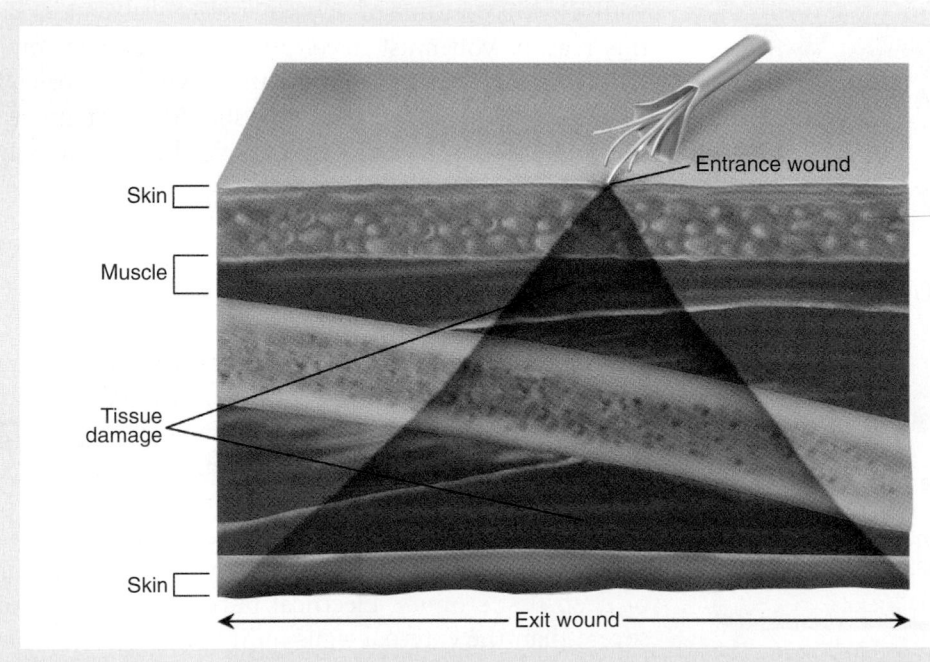

Skin

Muscle

Tissue damage

Skin

Entrance wound

Exit wound

Figure 24-25 External signs of an electrical burn may be deceiving. The entrance wound may be a small burn, whereas the damage to deeper tissue may be massive.

Another important source of thermal burns is the <u>flash burn</u> produced by an explosion, which may briefly expose a person to very intense heat. Lightning strikes can also cause a flash burn. These injuries are usually minor compared with the potential for trauma from whatever caused the flash.

Management of Thermal Burns

Management of thermal burns is largely the same as with any other burn. Stop the burning source, cool the burned area if appropriate, and remove all jewelry. You should maintain a high index of suspicion for inhalation injuries. Increased exposure time will increase damage to the patient. The larger the burn, the more likely the patient will be susceptible to hypothermia and/or hypovolemia. All patients with large surface burns should have a dry dressing applied to help maintain body temperature, prevent infection, and provide comfort.

■ Inhalation Burns

Inhalation injuries can occur when burning takes place in enclosed spaces without ventilation. When the upper airway is exposed to excessive heat, the patient can experience rapid and serious airway compromise. The heat can be an irritant to the lungs and the airway, causing coughing, wheezing, and rapid swelling or edema of the mucosa of the upper airway tissues, often evidenced by stridor. Upper airway damage is often associated with the inhalation of superheated gases. Lower airway damage is more often associated with the inhalation of

chemicals (eg, acids, aldehydes) and particulate matter. When treating a patient for inhalation injuries, you may encounter severe upper airway swelling, requiring intervention immediately after a severe burn, although this problem may not manifest itself until transport. You should consider requesting ALS backup if the patient has signs or symptoms of edema such as a hoarse voice, singed nasal hairs, singed facial hairs, burns of the face, or carbon particles in the sputum. Application of cool mist or aerosol therapy may help reduce some minor edema. Because most ambulances do not carry misters, apply an ice pack to the throat.

The combustion process produces a variety of toxic gases. The less efficient the combustion process, the more toxic the gases—such as carbon monoxide (CO) and carbon dioxide (CO_2)—that may be created. When furnaces, kerosene heaters, and other heating devices are in poor repair, they may emit unsafe levels of these toxic gases. Internal combustion engines may emit many of the same gases and, consequently, should always have their exhaust vented to the outdoors. A common cause of CO exposure is running a small engine in an enclosed space like a garage or basement. For this reason, many ambulance services and fire departments have added CO detectors to their garages or ambulance bays. Fire fighters who are performing an overhaul after a fire may be exposed to high levels of CO, as may people who are exposed to large amounts of car exhaust (such as toll takers and auto mechanics).

CO intoxication should be considered whenever a group of people in the same place all report a headache

You are the Provider: PART 5

You reassess the patient and then call your radio report to the receiving facility. The patient is still experiencing respiratory distress, but he is still moving air adequately. Your estimated time of arrival at the hospital is 6 minutes.

Recording Time: 17 Minutes	
Level of consciousness	Conscious and alert, but anxious
Respirations	22 breaths/min; labored; hoarse voice
Pulse	116 beats/min; strong and regular
Skin	Red, warm, and dry
Blood pressure	160/80 mm Hg
Sao_2	96% (on oxygen)

9. What, if any, additional treatment is indicated for this patient?

or nausea (a malfunctioning furnace or car exhaust being sucked into the air-handling system can cause CO intoxication in groups of people). Similarly, you should be suspicious when people complain of feeling sick at home but not when they go to work or school.

CO can displace oxygen from the alveolar air and the blood hemoglobin. Because CO binds to receptor sites on hemoglobin at least 250 times more easily than oxygen (O_2), the patient's hemoglobin may become saturated with the wrong chemical. Being exposed to relatively small concentrations of CO (such as in cigarette smoke) will result in progressively higher blood levels of CO. Most people have approximately 2% CO attached to their hemoglobin, but these levels may be as high as 4% to 8% in heavy smokers. Levels of 50% or higher may be fatal.

Traditional wisdom tells us that patients with CO intoxication will appear "cherry red." Most practitioners agree that this cherry red skin is most commonly seen in people who have died, not living people. So, never rule out CO intoxication because the patient's skin is not cherry red.

Patients with severe CO intoxication usually have an O_2 saturation level that is normal or better. For this reason, you should be suspicious of pulse oximeter readings when you are dealing with a patient who is suspected to have CO poisoning. New devices that can measure a patient's CO levels will soon be common in prehospital care; they will allow you to find and treat low-level CO intoxication far more readily than you can today.

The gaseous form of cyanide is hydrogen cyanide (HCN). It is generated by the combustion of hydrogen and contains substances such as paper, cotton, and wool. HCN is colorless and has the smell of bitter almonds; however, it can be difficult to detect at the scene of a fire. Prehospital diagnosis of HCN poisoning is difficult because laboratory studies are necessary. Signs and symptoms involve the central nervous, respiratory, and cardiovascular systems of the body and include faintness, anxiety, abnormal vital signs, headache, seizures, paralysis, and coma.

Management of Inhalation Burns

In situations where you have patients who have sustained inhalation injuries, you must first ensure your own safety and the safety of your coworkers. Once you have taken precautions, prehospital treatment of a patient with suspected HCN poisoning includes decontamination and supportive care according to signs and symptoms displayed by the patient.

Exposure to other toxic gases can also cause damage to organs and systems and may cause death. Care for any toxic gas exposure includes recognition, identification, and supportive treatment as necessary according to the patient's signs and symptoms.

■ Radiation Burns

Acute radiation exposure has become more than a theoretical issue because the use of radioactive materials has increased in industry and medicine; therefore, you must understand it to effectively manage patients exposed to radiation. Since 1944, there have been more than 400 radiation accidents involving significant radiation exposure to more than 3,000 people. Potential threats include incidents related to the use and transportation of radioactive isotopes and intentionally released radioactivity in terrorist attacks. To be effective, you must first determine if there has been a radiation exposure and then attempt to determine whether ongoing exposure continues to exist. Increasingly, special response units are equipped with pager-sized radiation detectors, or such detection may be provided by other public safety services.

There are three types of ionizing radiation: alpha, beta, and gamma. Alpha particles have little penetrating energy and are easily stopped by the skin. Beta particles have greater penetrating power and can travel much farther in air than alpha particles. They can penetrate the skin but can be blocked by simple protective clothing designed for this purpose. The threat from gamma radiation is directly proportional to its wavelength. This type of radiation is very penetrating and easily passes through the body and solid materials.

Radiation is measured in units of radiation absorbed dose (rad) or radiation equivalent in man (rem): 100 rad = 1 gray (Gy). Small amounts of everyday background radiation are measured in rad; the amount of radiation released in a major incident may be measured in gray. The average human exposure from background radiation is 0.36 rem per year. Mild radiation sickness can be expected with exposures of 1 to 2 Gy (100 to 200 rad), moderate sickness at 2 to 5 Gy, and severe sickness at 4 to 6 Gy. Exposure to more than 8 Gy is immediately fatal.

Most ionizing radiation accidents involve gamma radiation, or x-rays. People who have suffered a radiation exposure generally pose no risk to the people around them. However, in some types of incidents—particularly those involving explosions—patients may be contaminated with radioactive particulate matter. It is speculated that after a nuclear explosion, most patients will have sustained some type of trauma in addition to the radiation exposure.

Radiation burns require special rescue techniques beyond the initial training of the EMT. If not properly trained, maintain a safe distance and wait for the HazMat team to decontaminate the victim before initiating patient care.

Management of Radiation Burns

Being exposed to a radiation source does not make a patient contaminated or radioactive. However, when patients have a radioactive source on their body (such as debris from a bomb that dispersed radioactive material), they are contaminated and must be initially cared for by a HazMat responder. Once decontaminated by the HazMat team, care is often transferred to the EMT. Most contaminants can be removed by simply removing the patient's clothes. You will need to call for additional resources to manage this situation. Once the patient is decontaminated and there is no threat to you, you may begin treatment with the ABCs and treat the patient for any burns or trauma.

Irrigate open wounds. Washing should be gentle to avoid further damage to the skin, which could result in additional internal radiation absorption. The head and scalp should be irrigated the same way. The ED should be notified as soon as practical if you are transporting a potentially contaminated patient. In contrast with other types of contamination, radioactive particulate matter probably poses a relatively small risk to the rescuer. Consider providing basic care to the patient before decontamination if you are wearing protective clothing.

Increasing your (and your patient's) distance from the source by even a few feet may dramatically decrease your exposure, so it is important to identify the radioactive source and the length of the patient's exposure to it, if this information is available without putting you or your patient at risk for exposure. If not readily available, rely on the HazMat team to obtain this information. You must try to limit your duration of exposure, increase your distance from the source, and attempt to place shielding between yourself and sources of gamma radiation.

With contact radiation burns, decontaminate the wound as if it were a chemical burn to remove any radioactive particulate matter. You may then treat it as a burn.

Many radioactive isotopes are used in medicine and industry, some of which can be absorbed or have their toxic effects blunted by another substance. Like their radioactive effects, the toxic effects of these isotopes vary. Antidotes may help bind an isotope, enhance its elimination from the body, or reduce the toxic effects on other organs. Such antidotal therapy should be considered only under the guidance of a knowledgeable physician or public health agency.

Patient Assessment of Burns

When you are assessing a burn, it is important for you to classify the victim's burns. Classification of burns involves determining the source of the burn, the depth of the burn, and its severity. Assessment of a burn patient is essentially the same as with any other trauma patient. Again, you must use caution to avoid being distracted by dramatic burn injuries and, thus, possibly overlooking other potential life threats that require treatment.

Scene Size-up

Scene Safety

As you arrive on scene, observe the scene for hazards and threats to the safety of you and your crew, bystanders, and the patient. Ensure that the factors that led to the patient's burn injury do not pose a hazard to you and your crew. Is the electricity turned off? Is the chemical leak secure? Has the fire been extinguished? Is there any potential for violence?

Mechanism of Injury/Nature of Illness

When possible, attempt to determine the type of burn that has been sustained and the MOI. Burn patients can be very difficult to manage both physically and emotionally. It is easy to become overwhelmed by the sights, sounds, and smells of burn victims.

Assess the scene for any environmental hazards. If the patient is the victim of a lightning strike, is the weather still a threat to your safety? Anticipate using gloves and eye protection with any burn patient and gowns when serious injuries are expected. Determine the number of patients; the possibility for multiple patients grows if you are responding to a lightning strike or a vehicle crash. At vehicle crashes, ensure that there are no energized electrical lines or leaking fuel in the area where you will be working. If you determine that the power company, the fire department, or ALS units are needed, call for additional resources early. Remember, the burn patient is a trauma patient. Consider the potential for spinal injuries, inhalation injuries, and other injuries.

Primary Assessment

The primary assessment includes a rapid scan of the patient to identify and manage life-threatening concerns and to assist with transport decisions. The primary assessment begins when you approach the patient and form a general impression.

Form a General Impression

As you approach the burn trauma patient, simple clues can help identify how serious the injuries are and how quickly you need to assess and treat them. If your patient greets you with a hoarse voice or is reported to have been in an enclosed space with a fire or intense heat source, these should be indications of a significant MOI. Similarly, if the patient has singed facial hair, eyebrows, or nasal hair your initial general impression might be that the patient has a potential airway and/or breathing problem.

Child abuse and elder abuse are unpleasant situations to handle. Unfortunately, they are often situations that involve burns. As you enter a scene where burns are involved, be suspicious of clues that may indicate abuse.

The burned patient you encounter may have graphic injuries; however, you must not be distracted from the primary assessment. As you begin the primary assessment always consider the need for manual spinal stabilization.

Check for responsiveness using the AVPU scale. Assessing a patient's mental status is generally easy and can be done by asking the patient about the chief complaint. If the patient is alert, this should help direct you to any apparent life threats. If the patient is not alert, determine if he or she responds to verbal or painful stimuli or if he or she is unresponsive. An unresponsive patient may indicate a life-threatening condition. In all patients whose level of consciousness is less than alert and oriented, you should administer high-flow oxygen via a nonrebreathing mask and provide immediate transport to the ED.

Airway and Breathing

Ensure that the patient has a clear and patent airway. If the patient is unresponsive or has a significantly altered level of consciousness, consider inserting a properly sized oropharyngeal or nasopharyngeal airway. Be alert to signs that the patient has inhaled hot gases or vapors, such as singed facial hair or soot present in or around the airway. Copious secretions and frequent coughing may also indicate a respiratory burn.

You must also quickly assess for adequate breathing. Inspect and palpate the chest wall for DCAP-BTLS. Check for clear and symmetrical breath sounds and provide high-flow oxygen or provide assisted ventilations using a bag-mask device as needed, depending on the level of consciousness and breathing rate/quality of your patient. Burn patients are trauma patients. Evaluate and treat them for spinal injuries and airway problems concurrently. How you open the airway depends on whether a neck injury is suspected. Could the patient have fallen? Do the circumstances surrounding the MOI suggest a possible spinal injury?

Circulation

You must quickly assess the pulse rate and quality and determine perfusion based on the patient's skin condition, color, temperature, and capillary refill time. If you see significant bleeding, you must take the necessary steps to control it. Significant bleeding is an immediate life threat. If the patient has obvious life-threatening bleeding it must be controlled quickly. Shock frequently develops in burn patients. Support their circulation by elevating the arms and legs as appropriate or placing the patient in a Trendelenburg position. You should also treat the shock by preventing heat loss. This is very important because the damaged skin has only a limited ability to regulate body temperature.

Transport Decision

If the patient you are treating has an airway or breathing problem, significant burn injuries, significant external bleeding, or signs and symptoms of internal bleeding, you must consider quickly transporting this patient to the hospital for treatment. A rendezvous with ALS providers may be appropriate for burn patients with moderate or severe burns and burns of the airway or lungs. ALS providers can treat these patients with endotracheal intubation and intravenous fluids to support airway, breathing, and circulation (shock) problems. These problems can progress so rapidly that immediate ALS help can make the difference between life and death.

History Taking

Investigate Chief Complaint

Investigate the chief complaint or history of present illness. Next, be alert for signs or symptoms of other injuries due to the MOI.

You should obtain a medical history and be alert for injury-specific signs and symptoms as well as any pertinent negatives such as no pain. Typical signs of a burn are pain, redness, swelling, blisters, or charring. Typically symptoms include pain and/or burning at the injury site. Regardless of the type of burn injury, it is important for you to stop the burning process, apply dressings to prevent contamination, and treat the patient for shock.

SAMPLE History

You will need to obtain a SAMPLE history from your patient. In addition, be sure to ask the following questions of a burn patient:

- Are you having any difficulty breathing?
- Are you having any difficulty swallowing?
- Are you having any pain?

When you are assessing a burn patient, check to see whether he or she has an emergency medical identification device—a wallet card, necklace, or bracelet—or ask the patient or a family member because preexisting conditions may increase the chances of a poor outcome. Remember that the environment, bystanders, and medical identification devices may provide important clues about your patient's condition.

Secondary Assessment

The secondary assessment is a more detailed, comprehensive or focused examination of the patient that is conducted to reveal injuries that may have been missed during the primary assessment. In some instances where the patient is critically injured or the transport time is short, you may not have time to conduct a secondary assessment. In other instances, the secondary assessment may occur en route to the ED.

Physical Examinations

After the primary assessment is complete, perform a full-body scan. Quickly assess the patient from head to toe looking for DCAP-BTLS to be sure that you have found all of the problems and injuries. Make a rough estimate, using the rule of nines, of the extent of the burned area to report to medical control. Determine what classification of burns the victim has sustained. Superficial burns involve only the epidermis and are characterized by reddening of the skin, swelling, and pain. An example is a sunburn. Partial-thickness burns involve the epidermis and the dermis and are characterized by severe pain, reddening, blisters, and a spotted or mottled appearance. Full-thickness burns involve all layers of the skin and are characterized by charred areas of the skin that are dry and white or dark brown. The patient may or may not complain of pain depending on the amount of nerve damage. Before packaging your patient, determine the severity of the burns the victim has sustained. Severity is calculated by considering what caused the burn, the body region that is burned, the depth and extent of the burn, the patient's age, and preexisting illness or injures. You should follow your local protocols for criteria for transport to a burn center. Package the patient for transport based on your findings. Remember to stabilize your patient for spinal injuries as appropriate.

Assessment of the respiratory system involves looking, listening, and feeling. A patient who is conscious, alert, and talking has no immediate airway or breathing problems. When assessing the respiratory system of a burn patient, look specifically for the following findings:

1. Soot around the mouth
2. Soot around the nose
3. Singed nasal hairs

Next, listen to breath sounds with a stethoscope. Breath sounds should be clear and equal bilaterally, anteriorly, and posteriorly. Determine the patient's rate and quality of respiration. Finally assess the chest for DCAP-BTLS and asymmetrical chest wall movement. Burn patients who present with any type of airway problems should be considered critical.

You must be able to quickly assess pulse rate and quality; determine the skin condition, color, and temperature; and check the capillary refill time. If visible significant bleeding is seen, you must begin the steps necessary to control bleeding. Significant bleeding, internal or external, is an immediate life threat. If the patient has obvious life-threatening bleeding, it must be controlled quickly and treatment of shock begun as quickly as possible. Non–life-threatening bleeding, such as in abrasions, can be bandaged later in your assessment as necessary.

Assess the patient's neurologic system to formulate baseline data for further decisions on patient management. This examination should include assessment of the following:

- level of consciousness—use AVPU
- pupil size and reactivity
- motor response
- sensory response

Assess the musculoskeletal system by performing a detailed head-to-toe examination. Assess all anatomic regions looking for DCAP-BTLS. Specifically look for the following features:

- In the head, be alert for raccoon eyes, Battle's sign, and/or drainage of blood or fluid from the ears or nose.
- In the neck, check for jugular vein distention and tracheal deviation. Be alert for patients with a stoma or tracheostomy.
- In the pelvis, check for stability.
- In the abdomen, feel all four quadrants for tenderness or rigidity. If the abdomen is tender, expect internal bleeding.
- In the extremities, record pulse and motor and sensory function.

Vital Signs

A systematic examination helps you to understand what has happened to the outside of your patient. Vital signs are a good indication of how your patient is doing on the inside. Determining an early set of vital signs will help you to know how your patient is tolerating his or her injuries while en route to the hospital. These can be obtained in the ambulance on the way to the hospital, decreasing the delay to definitive care in a patient with moderate to severe burns. Because shock is often pronounced in a burn patient, blood pressure, pulse, and skin assessment for perfusion are important signs to obtain.

Monitoring Devices

In addition to hands-on assessment, you should use monitoring devices to quantify oxygenation and circulatory status. You may also use noninvasive blood pressure measurement to monitor blood pressure. It is recommended that you always assess the patient's first blood pressure measurement manually with a sphygmomanometer and stethoscope.

Reassessment

Repeat the primary assessment, and reassess the patient's vital signs. Reassess the patient's chief complaint. Reevaluate interventions and treatment you have provided to the patient, particularly those used to treat shock. Identify and treat any changes in the patient's condition.

Interventions

The goals in treating patients with burns are to stop the burning process, assess and treat breathing, support circulation, and provide rapid transport. Because burn patients are also trauma patients, provide complete spinal stabilization if you suspect spinal injuries. Oxygen is mandatory for inhalation burns but is also helpful in patients with smaller burns. If the patient has signs of hypoperfusion, treat aggressively for shock and provide rapid transport to the appropriate hospital. Cover all burns according to your local protocols. The risk of infection is very high and can be reduced if you cover large areas that are burned with sterile burn sheets or clean linen. Do not delay transport of a seriously injured patient to complete nonlifesaving treatments in the field, such as splinting extremity fractures. Instead, complete these types of treatment en route to the hospital.

Communication and Documentation

Provide hospital personnel with a description of how the burn occurred. Many times the ED staff can determine the appropriate dilutant for chemical burns or calculate appropriate treatments for other types of burns with enough advanced notice. Your report and documentation should include the extent of the burns. This should include the amount of body surface area involved, the depth of the burn, and the location. For example, you may say 10% full-thickness burns, 15% partial-thickness burns, and 25% superficial burns to the chest, abdomen, and left lower extremity. If special areas are involved (genitalia, feet, hands, face, or circumferential), they should be specifically mentioned and documented.

Emergency Medical Care for Burns

Your first responsibility in caring for a patient with a burn is to stop the burning process and prevent additional injury. When caring for a burn patient, follow the steps in **Skill Drill 24-3** :

1. Follow standard precautions. Because a burn destroys the patient's protective skin layer, always wear gloves and eye protection when treating a burn patient.

Skill Drill 24-3

Caring for Burns

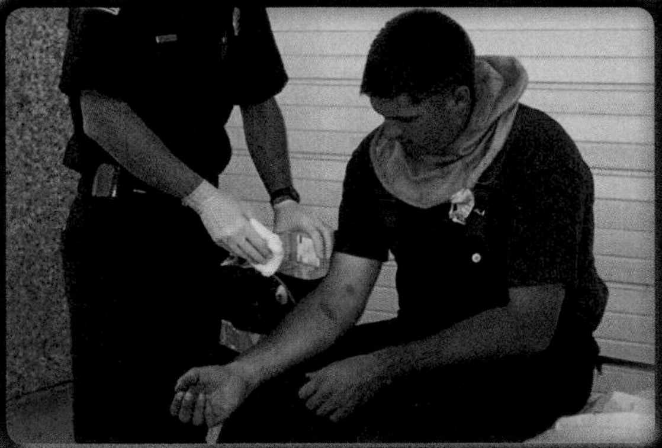

Step 1 Follow standard precautions to help prevent infection. If safe to do so, remove the patient from the burning area; extinguish or remove hot clothing and jewelry as necessary. If the wound(s) is still burning or hot, immerse the hot area in cool, sterile water, or cover with a wet, cool dressing.

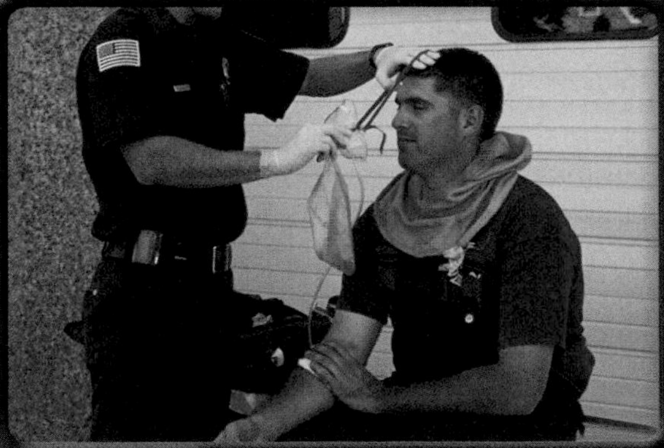

Step 2 Provide high-flow oxygen, and continue to assess the airway.

Skill Drill 24-3

Caring for Burns, continued

Step 3 Estimate the severity of the burn, and then cover the area with a dry, sterile dressing or clean sheet. Assess and treat the patient for any other injuries.

Step 4 Prepare for transport. Treat for shock.

Step 5 Cover the patient with blankets to prevent loss of body heat. Transport promptly.

2. Move the patient away from the burning area. If any clothing is on fire, wrap the patient in a blanket or follow specific guidelines outlined by your local fire department protocol to put out the flames, then remove any smoldering clothing and/or jewelry.

3. If allowed by local protocol, immerse the area in cool, sterile water or saline solution, or cover with a clean, wet, cool dressing if the skin or clothing is hot. This not only stops the burning, it also relieves pain. Prolonged immersion, however, may increase the risk of infection and hypothermia. For this reason, you should not keep the affected part submersed in water for more than 10 minutes. If the burning has stopped before you arrive, do not immerse the affected part at all. As an alternative to immersion, the burned area can be irrigated until the burning stops, followed by the application of a sterile dressing (Step 1).

4. Provide high-flow oxygen. Also remember that more fire victims die of smoke inhalation than of skin burns. A patient who has facial burns or has inhaled smoke or fumes may experience respiratory distress. Therefore, you should provide high-flow oxygen. Keep in mind that a patient who appears to

be breathing well at first may suddenly experience severe respiratory distress. Therefore, continually assess the airway for possible problems **Step 2**.

5. Rapidly estimate the burn's severity. Then cover the burned area with a dry, sterile dressing to prevent further contamination. Sterile gauze is best if the area is not too large. You may cover larger areas with a clean, white sheet. Most important, do not put anything else on the burned area. Never use ointments, lotions, or antiseptics of any kind. In addition, do not intentionally break any blisters.

6. Check for traumatic injuries or other medical conditions that may be more immediately life threatening. Most patients who have been burned have normal vital signs and can communicate at first, which will make your assessment easier **Step 3**.

7. Treat the patient for shock **Step 4**.

8. An extensive burn can produce hypothermia (loss of body heat). Prevent further heat loss by covering the patient with warm blankets.

9. Provide prompt transport by local protocol. Do not delay transport to do a prolonged assessment or to apply coverings to burns in a critical patient **Step 5**.

Dressing and Bandaging

All wounds require bandaging. In most instances, splints help to control bleeding and provide firm support for the dressing. There are many different types of dressings and bandages **Figure 24-26**. You should be familiar with the function and proper application of each.

In general, dressings and bandages have three primary functions:

- To control bleeding
- To protect the wound from further damage
- To prevent further contamination and infection

Sterile Dressings

Universal dressings, conventional 4″ × 4″ and 4″ × 8″ gauze pads, and assorted small adhesive-type dressings and soft self-adherent roller dressings will cover most wounds. Measuring 9″ × 36″ and made of thick, absorbent material, the universal dressing is ideal for covering large open wounds. It also makes an efficient pad for rigid splints. These dressings are available in compact, commercially sterilized packages.

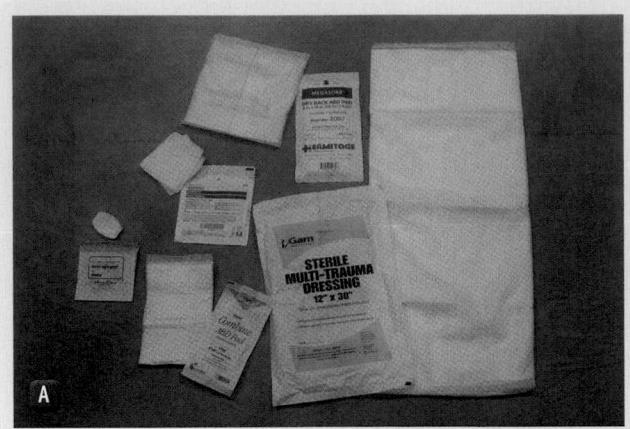

Figure 24-26   **A.** Many types of sterile dressings are used for covering open wounds, including universal dressings, gauze pads, adhesive dressings, and occlusive dressings. **B.** Bandages keep dressings in place and include soft roller bandages, triangular bandages, and adhesive tape. Splints may also be used to hold dressings in place.

Gauze pads are appropriate for smaller wounds, and adhesive-type dressings are useful for minor wounds. **Occlusive dressings**, made of Vaseline gauze, aluminum foil, or plastic, prevent air and liquids from entering (or exiting) the wound. They are used to cover sucking chest wounds, abdominal eviscerations, and neck injuries.

Bandages

To keep dressings in place during transport, you can use soft roller bandages, rolls of gauze, triangular bandages, or adhesive tape. The self-adherent, soft roller bandages are probably easiest to use. They are slightly elastic, which makes them easy to apply, and you can tuck the end of the roll into a deeper layer to secure it in place. The layers adhere somewhat but should not be applied too tightly to one another.

Adhesive tape holds small dressings in place and helps to secure larger dressings. Some people, however, are allergic to adhesive tape. If you know that a patient has this problem, use paper or plastic tape instead.

Do not use elastic bandages to secure dressings. If the injury swells, the bandage may become a tourniquet and cause further damage. Any improperly applied bandage that impairs circulation can result in additional tissue damage or even the loss of a limb. For this reason, you should always check a limb distal to a bandage for signs of impaired circulation and loss of sensation. Air splints are useful in stabilizing broken extremities, and they can be used with dressings to help control bleeding from soft-tissue injuries.

As discussed in Chapter 23, *Bleeding*, if a wound continues to bleed despite the use of direct pressure, quickly proceed to the use of a tourniquet. Research from the Iraq war has taught us that use of a tourniquet is rarely as harmful to the patient as it was once thought to be. If you cannot control bleeding from a major vessel in an extremity, a properly applied tourniquet may save a patient's life. Specifically, the tourniquet is useful if a patient is bleeding severely from a partial or complete amputation.

You are the Provider: SUMMARY

1. What should be your most immediate priority?

As with any patient, your first priority is to prevent further harm. The patient was brought to you wrapped in a blanket; this does not mean that his skin and/or clothing have stopped burning. You should remove the blanket to ensure that his clothes are not smoldering and that the burning process has stopped.

If the patient's clothes are still smoldering or if there is any other evidence indicating that the burning process is ongoing, pour sterile water or saline over the affected areas. Alternatively, you can apply moist, sterile dressings to extinguish the burning areas. These actions not only stop the burning process, they also help relieve pain.

It is important to note that you should use just enough water or saline to stop the burning process. Prolonged exposure to cold water or saline increases the risks of infection and hypothermia. If the burning process has already stopped, do not apply any water, saline, or moist dressings—at least until you have further assessed the patient.

2. What is a thermal burn? Are thermal burns always caused by fire?

A burn occurs when the body, or a body part, is exposed to more radiant energy than it can absorb without injury. Although a thermal burn is often referred to as "trauma by fire," it is any burn that is caused by heat energy. Heat energy can be transmitted in a variety of ways in addition to fire. Although thermal burns are all caused by heat—as opposed to radiation, chemicals, or electricity—many different situations can cause thermal burns.

Other sources of heat energy, other than fire, include scald burns, such as what occurs when a person opens a vehicle's hot radiator cap and is exposed to boiling liquids; contact burns, such as what occurs when a person comes in contact with a hot stovetop burner; steam burns, such as what occurs when the body is exposed to superheated gaseous water; and flash burns, such as what occurs when a person is briefly exposed to very intense heat (eg, explosion).

Most commonly, thermal burns are caused by an open flame (flame burn). The burn can range in severity from superficial damage to complete destruction of all the layers of the skin. Flame burns may also be associated with inhalation injuries that can result in airway and ventilation compromise secondary to airway swelling.

Thermal burns can occur when the skin is exposed to temperatures higher than 111ºF (44ºC). In general, the severity of a thermal burn correlates directly with the temperature of the heat source, the amount of heat energy possessed by the object or substance, and the duration of exposure. Burns are a progressive process; the greater the heat energy the patient is exposed to, the deeper the injury.

3. What additional information should be obtained from the fire fighters who rescued the patient?

It has already been established that the patient was trapped in an enclosed space because the fire fighters rescued him from the structure. However, you should try to determine an approximate length of exposure; this is usually a gross estimate at best.

As previously discussed, a factor to consider when determining the overall severity of a thermal burn is the duration of exposure. Victims who are trapped in a structural fire die more commonly of inhalation of toxic gases, such as carbon monoxide and cyanide, than of skin burns. The longer the patient is trapped, the greater the risk of significant carbon monoxide and/or cyanide toxicity.

It should also be determined if the patient was conscious or unconscious when he was found. Conscious patients are often able to extinguish themselves, unless they are completely engulfed in flames. They also reflexively limit the amount of time they are exposed to superheated air by holding their breath. If the patient is unconscious, however, he or she has

You are the Provider: SUMMARY, continued

no control over the duration of exposure to the fire itself or the quantity of superheated air that he or she inhales. This significantly increases the patient's risk of more severe skin burns and toxic gas inhalation.

It should also be determined how the patient was found. Was the patient in an open area of a room or was he or she trapped beneath a collapsed ceiling beam or other heavy structure? Although you should always assess the patient for traumatic injuries, information provided by the fire fighters regarding any mechanism of injury can help you focus on a particular area (or areas) of the body. Do not assume that the patient's problems are limited to skin burns and toxic gas exposure; the patient may have experienced other injuries (eg, blunt trauma with internal bleeding, head injury) that could be life threatening.

4. How are thermal burns classified? What are the characteristics of each type of burn?

Burns are classified according to their depth—that is, how far the burn injury extends through the layers of the skin (ie, epidermis, dermis). The types of burns you must be able to identify are superficial (first-degree), partial-thickness (second-degree), and full-thickness (third-degree).

Superficial (first-degree) burns involve only the outer layer of the skin—the epidermis. The skin turns red and is often painful, but it does not blister or burn through the epidermis. A sunburn is a good example of a superficial burn.

Partial-thickness (second-degree) burns involve the epidermis and some portion of the dermis. These burns do not destroy the entire thickness of the skin, nor is the subcutaneous (fatty) tissue injured. The skin is often moist, mottled, and white to red in appearance, and blister formation is typically present. Partial-thickness burns cause intense pain because portions of the nerve endings in the dermis are injured but not completely destroyed.

Full-thickness (third-degree) burns extend through all layers of the skin and may involve the subcutaneous layers, muscle, bone, or internal organs. The burned area is dry and leathery and may appear white, dark brown, or even charred. Some full-thickness burns feel hard to the touch. In many cases, the nerve endings have been completely destroyed; therefore, full-thickness burns are often painless. However, the surrounding, less severely burned areas may be extremely painful. Pure full-thickness burns are unusual. Severe burns are typically a combination of superficial, partial-thickness, and full-thickness burns.

5. What percentage of the patient's body surface area has been burned?

After identifying the depth of a burn, you must be able to rapidly estimate the extent of the burns—that is, the percentage of the patient's body surface area (BSA) that is burned.

The rule of nines is a quick way to estimate the amount of BSA that has been burned. It divides the body into sections, each representing approximately 9% of the total BSA.

In the adult, the entire head represents 9% of the BSA, the anterior torso (chest and abdomen) represents 18%, the posterior torso represents 18%, each upper extremity (anterior and posterior) represents 9%, each lower extremity (anterior and posterior) represents 18%, and the genitalia represents 1%.

Your patient has experienced burns to his anterior torso (chest and abdomen); this represents 18% of his BSA. Additionally, both upper extremities are burned, which represents 18% (9% per extremity) of his BSA. Therefore, *your patient has burns that cover approximately 36% of his BSA*.

Another method for estimating the percentage of BSA burned is to use the *rule of ones*, also known as the "rule of palm" or "palmar method." According to this method, the palm of the patient's hand represents approximately 1% of his or her BSA.

It is important to note that estimating the extent of a patient's burns is just that—an estimate. Do not take a long time to try to obtain an exact number; doing so delays patient care and may cause you to miss other injuries or conditions that present a greater threat to life.

6. What factors should you consider when determining the severity of a burn?

In most cases, burn patients are transported to an appropriate ED for stabilization, and are then transferred to a burn facility if needed. However, you must still determine the severity of a patient's burns in the prehospital setting for purposes of emergency care as well as communication with the receiving facility. Depending on local protocol and/or the proximity of a burn center in relation to an ED, you may be asked to transport a critically burned patient directly to a burn center.

When determining the severity of a burn, the two most important factors to consider initially are the depth and extent of the burn; what type of burns has the patient experienced (ie, superficial, partial-thickness, or full-thickness) and what percentage of the patient's body surface area (BSA) has been burned?

The American Burn Association classifies burns as being minor, moderate, and severe. However, your local protocols or regional burn center may have slightly different criteria.

According to the American Burn Association, minor burns in an adult include full-thickness burns that cover less than 2% of the BSA, partial-thickness burns that cover less than 15% of the BSA, and superficial burns that cover less than 50% of the BSA. Moderate burns include full-thickness burns that cover between 2% and 10% of the BSA (excluding burns to critical areas of the body), partial-thickness burns that cover 15% to 30% of the BSA, and superficial burns that cover more than 50% of the BSA. Severe burns include *any* full-thickness burn to a critical area of the body, full-thickness burns that cover more than 10% of the BSA to noncritical areas of the body, and partial-thickness burns that cover more than 30% of the BSA.

In addition to the depth and extent of the burn, you must also determine if the burns are located in any critical areas of the body. Critical areas of the body include the face, upper airway, hands, feet, and genitalia. Also included in the definition of critical areas is the presence of circumferential burns—burns that go completely around a body part. Circumferential burns of the chest, for example, can compromise breathing, and circumferential burns of an extremity can cause neurovascular compromise.

In some cases, burns that would otherwise be considered noncritical are complicated by the presence of an underlying injury (eg, fractures, blunt trauma) or a significant medical condition (eg, diabetes, heart disease, cancer). Therefore, it is important to perform a thorough assessment of the patient and to obtain information regarding his or her past medical history. You must also consider the patient's age; burns that would be considered moderate in younger adults are considered critical in patients younger than 5 years and older than 55 years. *The burn injury itself is not always what makes the patient's overall condition critical.*

7. What is the proper treatment for the patient's burns?

First and foremost, *do not* rupture any blisters that may have formed; they are the only barrier that is protecting the patient from infection. Furthermore, you should avoid the application of burn ointments, creams, or gels; these products also increase the risk of infection and will only have to be removed at the hospital so the burn can be thoroughly evaluated.

Large (greater than 10%) BSA burns should be covered with a dry, sterile, nonadherent dressing (eg, a sterile burn sheet); *your patient has burns that cover approximately 36% of his BSA.* Other than for the purpose of stopping the burning process, you should not apply water or saline to large surface area burns. The larger the burn area, the greater the risks of hypothermia, hypovolemia, and infection.

Dry, sterile dressings applied to large surface area burns help maintain body temperature (reduce the risk of hypothermia), prevent further contamination of the burn (reduce the risk of infection), and provide comfort.

Small (less than 10%) surface area burns may be cooled with sterile water or saline, depending on local protocol, and then covered with a dry, sterile dressing. Follow your local protocols regarding the treatment for burn injuries.

Further treatment should be aimed at preventing hypothermia (covering the patient with a blanket), closely monitoring the patient's airway and ventilation status, and monitoring for signs of shock.

8. How has the patient's condition changed? What should you do now?

Respiratory distress in a burn patient, especially in a patient without injuries that would cause breathing problems (eg, blunt chest trauma), indicates upper airway swelling secondary to inhaling excessive heat (inhalation injury).

Patients who are trapped in an enclosed space with poor ventilation, especially if they lost consciousness, are at highest risk for an inhalation injury. Although your patient denies losing consciousness, you should suspect that he has some degree of upper airway swelling because he was in an enclosed space.

Upper airway swelling is most often caused by the inhalation of superheated gas, which can cause rapidly progressing edema of the oral mucosa and complete airway obstruction. However, because the moist oral mucosa absorbs the heat and quickly cools it to body temperature, lower airway burns are uncommon.

Common early signs of an inhalation injury include a sore throat and hoarseness. The presence of stridor and respiratory distress indicates more severe upper airway swelling. Other signs and symptoms of upper airway edema include carbon particles in the sputum, singed nasal and facial hairs, and burns to the facial area.

Many patients with an inhalation injury present with rapidly progressing symptoms; others may not present with symptoms until you are en route to the hospital, as is the case with your patient.

Immediate treatment for a patient with signs of an inhalation injury includes a careful assessment of the patient's airway and breathing status. If the patient is breathing adequately, continue to administer high-flow oxygen and closely observe the patient. Cool mist or aerosol therapy may help reduce mild airway swelling; however, if this is not available, apply an ice pack to the throat area. If the patient is breathing inadequately (eg, shallow breathing [reduced tidal volume], labored respirations, falling oxygen saturation, decreasing level of consciousness), assist his or her ventilations with a bag-mask device.

Depending on your transport time to the closest appropriate hospital and the availability of ALS resources in your area, you should consider an intercept with an AEMT or paramedic unit. Some patients with inhalation injuries require advanced airway management, such as endotracheal intubation, to protect the airway before it closes completely.

9. What, if any, additional treatment is indicated for this patient?

Continuous, careful monitoring of this patient is essential. Although his condition does not seem to have worsened, it has not improved either. Continue to administer high-flow oxygen, closely monitor the adequacy of his breathing, and be prepared to assist his ventilations. You should also monitor him for signs of shock and treat accordingly.

Any patient who is experiencing respiratory distress will be anxious. Provide emotional support and let the patient assume a position of comfort; this is usually a full-Fowler's (90° angle) position.

You are the Provider: SUMMARY, continued

EMS Patient Care Report (PCR)					
Date: 10-14-09	**Incident No.:** 012309	**Nature of Call:** Burns		**Location:** 511 Bandera Rd.	
Dispatched: 1500	**En Route:** 1500	**At Scene:** 1500	**Transport:** 1514	**At Hospital:** 1530	**In Service:** 1541

Patient Information	
Age: 45 **Sex:** M **Weight (in kg [lb]):** 77 kg (170 lb)	**Allergies:** No known drug allergies **Medications:** None **Past Medical History:** None **Chief Complaint:** Burns to the torso and arms

Vital Signs				
Time: 1506	**BP:** 166/86	**Pulse:** 108	**Respirations:** 14	**Sao$_2$:** 98%
Time: 1512	**BP:** 158/84	**Pulse:** 120	**Respirations:** 22	**Sao$_2$:** 95%
Time: 1517	**BP:** 160/80	**Pulse:** 116	**Respirations:** 22	**Sao$_2$:** 96%

EMS Treatment
(circle all that apply)

Oxygen @ 15 L/min via (circle one): NC (NRM) Bag-Mask Device		**Assisted Ventilation**	**Airway Adjunct**	**CPR**
Defibrillation	**Bleeding Control**	**(Bandaging)**	**Splinting**	**Other:** (Thermal management, sterile burn sheet application)

Narrative

Medic 4 was standing by at the scene of a residential fire when fire fighters rescued a 45-year-old male from the burning structure. They presented the patient wrapped in a blanket; he was ambulatory and the burning process was stopped before EMS contact was made. The approximate time of exposure to the burning environment was 8 to 10 minutes. The patient was conscious and alert, his airway was patent, his breathing was adequate, and his face was covered with soot. He complained of severe pain to his chest, abdomen, and arms. Immediately moved patient into ambulance, applied high-flow oxygen via nonrebreathing mask, and assessed his burns. Assessment revealed burns that covered approximately 36% of his BSA. Superficial and partial-thickness burns were noted to the entire anterior torso, and partial- and full-thickness burns were noted to both of his upper extremities. Additional assessment revealed that his facial hair and the hair just above his hairline was singed; no facial skin burns were noted. The patient denied shortness of breath or any other symptoms other than severe pain from his burns. Secondary assessment did not reveal any other obvious injuries. Patient denies significant past medical history and states that he takes no medications. During the patient's entrapment, he stated that he did not lose consciousness. Applied dry, sterile burn sheets to patient's burns, covered him with a blanket for warmth, and began transport to the hospital. En route, notified medical control, who advised us to transport to the emergency department because the closest burn center was located 75 miles away. Reassessment revealed that the patient's respirations were becoming labored and his voice was becoming hoarse. Continued high-flow oxygen, applied ice pack to the patient's throat area, and continued to monitor his airway and breathing status. Notified the receiving facility of the patient's status and our impending arrival. The patient's vital signs remained stable throughout the duration of the transport, and his oxygen saturation never fell below 95%. Delivered patient to the emergency department without incident and gave verbal report to the attending physician. Medic 4 returned to service at 1541. *End of report*

Assessment and Emergency Care of Soft-Tissue Injuries

Scene Size-up

Scene Safety	Ensure scene safety by looking for threats, possible violence, and other hazards. Standard precautions should include a minimum of gloves and eye protection. Consider the number of patients, the need for additional resources/ALS, and cervical spine stabilization.
Mechanism of Injury (MOI)/ Nature of Illness (NOI)	Determine the MOI. Look for clues that may help you determine what happened to your patient.

Primary Assessment

Form a General Impression	Observe overall appearance of the patient and body position. Observe work of breathing and circulation. Determine level of consciousness. Perform a rapid scan to identify and manage immediate life threats. Determine priority of care based on the MOI. If the patient has a poor general impression, call for ALS assistance.
Airway and Breathing	If a cervical spine injury is suspected, open the airway using a modified jaw-thrust maneuver and ensure the airway is patent. Quickly assess the chest for DCAP-BTLS and treat any threats to life. Provide high-flow oxygen at 15 L/min, and evaluate depth and rate of the respiratory cycle, providing ventilatory support as needed.
Circulation	Evaluate pulse rate and quality; observe skin color, temperature, and condition; look for life-threatening bleeding, and treat accordingly by placing the patient in a supine or shock position. Be alert for signs and symptoms of internal bleeding.
Transport Decision	Significant MOI requires rapid transport.

History Taking

Investigate Chief Complaint	Investigate the chief complaint. Identify signs and symptoms and pertinent negatives. Ask pertinent OPQRST and SAMPLE questions. Be alert for pain or loss of sensation. Medications such as aspirin, blood thinners, and beta-blockers may alter your care plan.

NOTE: The order of the steps in this section differs depending on whether the patient is conscious or unconscious. The following order is for a conscious patient. For an unconscious patient, perform a primary assessment, perform a full-body scan, obtain vital signs, and obtain the past medical history from a family member, bystander, or emergency medical identification device.

Secondary Assessment

Physical Examinations	Perform a systematic full-body scan beginning with the head. Assess the pupils, and reassess the patient's mental status. If a spinal injury is suspected, apply a cervical immobilization device after assessing the neck. Inspect, palpate, and auscultate the chest, focusing on the respiratory effort and adequacy of ventilation. Assess the abdomen for signs of internal bleeding. Assess the musculoskeletal system for DCAP-BTLS. Log roll the patient, and assess the posterior regions.

Assessment and Emergency Care of Soft-Tissue Injuries, continued

| Vital Signs | Take vital signs, monitoring trends. Note skin color, temperature, and condition, as well as the patient's level of consciousness. Use pulse oximetry, if available, to assess the patient's perfusion status. This reading may not be accurate if peripheral blood flow is compromised. |

Reassessment

| Interventions | Consider the use of oxygen and proper positioning of the patient. Expose all wounds, cleanse the wound surface, control bleeding, and be prepared to treat the patient for shock. Reassess dressings and bandages. |
| Communication and Documentation | Contact medical control with a radio report. Include a thorough description of the MOI and the position in which the patient was found. Include injuries found, estimated blood loss, treatments performed, and patient response. Follow local protocols. Be sure to document any changes in patient status and the time. Document the reasoning for your treatment and the patient's response. |

NOTE: Although the following steps are widely accepted, be sure to consult and follow your local protocols.

Soft-Tissue Injuries

Closed Injuries

1. Ensure an open airway and adequate ventilations. Treat as required.
2. Be alert for and treat for shock (hypoperfusion) by raising the legs or backboard 6″ to 12″, maintain body temperature, and administer high-concentration oxygen.
3. Treat a closed soft-tissue injury by applying the mnemonic RICES:
   - Rest to keep patient quiet and comfortable
   - Ice to constrict blood vessels and reduce pain
   - Compression to compress blood vessels to slow bleeding
   - Elevation to raise injured part above level of the heart to decrease swelling
   - Splinting of extremity to decrease bleeding and pain

Open Injuries

1. Ensure you have followed standard precautions.
2. Ensure an open airway and adequate ventilations. Treat as required.
3. Apply an occlusive dressing to open chest injuries.
4. Apply direct pressure over the wound with a dry, sterile dressing.
5. Apply a pressure dressing.
6. If bleeding continues or recurs, apply a tourniquet to an extremity above the level of bleeding.
7. Be alert for and treat for shock (hypoperfusion).

Soft-Tissue Injuries, continued

Abdominal Wounds

1. Ensure you have followed standard precautions.
2. Ensure an open airway and adequate ventilations. Treat as required.
3. If organs are protruding (evisceration), do not attempt to replace.
4. Cover the wound with moist sterile gauze or dressing and an occlusive dressing.
5. Prevent heat loss.
6. Be alert for and treat for shock (hypoperfusion).

Impaled Objects

1. Ensure you have followed standard precautions.
2. Ensure an open airway and adequate ventilations. Treat as required.
3. Only remove object if it interferes with airway control or cardiopulmonary resuscitation.
4. Control bleeding with direct pressure.
5. Stabilize the object using bulky dressings to prevent movement during transport.
6. Tape a rigid object over the impaled item and its bandaging.
7. Be alert for and treat for shock (hypoperfusion).

Neck Injuries

1. Ensure you have followed standard precautions.
2. Ensure an open airway and adequate ventilations. Treat as required.
3. Cover the wound with an occlusive dressing first, then apply a pressure bandage, being careful not to compress both carotid arteries.
4. Be alert for and treat for shock (hypoperfusion).

Bites

1. Ensure you have followed standard precautions.
2. Ensure an open airway and adequate ventilations. Treat as required.
3. Apply a dry, sterile dressing.
4. Immobilize the area with a splint or bandage.
5. Transport patient to the emergency department for wound cleansing.

Assessment and Emergency Care of Burns

Scene Size-up

Scene Safety	Ensure scene safety by looking for threats, possible violence, and other hazards. Standard precautions should include a minimum of gloves and eye protection. Consider the number of patients, the need for additional resources/ALS, and cervical spine stabilization. Ensure that factors that led to the patient's burn injury do not pose a hazard to you.
Mechanism of Injury (MOI)/ Nature of Illness (NOI)	Determine the MOI/NOI. Observe the scene, and look for indicators of the MOI such as fire, electrical, chemical, or environmental.

▼   ▼

Primary Assessment

Form a General Impression	Observe overall appearance of patient and body position. Observe work of breathing and circulation. Determine level of consciousness. Perform a rapid scan to identify immediate life threats. Determine the priority of care based on the MOI. If the patient has a poor general impression, call for ALS assistance. Stop the burning process.
Airway and Breathing	If a cervical spine injury is suspected, open the airway using a modified jaw-thrust and ensure the airway is patent. Quickly assess the chest for DCAP-BTLS, and treat any threats to life. Provide high-flow oxygen at 15 L/min, and evaluate depth and rate of the respiratory cycle, providing ventilatory support as needed. Hoarsness and/or singed facial hair is an indicator of a potential airway/breathing problem; call for ALS. If the patient is unresponsive or has a significantly altered level of consciousness, consider inserting a properly sized oropharyngeal or nasopharyngeal airway.
Circulation	Evaluate pulse rate and quality; observe skin color, temperature, and condition; look for life-threatening bleeding and treat accordingly. Place patient in a supine or shock position. Prevent heat loss.
Transport Decision	If the patient has an airway or breathing problem, significant burn injuries, significant external bleeding, or signs and symptoms of internal bleeding, consider rapid transport or calling for ALS assistance. ALS providers can treat the patients with endotracheal intubation and intravenous fluids to support airway, breathing, and circulation (shock) problems. Consider the need for a burn center.

▼   ▼

History Taking

Investigate Chief Complaint	Investigate the chief complaint. Be alert for other injuries. If possible, ask OPQRST questions. Identify signs and symptoms and pertinent negatives. Ask the patient if he or she is having any difficulty breathing, difficulty swallowing, pain, or loss of sensation. Obtain a SAMPLE history from the patient or if unresponsive, from family, bystanders, or medical alert tags.

▼   ▼

NOTE: The order of the steps in this section differs depending on whether the patient is conscious or unconscious. The following order is for a conscious patient. For an unconscious patient, perform a primary assessment, perform a full-body scan, obtain vital signs, and obtain the past medical history from a family member, bystander, or emergency medical identification device.

Assessment and Emergency Care of Burns, continued

Secondary Assessment

Physical Examinations	Perform a systematic full-body scan beginning with the head, looking for DCAP-BTLS. Estimate extent of burned area using the rule of nines. Determine burn classification and severity. Inspect, palpate, and auscultate the chest, focusing on the respiratory effort and adequacy of ventilation. Perform a thorough neurologic examination. Assess the musculoskeletal system for DCAP-BTLS. Assess the abdomen for signs of internal bleeding.
Vital Signs	Take vital signs, monitoring trends. Because shock is often pronounced in a burn patient, blood pressure, pulse, and skin assessment for perfusion are important signs to obtain. Note the patient's level of consciousness. Use pulse oximetry, if available, to assess the patient's perfusion status.

Reassessment

Interventions	Stop the burning process. Manage airway, breathing, and circulation problems. Consider cervical spine precautions, and immobilize if needed. Provide high-concentration oxygen. Cover burn areas with sterile burn sheets following local protocols. Treat for shock. Do not delay transport.
Communication and Documentation	Contact medical control with a radio report. Include a thorough description of the MOI and the position the patient was found in. Include how the burn occurred, estimated body surface area burned, treatments performed, and patient response. Follow local protocols. Be sure to document any changes in patient status and the time. Document the reasoning for your treatment and the patient's response.

NOTE: Although the following steps are widely accepted, be sure to consult and follow your local protocols.

Burns

General Management of Burn Injuries

1. Stop the burning process.
2. Take appropriate standard precautions.
3. Treat life threats involving the ABCs.
4. Cool the burned area with sterile water or saline. Immerse or continuously irrigate the affected area, following local treatment protocols.
5. Cover the burned area with a sterile dressing. Dressing will be moist or dry depending on local protocol.
6. Maintain body temperature and treat for shock.

Chemical Burns

1. Stop the burning process by safely removing the chemical from the patient.
2. Remove patient's clothing and jewelry.
3. Flush burned area with large amounts of water for 15 to 20 minutes.

Assessment and Emergency Care of Burns, continued

Burns, continued

Electrical Burns

1. Ensure scene is safe and patient is not in contact with electrical source.
2. Treat respiratory and cardiac arrest. Defibrillate if necessary.
3. Administer high-concentration oxygen.
4. Cover burn wounds with dry, sterile dressings.
5. Splint suspected fractures.

Inhalation Burns

1. Ensure scene safety.
2. Treat life threats to airway, breathing, and circulation.
3. Consider requesting ALS.
4. Do not rely on pulse oximetry readings.

Radiation Burns

1. Ensure scene safety. Increase your distance from contaminated area.
2. Decontaminate the patient as needed.
3. Irrigate open wounds.
4. If a radiation burn is from a fission product, as found in nuclear power plants, contact medical control to find out if potassium iodide is available.

Prep Kit

Ready for Review

- The skin protects the body by keeping pathogens out, water in, and assisting in body temperature regulation.
- There are three types of soft-tissue injuries:
  - Closed injuries (Soft-tissue damage occurs beneath the skin or mucous membrane but the surface remains intact.)
  - Open injuries (There is a break in the surface of the skin or the mucous membrane, exposing deeper tissue to potential contamination.)
  - Burns (The soft tissue receives more energy than it can absorb without injury; the source of this energy can be thermal, toxic chemicals, electricity, or radiation.)
- Closed soft-tissue injuries are characterized by a history of blunt trauma, pain at the site of injury, swelling beneath the skin, and discoloration. Contusions, hematomas, and crushing injuries are classified as closed injuries. Treat a closed soft-tissue injury by applying the mnemonic RICES: *Rest*, *Ice*, *Compression*, *Elevation*, and *Splinting*.
- Open injuries differ from closed injuries in that the protective layer of skin is damaged. Abrasions, lacerations, avulsions, and penetrating wounds are classified as open injuries. Treat an open soft-tissue injury by applying direct pressure with a sterile bandage using a roller bandage, and splint the extremity.
- The assessment of an open injury is generally easier than the assessment of a closed injury because you can see the injury.
- Burns are serious and painful soft-tissue injuries caused by heat (thermal), chemicals, electricity, and radiation.
- Burns are classified primarily by the depth and extent of the burn injury and the body area involved.
- Burns are considered to be superficial, partial-thickness, or full-thickness based on the depth involved.
- When providing emergency care for burns, do the following:
  - Use standard precautions to protect yourself from potentially contaminated body fluid and to protect the patient from potential infection.
  - Ensure you have cooled the burned area to prevent further cellular damage.
  - Remove jewelry and constrictive clothing; never attempt to remove any synthetic material that may have melted into the burned skin.
  - Ensure an open and clear airway, provide high-flow oxygen, and be alert to signs and symptoms of inhalation injury such as difficulty breathing, stridor, or wheezing.
  - Place sterile dressings over the burned area(s); prevent hypothermia by covering the patient with a clean blanket. Provide prompt transport.
- Small animal and human bites can lead to serious infection and must be evaluated by a physician. Small animals can carry rabies.
- Dressings and bandages are designed to control bleeding, protect to the wound from further damage, prevent further contamination, and prevent infection.

Vital Vocabulary

abrasion Loss or damage of the superficial layer of skin as a result of a body part rubbing or scraping across a rough or hard surface.

amputation An injury in which part of the body is completely severed.

avulsion An injury in which soft tissue is torn completely loose or is hanging as a flap.

burns Injuries in which soft-tissue damage occurs as a result from thermal heat, frictional heat, toxic chemicals, electricity, or nuclear radiation.

closed injuries Injuries in which damage occurs beneath the skin or mucous membrane but the surface remains intact.

compartment syndrome Swelling in a confined space that produces dangerous pressure; may cut off blood flow or damage sensitive tissue.

contact burn A burn caused by direct contact with a hot object.

contamination The presence of infective organisms or foreign bodies such as dirt, gravel, or metal.

contusion A bruise from an injury that causes bleeding beneath the skin without breaking the skin.

crushing injury An injury that occurs when a great amount of force is applied to the body.

crush syndrome Significant metabolic derangement that develops when crushed extremities or body parts remain trapped for prolonged periods. This can lead to renal failure and death.

dermis The inner layer of the skin, containing hair follicles, sweat glands, nerve endings, and blood vessels.

ecchymosis Discoloration associated with a closed wound; signifies bleeding.

epidermis The outer layer of skin that acts as a watertight protective covering.

evisceration The displacement of organs outside the body.

fascia The fiberlike connective tissue that covers arteries, veins, tendons, and ligaments.

flame burn A burn caused by an open flame.

flash burn A burn caused by exposure to very intense heat, such as in an explosion.

full-thickness (third-degree) burns Burns that affects all skin layers and may affect the subcutaneous layers, muscle, bone, and internal organs, leaving the area dry, leathery, and white, dark brown, or charred.

hematoma Blood collected within the body's tissues or in a body cavity.

incision A sharp, smooth cut.

laceration A jagged, open wound.

mucous membranes The linings of body cavities and passages that are in direct contact with the outside environment.

occlusive dressings Dressings made of petrolatum (Vaseline) gauze, aluminum foil, or plastic that prevents air and liquids from entering or exiting a wound.

open injuries Injuries in which there is a break in the surface of the skin or the mucous membrane, exposing deeper tissue to potential contamination.

partial-thickness (second-degree) burns Burns affecting the epidermis and some portion of the dermis but not the subcutaneous tissue, characterized by blisters and skin that is white to red, moist, and mottled.

penetrating wound An injury resulting from a sharp, pointed object.

rabid Describes an animal that is infected with rabies.

rule of nines A system that assigns percentages to sections of the body, allowing calculation of the amount of skin surface involved in the burn area.

scald burn A burn caused by hot liquids.

steam burn A burn caused by exposure to hot steam.

superficial (first-degree) burns Burns affecting only the epidermis, characterized by skin that is red but not blistered or actually burned through.

thermal burns Burns caused by heat.

Assessment in Action

You are dispatched to a bar to respond to a fight. The police have cleared the scene and it is safe for you to enter. You see one male patient, conscious and alert. His face is mottled with blisters and abrasions, and he has blood on his shirt. He tells you that he was trying to stop the fight when he was hit in the face with scalding hot coffee and then fell backwards into a chair. Physical examination shows a jagged laceration measuring approximately 2″ on his abdomen. It is still bleeding, and you notice bruising on the right lateral chest. Vital signs are stable.

1. What is the classification of this burn?
   A. Superficial
   B. Partial-thickness
   C. Full-thickness
   D. Thermal

2. The priority in treating burns is to:
   A. clean any open wounds.
   B. take vital signs.
   C. stop the burning process.
   D. keep the airway open.

3. Is this a severe burn? Why or why not?

4. What kind of burn is this?
   A. Inhalation burn
   B. Thermal burn
   C. Radiation burn
   D. Chemical burn

5. What is the rule of nines?

6. According to the rule of nines, what percentage of skin surface is burned in this patient?
   A. 4½%
   B. 7½%
   C. 9%
   D. 18%

7. What is the priority for the open wound on the patient's abdomen?
   A. Clean it.
   B. Flush it with sterile saline.
   C. Probe it.
   D. Stop the bleeding.

8. What are the steps to stop bleeding?

9. How should you treat a closed injury?

10. Which of the patient's soft-tissue injuries is least likely to result in infection?
    A. Contusion on the right lateral chest
    B. Abdominal laceration
    C. Burns to the face
    D. Abrasions to the face

Face and Neck Injuries

National EMS Education Standard Competencies

Medicine

Applies fundamental knowledge to provide basic emergency care and transportation based on assessment findings for an acutely ill patient.

Diseases of the Eyes, Ears, Nose, and Throat

Recognition and management of

- Nosebleed (pp 869-870)

Trauma

Applies fundamental knowledge to provide basic emergency care and transportation based on assessment findings for an acutely injured patient.

Head, Facial, Neck, and Spine Trauma

- Recognition and management of:
  - Life threats (pp 856-857)
  - Spine trauma (Chapter 26, *Head and Spine Injuries*)
- Pathophysiology, assessment, and management of:
  - Penetrating neck trauma (pp 874-875)
  - Laryngotracheal injuries (p 875)
  - Spine trauma (Chapter 26, *Head and Spine Injuries*)
  - Facial fractures (p 872)
  - Skull fractures (Chapter 26, *Head and Spine Injuries*)
  - Foreign bodies in the eyes (pp 860-864)
  - Dental trauma (pp 855, 872-873)

Knowledge Objectives

1. Discuss the anatomy and physiology of the head, face, and neck, including major structures and specific important landmarks of which the EMT must be aware. (pp 851-854)

2. Describe the factors that may cause the obstruction of the upper airway following a facial injury. (pp 859-860)

3. Discuss the different types of facial injuries and patient care considerations related to each one. (pp 859-860)

4. Describe the process of providing emergency care to a patient who has sustained face and neck injuries, including assessment of the patient, review of signs and symptoms, and management of care. (pp 854-875)

5. List the steps in the emergency medical care of the patient with soft-tissue wounds of the face and neck. (pp 859-875)

6. List the steps in the emergency medical care of the patient with an eye injury based on the following scenarios: foreign object, impaled object, burns, lacerations, blunt trauma, closed head injuries, and blast injuries. (pp 860-869)

7. Describe the three different causes of a burn injury to the eye and patient management considerations related to each one. (pp 864-866)

8. List the steps in the emergency medical care of the patient with injuries of the nose. (pp 869-870)

9. List the steps in the emergency medical care of the patient with injuries of the ear, including lacerations and foreign body insertions. (pp 870-872)

10. Describe the physical findings of a patient with a facial fracture and list the steps related to providing emergency medical care to these patients. (p 872)

11. List the steps in the emergency medical care of the patient with dental and cheek injuries, including how to deal with an avulsed tooth. (pp 872-873)

12. List the steps in the emergency medical care of patient with an upper airway injury caused by blunt trauma. (pp 873-874)

13. List the steps in the emergency medical care of the patient with a penetrating injury to the neck, including how to control regular and life-threatening bleeding. (pp 874-875)

Skills Objectives

1. Demonstrate the removal of a foreign object from under a patient's upper eyelid. (pp 860-862, Skill Drill 25-1)

2. Demonstrate the stabilization of a foreign object that has been impaled in a patient's eye. (pp 862-864, Skill Drill 25-2)

3. Demonstrate irrigation of a patient's eye using a nasal cannula, bottle, or basin. (pp 864-865)

4. Demonstrate the care of a patient who has a penetrating eye injury. (pp 863-864)

5. Demonstrate how to control bleeding from a neck injury. (pp 874-875, Skill Drill 25-3)

Introduction

The face and neck are particularly vulnerable to injury because of their relatively unprotected positions on the body. Soft-tissue injuries and fractures to the bones of the face are common and vary greatly in severity. Some are potentially life threatening, and many leave disfiguring scars if not treated properly. Penetrating trauma to the neck may cause severe bleeding. An open injury may allow an air embolism to enter the circulatory system. If a hematoma forms in this area, it may stop or slow blood flow to the brain, causing a stroke. With appropriate prehospital and hospital care, a patient with a seemingly devastating injury can have a surprisingly good outcome.

As an EMT, your objectives when treating a patient with face and neck injuries include prevention of further injury, particularly to the cervical spine, managing any acute airway problems, and controlling bleeding. This chapter first reviews the anatomy of the head and neck and then examines the factors that can produce upper airway obstruction. A discussion follows that includes emergency medical care of soft-tissue wounds of the face, nose, and ear; facial fractures; penetrating injuries of the neck; and dental injuries.

Anatomy and Physiology

The head is divided into two parts: the cranium and the face. The cranium, or skull, contains the brain, which connects to the spinal cord through the foramen magnum,

a large opening at the base of the skull. The most posterior portion of the cranium is called the occiput. On each side of the cranium, the lateral portions are called the temples or temporal regions. Between the temporal regions and the occiput lie the parietal regions. The forehead is called the frontal region. Just anterior to the ear, in the temporal region, you can feel the pulse of the superficial temporal artery.

The face is composed of the eyes, ears, nose, mouth, cheeks, and jowls. Six bones—the nasal bone, the two maxillae (upper jawbones), the two zygomas (cheekbones), and the mandible (jawbone)—are the major bones of the face **Figure 25-1** .

The orbit of the eye is composed of the lower edge of the frontal bone of the skull, the zygoma, the maxilla, and the nasal bone. The bony orbit protects the eye from

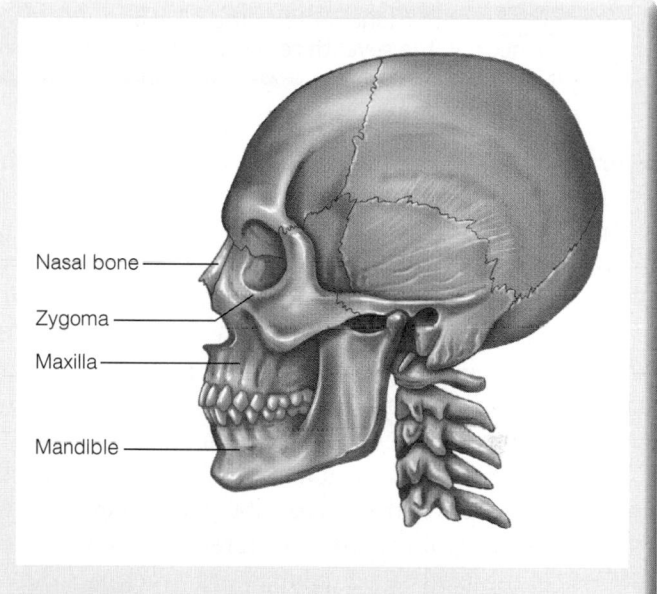

Nasal bone

Zygoma

Maxilla

Mandible

Figure 25-1 The face is composed of six bones: the nasal bone, two maxillae, two zygomas, and the mandible.

You are the Provider: PART 1

At 7:26 PM, you are dispatched to the parking lot of a convenience store at 1505 Eagle Rock Drive for a patient who was assaulted. You and your partner respond to the scene, which is located about 3 miles away. Law enforcement personnel, who are on scene, advise you that the scene has been secured, and that your patient, a young man, is conscious but has severe trauma to his face.

1. What should be your most immediate concern after receiving this initial patient information?
2. What should your initial actions consist of when you arrive at the scene?

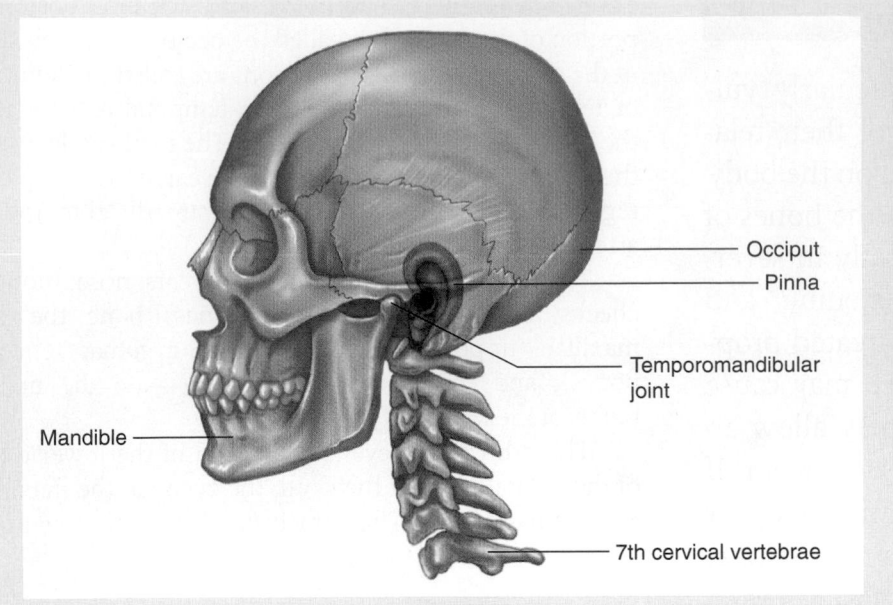

Occiput
Pinna
Temporomandibular joint
Mandible
7th cervical vertebrae

Figure 25-2 Specific landmarks of the head and neck include the pinna, the mandible, the occiput, the seventh cervical vertebra, and the temporomandibular joint.

injury. By viewing the face from the side, you can see the eyeball recessed in the orbit. Only the proximal third of the nose—the bridge—is formed by bone. The remaining two thirds are composed of cartilage.

The exposed portion of the ear is composed entirely of cartilage that is covered by skin. The external, visible part of the ear is called the **pinna** **Figure 25-2** . The earlobes are the fleshy portions at the bottom of each ear. The **tragus** is a small, rounded, fleshy bulge immediately anterior to the ear canal. The superficial temporal artery can be palpated just anterior to the tragus. About 1″ posterior to the external opening of the ear is a prominent bony mass at the base of the skull called the **mastoid process**.

The mandible forms the jaw and chin. The jaw is the lower border of the mouth, where the tongue and thirty-two teeth are located. Motion of the mandible occurs at the **temporomandibular joint**, which lies just in front of the ear on either side of the face. Below the ear and anterior to the mastoid process, the angle of the mandible is easily palpated.

The neck also contains many important structures. It is supported by the cervical spine, or the first seven vertebrae in the spinal column (C1 through C7). The spinal cord exits from the foramen magnum and lies within the spinal canal formed by the vertebrae. The upper part of the esophagus and the trachea lie in the midline of the neck. The carotid arteries are found on either side of the trachea, along with the jugular veins and several nerves.

Several useful landmarks can be palpated and seen in the neck **Figure 25-3** . The most obvious is the firm prominence in the center of the anterior surface,

commonly known as the Adam's apple. Specifically, this prominence is the upper part of the larynx, formed by the thyroid cartilage. It is more prominent in men than in women. The other portion of the larynx is the cricoid cartilage, a firm ridge of cartilage (the only complete circular cartilage structure of the trachea) below the thyroid cartilage, which is somewhat more difficult to palpate. Between the thyroid cartilage and the cricoid cartilage in the midline of the neck is a soft depression, the cricothyroid membrane. This is a thin sheet of connective tissue (fascia) that joins the two cartilages **Figure 25-4** . The cricothyroid membrane is covered at this point only by skin.

Below the larynx, several additional firm ridges are palpable in the anterior midline. These ridges are the cartilage rings of the trachea. The trachea connects the oropharynx and the larynx with the main air passages of the lungs (the bronchi). On either side of the lower larynx and the upper trachea lies the thyroid gland. Unless it is enlarged, this gland is usually not palpable.

Pulsations of the carotid arteries are easily palpable in a groove 1 to 2 cm lateral to the larynx. Lying immediately adjacent to these arteries, but not palpable, are the internal jugular veins and several important nerves. Lateral

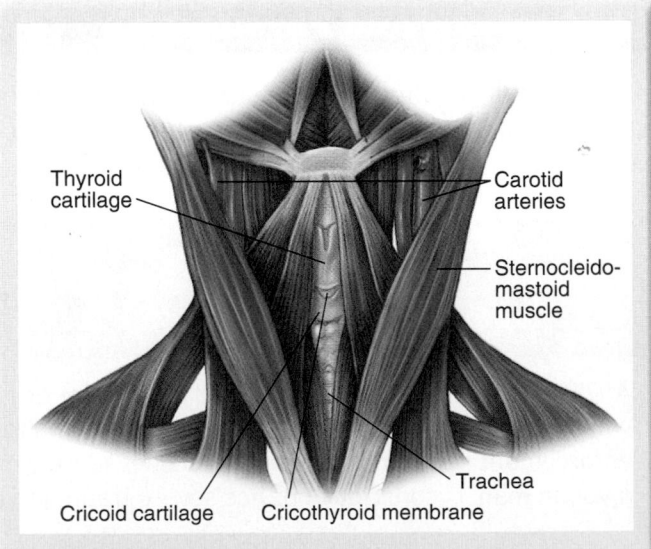

Thyroid cartilage
Carotid arteries
Sternocleido-mastoid muscle
Trachea
Cricoid cartilage
Cricothyroid membrane

Figure 25-3 Important landmarks in the neck include the cricoid cartilage, the thyroid cartilage, the carotid arteries, the cricothyroid membrane, and the sternocleidomastoid muscles.

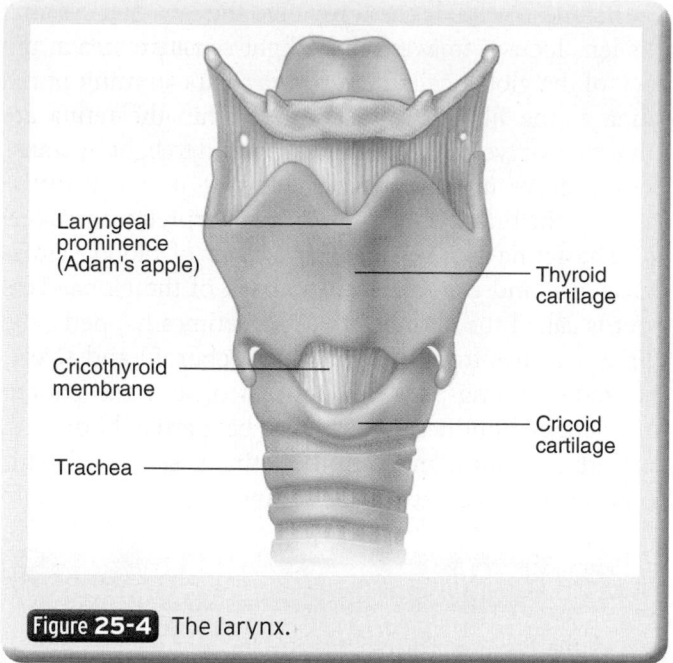

Figure 25-4 The larynx.

The Eye

The eye is globe-shaped, approximately 1″ in diameter, and located within a bony socket in the skull called the orbit **Figure 25-5**. The orbit is composed of the adjacent bones of the face and skull; the orbit forms the base of the floor of the cranial cavity, and directly above it are the frontal lobes of the brain. In the adult, more than 80% of the eyeball is protected within this bony orbit. Between and below the orbits are the nasal bone and the sinuses, respectively. Therefore, any severe injury to the face or head can potentially damage the eyeball or the muscles attached to the eyeball that cause the eye to move.

The eyeball, or **globe**, keeps its global shape as a result of the pressure of the fluid contained within its two chambers. The clear, jellylike fluid near the back of the eye is called the vitreous humor. In front of the lens is a clear fluid called the aqueous humor, named for its watery appearance; in Latin, aqua means water. In penetrating injuries of the eye, aqueous humor can also leak out, but with time and appropriate medical treatment, the body can make more.

The inner surface of the eyelids and the exposed surface of the eye itself, which are covered by a delicate membrane, the **conjunctiva**, are kept moist by fluid produced by the **lacrimal glands**, often called tear glands **Figure 25-6**. Humans blink unconsciously many times per minute. This action sweeps fluid from the lacrimal glands over the surface of the eye, cleaning it. The tears drain on the inner side of the eye through two lacrimal (tear) ducts into the nasal cavity. This is why, when people cry, they sometimes need to blow their nose.

The white of the eye, called the **sclera**, extends over the surface of the globe. This is extremely tough, fibrous tissue that helps maintain the eye's globular shape and protect the more delicate inner structures. On the front of the eye, the sclera is replaced by a clear, transparent membrane called the **cornea**, which allows light to enter the eye. A circular muscle lies behind the cornea with an opening in its center. Like the shutter in a camera, this muscle adjusts the size of the opening to regulate the amount of light that enters the eye. This circular muscle and surrounding tissue are called the **iris**. The iris is pigmented, giving the eye its characteristic brown, green, or blue color.

The opening in the center of the iris, which allows light to move

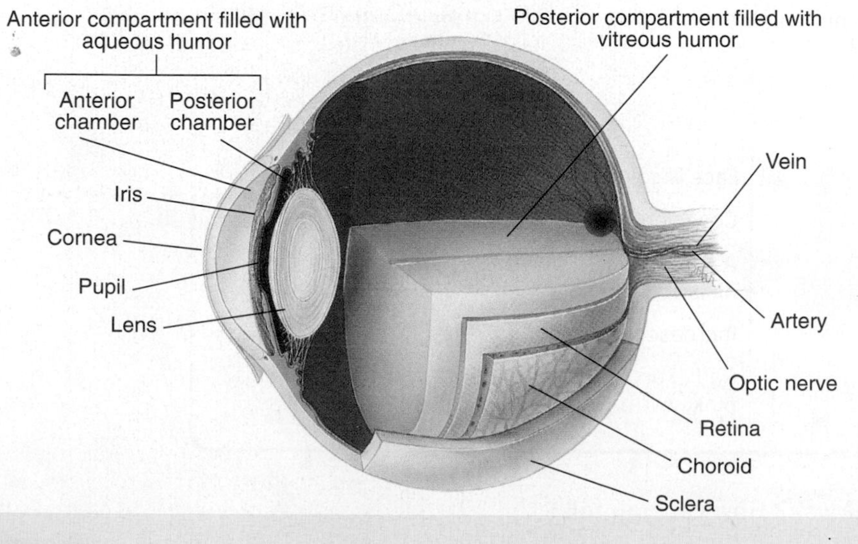

to these vessels and nerves lie the **sternocleidomastoid muscles**. These muscles originate from the mastoid process of the cranium and insert into the medial border of each collarbone and the sternum at the base of the neck. They allow movement of the head.

A series of bony prominences lie posteriorly, in the midline of the neck. They are the spines of the cervical vertebrae. The lower cervical spines are more prominent than the upper ones. They are more easily palpable when the neck is in flexion. At the base of the neck posteriorly, the most prominent spine is the seventh cervical vertebra.

Anterior compartment filled with aqueous humor

Posterior compartment filled with vitreous humor

Anterior chamber
Posterior chamber
Iris
Cornea
Pupil
Lens

Vein
Artery
Optic nerve
Retina
Choroid
Sclera

Figure 25-5 The major components of the eye.

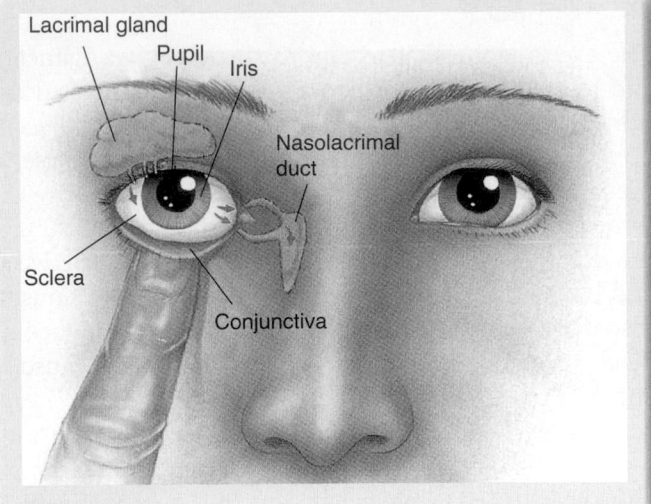

Lacrimal gland
Pupil
Iris
Nasolacrimal duct
Sclera
Conjunctiva

Figure 25-6 The lacrimal system consists of tear glands and ducts. Tears act as lubricants and keep the front of the eye from drying out.

Behind the iris is the **lens**. Like the lens of a camera, this lens focuses images on the light-sensitive area at the back of the globe, called the **retina**. You can think of the retina as the film in the camera. Within the retina are numerous nerve endings, which respond to light by transmitting nerve impulses through the **optic nerve** to the brain. In the brain, the impulses are interpreted as vision.

The retina is nourished by a layer of blood vessels between it and the sclera at the back of the globe. This layer is called the choroid. If, as sometimes happens, the retina detaches from the underlying choroid and sclera, the nerve endings are not nourished, and the patient experiences blindness. This may be partial blindness, depending on how much of the retina is separated. This condition is called **retinal detachment**.

Words of Wisdom

The eye can see objects directly in front (central line vision) and objects to the side (peripheral vision).

Injuries of the Face and Neck

Injuries about the face and neck can often lead to partial or complete obstruction of the upper airway. Several factors may contribute to the obstruction. Bleeding from facial injuries can be very heavy, producing large blood clots in the upper airway. These clots can lead to

to the back of the eye, is called the **pupil**. Normally, the pupil appears black. Like the opening in a camera, the pupil becomes smaller in bright light and larger in dim light. The pupil also becomes smaller and larger when the person is looking at objects near at hand and farther away; these adjustments occur almost instantaneously. Normally, the pupils in both eyes are equal in size. Some people are born with pupils that are not equal (**anisocoria**); however, particularly in unconscious patients, unequal pupil size may indicate serious injury or illness of the brain or eye.

You are the Provider: PART 2

When you arrive at the scene, you find the patient, a 30-year-old man, sitting on the ground. He is conscious and alert and tells you, "They hurt me bad!" His face is swollen and covered with blood, he keeps spitting out small amounts of blood from his mouth, and his voice is hoarse. As your partner manually stabilizes his head, you perform a primary assessment.

Recording Time: 0 Minutes	
Appearance	Face is covered with blood; appears anxious
Level of consciousness	Conscious and alert
Airway	A small amount of blood in the mouth, which he is spitting out
Breathing	Increased rate; adequate depth
Circulation	Radial pulses, increased rate and strong; skin is pink, warm, and dry

3. Does this patient have a patent airway? How can you tell?

4. What should be your initial treatment priorities for this patient?

complete obstruction, particularly in a patient who is not fully conscious. In particular, direct injuries to the nose and mouth, the larynx, or the trachea are often the source of significant bleeding and/or respiratory compromise. You may need to suction the airway if you are unable to control the bleeding. In addition, the injuries may cause loosened teeth or dentures to become dislodged into the throat where they may be swallowed or aspirated. The swelling that often accompanies direct and indirect injury to the soft tissues in these areas can also contribute to airway obstruction.

The airway may also be affected when the patient's head is turned to the side, as often is done when the patient has an altered level of consciousness or is unconscious. Other factors that interfere with normal respirations include possible injuries to the brain and/or cervical spine that may be associated with facial injuries. If the great vessels in the neck are injured, significant bleeding and pressure on the upper airway are common; these can result in airway obstruction as well.

Depending on the mechanism of injury there may be suspicion of a cervical spine injury. If there is significant impact to the face, suspect accompanying cervical spine injury and follow your agency's protocol for cervical injuries.

■ Soft-Tissue Injuries

Soft-tissue injuries of the face and neck are very common. Because the face and neck are extremely vascular, swelling from soft-tissue injuries in this area may be more severe than in other injured parts of the body. The skin and underlying tissues in these areas have a rich blood supply, so bleeding from penetrating injuries may be heavy. Indeed, even minor soft-tissue wounds of the face and neck may bleed profusely. A blunt injury that does not break the skin may cause a break in a blood vessel wall, leading blood to collect under the skin; this is called a hematoma **Figure 25-7** . In some situations, a flap of skin is peeled back, or avulsed, from the underlying muscle and fascia **Figure 25-8** .

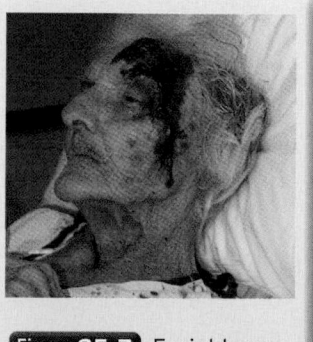

Figure 25-7 Facial hematoma.

■ Dental Injuries

Mandible (lower jaw) fractures are relatively common because of the prominence of the mandible itself. These fractures are second only to

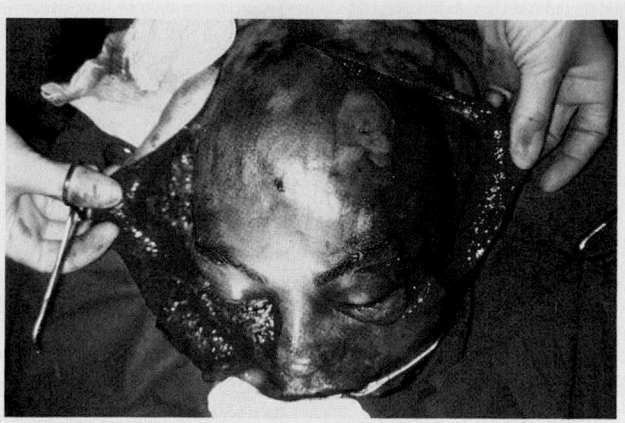

Figure 25-8 A major avulsion injury is characterized by a large flap of skin that is peeled back from the underlying muscle and tissue.

nasal fractures in frequency. Most of these fractures are the result of vehicle collisions and assaults. If your patient has a mandible fracture, then consider the major force necessary to cause that fracture—there is a strong probability your patient will have additional facial trauma and/or cervical injuries. Signs of a mandibular fracture include a misalignment of the teeth, numbness of the chin, and an inability to open the mouth. The patient will most likely have swelling, bruising, and loosened or missing teeth.

Maxillary fractures are predominantly found after blunt force high-energy impacts such as an unrestrained driver striking the steering wheel, a fall, or a direct blow from an object such as a pipe. The signs include massive facial swelling, instability of the facial bones, and misalignment of the teeth.

Fractured and avulsed teeth are common following facial trauma. Dental injuries may be associated with motor vehicle crashes or an assault. You should always assess the patient's mouth following a facial injury, especially if your examination reveals fractured or avulsed teeth. Teeth fragments (or even whole teeth) can become an airway obstruction and should be removed from the patient's mouth immediately.

▶ Patient Assessment

▶ Scene Size-up

Scene Safety

As you arrive on the scene, observe for hazards and threats to the safety of the crew, bystanders, and the patient. Assess the impact of hazards on patient care and address those hazards. Assess for the potential for violence and assess for environmental hazards.

Patients who are conscious and supine and have oral or facial bleeding may protect their airway by coughing, projecting the blood at you. Therefore, standard precautions require eye protection and a face mask. Also, put several pairs of gloves in your pocket for easy access in the event your gloves tear or there are multiple patients with bleeding.

If your response is to a motor vehicle crash, you may be confronted with more than one patient in a vehicle. Determine the number of patients and consider if you need additional or specialized resources on the scene.

Mechanism of Injury/Nature of Illness

As you observe the scene, look for indicators of the mechanism of injury (MOI). This assessment helps you develop an early index of suspicion for underlying injuries in the patient who has sustained a significant MOI. As you put together information from dispatch and your observations of the scene, consider how the MOI produced the injuries expected. Common MOI for face and neck injuries include motor vehicle accidents, sports, falls, penetrating trauma, and blunt trauma. In motor vehicle collisions, the probability of injury increases if the vehicle rolled over or came to an abrupt stop when striking an immovable object, such as a tree. Injuries sustained during sports participation may include a player without a helmet who was struck by a baseball or two players who sustained a helmet-to-helmet collision in football.

▶ Primary Assessment

The primary assessment focuses on identifying and managing life-threatening concerns. Threats to airway, breathing, or circulation must be treated immediately. Perform a rapid scan.

Form a General Impression

As you approach the patient, look for important indicators to alert you to the seriousness of the patient's condition. Is the patient interacting with the environment or lying still, making no sounds? Does the patient have any apparent life threats such as significant bleeding? How is the patient's skin color? Does he or she appear to be "sick" or "not so sick?" The general impression will help you develop an index of suspicion for serious injuries and determine your sense of urgency for medical intervention.

Injuries to the face and throat may be very obvious, such as bleeding and significant swelling, but may also be hidden under collars and hats. Because of the likelihood of respiratory distress with these injuries, these injuries should be recognized as early as possible.

As with any injury with life-threatening bleeding, control the blood loss with direct pressure. Always consider the need for manual spinal stabilization and check for responsiveness using the AVPU (*Alert* to person, place, and day; responsive to *Verbal* stimuli; responsive to *Pain*; *Unresponsive*) scale.

Airway and Breathing

Ensure that the patient has a clear and patent airway. If the patient is unresponsive or has a significant altered level of consciousness, consider inserting a properly sized oropharyngeal airway. The nasopharyngeal airway is contraindicated because of the possibility of insertion directly into the cranial vault and brain tissue if the patient has a basilar skull fracture.

Quickly assess for adequate breathing. Palpate the chest wall for DCAP-BTLS (Deformities, Contusions, Abrasions, Punctures/penetrations, Burns, Tenderness, Lacerations, and Swelling). If penetrating trauma is discovered, place an occlusive dressing on the wound. If a flail segment is discovered, stabilize the injury with a gloved hand or stabilize the injured chest wall with a bulky dressing. Check for clear and symmetric breath sounds and then provide high-flow oxygen, or provide assisted ventilation using a bag-mask device as needed, depending on the level of consciousness and your patient's breathing rate and quality. Face and throat injuries increase the need for airway and breathing maintenance, so do not hesitate to place a nonrebreathing mask over facial injuries. The seal may not be as easy to maintain, but airway and breathing take priority over soft-tissue injuries.

Circulation

You must quickly assess the pulse rate and quality; determine the skin condition, color, and temperature; and check the capillary refill time. Significant bleeding is an immediate life threat. If the patient has obvious life-threatening bleeding, you must control it quickly.

Transport Decision

If the patient you are treating has an airway or a breathing problem or significant bleeding, you must consider quickly transporting the patient to the hospital for treatment. Stabilization and maintenance of an airway and breathing and controlling bleeding can be very difficult in patients with facial or neck injuries, so delays in transport should be avoided and advanced life support backup considered if the transport time is long. A patient with signs and symptoms of internal bleeding must be transported quickly to the appropriate hospital for treatment by a physician. Internal bleeding in face and throat injuries often involves the brain or major vessels of the throat and can have a serious impact on the patient's airway. The condition of a patient with visible significant bleeding or signs of significant internal bleeding may quickly become unstable. Treatment is directed at quickly addressing life threats and providing rapid transport to the closest appropriate hospital. Signs such as tachycardia, tachypnea, low blood pressure, weak pulse, and cool, moist, pale skin are signs of hypoperfusion and imply the need for rapid transport. The patient who has a significant MOI but whose condition appears stable should also be transported promptly to the closest appropriate hospital. Remember that any significant blow to the face or throat should increase your suspicion of spinal or brain injury. You should be alert to these signs and reconsider your priority and transport decision if they develop.

Even if the patient has no signs of hypoperfusion or other life-threatening injuries, there is the possibility of eye injuries, which are considered serious; therefore, the patient should be transported to the hospital as quickly and as safely as possible. In some situations, surgery and/or restoration of circulation to the eye will need to be accomplished within 30 minutes or permanent blindness may result. For serious, isolated eye injuries, consideration should be given to transport to an eye care specialty center depending on local protocol. You should not delay transport of a seriously injured patient, particularly one with significant bleeding even if controlled, to take a patient's history or perform a secondary assessment. Further assessment can continue during transport.

History Taking

Investigate Chief Complaint
After the life threats have been managed during the primary assessment, investigate the chief complaint or history of present illness. You should obtain a medical history and be alert for injury-specific signs and symptoms as well as any pertinent negatives such as no pain or no loss of sensation.

SAMPLE History
Next, obtain a SAMPLE history from your patient. If the patient is not responsive, attempt to get the SAMPLE history from friends or family members who may be present.

In an unresponsive patient you will only be able to notice the signs of the patient's injuries; any other information will need to be obtained by someone who is knowledgeable about the patient. Keep in mind that the information you obtain may be or may not be accurate and may be incomplete. The person providing the information may not be able to give you the actual names of the patient's medications but might be able to provide some pertinent medical history and possibly known allergies.

Secondary Assessment

The secondary assessment is a more detailed, comprehensive examination of the patient that is used to uncover injuries that may have been missed during the primary assessment. In some instances, such as a critically injured patient or a short transport time, you may not have time to conduct a secondary assessment.

Physical Examinations
If there is significant trauma that likely affects multiple systems, start with a full-body scan looking for DCAP-BTLS to be sure that you have found all life threats and injuries. When this is completed, perform a detailed full-body scan. However, do not delay transport to complete a thorough physical examination.

In the responsive patient who has an isolated injury with a limited MOI, consider focusing your physical examination on the isolated injury, the patient's complaint, and the body region affected, which, in this case, is the face and throat. Ensure that control of bleeding is maintained and note the location of the injury. Inspect the open wound for any foreign matter or impaled object.

During the physical examination, use both your eyes and hands. Your eyes will be looking for swelling, deformities of the bones, contusions, and discoloration whereas your hands will be gently palpating the face, looking and feeling for any abnormalities such as deformity or tenderness. Ask yourself, do the facial bones seem to be in alignment? Does the nasal bone seem to deviate from the midline? You should make note of any variations from the normal facial examination. You should note any facial drooping. Does one eye appear to be lower than the other? If so, this is an indication of an orbital fracture. Does the mandible appear to deviate toward one side or the other?

If your patient is responsive, you should explain exactly what you are doing and what you are looking for. Your discovery of an abnormality may actually be an old injury that the patient can tell you more about.

Assess all underlying systems. This should include neurologic, including brain and major nerves; sensory organs, including the eyes and nose; respiratory system, including mouth, nose, sinuses, and airway; and circulatory system, particularly focusing on the carotid arteries and jugular veins.

When evaluating the eyes, start on the outer aspect of the eye and work your way in toward the pupils. Examine the eye for any obvious foreign matter. Your patient may relay this information to you also ("I have something in my eye."). In addition to discoloration of the eye, evaluate for clarity of the patient's vision, bleeding in the iris area, or redness. Look for eye symmetry because asymmetry is a possible indication of a brain injury.

Look at each pupil for equal size and reaction to light. If the pupils are not symmetrical, then ask the patient if he or she has had any previous eye surgeries or injuries. Previous surgery or injury, rather than brain injury, may be the root cause of the pupils not appearing the same. Cataract surgery can cause unequal pupils, but when you have a patient with a suspected head injury or ocular injury, anisocoria (unequal pupils in dim light) may be present. Determine whether the unequal pupils are caused by physiologic or pathologic issues. Use of over-the-counter eye drops can change pupil size, and certain asthma inhalers can have the same effect if inadvertently sprayed into the eye. Brain injury, nerve disease, glaucoma, and meningitis are all possible causes of unequal pupils.

Does the patient have the ability to follow your finger from side to side as well as up and down? Can the patient read normal print? Does the patient report blurry vision in either eye? Is there a new sensitivity to light?

Vital Signs

Assess vital signs to obtain a baseline so that you can observe any changes a patient may display during treatment. A systolic blood pressure reading of less than 100 mm Hg with a weak, rapid pulse and cool, moist skin that is pale or gray should alert you to the presence of hypoperfusion in a patient who may have significant bleeding. Remember, you must be concerned with visible bleeding and unseen bleeding inside a body cavity. With facial and throat injuries, baseline information about the rate and quality of respirations and pulse is very important, as is monitoring throughout patient care.

Monitoring Devices

In addition to hands-on assessment, use monitoring devices to quantify your patient's oxygenation and circulatory status. You may also use noninvasive methods to monitor the blood pressure. It is recommended that you always assess the patient's first blood pressure manually with a sphygmomanometer (blood pressure cuff) and stethoscope.

Reassessment

Repeat the primary assessment. Reassess vital signs and the chief complaint. You should continually reassess the adequacy of the patient's airway, breathing, and circulation. Recheck patient interventions. Are the treatments you provided for problems with the ABCs still effective? This is particularly important in patients with facial or neck injuries because of the ease in which injuries can affect associated systems, such as the respiratory (airway and breathing), circulatory, and nervous systems. The patient's condition should be reassessed at least every 5 minutes.

You are the Provider: PART 3

The patient is placed on high-flow oxygen via a nonrebreathing mask. As your partner continues to manually stabilize his head, you perform a rapid secondary assessment, which reveals bruising and mild swelling to the anterior neck, directly over the trachea. The remainder of your assessment does not reveal any obvious injuries. You apply a cervical collar and assess his vital signs. One of the police officers is asked to retrieve the long backboard and other spinal immobilization equipment.

Recording Time: 5 Minutes	
Respirations	22 breaths/min; adequate depth
Pulse	118 breaths/min; strong and regular
Skin	Pink, warm, and dry
Blood pressure	132/68 mm Hg
Oxygen saturation (Sao$_2$)	97% (on oxygen)

The patient tells you that he was struck in the face with a steel pipe. After he fell to the ground, he was kicked in the face and throat several times. His face is severely swollen, three of his front teeth are missing, and he tells you that it "doesn't feel right" when he closes his mouth.

5. On the basis of the mechanism of injury, what type of injuries should you suspect and assess for?

Interventions

Provide complete spinal immobilization to any patient with suspected spinal injuries. Spinal injuries should be suspected any time there is significant trauma to the face or neck. Maintain an open airway, be prepared to suction the patient, and consider an oropharyngeal airway. Whenever you suspect significant bleeding, provide high-flow oxygen. Oxygen and airway maintenance are important for all patients with face and neck injuries. If needed, provide assisted ventilation using a bag-mask device with high-flow oxygen.

Control any significant visible bleeding. If the patient has signs of hypoperfusion, treat the patient aggressively for shock and provide rapid transport to the appropriate hospital. Do not delay transport of a seriously injured trauma patient to complete nonlifesaving treatments in the field, such as splinting extremity fractures; instead, complete these types of treatment en route to the hospital. If there is no cervical spine injury suspected, the patient may be more comfortable in the sitting position during transport.

Communication and Documentation

Include a description of the MOI and the position in which you found the patient when you arrived at the scene. Document the method used to remove the patient from the vehicle, for example, "prolonged extrication." In patients with severe external bleeding, it is important to recognize, estimate, and report the amount of blood loss that has occurred and how rapidly or how much time has passed since the bleeding started. This can be a challenge for you, especially if the surface the patient is on is wet or absorbs fluids or the environment is dark. Inform the hospital personnel about all of the injuries involving the patient's head and neck. Specialists may need to be called to manage injuries involving the eyes, ears, teeth, mouth, sinuses, larynx, esophagus, or large vessels. These specialists are not always in the hospital, especially during the evening or night, or in smaller hospitals, so informing emergency department personnel of all injuries involving the face and throat can save valuable time.

■ Emergency Medical Care

The emergency care of soft-tissue injuries to the face and neck is the same as treatment of soft-tissue injuries elsewhere on the body. You should assess the ABCs and care for any life threats first. Remember also to follow standard precautions in all cases.

Your first step is to open and clear the airway. Securing and maintaining a patent airway is paramount. Remember that blood draining into the throat can produce vomiting and airway obstruction; therefore, the patient may need

frequent suctioning. Take appropriate precautions if you suspect that the patient has sustained a cervical spine injury; be sure to avoid moving the neck. Use the jaw-thrust maneuver to open the patient's airway, and then suction the mouth. Once the patient is immobilized in a cervical collar and on a backboard, you can turn the backboard to one side to allow any blood or vomitus to drain out of the mouth rather than pool in the pharynx and obstruct the airway.

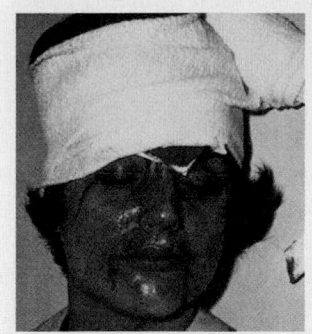

Figure 25-9 Use roller gauze, wrapped around the circumference of the head, to hold a pressure dressing in place.

Control bleeding by applying direct manual pressure with a dry, sterile dressing. Use roller gauze, wrapped around the circumference of the head, to hold a pressure dressing in place **Figure 25-9**. Do not apply excessive pressure if there is a possibility of an underlying skull fracture. When an injury exposes the brain, eye, or other structures, cover the exposed parts with a moist, sterile dressing to protect them from further damage. For injuries in which the skin is not broken, apply ice locally to help control the swelling of bruised tissues.

For soft-tissue injuries around the mouth, you should always check for bleeding inside the mouth. Broken teeth and lacerations to the tongue may cause profuse bleeding and obstruction of the upper airway **Figure 25-10**. Often, the patient will swallow the blood from lacerations inside the mouth, so the hemorrhage may not be apparent. You should also inspect the inside of the mouth for bleeding

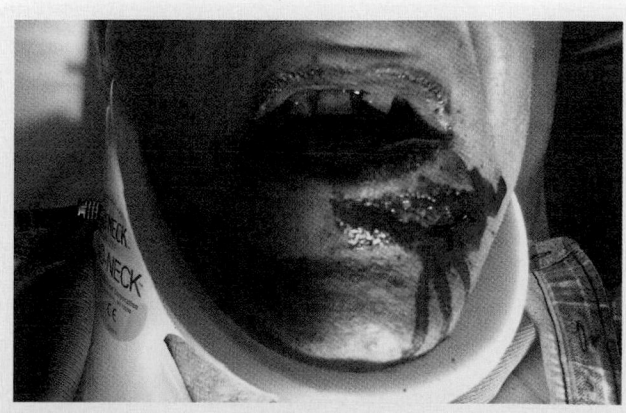

Figure 25-10 Soft-tissue injuries around the mouth can be associated with profuse bleeding inside the mouth and obstruction of the airway.

and hidden injuries in patients who have sustained facial trauma. Remember that patients who swallow blood are prone to vomiting.

Often, physicians will be able to graft a piece of avulsed skin back into the appropriate position. For this reason, if you find portions of avulsed skin that have become separated, you should wrap them in a sterile dressing, place them in a plastic bag, and keep them cool. Never place tissue directly on ice because freezing will destroy the tissue and make it unusable. Deliver the bag labeled with the patient's name to the emergency department along with the patient. In many avulsion injuries, the skin will still be attached in a loose flap **Figure 25-11** . Place the flap in a position that is as close to normal as possible, and hold it in place with a dry, sterile dressing. These steps will help to increase the patient's chances of having his or her normal appearance restored.

Emergency Medical Care for Specific Injuries

Injuries of the Eyes

Eye injuries are common, particularly in sports. An eye injury can produce severe lifelong complications, including blindness. Proper emergency treatment will minimize pain and may very well help to prevent a permanent loss of vision.

In a normal, uninjured eye, the entire circle of the iris is visible. The pupils are round, usually equal in size, and react equally when exposed to light **Figure 25-12** . Both eyes move together in the same direction when following your moving finger. After an injury, pupil reaction

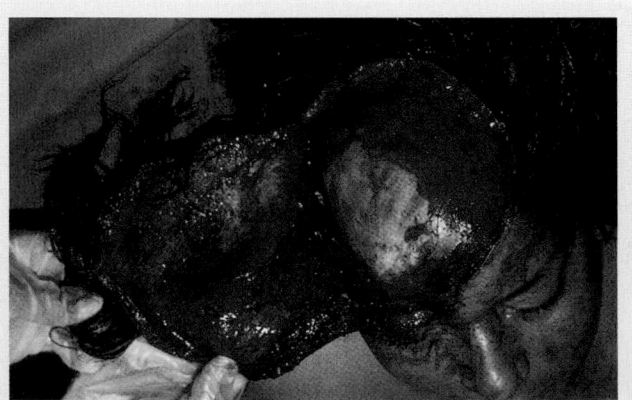

Figure 25-11 If avulsed skin is still attached, place the flap in a position that is as close to normal as possible, and hold it in place with a dry, sterile dressing.

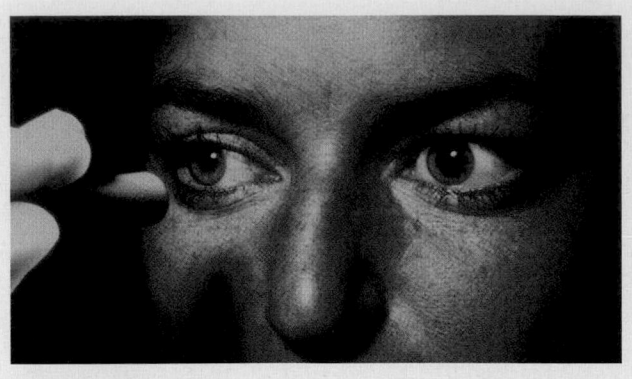

Figure 25-12 Normally, the pupils are round, equal in size, and equally reactive when exposed to light.

or shape and eye movement are often disturbed. Any of these conditions should cause you to suspect an injury of the globe or its associated tissues. Remember, though, that abnormal pupil reactions sometimes are a sign of brain injury rather than eye injury.

Treatment starts with a thorough examination to determine the extent and nature of any damage. Always perform your examination using standard precautions, taking great care to avoid aggravating any problems. You are looking for specific abnormalities or conditions that may suggest the nature of the injury **Figure 25-13** . For example, blunt or penetrating injuries can produce swollen or lacerated eyelids. Bleeding soon after irritation or injury can result in a bright red conjunctiva. A damaged cornea quickly loses its smooth, wet appearance.

Foreign Objects

Large objects are prevented from penetrating the eye by the protective orbit that surrounds it. However, moderately sized and smaller foreign objects of many different types can enter the eye and cause significant damage. Even a very small foreign object, such as a grain of sand lying on the surface of the conjunctiva, may produce severe irritation **Figure 25-14** . The conjunctiva becomes inflamed and red—a condition known as **conjunctivitis**—almost immediately, and the eye begins to produce tears in an attempt to flush out the object. Irritation of the cornea or conjunctiva causes intense pain. The patient may have difficulty keeping the eyelids open, because the irritation is further aggravated by bright light.

If a small foreign object is lying on the surface of the patient's eye, you should use a normal saline solution to gently irrigate the eye. Irrigation with a sterile saline solution will frequently flush away loose, small particles. If a small bulb syringe is available, you can use this, or a nasal

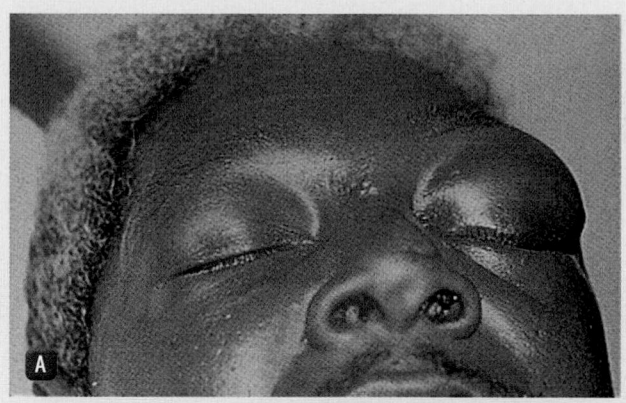

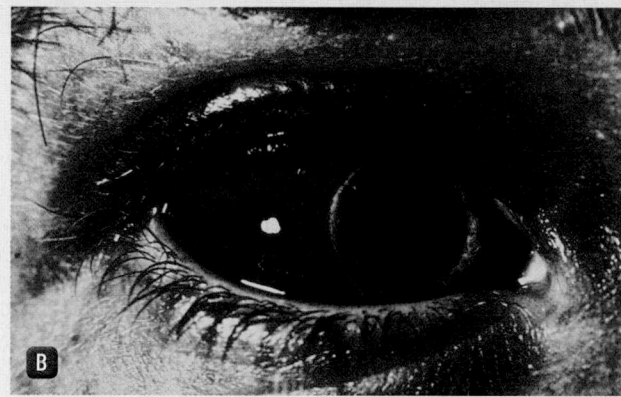

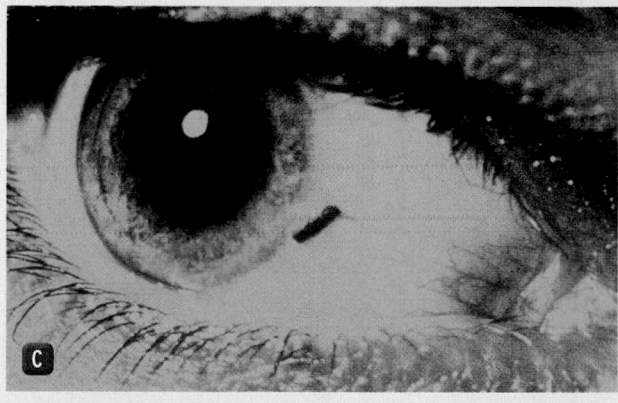

Figure 25-13 Injuries to the eyes are easily detected by **(A)** swelling, **(B)** bleeding, and **(C)** the presence of foreign objects in the eye.

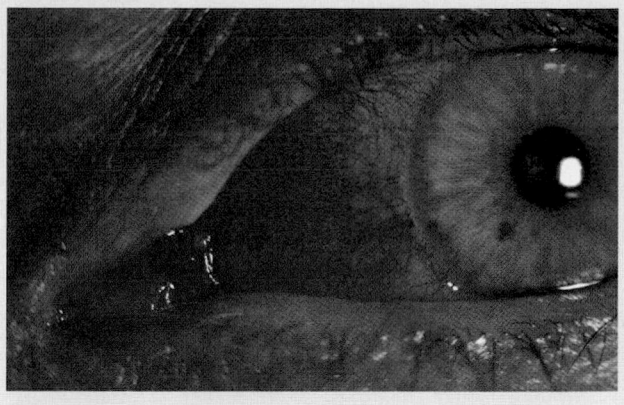

Figure 25-14 Conjunctivitis is often associated with the presence of a foreign object in the eye.

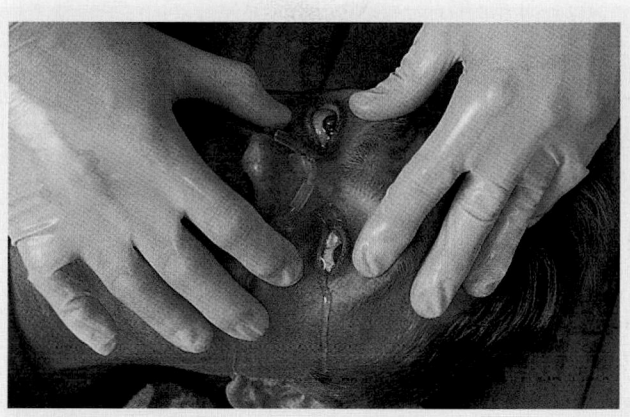

Figure 25-15 One method of irrigation is to direct saline into the injured eye using a round nasal airway or cannula. Always flush from the nose side of the eye toward the outside to avoid flushing material into the other eye.

airway or cannula, to direct the saline into the affected eye **Figure 25-15**. Always flush from the nose side of the eye toward the outside to avoid flushing material into the other eye. After it has been flushed away, a foreign body will often leave a small abrasion on the surface of the conjunctiva. For this reason, the patient will report irritation even when the particle itself is gone. It is always a good idea to transport the patient to the hospital for further assessment to ensure appropriate medical care to the affected eye.

Gentle irrigation usually will not wash out foreign bodies that are stuck to the cornea or lying under the upper eyelid. To examine the undersurface of the upper eyelid, pull the lid upward and forward. If you spot a foreign object on the surface of the eyelid, you may be able to remove it with a moist, sterile, cotton-tipped applicator **Skill Drill 25-1**. Never attempt to remove a foreign body that is stuck to the cornea.

1. Tell the patient to look down while you grasp the lashes of the upper eyelid with your thumb and index finger. Gently pull the eyelid away from the eyeball **Step 1**.
2. Gently place a cotton-tipped applicator horizontally along the center of the outer surface of the upper eyelid **Step 2**.

Skill Drill 25-1

Removing a Foreign Object From Under the Upper Eyelid

Step 1 Have the patient look down, grasp the upper lashes, and gently pull the lid away from the eye.

Step 2 Place a cotton-tipped applicator on the outer surface of the upper lid.

Step 3 Pull the lid forward and up, folding it back over the applicator.

Step 4 Gently remove the foreign object from the eyelid with a moistened, sterile, cotton-tipped applicator.

3. Pull the eyelid forward and up, which causes it to roll or fold back over the applicator, exposing the undersurface of the eyelid (Step 3).

4. If you see a foreign object on the surface of the eyelid, gently remove it with a moistened, sterile, cotton-tipped applicator (Step 4).

Foreign bodies ranging in size from a pencil to a sliver of metal may be impaled in the eye Figure 25-16. These objects must be removed by a physician. Your care involves stabilizing the object and preparing the patient for transport to definitive care. The greater the length of the foreign object you can see sticking out of the eye, the more important stabilization becomes in avoiding further

damage. Bandage the object in place to support it. Cover the eye with a moist, sterile dressing, and then surround the object with a doughnut-shaped collar made from roller gauze or a small gauze pack. Follow the steps in Skill Drill 25-2:

Skill Drill 25-2

1. Begin to prepare the doughnut ring by wrapping a 2″ gauze roll circumferentially around your fingers and thumb enough times to make a thick dressing layer. You can adjust the inner diameter

Skill Drill 25-2

Stabilizing a Foreign Object Impaled in the Eye

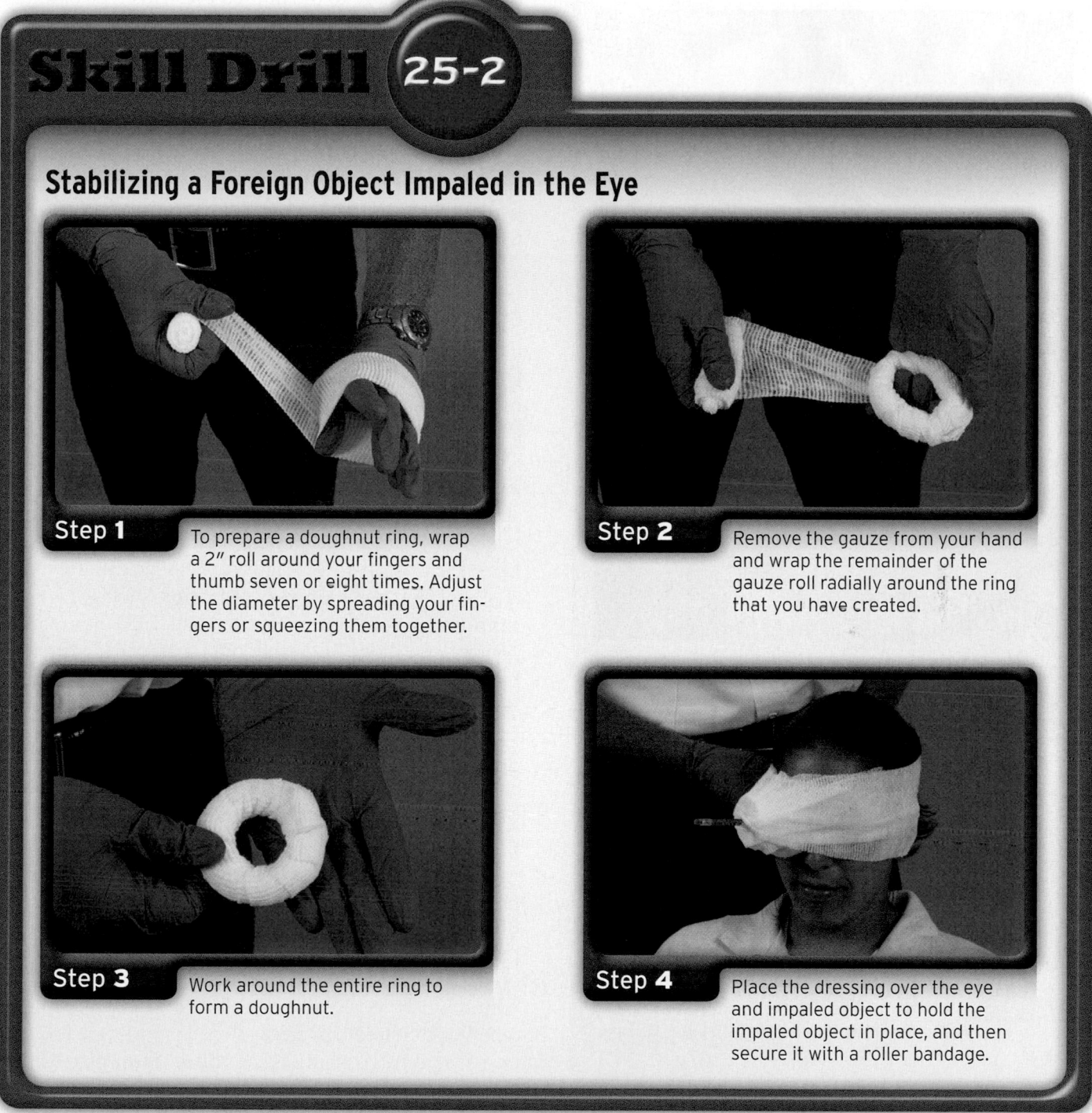

Step 1
To prepare a doughnut ring, wrap a 2″ roll around your fingers and thumb seven or eight times. Adjust the diameter by spreading your fingers or squeezing them together.

Step 2
Remove the gauze from your hand and wrap the remainder of the gauze roll radially around the ring that you have created.

Step 3
Work around the entire ring to form a doughnut.

Step 4
Place the dressing over the eye and impaled object to hold the impaled object in place, and then secure it with a roller bandage.

of what will become the ring by spreading your fingers or squeezing them together (Step 1).

2. Remove the gauze from your hand and wrap the remainder of the gauze roll radially around the ring that you have created (Step 2).

3. Work your way around the ring until you have wrapped all the way around it and finished the "doughnut" (Step 3).

4. Carefully place the ring over the eye and impaled object, without bumping the object. You can then

stabilize the object and the gauze collar with a roller bandage surrounding the head. Bandage both the injured and uninjured eyes to minimize eye movement and prevent further damage to the globe because when one eye moves, so does the other. Transport to an appropriate medical facility for treatment (Step 4).

Sometimes, a variety of types of large and small foreign bodies, particularly small metal fragments, become completely embedded within the eye itself. The

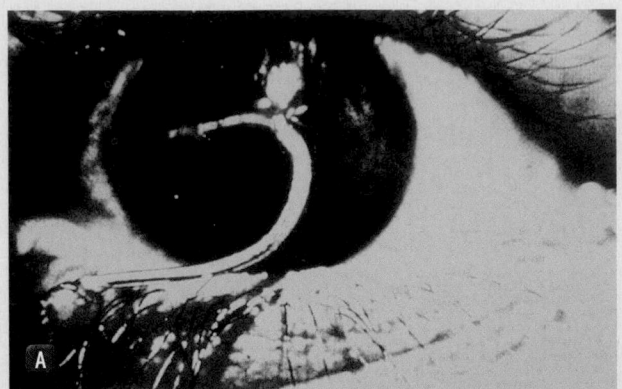

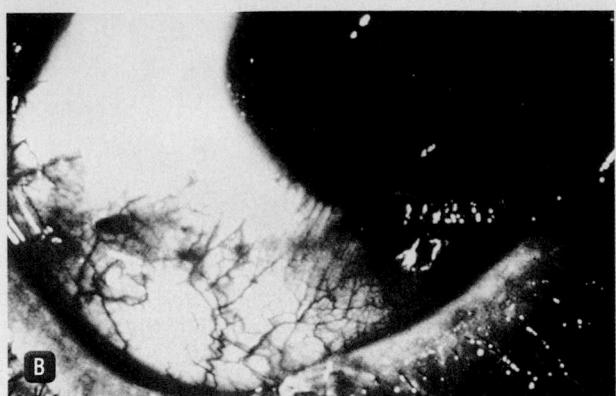

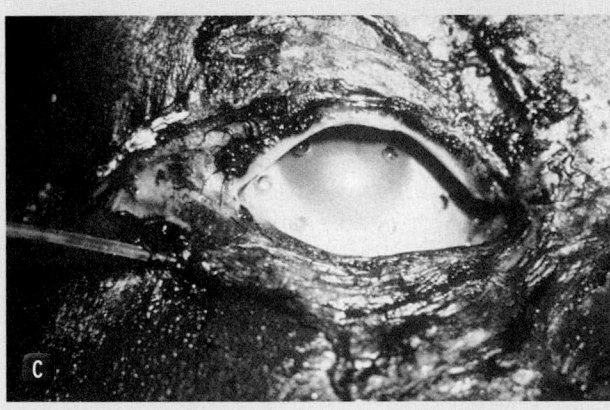

Figure 25-16 Any number of objects can become impaled in the eye. **A.** Fishhook. **B.** Sharp, metal sliver. **C.** Knife blade.

patient may not even be aware of the cause of the problem. Suspect such an injury when the history includes metal work (such as hammering, exposure to splinters, grinding, vigorous filing) and when there are other signs of ocular injury. When you see or suspect an impaled object in the eye, bandage both eyes with soft bulky dressings to prevent further injury to the affected eye. Your bandage should be loose enough to hold the eyelid closed but not cause pressure on the eye itself. Using this technique prevents sympathetic motion (the movement of one eye causing both eyes to move), which may cause

additional damage to the injured eye. This type of injury must be handled by an ophthalmologist on an urgent basis. X-rays and special equipment may be required to find the foreign body.

Burns of the Eye

Chemicals, heat, and light rays all can burn the delicate tissues, such as the cornea, often causing permanent damage. Your role is to stop the burn and prevent further damage.

Chemical Burns Chemical burns, usually caused by acid or alkaline solutions, require immediate emergency care **Figure 25-17**. This consists of flushing the eye with water or a sterile saline irrigation solution. If sterile saline is not available, you can use any clean water.

The idea is to direct the greatest amount of irrigating solution or water into the eye as gently as possible **Figure 25-18**. Because opening the eye spontaneously may cause the patient pain, you may have to force the lids open to irrigate the eye adequately. Ideally, you will use a bulb or irrigation syringe, a nasal cannula, or some other device that will allow you to control the flow. In some circumstances, you may have to resort to pouring water into the eye by holding the patient's head under a gently running faucet. You can even have the patient immerse his or her face in a large pan or basin of water and rapidly blink the affected eyelid. If only one eye is affected, care must be taken to avoid contaminated water from getting into the unaffected eye.

Be sure to flush from the inner corner of the affected eye toward the outside corner. Never flush from the outside corner as this may cause the substance to contaminate the unaffected eye. If the burn was caused by an alkali or a strong acid, you should irrigate the eye continuously for 20 minutes. Follow local protocols on whether to try to irrigate while transporting or to stay on scene until flushing is complete. Strong acids and all alkaline solutions can penetrate deeply, requiring a prolonged flush. Again, always take care to protect the uninjured eye and prevent irrigation fluid from running into it.

After you have completed irrigation, apply a clean, dry dressing to cover the eye, and transport the patient promptly to the hospital for further care **Figure 25-19**. If the irrigation can be carried out satisfactorily in the ambulance, it should be done during transport to save time.

Thermal Burns When a patient is burned in the face during a fire, the eyes usually close rapidly because of the heat. This reaction is a natural reflex to protect the eye from further injury. However, the eyelids remain exposed and are frequently burned Figure 25-20 . Burns of the eyelids require very specialized care. It is best to provide prompt transport for these patients without further examination. First, however, you should cover both eyes with a sterile dressing moistened with sterile saline. You may apply eye shields over the dressing.

Light Burns Infrared rays, eclipse light (if the patient has looked directly at the sun), and laser burns all can cause significant damage to the sensory cells of the eye when rays of light become focused on the retina. Retinal injuries that are caused by exposure to extremely bright light are generally not painful but may result in permanent damage to vision.

Superficial burns of the eye can result from ultraviolet rays from an arc welding unit, light from prolonged exposure to a sunlamp, or reflected light from a bright snow-covered area (snow blindness). This kind of burn often is not painful at first but may become so

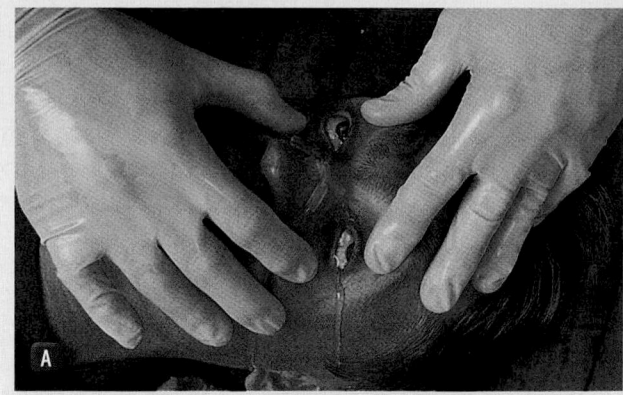

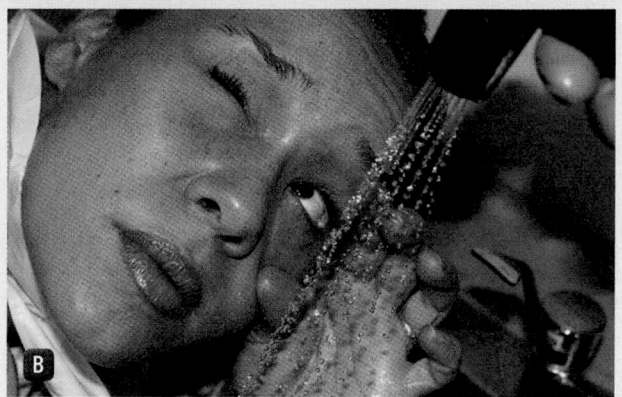

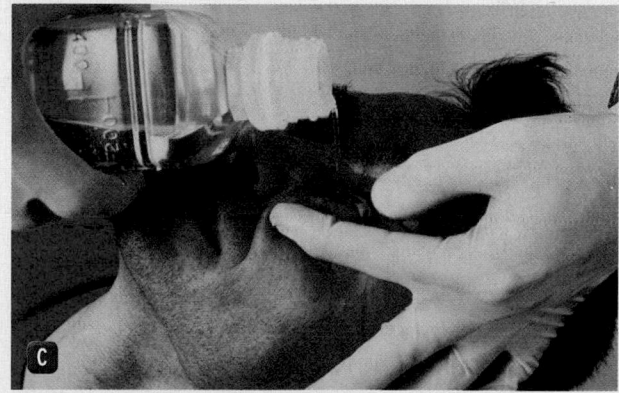

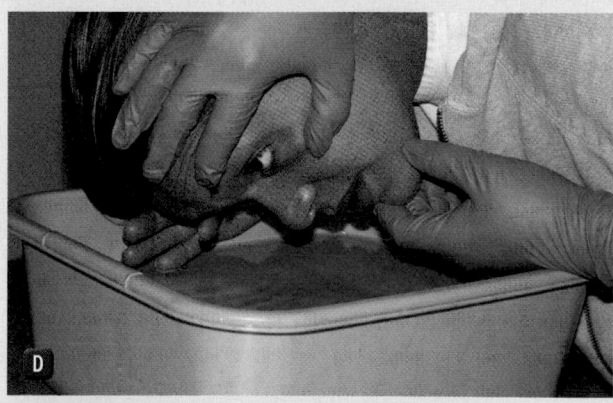

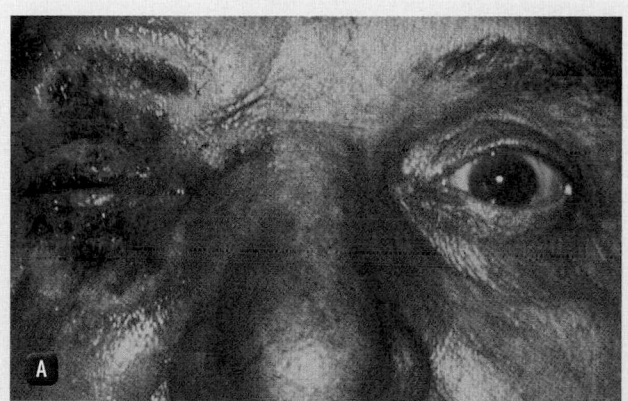

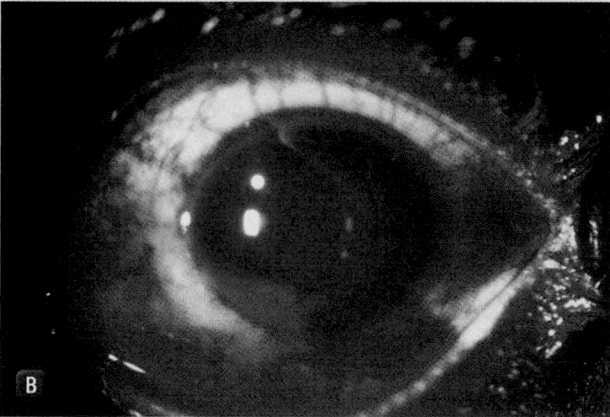

Figure 25-17 **A.** Chemical burns typically occur when an acid or alkali is splashed into the eye. **B.** This figure shows a chemical burn from lye, an alkaline solution. Because lye can continue to damage the eye even when diluted, fast action is needed.

Figure 25-18 The following are four ways to effectively irrigate the eye. **A.** Nasal cannula. **B.** Shower. **C.** Bottle. **D.** Basin. Remember, you must protect the uninjured eye from the irrigating solution to prevent exposure of the unaffected eye to the substance.

3 to 5 hours later, when the damaged cornea responds to the injury. Severe conjunctivitis usually develops, with redness, swelling, and excessive tear production. You can ease the pain from these corneal burns by covering each eye with a sterile, moist pad and an eye shield. Have the patient lie down during transport to the hospital, and protect him or her from further exposure to bright light.

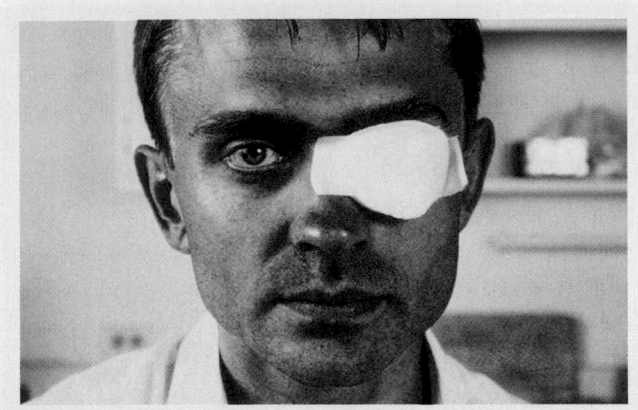

Figure 25-19 Apply a clean, dry dressing to cover the eye after you have finished irrigation.

The patient should be examined by a physician as soon as possible.

Lacerations

Lacerations of the eyelids require very careful repair to restore appearance and function **Figure 25-21**. Bleeding may be heavy, but it usually can be controlled by gentle, manual pressure. If there is a laceration of the globe itself, apply no pressure to the eye; compression can interfere with the blood supply to the back of the eye and result in loss of vision from damage to the retina. Furthermore, pressure may squeeze the vitreous humor, iris, lens, or even the retina out of the eye and cause irreparable damage or blindness.

Follow these three important guidelines in treating penetrating injuries of the eye:

1. Never exert pressure on or manipulate the injured eye (globe) in any way.
2. If part of the eyeball is exposed, gently apply a moist, sterile dressing to prevent drying.
3. Cover the injured eye with a protective metal eye shield, cup, or sterile dressing. Apply soft dressings to both eyes, and provide prompt transport to the hospital.

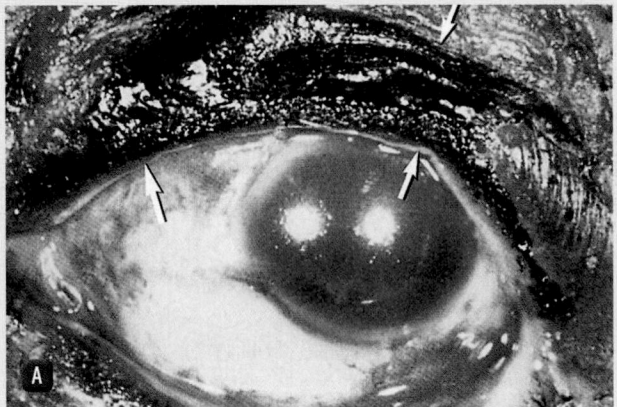

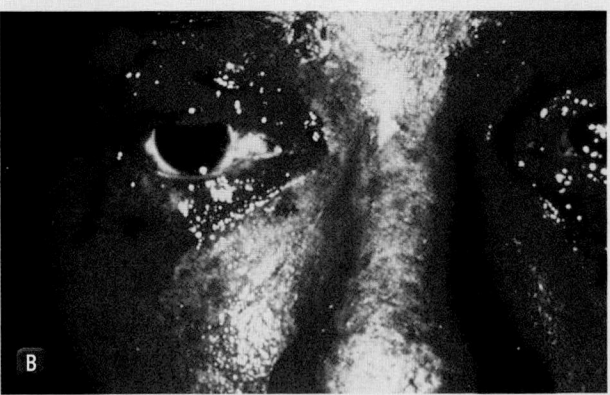

Figure 25-20 Thermal burns occasionally cause significant damage to the eyelids. **A.** Arrows show some full-thickness burns. **B.** Burns of the eyelids require immediate hospital care.

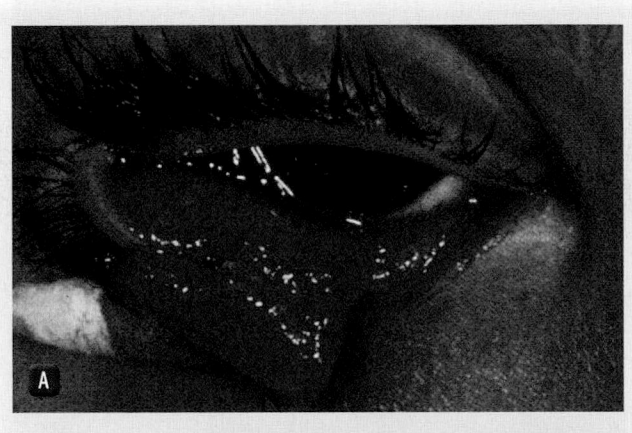

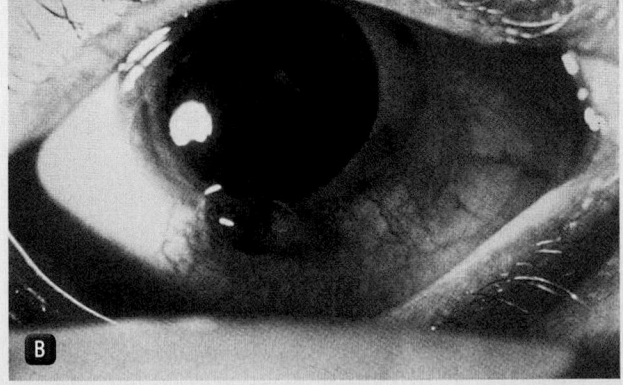

Figure 25-21 Lacerations are serious injuries that require prompt transport. **A.** Although bleeding can be heavy, never exert pressure on the eye. **B.** Pressure may squeeze the vitreous humor, iris, lens, or even the retina out of the eye.

On rare occasions following a serious injury, the eyeball may be displaced out of its socket. Do not attempt to reposition it. Simply cover the eye and stabilize it with a moist sterile dressing **Figure 25-22**; remember to cover both eyes to prevent further injury because of sympathetic movement. Have the patient lie in a supine position en route to the hospital to prevent further loss of fluid from the eye.

Blunt Trauma

Blunt trauma can cause a number of serious eye injuries. These range from the ordinary "black eye," a result of bleeding into the tissue around the orbit, to a severely damaged globe **Figure 25-23**. You may see an injury called hyphema, or bleeding into the anterior chamber of the eye, that obscures part or all of the iris **Figure 25-24**. This injury is common in blunt trauma and may seriously impair vision. Twenty-five percent of hyphemas are globe injuries, a serious injury to the eye. Cover the eye to protect it from further injury and provide transportation to the hospital for further medical evaluation.

Blunt trauma can also cause a fracture of the orbit, particularly of the bones that form its floor and support the globe. This injury is called a **blowout fracture**. The fragments of fractured bone can entrap some of the muscles that control eye movement, causing double vision **Figure 25-25**. Any patient who reports pain, double

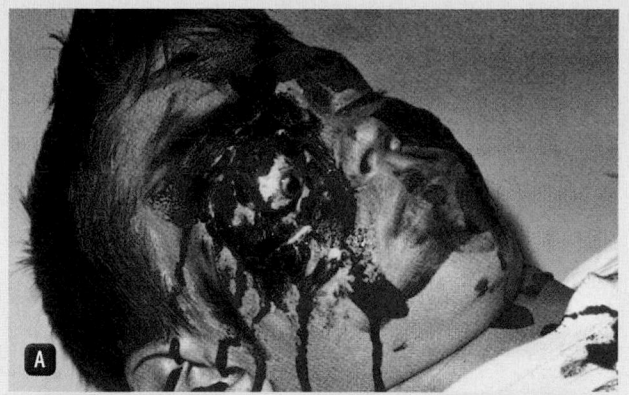

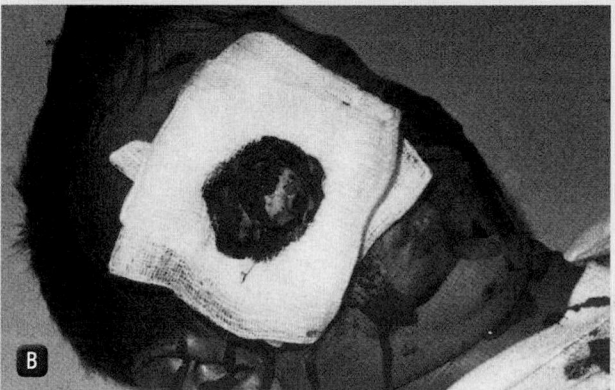

Figure 25-22 An injury that exposes the brain, eye, or other structures (**A**) should be covered with a moist, sterile dressing to prevent further damage (**B**).

You are the Provider: PART 4

After applying full spinal precautions, you place the patient onto the stretcher and load him into the ambulance. You recovered his teeth and placed them in a commercial tooth-saver container. While you reassess the interventions you have performed thus far, your partner reassesses the patient's vital signs.

Recording Time: 11 Minutes	
Level of consciousness	Conscious and alert
Respirations	20 breaths/min; adequate depth
Pulse	108 beats/min; strong and regular
Skin	Pink, warm, and dry
Blood pressure	128/62 mm Hg
Sao₂	98% (on oxygen)

Reassessment of the patient's mouth reveals that it is clear of blood. You begin transport to a trauma center, which is located 15 miles away. En route, the patient tells you that he is becoming nauseated. You call in your radio report and give an estimated time of arrival of 18 to 20 minutes.

6. What should you do if this patient begins to vomit?

7. How should you treat a patient with active oral bleeding *and* inadequate ventilation?

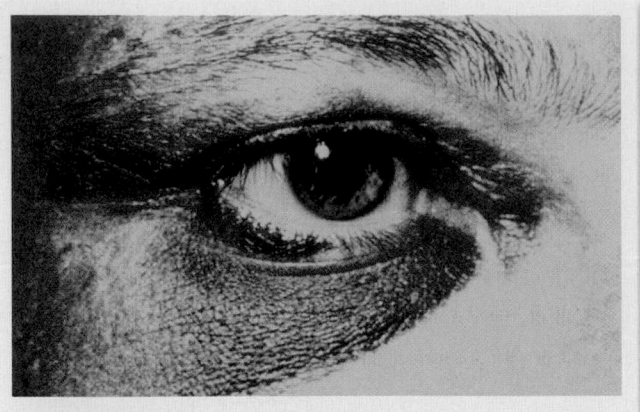

Figure 25-23 The typical "black eye" is caused by bleeding into the tissue around the orbit.

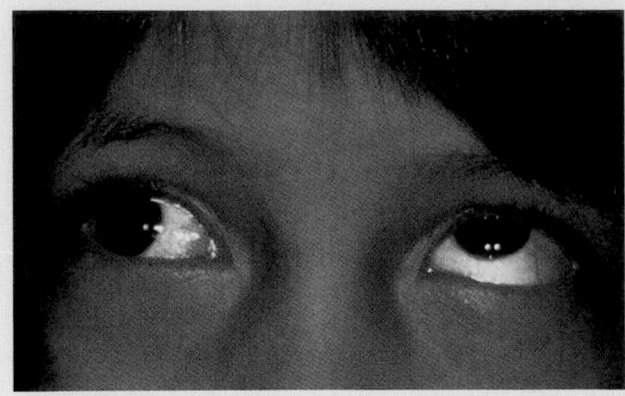

Figure 25-25 A patient with a blowout fracture may not move his or her eyes together because of muscle entrapment. Therefore, the patient sees double images of any object.

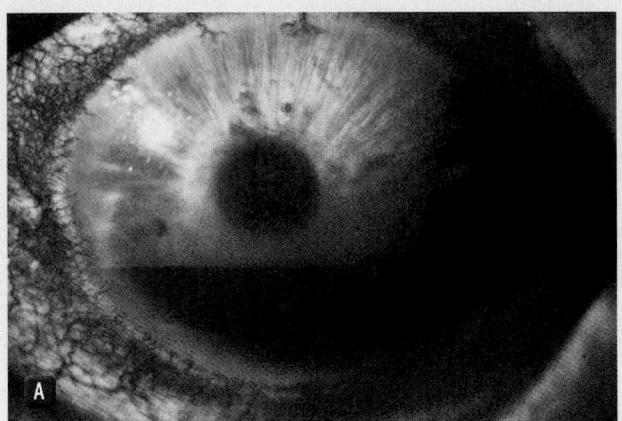

A

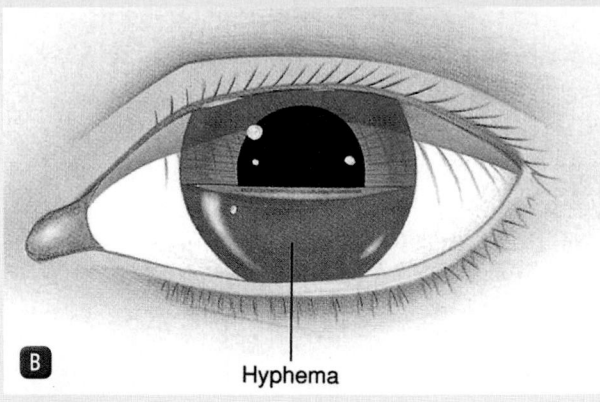

B

Hyphema

Figure 25-24 **A.** A hyphema, characterized by bleeding into the anterior chamber of the eye, is common following blunt trauma to the eye. This condition may seriously impair vision and should be considered a sight-threatening emergency. **B.** Illustration of hyphema.

Another possible result of blunt eye injury is retinal detachment. This injury is often seen in sports, especially boxing. It is painless but produces flashing lights, specks, or "floaters" in the field of vision and a cloud or shade over the patient's vision. Because the retina is separated from the nourishing choroid, this injury requires prompt medical attention to preserve vision in the eye.

Eye Injuries Following Head Injury

Abnormalities in the appearance or function of the eyes often occur following a closed head injury. Any of the following eye findings should alert you to the possibility of a head injury:

- One pupil larger than the other **Figure 25-26**
- The eyes not moving together or pointing in different directions
- Failure of the eyes to follow the movement of your finger as instructed

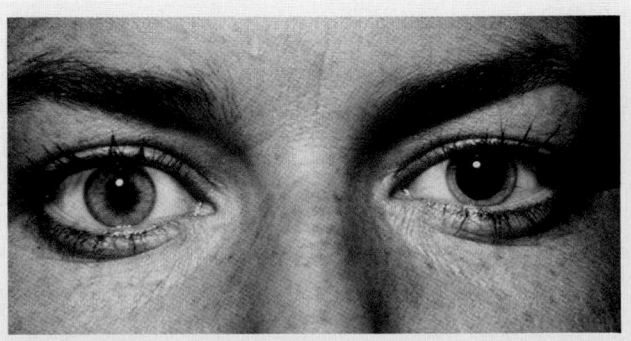

Figure 25-26 Variation of pupil size may indicate a head injury.

vision, or decreased vision following a blunt injury about the eye should be placed on a stretcher and transported promptly to the emergency department. Protect the eye from further injury with a metal shield; cover the other eye to minimize movement on the injured side.

- Bleeding under the conjunctiva, which obscures the sclera (white portion) of the eye
- Protrusion or bulging of one eye

Record any of these observations, along with the time that you make them. For an unconscious patient, remember to keep the eyelids closed; drying of the ocular tissue can cause permanent injury and may result in blindness. Cover the lids with moist gauze, or hold them closed with clear tape. Normal tears will then keep the tissues moist.

Blast Injuries

The signs and symptoms of blast injuries range from severe pain and loss of vision to foreign bodies within the globe. Before responding to patients after the blast, first ensure that the scene is safe.

Management of blast injuries to the eye depends on the severity of the injury. If there is a foreign body within the globe, do not attempt to remove it. Use a clean cup or similar item to protect the area. If only one eye is injured, then follow local protocol, which may include covering the other eye to eliminate sympathetic motion. Patients with a sudden loss or decrease of vision will need to be verbally instructed on what actions are taking place around them. If the patient has severe swelling or a hematoma to the eyelid, do not attempt to force the eyelid open to examine the eye because this increases the pressure already within the globe itself.

Contact Lenses and Artificial Eyes

Small, hard contact lenses usually are tinted, making them relatively easy to see. Large, soft contact lenses are clear and can be very difficult to see. In general, you should not attempt to remove either kind of lens from a patient. You should never attempt to remove a lens from an eye that has been—or may have been—injured because manipulating the lens can aggravate the problem. The only time that contact lenses should be removed immediately in the field is in the case of a chemical burn of the eye. In this situation, the lens can trap the chemical and make irrigation difficult.

If it is necessary to remove a hard contact lens, use a small suction cup, moistening the end with saline **Figure 25-27A**. To remove soft lenses, place one to two drops of saline in the eye **Figure 25-27B**, gently pinch it between your gloved thumb and index finger, and lift it off the surface of the eye **Figure 25-27C**. Place the contact lens in a container filled with sterile saline solution to prevent damage to the contact lens. Always advise the emergency department staff if a patient is wearing contact lenses.

Occasionally, you may find yourself caring for a patient who is wearing an eye prosthesis (an artificial eye). Many people are surprised to find that it can be difficult to distinguish a prosthesis from a natural eye. You should suspect an eye of being artificial when it does not respond to light, move in concert with the opposite eye, or appear quite the same as the opposite eye. If you think that a patient may have an artificial eye but you are not sure, go ahead and ask about it. Although no harm will be done if you care for an artificial eye as you would a normal one, you need to clearly understand the patient's eye function.

■ Injuries of the Nose

Nosebleeds (epistaxis) are a common problem that can occur spontaneously or from trauma. One of the most common causes of nosebleeds is digital trauma (picking the nose with a finger). Nosebleeds are further classified

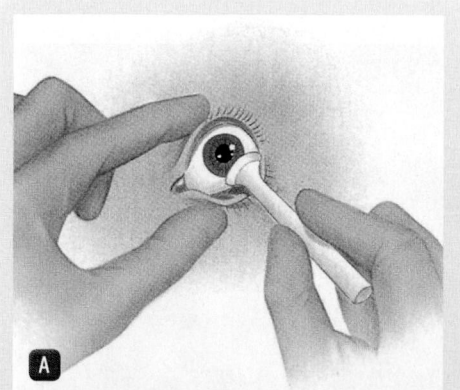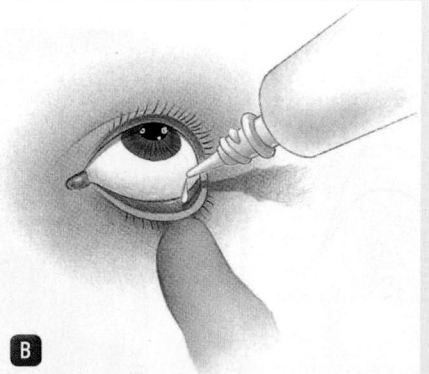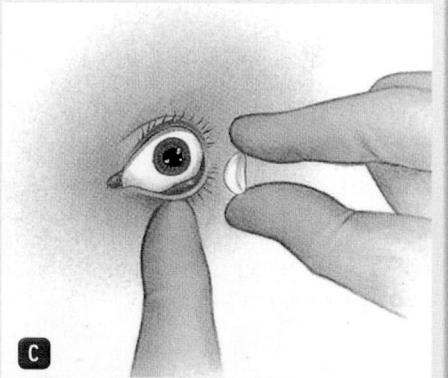

Figure 25-27 Removing contact lenses should be limited to patients with chemical burn injuries to the eye. **A.** To remove hard contact lenses, use a specialized suction cup moistened with sterile saline solution. **B.** To remove soft contact lenses, instill one or two drops of saline or irrigating solution. **C.** Next, pinch off the lens with your gloved thumb and index finger.

into anterior and posterior epistaxis. Anterior nosebleeds usually originate from the area of the septum and bleed fairly slowly. These are usually self-limiting and resolve quickly. Posterior nosebleeds are usually more severe and often cause blood to drain into the patient's throat, causing nausea and vomiting. Trauma to the face and skull that results in a basilar skull fracture often will cause the posterior wall of the nasal cavity to become unstable. You should not attempt to place a nasopharyngeal airway in a patient with a suspected basilar skull fracture or with facial injuries because insertion may permit the airway to enter through the unstable wall of the nasal cavity into the cranial vault.

The nose often takes the brunt of deliberate physical assaults and car crashes. Blunt injuries to the nose caused by a fist or a dashboard may be associated with fractures and soft-tissue injuries of the face, head injuries, and/or injuries to the cervical spine. Penetrating injuries to the nose can be seen when air guns and BB pellets are fired from a close range, resulting in pellets lodging in the nasal septum and sinuses. Another type of penetrating injury to the nose is a self-inflicted one that occurs when a person attempts to insert a foreign body into the nose, such as a pencil.

When you are assessing injuries involving the nose, it helps to picture the inside of the nose itself **Figure 25-28**.

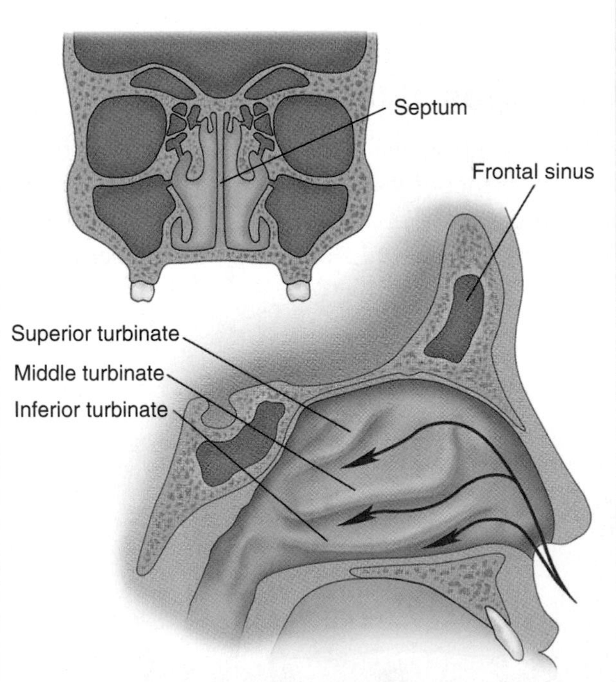

The nasal cavity is divided into two sections or chambers by the nasal septum, which is made of cartilage. Within each nasal chamber, there are layers of bone called the **turbinates**, which are covered with a moist lining. Both chambers have a superior turbinate, a middle turbinate, and an inferior turbinate. As a person breathes, air moves through the nasal chambers and is humidified as it passes over the turbinates. Directly above the nose are the frontal sinuses and, on either side, the orbit of the eye.

All of these structures should be assessed for injury. In patients with severe injury, there may also be injury to the cervical spine. Keep in mind that cerebrospinal fluid (CSF) may escape down through the nose (or ears) following a fracture at the base of the skull. If blood or drainage contains CSF, a characteristic staining of the dressing will occur. This can be seen by using a piece of gauze to absorb blood that is flowing from the nose or ears. If CSF is present, the blood will be surrounded by a lighter ring of fluid. This is often called the halo test.

You can control bleeding from abrasions and lacerations to the nose by applying a sterile dressing. If the patient is bleeding heavily from the nose, this is most likely caused by significant trauma and you must be concerned with cervical spine injury. The patient should not be moved if the airway can be managed in the patient's present position. For a nontrauma patient who is bleeding from the nose, you should place the patient in a sitting position, leaning forward, and pinch his or her nostrils together **Figure 25-29**. For a detailed discussion of the care for epistaxis, see Skill Drill 23-3 in Chapter 23, *Bleeding*.

■ Injuries of the Ear

The ear is a complex organ that is associated with hearing and balance. The ear is divided into three parts **Figure 25-30**. The external ear is composed of the pinna,

Figure 25-28 The nose has two chambers, divided by the septum. Each chamber is composed of layers of bone called turbinates. Above the nose are the frontal sinuses and, on either side, the orbit of the eye.

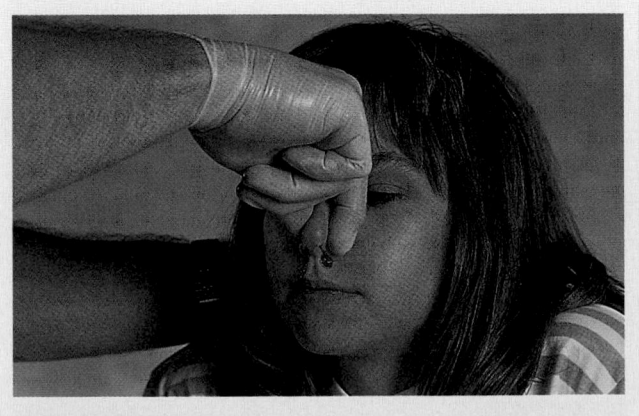

Figure 25-29 Control bleeding from the nose by pinching the nostrils together.

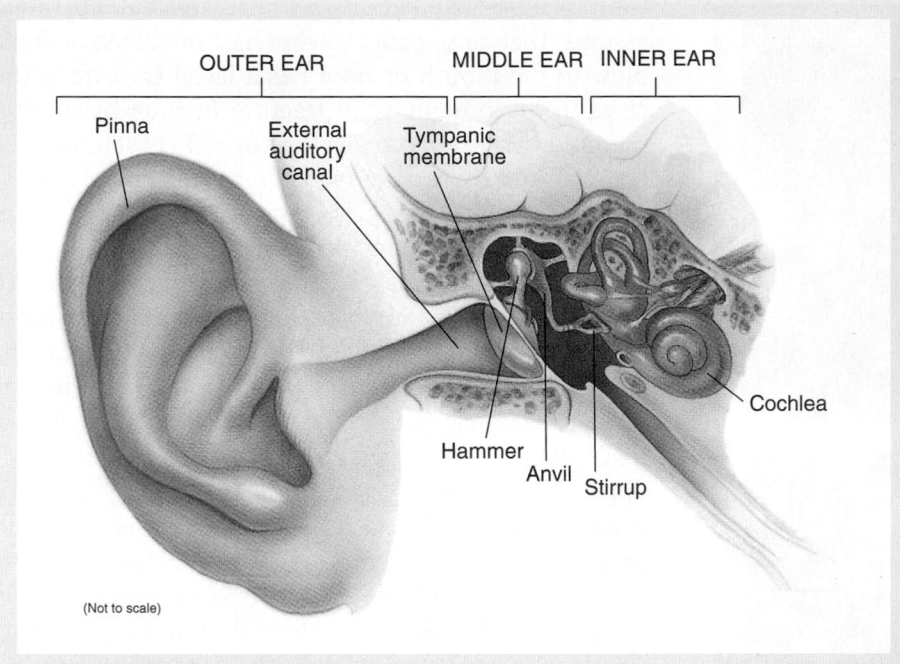

OUTER EAR   MIDDLE EAR   INNER EAR

Pinna

External
auditory
canal

Tympanic
membrane

Hammer

Anvil

Stirrup

Cochlea

(Not to scale)

Figure 25-30 The ear has three principal parts: the external ear, composed of the pinna, external auditory canal, and tympanic membrane; the middle ear, including the hammer, anvil, and stirrup; and the inner ear, composed of bony chambers filled with fluid.

contains three small bones (the hammer, anvil, and stirrup) that move in response to sound waves hitting the tympanic membrane. This is the mechanism by which sounds are heard and differentiated. The middle ear is connected to the nasal cavity by the **eustachian tube**, which is the internal auditory canal. This connection permits equalization of pressure in the middle ear when external atmospheric pressure changes. The inner ear is composed of bony chambers filled with fluid. As the head moves, so does the fluid. In response, fine nerve endings within the fluid send impulses to the brain indicating the position of the head and the rate of change of position.

Ears are often injured, but they usually do not bleed very much. If local pressure does not control the bleeding, you can apply a roller dressing **Figure 25-31**. First, however, you should place a soft, padded dressing between the back of the ear and the scalp because bandaging the ear against the tender underlying scalp can

or auricle, which is the part lying outside of the head, and the **external auditory canal**, which leads in toward the **tympanic membrane**, or eardrum. The middle ear

You are the Provider: PART 5

You reassess the patient en route, including his vital signs, and note that his breathing is becoming increasingly labored. His airway remains clear of secretions and blood, but his neck appears to be more swollen than it was during previous assessments.

Recording Time: 16 Minutes	
Level of consciousness	Conscious and alert; anxious
Respirations	26 breaths/min; labored
Pulse	124 beats/min; strong and regular
Skin	Cool and moist; perioral cyanosis is developing
Blood pressure	122/58 mm Hg
Sao$_2$	88% (on oxygen)

The hospital staff are notified of the change in the patient's condition. The nurse tells you to bring him straight to the trauma room on your arrival.

8. What is the most likely cause of your patient's labored breathing?

9. What adjustments, if any, should you make to your current treatment?

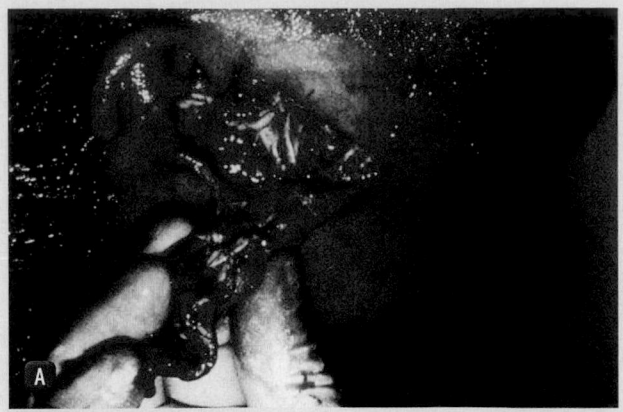

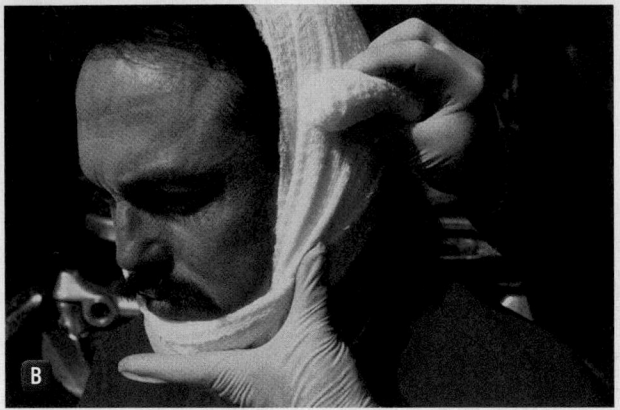

Figure 25-31 **A.** A major laceration of the ear. **B.** Proper treatment includes use of a soft, sterile pad behind the ear, between it and the scalp. Then wrap a roller gauze dressing around the head to include the entire ear.

be extremely painful for the patient. In the case of an ear avulsion, you should wrap the avulsed part in a moist, sterile dressing and put it in a plastic bag labeled with the patient's name. Keep the avulsed part cool and transport to the hospital with the patient. Often, avulsed tissue from the ear can be reattached.

The external auditory canal is a favorite place for children to place foreign bodies such as peanuts or candy. All such items should be removed by a physician in the emergency department. Never try to manipulate the foreign body because you may press it further into the auditory canal and cause permanent damage to the tympanic membrane.

Again, you should note any clear fluid coming from the ear of a severely injured patient because this may indicate a fracture at the base of the skull.

Facial Fractures

Fractures of the facial bones typically result from blunt impact. For example, the patient's head collides with a steering wheel or windshield in an automobile crash or

is hit by a baseball bat or pipe in an assault. You should assume that any patient who has sustained a direct blow to the mouth or nose has a facial fracture. Other clues to the possibility of fracture include bleeding in the mouth, inability to swallow or talk, absent or loose teeth, and/or loose or movable bone fragments. Patients may also report that "it doesn't feel right" when they close their jaw, signaling an irregularity of bite.

Facial fractures alone are not acute emergencies unless there is serious bleeding; however, they are an indication of significant blunt force trauma applied to that region of the body. Serious bleeding from a facial fracture can be life threatening. In addition to external hemorrhage, there is the danger of blood clots lodging in the upper airway and causing an obstruction **Figure 25-32**. Fractures around the face and mouth can also produce deformity and loose bone fragments. However, plastic surgeons can repair the damage if the injuries are treated within 7 to 10 days of the injury. Be sure to remove and save loose teeth or bone fragments from the mouth; it is often possible to reimplant them **Figure 25-33**. Remove any loose dentures or dental bridges to protect against airway obstruction. The removal of dentures will affect the shape of the patient's jaw.

Another source of potential airway obstruction is swelling, which can be extreme within the first 24 hours after injury. If you notice swelling during assessment or at any time while the patient is in your care, you should check for airway obstruction.

■ Dental Injuries

Dental injuries can be traumatic to a patient. Not only is the injury itself traumatic, but the patient's permanent teeth may also be lost—affecting everything from eating to smiling. Keep this in mind when providing care.

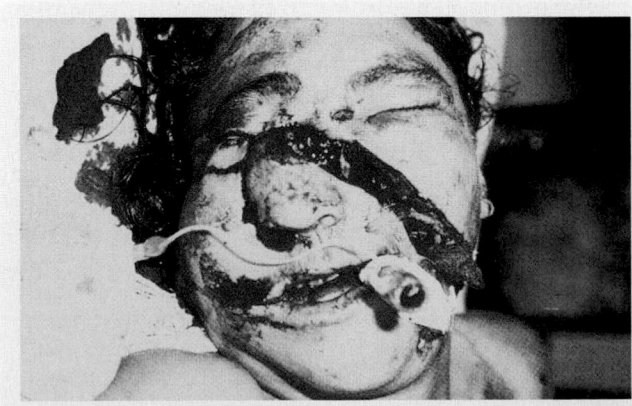

Figure 25-32 Bleeding following a crush injury to the face can be life threatening because, in addition to the external hemorrhage, blood clots in the airway can cause a complete obstruction.

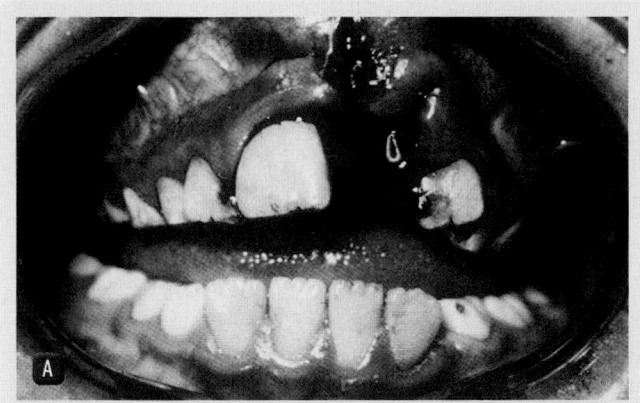

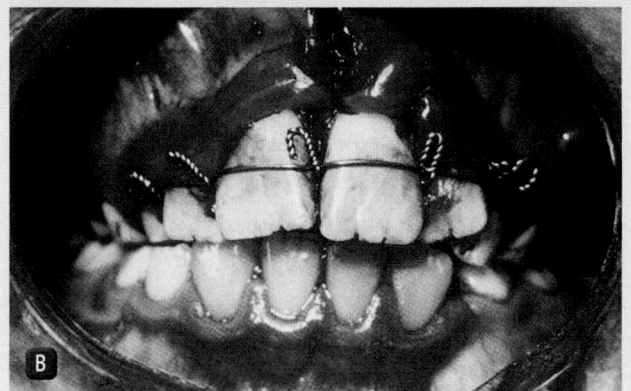

Figure 25-33 **A.** Save any lost teeth or bone fragments following an injury to the mouth. **B.** Even with traumatic loss of a tooth, the possibility of successful reimplantation is very good.

Bleeding will occur whenever a tooth is violently displaced out of its socket; therefore, apply direct pressure to stop the bleeding. To keep the airway patent, perform suctioning if needed. Also keep in mind that cracked or loose teeth are possible airway obstructions; therefore, suctioning may be necessary.

When dealing with an avulsed tooth, handle it by its crown and not by the root. When transporting the patient, bring along the tooth, placing it in either cold milk or sterile saline. There are also commercially available kits that may be used by your agency. Be familiar with how the kit is used before you encounter a patient with dental trauma. Notify the receiving facility about the avulsed tooth because reimplantation is recommended within 20 minutes to 1 hour after the trauma.

■ Injuries of the Cheek

You may encounter an object that is impaled in the patient's cheek. If you are unable to control the bleeding and it is compromising the patient's airway, consider removing the impaled object if possible and provide direct pressure on the inside and outside of the cheek. The amount of bandaging should not be so overwhelming that it occludes the mouth and makes it difficult for the patient to breathe.

■ Injuries of the Neck

The neck contains many structures that are vulnerable to injury by blunt trauma, such as from a steering wheel in a car crash, or by penetrating injury, such as a stab or gunshot wound. These structures include the upper airway, the esophagus, the carotid arteries and jugular veins, the thyroid cartilage or Adam's apple, the cricoid cartilage, and the upper part of the trachea. Any injury to the neck is serious and should be considered life threatening until proven otherwise in the emergency department.

Blunt Injuries

Any crushing injury of the upper part of the neck is likely to involve the larynx or trachea. Examples include a collision with a steering wheel, an attempted suicide by hanging, and a clothesline injury sustained while riding a bicycle. Once the cartilages of the upper airway and larynx are fractured, they do not spring back to their normal position. This type of fracture can lead to loss of voice, difficulty swallowing, severe and sometimes fatal airway obstruction, and leakage of air into the soft tissues of the neck **Figure 25-34**. The presence of air in the soft tissues produces a characteristic crackling sensation called **subcutaneous emphysema**. If you feel this sensation when you palpate the neck, you should maintain the airway as best you can and provide immediate transport. Be aware that complete airway obstruction can develop very rapidly in these patients as a result of swelling or bleeding into the underlying tissues. It may be very difficult to manage the airway in patients with these injuries; therefore, ALS support either by air or during an intercept may be necessary. Some patients will require a surgical

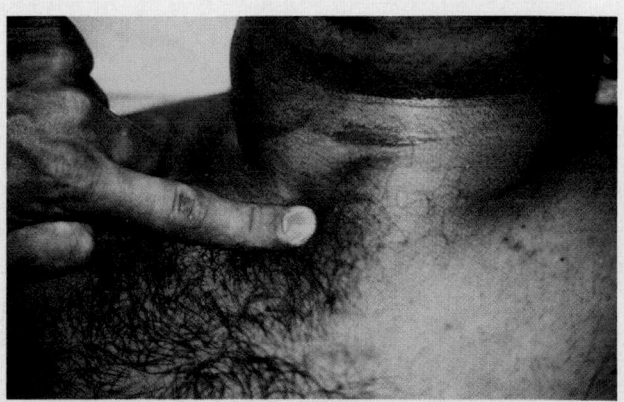

Figure 25-34 Fractures of the larynx or trachea can cause air to leak from the airway into the subcutaneous tissues. The presence of air in the soft tissues produces a crackling sensation called subcutaneous emphysema.

airway at the hospital. It is also possible that an incident involving an injury to the throat may also have caused a cervical spinal injury; therefore, spinal stabilization may be needed.

Penetrating Injuries

Penetrating injuries to the neck can cause profuse bleeding from laceration of the great vessels in the neck—the carotid arteries or the jugular veins **Figure 25-35**. Injuries to the carotid and jugular vessels in the neck can cause the body to bleed out, also known as exsanguination. Injuries to these large vessels may also allow air to enter the circulatory system and cause a pulmonary embolism. The airway, the esophagus, and even the spinal cord can be damaged by a penetrating injury.

Direct pressure over the bleeding site will control most neck bleeding. Follow the steps in **Skill Drill 25-3**:

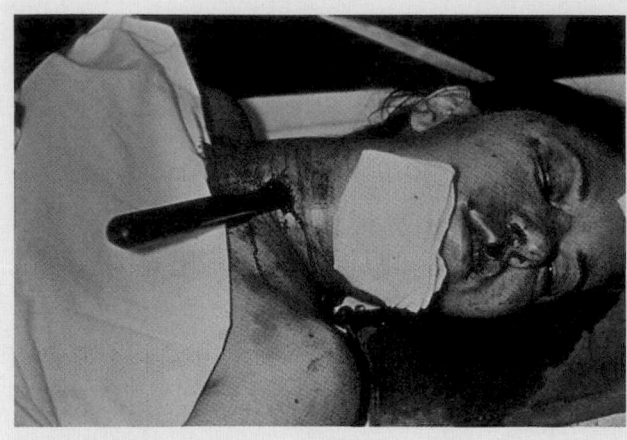

Figure 25-35 Penetrating injuries to the neck can result in profuse bleeding if a carotid artery or jugular vein is damaged.

Skill Drill 25-3

1. Apply direct pressure to the bleeding site using a gloved fingertip if necessary to control bleeding **Step 1**.

2. Apply a sterile occlusive dressing to ensure that air does not enter a vein or artery **Step 2**.

3. Secure the dressing in place with roller gauze, adding more dressings if needed.

4. Wrap the gauze around and under the patient's shoulder. To avoid possible airway and circulation problems, do not wrap the gauze around the neck.

Skill Drill 25-3

Controlling Bleeding From a Neck Injury

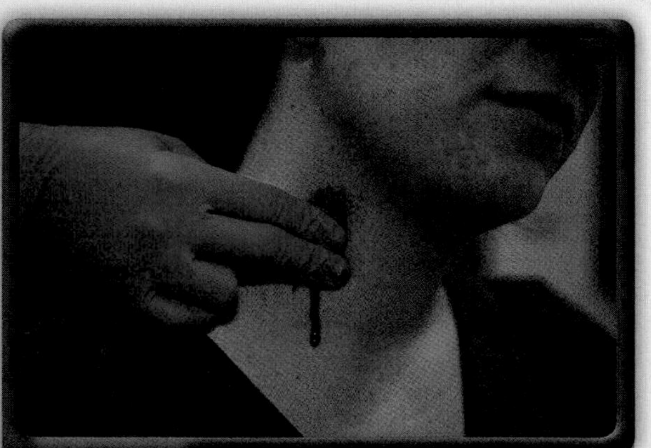

Step 1 Apply direct pressure to the bleeding site using a gloved fingertip if necessary to control bleeding.

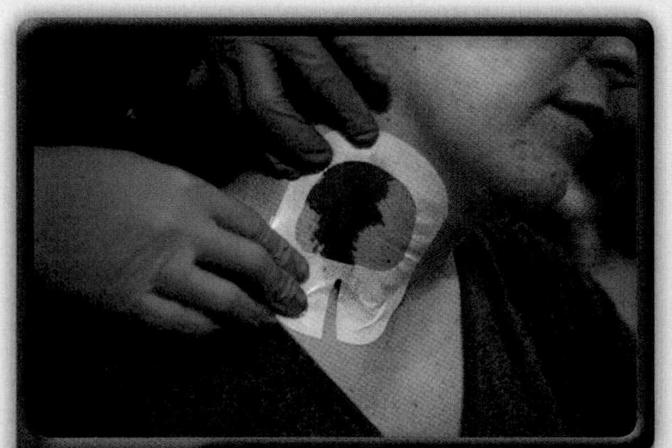

Step 2 Apply a sterile occlusive dressing to ensure that air does not enter a vein or artery.

Despite the use of these measures, the tissues within the neck may still continue to bleed and compress the upper airway, so you should look for signs of airway obstruction. If a vein has been punctured, air may be sucked through it to the heart, a clinical situation called **air embolism**. A large amount of air in the right atrium and right ventricle of the heart can lead to cardiac arrest.

You might find it necessary to apply pressure both above and below the penetrating wound to control life-threatening bleeding from the carotid artery (above) and the jugular vein (below). You may also need to treat the patient for shock.

Always maintain cervical spine stabilization, and with the patient fully immobilized to a backboard, provide prompt transport. Ensure that the airway remains open en route, and apply high-flow oxygen.

■ Laryngeal Injuries

Blunt force trauma to the larynx can occur when an unrestrained driver strikes the steering wheel or when a snowmobile rider or off-road biker strikes a clothesline or a fixed wire strung across a property line. The larynx becomes crushed against the cervical spine, resulting in soft-tissue injury, fractures, and/or separation of the fascia that connects the thyroid and cricoid cartilages. These strangulation injuries can also be found in either intentional or unintentional hangings. Any time there is suspected injury to the larynx, you should suspect possible cervical spine injury.

Open injuries to the larynx can occur as a result of a stabbing or penetration by a similar object. Penetrating and impaled objects should not be removed unless they interfere with cardiopulmonary resuscitation. Stabilize all impaled objects if they are not obstructing the airway (see Skill Drill 24-2, Chapter 24, *Soft-Tissue Injuries*).

Significant injuries to the larynx pose an immediate risk of airway compromise because of disruption of the normal passage of air, soft-tissue swelling, or aspiration of blood. The signs and symptoms of larynx injuries include respiratory distress, hoarseness, pain, difficulty swallowing (dysphagia), cyanosis, pale skin, sputum in the wound, subcutaneous emphysema, bruising on the neck, hematoma, or bleeding.

To manage a laryngeal injury, provide oxygenation and ventilation. Apply cervical immobilization but avoid the use of rigid collars because they may cause further damage to the soft tissues.

You are the Provider: SUMMARY

1. What should be your most immediate concern after receiving this initial patient information?

Although the patient is conscious at the present time, the presence of severe facial trauma should immediately increase your index of suspicion for potential airway and ventilation compromise.

Several factors may contribute to partial or complete upper airway obstruction in patients with massive facial trauma. Bleeding can be very heavy, producing large blood clots that can obstruct the upper airway; this risk is significantly higher if the patient has a decreased level of consciousness and is unable to expel blood from the mouth on his or her own.

In particular, direct trauma to the nose and mouth can be a source of significant bleeding. Direct trauma to the mouth can result in severe swelling and bleeding into the oropharynx. Direct nasal trauma can cause posterior epistaxis (nosebleed), in which case the patient may swallow large amounts of blood; because blood is a gastric irritant, the risks of vomiting and subsequent aspiration are high.

Direct facial trauma may also be associated with dislodged teeth that can fall to the back of the throat and be aspirated into the lungs or complicate obstruction of the upper airway by large clots of blood.

You should also be concerned about the potential for direct trauma and crushing injuries to the anterior part of the neck. Crush injuries to the larynx and trachea are associated with a high incidence of airway and ventilation compromise. You must also consider the possibility of a spinal injury; trauma that is severe enough to fracture facial bones, for example, could just as easily fracture spinal vertebrae in the neck; crush injuries to the throat should increase your index of suspicion for a spinal injury even further. In addition, the possibility of an underlying closed head injury cannot be ruled out and should be considered as well.

2. What should your initial actions consist of when you arrive at the scene?

Your initial actions on arriving at the scene of this call should be to conduct a scene size-up. Although law enforcement personnel have advised you that the scene is secure, you must still remain aware of your surroundings at all times; your patient has been assaulted, which means that the perpetrator, if he or she is not in custody, could return to the scene. Ascertain the total number of patients and call for additional resources if you think you will need them.

As you are approaching the patient, form a general impression. Although many life-threatening injuries or conditions can be discovered by simply looking at the patient, be careful to avoid getting distracted by visually impressive injuries (eg, fractures), which are usually not life threatening by themselves.

As you make physical contact with the patient, introduce yourself and manually stabilize his head. Remember, severe facial trauma may be associated with an injury to the cervical spine. Next, assess the status of the patient's airway; if he can talk to you, he has no *immediate* airway problems, although this can quickly change. If there is any blood in his oropharynx, remove it with suction. If he is conscious and alert, consider allowing him to suction his own mouth. Remember to follow standard precautions, including protective facial wear. When people have blood in their mouth, they have a tendency to spit it out; wearing a protective face shield minimizes the risk of having blood splattered in your face, eyes, and mouth.

After ensuring a patent airway, assess the quality of the patient's breathing and intervene at once. If he is breathing adequately, administer high-flow oxygen via a nonrebreathing mask. If he has signs of inadequate breathing (eg, fast or slow rate, shallow breathing [reduced tidal volume]), you should assist his ventilations.

Complete your primary assessment by assessing the rate and quality of his radial and carotid pulses and looking for and immediately controlling any severe external bleeding. If signs of shock are present, begin immediate shock treatment.

How you proceed after the primary assessment depends on the criticality of the patient's injuries and the presence or absence of abnormalities with the ABCs. The secondary assessment should include a full-body scan if he has multiple injuries or a focused examination if his injury is isolated.

3. Does this patient have a patent airway? How can you tell?

A patient who is conscious, alert, and talking has a patent airway—at least, for the time being. Although the patient has blood in his mouth, he is able to keep his own airway clear by spitting it out. The patient's voice is hoarse; this should concern you because it could indicate upper airway swelling or a crush injury to the structures of the anterior neck (ie, trachea, larynx). Closely monitor his airway and ventilation status!

At this point in the primary assessment, you have not yet determined if he has any injuries that could affect his level of consciousness, such as a closed head injury. Therefore, you must pay meticulous attention to his mental status and be prepared to suction the blood from his mouth. If he loses consciousness, he loses the ability to maintain his airway and you must intervene immediately. *Regardless of the situation, a patient's airway must remain patent at all times!*

4. What should be your initial treatment priorities for this patient?

The goal of the primary assessment is to *find and quickly treat* problems associated with the ABCs. Further assessment and treatment must focus on injuries or conditions that will kill your patient *first*.

Your patient has severe facial trauma and his voice is hoarse. Although he is able to maintain his own airway right now, do not assume that he will be able to do this until you get him to a hospital. Closely monitor his airway and ventilation status!

Your partner is providing manual stabilization of the patient's head based on the assumption that trauma significant enough to cause massive facial trauma can just as easily fracture or dislocate a spinal vertebra.

Apply high-flow oxygen via a nonrebreathing mask, but closely monitor his airway for continued oral bleeding. It may be necessary to occasionally remove the oxygen mask, allow him to expel any blood from his mouth, and then reapply the mask. If he remains conscious and can easily follow commands, you can allow him to hold the rigid suction catheter and suction the blood from his mouth himself.

Carefully monitor the patient's level of consciousness and breathing adequacy, keep his airway clear of blood, and be prepared to assist his ventilations.

5. On the basis of the mechanism of injury, what type of injuries should you suspect and assess for?

On the basis of your assessment, the patient's injuries appear to be isolated to his face and anterior part of the neck. However, a more in-depth assessment by a physician is needed to rule out occult injuries. Whereas lacerations and other open injuries may be obvious, facial swelling, which is common following blunt force facial trauma, can make assessment of the face difficult.

Your patient experienced significant blunt force trauma to the face when he was struck with the steel pipe and he experienced blunt trauma to the throat. On the basis of the mechanism of injury, you should be suspicious for facial bone fractures and injury to his trachea and/or larynx.

Indicators of facial fractures include oropharyngeal bleeding, loose or absent teeth, difficulty talking or swallowing, and loose or obviously movable bone fragments. Your patient reports that it "doesn't feel right" when he tries to close his mouth; this signals an irregular bite (dental malocclusion) and is a sign of a mandibular fracture. This is a significant finding because it takes a *lot* of force to fracture the mandible!

Any crushing injury to the anterior part of the neck is likely to involve the larynx or trachea. When the cartilages of the upper airway and larynx are fractured, they do not return to their normal position. During your assessment, you detected bruising and swelling of the throat, and the patient's voice is hoarse; these are red flag indicators of significant anterior neck trauma. Fractures of the larynx or trachea can cause a severe airway obstruction because of swelling. If the trachea has been fractured, air may leak into the soft tissues of the neck; suspect this if you detect subcutaneous emphysema, a crackling sensation that is felt when palpating the soft tissues of the neck.

Fractures of the zygomas (cheek bones) often present with a flattened appearance of the cheek bones; however, if the face is severely swollen, this may not be grossly apparent.

When caring for a patient with blunt facial trauma, you should assess extraocular movement, that is, the ability of the patient to move his or her eyes in all directions. Inability of the patient to look up (paralysis of upward gaze) suggests an orbital (blowout) fracture, in which case a bone fragment has entrapped one of the oculomotor nerves.

Severe periorbital swelling and bruising (raccoon eyes), in which case the patient's eyelids are swollen shut, can make extraocular movement and pupillary assessment extremely difficult or impossible.

Although the patient is conscious and alert, a closed head injury cannot be ruled out. During your assessment, you should look in and behind the ears. Bruising over the mastoid bone (Battle's sign) indicates a basilar skull fracture. Blood that is draining from the ears may contain cerebrospinal fluid; this is also an indicator of a basilar skull fracture. Blood or fluid drainage from the nose is also a sign of an underlying skull fracture.

When a person experiences significant blunt facial trauma, the head is suddenly thrust backwards (hyperextended). This sudden hyperextension can cause fractures or dislocations of the vertebrae of the cervical spine. Use spinal precautions when caring for a patient with significant blunt facial trauma.

The single most significant complication associated with facial fractures is airway compromise because of swelling, bleeding, or both, and while it is important to perform a thorough yet expedient secondary assessment, your main focus must be to ensure that the patient's airway remains patent.

6. What should you do if this patient begins to vomit?

You must be alert to the potential for vomiting when caring for any patient who has experienced facial trauma. Although your patient was spitting blood from his mouth, it is likely that he swallowed some as well. Blood is a gastric irritant that can cause nausea.

If aspiration occurs, the risk of mortality increases significantly so you *must* have a plan of action to prevent aspiration while transporting a patient who is supine and immobilized on a backboard.

If the patient begins to vomit, *immediately* place him on his side to allow vomitus to drain from his mouth. It is imperative that you protect the patient's airway, but as you turn the entire backboard onto its side, you must also avoid aggravating any possible spinal injury by ensuring that the patient is properly immobilized.

While the patient is on his side, suction his mouth to remove any remaining vomitus that did not drain with gravity. Before returning the patient to a supine position, make sure that *all* vomitus and other secretions are removed from his mouth!

7. How should you treat a patient with active oral bleeding *and* inadequate ventilation?

When you are transporting an immobilized patient with severe facial trauma, you must be prepared for a variety of situations. What if your patient's level of consciousness suddenly decreases, he begins bleeding into his mouth again, and his breathing becomes inadequate?

Patients with inadequate ventilation *and* active oral bleeding are a challenge to treat. Not only is the patient's airway in immediate jeopardy from obstruction from blood clots and aspiration, but inadequate ventilation will result in hypoxia and may lead to respiratory or cardiopulmonary arrest; therefore, you must treat both problems simultaneously.

If blood has pooled in the patient's mouth, immediately place the patient onto his or her side to allow the blood to drain from the mouth. Suction the airway for up to 15 seconds. You should then assist the patient's ventilations for 2 minutes.

This alternating pattern of suctioning for 15 seconds and assisting ventilations for 2 minutes should continue as needed, until the bleeding in the oropharynx is minimal or stops altogether. Regardless of the situation, the patient's airway must remain patent at all times and adequate ventilation must be ensured.

Patients with this type of airway and ventilation predicament would benefit from advanced airway management, especially if they are unconscious. Paramedics can intubate the patient, thus isolating the trachea and preventing aspiration, while manually ventilating the patient with a bag-mask device attached to the endotracheal tube. Request an ALS ambulance, if possible, when caring for patients with this type of complex airway and breathing problem.

8. What is the most likely cause of your patient's labored breathing?

This patient's status change is obvious. He is now laboring to breathe, his oxygen saturation is falling, and he is developing cyanosis and anxiety.

Reassessment reveals that his airway remains clear of blood; however, the anterior part of his neck is more swollen so you should suspect that his upper airway is swelling—most likely because of injury to his trachea or larynx, thus making it increasingly difficult for him to breathe. His oxygen saturation of 88%, even with high-flow oxygen, reflects significant hypoxemia, and perioral cyanosis (cyanosis around the mouth) is developing.

It is important to note that not all patients with significant injuries deteriorate at the scene; many of them deteriorate en route to the hospital or shortly after you arrive at the hospital. Therefore, it is critical to *frequently reassess* any patient with injuries that could jeopardize airway patency and impair ventilation.

9. What adjustments, if any, should you make to your current treatment?

Your patient's ventilation status has clearly deteriorated. You should consider assisting his ventilations with a bag-mask device and high-flow oxygen. Try to assist the patient's breathing, but do not be too aggressive. If he becomes combative and pushes the bag-mask device away from his face, reapply the nonrebreathing mask and carefully monitor his breathing.

If he tolerates assisted ventilation, use extreme caution. Although you must maintain adequate minute volume, a tracheal or laryngeal injury can be exacerbated by aggressive positive-pressure ventilation. Squeeze the bag-mask device just enough to improve the amount of tidal volume with each breath; observe for visible chest rise.

You should *not* use any type of mechanical ventilation device, such as a flow-restricted, oxygen-powered ventilation device, when ventilating a patient with tracheal or laryngeal trauma. These devices deliver oxygen under high pressure and can cause further injury to patients with fractures of the trachea or larynx.

You are the Provider: SUMMARY, continued

EMS Patient Care Report (PCR)

Date: 12-3-09	Incident No.: 012509	Nature of Call: Assault		Location: 1505 Eagle Rock Dr.

Dispatched: 1926	En Route: 1927	At Scene: 1930	Transport: 1941	At Hospital: 2001	In Service: 2021

Patient Information

Age: 30 Sex: M Weight (in kg [lb]): 80 kg (175 lb)	Allergies: Penicillin Medications: None Past Medical History: None Chief Complaint: Face and neck injury secondary to assault

Vital Signs

Time: 1935	BP: 132/68	Pulse: 118	Respirations: 22	Sao$_2$: 97%
Time: 1941	BP: 128/62	Pulse: 108	Respirations: 20	Sao$_2$: 98%
Time: 1946	BP: 122/58	Pulse: 124	Respirations: 26	Sao$_2$: 88%

EMS Treatment
(circle all that apply)

Oxygen @ 15 L/min via (circle one): NC (NRM) (Bag-Mask Device)	(Assisted Ventilation)	Airway Adjunct	CPR
Defibrillation	(Bleeding Control)	Bandaging	Splinting	Other: (Suctioning, spinal precautions)

Narrative

Ambulance 12 dispatched to the scene of an assault. The scene was secured by law enforcement prior to EMS arrival. On arrival at the scene, found the patient, a 30-year-old male, sitting on the ground; his face was covered with blood and his face was swollen. He was conscious but appeared anxious. He was spitting out small amounts of blood from his mouth, but was maintaining his own airway; his breathing was adequate. Manual c-spine stabilization was applied as patient was assessed further. He stated that he was struck across the face with a steel pipe, fell to the ground, and was kicked in the face and throat. He denies loss of consciousness. Applied high-flow oxygen via nonrebreathing mask and suctioned the patient's mouth as needed to remove blood. Rapid secondary assessment revealed severe swelling of the entire face, and bruising and mild swelling of the anterior part of the neck. Patient states that it "doesn't feel right" when he tries to close his mouth. No gross evidence of head or c-spine injury was noted during secondary assessment. Applied full spinal precautions, placed patient onto stretcher, loaded him into the ambulance, and began transport to the hospital. Reassessment en route revealed that his airway was free of blood or other secretions and his vital signs were stable. After radio report was called to receiving facility, patient became more anxious and began experiencing respiratory distress. Immediate reassessment revealed that his airway remained free of blood or other secretions, but the anterior part of his neck appeared more swollen than in previous assessments. Further reassessment revealed that he was developing perioral cyanosis, and his oxygen saturation decreased significantly. Began assisting the patient's ventilations with a bag-mask device and high-flow oxygen; he was initially resistant, but became more compliant with coaching. Notified hospital of patient status change. Continued to assist patient's ventilations and monitor his airway status for the duration of the transport. Noted improvement in patient's oxygen saturation and skin condition with assisted ventilation. Delivered patient to the emergency department staff and gave verbal report to the attending physician. Ambulance 12 cleared the hospital and returned to service at 2021 hrs. *End of report*

Assessment and Emergency Care of Face and Neck Injuries

Scene Size-up

Scene Safety	Ensure scene safety and safe access to the patient. Standard precautions should include a minimum of gloves and eye protection if there is vomiting. Consider the possibility that facial injuries can cause bleeding into the oropharynx, producing coughing; therefore, consider the use of face shields. Consider donning a gown and shoe covers if other bodily fluids are involved. Determine the number of patients. Assess the need for additional resources.
Mechanism of Injury (MOI)/ Nature of Illness (NOI)	Determine the MOI. Interview the patient, family, and/or bystanders to determine the exact nature of the traumatic forces applied. Maintain a high index of suspicion for associated spinal injuries, especially with rapid acceleration-deceleration MOIs.

▼   ▼

Primary Assessment

Form a General Impression	Inquire about the chief complaint and observe the patient's overall body position. Observe the work of breathing and circulation. Determine the level of consciousness using the AVPU scale. Identify immediate threats to life. Determine the priority of care based on the MOI. If the patient has a poor general impression, call for ALS assistance. A rapid scan will help you identify and manage life threats. Maintain a high index of suspicion for airway or respiratory compromise.
Airway and Breathing	Ensure the airway is open, clear, and self-maintained. Evaluate the patient's ventilatory status for rate and depth of breathing, respiratory effort, and tidal volume. Administer high-flow oxygen at 15 L/min, providing ventilatory support as needed. Hypoxia may cause changes in the patient's mental state. If vomiting or bleeding into the oropharynx is a possibility, tilt the backboard to the side after spinal immobilization has been performed and have suction ready.
Circulation	Observe skin color, temperature, and condition; look for life-threatening bleeding and treat accordingly. Evaluate distal pulse rate, quality (strength), and rhythm. Observe for significant oropharyngeal bleeding.
Transport Decision	If the patient has an airway or breathing problem, signs and symptoms of bleeding, or other life threats, manage them immediately and consider rapid transport, performing the secondary assessment en route to the hospital. Consider rapid transport to an appropriate trauma center.

▼   ▼

NOTE: The order of the steps in this section differs depending on whether the patient is conscious or unconscious. The following order is for a conscious patient. For an unconscious patient, perform a primary assessment, perform a full-body scan, obtain vital signs, and obtain the past medical history from a family member, bystander, or emergency medical identification device.

History Taking

Investigate Chief Complaint	Investigate the chief complaint. Monitor the patient for changes in mental status. Ask SAMPLE questions. SAMPLE can also be obtained from family, bystanders, and medical alert tags.

▼   ▼

Assessment and Emergency Care of Face and Neck Injuries, continued

Secondary Assessment

Physical Examinations	Perform a systematic full-body examination or a focused examination on the face and/or neck. Rule out any potential life threats. Advise the patient prior to performing any examination. Do not delay transport to perform the physical examination at the scene. Look for DCAP-BTLS and asymmetry in the face and neck. Pay close attention to injuries that could potentially obstruct the airway or occlude bloodflow to the brain.
Vital Signs	Obtain baseline vital signs as soon as practical. Vital signs should include blood pressure by auscultation, pulse rate and quality, respiration rate and quality, and skin assessment for perfusion. Note the patient's level of consciousness. Use pulse oximetry, if available, to assess the patient's perfusion status.

Reassessment

Interventions	Reassess the primary assessment, vital signs, chief complaint, and any interventions already performed. Assist breathing as required, administering high-flow oxygen.
Communication and Documentation	Contact medical control/receiving hospital with a radio report; many hospitals require additional personnel and a separate treatment area. Include a thorough description of the MOI and the position the patient was found in. Include treatments performed and patient response. Be sure to document the patient's distress, answers to your questions, and any changes in patient status and the time. Follow local protocols. Document the reasoning for your treatment and the patient's response.

NOTE: Although the following steps are widely accepted, be sure to consult and follow your local protocols. Take appropriate standard precautions when treating all patients.

Face and Neck Injuries

General Management of Face and Neck Injuries

1. Establish and maintain a patent airway. Provide oxygen. Monitor for vomiting/bleeding into the oropharynx and protect against aspiration.
2. Perform spinal immobilization if the MOI suggests the possibility of spinal injury.
3. Obtain SAMPLE history and vital signs.
4. Bleeding from soft-tissue injuries of the face or neck can be controlled with gentle direct pressure.
5. Injuries of the eye require specialized handling and definitive care.
6. Avulsed tissue should be kept cool and placed in a sealed container with a moist dressing.
7. Penetrating injuries of the neck could result in an air embolism if there is damage to the large blood vessels of the neck. Cover holes with occlusive dressings.
8. Request ALS when necessary.

Removal of a Foreign Object From the Eye

1. Tell the patient to look down while you grasp the lashes of the upper eyelid with your thumb and index finger. Gently pull the eyelid away from the eyeball.
2. Gently place a cotton-tipped applicator horizontally along the center of the outer surface of the upper eyelid.

Assessment and Emergency Care of Face and Neck Injuries, continued

Face and Neck Injuries

Removal of a Foreign Object From the Eye, continued

3. Pull the eyelid forward and up, which causes it to roll or fold back over the applicator, exposing the undersurface of the eyelid.
4. If you see a foreign object on the surface of the eyelid, gently remove it with a moistened, sterile, cotton-tipped applicator.

Stabilizing a Foreign Object Impaled in the Eye

1. To prepare a doughnut ring, wrap a 2" roll around your fingers and thumb seven or eight times. Adjust the diameter by spreading your fingers or squeezing them together.
2. Wrap the remainder of the roll, working around the ring to form a doughnut.
3. Place the dressing over the eye and the impaled object to hold the impaled object in place, and then secure it with a gauze dressing.

Chemical Burns of the Eye

Hold the patient's eyelid open. If only one eye is affected, take care to avoid contaminating the unaffected eye. Flush from the inner corner of the affected eye toward the outside corner. Irrigate the eye for 20 minutes. Apply a clean, dry dressing to cover the eye after irrigation. Transport the patient promptly to the hospital for further care.

Thermal Burns to the Eye

Cover both eyes with a sterile dressing moistened with sterile saline. Apply an eye shield over the dressing. Provide prompt transport.

Light Burns to the Eye

Cover each eye with a sterile, moist pad and an eye shield. Have the patient lie down during transport to the hospital. Protect the patient from further exposure to bright light.

Lacerations to the Eye

1. Never exert pressure on or manipulate the injured eye (globe) in any way.
2. If part of the eyeball is exposed, gently apply a moist, sterile dressing to prevent drying.
3. Cover the injured eye with a protective metal eye shield, cup, or sterile dressing. Apply soft dressings to both eyes, and provide prompt transport to the hospital.

Blunt Trauma to the Eye

Place the patient on a stretcher and transport promptly. Protect the eye from further injury with a metal shield. Cover the other eye to minimize movement on the injured side.

Assessment and Emergency Care of Face and Neck Injuries, continued

Face and Neck Emergencies

Blast Injuries to the Eye

First ensure that the scene is safe. Management depends on the severity of the injury. Do not attempt to remove a foreign body within the globe. Use a clean cup or similar item to protect the area. If only one eye is injured, follow local protocol, which may include covering the other eye to eliminate sympathetic motion. Patients with a sudden loss or decrease of vision will need to be verbally instructed on what actions are taking place around them. If the patient has severe swelling or a hematoma to the eyelid, do not attempt to force the eyelid open to examine the eye.

Nose Injuries

For a nontrauma patient who is bleeding from the nose, place the patient in a sitting position, leaning forward, and pinch his or her nostrils together. For a detailed discussion of the care for epistaxis, see Skill Drill 23-3 in Chapter 23, *Bleeding*.

Ear Injuries

Place a soft, padded dressing between the ear and the scalp. If the ear is avulsed, wrap it in a moist, sterile dressing and place it in a plastic bag. Keep the avulsed tissue cool and transport to the hospital with the patient. Leave any foreign object within the ear for the physician to remove. Note any clear fluid coming from the ear.

Facial Fractures

Remove and save loose teeth or bone fragments from the mouth and transport them with you. Remove any loose dentures or dental bridges to protect against airway obstruction. Maintain an open airway.

Dental Injuries

Apply direct pressure to stop the bleeding. Keep the airway open. Perform suctioning if needed. Handle the tooth by its crown and not by the root. Transport the patient. Bring along the tooth, placing it in either cold milk or sterile saline. Notify the receiving facility about the avulsed tooth.

Injuries of the Cheek

If bleeding is uncontrollable and compromising the patient's airway, consider removing the impaled object if possible. Provide direct pressure on the inside and outside of the cheek. Bandaging should not occlude the mouth or make it difficult for the patient to breath.

Injuries to the Neck

1. Apply direct pressure to the bleeding site using a gloved fingertip if necessary to control bleeding.
2. Apply a sterile occlusive dressing to ensure that air does not enter a vein or artery.
3. Use roller gauze to secure a dressing in place.
4. Wrap the bandage around and under the patient's shoulder.

Prep Kit

Ready for Review

- Soft-tissue injuries and fractures of the bones of the face and neck are common and vary in severity.

- In face and neck injuries, your priorities are to prevent further injury to the cervical spine, manage the airway and ventilation of the patient, and control bleeding.

- Airway compromise may be caused by heavy bleeding into the airway, swelling in and around the structures of the airway located in the face and neck, and injuries to the central nervous system that interfere with normal respiration.

- To control heavy bleeding from soft-tissue injuries to the face, use direct pressure with a dry, sterile dressing. If brain tissue is exposed, use a moist, sterile dressing.

- Always check for bleeding inside the mouth because this may produce airway obstruction.

- Open the airway using the modified jaw-thrust maneuver (when indicated), and clear the airway in all patients with facial injuries.

- Save avulsed pieces of skin and tissue, and transport them with the patient for possible reattachment at the hospital.

- Maintain a high index of suspicion for patients with unequal pupils—this sign may indicate an illness or an injury to the brain. Remember, some people are born with one pupil larger than the other. During your assessment, ask your patient whether he or she normally has unequal pupils.

- Foreign bodies on the surface of the eye should be irrigated gently with normal saline solution. Always flush from the region of the eye closest to the nose toward the outside, away from the midline.

- If a foreign body is on the underside of the eyelid, remove it gently with a cotton-tipped applicator. Never remove foreign bodies stuck to the cornea.

- Chemicals, heat, and light rays can all cause burn injury to the eyes, resulting in permanent damage.

- Be alert to clear fluid draining from the ears or nose. This may indicate a basilar skull fracture.

- Blunt and penetrating trauma to the neck can produce life-threatening injuries. Palpate the neck for signs of subcutaneous emphysema. In patients with this sign, complete airway obstruction may develop in minutes.

- If bleeding is present from a penetrating injury, direct pressure over the site will usually control most forms of bleeding.

- Be alert to the possibility of an air embolism from an open neck injury. Place an occlusive dressing over the site, and provide direct pressure.

Vital Vocabulary

air embolism The presence of air in the veins, which can lead to cardiac arrest if it enters the heart.

anisocoria Naturally occurring uneven pupil size.

blowout fracture A fracture of the orbit or of the bones that support the floor of the orbit.

conjunctiva The delicate membrane that lines the eyelids and covers the exposed surface of the eye.

conjunctivitis Inflammation of the conjunctiva.

cornea The transparent tissue layer in front of the pupil and iris of the eye.

eustachian tube A branch of the internal auditory canal that connects the middle ear to the oropharynx.

external auditory canal The ear canal; leads to the tympanic membrane.

globe The eyeball.

iris The muscle and surrounding tissue behind the cornea that dilate and constrict the pupil, regulating the amount of light that enters the eye; pigment in this tissue gives the eye its color.

lacrimal glands The glands that produce fluids to keep the eye moist; also called tear glands.

lens The transparent part of the eye through which images are focused on the retina.

mastoid process The prominent bony mass at the base of the skull about 1″ posterior to the external opening of the ear.

optic nerve A cranial nerve that transmits visual information to the brain.

pinna The external, visible part of the ear.

pupil The circular opening in the middle of the iris that admits light to the back of the eye.

retina The light-sensitive area of the eye where images are projected; a layer of cells at the back of the eye that changes the light image into electrical impulses, which are carried by the optic nerve to the brain.

retinal detachment Separation of the retina from its attachments at the back of the eye.

sclera The tough, fibrous, white portion of the eye that protects the more delicate inner structures.

sternocleidomastoid muscles The muscles on either side of the neck that allow movement of the head.

subcutaneous emphysema A characteristic crackling sensation felt on palpation of the skin, caused by the presence of air in soft tissues.

temporomandibular joint The joint formed where the mandible and cranium meet, just in front of the ear.

tragus The small, rounded, fleshy bulge that lies immediately anterior to the ear canal.

turbinates Layers of bone within the nasal cavity.

tympanic membrane The eardrum, which lies between the external and middle ear.

Assessment in Action

You are dispatched to a motor vehicle crash on a rural road where a single vehicle is up on an embankment. The vehicle has front-end damage and a starred windshield. The unrestrained driver is lying supine in the road; the police report that they found the patient outside of the vehicle. He is unconscious and has obvious facial bleeding. Examination reveals a hematoma and depression of the left temporal area, both eyes are black and blue, and one pupil is dilated, whereas the other is normal. The patient has blood coming from the nose and mouth, and palpation reveals subcutaneous emphysema around the clavicles and chest area. The patient's breathing is labored, and the pulse is rapid and thready.

1. What is the priority for this patient?
   A. Perform spinal immobilization.
   B. Assess the vital signs.
   C. Provide rapid transport.
   D. Assess the airway, breathing, and circulation.

2. How should you manage this patient's airway?
   A. Give oxygen via a nasal cannula at 4 L/min.
   B. Give oxygen via a nonrebreathing mask at 15 L/min.
   C. Suction the airway and assist with ventilations with a bag-mask device.
   D. Suction the airway and give 100% oxygen via a nonrebreathing mask.

3. The unequal pupils could indicate what type of injury?
   A. Head
   B. Eye
   C. Chest
   D. Spinal

4. What type of airway adjunct should be used to maintain a patent airway in this patient?
   A. Nasopharyngeal airway
   B. Gastric airway
   C. Oropharyngeal airway
   D. Endotracheal airway

5. When subcutaneous emphysema is found, what substance has accumulated under the skin?
   A. Blood
   B. Air
   C. Both blood and air
   D. Cerebrospinal fluid

6. Which of the following best describes hyphema?
   A. Bleeding in the interior chamber of the brain
   B. Bleeding in the oropharynx
   C. Bleeding in the posterior thorax
   D. Bleeding in the anterior chamber of the eye

7. Which of the following fractures is associated with bruising around the ears and blood coming from the nose?
   A. Basilar skull fracture
   B. Orbit fracture
   C. Mandible fracture
   D. Maxilla fracture

8. A hematoma is a collection of blood:
   A. in the bones.
   B. in the thorax.
   C. under the skin.
   D. under the eyes.

9. Why does this patient need rapid transport?

10. In patients with head injuries, cerebrospinal fluid (CSF) may escape from the skull. What test can you use to determine if there is leakage of CSF?

Head and Spine Injuries

Knowledge Objectives

1. Describe the anatomy and physiology of the nervous system, including its divisions into the central nervous system (CNS) and peripheral nervous system (PNS) and the structures and functions of each. (pp 887-890)
2. Explain the functions of both the somatic and autonomic nervous systems. (pp 887-890)
3. List the major bones of the skull and spinal column and their related structures, and describe their functions as related to the nervous system. (pp 890-891)
4. Discuss age-related variations that are required when providing emergency care to a pediatric patient who has a suspected head or spine injury. (pp 890, 895, 903, 905, 923-924)
5. Discuss the different types of head injuries, their potential mechanism of injury (MOI), and general signs and symptoms of a head injury that the EMT should consider when performing a patient assessment. (pp 891-897)
6. Define traumatic brain injury (TBI) and explain the difference between a primary (direct) injury and a secondary (indirect) injury, providing examples of possible mechanisms of injury that may cause each one. (pp 893-894)

7. Discuss the different types of brain injuries and their corresponding signs and symptoms, including increased intracranial pressure (ICP), concussion, contusion, and injuries caused by medical conditions. (pp 893-897)
8. Discuss the different types of injuries that may damage the cervical, thoracic, or lumbar spine, providing examples of possible mechanisms of injury that may cause each one. (pp 897-898)
9. List the mechanisms of injury that cause a high index of suspicion for the possibility of a head or spinal injury. (p 898)
10. Describe the steps in the patient assessment process for a person who has a suspected head or spine injury, including specific variations that may be required as related to the type of injury. (pp 898-905)
11. Describe the process of providing emergency medical care to a patient with a head injury, including the three general principles designed to protect and maintain the critical functions of the central nervous system and ways to determine if the patient has a traumatic brain injury. (pp 905-907)
12. Describe the process of providing emergency medical care to a patient with a spinal injury, including the implications of not properly caring for patients with injuries of this nature, the steps for performing manual in-line stabilization, implications for sizing and using a cervical spine immobilization device, and key symptoms that contraindicate in-line stabilization. (pp 907-924)
13. Describe the process of preparing patients who have suspected head or spinal injuries for transport, including the use and functions of a long backboard, short backboard, and other short spinal extrication devices to immobilize the patient's cervical and thoracic spine. (pp 909-924)
14. Explain the different circumstances in which a helmet should be either left on or taken off a patient with a possible head or spinal injury, and then list the steps EMTs must follow to remove a helmet, including the alternate method for removing a football helmet. (pp 919-924)

Skills Objectives

1. Demonstrate how to perform a jaw-thrust maneuver on a patient with a suspected spinal injury. (pp 907-908)
2. Demonstrate how to perform manual in-line stabilization on a patient with a suspected spinal injury. (pp 908-909, Skill Drill 26-1)
3. Demonstrate how to immobilize a patient with a suspected spinal injury to a long backboard. (pp 909-911, Skill Drill 26-2)
4. Demonstrate how to immobilize a patient with a suspected spinal injury who was found in a sitting position. (pp 911-914, Skill Drill 26-3)
5. Demonstrate how to immobilize a patient with a suspected spinal injury who was found in a standing position. (pp 915-916, Skill Drill 26-4)
6. Demonstrate how to apply a cervical collar to a patient with a suspected spinal injury. (pp 917-918, Skill Drill 26-5)
7. Demonstrate how to immobilize a patient with a suspected spinal injury to a short backboard. (pp 918-919)
8. Demonstrate how to remove a helmet from a patient with a suspected head or spinal injury. (pp 920-922, Skill Drill 26-6)
9. Demonstrate the alternate method for removal of a football helmet from a patient with a suspected head or spinal injury. (p 923)

Introduction

The nervous system is a complex network of nerve cells that enables all parts of the body to function. It includes the brain, the spinal cord, and several billion nerve fibers that carry information to and from all parts of the body. Because the nervous system is so vital, it is well protected. The brain lies within the skull, and the spinal cord is inside the bony spinal canal. Despite this protection, serious injuries can damage the nervous system.

This chapter briefly reviews the anatomy and function of the central and peripheral nervous systems and of the skeletal system. Discussion of specific head, brain, and spinal injuries follows, including signs, symptoms, assessment, and treatment. Extrication of patients with possible spinal injuries and removal of helmets are also described.

Anatomy and Physiology

Nervous System

The nervous system is divided into two anatomic parts: the central nervous system and the peripheral nervous system Figure 26-1 . The central nervous system (CNS) includes the brain and the spinal cord, including the nuclei and cell bodies of most nerve cells. Long nerve fibers link these cells to the body's various organs through openings in the spinal column. These cables of nerve fibers make up the peripheral nervous system.

Central Nervous System

The CNS is composed of the brain and spinal cord. The brain is the organ that controls the body; it is also the center of consciousness. It is divided into three major areas: the cerebrum, the cerebellum, and the brain stem Figure 26-2 .

The cerebrum, which contains about 75% of the brain's total volume, controls a wide variety of activities, including most voluntary motor function and conscious

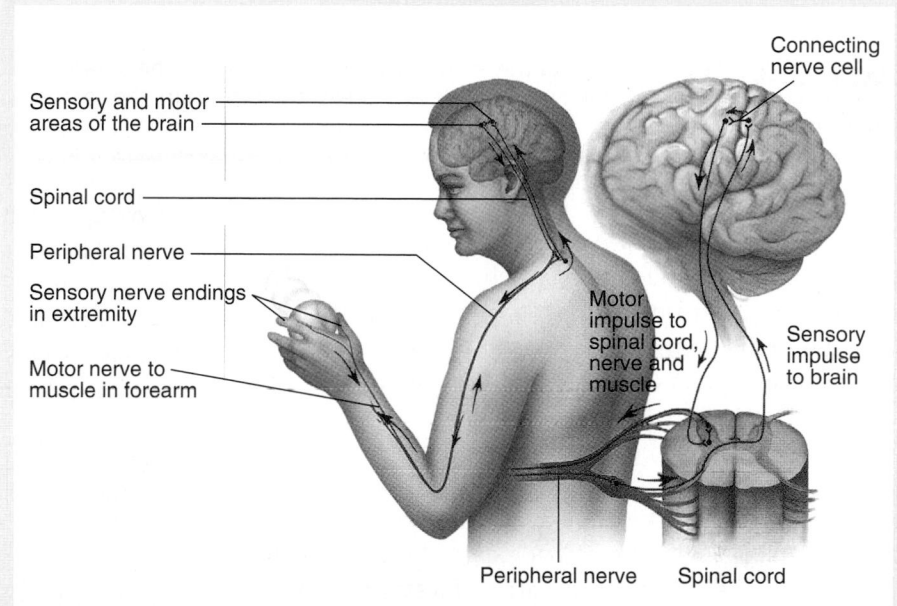

Figure 26-1 The nervous system has two anatomic components: the central nervous system and the peripheral nervous system. The central nervous system is composed of the brain and the spinal cord. The peripheral nervous system conducts sensory and motor impulses from the skin and other organs to the spinal cord.

Labels in figure:
- Sensory and motor areas of the brain
- Spinal cord
- Peripheral nerve
- Sensory nerve endings in extremity
- Motor nerve to muscle in forearm
- Connecting nerve cell
- Motor impulse to spinal cord, nerve and muscle
- Sensory impulse to brain
- Peripheral nerve
- Spinal cord

You are the Provider: PART 1

At 2:20 AM, you receive a call for a man who was assaulted outside a nightclub. Law enforcement personnel are present and have secured the scene. While you are en route, one of the police officers radios you and advises that the patient was struck in the side of the head with a baseball bat and is unconscious. Your response time to the scene is less than 5 minutes.

1. Why is the structure of the cranium considered to be a "mixed blessing?"
2. What is the difference between a primary and a secondary brain injury?

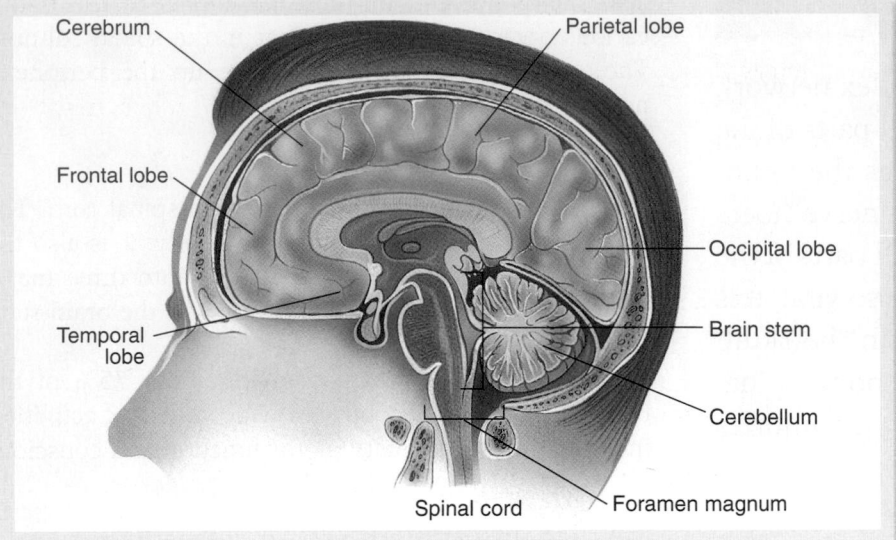

Figure 26-2 The brain is part of the central nervous system and is the organ that controls the body. It is divided into three major areas: the cerebrum, the cerebellum, and the brain stem.

thought. It is the main part of the brain and is divided into two hemispheres with four lobes. Underneath the cerebrum lies the cerebellum, which coordinates balance and body movements. The most primitive part of the CNS, the brain stem, controls virtually all the functions that are necessary for life, including the cardiac and respiratory systems and nerve function transmissions. Deep within the cranium, the brain stem is the best-protected part of the CNS.

The spinal cord, the other major portion of the CNS, is mostly made up of fibers that extend from the brain's nerve cells. The spinal cord carries messages between the brain and the body via the gray and white matter of the spinal cord. Grey matter is composed of neural cell bodies and synapses, which are connections between nerve cells. White matter consists of fiber pathways.

Protective Coverings The cells of the brain and spinal cord are soft and easily injured. Once damaged, they cannot be regenerated or reproduced. Therefore, the entire CNS is contained within a protective framework.

The thick, bony structures of the skull and spinal canal withstand injury very well. The skull is covered by layers of muscle, superficial fascia, and thick skin, which usually bears hair. Superficial fascia connects the muscle to the skin and contains white blood cells that are used to destroy pathogens when there is an open wound. The spinal canal is also surrounded by a thick layer of skin and muscles.

The CNS is further protected by the **meninges**, three distinct layers of tissue that suspend the brain and the spinal cord within the skull and the spinal canal **Figure 26-3**. The outer layer, the dura mater, is a tough, fibrous layer that

closely resembles leather. This layer forms a sac to contain the CNS, with small openings through which the peripheral nerves exit.

The inner two layers of the meninges, called the arachnoid and the pia mater, are much thinner than the dura mater. They contain the blood vessels that nourish the brain and spinal cord. Cerebral spinal fluid (CSF) is produced in a chamber inside the brain, called the third ventricle. CSF is located in the subarachnoid space below the arachnoid, which is a weblike structure. There is approximately 125 to 150 mL of CSF in the brain at any one time. CSF primarily acts as a shock absorber. The brain and spinal cord essentially float in this fluid, buffered from injury. The brain depends on a rich supply of oxygenated blood to function properly. When this supply is interrupted, even for short periods of time, serious damage to the brain tissue may occur.

When an injury does penetrate all of these protective layers, clear, watery CSF may leak from the nose, the ears, or an open skull fracture. Therefore, if a patient with a head injury has what looks like a runny nose or reports a salty taste at the back of the throat, you should assume that the fluid is CSF.

Ironically, the closed bony structure of the skull (which is similar to a vault) and the meninges, the very layers of tissue that isolate and protect the CNS, can

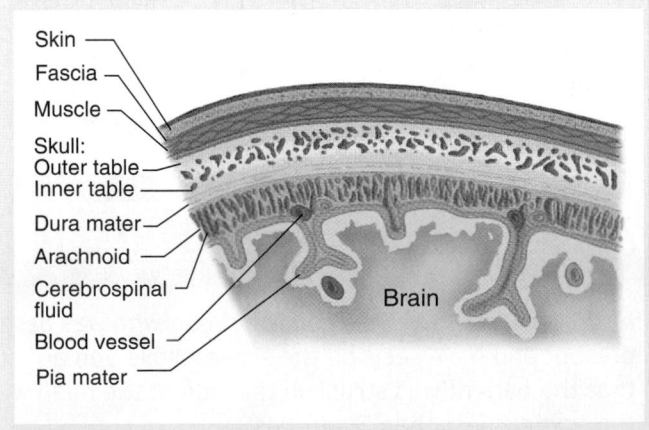

Figure 26-3 The central nervous system has several layers of protective coverings: the skin, muscles and their fascia, bone, and the meninges. The three layers of the meninges are the dura mater, the arachnoid, and the pia mater.

lead to serious problems in closed head injuries. Severe injury may cause bleeding within the skull, referred to as intracranial hemorrhage. Such bleeding causes increased pressure inside the skull and compresses softer brain tissue. In many cases, only prompt surgery can prevent permanent brain damage.

Peripheral Nervous System

The peripheral nervous system has two anatomic parts: 31 pairs of spinal nerves and 12 pairs of cranial nerves **Figure 26-4**.

The 31 pairs of spinal nerves conduct sensory impulses from the skin and other organs to the spinal cord. They also conduct motor impulses from the spinal cord to the muscles. Because the arms and legs have so many muscles, the spinal nerves serving the extremities are arranged in complex networks. The brachial plexus controls the arms, and the lumbosacral plexus controls the legs.

Cranial nerves are the 12 pairs of nerves that emerge from the brain stem and transmit information directly to or from the brain. For the most part, they perform special functions in the head and face, including sight, smell, taste, hearing, and facial expressions.

There are two major types of peripheral nerves. The sensory nerves, with endings that can perceive only one type of information, carry that information from the body to the brain via the spinal cord. The motor nerves, one for each muscle, carry information from the CNS to the muscles. The **connecting nerves**, found only in the brain and spinal cord, connect the sensory and motor nerves with short fibers, which allow the cells on either end to exchange simple messages.

■ How the Nervous System Works

The nervous system controls virtually all of the body's activities, including reflex, voluntary, and involuntary activities.

In connecting the sensory and motor nerves of the limbs, the connecting nerves in the spinal cord form a reflex arc. If a sensory nerve in this arc detects an irritating stimulus, such as heat, it will bypass the brain and send a message directly to a motor nerve **Figure 26-5**.

Voluntary activities are the actions that we consciously perform, in which sensory input determines the specific muscular activity—for example, reaching across the table for a salt shaker or to pass a dish. **Involuntary**

activities are the actions that are not under a person's conscious control, such as breathing; in most instances, we inhale and exhale without consciously thinking about it. Many of our body's functions occur independently of thought, or involuntarily.

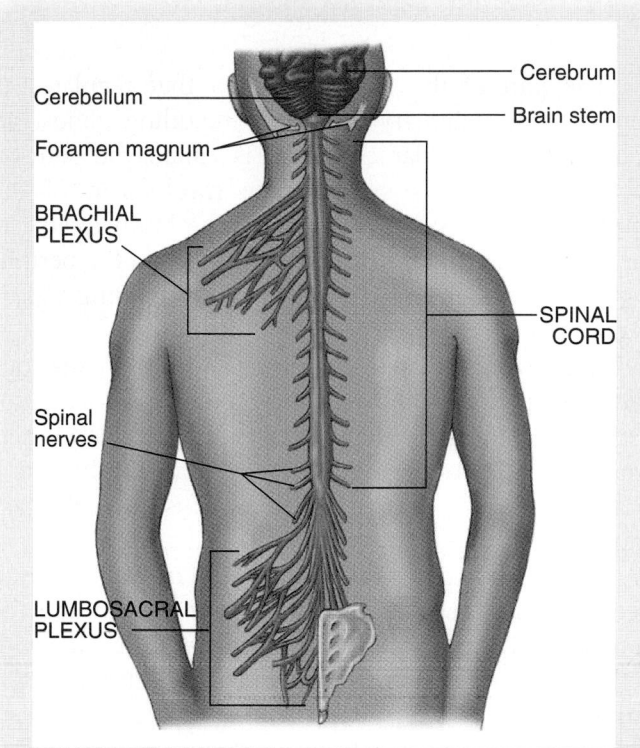

Figure 26-4 The peripheral nervous system is a complex network of motor and sensory nerves. The brachial plexus controls the arms, and the lumbosacral plexus controls the legs.

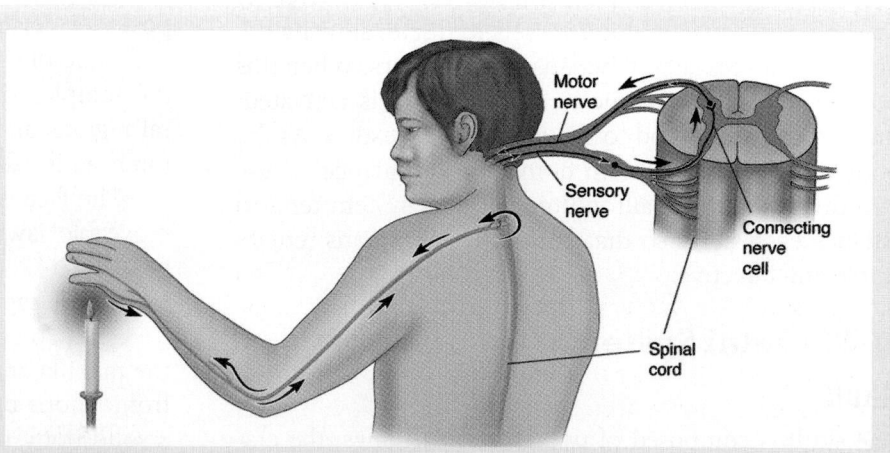

Figure 26-5 The connecting nerves in the spinal cord form a reflex arc. If a sensory nerve in this arc detects an irritating stimulus, it will bypass the brain and send a direct message to a motor nerve.

Words of Wisdom

Central nervous system structures, whose bony enclosures protect them quite well, are also very fragile. Protecting them from further damage is vital to the patient's future ability to live a normal life. Lean toward caution and overprotection in assessing and treating possible brain and spinal cord injuries.

The part of the nervous system that regulates or controls our voluntary activities, including almost all coordinated muscular activities, is called the somatic (voluntary) nervous system. The mechanism of the somatic nervous system is simple. The brain interprets the sensory information that it receives from the peripheral and cranial nerves and responds by sending signals to the voluntary muscles.

The body functions that occur without conscious effort are regulated by the much more primitive autonomic (involuntary) nervous system. The autonomic nervous system controls the functions of many of the body's vital organs, over which the brain has no voluntary control.

The autonomic nervous system is divided into two sections: the sympathetic nervous system and the parasympathetic nervous system. When confronted with a threatening situation, the sympathetic nervous system reacts to the stress with the fight-or-flight response. This response causes the pupils to dilate, smooth muscle in the lungs to dilate, heart rate to increase, and blood pressure to rise. This response also causes the body to shunt blood to vital organs and to skeletal muscle. During this time of stress, a hormone called epinephrine (also known as adrenaline) is released, which is responsible for much of these activities inside the body. The parasympathetic nervous system has the opposite effect on the body, causing blood vessels to dilate, slowing the heart rate, and relaxing the muscle sphincters. When this portion of the autonomic nervous system is activated, the body shunts blood to the organs of digestion. As the body attempts to maintain homeostasis (balance), these two divisions of the autonomic nervous system tend to balance each other so that basic body functions remain stable and effective.

■ Skeletal System

Skull

The skull is composed of two groups of bones: the cranium, which protects the brain, and the facial bones **Figure 26-6**. The cranium is composed of a number of thick bones that fuse together to form a shell above the eyes and ears that holds and protects the brain. It is

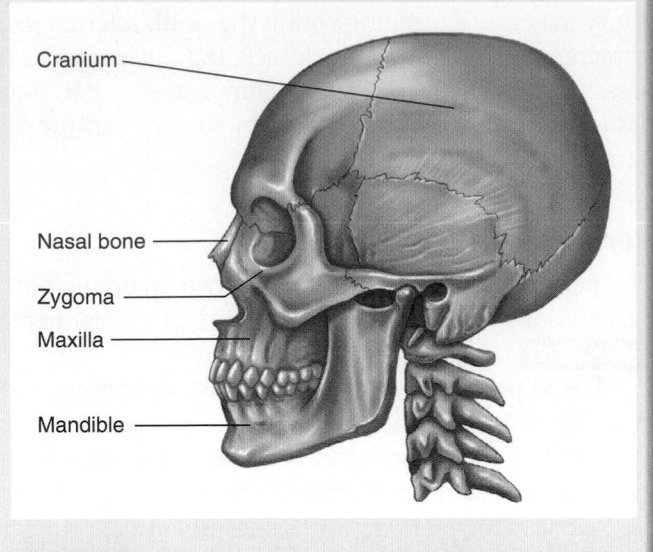

Figure 26-6 The skull includes two large structures: the cranium and the face.

Special Populations

The spinal canal is closed by birth and must grow and expand as the child grows. Neural tube deformities are common and serious birth defects. The most discussed neural tube deformity is spina bifida, in which the lower portion of the spine does not close prior to birth. As an EMT, you may be called on to treat or transport a child with one of these birth defects. Chapter 34, *Patients With Special Challenges*, covers spina bifida in detail.

occupied by 80% brain tissue, 10% blood supply, and 10% CSF. The brain connects to the spinal cord through a large opening at the base of the skull called the foramen magnum.

Four major bones make up the cranium. The most posterior portion of the cranium is called the occiput. On each side of the cranium, the lateral portions are called the temples or temporal regions. Between the temporal regions and the occiput lie the parietal regions. The forehead is called the frontal region.

The face is composed of 14 bones. The upper, nonmoveable jawbones are called the maxillae, the cheek bones are called the zygomas, and the mandible is the lower, moveable portion of the jaw.

The orbit (eye socket) is made up of two facial bones: the maxilla and the zygoma. The orbit also includes the frontal bone of the cranium. Together, these bones form a solid bony rim that protrudes around the eye to protect it. The nose mostly consists of flexible cartilage; in fact, only the proximal one third of the nose is formed by bone with very short bones forming the bridge of the nose.

Spinal Column

The spinal column is the body's central supporting structure. It has 33 bones, called vertebrae, and is divided into five sections: cervical, thoracic, lumbar, sacral, and coccygeal **Figure 26-7**. Injury to the vertebrae, depending on the level at which the injury occurs, can result in paralysis if the underlying spinal cord or nervous structures are also damaged.

The front part of each vertebra consists of a round, solid block of bone called the vertebral body; the back part forms a bony arch. From one vertebra to the next, the series of arches form a tunnel running the length of the spinal column. This tunnel is the spinal canal, which encases and protects the spinal cord **Figure 26-8**.

The vertebrae are connected by ligaments and separated by cushions, called **intervertebral disks**. These ligaments and disks allow the trunk to bend forward and back, but they also limit motion so that the spinal cord is not injured. When the spine is injured or fractured, the spinal cord and its nerves are left unprotected. Therefore, until the spine is stabilized, you must keep it aligned as best you can to prevent further injury to the spinal cord.

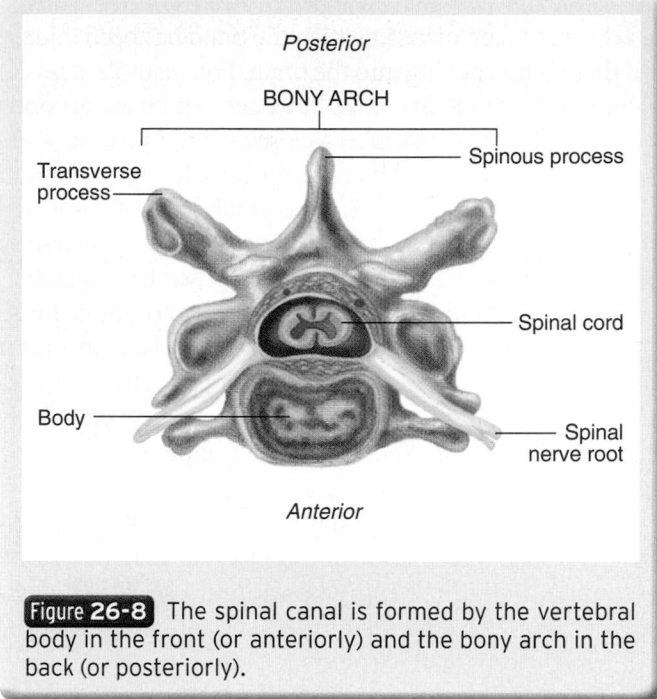

Figure 26-8 The spinal canal is formed by the vertebral body in the front (or anteriorly) and the bony arch in the back (or posteriorly).

The spinal column itself is almost entirely surrounded by muscles. However, you can usually palpate the posterior spinous process of each vertebra, which lies just under the skin in the midline of the back. The most prominent and most easily palpable spinous process is at the seventh cervical vertebra at the base of the neck.

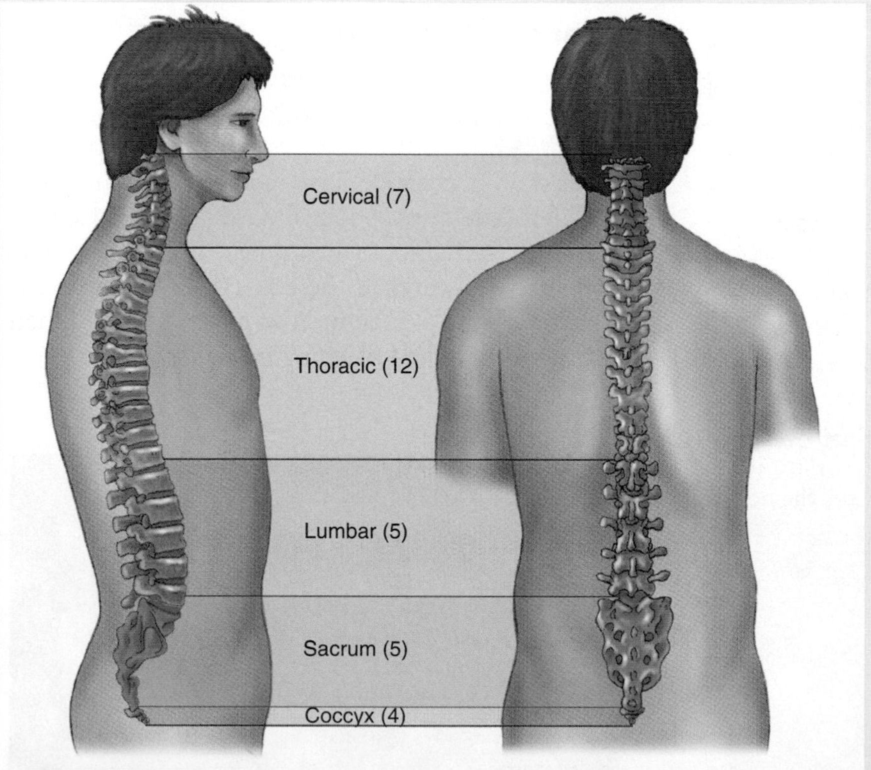

Head Injuries

A head injury is a traumatic insult to the head that may result in injury to soft tissue, bony structures, or the brain. Approximately 4 million people experience head injuries of varying severity in the United States each year. According to the Brain Trauma Foundation, 52,000 deaths occur annually as the result of severe head injury. More than 50% of all traumatic deaths result from a head injury. When head injuries are fatal, the cause is invariably associated injury to the brain. In addition to the head injury, and dependent on the mechanism of injury (MOI), you should be alert to the fact that the patient may have sustained additional trauma such as cervical spine injuries, pelvic injuries, and chest injuries.

Figure 26-7 The spinal column is the body's central supporting system and consists of 33 bones divided into five sections. Injury to the vertebrae can cause paralysis.

There are two general types of head injuries. **Closed head injuries** are those in which the brain has been injured but there is no opening into the brain. For example, a severe blow that fractures the skull but does not create an open wound would be considered a closed head injury. An **open head injury** is one in which an opening from the brain to the outside world exists. Obvious skull deformity is a sign of an open head injury, which is often caused by penetrating trauma. There may be bleeding and exposed brain tissue.

Motor vehicle crashes are the most common MOI, with more than two thirds of people involved in motor vehicle crashes experiencing a head injury. Head injuries also occur commonly in victims of assault, when elderly people fall, during sports-related incidents, and in a variety of incidents involving children.

Any head injury is potentially serious. If not properly treated, those that at first seem minor may end up becoming a life-threatening brain injury Table 26-1 . Conversely, severe lacerations of the scalp or fractures of the skull may occur with little or no brain injury and may lead to minimal or no long-term consequences.

■ Scalp Lacerations

Scalp lacerations can be minor or very serious. Because both the face and the scalp have unusually rich blood supplies, even small lacerations can quickly lead to significant blood loss Figure 26-9 . Occasionally, this blood loss may be severe enough to cause hypovolemic shock, particularly in children. In any patient with multiple injuries, bleeding from scalp or facial lacerations may contribute to hypovolemia. In addition, because scalp lacerations are usually the result of direct blows to the head, they are often an indicator of deeper, more serious injuries.

■ Skull Fracture

Significant force applied to the head may cause a skull fracture. As with any fracture, a skull fracture may be open or closed, depending on whether there is an overlying laceration of the scalp. Injuries from bullets or other penetrating weapons frequently result in fracture of the skull. The diagnosis of a skull fracture is usually made in the hospital with a computed tomography (CT) scan, but you should maintain a high index of suspicion that a fracture is present if the patient's head appears deformed or if there is a visible crack in the skull within a scalp laceration. Additional signs of skull fracture that you may see include ecchymosis (bruising) that develops under the eyes (**raccoon eyes**) Figure 26-10A or behind one ear over the mastoid process (**Battle's sign**) Figure 26-10B .

Linear Skull Fractures

Linear skull fractures (nondisplaced skull fractures) account for approximately 80% of all fractures to the skull Figure 26-11A . Radiographs are required to diagnose a linear skull fracture because there are often no physical signs such as deformity. If the brain is uninjured and there are no scalp lacerations, then linear fractures are not life threatening. However, if there is a scalp laceration with the linear fracture—making it an open fracture—there is a risk of infection and bleeding inside the brain.

Table 26-1	General Signs and Symptoms of a Head Injury

Following a head injury, any patient who exhibits one or more of these signs or symptoms has potentially sustained a very serious underlying brain injury:

- Lacerations, contusions, or hematomas to the scalp
- Soft area or depression on palpation
- Visible fractures or deformities of the skull
- Decreased mentation
- Irregular breathing pattern
- Widening pulse pressure
- Slow heart rate
- Ecchymosis about the eyes or behind the ear over the mastoid process
- Clear or pink CSF leakage from a scalp wound, the nose, or the ear
- Failure of the pupils to respond to light
- Unequal pupil size
- Loss of sensation and/or motor function
- A period of unconsciousness
- Amnesia
- Seizures
- Numbness or tingling in the extremities
- Irregular respirations
- Dizziness
- Visual complaints
- Combative or other abnormal behavior
- Nausea or vomiting
- Posturing (decorticate or decerebrate)

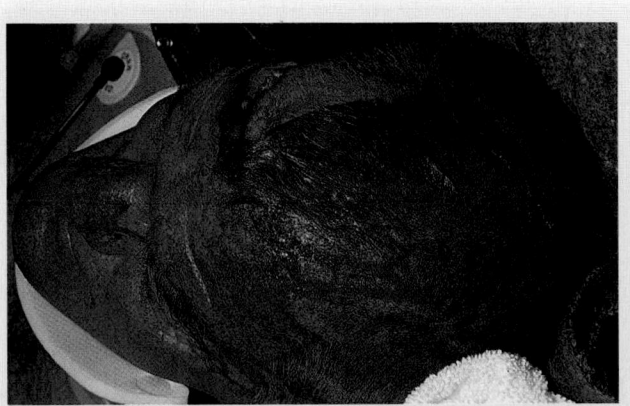

Figure 26-9 The scalp has an unusually rich blood supply; therefore, even small lacerations can result in significant blood loss.

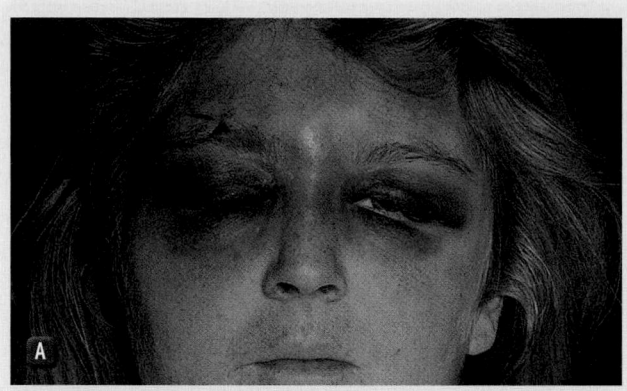

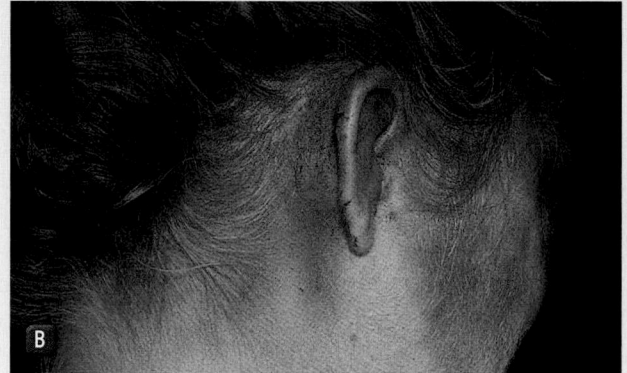

Figure 26-10 Signs of skull fracture include ecchymosis (**A**) under the eyes (raccoon eyes) or (**B**) behind one ear over the mastoid process (Battle's sign).

Compressed Skull Fractures

Compressed skull fractures result from high-energy direct trauma to the head with a blunt object (such as a baseball bat to the head) **Figure 26-11B**. The frontal and parietal bones of the skull are most susceptible to these types of fractures because the bones in these areas are relatively thin. As a consequence, bony fragments may be driven into the brain, resulting in injury. The scalp may or may not be lacerated. Patients with compressed skull fractures often present with neurologic signs (such as loss of consciousness).

Basilar Skull Fractures

Basilar skull fractures also are associated with high-energy trauma, but they usually occur following diffuse impact to the head (eg, falls, motor vehicle crashes). These injuries generally result from extension of a linear fracture to the base of the skull and can be difficult to diagnose with radiography (x-ray) **Figure 26-11C**.

Signs of a basilar skull fracture include CSF drainage from the ears, which indicates rupture of the tympanic membrane in the ear and freely flowing CSF through the ear. Patients with leaking CSF are at risk for bacterial meningitis.

Other signs of a basilar skull fracture include raccoon eyes or Battle's sign. Depending on the extent of

the damage, raccoon eyes and Battle's sign may appear relatively quickly, but in many patients, they may not appear until up to 24 hours following the injury, so their absence in the field does not rule out a basilar skull fracture.

Open Skull Fractures

Open fractures of the cranial vault result when severe forces are applied to the head and are often associated with trauma to multiple body systems **Figure 26-11D**. Brain tissue may be exposed to the environment, which significantly increases the risk of a bacterial infection (such as bacterial meningitis). Open cranial vault fractures have a high mortality rate.

■ Traumatic Brain Injuries

The National Head Injury Foundation defines a **traumatic brain injury (TBI)** as "a traumatic insult to the brain capable of producing physical, intellectual, emotional, social, and vocational changes." Traumatic brain injuries are the most serious of all head injuries. Traumatic brain injuries are classified into two broad categories: **primary (direct) injury** and **secondary (indirect) injury**. Primary brain injury is injury to the brain and its associated structures that results instantaneously from impact to the head. Secondary brain injury refers to a multitude of processes that increase the severity of a primary brain injury and therefore, negatively impact the outcome. Secondary injuries may be caused by cerebral edema, intracranial hemorrhage, increased intracranial pressure, cerebral ischemia, and infection however; hypoxia and hypotension are the two most common causes. According to the Brain Trauma Foundation, hypoxia or hypotension will increase death and disability significantly in a patient with a head injury. It is important to monitor and address hypoxia and hypotension when identified. Secondary brain injury can occur anywhere from a few minutes to several days following the initial head injury.

The brain can be injured directly by a penetrating object, such as a bullet, knife, or other sharp object. More commonly, such injuries occur indirectly, as a result of external forces exerted on the skull. Consider the most common cause of brain injury, the motor vehicle crash. When the passenger's head hits the windshield on impact with a fixed object, the brain continues to move forward until it comes to an abrupt stop by striking the inside of the skull. This rapid deceleration results in compression injury (or bruising) to the anterior portion of the brain along with stretching or tearing of the posterior portion of the brain **Figure 26-12**. As the brain strikes the front of the skull, the body begins its path of moving backward. The head falls back against the headrest and/or seat, and the brain slams into the rear of the skull. This

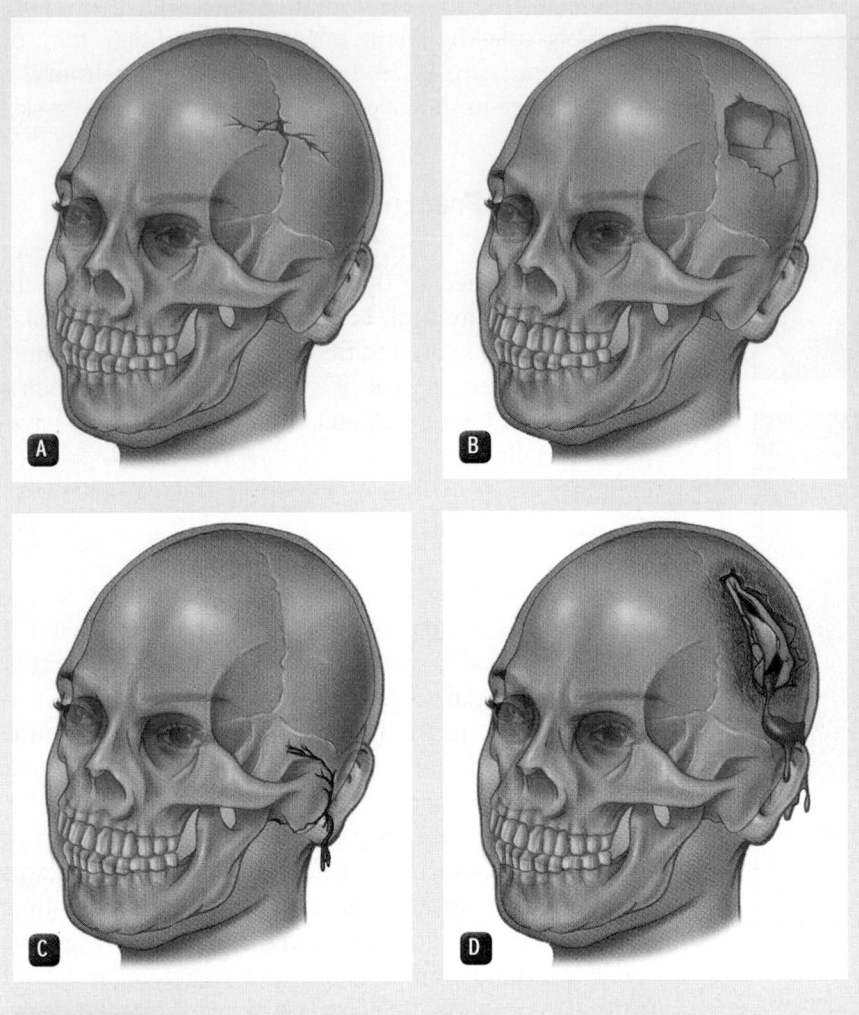

Figure 26-11 Types of skull fracture. **A.** Linear. **B.** Compressed. **C.** Basilar. **D.** Open.

type of front-and-rear injury is known as a <u>**coup-contrecoup injury**</u>. The same type of injury may occur on opposite sides of the brain in a lateral collision.

The injured brain starts to swell, initially because of cerebral vasodilation. An increase in cerebral water (cerebral edema) then contributes to further brain swelling. <u>**Cerebral edema**</u> (swelling of the brain) may not develop until several hours following the initial injury, however.

Cerebral edema is aggravated by low oxygen levels in the blood and improved by high ones. In fact, the brain consumes more oxygen than any other organ in the body. For this reason, you must make sure that the airway is open and that adequate ventilations and high-flow oxygen are given to any patient with a head injury. This is especially true if the patient is unconscious. Do not wait for cyanosis or other obvious signs of hypoxia to develop.

It is not uncommon for the patient with a head injury to have a convulsion, or seizure. This is the result of excessive excitability of the brain, caused by direct injury or the accumulation of fluid within the brain (edema). You should be prepared to manage seizures in all patients who have had a head injury because the brain may have sustained an injury as well.

■ Intracranial Pressure

For adults, the skull is a rigid, unyielding globe that allows little, if any, expansion of the intracranial contents. It also provides a hard and somewhat irregular surface against which brain tissue and its blood vessels can be injured when the head sustains trauma.

Accumulations of blood within the skull or swelling of the brain can rapidly lead to an increase in <u>**intracranial pressure (ICP)**</u>, the pressure within the cranial vault. Increased ICP squeezes the brain against bony prominences within the cranium. **Table 26-2** lists the levels of ICP and the corresponding signs and symptoms.

Other effects of cerebral edema and increased intracranial pressure may be increased systolic blood pressure, decreased pulse rate, and irregular respirations. This triad of signs is called Cushing's reflex.

Intracranial Hemorrhage

The closed box of the skull has no extra room for an accumulation of blood, so bleeding inside the skull also

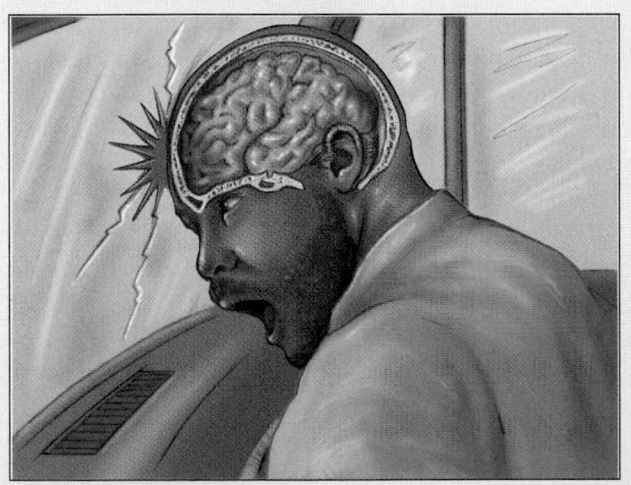

Figure 26-12 For the unrestrained person in a motor vehicle crash, the brain continues its forward motion and strikes the inside of the skull, resulting in compression injury to the anterior portion of the brain and stretching of the posterior portion.

Table 26-2 Levels of Intracranial Pressure

Mild elevation	■ Increased blood pressure; decreased pulse rate ■ Pupils still reactive ■ Cheyne-Stokes respirations (respirations that are fast and then become slow, with intervening periods of apnea) ■ Patient initially attempts to localize and remove painful stimuli; this is followed by withdrawal and extension ■ Effects are reversible with prompt treatment
Moderate elevation (indicates that the middle brain stem is involved)	■ Widened pulse pressure and bradycardia ■ Pupils are sluggish or nonreactive ■ Central neurogenic hyperventilation (deep, rapid respirations) ■ Decerebrate posturing ■ Survival possible but not without permanent neurologic deficit
Marked elevation (indicates that lower portion of brain stem involved/medulla)	■ Unilateral fixed and dilated pupil ■ Ataxic respirations (characterized by irregular rate, pattern, and volume of breathing with intermittent periods of apnea) or absent respirations ■ Flaccid response to painful stimuli ■ Irregular pulse rate ■ Diminished blood pressure ■ Most patients do not survive this level of intracranial pressure

increases the ICP. Bleeding can occur between the skull and dura mater, beneath the dura mater but outside the brain, or within the tissue of the brain itself.

Epidural Hematoma

An **epidural hematoma** is an accumulation of blood between the skull and dura mater **Figure 26-13**. An epidural hematoma is nearly always the result of a blow to the head that produces a linear fracture of the thin temporal bone. The middle meningeal artery runs along a groove in that bone; therefore, it is vulnerable when the temporal bone is fractured. Arterial bleeding into the epidural space will result in rapidly progressing symptoms.

Often, the patient loses consciousness immediately following the injury; this is often followed by a brief period of consciousness (lucid interval), after which the patient lapses back into unconsciousness. Meanwhile, as the ICP increases, the pupil on the side of the hematoma becomes fixed and dilated. Death will follow very rapidly without surgery to evacuate the hematoma.

Subdural Hematoma

A **subdural hematoma** is an accumulation of blood beneath the dura mater but outside the brain **Figure 26-14**. It usually occurs after falls or injuries involving strong deceleration forces. Subdural hematomas are more common than epidural hematomas and may or may not be associated with a skull fracture. Bleeding within the subdural space typically results from rupture of the veins that bridge the cerebral cortex and dura.

A subdural hematoma is associated with venous bleeding, so this type of hematoma and the signs of increased ICP typically develop more gradually than with an epidural hematoma. The patient with a subdural hematoma often experiences a fluctuating level of consciousness or slurred speech.

Special Populations

The elderly and those with a history of alcohol use are at higher risk for developing a subdural hematoma. This is caused by atrophy of the brain tissue that increases stretching of the bridging veins. Signs and symptoms of the condition may not occur for several hours, days, or weeks. Be sure to get a thorough history of any previous trauma.

Intracerebral Hematoma

An **intracerebral hematoma** involves bleeding within the brain tissue itself **Figure 26-15**. This type of injury can occur following a penetrating injury to the head or because of rapid deceleration forces.

Many small, deep intracerebral hemorrhages are associated with other brain injuries. The progression of increased ICP depends on several factors, including the presence of other brain injuries, the region of the brain involved (frontal and temporal lobes are the most common locations), and the size of the hemorrhage. Once symptoms appear, the patient's condition often deteriorates quickly. Intracerebral hematomas have a high mortality rate, even if the hematoma is surgically evacuated.

Subarachnoid Hemorrhage

In a **subarachnoid hemorrhage**, bleeding occurs into the subarachnoid space, where the CSF circulates. It results in bloody CSF and signs of meningeal irritation (such as neck

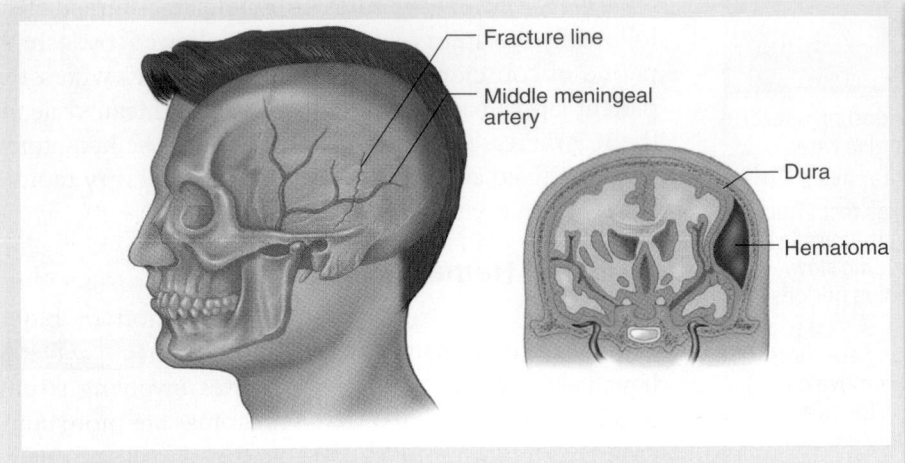

Figure 26-13 An epidural hematoma is usually the result of a blow to the head that produces a linear fracture of the temporal bone and damages the middle meningeal artery. Blood accumulates between the dura mater and the skull.

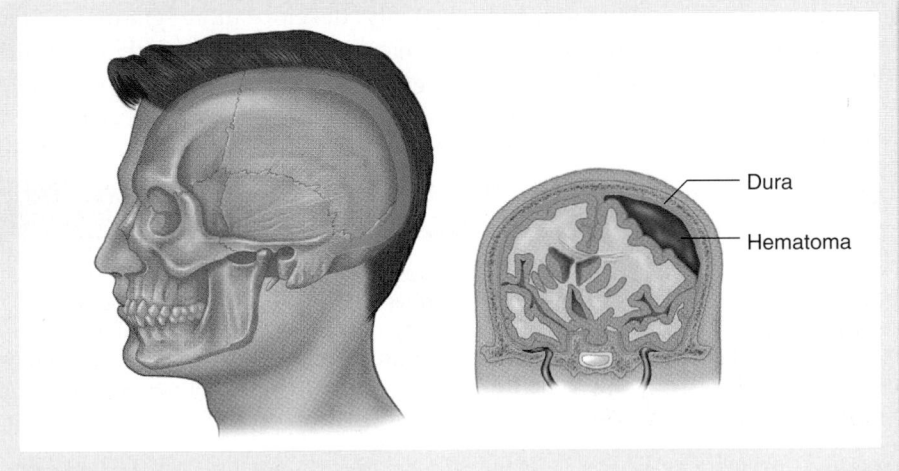

Figure 26-14 In a subdural hematoma, venous bleeding occurs beneath the dura mater but outside the brain.

rigidity, headache). Common causes of a subarachnoid hematoma include trauma or rupture of an aneurysm.

The patient with a subarachnoid hematoma reports a sudden, severe headache. As bleeding into the subarachnoid space increases, the patient experiences the signs and symptoms of increased ICP: decreased level of consciousness, changes in the pupils, vomiting, and seizures.

A sudden, severe subarachnoid hematoma usually results in death. People who survive often have permanent neurologic impairment.

■ Concussion

A blow to the head or face may cause **concussion** of the brain. Concussions are also known as mild traumatic brain injuries. There is no universal agreement on the exact definition of a concussion, but in general, it is a closed injury with a temporary loss or alteration of part or all of the brain's abilities to function without demonstrable physical damage to the brain. For example, a person who "sees stars" after being struck in the head has sustained a concussion that affects the occipital portion of the brain. A concussion may result in unconsciousness and even the inability to breathe for short periods of time; however, approximately 90% of patients who sustain a concussion do not experience a loss of consciousness.

A patient with a concussion may be confused or have amnesia (loss of memory). Occasionally, the patient can remember everything but the events leading up to the injury; this is called **retrograde amnesia**. Inability to remember events after the injury is called **anterograde (posttraumatic) amnesia**.

Usually, a concussion lasts only a short time. In fact, it has often resolved by the time you arrive. Nevertheless, you should ask about symptoms of concussion in any patient who has sustained an injury to the head; these symptoms include dizziness, weakness, or visual changes. Additional signs and symptoms you may encounter with a patient who has sustained a concussion may include nausea or vomiting and the patient may report ringing in the ears. Slurred speech and the inability to focus may also be present. Dependent on the severity of the concussion, you may also notice that the patient has a lack of coordination, a delay of motor functions, or displays inappropriate emotional responses. Patients may also report a temporary headache and may appear to be disoriented at times.

Patients with symptoms consistent with concussion can also have more serious underlying brain injury. A CT scan is necessary to differentiate between these conditions. You should always assume that a patient with signs or symptoms of concussion has a more serious injury until proven otherwise by a CT scan at the hospital or by evaluation by a physician.

■ Contusion

Like any other soft tissue in the body, the brain can sustain a contusion, or bruise, when the skull is struck.

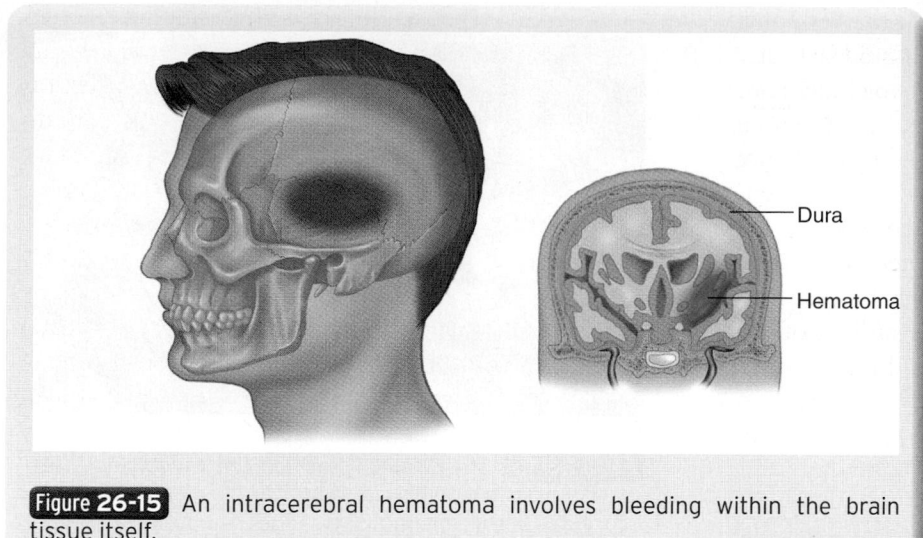

Figure 26-15 An intracerebral hematoma involves bleeding within the brain tissue itself.

Dura

Hematoma

can also cause brain injuries that produce significant bleeding or swelling. Problems with the blood vessels themselves, high blood pressure, or any number of other problems may cause spontaneous bleeding into the brain, affecting the patient's level of consciousness. This is known as altered mental status. The signs and symptoms of nontraumatic injuries are often the same as those of traumatic brain injuries, except that there is no obvious history of MOI or any external evidence of trauma. Altered mental status is discussed in Chapter 15, *Neurologic Emergencies*.

A contusion is far more serious than a concussion because it involves physical injury to the brain tissue, which may sustain long-lasting and even permanent damage. As with contusions that occur elsewhere in the body, there is associated bleeding and swelling from injured blood vessels. Injury of brain tissue or bleeding inside the skull causes an increase of pressure within the skull. A patient who has sustained a brain contusion may exhibit any or all of the signs of brain injury.

Other Brain Injuries

Brain injuries are not always a result of trauma. Certain medical conditions, such as blood clots or hemorrhages,

Spine Injuries

The cervical, thoracic, and lumbar portions of the spine can be injured in a variety of ways. Compression injuries can occur as a result of a fall, regardless of whether the patient landed on his or her feet, coccyx, or on the top of the head. Motor vehicle crashes or other types of trauma can overextend, flex, or rotate the spine. Any one of these unnatural motions, as well as excessive lateral bending, can result in fractures or neurologic deficit.

When the spine is pulled along its length, this is called **distraction** and it can cause injuries. For example, hangings often result in fracture of the vertebrae in the upper portion of the cervical spine.

You are the Provider: PART 2

When you arrive at the scene you find a 22-year-old man lying supine on the ground; he is motionless and his head is lying in a large pool of blood. Your partner manually stabilizes his head in a neutral position and opens his airway with the jaw-thrust maneuver as you perform a primary assessment.

Recording Time: 0 Minutes	
Appearance	Motionless; large pool of blood under his head
Level of consciousness	Responsive only to deep painful stimuli
Airway	Open; clear of secretions or foreign bodies
Breathing	Slow and irregular
Circulation	Radial pulses; slow and bounding; skin, warm and dry; bleeding from a large laceration to the right side of his head

3. What are your most immediate treatment priorities for this patient?

4. Where should you focus your secondary assessment of this patient?

Subluxation of the spine occurs when the vertebrae are no longer aligned. Think of blocks that should be neatly stacked one on top of another, with one or two of them out of alignment. This type of injury pattern can occur with a hyperextension mechanism or can be caused by a fracture or a dislocation. Common findings include pain and tenderness on palpation of the region. Less commonly you may feel or observe a deformity of the spine, sometimes referred to as a "step-off" where the spinous process may be palpable on physical examination. Regardless of the cause, subluxation is a dangerous injury and it can evolve into a full debilitating spinal cord injury. If you suspect a subluxation you should take extra precautions when stabilizing the spine, both manually and with adjuncts.

Words of Wisdom

When assessing the spine, be aware of the possibility of open wounds from the associated trauma. These open wounds can be penetrating injuries or lacerations. If you follow the mnemonic DCAP-BTLS, you will discover any open wounds prior to securing the patient to a backboard.

Patient Assessment

You should always suspect a possible head or spinal injury any time you encounter one of the following mechanisms of injury:

- Motor vehicle collisions
- Pedestrian–motor vehicle collisions
- Falls
- Blunt trauma
- Penetrating trauma to the head, neck, back, or torso
- Motorcycle crashes
- Rapid deceleration injuries
- Hangings
- Diving accidents
- Recreational accidents

Motor vehicle collisions, direct blows, falls from heights, assault, and sports injuries are common causes of head and spinal injury. A deformed windshield or dented helmet may indicate a major blow to the head, which is likely to have caused traumatic brain injury **Figure 26-16**. It is especially important to evaluate and monitor the level of consciousness in patients with suspected head injuries, paying particular attention to any changes that may occur.

Scene Size-up

Scene Safety
Evaluate every scene for hazards to your health and the health of your team or bystanders. Motor vehicle crashes are a common cause of head and spinal injuries. These

Figure 26-16 The classic "star" on the windshield after an automobile crash is a significant indicator of head injury. Be alert for the signs and symptoms of head injury.

situations have the potential to cause injury to rescuers and bystanders as well. Be prepared with appropriate standard precautions before you approach the patient. You will be spending a great deal of time at the head of the patient. Gloves, a mask, and eye protection should be the minimum standard precautions that you use. Because these patients can have very complicated injuries, call for ALS as soon as possible when a serious MOI or complicated presentation is evident. Law enforcement may be needed to control traffic or unruly people.

Words of Wisdom

Many mechanisms of injury that cause head and spine injuries can also entail risk to EMTs. Before you approach the patient, get the "big picture" of scene safety and take any actions necessary to ensure your own well-being. Do not rely entirely on assistance from fire or police personnel; maintain your own awareness of the scene.

When assessing a patient with a possible closed head injury, consider the MOI. Did the patient fall? Was he or she in an automobile crash or the victim of an assault? Was there deformity of the windshield or deformity of the helmet?

Words of Wisdom

Proper care of a patient with a possible spinal injury requires assessment of motor and sensory functions both before and after immobilizing the patient. Likewise, careful observation of level of consciousness at different stages of your care for a head-injured patient can provide crucial information. Document your detailed findings of these repeated neurologic examinations to make the information available to hospital personnel and to help establish that your care has been thorough and appropriate.

Mechanism of Injury/Nature of Illness

As you observe the scene, look for indicators of the MOI. This helps you develop an early index of suspicion for underlying injuries in the patient who has sustained a significant MOI. As you put together information from dispatch and your observations of the scene, consider how the MOI produced the injuries expected. For example, if you respond to a baseball field for a patient who was knocked unconscious by a foul ball, you may begin to suspect that the patient may have a compressed skull fracture and perform a neurologic assessment during the physical examination.

> ## Primary Assessment

The primary assessment should focus on identifying and managing life-threatening concerns. Threats to airway, breathing, or circulation are considered life threatening and must be treated immediately to prevent mortality. When assessing a patient with suspected head and/or spine injuries, be aware that any unnecessary movement of the patient can cause additional injury. Use the AVPU scale (*Alert* to person, place, and day; responsive to *Verbal* stimuli; responsive to *Pain*; *Unresponsive*) to assess the patient's level of consciousness.

Form a General Impression

Begin by forming a general impression of your patient based on his or her level of consciousness and ask about the chief complaint. This will give you a focal point to begin your assessment. Patients with head injuries frequently have spinal injuries and vice versa. When assessing a patient for possible head or spinal injury, you should begin by asking the responsive patient the following questions to determine his or her chief complaint:

- What happened?
- Where does it hurt?
- Does your neck or back hurt?
- Can you move your hands and feet?
- Did you hit your head?

Confused or slurred speech, repetitive questioning, or amnesia in responsive patients is a good indication of a head injury. Whereas other problems may cause similar symptoms, in the setting of trauma, assume your patient has a head injury until your assessment proves otherwise.

If the patient is found unresponsive, emergency responders, family members, or bystanders may have helpful information, including when the patient lost consciousness or what his or her previous level of consciousness was. Unresponsive patients with any trauma should be assumed to have a spinal injury. Patients with a decreased level of responsiveness on the AVPU scale (responds to verbal stimulus or responds to painful stimulus) should also be considered to have a spinal injury based on their chief complaint.

Unless the patient is absolutely clear in his or her thinking and does not have any other illnesses or injuries that may constitute a distraction, an MOI that suggests a potential spine injury should lead you to provide complete spinal motion restriction. A physician is widely considered to be the appropriate person to clear patients with potential spinal injuries. Some jurisdictions allow their EMTs to screen patients and to refrain from providing spinal immobilization on the basis of specific criteria in specific patients, although this is not a common practice. Other jurisdictions require that spinal immobilization be provided for every patient with an MOI that suggests potential spinal injury regardless of the patient's signs or symptoms. Understand and follow your local protocols.

Airway and Breathing

In patients with head and spinal injuries, airway and breathing problems are common and may result in death if not recognized and treated immediately. When a spinal injury is suspected, how you open and assess the airway is important. Begin by manually holding the patient's head still while you assess the airway. Use a jaw-thrust maneuver to open the airway. When performed correctly, this will prevent movement of the cervical spine. If, however, you are unable to provide a patent and open airway using the jaw-thrust maneuver, it is acceptable to use the head tilt–chin lift maneuver. The patient cannot survive if the airway is not functioning and even though this maneuver may cause further injury to the spine, it is considered the last resort to provide an airway for your patient. An oropharyngeal or nasopharyngeal airway may assist in maintaining an airway; however, the best way to adequately protect the patient's airway is to use advanced airway techniques, usually employed by AEMTs and paramedics. The decision to use an oropharyngeal or nasopharyngeal airway is based on the patient's ability to maintain his or her own airway, the presence of a gag reflex, and the extent of facial injuries. Review the indications and contraindications for these airway adjuncts in Chapter 9, *Airway Management*, and the use of advanced airway techniques in Chapter 40, *ALS Assist*.

Vomiting may occur in the patient with a head injury. With large amounts of emesis, the patient may need to be log rolled to the side and the mouth swept of secretions. When it is necessary to log roll the patient to clear the airway, roll the patient keeping the body in as straight a line as possible to minimize spinal injuries. Suctioning should be performed immediately to remove smaller amounts of secretions.

Apply a cervical spine immobilization device as soon as you have assessed the airway and breathing and provided necessary treatments. A cervical collar may help maintain spinal motion restriction as you treat the airway and breathing. The best time to apply the cervical collar depends on the patient's injuries and the seriousness of his or her condition. For some patients, you may have to apply the collar early on, while managing the ABCs; in other patients, manual stabilization may be adequate until you are ready to place the patient on a backboard. The key to managing spinal injuries and airway and breathing problems is to move the patient as little as possible and as carefully as possible, maintaining spinal alignment throughout. Place an appropriately sized cervical spine immobilization device on the patient when appropriate. Once the device is on, do not remove it unless it causes a problem with maintaining the ABCs. If you must remove the device, you will have to maintain manual stabilization of the cervical spine until the device is replaced and the patient has been once again secured to the backboard.

Irregular breathing, such as Cheyne-Stokes respirations, may result from increased pressure on the brain because of bleeding or swelling in the cranium.

If the ICP increases, there will be more periods of apnea. In either situation, determine whether breathing is present and adequate and continue to monitor the patient's respiratory rate and depth. Oxygen, delivered at a rate of 15 L/min by a nonrebreathing mask or via a bag-mask device, is always indicated for patients with head and spinal injuries. A single episode of hypoxia in a patient with a head injury increases the risk of death or permanent disability significantly. Pulse oximeter values should be maintained above 90%. Positive-pressure ventilations are not always necessary; however, if the patient's breathing rate is too slow or too fast and shallow, provide positive-pressure ventilations using a bag-mask device or a manually triggered ventilation device (see Chapter 9, *Airway Management*). The rate of ventilations should be based on the age of the patient and established BLS guidelines. Do not panic and hyperventilate the patient because his or her condition appears severe. Hyperventilation should be reserved for specific conditions and performed under specific guidelines. Be sure to know your local protocols on this subject.

Circulation

When approaching a patient who is unconscious, the obvious question is, "Is this person alive?" Whereas checking immediately to determine whether a pulse is present is tempting, it is more important for you to remember the ABCs. Always assess airway and breathing prior to moving on to assessment of circulation by checking a pulse. Patients who are responsive and moving obviously have a pulse; however, you should still check to see if the pulse is weak or strong and if it is generally too fast or too slow. A pulse that is too slow in the setting of a head injury can indicate a serious condition in your patient. If the pulse is present and adequate you can continue to evaluate your patient further.

A single episode of hypoperfusion in a patient with a head injury can lead to significant brain damage and even death. Assess for signs and symptoms of shock and treat appropriately. Bleeding may also be present from the same injury that caused the spine and/or head injury. That injury may involve blunt or penetrating forces. Consider again the MOI and the effects it has had on your patient. Control bleeding as previously discussed. When bandaging the head, be careful that you do not move the neck if spinal injuries are suspected. Remember that head and spine injuries often occur together.

Transport Decision

Most head injuries are considered mild and result in no or limited permanent disability. A smaller percentage of head injuries are considered moderate, and the patient is left with some permanent disabilities. A still smaller percentage of head injuries are considered severe, and many patients with a severe head injury die before ever reaching the hospital or are left in a comatose state despite hospital intervention. There will be a number of patients with head or spine injuries that will not require much intervention other than a thorough assessment and continued observation while being transported to the hospital. In these patients you may choose to take some time at the scene to provide careful spine stabilization before transport. In patients who have problems with ABCs or have other conditions for which you decide a rapid transport to the hospital is needed, rapid stabilization of the spine and quick loading into the ambulance may be indicated. Reduction of on scene time and recognition of a critical patient increases the patient's chances for survival or a reduction in the amount of irreversible damage.

Several transport considerations should be kept in mind for patients with head trauma. Patients with impaired airways, open head wounds, abnormal vital signs, or those patients who do not respond to painful stimuli may need to be rapidly extracted from a motor vehicle and transported. During transport, providing the patient with a patent airway and high-flow oxygen is paramount. Because of the risk of increasing ICP, there is a probability of vomiting and seizures, so suction should be readily available. A patient with head trauma may deteriorate rapidly, thus requiring aeromedical transport depending on your local protocols. In supine patients, the head should be elevated 30° to help reduce ICP if possible. Remember to maintain stabilization of the spine.

Studies have shown that the use of lights and sirens for transportation of patients does not significantly reduce transport time. In fact, the use of lights and sirens may increase the patient's level of distress. Patients who are conscious and aware of the inability to move their limbs need to be offered psychological support. Remember that it can be very traumatizing for a patient to realize that he or she may now have a debilitating and life-altering injury because of an accident; therefore, you need to be careful in your choice of words. A patient may ask you difficult questions, "Will I be able to walk?" It is best to tell the patient that you are providing immediate care and you cannot predict the outcome.

History Taking

Investigate Chief Complaint

After the life threats have been managed during the primary assessment, investigate the chief complaint. You should obtain a medical history and be alert for injury-specific signs and symptoms as well as any pertinent negatives such as no pain or no loss of sensation.

Using OPQRST (Onset, Provocation or palliation, Quality, Region/radiation, Severity, and Timing of pain) may provide some background on isolated extremity injuries. Does the patient have any recall of the incident? Inability to recall events is an important finding in patients with head injuries. You have the opportunity to interview the patient well in advance of the emergency physician. Any information you receive will be very valuable if the patient loses consciousness.

If the patient is not responsive, attempt to obtain the history from other sources, such as friends or family members. Medical identification jewelry and cards in wallets may also provide information about the patient's medical history. Does the patient have a recent or previous history of unresponsiveness? These key indicators may lead you to suspect a developing traumatic brain injury.

SAMPLE History

Make every attempt to obtain a SAMPLE history from your patient. History may be difficult to obtain when a person is confused from a head injury or frightened from a spinal injury. Whereas the prehospital environment is an excellent place to obtain important history, do not delay rapid transport for patients who need rapid hospital intervention. Gather as much SAMPLE history as you can while preparing for transport. In less urgent situations, you should have enough time to gather a complete SAMPLE history without compromising patient care.

Secondary Assessment

Remember that the ability to walk, move the extremities, or feel sensation does not necessarily rule out a spinal cord injury. Similarly, the absence of pain does not always indicate that a spinal injury has not occurred. Do not ask patients with possible spinal injuries to move their necks as a test for pain. Instead, you should instruct the patient to keep still and not to move the head or neck.

Physical Examinations

The physical examination may be a systematic head-to-toe, full-body scan or a systematic assessment that focuses on a certain area or region of the body, often determined through the chief complaint.

Patients with moderate or severe head injuries associated with a significant MOI should receive life-saving medical or surgical intervention at the hospital without delay. If time allows, perform a secondary assessment to identify and treat injuries that may have been missed during the primary assessment en route to the emergency department. Extremities can be stabilized using a long backboard and splinted individually while in the back of the ambulance as time and conditions permit.

Perform a full-body scan using DCAP-BTLS and examine the head, chest, abdomen, extremities, and back. Check perfusion, motor function, and sensation in all extremities prior to moving the patient. Make sure that you do not move any body parts excessively. Determine whether the strength in each extremity is equal by asking the patient to squeeze your hands and to gently push each foot against your hands **Figure 26-17**.

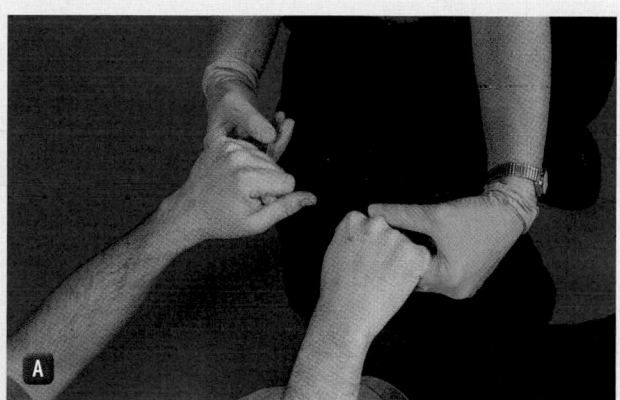

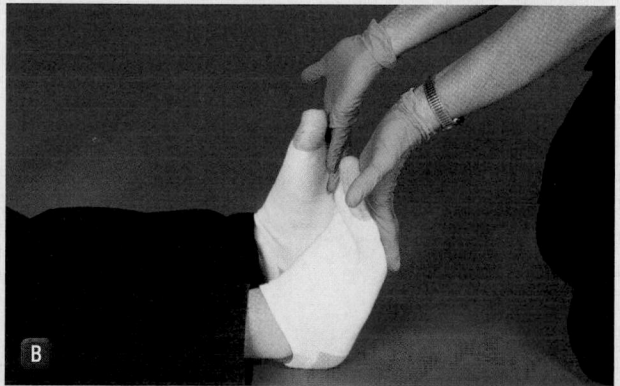

Figure 26-17 **A.** Assess the equality of strength in each extremity by asking the patient to squeeze your hands. **B.** Next, ask the patient to gently push each foot against your hands.

A decreased level of consciousness is the most reliable sign of a head injury. Monitor the patient for changes in level of consciousness, including signs of confusion, disorientation, or deteriorating mental status. Is the patient unresponsive or repeating questions? Experiencing seizures? Nauseous or vomiting?

Determine whether there is decreased movement and/or numbness and tingling in the extremities. Is there any spinal cord posturing? Is the patient able to perform motor function appropriately and equally such as squeezing your hands? Can the patient smile? An inability to smile is a sign that the cervical vertebra may be injured. Part the patient's hair and inspect the scalp for bruising. Look for blood or CSF leaking from the ears, nose, or mouth and for bruising around the eyes and behind the ears.

Evaluate the patient's pupils to see if they are equal and reactive to light, especially if he or she has a decreased level of consciousness. Unequal pupil size after a head injury in an unconscious patient often signals a serious problem. Developing blood clots may be compressing the brain, causing one pupil to dilate and indicating that the brain is at extreme risk of sustaining catastrophic damage **Figure 26-18** .

Do not probe open scalp lacerations with your gloved finger because this may push bone fragments into the brain. Do not remove an impaled object from an open head injury.

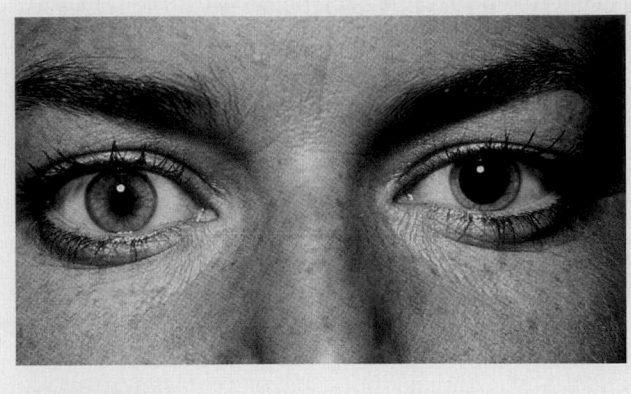

Figure 26-18 Assess pupil size if you suspect a head injury.

Head Injury

Depending on the chief complaint, you may focus your physical examination on the site of injury. For a patient with a head injury, perform a neurologic examination. Perform a baseline assessment using the Glasgow Coma Scale (GCS) and record the time **Table 26-3** . The GCS helps you to identify the patient's speech and ability to follow commands. Always use simple, easily understood terms when reporting the level of consciousness, such as "does not remember events immediately before the injury" or "confused about date and time." Terms such as "obtunded" or "dazed" have different meanings to

You are the Provider: PART 3

Your secondary assessment reveals an area of depression to the right side of his head over the temporal bone and dilated and sluggishly reactive pupils. The rest of his body is unremarkable for gross injury. He opens his eyes in response to pain, is making unrecognizable sounds, and his arms are flexed and drawn in toward his body. An engine company arrives at the scene to provide assistance. You ask them to prepare the backboard and straps while you quickly assess the patient's vital signs.

Recording Time: 5 Minutes	
Respirations	6 breaths/min and irregular (baseline); ventilations are being assisted
Pulse	60 beats/min; regular and bounding
Skin	Pink, warm, and dry
Blood pressure	190/104 mm Hg
Oxygen saturation (Sao$_2$)	94% (on oxygen)

There are numerous bystanders present; however, no one knows the patient. During your physical assessment, you did not find any medical alert bracelets or any other evidence of a past medical history.

5. What is this patient's Glasgow Coma Scale score?

6. What is the likely explanation for the patient's vital signs?

Table 26-3 Glasgow Coma Scale

Eye Opening		Best Verbal Response		Best Motor Response	
Spontaneous	4	Oriented conversation	5	Obeys commands	6
In response to speech	3	Confused conversation	4	Localizes pain	5
In response to pain	2	Inappropriate words	3	Withdraws to pain	4
None	1	Incomprehensible sounds	2	Abnormal flexion	3
		None	1	Abnormal extension	2
				None	1

Score: 13-15 may indicate mild dysfunction, although 15 is the score a person with no neurologic disabilities would receive.

Score: 9-12 may indicate moderate dysfunction.

Score: 8 or less is indicative of severe dysfunction.

different people and should not be used in either written or verbal reports.

If your jurisdiction uses the Rapid Trauma Score (RTS), then the findings from the Glasgow Coma Scale will be used in determining the RTS value. See Chapter 22, *Trauma Overview*, for a discussion of this scoring system.

Frequently, the levels of consciousness will fluctuate—improving, deteriorating, then improving again over time. On other occasions, there may be a gradual, progressive deterioration in the patient's response to stimuli; this usually indicates serious brain injury that may need aggressive medical and/or surgical treatment. The physicians who treat the patient will need to know when a loss of consciousness occurred. They will want to compare their neurologic evaluation with the one you performed in the field.

Special Populations

A modified Glasgow Coma Scale for pediatric and non-verbal patients assesses eye opening, verbal response, and motor response. The scoring indicators are the same as the Glasgow Coma Scale but the modified scale takes into consideration responses of coos and babbling, scoring these responses as oriented and appropriate.

Words of Wisdom

Change in the level of consciousness is the single most important observation that you can make in assessing the severity of brain injury. Level of consciousness usually corresponds to the extent of loss of brain function.

As you proceed with your assessment, ask yourself these questions: Is the patient's speech clear and appropriate? Does the patient answer in a logical manner and is the patient able to make decisions? Is the patient aware of his or her current location? Is the patient alert to person, place, time, and why you are at the scene? Can the patient recall the events leading up to the incident or is there a period of memory lapse? Can the patient recall major current events?

Spinal Injuries

Depending on the chief complaint, you may focus your physical examination on the site of injury. To examine the spine, first inspect for DCAP-BTLS and check the extremities for circulation, motor, or sensory problems. Sensation may be present throughout the body. If there is impairment, note the level. You do not need to know the exact nerve impairment because this will not change your treatment.

Pain or tenderness when you palpate the spinal area is certainly a warning sign that a spinal injury may exist. Patients with spinal injuries may report constant or intermittent pain along the spinal column or in the extremities. A spinal cord injury may also produce pain independent of movement or palpation. There may be altered sensation such as tingling or numbness distal to the injury.

Other signs and symptoms of spinal injury include an obvious deformity as you gently palpate the spine; numbness, weakness, or tingling in the extremities; and soft-tissue injuries in the spinal region. Patients with severe spinal injury may lose sensation or experience paralysis below the suspected level of injury or be incontinent (loss of urinary or bowel control) **Figure 26-19**. Obvious injury to the head and neck may indicate injury to the cervical spine.

Injuries to the cervical area can limit the ability of the diaphragm to function fully and minimize the

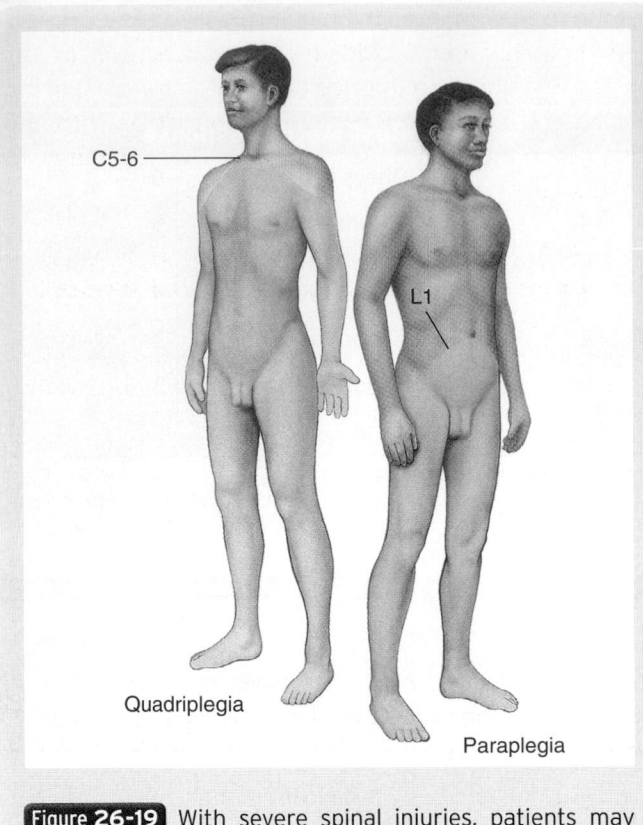

C5-6

L1

Quadriplegia

Paraplegia

Figure 26-19 With severe spinal injuries, patients may lose sensation or experience paralysis below the suspected level of injury.

As soon as you have assessed the patient's level of consciousness, determine the reaction of each pupil to light. Sketch the size of both pupils on the ambulance report to indicate any difference between the two eyes. Continue to monitor the pupils. Any change in their reactions over time may indicate progressive brain injury.

Monitoring Devices

In addition to hands-on assessment, you should use monitoring devices to quantify your patient's oxygenation and circulatory status. If available, CO_2 monitoring should be utilized on all patients suspected of having a head injury to ensure the patient is not hypo- or hyperventilating. You may also use noninvasive methods to monitor the blood pressure. It is recommended that you always assess the patient's first blood pressure manually with a sphygmomanometer (blood pressure cuff) and stethoscope.

Reassessment

Repeat the primary assessment. Reassess vital signs and the chief complaint. Are the airway, breathing, and circulation still adequate? Recheck patient interventions. Are the treatments you provided for problems with the ABCs still effective? This is particularly important in patients with head or spinal injuries because these injuries can suddenly affect the respiratory, circulatory, and nervous systems. The patient's condition should be reassessed at least every 5 minutes.

Multiple interventions may be necessary in patients with head and spinal injuries. The effectiveness of positive-pressure ventilations, spinal motion restriction, and treatments for shock can only be determined with both immediate and continuous observation after providing the intervention. If something is not working, try something else.

You have already established baseline vital signs as part of your assessment. Now is the time to compare those baseline vital signs with repeated vital signs. These changes will often tell you if treatments have been effective. For example, a dilated pupil may constrict with effective positive-pressure ventilations in an apneic head injury patient. Watch carefully for changes in the pulse, blood pressure, and respirations. If the ICP increases, the pulse may slow, blood pressure may rise, and respirations may become irregular. Document changes in the level of consciousness.

Interventions

Rapid deterioration of neurologic signs following a head injury is a sign of an expanding intracranial hematoma or rapidly progressing brain swelling. You must act quickly to evaluate and treat these patients. The trauma patient with signs and symptoms of head injury who also

ability of the chest wall to fully expand. Another sign of spinal injury is abdominal excursion—when the patient is unable to breathe without the assistance of the abdomen.

Additional signs of spinal cord trauma are an inability to maintain body temperature, priapism (a persistent erection lasting more than 4 hours), and a loss of bowel or bladder control.

Vital Signs

Obtaining a complete set of baseline vital signs is essential in patients with head and spine injuries. Significant head injuries may cause the pulse to slow and the blood pressure to rise. With neurogenic shock, the blood pressure may drop and the heart rate may increase to compensate. Respirations will become erratic with complications from both head and spine problems. Hypotension may be present with cervical or high thoracic spine injuries. The heart rate may become slow or fail to increase in response to hypotension.

Assess pupil size and reaction to light. The brain controls the diameter of pupils and how quickly they react. If an injury has occurred on one side of the brain, just one pupil will dilate. The pupils are windows to the brain and should be assessed as soon as possible to establish a baseline from which to monitor changes.

displays signs of shock has lost blood into another body cavity if hemorrhage is not seen externally.

As discussed earlier, the appearance of clear or pink watery CSF from the nose, the ear, or an open scalp wound indicates that the dura and the skull have both been penetrated. You should make no attempt to pack the wound, ear, or nose in this situation. Cover the scalp wound, if there is one, with sterile gauze to prevent further contamination, but do not bandage it tightly.

Hyperventilation is controversial because it can increase the severity of head injuries; it should therefore be avoided except in cases where signs of herniation have been identified. Even when employed, hyperventilation should be used with caution and only when capnography is available to assure an end-tidal carbon dioxide level of no less than 30 mm Hg.

Your local protocol for treatment of a suspected head injury should include the administration of high-flow oxygen and the application of a cervical collar as part of spinal immobilization. Reassessment should take place as the patient is transported to an appropriate facility. Monitor the patient's condition and vital signs and relay this information to the receiving facility, especially if there is a significant or noteworthy change.

Communication and Documentation

When providing care for patients with suspected head and spinal injuries, it is essential to maintain good communication between other providers and provide complete and detailed information to the destination facility. Key observations you relay help in the assessment and eventual treatment of your patient. Hospitals may better prepare for seriously injured patients with more advanced warning and a description of the most serious problems found during your assessment, and additional resources can be made available when you arrive. For example, a helicopter may be standing by for transport from a smaller hospital to a Level I trauma center. Larger hospitals may have trauma specialists or neurosurgeons available to meet you on arrival.

Your documentation should include the history you were able to obtain at the scene, your findings during your assessment, treatments you provided, and how the patient responded to them. How frequently you document repeat vital signs depends on the condition of your patient. More seriously injured patients should have documented vital signs every 5 minutes, whereas more stable patients should have documented vital signs every 15 minutes. Take time after your verbal report to hospital staff to sit and make a complete and accurate record of the situation. This will be your only accepted legal memory of the call.

Many events that cause spinal or head injuries may eventually result in some type of litigation. As with all responses, proper documentation of what you observed and the treatment provided will be beneficial as time passes. You may be requested to testify as a witness at incidents years later and proper and complete documentation recorded at the time of the incident will lay the framework for answering any questions that may be asked of you.

Special Populations

Infants may lose enough blood into the skull region to produce shock, but this is not the case with the older child or the adult patient. Provide oxygen, monitor the airway, treat for shock, and provide immediate transport.

A common response to head injuries, even among children with only very slight head injuries, is vomiting. This is sometimes the result of increased intracranial pressure. In managing such vomiting, you should pay particular attention to protecting the patient's airway.

Emergency Medical Care of Head Injuries

Treat the patient with a head injury according to three general principles that are designed to protect and maintain the critical functions of the central nervous system:

1. Establish an adequate airway. If necessary, begin and maintain ventilation and always provide high-flow supplemental oxygen.
2. Control bleeding, and provide adequate circulation to maintain cerebral perfusion. Begin cardiopulmonary resuscitation (CPR), if necessary. Be sure to follow standard precautions.
3. Assess the patient's baseline level of consciousness, and continuously monitor it.

As you continue to treat the patient, do not apply pressure to an open or compressed skull injury. In addition, you must assess and treat other injuries, dress and bandage open wounds as indicated in the treatment of soft-tissue injuries, splint fractures, anticipate and manage vomiting to prevent aspiration, be prepared for convulsions and changes in the patient's condition, and transport the patient promptly and with extreme care.

■ Managing the Airway

The most important step in the treatment of patients with head injury, regardless of the severity, is to establish an adequate airway. If the patient has an airway obstruction, you should perform the jaw-thrust maneuver to open the airway. Once the airway is open, maintain the head and cervical spine in a neutral, in-line position until the patient can be fully immobilized with a cervical collar and backboard **Figure 26-20**. Remove any foreign

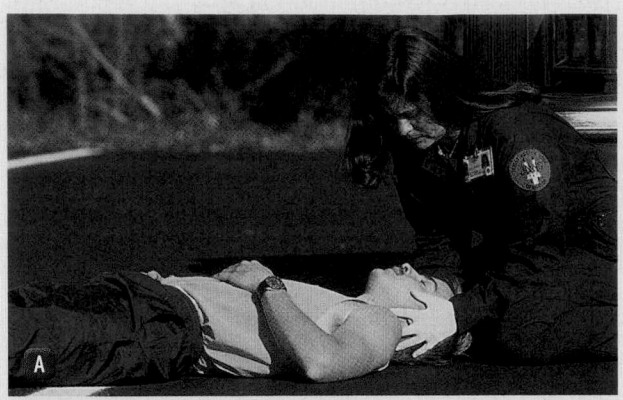

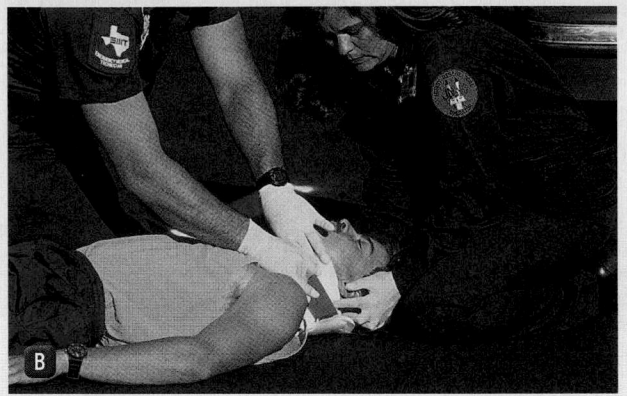

Figure 26-20 **A.** Maintain the head and cervical spine in a neutral in-line position. **B.** Apply a cervical collar as you finish the primary assessment.

bodies, secretions, or vomitus from the airway. Make sure a suctioning unit is available, because you will often need to clear blood, saliva, or vomitus from the airway.

Once you have cleared the airway, check ventilation. If the respiratory control center of the brain has been injured, the rate and/or depth of breathing may be ineffective. Ventilation may also be limited by chest injuries or, if the spinal cord is injured, by paralysis of some or all of the muscles of respiration. Give high-flow oxygen to any patient with suspected head injury, particularly anyone who is having trouble breathing. This reduces hypoxia and possible cerebral edema. An injured brain is even less tolerant of hypoxia than a healthy brain, and studies have shown that supplemental oxygen can reduce brain damage; to be effective, however, it must be started as soon as possible. Do not wait until the patient becomes cyanotic. Continue to assist ventilations and administer supplemental oxygen until the patient reaches the hospital.

■ Circulation

If the heart is not beating, providing airway maintenance, ventilation, and oxygen accomplishes nothing. You must also begin CPR if the patient is in cardiac arrest.

Active blood loss aggravates hypoxia by reducing the available number of oxygen-carrying red blood cells. Although scalp lacerations rarely cause shock except in infants and children, they often cause the loss of large volumes of blood, which must be controlled. Bleeding inside the skull may cause the ICP to rise to life-threatening levels, even though the actual volume of blood lost inside the skull is relatively small.

You can almost always control bleeding from a scalp laceration by applying direct pressure over the wound. Remember to follow standard precautions. Use a dry, sterile dressing, folding any torn skin flaps back down onto the skin bed before applying pressure **Figure 26-21A**. In some instances, you will have to apply firm compression for several minutes to control bleeding **Figure 26-21B**. If you suspect a skull fracture, do not apply excessive pressure to the open wound. Otherwise, you may

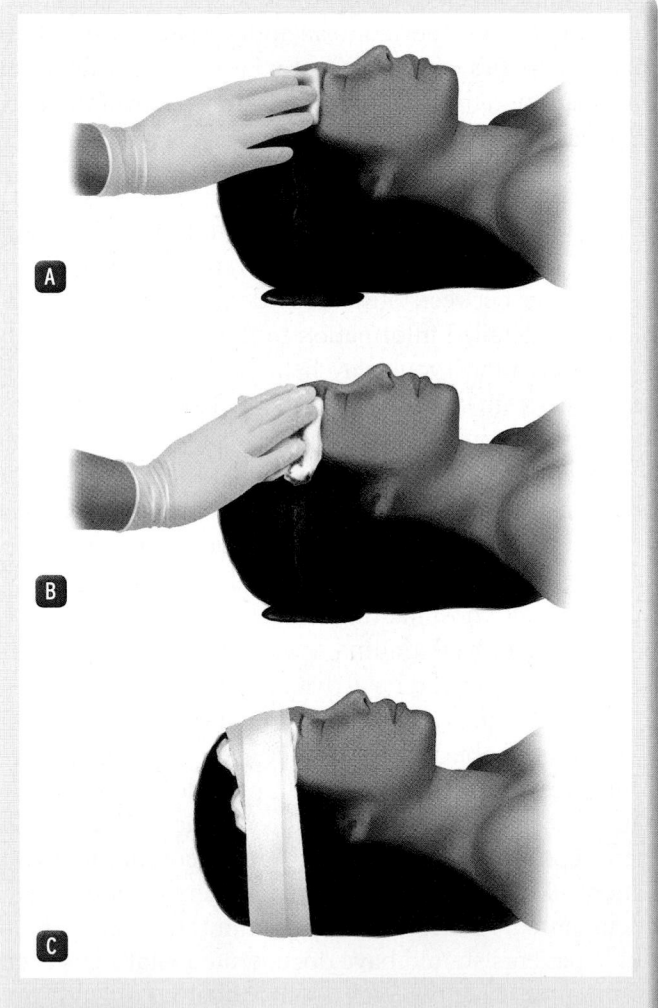

Figure 26-21 **A.** Use a dry sterile dressing to fold any flaps back down onto the skin bed before applying pressure. **B.** Apply firm compression for several minutes to control the bleeding. **C.** Secure the compression dressing in place with a soft, self-adhering roller bandage.

increase the ICP or push bone fragments into the brain. Bandages should not cover the mastoid process so that any apparent or developing sign of a basilar skull fracture still can be seen. If the bandage covers the patient's ears, remember that communication may become difficult because the patient's ability to hear will be decreased. To avoid limiting access to the patient's airway, do not cover the patient's mouth, nose, or jaw.

If the dressing becomes soaked, do not remove it. Instead, place a second dressing over the first. Continue applying manual pressure until the bleeding has been controlled, then secure the dressing in place with a soft, self-adhering roller bandage **Figure 26-21C**.

Shock that develops in a patient with a head injury is usually the result of hypovolemia caused by bleeding from other injuries. As with other trauma patients, shock in these cases indicates that the situation is critical. Such patients must be transported immediately to a trauma center. Maintain the airway while you protect the patient's cervical spine, ensure adequate ventilation, administer 100% oxygen, control obvious sites of bleeding with direct pressure, place the patient supine on a backboard, keep the patient warm, and provide immediate transport.

If the patient becomes nauseated or begins to vomit, elevate the right side of the backboard to prevent aspiration. Be sure to maintain the head in the in-line neutral position, with the cervical collar in place. You should also have a suctioning unit available.

Cushing's Triad

If the patient's head injuries are significant enough to cause a traumatic brain injury, the patient may begin to exhibit the signs of Cushing's triad: increased blood pressure (hypertension), decreased heart rate (bradycardia), and irregular respirations such as Cheyne-Stokes respirations, central neurogenic hyperventilation, and Biot respirations (irregular rate, pattern, and depth of breathing). Cushing's triad is also referred to as a herniation syndrome where the intracranial pressure is so great that it forces the brain stem and the midbrain through the foramen magnum, the hole at the base of the skull. If this process is allowed to continue, it is a fatal injury. If the patient exhibits these signs, it is commonly acceptable to hyperventilate your patient via positive-pressure ventilations. Follow local protocols and your medical direction in regards to hyperventilation in the presence of herniation.

Words of Wisdom

Hypoxemia is one of the key indicators along with hypotension of a poor outcome in patients with traumatic brain injuries.

Emergency Medical Care of Spinal Injuries

Emergency medical care of a patient with a possible spinal injury begins, as does all patient care, with your protection; therefore, you must remember to follow standard precautions. Next, you must maintain the patient's airway while manually keeping the spine in the proper position, assess respirations, and give supplemental oxygen.

Managing the Airway

Knowing that improper handling of a spinal injury can leave a patient permanently paralyzed must not prevent you from properly addressing an airway obstruction. Remember, all patients without an airway will die. If a patient with a spinal injury has an airway obstruction, you should perform the jaw-thrust maneuver to open the airway **Figure 26-22**. Do not use the head tilt–chin lift maneuver because it extends the neck and may further damage the cervical spine. If the patient is unconscious, you can lift or pull the tongue forward so that you do not have to move the neck. Once the airway is open, hold the

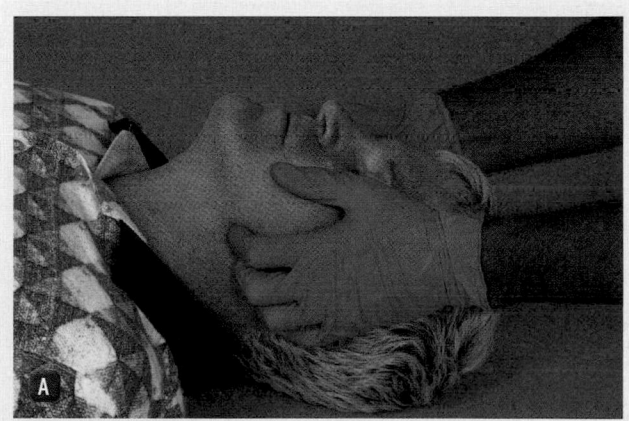

Figure 26-22 Jaw-thrust maneuver **A.** Stabilize the neck in a neutral, in-line position. **B.** Push the angle of the lower jaw upward.

head still in a neutral, in-line position until it can be fully immobilized.

After you open the airway, consider inserting an oropharyngeal airway. If your patient accepts an oropharyngeal airway, be sure to monitor the airway closely. Have a suctioning unit available because you will often need to clear away blood, saliva, or vomitus. Provide high-flow oxygen to any patient with suspected head or spine injury, especially those having trouble breathing.

Continuously monitor the patient's airway and be prepared for any changes in the patient's condition based on your treatment.

■ Stabilization of the Cervical Spine

Stabilizing the airway is your first priority. You must immobilize the head and trunk so that bone fragments do not cause further damage. Even small movements cause significant injury to the spinal cord. Follow the steps in Skill Drill 26-1 :

Skill Drill 26-1

Performing Manual In-Line Stabilization

Step 1 Kneel behind the patient and place your hands firmly around the base of the skull on either side.

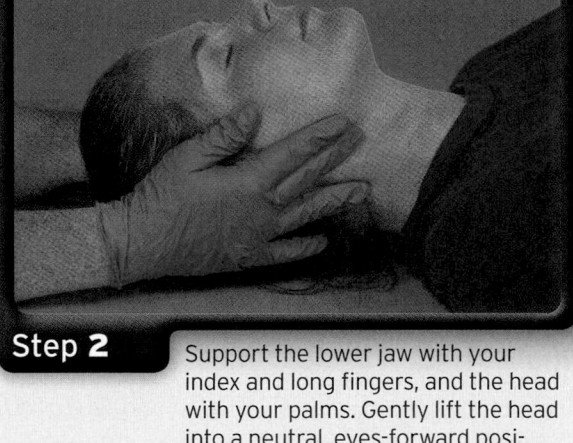

Step 2 Support the lower jaw with your index and long fingers, and the head with your palms. Gently lift the head into a neutral, eyes-forward position, aligned with the torso. Do not move the head or neck excessively, forcefully, or rapidly.

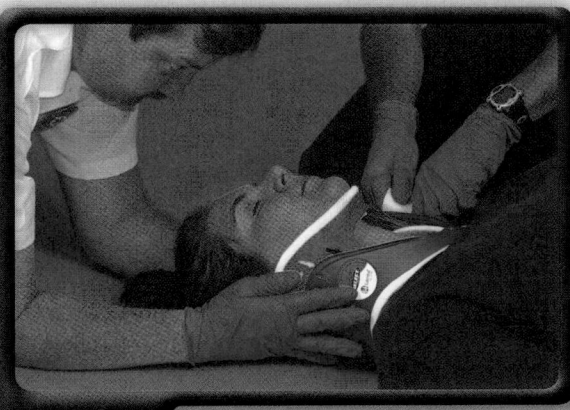

Step 3 Continue to support the head manually while your partner places a rigid cervical collar around the neck. Maintain manual support until you have completely secured the patient to a backboard.

1. Begin manual in-line stabilization by holding the head firmly with both hands. Whenever possible, kneel behind the patient, and place your hands around the base of the skull on either side **Step 1**.

2. Support the lower jaw with your index and long fingers, while you are supporting the head with your palms. Then gently lift the head until the patient's eyes are looking straight ahead and the head and torso are in line. This neutral **eyes-forward position** makes motion restriction easier. Align the nose with the navel. Never twist, flex, or extend the head or neck excessively **Step 2**.

3. Manually maintain this position as you continue to maintain the airway. Have your partner place a rigid cervical collar around the neck to provide more stability. Do not remove your hands from the patient's head until the patient has been completely secured to a backboard and the head has been immobilized. The patient must remain immobilized until he or she has been examined at the hospital **Step 3**.

Once the patient's head and neck have been manually immobilized, assess the pulse, motor functions, and sensation in all extremities. Then assess the cervical spine area and neck. Keep in mind that the cervical collar is used to provide increased stability to the neck. It is used in addition to, not instead of, manual cervical spine motion restriction. An improperly fitting collar will do more harm than good. If you do not have the proper size, place a rolled towel around the head, and tape it to the backboard as you immobilize the patient on the board. In any case, maintain manual support until the patient has been fully secured to a backboard.

Another method recommended for obtaining spinal immobilization is to place the backboard beside and approximately 3″ to 6″ higher than the patient. While maintaining cervical spine immobilization, the patient should be moved in an upward diagonal direction. This assists in maintaining axial alignment rather than a strictly horizontal movement.

You should never force the head into a neutral, in-line position. Do not move the head any farther if the patient reports any of the following symptoms:

- Muscle spasms in the neck
- Substantial increased pain
- Numbness, tingling, or weakness in the arms or legs
- Compromised airway or ventilations

In these situations, stabilize the patient in his or her current position.

Preparation for Transport

■ Supine Patients

A patient who is supine can be effectively immobilized by securing him or her to a long backboard. The ideal procedure for moving a patient from the ground to a backboard is the **four-person log roll**. This procedure is recommended any time you suspect a spinal injury. In other cases, you may choose instead to slide the patient onto a backboard or use a scoop stretcher. The patient's condition, the scene, and the available resources will dictate the method you choose.

You should first take the necessary precautions and then direct the team from a kneeling position at the patient's head so that you can maintain manual in-line immobilization. Your job is to ensure that the head, torso, and pelvis move as a unit, with your teammates controlling the movement of the body. If necessary, you may recruit bystanders to assist the team, but be sure to instruct them fully before moving the patient. To immobilize a patient to a backboard, follow the steps in **Skill Drill 26-2**:

1. Maintain in-line stabilization from a kneeling position at the patient's head. The EMT at the head will direct the log roll.

2. Assess pulse, motor, and sensory function in each extremity **Step 1**.

3. Apply an appropriately sized cervical collar **Step 2**.

4. The other team members should position the immobilization device (backboard) and place their hands on the far side of the patient to increase their leverage. Instruct them to use their body weight and their shoulder and back muscles to ensure a smooth, coordinated pull, concentrating their pull on the heavier portions of the patient's body **Step 3**.

5. On command from the EMT at the head, the rescuers roll the patient toward themselves. One rescuer quickly examines the back while the patient is rolled on the side, and then slides the backboard behind and under the patient. The team rolls the patient back onto the board, avoiding independent rotation of the head, shoulders, or pelvis **Step 4**.

6. Ensure the patient is centered on the board **Step 5**.

7. Secure the upper torso to the board once the patient is centered on the backboard **Step 6**. Consider padding voids between the patient and the

Skill Drill 26-2

Immobilizing a Patient to a Long Backboard

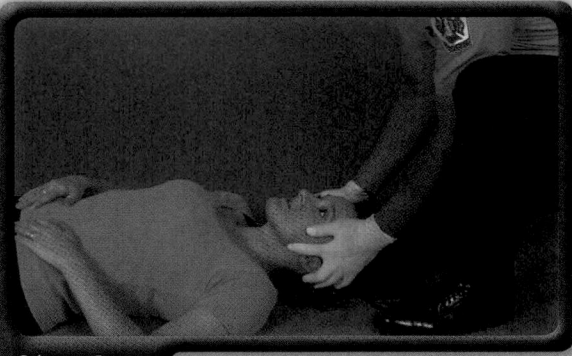

Step 1 Apply and maintain cervical motion restriction. Assess distal functions in all extremities.

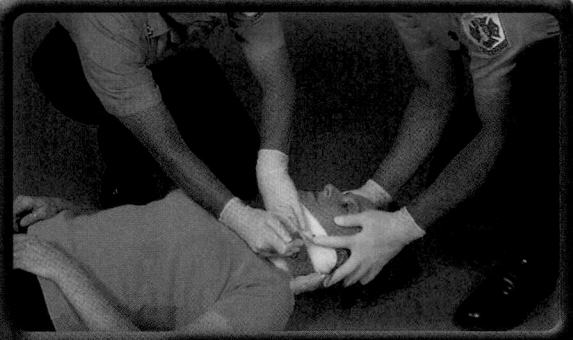

Step 2 Apply a cervical collar.

Step 3 Rescuers kneel on one side of the patient and place hands on the far side of the patient.

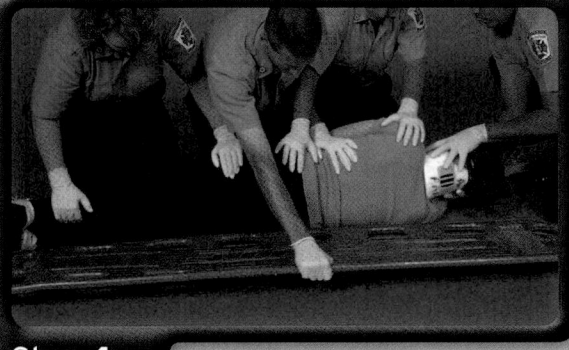

Step 4 On command, rescuers roll the patient toward themselves, quickly examine the back, slide the backboard under the patient, and roll the patient onto the board.

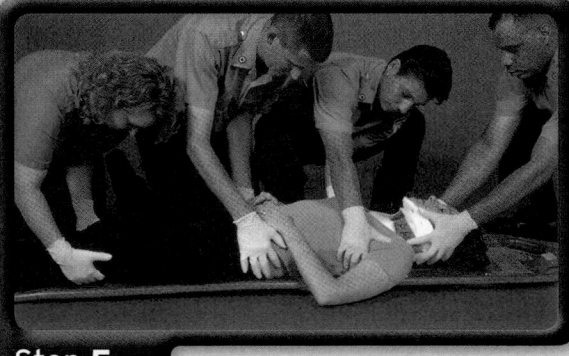

Step 5 Center the patient on the board.

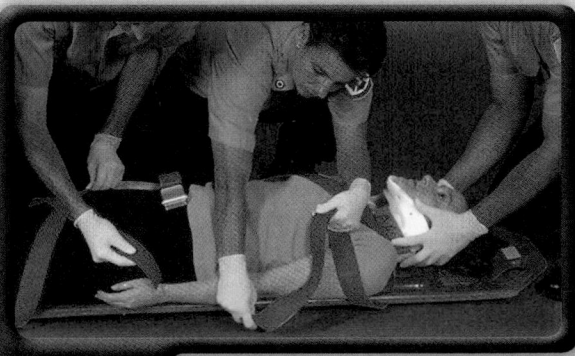

Step 6 Secure the upper torso first.

Skill Drill 26-2

Immobilizing a Patient to a Long Backboard, continued

Step 7 Secure the pelvis and upper legs.

Step 8 Begin to secure the patient's head using a commercial immobilization device or rolled towels.

Step 9 Place tape across the patient's forehead to secure the immobilization device.

Step 10 Check all straps and readjust as needed. Reassess distal functions in all extremities.

backboard to make transport more comfortable and protect the patient.

8. Secure the pelvis and upper legs, using padding as needed. For the pelvis, use straps over the iliac crests and/or groin loops (Step 7).

9. Begin to immobilize the head to the board by positioning a commercial immobilization device or towel rolls (Step 8).

10. Secure the head by taping the head immobilization device, or towels, across the forehead. To prevent airway problems and leave access to the airway, do not tape over the throat or chin (Step 9).

11. Check and readjust straps as needed to ensure that the entire body is snugly secured and will not slide during patient movement.

12. Reassess pulse, motor, and sensory function in each extremity, and continue to do so periodically (Step 10).

■ Sitting Patients

Some patients with a possible spinal injury will be in a sitting position, such as after an automobile crash. With these patients, you should use a short backboard or other short spinal extrication device to immobilize the cervical and thoracic spine. The short board is then secured to the long board.

The exceptions to this rule are situations in which you do not have time to first secure the patient to the short board, including the following situations:

- You or the patient is in danger.
- You need to gain immediate access to other patients.
- The patient's injuries justify urgent removal.

In these situations, your team should lower the patient directly onto a long backboard, using the rapid extrication technique as described in Chapter 35, *Lifting and Moving Patients*. Be sure that you provide manual motion restriction of the cervical spine as you move the patient. Rapid extrication is indicated only in cases of life-threatening or limb-threatening injury. In all other cases, follow the steps in **Skill Drill 26-3** to immobilize a sitting patient:

1. As with the supine patient, you must first stabilize the head and then maintain manual in-line motion restriction until the patient has been secured to the long backboard.

2. Assess pulse, motor, and sensory function in each extremity.

3. Apply the cervical collar (Step 1).
4. Insert a short spine immobilization device between the patient's upper back and the seat back (Step 2).
5. Open the board's side flaps (if present), and position them around the patient's torso and snug to the armpits (Step 3).
6. Once the board has been properly positioned, secure the upper torso straps and then the mid-torso straps (Step 4).
7. Position and fasten both groin (leg) straps. Check all torso straps to make sure they are secure. Make any adjustments necessary without excessive movement of the patient (Step 5).
8. Pad any space between the patient's head and the board as necessary.
9. Secure the forehead strap, and then fasten the lower head strap around the cervical collar (Step 6).
10. Place the long backboard next to the patient's buttocks, perpendicular to the trunk (Step 7).
11. Turn the patient parallel to the long board, and slowly lower him or her onto it.
12. Lift the patient (without rotating him or her), and slip the long board under the short board (Step 8).
13. Secure the short and long boards together.
14. Reassess the pulse, motor function, and sensation in all four extremities. Note your findings, and prepare for immediate transport (Step 9).

Immobilizing a Patient Found in a Sitting Position

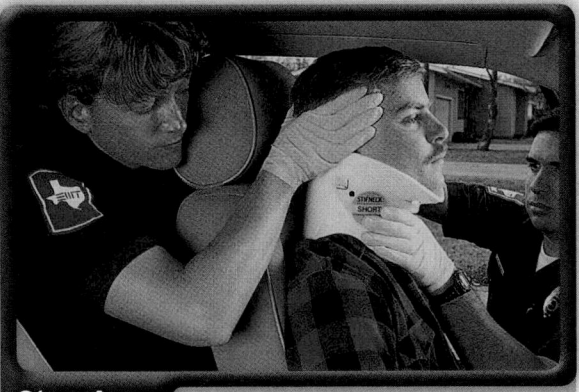

Step 1 Stabilize the head and neck in a neutral, in-line position. Assess pulse, motor, and sensory function in each extremity. Apply a cervical collar.

Step 2 Insert a short spine immobilization device between the patient's upper back and the seat.

Skill Drill 26-3

Immobilizing a Patient Found in a Sitting Position, continued

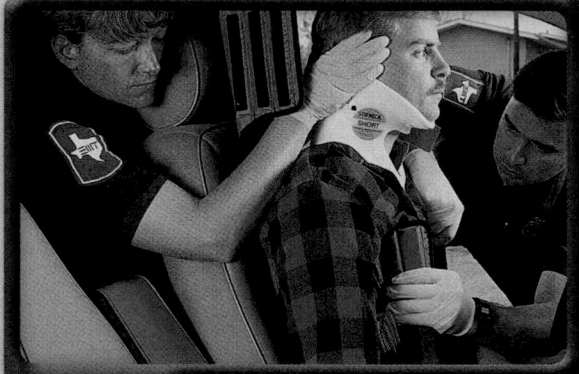

Step 3 Open the side flaps, and position them around the patient's torso, snug around the armpits.

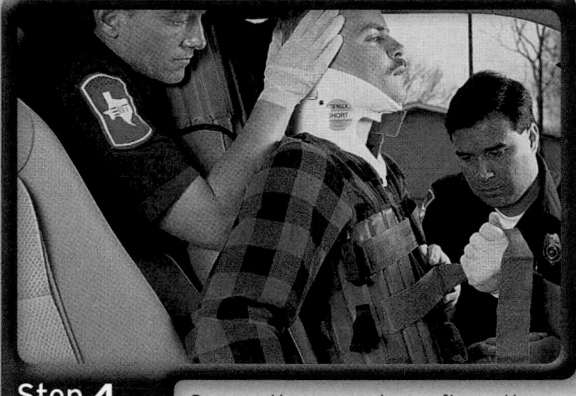

Step 4 Secure the upper torso flaps, then the midtorso flaps.

Step 5 Secure the groin (leg) straps. Check and adjust the torso straps.

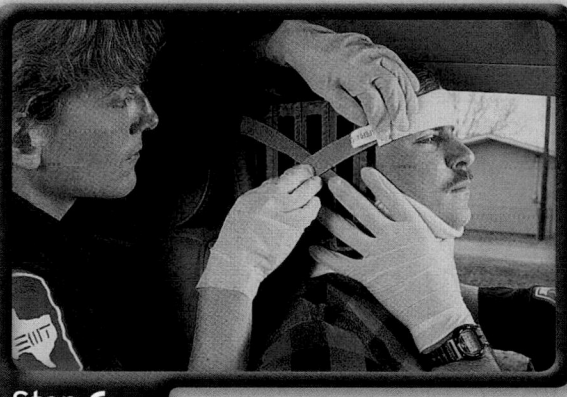

Step 6 Pad between the head and the device as needed. Secure the forehead strap and fasten the lower head strap around the cervical collar.

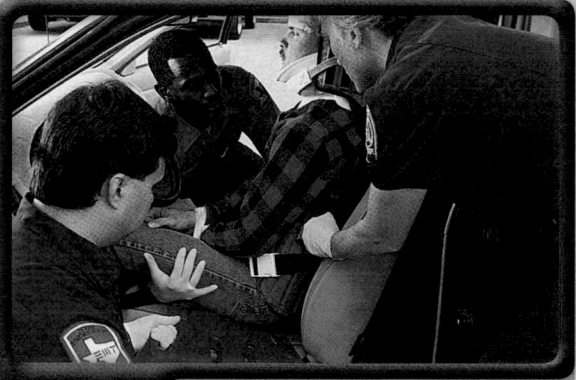

Step 7 Place a long backboard next to the patient's buttocks, perpendicular to the trunk.

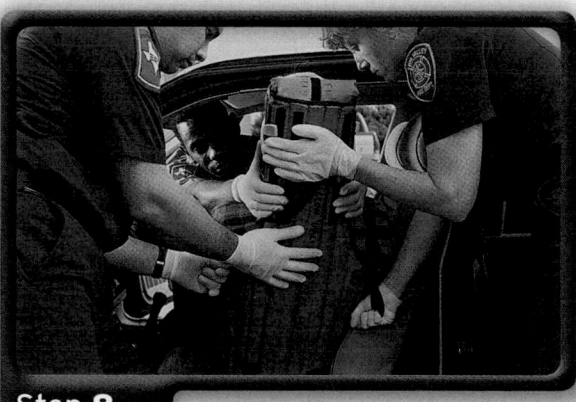

Step 8 Turn and lower the patient onto the long board. Lift the patient, and slip the long board under the short board.

Skill Drill 26-3

Immobilizing a Patient Found in a Sitting Position, continued

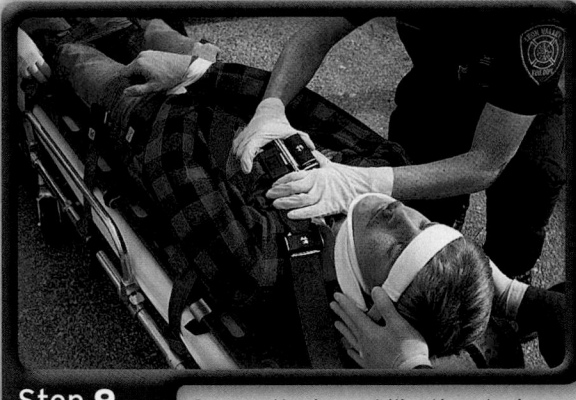

Step 9 Secure the immobilization devices to each other. Reassess pulse, motor, and sensory functions in each extremity.

You are the Provider: PART 4

Full spinal precautions are applied, assisted ventilations are continued, and the patient is loaded into the ambulance. An EMR from the engine company drives the ambulance so your partner can help you in the back with the patient. You begin transport to a local trauma center and reassess the patient's vital signs en route.

Recording Time: 11 Minutes	
Level of consciousness	Responsive only to deep painful stimuli
Respirations	6 breaths/min and irregular (baseline); ventilations are being assisted
Pulse	64 beats/min; regular and bounding
Skin	Pink, warm, and dry
Blood pressure	192/100 mm Hg
Sao₂	96% (on oxygen)

7. What further treatment is indicated for this patient?

8. What should you specifically monitor this patient for during transport?

■ Standing Patients

You may arrive at a scene in which you find a patient standing or wandering around after an accident or injury. If you suspect that there may be underlying head, neck, or spinal injuries, you should immobilize the patient to a long backboard before proceeding with assessment. This process will require three EMTs. To immobilize a standing patient, follow the steps in Skill Drill 26-4:

1. Establish manual, in-line motion restriction, apply a cervical collar, and instruct the patient to remain still.

2. Position the board upright directly behind the patient (Step 1).

3. Two EMTs stand on either side of the patient and the third is directly behind the patient, maintaining immobilization.

4. The two EMTs grasp the handholds at shoulder level or slightly above by reaching under the patient's arms while standing at either side (Step 2).

5. Prepare to lower the patient to the ground (Step 3).

6. Carefully lower the patient as a unit under the direction of the EMT at the head. The EMT at the head will have to make sure the patient's head stays against the board and carefully rotate his or her hands as the patient is being lowered to maintain in-line motion restriction (Step 4).

Skill Drill   26-4

Immobilizing a Patient Found in a Standing Position

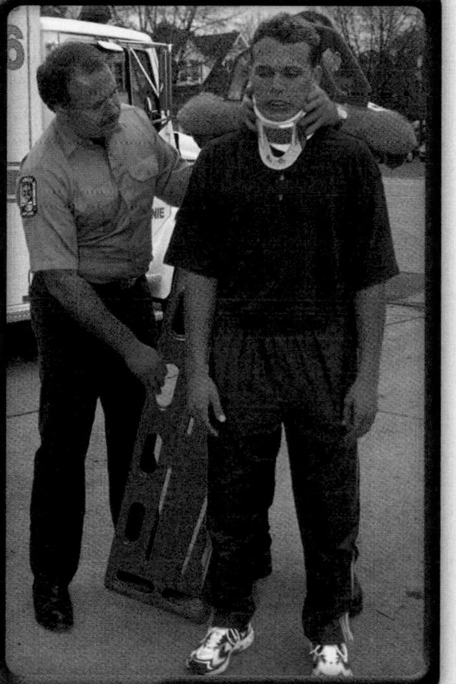

Step 1 While manually stabilizing the head and neck, apply a cervical collar. Position the board behind the patient.

Step 2 Position EMTs at sides and behind the patient. Side EMTs reach under the patient's arms and grasp handholds at or slightly above shoulder level.

Skill Drill 26-4

Immobilizing a Patient Found in a Standing Position, continued

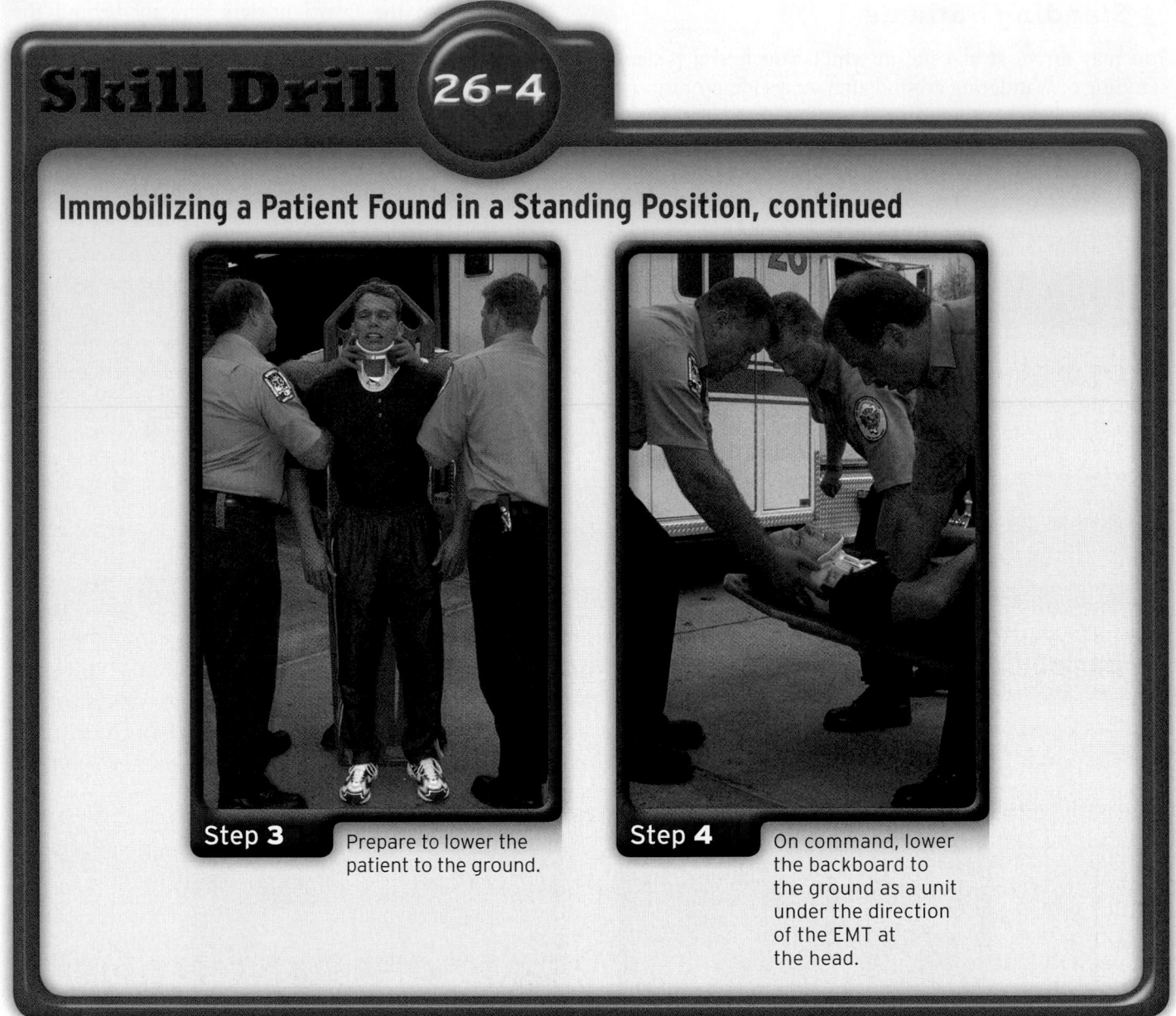

Step 3 Prepare to lower the patient to the ground.

Step 4 On command, lower the backboard to the ground as a unit under the direction of the EMT at the head.

■ Immobilization Devices

An injured spine is often very difficult to evaluate in a patient with a head injury. Sometimes, the patient has no neurologic loss. During assessment, pain in the spine may be missed because of shock or because the patient's attention is directed to more painful injuries. Evaluation is even more difficult if the patient is unconscious. Because any manipulation of the unstable cervical spine may cause permanent damage to the spinal cord, you must assume the presence of spinal injury in all patients who have sustained head injuries. Use manual in-line immobilization or a cervical collar and long backboard.

Cervical Collars

Rigid cervical immobilization devices, or cervical collars, provide preliminary, partial support. A cervical collar should be applied to every patient who has a possible spinal injury based on the MOI, history, or signs and symptoms. Keep in mind, however, that cervical collars do not fully immobilize the cervical spine. Therefore, you must maintain manual support until the patient has been completely secured to a spinal immobilization device, such as a long or short backboard.

To be effective, a rigid cervical collar must be the correct size for the patient. It should rest on the shoulder girdle and provide firm support under both sides of the

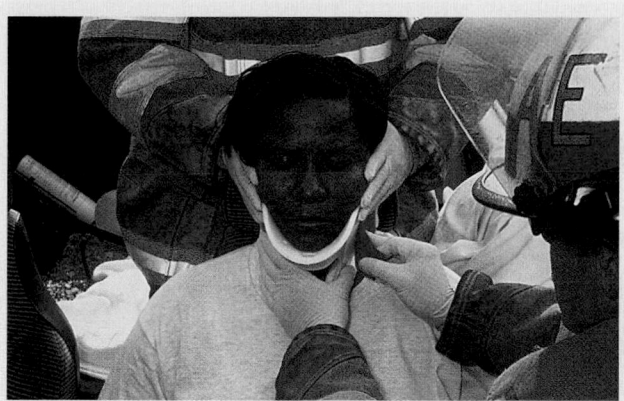

Figure 26-23 Proper fit is essential in applying a cervical collar. The collar should rest on the shoulder girdle and provide firm support under both sides of the mandible without obstructing the airway or any ventilation efforts.

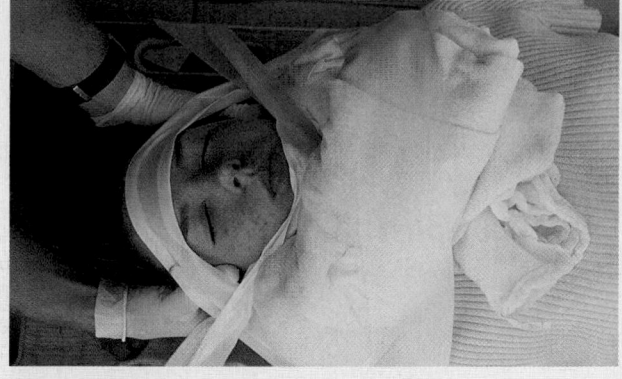

Figure 26-24 If you do not have an appropriately sized cervical collar, you may use a rolled towel. Tape it to the backboard around the patient's head, and provide continuous manual support.

mandible, without obstructing the airway or ventilation efforts in any way **Figure 26-23**. To apply a cervical collar, follow the steps in **Skill Drill 26-5**:

Skill Drill 26-5

1. One EMT provides continuous manual in-line support of the head while the other EMT prepares the collar (**Step 1**).

2. Measure the proper size collar according to the manufacturer's specifications. It is essential that the cervical collar fits properly. An improperly sized immobilization device may allow further injury to occur. If you do not have the correct size collar, use a rolled towel; tape it to the backboard around the patient's head, and provide continuous manual support **Figure 26-24** (**Step 2**).

3. Begin by placing the chin support snugly underneath the chin (**Step 3**).

4. Maintaining head stabilization and neutral neck alignment, wrap the collar around the neck and secure the collar to the far side of the chin support (**Step 4**).

Skill Drill 26-5

Application of a Cervical Collar

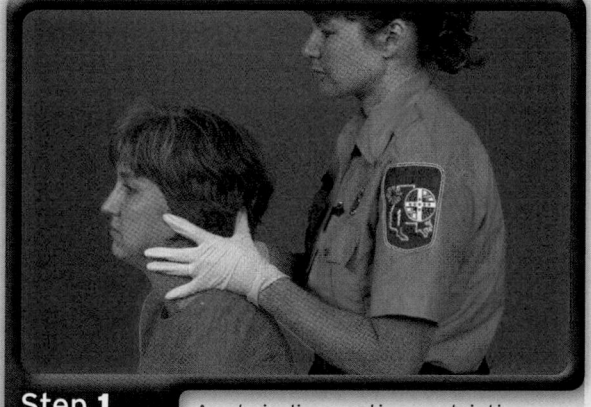

Step 1 Apply in-line motion restriction.

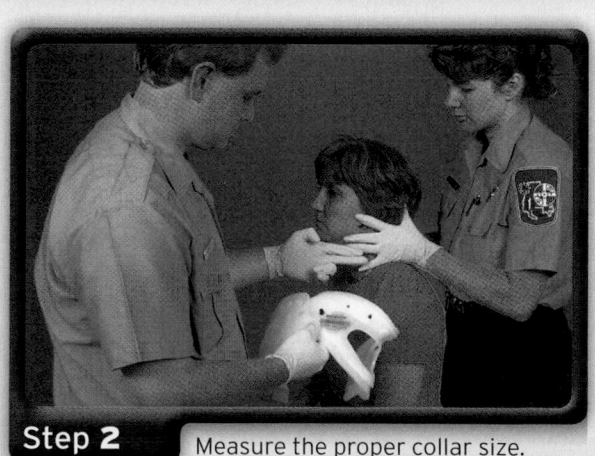

Step 2 Measure the proper collar size.

Skill Drill 26-5

Application of a Cervical Collar, continued

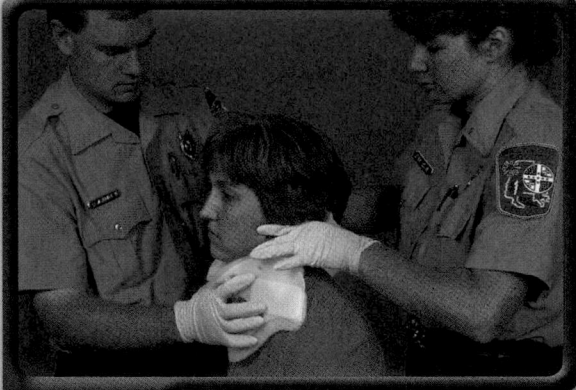

Step 3   Place the chin support first.

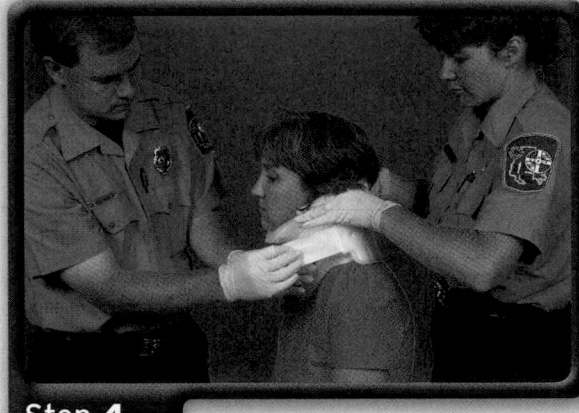

Step 4   Wrap the collar around the neck and secure the collar.

Step 5   Ensure proper fit and maintain neutral, in-line motion restriction until the patient is secured to a backboard.

5. Ensure that the collar fits properly and recheck that the patient is in a neutral, in-line position. Maintain in-line motion restriction until the patient has been completely secured to the backboard (Step 5).

Short Backboards

There are several types of short-board immobilization devices. The most common are the vest-type device Figure 26-25 and the rigid short board. These devices are designed to stabilize and immobilize the head, neck, and torso. They are used to immobilize noncritical patients who are found in a sitting position and have possible spinal injuries.

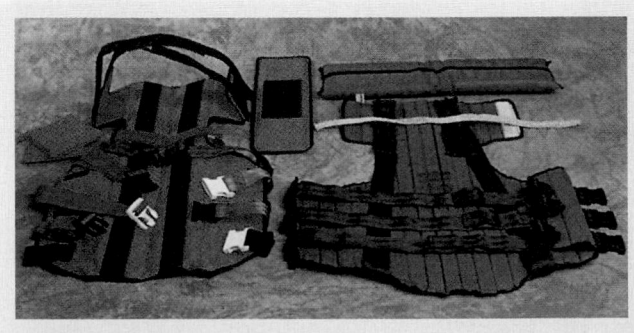

Figure 26-25 Common short-board immobilization devices are vest-type devices.

As described earlier in this chapter, the first step in securing a patient to a short board or device is to provide manual, in-line support of the cervical spine. Assess the pulse, motor function, and sensation in all extremities; next assess the cervical area; and then apply an appropriately sized cervical collar.

Position the device behind the patient, and secure it to the torso. Evaluate how well the torso and groin are secured, and make adjustments as necessary. Avoid excessive movement of the patient. Next, evaluate the position of the patient's head. Pad behind the head as needed to maintain neutral, in-line immobilization.

Now secure the patient's head to the device. Once the head is secured, you may release manual support of the head. Rotate or lift the patient to the long backboard. At this point, you must reassess the pulse, motor function, and sensation in all four extremities to determine whether the change in position has affected the patient's vital signs or neurologic status. Finally, you should immobilize the patient to the long backboard.

Long Backboards

There are several types of long-board immobilization devices that provide full body spinal immobilization Figure 26-26 . These devices also provide motion restriction and immobilization to the head, neck, torso, pelvis, and extremities. Long backboards are used to immobilize patients who are found in any position (standing, sitting, supine), sometimes in conjunction with short backboards.

Securing a patient to a long board was described in detail earlier in this chapter. Briefly, you should begin by providing manual, in-line support of the head. Assess pulse, motor function, and sensation in all extremities, and assess the cervical area. Then apply an appropriately sized cervical collar, and proceed as follows:

1. Position the device.
2. Log roll the patient onto the device. You may also move the patient onto the device using a suitable lift or slide or by using a scoop stretcher. As you maintain in-line support, your partner should kneel by the patient's head and direct the other two EMTs as you roll the patient. Your partner's job is to make sure that the head, torso, and pelvis move as a unit. As the patient's back comes into view, quickly assess its condition if you did not do so during initial assessment. One EMT should position the device under the patient. Then, at your partner's command, roll the patient onto the board.
3. If there are spaces between the patient's head and torso and the board, fill them with padding.
4. Secure the torso to the device by applying straps across the chest, pelvis, and legs. Adjust these straps as needed. Then secure the patient's head to the board.
5. Reassess pulse, motor function, and sensation in all extremities.
6. When the patient has been properly secured, you can safely lift the board or turn it on its side, if necessary.

Helmet Removal

As you plan your care of a patient wearing a helmet, ask yourself the following questions:

- Is the patient's airway clear?
- Is the patient breathing adequately?
- Can you maintain the airway and assist ventilations if the helmet remains in place?
- Can the face guard be easily removed to allow access to the airway without removing the helmet?
- How well does the helmet fit?
- Can the patient move within the helmet?
- Can the spine be immobilized in a neutral position with the helmet on?

A helmet that fits well prevents the patient's head from moving and should be left on, provided (1) there are no impending airway or breathing problems, (2) it does not interfere with assessment and treatment of airway or ventilation problems, and (3) you can properly immobilize the spine. You should also leave the helmet on if there is any chance that removing it will further injure the patient.

Remove a helmet if (1) it makes assessing or managing airway problems difficult and removal of a face guard

Figure 26-26 Long-board immobilization devices provide full body spinal immobilization, including stabilization of the head, neck, torso, pelvis, and extremities.

to improve airway access is not possible, (2) it prevents you from properly immobilizing the spine, or (3) it allows excessive head movement. Finally, always remove a helmet from a patient who is in cardiac arrest.

Sports helmets are typically open in the front and may or may not include an attached face mask. The mask can be removed without affecting helmet position or function by simply removing or cutting the straps that hold it to the helmet, thus allowing easy access to the airway Figure 26-27. A patient who is involved in full contact sports may be wearing bulky pads to protect various body regions, such as shoulder pads. Leaving a helmet in place whenever possible is preferred because it helps the body maintain an in-line neutral position. If the helmet must be removed, be sure to provide padding to compensate for the shoulder pads and maintain in-line positioning of the body. Motorcycle helmets often have a shield covering the face. This, too, can be unbuckled to allow access to the airway Figure 26-28. If a shield cannot be removed, then the helmet must be removed.

■ Preferred Method

Removing a helmet should always be at least a two-person job; however, the technique for helmet removal depends on the actual type of helmet worn by the patient. One EMT provides constant in-line support as the other EMT performs the various moves; you and your partner should not move at the same time. You should first consult with medical control, if possible, about your decision to remove a helmet. When you decide to do so, follow the steps in Skill Drill 26-6:

1. Begin by kneeling at the patient's head. Your partner should kneel on one side of the patient, at the shoulder area.
2. Open the face shield, if there is one, and assess the patient's airway and breathing. Remove eyeglasses if the patient is wearing them Step 1.
3. Stabilize the helmet by placing your hands on either side of it, with your fingers on the patient's lower jaw to prevent movement of the head. Once your hands are in position, your partner can loosen the face strap Step 2.
4. Once the strap has been loosened, your partner should place one hand on the patient's lower jaw at the angle of the jaw and the other behind the head at the occipital region. Once your partner's hands are in position, you may pull the sides of the helmet away from the patient's head Step 3.

You are the Provider: PART 5

Your partner continues to assist the patient's ventilations while you reassess his condition and vital signs. His Glasgow Coma Scale score remains unchanged from the previous readings. You call the trauma center to give your radio report and advise them of your estimated time of arrival.

Recording Time: 16 Minutes

Level of consciousness	Responsive only to deep painful stimuli
Respirations	6 breaths/min and irregular (baseline); ventilations are being assisted
Pulse	70 beats/min; regular and bounding
Skin	Pink, warm, and dry
Blood pressure	188/98 mm Hg
Sao$_2$	98% (on oxygen)

You deliver the patient to the emergency department and give your verbal report to the attending physician. After further assessment and treatment in the emergency department, the patient is taken to radiology for a CT scan, which reveals an epidural hematoma.

9. What is an epidural hematoma?

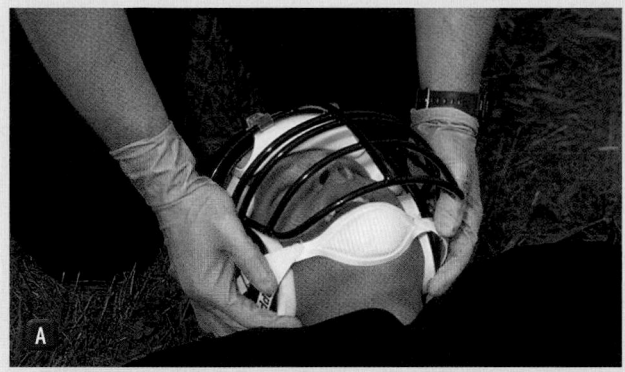

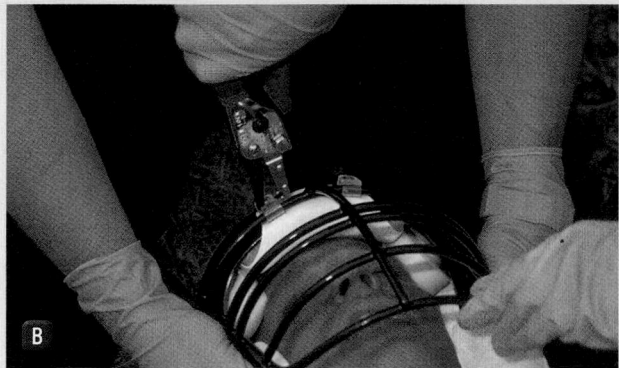

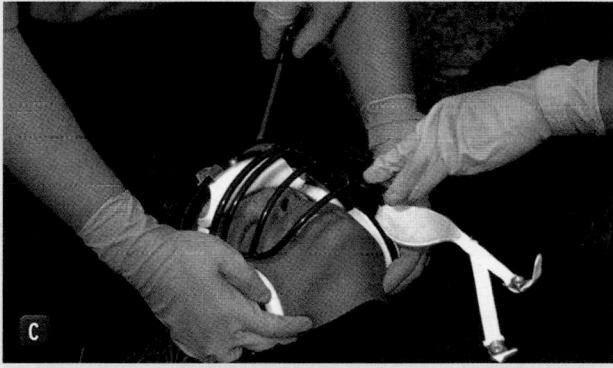

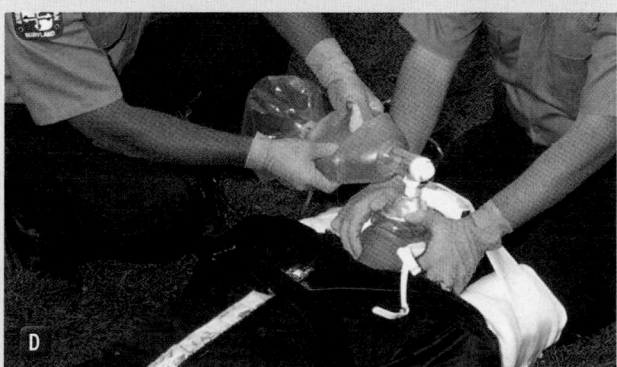

Figure 26-27 The mask on most sports helmets can be removed without affecting helmet position or function. **A.** Stabilize the patient's head and helmet. Then remove the face mask in one of two ways: **B.** Use a trainer's tool designed for cutting retaining clips, or, **C.** Unscrew the retaining clips for the face mask. **D.** Once the face mask has been removed, the helmet can be immobilized against the backboard and a bag-mask device can be used effectively.

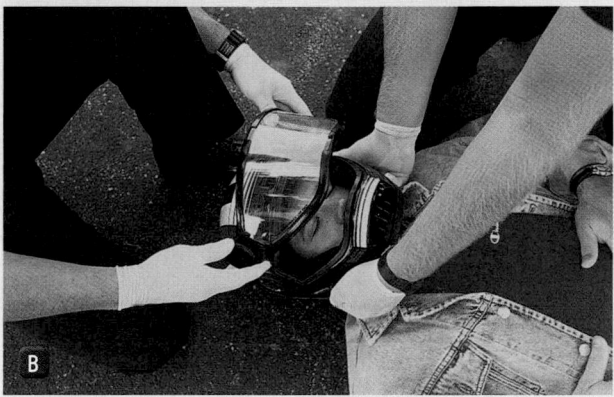

Figure 26-28 Motorcycle helmets often have a shield covering the face that can be removed. **A.** Stabilize the neck in a neutral, in-line position. **B.** Unbuckle or snap off the face shield to access the airway.

5. Gently slip the helmet halfway off the patient's head, stopping when the helmet reaches the halfway point (Step 4).
6. Your partner then slides his or her hand from the occiput to the back of the head. This will prevent the head from snapping back once the helmet has been completely removed (Step 5).
7. With your partner's hand in place, remove the helmet, and immobilize the cervical spine.
8. Apply the cervical collar, and then secure the patient to the backboard.
9. With large helmets or small patients, you may need to pad under the shoulders to prevent flexion of the neck. If shoulder pads or heavy clothing are in place, you may need to pad behind the patient's head to prevent extension of the neck (Step 6).

Remember, you do not need to remove a helmet if you can access the patient's airway, the head is snug inside the helmet, and the helmet can be secured to an immobilization device.

Skill Drill 26-6

Removing a Helmet

Step 1 Kneel at the patient's head with your partner at one side. Open the face shield to assess airway and breathing. Remove eyeglasses if present.

Step 2 Prevent head movement by placing your hands on either side of the helmet and fingers on the lower jaw. Have your partner loosen the strap.

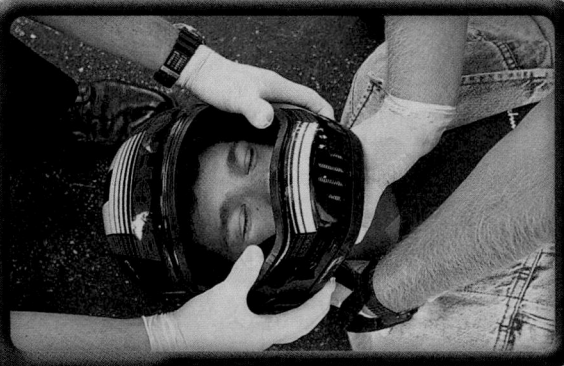

Step 3 Have your partner place one hand at the angle of the lower jaw and the other at the occiput.

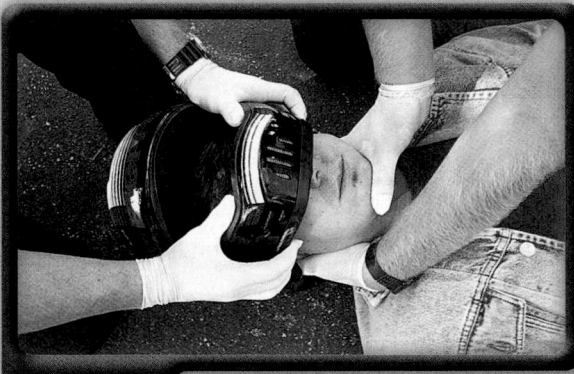

Step 4 Gently slip the helmet about halfway off, then stop.

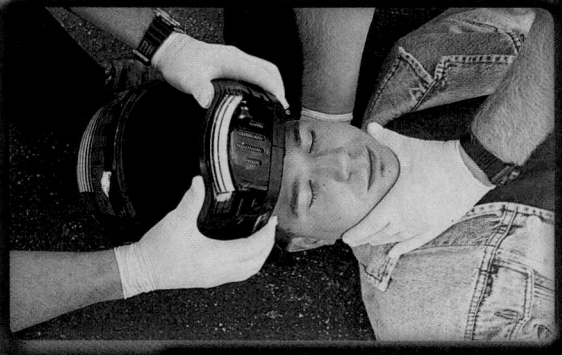

Step 5 Have your partner slide the hand from the occiput to the back of the head to prevent the head from snapping back.

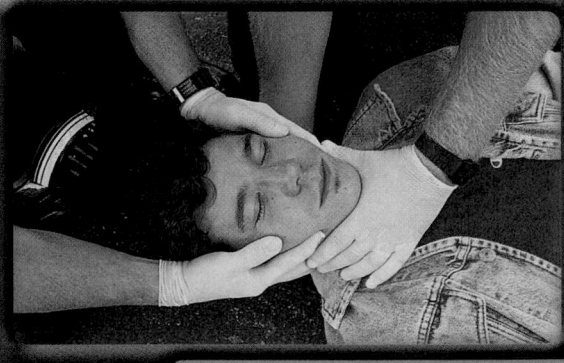

Step 6 Remove the helmet and stabilize the cervical spine. Apply a cervical collar and secure the patient to a long backboard. Pad as needed to prevent neck flexion or extension.

Alternate Method

An alternate method for removal of football helmets has also been used. The advantage of this method is that it allows the helmet to be removed with the application of less force, therefore reducing the likelihood of motion occurring at the neck. The disadvantage of this method is that it is slightly more time consuming. The first step involves removal of the chinstrap. This can be cut or carefully unsnapped. Be careful during removal of the chinstrap to avoid jarring the neck or head and causing excessive motion. Next, remove the face mask. The face mask is anchored to the helmet by plastic clips (loop straps) secured by screws. These can be removed with a screwdriver or cut with a knife. After the face mask has been removed, the jaw pads can be popped out of place. This can be accomplished with the use of a tongue depressor **Figure 26-29A** . You can then place your fingers inside the helmet, allowing greater control of the helmet during removal as the helmet is gently rocked back off the top of the head. The person at the side of the patient controls the head by holding the jaw with one hand and the occiput with the other **Figure 26-29B** . Padding is inserted behind the occiput to prevent neck extension. If the shoulder pads are in place, appropriate padding must be placed behind the head to prevent hyperextension. As with the previously described method, the person at the side of the patient's chest is responsible for making sure that the head and neck do not move during removal of the helmet.

Remember that small children may require additional padding to maintain the in-line neutral position. Children are not small adults. They have smaller airways and proportionally larger heads, so padding is important to maintain the airway. Pad under the shoulders to the toes, as needed, to avoid excessive neck flexion **Figure 26-30** . In addition, place blanket rolls between the child and the sides of an adult-sized board to prevent the child from slipping to one side or the other **Figure 26-31** . Appropriately sized backboards are available for children.

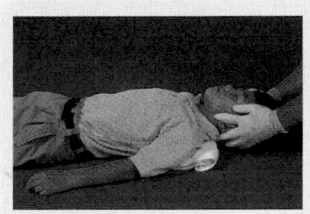

Figure 26-30 Children have proportionately larger heads than adults, so you may need to place padding under the shoulders to avoid excessive flexion of the head.

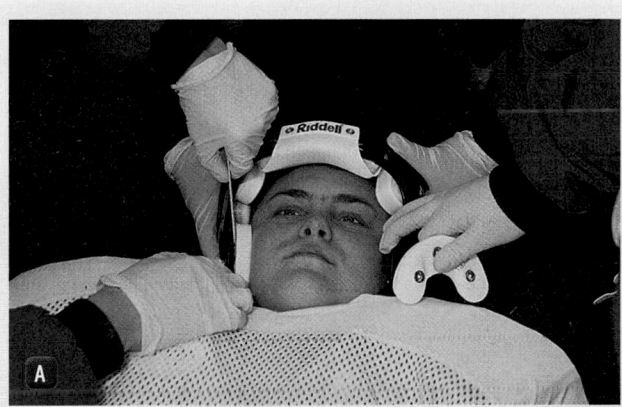

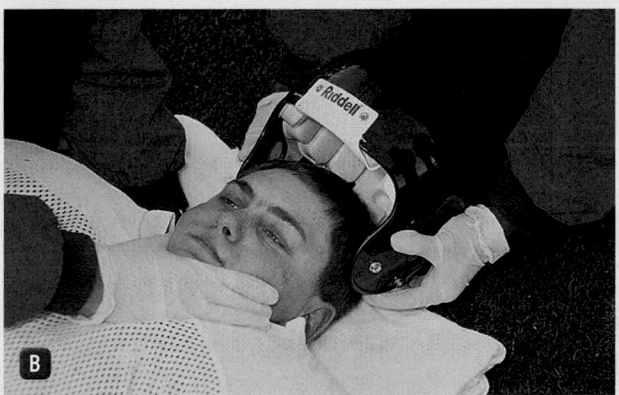

Figure 26-29 **A.** The jaw pads can be removed from the inside of a football helmet with the aid of a tongue depressor. **B.** Place the fingers inside the helmet and gently rock it out of place. The person at the side controls the lower jaw with one hand and the occiput with the other. Insert padding behind the occiput to prevent neck extension.

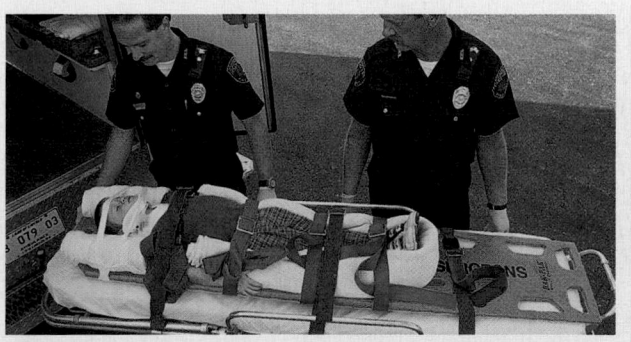

Figure 26-31 Place blanket rolls between the child and the sides of an adult-sized board to prevent the child from slipping to one side or the other.

Special Populations

You are likely to find infants and children who have been in automobile crashes and are still in their car seats. Your best course of action is to leave the child in the car seat and immobilize the child using an appropriately sized pediatric immobilization device, or to consider using a rigid short board device. These devices allow you to completely assess and conduct the ongoing assessment of the injured child while transporting to the hospital. Whenever you apply a cervical collar, make sure it is properly sized. If a properly fitting collar is not available, a rolled towel may be used as a substitution **Figure 26-32**. If the child is not in a car seat or was removed before your arrival, use an appropriately sized immobilization device. If the cervical immobilization device does not fit, use a rolled towel, and tape it to the board and manually support the head.

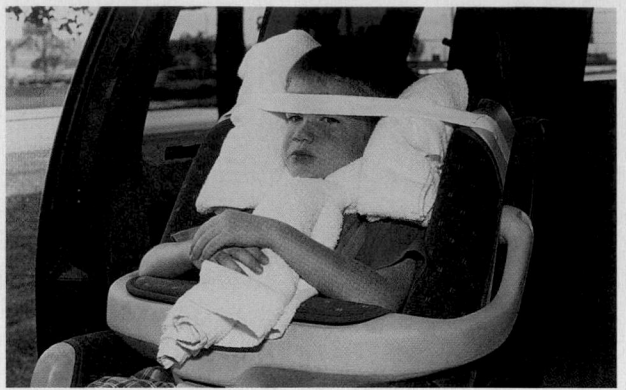

Figure 26-32 If you do not have an appropriately sized cervical collar for a child, you may use a rolled towel and tape it to the car seat. Pad the sides of the car seat, if needed, to prevent lateral movement.

You are the Provider: SUMMARY

1. Why is the structure of the cranium considered to be a "mixed blessing?"

The skull is divided into two large structures: the cranium and the face. The cranium, or cranial vault, is composed of a number of thick bones that fuse together to form a shell above the eyes and ears that holds and protects the brain. About 75% of the cranium is occupied by the cerebrum, the largest part of the brain (also called the gray matter); 10% is occupied by cerebrospinal fluid, and the remaining 10% is occupied by blood.

The cranium does an excellent job of protecting the brain from direct trauma following minor injuries, such as striking your head during a minor fall. The hard shell of the cranium provides protection for the cerebrum and other intracranial contents from the bumps and blows that occur as a part of everyday life. However, the very fact that the cranium is a hard shell makes it a mixed blessing following a significant head injury. The cranium's rigid, unyielding structure allows little, if any, expansion of the brain. Furthermore, its hard and somewhat irregular internal surface can injure the brain and its blood vessels following significant head trauma.

2. What is the difference between a primary and a secondary brain injury?

A traumatic brain injury (TBI)—the most serious type of head injury—is classified into two categories: primary (direct) injury and secondary (indirect) injury.

Primary brain injury is injury to the brain and its associated structures that occurs immediately on impact to the head. It can occur following a penetrating injury, such as a stabbing, gunshot wound, or if a bone fragment is driven into the brain following a skull fracture; however, it more commonly occurs following blunt force trauma.

Secondary brain injury refers to the "after effects" of the primary head injury; it includes abnormal processes such as cerebral edema, increased intracranial pressure (ICP), cerebral ischemia and hypoxia, and infection. Secondary brain injury can occur anywhere from a few minutes to several days following the initial head injury.

3. What are your most immediate treatment priorities for this patient?

As with any patient, your initial treatment must focus on what will kill him or her *first*. Your patient's breathing rate and quality—slow and irregular—is not adequate and requires immediate treatment. You should instruct your partner to stabilize the patient's head with his knees while he assists the patient's ventilations with a bag-mask device and high-flow oxygen. If a law enforcement officer is available, ask him or her to stabilize the patient's head while your partner assists his ventilations.

Consider inserting a simple airway adjunct. If the patient does not have a gag reflex, insert an oropharyngeal (oral) airway. Use of a nasopharyngeal (nasal) airway should be avoided in patients with a head injury, especially if you observe fluid drainage from the nose. This sign could indicate a midface fracture, in which case a nasal airway could inadvertently penetrate the brain. If a simple airway adjunct is not an option, keep the airway open with the jaw-thrust maneuver and maintain the patient's head in a neutral position.

While your partner is managing the patient's airway, you should control the bleeding from the laceration to his head. Use just enough pressure to control the bleeding; if a skull fracture is present, too much pressure could drive fractured bone fragments into his brain. Quickly scan the rest of the patient's body for any other external bleeding and control it as well. Remember the objective of the primary assessment: find it, fix it, and move on.

4. Where should you focus your secondary assessment of this patient?

Unconscious patients and those who experienced a significant mechanism of injury should be assessed using a full-body scan to look for life-threatening injuries that were not grossly apparent in the primary assessment. If immediate threats to the patient's life are found, they should be treated immediately.

Emphasis should be placed on the patient's head and face because this area appears to be where the patient experienced the most injury. Assess the integrity of the skull by *gently* palpating it and noting any areas of deformity, crepitus, or instability. Although you have already bandaged the laceration on the patient's head, you should reassess the bandage to ensure that the bleeding is controlled.

Look in the ears for fluid drainage and look behind the ears for Battle's sign. Fluid or blood drainage from the ears may contain cerebrospinal fluid and indicates a basilar skull fracture.

Assess the size, equality, and reactivity of the patient's pupils. The nerves that control the dilation and constriction of the pupils are very sensitive to intracranial pressure (ICP). Normally, both pupils should briskly constrict when a light is shone into either of the eyes. Pupils that are sluggish (slow) to react could indicate early increased ICP and/or cerebral hypoxia. Unequal or bilaterally fixed and dilated (blown) pupils are later, more ominous signs of increased ICP and indicate pressure on one or both oculomotor nerves.

Palpate the facial bones for stability and note any deformities or crepitus.

Patients with a significant head injury should also be assumed to have a cervical spine injury until proven otherwise. Blunt force trauma that is significant enough to render the patient unconscious could easily fracture a cervical vertebra. Palpate the cervical spine for obvious deformities and then apply a cervical collar. Manually stabilize the patient's head until *full* spinal precautions have been applied (cervical collar, backboard, straps, head blocks).

Your secondary assessment of *any* critically injured patient should not take an exorbitant amount of time. Address only life-threatening injuries and remain focused on preparing the patient for immediate transport.

5. What is this patient's Glasgow Coma Scale score?

The Glasgow Coma Scale (GCS) is a widely used method for assessing the level of consciousness in patients with neurologic trauma, and consists of three independent measurements: eye opening, verbal response, and motor response. The GCS score is used to classify the severity of a traumatic brain injury and has been shown to be a reliable predictor of the brain-injured patient's outcome.

The highest GCS score a patient can be assigned is 15; the lowest score is 3. A numeric value is assigned for each of the three parameters assessed. It is important to note that a single score using the GCS cannot reliably capture the patient's clinical progression. You should obtain a baseline GCS score and frequently (every 3 to 5 minutes) reassess it in the head-injured patient. Document all GCS scores and the times they were obtained and report this information to the emergency department physician.

The severity of a patient's brain injury can be classified based on his or her GCS score; a score of 13 to 15 indicates a mild brain injury, 9 to 12 indicates a moderate brain injury, 8 or less indicates a severe brain injury.

Your patient opens his eyes in response to painful stimuli; therefore, you should assign a score of *2* for eye opening. He is making unrecognizable sounds; therefore, you should assign a score of *2* for verbal response. You noted that his arms were flexed and drawn in toward his body (decorticate posturing); therefore, you should assign a score of 3 for motor response. Based on these findings, the patient's present GCS score is 7, which indicates a severe traumatic brain injury.

6. What is the likely explanation for the patient's vital signs?

Your patient's current vital signs represent a classic trio of findings in patients with a traumatic brain injury and increased ICP. Hypertension, bradycardia, and irregular respirations–called Cushing's triad–indicate significant cerebral edema and increased ICP.

A predictable response of the injured brain is swelling; this causes cerebral edema and a decrease in cerebral perfusion pressure (CPP) because there is little room in the cranium for the brain to swell. In an attempt to maintain cerebral blood flow, and thus CPP, arterial blood pressure increases, which presents with hypertension, and the cerebral blood vessels dilate. Bradycardia occurs as a reflex response to the increase in the patient's blood pressure.

Pressure on the respiratory centers of the brain stem causes a variety of abnormal respiratory patterns. Irregular respirations–slow or fast–are the third component of Cushing's triad. Cheyne-Stokes respirations are characterized by a pattern of rapid breathing (tachypnea), followed by slow breathing (bradypnea), and periods of apnea. Central neurogenic hyperventilation is characterized by deep, rapid breathing; this pattern is similar to Kussmaul respirations, but without an acetone breath odor. Biot respirations, also called ataxic respirations, are characterized by an irregular rate, pattern, and depth of breathing with intermittent periods of apnea.

7. What further treatment is indicated for this patient?

Continue to ensure adequate oxygenation and ventilation and take steps to decrease the ICP and maximize cerebral blood flow.

Continue to assist the patient's ventilations; however, do *not* hyperventilate him. According to the Brain Trauma Foundation, you should ventilate the brain-injured adult at a rate of 10 breaths/min. Hyperventilation with high-flow oxygen constricts the blood vessels in the brain, and while this *may* cause a slight decrease in the ICP, it also pushes oxygenated blood away from the brain, potentially causing a decrease in cerebral perfusion pressure and further brain injury.

Routine hyperventilation is *not indicated* for the brain-injured patient unless signs of brain herniation are present. Herniation occurs when severely increased ICP attempts to force the brain from the cranial vault, usually through the foramen magnum. Suspect brain herniation if the patient is unresponsive, has bilaterally fixed and dilated (blown) *or* unequal pupils, and decerebrate (extensor) posturing *or* no motor response to painful stimuli. If these signs are observed, medical control or your local protocols may call for a ventilation rate of 20 breaths/min.

Consider elevating the head of the backboard 30° in an attempt to reduce the ICP. Elevating the backboard greater than 30°, however, may cause blood (and oxygen) to leave the brain by gravity, thus causing a decrease in cerebral perfusion, and should be avoided. Follow your local protocols or contact medical control as needed regarding further treatment for the patient with a traumatic brain injury.

Early notification of the receiving facility is critical in the treatment of a patient with a traumatic brain injury. Report your findings, any treatment that you provided, the patient's response to your treatment, and your estimated time of arrival. This will allow the receiving facility adequate time to prepare to receive the patient.

8. What should you specifically monitor this patient for during transport?

The importance of reassessing the brain-injured patient cannot be overemphasized. It is important to make frequent observations of the patient's clinical status and report these to the emergency department physician.

Patients with increased ICP commonly vomit and experience seizures. You must be prepared to turn the backboard to the side and suction the patient's airway if vomiting occurs. If the patient experiences a seizure, continue to assist his ventilations and do not attempt to restrain him.

Carefully and frequently monitor the patient's vital signs, specifically, the blood pressure and oxygen saturation. A *single* episode of hypotension (a systolic BP of less than 90 mm Hg) in the adult with a traumatic brain injury is associated with a significant increase in mortality because it causes a decrease in cerebral perfusion. If the patient experiences hypotension, elevate the foot end of the backboard 6" to 12". A *single* drop in the patient's oxygen saturation to below 90% is also associated with a significant increase in mortality; ensure the continual delivery of high-flow oxygen and adequate ventilation!

Frequently reassess the patient's GCS score and pupils, and observe for signs of brain herniation. If directed by local protocol or online medical control, ventilate the patient at a rate of 20 breaths/min if signs of brain herniation are observed.

If possible, you should request an ALS ambulance at the scene, if it does not delay your scene time, or consider an ALS intercept during transport, provided that it does not cause a delay in transport. AEMTs and paramedics can provide IV therapy to treat hypotension if it occurs, and paramedics can perform advanced airway management (eg, intubation) and can give medications to stop seizures. The most important intervention for the patient with a traumatic brain injury, however, is to rapidly transport him or her to a definitive care facility as soon as possible.

9. What is an epidural hematoma?

As previously discussed, the closed box of the cranium has no extra room for the accumulation of blood, so intracranial hemorrhage also increases the ICP. If a blood vessel inside the brain or in the meninges that cover the brain is lacerated or ruptured, bleeding can occur between the skull and the dura mater (epidural hematoma), beneath the dura mater but outside the brain (subdural hematoma), or within the brain tissue itself (intracerebral hemorrhage).

Although the diagnosis of an intracranial hemorrhage can only be made at the hospital with the use of a computerized tomographic (CT) scan, the patient's clinical presentation and progression may provide clues as to the type of injury he or she has experienced, which you can relay to the receiving facility.

Your patient was diagnosed with an epidural hematoma, which is an accumulation of blood between the skull and dura mater. An epidural hematoma is almost always the result of a blow to the head that produces a fracture of the thin temporal bone (recall that the patient had a compressed area over the temporal region of his skull). The middle meningeal artery courses along the groove in the temporal bone, so it is prone to laceration or rupture when the temporal bone is fractured. When this occurs, brisk arterial bleeding will result in rapidly progressing symptoms. The patient with an epidural hematoma typically loses consciousness immediately; this may or may not be followed by a brief return of consciousness (lucid interval), after which the patient's level of consciousness rapidly declines and he or she manifests with signs and symptoms of increasing ICP. *This is consistent with how your patient presented.*

You are the Provider: SUMMARY, continued

EMS Patient Care Report (PCR)

Date: 1-13-09	Incident No.: 012609	Nature of Call: Head injury		Location: 147 Scottie Dr.	
Dispatched: 0220	En Route: 0221	At Scene: 0225	Transport: 0236	At Hospital: 0245	In Service: 0256

Patient Information

Age: 22 Sex: M Weight (in kg [lb]): 70 kg (155 lb)	Allergies: Unknown Medications: Unknown Past Medical History: Unknown Chief Complaint: Head injury; decreased LOC

Vital Signs

Time: 0230	BP: 190/104	Pulse: 60	Respirations: 6	Sao₂: 94%
Time: 0236	BP: 192/100	Pulse: 64	Respirations: 6	Sao₂: 96%
Time: 0241	BP: 188/98	Pulse: 70	Respirations: 6	Sao₂: 98%

EMS Treatment
(circle all that apply)

Oxygen @ 15 L/min via (circle one): NC NRM (Bag-Mask Device)	(Assisted Ventilation)	Airway Adjunct		CPR
Defibrillation	(Bleeding Control)	(Bandaging)	Splinting	Other: (Full spinal precautions)

Narrative

Medic 11 dispatched to a nightclub for a male patient who was assaulted. En route, law enforcement personnel, who were present at the scene, advised that the patient was struck in the side of the head with a baseball bat and was unconscious. On arrival at the scene, found the patient, a 22-year-old male, lying supine on the ground; a large pool of blood was under his head. He was motionless and his breathing appeared slow and irregular. Manual c-spine stabilization was initiated immediately and the patient's airway was opened with the jaw-thrust maneuver. Primary assessment revealed that he was responsive only to deep painful stimuli. His airway was clear of secretions or foreign bodies, his breathing was slow and irregular, and he was bleeding from a large laceration to the temporal region of his skull. The patient's gag reflex was intact, so an oral airway was not inserted. A nasal airway was avoided because of the potential for occult skull fracture. Patient's ventilations were assisted with a bag-mask device and high-flow oxygen and bleeding from the scalp laceration was controlled with a light pressure dressing. Secondary assessment revealed no gross trauma to the rest of the body. Assessment of the head revealed a depression over the area of the laceration. Pupils were bilaterally dilated and sluggish to react to light. There was no evidence of Battle's sign or fluid drainage from the ears or nose. Facial bones were stable. Initial GCS score of 7 was assigned (eye opening, 2; verbal response, 2; motor response, 3). Engine company 4 arrived to provide assistance; as they retrieved spinal immobilization equipment, initial vital signs were obtained. Applied full spinal precautions, loaded patient into ambulance, and began transport. EMR from assisting engine company drove the ambulance because of patient care demands that required two EMTs. En route to the hospital, reassessed vital signs and elevated the head end of the backboard 30°. Continued to assist ventilations at a rate of 10 breaths/min and noted that ventilations consistently produced adequate chest rise. Oxygen saturation remained greater than 95%. Notified trauma center of the patient's condition and our estimated time of arrival. Reassessment revealed no change in patient's condition; he remained responsive only to deep painful stimuli and his GCS score remained at 7. Pupils remained bilaterally dilated and sluggish to react to light. Delivered patient to emergency department and gave verbal report to attending physician. Medic 11 cleared the hospital and returned to service at 0256. *End of report*

Assessment and Emergency Care of Head and Spine Injuries

Scene Size-up

Scene Safety	Ensure scene safety and safe access to the patient. Standard precautions should include a minimum of gloves and eye protection if there is vomiting. Consider a gown and shoe covers if other bodily fluids are involved. Determine the number of patients. Assess the need for additional resources.
Mechanism of Injury (MOI)/ Nature of Illness (NOI)	Determine the MOI. Interview the patient, family, and/or bystanders to determine the exact nature of the traumatic forces applied. Maintain a high index of suspicion for associated spinal injuries, especially with rapid acceleration-deceleration MOIs. Blunt forces applied directly to the head can produce skull fractures with potential depression. With motor vehicle crashes, note the position of the patient in the vehicle in relation to the forces applied. Determine if the patient might have been ejected from the vehicle.

▼   ▼

Primary Assessment

Form a General Impression	Inquire about the chief complaint and observe the patient's overall body position. Observe the work of breathing and circulation. Determine the level of consciousness using the AVPU scale. Identify immediate threats to life. Determine the priority of care based on the MOI. If the patient has a poor general impression, call for ALS assistance. A rapid scan will help you identify and manage life threats. Maintain a high index of suspicion for airway or respiratory compromise. Alterations in mental status will be the most sensitive indicator of subtle or impending problems.
Airway and Breathing	Ensure the airway is open, clear, and self maintained. With potential spinal injuries, use the jaw-thrust maneuver to open the airway. Evaluate the patient's ventilatory status for rate and depth of breathing, respiratory effort, and tidal volume. Administer high-flow oxygen at 15 L/min, providing ventilatory support as needed; *do not* hyperventilate the patient. Hypoxia may cause changes in the patient's mental state. If vomiting or bleeding into the oropharynx is a possibility, tilt the backboard to the side after spinal immobilization has been performed and have suction ready.
Circulation	Observe skin color, temperature, and condition; look for life-threatening bleeding and treat accordingly. Evaluate distal pulse rate, quality (strength), and rhythm. Assess and treat for shock. Control bleeding.
Transport Decision	If the patient has an airway or breathing problem, signs and symptoms of bleeding, or other life threats, manage them immediately and provide rapid transport to an appropriate trauma center, performing the secondary assessment en route to the hospital.

▼   ▼

NOTE: The order of the steps in this section differs depending on whether the patient is conscious or unconscious. The following order is for a conscious patient. For an unconscious patient, perform a primary assessment, perform a full-body scan, obtain vital signs, and obtain the past medical history from a family member, bystander, or emergency medical identification device.

Assessment and Emergency Care of Head and Spine Injuries, continued

History Taking

| Investigate Chief Complaint | Investigate the chief complaint. Closely monitor the patient for changes in mental status. Ask SAMPLE questions. SAMPLE can also be obtained from family, bystanders, and medical alert tags. |

Secondary Assessment

| Physical Examinations | Perform a sytematic full-body physical examination or an examination focused on the head and spine. Rule out any potiental life threats. Advise the patient prior to performing any examination. Do not delay transport to perform the physical examination at the scene. Look for DCAP-BTLS and deformities in the head and neck. Palpate the head and neck, feeling for any subtle deformities or "step-offs" that could indicate a more serious underlying problem. |
| Vital Signs | Obtain baseline vital signs as soon as practical. Vitals signs should include blood pressure by auscultation, pulse rate and quality, respiration rate and quality, and skin assessment for perfusion. Note the patient's level of consciousness. Use pulse oximetry, if available, to assess the patient's perfusion status. Be alert for Cushing's triad with possible head injuries. |

Reassessment

| Interventions | Repeat the primary assessment, vital signs, and chief complaint. Assist breathing as required, administering high-flow oxygen. |
| Communication and Documentation | Contact medical control/receiving hospital with a radio report; many hospitals require additional personnel and a separate treatment area. Include a thorough description of the MOI and the position the patient was found in. Include treatments performed and patient response. Be sure to document the patient's distress, answers to your questions, and any changes in patient status and the time. Follow local protocols. Document the reasoning for your treatment and the patient's response. |

NOTE: Although the following steps are widely accepted, be sure to consult and follow your local protocols. Take appropriate standard precautions when treating all patients.

Assessment and Emergency Care of Head and Spine Injuries, continued

Head and Spine Injuries

General Management of Head Injuries

1. Establish and maintain a patent airway. Provide high-flow supplemental oxygen and provide ventilatory assistance if needed.
2. Control bleeding. Do not apply pressure to an open or compressed skull injury. Begin cardio-pulmonary resuscitation, if necessary.
3. Assess the patient's baseline level of consciousness, and continuously monitor it.
4. Assess and treat other injuries.
5. Anticipate and manage vomiting to prevent aspiration.
6. Be prepared for convulsions and changes in the patient's condition.
7. Transport the patient promptly and with extreme care.

General Management of Spine Injuries

1. Open and maintain a patent airway with the jaw-thrust maneuver.
2. Hold the head still in a neutral, in-line position.
3. Consider inserting an oropharyngeal airway.
4. Have a suctioning unit available.
5. Provide high-flow oxygen.
6. Continuously monitor the patient's airway.
7. Perform manual in-line stabilization to protect the cervical spine.
8. Prepare the patient for transport according to patient's position.
9. Transport to the appropriate trauma center.

Prep Kit

Ready for Review

- The nervous system of the human can be divided into two parts: the central nervous system and the peripheral nervous system.
- The central nervous system consists of the brain and the spinal cord; the peripheral nervous system consists of a network of nerve fibers, like cables, that transmit information to and from the body's organs to and from the brain.
- The central nervous system is well protected by bony structures; the brain is protected by the skull and the spinal cord is protected by the bones of the spinal column.
- The central nervous system is also covered and protected by three layers of tissue called the meninges. The layers are called the dura mater, the arachnoid, and the pia mater.
- A head injury is a traumatic injury to the head that may result in injury to soft tissue, bony structures, or the brain.
- A traumatic brain injury is a severe head injury that can be a life threat or leave the patient with life-altering injuries.
- The cervical, thoracic, and lumbar portions of the spinal column can be injured through compression such as in a fall, unnatural motions such as overextension from trauma, distraction such as from a hanging, or a combination of mechanisms. Each of these can also cause injury to the spinal cord encased in these regions of bone, causing permanent neurologic injury or death.
- Motor vehicle crashes, direct blows, falls from heights, assault, and sports injuries are common causes of spinal injury. A patient who has experienced any of these events may have also sustained a head injury.
- Treat the patient with a head injury according to three general principles that are designed to protect and maintain the critical functions of the central nervous system: establish an adequate airway, control bleeding, and reassess the patient's baseline level of consciousness.
- Treat the patient with a spinal injury by maintaining the airway while keeping the spine in proper alignment, assess respirations, and give supplemental oxygen.
- In those situations in which your patient has problems with the ABCs or has other conditions for which you decide a rapid transport to the hospital is needed, rapid stabilization of the spine and quick loading into the ambulance may be indicated. Reduction of on-scene time and recognition of a critical patient increases the patient's chances for survival or a reduction in the amount of irreversible damage.

Vital Vocabulary

anterograde (posttraumatic) amnesia Inability to remember events after an injury.

basilar skull fractures Usually occur following diffuse impact to the head (such as falls, motor vehicle crashes); generally result from extension of a linear fracture to the base of the skull and can be difficult to diagnose with a radiograph (x-ray).

Battle's sign Bruising behind an ear over the mastoid process that may indicate a skull fracture.

cerebral edema Swelling of the brain.

closed head injury Injury in which the brain has been injured but the skin has not been broken and there is no obvious bleeding.

concussion A temporary loss or alteration of part or all of the brain's abilities to function without actual physical damage to the brain.

connecting nerves Nerves in the spinal cord that connect the motor and sensory nerves.

coup-contrecoup injury Dual impacting of the brain into the skull; coup injury occurs at the point of impact; contrecoup injury occurs on the opposite side of impact, as the brain rebounds.

distraction The action of pulling the spine along its length.

epidural hematoma An accumulation of blood between the skull and the dura mater.

eyes-forward position A head position in which the patient's eyes are looking straight ahead and the head and torso are in line.

four-person log roll The recommended procedure for moving a patient with a suspected spinal injury from the ground to a long backboard.

intervertebral disk The cushion that lies between two vertebrae.

intracerebral hematoma Bleeding within the brain tissue (parenchyma) itself; also referred to as an intraparenchymal hematoma.

intracranial pressure (ICP) The pressure within the cranial vault.

involuntary activities Actions of the body that are not under a person's conscious control.

linear skull fractures Account for 80% of skull fractures; also referred to as nondisplaced skull fractures; commonly occur in the temporal-parietal region of the skull; not associated with deformities to the skull.

meninges Three distinct layers of tissue that surround and protect the brain and the spinal cord within the skull and the spinal canal.

open head injury Injury to the head often caused by a penetrating object in which there may be bleeding and exposed brain tissue.

primary (direct) injury An injury to the brain and its associated structures that is a direct result of impact to the head.

raccoon eyes Bruising under the eyes that may indicate a skull fracture.

retrograde amnesia The inability to remember events leading up to a head injury.

secondary (indirect) injury The "after effects" of the primary injury; includes abnormal processes such as cerebral edema, increased intracranial pressure, cerebral ischemia and hypoxia, and infection; onset is often delayed following the primary brain injury.

subarachnoid hemorrhage Bleeding into the subarachnoid space, where the cerebrospinal fluid circulates.

subdural hematoma An accumulation of blood beneath the dura mater but outside the brain.

subluxation A partial or incomplete dislocation.

traumatic brain injury (TBI) A traumatic insult to the brain capable of producing physical, intellectual, emotional, social, and vocational changes.

voluntary activities Actions that we consciously perform, in which sensory input or conscious thought determines a specific muscular activity.

Assessment in Action

You are called to an automotive body shop for an assault patient. Law enforcement personnel have arrived and declared the scene safe. You and your partner find a 44-year-old man seated in an office holding the left side of his head. According to bystanders, the patient was struck on the side of his head with a tire iron during a dispute with another employee. He was unconscious for approximately 3 minutes.

1. What is the most reliable sign of a head injury?
   A. Seizures
   B. Decreased level of consciousness
   C. Unequal pupils
   D. Vomiting

2. While performing a full-body scan of the patient you find a compressed area above the patient's left ear. An accumulation of blood between the skull and the dura mater is known as:
   A. an intracerebral hemorrhage.
   B. a subdural hematoma.
   C. a subarachnoid hemorrhage.
   D. an epidural hematoma.

3. An epidural hematoma is nearly always the result of a blow to the head that produces a linear fracture in what region of the cranium?
   A. Frontal
   B. Parietal
   C. Occipital
   D. Temporal

4. Which of the following arteries runs along a groove in the temporal bone?
   A. Cerebral
   B. Basilar
   C. Middle meningeal
   D. Craniofacial

5. The patient is unable to recall the events leading up to the injury. This is called _____ amnesia.
   A. anterograde
   B. hysterical
   C. retrograde
   D. dissociative

6. When immobilizing a patient to a long spine board, what area of the body should you secure first?
   A. Head
   B. Upper torso
   C. Pelvis
   D. Lower legs

7. During transport, the patient loses consciousness. The time between the two periods of unconsciousness is referred to as the:
   A. lucid interval.
   B. recognition period.
   C. danger zone.
   D. coherent stage.

8. Cushing's triad is a sign of increased intracranial pressure and impending brain stem herniation. Which of the following changes in vital signs is seen in patients demonstrating Cushing's triad?
   A. Decreased blood pressure, increased pulse, irregular respirations
   B. Increased blood pressure, decreased pulse, irregular respirations
   C. Increased blood pressure, increased pulse, irregular respirations
   D. Decreased blood pressure, decreased pulse, irregular respirations

9. Explain the difference between a primary (direct) brain injury and a secondary (indirect) brain injury.

10. What actions should you take to help prevent secondary brain injuries?

Chest Injuries

National EMS Education Standard Competencies

Trauma

Applies fundamental knowledge to provide basic emergency care and transportation based on assessment findings for an acutely injured patient.

Chest Trauma

Recognition and management of
- Blunt versus penetrating mechanisms (pp 938-940)
- Open chest wound (pp 938-940)
- Impaled object (pp 938-940)

Pathophysiology, assessment, and management of
- Blunt versus penetrating mechanisms (pp 938-940)
- Hemothorax (p 948)
- Pneumothorax (pp 945-948)
  - Open (pp 945-946)
  - Simple (pp 946-947)
  - Tension (pp 947-948)
- Cardiac tamponade (pp 948-949)
- Rib fractures (p 949)
- Flail chest (pp 949-950)
- Commotio cordis (pp 951-952)

Knowledge Objectives

1. Understand the mechanics of ventilation in relation to chest injuries. (pp 937-938)
2. Describe the differences between an open and closed chest injury. (pp 938-940)
3. Recognize the signs of chest injury. (pp 939-940)
4. Differentiate between a pneumothorax (open, simple, and tension) and hemothorax. (pp 945-948)
5. Describe the complications of cardiac tamponade. (pp 948-949)
6. Describe the complications of rib fractures. (p 949)
7. Describe the complications of a patient with a flail chest. (pp 949-950)
8. Explain the complications of a patient with an open pneumothorax (sucking chest wound). (pp 945-948)
9. Recognize the complications that can accompany chest injuries. (pp 945-952)
10. Describe the management of a patient with a suspected chest injury, including pneumothorax, hemothorax, cardiac tamponade, rib fractures, flail chest, pulmonary contusion, traumatic asphyxia, blunt myocardial injury, commotio cordis, and laceration of the great vessels. (pp 945-952)

Skills Objectives

1. Describe the steps to take in the assessment of a patient with a suspected chest injury. (pp 940-944)
2. Demonstrate the management of a patient with a sucking chest wound. (pp 945-948)
3. Demonstrate the management of a patient with a flail chest. (pp 949-950)

Introduction

Chest injuries are commonly encountered by EMTs. According to the Centers for Disease Control and Prevention, chest trauma causes more than 700,000 emergency department visits and more than 18,000 deaths in the United States annually. Given the location of the heart, lungs, and great blood vessels within the chest cavity, potentially serious injuries may occur. Any injury that interferes with the body's mechanics of normal breathing must be treated without delay to minimize or prevent permanent damage to tissues that depend on a continuous supply of oxygen. Another major problem with chest injuries may be internal bleeding. Blood from lacerations of the thoracic organs or major blood vessels can collect in the chest cavity, compressing the lungs or heart. This may also occur when air collects in the chest and prevents the lungs from expanding. Your ability to act quickly to care for patients with these injuries can make the difference between a successful outcome and death.

This chapter begins with a review of the anatomy of the chest and the physiology of respiration. It then describes the common signs and symptoms of chest injuries and the proper emergency medical treatment for specific injuries.

Anatomy and Physiology

To understand and evaluate chest injuries in the prehospital setting, you must first understand the anatomy of the chest and the mechanism by which gases are exchanged during breathing. A quick review will help you understand the logic in the emergency treatment of chest injuries and the potential complications of that treatment.

A key point to remember is the difference between ventilation and respiration. Ventilation is the body's ability to move air in and out of the chest and lung tissue. This is described later in the section on mechanics of ventilation. Any injury that affects the patient's ability to move air in and out of the chest is serious and may be life threatening. Respiration is the exchange of gases in the alveoli of the lung tissue. This is the terminal point of the pulmonary system. Oxygen must be delivered to the cells, and carbon dioxide (a waste product of cell function) must be removed from the body for proper organ system function.

The chest (thoracic cage) extends from the lower end of the neck to the diaphragm **Figure 27-1**. In an individual who is lying down or who has just completed exhalation, the diaphragm may rise as high as the nipple line. Thus, a penetrating injury to the chest, such as a gunshot or stab wound, may also penetrate the lung and diaphragm and injure the liver or stomach.

The skin, muscle, and bones of the thoracic region have similarities to other regions but also provide some unique features to allow for the ventilation process. Just under the normal three layers of skin, the epidermis, dermis, and subcutaneous layers, lies striated or skeletal muscle. This muscle extends between the ribs, forming the intercostal muscles. These muscles, innervated from the spinal nerves originating in the cervical region C6 and C7, allow the chest to expand on contraction and allow for the active portion of ventilation to occur. In patients who have sustained a spinal cord injury in that region, it is important for you to note that the patients may be unable to move these muscles and will have to breathe entirely with the diaphragm. This is often called

You are the Provider: PART 1

At 10:20 AM, you are dispatched to a construction site for a man with a chest injury. The caller, whose supervisor asked him to call 9-1-1, was unable to provide the dispatcher with the exact mechanism of injury. You respond to the scene, which is located about 5 miles away. The weather is clear, the traffic is moderate, and the temperature is 92°F.

1. What major organs and structures lie within the chest cavity?
2. What injuries commonly result from blunt chest trauma? Penetrating chest trauma?

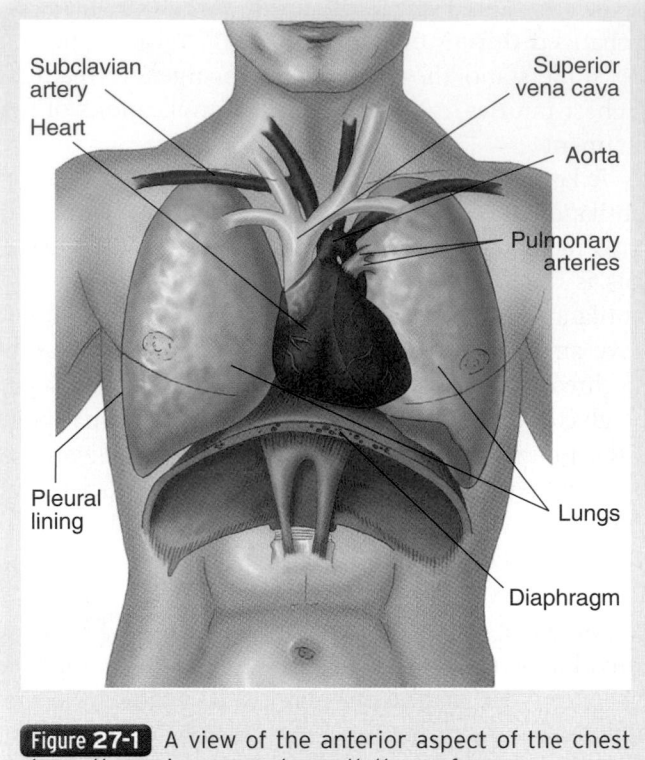

Figure 27-1 A view of the anterior aspect of the chest shows the major organs beneath the surface.

is the neurovascular bundle, composed of a network of nerves, arteries, and veins. Take this structure into consideration when evaluating patients who have sustained rib fractures because this may be a source of significant bleeding into the pleural space, creating a hemothorax. The ribs themselves create a protective and functional cage around the vital organs. Each side of the chest contains lung tissue that is separated into lobes. The right lung has three lobes, and the left lung has two lobes. The left lobe formation allows space for the heart to reside; this is called the cardiac notch. A thin membrane called the pleura covers each of the lungs and the thoracic cavity. The inner chest wall has a lining called the parietal pleura, and a lining called the visceral pleura covers the lung. Between these two linings is a small amount of fluid called surfactant that allows the lungs to move freely against the inner chest wall as a person breathes. Surfactant also creates surface tension to allow the lungs to adhere to the rib cage, thus allowing the mechanics of ventilation to occur.

The contents of the chest are partially protected by the ribs, which are connected in the back to the vertebrae and in the front, through the costal cartilages, to the sternum **Figure 27-2**. The trachea, which is in the middle of the neck, divides into the left and right mainstem bronchi, which supply air to the lungs. The thoracic cage also contains the heart and the great vessels: the aorta, the right and left subclavian arteries and their branches, the pulmonary arteries, and the superior and inferior venae cavae. The esophagus runs through the back of the chest, connecting the pharynx above with the stomach and the abdomen below. The esophagus, trachea, and great vessels lie in a specific cavity or space centrally located in the thorax. This is referred to as the mediastinum. Often

"belly breathing," and it is considered a clinical or positive diagnostic finding indicating cord damage at or above the level of C6 and C7. In very young children, the intercostal muscles are not yet developed. Children therefore have a tendency to breathe with their diaphragms. This is considered normal for this age group and does not typically indicate spinal cord injury.

Lying close to each rib along the bottom or inferior and slightly posterior to the lowest margin of each rib

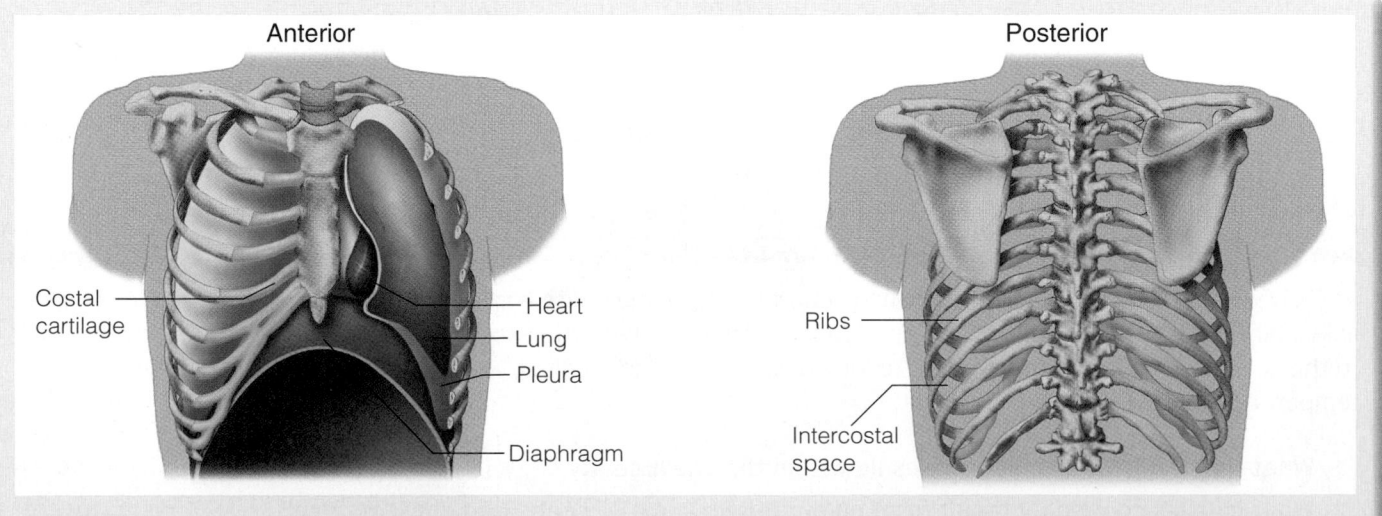

Figure 27-2 The organs within the chest are protected by the ribs, which are connected in back to the vertebrae and in the front, through the costal cartilages, to the sternum.

this is the location for the development of a pathologic condition called thoracic aortic aneurysm, which occurs when a weakened portion of the wall of the aorta develops a false lumen, or a bubble, that can eventually rupture and cause a fatal hemorrhage in the thorax. At the bottom of the chest, the diaphragm is a muscle that separates the thoracic cavity from the abdominal cavity.

Mechanics of Ventilation

When you inhale, the intercostal muscles between the ribs contract, elevating and expanding the rib cage. At the same time, the diaphragm contracts or flattens and pushes the contents of the abdomen down. The intrathoracic pressure inside the chest decreases, creating a negative pressure differential. Air then enters the lungs through the nose and mouth, which is the path of least resistance from the ambient air space to the upper and lower airway. When you exhale, the intercostal muscles and diaphragm relax, and the tissues move back to their normal positions, allowing air to be exhaled **Figure 27-3**. In a normal respiratory system, relaxation of the thoracic muscles and the diaphragm is a relatively passive function. Normal physiology dictates that the body should not have to work to breathe when in a resting state. When you are assessing the patient, you should be able to recognize when there is an increase in the work of breathing and equate that with respiratory distress and a life threat.

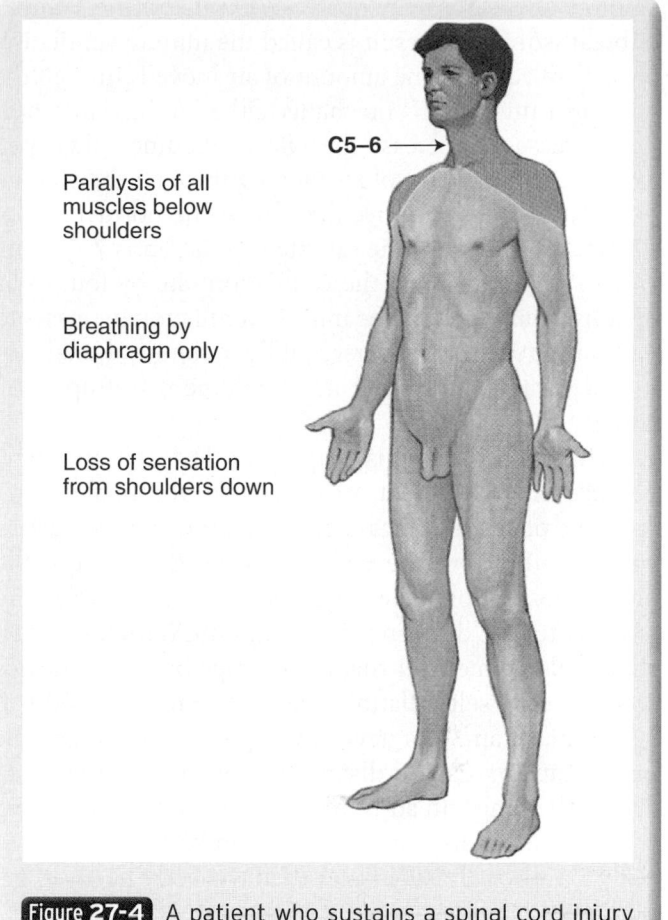

C5–6

Paralysis of all muscles below shoulders

Breathing by diaphragm only

Loss of sensation from shoulders down

Figure 27-4 A patient who sustains a spinal cord injury below the level of C5 and is paralyzed can still breathe spontaneously because the phrenic nerves originate at the C3, C4, and C5 levels.

Note that the nerves supplying the diaphragm (the phrenic nerves) exit the spinal cord at C3, C4, and C5. Refer back to Chapter 5, *The Human Body*, for a review of the spinal column. A patient whose spinal cord is injured below the C5 level will lose the power to move the intercostal muscles, but the diaphragm will still contract. The patient will still be able to breathe because the phrenic nerves remain intact. Patients with spinal cord injuries at C3 or above can lose their ability to breathe entirely **Figure 27-4**.

As discussed in Chapter 9, *Airway Management*, tidal volume is the amount of air in milliliters (mL) that is moved into or out of the lungs during a single breath. The average tidal volume for a male is approximately 500 mL. If you multiply this

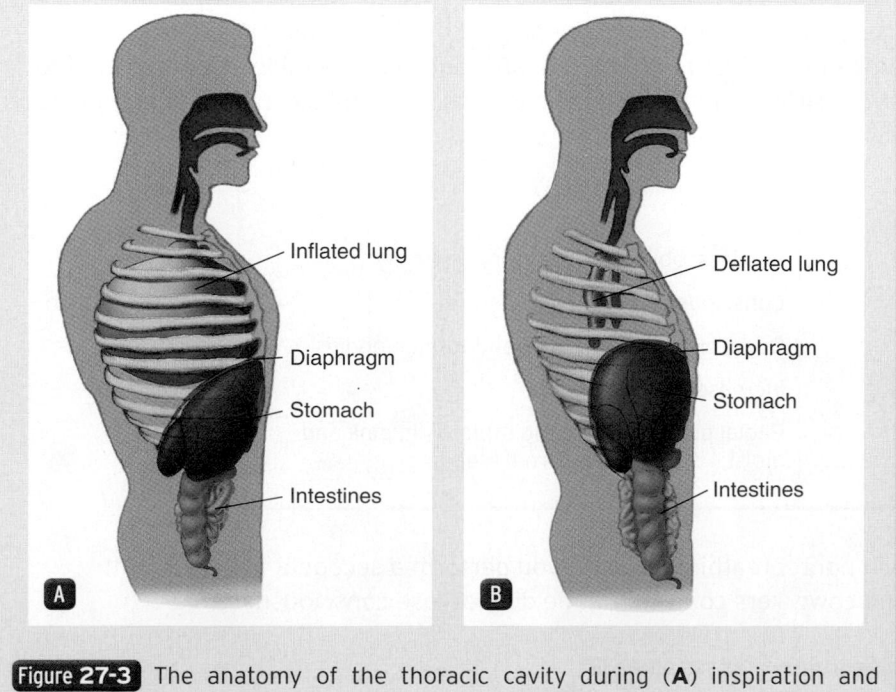

Inflated lung

Diaphragm

Stomach

Intestines

Deflated lung

Diaphragm

Stomach

Intestines

A

B

Figure 27-3 The anatomy of the thoracic cavity during **(A)** inspiration and **(B)** expiration.

amount of air, normally 500 to 700 mL, by the number of breaths/min, the result is called the minute ventilation or minute volume (the amount of air moved through the lungs in 1 minute). If you change either of these numbers (ie, increase or decrease the rate or volume), then you can affect the amount of air moving through the system. For example, if you move 600 mL at the normal rate of 12 breaths/min, then the minute ventilation is 7,200 mL (7.2 L). If you increase the ventilation rate by four extra breaths a minute, then the minute ventilation increases to 9,600 mL (9.6 L). Conversely, if the amount of tidal volume decreases, then the minute volume will drop along with it.

This information is important for you to know because if the patient is only able to inhale small amounts of air (in the case of a chest injury or a reactive airway pathology), the patient will need to exceed the normal respiratory rate range of 12 to 20 breaths/min to make up the difference in the minute ventilation. You should also remember that the average bag-mask device consists of a self-inflating bag that contains 1,000 to 1,500 mL of air. This device can quickly overinflate the lungs, causing gastric distention and impair the function of the lungs. In addition, there is the risk of causing acid-base imbalance and blood gas imbalance by "blowing off" carbon dioxide faster than the body needs to get rid of it.

Injuries of the Chest

There are two basic types of chest injuries: open and closed. As the name implies, a **closed chest injury** is one in which the skin is not broken. This type of injury is generally caused by blunt trauma, such as when a driver strikes a steering wheel or an air bag in a motor vehicle crash, is struck by a falling object, or is struck in the chest by some object during a fight or physical assault **Figure 27-5**. These types of injuries often cause significant contusions in both the cardiac muscle (cardiac contusion) and the lung tissue (pulmonary contusion), thus impairing the function of those organs.

If the heart is damaged in this manner, it may not be able to refill with blood or blood may not be able to

Words of Wisdom

The ability to pump blood depends on having a functional pump (the heart), an adequate volume of blood to be pumped, and an appropriate amount of resistance to the pumping mechanism. Collectively, these properties help determine cardiac output. Cardiac output is the volume of blood delivered to the body in 1 minute. Any injury that limits the heart's pumping ability, the delivery of blood to the heart, the blood's ability to leave the heart, or the heart rate will affect cardiac output.

You are the Provider: PART 2

You arrive at the scene and ensure it is safe. You are directed to the patient by a coworker. The patient, a 19-year-old man, was struck in the left side of the chest by a 2-by-4 board that slid off a table saw when another coworker was cutting it. You find the patient sitting on the ground; he is conscious and alert but restless and is experiencing respiratory distress. He tells you that it hurts to breathe.

Recording Time: 0 Minutes	
Appearance	Restless; obvious respiratory distress
Level of consciousness	Conscious and alert, but restless
Airway	Open; clear of secretions and foreign bodies
Breathing	Increased rate; labored
Circulation	Radial pulses, strong and rapid; skin, pink and moist; no obvious external bleeding

Your partner applies high-flow oxygen via nonrebreathing mask as you perform a secondary assessment. The patient denies any other injuries, and coworkers confirm that he did not lose consciousness.

3. How should you proceed with your secondary assessment?

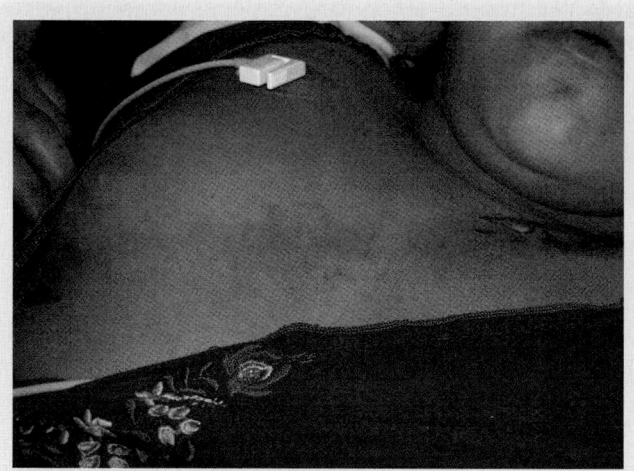

Figure 27-5 Closed chest injuries usually result from blunt trauma, such as when a patient strikes the steering wheel or an air bag in a motor vehicle crash or is struck by a falling object. A closed chest injury can occur even when a seatbelt is worn.

return to the heart, creating a form of inadequate tissue oxygenation called cardiogenic shock. Any bruising of the lung tissue can result in exponential loss of the surface area where oxygen and carbon dioxide exchange occurs. This impairment can cause hypoxic and hypercarbic states, both leading to alterations of consciousness and possible death if not recognized and treated. Rib fractures create sharp broken bone ends that can lacerate lung tissue and cause further vessel damage with every movement of the chest wall. This type of bleeding can be hidden from external view and rapidly lead to hypovolemic shock.

In an **open chest injury**, some object, such as a knife, a bullet, a piece of metal, or the broken end of a fractured rib, penetrates the chest wall itself **Figure 27-6**. The damage occurring from this type of trauma typically

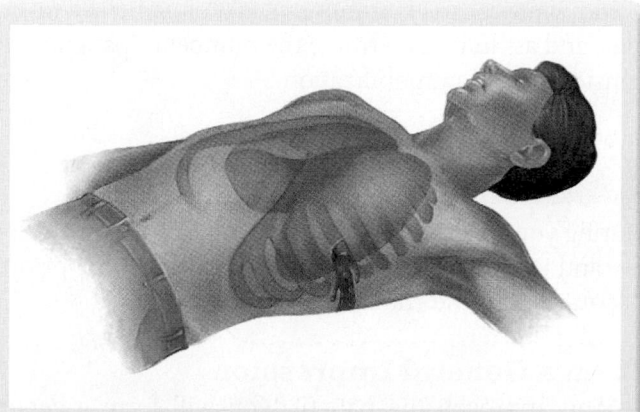

Figure 27-6 Open chest injuries occur when the chest wall is penetrated by an object or the broken end of a fractured rib.

is instant. However, the symptoms of these injuries may take time to develop as the damaged vessels continue to bleed or the lung collapses from a puncture. Occasionally, the object that penetrates and creates an open chest injury remains in place. This is referred to as an impaled object. When you have a patient with an impaled object, do not attempt to move or remove the object because it may be occluding the hole in the vessel that has been punctured. If you were to remove the object, the patient would bleed heavily. Another reason to not remove the impaled object from the chest is the fact that the objects that cause tissue damage on entry will likely cause damage on removal, resulting in further injury. The removal is best left for the surgeon. Any alteration from this standard should come directly from online medical control.

In blunt trauma, a blow to the chest may fracture the ribs, the sternum, or whole areas of the chest wall, bruise the lungs and the heart, and even damage the aorta. Almost one third of people who are killed immediately in car crashes die as a result of traumatic rupture of the aorta. Although the skin and chest wall are not penetrated in a closed injury, the contents of the chest may be lacerated by broken ribs. Damage to the chest wall structures may result in decreased ability of patients to ventilate on their own. Also, vital organs can actually be torn from their attachment in the chest cavity without any break in the skin; this condition can cause serious and life-threatening bleeding that is unseen outside the body.

■ Signs and Symptoms of Chest Injury

Important signs and symptoms of chest injury include the following:

- Pain at the site of injury
- Pain localized at the site of injury that is aggravated by or increased with breathing
- Bruising to the chest wall
- Crepitus with palpation of the chest
- Any penetrating injury to the chest
- Dyspnea (difficulty breathing, shortness of breath)
- Hemoptysis (coughing up blood)
- Failure of one or both sides of the chest to expand normally with inspiration
- Rapid, weak pulse and low blood pressure
- Cyanosis around the lips or fingernails

After a chest injury, any change in normal breathing is a particularly important sign. A healthy, uninjured adult usually breathes at a rate ranging from 12 to 20 breaths/min without difficulty and without pain. The chest should rise and fall in a symmetric pattern with each breath. Respirations of fewer than 12 breaths/min or of more than 20 breaths/min may indicate inadequate breathing. Patients with chest injuries often have **tachypnea** (rapid respirations) and shallow respirations because it hurts to take a deep breath. Note that the patient may be making breathing attempts but may not actually be moving air. Chest wall trauma may interfere with the ability to actually move air. Check the respiration rate and see if there is actual air movement from the mouth and/or nose. This is best accomplished through the use of auscultation of multiple locations on the chest wall for adequate breath sounds.

As with any other injury, pain and tenderness are common at the point of impact as a result of a bruise or fracture. Pain is usually aggravated by the normal process of breathing. Irritation of or damage to the pleural surfaces causes a characteristic sharp or sticking pain with each breath when these normally smooth surfaces slide on one another. This sharp pain is called *pleuritic pain*, or *pleurisy*.

In an injured patient, dyspnea, or difficulty breathing, has many causes, including airway obstruction, damage to the chest wall, improper chest expansion because of the loss of normal control of breathing, or lung compression because of accumulated blood or air in the chest cavity. Dyspnea in an injured patient indicates potential compromise of lung function; prompt, vigorous support of oxygenation and ventilation with prompt transport are required.

Hemoptysis, the spitting or coughing up of blood, usually indicates that the lung itself or the air passages have been damaged. With a laceration of the lung tissue, blood can enter the bronchial passages and is coughed up as the patient tries to clear the airway.

A rapid, weak pulse and low blood pressure are the principal signs of hypovolemic shock, which can result from extensive bleeding from lacerated structures within the chest cavity, where the great vessels and heart are located. Shock following a chest injury may also result from insufficient oxygenation of the blood by the poorly functioning lungs.

Cyanosis in a patient with a chest injury is a sign of inadequate respiration. The classic blue or ashen gray appearance around the lips and fingernails indicates that blood is not being oxygenated sufficiently. Patients with cyanosis are unable to provide a sufficient supply of oxygen to the blood through the lungs and require immediate ventilation and oxygenation.

Many of these signs and symptoms occur simultaneously. When any one of them develops as a result of a chest injury, the patient requires prompt hospital care. Remember that the principal reason for concern about a patient who has a chest injury is that his or her body has no means of storing oxygen; it is supplied and used continuously, even during sleep. Any interruption in this supply can be rapidly lethal and must be treated aggressively.

Patient Assessment

Scene Size-up

Scene Safety
As you arrive on the scene, observe for hazards and threats to the safety of the crew, bystanders, and the patient. Consider the possibility that the area where the patient is located may be a crime scene; therefore, make every attempt to not disturb potential evidence. Ensure that the police are on scene at incidents involving violence, such as assaults or gunshot wounds. Begin the encounter with scene safety as the highest priority. If you determine that power company, fire department, or advanced life support (ALS) units are needed, call for them early.

Ensure that you and your crew follow standard precautions, and put on a minimum of gloves and eye protection. Put several pairs of gloves in your pocket for easy access in case your gloves tear or there are multiple patients with bleeding. Because of the color of blood and the fact that it easily soaks through clothing, you can often identify patients with bleeding as you approach the scene. However, darker clothing may mask signs of bleeding, so you must remain vigilant when the mechanism of injury suggests the patient may be bleeding.

Mechanism of Injury/Nature of Illness
As you observe the scene, look for indicators and significance of the mechanism of injury (MOI). This helps you develop an early index of suspicion for underlying injuries in a patient who has sustained a significant MOI. Chest injuries are common in motor vehicle crashes, falls, and assaults. Determine the number of patients, and consider spinal immobilization.

Primary Assessment

During your primary assessment, you must quickly identify and treat potential life threats and determine priority of patient care and transport.

Form a General Impression
As you approach the patient, you will form a general impression of the patient's condition. It is important to note the patient's level of consciousness. Responsive patients may be able to tell you their chief complaint. Note not only what they say, but also how they say it.

Difficulty speaking may indicate several problems, and chest injury is an important one. Perform a rapid scan of the patient. Look for obvious injuries, the appearance of blood, and difficulty breathing. Look for cyanosis, irregular breathing, and chest rise and fall on only one side. Observe the neck, looking for accessory muscle use while breathing; also look for extended or engorged external jugular veins. If no obvious problems are seen, begin looking for them by focusing on the ABCs. The initial general impression will help you develop an index of suspicion for serious injuries and determine your sense of urgency for medical intervention. A good question to ask yourself is "How sick is this patient?" Patients with significant chest injuries will "look" sick and are often frightened or anxious. Keep in mind that you are rapidly searching for life threats and you will repeat the physical examination in a more detailed manner later in the assessment if time and patient condition allow.

Airway and Breathing

Addressing life threats begins with the assessment of airway and breathing. Ensure that the patient has a clear and patent airway. Normal breathing should be effortless, and any deviation from this pattern should be cause for concern. How you assess and manage the airway depends a great deal on whether you suspect a spinal injury. A significant number of patients with traumatic chest injuries also have spinal injuries, and proper precautions should be taken. Be suspicious, and protect the spine early in your care, even if your assessment later confirms that there is no spinal injury. While you are considering stabilization of the cervical spine, note whether the jugular veins are distended. If you note this finding, it is a sign of pressure or tamponade on the heart that is not allowing the right atrium to empty into the ventricle. This can be the result of a tension pnuemothorax (significant ongoing air accumulation in the pleural space) or injury to the heart that allows bleeding into the pericardium, creating a cardiac tamponade, otherwise referred to as a pericardial tamponade. Once you have determined the patient has a patent airway, determine whether breathing is present and adequate. With chest injuries, begin by inspecting for DCAP-BTLS (Deformities, Contusions, Abrasions, Punctures/penetrations, Burns, Tenderness, Lacerations, and Swelling), and look for equal expansion of the chest wall. Listen with a stethoscope to each side of the chest. Absent or decreased breath sounds on one side usually indicate significant damage to a lung, preventing it from expanding properly. Be alert to the pattern of symmetric rise and fall of the patient's chest wall. If the chest wall does not expand on each side when the patient inhales, the chest muscles may have lost their ability to work appropriately. Loss of muscle function may be the result of a direct injury to the chest wall, or it may be related to an injury of the nerves that control those muscles. Check also for **paradoxical motion**, an abnormality associated with multiple fractured ribs, in which one segment (often referred to as a flail segment) of the chest wall moves opposite the remainder of the chest; that is, out with expiration and in with inspiration. If you determine the patient has paradoxical movement of the chest wall or penetrating trauma, address this life threat at once. These conditions may interfere with the normal mechanics of breathing and can cause the patient's condition to worsen quickly. For quick initial care, you can use your gloved hand to stabilize a flail segment or to occlude an open chest wound. When further dressings can be applied, you should apply an **occlusive dressing** to all penetrating injuries to the chest. Depending on local protocol, the dressing may be taped on three sides to allow air to escape during exhalation. Stabilize paradoxical motion with a large, bulky dressing. Use a roller bandage, cravat, or tape to hold the dressing in place. Apply oxygen with a non-rebreathing mask at 15 L/min. Provide positive-pressure ventilations with 100% oxygen if breathing is inadequate based on the patient's level of consciousness and breathing rate and quality. As a note of caution, remember that when you are providing positive-pressure ventilation, you are overcoming the normal physiologic functions, and, if your patient has a pneumothorax (collapsed lung), you can quickly exacerbate the injury. Be diligent with auscultation of breath sounds, and evaluate the effectiveness of your ventilatory support with signs of circulation to the skin. Be aware of decreasing oxygen saturation (Sao_2) values because they may indicate the development of hypoxia. Watch for signs of an impending tension pneumothorax, such as increasingly poor compliance during ventilation.

Circulation

Assess the patient's pulse. Determine whether it is present and adequate. If the pulse is too fast or too slow, or if the skin is pale, cool, or clammy, consider your patient to be in shock. You need to treat aggressively to eliminate the cause and support the patient's circulatory system. Note that in the early stage of shock, the body compensates for blood loss by increasing the heart rate. Be alert for this change, especially if tachycardia is still present beyond a few minutes after the initial adrenaline rush from the incident or injury. External bleeding may or may not be significant, but if it is considered life threatening, address this threat immediately. Bleeding inside the chest can be significant and, as discussed earlier, can be a quick cause of death. Control bleeding with direct pressure and a bulky trauma dressing.

Transport Decision

Priority patients are considered patients who have a problem with their airway, breathing, and/or circulation. Sometimes the priority is obvious, and the decision to transport quickly is also easy. At other times, what is happening outside the body may not provide obvious clues to the seriousness of what is happening inside the body. Pay attention to subtle clues such as the appearance of the skin, level of consciousness, or a sense of impending doom in the patient. These symptoms are not as grand as a large gash across the chest or air being sucked into the chest; however, they can be equally important indicators of a life-threatening condition. When you find signs of poor perfusion or inadequate breathing, transport quickly and perform the remainder of the assessment en route to the emergency department. A delay on the scene to perform a lengthy assessment will reduce the chances of survival for your patient. With chest injuries, when in doubt, transport rapidly to a hospital. **Table 27-1** lists the "deadly dozen" chest injuries.

Table 27-1 **Deadly Dozen Chest Injuries**

Immediately life-threatening chest injuries that must be detected and managed during the primary assessment:

1. Airway obstruction
2. Bronchial disruption
3. Diaphragmatic tear
4. Esophageal injury
5. Open pneumothorax
6. Tension pneumothorax
7. Massive hemothorax
8. Flail chest
9. Cardiac tamponade

Potentially lethal chest injuries that may be identified during the secondary assessment:

10. Thoracic aortic dissection
11. Myocardial contusion
12. Pulmonary contusion

▶ History Taking

Investigate Chief Complaint

Once you have identified and treated life threats, you can move on to gathering a history from the patient. If you have not yet done so, you should determine and investigate the patient's chief complaint and further investigate the MOI. You will also identify any associated signs and symptoms and pertinent negatives. If the patient was assaulted with a blunt object such as a bat, further evaluate the spinal region for injury because the force may have been transferred through the body from the point of impact. If the patient fell from a great height and is complaining of chest discomfort or dyspnea, this

You are the Provider: PART 3

Your secondary assessment reveals bruising and crepitus to the left side of the chest and diminished breath sounds over that same side; no paradoxical chest wall movement is noted. The trachea is midline, and the jugular veins are nondistended. Your partner takes the patient's vital signs and reports them to you.

Recording Time: 5 Minutes

Respirations	24 breaths/min; labored
Pulse	110 beats/min; strong and regular
Skin	Pink, warm, and moist
Blood pressure	138/88 mm Hg
Oxygen saturation (Sao_2)	95% (on oxygen)

4. On the basis of your assessment findings, what injury or injuries should you suspect?

5. How should you proceed with your treatment of this patient?

may distract the patient from recognizing that he or she has fractures or is bleeding from the extremities. Palpation of the chest will typically cause direct pain at the site of the fracture. When a patient reacts to the pain, make certain to verify where the pain was located in relationship to the area being touched.

Pertinent negatives when examining the chest include no associated shortness of breath, no rapid breathing, no absent or abnormal breath sounds, and no areas of deformity or abnormal movement. In a patient with a suspected spinal cord injury, equal expansion of the chest and movement of the rib cage and the diaphragm can confirm to you that there is nerve conduction to that region of the body.

SAMPLE History

Obtaining a SAMPLE history from a patient with a chest injury may not seem very important. Regardless, a basic evaluation of signs and symptoms, allergies, medications, pertinent medical problems, including respiratory or cardiovascular disease, and last oral intake should be completed. The events leading to the emergency should also be identified. Questions about the events surrounding the incident should focus on the MOI: the speed of the vehicle or height of the fall, the use of safety equipment (eg, helmet, air bag, seatbelt, life jacket), the type of weapon used, the number of penetrating wounds, and so on. A SAMPLE history can be obtained quickly in most situations and can certainly be obtained while accomplishing other tasks. However, if the patient loses consciousness, it will no longer be possible to obtain the information.

Secondary Assessment

Physical Examinations

In a patient who has an isolated injury to the chest with a limited MOI, such as in a stabbing, you should focus your assessment on the isolated injury, the patient's complaint, and the body region affected. Ensure that wounds are identified and control of the bleeding has been established. Note the location and extent of the injury. Assess all underlying systems. Examine the anterior and posterior aspects of the chest wall, and be alert to changes in the patient's ability to maintain adequate respirations.

If there is significant trauma (such as a blunt trauma or gunshot wound) likely affecting multiple systems, start with a full-body scan looking for DCAP-BTLS to determine the nature and extent of thoracic injury. This examination will help to determine all of the injuries and the extent of the injuries. Inspection or visualization of

the region looking for deformities, such as asymmetry of the left and right sides of the chest or shoulder girdle, may reveal the presence of multiple rib fractures, crush injuries, or significant chest wall injury. Identification of discrete areas of contusion or abrasion may pinpoint a specific point of impact. The presence of puncture wounds or other penetrating injuries indicates a possible open chest injury that should be managed accordingly. Be alert for associated burns, which may alter respiratory mechanics. Palpate for tenderness to localize the injury and the presence of fractures. Look for lacerations and local swelling. Application of this systematic approach to patient assessment minimizes the chance of missing significant injury.

It is important in patients with a chest injury not to focus only on a chest wound. With significant trauma, you should quickly assess the entire patient from head to toe.

Special Populations

In elderly patients with reduced bone density or more fragile bones, even minor trauma to the chest wall can cause signficant injury to the underlying tissues and organs. Elderly patients may have also sustained a number of fractures to the rib cage. Be alert for these injuries and for signs and sypmtoms of respiratory compromise, even in lower energy mechanisms of injury.

Vital Signs

Once you have stabilized airway, breathing, and circulation problems and have checked the patient from head to toe to identify injuries, obtain a baseline set of vital signs. This activity should include assessment of pulse, respirations, blood pressure, skin condition, and pupils. Each of these is considered a sign indicating how your patient is tolerating the injuries. Consider these signs as a window to the functioning of the vital organs. This baseline set of vital signs will be used to evaluate changes in the patient's condition. Because patients with chest injury have so many risks of mortality, they should be reevaluated every 5 minutes or less. This will allow you to quickly recognize changes in the vital sign numbers or trends.

If you find an accelerated pulse rate or respiratory rate, the chest injury may be causing either a decrease in available oxygen (hypoxia) or blood loss that results in a decreased number of red blood cells that can carry oxygen (hypoxemia). The increased respiratory rate is often associated with an obvious increase in work of breathing. This can be identified by noting increased use of the accessory muscles in the face, neck, and chest to assist in the movement of air. In the later stages of injuries, the pulse rate can slow as the myocardium becomes starved for oxygen and the body is no longer able to keep up with the demands. The respiratory rate may drop as the brain becomes starved for oxygen and overloaded with carbon dioxide and other waste products. These are usually signs of impending cardiopulmonary arrest. In the case of increasing pressure on the heart from the pleural space or the pericardial space, the blood pressure may exhibit a narrowing pulse pressure as the systolic and diastolic pressures come closer together. This is a result of the inability of the heart to beat normally and effectively.

Monitoring Devices

Using diagnostic tools to evaluate the effectiveness of the respiratory system is especially important with an injury or insult to the thorax because the injury can create a decrease in the amount of available oxygen within the blood. The most commonly used device for this assessment is the pulse oximeter. The pulse oximeter monitors the oxygen saturation of hemoglobin by directing a beam of infrared light through the capillary beds of an area (such as a finger) between two probes. The light received after passing though the capillary bed is measured as a percentage that indicates how well the hemoglobin in the red blood cells is coated or saturated with oxygen. In normal circumstances, the hemoglobin is saturated with oxygen, and the number represents the percentage of oxygen saturation (Sao_2). However, if your patient has a decreased number of red blood cells, a damaged pump, or a pulmonary contusion, less oxygen is available to be absorbed by the hemoglobin, resulting in a decreased Sao_2. Another situation to note is when there is carbon monoxide in the patient's bloodstream. The hemoglobin is more likely to be saturated with carbon monoxide than with oxygen. Since the oximeter presumes that oxygen is saturating hemoglobin, the values produced are misleading, possibly resulting in your undertreating the victim. Because all of these situations can result in a decreased Sao_2, it is advisable to use the pulse oximeter on any patient with a chest injury to establish a baseline measurement and to help you recognize any downward trends that indicate the patient's condition is worsening.

Words of Wisdom

Pulse oximetry is an assessment tool used to evaluate the effectiveness of oxygenation. The pulse oximeter does not take the place of good assessment skills.

Reassessment

The reassessment identifies how your patient's condition is changing. It should focus on repeating the primary assessment, reassessing the chief complaint, and reassessing interventions performed. Reevaluate the patient's airway, breathing, pulse, perfusion, and bleeding. Has breathing improved now that the wound is sealed, or has it become more difficult and associated with the trachea deviating to one side? Is the splint providing stability, or is it too loose? Other interventions should also be assessed to determine if they are effective. For example, are pulse oximeter values rising now that the patient is receiving oxygen? Vital signs need to be reassessed and compared with vital signs taken earlier. Does a drop in blood pressure and tachycardia indicate increasing tension in the chest? Many chest injuries will worsen during transport to the hospital because of the seriousness of the injuries. An astute reassessment will help identify worsening conditions in a timely manner so that they can be addressed.

Interventions

Provide complete spinal immobilization of any patient with suspected spinal injuries. Maintain an open airway, be prepared to suction the patient, and consider an oropharyngeal or nasopharyngeal airway. Whenever you suspect significant bleeding, provide high-flow oxygen. If needed, provide assisted ventilation using a bag-mask device with high-flow oxygen. If significant bleeding is visible, you must control the bleeding. If you find penetrating trauma to the chest wall, place an occlusive dressing over the wound; if you find a flail segment, you should manually stabilize it using a bulky dressing. Use caution to avoid increasing the work of breathing and pain. Be prepared to provide positive-pressure ventilation if the patient's efforts are not effective. If the patient has signs of hypoperfusion, treat aggressively for shock and provide rapid transport to the appropriate hospital. Do not delay transport of a seriously injured trauma patient to complete nonlifesaving treatments such as splinting extremity fractures; instead, complete these types of treatments en route to the hospital.

Communication and Documentation

Communicating with hospital staff early when your patient has a significant MOI to the chest can help them be prepared with appropriate equipment and personnel when you arrive. If a penetrating injury is present, describe it in your report, along with what you have done to care for it. If a flail segment is present, hospital staff may be able to offer assistance on how to manage it. Your documentation should be complete and thorough. Describe all injuries and the treatment given. Remember, your documentation is your legal record of what happened.

Special Populations

In young children, the rib cage is very flexible and does not provide the same level of protection that the adult rib cage provides. This flexibility can allow any significant injury or compression of the rib cage to be masked because the ribs give way to the pressure and do not fracture. However, it is important to remember that the organs that underlie the rib cage have been exposed to that force and are likely injured. This flexibility of the ribs may result in fractures that are hidden on examination, and the only indication you may have of compromise is increased work of breathing or alterations in vital signs. This age group is often injured in walking or bicycle accidents that involve automobiles. In auto-pedestrian accidents, children often turn toward the vehicle instead of away as adults do, thus resulting in direct impact to the chest by the bumper or hood. Children also may not be cognizant of height or distances and, therefore, may be prone to falls from distances greater than twice their height, resulting in severe trauma.

Complications and Management of Chest Injuries

Pneumothorax

In any chest injury, damage to the heart, lungs, great vessels, and other organs in the chest can be complicated by the accumulation of air in the pleural space. This is a dangerous condition called a **pneumothorax** (commonly called a collapsed lung). In this condition, air enters through a hole in the chest wall or the surface of the lung as the patient attempts to breathe, causing the lung on that side to collapse **Figure 27-7**. As a result, any blood that passes through the collapsed portion of the lung is not oxygenated, and hypoxia can develop. If the lung is collapsed past 30% to 40% you may hear diminished breath sounds on that side of the chest. Absent breath sounds are a significant finding in chest trauma and may indicate the development of a tension pneumothorax, discussed later. Depending on the size of the hole and the rate at which air fills the cavity, the lung may collapse in a few seconds or a few hours. In the uncommon situation when the hole is in the chest wall, you can actually hear a sucking sound as the patient inhales and the sound of rushing air as he or she exhales. For this reason, an open or penetrating wound to the chest wall is often called an **open pneumothorax** or a **sucking chest wound** **Figure 27-8**.

This type of injury is a true emergency requiring immediate emergency medical care and transport. Initial emergency care, after clearing and maintaining the airway

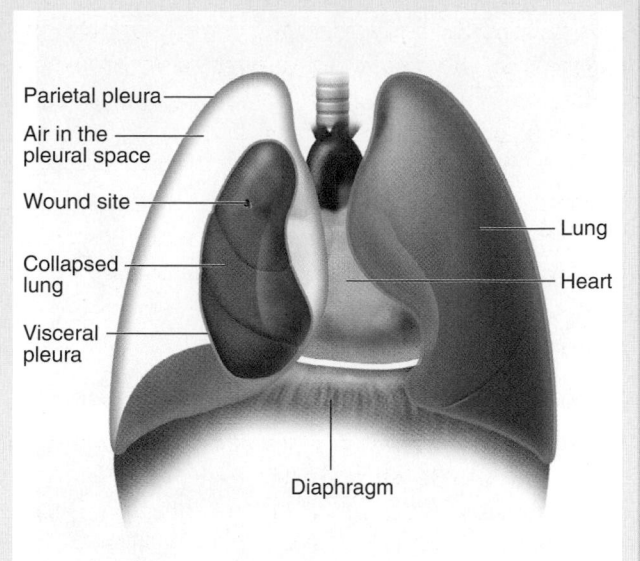

Figure 27-7 Pneumothorax occurs when air leaks into the space between the pleural surfaces from an opening in the chest wall or the surface of the lung. The lung collapses as air fills the pleural space.

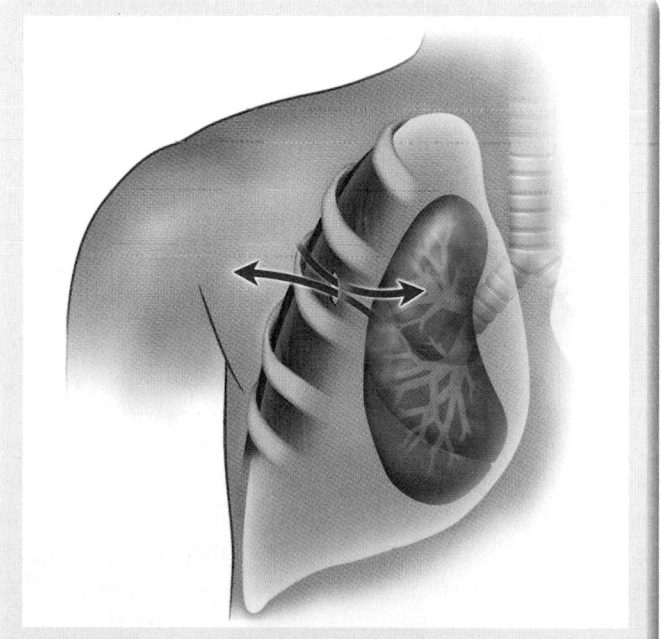

Figure 27-8 With a sucking chest wound, air passes from the outside into the pleural space and back out with each breath, creating a sucking sound.

and then providing oxygen, is to rapidly seal the open wound with a sterile occlusive dressing **Figure 27-9**. The purpose of the dressing is to seal the wound and prevent air from being sucked into the chest through the wound. Several sterile materials, including Vaseline gauze and aluminum foil, may be used to seal the wound. Use a

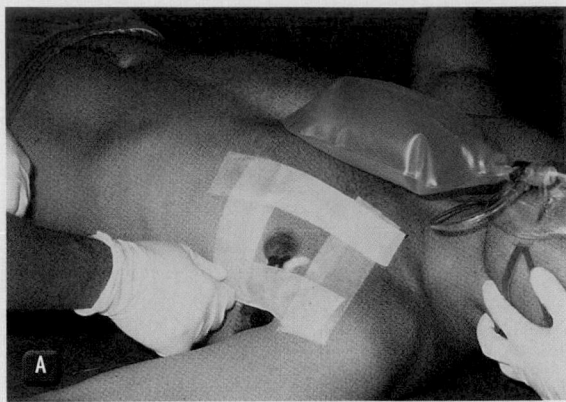

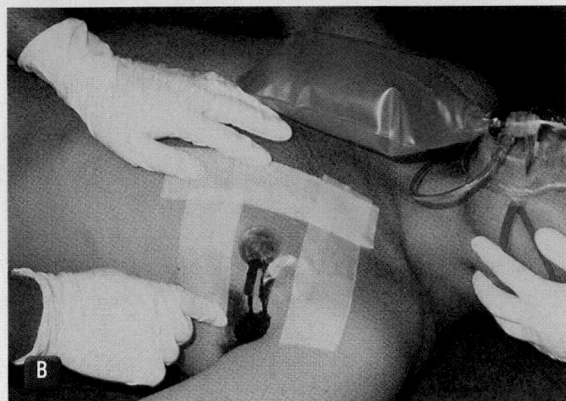

Figure 27-9   A sucking chest wound can be sealed with a large airtight dressing that seals all four sides **(A)** or seals three sides with the fourth left open as a flutter valve **(B)**. Your local protocol will dictate the way you are to care for this injury.

dressing that is large enough so that it is not pulled or sucked into the chest cavity. Depending on your local protocol, you may tape the dressing down on all four sides, or you may create a **flutter valve**, a one-way valve that allows air to leave the chest cavity but not return, by taping only three sides of the dressing. Careful observation is required after the placement of an occlusive dressing. The occlusive seal or a clot in the injury may allow a tension pneumothorax to develop. If signs of a tension pnemothorax develop, it is suggested that the occlusive dressing be partially removed to allow the chest to vent. Sometimes you may hear a sudden release of air pressure when you remove the dressing. Adhere to standard precautions, and be aware that if there is clotting present, it may be expelled with the force of the buildup of pressure. This situation can develop even after a flutter valve has been applied. There are several commercially available products that perform this seal, and some incorporate the one-way valve that is simulated by the three-sided method. Consult local protocols and manufacturers' guidelines for the use of these devices or for other methods that can be used to care for this condition.

Spontaneous Pneumothorax

Some people are born with or develop weak areas on the surface of the lungs. This weakened area of the lung is called a "bleb." Occasionally, such a weak area will rupture spontaneously, allowing air to leak into the pleural space. Usually, this event, called **spontaneous pneumothorax**, is not related to any major injury but simply happens with normal breathing or may occur during times of strenuous physical activity such as exercise or coughing forcefully. The patient experiences sudden sharp chest pain and increasing difficulty breathing. A portion of the affected lung collapses, losing its ability to ventilate normally. The amount of pneumothorax that develops varies, as does the amount of respiratory distress the patient experiences.

You should suspect a spontaneous pneumothorax in a patient who experiences sudden chest pain and shortness of breath without a specific known cause. You may find a decrease in breath sounds on the affected side; however, it is important to note that it takes a loss of greater than 40% of the lung surface to be able to hear this decrease. Instead, it is more prudent to maintain a high index of suspicion in these cases and assume that a pneumothorax is developing. You will see an increase in the respiratory rate and work of breathing long before you may hear the decreased breath sounds. There may also be hyperinflation of the injured side or hemothorax as a result of the air trapped in the pleural space. The prehospital treatment that you can provide for this type of pneumothorax is to administer oxygen and transport. Be alert for signs of developing tension pneumothorax, and consider ALS intercept or rendezvous for this life-threatening change in status.

Simple Pneumothorax

Any pneumothorax that does not result in major changes in the patient's physiology is referred to as a **simple pneumothorax**. These are commonly the result of blunt trauma that results in fractured ribs. As in the spontaneous pneumothorax, the simple pneumothorax is often difficult to diagnose. The lung has to collapse a significant amount before the effects will be heard as decreased breath sounds. The more common findings are similar to

Words of Wisdom

Be cautious about providing positive-pressure ventilation to a patient with a suspected pneumothorax. The positive pressure may cause the size of the pneumothorax to increase. If the patient requires positive-pressure ventilation, you must provide it, but be certain to communicate your concerns about the underlying pneumothorax at the receiving hospital.

other types of pneumothoraces: dyspnea increased work of breathing exhibited as increased rate; tachypnea, and accessory muscle use; and decreasing oxygen saturation on the pulse oximeter. Late findings can be decreased breath sounds on the injured side as well as lethargy and cyanosis. Be vigilant because the simple pneumothorax can often worsen or deteriorate into a tension pneumothorax or develop complications like bleeding or hemothorax. The treatment for a simple pneumothorax is much like any treatment for respiratory compromise; provide a high flow of oxygen. Monitor oximeter readings and breath sounds, and treat underlying causes of the injury. As in all pneumothorax treatment, adding positive-pressure ventilation will cause the pathology to advance rapidly and possibly cause a tension pneumothorax to develop. However, you should not withhold positive-pressure ventilation if the patient needs the support. Simply be aware of the risk, and plan on how to

resolve complications. Most patients with this problem require ALS intervention, so call for it early.

Tension Pneumothorax

A potential complication that may develop following chest injuries with pneumothorax is a **tension pneumothorax** Figure 27-10 . This can occur when there is significant ongoing air accumulation in the pleural space. This air gradually increases the pressure in the chest, first causing the complete collapse of the affected lung and then pushing the mediastinum (the central part of the chest containing the heart and great vessels) into the opposite pleural cavity. This prevents blood from returning through the venae cavae to the heart, decreasing cardiac output, causing shock, and ultimately leading to death.

Tension pneumothorax occurs more commonly as a result of closed, blunt injury to the chest in which a fractured rib lacerates a lung or bronchus. Only very rarely does a tension pneumothorax arise spontaneously.

The common signs and symptoms of tension pneumothorax include increasing respiratory distress, altered level of consciousness, distended neck veins, deviation of the trachea to the side of the chest opposite the tension pneumothorax, tachycardia, low blood pressure, cyanosis, and decreased breath sounds on the side of the pneumothorax. It should be noted that tracheal deviation is often a late sign and, if seen, the situation is grave and the patient requires intervention immediately.

You are the Provider: PART 4

The patient is placed onto the stretcher and loaded into the ambulance. While continuing to provide high-flow oxygen via nonrebreathing mask, you reassess his condition and vital signs shortly before departing the scene.

Recording Time: 10 Minutes	
Level of consciousness	Conscious, but confused and restless
Respirations	28 breaths/min; labored and shallow
Pulse	124 beats/min; weak at the radial artery
Skin	Cool, clammy, and pale; cyanosis around the mouth
Blood pressure	104/58 mm Hg
Sao$_2$	88% (on oxygen)

Breath sounds are now inaudible on the entire left side of his chest, and you note that cyanosis is developing around his mouth. His jugular veins appear somewhat distended, and his trachea is midline. The closest appropriate facility is about 10 minutes away, so you instruct your partner to begin transport at once.

6. What is most likely happening to your patient?

7. Should you adjust your current treatment? If so, how?

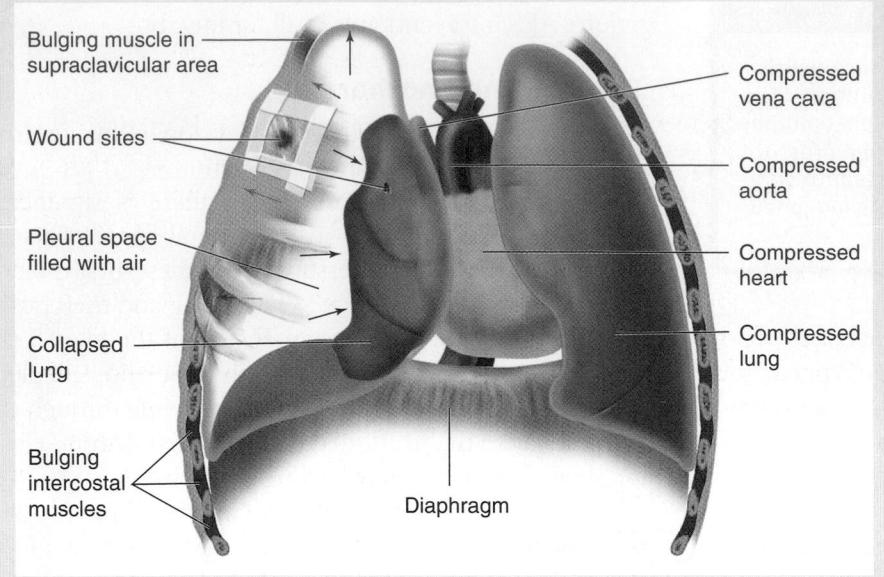

Bulging muscle in supraclavicular area

Wound sites

Pleural space filled with air

Collapsed lung

Bulging intercostal muscles

Compressed vena cava

Compressed aorta

Compressed heart

Compressed lung

Diaphragm

Figure 27-10 A tension pneumothorax can develop if a penetrating chest wound is bandaged tightly and air from a damaged lung cannot escape. The air then accumulates in the pleural space, eventually causing compression of the heart and great vessels.

Relieving a tension pneumothorax that is the result of blunt trauma is often done by inserting a needle through the rib cage into the pleural space; however, this procedure typically is performed by ALS personnel or emergency department staff depending on local protocols. A tension pneumothorax is a life-threatening condition. Be prepared to support ventilation with high-flow oxygen, and request ALS support or transport immediately to the closest hospital. Needle decompression may be performed at the hospital.

■ Hemothorax

In blunt and penetrating chest injuries, blood can collect in the pleural space from bleeding around the rib cage or from a lung or great vessel. This condition is called a **hemothorax** **Figure 27-11**. You should suspect a hemothorax if the patient has signs and symptoms of shock or decreased breath sounds on the affected side, an indication that the lung is being compressed by the blood. Because the bleeding is typically caused from severe damage within the chest cavity, there is virtually no way to control the bleeding in the prehospital setting. The only person who can treat this condition is a surgeon. The presence of air and blood in the pleural space is known as a **hemopneumothorax**. Again, because the injury has occurred within the walls of the chest, the treatment involves providing rapid transport to the nearest facility capable of performing surgery.

■ Cardiac Tamponade

Cardiac tamponade (pericardial tamponade) occurs more commonly in the presence of penetrating chest trauma, although it may occur in blunt trauma. Cardiac tamponade occurs when the protective membrane around the heart (**pericardium**), the pericardial sac, fills with blood or fluid, perhaps from a ruptured, torn, or lacerated coronary artery or vein **Figure 27-12**. It can also occur as a result of infection filling the sac with pus. As the fluid amount increases, the heart is less able to fill with blood during each relaxation phase. As a result, the heart cannot pump an adequate amount of blood and the patient experiences a decrease in systemic blood flow, or cardiac output. The signs of this condition are often subtle until the situation is dire. The signs and symptoms, referred to as Beck's triad, include distended or engorged jugular veins seen on both sides of the trachea, a narrowing pulse pressure

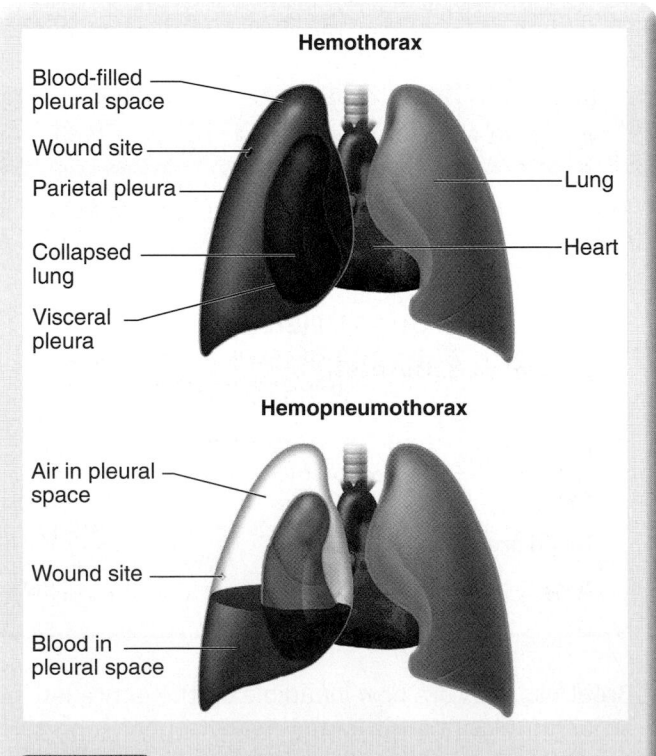

Hemothorax

Blood-filled pleural space

Wound site

Parietal pleura

Collapsed lung

Visceral pleura

Lung

Heart

Hemopneumothorax

Air in pleural space

Wound site

Blood in pleural space

Figure 27-11 **A.** A hemothorax is a collection of blood in the pleural space produced by bleeding within the chest. **B.** When both blood and air are present, the condition is a hemopneumothorax.

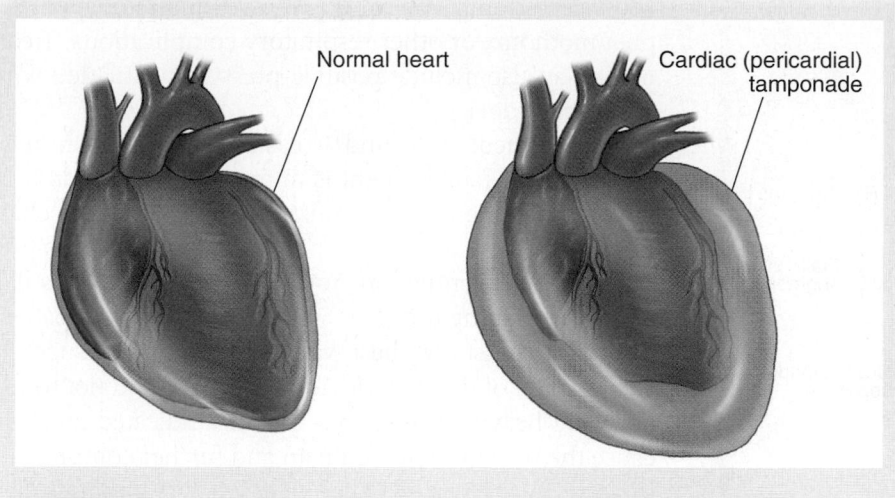

Figure 27-12 Cardiac pericardial tamponade is a potentially fatal condition in which fluid builds up within the pericardial sac, causing compression of the heart's chambers and dramatically impairing its ability to pump blood to the body.

(the difference between the systolic and diastolic blood pressure numbers), and muffled heart sounds. An associated and more commonly noticed sign is a decrease in mental status as blood flow decreases to the brain. The heart muscle is unique in that it needs to be stretched to create a good contraction to pump blood out of the ventricles. This mechanism, known as the Frank-Starling mechanism, can fail because of tamponade and can be directly related to a decrease in blood returning to the heart from a loss of blood or from some blockage of the returning veins. Because this injury is inaccessible, your role in treatment is supportive. Oxygen should never be withheld from a patient who needs it; however, you must weigh the need for positive-pressure ventilations against the possibility of hypoventilation. Rapidly transport the patient to a facility that is capable of intervention.

Words of Wisdom

In a trauma situation, even a small amount of fluid in the pericardial sac is enough to cause fatal pericardial tamponade. (Occasionally, fluid in surprisingly large amounts may collect in the pericardial sac as a chronic condition.)

■ Rib Fractures

Rib fractures are very common, particularly in older people, whose bones are brittle. Because the upper four ribs are well protected by the bony girdle of the clavicle and scapula, a fracture of one of these upper ribs is a sign of a very substantial MOI.

Be aware that a fractured rib that penetrates into the pleural space may lacerate the surface of the lung, causing a pneumothorax, a tension pneumothorax, a hemothorax, or a hemopneumothorax. One sign of this development can be a crackly feeling to the skin in the area (also called *crepitus* or *subcutaneous emphysema*), which indicates that air escaping from a lacerated lung is leaking into the chest wall. Be sure to relay this finding to hospital personnel.

Patients with one or more cracked ribs will report localized tenderness and pain when breathing. The pain is the result of broken ends of the fracture rubbing against each other with each inspiration and expiration. Patients will tend to avoid taking deep breaths and their breathing will be rapid and shallow instead. They will often hold the affected portion of the rib cage in an effort to minimize the discomfort. Patients with rib fractures should receive supplemental oxygen during assessment and transport. Note that in young children, the rib cage is very flexible and does not provide the same level of protection that the adult rib cage provides. This flexibility can allow any significant injury or compression of the rib cage to be masked as the ribs give way to the pressure and do not fracture. However, it is important to remember that the organs that underlie the rib cage have been exposed to that force and are likely injured.

■ Flail Chest

Ribs may be fractured in more than one place. If three or more ribs are fractured in two or more places or if the sternum is fractured along with several ribs, a segment of chest wall may be detached from the rest of the thoracic cage **Figure 27-13**. This condition is known as **flail chest**. In what is called paradoxical motion, the detached portion of the chest wall moves opposite of normal: it moves in instead of out during inhalation and out instead of in during exhalation. This occurs because of negative pressure that has built up in the thorax. Breathing with a flail chest can be painful and ineffective, and hypoxemia easily results. A flail segment seriously interferes with the body's normal mechanics of ventilation and must be addressed quickly.

Your treatment of a patient with a flail chest should include maintaining the airway, providing respiratory support if necessary, giving supplemental oxygen,

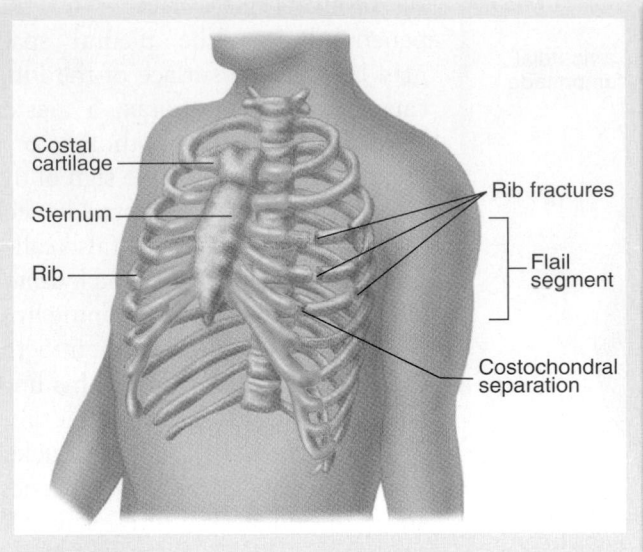

Figure 27-13 When two or more adjacent ribs are fractured in two or more places, a flail chest results. A flail segment will move paradoxically when the patient breathes.

and performing ongoing assessments for possible pneumothorax or other respiratory complications. Treatment may also include positive-pressure ventilation with a bag-mask device.

The patient may find it easier and less painful to breathe if the flail segment is immobilized. You can tape a bulky dressing or pad against that segment of the chest for this purpose, although taping too tightly will also prevent adequate ventilation. You can also immobilize a flail chest by splinting the chest with a pillow that the patient can hold against the chest wall. If your protocol suggests stabilization of the flail segment, keep in mind not to use anything heavy because this is contraindicated and will cause the patient increased pain and further compromise the respiratory system. Keep in mind that although flail chest itself is a serious condition, it suggests an injury that was forceful enough to cause other serious internal damage and possible spinal injury. Often the flail chest contributes less to the patient's ventilation difficulties than does the underlying pulmonary contusion (bruised lung segment).

You are the Provider: PART 5

You call for ALS support but find out that the closest unit is 15 miles away, so you elect to continue treatment and transport. After reassessing the patient's condition, you have your partner call a radio report to the receiving facility.

Recording Time: 15 Minutes	
Level of consciousness	Conscious, but confused and restless
Respirations	28 breaths/min; labored and shallow
Pulse	120 beats/min; weak at the radial artery
Skin	Cool and clammy; cyanosis around the mouth
Blood pressure	90/60 mm Hg
Sao$_2$	92% (on oxygen)

You arrive at the hospital and find that a physician and nurse are waiting for you in the ambulance bay. The patient is quickly taken into a treatment room, where further assessment is performed. After the physician performs a needle thoracentesis, the patient's condition improves. After returning to service, you follow-up with the hospital and learn that the patient had a tension hemopneumothorax.

8. What is a tension hemopneumothorax?

9. Should the EMT attempt to distinguish a tension pneumothorax from a tension hemopneumothorax? Why or why not?

Other Chest Injuries

Pulmonary Contusion

In addition to fracturing ribs, any severe blunt trauma to the chest can also injure or bruise the lung. The pulmonary alveoli become filled with blood, and fluid accumulates in the injured area, leaving the patient hypoxic. Severe **pulmonary contusion** should always be suspected in patients with a flail chest and usually develops during a period of hours. If you believe that a patient may have a pulmonary contusion, you should provide respiratory support and supplemental oxygen to ensure adequate ventilation.

Other Fractures

In addition to the rib fractures you have already learned about, there are other types of fractures that should be discussed.

Sternal Fractures

Any suspected fracture of the sternum should increase your index of suspicion for injuries to the underlying organs because the amount of force required to break the sternum is significant. There may be involvement of the lungs, great vessels, and the heart itself.

Clavicle Fractures

Whereas this fracture is also covered under skeletal injuries, it is important to mention here that the clavicle overlies and protects a large neurovascular bundle (nerve, artery, and vein) that can be significantly damaged or disrupted should injury to the clavicle occur. The pain, deformity, and swelling that accompany a clavicle fracture can also detract from assessment of the first and second ribs in proximity to the fracture. Suspect upper rib fractures in medial clavicle fractures, and be alert to possible signs of pneumothorax development.

Traumatic Asphyxia

Sometimes a patient will experience a sudden, severe compression of the chest, which produces a rapid increase in pressure within the chest. This may occur in an unrestrained driver who hits a steering wheel or a pedestrian who is compressed between a vehicle and a wall. The sudden increase in intrathoracic pressure results in a characteristic appearance, including distended neck veins, cyanosis in the face and neck, and hemorrhage into the sclera of the eye, signaling the bursting of small blood vessels Figure 27-14 . This is called **traumatic**

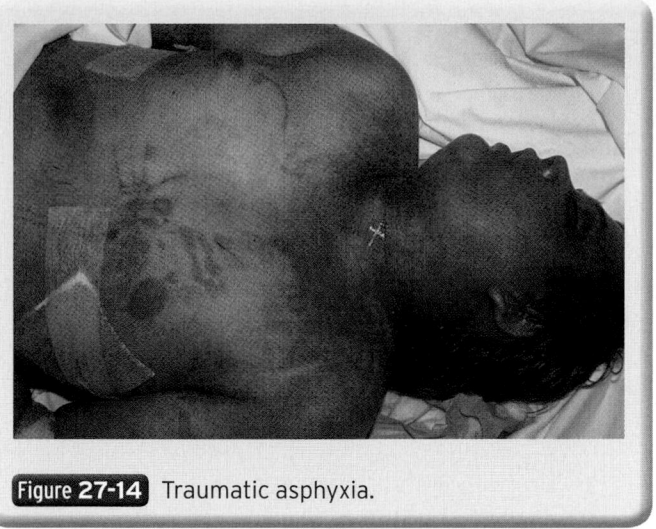

Figure 27-14 Traumatic asphyxia.

asphyxia. These findings suggest an underlying injury to the heart and possibly a pulmonary contusion. You should provide ventilatory support with supplemental oxygen and monitor the patient's vital signs as you provide immediate transport.

Blunt Myocardial Injury

Blunt trauma to the chest may injure the heart itself, making it unable to maintain adequate blood pressure. There is much debate in the medical literature about how to assess **myocardial contusion**, or bruising of the heart muscle. Often the pulse rate is irregular, but dangerous rhythms such as ventricular tachycardia and ventricular fibrillation are uncommon. There is no specific diagnostic test at this time, and there is no prehospital treatment for the condition. Still, you should suspect myocardial contusion in all cases of severe blunt injury to the chest. Check the patient's pulse carefully, and note any irregularities. Also note any change in blood pressure because this can be a direct result of the injury to the myocardium. Often the patient's signs and symptoms can mimic a heart attack in which the patient may complain of chest pain or discomfort that is similar in nature to cardiac symptoms. Provide supplemental oxygen, and transport immediately.

Commotio Cordis

Commotio cordis is a blunt chest injury caused by a sudden, direct blow to the chest (over the heart) that occurs only during the critical portion of a person's heartbeat. The result may be immediate cardiac arrest. This phenomenon has been documented to have occurred after patients were struck with softballs, baseballs, bats, snowballs, fists, and even kicks during kickboxing. The force of the blow to the chest is commonly at speeds of 35 to 40 mph. The blunt force causes a lethal abnormal heart rhythm called ventricular fibrillation. The ventricular fibrillation responds positively to early defibrillation if provided within the first 2 minutes after the injury. Commotio cordis is more commonly associated with sports-related injuries, although you should maintain a high index of suspicion for this condition in all cases in which the person is unconscious and unresponsive after a blow to the chest.

■ Laceration of the Great Vessels

The chest contains several large blood vessels: the superior vena cava, the inferior vena cava, the pulmonary arteries, four main pulmonary veins, and the aorta, with its major branches distributing blood throughout the body. Injury to any of these vessels may be accompanied by massive, rapidly fatal hemorrhage. Any patient with a chest wound who shows signs of shock may have an injury to one or more of these vessels. Frequently, significant blood loss is unseen because it remains within the chest cavity. You must remain alert to signs and symptoms of shock and to changes in the baseline vital signs, such as tachycardia and hypotension.

Emergency treatment in these cases includes cardiopulmonary resuscitation, if appropriate, ventilatory support, and supplemental oxygen. Here, particularly, immediate transport to the hospital may be critical. Occasionally, some of the patients can be treated. The overwhelming majority of injuries to the great vessels in the chest are rapidly fatal.

Words of Wisdom

One in every five deaths due to blunt trauma includes a transection of the aorta; the most common causes are high-speed motor vehicle crashes and falls from a height. Each year, 5,000 to 8,000 people in the United States die as a result of aortic or great vessel rupture. Given that the body's entire blood volume passes through this vessel, the high mortality associated with such an injury comes as no surprise. Of those patients who experience an aortic injury, only a few will survive until EMS units arrive; most of the individuals reached by EMS personnel can survive with prompt management including surgical intervention.

The most widely accepted theory of how this injury evolves suggests that the aorta is injured at its fixed points due to shearing forces. The high-velocity, high-energy impacts that result in these injuries cause the aortic arch to swing forward. The resulting tension on the area causes the descending aorta to rupture at its point of attachment to the posterior thoracic wall.

You are the Provider: SUMMARY

1. **What major organs and structures lie within the chest cavity?**

   To adequately assess and treat a patient with a chest injury, it is important for you to know what anatomic structures reside within the chest (thoracic) cavity and how they function. Chest injuries can range in severity from simple rib fractures to impairment in ventilation and internal bleeding with shock.

   The chest (thoracic cage) extends from the clavicles (superiorly) to the diaphragm (inferiorly). The organs and structures in the chest are protected by the ribs, which connect to the spinal vertebrae posteriorly and to the sternum (via the costal cartilages) anteriorly.

   It is important to note that when a person is lying down or has just completed exhalation, the diaphragm may rise as high as the nipple line. Thus, a penetrating chest injury (eg, gunshot, stab wound) may also penetrate the diaphragm, liver, or stomach, in addition to the thoracic organs and structures. Anatomically, the diaphragm separates the thoracic and abdominal cavities.

   Critical organs and structures in the chest cavity include the trachea, large bronchi, lungs, great vessels (eg, aorta, venae cavae), heart, esophagus, and the left and right subclavian arteries and their branches. The heart, trachea, great

   vessels, and a portion of the esophagus reside within the mediastinum—the space between the lungs.

   Each side of the chest (hemithorax) contains lung tissue that is separated into lobes, the right lung has three lobes, and the left lung has two lobes. The left lung has only two lobes to leave space for the heart. The space in the left hemithorax where the heart resides is called the cardiac notch.

   The lungs are covered by a thin lining called the visceral pleura; the inner chest wall is covered by the parietal pleura. A small amount of pleural fluid is found between the pleurae; it allows the lungs to move freely against the inner chest wall during breathing, thus facilitating the mechanics of ventilation. The space in between the visceral and parietal pleurae—the pleural space—is a potential space because the pleurae are in direct contact with each other. However, if trauma occurs to the lungs, blood or air could accumulate between the two pleurae, creating an actual space.

   The trachea, which is located in the midline of the neck, extends into the chest cavity and divides into the left and right mainstem bronchi, which allows air to move into and out of the lungs.

   The esophagus extends from the pharynx (throat) to the stomach. A portion of the esophagus lies posterior to the heart, trachea, and great vessels in the mediastinum.

You are the Provider: SUMMARY, continued

The heart is a double pump (the atria and ventricles contract independently of each other) that receives deoxygenated blood from the body via the superior and inferior venae cavae. Blood is then sent to the lungs, where carbon dioxide is exchanged for oxygen (pulmonary respiration). After being reoxygenated in the lungs, blood is sent to the left side of the heart where it is pumped throughout the entire body, beginning with the aorta, the largest artery in the body.

---

2. What injuries commonly result from blunt chest trauma? Penetrating chest trauma?

A critical component of any scene size-up is noting the mechanism of injury. With this information, you are often able to predict—to a certain degree—the type of injury (or injuries) the patient may have experienced. Blunt and penetrating chest trauma can cause a variety of injuries and can result in ventilation impairment, internal bleeding, damage to the heart, or a combination of any of these.

Blunt (closed) chest trauma occurs when an object strikes the chest (eg, a steel pipe during an assault) or when the chest strikes an object (eg, chest impacts the steering wheel during a car crash). The severity of the injury depends on the amount of energy that the chest wall sustained. Although the skin and chest wall are not penetrated by blunt force trauma, injury to the intrathoracic organs may be severe.

Rib fractures are a common injury associated with blunt chest trauma. In some cases, a single rib is fractured; in other cases, several ribs are fractured in more than one place (flail chest). A flail chest can impair ventilation because the section of fractured ribs collapses and puts pressure on the lung during inhalation. A fractured rib can also perforate an internal chest organ, such as the lung, causing air to leak out of the lung and into the pleural space (pneumothorax). If air freely moves between the pleural space and lung, the injury is called a simple pneumothorax. However, if the perforation in the lung acts as a one-way valve, air accumulates in the pleural space but cannot escape; therefore, pressure within the pleural space increases significantly, causing total collapse of the lung and shifting of pressure to the opposite side of the chest (tension pneumothorax). In a tension pneumothorax, the lung is completely collapsed, and the heart and great vessels (eg, aorta, venae cavae) are compressed as pressure shifts across the mediastinum; this causes ventilatory compromise, decreased cardiac output, and shock. A hemothorax, which may be caused by blunt or penetrating chest trauma, occurs when blood fills the pleural space rather than air.

Blunt force chest trauma can also cause bruising of the lung (pulmonary contusion) or the heart muscle (myocardial contusion). Pulmonary contusions are often associated with flail chest injuries; if the contusion is large enough, oxygenation and ventilation can become impaired. In a myocardial contusion, the heart itself is damaged, potentially making it unable to maintain adequate cardiac output.

Other injuries that may result from blunt chest trauma include shearing injuries of the aorta, which causes profound internal bleeding, and traumatic asphyxia, which occurs when the chest is suddenly compressed (eg, a person is pinned between two solid objects) and often results in injury to multiple organs in the chest cavity.

Commotio cordis is a unique blunt chest injury caused by a sudden, direct blow to the chest during a critical portion of the heartbeat; the result could be immediate cardiac arrest because of a lethal cardiac arrhythmia (ie, ventricular fibrillation). Patients who are struck in the chest by baseballs, snowballs, fists, or even kicks during kickboxing may experience this phenomenon.

In a penetrating (open) chest injury, the chest wall itself is penetrated by an object such as a knife or bullet, resulting in injury to vital organs in the thoracic cavity. Two common injuries caused by penetrating chest trauma are sucking chest wounds (open pneumothorax) and cardiac (pericardial) tamponade. In a sucking chest wound, air is drawn into the pleural space from an opening in the chest wall—unlike a simple or tension pneumothorax, in which air leaks into the pleural space from the lung. Sucking chest wounds cause various degrees of ventilation compromise, depending on the size of the hole in the chest wall. Cardiac tamponade occurs when blood accumulates in the pericardium—the fibrous sac that surrounds the heart—following penetration of the heart itself or one of the coronary arteries or veins. As the amount of blood in the pericardial sac increases, the heart is less able to fill with blood during diastole (the relaxation phase of the heart); as a result, the amount of blood pumped decreases and the patient will experience shock from a decrease in systemic blood flow.

---

3. How should you proceed with your secondary assessment?

Your secondary assessment should focus on areas of the body affected by the patient's chief complaint. For example, a patient who was ejected from his or her vehicle during a crash requires a full-body scan because the potential for multiple life-threatening injuries to more than one part of the body is high. For patients with an isolated injury, however, your secondary assessment should focus on that area of the body—in this case, the chest and adjacent structures.

Expose the patient's chest, and assess for obvious signs of injury, such as bruising, lacerations, or abrasions. Although bruising is an obvious indicator of injury, it may not appear immediately; you may only see a red mark where the object struck the patient.

Observe the chest wall for symmetry; Do both sides of the chest wall rise and fall equally (symmetrical chest movement)? Or, does one side of the chest move less than the other (asymmetrical chest movement)? Asymmetrical chest movement indicates decreased airflow into one lung. Look for any sections of the rib cage that collapse during inhalation and bulge during exhalation (paradoxical chest movement); this indicates a flail chest and should be stabilized immediately.

Palpate the chest wall to determine if it is stable, if there are any deformities, or if you feel crepitus—the sensation felt when broken bone ends grind together. Chest wall crepitus is a clear indicator of one or more fractured ribs.

Auscultate the apices (top) and bases (bottom) of both lungs to determine if breath sounds are clear and equal bilaterally

You are the Provider: SUMMARY, continued

or if there is a decrease in or absence of breath sounds on one side or the other. Decreased or absent breath sounds over the injured side of the chest indicate decreased or absent airflow into that lung and should immediately increase your index of suspicion for a pneumothorax.

Based on the mechanism of injury—blunt chest trauma— you should also assess the trachea and jugular veins. Note whether the trachea is in the midline position or if it appears to be deviated to one side or the other. Bear in mind that if the patient has a tension pneumothorax, tracheal deviation is an extremely late sign; a midline trachea does *not* rule out such an injury. Observe the jugular veins to determine if they appear to be normal, distended, or flat. Jugular vein distention is best assessed for with the patient sitting at a 45° angle. The presence of jugular vein distention in the context of chest trauma suggests a tension pneumothorax or cardiac tamponade, although it is often not present until the injury is well progressed. Conversely, collapsed jugular veins suggest a hemothorax.

As previously mentioned, further assessment beyond that discussed is based on any other complaints the patient may have and any obvious injuries that you observe.

4. On the basis of your assessment findings, what injury or injuries should you suspect?

On the basis of the mechanism of injury—blunt trauma to the chest—and the findings of your primary and secondary assessments, you should suspect that your patient has rib fractures and a pneumothorax. He has obvious chest wall bruising, labored breathing, crepitus to palpation (indicates at least one fractured rib), and diminished breath sounds on the same (ipsilateral) side that the injury occurred.

A pneumothorax—commonly referred to as a collapsed lung— occurs when air accumulates in the pleural space. Following blunt chest trauma, this is often caused by a fractured rib that penetrates the lung, resulting in air leakage out of the lung and into the pleural space. As the patient breathes, pressure in the pleural space increases and the lung begins to collapse. As a result, any blood that passes through the collapsed portion of the lung is not reoxygenated, and hypoxemia can develop.

Depending on the size of the lung perforation and the rate at which air fills the pleural space, the lung may collapse in a few seconds or a few hours. Breath sounds on the injured side are diminished because the collapsing lung is not fully expanding during inhalation.

Because a pneumothorax impairs oxygenation and ventilation, you should begin treatment immediately and prepare for rapid transport to the hospital. If the lung on the injured side collapses totally, pressure will shift to the opposite (contralateral) side of the chest—compressing the heart, aorta, and venae cavae in the process—and begin collapsing the unaffected lung; this condition is called a tension pneumothorax and is an *immediately* life-threatening condition that can result in shock and cardiac arrest.

5. How should you proceed with your treatment of this patient?

Patients with a pneumothorax need high-flow oxygen, continual close monitoring, and prompt transport to the hospital. Pneumothoraces do not heal spontaneously and require definitive treatment.

You should be especially concerned about the adequacy of the patient's breathing and should continuously monitor it. Your patient is tachypneic, which is common in patients with chest injuries; his respirations are labored, and it hurts when he breathes. Patients with chest injuries often breathe shallowly on purpose in an attempt to minimize the pain (respiratory splinting). It may be helpful to allow the patient to hold a pillow or similar object to his chest; this will help stabilize any rib fractures and may make it easier for him to breathe.

If his respirations become too shallow (indicates a marked tidal volume reduction), he will not move adequate amounts of air into his lungs during inhalation; this will only worsen any hypoxemia that already exists from the pneumothorax itself. If this occurs, begin assisting his ventilations with a bag-mask device attached to high-flow oxygen. Other signs of inadequate breathing, which may also necessitate assisted ventilations, include a falling oxygen saturation (despite administration of high-flow oxygen), cyanosis, and a decreasing level of consciousness.

Continue to monitor the patient's level of consciousness and vital signs. If he begins experiencing signs of shock (ie, decreased level of consciousness, pallor, weak pulses, hypotension), you should suspect that a tension pneumothorax is developing. If this occurs, you should consider an ALS intercept, if possible. Advanced EMTs and paramedics can perform a needle thoracentesis (also called a chest decompression); in this procedure, a needle is inserted through the chest wall to remove air from the pleural space and allow the lung to reexpand. If an ALS ambulance is not available, continue rapid transport, frequently reassess the patient, and notify the receiving facility as early as possible.

6. What is most likely happening to your patient?

Compared with previous assessments, your patient's clinical condition has obviously deteriorated. He is now confused; cyanosis is more tachypneic; his respirations are still labored, but are now shallow; cyanosis is developing around his mouth (perioral cyanosis); his radial pulses are weak; and his oxygen saturation has fallen to 88%, despite high-flow oxygen via nonrebreathing mask. Although his systolic blood pressure is still above 100 mm Hg, compared with your previous reading of 138/88 mm Hg, this is a significant decrease. The patient's signs and symptoms indicate worsened hypoxemia and hypoperfusion (shock).

The additional signs of inaudible (absent) breath sounds on the injured side of his chest and the appearance of jugular vein distention indicate that the entire left lung has collapsed, and pressure is now shifting across the mediastinum toward the unaffected lung. Your patient now has a tension pneumothorax!

You are the Provider: SUMMARY, continued

As pressure shifts across the mediastinum, the venae cavae are compressed, which is reducing blood return to the heart (preload) and is causing it to back up into the systemic circulation; clinically, this manifests as jugular vein distention. The heart and aorta are also compressed, which is reducing cardiac output; clinically, this manifests with a falling blood pressure and other signs of shock.

7. Should you adjust your current treatment? If so, how?

Treatment for a simple pneumothorax without signs of shock is supportive (ie, monitor the ABCs, give oxygen, allow the patient to assume a position of comfort, and transport). However, a tension pneumothorax is an immediate life threat and requires more aggressive treatment.

The patient now has clear evidence of inadequate breathing (ie, decreased level of consciousness [confusion]; falling oxygen saturation; labored, shallow breathing; and cyanosis), which is causing inadequate oxygenation and ventilation. You should begin assisting his ventilations with a bag-mask device. In doing so, however, you must exercise caution; ventilating too rapidly or with too much force could worsen his condition.

You must also initiate shock treatment. At a minimum, cover the patient with a blanket to keep him warm. Patients with any injury or condition that impairs their ability to breathe are often resistant to being placed in a supine position; this is especially true if they are conscious. If the patient becomes unconscious, however, you should place him in a supine position and continue to assist his ventilations.

Patients with a tension pneumothorax need an immediate needle thoracentesis (chest decompression). If it will not delay your transport time (you are only 10 minutes away from the hospital), intercept an ALS unit, if possible; AEMTs and paramedics can perform this lifesaving intervention. Otherwise, notify the receiving facility, continue ventilation assistance and shock treatment, and get the patient to the emergency department as soon as possible.

8. What is a tension hemopneumothorax?

In some cases, when a fractured rib perforates a lung, blood from the injured lung also accumulates in the pleural space; this is called a hemopneumothorax because the pleural space contains *both* blood and air. The amount of blood that accumulates in the pleural space depends on the size and severity of the lung injury. In some cases, blood accumulation is minimal; in other cases, it is significant.

As blood and air continue to accumulate in the pleural space, the lung on the injured side collapses and pressure shifts across the mediastinum toward the uninjured lung. In a tension hemopneumothorax, the patient experiences respiratory impairment from both blood and air in the pleural space, but also experiences internal blood loss of varying severity.

9. Should the EMT attempt to distinguish a tension pneumothorax from a tension hemopneumothorax? Why or why not?

It is impractical for the EMT to attempt to distinguish a tension pneumothorax from a tension hemopneumothorax in the prehospital setting. Both conditions cause impaired ventilation and perfusion; this is what the EMT should focus on when he or she is treating the patient.

In some cases, large amounts of blood *and* air accumulate in the pleural space. If this happens, the patient's total blood volume decreases because of severe intrathoracic hemorrhage and the jugular veins may appear less distended than you would expect if the patient were experiencing a tension pneumothorax. If the jugular veins are flat, you should suspect a massive hemothorax.

Although there is an anatomic difference between a tension pneumothorax and a tension hemopneumothorax, treatment at the EMT level is the same for both conditions—provide high-flow oxygen, assist ventilations as needed, initiate shock treatment, and transport without delay. Attempting to distinguish one injury from the other may only delay treatment and transport, thereby increasing the chance of a negative outcome.

EMS Patient Care Report (PCR)

Date: 11-5-09	Incident No.: 012709	Nature of Call: Chest injury		Location: 233 Indian Hills Dr.	
Dispatched: 1020	En Route: 1020	At Scene: 1028	Transport: 1035	At Hospital: 1045	In Service: 1053

Patient Information

Age: 19 Sex: M Weight (in kg [lb]): 66 kg (145 lb)	Allergies: Penicillin Medications: None Past Medical History: None Chief Complaint: Difficulty breathing secondary to chest injury

Vital Signs

Time: 1030	BP: 138/88	Pulse: 110	Respirations: 24	Sao$_2$: 95%
Time: 1035	BP: 104/58	Pulse: 124	Respirations: 28	Sao$_2$: 88%
Time: 1040	BP: 90/60	Pulse: 120	Respirations: 28	Sao$_2$: 92%

You are the Provider: SUMMARY, continued

EMS Treatment (circle all that apply)				
Oxygen @ _15_ L/min via (circle one): NC (NRM) (Bag-Mask Device)		(Assisted Ventilation)	**Airway Adjunct**	**CPR**
Defibrillation	**Bleeding Control**	**Bandaging**	**Splinting**	**Other:** (Shock treatment, blanket for warmth, pillow to chest for pain relief)

Narrative
Medic 30 dispatched to a construction site for a patient with a "chest injury." On arrival at the scene, a coworker directed the EMS crew to the patient, a 19-year-old man. He was found sitting on the ground. He was conscious, but restless; his airway was patent; his breathing was labored; and no gross external bleeding was noted. Patient complains of chest pain and difficulty breathing after being struck in the left anterior part of the chest by a 2 × 4 board, which slid off a table saw when another coworker was cutting it. He denies any other injuries, and coworkers confirm that there was no loss of consciousness. Applied oxygen at 15 L/min via nonrebreathing mask and performed further assessment. Obvious bruising was noted to left anterior part of the chest, as was crepitus on palpation. Remainder of chest wall was stable. Breath sounds were diminished to left side of the chest. No paradoxical chest wall movement was noted, jugular veins were nondistended, and trachea was midline. Gave patient a pillow to place against his chest because he stated that it hurt to take a breath. Obtained vital signs, placed patient onto stretcher, and loaded him into the ambulance. Reassessed patient's condition and vital signs and noted obvious deterioration. He was now confused, more tachypneic, and his respirations were markedly shallow. Oxygen saturation read 88%, and skin was now cool and clammy, with perioral cyanosis. Breath sounds were now absent on the left side of the chest. Began assisting ventilations with bag-mask device attached to high-flow oxygen and initiated transport. Intercept with ALS unit was not possible because it would have caused a significant delay in transport. Covered patient with a blanket for warmth, continued ventilation assistance, and reassessed patient every 5 minutes throughout transport. Delivered patient to hospital, gave verbal report to attending physician, and transferred patient care to hospital staff. Medic 30 returned to service at 1053. *End of report*

Assessment and Emergency Care of Chest Injuries

Scene Size-up

Scene Safety	Ensure scene safety and safe access to the patient. Be aware of potential violence or the possibility of a crime scene, and call for law enforcement if needed; disturb potential evidence as little as possible. Follow standard precautions, putting on a minimum of gloves and eye protection. Scenes with multiple patients may require the EMT to carry additional pairs of gloves. Determine the number of patients. Assess the need for additional resources such as utility services, fire department, or advanced life support (ALS). The mechanism of injury and presence of bleeding may suggest the injury extent and type. Dark clothing may mask signs of bleeding.
Mechanism of Injury (MOI)/ Nature of Illness (NOI)	Determine the MOI. Observe the scene, and look for indicators that will assist you with this. The nature of the problem may not be readily apparent until more information is gathered. Falls, assaults, and motor vehicle crashes are common mechanisms in chest trauma. Consider spinal immobilization in any trauma patient with a significant MOI.

Primary Assessment

Form a General Impression	Identify immediate threats to life. Determine the priority of care based on the MOI. If the patient has a poor general impression, call for ALS assistance. A rapid scan of the patient will help you identify and manage life threats. Keep alert for signs of shock. Assess the patient's level of consciousness using the AVPU scale. If the patient is able to communicate, obtain the chief complaint. Difficulty speaking, cyanosis, and hemoptysis may be signs of a chest injury.
Airway and Breathing	Ensure the airway is open, clear, and self-maintained. Unresponsive patients will need the airway opened and maintained using a modified jaw-thrust maneuver if spinal injury is suspected. A patient with an altered level of consciousness may need emergency airway management; consider inserting a properly sized oropharyngeal or nasopharyngeal airway. Evaluate the patient's ventilatory status for rate and depth of breathing, respiratory effort, and tidal volume. Patients breathing at a rate of less than 12 breaths/min or more than 20 breaths/min may have inadequate breathing and require assistance. Jugular vein distention may be a sign of a tension pneumothorax or cardiac tamponade. Inspect for DCAP-BTLS, observe chest for equal rise, and auscultate lung sounds. Treat life threats such as sucking chest wounds and flail chest immediately. Injuries to the chest may alter the mechanics of respiration; breathing problems should be observed early and managed aggressively. Paradoxical chest wall motion is a sign of a flail chest and must be managed immediately. Continuously monitor oxygen saturation and for additional signs of hypoxia. Administer high-flow oxygen at 15 L/min, providing ventilatory support as needed.
Circulation	Observe skin color, temperature, and condition. Assess capillary refill time; if greater than 2 seconds, treat aggressively for shock. Look for life-threatening bleeding, and treat accordingly. Consider that the patient may have significant internal bleeding with chest trauma. Evaluate distal pulse rate, quality (strength), and rhythm. Tachycardia may be an early indicator of shock.
Transport Decision	If the patient has an airway or breathing problem, signs and symptoms of bleeding, or other life threats, manage them immediately and consider rapid transport, performing the secondary assessment en route to the hospital. Serious chest injuries may be present without obvious signs; do not delay transport to perform a lengthy assessment.

NOTE: The order of the steps in this section differs depending on whether the patient is conscious or unconscious. The following order is for a conscious patient. For an unconscious patient, perform a primary assessment, perform a full-body scan, obtain vital signs, and obtain the past medical history from a family member, bystander, or an emergency medical identification device. Transport immediately.

Assessment and Emergency Care of Chest Injuries, continued

History Taking

Investigate Chief Complaint	Investigate the chief complaint and gather a history once you have identified and treated life threats. Identify associated symptoms and pertinent negatives. Ask SAMPLE questions, focusing on the events surrounding the incident and the mechanism of injury. SAMPLE can also be obtained from family, bystanders, and medical alert tags if the patient is not able to provide the information.

Secondary Assessment

Physical Examinations	If the patient is unconscious, perform a systematic full-body scan beginning with the head, looking for DCAP-BTLS. Assessment should be rapid if the patient has a poor general impression. Focus the assessment on an isolated injury once all the systems have been examined. Significant trauma requires a full-body scan. If a thoracic injury is noted, examine the anterior and posterior aspects of the thoracic cavity. Look for deformities, asymmetry, and contusions while palpating for tenderness. Note: The rib cage of children is very flexible. This flexibility can allow for compression and internal injury without fracture to the ribs. MOI, work of breathing, and changes to the vital signs may be the only indicators of chest trauma in children. Note: Elderly patients with trauma to the chest may have significant underlying injuries as a result of reduced bone density.
Vital Signs	Obtain baseline vital signs. Vital signs should include blood pressure by auscultation, pulse rate and quality, respiration rate and quality, pupils, and skin assessment for perfusion. Note the patient's level of consciousness. Use pulse oximetry, if available, to assess the patient's perfusion status. Tachycardia or hypotension may indicate hypoperfusion. Bradypnea may indicate cerebral hypoxia and impending cardiopulmonary arrest. A narrowing pulse pressure is an indicator of possible cardiac tamponade. Reassess the patient's vital signs every 5 minutes to observe trends.

Reassessment

Interventions	If spinal injury is suspected, stabilize and immobilize the spine as needed. Ensure an open airway, using an oropharyngeal or nasopharyngeal airway if necessary. Provide oxygen via a nonrebreathing mask or bag-mask device as required, and manage any chest wall injuries. Control external bleeding, and treat for shock. Reassess the chief complaint, primary assessment, vital signs, and any interventions already performed. If an occlusive dressing was placed, is the wound still sealed? Is the oxygen delivery sufficient and is the patient breathing adequately? Are pulse oximetry values improving? Vital signs should be repeated every 5 minutes and compared with those taken earlier. Place the patient in a position of comfort unless shock is suspected, then place the patient supine and treat accordingly. Continuously observe and reassess the patient during transport so worsening conditions can be managed.
Communication and Documentation	Contact medical control/receiving hospital with a radio report including notification of chest trauma so appropriate staff and equipment will be ready when you arrive. Include a thorough description of the MOI and the position the patient was found in. Include treatments performed and patient response. Be sure to document the chief complaint, physical findings, history, and any changes in patient status and the time. Document the scene observations on your arrival. Follow local treatment protocols.

NOTE: Although the following steps are widely accepted, be sure to consult and follow your local protocols. Take appropriate standard precautions when treating all patients.

Chest Injuries

General Management of Chest Injuries

Managing life threats to the patient's ABCs is the primary concern with any traumatic emergency. The MOI that caused the chest injury may also have caused a spinal injury or other fracture, and these must be managed at the appropriate time following local protocols. Breathing difficulties are often present with chest injuries, as are injuries involving bleeding. With all chest injuries, once crew safety has been established, management of airway, breathing, and circulatory problems is a priority. For all of the following specific chest injuries, begin with the following steps:

1. Ensure scene safety.
2. Determine the MOI.
3. Consider spinal immobilization.
4. Open, clear, and maintain the patient's airway.
5. Inspect, palpate, and auscultate the chest.
6. Administer high concentration oxygen via a nonrebreathing mask or bag-mask device as appropriate.
7. Control bleeding, and treat for shock.
8. Transport to the appropriate treatment facility.

Pneumothorax

A pneumothorax occurs when air accumulates in the pleural space during breathing. The different types of pneumothorax consist of spontaneous, simple, open, and tension. Treatment of a spontaneous or simple pneumothorax consists of supporting the patient's ABCs and transporting to an appropriate hospital. An open pneumothorax, often associated with a sucking chest wound, requires sealing the opening in the chest wall with an occlusive dressing, managing the ABCs, and transporting to a trauma center. An occlusive dressing should cover the wound sufficiently to avoid the danger of being pulled into the wound. The dressing can be taped on all four sides or on three sides to create a flutter valve allowing the release of air. Follow local protocol when deciding which method to use. A pneumothorax may worsen, causing the lung on the affected side to collapse because of an increased amount of trapped air in the pleural space. As air continues to enter the pleural space, the mediastinum and opposite lung are compromised. This is termed a tension pneumothorax. Treatment for a tension pneumothorax includes high-flow oxygen via nonrebreathing mask and the insertion of a needle between the second and third rib to release the trapped air. This is an ALS intervention.

Hemothorax

A hemothorax is similar to a pneumothorax except that blood, not air, is collecting in the pleural space. If blood and air are present, it is called a hemopneumothorax. The MOI, dyspnea, decreased breath sounds on the affected side, and signs of shock are indicators of a possible hemothorax. Definitive treatment for a hemothorax requires surgical intervention. Prehospital treatment consists of supporting the ABCs and rapid transport to a trauma center.

Assessment and Emergency Care of Chest Injuries, continued

Chest Injuries

Cardiac Tamponade/Pericardial Tamponade

Blunt or penetrating trauma to the chest can compromise the protective membrane surrounding the heart, allowing blood or fluid to collect. Blood or fluid collecting in the pericardial sac decreases cardiac output and must be managed quickly. Jugular vein distention, muffled heart sounds, and a narrowing pulse pressure, along with a mechanism suggestive of chest trauma, are signs of possible cardiac tamponade. Rapid surgical intervention is required. Provide high-flow oxygen and support the patient's breathing while transporting rapidly to the appropriate trauma center. Notify the hospital while en route that you suspect a cardiac tamponade.

Rib Fractures/Flail Chest

Patients with suspected rib fractures will often report pain with breathing. Patients tend to avoid taking deep breaths and will guard or attempt to self-splint the affected area with their arm. Although common, rib fractures can be serious because the broken ends may lacerate the lung surface, causing air to escape. Patients with rib fractures should receive high-flow oxygen via a nonrebreathing mask.

When three or more ribs are fractured in two or more places, a free-floating segment, called a flail chest, is created. Flail chest is a serious life threat requiring rapid intervention. The detached portion of the ribs moves in the opposite direction of the rest of the chest (paradoxical motion), resulting in painful and ineffective breathing. Provide support of the airway and breathing, and administer supplemental oxygen via a nonrebreathing mask or bag-mask device if breathing becomes inadequate. If local protocol allows, stabilize the flail section with a bulky dressing or pillow and transport to a trauma center. Be alert for signs of increasing respiratory distress because a pulmonary contusion often accompanies flail chest.

Pulmonary Contusion

Injury to lung tissue causes fluid accumulation in the injured area of the lung and blood filling the alveoli; this causes the patient to become hypoxic. Support the ABCs by providing high-concentration oxygen and respiratory assistance as needed. Lung contusions are often associated with a significant MOI. Transport to a trauma center for further evaluation.

Sternal Fractures/Traumatic Asphyxia

Significant force is required to fracture a patient's sternum. Airway maintenance, supplemental oxygen, and ventilatory support are required treatments. The severe compression forces that cause sternal fractures may produce a sudden increase in intrathoracic pressure, termed traumatic asphyxia. Ensure an open airway, assist with ventilations as required, administer supplemental oxygen, and provide rapid transport to a trauma center.

Assessment and Emergency Care of Chest Injuries, continued

Chest Injuries

Myocardial Contusion

In addition to lung contusions, cardiac contusions are often present when there has been blunt trauma to the chest. The patient's signs and symptoms often mimic those of a heart attack. Treatment is mostly supportive; administer oxygen, and provide rapid transport immediately on recognizing this condition.

Commotio Cordis

Suspect commotio cordis if your patient is unconscious and unresponsive immediately after a sudden impact to the chest. Ensure the airway is open, the patient is breathing, and a pulse is present. Begin cardiopulmonary resuscitation, and provide early defibrillation if the patient is pulseless. Transport to the nearest emergency department.

Laceration of the Great Vessels

Signs of shock in patients with chest trauma without signs of obvious bleeding may indicate an injury to one or more of the great vessels. This unseen bleeding may be rapid and fatal. Support the ABCs, provide cardiopulmonary resuscitation if necessary, and be alert for rapidly changing conditions. Surgical intervention is required. Monitor vital signs every 2 to 3 minutes en route to the hospital. Do not delay transport if an internal chest injury is suspected.

Prep Kit

- A penetrating chest injury has the potential to penetrate the lung and diaphragm and injure the liver or stomach.

- Chest injuries are classified as closed or open. Closed injuries are often the result of blunt force trauma, and open injuries are the result of some object penetrating the skin and/or chest wall.

- Blunt trauma may result in fractures to the ribs and the sternum.

- During the primary assessment, if an injury is encountered that interferes with the ability of the patient to ventilate or oxygenate, the injury must be addressed quickly.

- Any penetrating injury to the chest may result in air entering the pleural space and may cause pneumothorax. An occlusive dressing should be placed on this injury as soon as it is identified.

- When a penetrating injury creates a hole in the chest wall, you may hear a sucking sound as the patient inhales. This is called an open pneumothorax.

- A simple pneumothorax is a result of blunt trauma, such as fractured ribs.

- A spontaneous pneumothorax may be the result of rupture of a weak spot on the lung, allowing air to enter the pleural space and accumulate. This often results from nontraumatic injuries and may occur during times of physical activity such as exercise.

- A pneumothorax may progress to a tension pneumothorax and cause cardiac arrest.

- Hemothorax is the result of blood accumulating in the pleural space after a traumatic injury when the vessels of the lung are lacerated and leak blood.

- A flail chest segment is two or more ribs broken in two or more places.

- A flail chest segment should be secured with a large bulky dressing that is secured with a roller bandage, cravat, or tape.

- All patients with chest injuries should receive high-flow oxygen or ventilation with a bag-mask device.

- Pulmonary contusion, which is bruising of or injury to lung tissue after traumatic injury, may interfere with oxygen exchange in the lung tissue.

- Traumatic asphyxia is sudden, severe compression of the chest.

- Myocardial contusion is bruising of the heart muscle after traumatic injury. This condition may have the same signs and symptoms as a heart attack, including an irregular pulse. Remember that this is an injury to the heart muscle from trauma, not from a heart attack.

- Commotio cordis occurs from a direct blow to the chest during a critical portion of the patient's heartbeat. It may result in immediate cardiac arrest.

- Cardiac tamponade is when blood collects in the space between the pericardial sac and the heart. This condition results in pressure building up inside the pericardial sac until the heart cannot pump effectively; cardiac arrest may occur quickly.

- The great vessels of the body are located in the mediastinum. These large vessels may be lacerated or tear after traumatic injury and cause heavy, unseen bleeding inside the patient's chest cavity.

- Any patient who has signs of shock with a chest injury, even with unseen bleeding, should make you suspicious of unseen, life-threatening bleeding inside the chest cavity.

■ Vital Vocabulary

cardiac tamponade (pericardial tamponade) Compression of the heart as the result of buildup of blood or other fluid in the pericardial sac, leading to decreased cardiac output.

closed chest injury An injury to the chest in which the skin is not broken, usually caused by blunt trauma.

commotio cordis A blunt chest injury caused by a sudden, direct blow to the chest that occurs only during the critical portion of a person's heartbeat.

flail chest A condition in which two or more ribs are fractured in two or more places or in association with a fracture of the sternum so that a segment of the chest wall is effectively detached from the rest of the thoracic cage.

flutter valve A one-way valve that allows air to leave the chest cavity but not return; formed by taping three sides of an occlusive dressing to the chest wall, leaving the fourth side open as a valve.

hemopneumothorax The accumulation of blood and air in the pleural space of the chest.

hemothorax A collection of blood in the pleural cavity.

myocardial contusion A bruise of the heart muscle.

occlusive dressing A dressing made of Vaseline-impregnated gauze, aluminum foil, or plastic that protects a wound from air and bacteria.

open chest injury An injury to the chest in which the chest wall itself is penetrated by a fractured rib or, more frequently, by an external object such as a bullet or knife.

open pneumothorax An open or penetrating chest wall wound through which air passes during inspiration and expiration, creating a sucking sound; also referred to as a sucking chest wound.

paradoxical motion The motion of the portion of the chest wall that is detached in a flail chest; the motion—in during inhalation, out during exhalation—is exactly the opposite of normal chest wall motion during breathing.

pericardium The fibrous sac that surrounds the heart.

pneumothorax An accumulation of air or gas in the pleural cavity.

pulmonary contusion Injury or bruising of lung tissue that results in hemorrhage.

simple pneumothorax Any pneumothorax that is free from significant physiologic changes and does not cause drastic changes in the vital signs of the patient.

spontaneous pneumothorax A pneumothorax that occurs when a weak area on the lung ruptures in the absence of major injury, allowing air to leak into the pleural space.

sucking chest wound An open or penetrating chest wall wound through which air passes during inspiration and expiration, creating a sucking sound. See also open pneumothorax.

tachypnea Rapid respirations.

tension pneumothorax An accumulation of air or gas in the pleural cavity that progressively increases pressure in the chest that interferes with cardiac function with potentially fatal results.

traumatic asphyxia A pattern of injuries seen after a severe force is applied to the chest, forcing blood from the great vessels back into the head and neck.

Assessment in Action

You and your partner are dispatched to the rodeo arena for a person complaining of shortness of breath and chest pain. On arrival you are escorted to a first aid trailer where you find a man in his mid 30s seated in a chair and leaning forward clutching his chest. He tells you that he was riding a bull when he was thrown off. He states that he landed hard on his right side and heard a "loud pop."

1. He finds it difficult to take a deep breath and speak. On the basis of this information, what should be your first intervention?
   - **A.** Take vital signs.
   - **B.** Apply a bulky dressing to the right side of the chest.
   - **C.** Administer high-flow oxygen.
   - **D.** Immobilize the patient on a long spine board.

2. Exposure of the patient's chest reveals a large bruise on the lateral aspect of the right side of the chest. When you palpate the area, the patient yells out in extreme pain and states that he cannot take a deep breath. What condition should you be suspicious for?
   - **A.** Cardiac tamponade
   - **B.** Rib fractures
   - **C.** Spontaneous pneumothorax
   - **D.** Pulmonary contusion

3. Complications associated with fractured ribs can include:
   - **A.** pneumothorax.
   - **B.** tension pneumothorax.
   - **C.** hemothorax.
   - **D.** all of the above.

4. Which of the following blood vessels can be lacerated by a fractured rib?
   - **A.** Aorta
   - **B.** Brachial artery
   - **C.** Intercostal artery or vein
   - **D.** Jugular vein

5. Patients with chest injuries should be reevaluated every _____ minutes.
   - **A.** 5
   - **B.** 10
   - **C.** 15
   - **D.** 20

6. Diagnostic tools are used to assist you in assessing the severity of your patient's condition. What diagnostic tool is most commonly used to evaluate the effectiveness of the respiratory system?
   - **A.** End-tidal carbon dioxide detector
   - **B.** Peak flow meter
   - **C.** Pulse oximeter
   - **D.** Automated blood pressure

7. While en route to the hospital, the patient suddenly grabs your arm and states that he can't breathe. He appears pale, diaphoretic, and extremely anxious. Reassessment of the chest reveals diminished breath sounds on the right side and slight tracheal deviation. You suspect the patient has developed which of the following conditions?
   - **A.** Cardiac tamponade
   - **B.** Tension pneumothorax
   - **C.** Myocardial contusion
   - **D.** Pulmonary contusion

8. You are still 30 minutes away from the hospital. On the basis of the patient's current signs and symptoms, what is the most appropriate management?
   - **A.** Assist ventilations with a bag-mask device and continue transport.
   - **B.** Call for ALS assistance and continue transport.
   - **C.** Reassure the patient and continue transport.
   - **D.** Attempt to place a needle in the right side of the chest.

9. How can a tension pneumothorax develop as the result of fractured ribs?

10. Spinal cord injury should be suspected in patients with chest injury. The intercostal muscles receive nerve impulses from specific spinal nerves. Where are these nerves located?

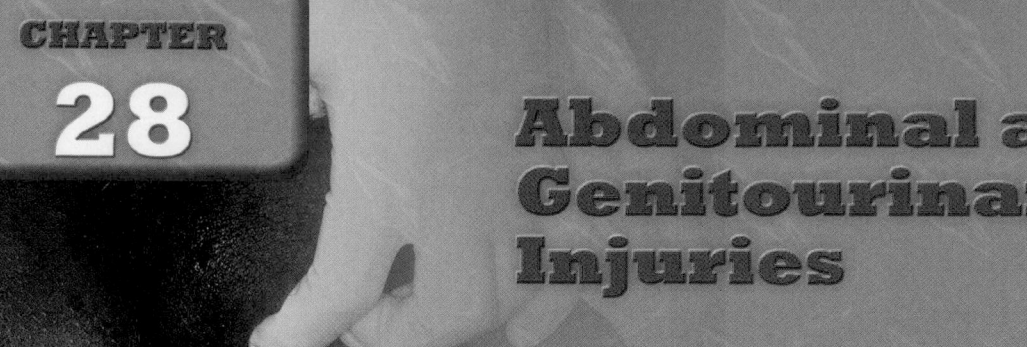

Abdominal and Genitourinary Injuries

National EMS Education Standard Competencies

Trauma

Applies fundamental knowledge to provide basic emergency care and transportation based on assessment findings for an acutely injured patient.

Abdominal and Genitourinary Trauma

- Recognition and management of
  - Blunt versus penetrating mechanisms (pp 969-972, 979-981)
  - Evisceration (pp 971-972, 980-981)
  - Impaled object (pp 971-972, 979-980)
- Pathophysiology, assessment, and management of
  - Solid and hollow organ injuries (pp 968-969, 972-974)
  - Blunt versus penetrating mechanisms (pp 969-972, 979-981)
  - Evisceration (pp 971-972, 980-981)
  - Injuries to the external genitalia (pp 983-985, 987-988)
  - Vaginal bleeding due to trauma (pp 983, 988)
  - Sexual assault (p 988)

Knowledge Objectives

1. Describe the anatomy and physiology of the abdomen, including an explanation of abdominal quadrants and boundaries and the difference between hollow and solid organs. (pp 967-969)

2. Describe some special considerations related to the care of pediatric patients and geriatric patients who have experienced abdominal trauma. (pp 968, 969, 974)

3. Define and discuss closed abdominal injuries, providing examples of the mechanisms of injury that are likely to cause this type of trauma in a patient, as well as key signs and symptoms. (pp 969-971)

4. Define and discuss open abdominal injuries, including ways to distinguish low-velocity, medium-velocity, and high-velocity injuries, examples of the mechanisms of injury that would cause each, and signs and symptoms exhibited by a patient who has experienced this type of injury. (pp 971-972)

5. Describe the different ways hollow and solids organs of the abdomen can be injured and include the signs and symptoms a patient might exhibit depending on the organ(s) involved. (pp 972-974)

6. Discuss assessment of a patient who has experienced an abdominal injury, including key indicators that will help determine the mechanism of injury (MOI) and whether it is significant or nonsignificant. (pp 974-979)

7. Discuss the emergency medical care of a patient who has sustained a closed abdominal injury, including blunt trauma caused by a seatbelt or air bag. (p 979)

8. Discuss the emergency medical care of a patient who has sustained an open abdominal injury, including penetrating injuries and abdominal evisceration, and considerations related to the use of a pneumatic antishock garment (PASG) when caring for these patients. (pp 979-981)

9. Describe the anatomy and physiology of the female and male genitourinary systems and distinguish between hollow and solid organs. (pp 981-982)

10. Discuss the types of traumatic injuries that may be sustained by the organs of the male and female genitourinary system, including the kidneys, urinary bladder, and internal and external genitalia. (pp 982-985)

11. Discuss assessment of a patient who has experienced a genitourinary injury, including special considerations related to patient privacy and determining the MOI. (pp 985-987)

12. Discuss the emergency medical care of a patient who has sustained a genitourinary injury related to the kidneys, bladder, external male genitalia, female genitalia, and rectum. (pp 987-988)

13. Explain special considerations related to a patient who has experienced a genitourinary injury caused by a sexual assault, including patient treatment, criminal implications, and evidence management. (p 988)

Skills Objectives

1. Demonstrate proper emergency medical care of a patient who has experienced a blunt abdominal injury. (p 979)

2. Demonstrate proper emergency medical care of a patient who has a penetrating abdominal injury with an impaled object. (pp 979-980)

3. Demonstrate how to apply a dressing to an abdominal evisceration wound. (pp 980-981)

Introduction

The abdomen is the major body cavity extending from the diaphragm to the pelvis. It contains organs that make up the digestive, urinary, and genitourinary systems. Although any of these organs can be injured, some organs are better protected than others. It is important for you to know the anatomy of the abdominal and pelvic cavities and where the organs are located. You must also understand the functions of the organs so that if an injury occurs, you can assess its seriousness.

Eight percent of all significant trauma involves the abdomen. Injuries to the abdomen that go unrecognized and are not repaired in surgery are a leading cause of traumatic death. In fact, 10% to 20% of all trauma patients have some form of genitourinary tract injury. This pathology is one of the most overlooked areas of trauma, which is unfortunate because these injuries can result in life-altering consequences for the patient, such as impotence or incontinence. It is paramount that you maintain a high index of suspicion if there is a mechanism of injury that suggests abdominal injury and share those findings with the receiving hospital prior to and on arrival.

Anatomy and Physiology of the Abdomen

Abdominal Quadrants

The abdomen is divided into four general quadrants Figure 28-1 . Imagine two lines intersecting at the umbilicus, dividing the abdomen into four equal areas. These areas are referred to as the right upper quadrant, left upper quadrant, right lower quadrant, and left lower quadrant. Remember that here, too, right and left refer to the patient's right and left, not yours.

The quadrant location of bruising or pain can delineate which organs are possibly involved in a traumatic injury. Organs commonly found in the right upper quadrant are the liver, gallbladder, and duodenum of the intestines and a small portion of the pancreas. The stomach occupies most of the left upper quadrant but it shares this space with the spleen. The pancreas occupies some of this space but is mostly posterior to the region. The left lower quadrant holds both the large and small bowel, notably the descending colon and the left half of the transverse colon. The right lower quadrant also holds portions of the large and small intestines that include the ascending colon and the right half of the transverse

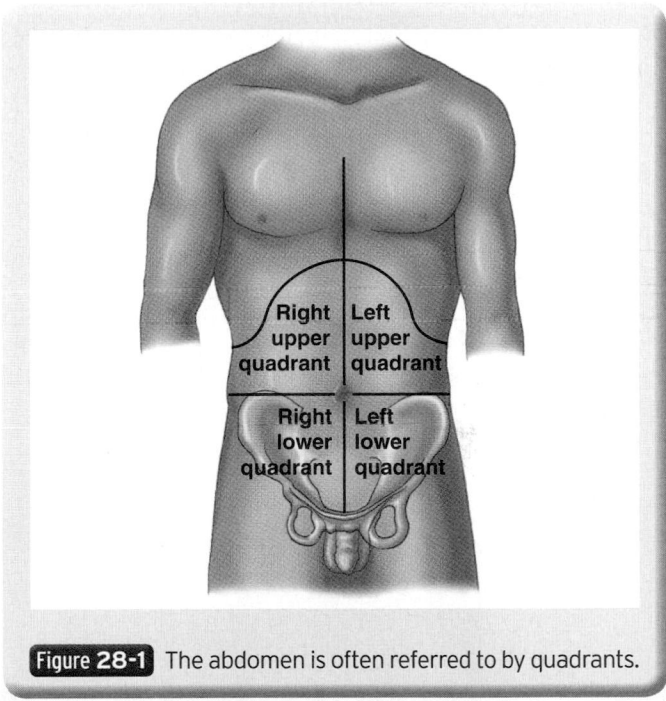

Figure 28-1 The abdomen is often referred to by quadrants.

You are the Provider: PART 1

You and your partner are working a special event—the annual rodeo. While on standby, you witness a rider being thrown and trampled by a bull. After a rodeo clown distracts and corrals the bull, the rider, a 20-year-old man, slowly gets up and begins to walk. He is clutching his abdomen with one hand and rubbing his lower back with the other hand.

1. How do hollow organ injuries differ from solid organ injuries?
2. Should you focus on trying to determine the origin of any intra-abdominal bleeding? Why or why not?

colon. The distal end of the descending colon, called the appendix, is located in this region. The region is a common location for swelling and inflammation, and the appendix is a source of infection if it ruptures.

■ Hollow and Solid Organs

The abdomen contains both hollow and solid organs, any of which may be damaged. __Hollow organs__, including the stomach, intestines, ureters, and bladder, are actually structures through which materials pass **Figure 28-2**. Most of these organs will contain food that is in the process of being digested, urine that is being passed to the bladder for release, or bile. When ruptured or lacerated, these organs spill their contents into the __peritoneal cavity__ (the abdominal cavity), causing an intense inflammatory reaction and possible infection. Peritonitis is an inflammation of the peritoneum that may be caused by this type of infection. The intestines and stomach contain acid-like substances that aid in the digestive process. When they spill or leak into the peritoneal cavity, pain and irritation of the peritoneum often follow. The first signs of peritonitis are severe abdominal pain, tenderness, and muscular

spasm. Later, bowel sounds diminish or disappear as the bowel stops functioning. A patient may feel nauseous and may vomit; the abdomen may become distended and firm to touch, and infection may occur. Peritonitis is serious and may become life threatening.

The small intestine is composed of the duodenum, the jejunum, and the ileum. The large intestine consists of the cecum, the colon, and the rectum. The intestinal blood supply comes from the mesentery. The term mesentery refers to any fold of tissue that attaches an organ to the body wall. However, the majority of time when the term is used it is in reference to the intestinal mesentery: a fold of tissue that contains a web of vessels, both arteries and veins, as well as nerves and lymphatic tissues. It connects the small intestine to the posterior of the abdominal wall. Both blunt and penetrating abdominal injuries affect this vasculature, and patients with injuries to the mesentery can bleed significantly into the peritoneal cavity. A common sign of bleeding in the abdomen is rigidity, with an almost boardlike feeling to the abdomen. Occasionally you will find periumbilical bruising or ecchymosis, referred to as the Cullen sign.

The __solid organs__, as their name suggests, are solid masses of tissue. They include the liver, spleen, pancreas, and kidneys **Figure 28-3**. It is here that much of the chemical work of the body—enzyme production, blood cleansing, and energy production—takes place. Solid organs have a rich blood supply, so injury can cause severe and unseen hemorrhage. The same is true of the aorta or inferior vena cava, whether the injury is open or closed. Blood may irritate the peritoneal cavity and cause

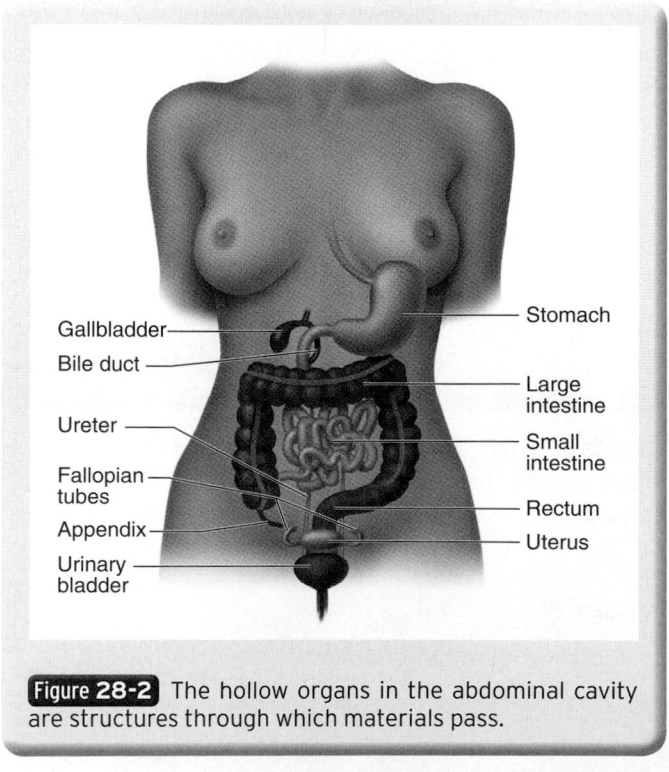

Figure 28-2 The hollow organs in the abdominal cavity are structures through which materials pass.

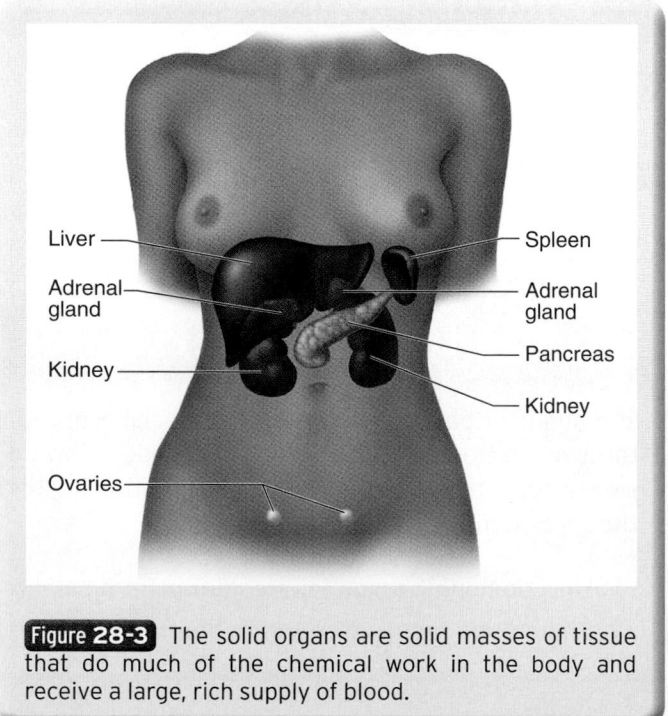

Figure 28-3 The solid organs are solid masses of tissue that do much of the chemical work in the body and receive a large, rich supply of blood.

the patient to report abdominal pain; however, this may not always occur. Therefore, the absence of pain and tenderness does not necessarily mean the absence of major bleeding in the abdomen.

Many solid organs are found in the retroperitoneal region, in addition to the great vessels, the abdominal aorta, and the inferior vena cava. This area also houses the kidneys, ureters, and bladder. The majority of the pancreas is located in this region, which is why the pancreas is referred to as a retroperitoneal organ. The last portion of the hollow organ, the colon, occupies the lowest portion of the retroperitoneal space.

Special Populations

Falls are the most common mechanism of injury in geriatric patients. In addition to the typical orthopaedic injuries that a geriatric patient sustains in a fall, the abdominal organs that have lost some elasticity over time are exposed to forces that can damage them. Specifically, the aorta, liver, and spleen are at risk of injury from falls. A geriatric patient's brittle bones can fracture in a fall, creating dangerously sharp edges that can puncture internal organs.

Injuries to the Abdomen

Abdominal injuries may be as obvious as loops of intestines protruding from a penetrating injury, or as occult as an unseen injury such as a laceration to the liver or spleen. Traumatic injuries to the abdomen are considered as either open or closed, and can involve hollow and/or solid organs.

Closed Abdominal Injuries

Closed abdominal injuries are those in which blunt force trauma, some type of impact to the body, results in injury to the abdomen without breaking the skin. Such a blow might come from the patient striking the handlebar of a bicycle or the steering wheel of a car, or when the patient is struck by an item such as a board or baseball bat during a fight or assault **Figure 28-4**. Other mechanisms of injury include the following:

- Motorcycle collisions
- Falls
- Blast injuries
- Pedestrian injuries
- Compression
- Deceleration

Compression injuries are typically caused by a poorly placed lap belt. This creates an injury pattern called a clasp-knife injury. A compression injury can also be

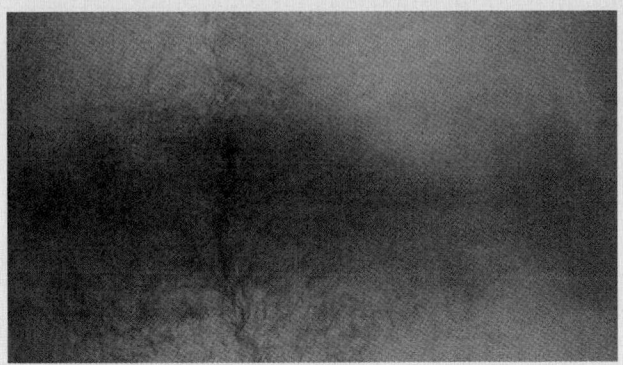

Figure 28-4 Blunt trauma to the abdomen can occur when a patient strikes the steering wheel of an automobile as a result of a crash.

caused when a person is run or rolled over by vehicles or objects. Deceleration injuries commonly occur when a person or the vehicle that he or she is traveling in strikes a large immovable mass such as a larger vehicle, a bridge abutment, or the ground.

Signs and Symptoms of a Closed Injury

In the abdomen, pain can often be deceiving because it is often diffuse in nature and may be referred from the site of injury to another location in the body. Most injured organs irritate the surrounding tissues. This commonly predictable radiation pattern can help you determine the source of the pain and possibly the site of the injury. In patients with liver and spleen injuries, in whom there is bleeding into the peritoneal space, pain is referred to the shoulder. This finding is called the Kehr sign when it involves injury to the spleen and pain in the tip of the left shoulder. However, shoulder pain can be misleading and injury to the liver or spleen could possibly be overlooked if the shoulder is also injured or if the mechanism of injury (MOI) suggests that an impact or injury may have occurred in the shoulder girdle.

When a patient reports pain that is tearing and describes it as going from the abdomen posteriorly, he or she is often describing symptoms of an abdominal aneurysm that is undergoing dissection. Pain that is following the angle from the lateral hip to the midline of the groin can be the result of damage to the kidneys or the ureters. Pain primarily located in the right lower quadrant can indicate an inflamed or ruptured appendix. Pain from the gallbladder due to direct injury or inflammation can be found just under the margin of the ribs on the right side or between the shoulder blades.

As blood and fluid from damaged organs flows into the peritoneal cavity, the common response is acute pain in the entire abdomen, which spreads as the blood or contaminant seeks out the voids in the peritoneal cavity. The resulting peritonitis or inflammation of the peritoneum can produce pain if the affected area is exposed to any jarring motion. This is commonly referred to as rebound tenderness or the Blumberg sign. As an EMT, you do not need to produce rebound tenderness intentionally when examining the patient. It is often discovered when you are moving the patient onto the stretcher or into the ambulance.

Determining the location of the pain or referred pain can be more difficult when the patient has voluntary or involuntary guarding. In **guarding**, the patient either consciously or unintentionally stiffens the muscles of the surface of the abdomen. Most often it is the rectus abdominis muscles that are held tight, and the tightness can be mistaken for abdominal rigidity. This stiffening is a natural response to abdominal pain; the body is attempting to splint the area to prevent unnecessary movement and to avoid further pain.

Abdominal distention or swelling that occurs between the xiphoid process and the groin is often the result of free fluid, blood, or organ contents spilling into the peritoneal cavity. Swelling can also be the result of air in the form of gases from the bowel or from infection. Tenderness is another sign of a closed abdominal injury.

Additional signs of abdominal injury are bruising and discoloration. Another likely injury is lower rib fractures—a trauma that was forceful enough to break the ribs may also have damaged internal organs.

Closed abdominal injuries may initially appear as abrasions to the surface of the skin depending on the MOI, such as an assault or auto versus pedestrian accident. In some circumstances, depending on how deep in the abdomen the injury occurs, it may take several minutes to hours for the contusion or hematoma to become present on the surface. Therefore, it is not prudent for you to rule out injury simply on the basis of absence of these findings.

Injuries From Seatbelts and Air Bags

Seatbelts have prevented many thousands of injuries and saved many lives, including those of people who otherwise would have been ejected from a smashed car. However, seatbelts occasionally cause blunt injuries of the abdominal organs. When worn properly, a seatbelt lies below the anterior superior iliac spines of the pelvis and against the hip joints. If the belt lies too high, it can squeeze abdominal organs or great vessels against the spine when the car suddenly decelerates or stops Figure 28-5. Occasionally, fractures of the lumbar spine have been reported. If you are called to the scene of such an accident, keep in mind that the use of seatbelts in many cases turns what could have been a fatal injury into a manageable one. In later stages of pregnancy the

gravid uterus displaces the bladder to the anterior. This anatomic change allows the normally protected bladder to become more susceptible to injuries from impacts and the seatbelt. Pregnant patients who adjust the lap belt portion for comfort as opposed to functionality can sustain further injuries.

In all current-model automobiles, the lap and diagonal (shoulder) safety belts are combined into one so that they may not be used independently. Of course, people can still place the diagonal portion of the belt behind the back, significantly reducing the effectiveness of this design. In some older cars, only lap belts or two separate belts are provided. Used alone, diagonal shoulder safety belts can cause injuries of the upper part of the trunk, such as a bruised chest, fractured ribs, lacerated liver, or even decapitation. Far fewer head and neck injuries are seen when this belt is used in combination with a lap belt and a headrest.

The air bag, which is standard in today's vehicles, represents a great advance in automotive safety. In head-on collisions, it can be a genuine lifesaver. However, because frontal air bags provide no protection in a side impact or rollover, they must be used in combination with safety belts. Small children and short stature individuals who

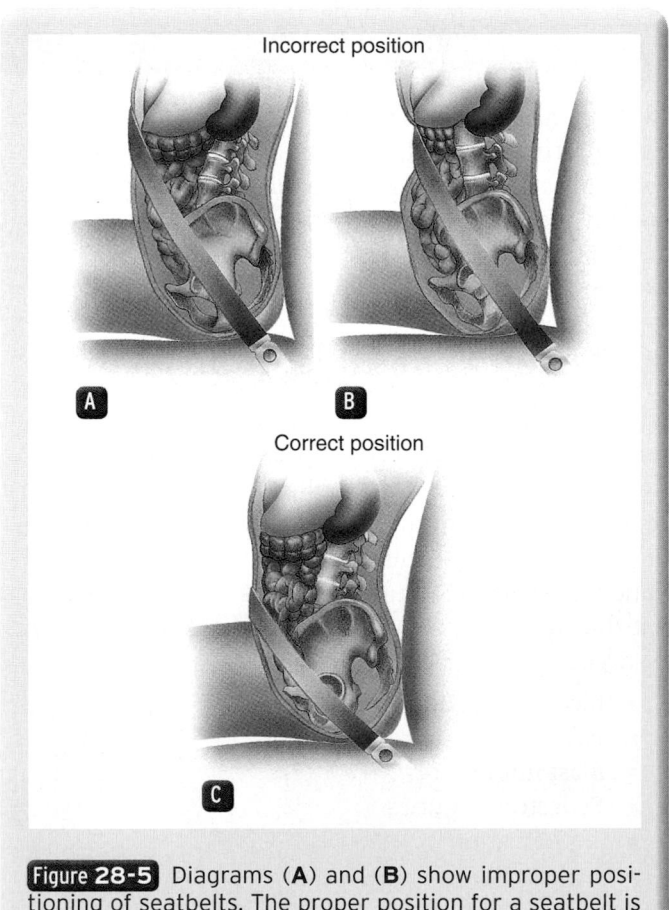

Figure 28-5 Diagrams (**A**) and (**B**) show improper positioning of seatbelts. The proper position for a seatbelt is below the anterior superior iliac spines of the pelvis and against the hip joints, as shown in diagram (**C**).

are in the front seat of the automobile may be at risk of injury if an air bag is deployed. Special attention should be used in evaluating these patients when a deployed air bag is noted. Remember to inspect beneath the air bag for signs of damage to the steering column.

> ## Words of Wisdom
>
> Hospital personnel will depend on you to record scene findings that explain the mechanism of injury. For example, be thorough in documenting your observations about the vehicle in which a patient rode. Notes about deployment of air bags and the condition of the exterior and the steering column will help in the assessment of possible internal injuries.

■ Open Abdominal Injuries

<u>Open abdominal injuries</u> are those in which a foreign object enters the abdomen and opens the peritoneal cavity to the outside; these are also known as penetrating injuries **Figure 28-6**. Stab wounds and gunshot wounds are examples of open injuries, or penetrating trauma. Open wounds may not be deeper than the muscular wall of the abdomen; however, this cannot be determined in the prehospital setting. Therefore, you should maintain a high index of suspicion for unseen injuries, internal damage to organs, and potential life-threatening injuries and provide prompt transport. Patients with open abdominal injuries must be assessed and evaluated at the hospital.

When a patient has sustained a penetrating injury to the abdomen, it is important to attempt to determine the velocity of the object that penetrated the abdominal wall because this can predict the amount of damage to tissue that has occurred. There are commonly three levels of velocity that are discussed in traumatic injuries.

- **Low-velocity injuries.** Caused by handheld or hand-powered objects such as knives and other edged weapons
- **Medium-velocity penetrating wounds.** Caused by smaller caliber handguns and shotguns
- **High-velocity injuries.** Caused by larger weapons such as high-powered rifles and the higher-powered handguns

High- and medium-velocity injuries have temporary wound channels in addition to the exit and entrance wounds. This temporary channel is caused by cavitation. A cavity forms as the pressure wave from the projectile is transferred to the tissues. This causes microscopic tears to the blood vessels and nerves, expanding the width and length of the wound beyond what you can see during physical examination. Cavitation can produce a large amount of bleeding depending on the speed or velocity

Figure 28-6 Because it is difficult to know how deep a penetrating injury is, assume organ damage and transport promptly.

of the penetrating object. The higher the velocity of the projectile, the larger the cavity it produces, typically resulting in a larger amount of tissue damage.

Low-velocity penetrations also have the capacity to damage underlying organs. This internal injury may not be apparent during the physical examination. The bleeding entrance wound may hide the fact that the object went farther and deeper into the peritoneal cavity and injured other organs and tissues. This is especially important information to remember when an injury occurs in the region where the thoracic cavity and the peritoneal cavity are separated by the diaphragm. Any time your patient has an injury at or below the xiphoid process, it should be assumed that both cavities have been violated.

An open abdominal injury that goes all the way through the skin and muscle layer and through the fascia or the interior covering of the abdomen and bowel now protrudes from the peritoneum, is an **evisceration**. This visually shocking injury can be extremely painful. Remember to not push down on the patient's abdomen and perform only a visual assessment when there is any suspicion of this type of injury. If there is clothing close to the wound, carefully cut the clothing around the wound, leaving a border of intact cloth outside the injured area. Never pull, even gently, on any clothing stuck to or in the wound channel because this may remove even more of the bowel.

Signs and Symptoms of an Open Injury

Patients with any type of abdominal injury generally have one principal complaint: pain. But other significant injuries may mask the pain at first, and some patients may not be able to tell you about pain because they are unconscious or unresponsive, such as after a head injury or a drug or alcohol overdose. A very common sign of significant abdominal injury is tachycardia because the heart is increasing its pumping action to compensate for blood loss, an early indication of compensated blood loss and shock. Later signs include evidence of shock, such as decreased blood pressure and pale, cool, moist skin, or changes in the patient's mental status, combined with trauma to the abdomen. In some cases, the abdomen may become distended from the accumulation of blood and fluid.

As an EMT, you must look for other signs and symptoms of potential problems and injuries to the abdomen. A patient may have both closed and open injuries. Blunt injuries include bruises (often indicated by red areas of skin at this early stage) or other visible marks, whose location should guide your attention to underlying structures Figure 28-7 . For example, bruises in the

right upper quadrant, left upper quadrant, or **flank** (the posterior region below the lower margin of the rib cage), called the Grey Turner sign, might suggest an injury to the liver, spleen, or kidney, respectively Figure 28-8 . Bruises around the umbilicus, called the Cullen sign, are predictive of significant internal abdominal bleeding Figure 28-9 .

Words of Wisdom

The signs of abdominal injury are usually more definite than the symptoms, including firmness on palpation of the abdomen, obvious penetrating wounds, bruises, and altered vital signs such as increased pulse rate, increased respiratory rate, decreased blood pressure, and shallow respirations (although these signs might not appear until later). Common symptoms include abdominal tenderness, particularly localized tenderness and difficulty with movement because of pain.

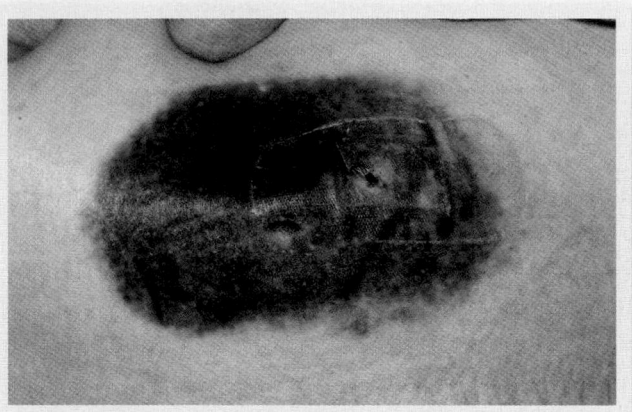

Figure 28-7 Bruising on the abdomen can provide clues to the possible injury of underlying organs.

■ Hollow Organ Injuries

Injuries that involve the hollow organs often have delayed signs and symptoms. The hollow organs commonly spill their contents into the abdomen and then an infection develops, which can take a few hours to days to develop. When the stomach and the intestines are injured, they can spill gastrointestinal contents such as food, waste, and digestive liquids that are highly toxic and acidic. These substances cause significant tissue damage to the entire peritoneum.

You are the Provider: PART 2

When you reach the patient, he is lying supine on the ground outside of the arena area. The scene is safe and the bull has been penned up. The patient is conscious, but restless, and reports pain over his entire abdomen, both flank areas, and the back of his head. Your partner manually stabilizes his head while you perform a primary assessment.

Recording Time: 0 Minutes	
Appearance	In obvious pain; restless
Level of consciousness	Conscious and alert; restless
Airway	Open, clear of secretions or foreign bodies
Breathing	Increased rate; adequate depth
Circulation	Radial pulses, rapid and weak; skin is pale, cool, and moist; no obvious bleeding

3. How should you interpret your primary assessment findings?

4. What immediate treatment is indicated for this patient?

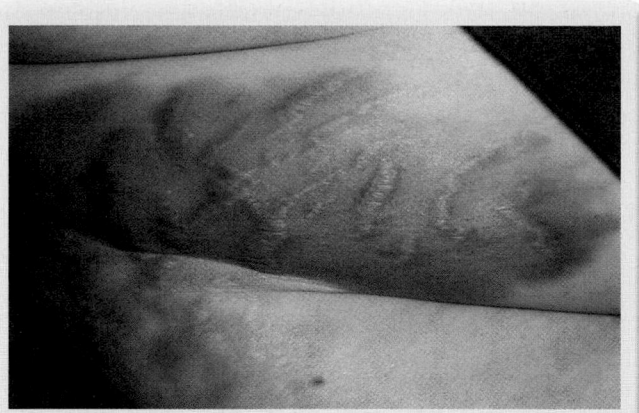

Figure 28-8 Bruises in the right upper quadrant, left upper quadrant, or flank, called the Grey Turner sign, suggest an injury to the liver, spleen, or kidney, respectively.

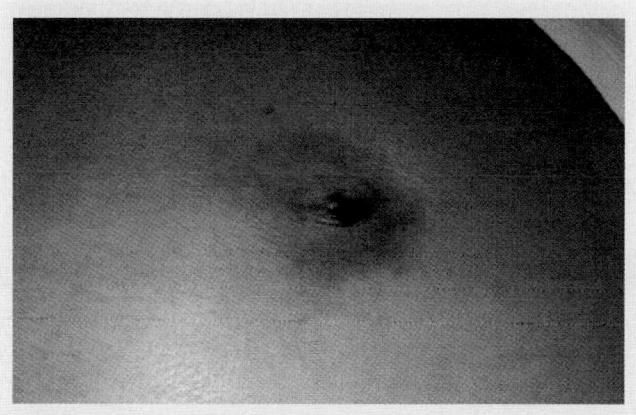

Figure 28-9 Bruises around the umbilicus, called the Cullen sign, are predictive of significant internal abdominal bleeding.

Both blunt and penetrating trauma can cause injuries to the hollow organs. Blunt trauma causes the organ to "pop," thus releasing fluids or air. Penetrating trauma causes direct injury such as laceration and punctures. In open wounds, patients typically report an intense pain that can be out of character for the size of the injury. Patients may also report intense pain with open wounds of the stomach or small bowel.

The gallbladder, which is filled with bile, and the urinary bladder, which is filled with urine, are two additional hollow organs whose contents are potentially irritating and damaging to the tissues of the abdomen if ruptured by injury. These fluids move via gravity into the loose spaces and voids in the peritoneal cavity, eventually leading to infection.

Air in the peritoneal cavity produces pain. It can irritate the tissues it contacts and can cause tissue ischemia and infarction. It can also cause a potentially fatal

infection. Any air in the peritoneal cavity seeks the most superior space or void, thus the location of the air can change with positioning of the patient.

■ Solid Organ Injuries

Solid organs (liver, spleen, diaphragm, kidneys, pancreas) can bleed significantly and cause rapid blood loss that can be hard to identify from a physical examination because the patient is not experiencing significant pain. Conversely, solid organs can slowly ooze blood into the peritoneal cavity, causing pain to increase slowly over time and increasing the chance for toxicity to develop. Blood in the peritoneal cavity irritates tissue and fills any voids or spaces, which can make it difficult for you to determine the exact source of the bleeding. Because of the structures in the retroperitoneal space and the spaces in the abdominal cavity, the peritoneal cavity can hold a large volume of blood following traumatic injuries of solid organs and major blood vessels.

The liver is the largest organ in the abdomen. It is very vascular; therefore, it can contribute to hypoperfusion if it is injured. It is often injured by a fractured lower right rib or a penetrating trauma, such as a knife wound. A common finding during assessment of patients with an injured liver is the **Kehr sign**.

Like the liver, the pancreas and spleen are organs responsible for filtering blood and are therefore very vascular. Both are prone to heavy bleeding when fractured by blunt force or lacerated or punctured by penetrating injury. The spleen is often injured during motor vehicle collisions, especially in the cases of improperly placed seatbelts or from the steering wheel, falls from heights or onto sharp objects, and bicycle and motorcycle accidents where the patient hits the handlebars on impact.

If the diaphragm is penetrated or ruptured, loops of bowel are likely to invade the thoracic cavity and can cause bowel sounds to be present during auscultation of the lungs. Because the bowel will now be displacing lung tissue and vital capacity, patients will exhibit dyspnea or feel short of breath.

In the retroperitoneal space, the kidneys can be impacted or penetrated by trauma. The kidneys are filtration organs; therefore, they are supplied with large quantities of blood. They can be sheared from their base, crushed, or fractured—causing significant amounts of blood loss. If the kidney is injured, a common finding is hematuria, or blood in the urine. This may be obvious to the naked eye or impossible to detect in the field. You may find drops of blood or blood-tinged urine on the patient's undergarments, leading you to inspect the exterior of the genitals. Blood visible on inspection of the urinary meatus (opening of the urethra situated on the glans penis in males and in the vulva in females) indicates significant trauma to the genitourinary system. If there is no blood present, do not take this as a sign that the patient is free from injury; it may be that the clinical manifestation of blood is not yet visible.

Patient Assessment of Abdominal Injuries

The assessment of abdominal injuries is one of the more difficult assessments that you will perform. The causes of the injury may be readily apparent as a result of the MOI or the visibility of a penetrating wound, but the resulting tissue damage may not be so apparent. Often other injuries, such as a fractured bone, may be painful and distracting for the patient. The patient may not tell you about more subtle pain that could indicate an abdominal injury. Additionally, some abdominal injuries develop and worsen over time, making reassessment critical.

Scene Size-up

Scene Safety

Your scene size-up begins with the information reported from dispatch. This information will help you to prepare for the call. Often the information will be sketchy or even inaccurate as reported to the dispatcher, but it will still provide some information to consider as you respond to the call. For example, is the patient injured or ill? Could one have led to the other? What equipment might you need to assess and treat the patient? Standard precautions should be taken prior to arrival at the scene; gloves and eye protection should be a minimum.

When you arrive at the scene, you will continue to gather information that will help to manage the incident. Observe the scene for hazards and threats to your safety. If dispatch information indicates a possible assault, domestic dispute, or drive-by shooting, all of which commonly lead to abdominal injuries, be sure that law enforcement personnel have controlled the scene. How many victims might be involved in the incident? If you determine additional resources are needed, call for them early in your assessment. Consider early ALS intercept or rendezvous for intravenous fluid therapy and pain management.

Mechanism of Injury/Nature of Illness

As you observe the scene, look for indicators of the MOI, and consider early spinal precautions. This helps you develop an early index of suspicion for underlying injuries in the patient who has sustained a significant MOI. As you put together information from dispatch and your observations of the scene, consider the possible injuries the MOI could have produced. As you inspect a vehicle,

look at the damage. Could this damage result in an abdominal injury? In the case of an assault, think about how many times the patient was struck, with what object, and where the patient was struck. Information from the scene will help to determine your index of suspicion.

If the wound is penetrating, inspect the object of penetration. Is the object's edge serrated, smooth, or jagged? Is it clean or dirty? The MOI may also provide indications of potential safety threats. For example, a knife wound may indicate the presence of a violent individual.

Special Populations

In pediatric patients, a common mechanism of injury is a motor vehicle versus pedestrian or motor vehicle versus bicycle accident. In the pediatric patient, the chest and abdomen are less protected by bony structures than in the adult. The pediatric patient may experience significant transfer of energy on impact. In the pediatric patient, the rib cage is so flexible that the chest can be flattened almost to the spine before rib fractures occur. This extensive compression can involve not only the organs of the chest, but also the abdomen as well. The ribs then recoil to their normal position, and the patient is left with very few outward signs that an injury has occurred.

Primary Assessment

Your goal in the primary assessment is to evaluate the patient's ABCs and then immediately care for any life threats. First perform a rapid scan. The general impression, including an evaluation of the level of consciousness, will help you establish the seriousness of the patient's condition. Some abdominal injuries will be obvious and graphic; however, most will be very subtle and may go unnoticed. Considering the MOI together with the general impression will help you focus on the immediate problem. Remember, in some cases of trauma or blows to the abdomen, the injury may have occurred hours or even days earlier and the pain has now reached a point where it is severe enough for the patient to seek help.

Form a General Impression

As you approach the trauma patient with a suspected closed abdominal injury, important indicators will alert you to the seriousness of the patient's condition. Is the patient awake and interacting with his or her surroundings, or is he or she lying still, making no sounds? Does the patient have any apparent life threats? What color is the patient's skin? Is he or she responding to you appropriately or inappropriately? Your general impression will help you develop an index of suspicion for serious injuries and determine how urgently your patient needs care.

Trauma patients with closed abdominal injuries may have what appear to be minor injuries; however, you must not be distracted from looking for more serious hidden injuries. For example, an abrasion to the abdomen may appear to be a superficial injury when in actuality it may be the only outward clue that abdominal organs are injured.

Check for responsiveness using the AVPU scale (*Alert* to person, place, and day; responsive to *Verbal* stimuli; responsive to *Pain; Unresponsive*). Assessing a patient's mental status is generally easy and can be done by asking the patient about his or her chief complaint. If the patient is alert, this should help direct you to any apparent life threats. If the patient is not alert, determine if he or she responds to verbal or painful stimuli or if he or she is unresponsive. An unresponsive patient may indicate a life-threatening condition. You should administer high-flow oxygen via a nonrebreathing mask to all patients whose level of consciousness is less than alert and oriented and provide immediate transport to the emergency department.

Trauma patients with open injuries may present with obvious significant injuries that indicate a serious condition. However, other injuries may not be as obvious but may still indicate a very serious condition. Your general impression of how the patient is doing is based on information as simple as the MOI and the patient's level of consciousness. Observations such as bleeding from open injuries, skin color and condition, and gasping respirations also contribute to your general impression and help to determine your treatment priorities and the urgency of care needed. A good question to ask yourself is, "How sick is my patient based on what I know right now?"

Airway and Breathing

Next, ensure that the patient has a clear and patent airway. If a spinal injury is suspected, prevent the patient from moving by having a team member hold the patient's head still and verbally remind the patient not to move. Patients may report that they feel nauseous, and they may vomit. Remember to keep the airway clear of vomitus so that it is not aspirated into the lungs, especially in a patient who is unconscious or has an altered level of consciousness. Turn the patient to one side, using spinal precautions if necessary, and try to clear any material from the throat and mouth. Note the nature of the vomitus: undigested food, blood, mucus, or bile.

You must also quickly assess the patient for adequate breathing. A distended abdomen or pain may prevent adequate inhalation. When these guarded respirations decrease the effectiveness of the patient's breathing, providing supplemental oxygen with a nonrebreathing mask will help improve oxygenation. If the patient's level of consciousness is decreased and respirations are shallow, consider supplementing respirations with a bag-mask device. Use airway adjuncts as necessary to ensure a patent airway and assist with breathing.

Circulation

Superficial abdominal injuries usually do not produce significant external bleeding. Internal bleeding from open or closed abdominal injuries, however, can be profound. Trauma to the kidneys, liver, and spleen can cause significant internal bleeding. Evaluate the patient's pulse and skin color, temperature, and condition to determine the stage of shock. If you suspect shock, treat the patient aggressively by providing oxygen, positioning the patient in a modified shock position, and keeping the patient warm. Wounds should be covered and bleeding controlled as quickly as possible.

Transport Decision

Because of the nature of abdominal injuries, a short on-scene time and quick transport to the hospital are generally indicated. Abdominal pain together with an MOI that suggests injury to the abdomen or flank is a good indication for rapid transport. In the prehospital environment, it is difficult to determine whether the liver, spleen, or kidney has been injured. Hollow organs that have ruptured are also difficult to identify without more advanced diagnostic equipment. A delay in medical evaluation may result in an unnecessary and dangerous progression of shock. Patients who have visible significant bleeding or signs of significant internal bleeding may quickly become unstable. Treatment should be directed at quickly addressing life threats and providing rapid transportation to the closest appropriate hospital.

Patients with abdominal injuries should be evaluated at the highest level of trauma center available because of the hidden or occult nature of most abdominal injuries. Transport to a trauma center is indicated for any patient who has an MOI that produces a high index of suspicion and who has any visible significant trauma, blunt or penetrating. Follow local protocols when considering a lower level of care such as acute care sites and clinics. Only the lowest levels of MOI should be considered eligible for these types of facilities.

History Taking

Investigate Chief Complaint

Once you have identified and treated life threats, you can then move on to gathering a history from the patient. If you have not yet done so, you should determine and investigate the patient's chief complaint and further investigate the MOI. You will also identify any associated signs and symptoms and pertinent negatives. You can quickly assess the patient's chief complaint with a simple inspection, noting the position in which he or she is lying. Movement of the body or the abdominal organs irritates the inflamed peritoneum, causing additional pain. To minimize this pain, patients will lie still, usually with their knees drawn up, and their breathing will be rapid and shallow. For the same reason, they will contract their abdominal muscles, a sign called guarding. Ask about previous injuries associated with a chief complaint of abdominal pain.

SAMPLE History

Next, obtain a SAMPLE history from your patient. Using OPQRST (Onset, Provocation or palliation, Quality, Region/radiation, Severity, and Timing of pain) to help explain an abdominal injury may provide some helpful information such as the description of the pain and if the pain is radiating. Take this time to confirm that you have all the necessary history to inform the hospital staff. If the patient is not responsive, attempt to obtain the SAMPLE history from friends or family members.

When investigating the history of the current injury or the details of the injury, make sure to ask if the patient has experienced any nausea, vomiting, or diarrhea. If the patient has experienced any of these symptoms, ask how many times and over what time period. Ask about the appearance of any bowel movements and urinary output to determine if there was any blood in the urine or black tarry stools (melena). This can help to determine if the patient has gastrointestinal bleeding and if there is bleeding in the lower gastrointestinal tract.

Secondary Assessment

The secondary assessment is a more detailed, comprehensive examination of the patient that is used to uncover injuries that may have been missed during the primary assessment. In some instances, such as with a critically injured patient or a short transport time, you may not have time to conduct a secondary assessment.

Physical Examinations

Usually, you will perform the physical examination on all patients with abdominal injuries in the same manner. Remove or loosen clothes to expose the injured regions of the body for the focused physical assessment. Inspect the patient for bleeding before removing the patient's clothing to prevent damaging any exposed tissues, such as in the case of an evisceration. Provide privacy as needed or wait until you are in the back of the ambulance. The patient without suspected spinal injury should be allowed to stay in the position of comfort—with the legs pulled up

You are the Provider: PART 3

After applying high-flow oxygen to the patient, you perform a full-body scan during the secondary assessment. Your partner continues to manually stabilize the patient's head in a neutral position. During your physical examination, you find numerous abrasions to his anterior abdomen and flanks, and he has pain to palpation of his right upper abdominal quadrant. The rest of your secondary assessment is unremarkable for gross injuries. An off-duty EMT from your agency, who was a spectator at the rodeo, provides assistance and obtains a set of vital signs.

Recording Time: 6 Minutes	
Respirations	24 breaths/min; adequate depth
Pulse	120 beats/min; weak and regular
Skin	Pale, cool, and clammy
Blood pressure	104/54 mm Hg
Oxygen saturation (Sao$_2$)	97% (on oxygen)

The patient's spine is fully immobilized and he is loaded into the ambulance. He is still conscious and alert, but is restless and complains of being very thirsty. You begin transport to the hospital while continuing to assess and treat the patient en route.

5. What are some common bruising patterns and clinical signs associated with intra-abdominal bleeding?

toward the abdomen. This position will relieve some of the tension on the abdomen and thus provide pain relief. For patients with spinal injury, place padding such as blankets or pillows under the patient's knees to help alleviate tension on the abdominal wall. Keep in mind that you can worsen the spinal injury if you are too aggressive when placing these items.

The patient without suspected spinal injury should not be forced to lie flat for the physical examination or transport. The fetal position may provide the patient with the most comfort during the physical examination or transport.

Examine the entire abdomen including all posterior, anterior, and lateral surfaces. This is a critical step when patients have an injury with an entrance wound. Examine the axillae (armpits) for entrance wounds.

When you auscultate the bowel, keep in mind that in the prehospital setting bowel sounds can be difficult to hear. Even if you are in a quiet environment, you may not have enough time to wait to hear them. If you hear nothing, do not state that bowels sounds are absent, use the term hypoactive. If you hear a lot of gurgling and the sounds of gas moving about frequently, the term hyperactive may be used. Most bowel sounds can be difficult to interpret and diverse in cause, so they are considered to be of limited value to you in your assessment.

Use DCAP-BTLS to help identify specific signs and symptoms of injury. Inspect and palpate the abdomen for the presence of deformity, which may be subtle in abdominal injuries. Look for the presence of contusions and abrasions, which can help localize focal points of impact and may indicate significant internal injury. Puncture wounds and other penetrating injuries must not be overlooked because the intra-abdominal extent of these injuries may be life threatening. The presence of burns, as in the case of flash burns or scalding fluids spilled onto the abdomen, must be noted and managed appropriately. Palpate for tenderness and attempt to localize to a specific quadrant of the abdomen. Identify and treat any lacerations with appropriate dressings. Swelling may involve the abdomen globally and indicate significant intra-abdominal injury.

Remember to also palpate the abdomen when examining the region. Palpation is typically performed first with a light touch, progressing to applying gentle increasing pressure deeper into the tissues to elicit a pain response for injuries. The object is to not cause further pain to the patient but to identify the location of the pain. Start by palpating the quadrant farthest away from the quadrant that is exhibiting signs and symptoms of injury and pain. This technique allows you to investigate the possibility of radiation and extension of the pain into other quadrants without causing the patient to guard the rest of the abdomen. If light touch elicits pain, deep palpation is not required or recommended.

If the patient has been subjected to a significant MOI, a full-body scan will help you to quickly identify any injuries your patient may have, not just abdominal injuries. Begin with the head and finish with the lower extremities, moving in a systematic manner. Your goal is not to identify the extent of all the injuries, but to determine whether other injuries are present. This requires you to work quickly but thoroughly. If you find a life-threatening problem, stop and treat it immediately, otherwise move on. The injuries you find will help you in packaging your patient for transport. Up to this point in the patient assessment process you may have been stabilizing the patient's spine by simply holding the head still and telling the patient not to move. If a cervical collar has not been applied, place one on the patient now before you log roll the patient to inspect the posterior part of the body and place the patient on a backboard.

If the MOI suggests an isolated injury to the abdomen, focus your physical examination on the injured area only. Inspect the skin of the abdomen for wounds through which bullets, knives, or other missile-type foreign bodies may have passed. Keep in mind that the size of the wound does not necessarily indicate the extent of the underlying injuries. If you find an entry wound, you must always check for a corresponding exit wound in the patient's back or sides. If the injury was caused by a very high-velocity missile from a rifle, you may see a small, harmless-looking entrance wound with a large, gaping exit wound. Do not attempt to remove a knife or other object that is impaled in the patient. Instead, stabilize the object with supportive bandaging. Bruises or other visible marks are important clues to the cause and severity of any blunt injury. Steering wheels and seatbelts produce characteristic patterns of bruising on the abdomen or chest.

The kidneys are located in the flank region of the back. Inspect and palpate this area for tenderness, bruising, swelling, or other signs of trauma. Remember that you may not be able to elicit pain from the specific organ, but the tissues around it may exhibit symptoms of pain. Hollow organs will spill their contents into the peritoneal cavity and will typically produce a significant peritonitis, which may be seen as diffuse pain with guarding, and reaction to sudden jarring movements. Bowel sounds may help to confirm these findings, but you should not depend on using these sounds to help rule out a specific injury.

Vital Signs

Quickly obtain the patient's vital signs. Many abdominal emergencies, in addition to those that cause severe bleeding, can cause a rapid pulse and low blood pressure.

Your record of vital signs, made as early as possible and periodically thereafter (every 5 minutes in the patient whom you suspect has a serious injury), will help you to identify changes in the patient's condition and be alert to signs of decompensation from blood loss. If the patient is experiencing external or internal hemorrhaging, as in the case of a stab wound or a direct blow to the abdomen, monitor the vital signs closely with a degree of suspicion and pay close attention to shifts in the vital signs.

Monitoring Devices

Use pulse oximetry and noninvasive blood pressure devices when these monitoring devices are available. It is recommended that you always assess the patient's first blood pressure manually with a sphygmomanometer (blood pressure cuff) and stethoscope.

Words of Wisdom

Occasionally you will have a patient who is extremely sensitive to palpation or is "ticklish." This can make the physical examination process more difficult. Because it is difficult for a patient to tickle himself or herself, use the technique of placing the patient's hand on the surface of his or her abdomen and then palpate and compress the abdomen with the patient's hand between your hand and the patient's skin.

Words of Wisdom

Log rolling the patient onto a backboard always provides a valuable chance to examine the back for signs of injury. Instruct and position helpers to ensure your ability to inspect and palpate the back briefly while the patient is rolled onto his or her side.

If possible, pad the long spine board before returning the patient to a supine position. This action helps reduce discomfort and prevents soft-tissue injury. It is best to avoid log rolling patients with an evisceration because this can cause more of the intestines to protrude from the wound. Instead, keep the patient in the supine position and allow him or her to flex the knees when possible to help relieve tension on the abdomen.

Reassessment

Repeat the patient's primary assessment and vital signs. Reassess the interventions and treatment you have provided to the patient. Identifying trends in pain, vital signs, and the progress of treatments will help determine whether the patient's condition is improving or getting worse. Adjustments in care can be based on these objective findings.

Interventions

Manage airway and breathing problems based on signs and symptoms found during the primary assessment. Provide complete spinal stabilization to the patient with suspected spinal injuries. If the patient has signs of hypoperfusion, provide aggressive treatment for shock and rapid transport to the appropriate hospital. If an evisceration is discovered, place a dressing moistened with normal saline over the wound, apply a bandage, and transport. Never attempt to push eviscerated tissue or organs back into the abdominal cavity. A patient who has a ruptured diaphragm may have an abdomen with a sunken anterior wall (scaphoid abdomen) and difficulty breathing because of bowel contents in the chest cavity. These patients should receive positive-pressure ventilation with a bag-mask device. Do not delay transport of the seriously injured trauma patient to complete nonlife-saving treatments such as splinting extremity fractures. Instead, complete these types of treatments en route to the hospital.

Communication and Documentation

Communicate the MOI and injuries found during your assessment. Use of appropriate medical and anatomic terminology is important; however, when in doubt just describe what you see. The content of your radio report will depend on your local protocols. The information you provide will help the hospital staff prepare for the patient.

Documentation of your assessment and trends in vital signs is a tremendous help to physicians in evaluating the problem when the patient arrives in the emergency department. Document the results of the physical examination and any pertinent negatives such as no blood loss noted in bowel movements. Also document if you passed over any step of the physical examination such as with a patient with acute abdominal pain in whom you opted to not perform palpation. Continuity of care is maintained when the emergency department that has an accurate record of your findings at the scene as well as the treatments you have provided. It is imperative that you be able to describe the scene in enough detail so the trauma team has a clear idea of the circumstances. Some services and departments now carry digital or other instant cameras to be able to show the trauma team the mechanism that the patient was exposed to. Remember that your written report is also a legal record of your care. If assault is suspected, you may have a legal requirement to inform the hospital staff of your suspicions; however, this information can wait until you have delivered the patient to the hospital and have a chance to discuss it privately with appropriate hospital personnel.

Be cautious and diligent when dealing with patients who refuse transport to the hospital after sustaining an injury to the abdomen or genitourinary system. These patients are at high risk for complications; therefore, that information should be explained to them in great detail. Contacting medical control for assistance to convince the patient of the need for transport can be very useful. Always document in detail the information you provide the patient and, if the patient continues to refuse transport, have the patient sign a document of refusal or an "against medical advice" form.

Emergency Medical Care of Abdominal Injuries

Closed Abdominal Injuries

The biggest concern in patients with closed abdominal injuries is the fact that you do not know the true extent of the injury. Because of this, the patient requires expedient transport to the nearest and highest level of care available, primarily a trauma center with a surgeon. If possible, position the patient for optimal comfort and apply high-flow oxygen. Treat for shock.

Blunt Abdominal Injuries

A patient with a blunt abdominal injury may have one or more of the following injuries:

- Severe bruising of the abdominal wall
- Laceration of the liver and spleen
- Rupture of the intestine

- Tears in the mesentery, the membranous folds that attach the intestines to the walls of the body, and injury to blood vessels within them
- Rupture of the kidneys or avulsion of the kidneys from their arteries and veins
- Rupture of the bladder, especially in a patient who had a full and distended bladder at the time of the injury
- Severe intra-abdominal hemorrhage
- Peritoneal irritation and inflammation in response to the rupture of hollow organs

A patient who has sustained a blunt abdominal injury should be log rolled to a supine position on a backboard. Ensure that you protect the spine while you roll him or her. If the patient vomits, turn him or her to one side and clear the mouth and throat of vomitus. Monitor the patient's vital signs for any indication of shock such as pallor; cold sweat; rapid, thready pulse; or low blood pressure. If you see any of these signs, administer high-flow supplemental oxygen via a non-rebreathing mask, or a bag-mask device if needed, and take all the appropriate measures to treat for shock. Patients with dyspnea due to a diaphragmatic rupture may require assistance with a bag-mask device. Keep the patient warm with blankets, and provide prompt transport to the emergency department.

Open Abdominal Injuries

Penetrating Abdominal Injuries

Patients with penetrating injuries generally have obvious wounds and external bleeding **Figure 28-10A**; however,

---

You are the Provider: PART 4

You reassess the patient en route and note that his clinical status has deteriorated. You ask your driver to notify the receiving facility as you continue to treat the patient. There are no ALS units available to rendezvous with and assist you.

Recording Time: 12 Minutes	
Level of consciousness	Responsive to pain only
Respirations	28 breaths/min; shallow
Pulse	134 beats/min; absent radial pulses
Skin	Cool, pale, and clammy
Blood pressure	82/54 mm Hg
Sao$_2$	88% (on oxygen)

6. Why is your patient's condition deteriorating? How should you modify your treatment?

7. Would a pneumatic antishock garment (PASG) benefit this patient? Why or why not?

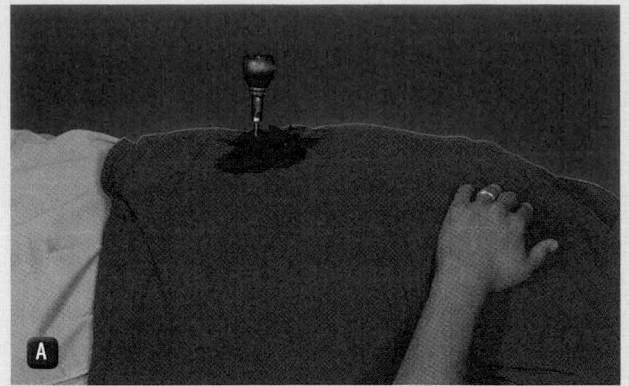

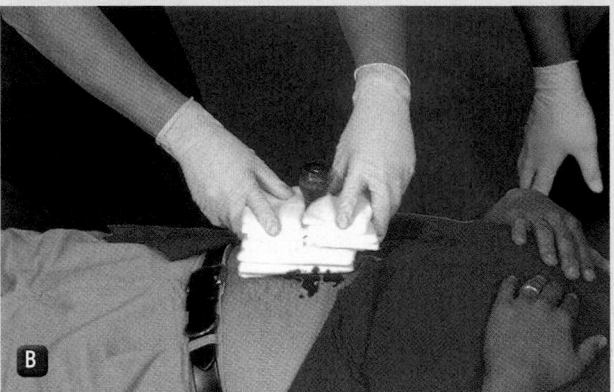

Figure 28-10 **A.** Penetrating injuries have obvious wounds and may also have external bleeding. **B.** If the penetrating object is still in place, use a roller bandage to stabilize the object and to control bleeding.

large amounts of external bleeding may not be present. As an EMT, you should have a high index of suspicion that the patient has serious unseen blood loss occurring inside the body. A large wound may have protrusions of bowel, fat, or other structures. In addition to pain, these patients often report nausea and vomiting. Patients with peritonitis generally prefer to lie very still with their legs drawn up because it hurts to move or straighten their legs. They may complain about every bump in the road during transport.

Some penetrating injuries go no deeper than the abdominal wall, but the severity of the injury often cannot be determined in the prehospital setting. Only a surgeon can accurately assess the damage. Therefore, as you care for a patient with this type of wound, you should assume that the object has penetrated the peritoneum, entered the abdominal cavity, and possibly injured one or more organs, even if there are no immediate obvious signs.

If major blood vessels are cut or solid organs are lacerated, bleeding may be rapid and severe. Other signs of intra-abdominal injuries may develop slowly, particularly

in penetrating wounds to hollow organs. Once such an organ is punctured and its contents are discharged into the abdominal cavity, peritonitis may develop, but this may take several hours.

In caring for a patient with a penetrating wound to the abdomen, follow the general procedures described previously for care of a blunt abdominal injury as well as the following specific steps for the penetrating wound. Inspect the patient's back and sides for exit wounds, and apply a dry, sterile dressing to all open wounds. If the penetrating object is still in place, apply a stabilizing bandage around it to control external bleeding and to minimize movement of the object **Figure 28-10B**.

Abdominal Evisceration

Severe lacerations of the abdominal wall may result in an evisceration, in which internal organs or fat protrude through the wound **Figure 28-11**. Never try to replace an organ that is protruding from an abdominal laceration, whether it is a small fold of peritoneum or nearly all of the intestines. Instead, cover it with sterile gauze compresses moistened with sterile saline solution and secure with a sterile dressing. (Protocols in some EMS systems call for an occlusive dressing over the organs, secured by trauma dressings.) Because the open abdomen radiates body heat very effectively, and because exposed organs lose fluid rapidly, you must keep the organs moist and warm. If you do not have gauze compresses, you may use moist, sterile dressings, covered and secured in place with a bandage and tape **Figure 28-12**. Do not use any material that is adherent or loses its substance when wet, such as toilet paper, facial tissue, paper towels, or absorbent cotton.

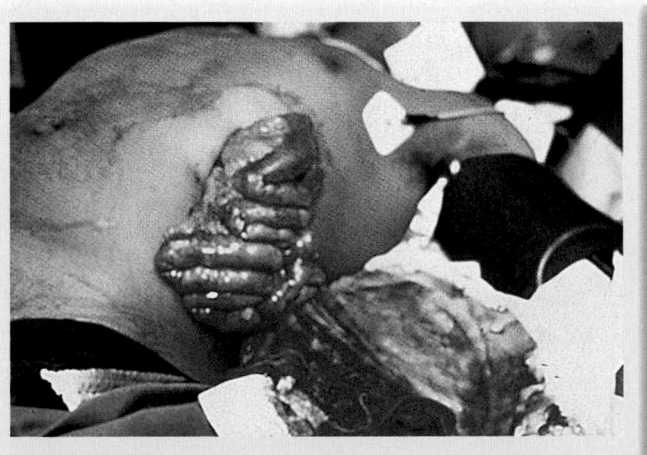

Figure 28-11 An abdominal evisceration is an open abdominal wound from which internal organs or fat protrude.

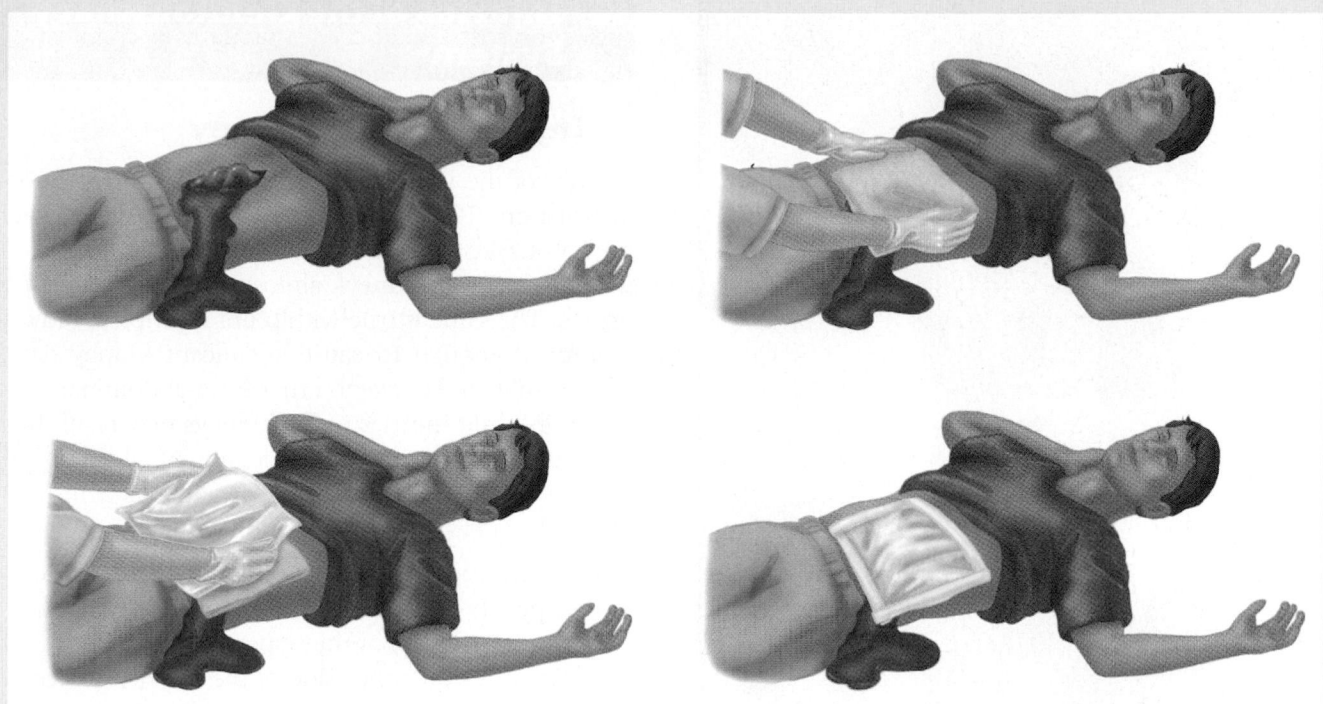

Figure 28-12 **A.** The open abdomen radiates body heat rapidly and must be covered. **B.** Cover the wound with moistened, sterile gauze or with an occlusive dressing, depending on local protocol. **C.** Secure the dressing with a bandage. **D.** Secure the bandage with tape.

Words of Wisdom

The pneumatic antishock garment (PASG) has a long and varied history of falling in and out of favor for use in the treatment of trauma patients. The technology behind the PASG was originally developed by the military for fighter pilots who wore them as "g-suits." The military needed to increase the pilot's systolic blood pressure to keep the pilot from losing consciousness at high altitudes or while performing certain maneuvers. The developers originally believed that the g-suits pushed several units of blood from the lower body up into the torso, making more blood available for circulation. Many studies have now shown that while the g-suit works very efficiently on healthy fighter pilots, it does not resolve shock in the typical trauma patient. Consequently many services have pulled the PASG from service and their protocols.

Currently the recommended practice is that if a PASG is used, it is used as a circumferential air splint for the pelvis, to produce external tamponade, or to decrease an internal abdominal hemorrhage. This use is further restricted by state and local protocols and by individual medical directors. Consult your local protocols and medical control in regard to the application of the PASG.

Treat the patient for shock by keeping the patient warm and, if possible, elevate the legs into the modified shock position. Provide high-flow oxygen and transport according to local protocols and destination policy. Transport the patient to the highest level trauma center available.

Anatomy of the Genitourinary System

The genitourinary system controls both the reproductive functions and the waste discharge system, which are generally considered together. The organs of the genitourinary system, such as the kidneys and bladder, are located in the abdomen.

The urinary system controls the discharge of certain waste materials filtered from the blood by the kidneys. In the urinary system, the kidneys are solid organs; the ureters, bladder, and urethra are hollow organs **Figure 28-13**.

The genital system is also important to reproductive processes. The male genitalia, except for the prostate gland and the seminal vesicles, lie outside the pelvic cavity **Figure 28-14**. The female genitalia, except for the vulva, clitoris, and labia, are contained entirely within the pelvis **Figure 28-15**. The male and female reproductive organs have certain similarities and, of course, basic differences. They allow for the production of sperm and egg cells and appropriate hormones, the act of intercourse, and, ultimately, reproduction.

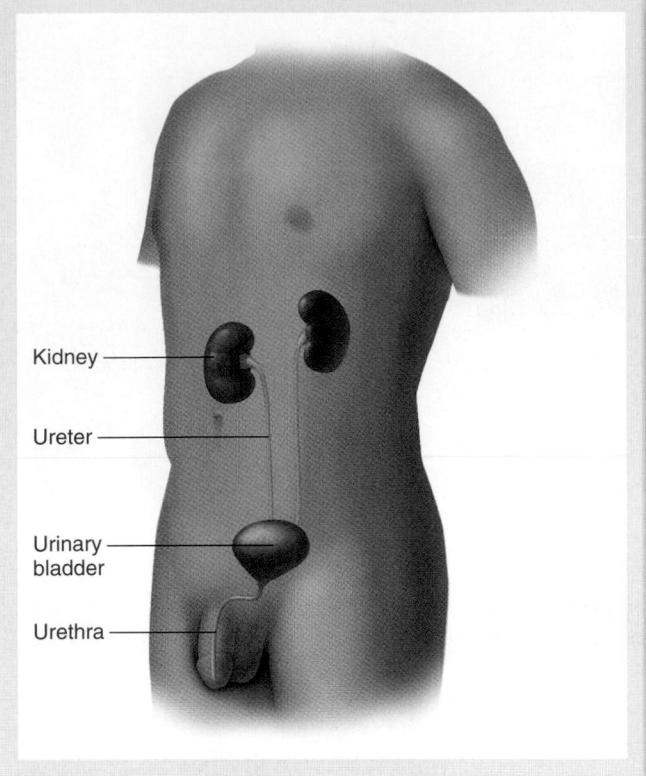

Figure 28-13 The urinary system lies behind the digestive tract. The kidneys are solid organs; the ureter, bladder, and urethra are hollow organs.

Injuries of the Genitourinary System

Injuries of the Kidney

Injuries of the kidney are not unusual and rarely occur in isolation. This is because the kidneys lie in such a well-protected area of the body. A penetrating wound that reaches the kidneys almost always involves other organs. The same is true with blunt injuries. A blow that is forceful enough to cause significant kidney damage often results in damage to other intra-abdominal organs. Less significant injuries to the kidneys may result from a direct blow or even from a tackle in football **Figure 28-16**. Suspect kidney damage if the patient has a history or physical evidence of any of the following:

- An abrasion, laceration, or contusion in the flank
- A penetrating wound in the region of the lower rib cage (the flank) or the upper abdomen
- Fractures on either side of the lower rib cage or of the lower thoracic or upper lumbar vertebrae
- A hematoma in the flank region

Injuries to the Urinary Bladder

Injury to the urinary bladder, either blunt or penetrating, may result in its rupture. When this happens, urine

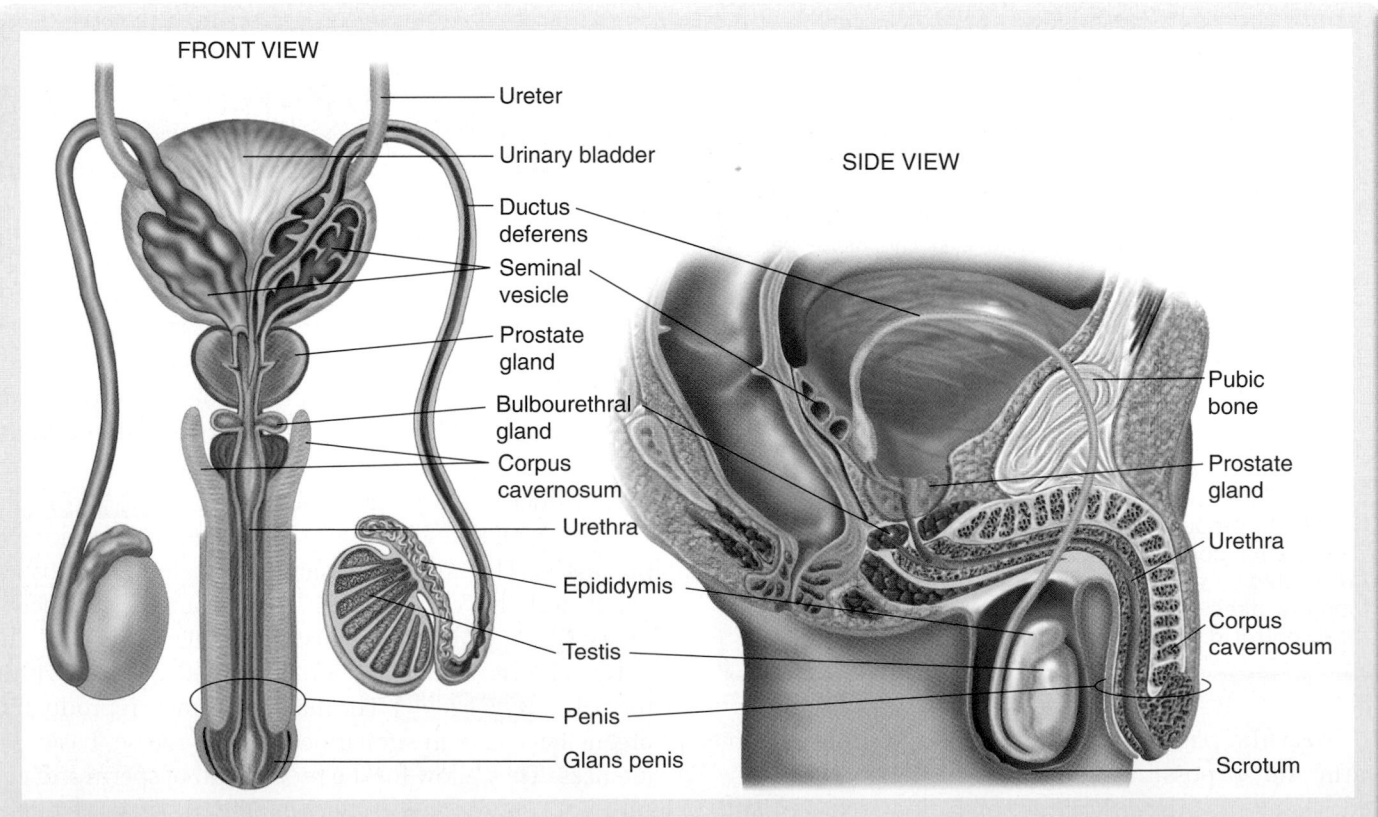

Figure 28-14 The male reproductive system includes the testicles, vasa deferentia, seminal vesicles, prostate gland, urethra, and penis.

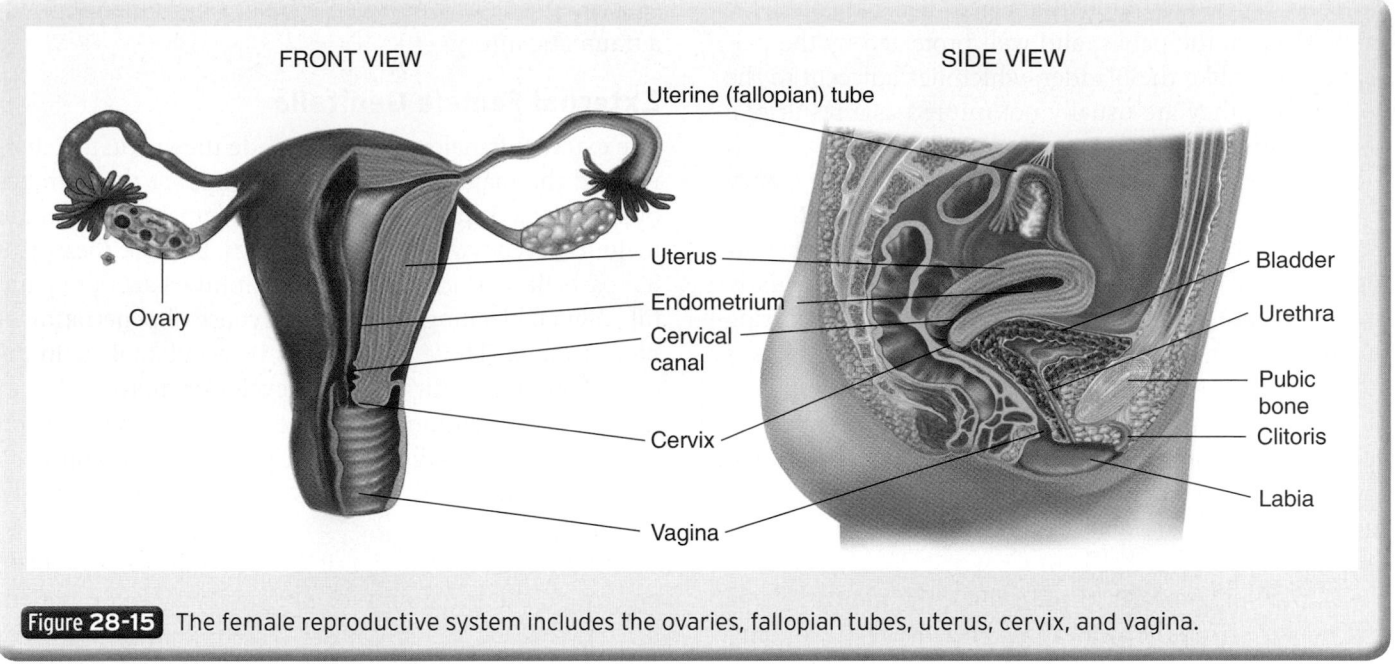

FRONT VIEW

SIDE VIEW

Uterine (fallopian) tube

Uterus

Endometrium

Cervical canal

Cervix

Vagina

Ovary

Bladder

Urethra

Pubic bone

Clitoris

Labia

Figure 28-15 The female reproductive system includes the ovaries, fallopian tubes, uterus, cervix, and vagina.

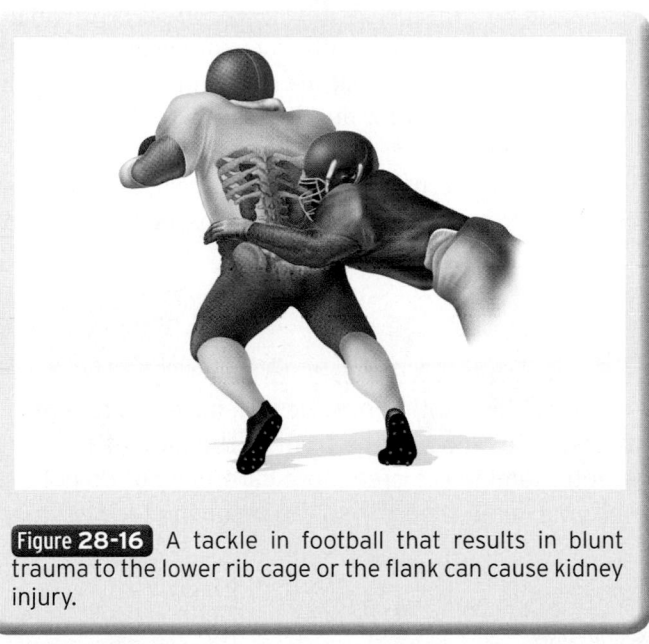

Figure 28-16 A tackle in football that results in blunt trauma to the lower rib cage or the flank can cause kidney injury.

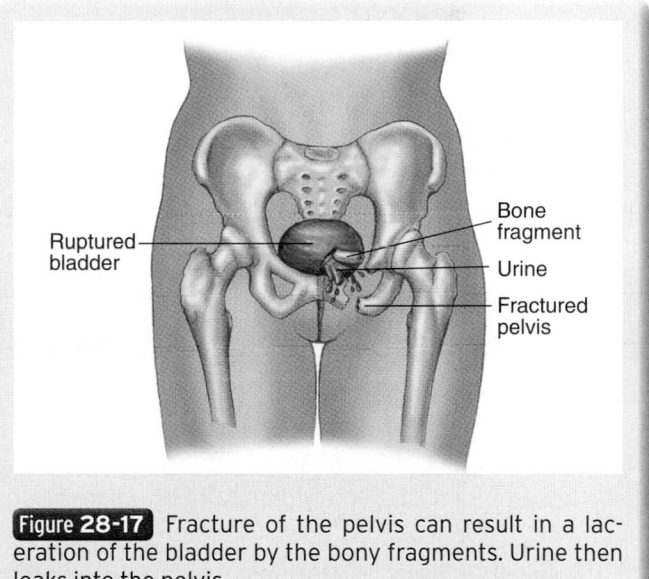

Ruptured bladder

Bone fragment

Urine

Fractured pelvis

Figure 28-17 Fracture of the pelvis can result in a laceration of the bladder by the bony fragments. Urine then leaks into the pelvis.

spills into the surrounding tissues, and any urine that passes through the urethra is likely to be bloody. Blunt injuries of the lower abdomen or pelvis often cause rupture of the urinary bladder, particularly when the bladder is full and distended. Sharp, bony fragments from a fracture of the pelvis often perforate the urinary bladder **Figure 28-17**. Penetrating wounds of the lower midabdomen or the perineum (the pelvic floor and associated structures that occupy the pelvic outlet) can directly involve the bladder. In the male, sudden deceleration from a motor vehicle or motorcycle crash can literally shear the bladder from the urethra. Remember that in the later trimesters of pregnancy, the incidence of injury to the bladder is increased by displacement of the uterus.

■ Injuries of the External Male Genitalia

Injuries of the external male genitalia include all types of soft-tissue wounds. Although these injuries are uniformly painful and generally a source of great concern to the patient, they are rarely considered life threatening and should not be given priority over other, more severe wounds.

■ Injuries of the Female Genitalia

Internal Female Genitalia

The uterus, ovaries, and fallopian tubes are subject to the same kinds of injuries as any other internal organ.

However, they are rarely damaged because they are small, deep in the pelvis, and well protected by the pelvic bones. Unlike the bladder, which lies adjacent to the bony pelvis, they are usually not injured as a result of a pelvic fracture.

An exception is the pregnant uterus. As pregnancy progresses, the uterus enlarges substantially and rises out of the pelvis, becoming vulnerable to both penetrating and blunt injuries. These injuries can be particularly severe because the uterus has a rich blood supply during pregnancy. You must also keep in mind that the fetus is at risk. You can expect to see the signs and symptoms of shock with these patients; be prepared to provide all necessary support and prompt transport. Note also that contractions may begin. If possible, ask the patient when she is due to deliver, and report this information to the hospital staff.

In the last trimester of pregnancy, the uterus is large and may obstruct the vena cava, decreasing the amount of blood returning to the heart if the patient is placed in a supine position (supine hypotensive syndrome). As a result, blood pressure may decrease. The patient should be carefully placed on her left side so that the uterus will not lie on the vena cava. If the patient is secured to a backboard, tilt the board to the left. Chapter 21, *Gynecologic Emergencies* and Chapter 31, *Obstetrics and Neonatal Care* cover in detail gynecologic emergencies and the special considerations when a pregnant woman sustains a traumatic injury.

External Female Genitalia

The external female genitalia include the vulva, the clitoris, and the major and minor labia (lips) at the entrance of the vagina. Injuries of the external female genitalia can include all types of soft-tissue injuries. Because these genital parts have a rich nerve supply, injuries are very painful. Vaginal bleeding may occur because of penetrating or blunt trauma. These injuries can be accidental, as in the case of straddle injuries from bicycles or motorcycles; or they can be intentional as in the case of assaults. Determining the MOI will assist you in deciding if you need to call for additional resources, as in the case of sexual assault.

In any case of trauma, it is important to attempt to determine the possibility of pregnancy. Ask the patient for the date of her last known menstrual period or if she has been sexually active. The assumption is that all women of childbearing age are possibly pregnant. This information is medically relevant because there are medications and tests that are harmful for a fetus and there is the potential for another source of blood loss in the gravid uterus.

In cases of external bleeding and trauma, a sterile absorbent sanitary napkin or pad may be applied to the labia.

You are the Provider: PART 5

The patient is becoming combative and resists your attempts to assist his ventilations. He will, however, tolerate oxygen via a nonrebreathing mask. During your reassessment, you note blood on the front of the patient's underwear, which was not there during previous assessments. You examine the area more carefully but do not see any open injuries to the genitalia or adjacent areas. You call your radio report in to the receiving facility.

Recording Time: 17 Minutes	
Level of consciousness	Responsive to pain only
Respirations	28 breaths/min; shallow
Pulse	128 beats/min; weak and regular
Skin	Cool, pale, and clammy
Blood pressure	90/58 mm Hg
Sao$_2$	93% (on oxygen)

The patient begins vomiting bright red blood. You quickly turn the backboard to the side to allow the vomitus to drain and then suction his mouth to ensure that his airway remains clear.

8. On the basis of your reassessment, what additional injuries should you suspect?

9. Will your reassessment findings change your current treatment plan?

Do not insert instruments, gloved fingers, or a tampon in the vagina because this can cause further damage.

Patient Assessment of the Genitourinary System

When assessing a genitourinary injury, there is a potential for embarrassment on the part of the patient. This is why is critical for you to maintain a professional presence at all times when dealing with these injuries. Remember to provide privacy for the patient during the assessment process. Whenever possible, have an EMT of the same gender as the patient perform the assessment. Look for blood first on the patient's undergarments and only inspect the external genitalia when there are complaints of pain or external signs of injury.

Scene Size-up

Scene Safety

As you arrive at the scene, observe it for hazards and threats to the safety of the crew, bystanders, and the patient. Assess the impact of hazards on patient care and address those hazards. Assess for the potential for violence and assess for environmental hazards.

Ensure that you and your crew have taken standard precautions—a minimum of gloves and eye protection. Control of blood and bloody contaminants can be difficult unless you are careful about what you touch and where. Apply standard precautions before you approach the scene to minimize your direct exposure to body fluids. Because of the color of blood and how well it soaks through clothing, you can often identify patients with an open injury as you approach the scene. However, blood can be hidden under thick clothing such as denim and leather. Eye protection is required when managing open injuries. Determine the number of patients and consider if you need additional or specialized resources on the scene.

Mechanism of Injury/Nature of Illness

As you observe the scene, look for indicators of the MOI. As you put together information from dispatch and your observations of the scene, consider how the MOI produced the injuries expected. Be aware that the patient may avoid discussing the injury to avoid undergoing a physical examination. Also, the patient may provide an MOI that seems "less embarrassing" than the actual MOI. By maintaining a professional demeanor, respecting the patient's privacy, and maintaining the patient's dignity, you will earn the patient's trust. If the patient trusts you, you are more likely to discover the true facts behind the injury.

Primary Assessment

During the primary assessment, you must quickly scan the patient to identify and treat potential life threats and determine the priority of patient care and transport. The genitourinary system is very vascular, and injuries to it can produce a significant volume of blood loss. Do not avoid this area during the rapid scan for life threats. Look externally at the patient's undergarments for signs of bleeding and injury. If bleeding is present, maintain privacy for the patient and inspect the exterior genitals for visible injury.

Form a General Impression

As you approach the trauma patient, important indicators will alert you to the seriousness of the patient's condition. Is the patient awake and interacting with his or her surroundings, or lying still, making no sounds? Does the patient have any apparent life threats? What color is the patient's skin? Is he or she responding to you appropriately or inappropriately? Your general impression will help you develop an index of suspicion for serious injuries and determine how urgently your patient needs care.

Airway and Breathing

Ensure that the patient has a clear and patent airway. Because trauma was involved, protect the patient from further spinal injury as you manage the airway. If the patient is unresponsive or has a significant altered level of consciousness, consider inserting an oropharyngeal airway or nasopharyngeal airway. Quickly assess the patient for adequate breathing. Provide assisted ventilations using a bag-mask device as needed, depending on the level of consciousness and if your patient is breathing inadequately.

Circulation

As stated previously, the genitourinary system is very vascular and can be a significant source of bleeding. Quickly assess the patient's pulse rate and quality; determine the skin condition, color, and temperature; and check the capillary refill time. These assessments will help you determine the presence of circulatory problems or shock. Closed injuries do not have visible signs of bleeding. Because the bleeding is occurring inside the body, shock may be present. Your assessment of the pulse and skin will give you an indication as to how aggressively you need to treat your patient for shock.

If visible significant bleeding is seen, you must begin the steps necessary to control bleeding. Significant bleeding is an immediate life threat and must be controlled quickly using appropriate methods. In dark environments, bleeding can be hard to see because of its color. Thick clothing may also hide bleeding. After you consider the MOI and form suspicions as to where bleeding may occur, expose that part of the body.

Transport Decision

A patient with a genitourinary system injury should be taken to a trauma center for evaluation and treatment. Any injury to this system can prove to be life altering and often requires a medical specialist to provide specialized care. When possible and protocols allow, transport the patient to a facility capable of treating this subset of injuries.

History Taking

Investigate Chief Complaint

When determining the chief complaint, you are seeking the primary reason that the patient called for assistance. Begin your interview by establishing why the patient called 9-1-1. Ask about associated complaints, but be cautious not to put words in the patient's mouth, such as when describing pain. This can be avoided by asking, "What else is wrong?" or "Is anything else bothering you?" Common associated complaints with genitourinary injuries are nausea, vomiting, diarrhea, blood in urine (hematuria), vomiting blood (hematemesis), or abnormal bowel and bladder habits such as an increase in frequency or the absence of the need to void. You can use the SAMPLE history to further elicit more facts and specifics about the chief complaint.

SAMPLE History

Use the SAMPLE mnemonic device to help determine the patient's baseline. Establish the signs and symptoms of the injury. Use OPQRST (Onset, Provocation or palliation, Quality, Region/radiation, Severity, and Timing of pain) to learn more about any pain the patient reports. Ask the patient about output from the genitourinary system, specifically the presence of blood in the urine. This may or may not be visible and the simple lack of it does not preclude your patient from having internal genitourinary injuries. Ask your patient about any allergies to medications or environmental triggers. Medications mask the signs and symptoms of injuries or make them more severe, so it important to determine what your patient is prescribed and what over-the-counter or herbal remedies the patient may have taken. The importance of past medical history cannot be overstated. Incidence of repeated or previous injury or illness involving the genitourinary system can help determine the extent of the current injury and possibly the MOI. The last intake of both food and fluids is important because it can help to predict what is contained in the genitourinary system and if the symptoms are related to the ingestion of those foods and fluids. Finally, addressing the events that led to the injury help to determine the MOI and help you to draw conclusions and develop an index of suspicion.

Secondary Assessment

The secondary assessment is a more detailed, comprehensive examination of the patient that is used to uncover injuries that may have been missed during the primary assessment. In some instances, such as with a critically injured patient or a short transport time, you may not have time to conduct a secondary assessment.

Physical Examinations

Genitourinary system injuries can be awkward to evaluate and can be even more awkward to treat. Privacy is a genuine concern. When examining the patient, expose only what is needed and cover what has been exposed. Being professional helps reduce anxiety for both you and your patient.

When your patient has an isolated injury to the genitourinary system with a limited MOI, focus your assessment on the isolated injury, the patient's complaint, and the body region affected. Look for DCAP-BTLS. Ensure that wounds are identified and control of the bleeding has been established. Note the location and extent of the injury.

If there is significant trauma (such as a blunt trauma) likely affecting multiple systems, start with a full-body scan looking for DCAP-BTLS to determine the nature and extent of genitourinary injury. This examination will help to determine all the injuries and the extent of those injuries. Inspection or visualization of the region looking for deformities may reveal the presence of multiple rib fractures (that could injure the kidneys). Identification of small areas of contusions or abrasions may pinpoint a specific point of impact. The presence of penetrating injuries indicates a possible internal injury that should be managed accordingly. The presence of burns must be noted and managed appropriately. Palpate for tenderness to localize the injury and the presence of fractures. Look for lacerations and local swelling. Application of this systematic approach to patient assessment minimizes the chance of missing significant injury.

With genitourinary injuries, it is important to not focus only on one area of the body. With significant trauma, you should quickly assess the entire patient from head to toe.

Vital Signs

Obtain the patient's vital signs. Patients who have hidden injuries may have internal bleeding and their condition may rapidly become unstable. It is important to reassess the vital signs to identify how quickly the patient's condition is changing. Signs such as tachycardia, tachypnea, low blood pressure, weak pulse, and cool, moist, and pale skin indicate hypoperfusion and imply the need for rapid treatment at the hospital. The reassessment of your patient's vital signs will give you a good understanding of how well or how poorly your patient is tolerating the injury.

Monitoring Devices

Use pulse oximetry and noninvasive blood pressure devices when these monitoring devices are available. It is recommended that you always assess the patient's first blood pressure manually with a sphygmomanometer (blood pressure cuff) and stethoscope.

Reassessment

Repeat the patient's primary assessment and vital signs. Reassess the interventions and treatment you have provided to the patient. Identifying trends in pain, vital signs, and the progress of treatments will help you determine whether the patient's condition is improving or getting worse. Adjustments in care should be based on these objective findings.

Interventions

When treating patients with trauma to the genitourinary system, the concerns are similar to those for other injuries to other body systems. Provide oxygen and maintain a patent airway. Attempt to control bleeding and treat for shock. Place the patient in a position of comfort and transport to the appropriate facility.

Communication and Documentation

Communicate early with the receiving facility staff your suspicions and concerns so they can be adequately prepared and, if required, have a specialist en route to evaluate and treat the patient. Your documentation should be complete and thorough. Describe all injuries and the treatment given. Remember, your documentation is your legal record of what happened.

Emergency Medical Care of Genitourinary Injuries

Kidneys

Damage to the kidneys may not be obvious on inspection of the patient. You may or may not see bruises or lacerations on the overlying skin. However, you will see signs of shock if the injury is associated with significant blood loss. Because one of the functions of the kidney is the formation of urine, another sign of kidney damage is blood in the urine (hematuria). Treat shock and associated injuries in the appropriate manner. Provide prompt transport to the hospital, monitoring the patient's vital signs carefully en route.

Urinary Bladder

Suspect a possible injury of the urinary bladder if you see blood at the urethral opening or physical signs of trauma on the lower abdomen, pelvis, or perineum. There may be blood at the tip of the penis or a stain on the patient's underwear.

The presence of associated injuries or of shock will dictate the urgency of transport. In most instances, provide prompt transport, and monitor the patient's vital signs en route.

External Male Genitalia

A few general rules apply to the treatment of injuries involving the external male genitalia:

- These injuries are very painful. Make the patient as comfortable as possible.
- Use sterile, moist compresses to cover areas that have been stripped of skin.
- Apply direct pressure with dry, sterile gauze dressings to control bleeding.
- Never move or manipulate impaled instruments or foreign bodies in the urethra.
- If possible, always identify and take avulsed parts to the hospital with the patient. Label the bag with the patient's name.

If you encounter a patient with an avulsion (tearing away) of skin of the penis, wrap the penis in a soft, sterile dressing moistened with sterile saline solution, and transport the patient promptly. Use direct pressure to control any bleeding. You should try to save and preserve the avulsed skin, but do not delay treatment or transport for more than a few minutes to do so.

Managing blood loss is your top priority in amputation of the penile shaft, whether partial or complete. You should use local pressure with a sterile dressing on the remaining stump. Never apply a constricting device to the penis to control bleeding. Surgical reconstruction of even a completely amputated penis is possible if you can locate the amputated part. Wrap it in a moist, sterile dressing; place it in a plastic bag; and transport it in a cooled container without allowing it to come in direct contact with ice.

If the connective tissue surrounding the erectile tissue in the penis is severely damaged, the shaft of the penis can be fractured or severely angled, sometimes requiring surgical repair. The injury may occur during particularly active sexual intercourse. It is associated with intense pain, bleeding into the tissues, and fear. Provide prompt transport to the emergency department.

Accidental laceration of the skin about the head of the penis usually occurs when the penis is erect and is associated with heavy bleeding. Local pressure with a sterile dressing is usually sufficient to stop the hemorrhage.

It is not uncommon for the skin of the shaft of the penis or the foreskin to get caught in the zipper of pants. If a small segment of the zipper is involved (one or two teeth), you can try to unzip the pants. If a longer segment is involved or the patient is agitated, use heavy scissors to cut the zipper out of the pants to make the patient more comfortable during transport. Be sure to explain how you are going to use the scissors before you begin cutting. Be particularly careful not to cause injury to the scrotum while cutting the zipper away from the penis.

Urethral injuries in the male are not uncommon. Lacerations of the urethra can result from straddle injuries, pelvic fractures, or penetrating wounds of the perineum. These injuries may bleed profusely, although bleeding may not be evident externally. Direct pressure with a dry, sterile dressing usually controls any external hemorrhage. Because the urethra is the channel for urine, it is very important to know whether the patient can urinate and whether hematuria is present. For this reason, you should save any voided urine for later examination at the hospital. Any foreign bodies that may be protruding from the urethra will have to be removed in a surgical setting.

Avulsion of the skin of the scrotum may damage the scrotal contents. If possible, preserve the avulsed skin in a moist, sterile dressing for possible use in reconstruction. Wrap the scrotal contents or the perineal area with a sterile, moist compress, and use a local pressure dressing to control bleeding. Transport this patient promptly to the emergency department.

Direct blows to the scrotum can result in the rupture of a testicle or significant accumulation of blood around the testes. In either case, you should apply an ice pack to the scrotal area while transporting the patient.

■ Female Genitalia

Lacerations, abrasions, and avulsions should be treated with moist, sterile compresses. Use local pressure to control bleeding and a diaper-type bandage to hold dressings in place. Under no circumstances should you pack or place dressings into the vagina. Leave any foreign bodies in place after you stabilize them with bandages.

In general, although these injuries are painful, they are not life threatening. Bleeding may be heavy, but it can usually be controlled by local compression. Contusions and other blunt injuries all require careful in-hospital evaluation. However, the urgency of the need for transport will be determined by associated injuries, the amount of hemorrhage, and the presence of shock.

■ Rectal Bleeding

Rectal bleeding is a common complaint and something that you may hear as a chief complaint or secondary to abdominal or pelvic complaints. Bleeding from the rectum may present as blood in or soaking through undergarments, or patients may report blood passed into the toilet associated with a bowel movement or attempted

bowel movement. Rectal bleeding can be caused by sexual assault, hemorrhoids, colitis, or ulcers of the digestive track. Significant rectal bleeding can occur after hemorrhoid surgery and can lead to a large amount of blood loss and shock. Acute rectal bleeding should never be passed off as something minor. Pack the crease between the buttocks with compresses, and consult with medical control to determine the need for transport.

■ Sexual Assault

Sexual assault and rape are all too common. Although most victims are women, men and children are also victims. Often, you can do little beyond providing compassion and transportation to the emergency department. On some occasions, these patients will have sustained multiple-system trauma and will also need treatment for shock.

Do not examine the genitalia of a victim of sexual assault unless obvious bleeding requires you to apply a dressing. Treat all other injuries according to appropriate procedures and protocols for your EMS system. Observe standard precautions. Take care to shield the patient from curious onlookers. Because you may have to appear in court as much as 2 or 3 years later, you must document the patient's history, assessment, treatment, and response to treatment in detail. Do not speculate. Record only the facts.

Follow any crime scene policy established by your system to protect the scene and any potential evidence for police. Advise the patient not to wash, douche, urinate, or defecate until after a physician has examined him or her; this will help to preserve any evidence of a crime. If oral penetration has occurred, advise the patient not to eat, drink, brush the teeth, or use mouthwash until he or she has been examined. If the patient will tolerate being wrapped in a sterile burn sheet, this may help investigators to find any hair, fluid, or fiber from the alleged offender. Handle the patient's clothes as little as possible, placing articles and any other evidence in paper bags. Do not use plastic bags. If the female patient insists on urinating, have her do so in a sterile urine container (if available). Also, have her deposit the toilet paper in a paper bag. Seal and mark the bag for the police because these items can be critical evidence.

To reduce the patient's anxiety, make sure the EMT who is caring for the patient is the same gender as the patient whenever possible. Remember that victims of sexual assault, whether they are male or female, may need medical assistance. In these cases, you must treat the medical injuries but also provide privacy, support, and reassurance. Chapter 21, *Gynecologic Emergencies*, also covers this topic in detail.

You are the Provider: SUMMARY

1. How do hollow organ injuries differ from solid organ injuries?

Although death can occur following significant injury to the hollow or solid abdominal organs, the manner and speed in which these injuries cause death is different. Furthermore, the clinical presentation may be somewhat different in a patient with a hollow organ injury versus a solid organ injury.

Hollow organs—the stomach, intestines, and urinary bladder—are structures through which materials pass. When hollow organs are ruptured or lacerated, they release their contents into the peritoneal (abdominal) cavity. The release of hollow organ contents into the peritoneal cavity causes an intense inflammatory response (peritonitis), which can result in a life-threatening infection. Hollow organ injury may be associated with some internal bleeding; however, the major cause of death is sepsis, which typically occurs later in the hospital.

Solid organs—the liver, spleen, pancreas, and kidneys—are highly vascular and tend to bleed profusely when injured by blunt or penetrating trauma. Unlike hollow organ injury, the major cause of death following injury to the solid organs is internal hemorrhage, which can lead to death very quickly.

Clinically, hollow and solid organ injury present similarly. Both types of injury commonly present with abdominal distention and rigidity and external evidence of trauma (ie, bruising, abrasions). The development and severity of pain differ, however, in the two types of injuries. When solid organ injury causes intra-abdominal hemorrhage, pain and tenderness may be delayed. Peritonitis, however, is almost always associated with intense abdominal pain because the acidic contents of the stomach and intestines are more irritating to the peritoneal cavity than is blood. The absence of abdominal pain and tenderness, however, does not rule out intra-abdominal bleeding.

2. Should you focus on trying to determine the origin of any intra-abdominal bleeding? Why or why not?

When you are caring for a patient with blunt or penetrating abdominal trauma, you should focus on recognizing signs and symptoms of shock, initiating treatment without delay, and providing prompt transport to the hospital.

If your patient sustained blunt abdominal trauma, has external evidence of injury (ie, bruising, distention, rigidity), and signs of shock—all signs that clearly suggest intra-abdominal bleeding—does the origin of the hemorrhage affect the treatment that you provide in the field? The answer, of course, is no. What *does* matter is that *the patient is bleeding from a source that you cannot control*, and that his or her outcome is dependent on your recognition of the situation, the initiation of prompt treatment, and rapid transport.

3. How should you interpret your primary assessment findings?

Rarely do EMS providers actually witness the injury when it occurs; this is why it is so important for you to pay attention to clues that suggest a particular mechanism of injury (MOI). In this situation, however, you witnessed the patient being trampled by a bull and were able to see that his injuries appear to be to the abdomen and flanks.

On the basis of the MOI and your primary assessment findings, your initial impression should be that your patient is in shock, which is likely the result of intra-abdominal bleeding. Although you will need to perform a head-to-toe secondary assessment to identify any other injuries, intra-abdominal bleeding is the most plausible field impression given the information that you have.

Although blood pressure is an important measurement, early assessment of a patient's blood pressure is not needed to recognize shock. Your patient is restless, which indicates that oxygen to his brain is decreased. His respirations are rapid (tachypnea) and his heart rate is rapid (tachycardia), which are other compensatory mechanisms of the body. The fact that your patient's radial pulses are weak indicates decreased peripheral perfusion.

On the basis of the MOI and the information obtained from the primary assessment, you should quickly identify this patient as a "load and go," begin immediate treatment, and make arrangements for prompt transport to the hospital.

4. What immediate treatment is indicated for this patient?

Your partner is already manually stabilizing the patient's head in a neutral position; this is necessary because of the MOI and because the patient reports pain in the back of his head.

Apply high-flow oxygen via a nonrebreathing mask. Oxygen is a critical treatment for any patient with signs and symptoms of shock and should be administered as soon as possible. Carefully monitor the patient's breathing and be ready to assist his ventilations if signs of inadequate breathing (ie, shallow breaths [reduced tidal volume], decreased mental status) are observed.

Cover the patient with a blanket to keep him warm. Patients in shock are less able to maintain body temperature because heat production requires energy, and energy requires oxygen; shock is a problem caused by a lack of oxygen!

5. What are some common bruising patterns and clinical signs associated with intra-abdominal bleeding?

In many patients, the only indicators of intra-abdominal injury are diffuse or localized pain, redness, and abrasions of the abdomen. Bruising of the abdomen is often a later manifestation and may not even be seen in the prehospital setting. For this reason, you must *not* rule out intra-abdominal bleeding based on the absence of bruising.

As blood accumulates in the abdominal cavity, the abdomen typically becomes distended and rigid. Palpation of the patient's abdomen can be challenging in the presence of abdominal guarding. Guarding is a conscious (voluntary) or unintentional (involuntary) response to abdominal trauma, and it is characterized by stiffening of the rectus abdominis muscles in an attempt to minimize the pain. Although guarding may be seen in patients who do not have significant intra-abdominal injury, *any abdominal rigidity following trauma should be assumed to be the result of internal bleeding.*

If bruising is observed following abdominal trauma, there are several patterns that you should look for during your

You are the Provider: SUMMARY, continued

assessment of the abdomen. The Cullen sign, or periumbilical (around the umbilicus) bruising, is an indicator of blood in the peritoneal cavity. The Grey Turner sign is characterized by bruising to the flank area; it is also an indicator of blood in the peritoneal cavity and suggests injury to the liver, spleen, or kidney.

Injury to the liver or spleen may present with referred pain to the shoulders. Unlike radiating pain, which is characterized by pain that "moves" from one area of the body to another, referred pain is characterized by pain in two separate locations. The Kehr sign, for example, is characterized by referred pain to the left shoulder following injury to the spleen.

It is important to note that some patients with intra-abdominal bleeding may present with no external signs of injury. The retroperitoneal space is a common location for hidden bleeding and can accommodate a large volume of blood.

In some patients with blunt abdominal trauma, your assessment will reveal obvious external injuries; in others, your abdominal examination findings will be grossly unremarkable. This underscores the importance of noting the mechanism of injury and closely monitoring the patient's clinical status (eg, mental status, vital signs).

6. Why is your patient's condition deteriorating? How should you modify your treatment?

Your patient was exhibiting signs of shock on initial contact with him. His level of consciousness has decreased, he is hypotensive, his breathing is inadequate, and his oxygen saturation is falling despite the use of high-flow oxygen. This marked clinical deterioration indicates that he is now in decompensated shock; the compensatory mechanisms that help maintain adequate perfusion to the tissues and cells of the body are failing.

Any deterioration in a patient's clinical status should prompt you to immediately repeat the primary assessment. His oxygen saturation is falling, which is likely the result of the combined effects of internal bleeding (you must have blood to carry oxygen) and inadequate breathing (you must be able to bring oxygen into the body to get it into the blood). At this point, you should begin assisting his ventilations with a bag-mask device attached to high-flow oxygen. Consider inserting a nasopharyngeal airway; his level of consciousness has decreased to the point that he may not be able to completely maintain his own airway.

In an attempt to improve perfusion to the vital organs (ie, the heart, brain, and lungs) you should elevate the foot end of the backboard 6″ to 12″, thus helping keep the vital organs perfused until the patient receives definitive treatment at the hospital.

As you continue to treat the patient, you must vigilantly monitor his ABCs. His condition is critical, and he is at high risk for cardiac arrest. If he becomes apneic and pulseless, begin cardiopulmonary resuscitation and ask your partner to update the receiving facility.

7. Would a pneumatic antishock garment (PASG) benefit this patient? Why or why not?

The pneumatic antishock garment (PASG) was originally designed to treat patients with severe trauma by shunting blood to the heart, lungs, brain, and other vital organs by applying circumferential pressure to the lower extremities and abdomen. When treating patients with internal bleeding–again, bleeding that cannot be controlled in the field–the PASG increases the patient's blood pressure. Whereas this may sound favorable, it has actually been shown to be detrimental to the patient by increasing the amount of internal bleeding that is already occurring.

Current shock treatment consists of high-flow oxygen, thermal management, and elevation of the lower extremities. Follow your state or local protocols regarding use of the PASG.

8. On the basis of your reassessment, what additional injuries should you suspect?

Blood stains on the front of the patient's underwear without evidence of any open genitalia injuries indicates that he has blood in his urine (hematuria) and suggests injury to his kidneys, urinary bladder, or both. Remember, the patient had diffuse abdominal pain and abrasions to *both* the anterior part of his abdomen and flanks. In addition to injury to his liver or spleen, which is likely what is causing his shock, it is clearly possible that he experienced injury to his genitourinary organs as well.

Kidney injuries are not uncommon and rarely occur in isolation because the kidneys lie in such a well-protected area of the body. However, blunt force trauma that is significant enough to cause injury to the kidney (ie, being trampled by a bull) almost always results in injury to other intra-abdominal organs. If you will recall from your full-body scan of the patient, he had pain in and abrasions of his flanks; this indicates direct trauma to that area, which overlies the kidneys. Injury to the kidneys may not be obvious during your assessment. If anything, you may only see abrasions or redness over the flanks; flank bruising (Grey Turner sign) typically does not manifest until later. Because one of the functions of the kidney is the formation of urine, another sign of kidney injury is hematuria.

Hematuria can also indicate rupture of the urinary bladder. Blood at the urethral opening should also make you suspicious for urinary bladder rupture. When the urinary bladder ruptures, urine spills into the surrounding tissues, and any urine that passes through the urethra will likely contain blood. Blunt injuries to the lower abdomen or pelvis can cause rupture of the urinary bladder, especially if the bladder is full and distended at the time of the injury. In isolation, bladder injuries typically do not cause shock; however, they can cause peritonitis and infection, thereby complicating the patient's condition.

Hematemesis, the vomiting of blood, indicates bleeding within the gastrointestinal tract. More specifically, vomiting bright red blood indicates injury to some part of the upper gastrointestinal tract. You should suspect injury to the patient's stomach. Although rupture of his stomach would have likely presented with hematemesis earlier, at a minimum, you should suspect a stomach laceration. As with any hollow organ injury, a lacerated stomach can cause gastric contents to leak into the abdominal cavity, potentially resulting in peritonitis and infection.

You are the Provider: SUMMARY, continued

9. Will your reassessment findings change your current treatment plan?

Despite the presence of indicators that suggest injury to both solid and hollow abdominal and genitourinary organs, you should maintain the patient's airway, ensure adequate oxygenation and ventilation, and treat for shock. Rapid transport to a trauma center is critical!

Blunt trauma with internal bleeding is often more fatal than penetrating abdominal trauma with external bleeding. Open injuries are obvious, whereas internal injuries are often hidden and may be missed. In the absence of obvious external injury, a trauma patient with signs of shock should be assumed to be bleeding into his or her abdomen.

EMS Patient Care Report (PCR)

Date: 12-16-09	**Incident No.:** 012809	**Nature of Call:** Traumatic injury		**Location:** 1333 Rodeo Blvd.	
Dispatched: 1904	**En Route:** 1904	**At Scene:** 1904	**Transport:** 1914	**At Hospital:** 1925	**In Service:** 1934

Patient Information

Age: 20 **Sex:** M **Weight (in kg [lb]):** 61 kg (135 lb)	**Allergies:** No known drug allergies **Medications:** None **Past Medical History:** None **Chief Complaint:** Pain in abdomen, flanks, and back of head

Vital Signs

Time: 1910	**BP:** 104/54	**Pulse:** 120	**Respirations:** 24	**Sao$_2$:** 97%
Time: 1916	**BP:** 82/54	**Pulse:** 134	**Respirations:** 28	**Sao$_2$:** 88%
Time: 1921	**BP:** 90/58	**Pulse:** 128	**Respirations:** 28	**Sao$_2$:** 93%

EMS Treatment
(circle all that apply)

Oxygen @ 15 L/min via (circle one): NC (NRM) (Bag-Mask Device)		(Assisted Ventilation)	**Airway Adjunct**	**CPR**
Defibrillation	**Bleeding Control**	**Bandaging**	**Splinting**	**Other:** (Thermal management, shock treatment, spinal precautions, suctioning)

Narrative

Medic 4 was standing by at a rodeo event when a bull rider was thrown from a bull and trampled. On contact with the patient, a 20-year-old male, he was found to be conscious, but restless, and in severe pain. His airway was patent and his breathing was adequate; radial pulses were rapid and weak. Manual c-spine stabilization was initiated immediately. Patient complains of pain in the RUQ of his abdomen, his flank areas, and the back of his head. Administered high-flow oxygen via nonrebreathing mask and performed secondary assessment. Assessment revealed abrasions of the anterior part of the abdomen and both flanks, and palpable tenderness in the RUQ. Breath sounds were clear to auscultation bilaterally, and the remainder of the assessment was unremarkable. Applied full spinal precautions, covered patient with a blanket, loaded him into the ambulance, obtained vital signs, and began transport to the hospital. En route, patient's mental status and vital signs deteriorated; he was now responsive to pain only; his respirations increased in rate, but decreased in depth; a marked decrease in his BP was observed; and his oxygen saturation decreased. Elevated the foot end of the backboard and began assisting the patient's ventilations with a bag-mask device attached to high-flow oxygen. Patient then become somewhat combative and would no longer tolerate assisted ventilation. However, he would tolerate oxygen via nonrebreathing mask. Reassessment of patient revealed little improvement in his vital signs. Noted blood on patient's underwear, which was previously not present. Assessment of genitalia and adjacent areas revealed no signs of open injury. Continued to monitor patient and contacted receiving facility. Patient then began vomiting bright red blood, so the backboard was immediately turned to the side and the patient's mouth was suctioned to ensure airway patency. Suctioned remaining vomitus from the patient's mouth and returned the backboard to a supine position. Called radio report to receiving facility to provide update on patient's status, and closely monitored his condition throughout the duration of the transport. Patient was delivered to emergency department staff without incident, and verbal report was given to attending physician. Medic 4 cleared the hospital and returned to service at 1934. *End of report*

Assessment and Emergency Care of Abdominal Injuries

Scene Size-up

Scene Safety	Review dispatch data en route to the call. Ensure scene safety and safe access to the patient. Be aware of potential violence or the possibility of a crime scene. Follow standard precautions, putting on a minimum of gloves and eye protection. Determine the number of patients. Assess the need for additional resources such as law enforcement, ALS, or additional ambulances. The mechanism of injury may suggest the injury extent and type; consider the possibility that the patient may have internal bleeding.
Mechanism of Injury (MOI)/ Nature of Illness (NOI)	Determine the MOI. Observe the scene and look for indicators that will assist you with this. The nature of the problem may not be readily apparent until more information is gathered. Observe the surroundings. Pediatric patients may have hidden internal injuries. Blunt force to the abdomen from bicycle handlebars is a common MOI in pediatric patients. Carefully review the mechanism of injury; could the mechanism result in an abdominal or genitourinary injury? If penetration is the MOI, inspect for cleanliness of the wound and whether the edges are smooth or jagged. Consider spinal immobilization in any trauma patient with a significant MOI.

▼

Primary Assessment

Form a General Impression	A rapid scan of the patient will help you identify and manage life threats. Determine the priority of care based on the general impression and the MOI. Assess the patient's level of consciousness using the AVPU scale. If the patient is able to communicate, obtain the chief complaint, the type of injury that occurred, and when it occurred. Closed abdominal injuries may be more severe than they appear.
Airway and Breathing	Ensure that the airway is clear and patent. If a spinal injury is suspected, have a team member hold the patient's head still and remind the patient not to move. Patients may feel nauseous and vomit; ensure the airway is clear through suction and by turning the patient to one side. Quickly assess the patient for adequate breathing. Abdominal pain or distention may prevent adequate inhalation. When the patient has guarded respirations that decrease the effectiveness of his or her breathing, provide supplemental oxygen with a nonrebreathing mask. If the patient's level of consciousness is decreased and respirations are shallow, consider using a bag-mask device. Use airway adjuncts as necessary to ensure a patent airway and assist with breathing.
Circulation	Abdominal injuries may cause internal bleeding leading to shock; be alert for signs and symptoms. Evaluate the patient's pulse and skin color, temperature, and condition to determine the stage of shock. If you suspect shock, provide oxygen, position the patient in a modified shock position, and keep the patient warm. Wounds should be covered and bleeding controlled as quickly as possible.
Transport Decision	Abdominal injuries require rapid assessment and quick transport to a trauma center. Do not delay transport to perform a lengthy assessment. A delay may result in unnecessary and dangerous progression of shock. Patients with abdominal injuries should be evaluated at the highest level of trauma center available.

▼

NOTE: The order of the steps in this section differs depending on whether the patient is conscious or unconscious. The following order is for a conscious patient. For an unconscious patient, perform a primary assessment, perform a full-body scan, obtain vital signs, and obtain the past medical history from a family member, bystander, or emergency medical identification device.

Assessment and Emergency Care of Abdominal Injuries, continued

History Taking

Investigate Chief Complaint	Once you have identified and treated life threats, investigate the chief complaint and gather a history. Identify associated symptoms and pertinent negatives. Note the patient's position and movement. Ask SAMPLE questions, focusing on the events surrounding the incident and the MOI. Ask OPQRST questions. SAMPLE can also be obtained from family, bystanders, and medical alert tags if the patient is not able to provide the information. Specifically ask about nausea, vomiting, diarrhea, blood in the urine (hematuria), and tarry stools (melena).

Secondary Assessment

Physical Examinations	If the patient has a significant MOI, perform a full-body scan. Begin with the head, using DCAP-BTLS to identify hidden and potentially life-threatening injuries. If the MOI suggests an isolated injury to the abdomen, focus your physical examination on the injured area only. Remove or loosen clothes to expose the injured area. Provide privacy as needed or wait until you are in the back of the ambulance. A patient without suspected spinal injury should be allowed to stay in a position of comfort. For patients with spinal injury, place padding under the patient's knees to help alleviate tension on the abdominal wall. Examine the entire abdomen including all posterior, anterior, and lateral surfaces for entrance and exit wounds. Auscultate the bowel. Use DCAP-BTLS. Palpate the abdomen, beginning with the quadrant farthest from the injured site. Inspect and palpate the kidneys. Ruptured or lacerated hollow organs will cause diffuse pain with guarding, and the patient will react to sudden jarring movements.
Vital Signs	Obtain the patient's vital signs. Tachycardia or hypotension may indicate hypoperfusion. Reassess the patient's vital signs every 5 minutes to observe trends. Pay close attention to shifts in the vital signs.

Reassessment

Interventions	Ensure a patent airway and provide oxygen as necessary. Stabilize and immobilize the spine as needed. Control external bleeding and treat for shock. Patients with a ruptured diaphragm may require positive-pressure ventilation to maintain adequate ventilation. Eviscerations and injured genitalia should be covered with a moist sterile dressing. Do not delay transport to complete nonlifesaving treatments. Complete these types of treatments en route to the hospital. Continuously observe and reassess the patient during transport so worsening conditions can be managed.
Communication and Documentation	Contact medical control/receiving hospital and communicate a thorough description of the MOI and the position the patient was found in using proper medical terminology. Document your assessment, trends in vital signs, results of the physical examination, and any pertinent negatives. Also document if you passed over any step of the physical examination. Document the scene observations you made on your arrival and be prepared to describe the scene for the trauma team. Refusals of medical aid or transport should documented completely.

NOTE: Although the following steps are widely accepted, be sure to consult and follow your local protocols. Take appropriate standard precautions when treating all patients.

Assessment and Emergency Care of Abdominal Injuries, continued

Abdominal Trauma

Blunt Abdominal Injuries

Log roll the patient to a supine position on a backboard. If the patient vomits, turn him or her to one side and clear the mouth and throat of vomitus. Monitor the patient's vital signs for any indication of shock. If shock is present, administer high-flow supplemental oxygen via a nonrebreathing mask and treat for shock. Keep the patient warm. Provide prompt transport to the emergency department.

Penetrating Abdominal Injuries

The injuries from penetrating trauma may not be so obvious. Large amounts of external bleeding may not be present, yet there can be significant blood loss internally. Accurate assessment of a penetrating injury takes place in the operating room. Prehospital treatment consists of controlling any external bleeding, immobilizing the patient on a long spine board, treating for shock, and transporting the patient to the appropriate specialty center. If the penetrating object is still in place (impaled), stabilize the object in place.

Abdominal Evisceration

Never attempt to replace protruding organs. Cover the exposed organs with a moist, sterile dressing. If local protocol allows, cover the sterile dressing with an occlusive dressing. Maintain body temperature, treat for shock, and transport to the highest level trauma center available.

Assessment and Emergency Care of Genitourinary System

Scene Size-up

Scene Safety	Ensure scene safety. Follow standard precautions, putting on a minimum of gloves and eye protection. Control of blood and bloody contaminants can be difficult unless you are careful about what you touch and where. Determine the number of patients. Assess the need for additional resources.
Mechanism of Injury (MOI)/ Nature of Illness (NOI)	Determine the MOI. Observe the scene and look for indicators that will assist you with this. The patient may avoid discussing the injury to avoid undergoing a physical examination. The patient may provide an MOI that seems "less embarrassing" than the actual MOI. Maintain a professional demeanor, respect the patient's privacy, and maintain the patient's dignity to earn the patient's trust.

Primary Assessment

Form a General Impression	Important indicators will alert you to the seriousness of the patient's condition. Perform a rapid scan of the patient. Is the patient awake and interacting with his or her surroundings, or lying still, making no sounds? Does the patient have any apparent life threats? What color is the patient's skin? The general impression will help you develop an index of suspicion for serious injuries and determine how urgently the patient needs care.
Airway and Breathing	Ensure that the airway is clear and patent. Protect the patient from further spinal injury. Consider inserting an oropharyngeal airway or nasopharyngeal airway if the patient is unresponsive or has a significantly altered level of consciousness. Quickly assess for adequate breathing. Provide assisted ventilations using a bag-mask device as needed.
Circulation	Quickly assess the pulse rate and quality; determine the skin condition, color, and temperature; and check the capillary refill time to determine the presence of circulatory problems or shock. Begin the steps to control visible bleeding. Significant bleeding is an immediate life threat and must be controlled quickly using appropriate methods.
Transport Decision	The patient should be taken to a trauma center for evaluation and treatment. Any injury to this system can prove to be life altering and often requires a medical specialist to provide specialized care.

NOTE: The order of the steps in this section differs depending on whether the patient is conscious or unconscious. The following order is for a conscious patient. For an unconscious patient, perform a primary assessment, perform a full-body scan, obtain vital signs, and obtain the past medical history from a family member, bystander, or emergency medical identification device.

Assessment and Emergency Care of Genitourinary System, continued

History Taking

Investigate Chief Complaint	Once you have identified and treated life threats, investigate the chief complaint and gather a history. Begin your interview by establishing why the patient called 9-1-1. Use the SAMPLE history to further elicit more facts and specifics about the chief complaint. Common associated complaints are nausea, vomiting, diarrhea, blood in the urine (hematuria), vomiting blood (hematemesis), or abnormal bowel and bladder habits such as an increase in frequency or the absence of the need to void.

Secondary Assessment

Physical Examinations	When examining the patient, expose only what is needed. For a patient with an isolated injury and a limited MOI, focus on the isolated injury, the patient's complaint, and the body region affected. Identify wounds and control bleeding. Note the location and extent of the injury. For a patient with significant trauma, start with a full-body scan looking for DCAP-BTLS from head to toe. With genitourinary injuries, it is important to not focus only on one area of the body.
Vital Signs	Obtain the patient's vital signs. Reassess the vital signs to identify how quickly the patient's condition is changing. Signs such as tachycardia, tachypnea, low blood pressure, weak pulse, and cool, moist, and pale skin indicate hypoperfusion and imply the need for rapid treatment at the hospital.

Reassessment

Interventions	Provide oxygen and maintain a patent airway. Attempt to control bleeding and treat for shock. Place the patient in a position of comfort and transport to the appropriate facility.
Communication and Documentation	Communicate early with the receiving facility so the staff can be adequately prepared and, if required, have a specialist en route to evaluate and treat the patient. Your documentation should be complete and thorough. Describe all injuries and the treatment given.

NOTE: Although the following steps are widely accepted, be sure to consult and follow your local protocols. Take appropriate standard precautions when treating all patients.

Assessment and Emergency Care of Genitourinary System, continued

Genitourinary Trauma

Kidney Injuries

Damage to the kidneys may not be obvious on inspection of the patient. You may or may not see bruises or lacerations on the overlying skin. You will see signs of shock if the injury is associated with significant blood loss. Another sign of kidney damage is blood in the urine (hematuria). Treat shock and associated injuries in the appropriate manner. Provide prompt transport to the hospital, monitoring the patient's vital signs carefully en route.

Urinary Bladder Injuries

Suspect a possible injury of the urinary bladder if you see blood at the urethral opening or physical signs of trauma on the lower abdomen, pelvis, or perineum. There may be blood at the tip of the penis or a stain on the patient's underwear. The presence of associated injuries or of shock will dictate the urgency of transport. In most instances, provide prompt transport, and monitor the patient's vital signs en route.

Genitalia Injuries

Soft-tissue injuries to the external genitalia should be treated like any other soft-tissue injury once all life threats have been assessed and managed. Female patients with external genitalia trauma should be questioned about the possibility of pregnancy. Use only external dressings, never place anything into the vagina.

Trauma to the abdomen or genitourinary area may produce injury to the female patient's uterus, fallopian tubes, and ovaries. Treat for shock because injuries to these organs may be hidden.

Rectal Bleeding

Rectal bleeding can be significant and lead to shock. Place dressings in the crease between the buttocks to manage bleeding. Contact medical control to determine the need for transport.

Sexual Assault

Follow local protocol for crime scene management and evidence preservation. If available, an EMT of the same sex as the patient should perform the assessment and treatment. Advise the patient not to change clothes, shower, drink, or eat. Maintain patient privacy at all times. Sexual assault victims may have serious multisystem trauma. Assessment and treatment should consist of managing life-threatening injuries. Do not examine the genitalia unless obvious bleeding must be managed.

Prep Kit

Ready for Review

- Abdominal injuries are categorized as either open (penetrating trauma) or closed (blunt force trauma).

- Either classification of injury can result in injury to the hollow or solid organs of the abdomen and cause significant life-threatening bleeding.

- Blunt force trauma that causes closed injuries results from an object striking the body without breaking the skin, such as being hit with a baseball bat or when the patient's body strikes the steering wheel during a motor vehicle crash.

- Penetrating trauma is often a result of a gunshot wound or stab wound. Other mechanisms of injury such as a fall on an object can also cause penetrating trauma to the abdomen.

- Injury to the solid internal organs often causes significant unseen bleeding that can be life threatening.

- Injury to the hollow organs of the abdomen may cause irritation and inflammation to the peritoneum as caustic digestive juices leak into the peritoneum. A serious infection may also occur over several hours.

- Always maintain a high index of suspicion for serious intra-abdominal injury in the trauma patient, particularly in the patient who exhibits signs of shock.

- Assess the abdomen for signs of bruising, rigidity, penetrating injuries, and complaints of pain.

- Never remove an impaled object from the abdominal region. Secure it in place with a large bulky dressing and provide prompt transport.

- Be prepared to treat the patient for shock. Place the patient in the modified shock position, keep the patient warm, and provide high-flow oxygen.

- Never replace an organ that protrudes from an open injury to the abdomen (evisceration). Instead, keep the organ moist and warm. Cover the injury site with a large sterile, moist, bulky dressing.

- Injuries to the kidneys may be difficult to detect because of the well-protected region of the body where they are located. Be alert to bruising or a hematoma in the flank region.

- Injury to the external genitalia of male and female patients is very painful but not usually life threatening.

- In the case of sexual assault or rape, treat for shock if necessary, and record all the facts in detail. Follow any crime scene policy established by your system to protect the scene and any potential evidence. Advise the patient not to wash, douche, or void until after a physician has examined him or her.

Vital Vocabulary

closed abdominal injury An injury in which there is soft-tissue damage inside the body but the skin remains intact.

evisceration The displacement of organs outside of the body.

flank The posterior region below the margin of the lower rib cage.

guarding Contracting the stomach muscles to minimize the pain of abdominal movement; a sign of peritonitis.

hollow organs Structures through which materials pass, such as the stomach, small intestines, large intestines, ureters, and bladder.

Kehr sign Left shoulder pain caused by blood in the peritoneal cavity.

open abdominal injury An injury in which there is a break in the surface of the skin or mucous membrane, exposing deeper tissue to potential contamination.

peritoneal cavity The abdominal cavity.

solid organs Solid masses of tissue where much of the chemical work of the body takes place (eg, the liver, spleen, pancreas, and kidneys).

Assessment in Action

You are dispatched to a private residence for a person who is reporting abdominal pain. You arrive to find a 25-year-old man lying on the couch in the fetal position. The patient states he was tackled while playing football 3 days ago and he now has pain that is becoming progressively worse. He is pale and diaphoretic. Physical examination of the abdomen shows bruising over the right and left upper quadrants, rebound tenderness, and guarding. The patient states the pain radiates to his shoulder blade. Assessment of his vital signs shows a pulse rate of 130 beats/min, a blood pressure of 90/60 mm Hg, and respirations of 24 breaths/min.

1. Bruising over the right and left upper quadrants is known as the:
   A. Cullen sign.
   B. Kehr sign.
   C. Blumberg sign.
   D. Grey Turner sign.

2. Peritonitis is an inflammation of the:
   A. peritoneum.
   B. pancreas.
   C. liver.
   D. appendix.

3. On the basis of your findings, the patient has what type of injury?
   A. Penetrating
   B. Crushing
   C. Blunt
   D. Compression

4. The pain radiating to the shoulder blade is known as the:
   A. Cullen sign.
   B. Kehr sign.
   C. Blumberg sign.
   D. Grey Turner sign.

5. What organs are found in the left upper quadrant?
   A. Spleen and liver
   B. Spleen and ascending colon
   C. Spleen and gallbladder
   D. Spleen and stomach

6. Which of the following are considered solid organs?
   A. Liver and intestines
   B. Kidneys and bladder
   C. Spleen and stomach
   D. Pancreas and spleen

7. What is the cause of peritonitis?

8. Rebound tenderness is known as the:
   A. Cullen sign.
   B. Kehr sign.
   C. Blumberg sign.
   D. Grey Turner sign.

9. Which of the following signs would indicate that an injury to the kidney has occurred?
   A. Hemoptysis
   B. Hematuria
   C. Hematoma
   D. Hematemesis

10. What is guarding?

National EMS Education Standard Competencies

Trauma

Applies fundamental knowledge to provide basic emergency care and transportation based on assessment findings for an acutely injured patient.

Orthopaedic Trauma

- Recognition and management of
  - Open fractures (pp 1007-1009, 1018-1047)
  - Closed fractures (pp 1007-1009, 1018-1047)
  - Dislocations (pp 1010, 1018-1047)
  - Amputations (pp 1006, 1011, 1047-1048)
- Pathophysiology, assessment, and management of
  - Upper and lower extremity orthopaedic trauma (pp 1005-1047)
  - Open fractures (pp 1007-1009, 1018-1047)
  - Closed fractures (pp 1007-1009, 1018-1047)
  - Dislocations (pp 1010, 1018-1047)
  - Sprains/strains (pp 1010-1011, 1048)
  - Pelvic fractures (1040-1041)
  - Amputations/replantation (pp 1006, 1011, 1047-1048)

Medicine

Applies fundamental knowledge to provide basic emergency care and transportation based on assessment findings for an acutely ill patient.

Nontraumatic Musculoskeletal Disorders

Anatomy, physiology, pathophysiology, assessment, and management of
- Nontraumatic fractures (pp 1007-1009, 1018-1047)

Knowledge Objectives

1. Describe the function of the musculoskeletal system. (pp 1001-1005)
2. Understand the anatomy and physiology of the musculoskeletal system. (pp 1001-1005)
3. Describe the different types of musculoskeletal injuries, including fractures, dislocations, amputations, sprains, and strains. (pp 1005-1011)
4. Name the four mechanisms of injury. (pp 1006-1007)
5. Differentiate between open and closed fractures. (pp 1007-1009)
6. Explain how to assess the severity of an injury. (pp 1012)
7. Understand the emergency medical care of the patient with an orthopaedic injury. (pp 1018-1048)
8. Describe the emergency medical care of the patient with a swollen, painful, deformed extremity (fracture). (pp 1018-1046)
9. Understand the need for, general rules of, and possible complications of splinting. (pp 1020-1021)
10. Explain the reasons for splinting fractures, dislocations, and sprains at the scene versus transporting the patient immediately. (p 1020)
11. Recognize the characteristics of specific types of musculoskeletal injuries. (pp 1005-1012, 1031-1048)
12. Describe the emergency medical care of the patient with an amputation. (pp 1047-1048)

Skills Objectives

1. Demonstrate the assessment of neurovascular status. (pp 1015-1017, Skill Drill 29-1)
2. Demonstrate the care of musculoskeletal injuries. (p 1018-1020, Skill Drill 29-2)
3. Demonstrate how to apply a rigid splint. (p 1021-1022, Skill Drill 29-3)
4. Demonstrate how to apply a zippered air splint. (pp 1023, Skill Drill 29-4)
5. Demonstrate how to apply an unzippered air splint. (pp 1023-1024, Skill Drill 29-5)
6. Demonstrate how to apply a vacuum splint. (pp 1024-1025, Skill Drill 29-6)
7. Demonstrate how to apply a Hare traction splint. (pp 1026-1028, Skill Drill 29-7)
8. Demonstrate how to apply a Sager traction splint. (p 1028-1030, Skill Drill 29-8)
9. Demonstrate how to apply a pneumatic antishock garment (PASG). (p 1031-1032, Skill Drill 29-9)
10. Demonstrate how to splint the hand and wrist. (p 1039-1040, Skill Drill 29-10)
11. Demonstrate how to splint the clavicle, the scapula, the shoulder, the humerus, the elbow, and the forearm. (pp 1031-1038)
12. Demonstrate how to care for a patient with an amputation. (pp 1047-1048)

Introduction

The human body is a well-designed system in which form, upright posture, and movement are provided by the musculoskeletal system. This system also protects the vital internal organs of the body. The term musculoskeletal refers to the bones and voluntary muscles of the body. However, the bones and muscles are susceptible to external forces that can cause injury. Also at risk are the tendons, the joints, and the ligaments.

Musculoskeletal injuries are among the most common reasons why patients seek medical attention. Complaints related to the musculoskeletal system result in almost 60 million visits to physicians annually in the United States. Approximately one in seven Americans will experience some type of musculoskeletal impairment, leading to millions of missed days of work or school and costing hundreds of billions of dollars yearly. An estimated 70% to 80% of all patients with multiple system trauma have one or more musculoskeletal injuries.

Musculoskeletal system injuries are often easily identified because of associated pain, swelling, and deformity. Although these injuries are rarely fatal, they often result in short- or long-term disability. By providing prompt assessment and treatment, such as splinting, EMTs may help reduce the disability period for patients. Despite the sometimes dramatic appearance of these injuries, you should not focus solely on a musculoskeletal injury without first determining that no life-threatening injuries exist. Never forget the ABCs!

As an EMT, you must be familiar with the basic anatomy of the musculoskeletal system. Although muscles are technically soft tissue, they are discussed in this chapter because of their close relationship with the skeleton. Therefore, the chapter begins with a review of the musculoskeletal anatomy. Various types and causes of musculoskeletal injuries in general are identified, and the assessment and treatment process for each is explained, followed by a detailed discussion of splinting. The chapter then focuses on specific musculoskeletal injuries, beginning at the clavicle and ending at the feet.

Anatomy and Physiology of the Musculoskeletal System

Muscles

The muscular system includes three types of muscles: skeletal, smooth, and cardiac. Skeletal muscle, also called striated muscle because of its characteristic stripes, attaches to the bones and usually crosses at least one joint, forming the major muscle mass of the body. This type of muscle is also called voluntary muscle, because it is under direct voluntary control of the brain, responding to commands to move specific body parts Figure 29-1. Usually, movement is the result of several muscles contracting and relaxing simultaneously. Skeletal muscle is the component of the muscular system that is included in the overall musculoskeletal system. Cardiac muscle contributes to the cardiovascular system, and smooth muscle is a component of other body systems, including the digestive system and the cardiovascular system.

All skeletal muscles are supplied with arteries, veins, and nerves. Blood from the arteries brings oxygen and nutrients to the muscles Figure 29-2. Waste products, including carbon dioxide and lactic acid, are carried away in the veins. Disease or trauma can result in the loss of a muscle's nervous supply; this, in turn, can lead to weakness and eventually atrophy, or a decrease in the size of the muscle and its inherent ability to function. Skeletal muscle tissue is directly attached to the bone by tough, ropelike fibrous structures known as tendons, which are extensions of the fascia that covers all skeletal muscle.

You are the Provider: PART 1

At 4:20 PM, you are dispatched to a soccer field for a player with a possible broken leg. You and your partner proceed to the scene, with a response time of about 5 minutes. En route, dispatch advises you that the patient is conscious, alert, and breathing. The weather is overcast, the temperature is 88°F, and the traffic is moderate.

1. Under which circumstances can orthopaedic injuries pose a threat to a patient's life?
2. Given the information you have, can you rule out a critical injury?

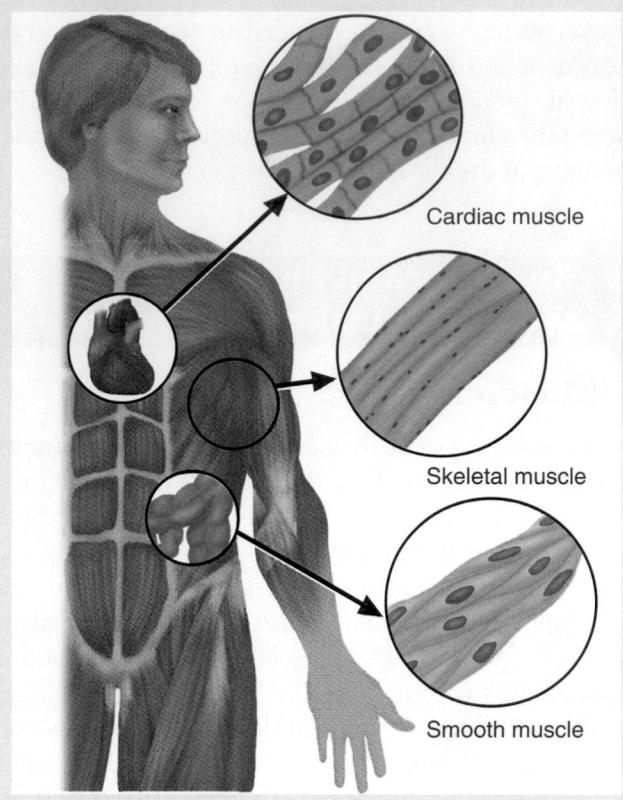

Figure 29-1 The major muscle type of concern for musculoskeletal injuries is skeletal, or voluntary, muscle.

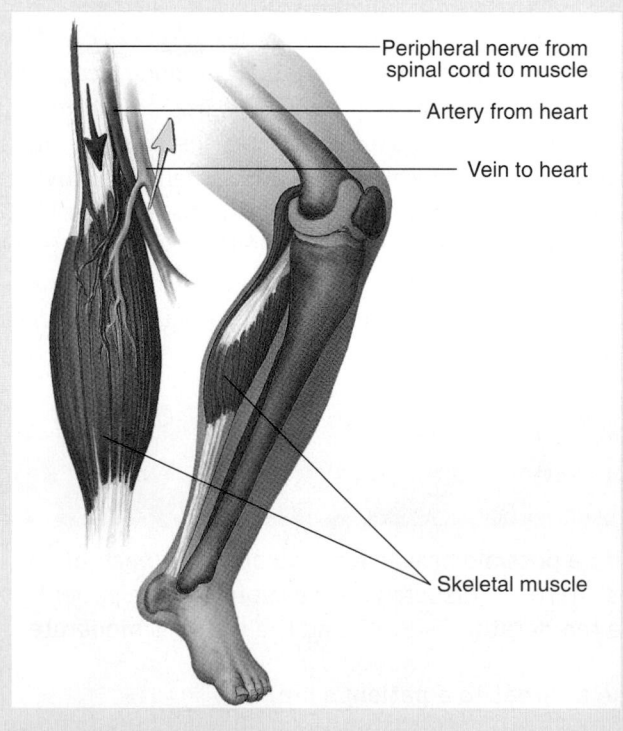

Figure 29-2 Skeletal muscles are supplied with arteries, veins, and nerves that bring oxygen and nutrients, carry away waste products, and supply nervous stimuli.

Smooth muscle, also called involuntary muscle because it is not under voluntary control of the brain, performs much of the automatic work of the body. This type of muscle is found in the walls of most tubular structures of the body, such as the gastrointestinal tract and the blood vessels. Smooth muscle contracts and relaxes to control the movement of the contents within these structures **Figure 29-3** .

The heart neither looks nor acts like skeletal or smooth muscle. It is composed largely of cardiac muscle, a specially adapted involuntary muscle with its own regulatory system. The remainder of this chapter is concerned exclusively with skeletal muscle.

■ The Skeleton

The skeleton, which gives us our recognizable human form, protects our vital internal organs, and allows us to move, is made up of approximately 206 bones **Figure 29-4** . The bones in the skeleton also produce blood cells (in the bone marrow) and serve as a reservoir for important minerals and electrolytes.

The skull is a solid vaultlike structure that surrounds and protects the brain. The thoracic cage protects the heart, lungs, and great vessels; the lower ribs protect the liver and spleen. The bony spinal canal encases and protects the spinal cord.

The pectoral girdle, also referred to as the shoulder girdle, consists of two scapulae and two clavicles **Figure 29-5** . The scapula (shoulder blade) is a flat, triangular bone held to the rib cage by powerful muscles that buffer it against injury. The clavicle (collarbone) is a slender, S-shaped bone attached by ligaments to the sternum on one end and to the acromion process on the other.

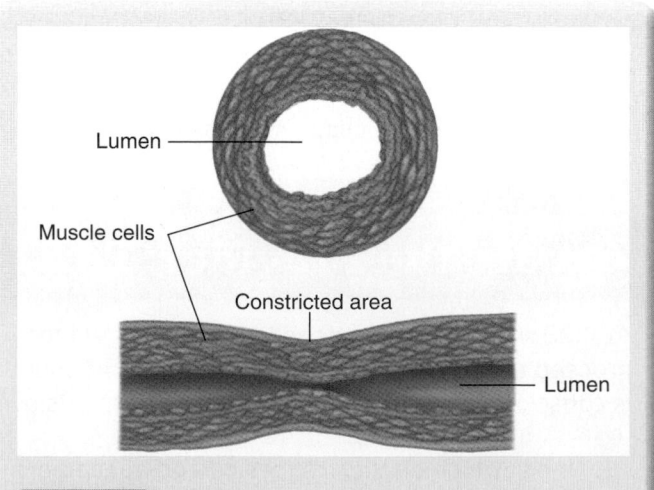

Figure 29-3 Smooth muscle is found in the walls of most tubular structures in the body. These muscles contract and relax to control the movement of the contents within these structures.

The clavicle acts as a strut to keep the shoulder propped up; however, because it is slender and very exposed, this bone is vulnerable to injury.

The upper extremity extends from the shoulder to the fingertips and is composed of the arm (humerus),

elbow, forearm (radius and ulna), wrist, hand, and fingers **Figure 29-6**. The arm extends from the shoulder to the elbow. The upper extremity joins the shoulder girdle at the glenohumeral joint. At the other end of the glenohumeral joint is the humerus. The humerus connects with the bones of the forearm—the radius and ulna—to form the hinged elbow joint.

The radius and ulna make up the forearm. The radius, the larger of the two forearm bones, lies on the *thumb* side of the forearm. The ulna is narrow and is on the little-finger side of the forearm. Because the radius and the ulna are parallel, when one is broken, the other is often broken as well.

The hand contains three sets of bones: *wrist bones* (carpals), *hand bones* (metacarpals), and *finger bones*

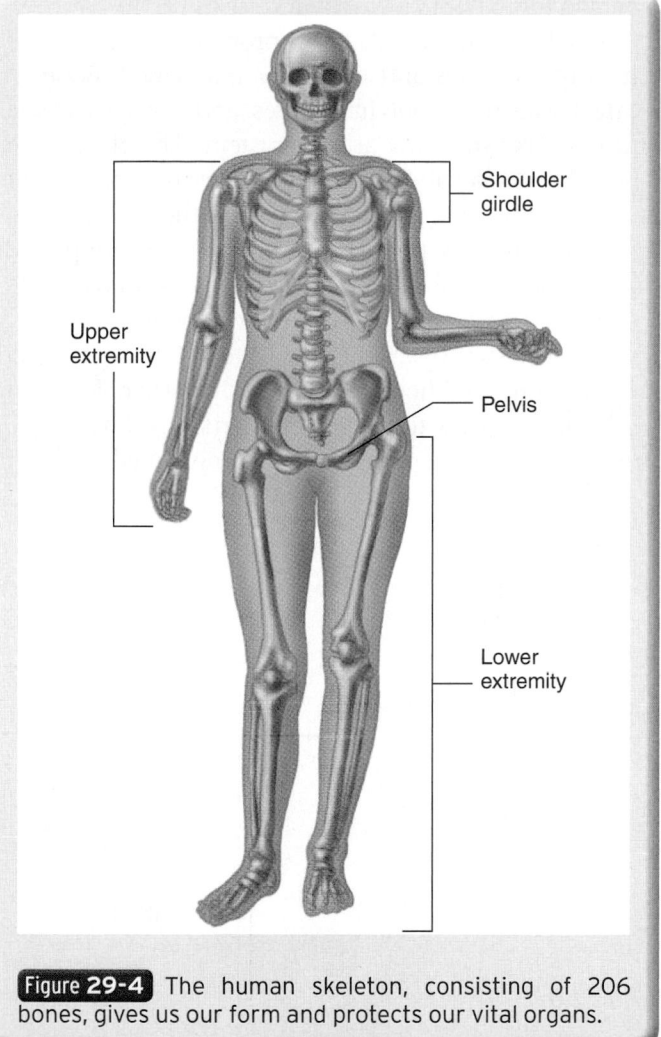

Figure 29-4 The human skeleton, consisting of 206 bones, gives us our form and protects our vital organs.

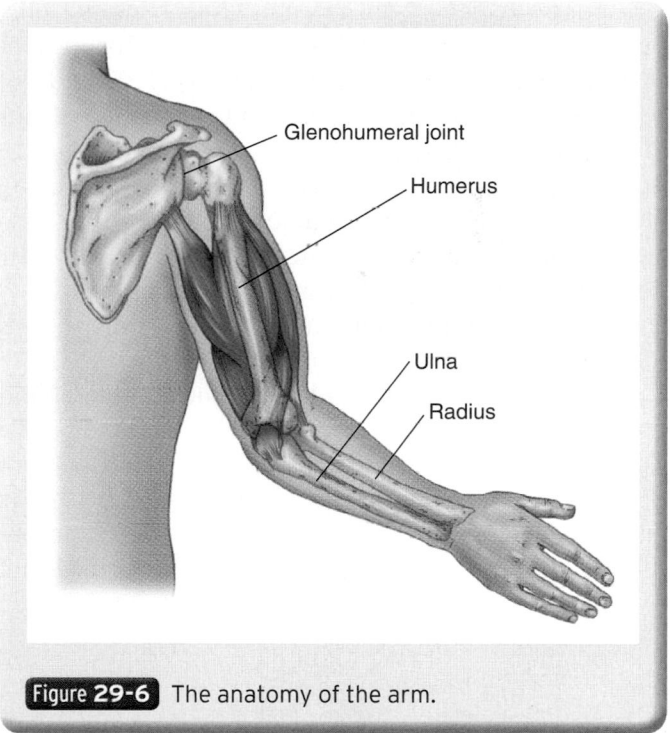

Figure 29-6 The anatomy of the arm.

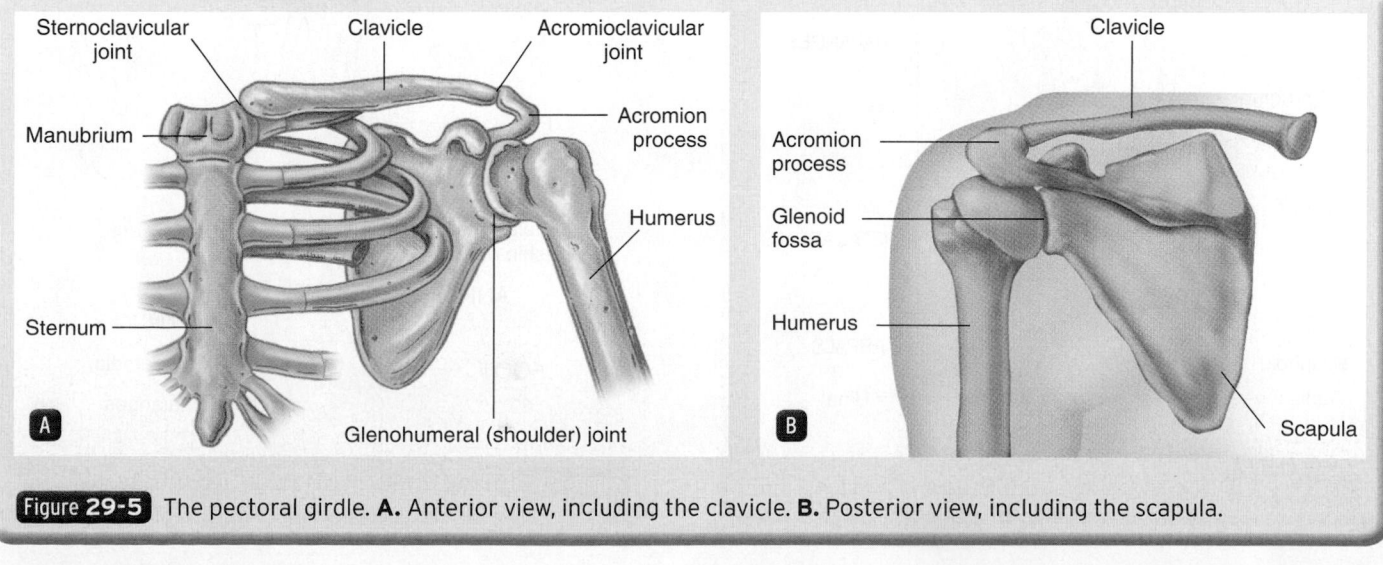

Figure 29-5 The pectoral girdle. **A.** Anterior view, including the clavicle. **B.** Posterior view, including the scapula.

(phalanges) Figure 29-7 . The carpals are vulnerable to fracture when a person falls on an outstretched hand. Phalanges are more apt to be injured by a crushing injury, such as being slammed in a car door.

The pelvis supports the body weight and protects the structures within the pelvis: the bladder, rectum, and female reproductive organs. The pelvic girdle is actually three separate bones—the ischium, ilium, and pubis—fused together to form the innominate bone. The two iliac bones are joined posteriorly by tough ligaments to the sacrum at the sacroiliac joints; the two pubic bones are connected anteriorly by equally tough ligaments to one another at the pubic symphysis. These joints allow very little motion, so the pelvic ring is strong and stable.

The lower extremity consists of the bones of the thigh, leg, and foot Figure 29-8 . The femur (thigh bone) is a long, powerful bone that connects in the ball-and-socket joint of the pelvis and in the hinge joint of the knee. The *head* of the femur is the ball-shaped part that fits into the acetabulum and is a common fracture point. It is connected to the *shaft,* or long tubular portion of the femur, by the femoral *neck.* The femoral neck is a common site for fractures, generally referred to as hip fractures, especially in the older population. The greater trochanter is the name given to the upper part of the femur, and the lesser trochanter is the name given to the lower part of the femur.

The lower leg consists of two bones, the tibia and the fibula. The tibia (shin bone) connects to the patella (kneecap) to form the knee joint and runs down the front of the lower leg. The tibia is vulnerable to direct blows and can be felt just beneath the skin. The much smaller fibula runs behind and beside the tibia. The fibula is not a component of the knee joint, but it does make up the outer knob of the ankle joint.

The foot consists of three classes of bones: *ankle bones* (tarsals), *foot bones* (metatarsals), and *toe bones* (phalanges) Figure 29-9 . The largest of the tarsal bones is the heel bone, or **calcaneus**, which is subject to injury when a person jumps from a height and lands on the feet.

The bones of the skeleton provide a framework to which the muscles and tendons are attached. Bone is a living tissue that contains nerves and receives oxygen and nutrients from the arterial system. Therefore, when a bone breaks, a patient typically experiences severe pain and bleeding. Bone marrow, located in the center of each bone, is constantly producing red blood cells to provide oxygen and nourishment to the body and remove waste.

A **joint** is formed wherever two bones come into contact. The sternoclavicular joint, for example, is where the sternum and the clavicle come together. Joints are held together in a tough fibrous structure known as a capsule, which is supported and strengthened in certain

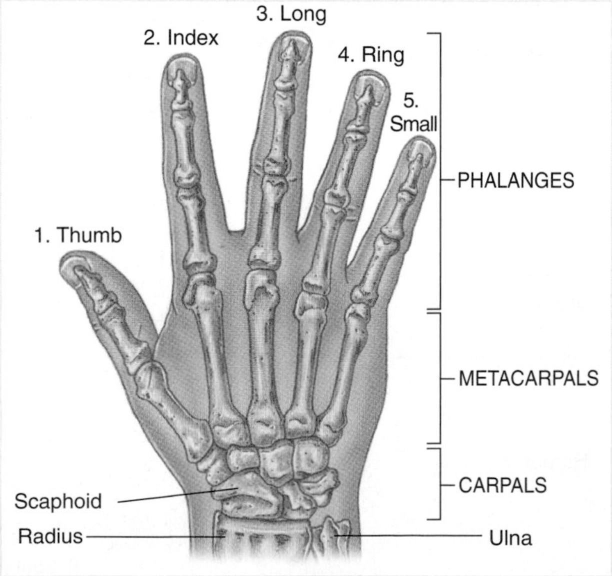

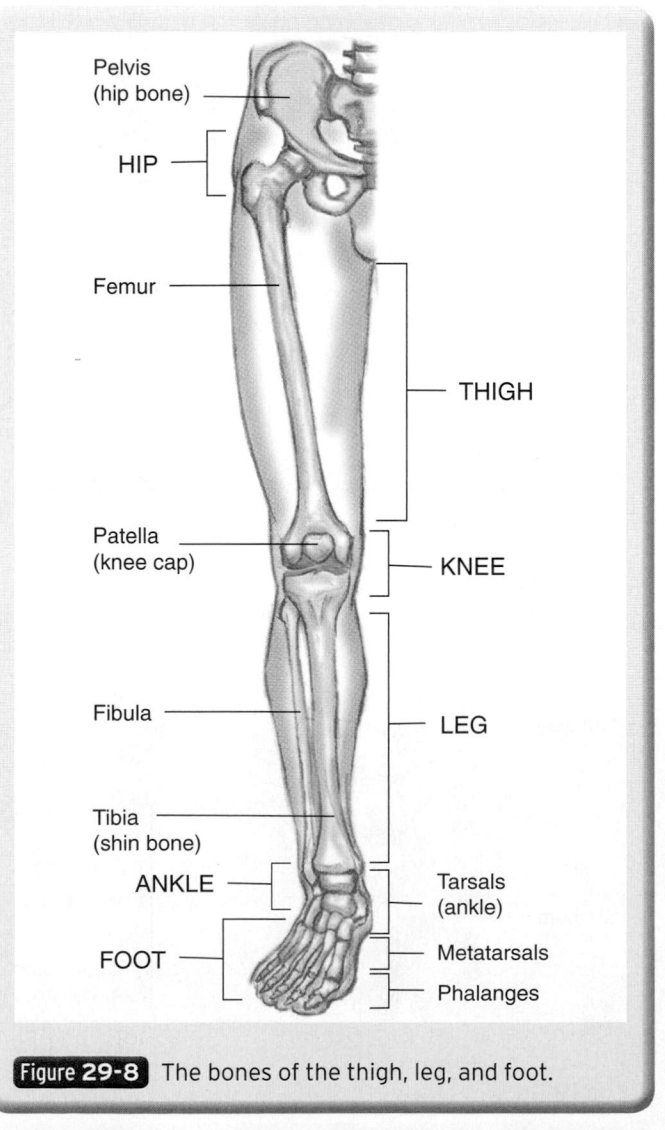

Figure 29-8 The bones of the thigh, leg, and foot.

Figure 29-7 The anatomy of the wrist and hand.

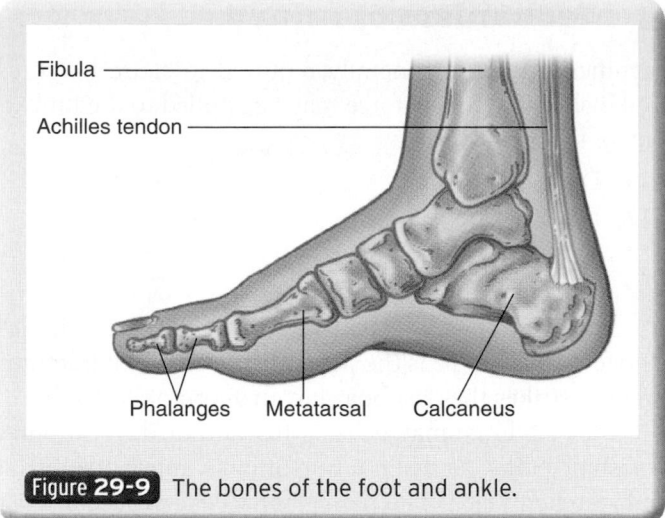

Figure 29-9 The bones of the foot and ankle.

key areas by bands of fibrous tissue called ligaments. In moving joints, the ends of the bones are covered with a thin layer of cartilage known as **articular cartilage**. This cartilage is a pearly substance that allows the ends of the bones to glide easily. Joints are bathed and lubricated by synovial (joint) fluid.

Some joints, such as the shoulder, allow motion to occur in a circular manner. Other joints, such as the knee and elbow, act as hinges. Still other joints, including the sacroiliac joint in the lower back and the sternoclavicular joints, allow only a minimum amount of motion. Certain joints, such as the sutures in the skull (present until about 18 months of life), fuse together during growth to create a solid, immobile, bony structure **Figure 29-10**.

Musculoskeletal Injuries

A **fracture** is a broken bone. More precisely, it is a break in the continuity of the bone, often occurring as a result of an external force **Figure 29-11**. The break can occur anywhere on the surface of the bone and in many different types of patterns. Contrary to a common misconception, there is no difference between a broken bone and a fractured bone. A potential complication of fractures is compartment syndrome. The name **compartment syndrome** refers to elevated pressure within a fascial compartment. Fascia is the fibrous tissue that surrounds and supports the muscles and neurovascular structures.

A **dislocation** is a disruption of a joint in which the bone ends are no longer in contact. The supporting ligaments are often torn, usually completely, allowing the bone ends to separate completely from each other **Figure 29-12**. A subluxation is similar to a dislocation except the disruption of the joints is not complete. Therefore, a **subluxation** is an incomplete dislocation of a joint. A fracture-dislocation is a combination injury at

the joint in which the joint is dislocated and there is a fracture of the end of one or more of the bones.

A **sprain** is an injury to ligaments, articular capsule, synovial membrane, and tendons crossing the joint.

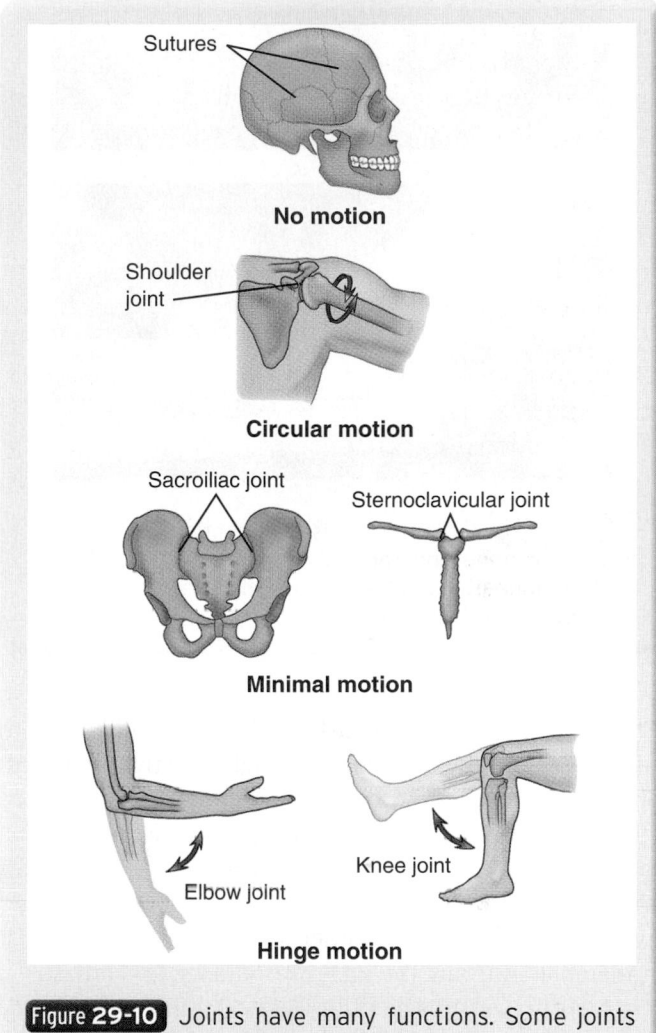

Figure 29-10 Joints have many functions. Some joints allow for motion to occur in a circular manner; others act as hinges. Still others allow only a minimum amount of motion, or none at all.

Figure 29-11 A fracture can occur anywhere on the surface of a bone and may or may not break the skin.

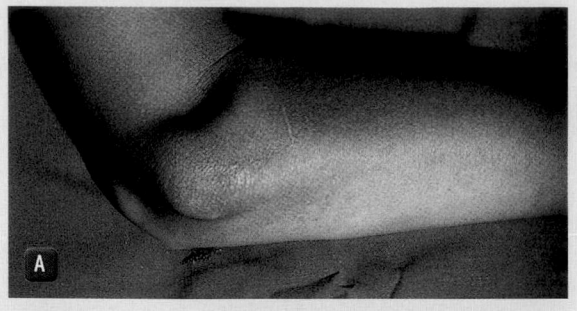

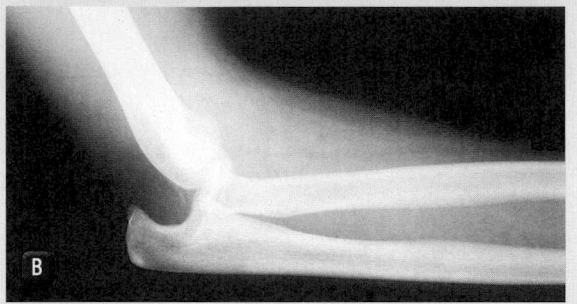

Figure 29-12 A dislocation is a disruption of a joint in which the bone ends are no longer in contact. **A.** The clinical appearance of an elbow dislocation. **B.** X-ray appearance of the same elbow.

After the injury, the joint surfaces generally fall back into alignment, so the joint is not significantly displaced. Sprains can range from mild to severe, depending on the amount of damage done to the supporting ligaments. The most severe sprains involve actual tearing of the ligament and may allow joint dislocation. Mild sprains are caused by ligament stretching rather than tearing. The most vulnerable joints are the knees, shoulders, and ankles.

A **strain**, or muscle pull, is a stretching or tearing of the muscle, causing pain, swelling, and bruising of the soft tissues in the area. It occurs because of an abnormal contraction. Strains may range from minute separation to complete rupture. Unlike a sprain, no ligament or joint damage typically occurs.

An amputation is an injury in which an extremity is completely severed from the body. This injury can potentially damage every aspect of the musculoskeletal system—from bone to ligament to muscle.

Injury to bones and joints is often associated with injury to the surrounding soft tissues, especially to the adjacent nerves and blood vessels. The entire area is known as the **zone of injury** Figure 29-13 . Depending on the amount of kinetic energy the tissues absorb from forces acting on the body, the zone may extend to a distant point. For this reason, you should not focus on a patient's obvious injury without first completing a rapid scan to check for associated injuries, which may be even more serious. This is especially true in assessing damage from high-energy trauma or gunshots.

Mechanism of Injury

Significant force is generally required to cause fractures and dislocations. This force may be applied to the limb in any of the following ways Figure 29-14 :

- Direct blows
- Indirect forces
- Twisting forces
- High-energy injury

A direct blow fractures the bone at the point of impact. An example is the patella (kneecap) that fractures when it strikes the dashboard in an automobile crash.

Indirect force may cause a fracture or dislocation at a distant point, as when a person falls and lands on an outstretched hand. The direct impact may cause a wrist fracture, but the indirect force can cause dislocation of the elbow or a fracture of the forearm, humerus, or even clavicle. Therefore, when you are caring for patients who have fallen, you must identify the point of contact and the mechanism of injury (MOI) so that you will not overlook associated injuries.

Twisting forces are a common cause of musculoskeletal injury, especially to the anterior cruciate ligament in the knee. Skiing injuries often happen because of

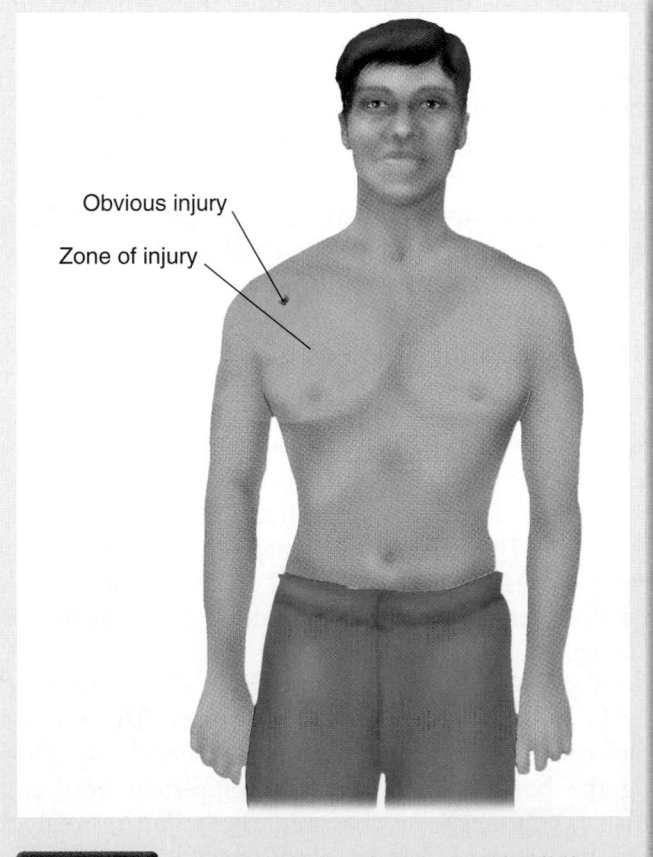

Figure 29-13 The zone of injury is the area of soft tissue, including the adjacent nerves and blood vessels, that surrounds the obvious injury of a bone or joint.

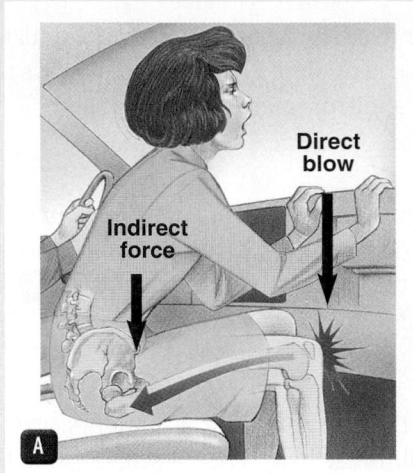  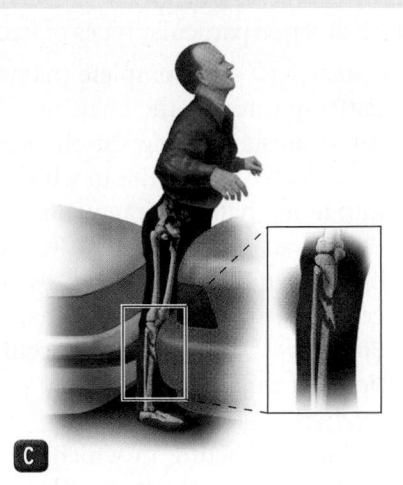

Figure 29-14 Significant force is required to cause fractures or dislocations. Among these are (**A**) Direct blows and indirect forces, (**B**) Twisting forces, and (**C**) High-energy crushing injuries.

twisting. A ski becomes caught, and the skier falls, applying a twisting force to the lower extremity.

High-energy injuries, such as those that occur in automobile crashes, falls from heights, gunshot wounds, and other extreme forces, produce severe damage to the skeleton, surrounding soft tissues, and vital internal organs. A patient may have multiple injuries to many body parts, including more than one fracture or dislocation in a single limb.

A significant MOI is not necessary to fracture a bone. A slight force can easily fracture a bone that is weakened by a tumor or osteoporosis, a generalized bone disease that is common among postmenopausal women. In geriatric patients with osteoporosis, minor falls, simple twisting injuries, or even a muscle contraction can cause a fracture, most often of the wrist, spine, or hip. You should suspect the presence of a fracture in any older patient who has sustained even a mild injury.

■ Fractures

Fractures are classified as either closed or open. In assessing and treating patients with possible fractures or dislocations, your first priority is to determine whether the overlying skin is damaged. If it is not, the patient has a **closed fracture**. However, making this determination is not always as easy as it sounds. With an **open fracture**, there is an external wound, caused either by the same blow that fractured the bone or by the broken bone ends lacerating the skin. The wound may vary in size from a very small puncture to a gaping tear that exposes bone and soft tissue. Regardless of the extent and severity of the damage to the skin, you should treat any injury that breaks the skin as a possible open fracture. Greater blood loss and a higher likelihood of infection are complications

that you must try to limit; these tend to occur with open fractures.

Fractures are also described by whether the bone is moved from its normal position. A **nondisplaced fracture** (also known as a hairline fracture) is a simple crack of the bone that may be difficult to distinguish from a sprain or simple contusion. X-ray examinations are required for hospital personnel to diagnose a nondisplaced fracture. A **displaced fracture** produces actual deformity, or distortion, of the limb by shortening, rotating, or angulating it. Often, the deformity is very obvious and can be associated with crepitus. However, in some cases the deformity is minimal. Be sure to look for differences between the injured limb and the opposite uninjured limb in any patient with a suspected fracture of an extremity **Figure 29-15** .

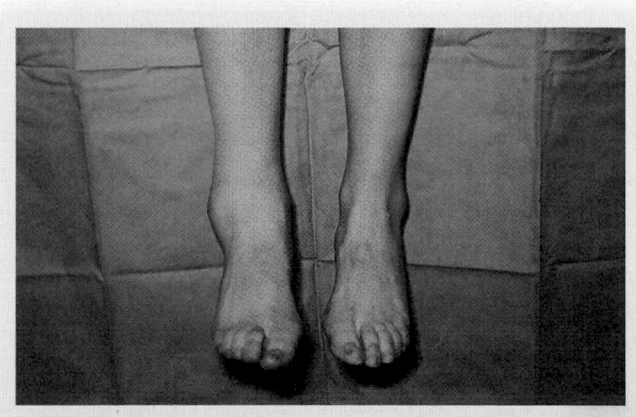

Figure 29-15 You should always compare the injured limb with the uninjured limb when checking for deformity.

Medical personnel often use the following special terms to describe particular types of fractures Figure 29-16 :

- **Greenstick.** An incomplete fracture that passes only partway through the shaft of a bone but may still cause substantial angulation; occurs in children.
- **Comminuted.** A fracture in which the bone is broken into more than two fragments.
- **Pathologic.** A fracture of weakened or diseased bone, seen in patients with osteoporosis or cancer, generally produced by minimal force.
- **Epiphyseal.** A fracture that occurs in a growth section of a child's bone and may lead to growth abnormalities.
- **Oblique.** A fracture in which the bone is broken at an angle across the bone. This is usually the result of a sharp angled blow to the bone.
- **Transverse.** A fracture that occurs straight across the bone. This is usually the result of a direct blow or stress fracture caused by prolonged running.
- **Spiral.** A fracture caused by a twisting force, causing an oblique fracture around the bone and through the bone. This is often the result of abuse in very young children.
- **Incomplete.** A fracture that does not run completely through the bone; a nondisplaced partial crack.

You should suspect a fracture if one or more of the following signs is present in any patient who has a history of injury and reports pain.

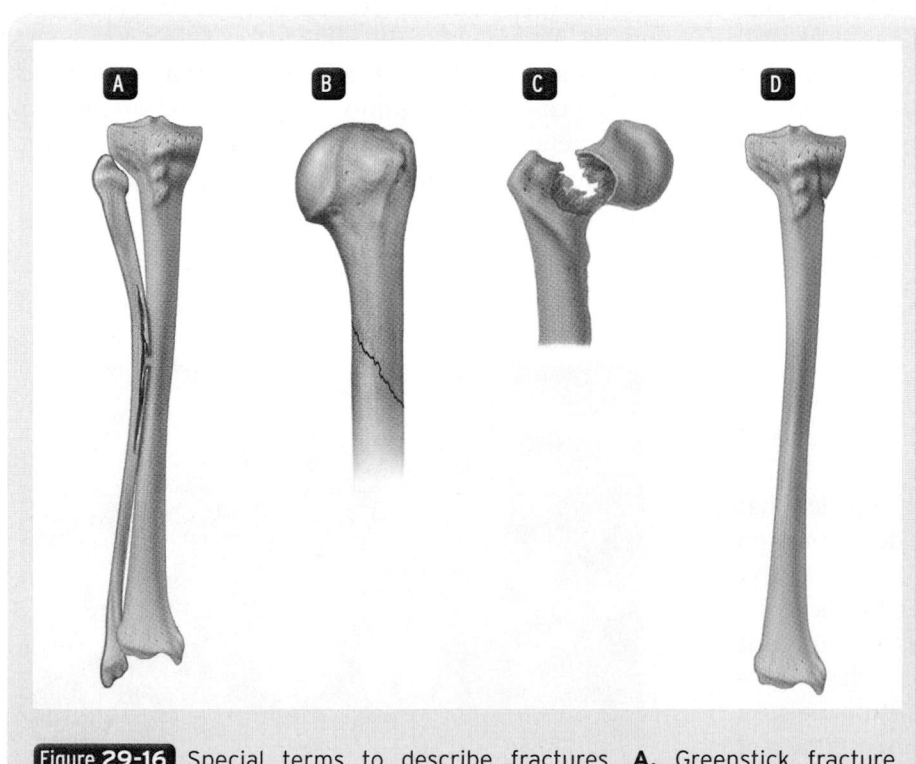

Figure 29-16 Special terms to describe fractures. **A.** Greenstick fracture. **B.** Oblique fracture. **C.** Pathologic fracture. **D.** Incomplete fracture.

Deformity

The limb may appear to be shortened, rotated, or angulated at a point where there is no joint Figure 29-17 . Always use the opposite limb as a mirror image for comparison.

Tenderness

Point tenderness on palpation in the zone of injury is the most reliable indicator of an underlying fracture, although it does not tell you the type of fracture Figure 29-18 . Be sure to wear gloves if there are any open wounds.

Guarding

An inability to use the extremity is the patient's way of immobilizing it to minimize pain. The muscles around the fracture contract in an attempt to prevent any movement of the broken bone. Guarding does not occur with all fractures; some patients may continue to use the injured part for a time. Occasionally, nondisplaced fractures are less painful, and there is minimal soft-tissue damage.

Swelling

Rapid swelling usually indicates bleeding from a fracture and is typically followed by substantial pain. Often, if the swelling is severe, it may mask deformity of the limb Figure 29-19 . Generalized swelling from fluid build-up may occur several hours after an injury.

Bruising

Fractures are almost always associated with **ecchymosis** (discoloration) of the surrounding soft tissues Figure 29-20 . Bruising may be present after almost any injury and may take hours to develop; it is not specific to bone or joint injuries. The discoloration associated with acute injuries is usually redness, as you may have seen with someone who has been punched. Within hours or days, blue, purple, and black will appear, followed by yellows and greens.

Crepitus

A grating or grinding sensation known as **crepitus** can be felt and sometimes even heard when fractured bone ends rub together.

False Motion

Also called free movement, this is motion at a point in the limb where there is no joint. It is a positive indication of a fracture.

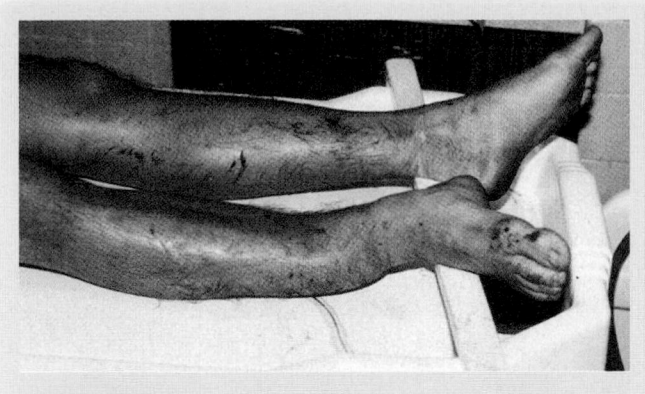

Figure 29-17 Obvious deformity, shortening, rotation, or angulation should increase your index of suspicion for a fracture. Remember to compare the injured limb with the opposite, uninjured limb.

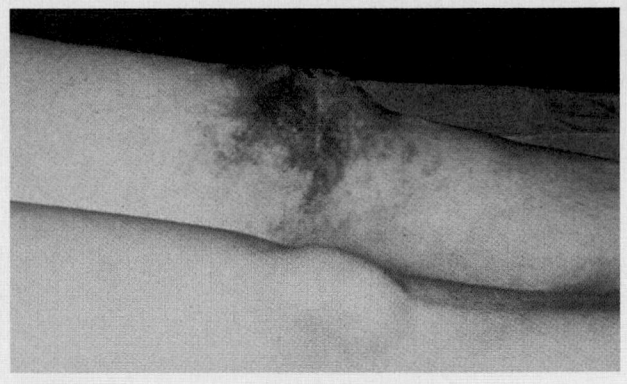

Figure 29-19 Swelling that occurs in association with a fracture can often mask deformity of the limb.

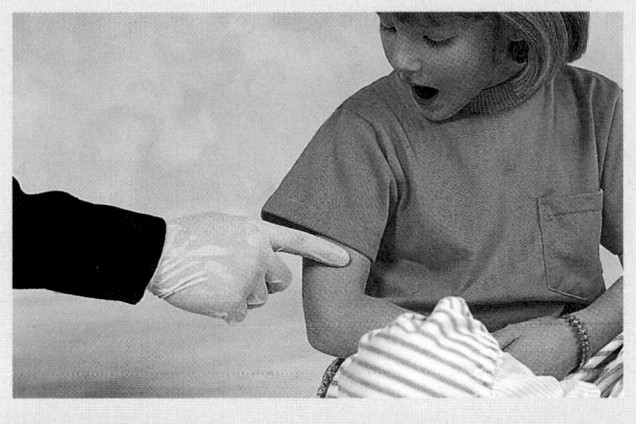

Figure 29-18 Point tenderness is the sensitive spot at the site of injury that can be located by palpation along the bone with the tip of your finger.

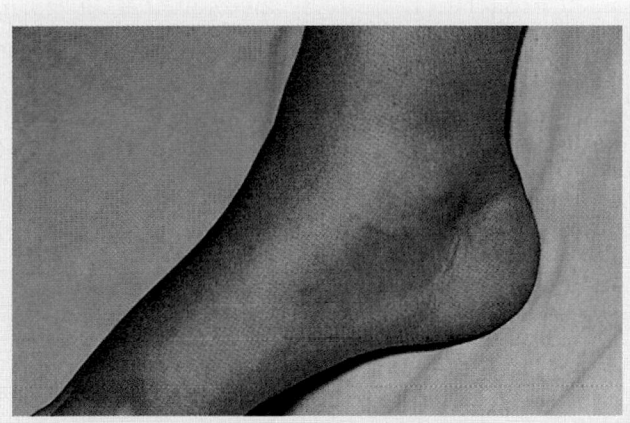

Figure 29-20 Fractures almost always have associated bruising into the surrounding soft tissue.

Exposed Fragments

In open fractures, bone ends may protrude through the skin or be visible within the wound **Figure 29-21**. Never attempt to push the end of a protruding bone back into place. This will increase the risk for infection.

Pain

Pain, along with tenderness and bruising, commonly occurs in association with fractures.

Locked Joint

A joint that is locked into position is difficult and painful to move. Keep in mind that crepitus and false motion appear only when a limb is moved or manipulated and are associated with injuries that are extremely painful. Do not manipulate the limb excessively in an effort to elicit these signs.

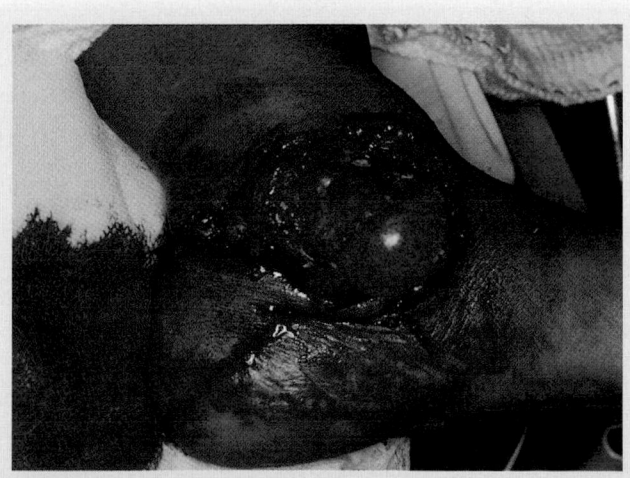

Figure 29-21 Bone ends may protrude through the skin or be visible within the wound of an open fracture.

Dislocations

A dislocated joint sometimes will spontaneously **reduce**, or return to its normal position, before your assessment. In this situation, you will be able to confirm the dislocation only by taking a patient history. Often, however, injury to the supporting ligaments and capsule is so severe that the joint surfaces remain completely separated from one another. A dislocation that does not spontaneously reduce is a serious problem. The ends of the bone can be locked in a displaced position, making any attempt at motion of the joint very difficult and very painful. Commonly dislocated joints include the fingers, shoulder, elbow, and knee.

The signs and symptoms of a dislocated joint are similar to those of a fracture **Figure 29-22** :

- Marked deformity
- Swelling
- Pain that is aggravated by any attempt at movement
- Tenderness on palpation
- Virtually complete loss of normal joint motion (locked joint)
- Numbness or impaired circulation to the limb or digit

Sprains

A sprain occurs when a joint is twisted or stretched beyond its normal range of motion. As a result, the supporting capsule and ligaments are stretched or torn. A sprain should be considered a partial dislocation or

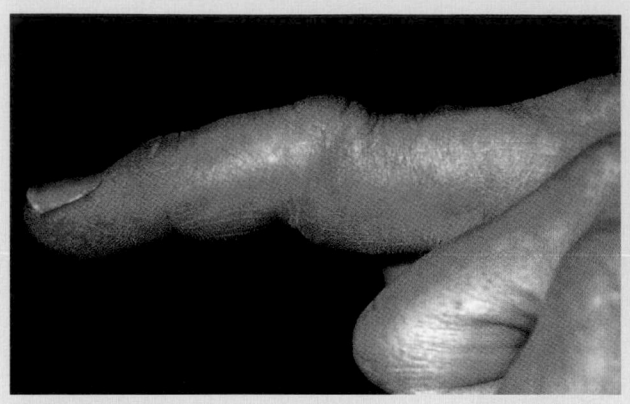

Figure 29-22 Joint dislocations, such as this finger, are characterized by deformity, swelling, pain with any movement, tenderness, locking, and impaired circulation.

subluxation. The alignment generally returns to a fairly normal position, although there may be some displacement. Note that severe deformity does not typically occur with a sprain. Sprains most often occur in the knee and the ankle, but a sprain can occur in any joint. The following signs and symptoms often indicate that the patient may have a sprain **Figure 29-23** :

- Point tenderness can be elicited over the injured ligaments.
- Swelling and ecchymosis appear at the point of injury to the ligament as a result of torn blood vessels.

You are the Provider: PART 2

When you arrive at the scene, you find the patient, a 21-year-old woman, sitting on the ground with an ice pack over her left tibial area. She is conscious and alert and tells you that another player fell against her leg.

Recording Time: 0 Minutes	
Appearance	Anxious; in obvious pain
Level of consciousness	Conscious and alert
Airway	Open; clear of secretions and foreign bodies
Breathing	Increased rate; adequate depth
Circulation	Radial pulses, increased rate; strong and regular

The patient denies having any other injuries, and tells you that she heard a "snap" when the other player fell against her leg. She is in severe pain.

3. What initial treatment should you provide to this patient?

4. What are some indicators of a fractured bone?

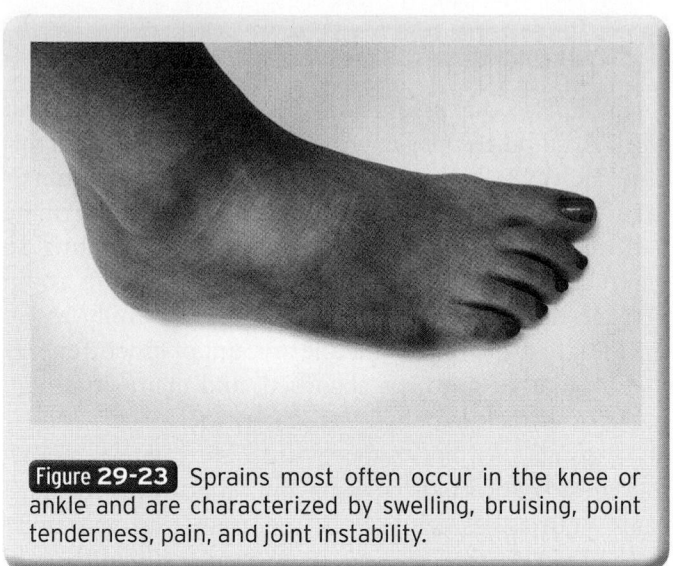

Figure 29-23 Sprains most often occur in the knee or ankle and are characterized by swelling, bruising, point tenderness, pain, and joint instability.

- Pain prevents the patient from moving or using the limb normally.
- Instability of the joint is indicated by increased motion, especially at the knee; however, this may be masked by severe swelling and guarding.

A fracture can look like a sprain, and vice versa. You will frequently not be able to distinguish a nondisplaced fracture from a sprain. Therefore, remember to document the MOI, because certain sprains and fractures occur more consistently with certain mechanisms. This is especially true at the ankle. In general, your approach should always be to determine the MOI. The basic principles of field management for sprains, dislocations, and fractures are essentially the same.

Strain

A strain (pulled muscle) is an injury to a muscle and/or tendon that results from a violent muscle contraction or from excessive stretching. Often no deformity is present and only minor swelling is noted at the site of the injury. Some patients may complain of increased pain with passive movement of the injured extremity. The general treatment of strains is similar to the field management for sprains, dislocations, and fractures.

Compartment Syndrome

Be on the alert for compartment syndrome, which most commonly occurs with a fractured tibia or forearm of children and is often overlooked, especially in patients with an altered level of consciousness. Compartment syndrome typically develops within 6 to 12 hours after injury, usually as a result of excessive bleeding, a severely crushed extremity, or the rapid return of blood to an ischemic limb. This syndrome is characterized by pain that is out of proportion to the injury, pain on passive stretch of muscles within the compartment, pallor, decreased sensation, and decreased power (ranging from decreased strength and movement of the limb to complete paralysis).

Amputations

Amputations can occur as a result of trauma or a surgical intervention. You must control bleeding and treat for shock when dealing with traumatic amputations. Complete traumatic amputations may not bleed much because the cut vessels may spasm, preventing the bleeding. You must also be aware of the victim's emotional stress, which can lead to psychogenic shock. Amputations are covered in detail in Chapter 24, *Soft-Tissue Injuries*.

Complications

Orthopaedic injuries can lead to numerous complications—not just those involving the skeletal system, but also systemic changes or illness. It is essential that you do not focus all of your attention on the skeletal injury: Keep in mind that there is a patient attached to the injured extremity! For example, pregnant women who sustain pelvic fractures tend to have higher mortality rates. Therefore, it is imperative to treat not only the fracture, but also the other needs of the woman and fetus.

The likelihood of having a complication is often related to the strength of the force that caused the injury, the injury's location, and the patient's overall health. Any injury to a bony structure is likely to be accompanied by bleeding. In general, the greater the force that caused the injury, the greater the hemorrhage will be. Following a fracture, the sharp ends of the bone may damage muscles, blood vessels, arteries, and nerves, or the ends may penetrate the skin and produce an open fracture. A significant loss of tissue may occur at the fracture site if the muscle is severely damaged or if the penetration of the bone into the skin causes a large deformity.

To prevent contamination following an open fracture, you should brush away any obvious debris on the skin surrounding an open fracture before applying a dressing. Do not enter or probe the open fracture site in an attempt to retrieve debris because this may lead to further contamination.

Long-term disability is one of the most devastating consequences of an orthopaedic injury. In many cases, a severely injured limb can be repaired and made to look almost normal. Unfortunately, many patients cannot return to work for long periods because of the extensive rehabilitation required and because of chronic pain. As an EMT, you have a critical role in mitigating the risk of long-term disability. You can help reduce the risk or duration of long-term disability by preventing further injury, reducing the risk of wound infection, minimizing pain by the use of cold and analgesia, and transporting patients with orthopaedic injuries to an appropriate medical facility.

Assessing the Severity of Injury

You must become skilled at quickly and accurately assessing the severity of an injury. The Golden Period is critical not only for life, but also for preserving limb viability. In an extremity with anything less than complete circulation, prolonged hypoperfusion can cause significant damage. For this reason, any suspected open fracture or vascular injury is considered a medical emergency. In a patient who has multisystem trauma, any additional bleeding can increase problems with underlying injuries or overall perfusion.

Remember that most injuries are not critical; you can identify critical injuries by using the musculoskeletal injury grading system shown in **Table 29-1**.

Table 29-1   Musculoskeletal Injury Grading System

Minor Injuries
- Minor sprains
- Fractures or dislocations of digits

Moderate Injuries
- Open fractures of digits
- Nondisplaced long-bone fractures
- Nondisplaced pelvic fractures
- Major sprains of a major joint

Serious Injuries
- Displaced long-bone fractures
- Multiple hand and foot fractures
- Open long-bone fractures
- Displaced pelvic fractures
- Dislocations of major joints
- Multiple digit amputations
- Laceration of major nerves or blood vessels

Severe, Life-Threatening Injuries (survival is probable)
- Multiple closed fractures
- Limb amputations
- Fractures of both long bones of the legs (bilateral femur fractures)

Critical Injuries (survival is uncertain)
- Multiple open fractures of the limbs
- Suspected pelvic fractures with hemodynamic instability

Patient Assessment

As an EMT, your assessments, attempts to splint, and work to stabilize the patient's condition are very important. However, always look at the big picture, evaluating the overall complexity of the situation to determine and treat any life threats. For example, overlooking an obstructed airway to splint a lower leg fracture could become deadly for the patient. Always carefully assess the MOI to try to determine the amount of kinetic energy that an injured limb has absorbed, and maintain a high index of suspicion for associated injuries.

Again, it is not important to distinguish among fractures, dislocations, sprains, and contusions. In most cases, your assessment will be reported as an "extremity injury." However, you must be able to distinguish mild injuries from severe injuries because some severe injuries may compromise neurovascular function, which could be limb threatening.

Scene Size-up

Scene Safety

Information from dispatch may indicate the MOI, the number of patients involved, and any first aid procedures used prior to your arrival. This will be useful information for you to think about as you travel to the scene. Remember, the information given by the dispatcher is only as accurate as the patient's or bystander's report. In addition, the situation may change prior to your arrival at the incident. Dispatch information can still be used to help you consider whether spinal stabilization will be needed, the equipment you may need, and whether hazards might be present.

As you arrive at the scene, observe the scene for hazards and threats to the safety of the crew, bystanders, and the patient. Try to identify the forces associated with the MOI. Could they have produced injuries other than the musculoskeletal injuries reported by dispatch? Standard precautions may be as simple as gloves. With a severe MOI or other risk factors, a mask and gown may be necessary. Consider the possibility that there may be hidden bleeding. Eye protection may also be indicated. Evaluate the need for law enforcement support, advanced life support, or additional ambulances and request them early based on your initial scene assessment.

Mechanism of Injury/Nature of Illness

As you observe the scene, look for indicators of the MOI. When you are assessing a patient who has experienced a significant MOI, be alert for both primary and secondary injuries. Primary injuries occur as a result of the MOI, whereas secondary injuries are the result of what happens after the initial injury. For example, being hit by a car will often result in a primary pelvic injury and often a secondary head injury when the patient rolls onto the hood of the car. As you put together information from dispatch and your observations of the scene, consider how the MOI produced the injuries expected. For example, when you are approaching a rear-end motor vehicle collision, you should suspect head, neck, and chest injuries.

Primary Assessment

The primary assessment should focus on identifying and managing life threats. Treating the patient according to his or her level of consciousness and ABCs is always a good choice. Threats to airway, breathing, and circulation are considered life threatening and must be treated immediately to prevent mortality. In some situations, significant bleeding may require management before applying oxygen for a person with adequate breathing. Significant bleeding, internal or external, is an immediate life threat. If the patient has obvious life-threatening bleeding, control it quickly and begin treating the patient for shock as quickly as possible. The decision on what to treat first will come with experience. For example, arterial bleeding from a compound fracture should be treated prior to giving oxygen.

Form a General Impression

Introduce yourself, and ask the patient his or her name. This helps you to evaluate the patient's level of consciousness and orientation. Check for responsiveness using the AVPU scale (*Alert* to person, place, and day; responsive to *Verbal* stimuli; responsive to *Pain*; *Unresponsive*). Generally you can assess a patient's mental status by asking the patient about his or her chief complaint. If the patient is alert, this should help direct you to any apparent life threats. If the patient is not alert, determine whether he or she responds to verbal or painful stimuli or whether he or she is unresponsive. An unresponsive patient may indicate a life-threatening condition. You should administer high-flow oxygen via a nonrebreathing mask (or a bag-mask device, if indicated) to all patients whose level of consciousness is less than alert and oriented and provide immediate transport to the emergency department.

Perform a rapid scan of the patient and ask about the MOI. Was it a direct blow, indirect force, twisting force, or high-energy injury? In many situations, the musculoskeletal complaints will be simple and usually not life threatening; however, some situations, such as those with a significant MOI, will include multiple problems that include musculoskeletal injuries. The initial interaction with your patient will provide you with a starting point and help you to distinguish the simple from the complex injuries. If there was significant trauma and multiple body systems are affected, the musculoskeletal injuries may be a lower priority. Scene time should not be wasted on prolonged musculoskeletal assessment or splinting.

Airway and Breathing

Fractures and sprains usually do not create airway and breathing problems. Other problems, such as injuries to the head, intoxication, or other related illnesses and injuries may cause inadequate breathing. Evaluating the chief complaint and MOI will help you to identify whether the patient has an open airway and whether breathing is present and adequate. In a conscious patient, this is as simple as noting whether the patient can speak normally. In an unconscious patient, it is as simple as opening the airway using the appropriate technique to look, listen, and feel for breathing. If a spinal injury is suspected, take the appropriate precautions and prepare for stabilization. Oxygen may be given to relieve anxiety and improve perfusion. Even though an injury to the arm or leg may be obvious, take the time to evaluate the adequacy of the airway and breathing. Very little else matters if the patient's airway and breathing are inadequate.

Circulation

Your circulatory assessment should focus on determining whether the patient has a pulse, has adequate perfusion, or is bleeding. If your patient is conscious, as most patients with fractures and dislocations are, he or she will have a pulse. If the patient is unconscious, make sure there is a pulse by palpating the carotid artery. Hypoperfusion (shock) and bleeding problems will most likely be your primary concern. If the skin is pale, cool, or clammy and capillary refill time is slow, treat your patient for shock immediately. Maintain a normal body temperature, and improve perfusion with oxygen. If musculoskeletal injuries in the extremities are suspected, they must be at least initially stabilized, if not splinted, prior to moving. Eliminating this cause of shock may need to be done later in your assessment.

Fractures can break through the skin and cause external bleeding. This may occur during the initial injury or during manipulation of the extremity while preparing for splinting or transport. Careful handling of the extremity minimizes this risk. If external bleeding is present, bandage the extremity quickly to control bleeding. The dressings that cover the wound and bone should be kept sterile to reduce the potential for bone infection. The bandage should be secure enough to control bleeding without restricting circulation distal to the injury. Monitor bandage tightness by assessing the circulation, sensation, and movement distal to the bandage. Swelling from fractures and internal bleeding may cause bandages to become too tight. If bleeding cannot be controlled, you should quickly proceed to apply a **tourniquet**.

Transport Decision

If the patient you are treating has an airway or breathing problem, or significant bleeding, provide rapid transport to the hospital for treatment. A patient who has a significant MOI but whose condition appears otherwise stable should also be transported promptly to the closest appropriate hospital. Patients with bilateral fractures of the long bones (humerus, femur, or tibia) have been subjected to a high amount of kinetic energy, which should dramatically increase your index of suspicion for serious unseen injuries. When a decision for rapid transport is made, you can use a backboard as a splinting device to splint the whole body rather than splinting each extremity individually. If you take time to splint the patient's arms and legs individually, you may delay the prompt surgical intervention that may be needed for other injuries when a significant MOI has occurred. Individual splints should be applied en route if the ABCs are stable and time permits.

Patients with a simple MOI, such as twisting of an ankle or dislocating a shoulder, may be further assessed and their condition stabilized on scene prior to transport if no other problems exist. Handle fractures carefully while preparing for transport. Careful handling is necessary to limit pain and prevent sharp bone ends from breaking through the skin or damaging nerves and blood vessels in the extremity.

History Taking

Investigate Chief Complaint

After the life threats have been managed during the primary assessment, investigate the chief complaint. You should obtain a medical history and be alert for injury-specific signs and symptoms and for any pertinent negatives such as no pain or loss of sensation.

SAMPLE History

A SAMPLE history should be obtained for all trauma patients. How much and in what detail you explore this history depends on the seriousness of the patient's condition and how quickly you need to transport the patient to the hospital. For example, if your patient has an isolated arm fracture, you do not need to worry about the patient's medical history of a hysterectomy. For patients with simple fractures, dislocations, or sprains, it is easier to obtain a SAMPLE history. At the scene you may have access to family members and others who have information about

the patient's history. Make an attempt to obtain this history without delaying time to definitive care.

OPQRST (Onset, Provocation or palliation, Quality, Region/radiation, Severity, Timing) can be of limited use in cases of severe injury and is usually too lengthy when matters of airway, breathing, circulation, and rapid transport require immediate attention. However, OPQRST may be useful when the MOI is unclear, the patient's condition is stable, or details of the injury are uncertain. This more detailed questioning for simple trauma may help you and the hospital staff to understand the specific injury better.

Secondary Assessment

The secondary assessment is a more detailed, comprehensive examination of the patient that can reveal injuries that may have been missed during the primary assessment. In some cases, such as a critically injured patient or a short transport time, you may not have time to conduct a secondary assessment.

Physical Examinations

If significant trauma has likely affected multiple systems, start with a full-body scan to be sure that you have found all of the problems and injuries. Begin with the head and work systematically toward the feet, checking the head, chest, abdomen, extremities, and back. The goal here is to identify hidden and potentially life-threatening injuries. This full-body scan will also help you to prepare for packaging and rapid transport. Knowing if an arm or leg is broken will be important when logrolling the patient onto a backboard and securing the patient to the board.

Assess the musculoskeletal system by performing a detailed full-body scan. Use the DCAP-BTLS (Deformities, Confusions, Abrasions, Punctures/penetrations, Burns, Tenderness, Lacerations, and Swelling) approach. Identify any extremity deformities that likely represent significant musculoskeletal injury, and stabilize them appropriately. Contusions and abrasions may overlie more subtle injuries and should prompt you to carefully evaluate the stability and neurovascular status of the limb. The presence of puncture wounds or other signs of penetrating injury should alert you to the possibility of an open fracture. Associated burns must be identified and treated appropriately. Palpate for tenderness, which, like contusions or abrasions, may be the only significant sign of an underlying musculoskeletal injury.

When lacerations are present in an extremity, an open fracture must be considered, bleeding controlled, and dressings applied. Careful inspection for swelling with comparison with the opposite limb may also reveal otherwise occult musculoskeletal injury. You may find a hematoma in the zone of injury during the assessment.

If your assessment finds no external signs of injury, ask the patient to move each limb carefully, stopping immediately if a movement causes pain. Skip this step in your evaluation if the patient reports neck or back pain; even slight motion could cause permanent damage to the spinal cord.

When nonsignificant trauma has occurred and your patient has a simple strain, sprain, dislocation, or fracture, you can take the time to focus your physical examination on that particular injury. Look for DCAP-BTLS. Evaluate the circulation, motor function, and abnormal sensations distal to the injury. If the patient has two or more extremities injured, treat the patient as a significant trauma patient and provide rapid transport to the hospital. The likelihood of other more severe injuries is greater when two or more bones have been broken. Be sure to assess the entire zone of injury by removing clothing from the area and looking and palpating for injuries. In musculoskeletal injuries, this zone generally extends from the joint above (proximal) to the joint below (distal), front and back. Do not forget to check perfusion, motion, and sensation.

Many important blood vessels and nerves lie close to the bone, especially around the major joints. Therefore, any injury or deformity of the bone may be associated with vessel or nerve injury. For this reason, you must assess neurovascular function every 5 to 10 minutes during the assessment, depending on the patient's condition, until the patient is at the hospital. Always recheck the neurovascular function before and after you splint or otherwise manipulate the limb. Manipulation can cause a bone fragment to press against or impale a nerve or vessel. Failure to restore circulation in this situation can lead to death of the limb. Always give priority to patients with impaired circulation resulting from bone fragments.

Examination of the injured limb should include the 6 Ps of musculoskeletal assessment—pain, paralysis, paresthesias (numbness or tingling), pulselessness, pallor (pale or delayed capillary refill in children), and pressure. To assess neurovascular status, follow the steps in Skill Drill 29-1:

Skill Drill 29-1

1. **Pulse.** Palpate the pulse distal to point of injury. First, palpate the radial pulse in the upper extremity (Step 1). Second, in the lower extremity, palpate the posterior tibial and dorsalis pedis pulses (Step 2).
2. **Capillary refill.** Note and record the skin color, identifying any pallor or cyanosis. Then apply firm pressure to the tip of the fingernail or toenail, which will cause the skin to blanch (turn white). If normal color does not return within 2 seconds

after you release the nail, you can assume that circulation is impaired. This test is typically recommended for use in children, although it can be used in adults also (Step 3).
3. **Sensation.** In the hand, check the feeling on the flesh near the tip of the index finger and thumb, as well as the little finger (Step 4). In the foot, check the sensation on the flesh of the big toe (Step 5) and on the lateral side of the foot (Step 6). The patient's ability to sense light touch in the fingers or toes distal to the site of a fracture is a good indication that the nerve supply is intact.
4. **Motor function.** Evaluate muscular activity when the injury is proximal to the patient's hand or foot. Ask the patient to open and close a fist for an upper extremity injury and to wiggle the toes and move the foot up and down for a lower extremity injury. Sometimes, an attempt at motion will produce pain at the injury site. If this happens, do not continue this part of the examination. To avoid causing pain, do not perform this test at all if the injury involves the hand or foot itself (Steps 7 through 10).

Words of Wisdom

Extremity injuries that impair circulation or nerve function in distal tissues are urgent conditions. Patients with these injuries need careful assessment, prompt transport, and frequent reassessment of distal functions. It is also crucial to report this information in your initial radio contact with the hospital to allow personnel to prepare for a condition in which prompt surgery may be necessary to save the limb.

Because many of the steps require patient cooperation, you will not be able to assess sensory and motor functions in an unconscious patient, but you can evaluate the limb for deformity, swelling, ecchymosis, false motion, and crepitus.

Vital Signs

Determine a baseline set of vital signs, including pulse rate, rhythm, and quality; respiratory rate, rhythm, and quality; blood pressure; skin condition; and pupil size and reaction to light. These baseline indicators need to be obtained as soon as possible. Your patient may appear to be tolerating the injury well until you reassess these vital signs and they indicate otherwise. Trending these vital signs helps you to understand whether your patient's condition is improving or getting worse over time, particularly during long transports. Shock or hypoperfusion is common in musculoskeletal injuries, and this baseline information is very important in assessing your patient's condition.

Skill Drill 29-1

Assessing Neurovascular Status

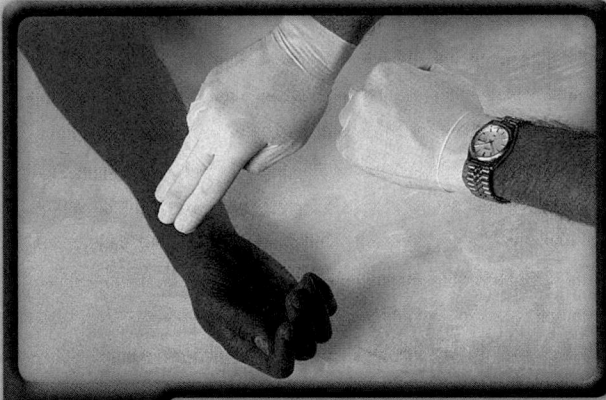

Step 1 Palpate the radial pulse in the upper extremity.

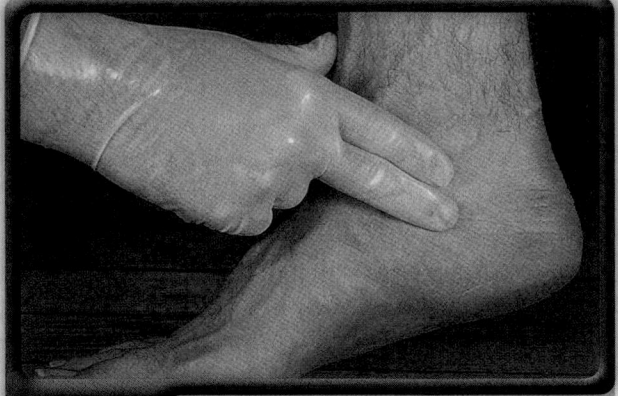

Step 2 Palpate the posterior tibial and dorsalis pedis pulse in the lower extremity.

Step 3 Assess capillary refill by blanching a fingernail or toenail.

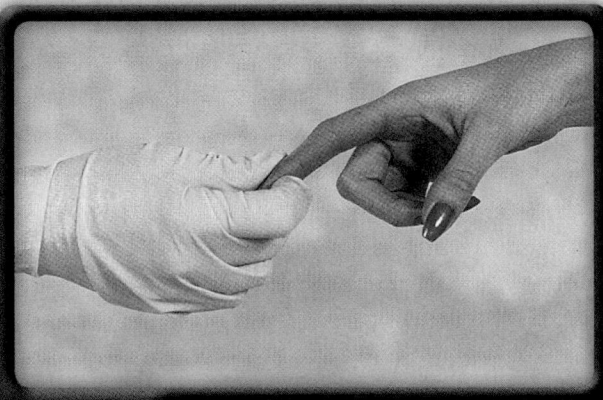

Step 4 Assess sensation on the flesh near the tip of the index finger and thumb, as well as the little finger.

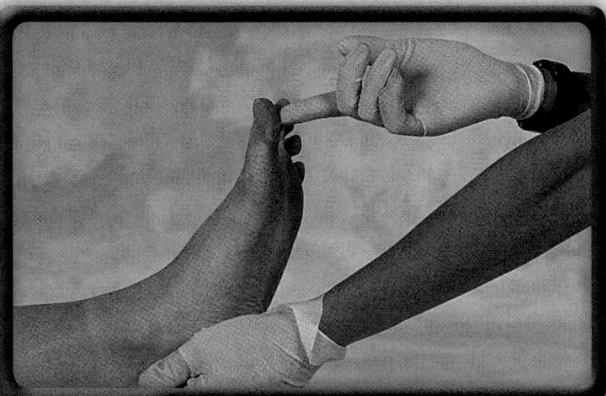

Step 5 On the foot, first check sensation on the flesh near the tip of the big toe.

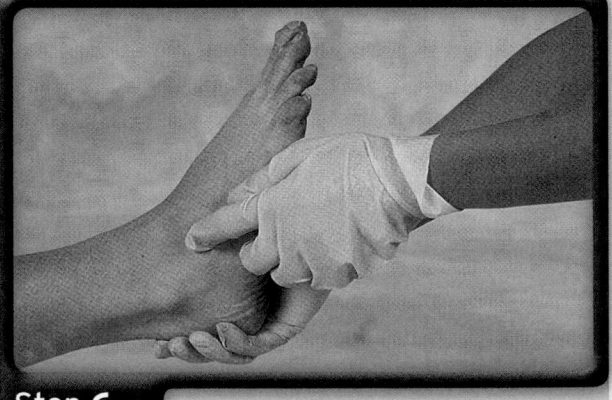

Step 6 Also check sensation on the lateral side of the foot.

Skill Drill 29-1

Assessing Neurovascular Status, continued

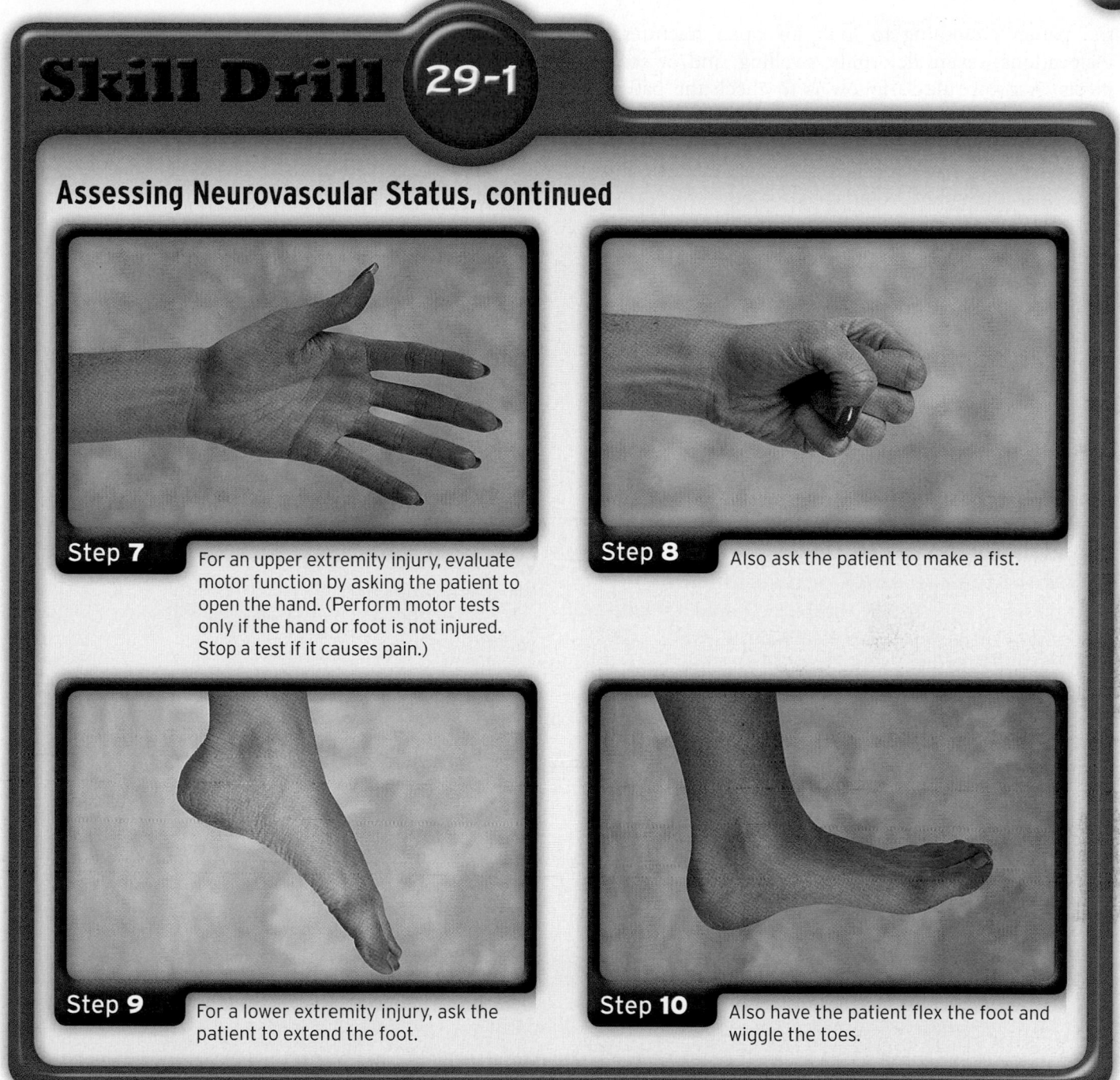

Step 7 For an upper extremity injury, evaluate motor function by asking the patient to open the hand. (Perform motor tests only if the hand or foot is not injured. Stop a test if it causes pain.)

Step 8 Also ask the patient to make a fist.

Step 9 For a lower extremity injury, ask the patient to extend the foot.

Step 10 Also have the patient flex the foot and wiggle the toes.

Reassessment

Repeat the primary assessment to ensure your interventions are working as they should. A reassessment should be performed every 5 minutes for an unstable patient and every 15 minutes for a stable patient.

Interventions

Because trauma patients often have multiple injuries, you must assess their overall condition, stabilize the ABCs, and control any serious bleeding before further treating the injured area. In a critically injured patient, you should secure the patient to a long backboard to stabilize the spine, pelvis, and extremities and provide prompt transport to a trauma center. In this situation, a secondary assessment with extensive evaluation and splinting of limb injuries in the field is a waste of valuable time. Perform the primary assessment and transport, reassessing the patient en route to the emergency department.

If the patient has no life-threatening injuries, you may take extra time at the scene to stabilize the patient's overall condition and more completely evaluate the injury. If possible, gently and carefully remove the patient's clothing to look for open fractures or dislocations, severe deformity, swelling, and/or ecchymosis. A good rule to follow is to check the patient's circulation, motor function, and sensation prior to and after splinting.

When you have finished assessing the extremity, apply a secure splint, commercial or otherwise, to stabilize the injury prior to transport. The joint above and below the site of injury should be included in the splint. To minimize the potential for problems, the splint should be well padded. A comfortable and secure splint will reduce pain, reduce shock, and minimize compromised circulation.

The main goal in providing care for musculoskeletal injuries is stabilization in the most comfortable position that allows for maintenance of good circulation distal to the injury. This should be done whether you are preparing the patient for rapid transport or you have as much time as you need to assess and treat the patient.

Communication and Documentation

Your radio report to the hospital should include a description of the problems found during your assessment. In particular, you should report problems with the patient's ABCs, open fractures, and compromised circulation that occurred before or after splinting. Many times the hospital staff can arrange for specialists or consider antibiotics early if they are aware of problems. How much you include in your radio report will depend on your local protocols. Additional details, such as the mandated reporting of situations involving elder or child abuse, can be given during your verbal report at the hospital when you transfer care to the nursing staff or physician.

Document complete descriptions of injuries and the MOIs associated with them. It is important to assess and document the presence or absence of circulation, motor function, and sensation distal to the injury before you move an extremity, after manipulation or splinting of the injury, and on arrival at the hospital. Hospital staff may later refer to your notes to clarify confusing situations or communication problems. Your careful documentation may protect you from legal action that patients may take later. Do not rely on your memory for details from situations; your memory is unreliable and will not hold up in a court of law.

Emergency Medical Care

Your first steps in providing care for any patient are the primary assessment and stabilizing the patient's ABCs. If needed, perform a rapid scan or focus on a specific injury. Remember to always follow standard precautions and be alert for signs and symptoms of internal bleeding. Internal bleeding should be suspected whenever the MOI suggests that severe forces have affected the body.

Follow the steps in **Skill Drill 29-2** when caring for patients with musculoskeletal injuries:

1. Remove any jewelry. Completely cover open wounds with a dry, sterile dressing, and apply direct pressure to control bleeding. If bleeding cannot be controlled, you should quickly proceed to the use of a tourniquet. Once you have applied

You are the Provider: PART 3

A nurse present at the scene assists by stabilizing the leg above the ankle and below the knee while you expose the injury. The patient has an obvious deformity in the midshaft area of her tibia/fibula; however, there are no open wounds. As you further assess the injury, your partner obtains the patient's vital signs.

Recording Time: 5 Minutes	
Respirations	22 breaths/min; adequate depth
Pulse	112 beats/min; strong and regular
Skin	Pink, warm, and moist
Blood pressure	130/78 mm Hg
Oxygen saturation (Sao$_2$)	98% (on room air)

5. How should you proceed with your assessment of this patient's injury?
6. How should you treat an injured extremity in which distal perfusion is absent?

a sterile dressing, treat an open fracture in the same way as a closed fracture (Step 1).

2. Apply the appropriate splint, and elevate the extremity. It is essential to splint the joint above and below the injury to ensure immobilization. Patients with lower extremity injuries should lie supine with the limb elevated about 6″ to minimize swelling. For any patient, be sure to position the injured limb slightly above the level of the heart. Never allow the injured limb to flop about or dangle from the edge of the backboard. Always assess pulse and motor and sensory functions before and after the application of splints. Assess the pulse by palpation, evaluate

motor function by asking the patient to open his or her hand or flex his or her foot and assess sensation and capillary refill on the flesh near the tip of the index finger. Finally, assess the skin color and condition, and evaluate sensory function by touch (Step 2).

3. If swelling is present, apply cold packs to the area; however, avoid placing cold packs directly on the skin or other exposed tissues (Step 3).

4. Prepare the patient for transport. A patient with an isolated upper extremity injury will most likely be more comfortable in a semiseated position rather than in a supine position; however, assuming there is no risk of spinal injury, either position is acceptable. Ensure that the extremity is elevated above the level of the heart and secured so that it does not dangle from the edge of the backboard (Step 4).

Skill Drill 29-2

Caring for Musculoskeletal Injuries

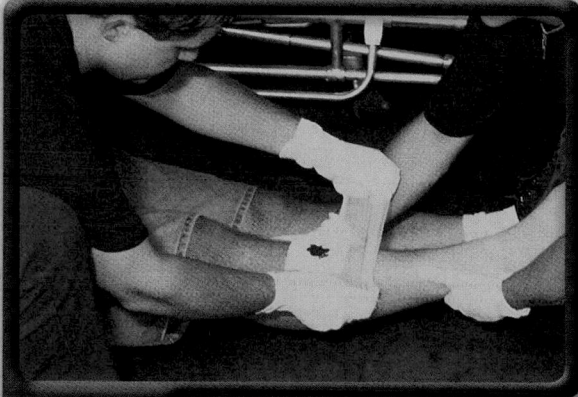

Step 1 Cover open wounds with a dry, sterile dressing, and apply pressure to control bleeding. If bleeding cannot be controlled, proceed to the use of a tourniquet.

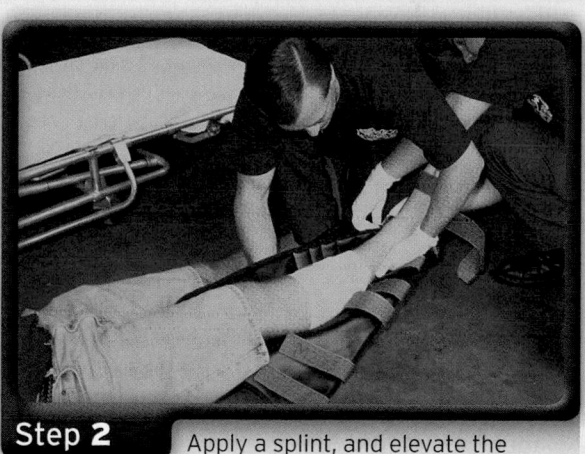

Step 2 Apply a splint, and elevate the extremity about 6″ (slightly above the level of the heart).

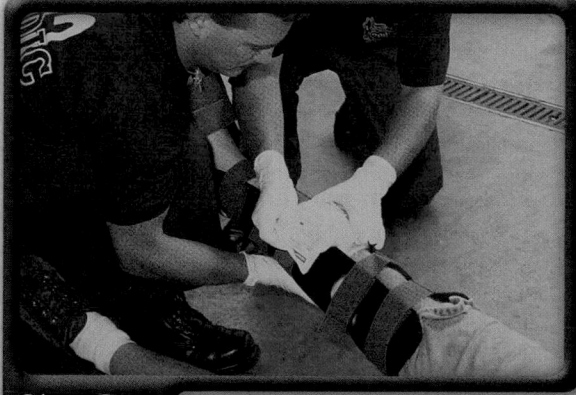

Step 3 Apply cold packs if there is swelling, but do not place them directly on the skin.

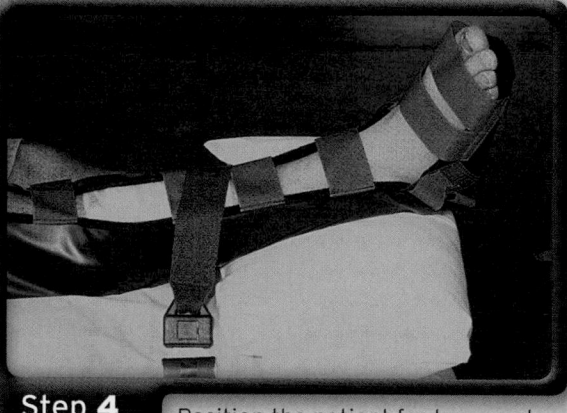

Step 4 Position the patient for transport, and secure the injured area.

5. Always transport your patient to the most appropriate facility, and consider the use of advanced life support backup for pain management.

6. Always inform hospital personnel about all wounds that have been dressed and splinted and any associated injuries treated by the EMS unit.

■ Splinting

A **splint** is a flexible or rigid device that is used to protect and maintain the position of an injured extremity Figure 29-24. Unless the patient's life is in immediate danger, you should splint all fractures, dislocations, and sprains before moving the patient. By preventing movement of fracture fragments, bone ends, a dislocated joint, or damaged soft tissues, splinting reduces pain and makes it easier to transfer and transport the patient. In addition, splinting will help to prevent the following:

- Further damage to muscles, the spinal cord, peripheral nerves, and blood vessels from broken bone ends.
- Laceration of the skin by broken bone ends. One of the primary indications for splinting is to prevent a closed fracture from becoming an open fracture (conversion).
- Restriction of distal blood flow resulting from pressure of the bone ends on blood vessels.
- Excessive bleeding of the tissues at the injury site caused by broken bone ends.
- Increased pain from movement of bone ends.
- Paralysis of extremities resulting from a damaged spine.

A splint is simply a device to prevent motion of the injured part. It can be made from any material on occasions when you need to improvise. However, you should have an adequate supply of standard commercial splints on hand.

Figure 29-24 Splinting reduces pain and prevents additional damage to the injured extremity.

Straightening or splinting an injured limb can compromise distal functions, just as the initial injury can. Record the status of distal circulation and nervous function (neurovascular status) before and after straightening or splinting. At a minimum, your written record should describe these functions before splinting and confirm that they were normal immediately after splinting and on hospital arrival. For any but the shortest transports, also indicate the results of reassessments while en route.

General Principles of Splinting

The following principles of splinting apply to most situations:

1. Remove clothing from the area of any suspected fracture or dislocation so that you can inspect the extremity for DCAP-BTLS.

2. Note and record the patient's neurovascular status distal to the site of the injury, including pulse, sensation, and movement. Continue to monitor the neurovascular status until the patient reaches the hospital.

3. Cover all wounds with a dry, sterile dressing before splinting. Be sure to follow standard precautions. Do not intentionally replace protruding bones. Notify the receiving hospital of all open wounds.

4. Do not move the patient before splinting an extremity unless there is an immediate danger to the patient or you.

5. In a suspected fracture of the shaft of any bone, be sure to stabilize the joints above and below the fracture.

6. With injuries in and around the joint, be sure to stabilize the bones above and below the injured joint.

7. Pad all rigid splints to prevent local pressure and discomfort to the patient.

8. While applying the splint, maintain manual stabilization to minimize movement of the limb and to support the injury site.

9. If fracture of a long-bone shaft has resulted in severe deformity, use constant, gentle manual traction to align the limb so that it can be splinted. This is especially important if the distal part of the extremity is cyanotic or pulseless.

10. If you encounter resistance to limb alignment, splint the limb in its deformed position.

11. Stabilize all suspected spinal injuries in a neutral in-line position on a backboard.

12. If the patient has signs of shock (hypoperfusion), align the limb in the normal anatomic position, and provide transport (total body stabilization).

13. When in doubt, splint.

General Principles of In-line Traction Splinting

Application of in-line **traction** is the act of pulling on a body structure in the direction of its normal alignment. It is the most effective way to realign a fracture of the shaft of a long bone so that the limb can be splinted more effectively. Excessive traction can be harmful to an injured limb. When applied correctly, however, traction stabilizes the bone fragments and improves the overall alignment of the limb. You should not attempt to reduce the fracture or force all of the bone fragments back into alignment. This is the physician's responsibility. In the field, the goals of in-line traction are as follows:

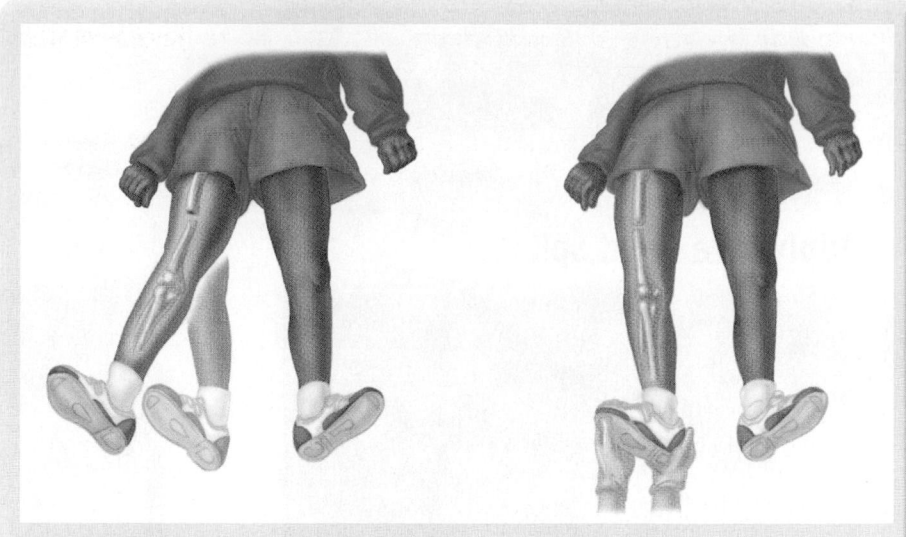

Figure 29-25 To apply traction, imagine the position where the uninjured limb would lie, and then gently pull along that line until the injured limb is in that position. Do not release traction once you have applied it.

1. To stabilize the fracture fragments to prevent excessive movement
2. To align the limb sufficiently to allow it to be placed in a splint
3. To avoid potential neurovascular compromise

Before you apply a traction splint, be sure to control any external bleeding. The amount of traction that is required varies but often does not exceed 15 lb. You should use the least amount of force necessary. Grasp the foot or hand at the end of the injured limb firmly; once you start pulling, you should not stop until the limb is fully splinted. Always apply the direction of traction along the long axis of the limb. Imagine where the uninjured limb would lie, and pull gently along the line of that imaginary limb until the injured limb is in approximately that position **Figure 29-25**. Grasping the foot or hand and the initial pull of traction usually causes some discomfort as the bone fragments move. It helps if a second person can support the injured limb directly under the site of the fracture. This initial discomfort quickly subsides, and you can then apply further gentle traction. However, if the patient strongly resists the traction or if it causes more pain that persists, you must stop and splint the limb in the deformed position.

Remember that many different materials can be used as splints if necessary. When no splinting materials are available, the arm can be bound to the chest wall, and an injured leg can be bound to the uninjured leg to provide at least temporary stability. The three basic types of splints are rigid, formable, and traction splints.

Rigid Splints

Rigid (nonformable) splints are made from firm material and are applied to the sides, front, and/or back of an injured extremity to prevent motion at the injury site. Common examples of rigid splints include padded board splints, molded plastic and metal splints, padded wire ladder splints, and folded cardboard splints. As always, be sure to follow standard precautions. It takes two EMTs to apply a rigid splint. Follow the steps in **Skill Drill 29-3**:

1. Gently support the limb at the site of injury as others prepare and begin to position the equipment. Apply steady, in-line traction if necessary. Maintain this support until the splint is completely applied (Step 1).
2. Place the rigid splint under or alongside the limb.
3. Place padding between the limb and the splint to make sure there is even pressure and even contact. Look for bony prominences, and pad them (Step 2).
4. Apply bindings to hold the splint securely to the limb (Step 3).
5. Check and record the distal nervous and circulatory (neurovascular) function (Step 4).

There are two situations in which you must splint the limb in the position of deformity—when the deformity is severe, as is the case with many dislocations, and when you encounter resistance or extreme pain when applying gentle traction to the fracture of a shaft of a

Skill Drill 29-3

Applying a Rigid Splint

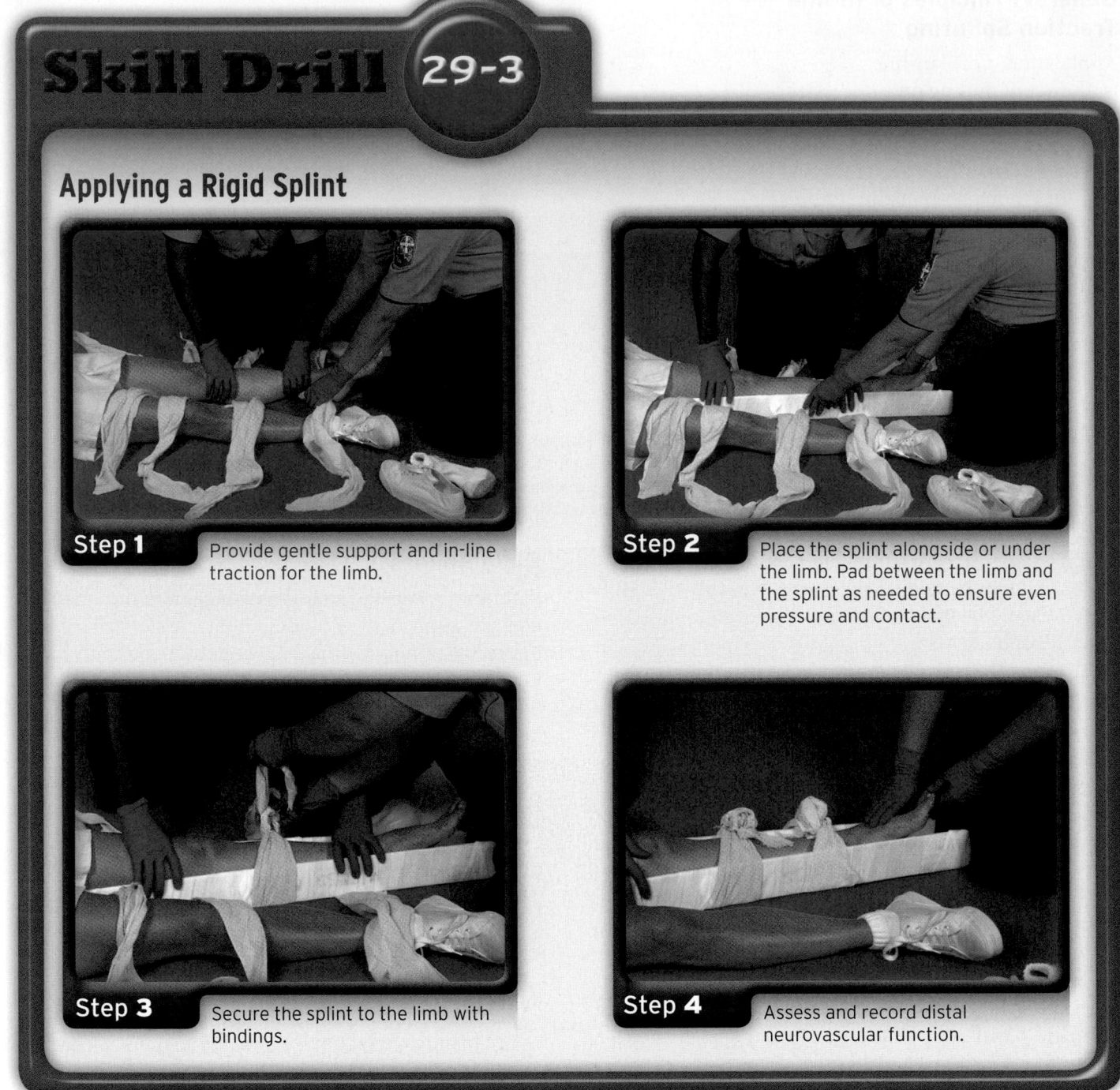

Step 1 Provide gentle support and in-line traction for the limb.

Step 2 Place the splint alongside or under the limb. Pad between the limb and the splint as needed to ensure even pressure and contact.

Step 3 Secure the splint to the limb with bindings.

Step 4 Assess and record distal neurovascular function.

long bone. In either situation, you should apply padded board splints to each side of the limb and secure them with soft roller bandages **Figure 29-26**. Most dislocations should be splinted as found, but follow local protocols. Attempts to realign or reduce dislocations can lead to more damage.

Formable Splints

The most commonly used formable or soft splint is the precontoured, inflatable, clear plastic air splint. These splints are available in a variety of sizes and shapes, with or without a zipper that runs the length of the splint.

Always inflate the splint after applying it. The air splint is comfortable, provides uniform contact, and has the added advantage of applying firm pressure to a bleeding wound. Air splints are used to stabilize injuries below the elbow or below the knee.

Air splints have some drawbacks, particularly in cold weather areas. The zipper can stick, clog with dirt, or freeze. Significant changes in the weather affect the pressure of the air in the splint, which decreases as the environment grows colder and increases as the environment grows warmer. The same thing happens when there are changes in altitude, which can be a problem with

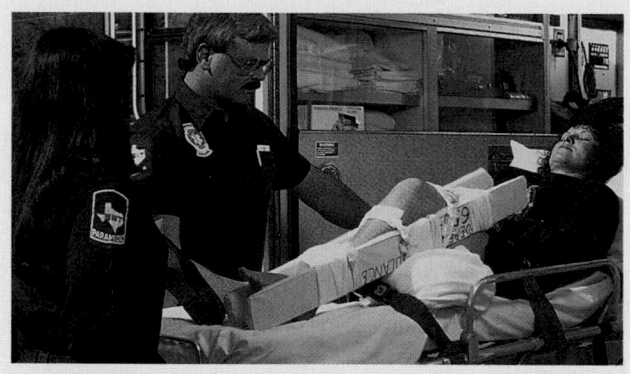

Figure 29-26 If you encounter resistance or extreme pain when applying traction to a long bone, apply padded board splints to each side of the limb, and secure them with soft roller bandages, stabilizing the limb in its deformed position.

helicopter transport of patients. Therefore, you should carefully monitor the splint and let air out if the splint becomes overinflated.

The method of applying an air splint depends on whether it has a zipper. With either type, you must first cover all wounds with a dry, sterile dressing, making sure that you use standard precautions. For a splint that has a zipper, follow the steps in **Skill Drill 29-4**:

1. Hold the injured limb slightly off the ground, applying gentle traction and supporting the site of injury. Have your partner place the open, deflated splint around the limb (Step 1).
2. Zip up the splint, and inflate it by pump or by mouth. When this is done, test the pressure in the splint. With proper inflation, you should just be able to compress the walls of the splint together with a firm pinch between the thumb and index finger near the edge of the splint.
3. Check and record pulse and motor and sensory functions, and monitor them periodically until the patient reaches the hospital (Step 2).

If you use an unzipped or partially zipped type of air splint, follow the steps in **Skill Drill 29-5**:

Skill Drill 29-5

1. Your partner supports the patient's injured limb until splinting is accomplished.
2. Place your arm through the splint. Extend your hand beyond the splint, and grasp the hand or foot of the injured limb (Step 1).

Skill Drill 29-4

Applying a Zippered Air Splint

Step 1 Support the injured limb, and apply gentle traction as your partner applies the open, deflated splint.

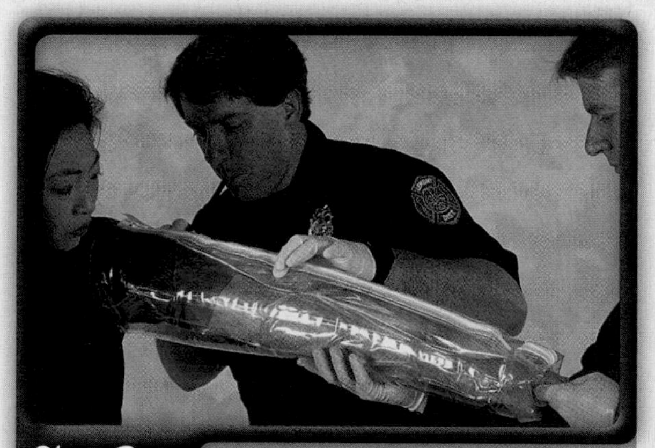

Step 2 Zip up the splint, inflate it by pump or by mouth, and test the pressure. Check and record distal neurovascular function.

Skill Drill 29-5

Applying an Unzipped Air Splint

Step 1 Your partner supports the injured limb. Place your arm through the splint to grasp the patient's hand or foot.

Step 2 Apply gentle traction while sliding the splint onto the injured limb.

Step 3 Your partner inflates the splint by pump or by mouth.

3. Apply gentle traction to the hand or foot while sliding the splint onto the injured limb. The hand or foot of the injured limb should always be included in the splint (Step 2).

4. Your partner inflates the splint by pump or by mouth (Step 3).

5. Test the pressure in the splint. This is something that you must do with either type of air splint.

6. Check and record pulse and motor and sensory functions, and monitor them en route.

Other formable splints include vacuum splints, pillow splints, structural aluminum malleable (SAM)

splints, a sling and swathe, and pelvic binders for pelvic fractures. Just like an air splint, a vacuum splint can be easily shaped to fit around a deformed limb. Instead of pumping air in, however, you can use a hand pump to pull the air out through a valve. Follow the steps in Skill Drill 29-6 to apply a vacuum splint:

Skill Drill 29-6

1. Your partner supports and stabilizes the injured limb, applying traction if needed (Step 1).

2. Gently place the injured limb onto the vacuum splint, and wrap the splint around the limb (Step 2).

3. Draw the air out of the splint through the suction valve, and then seal the valve. Once the valve is sealed, the vacuum splint becomes rigid, conforming to the shape of the deformed limb and stabilizing it (Step 3).

4. Check distal circulation and nervous functions, and monitor them en route.

Traction Splints

Traction splints are used primarily to secure fractures of the shaft of the femur, which are characterized by pain, swelling, and deformity of the midthigh. A traction splint should not be used if the patient has an obvious injury of the knee or ankle joint, foot, or lower leg. Several different types of lower extremity traction splints are commercially available, such as the Hare traction splint, the Sager splint, the Reel splint, and the Kendrick splint. Each has its own unique method of application; therefore, it is important to be familiar with each method. Consult with your local agency on which traction splint you will use in the field, and make sure that you are comfortable applying this device to a patient.

Traction splints are not suitable for use on the upper extremity because the major nerves and blood vessels in the patient's axilla cannot tolerate countertraction forces.

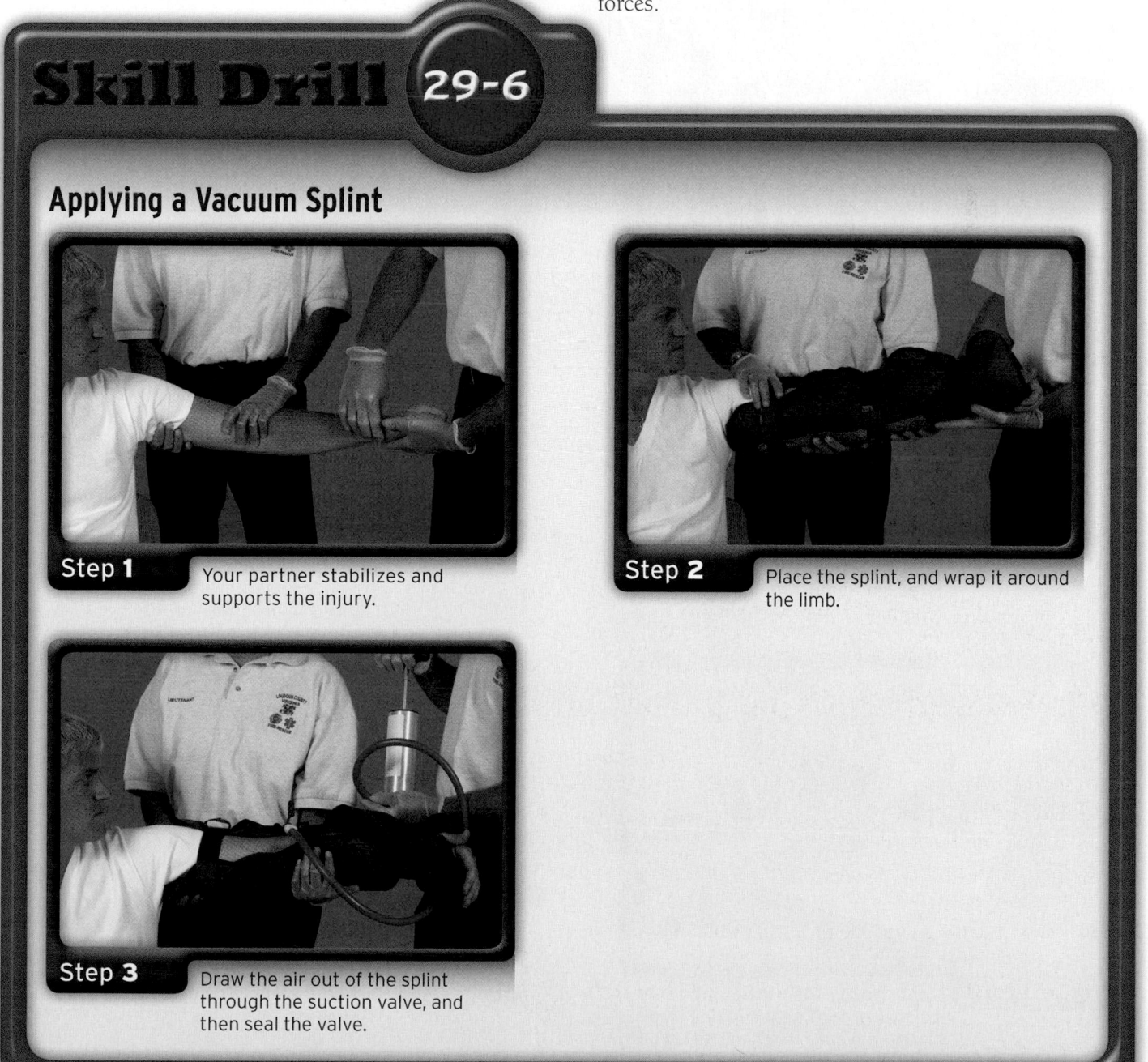

Skill Drill 29-6

Applying a Vacuum Splint

Step 1 Your partner stabilizes and supports the injury.

Step 2 Place the splint, and wrap it around the limb.

Step 3 Draw the air out of the splint through the suction valve, and then seal the valve.

Do not use traction splints for any of the following conditions:

- Injuries of the upper extremity
- Injuries close to or involving the knee
- Injuries of the hip
- Injuries of the pelvis
- Partial amputations or avulsions with bone separation
- Lower leg, foot, or ankle injury

Proper application of a traction splint requires two well-trained EMTs working together. To apply a Hare traction splint, follow the steps in Skill Drill 29-7:

Skill Drill 29-7

1. Cut open the patient's pant leg, or otherwise expose the injured lower extremity. Follow standard precautions as needed. Be sure to assess and record the pulse and motor function and sensation distal to the injury.

2. Place the splint beside the patient's uninjured leg, and adjust it to the proper length, with the ring at the ischial tuberosity and the splint extending 12″ beyond the foot. Open and adjust the four

Skill Drill 29-7

Applying a Hare Traction Splint

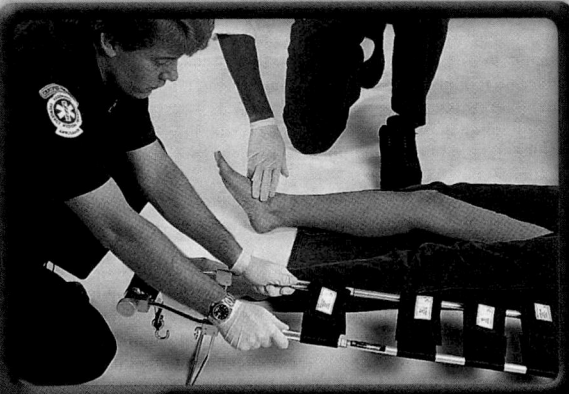

Step 1 Expose the injured limb and check pulse and motor and sensory function. Place the splint beside the uninjured limb, adjust the splint to proper length, and prepare the straps.

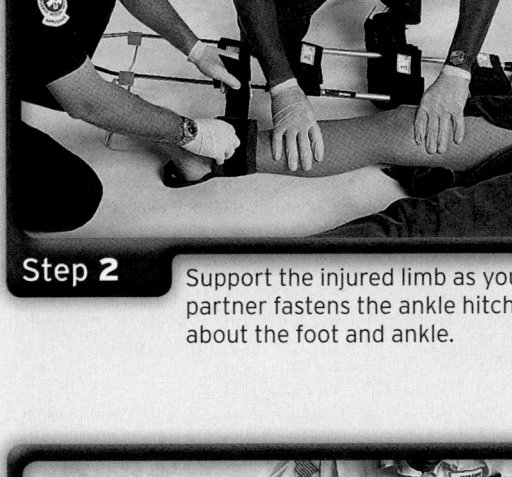

Step 2 Support the injured limb as your partner fastens the ankle hitch about the foot and ankle.

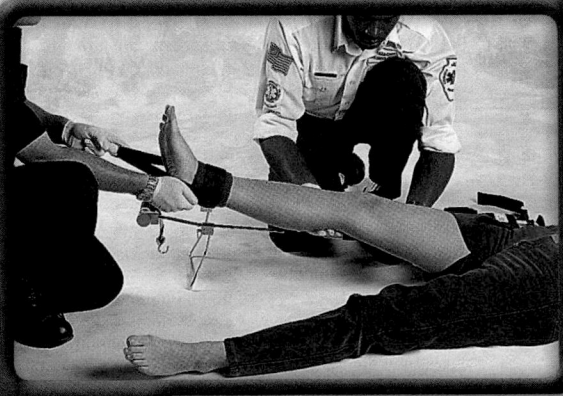

Step 3 Continue to support the limb as your partner applies gentle in-line traction to the ankle hitch and foot.

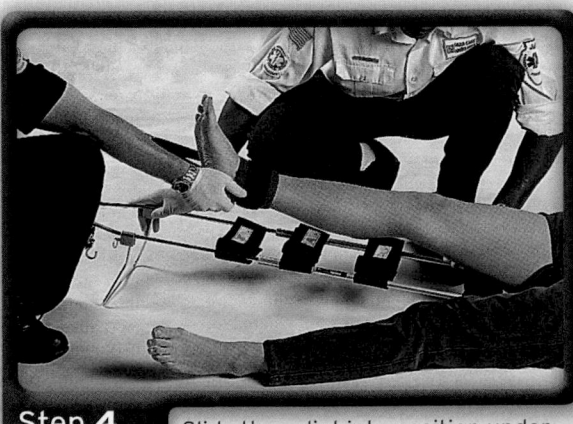

Step 4 Slide the splint into position under the injured limb.

Skill Drill 29-7

Applying a Hare Traction Splint, continued

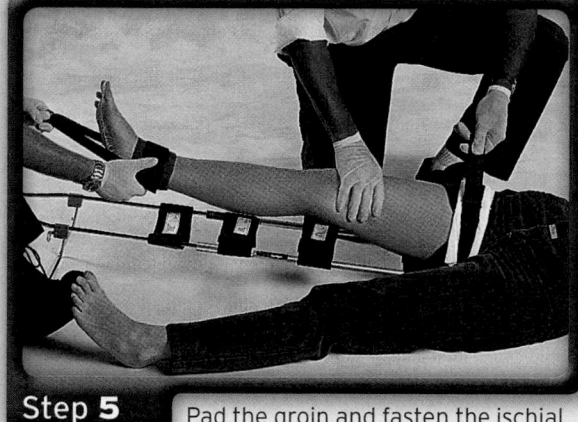

Step 5 Pad the groin and fasten the ischial strap.

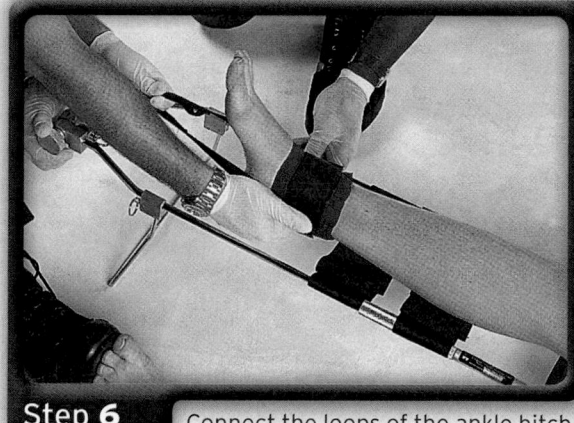

Step 6 Connect the loops of the ankle hitch to the end of the splint as your partner continues to maintain traction. Carefully tighten the ratchet to the point that the splint holds adequate traction.

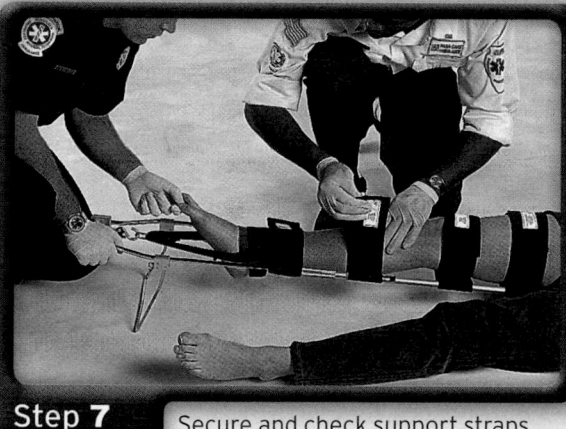

Step 7 Secure and check support straps. Assess pulse and motor and sensory functions.

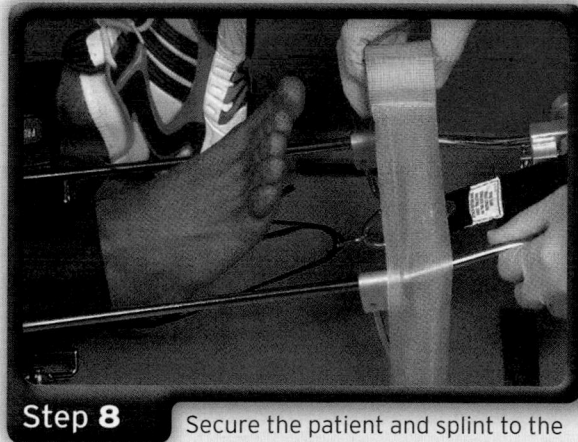

Step 8 Secure the patient and splint to the backboard in a way that will prevent movement of the splint during patient movement and transport.

Velcro support straps, which should be positioned at the midthigh, above the knee, below the knee, and above the ankle (Step 1).

3. Manually support and stabilize the injured limb so that no motion will occur at the fracture site while your partner fastens the appropriate-sized ankle hitch about the patient's ankle and foot. Normally, the patient's shoe is removed for this procedure (Step 2).

4. Support the leg at the site of the suspected injury while the your partner manually applies gentle longitudinal traction to the ankle hitch and foot. Use only enough force to align (reposition) the limb so that it will fit into the splint; do not attempt to align the fracture fragments anatomically (Step 3).

5. Slide the splint into position under the patient's injured limb, making certain that the ring is seated well on the ischial tuberosity (Step 4).

6. Pad the groin area, and gently apply the ischial strap (Step 5).

7. While the your partner continues to maintain traction, connect the loops of the ankle hitch to the end of the splint. Then apply gentle traction to the connecting strap between the ankle hitch and the splint, just strongly enough to maintain limb alignment. Use caution. This splint comes with a ratchet mechanism to tighten the strap, which can overstretch the limb and further injure the patient. Adequate traction has been applied when the leg is the same length as the other leg or the patient feels relief (Step 6).

8. Once proper traction has been applied, fasten the support straps so that the limb is securely held in the splint. Check all proximal and distal support straps to make sure they are secure (Step 7).

9. At this point, reassess distal pulses and motor function and sensation.

10. Place the patient securely on a long backboard for transport to the emergency department. You may need to load the patient feet first into the ambulance so that you do not shut the door against the splint (Step 8).

Because this traction splint stabilizes the limb by producing countertraction on the ischium and in the groin, use care to pad these areas well. You must avoid excessive pressure on the external genitalia. Always use commercially available padded ankle hitches rather than pieces of rope, cord, or tape. Such improvised hitches can sometimes be painful and can potentially obstruct circulation in the foot.

The Sager splint is lightweight and easy to store and applies a measurable amount of traction. Best of all, you can apply it by yourself when necessary. As with any splint, in addition to knowing the precise sequence of steps to apply the splint properly, you must practice the splinting technique frequently to maintain the necessary skills. Follow the steps below to apply a Sager splint (Skill Drill 29-8):

Skill Drill 29-8

1. Expose the injured extremity. Using standard precautions as needed, assess and record the pulse, motor function, and sensation distal to the injury.

2. Before applying the splint, adjust the thigh strap so that it will lie anteriorly when secured in place (Step 1).

3. Estimate the proper splint length by placing it alongside the uninjured limb, so that the wheel is at the level of the heel.

Words of Wisdom

Reel Splint
The Reel splint is a new traction splint that is being used by the military. Many devices used in the battlefield eventually appear in the ambulance and are used by EMTs in the field. This splint is designed to be used on a lower extremity (Figure 29-27).

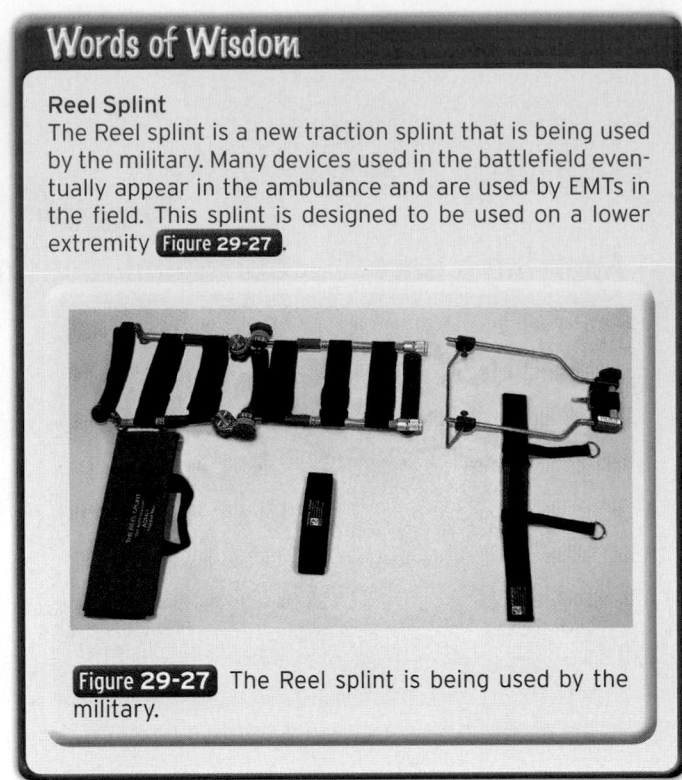

Figure 29-27 The Reel splint is being used by the military.

4. Arrange the ankle pads to fit the size of the patient's ankle (Step 2).

5. Place the splint along the inner aspect of the limb, and slide the thigh strap around the upper thigh so that the perineal cushion is snug against the groin and the ischial tuberosity. Tighten the thigh strap snugly (Step 3).

6. Secure the ankle harness tightly around the patient's ankle just above the malleoli.

7. Pull the cable ring snugly up against the bottom of the foot (Step 4).

8. Pull out the inner shaft of the splint to apply traction of approximately 10% of body weight, using a maximum of 15 lb (Step 5).

9. Secure the limb to the splint using elasticized cravat bandages (Step 6).

10. Secure the patient to a long backboard.

11. Check pulse and motor and sensory functions (Step 7).

Pelvic Binder

Pelvic binders are used to splint the bony pelvis to reduce hemorrhage from bone ends, venous disruption, and pain (Figure 29-28). Pelvic binders are meant to provide temporary stabilization until definitive stabilization can be achieved. Generally, pelvic binders should be light, made of soft material and easily applied by one person, and should allow access to the abdomen, perineum,

Skill Drill 29-8

Applying a Sager Traction Splint

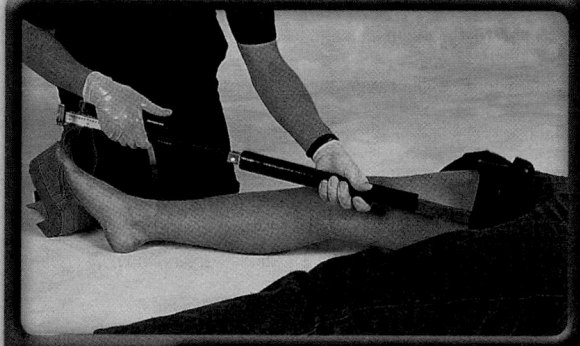

Step 1
After exposing the injured area, check the patient's pulse and motor and sensory functions. Adjust the thigh strap so that it lies anteriorly when secured.

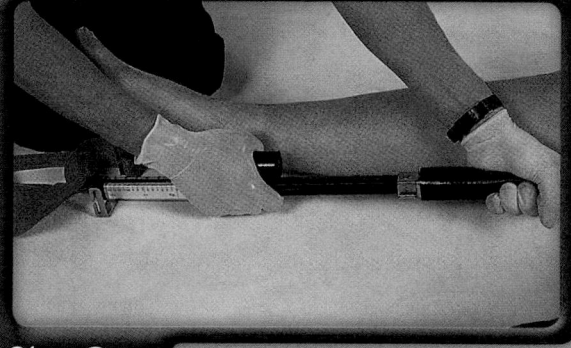

Step 2
Estimate the proper length of the splint by placing it next to the uninjured limb. Fit the ankle pads to the ankle.

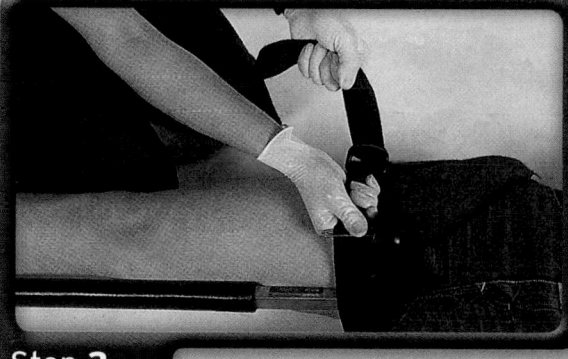

Step 3
Place the splint at the inner thigh, apply the thigh strap at the upper thigh, and secure snugly.

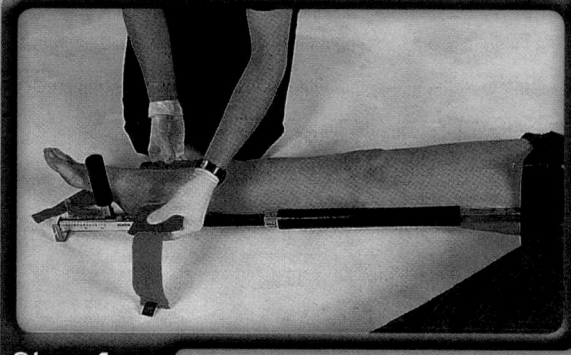

Step 4
Tighten the ankle harness just above the malleoli. Secure the cable ring against the bottom of the foot.

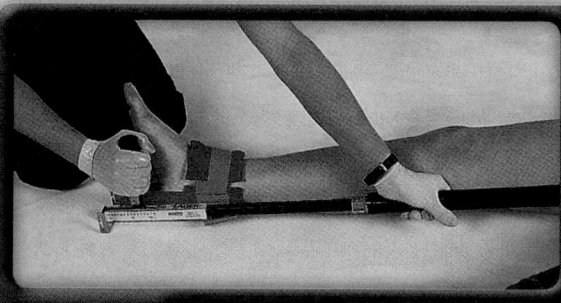

Step 5
Extend the splint's inner shaft to apply traction of about 10% of body weight.

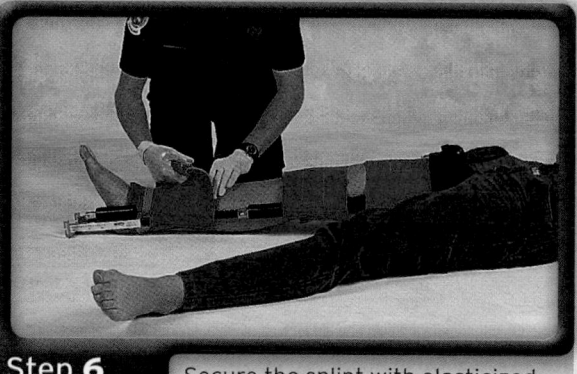

Step 6
Secure the splint with elasticized cravat bandages.

Skill Drill 29-8

Applying a Sager Traction Splint, continued

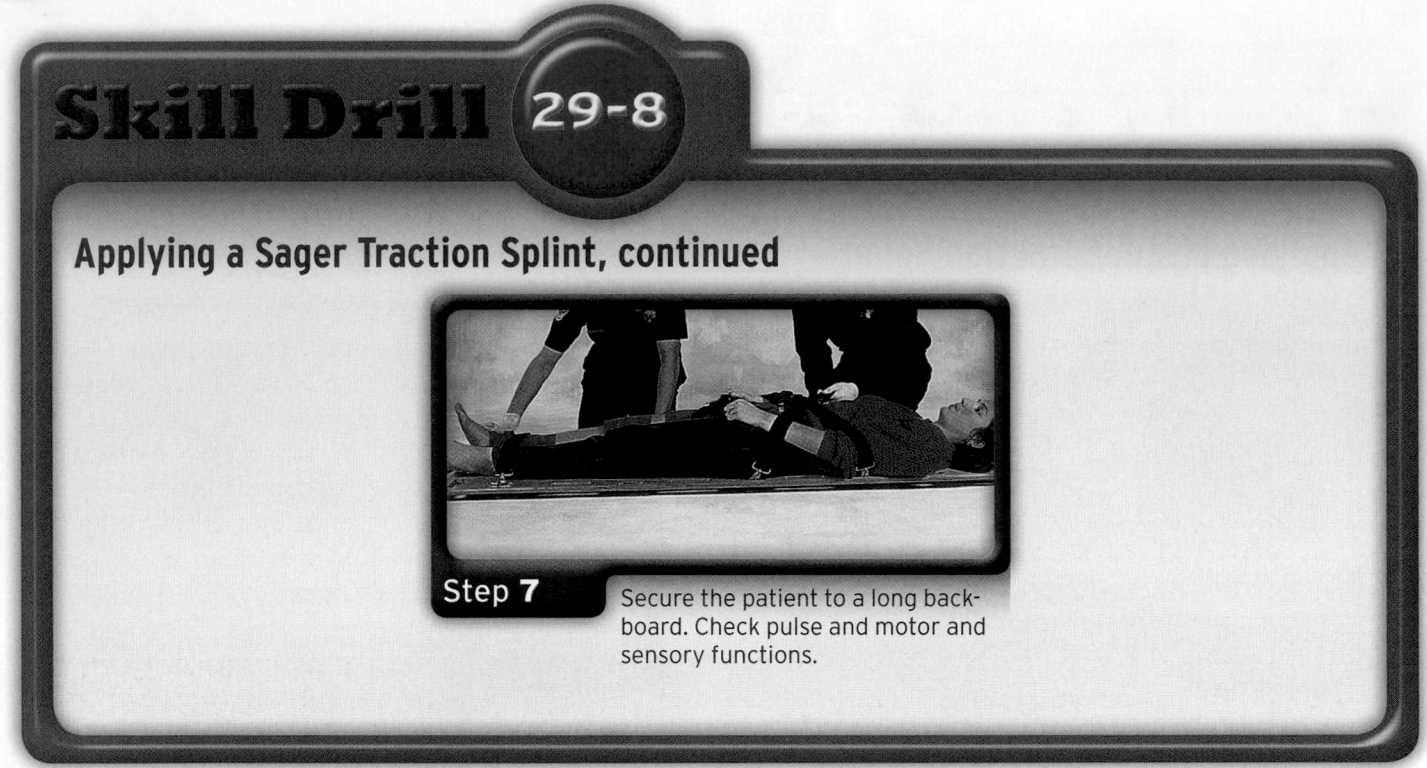

Step 7 Secure the patient to a long backboard. Check pulse and motor and sensory functions.

Figure 29-28 Pelvic binders are meant to provide temporary stabilization until definitive stabilization can be achieved.

anus, and groin for examination and diagnostic testing. Because there are various manufacturers of pelvic binder devices, you should be familiar with the manufacturer's instructions for your specific device.

Pneumatic Antishock Garments

If a patient has injuries to the lower extremities or pelvis, you may be able to use a pneumatic antishock garment (PASG) as a splinting device, if local protocol allows. Situations in which use of a PASG is allowed vary widely by locale. Many EMS systems no longer use this device because of problems reported with its use. Be sure to check with medical control in every case. The PASG is relatively contraindicated for treatment of shock but may have some value as a splinting device in rare circumstances.

Do not use the PASG if any of the following conditions exist:

- Pregnancy
- Pulmonary edema
- Acute heart failure
- Penetrating chest injuries
- Groin injuries
- Major head injuries
- A transport time of less than 30 minutes

Words of Wisdom

With a controversial therapy such as the pneumatic antishock garment, it is particularly important to seek direction from medical control if you have any doubts about a situation. EMS teams can learn how to maintain the flow of care while contacting medical control. Decide before your arrival on the scene who will direct patient care and who will handle radio communications.

In these situations, the PASG may worsen or complicate the patient's condition. Consult with medical control if you think prolonged use or use in unusual circumstances may be necessary. When applying the PASG, you should carefully inflate the device in increments. As a general rule, gradually inflate the legs of the PASG before inflating the abdominal portion. If you are using the device to stabilize a possible pelvic fracture, you must inflate all compartments. Always document all obvious injuries and deformities before application of the PASG. Follow these steps to apply the PASG **Skill Drill 29-9** :

Skill Drill 29-9

1. Apply the garment. If you will immobilize or move the patient on a backboard, lay the PASG out on the board before rolling the patient onto it. Position the top of the abdominal section of the PASG below the lowest rib to ensure that it does not compromise chest expansion **Step 1**.

2. Close and fasten both leg compartments and the abdominal compartment **Step 2**.

3. Open the stopcocks (valves) to the compartments you are preparing to inflate. Inflate all three compartments together **Step 3**.

4. Inflate the compartments with the foot pump. Do not increase the garment's pressure any more than necessary. A PASG is adequately inflated for stabilization when the splint feels firm **Step 4**.

5. Check the patient's blood pressure during inflation, and continue to monitor vital signs at least every 5 minutes afterward. Remember that the pressure gauges of the PASG measure the air pressure in the device. They do not reflect the patient's blood pressure. Be aware of temperature extremes and external pressure changes that can significantly affect the pressure exerted by the PASG, thus requiring frequent monitoring and adjustment **Step 5**.

Do not remove a PASG in the field. It must be deflated gradually in the hospital under careful supervision by a physician. Before turning your patient over to hospital personnel, report the patient's blood pressure, the time you applied the PASG, and the results.

Hazards of Improper Splinting

You must be aware of the hazards associated with the improper application of splints, including the following:

- Compression of nerves, tissues, and blood vessels
- Delay in transport of a patient with a life-threatening injury
- Reduction of distal circulation
- Aggravation of the injury
- Injury to tissue, nerves, blood vessels, or muscles as a result of excessive movement of the bone or joint

■ Transportation

Once an injured limb is adequately splinted, the patient is ready to be transferred to a backboard or stretcher and transported.

Very few, if any, musculoskeletal injuries justify the use of excessive speed during transport. The limb will be stable once a dressing and splint have been applied. However, a patient with a pulseless limb must be given a higher priority. Still, if the hospital is only a few minutes away, speeding to the emergency department will make little or no difference to the patient's eventual outcome. If the treatment facility is an hour or more away, a patient with a pulseless limb should be transported by helicopter or immediate ground transportation. If circulation in the distal limb is impaired, always notify medical control so that proper steps can be taken quickly once the patient arrives in the emergency department.

Specific Musculoskeletal Injuries

■ Injuries of the Clavicle and Scapula

The clavicle, or collarbone, is one of the most commonly fractured bones in the body. Fractures of the clavicle occur most often in children when they fall on an outstretched hand. They can also occur with crushing injuries of the chest. A patient with a fracture of the clavicle will report pain in the shoulder and will usually hold the arm across the front of his or her body **Figure 29-29**. A young child often reports pain throughout the entire

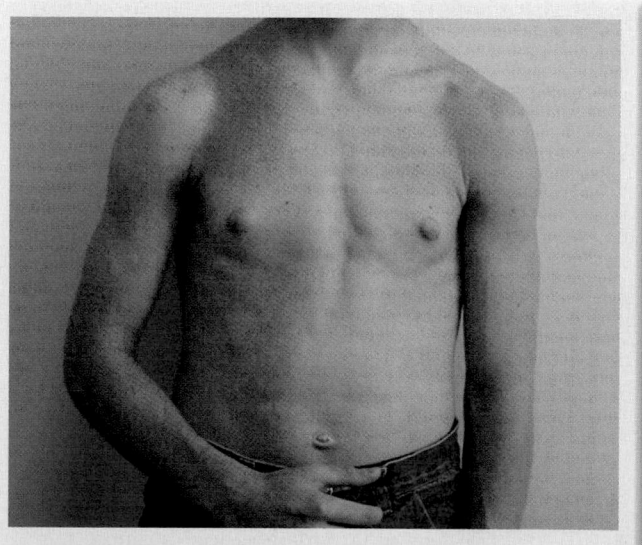

Figure 29-29 A patient with a fracture of the clavicle will usually hold the arm across the front of his or her body.

Skill Drill 29-9

Applying a Pneumatic Antishock Garment (PASG)

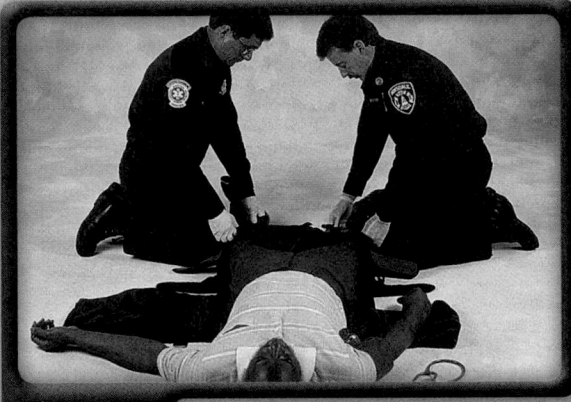

Step 1 Apply the garment so that the top is below the lowest rib.

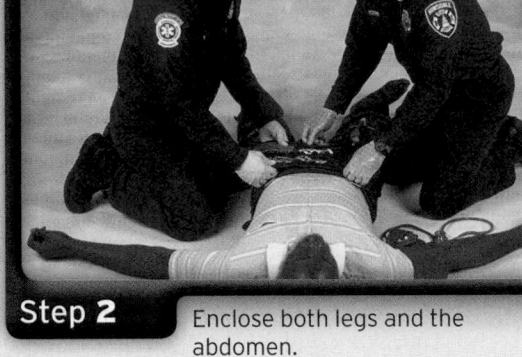

Step 2 Enclose both legs and the abdomen.

Step 3 Open the stopcocks to the compartments you are preparing to inflate.

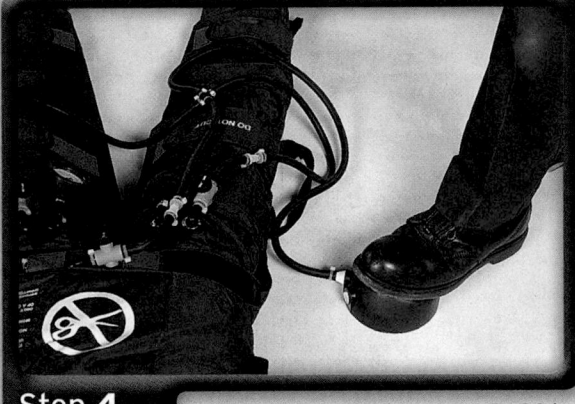

Step 4 Inflate with the foot pump. A PASG is adequately inflated for stabilization when the splint feels firm.

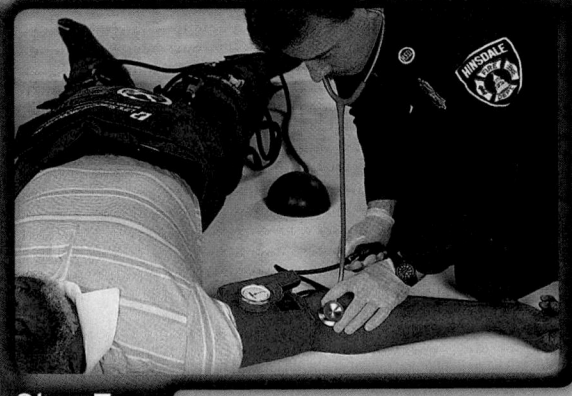

Step 5 Check the patient's blood pressure again. Monitor the vital signs.

arm and is unwilling to use any part of that limb. These complaints may make it difficult to localize the point of injury, but, generally, swelling and point tenderness occur over the clavicle. Because the clavicle is subcutaneous (just beneath the skin), the skin will occasionally "tent" over the fracture fragment. The clavicle lies directly over major arteries, veins, and nerves; therefore, fracture of the clavicle may lead to neurovascular compromise.

> ### Words of Wisdom
>
> Point tenderness is the most reliable indicator of an underlying fracture.

Fractures of the scapula, or shoulder blade, occur much less frequently because this bone is well protected by many large muscles. Fractures of the scapula are almost always the result of a forceful, direct blow to the back, directly over the scapula, which may also injure the thoracic cage, lungs, and heart. For this reason, you must carefully assess the patient for signs of breathing problems. Provide supplemental oxygen and prompt transport for patients who are having difficulty breathing. Remember, it is the associated chest injuries, not the fractured scapula itself, that pose the greatest threat of long-term disability.

Abrasions, contusions, and significant swelling may also occur, and the patient will often limit use of the arm because of pain at the fracture site. The scapula also has bony projections that may be fractured with a lesser degree of force.

The joint between the outer end of the clavicle and the acromion process of the scapula is called the **acromioclavicular (AC) joint**. This joint is frequently separated during football and hockey when a player falls and lands on the point of the shoulder, driving the scapula away from the outer end of the clavicle. This dislocation is often called an AC separation. The distal end of the clavicle will often stick out, and the patient will complain of pain, including point tenderness over the AC joint **Figure 29-30**.

Fractures of the clavicle and scapula and AC separations can all be splinted effectively with a sling and swathe. A **sling** is any bandage or material that helps support the weight of an injured upper extremity, relieving the downward pull of gravity on the injured site. To be effective, a sling must apply gentle upward support to the olecranon process of the ulna. The knot of the sling should be tied to one side of the neck so that it does not press uncomfortably on the cervical spine **Figure 29-31A**.

To fully stabilize the shoulder region, a **swathe**, a bandage that passes completely around the chest, must be used to bind the arm to the chest wall. The swathe should be tight enough to prevent the arm from swinging freely but not so tight as to compress the chest and compromise breathing. Leave the patient's fingers exposed so that you can assess neurovascular function at regular intervals **Figure 29-31B**.

Commercially available shoulder stabilizers or slings will provide adequate splinting for injuries of the shoulder region, as will triangular bandage slings.

■ Dislocation of the Shoulder

The glenohumeral joint (shoulder joint) is where the head of the humerus, the supporting bone of the upper arm, meets the **glenoid fossa** of the scapula. The glenoid fossa joins with the humeral head to form the glenohumeral joint. In shoulder dislocations, the humeral head most commonly dislocates anteriorly, coming to lie in front of the scapula as a result of forced abduction (away from the midline) and external rotation of the arm **Figure 29-32**.

Shoulder dislocations are extremely painful. The patient will guard the shoulder and try to protect it by holding the dislocated arm in a fixed position away from the chest wall **Figure 29-33**. The shoulder joint will usually be locked, and the shoulder will appear squared off or flattened. The humeral head will protrude anteriorly underneath the pectoralis major on the anterior chest wall. As a result, the axillary nerve may be compressed, causing a numb patch on the outer aspect of the shoulder. Be sure to document this finding. Some patients may

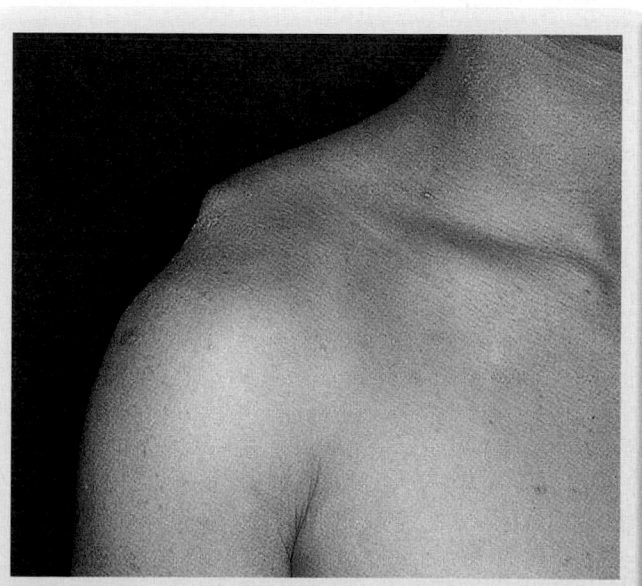

Figure 29-30 With acromioclavicular separations, the distal end of the clavicle usually sticks out.

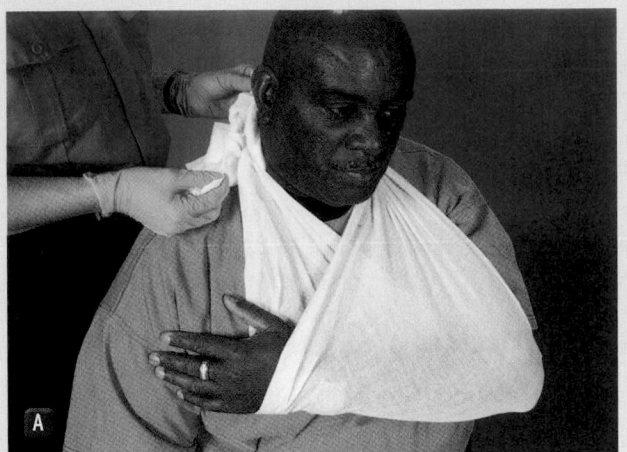

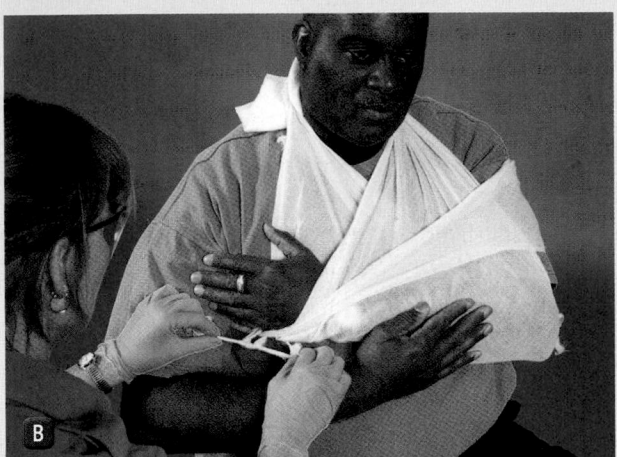

Figure 29-31 **A.** Apply the sling so that the knot is tied to one side of the neck. **B.** Bind the arm to the chest wall with a swathe so that the arm cannot swing freely. Leave the patient's fingers exposed so that you can assess distal circulation.

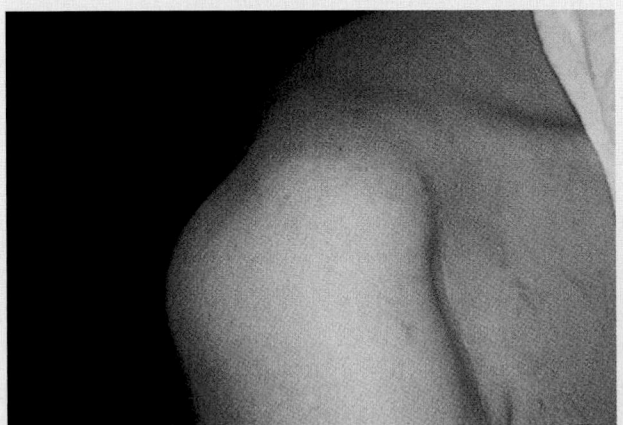

Figure 29-32 Most shoulder dislocations are anterior. Note the absence of the normal rounded appearance of the shoulder.

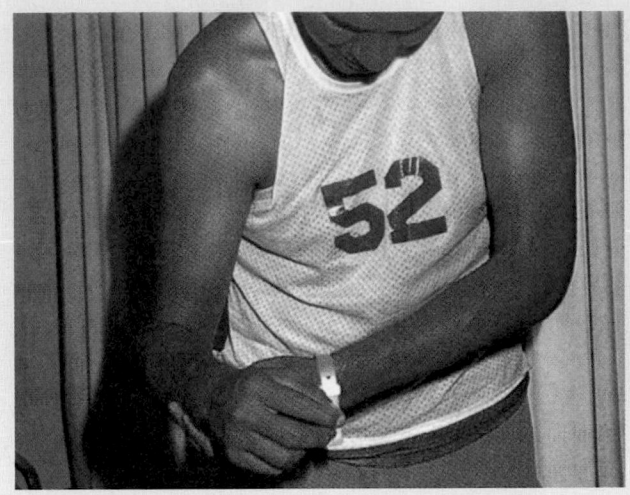

Figure 29-33 A patient with a dislocated shoulder will guard the shoulder, trying to protect it by holding the arm in a fixed position away from the chest wall.

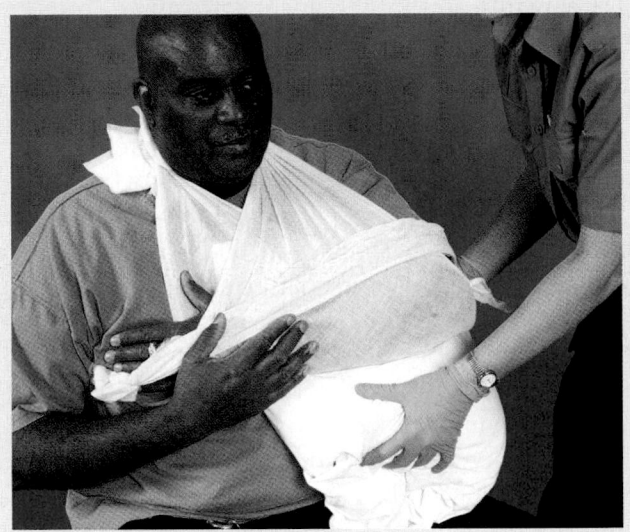

Figure 29-34 Splint the shoulder joint in a position of comfort, and place a pillow or towel between the arm and the chest wall to stabilize the arm, after which the elbow can be flexed to 90°. Apply a sling, and secure the arm to the chest with a swathe.

also report some numbness in the hand because of either nervous or circulatory compromise.

Stabilizing an anterior shoulder dislocation is difficult because any attempt to bring the arm in toward the chest will produce pain. You must splint the joint in whatever position is most comfortable for the patient. If necessary, place a pillow or rolled blankets or towels between the arm and chest to fill up the space between them **Figure 29-34**. Once the arm has been stabilized in this way, the elbow can usually be flexed to 90° without causing further pain. At this point, you can apply a sling to the forearm and wrist to support the weight of the arm. Finally, secure the arm in the sling to the pillow and chest with a swathe. Transport the patient in a seated or semiseated position.

Dislocation of the shoulder disrupts the supporting ligaments of the anterior aspect of the shoulder. Often, these ligaments fail to heal properly, so dislocation recurs, each time causing further neurovascular compromise and joint injury. In certain cases, surgical repair may be required. Some patients are able to reduce (set) their own dislocated shoulders. Generally, however, this maneuver must be done in a hospital setting and only after x-ray films have been obtained.

Words of Wisdom

When you are assessing a patient with a possible shoulder dislocation, position yourself behind the patient and compare the shoulders. The dislocated side is usually lower than the uninjured side.

Posterior dislocation is less common than anterior shoulder dislocation. Football players, especially linemen, are susceptible to this injury. The arm will often be locked in adduction (toward the midline), so it cannot be rotated. Reducing the dislocation usually requires medical supervision.

■ Fracture of the Humerus

Fractures of the humerus occur either proximally, in the midshaft, or distally at the elbow **Table 29-2**. Fractures of the proximal humerus resulting from falls are common among older people. Fractures of the midshaft occur more often in young patients, usually as the result of a violent injury.

With any severely angulated fracture, you should consider applying traction to realign the fracture fragments before splinting them. Check your local protocols for indications and techniques for applying traction to a severely angulated fracture. Support the site of the fracture with one hand, and with the other hand, grasp the two humeral condyles (its lateral and medial protrusions) just above the elbow. Pull gently in line with the normal axis of the limb **Figure 29-35**. Once you achieve gross realignment of the limb, splint the arm with a sling and swathe, supplemented by a padded board splint on the lateral aspect of the arm **Figure 29-36**. If the patient reports significant pain or resists gentle traction, splint

Table 29-2 Characteristics and Treatment of Fractures of the Humerus

Type	Characteristics	Treatment
Proximal humeral fractures	■ Significant swelling, but no deformity of the upper arm ■ Neurovascular compromise ■ Any or all of the brachial plexus affected, depending on the degree of displacement ■ Concurrent soft-tissue injuries ■ Possible rotator cuff injury (If x-ray films show no fracture, a tear of the rotator cuff is possible, especially if the patient cannot move the arm toward the medial plane.)	■ Stabilize in a sling and swathe or a shoulder stabilizer. ■ Use the chest wall as a splint, and secure the injured arm to the chest wall. ■ Place a short, padded board splint on the lateral side of the arm under the sling and swathe for additional support.
Midshaft fractures	■ Gross angulation of the arm ■ Marked instability and crepitus of fracture fragments ■ Possible neurovascular compromise ■ Possible entrapment of the radial nerve (The patient cannot extend or dorsiflex the wrist or fingers and may report numbness on the dorsum of the hand; classic "wrist drop.")	■ Stabilize with a sling and swathe or a shoulder stabilizer. ■ Use the chest wall as a splint, and secure the injured arm to the chest wall. ■ Place a short, padded board splint on the lateral side of the arm under the sling and swathe for additional support.
Distal humeral fractures	■ Significant swelling at the elbow ■ Possible neurovascular compromise ■ Possible injury to the ulnar or median nerve (Document nerve status before and after any attempt to reduce the fracture.)	■ Stabilize in a splint, in addition to a sling and swathe or a shoulder stabilizer.

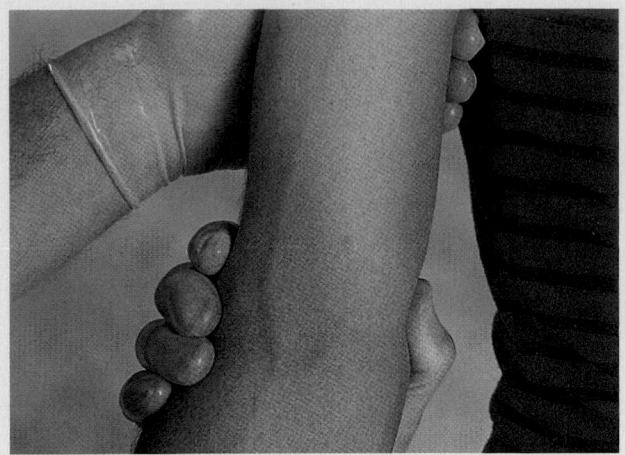

Figure 29-35 To align a severe deformity associated with a humeral shaft fracture, apply gentle pressure to the humeral condyles, as shown in this uninjured arm.

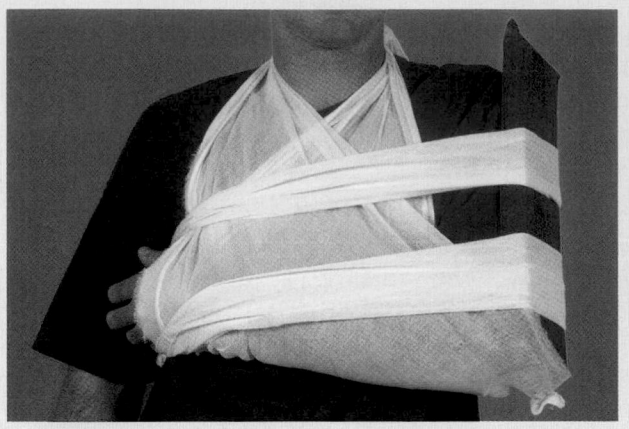

Figure 29-36 Splint a humeral shaft fracture with a sling and swathe supplemented by a padded board splint on the lateral aspect of the arm.

the fracture in the deformed position with a padded wire ladder or a padded board splint, using pillows to support the injured limb. Note that compartment syndrome can develop in the forearm in children with these fractures.

■ Elbow Injuries

Fractures and dislocations often occur around the elbow, and the different types of injuries are difficult to distinguish without x-ray examinations. However, they all produce similar limb deformities and require the same

emergency care. Injuries to nerves and blood vessels are quite common in this region. Such injuries can be caused or worsened by inappropriate emergency care, particularly by excessive manipulation of the injured joint.

Fracture of the Distal Humerus

This type of fracture, also known as a supracondylar or intercondylar fracture, is common in children. Frequently, the fracture fragments rotate significantly, producing deformity and causing injuries to nearby vessels and nerves. Swelling occurs rapidly and is often severe.

You are the Provider: PART 4

Your partner offers supplemental oxygen to the patient, but she does not want it; she only wants you to make her leg feel better. As the nurse continues to manually stabilize the patient's leg, you retrieve the splinting supplies from the ambulance, and your partner reassesses her vital signs.

Recording Time: 13 Minutes	
Level of consciousness	Conscious and alert
Respirations	22 breaths/min; adequate depth
Pulse	110 beats/min; strong and regular
Skin	Pink, warm, and moist
Blood pressure	134/80 mm Hg
Sao$_2$	99% (on room air)

7. How should you splint this patient's injury?

8. What are some methods for providing pain relief from orthopaedic trauma?

Dislocation of the Elbow

This type of injury typically occurs in athletes and rarely in young children. It can occur in toddlers when they are lifted or pulled by the arm. The ulna and radius are most often displaced posteriorly. The ulna, the bone on the small finger side of the forearm, and the radius, the bone on the thumb side of the forearm, both join the distal humerus. The posterior displacement makes the olecranon process of the ulna much more prominent Figure 29-37 . The joint is usually locked, with the forearm moderately flexed on the arm; this position makes any attempt at motion extremely painful. As with a fracture of the distal humerus, there is swelling and significant potential for vessel or nerve injury.

Elbow Joint Sprain

This diagnosis is often mistakenly applied to an occult, nondisplaced fracture.

Fracture of the Olecranon Process of the Ulna

This fracture can result from direct or indirect forces and is often associated with lacerations and abrasions. The patient will be unable to actively extend the elbow.

Fractures of the Radial Head

Often missed during diagnosis, this fracture generally occurs as a result of a fall on an outstretched arm or a direct blow to the lateral aspect of the elbow. Attempts to rotate the elbow or wrist cause discomfort.

Care of Elbow Injuries

All elbow injuries are potentially serious and require careful management. Always assess distal neurovascular functions periodically in patients with elbow injuries. If you find strong pulses and good capillary refill, splint the elbow injury in the position in which you found it, adding a wrist sling if this seems helpful. Two padded board splints, one applied to each side of the limb and secured with soft roller bandages, usually are enough to stabilize the arm Figure 29-38A . Make sure the board extends from the shoulder joint to the wrist joint, stabilizing the entire bone above and below the injured joint. Alternatively, you can mold a padded wire ladder splint or a SAM splint to the shape of the limb Figure 29-38B . If necessary, you may add further support to the limb with a pillow.

A cold, pale hand or a weak or absent pulse and poor capillary refill indicate that the blood vessels have likely been injured. Further care of this patient must be dictated by a physician. Notify medical control immediately.

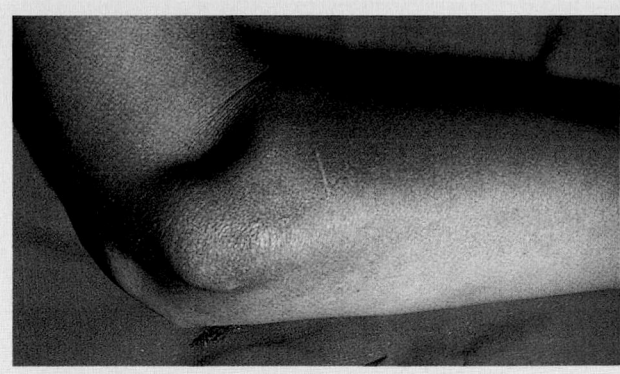

Figure 29-37 Posterior dislocation of the elbow makes the olecranon process of the ulna much more prominent.

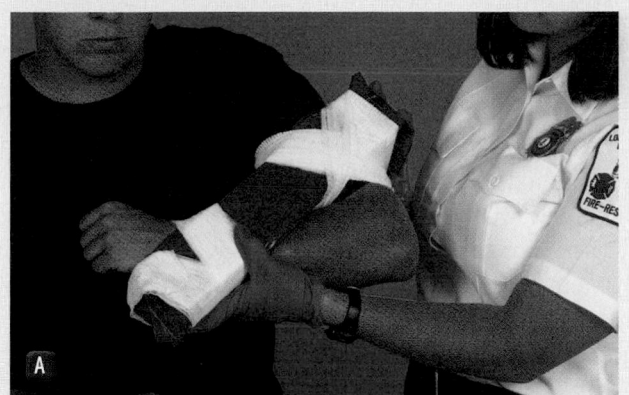

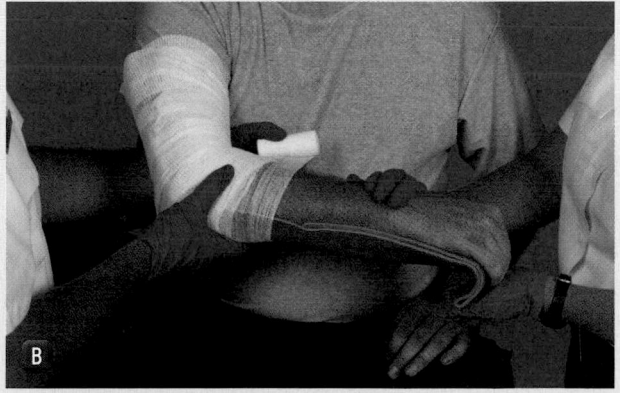

Figure 29-38 A. Two padded board splints provide adequate stabilization for an injured elbow. B. A structural aluminum malleable splint can be molded to the shape of the limb so that you can splint it in the position in which it was found.

If you are within 10 to 15 minutes of the hospital, splint the limb in the position in which you found it, and provide prompt transport. Otherwise, medical control may direct you to try to realign the limb to improve circulation in the hand.

Special Populations

Growth plate injuries in children are common, especially around the wrist, elbow, knee, and ankle. Injuries tend to occur through these cartilaginous growth centers because they are inherently weaker than the surrounding bone. Since longitudinal growth of the limb is dependent upon the function of the growth plate, it is extremely important to recognize the possibility of growth plate injuries, stabilize the injured limb, and transport the patient in timely manner to an appropriate center with pediatric, pediatric orthopaedic, and pediatric surgical coverage. Proper functioning of the injured growth plate throughout the remainder of skeletal growth may depend on timely anatomic reduction of the fracture and close follow up by an orthopaedist.

Any deformity close to a joint in children younger than 16 years should be assumed to be a growth plate injury, and the patient should be transported and treated appropriately.

If the limb is pulseless and significantly deformed at the elbow, apply gentle manual traction in line with the long axis of the limb to decrease the deformity. This maneuver may restore the pulse. Be careful, because excessive manipulation may only worsen the vascular problem. If no pulse returns after one attempt, splint the limb in the most comfortable position for the patient. If the pulse is restored by gentle longitudinal traction, splint the limb in whatever position allows the strongest pulse. Provide prompt transport for all patients with impaired distal circulation.

■ Fractures of the Forearm

Fractures of the shaft of the radius and ulna are common in people of all age groups but are seen most often in children and older people. Usually, both bones break at the same time when the injury is the result of a fall on an outstretched hand **Figure 29-39**. An isolated fracture

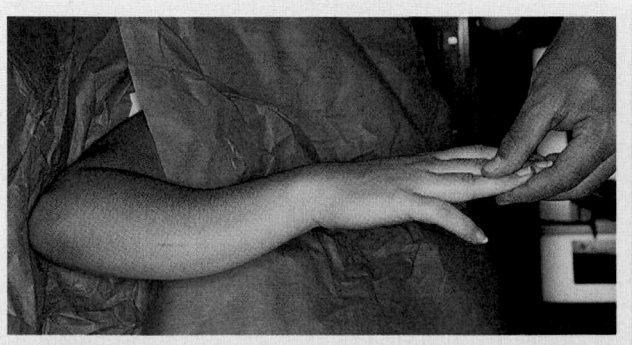

Figure 29-39 Fractures of the forearm often occur in children as a result of a fall on an outstretched hand.

of the shaft of the ulna may occur as the result of a direct blow to it; this is known as a nightstick fracture.

Fractures of the distal radius, which are especially common in elderly patients with osteoporosis, are often known as Colles fractures. The term "silver fork deformity" is used to describe the distinctive appearance of the patient's arm **Figure 29-40**. In children, this fracture may occur through the growth plate and can have long-term consequences.

To stabilize fractures of the forearm or wrist, you can use a padded board, air, vacuum, or pillow splint. If the shaft of the bone has been fractured, be sure to include the elbow joint in the splint. Splinting of the elbow joint is not essential with fractures near the wrist; however, the patient will be more comfortable if you add a sling or pillow for more support. If possible, elevate the injured extremity above the heart to help alleviate swelling.

■ Injuries of the Wrist and Hand

Injuries of the wrist, ranging from dislocations to sprains, must be confirmed by x-ray examination. Dislocations are usually associated with a fracture, resulting in a fracture dislocation. Another common wrist injury is the isolated, nondisplaced fracture of a carpal bone, especially

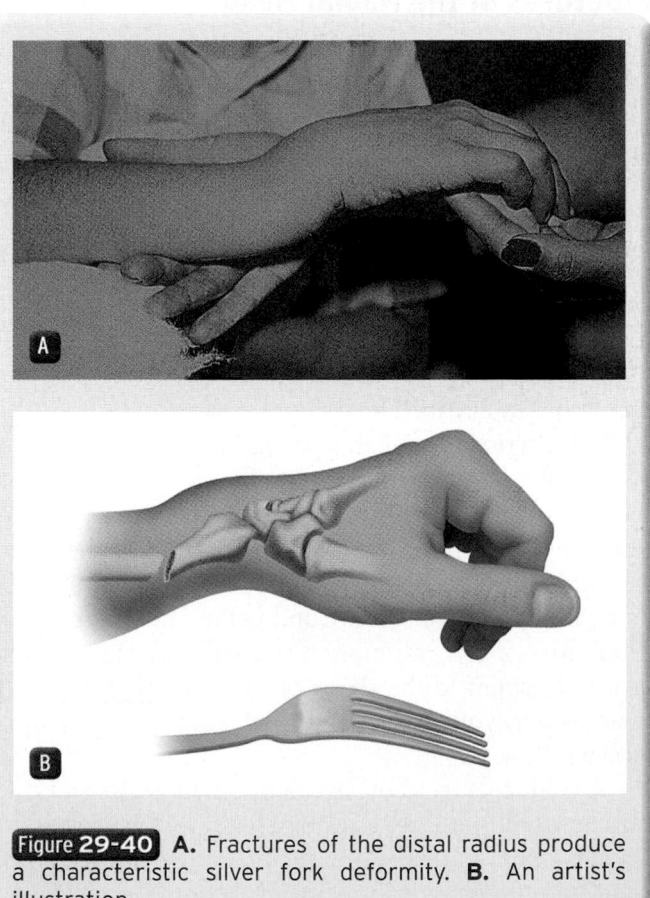

Figure 29-40 **A.** Fractures of the distal radius produce a characteristic silver fork deformity. **B.** An artist's illustration.

the scaphoid. Any questionable wrist sprain or fracture should be splinted and evaluated in the emergency department or an orthopaedic surgeon's office.

Hand injuries vary widely, some with potentially serious consequences. Industrial, recreational, and home accidents often result in dislocations, fractures, lacerations, burns, and amputations. Because the fingers and hands are required to function in such intricate ways, any injury that is not treated properly may result in permanent disability, as well as deformity. For this reason, all injuries to the hand, including simple lacerations, should be evaluated by a physician. For example, you should not attempt to "pop" a dislocated finger joint back in place

 **Figure 29-41**. Always take any amputated parts to the hospital with the patient. Be sure to wrap the amputated part in a dry or moist sterile dressing, depending on your local protocol, and place it in a dry plastic bag. Put the bag in a cooled container; do not soak the part in water or allow it to freeze.

A bulky forearm dressing makes an effective splint for any hand or wrist injury. Follow the steps in **Skill Drill 29-10**:

Skill Drill 29-10

1. Follow standard precautions.
2. Cover all wounds with a dry, sterile dressing.

Skill Drill 29-10

Splinting the Hand and Wrist

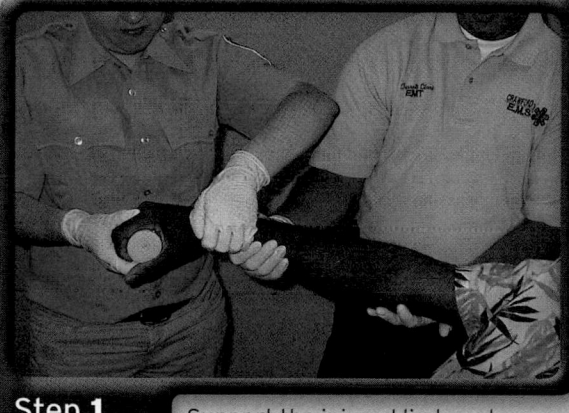

Step 1 Support the injured limb and move the hand into the position of function. Place a soft roller bandage in the palm.

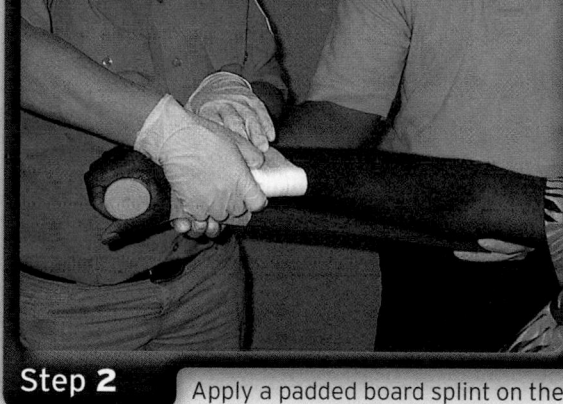

Step 2 Apply a padded board splint on the palmar side with fingers exposed.

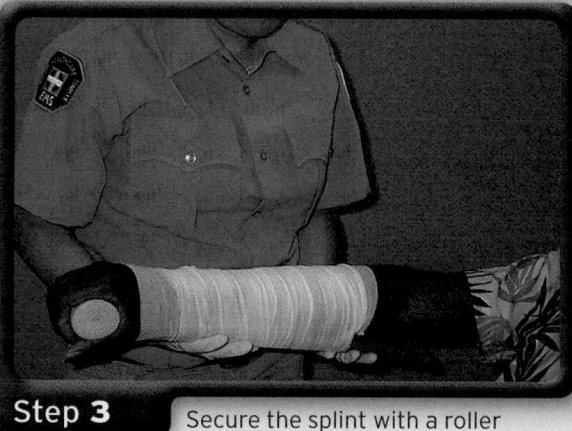

Step 3 Secure the splint with a roller bandage.

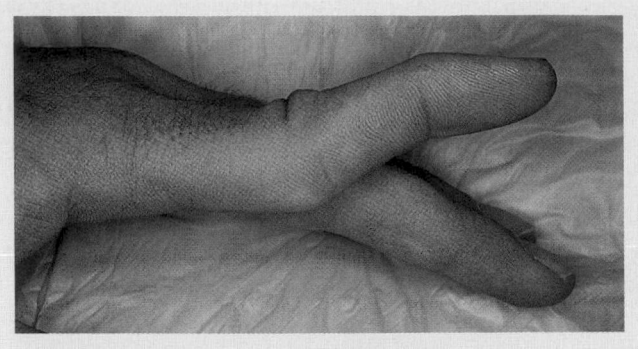

Figure 29-41 Dislocation of the finger joint. Do not be tempted to try to "pop" the joint back into place.

3. Supporting the injured limb, form the injured hand into the **position of function**, with the wrist slightly bent down and all finger joints moderately flexed. This is the position that is used to hold a can most comfortably.
4. Place a soft roller bandage into the palm of the hand (Step 1).
5. Apply a padded board splint to the palmar side of the wrist, leaving the fingers exposed (Step 2).
6. Secure the entire length of the splint with a soft roller bandage (Step 3).

7. Apply a sling and swathe, or prop the splinted hand and wrist on a pillow or on the patient's chest during transport to the hospital.

■ Fractures of the Pelvis

Fracture of the pelvis often results from direct compression in the form of a heavy blow that literally crushes the pelvis. The blow may be from a motor vehicle crash, a weapon, a falling object, or a fall from a height. Injuries to the pelvis can also be caused by indirect forces. For example, when the knee strikes the dashboard in an automobile crash, the impact of the force is transmitted along the line of the femur (the thigh bone), which is the longest and largest bone in the body. The head of the femur is driven into the pelvis, causing it to fracture. However, not all pelvic fractures result from violent trauma. Even a simple fall can produce a fracture of the pelvis, especially in older people with osteoporosis.

Fractures of the pelvis may be accompanied by life-threatening loss of blood from the laceration of blood vessels affixed to the pelvis at certain key points. Up to several liters of blood may drain into the pelvic space and the **retroperitoneal space**, which lies between the abdominal cavity and the posterior abdominal wall. The result is significant hypotension, shock, and sometimes death. For this reason, you must take immediate steps to treat shock, even if there is only minimal swelling. Often,

You are the Provider: PART 5

After properly splinting the patient's leg, you place her onto the stretcher, load her into the ambulance, and begin transport to the hospital. You reassess her condition and vital signs en route and note that her condition remains stable. You call your radio report in to the receiving facility; your estimated time of arrival is 8 minutes.

Recording Time: 23 Minutes	
Level of consciousness	Conscious and alert
Respirations	20 breaths/min; adequate depth
Pulse	115 beats/min; strong and regular
Skin	Pink, warm, and dry
Blood pressure	128/76 mm Hg
Sao$_2$	98% (on room air)

During transport, the patient begins to complain of numbness and tingling in her left foot. Your reassessment reveals that her pedal pulse is weaker than it was before and that her foot looks pale and feels cool.

9. **What is the most likely cause of the patient's complaint? Is there anything you can do to remedy the situation?**

10. **What factors increase the risk of complications following orthopaedic trauma?**

there are no visible signs of bleeding until severe blood loss has occurred. You should be prepared to resuscitate the patient rapidly if this becomes necessary.

Because the pelvis is surrounded by heavy muscle, open fractures of the pelvis are quite uncommon. However, pelvis fracture fragments can lacerate the rectum and vagina, creating an open fracture that is often overlooked. Once the protective pelvic ring is broken, the structures it is designed to protect, including the urinary bladder, are more susceptible to injury. The bladder may be lacerated by pelvic bone fragments, but more often, it tears or ruptures as a result of tension on either the bladder or the urethra.

You should suspect a fracture of the pelvis in any patient who has sustained a high-velocity injury and complains of discomfort in the lower back or abdomen. Because the area is covered by heavy muscle and other soft tissue, deformity or swelling may be very difficult to see. The most reliable sign of fracture of the pelvis is simple tenderness or instability on firm compression and palpation. Firm compression on the two iliac crests will produce pain at a fracture site in the pelvic ring. Assess for tenderness by taking the following steps **Figure 29-42**:

1. Place the palms of your hands over the lateral aspect of each iliac crest, and apply firm but gentle inward pressure on the pelvic ring.
2. With the patient lying supine, place a palm over the anterior aspect of each iliac crest, and apply firm downward pressure.
3. Use the palm of your hand to firmly but gently palpate the pubic symphysis, the firm cartilaginous joint between the two pubic bones. This area will

be tender if there is injury to the anterior portion of the pelvic ring.

If there has been injury to the bladder or the urethra, the patient will have lower abdominal tenderness and may have evidence of **hematuria** (blood in the urine) or blood at the urethral opening.

Perform the primary assessment, and carefully monitor the general condition of any patient whom you suspect has a pelvic fracture, because he or she is at high risk for hypovolemic shock. Patients in stable condition can be secured to a long backboard or a scoop stretcher to stabilize isolated fractures of the pelvis.

■ Dislocation of the Hip

The hip joint is a very stable ball-and-socket joint that dislocates only after significant injury. Most dislocations of the hip are posterior. The femoral head is displaced posteriorly to lie in the muscles of the buttock. Posterior dislocation of the hip most commonly occurs as a result of automobile accidents in which the knee meets with a direct force, such as the dashboard, and the entire femur is driven posteriorly, dislocating the hip joint **Figure 29-43**. Thus, you should suspect a hip dislocation in any patient who has been in an automobile crash and has a contusion, laceration, or obvious fracture in the knee region. Very rarely does the femoral head dislocate anteriorly; in this circumstance, the legs are suddenly and forcibly spread wide apart and locked in this position.

Posterior dislocation of the hip is frequently complicated by injury to the sciatic nerve, which is located

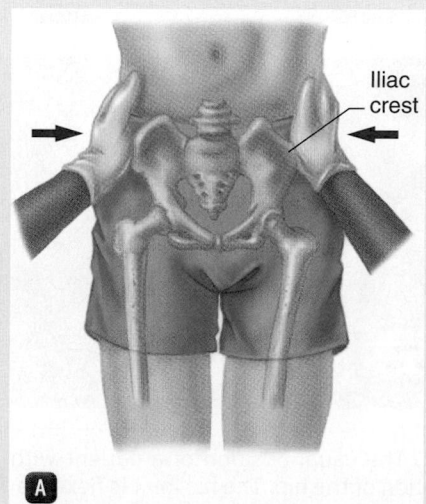
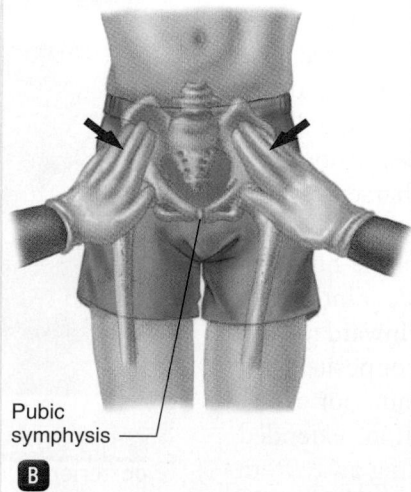
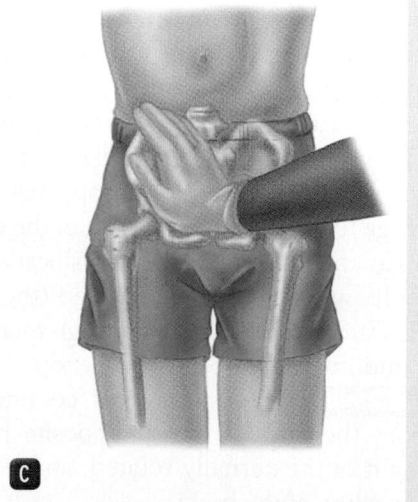

Figure 29-42 **A.** To assess for tenderness or instability in the pelvic region, place your hands over the lateral aspect of each iliac crest, and gently compress the pelvis. **B.** With the patient in a supine position, place your palms over the anterior aspect of each iliac crest, and apply firm but gentle downward pressure. **C.** Palpate the pubic symphysis with the palm of your hand.

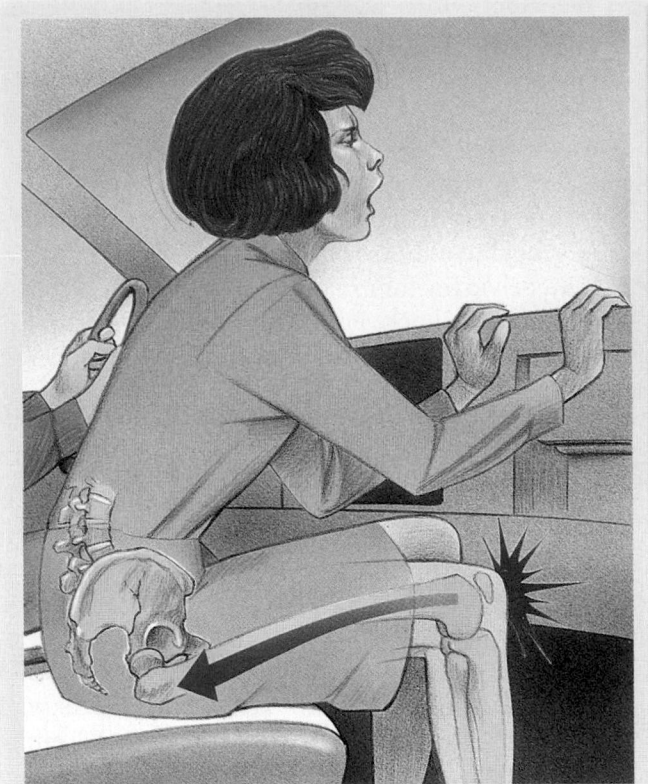

Figure 29-43 Posterior dislocation of the hip can occur as a result of the knee hitting the dashboard in an automobile crash. The impact drives the femur posteriorly (see arrow), dislocating the joint.

directly behind the hip joint. The **sciatic nerve** is the most important nerve in the lower extremity; it controls the activity of muscles in the posterior thigh and below the knee and the sensation in most of the leg and foot. When the head of the femur is forced out of the hip socket, it may compress or stretch the sciatic nerve, leading to partial or complete paralysis of the nerve. The result is decreased sensation in the leg and foot and frequently weakness in the foot muscles. Generally, only the dorsiflexors, the muscles that raise the toes or foot, are involved, causing the "foot drop" that is characteristic of damage to the peroneal portion of the sciatic nerve.

Patients with a posterior dislocation of the hip typically lie with the hip joint flexed (the knee joint drawn up toward the chest) and the thigh rotated inward toward the midline of the body over the top of the opposite thigh **Figure 29-44A**. With the less common anterior dislocation, the limb is in the opposite position, extended straight out, externally rotated, and pointing away from the midline of the body.

Dislocation of the hip is associated with very distinctive signs. The patient will have severe pain in the hip and will strongly resist any attempt to move the joint. The lateral and posterior aspects of the hip region will be tender

on palpation. With some thin patients, you can palpate the femoral head deep within the muscles of the buttock. Check for a sciatic nerve injury by carefully assessing sensation and motor function in the lower extremity. Occasionally, sciatic nerve function will be normal at first and then slowly diminish.

As with any other extremity injury, you should make no attempt to reduce the dislocated hip in the field. Splint the dislocation in the position of the deformity, and place the patient supine on a long backboard. Support the affected limb with pillows and rolled blankets, particularly under the flexed knee **Figure 29-44B**. Then secure the entire limb to the backboard with long straps so that the hip region will not move. Be sure to provide prompt transport.

■ Fractures of the Proximal Femur

Fractures of the proximal (upper) end of the femur are common fractures, especially in older people. Although they are usually called hip fractures, they rarely involve the hip joint. Instead, the break goes through the neck of the femur, the intertrochanteric (middle) region, or

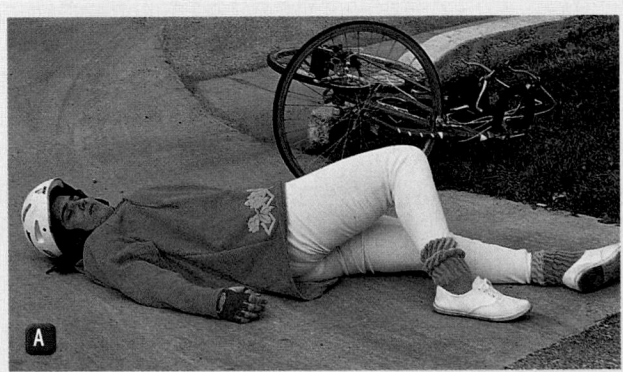

Figure 29-44 **A.** The usual position of a patient with a posterior dislocation of the hip. The hip joint is flexed, and the thigh is rotated inward and adducted across the midline of the body. **B.** Support the affected limb with pillows and blankets, particularly under the flexed knee. Secure the entire limb to a long board with long straps to prevent movement during transport.

across the proximal shaft of the femur (subtrochanteric fractures). Although these three fracture types occur most often in older patients, particularly patients with osteoporosis, they may also be seen as a result of high-energy injuries in younger patients.

Patients with displaced fractures of the proximal femur display a very characteristic deformity. They lie with the leg externally rotated, and the injured leg is usually shorter than the opposite, uninjured limb. When the fracture is not displaced, this deformity is not present. With any kind of hip fracture, patients typically are unable to walk or move the leg because of pain in the hip region or in the groin or inner aspect of the thigh. The hip region is usually tender on palpation, and gentle rolling of the leg will cause pain but will not do further damage. On occasion, the pain is referred to the knee, and it is not uncommon for a geriatric patient with a hip fracture to complain of knee pain after a fall. You should assess the pelvis for any soft-tissue injury and bandage appropriately. In addition, assess pulses and motor and sensory functions, looking for signs of vascular and nerve damage. Once your assessment is complete, you should splint the lower extremity of an older patient who has fallen and complains of pain in either the hip or the knee, even if there is no deformity, and then transport the patient to the emergency department.

The age of the patient and the severity of the injury will dictate how you splint the fracture. A geriatric patient with an isolated hip fracture does not require a traction splint. You can effectively stabilize such a fracture by placing the patient on a long backboard or scoop stretcher, using pillows or rolled blankets to support the injured limb in the deformed position. Then secure the injured limb carefully to the stretcher with long straps.

All patients with hip fractures may lose significant amounts of blood. Therefore, you should treat with high-flow oxygen and monitor vital signs frequently, being alert for signs of shock.

■ Femoral Shaft Fractures

Fractures of the femur can occur in any part of the shaft, from the hip region to the femoral condyles just above the knee joint. Following a fracture, the large muscles of the thigh spasm in an attempt to "splint" the unstable limb. The muscle spasm often produces significant deformity of the limb, with severe angulation or external rotation at the fracture site. Usually, the limb also shortens significantly. Fractures of the femoral shaft may be open, and fragments of bone may protrude through the skin. As with any other open fracture, never attempt to push the bone(s) back into the skin.

There is often a significant amount of blood loss, as much as 500 to 1,000 mL, after a fracture of the shaft of the femur. With open fractures, the amount of blood loss may be even greater. Thus, it is not unusual for hypovolemic shock to develop. Handle patients with these fractures with extreme care because any extra movement or fracture manipulation may increase the amount of blood loss.

Because of the severe deformity that occurs with these fractures, bone fragments may penetrate or press on important nerves and vessels and produce significant damage. For this reason, you must carefully and periodically assess the distal neurovascular function in patients who have sustained a fracture of the femoral shaft. Remove the clothing from the affected limb so that you can adequately inspect the injury site for any open wounds. Remember to follow standard precautions when any blood or body fluids are present. Monitor the patient's vital signs closely, and continue to watch for the onset of hypovolemic shock. You must provide immediate transport in this situation.

Cover any wound with a dry, sterile dressing. If the foot or leg below the level of the fracture shows signs of impaired circulation (is pale, cold, or pulseless), apply gentle longitudinal traction to the deformed limb in line with the long axis of the limb. Gradually turn the leg from the deformed position to restore the limb's overall alignment. Often, this restores or improves circulation to the foot. If it does not, the patient may have sustained a serious vascular injury and may be in need of prompt medical attention.

A fracture of the femoral shaft is best stabilized with a traction splint, such as a Sager splint.

■ Injuries of Knee Ligaments

The knee is very vulnerable to injury; therefore, many different types of injuries occur in this region. Ligament injuries, for example, range from mild sprains to complete dislocation of the joint. The patella can also dislocate. In addition, all the bony elements of the knee (distal femur, upper tibia, and patella) can fracture.

The knee is especially susceptible to ligament injuries, which occur when abnormal bending or twisting forces are applied to the joint. Such injuries are often seen in both recreational and competitive athletes. The ligaments on the medial side of the knee are the ones that are most frequently injured, typically when the foot is fixed to the ground and the lateral aspect of the knee is struck by a heavy object, such as when a football player is clipped or tackled from the side.

Usually, a patient with a knee ligament injury will report pain in the joint and be unable to use the extremity normally. When you examine the patient, you will generally find swelling, occasional ecchymosis, point tenderness at the injury site, and a joint effusion (excess fluid in the joint).

You should splint all suspected knee ligament injuries. The splint should extend from the hip joint to the foot, stabilizing the bone above the injured joint (the femur) and the bone below it (the tibia). A variety of splints can be used, including a padded, rigid, long leg splint or two padded board splints securely applied to the medial and lateral aspects of the limb. A long backboard, a pillow splint, or simply binding the injured limb to its uninjured mate are acceptable but less effective splinting techniques. The patient will usually be able to straighten the knee to allow you to apply the splint. However, if you encounter resistance or pain when trying to straighten the knee, splint it in the flexed position. Then continue to monitor the distal neurovascular function until the patient reaches the hospital.

■ Dislocation of the Knee

Dislocations of the knee are true emergencies that may threaten the limb. When the knee is dislocated, the ligaments that provide support to it may be damaged or torn. When this happens, the proximal end of the tibia completely displaces from its juncture with the lower end of the femur, usually producing a significant deformity. Although substantial ligament damage always occurs with a knee dislocation, the more urgent injury is often to the popliteal artery, which is frequently lacerated or compressed by the displaced tibia. When gross deformity, severe pain, and an inability to move the joint cause you to suspect a dislocation of the knee, always check the distal circulation carefully before taking any other step. If the distal pulses are absent, contact medical control immediately for further stabilization instructions.

The direction of dislocation refers to the position of the tibia with respect to the femur. Anterior knee dislocations, which result from extreme hyperextension of the knee, are the most common, occurring in almost half of all cases. Commonly, the anterior and posterior ligaments are damaged, but there is also a high risk of injury to the popliteal artery.

In posterior dislocations, a direct blow to the knee forces the tibia to shift posteriorly. There is also the possibility of damage to the ligaments and injury to the popliteal artery.

Medial dislocations result from a direct blow to the lateral part of the leg. Because the deforming force causes the medial aspect of the knee to stretch apart, there is a high likelihood of injury to the medial ligaments. When the force is applied from the medial direction, a lateral dislocation occurs and the lateral part of the knee is stretched apart, injuring the lateral ligament. Lateral and medial dislocations happen less commonly and are less likely to injure the popliteal artery.

Patients with a knee dislocation will typically complain of pain in the knee and report that the knee "gave out." If the knee did not spontaneously reduce, there may be evidence of significant deformity and decreased range of motion. Complications may include limb-threatening popliteal artery disruption, injuries to the nerves, and joint instability. Do not confuse this injury with a relatively minor patella dislocation.

If adequate distal pulses are present, splint the knee in the position in which you found it, and transport the patient promptly. Do not attempt to manipulate or straighten any severe knee injury if there are good distal pulses. If the limb is straight, apply standard rigid long leg splints to at least two sides of the limb to stabilize it **Figure 29-45A** . If the knee is bent and the foot has a good pulse, splint the joint in the bent position, using parallel padded board splints secured at the hip and ankle joint to provide a stable A-frame **Figure 29-45B** . Secure the limb to a backboard or stretcher with pillows and straps to eliminate any motion during transport.

On rare occasions, medical control may instruct you to realign a deformed, pulseless limb to reduce compression of the popliteal artery and, thus, restore distal circulation. You should make only one attempt to do this. First, straighten the limb by applying gentle longitudinal traction in the axis of the limb. Once you apply manual traction, maintain it until the limb is fully splinted; otherwise, the limb will return to its deformed position. If traction significantly increases the patient's pain, do not continue. As you apply traction, monitor the posterior tibial pulse to see whether it returns. Splint the limb in the position in which you feel the strongest pulse. If you are unable to restore the distal pulse, splint the limb in the position that is most comfortable for the patient, and then provide prompt transport to the hospital. Notify medical control of the status of the distal pulse so that arrangements to treat the patient can be made in advance.

■ Fractures About the Knee

Fractures about the knee may occur at the distal end of the femur, at the proximal end of the tibia, or in the patella. Because of local tenderness and swelling, it is easy to confuse a nondisplaced or minimally displaced fracture about the knee with a ligament injury. Likewise, a displaced fracture about the knee may produce significant deformity that makes it look like a dislocation. Management of the two types of injuries is as follows:

- If there is an adequate distal pulse and no significant deformity, splint the limb with the knee straight.
- If there is an adequate pulse and significant deformity, splint the joint in the position of deformity.

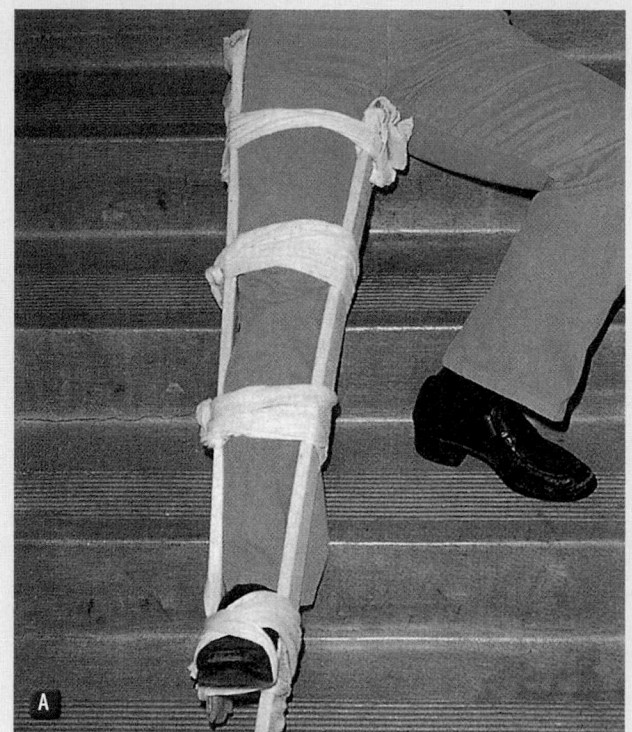

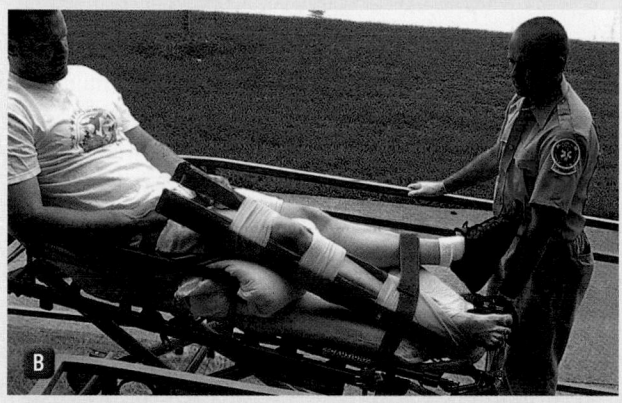

Figure 29-45 **A.** When the injured knee is straight, apply padded board splints extending from the hip to the ankle. **B.** If the knee is flexed and the foot has good pulses, apply padded board splints with the knee in the flexed position.

- If the pulse is absent below the level of the injury, suspect possible vascular and nerve damage, and contact medical control immediately for further instructions.
- Never use a traction splint if you suspect a fractured knee.

Dislocation of the Patella

A dislocated patella most commonly occurs in teenagers and young adults who are engaged in athletic activities. Some patients have recurrent dislocations of the patella. As with recurrent dislocation of the shoulder, a minor twisting may be enough to produce the problem. Usually, the dislocated patella displaces to the lateral side. The displacement of the patella produces a significant deformity in which the knee is held in a moderately flexed position, and the patella is displaced to the lateral side of the knee **Figure 29-46**.

Splint the knee in the position in which you found it; most often, this is with the knee flexed to a moderate degree. To stabilize the knee, apply padded board splints to the medial and lateral aspects of the joint, extending from the hip to the ankle. Use pillows to support the limb on the stretcher.

Occasionally, as you apply the splint, the patella will return to its normal position spontaneously. When this occurs, stabilize the limb as for a knee ligament injury in a padded long leg splint. The patient still needs to be transported to the emergency department. Report the spontaneous reduction as soon as you arrive at the hospital so that the medical staff is aware of the severity of the injury.

■ Injuries of the Tibia and Fibula

The <u>tibia</u> (shinbone) is the larger of the two leg bones that are responsible for supporting the major weight-bearing surface of the knee and ankle; the <u>fibula</u> is the smaller of them. Fracture of the shaft of the tibia or the fibula may occur at any place between the knee joint and the ankle joint. Usually, both bones fracture at the same time. Even a single fracture may result in severe deformity, with significant angulation or rotation. Because the tibia is located just beneath the skin, open fractures of this bone are quite common **Figure 29-47**.

Fractures of the tibia and fibula should be stabilized with a padded, rigid long leg splint or an air splint that extends from the foot to the upper thigh. Once splinted, the affected leg should be secured to the opposite leg. Traction splints are not indicated for isolated tibial

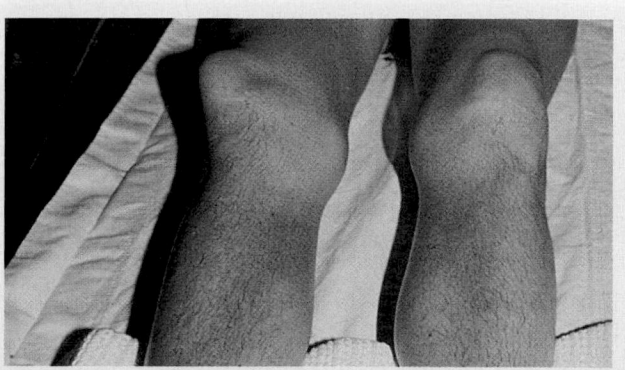

Figure 29-46 Usually, the dislocated patella displaces to the lateral side, and the knee is held in a partially flexed position.

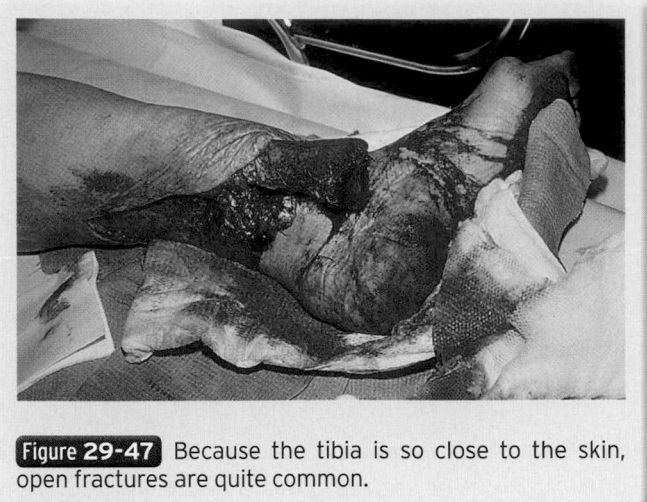

Figure 29-47 Because the tibia is so close to the skin, open fractures are quite common.

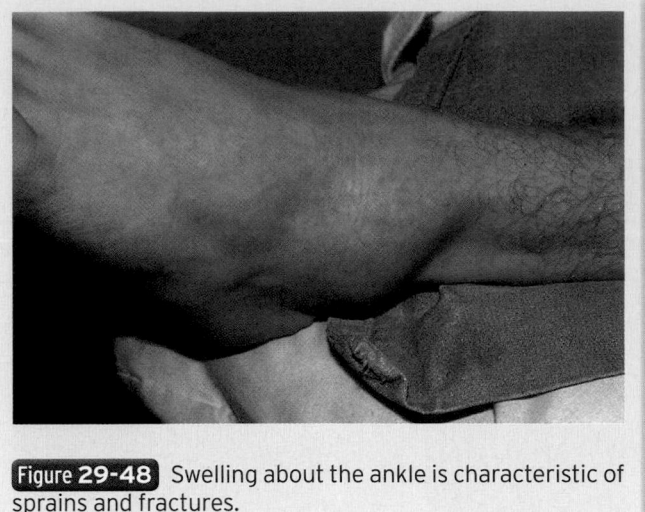

Figure 29-48 Swelling about the ankle is characteristic of sprains and fractures.

fractures. As with most other fractures of the shaft of long bones, you should correct severe deformity before splinting by applying gentle longitudinal traction. The goal here is to restore a position that will take a standard splint; it is not necessary to replace the fracture fragments in their anatomic position.

Fractures of the tibia and fibula are sometimes associated with vascular injury as a result of the distorted position of the limb following injury. Realigning the limb frequently restores an adequate blood supply to the foot. If it does not, transport the patient promptly and notify medical control while you are en route.

■ Ankle Injuries

The ankle is a very commonly injured joint. Ankle injuries occur in people of all ages and range in severity from a simple sprain, which heals after a few days' rest, to severe fracture-dislocations. As with other joints, it is sometimes difficult to tell a nondisplaced ankle fracture from a simple sprain without x-ray examination **Figure 29-48**. Therefore, any ankle injury that produces pain, swelling, localized tenderness, or the inability to bear weight must be evaluated by a physician. The most frequent mechanism of ankle injury is twisting, which stretches or tears the supporting ligaments. A more extensive twisting force may result in fracture of one or both malleoli. Dislocation of the ankle is usually associated with fractures of one or both malleoli.

You can manage the wide spectrum of injuries to the ankle in the same way, as follows:

1. Dress all open wounds.
2. Assess distal neurovascular function.
3. Correct any gross deformity by applying gentle longitudinal traction to the heel.
4. Before releasing traction, apply a splint.

You can use a padded rigid splint, an air splint, or a pillow splint. Just make sure it includes the entire foot and extends up the leg to the level of the knee joint.

■ Foot Injuries

Injuries to the foot can result in the dislocation or fracture of one or more of the tarsals, metatarsals, or phalanges of the toes. Toe fractures are especially common.

Of the tarsal bones, the calcaneus, or heel bone, is the most frequently fractured. Injury often occurs when the patient falls or jumps from a height and lands directly on the heel. The force of injury compresses the calcaneus, producing immediate swelling and ecchymosis. If the force of impact is great enough, as from a fall from a roof or tree, there may also be other fractures.

Frequently, the force of injury is transmitted up the legs to the spine, producing a fracture of the lumbar spine **Figure 29-49**. When a patient who has jumped or fallen from a height complains of heel pain, be sure to ask him or her about back pain and carefully check the spine for tenderness and deformity.

If you suspect that the foot is dislocated, immediately assess for pulses and motor and sensory functions. If pulses are present, immobilize the extremity using a commercially available splint or a pillow splint leaving the toes exposed so that you can periodically assess neurovascular function. If pulses are absent, contact medical control and discuss reduction of dislocation if the local scope of practice permits.

Injuries of the foot are associated with significant swelling but rarely with gross deformity. Vascular injuries are not common. As in the hand, lacerations about the ankle and foot may damage important underlying nerves and tendons. Puncture wounds of the foot are common and may cause serious infection if not treated

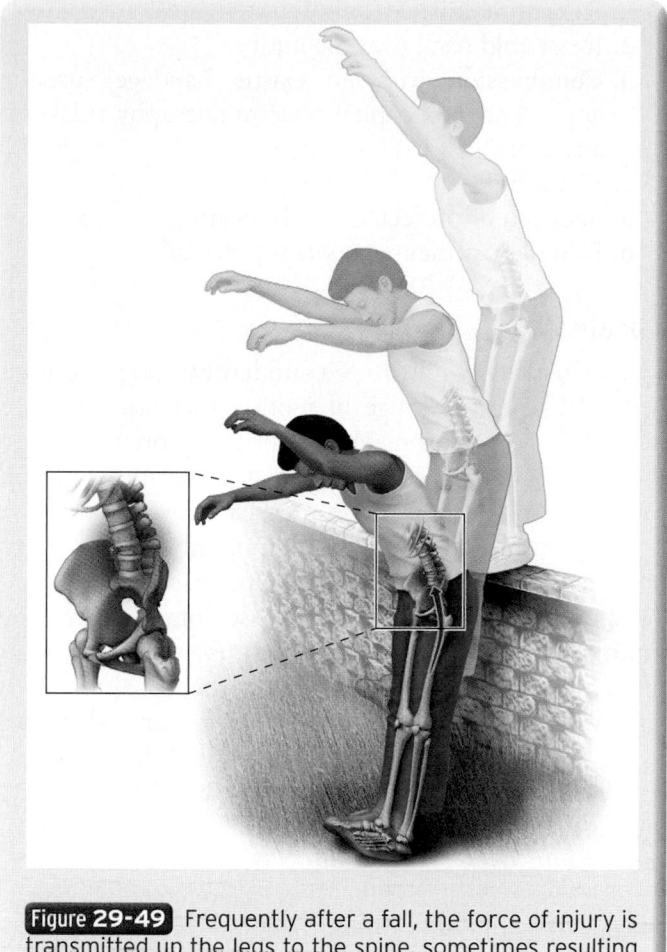

Figure 29-49 Frequently after a fall, the force of injury is transmitted up the legs to the spine, sometimes resulting in a fracture of the lumbar spine.

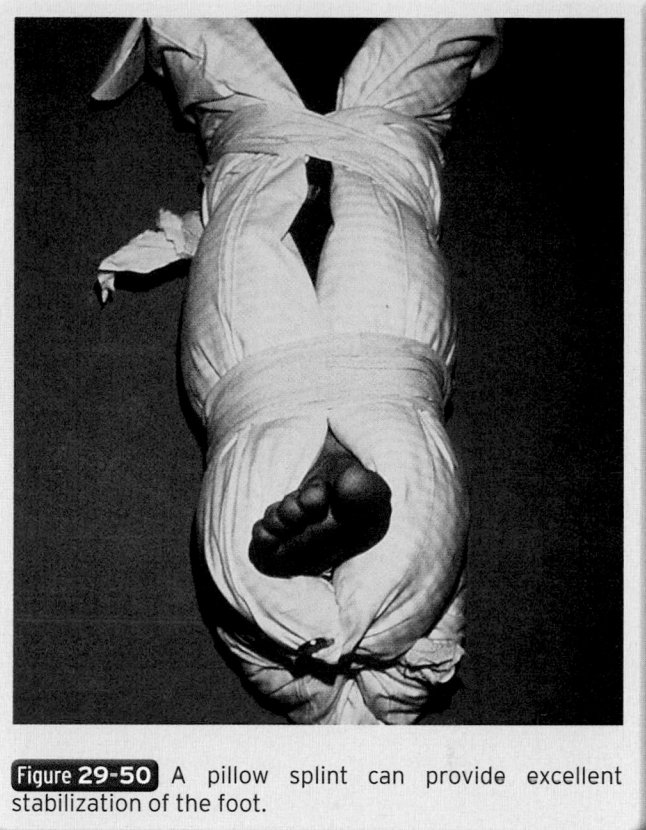

Figure 29-50 A pillow splint can provide excellent stabilization of the foot.

early. All of these injuries must be evaluated and treated by a physician.

To splint the foot, apply a rigid padded board splint, an air splint, or a pillow splint, stabilizing the ankle joint and the foot Figure 29-50. Leave the toes exposed so that you can periodically assess neurovascular function.

When the patient is lying on the stretcher, elevate the foot approximately 6″ to minimize swelling. All patients with lower extremity injuries should be transported in the supine position to allow for elevation of the limb. Never allow the foot and leg to dangle off the stretcher onto the floor or ground.

If a patient has fallen from a height and complains of heel pain, use a long backboard to stabilize any possible spinal injury in addition to splinting the foot.

■ Compartment Syndrome

If you have a pediatric patient with a fracture below the elbow or the knee, be on the lookout for these signs and symptoms: extreme pain, decreased pain sensation, pain on stretching of affected muscles, and decreased power. These are indicators that the pressure within a fascial

compartment is elevated. If you suspect that a patient has compartment syndrome, splint the affected limb, keeping it at the level of the heart, and provide immediate transport, reassessing neurovascular status frequently during transport. Compartment syndrome must be managed surgically.

■ Amputations

Surgeons today can occasionally reattach amputated parts Figure 29-51. However, correct prehospital care of the amputated part is vital to successful reattachment. With partial amputations, make sure to immobilize the part with bulky compression dressings and a splint to prevent further injury. Do not sever any partial amputations; this may complicate later reattachment. Hemorrhage from complete or incomplete amputations can be severe and life threatening. Control any bleeding to the stump. If bleeding cannot be controlled, you should quickly apply a tourniquet.

With a complete amputation, make sure to wrap the clean part in a sterile dressing and place it in a plastic bag. Follow your local protocols regarding how to preserve amputated parts. In some areas, dry sterile dressings are recommended for wrapping amputated parts; in other areas, dressings moistened with sterile saline are recommended. Put the bag in a cool container filled with

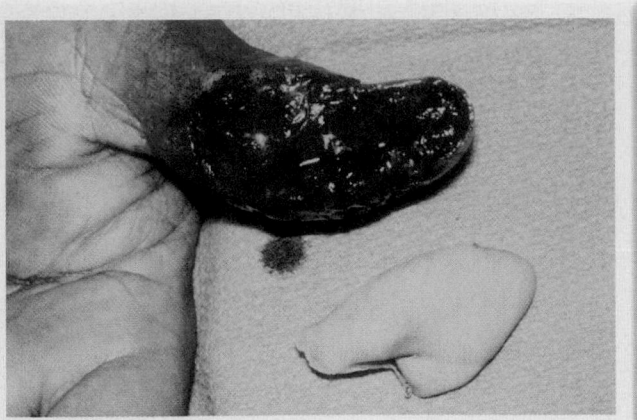

Figure 29-51 Amputated parts can occasionally be reimplanted, so you should make every attempt to find the part and transport it to the emergency department along with the patient.

ice. Lay the wrapped part on a bed of ice; do not pack it in ice. The goal is to keep the part cool without allowing it to freeze or develop frostbite. The amputated part should be transported with the patient to the appropriate resource hospital. Remember to care for the wound on the body side of the amputation also; control the bleeding, and apply an appropriate bandage.

■ Sprains and Strains

Strain

With a strain, often no deformity is present and only minor swelling is noted at the site of injury. Some patients may report a "snap" when a muscle tears and complain of increased sharp pain with passive movement of the injured extremity. Patients may complain of severe weakness of the muscle. Most patients also have extreme point tenderness. General treatment of strains is similar to that of fractures and includes the following (numbers 1 through 4 form the mnemonic RICE):

1. Rest; immobilize or splint injured area
2. Ice or cold pack over the injury
3. Compression with an elastic bandage (usually applied at the hospital once radiography rules out a fracture)
4. Elevation
5. Reduced or protected weight bearing
6. Pain management as soon as practical

Sprains

Sprains usually result from a sudden twisting of a joint beyond its normal range of motion that also causes a temporary subluxation. The majority of sprains involve the ankle or the knee because most occur after a person misjudges a step or landing. Evasive moves, like those done during a sporting event, commonly cause sprains in athletes. Sprains are typically characterized by pain, swelling (edema) at the joint, discoloration over the injured joint, and unwillingness to use the limb (point tenderness). Some patients might report hearing a snap when the injury occurred. Most patients also have extreme point tenderness. In contrast with fractures and dislocations, sprains usually do not involve deformity and joint mobility is usually limited by pain, not by joint incongruity.

Because it may be difficult to differentiate among the various types of injuries in the field, it is best to err on the side of caution and treat every severe sprain as if it is a fracture. General treatment of sprains is similar to that of fractures and includes the following:

1. Rest; immobilize or splint injured area
2. Ice or cold pack over the injury
3. Compression with an elastic bandage (usually applied at the hospital once radiography rules out a fracture)
4. Elevation
5. Reduced or protected weight bearing
6. Pain management as soon as practical

You are the Provider: SUMMARY

1. Under which circumstances can orthopaedic injuries pose a threat to a patient's life?

Although orthopaedic injuries can be extremely painful for the patient and can lead to varying degrees of temporary or permanent disability, most are not life threatening. Sprains and strains, which result in injury to the ligaments and muscles, respectively, are rarely life threatening. Dislocations—injuries that occur when two bones are completely separated from the joint capsule as a result of severe injury to the supporting ligaments—can cause neurovascular damage and permanent disability if not treated promptly, but typically do not pose a threat to life. Isolated closed fractures, such as fractures of the tibia or fibula or of the radius or ulna, can also result in neurovascular damage but are usually not life threatening; they are typically treated successfully, and the patient usually regains full use of the extremity.

There are certain orthopaedic injuries, usually fractures, that can pose a threat to the patient's life, and it is important for the EMT to recognize these types of injuries during patient assessment.

Multiple closed long-bone fractures, which can cause severe internal bleeding if bone fragments lacerate major blood vessels, can result in hypovolemic shock and death. Multiple long-bone fractures typically occur in the context of a significant mechanism of injury (ie, motor vehicle crash with ejection of the occupant). Bilateral femur fractures are an example of this type of injury.

Extremity amputations (excluding fingers and toes) can also result in hypovolemic shock due to severe external bleeding. Amputation of a leg or arm, above or below the joint, can quickly lead to exsanguination (bleeding to death) if not promptly treated. Limb amputations may require rapid application of a tourniquet.

Open fractures—especially to multiple long bones—can pose a threat to the patient's life for several reasons. The most immediate threat is severe external bleeding. Contamination of the wound could also lead to an infection of the underlying bone in muscle. In some patients, this infection could become systemic (sepsis), potentially resulting in septic shock and death. Infection is not seen during the acute phase of the fracture; it occurs within a few days after the injury.

Pelvic fractures are potentially life-threatening orthopaedic injuries because the pelvic cavity can accommodate a large volume of blood. Patients who die of pelvic fractures do so because hypovolemic shock occurs secondary to severe internal bleeding when a fractured bone fragment lacerates or severs a femoral artery or vein.

Although orthopaedic injuries are often obvious and quickly catch your eye, it is critical to remain focused on treating the most immediate life threat. Do not stay at the scene with a patient who has critical injuries (ie, of the head, chest, or abdomen) for the purpose of splinting fractures that are otherwise not life threatening. The saying "splinting to death" is used to describe such a situation; you are so busy splinting fractures that the patient dies of other injuries.

2. Given the information you have, can you rule out a critical injury?

As concise as the dispatch information can sometimes be, you will not know the extent of a patient's injury (or injuries) until you arrive at the scene and perform a patient assessment. In this case, all you know is that the patient is conscious and alert and breathing and has a possible leg fracture. While the dispatch information infers that this is an isolated injury secondary to a sports-related incident, you must keep an open mind and avoid approaching the patient with the preconceived notion that the leg injury is the only injury. Soccer can be a vicious sport, and, as with any other call, the possibility of other injuries cannot be ruled out by the dispatch information alone.

When you approach any patient, regardless of the nature of the call, you must perform a primary assessment in order to detect and correct immediate threats to airway, breathing, and circulation. Avoid tunnel vision when assessing and treating patients with orthopaedic trauma. A fractured leg may be the most obvious injury; however, it may not be the only injury. Furthermore, it may not be the most life-threatening injury.

3. What initial treatment should you provide to this patient?

Because the patient denies having any other injuries—for example, a neck injury, which may require spinal immobilization—your initial action, after taking standard precautions, should be to expose the injury site and then perform manual stabilization. With an injury in the tibia-fibula area, you or your partner should manually stabilize above the ankle and below the knee. Manual stabilization will help minimize the potential for further injury by preventing movement of the leg.

After you expose and manually stabilize the injury site, assess the patient for obvious signs of injury, such as swelling, deformity, bruising, and open wounds. The absence of deformity does not rule out an underlying fracture. Furthermore, swelling often masks underlying deformity. Treat any extremity injury as though an underlying fracture is present, and stabilize it appropriately.

4. What are some indicators of a fractured bone?

A fracture is defined as *any* break in the continuity of a bone and is classified as being open or closed. A closed fracture occurs when the bone is broken and the overlying skin remains intact. By contrast, an open fracture occurs when the overlying skin is not intact; this may be as obvious as broken bone ends protruding through the wound (compound fracture) or as subtle as an abrasion overlying the injury site.

Fractures are also described by whether the bone is moved from its normal position. A nondisplaced fracture (hairline fracture) is a simple crack in the bone that does not produce deformity, although swelling is often present. Nondisplaced fractures cannot be diagnosed in the field and require radiologic (eg, x-ray) confirmation. A displaced fracture occurs when the bone ends are no longer in alignment and produces actual deformity, or distortion, by shortening, rotating, or

You are the Provider: SUMMARY, continued

angulating the limb. In many cases, the deformity is grossly obvious; in other cases, it is very subtle.

Although the prehospital treatment for orthopaedic injuries—sprains, strains, fractures, and dislocations—is essentially the same, it is still important to conduct an adequate assessment of any patient with an extremity injury. Signs of a fracture include deformity, point tenderness, swelling, bruising, crepitus, and false motion.

A deformed limb may appear to be shortened, rotated, or angulated at a point where there is no joint (false motion). The presence of deformity and false motion are obvious indicators of a fracture. When you assess a limb for deformity, use the opposite limb for comparison, if possible.

Point tenderness on palpation in the zone of injury is the most reliable indicator of an underlying fracture, although it does not indicate the type of fracture. Point tenderness is the sensitive spot at the site of injury that can be located by palpation along the bone with the tip of your finger. The zone of injury is the area of soft tissue, including adjacent nerves and blood vessels, that surrounds the obvious injury of a bone or joint.

Rapid swelling usually indicates bleeding from a fracture and is typically followed by substantial pain. Often, if the swelling is severe, it may mask deformity of the limb.

Fractures are often associated with bruising (ecchymosis) of the surrounding soft tissue. Bruising may be present after almost any injury, although it may take several hours to develop. Redness is usually associated with acute injuries.

Crepitus, a grating or grinding sensation that can be felt or heard when fractured bone ends rub together, is another reliable indicator of an underlying fracture. However, you should not intentionally assess for crepitus while assessing an injured extremity, although crepitus is often noted if you must carefully manipulate the limb to facilitate splinting.

5. How should you proceed with your assessment of this patient's injury?

This patient did not experience multisystem trauma or a significant mechanism of injury; therefore, a full-body scan is not indicated. Instead, you should focus on evaluating perfusion and sensory and motor functions distal to the injury. When you are assessing an extremity injury, remember the 6 Ps of musculoskeletal assessment: pain, paralysis, paresthesias, pulselessness, pallor, and pressure.

Assess the patient's level of pain using a scale of 0 to 10, with 0 indicating no pain and 10 indicating the worst pain the patient has ever experienced. Pain assessment is important and should be repeated frequently, especially after splinting the injury and providing pain relief (eg, ice packs).

In some cases, a fractured bone end may compress or sever a blood vessel, resulting in inadequate or absent perfusion distal to the fracture. To assess perfusion, look at the area distal to the injury and compare the color of the skin with that of the uninjured limb. If perfusion is adequate, the skin should be pink and warm. Skin that is pale and/or cold suggests compromised perfusion. Palpate the dorsalis pedis pulse

(on the top of the foot) and the posterior tibial pulse (on the posterior aspect of the ankle). Pulses that are weak or absent in comparison with pulses in the uninjured limb also suggest compromised perfusion.

Use a blunt object, and stroke it up the bottom and sides of her foot. If she is unable to feel you touching her foot, you should suspect that a nerve has been compressed or possibly severed by a fractured bone end. The patient's ability to feel you touching the bottom and sides of her foot is a good indication that the nerve supply is intact. To test motor function, simply ask her to wiggle her toes; however, if this increases her pain, you should discontinue this part of the examination.

Paresthesias—a feeling of numbness or tingling—could indicate compromised perfusion and/or nerve injury. A feeling of pressure distal to the injury site could indicate elevated pressure within a fascial compartment due to internal bleeding; if this continues, it could lead to a condition called compartment syndrome. Although compartment syndrome typically develops within 6 to 12 hours after the injury, it is most common in fractures involving the tibia or fibula.

In addition to assessing perfusion and sensory and motor functions, you should assess the areas above and below the injury (the injury zone). On the basis of the location of the patient's injury, the injury zone extends above her knee and below her ankle. Remember, her leg may be the most obvious injury, but it may not be the only injury.

6. How should you treat an injured extremity in which distal perfusion is absent?

As a general rule, you should splint an orthopaedic injury in the position it was found, provided that distal perfusion is intact. In some cases, an injured extremity may be so severely angulated that gentle longitudinal traction may be required in order to splint the injury effectively, even if distal perfusion is adequate.

If your assessment reveals that perfusion distal to the injury is compromised or absent (ie, pallor, absent distal pulses, cold skin), apply gentle longitudinal traction in order to realign the limb until perfusion is restored. The goal is *not* to return the extremity to its normal anatomic position. The goal is to restore distal circulation. In many cases, gentle realignment of the limb restores adequate perfusion; however, if one attempt (local protocol may dictate more than one attempt) at realignment is unsuccessful, splint the injury, transport the patient as soon as possible, and notify the receiving facility early.

7. How should you splint this patient's injury?

Numerous techniques and devices can be used to splint an injured leg, and regardless of the technique or device used, the splint must protect and maintain the position of the injured extremity.

Splinting helps reduce pain and facilitates transfer and transport of the patient by preventing movement of fractured bone fragments, bone ends, and injured soft tissues.

You are the Provider: SUMMARY, continued

In addition, an effectively applied splint will help prevent further injury to muscles, nerves, and blood vessels from fractured bone ends; laceration of the skin by broken bone ends (a primary indication for splinting is to prevent a closed fracture from becoming an open one); and excessive bleeding of the tissues at the injury site caused by fractured bone ends.

Prior to splinting any extremity injury, assess the patient's distal perfusion and sensory and motor functions. Fractures of the tibia and fibula can be stabilized with a padded, rigid leg splint; a pneumatic (air) splint; or a vacuum splint that stabilizes the joints above and below the fracture site. In this case, you should stabilize the knee and ankle. As with most other fractures of the shaft of long bones, you should correct severe deformity before applying the splint by applying gentle longitudinal traction. Restore the deformed limb to a position that will accommodate a splint—not to its normal anatomic position. The affected leg, once splinted, should be secured to the opposite leg.

Immediately after the splint is secured in place, reassess distal perfusion and sensory and motor functions. If perfusion is found to be inadequate, the splint should be loosened or reapplied as necessary to restore this vital function.

---

8. What are some methods for providing pain relief from orthopaedic trauma?

Pain relief is an important aspect in the overall care of a patient with orthopaedic trauma. Whether the injury involves a sprain, strain, fracture, or dislocation, certain interventions can be performed to help reduce the amount of pain the patient experiences. Pain increases anxiety, which only adds to the patient's problems.

As previously discussed, proper application of a splint can help reduce the pain associated with extremity injuries. However, many patients may still complain of intense pain after the splint has been applied.

After you apply the splint, which should be padded for comfort, elevate the injured extremity above the level of the heart. This will help reduce pain and swelling by allowing blood to drain from the extremity. Of course, if the patient has critical injuries and is secured to a long backboard, this would not be practical.

Chemical cold packs can also be applied over the injury site; however, they should not come in direct contact with the skin. Wrap the cold pack with gauze or some other type of insulating material, and apply it directly over the injury site. A cold stimulus applied to the skin constricts the blood vessels; if the injured extremity is swollen and painful, this vasoconstrictive effect can help reduce swelling, thereby reducing pain.

Certain medications can be administered to the patient by a paramedic. In some cases, especially during a prolonged transport, it may be necessary to request a paramedic intercept so that he or she can start an intravenous line and administer analgesia. Common medications given for orthopaedic trauma include fentanyl (Sublimaze), meperidine (Demerol), and morphine.

9. What is the most likely cause of the patient's complaint? Is there anything you can do to remedy the situation?

On the basis of the patient's complaint of numbness and tingling and your findings of pallor and weak pedal pulses, you should suspect that you applied the splint too tightly and that it is now impairing distal circulation.

Although the goal of splinting is to stabilize a fracture, the splint must not be applied so tightly that it impedes blood flow to areas distal to the injury. One of the hazards of improper splinting is compression of nerves and blood vessels, which could cause a reduction in distal circulation. If broken bone ends injured a blood vessel and the splint is applied too tightly, compartment syndrome could develop because of increased pressure within the fascial compartment. Untreated compartment syndrome can result in necrosis of the tissues and subsequent loss of the limb.

Simply loosen the splint if it was applied too tightly. If you used padded board splints and triangular bandages (cravats), loosen the cravats. If you applied a pneumatic (air) splint, release some of the air from the splint. If you applied a vacuum splint, gently attempt to spread the edges of the splint apart to the point at which the patient feels relief. If this is not possible, it would not be unreasonable to stop the ambulance and apply a different type of splint, provided that the patient's condition is stable. Regardless of the splint that you used, immediately reassess distal circulation after making any adjustments, and ask the patient whether the numbness and/or tingling has subsided. In most cases, these adjustments will resolve the problem.

---

10. What factors increase the risk of complications following orthopaedic trauma?

Orthopaedic injuries can lead to numerous complications—not just those involving the skeletal system, but systemic problems as well. Do not focus all of your attention on the skeletal injury; after all, there is a patient attached to the injured extremity!

The risk of complications following orthopaedic trauma is increased by a variety of factors, such as the amount of force that caused the injury, the injury location, and the patient's overall health. Injuries in patients with diabetes, for example, tend to heal poorly owing to decreased peripheral perfusion; healing would be especially problematic following an open fracture.

Any fracture—open or closed—is accompanied by the risk of bleeding. In general, the severity of bleeding is directly related to the force that caused the injury. Sharp bone ends may damage muscles and blood vessels or may penetrate through the skin, resulting in an open fracture. A significant loss of tissue may occur at the fracture site if the muscle is severely damaged or if the bone's penetration of the skin causes a large deformity.

Infection is another potential complication associated with orthopaedic trauma especially in patients with open fractures or patients with weakened immune systems (ie, older adults or persons with human immunodeficiency infection). To prevent

You are the Provider: SUMMARY, continued

contaminating an open fracture, and minimize the risk of infection, you should brush away any obvious debris on the skin surrounding the fracture before covering it with a sterile dressing. Do not "probe" into an open fracture in an attempt to retrieve debris; probing may lead to further contamination, and it further increases the risk of infection.

Long-term disability is one of the most devastating complications of orthopaedic trauma. In many cases, a severely injured limb can be successfully repaired;

however, many patients may not be able to work for long periods because of severe, chronic pain and the extensive rehabilitation that is often required.

EMTs can help reduce the risk of complications, thus reducing the risk or duration of long-term disability following orthopaedic trauma, by preventing further injury, properly splinting orthopaedic injuries, reducing the risk of wound infection, and transporting patients to an appropriate medical facility.

EMS Patient Care Report (PCR)

Date: 12-23-09	Incident No.: 012909	Nature of Call: Leg injury		Location: 404 Field Drive	
Dispatched: 1620	En Route: 1620	At Scene: 1625	Transport: 1644	At Hospital: 1656	In Service: 1707

Patient Information

Age: 21	Allergies: Codeine
Sex: F	Medications: Birth control pills
Weight (in kg [lb]): 50 kg (110 lb)	Past Medical History: None
	Chief Complaint: Left leg pain

Vital Signs

Time: 1630	BP: 130/78	Pulse: 112	Respirations: 22	Sao₂: 98%
Time: 1638	BP: 134/80	Pulse: 110	Respirations: 22	Sao₂: 99%
Time: 1648	BP: 128/76	Pulse: 115	Respirations: 20	Sao₂: 98%

EMS Treatment
(circle all that apply)

Oxygen @ _ L/min via (circle one): NC NRM Bag-Mask Device		Assisted Ventilation	Airway Adjunct	CPR
Defibrillation	Bleeding Control	Bandaging	(Splinting)	Other (Cold pack; elevated injured leg)

Narrative

Medic 8 dispatched to a soccer field for a patient with a "possible broken leg." Arrived on scene and found the patient, a 21-year-old woman, sitting on the ground with her left leg extended and covered with an ice pack. She was conscious and alert; her airway was patent, and her breathing was adequate. Patient states that she injured her left leg when another player fell against it during the game. Assessment of her leg revealed obvious deformity to the midshaft tibial area. Patient describes pain severity as a "9" on a 0 to 10 scale. No open injuries were noted. Pulse and sensory and motor functions were grossly intact distal to the injury. A nurse was present at the scene and assisted EMS by manually stabilizing the injury site. The patient denies any other injuries; she further denies any past medical history. Secondary assessment was performed and revealed no gross evidence of injury to the areas above her knee and below her ankle on the injured extremity. Her right lower extremity was also unremarkable for gross injury. Patient was offered supplemental oxygen but refused to accept it. Vital signs were obtained and noted above. Splinted injured extremity with padded board splints; pulse and sensory and motor functions were assessed after splinting and were found to be grossly intact. Secured patient onto stretcher, elevated her left leg with pillows, loaded her into the ambulance, and began transport. Applied cold pack to injury site for pain relief. Patient stated that elevation of her leg and the cold pack reduced her pain to a 5/10. Monitored patient's vital signs en route and noted that they remained stable. Patient began complaining of paresthesias to her left foot during transport. Reassessment of area distal to the injury revealed that her foot was cool and pale and her pedal pulse was weaker than before. Loosened bandages that were securing splints in place, after which patient stated that the paresthesias resolved; her foot regained a pink color and became warm, and her pedal pulse was stronger following this intervention. Remainder of transport was uneventful. Delivered patient to emergency department, and gave verbal report to staff nurse. Medic 8 cleared the hospital and returned to service at 1707. *End of report*

Assessment and Emergency Care of Orthopaedic Injuries

Scene Size-up

Scene Safety	Ensure scene safety and safe access to the patient. Be aware of potential violence and the possibility of a crime scene. Follow standard precautions, putting on a minimum of gloves and eye protection. Scenes with multiple patients may require you to carry additional pairs of gloves in the event of tears. Determine the number of patients. Assess the need for additional resources such as utility services, fire department, ALS, or additional ambulances. The mechanism of injury may suggest the injury extent and type; consider the possibility that the patient may have internal bleeding.
Mechanism of Injury (MOI)/ Nature of Illness (NOI)	Determine the MOI. Observe the scene, and look for indicators that will assist you with this. The nature of the problem may not be readily apparent until more information is gathered. Falls, assaults, and motor vehicle crashes are common mechanisms in skeletal trauma. Be alert for primary and secondary injuries. Consider spinal immobilization in any trauma patient with a significant MOI.

Primary Assessment

Form a General Impression	Identify and manage immediate threats to life. Determine the priority of care based on the MOI. If the patient has a poor general impression, call for ALS assistance. A rapid scan of the patient will help you to identify and manage life threats. Keep alert for signs of shock. Assess the patient's level of consciousness using the AVPU scale. If the patient is able to communicate, obtain the chief complaint and type of injury that occurred. Do not let a nonlethal orthopaedic injury distract you from the ABCs.
Airway and Breathing	Ensure the airway is open, clear, and self-maintained. Unresponsive patients will need the airway opened and maintained using a modified jaw-thrust maneuver if a spinal injury is suspected. A patient with an altered level of consciousness may need emergency airway management; consider inserting a properly sized oropharyngeal or nasopharyngeal airway. Evaluate the patient's ventilatory status for rate and depth of breathing, respiratory effort, and tidal volume. Patients breathing at a rate of less than 12 breaths/min or more than 20 breaths/min may have inadequate breathing that requires assistance. Orthopaedic injuries are not common causes of breathing problems; if a breathing problem exists assess, the patient for other injuries. Continuously monitor the patient's oxygen saturation levels and for additional signs of hypoxia. Administer high-flow oxygen at 15 L/min, providing ventilatory support as needed.
Circulation	Observe skin color, temperature, and condition. Assess capillary refill time; if greater than 2 seconds, treat aggressively for shock. Open fractures may cause bone ends to protrude through the skin; therefore, look for life-threatening bleeding and treat accordingly. If you are not able to control bleeding in an extremity with a pressure dressing, apply a tourniquet. Fractures may cause internal bleeding leading to shock. Be alert for signs and symptoms. Evaluate the distal pulse rate, quality (strength), and rhythm. Tachycardia may be an early indicator of shock.
Transport Decision	If the patient has an airway or breathing problem, signs and symptoms of bleeding, or other life threats, manage them immediately and consider rapid transport, performing the secondary assessment en route to the hospital. Do not delay transport to perform a lengthy assessment or splint simple long-bone fractures. Pelvic and femoral fractures are indicators of severe external forces. Patients suspected of having pelvic, femoral, or bilateral fractures of any long bone should be packaged using a backboard and transported without delay. Simple fractures should be splinted to limit pain and blood vessel and nerve damage.

NOTE: The order of the steps in this section differs depending on whether the patient is conscious or unconscious. The following order is for a conscious patient. For an unconscious patient, perform a primary assessment, perform a full-body scan, obtain vital signs, and obtain the past medical history from a family member, bystander, or emergency medical identification device.

Assessment and Emergency Care of Orthopaedic Injuries, continued

History Taking

Investigate Chief Complaint

Investigate the chief complaint, and gather a history once you have identified and treated life threats. Identify associated symptoms and pertinent negatives. Ask SAMPLE questions, focusing on the events surrounding the incident and the mechanism of injury. SAMPLE can also be obtained from family, bystanders, and medical alert tags if the patient is not able to provide the information. Ask OPQRST questions when the MOI is unclear to help understand the injury better.

Secondary Assessment

Physical Examinations

If the patient is unconscious or multiple systems are affected, perform a full-body scan beginning with the head, using DCAP-BTLS to identify hidden and potentially life-threatening injuries. Assessment should be rapid if the patient has a poor general impression. Focus the assessment on an isolated injury once all the systems have been examined. Significant trauma requires a full-body scan. Look for swelling, deformities, asymmetry (compare the injured extremity with the opposite uninjured extremity), and contusions while palpating for tenderness. Look for shortening, rotation, and angulation of the limb. If no external signs of injury are present, and the patient is not reporting pain, you might ask the patient to move each extremity carefully to assess motor and neurologic status. Perform a focused examination when the patient has non-significant trauma. Assess the injured area, including the distal and proximal joints. Check for perfusion, motor, and sensory function. Look for the 6 Ps during your musculoskeletal assessment (pain, paralysis, paresthesias, pulselessness, pallor, and pressure).

Vital Signs

Obtain baseline vital signs. Vital signs should include blood pressure by auscultation, pulse rate and quality, respiration rate and quality, pupils, and skin assessment for perfusion. Note the patient's level of consciousness. Use pulse oximetry, if available, to assess the patient's perfusion status. Tachycardia or hypotension may indicate hypoperfusion. Reassess the patient's vital signs every 5 minutes to observe trends.

Reassessment

Interventions

If spinal injury is suspected, stabilize and immobilize the spine as needed. Ensure an open airway, using an oropharyngeal or nasopharyngeal airway if necessary. Provide oxygen via a nonrebreathing mask or bag-mask device as required, and manage any life-threatening injuries. Control external bleeding and treat for shock. In patients with non–life-threatening injuries, splint the affected area in a position that allows for good circulation distal to the injury. Reassess the chief complaint, primary assessment, vital signs, and any interventions already performed. Vital signs should be obtained every 5 minutes and results compared with those obtained earlier. Place the patient in a position of comfort unless shock is suspected, then place the patient supine and treat accordingly. Continuously observe and reassess the patient during transport so worsening conditions can be managed.

Assessment and Emergency Care of Orthopaedic Injuries, continued

Communication and Documentation	Contact medical control/receiving hospital with a radio report including notification of an orthopaedic emergency so appropriate staff and equipment will be ready when you arrive. Include a thorough description of the MOI and position in which the patient was found. Include treatments performed and patient's response. Let hospital staff know if there are open fractures and if circulation is compromised. Be sure to document the patient's chief complaint, physical findings, history, and any changes in patient status and the time. Document the scene observations on your arrival. If abuse is suspected, notify the hospital staff of your suspicions and complete any mandatory report forms. Follow local treatment protocols.

NOTE: Although the following steps are widely accepted, be sure to consult and follow your local protocols. Take appropriate standard precautions when treating all patients.

Orthopaedic Injuries

General Management of Orthopaedic Injuries

Managing life threats to the patient's ABCs is the primary concern with any traumatic emergency. The MOI that caused the injury may also have caused a spinal injury or other fracture, and these must be managed at the appropriate time following local protocol. Remove any jewelry the patient is wearing on injured extremities. Choose the correct type and size splint for the injury. Splints for long-bone fractures should be long enough to stabilize the injured bone and the joint above and below the injury. If swelling is present, a cold pack can be applied to the injured area. In all of the following specific injuries, perform the following:

1. Ensure scene safety.
2. Determine the MOI.
3. Consider the need for spinal stabilization.
4. Open, clear, and maintain the patient's airway.
5. Ensure adequate ventilation.
6. Administer high-concentration oxygen via a nonrebreathing mask or bag-mask device as appropriate.
7. Control bleeding, and treat for shock. Cover open wounds with a dry, sterile dressing and apply pressure to control bleeding.
8. Apply a splint, and elevate the extremity about 6" (slightly above the level of the heart).
9. Apply cold packs if there is swelling, but do not place them directly on the skin.
10. Position the patient for transport, and transport to the appropriate treatment facility.

Fractures, Dislocations, Sprains, and Strains

Field management for fractures, dislocations, sprains, and strains is essentially the same: Prevent further injury, reduce the risk of infection, minimize pain, and reduce the risk of long-term disability. Because it can be difficult to differentiate between the different types of orthopaedic injuries, it is necessary to provide the same emergency care to all, which includes control of bleeding, followed by splinting. Long-bone injuries can be immobilized with a padded board splint, an air splint, or other similar device. Isolated femoral fractures should be managed using a traction splint. If a pelvic injury is suspected, a pelvic binder should be applied. Clavicle and shoulder injuries can be immobilized using a sling and swathe.

Assessment and Emergency Care of Orthopaedic Injuries, continued

Orthopaedic Injuries

Compartment Syndrome

If you suspect compartment syndrome (pain out of proportion to the injury, pallor, decreased sensation, decreased power), splint the affected limb, keep it at or above the level of the heart, and transport immediately. Surgical intervention is required to manage this injury.

Amputation

Ensure that bleeding is controlled at the stump using a tourniquet, if necessary and if protocols allow. Manage life threats first; do not focus only on trying to save an amputated part. If life threats are under control and if the amputation is complete, you may wrap the part in a sterile dressing and place it in a plastic bag. You should then place the bag containing the amputated part on top of ice. Do not pack the amputated part in ice. A frozen part is useless to a surgeon. Transport the amputated part with the patient, but do not delay transport of a seriously injured patient to do so.

Prep Kit

Ready for Review

- Skeletal or voluntary muscle attaches to bone and forms the major muscle mass of the body. This muscle contains veins, arteries, and nerves.

- There are 206 bones in the human body. When this living tissue is fractured, it can produce bleeding and significant pain.

- A joint is a junction where two bones come into contact. Joints are stabilized in key areas by ligaments.

- A fracture is a broken bone, a dislocation is a disruption of a joint, a sprain is a stretching injury to the ligaments around a joint, and strain is a stretching of the muscle.

- Depending on the amount of kinetic energy absorbed by tissues, the zone of injury may extend beyond the point of contact. Always maintain a high index of suspicion for associated injuries.

- Fractures of the bones are classified as open or closed. Both are splinted in a similar manner, but remember to control bleeding and apply a sterile dressing to the open extremity injury before splinting.

- Fractures and dislocations are often difficult to diagnose without an x-ray examination. You will treat these injuries similarly. Stabilize the injury with a splint, and transport the patient.

- Signs of fractures and dislocations include pain, deformity, point tenderness, false movement, crepitus, swelling, and bruising.

- Signs of sprain include bruising, swelling, and an unstable joint.

- Compare the unaffected extremity with the injured extremity for differences whenever possible.

- There are three main types of splints used by EMTs: rigid splints, traction splints, and formable splints.

- Remember to splint the injured extremity from the joint above to the joint below the injury site for complete stabilization.

- A sling and swathe is used commonly to treat shoulder dislocations and to secure injured upper extremities to the body. Lower extremities can be secured to the unaffected limb or to a long backboard.

- The most common life-threatening musculoskeletal injuries are multiple fractures, open fractures with arterial bleeding, pelvic fractures, bilateral femur fractures, and limb amputations.

Vital Vocabulary

acromioclavicular (AC) joint A simple joint where the bony projections of the scapula and the clavicle meet at the top of the shoulder.

articular cartilage A pearly layer of specialized cartilage covering the articular surfaces (contact surfaces on the ends) of bones in synovial joints.

calcaneus The heel bone.

closed fracture A fracture in which the skin is not broken.

compartment syndrome An elevation of pressure within a closed fascial compartment, characterized by extreme pain, decreased pain sensation, pain on stretching of affected muscles, and decreased power; frequently seen in fractures below the elbow or knee in children.

crepitus A grating or grinding sensation or sound caused by fractured bone ends or joints rubbing together.

dislocation Disruption of a joint in which ligaments are damaged and the bone ends are completely displaced.

displaced fracture A fracture in which bone fragments are separated from one another and not in anatomic alignment.

ecchymosis Bruising or discoloration associated with bleeding within or under the skin.

fibula The outer and smaller bone of the two bones of the lower leg.

fracture A break in the continuity of a bone.

glenoid fossa The part of the scapula that joins with the humeral head to form the glenohumeral joint.

hematuria Blood in the urine.

joint The place where two bones come into contact.

nondisplaced fracture A simple crack in the bone that has not caused the bone to move from its normal anatomic position; also called a hairline fracture.

open fracture Any break in a bone in which the overlying skin has been damaged.

pelvic binders Used to splint the bony pelvis to reduce hemorrhage from bone ends, venous disruption, and pain.

point tenderness Tenderness that is sharply localized at the site of the injury, found by gently palpating along the bone with the tip of one finger.

position of function A hand position in which the wrist is slightly dorsiflexed and all finger joints are moderately flexed.

reduce Return a dislocated joint or fractured bone to its normal position; set.

retroperitoneal space The space between the abdominal cavity and the posterior abdominal wall, containing the kidneys, certain large vessels, and parts of the gastrointestinal tract.

sciatic nerve The major nerve to the lower extremities; controls much of muscle function in the leg and sensation in most of the leg and foot.

sling A bandage or material that helps to support the weight of an injured upper extremity.

splint A flexible or rigid appliance used to protect and maintain the position of an injured extremity.

sprain A joint injury involving damage to supporting ligaments, and sometimes partial or temporary dislocation of bone ends.

strain Stretching or tearing of a muscle; also called a muscle pull.

subluxation A partial or incomplete dislocation.

swathe A bandage that passes around the chest to secure an injured arm to the chest.

tibia The larger of the two lower leg bones responsible for supporting the major weight-bearing surface of the knee and the ankle; the shinbone.

tourniquet The bleeding control method used when a wound continues to bleed despite the use of direct pressure and elevation; useful if a patient is bleeding severely from a partial or complete amputation.

traction Longitudinal force applied to a structure.

zone of injury The area of potentially damaged soft tissue, adjacent nerves, and blood vessels surrounding an injury to a bone or a joint.

Assessment in Action

You and your partner are assigned to a rescue post at an extreme sports competition. You are called to assess a 21-year-old man who was performing a midair trick on his bike when he lost control and landed on the bottom of the concrete ramp. When you arrive on scene, the patient is awake and alert and in extreme pain. He tells you that he flew over his handlebars and felt both of his upper legs snap. He denies experiencing any loss of consciousness.

1. You perform a full-body scan and find instability in the pelvis and deformity to the midshaft area of the femur in both legs. The patient's injuries were the result of a/an:
   A. high-energy impact.
   B. indirect force.
   C. direct blow.
   D. twisting force.

2. Which of the following signs should make you suspicious of the presence of a fracture?
   A. Deformity
   B. Rotation
   C. Shortening
   D. All of the above

3. The most reliable indicator of an underlying fracture is:
   A. guarding.
   B. point tenderness.
   C. swelling.
   D. bruising.

4. When you assess distal circulation in the lower extremities, you should palpate the _____ pulse.
   A. femoral
   B. dorsalis pedis
   C. popliteal
   D. iliac

5. Further assessment of the patient reveals no other injuries or life threats. His vital signs include the following: pulse rate, 104 beats/min; blood pressure, 118/72 mm Hg; and respirations, 20 breaths/min. Using the musculoskeletal grading system, you would classify this patient's injuries as:
   A. minor.
   B. moderate.
   C. severe.
   D. critical.

6. In the field, the goal of in-line traction includes:
   A. minimizing pain.
   B. avoiding neurovascular compromise.
   C. reducing swelling.
   D. preventing permanent disability.

7. The best splinting option for this patient would be a:
   A. pneumatic antishock garment.
   B. Sager traction splint.
   C. vacuum splint.
   D. zippered air splint.

8. Neurovascular function should be reassessed every _____ minutes.
   A. 5 to 10
   B. 10 to 15
   C. 15 to 20
   D. 20 to 30

9. Why would this patient be at risk for shock?

10. How can you help minimize the risk or duration of long-term disability in patients with musculoskeletal injuries?

National EMS Education Standard Competencies

Trauma

Applies fundamental knowledge to provide basic emergency care and transportation based on assessment findings for an acutely injured patient.

Environmental Emergencies

Recognition and management of

- Submersion incidents (pp 1076-1079, 1083-1084)
- Temperature-related illness (pp 1063-1076)

Pathophysiology, assessment, and management of

- Near drowning (pp 1076-1079, 1081-1084)
- Temperature-related illness (pp 1063-1076)
- Bites and envenomations (pp 1086-1091)
- Dysbarism (pp 1079-1085)
  - High altitude (pp 1084-1085)
  - Diving injuries (pp 1079-1084)
- Electrical injury (pp 1085-1086)
- Radiation exposure (p 1076)

Knowledge Objectives

1. Describe four factors that affect how a person deals with exposure to a cold or hot environment and how each one relates to emergency medical care. (pp 1061-1062)
2. Explain the five different ways a body can lose heat and ways the rate and amount of heat loss or gain can be modified in an emergency situation. (pp 1062-1063)
3. Define and discuss hypothermia, including the signs and symptoms of its four different stages and the risk factors for developing it. (pp 1063-1064)
4. Explain local cold injuries and their underlying causes. (pp 1064-1066)
5. Describe the process of providing emergency care to a patient who has sustained a cold injury, including assessment of the patient, review of signs and symptoms, and management of care. (pp 1063-1069)
6. Explain the importance of following regional and state protocols when rewarming a patient who is experiencing moderate or severe hypothermia. (pp 1068-1069)
7. Describe the three forms of illness that are caused by heat exposure, including their signs and symptoms, and give examples of persons who are at the greatest risk of developing one of them. (pp 1070-1071)
8. Describe the process of providing emergency care to a patient who has sustained a heat injury, including assessment of the patient, review of signs and symptoms, and management of care. (pp 1070-1076)
9. Define drowning and discuss its incidence, risk factors, and prevention. (pp 1076-1079, 1084)
10. List the basic rules of performing a water rescue and discuss why rescue personnel should have a prearranged water rescue plan based on the environment in which they work. (p 1077)

11. List five conditions that may result in a spinal injury following a submersion incident and the steps for stabilizing a patient with a suspected spinal injury in the water. (pp 1076-1079)
12. Discuss recovery techniques and resuscitation efforts EMTs may need to follow when managing a patient who has been involved in a submersion incident. (p 1079)
13. Describe the three different types of diving emergencies, how they may occur, and their signs and symptoms. (pp 1079-1081)
14. Describe the process of providing emergency care to a patient who has been involved in a drowning or diving emergency, including assessment of the patient, review of signs and symptoms, and management of care. (pp 1079-1084)
15. Discuss the types of dysbarism injuries that may be caused by high altitudes, including their signs and symptoms and emergency medical treatment in the field. (pp 1084-1085)
16. Discuss lightning injuries, including their incidence, risk factors, assessment, and emergency medical treatment. (pp 1085-1086)
17. Identify the species of spiders found in the United States that may cause life-threatening injuries, and then describe the process of providing emergency care to patients who have been bitten by each type. (pp 1086-1087)
18. Discuss the emergency medical care of patients who have been stung by hymenoptera and scorpions, and bitten by ticks, including steps the EMT should follow if a patient develops a severe reaction to the sting or bite. (pp 1087, 1090-1091)
19. Identify the species of snakes found in the United States that are venomous, and then describe the process of providing emergency care to patients who have been bitten by each type and are showing signs of envenomation. (pp 1087-1090)
20. Discuss the emergency medical care of patients who have been stung by a coelenterate or other marine animal. (pp 1091-1092)

Skills Objectives

1. Demonstrate the emergency medical treatment of local cold injuries in the field. (p 1069)
2. Demonstrate using a warm-water bath to rewarm the limb of a patient who has sustained a local cold injury. (p 1069)
3. Demonstrate how to treat a patient with heat cramps. (pp 1073-1074)
4. Demonstrate how to treat a patient with heat exhaustion. (pp 1074-1075, Skill Drill 30-1)
5. Demonstrate how to treat a patient with heatstroke. (p 1075)
6. Demonstrate how to stabilize a patient with a suspected spinal injury in the water. (pp 1078-1079, Skill Drill 30-2)
7. Demonstrate how to care for a patient who is suspected of having an air embolism or decompression sickness following a drowning or diving emergency. (pp 1083-1084)
8. Demonstrate how to care for a patient who has been bitten by a pit viper and is showing signs of envenomation. (pp 1088-1090)
9. Demonstrate how to care for a patient who has been bitten by a coral snake and is showing signs of envenomation. (p 1090)
10. Demonstrate how to care for a patient who has sustained a coelenterate envenomation. (p 1092)

Introduction

Heat and cold can both overwhelm the body's mechanisms for regulating temperature, including sweating and radiation of body heat into the atmosphere. A variety of medical emergencies can result from exposure to heat or cold, particularly in children, older people, people with chronic illnesses, and young adults who overexert themselves. There is also a range of medical emergencies that arise from water recreation, and these can sometimes be complicated by the cold. These emergencies include localized injuries and systemic illnesses. As an EMT, you can save lives by recognizing and responding properly to these emergencies, most of which require prompt treatment in the hospital.

In this chapter you will learn how the body regulates core temperature, and the ways in which body heat is lost to the environment. The various forms of heat-, cold-, and water-related emergencies are described, including how to diagnose and treat hypothermia, frostbite, and hyperthermia. Other environmental medical emergencies include **dysbarism injuries** (the signs and symptoms related to changes in barometric pressure), caused by diving and high-altitude climbing; injuries caused by lightning; and envenomation, caused by bites and stings.

Factors Affecting Exposure

A number of factors will affect how a person deals with a cold or hot environment. These can certainly be used as prevention strategies for those who work or play in extreme environmental temperatures. They can also be useful during the assessment of your patient to determine how prepared he or she was for a cold or hot environment. A hiker prepared for a warm summer hike in the foothills will present and respond to treatment differently than a traveler stranded in a hot car because the radiator boiled over.

1. **Physical condition.** Patients who are already ill or in poor physical condition will not be able to tolerate extreme temperatures as well as those whose cardiovascular system, metabolic system, and nervous systems are all functioning well. A well-trained athlete performs much better and is less likely to experience injury or illness than the "weekend warrior" who is not in peak physical condition. Increasing your activity will generate more heat when out in the cold but will also produce more heat when it is not needed, as in walking on a hot asphalt road because you ran out of gas.

2. **Age.** Those who are at the extremes of age are more likely to experience illness as a result of temperatures. Small infants have poor thermoregulation at birth and do not have the ability to shiver and generate heat when needed until about 12 to 18 months of age. Their larger surface area and smaller mass contribute to increased heat loss and heat gain. When you get cold you put on a sweater; a small child may not think to do this or may have difficulty finding and putting one on. On the other end of the age spectrum, older adults lose subcutaneous tissues, reducing the amount of insulation they have. Poor circulation contributes to increased heat loss and gain in either environment. This is why older people often wear extra layers of clothing. Medications taken by older persons can also affect their body's thermostat, putting them at more risk to hot or cold problems.

3. **Nutrition and hydration.** Your body needs calories for your metabolism to function. Staying well

hydrated provides water as a catalyst for much of this metabolism. A decrease in either will aggravate both hot and cold stress. Calories provide fuel to burn, creating heat during the cold, and water provides sweat for evaporation and removing heat. Alcohol use may increase fluid loss and place the patient at greater risk for temperature-related problems.

4. **Environmental conditions.** Conditions such as air temperature, humidity levels, and wind can complicate or improve environmental situations. A cool breeze helps when it is hot outside, but a cold wind when it is cold outside can be uncomfortable. Extremes in temperature and humidity are not needed to produce hot or cold injuries. Many hypothermia cases occur at temperatures between 30°F and 50°F. Most heat stroke cases occur when the temperature is 80°F and the humidity is 80%. Be sure to examine the environmental temperature of your patient. Older patients may turn the heat down in the winter or neglect to use air conditioning in the summer because of cost concerns. Some people may not open windows in a heat wave for fear of burglars. When evaluating your patient's condition, consider the environment and whether your patient is prepared for that situation. It may help in your treatment decisions and give you an idea about how the patient will respond to your care.

Cold Exposure

Normal body temperature must be maintained within a very narrow range for the body's chemistry to work efficiently. If the body, or any part of it, is exposed to cold environments, these mechanisms may be overwhelmed. Cold exposure may cause injury to individual parts of the body, such as the feet, hands, ears, or nose, or to the body as a whole. When the entire body temperature falls, the condition is called <u>hypothermia</u>.

Because heat always travels from a warmer place to a cooler place, the body tends to lose heat to the environment. The body can lose heat in the following five ways:

- <u>Conduction</u> is the direct transfer of heat from a part of the body to a colder object by direct contact, such as when a warm hand touches cold metal or ice, or is immersed in water with a temperature of less than 98°F (37°C). Heat passes directly from the body to the colder object. Heat can also be gained if the substance being touched is warm. This is why people with chronic medical problems are advised to limit time in hot tubs.

- <u>Convection</u> occurs when heat is transferred to circulating air, such as when cool air moves across the body surface. A person standing outside in windy winter weather, wearing lightweight clothing, is losing heat to the environment mostly by convection. A person can gain heat if the air moving across the person's body is hotter than the environment's temperature such as in deserts or industrial settings like foundries, but it is more common to see rapid heat gain in spas and hot tubs where the water temperature may be well above body temperature.

- <u>Evaporation</u> is the conversion of any liquid to a gas, a process that requires energy, or heat. Evaporation is the natural mechanism by which sweating cools the body. This is why swimmers coming out of the water feel a sensation of cold as the water evaporates from their skin. Individuals who exercise vigorously in a cool environment may sweat and feel warm at first, but later, as their sweat evaporates, they can become exceedingly cool. Measures should be taken to keep a person dry if he or she is too cold.

- <u>Radiation</u> is the transfer of heat by radiant energy. Radiant energy is a type of invisible light that transfers heat. The body can lose heat by radiation, such as when a person stands in a cold room. Heat can also be gained by radiation; for example, when a person stands by a fire.

- <u>Respiration</u> causes body heat to be lost as warm air in the lungs is exhaled into the atmosphere and cooler air is inhaled. In warm climates, the air temperature can be well above body temperature, causing an individual to gain heat with each breath.

The rate and amount of heat loss or gain by the body can be modified in three ways:

1. **Increase or decrease heat production.** One way for the body to increase its heat production is to increase the rate of metabolism of its cells; the body can accomplish this through shivering. Also, people often have a natural urge to move around when they are cold. When a person is hot, he or she tends to reduce their level of activity, thus reducing heat production.

2. **Move to an area where heat loss is decreased or increased.** The most obvious way to decrease heat loss from radiation and convection is to move out of a cold environment and seek shelter from the wind. Just covering the head will minimize radiation heat loss by up to 70%. The same holds true for a patient who is too hot. Simply moving the patient into the shade can reduce the ambient temperature by 10° or more. If you cannot move the patient, create shade and increase air movement by fanning the patient.

3. **Wear insulated clothing, which helps to decrease heat loss in several ways.** Insulators, such as specific

materials or dry, still air, do not conduct heat. Thus, layers of clothing that trap air provide good insulation, as do wool, down, and synthetic fabrics that have small pockets of trapped air. Protective clothing also traps perspiration and prevents evaporation. Sweating without evaporation will not result in cooling. To encourage heat loss, loosen or remove clothing, particularly around the head and neck.

■ Hypothermia

Hypothermia literally means "low temperature." It is diagnosed when the **core temperature** of the body—the temperature of the heart, lungs, and vital organs—falls below 95°F (35°C). The body can usually tolerate a drop in core temperature of a few degrees. However, below this critical point, the body loses the ability to regulate its temperature and to generate body heat. Progressive loss of body heat then begins.

To protect itself against heat loss, the body normally constricts blood vessels in the skin; this results in the characteristic appearance of blue lips and/or fingertips. As a secondary precaution against heat loss, the body tends to create additional heat by shivering, which is the active moving of many muscles to generate heat. As cold exposure worsens and these mechanisms are overwhelmed, many body functions begin to slow down. Eventually, the functioning of key organs such as the heart begins to slow. Untreated, this can lead to death.

Hypothermia can develop either quickly, as when someone is immersed in cold water, or more gradually, as when a lost person is exposed to the cold environment for several hours or more. The temperature does not have to be below freezing for hypothermia to occur. In winter, homeless people and those whose homes lack heating may develop hypothermia at higher temperatures. Even in summer, swimmers who remain in the water for a long time are at risk of hypothermia. Like all heat- and cold-related injuries, hypothermia is more common among geriatric, pediatric, and ill individuals, who are less able to adjust to temperature extremes. Hypothermia is also common among the very young, who are unable to put on clothes to protect themselves against the cold. Infants and children are small, with a relatively large surface area, and have less body fat than do adults. Also, because of their small muscle mass, children may not be able to shiver as effectively as adults, and infants do not shiver at all.

Patients with injuries or illness, such as burns, shock, head injury, stroke, generalized infection, injuries to the spinal cord, diabetes, and hypoglycemia, are more prone to hypothermia, as are patients who have taken certain drugs or poisons.

Signs and Symptoms

Signs and symptoms of hypothermia generally become progressively more severe as the core temperature falls. Hypothermia generally progresses through four general stages, as shown in Table 30-1. Although there is no clear distinction among the stages, the different signs and symptoms of each will help you estimate the severity of the problem. When you assess a patient in the field, you should be able to distinguish between mild and severe hypothermia.

To assess the patient's general temperature, pull back on your glove and place the back of your hand on the patient's skin at the abdomen Figure 30-1. If the skin feels cool, the patient is likely experiencing a generalized cold emergency.

If you work in a cold environment, you may carry a hypothermia thermometer, which registers lower core temperatures Figure 30-2. It must be inserted in the rectum for an accurate reading. Regular thermometers will not register the temperature of a patient who has significant hypothermia.

Table 30-1 Characteristics of Systemic Hypothermia

Core temperature	93° to 95°F (34° to 35°C)	89° to 92°F (32° to 33°C)	80° to 88°F (27° to 31°C)	< 80°F (< 27°C)
Signs and symptoms	Shivering, foot stamping	Loss of coordination, muscle stiffness	Coma	Apparent death
Cardiorespiratory response	Constricted blood vessels, rapid breathing	Slowing respirations, slow pulse	Weak pulse, arrhythmias, very slow respirations	Cardiac arrest
Level of consciousness	Withdrawn	Confused, lethargic, sleepy	Unresponsive	Unresponsive

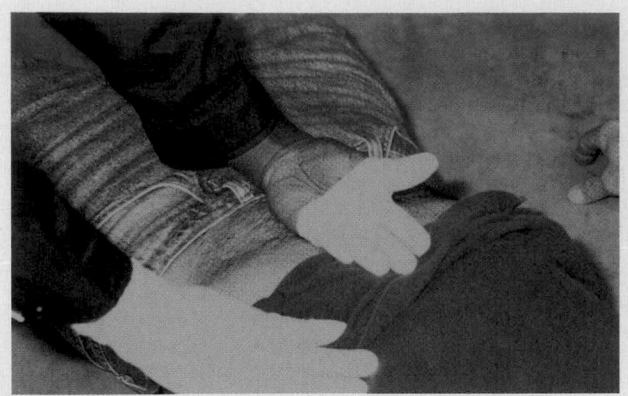

Figure 30-1 To assess a patient's temperature, pull back your glove and place the back of your hand on the patient's skin.

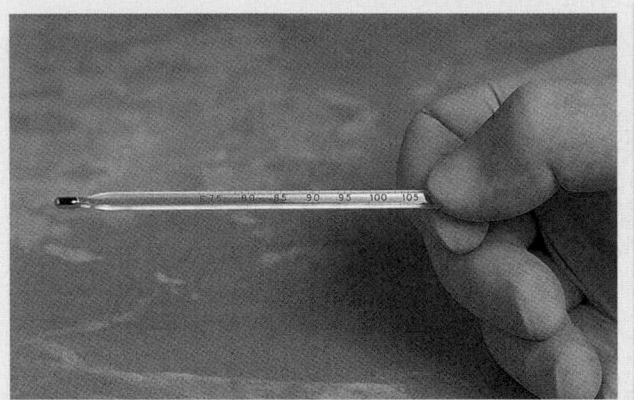

Figure 30-2 A special rectal hypothermia thermometer registers temperatures well below that of a regular thermometer.

Mild hypothermia occurs when the core temperature is between 90° and 95°F (32° and 35°C). The patient is usually alert and shivering in an attempt to generate more heat through muscular activity. The patient may jump up and down and stamp his or her feet. Pulse rate and respirations are usually rapid. The skin in light-skinned individuals can be red, but may eventually appear pale, then cyanotic. Individuals in a cold environment may have blue lips or fingertips because of the body's constriction of blood vessels at the skin to retain heat.

More severe hypothermia occurs when the core temperature is less than 90°F (32°C). Shivering stops and muscular activity decreases. At first, small, fine muscle activity such as coordinated finger motion ceases. Eventually, as the temperature falls further, all muscle activity stops.

As the core temperature drops toward 85°F (29°C), the patient becomes lethargic, usually losing interest in continuing to fight the cold. The level of consciousness decreases, and the patient may try to remove his or her own clothes. Poor coordination and memory loss follow, along with reduced or complete loss of sensation to touch, mood changes, and impaired judgment. The patient becomes less communicative, experiences joint or muscle stiffness, and has trouble speaking. The muscles eventually become rigid, and the patient begins to appear stiff or rigid.

If the temperature continues to fall to 80°F (27°C), vital signs slow; the pulse becomes weaker, and respirations slow to shallow or become absent. Cardiac arrhythmias may occur as the blood pressure decreases or disappears.

At a core temperature of less than 80°F (27°C), all cardiorespiratory activity may cease, pupillary reaction is slow, and the patient may appear dead.

Never assume that a cold, pulseless patient is dead. Patients may survive even severe hypothermia, if proper emergency measures are carried out.

Words of Wisdom

The stress of the cold environment, a remote terrain, a feeling of impending doom, and impaired judgment may lead to suicidal tendencies in some patients.

■ Local Cold Injuries

Most injuries from cold are confined to exposed parts of the body. The extremities, particularly the feet, and the exposed ears, nose, and face, are especially vulnerable to cold injury **Figure 30-3** . When exposed parts of the body become very cold but not frozen, the condition is called frostnip, chilblains, or immersion foot (trench foot). When the parts become frozen, the injury is called **frostbite**.

You should try to find out the duration of the exposure, the temperature to which the body part was exposed, and the wind velocity during exposure. These are important factors in determining the severity of a local cold injury. You should also investigate a number of underlying factors:

- Exposure to wet conditions
- Inadequate insulation from cold or wind
- Restricted circulation from tight clothing or shoes or circulatory disease
- Fatigue
- Poor nutrition
- Alcohol or drug abuse
- Hypothermia
- Diabetes
- Cardiovascular disease
- Older age

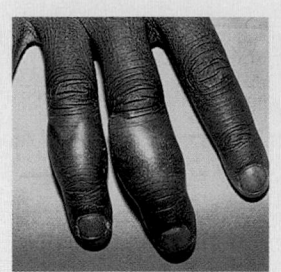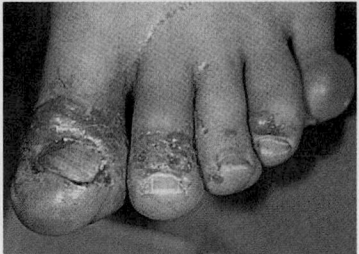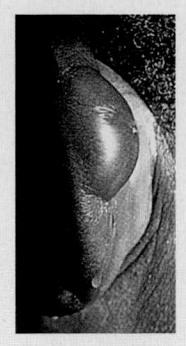

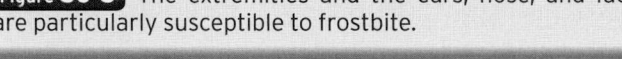
Figure 30-3 The extremities and the ears, nose, and face are particularly susceptible to frostbite.

- Alcohol or drug abuse
- Hypothermia
- Diabetes
- Cardiovascular disease
- Older age

In hypothermia, blood is shunted away from the extremities in an attempt to maintain the core temperature. This shunting of blood increases the risk of local cold injury to the extremities, ears, nose, and face. Thus, the patient with hypothermia should also be assessed for frostbite or other local cold injury. The reverse is also true. You must remember that both local and systemic cold exposure problems can occur in the same patient.

Frostnip and Immersion Foot

After prolonged exposure to the cold, the skin may be freezing whereas the deeper tissues are unaffected. This condition, which often affects the ears, nose, and fingers, is called frostnip. Because frostnip is usually not painful, the patient often is unaware that a cold injury has occurred. Immersion foot, also called trench foot, occurs after prolonged exposure to cold water. It is particularly common in hikers or hunters who stand for a long time in a river or lake. With both frostnip and immersion foot, the skin is pale (blanched) and cold to the touch; normal color does not return after palpation of the skin. In some cases, the skin of the foot will be wrinkled, but it can also remain soft. The patient reports loss of feeling and sensation in the injured area.

Frostbite

Frostbite is the most serious local cold injury because the tissues are actually frozen. Freezing permanently damages cells, although the exact mechanism by which damage occurs is not known. The presence of ice crystals within the cells may cause physical damage. The change in the water content in the cells may also cause changes in the concentration of critical electrolytes, producing

permanent changes in the chemistry of the cell. When the ice thaws, further chemical changes occur in the cell, causing permanent damage or cell death, called gangrene **Figure 30-4** . If gangrene occurs, the dead tissue must be surgically removed, sometimes by amputation. Following less severe damage, the exposed part will become inflamed, tender to touch, and unable to tolerate exposure to cold.

Frostbite can be identified by the hard, frozen feel of the affected tissues. Most frostbitten parts are hard and waxy **Figure 30-5** . The injured part feels firm to frozen as you gently touch it. If the frostbite is only skin deep it will feel leathery or thick, not hard and frozen through. Blisters and swelling

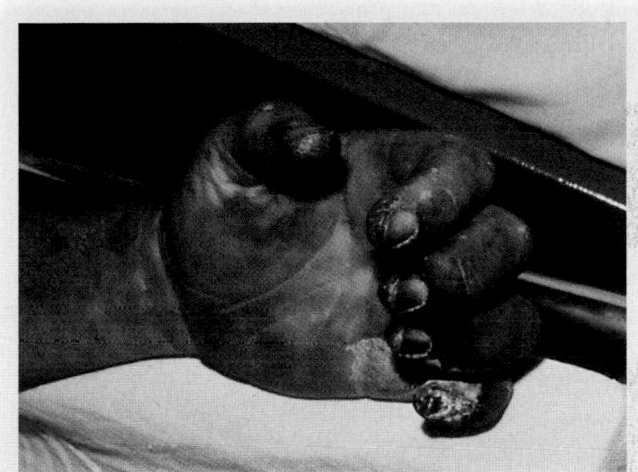
Figure 30-4 Gangrene, or permanent cell death, can occur when tissue is frozen and certain chemical changes occur in the cells.

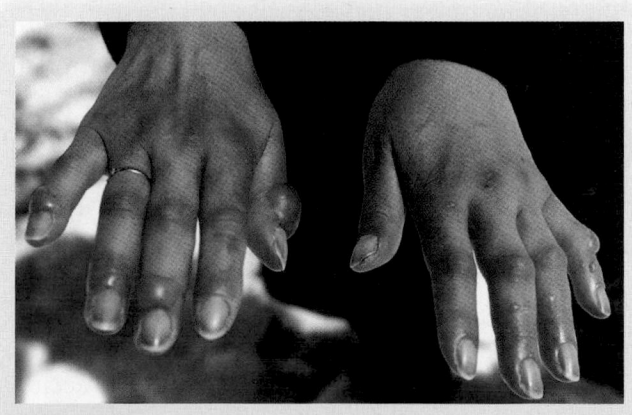
Figure 30-5 Frostbitten parts are hard and usually waxy to touch.

may be present. In light-skinned individuals with a deep injury that has thawed or partially thawed, the skin may appear red with purple and white, or it may be mottled and cyanotic.

As with a burn, the depth of skin damage will vary. With superficial frostbite, only the skin is frozen; with deep frostbite, the deeper tissues are frozen as well. You may not be able to tell superficial from deep frostbite in the field. Even an experienced surgeon in a hospital setting may not be able to tell until several days have gone by.

Assessment of Cold Injuries

Management of hypothermia in the field, regardless of the severity of the exposure, consists of stabilizing the ABCs and preventing further heat loss.

Scene Size-up

Scene Safety
Typically, your scene assessment begins with information provided by dispatch. This information helps you consider the mechanism of injury (MOI) and prepare for the problems your patient may have. Note environmental conditions. Air temperature, wind chill, and whether it is wet or dry are important aspects of scene size-up and will likely affect the patient.

Ensure that the scene is safe for you and other responders. Identify potential safety hazards, such as wet grass, mud, or icy streets. Consider special hazards such as avalanches. Cold environments may present special problems both for you and your patient. Use appropriate standard precautions and consider the number of patients you may have. Summon additional help, such as a search and rescue team, as quickly as possible.

Mechanism of Injury/Nature of Illness
As you observe the scene, look for indicators of the MOI. This helps you develop an early index of suspicion for underlying injuries in the patient who has sustained a significant MOI. As you put together information from dispatch and your observations of the scene, consider how the MOI produced the injuries expected. For example, if you find a vehicle in a secluded ditch off the highway and the vehicle's roof and hood are covered with fresh snow, then you may assume that the patient was in a motor vehicle accident and has been exposed to the cold for a long period of time.

Primary Assessment

Form a General Impression
In a cold emergency, your patient's chief complaint may be only that he or she is cold, or the cold may be an additional complication of an existing medical or trauma problem. Perform a rapid scan to determine whether a life threat exists, and if so, treat it. If the chief complaint is simply being cold, quickly assess how cold the patient actually is. This is done by feeling the patient's skin on the abdomen. This area of the body is usually well protected and insulated and will give you a quick and general idea of the patient's core body temperature. Evaluate the patient's mental status quickly using the AVPU scale (*Alert* to person, place, and day; responsive to *Verbal* stimuli; responsive to *Pain*; *Unresponsive*). An altered mental status indicates the intensity of the cold injury. Consider spinal precautions based on your scene size-up and the chief complaint.

Airway and Breathing
Your assessment should take into account the physiologic changes that occur as a result of hypothermia. If you believe the patient is in cardiac arrest proceed directly to the circulation ("C") step by providing high-quality chest compressions, then address airway and breathing ("A and B") after. Ensure that the patient has an adequate airway and is breathing. If your patient's breathing is slow or shallow, ventilation with a bag-mask device may be necessary. If warmed and humidified oxygen is available, use it because it helps to warm the patient from the inside out.

Circulation
If you cannot feel a radial pulse, gently palpate for a carotid pulse and wait for 30 to 45 seconds before you decide that the patient is pulseless. Some physicians disagree about performing cardiopulmonary resuscitation (CPR) on a patient with hypothermia who appears to be pulseless. Such a patient actually may be in a kind of "metabolic ice box," having achieved a metabolic balance that CPR may upset. Even a pulse rate of 1 or 2 beats/min indicates cardiac activity, and cardiac activity may spontaneously recover once the body core is warmed. However, there is evidence that CPR, when correctly done, will increase blood flow to the critical parts of the body. For this reason, some authorities recommend starting CPR on a patient with hypothermia and no pulse. The American Heart Association recommends that CPR be started if the patient has no detectable pulse or breathing. Again, for a patient with hypothermia, this may require a prolonged pulse check.

Perfusion will be compromised based on the degree of cold the patient is experiencing. Your assessment

Transport Decision

Even mild degrees of hypothermia can have serious consequences and complications, including cardiac arrhythmia and blood clotting abnormalities. Therefore, all patients with hypothermia require immediate transport for evaluation and treatment. Assess the scene for the safest way to quickly move your patient from the cold environment. As you package your patient for transport, work quickly, safely, and gently. Rough handling of a hypothermic patient may cause a cold, slow, weak heart to fibrillate and the patient to lose any pulse that may have existed. If transportation is delayed, protect the patient from further heat loss.

▶ History Taking

Investigate Chief Complaint

After the life threats have been managed during the primary assessment, investigate the chief complaint. You should obtain a medical history and be alert for injury-specific signs and symptoms as well as any pertinent negatives.

SAMPLE History

Obtaining a patient's history in these situations may be difficult but should be attempted. If possible, find out how long your patient has been exposed to the cold environment, either from the patient or bystanders. Exposures may be acute or chronic. Your SAMPLE history can provide important information affecting both your treatment in the field and the treatment your patient will receive in the hospital. Medications your patient has taken and underlying medical problems may have an impact on the way cold affects his or her metabolism. The patient's last oral intake and what the patient was doing prior to the exposure will help to determine the severity of the cold problem.

▶ Secondary Assessment

The secondary assessment is a more detailed, comprehensive examination of the patient that is used to uncover injuries that may have been missed during the primary assessment. In some instances, such as a critically injured patient or a short transport time, you may not have time to conduct a secondary assessment.

Physical Examinations

You should focus your physical examination on the severity of hypothermia, assessing the areas of the body directly affected by cold exposure, and the degree of damage. Is the whole body cold (hypothermia) or just parts

You are the Provider: PART 2

You arrive at the scene and find the patient, a 55-year-old man, sitting under a tree in his garden; he is conscious, but confused. His wife tells you that he has been working outside all day. She further states that despite her efforts to get him to take a break and drink some water, he would not. As your partner opens the jump kit, you perform a primary assessment.

Recording Time: 0 Minutes	
Appearance	Flushed appearance; confused
Level of consciousness	Conscious, but confused
Airway	Open, clear of secretions or foreign bodies
Breathing	Increased rate and depth
Circulation	Radial pulses, weak and rapid; skin is flushed, hot and moist; no gross bleeding

Your partner applies high-flow oxygen via a nonrebreathing mask and the patient's wife tells you that when she went to check on him, she found him sitting under a tree; initially, he would not respond to her. She further tells you that he has hypertension and angina, for which he takes furosemide (Lasix), potassium chloride (K-Dur), lisinopril (Prinivil), and nitroglycerin as needed.

3. What risk factors does this patient have that predispose him to heat illness?

4. What type of heat illness should you suspect that he is experiencing? Why?

(frostbite)? These determinations will have important consequences for your treatment decisions. For example, shivering indicates a protective mechanism to produce more heat because the body is cold. When shivering stops and the patient remains in a cold environment, the cold injury is more severe.

Determine the degree and extent of cold injury, as well as any other injuries or conditions that may not have been initially detected. The numbing effect of cold, both on the brain and on the body, may impair your patient's ability to tell you about other injuries or illnesses. Therefore, a careful examination of your patient's entire body, with special attention to skin temperatures, textures, and turgor, will help you avoid missing important clues to your patient's condition.

Vital Signs

Keep in mind that vital signs may be altered by the effects of hypothermia and can be an indicator of its severity. Respirations may be slow and shallow, resulting in low oxygen levels in the body. Low blood pressure and a slow pulse also indicate moderate to severe hypothermia. Carefully evaluate your patient for changes in mental status.

Monitoring Devices

Determine a core body temperature using a thermometer based on local protocol. A special low-temperature thermometer is required to take a hypothermic patient's temperature, generally done through the rectum. Pulse oximetry will often be inaccurate due to the lack of perfusion in the extremities.

Reassessment

Repeat the primary assessment. Reassess vital signs and the chief complaint. How is the patient's condition improving with the interventions? Identify and treat changes in the patient's condition. Keep a very close eye on your patient's level of consciousness and vital signs. As the body rewarms, the sudden redistribution of fluids and the release of built-up chemicals can have harmful effects, including cardiac arrhythmias. Be vigilant and monitor your patient closely, even if his or her condition appears to be improving.

Interventions

Review all treatments that have been performed. In a cold-related emergency, depending on your local and state protocols, your treatment may only include oxygen delivery. Reassess oxygen delivery and continue to provide for a warm environment by removing any wet or frozen clothing. Do not remove any clothing frozen to the patient's skin.

Communication and Documentation

Communicate all of the information you have gathered to the receiving facility. The conditions you found at the scene, what your patient was wearing, and information gathered from bystanders may be essential in evaluating and treating your patient in the hospital. Your documentation should always include the patient's physical status, the conditions at the scene, and any changes in the patient's mental status during treatment and transport.

General Management of Cold Emergencies

In most cases, you should move the patient from the cold environment to prevent further heat loss. To prevent further damage to the feet, do not allow the patient to walk. Remove any wet clothing, and place dry blankets over and under the patient Figure 30-6 . If available, give the patient warm, humidified oxygen if you have not already done so as part of the primary assessment.

Always make sure to handle the patient gently so that you do not cause any pain or further injury to the skin. Do not massage the extremities. Do not allow the patient to eat, to use any stimulants, such as coffee, tea, or cola, or to smoke or chew tobacco.

If the patient is alert, shivering, responds appropriately, and the core body temperature is between 90°F to 95°F, then the hypothermia is mild. Begin active rewarming, which includes applying heat packs or hot water bottles to the groin, axillary, and cervical regions. Turn the heat up high in the patient compartment of the ambulance. Use caution to avoid burns and rewarm the patient slowly. If possible, you can give warm fluids by mouth, assuming that the patient can swallow without a problem.

However, when the patient has moderate or severe hypothermia, you should never try to actively rewarm

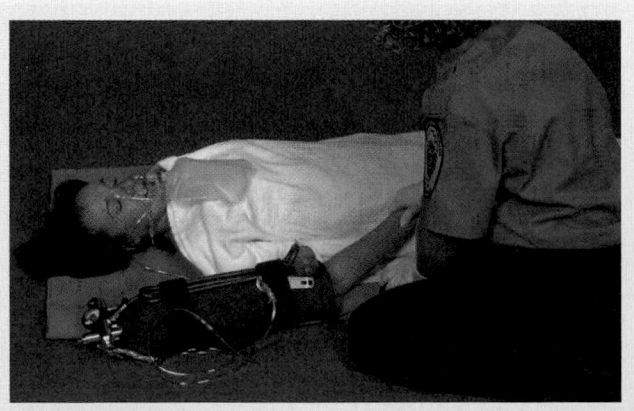

Figure 30-6 Place dry blankets over and under the patient with hypothermia; give warm, humidified oxygen, if available; assess the pulse before considering CPR.

the patient (placing heat on or into the body). Rewarming the patient too quickly may cause a fatal cardiac arrhythmia that requires defibrillation. For this reason, passive rewarming (high indoor heat) is also best delivered at an appropriate facility. Many regional and state protocols include a passive rewarming protocol that is based on determining the patient's body temperature. Follow your local protocols.

Your goal with the patient with moderate or severe hypothermia is to prevent further heat loss. Remove the patient immediately from the cold environment, place the patient in the ambulance, remove wet clothing, cover the patient with a blanket, and transport. Remember to handle the patient gently to decrease the risk of ventricular fibrillation.

If you cannot get the patient out of the cold immediately, move the patient out of the wind and away from contact with any object that will conduct heat away from the body. Place a protective cover on the patient. Remember that most body heat is lost around the head and neck.

Regardless of the nature or severity of the cold injury, remember that even an unresponsive patient may be able to hear you. Some patients have told of hearing themselves pronounced dead by someone who had forgotten the saying: "No one is dead unless he is warm and dead." If you carry an automated external defibrillator (AED), you should consider defibrillation. Although ventricular fibrillation is unlikely in patients with hypothermia, it can occur in patients who are rewarmed too rapidly.

■ Emergency Care of Local Cold Injuries

The emergency treatment of local cold injuries in the field should include the following steps:

1. Remove the patient from further exposure to the cold.
2. Handle the injured part gently, and protect it from further injury.
3. Administer oxygen, if this was not already done as part of the primary assessment.
4. Remove any wet or restricting clothing over the injured part.

If there is no chance of reinjury to a more superficial local cold injury, consider active rewarming. With frostnip, contact with a warm object may be all that is needed; you can use your hands, your breath, or the patient's own body. During rewarming, the affected part will often tingle and become red in light-skinned individuals. With immersion foot, remove wet shoes, boots, and socks, and rewarm the foot gradually, protecting it from further cold exposure. Next splint the extremity, and cover it loosely with a dry, sterile dressing. Never rub injured tissues with anything; rubbing causes further damage. Do not reexpose the injury to cold.

With a late or deep cold injury, such as frostbite, be sure to remove any jewelry from the injured part and cover the injury loosely with a dry, sterile dressing. Do not break blisters or rub or massage the area. Do not apply heat or rewarm the part. Unlike frostnip and trenchfoot, rewarming of the frostbitten extremity is best accomplished under controlled circumstances in the emergency department. You can cause a great deal of further injury to fragile tissues by attempting to rewarm a frostbitten part. Never apply something warm or hot, such as the exhaust from the ambulance engine or, even worse, an open flame. Do not allow the patient to stand or walk on a frostbitten foot.

Evaluate the patient's general condition for the signs or symptoms of systemic hypothermia. Support the vital functions as necessary, and transport the patient promptly to the hospital.

If prompt hospital care is not available and medical control instructs you to begin rewarming in the field, use a warm-water bath. Immerse the frostbitten part in water with a temperature of between 100°F and 105°F (38°C and 40.5°C). Check the water temperature with a thermometer before immersing the limb, and recheck it frequently during the rewarming process. The water temperature should never exceed 105°F (40.5°C). Stir the water continuously. Keep the frostbitten part in the water until it feels warm and sensation has returned to the skin. Dress the area with dry, sterile dressings, placing them also between injured fingers or toes. Expect the patient to report severe pain.

Never attempt rewarming if there is any chance that the part may freeze again before the patient reaches the hospital. Some of the most severe consequences of frostbite, including gangrene and amputation, have occurred when parts were thawed and then refrozen.

Cover the frostbitten part with soft, padded, sterile cotton dressings. If blisters have formed, do not break them. Remember, you cannot accurately predict the outcome of a case of frostbite early in its course. Even body parts that appear gangrenous may recover following proper emergency and hospital treatment.

■ Cold Exposure and You

As an EMT, you are also at risk for hypothermia if you work in a cold environment. If cold weather search-and-rescue operations are a possibility in your assigned areas, you should receive survival training and precautionary tips. You should be thoroughly familiar with local conditions. Be aware of existing and potential weather conditions, and stay on top of changes that are forecast for the area. Make sure proper clothing is available, and wear it whenever appropriate. Your vehicle, too, must be properly equipped and maintained for a cold environment. You cannot help others if you do not protect yourself. Never allow yourself to become a casualty!

Heat Exposure

Normal body temperature is 98.6°F (37°C). Complicated regulatory mechanisms keep this internal temperature constant, regardless of the **ambient temperature**, the temperature of the surrounding environment. In a hot environment or during vigorous physical activity, when the body itself produces excess heat, the body will try to rid itself of the excess heat. There are several ways of doing this. The two most efficient methods are sweating (and evaporation of the sweat) and dilation of skin blood vessels, which brings blood to the skin surface to increase the rate of heat radiation. In addition, a person who becomes overheated can remove clothing and try to find a cooler environment.

Ordinarily, the heat-regulating mechanisms of the body work very well, and individuals are able to tolerate significant temperature changes. When the body is exposed to more heat energy than it loses or it generates more heat than it can lose, hyperthermia results. **Hyperthermia** is a high core temperature, usually 101°F (38.3°C) or higher.

Words of Wisdom

Keeping yourself hydrated while on duty is very important, especially during periods of heavy exertion or when working in the heat. Drink at least 3 L of water a day and more when exertion or heat is involved. The color of urine (usually darker with dehydration) and frequency of urination correlate directly with the body's fluid level.

When the body's mechanisms to decrease body heat are overwhelmed and the body is unable to tolerate the excessive heat, illness develops. High air temperature can reduce the body's ability to lose heat by radiation; high humidity reduces the ability to lose heat through evaporation. A lack of acclimation to the heat is a risk factor. Another risk factor is vigorous exercise, during which the body can lose more than 1 L of sweat an hour, causing loss of fluid and electrolytes.

Illness from heat exposure can take the following three forms:

- Heat cramps
- Heat exhaustion
- Heatstroke

All three forms of heat illness may be present in the same patient because untreated heat exhaustion may progress to heatstroke. Heatstroke is a life-threatening emergency.

Persons at greatest risk for heat illnesses are children; geriatric patients; patients with heart disease, COPD, diabetes, dehydration, and obesity; and those with limited mobility. Older people, newborns, and infants exhibit poor thermoregulation. Newborns and infants often wear too much clothing. Alcohol and certain drugs, including medications that dehydrate the body or decrease the ability of the body to sweat, also make a person more susceptible to heat illnesses. When you are treating someone for a heat illness, always obtain a medication history.

Heat Cramps

Heat cramps are painful muscle spasms that occur after vigorous exercise. They do not occur only when it is hot outdoors. They may be seen in factory workers and even well-conditioned athletes. The exact cause of heat cramps is not well understood. It is known that sweat produced during strenuous exercise, particularly in a warm environment, causes a change in the body's electrolyte balance. The result may be a loss of essential electrolytes from the cells. Dehydration may also play a role in the development of muscle cramps. Large amounts of water can be lost from the body as a result of excessive sweating. This loss of water may affect muscles that are being stressed and cause them to go into spasm.

Heat cramps usually occur in the leg or abdominal muscles. When the abdominal muscles are involved, the pain and muscle spasm may be so severe that the patient appears to have an acute abdominal problem. If a patient with a sudden onset of abdominal cramps has been exercising vigorously in a hot environment, you should suspect heat cramps.

Heat Exhaustion

Heat exhaustion, also called heat prostration or heat collapse, is the most common serious illness caused by heat. Heat exposure, stress, and fatigue are causes of heat exhaustion, which is caused by hypovolemia as the result of the loss of water and electrolytes from heavy sweating. For sweating to be an effective cooling mechanism, the sweat must be able to evaporate from the body. Otherwise, the body will continue to produce sweat, with further loss of body water. People standing in the hot sun and particularly those wearing several layers of clothing, such as football fans or parade watchers, may sweat profusely but experience little body cooling. High humidity will also decrease the amount of evaporation that can occur.

Individuals working or exerting themselves in poorly ventilated areas are unable to release heat through convection. Thus, people who work or exercise vigorously and those who wear heavy clothing in a warm, humid, or poorly ventilated environment are particularly prone to heat exhaustion.

The signs and symptoms of heat exhaustion and those of associated hypovolemia are as follows:

- Dizziness, weakness, or faintness signifying a change in level of consciousness with accompanying nausea, vomiting, or headache. Muscle cramping may also be present, including abdominal cramping.
- Onset while working hard or exercising in a hot, humid, or poorly ventilated environment and sweating heavily.
- Onset, even at rest, in the older and infant age groups in hot, humid, and poorly ventilated environments or extended time in hot, humid environments. Individuals not acclimatized to the environment may also experience onset at rest.
- Cold, clammy skin with ashen pallor.
- Dry tongue and thirst.
- Normal vital signs, although the pulse is often rapid and weak (an indication for use of pulse oximetry) and the diastolic blood pressure may be low.
- Normal or slightly elevated body temperature; on rare occasions, as high as 104°F (40°C).

■ Heatstroke

Heatstroke, the least common but most serious illness caused by heat exposure, occurs when the body is subjected to more heat than it can handle and normal mechanisms for getting rid of the excess heat are overwhelmed. The body temperature then rises rapidly to the level at which tissues are destroyed. Untreated heatstroke always results in death.

Heatstroke can develop in patients during vigorous physical activity or when they are outdoors or in a closed, poorly ventilated, humid space. It also occurs during heat waves among individuals (particularly in geriatric patients) who live in buildings with no air conditioning or with poor ventilation. It may also develop in children who are left unattended in a locked car on a hot day.

Many patients with heatstroke have hot, dry, flushed skin because their sweating mechanism has been overwhelmed. However, early in the course of exertional heatstroke, the skin may be moist or wet. Keep in mind that a patient can have heatstroke even if he or she is still sweating. The body temperature rises rapidly in patients with heatstroke. It may rise to 106°F (41°C) or more. As the body core temperature rises, the patient's level of consciousness falls, resulting in unconsciousness.

Often, the first sign of heatstroke is a change in behavior. However, the patient then becomes unresponsive very quickly and seizures may occur. The pulse is usually rapid and strong at first, but as the patient becomes increasingly unresponsive, the pulse becomes weaker and the blood pressure falls. The respiratory rate increases as the body is attempting to compensate. One of the telltale signs you should be acutely aware of is when your patient no longer perspires, which means the body has lost its thermoregulatory mechanisms. If you are perspiring in the environment, your patient should also be perspiring.

■ Assessment of Heat Injuries

▷ Scene Size-up

Scene Safety

As part of your scene size-up, perform an environmental assessment. How hot is it outside? How hot is it in the room where your patient is? How well is the patient tolerating the heat? Dispatch may report the call initially as a medical or trauma emergency. The heat illness may only be secondary. Approach the scene looking for hazards as well as clues as to what may have caused your patient's problem. If you anticipate a prolonged scene time, protect yourself from the heat. Use appropriate standard precautions, including gloves and eye protection. Long-sleeved shirts and long pants may not be comfortable in warm weather; however, they can help protect you from being splashed by blood or other fluids. Consider whether you need ALS backup because intravenous fluids may need to be administered to treat the patient.

Mechanism of Injury/Nature of Illness

As you observe the scene, look for indicators of the MOI. This helps you develop an early index of suspicion for underlying injuries in the patient who has sustained a significant MOI. For example, you arrive on the scene at a shopping mall to find an older man with a decreased level of consciousness inside a parked vehicle on a warm, humid, sunny day. The MOI for this patient is sitting in a warm environment under direct sunlight with no ventilation.

▷ Primary Assessment

Form a General Impression

As you approach your patient, observe how the patient interacts with you and the environment. This will help identify the patient's degree of distress. Introduce yourself and ask about the chief complaint. A heat illness may be the primary problem or it may simply be aggravating a medical or trauma condition. Remember, prolonged heat exposure may stress the heart, causing a heart attack. Use this initial interaction to guide you in assessing for immediate life threats and related problems. Perform a rapid scan and avoid tunnel vision.

Assess the patient's mental status using the AVPU scale (*Alert* to person, place, and day; responsive to *Verbal* stimuli; responsive to *Pain*; *Unresponsive*). Heat stroke is a true life-threatening emergency. The severity of your patient's condition may be identified by gathering clues about his or her mental status. The more altered the patient's mental status is, the more serious the heat problem.

Airway and Breathing

Assess the patient's airway and breathing and treat any life-threatening problems found. Unless the patient is unresponsive, the airway should be patent. Nausea and vomiting, however, may occur. Position the patient to protect the airway as necessary. If the patient is unresponsive, be cautious how you open the airway; consider spinal precautions. Breathing will be fast depending on the patient's core temperature but should otherwise be adequate. Providing oxygen to the patient will assist with the perfusion of body tissues and may decrease nausea. If your patient is unresponsive, insert an airway and provide bag-mask device ventilations according to protocol.

Circulation

Circulation is assessed by palpating a pulse. If it is adequate, assess the patient for perfusion and bleeding. Assess the patient's skin condition carefully **Table 30-2**. Treat the patient aggressively for shock by removing the patient from the heat and positioning the patient to improve circulation. If the patient is bleeding, bandage according to protocol.

Table 30-2 Skin Condition

Skin Condition	Indicates
Moist, pale, cool skin	Excessive fluid and salt loss
Hot, dry skin	Body is unable to regulate core temperature
Hot, moist skin	Body is unable to regulate core temperature

Transport Decision

If your patient has any signs of heatstroke (high temperature; red, dry skin; altered mental status; tachycardia; poor perfusion), then transport without delay.

History Taking

Investigate Chief Complaint

After the life threats have been managed during the primary assessment, investigate the chief complaint. Obtain a medical history and be alert for injury-specific signs and symptoms such as the absence of perspiration, decreased level of consciousness, confusion, muscle cramping, nausea, and vomiting.

SAMPLE History

Obtain a SAMPLE history with an eye to noting any activities, conditions, or medications that may predispose a patient to dehydration or heat-related problems. Patients with inadequate oral intake, or who are taking diuretics, may have difficulty tolerating exposure to heat. Many psychiatric medications used by geriatric patients affect how well they tolerate heat. Be thorough in your questioning. Determine your patient's exposure to heat and humidity and activities prior to the onset of symptoms.

Secondary Assessment

The secondary assessment is a more detailed, comprehensive examination of the patient that is used to uncover injuries that may have been missed during the primary assessment. In some instances, such as a critically injured patient or a short transport time, you may not have time to conduct a secondary assessment.

Physical Examinations

If your patient is unresponsive, perform a full-body scan looking for problems or explanations as to what is wrong. Obtain the patient's vital signs to help understand how serious the problem is.

If the patient is conscious, perform a focused assessment. Exposure to heat has significant effects on the metabolism, muscles, and cardiovascular system. Assess the patient for muscle cramps or confusion. Examine the patient's mental status and skin temperature and wetness. Take the patient's vital signs, including body temperature.

Perform a detailed full-body scan if circumstances and time permit. Pay special attention to the patient's skin temperature, **turgor**, and wetness. Skin turgor is the ability of the skin to resist deformation. It is tested by gently pinching skin on the forehead or back of the hand. Normally the skin will quickly flatten out. In dehydration, with poor skin turgor, the skin will remain tented. Perform a careful neurologic examination.

Vital Signs

Patients who are hyperthermic will be tachycardic and tachypneic. As long as they maintain a normal blood pressure, their bodies are compensating for the fluid loss. Once their blood pressure begins to fall, it indicates they are no longer able to compensate for fluid loss and are going into shock. Your assessment of the patient's skin will help determine how serious the heat problem is. For example, in heat exhaustion, the skin temperature may be normal or may even be cool and clammy; however, in heatstroke, the skin is hot.

Monitoring Devices

Check the patient's temperature with a thermometer, depending on protocol. Your unit equipment may include disposable or oral thermometers with disposable covers. Some agencies provide tympanic/ear thermometers. You may not use these devices routinely so become familiar with how they work. In patients with a heat-related illness, pulse oximetry is also indicated.

Reassessment

Watch your patient's condition carefully for deterioration. Any decline in level of consciousness is an ominous sign. Monitor the patient's vital signs at least every 5 minutes. Evaluate the effectiveness of your interventions. Be careful not to cause shivering when cooling down a patient with heat problems. Shivering generates more heat and can occur when cooling is not monitored closely.

Interventions

Remove your patient as quickly as possible from the hot environment. Patients with heat cramps or exhaustion usually respond well to passive cooling and fluids by mouth. Patients with symptoms of heatstroke should be transported immediately and actively cooled.

Communication and Documentation

Inform the staff at the receiving facility early on that your patient is experiencing a heatstroke because additional resources may be required. Document the weather conditions and the activities the patient was performing prior to the emergency in your patient care report.

Management of Heat Emergencies

Heat Cramps

Take the following steps to treat heat cramps in the field Figure 30-7 :

1. Remove the patient from the hot environment, including sunlight, a source of radiant heat gain. Loosen any tight clothing.
2. Administer high-flow oxygen.
3. Rest the cramping muscles. Have the patient sit or lie down until the cramps subside.
4. Replace fluids by mouth. Use water or a diluted (half-strength) balanced electrolyte solution, such as Gatorade. In most cases, plain water is the most useful. Do not give salt tablets or solutions that have a high salt concentration. The patient already has an adequate amount of electrolytes circulating; they are just not distributed properly. With adequate rest and fluid replacement, the body will adjust the distribution of electrolytes, and the cramps will disappear.
5. Cool the patient with cool water spray or mist and add convection to the cooling method by manually or mechanically fanning the patient.

When the heat cramps are gone, the patient may resume activity. For example, an athlete can return to play once the heat cramps have disappeared. However, heavy sweating may cause the cramps to recur. The best preventive and treatment strategy is hydration by drinking sufficient quantities of water.

If the cramps do not go away after these measures, transport the patient to the hospital. If you are uncertain

Figure 30-7 A patient with heat cramps should be moved to a cool environment as you begin your assessment and treatment.

that the patient's cramps were caused by the heat or you note anything out of the ordinary, contact medical control or transport the patient to the hospital.

■ Heat Exhaustion

To treat the patient with heat exhaustion, follow the steps in Skill Drill 30-1:

1. Remove any excessive layers of clothing, particularly around the head and neck.

2. Move the patient promptly from the hot environment, preferably into the back of the air-conditioned ambulance. If outdoors, move out of the sun. (Step 1).

3. Give the patient oxygen if this was not already done as part of the primary assessment.

4. Splash the patient with cool water if his or her body temperature is elevated. Do not use ice water.

5. Encourage the patient to lie down and elevate the legs. Loosen any tight clothing and fan the patient for cooling (Step 2).

6. If the patient is fully alert, encourage him or her to sit up and slowly drink up to a liter of water,

Skill Drill 30-1

Treating for Heat Exhaustion

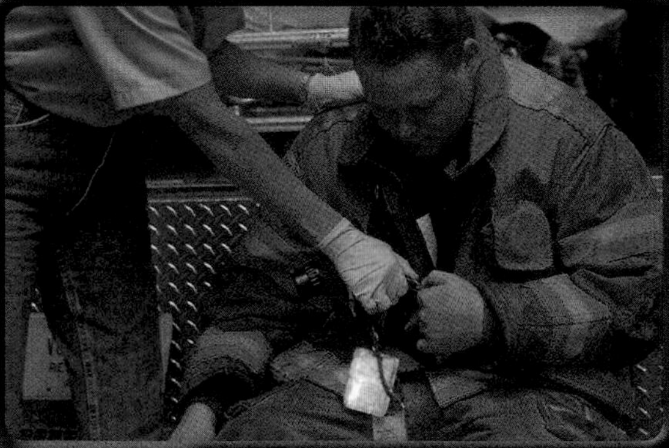

Step 1 Move the patient to a cooler environment. Remove extra clothing.

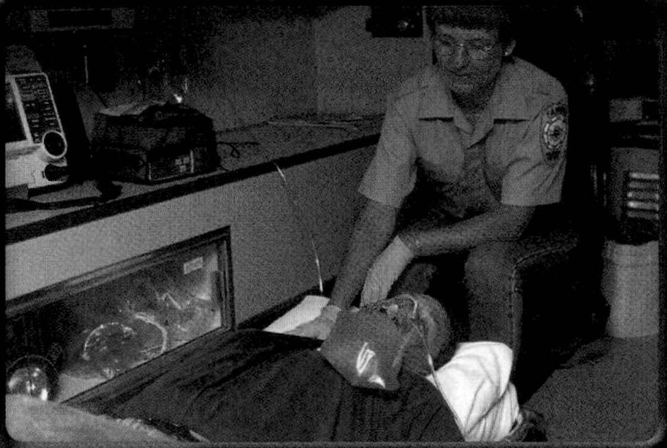

Step 2 Give oxygen. Place the patient in a supine position, elevate the legs, and fan the patient.

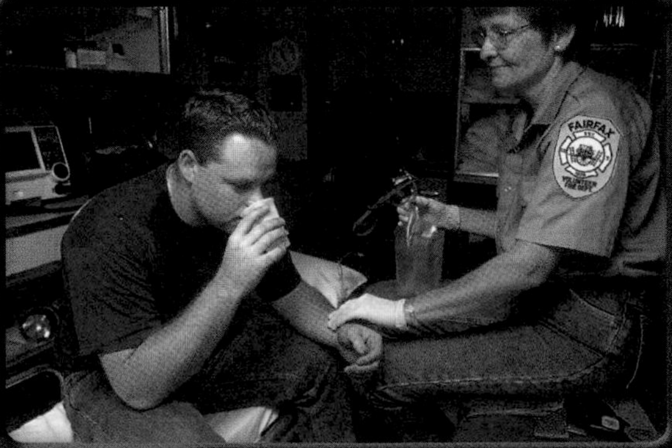

Step 3 If the patient is fully alert, give water by mouth.

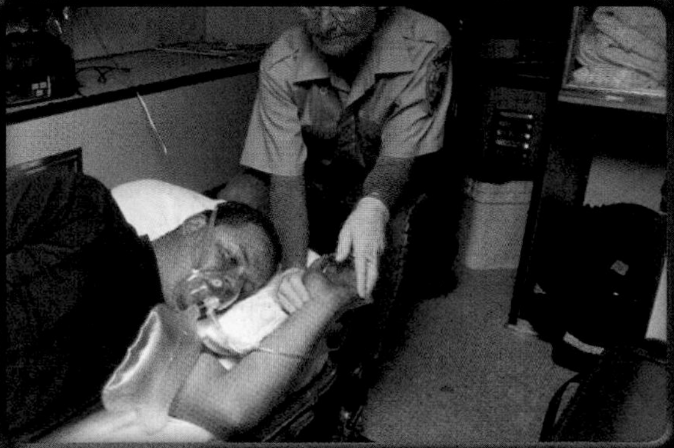

Step 4 If nausea develops, secure and transport the patient on his or her side.

as long as nausea does not develop. Never force fluids by mouth on a patient who is not fully alert, or allow drinking while supine, because the patient could aspirate the fluid into the lungs. If the patient does become nauseated, transport on the side to prevent aspiration (Step 3).

In most cases, these measures will reverse the symptoms, causing the patient to feel better within 30 minutes. But you should prepare to transport the patient to the hospital for more aggressive treatment, such as IV fluid therapy and close monitoring, especially in the following circumstances:

- The symptoms do not clear up promptly.
- The level of consciousness decreases.
- The body temperature remains elevated.
- The person is very young, older, or has any underlying medical condition, such as diabetes or cardiovascular disease.

7. Transport the patient on his or her side if you think the patient may be nauseated and ready to vomit, but make certain that the patient is secured (Step 4).

■ Heatstroke

Recovery from heatstroke depends on the speed with which treatment is administered, so you must be able to identify this patient quickly. Emergency treatment has one objective: Get the body temperature down by any means available. Take the following steps when treating a patient with heatstroke:

1. Move the patient out of the hot environment and into the ambulance.
2. Set the air conditioning to maximum cooling.

3. Remove the patient's clothing.
4. Give the patient 100% oxygen if this was not done as part of the primary assessment. If needed, assist the patient's ventilations with a bag-mask device and appropriate airway adjuncts as per your protocol.
5. Apply cool packs to the patient's neck, groin, and armpits Figure 30-8 .
6. Cover the patient with wet towels or sheets, or spray the patient with cool water and fan him or her to quickly evaporate the moisture on the skin.
7. Aggressively and repeatedly fan the patient with or without dampening the skin.
8. Provide immediate transport to the hospital.
9. Notify the hospital as soon as possible so that the staff can prepare to treat the patient immediately on arrival.

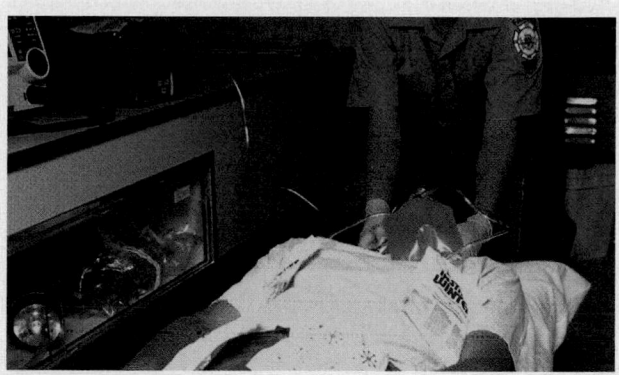

Figure 30-8 As part of treatment of heatstroke, give oxygen and place cool packs about the patient's neck, groin, and armpits.

You are the Provider: PART 3

You quickly place the patient onto the stretcher and load him into the ambulance. The air conditioner is on and set to high. You perform a secondary assessment, which does not reveal any gross signs of injury, while your partner assesses the patient's vital signs. You then depart the scene and begin further treatment en route to the hospital.

Recording Time: 6 Minutes	
Respirations	24 breaths/min; adequate depth
Pulse	130 beats/min; weak and regular
Skin	Hot, flushed, and moist
Blood pressure	88/66 mm Hg
Oxygen saturation (Sao$_2$)	95% (on oxygen)

5. What specific treatment is required for this patient?
6. What is the most likely explanation for this patient's vital signs?

Special Populations

As a person ages, the body can lose the ability to respond to the environment. Older adults undergo changes in their ability to compensate for low or high ambient temperatures. For example, if the ambient temperature rises from 85°F to 94°F (29.5°C to 34.5°C), the older adult may not recognize the change or be able to compensate for it. Therefore, unless the person is accustomed to the heat, heatstroke can develop relatively quickly.

Shivering, a common effect of hypothermia, is the body's attempt to maintain heat. However, because of a decrease in muscle mass or tone, the hypothermic geriatric patient may not shiver. Furthermore, a decrease in muscle mass and body fat means that there is less insulation and protection from the cold. Because of the body's altered response to heat loss and its ability to gain heat, the health care provider might not suspect or report hypothermia. In caring for the geriatric patient in cold climates, be sure to protect the patient against unwanted heat loss. Cover all exposed areas with loose-fitting blankets. Pay particular attention to protecting the patient's head because heat loss from the head and neck is substantial.

Because of reduced circulation to the skin, heat loss via conduction, convection, and radiation is significantly lower. Additionally, the aging process alters the patient's ability to perspire; therefore, heat loss through evaporation is reduced. Because the elderly patient cannot disperse heat effectively, classic heatstroke can develop rapidly. Typically, the older adult will not go through an initial stage of heat exhaustion. During the summer, you should be acutely aware of the potential for heatstroke and factors that can predispose a patient to heat illness. Factors that increase the possibility of heatstroke include medications, diabetes, alcohol abuse, malnutrition, parkinsonism, hyperthyroidism, and obesity.

Both hypothermia and hyperthermia can appear in older patients in environmental settings that are subtle. These problems are commonly found when an older person has cost concerns about heating or cooling their home and thus keeps the heat turned down in the winter or does not use air conditioning in hot weather. Thermal emergencies can develop over a period of time for older persons in these indoor, urban environments that may not seem uncomfortable to you.

Radiation Exposure

Exposure to non-ionized radiation occurs on a daily basis. Examples of non-ionized radiation are cell phones, microwave ovens, and ultraviolet (UV) light from the sun. Long-term exposure to UV light from the sun is one of the main risk factors for skin cancer. Exposure to sunlight can also burn unprotected skin in as little as 30 minutes. As an EMT, you will be working outside as well as inside, on cloudy and sunny days. To protect your skin and lower your risk for skin cancer, wear a sunscreen with SPF of at least 15 or higher.

A sunburn usually can be classified as a superficial burn. Prehospital treatment of sunburn includes removing the patient from the sun and treating any additional symptoms. If the sunburn is severe and dehydration is present, then ALS intervention may be needed for intravenous fluid replacement.

Drowning

Drowning is the process of experiencing respiratory impairment from submersion/immersion in liquid. Some agencies may still use the term "near drowning" to refer to a patient who survives at least temporarily (24 hours) after suffocation in water. According to the Centers for Disease Control, an average of 10 unintentional drownings occur per day. More than 25% are children younger than 14 years. Alcohol consumption, preexisting seizure disorders, geriatric patients with cardiovascular disease, and unsupervised access to water are among the major risk factors.

Drowning is often the last in a cycle of events caused by panic in the water. It can happen to anyone who is submerged in water for even a short period of time. Struggling toward the surface or the shore, the person becomes fatigued or exhausted, which leads him or her to sink even deeper. However, drowning also occurs in mop buckets, puddles, bathtubs, and other places where the person is not completely submerged. Small children can drown in only a few inches of water if left unattended.

Inhaling very small amounts of either fresh water or salt water can severely irritate the larynx, sending the muscles of the larynx and the vocal cords into spasm, called laryngospasm. The average person experiences this to a mild degree when a small amount of liquid is inhaled and the patient coughs and seems to be choking for a few seconds. This is the body's attempt at self-preservation; laryngospasm prevents more water from entering the lungs. In severe cases such as water submersion, however, the patient's lungs cannot be ventilated because significant laryngospasm is present. Instead, progressive hypoxia occurs until the patient becomes unconscious. At this point, the spasm relaxes, making rescue breathing possible. Of course, if the patient has not already been removed from the water, the patient may now inhale deeply, and more water may enter the lungs.

In 85% to 90% of cases, significant amounts of water enter the lungs of the drowning victim.

Spinal Injuries in Submersion Incidents

Submersion incidents may be complicated by spinal fractures and spinal cord injuries. You must assume that spinal injury exists with the following conditions:

- The submersion has resulted from a diving mishap or long fall.

Safety

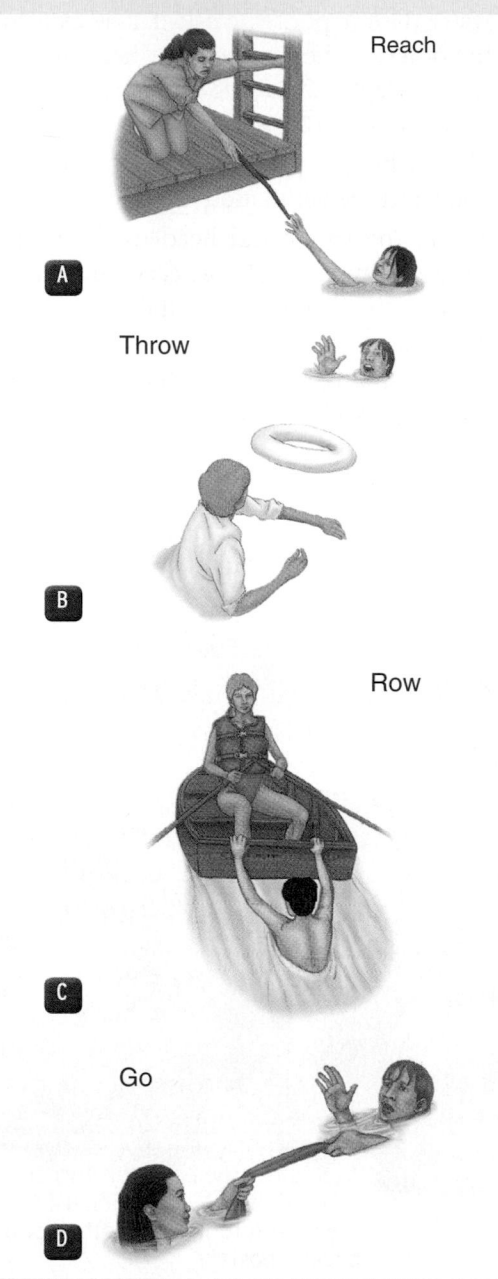

Reach

A

Throw

B

Row

C

Go

D

Figure 30-9 Basic rules of water rescue. **A.** Reach for the person from shore. If you cannot reach the person from shore, wade closer. **B.** If an object that floats is available, throw it to the person. **C.** Use a boat if one is available. **D.** If you must swim to the person, use a towel or board for him or her to hold onto. Do not let the person grab you.

You must ensure the safety of rescue personnel before a water rescue can begin. If the patient is conscious and still in the water, you should perform a water rescue. The saying: "Reach, throw, and row, and only then go" **Figure 30-9** sums up the basic rule of water rescue. First, try to reach for the patient. If that does not work, throw the patient a rope, a life preserver, or any floatable object that is available. For example, an inflated spare tire, rim and all, will float well enough to support two people in the water. Next, use a boat if one is available. Do not attempt a swimming rescue unless you are trained and experienced in the proper techniques. Even then, you should always wear a helmet and a personal flotation device **Figure 30-10**. Too many well-meaning individuals have themselves become victims while attempting a swimming rescue. In cold climates or cold-water locations, rapid hypothermia is also a concern for rescuers. Be prepared for this potential event.

If you work in a recreation area near lakes, rivers, or the ocean, you should have a prearranged plan for water rescue. This plan should include access to and cooperation with local personnel who are trained and skilled in water rescue; these personnel should help to develop the protocol for water rescue. Because the success of any water rescue depends on how rapidly the patient is removed from the water and ventilated, make sure you always have immediate access to personal flotation devices and other rescue equipment. Survival rates drastically decline the longer a victim is immersed. Cold water drowning survival rates are somewhat higher.

Figure 30-10 When performing a water rescue, you must wear proper personal protective equipment, including a personal flotation device.

- The patient is unconscious, and no information is available to rule out the possibility of a mechanism causing neck injury.
- The patient is conscious but complains of weakness, paralysis, or numbness in the arms or legs.

- You suspect the possibility of spinal injury despite what witnesses say.

Most spinal injuries in diving incidents affect the cervical spine. When spinal injury is suspected, the neck must be protected from further injury. This means that you will have to stabilize the suspected injury while the patient is still in the water. To stabilize a suspected spinal injury in water, follow the steps in Skill Drill 30-2:

1. Turn the patient supine. Two rescuers are usually required to turn the patient safely, although in some cases one rescuer will suffice. Always rotate the entire upper half of the patient's body as a single unit. Twisting only the head, for example, may aggravate any injury to the cervical spine (Step 1).

2. Restore the airway and begin ventilation. Immediate ventilation is the primary treatment of all drowning patients as soon as the patient is face up in the water. Use a pocket mask if it is available. Have the other rescuer support the head and trunk as a unit while you open the airway and begin artificial ventilation (Step 2).

3. Float a buoyant backboard under the patient as you continue ventilation (Step 3).

4. Secure the trunk and head to the backboard to eliminate motion of the cervical spine. Do not remove the patient from the water until this is done (Step 4).

5. Remove the patient from the water, on the backboard (Step 5).

Skill Drill 30-2

Stabilizing a Suspected Spinal Injury in the Water

Step 1 Turn the patient to a supine position by rotating the entire upper half of the body as a single unit.

Step 2 As soon as the patient is turned, begin artificial ventilation using the mouth-to-mouth method or a pocket mask.

Step 3 Float a buoyant backboard under the patient.

Step 4 Secure the patient to the backboard.

Skill Drill 30-2

Stabilizing a Suspected Spinal Injury in the Water, continued

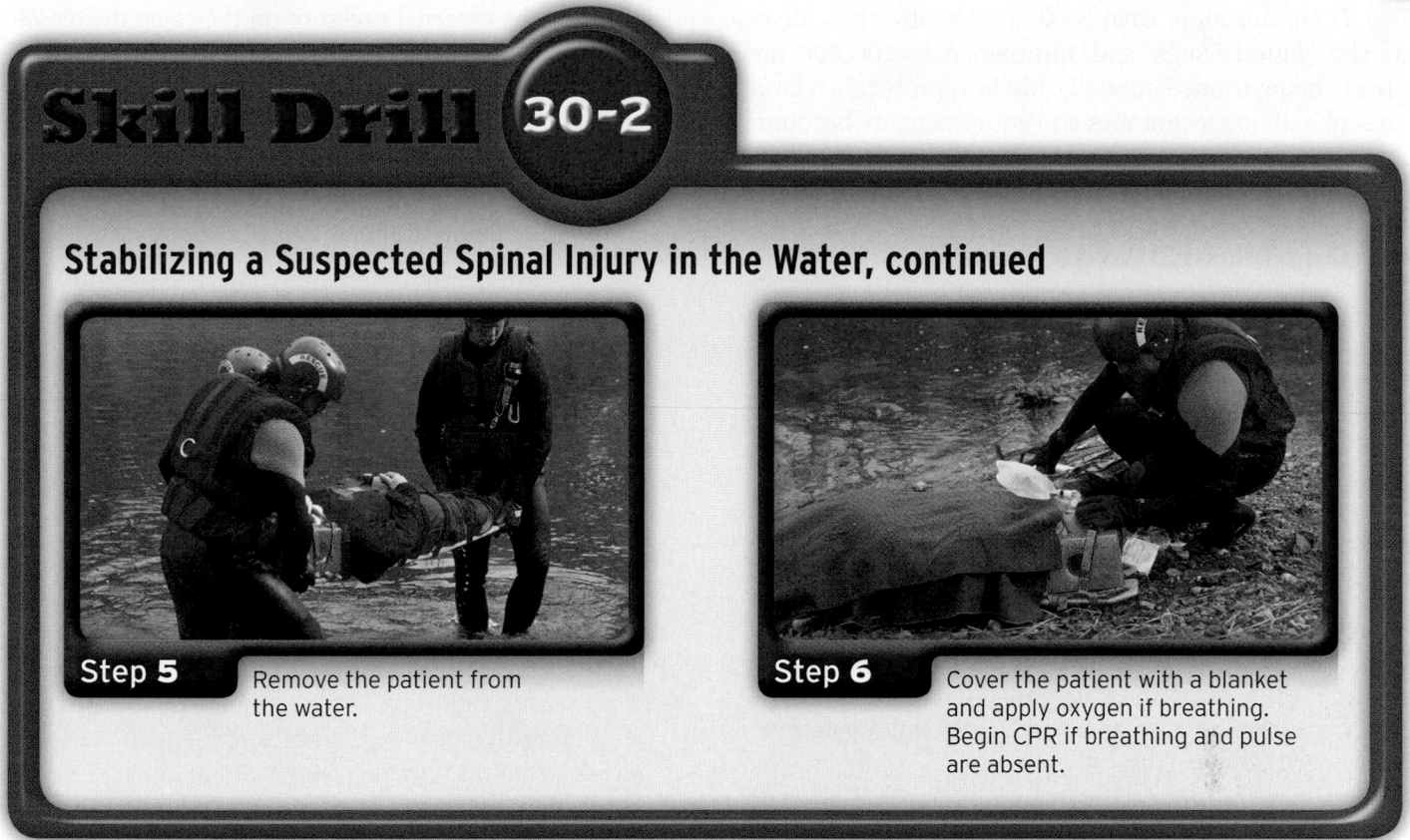

Step 5 Remove the patient from the water.

Step 6 Cover the patient with a blanket and apply oxygen if breathing. Begin CPR if breathing and pulse are absent.

6. Cover the patient with a blanket. Give oxygen if the patient is breathing spontaneously. Begin CPR if there is no pulse. Effective cardiac compression or CPR is extremely difficult to perform when the patient is still in the water (Step 6).

■ Recovery Techniques

On occasion, you may be called to the scene of a drowning and find that the patient is not floating or visible in the water. An organized rescue effort in these circumstances calls for personnel who are experienced with recovery techniques and equipment, including snorkel, mask, and scuba gear. <u>SCUBA</u> (self-contained underwater breathing apparatus) gear is a system that delivers air to the mouth and lungs at atmospheric pressures that increase with the depth of the dive.

As a last resort, when standard procedures for recovery are unsuccessful, you may have to use a grappling iron or large hook to drag the bottom for the victim. Although the hook could seriously wound the patient, it may be the only effective way to bring him or her to the surface for resuscitation efforts.

■ Resuscitation Efforts

You should never give up on resuscitating a cold-water drowning victim. When a person is submerged in water that is colder than body temperature, heat will be conducted from the body to the water. The resulting hypothermia can protect vital organs from the lack of oxygen. In addition, exposure to cold water will occasionally activate certain primitive reflexes, which may preserve basic body functions for prolonged periods.

In one case, a 2½-year-old girl recovered after being submerged in cold water for at least 66 minutes. Continue full resuscitation efforts until the patient recovers or is pronounced dead by a physician.

Also, whenever a person dives or jumps into very cold water, the **diving reflex**, slowing of the heart rate caused by submersion in cold water, may cause immediate bradycardia, a slow heart rhythm. Loss of consciousness and drowning may follow. However, the person may be able to survive for an extended period of time under water, thanks to a lowering of the metabolic rate associated with hypothermia. For this reason, you should continue full resuscitation efforts no matter how long the patient has been submerged.

Diving Emergencies

Most serious water-related injuries are associated with dives, with or without scuba gear. Some of these problems are related to the nature of the dive; others result from panic. Panic is not restricted to the person who is frightened by water. It can happen even to the experienced diver or swimmer.

There are more than 3,000,000 scuba sport divers in the United States, and approximately 200,000 new divers being trained annually. Medical problems relating to scuba diving techniques and equipment are becoming increasingly common. These problems are separated into three phases of the dive: descent, bottom, and ascent.

Descent Emergencies

Descent problems are usually caused by the sudden increase in pressure on the body as the person dives deeper into the water. Some body cavities cannot adjust to the increased external pressure of the water; the result is severe pain. The usual areas affected are the lungs, the sinus cavities, the middle ear, the teeth, and the area of the face surrounded by the diving mask. Usually, the pain caused by these "squeeze problems" forces the diver to return to the surface to equalize the pressures, and the problem clears up by itself. A diver who continues to complain of pain, particularly in the ear, after returning to the surface should be transported to the hospital.

A person with a perforated tympanic membrane (ruptured eardrum) may develop a special problem while diving. If cold water enters the middle ear through a ruptured eardrum, the diver may lose his or her balance and orientation. The diver may then shoot to the surface and run into ascent problems.

Emergencies at the Bottom

Problems related to the bottom of the dive are rarely seen. They include inadequate mixing of oxygen and carbon dioxide in the air the diver breathes and accidental feeding of poisonous carbon monoxide into the breathing apparatus. Both are the result of faulty connections in the diving gear. These situations can cause drowning or rapid ascent; they require emergency resuscitation and transport of the patient.

Ascent Emergencies

Most of the serious injuries associated with diving are related to ascending from the bottom and are referred to as ascent problems. These emergencies usually require aggressive resuscitation. Two particularly dangerous medical emergencies are air embolism and decompression sickness (also called "the bends").

Air Embolism

The most dangerous, and most common, emergency in scuba diving is **air embolism**, a condition involving bubbles of air in the blood vessels. Air embolism may occur on a dive as shallow as 6'. The problem starts when the diver holds his or her breath during a rapid ascent. The air pressure in the lungs remains at a high level while the external pressure on the chest decreases. As a result, the air inside the lungs expands rapidly, causing the alveoli in the lungs to rupture. The air released from this rupture can cause the following injuries:

- Air may enter the pleural space and compress the lungs (a pneumothorax).
- Air may enter the mediastinum (the space within the thorax that contains the heart and great vessels), causing a condition called pneumomediastinum.
- Air may enter the bloodstream and create bubbles of air in the vessels called air emboli.

Pneumothorax and pneumomediastinum both result in pain and severe dyspnea. An air embolus will act as a plug and prevent the normal flow of blood and oxygen to a specific part of the body. The brain and spinal cord are the organs most severely affected by air embolism because they require a constant supply of oxygen.

The following are potential signs and symptoms of air embolism:

- Blotching (mottling of the skin)
- Froth (often pink or bloody) at the nose and mouth
- Severe pain in muscles, joints, or abdomen
- Dyspnea and/or chest pain
- Dizziness, nausea, and vomiting
- Dysphasia (difficulty speaking)
- Cough
- Cyanosis
- Difficulty with vision
- Paralysis and/or coma
- Irregular pulse and even cardiac arrest

Decompression Sickness

Decompression sickness, commonly called the **bends**, occurs when bubbles of gas, especially nitrogen, obstruct the blood vessels. This condition results from too rapid an ascent from a dive, too long of a dive at too deep a depth, or repeated dives on the same day without the proper time intervals. During the dive, nitrogen that is being breathed dissolves in the blood and tissues because it is under pressure. When the diver ascends, the external pressure is decreased, and the dissolved nitrogen forms small bubbles within those tissues. These bubbles can lead to problems similar to those that occur in air embolism (blockage of tiny blood vessels, depriving parts of the body of their normal blood supply), but severe pain in certain tissues or spaces in the body is the most common problem.

The most striking symptom is abdominal and/or joint pain so severe that the patient literally doubles up or "bends." Dive tables and computers are available to show the proper rate of ascent from a dive, including the number and length of pauses that a diver should make

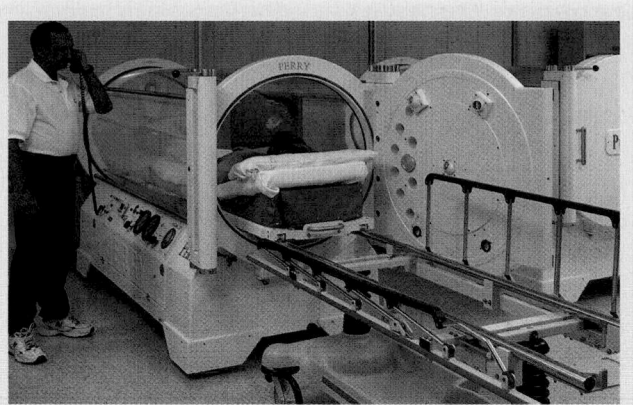

Figure 30-11 A hyperbaric chamber, usually a small room, is pressurized to more than atmospheric pressure and used in the treatment of decompression sickness and air embolism.

on the way up. However, even divers who stay within these limits can experience the bends.

Even after a "safe dive," decompression sickness can occur from driving a car up a mountain or flying in an unpressurized airplane that climbs too rapidly to a great height. However, the risk of this diminishes after 24 to 48 hours. The problem is exactly the same as ascent from a deep dive: a sudden decrease of external pressure on the body and release of dissolved nitrogen from the blood that forms bubbles of nitrogen gas within the blood vessels.

You may find it difficult to distinguish between air embolism and decompression sickness. As a general rule, air embolism occurs immediately on return to the surface, whereas the symptoms of decompression sickness may not occur for several hours. The emergency treatment is the same for both. It consists of basic life support (BLS) followed by recompression in a hyperbaric chamber, a chamber or a small room that is pressurized to more than atmospheric pressure Figure 30-11 . Recompression treatment allows the bubbles of gas to dissolve into the blood and equalizes the pressures inside and outside the lungs. Once these pressures are equalized, gradual decompression can be accomplished under controlled conditions to prevent the bubbles from reforming.

Assessment of Drowning and Diving Emergencies

Scene Size-up

Scene Safety
In managing water emergencies, your standard precautions should include gloves and eye protection at a minimum. Check for hazards to your crew. Never drive through moving water—a small amount can push the vehicle. Use extreme caution when driving through standing water. Never attempt a water rescue without proper training and equipment. Call for additional resources early.

If your patient is still in the water, look for the best, safest means of removal. This may require additional help from search and rescue teams or special extrication equipment. Trauma and spinal stabilization must be considered when the scene is a recreational setting. Check for additional patients based on where and how the problem occurred.

Mechanism of Injury/Nature of Illness
As you observe the scene, look for indicators of the MOI. This helps you develop an early index of suspicion for underlying injuries in the patient who has sustained a significant MOI. As you put together information from dispatch and your observations of the scene, consider how the MOI produced the injuries expected.

Primary Assessment

Form a General Impression
Use your evaluation of the patient's chief complaint to guide you in your assessment of life threats and determine whether spinal precautions are necessary. Pay particular attention to chest pain, dyspnea, and complaints related to sensory changes when a diving emergency is suspected. Determine the patient's level of consciousness using the AVPU scale (*Alert* to person, place, and day; responsive to *Verbal* stimuli; responsive to *Pain*; *Unresponsive*). Be suspicious of alcohol use and its effects on the patient's level of consciousness.

Airway and Breathing
Usual standard measures should be employed for any patient found or injured while in water. Begin with opening the airway and assessing breathing in unresponsive patients. Take into consideration the possibility of spinal trauma and take appropriate actions. The airway may be obstructed with water. Suction according to protocol if the patient has vomited or pink, frothy secretions are found in the airway. Provide ventilations with a bag-mask device for breathing that is inadequate. Use an airway adjunct to facilitate bag-mask device ventilations as necessary.

If the patient is responsive, provide high-flow oxygen with a nonrebreathing mask and if there is no risk of spinal injury, position the patient to protect the airway from aspiration in the event of vomiting.

Obtaining and continual monitoring of breath sounds in drowning patients is a key part of your assessment. Through auscultation, you may hear diminished sounds or even gurgling sounds from the water that has been inhaled. This information and any changes in the patient's lung sounds are important to relay to ALS providers who may rendezvous with your unit as well as to the receiving facility.

Words of Wisdom

The patient's breathing or lack of breathing can be a determining factor in submersion time. A patient who was briefly submerged may be coughing if rescued early. The longer the patient is under water, the more water enters the lungs. Agonal-type breathing can indicate prolonged submersion. Respiratory arrest occurs from very prolonged submersion.

Circulation

Check for a pulse. It may be difficult to find a pulse because of constriction of the peripheral blood vessels and low cardiac output resulting in cyanosis. Nevertheless, if the pulse is unmeasurable, then the patient may be in cardiac arrest. Begin CPR and apply your automated external defibrillator (AED) according to BLS and American Heart Association guidelines.

Evaluate the patient for adequate perfusion and treat for shock by maintaining normal body temperature and improving circulation through positioning. The patient's skin may be cold to the touch. If the MOI suggests trauma, assess for bleeding and treat appropriately.

Transport Decision

Even if resuscitation in the field appears completely successful, you must always transport patients to the hospital. Inhalation of any amount of fluid can lead to delayed complications lasting for days or weeks. Patients with decompression sickness and air embolism must be treated in a recompression chamber. If you live in an area with a significant amount of diving activity, you will have transport protocols in this regard. Usually, the patient will be stabilized in the nearest emergency department. Perform all interventions en route.

History Taking

Investigate Chief Complaint

After the life threats have been managed during the primary assessment, investigate the chief complaint. Obtain a medical history and be alert for injury-specific signs and symptoms as well as any pertinent negatives.

SAMPLE History

Obtain a SAMPLE history with special attention to the length of time the drowning victim was under water or the time of onset of symptoms in relation to the last dive. Note any physical activity, alcohol or drug consumption, and other medical conditions. All of these factors may have an effect on the diving or drowning emergency. In diving emergencies, it is important to determine the dive parameters in your history, including depth, time, and previous diving activity.

You are the Provider: PART 4

Your partner has notified the hospital and given them a patient report. You continue active cooling measures and initiate the appropriate shock treatment. When you reassess the patient, you note that his level of consciousness has decreased; he is responsive only to pain. His respirations are still rapid, but are now markedly shallow.

Recording Time: 11 Minutes	
Level of consciousness	Responsive only to pain
Respirations	26 breaths/min; shallow
Pulse	126 beats/min; weak and regular
Skin	Flushed, hot, and moist
Blood pressure	90/70 mm Hg
Sao_2	89% (on oxygen)

7. How should you adjust your treatment for this patient?

8. How will you know when you have adequately cooled the patient?

Secondary Assessment

The secondary assessment is a more detailed, comprehensive examination of the patient that is used to uncover injuries that may have been missed during the primary assessment. In some instances, such as a critically injured patient or a short transport time, you may not have time to conduct a secondary assessment.

Physical Examinations

If the patient is responsive, focus your physical examination on the basis of the chief complaint and the history obtained. This should include a thorough examination of the patient's lungs, including breath sounds.

Serious drowning situations typically result in an unresponsive patient. It is important to begin with a full-body scan in these situations to look for hidden life threats and potential trauma, even if trauma is not suspected. Look for signs of trauma or complications with the drowning. A diver with problems should be given a full-body scan for indications of the bends or an air embolism. Focus on pain in the joints and the abdomen. Pay attention to whether your patient is getting adequate ventilation and oxygenation, and check for signs of hypothermia. Obtain a Glasgow Coma Scale score to assess the patient's neurologic status and thinking.

Time and personnel permitting, complete a detailed full-body scan en route to the hospital. A careful examination may reveal additional injuries not initially observable. Examine the patient for respiratory, circulatory, and neurologic compromise. A careful distal circulatory, sensory, and motor function examination will be helpful in assessing the extent of the injury. Assess for peripheral pulses, skin color and discoloration, itching, pain, and paresthesia (numbness and tingling).

Vital Signs

The vital signs are a good indicator of how your patient is tolerating the effects of drowning or diving complications. Check the patient's pulse rate, quality, and rhythm. Pulse and blood pressure may be difficult to palpate in the hypothermic patient. Check carefully for both peripheral and central pulses, and listen over the chest for a heartbeat if pulses are weak. Check the respiratory rate, quality, and rhythm. Assess and document pupil size and reactivity.

Monitoring Devices

Although it is a valuable tool, oxygen saturation readings may produce a false low reading because of hypoperfusion of the patient's monitoring finger. Shivering also can interfere with obtaining an accurate reading because of excessive movement.

Reassessment

Repeat the primary assessment. Reassess vital signs and the chief complaint. Are the airway, breathing, and circulation still adequate? Recheck patient interventions. Are your treatments for problems with the ABCs still effective?

The condition of patients who have experienced submersion in water may deteriorate rapidly because of pulmonary injury, fluid shifts in the body, cerebral hypoxia, and hypothermia. Patients with air embolism or decompression sickness may decompensate quickly. Assess your patient's mental status constantly, and assess vital signs at least every 5 minutes, paying particular attention to respirations and breath sounds.

Interventions

Treatment for drowning begins with rescue and removal from the water. When necessary, artificial ventilation should begin as soon as possible, even before the victim is removed from the water. At the same time, you must take care to stabilize and protect the patient's spine when a long fall or dive has occurred (or if this is a possibility when no information is provided). Associated cervical spine injuries are possible, especially in diving accidents.

Communication and Documentation

Document the circumstances of the drowning and extrication. The receiving facility personnel will need to know how long the patient was submerged, the temperature of the water, the clarity of the water, and whether there was any possibility of cervical spine injury.

If you respond to a diving accident, the receiving facility personnel will also need a complete dive profile to properly treat your patient. This may be available in a dive log or from diving partners. Small diving computers have become standard equipment for most divers, and they record information from the current as well as previous dives. Be sure the computer is brought to the hospital with the patient. If possible, have all of the diver's equipment brought to the hospital. It will be helpful in determining the cause of the accident. Be sure to document the disposition of this equipment.

Emergency Care for Drowning or Diving Emergencies

If the patient does not have a possible spinal injury, turn the patient quickly to the left side to allow substances to drain from the upper airway. Note that water will not drain from the lungs. If there is evidence of upper airway obstruction by foreign matter, remove the obstruction manually or, if available, by suction. If necessary, use abdominal thrusts, followed by assisted ventilations.

Administer oxygen if this was not done as part of the primary assessment, either by mask for patients who are breathing spontaneously or via a bag-mask device for those requiring assisted ventilation.

Make sure that the patient is kept warm, especially after cold-water immersion. Make sure blankets and protection from the environment are provided as needed. If ventilation equipment is not available but oxygen is, you can breathe the oxygen in yourself and give mouth-to-mask ventilation until rescue equipment arrives. In this method, your expired air will have a higher percentage of oxygen.

When treating conscious patients who are suspected of having air embolism or decompression sickness, you should follow these accepted treatment steps:

1. Remove the patient from the water. Try to keep the patient calm.
2. Administer oxygen.
3. Place the patient in a left lateral recumbent position with the head down.
4. Provide prompt transport to the nearest recompression facility for treatment.

Injury from decompression sickness is usually reversible with proper treatment. However, if the bubbles block critical blood vessels that supply the brain or spinal cord, permanent central nervous system injury may result. Therefore, the key in emergency management of these serious ascent problems is to recognize that an emergency exists and treat as soon as possible. Administer oxygen and provide rapid transport.

■ Other Water Hazards

You must pay close attention to the body temperature of a person who is rescued from cold water. Treat hypothermia caused by immersion in cold water the same way you treat hypothermia caused by cold exposure. Prevent further heat loss from contact with the ground, stretcher, or air, and transport the patient promptly.

A person swimming in shallow water may experience **breath-holding syncope**, a loss of consciousness caused by a decreased stimulus for breathing. This happens to swimmers who breathe in and out rapidly and deeply before entering the water in an effort to expand their capacity to stay underwater. Whereas this technique increases the swimmer's oxygen level, the hyperventilation involved lowers the carbon dioxide level. Because an elevated level of carbon dioxide in the blood is the strongest stimulus for breathing, the swimmer may not feel the need to breathe even after using up all the oxygen in his or her lungs. The emergency treatment for a patient with breath-holding syncope is the same as that for a drowning patient.

Injuries caused by boat propellers, sharp rocks, water skis, or dangerous marine life may be complicated by immersion in cold water. In these cases, remove the patient from the water, take care to protect the spine, and administer oxygen. Apply dressings and splints if indicated, and monitor the patient closely for any signs of immersion or cold injury.

You should be aware that a child who is involved in a drowning may be the victim of child abuse. Although it may be difficult to prove, such incidents should be handled according to the rules set up for suspected child abuse.

■ Prevention

Appropriate precautions can prevent most immersion incidents. Each year, many small children drown in residential pools. All pools should be surrounded by a fence that is at least 6′ high, with slats no farther apart than 3″ and self-closing, self-locking gates. The most common problem is lack of adult supervision, even when attention is not given for a few seconds. Half of all teenage and adult drownings are associated with the use of alcohol. As a health care professional, you should be involved in public education efforts to make people aware of the hazards of swimming pools and water recreation.

High Altitude

High altitudes can cause dysbarism injuries. Dysbarism injuries are any signs and symptoms caused by the difference between the surrounding atmospheric pressure and the total gas pressure in various tissues, fluids, and cavities of the body. Altitude illnesses are illnesses caused by diminished oxygen pressure in the air at high altitudes on the central nervous system and pulmonary system as a result of unacclimatized people ascending to a high altitude. It runs the gamut from the common acute mountain sickness to the rare deaths from high-altitude cerebral edema (HACE) and high-altitude pulmonary edema (HAPE).

Acute mountain sickness is caused by diminished oxygen pressure in the air at altitudes above 8,000′, resulting in diminished oxygen in the blood (hypoxia). It strikes those who ascend too high too fast and those who have not acclimatized to high altitudes. The signs and symptoms include a headache, lightheadedness, fatigue, loss of appetite, nausea, difficulty sleeping, shortness of breath during physical exertion, and a swollen face.

With HAPE, fluid collects in the lungs, hindering the passage of oxygen into the bloodstream. It can occur at altitudes of 10,000′. The signs and symptoms include shortness of breath, cough with pink sputum, cyanosis, and a rapid pulse.

HACE usually occurs in climbers who climb above 12,000′. It may accompany HAPE and can quickly become life threatening. The symptoms of HACE and HAPE may overlap. The signs and symptoms include a severe constant throbbing headache, ataxia (lack of muscle coordination), extreme fatigue, vomiting, and loss of consciousness.

In the field, treatment for altitude illness consists of providing oxygen, descending from the height, and transporting the patient.

Lightning

According to the National Weather Service, there are an estimated 25 million cloud-to-ground lightning flashes in the United States each year. On average, lightning kills between 60 and 70 people per year in the United States based on documented cases. While documented lightning injuries in the United States average about 300 per year, undocumented lightning injuries are likely much higher. Lightning is the third most common cause of death from isolated environmental phenomena.

The energy associated with lightning is comprised of direct current of up to 200,000 amps and a potential of 100 million volts or more. Temperatures generated from lightning vary between 20,000°F and 60,000°F.

Most deaths and injuries caused by lightning occur during the summer months when people are enjoying outdoor activities, despite an approaching thunderstorm. Those most commonly struck by lightning include boaters, swimmers, and golfers. Any type of activity that exposes the person to a large, open area increases the risk of being struck by lightning.

Whether or not lightning injures or kills depends on whether a person is in the path of the lightning discharge. The current associated with the lightning discharge travels along the ground. Although some persons are injured or killed by a direct lightning strike, many individuals are indirectly struck when standing near an object that has been struck by lightning, such as a tree (splash effect).

The cardiovascular and nervous systems are most commonly injured during a lightning strike; therefore, respiratory or cardiac arrest is the most common cause of lightning-related deaths. The tissue damage caused by lightning is different from that caused by other electrical-related injuries (ie, high-power line injuries) because the tissue damage pathway usually occurs over the skin, rather than through it. During your assessment you should look for not only the entrance wound but also the exit wound. The exit wound does not necessarily occur on the same side of the body. Additionally, because the duration of a lightning strike is short, skin burns are usually superficial; full-thickness (third-degree) burns are rare. Lightning injuries are categorized as being mild, moderate, or severe:

- **Mild:** loss of consciousness, amnesia, confusion, tingling, and other nonspecific signs and symptoms. Burns, if present, are typically superficial.
- **Moderate:** seizures, respiratory arrest, cardiac standstill (asystole) that spontaneously resolves, and superficial burns.
- **Severe:** cardiopulmonary arrest. Because of the delay in resuscitation, often the result of occurrence in a remote location, many of these patients do not survive.

■ Emergency Medical Care

As with any scene response, the safety of you and your partner has priority. Take measures to protect yourself from being struck by lightning, especially if the thunderstorm is still in progress. Contrary to popular belief, lightning can, and does, strike in the same place twice. Move the patient to a place of safety, preferably in a sheltered area.

If you are in an open area and adequate shelter is not available, it is important to recognize the signs of an impending lightning strike and take immediate action to protect yourself. If you suddenly feel a tingling sensation or your hair stands on end, the area around you has become charged—a sure sign of an imminent lightning strike. Make yourself as small a target as possible by squatting down into a ball, close to but not touching the ground. If you are standing near a tree or other tall object, move away as fast as possible, preferably to a low-lying area. Lightning has an affinity for objects that project from the ground (ie, trees, fences, buildings).

The process of triaging multiple victims of a lightning strike is different than the conventional triage methods used during a mass-casualty incident. When a person is struck by lightning, respiratory or cardiac arrest, if it occurs, usually occurs immediately. Those who are conscious following a lightning strike are much less likely to develop delayed respiratory or cardiac arrest; most of these persons will survive. Therefore, you should focus your efforts on those who are in respiratory or cardiac arrest. This process, called **reverse triage**, differs from conventional triage, where such patients would ordinarily be classified as deceased.

When a person is struck by lightning, it causes massive direct current shock, with the patient experiencing massive muscle spasms (tetany) that can result in fractures of long bones and spinal vertebrae. Therefore, you should manually stabilize the patient's head in a neutral in-line position and open the airway with the jaw-thrust maneuver. If the patient is in respiratory arrest with a pulse, begin immediate bag-mask device ventilations with 100% oxygen. If the patient is in cardiac arrest, attach an AED as soon as possible and provide immediate defibrillation if indicated. If severe bleeding is present, control it immediately.

Provide full spinal stabilization and transport the patient to the closest appropriate facility. If CPR or ventilations are not required, address other injuries (ie, splint fractures, dress and bandage burns) and provide continuous monitoring while en route to the hospital.

Bites and Envenomations

This section discusses bites and stings from spiders, hymenoptera, snakes, scorpions, ticks, and injuries from marine animals.

Spider Bites

Spiders are numerous and widespread in the United States. Many species of spiders bite. However, only two, the female black widow spider and the brown recluse spider, are able to deliver serious, even life-threatening bites. When you care for a patient who has had some type of bite, be alert to the possibility that the spider may still be in the area, although it is not likely. Remember that your safety is of paramount importance.

Black Widow Spider

The female black widow spider (*Latrodectus*) is fairly large, measuring approximately 2″ long with its legs extended. It is usually black and has a distinctive, bright red-orange marking in the shape of an hourglass on its abdomen **Figure 30-12**. The female black widow spider is larger and more toxic than the male. Black widow spiders are found in every state except Alaska. They prefer dry, dim places around buildings, in woodpiles, and among debris.

The bite of the black widow spider is sometimes overlooked. If the site becomes numb right away, the patient may not even recall being bit. However, most black widow spider bites cause localized pain and symptoms, including agonizing muscle spasms. In some cases, a bite on the abdomen causes muscle spasms so severe that the patient may be thought to have an acute abdominal condition, possibly peritonitis. The main danger with this type of bite, however, is that the black widow's venom is poisonous to nerve tissues (neurotoxic). Other systemic symptoms include dizziness, sweating, nausea, vomiting, and rashes. Tightness in the chest and difficulty breathing develop within 24 hours, as well as severe cramps, with board-like rigidity of the abdominal muscles. Generally, these signs and symptoms subside over 48 hours.

If necessary, a physician can administer a specific **antivenin**, a serum containing antibodies that counteract the venom, but because of a high incidence of side effects, its use is reserved for very severe bites, for the aged or very feeble, and for children younger than 5 years. In children, these bites can be fatal. The severe muscle spasms are usually treated in the hospital with IV benzodiazepines such as diazepam (Valium) or lorazepam (Ativan). In general, emergency treatment for a black widow spider bite consists of BLS for the patient in respiratory distress. Much more often, the patient will only require relief from pain. If time permits, apply an ice pack to the bite area and clean the wound with soap and water. Transport the patient to the emergency department as soon as possible for treatment of both pain and muscle rigidity. If possible, bring the spider to the hospital.

Brown Recluse Spider

The brown recluse spider (*Loxosceles*) is dull brown and, at 1″, smaller than the black widow **Figure 30-13**. The short-haired body has a violin-shaped mark, brown to yellow in color, on its back. Although the brown recluse spider lives mostly in the southern and central parts of the country, it may be found throughout the continental United States. The spider takes its name from the fact that it tends to live in dark areas—in corners of old, unused buildings, under rocks, and in woodpiles. In cooler areas, it moves indoors to closets, drawers, cellars, and clothing.

In contrast to the venom of the black widow spider, the venom of the brown recluse spider is not neurotoxic but cytotoxic; that is, it causes severe local tissue damage. Typically, the bite is not painful at first but becomes so within hours. The area becomes swollen and

Figure 30-12 Black widow spiders are distinguished by their glossy black color and bright red-orange hourglass marking on the abdomen.

Figure 30-13 Brown recluse spiders are dull brown and have a dark, violin-shaped mark on the back.

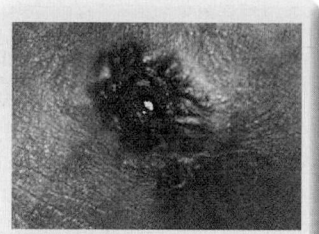

Figure 30-14 The bite of a brown recluse spider is characterized by swelling, tenderness, and a pale, mottled, cyanotic center. There may also be a small blister on the bite.

tender, developing a pale, mottled, cyanotic center and possibly a small blister Figure 30-14 . Over the next several days, a scab of dead skin, fat, and debris forms and digs down into the skin, producing a large ulcer that may not heal unless treated promptly. Transport patients with such symptoms as soon as possible.

Brown recluse spider bites rarely cause systemic symptoms and signs. When they do, the initial treatment is BLS and transportation to the emergency department. Again, it is helpful if you can identify the spider and bring it to the hospital with the patient.

Hymenoptera Stings

Typically **hymenoptera** (bees, wasps, ants, and yellow jackets) stings are painful but are not a medical emergency. If the patient is allergic to the venom, then anaphylaxis may occur. The signs and symptoms of anaphylaxis are flushed skin, low blood pressure, difficulty breathing usually associated with reactive airway sounds such as wheezes, or in severe cases diminished or absent breath sounds. The patient can also have swelling to the throat and tongue. This is a dire emergency and can be fatal if not recognized and treated quickly. The patient may develop hives (urticaria) near the site of envenomation or centrally on the body.

Remove the stinger and, if still present, the venom sac. This is best done by using a firm-edged item such as a credit card to scrape the stinger and sac off the skin. If you inadvertently squeeze the venom sac while trying to grasp the stinger with tweezers or forceps you will worsen the patient's exposure by increasing the amount of envenomation.

If anaphylaxis develops, be prepared to assist the patient in administering an EpiPen auto-injector. Also be prepared to support the airway and breathing should the patient experience significant respiratory compromise. Chapter 18, *Immunologic Emergencies*, has a detailed discussion on the treatment of anaphylaxis.

Snake Bites

Snake bites are a worldwide problem. More than 300,000 injuries from snake bites occur annually, including 30,000 to 40,000 deaths. The greatest number of fatalities occur in Southeast Asia and India (25,000 to 30,000) and in South America (3,000 to 4,000). In the United States, 40,000 to 50,000 snake bites are reported annually, with about 7,000 caused by poisonous snakes. However, snake bite fatalities in the United States are extremely rare, about 15 a year for the entire country.

Of the approximately 115 different species of snakes in the United States, only 19 are venomous. These include the rattlesnake (*Crotalus*), the copperhead (*Agkistrodon contortrix*), the cottonmouth, or water moccasin (*Agkistrodon piscivorus*), and the coral snakes (*Micrurus* and *Micruroides*) Figure 30-15 . At least one of these poisonous species is found in every state except Alaska, Hawaii, and Maine. As a general rule, these snakes are timid. They usually do not bite unless provoked or accidentally injured, as when they are stepped on. There are a few exceptions to these rules. Cottonmouths are

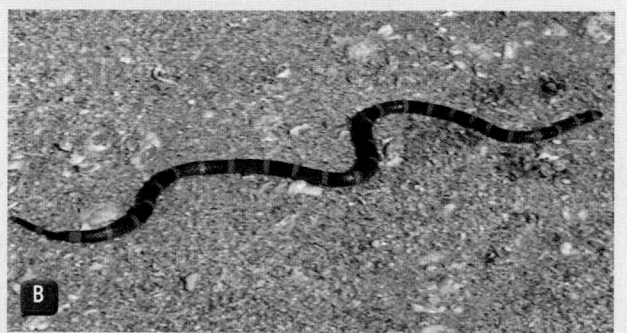

Figure 30-15 A. Copperhead. B. Coral snake. C. Rattlesnake. D. Cottonmouth.

often aggressive, and rattlesnakes are easily provoked. Coral snakes, in contrast, usually bite only when they are being handled.

Most snake bites occur between April and October, when the animals are active, and tend to involve young men who have been drinking alcohol. Texas reports the largest number of bites. Other states with a major concentration of snake bites are Louisiana, Georgia, Oklahoma, North Carolina, Arkansas, West Virginia, and Mississippi. If you work in one of these areas, you should be thoroughly familiar with the emergency handling of snake bites. Remember, almost any time you are caring for a patient with a snake bite, another snake may be in the area and create a second victim—you. Therefore, use extreme caution on these calls and be sure to wear the proper protective equipment for the area.

In general, only a third of snake bites result in significant local or systemic injuries. Often, envenomation does not occur because the snake has recently struck another animal and exhausted its supply of venom for the time being.

Poisonous snakes native to the United States all have hollow fangs in the roof of the mouth that inject the poison from two sacs at the back of the head. The classic appearance of the poisonous snake bite, therefore, is two small puncture wounds, usually about ½" apart, with discoloration and swelling, and the patient usually reports pain surrounding the bite **Figure 30-16**. Nonpoisonous snakes can also bite, usually leaving a horseshoe of tooth marks. However, some poisonous snakes have teeth as well as fangs, making it impossible to say which kind is responsible for a given set of tooth marks. Conversely, fang marks are a clear indication of a poisonous snake bite.

Pit Vipers

Rattlesnakes, copperheads, and cottonmouths are all pit vipers, with triangular-shaped, flat heads **Figure 30-17**. They take their name from the small pits located just behind each nostril and in front of each eye. The pit is a heat-sensing organ that allows the snake to strike accurately at any warm target, especially in the dark, when it cannot see through its vertical, slit-like pupils.

The fangs of the pit viper normally lie flat against the roof of the mouth and are hinged to swing back and forth as the mouth opens. When the snake is striking, the mouth opens wide and the fangs extend; in this way, the fangs penetrate whatever the mouth strikes. The fangs are actually special hollow teeth that act much like hypodermic needles. They are connected to a sac containing a reservoir of venom, which in turn is attached to a poison gland. The gland itself is a specially adapted salivary gland, which produces enzymes that digest and destroy tissue. The primary purpose of the venom is to kill small animals and to start the digestive process prior to their being eaten.

In the United States, the most common form of pit viper is the rattlesnake. Several different species of rattlesnake can be identified by the rattle on the tail. The rattle is actually numerous layers of dried skin that were shed but failed to fall off, coming to rest against a small knob on the end of the tail. Rattlesnakes have many patterns of color, often with a diamond pattern. They can grow to 6′ or more in length.

Copperheads are smaller than rattlesnakes, usually 2′ to 3′ long, with a reddish coppery color crossed with brown or red bands. These snakes typically inhabit woodpiles and abandoned dwellings, often close to areas of habitation. Although they account for most of

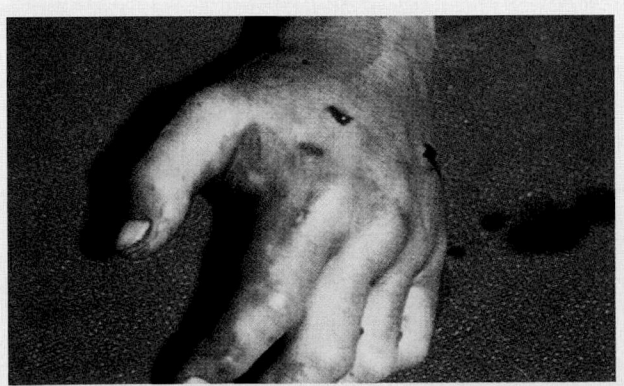

Figure 30-16 A snake bite wound from a poisonous snake has characteristic markings: two small puncture wounds about ¹/₂″ apart, discoloration, and swelling.

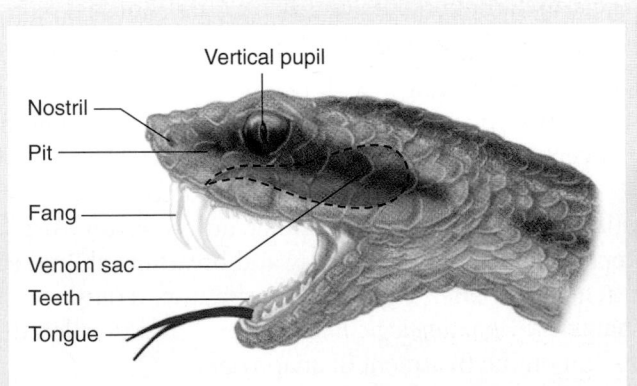

Figure 30-17 Pit vipers have small, heat-sensing organs (pits) located in front of their eyes that allow them to strike at warm targets, even in the dark.

the venomous snake bites in the eastern United States, copperhead bites are almost never fatal; however, note that the venom can destroy extremities.

Cottonmouths grow to about 4′ in length. Also called water moccasins, these snakes are olive or brown, with black cross-bands and a yellow undersurface. They are water snakes and have a particularly aggressive pattern of behavior. Although fatalities from these snake bites are rare, tissue destruction from the venom may be severe.

The signs of envenomation by a pit viper are severe burning pain at the site of the injury, followed by swelling and a bluish discoloration (ecchymosis) in light-skinned individuals that signals bleeding under the skin. These signs are evident within 5 to 10 minutes after the bite has occurred and spread over the next 36 hours. In addition to destroying tissues locally, the venom of the pit viper can also interfere with the body's clotting mechanism and cause bleeding at various distant sites. This toxin affects the entire nervous system. Other systemic signs, which may or may not occur, include weakness, nausea, vomiting, sweating, seizures, fainting, vision problems, changes in level of consciousness, and shock. If swelling has occurred, you should mark its edges on the skin. This will allow physicians to assess what has happened and when it happened with greater accuracy. If the patient has no local signs an hour after being bitten, it is safe to assume that envenomation did not take place.

The toxicity is related to the amount of toxin injected. A bite will affect children more than adults because there is less body mass to absorb the toxin. The same principle holds true for a small-statured adult.

Occasionally, a patient bitten by a snake will faint from fright. The patient will usually regain consciousness promptly when placed in a supine position. Do not confuse a fainting spell with shock. If shock occurs, it will happen much later.

In treating a snake bite from a pit viper, follow these steps to get the patient to the hospital in a timely manner:

1. Calm the patient; assure him or her that poisonous snake bites are rarely fatal. Place the patient in a supine position and explain that staying quiet will slow the spread of any venom through the system. Determine the approximate time of the bite and document your time en route to a receiving facility. This time from onset to evaluation at the facility is one of the criteria used in grading the severity of the incident and in determining the amount of antivenin to be used.

2. Locate the bite area; clean it gently with soap and water or a mild antiseptic. Do not apply ice to the area. If the patient is hypotensive, then a constricting band may be applied 4″ to 6″ above the bite site if called for by your local protocols. You should be able to slide two fingers underneath the band. Some local protocols do not allow the use of a constricting band in treating snake bites because of the potential

You are the Provider: PART 5

The patient's level of consciousness appears to have improved and he is now resisting your attempts to assist his ventilations. After reapplying the nonrebreathing mask, you reassess his vital signs and clinical condition. His skin, although still very warm, does not feel as hot as it did initially, and his skin appears less flushed. You will arrive at the hospital in approximately 5 minutes.

Recording Time: 16 Minutes	
Level of consciousness	Confused; somewhat combative
Respirations	22 breaths/min; depth has improved
Pulse	120 beats/min and regular; appears to be stronger
Skin	Less flushed, very warm to the touch, moist
Blood pressure	98/58 mm Hg
Sao$_2$	94% (on oxygen)

9. What other conditions should you consider as potential causes of the patient's altered mental status?

of the venom pooling into the localized bite area. This could potentially cause greater damage to the localized area.

3. If the bite occurred on an arm or leg, consider the use of a properly performed pressure immobilization bandage of the extremity (eg, 40 to 70 mm Hg in the arms and 55 to 70 mm Hg in the legs) and then place the affected extremity below the level of the heart.

4. Be alert for vomiting, which may be a sign of anxiety rather than the toxin itself.

5. Do not give anything by mouth.

6. If, as rarely happens, the patient was bitten on the trunk, keep him or her supine and quiet and transport as quickly as possible.

7. Monitor the patient's vital signs and mark the skin with a pen over the area that is swollen, proximal to the swelling, to note whether swelling is spreading.

8. If there are any signs of shock, place the patient in the shock position and administer oxygen.

9. If the snake has been killed, as is often the case, be sure to bring it with you in a secure container so that physicians can identify it and administer the proper antivenin.

10. Notify the hospital that you are bringing in a patient who has a snake bite; if possible, describe the snake.

11. Transport the patient promptly to the hospital.

If the patient shows no sign of envenomation, provide BLS as needed, place a sterile dressing over the suspected bite area, and immobilize the injury site. All patients with a suspected snake bite should be taken to the emergency department, whether they show signs of envenomation or not. Treat the wound as you would any deep puncture wound to prevent infection.

If you work in an area where poisonous snakes are known to live, you should know your local protocol for handling snake bites. You should also know the address of the nearest facility where antivenin is available. This may be a nearby zoo, the local or public state health department, or a local community hospital.

Coral Snakes

The coral snake is a small reptile with a series of bright red, yellow, and black bands completely encircling the body. Many harmless snakes have similar coloring, but only the coral snake has red and yellow bands next to one another, as this helpful rhyme suggests: "Red on yellow will kill a fellow; red on black, venom will lack."

A rare creature that lives in most southern states and in the Southwest, the coral snake is a relative of the cobra. It has tiny fangs and injects the venom with its teeth by a chewing motion, leaving behind one or more puncture or scratch-like wounds. Because of its small mouth and teeth and limited jaw expansion, the coral snake usually bites its victims on a small part of the body, such as a finger or toe.

Coral snake venom is a powerful toxin that causes paralysis of the nervous system. Within a few hours of being bitten, a patient will exhibit bizarre behavior, followed by progressive paralysis of eye movements and respiration. Often, there are limited or no local symptoms.

Successful treatment, either emergency or long term, depends on positive identification of the snake and support of respiration. Antivenin is available, but most hospitals do not stock it. Therefore, you should notify the hospital of the need for it as soon as possible. The steps for emergency care of a coral snake bite are as follows:

1. Immediately quiet and reassure the patient.

2. Flush the area of the bite with 1 to 2 quarts of warm, soapy water to wash away any poison left on the surface of the skin. Do not apply ice to the region.

3. If the bite occurred on an arm or leg, consider the use of a properly performed pressure immobilization bandage of the extremity (eg, 40 to 70 mm Hg in the arms and 55 to 70 mm Hg in the legs) and then place the affected extremity below the level of the heart.

4. Check the patient's vital signs and continue to monitor them.

5. Keep the patient warm and to help prevent shock, place the patient in the position dictated by local protocol for shock patients.

6. Give supplemental oxygen if needed.

7. Transport the patient promptly to the emergency department, giving advance notice that the patient has been bitten by a coral snake.

8. Give the patient nothing by mouth.

■ Scorpion Stings

Scorpions are eight-legged arachnids from the biologic group Arachnida with a venom gland and a stinger at the end of their tail **Figure 30-18**. Scorpions are rare; they live primarily in the southwestern United States and in deserts. With one exception, a scorpion's sting is usually very painful but not dangerous, causing localized swelling and discoloration. The exception is the *Centruroides sculpturatus*. Although it is found naturally in Arizona and New Mexico, as well as parts of Texas, California, and Nevada, it may be kept as a pet by anyone. The venom of this particular species may produce a severe systemic reaction that brings about circulatory collapse, severe

Figure 30-18 The sting of a scorpion is usually more painful than it is dangerous, causing localized swelling and discoloration.

muscle contractions, excessive salivation, hypertension, convulsions, and cardiac failure. Antivenin is available but must be administered by a physician. If you are called to care for a patient with a suspected sting from *C sculpturatus*, you should notify medical control as soon as possible. Administer BLS and transport the patient to the emergency department as rapidly as possible.

■ Tick Bites

Found most often on brush, shrubs, trees, sand dunes, or other animals, ticks usually attach themselves directly to the skin Figure 30-19 . Only a fraction of an inch long, they can easily be mistaken for a freckle, especially since their bite is not painful. Indeed, the danger with a tick bite is not from the bite itself, but from the infecting organisms that the tick carries. Ticks commonly carry two infectious diseases, Rocky Mountain spotted fever and Lyme disease. Both are spread through the tick's saliva, which is injected into the skin when the tick attaches itself.

Rocky Mountain spotted fever, which is not limited to the Rocky Mountains, occurs within 7 to 10 days after a bite by an infected tick. Its symptoms include nausea, vomiting, headache, weakness, paralysis, and possibly cardiorespiratory collapse.

Lyme disease has received extensive publicity. Originally seen only in Connecticut, Lyme disease has now been reported in 35 states. It occurs most commonly in the Northeast, the Great Lake states, and the Pacific Northwest; New York State reports the largest number of cases. The first symptom, a rash that may spread to several parts of the body, begins about 3 days after the bite of an infected tick. The rash may eventually resemble a target bull's-eye pattern in one third of patients Figure 30-20 . After a few more days or weeks, painful swelling of the joints, particularly the knees, occurs. Lyme disease may be confused with rheumatoid arthritis and, like that disease, may result in permanent disability. However, if it is recognized and treated promptly with antibiotics, the patient may recover completely.

Figure 30-19 Ticks typically attach themselves directly to the skin.

Tick bites occur most commonly during the summer months, when people are out in the woods wearing little protective clothing. Transmission of the infection from the tick to a person takes at least 12 hours, so if you are called on to remove a tick, you should proceed carefully and slowly. Do not attempt to suffocate

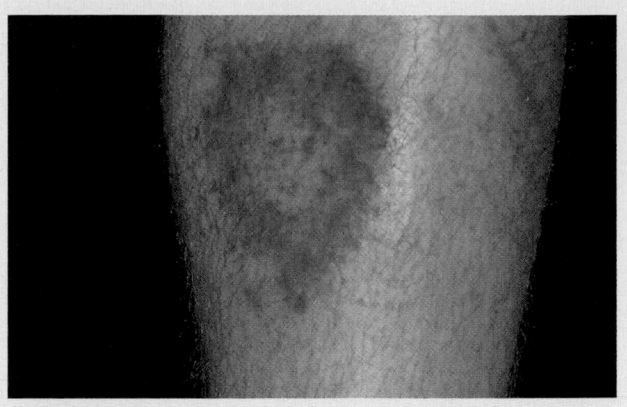

Figure 30-20 The rash associated with Lyme disease has a characteristic bull's-eye pattern.

the tick with gasoline or Vaseline or burn it with a lighted match; you will only burn the patient. Instead, using fine tweezers, grasp the tick by the body and pull gently but firmly straight up so that the skin is tented. Hold this position until the tick releases. Special tweezers are available for this, but are not necessary. This method will usually remove the whole tick. Even if part of the tick is left embedded in the skin, the part containing the infecting organisms has been removed. Cleanse the area with disinfectant and save the tick in a glass jar or other container so that it can be identified. Do not handle the tick with your fingers. Provide any necessary supportive emergency care, and transport the patient to the hospital.

■ Injuries From Marine Animals

Coelenterates, including the fire coral, Portuguese man-of-war, sea wasp, sea nettles, true jellyfish, sea anemones, true coral, and soft coral, are responsible for more envenomations than any other marine animals Figure 30-21 . The stinging cells of the coelenterate are called nematocysts, and large animals may discharge hundreds of thousands of them. Envenomation causes very painful, reddish lesions in light-skinned individuals extending in a line from the site of the sting. Systemic symptoms include headache, dizziness, muscle cramps, and fainting.

To treat a sting from the tentacles of a jellyfish, a Portuguese man-of-war, various anemones, corals, or hydras, remove the patient from the water and pour acetic acid (vinegar) on the affected area. Unlike fresh water, vinegar will inactivate the nematocysts. Do not try to manipulate the remaining tentacles; this will only cause further discharge of the nematocysts. Remove the tentacles by scraping them off with the edge of a sharp, stiff object such as a credit card. Persistent pain may respond

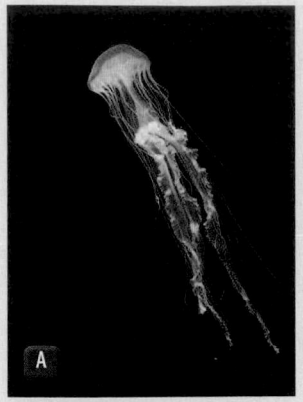
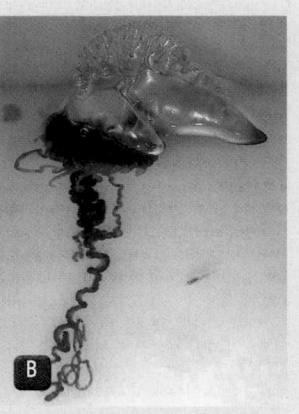

Figure 30-21 Coelenterates are responsible for many marine envenomations. **A.** Jellyfish. **B.** Portuguese man-of-war. **C.** Sea anemone.

Table 30-3	Common Marine Envenomations	
Dogfish	Marine snail	Starfish
Dragon fish	Portuguese man-of-war	Stingray
Fire coral	Ratfish	Stonefish
Hydroids	Scorpion fish	Tiger fish
Jellyfish	Sea anemone	Toadfish
Lionfish	Sea urchins	Weever fish

object such as a credit card. Persistent pain may respond to immersion of the area in hot water (110°F to 115°F, 43°C to 46°C) for 30 minutes. On very rare occasions, a patient may have a systemic allergic reaction to the sting of one of these animals. Treat such a patient for anaphylactic shock. Provide BLS and immediate transport to the hospital.

Toxins from the spines of urchins, stingrays, and certain spiny fish such as the lionfish, scorpion fish, or stonefish are heat sensitive **Table 30-3**. Therefore, the best treatment for such injuries is to immobilize the affected area and soak it in hot water for 30 minutes. This will often provide dramatic relief from local pain. However, the patient still needs to be transported to the emergency department because an allergic reaction or infection, including tetanus, could develop.

If you work near the ocean, you should be familiar with the marine life in your area. The emergency treatment of common coelenterate envenomations consists of the following steps:

1. **Limit further discharge** of nematocysts by avoiding fresh water, wet sand, showers, or careless manipulation of the tentacles. Keep the patient calm, and reduce motion of the affected extremity.
2. **Inactivate the nematocysts** by washing the site with vinegar as soon as possible and for at least 30 seconds. (Isopropyl alcohol may be used if vinegar is not available, but may not be as effective.)
3. **Remove the remaining tentacles** by scraping them off with the edge of a sharp, stiff object such as a credit card. Do not use your ungloved hand to remove the tentacles, because self-envenomation will occur. Persistent pain may respond to immersion in hot water (110°F to 115°F, 43°C to 46°C) for 30 minutes.
4. **Provide transport** to the emergency department.

You are the Provider: SUMMARY

1. **How does the body normally balance heat production and elimination?**

Although the normal core temperature—the temperature of the heart, lungs, and other vital organs—varies from person to person, it is usually around 98.6°F (37°C). A series of regulatory mechanisms keep this internal temperature constant, regardless of the ambient temperature (the temperature of the surrounding environment). However, heat elimination must balance heat production; if it does not, heat illness occurs.

In a hot environment or during vigorous physical activity, the body produces excess heat, which it must be able to remove. There are several ways the body does this, the most efficient of which are sweating (and evaporation of sweat) and dilation of the blood vessels, which brings warm blood to the surface of the skin and releases heat via radiation. Obvious ways of removing excess heat include removing the patient's clothing and moving him or her to a cooler environment.

Ordinarily, the heat-regulating mechanisms of the body work very well and people are able to tolerate significant temperature changes. However, if the body is exposed to or generates more heat energy than it can efficiently eliminate,

You are the Provider: SUMMARY, continued

hyperthermia—a core body temperature of 101°F (38.3°C) or higher—develops.

2. What factors can decrease the body's ability to eliminate excess heat?

A number of factors can decrease a person's ability to eliminate excess heat. High air temperature can reduce the body's ability to lose heat via radiation. Heat travels from a warmer place to a cooler place; if the ambient temperature is higher than body temperature, heat will move from the environment and into the body. If the relative humidity, the amount of moisture in the air, is high, the body's ability to lose heat by evaporation is reduced. Heat elimination is impaired the greatest when both the air temperature and relative humidity are high.

Vigorous exercise, during which the body can lose more than 1 liter of sweat per hour, causes a loss of fluids and electrolytes, resulting in dehydration. Dehydration decreases heat loss through sweating and evaporation; if the water content of the body is reduced, so is the ability to produce sweat.

Small children and older patients exhibit poor thermoregulation; therefore, they are less able to eliminate excess body heat. The body's water content decreases with age, which reduces the ability to sweat. Infants and small children often wear too much clothing, which they are physically unable to remove.

Certain medical conditions, such as heart disease, chronic obstructive pulmonary disease, diabetes, dehydration, and obesity interfere with the process of body heat elimination; if a patient has more than one medical condition, the ability to eliminate excess body heat is further reduced. In addition, alcohol and certain drugs, including medications that dehydrate the body (ie, diuretics) or decrease the ability of the body to sweat, also reduce heat elimination from the body.

When assessing a patient with a possible heat-related emergency, it is important to obtain a thorough medical history and determine which, if any, medications they are taking. Patients who are less able to eliminate excess body heat, such as those previously discussed, may experience heat illness even if the ambient temperature is not extremely high.

3. What risk factors does this patient have that predispose him to heat illness?

Your patient has several risk factors for heat illness, the single most significant of which is prolonged exertion in a hot, humid environment. Also, the patient has not been drinking any water. In combination with profuse sweating that occurs during exposure to a hot environment, you should suspect that he is dehydrated. When the body's water content is reduced, so is its ability to produce sweat. In conjunction with the high humidity, dehydration further reduces the ability to eliminate excess body heat by evaporation.

The patient's past medical history also predisposes him to heat illness. Diabetes is a systemic condition that impairs many body functions, including thermoregulation. The patient also

has hypertension, for which diuretics are often prescribed. Diuretics promote urination, further contributing to his dehydration.

4. What type of heat illness should you suspect that he is experiencing? Why?

There are several clinical findings that indicate your patient is experiencing the most serious heat illness, heatstroke. Heatstroke occurs when the body is exposed to more heat than it can handle and the normal compensatory mechanisms that eliminate excess heat are overwhelmed. When heat production or exposure exceeds heat elimination, body temperature rises rapidly and significantly; it is not uncommon to observe body temperatures of 106°F (41°C) or higher in patients with heatstroke. At a minimum, untreated heatstroke will cause permanent brain and tissue damage; in most cases, it causes death.

Heatstroke can be described as being exertional or classic. Exertional heatstroke often affects otherwise healthy people who vigorously exert themselves for prolonged periods of time in a hot, humid environment. Classic heatstroke commonly affects older patients who are confined to a closed, hot, poorly-ventilated environment, or children who are left unattended in a locked car on a hot day. Exertion is not a contributing factor in patients with classic heatstroke.

Unlike other less severe forms of heat illness (eg, heat cramps, heat exhaustion), patients with heatstroke have an altered level of consciousness, ranging from confusion to coma, and flushed, hot skin. *Your patient has both of these.*

Many patients with heatstroke have dry skin. In other cases, the patient's skin may be moist from residual perspiration, even though he or she is not actively sweating. Although hot, flushed, moist skin is more commonly seen in patients with exertional heatstroke, which is what your patient is experiencing, it may also be seen in patients with classic heatstroke, albeit less commonly. Common findings to both types of heatstroke are hot, flushed skin and an altered level of consciousness; these are the most reliable findings.

5. What specific treatment is required for this patient?

Immediate treatment for heatstroke includes moving the patient to a cool environment and administering oxygen. Further treatment is aimed at *actively* cooling the patient. Unless there are extenuating circumstances that will delay your transport, active cooling measures should be performed en route to the hospital. Heatstroke is a true emergency; it requires rapid cooling and immediate transport. Any delays in providing treatment increase the potential for permanent brain and tissue damage or death.

Remove the patient's clothes because they can trap heat and hamper your efforts to cool the patient. Place cold packs at the patient's groin, axillae, and behind the neck. Spray or pour saline on the patient and aggressively fan him; this measure, in conjunction with the air conditioner, will facilitate heat loss through convection and evaporation.

You are the Provider: SUMMARY, continued

Continue to actively cool the patient and notify the receiving facility early so they can continue treatment immediately on your arrival.

6. What is the most likely explanation for this patient's vital signs?

Your patient's vital signs—tachypnea, tachycardia, and hypotension—indicate shock. This is because heatstroke is associated with a severe loss of fluids and electrolytes, which results in hypovolemia. You should expect that patients with heatstroke will be tachypneic and tachycardic because of all the heat energy they have in their body. This response alone may enable patients to compensate for the severe fluid loss that occurs with heatstroke. The presence of hypotension, however, indicates that the body's compensatory mechanisms have failed (decompensated shock).

Patients with hypovolemic shock associated with heatstroke will need intravenous fluids and other treatment aimed at correcting electrolyte abnormalities at the hospital. Consider an ALS intercept, but do not delay transport to do this. AEMTs and paramedics are able to establish IV lines and administer fluids. Otherwise, elevate the patient's legs, continue high-flow oxygen, and closely monitor him.

7. How should you adjust your treatment for this patient?

Your patient's level of conscious has deteriorated and his respirations are now shallow. Furthermore, his oxygen saturation level has decreased to 89%. These clinical signs indicate that your patient is no longer breathing adequately.

Respiration is one of the body's mechanisms for removing heat, which explains why patients with heatstroke are usually tachypneic; in early heatstroke, their respirations may also be deep. However, when respirations become shallow, the patient has two problems: inadequate ventilation, which can result in worsened tissue hypoxia and a decreased ability to remove heat from the body.

Patients with inadequate breathing need assisted breathing with a bag-mask device and high-flow oxygen. In addition to actively cooling the patient, you must ensure that oxygenation and ventilation remain adequate. Consider inserting an airway adjunct to assist in maintaining airway patency. In this case, a nasal airway is the best choice because the patient is not completely unconscious and likely has an intact gag reflex.

8. How will you know when you have adequately cooled the patient?

Ideally, you should monitor the patient's core body temperature (CBT); this is most reliably obtained by assessing the patient's rectal temperature. It is important to note that the CBT increases rapidly in patients with heatstroke, but does not decrease as quickly, even with aggressive cooling measures. Therefore, you will likely have to actively cool the patient throughout the entire transport, unless the destination hospital is a great distance away.

If your protocols allow you to obtain a rectal temperature, you should assess it initially, and then actively cool the patient until his CBT reaches an acceptable level. Follow your local protocols or contact online medical control regarding the "target temperature" that you should attempt to achieve. If your protocols do not allow you to monitor a patient's CBT rectally, an axillary temperature should be assessed, although it is less accurate.

If you are unable to monitor the patient's CBT, frequently reassess his skin temperature during the cooling process. Does it feel as hot as it was initially, or does it seem to be cooler than it was before? It is important to frequently assess the effectiveness of your interventions.

When actively cooling a patient with heatstroke, you must not cool him or her to the point of shivering. Shivering generates more heat and can occur when cooling is not monitored closely. Although you must aggressively cool the patient with heatstroke, you do not want to overdo it; closely monitor your patient!

9. What other conditions should you consider as potential causes of the patient's altered mental status?

When assessing *any* patient, it is important to avoid tunnel vision. This is especially true when you are caring for a patient with an altered mental status. Whereas heatstroke obviously alters a patient's level of mentation, other factors should be considered. Altered mental status may be associated with heatstroke solely, or could be the result of a completely different problem.

An increase in heat energy, as with heatstroke, causes the body to expend a lot of glucose; therefore, you should consider the possibility of hypoglycemia. If you are trained and allowed by local protocol, assess the patient's blood glucose level. If it is found that he is hypoglycemic and he is not alert enough to swallow oral glucose, you should strongly consider an ALS intercept. Obviously, if the patient is conscious and alert enough to swallow (not likely during heatstroke), administer oral glucose per local protocol. If oral glucose administration and an ALS intercept are not options, continue rapid transport and report this finding in your radio report to the emergency department.

You should also consider the possibility of a head injury. If you will recall, the patient apparently fainted. When this occurred, he may have fallen and struck his head, resulting in a concussion or intracranial hemorrhage. Do not rule out an occult head injury in the absence of obvious signs of trauma.

The causes of altered mental status are numerous; heatstroke is just one. Although you may not be able to treat many of the causes of altered mental status, you should still consider and assess for them while aggressively cooling the patient and report any findings to the emergency department.

You are the Provider: SUMMARY, continued

EMS Patient Care Report (PCR)

Date: 8-11-10	Incident No.: 013010	Nature of Call: Fainting		Location: 1102 Rosewood Ave.
Dispatched: 1415	En Route: 1415	At Scene: 1422	Transport: 1430	At Hospital: 1443	In Service: 1450

Patient Information

Age: 55 Sex: M Weight (in kg [lb]): 75 kg (165 lb)	Allergies: PCN, Erythromycin Medications: Lasix, K-Dur, Prinivil, nitroglycerin Past Medical History: Hypertension, angina Chief Complaint: Heat exposure; confused

Vital Signs

Time: 1428	BP: 88/66	Pulse: 130	Respirations: 24	Sao₂: 95%
Time: 1433	BP: 90/70	Pulse: 126	Respirations: 26	Sao₂: 89%
Time: 1438	BP: 98/58	Pulse: 120	Respirations: 22	Sao₂: 94%

EMS Treatment (circle all that apply)

Oxygen @ 15 L/min via (circle one): NC (NRM) (Bag-Mask Device)	Assisted Ventilation	Airway Adjunct	CPR
Defibrillation	Bleeding Control	Bandaging	Splinting	Other: (Rapid cooling measures)

Narrative

Medic 4 was dispatched to a residence for a man who fainted after working outside in the heat for a prolonged period of time. Arrived on scene and found the patient, a 55-year-old male, sitting under a tree in his garden. He was conscious, but confused. His airway was patent and his breathing, although increased in rate, was producing adequate depth. Applied high-flow oxygen via nonrebreathing mask and quickly moved the patient to the cooled ambulance. Secondary assessment was performed, but did not reveal any gross signs of injury. Patient's skin was flushed, hot, and moist. According to the patient's wife, he would not come out of the heat to take a break and drink some water. When she found him, he would not respond to her initially. Patient's past medical history significant for hypertension and angina; medications listed above. Initial axillary temperature read 104.5°F. Removed patient's clothing and began rapid cooling measures by placing cold packs to his groin, axillae, and behind his neck. Began transport and continued cooling by spraying the patient with saline and fanning him. Vital signs indicated shock, so patient's legs were elevated and high-flow oxygen therapy was continued. Reassessment revealed that patient's mental status had markedly diminished; he was responsive only to pain. His respirations remained rapid, but were markedly decreased in depth. Inserted nasal airway and began assisting patient's ventilations with a bag-mask device and high-flow oxygen. After cooling measures, reassessment revealed that the patient's skin, although very warm, did not feel as hot as it was initially; he also appeared less flushed. Patient became somewhat combative and would no longer tolerate assisted ventilation. Reapplied nonrebreathing mask and reassessed his axillary temperature; it read 102.5°F. Continued to reassess patient's vital signs as indicated and monitored him for signs of overcooling. Remainder of transport was uneventful; patient was delivered to the emergency department and verbal report was given to attending physician. Medic 4 cleared the hospital and returned to service at 1450. *End of report*

Assessment and Emergency Care of Environmental Emergencies

Scene Size-up

Scene Safety	Ensure scene safety and safe access to the patient. Standard precautions should include a minimum of gloves and eye protection if there is a potential for vomiting. Consider a gown and shoe covers if other bodily fluids are involved. In situations of environmental extremes, ensure that you are wearing appropriate clothing. Determine the number of patients. Assess the need for additional resources, such as a search and rescue team, and summon additional help as quickly as possible.
Mechanism of Injury (MOI)/ Nature of Illness (NOI)	Determine the MOI. Observe the scene and look for indicators that will assist you with this. Interview the patient, family, and/or bystanders to determine the degree of environmental exposure.

Primary Assessment

Form a General Impression	Inquire about the chief complaint and observe the patient's overall condition. In an environmental emergency, your patient's chief complaint may be only that he or she is cold or hot. Determine the level of consciousness using the AVPU scale. The more altered the patient's mental status is, the more serious the problem. Identify immediate threats to life. Determine the priority of care based on the MOI. Pay particular attention to chest pain, dyspnea, and complaints related to sensory changes such as when a diving emergency is suspected. If the patient has a poor general impression, call for ALS assistance. A rapid scan of the patient will help you identify and manage life threats.
Airway and Breathing	Consider the possibility of spinal trauma. Ensure the airway is open, clear, and patent. Evaluate the patient's ventilatory status. Administer high-flow oxygen at 15 L/min, providing ventilatory support as needed. In patients with hypothermia, the oxygen should be warmed, if possible. Hypoxia may cause changes in the patient's mental status. If vomiting is a possibility, place the patient in the recovery position if no spinal injury is suspected.
Circulation	Observe skin color, temperature, and condition; look for life-threatening bleeding and treat accordingly. Evaluate the pulse rate, quality (strength), and rhythm. With severe hypothermia, heart function may be severely depressed, requiring an extended pulse check for accurate determination.
Transport Decision	If the patient has airway, breathing, or circulation problem, signs and symptoms of bleeding, or other life threats, manage them immediately and transport, performing the secondary assessment en route to the hospital. Consider the nature of the environmental emergency when making a transport decision. All patients with hypothermia require immediate transport for evaluation and treatment. If your patient has any signs of heatstroke (high temperature; red, dry skin; altered mental status; tachycardia; poor perfusion), then transport without delay. Patients with decompression sickness and air embolism must be treated in a recompression chamber.

NOTE: The order of the steps in this section differs depending on whether the patient is conscious or unconscious. The following order is for a conscious patient. For an unconscious patient, perform a primary assessment, perform a full-body scan, obtain vital signs, and obtain the past medical history from a family member, bystander, or emergency medical identification device.

History Taking

Investigate Chief Complaint	Investigate the chief complaint. Monitor the patient for changes in mental status. Ask OPQRST and SAMPLE questions. SAMPLE can also be obtained from family, bystanders, and medical alert tags. Ask probing questions specific to the nature of the environmental emergency. Certain medications can impair the body's natural ability to control its internal temperature.

Assessment and Emergency Care of Environmental Emergencies, continued

Secondary Assessment

Physical Examinations	Perform a systematic full-body examination or focused examination to rule out any potential life threats. Assess patients with suspected hypothermia by monitoring body temperature and focus on the areas of the body directly affected by cold exposure and assess the degree of damage. Frostbitten parts should be protected from further damage. Your assessment of the patient's skin will help determine the seriousness of a heat problem. For example, in heat exhaustion, the skin temperature may be normal or may even be cool and clammy; however, in heatstroke, the skin is hot. If a patient has been stung or bitten, locate any stingers or fangs and carefully remove them. Do not delay transport to perform the physical examination at the scene.
Vital Signs	Obtain baseline vital signs as soon as practical. Vitals signs should include blood pressure by auscultation, pulse rate and quality, respiration rate and quality, and skin assessment for perfusion. Note the patient's level of consciousness. Use pulse oximetry, if available, to assess the patient's perfusion status. Vital signs may be altered by the effects of hypothermia and can be an indicator of its severity. Patients who are hyperthermic will be tachycardic and tachypneic. Determine a core body temperature using a thermometer based on local protocol.

Reassessment

Interventions	Repeat the primary assessment and reassess vital signs and the chief complaint. Assist breathing as required, administering high-flow oxygen. Replace fluids by mouth for a heat emergency and cool the patient with a cool water spray or mist. For a cold emergency, reassess oxygen delivery and continue to provide a warm environment, removing any wet or frozen clothing.
Communication and Documentation	Contact medical control/receiving hospital with a radio report; many hospitals require additional personnel and a separate treatment area. Include a thorough description of the MOI and the position the patient was found in. Include treatments performed and patient response. Be sure to document the patient's distress, answers to your questions, and any changes in patient status and the time. Follow local protocols. Document the reasoning for your treatment and the patient's response.

NOTE: Although the following steps are widely accepted, be sure to consult and follow your local protocols. Take appropriate standard precautions when treating all patients.

Environmental Emergencies

Cold Exposure Emergency

1. Establish and maintain a patent airway. Provide oxygen. Monitor for vomiting and protect against aspiration.
2. Carefully move the patient to a protected environment. Remove any wet clothing. Place dry blankets over and under the patient.
3. Handle the patient gently to avoid further injury. With severe hypothermia, careful handling of the patient is necessary to prevent cardiac arrest; rough handling can cause ventricular fibrillation.
4. If the hypothermia is mild, begin active rewarming.
5. If the hypothermia is moderate or severe, prevent further heat loss and follow local protocols.

Assessment and Emergency Care of Environmental Emergencies, continued

Environmental Emergencies

Local Cold Injuries

1. Remove the patient from further exposure to the cold.
2. Handle the injured part gently, and protect it from further injury.
3. Administer oxygen.
4. Remove any wet or restricting clothing over the injured part.
5. For superficial local cold injury, consider active rewarming if there is no chance of reinjury. Splint the extremity, and cover it loosely with a dry, sterile dressing.
6. Do not rewarm a late or deep local cold injury unless specifically instructed to do so by medical control. Evaluate for hypothermia. Cover the injured part with soft, padded, sterile cotton dressings.
7. *Never attempt rewarming if there is any chance that the part may freeze again.*

Heat Cramps

1. Remove the patient from the hot environment. Loosen any tight clothing.
2. Administer high-flow oxygen.
3. Rest the cramping muscles. Have the patient sit or lie down until the cramps subside.
4. Replace fluids by mouth.
5. Cool the patient with cool water spray or mist and manually or mechanically fan the patient.

Heat Exhaustion

1. Remove extra clothing.
2. Move the patient to a cooler environment. Administer oxygen.
3. Place the patient in a supine position, elevate the legs, and fan the patient.
4. If the patient is fully alert, give water by mouth.
5. If nausea develops, transport the patient on his or her side.

Heatstroke

1. Move the patient out of the hot environment and into the ambulance.
2. Set the air conditioning to maximum cooling.
3. Remove the patient's clothing.
4. Administer high-flow oxygen and assist the patient's ventilations if needed.
5. Apply cool packs to the patient's neck, groin, and armpits.
6. Cover the patient with wet towels or sheets, or spray the patient with cool water and fan.
7. Aggressively and repeatedly fan the patient with or without dampening the skin.
8. Provide immediate transport to the hospital and notify the hospital of the patient's condition.

Drowning Injuries

1. Once the patient is removed from the water, ensure a patent and clear airway.
2. Begin CPR if pulse and breathing are absent.
3. If pulse and breathing are present, administer oxygen and assist ventilations if needed.
4. Keep the patient warm and transport.

Assessment and Emergency Care of Environmental Emergencies, continued

Environmental Emergencies

Diving Injuries

1. Remove the patient from the water.
2. Begin CPR if pulse and breathing are absent.
3. If pulse and breathing are present, administer oxygen.
4. Place the patient in a left lateral recumbent position with the head down.
5. Provide prompt transport to the nearest recompression facility for treatment.

Lightning Injuries

1. Move the patient to a sheltered area.
2. Stabilize the patient's head and open the airway with the jaw-thrust maneuver.
3. Those who are conscious following a lightning strike are much less likely to develop delayed respiratory or cardiac arrest; most of these persons will survive. Perform "reverse triage" by focusing your efforts on those who are in respiratory or cardiac arrest.
4. If the patient is in respiratory arrest with a pulse, begin ventilating with a bag-mask device with 100% oxygen. If the patient is in cardiac arrest, attach an automated external defibrillator. Control severe bleeding.
5. Provide full stabilization and transport.

Spider Bites

1. Provide basic life support for respiratory distress.
2. Apply ice to the bite area and clean the wound with soap and water.
3. Transport the patient and, if possible, the spider to the hospital.

Snake Bites

1. Calm the patient and minimize movement.
2. Clean the bite area gently with soap and water or a mild antiseptic. Do not apply ice.
3. Apply a pressure immobilization bandage if bite is located on an extremity.
4. Transport the patient and, if possible, the snake to the emergency department.
5. Notify the emergency department that you are bringing in a snake bite victim.

Marine Animal Injuries

1. Limit further discharge of nematocysts by avoiding fresh water, wet sand, showers, or careless manipulation of the tentacles. Keep the patient calm, and reduce motion of the affected extremity.
2. Inactivate the nematocysts by applying vinegar. Isopropyl alcohol may be used if vinegar is not available, but may not be as effective.
3. Remove the remaining tentacles by scraping them off with the edge of a sharp, stiff object such as a credit card. Do not use your ungloved hand to remove the tentacles because self-envenomation will occur. Persistent pain may respond to immersion in hot water (110°F to 115°F, 43°C to 46°C) for 30 minutes.
4. Provide transport to the emergency department.

Ready for Review

- Cold illness can be either a local or a systemic problem.

- Local cold injuries include frostbite, frostnip, and immersion foot. Frostbite is the most serious because tissues actually freeze. All patients with a local cold injury should be removed from the cold and protected from further exposure.

- If instructed to do so by medical control, rewarm frostbitten parts by immersing them in water at a temperature between 100°F and 112°F (38°C and 44.5°C).

- The key to treating hypothermic patients is to stabilize vital functions and prevent further heat loss. Do not attempt to rewarm patients who have moderate to severe hypothermia because they are prone to developing arrhythmias.

- Do not consider a patient dead until he or she is "warm and dead." Local protocol will dictate whether or not such patients receive cardiopulmonary resuscitation or defibrillation in the field.

- The body's regulatory mechanisms normally maintain body temperature within a very narrow range around 98.6°F (37°C). Body temperature is regulated by heat loss to the atmosphere via conduction, convection, evaporation, radiation, and respiration.

- Heat illness can take three forms: heat cramps, heat exhaustion, and heatstroke.
  - Heat cramps are painful muscle spasms that occur with vigorous exercise. Treatment includes removing the patient from the heat, resting the affected muscles, and replacing lost fluids.
  - Heat exhaustion is essentially a form of hypovolemic shock caused by dehydration. Symptoms include cold and clammy skin, weakness, confusion, headache, and rapid pulse. Body temperature can be high, and the patient may or may not still be sweating. Treatment includes removing the patient from the heat and treating for mild hypovolemic shock.
  - Heatstroke is a life-threatening emergency, usually fatal if untreated. Patients with heatstroke are usually dry and will have high body temperatures. Changes in mental status can include coma. Rapid lowering of the body temperature in the field is critical.

- The first rule in caring for drowning victims is to be sure not to become a victim yourself. Protect the spine when removing patients from the water because spinal cord injuries often occur in drownings. Be aware of the possibility of hypothermia.

- Injuries associated with scuba diving may be immediately apparent or may show up hours later. Patients with air embolism or decompression sickness may have pain, paralysis, or altered mental status. Be prepared to transport such patients to a recompression facility with a hyperbaric chamber.

- Poisonous spiders include the black widow spider and the brown recluse spider.

- Poisonous snakes include pit vipers and coral snakes.

- A person who has been bitten by a pit viper needs prompt transport; clean the bite area and keep the patient quiet to slow the spread of venom.

- Notify the hospital as soon as possible if a patient has been bitten by a coral snake; its venom can cause paralysis of the nervous system, and most hospitals do not have appropriate antivenin on hand.

- Patients who have been bitten by ticks may be infected with Rocky Mountain spotted fever or Lyme disease and should see a doctor within a day or two. Remove the tick using tweezers, and save it for identification.

- Always provide prompt transport to the hospital for any patient who has been bitten by a poisonous insect or animal. Remember that vital signs can deteriorate rapidly. Carefully monitor the patient's vital signs en route, especially for airway compromise.

Vital Vocabulary

air embolism Air bubbles in the blood vessels.

ambient temperature The temperature of the surrounding environment.

antivenin A serum that counteracts the effect of venom from an animal or insect.

bends Common name for decompression sickness.

breath-holding syncope Loss of consciousness caused by a decreased breathing stimulus.

conduction The loss of heat by direct contact (eg, when a body part comes into contact with a colder object).

convection The loss of body heat caused by air movement (eg, breeze blowing across the body).

core temperature The temperature of the central part of the body (eg, the heart, lungs, and vital organs).

decompression sickness A painful condition seen in divers who ascend too quickly, in which gas, especially nitrogen, forms bubbles in blood vessels and other tissues; also called "the bends."

diving reflex Slowing of the heart rate caused by submersion in cold water.

drowning The process of experiencing respiratory impairment from submersion or immersion in liquid.

dysbarism injuries Any signs and symptoms caused by the difference between the surrounding atmospheric pressure and the total gas pressure in various tissues, fluids, and cavities of the body.

evaporation Conversion of water or another fluid from a liquid to a gas.

frostbite Damage to tissues as the result of exposure to cold; frozen body parts.

heat cramps Painful muscle spasms usually associated with vigorous activity in a hot environment.

heat exhaustion A form of heat injury in which the body loses significant amounts of fluid and electrolytes because of heavy sweating; also called heat prostration or heat collapse.

heatstroke A life-threatening condition of severe hyperthermia caused by exposure to excessive natural or artificial heat, marked by warm, dry skin; severely altered mental status; and often irreversible coma.

hymenoptera A family of insects that includes bees, wasps, ants, and yellow jackets.

hyperthermia A condition in which the body core temperature rises to 101°F (38.3°C) or more.

hypothermia A condition in which the body core temperature falls below 95°F (35°C) after exposure to a cold environment.

radiation The transfer of heat to colder objects in the environment by radiant energy, for example heat gain from a fire.

respiration The loss of body heat as warm air in the lungs is exhaled into the atmosphere and cooler air is inhaled.

reverse triage A triage process in which efforts are focused on those who are in respiratory and cardiac arrest, and different from conventional triage where such patients would be classified as deceased. Used in triaging multiple victims of a lightning strike.

SCUBA A system that delivers air to the mouth and lungs at various atmospheric pressures, increasing with the depth of the dive; stands for self-contained underwater breathing apparatus.

turgor The ability of the skin to resist deformation; tested by gently pinching skin on the forehead or back of the hand.

Assessment in Action

You and your partner are called for a 25-year-old man who was found unresponsive by two hikers in a remote area of a national forest. On arrival you observe a young man lying supine on the ground with a liquor bottle nearby. Despite temperatures in the 30s, he is dressed in a t-shirt and jeans.

1. After establishing unresponsiveness, what should be your next step in patient management?
   A. Establish manual cervical spine immobilization.
   B. Open the airway.
   C. Check for a pulse.
   D. Measure the core body temperature.

2. Hypothermia is diagnosed when the core body temperature falls below what temperature?
   A. 98.6°F (37°C)
   B. 95.0°F (35°C)
   C. 92.0°F (33°C)
   D. 90.0°F (32°C)

3. The patient has a respiratory rate of 4 breaths per minute. Your partner is assisting ventilations with a bag-mask device while you perform a pulse check. How long should you assess for a carotid pulse?
   A. 5 to 10 seconds
   B. 15 to 30 seconds
   C. 30 to 45 seconds
   D. 45 to 60 seconds

4. The patient's core body temperature is 80°F (27°C). At this temperature, the patient's hypothermia would be classified as:
   A. mild.
   B. moderate.
   C. severe.
   D. extreme.

5. Your partner observes a medical alert bracelet on the patient's wrist. It says that the patient is diabetic. Other risk factors for hypothermia include:
   A. burns.
   B. head injury.
   C. shock.
   D. all of the above.

6. Shivering stops and muscle activity ceases once the body core temperature reaches:
   A. 95°F (35°C).
   B. 90°F (32°C).
   C. 85°F (29°C).
   D. 80°F (27°C).

7. Rough handling of a patient with severe hypothermia may cause which of the following arrhythmias?
   A. Ventricular fibrillation
   B. Sinus bradycardia
   C. Asystole
   D. Sinus tachycardia

8. Appropriate treatment for this patient includes:
   A. removing wet clothing.
   B. wrapping the patient in warm blankets.
   C. placing the patient in a heated ambulance.
   D. all of the above.

9. Discuss the controversy regarding performing cardiopulmonary resuscitation (CPR) in a patient with severe hypothermia.

10. Why are children and infants at a greater risk of developing hypothermia than adults?

National EMS Education Standard Competencies

Special Patient Populations

Applies a fundamental knowledge of growth, development, and aging and assessment findings to provide basic emergency care and transportation for a patient with special needs.

Obstetrics

- Recognition and management of:
  - Normal delivery (pp 1115–1124)
  - Vaginal bleeding in the pregnant patient (pp 1112–1113, 1132)
- Anatomy and physiology of normal pregnancy (pp 1107–1111)
- Pathophysiology of complications of pregnancy (pp 1111–1114)
- Assessment of the pregnant patient (pp 1115–1117)
- Management of
  - Normal delivery (pp 1117–1124)
  - Abnormal delivery (pp 1122–1123, 1128–1132)
    - Nuchal cord (pp 1122–1123)
    - Prolapsed cord (p 1129)
    - Breech delivery (p 1128)
  - Third trimester bleeding (pp 1112–1113)
    - Placenta previa (pp 1112–1113)
    - Abruptio placenta (pp 1112–1113)
  - Spontaneous abortion/miscarriage (pp 1129–1130)
  - Ectopic pregnancy (pp 1111–1112)
  - Preeclampsia/Eclampsia (pp 1111–1112)

Neonatal care

Assessment and management

- Newborn care (pp 1124–1128)
- Neonatal resuscitation (pp 1124–1128)

Trauma

Applies fundamental knowledge to provide basic emergency care and transportation based on assessment findings for an acutely injured patient.

Special Considerations in Trauma

- Recognition and management of trauma in the:
  - Pregnant patient (pp 1113–1114)
  - Pediatric patient (Chapter 32, *Pediatric Emergencies*)
  - Geriatric patient (Chapter 33, *Geriatric Emergencies*)
- Pathophysiology, assessment, and management of trauma in the:
  - Pregnant patient (pp 1112–1113)
  - Pediatric patient (Chapter 32, *Pediatric Emergencies*)
  - Geriatric patient (Chapter 33, *Geriatric Emergencies*)
- Cognitively impaired patient (Chapter 34, *Patients With Special Challenges*)

Knowledge Objectives

1. Be familiar with the anatomy and physiology of the female reproductive system. (pp 1107–1109)
2. Understand the normal changes that occur in the body during pregnancy. (pp 1109–1110)
3. Differentiate between the three stages of labor. (pp 1110–1111)
4. Recognize complications of pregnancy including hypertensive disorders, bleeding, and gestational diabetes. (pp 1111–1113)
5. Understand the need to consider two patients—the woman and the unborn fetus—when treating a pregnant trauma patient. (pp 1113–1114)
6. Be aware of special considerations involving pregnancy in different cultures and with teenage patients. (pp 1114–1115)
7. Describe assessment of the pregnant patient. (pp 1115–1117)
8. Explain the significance of meconium in the amniotic fluid. (pp 1116–1117)
9. Describe the indications of an imminent delivery. (pp 1116–1117)
10. Explain the steps involved in normal delivery management. (pp 1117–1123)
11. List the contents of an obstetrics kit. (p 1118)
12. Explain the necessary care of the baby as the head appears. (pp 1120–1123)
13. Describe the procedure followed to cut and tie the umbilical cord. (pp 1123–1124)
14. Describe delivery of the placenta. (p 1124)
15. Understand the steps to take in neonatal assessment and resuscitation. (pp 1124–1128)
16. Recognize complicated delivery emergencies including breech presentations, limb presentations, umbilical cord prolapse, spina bifida, abortion (miscarriage), multiple gestation, abuse, substance abuse, premature infants, postterm pregnancy, fetal demise, and delivery without sterile supplies. (pp 1128–1132)
17. Describe and know how to deal with postpartum complications. (p 1132)

Skills Objectives

1. Demonstrate the procedure to assist in a normal cephalic delivery. (pp 1117–1123, Skill Drill 31-1)
2. Demonstrate care procedures of the infant as the head appears. (pp 1120–1122)
3. Demonstrate the steps to follow in postdelivery care of the infant. (pp 1123–1124)
4. Demonstrate how to cut and tie the umbilical cord. (pp 1123–1124)
5. Demonstrate how to assist in delivery of the placenta. (p 1124)
6. Demonstrate the postdelivery care of the mother. (pp 1123–1124)
7. Demonstrate procedures to follow for complicated delivery emergencies including vaginal bleeding, breech presentation, limb presentation, and prolapsed umbilical cord. (pp 1112–1113, 1128–1129, 1132)

Introduction

Most infants in the United States are delivered in a hospital, with doctors and nurses in attendance to care for the mother and the newborn infant. Occasionally, the birth process moves faster than the mother expects, and you will find yourself with a decision to make: Should you stay on the scene and deliver the infant or transport the patient to the hospital? Are there other factors that would affect this decision, such as trauma, weather, and distance to the hospital? This chapter will tell you how to make this decision and how to proceed if on-scene delivery is necessary. It describes the anatomy and physiology of a normal pregnancy and the normal process of childbirth. Also discussed are common complications, including trauma in a pregnant patient, so that you will be prepared to handle normal and abnormal deliveries. Finally, the chapter discusses the evaluation and care of newborn infants and neonatal resuscitation.

Anatomy and Physiology of the Female Reproductive System

The female reproductive system includes the ovaries, the fallopian tubes, uterus, cervix, vagina, and the breasts. The ovaries are two glands, one on each side of the uterus, that are similar in function to the male testes. Each ovary contains thousands of follicles, and each follicle contains an egg (the female contribution to conception). Females are born with all the eggs they will release in their lifetime. Once puberty is reached, the monthly process of the menstrual cycle begins.

During each menstrual cycle, there will only be one follicle (out of 10 to 20 that attempt the process each month) that is successful at maturing and is able to release an egg. The remaining follicles die and are reabsorbed by the body. The processes that the follicle goes through and the actual release of the egg (ovulation) are stimulated by the release of specific hormones in the female body. Ovulation occurs approximately 2 weeks prior to menstruation. In the next step, the **endometrium**, or the lining of the inside of the uterus, begins to thicken and prepare for the potential fertilized egg to implant. If the egg is not fertilized within 36 to 48 hours after it has been released from the follicle, it will simply die. Eventually, the lining that has thickened inside the uterus will be shed because it is no longer needed. This shedding is the menstrual flow that occurs around the 28th day of a woman's cycle.

The fallopian tubes extend out laterally from the uterus, with one tube associated with each ovary. When the egg is released from the ovary, it travels through the fallopian tube to the uterus. Fertilization, when a sperm meets the egg, usually occurs when the egg is inside the fallopian tube. The fertilized egg then continues to the uterus where it continues to develop into an **embryo** (early stages of the fetus after fertilization) and implants in the wall of the uterus.

The uterus, or womb, is a muscular organ, and it is here that the **fetus** (the developing, unborn infant) grows

You are the Provider: PART 1

At 6:25 AM, you are dispatched to a residence at 2505 Landa Park Boulevard for a woman in labor. You and your partner proceed to the scene, which is located a short distance away. While en route, dispatch advises you that the patient is 38 weeks pregnant, and that her contractions are 3 minutes apart.

1. What anatomic and physiologic changes occur during pregnancy? How will they affect your assessment of the patient?

2. How will you determine if delivery is imminent or if there is enough time to transport a pregnant patient?

for approximately 9 months (40 weeks) Figure 31-1 . The uterus is responsible for contractions during labor and ultimately helps to push the infant through the **birth canal**. The birth canal is made up of the vagina and the lower third, or neck, of the uterus, called the **cervix**. During pregnancy, the cervix contains a mucous plug that seals the uterine opening, preventing contamination from the outside world. When the cervix begins to dilate, this plug is discharged into the vagina as pink-tinged mucus, or **bloody show**. This small amount of blood appears at the beginning of labor and may signal the first stage of labor.

The vagina is the outermost cavity of the female reproductive system and forms the lower part of the birth canal. It is about 8 to 12 cm in length, begins at the cervix, and ends as an external opening of the body. The vagina completes the passageway from the uterus to the outside world for the infant. The perineum is the area of skin between the vagina and the anus. During birth, as the infant moves through the birth canal, the perineum will begin to bulge significantly. If this occurs too quickly, the tissues may not be able to stretch enough, and this area may tear or rip to allow for the birth. This usually results in bleeding and increased pain for the mother. You should apply gentle pressure to the baby's head to allow the tissues to stretch and to prevent a precipitous birth (one that occurs too fast). In a hospital delivery, the physician may perform an episiotomy (incision to the perineum) to prevent the tissue from tearing or ripping.

The breasts (mammary glands) of the pregnant female are also a part of the female reproductive system. The breasts produce milk that is carried through small ducts to the nipple to provide nourishment to the infant once it is born. Signs of pregnancy in the breasts include increased size and tenderness.

As the fetus continues to develop, it requires increasingly more nourishment and support. The **placenta**, a disk-shaped structure, attaches to the inner lining of the wall of the uterus and is connected to the fetus by the umbilical cord. Blood normally does not mix between the fetus and the pregnant woman. The placental barrier Figure 31-2 consists of two layers of cells, keeping the circulation of the woman and fetus separated but allowing nutrients, oxygen, waste, carbon dioxide, and many toxins and most medications to pass between the fetus and woman. Anything ingested by a pregnant woman also affects the fetus. After delivery, the placenta, or afterbirth, separates from the uterus and is delivered. The **umbilical cord** is the lifeline of the fetus, connecting the woman and fetus through the placenta. The umbilical cord contains two arteries and one vein. These vessels supply blood to the fetus: The umbilical vein carries oxygenated blood from the woman to the heart of the fetus,

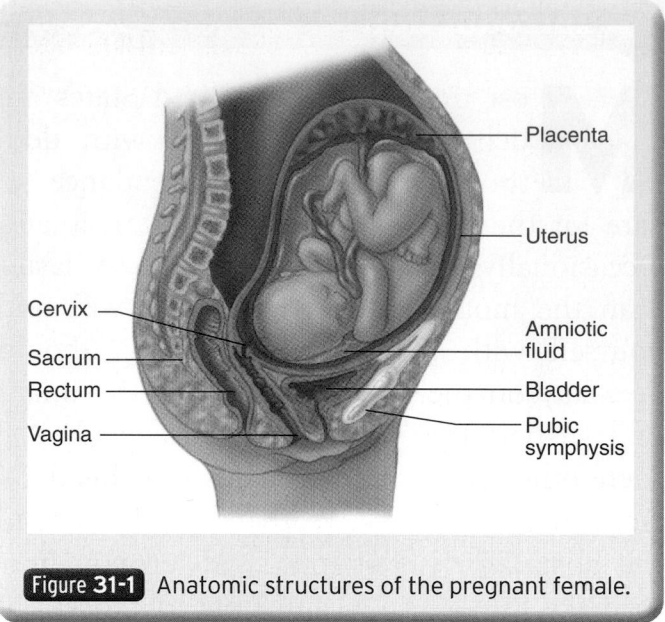

Figure 31-1 Anatomic structures of the pregnant female.

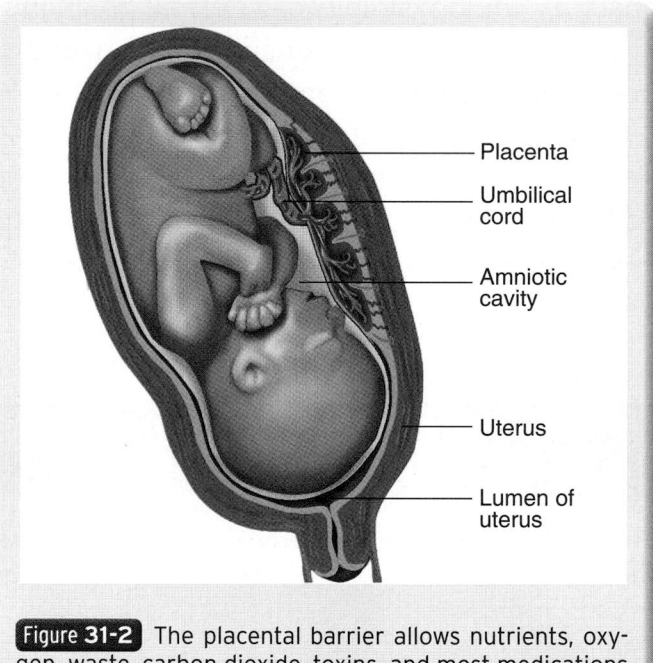

Figure 31-2 The placental barrier allows nutrients, oxygen, waste, carbon dioxide, toxins, and most medications to pass between fetus and pregnant woman.

and the umbilical arteries carry deoxygenated blood from the heart of the fetus to the woman. Oxygen and other nutrients cross from the woman's circulation through the placenta and then along the umbilical cord to support the fetus as it grows.

The fetus develops inside a fluid-filled, baglike membrane called the **amniotic sac**, or bag of waters. The sac contains about 500 to 1,000 mL of amniotic fluid, which helps insulate and protect the floating fetus. The amniotic fluid is released in a gush when the sac ruptures, usually

Words of Wisdom

Predicting the due date for an infant is not an exact science. Fewer than half of babies are born on the due date. Many factors influence when a baby is born, and neither the mother nor EMTs have much control over this. One confusing point is the expected date and how "far along" the pregnancy is. Most medical models base the due date on the first day of the last menstrual cycle. That adds approximately 2 weeks to the actual pregnancy because conception occurred sometime after ovulation, which occurred approximately 2 weeks after the beginning of the last menstrual cycle. Most women have a general idea of this date, but young women, women who have very irregular cycles, and women who did not think they were pregnant may not have a very accurate due date. Also, some women view their due date as the number of weeks from conception instead of menstruation. The important thing to remember is that a due date is not a firm number.

at the beginning of labor. Some women may experience a small leak rather than a gush of fluid. It is typical for the patient to tell you that her "water broke." This fluid helps to lubricate the birth canal and remove any bacteria.

Normal Changes in Pregnancy

Many normal changes occur in the body during pregnancy that are not all directly related to the reproductive system. It is important to understand these changes as you assess and treat a pregnant patient. The primary systems involved with these changes are the respiratory, cardiovascular, and musculoskeletal systems.

In the reproductive system, hormone levels increase to support fetal development and prepare the body for childbirth. However, these increased hormone levels also put pregnant patients at an increased risk for complications from trauma, bleeding, and some medical conditions. As the fetus develops and grows, the uterus also grows, stretching to accommodate a full-term infant. As the size of the uterus increases, so does the amount of fluid it must hold. These factors eventually result in displacement of the uterus out of its normally protected position and may expose it to injury. By the 20th week of pregnancy, the uterus is at or above the belly button. This increases the chance of direct fetal injury. The size of the uterus will also have a direct effect on the respiratory and cardiovascular systems because its enlarged size results in organs being shifted from their normal positions and an overall decreased area for all of the organs to fit.

Rapid uterine growth occurs in the second trimester of pregnancy. The increased size of the uterus directly affects the respiratory system because as the uterus grows,

it pushes up on the diaphragm and displaces it from its normal position. As the pregnancy continues, respiratory capacity changes, with increased respiratory rates and decreasing minute volumes. This is a normal change and you may observe a pregnant patient's increased breathing rate and a decreased ability for deep respirations. These changes result in a less than normal respiratory reserve. The pregnancy also increases the patient's overall demand for oxygen as her metabolic demands and workload for the fetus increase.

Changes also occur in the cardiovascular system. Overall blood volume gradually increases throughout the pregnancy to meet the increased needs of the fetus, to allow for adequate perfusion of the uterus, and to prepare for the blood loss that will occur during childbirth. During the third trimester of pregnancy, the woman's entire blood volume passes through the uterus every 8 to 11 minutes. Blood volume may eventually increase as much as 50% by the end of the pregnancy. The number of red blood cells also increases. This increases the woman's need for iron, which explains why pregnant women are advised to take iron supplements (prenatal vitamins) to avoid becoming anemic. Blood clotting factors also begin to change as the woman's body prepares for childbirth. These changes result in the woman's ability to clot faster to protect her from excessive bleeding during delivery. By the end of the pregnancy (third trimester), the pregnant patient's heart rate increases up to 20%, or about 20 beats more per minute, to accommodate the increase in blood volume. Cardiac output is significantly increased by the end of the pregnancy.

A patient in her third trimester of pregnancy has an increased risk of vomiting and potential aspiration following trauma because of changes that occur in the gastrointestinal tract. Gastrointestinal motility, or the filling and emptying of the stomach to the small intestine, is under the control of key hormones and the nervous system. Changes in these systems and the displacement of the stomach upward because of the increased size of the uterus significantly increase the chance that a pregnant trauma patient will vomit and aspirate if you are unable to clear her airway. You should be aware of this possibility when treating a pregnant trauma patient so that you will be prepared to quickly manage her airway if needed.

Changes in the cardiovascular system and the increased demands of the fetus significantly increase the workload of the heart. For healthy women, the increased workload and demand are able to be handled by the body. However, remember that not all pregnant women are healthy when they begin their pregnancy. Cardiac compromise is a life-threatening possibility.

Weight gain is expected in pregnant patients. The increase in body weight will eventually challenge the heart and will also have an impact on the musculoskeletal system. Increased hormones affect the musculoskeletal system by making the joints more "loose" or less stable. Patients in the third trimester of pregnancy also experience a change in the body's center of gravity, making them prone to slipping and falling.

Stages of Labor

There are three stages of labor: dilation of the cervix, delivery of the infant, and delivery of the placenta. The first stage begins with the onset of contractions as the fetus enters the birth canal and ends when the cervix is fully dilated. Because the cervix has to be stretched thin by uterine contractions until the opening is large enough for the infant to pass through into the vagina, the first stage of labor is usually the longest, lasting an average of 16 hours for a first delivery. You will usually have time to transport the mother during the first stage of labor.

The onset of labor starts with contractions of the uterus. Other signs of the beginning of labor are the bloody show and the rupture of the amniotic sac (water breaking). These events may occur before the first contraction or later in the first stage of labor. Initially, the uterine contractions may not occur at regular intervals. The mother may think that she simply has a nagging backache. The frequency and intensity of contractions in true labor increase with time. The uterine contractions become more regular and last about 30 to 60 seconds each. How long a woman will be in labor varies greatly. As a general rule, it is longer in a **primigravida**, a woman who is experiencing her first pregnancy, and shorter in a **multigravida**, a woman who has experienced previous pregnancies.

Table 31-1 discusses how to tell when true labor is occurring. During pregnancy, a woman may experience preterm or false labor, or Braxton-Hicks contractions. In this case, you should provide transport for the patient. However, if true labor is occurring, you may need to prepare for a delivery, depending on the patient's condition and transport time.

Some women experience a premature rupture of the membranes, in which the amniotic sac ruptures too early and the fetus is not developed or ready to be born. When this occurs, the patient may or may not go into labor. In this situation, you will need to provide supportive care and transport to the hospital. Some patients may experience a premature rupture of the membranes several months before they are due to deliver. They are usually placed on bed rest and followed up closely by an obstetrician.

You are the Provider: PART 2

When you arrive at the scene, you are greeted at the door by the patient's husband. He is obviously anxious and tells you, "She's having the baby! I thought I could get her to the hospital in time, but I was wrong." You find the patient, a 28-year-old woman, lying supine in her bed. You introduce yourself and your partner and perform a primary assessment.

Recording Time: 0 Minutes	
Appearance	Diaphoretic; in obvious pain
Level of consciousness	Conscious and alert
Airway	Open; clear of secretions and foreign bodies
Breathing	Increased rate; adequate depth
Circulation	Pulse rate is increased; strong and regular; no gross bleeding

The patient tells you that she feels like she needs to move her bowels and that her contractions are now about 2 minutes apart and last about 45 seconds. A brief visual examination of her perineum does not reveal crowning. According to the patient's husband, this is her third baby, and she has had gestational diabetes and preeclampsia with this pregnancy. Her amniotic sac ruptured 5 hours ago.

3. What are gestational diabetes and preeclampsia? How can they affect this delivery?

4. Is there time to transport this patient, or should you prepare for imminent delivery?

Table 31-1 False Labor Versus True Labor

False Labor or Braxton-Hicks Contractions	True Labor
Contractions are not regular and do not increase in intensity or frequency. Contractions come and go.	Contractions, once started, consistently get stronger and closer together. Change in position does not relieve contractions.
Pain is in the lower abdomen. Contractions start and stay in the lower abdomen.	Pains and contractions start in the lower back and "wrap around" to the lower abdomen.
Activity or changing position will alleviate the pain and contractions.	Activity may intensify the contractions. Pain and contractions are consistent in any position.
If there is any bloody show, it is brownish.	The bloody show will be pink or red and generally accompanied by mucus.
There may be some leakage of fluid, but it is usually urine and will be in small amounts and smell of ammonia.	The amniotic sac may have broken just before the contractions started or during contractions. A moderate amount of fluid will be present and may smell sweet, and fluid will continue to leak.

Toward the end of the third trimester of pregnancy, the head of the fetus normally descends into the woman's pelvis as the fetus positions for delivery. This movement down into the pelvis and the sensation that may accompany the descent is called **lightening**. Your patient may tell you that she has felt this sensation as the fetus moves into position. Some women who experience this feeling have described it as a "relief" because once the fetus has moved from under their rib cage, breathing becomes easier. Lightening may also be gradual and not as noticeable by some patients.

The second stage of labor begins when the fetus begins to enter the birth canal and ends when the infant is born (spontaneous birth). During this stage, you will have to make a decision about helping the mother to deliver at the scene or providing transport to the hospital. Because the fetus goes through positional changes as it moves through the birth canal during this stage, the uterine contractions are usually closer together and last longer. Pressure on the rectum may make the mother feel as if she needs to have a bowel movement. Under no circumstances should you let the mother sit on the toilet. She may also have the uncontrollable urge to push down. The perineum will begin to bulge significantly, and the top of the infant's head should begin to appear at the vaginal opening. This is called **crowning**.

The third stage of labor begins with the birth of the infant and ends with the delivery of the placenta. During this stage, the placenta must completely separate from the uterine wall. Contractions will continue to assist with the separation process and to clamp down and close the blood vessels that were connecting the placenta to the uterine lining. This may take up to 30 minutes.

Always follow standard precautions to protect yourself, the infant, and the mother from exposure to body fluids. A high potential for exposure exists because of the body fluids released during childbirth.

Complications of Pregnancy

Most pregnant women are healthy, but some may be ill when they conceive or become ill during pregnancy. You may safely use oxygen to treat any heart or lung disease in a pregnant patient without harm to the fetus.

Hypertensive Disorders

As delivery nears, complications can occur. One complication that occurs most commonly in patients who are pregnant for the first time is **preeclampsia**, or **pregnancy-induced hypertension**. This condition can develop after the 30th week of gestation and is characterized by the following signs and symptoms:

- Headache
- Seeing spots
- Swelling in the hands and feet (edema)
- Anxiety
- High blood pressure

Another condition, **eclampsia**, is characterized by seizures that occur as a result of hypertension. To treat eclampsia, lie the patient on her side—preferably her left side—maintain an airway, and provide supplemental oxygen; if vomiting occurs, suction the airway. Provide rapid transport for pregnant patients having seizures, and call for an advanced life support (ALS) intercept, if available.

Transporting the patient on her left side can also prevent **supine hypotensive syndrome**. This condition is caused by compression of the descending aorta and inferior vena cava by the pregnant uterus when the patient lies supine, which reduces the amount of blood that is returned to the heart. Hypotension (low blood pressure) results. Patients in the third trimester of pregnancy should always be transported on their side except during delivery.

◼ Bleeding

Internal bleeding may be the sign of an **ectopic pregnancy**, a pregnancy that develops outside the uterus, most often in a fallopian tube **Figure 31-3**. Ectopic pregnancy occurs about once in every 300 pregnancies. The leading cause of maternal death in the first trimester of pregnancy is internal hemorrhage into the abdomen following rupture of an ectopic pregnancy. For this reason, you should consider the possibility of an ectopic pregnancy in women who have missed a menstrual cycle and complain of sudden stabbing and usually unilateral pain in the lower abdomen. A history of pelvic inflammatory disease, tubal ligation, or previous ectopic pregnancies should heighten your suspicion of a possible ectopic pregnancy.

Hemorrhage from the vagina that occurs before labor begins may be very serious; call for ALS backup. In early pregnancy, it may be a sign of a spontaneous abortion, or **miscarriage**. In the later stages of pregnancy, vaginal hemorrhage may indicate a serious condition involving the placenta. In **abruptio placenta**, the placenta separates prematurely from the wall of the uterus **Figure 31-4**, most commonly from hypertension in the mother and as a result of trauma. In **placenta previa**, the placenta develops over and covers the cervix **Figure 31-5**.

In abruptio placenta, the patient often complains of severe pain and has vaginal bleeding. Regardless of

the cause of the bleeding, the pregnant patient will be emotional and very concerned about her infant. Your professional approach in communicating with the patient will be crucial in calming her emotions and gaining

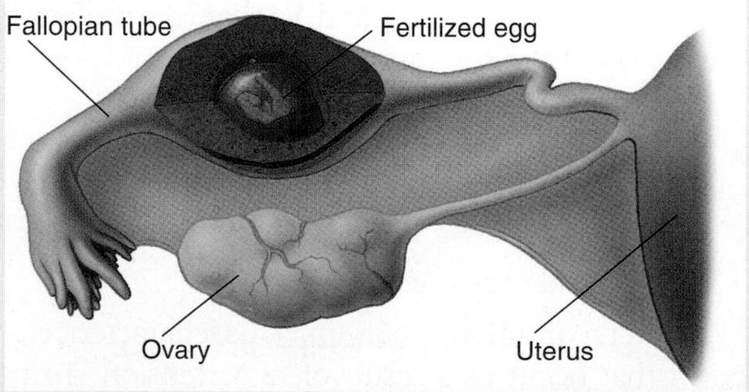

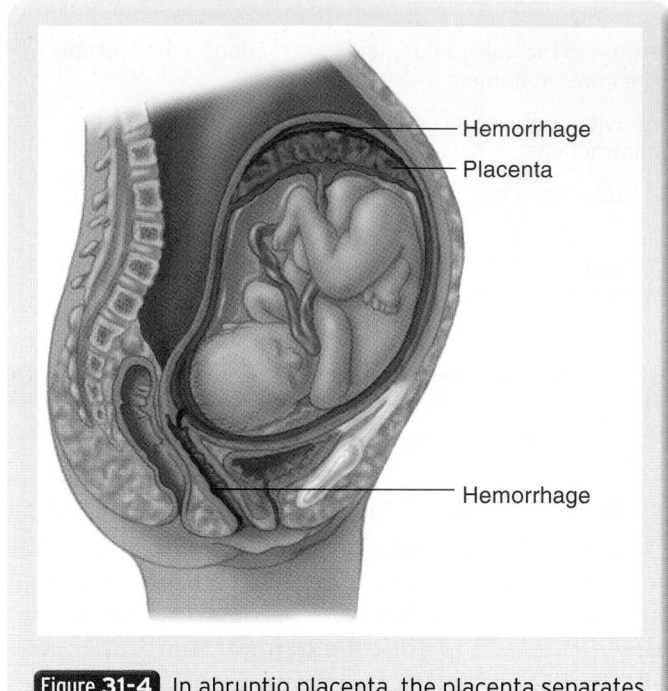

Figure 31-4 In abruptio placenta, the placenta separates prematurely from the wall of the uterus.

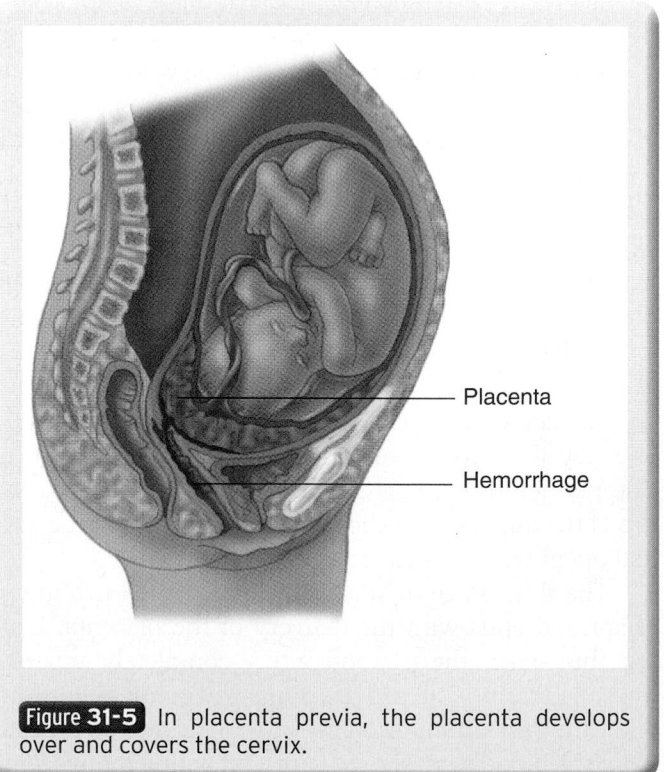

Figure 31-5 In placenta previa, the placenta develops over and covers the cervix.

Figure 31-3 In an ectopic pregnancy, a fertilized egg implants somewhere other than in the uterus. Here it is implanted in one of the fallopian tubes.

control of the situation. Decreasing the patient's anxiety will directly impact how she and the fetus may respond during this emergency.

Any bleeding from the vagina in a pregnant woman is a serious sign and should be treated in the hospital promptly. If the mother shows signs of shock, position her on her left side and administer high-flow oxygen. Place a sterile pad or sanitary napkin over the vagina, and replace it as often as necessary. Save the pads so that hospital personnel can estimate how much blood she has lost. Also save any tissue that may be passed from the vagina. Do not put anything into the vagina.

■ Diabetes

Diabetes develops during pregnancy in many women who have not had diabetes previously. This condition, called **gestational diabetes**, will clear up in most women after delivery. The treatment of a pregnant woman with diabetes is the same as the treatment for any patient who has diabetes. A pregnant woman may control her blood glucose level with diet and exercise or may take medication; in some cases, the woman will have to manage her condition with insulin injections. A pregnant woman experiencing hyperglycemia or hypoglycemia should be cared for in the same manner as any patient with diabetes. If a pregnant woman is found with an altered level of consciousness, your assessment should include determining if she has a history of diabetes, and you should check the blood glucose level if local protocols permit. Remember that childbirth labor is hard work. Many mothers experience nausea before labor and may not have eaten. These factors can lead to hypoglycemia and weakness in the mother and fetus. Consult with medical control if delivery is imminent.

■ Special Considerations for Trauma and Pregnancy

There are times when you will be dispatched to a trauma call that involves a pregnant woman. In these situations, you have two patients to consider—the woman and the unborn fetus. Any trauma to the woman has a direct effect on the condition of the fetus. Pregnant women may be victims of many types of trauma, including assaults, motor vehicle crashes, and shootings, and unfortunately, pregnant women are involved in cases of domestic abuse.

Pregnant women also have an increased risk of falls compared with nonpregnant women. Hormonal changes "loosen" up the joints in the musculoskeletal system, and the weight of the uterus and displacement of abdominal organs can change the patient's balance. These factors contribute to a pregnant woman's increased risk of falling.

Pregnant women have an increased amount of overall total blood volume and an approximate 20% increase in their heart rate by the third trimester of pregnancy. Therefore, a pregnant trauma patient may have a significant amount of blood loss before you will see or detect signs of shock. The fetus also may be in trouble well before signs of shock are present. Often, if the woman has sustained serious trauma, the blood supply to the fetus is reduced so that the body can supply an adequate amount of blood to the woman only.

Pregnant trauma patients need to have additional concerns considered and unique types of injuries assessed for and managed. As a pregnancy progresses, the uterus enlarges substantially and rises out of the pelvis. This makes the uterus especially vulnerable to penetrating trauma and blunt injuries. The fetus may be injured directly from penetrating types of trauma such as gunshot wounds and stabbings. Some data suggest that almost 70% of all penetrating abdominal trauma in pregnant patients results in fetal injury. A trauma injury to the pregnant uterus can be life threatening to the woman and fetus because the uterus has a rich blood supply. If the woman is hypoxic, is in shock, or has hypovolemia, the fetus will be in distress. In most cases, the only chance to save the fetus is to adequately resuscitate the woman.

When a pregnant woman is involved in a motor vehicle crash or a similar mechanism of injury (MOI), severe hemorrhage may occur from injuries to the pregnant uterus. Trauma is one of the leading causes of abruptio placenta, which is the premature separation of the placenta from the inside of the uterine wall. This condition results in significant intrauterine hemorrhage that can cause life-threatening hypovolemic shock in the woman and also increases the chance of fetal death. In a pregnant trauma patient, you should suspect abruptio placenta when the MOI is blunt trauma to the abdomen and the patient's signs and symptoms are suggestive of shock. Significant vaginal bleeding is common with severe abdominal pain. In this situation, quickly assess and transport the patient, support the airway, administer high-flow oxygen, place sanitary pads on the vagina, position the patient on her left side, and call for ALS backup.

Not all pregnant women properly position their seatbelts when in a vehicle. The lap belt should be placed under the abdomen and over the hip bones, and the shoulder belt should be positioned between the breasts. If a pregnant woman is involved in a motor vehicle crash with an improperly positioned seatbelt, the seatbelt can cause harm to the woman and fetus. Carefully assess a pregnant woman's abdomen and chest for seatbelt marks, bruising, and obvious trauma. Maintain a high index of suspicion for internal abdominal bleeding in the woman and possible direct injury to the fetus, regardless of seatbelt placement.

If a pregnant trauma patient goes into cardiac arrest, your focus is the same as with other patients in cardiac arrest. Remember that the only chance you have to save the infant is to do all you can to save the mother. Perform cardiopulmonary resuscitation (CPR) and provide transport to the hospital according to local protocol. You should notify the receiving facility personnel as soon as possible that you are en route with a pregnant trauma patient in cardiac arrest so they will have more time to prepare. In some situations, the hospital staff may decide to perform a cesarean section shortly after you arrive in an attempt to save the infant.

■ Assessment and Management

When your patient is pregnant, you have two patients to care for—the woman and the fetus. Your focus is on the assessment and the management of the woman. It is difficult to assess for or know the extent of internal blood loss in the patient. You should suspect shock based on the MOI because you will not see the typical signs and symptoms of shock as a result of the physiologic changes that occur with pregnancy. As you assess and treat the patient, be prepared for vomiting, and anticipate the need to manage the airway to protect the patient from aspirating. You should attempt to determine the gestational period (in number of weeks) to assist you with determining the size of the fetus and the position of the uterus in the patient's abdominal cavity. Since it is nearly impossible for you to accurately assess or determine the status of the fetus, you should aggressively provide emergency medical care to the woman to provide the best possible outcome for the fetus.

Follow these guidelines when treating a pregnant trauma patient:

1. **Maintain an open airway.** A pregnant patient has an increased risk of vomiting and aspiration compared with patients who are not pregnant. Be prepared for and anticipate vomiting; keep your suction unit readily available.

2. **Administer high-flow oxygen.** The patient is also supplying oxygen to the fetus. Keep the oxygen saturation level high, and administer high-flow, 100% oxygen by nonrebreathing mask.

3. **Ensure adequate ventilation.** Listen to lung sounds, and confirm that bilateral breath sounds are present. If the patient has inadequate ventilation, provide or assist ventilation with a bag-mask device and 100% oxygen.

4. **Assess circulation.** Control any external bleeding with direct pressure. Maintain a high index of suspicion for internal bleeding and shock based on the MOI because a pregnant patient will not always display the typical signs and symptoms of shock. Keep the patient warm.

5. **Transport considerations.** Transport the patient on her left side (to anticipate vomiting and to avoid supine hypotensive syndrome). If spinal injury is suspected, tilt the backboard to the left. Call early for ALS assistance or a medical helicopter for significant MOIs or major traumatic injuries. Transport the patient to a trauma center if one is available in your area; give early notification that you have a pregnant trauma patient in transport.

Words of Wisdom

By 20 weeks of gestation, the top of the uterus has grown to the level of the patient's belly button. This fact is important to remember when managing a pregnant patient who has sustained trauma (fetus is exposed and more prone to injury) and to aid in your assessment of a pregnant patient's abdomen.

■ Cultural Value Considerations

The United States is a culturally diverse nation. This diversity may be a factor when you are assessing and treating an obstetric patient from a culture different from yours. Women of some cultures may have a value system that will affect their pregnancy, the choice of how they care for themselves during pregnancy, and how they have planned the childbirth process. Some cultures may not permit a male health care provider, especially in the prehospital setting, to assess or examine a female patient. Different cultures may view pregnancy differently than you do in terms of social, psychological, and emotional issues. Some may see pregnancy as a means of achieving status and recognition within the family unit, whereas others may experience a drop in self-esteem. You should respect these differences and honor requests from the patients. Always remember that your responsibility is to the patient and is limited to providing care and transport, and keep in mind that a competent, rational adult has the right to refuse all or any part of your assessment or care.

■ Teenage Pregnancy

The United States has one of the highest teenage pregnancy rates compared with other developed countries. There is a good chance that during your career, you will respond to a pregnant teenager who may or may not be in labor. Adolescents present their own challenges to the EMS community in terms of physical and psychological development, even without the contributing factor of the female teenager being pregnant.

Pregnant teenagers may or may not know that they are pregnant or can be in denial about their pregnancy. As you begin to assess all female teenagers, you should remember that pregnancy is a possibility. The pregnancy itself may or may not be related to the nature of the call, but you should consider the possibility when assessing the patient, talking to the patient, obtaining a history, and providing treatment. Respect the teenager's privacy and need for independence. If possible, perform your assessment and obtain the history away from her parents.

Patient Assessment

Childbirth is seldom an unexpected event, but there are occasions when childbirth becomes an emergency. Dispatch protocols usually include the dispatcher asking simple questions to determine whether birth is imminent. Some of this information may be passed on to you to help you prepare for the situation. Contractions may be caused by trauma or medical conditions. It may just be "time" to deliver.

Scene Size-up

Scene Safety

As with every emergency call, your safety is a priority. Take standard precautions—gloves and eye protection are a minimum if delivery has already begun or is complete. If the call is going to result in a field delivery and if time allows, a mask and gown should also be used. Do not be complacent in your safety observations and precautions because a delivery is in progress or the family is anxious.

Rushing may not only hurt you, but it may also affect the fetus and pregnant woman. Remain calm and professional. Consider calling for additional or specialized resources.

Mechanism of Injury/Nature of Illness

You will encounter pregnant patients who are not in labor, so it is important to determine the MOI or nature of illness in a pregnant patient. Do not maintain tunnel vision during a call! Because a pregnant woman's balance may be altered by the weight and size of the fetus and hormones that relax the musculature, falls and spinal immobilization must be considered.

Primary Assessment

Form a General Impression

The general impression is a good across-the-room assessment that should tell you whether the patient is in active labor or if you have time to assess for imminent delivery and address other possible life threats. Perform a rapid scan of the patient to determine if there are airway, breathing, or circulation problems. The chief complaint may be, "The baby is coming!" Take a moment to confirm whether the infant will be delivered in the next few minutes or, again, whether you have time to continue to evaluate the situation. When trauma or other medical problems such as vaginal bleeding or seizures are the presenting complaint, evaluate these first and then assess the impact of these problems on the fetus. Use the AVPU (*Alert* to person, place, and day; responsive to *Verbal* stimuli; responsive to *Pain*; or *Unresponsive*) scale to determine the patient's level of consciousness.

You are the Provider: PART 3

Shortly after your partner assesses the patient's vital signs and administers supplemental oxygen to her, you observe the infant's head crowning at the vaginal opening. As the head delivers, you can feel the umbilical cord wrapped around the infant's neck.

Recording Time: 7 Minutes	
Respirations	24 breaths/min; adequate depth
Pulse	110 beats/min; strong and regular
Skin	Pink, warm, and moist
Blood pressure	122/82 mm Hg
Oxygen saturation (Sao$_2$)	98% (on oxygen)

5. How should you manage the umbilical cord situation?

6. Why is it important to suction the newborn's mouth before the nose?

Airway and Breathing

During an uncomplicated birth, life-threatening conditions with the mother's airway and breathing are not usually an issue. However, a motor vehicle crash, an assault, or any number of medical conditions in a pregnant woman may cause a life threat to exist and, sometimes, result in a complicated delivery. In these situations, assess the airway and breathing to ensure they are adequate. If needed, provide airway management and high-flow oxygen.

Circulation

External and internal bleeding are potential life threats to the patient and should be assessed early on. Blood loss after delivery is expected, but significant bleeding is not. Recall that normal changes in pregnancy result in increased overall blood volume, increased heart rate, and changes in blood clotting. These changes can have a significant impact on a pregnant patient who is bleeding, regardless of the cause. Quickly assess for any potential life-threatening bleeding, and begin treatment immediately. Assess the skin for color, temperature, and moisture, and check the pulse to determine if it is too fast or too slow. If there are signs of shock, control the bleeding, give oxygen, and keep the patient warm.

Transport Decision

If delivery is imminent, you must prepare to deliver at the scene. The ideal place to deliver an infant is in the security of your ambulance or the privacy of the mother's home. The area should be warm and private with plenty of room to move around.

If the delivery is not imminent, prepare the patient for transport and perform the remainder of the assessment en route to the emergency department. Administer oxygen. Pregnant women in the last two trimesters of pregnancy should be transported lying on the left side when possible. If spinal immobilization is indicated, secure the mother to the backboard and elevate the right side of the board with rolled towels or blankets to prevent supine hypotensive syndrome **Figure 31-6** . Provide rapid transport for pregnant patients who have significant bleeding and pain, are hypertensive, are having a seizure, or have an altered mental status.

▶ History Taking

Investigate Chief Complaint

You will encounter pregnant patients who are not in labor. Your thorough patient assessment and skill at obtaining a history will enhance your ability to determine the patient's primary problem. Determine a pregnant patient's chief complaint, and begin asking questions that will help you identify the cause of her complaint and identify all of the associated signs and symptoms. Regardless of whether the

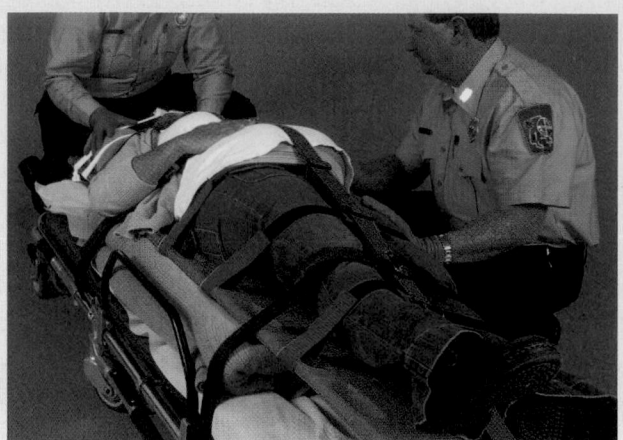

Figure 31-6 Place a blanket under the right side of the backboard to prevent supine hypotensive syndrome in pregnant patients.

patient is in active labor, is having an obstetric emergency, or is a pregnant patient with another complaint (eg, trauma), you should obtain a thorough obstetric history that should include her expected due date, any complications that she is aware of, if she has been receiving prenatal care, and her thorough medical history. Most pregnant patients in the late stages of pregnancy will be able to tell you if they feel anything different with the fetus, especially if they think the fetus is not moving as usual. If your patient is in labor, focus your questions to determine whether delivery is imminent. Find out how long the contractions have been occurring and how long they are lasting, whether the patient's water has broken, and whether the patient feels like having a bowel movement.

SAMPLE History

Obtain a SAMPLE history. Some pregnant women will have a history of medical problems and take prescription medications. Some women who have not experienced medical problems require medications when they become pregnant. Do not focus only on the pregnancy history; obtain a SAMPLE history as well. Pertinent past history should relate specifically to prenatal care. Identify any complications she may have had during the pregnancy or potential complications during delivery that her physician has noted. These complications may include the size or position of the fetus or the position and health of the placenta. Determine the due date, frequency of contractions, and history of previous pregnancies and deliveries and their complications, if any. Determine whether there is a possibility of twins and whether the mother has taken any drugs or medications during the pregnancy. If her water has broken, ask whether the fluid was green. Green fluid is due to **meconium** (fetal stool). The presence of meconium can indicate newborn distress, and it is possible for the fetus to aspirate meconium during delivery. Identifying potential complications prior to the delivery will better prepare you to treat the mother and newborn.

Secondary Assessment

Physical Examinations

A complete assessment of the major body systems should also be performed, with emphasis on the patient's chief complaint. You should assess for fetal movement by asking the patient whether she can feel the baby moving. For a pregnant patient who is in labor, your physical examination should be focused on contractions and possible delivery. Assess the length and frequency of contractions by asking the patient and by placing your hand on the abdomen. Compare what you feel with the patient's experience during each contraction. If at any point you suspect that delivery is imminent, you should check for crowning. This specific assessment should be performed only when appropriate and according to local protocol. If you do not suspect an imminent delivery and the patient has other complaints unrelated to delivery, you should not visually inspect the vaginal area. Be sure to protect the woman's privacy during the physical examination.

Vital Signs

The secondary assessment of an obstetric patient should include a complete set of vital signs and pulse oximetry, if available. Vital signs should include pulse; respirations; skin color, temperature, and condition; and blood pressure. Pay special attention to tachycardia and hypotension (which could mean hemorrhage or compression of the vena cava) or hypertension (possibly indicating preeclampsia). It is typical for a woman's blood pressure to drop slightly during the first two trimesters of pregnancy but return to normal during the third trimester. Compare your findings with previous blood pressure readings she may know of from prenatal visits. Hypertension, even mildly elevated blood pressure, may indicate more serious problems.

Reassessment

As time allows, repeat the primary assessment with a focus on the patient's ABCs and vaginal bleeding, particularly after delivery. Obtain another set of vital signs and compare the results with those obtained earlier. Frequent reassessment of vital signs may identify hypoperfusion from excessive blood loss as a result of delivery. Recheck interventions and treatments to see whether they were effective. Is the vaginal bleeding slowing with uterine massage? Uterine massage, discussed later in this chapter, can be used to slow vaginal bleeding after delivery.

Interventions

In most cases, childbirth is a natural process that does not require your assistance. When childbirth is complicated by trauma or other conditions, any interventions you provide for the patient will benefit the fetus. For example, if a pregnant patient has a low pulse oximetry reading, the fetus does as well. Applying oxygen to the patient also improves the oxygen level in the fetus.

Communication and Documentation

If your assessment determines that delivery is imminent, notify staff at the receiving hospital. Provide an update on the status of the mother and newborn after delivery. On the rare occasion that the delivery does not occur within 30 minutes or you determine that a complication is occurring that cannot be treated in the field, notify the hospital staff of your findings and provide rapid transport. Be sure to notify staff at the receiving hospital of all relevant information so there is time to prepare. The information you provide may help the hospital staff determine whether the patient will be seen in the emergency department or the labor and delivery unit. For a pregnant patient with complaints unrelated to childbirth (such as trauma or difficulty breathing), be sure to include the pregnancy status of your patient in your radio report. The hospital staff will want to know the number of weeks of gestation, her due date, and any known complications of the pregnancy. Thorough documentation is essential, especially the status of the newborn if delivery occurred in the field. You will have two patient care reports to complete. Obstetrics is among the most litigated specialties in medicine; therefore, scrupulous documentation is essential.

Normal Delivery Management

Preparing for Delivery

Consider delivering the infant at the scene when delivery can be expected within a few minutes or when a natural disaster, inclement weather, or other environmental factor makes it impossible to reach the hospital. To determine whether delivery is imminent (will occur within a few minutes), ask the patient the following questions:

- How long have you been pregnant?
- When are you due?
- Is this your first baby?
- Are you having contractions? How far apart are the contractions? How long do the contractions last?
- Do you feel as though you will have a bowel movement?
- Have you had any spotting or bleeding?
- Has your water broken?
- Were any of your previous children delivered by cesarean section?

Ask these questions to help determine any potential complications:

- Have you had any problems in a previous pregnancy?
- Do you use drugs, drink alcohol, or take any medications?
- Do you know if there is a chance of a multiple birth (having twins, more than one baby)?
- Does your doctor expect any complications?

If the patient has delivered before, she may be able to tell you whether she is about to deliver or not. If the patient says that she is about to deliver, you should immediately prepare for a delivery. Otherwise, does she have an extremely firm abdomen? Does she say that she has to move her bowels or feels the need to push? If so, the infant's head is probably pressing on the rectum, and delivery is about to occur. At this point, you should visually inspect the vagina to check for crowning. Crowning is an indication that the delivery is occurring. Do not touch the vaginal area until you have determined that delivery is imminent. In general, do not touch the vaginal area except during delivery (under certain circumstances) and when your partner is present. Spread the pregnant woman's legs apart gently, explaining that you are doing so to decide whether the baby should be delivered immediately or whether she should be transported to the hospital for the delivery.

Once labor has begun, there is no way it can be slowed or stopped. Never attempt to hold the woman's legs together. To do so would only complicate the delivery. Do not let her go to the bathroom. Instead, reassure her that the sensation of needing to move her bowels is normal and that it means she is about to deliver.

If your decision is to deliver at the scene, remember that you are only assisting the woman with the delivery. Your part is to help, guide, and support the baby as it is born. Use standard precautions at all times. Administer oxygen to the mother if indicated. Limit distractions for yourself and for the patient. You want to appear calm and reassuring while protecting the woman's modesty. Most important, recognize when the situation is beyond your level of training. If delivery is imminent with crowning, contact medical control for a decision to deliver on the scene or to transport. If there is any doubt, contact medical control for further guidance. Always recognize your own limitations, and when you are unsure about what to do, transport the patient even if delivery must occur during transport.

Your emergency vehicle should always be equipped with a sterile emergency obstetric (OB) kit containing the following items **Figure 31-7** :

- Surgical scissors or a scalpel
- Umbilical cord clamps
- Umbilical tape
- A small rubber bulb syringe
- Towels

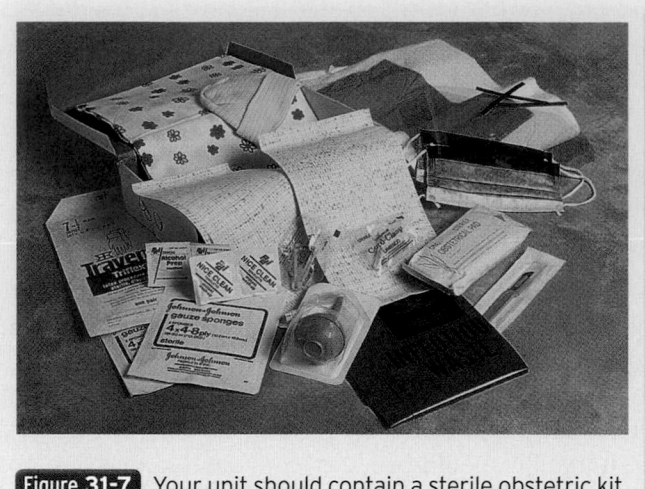

Figure 31-7 Your unit should contain a sterile obstetric kit.

- $4'' \times 4''$ gauze sponges and/or $2'' \times 10''$ gauze sponges
- Sterile gloves
- Infant blanket
- Sanitary napkins
- An infant-sized bag-mask device
- Goggles
- A plastic bag

Patient Position

The patient's clothing should be pushed up to her waist, and pants and undergarments should be removed. Remember to preserve the patient's modesty as much as you can while helping her to move into position. Place the patient on a firm surface that is padded with blankets, folded sheets, or towels. Put a pillow or blankets beneath her hips to elevate them about $2''$ to $4''$. It is sometimes better to put a pillow under one hip to allow the patient to turn to one side. Allow the patient to get comfortable. Support the patient's head, neck, and upper back with pillows and blankets. You should also begin preparing for the newborn's arrival. Communicate with your crew, and have a plan for where you will place the newborn after delivery, who will be responsible for drying off and keeping the infant warm, and who will be responsible for caring for the mother and the newborn after delivery.

If the emergency delivery is occurring at home, you should move the patient to a sturdy, flat surface or the floor if she will allow it. You will find it easier to work with the patient on a firm surface rather than on a bed. Elevate the patient's hips, and support her head with one or two pillows. Have her keep her legs and hips flexed, with her feet flat on the surface beneath her and her knees spread apart. Track the progression of the delivery closely at all times; you do not want an explosive delivery to occur, when the crowning head pops out uncontrollably and too quickly.

Preparing the Delivery Field

Take the following steps to prepare the area where the infant will be born:

1. As time allows, place towels or sheets on the floor around the delivery area to help soak up body fluids and to protect the mother and the infant. Elevate the patient's hips, and support her head and shoulders with folded blankets or pillows.
2. Open the OB kit carefully so that its contents remain sterile.
3. Put on the sterile gloves.
4. Use the sterile sheets and towels from the OB kit to make a sterile delivery field. Place one sheet or towel under the patient's buttocks, and unfold it toward her feet. Wrap another sheet behind the patient's back and drape over each thigh Figure 31-8A , and drape one sheet across the abdomen Figure 31-8B .

■ The Delivery

Your partner should be at the patient's head to comfort, soothe, and reassure her during the delivery. If she will allow it, apply oxygen. The patient may want to grip someone's hand. She may yell, cry, or say nothing at all. It is common for patients to become nauseated during delivery, and some may vomit. If this occurs, have your partner assist her and clear out her airway.

You must continually assess the patient for crowning. Some patients may experience precipitous (fast) labor and birth. This is more common in patients who have had children. Do not allow an abrupt or explosive delivery to occur. If the labor is too fast, the tissues do not have time to stretch, and the mother is at risk for rips and tears in the perineal area (see Delivering the Head, page 1120). Position yourself so that you can see the perineum at all times. Time the patient's contractions from the beginning of one to the beginning of the next to determine the frequency of the contractions. In addition, time the duration of each contraction. You do this by feeling the patient's abdomen from the moment the contraction begins (uterus and abdomen tightening) to the moment it ends (uterus and abdomen relaxing). Remind the patient to take quick, short breaths during each contraction but not to strain. Between contractions, encourage the mother to rest and breathe deeply through her mouth.

Follow the steps in Skill Drill 31-1 to deliver the infant:

1. Allow the mother to push the head out. Support it as it emerges, placing your gloved hand over its bony parts. Feel at the neck to see if the cord is wrapped around it. If it is, gently lift it over the infant's head

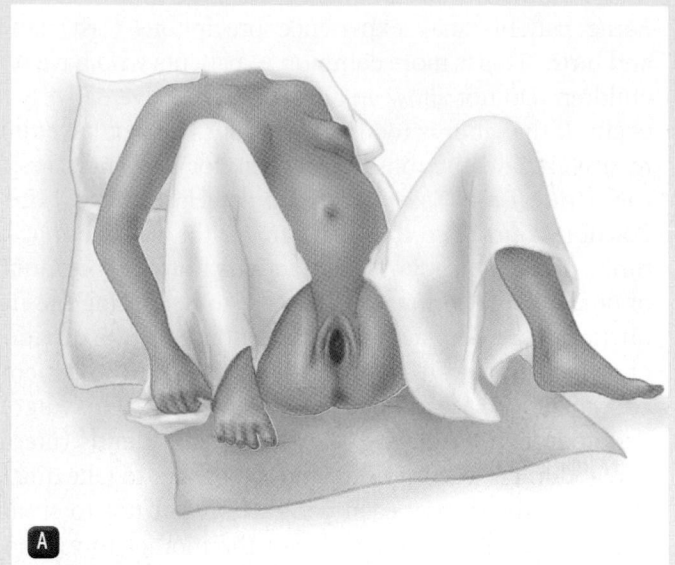
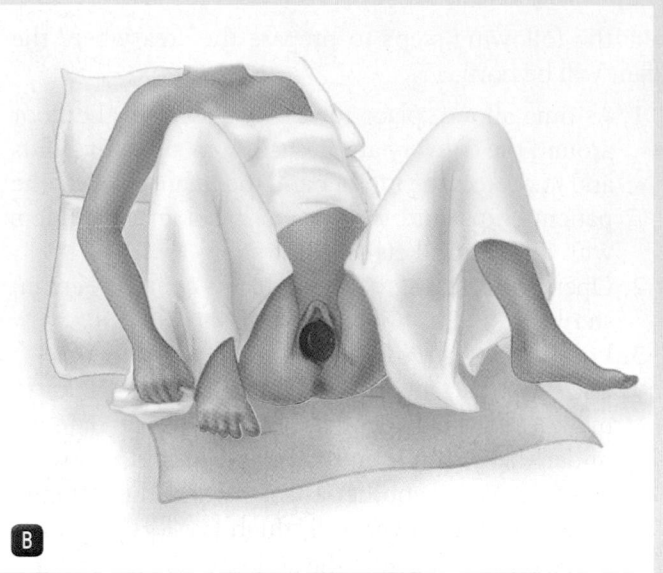

Figure 31-8 Preparing the delivery field. **A.** Use sterile sheets and towels from the obstetric kit to make a clean delivery field. Place one sheet under her buttocks. Wrap another sheet behind her back with either end draped over the thighs. **B.** Drape another sheet over her abdomen.

without pulling hard on the cord. Suction fluid from the mouth first, then the nostrils. Be sure to squeeze the bulb syringe before inserting it into the infant's mouth or nose (Step 1).

2. Once the head is delivered, the upper shoulder will be visible. Guide the head down slightly, if needed, to help the upper shoulder deliver (Step 2).

3. Support the head and upper body as the shoulders deliver. You may need to guide the head up slightly to deliver the lower shoulder (Step 3).

4. Once the body is delivered, handle the infant firmly but gently. It will be slippery. Make sure the infant's neck is in a neutral position to keep the airway open (Step 4).

5. Place the umbilical cord clamps about 2″ to 4″ apart, about four fingerbreadths from the infant's body. Depending on local protocol, cut between the clamps (Step 5).

6. The placenta delivers itself, usually within 30 minutes of birth. Never pull on the end of the umbilical cord in an attempt to speed delivery of the placenta (Step 6).

Words of Wisdom

When a fetus is positioned head first in the birth canal, this is called the cephalic presentation. Most births are cephalic presentation.

Delivering the Head

Observe the infant's head as it begins to exit the vagina so you can provide support as it emerges. It may take two, three, or more contractions for the delivery of the head to occur from the time it begins to crown. Once it is obvious that the head is coming out farther with each contraction, you should place your gloved hand over the emerging bony parts of the head and exert very gentle pressure on it, decreasing the pressure slightly between contractions. This will allow the head to come out smoothly and prevent it and the rest of the infant from suddenly popping out during a strong contraction, possibly causing injury to the mother's perineal area or to the infant. Continue to support the head as it rotates.

Methods of reducing the risk of perineal tearing during labor include applying gentle pressure across the perineum with a sterile gauze pad and applying gentle pressure to the head while gently stretching the perineum **Figure 31-9**. Consult your protocol regarding the methods used in your area. Also be prepared for the possibility of the mother having a bowel

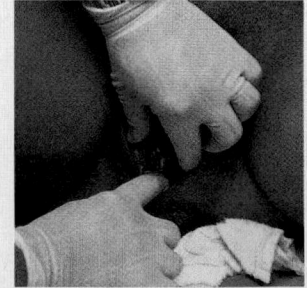

Figure 31-9 One method of reducing the risk of perineal tearing during labor is to apply gentle pressure on the infant's head while gently stretching the perineum.

Skill Drill 31-1

Delivering the Infant

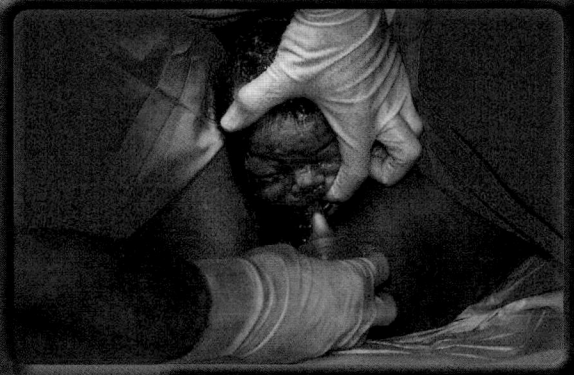

Step 1 Support the bony parts of the head with your hands as it emerges. Suction fluid from the mouth, then nostrils.

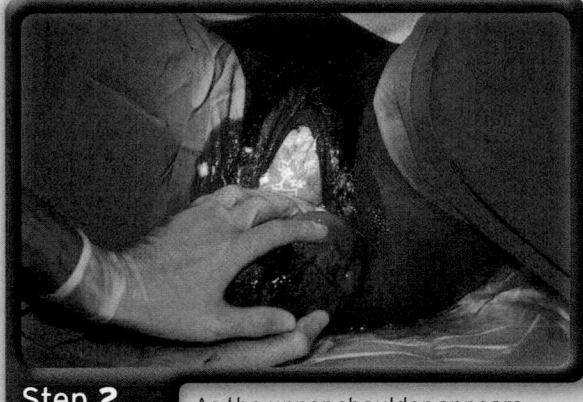

Step 2 As the upper shoulder appears, guide the head down slightly, if needed, to deliver the shoulder.

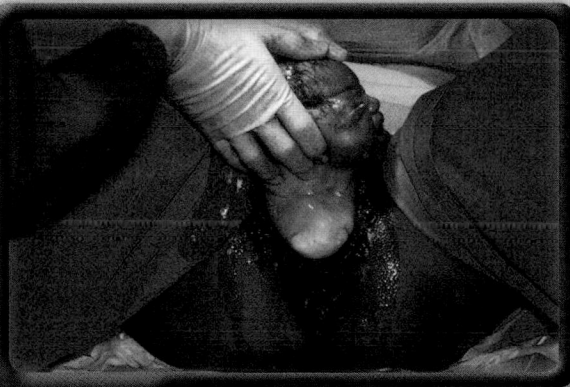

Step 3 Support the head and upper body as the lower shoulder delivers, guiding the head up if needed.

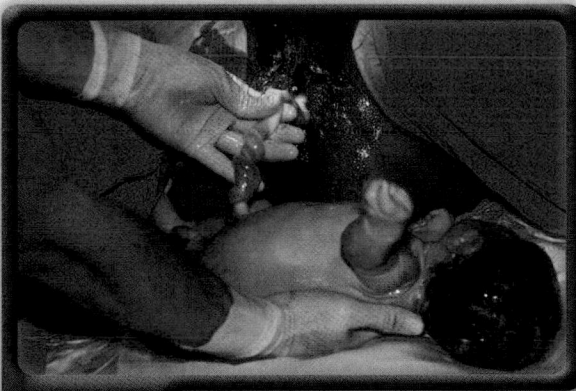

Step 4 Handle the infant firmly but gently, keeping the neck in neutral position to maintain the airway. Keep the infant approximately at the level of the vagina until the umbilical cord has been cut.

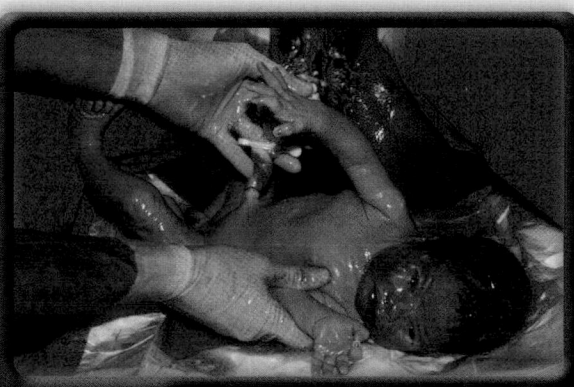

Step 5 Place the umbilical cord clamps 2" to 4" apart, and cut between them.

Skill Drill   31-1

Delivering the Infant, continued

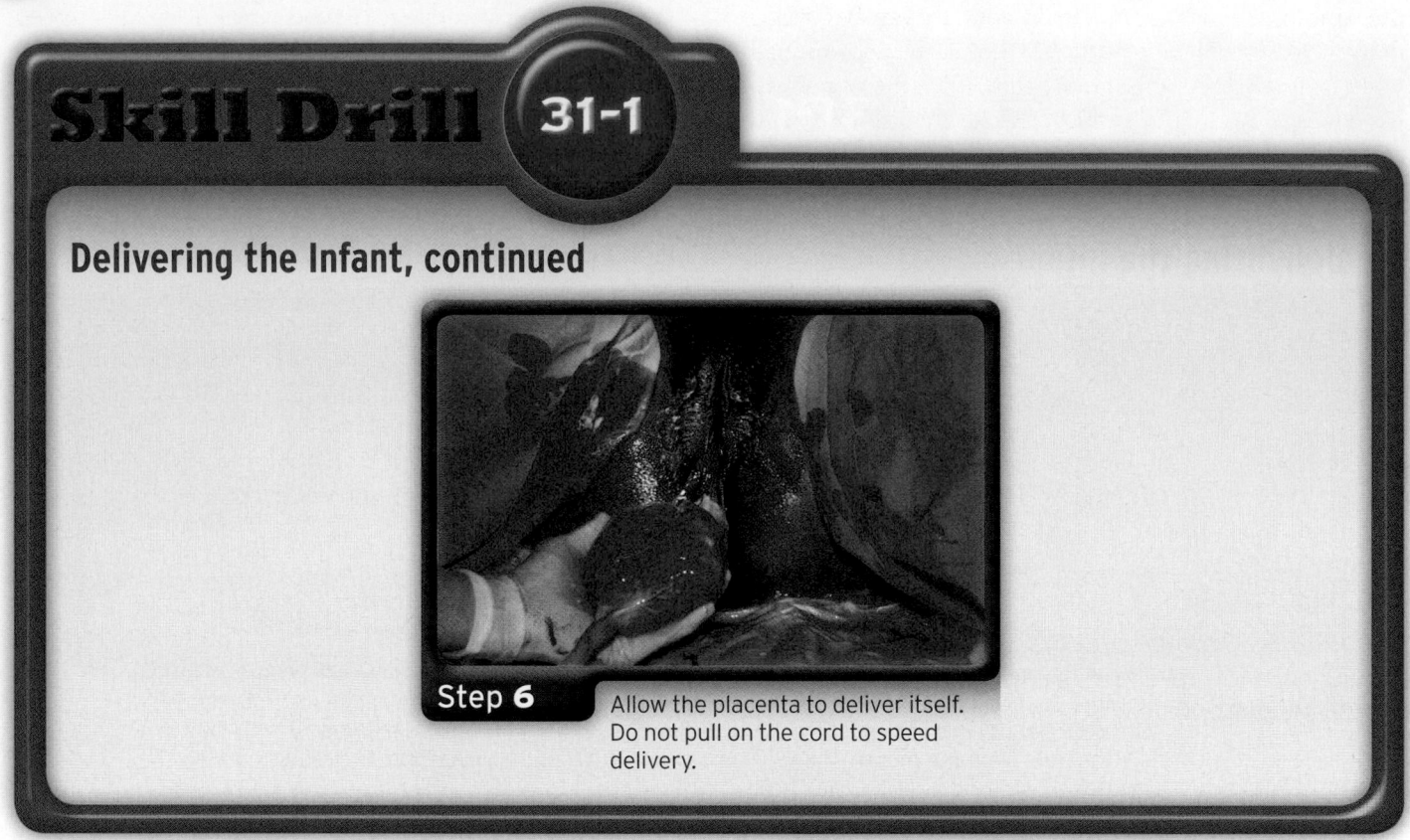

Step 6   Allow the placenta to deliver itself. Do not pull on the cord to speed delivery.

movement because of the increased pressure on the rectum.

As you are assisting with the delivery of the head, be careful that you do not poke your fingers into the infant's eyes or into the fontanelles. The fontanelles are soft spots on the newborn's skull that will eventually become covered with bone. At birth, the brain is covered only with skin and membranes at these areas. There are two primary fontanelles, one on the top of the head and one near the back of the head.

Unruptured Amniotic Sac   Usually, the amniotic sac will break or rupture at the beginning of labor. The sac may also rupture during contractions. If the amniotic sac has not ruptured by this point, it will appear as a fluid-filled sac (like a water balloon) emerging from the vagina. This situation is potentially life threatening for the infant because the sac will suffocate the infant if it is not removed. If the sac has not spontaneously ruptured, you may puncture it with a clamp. You should make sure that the puncture site is away from the infant's face and only perform this procedure as the head is crowning. Do not puncture the sac if the infant's head is not crowning. As the sac is punctured, amniotic fluid will gush out. Push the ruptured sac away from the infant's face as the head is delivered. Clear the infant's mouth and nose immediately, using the bulb syringe and gauze. If the amniotic fluid is greenish (meconium staining) instead of clear or has a foul odor,

make sure you notify the receiving hospital. Meconium is a sign of a newborn with respiratory distress or an airway obstruction. Thick meconium can clog the airway of the newborn. Aggressive suctioning of the infant's mouth and oropharynx before delivery of the body may prevent meconium aspiration and respiratory distress.

Umbilical Cord Around the Neck   As soon as the head is delivered, use one finger to feel whether the umbilical cord is wrapped around the infant's neck. This commonly is called a <u>nuchal cord</u>. A nuchal cord that is wound tightly around the neck could cause the infant to strangle. It must be released from the neck immediately. Usually, you can slip the cord gently over the infant's delivered head (or over the shoulder, if necessary). If not, you must cut it by placing two clamps about 2″ apart on the cord and cutting the cord between the clamps. In the rare case of the cord being wrapped more than once around the neck, you will need to clamp and cut only once; then you can unwrap the cord from around the neck. Handle the cord very carefully; it is fragile and easily torn. Do not let the clamps come off until the ends of the cord have been tied. Fortunately, the cord is usually not wrapped around the infant's neck and does not have to be cut until after the entire infant has been delivered. However, you must always check for a nuchal cord.

Once you have delivered the infant's head and verified that no nuchal cord is present, you will need to suction the amniotic fluids from the infant's airway before the delivery proceeds. You must ask the mother not to push while you are doing this, although her desire to do so will be very strong. While supporting the infant's head with one hand, quickly and efficiently suction the fluid from the mouth first and then the nostrils. If you suction the nostrils first, you may stimulate the infant to aspirate the fluid in the mouth or pharynx; because infants are nose breathers, any stimulation of the nose will cause a gasping response. In suctioning the airway, fully compress the bulb syringe before it is inserted 1″ to 1½″ into the infant's mouth, then release the bulb to suction fluids and mucus into the syringe. Make sure the syringe does not touch the back of the mouth. Discard the fluid into a towel, and repeat the procedure, suctioning the mouth and nostrils two or three times each, or until they are clear.

Delivering the Body

Once the head has been delivered, it usually rotates to one side or the other; this rotation places the infant in a better position to deliver the rest of the body. By the time you are finished suctioning, the mother will most likely be pushing again, and the upper shoulder will be visible in the vagina. The infant's head is the largest part of the body. Once it is born, the rest of the infant usually delivers easily. Support the head and upper body as the shoulders deliver. Do not pull the infant from the birth canal. The abdomen and hips will appear; once these deliver, support them with your other hand. Grasp the infant's feet as they are born; support and hold the infant with both of your hands. Handle the infant firmly but carefully. It will be slippery and covered with a white, cheesy substance, called vernix caseosa.

Words of Wisdom

As an infant delivers, you must divide your attention between two patients. This can keep two EMTs busy, even when things go well. To ensure that special care needs do not result in neglect of one of the patients, designate one member of the crew to pay primary attention to each patient. Call for additional help early if you suspect that both will need special care or that one will require resuscitation.

■ Postdelivery Care

As soon as the infant is born, he or she needs to be dried off and wrapped immediately in a blanket or towel and placed on one side, with the head slightly lower than the rest of the body. Wrap the infant so that only the face is exposed, making sure that the top of the head is covered. Also make sure that the neck is in a neutral position so

the airway remains open. Newborns are very sensitive to cold, so you should keep the blanket or towel warm, if possible, before you use it. A newborn's body temperature can drop very quickly, so dry and wrap the infant as soon as possible, Use a sterile gauze pad to wipe the infant's mouth, and once again suction the mouth and nose. Suctioning the nose is particularly important because newborns breathe through their noses. If you prefer, you can pick up and cradle the infant in your arm at the level of the mother's vagina, but always keep the head slightly downward to help prevent aspiration. After suctioning, keep the infant at the same level as the mother's vagina until the umbilical cord is cut. If the infant is higher than the vagina, blood will be siphoned from the infant through the umbilical cord back into the placenta.

Words of Wisdom

Recording the time of birth will ensure that the information is available for the birth certificate. It also provides you with a starting point from which to time the intervals for Apgar scores. This is even more important with multiple births. You will be busy; consider asking a family member to act as "timekeeper."

Once the infant is born, the umbilical cord is of no further use to the mother or infant. Postdelivery care of the umbilical cord is important because infection is easily transmitted through the cord to the infant. Using the two clamps in the OB kit, clamp the cord somewhere between the mother and the infant, preferably four fingerbreadths from the infant. Place the clamps about 2″ to 4″ apart. Once they are firmly in place, carefully cut the cord between them with sterile scissors or a scalpel. Remember, the cord is fragile; if handled roughly, it could be torn from the infant's abdomen, resulting in a fatal hemorrhage. Once the clamps are in place, there is no need to rush.

After you have cut the cord, tie the end coming from the infant. If it was a nuchal cord and cut during delivery, now is the time to tie it. Do not use ordinary string or twine, which will cut through the soft, fragile tissues of the cord. Place a loop of the special "umbilical tape" around the cord about 1″ nearer to the infant than to the clamp. Tighten the tape slowly so that it does not cut the cord, and then tie it firmly with a square knot. Cut the ends of the tape, but do not remove either clamp. The part of the cord that is coming out of the mother's vagina is attached to the placenta and will be delivered when the placenta delivers.

By now, the infant should be pink and breathing on his or her own. Give the infant, wrapped in a warm blanket, to your partner; he or she can monitor the infant and complete the infant's initial care. Alternatively, you can give the infant to the mother if she is alert and in stable condition, if allowed per protocol. The mother may want to begin breastfeeding at this time. You need to return your attention to the mother and the delivery of the placenta.

Delivery of the Placenta

The placenta is attached to the end of the umbilical cord that is coming out of the mother's vagina. Again, your job is only to assist. The placenta delivers itself, usually within a few minutes of the birth, although it may take as long as 30 minutes. Never pull on the end of the umbilical cord in an attempt to speed delivery of the placenta. You may tear the cord, the placenta, or both and cause serious, perhaps life-threatening hemorrhage in the mother.

The normal placenta is round, about 7″ in diameter, and about 1″ thick. One surface is smooth and covered with a shiny membrane; the other surface is rough and divided into lobes. Wrap the entire placenta and cord in a towel, place them into a plastic bag, and take them to the hospital. Hospital personnel will examine the placenta and the cord to make certain that the entire placenta has been delivered. If a piece of the placenta has been retained inside the mother, it could cause persistent bleeding or infection.

After delivery of the placenta and before transport, place a sterile pad or sanitary napkin over the vagina and straighten the mother's legs. You can help to slow bleeding by gently massaging the mother's abdomen with a firm, circular, "kneading" motion **Figure 31-10**. The abdominal skin will be wrinkled and very soft. You should be able to feel a firm, grapefruit-sized mass in the lower abdomen. This is called the **fundus**. As you massage the fundus, the uterus will contract and become firmer. This may sometimes be uncomfortable for the mother. Reassure her and explain that it is necessary to help control the bleeding. If the mother chooses to breastfeed, this will also stimulate the uterus to contract. Massaging the uterus and having the newborn stimulate the mother's nipples will cause a production of oxytocin, which is a hormone that will help to contract the uterus and slow bleeding. Take a minute to congratulate the mother and thank anyone who assisted. Be sure to record the time of birth in your patient care report.

Some bleeding, usually less than 500 mL, occurs before the placenta delivers and is normal and expected. The following are emergency situations:

- More than 30 minutes elapse, and the placenta has not delivered.
- There is more than 500 mL of bleeding before delivery of the placenta.
- There is significant bleeding after the delivery of the placenta.

If one or more of these events occur, transport the mother and infant to the hospital promptly. Never put anything into the vagina. Place a sterile pad or sanitary napkin over the mother's vagina, place her in the shock position, administer oxygen, keep her and the infant warm by preventing any heat loss, and monitor her vital signs closely.

■ Neonatal Assessment and Resuscitation

Follow standard precautions, and always put on gloves before handling a newborn infant. A newborn infant will usually begin breathing spontaneously within 15 to 30 seconds after birth, and the heart rate will be 120 beats/min or higher. This is the normal respiratory and cardiovascular physiologic response expected. If you do not observe these responses, gently tap or flick the soles of the infant's feet or rub the back to stimulate breathing. If the infant does not breathe after 10 to 15 seconds, you should begin resuscitation efforts. Many infants require some form of stimulation that will encourage them to breathe air and begin circulating blood through the lungs **Table 31-2**. These measures include positioning of the airway, drying, warming, suctioning, and tactile stimulation. To maximize the effects of these measures, follow these tips:

- Position the infant on his or her back with the head down and the neck slightly extended. Place a towel or blanket under the infant's shoulders to help maintain this position.

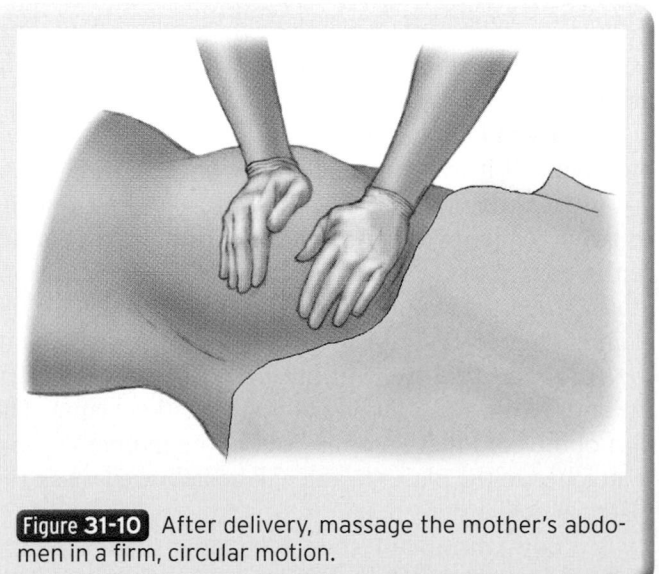

Figure 31-10 After delivery, massage the mother's abdomen in a firm, circular motion.

Table 31-2 Resuscitation for a Newborn Who Is Not Breathing

Assess and support	■ Temperature (warm and dry) ■ Airway (position and suction) ■ Breathing (stimulate to cry) ■ Circulation (heart rate and skin color)
Basic life support interventions	■ Dry and warm the infant. ■ Clear the airway with a bulb syringe. ■ Stimulate the infant if he or she is unresponsive. ■ Use a bag-mask device to ventilate the newborn if needed. This is seldom required. ■ Perform chest compressions if there is no pulse or if the heart rate is < 60 after 30 seconds of ventilation and heart rate is not increasing.

■ Suction the mouth and then the nose using a bulb syringe or suction device with an 8- or 10-French catheter. Suction both sides of the back of the mouth, where secretions tend to collect, but avoid deep suctioning of the mouth and throat; this can cause the heart rate to slow down. Aim blow-by oxygen at the infant's mouth and nose during resuscitation.

■ In addition to drying the infant's head, back, and body vigorously with dry towels, you may rub the infant's back and flick or slap the soles of his or her feet.

When a newborn is in distress, you should be properly equipped for resuscitation measures. Most of the equipment and supplies needed to resuscitate a newborn can be found in your OB kit. Other items you may need are clean, dry towels; an infant blanket; a bag-mask device with a 450-mL reservoir; clear masks in both newborn and premature sizes; and an oxygen source with tubing.

■ Additional Resuscitation Efforts

Observe the newborn for spontaneous respirations, skin color, and movement of the extremities. If the respiratory effort appears appropriate, evaluate the heart rate by palpating the pulse at the base of the umbilical cord or at the brachial artery. The heart rate is the most important

You are the Provider: PART 5

You reassess the newborn and find that she is breathing adequately and has a heart rate of 120 beats/min. Her hands and feet remain slightly cyanotic, but the central part of her body is pink. After clamping and cutting the umbilical cord, you allow the mother to hold her child. The placenta delivers and is appropriately cared for. You reassess the mother's vital signs and then prepare for transport.

Recording Time: 20 Minutes	
Level of consciousness	Conscious and alert
Respirations	20 breaths/min; adequate depth
Pulse	98 beats/min; strong and regular
Skin	Pink, warm, and moist
Blood pressure	126/60 mm Hg
Sao$_2$	97% (on oxygen)

En route to the hospital, you reassess the newborn and mother. The mother remains conscious and alert and has mild vaginal bleeding. The newborn's body is pink, but her hands and feet remain blue; her heart rate is 130 beats/min, and her respirations are rapid; she pulls her foot away when you flick the sole; and she resists your attempts to straighten her knees. You call the receiving hospital and give your radio report; your estimated time of arrival is 6 minutes.

9. What further treatment is indicated for the mother?

10. What Apgar score should you assign to this newborn?

measure in determining the need for further resuscitation Table 31-3.

If chest compressions are required, use either the hand-encircling technique or the two-finger technique Figure 31-11. Bag-mask ventilation is performed during a pause after every third compression. Avoid giving a compression and a ventilation simultaneously, since one will decrease the effectiveness of the other. Because cardiac arrest in neonates is nearly always the result of ventilation compromise, a compression to ventilation ratio of 3:1 should be used; this will yield a total of 120 "actions" per minute (90 compressions and 30 ventilations). If the newborn's cardiac arrest is believed to be of cardiac origin, however, consider a higher ratio (ie, 15:2). Remember that adequate ventilation is absolutely critical to the successful resuscitation of the neonate.

Any newborn who requires more than routine resuscitation requires transport, if available in your area, to a hospital with a level III neonatal intensive care unit. This type of unit is designed for newborns who require specialized care. If a level III neonatal intensive care until is not available in your area, provide rapid transport to the closest appropriate facility.

About 12% of deliveries are complicated by the presence of meconium, the dark green fecal material in the amniotic fluid. Meconium can be thick or thin. If the newborn aspirates thick meconium, significant lung disease and sometimes death can occur. If you see meconium in the amniotic fluid or meconium staining and the infant is not breathing adequately, you should continue vigorous suctioning of the infant after delivery.

■ The Apgar Score

The **Apgar score** is the standard scoring system used to assess the status of a newborn. This system assigns a number value (0, 1, or 2) to five areas of activity of the newborn infant:

- **Appearance.** Shortly after birth, the skin of a light-skinned newborn infant and the mucous membranes of a dark-skinned infant should turn pink. Newborn infants often have cyanosis of the extremities for a few minutes after birth, but hands and feet should "pink up" quickly. Blue skin all over or blue mucous membranes signal a central cyanosis.
- **Pulse.** If a stethoscope is unavailable, you can measure pulsations with your fingers in the umbilical cord or at the brachial pulse. An infant with no pulse requires immediate CPR.

Table 31-3 Additional Neonatal Resuscitation Efforts

If the Heart Rate Is...	More Than 100 Beats/Min	60 to 100 Beats/Min	Fewer Than 60 Beats/Min
Do this:	Keep the newborn warm. Transport the newborn. Assess the newborn continuously.	Begin assisted ventilation with a bag-mask device and 100% oxygen. Reassess the newborn every 30 seconds until heart rate and respirations are normal. Continue to reassess the infant. Call for ALS backup. Keep the newborn warm.	Begin assisted ventilation with a bag-mask device and 100% oxygen. Reassess the newborn every 30 seconds until heart rate and respirations are normal. Begin chest compressions. Call for ALS backup. If the heart rate does not increase, medication and ALS will be needed.

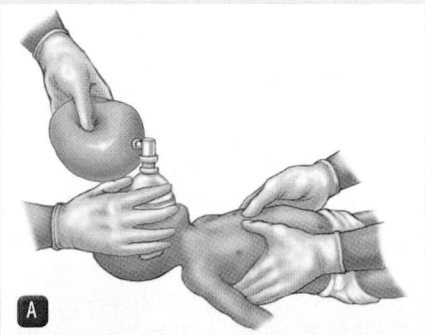
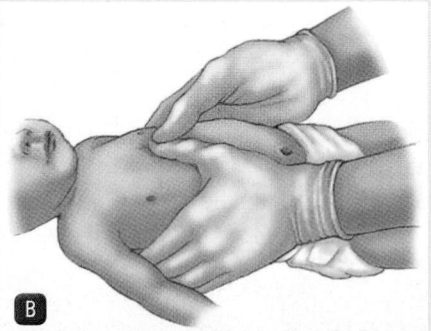
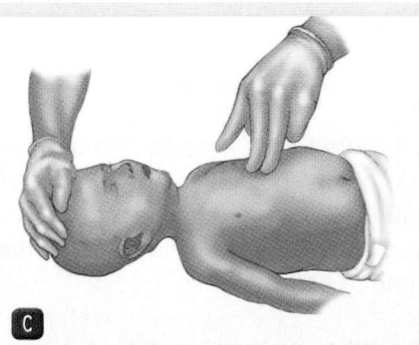

Figure 31-11 **A.** Chest compressions should be given with the hands encircling the infant and thumbs side by side. **B.** In very small infants, you may need to overlap the thumbs. **C.** In larger infants, you may use the two-finger technique, using the middle and ring fingers.

- **Grimace or irritability.** Grimacing, crying, or withdrawing in response to stimuli is normal in a newborn and indicates that the newborn infant is doing well. The way to test this is to snap a finger against the sole of the infant's foot.
- **Activity or muscle tone.** The degree of muscle tone indicates the oxygenation of the newborn infant's tissues. Normally, the hips and knees are flexed at birth, and, to some degree, the infant will resist attempts to straighten them out. A newborn should not be floppy or limp.
- **Respirations.** Normally, the newborn's respirations are regular and rapid, with a good strong cry. If the respirations are slow, shallow, or labored, or if the cry is weak, the newborn infant may have respiratory insufficiency and need assistance with ventilation. Complete absence of respirations or crying is obviously a very serious sign; in addition to assisted ventilation, CPR may be necessary.

The total of the five numbers is the Apgar score. A perfect score is 10. The Apgar score should be calculated at 1 minute and 5 minutes after birth. Most newborn infants will have a score of 7 or 8 at one minute and a score of 8 to 10 four minutes later. **Table 31-4** shows how to calculate an Apgar score.

Follow these steps in assessing a newborn infant:

1. Quickly calculate the Apgar score to establish a baseline on the newborn's status.
2. Suctioning and stimulation should result in an immediate increase in respirations. If they do not, you must begin ventilations with a bag-mask device. Unlike adults, who may have a sudden cardiac arrest, newborn infants who are in cardiac arrest usually have a respiratory arrest first. Therefore, it is essential to keep the infant ventilating and oxygenating well.

3. If the newborn is breathing well, you should next check the pulse rate by feeling the brachial pulse or the pulsations at the base of the umbilical cord. The pulse rate should be at least 100 beats/min. If it is not, begin ventilations with a bag-mask device. This alone may increase the newborn infant's heart rate. Reassess respirations and heart rate at least every 30 seconds to make sure that the pulse rate is increasing and that respirations are becoming spontaneous.
4. Assess the newborn's oxygenation via pulse oximetry and observe for central cyanosis. If central cyanosis is present or the newborn's oxygen saturation does not improve, administer blow-by oxygen by holding oxygen tubing or an oxygen mask close to the newborn's face. Set the oxygen flow rate at 5 L/min.
5. Remember, you now have two patients. You should request a second unit as soon as possible if you determine that the newborn infant is in any distress and will require resuscitation.

In situations in which assisted ventilation is required, you should use a newborn bag-mask device **Figure 31-12**. Cover the newborn's mouth and nose with the mask, and begin ventilation with high-flow oxygen at a rate of 40 to 60 breaths/min. Make sure you have a good mask-to-face seal. With gentle pressure, make the chest rise

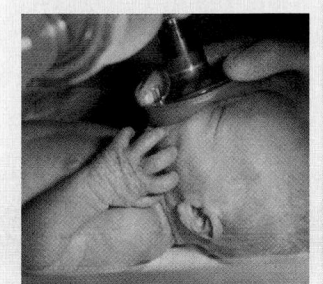

Figure 31-12 Use an infant bag and mask, and ensure that you cover the newborn's nose and mouth. Ventilate with high-flow oxygen at a rate of 40 to 60 breaths/min.

Table 31-4 Apgar Scoring System			
	Score		
Area of Activity	**2**	**1**	**0**
Appearance	Entire infant is pink.	Body is pink, but hands and feet remain blue.	Entire infant is blue or pale.
Pulse	More than 100 beats/min	Fewer than 100 beats/min	Absent pulse
Grimace or irritability	Infant cries and tries to move foot away from finger snapped against sole of foot.	Infant gives a weak cry in response to stimulus.	Infant does not cry or react to stimulus.
Activity or muscle tone	Infant resists attempts to straighten hips and knees.	Infant makes weak attempts to resist straightening.	Infant is completely limp, with no muscle tone.
Respiration	Rapid respirations	Slow respirations	Absent respirations

with each ventilation. It may be necessary to bypass the pop-off valve to accomplish this, especially during the first few breaths.

Special Populations

Current information on neonatal resuscitation varies from what you may have learned in your class on cardio-pulmonary resuscitation, which usually does not differ-entiate between an infant who needs cardiopulmonary resuscitation and a neonate (newborn). Be sure to know your specific local protocols on neonatal resuscitation.

If the infant does not begin breathing on his or her own or does not have an adequate heart rate, continue CPR and rapidly transport. Once CPR has been started, do not stop until the infant responds with adequate res-pirations and heart rates or is pronounced dead by a physician. Do not give up! Many infants have survived without brain damage after prolonged periods of effec-tive CPR. If the infant presents in distress, do not waste time on the assessing the Apgar score, begin resuscitation immediately.

Complicated Delivery Emergencies

■ Breech Delivery

The **presentation** is the position in which an infant is born or the body part that is delivered first. Most infants are born head first, in what is called a **vertex presentation**. Occasionally, the buttocks are delivered first. This is called a **breech presentation** Figure 31-13. With a breech presentation, the infant is at great risk for trauma from the delivery. In addition, prolapsed cords are more common in a breech delivery. Breech deliveries are usually longer than a normal delivery, so there is time to get the pregnant woman to the hospital. However, if the buttocks have already passed through the vagina, the delivery has begun. You should provide emergency care and call for ALS backup. In general, if the mother does not deliver within 10 minutes of the buttocks presenta-tion, provide prompt transport. Consult medical control to guide you in this difficult situation.

Preparing for a breech delivery is the same as for a normal childbirth. Position the pregnant woman, pre-pare the OB kit, and place yourself and your partner as you would for a normal delivery. Allow the buttocks and legs to deliver spontaneously, supporting them with your hand to prevent rapid expulsion. The but-tocks will usually come out easily. Let the legs dangle on either side of your arm while you support the trunk and

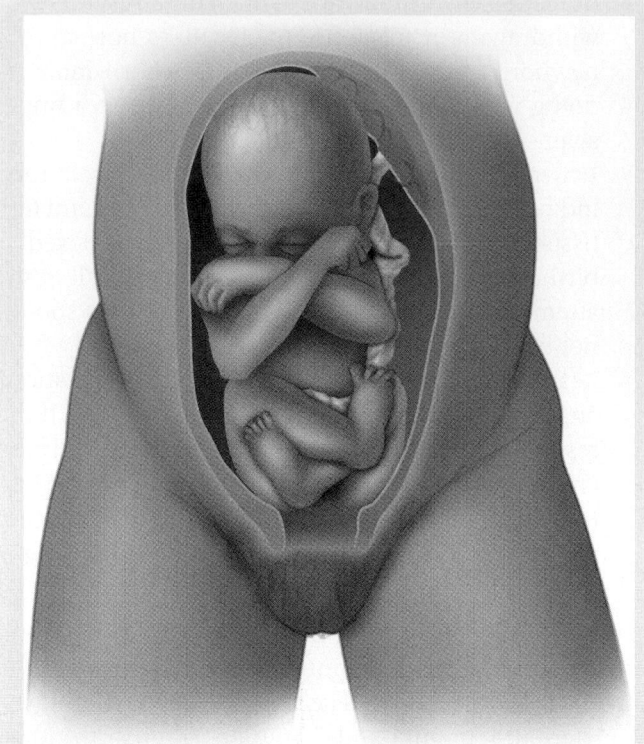

Figure 31-13 In a breech presentation, the buttocks are delivered first. Breech deliveries are usually slow, so you will often have time to transport the mother to the hospital.

chest as they are delivered. The head is almost always facedown and should be allowed to deliver spontane-ously. As the head is delivering, you will need to per-form a potentially lifesaving procedure to manage the newborn's airway. Make a "V" with your gloved fingers and position them in the vagina to keep the walls of the vagina from compressing the airway. This is one of only two circumstances in which you should insert your fingers into the vagina.

■ Presentation Complications

On rare occasions, the presenting part of the infant is neither the head nor the buttocks, but a single arm, leg, or foot. This is called a **limb presentation** Figure 31-14. You cannot successfully deliver an infant with limb presenta-tion in the field. These infants usually must be delivered surgically. If you are faced with a limb presentation, you must transport the patient to the hospital immediately. If a limb is protruding, cover it with a sterile towel. Never try to push it back in, and never pull on it. Place the patient on her back, with her head down and pel-vis elevated. Because the woman and fetus are likely to be physically stressed, remember to give the woman high-flow oxygen.

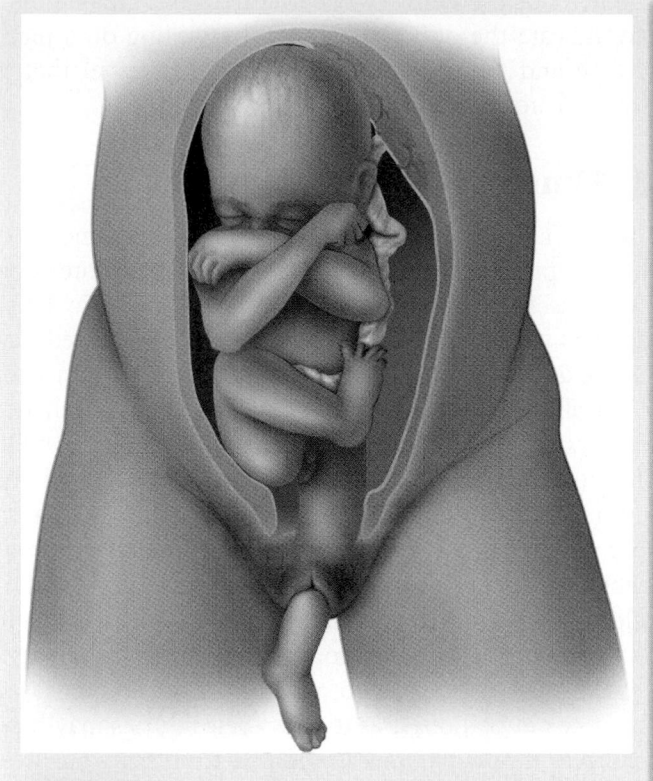

Figure 31-14 In rare cases, an infant's limb, usually a single arm or leg, presents first. This is a life-threatening situation, and you must provide prompt transport for hospital delivery.

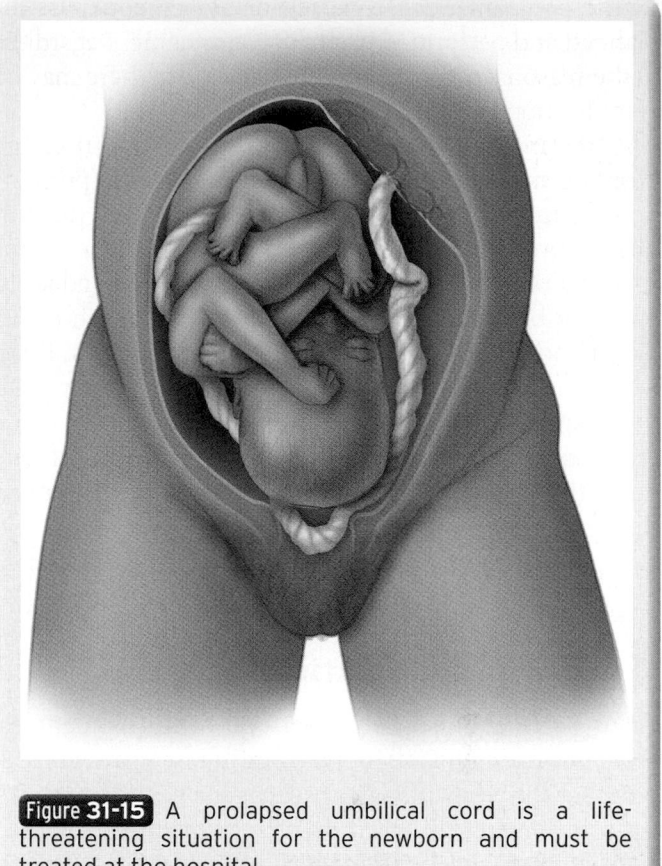

Figure 31-15 A prolapsed umbilical cord is a life-threatening situation for the newborn and must be treated at the hospital.

<u>**Prolapse of the umbilical cord**</u>, a situation in which the umbilical cord comes out of the vagina before the infant **Figure 31-15**, is another rare presentation that must be treated in the hospital. This situation is dangerous because the infant's head will compress the cord during birth and cut off circulation to the infant, depriving it of oxygenated blood. Do not attempt to push the cord back into the vagina. Prolapse of the umbilical cord usually occurs early in labor when the amniotic sac ruptures. There is usually time to get the patient to the hospital. Your job is to try to keep the infant's head from compressing the cord.

Place the pregnant woman on a backboard in Trendelenburg's position, with her hips elevated on a pillow or folded sheet. Alternatively, the mother may be placed in a knee-chest position: kneeling and bent forward, facedown. Either of these positions is meant to help keep the weight of the infant off the prolapsed cord. Carefully insert your sterile gloved hand into the vagina, and gently push the infant's head away from the umbilical cord. Note that this is the only other occasion on which you should actually place a hand into the vagina. You will maintain this position and continue to keep the pressure off of the cord continuously throughout the transport to the hospital and possibly until the operating room.

Wrap a sterile towel, moistened with saline, around the exposed cord. Give the patient high-flow oxygen, and transport rapidly.

■ Spina Bifida

<u>Spina bifida</u> is a developmental defect in which a portion of the spinal cord or meninges may protrude outside of the vertebrae and possibly outside of the body. This is easily seen on the newborn's back and usually occurs in the lower third of the back in the lumbar area. It is important to cover the open area of the spinal cord with a sterile, moist dressing immediately after birth to help prevent a potentially fatal infection. This treatment will have a positive impact on the newborn's outcome. However, maintenance of body temperature is important when applying moist dressings because the moisture can lower the newborn's body temperature. To prevent this, have someone hold the newborn against his or her body.

■ Abortion

Passage of the fetus and placenta before 20 weeks is called abortion. Abortions may be spontaneous (also called miscarriage), without any obvious known cause, or intentional. Deliberate abortions may be self-induced, by the pregnant woman herself or by someone else, or planned and performed in a hospital or clinic. Regardless of the reasons or the cause of the abortion, there may be complications.

The most serious complications of abortion are bleeding and infection. Bleeding can result from portions of the fetus or placenta being left in the uterus (incomplete abortion) or from injury to the wall of the uterus (perforation of the uterus and possibly the adjacent bowel or bladder). Infection can result from such perforation and from the use of nonsterile instruments. If the woman is in shock, treat and transport her promptly to the hospital. Collect and bring to the hospital any tissue that passes through the vagina. Never try to pull tissue out of the vagina. Place a sterile pad or sanitary napkin on the vagina. In rare cases, massive bleeding may occur and cause severe hypovolemic or hemorrhagic shock. In these cases, treat for shock and provide immediate transport.

Multiple Gestation

Twins occur about once in every 80 births. Sometimes, there is a family history of twins or the woman may suspect that she is having twins because she has an unusually large abdomen. Usually, however, twins are diagnosed early in pregnancy with modern ultrasound techniques. With twins, always be prepared for more than one resuscitation, and call for assistance.

Twins are smaller than single infants, and delivery is typically not difficult. Consider the possibility that you are dealing with twins any time the first infant is small or the mother's abdomen remains fairly large after the birth. You should also ask the patient about the possibility of multiples. If twins are present, the second one will usually be born within 45 minutes of the first. About 10 minutes after the first birth, contractions will begin again, and the birth process will repeat itself.

The procedure for delivering twins is the same as that for single infants. Clamp and cut the cord of the first infant as soon as it has been born and before the second infant is delivered. The second infant may deliver before or after the first placenta. There may be only one placenta, or there may be two. When the placenta has been delivered, check whether there is one umbilical cord or two. If you see only one umbilical cord coming out of the first placenta, there is still another placenta to be delivered. If both cords are attached to one placenta, the delivery is over. Identical twins are of the same sex; fraternal twins may be of different sexes, or they may be the same.

Record the time of birth of each twin separately. Twins may be so small that they look premature; handle them carefully, and keep them warm. In the case of twins, you should identify the first infant delivered as "Baby A" by loosely tying an extra length of umbilical tape around

a foot. With the delivery of more than two infants, you can indicate the order of delivery by writing on a piece of tape and placing it on the blanket or towel that is wrapped around each infant.

Abuse

There is an increased chance of domestic violence and abuse in pregnant women. Abuse is one of the more common causes of complications in pregnancy that may harm the woman or the fetus. Some studies have indicated that 15% to 25% of pregnant women are victims of physical or sexual abuse. Abuse during pregnancy increases the chance of miscarriage, premature delivery, and low birth weight. The woman is at risk from bleeding, infection, and uterine rupture. Use a calm, professional approach if you suspect your patient has been abused. Pay attention to the environment for any signs of abuse. Your attention to detail will be helpful in your documentation and to the physicians and staff who will be caring for your patient at the hospital.

Pregnant patients who are abused are often scared and may not be honest as to how their injuries may have occurred. If possible, talk to the patient in a private area, away from the potential abuser. As with other abuse situations, you may suspect abuse when the story of how an injury happened does not make sense to you. A pregnant abused patient will be concerned about her baby. Remember that the best way for you to care for the fetus is to treat the mother. Reassure the patient as you provide treatment. Remember that in trauma and pregnancy, your treatment should include supporting the ABCs, controlling any bleeding, immobilizing extremity injuries, treating for shock with oxygen, and keeping the patient warm.

Substance Abuse

Unfortunately, more and more infants are being born to women who are addicted to drugs or alcohol. The women often have had little or no prenatal care. The effects of the addiction on the fetus include prematurity, low birth weight, and severe respiratory depression. Some of the infants will die. **Fetal alcohol syndrome** is the term used to describe the condition of infants born to mothers who have abused alcohol.

If you are called to handle a delivery of a drug- or alcohol-addicted mother, pay special attention to your own safety. As with all other cases, follow standard precautions. Wear eye protection, a face mask, and gloves at all times. Clues that you are dealing with an addicted mother may include the presence of drug paraphernalia, empty wine or liquor bottles, and statements made by family or bystanders or by the patient herself. The newborn of an addicted mother will probably need immediate resuscitation. Assist with the delivery, and be prepared to support the infant's respirations and administer oxygen during transport. Do not judge or lecture the patient. Your job is to help deliver the infant, provide treatment to the mother and the newborn, and to transport both to the hospital.

■ Premature Infant

The usual gestational period is 9 calendar months, or 40 weeks. A normal, single infant will weigh approximately 7 lb at birth. Any infant who delivers before 8 months (36 weeks of gestation) or weighs less than 5 lb at birth is considered premature. This determination is not always easy to make. Often, the exact gestation time cannot be determined. A premature infant is smaller and thinner than a full-term infant, and the head is proportionately larger in comparison with the rest of the body Figure 31-16 . The vernix, a cheesy white coating on the skin that is found on a full-term infant, will be missing on the premature infant or will be very minimal. There will also be less body hair.

Premature infants need special care to survive. They often require resuscitation efforts, which should be performed unless it is physically impossible. With such care, infants as small as 1 lb have survived and developed normally.

■ Postterm Pregnancy

The normal pregnancy or gestation is 40 weeks. However, about 10% of the time, pregnancies are after term, meaning that the pregnancy lasts more than 42 weeks. This situation can be from an inaccurate calculation of the due date, especially if the date of the patient's last menstrual period was not accurately determined. The estimated due date is usually more reliable the earlier

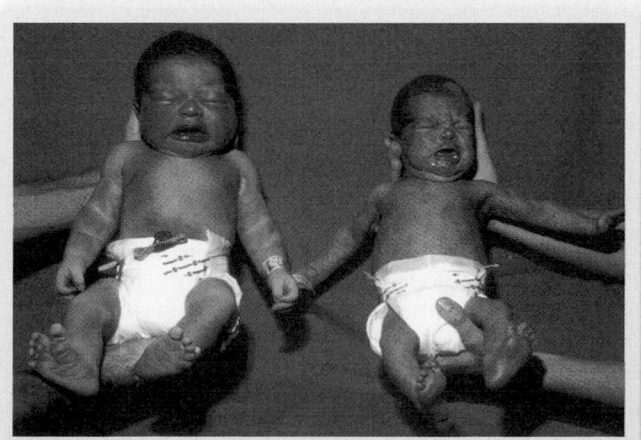

Figure 31-16 Premature infants (right) are smaller and thinner than full-term infants.

on in the pregnancy it is determined rather than later. Ultrasound examinations and measurements can be used to determine the due date in patients with an unknown last menstrual date, but ultrasound examinations are not typically as accurate. The patient may or may not know that her pregnancy has lasted more than 42 weeks.

This condition can lead to problems with the mother and infant. Infants can be larger than a typical 40-week infant, sometimes weighing 10 lb or more, which leads to a more difficult labor and delivery and an increased chance of injury to the fetus as it attempts to travel through the birth canal. There is an increased chance of cesarean section. The woman is also at risk for perineal tears and infection. Infants who are born after term have increased risks of meconium aspiration, infection, and being stillborn and may not have been able to develop normally because of the restricted size of the uterus. You should be prepared to resuscitate the newborn as respiratory and neurologic functions may have also been affected. The larger size of the infant causes it to take up more space inside the uterus, resulting in compression of the structures, including pressure on the blood vessels of the placenta and the umbilical cord.

■ Fetal Demise

Unfortunately, you may find yourself delivering an infant who died in the mother's uterus before labor. This will be a true test of your medical, emotional, and social abilities. Grieving parents will be emotionally distraught and perhaps even hostile, requiring all your professionalism and support skills.

The onset of labor may be premature, but labor will otherwise progress normally in most cases. If an intrauterine infection has caused the demise, you may note an extremely foul odor. The delivered infant may have skin blisters, skin sloughing, and a dark discoloration, depending on the stage of decomposition. The head will be soft and perhaps grossly deformed.

Do not attempt to resuscitate an obviously dead infant. However, do not confuse such an infant with infants who have had a cardiopulmonary arrest as a complication of the birthing process. You must attempt to resuscitate normal-appearing infants.

■ Delivery Without Sterile Supplies

On rare occasions, you may have to deliver an infant without a sterile OB kit. Even if you do not have an OB kit, you should always have eye protection, gloves, and a protective mask with you. This equipment is for your own protection and for that of the mother and infant. Carry out the delivery as if sterile supplies were available. If possible, use clean sheets and towels that have not been used since they were laundered. As soon as the infant is born, wipe the inside of the mouth with your finger to clear away blood and mucus. Without the OB kit, you should not cut or tie the umbilical cord. Instead, as soon as the placenta delivers, wrap it in a clean towel or put it in a plastic bag and transport it with the infant and mother to the hospital. Always keep the placenta and the infant at the same level, or elevate the placenta slightly if possible, so that blood does not drain from the infant into the placenta. Be sure to keep the infant warm.

Postpartum Complications

Some bleeding always occurs with delivery. However, bleeding that exceeds approximately 500 mL is considered excessive. Although up to 500 mL of blood loss is considered normal, you should continue to massage the uterus after delivery. Be sure to check your technique and hand placement if bleeding continues. If the mother appears to be in shock, treat her accordingly and transport, massaging the uterus en route. Excessive bleeding after birth is usually caused by the muscles of the uterus not fully contracting. This can be from delivering more than one infant, a long labor process so the uterus is too "tired" to contract, or parts of the placenta still being inside the uterus. This condition is potentially life-threatening for the mother. You should continue massaging the uterus and cover the vagina with a sterile pad, changing the pad as often as necessary. Do not discard any blood-soaked pads; hospital personnel will use

them to estimate the amount of blood that the mother has lost. Also save any tissue that may have passed from the vagina.

Place the woman in the shock position, administer oxygen, monitor vital signs frequently, and transport her immediately to the hospital. Never hold the woman's legs together in an effort to stop bleeding, and never pack the vagina with gauze pads in an attempt to control bleeding.

Postpartum patients are also at increased risk of an embolism, most commonly, a pulmonary embolism. As you recall, one of the normal changes in pregnancy that occurs is an increase in the woman's ability to clot in preparation for the bleeding that occurs with childbirth. Pregnant women who needed bed rest for any length of time are also more prone to forming clots. A pulmonary embolism results from a clot that travels through the blood stream and becomes lodged in the pulmonary circulation. This obstruction will block blood flow to the lungs and is potentially life threatening. Women have died of a postpartum pulmonary embolism from days to several weeks or months after childbirth. If you deliver a newborn in the field and the mother begins to report sudden difficulty breathing or shortness of breath, you should consider the possibility that she has a pulmonary embolism.

You should also suspect a pulmonary embolism in patients of childbearing age with respiratory complaints who have recently delivered, especially with the sudden onset of difficulty breathing or altered mental status. Provide supportive care of the ABCs with high-flow oxygen and rapid transport to the hospital.

You are the Provider: SUMMARY

1. What anatomic and physiologic changes occur during pregnancy? How will they affect your assessment of the patient?

It is important to understand the normal anatomic and physiologic changes that occur during the various stages of pregnancy; they explain deviations in your assessment findings that you would not encounter in a nonpregnant woman. Although these changes are normal, they place the mother at risk for certain injuries and medical conditions. Most pregnancy-related changes are observed in the respiratory, cardiovascular, musculoskeletal, and gastrointestinal systems.

Before a woman's first pregnancy, the uterus—a muscular, inverted pear-shaped organ—measures about 3" long, weighs about 0.07 oz (2 g), and has a fluid capacity of about 10 mL. During pregnancy, the uterus must grow many times its normal size to accommodate the developing fetus. By the end of pregnancy, the uterus weighs about 2.2 oz (1 kg) and has a fluid capacity of about 5,000 mL. As the uterus enlarges during pregnancy, it

is displaced upward from the pelvic cavity, where it is normally protected, and is therefore exposed to potential injury.

By the 20th week of pregnancy, the uterus is at or above the umbilicus, which increases the risk of direct fetal injury. Increased uterine size also has a direct effect on the woman's respiratory and cardiovascular systems because it causes upward shifting of organs toward the thoracic cavity.

During the second trimester of pregnancy, the uterus grows rapidly and pushes upward on the diaphragm, displacing it from its normal position. This decreases the ability to breathe deeply; therefore, the woman's respiratory rate increases slightly to maintain adequate minute volume. Pregnancy also increases maternal oxygen demand and consumption because of the increased metabolic needs of the woman and the fetus.

The cardiovascular system is impacted by pregnancy in several ways. Blood volume increases throughout pregnancy, which is necessary to meet the metabolic needs of the fetus, to adequately perfuse the woman's organs, especially the uterus and kidneys, and to help compensate for blood loss during delivery. By the end of the pregnancy, the woman's blood volume typically increases by as much as 50%, with about 15% of her total blood volume dedicated to the uterus. To accommodate the increase in total blood volume, the woman's heart rate increases up to 20%, or about 20 beats/min, by the third trimester of pregnancy.

As blood volume increases, so does the number of red blood cells (RBCs). This increased RBC count increases the woman's need for iron, which is why pregnant women are advised to take iron supplements (prenatal vitamins) to avoid becoming anemic. Blood clotting factors also increase during pregnancy, thus allowing the mother's blood to clot faster and protecting her from excessive bleeding during delivery.

During the latter part of pregnancy, the woman is at an increased risk for vomiting and potential aspiration because of changes in the gastrointestinal tract. Gastrointestinal motility, or the filling and emptying of the stomach to the small intestine, is under the control of key hormones and the nervous system. Pregnancy-related changes in these systems, in addition to upward displacement of the stomach because of increased uterine size, significantly increase the risks of vomiting and aspiration if the woman is in a supine position (eg, secured to a long backboard).

During pregnancy, a woman's weight typically increases by 25 to 27 lb. This increase in body weight will have an impact on the musculoskeletal system. Increased levels of hormones affect the musculoskeletal system by making the joints more "loose" or less stable. During the third trimester of pregnancy, the woman also experiences a change in her body's center of gravity, which increases her risk for falls.

Vital sign changes during pregnancy, such as an increase in heart rate and respiratory rate, should not be assumed to be pregnancy-related if the woman experiences an acute illness or injury. Instead, they should be assumed to be signs of shock until proven otherwise.

2. How will you determine if delivery is imminent or if there is enough time to transport a pregnant patient?

After conducting a primary assessment of the woman and providing any immediate care, you must determine if delivery is imminent (eg, within the next few minutes) or if there is enough time to safely transport the woman to the hospital. There are several questions that you should ask the patient when making this determination.

Begin by asking the patient how long she has been pregnant—preferably in weeks; normal gestation ranges between 37 and 42 weeks. Ask her when she is due to deliver. While most women will know their due date—which, of course, is an estimate—others will not; this is especially true if she has had no prenatal care.

Ask the patient if this is her first baby. As a general rule, labor is longer in women who are pregnant for the first time (primigravida) and shorter in women who have been pregnant more than once (multigravida). For example, the first stage of labor, which begins with the first contraction

and ends when the cervix is fully dilated (in the field, this is evidenced by the presence of crowning [presentation of the baby's head at the vaginal opening]), can last up to 16 hours in primigravidas and up to 8 hours in multigravidas. During the first stage of labor, there is generally enough time to transport the mother to the hospital.

If the patient is experiencing contractions, ask her how far apart they are and how long they last. True labor contractions consistently get stronger and closer together, are not relieved by a change in position, and begin in the lower back and "wrap around" to the lower abdomen. During early labor, contractions usually come at 5- to 15-minute intervals and may only last for 10 to 20 seconds. During the second stage of labor, contractions become more frequent and more intense; they typically occur every 2 to 3 minutes and can last up to 60 seconds. Regular and frequent contractions are an indicator that delivery is near.

Ask the patient if her bag of water (amniotic sac) has ruptured. This typically occurs toward the end of the first stage of labor, but may not occur until the delivery itself. Ask her if she is experiencing any vaginal spotting or bleeding; during the first stage of labor, a plug of mucus—sometimes mixed with blood (bloody show)—is expelled from the dilating cervix and discharged from the vagina.

Ask the patient if she feels the urge to push or move her bowels. During the second stage of labor, when the fetus is progressing through the birth canal, the head presses against the woman's rectum; this gives her the urge to move her bowels and is a sign of imminent delivery. The presence of crowning is an obvious indicator that delivery is in progress. The second stage of labor typically lasts about an hour in primigravidas women and 20 to 30 minutes in multigravidas.

Perhaps one of the most reliable indicators of imminent delivery is when the patient states, "I'm having this baby now!" This is especially true in women who have given birth in the past. This statement, coupled with the patient's urge to push or move her bowels, should prompt you to prepare for imminent delivery.

3. What are gestational diabetes and preeclampsia? How can they affect this delivery?

You should gather as much of the expectant mother's medical history as you can; this information will enable you to anticipate—to a certain degree—any complications that may occur before, during, or after delivery. It may also help you prepare for the possibility of resuscitating the infant after delivery.

Gestational diabetes (GD), or diabetes of pregnancy, is a condition that some women experience; it is more common in obese woman and women who have other risk factors for diabetes. Although GD typically resolves on its own after delivery, it can pose a threat to the mother and her infant. Gestational diabetes is caused by the increased production of the hormones progesterone and estrogen. During pregnancy, increased progesterone levels cause the pancreas to secrete insulin in greater amounts and at a faster pace, while the cells become less sensitive to insulin because of the effects

You are the Provider: SUMMARY, continued

of estrogen (insulin resistance). As a result, the cells are less able to uptake glucose, which accumulates in the woman's bloodstream. Much of this glucose is shunted to the fetus, where it is metabolized to fat; as a result, infants born to mothers with GD are often large for their gestational age, which could lead to shoulder dystocia—a condition in which the infant's shoulders are too broad to fit through the mother's pelvic opening—or other delivery complications caused by the infant's size. In addition, many women experience nausea before labor and have not eaten—factors that could lead to hypoglycemia and weakness in the woman and fetus.

Preeclampsia, or pregnancy-induced hypertension, is the most severe pregnancy-related hypertensive complication. It can occur as early as the 20th week of pregnancy, but is more common between the 20th and 30th weeks. Commonly called "toxemia of pregnancy," preeclampsia is characterized by hypertension; edema of the hands, feet, and face; and protein in the urine. Other symptoms may include visual disturbances (eg, seeing spots, blurred vision), headache, and anxiety.

Preeclampsia can cause damage to the mother's liver and kidneys and can cause bleeding disorders. Fetal complications include prematurity and growth retardation. Later in life, children born to mothers with preeclampsia may develop coronary artery disease, diabetes, and hypertension; they are also at an increased risk for stroke.

Untreated preeclampsia may lead to eclampsia, which is characterized by life-threatening seizures. Bear in mind that seizure deaths are caused by hypoxia; if the mother is hypoxic, her fetus will be as well. Many women with preeclampsia are treated with strict bed rest; others may require medications to control their blood pressure. The presence of preeclampsia, as with GD, increases the risk of fetal distress, which may necessitate resuscitation after the infant is born.

4. Is there time to transport this patient, or should you prepare for imminent delivery?

The patient is experiencing frequent (every 2 minutes) contractions that are lasting 45 seconds, and she has the urge to move her bowels (indicates that the infant is in the birth canal). On the basis of these factors, delivery will likely occur within the next few minutes; therefore, you and your partner should prepare for imminent delivery.

You should properly position the patient, take standard precautions, and open and prepare the obstetric kit. Your partner should be positioned at the patient's head to provide reassurance during delivery and to monitor her airway; it is not uncommon for women in labor to become nauseated, and some may vomit. If this occurs, your partner should be in a position so that he or she can turn the patient's head to the side to prevent aspiration.

The labor process is very exhausting and expends a lot of oxygen; therefore, supplemental oxygen should be given. Furthermore, the only way to oxygenate the fetus is to oxygenate the mother.

Position yourself so that you can see the vagina at all times and can continually assess for crowning. Time the frequency and duration of the mother's contractions. The frequency of contractions is determined from the beginning of one to the beginning of the next. To determine the duration of a contraction, feel the patient's abdomen from the moment the contraction begins (the uterus and abdomen become firm) to the moment that it ends (the uterus and abdomen relax). During each contraction, coach the patient to take quick, short breaths, but not to strain. Between contractions, encourage her to rest and breathe deeply through her mouth.

A friend or family member may want to be present during the delivery. He or she can help provide reassurance and help coach her breathing. Continue to monitor the patient for crowning, which will likely occur imminently.

5. How should you manage the umbilical cord situation?

As soon as the infant's head has delivered, you should instruct the mother to stop pushing while you stabilize the head with one hand and use the index finger of your other hand to feel whether the umbilical cord is wrapped around the neck (nuchal cord). A nuchal cord must be treated immediately; if the cord is wrapped tightly around the newborn's neck, the airway could be blocked.

Usually, you can slip the cord gently over the head during delivery (or shoulders, if necessary). If not, you must place two clamps about 2" apart and cut between the two clamps. In the rare event that the cord is wrapped more than once around the neck, clamp and cut only once and then unwrap it from around the neck. Handle the cord very carefully; it is fragile and easily torn. Do not let the clamps come off until the ends have been tied.

6. Why is it important to suction the newborn's mouth before the nose?

As soon as the infant's head has delivered and you have confirmed the absence of a nuchal cord (or treated appropriately, if present), you must suction amniotic fluid from the airway before the delivery proceeds. Although the mother's desire to continue pushing will be very strong, you must instruct her to stop.

While you are supporting the infant's head with one hand, quickly and efficiently *suction the fluid from the mouth first and then the nostrils*. Because infants are obligate nose breathers, suctioning the nostrils first may stimulate a gasping response, potentially causing the infant to aspirate the amniotic fluid that is in his or her mouth. Aspiration can lead to newborn distress, which increases the risk of complications (eg, hypoxia, aspiration pneumonia, sepsis) following delivery.

In addition to minimizing the risk of aspiration, the mouth should be suctioned first because more fluid volume can be retrieved from the mouth than from the nose. Infants have a large amount of amniotic fluid in their airway when they are born; this fluid must be removed quickly so that the infant can breathe effectively without distress. When it comes to suctioning a newborn's airway, remember that M (mouth) comes before N (nostrils).

You are the Provider: SUMMARY, continued

7. What is involved in the routine postdelivery care of a newborn?

Immediate postdelivery care of a newborn—regardless of his or her appearance—involves a series of actions aimed at keeping the infant warm and facilitating effective breathing. After these initial steps are performed, the need for further treatment is based on assessment of the newborn's respiratory effort, heart rate, and skin color.

The mnemonic "Do What Probably Seems Simple," which stands for Drying, Warming, Positioning, Suctioning, and Stimulation, will help you remember the immediate postdelivery care of newborns.

Newborns are covered with a substance called vernix; if this substance is not quickly removed, the newborn's body temperature will fall. Newborn hypothermia can result in depressed respirations and bradycardia, among other complications. After you have thoroughly dried off the newborn, wrap him or her in a clean, dry towel or blanket to help maintain the newborn's body temperature. Pay special attention to keeping the newborn's head covered; this can be a source of significant heat loss.

Position the newborn on his or her side, with the head slightly lower than the rest of his or her body, and make sure that the head is in a neutral position so the airway remains open. Resuction the mouth and nose with a bulb syringe or suction device with an 8- or 10-French catheter. Suction both sides of the back of the mouth, where secretions tend to collect, but avoid deep suctioning of the throat because this action can cause bradycardia. If you are using a mechanical suction device, ensure that the vacuum pressure is set to no more than 100 mm Hg.

In most cases, the actions of drying, warming, and suctioning will stimulate the newborn to breathe, if he or she is not already breathing. In fact, most newborns begin to breathe spontaneously within 15 to 30 seconds following birth. However, in some cases, you will need to perform some form of tactile stimulation to stimulate the newborn to breathe. Gently flick the soles of the feet or rub the back; these are the *only* acceptable methods for performing tactile stimulation. If the newborn does not breathe after 10 to 15 seconds of tactile stimulation, begin resuscitation, starting with positive-pressure ventilations.

8. What immediate treatment is indicated for this newborn?

As previously discussed, the need for and extent of newborn resuscitation is based on your assessment of the newborn's respiratory effort, heart rate, and skin color.

The newborn in this case study is breathing; however, her heart rate is 80 beats/min and she has cyanosis of her trunk and extremities. These clinical signs indicate that she is hypoxemic and will require additional resuscitative measures.

A healthy newborn will have a strong cry (which indicates adequate breathing), a heart rate of at least 100 beats/min (usually greater than 120 beats/min), and will quickly "pink up" in the central parts of the body (face, abdomen, and chest). Cyanosis of the hands and feet, called acrocyanosis, is a normal finding; in fact, the newborn's fingers and toes often remain cyanotic for up to 24 hours after birth.

If the newborn is found to be breathing adequately, assess his or her heart rate; an easy way to do this is to palpate the base of the umbilical cord or the brachial pulse. Because most newborns have very fast heart rates, count the number of pulsations you feel in 6 seconds and simply add a zero to that number. For example, if you feel 12 pulsations in 6 seconds, the newborn's heart rate is 120 beats/min. If the heart rate is found to be less than 100 beats/min—as is the case with the newborn in this case study—you should begin positive-pressure ventilations with a bag-mask device, even if he or she is breathing. Newborn bradycardia is defined as a heart rate of less than 100 beats/min and is almost always the result of hypoxemia.

Ventilate the newborn at a rate of 40 to 60 breaths/min for 30 seconds, and then reassess the heart rate. In most cases, only a brief period of positive-pressure ventilation is needed to increase the newborn's heart rate.

If the heart rate is still less than 100 beats/min after 30 seconds of positive-pressure ventilation, continue ventilations, keep the newborn warm, and transport at once. If the umbilical cord is still intact, keep the newborn at the same level as the mother's vagina.

9. What further treatment is indicated for the mother?

Some bleeding—usually less than 500 mL—occurs before the placenta delivers; this is normal and expected. After delivery and before transport, place a sterile pad or sanitary napkin over the vagina and straighten the mother's legs; *never place any pads or dressings into the vagina.*

In some cases, bleeding will continue after the placenta has delivered. To help slow this bleeding, gently massage the mother's abdomen with a firm, circular motion. You should be able to feel a firm, grapefruit-sized mass in the lower abdomen; this is the fundus, or the top part of the uterus. Fundal massage should cause the uterus to contract and become firmer. This can be uncomfortable for the mother, so reassure her and explain that this is necessary to help control bleeding. If the mother chooses to breastfeed, this will also stimulate the uterus to contract. Both uterine massage and stimulation of the mother's nipples by breastfeeding cause the pituitary gland to secrete oxytocin, a hormone that promotes uterine contraction, thus slowing any bleeding.

If there is severe bleeding after placental delivery, transport immediately, and treat the mother for shock. Place a sterile pad or sanitary napkin over the mother's vagina, elevate her legs 6" to 12", administer high-flow oxygen, and keep her warm with blankets. Closely monitor her vital signs en route.

10. What Apgar score should you assign to this newborn?

The Apgar score, which is calculated at 1 minute and 5 minutes after birth, is a numeric scoring system used to assess the status of an infant who does not require aggressive resuscitation, such as bag-mask ventilations or chest compressions. It is important to note that the need for and extent of resuscitation are based on assessment of the newborn's respiratory effort, heart rate, and oxygenation status—not the Apgar score.

If an infant is born with respiratory depression or apnea, you should begin resuscitation immediately. After you have stabilized the condition of the newborn, you may then calculate the Apgar score. In cases in which resuscitation is needed, you will likely only be able to calculate the 5-minute Apgar score.

The Apgar score assigns a numeric value (0, 1, or 2) to five areas of activity of the newborn: appearance, pulse, grimace or irritability, activity or muscle tone, and respirations. Refer to your EMT textbook for an in-depth discussion of the Apgar score.

The newborn in this case study has a pink body, but cyanotic hands and feet; therefore, you should assign a score of 1 for appearance. Her heart rate is 130 beats/min; therefore, a score of 2 should be assigned for the pulse. She moves her foot away when you flick the soles; therefore, you should assign a score of 2 for grimace. She resists your attempts to straighten her knees, which indicates good muscle tone; therefore, you should assign a score of 2 for activity. Finally, her respirations are rapid (40 to 60 breaths/min is a normal newborn respiratory rate); therefore, you should assign a score of 2 for respirations. On the basis of your assessment findings, you should assign the newborn an Apgar score of 9.

EMS Patient Care Report (PCR)

Date: 12-9-09	**Incident No.:** 013109	**Nature of Call:** Woman in labor		**Location:** 2505 Landa Park Blvd
Dispatched: 0625	**En Route:** 0627	**At Scene:** 0631	**Transport:** 0655	**At Hospital:** 0708	**In Service:** 0753

Patient Information

Age: 28 **Sex:** F **Weight (in kg [lb]):** 70 kg (155 lb)	**Allergies:** No known drug allergies **Medications:** Prenatal vitamins **Past Medical History:** Gestational diabetes, preeclampsia **Chief Complaint:** Active labor

Vital Signs

Time: 0636	**BP:** 122/82	**Pulse:** 110	**Respirations:** 24	**Sao₂:** 98%
Time: 0643	**BP:** 130/60	**Pulse:** 114	**Respirations:** 22	**Sao₂:** 98%
Time: 0651	**BP:** 126/60	**Pulse:** 98	**Respirations:** 20	**Sao₂:** 97%

EMS Treatment
(circle all that apply)

Oxygen @ _6_ L/min via (circle one): (NC) NRM **Bag-Mask Device**	**Assisted Ventilation**	**Airway Adjunct**	**CPR**
Defibrillation	(Bleeding Control)	**Bandaging**	**Splinting**	**Other:** (Assisted with delivery of baby; fundal massage)

Narrative

Medic 44 is dispatched to a residence for a woman in labor. Arrived on scene and found the patient, a 28-year-old woman, lying supine in her bed. She was conscious and alert; her airway was patent, and her breathing, although increased in rate, was of adequate depth. Patient states, "I am having my baby now!" She states that this is her third baby. Her husband, who was present at the scene, advised that she has had gestational diabetes and preeclampsia with this pregnancy and has had regular prenatal care. The patient states that her contractions are 2 minutes apart and are lasting about 45 seconds each; she further states that she has the urge to move her bowels and that her amniotic sac ruptured 5 hours ago. Initial visual inspection of the vaginal area did not reveal crowning. After initial set of vital signs were obtained, reassessment of vaginal area revealed crowning of the baby's head. Properly positioned patient, administered supplemental oxygen, and prepared for delivery. Husband remained present and provided emotional support and respiratory coaching to his wife. Delivery of infant's head revealed nuchal cord, which was easily corrected by sliding the cord over the shoulders. Suctioned the infant's mouth and nose and delivered the rest of the body without difficulty. After providing immediate postdelivery care to the newborn, a female, assessment revealed that she was breathing, had a heart rate of 80 beats/min, and had cyanosis to her trunk and extremities. Began assisted ventilation with a bag-mask device at 40 breaths/min; reassessment after 30 seconds revealed marked improvement in newborn's heart rate and skin color; her hands and feet remained slightly cyanotic, but her body was pink. Clamped and cut the umbilical cord, kept the newborn warm, and allowed mother to hold her child. Placental delivery occurred approximately 5 minutes after the birth of the baby. Reassessed mother's vital signs, and provided fundal massage for mild postpartum bleeding. Assigned 5-minute Apgar score of 9 (A, 1; P, 2; G, 2; A, 2; R, 2). Packaged mother and infant and began transport to the hospital. En route, continued to assess mother and infant; they both remained stable. Reassessment of vaginal bleeding revealed that it had stopped. Continued oxygen therapy for the mother, and delivered her and her baby to the hospital without incident. Provided verbal report to attending physician. Medic 44 returned to service at 0753. *End of report*

Assessment and Emergency Care of Obstetric Emergencies and Neonates

Scene Size-up

Scene Safety	Ensure scene safety and safe access to the patient. Standard precautions should be taken. If a field delivery appears imminent, gloves, gown, and eye protection are recommended. Determine the need for additional resources. Another ambulance or ALS may be needed if labor occurs, in order to continue care for the mother and newborn. Remain calm and professional.
Mechanism of Injury (MOI)/ Nature of Illness (NOI)	Assess the MOI/NOI. Observe the scene, and look for indicators that will assist you with this. The nature of the problem may not be readily apparent until more information is gathered. Avoid tunnel vision; the obstetric emergency may be related to a traumatic event. Spinal immobilization must be considered. Pay attention for signs of abuse. Abuse during pregnancy increases the chance of miscarriage, premature delivery, and low birth weight.

Primary Assessment

Form a General Impression	Your assessment of a patient with an obstetric emergency should begin at the door. Perform a rapid scan in order to ascertain if the patient is in active labor and whether delivery is imminent. Observe the scene for signs of trauma or other medical emergency. Look for signs of drug paraphenalia and for empty wine or liquor bottles that might suggest the possibility of fetal alcohol syndrome. Determine the level of consciousness using the AVPU scale. Identify and manage immediate threats to life. Determine priority of care based on the MOI/NOI. If the patient has a poor general impression, call for ALS assistance. Keep alert for signs of shock.
Airway and Breathing	Although airway problems usually are not an issue during uncomplicated childbirth, ensure the airway is open, clear, and self-maintained. Unresponsive patients will need their airway opened and maintained using a modified jaw-thrust if spinal injury is suspected or a head tilt-chin lift in nontrauma patients. A patient with an altered level of consciousness may need emergency airway management. Consider inserting a properly sized oropharyngeal or nasopharyngeal airway. Evaluate the patient's ventilatory status for rate and depth of breathing, respiratory effort, and tidal volume. Administer high-flow oxygen at 15 L/min, providing ventilatory support as needed.
Circulation	Observe skin color, temperature, and condition; look for life-threatening bleeding (including vaginal bleeding), and treat accordingly. Evaluate distal pulse rate, quality (strength), and rhythm. Tachycardia may be normal in pregnancy, but may also be an indicator of compensated shock. Treat for shock. If the patient is obviously pregnant or in the last two trimesters of her pregnancy, place her in a left lateral position to prevent supine hypotensive syndrome.
Transport Decision	If the patient has an airway or breathing problem, signs and symptoms of serious bleeding, hypertension, seizures, altered mental status, or other life threats, manage them immediately, and consider rapid transport, performing the secondary assessment en route to the hospital. Prevent supine hypotensive syndrome by transporting the pregnant patient on her left side. If the patient is immobilized, elevate the right side of the backboard. Occasionally the delivery will occur in the prehospital setting. If delivery is imminent, prepare to perform the delivery onscene.

NOTE: The order of the steps in this section differs depending on whether the patient is conscious or unconscious. The following order is for a conscious patient. For an unconscious patient, perform a primary assessment, perform a full-body scan, obtain vital signs, and obtain the past medical history from a family member, bystander, or emergency medical identification device.

Assessment and Emergency Care of Obstetric Emergencies and Neonates, continued

History Taking

Investigate Chief Complaint	Investigate the chief complaint to determine the patient's primary problem. Not all pregnant patients are in labor, instead there may be a medical or trauma emergency. Necessary questions may be very personal. Ensure patient privacy while remaining sensitive to the patient's feelings. Ask OPQRST and SAMPLE questions, but do not limit the history to only these. Obtaining the obstetric history should include questions about prenatal care, her expected due date, any complications she is aware of, and position of the fetus. Determine if contractions are occurring, how long they last, how far apart the contractions are, and if her amniotic sac (bag of water) has broken. If the amniotic sac has ruptured, ask about the color of the fluid; greenishbrown fluid indicates the presence of meconium. Call for ALS, and be prepared to suction prior to stimulating the newborn to breathe. SAMPLE can also be obtained from family, bystanders, and medical alert tags.

▼   ▼

Secondary Assessment

Physical Examinations	If the patient is unconscious, perform a systematic full-body or head-to-toe examination beginning with the head, looking for DCAP-BTLS. Assessment should be rapid if the patient had a poor general impression. Obstetric patients should receive a complete assessment of major body systems. Patients in labor should be assessed for fetal movement, contractions, and possible field delivery. If you believe that delivery of the infant is imminent, check for crowning, and prepare the mother and the scene for the delivery process. Protect patient privacy at all times to the best of your ability. Assess the vaginal area only if imminent delivery is suspected or the chief complaint warrants it.
Vital Signs	Obtain baseline vital signs. Vital signs should include blood pressure by auscultation; pulse rate, quality, and rhythm; respiration rate and quality; and skin assessment for perfusion. Note patient's level of consciousness. Use pulse oximetry, if available, to assess the patient's perfusion status. Tachycardia or hypotension may indicate hemorrhage or compression of the vena cava, leading to hypoperfusion. Hypertension may indicate preeclampsia or other serious problems. If possible, compare the patient's vital signs with those of previous prenatal visits.

▼   ▼

Reassessment

Interventions	Repeat the primary assessment and assessment of vital signs, and reassess the chief complaint. Identify any changes in the patient's condition. Vital signs should be repeated every 5 minutes if excessive blood loss is suspected to identify hypoperfusion. Place the patient in a position of comfort unless shock is suspected, then place patient supine and treat accordingly, monitoring the patient for possible supine hypotensive syndrome. If the patient is bleeding after delivery of the infant, provide uterine massage to slow it. Continue to provide high-concentration oxygen.
Communcation and Documentation	Contact medical control/receiving hospital with a radio report on the patient's condition. Advise the staff if delivery is imminent or has occurred so they can be prepared for your arrival. Often you will be asked to go directly to the labor and delivery floor, bypassing the emergency department. If you are treating a pregnant patient for something other than an obstetric emergency, be sure to notify the hospital staff that your patient is pregnant and give them her due date. When completing the patient care report for a pregnant patient or for an obstetric emergency, ensure it is thorough and complete.

▼   ▼

NOTE: Although the following steps are widely accepted, be sure to consult and follow your local protocols. Take approprite universal precautions when treating all patients.

Assessment and Emergency Care of Obstetric Emergencies and Neonates, continued

Obstetric Emergencies and Neonates

General Management of Obstetric Emergencies

Managing life threats to the patient's ABCs are primary concerns with any obstetric emergency. Avoid tunnel vision. Complete a full-body scan, remaining alert for signs and symptoms of shock. Manage as per local protocol. Request additional resources if delivery is imminent or has occurred. Provide high-concentration oxygen.

NOTE: Women of childbearing age, even if they are denying pregnancy, who are complaining of abdominal pain may have a life-threatening obstetric emergency.

Labor and Delivery

Once labor has begun, there is no way it can be slowed or stopped. If you suspect that delivery is imminent (abdomen is firm, she feels need to move her bowels or need to push, or crowning is present), explain to the patient that the infant will need to be delivered outside of the hospital. Calm and reassure her, protect her privacy, and prepare for the delivery. Place the patient in a position most comfortable for her, supporting her head, neck, and back with pillows and blankets. Sterile gloves should be worn, and sterile sheets should be placed to create a delivery area. As the infant's head emerges, be prepared to suction the mouth and nose. Ensure the cord is not wrapped around the neck (nuchal cord), and guide the head downward to assist with delivery of the shoulders. Support the infant as the body is delivered, being careful because the infant is slippery. Continue to suction the infant's mouth and nose with a bulb syringe. Warm and dry the infant as you vigorously stimulate the infant to breathe. Follow your local protocol regarding clamping and cutting of the cord. At all times, someone should be monitoring the mother. Document the sex of the infant and time of delivery. Once the infant is warmed, dry, and breathing, you can place the infant on the mother's chest. Encourage the mother to allow the infant to breastfeed to assist with uterine contraction. Locate the fundus, a grapefruit-sized mass in the lower abdomen, and begin uterine massage. After the infant has been delivered, you can begin the transportation process; remain alert as you might have to asssit with delivery of the placenta. At 1- and 5-minute intervals, assess and record the Apgar score.

Neonatal Assessment and Resuscitation

Following delivery, keep the infant warm and dry. Vigorously rubbing the infant's back, buttocks, and feet while drying the infant should stimulate the infant to breathe. Suction the infant's mouth first, then nose. Assess the newborn's respiratory effort and rate, as well as the heart rate. The newborn should begin breathing 15 to 30 seconds after delivery and have a heart rate of 120 beats/min or higher. If the heart rate is slow, begin by providing blow-by oxygen. If the rate does not increase, provide positive-pressure ventilation with an appropriately sized bag-mask device. If the heart rate is less than 60 beats/min, you should deliver compressions and ventilations in a 3:1 ratio, for a combined total of 120 "actions" per minute (90 compressions and 30 ventilations). If you suspect fetal alcohol syndrome, the newborn will probably need immediate resuscitation and transport. Do not attempt to resuscitate an obviously dead infant (one who has died in the mother's uterus before labor). An extremely foul odor, skin blisters, skin sloughing, dark discolorations, and gross deformities are indicators of fetal demise.

Assessment and Emergency Care of Obstetric Emergencies and Neonates, continued

Obstetric Emergencies and Neonates

Delivery Complications

Unruptured Amniotic Sac

Occasionaly the amniotic sac does not rupture during contractions or at the beginning of labor. This will be noted when you assess for crowning and see what appears to be a fluid-filled sac instead of the infant's head. If the head is crowning and the amniotic sac has not ruptured, you must rupture it. You can rupture the sac by pinching it and twisting. Amniotic fluid will rush out. Ensure you have taken standard precautions. Clear the ruptured sac from the infant's face, and be prepared to suction as soon as the head delivers. If meconium is noted, provide aggressive suctioning.

Nuchal Cord

When the umbilical cord is wrapped around the infant's neck it is called a nuchal cord. If it is wound tightly it will strangle the infant, so it must be removed. Attempt to slip the cord over the infant's head or shoulder. If you are unable to slip the cord over the head or shoulder, you will need to clamp the cord in two places about 2"apart, if possible, and cut the cord between the clamps. After cutting the cord, you can unwrap it and continue with the delivery as usual.

Breech Delivery

If the buttocks present first, the infant is at great risk of trauma from the delivery. If a breech birth is suspected, the patient should be transported to the hospital if time permits. Once the buttocks have passed through the vagina, the delivery process has begun. Call for ALS support, and contact medical control. Prepare the mother for delivery by placing her in a position of comfort. The buttocks and legs should deliver spontaneously; support them as they emerge. To keep the walls of the vagina from compressing the airway, make a "V" with your gloved fingers and insert them into the vagina to create an airway for the newborn. Your fingers will remain inserted until the infant's head is delivered. If delivery of the head does not occur within a few minutes, transport the mother to the hospital, keeping your fingers inserted to maintain an airway.

Limb Presentation

If the presenting part is the infant's arm, leg, or foot, you must prepare for immediate transport. You cannot succesfully deliver this infant in the field. Place the mother in a head-down, hips-elevated position, and cover the presenting limb with a sterile towel. Do not attempt to push or pull on the limb. Administer high-flow oxygen.

Prolapse of the Umbilical Cord

Presentation of the umbilical cord outside of the vagina before delivery of the infant is a very dangerous situation requiring immediate transport. Place the mother in a head-down, hips-elevated position to assist with preventing compression of the cord by the infant's head. Do not attempt to push the cord back into the vagina. Insert a sterile, gloved hand into the vagina, and gently push the infant's head away from the umbilical cord. Cover the exposed umbilical cord with a sterile, moistened towel. Administer high-flow oxygen.

Spina Bifida

Cover the exposed spinal cord with a moist, sterile dressing to prevent infection. Maintain the newborn's body temperature.

Spontaneous Abortion (Miscarriage)

If the delivery is occurring before the 20th week of gestation, be prepared to treat the patient for bleeding and shock. Place a sterile pad/dressing on the vagina. Collect any expelled tissue to take to the hospital, but never pull tissue out of the vagina. Transport immediately, continually monitoring the patient's ABCs while assessing for signs of shock.

Assessment and Emergency Care of Obstetric Emergencies and Neonates, continued

Obstetric Emergencies and Neonates

Delivery Complications, continued

Multiple Gestation

The procedure for delivering multiple infants is the same as that for a single newborn. If you suspect more than one infant, additional resources should be called for immediately. Record the time of birth for each infant separately, making sure to label them for identification after the delivery process is over. At 1- and 5-minute intervals, assess and record the Apgar score for each infant.

Postterm Pregnancy

Pregnancies lasting more than 42 weeks can lead to problems with the mother and infant. Infants can be larger, leading to a more difficult delivery and injury to the infant. Meconium aspiration risk increases, as does infection and stillborn birth. Respiratory and neurologic functions may be affected, so be prepared to resuscitate the infant.

Prep Kit

- Inside the uterus, the developing fetus floats in the amniotic sac. The umbilical cord connects the mother and fetus through the placenta. Eventually, contractions of the uterus will propel the neonate through the birth canal.

- Throughout pregnancy, the body changes to accommodate the fetus. The primary systems involved with these changes are the respiratory, cardiovascular, and musculoskeletal systems.

- As a result of enlargement of the uterus, a pregnant patient's respiratory capacity changes with increased respiratory rates and decreasing minute volumes.

- A pregnant patient's blood volume increases by as much as 50%, and the heart rate increases by 20%.

- Increased hormone levels affect the musculoskeletal system by making the joints more "loose" or less stable.

- The first stage of labor, dilation, begins with the onset of contractions and ends when the cervix is fully dilated. The second stage of labor, expulsion of the fetus, begins when the cervix is fully dilated and ends when the infant is born. The third stage of labor, delivery of the placenta, begins with the birth of the infant and ends with the delivery of the placenta.

- Once labor has begun, it cannot be slowed or stopped; however, there is usually time to transport the patient to the hospital during the first stage of labor. During the second stage of labor, you must decide whether to deliver the infant at the scene or transport the patient. During the third stage of labor, once the infant has been born, you will probably not transport the patient until the placenta has delivered.

- Complications of pregnancy include hypertensive disorders, bleeding, and diabetes.

- During a trauma call that involves a pregnant woman, you have two patients to consider—the woman and the unborn fetus. Any trauma to the woman will have a direct effect on the condition of the fetus.

- Abnormal or complicated deliveries include breech deliveries (buttocks first), limb presentations (arm, leg, or foot first), and prolapse of the umbilical cord (umbilical cord first). Quickly transport the patient with a limb presentation or prolapsed umbilical cord to the hospital.

- You should place a finger or hand into the vagina only to keep the walls of the vagina from compressing the infant's airway during a breech presentation or to push the infant's head away from the cord when the cord is prolapsed.

- Excessive bleeding is a serious emergency. Cover the vagina with a sterile pad; change the pad as often as necessary, and take all used pads to the hospital for examination.

Vital Vocabulary

abruptio placenta A premature separation of the placenta from the wall of the uterus.

amniotic sac The fluid-filled, baglike membrane in which the fetus develops.

Apgar score A scoring system for assessing the status of a newborn that assigns a number value to each of five areas of assessment.

birth canal The vagina and cervix.

bloody show A small amount of blood at the vagina that appears at the beginning of labor and may include a plug of pink-tinged mucus that is discharged when the cervix begins to dilate.

breech presentation A delivery in which the buttocks come out first.

cervix Narrowest portion of the uterus that opens into the vagina.

crowning The appearance of the infant's head at the vaginal opening during labor.

eclampsia Seizures (convulsions) resulting from severe hypertension in a pregnant woman.

ectopic pregnancy A pregnancy that develops outside the uterus, typically in a fallopian tube.

embryo The fertilized egg that is the early stages of a fetus.

endometrium The lining of the inside of the uterus.

fetal alcohol syndrome A condition of infants who are born to women who consume alcohol during pregnancy; characterized by growth and physical problems, mental retardation, and a variety of congenital abnormalities.

fetus The developing, unborn infant inside the uterus.

fundus The dome-shaped top of the uterus.

gestational diabetes Diabetes that develops during pregnancy in women who did not have diabetes before pregnancy.

lightening A sensation felt by a pregnant patient when the fetus positions itself for delivery.

limb presentation A delivery in which the presenting part is a single arm, leg, or foot.

meconium A dark green material in the amniotic fluid that can indicate distress or disease in the newborn; the meconium can be aspirated into the infant's lungs during delivery; the infant's first bowel movement.

miscarriage The passage of the fetus and placenta before 20 weeks; spontaneous abortion.

multigravida A woman who has had previous pregnancies.

nuchal cord An umbilical cord that is wrapped around the infant's neck.

placenta The tissue attached to the uterine wall that nourishes the fetus through the umbilical cord.

placenta previa A condition in which the placenta develops over and covers the cervix.

preeclampsia A condition of late pregnancy that involves headache, visual changes, and swelling of the hands and feet; also called pregnancy-induced hypertension.

pregnancy-induced hypertension A condition of late pregnancy that involves headache, visual changes, and swelling of the hands and feet; also called preeclampsia.

presentation The position in which an infant is born; the part of the infant that appears first.

primigravida A woman who is experiencing her first pregnancy.

prolapse of the umbilical cord A situation in which the umbilical cord comes out of the vagina before the infant.

spina bifida A developmental defect in which a portion of the spinal cord or meninges may protrude outside of the vertebrae and possibly even outside of the body, usually at the lower third of the spine in the lumbar area.

supine hypotensive syndrome Low blood pressure resulting from compression of the inferior vena cava by the weight of the pregnant uterus when the mother is supine.

umbilical cord The conduit connecting mother to infant via the placenta; contains two arteries and one vein.

vertex presentation A delivery in which the head comes out first.

Assessment in Action

You and your partner are called to a private residence for a woman in labor. On arrival, you find a 27-year-old woman lying on the living room couch, and she appears to be in the middle of a contraction.

1. The patient tells you that she is 38 weeks pregnant and her water broke when she was walking to the kitchen to get a drink. She says that she began experiencing contractions shortly afterwards. The onset of labor begins when the:
   A. amniotic sac ruptures.
   B. infant enters the birth canal.
   C. uterine contractions begin.
   D. woman has the urge to push.

2. When your partner asks the patient about any complications related to this pregnancy, she tells her that she has been on bed rest for the past 7 weeks because she was diagnosed with preeclampsia. Signs and symptoms of preeclampsia include:
   A. headache.
   B. swelling of the hands and feet.
   C. high blood pressure.
   D. all of the above.

3. Which of the following questions will help you determine whether delivery is imminent?
   A. Have you had a previous complicated pregnancy?
   B. Do you know if there is a chance of multiple births?
   C. Is this your first baby?
   D. Do you use drugs or drink alcohol?

4. The patient informs you that her amniotic water was green and she is concerned. The green color is caused by the presence of:
   A. premature rupture of the membranes.
   B. fetal stool.
   C. prolonged gestation.
   D. maternal infection.

5. As you deliver the infant's head, you observe the umbilical cord is wrapped once around the neck. What should you do?
   A. Ask the patient not to push, and prepare for immediate transport.
   B. Clamp and cut the cord, and remove the cord from around the infant's neck.
   C. Pull on the cord to deliver the infant faster.
   D. Try to slip the cord gently over the infant's head.

6. Following delivery, what is the correct way to stimulate the infant to breathe?
   A. Gently rub the infant's back.
   B. Blow oxygen into the infant's face.
   C. Smack the infant on the buttocks.
   D. Begin chest compressions.

7. You should evaluate the infant's heart rate by palpating which of the following arteries?
   A. Carotid
   B. Femoral
   C. Brachial
   D. Popliteal

8. Once an appropriate respiratory effort is present, what is the most important measure in determining the infant's need for resuscitation?
   A. Heart rate
   B. Blood pressure
   C. Pulse oximetry
   D. Apgar score

9. Discuss the proper management for an infant who needs ventilatory assistance.

10. What is the significance of meconium in the amniotic fluid?

Pediatric Emergencies

National EMS Education Standard Competencies

Special Patient Populations

Applies a fundamental knowledge of the growth, development, and aging and assessment findings to provide basic emergency care and transportation for a patient with special needs.

Patients With Special Challenges

- Recognizing and reporting abuse and neglect (pp 1194-1196, and Chapter 33, *Geriatric Emergencies*)

Health care implications of

- Abuse (pp 1194-1196 and Chapter 33, *Geriatric Emergencies*)
- Neglect (pp 1194-1196 and Chapter 33, *Geriatric Emergencies*)
- Homelessness (Chapter 34, *Patients With Special Challenges*)
- Poverty (Chapter 34, *Patients With Special Challenges*)
- Bariatrics (Chapter 34, *Patients With Special Challenges*)
- Technology dependent (Chapter 34, *Patients With Special Challenges*)
- Hospice/terminally ill (Chapter 34, *Patients With Special Challenges*)
- Tracheostomy care/dysfunction (Chapter 34, *Patients With Special Challenges*)
- Home care (Chapter 34, *Patients With Special Challenges*)
- Sensory deficit/loss (Chapter 34, *Patients With Special Challenges*)
- Developmental disability (Chapter 34, *Patients With Special Challenges*)

Pediatrics

Age-related assessment findings, and age-related assessment and treatment modifications for pediatric-specific major diseases and/or emergencies

- Upper airway obstruction (pp 1153-1162, 1168-1171)
- Lower airway reactive disease (pp 1153-1162, 1168-1173)
- Respiratory distress/failure/arrest (pp 1168-1180)
- Shock (pp 1180-1181)
- Seizures (pp 1181-1182)
- Sudden infant death syndrome (pp 1196-1199)

Age-related assessment findings, and developmental stage related assessment and treatment modifications for pediatric-specific major diseases and/or emergencies

- Upper airway obstruction (pp 1153-1162, 1168-1171)
- Lower airway reactive disease (pp 1153-1162, 1168-1173)
- Respiratory distress/failure/arrest (pp 1168-1180)
- Shock (pp 1180-1181)
- Seizures (pp 1181-1182)
- Sudden infant death syndrome (pp 1196-1199)
- Gastrointestinal disease (pp 1155-1156)

Trauma

Applies fundamental knowledge to provide basic emergency care and transportation based on assessment findings for an acutely injured patient.

Special Considerations in Trauma

Recognition and management of trauma in

- Pregnant patient (Chapter 31, *Obstetrics and Neonatal Care*)
- Pediatric patient (pp 1156-1168, 1186-1194)
- Geriatric patient (Chapter 33, *Geriatric Emergencies*)

Pathophysiology, assessment, and management of trauma in the

- Pregnant patient (Chapter 31, *Obstetrics and Neonatal Care*)
- Pediatric patient (pp 1156-1168, 1186-1194)
- Geriatric patient (Chapter 33, *Geriatric Emergencies*)
- Cognitively impaired patient (Chapter 34, *Patients With Special Challenges*)

Knowledge Objectives

1. Explain some of the challenges inherent in providing emergency care to pediatric patients and why effective communication with both the patient and his or her family members is critical to a successful outcome. (p 1148)

2. Discuss the physical and cognitive developmental stages of an infant, including signs that may indicate illness and patient assessment considerations when caring for an infant patient. (pp 1149-1150)

3. Discuss the physical and cognitive developmental stages of a toddler, including health risks, signs that may indicate illness, and patient assessment. (pp 1150-1151)

4. Discuss the physical and cognitive developmental stages of a preschool-age child, including health risks, signs that may indicate illness, and patient assessment. (p 1151)

5. Discuss the physical and cognitive developmental stages of a school-age child, including health risks, signs that may indicate illness, and patient assessment. (pp 1151-1152)

6. Discuss the physical and cognitive developmental stages of an adolescent, including health risks, patient assessment, and privacy issues. (pp 1152-1153)

7. Describe differences in the anatomy, physiology, and pathophysiology of the pediatric patient as compared to the adult patient and their implications for the health care provider, with a focus on the following body systems: respiratory, circulatory, nervous, gastrointestinal, musculoskeletal, and integumentary. (pp 1153-1156)

8. Describe the steps in the primary assessment for providing emergency care to a pediatric patient, including the elements of the pediatric assessment triangle (PAT), hands-on ABCs, transport decision considerations, and privacy issues. (pp 1156-1165)

9. Discuss the steps in the secondary assessment of a pediatric patient, describing what the EMT should look for related to different body areas and the method of injury. (pp 1166-1167)

10. Describe the different causes of pediatric respiratory emergencies, the signs and symptoms of increased work of breathing, the difference between respiratory distress and respiratory failure, and the emergency medical care strategies used in the management of each. (pp 1168-1180)

11. List the possible causes of an upper and a lower airway obstruction in a pediatric patient and the steps in the management of foreign body airway obstruction. (pp 1169-1171)

12. Describe asthma, its possible causes, signs and symptoms, and steps in the management of a patient who is experiencing an asthma attack. (pp 1171-1172)

13. Explain how to determine the correct size of an airway adjunct intended for a pediatric patient during an emergency. (pp 1173-1176)

14. List the different oxygen delivery device options that are available for providing oxygen to a pediatric patient, including the indications for the use of each and precautions the EMT must take to ensure the patient's safety. (pp 1176-1179)

15. Discuss the most common causes of shock (hypoperfusion) in a pediatric patient, its signs and symptoms, and emergency medical management in the field. (pp 1180-1181)

16. Discuss the most common causes of altered mental status (AMS) in a pediatric patient, its signs and symptoms, and emergency medical management in the field. (p 1181)

17. List the common causes of seizures in a pediatric patient, the different types of seizures, and their emergency medical management in the field. (pp 1181-1182)

18. List the common causes of meningitis, patient groups who are at the highest risk for contracting it, its signs and symptoms, special precautions, and emergency medical management in the field. (pp 1182-1183)

19. Discuss the types of gastrointestinal disease emergencies that might affect pediatric patients and their emergency medical management. (p 1183)

20. Discuss poisoning in pediatric patients, including common poison sources, signs and symptoms of poisoning, and its emergency medical management. (p 1184)

21. Discuss dehydration emergencies in pediatric patients, including how to gauge their severity based on key signs and symptoms, and emergency medical management. (p 1185)

22. Discuss the common causes of a fever emergency in a pediatric patient and the role of the EMT regarding patient management. (pp 1185-1186)

23. Discuss the common causes of drowning emergencies in pediatric patients, their signs and symptoms, and emergency medical management. (p 1186)

24. Discuss the common causes of pediatric trauma emergencies and differentiate between injury patterns in adults, infants, and children. (pp 1186-1194)

25. Discuss the significance of burns in pediatric patients, their most common causes, and general guidelines an EMT should follow when assessing patients who have sustained burns. (p 1192)

26. Explain the four triage categories used in the JumpSTART system for pediatric patients during disaster management. (p 1194)

27. Describe child abuse and neglect and its possible indicators, and then describe the medical and legal responsibilities of an EMT when caring for a pediatric patient who is a possible victim of child abuse. (pp 1194-1196)

28. Discuss sudden infant death syndrome (SIDS), including its risk factors, patient assessment, and special management considerations related to the death of an infant patient. (pp 1196-1198)

29. Discuss the responsibilities of the EMT when communicating with a family or loved ones following the death of a child. (pp 1197-1199)

30. Discuss some positive ways an EMT may cope with the death of a pediatric patient and why managing posttraumatic stress is important for all health care professionals. (pp 1197-1199)

Skills Objectives

1. Demonstrate how to position the airway in a pediatric patient. (pp 1160-1161, Skill Drill 32-1)

2. Demonstrate how to palpate the pulse and estimate the capillary refill time in a pediatric patient. (pp 1162-1163)

3. Demonstrate how to use a pediatric resuscitation tape measure to size equipment appropriately for a pediatric patient. (p 1173)

4. Demonstrate how to insert an oropharyngeal airway in a pediatric patient. (pp 1173-1174, Skill Drill 32-2)

5. Demonstrate how to insert a nasopharyngeal airway in a pediatric patient. (pp 1174-1176, Skill Drill 32-3)

6. Demonstrate how to administer blow-by oxygen to a pediatric patient. (p 1176)

7. Demonstrate how to apply a nasal cannula to a pediatric patient. (p 1177)

8. Demonstrate how to apply a nonrebreathing mask to a pediatric patient. (p 1177)

9. Demonstrate how to assist ventilation of an infant or child using a bag-mask device. (pp 1177-1179)

10. Demonstrate how to perform one-rescuer bag-mask device ventilation on a pediatric patient. (pp 1178-1179, Skill Drill 32-4)

11. Demonstrate how to perform two-rescuer bag-mask device ventilation on a pediatric patient. (p 1179)

12. Demonstrate how to immobilize a pediatric patient who has been involved in a trauma emergency. (pp 1187-1189, Skill Drill 32-5)

13. Demonstrate how to immobilize a pediatric patient who has been involved in a trauma emergency in a car seat. (pp 1189-1190, Skill Drill 32-6)

14. Demonstrate how to immobilize a pediatric patient who has been involved in a trauma emergency out of a car seat. (pp 1189-1191, Skill Drill 32-7)

Introduction

Pediatric patients have their own set of health-related problems that are unique to their population. Similarly, many problems that are common in adults do not occur in children and vice versa. It is important to remember that children are not small adults and their treatment can be a challenge for health care providers. Therefore, there is a specialized medical practice devoted to the care of the young, called **pediatrics**.

Most EMS providers have a level of discomfort when having to respond and care for a pediatric patient in distress. Pediatric patients differ in how they respond physiologically and emotionally to a stressful event. With the proper training and an understanding of this patient population, you will learn the tools necessary to form a baseline assessment and plan of care. In most situations, caring for an infant or child means that you must also care for the parents or caregivers as well Figure 32-1 . Therefore, it is imperative that you remain calm and professional, and remember to effectively communicate with your patient and the caregivers regarding your plan of care. However, once you learn how to approach children of different ages and what to expect while caring for them, you will find that treating children also offers some very special rewards. Not only are their innocence and openness appealing, they often respond to treatment much more rapidly than adults do.

Communication With the Patient and the Family

It is important to remember that when children are ill or injured, especially those with chronic illnesses, you may have more than one patient to treat rather than just one. Family members, especially the parent or primary caregiver, often need help or support when medical emergencies or problems develop. A calm parent usually helps to contribute to a calm child. An agitated parent usually means that the child will act the same way. Make sure that you are calm, efficient, professional, and sensitive as you deal with pediatric patients and their families.

Growth and Development

Adulthood begins at age 21, but when does childhood end? Many EMS systems use 18 years of age, others use 14, and still others use 12 or 16. Between birth and adulthood, many physical and emotional changes occur in children. Whereas each child is unique, the thoughts and behaviors of children as a whole are often grouped into stages: infancy, the toddler years, preschool years, school-age years, and adolescence. Children in each stage grapple with different developmental issues. Even though there are specific issues that are important to different age groups, there are also some general rules that apply when you care for children of any age.

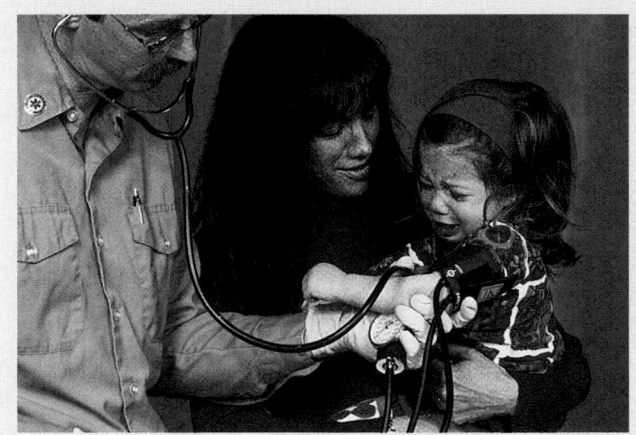

Figure 32-1 Treating a sick or injured child can be extremely challenging. A calm, professional demeanor is of utmost importance as you care for both the child and the parents.

You are the Provider: PART 1

At 11:23 PM, you are dispatched to 545 West San Antonio Street for a 4-year-old girl with respiratory distress. You and your partner proceed to the scene; your response time is 6 minutes. The weather is clear, the temperature is 58°F, and the traffic is light.

1. How does the pediatric patient's airway and respiratory system differ from an adult's?
2. What are some airway and breathing problems that are unique to pediatric patients?

■ The Infant

<u>Infancy</u> is usually defined as the first year of life; the first month after birth is called the neonatal or newborn period.

0 to 2 Months

Infants less than 2 months of age spend most of their time sleeping or eating. They respond mainly to physical stimuli such as light, warmth, hunger, and sound. It is important to remember that infants sleep for up to 16 hours a day between feeding times and caregiver interactions. An infant should be aroused easily from a sleeping state, and it should be considered an emergency if this is not the case.

Head control is limited, but infants can turn their heads and focus on faces. Infants at this stage have a sucking reflex for feeding. They also have a relatively large surface area that predisposes them to hypothermia, which is why you will often find infants bundled up in layers.

Crying is one of the main avenues of expression during this period. Infants may cry when hungry or if they require certain needs to be met. If all obvious needs have been addressed and the infant is still inconsolable, then this could be a sign of significant illness.

Infants are not able to tell the difference between parents and strangers. Their basic needs consist of being kept warm, dry, and fed. They experience the world through their bodies. Being held, cuddled, or rocked soothes the infant. Hearing is also well developed at birth, so calm and reassuring talk is often helpful in soothing the infant.

2 to 6 Months

Infants between 2 and 6 months of age are more active, which makes them easier to evaluate. They spend more time awake, they begin to smile and make eye contact, and they recognize caregivers. Healthy infants in this age group will have a strong sucking reflex, active extremity movement, and a vigorous cry. They may follow a bright light or toy with their eyes or turn their heads toward a loud sound or the caregiver's voice.

The infant now has an increased awareness of what is going on around him or her and will use both hands to examine objects and explore the world. About 70% of infants will sleep through the night by 6 months. At this point in development, infants will begin to roll over.

As with younger infants, persistent crying and irritability can be an indicator of serious illness. A lack of eye contact in a sick infant can also be a sign of significant illness, depressed mental status, or a delay in development.

6 to 12 months

During this stage, infants begin to babble and by their first year, infants are saying their first word. These infants also sit without support, progress to crawling, and finally begin to walk. This form of locomotion predisposes this age group to increased exposure to physical dangers. At this age, infants are also teething and prone to picking up anything and placing it in their mouths. The risk for foreign body aspirations and poisonings from toxic substances is yet another impending danger.

As with the younger infants, persistent crying or irritability can be a symptom of serious illness.

Infants are usually not afraid of strangers; however, by the end of their first year, they may show signs of preferring to be with their parents or caregivers and may cry if they are separated. This behavior is called separation anxiety, which is common among this age group **Figure 32-2** .

Assessment

Begin your assessment by observing the infant from a distance, preferably in a parent's or caregiver's arms. At 6 to 12 months, infants may begin to cry when touched or picked up by a stranger, so let the parent or caregiver continue to hold the baby as you start your assessment. Provide as much sensory comfort as you can: warm your hands and the end of the stethoscope and offer a pacifier if the parent or caregiver allows it. Have a parent or caregiver hold the infant, if possible, during procedures. Plan to complete any painful procedure in an efficient manner. If possible, plan to do any painful procedures at the end of the assessment process, so that the infant does

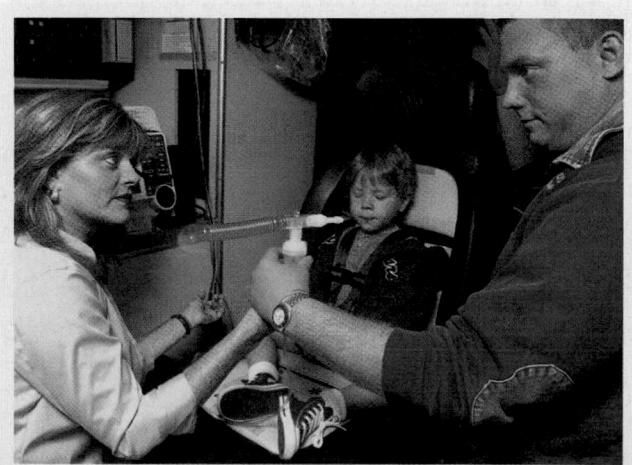

Figure 32-2 Infants are usually not afraid of strangers, but as they reach 6 months to 1 year, they may show signs that they prefer to be with their parents or caregivers.

not become agitated while you are trying to perform a full-body physical examination.

■ The Toddler

After infancy, until about 3 years of age, a child is called a **toddler**. Toddlers experience rapid changes in growth and development.

12 to 18 Months

During this period, toddlers begin to walk and to explore their environment. They are able to open doors, drawers, boxes, and bottles. Because they are explorers by nature and are not afraid, injuries in this age group increase. At 12 to 18 months, toddlers begin to imitate the behaviors of older children and parents and may express a desire to dress like mommy or daddy. The toddler knows major body parts when you point to them and may speak 4 to 6 words. Because of a lack of molars, toddlers may not be able to fully chew their food before swallowing, leading to an increased risk of food aspiration.

18 to 24 Months

The mind of the toddler is developing rapidly. At the beginning of this stage, the toddler may have a vocabulary of 10 to 15 words. By 2 years of age, a toddler should be able to pronounce approximately 100 words. When you point to a common object, toddlers should be able to name it. At this stage, toddlers begin to understand cause and effect with such activities as pop-up toys (jack-in-the-box) and turning on and off a light switch. The toddler's balance and gait also improve rapidly during this period. Running and climbing are two skills that develop. At this stage, toddlers tend to cling to their parents and caregivers and often have a special object such as a blanket or teddy bear that comforts them when they are separated.

Assessment

Stranger anxiety may still develop early in this period. Toddlers may resist separation from parents or caregivers and be afraid to let others come near them. Allow the toddler to hold any special object that brings the toddler comfort ("Would you like to hold your blankie while I listen to your tummy?"). Because of their newly found independence, they may also be very unhappy about being restrained or held for procedures **Figure 32-3** . Two-year-olds in particular have a well-deserved reputation for having their own ideas about almost everything, which is why these years are often called the "terrible twos."

Toddlers have a hard time describing or localizing pain because they do not have the verbal ability to be precise. Pain in the abdomen may be expressed as,

"My tummy hurts," and the physical examination may reveal tenderness throughout the body.

Toddlers can be curious and adventuresome, so you may be able to distract them **Figure 32-4** . For example, you might allow the toddler to play with a tongue depressor while assessing his or her vital signs. Restrain the toddler for as short a time as possible, and allow him or her to be comforted by the parent or caregiver immediately after a painful procedure. Begin your assessment at the feet to keep from upsetting the toddler whenever possible.

Like infants, persistent crying or irritability in a toddler can be a symptom of serious illness or injury. Painful procedures make a lasting impression at this stage. Older toddlers may remember earlier experiences with doctors or nurses, such as vaccinations or stitches, and be fearful about being treated. You should involve the parent or caregiver in any procedures because this not only provides you with an extra set of hands, but the

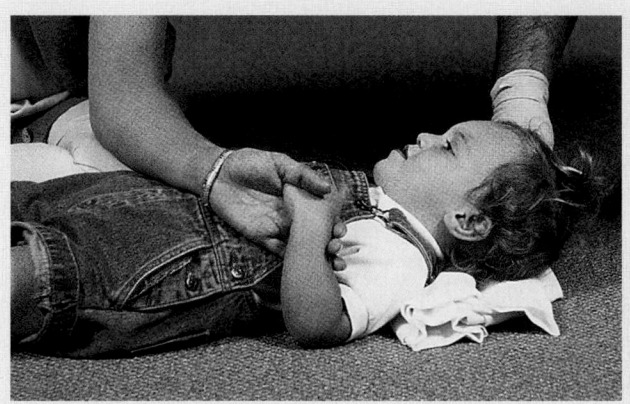

Figure 32-3 Because of their newly found independence, toddlers may be unhappy about being restrained or held for procedures.

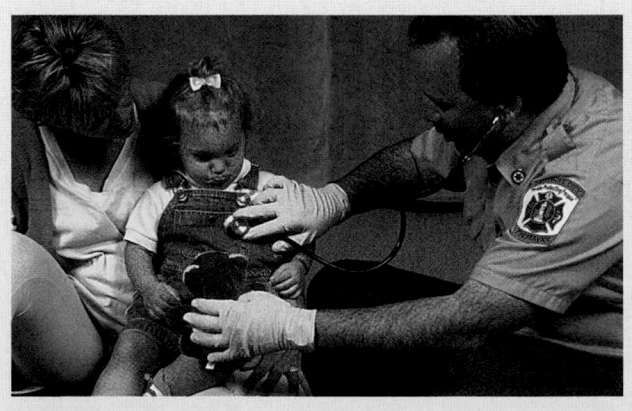

Figure 32-4 Leave a toddler on the caregiver's lap during your assessment, and use a toy to distract him or her.

presence of the parent or caregiver will provide comfort for the pediatric patient. If a parent or caregiver is not available, you should approach these pediatric patients using simple words to communicate and a calm, soothing voice while reassuring the toddler.

The Preschool-Age Child

Preschool-age children (ages 3 to 6 years) are able to use simple language quite effectively. The most rapid increase in language occurs during this stage of development. Children this age can walk and run well and begin throwing, catching, and kicking during play. Toilet training is mastered at this stage of development.

Preschool-age children have a rich fantasy life, which can make them particularly fearful about pain and change involving their bodies Figure 32-5 . At this age, they often believe that their thoughts or wishes can cause injury or harm to themselves or to others. They may believe that an injury is the result of a bad deed they did earlier in the day.

They are also learning which behaviors are appropriate and which behaviors will lead to a "time out." Tantrums may occur when preschool-age children feel they cannot control the situation or its outcomes.

The risk of foreign body airway obstruction continues to be high at this age.

Assessment

Preschool-age children can understand directions, be much more specific in describing their sensations, and identify painful areas when questioned. Much of their history must still be obtained from parents or caregivers, however. Tell the child what you are going to do immediately before you do it; this way, the child has no time to develop frightening fantasies. Appealing to the preschool-age child's imaginative thinking may allow treatment to go a bit smoother. For example, have the child pretend to be a dragon and "blow smoke" out her nose during a nebulizer treatment.

While caring for this age group and others within the pediatric population, you should never lie to the patient. Once you have lost your pediatric patient's trust, it will be challenging to regain it during your transport.

At this age, preschool-age children are easily distracted with counting games, small toys, or conversation Figure 32-6 . Be sure to adjust the level of game to the developmental level of the child; health care providers often assume that preschool-age children understand more than they actually do.

Begin your assessment with the feet and move toward the head, similar to assessing a toddler. Use adhesive bandages to cover the site of an injection or other small wound, because the preschool-age child might be worried about keeping his or her body together in one piece. Keep in mind that modesty is developing at this age, so keep the child covered when possible.

School-Age Years

School-age children (ages 6 to 12 years) are beginning to act more like adults. They can think in concrete terms, respond sensibly to direct questions, and help take care of themselves. School is important at this stage and concerns about popularity and peer pressure occupy a great deal of time and energy. Children with chronic illness or disabilities can become self-conscious because of concerns about fitting in with their peers. At this stage,

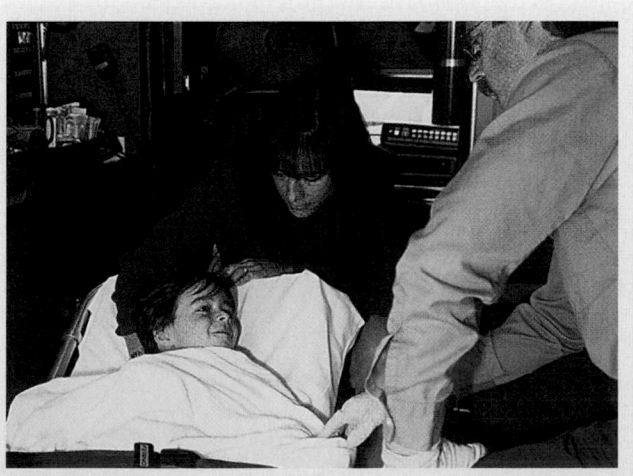

Figure 32-5 Preschool-age children have a vivid imagination, so much of the history must still be obtained from the caregiver.

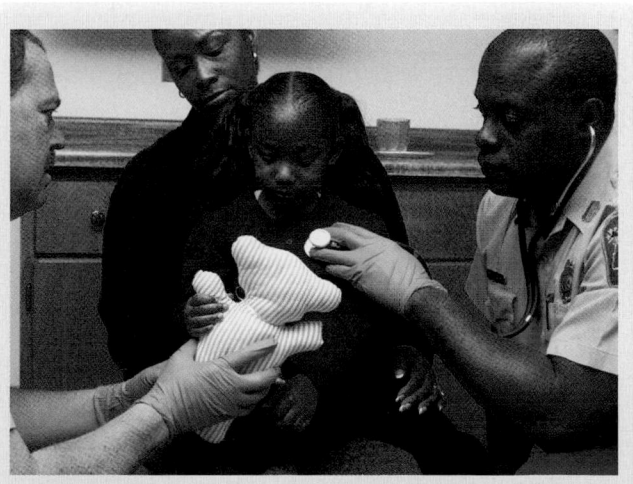

Figure 32-6 A preschool-age child can be easily distracted by games or conversation.

children begin to understand that death is final, which may increase their anxieties about illness or injury.

Assessment

Your assessment begins to be more like an adult assessment; talk to the child, not just the parent or caregiver, while taking the medical history **Figure 32-7**. At this stage, the child is usually familiar with the process of physical examination through check-ups and immunizations. You may begin at the head and move to the feet, in a process similar to assessing adolescent or adult patients.

Whenever possible, give the child appropriate choices: Would you like to sit up or lie down? Would you like to take off your clothes yourself? Only ask the type of questions that let you control the answer. For example, ask the child if you may find out the blood pressure on the child's right or left arm. Presenting a choice allows you to obtain assessment information and gives the child some control in a frightening situation. Also, asking if you may "take" the blood pressure may make younger patients think that you will not give it back. Encourage cooperation by allowing the child to listen to his or her own heartbeat through the stethoscope. Ensure the patient's modesty during the examination.

These children can understand the difference between emotional and physical pain, and have concerns about what pain means. Give them simple explanations about what is causing their pain and what will be done about it. Games and conversation may distract them. Ask them to describe their favorite place, their pets, school activities, or their toys. Ask the parent's or caregiver's advice in choosing the right distraction. Rewarding the child after a procedure can be very helpful in his or her future cooperation and recovery. Often, kind words and a smile make a good reward when stuffed toys or books are not available.

Adolescents

Most **adolescents** (ages 12 to 18 years) are able to think abstractly and can participate in decision making. This is also the stage when personal morals begin to develop. Adolescents are able to discriminate between what is right and wrong. They are now able to incorporate their own values and beliefs into their daily decision-making process. Even though this age group is physically similar to adults, adolescents are still children on an emotional level. They gradually shift from relying on family to relying on friends for psychological support, social development, and acceptance from their peers (especially the opposite sex). Interest in romantic relationships begins.

This is when puberty begins. This period of change makes the adolescent very concerned about body image and how he or she appears to peers. The adolescent may have very strong feelings against being observed during procedures.

Adolescence is a time of experimentation and risk-taking behaviors. Adolescents often feel that they are free from danger, and that they are "indestructible." Adolescents struggle with independence, loss of control, body image, sexuality, and peer pressure. They may have mood swings or depression and when ill or injured, may act younger than their age.

Assessment

Respect the adolescent's privacy at all times. Remember that adolescents can often understand very complex concepts and treatment options; you should provide them with information when they request it **Figure 32-8**. When the adolescent's condition is stable, discuss the situation and allow the adolescent to be involved in his or her care. Provide the adolescent with choices regarding his or her health, while also lending guidance if needed. You will

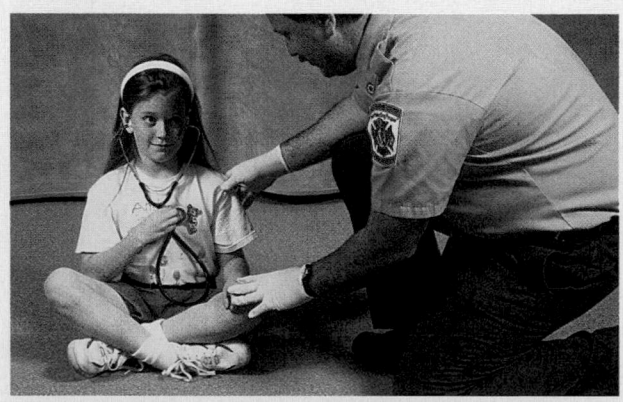

Figure 32-7 School-age children are more like adults in that they can answer your questions and can help to take care of themselves.

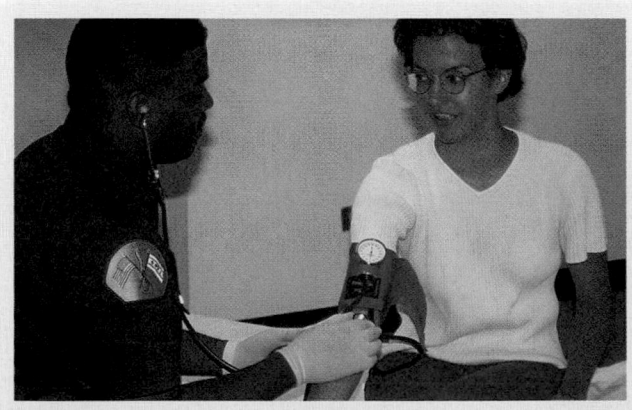

Figure 32-8 Respect the adolescent's privacy at all times; give the patient whatever information he or she requests.

find adolescents to be more helpful and understanding of necessary procedures than younger patients.

If the adolescent's condition requires him or her to be exposed or partially exposed to be assessed, you should take every measure to respect the patient's modesty and privacy. If an EMT of the same gender is available to perform the physical examination, it will lessen this stressful event. Adolescents are going through numerous body changes during puberty. An injury that could result in a scar from a laceration or burn will be challenging for you to address. The best practice is to be honest, tactful, and to reassure the adolescent that you are doing everything within your training to help in this situation. Allow the adolescent to speak openly about any thoughts and concerns. During this stage of development, an altered body image is a common concern.

Because the adolescent is under the influence of hormonal changes, peer pressure, and emotional highs and lows, risk-taking behaviors are common. Some of the risks that adolescents take can ultimately facilitate development and judgment, and help to shape their identity as an adult. However, the taking of risks in this age group can also result in unintentional trauma, dangerous sexual practices (unprotected sex), and teen pregnancy.

You must be aware that female adolescents may be pregnant, so you should ask, "Is there a chance you could be pregnant?" The answer is important to report to the receiving facility and on your patient care report. The adolescent might not want this information known to her parents and may fear facing the consequences of her actions. If you suspect that the patient might want to tell you something, but is silent in front of a caregiver, try to interview the adolescent without the caregiver present.

Adolescents have a clear understanding of the purpose and meaning of pain. Whenever possible, explain any necessary procedures well in advance. Assess their level of pain by observing facial and body expression as well as by asking questions; adolescents can be very stoic and may not request relief from pain even when they need it. To distract them, find out some of their interests, such as sports or movies, and get them talking.

Anatomy, Physiology, and Pathophysiology

There is no other time in a person's life that his or her body is growing and changing as fast as during childhood. Newborns have to quickly change to adapt to the world outside the mother's body (uterus). Toddlers learn to walk and talk. These changes can create difficulties during your assessment of the child if you do not expect them. For example, the child's head is proportionally larger than that of an adult patient. For

this reason, children are more susceptible to blunt head injuries from falls and other forms of trauma. In addition, excessive heat loss may occur if the child's head is bare.

■ The Respiratory System

To manage the pediatric airway effectively, you must understand the anatomic differences between the adult and pediatric airway. To start with, the pediatric airway is smaller in diameter and shorter in length, the lungs are smaller, and the heart is higher in a child's chest. The glottic opening (vocal cords) is higher and positioned more anteriorly (toward the front), and the neck appears to be nonexistent. As the child develops, the neck gets proportionally longer as the vocal cords and epiglottis achieve their anatomically correct adult position.

The anatomy of a pediatric airway and other important structures differs from that of an adult's in the following ways **Figure 32-9** :

- A larger, rounder occiput, or back of the head, which requires more careful positioning of the airway.
- A proportionately larger tongue relative to the size of the mouth and a more anterior location in the mouth. The child's tongue is also larger relative to the small mandible and can easily block the airway.

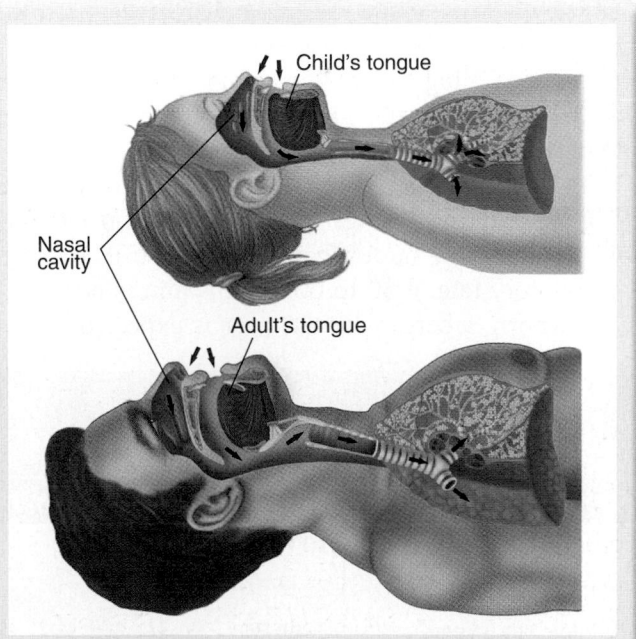

Figure 32-9 The anatomy of a child's airway differs from that of an adult's in several ways. The back of the head is larger in a child, so head positioning requires more care. The tongue is proportionately larger and more anterior in the mouth. The trachea is smaller in diameter and more flexible. The airway itself is lower and narrower (funnel shaped).

- A long, floppy, U-shaped epiglottis in infants and toddlers is larger than an adult's, relative to the size of the airway that extends at a 45° angle into the airway.
- Less well-developed rings of cartilage in the trachea that may easily collapse if the neck is flexed or hyperextended.
- A narrowing funnel-shaped (wide to narrow) upper airway compared to that of a cylinder-shaped (same width) lower airway.

These differences will influence the treatment decisions that you make about pediatric patients, including whether or not intervention is needed and, if so, what procedure to use.

Words of Wisdom

Because of the unique aspects of caring for children, it is wise to carry reference charts or measuring tools (Broselow tape) to assist you when assessing a child. Many services also carry copies of specialized pediatric protocols in their system. Refer to these resources during your care, and remember to make notes about your specific observations and treatment decisions. This "information-intensive" approach to pediatric care helps ensure both good care and thorough documentation.

Because of the smaller diameter of the trachea in infants, which is about the same diameter as a drinking straw, their airway is easily obstructed by secretions, blood, or swelling. Infants are obligate nose breathers, which may require diligent suctioning or reassessment and management to maintain a clear airway.

An infant needs to breathe faster than an older child **Table 32-1**. Children's lungs grow and develop increased abilities to handle the exchange of oxygen as they age. A respiratory rate of 30 to 60 breaths/min is normal for the newborn, whereas the teenager is expected to have

Table 32-1 Pediatric Respiratory Rates

Age	Respirations (breaths/min)
Newborn: 0 to 1 month	30 to 60
Infant: 1 month to 1 year	25 to 50
Toddler: 1 to 3 years	20 to 30
Preschool-age: 3 to 6 years	20 to 25
School-age: 6 to 12 years	15 to 20
Adolescent: 12 to 18 years	12 to 20
Older than 18 years	12 to 20

rates closer to the adult range (12 to 20 breaths/min). Children not only have a higher metabolic rate, but also a higher oxygen demand that is twice that of an adult. This in part is related to the actual size of the lung tissues and the volume that can be exchanged. Smaller lungs mean that the oxygen reserves are smaller. This higher oxygen demand combined with a smaller oxygen reserve increases the risk of hypoxia because of apnea or ineffective ventilation efforts.

Safety

In a pediatric patient, the lung tissues are prone to a simple or tension pneumothorax if excessive ventilatory pressures occur during assisted ventilations with a bag-mask device. To prevent hypoxia and to avoid damaging the lung tissues, use the appropriate size mask and reservoir bag to avoid administering excessive amounts of air volume. Only use enough force to make the chest rise slightly. Focus your attention on the rise and fall of the chest wall, versus just simply squeezing the reservoir bag. Ventilate with the patient's underlying respiratory rate and be careful not to ventilate against the patient's efforts.

Breathing also requires the use of the chest muscles and diaphragm. Because intercostal muscles are not well developed in children, movement of the diaphragm, their major muscle of respiration, dictates the amount of air that they inspire. Anything that places pressure on the abdomen of a young child can block the movement of the diaphragm and cause respiratory compromise. You must use caution when applying straps to the spinal immobilization device because this may hinder full symmetrical chest wall expansion and thus the tidal volume.

The breath sounds of the pediatric population are more easily heard because of their thinner chest walls. On another note, it may be difficult to assess for the absence of air movement because the air within the pediatric patient's lungs resonates with less surrounding muscle and fat to hinder the sounds.

Gastric distention can interfere with movement of the diaphragm and lead to hypoventilation. Young children also experience muscle fatigue much more quickly than older children. This can lead to respiratory failure if a child has to physically fight harder to breathe for long periods of time.

Pathophysiology

Respiratory problems are the leading cause of cardiopulmonary arrest in the pediatric population. Failure to recognize and treat declining respiratory status will lead to a certain death. A pediatric patient in respiratory distress still has the compensatory mechanisms and the ability to exchange oxygen and carbon dioxide. During respiratory distress, the pediatric patient is working harder to breathe and will eventually go into respiratory failure if left untreated. Respiratory failure occurs when the pediatric patient has exhausted all compensatory mechanisms and waste products begin to collect. If this is not treated, a total shutdown of the respiratory system will occur—respiratory arrest.

■ The Circulation System

It is important to know the normal pulse ranges when evaluating children **Table 32-2**. An infant's heart can beat as many as 160 times or more per minute if the body needs to compensate for injury or illness. This is the primary method the body uses to compensate for decreased perfusion.

The ability of children to constrict their blood vessels also helps them to compensate for decreased perfusion. Pale skin is an early sign that the pediatric patient may be compensating for decreased perfusion by constricting the vessels in the skin. Constriction of the blood vessels can be so profound that blood flow to the extremities can be diminished. Signs of vasoconstriction include weak distal (eg, radial or pedal) pulses in the extremities, delayed capillary refill, and cool hands or feet.

Pathophysiology

The pediatric cardiovascular system is not all that different than an adult's. Even though pediatric patients have a larger proportional amount of circulating blood volume than adults, they are more dependent on the actual cardiac output of the heart (amount of blood being pumped out of the heart in 1 minute). A pediatric patient actually may be in a state of shock while displaying a normal blood pressure. It may take only a small amount (one cup) of blood loss for the pediatric patient to go into shock.

■ The Nervous System

Compared with the adult nervous system, the pediatric nervous system is immature, underdeveloped, and not well protected. The head-to-body ratio of an infant and young child is disproportionally larger, making this population more prone to head injuries from falls or motor vehicle accidents. The occipital region of the head is larger, which increases the momentum of the head during a fall. The subarachnoid space is relatively smaller, leaving less cushioning for the brain. The brain tissue and the cerebral vasculature are fragile and prone to bleeding from shearing forces, such as during an incidence of **shaken baby syndrome**.

The pediatric brain also requires a higher amount of cerebral blood flow, oxygen, and glucose than adult brain tissue. Glucose stores are limited in the pediatric patients. These special needs mean that the pediatric brain is at risk for secondary brain damage from hypotension and hypoxic events.

Spinal cord injuries are less common in pediatric patients. If the cervical spine is injured, it is most likely to be an injury to the ligaments because of a rapid movement in the neck during a fall.

Pathophysiology

There are several causes of altered mental status in the pediatric population. Some of the most common causes are hypoglycemia, hypoxia, seizure, and drug or alcohol ingestion. The parent or caregiver is an important resource for you when you are gathering information regarding the baseline neurologic status of the pediatric patient. A pediatric patient with altered mental status may appear sleepy, lethargic, combative, or even unresponsive to tactile stimulus. Be diligent about assessing and managing the airway because pediatric patients may be prone to airway obstructions from their large tongues.

■ The Gastrointestinal System

The abdominal muscle structures are less developed in the pediatric patient, which results in less protection from blunt or penetrating trauma. The internal organs, such as the liver and the spleen, are proportionally larger and situated more anteriorly, so they are prone to bleeding and injury. Because the internal organs are positioned in a closer proximity to each other, there is a higher risk for multiple organ injury caused by minimal direct impact to this region, such as from automobile lap belts. The liver, spleen, and kidneys are more frequently injured in pediatric patients than in adults.

Pathophysiology

As with any injury or complaint in the abdominal region, the signs and symptoms may be vague in nature. The abdominal wall muscles are not as developed, which leaves this region more prone to injury. Pediatric patients may not be able to pinpoint the exact site where the pain

Table 32-2 Pediatric Pulse Rates	
Age	**Pulse Rate (beats/min)**
Infant	100 to 160
Toddler	90 to 150
Preschool-age	80 to 140
School-age	70 to 120
Adolescent	60 to 100

or discomfort originates, but will have complaints of diffuse tenderness. You should never take a complaint of abdominal pain and discomfort lightly because a large amount of bleeding may occur within the abdominal cavity without any outward signs of shock. Remember that liver and splenic injuries are common among this age group and may result in life-threatening consequences. The pediatric patient needs to be monitored for signs and symptoms of shock, which include an altered mental status, tachypnea, tachycardia, and bradycardia (late sign).

The Musculoskeletal System

Children's bones are softer than an adult's. The skeletal system contains open growth plates at the ends of long bones, which enable these bones to grow during childhood. As a result of the active growth plates, children's bones are weaker and more flexible, making them prone to fracture with stress. The open growth plates are also weaker than ligaments and tendons, leading to length discrepancies if there is an injury to the growth plate. Because of these factors, immobilize sprains or strains because they may actually be stress fractures.

The bones of the infant's head are flexible and soft, which allows the head to be delivered through the birth canal and for the growth of the brain during development. Located on the front (anterior) and back (posterior) portions of the head are soft spots, which are referred to as fontanelles. Each will close at particular stages of development, 18 months for the anterior suture and 6 months for the posterior suture. It is important to note that some bulging is a normal assessment finding when the infant is either crying, coughing, or lying on the back or stomach. The fontanelles of an infant can be a useful assessment tool for such issues as increased cranial pressure (bulging with a noncrying infant) or dehydration (a sunken appearance).

The thoracic cage in children is highly elastic and pliable because it is primarily composed of cartilaginous connective tissue. The ribs and vital organs are less protected by muscle and fat. The highly flexible ribs mean that fractures in pediatric patients are rare, unless a high energy force is encountered. You should assess for the potential of rib fractures caused by substantial motor vehicle crashes and accidents involving pedestrians. These mechanisms may result in a high-energy impact to the chest wall. Since this region is so pliable to impacts, underlying damage may still exist within the thoracic cavity without any exterior markings.

Pathophysiology

The muscles and bones of children continue to grow well into adolescence. For this reason, coupled with their risk-taking approach to activities, adolescents are prone to fractures of the extremities.

The younger the child, the more flexible the bone structures are to trauma. If a pediatric patient is unable to place weight on an extremity or favors an extremity, suspect injury until proven otherwise. Sprains are uncommon in this age group because the ligaments are more developed than the larger long bones.

A fracture of the femur is rare in pediatric patients, but when it does occur, it is a source of major blood loss. Older children and adolescents are prone to long bone fractures (femur and humerus) because they tend to take more risks during physical activities. The goal for care and treatment in this circumstance is to immobilize and stabilize the injured extremity, and to provide pain relief once advanced life support (ALS) has arrived on the scene.

The Integumentary System

The integumentary system of the pediatric population differs in a few ways. The skin is thinner with less subcutaneous fat than that of an adult. Infants and children also have a larger body surface area-to-body mass ratio, which can lead to large fluid and heat losses. The composition of the skin is thinner and tends to burn more deeply and easier with less exposure to the agent.

Pathophysiology

The thermoregulatory system in the pediatric body is immature. Paired with thinner skin and a lack of subcutaneous fat, this makes the pediatric population more prone to hypothermic events. Infants younger than 6 months lack the ability to shiver in response to a cold stimulus and therefore cannot generate heat from this protective mechanism. Newborns and infants less than 1 month are the most susceptible to hypothermia. However, newborns should not be overwarmed because this can worsen their neurologic outcomes.

Infants and young children should be kept warm during a transport or when the patient is exposed to assess or reassess an injury. The head should be covered, because up to 50% of heat loss can occur with a head that is larger in proportion to the rest of the body. Without recognition and treatment of a hypothermic event, the pediatric patient may progress to an unconscious state and lapse into convulsive seizure activity.

Patient Assessment

Because a young child might not be able to speak, your assessment of his or her condition must be based in large part on what you can see and hear yourself. Families may be helpful in providing vital information about an accident or illness. You should include families as part of the caregiving team and engage them to help comfort the infant or child during the assessment and during any interventions. Whenever possible, include them in all decisions about care and transportation.

Scene Size-up

The assessment begins at the time of initial dispatch. On the way to the scene, prepare mentally for approaching and treating an infant or child and interacting with the family. This means planning for a pediatric scene size-up, pediatric equipment, and the age-appropriate physical assessment. If possible, collect information from dispatch on the age and gender of the child, the location of the scene, and the nature of illness (NOI) or mechanism of injury (MOI).

Scene Safety

As with any EMS call, the scene size-up begins by ensuring that you and your partner have taken the appropriate safety precautions and standard precautions. As you enter the scene, note the position in which the pediatric patient is found. Look for any possible safety threats to the child, caregiver, bystanders, or EMS. Examples of safety threats include spilled toxins, open containers of alcohol, drug paraphernalia, weapons, or fire. Bring medications with you that could have been ingested by the pediatric patient. Keep in mind that the pediatric patient may be a safety threat if the child has an infectious disease.

Next, do an environmental assessment. The environmental assessment will give important information on the chief complaint, number of patients, MOI or NOI, and ongoing health risks. Evaluating the scene includes an inspection of the physical environment and watching the family–child and/or caregiver–child interactions. For example, documenting observations of dangerous scene conditions, and inappropriate statements from caregivers will greatly assist child protective services if the child is later determined to be a victim of an intentional injury. On the scene, be like a sponge; soak up as much useful information as possible to ensure scene safety and deliver timely care.

Mechanism of Injury/Nature of Illness

The MOI or the NOI refers to the circumstances in which the injury or illness occurred. As with the adult population, it is imperative for you to gather this information from the patient, parent, caregiver, or any bystanders who may have witnessed anything on the scene. At a traumatic scene when the child is unable to communicate because of his or her developmental age or is unresponsive, assume that the MOI was significant enough to cause head or neck injuries. Full spinal protocol with a cervical collar should be performed if you suspect the MOI to be severe. Remember the need to pad under the pediatric patient's head and/or shoulder to facilitate a neutral position for airway management.

Primary Assessment

As with the adult population, the objective of the primary assessment is to identify and treat immediate or potential threats to life.

You are the Provider: PART 2

You arrive at the scene, enter the residence, and find the child sitting on the couch next to her mother. She immediately makes eye contact with you, appears fearful of your presence, and starts clinging to her mother. You note that she is in obvious respiratory distress. As you approach the child, you make a visual assessment of her.

Recording Time: 0 Minutes	
Appearance	Obvious respiratory distress
Level of consciousness	Conscious; appears fearful
Airway	Open, no obvious obstructions
Breathing	Increased rate; moderate difficulty; nasal flaring; prominent supraclavicular retractions
Circulation	Skin, pink and dry; no gross bleeding

3. Why did you *not* immediately perform a hands-on assessment of this child?

4. On the basis of your initial observations, is this child experiencing respiratory distress or respiratory failure?

Form a General Impression

When assessing a pediatric patient, use the pediatric assessment triangle (PAT) to form a general impression.

Pediatric Assessment Triangle

The <u>**pediatric assessment triangle (PAT)**</u> is a structured assessment tool that allows you to rapidly form a general impression of the pediatric patient's condition without touching him or her. It provides a "first glance" assessment to identify the general category of the pediatric patient's physiologic problem and to establish urgency for treatment and/or transport. The PAT is a 15- to 30-second visual assessment of the pediatric patient.

The PAT **Figure 32-10** consists of three elements: appearance (muscle tone and mental status), **work of breathing**, and circulation to the skin. The only equipment required for the PAT are your own eyes and ears; no stethoscope, blood pressure cuff, cardiac monitor, or pulse oximeter is required.

Appearance Evaluating the pediatric patient's appearance involves noting the level of consciousness or interactiveness and muscle tone—signs that will provide you with information about the adequacy of the pediatric patient's cerebral perfusion (mentation) and overall function of the central nervous system.

Much of the information regarding the pediatric patient's level of consciousness can be obtained by using the PAT. In addition, you can evaluate the pediatric patient's level of consciousness by using the AVPU scale, modified as necessary for the pediatric patient's age **Table 32-3**.

An infant or child with a normal level of consciousness will act appropriately for his or her age, exhibiting good muscle tone and maintaining good eye contact

Table 32-3	The AVPU Scale
Alert	Normal interactiveness for age
Verbal	Appropriate: Responds to name Inappropriate: Nonspecific or confused
Painful	Appropriate: Withdraws from pain Inappropriate: Sound or movement without purpose or localization of pain
Unresponsive	No response to any stimulus

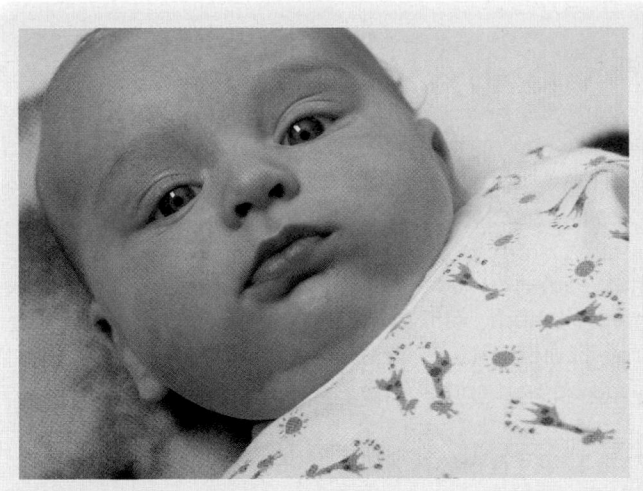

Figure 32-11 An infant or child making good eye contact is not very sick.

Figure 32-11. An abnormal level of consciousness is characterized by age-inappropriate behavior or interactiveness, poor muscle tone, or poor eye contact with the caregiver or with you **Figure 32-12**.

A helpful mnemonic called TICLS (or tickles) can also help to determine if the pediatric patient is sick or not sick. TICLS includes Tone, Interactiveness, Consolability, Look or gaze, and Speech or cry **Table 32-4**.

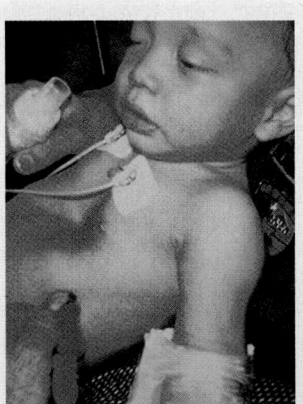

Figure 32-12 A limp child who is unable to maintain eye contact may be critically ill or injured.

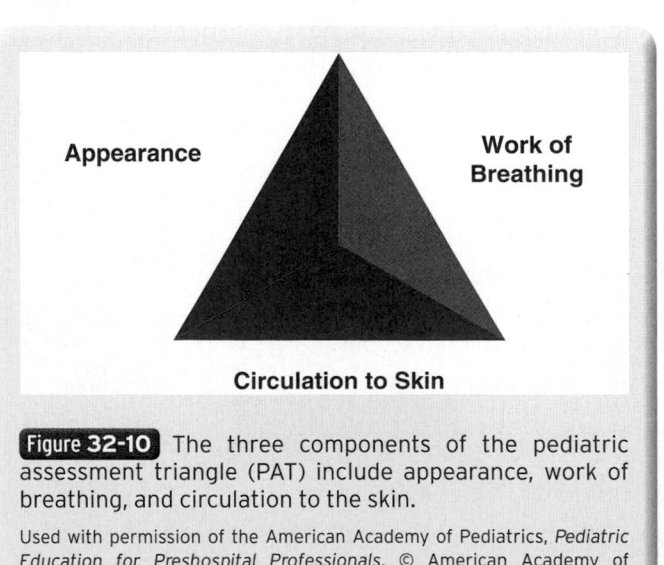

Figure 32-10 The three components of the pediatric assessment triangle (PAT) include appearance, work of breathing, and circulation to the skin.

Used with permission of the American Academy of Pediatrics, *Pediatric Education for Preshospital Professionals*, © American Academy of Pediatrics, 2000.

Work of Breathing A pediatric patient's work of breathing increases as the body attempts to compensate for abnormalities in oxygenation and ventilation. Increased work

	Table 32-4 Characteristics of Appearance: The TICLS Mnemonic	
Characteristic	**Features to Look For**
Tone	Is the child moving or resisting examination vigorously? Does the child have good muscle tone? Or is the child limp, listless, or flaccid?
Interactiveness	How alert is the child? How readily does a person, object, or sound distract the child or draw the child's attention? Will the child reach for, grasp, and play with a toy or exam instrument, like a penlight or tongue blade? Or is the child uninterested in playing or interacting with the caregiver or EMT?
Consolability	Can the child be consoled or comforted by the caregiver or by the EMT? Or is the child's crying or agitation unrelieved by gentle reassurance?
Look or gaze	Does the child fix his or her gaze on a face, or is there a "nobody home," glassy-eyed stare?
Speech or cry	Is the child's cry strong and spontaneous or weak or high-pitched? Is the content of speech age-appropriate or confused or garbled?

of breathing often manifests as **tachypnea**, abnormal airway noise (**grunting** or wheezing), retractions of the intercostal muscles or sternum Figure 32-13, or the way the pediatric patient positions himself or herself.

Circulation to the Skin An important sign of perfusion is circulation to the skin. When cardiac output falls, the body, through vasoconstriction, shunts blood from areas of lesser need (such as the skin) to areas of greater need (such as the brain, heart, and kidneys). The PAT is a valuable tool in the field when you are confronted with various etiologies, such as respiratory distress or failure, cardiovascular shock leading to cardiopulmonary failure or arrest, isolated head injury, ingestion of a toxic substance, neurologic injuries—or even as an approach to a stable pediatric patient.

Pallor of the skin and mucous membranes may be seen in compensated shock; it may also be a sign of anemia or hypoxia. Mottling is caused by constriction of peripheral blood vessels and is another sign of poor perfusion Figure 32-14.

Cyanosis, a blue discoloration of the skin and mucous membranes, reflects a decreased level of oxygen in the blood. Cyanosis is a late sign of respiratory failure or shock; absence of discoloration, however, does not rule out these conditions. Never wait for the development of cyanosis before administering oxygen!

Stay or Go On the basis of the findings of the PAT triangle, you will decide if the pediatric patient is stable or requires urgent care. If the pediatric patient is unstable, assess the ABCs, treat any life threats, and transport the pediatric patient immediately to an appropriate facility.

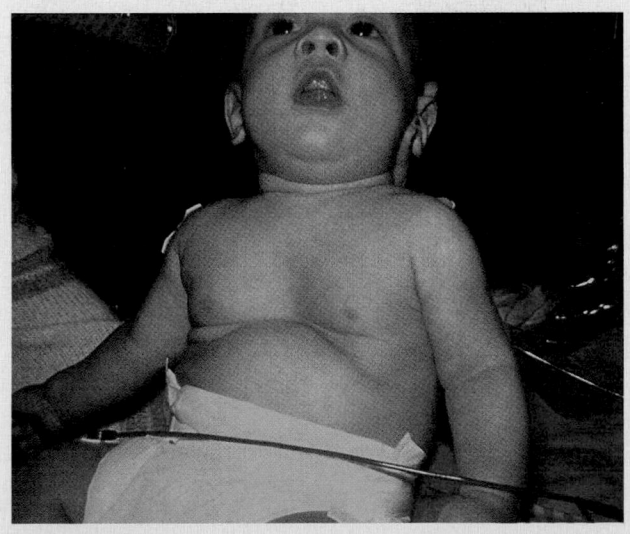

Figure 32-13 Retractions of the intercostal muscles or sternum indicate increased work of breathing.

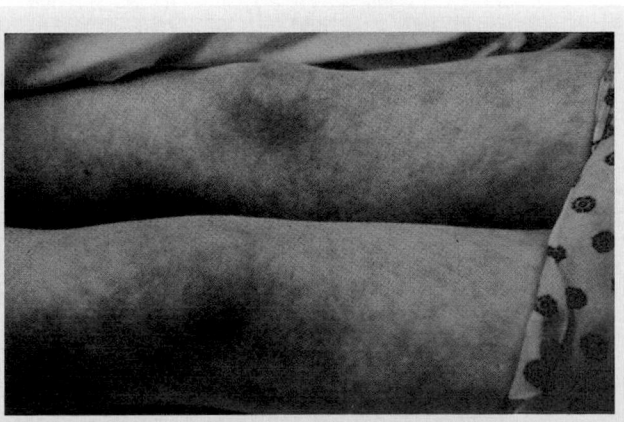

Figure 32-14 Mottling of the skin indicates poor perfusion and is the result of constriction of peripheral blood vessels.

If the pediatric patient is stable, then you have time to continue with the remainder of the patient assessment process.

Hands-on ABCs

For the pediatric patient, you will now perform a hands-on ABCs assessment. You will assess and treat any life threats to:

- **A**irway
- **B**reathing
- **C**irculation
- **D**isability
- **E**xposure

Airway

If the pediatric patient's airway is open and the patient can adequately keep it open (as is often the case in conscious pediatric patients), assess respiratory adequacy. However, if the pediatric patient is unresponsive or has difficulty keeping the airway clear, you must ensure that the airway is properly positioned and that it is clear of mucus, vomitus, blood, and foreign bodies.

If trauma has been ruled out, open the pediatric patient's airway with the head tilt–chin lift maneuver **Figure 32-15**. If the pediatric patient has been involved in trauma or trauma is suspected, use the jaw-thrust maneuver to open the airway **Figure 32-16**.

Positioning the airway correctly is critical in pediatric emergency care. Always position the airway in a neutral **sniffing position**. This accomplishes two goals at once, keeping the trachea from kinking and maintaining the proper alignment should you have to immobilize the spine. If the pediatric patient has been involved in trauma or trauma is suspected, use the jaw-thrust maneuver to open the airway.

Follow these steps to position the airway in a pediatric patient without trauma **Skill Drill 32-1**:

1. Place the pediatric patient on a firm surface such as a short backboard or pediatric immobilization device (Step 1).
2. Fold a small towel and place it under the pediatric patient's shoulders and back (Step 2).
3. Immobilize the pediatric patient's forehead to limit rolling of the head during transport. Use the head tilt–chin lift maneuver to open the airway (Step 3).

After the pediatric patient's airway has been opened, make sure that it is clear of potential obstructions such as mucus, blood, or foreign bodies. Next, establish whether the pediatric patient can maintain his or her own airway

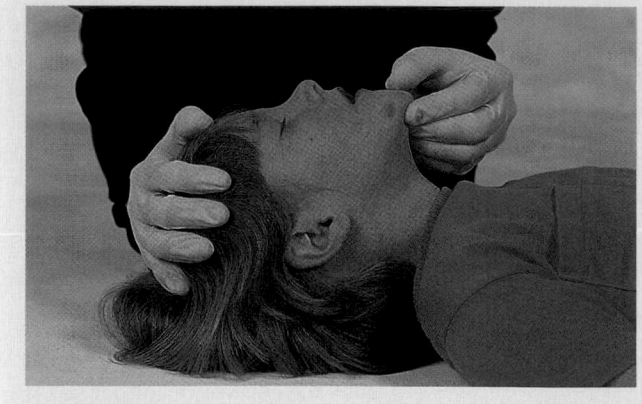

Figure 32-15 Use the head tilt–chin lift maneuver to open the airway of a pediatric patient without trauma.

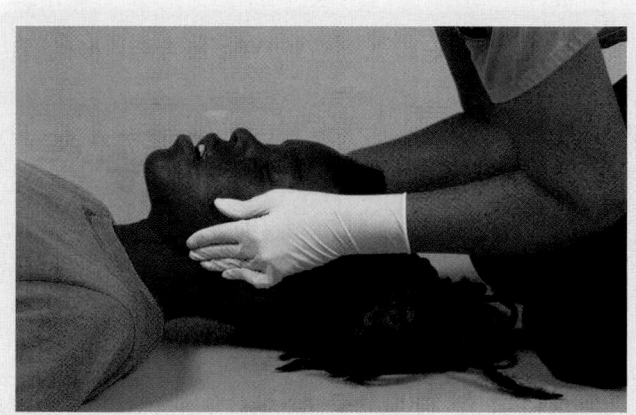

Figure 32-16 Use the jaw-thrust maneuver in a pediatric patient with possible spinal injury.

spontaneously (without the use of airway adjuncts) or whether adjuncts will be necessary to maintain airway patency. Techniques of airway management will be discussed later in this chapter.

Breathing

Assess the pediatric patient's breathing by using the look, listen, and feel technique, noting the degree of air movement at the nose and mouth and determining whether the chest is rising adequately and symmetrically. The respiratory rate and effort with which the pediatric patient is breathing should be assessed as well.

When you are assessing a pediatric patient, it is best to place both hands on the pediatric patient's chest to feel for the rise and fall of the chest wall. You will be able to count the actual respiratory rate and assess for symmetry. This assessment maneuver is especially helpful when your pediatric patient requires assisted ventilations with a bag-mask device or an endotracheal tube is in place. In infants, belly breathing is considered adequate because of the soft pliable bones of the chest and the strong muscular diaphragm.

Skill Drill 32-1

Positioning the Airway in a Pediatric Patient

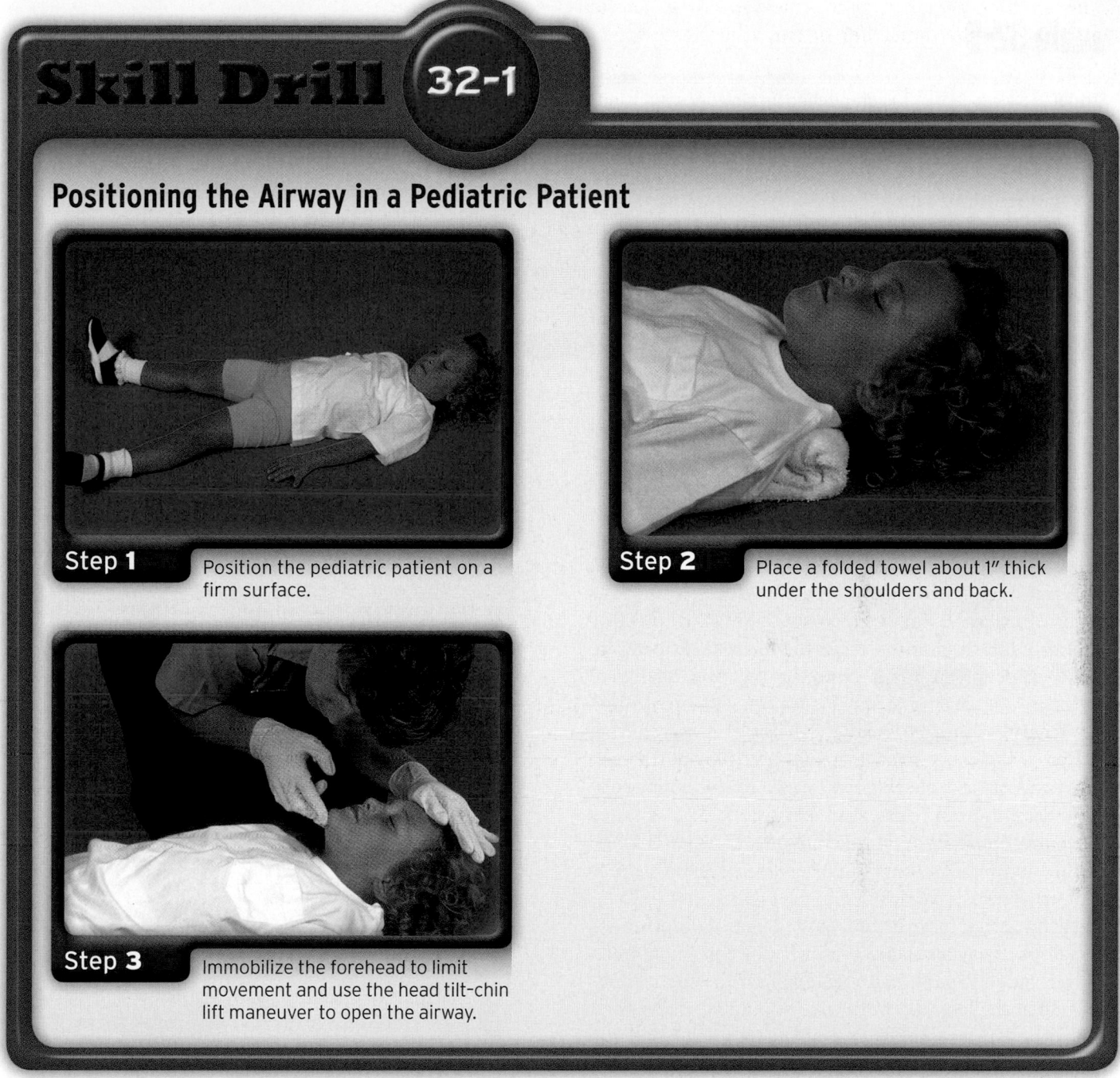

Step 1 Position the pediatric patient on a firm surface.

Step 2 Place a folded towel about 1" thick under the shoulders and back.

Step 3 Immobilize the forehead to limit movement and use the head tilt-chin lift maneuver to open the airway.

If the pediatric patient is conscious and not in need of immediate intervention (such as suctioning or assisted ventilation), assessing respirations is usually easier with the pediatric patient sitting on the caregiver's lap. Listen for abnormal respiratory sounds **Table 32-5**, and note any signs of increased respiratory effort.

When you are observing the pediatric patient's respiratory effort, note any signs of increased work of breathing, including:

- **Accessory muscle use:** Contractions of the muscles above the clavicles (supraclavicular)
- **Retractions:** Drawing in of the muscles between the ribs (intercostal retractions) or of the sternum (substernal retractions) during inspiration

- **Head bobbing:** The head lifts and tilts back during inspiration, then moves forward during expiration
- **Nasal flaring:** The **nares** (the external openings of the nose) widen; usually seen during inspiration
- **Tachypnea:** Increased respiratory rate

As the pediatric patient begins to tire, retractions often become weak and ineffective and the accessory muscles become less prominent during breathing. **Bradypnea**, a decrease in the respiratory rate, is an ominous sign and indicates impending respiratory arrest. Do not mistake bradypnea for a sign of improvement; it usually indicates that the pediatric patient's condition has deteriorated. Therefore, you must be prepared to begin ventilatory assistance.

Table 32-5 Abnormal Respiratory Sounds

Stridor	High-pitched inspiratory sound; indicates a partial upper airway obstruction (such as in croup or from a foreign body)
Wheezing	High- or low-pitched sound heard usually during expiration; indicates a partial lower airway obstruction (such as in asthma or bronchiolitis)
Grunting	An "uh" sound heard during exhalation; reflects the pediatric patient's attempt to keep the alveoli open by increasing pressure in the chest cavity; indicates inadequate oxygenation
Absent breath sounds (despite increased work of breathing)	Indicates a complete upper or lower airway obstruction (such as foreign body, severe asthma, or pneumothorax). This is also known as a silent chest, which is an ominous sign of impending respiratory distress; leading to respiratory arrest if not quickly addressed.

Circulation

When you are assessing circulation, you must determine if the pediatric patient has a pulse, is bleeding, or is in shock. Remember, infants and children can tolerate only small amounts of blood loss before circulatory compromise occurs. Assess and control any active bleeding early in your assessment.

A pulse may be difficult to palpate if it is weak, very fast, or very slow. In infants, palpate the brachial pulse or femoral pulse. In children older than 1 year, palpate the carotid pulse **Figure 32-17** . Note the rate and quality of the pulse: Is it weak or strong? Is it normal, slow, or fast? Strong **central pulses** usually indicate that the child is not hypotensive; however, this does not rule out the possibility of compensated shock. Weak or absent peripheral pulses indicate decreased perfusion. The absence of a central pulse (that is, brachial or femoral in infants, carotid in older children) indicates the need for cardiopulmonary resuscitation (CPR).

Tachycardia may be an early sign of hypoxia or shock, but it may also reflect less serious conditions such as fever, anxiety, pain, and excitement. Like the respiratory rate and effort, the pulse rate should be interpreted within the context of the overall history, PAT, and the entire primary assessment.

A trend of an increasing or decreasing pulse rate may be quite useful and may suggest worsening hypoxia or shock or improvement after treatment. When hypoxia or shock becomes critical, bradycardia occurs. As with slowing respirations, bradycardia in a pediatric patient is an ominous sign and often indicates impending cardiopulmonary arrest.

Feel the skin for temperature and moisture at the same time you assess the patient's pulse. Is the skin warm and dry, or cold and clammy? Estimate the capillary refill time by squeezing the end of a finger or toe for several seconds and then observing the return of blood to the area **Figure 32-18** . Color should return within 2 seconds after you let go. The capillary refill time is used to assess end-organ perfusion. It is most reliable in children younger

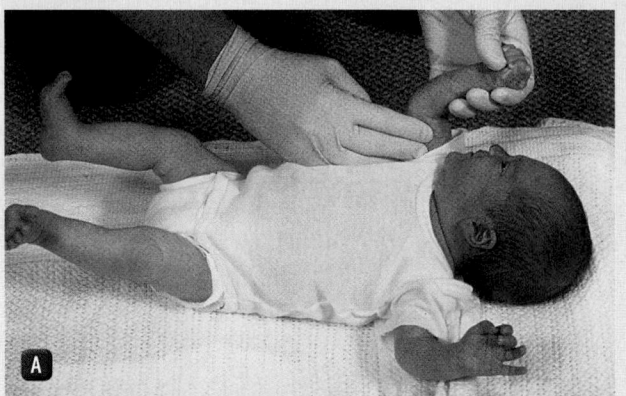
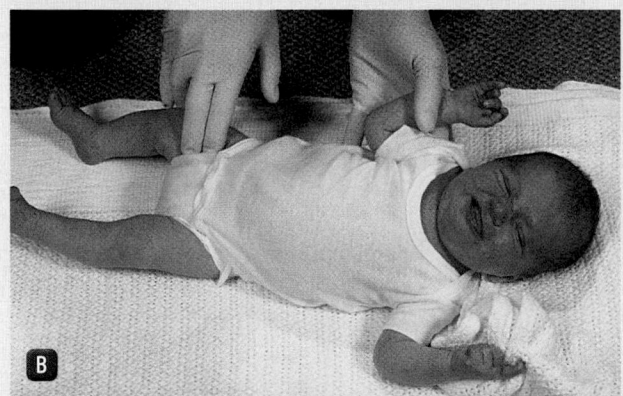
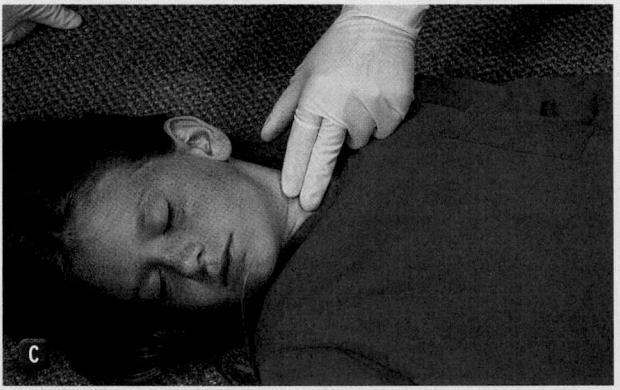

Figure 32-17 **A.** Palpate the brachial pulse in infants. **B.** Palpate the femoral pulse as a second choice. **C.** In children older than 1 year, palpate the carotid pulse.

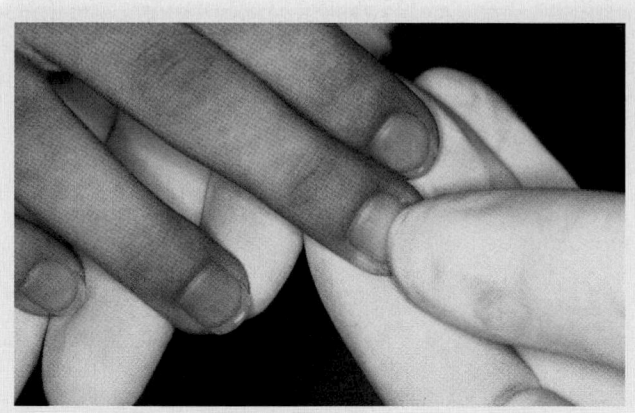

Figure 32-18 Estimate the capillary refill time by squeezing the end of a finger or toe for several seconds until the nailbed blanches. Normal color should return within 2 seconds after you let go.

than 6 years; however, factors such as cold temperatures may affect the capillary refill time.

Disability

The assessment of the pediatric patient's level of consciousness can be done using the AVPU scale or the Pediatric Glasgow Coma Scale **Table 32-6**.

Check the responses of each pupil to a direct beam of light. A normal pupil constricts after a light stimulus. Pupillary response may be abnormal in the presence of drugs, ongoing seizures, hypoxia, or brain injury. Note if the pupils are dilated, constricted, reactive, or fixed.

Next, look for symmetric movement of the extremities and note any neurologic motor deficit such as the inability to move the upper or lower extremities, an inability to communicate, weakness, or difficulty walking (gait).

Pain is present with most types of injury and many illnesses. Inadequate treatment of pain has many adverse effects on the pediatric patient and the family. Pain causes significant morbidity and misery for pediatric patients and caregivers and interferes with assessment.

Assessment of pain must take into consideration the developmental age of the patient. The ability to recognize pain will improve as patients become older. For example, crying and agitation in an infant may be the result of hunger or a dirty diaper. Meanwhile, a 3-year-old child can use words to say, "My tummy really hurts." In older children, pain scales using pictures of facial expressions (Wong-Baker FACES Scale) may be helpful in assessing the level of pain **Figure 32-19**.

Table 32-6   Pediatric Glasgow Coma Scale (GCS)

Activity	Score	Infant	Score	Child
Eye opening	4	Open spontaneously	4	Open spontaneously
	3	Open to speech or sound	3	Open to speech
	2	Open to painful stimuli	2	Open to painful stimuli
	1	No response	1	No response
Verbal	5	Coos, babbles	5	Oriented conversation
	4	Irritable cry	4	Confused conversation
	3	Cries to pain	3	Cries Inappropriate words
	2	Moans to pain	2	Moans Incomprehensible words/sounds
	1	No response	1	No response
Motor	6	Normal spontaneous movement	6	Obeys verbal commands
	5	Localizes pain	5	Localizes pain
	4	Withdraws to pain	4	Withdraws to pain
	3	Abnormal flexion (decorticate)	3	Abnormal flexion (decorticate)
	2	Abnormal extension (decerebrate)	2	Abnormal extension (decerebrate)
	1	No response (flaccid)	1	No response (flaccid)

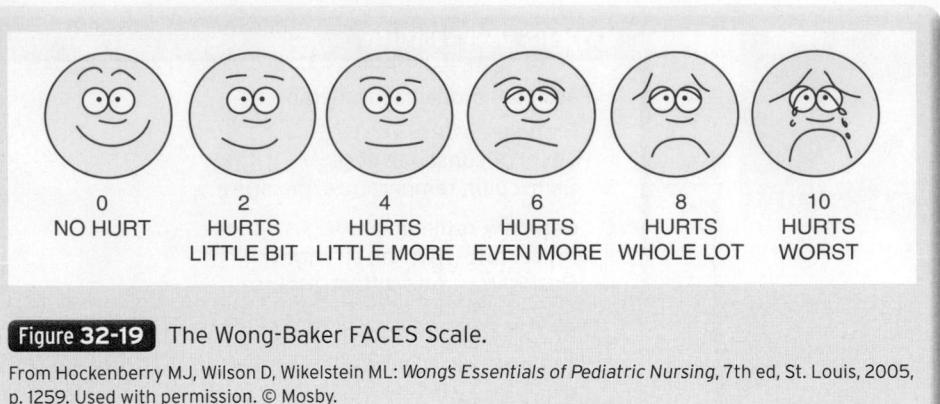

Figure 32-19 The Wong-Baker FACES Scale.

From Hockenberry MJ, Wilson D, Wikelstein ML: *Wong's Essentials of Pediatric Nursing*, 7th ed, St. Louis, 2005, p. 1259. Used with permission. © Mosby.

Exposure

Proper exposure of the pediatric patient is necessary for completing the hands-on ABCs. The PAT requires that the caregiver remove part of the pediatric patient's clothing to allow careful observation of the face, chest wall, and skin. Completing the components requires further exposure, as needed, to fully evaluate physiologic functions, anatomic abnormalities, and unsuspected injuries or rashes. Be careful to avoid heat loss, especially in infants, by covering the patient as soon as possible.

Transport Decision

After you have completed the primary assessment using hands-on ABCs and initiated any treatment, you must make a crucial decision: Is immediate transport to the hospital indicated? If the pediatric patient is in stable condition, you may elect to take a patient history and perform a secondary assessment at the scene.

However, immediate transport is indicated if the scene is unsafe for the pediatric patient or if any of the following conditions exist:

- A significant MOI—same MOIs as adults (Chapter 22, *Trauma Overview*), with the addition of:
  - Any fall from a height equal to or greater than a pediatric patient's height, especially with a head-first landing
  - Bicycle crash (when not wearing a helmet)
- A history compatible with a serious illness
- A physiologic abnormality noted during the primary assessment
- A potentially serious anatomic abnormality
- Significant pain
- Level of consciousness is not normal for the pediatric patient, altered mental status, and/or any signs or symptoms of shock

In addition to the preceding factors, you should also consider the following when making a transport decision:

- The type of clinical problem (injury versus illness)
- The expected benefits of ALS treatment in the field

- Local EMS system treatment and transport protocols
- The comfort level of the EMT
- Transport time to the hospital

If the pediatric patient's condition is urgent, then initiate immediate transport to the closest appropriate facility. Additional assessment and treatment should occur en route to the hospital.

If the pediatric patient's condition is nonurgent, obtain a history and perform a secondary assessment at the scene, transport, and provide additional treatment as needed.

Pediatric patients weighing less than 40 lb should be transported in a car seat as long as the situation allows. Many types of seats are available. A seat should be chosen to fit the appropriate weight of the pediatric patient and should meet the current applicable standards set by your governing agency. There are only a few locations to place a car seat in an ambulance. Seats are designed to be either forward-facing or rear-facing; they cannot be mounted sideways on a bench seat. Seats should not be mounted in the front of an ambulance, especially if the ambulance is equipped with air bags. To mount a car seat to the stretcher, place the head of the stretcher in an upright position. Place the seat so it is against the back of the stretcher. Secure one of the stretcher straps from the upper portion of the stretcher through the seatbelt positions on the seat and strap it tightly to the stretcher. Repeat on the lower portion of the stretcher. Push the seat into the stretcher tightly and retighten the straps.

To secure a seat to the captain's chair, follow the seat manufacturer's instructions. Remember that pediatric patients younger than 1 year must be transported in a rear-facing position because of the lack of mature neck muscles.

In some situations, it is not appropriate to secure a pediatric patient in a car seat, for example, if the pediatric patient has to be immobilized on a long board or requires splinting that does not fit in the seat. If the patient's condition is unstable and requires airway or ventilatory support, he or she should be positioned to maximize the airway and ventilatory requirements. Pediatric patients in cardiopulmonary arrest should likewise not be placed in a car seat.

You should use the pediatric patient's own car seat when feasible and when the pediatric patient is stable. The pediatric patient remains in the car seat, a cervical collar may be placed, and then the seat is properly secured to an EMS stretcher. Depending on the size of the patient, you may tape together additional towels and blankets to stabilize the head to the car seat. If the patient's position is unstable, the child should be removed immediately from the car seat and secured using a standard pediatric spinal board or an approved pediatric immobilizer.

History Taking

Investigate Chief Complaint

Your approach to the history will depend on the age of the pediatric patient. Historic information for an infant, toddler, or preschool-age child will need to be obtained from the parent or caregiver. When dealing with a young adolescent, you will usually be able to obtain most of the information from the patient.

Information about sexual activity, the possibility of pregnancy, or the use of illicit drugs or alcohol should be obtained from an adolescent patient in private. Most of these patients will be reluctant to provide this information in the presence of their parents or caregivers. When asking such questions, assure the adolescent that this information is important and is needed to provide the most appropriate care.

Questioning of the parent or child about the immediate illness or injury should be based on the child's chief complaint. Together with an evaluation of the child's medical history, this may provide clues to the underlying illness or injury and other conditions that may exist.

When interviewing the parent/caregiver or older child about the chief complaint, obtain the following pertinent information:

- Nature of the illness or mechanism of injury
- How long the pediatric patient has been sick or injured
- The key events leading to the injury or illness: Were there any witnesses to the injury? From what height did the pediatric patient fall? What surface did the pediatric patient land on (soft or hard)?
- Presence of fever
- Effects of the illness or injury on the pediatric patient's behavior
- Pediatric patient's activity level
- Recent eating, drinking, and urine output
- Change in bowel or bladder habits
- Presence of vomiting, diarrhea, abdominal pain
- Presence of rashes

If the parent or caregiver is unable to accompany you to the hospital, obtain a name and phone number so a staff person can call if there are questions. This might be the case when you respond to a daycare facility or babysitter's location. Most daycare facilities require emergency contact information, past medical history, and/or a list of current prescribed medications taken by the child in case of an emergency. Care may be delayed if this information is not discovered early.

SAMPLE History

Obtaining a SAMPLE history for a pediatric patient is the same as obtaining an adult's. However the questions should be based on the pediatric patient's age and developmental stage of life Table 32-7.

Table 32-7 Pediatric SAMPLE Components

Component	Explanation
Signs and symptoms	Onset and nature of symptoms of pain or fever Age-appropriate signs of distress
Allergies	Known drug reactions or other allergies
Medications	Exact names and doses of ongoing drugs (including over-the-counter, prescribed, herbal, and recreational drugs) Timing and amount of last dose Time and dose of analgesics or antipyretics
Past medical history	Previous illness or injuries Immunizations History of pregnancy, labor, delivery (infants and toddlers)
Last oral intake	Timing of the child's last food or drink, including bottle or breastfeeding
Events leading to illness or injury	Key events leading to the current incident Fever history

Secondary Assessment

The secondary assessment is a more detailed, comprehensive examination of the pediatric patient that is used to uncover illness or injuries that may have been missed during the primary assessment. In some instances, such as a critically ill or injured pediatric patient or a short transport time, you may not have time to conduct a secondary assessment.

Physical Examinations

A full-body scan should be used when pediatric patients have the potential for hidden illnesses or injuries; for example, unresponsive medical patients or trauma patients with a significant MOI. This examination may help to identify problems such as a distended abdomen or possible fractures.

The DCAP-BTLS mnemonic (Deformities, Contusions, Abrasions, Punctures/Penetrations, Burns, Tenderness, Lacerations, and Swelling) is a helpful way to remind you what to assess for on a pediatric patient involved in a traumatic event.

A focused assessment should be performed on pediatric patients without life-threatening illnesses or injuries who do not require a full-body scan. Focus your physical examination on the area(s) of the body affected by the illness or injury.

Infants, toddlers, and preschool-age children should be assessed starting at the feet and ending at the head; school-age children and adolescents can be assessed using the head-to-toe approach, as with adults. The extent of the physical examination will depend on the situation and may include the following:

- **Head.** The younger the infant or child, the larger the head is in proportion to the rest of the body, increasing the risk for head injury with deceleration (such as in motor vehicle crashes). Look for bruising, swelling, and hematomas. Significant blood can be lost between the skull and scalp of a small infant. Assessment of the fontanelle suggests elevated intracranial pressure caused by meningitis, encephalitis, or intracranial bleeding. A sunken fontanelle suggests dehydration.
- **Nose.** Young infants prefer to breathe through their nose, so nasal congestion with mucus can cause respiratory distress. Gentle bulb or catheter suction of the nostrils may bring relief.
- **Ears.** Look for any drainage from the ear canals. Leaking blood suggests a skull fracture. Check for

You are the Provider: PART 3

Your partner hands the child's mother a pediatric nonrebreathing mask with the oxygen flow rate set at 12 L/min, and he asks her to hold the mask near the child's face. Although the child becomes somewhat agitated by the oxygen, she is not pushing it away. After your partner talks to the child and explains what he is going to do, he assesses her vital signs. You gather additional information from the child's parents.

Recording Time: 5 Minutes	
Respirations	34 breaths/min; labored
Pulse	124 beats/min; strong and regular
Skin	Pink, warm, and dry; capillary refill time, 1 second
Blood pressure	86/56 mm Hg
Oxygen saturation (Sao$_2$)	95% (on oxygen)

The child's mother tells you that her daughter has had a cold for the past 2 days, and has slowly developed a low-grade fever and high-pitched cough, which she describes as a "barking seal" sound. She was going to take her to the doctor tomorrow, but called 9-1-1 when the child began experiencing trouble breathing. Further assessment of the child reveals that her breath sounds are clear and equal bilaterally and she has prominent intercostal retractions.

5. What is the most likely cause of this child's respiratory distress?

6. Should you separate this child from her parents to provide further treatment? Why or why not?

bruises behind the ear or Battle's sign, a late sign of skull fracture. The presence of pus may indicate an ear infection or perforation of the ear drum.

- **Mouth.** In the trauma patient, look for active bleeding and loose teeth. Note the smell of the breath. Some ingestions are associated with identifiable odors, such as hydrocarbons. Acidosis, as in diabetic ketoacidosis, may impart a sweet smell to the breath.

- **Neck.** Examine the trachea for swelling or bruising. Note if the pediatric patient cannot move his or her neck and has a high fever. This may indicate that the pediatric patient has bacterial or viral meningitis.

- **Chest.** Examine the chest for penetrating injuries, lacerations, bruises, or rashes. If the pediatric patient is injured, feel the clavicles and every rib for tenderness and/or deformity.

- **Back.** Inspect the back for lacerations, penetrating injuries, bruises, or rashes.

- **Abdomen.** Inspect the abdomen for distention. Gently palpate the abdomen and watch closely for guarding or tensing of the abdominal muscles, which may suggest infection, obstruction, or intra-abdominal injury. Note any tenderness or masses. Look for any seatbelt abrasions or bruising.

- **Extremities.** Assess for symmetry. Compare both sides for color, warmth, size of joints, swelling, and tenderness. Put each joint through full range of motion while watching the eyes of the pediatric patient for signs of pain, unless there is obvious deformity of the extremity suggesting a fracture.

Words of Wisdom

Because of the frequency of serious internal injuries in pediatric patients who show no external signs, it is especially important to investigate and thoroughly document the MOI. Do not let the rush at the scene distract you from determining the MOI or at least directing another reliable responder to do so. Hospital care providers need this information.

Vital Signs

Some of the guidelines used to assess adult circulatory status—heart rate and blood pressure—have important limitations in pediatric patients. First, normal heart rates vary with age in pediatric patients. Second, blood pressure is usually not assessed in pediatric patients younger than 3 years; it offers little information about the pediatric patient's circulatory status and is usually difficult to obtain. In these pediatric patients, assessment of the skin is a better indication of their circulatory status.

It is important to use appropriately sized equipment when you are assessing a pediatric patient's vital signs. To obtain an accurate reading of a pediatric patient's blood pressure, you must use a cuff that covers two thirds of the pediatric patient's upper arm. A blood pressure cuff that is too small may give you a falsely high reading, whereas a cuff that is too large may give you a falsely low reading. A useful tool to determine blood pressure in children 1 to 10 years of age (lower limits) is:

70 + (2 × child's age in years) = systolic blood pressure

Respiratory rates may be difficult to interpret. Rapid respiratory rates may simply reflect high fever, anxiety, pain, or excitement. Normal rates, on the other hand, may occur in a child who has been breathing rapidly with increased work of breathing for some time and is now becoming tired. Count the respirations for at least 30 seconds and then double that number (if counted for 30 seconds). In infants and children younger than 3 years, evaluate respirations by assessing the rise and fall of the abdomen. Assess the pulse rate by counting at least 1 minute, noting its quality and regularity.

Note that normal vital signs in pediatric patients vary with age **Table 32-8**. Remember that your approach to taking vital signs also varies with the age of the pediatric patient. Be gentle, talk to the pediatric patient, assess respirations and then pulse, and assess blood pressure last. Warm your stethoscope on your hands or a cloth before placing it on the skin. You may also want to let the pediatric patient hold the equipment first; this may help to reduce the pediatric patient's anxiety.

Evaluate pupils in the child using a small pen light. The response of pupils is a good indication of how well the brain is functioning, particularly when trauma has occurred. Be sure to compare the size of the pupils against each other.

Monitoring Devices

It is recommended that you always obtain the patient's first blood pressure reading manually with a sphygmomanometer (blood pressure cuff) and a stethoscope. In addition, a pulse oximeter is a valuable tool to measure the oxygen saturation in a pediatric patient with respiratory issues **Figure 32-20**.

Reassessment

Reassess the pediatric patient's condition as necessary—a general rule is to obtain vital signs every 15 minutes for a child in stable condition and at least every 5 minutes

Table 32-8 Vital Signs by Age

Age	Respirations (breaths/min)	Pulse (beats/min)	Systolic Blood Pressure (mm Hg)
Neonate: 0 to 1 month	30 to 60	90 to 180	50 to 70
Infant: 1 month to 1 year	25 to 50	100 to 160	70 to 95
Toddler: 1 to 3 years	20 to 30	90 to 150	80 to 100
Preschool age: 3 to 6 years	20 to 25	80 to 140	80 to 100
School age: 6 to 12 years	15 to 20	70 to 120	80 to 110
Adolescent: 12 to 18 years	12 to 20	60 to 100	90 to 110
Older than 18 years	12 to 20	60 to 100	90 to 140

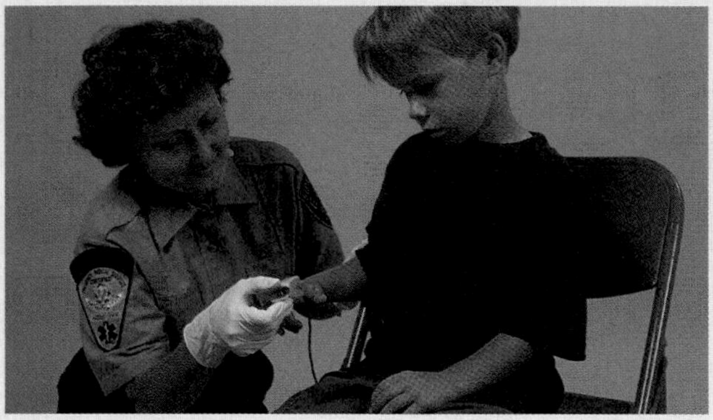

Figure 32-20 Pulse oximetry, which measures the pediatric patient's oxygen saturation, can be used to monitor the pediatric patient's status.

for a child in unstable condition. Remember that patient assessment is an ongoing process continuing until care is transferred to the receiving facility.

The physiologic safeguards in infants and children can decompensate with alarming unpredictability; therefore, continually monitor respiratory effort, skin color and condition, and level of consciousness or interactiveness. Frequently reassess vital signs and temperature. Repeat the primary assessment and adjust your treatment accordingly.

Interventions

When you are providing interventions for a pediatric patient, you should always consider getting help from the patient's parent or caregiver during these procedures. Parents are able to assist the EMS crew by calming and reassuring their child. They are also able to assist with administering oxygen or medications via a nebulizer. You should build a trusting environment and attempt to not frighten the pediatric patient who is already in a state of stress. Pediatric patients can sense fear from a provider and may not be willing to let you render care.

Communication and Documentation

Communicate with the hospital on your findings and the interventions you used to improve the pediatric patient's condition. Be sure that all of this information is documented and given to emergency department personnel. Remember, a patient care report is a legal document and you may be called to answer questions about this report for years to come.

Respiratory Emergencies and Management

According to a report by the Institute of Medicine in 2007, respiratory illnesses were among the top 10 reasons for emergency department visits in children younger than 17 years in the United States. Asthma is the most common cause of respiratory emergencies in children. Foreign bodies and trauma can also cause respiratory emergencies.

In the early stages of respiratory distress, you may note changes in the pediatric patient's behavior, such as combativeness, restlessness, and anxiety. As the body attempts to maximize the amount of air going into the lungs, the work of breathing increases. Signs and symptoms of increased work of breathing include:

- Nasal flaring, as the body tries to increase the size of the airway
- Grunting respirations, as the body attempts to keep the alveoli expanded at the end of expiration
- Wheezing, stridor, or other abnormal airway sounds

- Accessory (intercostal) muscle use; remember that in young children, the diaphragm is the major muscle of ventilation
- Retractions, or movements of the child's flexible rib cage
- The tripod position; in older children, this position will maximize the effectiveness of their airway **Figure 32-21**.

As the pediatric patient progresses to possible respiratory failure, efforts to breathe decrease; the chest rises less with inspiration. A definitive diagnosis of respiratory failure is made in the hospital. The body has used up its available energy stores and cannot continue to support the extra work of breathing under these conditions. At this point, cyanosis may develop (cyanosis is a late sign). Be aware that not all pediatric patients become cyanotic. You should be just as concerned about a pediatric patient with pale skin as one with bluish skin.

Changes in behavior will also occur until the pediatric patient demonstrates an altered level of consciousness. The pediatric patient may experience periods of apnea (absence of breathing). As the lack of oxygen becomes more serious, the heart muscle itself becomes hypoxic and slows down. This leads to bradycardia, a condition in which the heart rate is less than 80 beats/min in children or less than 100 beats/min in newborns. Bradycardia is almost always an ominous sign in pediatric patients. If the heart rate is fast, you need to investigate the cause. However, if the heart rate is slow (less than 60 beats/min) or absent, especially in an unconscious infant or child,

you must begin CPR immediately. Without aggressive airway management, bradycardia may quickly progress to cardiopulmonary arrest.

Of course, respiratory failure does not always indicate airway obstruction. It may indicate trauma, nervous system problems, dehydration (often caused by vomiting and diarrhea), or metabolic disturbances. For example, a pediatric patient with diabetes might have a blood glucose level that is too high or too low, or a pediatric patient might have a pH imbalance, as can happen with some rare pediatric diseases. Regardless of the cause, your first step is always to focus on ensuring adequate oxygenation and ventilation.

Never forget that a pediatric patient's condition can progress from respiratory distress to respiratory failure at any time. For this reason, you must reassess the pediatric patient frequently.

A child or infant in respiratory distress or possible respiratory failure needs supplemental oxygen. Anxiety, agitation, or crying may increase the effort or work of breathing, so use whichever method seems least upsetting to the pediatric patient—mask, blow-by, or nasal cannula. You may need to get creative by distracting the pediatric patient with games, a toy, or talking.

Allow the pediatric patient to remain in a comfortable position. For a small child, this may mean sitting on the caregiver's lap. Give nothing by mouth, in case the patient's condition deteriorates suddenly. If the patient's condition progresses to respiratory failure, you must begin assisted ventilation immediately and continue to provide supplemental oxygen.

■ Airway Obstruction

Children, especially those younger than 5 years, can (and do) obstruct their airway with any object that they can fit into their mouth: hot dogs, balloons, grapes, or coins **Figure 32-22**. In cases of trauma, a child's teeth may have been dislodged into the airway. Blood, vomitus, or other secretions can also cause mild or severe airway obstruction.

Airway obstructions can also be caused by infections, including pneumonia, croup, epiglottitis, and bacterial tracheitis **Figure 32-23**. Croup (laryngotracheobronchitis) is an infection of the airway below the level of the vocal cords, usually caused by a virus. Epiglottitis

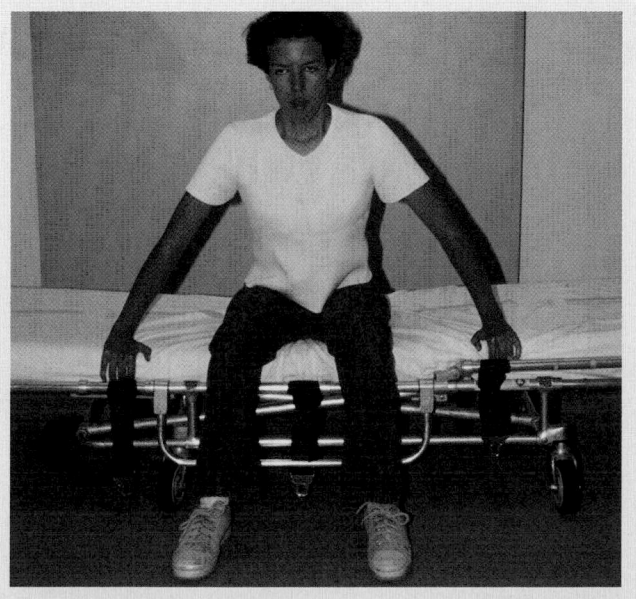

Figure 32-21 A patient in the tripod position will sit leaning forward on outstretched arms with the head and chin thrust slightly forward.

Special Populations

Pediatric patients who have a tracheostomy tube to assist in breathing are at risk of tracheostomy dysfunction. This is an airway obstruction that results from an accumulation of thick mucus at the opening of the tracheostomy tube. These pediatric patients require urgent care and transport.

Figure 32-22 Any number of objects can obstruct a child's airway. Some of the more common ones include batteries, coins, toys, buttons, and candy.

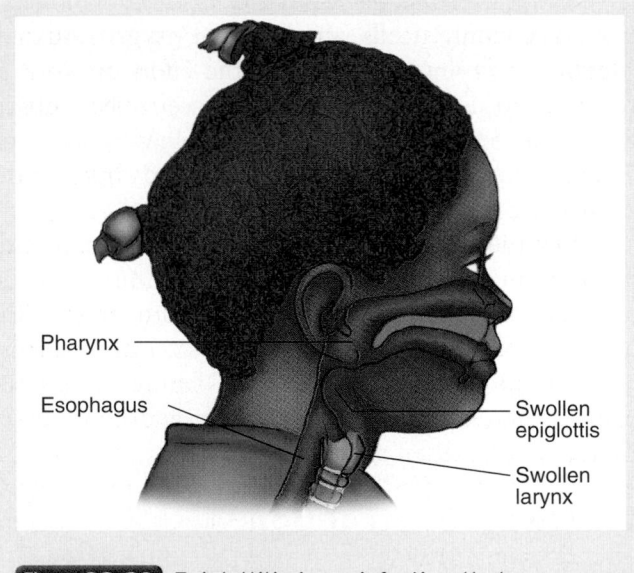

Pharynx

Esophagus

Swollen epiglottis

Swollen larynx

Figure 32-23 Epiglottitis is an infection that can cause airway obstruction in pediatric patients.

air through liquid, present in the air pouches and smaller airways in the lungs. They produce a crackling sound like that of blowing bubbles through a straw in a glass filled with liquid. The best way to auscultate breath sounds in a pediatric patient is to listen on both sides of the chest at the level of the armpit **Figure 32-24**.

Treatment of the pediatric patient with an airway obstruction must begin immediately. If the patient is conscious and coughing forcefully and you know for sure that there is a foreign body in the airway—that is, if someone actually saw the object go into the child's mouth—encourage the child to cough to clear the airway. If the material in the airway does not completely block the flow of air, the pediatric patient may be able to breathe adequately on his or her own without any intervention. In such cases, do not intervene except to provide supplemental oxygen **Figure 32-25**. Allow the

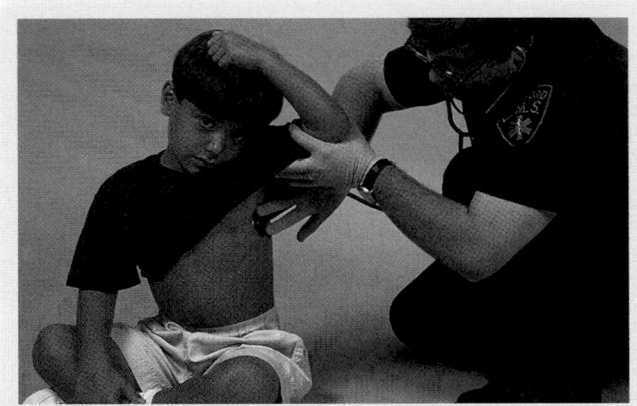

Figure 32-24 The best way to auscultate breath sounds in pediatric patients is to listen on both sides of the chest at the level of the armpit.

(supraglottitis) is an infection of the soft tissue in the area above the vocal cords. Infection should be considered as a possible cause of airway obstruction if a pediatric patient has congestion, fever, drooling, and cold symptoms.

Obstruction by a foreign object may involve the upper or the lower airway. Signs and symptoms that are frequently associated with an upper airway obstruction include decreased or absent breath sounds and stridor. Stridor, a high-pitched noise heard mainly on inspiration, is usually caused by swelling of the area surrounding the vocal cords or upper airway obstruction. In pediatric patients with croup, it resembles the bark of a seal.

Signs and symptoms of a lower airway obstruction include wheezing, a whistling sound caused by air traveling through narrowed air passages within the bronchioles, and/or crackles. Crackles are caused by the flow of

Figure 32-25 If a pediatric patient has a partial airway obstruction, do not intervene except to give supplemental oxygen and allow the child to remain in whatever position is most comfortable.

pediatric patient to remain in whatever position is most comfortable, and monitor his or her condition.

If you see signs of a severe airway obstruction, however, you must attempt to clear the airway immediately. The signs include the following:

- Ineffective cough (no sound)
- Inability to speak or cry
- Increasing respiratory difficulty, with stridor
- Cyanosis
- Loss of consciousness

If there is reason to believe that an unconscious child has a foreign body obstruction and there are no suspected spinal injuries, open the airway using the head tilt–chin lift maneuver and look inside the mouth to see whether the obstructing object is visible **Figure 32-26**. If the object is visible, try to remove it using a finger sweep motion. Never use finger sweeps if you cannot see the object because you may push it further into the airway.

Chest compressions are recommended to relieve a severe airway obstruction in an unconscious pediatric patient. Chest compressions increase the pressure in the chest, creating an artificial cough that may force a foreign body from the airway. Chapter 11, *BLS Resuscitation*, covers clearing a foreign body obstruction in an infant and child in detail.

■ Asthma

Asthma is an acute spasm of the smaller air passages, called bronchioles, associated with excessive mucous production and with swelling of the mucous lining of the respiratory passages. Asthma is one of the most common illnesses seen by EMS providers and is a true medical emergency if not promptly identified and treated. Almost 5 million children in the United States are affected by asthma. The emergency

department admission rate for children with asthma is more than twice the national average for all ages and the mortality rate is rising. Common causes for an asthma attack include upper respiratory infection and exercise. Exposure to cold air, emotional stress, and passive exposure to smoke may trigger attacks as well. Asthma is rare in children younger than 1 year.

Asthma produces a characteristic wheezing as patients attempt to exhale through partially obstructed lower air passages. These same air passages open easily during inspiration. This wheezing may be so loud that you can hear it without a stethoscope. In other cases, the airways are so blocked that no air movement is heard. In severe cases, the actual work of exhaling is very tiring, and cyanosis and/or respiratory arrest may quickly develop. Asthma patients in respiratory distress will typically assume a position of comfort to allow for maximum respiratory effort, such as the tripod position.

If possible, allow the pediatric patient to assume a position of comfort in a parent or caregiver's lap. Be cautious not to overexcite the pediatric patient because this may worsen the condition. Administer supplemental oxygen via a route that is tolerated by the child. Allow the parent or caregiver to assist the team by gathering any medications, calming the pediatric patient, or holding blow-by oxygen or a nonrebreather mask.

A bronchodilator (albuterol, a beta-2 agonist) via a metered-dose inhaler (MDI) with a spacer-mask device may be administered based on local agency protocols. Often the parents or caregivers have attempted multiple dosages of albuterol via the MDI or nebulizer. In this case, ALS providers should be dispatched immediately to meet you en route for additional medication administration and advanced care.

If you must assist ventilations in a pediatric patient who is having an asthma attack, use slow, gentle breaths. Remember, the problem in asthma is getting the air out of the lungs, not into them. Resist the temptation to squeeze the reservoir bag hard and fast.

A prolonged asthma attack that is unrelieved may progress into a condition known as *status asthmaticus*. The pediatric patient is likely to be frightened, frantically trying to breathe, while using all the accessory muscles. Status asthmaticus is a true emergency, and the pediatric patient must be given oxygen and transported immediately to the emergency department.

The effort to breathe during an asthma attack is very tiring, and the pediatric patient may be exhausted by the time you arrive. An exhausted pediatric patient may have stopped feeling anxious or even struggling to breathe. This patient is not recovering; he or she is at a very critical stage and is likely to stop breathing. Aggressive airway management, oxygen administration, and prompt

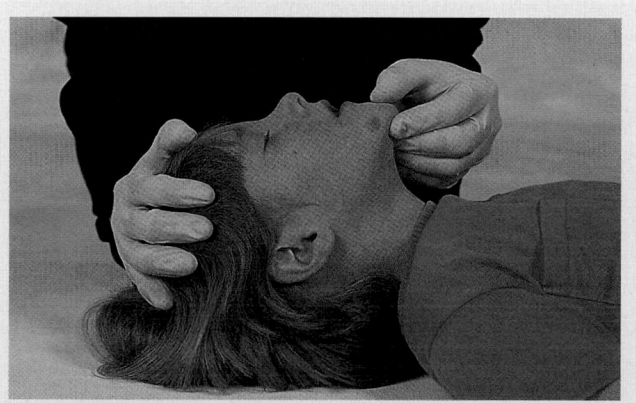

Figure 32-26 Open the airway and look inside the mouth of an unconscious pediatric patient with a possible airway obstruction.

transport are essential in this situation. ALS support should be considered. Follow local protocol.

Pneumonia

According to the World Health Organization, pneumonia is the leading cause of death in children worldwide. Pneumonia is a general term that refers to an infection of the lungs. Pneumonia is often a secondary infection, meaning it begins after an upper respiratory tract infection such as a cold or sore throat. It can also occur from chemical injury after an accidental ingestion, or a direct lung injury from a near drowning. Children with diseases causing immunodeficiency also increase the predisposition for pneumonia. You will notice that the incidence of this type of virus is greatest during the fall and winter months, affecting a large number of the pediatric population.

Often pediatric patients will present with unusually rapid breathing, or will breathe with grunting or wheezing sounds. Additional signs and symptoms include nasal flaring, tachypnea, crackles, and hypothermia or fever. The patient may also exhibit unilateral diminished breath sounds. Assess the work of breathing by observing for signs of accessory muscle usage. Pneumonia in the infant population may not be tolerated as well as in the older child or adult populations because infants have an increased oxygen demand and less reserve amounts.

For a pediatric patient with suspected pneumonia, your primary treatment will be supportive, consisting of monitoring the patient's airway and breathing status, and administering supplemental oxygen if required. A diagnosis of pneumonia must be confirmed in the hospital setting with a chest x-ray, followed by the administration of antibiotics.

Bronchiolitis

Bronchiolitis is a specific viral illness of newborns and toddlers, often caused by respiratory syncytial virus (RSV) that causes inflammation of the bronchioles. RSV is highly contagious and spread through droplets when the pediatric patient coughs or sneezes. RSV is more common in premature infants and results in copious secretions that may require suctioning. The virus can also survive on surfaces, including hands and clothing. The infection tends to spread rapidly through schools and in child care centers.

Bronchiolitis occurs during the first 2 years of life and is more common in males. These infections are most widespread in the winter and early spring. Bronchioles, the tiny airways that lead to alveoli in the lungs, become inflamed, swell, and fill with mucus. The airways of infants and young children can become easily blocked.

When assessing a pediatric patient, look for signs of dehydration—infants with RSV often refuse liquids.

You are the Provider: PART 4

The child's father carries her to the ambulance, where you appropriately secure her to the stretcher. You reassess her condition prior to transport and note that it has changed. On appearance, you note that her retractions have markedly weakened, she has a blank stare, and is listless. Your partner quickly secures the child's mother in the front seat of the ambulance and begins immediate transport.

Recording Time: 10 Minutes	
Level of consciousness	Decreased activity; blank stare; listless
Respirations	18 breaths/min; weak retractions
Pulse	90 beats/min; weak and regular
Skin	Perioral cyanosis; cool and dry; capillary refill time, 3 seconds
Blood pressure	76/56 mm Hg
Sao$_2$	85% (on oxygen)

7. How has this child's condition changed? What should you do next?

8. How could an ALS ambulance and crew benefit this child?

If the RSV has progressed to bronchiolitis, shortness of breath and fever may be present.

Approach the pediatric patient with a calm demeanor and allow for a position of comfort. Treat airway and breathing problems as appropriate. Humidified oxygen is helpful if available. Consider ALS backup for a higher level of care and transport to the appropriate hospital.

■ Airway Adjuncts

In children with inadequate ventilation, you should use an airway adjunct to maintain an open airway. Airway adjuncts are devices that help to maintain the airway or assist in providing artificial ventilation, including oral and nasal airways, bite blocks, and bag-mask devices. Placing the adjuncts correctly starts with choosing the appropriately sized equipment Table 32-9.

Words of Wisdom

Pertussis, also known as whooping cough, is a disease caused by a bacterium that is spread through respiratory droplets. As the result of vaccinations, this potentially deadly disease is less common in the United States. The typical signs and symptoms are similar to a common cold: coughing, sneezing, and a runny nose. As the disease progresses, the coughing becomes more severe and is characterized by the distinctive whoop sound heard during the inspiratory phase. To treat these pediatric patients, keep the airway patent and transport. Because pertussis is a communicable disease, practice standard precautions, including wearing a mask and eye protection.

Oropharyngeal Airway

An oropharyngeal airway is designed to keep the tongue from blocking the airway, and it makes suctioning the airway, if necessary, easier. An oropharyngeal airway should be used for pediatric patients who are unconscious and in possible respiratory failure. This adjunct should not be used in either conscious pediatric patients or those who have a gag reflex. Pediatric patients with a gag reflex do not tolerate an oropharyngeal airway. In addition, this adjunct should not be used in children who may have ingested a caustic or petroleum-based product because it may induce vomiting.

Skill Drill 32-2 shows the steps for inserting an oropharyngeal airway in a child:

1. Determine the appropriately sized airway by placing the airway next to the face with the flange

Table 32-9 Pediatric Equipment: Getting the Size Right

The best way to identify the appropriately sized equipment for a pediatric patient is to use the **pediatric resuscitation tape measure** (Broselow tape®), which can determine weight as well as height in pediatric patients weighing up to 75 lb (34 kg) Figure 32-27. The proper sequence for using the tape is the following:

1. Place the pediatric patient supine on a flat surface.
2. Lay the tape next to the pediatric patient with the multicolored side up.
3. Place the red end of the tape at the top of the pediatric patient's head (Red to Head).
4. Place one hand with its side down on top of the pediatric patient's head, covering the red box at the end of the tape.
5. Starting from the pediatric patient's head, run the side of your free hand down the tape.
6. Stretch the tape out the full length of the child, stopping at the heel. If the child is longer than the tape, stop here and use the appropriate adult technique.
7. Place your free hand, side down, at the bottom of the child's heel.
8. Note the color or letter block and weight range on the edge of the tape where your hand is. Say the color or letter out loud.
9. Select the appropriately sized equipment by matching the color or letter on the tape to the color or letter on the equipment.

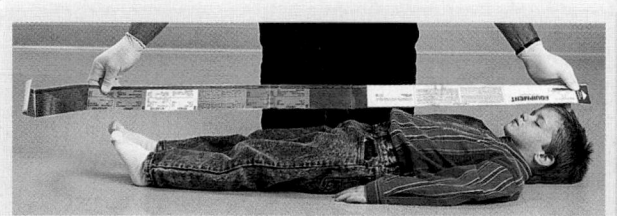

Figure 32-27 Use of a pediatric resuscitation tape measure is one way to identify the correct size for airway adjuncts.

at the level of the central incisors and the bite block segment parallel to the hard palate. The tip of the airway should reach the angle of the jaw Step 1. Or, use the length-based pediatric resuscitation tape to determine the appropriately sized airway.

2. Position the pediatric patient's airway. If the emergency is medical, use the head tilt–chin lift technique, avoiding hyperextension; you may place a towel under the pediatric patient's shoulders. If the pediatric patient has a traumatic injury, use the jaw-thrust maneuver and provide in-line spinal stabilization Step 2.

3. Open the mouth by applying pressure on the chin with your thumb.

4. Insert the airway by depressing the tongue with a tongue blade applied to the base of the tongue and inserting the airway directly over the tongue blade (Step 3). If a tongue blade is not available, point the airway tip toward the roof of the mouth to depress the tongue. Gently rotate the airway into position as it passes through the mouth toward the curve of the tongue. Insert the airway until the flange rests against the lips.

5. Reassess the airway after insertion. Take care to avoid injuring the hard palate as you insert the airway. Rough insertion can cause bleeding, which can aggravate airway problems and may even cause vomiting. Note also that if the pediatric patient's airway is too small, the tongue may be pushed back into the pharynx, obstructing the airway. If the airway is too large, it may obstruct the larynx.

Nasopharyngeal Airway

A nasopharyngeal airway is usually well tolerated and is not as likely as the oropharyngeal airway to

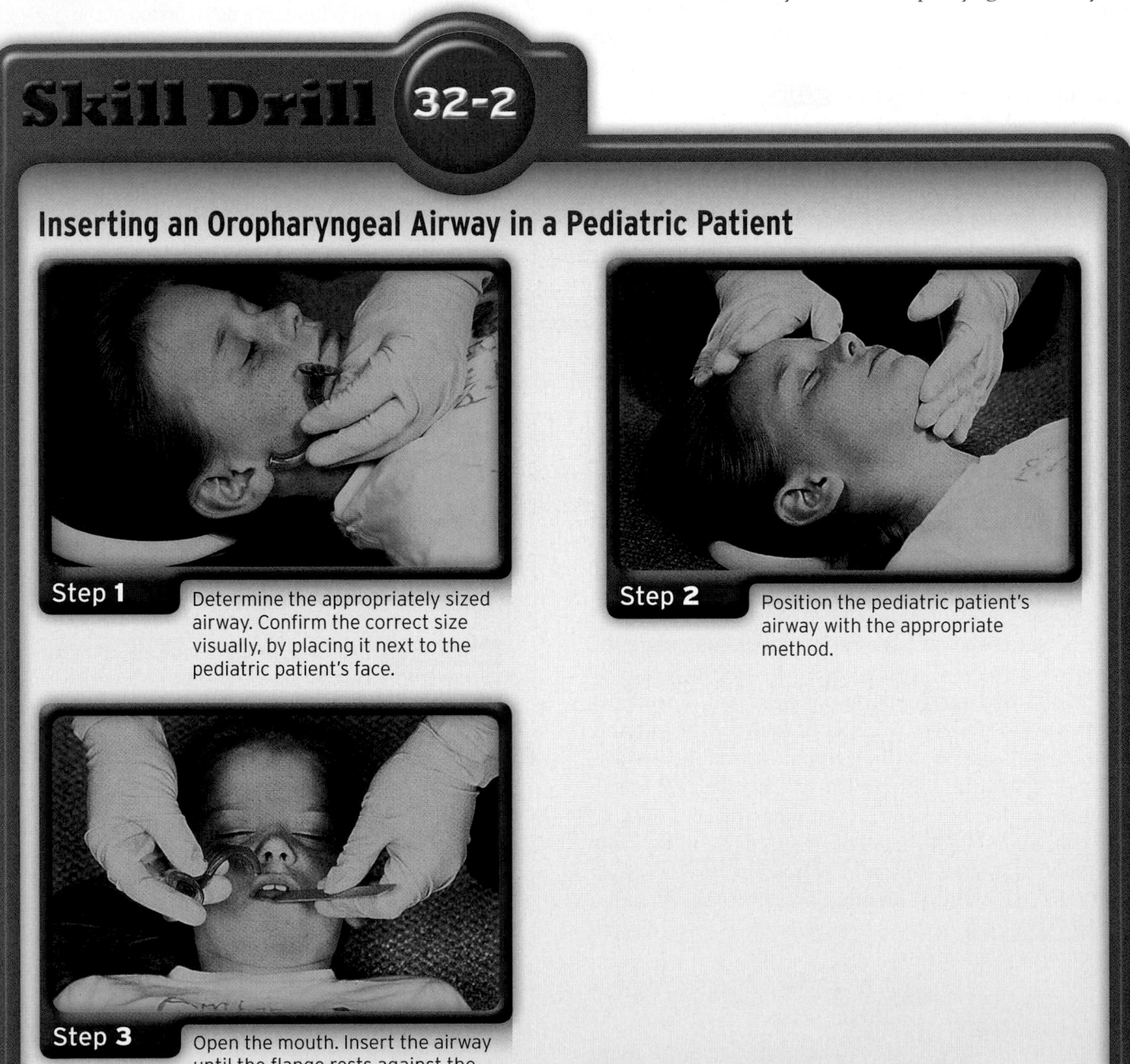

Skill Drill 32-2

Inserting an Oropharyngeal Airway in a Pediatric Patient

Step 1 Determine the appropriately sized airway. Confirm the correct size visually, by placing it next to the pediatric patient's face.

Step 2 Position the pediatric patient's airway with the appropriate method.

Step 3 Open the mouth. Insert the airway until the flange rests against the lips. Reassess the airway.

cause vomiting. Unlike the oropharyngeal airway, the nasopharyngeal airway is used for conscious pediatric patients or for pediatric patients with altered levels of consciousness. In pediatric patients, the nasopharyngeal airway is typically used in association with possible respiratory failure. It is rarely used in infants younger than 1 year.

A nasopharyngeal airway should not be used in pediatric patients with nasal obstruction or head trauma (possible basilar skull fracture), or in pediatric patients with moderate to severe head trauma because this adjunct could increase intracranial pressure.

Follow the steps in Skill Drill 32-3 to insert a nasopharyngeal airway in a pediatric patient:

Skill Drill 32-3

1. Determine the appropriately sized airway. The external diameter of the airway should not be larger than the diameter of the nares, and there should be no **blanching** of the nares after insertion.

Skill Drill 32-3

Inserting a Nasopharyngeal Airway in a Pediatric Patient

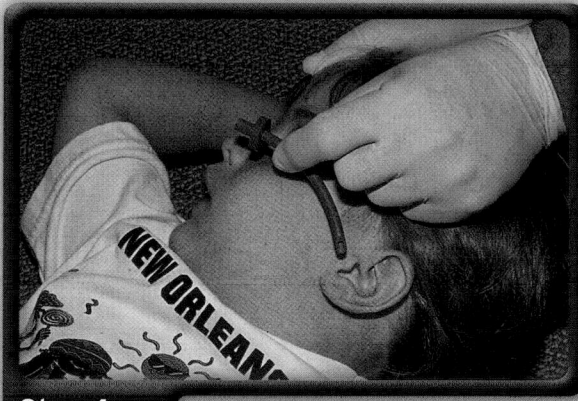

Step 1 Determine the correct airway size by comparing its diameter to the opening of the nostril (naris). Place the airway next to the pediatric patient's face to confirm correct length. Position the airway.

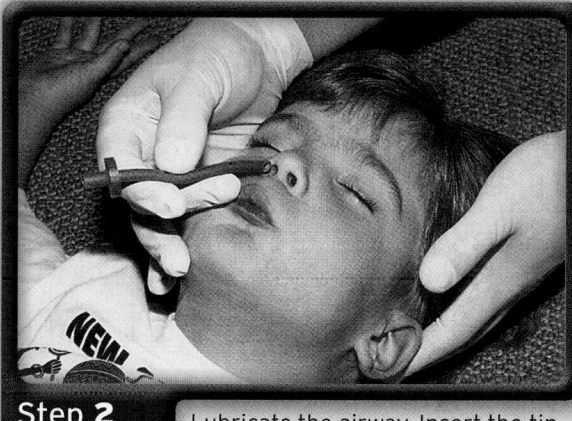

Step 2 Lubricate the airway. Insert the tip into the right naris with the bevel pointing toward the septum.

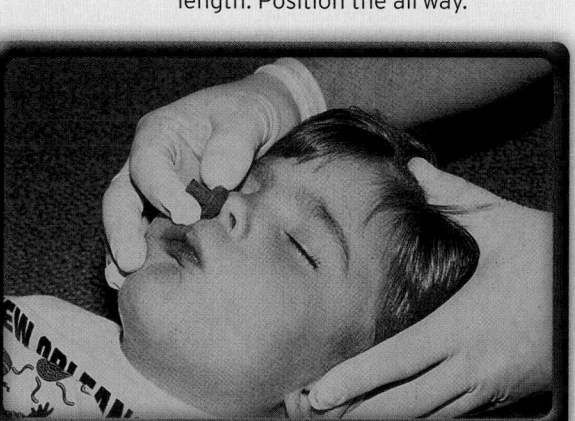

Step 3 Carefully move the tip forward until the flange rests against the outside of the nostril. Reassess the airway.

2. Place the airway next to the pediatric patient's face to make sure the length is correct. The airway should extend from the tip of the nose to the tragus of the ear. The tragus is the small cartilaginous projection in front of the opening of the ear.

3. Position the pediatric patient's airway, using the techniques described above for the oropharyngeal airway (Step 1).

4. Lubricate the airway with a water-soluble lubricant.

5. Insert the tip into the right naris (nostril opening) with the bevel pointing toward the septum, or central divider in the nose (Step 2). The right naris is commonly larger than the left naris in most patients.

6. Carefully move the tip forward, following the roof of the mouth, until the flange rests against the outside of the nostril (Step 3). If you are inserting the airway on the left side, insert the tip into the left naris upside down, with the bevel pointing toward the septum. Move the airway forward slowly about 1″ until you feel a slight resistance, and then rotate the airway 180°.

7. Reassess the airway after insertion.

As with the oropharyngeal airway, there can be problems with the nasopharyngeal airway. An airway with a small diameter may easily become obstructed by mucus, blood, vomitus, or the soft tissues of the pharynx. If the airway is too long, it may stimulate the vagus nerve and slow the heart rate or enter the esophagus, causing gastric distention. Inserting the airway in responsive patients may cause a spasm of the larynx and result in vomiting. Nasopharyngeal airways should not be used when pediatric patients have facial trauma because the airway may tear soft tissues and cause bleeding into the airway.

■ Oxygen Delivery Devices

In treating infants and children who require more than the usual 21% oxygen found in room air, you have several options:

- Blow-by technique at 6 L/min provides more than 21% oxygen concentration.
- Nasal cannula at 1 to 6 L/min provides 24% to 44% oxygen concentration.
- Nonrebreathing mask at 10 to 15 L/min provides up to 90% oxygen concentration (unassisted ventilations).
- Bag-mask device (with oxygen reservoir) at 15 L/min provides nearly 100% oxygen concentration (assisted ventilations).

Pediatric patients need enough air to be delivered for adequate gas exchange in the lungs. Therefore, use of a nonrebreathing mask, a nasal cannula, or a simple face mask is indicated only for pediatric patients who have adequate respirations and/or tidal volumes. The tidal volume is the amount of air that is delivered to the lungs and airways in one inhalation. Children with respirations of less than 12 breaths/min or more than 60 breaths/min, an altered level of consciousness, and/or an inadequate tidal volume should receive assisted ventilations with a bag-mask device.

Blow-by oxygen is not as effective as a face mask or nasal cannula for delivering oxygen. In the blow-by technique, an oxygen tube is held near the infant or child's nose and mouth. It is often used after childbirth to deliver a small amount of oxygen to the neonate. On rare occasions when other adjuncts cannot be used or the pediatric patient will not tolerate any other adjunct, this technique may be necessary. The blow-by technique does not provide a high concentration of oxygen but is better than no oxygen. To administer blow-by oxygen:

1. Place oxygen tubing through a small hole in the bottom of a 6- to 8-oz cup Figure 32-28. A cup is a familiar object that is less likely to frighten young children than an oxygen mask.

2. Connect tubing to an oxygen source set at 6 L/min.

3. Hold the cup approximately 1″ to 2″ away from the child's nose and mouth.

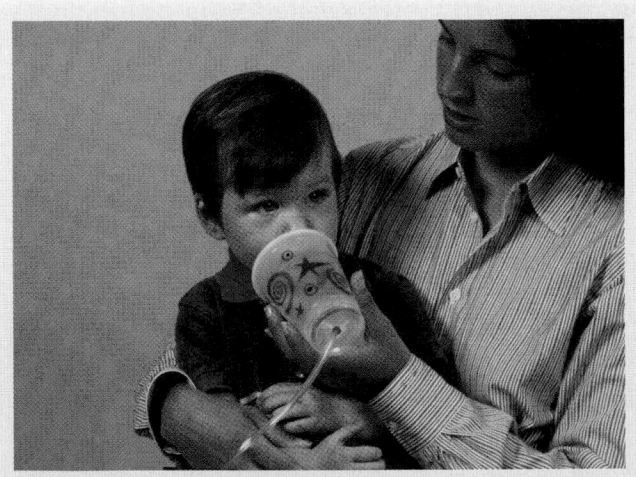

Figure 32-28 Blow-by techniques may be used when oxygen masks frighten children. Make a small hole in a 6- to 8-oz cup, or consider using a funnel inserted into the end of the oxygen tubing. Connect tubing to an oxygen source, and hold the cup about 1″ to 2″ from the child's face.

Nasal Cannula

Some pediatric patients prefer the nasal cannula whereas others find it uncomfortable. To apply a nasal cannula:

1. Choose the appropriately sized pediatric nasal cannula. The prongs should not fill the nares entirely **Figure 32-29**. If the nares blanch, select a smaller cannula.
2. Connect the tubing to an oxygen source set at 1 to 6 L/min.

Nonrebreathing Mask

A nonrebreathing mask delivers up to 90% oxygen to the pediatric patient and allows the pediatric patient to exhale all carbon dioxide without rebreathing it **Figure 32-30**. To apply a nonrebreathing mask:

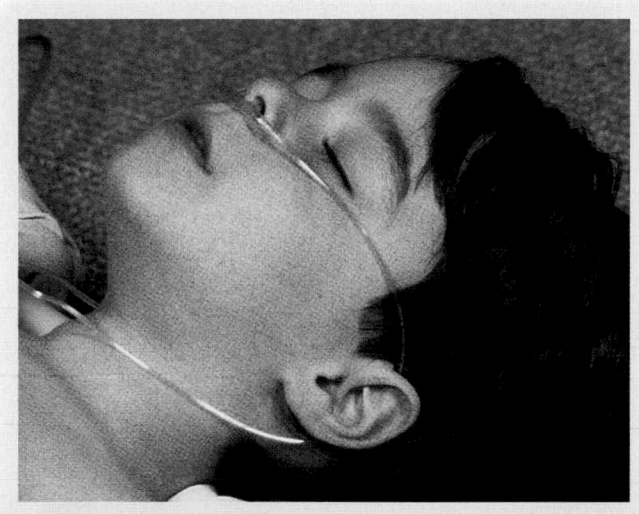

Figure 32-29 The prongs of a pediatric nasal cannula should not fill the nares entirely.

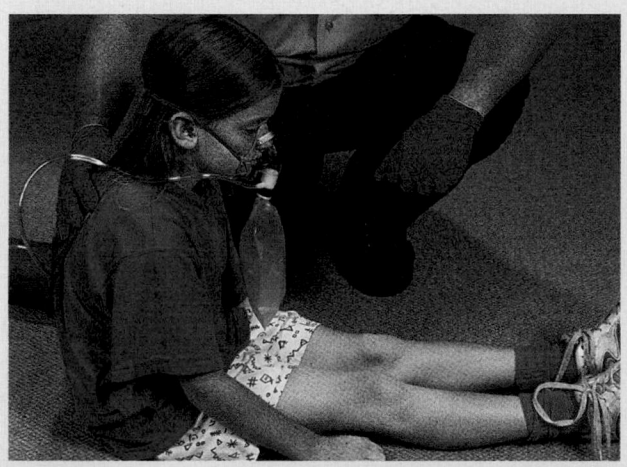

Figure 32-30 A pediatric nonrebreathing mask delivers up to 90% oxygen and allows the patient to exhale carbon dioxide without rebreathing it.

1. Select the appropriately sized pediatric nonrebreathing mask. The mask should extend from the bridge of the nose to the cleft of the chin.
2. Connect the tubing to an oxygen source set at 10 to 15 L/min.
3. Adjust oxygen flow as needed to match the pediatric patient's respiratory rate and depth. The reservoir bag should neither deflate completely nor fill to bulging during the respiratory cycle.

Bag-Mask Device

Assisting ventilations with a bag-mask device is indicated for pediatric patients who have respirations that are either too slow or too fast to provide an adequate volume of inhaled oxygen, who are unresponsive, or who do not respond in a purposeful way to painful stimuli.

Assist ventilation of an infant or child using a bag-mask device in the following way:

1. Ensure that you have the appropriate equipment in the right size. The proper size mask will extend from the bridge of the nose to the cleft of the chin, avoiding compression of the eyes **Figure 32-31**. The mask is transparent, so you can watch for cyanosis and vomiting. In addition, mask volume should be small to decrease dead space and avoid rebreathing; however, the bag should contain at least 450 mL of air. Use an infant bag, not a neonatal bag, for infants younger than 1 year; use a pediatric bag for children older than 1 year. Older children and adolescents may need an adult bag. Make sure that there is no pop-off valve on the bag; if the bag has a pop-off valve, make sure that you can hold it shut as necessary to achieve chest rise. Proper mask size for bag-mask device ventilation is critical.
2. Maintain a good seal with the mask on the face.

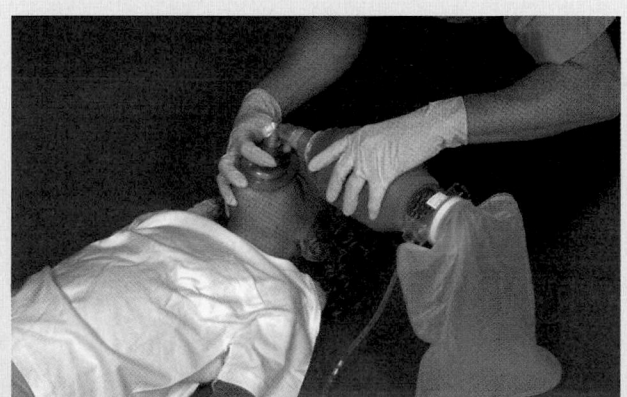

Figure 32-31 Proper mask size for bag-mask ventilation is critical. The mask should extend from the bridge of the nose to the cleft of the chin, avoiding compression of the eyes.

3. Ventilate at the appropriate rate and volume using a slow, gentle squeeze, not a sharp, quick one. Stop squeezing and begin to release the bag as soon as the chest wall begins to rise, indicating that the lungs are filled to capacity. To keep from ventilating too rapidly, use the phrase "squeeze, release, release." Say "squeeze" as you squeeze the bag; when you see the chest start to rise, release pressure on the bag and slowly say "release, release."

Special Populations

Errors in technique—providing too much volume with each breath, squeezing the bag too forcefully, or ventilating at a rate that is too fast—can result in gastric distention or overinflation of the lung resulting in a pneumothorax. An inadequate mask seal or improper head position can lead to hypoventilation or hypoxia.

Special Populations

One of the problems associated with abdominal injuries in children is the presence of air in the stomach. Pediatric patients, especially those who have had a traumatic injury, tend to swallow air. Air in the stomach can cause distention and interfere with your assessment. Air can also accumulate in the stomach with artificial ventilation, making it less effective.

One-Rescuer Bag-Mask Device Ventilation

Perform one-rescuer bag-mask device ventilation according to these steps Skill Drill 32-4 :

1. Open the airway, and insert the appropriate airway adjunct Step 1 .
2. Hold the mask on the pediatric patient's face with a one-handed head tilt–chin lift technique (E-C grip). Form a C with the thumb and index finger along the mask while the other three fingers form an E along the mandible. With infants and toddlers, support the jaw with only your third fingertip. Be careful not to compress the area under the chin because you may push the tongue into the back of the mouth and block the airway. Keep fingers on the mandible.
3. Make sure the mask forms an airtight seal on the face. Maintain the seal while checking that the airway is open Step 2 .
4. Squeeze the bag using the correct ventilation rate of 12 to 20 breaths/min.
5. Each ventilation (squeeze of the bag) should last 1 second Step 3 . Allow adequate time for exhalation.
6. Assess effectiveness of ventilation by watching for adequate bilateral rise and fall of the chest Step 4 .

You are the Provider: PART 5

The closest ALS ambulance is 8 miles across town in heavy traffic and your estimated transport time is about 5 minutes, so you make the decision to continue transport. Your partner notifies the hospital and informs them of your impending arrival. Following additional treatment, you reassess the child's condition.

Recording Time: 18 Minutes	
Level of consciousness	Eyes are open; still appears listless
Respirations	20 breaths/min and shallow (baseline); ventilations are being assisted
Pulse	110 beats/min; stronger
Skin	Cyanosis is resolving; capillary refill time, 2 seconds
Blood pressure	84/54 mm Hg
Sao$_2$	96% (on oxygen)

9. Are you providing adequate ventilations? How can you tell?
10. Why is it especially important to avoid hyperventilating infants and children?

Skill Drill 32-4

One-Rescuer Bag-Mask Device Ventilation on a Pediatric Patient

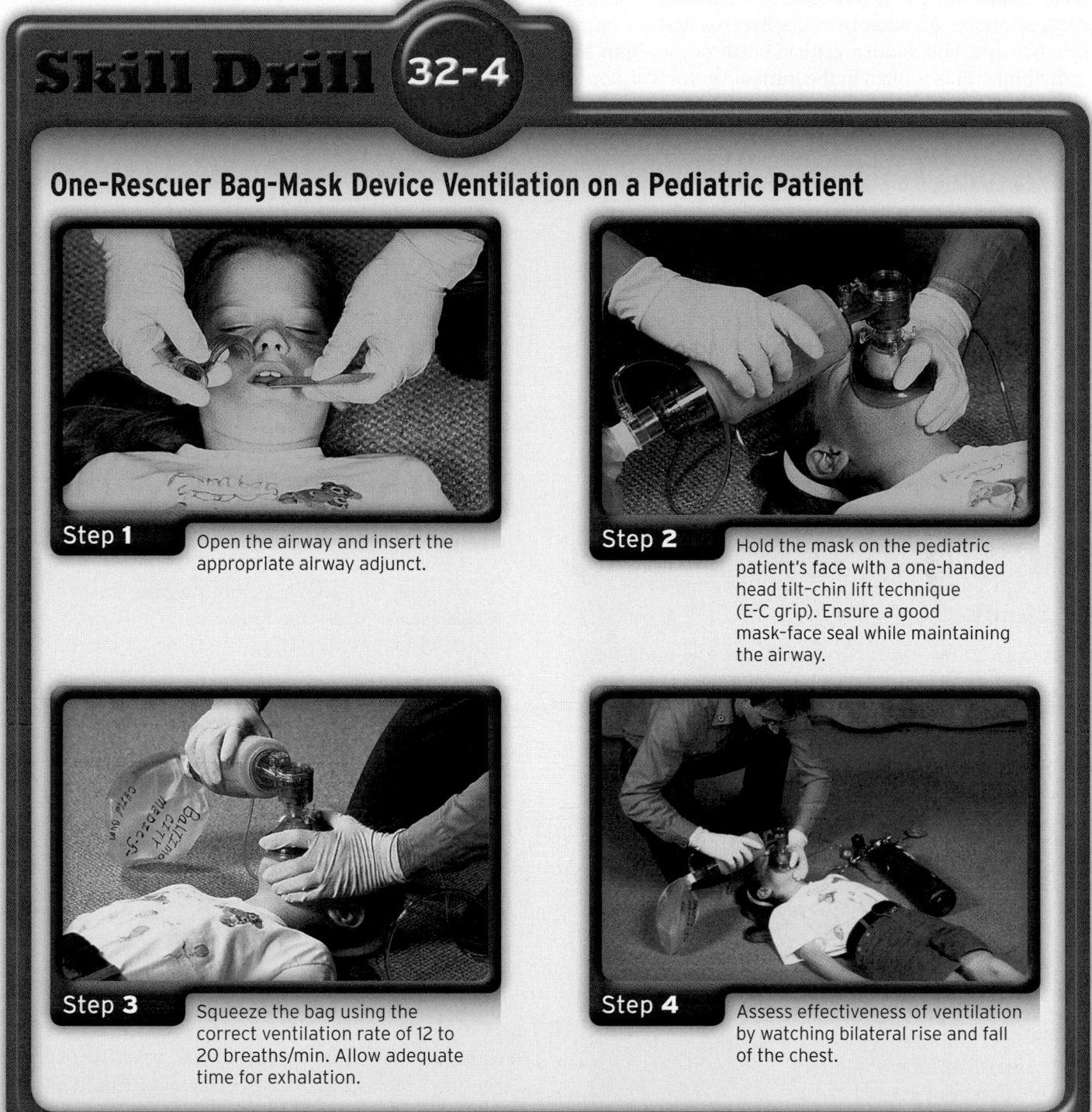

Step 1 Open the airway and insert the appropriate airway adjunct.

Step 2 Hold the mask on the pediatric patient's face with a one-handed head tilt-chin lift technique (E-C grip). Ensure a good mask-face seal while maintaining the airway.

Step 3 Squeeze the bag using the correct ventilation rate of 12 to 20 breaths/min. Allow adequate time for exhalation.

Step 4 Assess effectiveness of ventilation by watching bilateral rise and fall of the chest.

Two-Rescuer Bag-Mask Device Ventilation

This procedure is similar to one-rescuer ventilation except that it requires two rescuers—one to hold the mask to the pediatric patient's face and maintain the pediatric patient's head position, the other to ventilate the pediatric patient. This technique is usually more effective in maintaining a tight seal. When two rescuers are available, use your thumb and index finger to gently apply pressure over the area just below the Adam's apple (the

Sellick maneuver). This will decrease the risk of gastric distention and aspiration of vomitus by pushing the larynx back to compress and close off the esophagus.

■ Cardiopulmonary Arrest

As previously discussed, cardiac arrest in infants and children is most often associated with respiratory failure and respiratory arrest. Children are affected differently than adults when it comes to decreasing oxygen concentrations. An adult becomes hypoxic and the heart gets irritable, and sudden cardiac death occurs from an arrhythmia. This is often in the form of ventricular fibrillation, and is the reason that an automated external defibrillator (AED) is the treatment of choice. Children, on the other hand, become hypoxic and their hearts slow down, becoming more and more bradycardic. The heart will beat slower and become weaker with each beat until no pulse is felt. The survival rate from cardiac arrest in the prehospital setting is 3% to 5%. However, the survival rate from respiratory arrest is 75%. Therefore, a child who is breathing very poorly with a slowing heart rate must be ventilated with high concentrations of oxygen early to try to oxygenate the heart before cardiac arrest occurs. Chapter 11, *BLS Resuscitation*, covers providing CPR to pediatric patients in detail.

Circulation Emergencies and Management

Shock

As discussed in Chapter 10, *Shock*, shock is a condition that develops when the circulatory system is unable to deliver a sufficient amount of blood to the organs of the body. This results in organ failure and eventually cardiopulmonary arrest. The early stage of shock, while the body can still compensate for blood loss, is called compensated shock. The late stage, when blood pressure is falling, is called decompensated shock.

In pediatric patients, the most common causes include:

- Traumatic injury with blood loss (especially abdominal)
- Dehydration from diarrhea and vomiting
- Severe infection
- Neurologic injury, such as severe head trauma
- A severe allergic reaction/anaphylaxis to an allergen (insect bite or food allergy)
- Diseases of the heart
- A collapsed lung (pneumothorax)
- Blood or fluid around the heart (cardiac tamponade or pericarditis)

Infants and children have less blood circulating in their bodies than adults do, so the loss of even a small volume of fluid or blood may lead to shock. Pediatric patients also respond differently than adults to fluid loss. They may respond by increasing their heart rate, increasing respirations, and showing signs of pale or blue skin.

Greater than 25% blood volume loss significantly increases the risk of shock in children. Signs of shock in children are as follows:

- Tachycardia
- Poor capillary refill (> 2 seconds)
- Mental status changes

For comparison, signs of shock in adults are:

- Tachycardia
- Hypotension
- Mental status changes

Greater than 30% to 40% blood volume loss significantly increases risk of shock in adults.

Begin treating shock by assessing the ABCs, intervening immediately as required; do not wait until you have performed the complete assessment to take action. If cardiac arrest is suspected, the order becomes CAB because compressions are essential. Pediatric patients in shock often have increased respirations but do not demonstrate a fall in blood pressure until shock is severe.

In assessing circulation, you should pay particular attention to the following:

- **Pulse.** Assess both the rate and the quality of the pulse. A weak, "thready" pulse is a sign that there is a problem. The appropriate rate depends on age; anything over 160 beats/min suggests shock.
- **Skin signs.** Assess the temperature and moisture of the hands and feet. How does this compare with the temperature of the skin on the trunk of the body? Is the skin dry and warm, or cold and clammy?
- **Capillary refill time.** Squeeze a finger or toe for several seconds until the skin blanches, and then release it. Does the fingertip return to its normal color within 2 seconds, or is it delayed?
- **Color.** Assess the patient's skin color. Is it pink, pale, ashen, or blue?

Changes in pulse rate, color, skin signs, and capillary refill time are all important clues suggesting shock.

Blood pressure is the most difficult vital sign to measure in pediatric patients. The cuff must be the proper size—two thirds the length of the upper arm. The value for normal blood pressure is also age-specific. Remember that blood pressure may be normal with compensated shock. Low blood pressure is a sign of decompensated shock, requiring care from an ALS team and immediate transport.

Part of your assessment should also include talking with the parents or caregivers to determine when the signs and symptoms first appeared and whether any of the following has occurred:

- Decrease in urine output (with infants, are there fewer than 6 to 10 wet diapers?)
- Absence of tears, even when the child is crying
- A sunken or depressed fontanelle (infant patient)
- Changes in level of consciousness and behavior

Limit your management to these simple interventions. Time should not be wasted performing field procedures.

Ensure that the airway is open, prepare for artificial ventilation; control bleeding; and give supplemental oxygen by mask or blow-by method as tolerated. Continue to monitor airway and breathing. Place the pediatric patient in a position of comfort. Keep the pediatric patient warm with blankets and by turning up the heat in the patient compartment. Provide immediate transport to the nearest appropriate facility and continue monitoring vital signs en route. Contact ALS backup as needed. Allow a parent or caregiver to accompany the pediatric patient whenever possible.

Anaphylactic Shock

Anaphylaxis is a major allergic reaction that involves a generalized, multisystem response to an antigen (foreign substance). The airway and cardiovascular system are important sites of this potentially life-threatening reaction. Common causes are an insect sting or a food allergy.

A pediatric patient in anaphylactic shock will have hypoperfusion as well as additional signs such as stridor and/or wheezing, with increased work of breathing. The pediatric patient will also have an altered appearance with restlessness, agitation, and sometimes a sense of impending doom. Hives, an intensely itchy skin rash, are usually present.

Maintain the airway and administer oxygen via a route that is tolerated. If the pediatric patient is stable, allow the parent or caregiver to assist in the positioning of the patient, oxygen delivery, and to keep the pediatric patient calm. Increased agitation and crying, combined with an increased work of breathing, may lead to increased bronchoconstriction. Based on local protocol, assist the parent or caregiver with administering a prescribed epinephrine auto-injector, if available. Transport promptly.

■ Bleeding Disorders

Hemophilia is a congenital condition in which the patient lacks one or more of the normal clotting factors of blood. There are several forms of hemophilia, most of which are hereditary and some of which are severe. Hemophilia is predominantly found in the male population. Sometimes bleeding may occur spontaneously. Because the pediatric patient's blood does not clot, all injuries, no matter how minor, are potentially serious. A pediatric patient with hemophilia should be transported immediately.

Neurologic Emergencies and Management

■ Altered Mental Status

An altered mental status (AMS) is an abnormal neurologic state in which the pediatric patient is less alert and interactive than is age-appropriate. Sometimes, the concern of the parent or caregiver is vague, stating that the child is "not acting right." Understanding normal developmental or age-related changes in behavior, and listening carefully to the parent's or caregiver's opinion about a pediatric patient's normal behavior are key.

The mnemonic AEIOU-TIPPS reflects the major causes of AMS Table 32-10.

The signs and symptoms of AMS vary widely from simple confusion to coma. Management of AMS focuses on the ABCs and transport. If the pediatric patient's level of consciousness is low, then the pediatric patient may not be able to protect his or her airway. Ensure a patent airway and adequate breathing through a nonrebreathing mask or a bag-mask device. Pediatric patients with an altered level of consciousness may have inadequate breathing despite spontaneous respiratory effort, based on an inadequate respiratory rate or inadequate tidal volume. Transport to the hospital.

■ Seizures

A seizure is the result of disorganized electrical activity in the brain, causes of which are listed in Table 32-11. It can be very frightening to people around the pediatric patient. Therefore, it is important to reassure the family and to approach assessment and management in a calm, step-by-step manner.

Seizures in children may manifest in a wide variety of ways, depending on the age of the child. Seizures in

Table 32-10	AEIOU-TIPPS
A	Alcohol
E	Epilepsy, endocrine, electrolytes
I	Insulin
O	Opiates and other drugs
U	Uremia
T	Trauma, temperature
I	Infection
P	Psychogenic
P	Poison
S	Shock, stroke, space-occupying lesion, subarachnoid hemorrhage

Table 32-11	Common Causes of Seizures
■ Child abuse	
■ Electrolyte imbalance	
■ Fever	
■ Hypoglycemia (low blood glucose level)	
■ Infection	
■ Ingestion	
■ Lack of oxygen	
■ Medications	
■ Poisoning	
■ Seizure disorder	
■ Recreational drug use	
■ Head trauma	
■ Idiopathic (no cause can be found)	

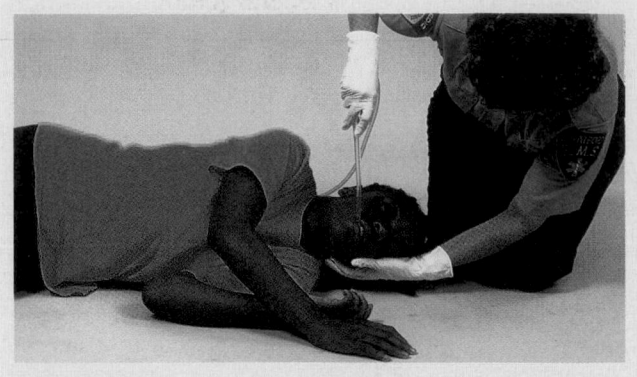

Figure 32-32 Position the head to open the airway and clear the airway with suction. If suction is inadequate or the patient is vomiting, consider placing the patient in the recovery position.

the longer it will take for this imbalance to correct itself. Likewise, longer and more severe seizures will result in longer postictal unresponsiveness and confusion. Once the pediatric patient regains a normal level of consciousness, the postictal state is over.

Seizures that continue every few minutes without regaining consciousness or last longer than 30 minutes are referred to as status epilepticus. Recurring or prolonged seizures should be considered potentially life-threatening situations in which pediatric patients need emergency medical care. If the pediatric patient does not regain consciousness or continues to seize, protect the pediatric patient from harming himself or herself and call for ALS backup. These pediatric patients need advanced airway management and medication to stop the seizure.

Securing and protecting the airway are your priorities. Position the head to open the airway. Clear the mouth with suction. Consider placing the pediatric patient in the recovery position if the pediatric patient is actively vomiting and suction is inadequate to control the airway **Figure 32-32**. Provide 100% oxygen by nonrebreathing mask or blow-by. If there are no signs of improvement, begin bag-mask device ventilation with appropriately sized equipment with supplemental oxygen. Transport the pediatric patient to the appropriate facility.

■ Febrile Seizures

Febrile seizures are common in children between the ages of 6 months and 6 years. Most pediatric seizures are the result of fever alone, which is why they are called febrile seizures.

These seizures typically occur on the first day of a febrile illness, are characterized by **generalized tonic-clonic seizure** activity, and last less than 15 minutes with

a short postictal phase or none at all. They may be a sign of a more serious problem, such as meningitis. Obtain a history from the parent or caregiver because these pediatric patients may have had a febrile seizure in the past.

If you are called to care for a pediatric patient who has had a febrile seizure, you often will find that the pediatric patient is awake, alert, and fully interactive when you arrive. Keep in mind that a persistent fever can lead to another seizure. Carefully assess the ABCs, begin cooling measures with tepid (not cold) water, and provide prompt transport. All pediatric patients with febrile seizures need to be seen in the hospital setting.

■ Meningitis

Meningitis is an inflammation of the tissue, called the meninges, that covers the spinal cord and brain. It is caused by an infection by bacteria, viruses, fungi, or parasites. If left untreated, meningitis can lead to permanent brain damage or death. Being able to recognize a pediatric patient who may have meningitis is an important skill for you to have.

Meningitis can occur in both children and adults, but some individuals are at greater risk than others, as follows:

- Males
- Newborn infants
- Geriatric population
- Compromised immune systems by AIDS or cancer
- People who have any history of brain, spinal cord, or back surgery
- Children who have had head trauma
- Children with shunts, pins, or other foreign bodies within their brain or spinal cord

At especially high risk are children with a ventriculo-peritoneal (VP) shunt. VP shunts drain excess fluids from around the brain into the abdomen. These special needs children have tubing that can usually be seen and felt just under the scalp.

The signs and symptoms of meningitis vary, depending on the age of the patient. Fever and altered level of consciousness are common symptoms of meningitis in patients of all ages. Changes in the level of consciousness can range from a mild or severe headache to confusion, lethargy, and/or an inability to understand commands or interact appropriately. The child may also experience a seizure, which may be the first sign of meningitis. Infants younger than 2 to 3 months can have apnea, cyanosis, fever, a distinct high-pitched cry, or hypothermia.

In describing children with meningitis, physicians often use the term "meningeal irritation" or "meningeal signs" to describe pain that accompanies movement. Bending the neck forward or back increases the tension within the spinal canal and stretches the meninges, causing a great deal of pain. This results in the characteristic stiff neck of children with meningitis, who will often refuse to move their neck, lift their legs, or curl into a "C" position, even if coached to do so. One sign of meningitis in an infant is increasing irritability, especially when being handled. Another sign is a bulging fontanel without crying.

One form of meningitis deserves special attention. **Neisseria meningitidis** is a bacterium that causes a rapid onset of meningitis symptoms, often leading to shock and death. Children with *N meningitidis* typically have small, pinpoint, cherry-red spots or a larger purple/black rash **Figure 32-33**. This rash may be on part of the face or body. These children are at serious risk of sepsis, shock, and death.

All pediatric patients with possible meningitis should be considered highly contagious and infectious. Therefore, you should use standard precautions whenever you suspect meningitis and follow up with the hospital to learn the patient's final diagnosis. If you have been exposed to saliva and respiratory secretions from a child with *N meningitidis*, you should receive antibiotics to protect yourself and others from the bacteria. This is particularly true if you managed the pediatric patient's airway. If you were not in close contact with the pediatric patient or his or her respiratory secretions, you do not need treatment.

Provide these pediatric patients with supplemental oxygen and assist with ventilations if needed. Reassess the pediatric patient's vital signs frequently as you transport the patient to the highest level of service available.

Gastrointestinal Emergencies and Management

Complaints of gastrointestinal origin are very common in the pediatric population. A common source of gastrointestinal upset is the ingestion of certain foods or unknown substances, such as milk or ice cream (lactose intolerance). In most cases, you will be faced with a pediatric patient who is experiencing abdominal discomfort with nausea, vomiting, and/or diarrhea. This can become a concern because both vomiting and diarrhea can cause dehydration in children.

Appendicitis is also common in pediatric patients and if untreated can lead to peritonitis (inflammation of the peritoneum, which lines the abdominal cavity) or shock. Appendicitis will typically present with a fever and pain on palpation of the right lower abdominal quadrant. Rebound tenderness is a common sign associated with appendicitis. Remember that constipation also can be a cause of abdominal pain in children. If you suspect appendicitis, immediately transport the pediatric patient to the hospital for further evaluation.

Because the pediatric population are very sensitive to fluid loss, obtain a thorough history from the primary caregiver. In particular, ask questions such as:

- How many wet diapers has the child had today?
- Is your child tolerating liquids and is he or she able to keep them down?
- How many times has your child had diarrhea and for how long?
- When he or she cries are there tears present?

These questions can help to determine just how dehydrated the pediatric patient may be. If the pediatric patient is dehydrated, transport to the hospital for further care.

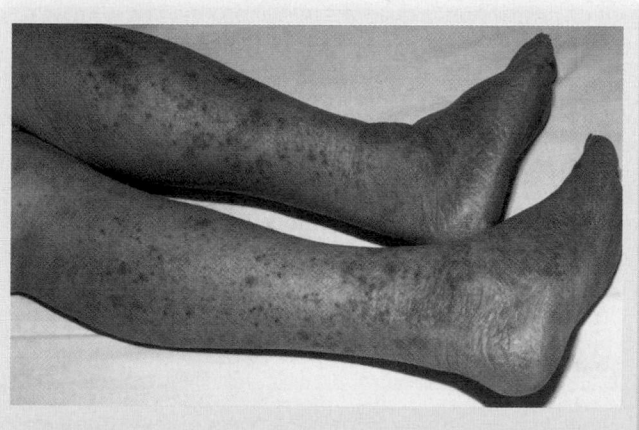

Figure 32-33 Children with *Neisseria meningitidis* typically have small, pinpoint, cherry-red spots or a larger purple/black rash.

Poisoning Emergencies and Management

Poisoning is common among children, and the common sources of poisoning in children are listed in Table 32-12 . It can occur by ingesting, inhaling, injecting, or absorbing a toxic substance.

The signs and symptoms of poisoning vary widely, depending on the substance and the age and weight of the pediatric patient. The pediatric patient may appear normal at first, even in serious cases, or he or she may be confused, sleepy, or unconscious.

Infants may be poisoned as a result of being fed a harmful substance by a sibling, parent, or caregiver, or as a result of child abuse. Infants can be exposed to drugs and poisons left on floors and carpeting. They can also be exposed in a room or automobile in which harmful drugs, such as crack, cocaine, or PCP, are being smoked. Toddlers are curious and often ingest poisons when they find them in the home or garage Figure 32-34 . For example, some people store petroleum products in soda bottles. Toddlers may believe the substance to be soda. Adolescents are more likely to have ingested alcohol and street drugs while partying or during a suicide attempt.

After you have completed your primary assessment, ask the parent or caregiver the following questions:

- What is the substance(s) involved?
- Approximately how much of the substance was ingested or involved in the exposure (eg, number of pills, amount of liquid)?
- What time did the incident occur?
- Are there any changes in behavior or level of consciousness?
- Was there any choking or coughing after the exposure? (These can be signs of airway involvement.)

To treat a pediatric patient exposed to a poisonous substance, first perform an external decontamination. Remove tablets or fragments from the patient's mouth,

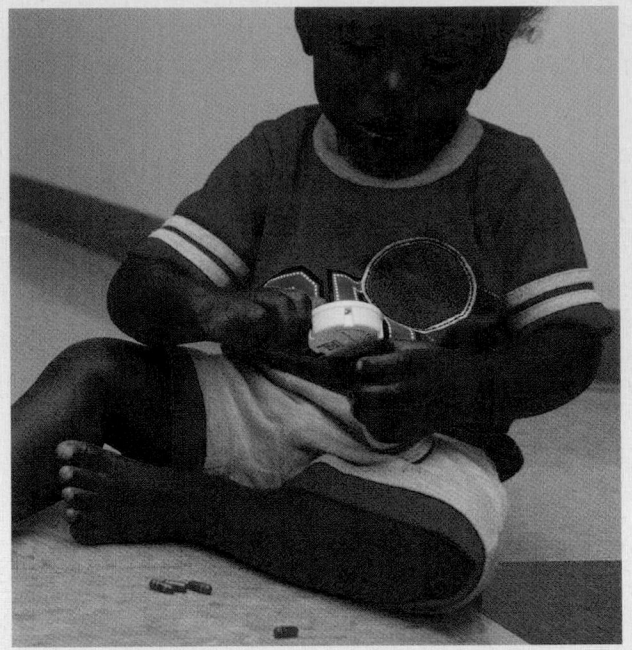

Figure 32-34 A curious child will try to taste or swallow almost any substance. A common victim of accidental ingestion of dangerous compounds is the unwatched toddler.

and wash or brush poison from the skin. Treatment focuses on support: assessing and maintaining the pediatric patient's ABCs and monitoring breathing. Provide oxygen and perform ventilations if necessary. If the patient demonstrates signs and symptoms of shock, position the child supine with the feet elevated, keep the child warm, and transport promptly to the nearest appropriate hospital.

In some cases, you will give activated charcoal to pediatric patients who have ingested poison, if approved by medical control or local protocol. Activated charcoal is not indicated for pediatric patients who have ingested an acid, an alkali, or a petroleum product; who have a decreased level of consciousness and cannot protect their airway; or who are unable to swallow. If local protocol permits, you will likely carry plastic bottles of premixed suspension, each containing up to 50 g of activated charcoal. Some common trade names for the suspension form are InstaChar, Actidose, and LiquiChar. The usual dose for a child is 1 g of activated charcoal per kilogram of body weight. The usual pediatric dose is 12.5 to 25 g. Chapter 19, *Toxicology*, discusses the administration of activated charcoal in detail.

Table 32-12 Common Sources of Poisoning in Children

- Alcohol
- Aspirin and acetaminophen
- Household cleaning products such as bleach and furniture polish
- Houseplants
- Iron
- Prescription medications of family members
- Street drugs
- Vitamins

Words of Wisdom

When you respond to a poisoning, remember to remain calm and control the scene. Poisonings can be very emotional for the parents or caregivers involved because of the potential for self-blame.

Dehydration Emergencies and Management

Dehydration occurs when fluid losses are greater than fluid intake. The most common cause of dehydration in pediatric patients is vomiting and diarrhea. If left untreated, dehydration can lead to shock and eventually death. Infants and children are at greater risk than adults for dehydration because their fluid reserves are smaller than those in adults. Life-threatening dehydration can overcome an infant in a matter of hours.

Dehydration can be mild, moderate, or severe. The severity of the dehydration can be gauged by looking at several clues **Table 32-13**. For example, an infant with mild dehydration may have dry lips and gums, decreased saliva, and fewer wet diapers throughout the day. As the dehydration grows more severe, the lips and gums may become very dry, the eyes may look sunken, and the infant may be sleepy and/or irritable, refusing bottles. The skin may be loose and have no elasticity; this is called poor skin turgor **Figure 32-35**. Also, infants may have sunken fontanelles.

Young children can compensate for fluid losses by decreasing blood flow to the extremities and directing blood flow to vital organs such as the brain and heart. Children who are moderately to severely dehydrated may have mottled, cool, clammy skin and delayed capillary response time. Respirations will usually be increased. Be aware that blood pressure may remain within a normal range while the pediatric patient is in shock, because the compensatory mechanisms are still in place.

Emergency medical care should include careful attention assessing the ABCs and obtaining baseline vital

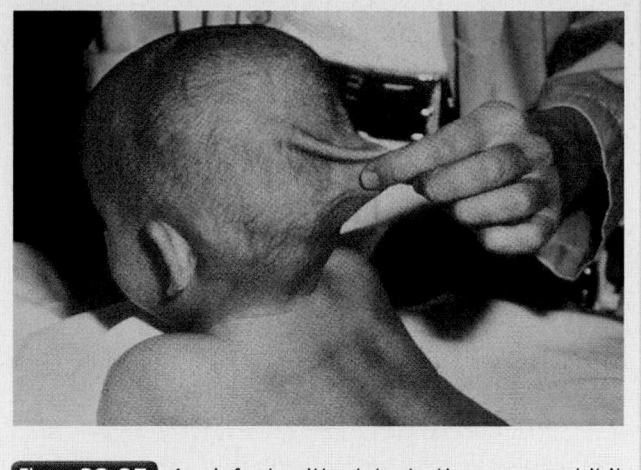

Figure 32-35 An infant with dehydration may exhibit "tenting" or poor skin turgor.

signs. However, if the dehydration is severe, ALS backup may be necessary so that IV access can be obtained and rehydration can begin. All pediatric patients with signs and symptoms of moderate to severe dehydration should be transported to the emergency department for further evaluation and treatment.

Fever Emergencies and Management

Fever is a common reason why parents or caregivers call 9-1-1. Simply defined, a fever is an increase in body temperature, usually in response to an infection. Body temperatures of 100.4°F (38°C) or higher are considered to be abnormal. A fever may have many causes and is rarely

Table 32-13 Vital Signs and Symptoms of Dehydration

	Mild Dehydration	Moderate Dehydration	Severe Dehydration
Pulse	Normal	Increased	Increased; 160+ is sign of impending shock
Level of activity	Normal or slowed	Slowed	Variable, weak to unresponsive
Urine output	Decreased	Decreased	No output
Skin	Normal	Cool, mottled; poor turgor	Cool, clammy; poor turgor; delayed capillary refill time
Mouth	Decreased saliva	Dry mucous membranes	Dry mucous membranes
Eyes	Normal	Tears	Sunken eyes
Anterior fontanelle	Normal to sunken	Sunken	Very sunken
Level of consciousness	Normal	Altered	Altered; lethargic
Blood pressure	Normal	Normal	Normal to low when shock sets in

life-threatening. However, you should not underestimate the potential seriousness of a fever that occurs in conjunction with a rash, which is a sign of serious illness, such as meningitis. Common causes of a fever in pediatric patients include the following:

- Infection, such as pneumonia, meningitis, or urinary tract infection
- Status epilepticus
- Neoplasm (cancer)
- Drug ingestion (aspirin)
- Arthritis and systemic lupus erythematosus (rash across nose)
- High environmental temperature

Fever is the result of an internal body mechanism in which heat generation is increased and heat loss is decreased. Note that there are other conditions in which the body temperature also increases. Hyperthermia differs from fever in that it is an increase in body temperature caused by an inability of the body to cool itself. Hyperthermia is typically seen in warm environments, such as a closed car on a hot day.

An accurate body temperature is an important vital sign for pediatric patients. A rectal temperature is the most accurate for infants to toddlers. Older children will be able to follow directions if placing a thermometer under the tongue or under the arm.

A fever can have several causes, such as a viral or a bacterial infection. Depending on the source of infection, the pediatric patient may present with additional signs of respiratory distress, shock, a stiff neck, a rash, skin that is hot to the touch, flushed cheeks, seizures, and bulging fontanelles in an infant. Assess the patient for other signs and symptoms such as nausea, vomiting, diarrhea, decreased feedings, and headache. A fever that is accompanied by a stiff neck, sensitivity to light, and a rash may be an indication that the patient has either bacterial or viral meningitis.

A pediatric patient with a fever may require only minimal interventions in the field. Provide immediate transport and manage the patient's ABCs. Follow standard precautions if you suspect that the patient may have a communicable disease such as meningitis.

Drowning Emergencies and Management

In drowning emergencies, you must always take steps to ensure your own safety when rescuing the patient from the water.

Drowning is the second most common cause of unintentional death among children in the United States; children younger than 5 years are at particular risk. At this age, children often fall into swimming pools and lakes, but many drown in bathtubs and even buckets. Older adolescents, who account for the most drowning after toddlers, drown when swimming or boating; alcohol is frequently a factor.

The principal condition that results from drowning is lack of oxygen. Even a few minutes (or less) without oxygen affects the heart, lungs, and brain, causing life-threatening problems such as cardiac arrest, respiratory difficulty, and coma. Submersion in icy water can rob the body of heat, causing hypothermia. Whereas a very few, very cold victims of submersion hypothermia have survived long periods in cardiac arrest in icy water, most people in this situation die. Diving into the water, of course, increases the risk of neck and spinal cord injuries.

Signs and symptoms of a drowning patient will vary based on the type and length of submersion. A pediatric patient involved in a drowning emergency may present with coughing; choking; airway obstruction; difficulty breathing; altered mental status; seizure activity; unresponsiveness; fast, slow, or no pulse noted; pale, cyanotic skin; and abdominal distention from ingestion of fluids.

Safety is critical when dealing with a drowning emergency. Do not become a victim yourself! Once the pediatric patient is successfully removed from the water, assess and manage the ABCs, and contact an ALS crew to intervene if needed. Oxygen should be administered at 100% via a nonrebreathing mask or bag-mask device if assisted ventilations are required. If trauma is suspected, apply a cervical collar and place the pediatric patient on a long board. Pad all open spaces under the pediatric patient before securing the patient onto the board. If the pediatric patient is unresponsive and in cardiopulmonary arrest, perform CPR.

Safety

Before using an automated external defibrillator (AED) on a pulseless drowning patient, ensure that the pediatric patient is dried off first. Use caution when operating an AED in this situation.

Pediatric Trauma Emergencies and Management

Trauma is the number one killer of children in the United States. More children die of injuries in 1 year than of all other causes combined. As an EMT, you will frequently treat injured children; therefore, you must have a thorough understanding of how trauma affects them. The quality of care in the first few minutes after a child has been injured can have an enormous impact on that child's chances for complete recovery.

Infants and toddlers are most commonly hurt as a result of falls or abuse. Older children and adolescents are usually injured as a result of mishaps involving automobiles. According to information collected by the National Pediatric Trauma Registry, automobile accidents, including those involving bicycles and pedestrians, are the most significant threat to the well-being of the child. Other common causes of traumatic injury and death include falls, gunshot wounds, blunt injuries, and sports activities. Another extremely serious and troublesome cause of injury is child abuse.

■ Physical Differences

Children are smaller than adults; therefore, when they are hurt in the same type of accident as an adult, the location of their injuries may differ from those in an adult. For example, the bumper of a car will strike an adult in the lower leg, whereas that same bumper will strike a child in the pelvis. In a crash involving sudden deceleration, an adult might injure a ligament in the knee; in that same accident, a child might injure the bones in the leg.

Children's bones and soft tissues are less well developed than those of adults; therefore, the force of an injury affects these structures somewhat differently than it does in an adult. Because a child's head is proportionately larger than an adult's, it exerts greater stress on the neck structures during a deceleration injury. Because of these anatomic differences, you should always carefully assess children for head and neck injuries.

■ Psychological Differences

Children are also less mature psychologically than adults; therefore, they are often injured because of their undeveloped judgment and their lack of experience. For example, children are more likely than adults to cross the street without looking for oncoming traffic. As a result, children are more likely than adults to be struck by cars. Children and adolescents are also more likely to sustain injuries from diving into shallow water because they forget to check the depth of the water before they dive. In such situations, you should always assume that the child has serious head and neck injuries.

■ Injury Patterns

Although you are not responsible for diagnosing injuries in children, your ability to recognize and report serious injuries will provide critical information to hospital staff. For this reason, it is important for you to understand the special physical and psychological characteristics of children and what makes them more likely to have certain kinds of injuries.

Vehicle Collisions

Children playing or riding a bicycle can dart out in front of motor vehicles without looking. In such a situation, the driver may have very little time to slow down or stop to prevent hitting the child. The area of greatest injury varies, depending on the size of the child and the height of the bumper at the time of impact. When vehicles slow down at the moment of impact, the bumper dips slightly, causing the point of impact with the child to be lowered. The exact area that is struck depends on the child's height and the final position of the bumper at the time of impact. Children who are injured in these situations often sustain high-energy injuries to the head, spine, abdomen, pelvis, or legs. In addition to differences in size and anatomy, children will often turn toward an oncoming vehicle when they see it approaching and, therefore, sustain different injuries than an adult who turns away.

Sports Activities

Children, especially those who are older or adolescents, are often injured in organized sports activities. Head and neck injuries can occur after high-speed collisions in contact sports such as football, wrestling, ice hockey, field hockey, soccer, or lacrosse. Remember to stabilize the cervical spine when caring for children with sports-related injuries. You should also be familiar with your local protocols related to helmet removal, and/or follow the guidelines presented in Chapter 26, *Head and Spine Injuries*.

■ Injuries to Specific Body Systems

Head Injuries

Head injuries are common in children. This is because the size of a child's head, in relation to the body, is larger than that of an adult. An infant also has a softer, thinner skull, which may result in injury to the underlying brain tissues. The scalp and facial vessels can bleed very easily and may cause a great deal of blood loss if the bleeding is not controlled. The signs and symptoms of head injury in a child are similar to those in an adult, but there are some important differences. Nausea and vomiting are common signs and symptoms of head injury in children; however, it is easy to mistake these for an abdominal injury or illness. You should suspect a serious head injury in any child who experiences nausea and vomiting after a traumatic event. Pediatric patients are managed in the same manner as adults. Chapter 26, *Head and Spine Injuries*, discusses head injuries in detail.

Immobilization Immobilization is necessary for all children who have possible head or spinal injuries after a traumatic event. Follow these steps **Skill Drill 32-5** :

Skill Drill 32-5

1. Maintain the child's head in a neutral position by placing a towel under the shoulders and torso **Step 1**.
2. Place an appropriately sized cervical collar on the pediatric patient **Step 2**.
3. Carefully log roll the child onto the immobilization device **Step 3**.

4. Secure the pediatric patient's torso to the immobilization device first **Step 4**.
5. Secure the child's head to the immobilization device **Step 5**.
6. Complete immobilization by ensuring that the child is strapped in properly **Step 6**.

Immobilization can be difficult to perform because of the child's body proportions. Young children require padding under the torso to maintain a neutral position. At around 8 to 10 years of age, children no longer require padding underneath the torso to create a neutral position. Instead, they can simply lie supine on the board. However, another complication may occur if a child is put onto an adult-sized long board. Because a child's body is narrower than an adult's, padding will be required along

Skill Drill 32-5

Immobilizing a Pediatric Patient

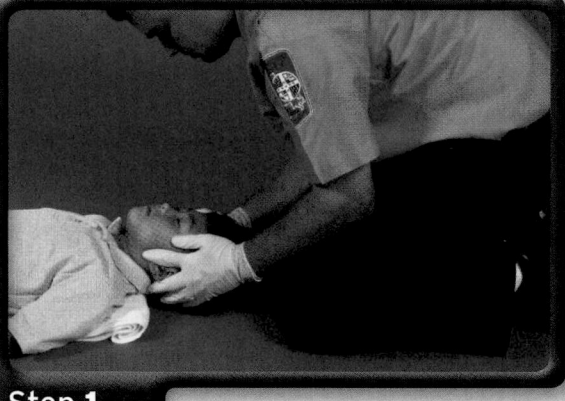

Step 1 Use a towel under the back, from the shoulders to the hips, to maintain the head in a neutral position.

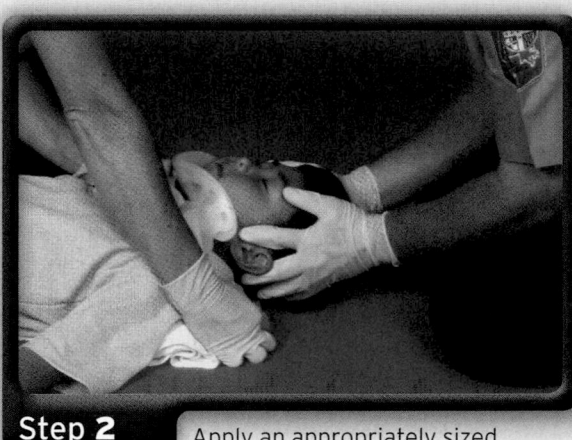

Step 2 Apply an appropriately sized cervical collar.

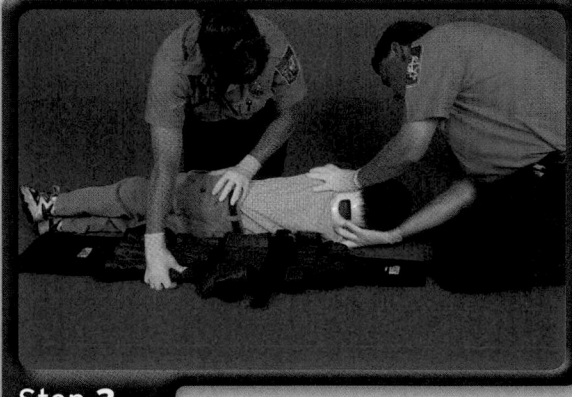

Step 3 Log roll the child onto the immobilization device.

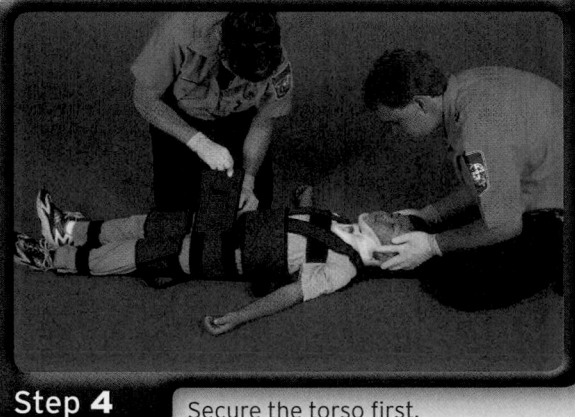

Step 4 Secure the torso first.

Skill Drill 32-5

Immobilizing a Pediatric Patient, continued

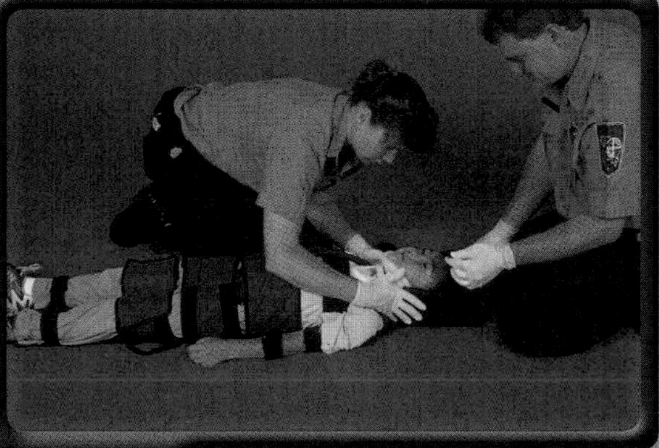

Step 5 Secure the head.

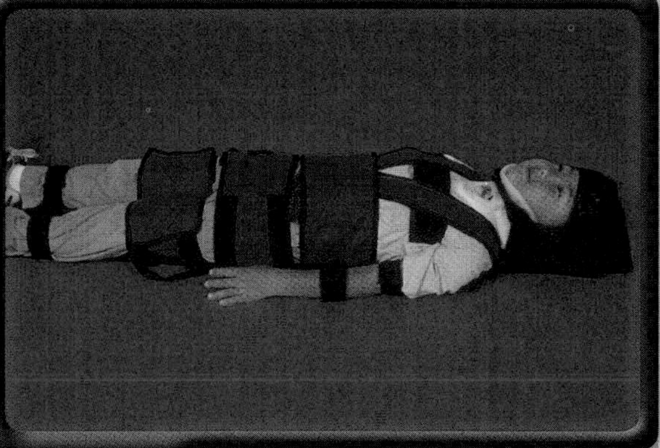

Step 6 Ensure that the child is strapped in properly.

the sides so that the child can be properly secured on an adult-sized long board.

Many infants and children will be in a car seat when you approach them. There are two methods of transportation that are determined by the severity of the pediatric patient's condition. If the pediatric patient has stable vital signs, minimal injury, and the car seat is visibly undamaged, the patient can be left in the seat and secured within it for transportation. If the pediatric patient is unstable, has injuries other than minor ones, or the car seat is visibly damaged, the patient must be removed to a board-type of device for immobilization and transportation.

Ideally, a cervical collar would be used when immobilizing an infant or toddler in a car seat; however, in most instances an appropriately sized cervical collar will not be available. In this case, place rolled towels on either side of the head to prevent side-to-side movement. Do not place a towel in the shape of an upside-down "U" over the pediatric patient's head; this may press down on the head and compromise the airway and spinal cord. The steps for immobilizing a patient in a car seat follow Skill Drill 32-6:

Skill Drill 32-6

1. Carefully stabilize the patient's head in a neutral position. Leave all car seat straps in place (Step 1).

2. Place an appropriately sized cervical collar on the patient if available. Otherwise, place rolled towels or padding alongside the patient to fill the voids in the car seat (Step 2).
3. Carefully secure the padding, using tape to keep it in place (Step 3).
4. Secure the car seat to the stretcher (Step 4).

Follow these steps to immobilize a patient out of a car seat Skill Drill 32-7:

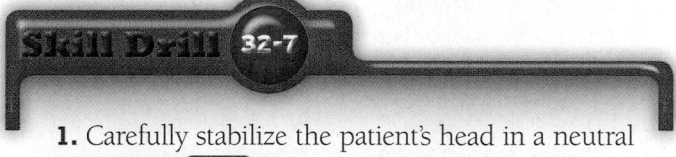

Skill Drill 32-7

1. Carefully stabilize the patient's head in a neutral position (Step 1).
2. Lay the seat down into a reclined position on a hard surface. Position a pediatric board or other similar device between the patient and the surface on which the patient is resting (Step 2).
3. Carefully slide the patient into position on the board (Step 3).
4. Make sure the patient's head is in a neutral position by placing a towel under the back, from the shoulders to the hips (Step 4).
5. Secure the torso first and place padding to fill any voids (Step 5).
6. Secure the patient's head to the board (Step 6).

Skill Drill 32-6

Immobilizing a Patient in a Car Seat

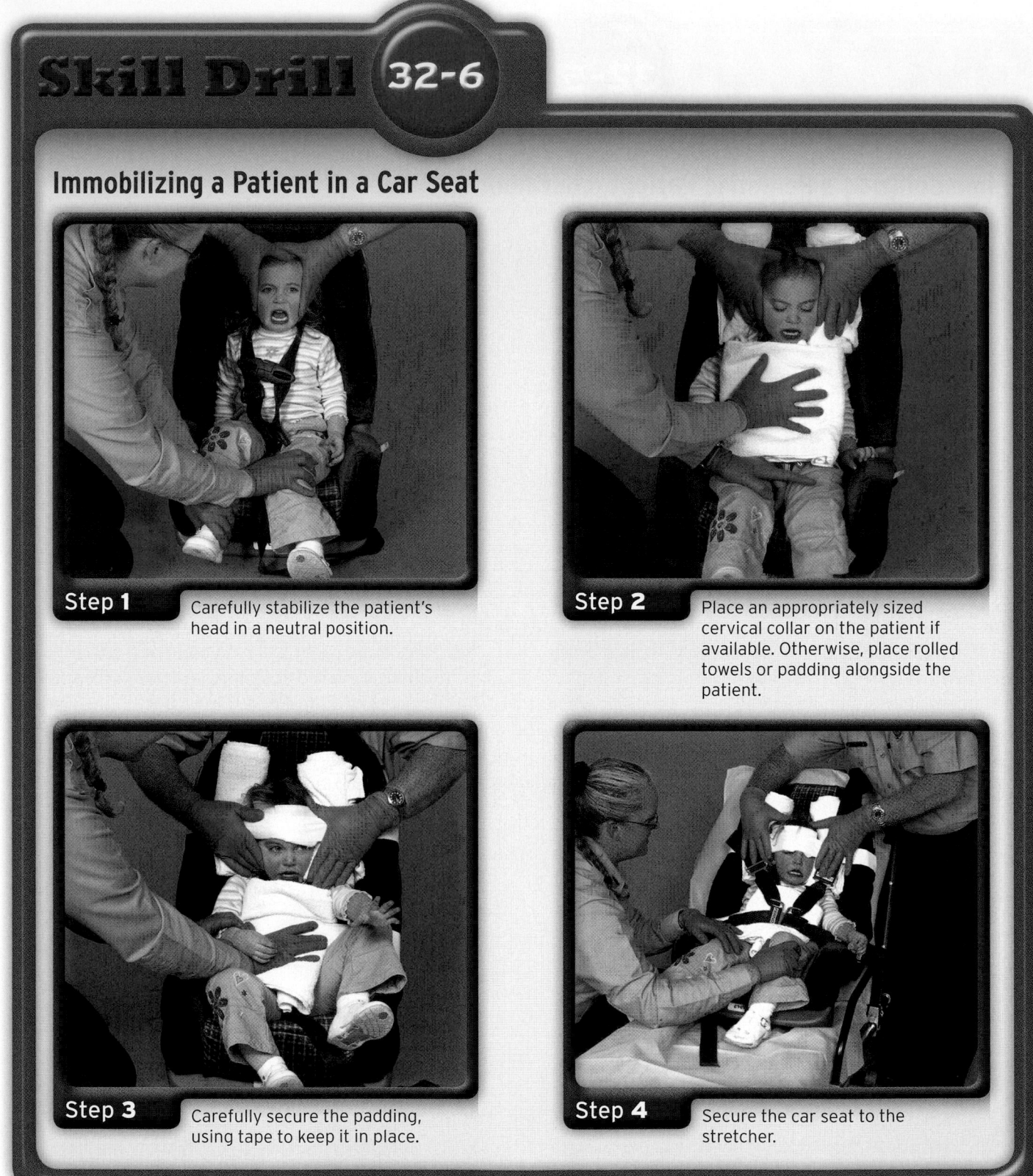

Step 1 Carefully stabilize the patient's head in a neutral position.

Step 2 Place an appropriately sized cervical collar on the patient if available. Otherwise, place rolled towels or padding alongside the patient.

Step 3 Carefully secure the padding, using tape to keep it in place.

Step 4 Secure the car seat to the stretcher.

Chest Injuries

Chest injuries in children are usually the result of blunt trauma rather than penetrating objects. Remember that children have very soft, flexible ribs that can be compressed a great deal without breaking. This chest wall flexibility can produce a flail chest. Keep this in mind as you assess a child who has sustained high-energy blunt trauma to the chest. Even though there may be no external sign of injury, such as broken ribs, contusions, or bleeding, there may be significant injuries within the

Skill Drill 32-7

Immobilizing a Patient Out of a Car Seat

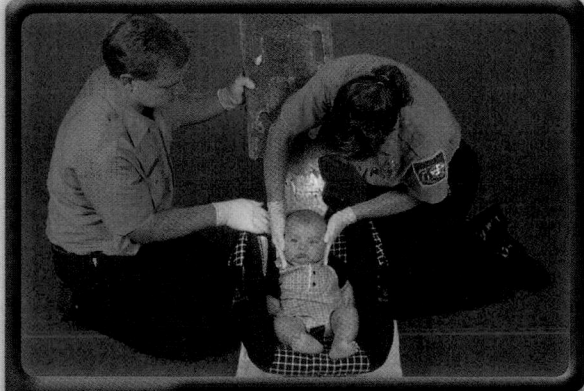

Step 1 Stabilize the head in neutral position.

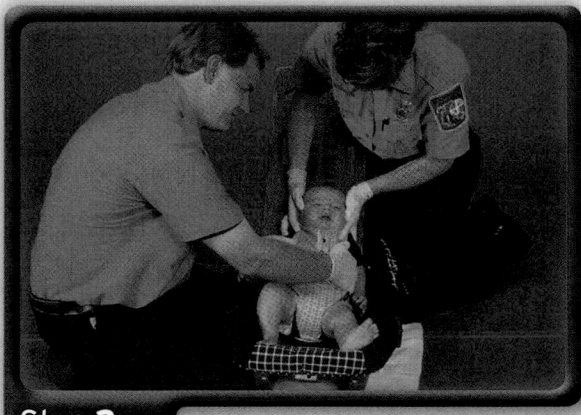

Step 2 Place an immobilization device between the patient and the surface he or she is resting on.

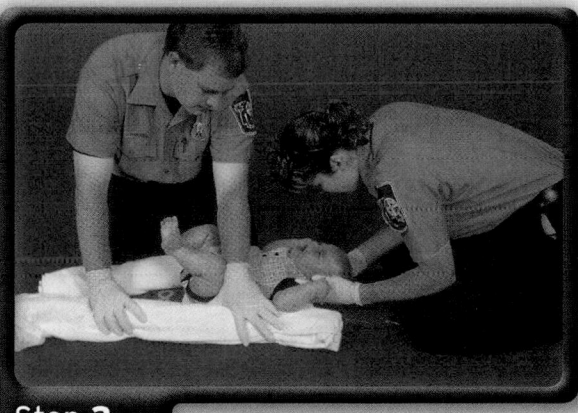

Step 3 Slide the patient onto the board.

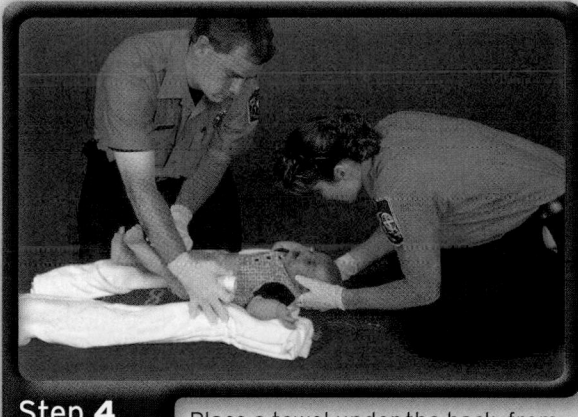

Step 4 Place a towel under the back, from the shoulders to the hips, to ensure neutral head position.

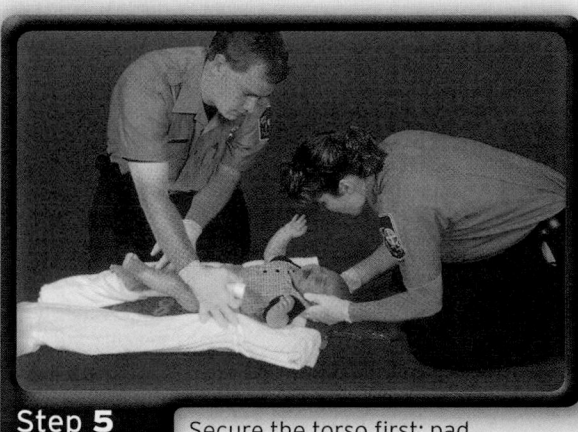

Step 5 Secure the torso first; pad any voids.

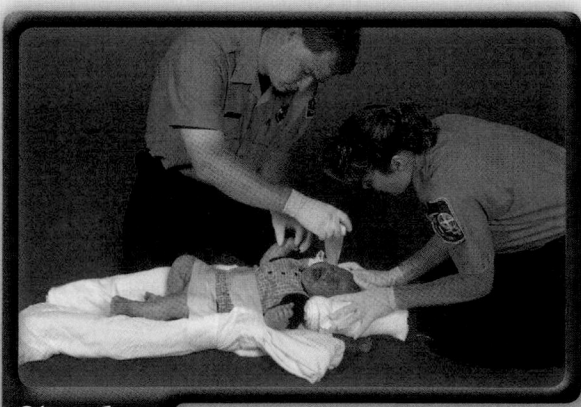

Step 6 Secure the head to the board.

chest **Figure 32-36** . Pediatric patients are managed in the same manner as adults. Chapter 27, *Chest Injuries*, discusses the treatment of chest injuries in detail.

Abdominal Injuries

Abdominal injuries are very common in children. Remember, though, that children can compensate for significant blood loss better than adults without signs or symptoms of shock developing **Figure 32-37** . They can also have a serious injury without early external evidence of a problem. All children with abdominal injuries should be monitored for signs and symptoms of shock, including a weak, rapid pulse; cold, clammy skin; decreased capillary refill (an early sign); confusion; and decreased systolic blood pressure (a late sign). Even in the absence of signs and symptoms of shock, or with only very few signs and symptoms, you should remain cautious about the possibility of internal injuries.

Pediatric patients are managed in the same manner as adults. Chapter 28, *Abdominal and Genitourinary Injuries*, discusses abdominal injuries in detail. If the patient shows signs and symptoms of shock, prevent hypothermia by keeping the patient warm with blankets. If the patient has bradycardia, ventilate. Monitor the patient's condition during transport.

■ Burns

Burns to children are generally considered more serious than burns to adults. This is because infants and children have more surface area relative to total body mass, which means greater fluid and heat loss. In addition, children do not tolerate burns as well as adults do. Children are also more likely to go into shock, develop hypothermia, and experience airway problems because of the unique differences of their ages and anatomy.

Children can be burned in a variety of ways. The most common involve exposure to hot substances such as scalding water in a bathtub, hot items on a stove, or exposure to caustic substances such as cleaning solvents or paint thinners **Figure 32-38** . You should suspect possible internal injuries from chemical ingestion when you see a child who has burns, particularly around the face and mouth.

One common problem following burn injuries in children is infection. Burned skin cannot resist infection as effectively as normal skin can. For this reason, sterile techniques should be used in handling the skin of children with burn wounds.

Table 32-14 provides some general guidelines to follow in assessing a pediatric patient who has been burned. These guidelines may help you to determine which pediatric patients should be treated primarily at specialized burn centers. Also note that you should consider the

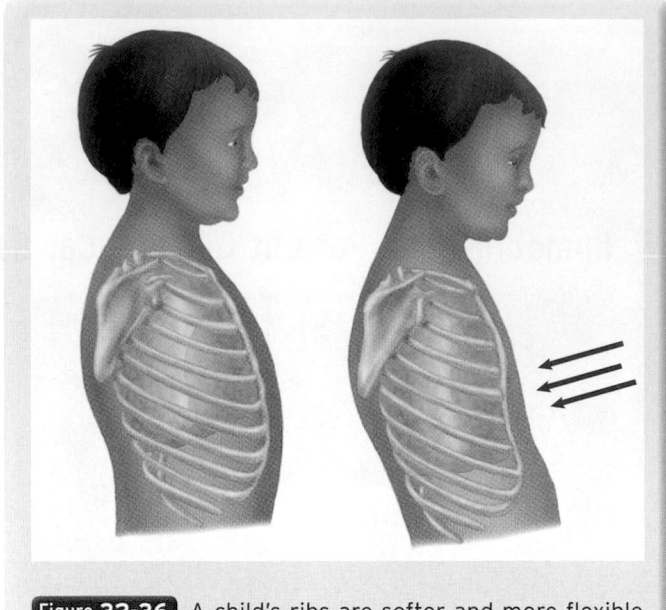

Figure 32-36 A child's ribs are softer and more flexible than an adult's. As a result, they may compress the lungs and heart, causing serious injury with no obvious external damage.

possibility of child abuse in any burn situation. Make sure you report any information about your suspicions to the appropriate authorities.

Pediatric patients are managed in the same manner as adults. Chapter 24, *Soft-Tissue Injuries*, discusses burn care in detail. If the patient shows signs and symptoms of shock, prevent hypothermia by keeping the pediatric patient warm with blankets. If the pediatric patient has bradycardia, ventilate. Monitor the pediatric patient's condition during transport.

■ Injuries of the Extremities

Children have immature bones with active growth centers. Growth of long bones occurs from the ends

Table 32-14	Severity of Burns in Pediatric Patients	
Severity of Burn	**Body Area Involved**
Minor	Partial-thickness burns involving less than 10% of the body surface
Moderate	Partial-thickness burns involving 10% to 20% of the body surface
Critical	Any full-thickness burn Any partial-thickness burn involving more than 20% of the body surface Any burn involving the hands, feet, face, airway, or genitalia

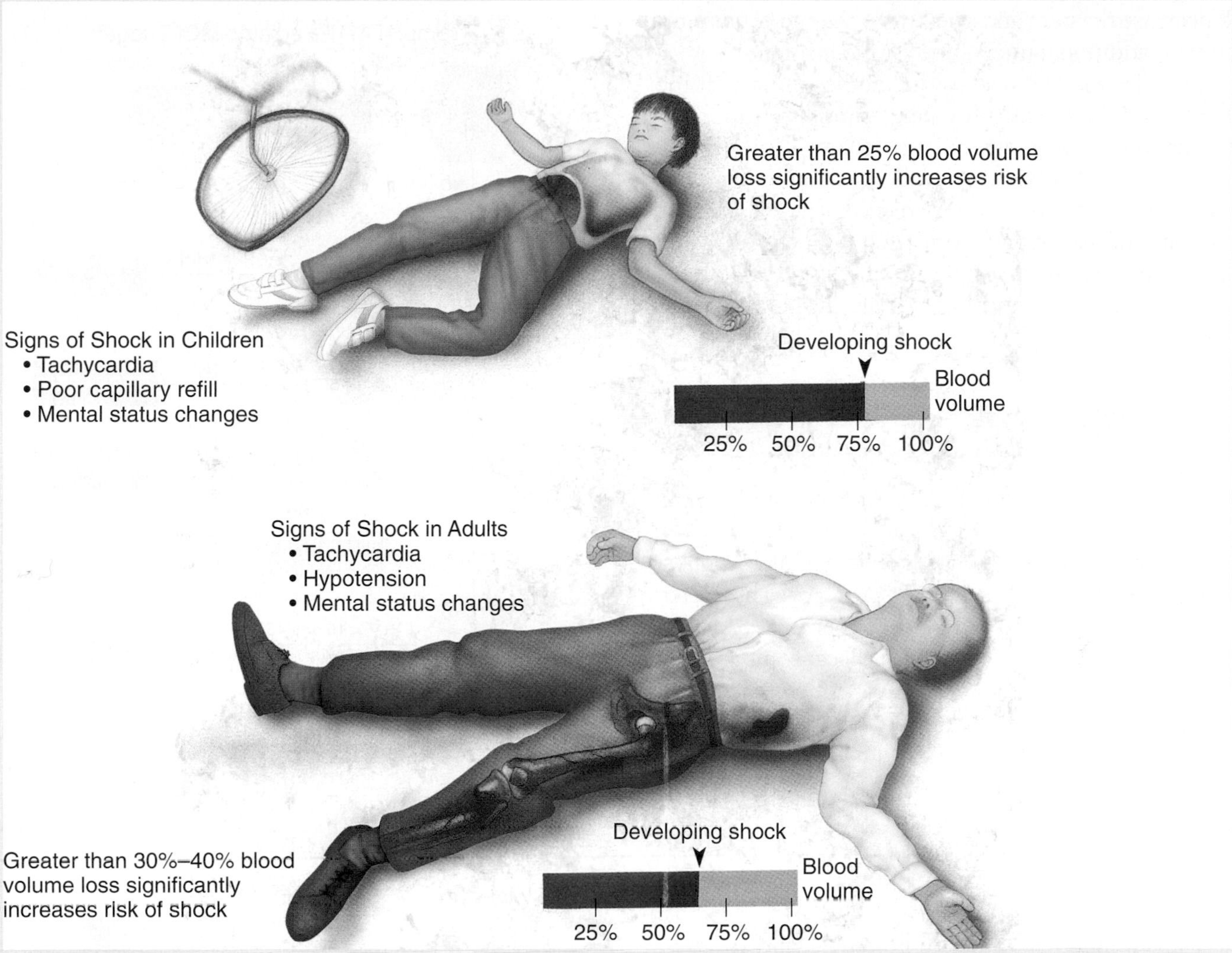

Greater than 25% blood volume loss significantly increases risk of shock

Signs of Shock in Children
• Tachycardia
• Poor capillary refill
• Mental status changes

Developing shock

Blood volume

25% 50% 75% 100%

Signs of Shock in Adults
• Tachycardia
• Hypotension
• Mental status changes

Greater than 30%–40% blood volume loss significantly increases risk of shock

Developing shock

Blood volume

25% 50% 75% 100%

Figure 32-37 All children with abdominal injuries should be monitored closely for signs of shock. Although children may compensate for significant blood loss better than adults, they develop shock after proportionally smaller blood losses.

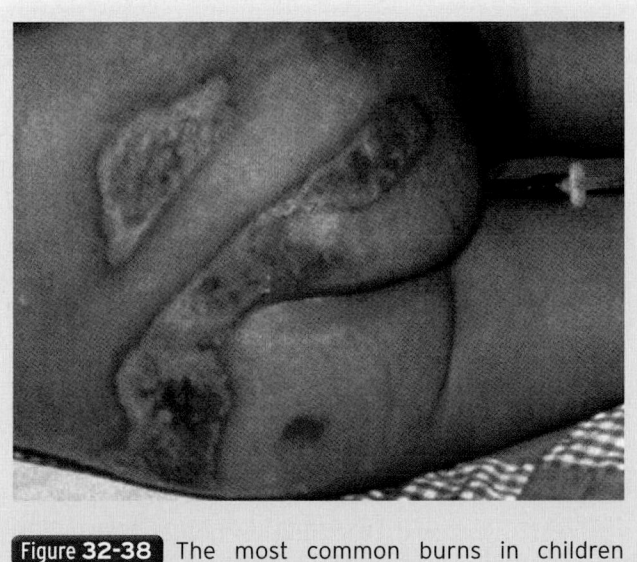

Figure 32-38 The most common burns in children involve exposure to hot surfaces. This child's buttocks were placed against a hot heating grate.

at specialized growth plates. These growth plates are potential weak spots in the bone and are often injured as a result of trauma. In general, children's bones bend more easily than adults' bones. As a result, incomplete or greenstick fractures can occur.

Extremity injuries in pediatric patients are generally managed in the same manner as those in adults. Painful deformed limbs with evidence of broken bones should be splinted. Specialized splinting equipment, such as a traction splint for fractures of the femur, should be used only if it fits the pediatric patient. You should not attempt to use adult immobilization devices on a pediatric patient unless the pediatric patient is large enough to properly fit in the device.

■ Pain Management

When dealing with pediatric pain management issues, you are limited to the following interventions: positioning, ice packs, and extremity elevation. These interventions will decrease the pain and swelling to the site of an injury. However, additional interventions (medications) from an ALS provider may be necessary. Another important tool is simple kindness and providing emotional support to the patient and the caregiver. This act alone can decrease pediatric patient anxiety and allow for a more soothing environment for all involved. Pediatric patients can sense fear and frustration from adults, so it is important to maintain a calm, professional, and trusting relationship with your patient and the family during the course of treatment.

Disaster Management

The JumpSTART triage system was developed for pediatric patients because the original START triage system did not take into account the developmental and physiological differences in children **Figure 32-39**. This system is intended for pediatric patients younger than 8 years of age and weighing less than 100 lb. Because infants and children may not be able to walk or follow commands during a disaster event, they must be considered for immediate delivery to the treatment area.

There are four triage categories in the JumpSTART system, designated by colors corresponding to different levels of urgency for treatment. Decision points include: able to walk (except in infants); presence of spontaneous breathing; respirations of less than 15 or of greater than 45 breaths/min; palpable peripheral pulse; and appropriate response to painful stimuli on the AVPU scale.

Pediatric patients who are able to walk are designated as Green for Minor and not in immediate need of treatment. Those patients breathing spontaneously, with a peripheral pulse and appropriately responsive to painful stimuli, are designated as Yellow for Delayed treatment. Pediatric patients who have apnea responsive to positioning or rescue breathing; respiratory failure; breathing but without a pulse; or inappropriate painful response, are designated as Red for Immediate response. Pediatric patients who are both apneic and without pulse, or apneic and unresponsive to rescue breathing are designated as Black and considered deceased or expectant deceased.

Child Abuse and Neglect

The term **child abuse** means any improper or excessive action that injures or otherwise harms a child or infant; it includes physical abuse, sexual abuse, neglect, and emotional abuse. The intentional injury of a child, whether physical or emotional, unfortunately is not rare

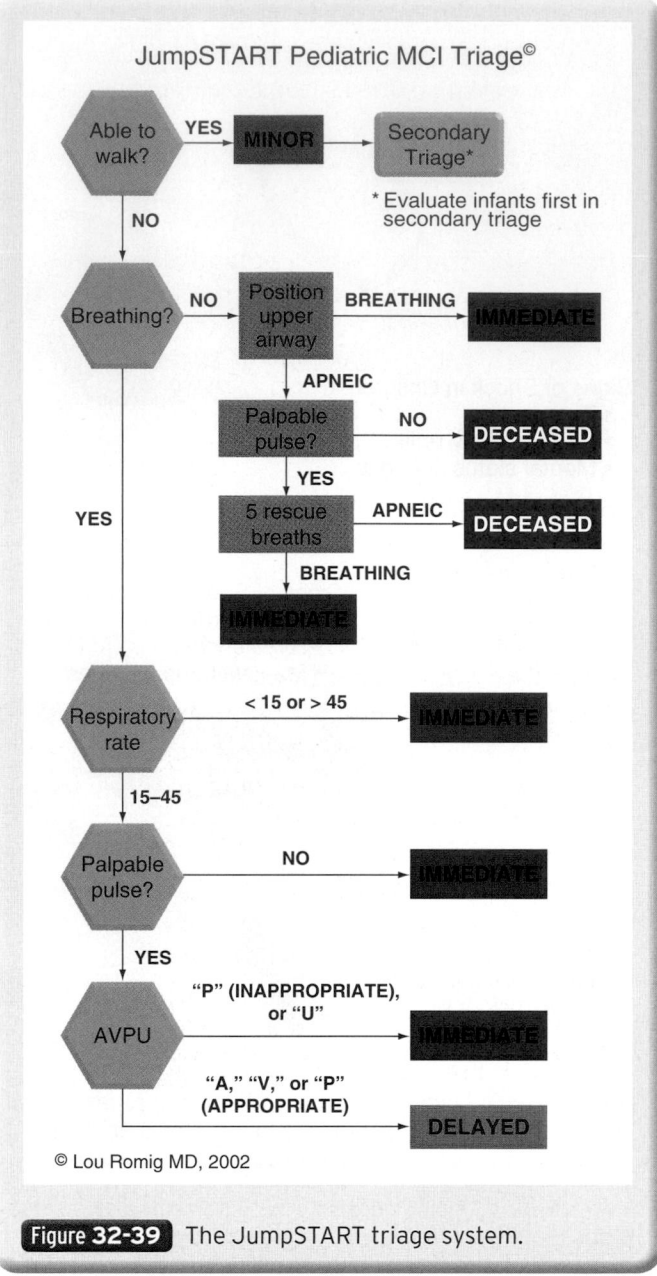

Figure 32-39 The JumpSTART triage system.

in our society. More than 2 million cases of child abuse are reported to child protection agencies annually. Many of these children suffer life-threatening injuries and some die. If suspected child abuse is not reported, the child is likely to be abused again and again, perhaps suffering permanent injuries or even dying. Therefore, you must be aware of the signs of child abuse and neglect, and it is your responsibility to report suspected abuse to law enforcement or child protection agencies.

■ Signs of Abuse

As an EMT, you will be called to homes because of a reported injury to a child. Child abuse occurs in every socioeconomic status, so you must be aware of the patient's surroundings and document your findings objectively. If you suspect that physical or sexual abuse is involved, you should ask yourself the following questions:

- Is the injury typical for the developmental level of the child?
- Is the method of injury reported by the parent or caregiver consistent with the pediatric patient's injury?
- Is the caregiver behaving appropriately (concerned about the child's well-being)?
- Is there evidence of drinking or drug use at the scene?
- Was there a delay in seeking care for the child?
- Is there a good relationship between the child and the caregiver?
- Does the child have multiple injuries at different stages of healing?
- Does the child have any unusual marks or bruises that may have been caused by cigarettes, grids, or branding injuries?
- Does the child have several types of injuries, such as burns, fractures, and bruises?
- Does the child have any burns on the hands or feet that involve a glove distribution (marks that encircle a hand or foot in a pattern that looks like a glove)?
- Is there an unexplained decreased level of consciousness?
- Is the child clean and an appropriate weight for his or her age?
- Is there any rectal or vaginal bleeding?
- What does the home look like? Clean or dirty? Is it warm or cold? Is there food?

Your assessment in the field will allow a better assessment by the medical staff later. An easy way to remember these points for the pediatric population is the mnemonic CHILD ABUSE, shown in **Table 32-15**.

As you assess the pediatric patient, look for and pay particular attention to the following signs **Figure 32-40**.

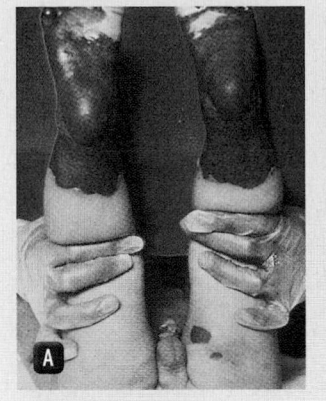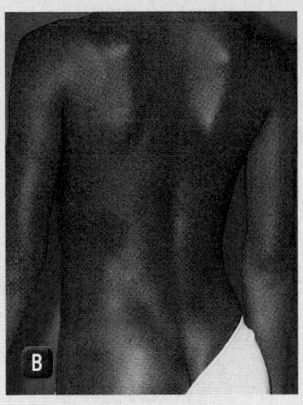

Figure 32-40 Signs of child abuse. **A.** Scald. **B.** Multiple injuries at different stages of healing.

Bruises

Observe the color and location of any bruises. New bruises are pink or red. Over time, bruises turn blue, then green, then yellow-brown and faded. Note the location. Bruises to the back, buttocks, or face are suspicious and are usually inflicted by a person.

Burns

Burns to the penis, testicles, vagina, or buttocks are usually inflicted by someone else, as are burns that encircle a hand or foot to look like a glove. You should suspect abuse if the child has cigarette burns or grid pattern burns.

Fractures

Fractures of the humerus or femur do not normally occur without major trauma, such as a fall from a high place or a motor vehicle crash. Falls from a bed are not usually associated with fractures. You should maintain some index of suspicion if an infant or young child sustains a femur fracture.

Shaken Baby Syndrome

Infants may sustain life-threatening head trauma by being shaken or struck on the head, a life-threatening condition called shaken baby syndrome. With this condition, there is bleeding within the head and damage to the cervical spine as a result of intentional, forceful shaking. The infant will be found unconscious, often without evidence of external trauma. The call for help may be for an infant who has stopped breathing or is unresponsive. The infant may appear to be in cardiopulmonary arrest, but what has likely occurred is that the shaking tore blood vessels in the brain, resulting in bleeding around the brain. The pressure from the blood results in an increased cranial pressure leading to coma, and/or death.

Table 32-15	Mnemonic for Assessing Possible Child Abuse
C	Consistency of the injury with the child's developmental age
H	History inconsistent with injury
I	Inappropriate parental concerns
L	Lack of supervision
D	Delay in seeking care
A	Affect
B	Bruises of varying ages
U	Unusual injury patterns
S	Suspicious circumstances
E	Environmental clues

Neglect

<u>Neglect</u> is refusal or failure on the part of the caregiver to provide life necessities, such as food, water, clothing, shelter, personal hygiene, medicine, comfort, and personal safety.

Children who are neglected are often dirty or too thin or appear developmentally delayed because of lack of stimulation. You may observe such children when you are making calls for unrelated problems. Report all cases of suspicious neglect.

■ Symptoms and Other Indicators of Abuse

An abused child may appear withdrawn, fearful, or hostile. You should be particularly concerned if the child refuses to discuss how an injury occurred. Occasionally, the parent or caregiver will reveal a history of several "accidents." Be alert for conflicting stories or a marked lack of concern from the parents or caregiver. Remember, the abuser may be a parent, caregiver, relative, or friend of the family. Sometimes the abuser is an acquaintance of a single parent.

EMTs in all states must report all cases of suspected abuse, even if the emergency department fails to do so. Most states have special forms for reporting. Supervisors are generally forbidden to interfere with the reporting of suspected abuse, even if they disagree with the assessment. You do not have to prove that there has been abuse. Law enforcement and child protection agencies are mandated to investigate all reported cases. You should take all necessary precautions to protect yourself, your crew, and the pediatric patient involved in this situation.

■ Sexual Abuse

Children of any age and either gender can be victims of sexual abuse. Most victims of rape are older than age 10 years, although younger children may be victims as well. This type of sexual abuse is often the result of long-standing abuse by relatives.

Your assessment of a child who has been sexually abused should be limited to determining the type of dressing any injuries require. Sometimes, a sexually abused child is also beaten. Therefore, you should treat any bruises or fractures as well. Do not examine the genitalia of a young child unless there is evidence of bleeding, or there is an injury that must be treated.

In addition, if you suspect that a child is a victim of sexual abuse, do not allow the child to wash, urinate, or defecate before a physician completes a physical examination. Although this step is difficult, it is important to preserve evidence. If the molested child is a girl, ensure that a female EMT or police officer remains with the child unless locating one will delay transport.

You must maintain professional composure the entire time you are assessing and caring for a sexually abused child. Assume a concerned, caring approach, and shield the child from onlookers and curious bystanders. Obtain as much information as possible from the child and any witnesses. The child may be hysterical or unwilling to say anything at all, especially if the abuser is a relative or family friend. You are in the best position to obtain the most accurate firsthand information about the incident. Therefore, you should record any information carefully and completely on the patient care report. Transport all children who are victims of sexual assault. Sexual abuse of a child is a crime. Cooperate with law enforcement officials in their investigations.

■ Sudden Infant Death Syndrome

The death of an infant or a young child is called <u>sudden infant death syndrome (SIDS)</u> when, after a complete autopsy, the cause of death remains unexplained. SIDS is the leading cause of death in infants younger than 1 year; most cases occur in infants younger than 6 months.

Although it is impossible to predict SIDS, there are several known risk factors:

- Mother younger than 20 years old
- Mother smoked during pregnancy
- Low birth weight

Deaths as the result of SIDS can occur at any time of the day; however, these children are often discovered in the morning when the parents go in to check on the infant. If you are the first provider at the scene of suspected SIDS, you will face three tasks: assessment of the scene, assessment and management of the patient, and communication and support of the family.

■ Patient Assessment and Management

SIDS is a diagnosis of exclusion. All other potential causes must first be ruled out, a process that may take physicians quite a while. An infant who has been a victim of SIDS will be pale or blue, not breathing, and unresponsive. Other causes for such a condition include the following:

- Overwhelming infection
- Child abuse
- Airway obstruction from a foreign object or as a result of infection
- Meningitis
- Accidental or intentional poisoning
- Hypoglycemia (low blood glucose level)
- Congenital metabolic defects

Regardless of the cause, assessment and management of the infant remain the same. Remember that what you find in assessing the infant and the scene may provide important diagnostic information.

Begin with an assessment of the ABCs, and provide interventions as necessary. Depending on how much time has passed since the child was discovered, he or she may show signs of postmortem changes. These include stiffening of the body, called rigor mortis, and dependent lividity, which is the pooling of blood in the lower parts of the body or those that are in contact with the floor or bed.

If the child shows such signs, call medical control. In some EMS systems, a victim of SIDS may be declared dead on the scene. Deciding whether to start CPR on a child who shows clear signs of rigor mortis or dependent lividity can be very difficult. Family members may consider anything less as withholding critical care. In this situation, the best course of action may be to initiate CPR and transport the patient and the family to the nearest emergency department, where the family can receive more extensive support (follow local protocols). If there is no evidence of postmortem changes, begin CPR immediately.

As you assess the infant, pay special attention to any marks or bruises on the child before performing any procedures, including CPR. Also note any intervention such as CPR that was done by the parents before you arrived.

■ Communication and Support of the Family

The sudden death of an infant is a very stressful event for a family; it also tends to evoke strong emotional responses among health care providers, including EMS personnel Table 32-16. Part of your job at this point is to allow the family to express their grief in ways that may differ from your own cultural, religious, and personal practices. Provide support in whatever ways you can.

Many times family members will ask specific questions about the event: Why did this happen? How did this happen? Let them know that their concerns will be addressed but that answers are not immediately available Table 32-17. Always use the infant's name in speaking to family members. If possible, allow the family to spend time with the infant and to ride in the ambulance to the hospital.

■ Scene Assessment

Carefully inspect the environment, following local protocols, noting the condition of the scene where the caregivers found the infant. Your assessment of the scene should concentrate on the following:

Table 32-16   How You Can Help the Family of a Deceased Child

When Arriving on Site
- Introduce yourself quickly.
- Obtain a brief history.
- When possible, one provider should stay with the family.

If Resuscitation Is Attempted
- Give brief, frequent updates and explanations.
- Allow family members to stay within viewing distance if they wish.
- Allow family members to accompany the child to the hospital when possible.

If No Resuscitation Is Performed
- Sit down with the family.
- Inform the family immediately.
- Explain why no resuscitation will be attempted.
- Offer to arrange for religious support, including baptism or last rites.

Beginning the Grieving Process
- Learn and use the child's name.
- Allow the family to express emotions; be nonjudgmental.
- Give brief explanations and answers.
- Explain to the family that the cause of death is still unknown.
- Allow time for questions.

DO
- Tell the family how sorry you are.
- Tell the family whom they can call if they have questions later.
- Give written instructions and referrals.

DON'T
- Say, "I know how you feel."
- Say, "You have other children" or "You can have other children."
- Attempt to answer the question "Why did this happen?"
- Try to tell the family that they will feel better in time.

- Signs of illness, including medications, humidifiers, or thermometers.
- The general condition of the house. (Note any signs of poor hygiene.)
- Family interaction. Do not allow yourself to be judgmental about family interactions at this time. Do note and report any behavior that is clearly not within the acceptable range, such as physical and verbal abuse.
- The site where the infant was discovered. Note all items in the infant's crib or bed, including pillows, stuffed animals, toys, and small objects.

The death of a child is difficult for everyone involved: parents, relatives, friends, and health care professionals.

Table 32-17 Common Questions Following Death of a Child

Q: Was there pain?

A: This often can be answered by a simple "No." If you are uncertain, you may give an indirect answer such as "We really don't know what patients feel in these circumstances."

Q: What did the child die of?

A: Do not answer this question; you would probably be guessing at this point.

Q: Why did this happen?

A: Do not attempt to answer this question either, as the answer depends on one's own individual philosophy or religion. "I wish I had an answer for you" is usually the most appropriate response.

Q: What happens now?

A: This question usually concerns the next few minutes or the next hour. If you know, you should give the family a general idea of what will happen. For example, if there is no history of illness, you can say that "a medical examination will be done, and then [child's name] will be taken to the mortuary."

You should arrange for a proper debriefing after your involvement with the case comes to a close. This can be a session with a trained counselor or a group discussion with your colleagues or the entire health care team.

■ Apparent Life-Threatening Event

Infants who are not breathing and are cyanotic and unresponsive when found by their families sometimes resume breathing and color with stimulation. These children have had what is called an **apparent life-threatening event (ALTE)**, called "near-miss SIDS" in the past. In addition to cyanosis and apnea, a classic ALTE is characterized by a distinct change in muscle tone (limpness) and choking or gagging. After the event, a child may appear healthy and show no signs of illness or distress. Nevertheless, you must complete a careful assessment and provide immediate transport to the emergency department.

Pay strict attention to management of the airway. Assess the infant's history and, if possible, the environment. Allow caregivers to ride in the ambulance. If asked, explain that you cannot say what caused the event, and that this is something that doctors will have to determine at the hospital.

■ Death of a Child

As with SIDS, the death of a child from any cause poses special challenges for EMS personnel. In addition to any medical treatment the child may require, you must be prepared to offer the family a high level of support and understanding as they begin the grieving process. First, the family may want you to initiate resuscitation efforts, which may or may not conflict with your EMS protocols. If the child is clearly deceased and, under protocol, can be declared dead in the field, but the family is so distraught

Words of Wisdom

Most parents of children who die suddenly will experience extremely strong emotional responses for a long time after the death. Counseling and support services begin with your care, including immediate referral to longer term services. You can usually make this referral through social services personnel in a hospital you work with, something you should know about in advance. Many communities also have support groups for families who lose children, including deaths caused by SIDS. Make sure the parents are aware of available services, offer to put them in touch while you are there, and leave the contact information in written form for their later reference even if you have helped them make the contact.

that they insist that resuscitation efforts be made, initiate CPR and transport the child.

The extent of your interaction with the family will depend, to some degree, on the number of providers available at the scene. Always introduce yourself to the child's caregivers, and ask about the child's date of birth and medical history. If and when the decision is made to start or stop resuscitation efforts, inform the family immediately. Find a place for family members from where they can watch resuscitation without being in the way. Do not, in any case, speculate on the cause of the child's death. The family will want to see the child and should be asked whether they want to hold the child and say good-bye. Parents may be experiencing strong feelings of denial.

The following interventions are helpful in caring for the family at this time:

- Learn and use the child's name rather than the impersonal "your child."
- Speak to family members at eye level, maintaining good eye contact with them.

- Use the word "dead" or "died" when informing the family of the child's death; euphemisms such as "passed away" or "gone" are not effective.
- Acknowledge the family's feelings ("I know this is devastating for you"), but never say "I know how you feel," even if you have experienced a similar event; the statement will anger many people.
- Offer to call other family members or clergy if the family wishes.
- Keep any instructions short, simple, and basic. Emotional distress may limit their ability to process information.
- Ask each adult family member individually whether he or she wants to hold the child.
- Wrap the dead child in a blanket, as you would if he or she were alive, and stay with the family while they hold the child. Ask them not to remove tubes or other equipment that was used in an attempted resuscitation.

Remember that each individual and each culture expresses grief in a different way, some more visibly than others. Some will require intervention; others will not. Most caregivers feel directly or indirectly responsible for the death of a child and may express this immediately; this does not mean that they actually are responsible. Parents often have questions that you should be prepared to answer. Although you should keep the possibility of abuse or neglect in mind, your role is not that of investigator. Any further inquiry is the responsibility of law enforcement.

Some EMS systems arrange for home visits after the death of a child so that EMS providers and family members can come to some sort of closure together. This also gives the family an opportunity to ask any remaining questions about the event. However, you need special training for such visits.

Again, coping with the death of a child can be very stressful for health care professionals. You may find yourself with unexpected feelings of pain and loss. It is helpful to take some time before going back on the job to work through your feelings and to talk about the event with your EMS colleagues. Be alert for signs of posttraumatic stress in yourself and others: nightmares, restlessness, difficulty sleeping, lack of appetite, a constant need for food, and the like. Consider the need for professional help if these signs or symptoms continue. All EMS programs should have critical incident stress management protocols and debriefing teams available for traumatic incidents.

Although you may consider the death of a child to be a failure, your skill at coping with this kind of emotional event can be a great comfort to the family, helping them to accept their loss and begin the long process of grieving.

You are the Provider: SUMMARY

1. How does the pediatric patient's airway and respiratory system differ from an adult's?

To effectively manage an infant or child's airway and ventilatory status, you must understand how the airway and respiratory system of an infant and child differs from that of the adult. Knowledge of these anatomic and physiologic differences is important because it often has a direct impact on your treatment decisions as well as the type and size of airway and ventilation equipment that you use.

Anatomically, the pediatric patient's airway differs from that of an adult in several ways. Proportionately, infants and children have a larger, rounder occiput (back of the head) when compared to an adult. This key anatomic difference directly affects how you manually position the child's airway; for example, towels or other padding are often placed in between the scapulae to maintain a neutral head position. The child's tongue is also proportionately larger than an adult's; because of this, oropharyngeal airways should be inserted by depressing the tongue and placing the device directly in the oropharynx, rather than rotating it in place as is done in adults. Furthermore, the child's tongue is also larger relative to the small mandible, which makes it easier to block the airway. As with adults, the tongue is the most common cause of upper airway obstruction in unresponsive infants and children.

In general, the pediatric airway is narrower at all levels when compared to an adult. The glottic opening, the space in between the vocal cords, is located higher in the neck and is positioned more anteriorly, and the neck itself is shorter. The trachea is easily collapsible because its cartilaginous rings are less well-developed. In addition, the smaller diameter of the trachea predisposes the infant or small child to airway obstruction from nonsolid substances (ie, secretions, blood) or swelling. The narrowest portion of the child's airway is the cricoid cartilage, the first ring of the trachea, whereas the narrowest portion in adults is the glottic opening. This means that foreign bodies tend to obstruct the airway at a lower level and can therefore be more difficult to remove.

Relative to the size of their airway, infants and children have a proportionately larger epiglottis than adults do, and it is floppier and U-shaped. Minimal swelling of the epiglottis can obstruct the glottic opening. The larger epiglottis can also cause problems during advanced airway management (ie, intubation).

Because of their relatively smaller lungs, infants and children need to breathe faster to adequately exchange oxygen and carbon dioxide. As the child ages, his or her lungs increase in size and the respiratory rate decreases accordingly. Children have both a higher metabolic rate and a higher demand for oxygen; this is due in part to the actual size of the lung tissues and the volume of air that can be exchanged. Smaller lungs mean smaller oxygen reserves; in conjunction with a higher demand for oxygen, this factor predisposes the infant or child to hypoxia following even brief periods of inadequate ventilation or apnea.

Breathing requires use of both the diaphragm and intercostal muscles. However, because the intercostal muscles are less developed in children, movement of the diaphragm, the major muscle of breathing, dictates the amount of air that is inhaled with each breath. Anything that puts pressure on the abdomen of an infant or small child can interfere with diaphragmatic movement. For example, gastric distention can cause significant diaphragmatic impairment, resulting in respiratory compromise. For this reason, it is critical to avoid ventilating the child too fast or with too much volume.

Infants and very young children have decreased oxygen reserves and increased oxygen demand; therefore, they experience muscle fatigue quicker than older children and adults. This can cause respiratory failure if the child has to physically fight hard to breathe for prolonged periods of time.

---

2. What are some airway and breathing problems that are unique to pediatric patients?

When you are assessing an infant or child with respiratory distress, it is important for you to remember that certain respiratory illnesses and conditions are unique to this population. Knowledge of these pediatric respiratory emergencies will help you when you are forming your field impression of the infant or child who presents with breathing difficulty.

Foreign body airway obstruction is common in both children and adults; however, the cause of the obstruction is usually different. Foreign body airway obstruction in adults typically occurs during a meal, whereas it more commonly occurs in children when they put a small toy or other object in their mouth.

A number of respiratory illnesses are unique to the pediatric population, and can be broadly classified by which part of the airway they affect—the upper or lower airway.

Croup (laryngotracheobronchitis) is a viral infection of the upper airway and the most common cause of upper airway emergencies in young children. Croup most commonly affects children between 6 months and 6 years of age, with most cases occurring in the fall and winter months. Most cases of croup are mild; however, in some cases, edema of the subglottic space can be severe enough to cause respiratory distress. A hallmark early sign of croup is a seallike barking cough. As airway swelling progresses and the area just beneath the vocal cords begins to narrow, the child begins experiencing varying degrees of respiratory distress. Significant airway narrowing produces stridor, a high-pitched sound that is typically heard during inhalation. Because croup is caused by a virus, fever is typically present.

Epiglottitis is an acute, life-threatening bacterial infection that results in a severely swollen and inflamed epiglottis—often to the point where it completely obstructs the tracheal opening. Classically, the disease presents with an acute onset of high fever, severe respiratory distress, a sore throat, and drooling. On appearance, the child appears sick; he or she is typically found in a tripod position with his or her head thrust forward in a sniffing position. Epiglottitis has become a rare condition because children are now vaccinated against the bacterium that was responsible for causing the disease. When it was most prevalent, it commonly affected children between 2 and 7 years of age.

Asthma, a lower airway disease that is caused by bronchospasm and inflammation of the bronchioles, is the most common chronic illness of childhood and the most common respiratory complaint encountered by prehospital care providers. Although asthma is not unique to children, children who experience an acute asthma attack may become hypoxic much more quickly than adults because children have small respiratory reserves and a propensity to fatigue much faster. Asthma is rarely seen in children younger than 1 year of age. Hallmarks of asthma include wheezing, which may or may not be audible without a stethoscope, and varying degrees of respiratory distress.

Bronchiolitis is an inflammation of the bronchioles in the lower respiratory tract and is caused by a viral infection. A number of viruses can cause bronchiolitis; however, the respiratory syncytial virus is the most common cause. Bronchiolitis most commonly occurs during the late fall and winter months, and primarily affects children younger than 2 years. The signs and symptoms of bronchiolitis can be difficult to distinguish from those of asthma; both illnesses typically present with wheezing and respiratory distress. One clue, however, is the child's age; asthma is rare in children younger than 1 year, whereas bronchiolitis is common in this age group. An infant who presents with a first-time episode of wheezing during the winter months likely has bronchiolitis. Bronchiolitis, being the result of a viral infection, commonly presents with fever, whereas asthma typically does not.

Pertussis (whooping cough) and bacterial tracheitis may also affect infants and children, although they are less common than the conditions previously discussed. Regardless of the underlying cause of a pediatric patient's respiratory distress, it is important to remember than children have a higher demand for oxygen and a smaller oxygen reserve; this means that they can decompensate very quickly. It is also important to remember that respiratory failure is the most common cause of cardiopulmonary arrest in the pediatric population.

---

3. Why did you *not* immediately perform a hands-on assessment of this child?

If an infant or child is clearly experiencing a life-threatening condition, such as unconsciousness or major trauma, then an immediate hands-on assessment is indicated. However, if the child is conscious, a slightly different approach to the assessment process is used. A key element in treating a conscious child, especially one with respiratory distress, is to avoid agitating the child; agitation breeds anxiety, and anxiety increases the child's respiratory and heart rates—functions that both require more oxygen. Recalling that children have smaller oxygen reserves, a sudden increase in anxiety could literally worsen their condition.

The pediatric assessment triangle (PAT) is a structured assessment tool that is used to quickly form a general impression of the child without touching him or her. It is a visual assessment that provides a "first glance" evaluation to identify the general category of the child's physiologic problem and to determine how urgently the child requires treatment and transport. The PAT should take no longer than 15 to 30 seconds.

The PAT focuses on three elements: appearance (muscle tone and mental status), work of breathing, and circulation to the skin. The only equipment required for the PAT is your own eyes and ears.

Evaluating the child's appearance involves noting the level of consciousness or interactiveness and muscle tone—signs that will provide information regarding the adequacy of the child's cerebral perfusion and overall function of the central nervous system. Much of the information regarding the child's level of consciousness can be obtained by using the PAT. An infant or child with a normal level of consciousness will act appropriately for his or her age, have good muscle tone, and maintain good eye contact. Age-inappropriate behavior or interactiveness, poor muscle tone, or poor eye contact with the caregiver or EMT indicates a decreased level of consciousness.

A child's work of breathing increases as the body attempts to compensate for abnormalities in oxygenation and ventilation. Signs of increased work of breathing include tachypnea, abnormal airway sounds, retractions of the intercostal muscles and sternum, nasal flaring, and preferential positioning (ie, tripod position), among others.

Circulation to the skin is an important indicator of perfusion but does not require a hands-on evaluation to assess. Abnormal skin signs include pallor or mottling, which indicates poor perfusion; cyanosis, which indicates a low level of oxygen in the blood (a late sign!); and flushing, which is commonly seen with a fever.

Based on the findings of the PAT, you should be able to determine whether the child is stable, in which case you can continue with your assessment, or unstable, in which case you should begin immediate treatment and prepare for transport.

4. On the basis of your observations so far, is this child experiencing respiratory distress or respiratory failure?

Although the child is fearful of your presence, a normal reaction for a young child, she is alert, is not exhibiting age-inappropriate behavior, is maintaining eye contact with you, and is clinging to her mother. Her skin is pink and appears to be dry; however, she is experiencing increased work of breathing, as noted by an increased respiratory rate, obvious breathing difficulty, nasal flaring, and prominent retractions. This clinical presentation is consistent with respiratory distress.

When assessing a child with a respiratory complaint, you must quickly determine the severity of his or her condition: Is the child in respiratory distress, respiratory failure, or respiratory arrest?

Respiratory distress is a compensated state in which increased work of breathing is able to maintain adequate oxygenation and ventilation. Signs of respiratory distress include intercostal

or suprasternal retractions, nasal flaring, and preferential positioning (ie, tripod position). Despite an increased work of breathing, however, the child with respiratory distress is alert or is otherwise exhibiting age-appropriate behavior; this indicates adequate cerebral perfusion. It is important to recognize the signs of respiratory distress and begin treatment immediately to prevent deterioration to respiratory failure.

Respiratory failure is a condition in which the child can no longer compensate for his or her condition by increased work of breathing; therefore, hypoxia and carbon dioxide retention occur. Respiratory failure is a decompensated state that requires immediate intervention to ensure adequate oxygenation and ventilation and prevent respiratory arrest. Signs include decreased or absent retractions owing to fatigue of the chest wall muscles; altered mental status, which indicates inadequate cerebral perfusion; and a decreased respiratory rate (bradypnea). Infants and children with respiratory failure require immediate ventilation assistance!

Respiratory arrest is characterized by absent breathing (apnea), and obviously requires positive-pressure ventilations with high-flow oxygen to prevent deterioration to cardiopulmonary arrest. Resuscitation of a child from respiratory arrest is often successful, whereas resuscitation of a child from cardiopulmonary arrest is often not.

Although your patient is currently in respiratory distress, it is critical for you to closely observe her for signs of deterioration and be prepared to assist her ventilations if they become inadequate.

5. What is the most likely cause of this child's respiratory distress?

As previously discussed, respiratory emergencies can be classified as those that affect the upper or lower airway. Assessment of the child's signs and symptoms as well as his or her past medical history will often enable you to determine where the problem resides.

Whereas both upper and lower airway problems present with respiratory distress, there are some clinical differences that you should observe for. Observe for the presence of abnormal airway sounds. Stridor, a high-pitched inspiratory sound, indicates upper airway swelling; it is common in children with severe croup, epiglottitis, and foreign body airway obstruction. In contrast, wheezing—a whistling sound that is heard while auscultating the chest—indicates narrowing of the bronchioles in the lungs; therefore, it is a sign of a lower airway problem (ie, asthma, bronchiolitis). In some cases, wheezing may be audible, meaning you can hear it without a stethoscope.

On the basis of the child's clinical presentation and the progression of her symptoms, you should suspect that she has croup. This field impression is further reinforced by the presence of a high-pitched cough, a low-grade fever, and preceding cold symptoms. Her breath sounds are clear to auscultation bilaterally; this makes a lower airway problem highly unlikely. As noted earlier, croup typically develops gradually and is most prevalent in children between 6 months and 6 years of age. The outside temperature is 58°F, which suggests that it is fall or early winter; this is when croup most commonly occurs.

6. Should you separate this child from her parents to provide further treatment? Why or why not?

An unconscious child or a child who is in respiratory or cardiac arrest obviously must be separated from his or her parents because immediate, aggressive treatment is required. If the child is stable, however, you will often find that parental presence is the best medicine for the child. It is important to keep a sick or injured child as calm as possible.

The decision to separate a sick or injured child from his or her parents is based on the child's condition and the reaction of the parent(s) to the situation. Although this child is experiencing respiratory distress, she is clinically stable and is clinging to her mother, who is showing no evidence of emotional distress. For these reasons, you should not separate the child from her mother. If you do, you will likely find that the child will become more anxious. Anxiety can easily worsen the child's condition by making her breathe harder and making her heart beat faster.

The presence of a parent or primary caregiver can assist you in your continued assessment of the child's mental status. It is expected that a young child will cling to his or her parent and express fear of your presence. However, failure of the child to recognize his or her parents is an ominous sign and indicates a significant decrease in cerebral perfusion.

7. How has this child's condition changed? What should you do next?

Compared with earlier assessments, the child's condition has obviously deteriorated; she is now in respiratory failure! As previously discussed, respiratory failure occurs when the body can no longer compensate for the underlying problem by increased work of breathing. When infants and children decompensate, they often do so with alarming speed. In this case, the child decompensated in the time it took to move her from the residence to the back of your ambulance!

Clinical signs of this child's deterioration include a decreased level of activity; she was previously conscious and maintaining eye contact. She is breathing slower and her retractions have markedly weakened; this indicates chest wall muscle fatigue from prolonged increased work of breathing. She is developing cyanosis around her mouth (perioral cyanosis), indicating a decreased level of oxygen in her blood. This sign is further substantiated by the marked decrease in her oxygen saturation level. Perhaps the most ominous sign is the marked decrease in her heart rate. Children rely heavily on their heart rate to adequately circulate oxygenated blood. When the body is no longer able to compensate, the heart rate begins to fall; this is a sign of impending cardiopulmonary arrest!

You must act immediately to prevent this child from developing respiratory or cardiac arrest. Begin assisting her ventilations with a bag-mask device and high-flow oxygen and transport immediately. If possible, coordinate an intercept with an ALS unit en route to the hospital. If her condition continues to deteriorate, she may require advanced airway management or medication therapy.

8. How could an ALS ambulance and crew benefit this child?

The decision to request an ALS ambulance to the scene or coordinate an ALS intercept en route to the hospital is based on several factors. First, you must ask yourself if you can effectively care for the patient by yourself; if you cannot, call for help. Secondly, you must take note of your transport distance to the hospital. If the closest appropriate facility is a lengthy distance away, you should call for help. Finally, you must ask yourself if the patient requires treatment that is above your level of training; if so, call for help.

If you are able to effectively care for the child by yourself and are a short distance away from the closest appropriate facility, you should not delay transport to wait for an ALS unit. Use reasonable judgment and common sense and weigh the risk versus benefits when making the decision to request ALS backup.

You are caring for a child who, at a minimum, requires assisted ventilation. If her condition continues to deteriorate, she may require cardiopulmonary resuscitation. AEMTs and paramedics are able to perform advanced airway management, and paramedics are able to administer cardiac medications and monitor the patient's cardiac rhythm. In some cases, children with respiratory failure require more than oxygen and assisted ventilations; certain medications may also be required. Because these interventions would clearly benefit the child—especially if she experiences cardiac arrest—you should consider an ALS intercept.

9. Are you providing adequate ventilations? How can you tell?

The child's level of consciousness (still listless) and slow, shallow respirations are clear indicators for the continued use of assisted ventilation. However, she is showing signs of improvement, indicating that your ventilations are adequate.

The child's heart rate has increased from 90 beats/min to 110 beats/min; as with any patient, an improvement in the heart rate is a sign of adequate artificial ventilation. The child's oxygen saturation has also improved significantly (from 85% to 96%); this indicates that you are adequately oxygenating her blood. Resolution of the cyanosis around her mouth is also an indicator that you have improved oxygenation of her blood with adequate ventilations.

Despite the fact that you have noted clinical improvement with assisted ventilation, this does not mean that you can stop. The child is still breathing at a slow rate and the depth of her breathing is shallow. Also, she is not resisting your treatment, thus indicating that her level of consciousness is still depressed and she is still fatigued. You must continue to assist her ventilations.

10. Why is it especially important to avoid hyperventilating infants and children?

You should avoid hyperventilating any patient! In the unprotected airway (ie, in the patient who is not intubated), hyperventilation forces excess air into the stomach, which causes gastric distention. Gastric distention increases the risk of aspiration if vomiting occurs.

You are the Provider: SUMMARY, continued

In infants and small children, there is the additional risk of pushing the diaphragm into the thoracic cavity, which may reduce the effectiveness of your ventilations. It takes much less air in the stomach of an infant or child to inhibit adequate positive-pressure ventilations than it does in an adult.

Hyperventilation has also been shown to impair blood return to the heart, thus impairing cardiac output (the amount of blood ejected from the left ventricle each minute). This occurs when hyperinflation of the lungs literally squeezes the heart, which reduces its ability to receive deoxygenated blood from the body.

Whether you are assisting ventilations of a spontaneously breathing patient or ventilating an apneic patient, you should squeeze the bag-mask device for no longer than 1 second; stop squeezing the bag as soon as you note visible chest rise.

EMS Patient Care Report (PCR)

Date: 1-15-10	**Incident No.:** 013210	**Nature of Call:** Respiratory distress		**Location:** 545 W. San Antonio St.	
Dispatched: 2323	**En Route:** 2324	**At Scene:** 2330	**Transport:** 2340	**At Hospital:** 2353	**In Service:** 2359

Patient Information

Age: 4 **Sex:** F **Weight (in kg [lb]):** 16 kg (35 lb)	**Allergies:** No known drug allergies **Medications:** None **Past Medical History:** Recent cold **Chief Complaint:** Respiratory distress

Vital Signs

Time: 2335	**BP:** 86/56	**Pulse:** 124	**Respirations:** 34	**Sao$_2$:** 95%
Time: 2340	**BP:** 76/56	**Pulse:** 90	**Respirations:** 18	**Sao$_2$:** 85%
Time: 2348	**BP:** 84/54	**Pulse:** 110	**Respirations:** 20	**Sao$_2$:** 96%

EMS Treatment
(circle all that apply)

Oxygen @ 12 **L/min via (circle one):** NC (NRM) (Bag-Mask Device)	(Assisted Ventilation)	**Airway Adjunct**	**CPR**
Defibrillation	**Bleeding Control**	**Bandaging**	**Splinting**	**Other**

Narrative

Medic 3 dispatched to a residence for a 4-year-old child with respiratory distress. Arrived on scene and found the patient sitting on the couch next to her mother. She was conscious, maintained eye contact, and began clinging to her mother. Her airway was patent; however, her breathing was obviously labored. Visual assessment revealed that her skin was pink and appeared dry. She had prominent supraclavicular retractions and nasal flaring during inhalation. Administered blow-by oxygen via pediatric nonrebreathing mask and performed further assessment. Patient's mother states that she has had a cold for the past 2 days, and slowly developed a low-grade fever and high-pitched cough, which she described as a "barking seal." Mother denies that her child has any significant past medical history; she has been giving her ibuprofen as needed for her low-grade fever. Auscultation of breath sounds revealed that they were clear and equal bilaterally, but prominent intercostal retractions were noted on exposure of the chest. Capillary refill time, 1 second. Continued blow-by oxygen as patient's father carried her to the ambulance. Shortly after securing the child to the stretcher, reassessment revealed that her level of consciousness had markedly decreased. Her respiratory rate also decreased, her retractions were weak, and her oxygen saturation level markedly decreased. Began assisting the child's ventilations with a bag-mask device and high-flow oxygen and began immediate transport. Attempted to coordinate ALS intercept; however, the closest ALS unit was too far away and would have resulted in unnecessary delay in transport. Notified receiving facility of the patient's condition and of our arrival and continued assisting ventilations. Reassessment revealed that the child was still listless and was not resistant to treatment; however, her heart rate, skin color, and oxygen saturation improved. Continued ventilatory assistance and delivered child to the emergency department. Verbal report was given to staff physician. Medic 3 cleared the hospital and returned to service at 2359. *End of report*

Assessment and Emergency Care of Pediatric Emergencies

Scene Size-up

Scene Safety	Take standard precautions. Note the position in which the pediatric patient is found. Look for any possible safety threats. Bring medications with you that could have been ingested by the pediatric patient. Perform an environmental assessment to obtain information on the chief complaint, number of patients, MOI or NOI, and ongoing health risks. Inspect the physical environment and watch the family–child and/or caregiver–child interactions.
Mechanism of Injury (MOI)/ Nature of Illness (NOI)	Determine the MOI/NOI. Observe the scene and look for indicators that will assist you with this. Gather information from the patient, parent, caregiver, or any bystanders. The nature of the problem may not be readily apparent until more information is gathered. Consider spinal immobilization at a traumatic scene in any pediatric patient with a significant MOI.

Primary Assessment

Form a General Impression	When assessing a pediatric patient, use the pediatric assessment triangle (PAT) to form a general impression.
PAT Triangle	Use the PAT to identify immediate threats to life. Decide if your patient is "sick" or "not sick." Perform a 15- to 30-second visual assessment of the patient. Evaluate the patient's appearance, work of breathing, and circulation to skin. Based on the findings of the PAT, determine if the patient is stable or requires urgent care.
Appearance	Note the pediatric patient's level of consciousness or interactiveness and muscle tone. This will provide information on the pediatric patient's cerebral perfusion and the overall function of the central nervous system. Use AVPU and TICLS.
Work of Breathing	Work of breathing increases as the body compensates for abnormalities in oxygenation and ventilation. Signs of increased work of breathing are tachypnea, grunting, wheezing, and retractions of the intercostal muscles or sternum.
Circulation to the Skin	Signs that indicate a sick pediatric patient include pallor of the skin and mucous membranes, mottling, and cyanosis.
Stay or Go	Assess the ABCs, treat any life threats, and transport immediately if the pediatric patient requires urgent care. Continue with the remainder of the pediatric patient assessment process if the patient is stable.
Hands-on ABCs	Perform a hands-on ABCs assessment. Assess and treat any life threats to the airway, breathing, and circulation and then assess the pediatric patient's central nervous system, and expose any injuries.
Airway	Ensure the airway is open, clear, and patent. Position the airway in a neutral sniffing position. You may have to place a small towel roll under the pediatric patient's shoulders to maintain the neutral airway. Unresponsive pediatric patients will need their airway opened and maintained using a jaw-thrust maneuver if spinal injury is suspected. Suction as needed.
Breathing	Assess breathing by using the look, listen, and feel technique. Note the degree of air movement at the nose and mouth and determine if the chest is rising adequately and symmetrically. Assess the respiratory rate and effort by placing both hands on the pediatric patient's chest to feel for the rise and fall of the chest wall. Listen for abnormal respiratory sounds. Note any signs of increased work of breathing, including accessory muscle use, retractions, head bobbing, nasal flaring, tachypnea. Bradypnea indicates impending respiratory arrest. Be prepared to begin ventilatory assistance.

Assessment and Emergency Care of Pediatric Emergencies, continued

Circulation	Determine if the pediatric patient has a pulse, is bleeding, or is in shock. Pediatric patients decompensate rapidly and without many early warning signs. Assess and control active bleeding early. Palpate the brachial or femoral pulse in an infant. Palpate a carotid pulse in children older than 1 year. Absence of a central pulse requires CPR starting with chest compressions. Bradycardia often indicates impending cardiopulmonary arrest. Feel the skin for temperature and moisture. Assess capillary refill time.
Disability	Assess the level of consciousness with the AVPU scale or Pediatric Glasgow Coma Scale. Check the pupils' response to light. Check for symmetric movement of the extremities. Assess the pediatric patient's pain level.
Exposure	Remove the pediatric patient's clothing and observe the face, chest wall, and skin. Evaluate physiologic functions, anatomic abnormalities, and unsuspected injuries or rashes. Avoid heat loss by covering the pediatric patient as soon as possible.
Transport Decision	Consider the clinical problem, the benefits of field management, transport time, and your comfort level. If the patient has an airway or breathing problem or other life threats, manage them immediately and consider rapid transport, performing the secondary assessment en route to the hospital. If the patient weighs less than 40 lb and the condition allows, transport in a car seat secured to the stretcher or captain's chair. If the parent is not present and the caregiver is unable to accompany the pediatric patient to the hospital, obtain emergency contact information and bring it to the hospital with you.

NOTE: The order of the steps in this section differs depending on whether the patient is conscious or unconscious. The following order is for a conscious patient. For an unconscious patient, perform a primary assessment, perform a full-body scan, obtain vital signs, and obtain the past medical history from a family member, bystander, or emergency medical identification device.

History Taking

Investigate Chief Complaint	Investigate the chief complaint and gather a history once you have identified and treated life threats. If the patient is an infant or toddler you will most likely need to get this information from the parent or caregiver. Note the patient's position and mannerism. Provide privacy for the patient, especially the adolescent patient when asking embarrassing questions such as sexual activity and the possibility of pregnancy. Identify associated symptoms and pertinent negatives. Ask SAMPLE questions, focusing on the events surrounding the incident and the MOI or NOI. SAMPLE can also be obtained from family, bystanders, and medical alert tags if the patient is not able to provide the information. Also ask about urine output, if the patient has diarrhea or vomiting, loss of appetite, fevers, or rashes. If the patient has a traumatic injury, try to determine what happened. If a fall occurred, determine the height of the fall and the surface the patient landed on.

Secondary Assessment

Physical Examinations	Remove or loosen clothes to expose the injured area. Significant trauma requires a full-body scan using DCAP-BTLS. Focus the assessment on an isolated injury once all the systems have been examined. Perform a focused assessment on patients without life-threatening illness or injury. The focused assessment for younger children should begin at the feet, ending at the head. Assess the fontanelles of children up to 18 to 24 months of age. Ensure the nasal passages are clear, suction as needed with a bulb syringe. Note any unusual breath odors. Fever and a stiff neck (nuchal rigidity) may indicate meningitis. Assess the extremities for symmetry and range of motion.

Assessment and Emergency Care of Pediatric Emergencies, continued

Vital Signs	Vital signs should include blood pressure by auscultation in children older than 3 years, pulse rate and quality, respiration rate and quality, pupils, and skin assessment for perfusion. Note patient's level of consciousness. Use pulse oximetry, if available, to assess the patient's perfusion status. Use appropriate sized equipment when assessing the vital signs to ensure accuracy; the use of a color-coded, length-based resuscitation tape (Broselow) can assist you in choosing the correct equipment.

Reassessment

Interventions	Reassess the chief complaint, repeat the primary assessment including vital signs, pain, and any interventions already performed. Vital signs and reassessment should be repeated every 5 minutes in the serious or unstable pediatric patient and every 15 minutes in the stable pediatric patient. Compare vital signs to those taken earlier. Continuously observe and reassess the pediatric patient during transport so worsening conditions can be managed. Pediatric patients decompensate quickly, patients may appear stable one minute and the next they are not. Whenever possible, include the parent or caregiver in the care of the pediatric patient; they can assist you with keeping the patient calm.
Communication and Documentation	Contact medical control/receiving hospital with a radio report including medical and anatomic terminology so appropriate staff and equipment will be ready when you arrive. Include a thorough description of the MOI/NOI and position the patient was found. Include treatments performed and pediatric patient response. Be sure to document the patient's chief complaint, parent or caregiver's description of the incident, physical findings, history, and any changes in patient status and the time. Document the scene observations on your arrival and be prepared to accurately describe the scene for the hospital staff. Follow local treatment protocols.

NOTE: Although the following steps are widely accepted, be sure to consult and follow your local protocols. Take appropriate standard precautions when treating all patients.

Pediatric Emergencies

General Management of Pediatric Emergencies

Managing life threats to the pediatric patient's ABCs are primary concerns with any emergency. Life-threatening illness or injury must be managed once discovered during the primary assessment. Remain calm. In all of the following specific situations, perform the following:

1. Ensure scene safety.
2. Determine the MOI/NOI.
3. Consider the need for spinal stabilization.
4. Determine the pediatric patient's level of consciousness and mental status using AEIOU-TIPPS.
5. Open, clear, and maintain the patient's airway.
6. Ensure adequate ventilation.
7. Administer high-concentration oxygen via a nonrebreathing mask or bag-mask device as appropriate. Use caution to avoid gastric distention, which can decrease lung volume; also avoid overinflation of the lung, which may cause a pneumothorax.
8. Control bleeding and treat for shock. Remember that internal injuries may be present without obvious bleeding or external injury.
9. Place the patient in a position of comfort.
10. Transport to the appropriate treatment facility.

Assessment and Emergency Care of Pediatric Emergencies, continued

Pediatric Emergencies

Respiratory Emergencies

Combativeness, restlessness, anxiety, nasal flaring, grunting, wheezing, stridor, accessory muscle use, retractions, and tripod positioning are all signs or symptoms of respiratory distress. Recognize and manage respiratory emergencies early and determine the cause.

Airway Obstruction

Determine the cause of the obstruction and whether it is involving the upper or lower airway. Upper airway obstructions may be from blood or vomit, a foreign body, croup, or epiglottitis. Fluid of any sort obstructing the airway must be suctioned as soon as possible. Conscious infants with a foreign body obstructing the airway should be managed with back slaps and chest thrusts. Conscious children with a foreign body obstructing the airway should have abdominal thrusts performed in an attempt to dislodge the object. All patients that become unconscious with a foreign body airway obstruction should be managed using cardiopulmonary resuscitation. Patients with croup should receive humidified oxygen and transport to the hospital. Patients with fever, lethargy, and drooling should be suspected of having epiglottitis. Do not agitate the patient because it may cause the airway to become completely obstructed. Allow the patient to stay with a parent or caregiver in the back of the ambulance and provide blow-by oxygen. Oxygen can be administered using a nonrebreathing mask or a paper cup and oxygen tubing taped through the bottom with the parent holding it near the child's face. A calm, rapid transport is required. Notify the hospital you are bringing a patient with suspected epiglottitis.

Asthma

Asthma may cause a lower airway obstruction and is characterized by the presence of wheezing and the pediatric patient sitting in a tripod position. Administer high-concentration oxygen with a nonrebreathing mask or blow-by oxygen as tolerated. If local protocol allows, assist with the pediatric patient's metered dose inhaler (MDI) or nebulized albuterol. Assist ventilations as required, using slow gentle breaths. Prolonged asthma attacks (status asthmaticus) require advanced life support intervention.

Shock (Hypoperfusion)

Common causes of shock include trauma with blood loss, burns, dehydration, infection, neurologic injury, severe allergic reaction, heart failure, and pneumothorax. Signs and symptoms may include tachycardia, tachypnea, delayed capillary refill, altered mental status, and pale or cyanotic skin. Treat for shock early, even if the pediatric patient is compensating; pediatric patients decompensate very quickly. Treatment for shock includes the administration of oxygen, placing the patient in a position of comfort, and maintaining body temperature. Pediatric patients in cardiogenic or anaphylactic shock may not want to be placed supine if they have an accompanying respiratory problem.

Anaphylaxis/Anaphylactic Shock

If you suspect the pediatric patient is having a severe allergic reaction (anaphylaxis) and the pediatric patient has hives, stridor, or wheezing, provide high-concentration oxygen, determine if the pediatric patient has a prescribed epinephrine auto-injector, and prepare for rapid transport. Request ALS assistance. Anaphylactic shock can develop rapidly. If you are authorized to assist or administer the epinephrine auto-injector, you should do so.

Assessment and Emergency Care of Pediatric Emergencies, continued

Pediatric Emergencies

Altered Mental Status

Use the AEIOU-TIPPS mnemonic to help you determine the cause of the pediatric patient's altered mental status. Manage any problems found with airway, breathing, and circulation immediately.

Seizures

If seizure is the cause of the altered mental status, look further into the cause. The pediatric patient is in status epilepticus if he or she has continuous seizures or seizures lasting more than 30 minutes without regaining consciousness. This is a life threat. Request ALS so medication can be administered to stop the seizure. Treatment for seizures consists of protecting the pediatric patient from injury, maintaining an open airway, suctioning as necessary, and administration of 100% oxygen via nonrebreathing or bag-mask device as required. Do not attempt to place anything in the pediatric patient's mouth. Transport to the appropriate facility. Place the pediatric patient in the recovery position once the seizure stops.

Febrile Seizures

Fever is a common reason for EMS response. Children ages 6 months to 6 years commonly have seizures related to fever. They usually are of short duration. Ensure there are no concerns with the ABCs and cool the child by placing towels moistened with tepid water on the chest and back. Transport the child for further evaluation. The concern is not the fever itself, but the cause of the fever. If meningitis is suspected, don an N-95 mask and inform the receiving hospital of your suspicions. The pediatric patient with meningitis may have a stiff neck, and might refuse to move the neck, lift the legs, or curl into a "C" position because of the pain. Infants may present with bulging fontanelles. Provide oxygen and airway support as required.

Gastrointestinal Emergencies

Abdominal discomfort, nausea, vomiting, and diarrhea are common complaints in the pediatric population. Often of minor concern, these complaints can be the byproduct of serious illness. Pediatric patients with nausea, fever, and right lower quadrant pain may have appendicitis. Dehydration and constipation with abdominal discomfort may be signs of a bowel obstruction. Both of these conditions require immediate transport to the hospital.

Poisoning

During your scene size-up, you should have scanned the area for household cleaning products, medicines, and other chemicals that may be the reason for EMS response. Perform a primary assessment and manage any life threats. Look for chemical burns in the mouth. If poisoning is suspected, in addition to the SAMPLE questions, ask about the substance, how much was on hand and how much was ingested, the time the event occurred, and any interventions taken prior to EMS arrival, such as inducing vomiting or administration of milk or water. Manage the ABCs. If the pediatric patient is conscious and able to swallow, medical control may ask that activated charcoal be administered. Transport to the appropriate hospital.

Assessment and Emergency Care of Pediatric Emergencies, continued

Pediatric Emergencies

Dehydration

Pediatric patients have less fluid volume than adults, so what may seem to be a simple case of vomiting or diarrhea is in fact a possible life threat. If your assessment reveals dry mucous membranes, decreased secretions, sunken eyes or fontanelles, loose skin (tenting), or the parent or caregiver states that there were less than normal the amount of diaper changes, suspect dehydration. Manage the ABCs and request ALS. Continuously monitor the vital signs and transport to the hospital for further evaluation.

Drowning Emergencies

Ensure personal and crew safety. Do not attempt a water rescue unless trained and authorized to do so. Attempt to determine the length of time the pediatric patient was submerged, the temperature of the water, and if a diving injury occurred. Shallow water diving may have caused a spinal injury that needs to be managed; always assume the child has neck and head injuries. Maintain the airway, taking spinal precautions, and manage any breathing or circulation issues. If trauma is suspected, immobilize the pediatric patient to a long backboard and pad under the back/shoulders to maintain a neutral airway. Consider requesting ALS. If cardiopulmonary resuscitation is required, ensure the pediatric patient is dry before attempting to use an automated external defibrillator. Cold water drowning victims should be treated aggressively no matter how long they were submerged.

Trauma Emergencies

Trauma is the number one killer of pediatric patients in the United States. Because of anatomic differences, a pediatric trauma victim may display differing injury patterns than an adult with a similar MOI. The pliable bones of a pediatric patient make internal injuries increasingly possible without the obvious outward signs. Investigate the MOI thoroughly. Head injuries are common; look for signs and symptoms of a head injury such as nausea and vomiting. Pediatric patients with head injuries need to be immobilized. Blunt chest trauma can cause significant internal injuries. Breathing needs to be monitored and assisted if necessary. Abdominal injuries may not be apparent early, but consider the possibility of internal bleeding and treat early for shock (hypoperfusion). Burns cause hypothermia and shock. Determine the extent of area burned using the rule of nines chart, the type of burn, and the depth. Inhalation burns need aggressive airway management; contact ALS to assist with airway management. Chemical burns as the result of ingestion may also affect the pediatric patient's ability to breath adequately. Extremity fractures should be splinted using the appropriate size equipment. Elevation and ice may be used to assist with pain management. If child abuse is suspected, remember that the patient is the priority; treat and transport the patient and do not confront the abuser. On arrival at the hospital, inform the hospital staff of your suspicions and complete the mandatory child abuse report form.

Prep Kit

- Children are not only smaller than adults and more vulnerable, they are also anatomically, physiologically, and psychologically different from adults in some important ways.

- Infancy is the first year of life. If possible, allow the parent or caregiver to hold the infant during the assessment.

- The toddler is 1 to 3 years of age. Toddlers may experience stranger anxiety but may be able to be distracted by a special object (blanket) or toy.

- Preschool-age children are 3 to 6 years of age. Preschool-age children can understand directions and can identify painful areas when questioned. Tell these children what you are going to do before you do it. This action can help prevent the development of frightening fantasies.

- School-age children are 6 to 12 years of age. These children are familiar with the physical examination process. Talk about their interests to distract them during a procedure.

- Adolescents are 12 to 18 years of age. Respect the adolescent's modesty. Remember that even though this age group is physically similar to adults, adolescents are still children on an emotional level.

- General rules for dealing with pediatric patients of all ages include appearing confident, being calm, remaining honest, and keeping parents or caregivers together with the pediatric patient as much as possible.

- The growing bodies of the pediatric patient create some special considerations.

- The tongue is large relative to other structures, so it poses a higher risk of airway obstruction than in an adult.

- An infant breathes faster than an older child.

- Breathing requires the use of chest muscles and the diaphragm.

- The airway in a child has a smaller diameter than the airway in an adult and is therefore more easily obstructed.

- A rapid heart beat and blood vessel constriction helps pediatric patients to compensate for decreased perfusion.

- Children's internal organs are not as insulated by fat and may be injured more severely, and children have less circulating blood, so that, although children exhibit the signs of shock more slowly, they go into shock more quickly, with less blood loss.

- Children's bones are more flexible and bend more with injury and the ends of the long bones, where growth occurs, are weaker and may be injured more easily.

- Because a young child might not be able to speak, your assessment of his or her condition must be based in large part on what you can see and hear yourself. Families may be helpful in providing vital information about an accident or illness.

- Use the pediatric assessment triangle to obtain a general impression of the infant or child.

- You will need to carry special sizes of airway equipment for pediatric patients.

- Use a pediatric resuscitation tape measure to determine the appropriately sized equipment for children.

- The three keys to successful use of the bag-mask device in a child are: (1) have the appropriate equipment in the right size; (2) maintain a good face-to-mask seal; and (3) ventilate at the appropriate rate and volume.

- Signs of shock in children are tachycardia, poor capillary refill time, and mental status changes. You must be very alert for signs of shock in a pediatric patient because they can decompensate rapidly.

- Febrile seizures may be a sign of a more serious problem such as meningitis.

- The most common cause of dehydration in children is vomiting and diarrhea. Life-threatening diarrhea can develop in an infant in hours.

- Fever is a common reason why parents or caregivers call 9-1-1. Body temperatures of 100.4°F (38°C) or higher are considered to be abnormal.

- Trauma is the number one killer of children in the United States.

- A victim of sudden infant death syndrome (SIDS) will be pale or blue, not breathing, and unresponsive. He or she may show signs of postmortem changes, including rigor mortis and dependent lividity; if so, call medical control to report the situation.

- Carefully inspect the environment where a SIDS victim was found, looking for signs of illness, abusive family interactions, and objects in the child's crib.

- Provide support for the family in whatever way you can, but do not make judgmental statements.

- Any death of a child is stressful for family members and for health care providers. In dealing with the family, acknowledge their feelings, keep any instructions short and simple, use the child's name, and maintain eye contact.

- Be prepared to respond to philosophical as well as medical questions, in most cases by indicating concern and understanding; do not be specific about the cause of death.

- Be alert for signs of posttraumatic stress in yourself and others after dealing with the death of a child. It can help to talk about the event and your feelings with your EMS colleagues.

Vital Vocabulary

adolescents Children between 12 to 18 years of age.

apparent life-threatening event (ALTE) An event that causes unresponsiveness, cyanosis, and apnea in an infant, who then resumes breathing with stimulation.

blanching Turning white.

bradypnea Slow respiratory rate; ominous sign in a child that indicates impending respiratory arrest.

central pulses Pulses that are closest to the core (central) part of the body where the vital organs are located; include the carotid, femoral, and apical pulses.

child abuse A general term applying to all forms of child abuse and neglect.

generalized tonic-clonic seizure A seizure that features rhythmic back-and-forth motion of an extremity and body stiffness.

infancy The first year of life.

grunting An "uh" sound heard during exhalation; reflects the child's attempt to keep the alveoli open; a sign of increased work of breathing.

nares The external openings of the nostrils. A single nostril opening is called a naris.

neglect Refusal or failure on the part of the caregiver to provide life necessities.

Neisseria meningitides A form of bacterial meningitis characterized by rapid onset of symptoms, often leading to shock and death.

pediatric assessment triangle (PAT) A structured assessment tool that allows you to rapidly form a general impression of the infant or child without touching him or her; consists of assessing appearance, work of breathing, and circulation to the skin.

pediatric resuscitation tape measure A tape used to estimate an infant or child's weight on the basis of length; appropriate drug doses and equipment sizes are listed on the tape.

pediatrics A specialized medical practice devoted to the care of the young.

pertussis An acute infectious disease characterized by a catarrhal stage, followed by a paroxysmal cough that ends in a whooping inspiration. Also called whooping cough.

preschool-age Children between 3 to 6 years of age.

school-age Children between 6 to 12 years of age.

shaken baby syndrome A syndrome seen in abused infants and children; the patient has been subjected to violent, whiplash-type shaking injuries inflicted by the abusing individual that may cause coma, seizures, and increased intracranial pressure due to tearing of the cerebral veins with consequent bleeding into the brain.

sniffing position Optimum neutral head position for the uninjured child who requires airway management.

sudden infant death syndrome (SIDS) Death of an infant or young child that remains unexplained after a complete autopsy.

toddler The period following infancy until 3 years of age.

tachypnea Increased respiratory rate.

work of breathing An indicator of oxygenation and ventilation. Work of breathing reflects the child's attempt to compensate for hypoxia.

Assessment in Action

Y ou and your partner are called to an elementary school cafeteria for an 8-year-old boy with an altered mental status. On arrival, you find the child lying supine on the cafeteria floor. He is responsive to verbal stimuli and is confused as to what happened and where he is.

1. You find out from the child's teacher that he has a history of seizures and recently had a medication change. What are other common causes of altered mental status in the pediatric population?
   A. Hypoglycemia
   B. Drug ingestion
   C. Hypoxia
   D. All of the above

2. The _____ is a structured assessment tool that is used before addressing the hands-on ABCs; it allows you to rapidly form a general impression of the pediatric patient's condition without touching him or her.
   A. Pediatric assessment triangle
   B. Glascow Coma Scale
   C. AVPU scale
   D. TICLS mnenomic

3. The three components of the pediatric assessment triangle are _____, work of breathing, and circulation to the skin.
   A. airway
   B. appearance
   C. activity
   D. ability

4. Evaluating the child's appearance involves noting the mental status and:
   A. skin color.
   B. pupil size.
   C. chest rise.
   D. muscle tone.

5. Once the seizure stops, a child's muscles relax and breathing becomes rapid. This is referred to as the:
   A. recovery state.
   B. postictal state.
   C. aural state.
   D. reversal state.

6. Status epilepticus is a condition in which seizures continue every few minutes without the patient regaining consciousness or seizures last longer than:
   A. 30 minutes.
   B. 45 minutes.
   C. 60 minutes.
   D. 120 minutes.

7. What is your priority in caring for a pediatric patient with seizures?
   A. Stop the seizure.
   B. Secure the airway.
   C. Provide rapid transport.
   D. Maintain the patient's body temperature.

8. Explain how seizure activity in an infant differs from that of an older child.

9. The postictal state is over when the patient's _____ returns to normal.
   A. level of consciousness
   B. respiratory rate
   C. muscle tone
   D. blood pressure

10. When you are using the pediatric assessment triangle, what information can you obtain by evaluating the appearance of a pediatric patient?

Geriatric Emergencies

National EMS Education Standard Competencies

Special Patient Populations

Applies a fundamental knowledge of growth, development, and aging and assessment findings to provide basic emergency care and transportation for a patient with special needs.

Geriatrics

- Impact of age-related changes on assessment and care (pp 1217-1223, 1238-1240)
- Changes associated with aging, psychosocial aspects of aging and age-related assessment and treatment modifications for the major or common geriatric diseases and/or emergencies
  - Cardiovascular diseases (pp 1217-1223, 1225-1228)
  - Respiratory diseases (pp 1217-1225)
  - Neurological diseases (pp 1217-1223, 1228-1231)
  - Endocrine diseases (pp 1217-1223, 1233-1234)
  - Alzheimer disease (pp 1217-1223, 1229)
  - Dementia (pp 1217-1223, 1229)

Patients With Special Challenges

- Recognizing and reporting abuse and neglect (pp 1243-1245 and Chapter 32, *Pediatric Emergencies*)

Health care implications of

- Abuse (pp 1243-1245 and Chapter 32, *Pediatric Emergencies*)
- Neglect (pp 1243-1245 Chapter 32, *Pediatric Emergencies*)
- Homelessness (Chapter 34, *Patients With Special Challenges*)
- Poverty (Chapter 34, *Patients With Special Challenges*)
- Bariatrics (Chapter 34, *Patients With Special Challenges*)
- Technology dependent (Chapter 34, *Patients With Special Challenges*)
- Hospice/terminally ill (Chapter 34, *Patients With Special Challenges*)
- Tracheostomy care/dysfunction (Chapter 34, *Patients With Special Challenges*)
- Homecare (Chapter 34, *Patients With Special Challenges*)
- Sensory deficit/loss (Chapter 34, *Patients With Special Challenges*)
- Developmental disability (*Chapter 34, Patients With Special Challenges*)

Trauma

Applies fundamental knowledge to provide basic emergency care and transportation based on assessment findings for an acutely injured patient.

Special Considerations in Trauma

- Recognition and management of trauma in the:
  - Pregnant patient (Chapter 31, *Obstetrics and Neonatal Care*)
  - Pediatric patient (Chapter 32, *Pediatric Emergencies*)
  - Geriatric patient (pp 1236-1241)
- Pathophysiology, assessment, and management of trauma in the:
  - Pregnant patient (Chapter 31, *Obstetrics and Neonatal Care*)
  - Pediatric patient (Chapter 32, *Pediatric Emergencies*)
  - Geriatric patient (pp 1236-1241)
  - Cognitively impaired patient (Chapter 34, *Patients With Special Challenges*)

Knowledge Objectives

1. Define the term "geriatrics." (p 1215)
2. Appreciate some of the special aspects of the lives of elderly persons. (pp 1215-1216)
3. Discuss generational considerations when communicating with a geriatric patient. (pp 1215-1216)
4. Describe the common complaints and the leading causes of death in the elderly. (pp 1216-1217)
5. Discuss special considerations when performing the patient assessment process on a geriatric patient with a medical condition. (pp 1217-1223)
6. Explain the GEMS Diamond and its role in the assessment and care of the geriatric patient. (pp 1221-1223)
7. Discuss the physiologic changes associated with the aging process and the age-related assessment and treatment modifications that result. (pp 1223-1236)
8. Define "polypharmacy," and explain the toxicity issues that can result. (p 1235)
9. Discuss the effect of aging on psychiatric emergencies. (p 1235-1236)
10. Explain special considerations when performing the patient assessment process on a geriatric patient with a traumatic injury. (pp 1238-1240)
11. Discuss the effects of aging on environmental emergencies. (p 1241)
12. Discuss the special considerations when responding to calls to nursing and skilled care facilities. (pp 1241-1242)
13. Define an advanced directive and explain its use with older patients. (pp 1242-1243)
14. Describe the causes of elder abuse and neglect, and explain why the extent of elder abuse is not well known. (pp 1243-1245)
15. Describe the assessment and care of a geriatric patient who has potentially been abused or neglected. (pp 1244-1245)

Skills Objectives

There are no skills objectives for this chapter.

Introduction

Geriatrics is the assessment and treatment of disease in a person who is 65 years or older. The baby boomer generation, people born between 1940 and 1960, are much more active today than people their age were in previous generations. In this chapter, we use 65 years as the threshold age to be consistent with the definition used by other medical groups and governmental agencies. How fast one ages, though, is a function of genetics, lifestyle, and, perhaps, attitude.

The process of aging is gradual and starts much earlier than most people realize. A decline in the systems of the human body starts as early as the late 20s and progresses slowly throughout life. The older population continues to grow to a larger percentage of the overall population. According to US census data, by 2010 more than 40 million people will be older than 65 years. It is projected that by the year 2025, the geriatric population will be almost 64 million, and in 2050, 88 million. This is a significant trend for EMTs because older people are major users of the EMS and health care systems in general.

Geriatric patients present as a special problem for caregivers because the classic presentation of disease is often altered by the presence of chronic conditions and the physiology of aging. In addition, geriatric patients may be taking multiple medications that can interact and have toxic effects.

Providing effective treatment for this growing number of patients requires that you understand the issues related to aging and that you modify some of your assessment and treatment approaches.

Being a patient advocate for an older patient involves much more than management of medical and traumatic emergencies. As the elderly population increases, communities, companies, and hospitals are encouraging awareness of geriatric issues through the media and creating programs that promote prevention of injuries. EMTs who respond to the homes of geriatric patients are in an ideal position to provide key information to others in health care and social services systems. Interventions for geriatric patients may include reviewing the home environment to ensure that safe and tolerable living conditions exist, providing information on preventing falls, and making referrals to appropriate social services agencies when your patient needs assistance. Often, simple preventive measures can help older people avoid further injury, costly medical treatment, and death. You are in a position to not only recognize and manage a serious emergency, but also to help to prevent problems before they occur.

Generational Considerations

It is important to understand and appreciate how the life of an elderly person might differ from yours. In their generation, many children referred to an adult as "Mr." or "Mrs." You will see elderly people who have recently lost a spouse and are struggling to fill the spouse's role, such as by managing the finances or doing the housework. Many elderly people also live on a fixed income, which can be very challenging. Some elderly people may not take all of their medications to save money. Many struggle to stay independent as long as possible and are living in the homes in which they raised their children—children who have now moved far away.

It takes time and patience to interact with an elderly person, which can sometimes be frustrating. Have patience and treat the patient with respect. Make every attempt to avoid "ageism," which is a stereotyping of elderly people that often leads to discrimination. Common stereotypes include assuming that the patient has

You are the Provider: PART 1

At 6:25 AM, you are dispatched to a residence at 644 Yolanda Street for an 82-year-old woman whose daughter called 9-1-1 because her mother is short of breath. The weather is cloudy, the temperature is 66°F, and the traffic is light. You and your partner respond to the scene, which is located only a short distance away.

1. Why is it important for the EMT to understand the anatomic and physiologic changes that occur with aging?
2. How does the process of aging affect a person's respiratory system?

Figure 33-1 Older people can stay fit and be active.

dementia, is hard of hearing, has a sedentary lifestyle, or is immobile. Elderly people can stay fit and be active, even though they are not able to perform at the same level as they did in their youth **Figure 33-1** . Many participate in the community, and are generally healthy.

Communication and Older Adults

Good verbal communication skills are essential to the successful assessment and treatment of older patients. Communication with elderly people can be challenging. The aging process brings with it changes in vision, hearing, taste, smell, and touch. Also, there are changes in communication abilities that accompany aging, dementia, and other diseases. These challenges are normal consequences of aging and should be expected, but not assumed.

Words of Wisdom

Elderly patients who learned English as a second language may revert back to their native language when stressed or in a crisis situation. Do not assume that the patient cannot understand you just because he or she is speaking another language.

■ Communication Techniques

Your first words to the patient and the attitude behind them can gain or lose a patient's trust. Speak respectfully when you introduce yourself. Using the patient's name shows respect and helps the patient focus on your questions. Older people may be insulted, however, if you use their first name. Introduce yourself with confidence and use their last name. Address the patient by using sir, ma'am, mister, missus, or miss. Asking patients how they

prefer to be addressed builds trust. Never use familiar or casual terms when referring to your patients unless they have told you to do so.

When you are interviewing an older patient, the following techniques should be used:

- Identify yourself. This helps to establish a relationship.
- Be aware of how you present yourself. Avoid showing frustration and impatience through body language.
- Look directly at the patient at eye level and ensure good lighting.
- Speak slowly and distinctly. Do not raise your voice. Try to talk in a lower tone to help facilitate communication.
- Have one person talk to the patient and ask only one question at a time. Do not answer questions for patients out of frustration or impatience.
- Do not assume that all older patients are hard of hearing. Ask the patient if he or she can hear you, and verify by asking him or her to confirm understanding of what you just said. Be aware of the presence of hearing aids. Ensure hearing aids are being worn and have working batteries.
- Give the patient time to respond unless the condition appears urgent.
- Listen to the answer the patient gives you.
- Explain what you are going to do before you do it. Use simple terms to explain the use of medical equipment and procedures; avoid using medical jargon or slang.
- Do not talk about the patient in front of him or her as though the patient was not there. This gives the patient the feeling that he or she has no say in decisions about his or her care. This is easy to forget when the patient has impaired cognitive (thought) processes or has difficulty communicating.

Words of Wisdom

Remember, as with patients of any age, older patients may have more difficulty communicating clearly when they are stressed by an emergency or personal crisis.

Common Complaints and the Leading Causes of Death in Elderly People

The changing physiology of geriatric patients can predispose this population to a host of problems not seen in youth. The acuity of a simple rib fracture in a 30- or 40-year-old patient can be inconsequential. The same injury in a geriatric patient who is 80 or 90 years old can result in pneumonia and even a deadly secondary lung infection. A hip fracture from a low-mechanism fall is common in elderly people and may have dire consequences. Hip fractures are more likely to occur when bones are weakened by osteoporosis or infection. Sedentary behavior while healing can predispose the patient to pneumonia and blood clots that may interfere with healing and can even cause death. Many patients who experience hip fractures do not return to their preinjury levels of activity. All of these factors make assessment and treatment decisions more complex and patient complaints potentially more serious.

The most frequently occurring conditions in older persons are hypertension, respiratory diseases, diagnosed arthritis, and heart disease. The leading causes of death in the geriatric population include heart disease, cancer, stroke, chronic obstructive pulmonary disease and other respiratory illnesses, diabetes, and trauma **Table 33-1**.

Special Considerations in Assessing a Geriatric Medical Patient

Assessing an elderly person can be challenging because of communication issues, hearing and vision deficits, alteration in consciousness, complicated medical histories, and

Table 33-1	**Common Conditions and the Leading Causes of Death in Geriatric Patients**

Common Conditions

- Hypertension (48%)
- Diagnosed arthritis (47%)
- Heart disease (32%)
- Cancer (20%)
- Diabetes (16%)
- Sinusitis (14%)

Leading Causes of Death

- Heart disease
- Cancer
- Stroke
- Chronic obstructive pulmonary disease
- Pneumonia and influenza
- Diabetes
- Trauma

Source: Sahyoun RN, Lentzner H, Hoyert D, Robinson KN. *Trends in Causes of Death Among the Elderly.* Hyattsville, MD: National Center of Health Statistics; March 2001. Aging Trends Series No. 1.

the effects of medications. Previous injury or illnesses that are not associated with the current problem may also alter the assessment findings. This may include medications that mask changes in vital signs that you might expect,

You are the Provider: PART 2

You arrive at the scene and are met at the door by the patient's daughter. She advises you that her mother has emphysema and that her respiratory distress has worsened during the past few days. You find the patient sitting in a chair in her living room; she has a blanket wrapped around her and is in mild respiratory distress. You introduce yourself to the patient and perform a primary assessment.

Recording Time: 0 Minutes	
Appearance	Obvious respiratory distress; chills
Level of consciousness	Conscious but confused
Airway	Open; clear of secretions and foreign bodies
Breathing	Increased rate; labored
Circulation	Radial pulse, irregular and weak; skin is pink, warm, and dry; no obvious bleeding

The patient is using home oxygen at 2 L/min via nasal cannula. She lives alone but is regularly checked on by her daughter and son-in-law. The patient's home appears to be well kept, and you can see numerous medication containers on a nearby table.

3. What is the GEMS diamond? How can it facilitate your overall care of an older patient?

such as tachycardia in shock. A previous stroke may have changed a patient's baseline level of consciousness and neurologic status.

Words of Wisdom

Medical and trauma conditions are often superimposed on each other. A simple fall may have been preceded by weakness and dizziness, suggesting a serious medical condition. Recovery from a hip fracture will be difficult in a patient with osteoporosis. Wounds in a patient with diabetes will take longer to heal.

Scene Size-up

Scene Safety

Every emergency call begins with a thorough scene size-up. Is the scene safe? Will you need assistance? How many patients do you have? What is the nature of illness (NOI)? Have you taken proper standard precautions?

Geriatric patients are commonly found in their own homes, retirement homes, or skilled nursing facilities, but calls for assistance can come from any location. Many older people live alone. Access to them may be hampered in cases in which the patients have conditions that prevent them from getting to the door to let you in such as chest pain, breathing difficulty, or an altered level of consciousness. Police or fire department assistance may be required.

Many older people try to maintain their independence as long as they can. They may or may not have someone who checks on their welfare. You will find some people living in conditions that are not safe or appropriate. You need to take note of negative or unsafe environmental conditions. Is the home well maintained and sanitary? Are the utilities working? Look for clues that might explain the patient's medical history or current problem: Is it too hot or too cold? A geriatric patient can have hypothermia or heat stroke in temperatures that are not considered extreme. Is there food available? Is there evidence of abuse or alcohol or illegal drug use? Are there medications on the nightstand in the bedroom?

In a nursing home or residential care facility, you will need to locate the patient's room and find a staff member who can explain why you were called. In any case in which the patient's mental status is altered, you need to find someone who can tell you the patient's history and whether the patient's behavior or level of consciousness is normal. The presence of a hospital bed, oxygen tanks, or therapeutic devices can give you a clue to the patient's medical history. The environment may give you the answer to questions when the patient cannot.

Mechanism of Injury/Nature of Illness

The NOI may be difficult to determine in older people who may have an altered mental status or dementia. Often it is someone other than the patient who called, so you must ask the family member, caregiver, or bystander why he or she called. Multiple and chronic disease processes may also complicate the determination of the NOI. Complaints from an elderly person may be vague, such as weakness, dizziness, or fatigue. These could be indicators of a more serious problem and require more assessment. You may need to ask specifically what is different *today* or specifically why the person called to determine acute versus chronic complaints. Chest pain, shortness of breath, and an altered level of consciousness should always be considered serious. You also may find that the patient's complaint is a symptom of something more serious. Sudden changes in the ability to talk could indicate a stroke, or the need to sleep on five pillows could suggest early congestive heart failure.

Primary Assessment

Once you have confirmed that the scene is safe, perform the primary assessment to address life threats such as problems with airway, breathing, and circulation. Determine the priority of your patient based on his or her condition. The priority of your patient may limit on-scene time and/or treatment.

Form a General Impression

The general impression is an important aspect of all patient assessment. As you approach the patient, you should be able to tell if the patient is generally in stable or unstable condition. You will use this information to help you with your further assessment. Use the AVPU (*Alert* to person, place, and day; responsive to *Verbal* stimuli; responsive to *Pain*; *Unresponsive*) scale to determine the patient's level of consciousness.

Airway and Breathing

Anatomic changes that occur as a person ages predisposes geriatric patients to airway problems. Aging and disease can compromise a patient's ability to protect his or her airway with loss of a gag reflex and normal swallowing mechanisms. Changes in level of consciousness, dementia, and poststroke weakness or paralysis can cause airway obstruction or aspiration. Ensure that the patient's airway is open and is not obstructed by dentures, vomitus, fluids, or blood. Suction may be necessary.

Anatomic changes with aging also affect a person's ability to breathe effectively. Increased chest wall stiffness, brittle bones, weakening of the airway musculature, and decreased muscle mass contribute to breathing problems. Loss of mechanisms that protect the upper airway, like cough and gag reflexes, cause a decreased ability to clear secretions. A decrease in the number of cilia that line the bronchial tree results in the inability of the patient to remove material from the lung, which can cause infection. In some patients, the alveoli are damaged, and a lack of elasticity results in a decreased ability to exchange oxygen and carbon dioxide. Superimposed on the physiologic changes are the chronic respiratory diseases common in elderly people that affect the ability of the patient to breathe effectively. Airway and breathing issues should be treated with oxygen as soon as possible.

Circulation

Poor perfusion is a serious issue in an older adult. People who normally live with compromised circulation have little in the way of reserves during a circulatory crisis. Physiologic changes may negatively affect circulation. Less responsive nerve stimulation may lower the rate and strength of the heart's contractions, so lower heart rates and weaker and irregular pulses are common in elderly patients. Vascular changes and circulatory compromise might make it difficult to feel a radial pulse on an older patient. If choosing an alternative pulse point like the carotid, press gently. Another option is to listen to the apical pulse right over the heart. The pulse may be irregular because of common heart rhythm problems. Circulation problems in older adults should be treated with oxygen as soon as possible.

Safety

Dentures should not be removed unless they obstruct the airway or interfere with ventilation when rescue breathing is needed.

Transport Decision

Patient assessment is more complicated in an elderly adult, and multiple problems can exist. Any complaints that compromise airway, breathing, or circulation should result in transportation of the patient as a priority patient. Your most important task is to determine conditions that are life threatening, treat them to the best of your ability, and provide transport to priority patients. Priority patients include patients who have a poor general impression, airway or breathing problems, acute altered level of consciousness, shock, any severe pain, or uncontrolled bleeding. Elderly people do not have the reserves that younger people do, and they will easily decompensate. Even a general complaint of weakness and dizziness can be an indication of something more serious like a heart problem. Consider early on in your call if advanced life support (ALS) treatment and immediate transport is appropriate and available. If possible, try to take the patient to a facility where the patient has been treated before and his or her medical records reside.

Words of Wisdom

Because of fear related to hospitalization, many older people will wait until their problem is life threatening and then finally call 9-1-1. A good example is a patient with congestive heart failure who will have significant pulmonary edema when you arrive. A cold could have become pneumonia. The elderly body does not have the capacity or flexibility to manage injury or illness like the younger body does. Complaints of shortness of breath and chest pain signify serious conditions that could easily become worse en route to the hospital. If a patient's complaint of chest pain indicates a myocardial infarction, the patient could go into cardiac arrest. Be alert and prepared for a worst-case scenario.

History Taking

Investigate Chief Complaint

Begin by inquiring about the chief complaint or history of the present illness. Find and account for all medications. If a patient lives alone, make sure to look for evidence that medication information may have been put in a specific place to provide a medication history for caregivers or EMS personnel. The "Vial of Life" is one program that recommends that the patient place a list of his or her medications on the refrigerator. There also may be similar local programs in your area. There should be a notice by the front door to alert you to this.

Communication may be more complicated with an older adult, but it is critical that you obtain a thorough patient history. Receiving information from an older adult may be hampered by an altered level of consciousness, a cognitive disorder, or vision or hearing impairments. It is best to obtain what information you can directly from the patient, but family or caregivers may need to assist.

The determination should be made early on as to whether the altered level of consciousness is acute or chronic. Chronic mental status impairment is not a normal process of aging but is caused by a pathologic or disease process. The elderly may not show severe symptoms even if they are very ill. It is important for you to determine your patient's "baseline" mental status. Is an acute change in level of consciousness the reason why you were called? Use family members, if available, to establish baseline mental status.

Multiple disease processes and multiple and/or vague complaints can make assessment complicated. Ask questions to assess the nature of the problem, and determine whether it may or may not be life threatening. Take a full set of vital signs, and ask what is "normal" for that patient. Use acronyms like OPQRST (Onset, Provocation/palliation, Quality, Radiation, Severity, Time of onset) for chest pain and PASTE (Progression, Associated chest pain, Sputum, Talking tiredness, Exercise tolerance) for shortness of breath. You may or may not be able to determine the exact nature of the problem and will need to use your general impression to guide you.

SAMPLE History

Getting an accurate SAMPLE history can be complicated, especially in a patient who has an altered mental status. You may have to depend on a relative or caregiver to help you in collecting a SAMPLE history. The chances are good that the chief complaint is related to a chronic medical problem and the patient may have experienced it before. Symptoms should be determined to be acute or chronic. Also note the signs you observe and your general impression. Allergies to food and medications are important. Is the patient taking any new medications? Make sure you have a list of the patient's medications or take the medications with you to the hospital. Patient medications can tell much about a patient's history. The hospital will also need to determine if the patient has been taking the medications as instructed or could have an overdose or underdose. If time and opportunity allow, check the patient's pill sorter. Has the patient taken every dose this week? Has the patient taken the medications that are scheduled for tomorrow or the next day?

The last meal is particularly important in a patient with diabetes, but lack of nutrition can have a negative effect on any patient. A history of last oral intake can indicate that the patient may be dehydrated. Last, what is the "event" that prompted the call? Again, it is advantageous to provide transport to a facility that "knows" the patient's medical history if the patient's condition and other factors allow. Another consideration is that the patient may go to more than one facility. Perhaps the patient goes to the local community hospital where the primary care doctor has admitting privileges for routine care but goes to the regional hospital for cardiology complaints.

Secondary Assessment

The secondary assessment may be performed on scene, en route to the emergency department, or in some cases, not at all. The priority of the patient will determine this for you.

Physical Examinations

Perform a physical examination when appropriate. You may find, though, that your elderly patient is not comfortable with being exposed. Protecting his or her modesty will help. The elderly are often cold, and you may have to remove several layers of clothing. Consider the need to keep your patient warm during your full-body scan.

Vital Signs

Vital signs may be different in elderly people because of the physiologic changes that come with aging, chronic disease, and the effects of medications. The heart rate should be in the normal adult range but may be compromised by medications such as beta-blockers. These medications keep the heart rate low and prevent the tachycardia that might be typically seen in dehydration or shock. Weaker and irregular pulses are common in elderly patients. The pulse may be irregular secondary to atrial fibrillation. Circulatory compromise may make it difficult to feel a radial pulse on an older patient, and other pulse points may need to be considered.

Blood pressure tends to be higher in elderly people. An elderly patient who has a blood pressure in a normal adult range could be hypotensive. Hypertension could signal impending stroke. Try to confirm if the patient has missed taking any medications for hypertension.

Capillary refill is not a good assessment tool in elderly adults because of skin changes and reduced circulation to the skin.

The respiratory rate should be in the same range as in a younger adult, but remember that chest rise will be compromised by increased chest wall stiffness. Be sure to auscultate breath sounds to listen for rales associated with pulmonary edema, rhonchi or rattles associated with pneumonia, and wheezes associated with asthma.

Monitoring Devices

Careful interpretation of pulse oximetry data is necessary in older adults because the pulse oximetry device requires adequate perfusion to get an accurate reading. Older adults may have poor circulation, vasoconstriction, hypotension, hypothermia, lack of red blood cells, or carbon monoxide poisoning that could result in an inaccurate reading. Adhesive temporal probes, if available and in local protocol, might help confirm accuracy of the data.

A blood pressure of 120/80 mm Hg, which may be normal for a younger adult, could be an indication of a significant problem in an elderly patient. Try to determine what the patient's normal blood pressure is. Your baseline blood pressure on this patient and any change from the patient's normal baseline can alert you to a potential problem.

Reassessment

Reassess the geriatric patient often because the condition of an older adult may deteriorate quickly. Repeat the primary assessment. Reassess the vital signs. Reassess the patient's complaint. Recheck interventions. Identify and treat changes in the patient's condition.

Interventions

Typical interventions include positioning, oxygenation, administration of glucose, and psychological support. In specific cases, you may also assist with nitroglycerin, aspirin, or inhalers. An elderly patient with a complaint of shortness of breath will want to sit up or assume the tripod position. Accommodations to these requests should be made except in cases in which you need to manage the patient's airway. The patient's position may be maintaining a patent airway. Forcing a patient supine who is short of breath may result in respiratory failure or arrest. Allow the patient to maintain a position of comfort, unless they are unable to maintain it. Provide ventilation as needed.

Oxygen is a useful therapy for many geriatric problems, including vague complaints of weakness or dizziness. When administering oxygen, be mindful of monitoring the level of consciousness in a patient with chronic obstructive pulmonary disease and the risks of providing a prolonged high concentration of oxygen. Be prepared to ventilate if the patient's hypoxic drive fails.

Words of Wisdom

In general, allow the patient to maintain a position of comfort unless contraindicated. If immobilization on a backboard is necessary, remember to pad the void spaces. You may have to weigh the risk of supine immobilization versus respiratory distress or failure. Be sure to document the reasons for your decision.

Diabetes is a common disease in the elderly. Consider glucose therapy for a patient with diabetes who has altered mental status but has a manageable airway. In a patient with a cardiac history, consider assisting with the patient's nitroglycerin if he or she is having chest pain, or assist a patient with medication for asthma when the patient is experiencing shortness of breath.

Last, and critically important in elderly adults, is providing psychological support. An older person is often fearful of what may be happening and that he or she may never return home from the hospital. Listen to your patient, respond to your patient, and provide reassurance.

Communication and Documentation

Communicate with the hospital staff on your findings and the interventions you used to improve the patient's condition. Be sure that all of this information is documented and given to emergency department personnel. Remember to document all history, medication, assessment, and intervention information.

A summary of the special considerations to keep in mind when assessing a geriatric patient can be found in **Table 33-2**.

■ The GEMS Diamond

When you are called on to care for older patients, it is important to remember certain key concepts. The GEMS diamond **Figure 33-2** was created to help you remember

Table 33-2 Geriatric Patient Assessment Guidelines

- When entering the home, take note of issues that would make it environmentally unsafe.
- Introduce yourself, show respect, and use patience to gain an older patient's confidence.
- Assessment of an elderly patient can be complicated by multiple medical or traumatic conditions, alterations in level of consciousness, and hearing and vision impairments.
- Airway, breathing, circulation, and vital signs are changed by the normal process of aging.
- Many older patients use multiple medications. Be aware of the possibility of overdose or underdose.
- An older person's body does not have the flexibility or reserves of a younger person's body when facing illness or injury.
- The elderly are more easily affected by poor nutrition.
- Older people cannot thermoregulate easily and tend to be cold.
- The memory and cognition of an older person may be impaired.
- The skin of an older adult may be fragile and can tear easily. Consider patient movement options that are safe and appropriate.

what is different about older patients. The GEMS diamond is not intended to be a format for the approach to geriatric patients, nor is it intended to replace the ABCs of care. Instead, it serves as an acronym for the issues to be considered when assessing every older patient.

The "G" of the GEMS diamond stands for "geriatric." When responding to an emergency involving an older patient, you should consider that older patients are different from younger patients and may present atypically.

The "E" of the GEMS diamond stands for an environmental assessment. Assessment of the environment can help give clues to the patient's condition and the cause of the emergency. Is the home too hot or too cold? Is the home well kept and secure? Are there hazardous conditions? Preventive care is also very important for a geriatric patient, who may not carefully study the environment or may not realize where risks exist.

The "M" of the GEMS diamond stands for medical assessment. Older patients tend to have a variety of medical problems and may be taking numerous prescription, over-the-counter (OTC), and herbal medications. Obtaining a thorough medical history is very important in older patients.

The "S" stands for social assessment. Older people may have less of a social network because of the death of a spouse, family members, and friends. Older people may also need assistance with activities of daily living, such as dressing and eating. There are numerous social agencies

G Geriatric Patients

- Present atypically
- Deserve respect
- Experience normal changes with age

E Environmental Assessment

- Check for hazardous conditions that may be present (eg, poor wiring, rotted floors, unventilated gas heaters, broken window glass, clutter that prevents adequate egress).
- Are smoke detectors present and working?
- Is the home too hot or too cold?
- Is there an odor of feces or urine in the home? Is bedding soiled or urine-soaked?
- Is food present in the home? Is it adequate and unspoiled?
- Are liquor bottles present? If so, are they lying empty?
- If the patient has a disability, are appropriate assistive devices (eg, ramps, rails, wheelchairs, or walkers) present?
- Does the patient have access to a telephone?
- Are medications out of date or unmarked, or are prescriptions for the same or similar medications from many physicians? Are any of the medications prescribed to other people?
- If living with others, is the patient confined to one part of the home?
- If the patient is residing in a nursing facility, does the care appear to be adequate to meet the patient's needs?

M Medical Assessment

- Older patients tend to have a variety of medical problems, making assessment more complex. Keep this in mind in all cases—both trauma and medical. A trauma patient may have an underlying medical condition that could have caused or may be exacerbated by the injury.
- Obtaining a medical history is important in older patients, regardless of the chief complaint.
- Primary assessment
- Reassessment

S Social Assessment

- Assess activities of daily living (eating, dressing, bathing, toileting).
- Are these activities being provided for the patient? If so, by whom?
- Are there delays in obtaining food, medication, or other necessary items? The patient may complain of this, or the environment may suggest this.
- If in an institutional setting, is the patient able to feed himself or herself? If not, is food still sitting on the food tray? Has the patient been lying in his or her own urine or feces for prolonged periods?
- Does the patient have a social network? Does the patient have a mechanism to interact socially with others on a daily basis?

Figure 33-2 The GEMS diamond provides a concise way to remember the important issues of older patients.

You are the Provider: PART 3

Further assessment of the woman reveals that she has a fever of 101°F and that she has not been too hungry lately. She has a weak cough, which is producing thick green sputum. Auscultation of her breath sounds reveals bilateral basilar rhonchi. Your partner takes her vital signs as you ask additional questions regarding her past medical history.

Recording Time: 5 Minutes	
Respirations	22 breaths/min; labored
Pulse	68 beats/min; weak and irregular
Skin	Warm to the touch; pink and dry
Blood pressure	158/88 mm Hg
Oxygen saturation (Sao$_2$)	92% (on oxygen)

The patient's daughter advises you that her mother has emphysema, hypertension, atrial fibrillation, rheumatoid arthritis, and Alzheimer disease. You inquire about her confusion, and her daughter tells you that this is normal for her. She hands you a list of medications, which are numerous, and states that she personally gives her mother her medications every day.

4. What are some common factors that affect an older patient's vital signs?

5. What should concern you about patients who take numerous medications?

that are readily available to help geriatric patients. Consider obtaining information pamphlets about some of the agencies for older people in your area. If you have these brochures with you and encounter a person in need, you can provide this valuable information. Social agencies that deal with the older population will be more than happy to share a listing of the services they provide.

The GEMS diamond provides a concise way to remember the important issues for older patients. Using this concept will help you make appropriate referrals, and, as a result, you will help older patients maintain their quality of life.

Changes in the Body

Human growth and development peaks in the late 20s and early 30s, at which point the aging process sets in. Aging is a linear process, that is, the rate at which a person loses functions does not increase with age. A 35-year-old is aging just as fast as an 85-year-old, but the older person exhibits the cumulative results of a longer process. Of course, the aging process can vary dramatically from one person to another. You can probably recall seeing 60-year-olds who look frail and elderly and 80-year-olds who run marathons Figure 33-3 .

The aging process is inevitably accompanied by changes in physiologic function, such as a decline in the function of the liver and kidneys. All tissues in the body undergo aging, albeit not at the same rate. The decrease in the functional capacity of various organ systems is normal but can affect the way in which a patient responds to illness.

Health professionals need to make sure they are knowledgeable about decreased function in organ systems in elderly people because it will enable them to correctly respond to a patient's illness. For example, if a health care provider is unaware of the normal changes of aging, he or she may mistake the changes for signs of illness and be tempted to give treatment when none is necessary. At the other end of the spectrum, there is a widespread—and unfortunate—tendency to attribute genuine disease symptoms to "just getting old" and to neglect their treatment.

Figure 33-3 Many older people, especially those who have hobbies and activities, are healthy and vital.

Changes in the Respiratory System

Anatomy and Physiology

Age-related changes in the respiratory system can predispose an older adult to respiratory illness. Even a minor lung infection can become a life-threatening event. One of the conditions contributing to breathing problems is weakening of the airway musculature that can cause decreased breathing capacity. This decreased muscle mass means that elderly patients have less help from muscles in the chest wall when they have trouble breathing. As one gets older, there is loss of elastic recoil in the chest wall, resulting in air trapping and an increase in the amount of air left in the lungs at the end of an exhalation. Loss of mechanisms that protect the upper airway include decreased cough and gag reflexes, resulting in a decreased ability to clear secretions. There is also a decrease in the number of cilia that line the bronchial tree, lessening a person's ability to cough and, therefore, increasing the chances of infection.

The alveoli in an older person's lung tissue can become enlarged and the elasticity decreases, making it harder to expel used air. This change in lung tissue quality is comparable to a balloon that has been expanded and then deflated; the balloon loses some of its ability to contract to its original state after inflation. The lack of elasticity results in a decreased ability to exchange oxygen and carbon dioxide. The body's chemoreceptors, which monitor the changes in oxygen and carbon dioxide levels in the blood, slow with age. This can present as lower pulse oximetry readings, even in healthy people.

Pathophysiology

Pneumonia

Although tobacco abuse seems to be decreasing among elderly people, chronic lower respiratory disease, influenza, and pneumonia remain in the top five causes of geriatric deaths. In fact, one of the most common causes of death in older patients is infection with *Pneumococcus* bacteria.

Pneumonia is an inflammation/infection of the lung from bacterial, viral, or fungal causes. Pneumonia is the leading cause of death from infection in Americans older than age 65 years. It especially affects people who are chronically and terminally ill. The infection issue creates problems for a person with lowered resistance, as does the fact that the products of inflammation in the lung interfere with the ability of the alveoli to oxygenate the blood. Some types of pneumonia are infectious, cause a cough that produces greenish or yellow sputum, and are accompanied by a high fever and sometimes shaking chills. Remember to wear respiratory protection when you are assessing a patient with a potentially infectious respiratory disease. You can also place a surgical mask on the patient. If the patient requires oxygen, you can place the surgical mask over the oxygen mask.

You will need to evaluate the pathophysiology through history and possible risk factors. Some factors that predispose a patient to pneumonia are whether the patient is institutionalized, has chronic disease processes, has immune system compromise, or has a history of chronic obstructive pulmonary disease or cancer. Other factors can include a history of inhaled toxins or aspiration of material into the lung.

Associated signs and symptoms of pneumonia include exertional dyspnea, a productive cough, chest discomfort and pain, wheezing, headache, nausea and vomiting, musculoskeletal pain, weight loss, and/or confusion.

During physical assessment, the patient may present with cyanosis and pallor; dry skin; possible fever; increased skin turgor; pale, dry mucosa; and a furrowed tongue. The patient's heart rate is tachycardic, the blood pressure is possibly hypotensive, and an assessment of the lungs will most likely reveal diminished breath sounds with the adventitious sounds of wheezing (the production of whistling sounds, usually during expiration, caused by a partial obstruction of the lower airway), rales (the sounds of air trying to pass through fluid in the alveoli), or rhonchi (lower pitched rattling sounds caused by secretions or mucus in the larger airway). Gentle percussion produces a dull sound. With the mucous consolidation seen in pneumonia, vocal resonance is increased. When the chest is auscultated, the patient's voice can be heard more clearly than in a healthy lung, where speech sounds are muffled.

If possible, the core body temperature should be obtained to determine the presence of fever. Temperatures taken orally or by the axillary method will not be as accurate as a rectal temperature. Pulse oximetry readings, if available, will most likely be low. Blood circulation may be impaired by a loss of fluid. When a blood pressure is taken, the values may be lower than normal. Fluid loss in pneumonia can also cause moderate to severe dehydration and orthostatic hypotension that is manifested by dizziness or fainting on standing.

The patient should be treated with airway, ventilatory, and circulatory support. Provide high-flow oxygen by nonrebreathing mask for a patient with a manageable airway. Use high-flow oxygen with an oropharyngeal or nasopharyngeal airway for a patient who needs ventilatory support with a bag-mask device. If necessary, provide circulatory support by placing the patient in the shock position or performing cardiopulmonary resuscitation. Follow the respiratory and cardiac arrest management standard of your agency. Evaluate patient treatment through reassessment, and prepare for possible deterioration in your patient's condition.

Pulmonary Embolism

Another condition that can cause respiratory distress in the elderly is a **pulmonary embolism**. Pulmonary embolism is a condition that causes a sudden blockage of an artery by a venous clot. Clots develop in the veins of the legs or pelvis and then break off and embolize (move) through the pulmonary artery or one of its branches, where they lodge. This potentially life-threatening condition can present as another disease. A patient with a pulmonary embolism will generally complain of symptoms of chest pain; thus, the pulmonary embolism can be confused with a cardiac, lung, or musculoskeletal problem. Risk factors for a pulmonary embolism include recent surgery (especially in a lower extremity), history of blood clots, obesity, recent long-distance travel, and sedentary behavior, especially after surgery. Other conditions that render the patient bedridden also increase the risk of a pulmonary embolism.

The presentation can be subtle or dramatic depending on how large the clot is and how much lung tissue is damaged. Patients present with tachycardia; sudden onset of dyspnea (which differentiates this from a secondary infection like pneumonia); shoulder, back, or chest pain; cough; syncope in patients in whom the clot is larger; anxiety, which may be communicated as a sense of impending doom; apprehension; and possibly a low-grade fever. Also look for leg pain, redness, and **unilateral pedal edema** (swelling in just one ankle and foot) for the source of the clot. The patient may present with profound fatigue and may go into cardiac arrest in a worst-case scenario.

Assessment of vital signs will reveal changes in circulation because the patient may have an unusually rapid and possibly irregular heartbeat and possibly hypotension. Adventitious lung sounds such as wheezing, rales, or decreased breath sounds may be heard. Consider patient history in a geriatric patient and whether other lung diseases like asthma, congestive heart failure, or chronic obstructive pulmonary disease might complicate the patient assessment. The assessment for pulmonary embolism may not differ considerably from that for any other condition, but remember that geriatric patients are more likely to be compromised because of changes in the respiratory system because of aging and a decreased ability to tolerate extreme injury or illness.

Patients with pulmonary embolism may present with a pulse oximetry reading of 70% or lower; therefore, a pulse oximetry machine may be useful when you suspect this condition.

Treatment should focus on airway, ventilatory, and circulatory support. Hemoptysis, or coughing up blood, is usually not severe, but any blood that has been coughed up should be cleared from the airway. Because a considerable amount of lung tissue may not be functioning, supplemental oxygen is mandatory in a patient with a pulmonary embolism. Place the patient in a comfortable position, usually sitting, and assist breathing using high-flow oxygen and a nonrebreathing mask. Aggressive airway management may be necessary because a large pulmonary embolus may cause significant impairment of the patient's ability to breathe and could result in cardiac arrest if not managed properly. In this situation, ventilate with a bag-mask device, using an oropharyngeal or nasopharyngeal airway.

When a patient is in respiratory and/or cardiac arrest, manage according to current Emergency Cardiovascular Care guidelines and area protocol. Continue to reevaluate patient treatment through reassessment.

Changes in the Cardiovascular System

Anatomy and Physiology

A variety of changes occur in the cardiovascular system as a person grows older, with the net effect of a decrease in the efficiency of the system. Specifically, the heart hypertrophies (enlarges) with age, probably in response to the chronically increased afterload imposed by stiffened blood vessels. Bigger is not better, however. Over time, cardiac output declines, mostly as a result of a decreasing stroke volume.

Arteriosclerosis—the stiffening of vessel walls—contributes to systolic hypertension in many older patients, which places an extra burden on the heart. This phenomenon may be a consequence of disease states such as diabetes, atherosclerosis, and renal compromise, and it is associated with an increased risk of cardiovascular disease, dementia, and death . Compliance of the vascular walls depends on the production of collagen and elastin, proteins that are the primary components of muscle and connective tissue. An increase in pressure (normal hypertension seen with aging) leads to overproduction of abnormal collagen and decreased quantities of elastin; these actions contribute to vascular stiffening. The result is a widening pulse pressure, decreased coronary artery perfusion, and changes in cardiac ejection efficiency.

Some changes in cardiovascular performance are probably not a direct consequence of aging, but rather reflect the deconditioning effect of a sedentary lifestyle. Whether because of other disabilities (such as arthritis) or for psychological reasons, many people tend to limit physical activity and exercise as they grow older. The phrase, "Use it or lose it," applies just as much to the cardiac muscle as to the biceps muscle.

Pathophysiology

Cardiac output is a measure of the workload of the heart. A younger person's body normally compensates for an increased demand on the cardiovascular system by increasing the heart rate, increasing the contraction of the heart, and constricting the blood vessels to nonvital organs. However, with aging, a person's ability to speed up contractions, increase contraction strength, and constrict or narrow blood vessels (vasoconstriction) is decreased because of stiffer vessels. When stroke volume is lowered (the amount of blood pumped out in one beat), it causes a corresponding decrease in heart rate and cardiac output, which is the amount of blood pumped in 1 minute. The heart may lose its ability to raise cardiac output to meet the needs of the body.

Geriatric patients are at risk for atherosclerosis, an accumulation of fatty material in the arteries Figure 33-4. Major complications of atherosclerosis include myocardial infarction (heart attack) and stroke. Atherosclerotic disease begins in the teen years and affects more than 60% of people older than 65 years. The presence of arteriosclerosis, a disease that causes the arteries to thicken, harden, and calcify, makes stroke, heart disease, hypertension, and bowel infarction more likely.

Older people are also at an increased risk for **aneurysm**, an abnormal, blood-filled dilation of the wall of a blood vessel. Severe blood loss can occur when an aneurysm bursts.

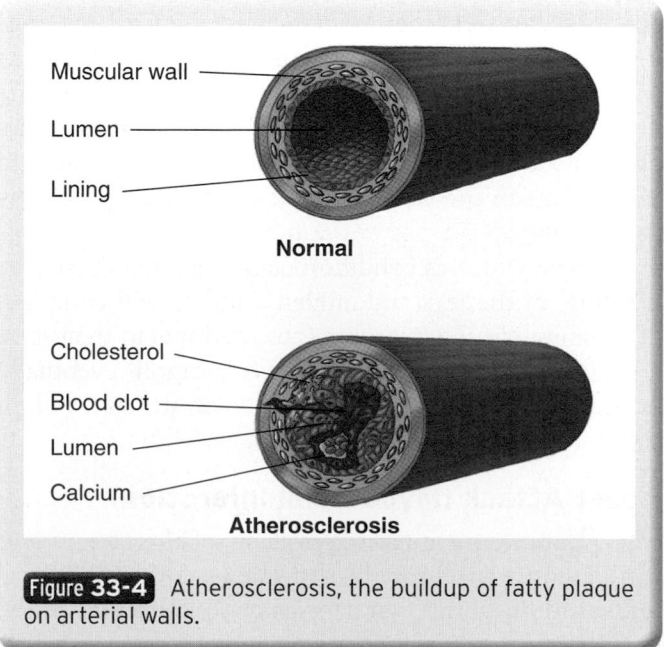

Figure 33-4 Atherosclerosis, the buildup of fatty plaque on arterial walls.

The blood vessels themselves become stiff, which results in a higher systolic blood pressure. As a result, the left ventricle, which pushes blood out into the body, becomes thicker and eventually loses elasticity, resulting in decreased filling of the left ventricle, which in turn causes decreased cardiac output.

Other anatomic changes include stiffening and degeneration of the heart valves, which may impede normal blood flow in and out of the heart. Aging also alters the heart's electrical conduction system. The sinoatrial node is the normal pacemaker of the heart, but by the age of 75 years, the number of the cells in the sinoatrial node will decrease by 90%. This event, combined with fibrosis and fatty deposits attaching to the electrical pathway, makes it likely that the patient will have some kind of heart rhythm disturbance, or arrhythmia. This can cause a heart rate that is too fast, too slow, or too erratic to provide effective blood flow to the body.

Another condition that affects older people is orthostatic hypotension (postural hypotension), which is a drop in blood pressure with a change in position. In younger people, the baroreceptors in the body sense changes in blood pressure and send a message to the adrenal glands to secrete the hormones that will compensate. In older people, you will see a drop in the systolic blood pressure of at least 20 mm Hg when an older patient is moved from a sitting position to a standing position. This happens because the baroreceptors have become less sensitive to the changes in blood pressure. The result is that the older heart takes longer to speed up and then return to normal in response to these hormones. The body, therefore, is less able to compensate for rapid postural changes.

Another vessel-related problem is called venous stasis. Stasis means motionless state and in this context refers to loss of proper function of the veins in the legs that would normally carry blood back to the heart. This condition creates problems such as blood clots in the superficial veins (superficial phlebitis) and blood clots in the deep veins, known as **deep venous thrombosis**.

People with this condition usually exhibit edema, or swelling, of the legs and ankles. Patients will complain of a feeling of fullness, aching, or tiredness in their legs, especially when standing. This condition eventually causes a reddish-brown discoloration on the skin and, in some cases, skin ulcers.

Heart Attack (Myocardial Infarction)

Chest pain is a common complaint of the elderly and can often mean heart-related issues like myocardial infarction. It is important to remember that the classic symptoms of a heart attack or myocardial infarction are often not present in geriatric patients. As many as one third of older patients have "silent" heart attacks in which the usual chest pain is not present. This is particularly common in women and people with diabetes. Do not assume that your patient is not having a myocardial infarction because he or she is not reporting the classic, pressure-type, substernal chest pain. With older people, treat associated symptoms such as dyspnea; epigastric and abdominal pain; nausea and vomiting; weakness, dizziness, light-headedness, and syncope; fatigue; and confusion as seriously as if the patient had chest pain.

Other signs and symptoms in elderly persons that can indicate a cardiovascular problem include issues with circulation; diaphoresis (profound sweating); pale, cyanotic (blue) mottled skin; adventitious (outside of normal) or decreased breath sounds; and increased peripheral edema (swelling).

It is important to obtain as much information as possible from the patient. Even if the patient has altered mental status or is having a hard time communicating, no one else can describe subjective symptoms like pain or shortness of breath as well as the patient can.

It is essential to obtain baseline vital signs because this information will provide you with a primary picture of the severity of your patient's condition, and you can use these findings to measure against in your ongoing assessment of the patient. Pulse rates can be too slow, too fast, weak, or irregular. A blood pressure can show hypertension or hypotension, either of which is significant in the presence of cardiac chest pain. The respiratory rate may be higher as the body attempts to take in and use more oxygen to aid an ailing heart.

At the EMT level, treatment for an elderly patient's cardiac problem mostly consists of airway, ventilatory, and circulatory support. Give oxygen with adjuncts appropriate to the patient's condition. Continue to evaluate your patient treatment through reassessment. Cardiac problems can be expected to worsen suddenly, so be prepared. **Table 33-3** shows signs and symptoms that are commonly noted in geriatric patients who are experiencing a heart attack.

Associated signs and symptoms such as weakness and dizziness, nausea and vomiting, loss of bladder and bowel control, and shortness of breath should be evaluated by ALS personnel as possible indicators of myocardial infarction.

Heart Failure

The signs and symptoms of heart failure will differ depending on the extent to which the right and/or left side of the heart is not functioning correctly. In heart failure, the heart is not able to maintain cardiac output that meets the needs of the body. The heart is failing as a pump. Patient risk factors include hypertension, a history of coronary artery disease, and/or atrial

Table 33-3	Common Signs and Symptoms of Heart Attack in Geriatric Patients
Dyspnea	<u>Dyspnea</u>, the feeling of shortness of breath or difficulty breathing, is a common complaint in geriatric patients and is sometimes associated with a heart attack. It is often combined with other symptoms, such as nausea, weakness, and sweating. Chest pain associated with angina typically has an onset during periods of stress or exertion. In geriatric patients, chest pain is often not present, but exertional dyspnea is. As the disease progresses, dyspnea may occur without exertion. Dyspnea in older people can be the equivalent of chest pain in younger patients who are having angina or a heart attack. In addition, congestive heart failure and acute pulmonary edema may result from a "silent" heart attack.
A weak feeling	Weakness can have many causes; however, you should suspect a heart attack in a patient with a sudden onset of weakness. Weakness is often associated with sweating.
Syncope, confusion, altered mental status	Syncope can have many causes, and in geriatric patients, none of these causes should be presumed to be minor. Major life-threatening causes of syncope are often cardiac in origin. Altered mental status is usually a signal of poor blood supply to the brain, often from cardiac arrhythmia and heart attack.

fibrillation, a condition in which the atria no longer contract normally.

With left-sided heart failure, fluid backs up into the lungs. The excess fluid in the lungs causes a condition called pulmonary edema, and the patient will have shortness of breath. Right-sided heart failure occurs when the fluid backs up into the body. You will see **jugular vein distention** and fluid in the abdomen (**ascites**), which will be seen as peripheral edema in the body tissues. An enlarged liver may also be present from blood backing up through the portal vein. This may be determined by palpation.

The associated signs and symptoms of congestive heart failure include dyspnea on exertion and paroxysmal nocturnal dyspnea. Dyspnea on exertion means the patient has difficulty breathing during exertion because the heart cannot keep up with the body's need for oxygen. In addition, the lungs cannot oxygenate the blood coming through them when fluid has accumulated in the lungs.

Paroxysmal nocturnal dyspnea is a condition that is characterized by a sudden attack of respiratory distress that wakes the person at night when the patient is in a reclining position. The respiratory distress is caused by fluid accumulation in the lungs. Patients complain of coughing, feeling suffocated, and cold sweats, and you will notice tachycardia. The term for not being able to breathe while lying down is orthopnea. If you suspect that your patient may have congestive heart failure, one question you can ask is, "How many pillows do you sleep on at night?"

People with pulmonary edema due to congestive heart failure also complain of tachypnea (fast breathing), use their accessory muscles to facilitate breathing, feel anxious and fatigued, and may complain of associated chest pain. On your assessment, you might find the patient's skin diaphoretic (sweating profusely) and cyanotic (blue from insufficient oxygen). You may hear adventitious

breath sounds like crackles, wheezing, or rales. The patient's pulse may be tachycardic. The patient may have hypertension (high blood pressure) early, which then deteriorates to hypotension (low blood pressure) as a late finding. Continue to assess the blood pressure, breathing, and skin for changes.

Treatment should consist of airway, ventilatory, and circulatory support. Provide oxygen with adjuncts appropriate to the patient's condition, and prepare for the next level of deterioration.

Stroke

Stroke (cerebrovascular accident, or CVA) is a leading cause of death in the elderly. The likelihood of having a stroke becomes greater as a person gets older. Causes of strokes are both preventable and nonpreventable. Preventable risk factors include smoking, obesity, and a sedentary lifestyle. Less preventable causes are high cholesterol and hypertension. Uncontrollable factors include cardiac disease and atrial fibrillation.

Signs and symptoms of stroke include acute altered level of consciousness; numbness, weakness or paralysis on one side of the body; slurred speech; difficulty speaking (aphasia); visual disturbances; headache and dizziness; incontinence; and, in the worst cases, seizure.

Hemorrhagic strokes, in which a broken blood vessel causes bleeding into the brain, are less common and more likely to be fatal. Ischemic strokes occur when a blood clot blocks the flow of blood to a portion of the brain. Brain tissue distal to this clot is deprived of oxygen and will die if the clot is not removed.

The treatment goal is to salvage as much of the surrounding brain tissue as possible. Many communities now have stroke centers that specialize in fast, effective treatment of stroke. Determining the onset of the symptoms of stroke is important. If the symptoms occurred within the past few hours, the patient will be a candidate for stroke center therapy and has a higher chance for recovery.

Changes in the Nervous System

Anatomy and Physiology

Aging produces changes in the nervous system that are reflected in the neurologic examination. Changes in thinking speed, memory, and posture stability are the most common normal findings in older people. Studies have documented age-associated declines in mental function, especially slower central processing of sensory stimuli and language, and longer retrieval times for short- and long-term memory. Common findings during the physical examination include slow responses to questions and requests to repeat a question.

The brain decreases in terms of weight (10% to 20%) and volume as a person ages. This increases the amount of space in the cranium, thus increasing the chance for head injuries. Head injuries with a minimal mechanism are commonly missed in elderly people. In addition, there is a 5% to 50% loss of neurons in older people. Neurons are responsible for transmission of impulses, so the motor and sensory neural networks slow down with age. This affects the control of the rate and depth of breathing, heart rate, blood pressure, hunger, thirst, and body temperature. However, the functional significance of these changes is not clear. The human brain has an enormous reserve capacity, and having a smaller and lighter brain does not necessarily interfere with the mental capabilities of all elderly people, as evidenced by older people who remain active and productive.

Undeniably, though, the performance of most of the sense organs declines with increasing age. The senses of taste and smell become diminished as a person ages. Visual changes may begin as early as age 40 years, such that as many as 50% of patients older than 65 years have vision problems. Causes of visual impairment in elderly people may include diabetic retinopathy and age-related macular degeneration.

Vision

Visual acuity, depth perception, and the ability of the eyes to accommodate to light change with age. The pupils require more time to adjust, which can make driving and even walking more hazardous **Figure 33-5**. **Cataracts**, clouding of the lenses or their surrounding membranes, interfere with vision and make it difficult to distinguish colors and see clearly, increasing the likelihood of falls and accidents and accounting for some mistakes in taking medications. Decreased tear production leads to drier eyes. Older people develop an inability to differentiate colors and have decreased night vision, affecting their

ability to drive. Vision changes, however, can occur earlier in life. Many people in their 40s lose the ability to see up close and may have to get glasses for the first time. This condition is called presbyopia and is caused by a loss of elasticity of the crystalline lens.

A number of other disease processes plague the vision of older adults. These can include glaucoma, macular degeneration, and retinal detachment. Increased intraocular pressure is a risk factor for glaucoma, which can cause damage to the optic nerve. It sometimes causes headache with nausea and vomiting and visual disturbances. Macular degeneration is a deterioration of the macula, which is in the central portion of the retina; this condition generally affects adults older than 50 years. It causes a vision loss in the central part of the visual field. Retinal detachment is a medical emergency requiring prompt surgical treatment to preserve vision. In retinal detachment, the retina is pulled away from the choroid, a thin layer of blood vessels that supply nutrients and oxygen to the retina. This condition leaves the retinal cells deprived of oxygen; therefore, there is a potential for permanent vision loss. The patient may complain of floaters, debris in the visual field, sudden flashes of light or shadow, or visual blurring.

Hearing

Hearing is the sensory change that affects the most elderly people. Typical hearing problems cause changes in the inner ear and make hearing high-frequency sounds difficult. Changes in the ear can also cause problems with balance and make falls more likely. **Presbycusis**, like presbyopia in the eye, is a gradual hearing loss that occurs as we age. Over time, the wear and tear on your ears from noise contributes to hearing loss by damaging your inner ear. Heredity and long-term exposure to loud noises are the main factors that contribute to hearing loss. When assessing your patient, check for the use of hearing aids. If the patient wears hearing aids, ensure that the aids are properly in place and are on. Some patients may want you to talk into their "good ear."

Figure 33-5 Changes in vision, hearing, posture, and motor ability predispose older people to a greater risk of being struck by a vehicle.

Taste

Even the sense of taste can be diminished for an older person because of a decrease in the number of taste buds. The negative result might be lessened interest in eating, which can lead to weight loss, malnutrition, and complaints of fatigue.

Touch

An older person may have a decreased sense of touch and pain perception from the loss of the end nerve fibers. This loss, in conjunction with the slowing of the peripheral nervous system, can create situations in which an older person may be injured and not know it. Specifically, there is a decreased sensation of hot and cold. An older person may be slow to react when touching something hot. This delayed response could result in a burn. This is an especially acute problem in people with diabetes who also lose sensation because of diabetic neuropathy, or nerve damage.

■ Pathophysiology

Dementia

You may come across an elderly patient who is exhibiting delusions, hallucinations, or aggressive behavior. This patient may have dementia. **Dementia** is the slow onset of progressive disorientation, shortened attention span, and loss of cognitive function. It is a chronic, generally irreversible condition that causes a progressive loss of cognitive abilities, psychomotor skills, and social skills. Dementia develops slowly over a period of years rather than a few days. Dementia is the result of many neurologic diseases. Alzheimer disease, cerebrovascular accidents, and genetic factors may cause dementia.

You may or may not have someone in attendance with the patient who can provide information you need to know about whether the patient's mental status is at baseline or abnormal. You should make an evaluation of pathophysiology through history, risk factors, and current medications.

On assessment, the patient might exhibit loss of cognitive function. It helps to determine if this was progressive over a period of time or acute, ie, it began suddenly. Patients with dementia may have short- and long-term memory problems and a decreased attention span, or they may be unable to perform their daily routines. They also may show a decreased ability to communicate and appear confused. Again, determine why you were called, and establish a baseline of the person's cognitive abilities and functioning.

Other aspects of dementia can complicate your ability to assess and manage the patient. Sometimes patients are not only confused, but angry as well. They will generally be poor historians and have impaired judgment. Patients may be unable to vocalize areas of pain

and current symptoms, or they may be unable to follow commands. The patients exhibit disorganized thoughts: inattention, memory loss, disorientation, hallucinations, delusions, and a reduced level of consciousness.

Patients with dementia may express anxiety over movement out of their current residence. They may not understand why they need to go to the hospital and often express anxiety and fear of treatment. Their level of tolerance to changes in their routine may be very low. You have to exercise extreme tolerance and patience with patients who have altered mental status or are experiencing dementia.

Special Populations

Alzheimer disease is a common cause of dementia and affects nearly 4 million Americans. Although its cause is unknown, the disease results in loss of brain tissue. Symptoms include memory loss, lack of spontaneity, subtle personality changes, disorientation to time and date, impaired thinking, restlessness, agitation, wandering, impaired judgment, and inappropriate social behavior. In late stages of the disease, the patient shows indifference to food, an inability to communicate, incontinence, and seizures. Patients with Alzheimer disease may live at home with a spouse or child who is also the caregiver or they may live in a specialized nursing facility. As with all patients with dementia, you must treat patients with Alzheimer disease with patience and respect.

Delirium

Delirium is a sudden change in mental status, consciousness, or cognitive processes, and is marked by the inability to focus, think logically, and maintain attention. According to *The Merck Manual of Medical Information, Second Edition*, delirium affects 15% to 50% of hospitalized people aged 70 years or older.

Acute anxiety may be present in addition to the other symptoms. Usually memory remains intact. Delirium is commonly marked by acute or recent onset and is a red flag for some type of new health problem. This condition is generally the result of a reversible physical ailment, such as tumors or fever. However, delirium also can be present from metabolic causes. Any time a patient has an acute onset of delirious behavior, consider the evaluation of pathophysiology through history, possible risk factors, and current medications.

Other important things to look for in the history are intoxication or withdrawal from alcohol; withdrawal from sedatives; medical conditions such as urinary tract infections, bowel obstructions, dehydration, fever, cardiovascular disease, and hyperglycemia or hypoglycemia; psychiatric disorders like depression; malnutrition/vitamin deficiencies; and environmental emergencies.

Assess the patient for the three specific conditions that can be managed at the prehospital level:

- Hypoxia
- Hypovolemia
- Hypoglycemia

Any of these three conditions, if unrecognized or untreated, can be rapidly fatal. With these conditions, delirium has a rapid onset and is usually curable if identified early. Geriatric patients will respond to oxygen for hypoxia, placement in the shock position, and ALS support with intravenous fluids given for hypovolemia and intravenous glucose given for hypoglycemia. The onset may be described in terms of minutes, hours, or days.

During physical examination, you may see changes in circulation, response of the pupils, and response to motor tests or find adventitious breath sounds. A low blood pressure can indicate hypovolemia. Dilated pupils could suggest hypoxia; wheezing, rales, and rhonchi are the result of disease processes that impair breathing and oxygenation.

Treatment will depend on the results of your assessment but should include airway, ventilatory, and circulatory support and oxygen with airway adjuncts appropriate to the patient's condition if tolerated by the patient. ALS personnel will attempt venous access to introduce fluids that will help correct hypovolemia.

Syncope

You should always assume that syncope or fainting in an older patient is a life-threatening problem until proven otherwise. **Syncope** is often caused by an interruption of blood flow to the brain. Syncope has many causes—some are serious and others are not. Regardless, an older person who has a period of unconsciousness should be examined to determine the cause of the syncope. **Table 33-4** shows some of the causes of syncope in geriatric patients.

Neuropathy

Your patient could be experiencing a **neuropathy**, a disorder of the nerves of the peripheral nervous system in which function and structure of the peripheral motor, sensory, and autonomic neurons are impaired. Symptoms depend on whether the nerves affected are motor, sensory, or autonomic and where the nerves are located.

- **Motor nerves:** muscle weakness, cramps, spasms, loss of balance, and loss of coordination
- **Sensory nerves:** tingling, numbness, itching, and pain; burning, freezing, or extreme sensitivity to touch

You are the Provider: PART 4

You advise the patient that she should be transported to the hospital for evaluation by a physician; however, she is reluctant to go. Her daughter reassures her that everything will be okay and that she will meet her at the hospital. After several minutes of deliberation, the patient consents to transport. Your partner increases the oxygen flow rate to 4 L/min as you reassess the patient's vital signs.

Recording Time: 12 Minutes	
Level of consciousness	Conscious but confused
Respirations	22 breaths/min; labored
Pulse	84 beats/min; weak and irregular
Skin	Warm to the touch; pink and dry
Blood pressure	152/90 mm Hg
Sao$_2$	94% (on oxygen)

The patient's blood glucose level is assessed and is noted to read 98 mg/dL. You place her onto the stretcher, place her in a position of comfort, load her into the ambulance, and begin transport to the hospital. Her daughter tells you that she will follow you to the hospital in her own vehicle.

6. **Why do older patients commonly refuse EMS transport?**

7. **On the basis of the patient's past medical history and signs and symptoms, what do you suspect as the cause of her problem?**

Table 33-4	Possible Causes of Syncope in Geriatric Patients
Arrhythmias and heart attack	The heart is beating too fast or too slowly, the cardiac output drops, and blood flow to the brain is interrupted. A heart attack can also cause syncope.
Vascular and volume changes	Medication interactions can cause venous pooling and vasodilation, the widening of a blood vessel that results in a drop in blood pressure and inadequate blood flow to the brain. Another cause of syncope can be a drop in blood volume because of hidden bleeding from a condition such as an aneurysm.
Neurologic cause	A transient ischemic attack or stroke can sometimes mimic syncope.

- **Autonomic nerves:** affect involuntary functions that could include changes in blood pressure and heart rate, constipation, bladder and sexual dysfunction

Neuropathies are treated with medication and other therapies not available in a field setting. You should make your patient as comfortable as possible and transport.

Changes in the Gastrointestinal System

Anatomy and Physiology

Changes in the mouth include a reduction in the volume of saliva, with a resulting dryness of the mouth. Dental loss is not a normal result of the aging process, but rather the result of disease of the teeth and gums; nevertheless, dental loss is widespread in the elderly population and contributes to nutritional and digestive problems.

Like oral secretions, gastric secretions are reduced as a person ages—although enough acid is still present to produce ulcers under certain conditions. Changes in gastric motility also occur, which may lead to slower gastric emptying—a factor of some importance when assessing the risk of aspiration.

Function of the small and large bowel changes little as a consequence of aging, although the incidence of certain diseases involving the bowel (such as diverticulosis) increases as a person grows older. In addition, nutrients from food are not as readily absorbed.

Blood flow to the liver declines. There are changes in hepatic enzyme systems, with some systems declining in activity and others increasing. Notably, the activity of the enzyme systems involved with the detoxification of drugs declines as a person ages.

Pathophysiology

Gastrointestinal issues in the elderly are attributable to changes related to age or to the diseases that come with advanced age. Age-related changes in the gastrointestinal system include issues with dental problems; decrease in saliva and sense of taste leading to poor nutrition; and poor muscle tone of the smooth muscle sphincter between the esophagus and stomach that can cause regurgitation and lead to heartburn and acid reflux. Other changes include a decrease in hydrochloric acid in the stomach and alterations in absorption of nutrients and slowing peristalsis (motion that moves feces through the colon), which can cause constipation. The rectal sphincter may also become weak, resulting in fecal incontinence, or lack of bowel control.

Changes in the liver predispose elderly patients to a number of problems. The liver, which is responsible for removing toxins and breaking down drugs in the body, shrinks with age. Blood flow to the liver declines, and there is decreased metabolism. This has a direct effect on how medications may affect the patient.

Serious gastrointestinal issues that affect elderly people are gastrointestinal bleeding caused by disease processes, inflammation, infection, and obstruction of the upper and lower gastrointestinal tract. Gastrointestinal bleeding is usually heralded by the vomiting of blood or coffee ground–like vomitus. Bleeding that travels through the lower digestive tract usually manifests as **melena** (black, tarry stools), whereas red blood usually means a local source of bleeding, such as hemorrhoids. A patient with gastrointestinal bleeding may experience weakness, dizziness, or syncope. Bleeding into the gastrointestinal system can be life threatening because of the potential for blood loss and shock.

Specific gastrointestinal problems that are more common in older patients include diverticulitis, bleeding in the upper and lower gastrointestinal system, peptic ulcer disease, gallbladder disease, and bowel obstruction. Diverticulosis is a condition in which small pouches protrude from the colon. When inflammation develops in one of these pouches, the condition is called diverticulitis. A geriatric patient with diverticulitis generally presents with left lower quadrant pain and fever. Fever suggests a condition that requires immediate attention.

Upper gastrointestinal bleeding occurs in the esophagus, stomach, or duodenum. These bleeding episodes are sometimes seen in people who are long-term users of nonsteroidal anti-inflammatory drugs (NSAIDs) like celecoxib (Celebrex), ibuprofen, and naproxen or people who are long-term alcohol users. Irritation of the lining of the stomach or ulcers can cause forceful vomiting that tears the esophagus. Hepatitis and cancer can also contribute to bleeding problems.

Lower gastrointestinal bleeding occurs in the colon or rectum. Lower gastrointestinal bleeding is not as serious as upper gastrointestinal bleeding unless the patient presents with tachycardia and hypotension.

Peptic ulcer disease is more common in older adults, especially people who use NSAIDs. The patient will complain of a gnawing, burning pain in the upper abdomen that improves after eating but returns later. Complications of peptic ulcer disease include bleeding, anemia, and bowel perforation, which is a medical emergency.

Gallbladder disease is more common in older adults and they have a higher risk of complications from gallstones. The risk of death from surgery to remove the gallbladder increases with age. Patients will complain of sharp right upper quadrant pain, and the pain may radiate to the back and right shoulder. Inflammation of the gallbladder, cholecystitis, will present with right upper quadrant pain and fever. Patients may also present with jaundice, which is a yellow appearance of the eyes and skin. This condition is dangerous because infection can spread to the blood, causing systemwide sepsis and shock.

Bowel obstructions occur frequently in the geriatric population. The ability of the gastrointestinal tract to move feces through the system slows with aging, and patients can experience problems having bowel movements. When patients are straining to have a bowel movement, they can stimulate the vagus nerve and produce a reaction called a vasovagal attack; this is a condition in which the heart rate drops dramatically and the patient becomes dizzy or passes out. The patient will usually be in stable condition on your arrival but requires transport to rule out other conditions.

In general, patients with gastrointestinal issues will present with hematemesis (bloody vomitus), melena (dark, tarry stool), dyspepsia (indigestion), hepatomegaly (enlarged liver), constipation, or diarrhea. The patients are agitated and are unable to find a comfortable position. Some will complain of dizziness or syncope when more serious bleeding is occurring.

When assessing patients with gastrointestinal problems, ask about NSAID and alcohol use. Presentation can include pale or yellow, thin skin; frail musculoskeletal system; peripheral, sacral, and periorbital edema; hypertension; fever; tachycardia; and dyspnea.

Orthostatic vital signs can determine if a patient is bleeding internally. Blood pressures are taken with the patient lying, sitting, and then standing; note any change of 10 mm Hg or more as the patient moves to an upright position. Pulses are also obtained with the patient lying, sitting, and standing, taking care to note any change of 10 beats/min or higher as the patient moves to an upright position. Auscultation of breath sounds should be performed to detect adventitious noises or possible foreign bodies.

Treatment consists of airway, ventilatory, and circulatory support. Oxygen should be delivered with adjuncts appropriate to the patient's condition.

The Acute Abdomen—Nongastrointestinal Complaints

Because of an aging nervous system, abdominal complaints in geriatric patients are extremely difficult to assess. A number of life-threatening problems are common in older patients. In the field, the most serious threat from abdominal complaints is blood loss, which can lead to shock and death. **Abdominal aortic aneurysm (AAA)** is one of the most rapidly fatal conditions. An AAA (triple A) tends to develop in people who have a history of hypertension and atherosclerosis. The walls of the aorta weaken, and blood begins to leak into the layers of the vessel, causing the aorta to bulge like a bubble on a tire. If enough blood is lost into the vessel wall itself, shock occurs. If the vessel wall bursts, it rapidly leads to fatal blood loss. When the problem is found early, there is a chance to repair the vessel before rupture, and fatal blood loss is less likely to occur.

A patient with an AAA most commonly reports abdominal pain radiating through to the back with occasional flank pain. If the AAA becomes large enough, it can be felt as a pulsating mass just above and slightly to the left of the navel during your physical examination. If you see or palpate a pulsating mass, do not continue manipulation or allow other providers to palpate the mass. Occasionally, the AAA causes a decrease in blood flow to one of the legs, and the patient complains of some discomfort in the affected extremity. Assessment may also reveal diminished or absent pulses in the extremity. Compensated shock (early shock) and decompensated shock (late shock) as a result of blood loss are common occurrences. Because of a decrease in blood volume and decreased blood flow to the brain, the patient may experience syncope. You should treat the patient for shock and provide prompt transport to the hospital.

Changes in the Renal System

Anatomy and Physiology

The genitourinary system includes the reproductive organs and the urinary system. The largest component of the urinary system is the kidneys, or renal organs. Age-related changes in the genitourinary system specific to the kidney include a reduction in renal function, a reduction in renal blood flow, and tubule degeneration. For the genitourinary system in general, there is decreased bladder capacity, decline in sphincter muscle control, decline in voiding senses, increase in nocturnal voiding, and, in men, benign prostatic hypertrophy (enlarged prostate).

Age brings changes in the kidneys as well. The kidneys are responsible for maintaining the body's fluid and electrolyte balance and have important roles in maintaining the body's long-term acid-base balance and eliminating drugs from the body. In a young adult, the kidneys weigh 250 to 270 g; in a healthy 70-year-old, they weigh 180 to 200 g. This decline in weight results from a loss of functioning nephron units, or tubule degeneration, translating into a smaller effective filtering surface. At the same time, renal blood flow decreases by as much as 50% as a person ages.

Pathophysiology

Although the kidneys of an elderly person may be capable of dealing with day-to-day demands, they may not be able to meet unusual challenges, such as those imposed by illness. For that reason, acute illness in elderly patients is often accompanied by derangements in fluid and electrolyte balance. Aging kidneys, for example, respond sluggishly to sodium deficiency. An elderly patient may lose a great deal of sodium before the kidneys halt urinary sodium excretion, a problem that is exacerbated by the markedly decreased thirst mechanism in elderly people. The net result may be a rapid development of severe dehydration.

Bowel and bladder continence require anatomically correct gastrointestinal and genitourinary tracts, functioning and intact sphincters, and properly working cognitive and physical functions. Urinary incontinence (involuntary loss of urine) can have significant social and emotional impact, but relatively few people admit to the problem and even fewer seek treatment. Incontinence is not a normal part of aging and can lead to skin irritation, skin breakdown, and urinary tract infections. As people age, the capacity of the bladder decreases. As a consequence, an older person may find it difficult to postpone voiding or may have involuntary bladder contractions. An increase in nocturnal voiding is common. Two major types of incontinence are distinguished: stress and urge. Stress incontinence occurs during activities such as coughing, laughing, sneezing, lifting, and exercise. Urge incontinence is triggered by hot or cold fluids, running water, and even thinking about going to the bathroom. Treatment of incontinence consists of medications, physical therapy, and, possibly, surgery.

The opposite of incontinence is urinary retention or difficulty urinating. Patients may have difficulty voiding or absence of voiding as a result of many medical causes. In men, enlargement of the prostate can place pressure on the urethra, making voiding difficult. Bladder and urinary tract infections can also cause inflammation.

In severe cases of urinary retention, patients may have acute or chronic renal failure.

Changes in the Endocrine System

Anatomy and Physiology

The endocrine system functions as the control center of the body. It uses hormones to control physiologic processes. A significant change that occurs in an elderly person is decreased metabolism of thyroxine. This is a thyroid hormone that has an effect on metabolism in the body. There is also decreased conversion of thyroxine to triiodothyronine. Triidothyronine is the most powerful thyroid hormone, and it affects almost every process in the body, including body temperature, growth, and heart rate. A reduction in these hormones can cause a condition called hypothyroidism. Most of the signs and symptoms people experience are attributed to the process of aging and include slower heart rate, fatigue, drier skin and hair, cold intolerance, and weight gain.

Other endocrine system changes include an increase in the secretion of antidiuretic hormone, causing fluid imbalance; increases in the levels of norepinephrine, possibly having a harmful effect on the cardiovascular system; and a reduction in pancreatic beta cell secretion, causing hyperglycemia.

Pathophysiology

Hyperosmolar hyperglycemic nonketotic coma (HHNC) is a type 2 diabetic complication in elderly people. Unlike diabetic ketoacidosis (DKA), which occurs in type 1 diabetes, the resulting high blood glucose level does not cause ketosis; instead, it leads to osmotic diuresis and a shift of fluid to the intravascular space that results in dehydration. The signs and symptoms of HHNC and DKA often overlap. Associated signs and symptoms include hyperglycemia, polydipsia (thirst), polyuria (urination), and polyphagia (hunger), as well as dizziness, confusion, altered mental status, and possibly seizures.

On assessment, you may see changes in circulation such as warm, flushed skin; poor skin turgor; pale, dry, oral mucosa; and a furrowed tongue. The patient may present with signs and symptoms of hypotension and shock, including tachycardia. The blood glucose level will be greater than 500 mg/dL in DKA, whereas in HHNC, the value is greater than 300 mg/dL. Another assessment difference is that DKA will present with Kussmaul respirations (deep and labored), and HHNC does not.

Assessment of the patient should include obtaining blood pressure, distal pulses, auscultation of breath sounds to detect adventitious noises, and determination of temperature.

Treatment should include airway, ventilatory, and circulatory support. Provide oxygen with adjuncts appropriate to the patient's condition.

Changes in the Immune System

Infections are commonly seen in elderly people because they generally have an increased risk of infection and are less able to fight infections once they occur. With age, systemic and cellular immune responses become less effective at fighting infection. Pneumonia and **urinary tract infection** are common in patients who are bedridden. When infection occurs, signs and symptoms may be decreased or minimized by the patient because of the loss of sensation, lack of awareness, or fear of being hospitalized.

Changes in the Musculoskeletal System

■ Anatomy and Physiology

Aging brings a widespread decrease in bone mass in men and women, but especially among postmenopausal women. Bones become more brittle and tend to break more easily. The disks between the vertebrae of the spine begin to narrow, and a decrease in height of between 2″ and 3″ may occur through the lifespan, along with changes in posture. Joints lose their flexibility and may be further immobilized by arthritic changes. In fact, more than half of all elderly people have some form of arthritis. A decrease in the amount of muscle mass often results in less strength.

■ Pathophysiology

Changes in physical abilities can affect older adults' confidence in their mobility. The muscle system atrophies and weakens with age. Muscle fibers become smaller and fewer, motor neurons decline in number, and strength declines. The ligaments and cartilage of the joints lose their elasticity. Cartilage also goes through degenerative changes with aging, contributing to arthritis.

The stooped posture of older people comes from atrophy of the supporting structures of the body. Two of every three older patients will show some degree of **kyphosis** (also called humpback or hunchback). Lost height in older adults generally results from compression in the spinal column, first in the disks and then from the process of **osteoporosis** in the vertebral bodies.

Osteoporosis, a condition that affects men and women, is characterized by a decrease in bone mass leading to reduction in bone strength and greater susceptibility to fracture. The extent of bone loss that a person undergoes is influenced by numerous factors, including genetics, smoking, level of activity, diet, alcohol consumption, hormonal factors, and body weight. The most rapid loss of bone occurs in women during the years following menopause, and many post-menopausal women use hormone replacement therapy as a means to reduce the loss of bone. Calcium and vitamin D supplementation is another treatment for the condition, and many other medications are available to improve bone strength. Older people should remain active and perform low-impact exercises to maintain bone and muscle strength.

Osteoarthritis is a progressive disease of the joints that destroys cartilage, promotes the formation of bone spurs in joints, and leads to joint stiffness. This type of arthritis is thought to result from "wear and tear" and, in some cases, from repetitive trauma to the joints. It affects 35% to 45% of the population older than 65 years. Typically, osteoarthritis affects several joints of the body, most commonly those in the hands, knees, hips, and spine. Patients complain of pain and stiffness that gets worse with exertion. The end result is often substantial disability and disfigurement. Patients are typically treated with anti-inflammatory medications and physical therapy to improve the range of motion.

Changes in Skin

Collagen is a protein that is the chief component of connective tissue and bones, and elastin is a protein that helps to make the skin pliable. Reproduction of these proteins slows as the body ages, bringing on a thinner and less robust appearance in older people. The layer of fat under the skin also becomes thinner because of the redistribution of fluids and proteins. As the elasticity of the skin declines, bruising becomes more common because the skin can tear more easily. Exocrine (sweat) glands do not respond as readily to heat because of atrophy and because of remodeling of the tissues of the dermal layer of the skin.

Another problem that affects the skin is pressure ulcers, sometimes referred to as bedsores or **decubitis ulcers**. Pressure ulcers form when a patient is lying or sitting in the same position for a long time. The pressure from the weight of the body cuts off the blood flow to the area of skin. With no blood flow to the skin, a sore develops. These sores can develop in as little as 45 minutes. To help prevent these ulcers, take special care to pad voids in a patient who may be on a backboard for an extended period.

You may see these ulcers in the following various stages of development:

- **Stage I:** Nonblanching redness with damage under the skin
- **Stage II:** Blister or ulcer that can affect the dermis and epidermis

Falls are some of the more common mechanisms of injury for elderly people. Safety and environment factors such as poor lighting, loose floor coverings, and lack of handrails are often responsible for falls. Physiologic factors include vision and balance issues, decreased visual acuity, and decreased strength. A critically important assessment process is to determine if the fall was mechanical or pathologic. Cardiac, neurologic, and metabolic issues can create weakness, dizziness, or near syncope, resulting in a fall. Traumatic and medical factors are much more interconnected in elderly people.

Elderly people are more likely to experience burns because of altered mental status, inattention, and a compromised neurologic status. Their risk of mortality from burns is increased when preexisting medical conditions exist, the immune system is weakened, and fluid replacement is complicated by renal compromise.

There is higher mortality from penetrating trauma in older adults, especially in the case of gunshot wounds. Penetrating trauma can easily cause serious internal bleeding. An older patient's limited physiologic reserves and more subtle presentation can affect proper management and transportation options.

Trauma in elderly people can also be caused by abuse. Abuse comes in many forms and may include physical assault. Be aware of the environment and conditions a patient lives in, and take note of soft-tissue injuries that cannot be explained by the person's lifestyle and physical condition. Patients may be reluctant to talk about it. If you have any doubt that abuse is a consideration, submit an elder abuse report.

■ Anatomic Changes and Trauma

Changes in pulmonary, cardiovascular, neurologic, and musculoskeletal systems make older patients more susceptible to trauma. The brain shrinks, leading to higher risk of cerebral bleeding following head trauma. Skeletal changes cause curvature of the upper spine that often requires additional padding during spinal immobilization. Loss of strength, sensory impairment, and medical illness all increase the risk of falls.

A geriatric patient's overall physical condition may lessen the ability of the patient's body to compensate for the effects of even simple injuries. For example, the aging body has a heart that no can longer beat faster when it needs to compensate for blood loss, vessels that cannot constrict due to atherosclerosis, and lungs that do not exchange oxygen as well. Additional changes in the circulatory system leave the geriatric patient's body unable to maintain normal vital signs during hemorrhage. Also, a geriatric patient's blood pressure drops sooner than in a younger adult patient during a traumatic emergency.

You are the Provider: PART 5

During transport, the patient remains conscious, although confused. You reassess her vital signs and then call your radio report to the hospital. During your radio report, the nurse asks you if the patient is normally confused. You inform her that the patient has Alzheimer disease and that this is her baseline mental status.

Recording Time: 22 Minutes	
Level of consciousness	Conscious, but confused
Respirations	22 breaths/min; labored
Pulse	70 beats/min; weak and irregular
Skin	Warm to the touch; pink and dry
Blood pressure	148/88 mm Hg
Sao$_2$	95% (on oxygen)

During the remainder of the transport, you reassure the patient that the hospital staff will take good care of her. Although confused, she looks at you and smiles. You deliver her to the hospital, transfer patient care to the emergency department staff, and return to service.

8. How does dementia differ from delirium? Is dementia a normal part of the aging process?

9. What strategies should you use when communicating with older patients?

Special Considerations in Assessing Geriatric Trauma Patients

Trauma is never isolated to a single issue when you are assessing and caring for a geriatric patient. An isolated hip fracture in a healthy 25-year-old adult is rarely associated with overall decline. However, the same injury in an 85-year-old patient can produce a wide-ranging, systemic impact that results in deterioration, shock, and life-threatening hypoxia, a dangerous condition in which the body tissues and cells do not have enough oxygen.

Scene Size-up

Scene Safety

As with all scenes, ensure your own safety first. Take standard precautions. Consider the number of patients, epecially in the case of a motor vehicle accident. Determine if you need additional or specialized resources.

Mechanism of Injury/Nature of Illness

Gather information on the mechanism of injury. Would you always assume that a geriatric driver involved in a vehicle collision is just a trauma patient? As with any call, look for clues that indicate your patient's traumatic incident may have been preceded by a medical incident like syncope, a cardiac problem, or a diabetic issue. Bystander information may help determine if a loss of consciousness occurred before the accident. You should use the same thought processes when you have a patient who has fallen. Was the fall mechanical, or was an episode of weakness or dizziness a factor?

Mechanism of injury is also important in establishing whether an injury is considered critical and it affects treatment and transport considerations.

Primary Assessment

During the primary assessment you will address life threats. A determination needs to be made on whether this is a priority patient and to which facilty the patient will be transported. The decision that the patient has a potentially critical or life-threatening condition would limit on-scene treatment to that which is absolutely necessary for patient stabilization. Be conservative in your thinking. A geriatric patient who sustained a minor fall could have intracranial bleeding, especially if the patient drinks alcohol or is taking blood thinners.

Form a General Impression

The general impression is an important aspect of all patient assessment. As you approach the patient, you should be able to tell if he or she is generally in stable or unstable condition. You will use this information to help you with your further assessment. Determining neurologic status may be difficult if you do not know the patient's baseline. Try to get information from someone familiar with the patient, if possible. Use the AVPU mnemonic to determine posttraumatic status. An important consideration with any patient is the inability to remember the event.

Airway and Breathing

If the patient is talking to you, the airway is patent. Patients who have noisy respirations have airway compromise. Older patients may have a diminished ability to cough, so suctioning is important. Suction any blood or foreign material. Dentures may cause an airway obstruction, so assess for the presence of dentures but do not remove them unless they are creating an airway patency problem. It is more difficult to ventilate a patient with no teeth.

In an unresponsive patient, open the airway with a modified jaw-thrust maneuver. Use an oropharyngeal or nasopharyngeal airway as appropriate, and ventilate with a bag-mask device if the patient's respiratory effort is inadequate or absent. Any curvature of the patient's spine will require padding to keep the patient supine and the airway open.

Breathing problems caused by trauma can be made worse by preexisting respiratory disease and the compromised respiratory effort that comes with aging. Remember that minor chest trauma can cause lung injury. Perform a thorough respiratory assessment and physical assessment of the chest, and treat accordingly. Use pulse oximetry to monitor oxygenation.

Circulation

Manage any external bleeding immediately. Be suspicious of signs and symptoms of internal bleeding. The bodies of older people do not compensate for blood loss as well as the bodies of younger people, and older patients can more easily go into shock. A head injury with minimal mechanism can cause cerebral bleeding. Many elderly people take blood-thinning medications that can make internal bleeding worse or external bleeding more difficult to control. Also, remember that patients who were hypertensive prior to injury may have a normal blood pressure when they are actually in shock.

Transport Decision

When determining your patient's priority status and making the destination decision, remember that physiologic changes secondary to aging can worsen the effects of trauma and that older people do not heal from trauma as easily as do younger adults. Consider trauma center transport for geriatric patients if there is the potential for a serious injury.

History Taking

Investigate Chief Complaint

Considerations in your assessment of the patient's condition and stability must include past medical conditions, even if they are not currently acute or symptomatic. For example, you respond to a call for a patient with a history of unstable angina who sustains a simple isolated fracture of the ankle. You must consider this patient to have the potential for an unstable condition and provide prompt transport before the stress of the simple trauma worsens the angina and an unstable overall scenario develops. In these cases, the remainder of the assessment can be performed en route to the emergency department.

Secondary Assessment

Physical Examinations

The physical examination should be performed on a geriatric trauma patient in the same manner as for any adult but with consideration of the higher likelihood of damage from trauma. Remember that any head injury can be life threatening in an older adult. When examining the chest, consider that breathing is normally impaired. Check lung sounds, and look to see if there is any evidence of pacemakers or previous cardiac surgery. Even though it may appear that the patient has only experienced trauma, keep in mind that this does not mean he or she may not also be having medical problems. When assessing the abdomen, remember that older patients have a flaccid abdominal wall and may not present with pain and rigidity in the abdomen when trauma has been sustained. Decreased muscle size in the abdomen may mask abdominal trauma. Look for bruising and other evidence of trauma. Injury to the liver or spleen may present with diffuse abdominal pain, or pain may refer to the left shoulder.

Vital Signs

Assess the pulse, blood pressure, and skin signs. Capillary refill is unreliable in elderly people because of compromised circulation. Remember that some elderly people take beta-blockers, which will inhibit their heart from becoming tachycardic as you would expect in shock. Even a heart rate in normal ranges may be high for someone taking beta-blockers. Try to determine if the patient's blood pressure is normal for him or her. Remember that a blood pressure that may be normal for an older adult could indicate shock in a younger patient.

Reassessment

Reassessment of primary assessment, level of consciousness, vital signs, and interventions should be performed and documented as with any patient, but remember that a geriatric patient has a higher likelihood of decompensating after trauma. Be prepared.

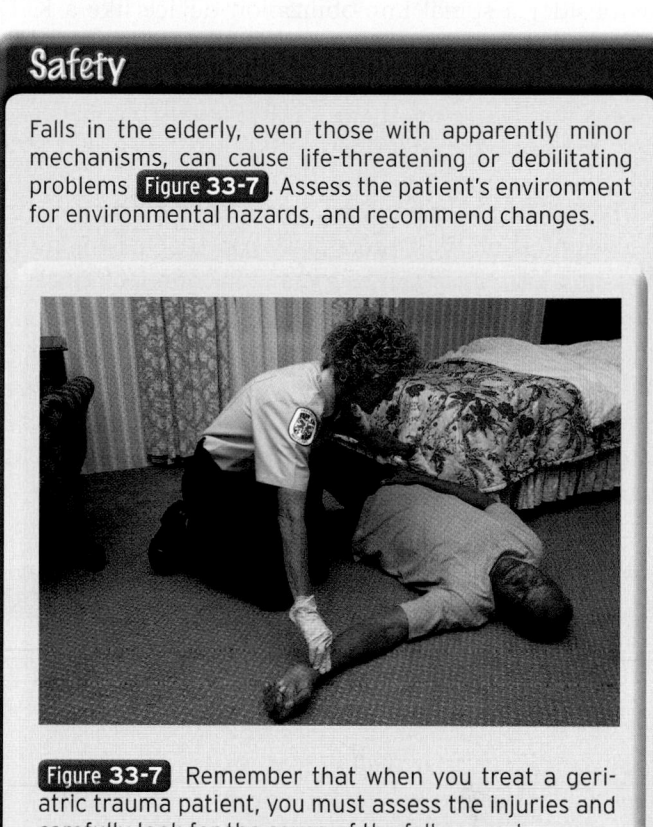

Safety

Falls in the elderly, even those with apparently minor mechanisms, can cause life-threatening or debilitating problems Figure 33-7. Assess the patient's environment for environmental hazards, and recommend changes.

Figure 33-7 Remember that when you treat a geriatric trauma patient, you must assess the injuries and carefully look for the cause of the fall or crash.

Interventions

Broken bones are common and should be splinted in a manner appropriate to the injury. Because of the amount of flexion that occurs in the spinal column, hips, and knees of older patients, effective application of conventional splints and backboards to immobilize them may be difficult or impossible unless a large amount of padding is used. What is considered a normal anatomic position for children and adults is often very abnormal for some geriatric trauma patients. Trying to force a patient with pronounced joint flexion or kyphosis into a "normal" anatomic position can be very painful for the patient and frustrating for you. Some devices, such as traction splints, simply do not work on patients with flexed hips and knees and should never be used to treat hip fractures. Splinting devices such as vacuum mattresses that conform to body contours may be a good choice for immobilization in these cases. In hip and pelvic fractures, you must remember not to log roll the patient because you run the risk of causing more damage. Patients with kyphosis will require padding to keep the patient supine. In general, padding should be done for comfort and to help decrease the likelihood of decubitis ulcers forming. Consider also that patients with chronic cardiac or respiratory disorders, particularly congestive heart failure, may have immense difficulty lying supine for immobilization. An alternative solution may be to consider a spinal immobilization device like a KED intended for a patient in a seated position.

Remember that elderly people do not have the mechanisms that help keep them warm. Provide blankets and heat to prevent hypothermia.

Communication and Documentation

Communication with the elderly can be challenging in any situation, but it can become even more complicated when the patient is in pain, or is experiencing fear from trauma. Older people also tend to fear that a trauma may end their mobility and independence. Remember to provide psychological support, as well as medical treatment. Document assessment, treatment, and reassessment, including any changes in the patient's status.

Falls and Trauma

A medical condition such as fainting, a cardiac rhythm disturbance, or a medication interaction may lead to a fall that injures the patient. Whenever you assess a geriatric patient who has fallen, it is important to find out why the fall occurred. Was it simply a mechanical fall, or was it preceded by a medical event? Was the patient dizzy before the fall? Does the patient remember the fall? Did a fainting episode cause the fall and injury, or did the patient trip on something and lose his or her balance? Sometimes, a recent history of starting or stopping blood pressure medication is enough to cause a patient to become dizzy and fall.

Consider that the fall may have been caused by a medical condition, and look carefully for clues from the patient, bystanders, and the environment. Although the trauma that the patient sustains from the fall can be serious in and of itself, you should also consider that if a medical condition caused the fall, it may be life threatening. In motor vehicle accidents, be alert to the possibility that a medical emergency may have caused the accident, especially in single vehicle collisions with no apparent cause.

As a result of bone loss from osteoporosis, a generalized bone disease that is commonly associated with postmenopausal women, older patients of both sexes are prone to fractures, especially of the hip. Hip fractures are much more common among women. Not all hip fractures are the result of trauma. In some people, a hip fracture may result from the stresses of ordinary activity such as getting in and out of a chair. People with osteoporosis may fracture a hip from a standing fall. This is considered a pathologic fracture because it would not occur in a person with normal bone density. Other contributing factors could include vitamin D and calcium deficiencies, metabolic bone diseases, and tumors. Injuries to the hip also tend to be recurring. A previous fracture increases the likelihood of a future injury.

Elderly people with osteoporosis are also at risk for pelvic fractures **Figure 33-8**. A pelvic fracture may also be caused by a "low-energy" mechanism or a standing fall. The person may sustain this injury when getting out of the bathtub or descending stairs. These injuries do not usually damage the structural integrity of the pelvic ring but may fracture an individual bone.

Recovering from these kinds of injuries can be complicated for an elderly person, especially one with a compromised immune system or diabetes. The fact that the person will be bedridden for a considerable amount of time may inhibit his or her ability to continue to live independently.

With age, the spine stiffens as a result of shrinkage of disk spaces, and vertebrae become brittle. Compression fractures of the spine are more likely to occur. As with a head injury, you must be suspicious of the possibility of other fractures and complicating issues.

Because brain tissue shrinks with age, older patients are more likely to sustain closed head injuries, such as subdural hematomas. This is where tiny veins between the surface of the brain and its outer covering (the dura) stretch and tear, allowing blood to collect. In elderly

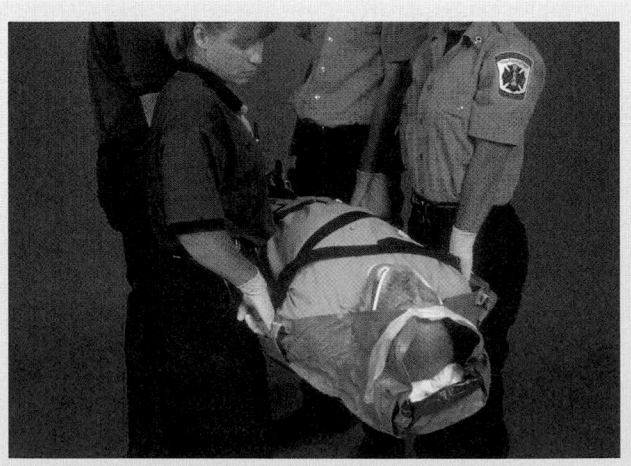

Figure 33-8 Vacuum mattresses that conform to body contours can be a good choice for immobilizing geriatric patients with pelvic fractures.

people, the veins are often already stretched because of brain atrophy (shrinkage) and are more easily injured.

Acute subdural hematomas are among the deadliest of all head injuries. Blood fills the skull very rapidly, compressing brain tissue, which often results in brain injury. This type of bleeding can go unnoticed initially because the blood has a void to fill before it can produce pressure in the skull; only then will the familiar signs of head trauma appear. Also, serious head injuries are often missed in elderly patients because the mechanism may seem relatively minor. These "chronic" subdural hematomas may go unnoticed for many days to weeks.

Other factors that predispose an elderly patient to a serious head injury include long-term abuse of alcohol, recurrent falls or repeated head injury, and/or anticoagulant medication (blood thinners, including aspirin). Be sure to get a history of these risk factors when you assess a geriatric patient with a potential head injury. If a patient shows signs and symptoms of increased intracranial pressure, ask about recent head trauma.

Words of Wisdom

Elderly Patients With Head Trauma
Elderly people are more predisposed to head trauma like subdural hematomas. The mechanism may be minor, thus the severity of these injuries is often underestimated. Consider that the signs and symptoms of head trauma can sometimes mimic the signs and symptoms of a stroke. Be sure to ask about these risk factors during your assessment of a patient with a potential head injury:

- Long-term abuse of alcohol
- Recurrent falls
- Repeated head injury
- Use of anticoagulant medication (blood thinners, including aspirin)

Environmental Injury

Internal temperature regulation is slowed in elderly people and gets slower with increasing age. The body's ability to recognize fluctuations in temperature becomes delayed owing to a slowed endocrine system. Heat gain or loss in response to environmental changes is delayed by slowed circulation and decreased sweat production in the skin. In addition, thermoregulation can be adversely affected by chronic disease, medication use, and alcohol use, all of which are more common in elderly people.

Not surprisingly, about half of all deaths from hypothermia occur in elderly people, and most indoor hypothermia deaths involve geriatric patients. Although living where harsh winters occur is a risk factor, hypothermia can develop at temperatures above freezing when an older person is exposed for a prolonged period.

The death rates from hyperthermia are more than doubled in elderly people compared with younger persons; people older than 85 years are at highest risk. Arizona has more heat-related deaths than all other states combined, reflecting its very long, hot summers and large geriatric population.

Response to Nursing and Skilled Care Facilities

Elderly patients who you will see, assess, and transfer are often found in convalescent homes, nursing homes, and other skilled care facilities. The kind of facility will depend on the type of care needed. Relatively healthy and active seniors live in age-restricted active adult communities that offer resort-type amenities. A less expensive option is age-restricted apartments that provide seniors with the physical and emotional security that comes with living with other seniors. A similar type of facility, but one that also provides communal meals, social events, and other types of support, is an independent living facility. People living in this type of facility are more likely to have minor health problems. The next level would be facilities that assist residents with activities of daily living and provide 24-hour assistance. These are assisted living residential facilities. Residents get assistance with daily medication administration, and some facilities address specialized patient issues like Alzheimer disease and dementia.

Nursing homes, also called convalescent or long-term care facilities, are facilities that serve patients who need 24-hour care and are sometimes a step down from an acute care hospital. Patients require assistance with daily living and need therapeutic or rehabilitation services. A larger number of the patients with whom you interact will be found in one of these facilities.

Calls to these types of facilities can sometimes be challenging. The staff is usually spread thin and may not be familiar with what needs to be done to assist you when transport is necessary. The most important piece of information you need to establish immediately is, "What is wrong with the patient?" Patients in these facilities often have an altered level of consciousness and may not be able you give you a nature of illness or mechanism of injury. The nurse who greets you may not be familiar with the patient. As soon as possible, establish what the "baseline" status of the patient is. Talk to the staff who directly care for the patient on a daily basis. They tend to have more interaction with the patient and may have a better understanding of the patient's baseline mental status.

With potentially limited information, you need to do an assessment to determine if the patient's problem is life threatening and/or requires ALS level care. Optimally, the facility will provide you with a transfer record that provides critical information on the patient's history, medications, allergies, and current complaint. This information is critical because the emergency department staff needs to know how to best manage the patient. Ideally, and when appropriate, transport the patient to the acute care facility where the patient has been treated before and his or her records are available.

Infection control needs to be a high priority for EMTs when visiting these facilities. You not only need to protect yourself, but you also need to inhibit the spread of pathogens from patient to patient. Good handwashing and standard precautions can inhibit the spread of infectious pathogens to people who already have compromised immune systems. An infection in an older patient can lead to life-threatening sepsis. There are many risks to the patients and the EMTs. Some of these risks are described here:

Methicillin-resistant _Staphylococcus aureus_ (MRSA) infections are common among people who are living in close quarters like nursing homes. The organism can be found in decubitus ulcers (bed sores), on feeding tubes, and on indwelling urinary catheters. The symptoms of MRSA depend on the type of infection. It can cause mild infections on the skin or invade the bloodstream, lungs, or the urinary tract. MRSA is primarily spread by broken skin-to-skin contact but is also acquired by touching objects that have the bacteria on them.

Similarly, many infections in hospitals are caused by vancomycin-resistant enterococci. Enteroccocci are bacteria that are normally present in the human intestines and the female genital tract. Under the right circumstances, these bacteria can cause infection. Some of the enterococci have become resistant to the antibiotic commonly used to treat these infections, which is vancomycin.

The **respiratory syncytial virus** causes an infection of the upper and lower respiratory tracts. Although more typically seen in children, the virus can also cause serious illness in elderly persons, especially those with lung disease or weakened immune systems. The symptoms are similar to the common cold but can be more severe and last longer. The virus is highly contagious and is found in discharges from the nose and throat of an infected person. Respiratory syncytial virus is also transmitted by direct contact with droplets from coughs or sneezes and by touching a contaminated surface.

MRSA and respiratory syncytial virus infections can be life threatening, especially in an immune-compromised patient. Look for "isolation" signs or ask about contagious disease when you approach a patient. Be sure to wear appropriate personal protective equipment and decontaminate your ambulance and diagnostic equipment after contact with nursing home residents whether a history of infectious disease is known or not. Be sure to document the infection control issue; advise the receiving facility; and, depending on local protocol, report an infectious disease to your company or the local health department.

Clostridium difficile is a bacterium responsible for the most common cause of hospital-acquired infectious diarrhea and regularly causes sporadic cases of diarrhea in nursing homes. It is a bacterium that normally grows in the intestines. Antibiotic use may account for the rapid increase in toxic strains that ultimately cause illness. Health care workers may carry this bacterium following contact with contaminated feces. It can also be found on environmental surfaces like furniture, floors, toilets, sinks, and bedding. The symptoms from the resultant colitis can range from minor diarrhea to a life-threatening inflammation of the colon.

You should also be cognizant of potential airborne pathogens. Something as simple as a cold or flu virus could result in a life-threatening pneumonia for a compromised older adult. Be sure to mask yourself if you have an upper respiratory infection, and mask the patient if the patient has one.

■ Dying Patients

As elderly patients are living longer, more terminally ill patients are choosing to die at home rather than in a hospital. Many have family support and/or hospice support. Often the patient comes to terms with his or her impending death before the family does. Dying patients receive what is called palliative, or comfort, care. Palliative care recognizes that death, as life, is normal. It neither hastens nor prolongs death and includes relief of pain and psychological care.

You may be called on to interact with a dying patient. One thing to remember is that this interaction will have a long-term effect on the family. Being understanding, sensitive, and compassionate is critical, although the situation may be uncomfortable for you as well. Determine if the family wishes the patient to go to the hospital or stay in the home.

Advance Directives

Many people today are making use of **advance directives**, specific legal papers that direct relatives and caregivers about what kind of medical treatment may be given to patients who cannot speak for themselves. An advance directive is also commonly called a "living will." Mentally competent adults and emancipated minors have the right to consent to or decline treatment, provided they are competent to do so. The definition of competence is often hotly debated, but a person who is older than 18 years, alert, and not intoxicated and who understands the consequences of his or her decision is generally deemed competent. Unfortunately, patients who are unconscious or in a medical crisis are not able to inform medical personnel about their wishes to consent to or decline treatment. It is dangerous to take someone else's word for what the patient's wishes are. Written advance directives have been developed for this reason.

Advance directives may also take the form of a "Do Not Resuscitate" (DNR) order **Figure 33-9**. A DNR order gives you permission not to attempt resuscitation for a patient in cardiac arrest. However, for a DNR order to be valid, the form must be signed by the patient or legal surrogate and by one or more physicians. In most states, the form must be dated within the preceding 12 months. In the presence of a DNR order, if the patient is still alive, you are obligated to provide supportive measures that may include oxygen delivery, pain relief, and comfort. DNR does not mean do not treat. Basic airway, breathing, and circulatory support should be provided; however, cardiopulmonary resuscitation may not. Learn and become familiar with your state laws regarding this issue.

A "health care power of attorney" is an advance directive that is exercised by a person who has been authorized by the patient to make medical decisions for him or her. Be sure to follow your service's protocol when faced with any advance directive.

Dealing with advance directives has become more common for EMS providers because more people are electing to use hospice services and spend their final days at home. Although advance directives may be in place, family members or caregivers who are faced with the final moments of life or when the patient's condition worsens often become alarmed and call 9-1-1. Family members and caregivers may then become upset when you take resuscitative action and begin transportation to the hospital.

Another common situation is the transportation of patients from nursing facilities. Specific guidelines vary from state to state; however, you should consider the following general guidelines:

- Patients have the right to refuse treatment, including resuscitative efforts, provided that they are able to communicate their wishes.
- A DNR order is valid in a health care facility only if it is in the form of a written order by a physician.
- You should periodically review state and local protocols and legislation regarding advance directives.
- When you are in doubt or when there are no written orders, you should try to resuscitate the patient.

Every service should also provide training on the actions you should take when presented with advance directives. When in doubt, your best course of action is to take resuscitative action that is appropriate to the situation and to practice sound medical treatment.

Elder Abuse and Neglect

Reports and complaints of abuse, neglect, and other related problems among the nation's older population are on the rise. **Elder abuse** is defined as any action on the part of an older person's family member, caregiver, or other associated person that takes advantage of the older person's person, property, or emotional state. It is also called "granny beating" and "parent battering."

The exact extent of elder abuse is not known for several reasons, including the following:

- Elder abuse is a problem that has been largely hidden from society.
- The definitions of abuse and neglect among the geriatric population vary.
- Victims of elder abuse are often hesitant to report the problem to law enforcement agencies or human and social welfare personnel.

A parent who feels ashamed or guilty because he or she raised the abuser is a typical victim of elder abuse. The abused person may also feel traumatized by the situation or be afraid that the abuser will try to get back at

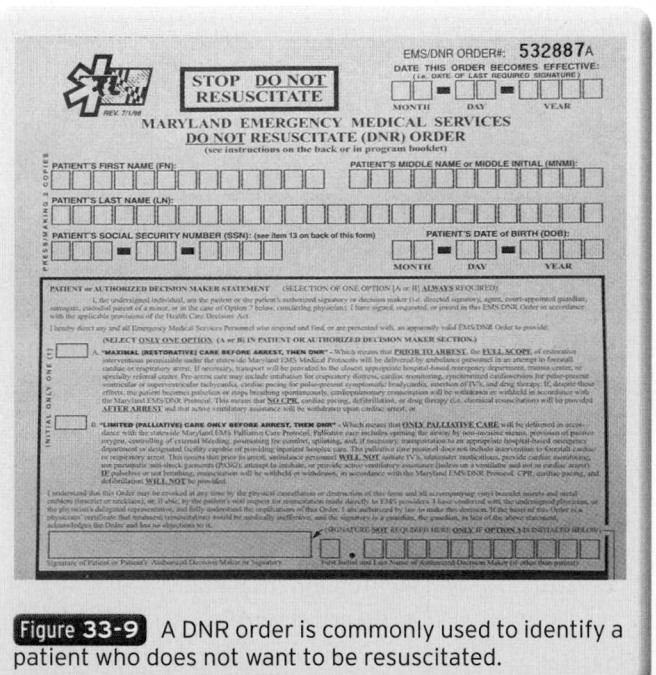

Figure 33-9 A DNR order is commonly used to identify a patient who does not want to be resuscitated.

him or her for reporting the abuse. In some areas of the country, there is a lack of formal reporting mechanisms, and some states lack statutory provisions that require that elder abuse be reported.

The physical and emotional signs of abuse, such as rape, spouse beating, and nutritional deprivation, are often overlooked or not accurately identified. Older women in particular are not likely to report incidents of sexual assault to law enforcement agencies. Patients with sensory deficits, dementia, and other forms of altered mental status, such as drug-induced depression, may not be able to report abuse.

Elder abuse occurs most often in women older than 75 years. The abused person is often frail and has multiple chronic medical conditions and dementia. The person may sleepwalk, have an impaired sleep cycle, and periodically shout at others. The person may also be incontinent and, in general, is dependent on others for activities of daily living.

Abusers of older people are often products of child abuse themselves, and the abuse that is inflicted on the older person may be retaliatory. Most of these abusers are not trained in the particular care that older people require and have little relief time from the constant care demands of their own family, children, and spouse. Their lives are now complicated by the constant, demanding needs of the older person they have to care for.

The abuser may also have marked fatigue, be unemployed with financial difficulties, and abuse one or more substances. With a careful eye, you can recognize the clues to these stressful situations and help guide the family toward programs in their community that are geared to helping the whole family. Programs such as adult day-care, Meals on Wheels, and many local individualized programs help to decrease the stress put on the family and lower the chances of abuse.

Abuse is not restricted to the home. Environments such as nursing, convalescent, and continuing care centers are also sites where older people sustain physical, psychological, financial, or pharmacologic harm. Often, care providers in these environments consider older people to be management problems or categorize them as obstinate and undesirable patients.

■ Assessment of Elder Abuse

While assessing the patient, you should try to obtain an explanation of what happened. You should suspect abuse when answers to questions about what caused the injury are concealed or avoided.

You must also suspect abuse when you are given unbelievable answers. You should be suspicious if you think "Does this make sense?" or "Do I really believe this story?" while reviewing the patient's history. As an EMT, you may be the first health care provider to observe the signs of possible abuse. Information that may be important in assessing possible abuse includes the following:

- Repeated visits to the emergency department or clinic
- A history of being accident-prone
- Soft-tissue injuries
- Unbelievable or vague explanations of injuries
- Psychosomatic complaints
- Chronic pain without medical explanation
- Self-destructive behavior
- Eating and sleep disorders
- Depression or a lack of energy
- Substance and/or sexual abuse history

You should remember that many patients who are being abused are so afraid of retribution that they make false statements. A geriatric patient who is being abused by family members may lie about the origin of abuse for fear of being thrown out of the home. In other cases of elder abuse, sensory deprivation or dementia may hinder adequate explanation.

Repeated abuse can lead to a high risk of death. A preventive measure in reducing additional maltreatment of the patient is identification of the abuse by emergency medical providers **Table 33-5**. This may allow for referral and protective services of human, social, and public safety agencies.

■ Signs of Physical Abuse

Signs of abuse may be quite obvious or subtle. Inflicted bruises are usually found on the buttocks and lower back, genitals and inner thighs, cheeks or earlobes, neck, upper lip, and inside the mouth. Pressure bruises caused by the human hand may be identified by oval grab marks, pinch marks, or handprints. Human bites

Table 33-5 Categories of Elder Abuse

Physical	• Assault • Neglect • Dietary • Poor maintenance of home • Poor personal hygiene
Psychological	• Benign neglect • Verbal • Treating the person as an infant • Deprivation of sensory stimulation
Financial	• Theft of valuables • Embezzlement

are typically inflicted on the upper extremities and can cause lacerations and infection. You should inspect the patient's ears for indications of twisting, pulling, or pinching and evidence of frequent blows to the outer ears. You should also investigate multiple bruises in various states of healing by asking the patient and reviewing the patient's activities of daily living.

Words of Wisdom

As with other legally complex and emotionally charged issues, the possibility of elder abuse demands particularly careful documentation. Be thorough, objective, and factual, avoiding unsupported opinions and personal judgments. You may be called on to explain your report in a legal proceeding.

Burns are a common form of abuse. If you see burns, especially cigarette burns or physical marks that indicate that certain parts of the patient's body have been scalded systematically, you must suspect abuse. Typical abuse from burns is caused by contact with cigarettes, matches, heated metal, forced immersion in hot liquids, chemicals, and electrical power sources.

It may be difficult to see a failure to thrive in an older patient who has been abused. You should observe the patient's weight and try to determine whether the patient appears undernourished or has been unable to gain weight in the current environment. Does the patient have a ravenous appetite? Has medication been withheld? Is money being withheld, so the patient cannot buy food or medicine? You should also check for signs of neglect, such as evidence of a lack of hygiene, poor dental hygiene, poor temperature regulation, or lack of reasonable amenities in the home **Figure 33-10**.

You must regard injuries to the genitals or rectum with no reported trauma as evidence of sexual abuse in any patient. Geriatric patients with altered mental status may never be able to report sexual abuse. In addition, many women do not report cases of sexual abuse because of shame and the pressure to forget.

Figure 33-10 Check for signs of neglect, such as evidence of a lack of hygiene, poor dental hygiene, poor temperature regulation, or lack of reasonable amenities in the home.

You are the Provider: SUMMARY

1. Why is it important for the EMT to understand the physiologic changes that occur with aging?

Aging is a linear process; ie, the rate at which physiologic changes occur does not increase with age. A 25-year-old is aging just as fast as a 75-year-old; however, the 75-year-old exhibits the cumulative results of a longer process.

The aging process is inevitably accompanied by changes in physiologic function, such as a decline in the function of the brain, respiratory system, cardiovascular system, and renal system, among others. All organs and tissues are affected by the process of aging, but not at the same rate. Although the decrease in the function of various organs is a normal process of aging, it can affect the way in which a patient responds to illness or injury.

Understanding the normal anatomic and physiologic changes that occur with aging is important when you are assessing and treating an older patient who is experiencing an illness or injury. You do not want to mistake these changes for signs of illness and provide treatment when it is not indicated, nor do you want to attribute them to "just getting old" and fail to provide treatment when it is indicated.

You must understand the normal physiologic processes in patients of all ages, not just older adults, and realize that the aging process can vary dramatically from one person to another. Base your treatment on the patient's clinical presentation, not his or her age.

You are the Provider: SUMMARY, continued

2. How does the process of aging affect a person's respiratory system?

Age-related changes in the respiratory system can predispose older adults to a variety of respiratory illnesses. An otherwise minor lung infection in a younger adult could be a life-threatening infection in an older adult. You must consider these age-related changes when assessing an older adult with a respiratory complaint.

Just like other muscles in the body, the respiratory muscles atrophy (decrease in mass) with age, causing decreased breathing capacity. The decreased muscle mass means that an older adult has less help from the muscles in the chest wall when he or she is experiencing respiratory distress. A loss of elastic recoil of the chest causes air trapping in the lungs, resulting in increases in lung capacity and residual volume.

Important upper airway mechanisms, such as the cough and gag reflexes, decrease with age, resulting in a decreased ability of older adults to clear secretions from the airway. In addition, there is a decrease in the number of functional cilia in the bronchial tree. These upper and lower airway changes decrease the patient's ability to cough, which increases the risk of a respiratory infection.

Although the alveoli in an older person become enlarged, their elasticity decreases; this makes it harder for the patient to expel used air. Just like a balloon that has been expanded and deflated, the balloon loses some of its ability to contract to its original state after inflation. As a result, pulmonary respiration—the exchange of oxygen and carbon dioxide in the lungs—decreases.

The body's chemoreceptors, which monitor changes in oxygen and carbon dioxide levels in the blood, become less sensitive with age. As a result, the respiratory centers in the brain may become less able to increase (or decrease) the patient's respiratory rate accordingly. This is an important age-related change because it affects an older person's ability to compensate for conditions such as hypoxia and shock.

3. What is the GEMS diamond? How can it facilitate your overall care of an older patient?

When caring for an older patient—especially one who lives alone—it is important to remember certain key concepts. The GEMS diamond was created to help you recall what is different about older patients. It is not intended to be a format for the approach to an older patient, nor is it intended to replace the processes of assessment and treatment. Instead, it serves as a useful acronym for the issues to consider when assessing every older patient.

The "G" of the GEMS diamond stands for "geriatric." Remember that older adults are different from younger adults and may present atypically. For example, older patients may not present with the classic signs and symptoms of a heart attack (eg, chest pain, diaphoresis); their presenting complaint may only be weakness.

The "E" of the GEMS diamond stands for an environmental assessment. Assessment of the environment may provide clues about the patient's condition or the cause of the emergency. Is the home too hot or too cold? Are there hazardous conditions (eg, rotted floors, poor wiring, unventilated gas heaters)? If the patient has a disability, are there appropriate asssitive devices (eg, wheelchair, walker) present? Are medications out-of-date or unmarked, or are there prescriptions for the same or similar medications prescribed by more than one physician?

The "M" of the GEMS diamond stands for medical assessment. Older patients tend to have a variety of medical problems and may be taking numerous prescription, over-the-counter (OTC), and herbal medications. Obtaining a thorough medical history is very important in older patients.

The "S" stands for social assessment. Older people may have an inadequate social network because of death of a spouse, a lack of friends, and family members who live far away or simply do not check in on the patient. Older patients often need assistance with activities of daily living, such as dressing, bathing, or eating. Is there someone who can assist the patient with these tasks?

The GEMS diamond provides a concise way to remember the important issues for older patients. Using this concept will help you make appropriate referrals, and, as a result, you will help older patients maintain their quality of life.

4. What are some common factors that affect an older patient's vital signs?

Vital signs may vary from normal values in older patients for a variety of reasons, such as age-related physiologic changes, chronic disease, and the effects of one or more medications. If the patient is taking beta-blockers (medications that are used to treat hypertension and abnormally fast heart rates) his or her pulse rate may be slower than expected. Furthermore, the use of beta-blockers may prevent the patient from developing a compensatory tachycardia, such as what is typically seen in dehydration and shock. When the patient has decreased peripheral perfusion, the radial pulse may be weak and difficult to feel; other pulse points should then be considered during assessment. Abnormal cardiac rhythms, such as atrial fibrillation, will cause the pulse to be irregular.

Capillary refill time is an unreliable indicator of perfusion in an older patient because of skin changes and decreased perfusion to the periphery.

Blood pressure tends to increase with age and is often the result of stiff blood vessels that are less able to dilate (arteriosclerosis). If an older patient has a history of hypertension, you should be aware that a blood pressure in a "normal" adult range could actually be hypotensive for him or her. If the patient presents with an elevated blood pressure, determine if he or she has hypertension; if so, try to ascertain if the patient is compliant with his or her antihypertensive medications.

An older patient's respiratory rate should be in the same range as that of a younger adult; however, increased chest wall stiffness may make it appear that chest rise is not adequate. Carefully assess the older patient's respiratory rate, depth, and regularity. Auscultate breath sounds to listen for adequate air movement and abnormal breath sounds, such as rales, rhonchi, or wheezing.

The pulse oximeter requires adequate perfusion to obtain an accurate reading. Older patients may have poor circulation, vasoconstriction, a lack of red blood cells (anemia), and hypothermia, among other conditions. These factors may influence the accuracy of the patient's oxygen saturation or your ability to obtain a reading.

5. What should concern you about patients who take numerous medications?

Older patients account for one eighth of the population, but use one quarter of the prescribed medications and one third of the over-the-counter medications sold in the United States.

The term "polypharmacy" refers to the use of multiple prescription medications by one patient. Older patients commonly have more than one physician: a primary care physician for everyday care, a cardiologist for the heart, and an endocrinologist for care of diabetes, among other physicians. Each of these physicians prescribes different medications to the patient for different medical conditions.

The risks of inadvertent overdosing or negative medication interactions increase when patients take numerous medications. Furthermore, the physiologic changes that occur with aging make an older patient susceptible to drug toxicity—even if the patient takes all of his or her medications as prescribed. Metabolism and excretion are affected because of altered gastrointestinal function, decreased kidney function, and decreased liver function. As a result, medications tend to stay in the body for longer periods and are not detoxified by the liver as fast; these factors increase the risk of drug toxicity.

Some patients may take as many as 10 (or more) different medications for multiple medical conditions. When this many medications are combined, it is impossible to predict how they may interact with each other. Some medication interactions can produce acute life-threatening emergencies; others may have negative effects but cause damage over an extended period. When caring for a patient who takes numerous medications, consider the possibility of an interaction as the underlying cause of, or a contributor to, his or her problem.

In many cases, an older patient is unable to tell you some or all of the medications he or she is taking. Statements such as, "I take a little white pill for my blood pressure and a water pill" are not at all uncommon. In other cases, the patient or a family member may present you with a medication list, which identifies what is taken, how much is taken, and when it is taken.

As an EMT, you are not expected to know the intricacies of every drug a patient takes; however, you should carry an EMT field guide or similar reference, which can be used to look up a particular drug to determine what it is commonly prescribed for. It is important to document all of the patient's medications and pass this information along to the physician at the emergency department.

6. Why do older patients commonly refuse EMS transport?

The single most common reason why older patients are resistant to EMS transport is fear. They often fear that when they are loaded into an ambulance and leave the safety of their own home, they will be moved to a nursing home or, even worse, never leave the hospital.

In other cases, patients simply do not realize that their condition has deteriorated to the point that they can no longer safely care for themselves. As an EMT, you are always obligated to treat the patient to the best of your abilities; but in the same context, you must acknowledge and respect the older patient's need for independence and the emotional attachment that many of them have to their homes.

Older patients are often concerned about a longtime companion. Stranger anxiety—a common problem in children—is also common in older patients. If possible, transport a patient's loved one in the front of the ambulance; if the patient's loved one does not wish to ride in the ambulance, encourage him or her to follow the ambulance to the hospital if he or she is able to drive. Pets can also be a source of anxiety; patients who live alone often do not want to go to the hospital and abandon their pet. If there are no family members present, ask a neighbor or friend if he or she will watch the patient's pet.

Money is a significant concern for older patients, especially patients living on a fixed income. Health care is expensive, and many older patients fear that being hospitalized will deplete their life savings or even cause them to lose their homes.

You must acknowledge the patient's fears, regardless of how insignificant they may seem, and recognize that they are very real to the patient. If the patient is fearful, ask specifically what he or she is fearful about; simply vocalizing these concerns may help the patient establish a sense of trust in your care. As with any patient, provide emotional support and reassurance; however, do not provide false reassurance. For example, if an older patient asks you if he or she will be placed in a nursing home, a suggested response would be, "I do not know, but I am going to take good care of you." Do not make promises that you cannot keep just to pacify the patient; it is unethical, unprofessional, and inhumane.

7. On the basis of the patient's past medical history and signs and symptoms, what do you suspect as the cause of her problem?

The EMT can often gather enough information through the processes of physical assessment and history taking to formulate a working field impression. It is important to note that the field impression is not a diagnosis; it is what you believe is the cause of the patient's problem.

As previously discussed, age-related changes to the respiratory system predispose older patients to a variety of respiratory diseases. A decrease in the number of functional cilia in the bronchial tree, which lessens the ability to cough, increases the chance of an infection.

On the basis of the patient's past medical history, which includes emphysema, and her current presentation—fever, chills, and worsened shortness of breath—you should suspect that she has pneumonia.

Pneumonia is an inflammation/infection of the lung from bacterial, viral, or fungal causes and is the leading cause of death from infection in Americans older than 65 years. Pneumococcal pneumonia, which is caused by the *Pneumococcus* bacterium, is the most common infection that affects older patients. Pneumonia creates a barrier to effective gas exchange in the lungs, resulting in varying degrees of hypoxemia and hypercarbia.

Older patients have a decreased cough mechanism, which lessens their ability to expel secretions from the lungs. As a result, secretions settle (consolidate) in the lungs—usually at the bases—and become infected. In addition, the immune system deteriorates with age, making older patients less able to fight off an infection.

The presence of a preexisting lower respiratory disease (eg, chronic obstructive pulmonary disease), coupled with age-related deterioration of the respiratory and immune systems, makes older patients especially prone to pneumonia.

Pneumonia is not an acute illness; it usually progresses over a period of days. The patient typically presents with fever and chills (indicators of infection), a cough that produces yellow or green sputum, and varying degrees of respiratory distress. If the patient has a preexisting lower respiratory disease that creates a baseline respiratory distress, the patient typically complains of worsened respiratory distress. Assessment of breath sounds typically reveals rales (fine, moist thin sounds) or rhonchi (coarse crackles), which indicate fluid in the small and large airways.

8. How does dementia differ from delirium? Is dementia a normal part of the aging process?

Any change in a patient's level of consciousness or mentation, regardless of his or her age, is abnormal. Confusion, abnormal behavior, and other changes in cognitive function do not come automatically with age; they should be regarded as an indicator of an underlying disease process.

When caring for an older patient who is confused, you should ascertain his or her baseline mental status—that is, how he or she normally acts. A family member or close friend who sees the patient on a regular basis should be able to provide you with this information. Even if the patient's confused state is "normal" for him or her, the possibility of a secondary event cannot be ruled out in the prehospital setting, and the patient should be transported to the hospital.

Dementia and delirium—two processes that alter a patient's mental status—are commonly seen in older patients; however, they are both abnormal processes and require evaluation by a physician.

Dementia is a slow onset of progressive disorientation, as well as a progressive loss of cognitive function, psychomotor skills, and social skills. It is a chronic, generally irreversible condition that develops slowly—typically over a period of years—and is the result of a progressive deterioration in cerebral function. There are a number of causes of dementia (eg, genetic disorders); however, Alzheimer disease is perhaps the most common form of dementia that you will encounter as an EMT. There are medications that are used to slow the progression of Alzheimer disease, such as donepezil (Aricept); however, no cure exists at the present time.

In contrast with dementia, delirium is an acute change in mental status or cognitive function. It is marked by the inability to focus, think logically, and maintain attention and signals the onset of a new health problem. Unlike dementia, many causes of delirium are reversible. In older patients, delirium is commonly the result of a urinary tract infection, hypoglycemia or hyperglycemia, dehydration, bowel obstruction, fever, and vitamin deficiencies. Other causes include acute alcohol intoxication, drug overdose, withdrawal from drugs or alcohol, hypoxia, and hypovolemia, among others.

Although dementia cannot be treated in the prehospital setting, there are certain causes of delirium that can be treated. Regardless of whether you suspect dementia or delirium, you should assess the patient for three specific conditions that can be treated in the prehospital setting: hypoxia, hypoglycemia, and hypovolemia. Any of these three conditions, if unrecognized or untreated, can be rapidly fatal. With prompt treatment, however, the patient's delirious behavior can often be reversed.

9. What strategies should you use when communicating with older patients?

Effective communication skills are an essential component when assessing and treating any patient but are especially important when caring for older patients. A number of factors can hamper communication with older patients; however, this does not mean that an older patient cannot be communicated with or is unable to communicate.

Age-related changes in vision, hearing, taste, smell, and touch can make effective communication with older patients a challenge, and although these challenges should be expected, they should not be assumed.

You are the Provider: SUMMARY, continued

Patients with processes such as dementia can be especially difficult to communicate with, and although you will likely obtain most of the patient's medical history from a family member or caretaker, you should not abandon attempts to talk to your patient. As with patients of any age, many older patients may have more difficulty communicating clearly when they are stressed by an emergency or personal crisis; therefore, patience on your part is essential.

Your first words to the patient and the attitude behind them can gain or lose a patient's trust. Speak respectfully when you introduce yourself. If you know the patient's name, use it; however, address him or her as "Mr," "Missus," or "Miss." If you do not know the patient's name, address him or her as "sir" or "ma'am." Some older patients may be insulted if you use their first name. Asking the patient how he or she prefers to be addressed is a trust-building technique. If the patient requests that you call him or her by first name, it is okay to do so. However, words such as "Hon" and "Dear" are unprofessional and inappropriate; they should be avoided.

When you are interviewing an older patient, identify yourself, your partner, and any other responders at the scene; this helps establish a rapport with the patient. Remain aware of how you present yourself; avoid showing frustration and impatience through body language.

Look directly at the patient and speak slowly and distinctly. Do not assume that the patient is hearing-impaired simply because of his or her age. Many older patients' hearing is just as acute as that of someone half their age. If the patient is hearing-impaired, try speaking in lower frequencies; most hearing-impaired patients have difficulty with high-frequency sounds. Do not yell in the patient's ear! If the patient has hearing aids, ensure that they are in and that the batteries are working.

Have one person talk to the patient, and ask only one question at a time. Do not answer questions for patients out of frustration or impatience. When you ask the patient a question, allow adequate time for him or her to respond, unless the condition is urgent or critical. When the patient responds, listen carefully to what he or she is saying.

Explain what you are going to do before you do it. Use simple terms to explain the use of medical equipment and procedures, avoiding medical jargon and slang. Plain English is always the best approach, regardless of the patient's age.

Do not talk about the patient in front of him or her as though he or she was not there. This gives the patient the feeling that he or she has no say in decisions about his or her care. This is easy to forget when the patient has impaired cognitive (thought) processes or has difficulty communicating.

EMS Patient Care Report (PCR)

Date: 11-28-09	**Incident No.:** 013309	**Nature of Call:** Shortness of breath		**Location:** 644 Yolanda St.	
Dispatched: 0625	**En Route:** 0626	**At Scene:** 0630	**Transport:** 0644	**At Hospital:** 0657	**In Service:** 0707

Patient Information

Age: 82 **Sex:** F **Weight (in kg [lb]):** 47 kg (105 lb)	**Allergies:** Erythromycin **Medications:** Plavix, Digoxin, Aricept, Verapamil, Celebrex, Albuterol, Flovent **Past Medical History:** Emphysema, atrial fibrillation, Alzheimer disease, rheumatoid arthritis, hypertension **Chief Complaint:** Difficulty breathing

Vital Signs

Time: 0635	**BP:** 158/88	**Pulse:** 68	**Respirations:** 22	**Sao$_2$:** 92%
Time: 0642	**BP:** 152/90	**Pulse:** 84	**Respirations:** 22	**Sao$_2$:** 94%
Time: 0652	**BP:** 148/88	**Pulse:** 70	**Respirations:** 22	**Sao$_2$:** 95%

EMS Treatment
(circle all that apply)

Oxygen @ 4 L/min via (circle one): (NC) NRM Bag-Mask Device		**Assisted Ventilation**	**Airway Adjunct**	**CPR**
Defibrillation	Bleeding Control	Bandaging	Splinting	**Other:** (Position of comfort; emotional support)

You are the Provider: SUMMARY, continued

Narrative
Medic 53 dispatched to a residence for a woman with respiratory distress. Arrived on scene and found the patient, an 82-year-old woman, sitting in a chair in her living room. She was conscious but confused. Her airway was patent, and her breathing was labored. The patient had a blanket wrapped around her and was shivering. Her daughter, who was present at the scene, advised that the patient has a history of emphysema and that her usual shortness of breath has worsened during the past few days. The patient was currently receiving home oxygen at 2 L/min via nasal cannula. The daughter further stated that the patient has been running a fever (101°F) and has had a cough, which is producing thick green sputum. Other past medical history significant for A-Fib, Alzheimer disease, hypertension, and rheumatoid arthritis. Inquired about patient's current mental status (confused) and was advised by the daughter that this is normal for her. Further assessment revealed bilateral basilar rhonchi on auscultation of breath sounds. Obtained vital signs, and increased oxygen flow rate to 4 L/min. Advised patient that transport to the hospital for evaluation was necessary; however, she stated that she did not want to go. Communication with the patient was difficult given her history of Alzheimer disease. The patient's daughter reassured her that she would be okay and that she would follow the ambulance to the hospital. After reassurance by the daughter and EMS, the patient consented to transport. Placed patient onto stretcher, placed her in a position of comfort, loaded her into the ambulance, and began transport. En route, patient's condition remained unchanged; she remained confused and short of breath, and her vital signs remained stable. Continued to provide emotional support and reassurance to the patient; she was able to answer some questions but had difficulty with others. Delivered patient to emergency department staff, gave verbal report to staff nurse, and transferred patient care. Medic 53 returned to service at 0707. *End of report*

Assessment and Emergency Care of Geriatric Emergencies

Scene Size-up

Scene Safety	Ensure scene safety and safe access to the patient. You may need assistance gaining access to patients with breathing problems, chest pain, or altered mental status who are not able to get to the door to let you in. If the call takes place in an adult care facility or a nursing home, locate a staff member to help provide you with the patient's medical history. Follow standard precautions, putting on a minimum of gloves and eye protection. Determine the number of patients. Assess the need for additional resources such as law enforcement, ALS, or additional ambulances. Take note of negative or unsafe environmental conditions. Consider which equipment is needed to treat the patient.
Mechanism of Injury (MOI)/ Nature of Illness (NOI)	Determine the MOI/NOI. Observe the scene and look for clues that may help you determine what happened to your patient. The NOI may be difficult to determine in older people with an altered mental status or dementia. Multiple and chronic disease processes may also complicate the determination of the NOI. Vague complaints may be indicators of a more serious illness or injury. Determine what is different today that resulted in the call to EMS. Observe the surroundings, noting the presence of pills, medicine bottles, medical equipment, and the overall condition of the living environment. Look for clues that indicate your patient's traumatic incident may have been preceded by a medical incident. Bystander information may help determine if a loss of consciousness occurred before the accident. Consider spinal immobilization in any trauma patient with a significant MOI.

▼ ▼

Primary Assessment

Form a General Impression	A rapid visual examination and full-body scan of the patient will help you identify and manage life threats. Determine the priority of care based on the general impression and the MOI. Assess the patient's level of consciousness using the AVPU scale. If the patient is able to communicate, obtain the chief complaint.
Airway and Breathing	Ensure the airway is open, clear, and patent. Geriatric patients are predisposed to airway problems. Be alert for airway obstructions, loss of a gag or cough reflex, and aspiration. Only remove dentures if they are creating an airway patency problem. It is more difficult to ventilate a patient with no teeth. Open the airway with a modified jaw-thrust maneuver in an unresponsive patient. Use an airway adjunct as appropriate and ventilate with a bag-mask device if needed. Suction as needed. Evaluate the patient's ventilatory status for rate and depth of breathing, respiratory effort, and tidal volume. Assess the patient's lungs sounds. Any curvature of the patient's spine will require padding to keep the patient supine and the airway open. Minor chest trauma can cause lung injury. Perform a thorough respiratory assessment and physical assessment of the chest and treat accordingly. Administer high-flow oxygen at 15 L/min.
Circulation	Lower heart rates and weaker and irregular pulses are common. It may be difficult to feel a radial pulse. Try an alternative pulse point like the carotid artery or listen to the apical pulse right over the heart. The pulse may be irregular because of common heart rhythm problems. Treat circulation issues with oxygen as soon as possible. Manage any external bleeding immediately. Be suspicious of signs and symptoms of internal bleeding. Geriatric patients can quickly go into shock from blood loss. Geriatric patients with hypertension may have a normal blood pressure when they are actually in shock. A head injury with minimal mechanism can cause cerebral bleeding. Blood-thinning medications can make internal bleeding worse or external bleeding more difficult to control.
Transport Decision	Any examination findings that cause compromises in airway, breathing, or circulation should result in transportation. This includes patients with a poor general impression, airway or breathing problems, acute altered level of consciousness, shock, any severe pain, or uncontrolled bleeding. A general complaint of weakness and dizziness can be an indication of something more serious. Consider early on in the call if ALS treatment and transport is appropriate and available. If possible, try to take the patient to a facility where the patient has been treated before and his or her medical records reside. Consider trauma center transport for geriatric patients if there is the potential for a serious injury.

▼ ▼

Assessment and Emergency Care of Geriatric Emergencies, continued

NOTE: The order of the steps in this section differs depending on whether the patient is conscious or unconscious. The following order is for a conscious patient. For an unconscious patient, perform a primary assessment, perform a full-body scan, obtain vital signs, and obtain the past medical history from a family member, bystander, or emergency medical identification device.

History Taking

Investigate Chief Complaint	Investigate the chief complaint and gather a history once you have identified and treated life threats. It is best to obtain the information directly from the patient, but family or caregivers may need to assist you. Find and account for all medications. Ask OPQRST and SAMPLE questions, focusing on the events surrounding the incident and the MOI or NOI. SAMPLE can also be obtained from family, bystanders, and medical alert tags if the patient is not able to provide the information. Make special note of the patient's medications. Is the patient taking them as prescribed? Does the patient have many medications from different physicians or pharmacies (polypharmacy)? The illness may be the result of medication interactions. If the patient has a traumatic injury, consider the geriatric patient's past medical conditions, even if the patient is not currently acute or symptomatic.

Secondary Assessment

Physical Examinations	Determine if an alteration of consciousness is acute or chronic. Chronic mental status impairment is not a normal process of aging. Determine your patient's baseline mental status using family members or caregivers, if available. Ask questions to assess the nature of the problem and determine whether it may or may not be life threatening. Take a full set of vital signs and ask what is considered "normal" for the patient. Perform a physical examination when appropriate. However, you may find that your elderly patient is not comfortable with being exposed. Protecting his or her modesty will help. The elderly are often cold and you may have to remove several layers of clothing. Consider the need to keep your patient warm during your full-body scan. Determine the conditions that are life threatening, treat them to the best of your ability, and provide transport to the appropriate facility.
Vital Signs	Obtain baseline vital signs. Assess pulse, blood pressure, and skin signs. Measurement of capillary refill may be unreliable because of compromised circulation in geriatric patients. Drugs such as beta-blockers inhibit the heart from becoming tachycardic as you would expect in shock; therefore, a heart rate in the normal range may be high for a person who is taking beta-blockers. Try to determine if the patient's blood pressure is normal. A blood pressure that may be normal for an older adult could indicate shock in a younger patient. Auscultate breath sounds to listen for rales or rattles. Careful interpretation of pulse oximetry data is necessary because the pulse oximetry device requires adequate perfusion to provide an accurate reading.

Reassessment

Interventions	Repeat the primary assessment, vital signs, chief complaint, and interventions. Vital signs and reassessment should be repeated every 5 minutes in the serious or unstable patient and every 15 minutes in the stable patient. Compare the findings to those measurements taken earlier. Continuously observe and reassess the patient during transport so worsening conditions can be managed. A geriatric patient who is experiencing shortness of breath will want to sit up or assume the tripod position. Patients with kyphosis will require padding to allow placement of the patient in a supine position. Provide blankets and heat to prevent hypothermia.

Assessment and Emergency Care of Geriatric Emergencies, continued

| Communication and Documentation | Contact medical control/receiving hospital with a radio report, using appropriate medical and anatomic terminology so appropriate staff and equipment will be ready when you arrive. Include a thorough description of the MOI/NOI and the position in which the patient was found. Include treatment performed and patient response. Be sure to document the patient's chief complaint, description of the incident, physical findings, history, and any changes in patient status and the time. Document the scene observations on your arrival and be prepared to describe your observations to the hospital staff. Follow local treatment protocols. Refusals of medical aid or transport should be documented completely and approved through medical control. Communicating with the geriatric patient during an emergency may be difficult. Do not assume that every elderly patient is hard of hearing; do not shout at the patient. |

NOTE: Although the following steps are widely accepted, be sure to consult and follow your local protocols. Take appropriate standard precautions when treating all patients.

Geriatric Emergencies

General Management of Geriatric Emergencies

Managing life threats to the patient's ABCs is the primary concern. Life-threatening illness or injury detected during the physical examination must be managed immediately. Remain calm and professional. Treat every geriatric patient with respect. Give the patient time to answer your questions.

1. Ensure scene safety.
2. Determine the MOI/NOI.
3. Consider the need for spinal stabilization.
4. Determine the patient's baseline mental status.
5. Open, clear, and maintain the patient's airway.
6. Administer high-concentration oxygen via a nonrebreathing mask or a bag-mask device as appropriate.
7. Control bleeding and treat for shock. Remember, internal injuries may be present without obvious bleeding or external injury.
8. Place the patient in a position of comfort.
9. Transport to the appropriate treatment facility.

Respiratory Emergencies

Age-related changes in the respiratory system predispose an older adult to respiratory illness. Even a minor lung infection can become a life-threatening event.

Pneumonia
Treat the patient with airway, ventilatory, and circulatory support. Provide high-flow oxygen via a nonrebreathing mask for a patient with a manageable airway. Use high-flow oxygen with an oropharyngeal or nasopharyngeal airway for the patient who needs ventilatory support with a bag-mask device. If necessary, provide circulatory support by placing the patient in the shock position or performing cardiopulmonary resuscitation. Follow the respiratory and cardiac arrest management standard of your agency. Evaluate patient treatment through reassessment and prepare for possible deterioration of your patient's condition.

Pulmonary Embolism
Focus on airway, ventilatory, and circulatory support. Suction any blood from the airway. Place the patient in a position of comfort and assist breathing using high-flow oxygen and a nonrebreathing mask. If aggressive airway management is necessary, ventilate with a bag-mask device using an oropharyngeal or nasopharyngeal airway. If a patient is in respiratory and/or cardiac arrest, manage according to current Emergency Cardiovascular Care (ECC) guidelines and area protocol. Reassess the patient during transport.

Assessment and Emergency Care of Geriatric Emergencies, continued

Geriatric Emergencies

Cardiovascular System

The cardiovascular system in a geriatric patient is at risk because of age-related changes such as enlargement of the heart with age, arteriosclerosis, and a more sedentary lifestyle.

Heart Attack (Myocardial Infarction)

Treatment should consist of airway, ventilatory, and circulatory support. Provide oxygen with adjuncts appropriate to the patient's condition. Continue to evaluate your patient's response to treatment through reassessment. Cardiac problems can worsen suddenly, so be prepared to act quickly. Many patients will not present with classic chest pain. Atypical presentations are seen mostly in women, the elderly, and patients with diabetes. Associated signs and symptoms such as weakness and dizziness, nausea and vomiting, loss of bladder and bowel control, and shortness of breath should be evaluated by ALS personnel as a possible myocardial infarction.

Heart Failure

Signs and symptoms that your patient may have heart failure include dyspnea on exertion, sleeping with multiple pillows or sitting upright, fatigue, and edema in the lower back and extremities. Treatment consists of airway, ventilatory, and circulatory support. Provide oxygen with adjuncts appropriate to the patient's condition and prepare for continued deterioration.

Stroke (Cerebrovascular Accident)

The signs of stroke include altered mental status; numbness, weakness, or paralysis on one side of the body; slurred speech; difficulty speaking; vision disturbances; headache; dizziness; or incontinence. The treatment goal is to make sure that surrounding brain tissue sustains as little damage as possible. If available, transport to a stroke center.

Nervous System

Delirium

Assess the patient for the three specific conditions that can be managed at the prehospital level: hypoxia, hypovolemia, and hypoglycemia. Treatment will depend on the results of your assessment but should include airway, ventilatory, and circulatory support and oxygen with airway adjuncts appropriate to patient's condition if tolerated by the patient. Contact ALS for support.

Gastrointestinal System

Geriatric patients are more likely to experience gastrointestinal bleeding. Causes include inflammation, infection, and intestinal obstruction. Patients with gastrointestinal problems will present with hematemesis, melena, dyspepsia, constipation, or diarrhea. Patients are usually agitated and are unable to find a comfortable position. Some patients will report dizziness or syncope. Treatment consists of airway, ventilatory, and circulatory support. Oxygen should be delivered with adjuncts appropriate to the patient's condition. Transport the patient to the hospital.

Abdominal Aortic Aneurysm (AAA)

The signs and symptoms include radiating abdominal pain, a pulsating mass near the navel, diminished or absent pulses in the extremity, shock, and syncope. Treat the patient for shock and provide prompt transport to the hospital.

Assessment and Emergency Care of Geriatric Emergencies, continued

Geriatric Emergencies

Endocrine System

With age, the endocrine system metabolizes hormones differently. Hypothyroidism can present with a slower heart rate, fatigue, drier skin and hair, cold intolerance, and weight gain. An increase in the secretion of the antidiuretic hormone can cause fluid imbalance; increases in the levels of norepinephrine possibly have a harmful effect on the cardiovascular system and cause hyperglycemia.

Hyperosmolar Hyperglycemic Nonketotic (HHNC) Coma

Signs and symptoms include hyperglycemia, polydipsia, dizziness, confusion, altered mental status, and possibly seizures. The patient may present with signs and symptoms of hypotension and shock, including tachycardia. Treatment should include airway, ventilatory, and circulatory support. Provide oxygen with adjuncts appropriate to the patient's condition.

Trauma

The risk of serious injury or death is more common in elderly patients who sustain trauma. Elderly pedestrians are more likely to have life-threatening complications after being struck by a vehicle because of changes in the body such as fragile bones. Elderly pedestrians commonly suffer injury to the legs and arms, and a secondary collision onto the street often involves the head, which can cause fractures, traumatic brain injury, spinal injury, and paralysis. Falls are a common MOI for the elderly because of poor vision and balance problems, decreased visual acuity, and decreased strength. A fall in a geriatric patient with osteoporosis can result in a pelvic fracture. Determine if the fall was the result of cardiac, neurologic, or metabolic issues. Burns are more common because of altered mental status, inability to focus, and a compromised neurologic status. Remember, trauma is never isolated to a single issue when you are assessing and caring for a geriatric patient. Trauma in the elderly can also be caused by abuse. Be aware of the patient's environment and living conditions a patient lives in and take note of soft-tissue injuries that cannot be explained by the person's lifestyle and physical condition.

Prep Kit

- Although assessment of geriatric patients involves the same basic approach as that for any other patient, you must take a more wary approach.

- Assessing an elderly person can be challenging because of communication issues, hearing and vision deficits, alteration in consciousness, complicated medical history, and the effects of multiple medications.

- To obtain an accurate history for a geriatric patient, patience and good communication skills are essential. A slow, deliberate approach to the patient history, with one EMT asking questions, is generally the best strategy.

- With changes in the respiratory system, such as a decreased ability to cough, geriatric patients are more likely to present with pneumonia.

- Changes in the cardiovascular system can lead to atherosclerosis, aneurysm, stiffening heart valves, orthostatic hypotension, venous stasis, deep venous thrombosis, heart attack, heart failure, and stroke.

- Many patients do not present with the classic symptom of chest pain when experiencing a heart attack. Atypical presentations are seen mostly in women, elderly patients, and patients with diabetes.

- Dementia and delirium must be carefully evaluated in geriatric patients.

- As the body ages, the bones become more fragile. This leads to a higher risk of fracture in geriatric patients.

- Polypharmacy and changes in medications can cause serious problems for geriatric patients.

- Depression is treatable with medication and therapy but is a risk factor for suicide if it remains untreated in geriatric patients.

- The risk of serious injury or death is more common in elderly patients who experience a traumatic injury.

- When you treat a geriatric trauma patient, assess the injuries and carefully look for the cause of the injury. A medical condition such as fainting could actually be the cause of a fall. The injuries from the fall and the medical condition will need to be addressed.

- When responding to nursing and skilled care facilities, you should determine the patient's chief complaint on that day and what initial problem caused the patient to be admitted to the facility.

Vital Vocabulary

abdominal aortic aneurysm (AAA) A condition in which the walls of the aorta in the abdomen weaken and blood leaks into the layers of the vessel, causing it to bulge.

advance directives Written documentation that specifies medical treatment for a competent patient should the patient become unable to make decisions; also called living wills.

aneurysm An enlargement of a part of an artery, resulting from weakening of the arterial wall.

ascites Fluid in the abdomen.

cataracts Clouding of the lens of the eye or its surrounding transparent membranes.

decubitus ulcers Also known as bedsores. These are sores caused by the pressure of skin against a surface for long periods. These sores can range from a pink discoloration of the skin to a deep wound that may invade into bone or organs.

deep venous thrombosis The formation of a blood clot within the larger veins of an extremity, typically following a period of prolonged immobilization.

delirium A more or less sudden change in mental status marked by the inability to focus, think logically, and maintain attention.

dementia The slow onset of progressive disorientation, shortened attention span, and loss of cognitive function.

dyspnea Shortness of breath or difficulty breathing.

elder abuse Any action on the part of an older person's family member, caregiver, or other associated person that takes advantage of the older person's person, property, or emotional state; also called granny beating and parent battering.

geriatrics The assessment and treatment of disease in someone who is 65 years or older.

jugular vein distention A visual bulging of the jugular veins in the neck that can be caused by fluid overload, pressure in the chest, cardiac tamponade, or tension pneumothorax.

kyphosis A forward curling of the back caused by an abnormal increase in the curvature of the spine.

melena Black, tarry stools.

methicillin-resistant *Staphylococcus aureus* (MRSA) A bacterium that causes infections in different parts of the body and is often resistant to commonly used antibiotics; can be found on the skin, in surgical wounds, in the bloodstream, lungs, and urinary tract.

neuropathy A group of conditions in which the nerves leaving the spinal cord are damaged, resulting in distortion of signals to or from the brain.

osteoporosis A generalized bone disease, commonly associated with postmenopausal women, in which there is a reduction in the amount of bone mass leading to fractures after minimal trauma in either sex.

peptic ulcer disease An abrasion of the stomach or small intestine.

pneumonia An inflammation/infection of the lung from a bacterial, viral, or fungal cause.

polypharmacy The use of multiple medications by a patient as typically seen in elderly people.

presbycusis An age-related condition of the ear that produces progressive bilateral hearing loss that is most noted at higher frequencies.

pulmonary embolism A condition that causes a sudden blockage of the pulmonary artery by a venous clot.

respiratory syncytial virus A highly contagious virus that causes an infection of the upper and lower respiratory system.

syncope A fainting spell or transient loss of consciousness, often caused by an interruption of blood flow to the brain.

unilateral pedal edema Pedal edema is a swelling of the foot and ankle caused by fluid overload; unilateral would present in only one extremity.

urinary tract infection A bacterial infection that affects the urinary tract.

Assessment in Action

It is 3:00 AM and you and your partner are called to a private residence for an 86-year-old woman complaining of a sudden onset of shortness of breath. On arrival, you observe an older home in need of repair. When you enter the home, you find that there are seven cats, and there is the strong odor of stale cat food and litter boxes.

1. What is one of the most frequently occurring conditions that affect older people?
   - **A.** Diabetes
   - **B.** Altered mental status
   - **C.** Respiratory distress
   - **D.** Cancer

2. You find your patient sitting upright in her bed with four pillows behind her. She is having difficulty speaking because of a previous stroke. Which of the following interview techniques should you use when addressing the patient?
   - **A.** Speak loudly and slowly.
   - **B.** Listen carefully to the answers the patient provides.
   - **C.** Refer to the patient by her first name.
   - **D.** Have both you and your partner ask questions.

3. Which of the following conditions is considered to be a risk factor for congestive heart failure?
   - **A.** Coronary artery disease
   - **B.** Emphysema
   - **C.** Dementia
   - **D.** Diabetes

4. The patient explains that she was awakened by a sudden feeling of suffocation and respiratory distress. What are her symptoms a characteristic of?
   - **A.** Orthopnea
   - **B.** Exertional dyspnea
   - **C.** Intermittent sleep apnea
   - **D.** Paroxysmal nocturnal dyspnea

5. What assessment question should you ask to help clarify her symptoms of a sudden feeling of suffocation and respiratory distress?
   - **A.** How many hours of sleep do you get each night?
   - **B.** What position do you normally sleep in?
   - **C.** How many pillows do you sleep on?
   - **D.** Do you take any medication to help you sleep at night?

6. On auscultation of the patient's lungs, your partner hears rales. These lung sounds are caused by air passing through:
   - **A.** constricted airways.
   - **B.** thick secretions in the airways.
   - **C.** fluid in the alveoli.
   - **D.** inflamed airways.

7. The pulse oximeter is unable to provide an accurate reading. What is a possible cause for this finding?
   - **A.** Poor circulation
   - **B.** Hypertension
   - **C.** Fever
   - **D.** Increased red blood cell count

8. Observing and documenting the condition in which you find the home is part of the GEMS _____ assessment.
   - **A.** general
   - **B.** environmental
   - **C.** medical
   - **D.** social

9. Explain the importance of performing a social assessment when caring for a geriatric patient.

10. Your patient tells you that she takes numerous medications every day for her heart and blood pressure; however, she cannot remember their names. She points to a plastic bag by the bedside that contains 11 prescription bottles with similar medications, and some of the medication were not prescribed to her. What problems might this situation pose?

National EMS Education Standard Competencies

Special Patient Populations

Applies a fundamental knowledge of growth, development, and aging and assessment findings to provide basic emergency care and transportation for a patient with special needs.

Patients With Special Challenges

Recognizing and reporting abuse and neglect (Covered in Chapter 32, *Pediatric Emergencies*, and Chapter 33, *Geriatric Emergencies*)

Health care implications of

- Abuse (Chapter 32, *Pediatric Emergencies*, and Chapter 33, *Geriatric Emergencies*)
- Neglect (Chapter 32, *Pediatric Emergencies*, and Chapter 33, *Geriatric Emergencies*)
- Homelessness (p 1275)
- Poverty (p 1275)
- Bariatrics (pp 1268-1269)
- Technology dependent (pp 1269-1273)
- Hospice/terminally ill (pp 1274-1275)
- Tracheostomy care/dysfunction (pp 1269-1270)
- Homecare (pp 1273-1274)
- Sensory deficit/loss (pp 1264-1266)
- Developmental disability (pp 1261-1264)

Trauma

Applies fundamental knowledge to provide basic emergency care and transportation based on assessment findings for an acutely injured patient.

Special Considerations in Trauma

Recognition and management of trauma in

- Pregnant patient (Chapter 31, *Obstetrics and Neonatal Care*)
- Pediatric patient (Chapter 32, *Pediatric Emergencies*)
- Geriatric patient (Chapter 33, *Geriatric Emergencies*)

Pathophysiology, assessment, and management of trauma in the

- Pregnant patient (Chapter 31, *Obstetrics and Neonatal Care*)
- Pediatric patient (Chapter 32, *Pediatric Emergencies*)
- Geriatric patient (Chapter 33, *Geriatric Emergencies*)
- Cognitively impaired patient (pp 1261-1264)

Knowledge Objectives

1. Give some examples of patients with special needs whom an EMT may encounter during a medical emergency. (pp 1261-1275)

2. Discuss the special patient care considerations that may be required when providing emergency medical care to patients with developmental disabilities, including patients with autism, Down syndrome, and prior brain injuries. (pp 1261-1264)

3. Discuss different types of visual impairments and the special patient care considerations that may be required when providing emergency medical care for these patients depending on the level of their disability. (p 1264)

4. Explain the various types of hearing impairments and the special patient care considerations that may be required when providing emergency medical care for these patients, including tips on effective communication. (pp 1265-1266)

5. List the various types of hearing aids that may be worn by patients and describe troubleshooting strategies that may help to fix a hearing aid that is not working. (pp 1265-1266)

6. Discuss the special patient care considerations that may be required when providing emergency medical care to patients who have cerebral palsy, spina bifida, and paralysis. (pp 1266-1268)

7. Define obesity and discuss the special patient care considerations, including the best way to move a morbidly obese patient, that may be required when providing emergency medical care to bariatric patients. (pp 1268-1269)

8. Discuss the special patient care considerations that may be required when providing emergency medical care to a patient who relies on a form of medical technological assistance, including a tracheostomy tube, mechanical ventilator, apnea monitor, internal cardiac pacemaker, left ventricular assist device, central venous catheter, gastrostomy tube, shunt, vagal nerve stimulator, colostomy, and ileostomy. (pp 1269-1273)

9. Describe home care, the types of patients it serves, and the services it encompasses. (pp 1273-1274)

10. Discuss hospice and palliative care and how they differ from curative care, and then explain the responsibilities of the EMT when responding to calls for terminally ill patients who have DNR orders. (pp 1274-1275)

11. Discuss the issues of poverty and homelessness in the US, its negative effects on a person's health, and the role of the EMT as a patient advocate. (p 1275)

Skills Objectives

1. Demonstrate different strategies to communicate effectively with a patient who has a hearing impairment. (pp 1265-1266)

Introduction

The approach to health care in our society continues to focus on decreasing the length of hospitalization. At the same time, medicine and medical technology continue to improve. As a result, the number of children and adults with chronic diseases and injuries who are living at home or in other environments outside of a hospital setting continues to grow. EMTs should be familiar with the special needs created by patients with chronic diseases and conditions.

Some examples of patients with special needs include:

- Children who were born prematurely and who have associated respiratory problems
- Infants or small children with congenital heart disease
- Patients with neurologic disease (occasionally caused by hypoxemia at the time of birth, as with cerebral palsy)
- Patients with congenital or acquired diseases resulting in altered body function that requires medical assistance for breathing, eating, urination, or bowel function
- Patients with sensory deficits such as hearing or visual impairments
- Geriatric patients with chronic diseases requiring visitation from a home health care service

You may be called on to treat children and adults who are living at home who depend on mechanical ventilators, intravenous pumps, or other devices to maintain their lives. You should assess and care for patients with special needs the same way you care for your other patients. Your focus on the assessment and treatment of the ABCs remains the priority. Do not allow yourself to be distracted by the noise and mechanics of the medical equipment—your focus needs to remain on the patient the medical equipment may be assisting. If the emergency is the result of medical equipment failure, use the equipment on the ambulance. In some cases, some families will have a "go bag," which is a collection of spare equipment and supplies for such situations.

Words of Wisdom

During stressful emergency events, it is imperative to use the TEAM approach (Trust, Every, Available, Member) to collaborate with others regarding your patient's treatment, which leads to a better standard of care and patient outcome.

Developmental Disability

A **developmental disability** (mental retardation) is caused by insufficient cognitive development of the brain, which results in a person's inability to learn and socially adapt at a normal developmental rate. A developmental disability may be caused by genetic factors, congenital infections, complications at birth, malnutrition, or environmental factors. Prenatal drug or alcohol use, as in fetal alcohol syndrome, may also cause developmental disability. Other causes that may occur after birth include traumatic brain injury and poisoning (eg, with lead or other toxins).

A person with slight impairment may appear slow to understand or have a limited vocabulary. Such patients will often behave immaturely in comparison to their

You are the Provider: PART 1

At 2:35 PM, you are dispatched to a residence at 575 Ranger Drive for a 19-year-old man with a fever. You recognize the address because you have responded to this patient on several occasions. He is a quadriplegic who is ventilator-dependent because of a spinal injury that occurred 2 years ago. You and your partner proceed to the scene; your response time is 5 minutes.

1. Will your assessment and treatment of this patient differ from those of any other patient?
2. What role do the parents or caregivers of patients with special health care needs have in the prehospital setting?

peers. Severely disabled persons may not have the ability to care for themselves, communicate, understand, or respond to their surroundings.

Speaking to patients and family members will give you a good idea of how well the patient can understand you and how the patient will interact with you. Family or friends of the patient also may be able to supply additional medical information regarding the patient.

Because patients with disabilities may have difficulty adjusting to change or a break in routine, an emergency call that generates a roomful of strangers can be overwhelming. A patient may become more difficult to interact with as his or her anxiety level increases. Make every effort to respect the patient's wishes and concerns; take as much time as necessary to explain in a calming, understandable way the treatment the patient is about to receive.

Patients with developmental disabilities are susceptible to the same disease processes as other patients, including diabetes, heart attack, and respiratory difficulties. Assess and treat the patient according to the chief complaint. Transport should be accomplished with as little stress as possible.

■ Autism

Autism is a term that is used widely in the general public. Autism is a pervasive developmental disorder characterized by impairment of social interaction. Other characteristics can include severe behavioral problems, repetitive motor activities, and impairment in verbal and nonverbal skills. The spectrum of disability is wide. Some children will grow up to be independent while others will be unable to care for themselves.

Patients with autism fail to use or understand nonverbal means of communicating messages. They frequently have difficulty making eye-to-eye contact and resist encouragement to do so. They have extreme difficulty with complex tasks that require many steps and do best with simple, one-step directions ("Please roll up your sleeve."). Patients with autism tend to get lost in long conversations and have trouble answering open-ended questions (eg, "What sorts of things do you enjoy doing?"). They tend to talk in robotic or monotone speech patterns and sometimes repeat phrases over and over again. Many patients with autism confuse pronouns and will say "you" when they really mean "I," as in "You are going to the hospital," when they really mean, "I am going to the hospital." A small percentage of patients with autism do not speak at all, but instead rely on pulling parents and caregivers around by the hand to get their needs met.

There is no simple explanation as to why autism develops in children. According to the Centers for Disease Control and Prevention, approximately 1 in every 150 American children is diagnosed with autism. Autism affects males four times greater than females and is typically diagnosed by 3 years of age. The parents or caregivers often report unique repetitive (hand-flapping, twirling objects) or isolated abnormal behaviors. Today children with autism-spectrum disorders receive special instruction and care in school-based settings. It is likely that some older adults with autism have never been diagnosed and have never received any assistance.

Patients with autism generally do not have other medical disorders and will have medical needs similar to their peers without autism. Rely on parents or caregivers for information and keep them involved in the treatment of the patient.

■ Down Syndrome

Down syndrome is characterized by a genetic chromosomal defect that can occur during fetal development, resulting in mild to severe mental retardation **Figure 34-1**. The normal human somatic cell contains 23 chromosomes. Down syndrome, which is also known as trisomy 21, occurs when chromosome 21 fails to separate, so that the ovum contains 24 chromosomes. When the ovum is fertilized by a normal sperm with 23 chromosomes, a triplication ("trisomy") of chromosome 21 occurs.

Increased maternal age and a family history of Down syndrome are known risk factors for this condition. A variety of abnormalities are associated with Down syndrome—a round head with a flat occiput; an enlarged, protruding tongue; slanted, wide-set eyes and

Figure 34-1 A child with Down syndrome.

folded skin on either side of the nose, covering the inner corners of the eye; short, wide hands; a small face and features; congenital heart defects; thyroid problems; and hearing and vision problems. Persons with Down syndrome do not usually have all of these signs, but a diagnosis is able to be made rapidly at birth because a combination of them can be seen. Persons with Down syndrome, depending on their level of mental disability, may lead more independent lives through employment, voting, and getting involved in the community.

Patients with Down syndrome are at increased risk for medical complications, including those that affect the cardiovascular, sensory, endocrine, orthopaedic, dental, gastrointestinal, and neurologic development. As many as 40% of these patients may have heart conditions and hearing and vision problems. Two thirds of children born with Down syndrome have congenital heart disease.

Because persons with Down syndrome often have large tongues and small oral and nasal cavities, intubation may be difficult. These patients may also have misalignment of teeth and other dental anomalies. The enlarged tongue and dental anomalies can lead to speech abnormalities as well. In an emergency situation, if airway management is necessary, mask ventilation can be challenging. In the case of airway obstruction, a jaw-thrust maneuver may be all that is needed to clear the airway. In an unconscious patient, either the jaw-thrust maneuver or a nasopharyngeal airway may be necessary.

Many persons with Down syndrome have epilepsy. Most of the seizures are tonic-clonic. Patient management is the same as with other patients with seizures. Chapter 15, *Neurologic Emergencies,* discusses the emergency management of seizures in detail.

■ Patient Interaction

It is normal to feel somewhat uncomfortable when initiating contact with a developmentally disabled patient, especially if you have not encountered such situations frequently. The best plan of action is to treat the patient as you would any other patient.

Approach the patient in a calm, friendly manner, watching for signs of increased anxiety or fear. Remember, you are a stranger and are approaching with a group of people. The patient may not understand your uniform or realize that you and your crew are there to help. It may be helpful to have the members of your team hold back slightly until you can establish a rapport with the patient. You can then introduce the team members and explain what they are going to do. This method will slowly bring forward the other providers.

You are the Provider: PART 2

You arrive at the scene and find the patient lying supine in a hospital-style bed in the living room. He immediately looks at you when you approach him but does not talk to you. The patient's mother tells you that he began running a fever earlier in the day. She further advises you that the patient's home health nurse was present earlier and contacted his physician, who requested that EMS transport him to the hospital. While your partner gathers additional information, you perform a primary assessment.

Recording Time: 0 Minutes	
Appearance	Eyes open
Level of consciousness	Conscious and alert; this is his baseline mental status
Airway	Tracheostomy tube in place; upper airway clear of secretions or foreign bodies
Breathing	14 breaths/min via mechanical ventilator
Circulation	Increased pulse rate (strong and regular); skin is pink, hot, and moist; no gross bleeding

3. What are some conditions that would cause a patient to become dependent on a mechanical ventilator?

4. Are patients with tracheostomy tubes able to speak? How can you determine if your patient is alert?

You might interact with a patient as follows: "Hello Mr. Pemberton. My name is Jerry Booker." Shake Mr. Pemberton's hand if he will allow it. "We're here to help you. Your sister called us. She says you're not feeling well today, and we're here to help you feel better. My partner Tim is going to take your blood pressure. Do you remember having that done before?" Allow Mr. Pemberton to see and touch the blood pressure cuff as your partner moves forward. Move slowly but deliberately, explaining beforehand what you are going to do, just like you would with any other patient. Watch carefully for signs of fear or reluctance from the patient. Make sure you are eye level with the patient. If the patient is sitting, kneel or sit down. This is important in communicating with all patients, however, it is even more important in making the patient with special challenges comfortable.

Do your best to soothe the patient's anxiety and discomfort as you work through your assessment and provide treatment. By initially establishing trust and communication, you will have much better chance for a successful outcome.

Brain Injury

Patients who previously experienced head injuries may be difficult to assess and treat. Chapter 26, *Head and Spine Injuries*, discusses head injuries and traumatic brain injuries in detail. Patients with brain injuries may face a complex array of challenges related to their injury. In such cases, gathering a complete medical history from the patient, family, and friends will assist you. Your interaction with patients with brain injuries will need to be tailored to their specific abilities. Take the time to speak with the patient and family to establish what is considered normal for the patient; for example, determine whether the patient has cognitive, sensory, communication, motor, behavioral, or psychological deficits.

When you are caring for a patient with a previous head injury, talk in a calm, soothing tone, and watch the patient closely for signs of anxiety or aggression. In some cases, the patient may need to be specially positioned or restrained to ensure your safety and the safety of the patient. Do not expect the patient to walk to the ambulance or stretcher. As always, treat the patient with respect, use his or her name, explain procedures, and reassure the patient throughout the process.

◾ Sensory Disabilities

◾ Sight

Visual impairments may result from many different causes—a congenital defect, disease, injury, and

degeneration of the eyeball, optic nerve, or nerve pathway (eg, with aging). The degree of blindness may range from partial to total. Some patients lose peripheral or central vision; others can distinguish light from dark or discern general shapes.

Figure 34-2 A service dog is easily identified by its special harness.

Visual impairments may be difficult to recognize. During your scene size-up, look for signs that indicate the patient is visually impaired, such as the presence of eyeglasses, a cane, or a service dog **Figure 34-2**. Make yourself known when you enter the room, and introduce yourself and others in the room or have them introduce themselves so that the patient can identify their placement and voice. In addition, retrieve any visual aids to make the interaction more comfortable for your patient.

A visually impaired patient may feel vulnerable, especially during the chaos of an accident scene. The patient may have learned to use other senses such as hearing, touch, and smell to compensate for the loss of sight, and the sounds and smells of the scene may be disorienting. Remember to tell the patient what is happening, identify noises, and describe the situation and surroundings, especially if you must move the patient.

To ambulate safely, the patient may use a cane or walker. Even if the patient will be carried on a gurney, do not forget to take the patient's cane or walker. Unless the patient is in critical condition, a service dog can remain in the room and will provide reassurance for the patient and prevent delays in transport; however, you may need to make arrangements for the care or accompaniment of the dog. A friend or animal control officer can be helpful in this situation.

An ambulatory patient may be led by a light touch on the arm or elbow. You may also allow the patient to rest his or her hand on your shoulder, as this may enhance the patient's sense of balance and security while moving. You may also ask patients which method they prefer to use. Patients should be gently guided but never pulled or pushed. Obstacles need to be communicated in advance. Statements such as "You're approaching the stairs," and instructions about how many stairs to expect, will allow the patient to anticipate and navigate the obstacles safely.

■ Hearing

Hearing impairment may range from a slight hearing loss to total deafness. Some patients may have difficulty with pitch, volume, and speaking distinctly. Some patients learn to speak even though they have never heard sounds. Other patients may have heard speech and learned to speak, but have since lost some or all of their hearing, leading them to speak too loudly. Parkinson disease or other disease processes may cause patients to slur words, speak very slowly, or speak in a monotone.

The two most common forms of hearing loss are known as sensorineural deafness and conductive hearing loss. **Sensorineural deafness**, or nerve damage, is the most common hearing loss you will encounter in the field. Sensorineural deafness occurs from a lesion or damage to the inner ear. Elderly persons will have some degree of sensorineural hearing loss because of advanced age. Conductive hearing loss is caused by a faulty transmission of sound waves, which can occur when a person has an accumulation of wax within the ear canal or a perforated eardrum.

Communication with a patient who is hearing impaired can be challenging at best without hearing aids. A piece of paper and a writing utensil may prove helpful until the hearing aid(s) can be located.

Words of Wisdom

As with all barriers to communication, remember to document whenever you enlist the help of an interpreter or a person who signs. Also remember that conclusions reached based on the information from interpreters may not be valid. Ask the interpreter to report exactly what the patient signs and not to add any commentary, however well intentioned.

Once a hearing impairment is noted, you should assist the patient with finding and inserting any hearing aids. Hearing aids can be either external or internal, depending on the type of hearing damage. Clues that a person could be hearing impaired include the presence of hearing aids, poor pronunciation of words, or failure to respond to your presence or questions. Face the patient while you communicate so that he or she can see your mouth; do not exaggerate your lip movements or look away. Position yourself approximately 18″ directly in front of the patient. Most people who are hearing impaired have learned to use body language, such as hand gestures and lip reading. Because hearing-impaired patients typically have more difficulty hearing higher frequency sounds, if the patient seems to have difficulty hearing you, do not speak louder—try lowering the pitch of your voice.

Ask the patient, "How would you like to communicate with me?" American sign language may be his or her preferred method of communication Figure 34-3. An interpreter, family member, or friend may prove to be a valuable teammate. If needed, take the interpreter with you to the hospital because this may decrease the stress of communication on the patient, EMS crew, and the hospital staff.

If an interpreter is not readily available, call your receiving facility early on to request one. Ideally, an interpreter will arrive before you begin the patient assessment. Other patients may prefer written communication or communication of concepts or procedures with gestures or pictures. Simply asking a team member to retrieve the patient's hearing aid or auditory electronic enhancement device may help a great deal.

Here are some helpful hints for working with patients with hearing impairments:

- Speak slowly and distinctly into a less-impaired ear, or position yourself on that side.

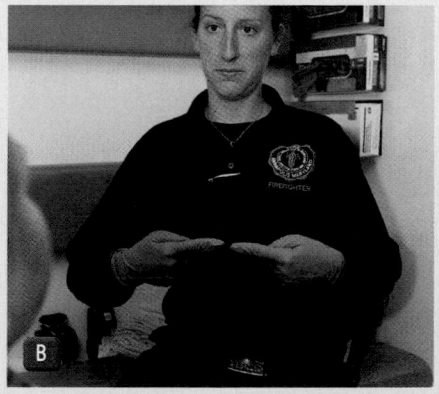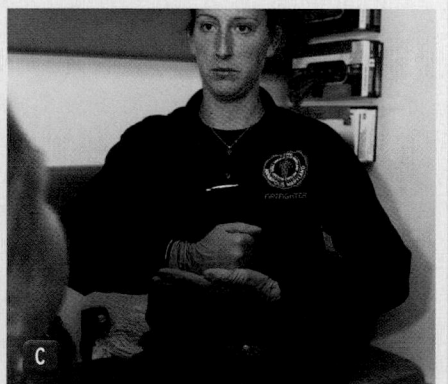

Figure 34-3 Consider learning American sign language for common terms related to illness and injury. **A.** Sick. **B.** Hurt. **C.** Help.

- Change speakers. Given that 80% of hearing loss is related to an inability to hear high-pitched sounds, look for a team member with a low-pitched voice if you think this may be the issue.
- Provide paper and a pencil so that you may write your questions and the patient may write responses.
- Only one person should ask interview questions, to avoid confusing the patient.
- Try the "reverse stethoscope" technique: put the earpieces of your stethoscope in the patient's ear and speak softly into the diaphragm of the stethoscope. This will amplify your voice.

Words of Wisdom

When you are caring for a hearing-impaired patient, one communication solution is to place the ear pieces of your stethoscope into the patient's ears while you speak softly into the bell of the stethoscope.

Hearing Aids

A hearing aid is essentially a device that makes sound louder. Hearing aids cannot restore hearing to normal, but they do improve hearing and listening ability. Several types of hearing aids are available Figure 34-4 :

- **Behind-the-ear type.** All parts are contained in a plastic case that rests behind the ear.
- **Conventional body type.** This older style is generally used by people with profound hearing loss.
- **In-the-canal and completely in-the-canal type.** These hearing aids are contained in a tiny case that fits partly or completely into the ear canal.
- **In-the-ear type.** All parts are contained in a shell that fits in the outer part of the ear.

Implantable hearing aids are also an option for patients with less profound hearing loss. To insert a hearing aid, follow the natural shape of the ear. The device needs to fit snugly without forcing. If you hear a whistling sound, the hearing aid may not be in far enough to create a seal or the volume may be too loud. Try repositioning the hearing aid, or remove it and turn down the volume. If you cannot insert the hearing aid after two tries, put it in the box, take it with you, and document the transport and transfer of hearing aids to hospital personnel. Never try to clean hearing aids, and do not get them wet.

If a patient's hearing aid is not working, try troubleshooting the problem. First, make sure the hearing aid is turned on. Try a fresh battery, and check the tubing to make sure it is not twisted or bent. Ensure that the switch is set on M (microphone), not T (telephone). For a conventional body type aid, try a spare cord; the old one may be broken or shorted. Finally, make sure the ear mold is not plugged with wax.

Words of Wisdom

Many patients with borderline hearing impairments may not be aware of the extent of their problem. The distracting and noisy EMS environment may worsen the situation. If a patient frequently asks you to repeat things, suspect a hearing impairment.

Words of Wisdom

The ears of some hearing-impaired patients are sensitive to very loud noises close to their ears. Remember to use a normal tone of voice when speaking to them.

Physical Disabilities

Cerebral Palsy

Cerebral palsy is a term for a group of disorders characterized by poorly controlled body movement Figure 34-5 . This disorder is a result of damage to the developing fetal brain while in utero, traumatic brain injury at birth or early during childhood, or from a postpartum infection such as meningitis. Patients with cerebral palsy can have symptoms that range from mild to severe, involving poor posture and uncontrolled, spastic movements of the limb.

This disorder is also associated with other conditions such as visual and hearing impairments, difficulty communicating, epilepsy (seizures), and mental retardation. A significant majority (75%) of patients with cerebral palsy possess some varying degrees of developmental delay, whereas others have a normal intelligence level and are able to live

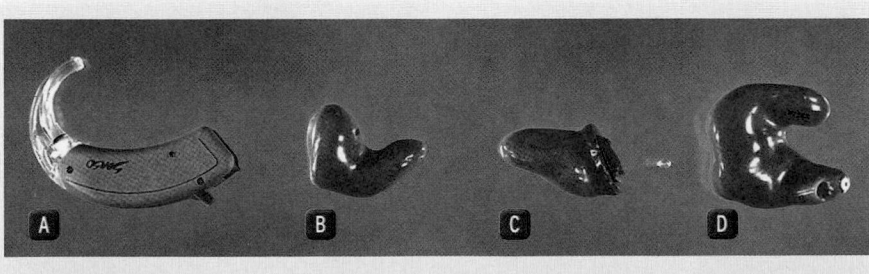

Figure 34-4 Different types of hearing aids. **A.** Behind-the-ear-type. **B.** In-the-canal type. **C.** Completely in-the-canal type. **D.** In-the-ear type.

Figure 34-5 A person with cerebral palsy.

independently with minimal support. Patients with cerebral palsy may have an unsteady gait (ataxia) and may require the assistance of a wheelchair or walker. This type of equipment should be transported with the patient, providing it can be secured properly in the ambulance. One in every four patients with cerebral palsy may also have a seizure disorder, which may require additional advanced life support (ALS).

As with all patients, assessing the ABCs is of the utmost importance. The airway status of a patient with cerebral palsy should be observed closely because a patient may have increased secretion production and difficulty swallowing (dysphagia), requiring aggressive suctioning to clear the airway.

When caring for a patient with cerebral palsy, note the following:

- Do not assume that patients with cerebral palsy are mentally disabled. Although 75% of patients have some developmental disability, many people with cerebral palsy have a normal IQ or only slight mental impairment.
- Limbs are often underdeveloped and are prone to injury (eg, from a fall from a wheelchair).
- Patients who have the ability to walk may have an ataxic or unsteady gait and are prone to falls.
- If the patient has a specially made pillow or chair (pediatric patients), the patient may prefer to use it during transport. Remember to pad the patient to ensure his or her comfort, and never force a patient's extremities into any position.
- Whenever possible, take walkers or wheelchairs along during transport.
- Approximately 25% of patients with cerebral palsy also have seizures. Be prepared to care for a seizure if one occurs, and keep suctioning available.

You are the Provider: PART 3

The patient's axillary temperature reads 101.2°F. The patient has a gastrostomy tube and his mother tells you that she administered an appropriate dose of acetaminophen through the tube about an hour ago. Your partner assesses the patient's vital signs while you perform a physical examination, which reveals no obvious abnormalities. You note that the patient has a colostomy bag and an indwelling urinary catheter.

Recording Time: 5 Minutes	
Respirations	14 breaths/min; provided by mechanical ventilator
Pulse	110 beats/min; strong and regular
Skin	Pink, hot, and moist
Blood pressure	118/62 mm Hg
Oxygen saturation (Sao$_2$)	99% (on oxygen)

5. What are some potential complications that may result from your patient's condition? Is there anything you can do to prevent them?

6. What do you suspect is the patient's underlying problem? Is there specific treatment that you should provide to him?

Spina Bifida

Spina bifida is a birth defect caused by the incomplete closure of the spinal column that results in an exposed spinal cord and undeveloped vertebrae **Figure 34-6**. The opening can be surgically closed, but the child is left with spinal damage. To reduce the occurrences of such disabling birth defects, pregnant women are advised to take vitamin B (folic acid). Unfortunately, spina bifida is still one of the most common disabling birth defects in the United States. Most patients with spina bifida also have hydrocephalus, which requires the placement of a shunt to drain excessive amounts of cerebrospinal fluid from the brain.

Be aware that patients with spina bifida will have partial or full paralysis of the lower extremities, loss of bowel and bladder control, and an extreme allergy to latex products. A supply of latex-free products should be kept on the ambulance to avoid a severe anaphylactic reaction in patients with spina bifida.

Patients with spina bifida will benefit from the same considerations that you offer when you treat a patient with paralysis or a patient who has difficulty moving. Ask patients how it is best to move them before you transport them. It is highly likely that a patient in your community will have spina bifida.

Paralysis

Paralysis is the inability to voluntarily move one or more body parts and may be caused by stroke, trauma, or birth defects. Paralysis does not always entail a loss of sensation, however. In some cases, the patient will have normal sensation or hyperesthesia (increased sensitivity), which may cause the patient to interpret touch as pain in the affected area. Paralysis of one side of the face may also cause communication challenges.

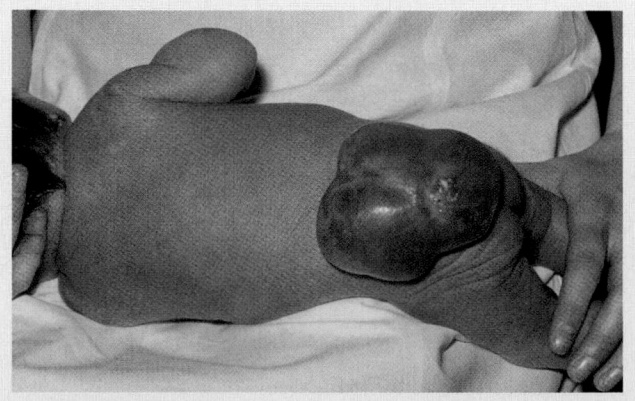

Figure 34-6 Spina bifida is still one of the most common disabling birth defects in the United States.

The diaphragm of some paralyzed patients may not function correctly, requiring the use of a ventilator. Patients may also rely on specialized equipment such as urinary catheters, tracheotomies, colostomies, or feeding tubes, which are discussed later in this chapter. Some patients may have difficulty swallowing, creating the need for suctioning. Each type of spinal cord paralysis requires its own equipment and may have its own complications.

If patients have lost some or all of the sensation in the affected limbs, they cannot tell you when you are hurting them. Take special care to use a gentle touch. Always take great care when lifting or moving a paralyzed patient. Ask patients how it is best to move them before you transport them.

Bariatric Patients

Obesity is a condition in which a person has an excessive amount of body fat and is the result of an imbalance between food eaten and calories used. The solution to the obesity problem may sound relatively simple—reestablish the balance and cure the problem. Unfortunately, obesity can be a much more complex situation. The causes of obesity are not fully understood. Oftentimes, this problem may be attributed to a low metabolic rate or genetic predisposition.

The term obese is used when someone is 20% to 30% over his or her ideal weight. In severe or morbid obesity, the person is 50 to 100 lb over the ideal weight. Severe obesity afflicts about 9 million adult Americans. Obese persons are often ridiculed publicly and sometimes are victims of discrimination. Mobility and the person's general quality of life are often negatively affected by their size, and the extra weight can cause a myriad of health problems, such as diabetes, hypertension, heart disease, and stroke.

Interaction With Obese Patients

Obese patients may be embarrassed by their condition or fearful of ridicule as a result of past experiences. Some of those negative interactions may have occurred with an insensitive health care professional. As with any patient, work hard to put these patients at ease. Establish the patient's chief complaint and then communicate your plan to help. Many severely obese patients have a complex and extensive medical history, so mastering the art of conducting a patient interview will serve you well in your interactions with obese patients.

If transport is necessary, plan early for extra help and do not be afraid to call for additional help if necessary. In particular, send a member of your team to find the easiest and safest exit to use. Remember, everyone's safety is at stake! You do not want to risk dropping the patient or injuring a team member by trying to lift too much weight. Moves, no matter how simple they may seem, become far more complex with an oversized patient.

Interaction With Morbidly Obese Patients

Morbidly obese patients may overcome mobility difficulties by pulling, rocking, or rolling into a position. The constant strain on their body's structures may leave them with chronic joint injuries or osteoarthritis. When you are moving a morbidly obese patient, follow these tips:

- Treat the patient with dignity and respect.
- Ask your patient how it is best to move him or her before attempting to do so.
- Avoid trying to lift the patient by only one limb, which would risk injury to overtaxed joints.
- Coordinate and communicate all moves to all team members prior to starting to lift.
- If the move becomes uncontrolled at any point, stop, reposition, and resume.
- Look for pinch or pressure points from equipment because they could cause a deep venous thrombosis.
- Very large patients may have difficulty breathing if you lay the patient in a supine position.
- Many manufacturers make specialized equipment for morbidly obese patients, and some areas have specially equipped bariatric ambulances for such patients. Become familiar with the resources available in your area.
- Plan egress routes to accommodate large patients, equipment, and the lifting crew members. Remember: Do no harm!
- Notify the receiving facility early to allow special arrangements to be made prior to your arrival to accommodate the patient's needs.

Patients With Medical Technology Assistance

Tracheostomy Tubes

A <u>tracheostomy tube</u> is a plastic tube placed in a surgical opening from the anterior part of the neck into the trachea. The tube can be temporary or permanent and passes from the neck directly into the major airways Figure 34-7 .

Patients who depend on home automatic ventilators or those who have chronic pulmonary medical conditions may breathe through a tracheostomy tube. Because these tubes bypass the nose and mouth, such devices are foreign to the respiratory tract. The body reacts to the tube by building up secretions in or around the tube. The tubes are prone to becoming obstructed by mucous plugs or foreign bodies. Routine care that is provided by caregivers includes keeping the stoma clean and dry and suctioning any secretions.

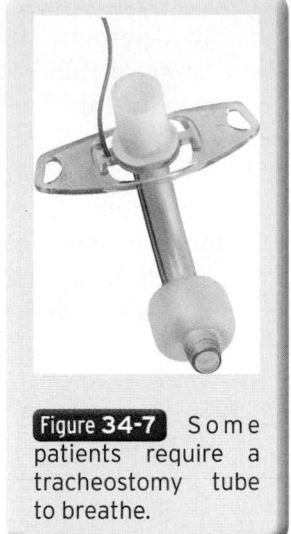

Figure 34-7 Some patients require a tracheostomy tube to breathe.

Obstructions of the tracheostomy tube are emergency events that require you to intervene immediately. This type of emergency can be stressful to deal with, so it is imperative to remember the ABCs and airway management. A useful mnemonic to remember regarding such an emergency is DOPE Table 34-1 . The DOPE mnemonic helps you to recognize the causes of an airway obstruction and alleviate the problem, which could lead to cardiopulmonary arrest.

There may be bleeding or air leaking around the tube, which usually happens with new tracheostomies, and the tube can become loose or dislodged. Occasionally, the opening around the tube may become infected. Your care of a patient with a tracheostomy tube includes maintaining an open airway. This management can include suctioning the tube if necessary to clear a mucous plug, while maintaining the patient in a position of comfort, administering supplemental oxygen, and providing transport to the hospital.

Some patients with special health care needs may have muscle contractions. You will not be able to place them in a "typical" semi-Fowler position; in these cases, suction the patient in a position of comfort. If trauma is involved, the cervical spine must be protected.

If suctioning of the tracheostomy tube is necessary, first attempt to use the patient's suction device. It is probably already sized correctly and readily available. If the size of the suction catheter is unknown, estimate the

Table 34-1 DOPE Mnemonic

D	Displacement, dislodged, or damaged tube
O	Obstruction of the tube (secretions, blood, mucus, vomitus)
P	Pneumothorax, pulmonary problems
E	Equipment failure (kinked tubing, ventilator malfunction, empty oxygen supply)

size by doubling the inner diameter of the tracheostomy tube. To determine the depth of the suction device, ask a family member, obtain the length from a spare tracheostomy tube, or insert the device to no more than 3- to 6-cm deep. The suction unit should be set to 100 mm Hg, and you may need to instill 2 to 3 mL of saline before suctioning thick tracheal secretions. Remember, do not suction for more than 10 seconds, do not force the suction catheter into the cannula, and oxygenate before and after such a procedure. Call for ALS backup.

■ Mechanical Ventilators

Patients who are on a mechanical ventilator at home cannot breathe without assistance Figure 34-8 . Patients requiring a mechanical ventilator may or may not have an underlying respiratory drive because of a congenital defect or a chronic lung disease process. Others patients may have a traumatic brain injury, muscular dystrophy, or another disease process that weakens their ability to breathe and requires a permanent tracheostomy and mechanical ventilator.

If the ventilator malfunctions, remove the patient from the ventilator and begin ventilations with a bag-valve device. To do this, remove the mask from a bag-valve device and directly attach the bag and valve to the tracheostomy tube, which will allow you to ventilate directly through the tracheostomy tube. Patients with

tracheostomies do not breathe through their mouth and nose. A face mask or nasal cannula therefore cannot be used to treat them. Masks designed specifically for patients with tracheostomies cover the tracheostomy hole and have a strap that goes around the neck. These masks are usually available in intensive care units, where many patients have tracheostomies, and may not be available in a prehospital setting. If you do not have a tracheostomy mask, you can improvise by placing a face mask over the stoma Figure 34-9 . Even though the mask is shaped to fit the face, you can usually achieve an adequate fit over the patient's neck by adjusting the strap.

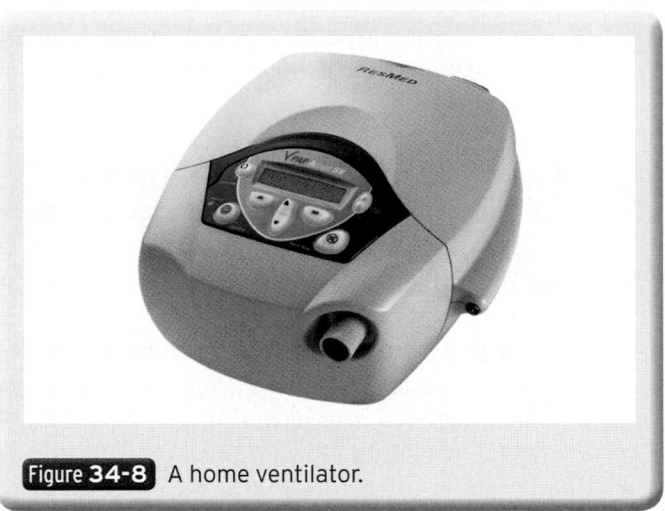

Figure 34-8 A home ventilator.

You are the Provider: PART 4

You consult with medical control and recieve instructions on how to proceed. The mechanical ventilator is too large to fit in your ambulance, so you prepare the bag-valve device, detach the ventilator circuit from the tracheostomy tube, and resume manual ventilations at 14 breaths/min. You carefully move the patient to your stretcher, secure him properly, and load him into the ambulance. Following medical control's direction, you connect your automatic transport ventilator to the patient's tracheostomy tube, reassess his vital signs, and begin transport to the hospital.

Recording Time: 15 Minutes	
Level of consciousness	Conscious and alert
Respirations	14 breaths/min; provided by ATV
Pulse	118 beats/min; strong and regular
Skin	Pink, hot, and dry
Blood pressure	120/60 mm Hg
Sao$_2$	98% (on oxygen)

7. Should the patient's mother be allowed to accompany her son to the hospital? Why or why not?

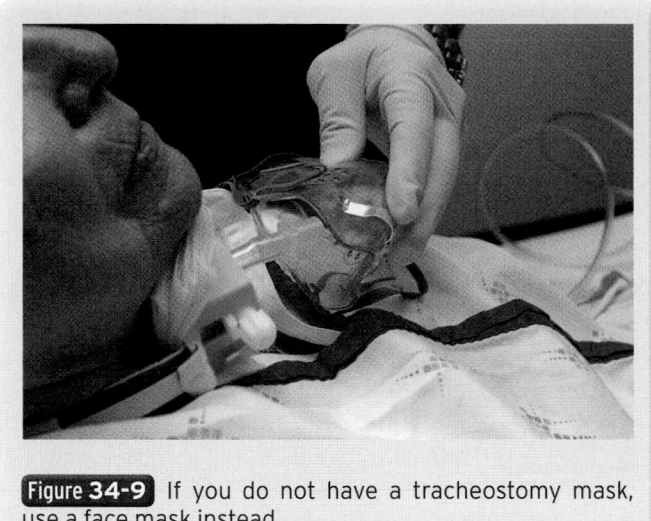

Figure 34-9 If you do not have a tracheostomy mask, use a face mask instead.

Patients on home mechanical ventilators require assisted ventilation throughout transport. Remember that the patient's caregivers will know how the mechanical ventilator works and will be of great help to you in attaching the bag and valve from a bag-valve-mask device to the tracheostomy tube in preparation for transport.

Words of Wisdom

Several states have adopted laws that require a backup generator or other devices to prevent the loss of electric supply to the homes of families or institutions that have persons using mechanical ventilators.

■ Apnea Monitors

While caring for infants with special challenges, you may come across an apnea monitor. The apnea monitor is typically used when an infant is born prematurely, has severe gastroesophageal reflux that causes choking episodes, or if there is a family history of sudden infant death syndrome or if the infant has experienced an apparent life-threatening event. Chapter 32, *Pediatric Emergencies*, discusses sudden infant death syndrome and apparent life-threatening events in detail. Because the central nervous system is not mature in pediatric patients with special challenges, the apnea monitor is used for 2 weeks to 2 months after birth to monitor the respiratory system. A typical episode of apnea may only last for approximately 15 to 20 seconds, during periods of sleep. The apnea monitor is designed to sound an alarm if the infant experiences bradycardia or if an episode of apnea occurs.

The apnea monitor is attached with electrodes or a belt wrapped around the infant's chest or stomach. A pulse oximeter may also be used, which measures the oxygenation of the infant's hemoglobin. The apnea monitor will provide a pulse oximetry reading that will assist you in assessing the patient's respiratory status.

The parents or caregivers of pediatric patients with special challenges will be a useful resource to obtain a patient history and the events leading to the call for assistance. Parents and caregivers become very knowledgeable regarding the use of the apnea monitors and may be able to provide you and your partner with a computerized printout to share with ALS providers or emergency department personnel. If possible, bring the apnea monitor to the receiving hospital with the pediatric patient so that it may be evaluated and any stored information may be retrieved for further analysis.

■ Internal Cardiac Pacemakers

An internal cardiac pacemaker is a device implanted under the patient's skin to regulate the heart rate. These devices are typically placed on the nondominant side of the patient's chest so that normal activities are not hindered. In patients who are small or extremely thin, the device may be implanted in the abdomen. In some cases, the pacemaker may also include an automated implanted cardioverter defibrillator, which monitors the patient's heart rhythm and is able to slow down or stop accelerated heart rates.

You should never place defibrillator paddles or pacing patches directly over the implanted device. When you are obtaining the patient's history during the patient assessment process, you may find it helpful to gather specific information for the hospital staff, such as the type of cardiac pacemaker **Table 34-2**.

■ Left Ventricular Assist Devices

A left ventricular assist device is a special piece of medical equipment that takes over the function of either one or both heart ventricles. These types of devices are used as

Table 34-2	Questions for Patients With Pacemakers

- What type of heart disorder does the patient have?
- How long has this device been implanted?
- What is the patient's normal baseline rhythm and heart rate?
- Is the patient's heart completely dependent on the pacemaker device?
- At what heart rate will the defibrillator fire?
- How many times has the defibrillator shocked the patient?

a bridge to heart transplantation while a donor heart is being located. To date, there is only one approved ventricular assist device designed for persons aged 5 to 16 years.

If you encounter a patient with this device, you will primarily provide support measures and basic care while using the caregiver as a resource during the transport. There are risk factors associated with the implantation of a left ventricular assist device, such as excessive bleeding following the surgery, infection, blood clots leading to strokes, and acute heart failure. Although medical equipment failure is rare in these cases, you must be prepared to provide cardiopulmonary resuscitation efforts if the situation arises. The ALS crew should be notified as soon as possible so that other supportive measures may be initiated.

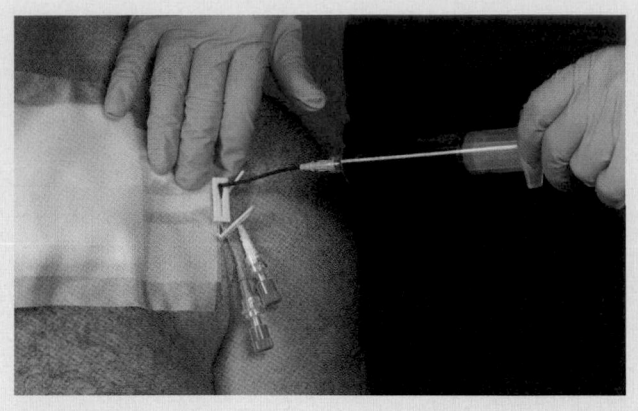

Figure 34-10 Patients who require frequent intravenous medications may have a central line in place.

Central Venous Catheter

A central venous catheter—a venous access device with the tip of the catheter in the vena cava—is used for many types of home care patients, including those receiving chemotherapy, long-term antibiotic or pain management, high-concentration glucose solutions, and hemodialysis **Figure 34-10** . Central venous catheters are often located in the chest, upper arm, or subclavicular area.

Problems associated with these devices may include broken lines, infections around the lines, clotted lines, and bleeding around the line or from the tubing attached to the line. If bleeding occurs, you should apply direct pressure to the tubing and provide immediate transport to the hospital.

Gastrostomy Tubes

Gastrostomy tubes are sometimes referred to as gastric tubes or G-tubes. Gastric tubes are placed directly into the stomach for feeding patients who cannot ingest fluids, food, or medication by mouth **Figure 34-11** . These tubes may be inserted through the nose or mouth into the stomach (using a nasogastric or orogastric tube).

In some cases, a gastric tube may be placed surgically. Gastric tubes are typically sutured in place and may become dislodged during the patient's normal daily activity. If such a situation arises, assess the patient for signs or symptoms of bleeding into the stomach such as vague abdominal discomfort, nausea, vomiting (especially "coffee ground" emesis), and blood in emesis.

Patients who have a gastric tube in place may still be at increased risk of aspiration. Always have suction readily available to clear any materials from the patient's mouth and to prevent airway problems. Patients with gastric tubes who have difficulty breathing should be transported while sitting or lying on the right side with the head elevated 30° to prevent the contents of the

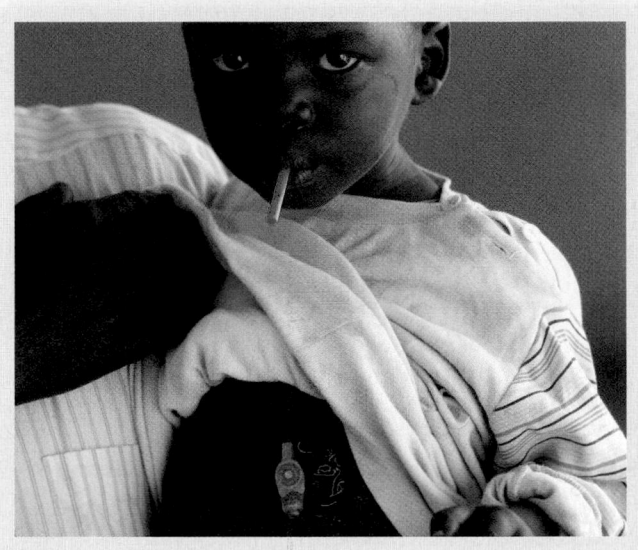

Figure 34-11 Gastric tubes are placed through the skin into the stomach for children or adults who cannot be fed by mouth.

stomach from passing into the lungs. Give supplemental oxygen if the patient has any difficulty breathing.

Patients with diabetes who receive insulin and gastric tube feedings may become hypoglycemic quickly if the gastric tube feedings are discontinued for any reason. Be alert for an altered mental status or a change in the baseline behavior of your patient.

Shunts

Some patients with chronic neurologic conditions may have shunts in place. For example, patients with hydrocephalus will have a shunt. **Shunts** are tubes that extend from the brain to the abdomen to drain excess cerebrospinal fluid that may accumulate near the brain.

There are a few different types of shunts, such as a ventricular peritoneum shunt and a ventricular atrium shunt. A ventricular peritoneum shunt drains excess fluid from the ventricles of the brain into the peritoneum of the abdomen. A ventricular atrium shunt drains excess fluid from the ventricles of the brain into the right atrium of the heart. These shunts keep pressure in the skull from building up.

If a shunt becomes blocked or becomes infected, changes in mental status and respiratory arrest may occur. Infections of shunts may occur within the first 2 months after the insertion. A blocked shunt may also present as a medical emergency. If the shunt is unable to drain properly, intracranial pressure may increase and the patient will experience an altered mental status.

During your patient assessment, you will likely feel a device beneath the skin on the side of the head, behind the ear. The device is a fluid reservoir, and the presence of the device should alert you to the possibility that the patient has an underlying shunt. Should the shunt become dysfunctional, the patient could be predisposed to respiratory arrest.

The signs that a patient is in distress include bulging fontanels (in infants), headache, projectile vomiting, altered mental status, irritability, high-pitched cry, fever, nausea, difficulty with coordination (walking), blurred vision, seizures, redness along the shunt track, bradycardia, and heart arrhythmias. Emergency medical care includes airway management and artificial ventilation during transport.

■ Vagal Nerve Stimulators

An alternative treatment to medication for patients with chronic seizures disorders may be provided by vagal nerve stimulators. These devices are implanted in patients when medications fail to resolve seizure activity or if the patient is not a good candidate for brain surgery. To date, approximately 32,000 patients have received this device, but further studies are being done regarding its effectiveness for seizure disorders.

Vagal nerve stimulators, which stimulate the vagus nerve to keep seizure activity from occurring, are used in children older than 12 years. The nerve stimulator is surgically implanted under the patient's skin and is about the size of a silver dollar. The stimulator can last for up to 6 years or until the battery begins to run out. If you encounter a patient with this device, contact medical control or follow your local protocols.

■ Colostomies and Ileostomies

A <u>colostomy</u> or <u>ileostomy</u> is a surgical procedure that creates an opening between the small or large intestine and the surface of the body that allows for elimination of waste products. The special opening is referred to as a stoma. Either urine or feces are expelled and collected into a clear external bag or pouch, which is emptied or changed frequently.

If you encounter a patient with a colostomy or ileostomy bag, assess for signs and symptoms of dehydration if the patient has been complaining of diarrhea or vomiting. The area around the stoma is prone to infection, so patients and caregivers must be diligent with daily hygiene. Signs of infection include redness, warm skin around the stoma, and tenderness with palpation over the colostomy or ileostomy site. Contact medical control or follow your local protocols on caring for a patient with a colostomy or ileostomy.

■ Patient Assessment Guidelines

Interaction with the caregiver of a child or adult with special needs will be an important part of the patient assessment process. Always speak with the caregiver or family members; they have become experts on the illness or disability. The parents, caregivers, or home health care staff members are trained to use and troubleshoot problems with medical equipment on a daily basis. Assess the patient's baseline vital signs, note any allergies (eg, latex), medications, and other pertinent medical history. You must first determine the patient's normal baseline status before an assessment of the current condition can be made. It is often helpful to ask, "What is different today?"

■ Home Care

Home care occurs within a patient's home environment. Patients requiring home services involve a spectrum of special health care needs, including infants and the elderly, patients with chronic illnesses, and patients with developmental disabilities. These services are commonly needed among patients older than 65 years.

Services offered by home care agencies include but are not limited to delivering prepared meals, house cleaning, washing laundry, yard maintenance, providing physical therapy, and providing personal hygiene, including bathing and wound care. Oftentimes, EMS is called to a residence when a home care provider has either found the patient injured or has recognized a change in the patient's health status. Home care personnel are an important resource for you when you are obtaining the patient's baseline health status and the history of the present illness or condition. Home care personnel are usually familiar with the patient's surroundings and can obtain any health care documentation or medications that need to be transported with the patient to the hospital.

Hospice Care and Terminally Ill Patients

Unfortunately, not all illnesses can be cured. As health care providers, you and your team may be called on to assist a patient who has a terminal illness. The patient may be receiving hospice care at a hospice facility or at home.

Patients receiving hospice care at a hospice facility or at home are terminally ill with diseases such as cancer, heart and lung failure, end-stage Alzheimer disease, or acquired immunodeficiency syndrome (AIDs). The patient's physician has determined the illness terminal and has completed a do not resuscitate (DNR) order or given medical orders for the scope of treatment, outlining the care agreed upon by the patient and/or the family. Hospice care provides comfort care (pain medications) during a person's last days. Comfort care, or palliative care, improves the patient's quality of life before the patient dies and allows the patient to be with family and friends. If you are called to a facility that provides hospice care or a home with a patient receiving hospice care, you will need to follow your local protocols, the patient's wishes, or legal documents such as a DNR order. All necessary documentation must be brought to the hospital if the patient is to be transported to the hospital and noted in the patient care report.

If you are called to the home of a terminally ill patient, the care you give will have a lasting impact on the family. This is a time when compassion, understanding, and sensitivity are most needed. Some homes with patients receiving hospice care may be chaotic. Family members may be having a difficult time coping with the situation, and they may act angry and hostile. Treat everyone with compassion and understanding. Members of your team may be able to separate family members to speak with them privately to diffuse intense emotions and restore order.

Some terminally ill patients at home may also be receiving outpatient care from a hospice or a home health nurse. You may be called to the home because of a delay in the arrival of the regular care provider or for transport so that a physician can address an immediate need, such as increasing pain. Because terminally ill patients may use a complex array of pain medications, transdermal patches, or self-administered pain management devices, you may need to consult medical control for guidance.

Even if a DNR order is in place, family members may not understand what to do and they may not be ready to face the death of a loved one. In such cases, obtain a thorough history and compassionately discuss the patient's wishes. Ask to review the DNR order and contact medical control.

Ascertain the family's wishes about having the patient remain in the home or having the patient transported to the hospital. If a family member requests to accompany the patient, he or she should be allowed to do so. If the

You are the Provider: PART 5

You call in your radio report to the emergency department. During your reassessment, you note that the patient appears to be fighting the ventilator by moving his head around. You also note an acute change in his vital signs. His mother, who is riding in the back of the ambulance with you, asks you if you know what to do.

Recording Time: 25 Minutes

Level of consciousness	Conscious; moving his head around; mother advises that he is agitated
Respirations	14 breaths/min; provided by ATV
Pulse	130 beats/min; strong and regular
Skin	Pink, hot, and moist
Blood pressure	134/74 mm Hg
Sao$_2$	87% (on oxygen)

8. What has most likely happened to your patient? What should you do next?

family wishes the patient to remain at home, this request should be honored provided it is in accordance with your local or state protocol.

Local protocols for handling the death of a patient vary, so be familiar with your local or state regulations. The protocols identify whether the coroner needs be to called to report the death and, if so, who is responsible for contacting the coroner. Also determine whether a pronouncement of death is required and, if so, who is responsible for the determination.

Poverty and Homelessness

According to a US Bureau of the Census report in 2007, 12.5% of the US population lives in poverty. People who live in poverty are unable to provide for all of their basic needs such as housing, food, child care, health insurance, and medication. An impoverished person or family may have housing but may go without food or medication in order to pay for housing. Disease prevention strategies such as dental care, good nutrition, and exercise are likely absent, which increases the probability of disease in people who live in poverty.

It is estimated that close to 4 million people are homeless. Of those 4 million, approximately 40% are women and children. The homeless population includes persons with mental illness, victims of domestic violence, persons with addiction disorders, and impoverished families.

You are an advocate for all patients. Your job is to provide emergency medical care and transport patients to the appropriate facility. Remember, all health care facilities *must* provide a medical assessment and required treatment, regardless of the patient's ability to pay under the Emergency Medical Treatment and Active Labor Act (referred to as EMTALA). You can also be an advocate by becoming familiar with the social services resources within your community so you can refer patients to these lifelines.

You are the Provider: SUMMARY

1. Will your assessment and treatment of this patient differ from those of any other patient?

The principles of patient assessment and treatment are the same, regardless of the patient's special health care needs and any medical equipment that he or she requires to function or live. As with any patient you encounter, your initial focus must be on ensuring a patent airway, adequate oxygenation and ventilation, and adequate circulation. Chronically ill patients can present with acute problems just like any other patient, and as such, your goal remains the same, maintaining the ABCs and safely transporting the patient to an appropriate medical facility.

2. What role do the parents or caregivers of patients with special health care needs have in the prehospital setting?

When you are caring for a patient with special health care needs, it is imperative to *listen* to the people who take care of the patient. Whereas your encounter with the patient will be brief, the parents or caregivers provide for the medical needs of the patient *every day*. They are aware of the patient's medical and/or surgical history, his or her baseline mental status, and the names and doses of any medications the patient may be taking.

Parents or caregivers of this patient population are also trained and experienced in the use of any special equipment the patient requires, such as tracheostomy tubes, mechanical home ventilators, urinary catheters, and gastrostomy tubes. Your exposure to this equipment will most likely be limited. In many cases, the parent or caregiver will have performed certain interventions prior to calling 9-1-1; this is because they have been educated and trained to troubleshoot certain problems that may occur. It is important to determine what interventions were performed, why they were performed, and what affect they had on the patient's condition.

You should always consider the parents or caregivers of a patient with special health care needs as experts in the care of the patient, so listen carefully to what they tell you.

3. What are some conditions that would cause a patient to become dependent on a mechanical ventilator?

A mechanical device (machine) is used to inflate and deflate the lungs. Spontaneous breathing normally occurs when the major muscles of breathing—the diaphragm and intercostal muscles—contract and relax. In addition, the respiratory centers in the brain stem, which send messages to the respiratory muscles, must be intact. When a person has any acute or chronic condition that impairs his or her respiratory muscles or injures the respiratory centers in the brain, he or she will require the use of a mechanical ventilator.

Your patient is a quadriplegic secondary to a spinal injury. Quadriplegia is paralysis of the arms, legs, and trunk. If the spinal cord is injured at or above the level of the fourth cervical vertebrae (C4), paralysis of the respiratory muscles will also occur. Without a mechanical ventilator, these patients are not able to breathe at all.

Other conditions that often require mechanical ventilation include traumatic brain injury, in which case the respiratory centers in the brain stem are not able to send messages to the respiratory muscles; muscular dystrophy, a slow but progressive weakness and deterioration of the skeletal muscles; cystic fibrosis, a genetic disease that causes the lungs to produce thick secretions that obstruct the air passages and can lead to life-threatening infections; and spina bifida, a birth defect caused by the incomplete closure of the spinal column, which results in an exposed spinal cord and undeveloped vertebrae.

Regardless of why a patient requires mechanical ventilation, the most important thing for you to remember is that without it, he or she is unable to breathe!

---

4. Are patients with tracheostomy tubes able to speak? How can you determine if your patient is alert?

A tracheostomy is a surgical procedure that forms an opening in the trachea, just below the Adam's apple. A tube is placed into the opening, and air moves in and out of the body through the tube instead of the nose and mouth. Some patients only require a tracheostomy tube for short periods; others require it for long periods–or in the case of your patient, permanently.

Because a tracheostomy tube is placed below the vocal cords, air no longer passes through the vocal cord folds (which make sound when they vibrate); therefore, most patients with tracheostomy tubes are unable to speak. This is especially true if the patient requires mechanical ventilation because he or she will be unable to breathe if the ventilator is detached from the tracheostomy tube. There are patients with tracheostomy tubes who are not dependent on ventilators. These patients may be able to speak–although not as clearly–if they occlude the opening of the tube.

If patients cannot speak, they may communicate in other ways. The head, face, and eyes can be used to communicate. Some patients may nod their head a certain way in response to a question; others may blink their eyes.

When you are assessing the mental status of a patient who is dependent on a ventilator, you must rely on the caregiver, who will know how the patient communicates. The caregiver should be able to tell you whether the patient is communicating as he or she typically does and whether he or she is responding to questions appropriately. The objective is to determine whether the patient's mental status has changed. If the patient responds to questions appropriately, it could be said that he or she is alert.

---

5. What are some potential complications that may result from your patient's condition? Is there anything you can do to prevent them?

When you are caring for any patient with a chronic condition, it is important to understand the potential complications that could be caused by their condition. Although you will not be able to prevent all of these potential complications, there are often some basic steps you can take to minimize the risk.

Patients with quadriplegia who are dependent on a ventilator are typically confined to a bed for prolonged periods. Prolonged immobilization can cause potentially serious complications, including pressure sores and pulmonary embolism.

In particular, patients who are paralyzed are susceptible to urinary tract infections and pneumonia. Urinary tract infections are typically caused by indwelling urinary catheters that are not changed regularly; however, infection can still occur with proper care of the catheter. Pneumonia often occurs because of decreased or absent cough reflexes and prolonged immobilization, which increases the risk for pulmonary secretions to settle in the lungs and become infected.

In patients with indwelling urinary (Foley) catheters, always maintain the catheter collection bag below the level of the bladder; this position will prevent urine from flowing back into the bladder and therefore minimizes the risk of infection.

Patients with tracheostomy tubes, such as your patient, can develop thick mucous plugs in the tube, which can impair oxygenation and ventilation. Ask the caregiver when the tube was last suctioned and observe for signs that indicate it may need to be suctioned (eg, restlessness, signs of hypoxemia). Many mechanical ventilators will sound an alarm if there is any obstruction in the ventilator circuit, such as a mucous plug in the tracheostomy tube.

After you have transferred the patient to the ambulance stretcher, ensure that there are no wrinkles or areas that are thicker than others in the sheet or blanket under the patient. This simple step can help prevent pressure sores.

---

6. What do you suspect is the patient's underlying problem? Is there specific treatment that you should provide to him?

The presence of fever suggests infection. In this patient, the source of infection could be a number of places. Furthermore, it may be the result of more than one underlying problem.

Fever is often the only presenting sign of pneumonia in paralyzed patients. This is especially true in patients with quadriplegia because their respiratory muscles are also paralyzed; therefore, outward signs of respiratory distress, such as retractions, are not present.

A urinary tract infection is also another possibility and is commonly caused by the long-term presence of an indwelling urinary catheter. Even with proper care of the catheter by the caregiver, urinary tract infections are common in paralyzed patients.

Infection requires antibiotic therapy, which can only be administered in a hospital setting. Treatment for patients with a possible infection is mainly supportive; monitor the patient's ABCs, observe standard precautions (eg, gloves, mask if necessary), and transport the patient to the hospital.

---

7. Should the patient's mother be allowed to accompany her son to the hospital? Why or why not?

The decision as to whether the patient's mother accompanies her son to the hospital is ultimately hers. By no means, however, should you not allow her to go. Remember, she is his primary caregiver and can continue to be a source of information en route to the hospital. Furthermore, she could alert you to any changes in the patient's mental status, which may be obvious only to her.

You are the Provider: SUMMARY, continued

The patient's mother can also bring supplies that the patient needs, such as a sterile suction catheter kit and comfort supplies or equipment, which you may or may not carry on your ambulance.

You must also consider the emotional needs of the patient. A patient and his or her caregiver, whether it is a parent, other relative, or home health nurse, often develop a special bond. Unnecessarily separating a patient with special health care needs from his or her primary caregiver can be a source of great emotional distress for the patient and the caregiver.

Although your patient is ill and in need of further assessment and treatment at the hospital, his condition is stable. The patient's mother should be allowed to accompany her son in the ambulance if she wishes. If she prefers to follow the ambulance in her own vehicle, reassure her that you will take good care of her son.

---

8. What has most likely happened to your patient? What should you do next?

Your patient's clinical condition has changed. He appears to be fighting the ventilator by moving his head around, which is a sign of agitation. Furthermore, his heart rate has increased and his oxygen saturation level has decreased. You should suspect that he is not receiving adequate ventilation.

Because tracheostomy tubes bypass the nose and mouth, they are foreign to the respiratory tract. As a result, the body reacts by building up secretions in or around the tube, resulting in obstruction of the tube by mucous plugs.

Tracheostomy tube obstruction, which may occur acutely, is an emergency that requires immediate intervention; you must suction the tube and reassess the patient for signs of adequate oxygenation and ventilation.

The DOPE mnemonic can be used to troubleshoot acute deterioration in a patient with a tracheostomy tube. The "D" stands for displacement, dislodgment, or damage to the tube; "O" stands for obstruction of the tube (eg, from secretions, blood, mucus); "P" stands for pneumothorax; and "E" stands for equipment failure (eg, ventilator malfunction, depleted oxygen supply). It is critical for you to recognize which of these problems you can correct and which problems require ALS. For example, if the tracheostomy tube has become dislodged, it must be replaced by a paramedic or physician; EMTs are typically not trained or permitted to perform this procedure. If it is determined that the patient has a pneumothorax, as evidenced by decreased or absent breath sounds on one side of the chest, the patient may require a needle chest decompression, which EMTs are not trained to perform. Follow your local protocols and request an ALS ambulance if necessary.

EMS Patient Care Report (PCR)

Date: 1-20-10	Incident No.: 013410	Nature of Call: Fever		Location: 575 Ranger Drive	
Dispatched: 1435	En Route: 1435	At Scene: 1440	Transport: 1501	At Hospital: 1515	In Service: 1523

Patient Information

Age: 19 Sex: M Weight (in kg [lb]): 52 kg (115 lb)	Allergies: Aspirin Medications: Tylenol (as needed) Past Medical History: Quadriplegia from spinal cord injury 2 years ago; ventilator-dependent Chief Complaint: Fever

Vital Signs

Time: 1445	BP: 118/62	Pulse: 110	Respirations: 14	Sao$_2$: 99%
Time: 1500	BP: 120/60	Pulse: 118	Respirations: 14	Sao$_2$: 98%
Time: 1510	BP: 134/74	Pulse: 130	Respirations: 14	Sao$_2$: 87%

EMS Treatment
(circle all that apply)

Oxygen @ 15 L/min via (circle one): NC NRM **Bag-Mask Device**	**Assisted Ventilation**	Airway Adjunct	CPR
Defibrillation	Bleeding Control	Bandaging	Splinting	Other: Emotional support, tracheostomy tube suctioning

You are the Provider: SUMMARY, continued

Narrative
Medic 53 dispatched to a residence for a 19-year-old male with a fever. On arrival, found the patient lying supine in a hospital-style bed in the living room of his home. The patient is a quadriplegic and is ventilator-dependent secondary to a spinal injury he experienced 2 years ago. On presentation, the patient's eyes were open and he acknowledged our presence by nodding his head. According to his mother, who is his primary caregiver, his present mental status is consistent with his baseline. The patient has a tracheostomy tube and is receiving mechanical ventilation at 14 breaths/min. According to the patient's mother, he began running a fever earlier in the day, and the last reading was 101.2°F. She administered 500 mg of Tylenol via his gastrostomy tube about an hour before EMS arrival. The patient's physician, who was contacted by the home health agency who assists the patient's mother in his care, requested transport via ambulance to the emergency department. Secondary assessment revealed no gross abnormalities. Breath sounds were auscultated and found to be clear and equal bilaterally. Patient also has an in-dwelling urinary catheter and a colostomy. Assessment of these devices revealed no bleeding, redness around the area of the devices, or any other abnormalities. After consulting with medical control, carefully moved the patient to the ambulance stretcher, disconnected him from the mechanical ventilator circuit, and resumed ventilations with a bag-valve device at 14 breaths/min. Moved patient to ambulance, resumed mechanical ventilations via automatic transport ventilator at the proper rate and depth, and reassessed his vital signs. Patient's mother accompanied her son in the back of the ambulance. Began transport to the hospital and continued to monitor the patient en route. Shortly after calling radio report to receiving facility, noted that patient became acutely agitated; his heart rate markedly increased and his oxygen saturation decreased. Suspected obstruction of the tracheostomy tube, provided suctioning one time and reassessed patient. He was now calm and his heart rate and oxygen saturation stabilized. Remainder of transport was uneventful. Delivered patient to emergency department and gave verbal report to attending physician. Medic 53 cleared the hospital and returned to service at 1523. *End of report*

Prep Kit

Ready for Review

- Medicine and medical technology continue to improve and the number of children and adults with chronic diseases or injuries who are living at home or in other environments outside of the hospital setting continues to grow.

- You may find children and adults who are living at home who depend on mechanical ventilators, intravenous pumps, or other medical devices to maintain their lives.

- Assess and care for patients with special needs in the same manner as all other patients.

- Developmental disability is caused by insufficient development of the brain, resulting in the inability to learn and socially adapt at a normal developmental rate.

- People with Down syndrome often have large tongues and small oral and nasal cavities, so intubation of these patients may be difficult.

- Visual impairments may be difficult to recognize. During your scene size-up, look for signs that indicate the patient is visually impaired, such as the presence of eyeglasses, a cane, or a service dog. Make yourself known when you enter the room, and introduce yourself and others in the room so that the patient can identify their placement and voices.

- Hearing impairment may range from a slight hearing loss to total deafness. Clues that a person could be hearing impaired include the presence of hearing aids, poor pronunciation of words, or failure to respond to your presence or questions.

- Cerebral palsy is associated with other conditions such as visual and hearing impairments, difficulty communicating, epilepsy, and mental retardation. Patients may also have an unsteady gait and may require the assistance of a wheelchair or walker.

- Patients with spina bifida will have either partial or full paralysis of the lower extremities, loss of bowel and bladder control, and an extreme allergy to latex products.

- Obese patients may be embarrassed by their condition or fearful of ridicule as a result of past experiences. If transport is necessary, plan early for extra help and do not be afraid to call for more help if necessary.

In particular, send a member of your team to find the easiest and safest exit.

- Patients who depend on home automatic ventilators or those who have chronic pulmonary medical conditions may breathe through a tracheostomy tube.

- Patients who are on a mechanical ventilator at home cannot breathe without assistance. If the ventilator malfunctions, remove the patient from the mechanical ventilator and begin ventilations with a bag-valve-mask device.

- An apnea monitor is typically used when an infant is born prematurely, has severe gastroesophageal reflux that causes episodes of choking, or if there is a family history of sudden infant death syndrome or if the infant has experienced an apparent life-threatening event. The apnea monitor is designed to sound an alarm if the infant experiences bradycardia or if apnea occurs.

- An internal cardiac pacemaker is a device implanted under the patient's skin to regulate the heart rate.

- A left ventricular assist device is a special piece of medical equipment that takes over the function of either one or both heart ventricles. These types of devices are used as a bridge to transplantation while a donor heart is being located.

- Gastrostomy tubes are placed directly into the stomach for feeding in patients who cannot ingest fluids, food, or medication by mouth. These tubes may be inserted through the nose, mouth, or placed through the abdominal wall surgically.

- Shunts are tubes that extend from the brain to the abdomen to drain excess cerebrospinal fluid that may accumulate near the brain.

- A colostomy or ileostomy is a section of small or large intestine that is surgically attached to the abdominal wall and allows for elimination of waste products. Urine and/or feces are expelled and collected into a clear external bag or pouch, which is emptied or changed frequently.

- You and your team may be called on to assist a patient who is terminally ill. Terminally ill patients may be in a hospice facility or at home.

Vital Vocabulary

<u>cerebral palsy</u> A term for a group of disorders characterized by poorly controlled body movement.

<u>colostomy</u> A surgical procedure to establish an opening between the colon and the surface of the body.

<u>developmental disability</u> Insufficient development of the brain, resulting in some level of dysfunction or impairment.

<u>Down syndrome</u> A genetic chromosomal defect that can occur during fetal development and that results in mental retardation as well as certain physical characteristics, such as a round head with a flat occiput and slanted, wide-set eyes.

<u>ileostomy</u> A surgical procedure to create an opening between the small intestine and the surface of the body.

<u>obesity</u> A condition in which a person has an excessive amount of body fat.

<u>sensorineural deafness</u> A permanent lack of hearing caused by a lesion or damage of the inner ear.

<u>shunts</u> Tubes that drain fluid from the brain to another part of the body outside of the brain, such as the abdomen; lowers pressure in the brain.

<u>spina bifida</u> A development defect in which a portion of the spinal cord or meninges may protrude outside of the vertebrae and possibly even outside of the body, usually at the lower third of the spine in the lumbar area.

<u>tracheostomy tube</u> Plastic tube placed within the tracheostomy site (stoma).

Assessment in Action

Today you and your partner are traveling to headquarters for your monthly in-service training session. This month's training is focused on special challenges in prehospital emergency care.

1. Which of the following developmental disorders is also known as trisomy 21?
   A. Autism
   B. Fetal alcohol syndrome
   C. Cerebral palsy
   D. Down syndrome

2. Which of the following techniques should you use when you are communicating with a patient who has a hearing disorder?
   A. Exaggerate your lip movements.
   B. Face the patient so he or she can see your mouth.
   C. Speak loudly into the patient's ear.
   D. Talk to the patient slowly and quietly.

3. In an effort to reduce birth defects such as spina bifida, pregnant women are advised to take vitamin:
   A. A.
   B. C.
   C. B.
   D. D.

4. The DOPE mnemonic helps you to recognize the causes of an airway obstruction and alleviate the problem. The "P" in DOPE stands for:
   A. pneumothorax.
   B. pneumonia.
   C. pulmonary embolism.
   D. pulmonary edema.

5. If your patient has a tracheostomy tube and requires suctioning, you should never suction for more than:
   A. 5 seconds.
   B. 10 seconds.
   C. 15 seconds.
   D. 20 seconds.

6. When lifting and moving a morbidly obese patient, you should:
   A. ask the patient the best way to move him or her.
   B. avoid placing the patient in a supine position.
   C. avoid lifting the patient by only one limb.
   D. all of the above.

7. An apnea monitor is used when an infant has a:
   A. premature birth.
   B. severe gastroesophageal reflux.
   C. family history of sudden infant death syndrome.
   D. all of the above.

8. Care that improves the patient's quality of life before the patient dies is known as:
   A. curative.
   B. palliative.
   C. terminal.
   D. holistic.

9. Explain what the "reverse stethoscope" technique is and when it is used.

10. Why might airway management be difficult in a patient with Down syndrome?

National EMS Education Standard Competencies

EMS Operations

Knowledge of operational roles and responsibilities to ensure patient, public, and personnel safety.

• •

Knowledge Objectives

1. Describe the technical skills and general considerations that are required of the EMTs during patient packaging and patient handling. (pp 1285-1313)

2. Define the term body mechanics and discuss how following proper patient lifting and moving techniques can help prevent work-related injuries. (pp 1286-1289)

3. Describe the guidelines and safety precautions the EMT should follow when lifting and carrying a patient on a stretcher or backboard and identify how to avoid common mistakes. (pp 1290-1299)

4. Describe the guidelines for lifting a patient, including using a power grip and using a sheet or blanket. (p 1286-1290)

5. Explain how to carry patients safely on stairs, including the selection of appropriate equipment to aid in the process. (pp 1292-1295)

6. Summarize the general considerations required to move patients safely without causing them further harm while simultaneously protecting the EMT from injury. (pp 1299-1300, 1322)

7. Describe specific situations in which an urgent move or rapid extrication may be necessary to move a patient and explain how each one is performed. (pp 1300-1306)

8. Describe specific situations in which a nonurgent move may be necessary to move a patient and explain how each one is performed. (pp 1306-1311)

9. Discuss special considerations related to moving and transporting geriatric patients and guidelines that must be followed during their lifting and moving. (pp 1311-1312)

10. Define the term bariatrics and discuss the guidelines for lifting and moving bariatric patients. (p 1312-1313)

11. Provide seven examples of patient-moving equipment and explain how each one is used to move a patient. (pp 1313-1321)

12. Explain the relationship between equipment decontamination and the prevention of disease transmission. (p 1321)

13. Discuss situations that may require the use of medical restraints on a patient and explain guidelines and safety considerations for their use. (pp 1321-1322)

Skills Objectives

1. Perform a power lift to lift a patient. (pp 1287-1289, Skill Drill 35-1)

2. Demonstrate using a power grip. (p 1289)

3. Perform the diamond carry to move a patient. (pp 1290-1291, Skill Drill 35-2)

4. Perform the one-handed carrying technique to move a patient. (pp 1291-1292, Skill Drill 35-3)

5. Perform a patient carry to move a patient down the stairs. (pp 1292-1294, Skill Drill 35-4)

6. Perform a patient carry using a stair chair to move a patient down the stairs. (pp 1296-1297, Skill Drill 35-5)

7. Demonstrate the body mechanics and principles required for safe reaching and pulling, including the safe reaching technique used for performing log rolls. (pp 1296-1299)

8. Demonstrate how to perform an emergency or urgent move. (pp 1300-1306)

9. Perform the rapid extrication technique to move a patient from a vehicle. (pp 1301-1306, Skill Drill 35-6)

10. Perform the direct ground lift to lift a patient. (pp 1306-1307, Skill Drill 35-7)

11. Perform the extremity lift to move a patient. (pp 1306-1308, Skill Drill 35-8)

12. Perform the direct carry to move a patient. (pp 1308-1309, Skill Drill 35-9)

13. Demonstrate how to use the draw sheet method to transfer a patient onto a stretcher. (pp 1309-1310)

14. Use a scoop stretcher to move a patient. (pp 1310-1311, Skill Drill 35-10)

15. Demonstrate how to load a stretcher into an ambulance. (pp 1315-1317, Skill Drill 35-11)

16. Demonstrate the correct use of medical restraints on a patient. (pp 1321-1322)

Introduction

In the course of a call, you will have to move a patient several times to provide emergency medical care in the field and transport the patient to the emergency department. Often, you will have to move the patient into a different position or location. Once you have assessed the patient and provided emergency care, you and your team may have to move the patient onto a long backboard or stretcher. Then you must move the patient to the waiting ambulance and load the patient into the patient compartment. After you arrive at the hospital, you must unload the patient, move him or her to the correct examining room, and transfer the patient from the stretcher to the emergency department bed. To avoid injury to the patient, yourself, or your partners, you need to learn how to lift and carry the patient properly, using proper body mechanics and a power grip.

To be able to move a patient safely and properly in the various situations that you may encounter in the field, it will be necessary to learn how to perform emergency body drags and lifts, rapidly extricate a patient from a car onto the stretcher, assist a patient from a chair or bed onto the stretcher, and lift a patient from the floor onto the stretcher. In addition, you may need to move a patient from the bed onto the stretcher or manually carry a patient up or down stairs. You and your team should know how to place a patient with a suspected spinal injury onto a long backboard and package patients with and without suspected spinal injury. At times, you and your team may need to move a patient who is very heavy or carry a patient on a trail or across rugged, uneven terrain. Special techniques for loading and unloading the stretcher and transferring the patient from the stretcher to an examining table or bed in the emergency department are necessary, and you and your team should be familiar with these techniques and practice them often.

Lifting and carrying are dynamic processes. To ensure that no individual suddenly bears unexpected, dangerous weight and to reduce the risk of injury to yourself and the patient, you must know where rescuers should be positioned and how to give and receive lifting commands so that all parties act simultaneously. You will also need to know how to prepare patient-moving devices, such as a wheeled ambulance stretcher (also called an ambulance stretcher, gurney, or simply "the stretcher"), stair chair, backboard, scoop stretcher, folding ambulance stretcher, basket stretcher, and flexible stretcher and when and how to use them. This chapter will cover lifting, carrying, and reaching techniques as well as principles of moving patients, including emergency, urgent, and nonurgent moves, and the use of physical restraints to protect the patient and your crew from further harm. In addition, different types of equipment and patient positioning will be discussed in detail.

Moving and Positioning the Patient

Every time you have to move a patient, special care must be taken so that neither you, your team, nor the patient is injured. Patient packaging and handling are technical skills that you will learn and perfect through repeated training and practice. Every year a significant number of EMTs are injured when they attempt to lift and move patients. Even when you are lifting, moving, or transferring relatively light patients, the need for proper body mechanics should remain paramount.

You are the Provider: PART 1

You and your partner arrive at the residence of a 59-year-old man who fell out of bed and needs assistance. You climb a flight of stairs to the second floor of the residence where you find the patient on the floor beside his bed. The patient, who weighs approximately 280 lb, complains of lower back pain and cannot move without extreme pain. He tells you that he does not want to go to the hospital; he simply needs someone to help move him back to his bed.

1. What factors should you consider before lifting or moving a patient?
2. Why is knowledge of body mechanics important when lifting and moving a patient?

Training and practice are required so that you will be able to properly use all the equipment that is described in this chapter. You must master the skills necessary for their use and understand the advantages and limitations of each device. Practice each technique with your team often so that when you must move a patient, you can perform the move quickly, safely, and efficiently. After each patient transfer, you and your team should evaluate the appropriateness of the technique that you used, as well as your technical skill in completing the transfer. You must also be sure to maintain your equipment according to the manufacturer's instructions. Using clean, well-maintained equipment is but one part of providing high-quality patient care.

After you deliver the patient to the emergency department, you and your team must begin preparing for your next call. Review the positive points about the transport. Discuss changes that would improve the next run. This process of review and evaluation should help you identify the following:

- Procedures that need more practice
- Equipment that needs to be cleaned or serviced
- Skills that you need to review or acquire

Most important, a critical review helps you and your team to become more confident and better skilled EMTs.

Certain patient conditions, such as head injury, shock, spinal injury, pregnancy, and obese patients call for special lifting and moving techniques. Patients with chest pain or who are having difficulty breathing should sit in a position of comfort, as long as they are not hypotensive. Patients with suspected spinal injuries must be immobilized in a supine position on a long backboard, or scoop stretcher. Patients who are in shock should be packaged and moved in a Trendelenburg's position or supine with their legs elevated 6″ to 12″. Pregnant patients should be positioned and transported on their left sides. Place an unresponsive patient with no suspected spinal injury into the recovery position by rolling the patient onto his or her side without twisting the body. Transport a patient who is nauseated or vomiting in a position of comfort, but be sure that you are positioned appropriately to manage and maintain a patent airway. Obese patients should be positioned the same as other patients; however, particular attention must be made to ensure their dignity is maintained.

Body Mechanics

Anatomy Review

The shoulder girdle rests on the rib cage and is supported by the vertebrae that lie inferior to it. The arms are connected to and hang from the shoulder girdle.

When a person is standing upright, the individual weight-bearing vertebrae are stacked on top of each other and aligned over the sacrum. The sacrum is both the mechanical weight-bearing base of the spinal column and the fused central posterior section of the pelvic girdle.

When a person is standing upright, the weight of anything being lifted and carried in the hands is reflected onto the shoulder girdle, the spinal column inferior to it, the pelvis, and then the legs **Figure 35-1**. In lifting, if the shoulder girdle is aligned over the pelvis and the hands are held close to the legs, the force that is exerted against the spine occurs in an essentially straight line down the vertebrae in the spinal column. Therefore, with the back properly maintained in an upright position, very little strain occurs against the muscles and ligaments that keep the spinal column in alignment, and significant weight can be lifted and carried without injury to

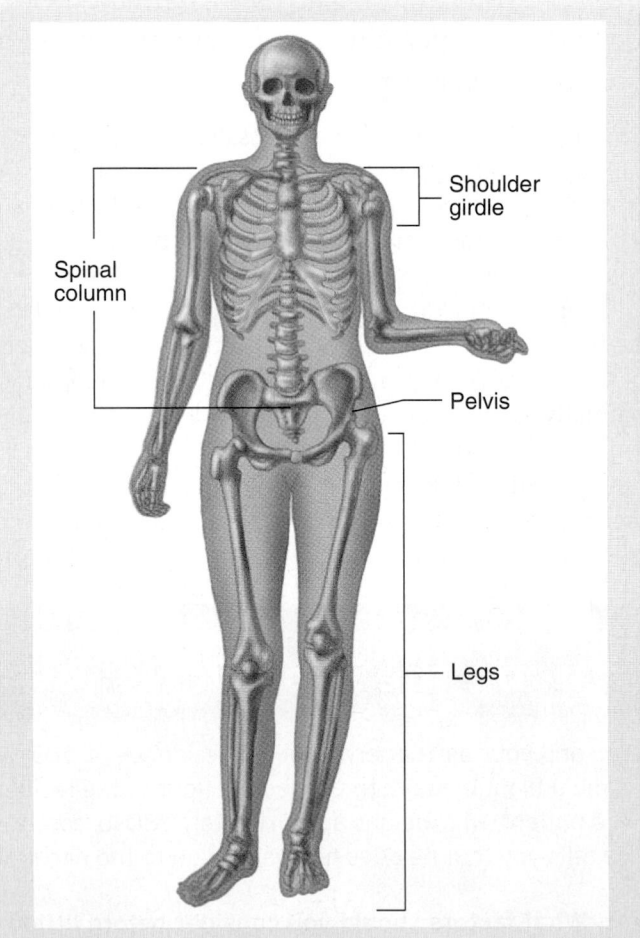

Figure 35-1 When you are standing upright, the weight of anything that you lift and carry in your hands is borne by the shoulder girdle, the spinal column, the pelvis, and the legs.

the back Figure 35-2 . However, you may injure your back if you lift with your back curved or, even if straight, bent significantly forward at the hips Figure 35-3 . With the back in either of these positions, the shoulder girdle lies significantly anterior to the pelvis, and the force of lifting is exerted primarily across, rather than down, the spinal column. When this occurs, the weight is supported by the muscles of the back and ligaments that run from the base of the skull to the pelvis, keeping the spinal column in alignment, rather than by each vertebral body and disk resting on those aligned below it. In addition, the upper spine and torso serve as a lever so that the force that is exerted against the muscles and ligaments in the lumbar and sacral regions, as a result of the mechanical advantage produced, is many times that of the combined weight of your upper body and the object you are lifting. Therefore, the first key rule of lifting is to always keep

Figure 35-3 This photo demonstrates an incorrect method of lifting. You may be injured if you lift with your back curved because the lifting force is exerted primarily across, rather than down, the spinal column. When this occurs, the muscles of the back, not the vertebrae, are supporting the lift.

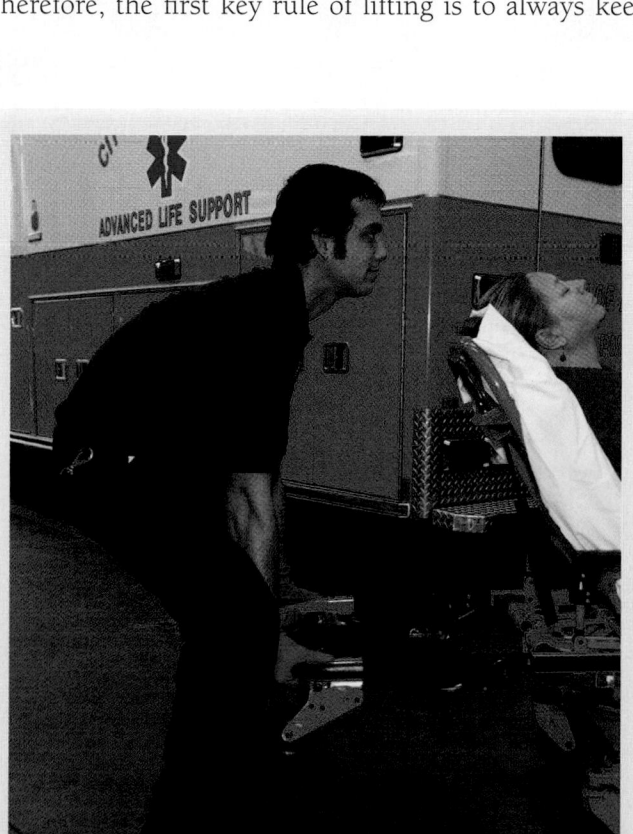

Figure 35-2 If your body is properly aligned when you lift, the line of force exerted against the spine occurs in an essentially straight line down the vertebrae. In this way, the vertebrae support the lift.

your back in a straight, upright (vertical) position and lift without twisting.

When lifting, you should spread your legs about 15″ apart (shoulder width) and place your feet so that your center of gravity is properly balanced between them. Then, with the back held upright, bring your upper body down by bending the legs. Once you have properly grasped the patient or stretcher and made any necessary adjustments in the location of your feet, lift the patient by raising your upper body and arms and by straightening your legs until you are again standing. Because the leg muscles are regularly exercised by walking, climbing stairs, or running, they are well developed and extremely strong. Therefore, as well as being the safest way to lift, lifting by extending the properly placed flexed legs is also the most powerful way to lift. This method is appropriately called a **power lift**. The power lift position is also useful for individuals who have weak knees or thighs.

One mistake you can make while performing a patient lift is to lift a patient or other heavy object while reaching any significant distance in front of your torso or face. Even if your back is held properly upright, adverse forces across the spinal column and leverage against the lower back will occur if you are lifting a heavy object with your arms outstretched so that your hands are significantly anterior to the plane described by the front of the torso (the plane consists of the anterior torso and imaginary lines extended vertically above and below it). Whenever you are lifting or carrying a patient, be sure to hold your arms so that your hands are almost immediately adjacent to the plane described by your anterior torso, and always keep the weight that you are lifting as close to your body as possible.

Another rule to remember when lifting is to avoid placing lateral force across the spine and sideways leverage against the lower back. If you lift with only one arm or with the arms extended more to one side than the other, more force will be exerted against one side of the shoulder girdle than the other, causing lateral force to be exerted across the spinal column. To prevent this, keep your arms approximately the same distance apart as when hanging at each side of the body, with the weight distributed equally and properly centered between them. If the weight is not balanced between both arms or properly centered between the shoulders when you are preparing to lift, turn your body and/or move to the left or right until the weight is properly balanced and centered. To lift safely and produce the maximal power lift, you should take the following steps Skill Drill 35-1 :

Skill Drill 35-1

1. Tighten your back in its normal upright position, and use your abdominal muscles to lock it in a slight curve.
2. Spread your legs apart about 15″, and bend your legs to lower your torso and arms.
3. With arms extended down each side of the body, grasp the stretcher or backboard with your hands held palm up and just in front of the plane described by the anterior torso and imaginary lines extending vertically from it to the ground.
4. Adjust your orientation and position until the weight is balanced and centered between both arms Step 1 .

Skill Drill 35-1

Performing the Power Lift

Step 1 Lock your back into an upright curve. Spread and bend your legs. Grasp the backboard, palms up and just in front of you. Balance and center the weight between your arms.

Step 2 Position your feet, straddle the object, and distribute weight.

Step 3 Lift by straightening your legs, keeping your back locked in.

5. Reposition your feet as necessary so that they are about 15″ apart with one slightly farther forward and rotated so that you and your center of gravity will be properly balanced between them. Be sure to straddle the object, keep your feet flat, and distribute your weight to the balls of the feet or just behind them (Step 2).

6. With the arms extended downward, lift by straightening your legs until you are fully standing. Make sure your back is locked in and that your upper body comes up before your hips (Step 3).

Reverse these steps whenever you are lowering the stretcher. Always remember to avoid bending at the waist and twisting as you stand.

Your safety, as well as that of the other EMTs and the patient, depends on the use of proper lifting techniques and having and maintaining a proper hold when lifting or carrying a patient. If you do not have proper hold of the stretcher or of the patient in a body lift, you will not be able to bear a proper share of the weight, and there is an increased chance that you can suddenly lose your grasp with one or both hands. If you temporarily lose your grasp with one or both hands, the position and weight distribution of the stretcher change suddenly, and the other members of the team must quickly reach beyond a safe distance to avoid dropping the patient. As a result, sudden excessive force may be placed across each one's spine, causing lower back injury.

You should use the **power grip** to get the maximum force from your hands whenever you are lifting a patient Figure 35-4 . The arm and hand have their greatest lifting strength when facing palm up. Whenever you grasp a stretcher or backboard, your hands should be at least 10″ apart. Each hand should be inserted under the handle with the palm facing up and the thumb extended upward. You should then advance the hand until the thumb prevents further insertion and the cylindrical handle lies firmly in the crease of your curved palm. Curl your fingers and thumb tightly over the top of the handle. All your fingers should be at the same angle. To have the proper power grip, make sure that the underside of the handle is fully supported on your curved palm with only the fingers and thumb preventing it from being pulled sideways or upward out of the palm.

If you must lift the object higher once you have lifted by extending your legs, you will be able to "curl" the object higher by using your biceps to flex the arms while maintaining the power grip and weight supported in the palms.

You should never grasp a stretcher or backboard with the hand placed palm down over the handle unless you are standing at the front end with your back to the stretcher, as when performing a diamond carry. When lifting with the palm down, the weight is supported by the fingers rather than the palm. This hand orientation places the tips of the fingers and thumb under the handle. If the weight forces them apart, your grasp on the handle will be lost.

When lifting a patient by a sheet or blanket, you should center the patient on the sheet and tightly roll up the excess fabric on each side. This produces a cylindrical handle that provides a strong, secure way to grasp the fabric Figure 35-5 .

When directly lifting a patient, you should tightly grip the patient in a place and manner that will ensure that you will not lose your grasp on the patient.

Figure 35-4 To perform the power grip, grasp the handle of the stretcher or backboard with your palms up and your thumbs extending up. Make sure your hands are about 10″ apart and that your fingers are all at the same angle. The underside of the handle should be fully supported by the palms of your hands.

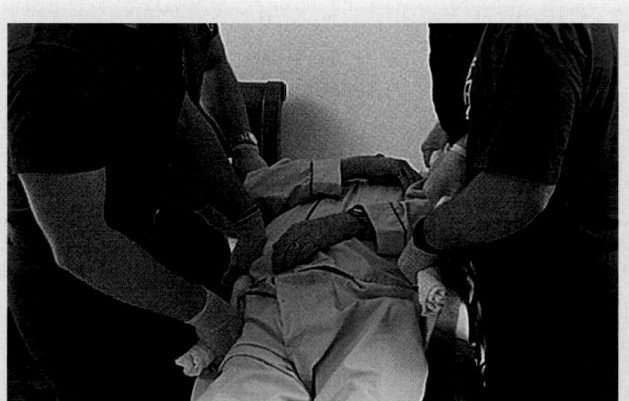

Figure 35-5 When lifting a patient by a bedsheet, you should center the patient on the sheet and tightly roll up the excess fabric on each side. This produces a cylindrical handle that provides a strong way to grasp the fabric.

Weight and Distribution

Whenever possible, you should use a device that can be rolled to move a patient. However, in a situation where a wheeled device is not available, you must make sure that you understand and follow certain guidelines for carrying a patient on a stretcher. Table 35-1 shows the guidelines.

If a patient is supine on a backboard or is lying or in a semi-Fowler's position on the stretcher, his or her weight is not equally distributed between the two ends of the device. Between 68% and 78% of the body weight of a patient in a horizontal position is in the torso. Therefore, more of the patient's weight rests on the head half of the device than on the foot half.

A patient on a backboard or stretcher can be lifted and carried by four rescuers in a **diamond carry**, with one EMT at the head end of the device, one at the foot end, and one at each side of the patient's torso Figure 35-6. Follow these steps to perform the diamond carry Skill Drill 35-2:

Table 35-1 Guidelines for Carrying a Patient on a Stretcher

- Be sure that you know or can find out the weight, both of the patient and the associated equipment, to be lifted and the limitations of the team's abilities.
- Coordinate your movements with those of the other team members while constantly communicating with them.
- Do not twist your body as you are carrying the patient.
- Keep the weight that you are carrying as close to your body as possible while keeping your back in a locked-in position.
- Be sure to flex at the hips, not at the waist, and bend at the knees, while making sure that you do not hyperextend your back by leaning back from your waist.

Figure 35-6 The diamond carry requires four rescuers, one each at the head of the backboard, the foot end, and each side of the patient's torso.

You are the Provider: PART 2

Because of the patient's weight, you ask dispatch to send an engine company to assist with the lifting and moving process. After performing a primary assessment of the patient, you explain to him that there will be a slight delay in moving him back to his bed.

Recording Time: 0 Minutes	
Appearance	Anxious; in obvious pain
Level of consciousness	Conscious and alert
Airway	Open; clear of secretions and foreign bodies
Breathing	Increased rate; adequate depth
Circulation	Pulse, increased and strong; skin, pink, warm, and moist; no obvious bleeding

The patient tells you that he has chronic back pain but has never been to see a physician. Although he denies injuring himself when he accidentally rolled out of bed during his sleep, he tells you that he is unable to move himself back to his bed.

3. Should you and your partner attempt to lift the patient and move him back to his bed? Why or why not?

should be facing the same direction and will be walking forward when carrying the patient (Step 3).

A patient on a backboard or stretcher should be carried feet first to place the lightest load on the EMT at the patient's feet, who, to walk forward, must turn and grasp the handles with his or her back to the device. Carrying the patient feet first will also allow a conscious patient to see in the direction of movement.

It is important that you and your team use the correct lifting techniques to lift the stretcher. One method of lifting and carrying a patient on a backboard is the one-handed carrying technique. With this method, four or more EMTs each use one hand to support the backboard so that they are able to face forward as they are walking. To perform the one-handed carrying technique, follow the steps in Skill Drill 35-3:

1. To best balance the weight, the EMTs at each side should be located so that they are able to grasp the backboard or stretcher with one hand adjacent to the distal edge of the patient's pelvis and the other midthorax. All four EMTs lift the device while facing toward the patient (Step 1).

2. The EMT at each side should grasp the backboard or stretcher with the head-end hand (Step 2).

3. The EMTs at the sides turn toward the patient's feet. The EMT at the foot turns to face forward. All four

Skill Drill 35-2

Performing the Diamond Carry

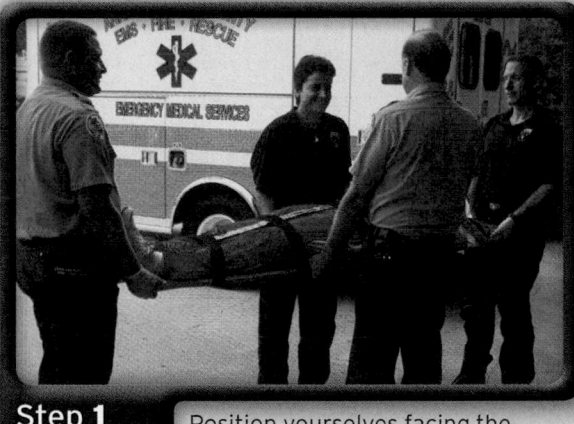

Step 1 Position yourselves facing the patient.

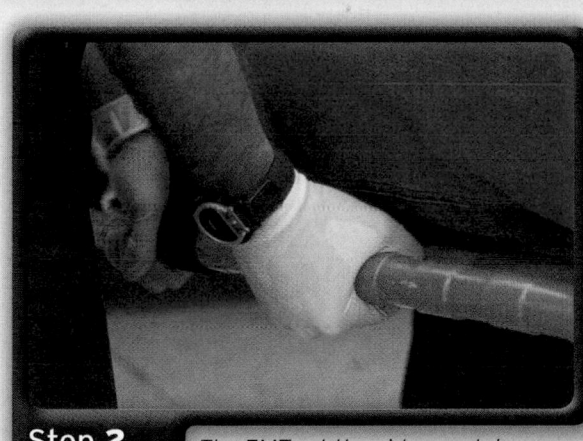

Step 2 The EMTs at the sides each turn the head-end hand palm down and release the other hand.

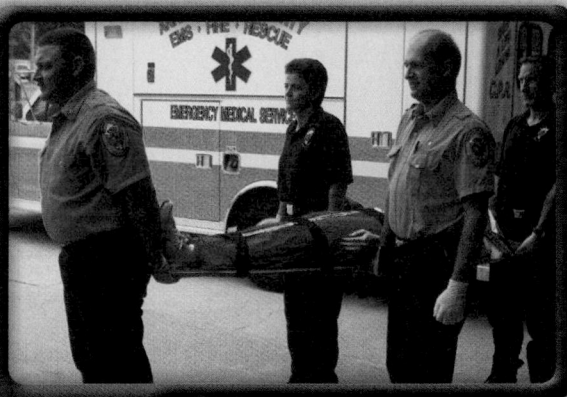

Step 3 The EMTs at the side turn toward the foot end. The EMT at the foot turns to face forward.

1. Before lifting the backboard, be sure that at least two EMTs are on each side of the backboard facing across from each other and using both hands (Step 1).
2. Lift the backboard to carrying height using correct lifting techniques, including a locked-in back (Step 2).

3. Once you have lifted the backboard to carrying height, you and your partners turn in the direction you will be walking and switch to using one hand (Step 3).

Be sure to pick up and carry the backboard with your back in the locked-in position. If you need to lean to either side to compensate for a weight imbalance, you have probably exceeded your weight limitation. If this occurs, you may need to add helpers or reevaluate the carry, or you might injure yourself or drop the patient.

When you must carry a patient up or down a flight of stairs or other significant incline, use a stair chair if

Skill Drill 35-3

Performing the One-Handed Carrying Technique

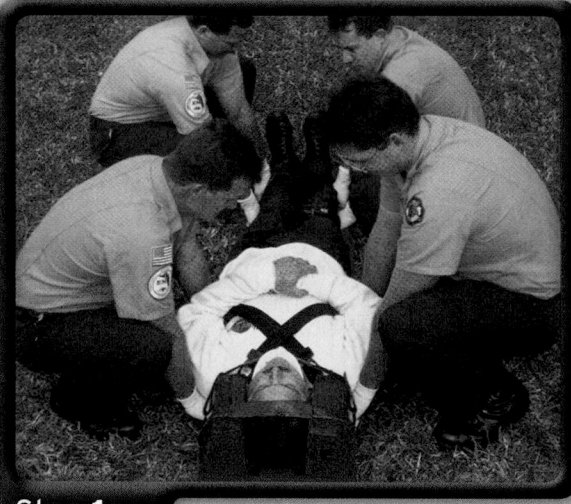

Step 1 Face each other and use both hands.

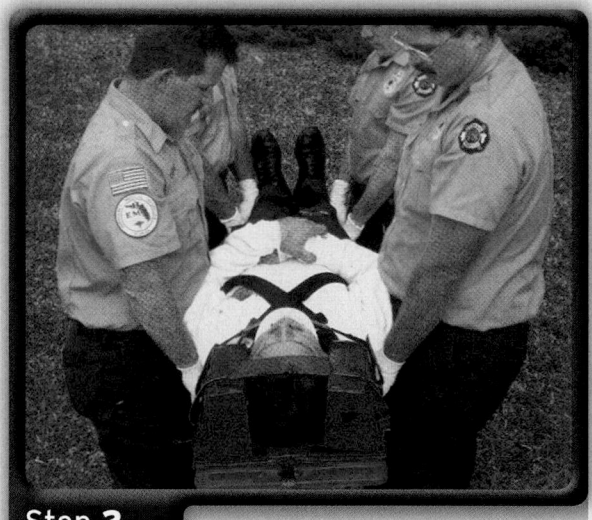

Step 2 Lift the backboard to carrying height.

Step 3 Turn in the direction you will walk, and switch to using one hand.

possible. When you must use a backboard or stretcher, be sure that the patient is anatomically secured to the device in such a way that he or she cannot slide significantly when the stretcher is at an angle. To carry a patient on stairs, follow the steps in Skill Drill 35-4:

Skill Drill 35-4

1. Apply a strap that passes tightly across the upper torso and through each armpit, but not over the arms, to hold the patient in place while leaving the arms free. The strap is secured to the handles at both sides of the backboard so that it cannot slide toward the foot end of the board. Strap the patient securely to the backboard (Step 1).
2. When you carry the patient down stairs or an incline, make sure the backboard or stretcher is carried with the foot end first so that the head end is elevated higher than the foot end. The straps will prevent the patient from sliding down or off the backboard (Step 2).
3. When you carry a patient up stairs or an incline, the elevated head end of the backboard or stretcher should go first (Step 3).

It is helpful to put taller rescuers at the foot of the stretcher when moving a patient up and down steps. This minimizes bending while lifting and moving the patient.

Safety

Since lifting and moving techniques require a team leader to coordinate and direct the process, it will save time and prevent confusion to establish either informal practices or formal procedures that tell all team members—in advance—who will be in charge of these activities.

The **wheeled ambulance stretcher**, which is a specially designed stretcher that can be rolled along the ground, weighs between 40 and 145 lb, depending on its design and features Figure 35-7. Because its weight must be added to that of the patient, it is generally not taken up or down stairs or to other locations where the patient must be carried for any significant distance. Moving a patient by rolling, using a stretcher or other wheeled device, is preferred when the situation allows and helps prevent injuries from carrying. When the patient is upstairs, you should take the wheeled ambulance stretcher to the ground floor landing and prepare it for the patient by lowering the side rails, turning down the cover sheet, and removing any equipment that you may have secured on the top. You should then take either a wheeled stair chair or a backboard upstairs. Both of these devices are considerably lighter than a wheeled stretcher and may be

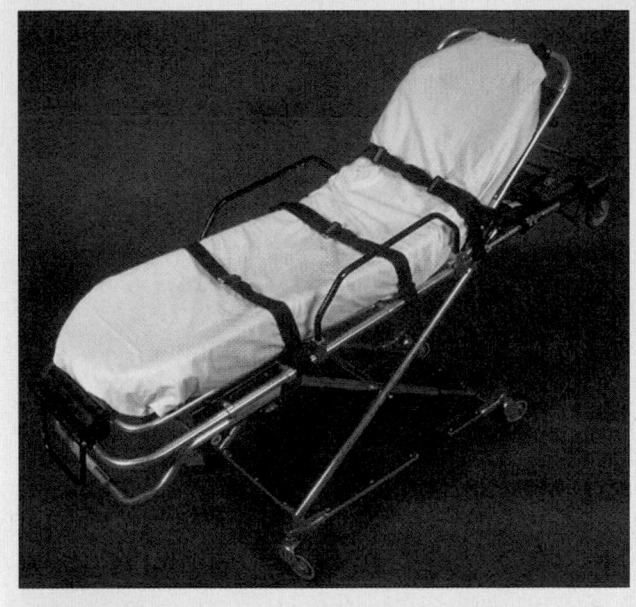

Figure 35-7 The wheeled ambulance stretcher is specially designed to roll along the ground.

used to carry the patient down to the waiting stretcher. Use a wheeled **stair chair** to bring a conscious patient down to the waiting stretcher if the patient's condition allows him or her to be placed in a sitting position Figure 35-8. Once the stretcher has been reached, transfer the patient from the stair chair onto the stretcher. When a patient is

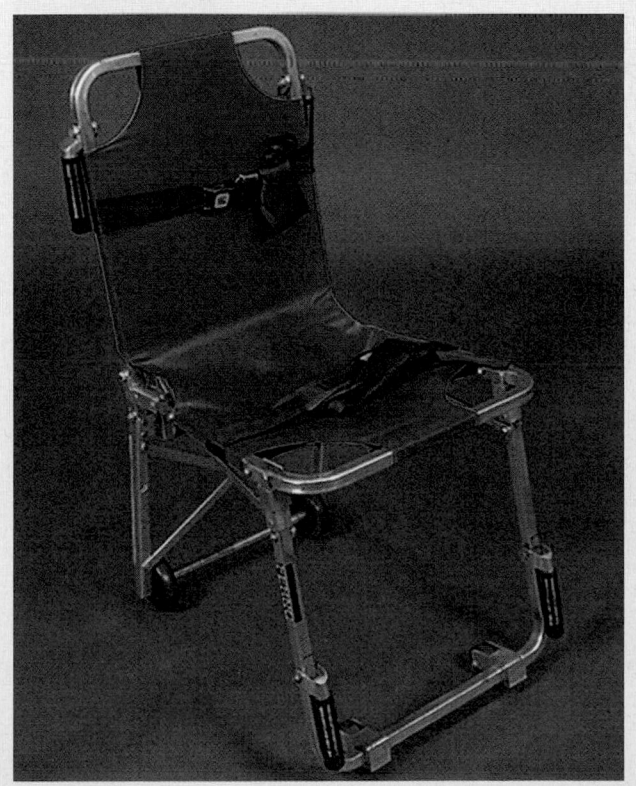

Figure 35-8 A wheeled stair chair can be used to transfer a conscious patient up or down a flight of stairs.

Skill Drill 35-4

Carrying a Patient on Stairs

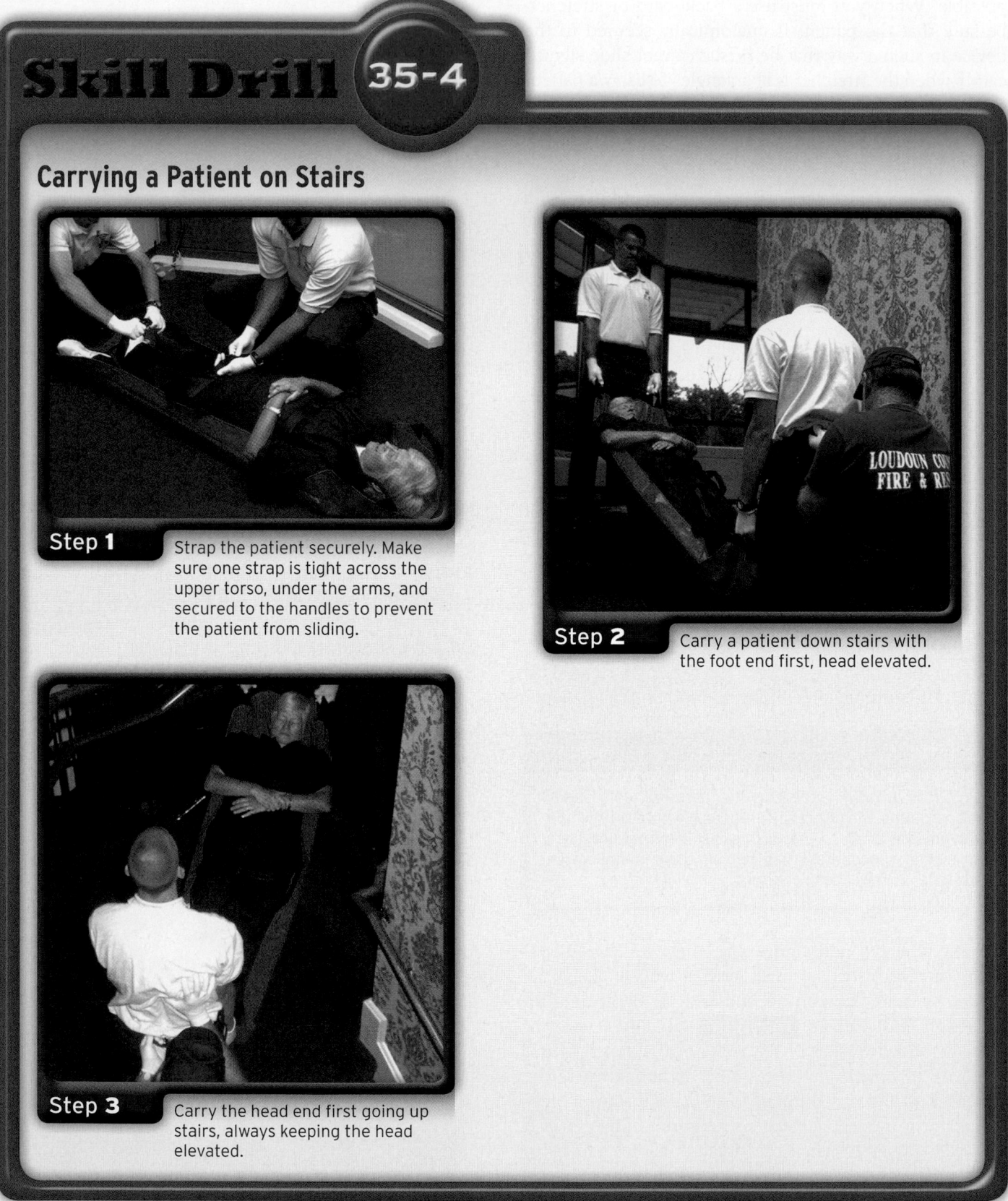

Step 1 Strap the patient securely. Make sure one strap is tight across the upper torso, under the arms, and secured to the handles to prevent the patient from sliding.

Step 2 Carry a patient down stairs with the foot end first, head elevated.

Step 3 Carry the head end first going up stairs, always keeping the head elevated.

in cardiac arrest, must be moved in a supine position, or must be immobilized, secure the patient onto a backboard. A **backboard**, which is a device that provides support to patients whom you suspect have hip, pelvic, spinal, or lower extremity injuries, is also called a spine board, trauma

board, or longboard **Figure 35-9**. You can then carry the patient on the backboard down the stairs to the prepared stretcher. Once you reach the stretcher, place both the backboard and patient on the stretcher; then secure both to the stretcher with additional straps.

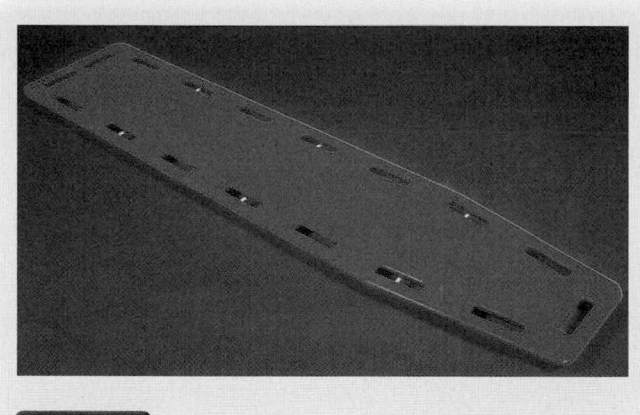

Figure 35-9 A backboard is used to transfer patients who must be moved in a supine or immobilized position.

■ Directions and Commands

To safely lift and carry a patient, you and your team must anticipate and understand every move, and each move must be executed in a coordinated manner. The team leader should indicate where each team member is to be located and rapidly describe the sequence of steps that will be performed to ensure that the team knows what is expected before any lifting is initiated. If you must lift and move the patient through a number of separate stages, the team leader should first give an abbreviated overview of the stages, followed by a more detailed explanation of each stage just before it will occur.

Orders that will initiate the actual lifting or moving or any significant changes in movement should be given in two parts: a preparatory command and a command of execution. For example, if the team leader says "All ready to stop. STOP!" the "All ready to stop" will get your attention, identify who should act, and prepare them to act; the declarative "STOP!" will indicate the exact moment for execution. Commands of execution should be delivered in a louder voice. Often, a countdown is helpful when you need to lift a patient. To avoid confusion in using a countdown, always clarify whether "three" is to be a part of the preparatory command or whether it is to serve as the order to execute. You can say "We're going to lift on three. One-two-THREE!" or "I'm going to count to three, and then we're going to lift. One-two-three-LIFT!"

■ Additional Lifting and Carrying Guidelines

You should estimate how much the patient weighs before attempting to lift him or her. Commonly, adult patients weigh between 120 and 220 lb. If you use the correct technique, you and one other EMT should be able to safely lift this weight. Depending on your individual strength, you and another EMT may be able to safely lift

an even heavier patient. However, because it is safer to have four rescuers lift, you should try to use four rescuers whenever the available resources will allow. You should know how much you can comfortably and safely lift and should not attempt to lift a proportional weight (the share of the weight that you will bear) that exceeds this amount. If you find that lifting the patient places a strain on you, call for the lifting to be stopped and the patient to be lowered. You should then obtain additional help before again attempting to lift the patient. Be sure to communicate clearly and frequently with your partner and other rescuers whenever you are lifting a patient.

You should not attempt to lift a patient who weighs more than 250 lb with fewer than four rescuers, regardless of individual strength. Protocols should include a method to rapidly summon additional help to lift and carry such a patient or, as in the case of a cardiac arrest, provide and maintain the necessary care in the field, as well as when moving and transporting the patient. In addition, you must know, or be able to find out, the weight limitations of the equipment you are using and how to handle patients who exceed the weight limitations. Special techniques, equipment, and resources generally are required to move any patient who weighs more than 350 lb to the ambulance. These resources should be summoned when you arrive.

Because more than half of a patient's weight is distributed to the head end of the backboard or stretcher, the strongest of the available EMTs should be located at the head end of the device. Even with four or more EMTs carrying the patient, the strain on the EMT carrying the head end of the device will be increased when you must negotiate a narrow area or a flight of stairs. In carrying a patient up or down a flight of stairs, proportionally greater weight will also be distributed to the EMT who is carrying the foot end when the backboard or stretcher becomes angled because of the incline or decline. You should anticipate this and, in such cases, make sure the two strongest EMTs are positioned at the head and foot ends of the board. Because of the incline of the stairway, if one of the two EMTs is considerably taller than the other, it will be easier if the shorter of the two is at the head end and the taller person is at the foot end.

The dynamics that are involved in carrying a patient down a flight of stairs or for any significant distance will not allow you to carry as much proportional weight as you can to safely lift or support the patient during a move onto a nearby backboard or stretcher. Therefore, if you feel that you are approaching your maximum lifting capacity as you are moving the patient onto a backboard or stretcher, you should not attempt to lift and carry the patient for any significant distance or down a flight of stairs. You can again attempt to lift and carry the patient after you have decreased the amount of proportional weight you will be carrying by changing your position on the device or that of the others on the team or have obtained additional help.

You should try to use a stair chair instead of a stretcher, whenever possible, to carry a patient down stairs. Follow these steps to use a stair chair **Skill Drill 35-5** :

Skill Drill 35-5

1. Secure the patient to the stair chair with straps. At a minimum, use a lap belt at the hips and a strap around the chest. You should also use some method to secure the arms and hands so the patient does not reach out to grasp something and throw the carrying team off balance. You can ask the patient to fold his or her hands on the chest or lap **Step 1**.

2. Rescuers take their places around the patient seated on the chair: one at the head and one at the foot. The rescuer at the head will give directions to coordinate the lift and carry **Step 2**.

3. The third rescuer precedes the two carrying the chair to open doors, with his or her hand on the back of the second rescuer, providing guidance and support. For lengthy carries, the third rescuer can also rotate into the carrying team to provide breaks for the other two **Step 3**.

4. When reaching landings and other flat intervals in the carry, lower the chair to the ground and roll it rather than carrying it. When reaching the level where the stretcher awaits, roll the chair into position next to the stretcher in preparation for transferring the patient **Step 4**.

As with other carries, always remember to keep your back in a locked-in position and to flex at the hips, not the waist. You should also bend at the knees and keep the patient's weight and your arms as close to your body as possible. Twisting while carrying or moving a patient will increase your risk of injury. Try to avoid any unnecessary lifting and carrying of the patient. You may find that a log roll or a body drag will aid you in moving your patient

Words of Wisdom

When you encounter a patient in a confined space, such as a bathroom, it can pose a unique set of problems. Prior to moving a patient in a confined space, it is important to discuss the process with your fellow team members. Ensure that everyone agrees with the extrication plan and understands his or her role. Remember that communication with the crew, as well as the patient, will assist in minimizing potential problems.

onto the backboard or the stretcher. If these techniques will not harm or jeopardize your patient's condition, use one of these moves.

Principles of Safe Reaching and Pulling

When you use a body drag to move a patient, the same basic body mechanics and principles apply as when lifting and carrying. Your back should always be locked and straight, not curved or bent laterally, and you should avoid any twisting so that the vertebrae remain in their normal alignment. When you are reaching overhead, avoid hyperextending your back. When you are pulling a patient who is on the ground, you should always kneel to minimize the distance that you will have to lean over **Figure 35-10A**. To keep your reach within the recommended distance, reach forward and grasp the patient so that your elbows are just beyond the anterior torso **Figure 35-10B**. When you are pulling a patient who is at a different height from you, bend your knees until your hips are just below the height of the plane across which you will be pulling the patient. During pulling, you should extend your arms no more than about 15″ to 20″ in front of your torso. Reposition your feet (or knees, if kneeling) so that the force of pull will be balanced equally between both arms and the line of pull will be centered between them **Figure 35-10C**. Pull the patient by slowly flexing your arms. When you

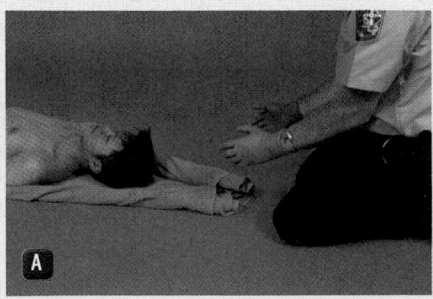 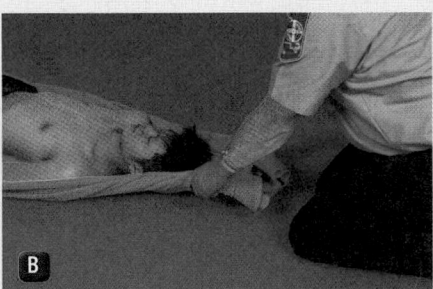 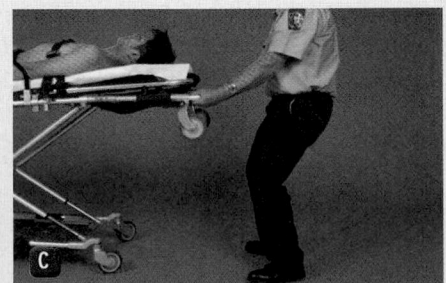

Figure 35-10 Reaching and pulling safely. **A.** Kneel to pull a patient who is on the ground. **B.** When pulling, your elbows should only extend just beyond the anterior torso. **C.** Bend your knees to pull a patient who is at a different height than you are. Position your feet or knees to balance the force of pull.

Skill Drill 35-5

Using a Stair Chair

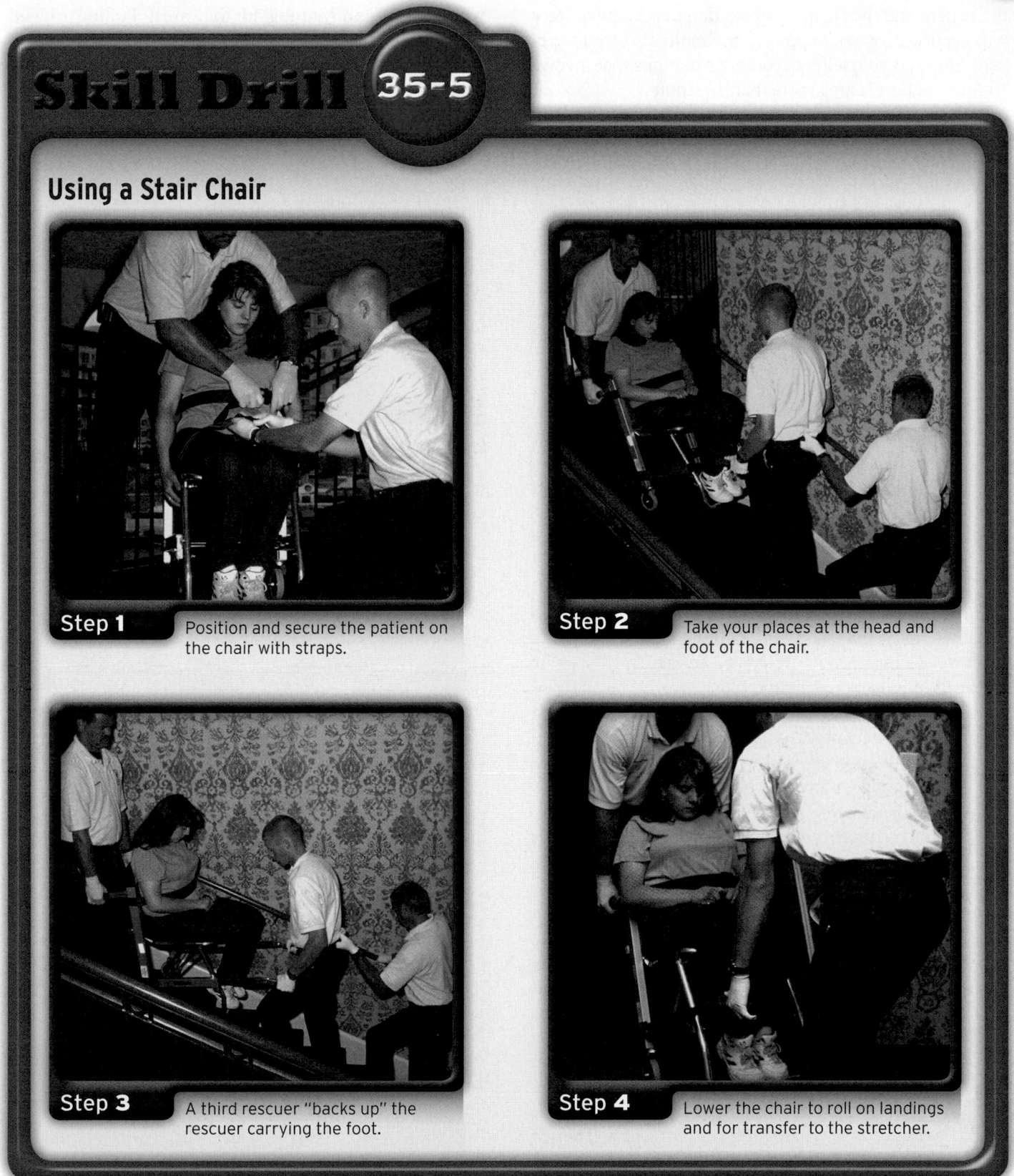

Step 1 Position and secure the patient on the chair with straps.

Step 2 Take your places at the head and foot of the chair.

Step 3 A third rescuer "backs up" the rescuer carrying the foot.

Step 4 Lower the chair to roll on landings and for transfer to the stretcher.

can pull no farther because your hands have reached the front of your torso, stop and move back another 15″ to 20″. Then, when properly positioned, repeat the steps. You should alternate between pulling the patient by flexing

your arms and then repositioning yourself so that your arms are again extended with your hands about 15″ in front of your torso. By not moving yourself and the patient simultaneously, you will prevent undesirable jostling of the patient and the chance that sudden unscheduled force will occur across your spine. You should also try to prevent injury to yourself by avoiding situations that involve strenuous effort lasting more than 1 minute.

If you must drag a patient across a bed, you will have to kneel on the bed to avoid reaching beyond the recommended distance. Then follow the steps described previously until the patient is within 15″ to 20″ of the bed's edge (see Figure 35-10). You can then complete the drag while standing at the side of the bed. Rather than dragging the patient by his or her clothing, use the sheet or blanket under the patient for this purpose. You can roll the bedding under the patient until it is about 6″ wider than the patient. Pull on the rolled bedding smoothly and evenly to glide the patient to the bedside.

Unless the patient is on a backboard, transfer the patient from the stretcher to a bed in the emergency department or the patient's hospital room with a body drag. With the stretcher at the same height as the bed or slightly higher and held firmly against the bed's side, you and another EMT should kneel on the hospital bed and, in the manner previously described, drag the patient in increments until he or she is properly centered on the bed. When transferring the patient onto a narrow examining table, rather than kneeling on the table, you can usually drag the patient while standing against the opposite side. A third person may need to take both sides of the head to move the patient safely.

Sometimes during a body drag, you and another EMT may have to pull the patient with one of you on each side of the patient. You will have to alter the usual pulling technique to prevent pulling sideways and producing adverse lateral leverage against your lower back. You should position yourself by kneeling just beyond the patient's shoulder and facing toward his or her groin **Figure 35-11A**. By extending one arm across and in front of your chest, you can grasp the armpit and, with your other arm extended in front and to the side of the patient's torso, the patient's belt. Then, by raising your elbows and flexing your arms, you can pull the patient with the line of force at the minimum angle possible **Figure 35-11B**.

Generally, when log rolling a patient onto his or her side, you will initially have to reach farther than 18″ **Figure 35-12**. To minimize this distance, kneel as close to the patient's side as possible, leaving only enough room so that your knees will not prevent the patient from being rolled. When you lean forward, keep your back straight and lean solely from the hips. Be sure to use your shoulder muscles to help with the roll. To minimize the amount of time you are extended like this and to support the patient's weight, roll the patient without stopping until the patient is resting on his or her side. Some EMS experts consider that, during a log roll, you should pull rather than push the patient. Local protocols

will guide your training in this area. Pulling toward you allows your legs to prevent the patient from rolling over completely and from rolling beyond the intended distance.

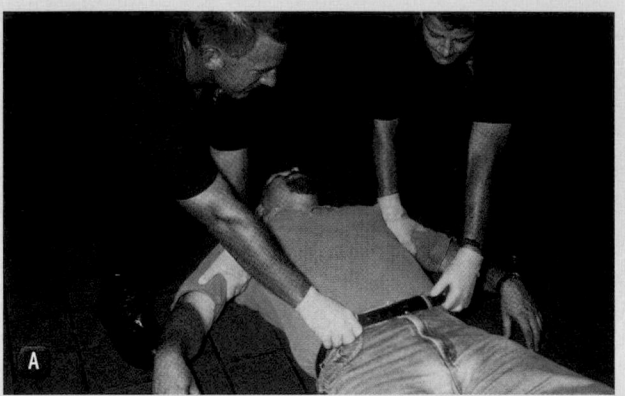

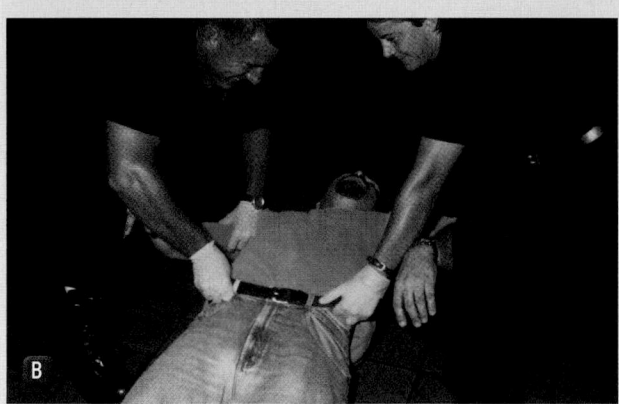

Figure 35-11 A body drag with an EMT on each side of the patient. **A.** Kneel just beyond the patient's shoulder facing his or her groin. Extend one arm across and in front of your chest, and grasp the armpit. Extend your other arm in front and to the side of the patient's torso, and grasp the patient's belt. **B.** Raise your elbows and flex your arms to pull the patient.

Figure 35-12 When placing a patient onto a backboard, roll the patient onto his or her side. Kneel as close to the patient's side as possible, leaving only enough room so that your knees will not prevent the patient from being rolled. Lean forward, keeping your back straight and leaning solely from the hips. Use your shoulder muscles to help with the roll.

When you are rolling the wheeled ambulance stretcher, make sure that it is in the fully elevated position and that the intended travel path is free from debris and potential obstacles Figure 35-13 . Push the stretcher from the head end. If you are guiding the stretcher from the foot end, make sure your arms are held close to your body, and be careful to avoid reaching significantly behind you or hyperextending your back. Your back should be locked, straight, and untwisted. While you are walking and guiding the stretcher, bend slightly forward at the hips. As you walk, your legs are pulled back with your feet on the ground, your pelvis is moved forward, and the movement of the pelvis is transferred to the stretcher through your straight torso and firmly held arms. You should try to keep the line of the pull through the center of your body by bending your knees.

The second EMT should guide the head end and assist you by pushing with his or her arms held with the elbows bent so that the hands are about 12″ to 15″ in front of the torso. To protect your elbows from injury, you should never push an object with your arms fully extended in a straight line and the elbows locked. When you push with the elbow bent but firmly held from bending further, the strong muscles of the arm serve as a shock absorber if the wheels or foot end of the stretcher strikes an obstacle that causes its progress to be suddenly slowed or stopped. You must be sure that you push from the area of your body that is between the waist and shoulder. If the weight you are pushing is lower than your waist, you should push

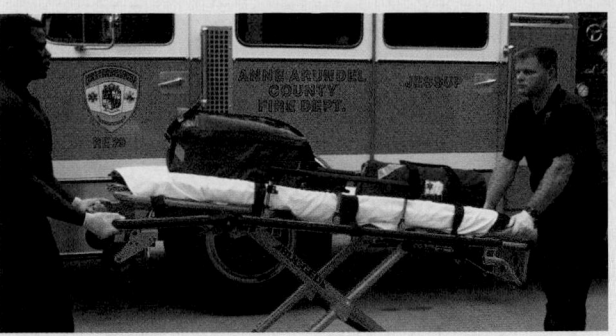

Figure 35-13 Push the stretcher from the head end. If you are guiding the stretcher from the foot end, make sure your arms are held close to your body, and be careful to avoid reaching significantly behind you or hyperextending your back. Your back should be locked, straight, and untwisted.

from a kneeling position. Be careful that you do not push or pull from an overhead position.

General Considerations

Moving a patient should normally be done in an orderly, planned, and unhurried manner. This approach will protect you and the patient from further injury and reduce the risk of worsening the patient's condition when he or she is moved. At a minimum, on most calls you will have to lift and carry the patient to the wheeled ambulance

You are the Provider: PART 3

The patient calls his daughter, who convinces him to go to the hospital to be evaluated for his chronic back pain. You perform a secondary assessment, which does not reveal any obvious injury, while your partner takes his vital signs. The patient denies having any past medical history, except for chronic back pain.

Recording Time: 5 Minutes	
Respirations	22 breaths/min; adequate depth
Pulse	104 beats/min; strong and regular
Skin	Pink, warm, and moist
Blood pressure	156/90 mm Hg
Oxygen saturation (Sao₂)	98% (on room air)

The engine company arrives, and you inform them that the patient will need to be moved downstairs and into the ambulance. You now have a total of six personnel at the scene. The patient tells you that he cannot sit up without experiencing severe pain.

4. What type of carrying device should you use for this patient?

5. What steps can you take to maximize safety while lifting a patient?

stretcher, move the stretcher and patient to the ambulance, and load the stretcher into the patient compartment.

You will often have to include several additional steps to place the patient onto a backboard and/or carry him or her down a flight of stairs. You will also have to add a stop at the top of the stairway so that everyone can reposition for carrying the patient down the stairs. Repositioning usually requires lowering the backboard to the ground and lifting it again when all EMTs are in their proper places. If you are carrying the patient in a stair chair, the additional step occurs after you have descended the stairs and reached the stretcher. At that point, you will have to assist or lift the patient from the stair chair onto the stretcher.

You should carefully plan ahead and select the methods that will involve the least amount of lifting and carrying. Remember to always consider whether there is an option that will cause less strain to you and the other EMTs.

Safety

Follow these rules to keep your back and your patient safe:
- Minimize the number of total body lifts you have to perform
- Coordinate every lift in advance
- Minimize the total amount of weight you have to lift
- Never lift with your back, not ever
- Don't carry what you can put on wheels
- Ask for help any time

Emergency Moves

You should use an **emergency move** to move a patient before assessment and care are provided when there is a potential for danger, and you and the patient must move to a safe place to avoid possible serious harm or death. The presence of fire, explosives, or hazardous materials, your inability to protect the patient from other hazards or gain access to others in a vehicle that need lifesaving care, or difficulty in performing lifesaving treatment because of patient positioning are all situations in which you should use an emergency move.

The only other time you should use an emergency move is if you cannot properly assess the patient or provide immediate critical emergency care because of the patient's location or position.

If you are alone and danger at the scene makes it necessary for you to use an emergency move, regardless of a patient's injuries, you should use a drag to pull the patient along the long axis of the body. This will help to keep the spinal column in line as much as possible. When performing an emergency move, one of your primary concerns is the danger of aggravating an existing spinal injury. Remember that it is impossible to remove a patient quickly from a vehicle while providing as much protection to the spine as you would give by using an immobilization device. However, if you follow certain guidelines during the move, you can usually move a patient from a life-threatening situation without causing further injury to the patient.

You can move a patient on his or her back along the floor or ground by using one of the following methods:

- Pull on the patient's clothing in the neck and shoulder area **Figure 35-14A**. If the shirt has buttons, the top two should be undone to prevent the patient from choking.
- Place the patient onto a blanket, coat, or other item that can be pulled **Figure 35-14B**.
- Rotate the patient's arms so that they are extended straight on the ground beyond his or her head, grasp the wrists, and, with the arms elevated above the ground, drag the patient **Figure 35-14C**.
- Place your arms under the patient's shoulders and through the armpits, and, while grasping your opposite wrist, drag the patient backward **Figure 35-14D**.

If you are alone and must remove an unconscious patient from a vehicle, you should first move the patient's legs so they are clear of the pedals and are against the seat. Then rotate the patient so that his or her back is positioned toward the open car door. Next, place your arms through the armpits and support the patient's head against your body **Figure 35-15A**. While supporting the patient's weight, drag the patient from the seat. If the legs and feet clear the car easily, you can rapidly drag the patient to a safe location by continuing this method **Figure 35-15B**. If the legs and feet do not clear the car easily, you can slowly lower the patient until he or she is lying on his or her back next to the vehicle, clear the legs from the vehicle, and, as previously described, use a long-axis body drag to move the patient a safe distance from the vehicle.

You should use one-person techniques to move a patient only if an immediately life-threatening danger exists and you are alone or, because of the pressing nature of the danger, your partner is moving a second patient simultaneously. Additional one-rescuer drags, carries, and lifts are shown in **Figure 35-16**.

Urgent Moves

An urgent move may be necessary for moving a patient with an altered level of consciousness, inadequate

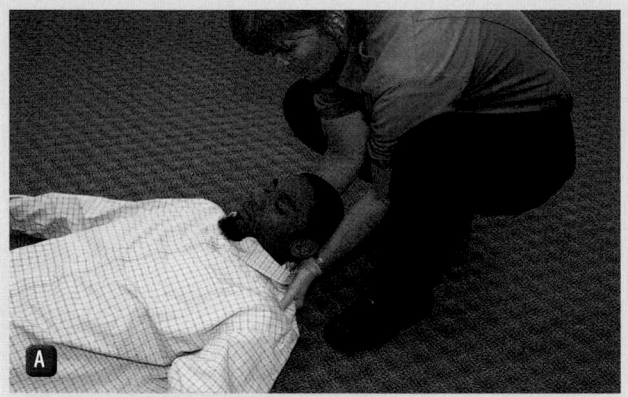

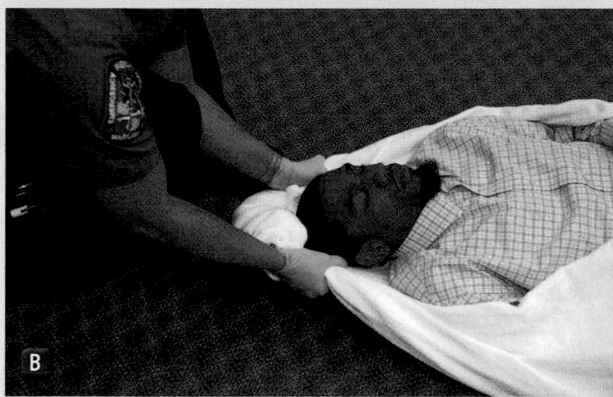

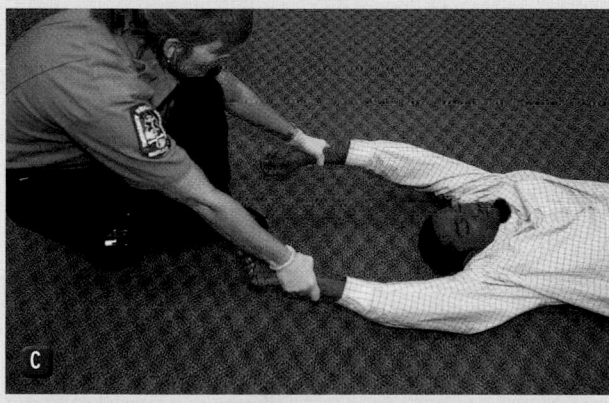

Figure 35-14 Dragging methods. **A.** Emergency clothes drag. **B.** Blanket drag. **C.** Arm drag. **D.** Arm-to-arm drag.

Figure 35-15 One-person technique for moving an unconscious patient from a vehicle. **A.** Grasp the patient under the arms. **B.** Lower the patient down into a supine position.

ventilation, or shock (hypoperfusion). An extreme weather condition may also make an urgent move necessary. In some cases, patients must be urgently moved from the location or position in which they are found. When a patient who is sitting in a car or truck must be urgently moved, you should use the rapid extrication technique.

■ Rapid Extrication Technique

The long backboard, short backboard, and vest-type devices are known as immobilization devices. Normally, you would use an extrication-type vest or half-backboard device to immobilize a seated patient with a suspected spinal injury before removing the patient from the vehicle. However, proper placement of either of these devices on the patient usually requires between 6 and 8 minutes, and in some cases even longer. By using the

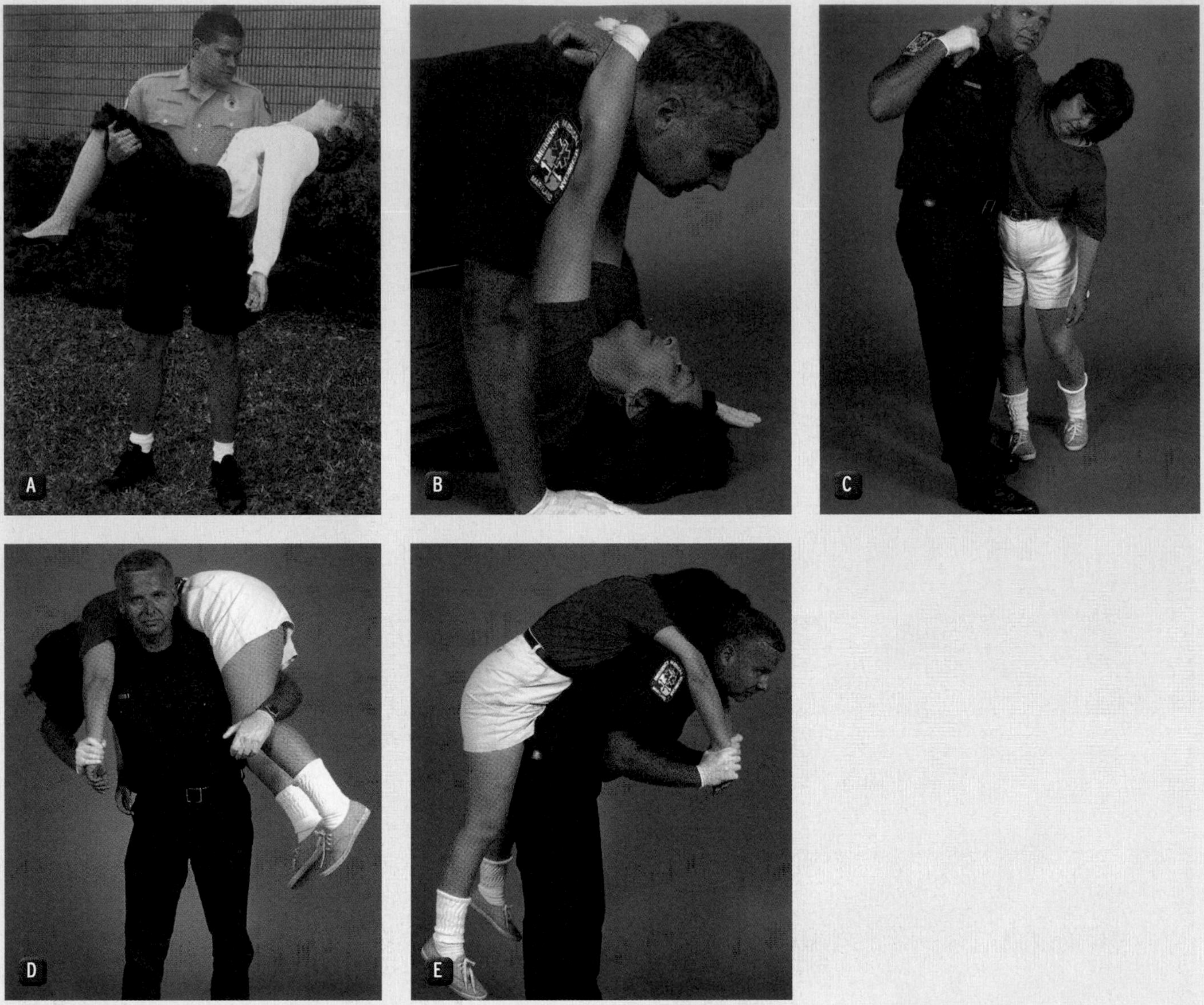

Figure 35-16 One-rescuer drags, carries, and lifts. **A.** Front cradle. **B.** Fire fighter's drag. **C.** One-person walking assist. **D.** Fire fighter's carry. **E.** Pack strap.

rapid extrication technique instead, the patient can be moved from sitting in the vehicle to supine on a backboard in 1 minute or less. However, the rapid nature of this type of extrication can potentially increase the risk of damage if the patient has a spinal injury. Because of this possible patient injury, all available options need to be looked at prior to performing a rapid extrication. **Table 35-2** describes the situations in which you should use the rapid extrication technique.

In such cases, the delay that occurs in applying an extrication-type vest or half-board makes its application contraindicated. However, the manual support and immobilization that you provide when using the rapid extrication technique produce a greater risk of spine movement. Because of this increased risk, you should not use the rapid extrication technique if no urgency exists.

The rapid extrication technique requires a team of three EMTs who are knowledgeable and practiced in the procedure. You should take the following steps when using the rapid extrication technique **Skill Drill 35-6**:

Skill Drill 35-6

1. The first EMT applies manual in-line support of the patient's head and cervical spine from behind. Support may be applied from the side, if necessary, by reaching through the driver's side doorway **Step 1**.

2. The second EMT serves as team leader and, as such, gives the commands until the patient is

Table 35-2 Situations in Which to Use the Rapid Extrication Technique

- The vehicle or scene is unsafe.
- Explosives or other hazardous materials are on the scene.
- There is a fire or a danger of fire.
- The patient cannot be properly assessed before being removed from the car.
- The patient needs immediate intervention that requires a supine position.
- The patient has a life-threatening condition that requires immediate transport to the hospital.
- The patient blocks your access to another seriously injured patient.

supine on the backboard. Because the second EMT lifts and turns the patient's torso, he or she must be physically capable of moving the patient. The second EMT works from the driver's side doorway. If the first EMT is also working from that doorway, the second EMT should stand closer to the door hinges toward the front of the vehicle. The second EMT applies a cervical immobilization device and may perform the primary assessment (Step 2).

3. The second EMT provides continuous support of the patient's torso until the patient is supine on the backboard. Once the second EMT takes control of the patient's torso, usually in the form of a body hug, he or she should not let go of the patient for any reason. Some type of cross-chest shoulder hug usually works well, but you will have to decide what method works best for you on any given patient. You must remember that you cannot simply reach into the car and grab the patient; this will only twist the patient's torso. You must rotate the patient as a complete unit.

4. The third EMT works from the front passenger's seat and is responsible for rotating the patient's legs and feet as the torso is turned, ensuring that they are free of the pedals and any other obstruction. With care, the third EMT should first move the patient's nearer leg laterally without rotating the patient's pelvis and lower spine. The pelvis and lower spine rotate only as the third EMT moves the patient's second leg during the next step. Moving the nearer leg early makes it much easier to move the second leg in concert with the rest of the body. After the third EMT moves the legs together, they should be moved as a unit (Step 3).

5. These initial steps of the rapid extrication technique direct the team to its starting positions and responsibilities. The first EMT applies in-line

support and immobilization of the head and neck. The second EMT gives orders and supports the torso. The third EMT moves and supports the patient's legs. The team is now ready to move the patient.

6. The patient is rotated 90° so that the patient's back is facing out the driver's door and the feet are on the front passenger's seat. This coordinated movement is done in three or four short, quick "eighth turns." The second EMT directs each quick turn by saying, "Ready, turn" or "Ready, move." Hand position changes should be made between moves.

7. In most cases, the first EMT will be working from the back seat and will have removed the headrest (if able). At some point, either because the doorpost is in the way or because he or she cannot reach farther from the back seat, the first EMT will be unable to follow the torso rotation. At that time, the third EMT should assume temporary in-line support of the head and neck until the first EMT can regain control of the head from outside the vehicle. If a fourth EMT is present, the fourth EMT stands next to the second EMT. The fourth EMT takes control of the patient's head and neck from outside the vehicle without involving the third EMT. As soon as the change has been made, the rotation can continue (Step 4).

8. Once the patient has been fully rotated, the backboard should be placed against the patient's buttocks on the seat. Do not try to wedge the backboard under the patient. If only three EMTs are present, be sure to place the backboard within arm's reach of the driver's door before the move so that the board can be pulled into place when needed. In such cases, the far end of the board can be left on the ground. When a fourth EMT is available, the first EMT exits the back seat of the vehicle, places the backboard against the patient's buttocks, and maintains pressure toward the interior of the vehicle from the far end of the board. (Note: When the door opening allows, some EMTs prefer to insert the backboard onto the seat before the patient is rotated.)

9. As soon as the patient has been rotated and the backboard is in place, the second EMT and the third EMT lower the patient onto the backboard while supporting the head and torso so that neutral alignment is maintained. The first EMT holds the backboard until the patient is secured (Step 5).

10. Next, the third EMT must move across the front seat to be in position at the patient's hips. If the third EMT stays at the patient's knees or feet, he or she will be ineffective in helping to move the body's weight. The knees and feet follow the hips.

11. The fourth EMT maintains manual in-line support of the head and now takes over giving the commands. If a fourth EMT is not present, you can direct a volunteer to assist you. The second EMT maintains the direction of the extrication. The second EMT stands with his or her back to the door, facing the rear of the vehicle. The backboard should be immediately in front of the third EMT. The second EMT grasps the patient's shoulders or armpits. Then, on command, the second EMT and the third EMT slide the patient 8″ to 12″ along the backboard, repeating this slide until the patient's hips are firmly on the backboard (Step 6).

Skill Drill 35-6

Performing the Rapid Extrication Technique

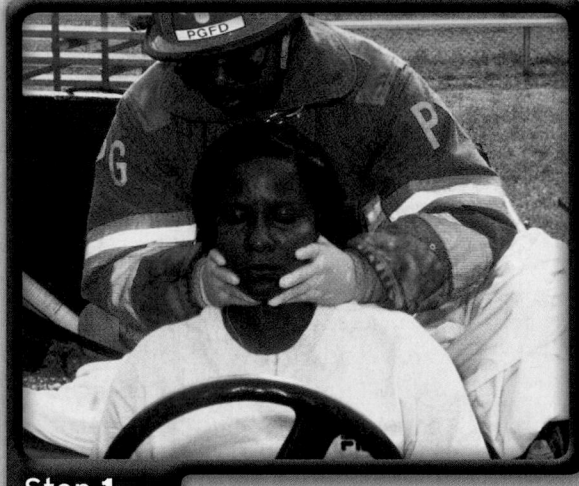

Step 1 The first EMT provides in-line manual support of the head and cervical spine.

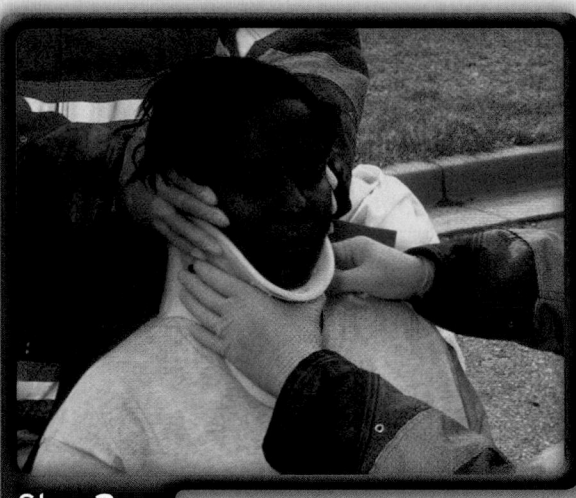

Step 2 The second EMT gives commands, applies a cervical collar, and performs the primary assessment.

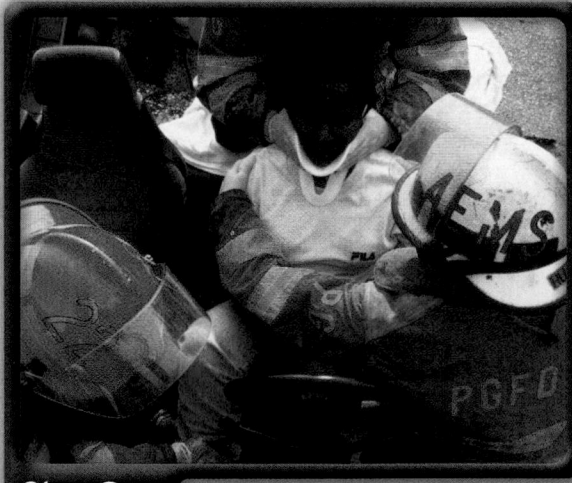

Step 3 The second EMT supports the torso. The third EMT frees the patient's legs from the pedals and moves the legs together, without moving the pelvis or spine.

Step 4 The second EMT and the third EMT rotate the patient as a unit in several short, coordinated moves. The first EMT (relieved by the fourth EMT or a bystander as needed) supports the patient's head and neck during rotation (and later steps).

12. At that time, the third EMT gets out of the vehicle and moves to the opposite side of the backboard, across from the second EMT. The third EMT now takes control at the shoulders, and the second EMT moves back to take control of the hips. On command, these two EMTs move the patient along the board in 8″ to 12″ slides until the patient is placed fully on the board (Step 7).

13. The first (or fourth) EMT continues to maintain manual in-line support of the patient's head. The second EMT and the third EMT now grasp their side of the board, and then carry it and the patient away from the vehicle onto the prepared stretcher nearby (Step 8).

Skill Drill 35-6

Performing the Rapid Extrication Technique, continued

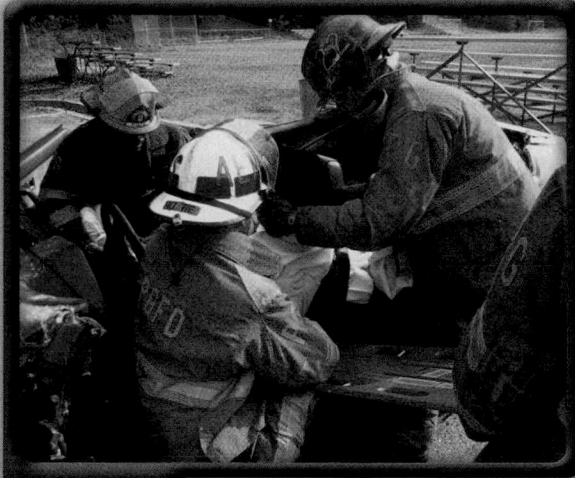

Step 5 The first (or fourth) EMT places the backboard on the seat against the patient's buttocks.

Step 6 The third EMT moves to an effective position for sliding the patient. The second and the third EMTs slide the patient along the backboard in coordinated 8″ to 12″ moves until the patient's hips rest on the backboard.

Step 7 The third EMT exits the vehicle, moves to the backboard opposite the second EMT, and they continue to slide the patient until the patient is fully on the board.

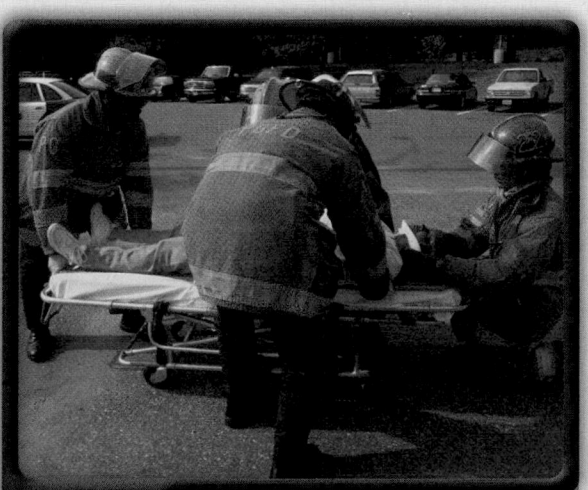

Step 8 The first (or fourth) EMT continues to stabilize the head and neck while the second EMT and the third EMT carry the patient away from the vehicle and onto the prepared stretcher.

In some cases, you will be able to rest the head end of the backboard on the stretcher while the patient is moved onto the backboard. In other situations, you will not be able to do this. Once the backboard and patient have been placed on the stretcher, you should begin lifesaving treatment immediately. If you used the rapid extrication technique because the scene was dangerous, you and your team should immediately move the stretcher a safe distance away from the vehicle, or scene, before you assess or treat the patient.

The steps of the rapid extrication technique must be considered a general procedure to be adapted as needed. Two-door cars differ from four-door models. Larger cars differ from smaller compact models, pickup trucks, and full-size sedans and four-wheel-drive vehicles. You will handle a large, heavy adult differently from a small adult or child. Every situation will be different—a different vehicle, a different patient, and different partners. Your resourcefulness and ability to adapt are necessary elements to successfully perform the rapid extrication technique.

Nonurgent Moves

When both the scene and the patient are stable, you should carefully plan how to move the patient. If your patient move is rushed or not well planned, it may result in discomfort or injury to the patient, you, and your team. Before you attempt any move, the team leader must be sure that there are enough personnel, any obstacles have been identified or removed, the proper equipment is available, and the procedure and path to be followed have been clearly identified and discussed.

In nonurgent situations, you and your team may choose one of several methods for lifting and carrying a patient. Three general methods are presented here, which may serve as a basis for your plan. You may adapt these procedures to meet your needs on a case-by-case basis.

Direct Ground Lift

The **direct ground lift** is used for patients with no suspected spinal injury who are found lying supine on the ground. You should use this lift when you have to lift and carry the patient some distance to be placed on the stretcher. If you find the patient semiprone or lying on his or her side, you should first roll the patient onto his or her back. Ideally, the direct ground lift should be performed by three EMTs; however, it can be done with only two. The direct ground lift is performed as follows **Skill Drill 35-7** :

1. Line up on one side of the patient with the first EMT at the patient's head, the second EMT at the patient's waist, and the third EMT at the patient's knees. All EMTs kneel on one knee, preferably the same knee.

2. The patient's arms should be placed on his or her chest if possible (Step 1).

3. The first EMT places one arm under the patient's neck and shoulders and cradles the patient's head. The first EMT then places the other arm under the patient's lower back.

4. The second EMT places one hand under the patient's waist, and the other under the knees.

5. The third EMT places one arm under the patient's knees and the other under the ankles.

6. On command, the team lifts the patient up to knee level as each EMT rests an arm on his or her knee (Step 2).

7. As a team and on command, each EMT rolls the patient in toward his or her chest. Again on command, the team stands and carries the patient to the stretcher (Step 3).

8. The steps are reversed to lower the patient onto the stretcher.

Extremity Lift

The **extremity lift** may also be used for patients with no suspected extremity or spinal injuries who are supine or in a sitting position. The extremity lift may be especially helpful when the patient is in a very narrow space or there is not enough room for the patient and a team of EMTs to stand side by side.

Communication is the key to success with this lift. You and your partner must coordinate your movements through direct verbal commands. You should perform the extremity lift as follows **Skill Drill 35-8** :

1. The first EMT kneels behind the patient's head as the second EMT kneels at the patient's feet. The two EMTs should be facing each other.

2. The patient's hands should be crossed over his or her chest.

3. The first EMT places one hand under each of the patient's armpits. The first EMT grasps the patient's

Skill Drill 35-7

The Direct Ground Lift

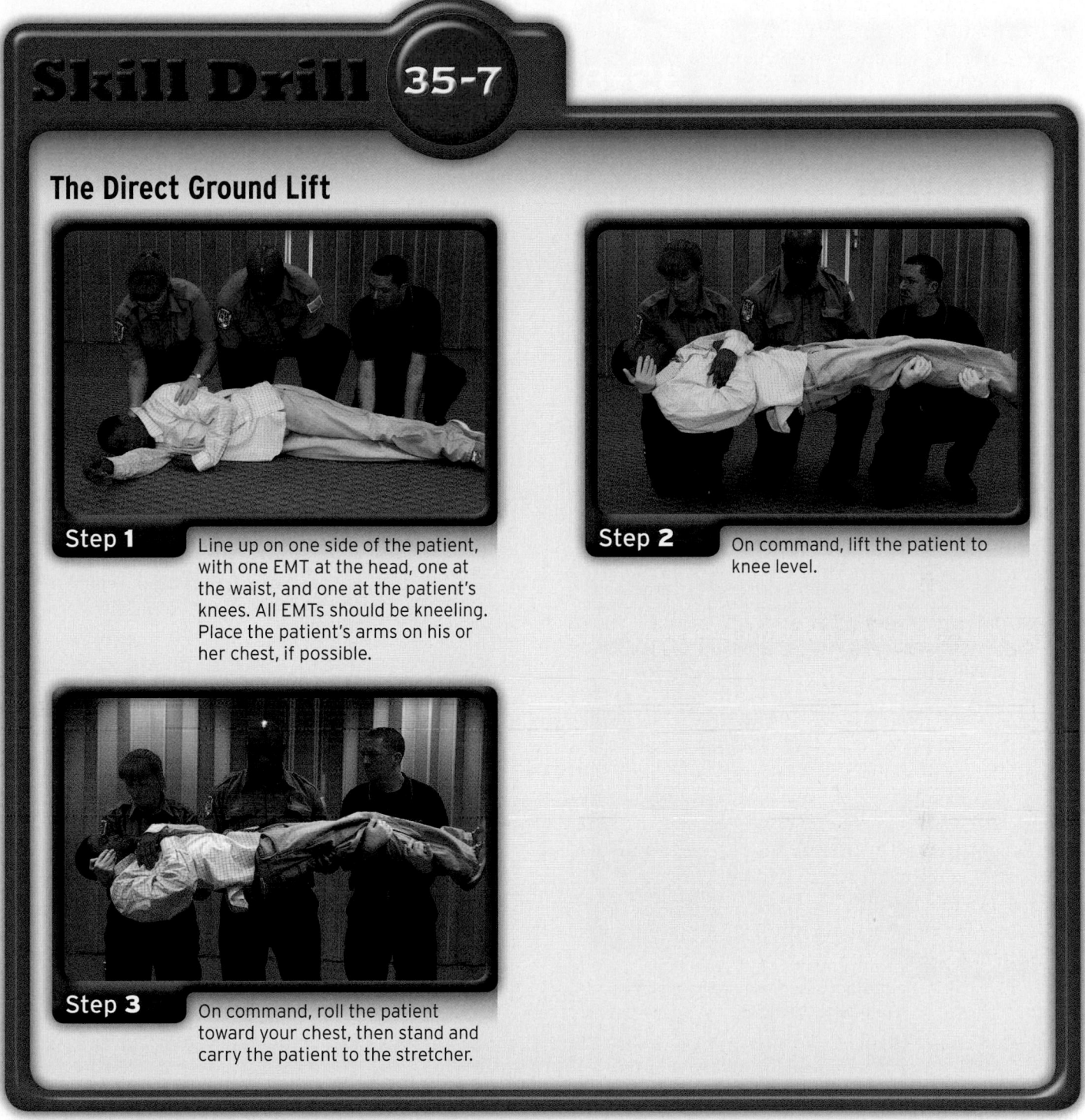

Step 1 Line up on one side of the patient, with one EMT at the head, one at the waist, and one at the patient's knees. All EMTs should be kneeling. Place the patient's arms on his or her chest, if possible.

Step 2 On command, lift the patient to knee level.

Step 3 On command, roll the patient toward your chest, then stand and carry the patient to the stretcher.

wrists or forearms and pulls the upper torso until the patient is in a sitting position (Step 1).

4. The second EMT moves to a position between the patient's legs, facing in the same direction as the patient, and slips his or her hands under the patient's knees (Step 2).

5. As the EMT at the head gives the command, both stand fully upright and move the patient to the stretcher (Step 3).

You will be less likely to injure yourself if you bend at the hips and knees and use your legs for lifting. However, this lift and carry method increases pressure on the patient's chest, so the patient may be uncomfortable in this position.

■ Transfer Moves

There are several ways to transfer the patient from a bed onto the stretcher.

Skill Drill 35-8

Extremity Lift

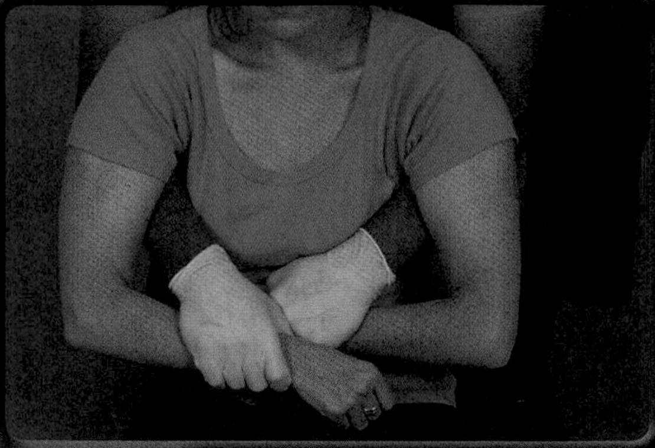

Step 1
The patient's hands are crossed over the chest. The first EMT grasps the patient's wrists or forearms and pulls the patient to a sitting position.

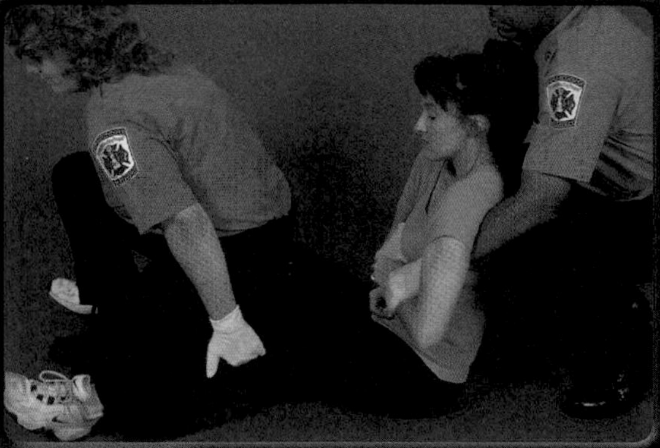

Step 2
The second EMT moves to a position between the patient's legs, facing in the same direction as the patient, and places his or her hands under the knees.

Step 3
Both EMTs rise to a crouching position. On command, both lift and begin to move.

Direct Carry

Transfer a supine patient from a bed to the stretcher using the direct carry method **Skill Drill 35-9**.

Skill Drill 35-9

1. Position the stretcher parallel to the bed, with the head of the stretcher at the foot of the bed. Be sure that you prepare the stretcher by unbuckling the straps and removing any other items from it.

Secure the stretcher to prevent movement. Both you and your partner should face the patient while standing between the bed and the stretcher.

2. You should slide one arm under the patient's neck and cup the patient's shoulder. Your partner should slide his or her hand under the patient's hip and lift slightly. You should then slide your other arm under the patient's back, and your partner should place both arms underneath the patient's hips and calves.

3. Slide the patient to the edge of the bed, and lift and curl the patient toward your chest (Step 1).

Skill Drill 35-9

Direct Carry

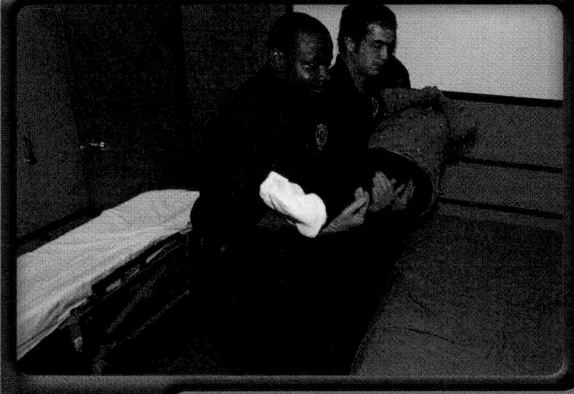

Step 1
Place the stretcher parallel to the bed with the patient's feet facing the head of the stretcher. Secure the stretcher to prevent movement. Face the patient while standing between the bed and the stretcher. Slide one arm under the patient's neck and cup the patient's shoulder. Your partner should slide his or her hand under the patient's hip and lift slightly. You should then slide your other arm under the patient's back, and your partner should place both arms underneath the patient's hips and calves.

Step 2
Lift the patient in a smooth, coordinated fashion. Slowly walk the patient around, and position him or her over the stretcher.

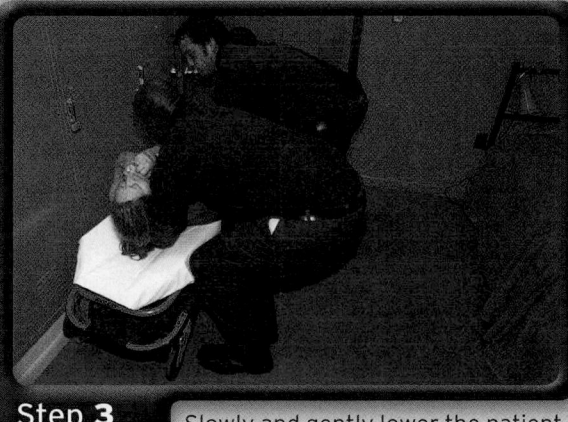

Step 3
Slowly and gently lower the patient onto the stretcher.

4. Slowly walk the patient around, and position him or her over the stretcher (Step 2).

5. Gently lower the patient to the stretcher (Step 3).

This carry can be performed more easily with three providers (as illustrated).

Draw Sheet Method

To move the patient from a bed onto a stretcher, use the draw sheet method. Place the stretcher next to the bed, making sure it is at the same height, or slightly higher than the bed and that the rails are lowered and straps are unbuckled. Be sure to hold or secure the stretcher to keep it from moving. Loosen the bottom sheet underneath the patient, or log roll the patient onto a blanket **Figure 35-17A**. Reach across the stretcher, and grasp the sheet or blanket firmly at the patient's head, chest, hips, and knees **Figure 35-17B**. Gently slide the patient onto the stretcher **Figure 35-17C**.

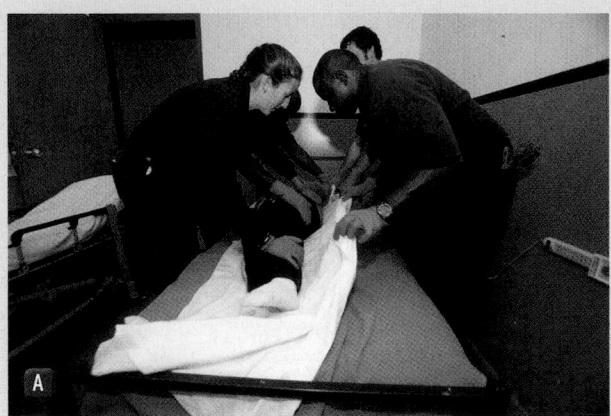

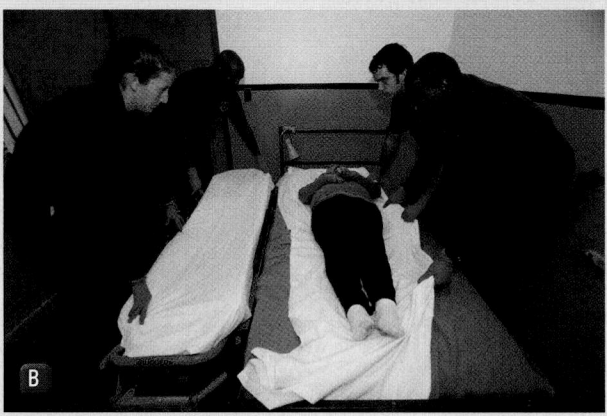

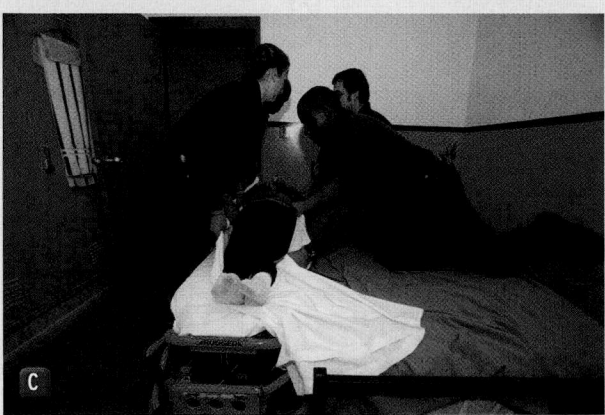

Figure 35-17 The draw sheet method. **A.** Log roll the patient onto a sheet or blanket. **B.** Bring the stretcher in parallel to the bed. Secure the stretcher. Gently pull the patient to the edge of the bed. **C.** Transfer the patient to the stretcher.

Other Carries

Other carries are performed in the following manner:

- Place a backboard next to the patient and, after using a log roll or slide to move the patient onto the backboard, secure the patient and lift and carry the backboard to the nearby prepared stretcher.
- Insert the halves of a scoop stretcher under each side of the patient, and fasten the two sides together. Lift and carry the patient to the nearby prepared stretcher. Follow the steps in **Skill Drill 35-10** to use a scoop stretcher. (Note that you can also log roll a patient onto a scoop stretcher that is already locked together.)

Skill Drill 35-10

1. With the scoop stretcher separated, measure the length of the scoop and adjust to the proper length **Step 1**.
2. Position the stretcher, one side at a time. One EMT lifts the patient's side slightly by pulling on the far hip and upper arm, while the other EMT slides the stretcher into place **Step 2**.
3. Lock the stretcher ends together by engaging their locking mechanisms one at a time and continue to lift the patient slightly as needed to avoid pinching **Step 3**.
4. Apply and tighten straps to secure the patient to the scoop stretcher before transferring to the stretcher **Step 4**.

- Assist an able patient to the edge of the bed, and place the patient's legs over the side, helping the patient to sit up. Move the stretcher so that its foot end touches the bed near the patient. Help the patient to stand and rotate so that he or she can sit down on the center of the stretcher. Lift the patient's legs, and rotate them onto the stretcher while your partner lowers the patient's torso onto the stretcher.

To avoid the strain of unnecessary lifting and carrying, you should use the draw sheet method or assist an able patient to the stretcher whenever possible.

To move a patient from the ground or the floor onto the stretcher, you should use one of the following methods:

- Lift and carry the patient to the nearby prepared stretcher using a direct body carry.
- Use a log roll or long-axis drag to place the patient onto a backboard, and then lift and carry the backboard to the stretcher. Place both the backboard and the patient onto the stretcher.

Skill Drill 35-10

Using a Scoop Stretcher

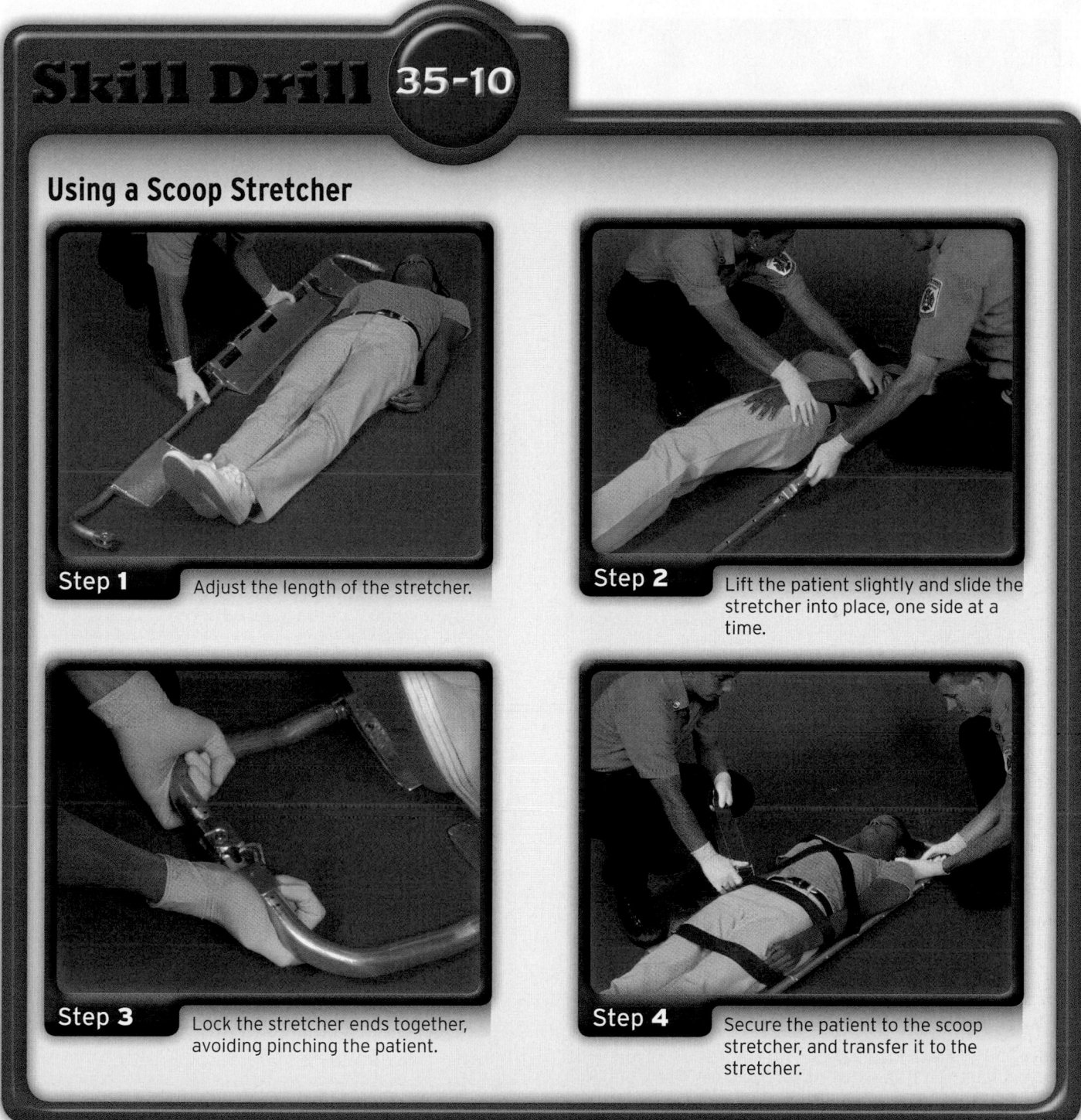

Step 1 Adjust the length of the stretcher.

Step 2 Lift the patient slightly and slide the stretcher into place, one side at a time.

Step 3 Lock the stretcher ends together, avoiding pinching the patient.

Step 4 Secure the patient to the scoop stretcher, and transfer it to the stretcher.

- Use a scoop stretcher.
- Log roll the patient onto a blanket, centering the patient on the blanket and rolling up the excess material on each side Figure 35-18A. Lift the patient by the blanket, and carry him or her to the nearby stretcher Figure 35-18B.

If a patient is sitting in a chair and cannot assist you, transfer the patient from the chair to a wheelchair Figure 35-19A and Figure 35-19B.

Geriatrics

Most patients transported by EMS are geriatric patients. For many older patients, the fear of illness and disability is ever present, and an emergency trip to the hospital can be a terrifying and disorienting experience. In addition, there are physiologic changes that occur with aging that require special attention on your part as an EMT.

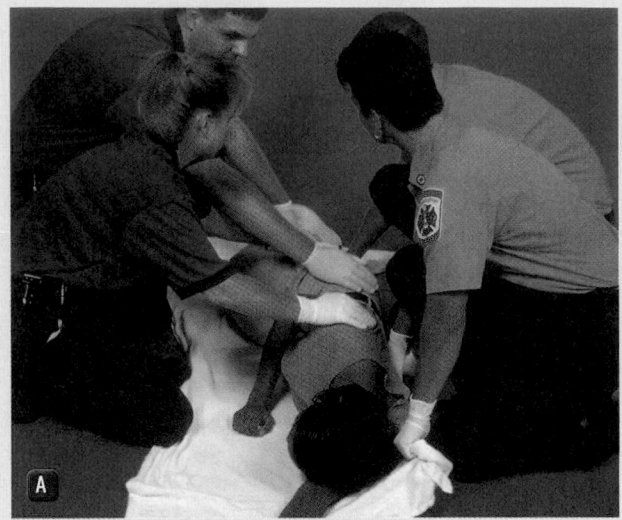

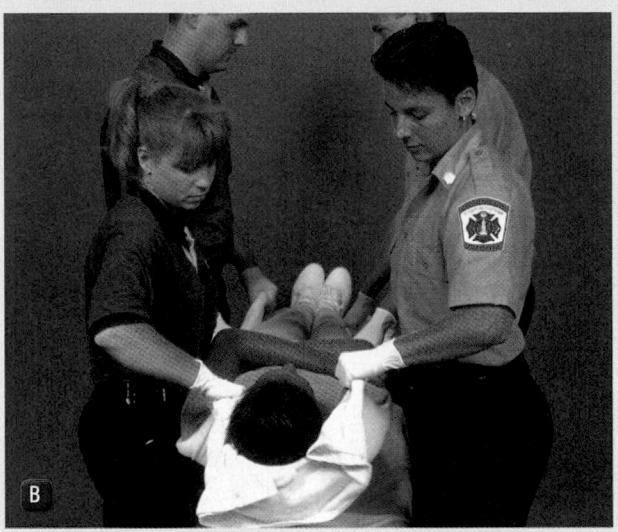

Figure 35-18 Log-rolling a patient on the ground. **A.** Log roll the patient onto a blanket. **B.** Lift the blanket, and transfer the patient to the stretcher.

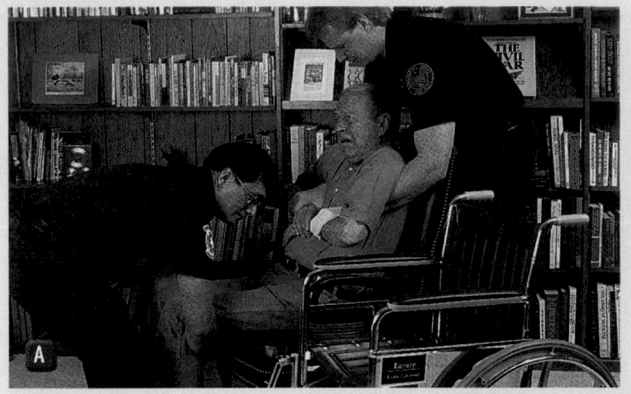

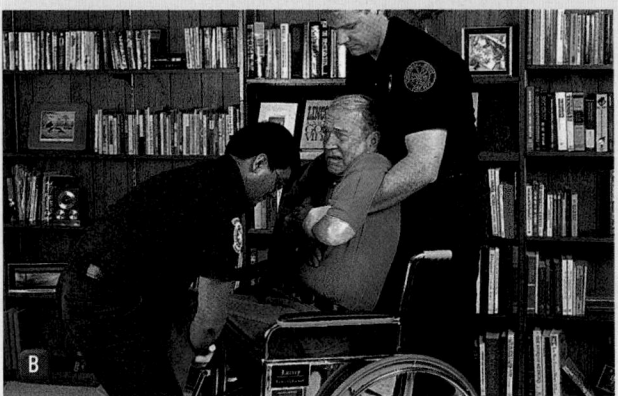

Figure 35-19 Moving a patient from a chair to a wheelchair. **A.** If present, any removable side pieces on the chair should be removed or placed in a position so as to not interfere. Slide your arms through the patient's armpits, and grasp the patient's crossed forearms. The second EMT grasps the patient's legs at the knees. **B.** Gently lift the patient into the locked wheelchair.

1. Skeletal changes: Brittle bones (osteoporosis), rigidity, and spinal curvatures (kyphosis and spondylosis) **Figure 35-20** present special challenges in packaging and moving older patients. Many patients cannot lie supine on a backboard without causing additional injury, such as fractures, pressure sores, and skin breakdown. Special care and creativity must be used in immobilizing such patients. For example, a patient with spinal curvature may have to be placed on his or her side and immobilized in place with towel and blanket rolls to prevent exacerbating his or her injuries. Be sure to consult your local protocols and medical director about alternative ways of immobilizing such patients.

2. Fear: A sympathetic and compassionate approach can go a long way in allaying the natural fears many older patients experience when interacting with

caregivers. Slow down, explain, and anticipate: these actions can help you gain an older patient's cooperation and take some of the anxiety out of the process of packaging and transportation. Imagine how frightening being strapped to a stretcher and carried down a flight of stairs can be to an individual who lives in constant fear of falls and broken bones.

Special Populations

Cover the patient with a blanket to protect privacy and keep the patient warm.

Bariatrics

Estimates suggest that approximately 100 million adults in the United States are at least overweight or are obese. Approximately 35% of women and 31% of men older than 19 years are obese or overweight. The numbers among children are even more imposing. The prevalence

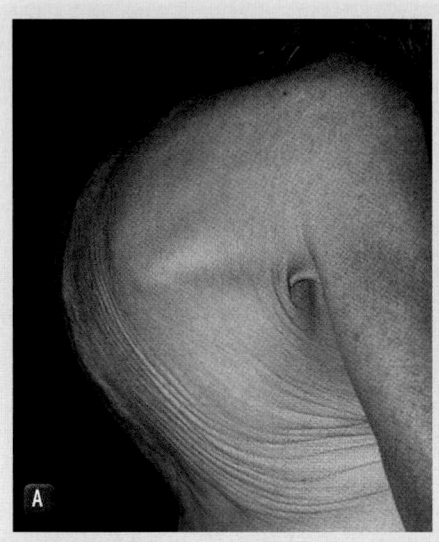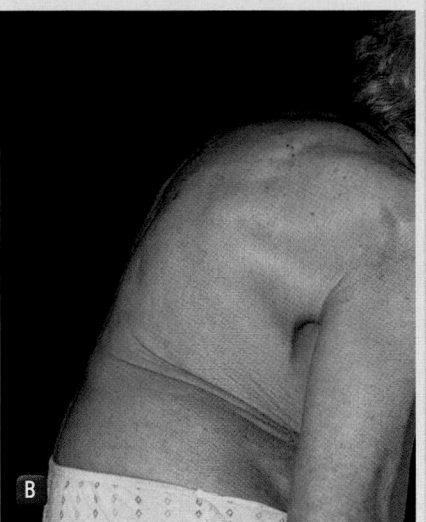

Figure 35-20 Skeletal changes. **A.** Kyphosis. **B.** Spondylosis.

of obesity in children in the United States has increased markedly. Approximately 20% to 25% of children are either overweight or obese, and the prevalence is even greater in some minority groups, including Pima Indians, Mexican Americans, and African Americans. Conservative estimates suggest that the management of obesity consumes approximately $100 billion yearly, without factoring in the costs of various commercial dietary and weight loss programs.

Americans are becoming so large that a new field of medicine has been named for the care of obese people. **Bariatrics** is the branch of medicine concerned with the management (prevention or control) of obesity and allied diseases. It comes from the Greek words *baros*, weight, and *iatreia*, medical treatment. Because there is a direct correlation between the degree of obesity and the frequency and severity of health problems, the larger the patient the more likely he or she is to need emergency treatment and transportation. This problem is taking an increasing toll on the health and functioning of EMTs because back injuries account for the largest number of missed days of work and both temporary and permanent disability.

Although ambulance stretcher and equipment manufacturers are producing equipment with ever higher capacities, this does not address the danger to the users of that equipment. Although European ambulance manufacturers regularly install mechanical lifts on their units, these are not as common in the United States.

■ Patient-Moving Equipment

The modern stretcher is available in a number of different models, which may include different features Figure 35-21 . Before going on a call, you should be fully familiar with the specific features of the stretcher that your ambulance carries. You must know where the controls to adjust and lock each feature are located and how each works.

You are the Provider: PART 4

As your team prepares to lift the patient, you reassess him. He is still conscious and alert but is in significant pain. The ambulance stretcher has already been prepared and is at the base of the front stairs.

Recording Time: 20 Minutes	
Level of consciousness	Conscious and alert
Respirations	20 breaths/min; adequate depth
Pulse	100 beats/min; strong and regular
Skin	Pink, warm, and moist
Blood pressure	150/88 mm Hg
Sao$_2$	97% (on room air)

6. How is a patient's weight distributed when he or she is on a carrying device? Why is it important to know this?

7. How will you and your team safely carry the patient down the stairs?

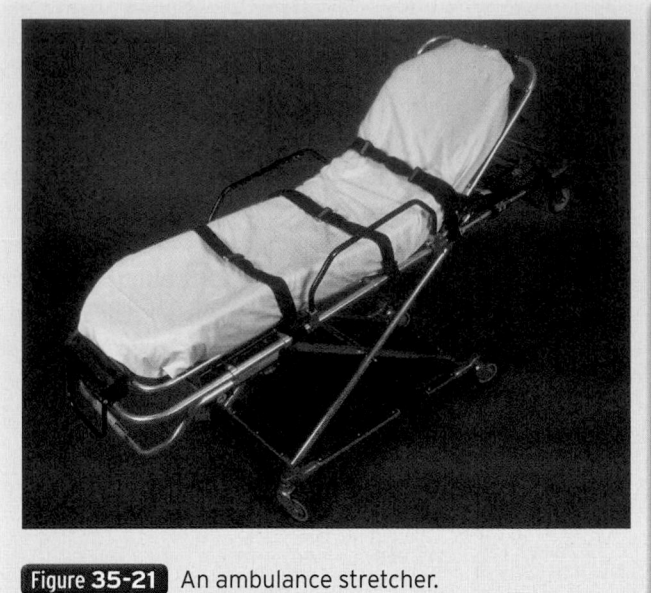

Figure 35-21 An ambulance stretcher.

The stretcher has a specific head end and foot end. The stretcher has a strong horizontal rectangular, tubular metal main frame to which all of its other parts are attached. The stretcher should be pulled, pushed, and lifted only by its main frame or handles, which are attached to the main frame specifically for this purpose.

On most models, a second tubular frame made up of three sections is attached within or above the main frame. A metal plate is fastened to each of the three sections between its sides. This plate serves as the platform on which the stretcher mattress and patient are supported. The head section runs from the head end of the stretcher to near the center of the stretcher, where the patient's hips will be. Hinges at the area where the hips will be allow the head end to be elevated and the patient's back to be positioned at any desired angle from flat to fully upright. The head end of the stretcher is designed to be elevated or moved down only when a tilt control is purposely released. At all other times, the back will remain locked at the position in which it was placed. The frame and plates that lie from the hips to the foot end of the stretcher are divided into two hinged sections. These sections may be connected so that the foot end can be drawn in toward the knees, causing the frame and plates to hinge upward under the patient's knees to elevate them as desired. This feature is not found in all models.

A retractable guardrail is attached along the central portion of the main frame of the stretcher at each side and is lowered out of the way when a patient is being loaded onto the stretcher. Once the patient has been properly placed on the stretcher, the handle is drawn up and locked in an elevated position perpendicular to the surface of the stretcher. The patient cannot roll off either side of the stretcher even if a securing strap becomes released. The guardrail at each side can be lowered only if its locking handle is released.

Words of Wisdom

Ensure a thorough patient care report by including details of how you moved the patient. For example "Moved patient to stretcher with draw-sheet lift."

The underside of the main frame of the stretcher is supported on a folding undercarriage that has a smaller horizontal rectangular frame and four large rubber casters at its bottom end. The folding undercarriage is designed so that the litter can be adjusted to any height from about 12″ above the ground, which is the desired height when the stretcher is secured in the ambulance, to 32″ to 36″ above the ground, which is the desired height when the stretcher is being rolled. Because you are able to lock the stretcher at any height between its lowest height and its fully extended height, it can be locked at the same height as any bed or examining table to allow the patient to be slid from one to the other. This permits you to transfer the patient without the need for any additional lifting. The controls for folding the undercarriage are designed so that the stretcher remains locked at its present height when the controls are not being activated. As an additional safety feature on most stretchers, the main frame must be slightly lifted so that the undercarriage becomes unweighted before it will fold, even if the control is pulled. Therefore, if the handle is accidentally pulled, the elevated stretcher will not suddenly drop. Controls for elevating and lowering most stretchers are located at the foot end and at one or both sides. You and your partner must use the proper lifting mechanics to lift the wheeled ambulance stretcher.

The mattress on a stretcher must be fluid resistant so that it does not absorb any type of potentially infectious material, including water, blood, or other body fluid.

■ The Wheeled Ambulance Stretcher

The wheeled ambulance stretcher, or stretcher or gurney, is the most commonly used device to move and transport patients. Only when you must transport two patients in the same ambulance should it be necessary to transport one patient on a folding stretcher or backboard placed on the long squad bench.

Most patients are placed directly on the stretcher. However, you will need to place and secure patients with a possible spinal injury or multiple system trauma onto a backboard. Patients who may need cardiopulmonary resuscitation or must be carried down (or up) a flight of stairs while supine should also be placed on a backboard. The backboard and patient are then secured onto the stretcher.

You can use a stair chair to carry a patient who can tolerate being in a sitting position down a flight of stairs to the prepared stretcher, which is waiting on the ground floor. You should then transfer the patient from the chair to the stretcher.

In most instances, it is best if you push the head of the stretcher while your partner guides the foot of the stretcher. When the stretcher must be carried, it is best if four rescuers are available to carry it. There is more stability with a four-person carry, and the carry requires less strength. One EMT should be positioned at each corner of the stretcher to provide an even lift. A four-person carry is much safer if the stretcher must be moved over rough ground. If only two EMTs are available, or if limited space allows room for only two EMTs to carry the stretcher, there is a risk that the stretcher will become unbalanced. In a two-person carry, the two EMTs should stand facing each other, with one person at the head end of the stretcher and the other at the foot end. With this type of carry, one EMT will have to walk backward.

■ Bariatric Stretchers

Because of the large girth of bariatric patients, they may not fit comfortably on the standard wheeled stretcher. As a result, a specialized type of wheeled stretcher has been developed, called the bariatric stretcher **Figure 35-22** . This type of stretcher is similar in design to the common wheeled stretcher; however, it has several differences. Bariatric stretchers typically have a wider patient surface area to allow for increased comfort, as well as increased dignity when transporting the patients. Bariatric stretchers also have a wider wheelbase, allowing for increased stability when rolling the patient over uneven terrain.

Bariatric stretchers are also sometimes equipped with optional features such as a tow package, which allows an ambulance-mounted winch to assist in loading the patient into the ambulance, decreasing the potential for EMT back injuries. Another optional feature is telescoping side lift handles, which allow for increased leverage when lifting with multiple responders. However, the most important feature of the bariatric stretcher is the increased weight-lifting capacity. Typical wheeled ambulance stretchers, depending on manufacturer ratings, are rated to a maximum weight of 650 lb. Bariatric stretchers are usually rated to 850 lb to 900 lb.

■ Pneumatic and Electronic Powered Wheeled Stretchers

In an effort to decrease the potential for back injuries to EMS providers, manufacturers have developed pneumatic and electronic stretchers. Similar in appearance to conventional wheeled stretchers, electronic stretchers are battery operated and have electronic controls to facilitate raising and lowering of the undercarriage at the touch of a button **Figure 35-23** . A drawback to the powered wheeled stretcher is that by adding the electronic controls, as well as the associated equipment, the weight of the stretcher is increased, typically 75 lb to 100 lb. Coupled with the weight of the patient on the loaded stretcher, this creates a potential hazard when transporting the patient over uneven terrain or down one to two steps in the front of a residence.

■ Loading a Wheeled Stretcher into an Ambulance

Whenever a patient has been placed onto the stretcher, one EMT must hold the main frame to make sure that it cannot roll. When the stretcher is elevated, the main frame and the patient extend considerably beyond the

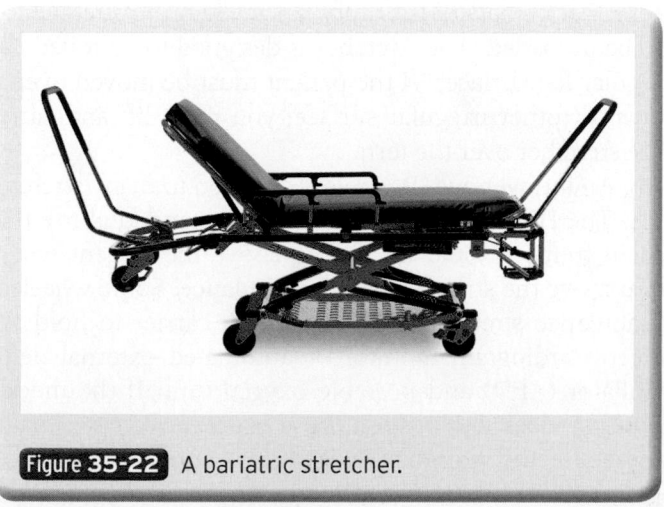

Figure 35-22 A bariatric stretcher.

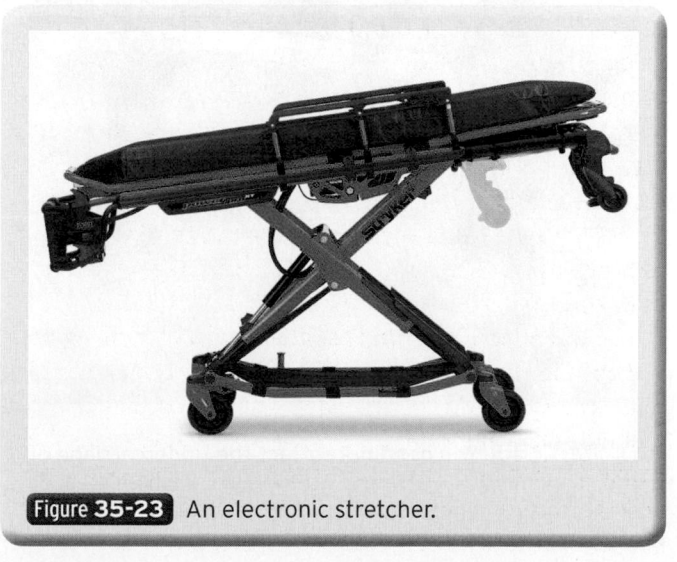

Figure 35-23 An electronic stretcher.

wheels at both the head end and foot end of the stretcher. Therefore, whenever a patient is on an elevated stretcher, you must ensure that it is held firmly between two hands at all times so that even if the patient moves, the stretcher cannot tip **Figure 35-24**.

If the loaded stretcher must be carried down a short flight of steps, be sure to first retract the undercarriage; however, this is not necessary when the stretcher must be lifted over a curb, a single step, or an obstacle of a similar height **Figure 35-25**. Remember, if the patient must

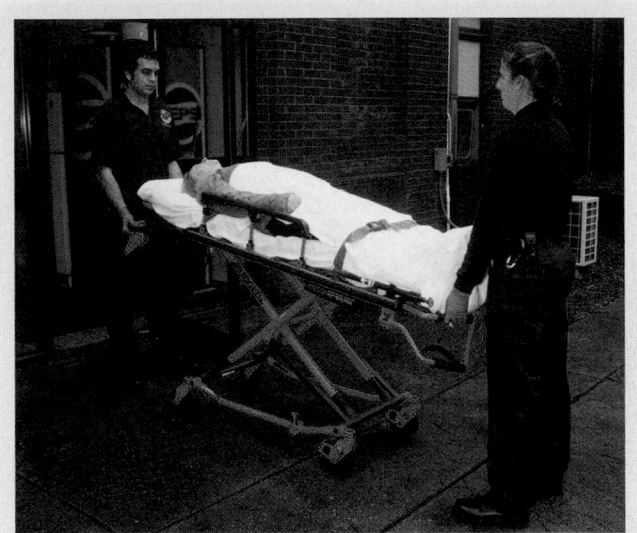

Figure 35-24 Make sure that you hold the main frame of the stretcher when it is elevated so that even when the patient moves, the stretcher does not tip.

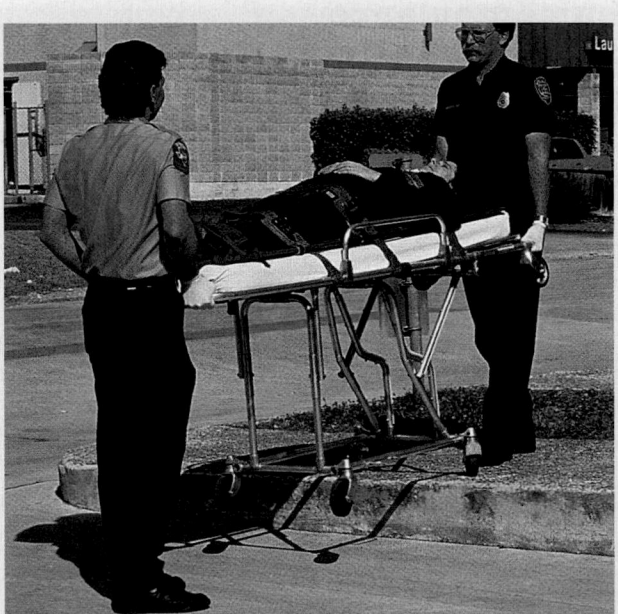

Figure 35-25 You need not retract the undercarriage of the stretcher when lifting it over a curb, a single step, or an obstacle of similar height.

be carried up or down a full flight or several flights of stairs, you should prepare the stretcher and leave it on the ground floor at the bottom (or top) of the stairs. Use a backboard or stair chair to carry the patient up or down the stairs to the waiting stretcher.

Follow these steps to load the stretcher into an ambulance **Skill Drill 35-11**:

Skill Drill 35-11

1. Tilt the head end of the main frame upward, and place it into the patient compartment with the wheels on the floor. The two additional wheels that extend just below the head end are attached to the main frame and will enable this movement **Step 1**.

2. With the patient's weight supported by these two head-end wheels and the EMT at the foot end of the stretcher, move to the side of the main frame and release the undercarriage lock to lift the undercarriage up to its fully retracted position. The wheels of the undercarriage and the two on the head end of the main frame will now be on the same level **Step 2**.

3. Simply roll the stretcher the rest of the way into the back of the ambulance, where it will rest on all six wheels **Step 3**.

4. Secure the stretcher in the ambulance with the strong clamps that fasten around the undercarriage when the stretcher is pushed into them. The clamps are located in a rack on the floor or side of the patient compartment **Step 4**.

The clamps will hold the stretcher in place until they are released at the hospital. You can control and release the clamps with a single handle that is positioned so that you can activate it when standing on the ground at the open back doors of the ambulance when the stretcher is to be unloaded. The stretcher is designed to be rolled on regular flat surfaces. If the patient must be moved over a lawn or other irregular surface, you must lift and carry the stretcher over the terrain.

An intravenous (IV) pole is attached to many stretchers. The IV pole can be unfolded or extended above the main frame to hold an IV bag above the patient while you move the stretcher to the ambulance. Some wheeled ambulance stretchers even include a carrier to hold an electrocardiogram monitor or automated external defibrillator (AED) and portable oxygen unit. If the model you use does not include these features, you will have to secure the portable oxygen unit and electrocardiogram monitor or AED to the top surface of the stretcher

Skill Drill 35-11

Loading a Stretcher into an Ambulance

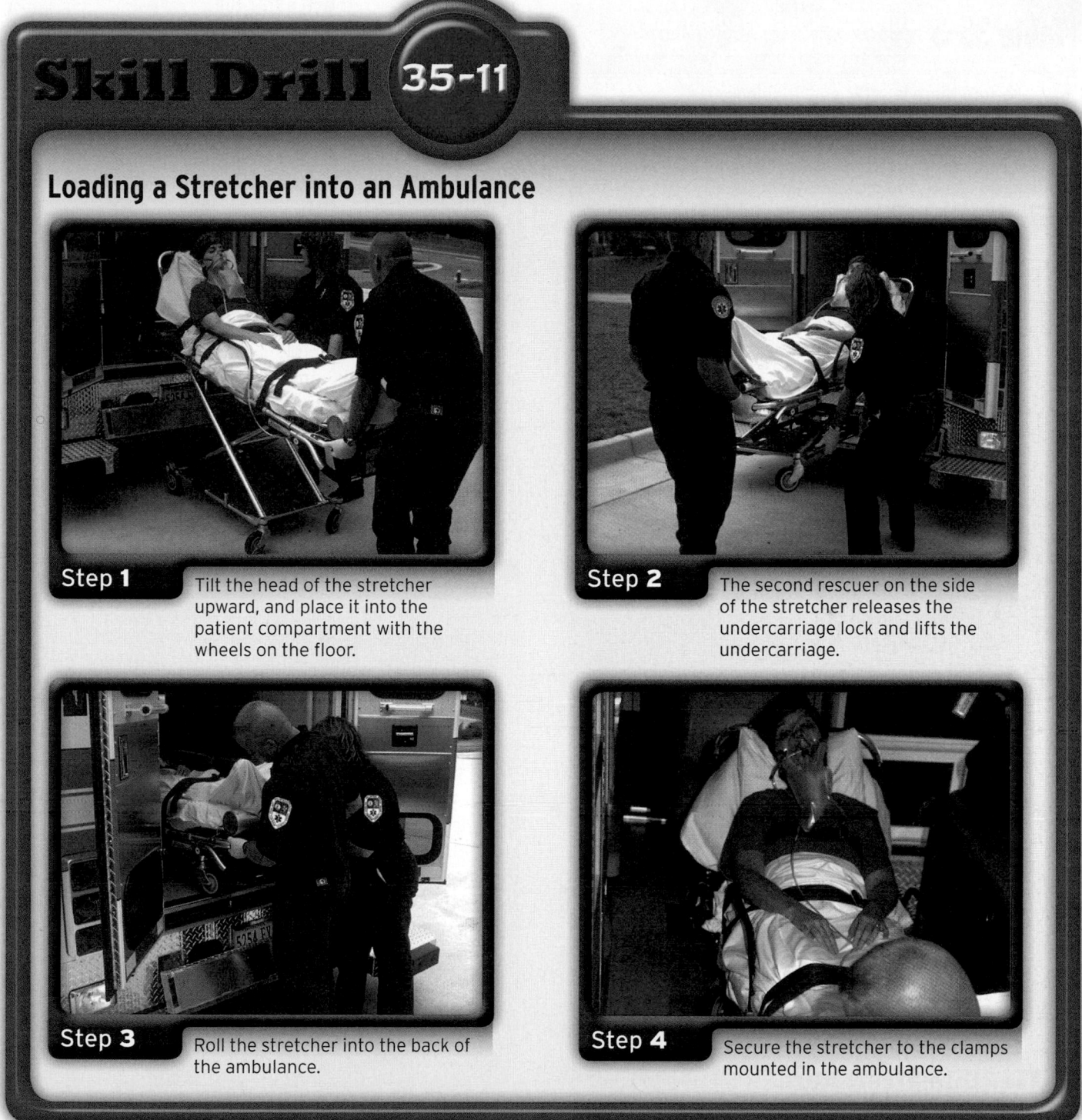

Step 1 Tilt the head of the stretcher upward, and place it into the patient compartment with the wheels on the floor.

Step 2 The second rescuer on the side of the stretcher releases the undercarriage lock and lifts the undercarriage.

Step 3 Roll the stretcher into the back of the ambulance.

Step 4 Secure the stretcher to the clamps mounted in the ambulance.

mattress at the patient's legs, remembering that these items will add excess weight when attempting to lift the stretcher.

The extra wheels below the head end of the main frame of the stretcher are not featured on some older or less expensive wheeled ambulance stretchers. These stretchers are not self-loading. When you reach the back of the ambulance with such a stretcher, you must lower it until the undercarriage is in its lowest retracted position and then, with you and your partner at each side of the stretcher, lift it to the height of the floor of the ambulance and roll it into the track that locks it into place. Table 35-3 shows the guidelines that you must follow to load the stretcher into the ambulance.

■ Portable/Folding Stretchers

A **portable stretcher** is a stretcher with a strong rectangular tubular metal frame and rigid fabric stretched across it Figure 35-26. Portable stretchers

Table 35-3	Guidelines for Loading the Stretcher into the Ambulance

- Make sure there is sufficient lifting power.
- Follow the manufacturer's directions for safe and proper use of the stretcher.
- Make sure that all stretchers and patients are fully secured before you move the ambulance.

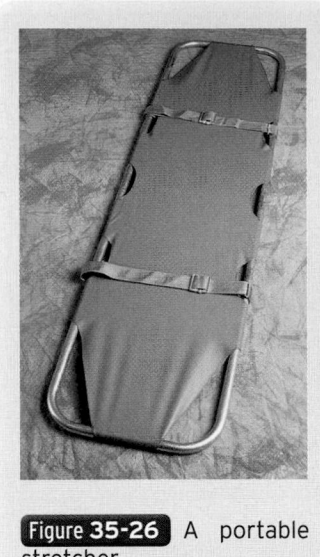

Figure 35-26 A portable stretcher.

do not have a second multipositioning frame or adjustable undercarriage. Some models have two wheels that fold down about 4″ underneath the foot end of the frame and legs of a similar length that fold down from the head end at each side. The wheels make it easier to move the loaded stretcher. The legs should not be used as handles.

Some portable stretchers can be folded in half across the center of each side so that the stretcher is only half its usual length during storage. Many ambulances carry a portable stretcher to use if a patient is in an area that is difficult to reach with a wheeled ambulance stretcher or a second patient must be transported on the squad bench of the ambulance.

A portable stretcher weighs much less than a wheeled stretcher and does not have a bulky undercarriage. However, because most models do not have wheels, you and your team must support all of the patient's weight and any equipment along with the weight of the stretcher.

■ Flexible Stretchers

Several types of **flexible stretchers**, such as the Sked®, Reeves®, and Navy stretcher, are available and can be rolled up across either the stretcher's width or, in the case of the Sked, its length, so that the stretcher becomes a smaller tubular package for storage and carrying **Figure 35-27** . When you must carry the equipment a considerable distance from the nearest place that the ambulance can be located, this is an important consideration. A flexible stretcher forms a rigid stretcher that conforms around the patient's sides and does not extend beyond them. When these stretchers are extended, they are particularly useful

when you must remove a patient from or through a confined space. The Sked stretcher can also be used if the patient must be belayed or rappelled by ropes.

The flexible stretcher is the most uncomfortable of all the various devices; however, it provides excellent support and immobilization. When the stretcher is wrapped around the patient and the straps are secured, the patient is completely immobilized. The stretcher can then be lowered by rope or slid down a flight of stairs by resting it on the front edge of each step.

Figure 35-27 A flexible stretcher.

■ Backboards

Backboards are long, flat boards made of rigid, rectangular material **Figure 35-28** . Backboards were originally made of wood but are now made of other materials as well, mostly plastic. They are used to carry patients and to immobilize supine patients with suspected spinal injury or other multiple trauma. Backboards can also be used to move patients out of awkward places. They are 6′ to 7′ long and are commonly used for patients who are found lying down. Parallel to the sides and ends of the backboard are a number of long holes that are about ½″ to 1″ from the outer edge. These holes form handles and handholds so that the board can be easily grasped, lifted, and carried. The handles and adjacent holes also allow straps used to secure and immobilize the patient to the backboard to be secured to each side and end of the backboard at any needed location.

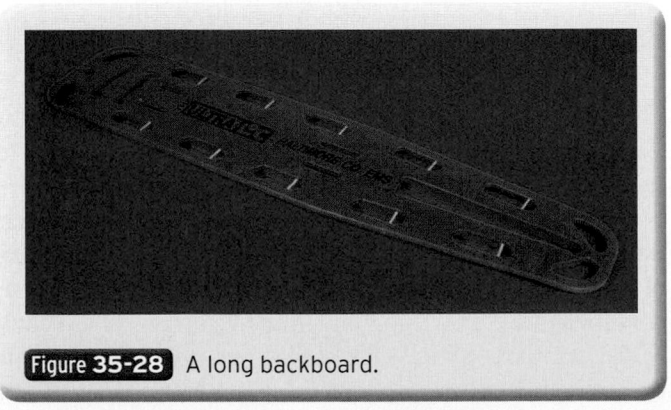

Figure 35-28 A long backboard.

For many years, backboards were made of thick marine plywood whose surface was sealed with polyurethane or another marine varnish. Wooden backboards are still used in some places. If your service uses wooden backboards, you must follow infection control procedures before you can reuse the backboards. Where wooden backboards are no longer used, they have generally been stored so that they will be available in the event of a mass-casualty situation. Newer backboards are made of plastic materials that will not absorb blood or other infectious substances.

You can use a short backboard, or half-board, to immobilize the torso, head, and neck of a seated patient with a suspected spinal injury until you can immobilize the patient on a long backboard. Short backboards are 3′ to 4′ long. The original short wooden backboard has generally been replaced with a vest-type device that is specifically designed to immobilize the patient until he or she is moved from a sitting position to a supine position on a backboard **Figure 35-29**. The vest-type devices are easier to use than the wooden backboard.

◼ Basket Stretchers

You should use a rigid **basket stretcher**, often called a Stokes litter, to carry a patient across uneven terrain from a remote location that is inaccessible by ambulance or other vehicle **Figure 35-30**. If you suspect that the patient has a spinal injury, you should first immobilize him or her on a backboard and then place the backboard into the basket stretcher. Once you have reached the ambulance and wheeled ambulance stretcher, you can remove the patient and backboard from the basket stretcher and place them on the stretcher.

Figure 35-30 A basket stretcher.

Basket stretchers are made of plastic with an aluminum frame or have a full steel frame that is connected by a woven wire mesh. The wire basket is very uncomfortable for the patient unless the wire is padded. Either type can be used to carry a patient across fields, rough terrain, or trails or on a toboggan, boat, or all-terrain vehicle. Basket stretchers surround and support the patient, yet their design allows water to drain through holes in the bottom. Basket stretchers are also used for technical rope rescues and some water rescues. Not all basket stretchers are rated or appropriate for each of these specialized rescue uses. The types of basket stretchers that are acceptable for specialized rescue must be determined by individuals with additional special training.

◼ Scoop Stretcher

The **scoop stretcher**, or orthopaedic stretcher, is designed to be split into two or four pieces **Figure 35-31**. These sections are fitted around a patient who is lying on the ground or another relatively flat surface. The parts are reconnected, and the patient is lifted and placed on a long

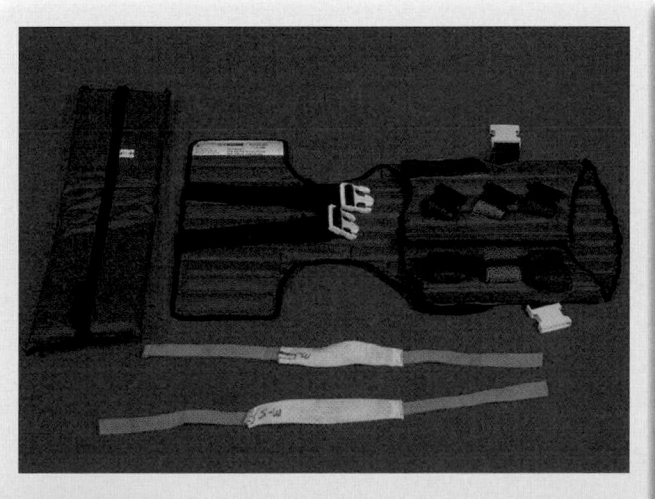

Figure 35-29 The KED device is a vest-type immobilization device.

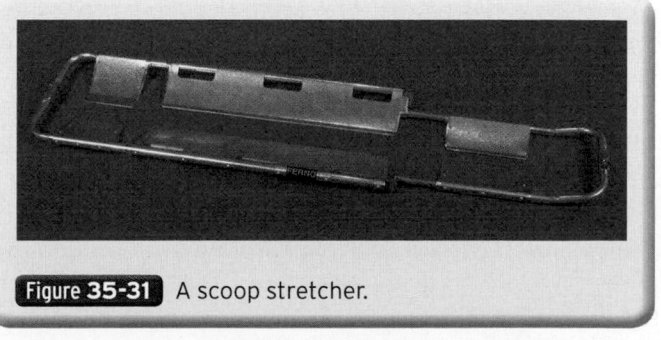

Figure 35-31 A scoop stretcher.

backboard or stretcher. A scoop stretcher may be used for patients who have been struck by a motor vehicle.

A scoop stretcher is efficient; however, both sides of the patient must be accessible. You must also pay special attention to the closure area beneath the patient so that clothing, skin, and other objects are not trapped. As with the long backboard, you must fully stabilize and secure the patient before moving him or her; however, you cannot slip a scoop stretcher under the long axis of the patient's body. Scoop stretchers are narrow, well constructed, and compact and have excellent body support features but are not adequate when used alone for standard immobilization of a spinal injury. You and your team should practice often with a scoop stretcher to be ready for using it with a patient. It is important to remember that a scoop stretcher has internal supports running throughout its length; this feature prohibits hospitals from being able to obtain an x-ray while the patient is secured to it, often mandating another move to a standard backboard prior to obtaining the x-ray.

■ Stair Chairs

Stair chairs are folding aluminum frame chairs with fabric stretched across them to form a seat and seat back **Figure 35-32** . They have fold-out handles to help

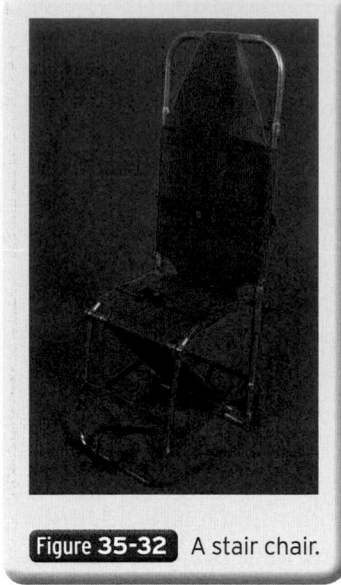

Figure 35-32 A stair chair.

you carry their head and foot ends up or down a flight of stairs, and most have rubber wheels in the back with casters in front so that they can be rolled along the floor and make turns. Stair chairs serve as an adjunct for moving a patient up or down stairs to the ground floor, where the prepared wheeled ambulance stretcher is waiting. You can roll the stair chair on the floor until you reach the stairwell, then carry it (rather than roll and bump it) up or down the stairs. Once you reach the ground floor, you can roll it to the waiting stretcher and assist or lift the patient onto the stretcher.

Be sure to follow the manufacturer's directions for maintenance, inspection, repair, and upkeep for any device that you use as patient-handling equipment.

You are the Provider: PART 5

You move the patient down the stairs, place him onto the stretcher, load him into the ambulance, and begin transport to the hospital. You place padding under the patient's back, after which he expresses some relief from the pain. You reassess the patient and then call in your radio report.

Recording Time: 27 Minutes	
Level of consciousness	Conscious and alert
Respirations	20 breaths/min; adequate depth
Pulse	98 beats/min; strong and regular
Skin	Pink, warm, and moist
Blood pressure	148/84 mm Hg
Sao$_2$	98% (on room air)

On arrival at the hospital, the patient tells you that he is still in pain, although it is not as severe. After moving him from your stretcher to the hospital bed, you give your verbal report to the staff nurse and return to service.

8. How can you minimize the risk of injury while moving a patient on a wheeled ambulance stretcher?

■ Neonatal Isolettes

When you are requested to transport a neonatal patient from one hospital to another, the common wheeled ambulance stretcher will not suffice. To safely transport a neonatal patient, the patient must be placed inside of an isolette, sometimes referred to as an incubator. The isolette keeps the neonatal patient warm with moistened air in a clean environment and helps to protect the infant from noise, drafts, infection, and excess handling. The specialized transport devices come in one of two forms: the isolette is placed directly on top of the wheeled stretcher and secured with seatbelts, or a freestanding type of isolette is secured into the back of the ambulance, taking the place of the standard stretcher Figure 35-33 .

■ Decontamination

It is essential that you decontaminate your equipment after use, for your own safety, the safety of the crew using the equipment after you, and the safety of your patients, to prevent the spread of disease. Just as we expect a hospital bed to be disinfected after the last patient, so too with your stretcher and other transport equipment. Know and follow your local standard operating procedures for disinfecting equipment after each call.

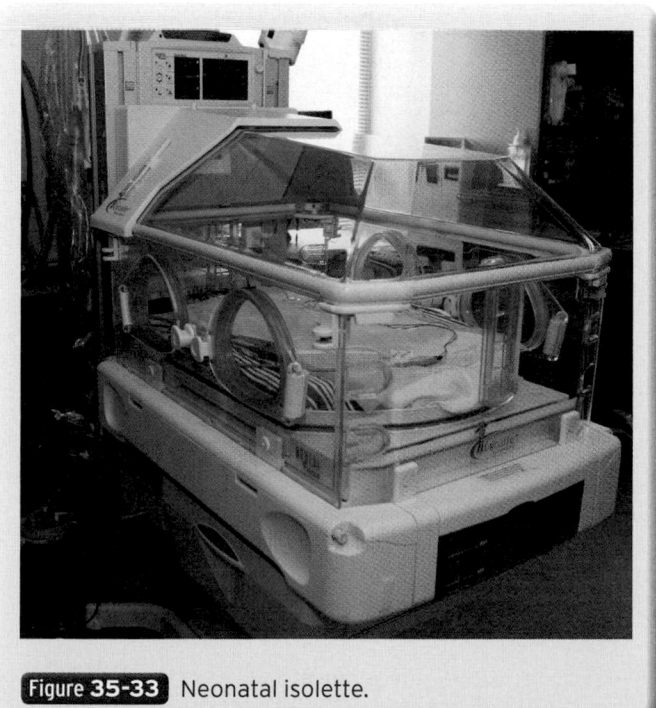

Figure 35-33 Neonatal isolette.

■ Medical Restraints

While not a common occurrence, there may be a time when you are called on to physically restrain a patient. After evaluating the patient for correctable causes of combativeness, such as head injury, hypoxia, or hypoglycemia, the decision needs to be made as to whether to restrain your patient. There may be consequences for either applying the restraints, or failing to restrain a patient that should have been restrained. Local protocols should be consulted prior to applying restraints, and in some jurisdictions, medical control authorization is needed before an EMT can apply restraints.

The decision to restrain a patient is not one to be taken lightly; however, if the patient is posing a danger to you, your crew, himself or herself, or to bystanders, the application of physical restraints needs to be considered. However, before you take action to restrain the patient, you should attempt to speak to the patient in a calming manner, while remaining firm in your requests. If that does not work and the patient continues to be combative, a plan needs to be developed among all responders present as to who will do what, when it will happen, and how you will accomplish the restraint.

There should be a minimum of five personnel present to assist in the restraint of a combative patient, one for each extremity, and one for the head. One EMT should be established as the team leader, the one who will give commands. A plan to restrain the patient should be developed and agreed on by all team members. A patient who is caught off guard and unsuspecting allows for a decreased likelihood of injury to the responders.

When preparing to secure the patient on the stretcher, it is of the utmost importance to have the patient in the supine position. If the patient is placed in a prone position, a condition called positional asphyxia could develop. In prone positioning, the increased weight on the patient's lungs and his or her inability to fully expand the thoracic cavity could render the patient unable to breathe, creating a preventable, life-threatening emergency.

If a patient for whom use of medical restraints is indicated is in the supine position, some type of humane restraint should be applied to each extremity, such as triangle bandages, roller gauze, soft commercially available disposable restraints, or leather restraints. Preferably the patient should be restrained onto a backboard, which allows for easy movement should the patient begin to vomit. However, if it is impractical or inadvisable to secure the patient to the backboard, then secure the patient to the stretcher. Regardless of whether the patient is secured to a backboard or the stretcher, one arm should be secured above the patient's head and one arm should be secured at the patient's side. This technique will not give the patient the leverage to break free from the restraints. After the upper extremities are secured, each leg should be secured as well.

After application of the restraints, it is imperative to assess the patient's distal circulation (pulse and motor and sensory function) **Figure 35-34**. Document your findings on the patient care report.

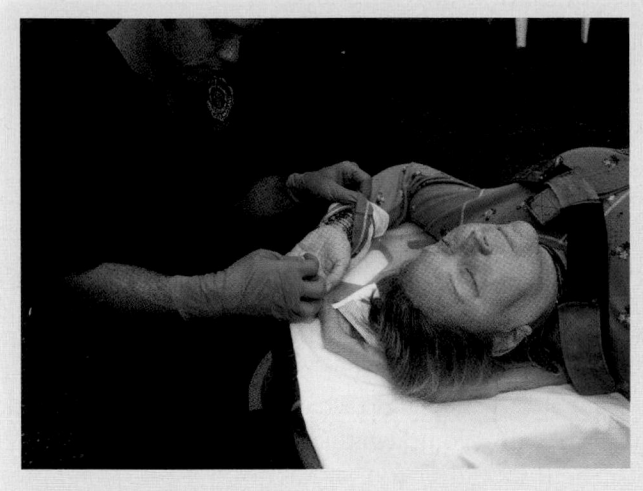

Figure 35-34 After application of the restraints, it is imperative to assess the patient's distal circulation.

Personnel Considerations

In an effort to minimize personnel injuries, prior to moving any patient, a complete plan needs to be developed and discussed among the crew. Some questions to ask are the following: Am I physically strong enough to lift/move this patient? Many back injuries are the result of poor physical condition. As an EMT, you will be required to assist in the movement of patients. Do your best to maintain a level of physical fitness. Other questions may include the following: Is there adequate room to get the proper stance to lift the patient? Do I need additional personnel for lifting assistance? The answers to these questions need to be evaluated prior to moving your patient. Remember that injured rescuers cannot help anyone.

You are the Provider: SUMMARY

1. What factors should you consider before lifting or moving a patient?

There are numerous factors to consider before lifting and/or moving any patient. A hasty patient move can result in injury to yourself, your partner, and the patient. The first factor to consider is the most obvious—the patient's weight. How much does the patient weigh? Does he or she weigh more than you and your partner can safely lift?

Adult patients commonly weigh between 100 and 210 lb. If you use proper technique, you and one other EMT should be able to safely lift this weight. However, you must be aware of your own lifting and moving abilities, as well as the abilities of your partner. You should know how much you can comfortably and safely lift and should not attempt to lift a proportional weight (the share of the weight that you will bear) that exceeds this amount.

You should not attempt to lift or move a patient who weighs more than 250 lb with fewer than four rescuers, regardless of individual strength. If the patient weighs more than 250 lb, or otherwise weighs more than you and your partner think you can safely lift, summon additional help to the scene first.

Another factor to consider is the distance that you will be moving the patient. In this case, the patient has requested that you simply assist him back to his bed. While this seems easy enough, *it only takes one wrong move to cause injury.*

Although you and your partner may be able to safely lift a 180-lb patient onto a stretcher, move him or her through a hallway, and load him or her into the ambulance, moving a 280-lb patient—even if the distance of the move is short (ie, from the ground to his or her bed)—should not be attempted until additional help arrives.

2. Why is knowledge of body mechanics important when lifting and moving a patient?

Body mechanics refers to the way you move your body. Posture is an important component in body mechanics. Good posture means the spine is in a neutral position. Knowing how your body moves will minimize the risk of a back injury when you are lifting or moving a patient.

When you are standing upright, the weight of anything being lifted and carried in the hands is reflected onto the shoulder girdle, the spinal column, the pelvis, and then the legs. With proper lifting technique, the shoulder girdle is aligned over the pelvis and the hands are held close to the legs, thus the force that is exerted against the spine occurs in a straight line down the vertebrae in the spinal column. Therefore, with the back properly maintained in an upright position, very little strain occurs against the muscles and ligaments that keep the spinal column in alignment, and significant weight can be lifted and carried without injury to the back.

Back injuries can occur if you lift with your back curved or, even if straight, bent significantly forward at the hips. With the back in either of these positions, the shoulder girdle lies significantly anterior to the pelvis, and the force of lifting is exerted primarily across, rather than down, the spinal column. When this situation occurs, the weight is supported by the muscles of the back and ligaments that run from the base of

the skull to the pelvis, rather than by each vertebral body and disk resting on those aligned below it. In addition, the upper spine and torso serve as a lever so that the force exerted against the muscles and ligaments in the lumbar and sacral regions, as a result of the mechanical advantage produced, is many times that of the combined weight of your upper body and the object you are lifting. Therefore, the first rule of lifting is to always keep your back in a straight, upright (vertical) position and lift without twisting.

3. Should you and your partner attempt to lift the patient and move him back to his bed? Why or why not?

There are several reasons why you and your partner should not attempt to lift and move this patient without additional resources. First and foremost, he weighs 280 lb; you should not attempt to lift a patient who weighs more than 250 lb with fewer than four rescuers—regardless of your or your partner's individual strength. Second, the patient is unable to provide any assistance, which means that you and your partner will be lifting dead weight. Finally, between you and your partner, one of you will have to lift a greater proportion of the patient's weight. Remember that the objective of a safe patient lift is to *equally distribute the weight among all of the rescuers*. Any attempt to lift him without assistance—even for a brief period—may be just enough to cause injury to you, your partner, or the patient.

There was a reason why you requested additional lifting help—you and your partner did not think that you could safely lift the patient. The patient's condition clearly does not warrant any type of urgent move; simply make him as comfortable as possible until help arrives.

4. What type of carrying device should you use for this patient?

The mechanism of injury (ie, rolling out of bed) is not significant enough to warrant full spinal precautions. Furthermore, the patient's back pain is chronic; it is not the result of this incident. There are numerous devices available for carrying a patient, and you must determine which of these will be the safest for the patient and your team. The scoop stretcher, or orthopedic stretcher, is probably the best choice for this patient. If he were able to flex at the waist and sit up without pain, the stair chair would be a good option.

The scoop stretcher is designed to be split into two sections. These sections are then fitted around a patient who is lying on a flat or relatively flat surface. The scoop stretcher can easily be applied to the patient by simply sliding each half under the left and right sides of his body and then reconnecting the device at the ends. Depending on the patient's size, you may have to slightly lift each side of his or her body to slide each half of the device in place. Overall, the scoop stretcher eliminates the need to roll the patient or lift him or her onto the device.

The scoop stretcher is commonly used for patients with hip fractures, for elderly patients with brittle bones, and for moving patients up or down stairs. The concave design of the scoop stretcher reduces the risk that the patient will slide while he or she is being carried. As with any carrying device, you must fully stabilize and secure the patient before moving him or her.

You are the Provider: SUMMARY, continued

Scoop stretchers are narrow, well constructed, and compact and have excellent body support features. They are ideal for carrying patients in a variety of different situations.

5. What steps can you take to maximize safety while lifting a patient?

Your safety, as well as the safety of the other team members and the patient, depends on the use of proper lifting techniques and having and maintaining a proper hold when lifting and carrying a patient.

To safely lift and carry a patient, you and your team must anticipate and understand every move, and each move must be executed in a coordinated manner. To avoid confusion, which may result in one or more rescuers suddenly bearing an unexpected amount of weight, one person should call out all of the lifting commands.

The following general guidelines should be followed to maximize safety when lifting any patient:

- Keep your legs shoulder width apart.
- Keep your back in a straight, locked-in position.
- Keep the weight as close to your body as possible.
- Bend at the knees, not the waist, when lifting.
- Avoid lifting and reaching at the same time.
- Avoid twisting your body as you are lifting.
- Lift with your palms facing up (power grip).
- Communicate with your partner (or team) at all times.

6. How is a patient's weight distributed when he or she is on a carrying device? Why is it important to know this?

To position your team accordingly, and thus minimize unnecessary rescuer strain and the potential for injury, it is important to know how a patient's weight is distributed when he or she is on a carrying device.

If a patient is supine on a backboard or a scoop stretcher or is in a semisitting position on an ambulance stretcher, his or her weight is not equally distributed between the two ends of the device. When a patient is in a horizontal position, between 68% and 78% of his or her weight is in the torso. Therefore, the strongest rescuer(s) should be positioned at the head end of the carrying device. However, you should still position the remaining rescuers so that each person—including the rescuer(s) at the patient's head—bears an equal amount of the patient's weight.

7. How will you and your team safely carry the patient down the stairs?

Carrying a patient down a flight of stairs or other incline poses a significant risk for injury to the rescuers and/or patient if not performed in a safe, coordinated manner. Unlike carrying a patient across flat terrain in an open area, you must now carry the patient down an incline and within the confines of the walls on either side of you.

The best device to use when carrying a patient down a flight of stairs is a stair chair. However, because the patient cannot sit up, this is not an option.

When carrying a patient on a backboard or scoop stretcher down a flight of stairs, you must first ensure that the patient is anatomically secured to the device in such a way that he or she will not slide when the carrying device is at an angle.

The scoop stretcher is adjustable in length; therefore, you should make it long enough so that the patient's heels abut against the device on either side of the foot-end locking mechanism; this will help prevent the patient from sliding.

Once the patient has been properly secured to the carrying device, position your team accordingly, and lift on the command of a single rescuer. Remember to keep your back in a straight, locked-in position and bend at the knees, not at the waist.

When you carry the patient down stairs on a backboard or a scoop stretcher, make sure the device is carried with the foot end first so that the head end is higher than the foot end. Position taller personnel at the foot end of the device to minimize bending while carrying the patient.

Considering the fact that most stairwells are relatively narrow, it is unlikely that you will be able to position rescuers at the sides of the carrying device. Therefore, you should position two rescuers at the patient's head and two rescuers at the feet. Do not, however, position the rescuers so close together that they risk tripping over their partner's feet!

In many cases, there simply will not be enough room to position two rescuers at both ends of the carrying device. However, you must still ensure that the patient's weight is equally distributed. If you must turn a corner while carrying the patient down stairs, it may be necessary to reposition personnel accordingly to ensure that no one person bears more weight than he or she can safely carry.

A rescuer should be waiting at the base of the stairs to stabilize the stretcher, and another rescuer should be used as a guide for the personnel carrying the foot end of the device. The purpose of a guide is to ensure safety by helping stabilize the rescuer(s) at the foot end of the carrying device by placing a hand on their back and making sure they are aware of when the next step is approaching.

Regardless of whether the stairs are a "straight shot" or if you must turn a corner, the key is to ensure that each move is coordinated and that all rescuers are aware of what is happening at all times. Communication among all rescuers is absolutely critical for a safe patient carry.

8. How can you minimize the risk of injury while moving a patient on a wheeled ambulance stretcher?

Although the safest approach to moving a patient is to use a device that can be rolled, such as a wheeled ambulance stretcher, there is still a risk of injury if improper technique is not observed. EMTs and patients have been injured during the seemingly minor task of transporting a patient on a wheeled ambulance stretcher; therefore, knowledge of safe ambulance stretcher operation is essential.

When you are moving a patient on a wheeled ambulance stretcher, make sure that it is elevated whenever possible, not lowered to the ground. If the stretcher is lowered to the

You are the Provider: SUMMARY, continued

ground, you will have to bend down and move the patient at the same time; this increases the potential for a back injury.

When the stretcher is elevated, the main frame and the patient are considerably higher than the wheels; this makes the stretcher top-heavy. Therefore, when you are moving a patient on an elevated stretcher, ensure that you hold it firmly between both of your hands at all times so that if the patient moves, the stretcher will not tip over.

If you are guiding the stretcher from the foot end, make sure your arms are held close to your body, and avoid reaching a great distance behind you or hyperextending your back. Your back should be locked, straight, and untwisted. To avoid

hyperextending your elbows or injuring your shoulder, keep your elbows slightly flexed and use the muscles of your arms to pull.

If you are guiding the stretcher from the head end, push with your arms and bend your elbows so that your hands are about 12″ to 15″ in front of your torso. To protect your elbows from injury, you should avoid pushing the stretcher with your arms fully extended and your elbows locked. When you push with your elbows bent but firmly held from bending further, the strong muscles of your arms serve as a shock absorber if the wheels of the stretcher strike an obstacle, causing the stretcher to come to an abrupt stop.

EMS Patient Care Report (PCR)

Date: 10-09-09	Incident No.: 013409	Nature of Call: Lift assist		Location: 125 Parkview Place	
Dispatched: 0822	En Route: 0822	At Scene: 0828	Transport: 0913	At Hospital: 0927	In Service: 0935

Patient Information

Age: 59 Sex: M Weight (in kg [lb]): 127 kg (280 lb)	Allergies: No known drug allergies Medications: Ibuprofen Past Medical History: Chronic back pain Chief Complaint: Back pain; request lift assist

Vital Signs

Time: 0833	BP: 156/90	Pulse: 104	Respirations: 22	Sao$_2$: 98%
Time: 0853	BP: 150/88	Pulse: 100	Respirations: 20	Sao$_2$: 97%
Time: 0918	BP: 148/84	Pulse: 98	Respirations: 20	Sao$_2$: 98%

EMS Treatment
(circle all that apply)

Oxygen @ ____ L/min via (circle one): NC NRM Bag-Mask Device	Assisted Ventilation	Airway Adjunct	CPR
Defibrillation	Bleeding Control	Bandaging	Splinting	Other: Transport only; patient comfort measures en route

Narrative

Dispatched to the residence of a 59-year-old male who rolled out of bed and requested assistance getting back into bed. On arrival at the scene, found the patient lying supine next to his bed; his bedroom was located on the second story of his home. He was conscious and alert; his airway was patent, and his breathing was adequate. Patient states that he suffers from chronic back pain but has never been evaluated by a physician. He did not wish to go to the hospital; he only wanted to be moved back to his bed. The patient weighed approximately 280 lb; therefore, requested Engine Company 44 to respond to the scene to assist with lifting and moving. Obtained vital signs and performed secondary assessment, which did not reveal any obvious injury. The patient denies striking his head when he rolled out of bed and further denies any other injury or pain. He states that the pain in his back is severe, as it usually is, and that he cannot sit up. Other than chronic back pain, he denied any significant past medical history. While waiting for the engine company to arrive, the patient called his daughter, who convinced him to go to the hospital for physician evaluation; he agreed to this. Engine Company 44 arrived; with their assistance, the patient was placed onto a scoop stretcher, secured appropriately, and moved down a flight of stairs to the stretcher. Reassessed the patient, he was still conscious and alert, but still experiencing pain. Loaded patient into the ambulance, placed padding under his back for pain relief, and began transport to the hospital. Continued to monitor patient en route; his mental status remained unchanged and vital signs indicated that he was stable. He expressed some relief of his pain after padding was placed under his back. Further treatment involved supportive care only. Patient was delivered to emergency department staff without incident. Verbal report was given to staff nurse. *End of report*

Prep Kit

- The first key rule of lifting is to always keep your back in an upright position and lift without twisting. You can lift and carry significant weight without injury as long as your back is in the proper upright position.

- The power lift is the safest and most powerful way to lift.

- The safety of you, your team, and the patient depends on the use of proper lifting techniques and maintaining a proper hold when lifting or carrying a patient.

- Pushing is better than pulling.

- If you do not have a proper hold, you will not be able to bear your share of the weight, or you may lose your grasp with one or both hands and possibly cause a lower back injury to one or more EMTs.

- It is always best to move a patient on a device that can be rolled. However, if a wheeled device is not available, you must understand and follow certain guidelines for carrying a patient on a stretcher.

- You must constantly coordinate your movements with those of the other team members and make sure that you communicate with them.

- When lifting a stretcher, you must make sure that you and your team use correct lifting techniques.

- Ideally, members of the lifting team should also be of similar height and strength.

- If you must carry a loaded backboard or stretcher up or down stairs or other inclines, be sure that the patient is tightly secured to the device to prevent sliding.

- Be sure to carry the backboard or stretcher foot end first so that the patient's head is elevated higher than the feet.

- Directions and commands are an important part of safe lifting and carrying.

- You and your team must anticipate and understand every move and execute it in a coordinated manner.

- The team leader is responsible for coordinating the moves.

- You should try to use four rescuers whenever resources allow.

- You should know how much you can comfortably and safely lift and not attempt to lift more than this amount.

- Rapidly summon additional help to lift and carry a weight that is greater than you are able to lift.

- The same basic body mechanics apply for safe reaching and pulling as for lifting and carrying.

- Keep your back locked and straight, and avoid twisting.

- Do not hyperextend your back when reaching overhead.

- You should normally move a patient with nonurgent moves, in an orderly, planned, and unhurried manner, selecting methods that involve the least amount of lifting and carrying.

- At times, you may have to use an emergency move to maneuver a patient before providing assessment and care.

- You should perform an urgent move if a patient has an altered level of consciousness, inadequate ventilation, or shock or in extreme weather conditions.

- The wheeled ambulance stretcher is the most commonly used device to move and transport patients.

- Other devices that are used to lift and carry patients include portable stretchers, flexible stretchers, backboards, basket stretchers (Stokes litters), scoop stretchers, and stair chairs.

- Whenever you are moving a patient, you must take special care so that neither you, your team, nor the patient is injured.

- You will learn the technical skills of patient packaging and handling through practice and training.

- Training and practice are required to use all the equipment that is available to you.

- You must practice each technique with your team often so that you are able to perform the move quickly, safely, and efficiently.

Vital Vocabulary

backboard A device that is used to provide support to a patient who is suspected of having a hip, pelvic, spinal, or lower extremity injury. Also called a spine board, trauma board, and longboard.

bariatrics A branch of medicine concerned with the management (prevention or control) of obesity and allied diseases.

basket stretcher A rigid stretcher commonly used in technical and water rescues that surrounds and supports the patient yet allows water to drain through holes in the bottom. Also called a Stokes litter.

diamond carry A carrying technique in which one EMT is located at the head end of the stretcher or backboard, one at the foot end, and one at each side of the patient; each of the two EMTs at the sides uses one hand to support the stretcher/backboard so that all are able to face forward as they walk.

direct ground lift A lifting technique that is used for patients who are found lying supine on the ground with no suspected spinal injury.

emergency move A move in which the patient is dragged or pulled from a dangerous scene before assessment and care are provided.

extremity lift A lifting technique that is used for patients who are supine or in a sitting position with no suspected extremity or spinal injuries.

flexible stretcher A stretcher that is a rigid carrying device when secured around a patient but can be folded or rolled when not in use.

portable stretcher A stretcher with a strong rectangular tubular metal frame and rigid fabric stretched across it.

power grip A technique in which the litter or backboard is gripped by inserting each hand under the handle with the palm facing up and the thumb extended, fully supporting the underside of the handle on the curved palm with the fingers and thumb.

power lift A lifting technique in which the EMT's back is held upright, with legs bent, and the patient is lifted when the EMT straightens the legs to raise the upper body and arms.

rapid extrication technique A technique to move a patient from a sitting position inside a vehicle to supine on a backboard in less than 1 minute when conditions do not allow for standard immobilization.

scoop stretcher A stretcher that is designed to be split into two or four sections that can be fitted around a patient who is lying on the ground or other relatively flat surface; also called an orthopedic stretcher.

stair chair A lightweight folding device that is used to carry a conscious, seated patient up or down stairs.

wheeled ambulance stretcher A specially designed stretcher that can be rolled along the ground. A collapsible undercarriage allows it to be loaded into the ambulance. Also called the stretcher or an ambulance stretcher.

Assessment in Action

You are dispatched to a residence to find a 60-year-old man lying supine at the bottom of the basement steps. He is conscious and complaining of pain in his hip. Your assessment reveals pain and tenderness in the left pelvic area. Vital signs are stable, and the patient has no medical history. You and your partner decide to use the scoop stretcher to carry him up the steps and place him on the wheeled stretcher. The patient weighs approximately 170 lb.

1. When lifting the patient, you should use the:
   A. power lift.
   B. power grip.
   C. lateral lift.
   D. diamond lift.

2. What is the best method for carrying this patient up the steps?
   A. Feet first and head elevated
   B. Feet first and elevated
   C. Head first and feet elevated
   D. Head first and slightly elevated

3. Where should you place the wheeled stretcher for easy access?
   A. Outside by the front door
   B. To the side at the top of the stairs
   C. Right at the top of the stairs
   D. Outside at the back of the ambulance

4. The power lift should be done with your body in what position?
   A. Legs straight and back bent
   B. Back bent and knees bent
   C. Knees bent and back straight
   D. Back straight and legs locked

5. What is the best carry for this patient?
   A. Diamond carry
   B. One-handed carry
   C. End-to-end carry
   D. Direct carry

6. To perform the power grip, your palms should be facing:
   A. down with your thumbs pointing sideways.
   B. down with your thumbs curled over your fingers.
   C. up with your thumbs curled over your fingers.
   D. up with your thumbs pointing up.

7. The scoop stretcher is also known as a(n):
   A. orthopedic stretcher.
   B. flexible litter.
   C. basket litter.
   D. ambulance stretcher.

8. What is the safest way to lift a patient?
   A. Extremity lift
   B. Power lift
   C. Direct ground lift
   D. Emergency lift

9. Describe how constant communication between team members can avoid rescuer injury.

10. What are three questions you should ask yourself before lifting a patient?

National EMS Education Standard Competencies

EMS Operations

Knowledge of operational roles and responsibilities to ensure patient, public, and personnel safety.

Principles of Safely Operating a Ground Ambulance

- Risks and responsibilities of emergency response (pp 1341-1353)
- Risks and responsibilities of transport (pp 1341-1353)

Air Medical

- Safe air medical operations (pp 1353-1358)
- Criteria for utilizing air medical response (pp 1353-1355)

Medicine

Applies fundamental knowledge to provide basic emergency care and transportation based on assessment findings for an acutely ill patient.

Infectious Diseases

Awareness of

- How to decontaminate equipment after treating a patient (pp 1345-1346)
- How to decontaminate the ambulance and equipment after treating a patient (pp 1345-1346)

....

Knowledge Objectives

1. Describe the nine phases of an ambulance call and provide examples of key tasks the EMT performs during each phase. (pp 1332-1346)
2. Describe the medical equipment carried on an ambulance and provide examples of supplies that are included in each main category of the ambulance equipment checklist. (pp 1332-1338)
3. Provide examples of the safety and operations equipment carried on an ambulance and explain how each item might be used in an emergency by EMTs. (pp 1338-1340)
4. Discuss the importance of performing regular vehicle inspections and list the specific parts of an ambulance that should be inspected daily. (p 1341)
5. Describe the minimum dispatch information required by EMS to respond to an emergency call. (pp 1341-1342)

6. Provide examples of some high-risk situations and hazards that may affect the safety of the ambulance and its passengers during both pretransport and transport. (pp 1342-1343, 1349-1353)
7. Discuss specific considerations that are required for ensuring scene safety, including personal safety, patient safety, and traffic control. (pp 1342-1344)
8. Describe the key elements related to patient information that must be included in the written patient report upon patient delivery to the hospital. (pp 1344-1345)
9. Summarize the tasks that must be completed by EMS at the completion of an ambulance call. (pp 1345-1346)
10. Define the terms cleaning, disinfection, high-level disinfection, and sterilization and explain how they differ. (pp 1345-1346)
11. Discuss the guidelines for driving an ambulance safely and defensively and identify key steps EMS personnel can take to improve safety while en route to the scene, the hospital, and the station. (pp 1346-1353)
12. Describe the elements that dictate the use of lights and siren to the scene and to the hospital and the factors required to perform a risk-benefit analysis regarding their use. (p 1348)
13. Give examples of the specific, limited privileges that are provided to emergency vehicle drivers by most state laws and regulations. (pp 1351-1353)
14. Explain why using police escorts and crossing intersections pose additional risks to EMS personnel during transport and discuss special considerations related to each. (pp 1352-1353)
15. Describe the capabilities, protocols, and methods for accessing air ambulances. (pp 1353-1355)
16. Describe key scene safety considerations when preparing for a helicopter medivac, including establishing a landing zone, securing loose objects, mitigating onsite hazards, and approaching the aircraft. (pp 1355-1358)

Skills Objectives

1. Demonstrate how to perform a daily inspection of an ambulance. (p 1341)
2. Demonstrate how to present a verbal report that would be given to arrival personnel at the hospital upon patient transfer. (pp 1344-1345)
3. Demonstrate how to write a written report that includes all pertinent patient information following patient transfer to the hospital. (p 1345)
4. Demonstrate how to clean and disinfect the ambulance and equipment during the postrun phase. (pp 1345-1346)

Introduction

During the late 1700s, Napoleon Bonaparte commissioned one of the more advanced professional emergency medical patient care systems in the world. By that time, horse-drawn ambulances already were in use in major cities throughout the United States Figure 36-1 . American hospitals initiated their own professional ambulance services during the late 1860s. Ambulance attendants traveled with limited medical supplies, including brandy, a few tourniquets, several assorted bandages and sponges, basic splinting material, and blankets.

Today's ambulances are stocked with standard medical supplies. Many are equipped with state-of-the-art technology, including defibrillators and monitors that can transmit information directly to the emergency department, blood and oxygen testing equipment, automatic

Figure 36-1 Horse-drawn ambulances were used in major cities throughout the United States during the 1800s.

ventilators, automated cardiopulmonary resuscitation (CPR) machines, global positioning systems, and computer-aided dispatch consoles. Even when following all safety guidelines, today's emphasis on rapid response places the EMT in great danger while driving to calls.

This chapter discusses ambulance design and how to equip and maintain an ambulance. It also focuses on the techniques and judgment that you will need to learn to drive an ambulance or ambulance service vehicle, which includes parking considerations, emergency vehicle control and operation, the effects of weather on driving, and common hazards that are encountered in driving an ambulance. Finally, the chapter describes how to work safely with air ambulances.

Emergency Vehicle Design

An **ambulance** is a vehicle that is used for treating and transporting patients who need emergency medical care to a hospital. The first motor-powered ambulance was introduced in 1906. For many decades after that, a hearse was the vehicle that was most often used as an ambulance because it was the only vehicle with enough room for a person to lie down. Few supplies were carried on board, and there was little space for attendants.

The hearse-ambulance has gone the way of its horse-drawn predecessor. Ambulances today are designed according to strict government regulations based on national standards. The standards themselves are based in large part on suggestions from the ambulance industry and from EMS personnel. One of the most significant developments in ambulance design has been the enlargement of the patient compartment. Another development is the use of **first-responder vehicles** Figure 36-2 , which respond initially to the scene with personnel and equipment to treat the sick and injured until an ambulance can arrive.

You are the Provider: PART 1

At 3:15 PM, you are dispatched to 37525 Wells Avenue for a "sick person." It is raining, the temperature is 84°F, and the traffic is heavy. You have completed an emergency vehicle operator course and have been cleared by your supervisor to drive the ambulance in emergency mode. You and your partner respond to the scene, which is located a short distance away.

1. What attributes should an emergency vehicle operator possess?
2. What factors should you consider before responding to the scene?

Figure 36-2 First responders, such as fire fighters and law enforcement personnel, are often the first to arrive at a scene.

Each state establishes its own standards for licensing or certifying ambulances. Many agencies use the federal specifications (KKK-A-1822F, August 2008) that cover the following three types of basic ambulance designs **Figure 36-3** and **Table 36-1**.

The six-pointed **Star of Life**® emblem **Figure 36-4** identifies vehicles as ambulances. It is often affixed to the sides, rear, and roof of the ambulance. Local regulatory authorities determine what emblems may be displayed on the side of a prehospital care ambulance. **Figure 36-5** illustrates some of the required features of a licensed or certified ambulance.

As defined by the National Research Council of the National Academy of Sciences, the modern ambulance is a vehicle for emergency medical care that has the following features:

- A driver's compartment
- A patient compartment that can accommodate two EMTs and two supine patients (one on the stretcher, one on the bench) positioned so that at least one of the patients can receive CPR during transport
- Equipment and supplies to provide emergency medical care at the scene and during transport, to safeguard personnel and patients from hazardous conditions, and to carry out light extrication procedures
- Two-way radio communication so that ambulance personnel can speak with the dispatcher, the hospital, public safety authorities, and online medical control
- Design and construction that ensure maximum safety and comfort

Phases of an Ambulance Call

An ambulance call has nine phases: preparation, dispatch, en route, arrival at scene, transfer of patient to ambulance, en route to receiving facility (transport), at receiving facility (delivery), en route to station, and postrun, as shown in **Table 36-2**. These nine phases address the vehicle and its crew and their roles in a response to a medical emergency. The details of patient care are not included in these nine phases.

The Preparation Phase

Making sure that equipment and supplies are in their proper places and ready for use is an important part of preparing for the call. Items that are missing or that do not work are of no use to you or the patient. As a general rule, the more complex a piece of equipment is and the harder it is to learn to use, the more likely it is to

Table 36-1 Basic Ambulance Designs

Type I	Conventional, truck cab-chassis with a modular ambulance body that can be transferred to a newer chassis as needed
Type II	Standard van, forward-control integral cab-body ambulance
Type III	Specialty van, forward-control integral cab-body ambulance

Table 36-2 Phases of an Ambulance Call

1. Preparation for the call
2. Dispatch
3. En route
4. Arrival at scene
5. Transfer of the patient to the ambulance
6. En route to the receiving facility (transport)
7. At the receiving facility (delivery)
8. En route to the station
9. Postrun

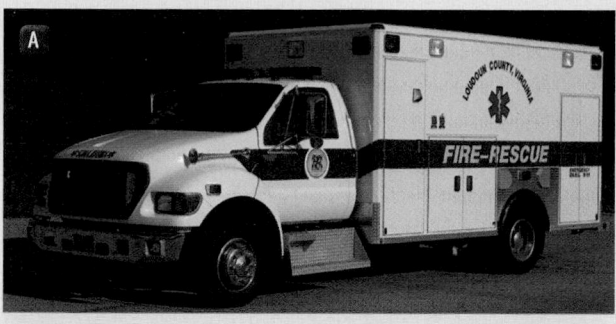

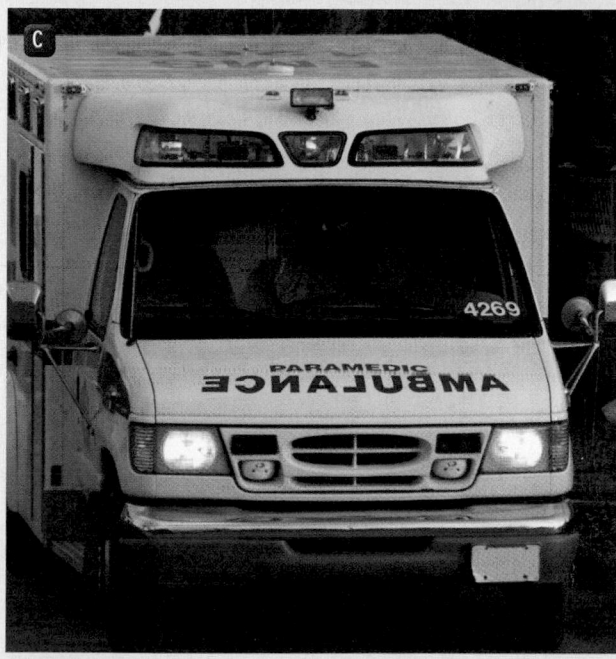

Figure 36-3 **A.** The conventional, truck cab-chassis has a modular ambulance body that can be transferred to a newer chassis (type I). **B.** The standard van ambulance has a forward-control integral cab body (type II). **C.** The specialty van ambulance has a forward-control integral cab body (type III).

malfunction during an emergency. Many EMS items have never been rigorously tested under field conditions and could turn out to be expensive mistakes. For this reason, new equipment should be placed on an ambulance only after proper instruction on its use and consulting with the medical director.

Figure 36-4 The Star of Life®.

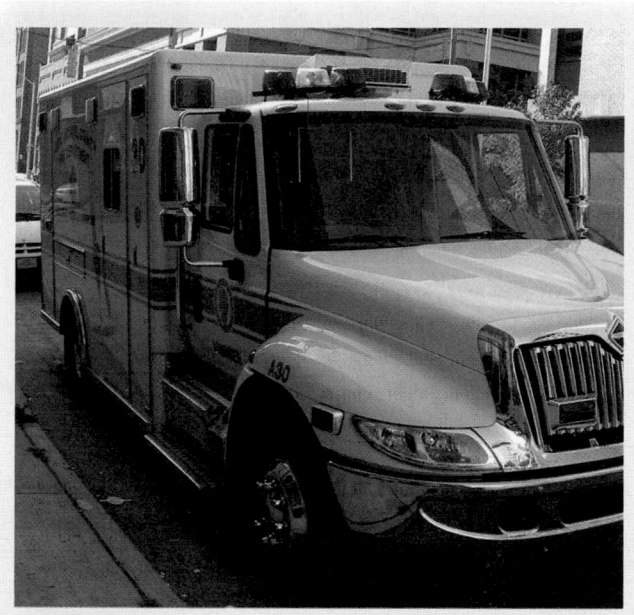

Figure 36-5 Warning lights and public address systems are necessary on licensed or certified ambulances.

Equipment and supplies should be durable and, to the extent possible, standardized. This makes it easy to quickly exchange equipment with other ambulances or with the emergency department, thus saving time during patient transfer.

Store equipment and supplies in the ambulance according to how urgently and how often they are used **Figure 36-6** . Give priority to items that are needed to care for life-threatening conditions. These include equipment for airway management, artificial ventilation, and oxygen delivery. Place these items within easy reach, at the head of the primary stretcher. Place items for cardiac care, control of external bleeding, and monitoring blood pressure at the side of the stretcher.

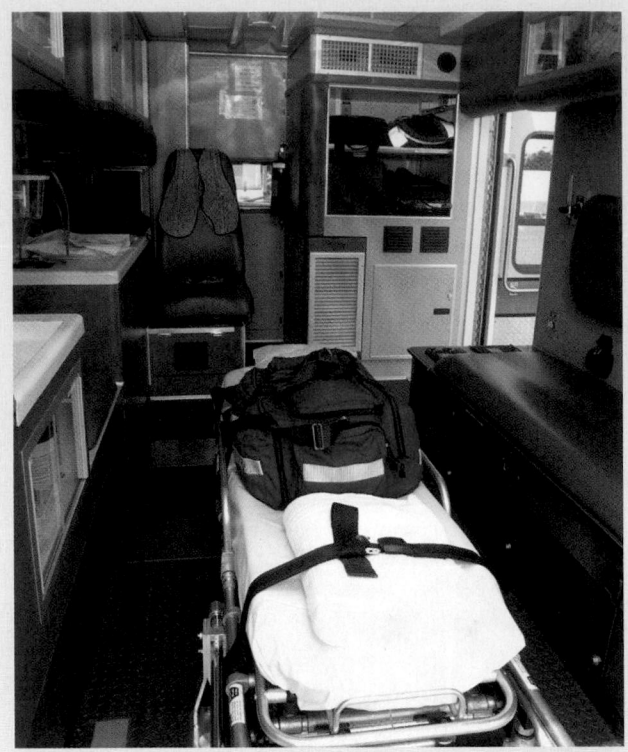

Figure 36-6 Store equipment and supplies in the ambulance according to how urgently and how often they are used.

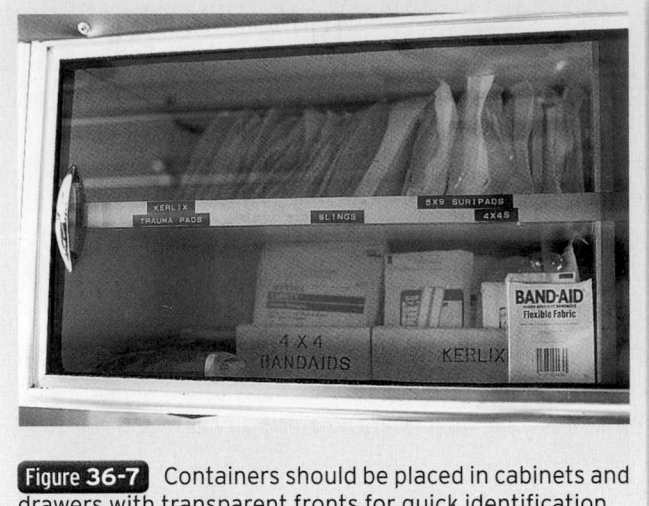

Figure 36-7 Containers should be placed in cabinets and drawers with transparent fronts for quick identification.

Storage cabinets and kits should open easily. They should also close securely so that they do not fly open while the ambulance is in motion. Cabinet and drawer fronts should be transparent so that you can quickly

identify their contents; if they are not, be sure to label each container **Figure 36-7**.

Medical Equipment

As an EMT, you have access to a large variety of medical equipment and supplies, far more than can be described here. Certain items must be available on the ambulance at all times, as dictated by state and jurisdictional requirements.

Basic Supplies **Table 36-3** lists the common supplies carried on ambulances. These include basic items such as disposable gloves and sharps, airway and ventilation equipment, basic wound care supplies, splinting supplies, childbirth supplies, an automated external

You are the Provider: PART 2

You arrive at the scene and knock on the door of the residence. The patient, a 70-year-old woman, answers the door and tells you that she began feeling very weak earlier in the day. You perform a primary assessment as your partner opens the jump kit.

Recording Time: 0 Minutes	
Appearance	Appears ill; skin slightly pale
Level of consciousness	Conscious and alert
Airway	Open; clear of secretions or foreign bodies
Breathing	Normal rate; adequate depth
Circulation	Increased pulse rate, but strong and regular; skin is pink, warm, and dry

3. What equipment and supplies are typically taken from the ambulance to the patient's side?

Table 36-3 Ambulance Equipment Checklist

Basic Supplies

Pillows and pillowcases

Sterile sheets

Blankets

Towels

Disposable emesis bags or basins

Boxes of disposable tissue

Bedpan (optional)

Urinals (one male, one female; optional)

Blood pressure cuffs (pediatric, adult, large adult)

Stethoscope

Disposable drinking cups

Unbreakable container of water

Wet wipes

Chemical cold/hot packs

Sterile irrigation fluid

Restraining devices

Plastic bags for waste or severed parts

Hypoallergenic latex, vinyl, or other disposable gloves (various sizes)

Sharps container

Set of hearing protectors

Airway and Ventilation Equipment

Infection control kits (goggles, masks, waterproof gowns)

Oropharyngeal airways and nasopharyngeal airways of various sizes

Advanced airway supplies, if local protocol permits (laryngeal mask airway, Combitube, King, endotracheal intubation equipment), with secondary placement confirmation devices

Bag-mask devices (adult and pediatric)

Mounted suction unit and a portable suction unit

Assorted oxygen delivery devices (adult and pediatric)

Oxygen supply units (both portable and installed)

Disposable humidifier (for mounted oxygen system)

Basic Wound Care Supplies

Trauma shears

Sterile sheets

Sterile burn sheets

Adhesive tape in several widths

Self-adhering, soft roller bandages, 4″ × 5 yd

Self-adhering, soft roller bandages, 2″ × 5 yd

Sterile dressings, gauze, 4″ × 4″

Sterile dressings, abdominal or laparotomy pads, usually 6″ × 9″ or 8″ × 10″

Sterile universal trauma dressings, usually 10″ × 36″, folded into 9″ × 10″ packages

Sterile, occlusive, nonadherent dressings (aluminum foil sterilized in original package)

Occlusive dressings, or chest seals

Assortment of adhesive bandages

Tourniquet

Adult-size pneumatic antishock garment, previously called military antishock trousers (depending on local protocols)

Splinting Supplies

Adult-size traction splint

Child-size traction splint

A variety of arm and leg splints, such as inflatable, vacuum, cardboard, plastic, foam wire-ladder, or padded board (the number and type of splints should be determined by state regulations and your medical director)

A variety of triangular bandages and roller bandages

Short backboard device

Long backboard

Cervical collars in an adjustable size or a variety of sizes

Head stabilization devices

Childbirth Supplies

Emergency obstetric kit

Surgical scissors

Hemostats or special cord clamps

Umbilical tape or sterilized cord

Small rubber bulb syringe

Towels

Gauze sponges

Sterile gloves

Sanitary napkins

Plastic bag

Baby blanket

Baby stocking cap

Automated External Defibrillator

Semiautomated defibrillation equipment

Patient Transfer Equipment

Wheeled ambulance stretcher

Wheeled stair chair

Long backboard

Short backboard/short immobilization device

Other devices also carried on ambulances include:

    Scoop stretcher

    Portable/folding stretcher

    Flexible stretcher

    Basket stretcher

Medications and Other Supplies

Activated charcoal

Drinkable water and cups

Oral glucose

Oxygen

Supplies for irrigating the skin and eyes

Snakebite kit or other regional equipment, depending on the area and local protocol defibrillator, patient transfer equipment, medications, and other supplies such as a snakebite kit or regional supplies.

Airway and Ventilation Equipment Airway management equipment that should be carried on ambulances includes the following:

- Oropharyngeal airways for adults, children, and infants
- Nasopharyngeal airways for adults and children
- Two sets of equipment for advanced airway procedures if your service is authorized by state regulation and the medical director to perform these: one in the ambulance and one in the jump kit that you carry to the patient

It is important that two portable artificial ventilation devices that operate independently of an oxygen supply are carried on the ambulance: one for use in the ambulance and one for use outside the ambulance or as a spare. These devices include disposable pocket masks and bag-mask devices. In addition, bag-mask devices capable of oxygen enrichment that, when attached to an oxygen supply with the oxygen reservoir in place, are able to supply almost 100% oxygen, should also be carried on the ambulance. Masks for these devices come in a variety of sizes, from neonatal to adult, and are necessary materials to carry on the ambulance. Oxygen-powered devices are also available to provide ventilation to a patient but may quickly deplete available oxygen sources. You should follow local guidelines in identifying the specific ventilation equipment carried on the ambulance.

The ambulance should carry both portable and mounted suctioning units **Figure 36-8**. These units must be powerful enough to generate a vacuum of 300 mm Hg when the tube is clamped. The suctioning force must be adjustable for use on infants and children. The units should include large-bore, nonkinking suction tubing with a semirigid pharyngeal tip, with additional semirigid tips available. The installed unit should include a suction yoke, an unbreakable collection cannister, water for rinsing the suction tips, and suction tubing, all easily accessible when you are sitting at the head of the stretcher. The tubing must reach the patient's airway, regardless of the patient's position. All components of the suctioning unit must be disposable or made of material that is easily cleaned and **decontaminated**.

The ambulance should carry at least two oxygen supply units: one portable and one installed. The portable unit should be located near a door or in the jump kit, for easy use outside the ambulance. It should have a capacity of minimum 500 L of oxygen and be equipped with a yoke, pressure gauge, flowmeter, oxygen supply tubing,

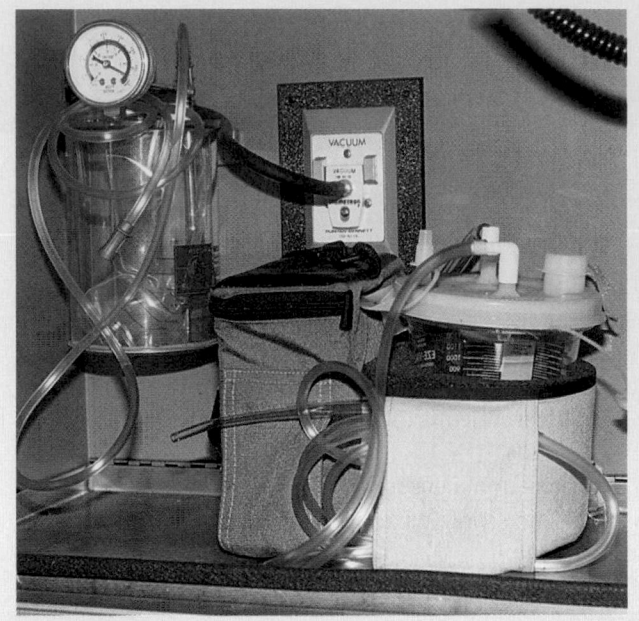

Figure 36-8 The ambulance should carry both a mounted suctioning unit and a portable unit.

nonrebreathing mask, and nasal cannula. This unit must be able to deliver oxygen at a variable rate between 1 and 15 L/min. At least one extra portable 500-L cylinder should be kept on the ambulance. Many services equip the backup cylinder with its own yoke, gauge, regulator, and tubing so that it can be used for a second patient.

The mounted oxygen unit should have a capacity of 3,000 L of oxygen **Figure 36-9**. It should also be equipped with visible flowmeters that are capable of delivering 1 to 15 L/min that are accessible when you are at the head of the stretcher. Oxygen masks, with and without nonbreathing bags, should be transparent, disposable, and in sizes for adults, children, and infants.

Ambulance services that often transport patients on runs lasting longer than 1 hour should consider using a disposable, single-use humidifier for the mounted oxygen system. On runs of less than 1 hour, humidification is not usually necessary. Humidification may increase a patient's risk of infection unless the equipment is rigorously maintained.

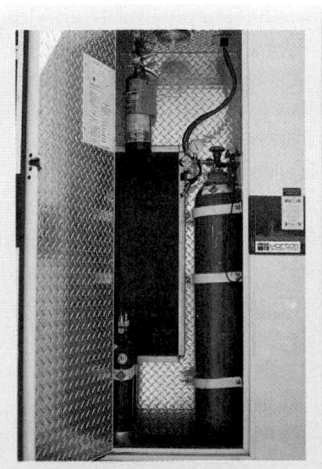

Figure 36-9 An oxygen unit with a capacity of 3,000 L of oxygen should be mounted on the ambulance.

CPR Equipment A **CPR board** provides a firm surface under the patient's torso so that you can give effective chest compressions **Figure 36-10A**. It also establishes an appropriate degree of head tilt **Figure 36-10B**. Today this item is carried in only a few ambulances across the country. If you do not have a special CPR board, you can place a long or short backboard under the patient on the stretcher. Use a tightly rolled sheet or towel to raise the patient's shoulders 3″ to 4″; this will also keep the patient's head in a position of maximum backward tilt and keep the shoulders and chest in a straight position. Caution: Do not use this roll to hyperextend the neck if you suspect a spinal injury.

Mechanical devices that operate on compressed gas and deliver chest compressions and ventilations are also available.

Basic Wound Care Supplies Basic supplies for dressing open wounds should be included on the ambulance. These include a pair of trauma shears; sterile sheets; sterile burn sheets; adhesive tape in several widths; self-adhering, soft roller bandages; sterile dressings; gauze; abdominal or laparotomy pads; sterile universal trauma dressings; sterile, occlusive, nonadherent dressings (aluminum foil sterilized in original package); an assortment of adhesive bandages; a tourniquet; and an adult-size pneumatic antishock garment, previously called military antishock trousers (depending on local protocols).

Splinting Supplies Examples of supplies for splinting fractures and dislocations that may be carried on ambulances are shown in **Figure 36-11**. These include an adult-size and a child-size traction splint; a variety of arm and leg splints, such as inflatable, vacuum, cardboard, plastic, foam wire-ladder, or padded board; a variety of triangular bandages and roller bandages; a short backboard device; a long backboard; head immobilization devices; and cervical collars in an adjustable size or a variety of sizes.

Childbirth Supplies You must carry at least one sterile emergency obstetric kit **Figure 36-12** that includes the supplies listed in Table 36-3, including a pair of surgical scissors, hemostats or special cord clamps, umbilical tape or sterilized cord, a small rubber bulb syringe, towels, gauze sponges, pairs of sterile gloves, sanitary napkins, a plastic bag, a baby stocking cap, and a baby blanket.

Automated External Defibrillator Modern-day EMS was ushered in by the first-ever prehospital use of the defibrillator by a St. Vincent's Hospital ambulance in New York City under the direction of Dr. William Grace in the

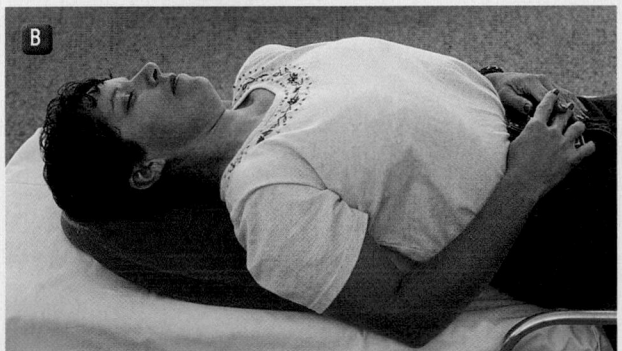

Figure 36-10 **A.** A CPR board may be carried on the ambulance. **B.** A patient on a CPR board has the appropriate degree of head tilt for effective artificial ventilation.

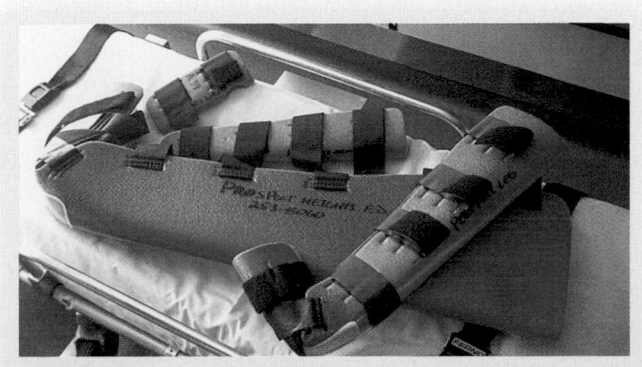

Figure 36-11 Supplies for splinting fractures and dislocations should be carried on the ambulance.

Figure 36-12 A sterile emergency obstetric kit must be carried on the ambulance.

early 1970s. Now a prehospital standard of care, semi-automated defibrillation equipment or manual monitor/defibrillators that have automated external defibrillation capability, as permitted by regulation and the local medical director, should always be carried on the ambulance Figure 36-13.

Patient Transfer Equipment Each ambulance should carry the following patient transfer equipment:

- A primary wheeled ambulance stretcher
- A wheeled stair chair for use in narrow spaces
- A long backboard
- A short backboard or short immobilization device

You should be able to tilt the head of the stretcher upward to at least a 60° semisitting position and tilt the entire stretcher into a 10° to 15° Trendelenburg's position (feet elevated 6″ to 10″) for airway care and treatment of shock Figure 36-14. Stretchers must be provided with fasteners to secure them firmly to the floor or side of the ambulance during transport. Stretcher restraints should be capable of holding the stretcher in place in case the vehicle rolls over. Make certain that

the wheeled stretcher is locked into position properly, because injuries can occur to the patient and you if the stretcher becomes loose while the ambulance is in motion. Make sure there are at least three restraining devices for the patient, such as deceleration or stopping straps over the shoulders, to prevent the patient from continuing to move forward in case the ambulance suddenly slows or stops. Other devices that can be used include the following:

- A scoop stretcher
- A portable/folding stretcher
- A flexible stretcher
- A basket stretcher

Medications It is important that the ambulance carry appropriate medications and that they have not expired. Be certain that you have the telephone number and radio frequency of online medical control or the local poison control center with you on the ambulance. The back of your clipboard is a good place to keep this information.

The Jump Kit The ambulance must be equipped with a portable, durable, and waterproof jump kit that you can carry to the patient Figure 36-15. Think of the **jump kit** as the "5-minute kit," containing anything you might need in the first 5 minutes with the patient except for the semiautomated external defibrillator, possibly the oxygen cylinder, and portable suctioning unit. The jump kit must be easy to open and secure. Table 36-4 lists the items that are typically contained in a jump kit.

Safety and Operations Equipment

In addition to medical equipment, a properly stocked ambulance carries several kinds of equipment for

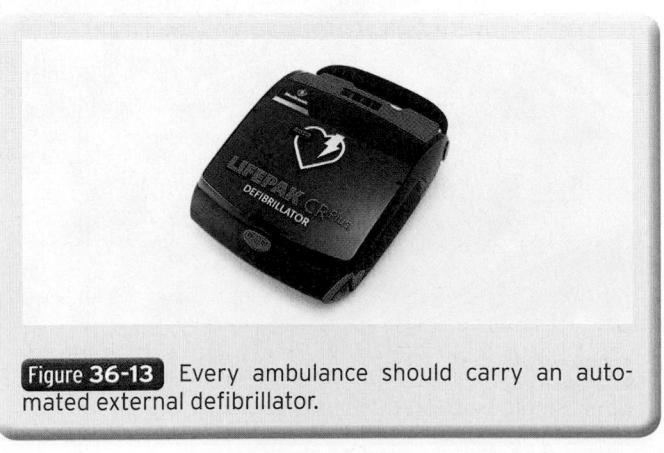

Figure 36-13 Every ambulance should carry an automated external defibrillator.

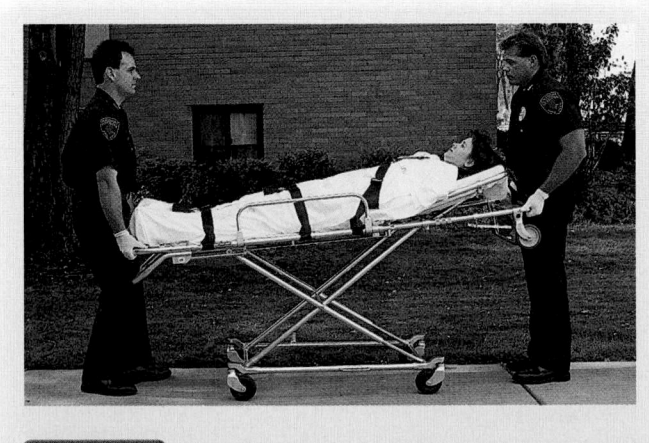

Figure 36-14 The wheeled ambulance stretcher should be locked into place at an appropriate height.

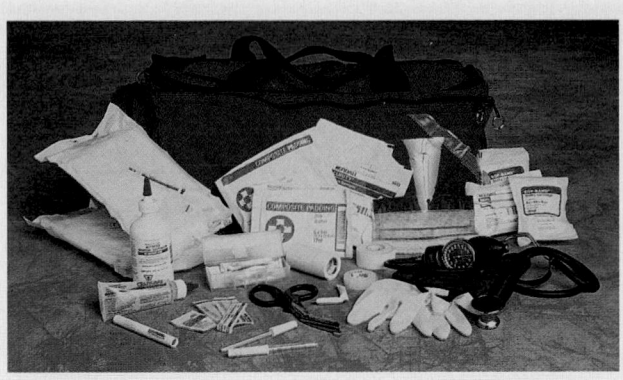

Figure 36-15 A portable jump kit should contain practically anything you will need during the first 5 minutes with the patient.

Table 36-4 Items Carried in a Jump Kit

- Latex, vinyl, or other disposable gloves
- Triangular bandages
- Trauma shears
- Adhesive tape in various widths
- Universal trauma dressings
- Self-adhering soft roller bandages, 4" × 5 yd and 2" × 5 yd
- Oropharyngeal airways in adult, child, and infant sizes*
- Bag-mask device with masks for adults, children, and infants*
- Blood pressure cuff
- Stethoscope
- Penlight
- Sterile gauze dressings, 4" × 4"
- Sterile dressings (abdominal pads), 6" × 9" or 8" × 10"
- Adhesive strips
- Oral glucose
- Activated charcoal

*These might be carried in a separate airway kit, along with the portable oxygen cylinder.

responder safety, rescue operations, and locating emergency scenes. To do the job effectively, the EMT team will need the following equipment:

- Personal protective equipment
- Equipment for work areas
- Preplanning/navigation guides
- Extrication equipment

Personal Safety Equipment You should always carry personal protective equipment that allows you to work safely in a limited variety of hazardous or contaminated situations. These situations include the edges of a structural

fire or explosion, vehicle extrication, and in crowds. The equipment should protect you from exposure to blood and other potentially infectious body fluids. Note that you will not be equipped to face all hazardous materials (HazMat) and other exposure situations that you may encounter; this is the job of specially trained HazMat technicians and response teams. Your equipment might include the following:

- Face shields
- Gowns, shoe covers, caps
- Turnout gear
- Helmets with face shields or safety goggles
- Safety shoes or boots

Equipment for Work Areas A weatherproof compartment that you can reach from outside the patient compartment should hold equipment for safeguarding patients and EMTs, controlling traffic and bystanders, and illuminating work areas **Figure 36-16**. The following items are recommended:

- Warning devices that flash intermittently or have reflectors (Road flares can pose an additional hazard, such as ignition of flammable liquids or gases.)
- Two high-intensity halogen 20,000 candlepower flashlights of the recharging battery-powered, stand-up type
- Fire extinguisher, type BC, dry powder, 5-lb minimum
- Hard hats or helmets with face shields or safety goggles
- Portable floodlights

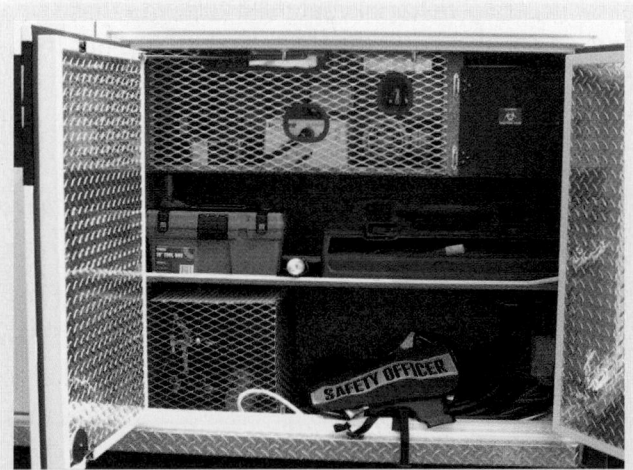

Figure 36-16 The ambulance should have a weatherproof compartment that can be reached from outside the patient compartment. It should hold equipment for safeguarding patients and EMTs, controlling traffic, and illuminating work areas.

Preplanning and Navigation Equipment Make sure you have detailed street and area maps in the driver's compartment of the ambulance, along with directions to key locations, such as local hospitals. Become familiar with the roads and traffic patterns in your town or city so that you can plan alternative routes to common destinations. Pay particular attention to ways around frequently opened bridges, congested traffic, and blocked railroad crossings. Often, switching to an alternative route will save more time than driving faster. Also become familiar with special facilities and locations within your regional operating area, such as other medical facilities, airports, arenas and stadiums, detention facilities, and chemical or research facilities that might pose unusual problems (staging areas may be predefined for emergency operations).

Extrication Equipment A weatherproof compartment outside the patient compartment should contain equipment that is needed for simple, light extrication, even if an extrication and rescue unit is readily available. **Table 36-5** lists the items that should be included in the compartment.

If rescue and extrication services are not readily available, additional equipment may be needed.

Personnel

Every ambulance must be staffed with at least one EMT in the patient compartment whenever a patient is being transported; two EMTs are strongly recommended. Some

Table 36-5 Extrication Equipment

- 12″ wrench, adjustable, open-end
- 12″ screwdriver, standard square bar
- 8″ screwdriver, Phillips head #2
- Hacksaw with 12″ carbide wire blades
- Vise-grip pliers, 10″
- 5-lb hammer with 15″ handle
- Fire ax, butt, 24″ handle
- Wrecking bar with 24″ handle. This may be a combination tool with a hammer and ax.
- 51″ crowbar, pinch point
- Bolt cutter with 1″ to 1¼″ jaw opening
- Folding shovel, pointed blade
- Tin snips, double action, 8″ minimum
- Gauntlets, reinforced, leather covering past midforearm, one pair per crew member
- Rescue blanket
- Ropes, 5,400-lb tensile strength in 50′ lengths in protective bags
- Mastic knife (able to cut seatbelt webbing)
- Spring-load center punch
- Roll of duct tape (for window application prior to center punch use)
- Pruning saw
- Heavy-duty 2″ × 4″ and 4″ × 4″ shoring (cribbing) blocks, various lengths

You are the Provider: PART 3

After performing a secondary assessment and obtaining the patient's vital signs, you ask the patient if she will allow you to give her oxygen. She tells you that she does not want a mask covering her face but will allow you to apply a nasal cannula. She asks you to take her to the hospital because she feels too weak to drive herself. You assess her blood glucose level and note that it reads 90 mg/dL.

Recording Time: 5 Minutes	
Respirations	14 breaths/min; adequate depth
Pulse	72 beats/min; strong and regular
Skin	Pink, warm, and dry
Blood pressure	132/70 mm Hg
Oxygen saturation (Sao$_2$)	99% (on oxygen)

You place the patient onto the stretcher and load her into the ambulance. The hospital you will be transporting to is 18 miles away.

4. What actions should be taken before you depart the scene?
5. Should lights and siren be used while transporting this patient? Why or why not?

services may operate with a non-EMT driver and a single EMT in the patient compartment.

Daily Inspections

Being fully prepared means that you and your team must inspect both the ambulance and equipment daily to ensure that all items are in proper working order. The ambulance inspection should include the following:

- Fuel level
- Oil level
- Transmission fluid level
- Engine cooling system and fluid levels
- Batteries
- Brake fluid
- Engine belts
- Wheels and tires, including the spare, if there is one. Check inflation pressure and look for signs of unusual or uneven wear.
- All interior and exterior lights
- Windshield wipers and fluid
- Horn
- Siren
- Air conditioners and heaters
- Ventilating system
- Doors. Make sure they open, close, latch, and lock properly.
- Communication systems, vehicle and portable
- All windows and mirrors. Check for cleanliness and position.

Check all medical equipment and supplies at least daily, including all the oxygen supplies, the jump kit, splints, dressings and bandages, backboards and other stabilization equipment, and the emergency obstetrics kit. Is the equipment functioning properly? Are the supplies clean? Are there enough of them? All battery-operated equipment, including the defibrillator, should be operated and checked each day **Figure 36-17**. Rotate the batteries according to an established schedule.

Safety Precautions

A final part of the preparation phase is reviewing safety precautions. These precautions, which include standard traffic safety rules and regulations, should be followed on every call. Check to make sure that safety devices, such as seatbelts, are in proper working order. Regardless of their location, portable oxygen tanks must always be secured by fixed clasps or housings. Never attempt to secure a tank to the stretcher or bench, unless using a commercially manufactured device specifically designed for such purpose; tanks may become projectiles if the ambulance is involved in a motor vehicle crash. In fact, all equipment in the cab, the rear, and in compartments needs to be secured appropriately.

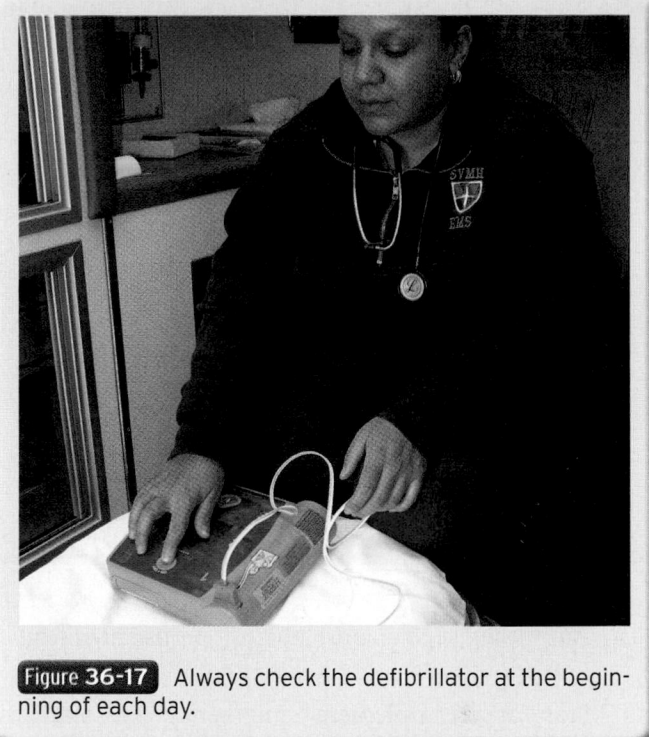

Figure 36-17 Always check the defibrillator at the beginning of each day.

Words of Wisdom

Because mechanical aspects of emergency work such as driving and moving patients strongly impact your safety and that of others, your service should have specific procedures for daily inspections. Following them protects you physically, and documenting your compliance is an important legal protection. Procedures should call for dating and either signing or initialing the check sheets and for storing them where they can be found later if needed.

■ The Dispatch Phase

Dispatch must be easy to access and in service 24 hours a day **Figure 36-18**. It may be operated by the local EMS or by a shared service that also covers law enforcement and the fire department. The dispatch center might serve only one jurisdiction, such as a single city or town, or it might be an area or regional center serving several communities or an entire county. In either case, it should be staffed by trained personnel who are familiar with the agencies they are dispatching and the geography of the service area. For every emergency request, the dispatcher should gather and record the following minimum information:

- The nature of the call
- The name, present location, and call-back telephone number of the caller
- The location of the patient(s)
- The number of patients and some idea of the severity of their conditions

Figure 36-18 The dispatcher is the key communications link throughout all phases of the ambulance run.

- Any other special problems or pertinent information about hazards or weather conditions

Many areas implement emergency medical dispatching, which provides the caller with instructions for patient care before the ambulance arrives.

■ En Route to the Scene

In many ways, the en route or response phase of the call is the most dangerous for you. Collisions between automobiles and emergency vehicles cause many serious injuries among EMS personnel. Techniques to make vehicle operation safer will be discussed later in this chapter. As you and your partner prepare to respond to the scene, make sure you fasten your seatbelts and shoulder harnesses before you move the ambulance. At this point, you should inform dispatch that your unit is responding and confirm the nature and location of the call. This is also an excellent time to ask for any other available information about the location. For example, you might learn that the patient is on the third floor or that the best door to use is around the side of the house.

While en route to the call, the team should prepare to assess and care for the patient. Review dispatch information about the nature of the call and the location of the patient. Assign specific initial duties and scene management tasks to each team member, and decide what type of equipment to take initially. Depending on your operating procedures, you may also decide which stretcher to take to the patient. Arriving at the scene safely and transporting the patient safely are two of the most challenging aspects of being an EMT. For more on how to operate the ambulance safely, refer to the defensive ambulance driving section.

■ Arrival at the Scene

On arrival at the incident, you will perform a scene size-up. After you complete your size-up, report to dispatch the nature of the incident if this is part of your local protocol. If other units are en route, providing dispatch with your size-up information will help to determine whether the units should continue to the scene. For example, if your size-up determines that the patient is potentially violent, the police unit should continue to the incident scene.

If you are the first to arrive on the scene of a mass-casualty incident, you should inform dispatch that you have arrived and give a brief report of what you see. Also report any unexpected situations, such as the need for additional units, a heavy rescue unit, or a HazMat team **Figure 36-19**. Do not enter the scene if there are any hazards to you. If there are hazards at the scene, the patient should be moved somewhere safe before you begin care. The patient may have to be moved by others if you are not appropriately equipped.

Immediately size up the scene by using the following guidelines:

- Look for safety hazards to yourself, your partner, and your patient(s).
- Evaluate the need for additional units or other assistance.
- Determine the mechanism of injury in trauma patients or the nature of the illness on medical calls.
- Evaluate the need to stabilize the spine.
- Make sure that you follow standard precautions. The type of care that you expect to give will dictate the personal protective equipment you should wear.

If you are the first EMT at the scene of a mass-casualty incident, quickly estimate the number of patients, and communicate with the incident commander **Figure 36-20**.

Figure 36-19 If you are the first to arrive on the scene of a mass-casualty incident, you should report to dispatch and ask for additional units, such as rescue, or HazMat units as needed.

Figure 36-20 At a mass-casualty incident, follow instructions from the incident commander assigning your roles, which may include assisting with triage, treating patients, or loading patients for transportation to the hospital.

Inform dispatch that additional units are needed at the scene. Mass-casualty incidents involve complex organization of personnel under the incident command system (see Chapter 38, *Incident Management*). In this system, individual EMTs may be assigned roles, for example, to begin the triage process, assist in treating patients, and load patients for transportation to a hospital.

Safe Parking

In assessing the situation, you must decide where to park the ambulance. Pick a position that will allow for efficient traffic control and flow around a crash scene. Do not park alongside the scene, as you may block the movement of other emergency vehicles. Instead, park about 100′ past the scene on the same side of the road **Figure 36-21**. Parking about 100′ before the scene to create a barrier between you and oncoming traffic may also be a good idea. It is best to park uphill and/or upwind of the scene if smoke or hazardous materials are present. Always leave your warning lights or devices on, and use extra caution if you must park on the backside of a hill or curve. Do the same when parking at night. Always park so as to provide a cushion of space between your vehicle and operations at the scene. Assume that someone may collide with your vehicle and strike personnel on the scene **Figure 36-22**.

Stay away from any fires, explosive hazards, downed wires, and structures that might collapse. Be sure to set the parking brake. If your vehicle is blocking part of the roadway, leave the emergency warning lights on. Leave only the flashing yellow lights on if your vehicle has them. Other drivers tend to drive toward emergency vehicles with flashing red or red and white lights. Within these safety guidelines, you should try to park your ambulance as close to the scene as possible to facilitate emergency medical care. If necessary, you can temporarily block traffic to unload equipment and to load patients quickly and safely. If you must do this, try to do it quickly so that traffic is not blocked any longer than necessary. Also, park in a location that will not hamper leaving the scene.

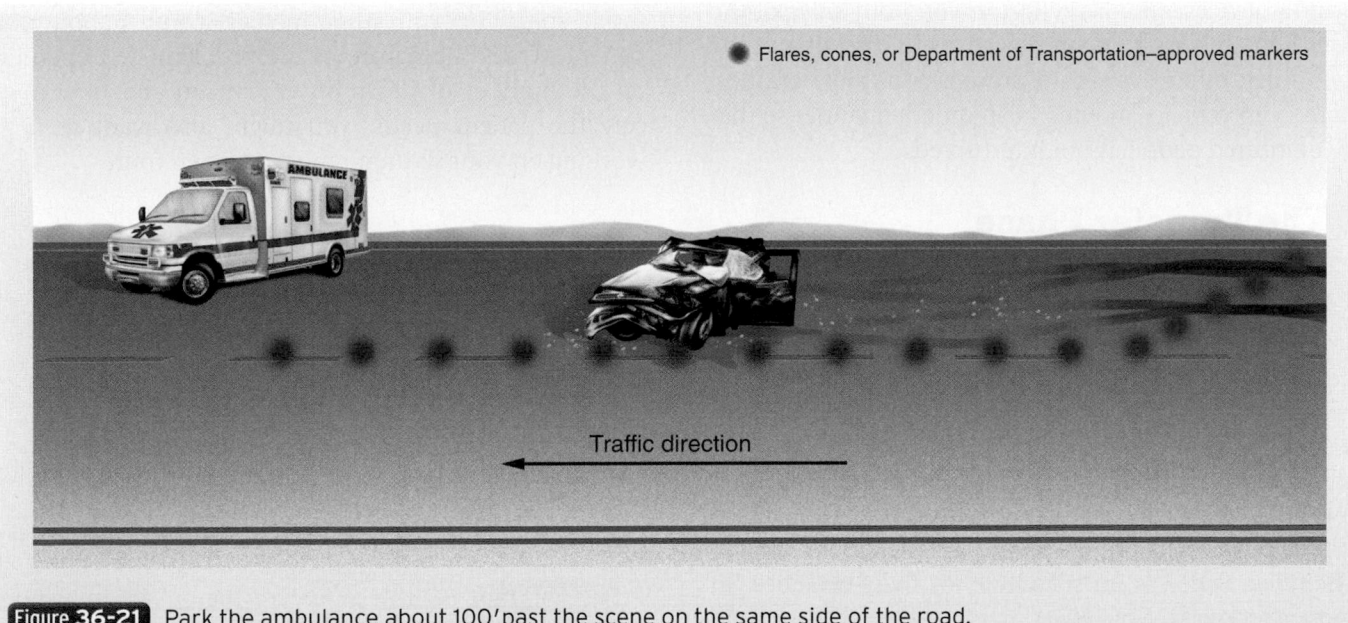

● Flares, cones, or Department of Transportation–approved markers

Traffic direction

Figure 36-21 Park the ambulance about 100′ past the scene on the same side of the road.

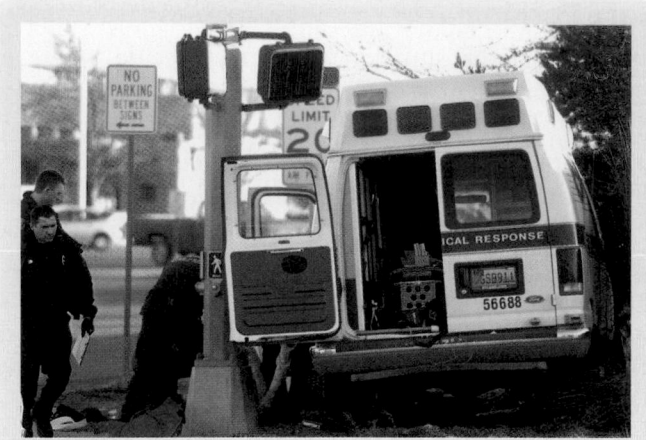

Figure 36-22 Unsafe parking of an ambulance can result in collision with other vehicles and injury of personnel.

Traffic Control

After ensuring your own safety, your first responsibility at a crash scene is to care for the patients. Only when all the patients have been treated and the emergency situation is under control should you be concerned with restoring the flow of traffic. If the police are slow to arrive at the scene, you might then need to take action.

The purposes of traffic control are to ensure an orderly traffic flow and to prevent another crash. Under ordinary circumstances, traffic control is difficult. A crash or disaster scene presents serious additional problems. Passing motorists often slow down and stare, paying little attention to the roadway in front of them. Some curiosity seekers may park down the road and return on foot, creating additional hazards. As soon as possible, place appropriate warning devices, such as reflectors, on both sides of the crash. Remember, the main objectives in directing traffic are to warn other drivers, to prevent additional crashes, and to keep vehicles moving in an orderly manner so that care of injured people is not interrupted.

■ The Transfer Phase

Many patients have said that one of the most frightening parts of being suddenly ill or injured is the ambulance ride to the hospital. Already anxious, a patient may be made more so by a fast, bumpy ride with a siren blaring. Sometimes, such a ride is truly lifesaving. However, in most cases, excessive speed is unnecessary and dangerous. What is necessary is that the patient be safely transported to an appropriate medical care facility in the shortest practical time. This takes common sense and defensive driving techniques. Speed is no substitute for these qualities. In almost every case, you will provide lifesaving care right where you find the patient, before moving the patient to

the ambulance. You may then begin less critical measures, such as bandaging and splinting. Next, you must package the patient for transport, securing him or her to a device such as a backboard, a scoop stretcher, or the wheeled ambulance stretcher. Then move to the ambulance, and properly lift the patient into the patient compartment.

No matter how careful the ambulance driver may be, riding to the hospital while lying down on a stretcher can be uncomfortable and even dangerous. So be sure to secure the patient with at least three straps across the body **Figure 36-23**. Use deceleration or stopping straps over the shoulders to prevent the patient from continuing to move forward in case the ambulance suddenly slows or stops. This is especially important if the patient is lying flat or secured to a backboard.

■ The Transport Phase

Inform dispatch when you are ready to leave with the patient. Report the number of patients you have, the name of the receiving hospital, and, in some jurisdictions, the beginning mileage of the ambulance. In most cases, even though you have already assessed and treated the patient, you should continue to monitor the patient's condition en route. These ongoing assessments may reveal changes in the patient's vital signs and overall condition. Be sure to recheck the patient's vital signs en route. The frequency of checking vital signs depends on the situation, but checking them every 15 minutes for a stable patient and every 5 minutes for an unstable patient is a practice that many services use. In addition, it is important that you continually reassess the patient's clinical situation and record and address new problems and the patient's responses to earlier treatment.

At this time, you should also contact the receiving hospital. Inform online medical control about your patient(s) and the nature of the problem(s). Depending on the number of EMTs on your team and how much care the patient needs, you might also want to begin working on your written report while en route.

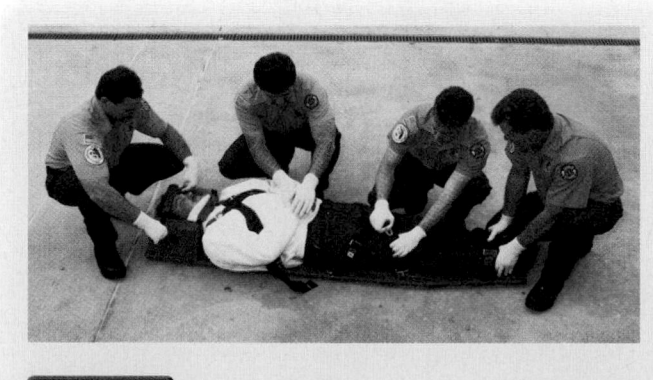

Figure 36-23 Be sure to secure the patient appropriately for protection during transport.

Finally, and most important, do not abandon the patient emotionally. Do not become so involved in paperwork and ongoing assessments that you ignore the patient's fears. You are there to help the patient as a person, so use this time to reassure him or her. Some patients, such as very young or older people, may benefit from added attention during transport. Be aware of the differing levels of need of different patients.

The Delivery Phase

Inform dispatch as soon as you arrive at the hospital and, depending on your jurisdiction, your ending mileage as well. Then follow these steps to transfer the patient to the receiving hospital:

1. Report your arrival to the triage nurse or other arrival personnel.
2. Physically transfer the patient from the stretcher to the bed directed for your patient.
3. Present a complete verbal report at the bedside to the nurse or physician who is taking over the patient's care.
4. Complete a detailed written report, obtain the required signatures, and leave a copy with an appropriate staff member.

The written report should include a summary of the history of the patient's current illness or injury with pertinent positives and negatives, mechanism of injury, and findings on your arrival. In addition, you should list vital signs and briefly mention relevant past medical or surgical history, as well as information regarding medication and allergies. Also, be sure to include any treatment and its effect during the prehospital setting.

While at the hospital, you may be able to restock any items that were used during the call, such as oxygen masks or dressings and bandages **Figure 36-24**. Remember, though, that your priority is transfer of the patient and patient information to the hospital staff. Restocking the ambulance comes second.

En Route to the Station

Once you leave the hospital, inform dispatch whether you are in service and where you are going. As soon as you are back at the station, you should do the following:

- Clean and disinfect the ambulance and any equipment that was used, if you did not do so before leaving the hospital **Figure 36-25**.
- Restock any supplies you did not get at the hospital.

The Postrun Phase

During the postrun phase, you should complete and file any additional written reports and again inform dispatch of your status, location, and availability.

You are also responsible for maintaining the ambulance so that it is safe and available on a moment's notice. This means you should perform routine inspections and refuel the vehicle. Use a written checklist to document needed repairs or replacement of equipment and supplies.

It is important that you know the meanings of the terms "cleaning," "disinfection," "high-level disinfection," and "sterilization," as follows:

- **Cleaning**. The process of removing dirt, dust, blood, or other visible contaminants from a surface or equipment.
- **Disinfection**. The killing of pathogenic agents by directly applying a chemical made for that purpose to a surface or equipment.
- **High-level disinfection**. The killing of pathogenic agents by the use of potent means of disinfection.

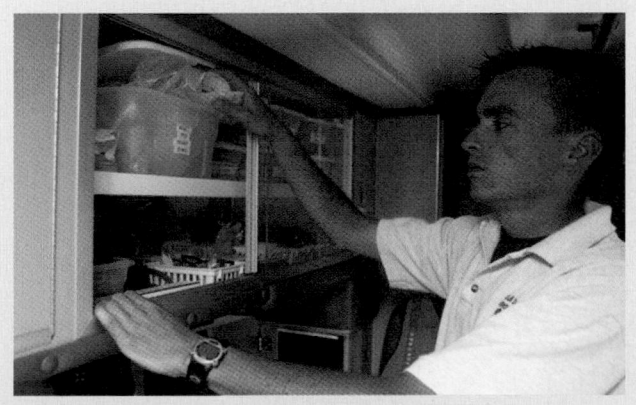

Figure 36-24 After transferring the patient and relating patient information to the hospital staff, you should restock any items that were used during the run.

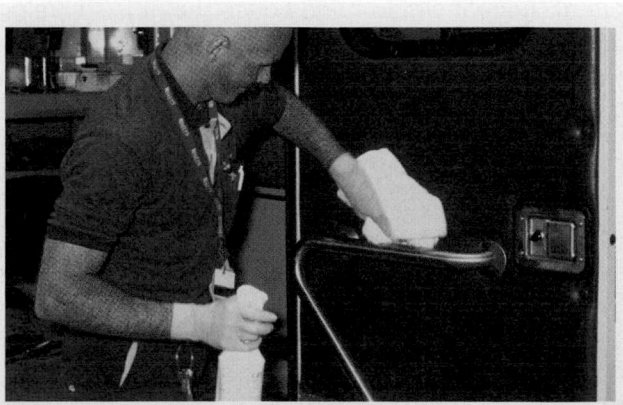

Figure 36-25 Be sure to clean and disinfect the ambulance and equipment at the station if you did not do so at the hospital.

- **Sterilization**. A process, such as the use of heat, that removes all microbial contamination.

A basic rule is to do the following after every call:

1. Strip used linens from the stretcher immediately after use, and place them in a plastic bag or in the designated receptacle in the emergency department.

2. In an appropriate receptacle, discard all disposable equipment used for care of the patient that meets your state's definition of medical waste. Most items will be considered general trash.

3. Wash contaminated areas with soap and water. For disinfection to be effective, cleaning must be done first.

4. Disinfect all nondisposable equipment used in the care of the patient. For example, disassemble the bag-mask device and place the components in a liquid sterilization solution as recommended by the manufacturer.

5. Clean the stretcher with an EPA-registered germicidal/virucidal solution or bleach and water at 1:100 dilution.

6. If any spillage or other contamination occurred in the ambulance, clean it up with the same germicidal/virucidal or bleach/water solution.

7. Create a schedule for routine full cleaning for the vehicle.

8. Have a written policy/procedure for cleaning each piece of equipment. Refer to the manufacturer's recommendations as a guide.

Defensive Ambulance Driving Techniques

Every year there are more than 6,000 ambulances involved in crashes, some of them fatal. Between 1991 and 2001, the Centers for Disease Control and Prevention found that there were 300 fatal ambulance accidents with 275 pedestrians and motorists killed. Of the passengers onboard the ambulances, there were 82 fatalities, of which 27 were EMS personnel. These statistics do not include the thousands of injured pedestrians, motorists, ambulance passengers, and EMS personnel Figure 36-26 . Learning how to properly operate your vehicle is just as important as learning how to care for patients when you arrive on the scene. An ambulance that is involved in a crash delays patient care, at a minimum, and may take the lives of the EMTs, or other motorists, or pedestrians at worst. The following section is provided to introduce you to safe driving techniques; however, you cannot become a proficient and safe ambulance driver without specialized training and practice. You are strongly encouraged to participate in a certified defensive driving program, such as those offered through your EMS organization, before attempting to operate an emergency vehicle.

■ Driver Characteristics

Not everyone who drives an automobile is qualified to drive an emergency vehicle. In some states, you must

You are the Provider: PART 4

While en route to the hospital, you reassess the patient, including her vital signs. She remains stable. You look out of the rear window and notice that it is still raining and that a motorist is following very closely behind the ambulance.

Recording Time: 12 Minutes	
Level of consciousness	Conscious and alert
Respirations	14 breaths/min; adequate depth
Pulse	80 beats/min; strong and regular
Skin	Pink, warm, and dry
Blood pressure	128/68 mm Hg
Sao$_2$	98% (on oxygen)

6. What factors should be considered when operating an ambulance in inclement weather?

7. What should the emergency vehicle operator do if he or she is being tailgated?

Figure 36-26 Each year, ambulance crashes are the cause of thousands of injuries to pedestrians, motorists, ambulance passengers, and EMS personnel.

successfully complete an approved emergency vehicle operations course before you are allowed to drive the ambulance on emergency calls. In any state, due diligence and caution are important characteristics, as are a positive attitude about your ability and tolerance of other drivers.

One basic requirement is physical fitness. Many crashes occur as a result of physical impairment of the driver. You should not be driving if you are taking medications that may cause drowsiness or slow your reaction times. These include cold remedies, analgesics, and tranquilizers. And, of course, you should never drive or provide medical care after drinking alcohol. Working long shifts or multiple consecutive shifts also puts drivers at risk for delayed reaction time and falling asleep behind the wheel. While many services have regulations against working beyond a prescribed number of hours, most do not take into account EMTs who may work for more than one service. It is your responsibility to notify your employer if you have worked a shift previously and feel unable to safely operate an emergency vehicle.

Words of Wisdom

The proper attitude is very important for an ambulance driver. The good judgment needed to drive an ambulance requires practice—even for the best drivers.

Another requirement is emotional fitness. Emotions should not be taken lightly. A person's personality often changes once he or she gets behind a steering wheel. Emotional maturity and stability are closely related to the ability to operate under stress. In addition to knowing exactly what to do, you must be able to do it under difficult conditions.

Having the proper attitude is very important for an ambulance driver. Never get behind the wheel of an emergency vehicle thinking that you can drive in any manner that pleases you simply because you have lights and sirens on. A greater responsibility is placed on the driver of an ambulance, and generally a lower burden of proof is needed to find that an EMT has caused a crash. As a rule, whenever lights and sirens are used on an emergency call and there is a crash, the actions of the emergency vehicle operator fall under the most scrutiny.

Words of Wisdom

Ambulance crashes that kill EMTs, patients, or occupants of other vehicles are disturbingly common. Most of them could be prevented by the driver of the ambulance. Attending thoroughly to your own driving skills, driving according to established standards, and dealing with any obvious deficiencies in your partner's driving skills are all crucial to your safety on the job.

■ Safe Driving Practices

The first rule of safe driving in an emergency vehicle is that speed does not save lives; good care does. The second rule is that the driver and all passengers must wear seatbelts and shoulder restraints at all times. These are the most important items of safety equipment on every ambulance. All EMTs should wear restraints en route to the scene and whenever they are not performing direct patient care. Patients should also be properly restrained. Studies show that fewer than half of all EMTs wear seatbelts while the vehicle is in emergency mode, and few wear lap belts in the rear compartment while patient care is being rendered. If you must remove your seatbelt to care for the patient, fasten the belt again as soon as possible. Also, unrestrained or improperly restrained patients and medical equipment (especially portable oxygen tanks) may become airborne during a collision and place you and your patient at an additional risk. All equipment and cabinets must be secured, as well as the patient and any passengers accompanying the patient.

Learn how your vehicle accelerates, corners, sways, and stops. You must become familiar with exactly how your particular vehicle will respond to steering, braking, and accelerating under various conditions.

Getting a feel for the proper brake pressure comes with experience and practice. Each vehicle has a different braking action. For example, the brakes on types I and III vehicles have a heavier feel than the brakes on a type II vehicle. Braking on a diesel-powered unit will be different from braking on an identically equipped gasoline-powered unit. Certain heavy vehicles use air brakes, which have yet another feel. Get to know each vehicle you drive, and be sure you understand its braking characteristics and the best downshifting techniques.

When you are driving an ambulance on a multilane highway, you should usually stay in the extreme left-hand (fast) lane. This allows other motorists to move over to the right when they see or hear you approach.

Table 36-6 lists further guidelines to follow when you are en route to a call.

Words of Wisdom

Centrifugal force is the tendency for objects to be pulled outward when rotating around a center. Vehicles are subject to this force when making a turn. If you must brake on a turn, brake gently while making the turn.

Siren Risk-Benefit Analysis

Whether responding to a call or transporting a patient from the scene to the hospital, the decision to activate the emergency lighting and sirens will depend on several factors such as local protocols, patient condition, and the anticipated clinical outcome of the patient. Some local protocols require that all responses to the scene use emergency lights and sirens, whereas other systems incorporate response modes based on the information received from dispatch. Regardless of your jurisdictional requirements, as the driver of the ambulance, you need to evaluate the risk versus benefit of your response mode. Numerous studies have been done to determine whether the emergency lights and sirens save time getting to the patient or getting the patient from the scene to the hospital. The findings of these studies show that while time is saved, the time that you do save is minimal.

As an EMT, you will also need to take into account the patient's condition before activating emergency lights and sirens. For example, patients who have experienced a seizure may have another seizure as a result of the rapid flash pattern of the emergency lighting. In cases such as this, it may be preferable to transport your patient without lights and sirens activated in an effort to minimize external stimuli and prevent the worsening of your patient's condition.

Driver Anticipation

Always assume that motorists around your vehicle have not heard your siren/public address (PA) system or seen you until proven otherwise by their actions. Ambulance drivers often make the mistake of assuming that motorists and pedestrians will do what is expected of them when an emergency vehicle is in the vicinity. Motorists may indeed pull over to the right and stop or drive as close to the curb as possible, but you cannot take this behavior for granted. At any time, a motorist might stop suddenly in front of the ambulance or pull to the left. Both of these motorist responses may result in a crash. Aggressive ambulance driving may have an opposite effect on motorists, as you may not allow for their reaction time to respond to your vehicle, or they may become nervous and not react in a rational manner. Whenever a motorist yields the right-of-way, the emergency vehicle operator should attempt to establish eye contact with the other driver. When anticipating how motorists may respond to your lights and sirens, always assume that they will react in a manner that may cause a collision. You can also look at the direction of the other vehicle's front tires to get an early indication of which way the vehicle will turn.

Table 36-6 Guidelines for Safe Ambulance Driving

1. Select the shortest and least congested route to the scene at the time of the dispatch.
2. Avoid routes with heavy traffic congestion; know alternative routes to each hospital during rush hours.
3. Avoid one-way streets; they may become clogged. Do not go against the flow of traffic on a one-way street, unless absolutely necessary.
4. Watch carefully for bystanders as you approach the scene. Curiosity seekers rarely move out of the way.
5. Park the ambulance in a safe place once you arrive at the scene. If you park facing into traffic, turn off your headlights so that they do not blind oncoming drivers unless they are needed to illuminate the scene. If the vehicle is blocking part of the road, keep your warning lights on to alert oncoming motorists; otherwise, turn them off.
6. Drive within the speed limit while transporting patients, except in the rare extreme emergency.
7. Go with the flow of the traffic.
8. Always drive defensively.
9. Always maintain a safe following distance. Use the "4-second rule": Stay at least 4 seconds behind another vehicle in the same lane.
10. Try to maintain an open space or cushion in the lane next to you as an escape route in case the vehicle in front of you stops suddenly.
11. Use your siren if you turn on the emergency lights, except when you are on a freeway.
12. Always assume that other drivers will not hear the siren or see your emergency lights.

It is often quite difficult for motorists to hear instructions called out over the vehicle's PA system, especially when their windows are rolled up. The PA system may actually make the situation worse because motorists may hesitate or make unexpected moves so that they can hear or follow instructions. Moreover, when the driver of the ambulance is shouting to motorists and pedestrians over the PA system, he or she is now distracted from the business of driving and forced to handle the microphone when both hands should be on the steering wheel. With this said, the ambulance's PA system should not be used often, if at all, during emergency driving.

Most important, you must always drive defensively. Never rely on what another motorist will do unless you get a clear visual signal. Even then, you must be prepared to take defensive action in the case of a misunderstanding, panic, or careless driving on the part of the other driver.

The Cushion of Safety

To operate an emergency vehicle safely, you must maintain a safe following distance from the vehicles in front of you and try to avoid being tailgated from behind. You also must ensure that the **blind spots** in your vehicle's mirrors do not prevent you from seeing vehicles or pedestrians on either side of the ambulance. Keeping a safe distance between your vehicle and the one in front of you, checking for tailgaters behind your ambulance, and keeping aware of vehicles potentially hiding in your mirror's blind spots are considered maintaining a **cushion of safety**. To ensure that you have enough reaction time and stopping distance from the vehicle in front of you, follow at a safe distance, allowing the motorist enough time to move over to the right. If the motorist does not move, you will need to allow for enough time to evade the vehicle. This entails driving about 4 or 5 seconds behind a vehicle traveling at an average speed.

While operating in emergency mode, tailgaters may follow your vehicle dangerously close in congested areas simply to use your ambulance to get through traffic. This poses a threat to the crew and patient. If the ambulance stops suddenly to avoid a collision (which should not happen if a cushion of safety was maintained), the tailgating vehicle could smash into the rear of the ambulance, possibly causing you to lose control and strike other vehicles or pedestrians. Always scan your rearview and side mirrors for cars following too closely. Instruct your partner to stay alert for such vehicles while he or she is in the rear compartment rendering care and to inform you about any tailgaters.

If you are being tailgated, never speed up to create more distance. The tailgater may, in turn, increase his or her speed to continue to follow you through traffic, thereby decreasing your cushion of safety and reaction time and increasing the time and distance needed to avoid a collision. Slamming on your brakes to scare the other driver usually does not work either and may cause a crash. The best method for distancing yourself from the vehicle is to slow down. Generally tailgaters are impatient and will speed up past you. You also can have your dispatcher contact the local police to let them know that someone is driving recklessly behind you.

Never, under any circumstance, get out of the ambulance to confront a driver. This will only delay your response to or transport of the patient and can lead to a dangerous situation. It is also unprofessional for you to become involved in a verbal altercation with any member of the public and may lead to disciplinary actions or termination, depending on your service's conduct regulations.

Finally, there are three blind spots around the ambulance that you cannot see with the mirrors:

- The mirror itself creates a blind spot, obstructing the view ahead and preventing the driver from seeing objects such as a pedestrian or car. Many new ambulance drivers will not be accustomed to the larger ambulance mirrors, which create a special hazard that the driver should be aware of. To eliminate this blind spot, you should lean forward in your seat so that the mirror does not obstruct the view, especially when making turns at intersections.
- The rear of the vehicle cannot be seen fully through the mirror and is therefore a blind spot. Because of the configuration of today's ambulances and the relative height of the vehicle, the rearview mirror generally gives the driver a view of the patient compartment at best and is not intended to be used for alerting the driver of a vehicle behind the ambulance. Because of this blind spot, many crashes occur when the ambulance is being backed up. It is highly recommended, and required in many jurisdictions, that a **spotter** be used to help you back up the vehicle. Rear-facing cameras are also helpful and much more common; however, they do not replace the use of a spotter if one is available.
- The side of the vehicle often cannot be seen through the driver and passenger mirrors at a certain angle. Entire cars may not be seen in the mirror, even though they are right next to the ambulance. To eliminate this problem, many EMS services place small rounded mirrors on the side mirrors to assist you in visualizing this blind spot. However, if these mirrors are not available, you need to lean forward or backward in the seat to help eliminate the blind spot. This is an especially important technique to use when shifting lanes or making turns. Remember,

just because you are turning from the appropriate lane does not mean that another motorist will not try to cut in beside the ambulance or that there is not a bicyclist riding on the side of the road next to you.

You should always scan your mirrors frequently for any new hazards, which may encroach on your cushion of safety. However, it is important to understand that your mirrors can give you false information and may hide people or cars. Adjust for blind spots in your mirrors by adjusting your position in the driver's seat. Always use a spotter whom you can see from the driver's side mirror and predetermined hand signals when backing up the ambulance.

The Problem of Excessive Speed

Even in extreme life-and-death emergencies, excessive speed is not indicated. In most cases, if you properly assess and stabilize the patient at the scene, speeding during transport is unnecessary and undesirable. No matter what the situation, you should never travel at a speed that is unsafe for the given road conditions.

Excessive speeds, in addition to being unnecessary, do not increase a patient's chance of survival. More often, using excessive speed while driving to and from the scene has resulted in crashes in which the EMT, the patient, and occupants of other vehicles are killed. It also makes it very difficult for the EMT attending to the patient to be able to provide care because of the rough ride typically created by the excessive speed and maneuvering. Excessive speed also cuts down on the driver's reaction time and increases the time and distance needed to stop the ambulance. While many state laws allow emergency vehicles to travel beyond the speed limits in emergencies, they offer little or no protection against prosecution should the driver become involved in a deadly collision. The legal ramifications of driving an emergency vehicle will be covered later in this section.

Recognition of Siren Syndrome

The siren may have a psychological effect on other drivers. Recognizing that the siren may increase the anxiety of other drivers will help you become aware of your or other drivers' tendencies to drive faster in the presence of sirens. Although a siren signifies a request for drivers to yield the right-of-way, drivers do not always do so.

Vehicle Size and Distance Judgment

Vehicle length and width are critical factors when maneuvering, driving, and parking an emergency vehicle. They are especially important with types I and III vehicles, which are wider than they look from behind the steering wheel. To brake and pass effectively, you must know the width and length of your vehicle. Crashes often occur when the vehicle is backing up. Always use someone outside the ambulance as a ground guide when you are backing up to avoid any incidents. Vehicle size and weight greatly influence braking and stopping distances. Good peripheral vision and depth perception will help you to judge distances, but they are no substitute for intensive training, experience, and frequent evaluation of the vehicle.

Road Positioning and Cornering

Road position means the position of the vehicle on the roadway relative to the inside or outside edge of the paved surface. To corner efficiently, you must know the vehicle's present position and its projected path. The aim is to take the corner at the speed that will put you in the proper road position as you exit the curve **Figure 36-27**. Whereas the fastest path through a curve is to enter high in the lane (positioned to the outside of the lane), apex low in the lane (to the inside of the lane), and exit high, these actions can result in misjudgment of speed and position, creating the danger of ending up in the opposing lane or off the road if you are traveling too fast. The safest path is to enter high in the lane (to the outside), and exit low (to the inside). This allows room for error if you enter the turn too fast.

Weather and Road Conditions

Whereas most ambulance collisions occur on clear days with dry roads, there are certain conditions that can limit your ability to control your vehicle. Ambulances do not handle the same as cars. Ambulances have a longer braking time and stopping distance. In addition, the

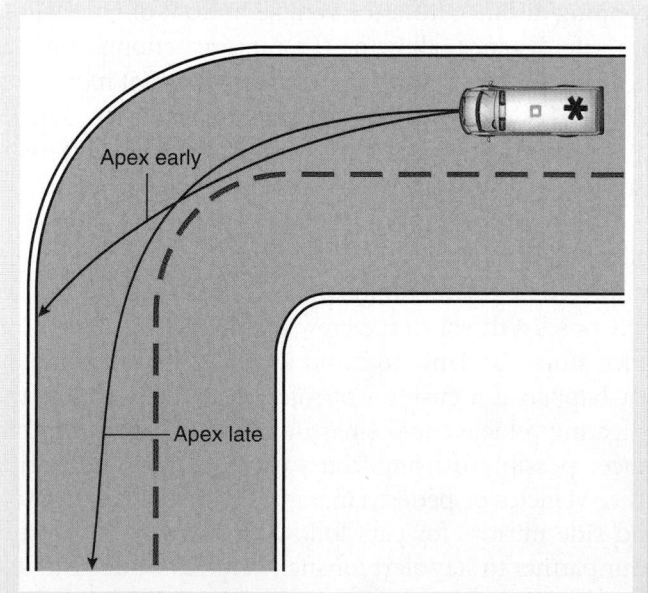

Figure 36-27 To keep the ambulance in the proper lane on a curve, you must know the vehicle's present position and projected path and take the corner at the correct speed.

weight of the ambulance is unevenly distributed, which makes it more prone to roll over. These factors, in addition to adverse environmental conditions, greatly increase the chance that an accident may occur. Therefore, you should be alert to changing weather, road, and driving conditions Figure 36-28 . Whether going to or coming from an emergency, you must modify your speed according to road conditions. Take warnings of ice or hazardous conditions seriously, and be prepared to take an alternative route, if necessary. During a major disaster, all public safety and emergency services should be coordinated. If you run into unexpected traffic congestion, notify the dispatcher so that other emergency vehicles can select alternative routes.

Even the most careful drivers will occasionally run into unexpected situations that may require special driving skills. However, if you drive at a speed that is appropriate for the weather and road conditions and maintain an adequate cushion of safety, you will minimize these situations. Therefore, it is safer if you decrease your speed in weather situations involving fog, rain, snow, or ice. The following are examples of conditions that require the emergency vehicle operator to decrease speed, increase following distance, and be alert.

Hydroplaning On a wet road, a tire usually displaces the water on the road surface and stays in direct contact with the road. However, at speeds of greater than 30 mph, the tire may be lifted off the road as water "piles up" under it; the vehicle may then feel as if it is floating. This problem is known as **hydroplaning**. At higher speeds on wet roadways, the front wheels may actually be riding on a sheet of water, robbing the driver of control of the vehicle. If hydroplaning occurs, you should gradually slow down without jamming on the brakes.

Water on the Roadway Wet brakes will not slow the vehicle as efficiently as dry brakes, and the vehicle may pull to one side or the other. If at all possible, avoid driving through large pools of standing water; often, you cannot tell how deep they are. If you must drive through standing water, make sure to slow down and turn on the windshield wipers. After driving out of the water, lightly tap the brakes several times until they are dry. If the vehicle is equipped with antilock brakes, apply a steady, light pressure to dry the brakes. Driving through moving water should be avoided at all times.

Decreased Visibility In areas where there is fog, smog, snow, or heavy rain, slow down after warning cars behind you. At night, use only low headlight beams for maximum visibility without reflection. You should always use headlights during the day to increase your visibility to other drivers. Also, watch carefully for stopped or slow-moving vehicles.

Ice and Slippery Surfaces A light mist on an oily, dusty road can be just as slippery as a patch of ice. Good all-weather tires and an appropriate speed will reduce traction problems significantly. If you are in an area that often has snowy or icy conditions, consider using studded snow tires or tire chains, if they are permitted by law. You should be especially careful on bridges and overpasses when temperatures are close to freezing. These road surfaces will freeze much faster than surrounding road surfaces because they lack the warming effect of the ground underneath.

Words of Wisdom

Although preventing skids and sliding is ideal, you are likely to skid or slide at least occasionally, especially if you live in climates with ice and snow. Your training should include the technique for correcting slides during turns. If you are likely to drive on ice and snow, you should practice control maneuvers until they become automatic—at low speeds in an area where there is no danger of collisions. Remember that four-wheel-drive and front-wheel-drive vehicles behave differently when sliding than rear-wheel-drive vehicles do. It is also important to remember that although four-wheel-drive vehicles have better traction for acceleration in slippery conditions, they do not stop any faster than two-wheel-drive vehicles.

Laws and Regulations

Regulations regarding vehicle operations vary from state to state and from city to city, but some regulations are the same regardless of location. Drivers of emergency vehicles have certain limited privileges in every state. However, these privileges do not lessen their liability in a crash. In fact, in most cases, the driver is presumed to be guilty if a collision occurs while the ambulance is operating with warning lights and a siren. Motor vehicle crashes are the single largest source of lawsuits against EMS personnel and services.

Figure 36-28 Modify your speed according to changing weather, road, and driving conditions.

While on an emergency call, emergency vehicles typically are exempt from normal vehicle operations. If you are on an emergency call and are using your warning lights and siren, you may be allowed to do the following:

- Park or stand in an otherwise illegal location
- Proceed through a red traffic light or stop sign, but never without stopping first.
- Drive faster than the posted speed limit
- Drive against the flow of traffic on a one-way street or make a turn that is normally illegal
- Travel left of center to make an otherwise illegal pass

Remember that these exemptions vary by state and local jurisdiction. Therefore, you should check your local statutes for regulations in your area.

An emergency vehicle is never allowed to pass a school bus that has stopped to load or unload children and is displaying its flashing red lights or extended "stop arm." If you approach a school bus that has its lights flashing, you should stop before reaching the bus, turn off your siren, and wait for the driver to make sure the children are safe, close the bus door, and turn off the warning lights. Only then may you carefully proceed past the stopped school bus.

Use of Warning Lights and Siren Three basic principles govern the use of warning lights and siren on an ambulance:

1. The unit must be on a true emergency call to the best of your knowledge.
2. Both audible and visual warning devices must be used simultaneously.
3. The unit must be operated with due regard for the safety of all others, on and off the roadway.

The siren is probably the most overused piece of equipment on an ambulance. In general, the siren does not help you as you drive, nor does it really help other motorists. Motorists who are driving at the speed limit with the windows up, the radio on, and the air conditioner or heater set on high cannot hear the siren until the ambulance is very close. If the radio is loud, they may not hear the siren at all.

If you do have to use the siren, be sure to tell the patient before you turn it on. Be especially mindful not to increase the speed of the ambulance just because the siren is in use. Always travel at a speed that will allow you to stop safely at all times, especially so that you are prepared for drivers who do not give you the right-of-way. Never assume that warning lights and sirens will allow you to drive through a congested area without stopping or slowing down. Slow down to ensure that all drivers are stopping as you approach an intersection, and proceed with caution. Remember, the siren is requesting that other drivers give you the right-of-way: It does not magically clear traffic.

Some ambulance headlights are equipped with a high-beam flasher unit. These are the most visible, effective warning devices for clearing traffic in front of the vehicle.

Right-of-Way Privileges State motor vehicle statutes or codes often grant an emergency vehicle, such as an ambulance, the right to disregard the rules of the road when responding to an emergency. However, in doing so, the operator of an emergency vehicle must not endanger people or property under any circumstances.

Consider this case: An ambulance is approaching an intersection that is controlled by a four-way stop sign. The ambulance, with lights and audible warning device functioning, proceeds through the intersection without slowing or stopping and crashes into a car coming from its right. Did the operator of the ambulance act appropriately by going through the intersection in this manner?

Right-of-way privileges for ambulances vary from state to state. Some states allow you to proceed through a red light or stop sign after you stop and make sure it is safe to go on. Other states allow you to proceed through a controlled intersection "with due regard," using flashing lights and siren. This means that you may proceed only if you consider the safety of all people who are using the highway. If you fail to use due regard, your service may be sued. If you are found to be at fault, you may personally have to pay punitive damages or face civil and criminal sanctions.

Get to know your local right-of-way privileges. Exercise them only when it is absolutely necessary for the patient's well-being. The use of lights and audible warning devices is a matter of state and local practice and protocol.

Use of Escorts Using a police escort is an extremely dangerous practice. When other motorists hear a siren and see a police car passing, they might assume that the police car is the only emergency vehicle and not see the ambulance. The only time an escort is justified is when you are in unfamiliar territory and truly need a guide more than an escort. In such cases, neither vehicle should use warning lights or sirens. If you are being guided, make sure that you follow at a safe distance.

Intersection Hazards Intersection crashes are the most common and usually the most serious type of collision in which ambulances are involved. Always be alert and careful when approaching an intersection. If you are on an urgent call and cannot wait for traffic lights to change, you should still come to a momentary stop at the light; look around for other motorists and pedestrians before proceeding into the intersection.

Motorists who "time the traffic lights" present a serious hazard. You may arrive at an intersection while the light is green. At the same time, a motorist who is timing the lights on the cross street arrives at the intersection. The motorist has a red light but knows that it is about to turn green and is expecting to go through. The stage is now set for a serious crash.

Another common intersection hazard occurs when the driver of one emergency vehicle follows another emergency vehicle through an intersection without assessing the situation carefully. A motorist who has yielded the right-of-way to the first vehicle may proceed into the intersection without expecting a second vehicle. You should exercise extreme caution in these situations. To signal motorists that a second unit is approaching, use a siren tone that is different from that of the first vehicle.

Highways When you are responding to an emergency call and you must travel on the highway, you should shut down your emergency lights and siren until you have reached the far left lane. Shutting down your emergency devices minimizes the possibility of confusion for drivers who might not know what to do or where to go.

When driving on a highway with your emergency devices activated, you should always travel in the far left-hand lane. Also known as the "passing lane," this allows the ambulance to safely pass vehicles, while still leaving a safety corridor on the left side of the ambulance in case of emergency or unexpected obstacles.

When you exit the highway, the same procedures should be followed as when you entered the highway: deactivate all emergency devices, move onto the off-ramp, and then reactivate the emergency lights and sirens if necessary.

Unpaved Roadways When you are required to drive the ambulance on an unpaved roadway, special care must be taken. Unpaved roadways often have uneven surfaces, as well as large potholes. While responding on this type of roadway, you must operate the vehicle at a lower speed and maintain a firm grip on the steering wheel in an effort to maintain complete control of the ambulance at all times.

School Zones When you respond through a school zone with your emergency lights activated, it is important to remember that lights and sirens tend to attract children to the roadway and create a potential hazard. In many states, it is unlawful for an emergency vehicle to exceed the speed limit in school zones regardless of the condition of the patient.

Distractions

As technology progresses, so do the distractions that you will face while operating the ambulance. Some ambulances have mobile dispatch terminals (MDT) and global positioning systems (GPS) that assist EMTs in determining the location of the call. However, these devices, along with using the vehicle's mounted mobile radio, listening to the vehicle's stereo, talking on the cell phone, and eating/drinking create additional driving hazards. While the ambulance is in motion, you should be focused on driving and anticipating roadway hazards, while your partner operates the MDT, GPS, and portable radios or activates the siren. Minimizing the potential distractions to the driver allows for a safer response and minimizes the potential for mishaps.

Driving Alone

Although driving alone is not a standard practice or even allowable in certain locations, you may be faced with responding to a scene by yourself in the ambulance and meeting your partner on the scene. When presented with this situation, you have additional duties and responsibilities, such as figuring out the safest route to the call, operating the radios and emergency warning devices, and preparing for the call mentally. Situations such as these demand your complete attention and focus.

Fatigue

Fatigue has many causes, such as stress, working the night shift, and lack of quality sleep in accordance with your body's circadian rhythms. As a result of these causes of fatigue, operating a large vehicle, such as an ambulance, creates a large risk. You must be able to recognize when you are fatigued. Do not be ashamed to admit it to yourself, your partner, or your supervisor. If you are feeling fatigued, you should be placed out of service for the remainder of the shift or until the fatigue has passed and you feel capable of operating the vehicle safely.

■ Air Medical Operations

Air ambulances are used to evacuate medical and trauma patients. They land at or near the scene and transport patients to trauma facilities every day in many areas. There are two basic types of air medical units: fixed-wing and rotary-wing, otherwise known as helicopters Figure 36-29. Fixed-wing aircraft generally are used for interhospital patient transfers over distances greater than 100 to 150 miles. For shorter distances, ground transport or rotary-wing aircraft are more efficient.

Specially trained medical flight crews accompany all air ambulance flights. Your role in fixed-wing aircraft transfers probably will be limited to providing ground transport for the patient and medical flight crew between the hospital and the airport.

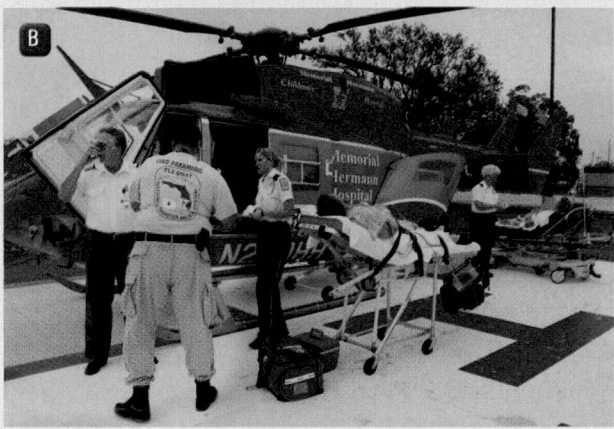

Figure 36-29 **A.** Fixed-wing aircraft are generally used to transfer patients from one hospital to another over distances greater than 100 to 150 miles. **B.** A rotary-wing aircraft, or helicopter, is used to help provide emergency medical care to patients who need to be transported quickly over shorter distances.

Words of Wisdom

You should be familiar with the capabilities, protocols, and methods for accessing helicopters in your area. Helicopter services provide training for EMTs in ground operations and safety.

Rotary-wing aircraft have become an important tool in providing emergency medical care. Trauma patient survival is directly related to the time that elapses between injury and definitive treatment. Most helicopters that are used for emergency medical operations fly well in excess of 100 mph in a straight line, without road or traffic hazards, straight to a hospital helipad. The crew may include EMTs, paramedics, flight nurses, and/or physicians.

You should be familiar with the capabilities, protocols, and methods for accessing helicopters in your area. Helicopter services provide training for EMTs in ground operations and safety. The following discussion is an introduction to safe operations and is not intended to be substituted for the more extensive courses available locally.

■ Helicopter Medical Evacuation Operations

A medical evacuation is commonly known as a **medivac** and is generally performed exclusively by helicopters. Most rural and suburban EMS jurisdictions and many urban systems have the capability to perform helicopter medivacs or have a mutual aid agreement with another agency such as police or hospital-based medivac service to provide such service. You should become familiar with the medivac capabilities, protocols, and procedures of your particular EMS service because they vary from service to service. The following are some general guidelines that you should be familiar with when considering whether to initiate a medivac operation.

Calling For a Medivac

Every agency has specific criteria for the type of patient who may receive medical evacuation and how and when to call for a medivac. These basic guidelines will help you to understand the process better.

- *Why call for a medivac?* The transport time to the hospital by ground ambulance is too long considering the patient's condition. Road, traffic, or environmental conditions limit or completely prohibit the use of a ground ambulance. The patient requires advanced care that you are unable to provide, such as administering pain medications or other specialized medications and inserting advanced airways. There are multiple patients who will overwhelm resources at the hospital reachable by ground transport.

- *Who receives a medivac?* Medical evacuations should be utilized for patients with time-dependent injuries or illnesses. They are widely used for patients suspected of having a stroke, heart attack, or serious spinal cord injury, such as injuries sustained in a motor vehicle collision or while diving into a pool or horseback riding. Serious conditions that may require the use of helicopter medivacs may be found in remote areas and involve scuba diving accidents, near-drownings, or skiing and wilderness accidents. Other patients who may warrant the use of medical evacuation are trauma patients and candidates for limb replantation (for amputations), a burn center, a hyperbaric chamber, or a venomous bite center. Because specific criteria differ between services, you must be familiar with the criteria used to call for this lifesaving service.

- *Whom do you call?* Generally your dispatcher must be notified first. In some regions, after the medivac has been initiated, the ground EMS crew may be able to access the flight crew on a specially designated radio frequency for one-on-one communications. If available, it is important to keep this frequency clear of chatter and long, drawn-out communications. You may be asked to give a brief presentation or update on the patient's condition. In this case, you should gather your thoughts and speak clearly and concisely, avoiding information that is not immediately pertinent. Another important topic of communication between the ground and flight EMS crews will be where to land the helicopter. This will be covered in the next section.

■ Establishing a Landing Zone

Although a helicopter can fly straight up and down, this is the most dangerous mode of operation. The safest and most effective way to land and take off is similar to that used by fixed-wing aircraft. Landing at a slight angle allows for safer operations. Takeoff combines a gradual lift and forward motion to travel up and out on a slight angle.

An important part of conducting a medivac is choosing the best location. Establishing a landing zone is the responsibility of the ground EMS crew. It involves more than simply looking for a clear space. You must be prepared to take action to make certain that the flight crew is able to land and take off safely. Things to do and considerations when selecting and establishing a landing zone include the following:

- The area should be a hard or grassy level surface that measures $100' \times 100'$ (recommended) and no less than $60' \times 60'$ **Figure 36-30**. If the site is not level, the flight crew must be notified of the steepness and direction of the slope.
- The area must be cleared of any loose debris that could become airborne and strike the helicopter or the patient and crew. This includes branches, trash bins, flares, accident tape, and medical equipment and supplies.
- You must survey the immediate area for any overhead or tall hazards such as power lines or telephone cables, antennas, and tall or leaning trees. The presence of these must be relayed immediately

You are the Provider: PART 5

You reassess the patient and determine that her condition has deteriorated significantly. She has experienced a significant decrease in level of consciousness, is unable to talk, and cannot move the entire left side of her body. Her right pupil is now dilated and minimally reactive.

Recording Time: 22 Minutes	
Level of consciousness	Confused; unable to speak
Respirations	16 breaths/min; adequate depth
Pulse	90 beats/min; strong and regular
Skin	Pink, warm, and dry
Blood pressure	132/74 mm Hg
Sao$_2$	97% (on oxygen)

You ask the driver to upgrade your transport mode. After treating the patient accordingly, you call in your radio report. You are approximately 9 miles away from the hospital, and the driver must proceed through a school zone to get to the hospital.

8. **Where do most serious ambulance crashes occur, and what should the emergency vehicle operator do to help avoid a collision?**

9. **Are there any special privileges given to emergency vehicles when driving through a school zone?**

Figure 36-30 A landing area should be a level surface measuring 100' × 100'.

to the flight crew because an alternative landing site may be required. The flight crew may request that the hazard be marked or illuminated by weighted cones or by positioning an emergency vehicle with its lights turned on next to or under the potential hazard.

- To mark the landing site, use weighted cones or position emergency vehicles at the corners of the landing zone with headlights facing inward to form an X. This procedure is essential during night landings as well. Never use accident tape or people to mark the site. The use of flares is also not recommended, because not only can they become airborne, but they also have the potential to start a fire or cause an explosion.

- Make sure that all nonessential persons and vehicles are moved to a safe distance outside of the landing zone.

- If the wind is strong, radio the direction of the wind to the flight crew. They may request that you improvise some form of wind directional device to aid their approach. A bed sheet tightly secured to a tree or pole may be used to help the crew determine wind direction and strength. Never use tape.

■ Landing Zone Safety and Patient Transfer

Helicopter safety is a combination of good sense and a constant awareness of the need for personal safety. You should stay away from the helicopter and go only where the pilot or crewmember directs you. The most important rule is to keep a safe distance from the aircraft whenever it is on the ground and "hot," which means when the helicopter blades are spinning. Most of the time, the rotor blades will remain running because the flight crew

does not generally expect to remain on the ground for a long time. This means that all EMTs should stay outside the landing zone perimeter unless directed to come to the aircraft by the pilot or a member of the flight crew. Usually, the flight crew will come to the EMTs carrying their own equipment and not require any assistance inside the landing zone. If you are asked to enter the landing zone, stay away from the tail rotor; the tips of its blades move so rapidly that they are invisible. Never approach the helicopter from the rear, even if it is not running. If you must move from one side of the helicopter to another, go around the front. Never duck under the body, the tail boom, or the rear section of the helicopter. The pilot cannot see you in these areas.

Another area of concern is the height of the main rotor blade. On many aircraft, it is flexible and may dip as low as 4' off the ground **Figure 36-31** . When you approach the aircraft, walk in a crouched position. Wind gusts can alter the blade height without warning, so be sure to protect equipment as you carry it under the blades. Air turbulence created by the rotor blades can blow off hats and loose equipment. These objects, in turn, can become a danger to the aircraft and personnel in the area.

When accompanying a flight crew member, you must follow directions exactly. Never try to open any aircraft door or move equipment unless a crew member tells you to. When told to approach the aircraft, use extreme caution and pay constant attention to hazards.

Keep the following guidelines in mind when operating at a landing zone:

- Become familiar with helicopter hand signals used within your jurisdiction **Figure 36-32** .
- Do not approach the helicopter unless instructed and accompanied by flight crew.
- Make certain that all patient care equipment is properly secured to the stretcher and that the

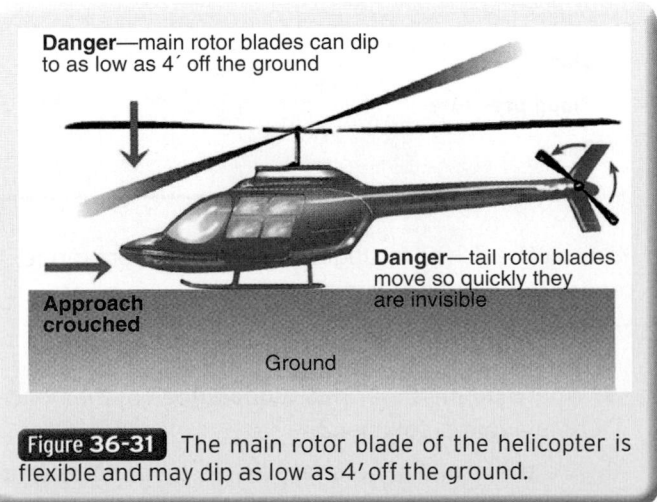

Figure 36-31 The main rotor blade of the helicopter is flexible and may dip as low as 4' off the ground.

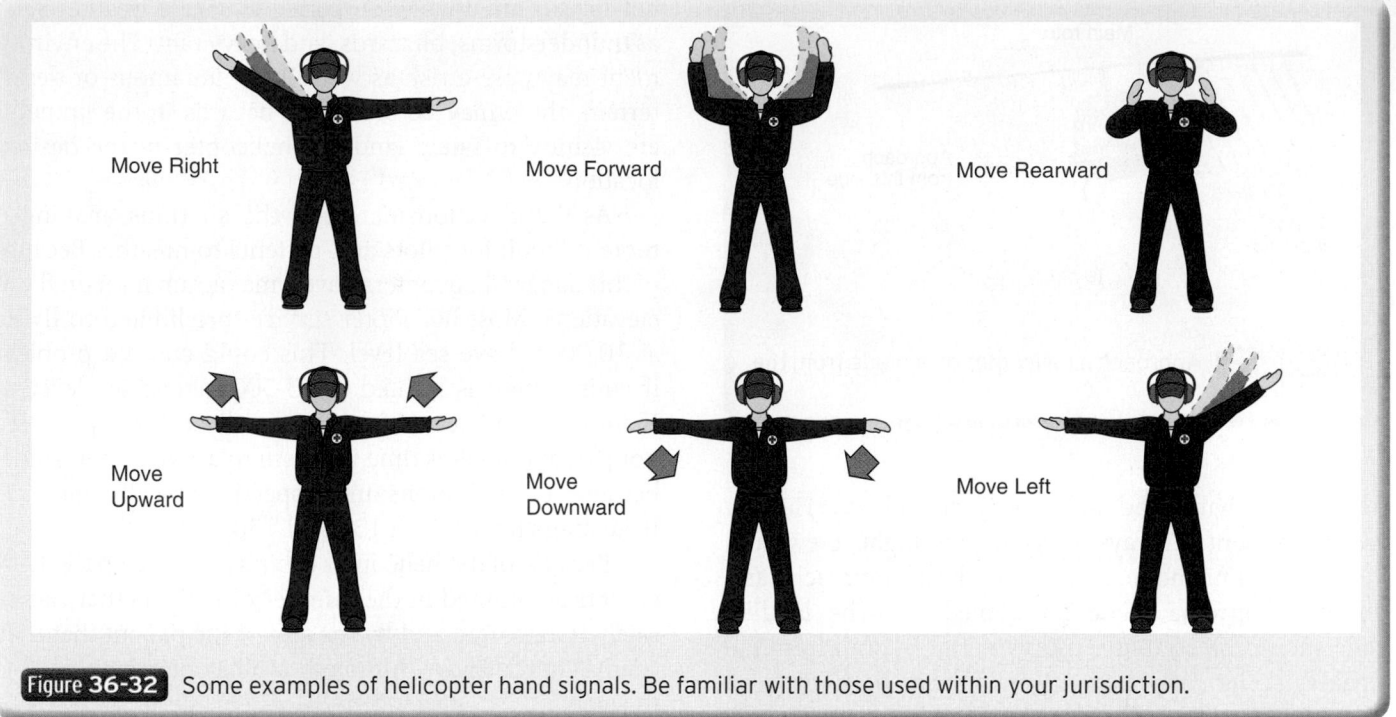

Figure 36-32 Some examples of helicopter hand signals. Be familiar with those used within your jurisdiction.

patient is fastened as well. This includes oxygen tanks, cervical collars, and head immobilizers. Any loose articles or belongings such as hats, coats, or bags that belong to the patient or crew should not be brought into the landing zone and will likely need to be transported to the hospital by ground.

- Be mindful that some helicopters may load patients from the side, whereas others have rear-loading doors. Regardless of where the patient is being loaded, always approach the aircraft from the front unless otherwise instructed by the flight crew. Always take the same path when exiting away from the helicopter, moving the patient headfirst.
- Smoking, open lights or flames, and flares are prohibited within 50′ of the aircraft at all times.

Communication Issues

When interacting with other agencies, there is always the possibility of communication issues. Medivacs are no exception. While the typical EMS service has its specific and well-defined jurisdiction, medivacs respond to service requests throughout a large, multijurisdictional area. Because of this large area with numerous jurisdictions, the medivac interacts with many services on a multitude of different radio frequencies.

To prevent any miscommunication, when the request is made for a medivac response, the request should include a ground contact radio channel (typically a pre-established mutual aid channel), as well as a call sign of the unit that the medivac should make contact with.

■ Special Considerations

Night Landings

Nighttime operations are considerably more hazardous than daytime operations because of the darkness. The pilot may fly over the area with the helicopter's lights on in order to spot obstacles and overhead wires, which can be hard to see. Do not shine spotlights, flashlights, or any other lights in the air to help the pilot; they may temporarily blind the pilot. Instead, direct low-intensity headlights or lanterns toward the ground at the landing site from opposite corners to form an X at the center of the landing zone. Turn off all headlights or lanterns that are facing in the direction of the aircraft once it has landed; after the helicopter has landed, you should not aim lights near the aircraft. Always make certain that the flight crew is aware of any overhead hazards or obstructions, and illuminate these if possible.

Landing on Uneven Ground

If the helicopter must land on a grade (uneven surface), extra caution is advised. The main rotor blade will be closer to the ground on the uphill side. In this situation, approach the aircraft from the downhill side only or as directed by the flight crew **Figure 36-33**. Do not move the patient to the helicopter until the crew has signaled that they are ready to receive you.

Medivacs at Hazardous Materials Incidents

The flight crew must be notified immediately of the presence of HazMat at the scene. The aircraft generates

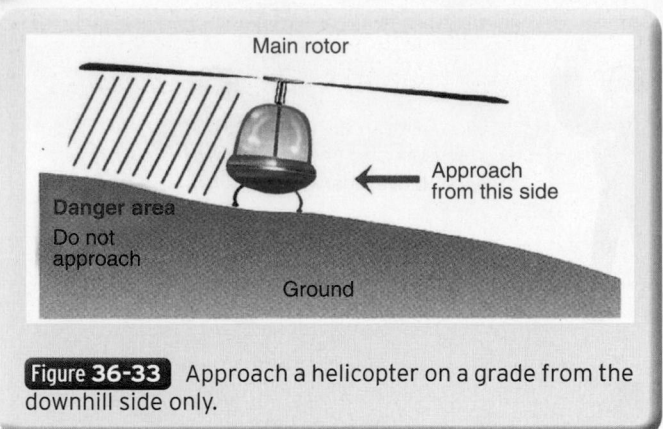

Main rotor

Approach from this side

Danger area
Do not approach

Ground

Figure 36-33 Approach a helicopter on a grade from the downhill side only.

tremendous wind and may easily spread any HazMat vapors present. Always consult the flight crew and incident commander about the best approach and distance from the scene for a medivac. The landing zone should be established upwind and uphill from the HazMat scene. Any patients who have been exposed to a HazMat must be properly decontaminated before they can be loaded into the aircraft. For proper procedures at HazMat incidents, refer to Chapter 38, *Incident Management*.

■ Medivac Issues

While making the decision to request medivac, several important factors need to be taken into consideration. These factors are weather, the environment/terrain, altitude, airspeed limitations, cabin size, and cost. Typically, helicopters are unable to operate in severe weather such as thunderstorms, blizzards, and heavy rain. The environment may pose a risk as well. In mountainous or desert terrain, there may be too many hazards in the immediate vicinity to safely land the helicopter in the desired location.

As the elevation increases, the air thins, making it more difficult for pilots and patients to breathe. Because of this danger, helicopters have a maximum limit on flight elevations. Most helicopter services are limited to flying at 10,000′ above sea level. This could create a problem if your patient is located at 13,500′ above sea level. It is important to remember that medivac helicopters are not jets, and it takes time for them to arrive on the scene, because of limitations in airspeed. Typically medivac helicopters fly between 130 and 150 mph.

Because of the helicopter cabin's confined space, helicopters are limited in the number of patients that can be safely transported and by the size of the patient that they can safely transport. Although a helicopter may be able to safely lift off with a 500-lb patient, because of his or her size and girth, it may be impossible to safely fit and secure the patient into the cabin area.

Typical medivac flights cost in the range of $8,000 to $10,000, whereas the typical ambulance transport costs $400 to $1,000. The decision to request a medivac should not be based on the perceived ability of the patient to pay the bill, but rather on the medical necessity. However, the cost factor should be considered so as not to create any unnecessary financial hardship, while still providing the best care for the patient.

You are the Provider: SUMMARY

1. What attributes should an emergency vehicle operator possess?

Clearly, not everyone who drives an automobile is qualified to operate an emergency vehicle. In many states, EMTs are required to successfully complete an emergency vehicle operator course before they are allowed to drive the ambulance in emergency mode. Many EMS systems require completion of an emergency vehicle operator course as well as clearance by a supervisor or field training officer.

All the training in the world cannot replace reasonable judgment and common sense; these are the most crucial attributes that *any* vehicle operator must possess, not just an emergency vehicle operator.

Regardless of your systems' requirements for operating an emergency vehicle, diligence and caution are important characteristics, as is a positive attitude about your driving abilities and tolerance of other motorists.

Many crashes involving emergency vehicles occur because the operator was physically impaired; therefore, a basic requirement is physical fitness. Avoid operating an emergency vehicle if you are taking prescription or over-the-counter medications that may cause drowsiness or slow your reaction times (eg, sedatives, cold remedies, pain relievers). Another hard and fast rule is that you should *never* operate an emergency vehicle or provide patient care after drinking alcohol.

Long shifts or multiple consecutive shifts also affect a vehicle operator's physical ability because fatigue and exhaustion increase reaction times and increase the risk of literally falling asleep behind the wheel. It is *your* responsibility to notify your supervisor if you do not feel that you can safely operate an emergency vehicle because of fatigue or exhaustion.

Just as important as physical fitness is emotional fitness. Do not take your emotions lightly; attitudes and personalities often change when getting behind a steering wheel. The ability to safely operate an emergency vehicle under stress relies heavily on emotional maturity and stability. Not only must you know exactly what to do, you must also be able to do it under difficult conditions.

Your attitude when operating an emergency vehicle is absolutely critical. You must be able to safely get to the scene or receiving hospital, while driving with due regard for the safety of other motorists on the road. When operating an emergency vehicle, you do not "own the road," nor can you do as you please simply because your lights are on and your siren is blaring! This type of attitude places your own safety in jeopardy, as well as the safety of your partner, the patient, and other motorists on the road.

2. What factors should you consider before responding to the scene?

When you receive a call, regardless of its nature, do not simply jump in the ambulance, buckle up, and drive. Before you depart your station, you need to consider a number of factors that will facilitate an expedient and safe response. Some EMS systems use global positioning systems to guide them to the scene; others use a map book. Regardless of what method you use for directions to the location, you must know exactly where you are going and then notify the dispatcher that you are en route.

If there is any question or confusion regarding the location of the call, ask the dispatcher for confirmation. If you are unfamiliar with the location, ask if there is a cross street that you can use as a point of reference. Choose the shortest and least congested route to the scene, taking into account the time of day because this often influences the number of other motorists on the road. In this case, it is 3:15 PM, the traffic is heavy, and it is raining; therefore, the most obvious route to the scene may not be the quickest and safest.

Another factor to consider before you respond is the weather. In this case, it is raining; therefore, you know that the roads will be slippery. Although you must go into "safety mode" prior to responding—regardless of the weather conditions—you must use extra caution during inclement weather. If it is raining hard, you should be aware of physical barriers, such as low water crossings, that may impede your ability to get to the scene.

There are numerous other factors to consider before responding—some of which may be unique to your location. However, if safety is at the forefront of your mind, your ability to get to the scene *quickly and safely* will be maximized.

3. What equipment and supplies are typically taken from the ambulance to the patient's side?

The equipment and supplies that you take from the ambulance to the patient's side are largely determined by the nature of the call and the location of the patient in relation to the ambulance.

You should wear the appropriate personal protective equipment (PPE) on *every* call, and it should be donned *before* making contact with the patient. At a minimum, this includes exam gloves; however, depending on the situation, you may need protective eyewear, a mask, or other PPE. Additionally, you should take a portable radio with you in the event that you need to quickly summon additional resources to the scene.

You should carry a portable treatment bag (often referred to as a jump kit) to the patient on every call. The jump kit should contain anything that you might need within the first 5 minutes after making contact with the patient—regardless of his or her chief complaint. A typical jump kit contains trauma shears, bleeding control supplies (ie, sterile gauze, tape, trauma dressings), airway and ventilation equipment (ie, oral/nasal airways, bag-mask device), vital sign equipment (ie, blood pressure cuff, stethoscope, penlight, pulse oximeter), and certain emergency medications (ie, oral glucose, activated charcoal, aspirin, epinephrine).

Ideally, the jump kit should also contain a portable oxygen cylinder; however, this may be contained in a separate airway kit, along with oral/nasal airways and a bag-mask device, if not carried in the jump kit.

Some EMS systems have a separate jump kit for pediatric patients that contains equipment tand supplies appropriate for infants and children, such as a pediatric blood pressure cuff, pediatric nonrebreathing masks, and infant and child bag-mask devices.

The decision to bring the wheeled ambulance stretcher to the patient is determined by the nature of the call and whether other personnel arrive at the scene before you. For example, if the call is for a patient in cardiac arrest and you and your partner are the first to arrive at the scene, you would take in the jump kit, automated external defibrillator, airway kit (if not a part of your jump kit), and portable suction unit. As additional responders and/or law enforcement personnel arrive, one of those personnel can retrieve the stretcher.

By contrast, if you arrive at the scene and quickly determine that the patient is conscious, alert, and requesting transport to the hospital, you or your partner could begin assessing the patient while the other retrieves the stretcher.

Whereas it is clearly not practical, not to mention time-consuming, to take every piece of portable equipment to the patient's side, you do not want to make repeated trips back and forth to the ambulance. Use the information provided to you during the initial dispatch, as well as additional information that may be provided while you are en route to the scene, to determine which equipment and supplies are appropriate to take to the patient's side.

4. What actions should be taken before you depart the scene?

Prior to loading the patient into the ambulance, you must ensure that he or she is properly and safely secured to the stretcher. Place at least three straps across the patient's body: one under the arms and across the chest, another across the pelvis, and a third across the lower extremities. Use deceleration straps (ie, a shoulder harness) to prevent the patient from continuing to move forward in the event that the ambulance suddenly slows or stops. After loading the patient into the ambulance, recheck these straps to ensure that they are securely fastened before the ambulance moves.

If anyone will be accompanying the patient, ensure that the person is properly secured with a seatbelt. Unless the patient is a child, family members or friends should ride in the front seat of the ambulance.

In most cases, you typically will not simply load the patient into the ambulance and immediately depart the scene. Certain actions are usually carried out before transport begins. Once the ambulance begins to move, you want to minimize your movement in the back of the ambulance as much as possible. If the patient is attached to portable oxygen, for example, transfer him or her to the onboard oxygen.

If the patient requires additional assessment or treatment that is easier to perform when the ambulance is stationary, and his or her condition permits, do so before the ambulance begins to move. Examples of these actions include reassessing breath sounds and obtaining a manual blood pressure reading.

Secure any items or pieces of equipment to prevent them from becoming projectiles in the event of a crash. The jump kit, for example, should not be placed on the ambulance floor; place it in an outside compartment or otherwise secure

it—either to the bench seat or the seat in front of the patient. If a portable oxygen cylinder was used at the scene, secure it properly after transferring the patient to the onboard oxygen.

After you have carried out any additional assessment and treatment the patient may require, you have ensured that he or she is safely fastened to the stretcher, and any equipment has been safely secured, you should position yourself in such a way that you can monitor the patient. After doing so, ensure that you are securely fastened. Some ambulances have a lap belt on the bench seat; others have a shoulder harness.

Finally, the driver should notify the dispatcher when you are ready to depart the scene. He or she should report the number of patients being transported, the name of the destination facility, and, in some jurisdictions, the beginning mileage and transport mode.

5. Should lights and siren be used while transporting this patient? Why or why not?

The transport mode (eg, lights and siren [emergency mode] versus no lights and siren [nonemergency mode]) is based on the patient's present condition and his or her anticipated clinical outcome. In this case, the patient is stable; she is conscious and alert, her vital signs are stable, and she has no life-threatening conditions (eg, airway problems, uncontrolled bleeding). Therefore, the use of lights and siren during transport is not indicated.

Not only does the siren have a psychological effect on the patient, which often causes unnecessary anxiety, it also has a psychological effect on the emergency vehicle operator and other drivers on the road. Other motorists often drive faster to get out of the way, and the emergency vehicle operator often increases the speed of the ambulance. When the speed of the ambulance increases, so does the risk of an ambulance crash.

Use of lights, siren, and speed does not save lives; good patient care does. If the patient has been properly assessed and determined to be stable, lights and siren are not indicated. If the patient's condition deteriorates en route, you can always upgrade your transport mode.

In most states, emergency vehicle operators are given certain limited privileges when lights and siren are being used. However, these privileges do not lessen their legal liability in a crash. In fact, in most cases, the driver is presumed to be at fault if a crash occurs while the ambulance is operating with warning lights and siren. If you are using lights and siren, you may be able to exceed the posted speed limit (within reason, and with safety in mind), and proceed through red lights and stop signs (after stopping first!).

If lights and siren are being used, you must remember that they signify a request for drivers to yield the right-of-way. They are not used to "clear the traffic," nor do they replace your responsibility to safely operate the vehicle and drive with due regard for the safety of others on the road.

You must always evaluate the risk versus benefit of your response or transport mode. Numerous studies have been conducted to determine whether the emergency lights and siren save time getting to the patient or getting the patient to the hospital. The findings of these studies show that while time is saved, the amount of savings is minimal.

6. What factors should be considered when operating an ambulance in inclement weather?

Inclement weather, such as fog, rain, snow, and ice, creates special hazards that require the emergency vehicle operator to adjust the way that he or she is operating the vehicle to maintain a cushion of safety. Maintaining a cushion of safety involves keeping a safe distance between your vehicle and the one in front of you, checking for tailgaters behind the ambulance, and looking for vehicles that are potentially hidden in your side mirror's blind spots.

Heavy rain and wet roads can result in hydroplaning of the vehicle and decreased visibility. The vehicle's brakes also become wet, affecting the ability of the operator to safely slow down or come to a complete stop.

In general, when operating the ambulance on wet roads, the emergency vehicle operator should decrease the speed of the vehicle accordingly and increase the distance between the ambulance and the vehicle it is following.

On wet roads, a tire usually displaces the water on the road surface and remains in direct contact with the road. However, at speeds of greater than 30 mph, the tire may be lifted off the road as the water "piles up" under it; the vehicle feels as if it is floating (hydroplaning). At higher speeds on wet roadways, the front tires may actually be riding on a sheet of water, decreasing the operator's control of the vehicle. If hydroplaning occurs, you should *gradually slow down without slamming on the brakes*.

Because you can often not tell how deep large pools of standing water are, you should avoid driving through them if at all possible. However, if you must drive through standing water, slow the vehicle down and proceed with caution. After driving out of the pool of water, lightly tap the brakes several times until they are dry. If the vehicle is equipped with antilock brakes, apply a steady, light pressure to dry the brakes. Never attempt to drive through moving water; if you encounter a flooded crossing with fast-moving water; notify the dispatcher of the situation and seek an alternative route.

Never assume that other motorists on the road will take the same safety measures that you will when driving on wet roads. Safe vehicle operation relies on little more than reasonable judgment and common sense; however, not everyone practices (or possesses) these skills.

7. What should the emergency vehicle operator do if he or she is being tailgated?

If you are being tailgated, never speed up to create more distance. The tailgater may, in turn, increase his or her speed to continue to follow you through traffic, thereby decreasing your cushion of safety and reaction time and increasing the time and distance needed to avoid a collision with any vehicles that you may be following.

Slamming on your brakes should also be avoided. In addition to increasing the risk of injury to your partner and patient in

the back of the ambulance, it usually does not result in getting the tailgater to back off and may cause him or her to collide with the rear of the ambulance.

If you are being tailgated, the best method for distancing yourself from the tailgater is to slow down. Tailgaters are generally impatient people and they will usually speed up and pass you. If another vehicle is persistently tailgating you, you should have the dispatcher contact law enforcement personnel and let them know that someone is driving recklessly behind you.

Never, under *any* circumstances, stop the ambulance and get out to confront the person who is tailgating you! Not only will this cause a delay in getting your patient to the hospital, it can lead to a dangerous situation. Remember, tailgaters are usually impatient people so there is always the risk that they may become violent if confronted. It is also unprofessional for you to engage in a verbal altercation with any member of the public—regardless of what the person is doing.

8. Where do most serious ambulance crashes occur, and what should the emergency vehicle operator do to help avoid a collision?

Intersection crashes are the most common—and usually the most serious—type of collision in which ambulances are involved. Therefore, the emergency vehicle operator must be especially careful and alert when approaching an intersection. If you are operating the vehicle in emergency mode and cannot wait for traffic lights to change, *you should still come to a complete stop*, look around for other motorists and pedestrians, and then cautiously proceed. The same applies if you approach an intersection with stop signs.

Motorists who "time the traffic lights" present a serious hazard. You may arrive at an intersection while the light is green (or yellow). At the same time, a motorist who is timing the lights on the cross street arrives at the intersection. Although the motorist has a red light, he or she knows that it is about to turn green and is expecting to go through. The stage is now set for a serious collision!

A common scenario occurs when an ambulance is approaching an intersection that is controlled by stop signs or intersection lights. The ambulance, with warning lights and siren in use, proceeds through the intersection without slowing down or stopping and broadsides or is broadsided by a car that is coming from the left or right. Instead of taking the extra few seconds to ensure that traffic is clear in all directions, the operator hastily proceeded through the intersection.

The best way to help avoid a collision at an intersection is to approach it as though you were any other motorist. If the traffic is clear (or has yielded), carefully proceed through the intersection; if it has not cleared or yielded, stop and wait until it is safe to proceed.

9. Are there any special privileges given to emergency vehicles when driving through a school zone?

Although state motor vehicle statutes or codes often grant an emergency vehicle certain privileges when responding

You are the Provider: SUMMARY, continued

to an emergency, there are *no special privileges* given to emergency vehicles when proceeding through a school zone or approaching a stopped school bus that is loading or unloading children. In these situations, the emergency vehicle operator is required by law to act no differently than any other motorist.

If you encounter a school zone during school hours and the orange flashing lights, which indicate the posted school zone speed limit is in effect, you should keep your warning lights and siren on to alert others, but reduce your speed to the posted speed limit. Never exceed the posted speed limit when

driving through a school zone during school hours—regardless of the severity of your patient's condition.

An emergency vehicle is *never* allowed to pass a school bus that has stopped to load or unload children and is displaying its red warning lights and extended "stop arm." If you approach a school bus that is loading or unloading children, you should stop before reaching the bus and turn off your siren. Wait for the driver to make sure the children are safe, that the bus door has been closed, and its red warning lights are off. *Only then may you cautiously proceed past the stopped school bus.*

EMS Patient Care Report (PCR)

Date: 10-19-09	**Incident No.:** 013509	**Nature of Call:** Sick person		**Location:** 37525 Wells Ave	
Dispatched: 1515	**En Route:** 1516	**At Scene:** 1522	**Transport:** 1530	**At Hospital:** 1552	**In Service:** 1601

Patient Information

Age: 70 **Sex:** F **Weight (in kg [lb]):** 50 kg (110 lb)	**Allergies:** Sulfa, morphine, penicillin **Medications:** Prinivil, Os-Cal, vitamins **Past Medical History:** Hypertension, osteoporosis **Chief Complaint:** Generalized weakness

Vital Signs

Time: 1527	**BP:** 132/70	**Pulse:** 72	**Respirations:** 14	**Sao$_2$:** 99%
Time: 1534	**BP:** 128/68	**Pulse:** 80	**Respirations:** 14	**Sao$_2$:** 98%
Time: 1544	**BP:** 132/74	**Pulse:** 90	**Respirations:** 16	**Sao$_2$:** 97%

EMS Treatment
(circle all that apply)

Oxygen @ 15 L/min via (circle one): NC **(NRM)**		**Assisted Ventilation**	**Airway Adjunct**	**CPR**
Defibrillation	**Bleeding Control**	**Bandaging**	**Splinting**	**Other:** (Blood glucose assessment; emotional support)

Narrative

Dispatched to a residence for a "sick person." Arrived on scene and was greeted at the door of the residence by the patient, a 70-year-old female. She was conscious and alert; her airway was patent, and her breathing was adequate. Patient states that she began feeling very weak earlier in the day and that she would like to be transported to the hospital. Past medical history significant for hypertension and osteoporosis. Patient denied chest pain, shortness of breath, abdominal pain, or any other symptoms; her only complaint was weakness. Oxygen was applied via nasal cannula at 4 L/min. Secondary assessment was performed and revealed that her breath sounds were clear and equal bilaterally, her pupils were noted to be equal and reactive to light, and she was able to move all extremities equally. Obtained vital signs and assessed blood glucose level, which read 90 mg/dL. Remainder of secondary assessment was unremarkable. Placed patient onto the stretcher, covered her with a blanket to protect her from the rain, and loaded her into the ambulance. Began transport to the hospital and continued to monitor the patient's level of consciousness and vital signs en route. On reassessment, noted that patient was now confused, was unable to speak, and could not move the entire left side of her body. Her airway remained patent and her breathing remained adequate. Replaced nasal cannula with nonrebreathing mask and set the flow rate at 15 L/min. Reassessment of her pupils revealed that her right pupil was dilated and minimally reactive. Instructed driver to upgrade transport mode, and then contacted the receiving facility to give a radio report. Closely monitored the patient's ABCs for the duration of the transport and provided emotional support as needed. Her condition remained unchanged. Arrived at the hospital, gave verbal report to the staff nurse, and transferred patient care. Medic 527 cleared the hospital and returned to service at 1601. *End of report**

Prep Kit

Ready for Review

- Today's ambulances are designed according to strict government regulations based on national standards.

- The six-pointed Star of Life® emblem identifies vehicles that meet federal specifications as licensed or certified ambulances.

- An ambulance call has nine phases:
  - Preparation for the call
  - Dispatch
  - En route
  - Arrival at scene
  - Transfer of the patient to the ambulance
  - En route to the receiving facility (transport)
  - At the receiving facility (delivery)
  - En route to the station
  - Postrun

- Certain items, like sterile gloves, must be available on the ambulance at all times, as dictated by state and jurisdictional requirements.

- Every ambulance must be staffed with at least one EMT in the patient compartment whenever a patient is being transported. However, two EMTs are strongly recommended. Some services may operate with a non-EMT driver and a single EMT in the patient compartment.

- Check all medical equipment and supplies at least daily, including all the oxygen supplies, the jump kit, splints, dressings and bandages, backboards and other stabilization equipment, and the emergency obstetric kit.

- During the postrun phase, you should complete and file any additional written reports and inform dispatch of your status, location, and availability. Perform a routine inspection to ensure that the ambulance is ready to respond to the next call.

- Learning how to properly operate your vehicle is just as important as learning how to care for patients when you arrive on the scene.
  - The first rule of safe driving in an emergency vehicle is that speed does not save lives; good care does.
  - The second rule is that the driver and all passengers must wear seatbelts and shoulder restraints at all times.

- Air ambulances are used to evacuate medical and trauma patients.
  - There are two basic types of air medical units: fixed-wing and rotary-wing, otherwise known as helicopters.
  - A medical evacuation is commonly known as a medivac and is generally performed exclusively by helicopters.

Prep Kit, continued

Vital Vocabulary

air ambulances Fixed-wing aircraft and helicopters that have been modified for medical care; used to evacuate and transport patients with life-threatening injuries to treatment facilities.

ambulance A specialized vehicle for treating and transporting sick and injured patients.

blind spots Areas of the road that are blocked from your sight by your own vehicle or mirrors.

cleaning The process of removing dirt, dust, blood, or other visible contaminants from a surface.

CPR board A device that provides a firm surface under the patient's torso.

cushion of safety Keeping a safe distance between your vehicle and other vehicles on any side of you.

decontaminate To remove or neutralize radiation, chemical, or other hazardous material from clothing, equipment, vehicles, and personnel.

disinfection The killing of pathogenic agents by direct application of chemicals.

first-responder vehicles Specialized vehicles used to transport EMS equipment and personnel to the scenes of medical emergencies.

high-level disinfection The killing of pathogenic agents by using potent means of disinfection.

hydroplaning A condition in which the tires of a vehicle may be lifted off the road surface as water "piles up" under them, making the vehicle feel as though it is floating.

jump kit A portable kit containing items that are used in the initial care of the patient.

medivac Medical evacuation of a patient by helicopter.

spotter A person who assists a driver in backing up an ambulance to compensate for blind spots at the back of the vehicle.

Star of Life® The six-pointed star that identifies vehicles that meet federal specifications as licensed or certified ambulances.

sterilization A process, such as heating, that removes microbial contamination.

Assessment in Action

At 7:30 AM you arrive at the station and begin preparing for your 24-hour shift. As you inspect your ambulance inside and out, you reflect on the equipment in your ambulance, the various situations you may have to respond to that day, and the risks and responsibilities of emergency transport.

1. The features found in a modern ambulance are defined by which of the following agencies?
   - A. Department of Transportation
   - B. National Association of EMS Directors
   - C. American Ambulance Association
   - D. National Research Council of the National Academy of Sciences

2. Equipment and supplies in the ambulance should be stored according to:
   - A. size and space availability.
   - B. alphabetical order.
   - C. frequency of use.
   - D. ease of access.

3. Items contained in a jump kit should include:
   - A. trauma dressings.
   - B. childbirth supplies.
   - C. portable suction.
   - D. cardiopulmonary resuscitation equipment.

4. Portable floodlights, hard hats, and fire extinguishers are examples of:
   - A. preplanning and navigation equipment.
   - B. work area equipment.
   - C. extrication equipment.
   - D. personal safety equipment.

5. For every emergency request, the minimum information that should be gathered and recorded by the dispatcher includes:
   - A. the patient's social security number.
   - B. the patient's past medical history.
   - C. insurance information.
   - D. the location of call.

6. If you arrive on the scene of a mass-casualty incident, what is the first thing you should do?
   - A. Declare the area a crime scene.
   - B. Direct traffic until law enforcement arrives.
   - C. Ask for additional resources.
   - D. Begin treating patients.

7. A heat process that removes all microbial contamination is called:
   - A. sterilization.
   - B. disinfection.
   - C. cleaning.
   - D. high-level disinfection.

8. What is the minimum size required for a helicopter landing zone?
   - A. 30′ × 30′
   - B. 60′ × 60′
   - C. 50′ × 50′
   - D. 20′ × 20′

9. What should you do to properly position a patient if you do not have a CPR board available?

10. It is now the end of your 24-hour shift in which you were extremely busy because you worked a pediatric mass-casualty incident that involved an elementary school bus. The supervisor approaches you and asks you to work a second 24-hour shift. Knowing that you had only 2 hours of sleep the night before, already worked an overtime shift that week, and are fatigued both physically and mentally, how should you answer?

Vehicle Extrication and Special Rescue

National EMS Education Standard Competencies

EMS Operations

Knowledge of operational roles and responsibilities to ensure patient, public, and personnel safety.

Vehicle Extrication

- Safe vehicle extrication (pp 1368-1376)
- Use of simple hand tools (pp 1372-1376)

Knowledge Objectives

1. Explain the responsibilities of an EMT in patient rescue and vehicle extrication. (pp 1368-1370)
2. Discuss how to ensure safety at the scene of a rescue incident, including scene size-up and the selection of the proper personal protective equipment and additional necessary gear. (pp 1367-1372)
3. Provide examples of vehicle safety components that may be hazardous to both EMTs and patients following a collision and explain how to mitigate their dangers. (pp 1367-1368)
4. Define the terms extrication and entrapment and explain how they differ. (p 1368)

5. Describe the ten phases of vehicle extrication and the role of the EMT during each one. (pp 1368-1376)
6. Discuss the various factors related to ensuring situational safety at the site of a vehicle extrication, including controlling traffic flow, performing a 360° assessment, stabilizing the vehicle, dealing with unique hazards, and evaluating the need for additional resources. (pp 1368-1371)
7. Describe the special precautions the EMT should follow to protect the patient during a vehicle extrication. (pp 1371-1376)
8. Explain the different factors that must be considered before attempting to gain access to the patient during an incident that requires extrication. (pp 1371-1373)
9. Discuss patient care considerations related to assisting with rapid extrication, providing emergency care to a trapped patient, and removing and transferring a patient. (pp 1373-1376)
10. Explain the difference between simple access and complex access in vehicle extrication. (pp 1372-1373)
11. Give examples of situations that would require special technical rescue teams and describe the EMT's role in these situations. (pp 1376-1380)

Skills Objectives

There are no skills objectives for this chapter.

Introduction

As an EMT, you will usually not be responsible for rescue and extrication. Rescue involves many different processes and environments. It also requires training beyond the level of the EMT. In this chapter, you will learn basic concepts of extrication.

The chapter begins with a discussion of safety at the scene of a rescue incident, followed by the 10 phases of extrication. Gaining access is one of the phases examined. This includes how to gain access to patients and how to keep yourself, patients, and bystanders safe in the process. Your main concern is reaching the patient so that you can begin providing care. In most cases, once you have reached the patient, extrication will occur around you and the patient. Communication between the EMT caring for the patient and personnel performing the extrication is vital.

Safety

You must always be prepared, mentally and physically, for any incident that requires rescue or extrication. The most important part of this preparation is thinking about your safety and the safety of your team. Safety begins with the proper mind-set and the proper protective equipment.

The equipment that you use and the gear that you wear will depend on the hazards you expect to encounter, as well as what you observe during your scene size-up **Figure 37-1**. Such protective gear may include turnout gear, helmets, hearing protection, and a fire extinguisher. However, the importance of wearing

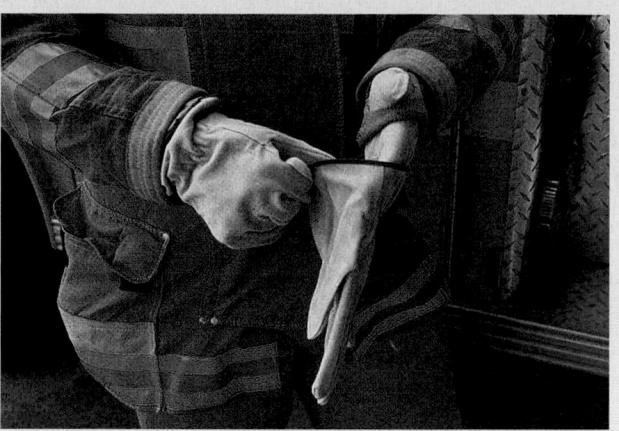

Figure 37-1 Proper protective equipment varies depending on the hazards expected to be encountered.

blood- and fluid-impermeable gloves at all times during patient contact cannot be emphasized enough. If you will be involved with extrication, you should wear a pair of leather gloves over your disposable gloves to protect you from injury when handling ropes, tools, broken glass, hot or cold objects, or sharp metal. Additional information on protective clothing is given later in this chapter.

Vehicle Safety Systems

A variety of safety systems are used in modern vehicles. Although many of these devices are useful when the vehicle is in motion, they can become hazards after the vehicle has been involved in a collision.

Shock-absorbing bumpers provide vehicle protection from low-speed impact. Following a front or rear-end collision, the shock absorbers within these bumpers may be compressed or "loaded." You should avoid standing

You are the Provider: PART 1

Your unit arrives at the scene of a motor vehicle crash. The vehicle involved is a 4-door sedan that ran off the road, slid down a small hill, and struck a tree. Law enforcement personnel are diverting traffic around the incident, and a rescue team is in the process of stabilizing the vehicle. A rescue team member advises you that a 360° assessment of the scene has already been performed and that the single occupant of the vehicle is trapped.

1. **What information is obtained from a 360° assessment of a motor vehicle crash?**
2. **How does the rescue team's role differ from that of EMS personnel?**

directly in front of such bumpers, and always approach vehicles from the side, because the shock absorbers can release and injure your knees and legs.

Manufacturers are now mandated to incorporate supplemental restraint systems or air bags into their vehicles. These air bags fill with a nonharmful gas on impact and quickly deflate after the collision. Air bags are located in the steering wheel and the dash in front of the passenger, and they deploy when the vehicle is struck from the front or rear. Additional bags may be present to protect the driver and passengers from side impacts. These bags may be located in the doors or seats. Air bags should normally deploy and deflate before your arrival on the scene. Air bags have, however, inflated while EMTs were providing patient care, causing injury to the EMT. Use caution when working in damaged vehicles in which air bags have not inflated. Generally, you should maintain at least a 5" clearance around side-impact air bags that have not deployed, 10" around driver air bags that have not deployed, and 20" around passenger-side air bags that have not deployed.

Words of Wisdom

A vehicle crash scene can present many hazards to rescuers and patients, including fuel spills that pose fire and explosion risks, downed electrical lines that pose electrical hazards, broken glass and torn metal, and exposure to potentially infectious body fluids. Your safety at every type of emergency scene begins with, and depends on, your scene size-up. What you see at the scene helps you determine which personal protective equipment to use and whether to call for additional or specialized assistance.

You may notice a haze similar to smoke inside vehicles in which air bags have deployed. Manufacturers use cornstarch or talc on the air bags to reduce friction that may cause a minor skin irritation. Appropriate protective gear, including eye protection, will reduce the potential for such irritation.

Fundamentals of Extrication

During all phases of rescue, your primary concern is safety, and your primary roles are to provide emergency medical care and prevent further injury to the patient. You will provide care as extrication goes on around you unless this proves to be too dangerous for you or the patient. **Extrication** is the removal from entrapment or from a dangerous situation or position. **Entrapment** is the term used when a person is caught within a closed area with no way out or has a limb or other body part trapped. In the context of this chapter, extrication means

removal of a patient from a wrecked vehicle. However, the same principles and concepts apply to other situations.

There are 10 phases to the extrication process **Table 37-1**. Many are similar to the phases of an ambulance call (discussed in Chapter 36, *Transport Operations*). Each will be discussed, with emphasis on the phases in which you will participate.

Safety

Just as in patient care, the first priority in rescue is *rescuer safety*.

■ Preparation

Preparing for an incident requiring extrication involves training for the various types of rescue situations your team might face. Some are discussed later in this chapter. Just as you must check the equipment carried on the ambulance, rescue personnel must also routinely check the extrication tools and their response vehicle to ensure its proper operation. Such preparations reduce the possibility of equipment failure at an emergency scene.

■ En Route to the Scene

Procedures and safety precautions similar to those discussed in the phases of an ambulance call are used when responding to a rescue call.

■ Arrival and Scene Size-up

When you arrive on the scene, you should position your unit in a safe location, with the emergency lights activated. You must avoid adding a hazard to the scene. When a hazardous materials incident is encountered, park uphill and upwind from the hazard. Before proceeding, make sure that the scene is properly marked and protected and that the road is closed or traffic flow is diverted safely around

Table 37-1 Ten Phases of Extrication

1. Preparation
2. En route to the scene
3. Arrival and scene size-up
4. Hazard control
5. Support operations
6. Gaining access
7. Emergency care
8. Removal of the patient
9. Transfer of the patient
10. Termination

the scene using cones, flares, or taping; it is also a good idea to designate a traffic control person Figure 37-2 . Before exiting your vehicle at an emergency scene, be alert for any vehicles that might cause injury to you. Do not assume that motorists will always heed the warning lights.

<u>Size-up</u> is the ongoing process of information gathering and scene evaluation to determine appropriate strategies and tactics to manage an emergency, while paying attention to downed electrical lines, leaking fluids, fire, and broken glass. One of the important responsibilities of scene size-up is to determine what, if any, additional resources will be needed. These resources may include additional EMS units and personnel. If you are first on the scene, you may need to initiate a rescue response or call for law enforcement or specialized crews, such as Hazardous Materials (HazMat) or utility departments.

Situational awareness is the ability to recognize any possible issues once you arrive on the scene and act proactively to avoid a negative impact. For example, upon arriving at the scene of a motor vehicle collision, you should pay attention to the traffic flow. Park the ambulance in a safe position to protect the scene. Make sure you are uphill and upwind if a HazMat scenario is possible. Use emergency lights to warn oncoming

Figure 37-2 The scene of a crash should be marked properly, and traffic should be diverted so that responders have enough room to work.

vehicles of your presence. Assign a person to control traffic or make sure law enforcement is present to assist with traffic control. Take weather and environmental conditions into account and always be aware of all patients and bystanders.

A 360° walk-around of the scene will allow you to evaluate the hazards present and determine the number

You are the Provider: PART 2

You proceed to the vehicle and note that it struck the tree head-on; there is significant damage to the front of the vehicle. After the rescue team properly stabilizes the vehicle, your partner accesses the patient from the back seat and manually stabilizes his head. You access the patient through the broken driver's side window and perform a primary assessment. The patient is a 22-year-old man who is conscious, but he is screaming in pain. His legs are trapped by the steering column, and he tells you that his feet are stuck. You note that the driver's side air bag deployed. The patient tells you that he was wearing his seatbelt, but he removed it after the crash.

Recording Time: 0 Minutes	
Appearance	Conscious; restless; in obvious pain
Level of consciousness	Conscious, but restless
Airway	Open; clear of secretions or foreign bodies
Breathing	Increased rate; shallow depth
Circulation	Radial pulses, weak and rapid; skin is cool, clammy, and pale; laceration to the forehead with active bleeding

Because of the mechanism of injury, the patient's clinical condition, and the possibility of a prolonged extrication, you request air medical transport to respond to the scene and transport him to a trauma center located 35 miles away. You apply high-flow oxygen, cover the laceration on his forehead, and apply a cervical collar.

3. How should you initially attempt to gain access to a crash victim?

4. What treatment can and should you provide to a patient who is entrapped in his or her vehicle?

of patients. If there is a large group of patients, implement local mass-casualty incident protocols as necessary. During your walk, look for the following:

- The mechanism of injury
- Downed electrical lines
- Leaking fuels or fluids
- Smoke or fire
- Broken glass
- Trapped or ejected patients

Evaluate the need for additional resources such as:

- Extrication equipment
- Fire suppression
- Law enforcement
- HazMat units
- Utility companies
- Advanced life support units
- Aeromedical transport

Look for spilled fuel and other flammable substances. Motor vehicles carry a variety of fuels and lubricants that pose a fire hazard. Sometimes postcrash fires are started when sparks created during the crash ignite spilled fuel. A short in a vehicle's electrical system or a damaged battery may also cause a postcrash fire. These fires may trap the occupants of the vehicle and require fire suppression.

Environmental conditions can lead to unique hazards at a crash scene. Crashes that occur in rain, sleet, or snow, for example, present an added hazard for rescue personnel and patients. Crashes that occur on hills are harder to handle than those that occur on level ground.

Some crash scenes may present threats of violence. Intoxicated people or people who are upset with other motorists may pose a threat to you or to other people present at the scene. Be alert for weapons that are carried in civilian vehicles.

You will need to coordinate your efforts with the rescue teams and with law enforcement officials. If you respect their job, they will respect yours. You should communicate with members of the rescue team throughout the extrication process. Start talking to the rescue team leader as soon as you arrive at the scene. Under the incident command system (described in Chapter 38, *Incident Management*), rescue operations are integrated as a separate group. You become a member of this group and will enter the vehicle and provide care for the patient(s) when approved by the extrication leader.

The rescue team is responsible for properly securing and stabilizing the vehicle, providing safe entrance and **access** to the patients (the ability to reach the patient), extricating any patients, ensuring that patients are properly protected during extrication or other rescue activities, and providing adequate room so that patients can be removed properly.

EMS personnel are responsible for assessing and providing immediate medical care, triaging and assigning priority to patients, packaging patients, providing additional assessment and care as needed once patients are removed, and providing transport to the emergency department.

> **Safety**
>
> The EMT's job at the scene of an accident is to care for the patients.

■ Hazard Control

A variety of hazards may be present at the extrication scene. Law enforcement personnel are responsible for traffic control and direction, maintaining order at the scene, investigating the crash or crime scene, and establishing and maintaining lines so that bystanders are kept at a safe distance and out of the way of rescuers. Fire fighters are responsible for extinguishing any fire, preventing additional ignition, ensuring that the scene is safe, and removing any spilled fuel Figure 37-3 .

Downed electrical lines are a common hazard at vehicle crash scenes. You should never attempt to move downed electrical lines. If power lines are touching or located in proximity to a vehicle involved in the crash, patients should be instructed to remain in the vehicle until power is removed. In most incidents, there will be an area designated as the **safe zone**. You and the

Figure 37-3 Every crash requires cooperation, as each responder has a specific role at the scene. Fire fighters, law enforcement personnel, the rescue team, and EMS personnel all have distinct responsibilities.

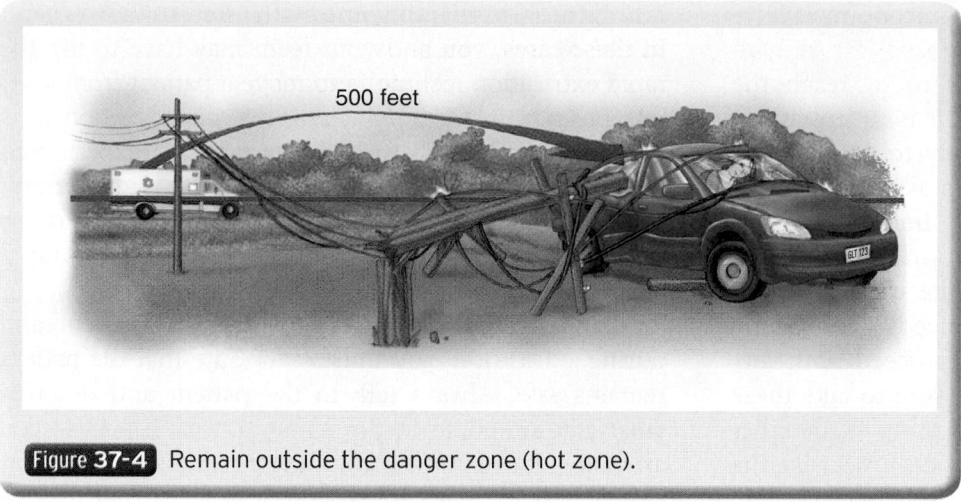

Figure 37-4 Remain outside the danger zone (hot zone).

ambulance should remain in that area, outside of the danger zone (hot zone) **Figure 37-4** . A **danger zone (hot zone)** is an area where individuals can be exposed to sharp metal edges, broken glass, toxic substances, lethal rays, or ignition or explosion of hazardous materials.

Bystanders and family members can be hazards themselves. If they are allowed to get too close, they are at risk of injury and may also interfere with the overall management of the incident. For these reasons, the rescue team will set up a danger zone that is off-limits to bystanders **Figure 37-5** . You should help to set up and enforce this zone. If you arrive before the rescue team, you should coordinate crowd control with law enforcement officials.

The vehicle also can be a hazard. An unstable automobile on its side or roof can be a danger to you. Rescue personnel can stabilize the vehicle with a variety of jacks or cribbing (wooden blocks). Prior to attempting to gain access to a vehicle involved in a crash, you should ensure that the vehicle is in "park" with the parking brake set and the ignition is turned off. The battery should also be disconnected, negative side first, to minimize the possibility of sparks or fire.

Safety

Always assume that oncoming traffic cannot see you, and take appropriate steps to keep yourself safe.

Alternative Fuel Vehicles

Today, with advances in automotive technology, EMS responders should keep in mind that some vehicles on the road are powered by alternative fuel. Vehicles may be powered by electricity and electricity/gasoline hybrids, or fuels such as propane, natural gas, methanol, or hydrogen. Although each type of vehicle has its own unique features, one feature is common throughout—the need for responders to disconnect the battery to prevent further fire or explosion. In more than 40% of today's alternative fuel vehicles, the batteries are not located in the engine compartment, but in other areas, such as the trunk or under the seats. Furthermore, there also may be more than one battery present. You must remain vigilant when presented with these alternative types of vehicles and their inherent dangers. For example, hybrid batteries have higher amperes than a traditional vehicle's battery, and these amperes can injure you.

■ Support Operations

Support operations include lighting the scene, establishing tool and equipment staging areas, and marking helicopter landing zones. Fire and rescue personnel will work together on these functions.

■ Gaining Access

A critical phase of extrication is gaining access to the patient. Remember, you should not attempt to gain access to the patient or enter the vehicle until you are sure that the vehicle is stable and that any hazards have been identified and properly controlled or eliminated. When there is a rescue leader present, you will be authorized to enter

Figure 37-5 A danger zone should be established to prevent bystanders from entering the area around an incident.

the scene or the vehicle only when these conditions have been met.

The exact way you gain access to or reach the patient(s) depends on the situation. It is up to you to identify the safest, most efficient way to gain access. Darkness, uneven terrain, tall grass, shrubbery, and wreckage may make patients hard to find Figure 37-6. Multiple vehicles with multiple patients may be involved. If this is the situation, you should locate and rapidly triage each patient to determine who needs urgent care. This step is important before you proceed with any treatment and patient packaging. Be sure to take these factors into account in your scene size-up. Remember that scene size-up is a continuing process, because the situation often changes. As a result, you may need to change your plans for gaining access and providing treatment.

To determine the exact location and position of the patient, you and your team should consider the following questions:

- Is the patient in a vehicle or in some other structure?
- Is the vehicle or structure severely damaged?
- What hazards exist that pose a risk to the patient and rescuers?
- In what position is the vehicle? On what type of surface? Is the vehicle stable or is it apt to roll or tip?

You must also take into account the patient's injuries and their severity. You may have to change your course of action as you learn more about the patient's condition. Do not try to access the patient until you are sure that the vehicle is stable and that hazards have been identified and rendered safe. Hazards might include electrical or gas lines.

What should you do if you have to remove a patient quickly because the environment is threatening or you need to perform cardiopulmonary resuscitation? Cardiopulmonary resuscitation is not effective when the patient is in a sitting position or lying on the soft seat of a vehicle. In these cases, you and your team may have to use the rapid extrication technique to move a patient from a sitting position inside a vehicle to a supine position on a long backboard. A team of EMTs who are experienced in using this technique should be able to rapidly remove a patient who is not entrapped, keeping in mind the patient's condition and the group's safety. Use the rapid extrication technique only as a last resort.

While you are gaining access to the patient and during extrication, you must make sure that the patient remains safe. Always talk to the patient and describe what you are going to do before you do it and as you are doing it, even if you think the patient is unconscious Figure 37-7. In many cases, you or your partner may be providing cervical spine immobilization or other care during extrication. All EMS personnel should wear proper protective gear while in the working area. A heavy, nonflammable blanket can be used to cover the patient and EMS personnel to protect them from flying glass or other objects. A long backboard may also be used as a protective shield. Try to keep heat, noise, and force to a minimum. Use only what is necessary to extricate the patient safely.

Simple Access

Your first step is <u>simple access</u>, trying to get to the patient as quickly and simply as possible without using any tools or breaking any glass. Automobiles are built for easy entry and exit; however, it may be necessary to use tools or other forcible entry methods. Enter through the doors when there is no danger to the patient. Whenever possible, you should first try to unlock the doors (or ask the patient to unlock them) or roll down the windows. Try to open every door using the door handles to gain access before breaking any windows or using other

Figure 37-6 The exact way to gain access depends on many factors, including the terrain, the way in which the vehicle is situated, and the weather.

Figure 37-7 Always explain to the patient why you are there and what you are doing.

methods of forced entry Figure 37-8 . The rescue team should provide the entrance you need to gain access to the patient. In situations where the rescue team has not yet arrived, and delayed access to the patient could be life threatening, simple hand tools, such as hammers, center punches, pry bars, and hacksaws, should be available on the ambulance for you to use.

Safety

Always try before you pry.

Complex Access

Complex access requires the use of special tools, such as hand, pneumatic, and hydraulic devices, and special training and includes breaking windows or other means of forcible entry. These skills may be too advanced for the EMT course and are not covered in this text Figure 37-9 .

■ Emergency Care

Providing medical care to a patient who is trapped in a vehicle is principally the same as for any other patient. Unless there is an immediate threat of fire, explosion, or other danger, once entrance and access to the patient

Figure 37-8 Get to the patient as quickly and simply as possible by opening the door without using tools or breaking any glass.

have been provided and the scene is safe, you should perform a primary assessment and provide care before further extrication begins, as follows:

1. Provide manual stabilization to protect the cervical spine, as needed.
2. Open the airway.
3. Provide high-flow oxygen.
4. Assist or provide for adequate ventilation.

You are the Provider: PART 3

The driver's door is badly damaged and cannot be opened manually. However, a rescue team member is able to open the front passenger door with a pry bar. As your partner continues to manually stabilize the patient's head, you access the patient through the front passenger door. As the extrication team prepares to remove the driver's side door, you quickly assess the patient's vital signs.

Recording Time: 7 Minutes	
Respirations	26 breaths/min; shallow
Pulse	120 beats/min; weak at the radial artery
Skin	Cool, clammy, and pale
Blood pressure	100/50 mm Hg
Oxygen saturation (Sao$_2$)	96% (on oxygen)

A rescue team member informs you that an air medical helicopter is en route to the scene and will arrive in approximately 12 minutes. A landing zone has been established about 500 yards away.

5. Is it always appropriate to obtain the vital signs of a patient who is trapped in his or her vehicle? Why or why not?
6. What should you do as the patient is being extricated from the vehicle?

Figure 37-9 Complex access requires the use of hydraulic devices.

5. Control any significant external bleeding.
6. Treat all critical injuries.

Good communication among team members and clear leadership are essential to safe, efficient provision of proper emergency care. Although your input at the scene is important, one member of your team must be clearly in charge. The team leader's assessment of the patient and the situation will dictate the way in which medical care, packaging, and transport will proceed. Customarily, the senior medical person is responsible for this role. If a team leader has not been identified, this must be decided and agreed on before you arrive at the scene. A lack of identifiable leadership at the scene hinders the rescue effort and patient care. Leaders should be identified as part of a larger incident command system. They should be medically trained and qualified to judge the priorities of patient care, and they must also be experienced in extrication.

■ Removal of the Patient

Extrication, also called disentanglement, involves the removal of the patient from a dangerous situation or position. In the case of a vechicle extrication, rescue personnel should coordinate with you to determine the best route for removal `Table 37-2`. Whereas one accident may require removal of the patient through the driver's door, a similar accident may require complete removal of the vehicle's roof. Removal of a patient from a motor vehicle is a multistep process that is intensive in terms of the number of rescuers involved, the equipment used, and the time required to prevent further injury or harm.

Extrication requires the use of a variety of complex hand and power tools. Specialized education is required for their safe operation.

As a part of your assessment, you should participate in the preparation for patient removal. Determine how urgently the patient must be extricated, where you should be positioned to best protect the patient during extrication, and, once the patient has been freed, how you will best move the patient from within the vehicle onto the long backboard and onto the stretcher. Carefully examine the exposed area of the limb or other part of the patient that is trapped to determine the extent of injury and whether there is a possibility of hidden bleeding. If possible, you should also evaluate sensation in the trapped area so that you will know whether increased pain indicates that an object is pressing on or impaled in the patient during extrication.

During this time, the rescue team is assessing exactly how the patient is trapped and determining the safest, easiest way to extricate him or her. Your input is essential so that the patient's injuries are considered as the rescue team plans a move that protects the patient from further harm. Reevaluate whether the patient needs to be immediately removed by using manual stabilization and the rapid extrication technique or whether the patient's condition and the scene allow for immobilization using an extrication vest or short backboard before he or she is moved further. In most cases, it is impractical and difficult to properly apply extremity splints within the vehicle. Extremity injuries can generally be rapidly supported and immobilized while the patient is being removed by securing an injured arm to the body and, if a leg is injured, securing one leg to the other. This will be adequate until the patient is secured to the backboard or time permits a more detailed assessment and splinting of each injury.

Once the plan has been devised and everyone understands what will be done, you should determine how best to protect the patient. Often, you or another EMT will be placed in the vehicle alongside the patient to monitor his

Table 37-2 Vehicle Extraction Techniques

- Brake and gas pedal displacement
- Dash roll-up
- Door removal
- Roof opening and removal
- Seat displacement
- Steering column displacement
- Steering wheel cutting

or her condition and well-being as the vehicle is being forcibly cut, bent, or disassembled. Be sure to wear proper protective clothing.

Naturally, your safety and that of the patient are paramount during this process. Both you and the patient should be covered by a thick, fire-resistant canvas or blanket for protection from broken glass, flying particles, tools, or other hazards during any cutting or forceful extrication maneuvers. Extrication is often extremely noisy, and appropriate hearing protection should be worn by the EMTs and the patient. You also must be sure that you can communicate effectively with the patient and the rescue group so that you can instantly let the rescuers know if it is necessary that they stop.

■ Transfer of the Patient

Once the patient has been freed, rapidly assess any other patients who were previously inaccessible, and then perform a complete primary assessment, providing critical interventions, if required. Make sure that the patient's spine is manually stabilized, and apply a cervical collar if this was not previously done Figure 37-10 .

Moving the patient in one fast, continuous step increases the risk of harm and confusion. To ensure

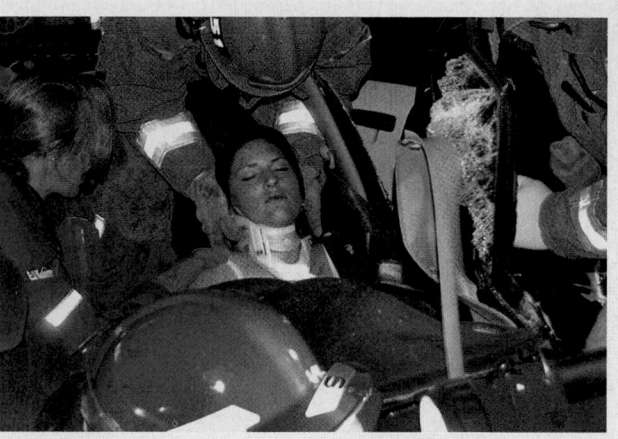

Figure 37-10 Once the patient has been accessed, rapidly assess the patient and make sure that the spine is manually stabilized. Apply a cervical collar if this was not previously done.

that each EMT can be positioned so that he or she can lift and carry properly at all times, move the patient in a series of smooth, slow, controlled steps, with stops designated between them to allow for the repositioning and adjustments that are needed. One person should be

You are the Provider: PART 4

The driver's side door has been removed and the steering column has been lifted from the patient's legs. His feet were entangled in the pedals on the floorboard and have also been freed. After removing the patient from the vehicle using the rapid extrication technique, you perform a rapid secondary assessment as your partner reassesses his vital signs.

Recording Time: 18 Minutes	
Level of consciousness	Responsive to pain only
Respirations	28 breaths/min; shallow
Pulse	130 beats/min; weak radial pulses
Skin	Cool, clammy, and pale
Blood pressure	80/58 mm Hg
Sao$_2$	96% (on oxygen)

Noting deterioration in the patient's condition, your partner begins assisting his ventilations with a bag-mask device and high-flow oxygen while another rescue team member continues to stabilize the patient's head. After applying full spinal precautions, you cover the patient with a blanket and quickly load him into the ambulance.

7. How does the rapid extrication technique differ from other methods of patient removal? When is it indicated?

in charge of the move and plan and verbalize the exact steps and pathway that you will follow in moving the patient from sitting in the vehicle to lying supine on the backboard and prepared ambulance stretcher. Choose a path that requires the least manipulation of the patient and equipment. Make sure that sufficient personnel are available. Once you are sure that everyone understands the steps and is ready, you can transfer the patient safely. Make sure everyone knows to move on your command and to move the patient as a unit, resisting the temptation to move the immobilization device instead. While transferring the patient, continue to protect him or her from any hazards.

Once the patient has been placed on the stretcher, continue with any additional assessment and treatment that was deferred. If it is extremely cold or hot, raining, or snowing, you should load the stretcher and patient into the climate-controlled ambulance before continuing assessment and treatment. If the patient's condition requires that transport be initiated without further delay, you should provide only the additional care that is essential or necessary to package the patient. Leave the remaining steps to be performed en route to the hospital.

■ Termination

Termination involves returning the emergency units to service. For rescue units, this process may be quite involved. All equipment used on the scene, including hydraulic, electrical, and hand tools, must be checked before reloading them on the apparatus. While some tools require only generalized cleaning, others may need refueling and checking of the various fluid levels.

You will also be required to check the ambulance thoroughly, replacing used supplies and conforming to cleaning needs required by bloodborne pathogen standards.

Finally, rescue units and medical units will be required to complete all necessary reports.

Specialized Rescue Situations

On most calls, you can drive the ambulance to within a short distance of the patient's location, and, with simple or complex access, you can reach and treat the patient. However, in some situations, the patient can be reached only by teams trained in special technical rescues. Specialized skills of these teams include the following:

- Cave rescue
- Confined space rescue
- Cross-field and trail rescue (park rangers)
- Dive rescue
- Lost person search and rescue

- Mine rescue
- Mountain-, rock-, and ice-climbing rescue
- Ski slope and cross-country or trail snow rescue (ski patrol)
- Structural collapse rescue
- Special weapons and tactics (SWAT)
- Technical rope rescue (low- and high-angle rescue)
- Trench rescue
- Water and small craft rescue
- White-water rescue

■ Technical Rescue Situations

A **technical rescue situation** may contain hidden dangers, and personnel need special technical skills to safely enter and move around. It is not safe to include personnel who do not have the necessary special training and experience in such a rescue. A **technical rescue group** is made up of individuals from one or more departments in a region who are trained and on call for certain types of technical rescues. Many members of a technical rescue group are also trained as emergency medical responders (EMRs) or EMTs so that they can provide the necessary immediate care when only they can safely reach the patient. Even when the technical rescue group includes a paramedic or physician, generally nothing but essential simple care is provided until the rescuers can bring the patient to the nearest point where a safe, stable setting exists.

If a technical rescue group is necessary but is not present when you arrive, you should immediately check with the **incident commander** to make sure that the group has been summoned and is en route to your location. The incident commander is the individual who has overall command of the scene in the field **Figure 37-11**. If no incident commander is present, follow local guidelines. (Chapter 38, *Incident Management*, discusses this in more detail.)

Figure 37-11 The incident commander is the individual who has overall command of the scene.

When you arrive at a scene where a technical rescue is in progress, you will usually be met by a member of the technical rescue group and directed or led to the actual rescue site. If the rescue site is some distance from the road, you may need to leave the ambulance on the road. The use of the ambulance stretcher is impractical in these situations; you should instead take a long backboard and/or basket stretcher or similar rescue stretcher to carry the patient back to the waiting ambulance. Be sure that you take all of the carry-in kits and other equipment you may need to treat and immobilize the patient at the rescue site.

When you arrive at the rescue site, identify the stable location to which the technical rescue group will bring the patient, and set up your equipment there. As soon as the technical rescue group has brought the patient to this staging area, you should perform a primary assessment and, after providing the treatment indicated, package the patient without delay. Although you and the other EMTs who responded with the ambulance will assume the primary responsibility for the patient's care, at this point it usually requires a cooperative effort by the technical rescue group and the EMS team to carry the patient to the waiting ambulance. Consider using an air medical unit if the patient will need to be carried or transported an extensive distance.

Safety

Treat all downed wires as if they are charged (live) until you receive specific clearance from the electric company. Even if the lights are out along the street where the wires are down, never assume that the wires are dead. Be especially alert for downed wires after a storm that has blown down trees or tree limbs.

Lost Person Search and Rescue

When someone is lost outdoors and a search effort is initiated, an ambulance is usually summoned to the incident command post or a designated staging area. Each search team will be organized to include a member who is trained at the EMR or EMT level and who carries the essential equipment to provide simple immediate care. Your role, and that of the other EMTs who arrive with the ambulance, is to stand by at the command post until the lost person or people have been found.

As soon as you arrive at the scene and have been briefed on the situation, you should isolate and prepare the equipment you will need to carry to the patient's location so that no time is lost once the patient has been found or a member of the search team is injured. The prepared

You are the Provider: PART 5

A rescue team member drives the ambulance to the landing zone as you and your partner care for the patient. Further assessment reveals that he has bilateral femoral deformities and a rigid, bruised abdomen. Your partner continues assisting the patient's ventilations while you reassess his vital signs.

Recording Time: 23 Minutes	
Level of consciousness	Responsive to pain only
Respirations	28 breaths/min; shallow; ventilations are being assisted
Pulse	128 beats/min; weak radial pulse
Skin	Cool, clammy, and pale
Blood pressure	84/60 mm Hg
Sao$_2$	95% (on oxygen)

The air medical helicopter arrives, and you give a verbal report to the flight medic. After assisting in the loading of the patient onto the aircraft, you return to the scene to retrieve any equipment you may have left. Your partner informs you that he looked in the patient's vehicle and noticed that the passenger side air bag had not deployed.

8. What should you do when entering a vehicle in which an air bag has not deployed?

carry-in equipment, including a long backboard and other equipment you will need to immobilize the patient, should be left in the back of the ambulance so that it is protected from the weather. In addition, if the ambulance should need to be relocated, the equipment will not need to be reloaded and will not be left behind. You will usually be given a portable radio that is tuned to the search frequency so that you can monitor the progress of the search and communicate with and be contacted by those in charge of the search operation.

Sometimes, you may be asked to stay with relatives of the lost individual who are at the scene. Find out from relatives whether the lost person has any medical history that may need to be addressed, and pass this information on to those who are in charge of the search. Unless you have been instructed otherwise, only the incident commander should communicate any news or progress of the search to the family. For this reason, you must be sure that your radio is set at a discreet volume.

Once the lost person has been found, you will be guided by search personnel to that location or to a pre-arranged intersecting point where the patient will be carried to decrease the amount of time you need to reach the patient and begin treatment. You should be sure that the carry-in equipment is evenly distributed among personnel and that the pace is such that all can stay together easily. Sometimes the time and effort that are needed to reach and carry the patient out can be decreased by relocating the ambulance or, if one is available, by using a four-wheel drive or all-terrain vehicle. As with other specialized rescues, although the ambulance crew will assume the responsibility for patient care once they are at the patient's side, a cooperative effort of the EMS and search teams is necessary to safely carry the patient to the base and waiting ambulance.

■ Trench Rescue

Owing to the physical forces involved, many cave-ins and trench collapses have poor outcomes for victims. Collapses usually involve large areas of falling dirt that weigh approximately 100 lb per cubic foot. Victims with thousands of pounds of dirt resting on their chests cannot fully expand their lungs and may become hypoxic.

The risk of a secondary collapse during the rescue operation is of concern to rescue personnel and to the EMTs. Safety measures can reduce the potential for injury from this and other hazards. When arriving on the scene of a cave-in or trench collapse, response vehicles should be parked at least 500′ from the scene. Because vibration is a primary cause of secondary collapse, all vehicles, including on-scene construction equipment, should be turned off. In addition, all road traffic should be diverted

from the 500′ safety area. Other hazards include exposed or downed electric wires and broken gas or water lines. In addition, construction equipment at the collapse may be unstable and could fall into the cave-in or trench site.

Any witnesses to the incident should be identified. They may be valuable in providing information on the number of victims and their location within the collapsed area. Any nontrapped individuals should be assisted from the area. At no time should medical or rescue personnel enter a trench deeper than 4′ without proper shoring in place.

During the extrication of any live victims, medical personnel trained in cave-in and trench collapse rescue will provide most medical care. You should be prepared to receive patients once they have been extricated from the site.

■ Tactical Emergency Medical Support

A steady increase in violence throughout the country has resulted in EMTs taking precautions to ensure personal safety. Normally, when the potential for violence exists—as in shootings, stabbings, and attempted suicides—responding units should wait until the scene is secured by law enforcement personnel. However, some incidents pose an increased risk to EMTs and law enforcement personnel. Hostage incidents, barricaded subjects, and snipers require the use of specialized law enforcement tactical units or the **special weapons and tactics team (SWAT)**.

Words of Wisdom

Physical dangers such as fire, infectious disease, and electricity are not the only risks to your safety during emergency responses. Some calls involve the possibility of deliberate violence against rescuers. Formal tactical situations are obvious examples, but "simple" calls involving assaults, possible alcohol or drug use, and domestic disputes can be just as dangerous. Your training, your attitude when responding to calls, and your routine daily procedures should all help you take these risks into account. Never become complacent, and continually reassess your surroundings.

Owing to the high potential for injuries at these incidents, many communities have incorporated specially trained EMTs, paramedics, nurses, and even physicians into their police SWAT units. These EMS personnel provide a special level of care to the sick and injured at such volatile incidents. Their training goes well beyond the practices seen in standard emergency medical care. Thus, the techniques used may not seem appropriate or adequate. For example, spinal immobilization is not used within an

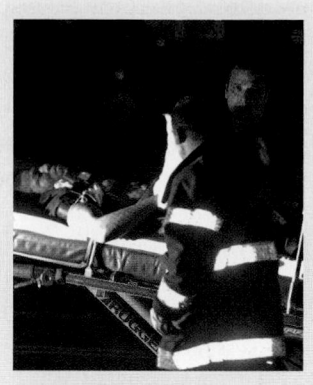

Figure 37-12 Tactical EMS providers move a downed officer; only the most basic medical care is provided in an unsecured area.

unsecured area where gunfire may still erupt. The time and manpower necessary to completely secure a victim to a backboard with a collar, straps, and head immobilization may expose EMS providers and SWAT officers to injury or death from gunfire **Figure 37-12** . Such altered standards of care are similar to those used by military EMS providers on the battlefield and are not used in "standard" situations encountered by EMTs.

When called to the scene of a law enforcement **tactical situation**, you should determine the location of the **command post** (location of the incident commander) and report to the incident commander for instructions. Lights and siren should be turned off when nearing the scene, and outside radio speakers should not be used. The command post is usually located in an area that cannot be seen by the suspect and is out of range of possible gunfire. You should remain in this area and not roam beyond this site. Nearby areas may be visible to the suspect, and you could be injured.

A number of planning measures should be started after checking with the incident commander. Such planning will reduce the potential for chaos should a mass-casualty incident occur at the scene. First, have the incident commander identify the specific location of the incident. The information should include the street address and the side of the street on which the house or building is located. The incident commander should determine a safe location where you can meet SWAT team members or tactical EMS providers should an injury occur. Tactical EMS providers or officers will remove the patient to this area for your continued treatment and transport to a medical facility. The incident commander should also determine a safe route to this meeting point.

Designate primary and secondary helicopter landing zones if your region uses aeromedical evacuation. Such preplanning will save valuable time in critical situations. The closest hospital, burn center, and trauma center should be identified. The route of travel to these facilities should also be noted. Many of these measures are incorporated into the operational plan used by tactical EMS providers. If tactical EMS providers are used in your jurisdiction, coordinate with them on your arrival at the command post.

Words of Wisdom

F-A-I-L-U-R-E

The reasons for rescue failure can be referred to by the mnemonic "FAILURE":

F   Failure to understand the environment or underestimating it
A   Additional medical problems not considered
I   Inadequate rescue skills
L   Lack of teamwork or experience
U   Underestimating the logistics of the incident
R   Rescue versus recovery mode not considered
E   Equipment not mastered

■ Structure Fires

In most areas, an ambulance is dispatched with the fire department apparatus to any structure fire, whether or not injuries are reported. A fire in a house, apartment building, office, school, plant, warehouse, or other building is considered a **structure fire**. When responding to a major fire scene, you should determine whether, because of the fire, any special route will be necessary. Once you arrive at the scene, you should ask the incident commander where the ambulance should be staged. It is essential that the ambulance be parked far enough from the fire to be safe from the fire itself or a collapsing building. You must also ensure that the ambulance will not block or hinder other arriving equipment or be blocked in by other equipment or hose lines. However, you must also make sure that the ambulance will be close enough to be visible and that patients can be brought to it easily. The fire officer who is the incident commander will determine this location.

Your next step is to determine whether there are any injured patients at the scene or whether you have been called to stand by. A number of ambulances may be dispatched to a major fire to ensure that one or more units will always remain immediately available at the scene if others leave to transport the injured.

As with other specialized rescue situations, search and rescue in a burning building requires special training and equipment. Search and rescue operations are performed by teams of fire fighters wearing full turnout gear and **self-contained breathing apparatus (SCBA)** and carrying tools and fully charged hose lines. These teams will bring patients out of the burning building to the area where the ambulance is standing by. Therefore, unless otherwise ordered, you should always stay with the ambulance. Do not leave the scene even after the fire is out because you may need to treat a fire fighter who has been injured during salvage and overhaul. The ambulance should leave the scene only if transporting a patient or if the incident commander has released it.

Sometimes the scene at a crash or fire is further complicated by the presence of **hazardous materials**. A hazardous material is any substance that is toxic, poisonous, radioactive, flammable, or explosive and can cause injury or death with exposure. In addition to posing a threat to you and others at the immediate scene, hazardous materials may pose a threat to a much larger area and population. Whenever there is a possibility that a hazardous material is involved, you will have to follow a number of additional special procedures. Chapter 38, *Incident Management*, covers the specifics of hazardous material procedures.

You are the Provider: SUMMARY

1. What information is obtained from a 360° assessment of a motor vehicle crash?

The 360° assessment is a component of the scene size-up. It is not unique to motor vehicle crashes; it is used by rescue and fire personnel at the scene of any rescue operation or major incident, such as structural fires, building collapses, and trench rescues, among others. The objective of this type of assessment is to look at the "big picture" with a focus on safety.

The 360° assessment of a motor vehicle crash focuses on hazards unique to this type of incident. Common safety hazards include downed electrical lines, leaking fuels or other fluids, smoke or fire, broken glass, and vehicle stability. Safety issues must be addressed by the appropriate personnel before anyone attempts to gain access to the patient(s).

Other information obtained during the 360° assessment of a motor vehicle crash includes the mechanism of injury, the position of the patient(s) in the vehicle, and whether the patient(s) are entrapped or have been ejected.

If you are the first unit to arrive at the scene, you should perform this assessment and then request the appropriate resources. Do not be hasty; look at the entire scene first, and then proceed accordingly. If fire or rescue personnel have performed this assessment before your arrival, obtain the pertinent information from them.

2. How does the rescue team's role differ from that of EMS personnel?

An integral part of any rescue or extrication operation is communication and cooperation among all personnel present at the scene. EMS, rescue, fire, and law enforcement personnel have duties that are unique to their training; in turn, they must all work together as a team to facilitate a seamless operation that is safe and effective.

At the scene of a motor vehicle crash, the rescue team is responsible for properly securing and stabilizing the vehicle, eliminating safety hazards (ie, disconnecting the vehicle's battery cables), providing complex access to the patient, extricating the patient with heavy rescue equipment, ensuring that the patient is properly protected during extrication or other rescue activities, and providing adequate room so the patient can be safely and properly removed.

The fire department and rescue team are often one in the same. They respond with an engine or pumper for fire suppression and a rescue vehicle that carries specialized equipment such as the Jaws of Life. In some cases, the engine serves as both a fire suppression apparatus and a rescue vehicle. It is important to be familiar with the capabilities of the fire department and rescue team that you will be working with.

EMS personnel are responsible for assessing and providing immediate medical care (after access to the patient has been achieved), triaging and assigning priorities to patients, packaging patients, providing additional assessment and care as needed once patients are removed from the vehicle, and providing transport to the emergency department.

3. How should you initially attempt to gain access to a crash victim?

A critical phase of extrication is gaining access to the patient. Do not attempt to gain access to the patient or enter the vehicle until you are sure that the vehicle is stable and that any hazards have been identified and properly controlled or eliminated. The rescue team leader will inform you when it is safe to gain access to the patient.

In the case of this patient, the driver's side window is already broken; therefore, you can access him this way. Additionally, your partner was able to access the patient through a rear door and manually stabilize his head. However, there will be times when this type of access will not be possible; in many situations, the windows are still intact and the doors are not easily opened. Although modern automobiles are built for easy entry and exit, at times it may be necessary to use tools or other forcible entry methods.

Whenever possible, you should use simple access techniques to get to the patient. Simple access involves getting to the patient quickly, but without using any tools or breaking any glass. The simplest way to gain access to a crash victim is to open a door. Try all of the doors first—even if they are badly damaged. It is a waste of time and energy to open a jammed door with heavy rescue equipment when another door can be opened easily and without any special equipment. Attempt to unlock and open the least damaged door first. Make sure the locking mechanism is released. Try the outside and inside handles at the same time if possible. If the window is rolled up and the door is locked, ask the patient, if he or she can, to unlock the door or roll down the window; this may or may not be possible, depending on the patient's location, the severity of his or her injuries, and the amount of damage to the vehicle.

If the patient's condition is serious enough to require immediate care and you cannot enter the vehicle through a door, it will be necessary to break a window. Do not try to break and enter through the windshield because it is made of laminated glass, which is difficult to break. When it does break, it creates large shards of glass that can easily injure you or the patient. The side and rear windows are made of tempered glass (safety glass); this type of glass will break easily into small pieces when hit with a sharp, pointed object, such as a spring-loaded

You are the Provider: SUMMARY, continued

center punch. These windows do not pose as great a safety threat; therefore, they should be your primary access route if a window must be broken.

If you must break a window to gain access, try to break one that is farthest away from the patient. If the patient's condition warrants immediate entry (ie, airway compromise, severe bleeding), however, do not hesitate to break the closest window. Small pieces of tempered glass generally do not pose a danger to people trapped in their vehicle.

If you cannot gain access to the patient by opening a door or breaking a window, heavier rescue tools will be necessary to gain access to the patient. The most common technique is door displacement; however, this usually requires tools such as pry bars or the Jaws of Life. Most ambulances do not carry hydraulic rescue equipment; therefore, the rescue team generally performs this task.

4. What treatment can and should you provide to a patient who is entrapped in his or her vehicle?

Unless there is an immediate threat of fire, explosion, or other danger, once entry and access to the patient have been provided, you should perform a primary assessment and begin immediate emergency care before the process of extrication begins. Once you are in the vehicle, you can often assess the degree of patient entrapment better than those who are not in the vehicle. The information that you obtain about the type or degree of entrapment should be passed on to the extrication team; it will help them when they are formulating the most appropriate extrication approach.

When you have a patient trapped in a vehicle, the assessment you perform and the treatment you provide should focus on identifying and correcting immediate life threats—just as you would for any other patient. A lengthy, detailed assessment is not appropriate because this delays the process of patient extrication. The ability to adequately assess and treat the patient depends on the severity of entrapment and how much of the patient you can access. In the case of this patient, you can access his head, torso, and upper extremities. You cannot, however, access his lower extremities.

Your partner has already gained access from the back seat and is manually stabilizing the patient's head to protect his cervical spine. The patient's airway is obviously patent because he is screaming in pain. You have already determined that his pulse is weak and rapid, and his skin is cool, clammy, and pale. At this point, the only obvious injury you have discovered is an actively bleeding laceration on the patient's forehead.

On the basis of your findings, you should apply high-flow oxygen, control the bleeding from his forehead, and look for and control any other bleeding to the areas of his body that you can access. It would be appropriate to apply a cervical collar—after assessing the back of the neck for any deformities—to help maintain cervical spine stabilization.

The patient is showing signs of shock; therefore, after providing immediate care aimed at correcting problems with airway, breathing, and circulation, he must be extricated so that you can complete your assessment and provide additional treatment.

5. Is it always appropriate to obtain the vital signs of a patient who is trapped in his or her vehicle? Why or why not?

Clearly, it is not always appropriate or practical to obtain a complete set of vital signs while a patient is still entrapped in his or her vehicle. Your primary focus should be to provide immediate lifesaving care and then have the patient extricated as quickly and safely as possible.

In some cases, there may be a delay in extricating the patient. The rescue team may still be determining the safest extrication approach and is in the process of setting up the necessary rescue tools. If this is the case, it would not be unreasonable to assess the patient's vital signs; doing so will provide you with additional information about the patient's condition. However, make sure that you have identified and corrected all immediate life threats *first*.

It should be noted that you should *not* delay the process of extrication to assess the patient's vital signs. The longer the patient remains trapped in the vehicle, the longer the delay in providing further assessment and treatment.

6. What should you do as the patient is being extricated from the vehicle?

After providing initial emergency care, the next step in the extrication process is extrication of the patient—a measure that involves removing those parts of the vehicle that are trapping the patient. The goal of extrication is to remove the vehicle from around the patient—not to "cut the patient out of the vehicle."

During the extrication process, your input is essential so that the patient's injuries are considered as the rescue team plans a move that is quick and effective, but also protects the patient from further harm. Once a plan has been devised and all team members understand what will be done, you should determine how best to protect the patient. As is the case with this patient, you or another EMT will often be placed in the vehicle alongside the patient to monitor his or her condition and well-being as the vehicle is being forcibly cut, bent, or disassembled. Be sure that you are wearing the appropriate protective clothing (ie, a bunker coat, heavy-duty gloves [over your exam gloves], a rescue helmet, and facial protection).

The safety of you and the patient are paramount. All persons in the vehicle (in this case: you, the patient, and your partner in the back seat) should be covered by a thick, fire-resistant canvas or blanket that provides adequate protection from flying glass, metal, or other hazards during the extrication process.

Extrication is a noisy, frightening experience for the patient. During the entire process, it is important to communicate effectively with the patient to let him or her know what is happening. It is also important to communicate with the extrication team. If the patient suddenly screams in pain, instruct the extrication team to stop while you determine where the patient is hurting; it may be necessary for the rescue team to take a different extrication approach. Remember, *the goal of any extrication procedure is to extricate the patient without causing him or her further harm.*

You are the Provider: SUMMARY, continued

7. How does the rapid extrication technique differ from other methods of patient removal? When is it indicated?

Before the patient is freed from the vehicle, a plan for patient removal should already be in place. Do not wait until the patient is freed from the wreckage before deciding on the appropriate removal technique; this is a clear waste of time.

The rapid extrication technique involves manually stabilizing the patient's head, and in most cases, moving the patient from a sitting position in a vehicle to a supine position on a backboard in 1 minute or less. The only resources and equipment needed are a cervical collar, a backboard, and adequate personnel.

The rapid extrication technique is indicated when the patient cannot be properly assessed in the vehicle, when he or she needs immediate treatment that requires a supine position, or if his or her condition requires immediate transport.

Although the goal of the rapid extrication technique is to quickly remove a patient from a vehicle, moving the patient in one fast, continuous step increases the risk of harm to the patient and confusion among the rescuers and, thus, should be avoided.

To ensure that each team member can be positioned so that he or she can properly lift and carry the patient at all times, move the patient in a series of smooth, slow, controlled steps, with stops designed in between to allow for repositioning and adjustments. Plan the exact steps and pathway that you will follow *before* moving the patient; the path you choose should require the least amount of manipulation of the patient and the equipment. Ensure that only one person calls out all of the move commands; this should be the person who is maintaining cervical spine control. When you begin moving the patient, move him or her as a unit; do not move the backboard. The ambulance stretcher should be ready so that the patient can be placed directly on the stretcher instead of the ground.

The main difference between the rapid extrication technique and other methods of patient removal (ie, short backboard, vest-style extrication device) is that it is fast and requires minimal preparation of the patient in the vehicle before he or she is removed. For example, it takes between 6 and 8 minutes—and in some cases, even longer—to properly apply a short backboard or vest-style immobilization device. This is clearly too long when your patient is critically injured and needs immediate treatment and transport.

8. What should you do when entering a vehicle in which an air bag has not deployed?

In this case, the air bag on the driver's side of the vehicle deployed, but the passenger-side air bag did not. While you were in the front seat with the patient during the extrication process, the air bag could have unexpectedly deployed, potentially causing serious injury to you.

If an air bag does not deploy during a collision, it presents a hazard for the passenger of the vehicle (if one is present) and for rescue personnel. It could potentially deploy if the wires are cut or if it becomes activated during the rescue operation.

If you gain access to a patient via the passenger-side front seat and notice that the air bag has not deployed, get out of the vehicle and seek alternative access. If the driver's side air bag has not deployed, rescue personnel should not attempt to cut the steering wheel, nor should you place a hard object, such as a short backboard, in between the patient (or yourself) and an undeployed air bag. The rescue team or fire department should be informed about this situation so they can disconnect the battery cables on the vehicle. This action should allow the air bag capacitor to discharge. However, the time required to discharge the capacitor varies from one vehicle model to another. Some newer vehicles have a switch mounted under the dash that allows you to turn off the passenger-side air bag.

EMS Patient Care Report (PCR)

Date: 10-12-09	Incident No.: 013609	Nature of Call: Motor vehicle crash		Location: 4700 Block Hwy 46 W
Dispatched: 0422	En Route: 0423	At Scene: 0428	Transport: 0450	At Landing Zone: 0451	In Service: 0458

Patient Information

Age: 22 Sex: M Weight (in kg [lb]): 68 kg (150 lb)	Allergies: No known drug allergies Medications: None Past Medical History: None Chief Complaint: Multiple injuries

Vital Signs

Time: 0435	BP: 100/50	Pulse: 120	Respirations: 26	Sao$_2$: 96%
Time: 0446	BP: 80/58	Pulse: 130	Respirations: 28	Sao$_2$: 96%
Time: 0451	BP: 84/60	Pulse: 128	Respirations: 28	Sao$_2$: 95%

You are the Provider: SUMMARY, continued

EMS Treatment (circle all that apply)				
Oxygen @ 15 **L/min via (circle one):** NC **(NRM) (Bag-Mask Device)**	**(Assisted Ventilation)**		**Airway Adjunct**	**CPR**
Defibrillation	**(Bleeding Control)**	**Bandaging**	**Splinting**	**Other:** (Full spinal precautions, thermal management)

Narrative
Dispatched to the scene of a one-vehicle crash in which a passenger car ran off the road, slid down a small hill, and struck a tree head-on. On arrival at the scene, Rescue 33 was present and advised that the scene was safe, the vehicle was being stabilized, and that the single patient was trapped. After the vehicle was properly stabilized, made contact with the patient, a 22-year-old male. He was still in the vehicle; his legs were trapped by the steering column, and his feet were stuck. The patient, who was conscious but anxious, complained of severe pain to his legs. As medic partner accessed the patient from the back seat and manually stabilized his head, a primary assessment was performed via access to the patient by way of the broken driver's side window. The patient's airway was patent, his breathing was rapid and shallow, and he had an actively bleeding laceration to his forehead. Covered the wound on the patient's forehead and applied high-flow oxygen via nonrebreathing mask. No other obvious bleeding was noted. Skin was cool, clammy, and pale. Driver-side air bag had deployed, and the patient stated that he was wearing his seatbelt at the time of the crash, but removed it shortly thereafter. Rescue personnel prepared to extricate patient using the Jaws of Life as better access was gained via the front-side passenger door. Due to the mechanism of injury, the patient's clinical condition, and a potentially lengthy extrication, air medical transport was requested. Was able to obtain vital signs shortly before extrication began. Rescue leader advised that extrication was ready to commence, and covered medics and patient with heavy-duty blanket. After removing the door and lifting the steering wheel from the patient's legs, removed patient from the vehicle and performed rapid secondary assessment. Patient was now responsive to pain only, and his breathing remained rapid and shallow, so ventilation assistance was initiated. Additional injuries included bilateral femoral deformities and a rigid, bruised abdomen. Applied full spinal precautions, covered patient with a blanket, loaded him into the ambulance, and began transport to the landing zone. En route, reassessed patient, including his vital signs, and continued to assist his ventilations. Air medical transport arrived, and verbal report was given to flight medic. Medic 8 returned to service at 0458. *End of report*

Prep Kit

- You must always be prepared, mentally and physically, for any incident that requires rescue or extrication.
- Vehicle safety systems, such as shock-absorbing bumpers and air bags, protect your patients but also have the potential to injure rescuers.
- The ten phases of extrication are:
  - Preparation
  - En route to the scene
  - Arrival and scene size-up
  - Hazard control
  - Support operations
  - Gaining access
  - Emergency care
  - Removal of the patient
  - Transfer of the patient
  - Termination
- The rescue team is responsible for securing and stabilizing vehicles, providing safe entrance and access to patients, extricating patients, and protecting patients during extrication.
- EMS personnel are responsible for assessment, medical care, triage, packaging, and transport of patients.
- In some situations, the patient can only be reached by teams trained in special technical rescues. Situations requiring specialized teams include:
  - Cave rescue
  - Confined space rescue
  - Cross-field and trail rescue (park rangers)
  - Dive rescue
  - Lost person search and rescue
  - Mine rescue
  - Mountain-, rock-, and ice-climbing rescue
  - Ski slope and cross-country or trail snow rescue (ski patrol)
  - Structural collapse rescue
  - Special weapons and tactics (SWAT)
  - Technical rope rescue (low- and high-angle rescue)
  - Trench rescue
  - Water and small craft rescue
  - White-water rescue

Vital Vocabulary

access Gaining entry to an enclosed area and reaching a patient.

command post The location of the incident commander at the scene of an emergency and where command, coordination, control, and communication are centralized.

complex access Complicated entry that requires special tools and training and includes breaking windows or using other force.

danger zone (hot zone) An area where individuals can be exposed to electrical hazards such as sharp metal edges, broken glass, toxic substances, lethal rays, or ignition or explosion of hazardous materials.

entrapment To be caught (trapped) within a vehicle, room, or container with no way out or to have a limb or other body part trapped.

extrication Removal of a patient from entrapment or a dangerous situation or position, such as removal from a wrecked vehicle, industrial accident, or building collapse.

hazardous materials Any substances that are toxic, poisonous, radioactive, flammable, or explosive and cause injury or death with exposure.

incident commander The individual who has overall command of the incident in the field.

safe zone An area of protection providing safety from the danger zone (hot zone).

self-contained breathing apparatus (SCBA) Respirator with independent air supply used by fire fighters to enter toxic and otherwise dangerous atmospheres.

simple access Access that is easily achieved without the use of tools or force.

size-up The ongoing process of information gathering and scene evaluation to determine appropriate strategies and tactics to manage an emergency.

special weapons and tactics team (SWAT) A specialized law enforcement tactical unit.

structure fire A fire in a house, apartment building, office, school, plant, warehouse, or other building.

tactical situation A hostage, robbery, or other situation in which armed conflict is threatened or shots have been fired and the threat of violence remains.

technical rescue group A team of individuals from one or more departments in a region who are trained and on call for certain types of technical rescue.

technical rescue situation A rescue that requires special technical skills and equipment in one of many specialized rescue areas, such as technical rope rescue, cave rescue, and dive rescue.

Assessment in Action

You are dispatched to a motor vehicle accident. On arrival you find two vehicles have been involved in a lateral collision. One vehicle is on its side and the other remains upright. You notice extensive damage to both vehicles. Not all of the air bags have deployed.

1. What is your first concern?
   - **A.** Mechanism of injury
   - **B.** Scene safety
   - **C.** Physical assessment
   - **D.** Extrication

2. What is the standard clearance for a side air bag that has not deployed?
   - **A.** 5″
   - **B.** 10″
   - **C.** 15″
   - **D.** 20″

3. List the 10 phases of extrication.

4. Define extrication.

5. To gain access to the passengers, what should be done first?
   - **A.** Use a hand tool to gain access.
   - **B.** Check to see whether the doors are unlocked.
   - **C.** Use hydraulic devices to pry open the vehicle.
   - **D.** Use a crow bar to open the doors.

6. Evaluating the scene for hazardous materials and downed electrical wires and determining the number of patients is known as:
   - **A.** hazard control.
   - **B.** support operations.
   - **C.** scene size-up.
   - **D.** emergency care.

7. An area where individuals can be exposed to sharp metal edges, broken glass, toxic substances, lethal rays, or ignition or explosion of hazardous materials is known as a:
   - **A.** danger zone.
   - **B.** control zone.
   - **C.** cold zone.
   - **D.** warm zone.

8. Before gaining access to each vehicle, you should:
   - **A.** turn the car off.
   - **B.** disconnect the battery.
   - **C.** empty the gas tank.
   - **D.** stabilize the vehicle.

9. Define complex access.

10. When you are at the scene of a motor vehicle accident, what is one of the most important things you should do?
    - **A.** Gain access to the vehicles.
    - **B.** Communicate with the patient and your rescue team.
    - **C.** Clean up fuel spills.
    - **D.** Provide transport to the most appropriate facility.

National EMS Education Standard Competencies

EMS Operations

Knowledge of operational roles and responsibilities to ensure patient, public, and personnel safety.

Incident Management

- Establish and work within the incident management system. (pp 1387-1394)

Multiple Casualty Incidents*

- Triage principles (pp 1394-1400)
- Resource management (p 1388)
- Triage (pp 1396-1400)
  - Performing (pp 1396-1399)
  - Retriage (p 1396)
  - Destination decisions (pp 1399-1400)
  - Posttraumatic and cumulative stress (p 1394)

*This text uses the term mass-casualty incident.

Hazardous Materials Awareness

- Risks and responsibilities of operating in a cold zone at a hazardous material or other special incident. (pp 1409-1411)

....

Knowledge Objectives

1. Describe the National Incident Management System (NIMS) and its major components. (pp 1387-1388)
2. Describe the purpose of the incident command system (ICS) and its organizational structure and explain the role of EMS response within it. (pp 1388-1392)
3. Describe how the ICS assists the EMS in ensuring both personal safety and the safety of bystanders, health care professionals, and patients during an emergency. (pp 1392-1393)
4. Describe the role of the EMT in establishing command under the ICS. (p 1392)
5. Explain the purpose of medical incident command within the incident management system and describe its organizational structure within ICS. (pp 1393-1394)

6. Describe the specific conditions that would define a situation as a mass-casualty incident (MCI) and give some examples. (pp 1394-1396)
7. Describe what occurs during primary and secondary triage, how the four triage categories are assigned to patients on the scene, and how destination decisions regarding triaged patients are made. (pp 1396-1400)
8. Describe how the START and JumpSTART triage methods are performed. (pp 1398-1399)
9. Explain how a disaster differs from a mass-casualty incident and describe the role of the EMT during a disaster operation. (p 1400)
10. Recognize the entry-level training or experience requirements identified by the HAZWOPER regulation for an EMT to respond to a HazMat incident. (pp 1400-1401)
11. Define the term hazardous material, including the classification system used by the NFPA, and discuss the specific types of information and resources an EMT can use to recognize a HazMat incident. (pp 1400-1412)
12. List the different reference materials that may assist personnel who respond to a HazMat incident. (pp 1406-1408)
13. Explain the role of the EMT during a hazardous materials incident both before and after the HazMat team arrives, including precautions required to ensure the safety of civilians and public service personnel. (pp 1409-1411)
14. Explain how the three control zones are established at a HazMat incident and discuss the characteristics of each zone, including the personnel who work within each one. (pp 1409-1411)
15. Describe patient care at a HazMat incident and explain special requirements that are necessary for those patients who require immediate treatment and transport prior to full decontamination. (pp 1411-1413)
16. Describe the four levels of personal protective equipment (PPE) that may be required at a HazMat incident to protect personnel from injury by or contamination from a particular substance. (pp 1413-1414)

Skills Objectives

1. Demonstrate how to perform triage based on a fictitious scenario that involves a mass-casualty incident. (pp 1394-1399)
2. Correctly identify DOT labels, placards, and markings that are used to designate hazardous materials. (pp 1404-1406)
3. Demonstrate the ability to use a variety of reference materials to identify a hazardous material. (pp 1404-1408)

Introduction

The most challenging situations you can be called to are disasters and mass-casualty incidents (MCIs). These incidents, also referred to as multiple-casualty incidents, can be overwhelming because you will find a large number of patients and a lack of specialized equipment and/or adequate help. When you respond to an event with a large number of patients, you must use a systematic approach to manage the incident most efficiently. By learning to use the principles of the incident command system (ICS) (also called the Incident Management System in some agencies), you will be able to do the greatest good for the greatest number. As an EMT, you will typically be assigned to work within the EMS/medical branch under an ICS, but you may be asked to function in other areas, which will be discussed later in this chapter. The National Incident Management System (NIMS) was developed to promote more efficient coordination of emergency incidents at the regional, state, and national levels. To reduce on-scene problems and to increase your efficiency, you should attend training sessions and have a solid understanding of the basics of the NIMS.

National Incident Management System

Although most incidents are handled at the local level, the president directed the Secretary of Homeland Security to implement the **National Incident Management System (NIMS)** in March 2004. Major incidents require the involvement and coordination of multiple jurisdictions, functional agencies, and emergency response disciplines. The NIMS provides a consistent nationwide template to enable federal, state, and local governments, as well as private-sector and nongovernmental organizations, to work together effectively and efficiently. The NIMS is used to prepare for, prevent, respond to, and recover from domestic incidents, regardless of cause, size, or complexity, including acts of catastrophic terrorism and hazardous materials (HazMat) incidents.

Two important underlying principles of the NIMS are flexibility and standardization. The organizational structure must be flexible enough to be rapidly adapted for use in any situation. The NIMS provides standardization in terminology, resource classification, personnel training, certification, and more. Another important feature of the NIMS is the concept of interoperability, which refers to the ability of agencies of different types or from different jurisdictions to communicate with each other.

The ICS is one component of the NIMS. The major NIMS components are as follows:

- **Command and management.** The NIMS standardizes incident management for all hazards and across all levels of government. The NIMS standard incident command structures are based on three key constructs: ICS, multiagency coordination systems, and public information systems.

You are the Provider: PART 1

At 3:05 AM, you are dispatched to the 539 mile marker of Interstate 10 where a van crossed the median and struck a passenger car head-on. Law enforcement and an engine company are en route as well. Dispatch advises you that the caller told her there were at least three patients; however, the severity of their injuries is unknown. Of the three ambulances in your system, two of them are out on other calls and are unavailable.

1. **Do you have enough information to declare this a mass-casualty incident?**
2. **How will the incident command system facilitate operations at this scene?**

- **Preparedness.** The NIMS establishes measures for all responders to incorporate into their systems to prepare for their response to all incidents at any time.
- **Resource management.** The NIMS sets up mechanisms to describe, inventory, track, and dispatch resources before, during, and after an incident. The NIMS also defines standard procedures to recover equipment used during the incident.
- **Communications and information management.** Effective communications, information management, and sharing are critical aspects of domestic incident management. The NIMS communications and information systems enable the essential functions needed to provide interoperability.
- **Supporting technologies.** The NIMS promotes national standards and interoperability for supporting technologies to successfully implement the NIMS and standard technologies for professions or incidents. It provides structure for the science and technology used in incident management.
- **Ongoing management and maintenance.** The US Department of Homeland Security will establish a multijurisdictional, multidisciplinary NIMS Integration Center. This center will provide strategic direction for and oversight of the NIMS, supporting routine maintenance and continuous improvement of the system in the long term.

Incident Command System

It is important for you to be familiar with the terminology and concepts of the **incident command system (ICS)**. Some agencies refer to the incident command system as the incident management system. However, the terminology under NIMS is incident command system. The purpose of the ICS is ensuring responder and public safety, achieving incident management goals, and ensuring the efficient use of resources.

As you know, communication is the building block of good patient care. Common terminology and the use of "clear text" communications (plain English as opposed to 10-codes) help responders from multiple agencies work efficiently together.

Using the ICS gives you a modular organizational structure that is built on the size and complexity of the incident. The goal of the ICS is to make the best use of your resources to manage the environment around the incident and to treat patients during

an emergency. The ICS is designed to control duplication of effort and **freelancing**, in which individual units or different organizations make independent and often inefficient decisions about the next appropriate action. Follow your local standard operating procedures for establishing the ICS.

One of the organizing principles of the ICS is limiting the **span of control** of any one individual. This principle refers to keeping the supervisor/worker ratio at one supervisor for three to seven workers. A supervisor who has more than seven people reporting to him or her is exceeding an effective span of control and needs to divide tasks and delegate the supervision of some tasks to another person.

Organizational divisions may include sections, branches, divisions, and groups Figure 38-1 . In some regions, emergency operations centers may exist. The centers are usually operated by the city, state, or federal government. These centers will usually only be activated in a large catastrophic event that may go on for days, involves hundreds of patients, and taxes the whole system.

The individuals who will participate in the many tasks in an MCI or a disaster should use the ICS. You should find out from your service if one exists, who is in charge, how it is activated, and what your expected role will be.

Incident Command System Roles and Responsibilities

There are many roles defined in the ICS. The general staff includes command, finance, logistics, operations, and planning. It is important for you to understand the specific duties of each and how they work in coordinating the response. **Command** functions include the public information officer (PIO), safety officer, and liaison officer.

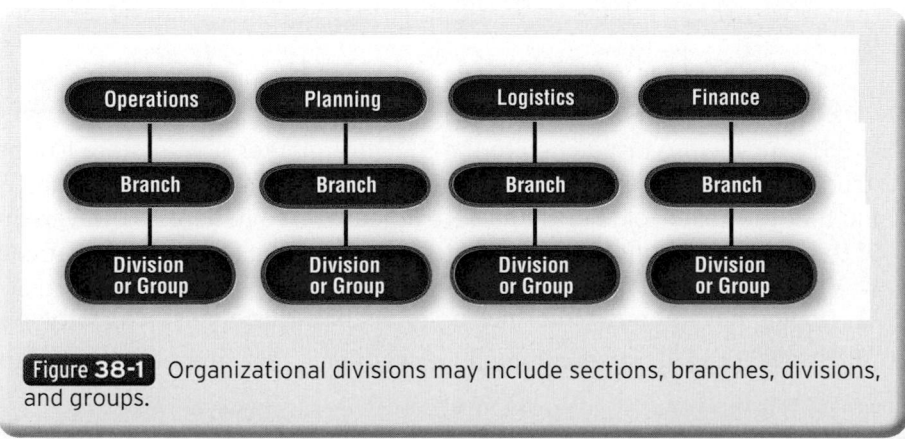

Figure 38-1 Organizational divisions may include sections, branches, divisions, and groups.

Command

The **incident commander (IC)** is the person in charge of the overall incident. The IC will assess the incident, establish the strategic objectives and priorities, and develop a plan to manage the incident **Figure 38-2**. The number of command duties (public information, safety, and liaison) the IC takes on often varies by the size of the incident. Small incidents often mean the IC will do it all. In an incident of medium size or complexity, the IC may delegate some functions but retain others. For example, at a motor vehicle collision with multiple patients, the IC may designate a safety officer or assign a PIO but maintain responsibility for the other command functions. In a complex situation, the IC may appoint team members to all of the command roles.

Large MCIs, such as a HazMat incident, require a multiagency or multijurisdiction response and need to use a **unified command system**. In this case, plans are drawn up in advance by all cooperating agencies that assume a shared responsibility for decision making. The response plan should designate the lead and support agencies in several kinds of MCIs. (For example, the HazMat team will take the lead in a chemical leak. However, the medical team might take the lead in a multivehicle car crash.) Agencies bordering each other should train often with each other to ensure that a unified command system will function well and that communication among the people involved is well established before a real incident occurs.

A **single command system** is one in which one person is in charge, even if multiple agencies respond. It is generally used with incidents in which one agency has the majority of responsibility for incident management.

Ideally, it is used for short-duration, limited incidents that require the services of a single agency.

Your IC should be on or near the scene, where he or she can easily communicate with all emergency responders operating at the scene. It is important that you know who the IC is, where the **command post** is located, and how to communicate with your supervisor. If the incident is very large, you will be reporting to a supervisor working under the IC. (Remember the rule of span of control? The number of people who can be effectively supervised is between three and seven.) To make the IC easily identifiable, some type of garment can be worn, such as a brightly colored vest emblazoned with the word COMMAND. If the command post is set up in a vehicle, it should be well marked, and you should know its location. Make sure that your supervisor or the IC knows of any plans or operations before they are initiated.

This communication is particularly important if a transfer of command takes place. Because an MCI can be ever changing and ever increasing in scope, an IC may turn over command to someone with more experience in a critical area. This change, or transfer of command, must take place in an orderly manner and, if possible, face to face. In extreme situations, it could be done by phone, radio, or e-mail, although this method is not recommended. Your agency should have standard operating procedures that govern the transfer of command. Make certain to follow the standard operating procedures. When an incident draws to a close, there should be a **termination of command**. Your agency should have **demobilization** procedures to implement as the situation deescalates or comes to an end.

Finance

The **finance** section chief is responsible for documenting all expenditures at an incident for reimbursement. A financial person is not usually needed at smaller incidents, but larger incidents demand keeping track of personnel hours and expenditures for materials and supplies and reporting at meetings of the general staff. Responding agencies and organizations may be eligible for some types of reimbursement after the incident, and an efficient finance section chief will help your agency to succeed in the reimbursement process. Finance personnel should be trained in the process of assessing expenditures with an eye to reimbursement long before an actual event.

The various functions within the finance section are the time unit, the procurement unit, the compensation/claims unit, and the cost unit. The time unit is responsible for ensuring the daily recording of personnel time and equipment use. The procurement unit deals with all matters concerning vendor contracts. The

Figure 38-2 The person in command at a mass-casualty incident oversees the incident and develops a plan for the response.

compensation and claims unit has two major purposes: dealing with claims as a result of the incident and injury compensation. Finally, the cost unit is responsible for collecting, analyzing, and reporting the costs related to an incident.

Logistics

The **logistics** section or section chief has responsibility for communications equipment, facilities, food and water, fuel, lighting, and medical equipment and supplies for patients and emergency responders. Local standard operating procedures will list the medical equipment needed for the incident, depending on the type of incident. Logistics personnel are trained to find food, shelter, and health care for you and the other responders at the scene of an MCI. In a large incident, it is often necessary for many people to handle logistics, even though only one person will report to the IC.

Operations

At a very large incident, the **operations** section is responsible for managing the tactical operations usually handled by the IC on routine EMS calls. In a complex incident, however, the IC must coordinate with other agencies and the media, engage in strategic planning, and ensure that logistics are functioning effectively. In these cases, the IC should appoint an operations section chief. The operations section chief will supervise the people working at the scene of the incident, who will be assigned to branches, divisions, and groups. Operations personnel often have experience in management within EMS.

Planning

The **planning** section solves problems as they arise during the MCI. Planners obtain data about the problem, analyze the previous incident plan, and predict what or who is needed to make the new plan work. They need to work closely with the operations, finance, and, especially, logistics sections. Planners can and should call on technical experts to help with the planning process. They should document their decisions and what they learned from the incident and also set out a course for demobilizing the response when necessary.

Another function of the planning section is the development of an **incident action plan**, which is the central tool for planning during a response to a disaster emergency. The incident action plan is prepared by the planning section chief with input from the appropriate sections and units of the incident management team. It should be written at the outset of the response and revised continually throughout the response. In an initial response for an incident that is readily controlled, a written plan may not be necessary. Larger, more complex incidents will require an incident action plan to coordinate activities. The level of detail required in an incident action plan will vary according to the size and complexity of the response.

Command Staff

Three important positions that help the general staff (all staff described previously) and the IC are the safety officer, the public information officer, and the liaison officer. The **safety officer** monitors the scene for conditions or operations that may present a hazard to responders and patients. The safety officer may need to work with environmental health and HazMat specialists. The importance of the safety officer cannot be underestimated—he or she has the authority to stop an emergency operation whenever a rescuer is in danger. A safety officer should remove hazards to EMS personnel and patients before the hazards cause injury.

The **public information officer (PIO)** provides the public and media with clear and understandable information. A wise PIO positions his or her headquarters well away from the incident command post and, most important, away from the incident, to minimize distractions. Also, the PIO must keep the media safe and from becoming part of the incident. The designated PIO may work in cooperation with PIOs from other agencies in a **joint information center (JIC)**. In some circumstances, the PIO/JIC may be responsible for disseminating a message designed to help a situation, prevent panic, and provide evacuation directions.

The **liaison officer** relays information and concerns among command, the general staff, and other agencies. If an agency is not represented in the command structure, questions and input should be given through the liaison officer.

■ Communications and Information Management

Communication has historically been the weak point at most major incidents. To minimize the effects of communications problems, it is recommended that communications be integrated. This means that all agencies involved should be able to communicate quickly and effortlessly via radios. Communications allow for accountability throughout the incident, as well as instant communication between recipients. As always, and more so during a large incident, it is important to maintain professionalism on all radio communications, remembering to communicate clearly, concisely, and using clear text (no codes).

■ Mobilization and Deployment

When an incident has been declared and the need for additional resources has been identified, a request is made for additional resources. Once a request is made, these resources are mobilized and deployed to the scene. It is important to wait until the request is made, to minimize the potential for freelancing.

Check-In at the Incident

On arrival at an incident, you should check in with the finance section. Checking in accomplishes many different functions. It allows you to be assigned to a supervisor for job tasking and allows for personnel tracking throughout the incident. Checking in also ensures that costs, pay, and reimbursement can be calculated accurately.

Initial Incident Briefing

After the check-in process is complete, you should report to your supervisor for an initial briefing that will allow you to get information regarding the incident, as well as specific job functions and responsibilities.

Incident Record Keeping

Record keeping is important for financial reasons and for documentation purposes. If a large piece of equipment becomes inoperable, it may be possible for replacement costs to come from the incident. Record keeping also allows for tracking of time spent on the actual incident for reimbursement purposes.

Accountability

Because of the large number of responders at a large incident, accountability is important. Accountability means keeping your supervisor advised of your location, actions, and completed tasks. It also includes advising your supervisor of the tasks that you have been unable to complete and what tools you need to complete them.

Incident Demobilization

Once the incident has been stabilized and all of the hazards mitigated, the IC will determine which resources are needed or not needed and when to begin demobilization. This process allows for an expeditious return of resources to their parent organizations to be placed back in service.

■ EMS Response Within the Incident Command System

■ Preparedness

Preparedness involves the decisions made and basic planning done before an incident occurs. Every area is prone to natural disasters, such as hurricanes, tornadoes, earthquakes, and wildfires. Therefore, preparedness in a given area involves decisions and planning about the most likely natural disasters for the area, among other disasters.

Your EMS agency should have written disaster plans that you are regularly trained to carry out. A copy of the disaster plan should be kept in each EMS vehicle. EMS facilities should have disaster supplies for at least a 72-hour period of self-sufficiency. Your EMS service should have mutual aid agreements with surrounding organizations so that requests for help can be expedited in an emergency. All groups with mutual aid agreements should practice using the plans frequently. Organizations should share a list of resources with each other so they will know early on what they can access. Also, your local EMS organizations should develop an assistance program for the families of EMS responders. If EMS responders have concerns about their families during a disaster, their effectiveness on the job could be diminished.

You are the Provider: PART 2

When you arrive at the scene, the engine company captain approaches you and identifies himself as the incident commander; his lieutenant is acting as the safety officer. He tells you that no immediate safety hazards have been identified. There are a total of five patients, two of whom are still in their vehicles. The interstate has been closed, and law enforcement personnel are rerouting traffic. Three additional ambulances are en route and will arrive in approximately 10 minutes. Both of the vehicles have sustained massive damage.

3. What functions does the safety officer perform?

4. How should you and your partner proceed?

Of course, you should have a personal disaster plan for your family. Families need to be prepared and know what to expect should you be required to be a disaster responder. You should be up to date on immunizations for influenza, hepatitis A and B, and tetanus.

■ Scene Size-up

Remember that sizing up a scene starts with dispatch. If dispatch information indicates a possible unsafe scene, you should stay away from the scene or get only close enough to make an assessment without putting yourself in harm's way. When you arrive first on the scene of an MCI, you will make an initial assessment and some preliminary decisions. The size-up will be driven by three basic questions that responders must ask themselves:

- *What do I have?*
- *What do I need to do?*
- *What resources do I need?*

These questions have a symbiotic relationship. The answer from one helps answer the others, and each answer represents a piece to the puzzle. Work as a team when you answer these questions because over looking just one safety issue early on can start a chain reaction of problems.

What Do I Have?

Start with scene safety. First, assess the scene for hazards. Warn all other responders about hazardous materials, fuel spills, electrical hazards, or other safety concerns as soon as possible. Confirm the incident location. Establish whether the incident is open or closed. Estimate the number of casualties. Report immediately to dispatch. An example of such a report would say: "EMT unit number one arriving on scene, multiple vehicles involved, full road blockage, no apparent hazards at this time, EMT unit number one is assuming command."

What Do I Need to Do?

You should keep the following priorities in mind:

- Safety
- Incident stabilization
- Preservation of property and the environment

You need to consider these priorities in the order they are given. Safety is paramount. Safety includes your life, your partner's life, and other rescuers' lives. Then, consider the safety of the patient and any bystanders. This will be difficult for anyone dedicated to saving lives, but it is important to put yourself and your partner first—you have the skills, and bystanders usually do not; the situation can be far worse if you do not put yourself first. Often, if a responder is injured, other responders will focus on "their own," removing available resources from the incident.

You may have to initially work to isolate or stabilize the incident before providing care to injured persons—this is another difficult concept for all emergency workers. Remember, you cannot help the injured if the scene is unstable. An unstable scene can lead to an injured EMT.

What Do I Need?

Decide what resources are needed. You may need more EMS responders, ambulances, or other forms of transportation. If extrication is required, a rescue unit and fire department response may be needed. If there are hazardous materials, get a HazMat team immediately. Many large EMS systems deploy specialized MCI units or mobile emergency room vehicles that are able to treat dozens of patients on the scene **Figure 38-3**.

■ Establishing Command

Once you have performed a good scene size-up and answered the three basic questions, command should be established by the most senior official, notification to other responders should go out, and necessary resources should be requested. A command system ensures that resources are effectively and efficiently coordinated. Command must be established early, preferably by the first-arriving, most experienced public safety official. These officials may include police, fire, or EMS personnel.

■ Communications

Communications is often the key problem at an MCI or a disaster. The infrastructure may be damaged, or communications capabilities may be overwhelmed. If possible, use face-to-face communications to limit radio traffic. Some organizations responding to a disaster might not know how to use a radio. If you communicate via radio, do not use codes or signals. Most communications problems should be worked out before a disaster happens

Figure 38-3 This mobile emergency room is staffed by EMTs, paramedics, and physicians who are able to provide advanced life support to multiple patients simultaneously on the scene of a mass-casualty incident.

by designating channels strictly for command during a disaster. Whatever form of communications equipment is used, it must be reliable, durable, and field-tested. Be sure there are backups in place if the primary communications system does not work. Some regions have mobile self-contained communications centers, whereas others use local radio groups such as ham radio operators to assist with communications. Most important, your plan should include a "Plan B" in case of communications failure.

Medical Incident Command

What has traditionally been referred to as **medical incident command** is also known as the medical (or EMS) branch of the ICS **Figure 38-4** . At incidents that have a significant medical factor, the IC should appoint someone as the medical branch director. This person will supervise the primary roles of the medical branch—triage, treatment, and transport of injured people. The medical branch director should help ensure that EMS units responding to the scene are working within the ICS, each medical division or group receives a clear assignment before beginning work at the scene, and personnel remain with their vehicle in the staging area until they are assigned their duties. Depending on the scale of the incident, EMS may be a branch or may fall under the logistics section as a unit.

Triage Supervisor

The **triage supervisor** is ultimately in charge of counting and prioritizing patients. During large incidents, a number of triage personnel may be needed. The primary duty of the triage division or group is to ensure that every patient receives initial assessment of his or her condition. EMTs doing triage will help move patients to the appropriate treatment sector. One of the most difficult parts of being a triage supervisor is that you must not begin treatment until all patients are triaged, or you will compromise your triage efforts.

Treatment Supervisor

The **treatment supervisor** will locate and set up the **treatment area** with a tier for each priority of patient. Treatment supervisors ensure that secondary triage of patients is performed and that adequate patient care is given as resources allow. Treatment supervisors also have a responsibility to assist with moving patients to the **transportation area**. As treatment supervisors supervise the responders, they must communicate with the medical branch director to request sufficient quantities of supplies, including bandages, burn supplies, airway and respiratory supplies, and patient packaging equipment.

Transportation Supervisor

The **transportation supervisor** coordinates the transportation and distribution of patients to appropriate receiving hospitals and helps to ensure that hospitals do not become overwhelmed by a patient surge. Transportation requires coordination with incident command to help ensure that enough personnel and ambulances are in staging or have been requested. A key role of the transportation supervisor is to communicate with the area hospitals to determine where to transport the patients. Some regions may have planned for a designated hospital within a region to perform the coordination between hospitals on destination decisions. An MCI typically disrupts the everyday functioning of the region's trauma system, so good coordination is needed. The transportation supervisor documents and tracks the number of vehicles transporting, patients transported, and the facility destination of each vehicle and patient.

Staging Supervisor

A **staging supervisor** should be assigned when MCIs or scenes require response by numerous emergency vehicles or agencies. The vehicles cannot and should not drive into the scene of the MCI without direction from the

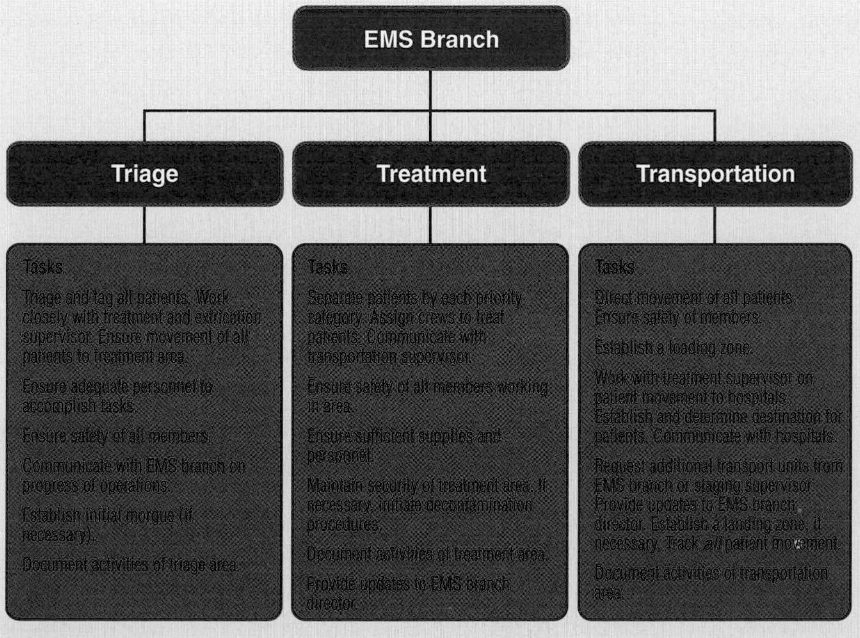

Figure 38-4 Components of the EMS branch within the incident command system.

EMS Branch

Triage | Treatment | Transportation

Tasks (Triage): Triage and tag all patients. Work closely with treatment and extrication supervisor. Ensure movement of all patients to treatment area. Ensure adequate personnel to accomplish tasks. Ensure safety of all members. Communicate with EMS branch on progress of operations. Establish initial morgue (if necessary). Document activities of triage area.

Tasks (Treatment): Separate patients by each priority category. Assign crews to treat patients. Communicate with transportation supervisor. Ensure safety of all members working in area. Ensure sufficient supplies and personnel. Maintain security of treatment area. If necessary, initiate decontamination procedures. Document activities of treatment area. Provide updates to EMS branch director.

Tasks (Transportation): Direct movement of all patients. Ensure safety of members. Establish a loading zone. Work with treatment supervisor on patient movement to hospitals. Establish and determine destination for patients. Communicate with hospitals. Request additional transport units from EMS branch or staging supervisor. Provide updates to EMS branch director. Establish a landing zone, if necessary. Track all patient movement. Document activities of transportation area.

staging supervisor. The staging area should be established away from the scene because the parked vehicles can be in the way. The staging supervisor locates an area to stage equipment and responders, tracks unit arrivals, and sends out vehicles as needed. This position plans for efficient access to and exit from the disaster site and prevents traffic congestion among responding vehicles. The staging supervisor releases vehicles and supplies when ordered by command.

■ Physicians on Scene

In an MCI, some areas have plans in place for physicians on scene. Sometimes, even without a plan, the enormity of the situation may require that physicians be sent to the scene. Emergency physicians, especially, will have the ability to make difficult triage decisions. They also provide secondary triage decisions in the treatment sector, deciding which priority patients are to be transported first. Physicians can provide on-scene medical direction for EMTs, and they can provide care in the treatment sector as appropriate.

■ Rehabilitation Supervisor

In disasters or situations that will last for extended periods, a rehabilitation section for the responders should be established. The **rehabilitation supervisor** should establish an area that provides protection for responders from the elements and the situation. The **rehabilitation area** should be located away from exhaust fumes and crowds (especially members of the media) and out of view of the scene itself. Rehabilitation is where a responder's needs for rest, fluids, food, and protection from the elements are met. The rehabilitation supervisor must also monitor responders for signs of stress. These signs may include fatigue, altered thinking patterns, and complete collapse. You should remember that all EMS personnel should be responsible to be aware of signs of stress. Your service might consider having a defusing or debriefing team in this area. Responders should be encouraged to take advantage of these services but should never be forced to participate.

■ Extrication and Special Rescue

Some disasters require search and rescue or extrication of patients **Figure 38-5**. An **extrication supervisor** or **rescue supervisor** may need to be appointed. These officers determine the type of equipment and resources needed for the situation. In some incidents, victims may need to be extricated or rescued before they can be triaged and treated. Because extrication and rescue are medically complex, the supervisors will usually function under the EMS branch of the ICS. The extrication and rescue supervisors identify the special equipment and personnel

Figure 38-5 Some disasters will involve search and rescue or extrication.

needed for the rescue. Extrication and rescue can be dangerous, so crew safety is of utmost importance.

■ Morgue Supervisor

In some disasters, there will be many dead patients. The **morgue supervisor** will work with area medical examiners, coroners, disaster mortuary assistance teams, and law enforcement agencies to coordinate removal of the bodies and even, possibly, body parts. The morgue supervisor should attempt to leave the dead victims in the location found, if possible, until a removal and storage plan can be determined. The location of victims may help in the identification of the dead victims in mass-fatality situations, or there may be crime scene considerations. If it is determined that a morgue area is needed, the morgue supervisor should ensure that the morgue is out of view of the living patients and other responders because the psychological impact could worsen the situation. In addition, the morgue should be secured from the public to prevent theft of any personal effects of the dead victims.

■ Mass-Casualty Incidents

In this text, a **mass-casualty incident (MCI)** refers to any call that involves three or more patients, any situation that places such a great demand on available equipment or personnel that the system would require a **mutual aid response** (an agreement between neighboring EMS systems to respond to MCIs or disasters in each other's

Figure 38-6 Mass-casualty incidents can be large, such as the attack on September 11, 2001, or they can be much smaller in scope.

the various roles and responsibilities of each position, the responders and/or IC can manage the incident in a smooth, organized manner.

All systems have different protocols for when to declare an MCI and initiate the ICS; however, as the EMT, ask yourself the following questions when considering whether the call is an MCI:

- How many seriously injured or ill patients can you care for effectively and transport in your ambulance? One? Two?
- What happens when you have three patients to deal with?
- How long will it take for additional help to arrive?
- What do you do when a school bus crashes, resulting in eight critically injured patients, and you have only three ambulances available?

region when local resources are insufficient to handle the response), or any incident that has the potential to create one of the previously mentioned situations **Figure 38-6**. Bus or train crashes and earthquakes are obvious examples of MCIs. However, other causes of MCIs are far more common than such disasters and are usually much smaller in scope. **Figure 38-7** is a diagrammed example of a residential building fire confined to one apartment that may only produce one patient but that has the potential to generate dozens of patients from among the rescuers and residents. Loss of power to a hospital or nursing home with ventilator-dependent and nonambulatory victims is considered an MCI, although no one is injured. By using the ICS and the NIMS and understanding

Words of Wisdom

The terminology used to describe an incident with multiple patients varies in different communities. Many communities use the term *mass-casualty situation* to describe an emergency that involves more than one patient but use the term *mass-casualty incident* to describe larger scale events, such as those with more than 20 patients. In this text, the term *mass-casualty incident* is used to describe any call that involves three or more patients.

Obviously, you and your team cannot treat and transport all injured patients at the same time. At an MCI, you will often experience an increased demand for equipment and personnel. For example, you may realize

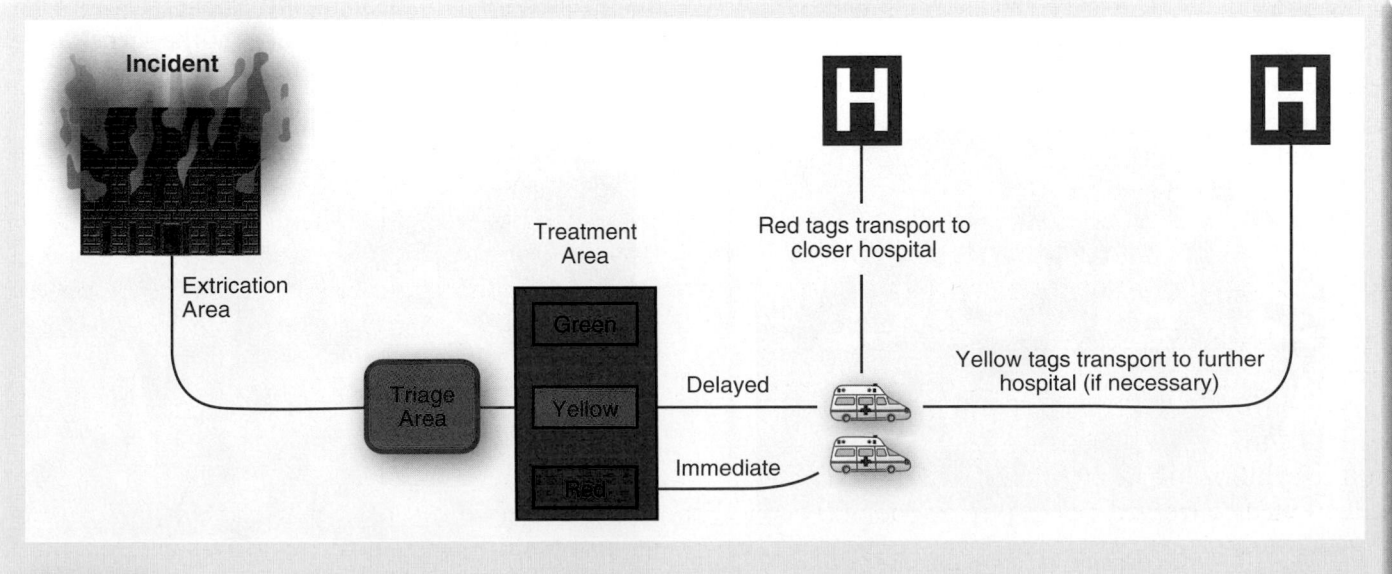

Figure 38-7 Diagram of a mass-casualty incident. The incident command system established at the scene of a building fire may look similar to this diagram.

that you are the only ambulance crew currently at the scene and there is a wait of 15 or more minutes before the next ambulance arrives. You should never leave the scene with patients who are loaded if there are still other patients present who are sick or wounded. This would leave patients at the scene without medical care and can be considered abandonment. If there are multiple patients and not enough resources to handle them without abandoning victims, you should declare an MCI (at least for the present time), request additional resources, and initiate the ICS and triage procedures (described below) Figure 38-8 . Although this may cause some delay in initiating treatment to all patients, it will not adversely affect the patient care. Always follow your local protocol. Many large EMS systems deploy specialized MCI units or mobile emergency room vehicles that are able to treat dozens of patients on the scene.

Triage

Triage simply means "to sort" your patients based on the severity of their injuries Figure 38-9 . The goal of doing the greatest good for the greatest number means that the triage assessment is brief and the patient condition categories are basic. **Primary triage** is the initial triage done in the field, allowing the EMT to quickly and accurately categorize the patient's condition and transport needs, whereas **secondary triage** is done as patients are brought to the treatment area. During primary triage, patients are briefly assessed and then identified in some way, such as by attaching a triage tag or triage tape. The main information needed on the tag is a unique number and a triage category. Rapid and accurate triage will help bring order to the chaos of the MCI scene and allow the most critical patients to be transported first. After the primary triage,

the triage supervisor should communicate the following information to the medical branch director:

- The total number of patients
- The number of patients in each of the triage categories
- Recommendations for extrication and movement of patients to the treatment area
- Resources needed to complete triage and begin movement of patients

When the initial triage has been completed, secondary triage, or retriage, can occur, allowing for the EMT to reassess all remaining patients and to upgrade the triage category, if necessary. In smaller MCI events, this step may not be necessary, if enough resources have arrived on the scene at this point.

■ Triage Categories

There are four common triage categories. They can be remembered using the mnemonic IDME, which stands for Immediate (red), Delayed (yellow), Minor or Minimal (green; hold), and Expectant (black; likely to die or dead) Table 38-1 . This is the order of priority for treatment and transport of the patients at an MCI.

Immediate (red-tag) patients are your first priority. They will need immediate care and transport. They usually have problems with the ABCs, head trauma, or signs and symptoms of shock.

Delayed (yellow-tag) patients are the second priority and will need treatment and transport, but it can be delayed. Patients usually have multiple injuries to bones or joints, including back injuries with or without spinal cord injury.

Minimal (green-tag) patients are the third priority. Patients may require no field or only "minimal" treatment. In some parts of the world, this is the hold category. These patients are the "walking wounded" at the scene. If they

Figure 38-8 Mass-casualty incidents require additional ambulances and EMS providers from the immediate region.

Figure 38-9 Triage is the process of sorting and prioritizing patients based on severity of conditions.

Table 38-1 Triage Priorities

Triage Category	Typical Injuries
Red tag: first priority (immediate) Patients who need immediate care and transport Treat these patients first, and transport as soon as possible	• Airway and breathing difficulties • Uncontrolled or severe bleeding • Severe medical problems • Signs of shock (hypoperfusion) • Severe burns • Open chest or abdominal injuries
Yellow tag: second priority (delayed) Patients whose treatment and transport can be temporarily delayed	• Burns without airway problems • Major or multiple bone or joint injuries • Back injuries with or without spinal cord damage
Green tag: third priority, minimal (walking wounded) Patients who require minimal or no treatment and transport can be delayed until last	• Minor fractures • Minor soft-tissue injuries
Black tag: fourth priority (expectant) Patients who are already dead or have little chance for survival; treat salvageable patients before treating these patients	• Obvious death • Obviously nonsurvivable injury, such as major open brain trauma • Respiratory arrest (if limited resources) • Cardiac arrest

have any apparent injuries, they are usually soft-tissue injuries such as contusions, abrasions, and lacerations.

The last priority is the expectant (black-tag) patients who are dead or whose injuries are so severe that they have, at best, a minimal chance of survival. This category may include patients who are in cardiac arrest or who have an open head injury, for example. If you have limited resources, this category may also include patients in respiratory arrest. Patients in this category receive treatment and transport only after patients in the other three categories have received care.

■ Triage Tags

Whatever triage system is used, it is vital that a patient has a tag or some type of label. Tagging patients early assists in tracking them and can help keep an accurate record of their condition. Triage tags should be weatherproof and easily read Figure 38-10 . The patient tags or tape should be color-coded and should clearly show the category of the patients. The use of symbols and colors to indicate the triage categories is important in case some rescuers are color blind.

The tags will become part of the patient's medical record. Most have a tear-off receipt with a number correlating with the number on the tag. When torn off by the transportation officer, it will assist him or her in tracking a patient. If the patient is unconscious and cannot be identified at the scene, the tag will be an identifier for tracking purposes. Some areas use digital photography of patients to assist in later identification. The photograph is catalogued with the patient's tag number, and the patient's location is tracked with this information. When family members are brought to crisis centers to help locate loved ones, the pictures may be of assistance. This technique has been used quite effectively in Europe and Israel with Polaroid or digital pictures. Another way of tracking and accounting for patients is to only issue 20 to 25 cards or tags at a time with a scorecard to mark how patients are triaged and their priority. When the responder returns for more tags, the scorecard will provide a patient count to help command

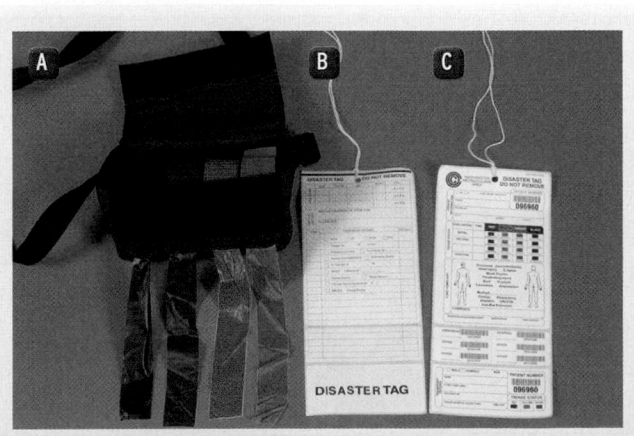

Figure 38-10 Triage tags (from left to right). **A.** Waterproof weapons of mass destruction tape. **B.** Triage tag: back. **C.** Triage tag: Front.

and the staff to develop a plan to respond and ensure that appropriate resources are either available or summoned. Whatever labeling system is used, it is imperative for the transportation officer to be able to identify which patient was transported by which unit and to which destination, and the priority of the patient's condition.

■ START Triage

START triage is one of the easiest methods of triage. START stands for Simple Triage And Rapid Treatment. The staff members at Hoag Memorial Hospital, Newport Beach, CA, are responsible for developing this method of triage. It is easily mastered with practice and will give you the ability to rapidly categorize patients at an MCI. START triage uses a limited assessment of the patient's ability to walk, respiratory status, hemodynamic status (pulse), and neurologic status.

The first step of the START triage system is performed on arrival at the scene by calling out to patients at the disaster site, "If you can hear my voice and are able to walk . . ." and then directing patients to an easily identifiable landmark. The injured persons in this group are the walking wounded and are considered minimal (green) priority, or third-priority patients.

The second step in the START process is directed toward nonwalking patients. You move to the first nonambulatory patient and assess the respiratory status. If the patient is not breathing, you should open the airway by using a simple manual maneuver. A patient who still does not begin to breathe is triaged as expectant (black). If the patient begins to breathe, tag him or her as immediate (red) and place in the recovery position and move on to the next patient.

If the patient is breathing, a quick estimation of the respiratory rate should be made. A patient who is breathing faster than 30 breaths/min or slower than 10 breaths/min is triaged as an immediate priority (red). If the patient is breathing from 10 to 29 breaths/min, move to the next step of the assessment.

The next step is to assess the hemodynamic status of the patient by checking for bilateral radial pulses. An absent radial pulse implies the patient is hypotensive and should be triaged as an immediate priority. If the radial pulse is present, go to the next assessment.

The final assessment in START triage is to assess the patient's neurologic status, which simply means to assess the patient's ability to follow simple commands, such as "show me three fingers." This assessment establishes that the patient can understand and follow commands. A patient who is unconscious or cannot follow simple commands is an immediate priority patient. A patient who complies with a simple command should be triaged in the delayed category.

Words of Wisdom

Another triage method is the Sort, Assess, Lifesaving interventions, and Treatment and/or Transport (SALT) triage system. This triage system begins by using a global sorting of patients. This first step identifies the patients who are able to understand verbal instructions and are therefore likely to have good perfusion. These patients are given a collection point to move to for further instructions. This is an attempt to decrease the number of patients leaving the scene and overwhelming local hospital resources before EMS can begin to move highest priority patients. The SALT method differs from others in its lifesaving intervention steps, which include bleeding control, opening the airway, two rescue breaths for children, needle decompression for tension pneumothorax, and auto-injector antidotes. The START method uses respirations, pulse, and neurologic status to assign priority.

■ JumpSTART Triage for Pediatric Patients

Lou Romig, MD, recognized that the START triage system does not take into account the physiologic and developmental differences of pediatric patients. She developed the JumpSTART triage system for pediatric patients. JumpSTART is intended for use in children younger than 8 years or who appear to weigh less than 100 lb. As in START, the JumpSTART system begins by identifying the walking wounded. Infants or children not developed enough to walk or follow commands (including children with special needs) should be taken as soon as possible to the treatment sector for immediate secondary triage. This action assists in getting children who cannot take care of their own basic needs into a caregiver's hands. There are several differences within the respiratory status assessment compared with that in START. First, if you find that a pediatric patient is not breathing, immediately check the pulse. If there is no pulse, label the patient as expectant. If the patient is not breathing but has a pulse, open the airway with a manual maneuver. If the patient does not begin to breathe, give five rescue breaths and check respirations again. A child who does not begin to breathe should be labeled expectant. The primary reason for this difference is that the most common cause of cardiac arrest in children is respiratory arrest.

The next step of the JumpSTART process is to assess the approximate rate of respirations. A patient who is breathing fewer than 15 breaths/min or more than 45 breaths/min is tagged as immediate priority, and you move on to the next patient. If the respirations are within the range of 15 to 45 breaths/min, the patient is assessed further.

The next assessment in JumpSTART triage is also the hemodynamic status of the patient. Just like in START, you are simply checking for a distal pulse. This does not need to be the brachial pulse; assess the pulse that you feel the most competent and comfortable checking. If there is an absence of a distal pulse, label the child as an immediate priority and move to the next patient. If the child has a distal pulse, move on to the next assessment.

The final assessment is for neurologic status. Because of the developmental differences in children, their responses will vary. For JumpSTART, a modified AVPU (*Alert* to person, place, and day; responsive to *Verbal* stimuli; responsive to *Pain*; or *Unresponsive*) score is used. A child who is unresponsive or responds to pain by posturing or with incomprehensible sounds or is unable to localize pain is considered an immediate priority and tagged as such. A child who responds to pain by localizing it or withdrawing from it or is alert is considered a delayed-priority patient.

■ Triage Special Considerations

There are a few special situations in triage. Patients who are hysterical and disruptive to rescue efforts may need to be made an immediate priority and transported out of the disaster site, even if they are not seriously injured. Panic breeds panic, and this type of behavior could have a detrimental impact on other patients and on the rescuers.

A rescuer who becomes sick or injured during the rescue effort should be handled as an immediate priority and be transported off the site as soon as possible to avoid negative impact to the morale of remaining rescuers.

HazMat and weapons of mass destruction incidents force the HazMat team to identify patients as contaminated or decontaminated before the regular triage process. Contamination by chemicals or biologic weapons in a treatment area, a hospital, or trauma center could obstruct all systems and organizations coping with the MCI. Bear in mind that some incidents may require multiple triage areas or teams because the victims are located far apart.

■ Destination Decisions

All patients triaged as immediate (red) or delayed (yellow) should preferably be transported by ground ambulance or air ambulance, if available. In extremely large situations, a bus may transport the walking wounded. If a bus is used for minimal-priority patients, it is strongly suggested that they be transported to a hospital or clinic distant from the MCI or disaster site to avoid overwhelming the local area hospital resources. It is advisable when using a bus to plan for at least one EMT or paramedic to ride on the bus and to have an ambulance follow the bus. If a minimal-priority patient's condition worsens, the patient could be moved to the ambulance and transported to a closer facility. The EMT or paramedic can stay with the patients triaged as needing minimal care until their arrival at the designated hospital. Any worsening of a patient's

You are the Provider: PART 3

There were two occupants in the van and three in the passenger car. You and your partner split up and begin triaging the patients.

Patient No. 1, a young male, was the passenger in the van. He is ambulatory and is talking on his cell phone. He has an obvious closed deformity to his left forearm and abrasions to his face.

Patient No. 2, a middle-aged male, was the driver of the van and was ejected. He is lying motionless on the ground; is unconscious and not breathing; and has a slow, weak carotid pulse. Both of his femurs are severely angulated and he has massive facial trauma.

Patient No. 3, a young female, was the driver of the passenger car, and is still in her vehicle. A fire fighter is in the rear seat manually stabilizing her head. She is conscious but confused, is experiencing obvious respiratory distress, and has a large hematoma to her forehead.

Patient No. 4, an older female, was the back seat passenger of the car. She is conscious and alert and has small abrasions on her arms. She tells you that she has heart problems and is experiencing severe, crushing chest pain. She is breathing with difficulty and has a rapid, irregular pulse.

Patient No. 5, a young male, was the front seat passenger in the car, and is found lying supine about 10′ away from the car. He is conscious but restless, his abdomen is rigid, and his pulse is rapid and weak.

5. What triage category should be assigned to each of these patients?

condition must be relayed to the receiving hospital as soon as possible in whatever manner the incident dictates.

Immediate-priority patients should be transported two at a time until all are transported from the site. Then patients in the delayed category can be transported two or three at a time until all are at a hospital. Finally, the slightly injured are transported. Expectant patients who are still alive would receive treatment and transport at this time. Dead victims are handled or transported according to the standing operating procedure for the area.

It is important to remember that during an MCI, local hospitals may have their resources overwhelmed as well. Early notification to receiving facilities will allow for the hospitals to increase staffing and move patients within their facility as required. Typically EMS agencies will know a hospital's surge capacity, which will tell the agency how many patients of each category the hospital is able to safely handle and care for.

Disaster Management

A **disaster** is a widespread event that disrupts functions and resources of a community and threatens lives and property. Many disasters may not involve personal injuries. Droughts causing widespread crop damage are an example. On the other hand, many disasters such as floods, fires, and hurricanes also result in widespread injuries. Unlike an MCI, which generally lasts no longer than a few hours, emergency responders will generally be on the scene of a disaster for days to weeks and sometimes months (as in the events following Hurricane Katrina). Although you can "declare" an MCI, only an elected official can declare a disaster.

Your role in a disaster is to respond when requested and to report to the IC for assigned tasks. In a disaster with an overwhelming number of casualties, area hospitals may decide that they cannot treat all patients at their facility. In this case, they may mobilize medical and nursing teams with equipment. Using a facility such as a warehouse near the disaster scene, they will set up a **casualty collection area**. Once at the casualty collection area, the teams can perform triage, provide medical care, and transport patients to the hospital on a priority basis.

If a casualty collection area is established, it will be coordinated through the ICS in the same way as all other branches and areas of the operation. This is usually done only in a major disaster such as an earthquake when transportation to a hospital facility is impossible or involves prolonged delays. It may take several hours to establish a casualty collection area.

Words of Wisdom

Mass-casualty incidents and disasters take a physical and emotional toll on emergency responders. Make certain that you are medically evaluated if you have been injured, come into contact with any hazardous substance, or inhale any dust, fumes, or smoke. Often the health effects of such exposures do not manifest for years and are difficult to link back to a particular event. Also be aware of the signs of stress in yourself and in your coworkers. Consider taking advantage of stress debriefing opportunities after an incident if you feel they may be valuable.

Introduction to Hazardous Materials

Your training has taught you that rapid response to the scene of a crash can save lives. However, when you arrive at the scene of a possible HazMat incident, you must first step back and assess the situation. This can be very stressful for you, particularly if you can see a patient. However, rushing into such events can have catastrophic results. If you are overcome by a hazardous substance, not only will patients suffer because you will be unable to assist them, but also you will place a strain on the system because you will require emergency care.

Because of the unique aspects of responding to and working at a **hazardous materials (HazMat) incident**, the Occupational Safety and Health Administration, or OSHA, has set specific additional training requirements in publication "29 CFR 1910.120(q)(6)(i)— Hazardous Waste Operations and Emergency Response Standard," (or HAZWOPER) which all individuals, including EMTs, must meet before becoming involved in these situations. In addition, you need to meet training requirements published in "1910.120(q)(6)(i)—First Responder Awareness Level." This text does not include the skills and information to meet the requirements in training individuals to respond to HazMat incidents at the awareness level. You need to check with your agency for information about additional specific awareness level training.

On the basis of the HAZWOPER regulation, first responders at the awareness level should have sufficient training or experience to objectively demonstrate competency in the following areas:

- An understanding of what hazardous substances are and the risks associated with them
- An understanding of the potential outcomes of an incident
- The ability to recognize the presence of hazardous substances

- The ability to identify the hazardous substances, if possible
- An understanding of the role of the first responder awareness individual in the emergency response plan
- The ability to determine the need for additional resources and to notify the communication center

Recognizing a Hazardous Material

A **hazardous material** is any material that poses an unreasonable risk of damage or injury to persons, property, or the environment if it is not properly controlled during handling, storage, manufacture, processing, packaging, use and disposal, and transportation. Recognizing a HazMat incident, determining the identity of the material(s), and understanding the hazards involved often require some detective work. You must train yourself to take the time to look at the whole scene so that you can identify the critical visual indicators and fit them into what is known about the problem.

Hazardous materials may be involved in any of the following situations Figure 38-11 :

- A truck or train crash in which a substance is leaking from a tank truck or railroad tank car
- A leak, fire, or other emergency at an industrial plant, refinery, or other complex where chemicals or explosives are produced, used, or stored
- A leak or rupture of an underground natural gas pipe
- Deterioration of underground fuel tanks and seepage of oil or gasoline into the surrounding ground
- Buildup of methane or other by-products of waste decomposition in sewers or sewage-processing plants
- A motor vehicle crash in which a gas tank has ruptured

Initially, it is important to approach the scene from a safe location and direction. The traditional rules of staying uphill and upwind are a good place to start. In addition, it may be wise to use binoculars and view the scene from a safe distance. Be sure to question anyone involved in the incident—a wealth of information may be available to you if you simply ask the right person. Take enough time to assess the scene and interpret other clues such as dead animals near the release, discolored pavement, dead grass, visible vapors or puddles, or labels that may help identify the presence of a hazardous material. Once you have a basic idea of what happened or determine that danger may be present, you can begin to formulate a plan for addressing the incident.

Figure 38-11 Two examples of hazardous materials incidents.

■ Occupancy and Location

A wide variety of chemicals are stored in warehouses, hospitals, laboratories, industrial occupancies, residential garages, bowling alleys, home improvement centers, garden supply stores, restaurants, and scores of other facilities or businesses in your response area. So many different chemicals exist in so many different locations that you could encounter almost anything during any type of emergency situation. The location and type of building are two good indicators of the possible presence of a hazardous material. For example, a biomedical laboratory is more likely to have chemicals that could be hazardous on site than a preschool.

■ Senses

Another way to detect the presence of hazardous materials is to use your senses, although this technique must be used carefully to avoid becoming contaminated or exposed. The senses that can be safely used are those of sight and sound. Initially, the farther you are from the incident when you notice a problem, the safer you will be. Using any of your senses that bring you in proximity to the chemical should be done with caution or avoided. When it comes to HazMat incidents, "leading with your nose" is not a good tactic—but using binoculars from a distance is.

Clues that are seen or heard may provide warning information from a distance, enabling you to take precautionary steps. Vapor clouds at the scene, for example, are a signal to move yourself and others away to a place of safety; the sound of an alarm from a toxic gas sensor in a chemical storage room or laboratory may also serve as a warning to retreat. Some highly vaporous and odorous chemicals—chlorine and ammonia, for example—may be detected by smell a long way from the actual point of release.

■ Containers

In basic terms, a **container** is any vessel or receptacle that holds a material. Often the container type, size, and material of construction provide important clues about the nature of the substance inside. Nevertheless, you should not rely solely on the type of container when making a determination about hazardous materials.

Red phosphorus from a drug laboratory, for example, might be found in an unmarked plastic container. In this case, there may be no legitimate markings to alert you to the possible contents. Gasoline or waste solvents may be stored in 55-gallon steel drums. Sulfuric acid, at 97% concentration, could be found in a polyethylene drum that might be colored black, red, white, or blue. In most cases, there is no correlation between the color of the drum and the possible contents. The same sulfuric acid might also be found in a 1-gallon

Figure 38-12 Drums may be constructed of many different types of materials, including cardboard, polyethylene, and stainless steel. The drum shown here is a polyethylene drum.

amber glass container. Steel or polyethylene drums, bags, high-pressure gas cylinders, railroad tank cars, plastic buckets, above-ground and underground storage tanks, cargo tanks, and pipelines are all representative examples of how hazardous materials are packaged, stored, and shipped **Figure 38-12**.

Some very recognizable chemical containers, such as 55-gallon drums and compressed gas cylinders, can be found in almost every type of manufacturing facility. Materials stored in a cardboard drum are usually in solid form. Stainless steel containers hold particularly dangerous chemicals, and cold liquids are kept in containers designed to maintain the appropriate temperature **Figure 38-13**.

One way to distinguish containers is to divide them into two categories based on their capacity: bulk and nonbulk storage containers.

Safety

When you consider locations for possible hazardous materials incidents, do not limit your thinking. You may be surprised at how many different kinds of containers you may find in your area.

Container Volume

Bulk storage containers include fixed tanks, highway cargo tanks, rail tank cars, totes, and intermodal tanks. In general, bulk storage containers are found in buildings that rely on and need to store large quantities of a particular chemical. Most manufacturing facilities have at least one type of bulk storage container. Often these bulk

Figure 38-13 A series of chemical storage containers.

storage containers are surrounded by a supplementary containment system to help control an accidental release. **Secondary containment** is an engineered method to control spilled or released product if the main containment vessel fails. A 5,000-gallon vertical storage tank, for example, may be surrounded by a series of short walls that form a catch basin around the tank.

Large-volume horizontal tanks are also common. When stored above ground, these tanks are referred to as above-ground storage tanks; if they are placed underground, they are known as underground storage tanks. These tanks can hold a few hundred gallons to several million gallons of product and are usually made of aluminum, steel, or plastic.

Another commonly encountered bulk storage vessel is the tote, also referred to as an intermediate bulk container. Totes have capacities ranging from 119 gallons to 703 gallons. These portable plastic tanks are surrounded by a stainless steel web that adds both structural stability and protection to the container. They can contain any type of chemical, including flammable liquids, corrosives, food-grade liquids, or oxidizers **Figure 38-14**.

Shipping and storing totes can be hazardous. These containers often are stacked atop one another and moved with a forklift, such that a mishap with the loading or moving process can compromise the tote. Because totes have no secondary containment system, any leak has the potential to create a large puddle. In addition, the steel webbing around the tote makes it difficult to access and patch leaks.

Intermodal tanks are both shipping and storage vessels. They hold between 5,000 and 6,000 gallons of product and can be pressurized or nonpressurized. Intermodal tanks can also be used to ship and store gaseous substances that have been chilled until they liquefy such as liquid nitrogen. In most cases, an intermodal tank is shipped to a facility, where it is stored and used and then returned to the shipper for refilling. Intermodal tanks can be shipped by all methods of transportation—air, sea, and land **Figure 38-15**.

Nonbulk Storage Vessels

Essentially, **nonbulk storage vessels** are all types of containers other than bulk containers. Nonbulk storage vessels can hold a few ounces to 119 gallons of product and include vessels such as drums, bags, compressed gas cylinders, cryogenic containers, and more. Nonbulk storage vessels hold commonly used commercial and industrial chemicals such as solvents, industrial cleaners, and compounds. This section describes the most commonly encountered types of nonbulk storage vessels.

Drums **Drums** are easily recognizable, barrel-like containers. They are used to store a wide variety of substances, including food-grade materials, corrosives, flammable liquids, and grease. Drums may be constructed of low-carbon steel, polyethylene, cardboard, stainless steel, nickel, or other materials. Generally, the nature of the chemical dictates the construction of the storage drum. Steel utility drums, for example, hold flammable liquids, cleaning fluids, oil, and other noncorrosive chemicals. Polyethylene drums are used for corrosives such as acids, bases, oxidizers, and other materials that cannot be stored in steel containers. Cardboard drums hold solid materials such as soap flakes, sodium hydroxide pellets, and food-grade materials. Stainless steel or other heavy-duty drums generally hold materials too aggressive (ie, too reactive) for either plain steel or polyethylene.

Bags Bags are commonly used to store solids and powders such as cement powder, sand, pesticides, soda ash, and slaked lime. Storage bags may be constructed of plastic, paper, or plastic-lined paper. Bags come in different sizes and weights, depending on their contents.

Figure 38-14 A tote is a commonly encountered bulk storage vessel.

Figure 38-15 An intermodel tank.

Pesticide bags must be labeled with specific information Figure 38-16 . You can learn a great deal from the label, including the following details:

- Name of the product
- Active ingredients
- Hazard statement
- The total amount of product in the container
- The manufacturer's name and address
- The Environmental Protection Agency (EPA) registration number, which provides proof that the product was registered with the EPA
- The EPA establishment number, which shows where the product was manufactured
- Signal words to indicate the relative toxicity of the material:
    - Danger—Poison: Highly toxic by all routes of entry
    - Danger: Severe eye damage or skin irritation
    - Warning: Moderately toxic
    - Caution: Minor toxicity and minor eye damage or skin irritation
- Practical first-aid treatment description
- Directions for use
- Agricultural use requirements
- Precautionary statements such as mixing directions or potential environmental hazards
- Storage and disposal information
- Classification statement on who may use the product

In addition, every pesticide label must carry the statement, "Keep out of reach of children."

Carboys Some corrosives and other types of chemicals are transported and stored in vessels called **carboys** Figure 38-17 . A carboy is a glass, plastic, or steel container that holds 5 to 15 gallons of product. Glass carboys are often placed in a protective wood, foam, fiberglass, or steel box to help prevent breakage. For example, nitric acid, sulfuric acid, and other strong acids are often transported and stored in thick glass carboys protected by a

Figure 38-17 A carboy is used to transport and store corrosive chemicals.

wooden or polystyrene (Styrofoam) crate to shield the glass container from damage during normal shipping.

Cylinders Several types of **cylinders** are used to hold liquids and gases. Uninsulated compressed gas cylinders are used to store substances such as nitrogen, argon, helium, and oxygen. They come in a range of sizes. As an EMT, you are very familiar with the shape of a cylinder, it holds the oxygen for your patients.

■ The Department of Transportation Marking System

The presence of labels, placards, and other markings on buildings, packages, boxes, and containers often enables EMTs to identify a released chemical. When used correctly, marking systems indicate the presence of a hazardous material from a safe distance and provide clues about the substance.

The US Department of Transportation (DOT) marking system is an identification system characterized by labels, placards, and markings Figure 38-18 .

This marking system is used when materials are being transported from one location to another in the United States. The same marking system is also used in Canada by Transport Canada.

Placards are diamond-shaped indicators (10¾″ on each side) that are placed on all four sides of highway transport vehicles, railroad tank cars, and other forms of transportation carrying hazardous materials Figure 38-19 . Labels are smaller versions (4″ diamond-shaped indicators) of placards; they are placed on the four sides of individual boxes and smaller packages being transported.

Figure 38-16 A pesticide bag must be labeled with the appropriate information.

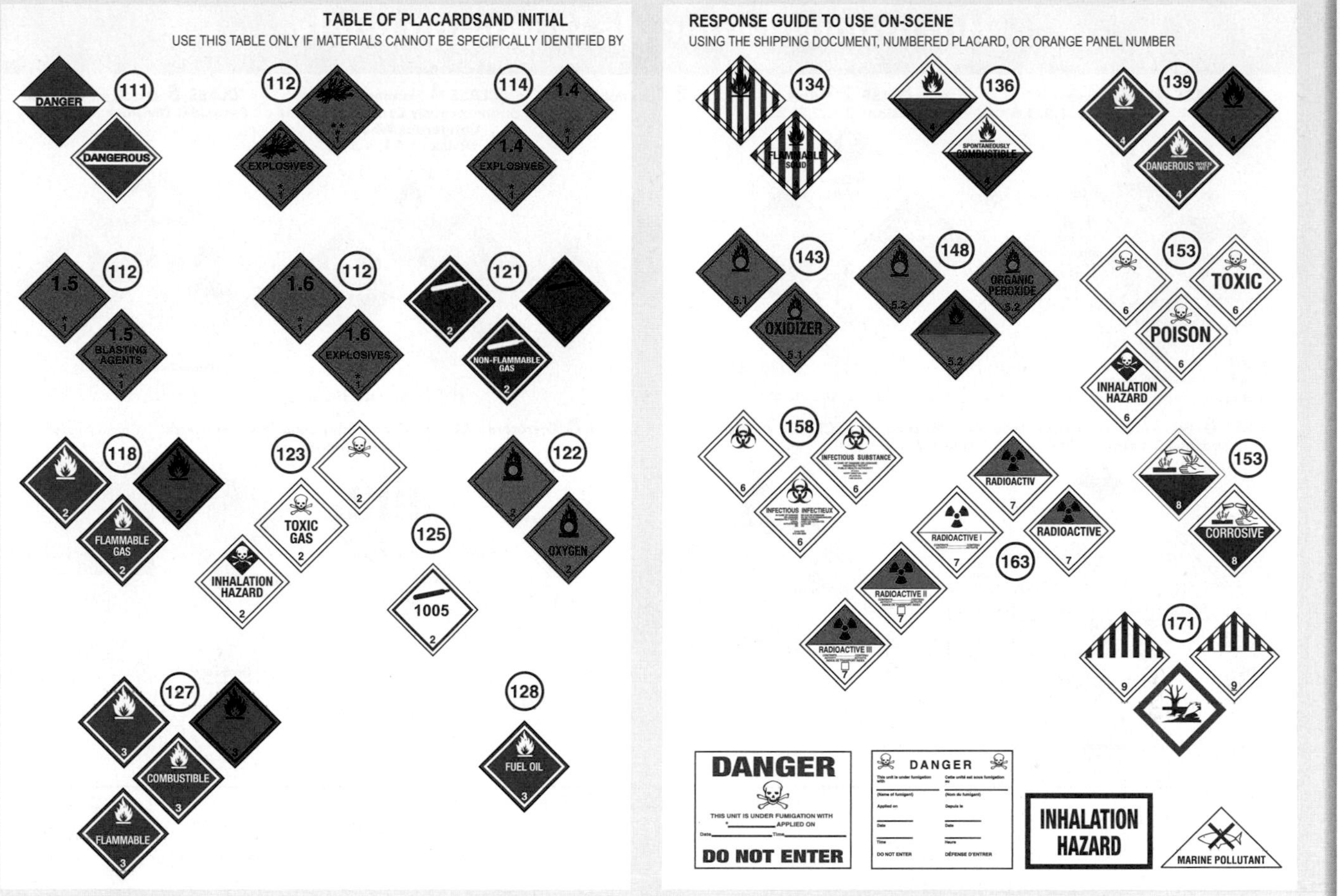

TABLE OF PLACARDSAND INITIAL
USE THIS TABLE ONLY IF MATERIALS CANNOT BE SPECIFICALLY IDENTIFIED BY

RESPONSE GUIDE TO USE ON-SCENE
USING THE SHIPPING DOCUMENT, NUMBERED PLACARD, OR ORANGE PANEL NUMBER

Figure 38-18 The Department of Transportation uses labels, placards, and markings (such as these found in the *Emergency Response Guidebook*) to give a general idea of the hazard inside a particular container or cargo tank.

Source: Courtesy of the US Department of Transportation.

Placards, labels, and markings are intended to give a general idea of the hazard inside a particular container or cargo tank. A placard identifies the broad hazard class (flammable, poison, corrosive) to which the material inside belongs. A label on a box inside a delivery truck, for example, relates only to the potential hazard inside that particular package **Figure 38-20**.

Figure 38-19 A placard is a large diamond-shaped indicator that is placed on all sides of transport vehicles that carry hazardous materials.

■ Other Considerations

The DOT system does not require that all chemical shipments be marked with placards or labels. In most cases, the package or cargo tank must contain a certain amount of hazardous material before a placard is required. For example, the "1,000-pound rule" applies to blasting agents, flammable and nonflammable gases, flammable/combustible liquids, flammable solids, air-reactive solids, oxidizers and organic peroxides, poison solids, corrosives, and miscellaneous (class 9) materials. Placards are required for these materials only when the shipment weighs more than 1,000 pounds.

Conversely, some chemicals are so hazardous that shipping any amount of them requires the use of labels or placards. These materials include explosives, poison gases, water-reactive solids, and high-level radioactive substances. A four-digit United Nations number may be required on some placards. This number identifies the specific material being shipped; a list of United Nations numbers is included in the *Emergency Response Guidebook*.

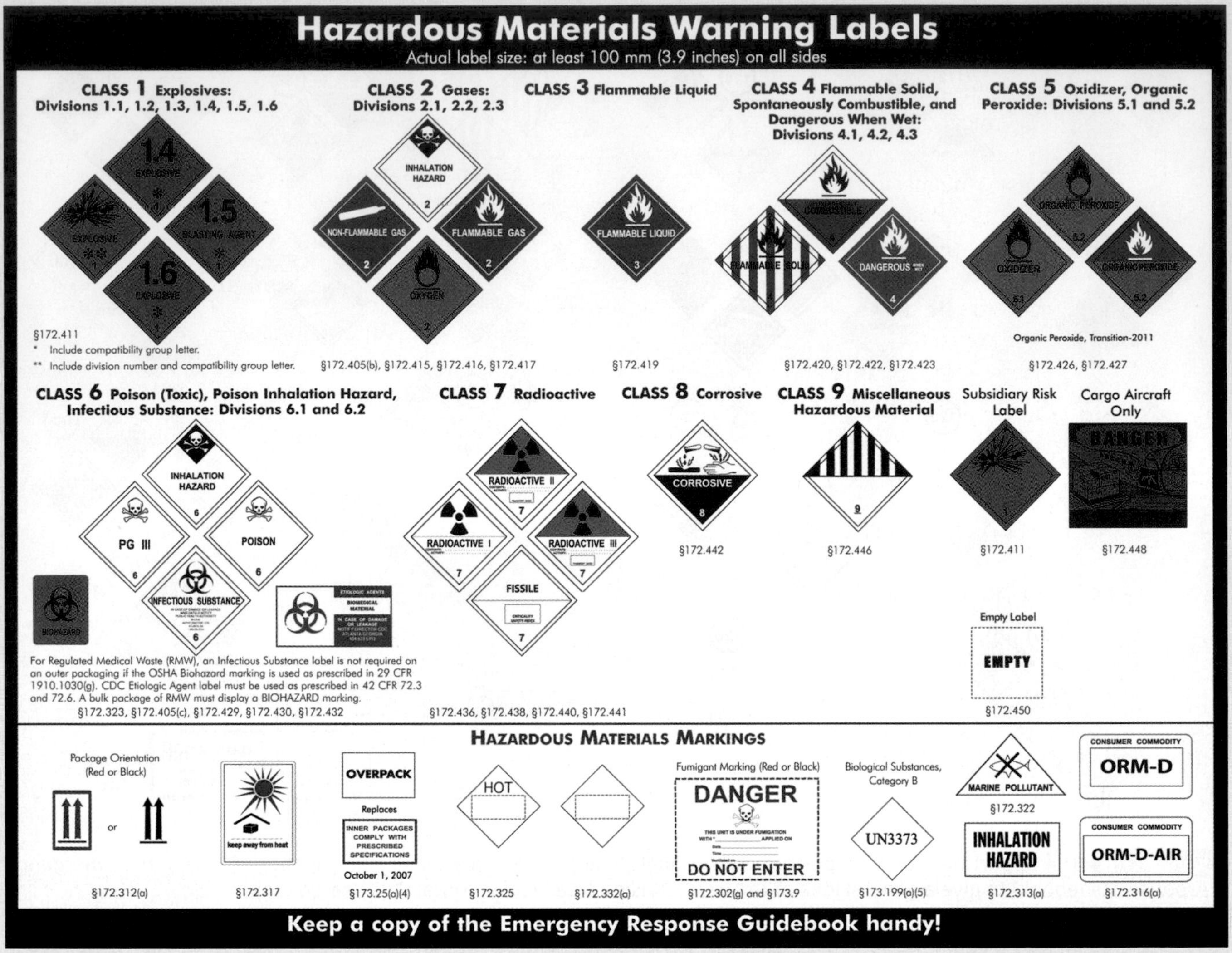

Hazardous Materials Warning Labels

Actual label size: at least 100 mm (3.9 inches) on all sides

CLASS 1 Explosives: Divisions 1.1, 1.2, 1.3, 1.4, 1.5, 1.6

§172.411
* Include compatibility group letter.
** Include division number and compatibility group letter.

CLASS 2 Gases: Divisions 2.1, 2.2, 2.3

§172.405(b), §172.415, §172.416, §172.417

CLASS 3 Flammable Liquid

§172.419

CLASS 4 Flammable Solid, Spontaneously Combustible, and Dangerous When Wet: Divisions 4.1, 4.2, 4.3

§172.420, §172.422, §172.423

CLASS 5 Oxidizer, Organic Peroxide: Divisions 5.1 and 5.2

Organic Peroxide, Transition-2011
§172.426, §172.427

CLASS 6 Poison (Toxic), Poison Inhalation Hazard, Infectious Substance: Divisions 6.1 and 6.2

For Regulated Medical Waste (RMW), an Infectious Substance label is not required on an outer packaging if the OSHA Biohazard marking is used as prescribed in 29 CFR 1910.1030(g). CDC Etiologic Agent label must be used as prescribed in 42 CFR 72.3 and 72.6. A bulk package of RMW must display a BIOHAZARD marking.
§172.323, §172.405(c), §172.429, §172.430, §172.432

CLASS 7 Radioactive

§172.436, §172.438, §172.440, §172.441

CLASS 8 Corrosive

§172.442

CLASS 9 Miscellaneous Hazardous Material

§172.446

Subsidiary Risk Label

§172.411

Cargo Aircraft Only

§172.448

Empty Label

EMPTY
§172.450

HAZARDOUS MATERIALS MARKINGS

Package Orientation (Red or Black)
§172.312(a)

§172.317

OVERPACK
Replaces
INNER PACKAGES COMPLY WITH PRESCRIBED SPECIFICATIONS
October 1, 2007
§173.25(a)(4)

HOT
§172.325

§172.332(a)

Fumigant Marking (Red or Black)
DANGER
THIS UNIT IS UNDER FUMIGATION WITH
DO NOT ENTER
§172.302(g) and §173.9

Biological Substances, Category B
UN3373
§173.199(o)(5)

MARINE POLLUTANT
§172.322

INHALATION HAZARD
§172.313(a)

CONSUMER COMMODITY
ORM-D
CONSUMER COMMODITY
ORM-D-AIR
§172.316(a)

Keep a copy of the Emergency Response Guidebook handy!

Figure 38-20 A label is a smaller version of the placard and is placed on boxes or smaller packages that contain hazardous materials.

Source: Courtesy of the US Department of Transportation.

References

Numerous reference materials are available to the responder, including the DOT's *Emergency Response Guidebook* and Jones and Bartlett Publishers' *Fire Fighter's Handbook of Hazardous Materials*. The following sections describe these resources.

The *Emergency Response Guidebook*

The DOT's *Emergency Response Guidebook* offers a certain amount of guidance for responders operating at a HazMat incident **Figure 38-21**. This guide is updated every 3 to 4 years and provides information on approximately 4,000 chemicals. The US DOT and the Secretariat of Communications and Transportation of Mexico, along with Transport Canada, jointly developed the *Emergency Response Guidebook*.

Material Safety Data Sheets

A common source of information about a particular chemical is the **material safety data sheet (MSDS)** specific to that substance **Figure 38-22**. Essentially, an MSDS provides basic information about the chemical makeup of a substance, the potential hazards it presents, appropriate first aid in the event of an exposure, and other pertinent data for safe handling of the

Figure 38-21 The *Emergency Response Guidebook* is a reference used as a base for your initial actions at a hazardous materials incident.

material. An MSDS will typically include the following details:

- The name of the chemical, including any synonyms for it
- Physical and chemical characteristics of the material
- Physical hazards of the material
- Health hazards of the material
- Signs and symptoms of exposure
- Routes of entry
- Permissible exposure limits
- Responsible-party contact
- Precautions for safe handling (including hygiene practices, protective measures, and procedures for cleaning up spills or leaks)
- Applicable control measures, including personal protective equipment
- Emergency and first-aid procedures
- Appropriate waste disposal

All facilities that use or store chemicals are required by law to have an MSDS on file for each chemical used or stored in the facility. Many sites, but especially those that stock many different chemicals, may keep this information archived on a computer database. Although the MSDS is not a definitive response tool, it is a key piece of the puzzle. An MSDS can also be obtained from the transporting vehicle.

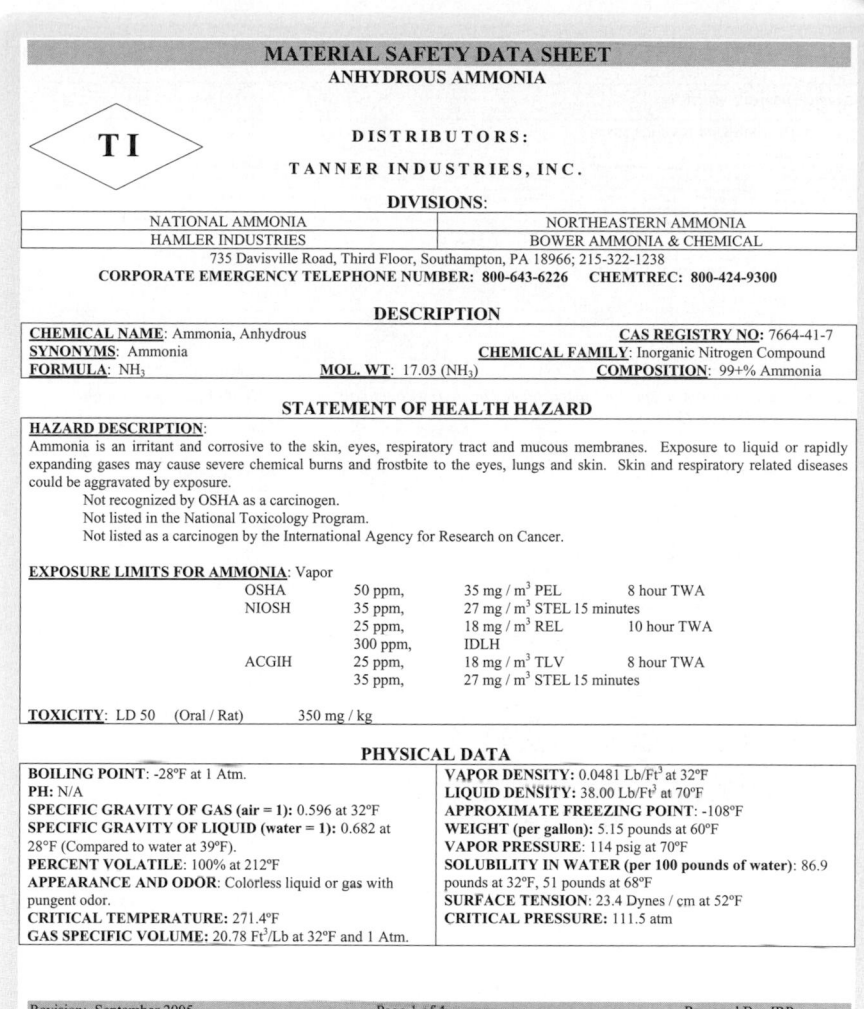

MATERIAL SAFETY DATA SHEET
ANHYDROUS AMMONIA

T I

DISTRIBUTORS:

TANNER INDUSTRIES, INC.

DIVISIONS:

| NATIONAL AMMONIA | NORTHEASTERN AMMONIA |
| HAMLER INDUSTRIES | BOWER AMMONIA & CHEMICAL |

735 Davisville Road, Third Floor, Southampton, PA 18966; 215-322-1238
CORPORATE EMERGENCY TELEPHONE NUMBER: 800-643-6226   CHEMTREC: 800-424-9300

DESCRIPTION

CHEMICAL NAME: Ammonia, Anhydrous   **CAS REGISTRY NO**: 7664-41-7
SYNONYMS: Ammonia   **CHEMICAL FAMILY**: Inorganic Nitrogen Compound
FORMULA: NH_3 **MOL. WT**: 17.03 (NH_3)   **COMPOSITION**: 99+% Ammonia

STATEMENT OF HEALTH HAZARD

HAZARD DESCRIPTION:
Ammonia is an irritant and corrosive to the skin, eyes, respiratory tract and mucous membranes. Exposure to liquid or rapidly expanding gases may cause severe chemical burns and frostbite to the eyes, lungs and skin. Skin and respiratory related diseases could be aggravated by exposure.
   Not recognized by OSHA as a carcinogen.
   Not listed in the National Toxicology Program.
   Not listed as a carcinogen by the International Agency for Research on Cancer.

EXPOSURE LIMITS FOR AMMONIA: Vapor
OSHA	50 ppm,	35 mg / m³ PEL	8 hour TWA
NIOSH	35 ppm,	27 mg / m³ STEL 15 minutes
	25 ppm,	18 mg / m³ REL	10 hour TWA
	300 ppm,	IDLH
ACGIH	25 ppm,	18 mg / m³ TLV	8 hour TWA
	35 ppm,	27 mg / m³ STEL 15 minutes

TOXICITY: LD 50   (Oral / Rat)   350 mg / kg

PHYSICAL DATA

BOILING POINT: -28°F at 1 Atm.
PH: N/A
SPECIFIC GRAVITY OF GAS (air = 1): 0.596 at 32°F
SPECIFIC GRAVITY OF LIQUID (water = 1): 0.682 at 28°F (Compared to water at 39°F).
PERCENT VOLATILE: 100% at 212°F
APPEARANCE AND ODOR: Colorless liquid or gas with pungent odor.
CRITICAL TEMPERATURE: 271.4°F
GAS SPECIFIC VOLUME: 20.78 Ft³/Lb at 32°F and 1 Atm.

VAPOR DENSITY: 0.0481 Lb/Ft³ at 32°F
LIQUID DENSITY: 38.00 Lb/Ft³ at 70°F
APPROXIMATE FREEZING POINT: -108°F
WEIGHT (per gallon): 5.15 pounds at 60°F
VAPOR PRESSURE: 114 psig at 70°F
SOLUBILITY IN WATER (per 100 pounds of water): 86.9 pounds at 32°F, 51 pounds at 68°F
SURFACE TENSION: 23.4 Dynes / cm at 52°F
CRITICAL PRESSURE: 111.5 atm

Revision: September 2005   Page 1 of 4   Prepared By: JRP

Figure 38-22 An example of a material safety data sheet for anhydrous ammonia.

Shipping Papers

Shipping papers are required whenever materials are transported from one place to another. They include the names and addresses of the shipper and the receiver, identify the material being shipped, and specify the quantity and weight of each part of the shipment.

Shipping papers for road and highway transportation are called **bills of lading** or **freight bills** and are located in the cab of the vehicle **Figure 38-23**. Drivers transporting chemicals are required by law to have a set of shipping papers on their person or within easy reach inside the cab at all times.

You are the Provider: PART 4

After triaging all the patients, you and your partner begin treating the most critically injured. A few minutes later, the backup ambulances arrive at the scene. You update the incident commander on the status of your EMS personnel and ambulances. Patient No. 1 is still talking on his cell phone. He is conscious and alert and tells you that his girlfriend is coming to pick him up and will take him to the hospital; he refuses to go via EMS.

6. Should your initial triage assignments remain the same?

7. Is it necessary to set up a treatment area at this incident?

Figure 38-23 A bill of lading or freight bill.

The figure contains the following form text:

STRAIGHT BILL OF LADING
ORIGINAL - NOT NEGOTIABLE

BOL/Reference No.
RSI82715

CARRIER: NORFOLK SOUTHERN   Date: 12/23/2008

Shipper: RSI LOGISTICS, INC (OKEMOS, MI US)

The property described below, in apparent good order, except as noted (contents and condition of packages unknown), marked, consigned, and destined as indicated below, which said carrier (the word carrier being understood throughout this contract as meaning any person or corporation in possession of the property under the contract) agrees to carry to its usual place of delivery at said destination, if on itsroute, otherwise to deliver to another carrier on the route to said destination. It is mutually agreed, as to each carrier of all or any said property, that every service to be performed hereunder shall be subject to all the terms and conditions of the Uniform Domestic Straight Bill of Lading set forth (1) in Official, Southern, Western and Illinois Freight Classificatio in effect on the date hereof, if this is rail or a rail-water shipment, or (2) in the applicable motor carrier classification or tariff if this is a motor carrier shipment
Shipper hereby certifies that he is familiar with all the terms and conditions of the said bill of lading, including those on the back thereof, set forth in the classification or tariff which governs the transportation of th shipment, and the said terms and conditions are hereby agreed to   by the shipper and accepted for himself and his assigns.

Consignee Information: CONSIGNEE DEER PARK, TX
Address:
City: DEER PARK, TX US

Route: NS-ESTL-BNSF

Origin Switch Route:

Destination Switch Route: HUSTN-PTRA   Rail Car No: GATX290861

For assistance in any transportation emergency involving chemicals, phone CHEMTREC, day or night, Toll Free 1-800-424-9300

DESCRIPTION	*WEIGHT
ONE TANK CAR	Contains: Methyl Esters STCC#2899415 BIODIESEL-15, Biodiesel	(Sub. To Correction) 204400 Lbs.

Sales Order Contract No: RSI82715
Sales Order Contract No: AAT122308-4
Purchase Order Contract No: AAT122308-4

SEAL NUMBERS: Gross

                       Tare

                       Net

                       Weighed By: _____

If charges are to be prepaid, write or stamp here, "To be Prepaid"
Prepaid

Subject to Section 7 of the conditions of applicable bill of lading, if this shipment is to be delivered to the consignee without recourse on the consignor, the consignor shall sign the following statement:: *The carrier shall not make delivery of this shipment without payment of freight and all other lawful charges.*

Not In Effect

* This is to certify that the above named materials are properly classified, described, packaged marked, and labeled, and are in proper condition for transportation, according to the applicable regulations of the Department of Transportation.

CHEMTREC

Located in Arlington, VA, the **Chemical Transportation Emergency Center (CHEMTREC)**, now operated by the American Chemistry Council, is a clearinghouse of technical chemical information. Since 1971, this emergency call center has served as an invaluable information resource for first responders of all disciplines who are called upon to respond to chemical incidents. The toll-free number for CHEMTREC is 1-800-262-8200. CHEMTREC has the ability to provide you with technical chemical information via telephone, fax, or other electronic media. It also offers a phone conferencing service to connect you with thousands of shippers, subject matter experts, and chemical manufacturers.

When you call CHEMTREC, be sure to have the following basic information ready:

- The name of the chemical(s) involved in the incident (if known)
- Name of the caller and callback telephone number
- Location of the actual incident or problem
- Shipper or manufacturer of the chemical (if known)

- Container type
- Railcar or vehicle markings or numbers
- The shipping carrier's name
- Recipient of material
- Local conditions and exact description of the situation

When you are speaking with CHEMTREC personnel, spell out all chemical names; if using a third party, such as a dispatcher, it is vital that you confirm all spellings to avoid misunderstandings. One number or letter out of place could throw off all subsequent research. When in doubt, be sure to obtain clarification.

■ Identification

Unfortunately, even with all of these resources, identifying materials can still be difficult. Little consistency is used on labels and placards, and sometimes dishonest transporters will not label containers or vessels appropriately. The laws and regulations that cover labeling of packages and transport vehicles can also be misleading. In most cases, the package or tank must contain a certain amount of a hazardous material before a placard is required. For example, because of the small quantities of hazardous materials that are involved, a truck carrying 99 lb of HazMat No. 1 and 99 lb of both HazMat No. 2 and HazMat No. 3 may not be required by law to display any labels or placards. The truck may show only a "Please drive carefully" placard, implying that it carries no hazardous materials. Therefore, a crash involving this truck is a serious situation, but you would not necessarily know this if you relied on labels and placards. Always maintain a high index of suspicion when approaching the scene of a truck or train tanker accident.

Some substances are not hazardous; however, when mixed with another substance, they may become highly toxic. There may be no regulations against carrying such substances together on one truck or railroad car (or adjacent tank cars). The driver of a commercial truck and the conductor of a train, however, must carry shipping papers that identify what is being transported in their care. These shipping papers may be your first clue that there is a possible HazMat problem, although, depending on the nature of the incident, the papers may not be available to you.

In the event of a leak or spill, a HazMat incident is often indicated by the presence of the following:

- A visible cloud or strange-looking smoke resulting from the escaping substance
- A leak or spill from a tank, container, truck, or railroad car with or without HazMat placards or labels
- An unusual, strong, noxious, acrid odor in the area

To indicate the presence of normally odorless toxic gases or fluids during a leak or spill, manufacturers may add a substance that produces a strong noxious odor. However, a large number of hazardous gases and fluids are essentially odorless (or do not have a distinctive unpleasant smell) even when a substantial leak or spill has occurred. In some incidents, a large number of people are exposed and may be injured or killed before the presence of a HazMat incident is identified. If you approach a scene where more than one person has collapsed or is unconscious or in respiratory distress, you should assume that there has been a HazMat leak or spill and that it is unsafe to enter the area.

It is important for you to understand the potential danger of hazardous materials and know how to operate safely at a HazMat incident. If you do not follow the proper safety measures, you and many others could end up needlessly injured or dead. The safety of you and your team, the other responders, and the public must be your most important concern.

Words of Wisdom

Safety considerations at HazMat scenes differ considerably from those involved in emergency response in general. A HazMat scene requires you to have an even higher degree of alertness than usual to avoid entering a dangerous environment and to help others avoid it. There is also a need to prevent the spread of contamination to yourself and your ambulance. Understanding these two concepts is a good start toward safe operations in the presence of hazardous materials.

There will be times when the ambulance is the first to arrive at the scene. If, as you approach, any signs suggest that a HazMat incident has occurred, you should stop at a safe distance and park upwind or uphill from the incident. After rapidly sizing up the scene, call for a HazMat team. If you do not recognize the danger until you are too close, immediately leave the **danger zone**. Once you have reached a safe place, try to rapidly assess the situation and provide as much information as possible when calling for the HazMat team, including your specific location, the size and shape of the containers of the hazardous material,

and what you have observed and have been told has occurred. Do not reenter the scene, and do not leave the area until you have been cleared by the HazMat team, or you may contribute to the situation by spreading hazardous materials. Finally, do not allow civilians to enter the scene, if possible. No one should enter the area without the proper protective equipment, respiratory protection, or training.

Above all, avoid all contact with the material!

■ HazMat Scene Operations

Once you have recognized the incident as one involving hazardous materials and have called for the HazMat team, you should focus your efforts on activities that will ensure the safety and survival of the greatest number of people. Use the ambulance's public address system to alert individuals who are near the scene and direct them to move to a location where they will be sufficiently far from danger. With the aid of others on your team, try to set up a perimeter to stop traffic and individuals from entering the danger zone.

Establishing Control Zones

Managing a HazMat incident by setting control zones and limiting access to the incident site helps reduce the number of civilians and public service personnel who may be exposed to the released substance. **Control zones** are established at a HazMat incident based on the chemical and physical properties of the released material, the environmental factors at the time of the release, and the general layout of the scene. Of course, isolating a city block in the busy downtown area of a large city presents far different challenges than isolating the area around a rolled-over cargo tank on an interstate highway. Each situation is different, requiring flexibility and thoughtfulness. Securing access to the incident helps ensure that no one will accidentally enter a contaminated area.

If the incident takes place inside a structure, the best place to control access is at the normal points of ingress and egress—doors. Once the doors are secured so that no unauthorized personnel can enter, appropriately trained emergency response crews can begin to isolate other areas as appropriate.

The same concept applies to outdoor incidents. The goal is to secure logical access points around the hazard. Begin by controlling intersections, on and off ramps, service roads, and other access routes to the scene. Police officers should assist by diverting traffic at a safe distance outside the hazard area. They should block off streets, close intersections, and redirect traffic as needed.

During a long-term incident, highway department or public works department employees may be called upon to set up traffic barriers. Whatever methods or devices are used to restrict access, they should not limit or prevent a rapid withdrawal from the area by personnel working inside the hot zone.

It is not uncommon to set large control zones at the onset of an incident, only to discover that the zones may have been established too liberally. At the same time, control zones should not be defined too narrowly **Figure 38-24**. As the IC gets more information about the specifics of the chemical or material involved, the control zones may be changed. Ideally, the control zones will be established in the right place, geographically, the first time. Nevertheless, you should be prepared to expand or contract them if necessary. Wind shifts are a common reason why control zones are modified during the incident. If there is a prevailing wind pattern in your area, factor that consideration into your decision making when it comes to control zones.

Typically, control zones at HazMat incidents are labeled as *hot, warm,* or *cold.* You may also discover that other terms are used, such as *exclusionary zone* (hot zone), *contamination reduction zone* (warm zone), and *outer perimeter* (cold zone). In any case, make sure you understand the terminology used in your jurisdiction. Be prepared to discover that different jurisdictions may use terminology and setup procedures unlike the ones used in your agency. As long as you understand the concepts behind the actions and remember that safety is the main focus, the act of setting up and naming zones can remain flexible.

The **hot zone** is the area immediately surrounding the release, which is also the most contaminated area. Its boundaries should be set large enough that adverse effects from the released substance will not affect people outside of the hot zone. An incident involving a gaseous substance or a vapor, for example, may require a larger hot zone than one involving a solid or nonvolatile liquid leak. In some cases, atmospheric monitoring, plume modeling, or reference sources such as the *Emergency Response Guidebook* may prove useful in helping to establish the parameters of a hot zone. Specially trained responders, in accordance with their level of training, should be tasked with using these tools. Keep in mind that the physical characteristics of the released substance will significantly affect the size and layout of the hot zone. In addition, all specially trained responders entering the hot zone should avoid contact with the product to the greatest extent possible—an important goal that should be clearly understood by those entering the hot zone. Adhering to this policy makes the job of decontamination easier and reduces the risk of cross-contamination.

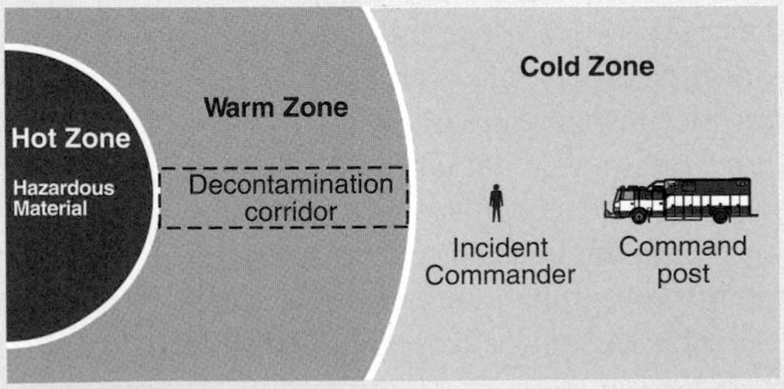

Personnel accountability is important, so access into the hot zone must be limited to only the persons necessary to control the incident. All personnel and equipment must be decontaminated when they leave the hot zone. This practice ensures that contamination is not inadvertently spread to "clean" areas of the scene.

The **warm zone** is where personnel and equipment transition into and out of the hot zone. It contains control points for access to the hot zone as well as the decontamination area. Only the minimal number of personnel and the equipment necessary to perform decontamination, or support those operating in the hot zone, should be permitted in the warm zone.

A patient's skin and clothing may contain hazardous material, so the **decontamination area** is set up in the warm zone. The decontamination area is the designated area where contaminants are removed before an individual can go to another area. **Decontamination** is the process of removing or neutralizing and properly disposing of hazardous materials from equipment, patients, and rescue personnel. The decontamination area must include special containers for contaminated clothing and special bags to isolate each patient's personal effects safely until they can be decontaminated **Figure 38-25**.

Figure 38-24 Control zones spread outward from the center of a hazardous materials incident.

Figure 38-25 Patients should be decontaminated before they are taken to treatment areas.

The area will also contain a number of special facilities to thoroughly wash and rinse patients and backboards. The water that is used must be captured and delivered into special sealable containers.

Anyone who leaves the hot zone must pass through the decontamination area. Fire fighters' and HazMat team members' outer protective gear is rinsed and washed in the decontamination area before it is removed **Figure 38-26**. To prevent needless contact and transmission of splash or residues, different personnel are used in the decontamination and treatment areas. You should not move into the decontamination area unless you are properly trained and equipped. You should wait for the patients to be brought to you.

Beyond the warm zone is the **cold zone**. The cold zone is a safe area where personnel do not need to wear any special protective clothing for safe operation. Personnel staging, the command post, EMS providers, and the area for medical monitoring, support, and/or treatment after decontamination are all located in the cold zone.

Role of the EMT

As an EMT, your job is to report to a designated area outside of the hot and warm zones and provide triage, treatment, transport, or rehabilitation when HazMat team members bring patients to you.

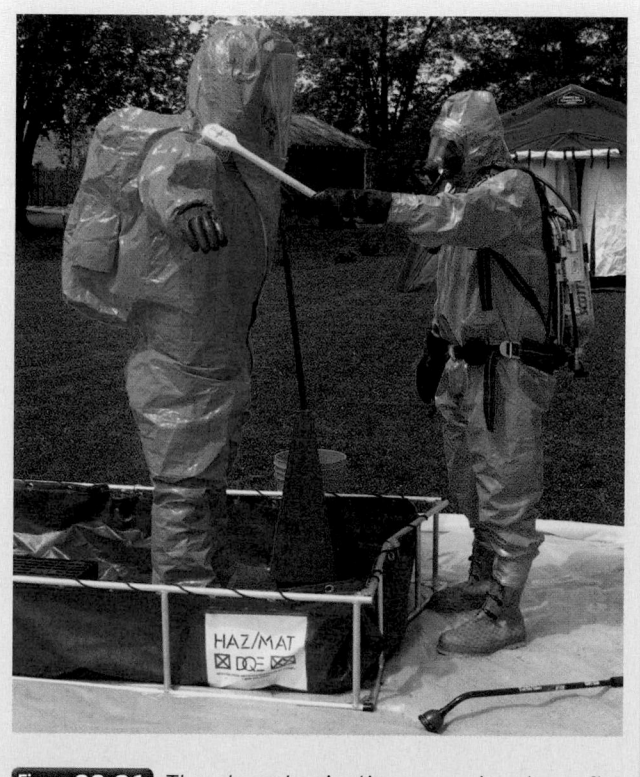

Figure 38-26 The decontamination zone is where fire fighters' and HazMat team members' outer protective gear is rinsed and washed before removal.

Classification of Hazardous Materials

The National Fire Protection Association (NFPA) 704 Hazardous Materials Classification standard classifies hazardous materials according to health hazard or toxicity levels, fire hazard, chemical reactive hazard, and special hazards (such as radiation and acids) for fixed facilities that store hazardous materials. Toxicity protection levels are also classified according to the level of personal protection required. For your safety, you must know the type and degree of health, fire, and reactive hazard protection you need to operate safely near these substances before you enter the scene.

Toxicity Level

Toxicity levels are measures of the health risk that a substance poses to someone who comes into contact with it. There are five toxicity levels: 0, 1, 2, 3, and 4. The higher the number, the greater the toxicity, as follows:

- **Level 0** includes materials that would cause little, if any, health hazard if you came into contact with them.

- **Level 1** includes materials that would cause irritation on contact but only mild residual injury, even without treatment.

- **Level 2** includes materials that could cause temporary damage or residual injury unless prompt medical treatment is provided. Both levels 1 and 2 are considered slightly hazardous but require use of self-contained breathing apparatus (SCBA) if you are going to come into contact with them.

- **Level 3** includes materials that are extremely hazardous to health. Contact with these materials requires full protective gear so that none of your skin surface is exposed.

- **Level 4** includes materials that are so hazardous that minimal contact will cause death. For level 4 substances, you need specialized gear that is designed for protection against that particular hazard.

You must note that all health hazard levels, with the exception of 0, require respiratory and chemical protective gear that is not standard on most ambulances and specialized training. **Table 38-2** further describes the four hazard classes.

Caring for Patients at a HazMat Incident

Generally, HazMat team members who are trained in prehospital emergency care will initiate emergency care for

Table 38-2	Toxicity Levels of Hazardous Materials	
Level	**Health Hazard**	**Protection Needed**
0	Little or no hazard	None
1	Slightly hazardous	SCBA (level C suit) only
2	Slightly hazardous	SCBA (level C suit) only
3	Extremely hazardous	Full protection, with no exposed skin (level A or B suit)
4	Minimal exposure causes death	Special HazMat gear (level A suit)

patients who have been exposed to a hazardous material. However, because of the dangers, time constraints, and bulky protective gear that team members wear, it is practical only to provide the simplest assessment and essential care in the hazard zone and the decontamination area. In addition, to avoid entrapment and spread of contaminants, no bandages or splints are applied—except pressure dressings that are needed to control bleeding—until the "clean" (decontaminated) patient has been moved to the treatment area. Therefore, the EMTs providing care in the treatment area should assess and treat the patient in the same way as they would a patient who has not been previously assessed or treated.

Your care of patients at a HazMat incident must address the following two issues:

- Any trauma that has resulted from other related mechanisms, such as vehicle collision, fire, or explosion
- The injury and harm that have resulted from exposure to the toxic hazardous substance

Most serious injuries and deaths from hazardous materials result from airway and breathing problems. Therefore, you should be sure to maintain the airway, and, if the patient appears to be in distress, give oxygen at 12 to 15 L/min with a nonrebreathing mask. Monitor the patient's breathing at all times. If you see signs that indicate that respiratory distress is increasing, you may need to provide assisted ventilation with a bag-mask device and high-flow oxygen.

You should treat the patient's injuries in the same way that you would treat any injury. There are few specific antidotes or treatments for exposure to most hazardous materials. Different people may respond differently to contact with the same hazardous material. Therefore, your treatment for the patient's exposure to the toxic substance should focus mainly on supportive care and initiating transport to the hospital with a minimum of additional delay.

If special antidotes or other special treatments need to be initiated in the field, they will be ordered by medical control and relayed to the officer in charge of EMS operations at the scene. If special treatment includes medications, intravenous fluids, or other advanced care, paramedics or other advanced personnel will be sent to work with you at the treatment area.

Special Care

In some cases, before the decontamination area has been completely set up, the HazMat team will find one or two patients who need immediate treatment and transport without further delay if they are to survive. Even after the decontamination area is set up and functioning, some patients may have such respiratory distress or other urgent critical condition that the time necessary for full decontamination may prove fatal. If additional delay for proper decontamination seems life threatening in nontoxic exposure situations, it may be necessary to simply cut away all of the patient's clothing and do a rapid rinse to remove the majority of the contaminating matter before transport.

If you are treating and transporting a patient who has not been fully and properly decontaminated, you

You are the Provider: PART 5

As the patients are being retriaged and treated, you contact the local trauma center personnel to determine if they can handle the critically injured patients; they advise you that they can. You reassess the middle-aged man whom you originally triaged as expectant. He is now apneic and pulseless. Your partner begins cardiopulmonary resuscitation (CPR) as you contact medical control. Based on the patient's injuries, medical control determines that resuscitation is not indicated.

8. What factors should be considered when determining the appropriate transport destination?

will need to increase the amount of protective clothing you wear, including the use of SCBA. At the least, this should include two pairs of gloves, goggles or a face shield, a protective coat, respiratory protection, and a disposable fluid-impervious apron or similar outfit. Many HazMat teams carry light, easy-to-use, disposable, fluid-impervious protective suits for such a purpose. Remember, however, that transporting a contaminated patient merely increases the scope of the event. The decision to transport even a patient with critical injuries rests with the IC, who bases his or her decision on recommendations made by the HazMat team.

To make decontaminating the ambulance easier, tape the cabinet doors shut. Any equipment kits, monitors, and other items that will not be used en route should be removed from the patient compartment and placed in the front of the ambulance or in outside compartments. Before loading the patient, you should turn on the power vent ceiling fan and patient compartment air-conditioning unit fan. Unless the weather is too severe, the windows in the driver's area and sliding side windows in the patient compartment should also be partially opened to prevent creating a "closed box" inside the ambulance and to ensure that it is properly ventilated for the safety of the patient and EMTs.

When you leave the scene, inform the hospital that you are transporting a critically injured patient who has not been fully decontaminated at the scene. This will allow the hospital to prepare to receive the patient. Many emergency departments have decontamination facilities and trained personnel for such an event. You may be diverted to a facility with these capabilities if the receiving hospital is not so equipped. Be sure that one EMT enters the emergency department and, after giving hospital staff the report and advising them again of the incomplete decontamination, obtains directions before the patient is unloaded and brought in. If there are enough ambulances at a HazMat scene, one may be isolated and used only to transport such patients. Remember, the ambulance needs to be decontaminated before transporting another patient.

■ Personal Protective Equipment Level

Personal protective equipment (PPE) levels indicate the amount and type of protective gear that you need to prevent injury from a particular substance. The four recognized protection levels, A, B, C, and D, are as follows Figure 38-27 :

- **Level A**, the most hazardous, requires fully encapsulated, chemical-resistant protective clothing that provides full body protection, as well as SCBA and special, sealed equipment.

- **Level B** requires nonencapsulated protective clothing or clothing that is designed to protect against a particular hazard Figure 38-28 . Usually, this clothing is made of material that will let only limited amounts of moisture and vapor pass through

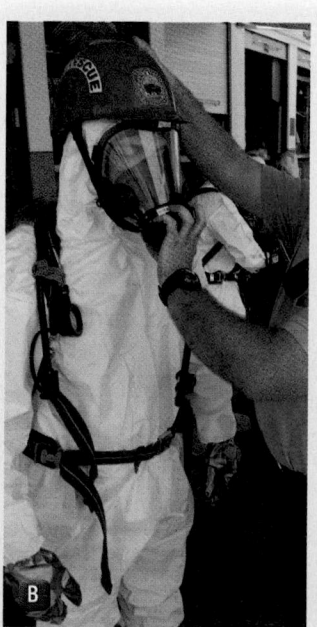

Figure 38-27 Four levels of protection. **A.** Level A protection. **B.** Level B protection. **C.** Level C protection. **D.** Level D protection. Most serious injuries and deaths from hazardous materials result from airway and breathing problems.

Figure 38-28 Workers in Level B protection.

(nonpermeable). Level B also requires breathing devices that contain their own air supply, such as SCBA, and eye protection.

- **Level C**, like Level B, requires the use of nonpermeable clothing and eye protection. In addition, face masks that filter all inhaled outside air must be used.

- **Level D** requires a work uniform, such as coveralls, that affords minimal protection.

- **All levels of protection require the use of gloves**. Two pairs of rubber gloves are needed for protection in case one pair must be removed because of heavy contamination.

You are the Provider: SUMMARY

1. Do you have enough information to declare this a mass-casualty incident?

A mass-casualty incident (MCI) refers to any call involving three or more patients, any situation that places such a great demand on available equipment or personnel that the system requires a mutual aid response (an agreement between neighboring EMS systems to respond to MCIs or disasters in each other's jurisdiction when local resources are insufficient to handle the incident), or any incident that has the potential to create one of the previously mentioned situations.

In this case, you are responding to an incident involving *at least* three patients, and the severity of their injuries is not known. Considering the fact that you are the only ambulance left in your jurisdiction, your resources have already been depleted! One ambulance and two EMTs can effectively treat and transport only two stable patients *or* one critical patient at a time.

Compared with a plane or bus crash, this is a small-scale MCI; however, it is still an MCI. On the basis of the information provided by dispatch, mutual aid assistance should be requested. It is arguably better to call for help earlier, rather than wait until you arrive at the scene and find yourself overwhelmed by patients—some of whom may be critically injured. The longer it takes to call for help, the longer it will take for help to arrive.

2. How will the incident command system facilitate operations at this scene?

Any time you respond to a scene involving multiple patients, you must use a systematic approach to manage the incident efficiently; the incident command system (ICS) was established for this very purpose. The ICS will enable you and your team to do the greatest good for the greatest number of people by providing a modular organizational structure that is built on the size and complexity of the incident.

The purpose of the ICS is to ensure responder and public safety, achieve incident management goals, and ensure the effective use of resources. The goal of the ICS is to make the best use of your resources to manage the environment around the incident and to treat patients during an emergency.

You have requested mutual aid assistance from other EMS systems because you have determined that there are more patients than you can effectively take care of. When you and the other units arrive, the ICS will facilitate the processes of triage, treatment, and transport; it will also help control the duplication of efforts and freelancing, in which individual units or different organizations make independent and often inefficient decisions that could compromise the effectiveness of the entire operation.

The most critical part of an effective ICS is the *early* establishment of an incident commander (IC)—the person in charge of the overall incident. The IC will assess the incident and develop an action plan to effectively manage it. The role of IC should be assumed by the first-arriving, most experienced public safety official; depending on the nature of the incident, that person may be a police, fire, or EMS official.

The IC will determine—based on the size and complexity of the incident—which positions of the ICS need to be filled. Although not all positions of the ICS will be filled during every incident, the IC is ultimately responsible for all activity at the scene.

The number of command duties the IC assumes varies by the size of the incident. In small incidents, the IC often assumes all of these duties. In an incident of medium size or complexity, the IC may delegate some functions but retain others. At a motor vehicle crash with multiple patients, for example, the IC may designate safety and triage officers but maintain responsibility for the other command functions.

The success of any ICS is dependent on cooperation and communication among all providers at the incident. Your EMS system should have protocols regarding use of the ICS, and it is important for you to become familiar with them.

3. What functions does the safety officer perform?

The safety officer monitors the entire scene for conditions or operations that may present a hazard to responders and patients. Depending on the situation, the safety officer may need to work with environmental health, hazardous materials (HazMat) teams, and other agencies.

The importance of the safety officer cannot be underestimated; he or she has the ultimate authority to stop an emergency operation whenever a rescuer or patient is in danger. The safety officer should remove hazards to EMS personnel and patients before the hazards cause injury.

Throughout the entire operation, the safety officer will constantly monitor the situation. He or she will perform ongoing 360° assessments of the entire incident to identify potential or actual threats to safety.

At an MCI, you will often be engrossed in the function that you were assigned and are often not aware of all that is going on around you. The safety officer ensures safe scene operations and should be listened to. If the safety officer gives the order to "STOP," you should take him or her seriously and stop what you are doing until the safety hazard has been removed.

4. How should you and your partner proceed?

An IC and safety officer have already been identified. You have five patients, and additional EMS personnel and ambulances will arrive in about 10 minutes. You and your partner should begin the process of locating the patients and triaging them.

Triage simply means "to sort" your patients, based on the severity of their injuries, to establish treatment priorities based on available resources. The goal of triage is to provide the greatest good for the greatest number; therefore, the triage assessment should be brief, and the patient condition categories should be basic.

Primary triage is the initial triage done in the field; it allows the EMT to quickly and accurately categorize the patient's condition and transport needs. During primary triage, patients are briefly assessed and identified in some way–usually by attaching a color-coded triage tag.

Secondary triage is performed as patients are brought to the treatment area, if one has been established, and is a function of the treatment officer. When primary triage has been completed, secondary triage, or retriage, can occur. Secondary triage allows the EMT to reassess the patients, upgrade their triage category (if necessary), and begin initial treatment.

In this situation, you and your partner are the only EMTs at the scene; therefore, you will need to function as the triage and treatment officer, at least initially, and the primary and secondary triage will likely occur in the same area. If you have triaged all patients and additional help has still not arrived, you should begin treating the most critically injured patients first. As additional EMTs and ambulances arrive, they should be assigned accordingly–again, with priority given to the most critically injured patients.

Rapid and accurate triage is an integral part of any MCI. It allows you to identify the most critically injured patients and ensures they are treated and transported first.

5. What triage category should be assigned to each of these patients?

There are four common triage categories–regardless of the triage method that is used–and each patient is identified with a color-coded triage tag. The four triage categories can be remembered using the mnemonic "IDME," which stands for Immediate (red), Delayed (yellow), Minor (green), and Expectant (black).

Triage tags are perforated, which allows a patient to be triaged only to a higher category, but never downgraded. For example, a patient initially triaged as delayed (yellow tag), but whose condition later deteriorates, the yellow portion of the tag can be torn off, leaving only the red and black sections.

Patients who are categorized as immediate (first priority; red tag) need immediate treatment and should be transported first. Examples of injuries that would classify a patient as immediate include airway or breathing problems, uncontrolled or severe bleeding, signs of shock, severe burns, a decreased level of consciousness, severe underlying medical problems, and open chest or abdominal injuries.

Patients who are categorized as delayed (second priority; yellow tag) need treatment and transport; however, it can be temporarily delayed. Examples of injuries that would classify a patient as delayed include burns *without airway problems*, major or multiple bone or joint injuries, and back injuries with or without spinal cord damage.

Patients who are categorized as minor (third priority; green tag) are commonly referred to as "walking wounded." These patients require minimal or no treatment, and transport and can wait until higher priority patients have been treated and transported first. Examples of injuries that would classify a patient as minor include minor fractures and soft-tissue injuries.

Patients who are categorized as expectant (fourth priority; black tag) include those who are obviously dead (ie, decapitation, burned beyond recognition), are in cardiac arrest, are in respiratory arrest (if your resources are limited), or who have injuries that are obviously nonsurvivable (ie, major open brain trauma).

Patients 3, 4, and 5 should be categorized as immediate (red tag); they require immediate treatment and prompt transport. Patient No. 3 is confused and is experiencing respiratory distress. Patient No. 4, although conscious and alert, has a history of heart problems; is breathing with difficulty; has severe, crushing chest pain; and has a rapid, irregular pulse. Patient No. 5 has signs of shock (eg, restlessness; rapid, weak pulse) and a rigid abdomen.

Patient No. 1 should be categorized as minor (green tag). He is ambulatory and is talking on his cell phone. Although he has a deformity to his arm and abrasions to his face, these injuries are not life threatening.

Patient No. 2 should be categorized as expectant (black tag). He is unconscious; is not breathing; has a slow, weak pulse; has bilaterally angulated femurs; and has massive facial trauma. If adequate help was available at the scene, this patient would be categorized as immediate; however, compared with patients 3, 4, and 5, he has the least likely chance of surviving. If you and your partner focused on treating him, the conditions of patients 3, 4, and 5 could deteriorate further, potentially resulting in the death of four patients instead of one.

As previously discussed, some patients who are initially triaged at a lower category may need their triage status upgraded if their condition deteriorates. For example, if a patient is initially conscious and alert but then experiences a decreased level of consciousness or other signs of deterioration (ie, signs of shock), you should upgrade his or her triage category.

During large-scale MCIs involving large numbers of patients, such as bus crashes or building collapses, patients are moved from the triage area to the treatment area, where they are retriaged (secondary triage), and, if necessary, upgraded to a higher triage category. During small-scale MCIs, however, patients are often retriaged in the same area where they were initially triaged–especially if available resources are limited.

6. Should your initial triage assignments remain the same?

Patients 3, 4, and 5 were initially triaged as immediate (red tag); their triage categories should stay the same. Remember, you should not downgrade a patient's initial triage category; you should only upgrade it if the patient's condition deteriorates.

When you and your partner were the only EMTs at the scene, you had to decide which patients were the most critical, yet the most likely to survive with immediate treatment and prompt transport. Patient No. 2, the middle-aged, nonbreathing male, was initially triaged as expectant (black tag) because he was the least likely to survive and your available resources were limited. However, now that you have a total of eight EMTs and four ambulances at the scene, you should reconsider his triage category. Patients 3, 4, and 5 should still receive the most immediate treatment and transport; however, you should reassess patient No. 2 now that your resources are no longer limited.

If patient No. 2 still has a pulse—even though he is not breathing—you should begin immediate treatment. If you leave him in an expectant (black tag) category, he will assuredly die. However, if you begin immediate treatment and transport promptly, there is a chance that he will survive. If it is found that he is now pulseless and apneic, you should begin CPR and contact medical control; treatment and transport may still be indicated.

7. Is it necessary to set up a treatment area at this incident?

During a large-scale MCI involving large numbers of patients, there must be a systematic flow between the triage area and treatment area because of the large number of patients that are generally associated with these incidents. Therefore, a treatment area is an important part of the ICS structure in large-scale MCIs.

During a small-scale MCI, such as the one in this case study, a treatment area would likely not be necessary. Initially, there were only two EMTs at the scene and three critically injured patients. It would be more practical and time-saving to triage and treat the patients in the same area. Taking the time to set up a treatment area would delay patient care and would require more personnel.

Now that there are ample EMS resources at the scene, the need for a designated treatment area is even less practical. It would clearly be more efficient to simply assign one critical patient to each EMT crew so they can reassess, treat, and transport.

Although all components of the ICS will not be needed during all MCIs, the concepts of the ICS—structure and efficiency—must remain the same and are crucial for overall operation effectiveness.

8. What factors should be considered when determining the appropriate transport destination?

During an MCI—especially one that involves large numbers of patients—it is important to remember that the resources of local hospitals may be overwhelmed as well. Not only do they have non-MCI patients to care for, they must now prepare for a potentially large number of additional patients.

As soon as you declare an MCI, notify area hospitals as early as possible, apprise them of the situation, and determine their surge capacity; this will tell you how many patients of each category they are able to safely and effectively care for. It will also allow the hospitals to increase their staffing, and if needed, move patients within the facility.

Do *not* begin transport until the intended destination facility has been notified and has accepted the patient(s). There must be a plan of action in place *before* any patients are transported from the scene.

During large-scale MCIs, a transportation officer is designated to coordinate the transportation and distribution of patients to appropriate hospitals and to help ensure that the hospitals do not become overwhelmed by a surge of patients. In small-scale MCIs (ie, a motor vehicle collision with four or five patients), a specific person may not always be designated as the transportation officer; therefore, the most senior EMT at the scene should ensure that local hospitals are contacted early. Regardless of how you communicate with receiving hospitals, make sure that you give them as much "heads up" time as possible and keep track of where each patient is transported.

In some cases, the EMS medical director will help determine the appropriate transport destination based on information that he or she receives from the EMTs in the field. However, if this is not possible, a rotation system can be used to distribute patients properly to each hospital on the basis of hospital capacity and capabilities. The rotation system allows distribution of patients throughout the hospitals of the system and, therefore, does not overburden any one facility.

The basic principles of transport that apply to any other patient also apply to MCIs; the most critically injured (red-tagged) patients should be transported to a designated trauma center, whereas yellow-tagged patients can be transported to hospitals that are located further away. In cases where the closest trauma center is located a great distance away, air medical transport should be considered.

Many EMS systems have preplanned for a designated hospital within a region to perform the coordination between hospitals regarding transport decisions. It is critical to remember that an MCI disrupts the everyday activities of EMS systems and local hospitals; good coordination and communication are essential.

You are the Provider: SUMMARY, continued

Patient No. 1

Triage Tag
No. 239351

(Move the Walking Wounded)	**MINIMAL**
No respirations after head tilt	**EXPECTANT**
☐ Respirations–over 30	IMMEDIATE
☐ Perfusion–capillary refill over 2 seconds	IMMEDIATE
☐ Mental status–unable to follow simple commands	IMMEDIATE
Otherwise	**DELAYED**

MAJOR INJURIES: None
HOSPITAL DESTINATION: No transport
ORIENTED × 4   DISORIENTED ☐   UNCONSCIOUS ☐

TIME	PULSE	B/P	RESPIRATION
N/A	N/A	N/A	N/A
N/A	N/A	N/A	N/A

PERSONAL INFORMATION:
NAME: Joshua Smith
MALE ☒   FEMALE ☐   AGE: 22   WEIGHT: 140 lb
MEDICAL COMPLAINTS/HISTORY
Closed deformity to left forearm, facial abrasions, no medical history

EXPECTANT   No 239351

IMMEDIATE   No 239351

DELAYED   No 239351

MINIMAL   No 239351

Patient No. 2

Triage Tag
No. 239352

Move the Walking Wounded	**MINIMAL**
(No respirations after head tilt)	**EXPECTANT**
☐ Respirations–over 30	IMMEDIATE
☐ Perfusion–capillary refill over 2 seconds	IMMEDIATE
☐ Mental status–unable to follow simple commands	IMMEDIATE
Otherwise	**DELAYED**

MAJOR INJURIES: Massive facial trauma, bilateral femur fractures
HOSPITAL DESTINATION: No transport
ORIENTED ×   DISORIENTED ☐   UNCONSCIOUS ☒

TIME	PULSE	B/P	RESPIRATION
N/A	N/A	N/A	N/A
N/A	N/A	N/A	N/A

PERSONAL INFORMATION:
NAME: Not available
MALE ☒   FEMALE ☐   AGE: N/A   WEIGHT: N/A
MEDICAL COMPLAINTS/HISTORY
Unconscious, respiratory arrest, unknown medical history

EXPECTANT   No 239352

Patient No. 3

Triage Tag
No. 239353

Move the Walking Wounded	MINIMAL
No respirations after head tilt	EXPECTANT
☒ Respirations–over 30	IMMEDIATE
☐ Perfusion–capillary refill over 2 seconds	IMMEDIATE
☐ Mental status–unable to follow simple commands	IMMEDIATE
Otherwise	DELAYED

MAJOR INJURIES: Chest and head
HOSPITAL DESTINATION: University Hospital
ORIENTED × 2   DISORIENTED ☐   UNCONSCIOUS ☐

TIME	PULSE	B/P	RESPIRATION
0325	120	90/60	34 (labored)
0330	118	92/62	30 (labored)

PERSONAL INFORMATION:
NAME: Anita Michaels
MALE ☐   FEMALE ☒   AGE: 26   WEIGHT: 110 lb
MEDICAL COMPLAINTS/HISTORY
Confused, respiratory distress, hematoma to head, no medical history

EXPECTANT	No	239353
IMMEDIATE	No	239353

Patient No. 4

Triage Tag
No. 239354

Move the Walking Wounded	MINIMAL
No respirations after head tilt	EXPECTANT
☒ Respirations–over 30	IMMEDIATE
☐ Perfusion–capillary refill over 2 seconds	IMMEDIATE
☐ Mental status–unable to follow simple commands	IMMEDIATE
Otherwise	DELAYED

MAJOR INJURIES: None
HOSPITAL DESTINATION: University Hospital
ORIENTED × 4   DISORIENTED ☐   UNCONSCIOUS ☐

TIME	PULSE	B/P	RESPIRATION
0329	116	106/60	32 (labored)
0334	110	110/64	28 (labored)

PERSONAL INFORMATION:
NAME: Gloria Michaels
MALE ☐   FEMALE ☒   AGE: 64   WEIGHT: 145 lb
MEDICAL COMPLAINTS/HISTORY
Abrasions to forearms, crushing chest pain, difficulty breathing,
history of heart problems

EXPECTANT	No	239354
IMMEDIATE	No	239354

You are the Provider: SUMMARY, continued

Patient No. 5

Triage Tag
No. 239355

Move the Walking Wounded	MINIMAL
No respirations after head tilt	EXPECTANT
☐ Respirations—over 30	IMMEDIATE
☒ Perfusion—capillary refill over 2 seconds	IMMEDIATE
☐ Mental status—unable to follow simple commands	IMMEDIATE
Otherwise	DELAYED

MAJOR INJURIES: _Possible internal bleeding, shock_

HOSPITAL DESTINATION: University Hospital

ORIENTED × 4 · DISORIENTED ☐   UNCONSCIOUS ☐

TIME	PULSE	B/P	RESPIRATION
0331	128	88/58	24 (shallow)
0336	120	92/60	24 (shallow)

PERSONAL INFORMATION:

NAME: James Michaels

MALE ☒   FEMALE ☐   AGE: 29   WEIGHT: 170 lb

MEDICAL COMPLAINTS/HISTORY

Abdominal rigidity; restless; weak, rapid pulse; no medical history

EXPECTANT   No 239355

IMMEDIATE   No 239355

Prep Kit

Ready for Review

- The National Incident Management System (NIMS) provides a consistent nationwide template to enable federal, state, and local governments, as well as private-sector and nongovernmental organizations, to work together effectively and efficiently. The NIMS is used to prepare for, prevent, respond to, and recover from domestic incidents, regardless of cause, size, or complexity, including acts of catastrophic terrorism and hazardous materials (HazMat) incidents.

- The major NIMS components are command and management, preparedness, resource management, communications and information management, supporting technologies, and ongoing management and maintenance.

- The purpose of the incident command system is ensuring responder and public safety; achieving incident management goals; and ensuring the efficient use of resources.

- Preparedness involves the decisions made and basic planning done before an incident occurs.

- Your agency should have written disaster plans that you are regularly trained to carry out.

- At incidents that have a significant medical factor, the incident commander should appoint someone as the medical group or branch leader. This person will supervise the primary roles of the medical group: triage, treatment, and transport of the injured.

- A mass-casualty incident refers to any call that involves three or more patients, any situation that places such a great demand on available equipment or personnel that the system would require a mutual aid response, or any incident that has a potential to create one of the previously mentioned situations.

- The goal of triage is to do the greatest good for the greatest number. This means that the triage assessment is brief and the patient condition categories are basic.

- There are four basic triage categories that can be recalled using the mnemonic IDME:
  - Immediate (red)
  - Delayed (yellow)
  - Minimal (green; hold)
  - Expectant (black; likely to die or dead)

- A disaster is a widespread event that disrupts functions and resources of a community and threatens lives and property.

- Many disasters, such as a drought, may not involve personal injuries.

- When you arrive at the scene of a HazMat incident, you must first step back and assess the situation. This can be very stressful, particularly if you see a patient.

- A valuable resource for determining what the hazardous material is and what you should do is CHEMTREC.

Vital Vocabulary

bills of lading The shipping papers used for transport of chemicals over roads and highways. Also referred to as freight bills.

carboys Glass, plastic, or steel containers, ranging in volume from 5 to 15 gallons.

casualty collection area An area set up by physicians, nurses, and other hospital staff near a major disaster scene where patients can receive further triage and medical care.

Chemical Transportation Emergency Center (CHEMTREC) An agency that assists emergency personnel in identifying and handling hazardous materials transport incidents.

cold zone A safe area at a hazardous materials incident for the agencies involved in the operations. The incident commander, the command post, EMS providers, and other support functions necessary to control the incident should be located in the cold zone. Also referred to as the clean zone or the support zone.

command In incident command, the position that oversees the incident, establishes the objectives and priorities, and from there develops a response plan.

command post The designated field command center where the incident commander and support personnel are located.

container Any vessel or receptacle that holds material, including storage vessels, pipelines, and packaging.

control zones Areas at a hazardous materials incident that are designated as hot, warm, or cold, based on safety issues and the degree of hazard found there.

cylinders Portable, compressed gas containers used to hold liquids and gases. Uninsulated compressed gas cylinders are used to store substances such as nitrogen, argon, helium, and oxygen. They have a range of sizes and internal pressures.

danger zone An area where individuals can be exposed to toxic substances, lethal rays, or ignition or explosion of hazardous materials.

decontamination The process of removing or neutralizing and properly disposing of hazardous materials from equipment, patients, and rescue personnel.

decontamination area The designated area in a hazardous materials incident where all patients and rescuers must be decontaminated before going to another area.

demobilization The process of directing responders to return to their facilities when work at a disaster or mass-casualty incident has finished, at least for those particular responders.

disaster A widespread event that disrupts community resources and functions, in turn threatening public safety, citizens' lives, and property.

drums Barrel-like containers used to store a wide variety of substances, including food-grade materials, corrosives, flammable liquids, and grease. Drums may be constructed of low-carbon steel, polyethylene, cardboard, stainless steel, nickel, or other materials.

Emergency Response Guidebook A preliminary action guide for first responders operating at a hazardous materials incident in coordination with the US Department of Transportation's labels and placards marking system. The ERG was jointly developed by the DOT, the Secretariat of Communications and Transportation of Mexico, and Transport Canada.

extrication supervisor In incident command, the person appointed to determine the type of equipment and resources needed for a situation involving extrication or special rescue; also called the rescue officer.

finance In incident command, the position in an incident responsible for accounting of all expenditures.

freelancing When individual units or different organizations make independent and often inefficient decisions about the next appropriate action.

freight bills The shipping papers used for transport of chemicals along roads and highways. Also referred to as bills of lading.

hazardous material Any substance that is toxic, poisonous, radioactive, flammable, or explosive and causes injury or death with exposure.

hazardous materials (HazMat) incident An incident in which a hazardous material is no longer properly contained and isolated.

hot zone The area immediately surrounding a hazardous materials spill/incident site that is directly dangerous to life and health. All personnel working in the hot zone must wear complete, appropriate protective clothing and equipment. Entry requires approval by the incident commander or other designated officer.

incident action plan An oral or written plan stating general objectives reflecting the overall strategy for managing an incident.

incident commander (IC) The overall leader of the incident command system to whom commanders or leaders of incident command system divisions report.

incident command system (ICS) A system implemented to manage disasters and mass-casualty incidents in which section chiefs, including finance, logistics, operations, and planning, report to the incident commander.

intermodal tanks Shipping and storage vessels that can be either pressurized or nonpressurized.

joint information center An area designated by the incident commander, or a designee, in which public information officers from multiple agencies disseminate information about the incident.

JumpSTART triage A sorting system for pediatric patients younger than 8 years or weighing less than 100 lb. There is a minor adaptation for infants since they cannot ambulate on their own.

liaison officer In incident command, the person who relays information, concerns, and requests among responding agencies.

logistics In incident command, the position that helps procure and stockpile equipment and supplies during an incident.

mass-casualty incident (MCI) An emergency situation involving three or more patients or that can place great demand on the equipment or personnel of the EMS system or has the potential to produce multiple casualties.

material safety data sheet (MSDS) A form, provided by manufacturers and compounders (blenders) of chemicals, containing information about chemical composition, physical and chemical properties, health and safety hazards, emergency response, and waste disposal of a specific material.

medical incident command A branch of operations in a unified command system, whose three designated sector positions are triage, treatment, and transport.

morgue supervisor In incident command, the person who works with area medical examiners, coroners, and law enforcement agencies to coordinate the disposition of dead victims.

mutual aid response An agreement between neighboring EMS systems to respond to mass-casualty incidents or disasters in each other's region when local resources are insufficient to handle the response.

National Incident Management System (NIMS) A Department of Homeland Security system designed to enable federal, state, and local governments and private-sector and nongovernmental organizations to effectively and efficiently prepare for, prevent, respond to, and recover from domestic incidents, regardless of cause, size, or complexity, including acts of catastrophic terrorism.

nonbulk storage vessels Any container other than bulk storage containers such as drums, bags, compressed gas cylinders, and cryogenic containers. Nonbulk storage vessels hold commonly used commercial and industrial chemicals such as solvents, industrial cleaners, and compounds.

operations In incident command, the position that carries out the orders of the commander to help resolve the incident.

personal protective equipment (PPE) levels Measures of the amount and type of protective equipment that an individual needs to avoid injury during contact with a hazardous material.

placards Signage required to be placed on all four sides of highway transport vehicles, railroad tank cars, and other forms of hazardous materials transportation; the sign identifies the hazardous contents of the vehicle, using a standardization system with 10¾-inch diamond-shaped indicators.

planning In incident command, the position that ultimately produces a plan to resolve any incident.

primary triage A type of patient sorting used to rapidly categorize patients; the focus is on speed in locating all patients and determining an initial priority as their conditions warrant.

public information officer (PIO) In incident command, the person who keeps the public informed and relates any information to the press.

rehabilitation area The area that provides protection and treatment to fire fighters and other personnel working at an emergency. Here, workers are medically monitored and receive any needed care as they enter and leave the scene.

rehabilitation supervisor In incident command, the person who establishes an area that provides protection for responders from the elements and the situation.

rescue supervisor In incident command, the person appointed to determine the type of equipment and resources needed for a situation involving extrication or special rescue; also called the extrication officer.

safety officer In incident command, the person who gives the "go ahead" to a plan or who may stop an operation when rescuer safety is an issue.

secondary containment An engineered method to control spilled or released product if the main containment vessel fails.

secondary triage A type of patient sorting used in the treatment sector that involves retriage of patients.

single command system A command system in which one person is in charge, generally used with small incidents that involve only one responding agency or one jurisdiction.

span of control In incident command, the subordinate positions under the commander's direction to which the workload is distributed; the supervisor/worker ratio.

staging supervisor In incident command, the person who locates an area to stage equipment and personnel and tracks unit arrival and deployment from the staging area.

START triage A patient sorting process that stands for Simple Triage And Rapid Treatment and uses a limited assessment of the patient's ability to walk, respiratory status, hemodynamic status, and neurologic status.

termination of command The end of the incident command structure when an incident draws to a close.

toxicity levels Measures of the risk that a hazardous material poses to the health of an individual who comes into contact with it.

transportation area The area in a mass-casualty incident where ambulances and crews are organized to transport patients from the treatment area to receiving hospitals.

transportation supervisor The individual in charge of the transportation sector in a mass-casualty incident who assigns patients from the treatment area to awaiting ambulances in the transportation area.

treatment area The location in a mass-casualty incident where patients are brought after being triaged and assigned a priority, where they are reassessed, treated, and monitored until transport to the hospital.

treatment supervisor The individual, usually a physician, who is in charge of and directs EMS personnel at the treatment area in a mass-casualty incident.

triage The process of sorting patients based on the severity of injury and medical need to establish treatment and transportation priorities.

triage supervisor The individual in charge of the incident command triage sector who directs the sorting of patients into triage categories in a mass-casualty incident.

unified command system A command system used in larger incidents in which there is a multiagency response or multiple jurisdictions are involved.

warm zone The area located between the hot zone and the cold zone at a hazardous materials incident. The decontamination corridor is located in the warm zone.

Assessment in Action

You and your partner have just heard a news report that a plane carrying 155 people lost use of its landing gear and has been granted permission for an emergency landing at the local airport. As you head to your unit, the news report announces that the plane has crashed and broke apart on impact.

1. Which of the following individuals will be in charge of the overall incident?
   A. Medical director
   B. Shift supervisor
   C. Incident commander
   D. Police chief

2. On arrival at the scene, you and your partner need to check in to receive your assignment. What section should you report to?
   A. Planning section
   B. Finance section
   C. Triage section
   D. Treatment section

3. You are assigned to the triage unit. What is the primary responsibility of the triage unit?
   A. Ensure that every patient receives initial assessment of his or her condition
   B. Begin immediate treatment of critically ill patients
   C. Determine which patients will require evacuation with a helicopter
   D. Decide which patients will be transported to each hospital

4. After the primary triage has been completed, the team leader should communicate what information to the medical team leader?
   A. The total number of patients
   B. The number of patients in each category
   C. Resources needed to complete triage
   D. All of the above

5. You are triaging a woman who has severe burns over the lower portion of her body and weak carotid pulses. What color tag should you apply to the patient?
   A. Green
   B. Yellow
   C. Red
   D. Black

6. The JumpSTART triage system is used for:
   A. children younger than 5 years who appear to weigh less than 50 lb.
   B. children younger than 8 years who appear to weigh less than 100 lb.
   C. children up to the onset of puberty who appear to weigh less than 150 lb.
   D. children up to 18 years who appear to weigh less than 150 lb.

7. When using the START triage system, a patient who is breathing between 10 and 29 breaths/min but does not have radial pulses should be triaged as:
   A. expectant.
   B. minimal.
   C. delayed.
   D. immediate.

8. Which of the following is often the key problem at a mass-casualty incident or a disaster?
   A. Communication
   B. Resources
   C. Finances
   D. Planning

9. The JumpSTART triage system is a sorting system for pediatric patients. Explain the differences in the respiratory assessment used in the JumpSTART triage system compared with the standard START triage system.

10. Discuss the role of the transport officer at a mass-casualty incident.

National EMS Education Standard Competencies

EMS Operations

Knowledge of operational roles and responsibilities to ensure patient, public, and personnel safety.

Mass-Casualty Incidents Due to Terrorism and Disaster

- Risks and responsibilities of operating on the scene of a natural or man-made disaster. (p 1427-1448)

Knowledge Objectives

1. Define the terms international terrorism and domestic terrorism and provide some examples of incidents that have been caused by each one. (pp 1427-1428)
2. Provide examples of four different types of goals that commonly motivate terrorist groups to stage a terrorist attack. (pp 1427-1428)
3. Define the terms weapon of mass destruction (WMD) and weapon of mass casualty (WMC), and list and give examples of the five categories of weapons that are considered WMDs. (pp 1428-1429)
4. Discuss the history of chemical agents, their four main classifications, routes of exposure, affects on the patient, and patient care. (pp 1429, 1432-1438)
5. Discuss three categories of biologic agents, their routes of exposure, effects on the patient, and patient care. (pp 1429, 1439-1444)
6. Describe the history of nuclear/radiologic devices, sources of radiologic materials and dispersal devices, medical management of the patient, and protective measures that can be taken by the EMT during a nuclear/radiologic incident. (pp 1429, 1445-1447)
7. Describe how the Department of Homeland Security (DHS) Homeland Security Advisory System relates to the daily activities of EMTs and their ability to respond to and survive a terrorist attack. (pp 1429-1430)
8. Describe key observations an EMT must make on each call to assist in the determination of whether an incident is related to terrorism. (pp 1429-1432)
9. Explain the colors and threat levels that are used by the DHS daily to heighten awareness of the current terrorist threat. (p 1430)
10. Describe the critical response actions related to establishing and reassessing scene safety, personnel protection, notification procedures, and establishing command an EMT must perform at a suspected terrorist event. (pp 1430-1432)
11. Explain the role of EMS in relation to syndromic surveillance and points of distribution (PODS) during a biologic event. (pp 1444-1445)
12. Describe the mechanisms of injury caused by incendiary and explosive devices, including the types of wounds and their severity. (pp 1447-1448)

Skills Objectives

1. Demonstrate the steps an EMT can take to establish and reassess scene safety based on a scenario of a terrorist event. (pp 1430-1432)
2. Demonstrate the steps an EMT can take for the management of a patient exposed to a chemical agent. (pp 1432-1438)
3. Demonstrate the use of the Mark 1 Nerve Agent Antidote Kit (NAAK) and/or the DuoDote Auto-Injector. (pp 1436-1437)

Introduction

As a result of the increase in terrorist activity, it is possible that you may be called on to respond to a terrorist event during your career. International terrorists and domestic groups have increased their targeting of civilian populations with acts of terror. The question is not will terrorists strike again, but rather when and where they will strike. You must be mentally and physically prepared for the possibility of a terrorist event.

The use of weapons of mass destruction, or weapons of mass casualty, further complicates the management of the terrorist incident and places you in greater danger. Although it is difficult to plan and anticipate a response to many terrorist events, there are several key principles that apply to every response. This chapter describes types of terrorist events, personnel safety, and patient management and how you can prepare to respond to these events. You will learn the signs, symptoms, and treatment of patients who have been exposed to nuclear, chemical, or biologic agents or an explosive attack. At the end of this chapter, you will be able to answer the following key questions:

- What are your initial actions?
- Whom should you notify, and what should you tell them?
- What type of additional resources might you require?
- How should you proceed to address the needs of the victims?
- How do you ensure your own and your partner's safety, as well as the safety of the victims?

- What is the clinical presentation of a victim exposed to a weapon of mass destruction (WMD)?
- How are WMD patients to be assessed and treated?
- How do you avoid becoming contaminated or cross-contaminated with a WMD agent?

What Is Terrorism?

No one is quite sure who the first terrorist was, but terrorist forces have been at work since early civilizations. Today, terrorists pose a threat to nations and cultures everywhere. **International terrorism** has brought a new fear into the lives of many American citizens.

Modern-day terrorism is common in the Middle East, where terrorist groups have frequently attacked civilian populations. In Colombia, political terrorist groups target oil resources as a means to instill fear.

In the United States, domestic terrorists have struck multiple times in previous years. The Centennial Park bombing during the 1996 Summer Olympics and the destruction of the Alfred P. Murrah Federal Building in Oklahoma City in 1995 are examples. Terrorist organizations are generally categorized. Only a small percentage of groups actually turn toward terrorism as a means to achieve their goals, such as the following:

1. **Violent religious groups/doomsday cults.** These include groups such as Aum Shinrikyo, who carried out chemical attacks in Tokyo in 1994 and 1995. Some of these groups may participate in apocalyptic violence.
2. **Extremist political groups.** They may include violent separatist groups and those who seek political, religious, economic, and social freedom Figure 39-1 .

You are the Provider: PART 1

At 10:05 AM, you are dispatched to an abortion clinic where a car crashed into the building and then exploded. According to dispatch, the frantic caller could not tell her how many victims were involved; he could only tell her that the building was on fire. Law enforcement and the fire department's hazardous materials (HazMat) team are en route to the scene as well. The weather is clear, and a light breeze is blowing from the northwest.

1. On the basis of the dispatch information, how should you approach this incident?
2. What indicators suggest that an incident is the result of terrorism?

Figure 39-1 Taliban militants have been associated with terrorism.

3. **Technology terrorists.** Those who attack a population's technological infrastructure as a means to draw attention to their cause, such as cyberterrorists.

4. **Single-issue groups.** These include antiabortion groups, animal rights groups, anarchists, racists, and even ecoterrorists who threaten or use violence as a means to protect the environment **Figure 39-2**.

Most terrorist attacks require the coordination of multiple terrorists or "actors" working together. Nineteen hijackers worked together to commit the worst act of terrorism in US history on September 11, 2001 **Figure 39-3**. At least four terrorists worked together to commit the London Subway bombings on July 7, 2005. However, in a few instances there has been a single terrorist who struck with devastating results. Terrorists who acted alone carried out all of the Atlanta abortion clinic attacks and the 1996 Summer Olympics attack.

Figure 39-2 Demonstrators being held back by police near the World Bank in Washington, DC.

Figure 39-3 The September 11, 2001 attacks on the World Trade Center in New York City accounted for the majority of the deaths caused by terrorists in 2001.

Weapons of Mass Destruction

A **weapon of mass destruction (WMD)**, or **weapon of mass casualty (WMC)**, is any agent designed to bring about mass death, casualties, and/or massive damage to property and infrastructure (bridges, tunnels, airports, and seaports). These instruments of death and destruction include biologic, nuclear, incendiary, chemical, and explosive weapons (**B-NICE**), or chemical, biologic, radiologic, nuclear, and explosive (CBRNE) weapons. B-NICE and CBRNE are helpful mnemonics that are commonly used to remember the kinds of weapons of mass destruction. To date, the preferred WMD for terrorists has been explosive devices. Terrorist groups have favored tactics that use truck bombs or car or pedestrian suicide bombers. Many previous terrorist attempts to use either chemical or biologic weapons to their full capacity have been unsuccessful. Nonetheless, as an EMT, you should understand the destructive potential of these weapons.

The motives and tactics of the new-age terrorist groups have begun to change. As with the doomsday cults, many terrorist groups participate in apocalyptic, indiscriminate killing. This doctrine of total carnage would make the use of WMDs highly desirable. WMDs are relatively easy to obtain or create and are specifically geared toward killing large numbers of people. Had the proper techniques been used during the 1995

Aum Shinrikyo attack on the Tokyo subway, there may have been tens of thousands of casualties. With the fall of the former Soviet Union, the technology and expertise to produce WMDs may be available to terrorist groups with sufficient funding. Moreover, the technical recipes for making B-NICE weapons can be found readily on the Internet; in fact, they have even been published on terrorist group Web sites.

> ## Words of Wisdom
>
> Chemical warfare may consist of agents in the form of a liquid, powder, or vapor.

■ Chemical Terrorism/Warfare

Chemical agents are manufactured substances that can have devastating effects on living organisms. They can be produced in liquid, powder, or vapor form depending on the desired route of exposure and dissemination technique. Developed during World War I, these agents have been implicated in thousands of deaths since being introduced on the battlefield and since then have been used to terrorize civilian populations. These agents consist of the following types:

- Vesicants (blister agents)
- Respiratory agents (choking agents)
- Nerve agents
- Metabolic agents (cyanides)

■ Biologic Terrorism/Warfare

Biologic agents are organisms that cause disease. They are generally found in nature; for terrorist use, however, they are cultivated, synthesized, and mutated in a laboratory. The **weaponization** of biologic agents is performed to artificially maximize the target population's exposure to the germ, thereby exposing the greatest number of people and achieving the desired result.

The primary types of biologic agents that you may come into contact with during a biologic event include the following:

- Viruses
- Bacteria
- Toxins

■ Nuclear/Radiologic Terrorism

There have been only two publicly known incidents involving the use of a nuclear device. During World War (WW) II, Hiroshima and Nagasaki were devastated when they were targeted with nuclear bombs. The awesome destructive power demonstrated by the attack ended WWII and has served as a deterrent to nuclear war.

There are also nations that hold close ties with terrorist groups (known as **state-sponsored terrorism**) and have obtained some degree of nuclear capability.

It is also possible for a terrorist to secure radioactive materials or waste to perpetrate an act of terror. These materials are far easier for a determined terrorist to acquire and require less expertise to use. The difficulties in developing a nuclear weapon are well documented. Radioactive materials, however, such as those in radiologic dispersal devices (RDDs), also known as "dirty bombs," can cause widespread panic and civil disturbances. More on these devices will be covered later in this chapter.

■ EMT Response to Terrorism

When you respond to a terrorist event, the basic foundations of patient care remain the same; however, the treatment can and will vary. Terrorist events can produce a single casualty, hundreds of casualties, or even thousands of casualties. When presented with widespread mass casualties, you must remember situational awareness. What you may do in one situation may not be appropriate for another situation. In large-scale terrorist events, it is important to use triage and base patient care on available resources.

■ Recognizing a Terrorist Event (Indicators)

Most acts of terror are **covert**, which means that the public safety community generally has no prior knowledge of the time, location, or nature of the attack. This element of surprise makes responding to an event more complex. You must constantly be aware of your surroundings and understand the possible risks for terrorism associated with certain locations, at certain times. It is therefore important that you know the current threat level issued by the federal government through the Department of Homeland Security (DHS).

The Homeland Security Advisory System alerts responders to the potential for an attack, although the specifics of the current threat will not be given. On the basis of the current threat level, EMTs should take appropriate actions and precautions while continuing to perform daily duties and responding to calls. The system of colors is used to inform the public safety community of the climate of terrorism (derived from intelligence gathering and the amount of terrorist communication) and to heighten the awareness of the potential for a terrorist attack. The system is designed to save lives, including yours.

The DHS has not issued specific recommendations for EMS personnel to follow in response to the alert system. Follow your local protocols, policies, and procedures.

It is your responsibility to make sure you know the advisory level at the start of your workday. Daily newspapers, television news programs, and multiple Web sites (including the DHS Web site) all give up-to-date information on the threat level. Many EMS organizations are starting to display the advisory system on boards where it can be seen by staff when they arrive for a shift.

Understanding and being aware of the current threat is only the beginning of responding safely to calls. Once you are on duty, you must be able to make appropriate decisions regarding the potential for a terrorist event. In determining the potential for a terrorist attack, on every call you should make the following observations:

- **Type of location.** Is the location a monument, infrastructure, government building, or a specific type of location such as a temple? Is there a large gathering? Is there a special event taking place?
- **Type of call.** Is there a report of an explosion or suspicious device nearby? Does the call come into dispatch as someone having unexplained coughing and difficulty breathing? Are there reports of people fleeing the scene?
- **Number of patients.** Are there multiple victims with similar signs and symptoms? This is probably the single most important clue that a terrorist attack or an incident involving a WMD has occurred.
- **Victims' statements.** This is probably the second best indication of a terrorist or WMD event. Are the victims fleeing the scene giving statements such as, "Everyone is passing out," "There was a loud explosion," or "There are a lot of people shaking on the ground." If so, something is occurring that you do not want to rush into, even if it is determined not to be a terrorist event.
- **Preincident indicators.** Is the terror alert level high (orange) or severe (red)? Has there been a recent increase in violent political activism? Are you aware of any credible threats made against the location, gathering, or occasion?

■ Response Actions

Once you suspect that a terrorist event has occurred or a WMD has been used, there are certain actions you must take to ensure that you will be safe and be in the proper position to help the community.

Scene Safety

Ensure that the scene is safe, remembering to stage your vehicle a safe distance (usually 1 to 2 blocks) from the incident, and wait for law enforcement personnel to

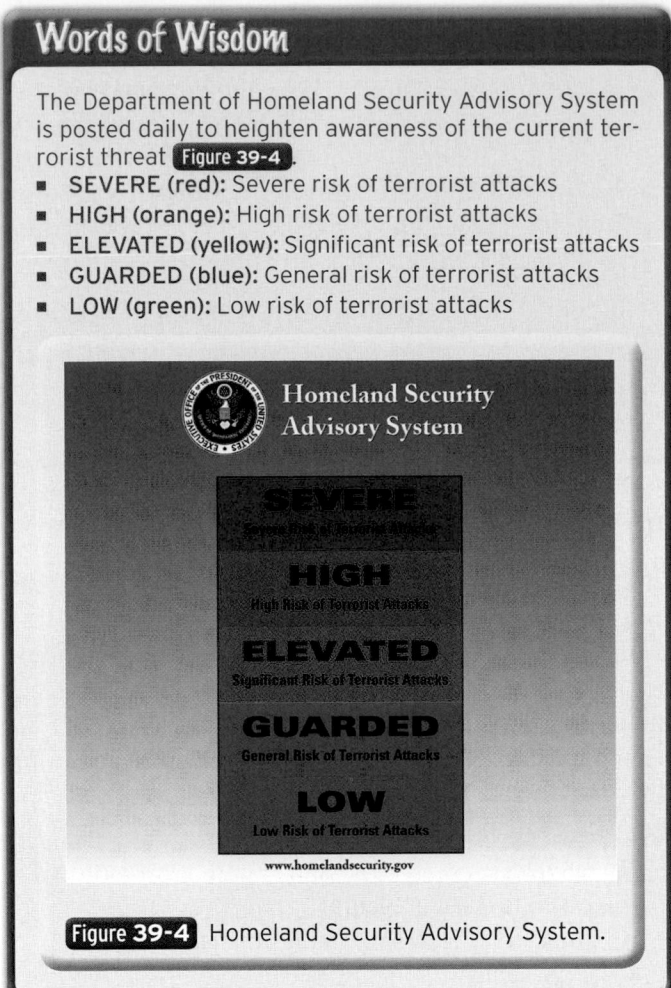

Words of Wisdom

The Department of Homeland Security Advisory System is posted daily to heighten awareness of the current terrorist threat **Figure 39-4**.

- **SEVERE (red):** Severe risk of terrorist attacks
- **HIGH (orange):** High risk of terrorist attacks
- **ELEVATED (yellow):** Significant risk of terrorist attacks
- **GUARDED (blue):** General risk of terrorist attacks
- **LOW (green):** Low risk of terrorist attacks

Figure 39-4 Homeland Security Advisory System.

advise you that the scene has been made secure. If you have any doubt that it may not be safe, do not enter. When dealing with a WMD scene, it is safe to assume that you will not be able to enter where the event has occurred—nor do you want to. The best location for staging is upwind and uphill from the incident. Wait for assistance from those who are trained in assessing and managing WMD scenes **Figure 39-5**. You should expect that a perimeter will be created, usually by law enforcement personnel, in an effort to isolate the scene, prevent further contamination of evidence, and protect rescuers and the public from further danger. Also remember the following rules:

- Failure to park your vehicle at a safe location can place you and your partner in danger **Figure 39-6**. It is important to remember to always have an escape plan determined beforehand, in case the scene becomes unsafe.
- If your vehicle is blocked in by other emergency vehicles or damaged by a secondary device (or event), you will be unable to provide victims with transportation **Figure 39-7** or escape yourself.

Figure 39-5 Improper staging of a mass-casualty scene could lead to injury or even death of EMS personnel. Wait for assistance from persons who are trained in assessing and managing such scenes.

Figure 39-7 Make sure that your vehicle is not blocked in by other emergency vehicles.

Responder Safety (Personnel Protection)

The best form of protection from a WMD agent is preventing yourself from coming into contact with the agent. The greatest threats facing you in a WMD attack are contamination and **cross-contamination**. Contamination with an agent occurs when you have direct contact with the WMD or are exposed to it. Cross-contamination occurs when you come into contact with a contaminated person who has not yet been decontaminated.

Words of Wisdom

One of the easiest ways to distinguish between a nonterrorist mass-casualty event and a terrorist event is that the intentional use of a WMD affects multiple persons. These casualties will generally exhibit the same signs and symptoms. It is highly unlikely for more than one person to experience a seizure at any given time. It is not uncommon to find multiple patients complaining of difficulty breathing at the scene of a fire. However, the same report in the subway at rush hour, when no smell of smoke has been reported, is certainly cause for suspicion. In these situations, you must use good judgment and resist the urge to "rush in and help," especially when there are multiple victims from an unknown cause.

Notification Procedures

When you suspect a terrorist or WMD event has taken place, notify the dispatcher, providing that communications function properly. Vital information needs to be communicated effectively if you are to receive the appropriate assistance (see Chapter 4, *Communications and Documentation*, for information on effective communication). Inform dispatch of the nature of the event, any additional resources that may be required, the estimated number of

Figure 39-6 Park your vehicle at a safe location.

You are the Provider: PART 2

The HazMat team has assessed the scene. On the basis of its assessment and observations, it has determined that radiation is present in the area of the incident. The driver of the vehicle is obviously deceased, and there are four patients who require emergency care. The severity of their injuries, however, is unknown.

3. On the basis of the HazMat team's findings, what has most likely occurred?

4. Is it safe for you and other EMS personnel to enter the scene? If not, what should be done first?

patients, and the upwind route of approach or optimal route of approach.

It is extremely important to establish a staging area, where other units will converge. Be mindful of access and exit routes when you direct units to respond to a location. It is unwise to have units respond to the front entrance of a hotel or apartment building that has had an explosion (see Chapter 36, *Transport Operations*, on vehicle positioning). Last, trained responders in the proper protective equipment are the only persons equipped to handle the WMD incident. These specialized units, traditionally hazardous materials (HazMat) teams, must be requested as early as possible because of the time required to assemble and dispatch the team and their equipment. Many jurisdictions share HazMat teams, and the team may have to travel a long distance to reach the location of the event. It is always better to be safe than sorry; call the team early, and the outcome of the call will be more favorable.

Keep in mind that there may be more than one type of device or agent present.

Words of Wisdom

Weapons of Mass Destruction
On September 11, 2001, communications were severely affected by the collapse of the World Trade Center. The primary communications repeater was situated on top of one of the towers. In addition, excess radio traffic made transmitting and receiving messages extremely difficult. Not only were radio communications affected, but also most cellular phones and the majority of radio and television stations were disabled. The lesson learned from this event is to have multiple backups to your ability to communicate with your dispatcher. In the event of a terrorist or WMD event, refrain from using the radio unless you have something important to transmit. If you do transmit, gather your thoughts and speak in as calm a tone as possible, avoiding unnecessary chatter. Remember, while you are transmitting, others may be unable to call for help.

Establishing Command

The first arriving provider on the scene must begin to sort out the chaos and define his or her responsibilities under the incident command system (ICS). As the first person on scene, the EMT may need to establish command until additional personnel arrive. Depending on the circumstances and stage of the operation, you and other EMTs may function as medical branch directors, triage supervisors, treatment supervisors, transportation supervisors, logistic officers, or command and general staff. If the initial ICS is already in place, then you should immediately seek out the medical staging officer to receive your assignment.

Secondary Device or Event (Reassessing Scene Safety)

Terrorists have been known to plant additional explosives that are set to explode after the initial bomb. This type of **secondary device** is intended primarily to injure responders and to secure media coverage because the media generally arrive on scene just after the initial response. Secondary devices may include various types of electronic equipment such as cell phones or pagers that are detonated when "answered." Do not rely on others to secure your safety. It is every EMT's responsibility to constantly assess and reassess the scene for safety. It is easy to overlook a suspicious package lying on the floor while you are treating casualties. Stay alert. Something as subtle as a change in the wind direction during a gas attack or an increase in the number of contaminated patients can place you in danger. Never become so involved with the tasks you are performing that you do not look around and make sure that the scene remains safe.

Words of Wisdom

Whereas it may be difficult for you because of ethical or moral reasons, when you are called on to treat a suspected criminal or suspected terrorist, it is important that this patient receive the same care as your other patients. Remember that you are not the judge or jury. It is up to the legal system to prove someone guilty in a court of law.

Chemical Agents

Chemical agents are liquids or gases that are dispersed to kill or injure. Modern-day chemicals were first developed during WWI and WWII. During the Cold War, many of these agents were perfected and stockpiled. Whereas the United States has long renounced the use of chemical weapons, many nations still develop and stockpile them. These agents are deadly and pose a threat if acquired by terrorists.

Chemical weapons have several classifications. The properties or characteristics of an agent can be described as liquid, gas, or solid material. **Persistency** and **volatility** are terms used to describe how long the agent will stay on a surface before it evaporates. Persistent or nonvolatile agents can remain on a surface for long periods, usually longer than 24 hours. Nonpersistent or volatile agents evaporate relatively fast when left on a surface in the optimal temperature range. An agent that is described as highly persistent (such as VX, a nerve agent) can remain in the environment for weeks to months, whereas an agent that is highly volatile (such as sarin, also a nerve agent) will turn from liquid to gas (evaporate) within minutes to seconds.

Route of exposure is a term used to describe how the agent most effectively enters the body. Chemical agents can have either a vapor or contact hazard. Agents with a **vapor hazard** enter the body through the respiratory tract in the form of vapors. Agents with a **contact hazard** (or skin hazard) give off very little vapor or no vapors and enter the body through the skin.

■ Vesicants (Blister Agents)

The primary route of exposure of blister agents, or **vesicants**, is the skin (contact); however, if vesicants are left on the skin or clothing long enough, they produce vapors that can enter the respiratory tract. Vesicants cause burn-like blisters to form on the victim's skin and in the respiratory tract. The vesicant agents consist of sulfur mustard (H), Lewisite (L), and phosgene oxime (CX) (the symbols H, L, and CX are military designations for these chemicals). The vesicants usually cause the most damage to damp or moist areas of the body, such as the armpits, groin, and respiratory tract. Signs of vesicant exposure on the skin include the following:

- Skin irritation, burning, and reddening
- Immediate, intense skin pain (with L and CX)
- Formation of large blisters
- Gray discoloration of skin (a sign of permanent damage seen with L and CX)
- Swollen and closed or irritated eyes
- Permanent eye injury (including blindness)

If vapors were inhaled, the patient may experience the following signs/symptoms:

- Hoarseness and stridor
- Severe cough
- Hemoptysis (coughing up blood)
- Severe dyspnea

Sulfur mustard (H) is a brownish, yellowish oily substance that is generally considered very persistent. When released, mustard has the distinct smell of garlic or mustard and is quickly absorbed into the skin and/ or mucous membranes. As the agent is absorbed into the skin, it begins an irreversible process of damage to the cells. Absorption through the skin or mucous membranes usually occurs within seconds, and damage to the underlying cells takes place within 1 to 2 minutes.

Mustard is considered a **mutagen**, which means that it mutates, damages, and changes the structures of cells.

Eventually, cellular death will occur. On the surface, the patient will generally not produce any signs or symptoms until 4 to 6 hours after exposure (depending on concentration and amount of exposure) **Figure 39-8**.

The patient will develop a progressive reddening of the affected area, which will gradually develop into large blisters. These blisters are very similar in shape and appearance to those associated with thermal second-degree burns. The fluid within the blisters does not contain any of the agent; however, the skin covering the area is considered to be contaminated until decontamination by trained personnel has been performed.

Mustard also attacks vulnerable cells within the bone marrow and depletes the body's ability to reproduce white blood cells. As with burns, the primary complication associated with vesicant blisters is secondary infection. If the patient survives the initial direct injury from the agent, the depletion of the white blood cells leaves the patient with a decreased resistance to infections. Although sulfur mustard is regarded as persistent, it releases enough vapors when dispersed to be inhaled. This creates upper and lower airway compromise. The result is damage and swelling of the airways. The airway compromise makes the patient's condition far more serious.

Lewisite (L) and **phosgene oxime (CX)** produce blister wounds very similar to those caused by mustard. They are highly volatile and have a rapid onset of symptoms, as opposed to the delayed onset seen with mustard. These agents produce immediate intense pain and discomfort when contact is made. The patient may have a grayish discoloration at the contaminated site. While tissue damage also occurs with exposure to these agents, they do not cause the secondary cellular injury that is associated with mustard.

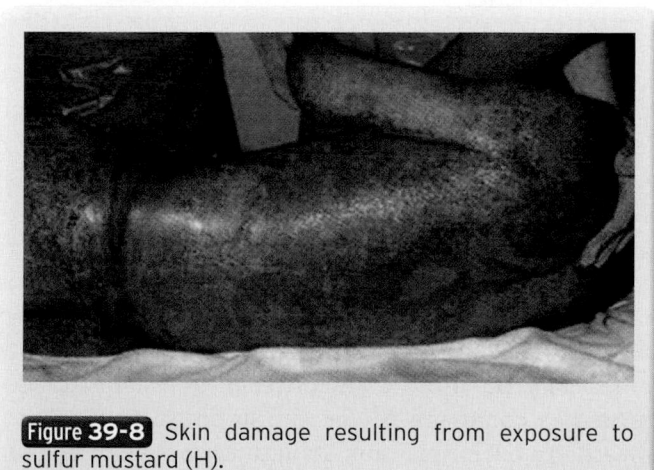

Figure 39-8 Skin damage resulting from exposure to sulfur mustard (H).

Vesicant Agent Treatment

There are no antidotes for mustard or CX exposure. British anti-Lewisite is the antidote for agent L; however, it is not carried by civilian EMS. You must ensure that the patient has been decontaminated before the ABCs are initiated. The patient may require prompt airway support if any agent has been inhaled, but this should not occur until after decontamination. Transport should be initiated as soon as possible. Generally, burn centers are best equipped to handle the wounds and subsequent infections produced by vesicants. Follow your local protocols when deciding the transport destination.

■ Pulmonary Agents (Choking Agents)

The pulmonary agents are gases that cause immediate harm to persons exposed to them. The primary route of exposure for these agents is through the respiratory tract, which makes them an inhalation or vapor hazard. Once inside the lungs, they damage the lung tissue and fluid leaks into the lungs. Pulmonary edema develops in the patient, resulting in difficulty breathing because of the inability for air exchange. These agents produce respiratory-related symptoms such as dyspnea, tachypnea, and pulmonary edema. This class of chemical agents consists of chlorine (CL) and phosgene.

Chlorine (CL) was the first chemical agent ever used in warfare. It has a distinct odor of bleach and creates a green haze when released as a gas. Initially it produces upper airway irritation and a choking sensation. The patient may later experience the following signs/symptoms:

- Shortness of breath
- Chest tightness
- Hoarseness and stridor as the result of upper airway constriction
- Gasping and coughing

With serious exposures, patients may experience pulmonary edema, complete airway constriction, and death. The fumes from a mixture of household bleach (CL) and ammonia create an acid gas that produces similar effects. Each year, such mixtures overcome hundreds of people when they try to mix household cleaners.

Phosgene should not be confused with phosgene oxime, a blistering agent, or vesicant. Not only has phosgene been produced for chemical warfare, but it is a product of combustion such as might be produced in a fire at a textile factory or house or from metalwork or burning Freon (a liquid chemical used in refrigeration). Therefore, you may encounter a victim of exposure to this gas during the course of a normal call or at a fire scene. Phosgene is a very potent agent that has a delayed onset of symptoms, usually hours. Unlike CL, when phosgene enters the body, it generally does not produce severe irritation that would possibly cause the victim to leave the area or hold his or her breath. In fact, the odor produced by the chemical is similar to that of freshly mown grass or hay. The result is that much more of the gas is allowed to enter the body unnoticed. Initially, a mild exposure may include the following signs/symptoms:

- Nausea
- Chest tightness
- Severe cough
- Dyspnea on exertion

The victim of a severe exposure may present with dyspnea at rest and excessive pulmonary edema. (The patient will actually expel large amounts of fluid from the pulmonary edema in the lungs.) A severe exposure produces such large amounts of fluid in the lungs that the patient may actually become hypovolemic and subsequently hypotensive.

Pulmonary Agent Treatment

The best initial treatment for any patient who has been exposed to a pulmonary agent is to remove the patient from the contaminated atmosphere. This should be done by trained personnel in the proper personal protective equipment. Aggressive management of the ABCs should be initiated, paying particular attention to oxygenation, ventilation, and suctioning if required. Do not allow the patient to be active because this will worsen the condition much faster. There are no antidotes to counteract the pulmonary agents. Performing the ABCs, allowing the patient to rest in a position of comfort with the head elevated, and initiating rapid transport are the primary goals for prehospital emergency care.

■ Nerve Agents

The **nerve agents** are among the most deadly chemicals developed. Designed to kill large numbers of people with small quantities, nerve agents can cause cardiac arrest within seconds to minutes of exposure. Nerve agents, discovered while in search of a superior pesticide, are a class of chemical called organophosphates, which are found in household bug sprays, agricultural pesticides, and some industrial chemicals, at far lower strengths than in nerve agents. Organophosphates block an essential enzyme in the nervous system, causing the body's organs to become overstimulated and burn out.

G agents came from the early nerve agents, the G series, which were developed by German scientists (hence the G) in the period after WWI and into WWII. There are three G series agents, which are all designed with the same basic chemical structure with slight variations to produce different properties. The two variations of these agents are lethality and volatility. The following G agents are listed from high volatility to low volatility:

- **Sarin (GB)**. Highly volatile colorless and odorless liquid. Turns from liquid to gas within seconds to minutes at room temperature. Highly lethal, with an **LD$_{50}$** of 1,700 mg/70 kg (about 1 drop, depending on the purity). The LD$_{50}$ is the amount that will kill 50% of people who are exposed to this level. Sarin is primarily a vapor hazard, with the respiratory tract as the main route of entry. This agent is especially dangerous in enclosed environments such as office buildings, shopping malls, and subway cars. When this agent comes into contact with the skin, it is quickly absorbed and evaporates. When sarin is on clothing, it has the effect of **off-gassing**, which means that the vapors are continuously released over a period of time (like perfume). This renders the victim and the victim's clothing contaminated.

- **Soman (GD)**. Twice as persistent as sarin and five times as lethal. It has a fruity odor as a result of the type of alcohol used in the agent and generally has no color. This agent is a contact and an inhalation hazard that can enter the body through skin absorption and through the respiratory tract. A unique additive in GD causes it to bind to the cells that it attacks faster than any other agent. This irreversible binding is called **aging**, which makes it more difficult to treat patients who have been exposed.

- **Tabun (GA)**. Approximately half as lethal as sarin and 36 times more persistent. Under the proper conditions it will remain present for several days.

It also has a fruity smell and an appearance similar to sarin. The components used to manufacture GA are easy to acquire, and the agent is easy to manufacture, which make it unique. GA is a contact and an inhalation hazard that can enter the body through skin absorption and through the respiratory tract.

- **V agent (VX)**. Clear oily agent that has no odor and looks like baby oil. V agent was developed by the British after WWII and has chemical properties similar to the G series agents. The difference is that VX is more than 100 times more lethal than sarin and is extremely persistent **Figure 39-9**. In fact, VX is so persistent that given the proper conditions, it will remain relatively unchanged for weeks to months. These properties make VX primarily a contact hazard because it lets off very little vapor. It is easily absorbed into the skin, and the oily residue that remains on the skin's surface is extremely difficult to decontaminate.

Nerve agents all produce similar symptoms but have varying routes of entry. Nerve agents differ slightly in lethal concentration or dose and also differ in their volatility. Some agents are designed to become a gas quickly (nonpersistent or highly volatile), whereas others remain liquid for a period of time (persistent or nonvolatile). These agents have been used successfully in warfare and to date represent the only type of chemical agent that has been used successfully in a terrorist act. Once the agent has

Figure 39-9 VX is the most toxic chemical ever created. The dot on the penny demonstrates the amount needed to achieve the lethal dose.

You are the Provider: PART 3

You have been informed that the patients have been properly decontaminated and have been moved to an area that has been designated as being safe for you to enter. After donning the appropriate personal protective equipment, you and the other responders report to the incident scene.

5. Given the situation, what are some unique concerns about this incident?

6. What types of injuries should you expect to encounter?

entered the body through skin contact or through the respiratory system, the patient will begin to exhibit a pattern of predictable symptoms. Like all chemical agents, the severity of the symptoms will depend on the route of exposure and the amount of agent to which the patient was exposed. The resulting symptoms are described below using the military mnemonic SLUDGEM and the medical mnemonic DUMBELS. The medical mnemonic is more useful to you because it lists the more dangerous symptoms associated with exposure to nerve agents.

There are only a handful of medical conditions that are associated with the bilateral pinpoint constricted pupils (**miosis**) seen with nerve agent exposure. Conditions such as a cerebrovascular accident, direct light to both eyes, and a drug overdose all can cause bilateral constricted pupils. You should therefore assess the patient for all of the SLUDGEM/DUMBELS signs and symptoms to determine whether the patient has been exposed to a nerve agent.

Miosis is the most common symptom of nerve agent exposure and can remain for days to weeks. This symptom, along with the others listed in `Table 39-1`, will help you recognize exposure to a nerve agent early. The seizures that are associated with nerve agent exposure are unlike those found in patients with a history of seizure. The seizure will continue until the patient dies or until treatment is given with a nerve agent antidote kit (Mark 1 NAAK).

Nerve Agent Treatment (Mark 1 NAAK)

Fatalities from severe nerve agent exposure occur as a result of respiratory complications, which lead to respiratory arrest. Once the patient has been decontaminated, the EMT should be prepared to treat aggressively, if the patient is to be saved. You can greatly increase the patient's chances of survival simply by providing airway and ventilatory support. As with all emergencies, securing the ABCs is the best and most important treatment that you can provide. Often in patients exposed to these agents, seizures will begin and will not stop. These patients will require administration of nerve agent antidote kits in addition to support of the ABCs.

In terms of medical treatment for nerve agent exposure, the most common treatment is the **Mark 1 Nerve Agent Antidote Kit (NAAK)**. The Mark 1 NAAK contains two medications—2 mg of atropine and 600 mg of pralidoxime chloride (2-PAM)—in two separate auto-injectors. An updated version of the Mark 1 is the **DuoDote Auto-Injector**. The DuoDote contains 2.1 mg of atropine and 600 mg of 2-PAM and is delivered as a single dose through one needle.

In some regions, EMTs may carry Mark 1 or DuoDote kits on the unit and will be called on to administer one or

Table 39-1	Symptoms of Persons Exposed to Nerve Agents
Military Mnemonic: SLUDGEM	**Medical Mnemonic: DUMBELS**
Salivation, Sweating	Diarrhea
Lacrimation (excessive tearing)	Urination
Urination	Miosis (pinpoint pupils)
Defecation, Drooling, Diarrhea	Bradycardia, Bronchospasm (spasm of the bronchioles)
Gastric upset and cramps	Emesis (vomiting)
Emesis (vomiting)	Lacrimation (excessive tearing)
Muscle twitching/Miosis (pinpoint pupils)	Seizures, Salivation, Sweating

Words of Wisdom

On March 20, 1995, members of Aum Shinrikyo, a Japanese cult, released sarin (GB) in the Tokyo subway. The first arriving medical responders were met with chaos as hundreds and then thousands of people fled the subway system `Figure 39-10`. Many were contaminated and showing signs and symptoms of nerve agent exposure. In the end, more than 5,000 people sought medical care for exposure to sarin, and 12 people died. None of the EMS personnel wore protective clothing, and most became cross-contaminated. Remember, you can avoid becoming exposed. Do not become a victim.

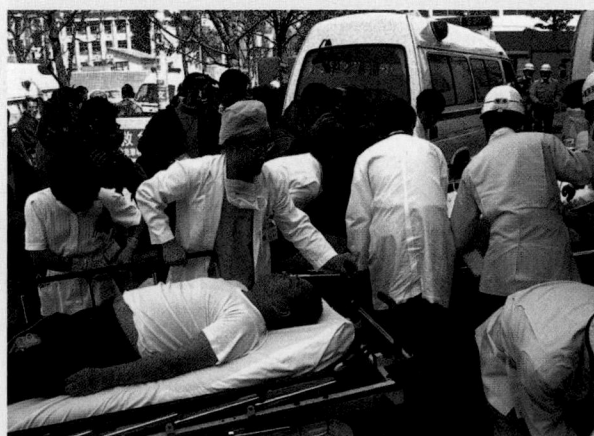

Figure 39-10 Medical professionals responding to an attack in 1995, where cult members released sarin in the Tokyo subway.

both of the antidotes to themselves or their patients. These medications are delivered using the same technique as the EpiPen auto-injector; however, multiple doses may need to be administered, remembering that activated NAAKs need to be disposed of properly, in a sharps container.

Atropine is used to block the nerve agent from affecting the body. However, because the nerve agent may remain in the body for long periods, 2-PAM is used to eliminate the agent from the body. Many of the symptoms described in the DUMBELS mnemonic will be reversed with the use of atropine; however, many doses may need to be administered to see these results. If your service carries a nerve agent antidote, refer to your local protocols for dose and use information.

Table 39-2 has been provided for quick reference and comparison of the nerve agents.

■ Metabolic Agents (Cyanides)

Hydrogen cyanide (AC) and cyanogen chloride (CK) are both agents that affect the body's ability to use oxygen. <u>Cyanide</u> is a colorless gas that has an odor similar to almonds. The effects of the cyanides begin on the cellular level and are very rapidly seen at the organ and system levels. Besides the nerve agents, metabolic agents are the only chemical weapons known to kill within seconds to minutes. Unlike nerve agents, however, these deadly gases are commonly found in many industrial settings. Cyanides are produced in massive quantities throughout the United States every year for industrial uses such as gold and silver mining, photography, and plastics processing. They are often present in fires associated with textile and plastic factories. In fact, cyanide is naturally found in the pits of many fruits in very low doses.

Safety

Industrial Chemicals/Insecticides

As previously mentioned, the basic chemical ingredient in nerve agents is organophosphate. This is a common chemical that is used in lesser concentrations for insecticides. Whereas industrial chemicals do not possess sufficient lethality to be effective WMDs, they are easy to acquire and inexpensive and would have similar effects as the nerve agents. Crop-duster planes could be used to disseminate these chemicals. You should be cautious when responding to calls where insecticide equipment is stored and used, such as a farm or supply store that sells these products. The symptoms and medical management of victims of organophosphate insecticide poisoning are identical to those of the nerve agents.

There is very little difference in the symptoms found between AC and CK. In low doses, these chemicals are associated with dizziness, light-headedness, headache, and vomiting. Higher doses will produce symptoms that include the following:

- Shortness of breath and gasping respirations
- Tachypnea
- Flushed skin
- Tachycardia
- Altered mental status
- Seizures
- Coma
- Apnea
- Cardiac arrest

The symptoms associated with the inhalation of a large amount of cyanide will all appear within several minutes. Death is likely unless the patient is treated promptly.

Table 39-2 The Nerve Agents

Name	Military Designation	Odor	Special Features	Onset of Symptoms	Volatility	Route of Exposure
Tabun	GA	Fruity	Easy to manufacture	Immediate	Low	Contact and vapor hazard
Sarin	GB	None (if pure) or strong	Will off-gas while on victim's clothing	Immediate	High	Primarily respiratory vapor hazard; extremely lethal if skin contact is made
Soman	GD	Fruity	Ages rapidly, making it difficult to treat	Immediate	Moderate	Contact with skin; minimal vapor hazard
V agent	VX	None	Most lethal chemical agent; difficult to decontaminate	Immediate	Very low	Contact with skin; no vapor hazard (unless aerosolized)

Cyanide Agent Treatment

Cyanide binds with the body's cells, preventing oxygen from being used. Several medications act as antidotes, but most services do not carry them. Once trained personnel wearing the proper personal protective equipment have removed the patient from the source of exposure, even if there is no liquid contamination, all of the patient's clothes must be removed to prevent off-gassing in the ambulance. Trained and protected personnel must decontaminate any patients who may have been exposed to liquid contamination before an EMT can initiate treatment. Then you should support the patient's ABCs. Mild effects of cyanide exposure will generally resolve by simply removing the victim from the source of contamination and administering supplemental oxygen. Severe exposure, however, will require aggressive oxygenation and perhaps ventilation with supplemental oxygen. Always use a bag-mask device or oxygen-powered ventilator device to ventilate a victim of a metabolic agent. The agent can easily be passed on from the patient to you through mouth-to-mouth or mouth-to-mask ventilations. If no antidote is available, initiate transport immediately.

Words of Wisdom

Always make sure that your patients have been thoroughly decontaminated by trained personnel before you come into contact with them. Chemical agents are primarily a vapor hazard, and all of the patient's clothing must be removed prior to you providing treatment to prevent off-gassing to you. Finally, never perform mouth-to-mouth or mouth-to-mask ventilation on a victim of a chemical agent exposure. Many of the vapors may linger in the patient's airway, and cross-contamination may occur.

Table 39-3 summarizes the chemical agents. The odors of the particular chemicals are provided for informational purposes only. The sense of smell is a poor tool to use to determine whether there is a chemical agent present. Many persons are unable to smell the agents, and the odor could be derived from another source. This information is useful to you if you receive reports from victims who claimed to smell bleach or garlic, for example. You should never enter a potentially hazardous area and "smell" to determine whether a chemical agent is present.

Table 39-3 Chemical Agents

Name	Military Designations	Odor	Lethality	Onset of Symptoms	Volatility	Primary Route of Exposure
Nerve agents	Tabun (GA) Sarin (GB) Soman (GD) VX	Fruity or none	Most lethal chemical agents; can kill within minutes; effects are reversible with antidotes	Immediate	Moderate (GA, GD) Very high (GB) Low (VX)	GA—both GB—vapor hazard GD—both VX—contact hazard
Vesicants	Mustard (H) Lewisite (L) Phosgene oxime (CX)	Garlic (H) Geranium (L)	Causes large blisters to form on victims; may severely damage upper airway if vapors are inhaled; severe, intense pain and grayish skin discoloration (L and CX)	Delayed (H) Immediate (L, CX)	Very low (H, L) Moderate (CX)	Primarily contact, with some vapor hazard
Pulmonary agents	Chlorine (CL) Phosgene (CG)	Bleach (CL) Cut grass (CG)	Causes irritation choking (CL); severe pulmonary edema (CG)	Immediate (CL) Delayed (CG)	Very high	Vapor hazard
Cyanide agents	Hydrogen cyanide (AC) Cyanogen chloride (CK)	Almonds (AC) Irritating (CK)	Highly lethal chemical gases; can kill within minutes; effects are reversible with antidotes	Immediate	Very high	Vapor hazard

Biologic Agents

Biologic agents pose many difficult issues when used as a WMD. Biologic agents can be almost completely undetectable. Also, most of the diseases caused by these agents will be similar to other minor illnesses commonly seen by EMS providers.

Biologic agents are grouped as viruses, bacteria, and neurotoxins and may be spread in various ways. **Dissemination** is the means by which a terrorist will spread the agent—for example, poisoning the water supply or aerosolizing the agent into the air or ventilation system of a building. A **disease vector** is an animal that spreads disease, once infected, to another animal. For example, bubonic plague can be spread by infected rats, smallpox by infected persons, and West Nile virus by infected mosquitoes. How easily the disease is able to spread from one human to another human is called communicability. Some diseases, such as those caused by human immunodeficiency virus, are difficult to spread by routine contact. Therefore, communicability is considered low. In other instances when communicability is high, such as with smallpox, the person is considered **contagious**. Typically, routine standard precautions are enough to prevent contamination from contagious biologic organisms.

Incubation describes the period of time between the person becoming exposed to the agent and when symptoms begin. The incubation period is especially important for the EMT to understand. Although your patient may not exhibit signs or symptoms, he or she may be contagious.

EMTs need to be aware of when they should suspect the use of biologic agents. If the agent is in the form of a powder, such as in the October 2001 Amerithrax attacks involving anthrax powder mailed in letters, the incident must be handled by HazMat specialists. Patients who have come into direct contact with the agent need to be decontaminated before there is any contact with EMS personnel or treatment is initiated.

■ Viruses

Viruses are germs that require a living host to multiply and survive. A virus is a simple organism and cannot thrive outside of a host (living body). Once in the body, the virus invades healthy cells and replicates itself to spread through the host. As the virus spreads, so does the disease that it carries. Viruses move from host to host by direct methods, such as respiratory droplets, or through vectors. A vector is any agent that acts as a carrier or transporter.

Viral agents that may be used during a biologic terrorist release pose an extraordinary problem for health care providers, especially those in EMS. Although some viral agents do have vaccines, there is no treatment for a viral infection other than antiviral medications for some agents. Because of this characteristic, the following viruses have the potential to be used as terrorism agents.

Smallpox

Smallpox is a highly contagious disease. All forms of standard precautions must be used to prevent cross-contamination to health care providers. Simply by wearing examination gloves, a HEPA-filtered respirator, and eye protection, you will greatly reduce your risk of contamination. The last natural case of smallpox in the world was seen in 1977. Before the rash and blisters show, the illness will start with a high fever and body aches and headaches. The patient's temperature is usually in the range of 101°F to 104°F.

An easy, quick way to differentiate the smallpox rash from other skin disorders is to observe the size, shape, and location of the lesions. In smallpox, all the lesions are identical in their development. In other skin disorders, the lesions will be in various stages of healing and development. Smallpox blisters also begin on the face and extremities and eventually move toward the chest and abdomen. The disease is in its most contagious phase when the blisters begin to form **Figure 39-11**. Unprotected contact with these blisters will promote transmission of the disease **Table 39-4**. There is a vaccine to prevent smallpox; however, it has been linked to medical complications and, in rare cases, death. Should an outbreak occur, the US government has enough vaccine to vaccinate every person in the United States.

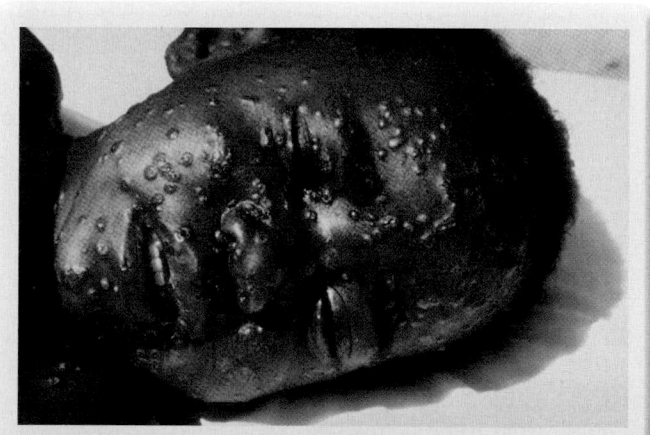

Figure 39-11 In smallpox, all the lesions are identical in their development. In other skin disorders, the lesions will be in various stages of healing and development.

Table 39-4 Characteristics of Smallpox

Dissemination	Aerosolized for warfare or terrorist uses
Communicability	High from infected individuals or items (such as blankets used by infected patients); person-to-person transmission possible
Route of entry	Inhalation of coughed droplets or direct skin contact with blisters
Signs and symptoms	Severe fever, malaise, body aches, headaches, small blisters on the skin, bleeding of the skin and mucous membranes; incubation period 10 to 12 days; duration of the illness, approximately 4 weeks
Medical management	Standard precautions; no specific treatment for smallpox, provide with supportive care (ABCs)

Viral Hemorrhagic Fevers

Viral hemorrhagic fevers (VHF) consist of a group of diseases caused by viruses that include the Ebola, Rift Valley, and yellow fever viruses, among others. This group of viruses causes the blood in the body to seep out from the tissues and blood vessels Figure 39-12 . Initially, the patient will have flulike symptoms, progressing to more serious symptoms such as internal and external hemorrhaging. Outbreaks are not uncommon in Africa and South America. Outbreaks in the United States, however, are extremely rare. All standard precautions must be taken when treating these illnesses. Mortality rates can range from 5% to 90%, depending on the strain of virus, the victim's age and health condition, and the availability of a modern health care system Table 39-5 .

■ Bacteria

Unlike viruses, **bacteria** do not require a host to multiply and live. Bacteria are much more complex and larger than viruses and can grow up to 100 times larger than the largest virus. Bacteria contain all the cellular structures of a normal cell and are completely self-sufficient. Most bacterial infections can be fought with antibiotics.

Most bacterial infections will generally begin with flulike symptoms, which can make it quite difficult for health care providers to identify whether the cause is a biologic attack or a natural epidemic.

Words of Wisdom

Because humans are acceptable hosts and vectors for many viruses and bacteria, it is important for you to use standard precautions at all times. If you fail to use standard precautions, you may not only become a host for a virus, but you may spread it as well. Remember, a virus moves from person to person to survive, and many infectious diseases present like common colds.

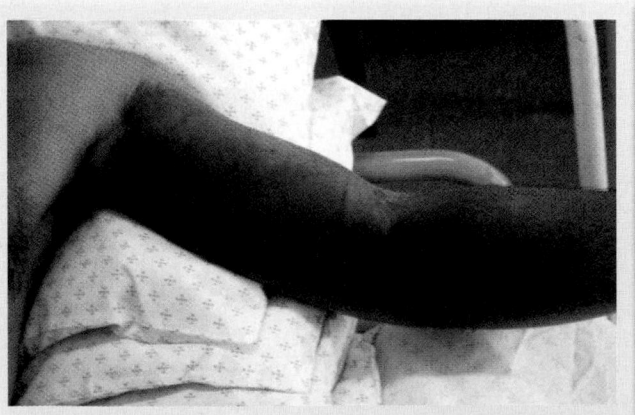

Figure 39-12 Viral hemorrhagic fevers cause the blood vessels and tissues to seep blood. The end result is ecchymosis, hemoptysis, and blood in the patient's stool. Notice the severe discoloration in this patient with Crimean Congo hemorrhagic fever, indicating internal bleeding.

Table 39-5 Characteristics of Viral Hemorrhagic Fevers

Dissemination	Direct contact with an infected person's body fluids; can also be aerosolized for use in an attack
Communicability	Moderate from person to person or contaminated items
Route of entry	Direct contact with an infected person's body fluids
Signs and symptoms	Sudden onset of fever, weakness, muscle pain, headache, and sore throat; all followed by vomiting and, as the virus runs it course, internal and external bleeding
Medical management	Standard precautions; no specific treatment for viral hemorrhagic fever; provide supportive care (ABCs) and treatment for shock and hypotension, if present

Inhalation and Cutaneous Anthrax (*Bacillus anthracis*)

<u>Anthrax</u> is caused by a deadly bacterium that lays dormant in a spore (protective shell). When exposed to the optimal temperature and moisture, the germ will be released from the spore. The routes of entry for anthrax bacteria are inhalation, cutaneous, and gastrointestinal (from consuming food that contains spores) Figure 39-13. The inhalational form, or pulmonary anthrax, is the most deadly and often presents as a severe cold. Pulmonary anthrax is associated with a 90% death rate if untreated. Antibiotics can be used to treat anthrax successfully. There is also a vaccine to prevent anthrax infections Table 39-6.

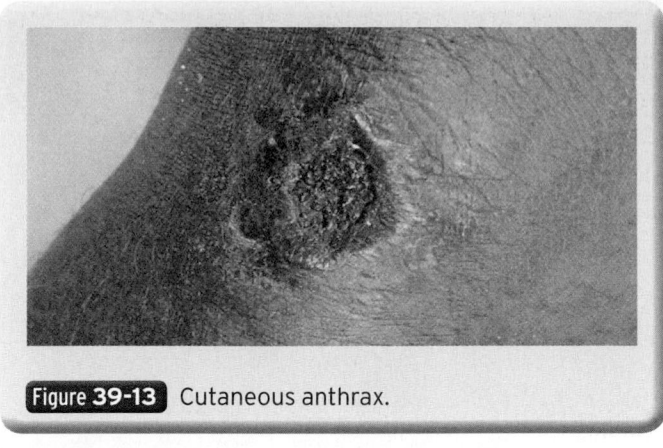

Figure 39-13 Cutaneous anthrax.

Plague (Bubonic/Pneumonic)

The 14th century plague that ravaged Asia, the Middle East, and finally Europe (the Black Death) killed an estimated 33 to 42 million people. Later on, in the early 19th century, almost 20 million people in India and China died due to plague. The plague's natural vectors are infected rodents and fleas. When a person is bitten by an infected flea or comes into contact with an infected rodent (or the waste of the rodent), the person can contract bubonic plague.

<u>Bubonic plague</u> infects the <u>lymphatic system</u> (a passive circulatory system in the body that bathes the tissues in lymph and works with the immune system). When this occurs, the patient's <u>lymph nodes</u> (area of the lymphatic system where infection-fighting cells are housed) become infected and grow. The glands of the nodes will grow large (up to the size of a tennis ball) and round, forming <u>buboes</u> Figure 39-14. If left untreated, the infection may spread through the body, leading to sepsis and possibly death. This form of plague is not contagious and is not likely to be seen in a bioterrorist incident.

Table 39-6 Characteristics of Anthrax

Dissemination	Aerosol
Communicability	Only in the cutaneous form (rare)
Route of entry	Through inhalation of spore or skin contact with spore or direct contact with skin wound (cutaneous)
Signs and symptoms	Flulike symptoms, fever, respiratory distress with tachycardia, shock, pulmonary edema, and respiratory failure after 3 to 5 days of flulike symptoms
Medical management	Pulmonary/inhalation: Standard precautions, oxygen, ventilatory support if in pulmonary edema or respiratory failure, and transport Cutaneous: Standard precautions, apply dry sterile dressing to prevent accidental contact with wound and fluids

You are the Provider: PART 4

On arriving at the scene, you are directed to the patients by the staging officer. They have been properly decontaminated and placed on a large tarp in a safe location. Patient No. 1 is conscious but disoriented; the front of her shirt is covered with blood and she is in obvious respiratory distress. Patient No. 2 is conscious and alert but is also experiencing respiratory distress; he has numerous lacerations and abrasions on his face. Patient No. 3 is conscious, alert, and ambulatory; he is holding his left arm against his body. Patient No. 4 is unconscious, breathing rapidly, and has burns to her face and neck.

7. On the basis of the number of patients and their apparent conditions, how many ambulances and EMTs should be present at the scene?

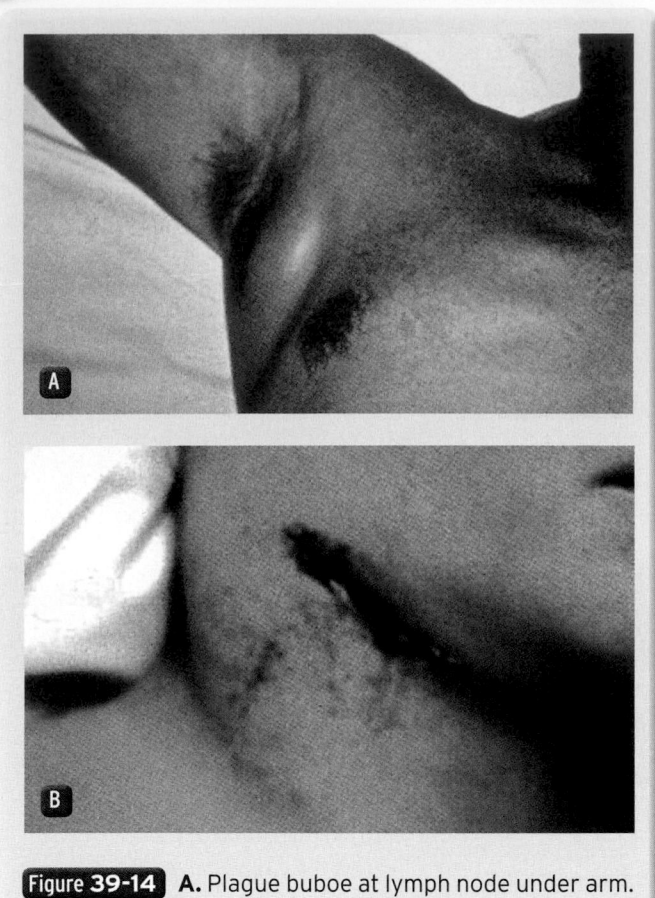

Figure 39-14   **A.** Plague buboe at lymph node under arm. **B.** Plague buboe at lymph node on neck.

Pneumonic plague is a lung infection, also known as plague pneumonia, that results from inhalation of plague bacteria. This form of the disease is contagious and has a much higher death rate than the bubonic form Table 39-7.

■ Neurotoxins

Neurotoxins are the most deadly substances known to humans. The strongest neurotoxin is 15,000 times more lethal than VX and 100,000 times more lethal than sarin. These toxins are produced from plants, marine animals, molds, and bacteria. The route of entry for these toxins is through ingestion, inhalation from aerosols, or injection. Unlike viruses and bacteria, neurotoxins are not contagious and have a faster onset of symptoms. Although these biologic toxins have immense destructive potential, they have not been used successfully as a WMD.

Botulinum Toxin

The most potent neurotoxin is **botulinum**, which is produced by bacteria. When introduced into the body,

this neurotoxin affects the nervous system's ability to function. Voluntary muscle control diminishes as the toxin spreads. Eventually the toxin causes muscle paralysis that begins at the head and face and travels downward throughout the body. The patient's accessory muscles and diaphragm will become paralyzed, and the patient will go into respiratory arrest Table 39-8.

Ricin

While not as deadly as botulinum, **ricin** is still five times more lethal than VX. This toxin is derived from mash that is left from the castor bean Figure 39-15. When introduced into the body, ricin causes pulmonary edema

Table 39-7   Characteristics of Plague

Dissemination	Aerosol
Communicability	Bubonic: low, only from contact with fluid in buboes Pneumonic: high, from person to person
Route of entry	Ingestion, inhalation, or cutaneous
Signs and symptoms	Fever, headache, muscle pain and tenderness, pneumonia, shortness of breath, extreme lymph node pain and enlargement (bubonic)
Medical management	Standard precautions, ABCs, provide oxygen, and transport

Table 39-8   Characteristics of Botulinum Toxin

Dissemination	Aerosol or food supply sabotage or injection
Communicability	None
Route of entry	Ingestion, inhalation
Signs and symptoms	Dry mouth, intestinal obstruction, urinary retention, constipation, nausea and vomiting, abnormal pupil dilation, blurred vision, double vision, drooping eyelids, difficulty swallowing, difficulty speaking, and respiratory failure as the result of paralysis
Medical management	ABCs, provide oxygen, and transport; ventilatory support in case of paralysis of the respiratory muscles; vaccine is available

Figure 39-15 These seemingly harmless castor beans contain the key ingredient for ricin, one of the most potent toxins known to humans.

and respiratory and circulatory failure leading to death **Table 39-9**.

The clinical picture depends on the route of exposure. The toxin is quite stable and extremely toxic by many routes of exposure, including inhalation. It is likely that 1 to 3 mg of ricin can kill an adult, and the ingestion of one seed can most likely kill a child.

Although all parts of the castor bean are actually poisonous, it is the seeds that are the most toxic. Castor bean ingestion causes a rapid onset of nausea, vomiting, abdominal cramps, and severe diarrhea, followed by vascular collapse. Death usually occurs on the third day in the absence of appropriate medical intervention.

Ricin is least toxic by the oral route. This is probably a result of poor absorption in the gastrointestinal tract, some digestion in the gut, and, possibly, some expulsion of the agent as caused by the rapid onset of vomiting. Ingestion causes local hemorrhage and necrosis of the liver, spleen, kidneys, and gastrointestinal tract. Signs and symptoms appear 4 to 8 hours after exposure.

Signs and symptoms of ricin ingestion are as follows:

- Fever
- Chills
- Headache
- Muscle aches
- Nausea
- Vomiting
- Diarrhea
- Severe abdominal cramping
- Dehydration
- Gastrointestinal bleeding
- Necrosis of the liver, spleen, kidneys, and gastrointestinal tract

Inhalation of ricin causes nonspecific weakness, cough, fever, hypothermia, and hypotension. Symptoms occur about 4 to 8 hours after inhalation, depending on the inhaled dose. The onset of profuse sweating some hours later signifies the termination of the symptoms.

Signs and symptoms of ricin inhalation are as follows:

- Fever
- Chills
- Nausea
- Local irritation of eyes, nose, and throat
- Profuse sweating
- Headache
- Muscle aches
- Nonproductive cough
- Chest pain
- Dyspnea
- Pulmonary edema
- Severe lung inflammation
- Cyanosis
- Seizures
- Respiratory failure

Treatment is supportive and includes both respiratory support and cardiovascular support as needed. Early intubation and ventilation, combined with treatment of pulmonary edema, are appropriate. Intravenous fluids and electrolyte replacement are useful for treating the dehydration caused by profound vomiting and diarrhea.

Table 39-10 summarizes the biologic agents.

Table 39-9	Characteristics of Ricin
Dissemination	Aerosol or contamination of a food or water supply by sabotage
Communicability	None
Route of entry	Inhalation, ingestion, injection
Signs and symptoms	Inhaled: cough, difficulty breathing, chest tightness, nausea, muscle aches, pulmonary edema, and hypoxia Ingested: nausea and vomiting, internal bleeding, and death Injection: no signs except swelling at the injection site and death
Medical management	ABCs, no treatment or vaccine available

Words of Wisdom

In a mass-casualty incident, it is important to frequently communicate with your patient. Remember that your patient is probably scared and does not know what is going on. By explaining to your patient any delays that are occurring, as well as the actions you are taking, you may alleviate the patient's fears.

It is also important to provide your patient with some type of protection in an effort to protect the patient from further harm. Whether it is building materials, backboards, or tarps, some type of material should be used to minimize the potential for further harm.

■ Other EMT Roles During a Biologic Event

Syndromic Surveillance

Syndromic surveillance is the monitoring, usually by local or state health departments, of patients presenting to emergency departments and alternative care facilities, the recording of EMS call volume, and monitoring the use of over-the-counter medications. Patients with signs and symptoms that resemble influenza are particularly important. Local and state health departments monitor for an unusual influx of patients with these symptoms in hopes of discovering an outbreak early. The EMS role in syndromic surveillance is a small one, yet is valuable in the overall tracking of a biologic terrorist event or infectious disease outbreak. Quality assurance and dispatch operations need to be aware of an unusual number of calls from patients with "unexplainable flu" coming from a particular region or community.

Figure 39-16 The Centers for Disease Control and Prevention Strategic National Stockpile can deliver one of many push packs to any location in the country within 12 hours of an emergency.

Points of Distribution (Strategic National Stockpile)

Points of distribution (PODs) are existing facilities that are established in a time of need, for the mass distribution of antibiotics, antidotes, vaccinations, and other medications and supplies. These medications may be delivered in large containers known as "push packs" by the Centers for Disease Control and Prevention National Pharmaceutical Stockpile **Figure 39-16**. These containers have a delivery time of 12 hours anywhere in the country and contain antibiotics, chemical antidotes, antitoxins, life-support medications, intravenous administration supplies, airway maintenance supplies, and medical/surgical items. In some regions, local and state municipalities have started to stockpile their own supplies to reduce the time delay.

EMTs, AEMTs, and paramedics may be called on to assist in the delivery of the medications to the public

Table 39-10 Biologic Agents

Disease	Transmission Person to Person	Incubation Period	Duration of Illness	Lethality (approximate case fatality rates)
Inhalation anthrax	No	1 to 6 d	3 to 5 d (usually fatal if untreated)	High
Pneumonic plague	High	2 to 3 d	1 to 6 d (usually fatal)	High unless treated within 12 to 24 h
Smallpox	High	7 to 17 d (average, 12 d)	4 wk	High to moderate
Viral hemorrhagic fevers	Moderate	4 to 21 d	Death between 7 and 16 d	High to moderate, depending on type of fever
Botulinum poisoning	No	1 to 5 d	Death in 24 to 72 h; lasts months if patient does not die	High without respiratory support
Ricin poisoning	No	18 to 24 h	Days; death within 10 to 12 d for ingestion	High

(depending on local emergency management planning). Your role may include triage, treatment of seriously ill patients, and patient transport to the hospital. Most plans for PODs include at least one ambulance on standby for the transport of seriously ill patients.

Radiologic/Nuclear Devices

What Is Radiation?

Ionizing radiation is energy that is emitted in the form of rays, or particles. This energy can be found in radioactive material, such as rocks and metals. Radioactive material is any material that emits radiation. This material is unstable, and it attempts to stabilize itself by changing its structure in a natural process called decay. As the substance decays, it gives off radiation, until it stabilizes. The process of radioactive decay can take from as little as minutes to billions of years; meanwhile, the substance remains radioactive.

The energy that is emitted from a strong radiologic source is alpha, beta, gamma (x-ray), or neutron radiation. Alpha is the least harmful penetrating type of radiation and cannot move through most objects. In fact, a sheet of paper or the body's skin easily stops it. Beta radiation is slightly more penetrating than alpha and requires a layer of clothing to stop it. Gamma rays are far faster and stronger than alpha and beta rays. These rays easily penetrate through the human body and require lead or several inches of concrete to prevent penetration. Neutron particles are among the most powerful forms of radiation. Neutrons easily penetrate through lead and require several feet of concrete to stop them Figure 39-17 .

Sources of Radiologic Material

There are thousands of radioactive materials found on the earth. These materials are generally used for purposes that benefit humankind, such as medicine, killing germs in food (irradiating), and construction work. Once radiologic material has been used for its purpose, the material remaining is called radiologic waste. Radiologic waste remains radioactive but has no more usefulness. These materials can be found at the following locations:

- Hospitals
- Colleges and universities
- Chemical and industrial sites

Not all radioactive material is tightly guarded, and the waste is often not guarded. This makes use of radioactive material and substances appealing to terrorists.

Radiologic Dispersal Devices (RDD)

A radiologic dispersal device (RDD) is any container that is designed to disperse radioactive material. This would generally require the use of a bomb, hence the nickname "dirty bomb." A dirty bomb carries the potential to injure victims with not only the radioactive material, but also the explosive material used to deliver it. Just the thought of an RDD creates fear in a population, and so the ultimate goal of some terrorists—fear—is accomplished. In reality, however, the destructive capability of a dirty bomb is limited to the explosives that are attached to it. Therefore, if the explosive is sufficient to kill 10 persons without radioactive material, it will also kill 10 persons with the radioactive material added. There may be long-term injuries and illness associated with the use of an RDD, yet not much more than the bomb by itself would create. In short, the dirty bomb is an ineffective WMD.

Nuclear Energy

Nuclear energy is artificially made by altering (splitting) radioactive atoms. The result is an immense amount of energy that usually takes the form of heat.

You are the Provider: PART 5

The patients have been triaged and appropriately treated. After the transportation officer notifies the receiving facilities about the patients' injuries and their exposure to radiation, they are transported. After decontaminating yourselves and the ambulance, you and your crew discuss the incident, including terrorism in general.

8. What type of terrorist group was most likely responsible for this incident?

9. What level of knowledge of terrorism and weapons of mass destruction should the EMT possess?

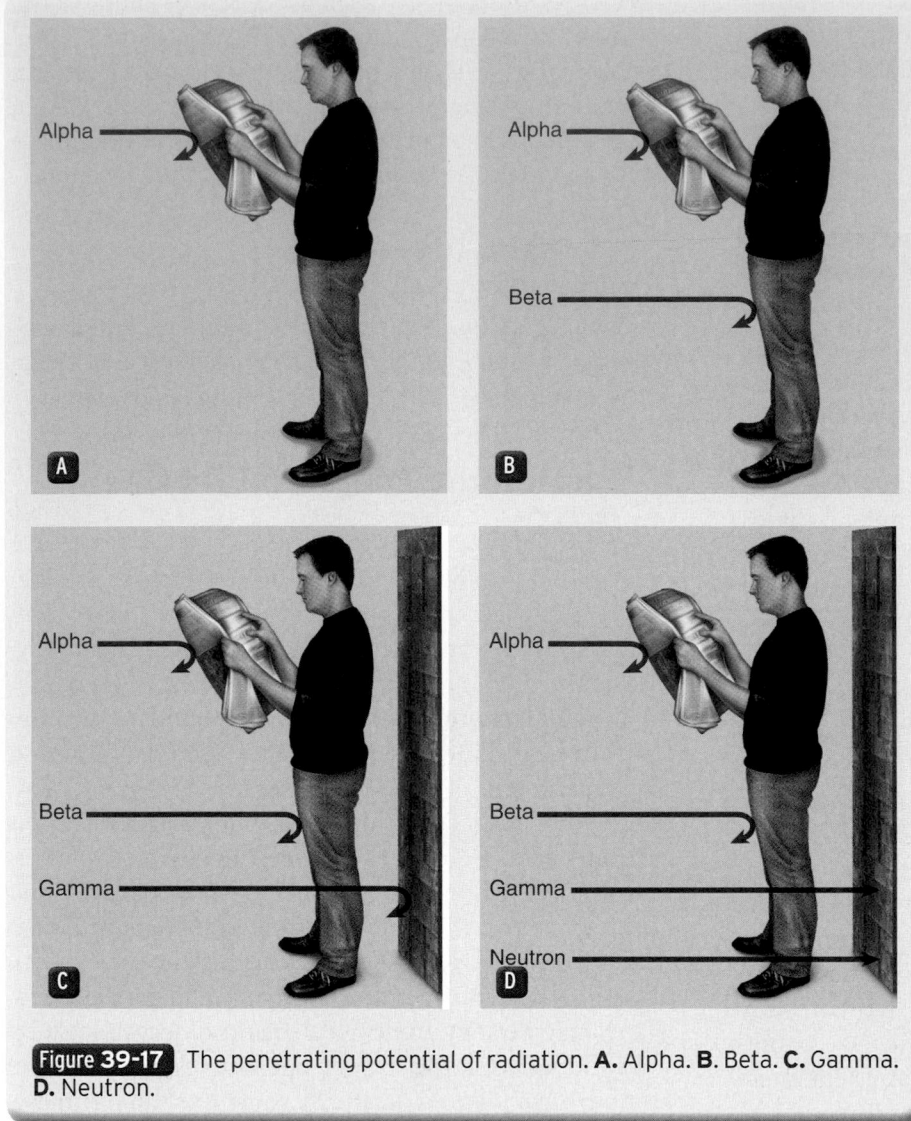

Figure 39-17 The penetrating potential of radiation. **A.** Alpha. **B.** Beta. **C.** Gamma. **D.** Neutron.

the likelihood of a nuclear attack is extremely remote.

Unfortunately, however, due to the collapse of the former Soviet Union, the whereabouts of many small nuclear devices is unknown. These small suitcase-sized nuclear weapons are called **Special Atomic Demolition Munitions (SADM)**. The SADM, or "suitcase nuke," was designed to destroy individual targets, such as important buildings, bridges, tunnels, and large ships. The estimate is that perhaps as many as 80 are missing as of 1998. No other information or updates on the whereabouts of these devices have been made public.

■ Symptomatology

The effects of radiation exposure will vary depending on the amount of radiation that a person receives and the route of entry. Radiation can be introduced into the body by all routes of entry as well as through the body (irradiation). The patient can inhale radioactive dust from nuclear fallout or from a dirty bomb or have radioactive liquid absorbed into the body through the skin. Once in the body, the radiation source will irradiate the person from within rather than from an external source (such as x-ray equipment). Some common signs of acute radiation sickness are listed in **Table 39-11**. Additional injuries will occur with a nuclear blast such as thermal and blast trauma, trauma from flying objects, and eye injuries.

Nuclear material is used in medicine, weapons, naval vessels, and power plants. Nuclear material gives off all forms of radiation, including neutrons (the most deadly type). Like radioactive material, when nuclear material is no longer useful it becomes waste that is still radioactive.

■ Nuclear Weapons

The destructive energy of a nuclear explosion is unlike any other weapon in the world. That is why nuclear weapons are kept only in secure facilities throughout the world. There are nations that have ties to terrorists and that have actively attempted to build nuclear weapons. Yet the ability of these nations to deliver a nuclear weapon, such as a missile or bomb, is as yet, incomplete. There is also the deterrent of complete mutual annihilation. Therefore,

Table 39-11	Common Signs of Acute Radiation Sickness
Low exposure	Nausea, vomiting, diarrhea
Moderate exposure	First-degree burns, hair loss, depletion of the immune system (death of white blood cells), and cancer
Severe exposure	Second- and third-degree burns, cancer, and death

Medical Management

Being exposed to a radiation source does not make a patient contaminated or radioactive. However, when patients have a radioactive source on their body (such as debris from a dirty bomb), they are contaminated and must be initially cared for by a HazMat responder. Once the patient is decontaminated and there is no threat to you, you may begin treatment with the ABCs and treat the patient for any burns or trauma.

Protective Measures

There are no suits or protective gear designed to completely shield you from radiation. The people who work in high-risk areas wear some protection (lead-lined suits); however, this equipment is not available to EMTs. The best ways to protect yourself from the effects of radiation are to use time and distance and shield yourself using buildings and walls for protection. Do not enter a HazMat area unless you are trained as a HazMat responder and have proper training in the use of self-contained breathing apparatus.

- **Time.** Radiation has a cumulative effect on the body. The less time that you are exposed to the source, the less the effects will be. If you realize that the patient is near a radiation source, leave the area immediately.
- **Distance.** Radiation is limited as to how far it can travel. Depending on the type of radiation, often moving only a few feet is enough to remove you from immediate danger. Alpha radiation cannot travel more than a few inches but gamma rays can travel hundreds or thousands of meters. You should take this into account when responding to a nuclear or radiologic incident and make certain that responders are stationed far enough from the incident.
- **Shielding.** As discussed earlier, the path of all radiation can be stopped by a specific object. It will be impossible for you to recognize the type of radiation being emitted or even from which direction it is coming. Therefore, you should always assume that you are dealing with the strongest form of radiation and use concrete shielding (such as buildings or walls) between yourself and the incident. The importance of shielding cannot be overemphasized.

Incendiary and Explosive Devices

Incendiary and explosive devices come in various shapes and sizes. Although you are not tasked with recognizing all of the possible types of explosive devices, including improvised explosive devices (IED), it is important for you to be able to identify an object you believe is a potential device, notify the proper authorities, and safely evacuate the area. Always remember that there is the possibility of a secondary device when you are responding to the scene of an incendiary or explosive device call.

Mechanisms of Injury

The type and severity of wounds sustained from incendiary and explosive devices primarily depend on the patient's distance from the epicenter of the explosion. Patients close to the epicenter of the explosion are likely to suffer from all wound-causing agents of the munitions. Patients who are farther away from the epicenter are likely to experience a combination of blast injuries from the explosion and penetrating trauma injuries from primary and secondary projectiles created by the explosion.

Blast injuries occur in a number of ways.

- **Primary blast injury**. Due solely to the direct effects of the pressure wave on the body. The injury from the primary blast is seen almost exclusively in the hollow organs of the body—the lungs, intestines, and inner ears. An injury to the lungs causes the greatest morbidity and mortality.
- **Secondary blast injury**. Penetrating or nonpenetrating injury that results from being struck by flying debris, such as ordnance projectiles or secondary missiles, that has been set in motion by the explosion. Objects are propelled by the force of the blast and strike the victim, causing injury.
- **Tertiary blast injury**. Results from whole body displacement and subsequent traumatic impact with environmental objects (eg, trees, buildings, and vehicles). Other indirect effects include crush injury from the collapse of structures (buildings, bunkers, or tunnels) and toxic effects from the inhalation of combustion gases.

The Physics of an Explosion

When a substance is detonated, a solid or liquid is chemically converted into large volumes of gas under high pressure with resultant explosive energy release. Propellants, like gunpowder, are explosives designed to release energy relatively slowly compared with high energy explosives, which are designed to detonate very quickly. This generates a pressure pulse in the shape of a spherical blast wave that expands in all directions from the point of explosion. Flying debris and high winds commonly cause conventional blunt and penetrating trauma.

Tissues at Risk

Hollow organs such as the middle ear, lung, and gastrointestinal tract are most susceptible to pressure changes. The junction between tissues of different densities and exposed tissues such as the head and neck are prone to injury as well. The ear is the organ system most sensitive to blast injuries. The patient may complain of ringing or pain in the ears or some loss of hearing, and blood may be visible in the ear canal. Permanent hearing loss is possible.

Primary **pulmonary blast injuries** occur as contusions and hemorrhages. When the explosion occurs in an open space, the patient's side that is toward the explosion is usually injured, but the injury can be bilateral when the patient is located in a confined space. The patient may complain of tightness or pain in the chest and may cough up blood and have tachypnea or other signs of respiratory distress. Subcutaneous emphysema (crackling under the skin) over the chest can be palpated, indicating air in the thorax. Pneumothorax is common and may require emergency decompression.

Solid organs are relatively protected from shock-wave injury but may be injured by secondary missiles or a hurled body. Hollow organs, however, may be injured by similar mechanisms as lung tissue. Petechiae, or pinpoint hemorrhages that show up on the skin, to large hematomas are the most visible sign.

Neurologic injuries and head trauma are the most common causes of death from blast injuries. Subarachnoid (beneath the arachnoid layer covering the brain) and subdural (beneath the outermost covering of the brain) hematomas are often seen. Permanent or transient neurologic deficits may be secondary to concussion, intracerebral bleeding, or air embolism. Instant but transient unconsciousness, with or without retrograde amnesia, may be initiated not only by head trauma, but also by cardiovascular problems. Bradycardia and hypotension are common after an intense pressure wave from an explosion. This is a vagal nerve–mediated form of cardiogenic shock without compensatory vasoconstriction (for example, vasovagal syncope).

Extremity injuries, including traumatic amputations, are common. Other injuries are often associated with tertiary blasts. Patients with traumatic amputation are likely to sustain fatal injuries secondary to the blast.

You are the Provider: SUMMARY

1. On the basis of the dispatch information, how should you approach this incident?

The fact that a vehicle has crashed into an abortion clinic and exploded should raise your index of suspicion that this incident is the result of terrorism because, in most cases, vehicles rarely "explode" when they crash into something. When explosions do occur, they are often the result of a ruptured gas line or gas tank combined with sparks created during the crash that cause the vehicle to catch fire. In this incident, the explosion could have been meant to disperse a weaponized chemical.

Many of the same fundamental principles that apply to an accidental hazardous materials incident apply to suspected acts of terrorism. If you suspect that a terrorist event has occurred—whether there is evidence that a weapon of mass destruction (WMD) has been used or not—there are certain actions you need to take to ensure that you and your team are safe and in the proper position to take care of any casualties.

During your response, take a route that is uphill and upwind from the incident scene, keeping in mind that wind direction and speed can change very quickly. If you do not have access to information about the wind direction and speed, ask the dispatcher for a current weather report.

Do not enter the scene where the incident has occurred; stage your ambulance in a safe location—again, upwind and uphill—and wait for further information from the personnel who are trained in assessing and managing WMD scenes.

The incident command system (ICS) is an integral part of any scene in which there are multiple casualties—whether the incident was natural or manmade. If the ICS has already been established, wait in the staging area until you receive further directions from the incident commander. At a disaster scene, EMS personnel are usually assigned duties such as medical branch officers, triage officers, treatment officers, transportation officers, or general staff.

Avoid unnecessary use of the radio; monitor the radio traffic between the dispatcher and responding fire, hazardous materials (HazMat), and law enforcement personnel to receive the most up-to-date information regarding the incident. When trained personnel have properly assessed the scene, they will let you know when to proceed and where to go. Communication and coordination among *all* responders is critical.

2. What indicators suggest that an incident is the result of terrorism?

Most terrorist attacks are covert, which means that public safety personnel and the general community have no prior knowledge of the time, location, or nature of the attack. This element of surprise creates a higher degree of danger and complexity when responding to an incident.

You are the Provider: SUMMARY, continued

EMS personnel must be constantly aware of their surroundings and understand the possible risks for terrorism. This process begins with you knowing the current terrorist threat level, as established by the Department of Homeland Security (DHS) color-coded advisory system. The DHS has not issued specific recommendations for EMS personnel to follow in response to the current threat level; therefore, you must follow your local protocols, policies, and procedures. Many EMS systems post the current threat level on a bulletin board at their facility; it is your responsibility to be aware of this level when you begin your shift. However, remember that while the DHS advisory system is designed to increase awareness, it should not be your sole source of information. There are other preincident indicators that suggest terrorist activity. For example, has there been an increase in recent political activism, domestic or abroad? Are you aware of any credible threats made against a particular location, gathering, or occasion?

As with any call, you must *carefully listen* to the dispatch information. In addition to preincident indicators, the location and type of incident, the number of patients, and victims' statements should be carefully noted.

For example, is the incident location a monument, government building, or a specific structure such as a temple, church, or abortion clinic? Is there a large gathering or special event taking place? Is there a report of an explosion or a suspicious item or device near the incident?

The number of patients and their symptoms is perhaps the single most important clue that a terrorist attack or an incident involving a WMD has occurred. Although multiple patients, In the same location, with similar symptoms could indicate carbon monoxide poisoning, it could just as easily signify the presence of a WMD.

Information provided to the dispatcher by the caller regarding victims' statements at the scene is another reliable indicator for you to take note of. Victims who are fleeing the scene and making statements, such as "Everyone is passing out," or "There was a loud explosion," should heighten your suspicion that a terrorist event has occurred.

The indicators discussed above represent some of the more common and reliable clues. You must remain alert at *any* scene and report any suspicious activity or findings to the appropriate authorities.

3. On the basis of the HazMat team's findings, what has most likely occurred?

The presence of radiation in the area following an explosion indicates that a radiologic dispersal device (RDD) was used. An RDD combines a conventional explosive, such as dynamite, with radioactive material; this is why an RDD is commonly referred to as a "dirty bomb."

An RDD, while harmful, does not compare to a nuclear bomb, which is capable of creating an explosion that is millions of times more powerful. Furthermore, a nuclear bomb is capable of spreading radiation over hundreds of square miles, whereas an RDD would likely only spread radiation within a few blocks or miles of the explosion. As radioactive material spreads, it becomes less concentrated and less harmful.

The destructive power of an RDD is limited to the explosive that is attached to it; a bomb that is powerful enough to kill 10 people without radioactive material will also kill 10 people when the radioactive material is added. In short, the explosive device used would be more harmful than the radioactive material.

Injury and death caused by an RDD are the result of the explosive pressure wave and blunt and penetrating trauma—just like any other explosion. The addition of radiation, although not likely to cause immediate health problems, could result in people experiencing long-term injury or illness depending on the type and amount of radiation involved, the means of exposure (eg, absorption, inhalation, or ingestion), and the length of exposure.

To date, explosive devices have been the preferred WMD for terrorists. Terrorist groups have favored tactics such as truck bombs or car or pedestrian suicide bombers. Compared with other WMDs (eg, chemical or biologic agents), explosive devices are easy to manufacture. Furthermore, not all radioactive material is tightly guarded, and radioactive waste, such as what may be found in hospitals, universities, and industrial plants, is often not guarded at all. This makes radioactive material quite easy for a terrorist to acquire.

The thought of an RDD incites panic and fear in a population, thus allowing some terrorists to achieve their ultimate goal. But to look at the event in another way, while contamination is the main threat posed by the amount of radiation released by an RDD, the actual bomb used to disperse it is far more deadly. The RDD is more a weapon of mass "disruption" than it is a weapon of mass destruction. To date, no dirty bombs have been detonated by terrorists in the United States.

4. Is it safe for you and other EMS personnel to enter the scene? If not, what should be done first?

The HazMat team has identified radiation in the area of the incident; however, you have not received information regarding what has been done to make the scene safe for you to enter. Therefore, *do not* enter the scene until you have been given specific instructions by the HazMat team!

Exposure to a radiation source does not make a patient contaminated or radioactive; however, if the patient has a radioactive source on his or her body, such as debris from a dirty bomb, the patient is contaminated and must be initially decontaminated by HazMat personnel.

HazMat teams use a Geiger counter to detect ionizing radiation—usually beta particles and gamma rays. However, some models can also detect alpha particles. After the HazMat team obtains this information, it will use it to determine the appropriate level of personal protective equipment (PPE) that should be donned; the team will also move the patients to an area that is safe for you to access and treat them.

You are the Provider: SUMMARY, continued

Time, distance, and shielding are the best ways to protect yourself from the effects of radiation. Radiation has a cumulative effect on the body; the less time you are exposed to the source, the less the effects will be. Radiation is also limited as to how far it can travel; thus, the type of radiation present will determine what is considered a safe distance. Depending on the type of radiation, its path can be stopped by a specific object. Alpha rays, the least harmful type of radiation, cannot travel through most objects; a sheet of paper or the skin easily stops them. Beta radiation is slightly more penetrating than alpha rays and usually require a layer of clothing to stop them. Gamma rays are far faster and stronger than alpha and beta rays; these rays easily penetrate the human body and require lead or several inches of concrete to prevent penetration of the body. Neutrons easily penetrate through lead and require several feet of concrete to stop them.

Although it is important to understand the different types of ionizing radiation and what it takes to protect yourself from exposure, it is far more important to rely on specialized teams to determine the type and amount of radiation present and to follow their instructions. Once the patients have been properly decontaminated and moved to a safe area, you will be allowed to begin the processes of assessment and treatment.

5. Given the situation, what are some unique concerns about this incident?

The size and complexity, as well as a variety of other factors, can rapidly change at any incident; however, you should always remain especially vigilant when functioning at the scene of a terrorist incident.

Terrorists have been known to plant additional explosives (secondary devices) that are set to detonate after the initial device. Secondary devices, which are primarily intended to injure emergency responders and secure media coverage, may not be in the same location as the primary device. The secondary device could be a package, briefcase, or other container that has been planted across the street or in an adjacent building, or it could be an electronic device, such as a pager or mobile phone, that is set to detonate when "answered."

Although the HazMat team has secured the scene initially, you must realize that the scene can easily become unsafe. Do not rely on others to keep you safe; it is *your* responsibility to constantly reassess the scene for indicators of danger. A subtle change in wind direction during an incident involving radiation or an increase in the number of patients who are contaminated can place you in danger.

Treat your patients accordingly, but do not get so absorbed in what you are doing that you do not look around to ensure that the scene remains safe. Retreat to a place of safety immediately if the scene becomes unsafe.

6. What types of injuries should you expect to encounter?

Although the victims may experience the effects of radiation exposure (eg, skin burns), depending on the amount of radiation they were exposed to, their most immediately life-threatening injuries will be the result of the explosion itself.

The severity of the injuries during an explosion depends primarily on the location of the victims relative to the epicenter of the blast; the closer they are to the explosion, the more severe their injuries will be.

The victims who were in the building and closest to the blast will experience all of the effects of the explosion—severe burns, blunt and penetrating trauma, crush injuries if the building collapsed, and inhalation injury, among others.

People who are injured during an explosion can by injured by any one of four mechanisms: primary blast injuries, secondary blast injuries, tertiary blast injuries, and miscellaneous injuries.

Primary blast injuries are caused exclusively by the blast itself. The pressure wave that is generated by the explosion causes injury to the hollow organs, such as the lungs, stomach, and intestines. Rupture of the tympanic membrane (eardrum) is also a common primary blast injury. Injury to the lung is usually the most life-threatening consequence of the primary blast. This injury occurs because of the sudden change in intrathoracic pressure (barotrauma), resulting in injuries such as a pneumothorax. Because primary blast injuries are usually internal, they can easily be missed if a careful patient assessment is not performed.

Secondary blast injuries include penetrating and nonpenetrating trauma that occurs when the victim is struck by shrapnel, glass, and other debris that is set in motion by the explosion. These objects can travel great distances and be propelled at tremendous speeds.

Tertiary blast injuries occur when the victim is propelled by the force of the explosion and strikes a stationary object, such as a tree, building, or vehicle. Injuries experienced during this phase can range from fractures to blunt trauma to multiple body systems.

Other injuries that may occur during an explosion include burns from hot gases or fires that are started by the blast and respiratory injury from inhaling toxic gases.

Most victims who survive an explosion will have some combination of the four types of injury patterns previously discussed. Furthermore, it is important to note that life-threatening injuries can occur during any phase of the explosion. For example, a person standing some distance from the explosion can easily sustain a penetrating injury to the chest, head, or abdomen when he or she is struck by flying debris.

7. On the basis of the number of patients and their apparent conditions, how many ambulances and EMTs should be present at the scene?

Your general impression of the scene has revealed that there are four patients—one of whom appears to be unconscious. Even though the other three patients are conscious, this does not mean that they do not have life-threatening injuries.

Therefore, you should assume that all four patients have critical injuries until they have been triaged.

One ambulance and two EMTs can effectively care for only one critically injured patient at a time. Therefore there should be a total of four ambulances and eight EMTs at the scene—two EMTs and one ambulance per patient.

As soon as you determine that you have more patients that you can effectively manage, your first priority is to call for additional resources. When you were staging at the scene, the HazMat team informed you that there were a total of four patients; however, the severity of their injuries was unknown at that time. Unless there were already ample resources at the staging area, this is when you should have summoned the additional teams.

Once you arrive at the scene and triage the patients, you can determine whether additional resources are needed. In some cases, two EMTs may be needed in the back of the ambulance to care for one critically injured patient; even if the additional ambulances are not needed, you may still need the additional EMTs.

8. What type of terrorist group was most likely responsible for this incident?

Not all terrorists resort to violence to further their agenda; however, they all have the same ultimate goal—to get their way through fear. Terrorist organizations are generally categorized as follows: violent religious groups/doomsday cults, extremist political groups, technology terrorists (eg, cyber-terrorists), and single-issue groups. Some terrorists may act alone, although most attacks require coordination of multiple terrorists or "actors" working together.

In this incident, a single actor crashed his vehicle into an abortion clinic and detonated a dirty bomb in the process. This behavior is indicative of a single-issue terrorist; he clearly felt strongly enough about his cause to destroy a clinic, release radiation to induce further injury, and take his own life in the process.

Single-issue terrorist groups threaten or use violence as a means to draw attention to their cause; they include antiabortion groups, animal rights groups, anarchists, racists, and ecoterrorists (use of terrorism as a means of protecting the environment).

Abortion has been a topic of great debate for many years, and terrorism has been used by antiabortion groups in the past. Several abortion clinics were bombed in Atlanta; other attacks have occurred when a gunman opened fire on clinic employees.

9. What level of knowledge of terrorism and weapons of mass destruction should the EMT possess?

EMS personnel are not expected to be experts in terrorism and WMDs. However, they are expected to be aware of the various types of threats—just like any other citizen—and recognize certain indicators of terrorism when responding to an incident, such as those discussed earlier in this case study.

As with any hazardous material, EMTs should possess a basic working knowledge of the different types of WMDs that terrorists could use; their respective signs and symptoms and treatment; measures to take to ensure personal safety and safety of the patient; and knowledge of the incident command system (ICS), which should be implemented at any disaster—natural or manmade.

The various types of agents that could be used as WMDs can be remembered by the mnemonic "B-NICE," which stands for Biologic, Nuclear, Incendiary, Chemical, and Explosive. Refer to the chapter for specific information regarding these potential WMDs.

Biologic agents are organisms—viruses, bacteria, toxins—that cause disease and are generally found in nature. However, terrorists may cultivate, synthesize, and mutate them in a laboratory to artificially maximize a target population's exposure (weaponization). Examples of biologic agents include smallpox, viral hemorrhagic fevers (eg, Ebola, Rift Valley, and yellow fever), the bacteria that causes anthrax and plague, and ricin.

Nuclear or radiologic agents and devices can range from an RDD or dirty bomb to a nuclear bomb capable of producing apocalyptic destruction and death.

Incendiary devices are used for the purpose of starting a fire—for example, a Molotov cocktail. Explosive devices are also used to start fires, while at the same time causing injury, death, and destruction from the explosive wave (eg, blast injuries).

Chemical agents are manufactured substances that can have devastating effects on living organisms. They can be produced in liquid, powder, or vapor form depending on the desired route of exposure and dissemination technique. Examples of chemical agents include vesicants, or blister agents (eg, sulfur mustard [agent H], Lewisite [L], phosgene oxime [CX]); pulmonary, or choking agents (eg, chlorine [CL], phosgene); nerve agents (eg, sarin [GB], soman [GD], tabun [GA], V agent [VX]); and metabolic agents (eg, cyanide).

You are the Provider: SUMMARY, continued

Patient No. 1

Triage Tag
No. 240351

Move the Walking Wounded	MINIMAL
No respirations after head tilt	EXPECTANT
☐ Respirations–over 30	IMMEDIATE
☐ Perfusion–capillary refill over 2 seconds	IMMEDIATE
☒ Mental status–unable to follow simple commands	IMMEDIATE
Otherwise	DELAYED

MAJOR INJURIES: _Open chest and abdominal injuries_
HOSPITAL DESTINATION: _Harbor Bay Trauma Center_
ORIENTED × DISORIENTED ☒ UNCONSCIOUS ☐

TIME	PULSE	B/P	RESPIRATION
1025	118	80/60	26 (labored)
1030	110	94/64	24 (labored)

PERSONAL INFORMATION:
NAME: _Lisa Malone_
MALE ☐ FEMALE ☒   AGE: 41   WEIGHT: 125 lb
MEDICAL COMPLAINTS/HISTORY
Penetrating injuries to chest and abdomen; shard of glass impaled in chest; abdominal evisceration; history of high blood pressure

EXPECTANT No 240351

IMMEDIATE No 240351

Patient No. 2

Triage Tag
No. 240352

Move the Walking Wounded	MINIMAL
No respirations after head tilt	EXPECTANT
☐ Respirations–over 30	IMMEDIATE
☒ Perfusion–capillary refill over 2 seconds	IMMEDIATE
☐ Mental status–unable to follow simple commands	IMMEDIATE
Otherwise	DELAYED

MAJOR INJURIES: _Possible pneumothorax, deformed left femur_
HOSPITAL DESTINATION: _Harbor Bay Trauma Center_
ORIENTED × 4 DISORIENTED ☐ UNCONSCIOUS ☐

TIME	PULSE	B/P	RESPIRATION
1026	112	100/58	28 (labored)
1031	118	98/60	28 (labored)

PERSONAL INFORMATION:
NAME: _Stanley Green_
MALE ☒ FEMALE ☐   AGE: 34   WEIGHT: 165 lb
MEDICAL COMPLAINTS/HISTORY
Difficulty breathing; multiple abrasions and lacerations; left femur deformity; no medical history

EXPECTANT No 240352

IMMEDIATE No 240352

Patient No. 3

Triage Tag
No. 240353

Move the Walking Wounded	**MINIMAL**
No respirations after head tilt	**EXPECTANT**
☐ Respirations–over 30	**IMMEDIATE**
☐ Perfusion–capillary refill over 2 seconds	**IMMEDIATE**
☐ Mental status–unable to follow simple commands	**IMMEDIATE**
(Otherwise)	**DELAYED**

MAJOR INJURIES: _Possible ruptured eardrums, left forearm deformity_
HOSPITAL DESTINATION: _Harbor Bay Trauma Center_
ORIENTED × 4   DISORIENTED ☐   UNCONSCIOUS ☐

TIME	PULSE	B/P	RESPIRATION
1030	100	132/92	14
1040	98	128/88	14

PERSONAL INFORMATION:
NAME: _Brett Lackey_
MALE ☒   FEMALE ☐   AGE: 50   WEIGHT: 180 lb
MEDICAL COMPLAINTS/HISTORY
Difficulty hearing; left arm deformity; minor abrasions to arms and face; history of depression

EXPECTANT   No 240353

IMMEDIATE   No 240353

DELAYED   No 240353

Patient No. 4

Triage Tag
No. 240354

Move the Walking Wounded	**MINIMAL**
No respirations after head tilt	**EXPECTANT**
☒ Respirations–over 30	**IMMEDIATE**
☐ Perfusion–capillary refill over 2 seconds	**IMMEDIATE**
☐ Mental status–unable to follow simple commands	**IMMEDIATE**
Otherwise	**DELAYED**

MAJOR INJURIES: _Closed head injury, facial burns_
HOSPITAL DESTINATION: _Harbor Bay Trauma Center_
ORIENTED ×   DISORIENTED ☐   UNCONSCIOUS ☒

TIME	PULSE	B/P	RESPIRATION
1027	60	166/100	34
1032	66	170/98	32

PERSONAL INFORMATION:
NAME: _Georgia Wayland_
MALE ☐   FEMALE ☒   AGE: 30   WEIGHT: 120 lb
MEDICAL COMPLAINTS/HISTORY
Large hematoma to back of head; partial-thickness burns to face and neck; unknown medical history

EXPECTANT   No 240354

IMMEDIATE   No 240354

Prep Kit

- As a result of the increase in terrorist activity, it is possible that you, the EMT, could witness a terrorist event. You must be mentally and physically prepared for the possibility of a terrorist event.

- Types of groups that tend to use terrorism include violent religious groups/doomsday cults, extremist political groups, technology terrorists, and single-issue groups.

- A weapon of mass destruction (WMD) is any agent designed to bring about mass death, casualties, and/or massive damage to property and infrastructure (bridges, tunnels, airports, and seaports). These can be biologic, nuclear, incendiary, chemical, and explosive weapons (B-NICE).

- Indicators that may give you clues as to whether the emergency is the result of a terrorist attack include the type of location, type of call, number of patients, victims' statements, and preincident indicators.

- If you suspect that a terrorist or a weapon of mass destruction event has occurred, ensure that the scene is safe. If you have any doubt that it may not be safe, do not enter. Wait for assistance.

- Terrorists may set secondary devices that are designed to explode after the initial bomb, thus injuring responders and media coverage. Constantly assess and reassess the scene for safety.

- Chemical agents are manufactural substances that can have devastating effects on living organisms.

- The route of exposure is how the agent most effectively enters the body.

- Biologic agents are organisms that cause disease.

- Biologic agents include viruses such as smallpox and those that cause viral hemorrhagic fevers; bacteria such as those that cause anthrax and plague; and neurotoxins such as botulinum toxin and ricin.

- Nuclear or radiologic weapons can create a massive amount of destruction.

- Ionizing radiation is energy that can enter the human body and cause damage.

- Explosive and incendiary devices come in various shapes and sizes. It is important to be able to identify an object you believe is a potential device and notify the proper authorities, while safely evacuating the area.

Vital Vocabulary

alpha A type of energy that is emitted from a strong radiologic source; it is the least harmful penetrating type of radiation and cannot travel fast or through most objects.

anthrax A disease caused by deadly bacteria (*Bacillus anthracis*) that lay dormant in a spore (protective shell); the germ is released from the spore when exposed to the optimal temperature and moisture. The routes of entry are inhalation, cutaneous, and gastrointestinal (from consuming food that contains spores).

aging Aging is the process by which the temporary bond between the organophosphate and acetylcholinesterase undergoes hydrolysis, resulting in a permanent covalent bond.

bacteria Microorganisms that reproduce by binary fission. These single-cell creatures reproduce rapidly. Some can form spores (encysted variants) when environmental conditions are harsh.

beta A type of energy that is emitted from a strong radiologic source; is slightly more penetrating than alpha and requires a layer of clothing to stop it.

B-NICE A memory device to recall the types of weapons of mass destruction: biologic, nuclear, incendiary, chemical, and explosive.

botulinum Produced by bacteria, this is a very potent neurotoxin. When introduced into the body, this neurotoxin affects the nervous system's ability to function and causes botulism.

buboes Enlarged lymph nodes (up to the size of a tennis ball) that were characteristic in people infected with the bubonic plague.

bubonic plague An epidemic that spread throughout Europe in the Middle Ages, causing more than 25 million deaths, also called the Black Death; transmitted by infected fleas and characterized by acute malaise, fever, and the formation of tender, enlarged, inflamed lymph nodes that appear as lesions, called buboes.

chlorine (CL) The first chemical agent ever used in warfare. It has a distinct odor of bleach and creates a green haze when released as a gas. Initially it produces upper airway irritation and a choking sensation.

contact hazard A hazardous agent that gives off very little or no vapors; the skin is the primary route for this type of chemical to enter the body; also called a skin hazard.

contagious An infectious disease that can be transmitted to another; communicable. A person who has a contagious disease and can transmit it to another person might be considered "contagious."

covert An act in which the public safety community generally has no prior knowledge of the time, location, or nature of the attack.

cross-contamination Occurs when a person is contaminated by an agent as a result of coming into contact with another contaminated person.

cyanide An agent that affects the body's ability to use oxygen. It is a colorless gas that has an odor similar to almonds. The effects begin on the cellular level and are very rapidly seen at the organ and system levels.

decay A natural process in which a material that is unstable attempts to stabilize itself by changing its structure.

dirty bomb Name given to a bomb that is used as a radiologic dispersal device.

disease vector An animal that spreads a disease, once infected, to another animal.

dissemination The means by which a terrorist will spread a disease, for example, by poisoning the water supply or aerosolizing the agent into the air or ventilation system of a building.

DuoDote Auto-Injector A nerve agent antidote kit containing atropine and pralidoxime chloride; delivered as a single dose through one needle.

G agents Early nerve agents that were developed by German scientists in the period after World War I and into World War II. There are three such agents: sarin, soman, and tabun.

gamma (x-ray) A type of energy that is emitted from a strong radiologic source that is far faster and stronger than alpha and beta rays. These rays easily penetrate through the human body and require lead or several inches of concrete to prevent penetration.

incubation The period of time from a person being exposed to a disease to the time when symptoms begin.

international terrorism Terrorism that is carried out by people in a country other than their own; also known as cross-border terrorism.

ionizing radiation Energy that is emitted in the form of rays, or particles.

LD_{50} The amount of an agent or substance that will kill 50% of people who are exposed to this level.

Lewisite (L) A blistering agent that has a rapid onset of symptoms and produces immediate, intense pain and discomfort on contact.

lymph nodes The area of the lymphatic system where infection-fighting cells are housed.

lymphatic system A passive circulatory system that transports a plasmalike liquid called lymph, a thin fluid that bathes the tissues of the body.

Mark 1 Nerve Agent Antidote Kit (NAAK) A nerve agent antidote kit containing two auto-injectors containing atropine and pralidoxime chloride.

miosis Excessively constricted pupil; often bilateral after exposure to nerve agents.

mutagen A substance that mutates, damages, and changes the structures of DNA in the body's cells.

nerve agents A class of chemical called organophosphates; they function by blocking an essential enzyme in the nervous system, which causes the body's organs to become overstimulated and burn out.

neurotoxins Biologic agents that are the most deadly substances known to humans; they include botulinum toxin and ricin.

neutron radiation The type of energy that is emitted from a strong radiologic source; neutron particles are among the most powerful forms of radiation. Neutrons easily penetrate through lead and require several feet of concrete to stop them.

off-gassing The emitting of an agent after exposure, for example from a person's clothes that have been exposed to the agent.

persistency Term used to describe how long a chemical agent will stay on a surface before it evaporates.

phosgene A pulmonary agent that is a product of combustion, such as might be produced in a fire at a textile factory or house or from metalwork or burning Freon. Phosgene is a very potent agent that has a delayed onset of symptoms, usually hours.

phosgene oxime (CX) A blistering agent that has a rapid onset of symptoms and produces immediate, intense pain and discomfort on contact.

pneumonic plague A lung infection, also known as plague pneumonia, that is the result of inhalation of plague-causing bacteria.

points of distribution (PODs) Existing facilities that are established in a time of need for the mass distribution of antibiotics, antidotes, vaccinations, and other medications and supplies.

primary blast injury Injuries caused by an explosive pressure wave on the hollow organs of the body.

pulmonary blast injuries Pulmonary trauma resulting from short-range exposure to the detonation of high energy explosives.

radioactive material Any material that emits radiation.

radiologic dispersal device (RDD) Any container that is designed to disperse radioactive material.

ricin A neurotoxin derived from mash that is left from the castor bean; causes pulmonary edema and respiratory and circulatory failure leading to death.

route of exposure The manner by which a toxic substance enters the body.

sarin (GB) A nerve agent that is one of the G agents; a highly volatile colorless and odorless liquid that turns from liquid to gas within seconds to minutes at room temperature.

secondary blast injury A penetrating or nonpenetrating injury caused by ordnance projectiles or secondary missiles.

secondary device An additional explosive used by terrorists, set to explode after the initial bomb.

smallpox A highly contagious disease; it is most contagious when blisters begin to form.

soman (GD) A nerve agent that is one of the G agents; twice as persistent as sarin and five times as lethal; it has a fruity odor, as a result of the type of alcohol used in the agent, and is a contact and an inhalation hazard that can enter the body through skin absorption and through the respiratory tract.

Special Atomic Demolition Munitions (SADM) Small suitcase-sized nuclear weapons that were designed to destroy individual targets, such as important buildings, bridges, tunnels, and large ships.

state-sponsored terrorism Terrorism that is funded and/or supported by nations that hold close ties with terrorist groups.

sulfur mustard (H) A vesicant; it is a brownish, yellowish oily substance that is generally considered very persistent; has the distinct smell of garlic or mustard and, when released, is quickly absorbed into the skin and/or mucous membranes and begins an irreversible process of damaging the cells.

syndromic surveillance The monitoring, usually by local or state health departments, of patients presenting to emergency departments and alternative care facilities, the recording of EMS call volume, and the use of over-the-counter medications.

tabun (GA) A nerve agent that is one of the G agents; is 36 times more persistent than sarin and approximately half as lethal; has a fruity smell and is unique because the components used to manufacture the agent are easy to acquire and the agent is easy to manufacture.

tertiary blast injury An injury from whole body displacement and subsequent traumatic impact with environmental objects.

V agent (VX) One of the G agents; it is a clear, oily agent that has no odor and looks like baby oil; more than 100 times more lethal than sarin and is extremely persistent.

vapor hazard An agent that enters the body through the respiratory tract.

vesicants Blister agents; the primary route of entry for vesicants is through the skin.

viral hemorrhagic fevers (VHF) A group of diseases caused by viruses that include the Ebola, Rift Valley, and yellow fevers, among others. This group of viruses causes the blood in the body to seep out from the tissues and blood vessels.

viruses Germs that require a living host to multiply and survive.

volatility A term used to describe how long a chemical agent will stay on a surface before it evaporates.

weapon of mass casualty (WMC) Any agent designed to bring about mass death, casualties, and/or massive damage to property and infrastructure (bridges, tunnels, airports, and seaports); also known as a weapon of mass destruction (WMD).

weapon of mass destruction (WMD) Any agent designed to bring about mass death, casualties, and/or massive damage to property and infrastructure (bridges, tunnels, airports, and seaports); also known as a weapon of mass casualty (WMC).

weaponization The creation of a weapon from a biologic agent generally found in nature and that causes disease; the agent is cultivated, synthesized, and/or mutated to maximize the target population's exposure to the germ.

Assessment in Action

You and your partner are dispatched to the federal courthouse for a possible mass-casualty incident secondary to a terrorist attack. While driving to the scene, you both review key concepts concerning weapons of mass destruction.

1. To date, the preferred weapon of mass destruction for terrorists has been what type of device?
   A. Biologic
   B. Chemical
   C. Explosive
   D. Incendiary

2. When determining the potential for a terrorist attack, you should consider the:
   A. type of call.
   B. number of patients.
   C. victims' statements.
   D. all of the above.

3. What is the best location for staging following a weapons of mass destruction incident?
   A. Upwind and uphill from the incident
   B. Upwind and downhill from the incident
   C. Downwind and uphill from the incident
   D. Downwind and downhill from the incident

4. What is the goal of terrorists when deploying a secondary device?
   A. Injure responders
   B. Maximize the destruction
   C. Cover up any evidence
   D. Minimize access to the scene

5. What is the primary complication associated with vesicant blisters?
   A. Hypothermia
   B. Secondary infection
   C. Volume loss
   D. Scar tissue

6. What is the most toxic chemical ever produced?
   A. Sarin
   B. Soman
   C. VX
   D. Tabun

7. When assessing patients exposed to a nerve agent, you should expect to find which of the following signs or symptoms?
   A. Tachycardia
   B. Hot, dry skin
   C. Confusion
   D. Miosis

8. Which of the following forms of anthrax is the most deadly?
   A. Cutaneous
   B. Inhalational
   C. Bloodborne
   D. Gastrointestinal

9. Describe the Department of Homeland Security Advisory System and the color-coded system.

10. Explain how pulmonary agents inflict their damage.

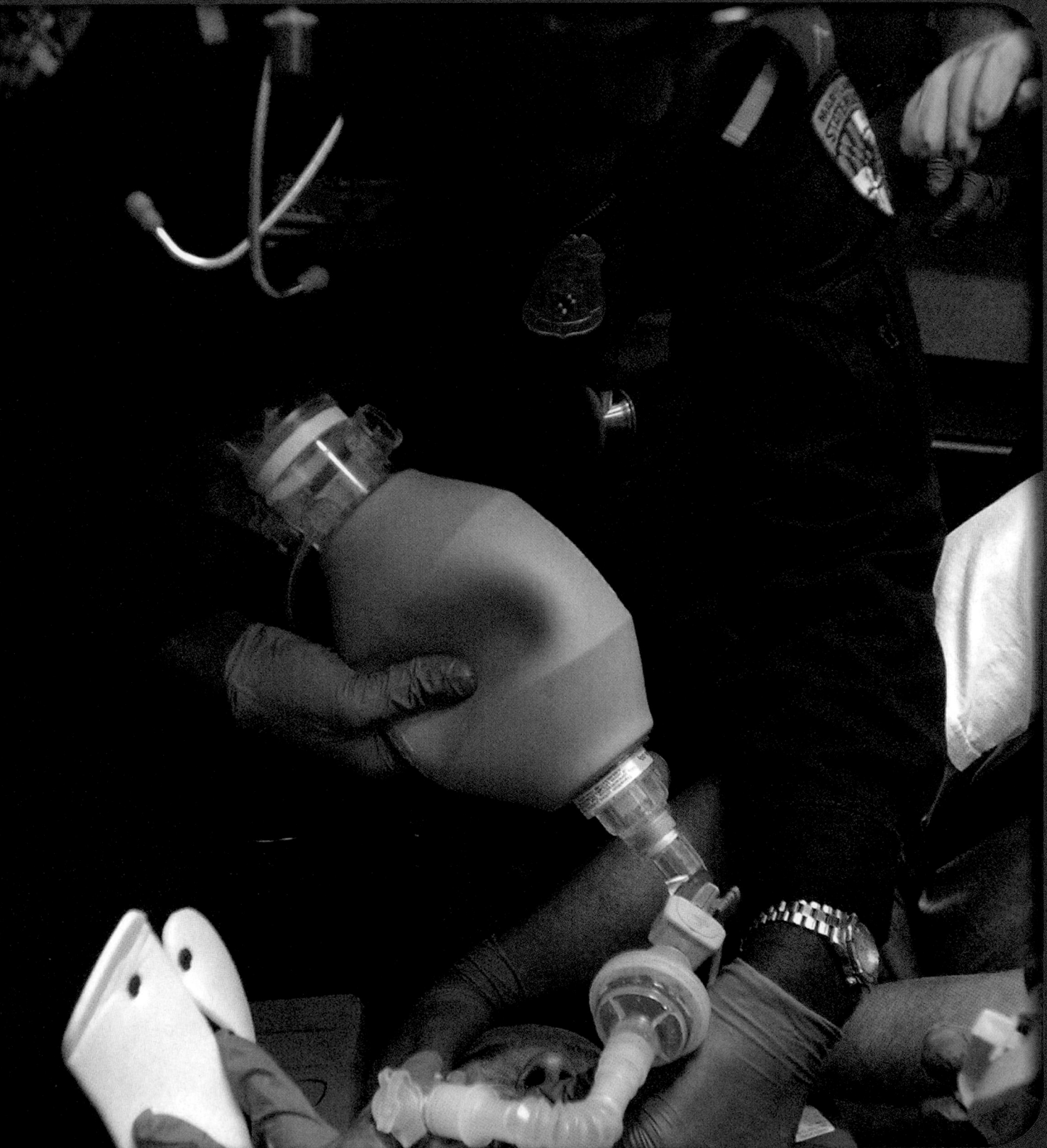

40 ALS Assist

National EMS Education Standard Competencies

There are no National EMS Education Standard Competencies for this chapter.

. . . .

Knowledge Objectives

1. Discuss advanced airway techniques. (p 1463)
2. Describe the basic anatomy and physiology of the airway. (pp 1463-1465)
3. Explain the principles of basic airway management. (p 1465)
4. Describe the equipment and techniques used in endotracheal intubation. (pp 1465-1475)
5. Discuss the benefits and disadvantages of multilumen and single-lumen airways. (pp 1475-1476)
6. Discuss placement of a gastric tube. (pp 1476-1477)
7. Understand the uses for continuous positive airway pressure (CPAP). (p 1477)
8. Be familiar with the equipment necessary to gain intravenous (IV) access. (pp 1477-1482)
9. Understand the techniques, alternative IV sites, and complications associated with IV access. (pp 1482-1490)
10. Describe age-specific considerations in the care of pediatric patients. (p 1489)
11. Describe age-specific considerations in the care of geriatric patients. (p 1489)
12. Understand the use and techniques of cardiac monitoring. (pp 1490-1497)

Skills Objectives

1. Demonstrate how to perform orotracheal intubation. (pp 1470-1475, Skill Drill 40-1)
2. Demonstrate the steps in assembling IV equipment. (p 1478)
3. Demonstrate how to spike the bag with an IV administration set. (pp 1481-1482, Skill Drill 40-2)
4. Demonstrate how to start an IV. (pp 1482-1486, Skill Drill 40-3)
5. Demonstrate appropriate lead placement for both 4-lead and 12-lead ECG monitoring. (pp 1496-1497)

Introduction

There may be cases in which you may find it necessary to be familiar with skills normally practiced at the AEMT and paramedic certification levels. Advanced airway techniques, intravenous (IV) therapy, and cardiac monitoring are among these skills, and they will be discussed in this chapter. All of these skills require additional training and practice beyond what is normally a part of the EMT curriculum and require oversight by the medical director. It is also necessary that you perform these skills on a regular basis so that you can maintain proficiency in these techniques. A team approach to patient care will positively affect the quality of that care and will foster a cooperative working relationship among your crew.

Advanced Airway Techniques

The single most important manipulative skill you will use as an EMT is establishing and maintaining a patient's airway. While the obviously broken leg or amputated finger may be eye-catching, the airway must be secured immediately or the patient will die. The vast majority of conscious patients with an intact gag reflex can maintain their own airway. Therefore, in managing a conscious patient, you may need only to provide oxygen and monitor the patient closely for any changes. Patients whose consciousness is altered may require an oropharyngeal or nasopharyngeal airway and suctioning after you make sure the patient's airway is opened properly by using the head tilt–chin lift or jaw-thrust maneuver.

However, patients who are unresponsive and not breathing on their own may fare better when advanced airway techniques are used to maintain their airway. The purpose of advanced airway management is to provide better airway protection and improve ventilation by using a tube to create a direct channel to the trachea. Endotracheal intubation is a difficult skill to master and requires additional training for EMTs. Additional options include a multilumen airway and a laryngeal mask airway. All of these techniques require additional training, appropriate approval for their use, and medical oversight. During advanced procedures, as with basic procedures, you need to recognize and respect the feelings of the patient and family.

Anatomy and Physiology of the Airway

As you learned in previous chapters, the respiratory system consists of all the structures in the body that are used for breathing **Figure 40-1**. The upper airway begins with the nose, mouth, throat (pharynx), and larynx (vocal cords). The larynx is typically considered to be the dividing line between the upper and lower airway. The lower airway includes the trachea, bronchi, and lungs. The epiglottis is a leaf-shaped structure located at the glottic opening (covering the larynx) that prevents food and liquid from entering the lower airway during swallowing. The bronchi and other air passages branch off from the trachea, extending into each lung, subdividing into bronchioles (smaller passages) down to the alveoli, where the exchange of oxygen and carbon dioxide occurs.

The mechanical process of breathing occurs through the use of the diaphragm and intercostal muscles (muscles between the ribs). The diaphragm is a thin, dome-shaped muscle that separates the thoracic cavity from the abdominal cavity. The diaphragm and intercostal

You are the Provider: PART 1

At 6:22 PM, you are dispatched to back up an advanced life support (ALS) unit for a "CPR in progress" call involving a male of unknown age. The address is 105 Oak Park Drive, which is located about 5 miles away from your station. The weather is cloudy, the temperature is 82°F, and the traffic is moderate.

1. What contributions can the EMT make to the overall resuscitation effort of a patient in cardiac arrest?
2. What tasks will most likely be assigned to you and your EMT partner when you arrive at the scene?

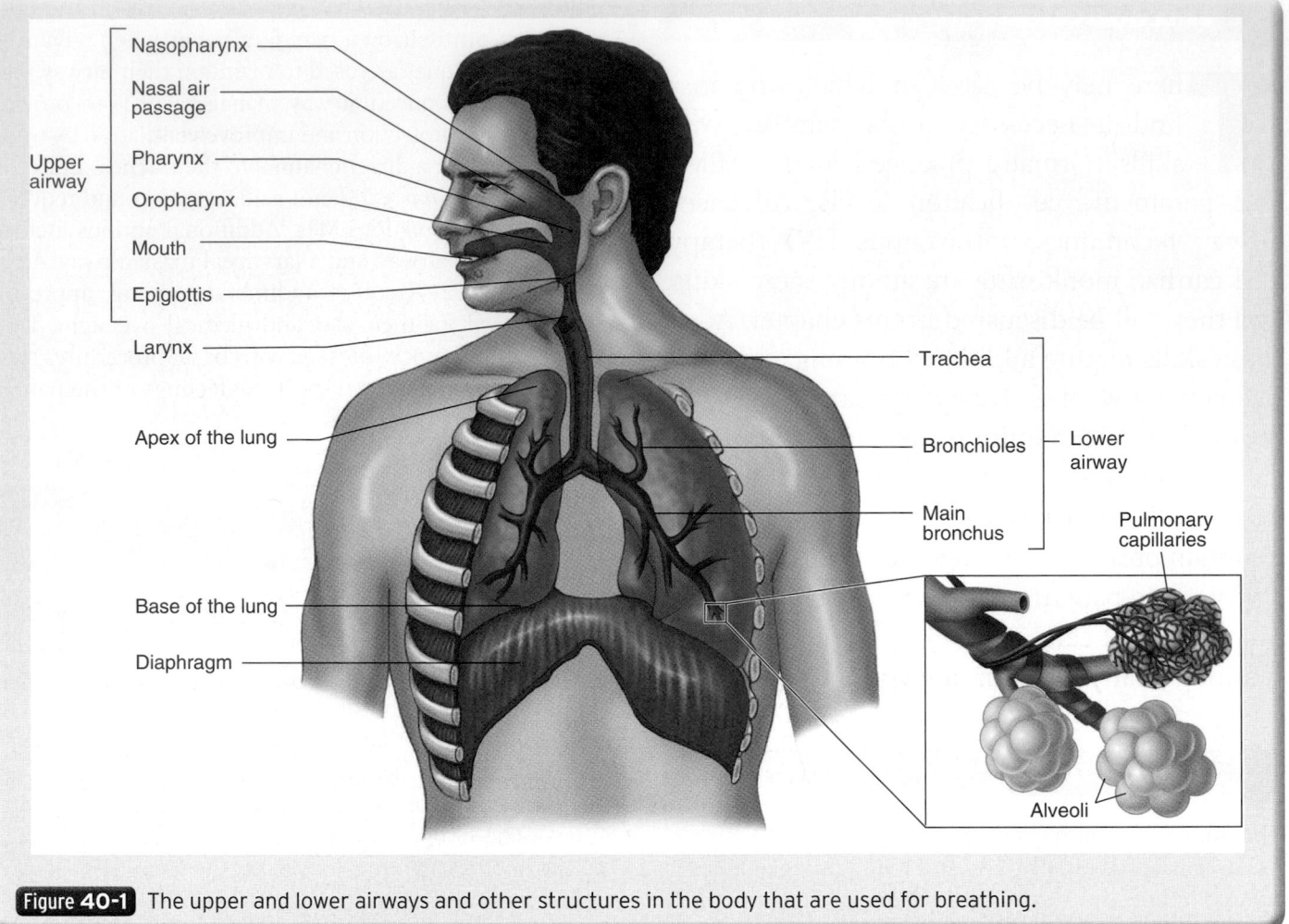

Figure 40-1 The upper and lower airways and other structures in the body that are used for breathing.

muscles contract during the active phase of breathing (inhalation), increasing the size of the chest cavity. Contraction of the diaphragm pulls the chest cavity down; contraction of the intercostal muscles pulls the rib cage up and out. The increased size of the chest cavity allows air to flow into the lungs. During the passive phase of breathing (exhalation), air flows out of the lungs. The diaphragm and intercostal muscles relax, and the size of the chest cavity decreases. The diaphragm moves up, and the ribs move in.

The respiratory system delivers oxygen to the body and removes carbon dioxide, a process that takes place on two levels: the alveolar-capillary exchange and the capillary-cellular exchange **Figure 40-2**.

The alveolar-capillary exchange works in the following way:

1. Air breathed in during inhalation travels through the airways to the alveoli.
2. As this oxygen-rich air enters the alveoli, oxygen-poor blood is circulated through the capillaries around each alveolus.
3. Oxygen in the alveoli crosses over into the bloodstream. Carbon dioxide in the blood from the capillaries crosses over into the alveoli, creating a shift of oxygen and carbon dioxide.

The capillary-cellular exchange occurs throughout the body's cells. Cells give up carbon dioxide into the capillaries, and capillaries give up oxygen to the cells.

Each living cell in the body requires a regular supply of oxygen; some cells, such as those in the heart, brain, and nervous system, need a constant supply of oxygen to survive. Cells in the heart will be damaged if the oxygen supply is interrupted for more than a few minutes. After 4 to 6 minutes without oxygen, cells in the brain and nervous system begin to die. Dead brain cells can never be replaced. Brain damage and other permanent changes in the body result from damage caused by a lack of oxygen.

Other cells in the body that are not as dependent on a constant oxygen supply can tolerate short periods without oxygen and still survive.

Words of Wisdom

Advanced airway techniques are begun only after proper basic airway management has been completed.

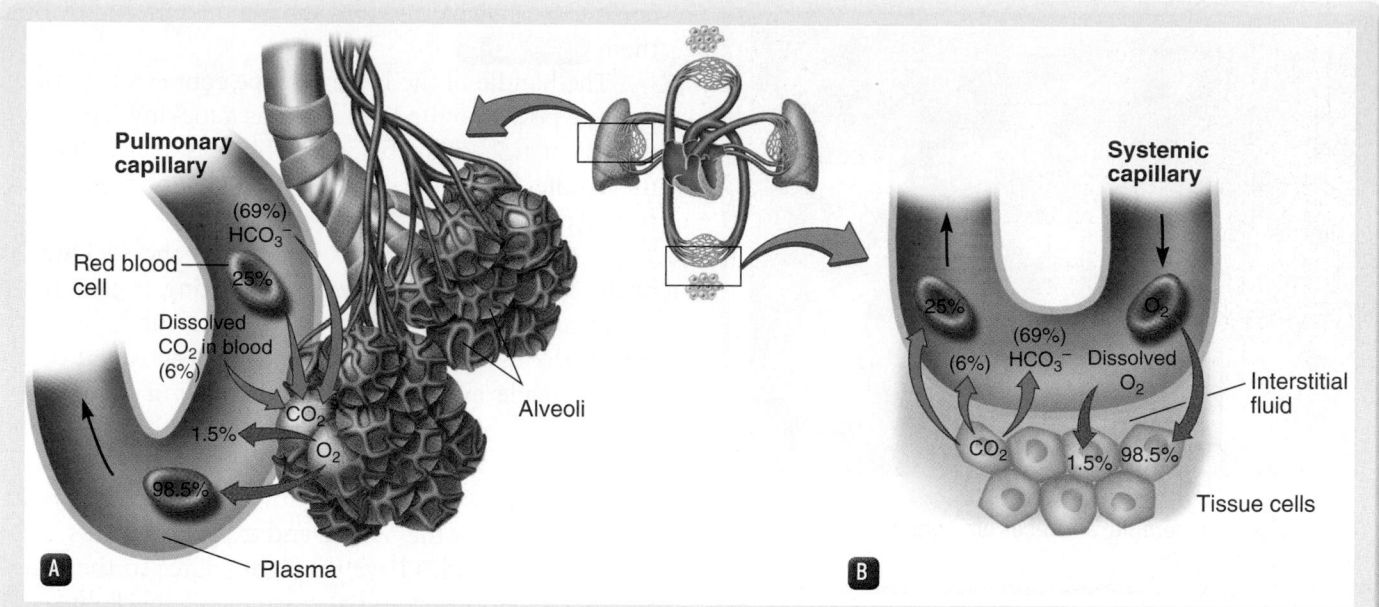

Figure 40-2 **A.** Alveolar-capillary exchange. Oxygen-rich air enters the alveoli, where it crosses into the bloodstream. Carbon dioxide leaves the blood and enters the alveoli. **B.** Capillary-cellular exchange. Throughout the body, carbon dioxide from the cells enters the capillaries, and oxygen leaves the capillaries and enters the cells.

Basic Airway Management

You should always assess the airway first in an injured or ill patient. This rule applies to the basic and advanced levels of airway management. Advanced airway techniques are begun only after proper basic airway management has been completed.

As you have already learned, the first step in airway management is opening a patient's airway. You should use the head tilt–chin lift maneuver in a patient with no suspected spinal injury and the jaw-thrust maneuver in a patient you suspect may have a spinal injury. After you have opened the airway, you should assess the airway and evaluate the need for suctioning to remove foreign bodies, liquid, and/or blood from the patient's mouth.

After the airway has been cleared, you need to determine whether the patient needs an airway adjunct. The basic airway adjuncts that are already available to you are oropharyngeal and nasopharyngeal airways. The more advanced airway adjuncts that may be available to you, with approval of your medical director, will be discussed in this chapter.

Endotracheal Intubation

Endotracheal intubation is the insertion of a tube into the trachea to maintain the airway. This can be done through the mouth (called **orotracheal intubation**) or through the nose (called nasotracheal intubation). In either case, the tube passes directly through the larynx between the vocal cords and then into the trachea. Endotracheal intubation is a very effective method for controlling a patient's airway and has many advantages over other airway management techniques. It is indicated for patients who cannot protect their own airway and also for patients who need prolonged artificial ventilation. Initial care of the airway is essential and should include basic airway care as outlined above, including ventilation with a bag-mask device if needed. If the patient is being adequately ventilated, endotracheal intubation may not be necessary. If advanced airway care is required, you should continue to effectively ventilate the patient while the equipment to perform the intubation is prepared.

■ Equipment

Before an orotracheal intubation attempt begins, it is vitally important to assemble all the equipment that you will need. The following is a list of equipment needed Figure 40-3 :

- Laryngoscope handle and blade
- Properly sized endotracheal (ET) tube
- Stylet
- 10-mL syringe
- Water-soluble lubricant for the ET tube
- Suction unit with rigid and soft-tip catheters
- Magill forceps
- Stethoscope
- Commercial securing device
- Secondary confirmation device

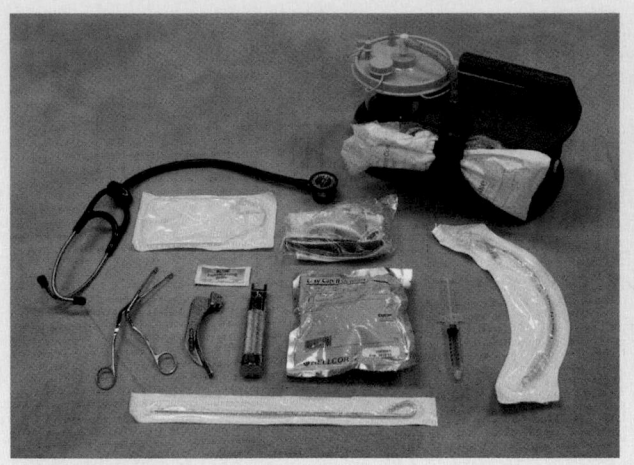

Figure 40-3 Assemble all necessary equipment before you begin intubation.

Words of Wisdom

While collecting the orotracheal intubation equipment, ensure that your partner provides the patient with adequate ventilation with airway adjuncts while oxygenating with a bag-mask device.

Laryngoscope

The purpose of a **laryngoscope** is to sweep the tongue out of the way and align the airway so that the vocal cords can be visualized and the ET tube passed through them **Figure 40-4**.

The handle of the laryngoscope contains batteries to provide power to the light and has a locking bar to connect the handle to the blade; the blade is detachable from the handle. Blades are curved or straight and range in size from 0 to 4 **Figure 40-5**.

A notch on the blade locks onto the locking bar of the handle. Because adequate lighting is essential to visualize the epiglottis and vocal cords, the light source is near the tip of the blade. The light is activated by lifting the blade away from the handle until it locks at a right angle. The light will not come on if the blade is not attached properly, the bulb is burned out or loose, or the batteries in the handle are low or dead. Always carry extra batteries for the handle and extra light bulbs. Some laryngoscope blades have the light source in the handle and use fiberoptic material to transmit the light to the end of the blade.

Endotracheal Tubes

ET tubes come in many sizes; the size is specified by the measurement of the inside diameter of the tube. Sizes range from 2.5 to 9 mm **Figure 40-6**. The length of the ET tube is marked on the outside of the tube in centimeters.

The proper-sized tube for adults ranges from 7.0 to 8.5 mm. For most efficient use of the tube, use the largest-diameter ET tube that will pass easily through

You are the Provider: PART 2

You and your partner arrive shortly after the ALS crew. When you enter the residence, you find an AEMT and a paramedic performing two-rescuer cardiopulmonary resuscitation (CPR) on the patient, a 74-year-old man. Following 2 minutes of CPR, the patient is reassessed, and then you and your partner resume two-rescuer CPR.

Recording Time: 2 Minutes	
Appearance	Motionless; cyanotic
Level of consciousness	Unconscious and unresponsive
Airway	Open; clear of secretions and foreign bodies
Breathing	Absent
Circulation	No pulse; face is cyanotic

A rescue squad with two EMRs arrives at the scene to provide assistance. The EMRs continue two-rescuer CPR as the paramedic asks you to apply the defibrillation pads and turn on the cardiac monitor. The AEMT begins looking for a site to insert an intravenous (IV) line and asks your partner to set up the equipment.

3. How does a manual cardiac monitor/defibrillator differ from an automated external defibrillator?

4. Would a 12-lead electrocardiogram (ECG) be of benefit to this patient? Why or why not?

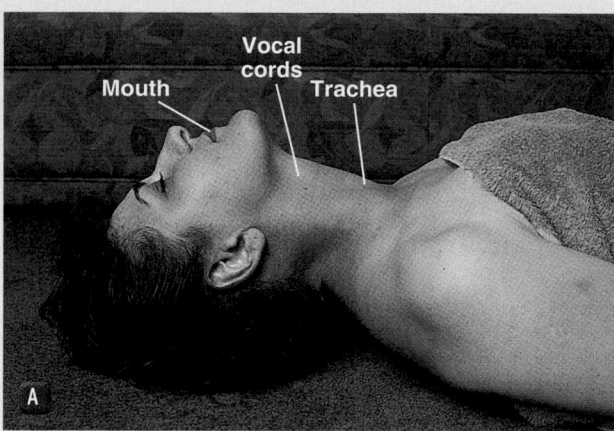
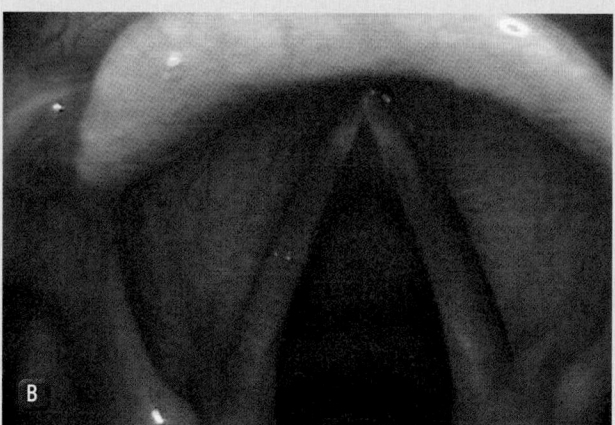

Mouth
Vocal cords
Trachea

A
B

Figure 40-4 **A.** You must see the vocal cords to pass an endotracheal tube through them. The vocal cords are located in the upper airway at the entrance to the larynx. **B.** A view of the vocal cords.

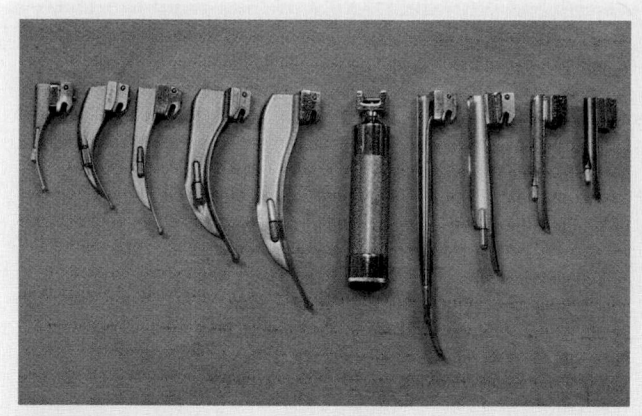

Figure 40-5 Laryngoscope blades can be curved or straight and come in different sizes.

of tube sizes to ensure that no matter what size you choose, you have one tube smaller and one tube larger, in case you need it.

For children, it is best to have a chart or length-based resuscitation tape device to help you with sizing the ET tube. Another method is to select a tube that roughly equals the size of the diameter of the patient's little finger across the nail bed **Figure 40-7**. No matter what size you decide to use, you should also have one tube larger and one tube smaller available in case you need it. Cuffed tubes should be used when intubating older children and adults. However, in younger children, the circular narrowing of the trachea at the level of the cricoid cartilage functions as a cuff. Therefore, uncuffed tubes are used in younger children.

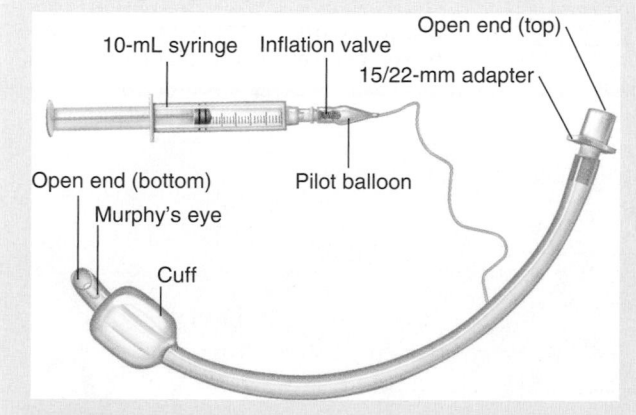

10-mL syringe   Inflation valve   Open end (top)
15/22-mm adapter
Open end (bottom)   Pilot balloon
Murphy's eye
Cuff

Figure 40-6 Endotracheal tubes come in many sizes; the size is specified by the measurement of the inside diameter of the tube. Sizes range from 2.5 to 9 mm.

the vocal cords. A good rule of thumb is to always have a 7.5-mm ET tube on hand; this size tube will fit most adults. However, you should carry a complete selection

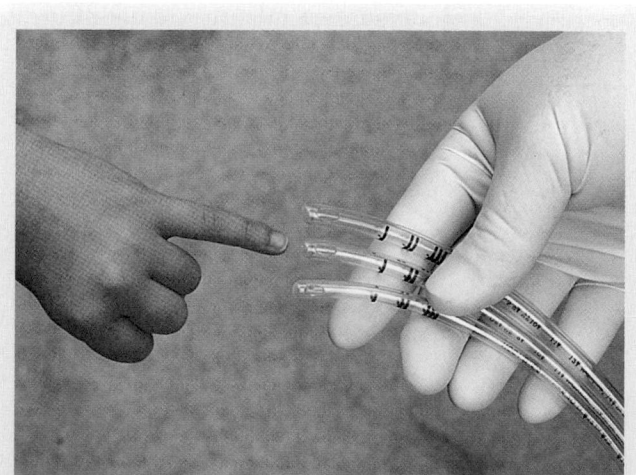

Figure 40-7 When selecting an endotracheal tube for a pediatric patient, one method of sizing is to select a tube that roughly equals the size of the diameter of the patient's little finger across the nail bed.

A standard 15/22-mm adapter attaches to any ventilation device, such as a bag-mask device or a mechanical ventilator. Make sure that the adapter is securely pushed into the tube so that it does not pull off as you ventilate the patient. A pilot balloon is attached to the tube to indicate how well the balloon cuff at the distal end of the ET tube is inflated (cuffed ET tubes only). The cuff at the end of the tube holds about 10 mL of air. The small hole at the distal end of the tube across from the bevel end, called Murphy's eye, helps to prevent tube obstruction by secretions **Figure 40-8**.

Note the centimeter mark for placement of the ET tube by looking at where the tube lines up with the teeth on an intubated patient. In average-sized adults, the tube-to-teeth mark is usually at around 22 cm. Once the patient is intubated, monitor the centimeter marking at the patient's teeth or lips to help determine whether the tube has moved from its original position, and make sure to document this information on the patient care report.

> ## Safety
>
> During the intubation attempt, your patient is not being oxygenated, so make sure to stop the procedure and ventilate the patient if the attempt is prolonged for any reason.

Stylet

A plastic-coated wire called a **stylet** may be inserted into the ET tube to add rigidity and shape to the tube during the intubation **Figure 40-9**. You should bend the tip of the stylet to form a gentle curve for adult intubations **Figure 40-10**. Because an infant's or a child's airway is more angular and less aligned than an adult's, the tip of the stylet should be bent into a hockey stick shape for use in an infant or child.

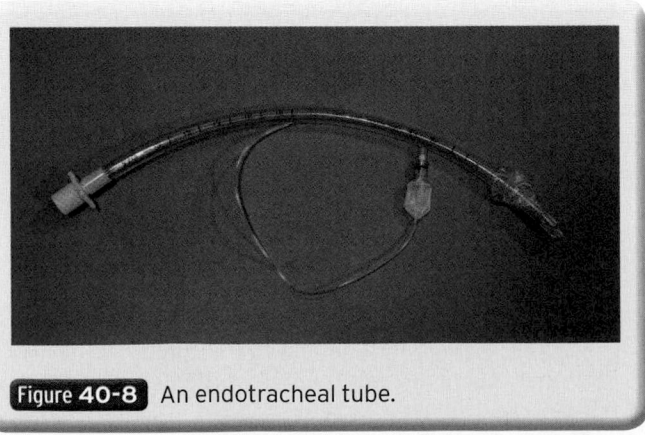

Figure 40-8 An endotracheal tube.

Do not insert the stylet past Murphy's eye because it could puncture or lacerate delicate airway tissues **Figure 40-11**. Make sure to keep the stylet about ¼″ proximal to the cuff in adults and 1″ from the end of the tube in infants and children. Before intubation is attempted, always confirm that the stylet is not sticking out past the end of the ET tube.

Syringe

Use a 10-mL syringe to test for air leaks in the ET tube before intubation and to fill the cuff when the tube is placed correctly in the trachea. As you are assembling and checking the equipment before intubation, attach the syringe to the pilot balloon, and test the cuff by inflating it with 5 to 10 mL of air **Figure 40-12**. Deflate the cuff after you have confirmed that there are no air leaks. Be sure that the syringe remains attached to the pilot balloon with the plunger pulled back to the 10-mL mark. After the ET tube has been properly placed in the

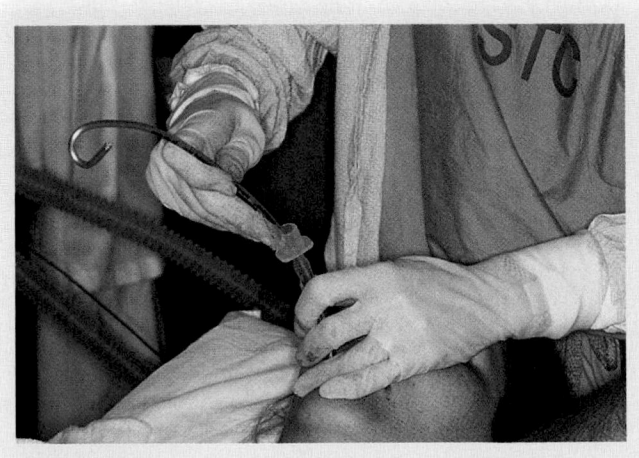

Figure 40-9 A wire stylet, which adds rigidity and shape to the tube, must be removed after tube placement.

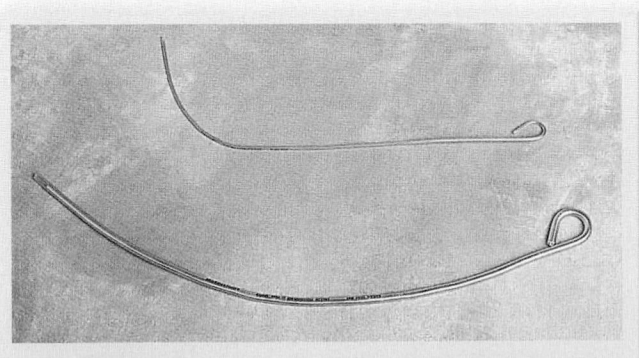

Figure 40-10 Bend the tip of the stylet into a hockey stick shape for a pediatric patient, as shown at the top. Bend the stylet to form a gentle curve for an adult patient, as shown at the bottom.

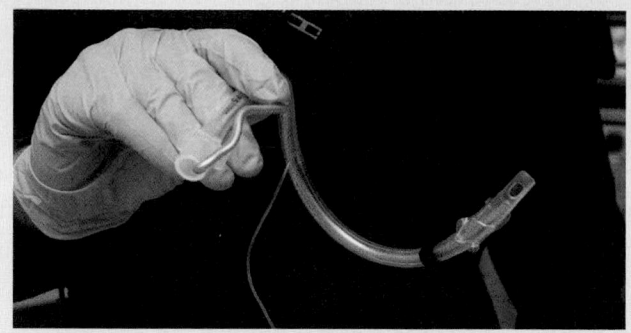

Figure 40-11 Do not insert the stylet past Murphy's eye because it could puncture or lacerate delicate airway tissues.

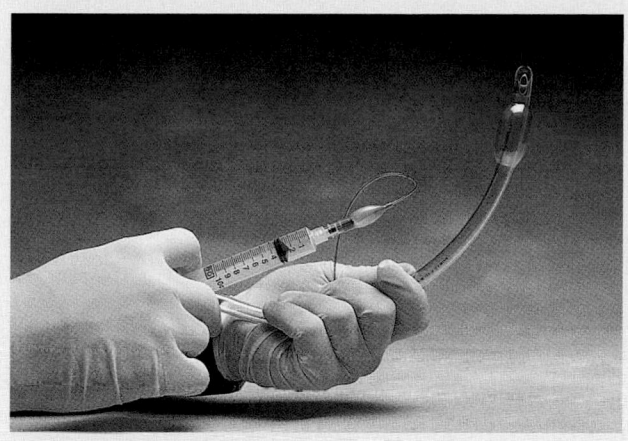

Figure 40-12 Inflate the cuff with 5 to 10 mL of air to check for air leaks.

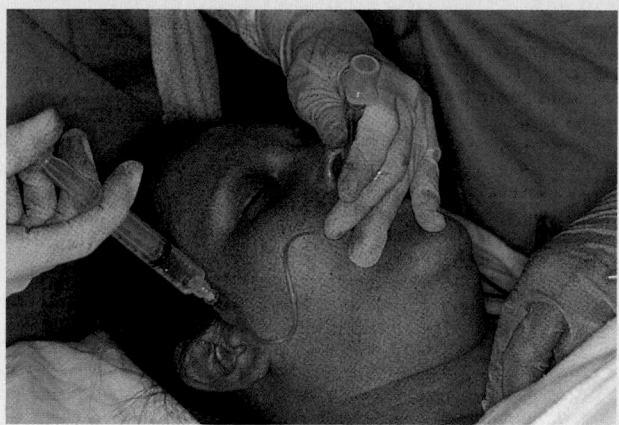

Figure 40-13 Inflate the cuff with 5 to 10 mL of air, and then immediately remove the syringe from the pilot balloon to prevent air from leaking back into the syringe.

Words of Wisdom

Water-soluble lubricant should be applied to the distal end of the endotracheal tube to make it easier to insert and to the end of the stylet to make it easier to remove once the tube is in place.

Words of Wisdom

Medical control may advise you to use an oral airway or similar device to prevent the patient from biting down on the endotracheal tube.

Words of Wisdom

The Sellick maneuver, also known as cricoid pressure, has been used to inhibit the flow of air into the stomach (and thus reduce gastric distention) and reduce the chance of aspiration by helping block the regurgitation of gastric contents from the esophagus. It has also been used to improve visualization of the vocal cords or positioning of the lighted stylet during intubation. In this maneuver, a rescuer applies cricoid pressure on the patient by placing the thumb and index finger on either side of the cricoid cartilage (at the inferior border of the larynx) and pressing down.

According to several studies cited in the 2010 American Heart Association Guidelines, cricoid pressure may actually *impede* ventilation and not completely prevent aspiration. For this reason, the procedure is generally not recommended. Be sure to follow your local protocol regarding the use of the Sellick maneuver.

patient, inflate the cuff with 5 to 10 mL of air and then remove the syringe from the pilot balloon to prevent air from leaking back into the syringe Figure 40-13.

Other Equipment

A suction unit may be needed to clear secretions or blood so the vocal cords can be visualized. During the intubation attempt, the patient may vomit, so make sure the suction unit is readily available.

The use of a commercial securing device is an effective means of ensuring the tube does not move once it is in the correct position. There are many manufacturers of securing devices; you should become familiar with the device used in your system. Some EMS systems use tape to secure the ET tube.

The Intubation Procedure: Visualized (Oral) Intubation

You may intubate only if authorized to do so by off-line or online medical control, according to your local medical protocols. Once you and medical control have made the decision to intubate, you must act quickly, carefully, and efficiently. Be sure to use standard precautions, including the use of gloves, eye protection, and a mask. Only when the patient is adequately oxygenated and all of the equipment is in place should the intubation attempt occur. An intubation attempt should not take more than 30 seconds. The 30-second time limit begins when ventilation stops and the laryngoscope blade is inserted into the patient's mouth; it ends when ventilation has begun again. If the attempt is not successful, stop, withdraw the tube, oxygenate the patient, and try again according to your local protocols.

Intubation is a multiple-person task, especially in a situation involving cardiac arrest and use of an automatic external defibrillator (AED). The following tasks should be divided among the EMTs who are present:

- First EMT applies and uses the AED.
- Second and third EMTs perform synchronous cardiopulmonary resuscitation (CPR) starting with chest compressions at a ratio of 30 compressions to 2 ventilations (CPR is performed asynchronously after the patient has been intubated).
- Fourth EMT prepares and intubates the patient.

You should note that effective CPR and defibrillation with an AED are critical to survival from cardiac arrest. Intubation of a patient in cardiac arrest should be performed during the 2-minute period of CPR that occurs between defibrillation attempts. If possible, intubation should be performed without interrupting CPR. Follow the steps in **Skill Drill 40-1** to perform visualized orotracheal intubation:

Skill Drill 40-1

1. Open the patient's airway with a basic life support maneuver, and clear the airway of any foreign material. Be sure to use standard precautions.
2. Insert an oral airway, and oxygenate the patient with a bag-mask device at the appropriate rate, which will vary depending on the age of the patient. You should oxygenate the patient at a rate of between 20 and 24 breaths/min for 1 to 2 minutes before attempting intubation (Step 1).
3. As your partner ventilates the patient, you should quickly assemble and test your equipment. Verify that the bulb on the laryngoscope or lighted stylet is working, select the proper-sized tube, and make sure the ET tube cuff has no leaks. If you are using a nonlighted stylet, insert it in the ET tube. Lubricate the tube and stylet as needed (Step 2).
4. Confirm that the patient has been properly preoxygenated. Stop ventilating the patient and remove the oral airway if it is in place.
5. Position the patient's head and neck to allow for the best visualization of the vocal cords. In a patient with no spinal cord injury, use the head tilt–chin lift maneuver to align the structures. When you are using the laryngoscope, place towels under the patient's shoulders, if necessary, to raise the head for a better view of the vocal cords. When you are using the lighted stylet, grasping the tongue and jaw and pulling upward will aid in insertion (Step 3).
6. To intubate a patient you suspect has a spinal cord injury, you should make sure that your partner maintains manual in-line stabilization of the head and neck in the neutral position with a cervical collar in place while you attempt the intubation (Step 4).
7. Grasp the laryngoscope handle in your left hand. Make sure the blade is locked into place and the bulb is illuminated. Open the patient's mouth with the gloved fingers of your right hand. Gently place the blade in the right side of the patient's mouth, then move it toward the center of the mouth, gently pushing the tongue to the left. The tongue must be displaced for you to visualize the vocal cords. Visualize the epiglottis. Advance a curved blade along the base of the tongue until its tip rests at the vallecula; advance a straight blade along the base of the tongue until you see it catch the epiglottis. Lift the laryngoscope away from the posterior pharynx so that you can see the vocal cords. The lifting force is directed straight up, parallel to the long axis of the laryngoscope handle, not back toward the patient's head. It should feel as if you are picking up the patient's head by the jaw. To avoid breaking the patient's teeth or lacerating the lips, never use the blade as a lever or fulcrum against the upper teeth. Do not lose sight of the vocal cords at any time after you have visualized them. Proper placement of the ET tube depends on your visualization of the tube as it is placed between the vocal cords.
8. Insert the ET tube with your right hand, keeping the vocal cords and the tip of the tube in sight at all times. Do not advance the ET tube down the

Skill Drill 40-1

Performing Orotracheal Intubation

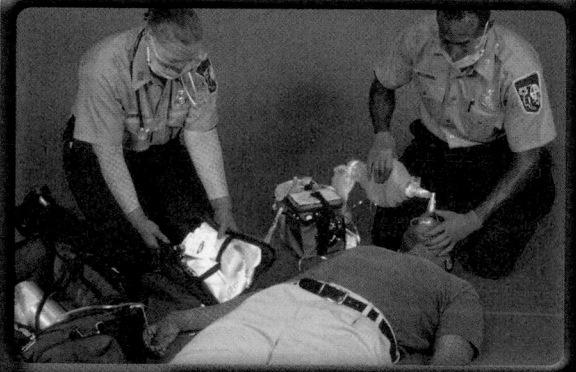

Step 1 Open and clear the airway. Insert an oral airway, and oxygenate with a bag-mask device.

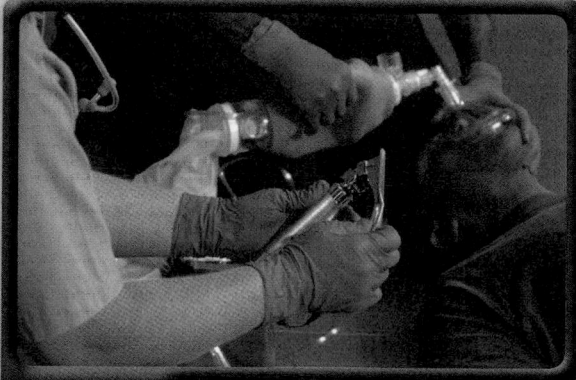

Step 2 Assemble and test the intubation equipment.

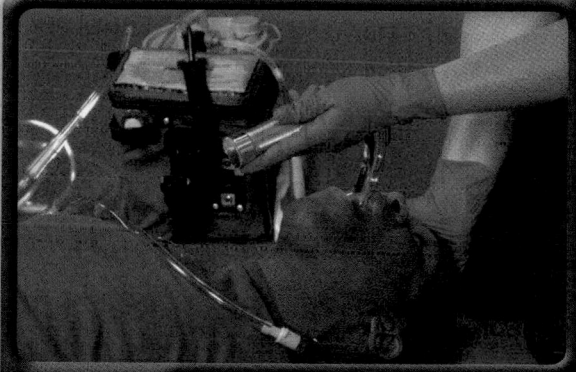

Step 3 Position the patient's head and remove the oral airway.

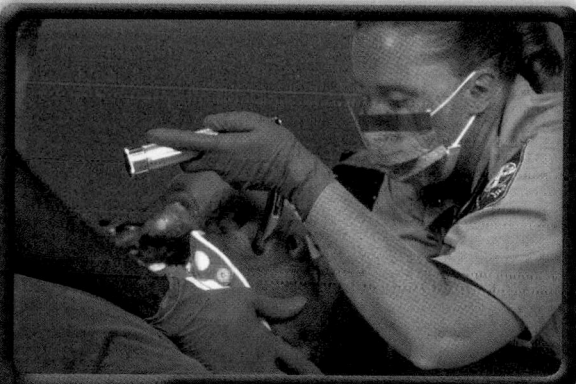

Step 4 For trauma patients, maintain the cervical spine in-line and neutral.

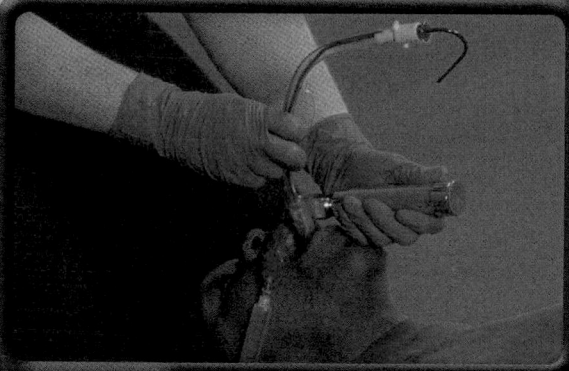

Step 5 Visualize the vocal cords and watch the ET tube pass between them. Remove the laryngoscope and stylet. Hold the tube carefully.

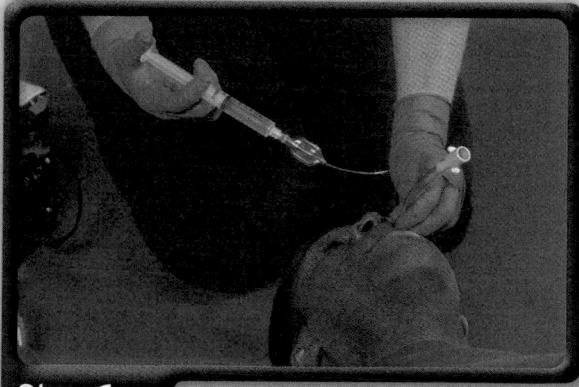

Step 6 Inflate the balloon cuff, and remove the syringe as your partner prepares to ventilate.

Skill Drill 40-1

Performing Orotracheal Intubation, continued

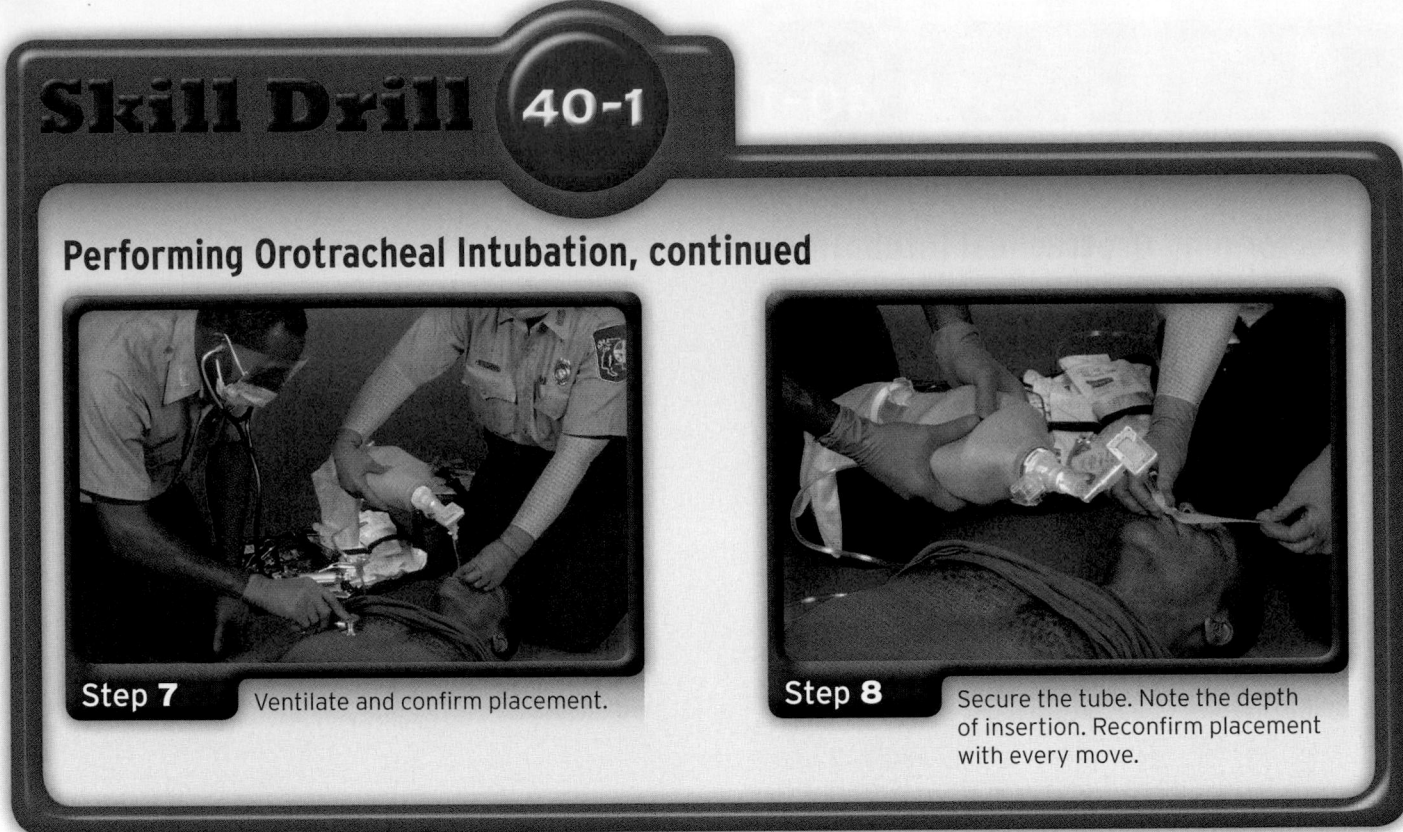

Step 7 Ventilate and confirm placement.

Step 8 Secure the tube. Note the depth of insertion. Reconfirm placement with every move.

center of the laryngoscope blade, or your view of the vocal cords will be obstructed. Advance the tube from the right side of the patient's mouth. Watch the uninflated cuff on the tube as it passes through the vocal cords, then advance the ET tube until the cuff is just past the vocal cords. Note and document the centimeter markings on the outside of the ET tube at the level of the teeth or lips. Once the tube has been inserted through the vocal cords into the trachea, gently remove the laryngoscope and stylet, if a stylet was used. Do not let go of the ET tube until it is secured Step 5 .

9. Inflate the soft balloon cuff on the end of the tube with 5 to 10 mL of air. This will seal the trachea so that air can be blown directly into the lungs. Gently squeeze the pilot balloon cuff to verify the amount of air you should use. The pilot balloon should be full but easily compressed between your fingers. Immediately detach the syringe so that the air in the cuff will not empty back into it Step 6 .

10. You or your partner (whoever is not holding the ET tube in place) should begin ventilating the patient with a bag-mask device attached to the ET tube. Confirm placement of the ET tube. Listen with a stethoscope over the stomach, then both lungs as you ventilate the patient through the tube. You should be able to hear equal breath sounds over the right and left lung fields and no

sounds over the stomach. Also listen at the sternal notch in children. You should see both sides of the chest rise and fall with each ventilation. This is especially important in children because breath sounds in children may be misleading. You may hear them even if the tube is in the esophagus. You should not be able to hear breath sounds in the stomach. If sounds are heard over the epigastrium, the tube should be removed, the patient ventilated, and intubation reattempted Step 7 .

11. Proper confirmation of ET tube placement is essential. A misplaced tube that goes undetected is a fatal error. The actual visualization of the ET tube as it passes through the vocal cords followed by chest rise with ventilation is the primary method to confirm proper placement. Auscultation of the stomach (silent) and lungs (equal breath sounds bilaterally) is also considered to be primary confirmation Step 8 .

It is essential that you use a secondary method of confirming proper tube placement. There are several devices available—esophageal detector devices, **end-tidal carbon dioxide detectors**, colorimetric detectors, and portable capnography monitors Figure 40-14 .

Esophageal detector devices are designed to connect directly to the ET tube adapter end (where the bag-mask device is attached) once the patient has been intubated. Two commonly used devices are a syringe with a plunger

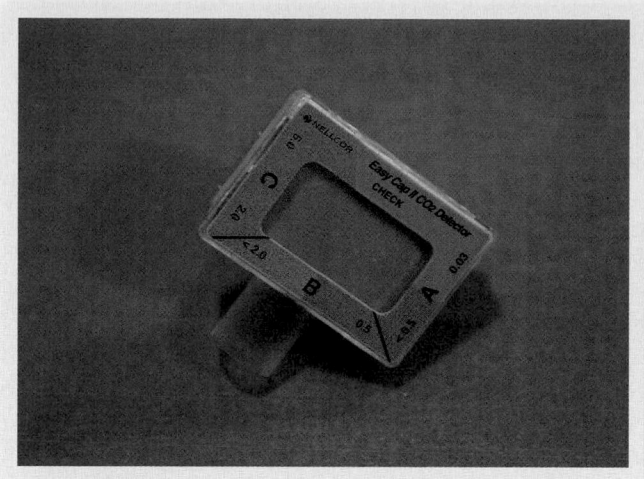

Figure 40-14 Secondary confirming devices include colorimetric detectors.

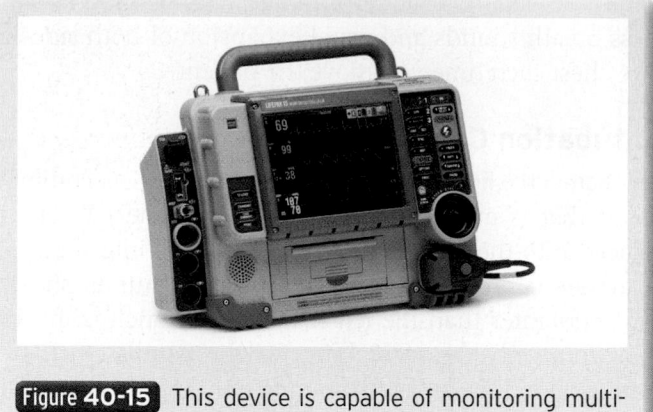

Figure 40-15 This device is capable of monitoring multiple functions simultaneously, including continuous capnography (bottom tracing).

and a bulb syringe. The devices work by attempting to withdraw air from the ET tube. If the tube is properly placed, air will freely withdraw through the syringe or will inflate the bulb because the trachea will not collapse around the end of the ET tube. However, if the ET tube is in the esophagus, the soft tissues of the esophagus will collapse around the end of the tube and there will be noticeable resistance to the withdrawal of air through the syringe or plunger or to the inflation of the bulb.

End-tidal carbon dioxide (sometimes abbreviated as ETCO$_2$) detectors and capnography monitors sense the amount of carbon dioxide during the exhalation phase of ventilation. Place the device between the ET tube adapter and the bag-mask device and ventilate the patient normally. After a minimum of six breaths, look for evidence of carbon dioxide production. Patients in cardiac arrest may not produce enough carbon dioxide for the device to measure. Therefore, accurate readings are not possible until effective ventilation and circulation are restored.

One example of an end-tidal carbon dioxide detector is a disposable plastic indicator with chemically treated paper that changes from purple to yellow in the presence of carbon dioxide. Secondary confirmation is made by verifying the appropriate color change. If carbon dioxide is present, it will change the color of the indicator to yellow. If the ET tube is in the esophagus, ventilation will not produce any carbon dioxide and there will be no color change on the indicator (may change to tan, but generally will stay purple).

Capnography monitors can also be used for secondary confirmation of ET tube placement **Figure 40-15**. These devices directly measure the presence of carbon dioxide. During ventilation, the amount of carbon dioxide present is displayed as a number or as a positive waveform on the monitor. Positive waveforms and/or

carbon dioxide readings provide secondary confirmation of proper placement. If the ET tube is in the esophagus, waveforms and readings will not be present.

Remember that these devices are used as secondary confirmation of ET tube placement. The devices do not give a 100% guarantee that the tube is in the correct location. Primary confirmation is direct visualization of the tube passing through the vocal cords, auscultating good bilateral breath sounds (no epigastric sounds), and seeing the patient's chest rise and fall with each ventilation.

Keep in mind that even when the ET tube is properly placed, it will move if it is not secured. For this reason, never let go of the ET tube until it is secured. Even then, you must continuously check the tube to make sure it is secure and in the correct place. Check the centimeter marking on the ET tube at the teeth or lips, and assess it frequently for tube movement **Figure 40-16**. Also, frequently reassess the epigastrium, breath sounds, and other confirmation methods. Continue to artificially ven-

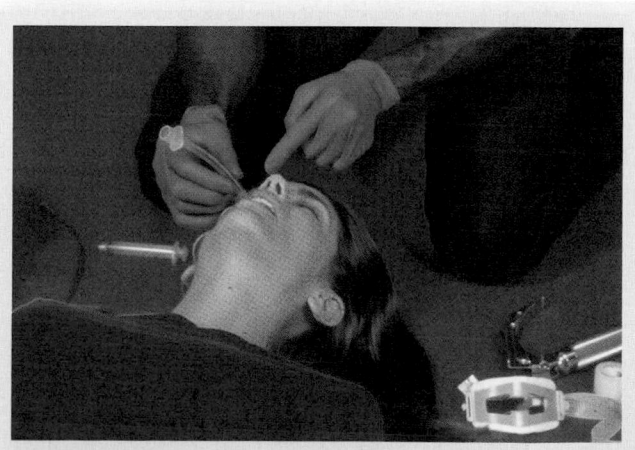

Figure 40-16 Before securing the tube, note the centimeter marking on the endotracheal tube. The tube typically lines up with the teeth on the intubated patient around the 22-cm mark.

tilate the patient at an appropriate rate. Be sure to reassess breath sounds and equal expansion of both sides of the chest each time you move the patient.

Intubation Complications

Intubating the Right Main-stem Bronchus This is a common error that is made during intubation. If the ET tube is placed into the trachea too far, it will pass into the right main-stem bronchus because that bronchus is shorter and straighter than the left main-stem bronchus. In this position, only the right lung will be ventilated; therefore, you will hear breath sounds only on the right side. Correct this problem by deflating the cuff and pulling the ET tube back slightly. Listen again for bilateral breath sounds. Make sure you do not pull the tube back so far that the cuff is above the level of the vocal cords.

Intubating the Esophagus If the ET tube is inserted without first seeing the vocal cords, it may easily be inserted into the esophagus rather than the trachea. The result is rapid inflation of the patient's stomach rather than ventilation of the lungs; this situation must be corrected immediately. Immediately after ventilation has begun, auscultate over the epigastrium and over the left and right apices and bases of the lungs. Watch for the rise

and fall of the chest. If there is any doubt, pull out the ET tube and oxygenate the patient with a bag-mask device before another attempt.

Aggravating a Spinal Injury Whenever there is concern about a spinal injury, intubation must occur without moving the patient's neck from the neutral, in-line position. Make sure to have another EMT hold the patient's head in position during the intubation attempt.

Increased Hypoxia: Taking Too Long to Intubate An intubation attempt should not take any longer than 30 seconds to avoid causing increased hypoxia. If the procedure cannot be completed within 30 seconds, stop and ventilate the patient with a bag-mask device and 100% oxygen for 2 to 3 minutes before trying again.

Patient Vomiting A patient who is not totally unresponsive may begin to gag or try to remove the tube. Gagging may cause the patient to vomit and aspirate stomach contents. To avoid this complication, always check for a gag reflex before intubation. Always have a suction unit ready in case the patient vomits during the intubation procedure. Vomiting is also common when the tube has been misplaced in the esophagus and is removed. Be prepared with immediate suctioning.

You are the Provider: PART 3

The paramedic interprets the patient's cardiac rhythm as ventricular tachycardia; however, the patient remains apneic and pulseless. After defibrillating the patient, the paramedic calls for the immediate resumption of CPR. The patient's wife advises you that he has a history of several heart attacks, hypertension, and type 2 diabetes. After 2 minutes of CPR, the patient is reassessed.

Recording Time: 4 Minutes	
Respirations	Absent
Pulse	Absent
Skin	Cyanotic
Blood pressure	Not obtainable; the patient is pulseless
Oxygen saturation (Sao$_2$)	Not obtainable; the patient is pulseless

An IV line has been established, and the appropriate cardiac drugs are given to the patient. The patient's blood glucose level is assessed and noted to be 104 mg/dL. The paramedic moves to the patient's airway and prepares to perform endotracheal intubation.

5. How does ventricular tachycardia differ from ventricular fibrillation?

6. How can the EMT assist the paramedic in intubating the patient?

Laryngospasm Trying to insert an ET tube through the vocal cords can cause the cords to spasm. If this occurs, the intubation attempt should be stopped, and the patient should be ventilated with a bag-mask device.

Trauma The laryngoscope and the tip of the ET tube can injure the lips, teeth, tongue, gums, and other airway structures. Used as a lever, the laryngoscope blade can easily break teeth, whereas a tube pushed blindly through the vocal cords can lacerate adjacent structures. Careful attention to technique will minimize the risk of trauma.

Mechanical Failure You may hear or feel air coming from the oropharynx when ventilating the patient. In an adult, this means that the cuff may not have enough air in it or that it has been torn and is leaking. If this occurs and ventilation is inadequate, you must get more air into the cuff (check the pilot balloon) or replace the tube. In a child, an air leak may mean that the uncuffed tube is too small or the child is large enough to need a cuffed tube.

Patient Intolerant of the Endotracheal Tube Because of the reversal of hypoxia from direct oxygenation, a patient may regain a gag reflex or regain consciousness and try to remove the ET tube. Before removal of the tube (**extubation**) becomes necessary, ensure that a suction unit is nearby and ready. Then deflate the cuff, and carefully withdraw the ET tube as the patient exhales. Provide immediate suctioning if the patient vomits, reassess the airway, and administer supplemental oxygen. Be aware that conscious patients are at high risk for laryngospasm immediately following extubation. Local advanced life support (ALS) protocol may allow sedation for patients who are intolerant of an ET tube.

Decrease in Heart Rate Be sure to monitor the patient's vital signs carefully and continuously, particularly the heart rate, especially in children. With endotracheal intubation, the heart rate may decrease when the airway is stimulated or if the tube is misplaced in the esophagus. If this occurs, it is important to reassess and confirm tube placement.

Multilumen Airways

In addition to endotracheal intubation, other advanced airway devices are available that do not require visualization of the vocal cords for placement Table 40-1 . **Multilumen airways** such as the **Combitube** and **pharyngeotracheal lumen airway**, which are inserted without direct visualization of the vocal cords, have been

Table 40-1   Benefits and Disadvantages of Multilumen Airways

Benefits	Disadvantages
Ease of proper placement	Loses effectiveness (cuff malfunction)
No mask seal necessary	Requires deeply comatose patient
Requires minimal skill and practice to maintain	Requires constant balloon observation
Easily used in spinal injury patients	Cannot be used on patients shorter than 5′
May be inserted blindly	Requires great care in listening for breath sounds
Protects the airway from upper airway secretions	Large balloon is easily broken and tends to push the PtL out of the mouth when inflated

designed to provide lung ventilation when placed in the trachea or the esophagus, thus making them much easier to insert than an ET tube Figure 40-17 . If the tube happens to go into the trachea, ventilation is provided directly into the lungs as with an ET tube. If the tube goes into the esophagus, as occurs most often, ventilation can still be provided to the patient. You do not need to maintain a constant face-mask seal with these airways because they have a balloon that inflates in the oropharynx, so the lungs can be inflated via a tube rather than a mask. Insertion of these devices requires additional training and practice. Local protocols and your medical director will determine when and how these devices may be used.

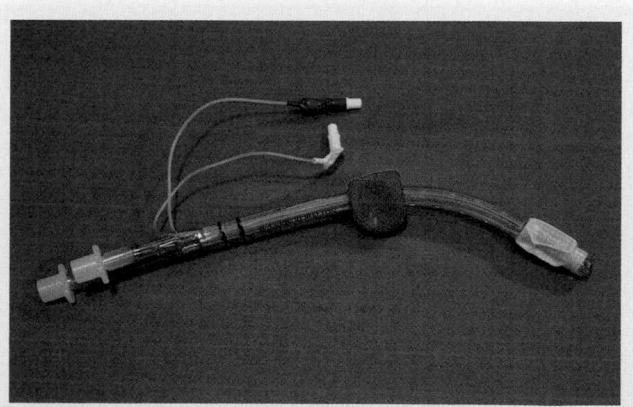

Figure 40-17 Multilumen airways such as the Combitube (shown here) and pharyngeotracheal lumen airway have been designed to provide lung ventilation when placed in the trachea or the esophagus, making them much easier to insert than an endotracheal tube.

■ Contraindications

Multilumen airways should not be used in the following patients:

- Conscious or semiconscious patients with a gag reflex
- Children younger than 14 years
- Adults shorter than 5′
- Patients who have ingested a caustic substance
- Patients who have a known esophageal disease

■ Removing the Multilumen Airway

If the patient will no longer tolerate the airway, it should be removed. Remember that the patient will likely vomit when the airway is removed, so a suction unit must be readily available. Be sure to turn the patient on his or her side to keep the airway clear of vomitus. When you are ready, simply deflate both balloon cuffs, and gently remove the tube.

Single Lumen Airway

■ King LT Airway

The **King LT** is a single lumen airway that is blindly inserted into the esophagus **Figure 40-18**. It consists of a curved tube with ventilation ports located between two inflatable cuffs. Both cuffs are inflated using a single valve/pilot balloon. When the airway is properly placed in the esophagus, one cuff is designed to seal the esophagus, while the other is intended to seal the oropharynx. Openings located between these two cuffs provide ventilation of the lungs.

Contraindications

The King LT is intended for airway management in patients who are taller than 4′. It does not protect the airway from the effects of vomiting and aspiration. High airway pressures may cause air to leak into the stomach or out of the mouth. If the trachea is intubated, the airway must be removed and another attempt made to place it in the esophagus. As with other advanced airway devices, confirm proper placement by observing chest rise, auscultating the epigastrium and lungs, and use of a secondary confirmation device.

■ Laryngeal Mask Airway

The **laryngeal mask airway** was originally developed for use in the operating room **Figure 40-19**. However, since its inception, its use has been expanded to the field, especially as an alternative for basic life support (BLS) providers.

The LMA consists of two parts: the tube and the mask or cuff. The device is made of silicone and is available in reusable (after proper sterilization) and disposable types. After blind insertion, the device molds and seals itself around the laryngeal opening by inflation of the mask. The epiglottis is contained within the mask or cuff. The device comes in seven sizes and can be used in children and in adults.

Gastric Tubes

There may be occasions, such as cardiac arrest, when an advanced airway has been placed in a patient, and the patient now requires placement of a tube through the nose or mouth that extends into the stomach. A nasal

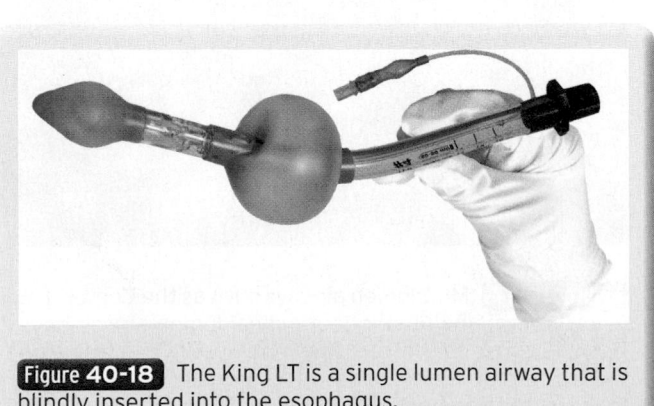

Figure 40-18 The King LT is a single lumen airway that is blindly inserted into the esophagus.

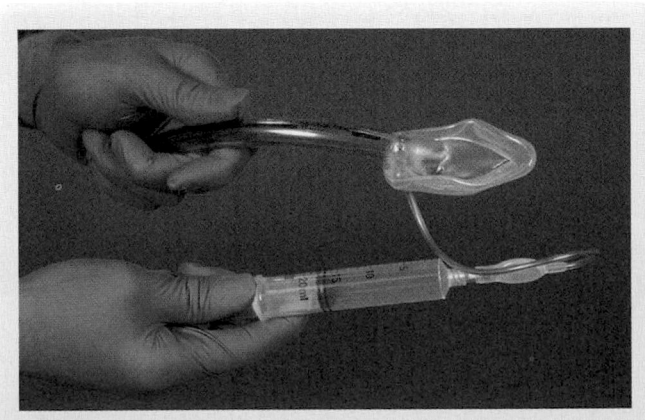

Figure 40-19 The laryngeal mask airway was originally developed for use in the operating room.

or oral **gastric tube** relieves gastric distention caused by the introduction of air into the stomach during positive-pressure ventilation prior to the advanced airway being placed **Figure 40-20**. Gastric distention causes a significant risk of passive regurgitation and aspiration in unconscious patients, which can result in pneumonia that may significantly complicate the potential recovery of your patient. The gastric tube may also be used by emergency department staff to lavage (wash out) the stomach in cases of accidental or intentional overdose.

After the advanced airway has been placed, the gastric tube is measured to determine the proper depth of insertion, lubricated, and then passed through the nostril or mouth and into the stomach. Stomach contents and air can then be aspirated or passively allowed to escape through the tube. Proper placement can be confirmed by aspiration of stomach contents with a syringe, listening with a stethoscope as air is introduced into the gastric tube with a syringe, and by x-ray on arrival at the emergency department.

Continuous Positive Airway Pressure

As discussed in Chapter 9, *Airway Management*, continuous positive airway pressure (CPAP) is used in breathing patients, who are alert and able to follow commands, who have reduced function of the alveoli because of congestive heart failure, chronic obstructive pulmonary disease, and possibly asthma. The devices for CPAP use a tight-fitting mask that is placed over the mouth and nose and connects to an oxygen source capable of delivering flow rates of at least 50 L/min **Figure 40-21**. During the inspiratory phase of respiration, oxygen-rich air is supplied at flow rates high enough to increase airway pressure. During the expiratory phase, the patient exhales against a resistance called positive end-expiratory pressure. The combination of positive inspiratory and positive expiratory pressure results in an increased volume of air in the lungs, increased alveolar surface available for gas exchange, and increased oxygen diffusion across the cell membrane.

In patients with severe respiratory distress when initial treatments of high-flow oxygen administration do not result in increases in oxygen saturation and/or a decrease in the work of breathing CPAP may be helpful. To use CPAP, place the mask tightly over the patient's mouth and nose (this may take some convincing in some patients) and adjust the pressure being delivered by the device according to local protocol. Because there is the potential for the patient's blood pressure to drop while CPAP is being used, make sure to monitor vital signs regularly.

Intravenous Therapy

This section is designed to familiarize you with setting up the equipment necessary to gain IV access. You will learn about the equipment used and understand the importance of early access and how to recognize complications when they happen.

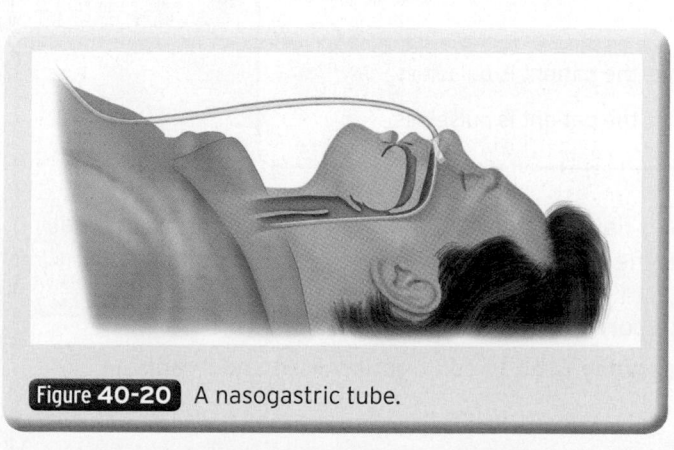

Figure 40-20 A nasogastric tube.

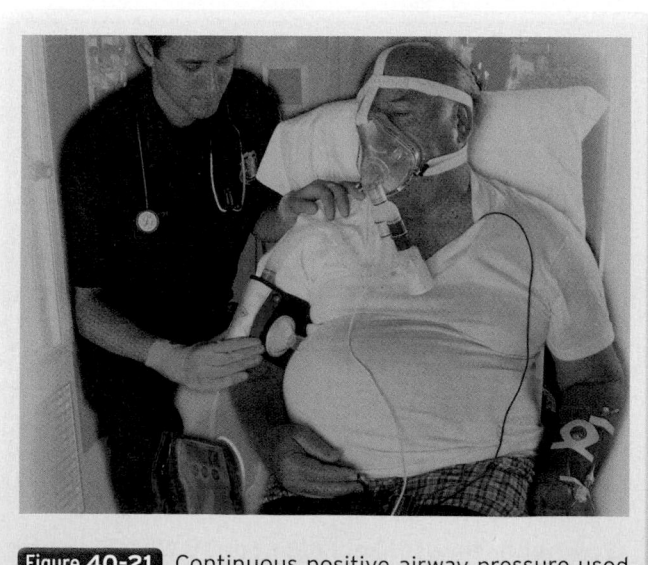

Figure 40-21 Continuous positive airway pressure used in the acute setting is usually administered via a face mask, which must make a tight seal to function properly.

One way to ensure proper technique is to develop a routine to follow as you assemble the appropriate equipment. A routine will help you keep track of your equipment and the steps necessary to complete successful IV administration.

■ Indications

Many medications used by ALS crews are given by the IV route. A fluid bolus may be indicated for patients who are unable to maintain adequate fluid balance because of dehydration caused by vomiting or excessive diarrhea or patients who have experienced blood loss because of hemorrhage.

■ Assembling the Equipment

To avoid delays and the possibility of IV site contamination, gather and prepare all of your equipment before the attempt to start IV administration. Sometimes the condition and presentation of the patient make full preparation difficult. In this situation, working as a team becomes critical. By anticipating the needs of your ALS partner, you can help make the IV equipment assembly possible. **Table 40-2** shows a logical sequence of steps for assembling your equipment. Depending on local protocols, your ALS partner will take care of inserting the catheter, but your assistance will be very helpful in other areas.

Words of Wisdom

Teamwork is critical to good patient care. As an EMT, you are a most valuable member of the team. Learning how to assemble the equipment is often done on the job. You may hear your ALS partner ask you to "Spike 1,000 LR with a macro and get me a 16-gauge catheter." "Spike" means attaching the tubing to an IV bag; "1,000" means the number of milliliters, or size, of the bag; "LR" means lactated Ringer's solution; "macro" means the size of drip chamber to choose for the tubing; and "16-gauge" refers to the size of catheter.

■ Choosing an IV Solution

While ALS providers are likely to select the solution to use based on medical direction or standard operating guidelines, this section discusses likely possibilities so that you may become familiar with them. In the prehospital setting, the choice of IV solution is limited to the **isotonic crystalloids**, normal saline and lactated Ringer's solution. D_5W (5% dextrose in water) is often reserved for administering medication.

You are the Provider: PART 4

After CPR and further defibrillation attempts, the cardiac monitor now shows asystole. Reassessment of the patient reveals that he remains apneic and pulseless. The patient has been intubated, CPR is ongoing, and the appropriate medications are administered.

Recording Time: 6 Minutes	
Level of consciousness	Unconscious and unresponsive
Respirations	Absent
Pulse	Absent
Skin	Cyanotic
Blood pressure	Not obtainable; the patient is pulseless
Sao_2	Not obtainable; the patient is pulseless

The paramedic auscultates the patient's epigastrium and lung fields to confirm continued correct endotracheal tube placement and notes that breath sounds are absent on the left side and present on the right side.

7. What has most likely happened? What must be done to correct the situation?

8. How are end-tidal carbon dioxide detectors and capnography used in conjunction with endotracheal intubation?

Table 40-2 EMT Steps in Assembling IV Equipment

1. Put your gloves on! Standard precautions cannot be emphasized strongly enough.
2. Obtain the solution requested by your ALS partner. Check the bag for clarity, expiration date, and correct solution.
3. Choose an appropriate administration set for the patient.
4. Obtain the catheter requested by your ALS partner. Have several catheters ready for insertion.
5. Spike the bag by inserting the administration set into the port in the fluid bag.
6. Make sure the tubing is clamped, invert the bag, and fill the drip chamber about halfway.
7. Unclamp the tubing to allow fluid to pass through the administration set and completely displace all of the air in the tubing.
8. Tear tape for securing the IV site, or prepare a commercial securing device.
9. Alcohol wipes or skin prep solutions should be ready to cleanse the skin at the puncture site.
10. Have 4″ × 4″ pieces of gauze ready for catching blood.
11. A tourniquet or constricting band will be used to occlude venous flow, resulting in engorgement of the veins distal to the tourniquet and making cannulation easier.
12. After your ALS partner has inserted the catheter, safely dispose of sharps in the appropriate container.
13. Hook up the IV tubing, and adjust the flow.

Each IV solution bag is wrapped in a protective sterile plastic bag and is guaranteed to remain sterile until the posted expiration date **Figure 40-22**. Once the protective wrap is torn and removed, the IV solution has a shelf life of 24 hours. The bottom of each IV bag has two ports: an injection port for medication and an **access port** for connecting the administration set. The sterile access port is protected by a removable pigtail **Figure 40-23**. Once this is removed, the bag must be used immediately or discarded.

Bags of IV solution come in different fluid volumes **Figure 40-24**. The more common prehospital volumes are 1,000 mL and 500 mL.

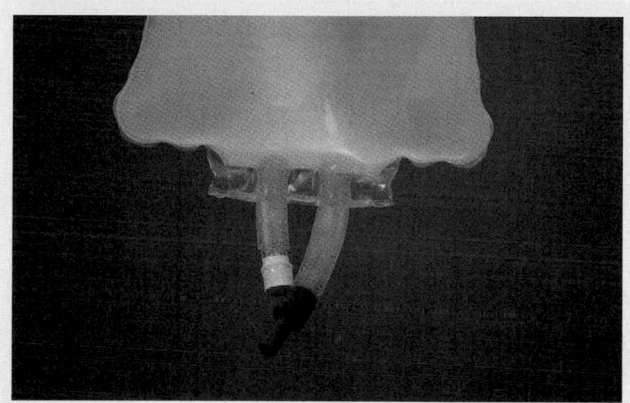

Figure 40-23 The sterile access port is protected by a removable pigtail.

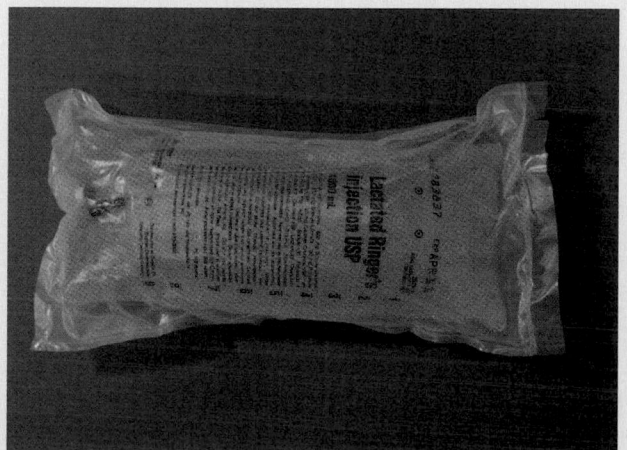

Figure 40-22 Each intravenous solution bag is wrapped in a protective sterile plastic bag and is guaranteed to remain sterile until the posted expiration date.

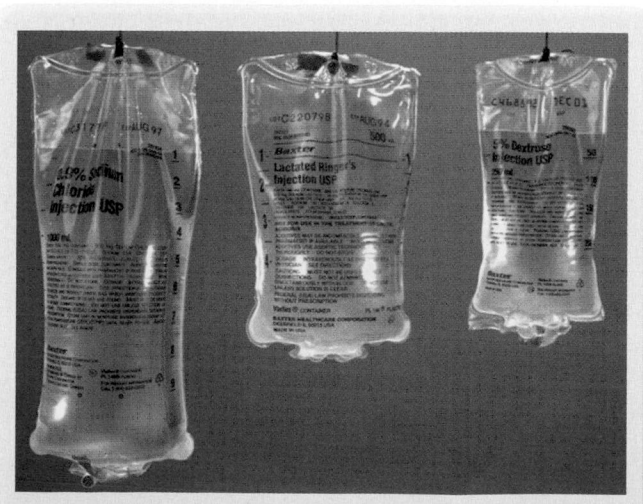

Figure 40-24 Examples of different intravenous bag sizes.

■ Choosing an Administration Set

An administration set moves fluid from the IV bag into the patient's vascular system. As with IV solution bags, IV administration sets are sterile as long as they remain in their protective packaging **Figure 40-25** . Once they are removed from the packaging, their sterility cannot be guaranteed. Each IV administration set has a **piercing spike** protected by a plastic cover. Again, once the piercing spike is exposed and the seal surrounding the cap is broken, the set must be used immediately or discarded.

There are different sizes of administration sets for different situations and patients. Most drip sets have a number visible on the package **Figure 40-26** , which indicates the number of drops it takes for a milliliter of fluid to pass through the orifice and into the **drip chamber**. **Drip sets** commonly used in the prehospital environment come in two primary sizes: microdrip and macrodrip. A **microdrip set** allows 60 gtt (drops)/mL through the small, needlelike orifice inside the drip chamber. Microdrips are ideal for medication administration or pediatric fluid delivery because it is easy to control their fluid flow. A **macrodrip set** allows 10 to 15 gtt/mL through a large opening between the piercing spike and the drip chamber **Figure 40-27** . Macrodrip sets are best used for rapid fluid replacement but can also be used for maintenance.

Words of Wisdom

To differentiate between macrodrip and microdrip sets, remember that the prefixes refer to the size of the drops, not the size of the tubing.

Macro means large. A 10-gtt set, which is a macrodrip set, has 10 drops that equal 1 mL of fluid. Micro means small. A 60-gtt set, which is a micro (or mini) drip, has 60 drops that equal 1 mL of fluid.

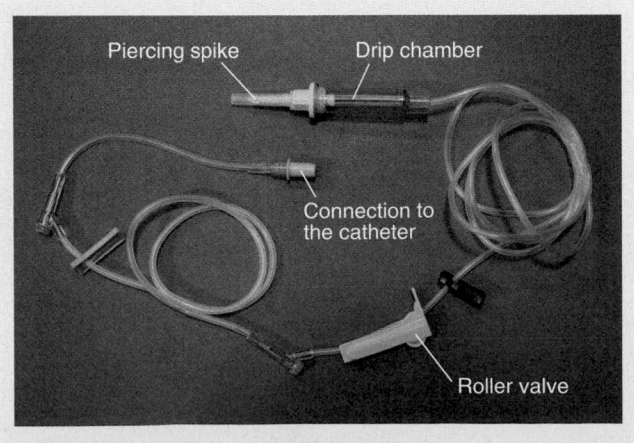

Figure 40-25 An administration set.

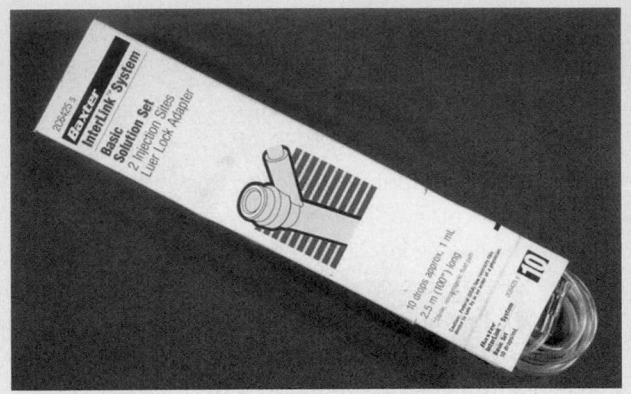

Figure 40-26 The number visible on the drip set refers to the number of drops it takes for a milliliter of fluid to pass through the orifice and into the drip chamber.

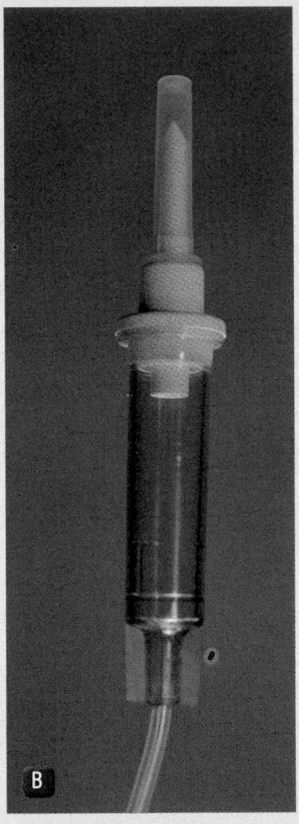

Figure 40-27 **A.** Microdrip sets allow 60 gtt (drops)/mL through the small, needlelike orifice inside the drip chamber. **B.** Macrodrip sets allow 10 to 15 gtt/mL through a large opening between the piercing spike and the drip chamber.

Preparing an Administration Set

After choosing the IV administration set and the IV solution bag, verify the expiration date of the solution and check for solution clarity. To spike the bag with the administration set, follow the steps in **Skill Drill 40-2** :

1. Remove the rubber pigtail found on the end of the IV bag by pulling on it. The bag is still sealed and will not leak until the piercing spike of the IV administration set punctures this port. Remove the protective cover from the piercing spike (remember, this spike is sterile!) (Step 1).

2. Slide the spike into the IV bag port until you see fluid enter the drip chamber (Step 2).

3. Invert the bag. Squeeze and release the drip chamber until about half full. Unclamp the tubing (to allow fluid into the tubing to prime the line and flush the air out of the tubing) (Step 3).

4. Twist the protective cover on the opposite end of the IV tubing to allow air to escape. Carefully remove the cover without breaching sterility. Let

Skill Drill 40-2

Spiking the Bag

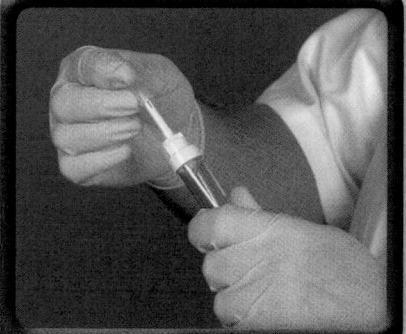

Step 1 Remove the pigtail from the port on the IV bag and the cover from the spike on the administration set.

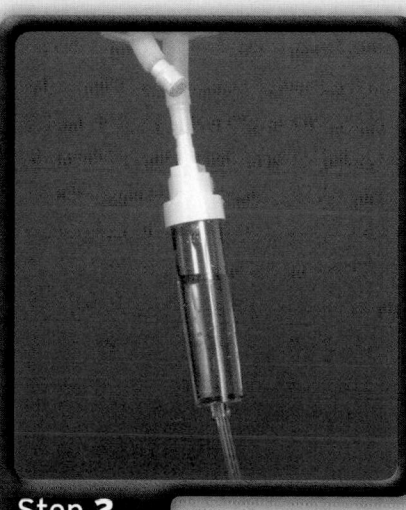

Step 2 Slide the spike into the IV bag port.

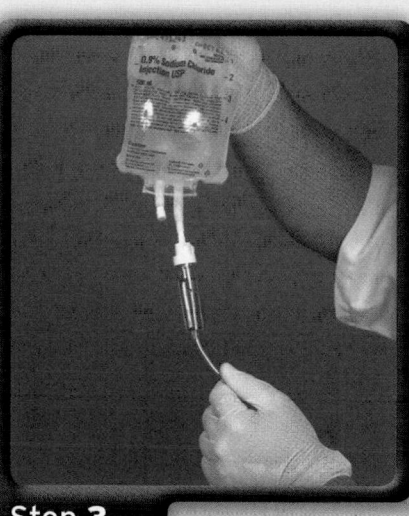

Step 3 Prime the chamber.

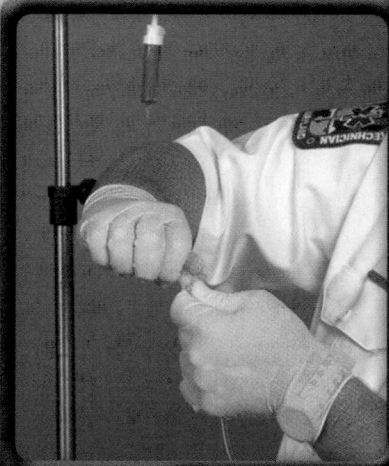

Step 4 Prime the line to remove air.

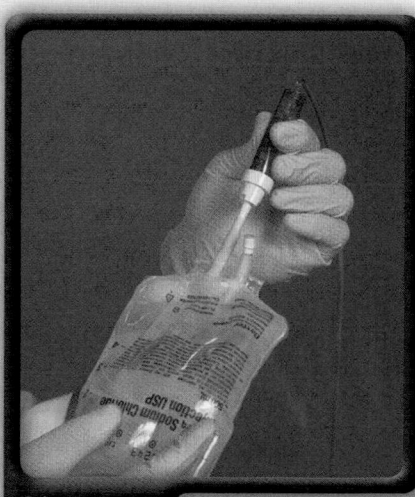

Step 5 Check the drip chamber. It should only be half filled. If the fluid level is too low, squeeze the chamber until it fills. If the drip chamber fills completely, invert the IV bag and squeeze the excess back into the bag.

the fluid flow until air bubbles are removed from the line before turning the roller clamp wheel to stop the flow (Step 4). Replace the cover.

5. Next, go back and check the drip chamber; it should be only half filled. The fluid level must be visible to calculate drip rates. If the fluid level is too low; squeeze the IV bag until it fills. If the chamber is too full, invert the IV bag and squeeze the chamber to empty the fluid back into the bag (Step 5).

6. Hang the bag in the appropriate location with the end of the IV tubing easily accessible.

<u>Saline locks (buff caps)</u> are a way to maintain an active IV site without running fluids through the vein (Figure 40-28). These access devices are used primarily for patients who do not need additional fluids but may need rapid medication delivery. A saline lock is attached to the end of an IV catheter and filled with a few milliliters of normal saline to keep blood from clotting at the end of the catheter. Saline remains in the port without entering the vein.

■ Catheters

An IV <u>catheter</u> is a hollow, laser-sharpened needle inside a hollow plastic tube that is inserted into a vein (Figure 40-29). Once the catheter is properly placed, the needle is removed, leaving the catheter in the vein. Select the catheter size based on the need for the IV, the condition of the patient's veins, and the location for the IV.

Catheters are sized by their diameter and referred to by the <u>gauge</u> of the catheter. A larger-diameter catheter

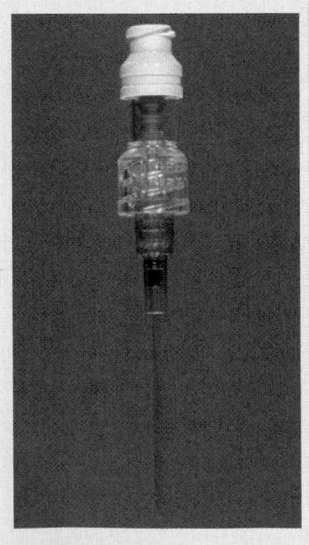

corresponds to a smaller gauge. Thus, a 14-gauge catheter has a greater diameter than a 22-gauge catheter. With larger-diameter catheters, more fluid can be delivered into the vein faster.

Figure 40-28 A saline lock is attached to the end of an intravenous catheter and filled with approximately 2 mL of normal saline.

Starting an IV

The first step in starting an IV is to apply a tourniquet proximal to the site where venipuncture is to be performed. This will engorge the veins with blood, allowing the catheter to be more easily placed in the vein. When a suitable vein is identified, the area should be cleaned according to local protocol

Figure 40-29 An intravenous catheter is a hollow, laser-sharpened needle inside a hollow plastic tube that is inserted into a vein.

to decrease the chance of introducing potentially infectious materials through the skin. The needle/catheter is then introduced into the vein, the needle withdrawn and disposed of properly, and IV tubing or lock placed. Use tape or a commercially available device to secure the catheter so it is not accidentally pulled out. The flow rate is then adjusted according to the purpose of the IV therapy. Remember to always wear gloves during the procedure to protect both you and your patient from being exposed to potentially infectious materials.

Skill Drill 40-3 covers how to start an IV:

1. Choose the fluid appropriate for the patient's condition. Choose the appropriate drip set, and attach it to the fluid bag. A macrodrip set (eg, 10 gtt/mL) should be used for a patient who needs volume replacement; a microdrip set (eg, 60 gtt/mL) should be used for a patient who needs a medication route. Fill the drip chamber halfway by squeezing it (Step 1).

2. Flush or "bleed" the tubing to remove any air bubbles by opening the roller clamp (Step 2). Make sure no errant bubbles are floating in the tubing.

3. Tear tape prior to venipuncture, or have a commercial device available (Step 3).

4. Apply gloves before making contact with the patient. Palpate a suitable vein (Step 4). Veins should be "springy" when palpated. Stay away from areas that are hard when palpated.

5. Apply the constricting band above the intended IV site (Step 5). It should be placed approximately 6″ to 10″ above the intended site.

6. Clean the area using aseptic technique. Use an alcohol pad to cleanse in a circular motion from

the inside out. Use a second alcohol pad to wipe straight down the center (Step 6).

7. Choose the appropriately sized catheter, and twist the catheter to break the seal. Do not advance

the catheter upward because this may cause the needle to shear the catheter. Examine the catheter and discard it if you discover any imperfections (Step 7). Occasionally you will find "burrs" on the edge of the catheter.

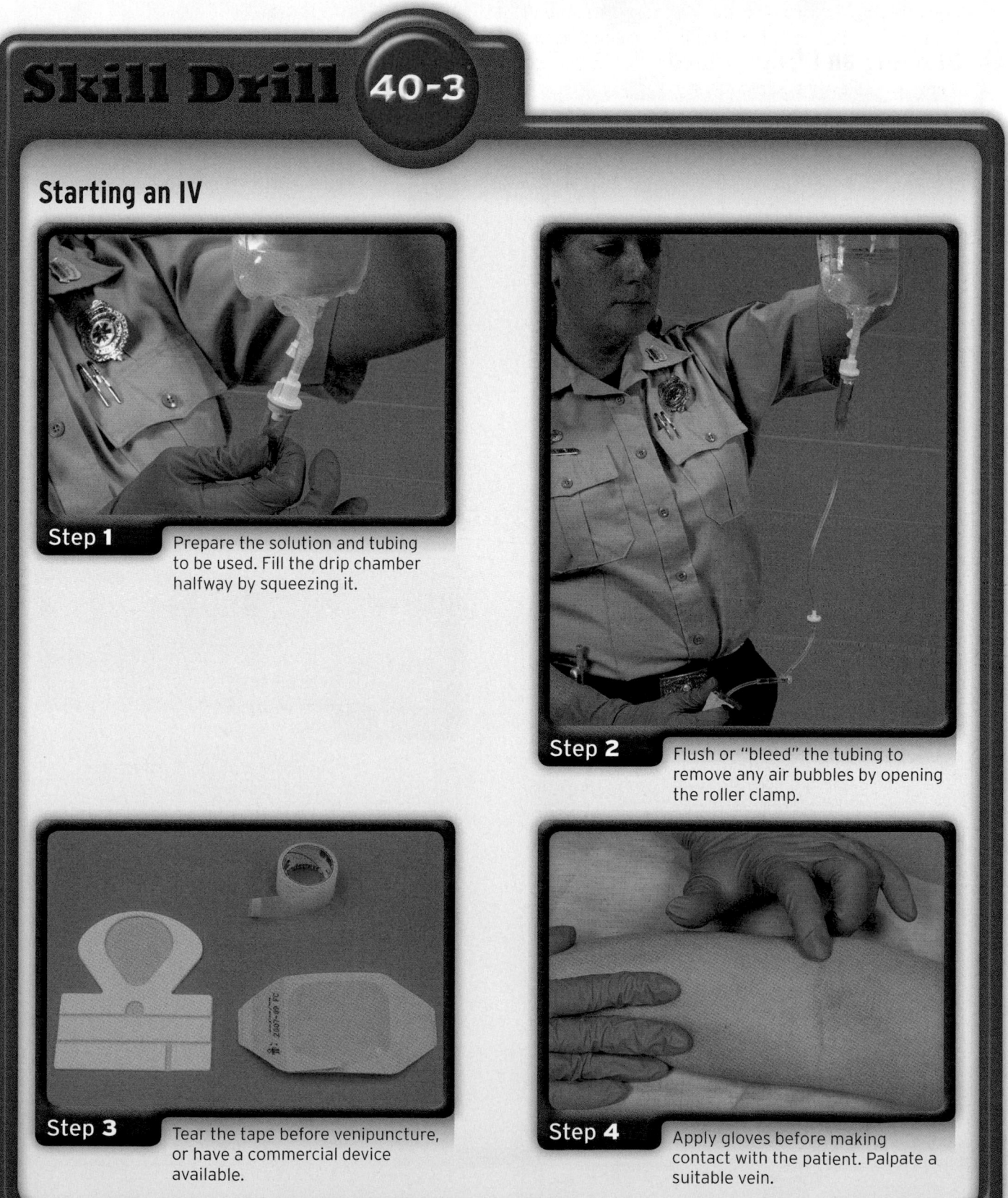

Skill Drill 40-3

Starting an IV

Step 1 Prepare the solution and tubing to be used. Fill the drip chamber halfway by squeezing it.

Step 2 Flush or "bleed" the tubing to remove any air bubbles by opening the roller clamp.

Step 3 Tear the tape before venipuncture, or have a commercial device available.

Step 4 Apply gloves before making contact with the patient. Palpate a suitable vein.

Skill Drill 40-3

Starting an IV, continued

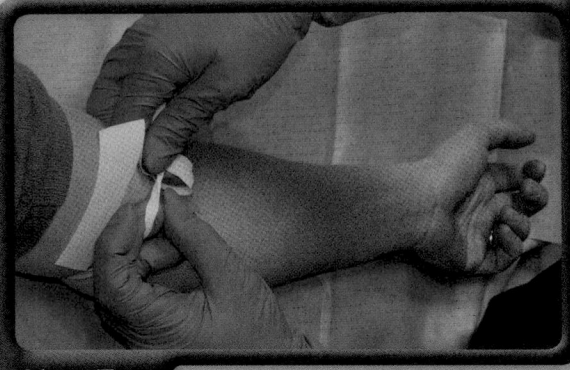

Step 5   Apply the constricting band above the intended IV site.

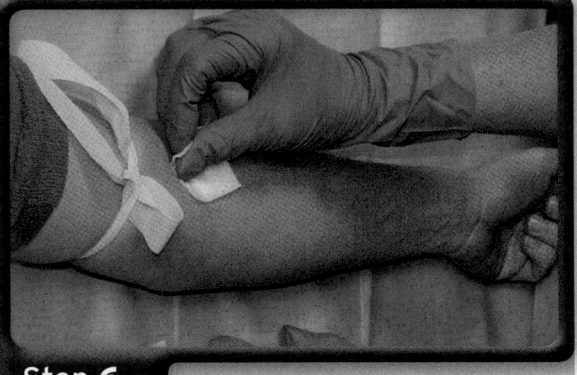

Step 6   Clean the area using aseptic technique.

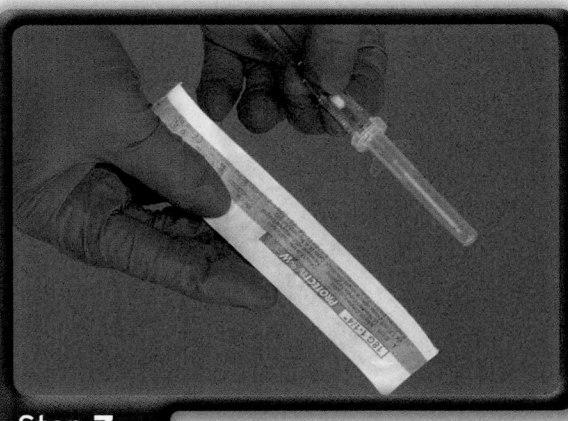

Step 7   Choose the appropriately sized catheter, and examine it for any imperfections.

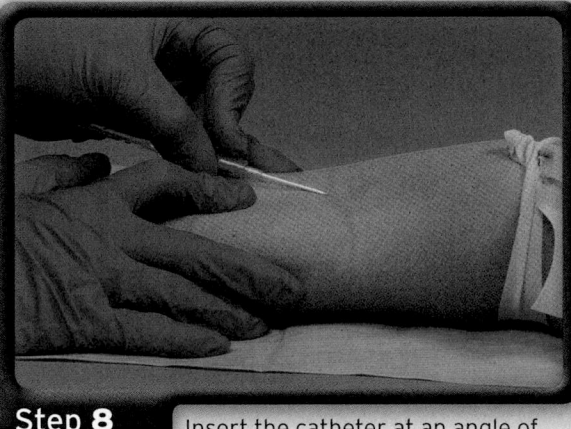

Step 8   Insert the catheter at an angle of approximately 45° with the bevel up while applying distal traction with the other hand.

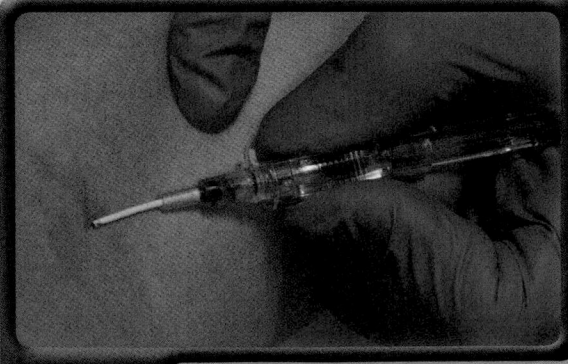

Step 9   Observe for "flashback" as blood enters the catheter.

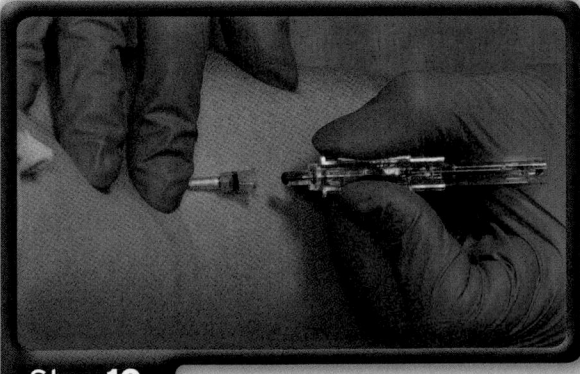

Step 10   Occlude the catheter to prevent blood leaking while removing the stylet.

Skill Drill 40-3

Starting an IV, continued

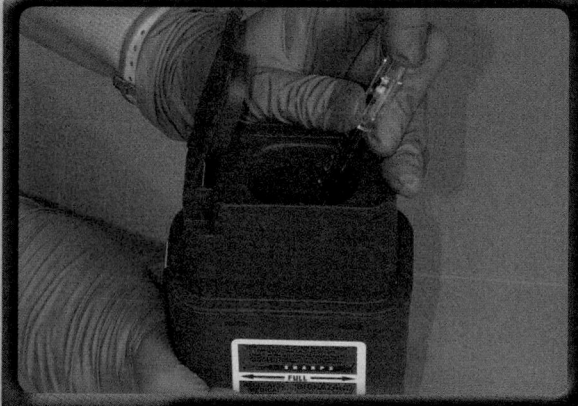

Step 11 Dispose of all sharps In the proper container.

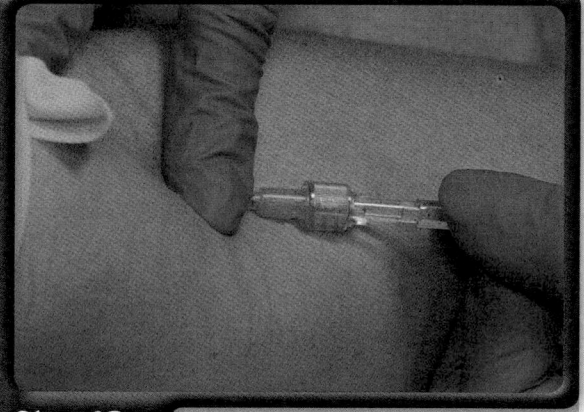

Step 12 Attach the prepared IV line.

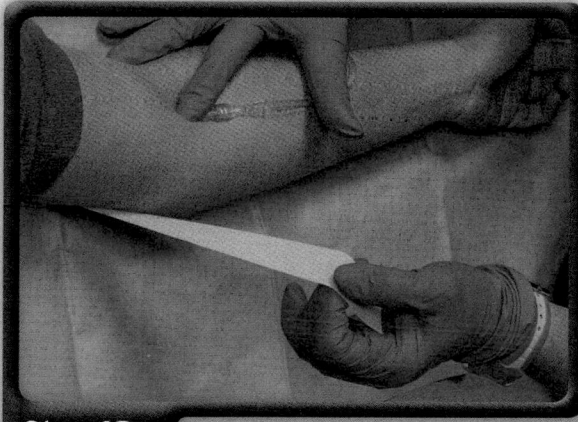

Step 13 Remove the constricting band.

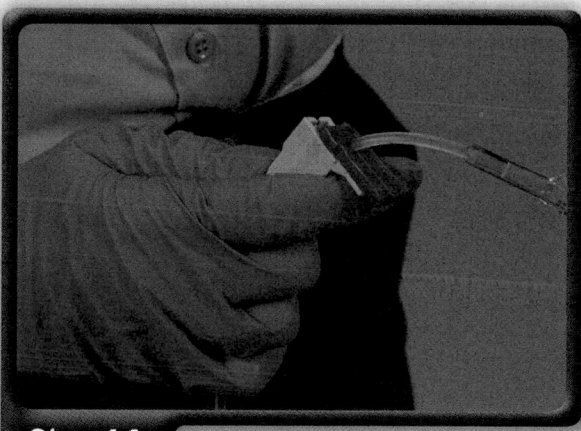

Step 14 Open the IV line to ensure fluid is flowing and the IV is patent. Observe for swelling and infiltration around the IV site.

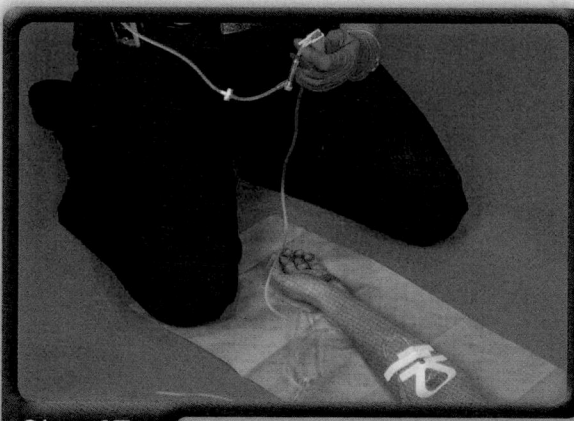

Step 15 Secure the catheter with tape or a commercial device. Secure IV tubing and adjust the flow rate.

8. Insert the catheter at an angle of approximately 45° with the bevel up while applying distal traction with the other hand (Step 8). This traction will stabilize the vein and help to keep it from "rolling" as you stick.

9. Observe for "flashback" as blood enters the catheter (Step 9). The clear chamber at the top of the catheter should fill with blood when the catheter enters the vein. If you note only a drop or two, you should gently advance the catheter farther into the vein.

10. Occlude the catheter to prevent blood leaking while removing the stylet (Step 10). Place the thumb of the hand not holding the catheter over the end of the catheter that is currently situated inside the vein to prevent blood running out when you remove the needle. With practice, you will be able to feel the catheter.

11. Immediately dispose of all sharps in the proper container (Step 11).

12. Attach the prepared IV line (Step 12).

13. Remove the constricting band (Step 13).

14. Open the IV line to ensure fluid is flowing and the IV is patent. Observe for any swelling or infiltration around the IV site (Step 14). If the fluid does not flow, check whether the constriction band has been released. If infiltration is noted, immediately stop the infusion and remove the catheter while holding pressure over the site to prevent bleeding.

15. Secure the catheter with the tape or the commercial device.

16. Secure IV tubing and adjust the flow rate while monitoring the patient (Step 15).

Special Populations

Bariatric and pediatric patients can offer special challenges for inserting an intravenous catheter. If protocols allow, consider using specialized equipment such as an intraosseous infusion.

Words of Wisdom

When inserting an intravenous catheter in a patient who is frightened of needles, make sure he or she is lying down before the paramedic begins catheter insertion. Advise the patient of ALS each step, even when you are only cleansing the site.

■ Securing the Line

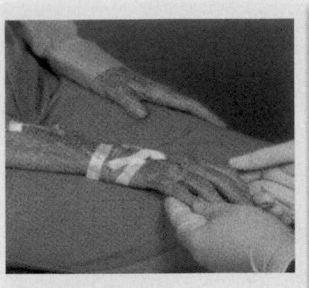

Figure 40-30 Tape the area so that the catheter and tubing are securely anchored.

Once the catheter is in position and the contents of the IV bag are flowing properly, the site must be secured. Tape the area so that the catheter and tubing are securely anchored in case of a sudden pull on the line (Figure 40-30). You should tear the tape before the IV catheter is inserted because you will need one hand to stabilize the site while you tape the IV catheter and tubing. Double back the tubing to create a loop that will act as a shock absorber if the line is pulled accidentally. Avoid circumferential taping around any extremity because circumferential taping can act like a constricting band and stop circulation.

■ Alternative Intravenous Sites and Techniques

Intraosseous (IO) needles are used for emergency venous access as defined by protocol when other IV access is difficult or impossible (Figure 40-31). Often patients are experiencing a life-threatening situation such as cardiac arrest, status epilepticus, or progressive shock. The IO needles are generally inserted in the proximal tibia. A double needle, consisting of a solid boring needle inside a sharpened hollow needle, is pushed or screwed into the bone.

An external jugular IV provides venous access through the external jugular veins of the neck. These are the same veins used to assess jugular vein distention. The vein is compressed by placing a finger on the vein above the clavicle, causing the vein to fill.

The catheter is inserted into the vein in the same manner as any other IV catheter, except the insertion point is very specific. The catheter is inserted midway between the angle of the jaw and the midclavicular line. These punctures can be difficult because these veins are surrounded by a very tough, fibrous sheath that may make access difficult. This is an advanced skill that can be performed only by properly trained providers with the permission of a medical director.

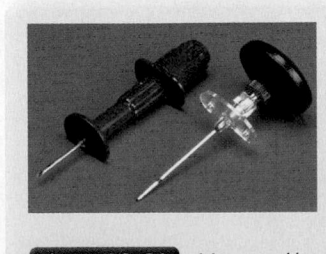

Figure 40-31 Manually inserted intraosseous needles.

Possible Complications of Intravenous Therapy

Peripheral IV insertion carries risks. The problems associated with IV administration can be categorized as local and systemic reactions. Local reactions include problems like infiltration and phlebitis. Systemic complications include allergic reactions and circulatory overload.

Local IV Site Reactions

Infiltration

Infiltration is the escape of fluid into the surrounding tissue when the IV catheter is not in the vein. This escape of fluid can cause a localized area of edema or swelling. Some of the more common reasons for infiltration include the following:

- The IV catheter has passed completely through the vein and out the other side.
- The IV catheter tip moves out of the vein because of patient movement or the tape securing the line coming loose.
- The catheter was inserted at too shallow an angle and has only entered the tissue surrounding the vein.
- Fluid is escaping from the vein because of prior venipuncture.

The following are some of the signs and symptoms of infiltration:

- Edema or swelling at the catheter site
- Extremely slow IV flow
- Patient complaint of tightness and pain around the IV site

To correct the infiltration, stop the flow, remove the IV catheter, and reinsert it at an alternative site. Apply direct pressure over the swollen area to reduce further swelling and/or bleeding into the tissue.

You are the Provider: PART 5

The patient remains apneic and pulseless, and the cardiac monitor continues to show asystole. The paramedic asks the AEMT to administer a bolus of IV fluids to rule out hypovolemia as an underlying cause of the patient's cardiac arrest.

Recording Time: 12 Minutes	
Level of consciousness	Unconscious and unresponsive
Respirations	Absent
Pulse	Absent
Skin	Cyanotic
Blood pressure	Not obtainable; the patient is pulseless
Sao$_2$	Not obtainable; the patient is pulseless

Despite the administration of a bolus of normal saline through a large-bore IV catheter, as well as additional treatment, the patient is still apneic, pulseless, and in asystole. After conferring with the patient's wife and contacting medical control, resuscitation efforts are stopped.

9. What is the role of vascular access in the resuscitation effort of a patient in cardiac arrest?

10. Why does asystole often not respond to even the most aggressive treatment?

Phlebitis

Phlebitis is inflammation of the vein. Phlebitis is not usually seen in emergency prehospital patients. Often phlebitis is associated with fever, tenderness, and red streaking along the course of the associated vein. Some of the more common causes of phlebitis include localized irritation and infection from nonsterile equipment, prolonged IV therapy, and irritating IV solutions.

Occlusion

In IV therapy, **occlusion** is the physical blockage of a vein or catheter. If the flow rate is not sufficient to keep fluid moving out of the catheter tip or if blood enters the catheter, a clot may form and occlude the flow. Proximity to a valve may often cause occlusion. Other causes can be related to patient movement that allows the line to become physically blocked, such as resting on the IV line or crossing the arms. Occlusion may also develop if the IV bag is nearly empty and the blood pressure overcomes the flow and causes blood to back up into the line.

Vein Irritation

Occasionally, a patient will experience vein irritation in reaction to the IV fluid. This is more common with IV medication administration and very uncommon with administration of pure IV fluids. Patients who have this problem often complain immediately that the IV is bothering them. It may tingle, sting, or itch. Note these complaints, and observe the patient closely in case a more serious allergic reaction develops.

Hematoma

A hematoma is an accumulation of blood in the tissues surrounding an IV site Figure 40-32. Hematomas result from vein perforation or catheter removal, which allows blood to accumulate in the surrounding tissues. If a hematoma develops when IV catheter insertion is attempted, the procedure should stop. Direct pressure should be applied to help minimize bleeding. Application of ice may help. If a hematoma develops after a successful catheter insertion, evaluate the IV flow and the hematoma. If the hematoma appears to be controlled and the flow is not affected, monitor the IV site and leave the line in place.

■ Systemic Complications

A **systemic complication** can evolve from reactions or complications associated with IV insertion. Systemic complications usually involve other body systems and can be life threatening.

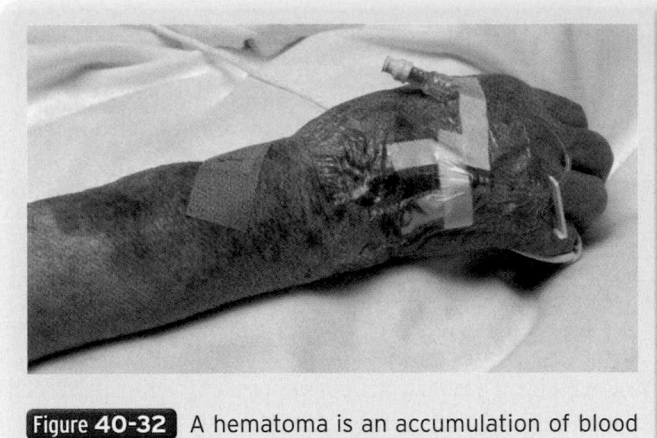

Figure 40-32 A hematoma is an accumulation of blood in the tissues surrounding an intravenous site.

Allergic Reactions

Often allergic reactions are minor, but true anaphylaxis is possible and must be treated aggressively. Allergic reactions can be related to a person's unexpected sensitivity to an IV fluid or (much more commonly) medication. Such a sensitivity could be an unknown condition to the patient; thus, vigilance must be maintained with any IV therapy for a possible reaction. Patient presentation depends on the extent of the reaction. Common signs and symptoms of an allergic reaction include the following:

- Itching
- Edema of face and hands
- Bronchospasm
- Wheezing
- Shortness of breath
- Hives

If an allergic reaction occurs, the ALS provider must discontinue the IV fluid and remove the solution. The catheter can be left in place with a saline lock as an emergency medication route. Maintain the airway, and monitor ABCs and vital signs.

Air Embolus

Healthy adults can tolerate as much as 200 mL of air introduced into the circulatory system, but patients who are already ill or injured can be adversely affected if any air is introduced. Properly flushing an IV line will help eliminate any potential of introducing air into a patient. IV bags are designed to collapse as they empty to help prevent this problem. Be sure to replace empty IV bags with full ones. If your patient begins developing unexplained respiratory distress, consider the possibility of an air embolus. Treat a patient with a suspected air embolus by placing the patient on his or her left side with the head down. Symptomatic air embolus is an extremely rare event and should be considered only after more common explanations for the patient's presenting symptoms have been excluded.

Circulatory Overload

An unmonitored IV bag can lead to circulatory overload. Healthy adults can handle as much as 2 to 3 extra liters of fluid without compromise. Problems occur more frequently when the patient has cardiac, pulmonary, or renal dysfunction. These types of conditions do not allow the patient to tolerate the additional demands associated with increased circulatory volume. The most common cause for circulatory overload in the prehospital setting is failure to readjust the drip rate after flushing an IV line immediately after insertion. Always monitor IV bags to ensure the proper drip rate. Patient presentation includes shortness of breath, jugular vein distention, and increased blood pressure. Crackles are often heard when evaluating breath sounds. Acute peripheral edema can also indicate circulatory overload.

To treat a patient with circulatory overload, slow the IV rate to keep the vein open, and raise the patient's head to ease respiratory distress. Administer high-flow oxygen, and monitor vital signs and shortness of breath.

Vasovagal Reactions

Some patients have anxiety concerning needles or in response to the sight of blood. Patients can present with anxiety, diaphoresis, nausea, and syncopal episodes. Treatment for patients with vasovagal reactions centers on providing supportive care:

1. Lower the head of the stretcher.
2. Administer oxygen if indicated.
3. Monitor vital signs.

Catheter Shear

Although uncommon, **catheter shear** is a potential complication when starting an IV that could have a devastating effect on your patient. Catheter shear may occur if you attempt to reinsert the needle through the catheter after the needle has been partially withdrawn. Once the needle has been even partially withdrawn, it should never be reintroduced through the catheter. The sharp tip of the needle can cut through the plastic of the catheter, resulting in a portion of the catheter being detached and released into the bloodstream.

■ Troubleshooting

Several factors can influence the IV flow rate. For example, if the IV bag is not hung high enough, the flow rate will not be sufficient. It is always helpful to perform the following checks after completing IV administration.

Also, if there is a flow problem, rechecking these items will help you determine the problem:

- Check your administration set. Macrodrips are used for rapid fluid delivery, whereas microdrips are designed to deliver a more controlled flow.
- Check the height of the IV bag. The IV bag must be hung high enough to overcome the patient's own blood pressure. Hang the bag as high as possible.
- Check the type of catheter used. The wider the catheter (the smaller the gauge), the more fluid can be delivered; 14 gauge is the widest, 27 gauge is the narrowest.
- Check the tourniquet. Leaving the tourniquet on the patient's arm after establishing IV access can prevent the IV fluid from flowing at the proper rate.

■ Age-Specific Considerations

Pediatric and geriatric populations warrant specific attention. What sets these populations apart are physical differences specific to the populations and communication barriers that might prevent the patients from expressing themselves.

IV Therapy for Pediatric Patients

The same IV solutions and equipment used for adults can be used for pediatric patients with a few exceptions. Physically, a child has smaller veins. A small-gauge catheter should be used so the vessel will be traumatized as little as possible during IV placement. A 22-gauge to 24-gauge catheter is appropriate for most pediatric patients. Volume control for pediatric patients is important. Using a special type of microdrip set called a Volutrol allows you to fill the large drip chamber with a specific amount of fluid and administer only this amount to avoid fluid overload. The 100-mL calibrated drip chamber can be shut off from the IV bag.

IV Therapy for Geriatric Patients

Smaller catheters may be preferable for geriatric patients unless rapid fluid replacement is needed. Some medications commonly used by older patients have the tendency to create fragile skin and veins. Often, simply puncturing the vein will cause a massive hematoma. The use of tape can lead to skin damage, so be careful when taping IV catheters and tubing on older patients.

The use of the smaller catheters may be more comfortable for the patient and can reduce the risk of fluid leaking into surrounding tissues. If fluid resuscitation is necessary, an appropriately sized catheter must be used.

Be careful when using macrodrips because they can allow rapid infusion of fluids, which may lead to fluid overload if they are not monitored closely. With both geriatric and pediatric patients, fluid overload is a real possibility and can happen quickly. If necessary, use the Volutrol IV set to prevent fluid overloads **Figure 40-33**.

Cardiac Monitoring

Within the past few years, a number of studies have examined the various aspects of the use of the 12-lead **ECG**, or electrocardiogram, in the prehospital setting. In 1993, the National Institutes of Health recommended that, "EMS systems should consider providing prehospital 12-lead ECGs to facilitate early identification of AMI [acute myocardial infarction]." In fact, studies have indicated a 95% or better accuracy rate in the diagnosis of myocardial infarction when a 12-lead ECG is used. Why is this so important? The evaluation and treatment of a patient experiencing the signs and symptoms of AMI has progressed to the point that use of a 12-lead ECG is rapidly becoming the norm when one considers the overall patient outcome. Early identification of an impending AMI allows the EMS provider to alert the receiving hospital, which allows the hospital to prepare before receiving the patient, drastically reducing the time to definitive care. Reduction in cell death results in a more positive patient outcome.

Although the interpretation of cardiac rhythm may not be an EMT skill, you will find it quite helpful to be able to place electrodes and leads on a patient in preparation for **cardiac monitoring**. Whether it be a 4- or 12-lead system, it may also be helpful to be able to recognize a normal ECG tracing and become familiar with basic rhythm disturbances.

The identification of cardiac conditions and correct management in the field requires that you have a good understanding of the basic anatomy and physiology of the heart and that you understand the basic terminology and techniques of cardiac monitoring.

Safety

Movement, such as moving down the road in the ambulance, may cause artifact on the cardiac monitor.

Electrical Conduction System of the Heart

Before discussing how to assist with cardiac monitoring, basic information on the electrical conduction system of

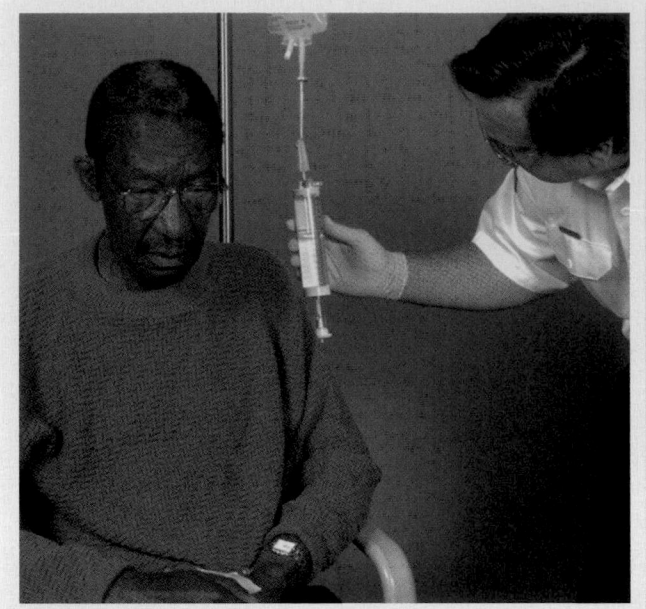

Figure 40-33 It may be necessary to use the Volutrol IV set to prevent fluid overload.

the heart is required to understand what is happening on an ECG.

As discussed in Chapter 5, *The Human Body*, the heart contains a network of specialized tissue that is capable of conducting electrical current throughout the heart called the **electrical conduction system**. The flow of electrical current through this network causes contractions of the heart that produce pumping of blood.

When the heart is working normally, the electrical impulse moves through the electrical conduction system and produces a coordinated contraction. If the heart is deprived of oxygen or is injured, the electrical system may not function normally and the heart may not beat properly. Blood pressure decreases, and the patient may lose consciousness.

The Process of Electrical Conduction

The main function of the electrical conduction system is to create an electrical impulse and transmit it through the heart in an organized manner. Electrical conduction in the heart occurs through this pathway of special cells **Figure 40-34**. As the impulse travels through this system, it is conducted to the "working" myocardial cells that contract in response to the stimulus and cause the heart to squeeze. The electrical conduction system contains the following components:

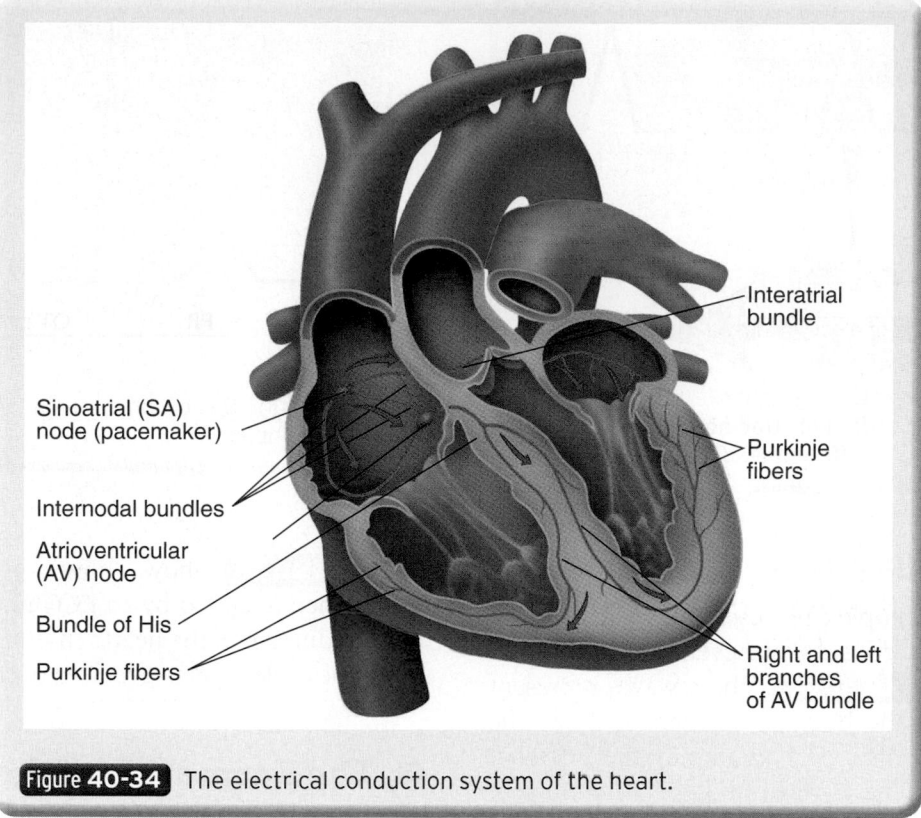

Figure 40-34 The electrical conduction system of the heart.

- The sinoatrial (SA) node, the heart's main pacemaker, which is located in the wall of the right atrium where it meets the superior vena cava
- Three internodal pathways that transmit the pacing impulse from the SA node to the atrioventricular (AV) node
- The AV node, which transmits the impulse from the atria to the ventricles
- The bundle of His, which starts at the AV node and then splits into the right and left bundle branches.
- The right and left bundle branches, which travel through the interventricular septum and lead to the Purkinje fibers in the ventricular walls

For the heart to pump, one of the parts of the electrical conduction system needs to act as the heart's pacemaker. In a normally functioning heart, the SA node performs this function. It paces at a rate of 60 to 100 beats/min. Every cell in the conduction system is capable of setting the pace. However, the rate of each segment of the conduction system is slower than the segment that precedes it. This means that the fastest pacer is the SA node, the next fastest is the AV node, and so on.

Words of Wisdom

When applying electrocardiographic electrodes to your patient, make sure there is good contact between the pad and skin.

■ Electrodes and Waves

The ECG electrodes pick up the electrical activity of the heart, and the ECG machine converts them to waves. When an electrical impulse is moving away from the positive electrode, the ECG machine converts it into a negative (downward) wave. When a wave moves toward a positive electrode, the ECG machine records a positive (upward) wave.

The way an ECG tracing looks depends on where the lead is placed **Figure 40-35**. For example, if an electrical impulse is moving toward the patient's left side, a lead on the right arm will create a negative wave on the ECG, and a lead on the left arm will create a positive wave.

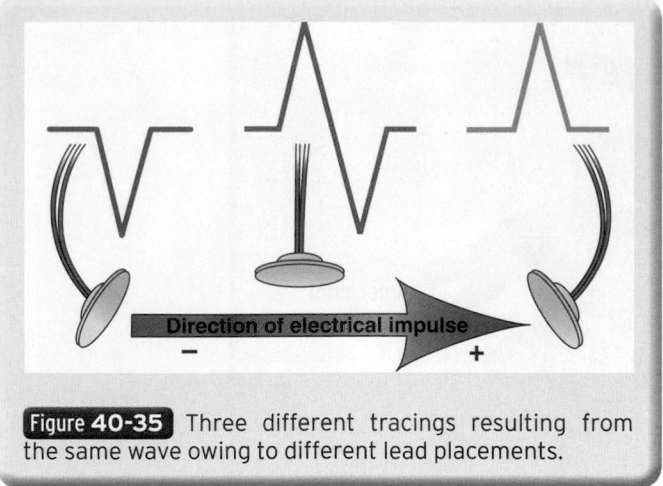

Figure 40-35 Three different tracings resulting from the same wave owing to different lead placements.

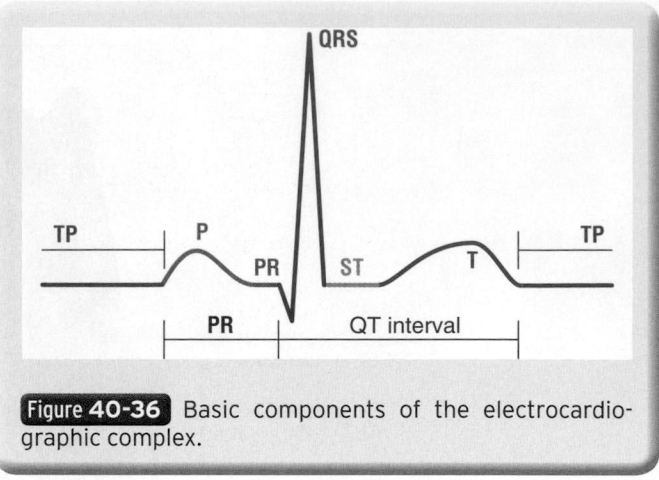

Figure 40-36 Basic components of the electrocardiographic complex.

The ECG Complex

On the ECG, one complex represents one beat in the heart. The complex consists of several waves: the P, QRS, and T waves **Figure 40-36**. These waves represent electrical activity in the heart. A segment is a specific portion of the complex. For example, the segment between the end of the P wave and the beginning of the Q wave is known as the P-R segment. An interval is the distance, measured as time, occurring between two cardiac events. The time between the beginning of the P wave and the beginning of the QRS complex is known as the P-R interval. Note that there is a P-R interval and a P-R segment.

ECG Paper

The paper on which an ECG is recorded contains a grid. Each little box on the ECG paper represents 1/25 of a second, or 0.04 seconds. Each bigger box on the paper is composed of five smaller boxes, making each big box 5 × 0.04 seconds, or 0.20 seconds. Finally, five big boxes equal 1 second. By knowing how much time each box on the grid represents, we can look for problematic waves or intervals that are slower or faster than normal.

Normal Sinus Rhythm

Sinus rhythm is a rhythm in which the SA node acts as the pacemaker **Figure 40-37**. All of the P waves on the patient's ECG should be the same. A normal rate for most people is from 60 to 100 beats/min. A rhythm strip with consistent P waves, consistent P-R intervals, and a regular heart rate of between 60 and 100 beats/min is showing normal sinus rhythm

Figure 40-38 shows how normal sinus rhythm looks when recorded by an ECG machine. Keep in mind that production of the heart's rhythm is a continuous process. The cycle repeats over and over again.

The Formation of the ECG

When you are viewing a normal sinus rhythm that has been recorded by an ECG machine, keep in mind that production of the heart's rhythm is a continuous process, with no actual period of rest or inactivity. The process is as follows:

- The baseline is a period when the majority of the cardiac muscle is at rest **Figure 40-39A**.
- The SA node produces an electrical impulse that is conducted through the internodal pathways and into the atrial cardiac muscle cells (P-wave), resulting in atrial depolarization **Figure 40-39B**, **Figure 40-39C**.
- The impulse passes through the AV node; the only route of communication between the atria and the ventricles (PR interval) **Figure 40-39D**.
- The impulse next travels through the bundle of His, right and left bundles, the fascicles, and the Purkinje system. As the impulse travels through the ventricular muscle cells, the ventricles contract, and the QRS complex is formed **Figure 40-39E-H**.
- As the ventricular muscle cells repolarize the T wave is formed. Cardiac cells may not respond to another electrical impulse until the cell is repolarized **Figure 40-39I-K**.

When the tracing returns to the baseline, the heart muscle is relaxing. If the heart is functioning normally, the above process will repeat over and over continuously.

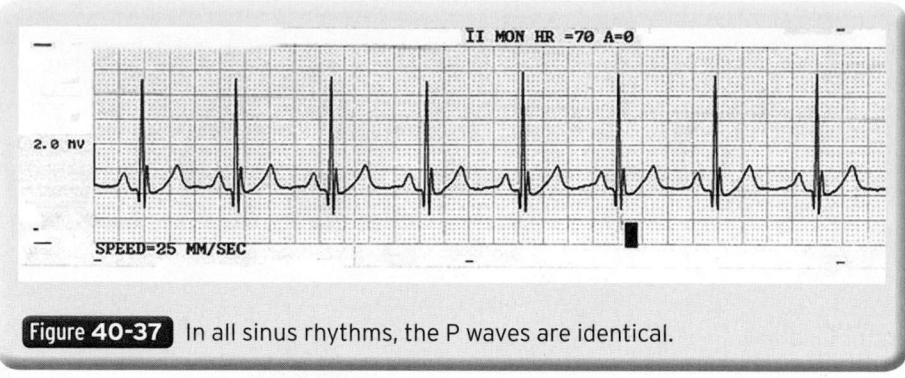

Figure 40-37 In all sinus rhythms, the P waves are identical.

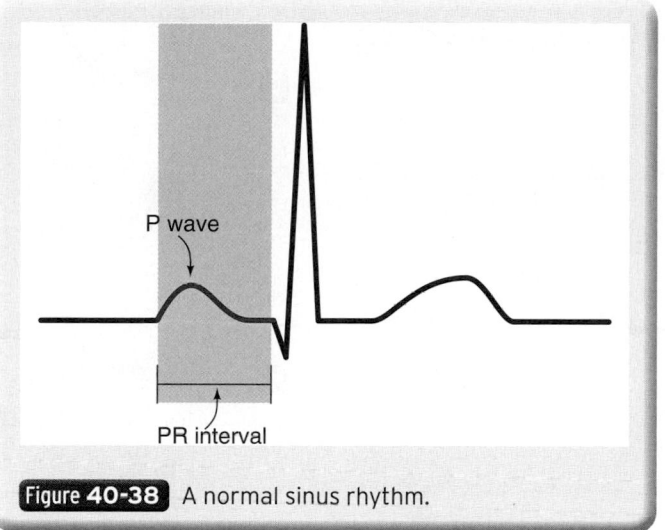

P wave

PR interval

Figure 40-38 A normal sinus rhythm.

■ Arrhythmias

An **arrhythmia** is an abnormal rhythm of the heart. The following section discusses some basic arrhythmias with which you should be familiar.

Sinus Bradycardia

Bradycardia refers to a slow heart rate, usually less than 60 beats/min. Therefore, **sinus bradycardia** is a rhythm that has consistent P waves, consistent P-R intervals, and a regular heart rate that is less than 60 beats/min **Figure 40-40**.

Most patients can tolerate heart rates of between 50 and 60 beats/min without much difficulty. Sinus brady-cardia typically becomes a problem when the heart rate drops to less than 50 beats/min. Under normal circumstances, however, heart rates as slow as the low 40s may be normal for very well-conditioned athletes and for some patients during sleep.

Sinus Tachycardia

Tachycardia refers to a fast heart rate, usually more than 100 beats/min. **Sinus tachycardia** is a rhythm that has consistent P waves, consistent P-R intervals, and a regular heart rate that is more than 100 beats/min **Figure 40-41**.

Tachycardia may cause a decrease in cardiac output when the rate becomes so high that the stroke volume is affected. Tachycardia decreases the amount of passive filling time. In other words, as the amount of time available to fill the ventricles decreases with increasing heart rates, the less the ventricles are filled, thereby decreasing cardiac output.

As mentioned, the heart rate in sinus tachycardia is more than 100 beats/min. At this rate, the tachycardia itself does not pose significant problems. The heart rate in sinus tachycardia, however, can go up to 160 or even 220 beats/min in rare circumstances. At these rates, the rhythm can pose clinical and diagnostic challenges.

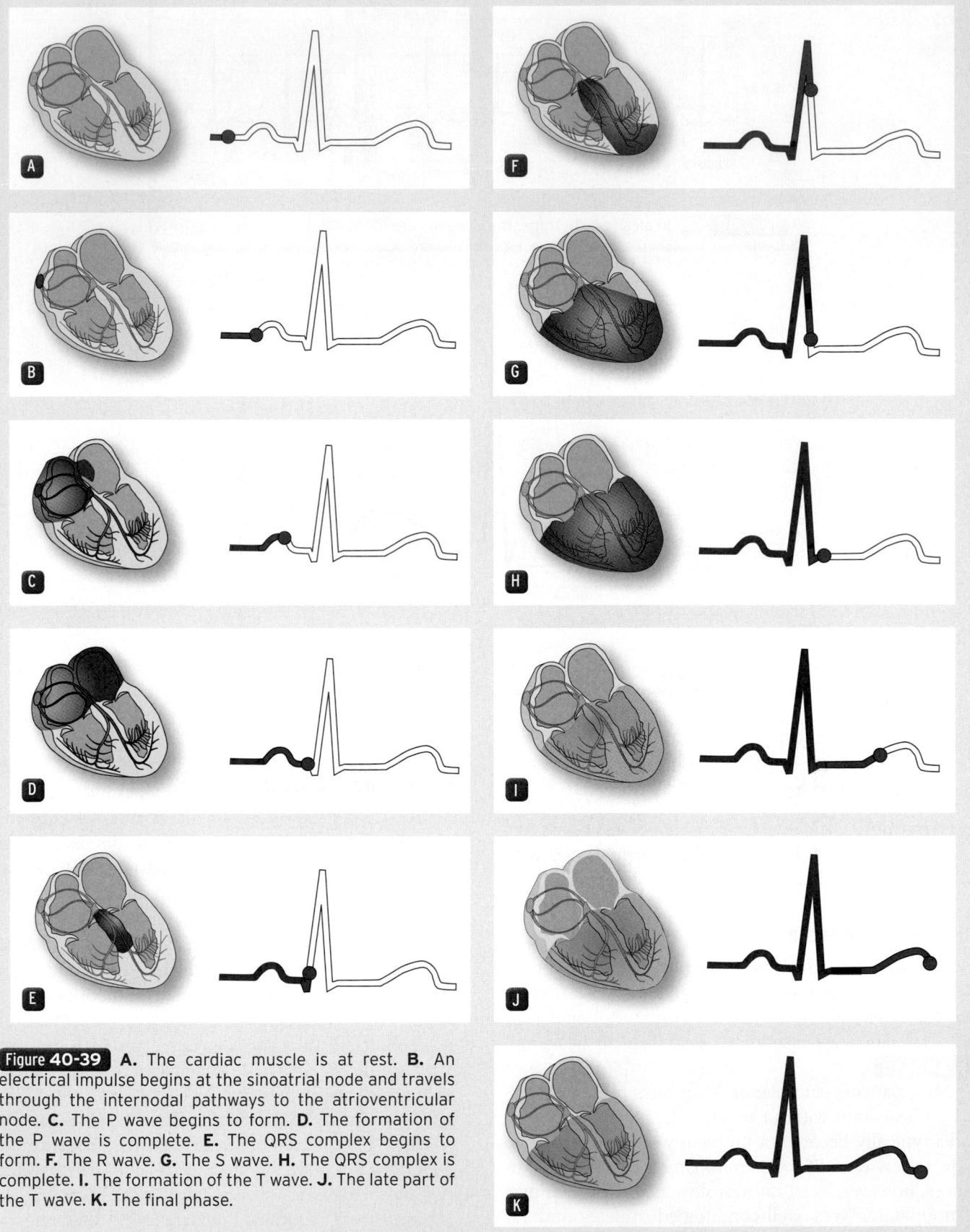

Figure 40-39 **A.** The cardiac muscle is at rest. **B.** An electrical impulse begins at the sinoatrial node and travels through the internodal pathways to the atrioventricular node. **C.** The P wave begins to form. **D.** The formation of the P wave is complete. **E.** The QRS complex begins to form. **F.** The R wave. **G.** The S wave. **H.** The QRS complex is complete. **I.** The formation of the T wave. **J.** The late part of the T wave. **K.** The final phase.

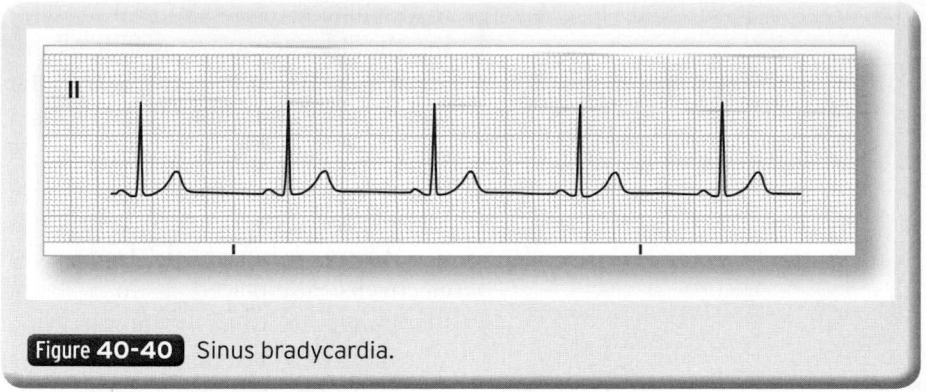

Figure 40-40 Sinus bradycardia.

Ventricular Tachycardia

Any impulse that has its origin in the ventricular conduction system generally has wide, abnormally shaped QRS complexes. A basic definition of <u>ventricular tachycardia</u> is simply the presence of three or more abnormal ventricular complexes in a row with a rate of more than 100 beats/min Figure 40-42. The rate for ventricular tachycardia is between 100 and 200 beats/min, but the rate most commonly is between 140 and 200 beats/min. Rates of more than 200 beats/min can occur. In general, ventricular tachycardia is a very regular rhythm.

Ventricular Fibrillation

<u>Ventricular fibrillation</u> is a rapid, completely disorganized ventricular rhythm with chaotic characteristics. The electrocardiographic characteristics of this arrhythmia are undulations of varying shapes and sizes with no specific pattern and no discernable P, QRS, or T waves Figure 40-43.

Notice that the word "beats" is not used to describe the undulations. This is because in ventricular fibrillation (also called V-fib), there is no organized beating of the heart. Ventricular fibrillation is a deadly arrhythmia and does not cease on its own. The best chance for patient survival is dependent on rapid defibrillation.

Asystole

<u>Asystole</u> refers to the complete absence of any electrical cardiac activity Figure 40-44. It looks like a straight or almost straight line on an ECG strip. There is a complete absence of any P, QRS, and T waves anywhere along the strip because no electrical activity is occurring. The patient is clinically dead at this point. There is no electrical or mechanical activity whatsoever.

■ Assisting With Cardiac Monitoring

Now that you have an understanding of how an ECG is formed, this section discusses the cardiac monitor and how to apply the leads. Depending on the patient and your local protocols, the number of leads on your cardiac monitor may vary. You may have a <u>4-lead ECG</u> or <u>12-lead ECG</u> system.

In recent years, manufacturers have developed a vast array of cardiac monitoring devices Figure 40-45. These cardiac monitors include several new features using modern technology. They are compact, light, and portable and combine defibrillation and monitoring capabilities. Many offer the capabilities of pulse oximetry, blood pressure monitoring capabilities, and manual and semiautomatic defibrillation functions. Some are capable of transmitting ECG tracings to a physician for remote diagnosis.

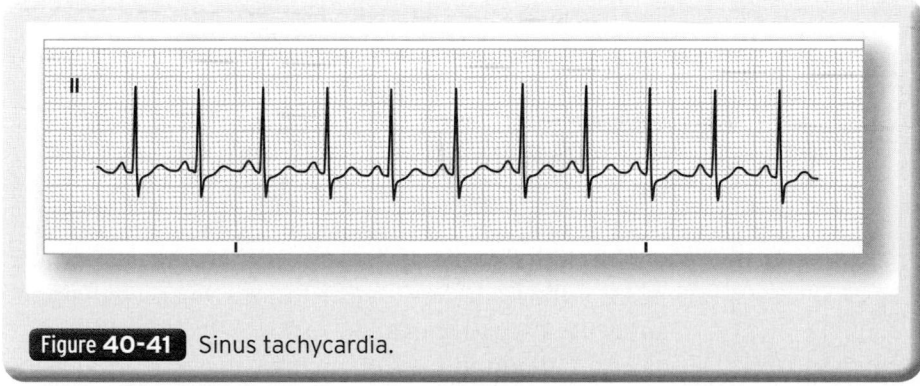

Figure 40-41 Sinus tachycardia.

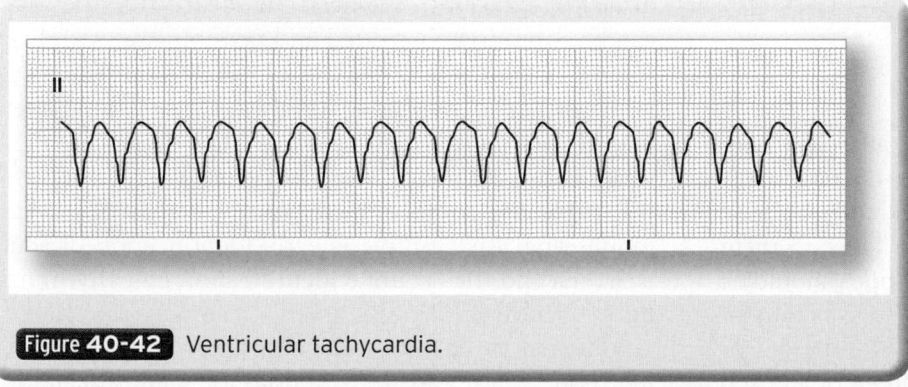

Figure 40-42 Ventricular tachycardia.

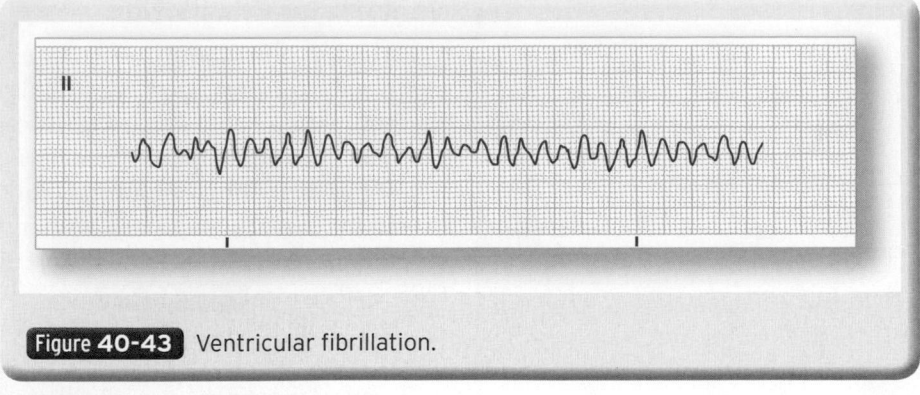

Figure 40-43 Ventricular fibrillation.

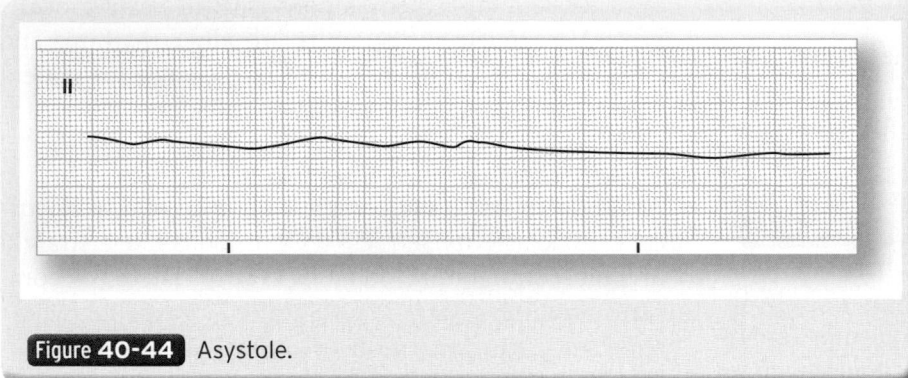

Figure 40-44 Asystole.

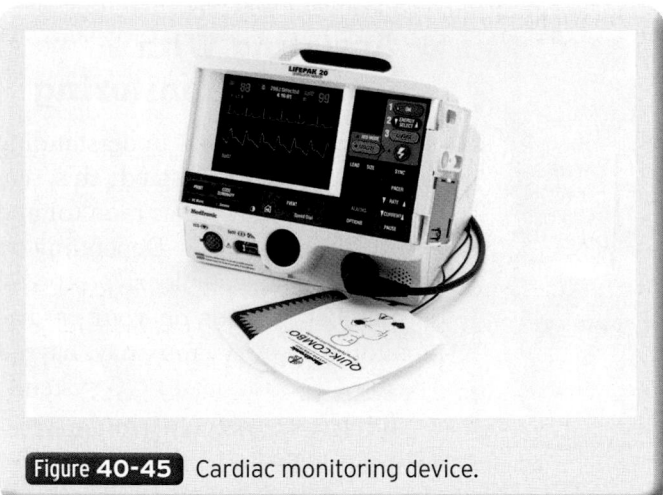

Figure 40-45 Cardiac monitoring device.

■ Lead Placement

A 4-lead ECG uses four leads, which are electrodes attached to wires that are attached to the cardiac monitor. These four leads are called the **limb leads** because they are placed on the patient's limbs, or close to them Figure 40-46. For a 4-lead ECG, the electrodes should be placed with the white lead on the patient's right shoulder or arm, the black lead on the patient's left shoulder or arm, the green lead on the right low abdomen or leg, and the red lead on the left low abdomen or leg. It does not

matter if you place the arm leads on the shoulders or arms, as long as they are at least 10 cm from the heart. Likewise, for the abdominal or leg leads, it does not matter whether these are on the abdomen or legs, as long as they are at least 10 cm from the heart.

When using a 12-lead ECG, electrodes are placed as in a 4-lead placement as well as in very specific locations on the patient's chest Figure 40-47. The **chest leads** must be placed exactly. Position the V_1 and V_2 leads on each side of the sternum at the fourth intercostal space. To find the space, first find the angle of Louis. This is a hump located near the top third of the sternum. Start feeling down the sternum from the top, and you will feel it. It is located next to the second rib. The space directly beneath it is the second intercostal space. Count down two more spaces, and place an electrode on each side of the sternum. V_4 is at the fifth intercostal space in the midclavicular line. V_3 is placed right between V_2 and V_4. V_5 is placed in the anterior axillary line and V_6 in the midaxillary line at the fifth intercostal space.

It is very important to have direct skin contact when obtaining an ECG. It may be difficult to place leads on patients in certain situations, for example if the patient's skin is diaphoretic (sweaty), oily, dirty, or hairy. Patients with cardiac emergencies may be sweating as a result of the situation. Wipe and clean the patient's skin thoroughly with a towel. If the patient's hair prevents attachment of the leads, use a razor to remove excess hair and apply the leads as above.

The importance of obtaining a 12-lead ECG is for early identification of potential myocardial ischemia, or lack of oxygen to the heart's tissue, so that the cause can be appropriately treated and possibly reversed. Even though obtaining a 12-lead ECG has distinct benefits, it is important to remember to always treat your patient first. If your patient is in severe distress, do not withhold treatment to obtain a 12-lead ECG.

There are some immediate advantages of 12-lead monitoring. An example is early identification of acute ischemia and the accurate identification of arrhythmias. Early identification and timely treatment can lead to reperfusion of valuable cardiac muscle cells, preventing tissue death and the potential for a life-threatening arrhythmia.

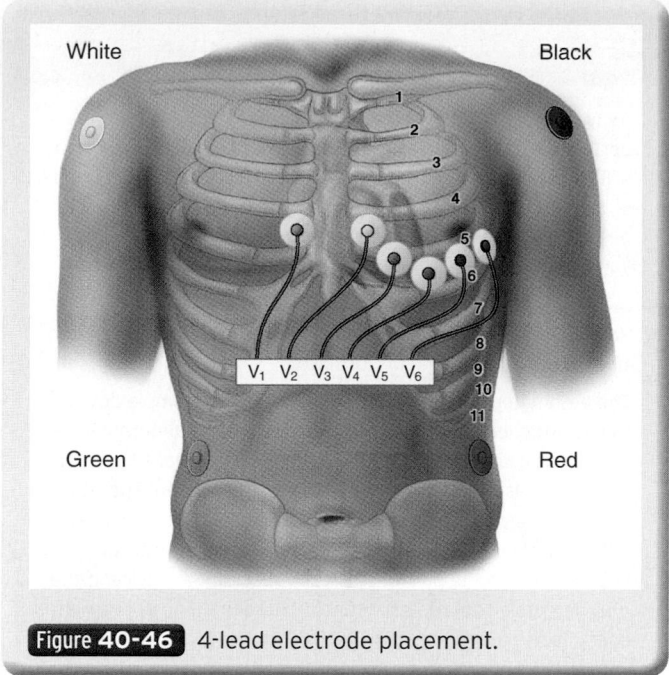

Figure 40-46 4-lead electrode placement.

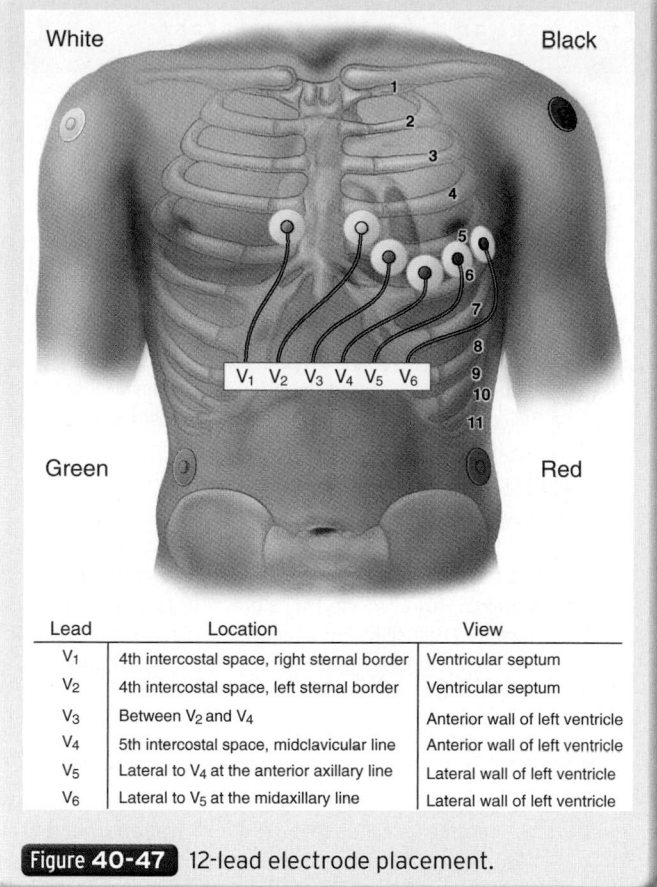

Lead	Location	View
V_1	4th intercostal space, right sternal border	Ventricular septum
V_2	4th intercostal space, left sternal border	Ventricular septum
V_3	Between V_2 and V_4	Anterior wall of left ventricle
V_4	5th intercostal space, midclavicular line	Anterior wall of left ventricle
V_5	Lateral to V_4 at the anterior axillary line	Lateral wall of left ventricle
V_6	Lateral to V_5 at the midaxillary line	Lateral wall of left ventricle

Figure 40-47 12-lead electrode placement.

■ ST-Segment Elevation Myocardial Infarction

A <u>STEMI (ST-segment elevation myocardial infarction)</u> is a specific type of myocardial infarction in which the ST segment of the cardiac cycle is elevated. A STEMI is treatable by techniques that rapidly restore perfusion to the coronary arteries. Recent studies have shown that prehospital recognition of STEMI, rapid transportation to a facility capable of these techniques, and early notification to activate AMI teams result in shorter reperfusion times. Shorter reperfusion times lead to improved patient outcomes.

Many cardiac monitors available are capable of reading a 12-lead ECG and indicating whether the patient may be having a STEMI. Others are capable of sending the ECG to a physician at a primary cardiac hospital.

Therefore, in the goal of achieving the shortest time to reperfusion, potential delays can be minimized. Remember, "Time is muscle."

Working as a team, regardless of certification level, to provide optimal care to your patient is essential. Understanding ALS procedures and recognizing your role in your service can greatly affect the treatment and, therefore, potential outcomes of your patients. Learning what to look for and how to anticipate the need for ALS procedures will go a long way in improving the team effort.

You are the Provider: SUMMARY

1. What contributions can the EMT make to the overall resuscitation effort of a patient in cardiac arrest?

As an EMT, you have an integral role on any call—regardless of its nature or severity. During the attempted resuscitation of a patient in cardiac arrest, you are a member of a team of health care providers; your contributions are just as important as any other team member—regardless of his or her level of training and certification.

It is critical to remember that good advanced life support (ALS) is built on a solid foundation of basic life support (BLS). As a BLS provider, you are in the optimum position to ensure that this foundation remains solid.

Many EMTs have been familiarized with and trained to assist ALS providers with various advanced level interventions. As such, the EMT can be of great benefit to the patient by assisting paramedics and AEMTs with these techniques and interventions. However, the EMT's greatest contribution during the attempted resuscitation of a patient in cardiac arrest is to focus on the techniques and interventions that are inherent to his or her training—high-quality BLS!

2. What tasks will most likely be assigned to you and your EMT partner when you arrive at the scene?

During a cardiac arrest call, you are *not* "just another pair of hands." You will be assigned tasks that are consistent with your level of training and certification. You will not be expected to perform, nor should you attempt to perform, any procedure that you are not trained and authorized by your medical director to do.

You will be asked to perform CPR, the most important treatment that any cardiac arrest patient should receive, and will be expected to ensure that your CPR is of consistent high quality. You will also be asked to ventilate the patient. Remember, good ALS stems from good BLS; use your training and experience and ensure that your ventilations are effective. If an advanced airway device (ie, endotracheal tube) has not already been inserted, you should ensure that a basic airway adjunct (eg, oral airway) is in place and that the bag-mask device is attached to high-flow oxygen.

Although EMTs are typically not trained to perform ALS interventions, they should be familiar with them. A basic working knowledge of the equipment and supplies that are used to perform advanced airway management, IV therapy, and cardiac monitoring will enable you to assist the ALS provider, thus making the overall resuscitation effort more effective. Furthermore, a basic working knowledge of when and why these procedures are performed will enable you to anticipate, to a certain degree, what will be done next.

Other tasks that you will likely be asked to perform include retrieving additional equipment and supplies from the ambulance, assisting with lifting and moving of the patient, and obtaining patient information from family members or bystanders. By no means are these menial tasks; they are critical to the overall resuscitation effort.

As with any critical call, there will be many actions being performed at the same time; if you know that a particular task will be carried out next and are able to assist the ALS provider in performing that task, the ultimate benefit will be to the patient. You are *not* there to make a paramedic's life easier; you are there to help take care of the patient!

3. How does a manual cardiac monitor/defibrillator differ from an automated external defibrillator?

The automated external defibrillator (AED) is a simple device that is attached to a patient in cardiac arrest to determine if he or she has a shockable rhythm, such as ventricular fibrillation (V-fib) or pulseless ventricular tachycardia (V-tach). The AED analyzes the patient's cardiac rhythm using an internal computer chip; thus, knowledge of cardiac rhythm interpretation is not required on the part of the rescuer using the device. If the AED detects a shockable rhythm, it charges to the appropriate energy setting and then advises the rescuer to deliver a shock.

A manual cardiac monitor/defibrillator features an oscilloscope (screen) that allows the paramedic to actually see (and interpret) the patient's cardiac rhythm. Because the paramedic must visually analyze the patient's cardiac rhythm instead of having it done by the internal computer chip of an AED, he or she must be knowledgeable at interpreting cardiac rhythms. On the basis of the paramedic's interpretation of the patient's cardiac rhythm, he or she can determine whether a shock is indicated.

Most manual cardiac monitors/defibrillators on the market are compact, light, and portable. Many offer additional monitoring capabilities, such as pulse oximetry, end-tidal carbon dioxide, and blood pressure, and semiautomatic and manual defibrillation functions.

Although EMTs are not trained to interpret cardiac rhythms, they can assist the paramedic by applying the leads or defibrillator pads as well as using the cardiac monitor's additional functions (eg, pulse oximetry, blood pressure monitoring).

4. Would a 12-lead electrocardiogram (ECG) be of benefit to this patient? Why or why not?

In a patient in cardiac arrest, the most urgent assessment is to determine if he or she needs to be defibrillated; it does *not* take a 12-lead ECG to do this. The defibrillation pads—also called "multipads"—are able to monitor, defibrillate, and perform other electrical interventions and can be more quickly applied than the leads from a 12-lead ECG; clearly, this is the most practical approach.

If multipads are not available, a 4-lead configuration can be used; this involves applying four leads to the patient: a negative (white) lead, a ground (black) lead, a positive (red) lead, and an augmented (green) lead. The negative lead is placed on the right arm (or right upper shoulder), the ground lead is placed on the left arm (or left upper shoulder), the positive lead is placed on the left leg (or left lower abdomen), and the augmented lead is placed on the right leg (or right lower abdomen).

Twelve-lead ECGs are typically used on patients who are not in cardiac arrest but who are suspected of experiencing myocardial ischemia or an acute myocardial infarction. The 12-lead ECG provides more "views" of the heart than does a standard 4-lead configuration, thus allowing it to provide more diagnostic information. However, if the multipads or standard 4-lead configuration shows that the patient is in ventricular fibrillation, the 12-lead ECG will show the same information. Furthermore, it takes about 15 to 20 seconds to acquire a 12-lead ECG reading, and the patient must be motionless; this is *not* acceptable in a patient in cardiac arrest because it would cause a delay in performing CPR.

If return of spontaneous circulation occurs, the paramedic may acquire a 12-lead ECG at that time; however, as long as the patient is in cardiac arrest, the 12-lead ECG is of no additional benefit for patient care.

5. How does ventricular tachycardia differ from ventricular fibrillation?

Ventricular tachycardia and ventricular fibrillation are cardiac arrhythmias that are commonly encountered in a patient in cardiac arrest. On the cardiac monitor, they are not difficult to recognize.

Ventricular tachycardia is a regular, rapid cardiac rhythm with wide, bizarre-looking complexes. Ventricular fibrillation, on the other hand, is a rapid, chaotic, disorganized rhythm; it is characterized by undulations of varying shapes and sizes with no identifiable pattern.

Whereas ventricular tachycardia can produce a pulse in some patients, ventricular fibrillation is never accompanied by a pulse. Some patients in cardiac arrest present with ventricular tachycardia but do not have a palpable pulse (pulseless ventricular tachycardia). The treatment for pulseless ventricular tachycardia is the same as it is for ventricular fibrillation—defibrillation.

It is important for the EMT to be aware of the patient's present cardiac rhythm, especially during cardiac arrest. If the paramedic states that the patient is in ventricular fibrillation or pulseless ventricular tachycardia, for example, the EMT will know that defibrillation is indicated; in some cases, he or she may even be familiar with the various cardiac medications that are given for these arrhythmias. Conversely, if the patient is in asystole—an absence of electrical and mechanical activity—the EMT will know that defibrillation is not indicated.

6. How can the EMT assist the paramedic in intubating the patient?

Endotracheal intubation is the insertion of an endotracheal tube into the trachea to maintain the airway. The tube passes directly through the larynx between the vocal cords and then into the trachea. Endotracheal intubation is a very effective method for controlling a patient's airway and facilitates the instillation of 100% oxygen directly into the patient's lungs.

Of course, advanced airway management techniques should be preceded by basic airway care, such as ventilating the patient with a bag-mask device with an oral airway inserted. As the paramedic is preparing the necessary equipment to carry out the intubation procedure, the EMT should continue *effective* bag-mask ventilations for at least 2 to 3 minutes; adequate preoxygenation of the patient is essential prior to any advanced airway management interventions. After the patient has been intubated, the EMT can resume ventilations at the appropriate rate, thus freeing up the paramedic to perform other advanced interventions.

Other devices are available to perform advanced airway management, such as the laryngeal mask airway, the King LT, and multilumen airway devices (eg, Combitube, pharyngeotracheal lumen airway. In some EMS systems, EMTs are trained and authorized to use these alternative airway devices.

It is important to note that if a patient's airway can be adequately maintained with an oral airway, and ventilations can be effectively performed with a bag-mask device, advanced airway management may not be necessary. In any patient in cardiac arrest, the goal is to ensure effective ventilations; if this can be accomplished with basic means, then you have achieved your goal.

7. What has most likely happened? What must be done to correct the situation?

If the endotracheal tube is inserted too far into the trachea, it will pass into the right main-stem bronchus, which is shorter and straighter than the left main-stem bronchus. If this occurs, only the right lung can be ventilated; therefore, breath sounds will be heard only on the right side of the chest. Right main-stem intubation is a relatively common error; however, it is easily correctable.

To correct a right main-stem bronchus intubation, the paramedic must first remove the tube-securing device and deflate the cuff at the distal end of the endotracheal tube. Then, he or she will *slightly* withdraw the tube while auscultating the lungs at the same time and stop when breath sounds are heard equally over both sides of the chest. The distal cuff is then reinflated, and the tube is resecured.

It is critical to *frequently* reconfirm proper placement of any advanced airway device, especially after any patient move. An endotracheal tube that becomes dislodged and ends up in the esophagus can result in death of the patient if it goes unrecognized.

8. How are end-tidal carbon dioxide detectors and capnography used in conjunction with endotracheal intubation?

Auscultation of breath sounds and observation for visible chest rise should not be the only methods used to help confirm proper endotracheal tube placement. The use of additional methods to confirm proper placement is *highly*

You are the Provider: SUMMARY, continued

recommended. Of these additional confirmation methods, end-tidal carbon dioxide detectors and capnography are perhaps the most commonly used.

End-tidal carbon dioxide detectors and capnography monitors sense the amount of carbon dioxide during the exhalation phase of ventilation. They are placed between the endotracheal tube adaptor and bag-mask device. After a minimum of six breaths, look for evidence of carbon dioxide production, depending on the type of device used. Patients in cardiac arrest may not produce enough carbon dioxide for the device to measure. Therefore, accurate readings are usually not possible until effective ventilation and circulation have been restored.

One example of an end-tidal carbon dioxide detector is a disposable plastic device that contains chemically treated paper (eg, EZ-Cap). During exhalation, the paper in the device should change from purple to yellow, indicating the presence of carbon dioxide in the patient's exhaled breath. Confirmation is made by observing the paper for the appropriate color change. If the endotracheal tube is in the esophagus, there will be no carbon dioxide in the patient's exhaled breath, and thus, the chemically treated paper will stay purple during exhalation.

Capnography monitors can also be used for secondary confirmation of endotracheal tube placement and are typically used to replace the disposable end-tidal carbon dioxide detector after initial confirmation has been made. Like the disposable end-tidal carbon dioxide detector, electronic capnography devices also measure the presence of carbon dioxide in the patient's exhaled breath. However, the amount of carbon dioxide present is displayed as a number or as a positive waveform on the monitor. Positive waveforms and/or carbon dioxide readings provide secondary confirmation of proper placement. If the endotracheal tube is in the esophagus, waveforms and readings will not be present.

Because no single method of confirming correct endotracheal tube placement is 100% accurate, it is important to use several confirmation methods and to reassess them frequently.

9. What is the role of vascular access in the resuscitation effort of a patient in cardiac arrest?

The primary reason that vascular access is established during cardiac arrest is to administer medications that are based on the patient's condition and presenting cardiac arrhythmia. Depending on the situation, IV fluids (ie, normal saline) and other drugs may also be given. Vascular access is but one component in the overall resuscitation effort; it must be established during CPR and without delaying defibrillation, if indicated.

There are two methods for establishing vascular access—intravenous (IV) and intraosseous (IO) access. IV access involves inserting a Teflon catheter directly into a vein; fluids and medications given by the IV route are returned back to the heart and then distributed throughout the body. IO access involves inserting a stainless steel catheter into the cavity of the bone. Fluids and medications given by the IO route are

quickly absorbed into the circulatory system; therefore, the IO route has shown to be just as fast as the IV route.

EMTs often receive on-the-job training in setting up an IV line or preparing the equipment needed to insert an IO catheter; at a scene, they are often asked to set up the IV or IO line while the ALS provider finds a suitable location to insert the catheter. Therefore, it is important for the EMT to be familiar with equipment and supplies used to establish vascular access.

The IV catheter size (bore), the type of IV fluid, and the administration set that is used depend on the reason that vascular access is being established. IV catheters range in size from 14-gauge to 24-gauge; the smaller the number, the larger the bore. For example, fluids and medications can be delivered more rapidly through a 16-gauge catheter than they can through a 20-gauge catheter. The administration set is designed to deliver a certain amount of fluid based on the number of drops; it takes 60 drops to deliver 1 mL of fluid with a microdrip set and 10 drops to deliver 1 mL with a macrodrip set. If large amounts of fluid need to be given, the macrodrip set is selected because it can deliver a greater volume of fluid with fewer drops. By contrast, a microdrip set is used if the purpose of the IV line is to administer a medication instead of giving large volumes of fluid. When vascular access is established in a patient in cardiac arrest, a macrodrip set is most commonly used because the patient may need fluid boluses—as was the case with the patient in this case study—as well as various cardiac medications. The IV fluid that is selected is also based on the reason for establishing vascular access. Normal saline solution is the most common IV fluid used because it helps expand circulating volume and is compatible with the drugs that are given during cardiac arrest. Other IV fluids may be used, and use is dependent on local protocol.

During cardiac arrest, it is important to note that the only circulation the patient has is what is being provided by CPR. This fact underscores the importance of high-quality CPR at all times, not only to ensure effective circulation of blood, but to maximize the ability of drugs and IV fluids to be circulated.

10. Why does asystole often not respond to even the most aggressive treatment?

Despite a well-coordinated resuscitation effort between EMTs and ALS providers, many patients in cardiac arrest do not survive. Although this is an unfortunate reality, it does *not* reflect on the resuscitation efforts of the team. A major determining factor in the survival of prehospital cardiac arrest is duration of the arrest; the longer the patient is in cardiac arrest, the less likely that he or she will survive.

Although patients have been resuscitated after being in asystole (a total absence of electrical and mechanical activity), it is a terminal rhythm in most patients. On the ECG, asystole manifests as a flat line; its presence signifies profound cardiac hypoxia and acidosis.

Asystole can be the result of a sudden, catastrophic event (eg, massive heart attack, ruptured cerebral aneurysm, massive pulmonary embolism) or the end result of another cardiac arrest rhythm (eg, ventricular fibrillation, pulseless ventricular

You are the Provider: SUMMARY, continued

tachycardia) that has not responded to CPR, defibrillation, medication therapy, and other treatment.

The patient in this case study presented with pulseless ventricular tachycardia; however, despite appropriate treatment—CPR, defibrillation, and drug therapy—his cardiac rhythm deteriorated to asystole. Patients who remain in asystole for more than 10 minutes rarely survive; it is important for both EMTs and ALS providers to understand this and to realize that it is *not a reflection of their resuscitation efforts.*

EMS Patient Care Report (PCR)

Date: 11-17-09	Incident No.: 400109	Nature of Call: Cardiac arrest		Location: 105 Oak Park Dr.	
Dispatched: 1822	En Route: 1822	At Scene: 1827	Transport: N/A	At Hospital: N/A	In Service: 1848

Patient Information

Age: 74 Sex: M Weight (in kg [lb]): 73 kg (160 lb)	Allergies: Sulfa, Contrast dye, Demerol Medications: Nitroglycerin, Prinivil, Plavix, Metformin Past Medical History: AMI, type 2 diabetes, hypertension Chief Complaint: Cardiopulmonary arrest

Vital Signs

Time: 1829	BP: not obtainable	Pulse: absent	Respirations: absent	Sao$_2$: not obtainable
Time: 1833	BP: not obtainable	Pulse: absent	Respirations: absent	Sao$_2$: not obtainable
Time: 1835	BP: not obtainable	Pulse: absent	Respirations: absent	Sao$_2$: not obtainable
Time: 1841	BP: not obtainable	Pulse: absent	Respirations: absent	Sao$_2$: not obtainable

EMS Treatment
(circle all that apply)

Oxygen @ 15 L/min via (circle one): NC NRM (Bag-Mask Device)		(Assisted Ventilation)	(Airway Adjunct)	(CPR)
(Defibrillation)	Bleeding Control	Bandaging	Splinting	Other: (Intubation, IV therapy, cardiac drug administration, blood glucose assessment)

Narrative

Medic 7 was dispatched to a residence to provide assistance to an ALS crew for a patient in cardiac arrest. Arrived on scene and found an AEMT and paramedic performing two-rescuer CPR on the patient, a 74-year-old man. After 2 minutes of CPR, the patient was reassessed; he was apneic, pulseless, and cyanotic. A rescue squad with two EMRs arrived at the scene to provide further assistance. Applied the ECG pads to the patient's chest; the paramedic noted that the patient's presenting cardiac rhythm was ventricular tachycardia without a pulse. Defibrillation was performed several times by the paramedic; CPR was immediately resumed following each defibrillation. Assisted the AEMT in establishing IV access, after which the appropriate medications were given to the patient by the paramedic. Also assisted the paramedic in performing endotracheal intubation; after successful intubation was confirmed, ventilations were resumed at the appropriate rate, and chest compressions were continuous. CPR was continued throughout the resuscitation effort, and rescuer roles were switched every 2 minutes. According to the patient's wife, he has a history of several heart attacks, type 2 diabetes, and hypertension. After further treatment, the patient's cardiac rhythm deteriorated to asystole, and he remained apneic and pulseless. BLS crew continued to assist the ALS crew in performing additional treatment. Patient's blood glucose level was assessed and noted to read 104 mg/dL. A normal saline bolus was administered by the AEMT to rule out hypovolemia as a potential cause of the patient's cardiac arrest. Additional assessment and treatment were performed by the paramedic; however, the patient remained in asystole. After approximately 12 minutes of adequate BLS and ALS, the patient's condition remained unchanged. After conferring with the patient's wife and contacting medical control, the decision was made to discontinue the resuscitation effort. Medic 7 remained on scene to help provide emotional support to the patient's wife and to assist the ALS crew in cleaning up the resuscitation area. Medic 7 returned to service at 1848. *End of report*

Prep Kit

Ready for Review

- There may be cases in which an EMT may find it necessary to be familiar with skills normally practiced at the AEMT and paramedic level. These skills include advanced airway techniques, intravenous therapy, and cardiac monitoring.

- An advanced airway technique is endotracheal intubation, the insertion of a tube into the trachea to maintain the airway.

- Additional advanced airway care devices include the Combitube, pharyngeotracheal lumen airway, the King LT, and the laryngeal mask airway.

- Intravenous therapy is used to replace fluids in a patient with shock or to administer medications.

- Cardiac monitoring with an electrocardiogram is an advanced skill that the EMT may provide assistance to the AEMT or paramedic.

Vital Vocabulary

4-lead ECG An ECG that uses 4 leads attached to the patient's skin; these include the limb leads.

12-lead ECG An ECG that uses 12 leads attached to the patient's skin; these include the limb leads and chest leads.

access port A sealed hub on an administration set designed for sterile access to the intravenous fluid.

arrhythmia An abnormal rhythm of the heart, sometimes called an arrhythmia.

asystole The complete absence of any electrical cardiac activity, appearing as a straight or almost straight line on an ECG strip.

cardiac monitoring The act of viewing the electrical activity of the heart through the use of an ECG machine or cardiac monitor.

catheter A flexible, hollow structure that drains or delivers fluids.

catheter shear The cutting of the catheter by the needle during improper rethreading of the catheter with the needle; the severed piece can then enter the circulatory system.

chest leads The leads that are used only with a 12-lead ECG and must be placed exactly; includes leads V_1, V_2, V_3, V_4, V_5, and V_6.

Combitube A multilumen airway device that consists of a single tube with two lumens, two balloons, and two ventilation ports; an alternative airway device if endotracheal intubation is not possible or has failed.

drip chamber The area of the administration set where fluid accumulates so that the tubing remains filled with fluid.

drip sets Another name for administration sets.

ECG Electrocardiogram; an electronic tracing of the heart's electrical activity through leads, which originate in the electrocardiograph machine and contain electrodes that attach to the patient's chest and/or limbs.

electrical conduction system A network of special cells in the heart through which an electrical current flows, causing contractions of the heart that produce pumping of blood.

endotracheal intubation Insertion of an endotracheal tube directly through the larynx between the vocal cords and into the trachea to maintain and protect an airway.

end-tidal carbon dioxide detectors Plastic, disposable indicators that signal by color change when an endotracheal tube is in the proper place.

external jugular IV IV access established in the external jugular vein of the neck.

extubation Removal of a tube after it has been placed.

gastric tube An advanced airway adjunct that provides a channel directly into a patient's stomach, allowing for removal of gas, blood, and toxins and for instilling medications and nutrition.

gauge A measure of the interior diameter of the catheter. It is inversely proportional to the true diameter of the catheter.

infiltration The escape of fluid into the surrounding tissue when the IV catheter is not in the vein.

intraosseous (IO) needles Rigid, boring catheters placed into a bone to provide intravenous fluids.

isotonic crystalloids Intravenous solutions that do not cause a fluid shift into or out of the cell; examples include normal saline and lactated Ringer's solutions.

King LT A disposal supraglottic airway used as an alternative to tracheal or mask ventilation.

laryngeal mask airway An advanced airway device that is blindly inserted into the mouth to isolate the larynx for direct ventilation; consists of a tube and a mask or cuff that inflates to seal around the laryngeal opening.

laryngoscope An instrument used to give a direct view of the patient's vocal cords during endotracheal intubation.

limb leads The four leads used with a 4-lead ECG; placed on or close to the right arm, left arm, right leg, and left leg.

macrodrip set An administration set named for the large orifice between the piercing spike and the drip chamber; allows for rapid fluid flow into the vascular system.

microdrip set An administration set named for the small orifice between the piercing spike and the drip chamber; allows for carefully controlled fluid flow and is ideally suited for medication administration.

multilumen airways Advanced airway devices, such as the esophageal tracheal Combitube and the pharyngeotracheal lumen airway, that have multiple tubes to aid in ventilation and will work whether placed in the trachea or esophagus.

normal sinus rhythm A rhythm that has consistent P waves, consistent P-R intervals, and a regular heart rate of between 60 and 100 beats/min.

occlusion Blockage, usually of a tubular structure such as a blood vessel.

orotracheal intubation Endotracheal intubation through the mouth.

pharyngeotracheal lumen airway A multilumen airway that consists of two tubes, two masks, and a bite block.

phlebitis Inflammation of a vein; often associated with a clot in the vein.

piercing spike The hard, sharpened plastic spike on the end of the administration set designed to pierce the sterile membrane of the intravenous bag.

proximal tibia Anatomic location for intraosseous catheter insertion; the wide portion of the tibia located directly below the knee.

saline locks (buff caps) Special types of intravenous apparatus, also called heparin caps and heparin locks.

sinus bradycardia A rhythm that has consistent P waves, consistent P-R intervals, and a regular heart rate that is less than 60 beats/min.

sinus rhythm A rhythm in which the sinoatrial node acts as the pacemaker.

sinus tachycardia A rhythm that has consistent P waves, consistent P-R intervals, and a regular heart rate that is more than 100 beats/min.

STEMI (ST-segment elevation myocardial infarction) Elevation of the ST segment of the 12-lead EKG that is likely evidence that the patient is having a heart attack.

stylet A plastic-coated wire that gives added rigidity and shape to the endotracheal tube.

systemic complication A moderate to severe complication affecting the systems of the body; after administration of medications, the reaction might be systemic.

ventricular fibrillation A rapid, completely disorganized ventricular rhythm with chaotic characteristics, no specific pattern, and no discernable P, QRS, or T waves.

ventricular tachycardia The presence of three or more abnormal ventricular complexes in a row with a rate of more than 100 beats/min.

Assessment in Action

You are dispatched to a residence where a 60-year-old man is lying supine on the kitchen floor. The patient is unresponsive, is not breathing, and has a weak pulse. His airway is open, and you secure the airway with an endotracheal tube. En route to the hospital you meet up with the paramedic unit. The paramedic asks you to place the patient on the cardiac monitor as he starts an IV. The monitor shows ventricular tachycardia. The paramedic then asks you to spike a second IV bag while he cardioverts the patient. The rhythm converts to a normal sinus rhythm with ST elevation.

1. Endotracheal intubation should be performed:
   A. immediately to secure the airway.
   B. after the patient has been properly ventilated.
   C. as the only method to ventilate a patient.
   D. as a last resort for ventilation.

2. On average, the tube-to-teeth mark should be how many centimeters?
   A. 20
   B. 21
   C. 22
   D. 23

3. To confirm tube placement you should:
   A. auscultate over the lungs.
   B. auscultate over the stomach.
   C. use an end-tidal carbon dioxide device.
   D. all of the above.

4. List the steps of intubation.

5. Which of the following describes the escape of fluid into the surrounding tissue when an IV catheter is not in the vein?
   A. Phlebitis
   B. Cellulitis
   C. Infiltration
   D. Hepatitis

6. Which of the following best describes infiltration?
   A. The escape of fluid into the surrounding tissues
   B. A blockage, usually of a tubular structure such as a blood vessel
   C. Inflammation of a vein, often associated with a clot in the vein
   D. Infection of the blood vessels

7. Describe the locations for placement of chest leads to obtain a 12-lead ECG.

8. Which of the following structures is the main pacemaker of the heart?
   A. AV node
   B. Bundle of His
   C. Purkinje fibers
   D. SA node

9. Which of the following best describes ventricular tachycardia?
   A. A disorganized rhythm with chaotic characteristics
   B. The presence of three or more ventricular complexes in a row
   C. A rhythm that has consistent P waves and QRS complexes and is regular
   D. A regular rhythm at a rate of more than 100 beats/min

Medical Terminology

National EMS Education Standard Competencies

Medical Terminology

Uses foundational anatomical and medical terms and abbreviations in written and oral communication with colleagues and other health care professionals.

Medical Terminology

It is critical that you have a strong working knowledge of medical terminology. The language of medicine is primarily derived from ancient Greek and Latin. Medical terminology is used in international language, and it is also necessary for communicating with other medical personnel. The wider your vocabulary base, the more competent you seem to the rest of the medical community and the better the patient care you will be able to provide. Understanding terminology involves breaking words down into their separate components of prefix, suffix, and root word and having a good working knowledge of those parts.

Prefixes

A prefix appears at the beginning of a word and generally describes location and intensity. Prefixes are frequently found in general language (ie, autopilot, submarine, tricycle), as well as in medical and scientific terminology. When a medical word (ventilation) contains a prefix (hyper), the meaning of the word is altered (hyperventilation). Not all medical terms have prefixes.

By learning to recognize a few of the more commonly used medical prefixes, you can figure out the meaning of terms that may not be immediately familiar to you. See **Table A-1** for a list of common prefixes.

Table A-1 Common Prefixes

Prefix	Meaning	Prefix	Meaning	Prefix	Meaning
a-	without, lack of	calc-	stone; also heel	dys-	difficult, painful, abnormal
ab-	away from	cardi(o)-	pertaining to the heart	ect(o)-	out from
abdomi(n)-	abdomen	cephal(o)-	pertaining to the head	electro-	pertaining to electricity
acr(o)-	pertaining to an extremity	cerebr(o)-	pertaining to the cerebrum, a part of the brain	end(o)-	within
ad-	to, toward			enter(o)-	pertaining to the intestines
aden(o)-	pertaining to a gland	cervic(o)-	pertaining to the neck or the uterine cervix	epi-	upon, on
an-	without, lack of	chole-	pertaining to bile	erythr(o)-	pertaining to red or to erythrocytes (red blood cells)
ana-	up, back, again	chondr(o)-	pertaining to cartilage		
angio-	vessel	circum-	around, about	eu-	easy, good, normal
ante-	before, forward	contra-	against, opposite	ex(o)-	outside
anti-	against, opposed to	cost(o)-	pertaining to a rib	extra-	outside, in addition
arteri(o)-	artery	cyan(o)-	blue	gastr(o)-	pertaining to the stomach
arthro-	pertaining to a joint	cyst(o)-	pertaining to the bladder or any fluid-containing sac	glyc(o)-	sugar
auto-	self			gynec(o)-	pertaining to females or the female reproductive organs
bi-	two	cyt(o)-	pertaining to a cell		
bi(o)-	pertaining to life	de-	down from	hemat(o)-	pertaining to blood
blast(o)-	germ or cell	dermat(o)-	pertaining to the skin	hemi-	half
blephar(o)-	pertaining to an eyelid	di-	twice, double	hem(o)-	pertaining to blood
brady-	slow	dia-	through, completely	hepat(o)-	pertaining to the liver
				hom-	same or like

▼ *Table A-1 continues*

Table A-1		Common Prefixes, continued			
Prefix	**Meaning**	**Prefix**	**Meaning**	**Prefix**	**Meaning**
hydr(o)-	water	nephr(o)-	pertaining to the kidney	pre-	before
hyper-	over, excessive	neur(o)-	pertaining to a nerve or the nervous system	pro-	before, in front of
hypo-	under, deficient	noct-	night	proct(o)-	pertaining to the rectum
hyster(o)-	pertaining to the uterus	olig(o)-	little, deficient	pseud(o)-	false
infra-	below	oophor(o)-	pertaining to the ovary	psych(o)-	pertaining to the mind
inter-	between	ophthalm(o)-	pertaining to the eye	pulm(o)-	pertaining to the lung
intra-	within	orchid(o)-	pertaining to the testicles	pyel(o)-	pertaining to the kidney or pelvis
iso-	equal	orchi(o)-	pertaining to the testicles	py(o)-	pertaining to pus
latero-	side	ortho-	straight or normal	quadr(i)-	four
leuk(o)-	pertaining to anything white or to leukocytes (white blood cells)	oste(o)-	pertaining to bone	retr(o)-	backward or behind
lith(o)-	pertaining to a stone	ot(o)-	pertaining to the ear	rhin(o)-	pertaining to the nose
macro-	large	para-	by the side of	salping(o)-	pertaining to a tube
mal-	bad or abnormal	path(o)-	pertaining to disease	scler(o)-	hard; also means pertaining to the sclera
medi-	middle	per-	through	semi-	half or partial
mega-	large	peri-	around	sub-	under, moderately
mening(o)-	pertaining to a membrane, particularly the meninges	phag(o)-	pertaining to eating, ingesting, or engulfing	super-	above, excessive, or more than normal
micro-	small	pharyng(o)-	pertaining to the throat, or pharynx	supra-	above
mono-	one			tachy-	fast
myel(o)-	pertaining to the spinal cord, the bone marrow, or myelin	phleb(o)-	pertaining to a vein	thorac(o)-	pertaining to the chest
my(o)-	pertaining to muscle	pneum(o)-	pertaining to respiration, the lungs, or air	trans-	across
nas(o)-	pertaining to the nose	poly-	many	tri-	three
ne(o)-	new	post-	after, behind	uni-	one
				vas(o)-	vessel

Suffixes

Suffixes are placed at the end of words to change the original meaning. In medical terminology, a suffix usually indicates a procedure, condition, disease, or part of speech. A commonly used suffix is -itis, which means "inflammation." When this suffix is paired with the prefix arthro-, meaning joint, the resulting word is arthritis, an inflammation of the joints. Sometimes it is necessary to change the last letter or letters of the root word or prefix when a suffix is added to make pronunciation easier. See **Table A-2** for a list of common suffixes.

Table A-2		Common Suffixes			
Suffix	**Meaning**	**Suffix**	**Meaning**	**Suffix**	**Meaning**
-algia	pertaining to pain	-centesis	pertaining to a procedure in which an organ or body cavity is punctured, often to drain excess fluid or obtain a sample for analysis	-emia	pertaining to the presence of a substance in the blood
-asthen(o)	weakness			esthesi(o)-	pertaining to sensation or perception
-blast	immature cell	-cyte	cell	-genic	causing
-cele	pertaining to a tumor or swelling	-ectomy	surgical removal of	-gram	record

▼ *Table A-2 continues*

Table A-2 — Common Suffixes, continued

Suffix	Meaning	Suffix	Meaning	Suffix	Meaning
-graph	a record or the instrument used to create the record	-pathy	disease or a system for treating disease	-rrhagia	abnormal or excessive flow or discharge
-itis	inflammation	-phagia	pertaining to eating or swallowing	-rrhaphy	suture of; repair of
-lysis	decline, disintegration, or destruction	-phasia	pertaining to speech	-rrhea	flow or discharge
-megaly	enlargement of	-phobia	pertaining to an irrational fear	-scope	instrument for examination
-ology	science of	-plasty	plastic surgery	-scopy	examination with an instrument
-oma	tumor	-plegia	paralysis	-sis	a process, action, or condition
-osis	pertaining to a disease process (see also -sis)	-pnea	pertaining to breathing	-taxis	order, arrangement of
-ostomy	surgical creation of an opening	-ptosis	drooping	-trophic	pertaining to nutrition
		-rrhage	abnormal or excessive flow or discharge	-uria	pertaining to a substance in the urine or the condition so indicated

Root Words

The main part or stem of a word is called a root word. A root word conveys the essential meaning of the word and frequently indicates a body part. Using a combining form, the root word and a combining vowel such as i, e, o, or a may be combined with another root word, a prefix, or a suffix to describe a particular structure or condition.

A frequently used term in EMS is CPR, which stands for cardiopulmonary resuscitation. When we break it down, cardio is a root word meaning "heart," and pulmonary is a root word meaning "lungs." By performing CPR we introduce air into the lungs and circulate blood by compressing the heart to resuscitate the patient. Some root words may also be used as prefixes or suffixes. See **Table A-3** for a list of common root words.

Table A-3 — Common Root Words

Root Word	Meaning	Root Word	Meaning	Root Word	Meaning
abdomin-	abdomen	bi-	life; also two	chondr-	cartilage
acou-	hear	bronch-	windpipe	cili-	eyelid
aden-	gland	bucc-	cheek	cleid(o)-	clavicle
adip-	fat	bursa	pouch or sac	cubitus	elbow
alb-	white	callus	hard, thick skin; also a meshwork of connective tissue that forms during the healing process after a fracture	cyan-	blue
alges-	pain			cycl-	circle or cycle
andr-	male			cyst-	bladder
angi-	vessel	carcin-	cancer	cyt-	cell
aorta	large artery exiting from the left ventricle of the heart	cardi-	heart	derm(at)-	skin
		carotid	great arteries of the neck	digit	finger or toe
aqua-	water	carpus	wrist	ede-	swelling
arteri-	artery	cent-	a fraction in the metric system; one hundredth or 100	enter-	intestine
arthr-	joint			erythr-	red
asphyxia	lack of oxygen or excess of carbon dioxide in the body that results in unconsciousness	cente-	to puncture (a body cavity)	esthe-	sensation or perception
		cephal-	head	febr-	fever
		cervic-	neck	flex	bend
asthen-	weak	chol-	bile	foramen	opening
audi-	to hear				

▼ *Table A-3 continues*

Table A-3 **Common Root Words, continued**

Root Word	Meaning	Root Word	Meaning	Root Word	Meaning
fract-	break	ov-	egg	sept-	wall, divider; also seven
gastr-	stomach	palpate	to examine by touch	serum	the clear portion of body fluids, including blood
gest-	carry, produce, congestion	path-	disease	sinus	cavity, channel or hollow space
glyc(y)-	sweet	ped-	child or foot	som(a)-	body
gno-	know	percuss	to examine by striking	spir-	coil
-gram	something written or recorded	phag-	eat	stasis	slowing or stopping of the normal flow of a fluid, such as blood
graph-	write, record	pharyng-	throat		
gyn(ec)-	female	phot-	light	stature	height
hem(at)-	blood	pleur-	rib, side	stern(o)-	sternum (breastbone)
hepat(ic)-	liver	pneum(at)-	breath	stoma	any small opening on the surface of the body, such as a pore; also, the opening created in the abdominal wall for the passage of urine or feces
heter-	other, different	pneumo(n)-	lung		
hom-	the same	pod-	foot		
humerus	the bone in the upper arm	pseud-	false		
hydr-	water	psych-	mind		
idi-	separate, distinct	pto-	fall	tach(y)-	rapid
iod(o)-	iodine	ptyal-	saliva	tact-	touch
lact-	milk	pur-, py-	pus	tetra-	four
leuk-	white	pyr-	fire-	thorac-	chest
lingu-	tongue	quadr, quar-, quat-	four	tom-	cut
mal-	abnormal			toxic	poisonous
medi-	middle	radius	the forearm bone on the thumb side; also a line from the center of a circle or sphere to the edge	trich-	hair
mega-	large			ur-	urine
melan-	black			varic-	varicose vein
men-	month	ren-	kidney	vas-	vessel
mening-	membrane, usually refers to the meninges	retina	inner nerve-containing layer of the eye	vertigo	a disordered sensation in which one's own body or the surroundings are perceived as moving
myel-	marrow or spinal cord	rhin-	nose		
my-	muscle	sangui(n)-	blood		
nephr-	kidney	scler-	hard	viscer-	internal organs
neur-	nerve	sebum	a fatty secretion of the sebaceous glands	viscous	sticky
ocul-	eye			xen-	foreign (material)
ophthalm-	eye	sect-	cut	xer-	dry
ost(e)-	bone	sepsis	the presence of microorganisms or their toxins in the blood; also the toxic condition caused by such presence		
ot-	ear				

Abbreviations

Abbreviations take the place of words to shorten notes or documentation. When using abbreviations on patient care reports, remember to use only standard, accepted abbreviations to avoid confusion and errors. See **Table A-4** for a list of commonly used abbreviations. This list is intended to help you decipher documents written by other health care professionals. Before using any abbreviations in your own reports, be familiar with accepted use of abbreviations in your local jurisdiction or service area.

Table A-4   Common Abbreviations*

Sometimes abbreviations are written with periods (for example, abd. and a.c.), and sometimes different capitalization might be used and might convey a different meaning. Not all possible meanings for the abbreviations in this table are given here. Unless you are certain about the meaning, ask the person who used the abbreviation.

Abbreviation	Meaning
A&P	anatomy and physiology
$\bar{a}$	before
$\bar{a}\bar{a}$	of each (used in writing prescriptions)
abd	abdomen
ABG	arterial blood gas
ac	before meals
ACLS	advanced cardiac life support
ADL	activity of daily living
ad lib	as much as desired
AED	automated external defibrillator
AF	atrial fibrillation
AIDS	acquired immunodeficiency syndrome
AK	above the knee
AKA	above the knee amputation
A-line	arterial line
AMA	against medical advice
amb	ambulatory
AMI	acute myocardial infarction
AMS	altered mental status
ant	anterior
AO × 4	alert and oriented to person, place, time, and self
AP	anteroposterior, front-to-back, action potential, angina pectoris, anterior pituitary, arterial pressure
APC	atrial premature complex, activated protein C, aspirin-phenacetin-caffeine
Aq	water
ARDS	adult respiratory distress syndrome
ASA	aspirin (acetylsalicylic acid)
ASAP	as soon as possible
ASHD	arteriosclerotic or atherosclerotic heart disease
AV, A-V	atrioventricular, arteriovenous
BBB	bundle branch block
bid	twice daily
BKA	below the knee amputation
BM	bowel movement
BP	blood pressure

Abbreviation	Meaning
BS	blood sugar, breath sounds, bowel sounds, bachelor of science (degree)
BSA	body surface area
BVM	bag-valve-mask
bx	biopsy
$\bar{c}$	with
°C	degrees Celsius (centigrade)
Ca	calcium
CA	cancer, cardiac arrest, chronologic age, coronary artery, cold agglutinin
CABG	coronary artery bypass graft
CAD	coronary artery disease
CBC	complete blood count
cc	cubic centimeter
CC or C/C	chief complaint
CCU	coronary care unit
CHF	congestive heart failure
Cl⁻	chloride
cm	centimeter
cm³	cubic centimeter
CNS	central nervous system
c/o	complaining of
CO	cardiac output, carbon monoxide
CO_2	carbon dioxide
COLD	chronic obstructive lung disease
COPD	chronic obstructive pulmonary disease
CP	chest pain, chemically pure, cerebral palsy
CPR	cardiopulmonary resuscitation
CRNA	certified registered nurse anesthetist
CRT	capillary refill time, cathode-ray tube
CSF	cerebrospinal fluid
CSM	carotid sinus massage, cerebrospinal meningitis
CVA	cerebrovascular accident
CVP	central venous pressure
CXR	chest x-ray

Abbreviation	Meaning
D&C	dilation and curettage
D/C	discontinue
diff	differential
dig	digoxin
DM	diabetes mellitus
DOA	dead on arrival
DOE	dyspnea on exertion
DON	director of nursing
DOS	dead on scene
DPT	diphtheria and tetanus toxoids and pertussis vaccine
DSD	dry sterile dressing
DtaP	diphtheria and tetanus toxoids and acellular pertussis vaccine
DTP	diphtheria and tetanus toxoids and pertussis vaccine
DTs	delirium tremens
DVT	deep venous thrombosis
D_5W	dextrose 5% in water
Dx	diagnosis
ECG	electrocardiogram
ED	emergency department
EDC	estimated date of confinement
EEG	electroencephalogram
eg	for example
EKG	electrocardiogram
ENT	ears, nose, and throat
ER	emergency room
ET	endotracheal tube, endotracheal
ETA	estimated time of arrival
ETOH	ethyl alcohol
ETT	endotracheal tube
°F	degrees Fahrenheit
F_{IO_2}	fraction of inspired oxygen
FBS	fasting blood sugar
Fe	iron
FHR	fetal heart rate
FHT	fetal heart tones
FHx	family history
fL	femtoliter

▼ *Table A-4 continues*

Table A-4   Common Abbreviations, continued

Abbreviation	Meaning	Abbreviation	Meaning	Abbreviation	Meaning
fl or fld	fluid	K^+	potassium	MVP	mitral valve prolapse
FSH	follicle-stimulating hormone	KCl	potassium chloride	Na	sodium
fx	fracture	kg	kilogram	NA, N/A	not applicable
g	gram	KUB	kidneys, ureters, and bladder	NaCl	sodium chloride
GB	gallbladder	KVO	keep vein open	NAD	no apparent distress, no appreciable disease
GI	gastrointestinal	L	liter		
gm	gram	LAC	laceration, laparoscopic-assisted colectomy	$NaHco_3$	sodium bicarbonate
gr	grain			NC	nasal cannula
GSW	gunshot wound	lb	pound	NG	nasogastric
gtt	drop(s)	LE	lower extremity, left eye, lupus erythematosus	NICU	neonatal intensive care unit
GTT	glucose tolerance test			NIDDM	non-insulin-dependent diabetes mellitus
GU	genitourinary	LLL	left lower lobe of the lung		
gyn	gynecology	LLQ	left lower quadrant of the abdomen	NKA	no known allergies
h	hour	L/M	liters per minute	NKDA	no known drug allergies
H, (H)	hypodermic	LMP	last menstrual period	NPA	nasopharyngeal airway
H&H	hemoglobin and hematocrit	LOC	level of consciousness, loss of consciousness	NPO	nil per os (nothing by mouth)
H&P	history and physical			NS	normal saline
H/A	headache	LPM	liters per minute	NSR	normal sinus rhythm
Hb, Hgb	hemoglobin	LPN	licensed practical nurse	NTG	nitroglycerin
Hct	hematocrit	LR	lactated Ringer's	N/V	nausea and vomiting
Hg	mercury	LSD	lysergic acid diethylamide	N/V/D	nausea, vomiting, and diarrhea
HH	hiatal hernia	LUL	left upper lobe of the lung	NVD	neck vein distention
HIV	human immunodeficiency virus	LUQ	left upper quadrant of the abdomen	O_2	oxygen
H_2O	water	LVN	licensed vocational nurse	OB	obstetrics
H_2O_2	hydrogen peroxide	m	meter	OBS	organic brain syndrome
HPI	history of present illness	MAE	moves all extremities	OD	overdose, right eye, optical density, outside diameter, doctor of optometry
hr	hour	MAEW	moves all extremities well		
hs	at bedtime	MAP	mean arterial pressure		
HTN	hypertension	mcg	microgram	OP	outpatient
Hx	history	MCL	midclavicular line, modified chest lead	OPA	oropharyngeal airway
Hz	hertz			OR	operating room
I&O	intake and output	mEq	milliequivalent	OS	left eye
IC	intracardiac, inspiratory capacity, irritable colon	mg	milligram (mgm is a former symbol)	OU	both eyes
		MI	myocardial infarction	oz	ounce
ICP	intracranial pressure	MICU	mobile intensive care unit; medical intensive care unit	$\bar{p}$	after
ICU	intensive care unit			pc	after meals
IDDM	insulin-dependent diabetes mellitus	min	minute	Pco_2	partial pressure of carbon dioxide
IM	intramuscular	mL	milliliter	PDR	Physician's Desk Reference
IO	intraosseous	mm	millimeter	PE	pulmonary embolism, physical examination
IPPB	intermittent positive pressure breathing	mm Hg	millimeters of mercury		
		MRI	magnetic resonance imaging	PEA	pulseless electrical activity
IUD	intrauterine (contraceptive) device	MS	morphine sulfate, multiple sclerosis	PEARL or PERL	pupils equal and reactive to light
IV	intravenous	MSO_4	morphine sulfate		
JVD	jugular venous distention	MVA	motor vehicle accident	ped or peds	pediatric
		MVC	motor vehicle crash	PEEP	positive end-expiratory pressure

▼ *Table A-4 continues*

Table I-4   Common Abbreviations, continued

Abbreviation	Meaning	Abbreviation	Meaning	Abbreviation	Meaning
PERRL	pupils equal, round, and reactive to light	ROM	range of motion, rupture of membranes	URI	upper respiratory infection
pH	hydrogen ion concentration	RUL	right upper lobe of the lung	USP	United States Pharmacopeia
PID	pelvic inflammatory disease	RUQ	right upper quadrant of the abdomen	UTI	urinary tract infection
PND	paroxysmal nocturnal dyspnea			VD	venereal disease
po	per os (by mouth)	Rx	prescription	vol	volume
PO	postoperative, "post op"	$\bar{s}$	without	VS	vital signs
Po_2	partial pressure of oxygen	SC	subcutaneous, secretory component	W/	with
PRN	pro re nata (as needed)	SICU	surgical intensive care unit	WBC	white blood cell
psi	pounds per square inch	SIDS	sudden infant death syndrome	WNL	within normal limits
PSVT	paroxysmal supraventricular tachycardia	SL	sublingual	wt	weight
pt	patient	SOB	shortness of breath	yo	year old
PT	physical therapy	SQ	subcutaneous	$\bar{x}$	except
PTA	prior to admission, plasma thromboplastin antecedent	ss	half	1°	first, first degree, primary
PTT	partial thromboplastin time	S/S	signs and symptoms	2°	secondary, second degree
PVC	premature ventricular complex, polyvinyl chloride	stat	immediately	↑	increase(d)
		STD	sexually transmitted disease	↓	decrease(d)
PVD	peripheral vascular disease	Sub Q	subcutaneous	Ø	no, not, none
q	every	SVT	supraventricular tachycardia	®	right
qd	every day	sym or Sx	symptoms	Ⓛ	left
qh	every hour	tab	tablet	μ	micro
qid	four times a day	TB	tuberculosis	α	alpha
qod	every other day	TBA	to be admitted, to be announced	β	beta
RA	rheumatoid arthritis, right atrium	tbsp	tablespoon	~	approximately
RAD	reactive airway disease, right axis deviation	tech	technician, technologist	N	normal
RBC	red blood cell	TIA	transient ischemic attack	×2	times two
Rh	Rhesus blood factor, rhodium	tid	three times a day	/	per
RHD	rheumatic heart disease	TKO	to keep open	≠	not equal
RL	Ringer's lactate	TPR	temperature, pulse, respiration	>	greater than
RLL	right lower lobe of the lung	tsp	teaspoon	<	less than
RLQ	right lower quadrant of the abdomen	Tx	treatment	?	questionable, possible
RN	registered nurse	U	unit	Δ	change
R/O	rule out	UA	urinalysis	—	negative
		UE	upper extremity	♀	female
				♂	male

12-lead ECG An ECG that uses 12 leads attached to the patient's skin; these include the limb leads and chest leads.

4-lead ECG An ECG that uses 4 leads attached to the patient's skin; these include the limb leads.

abandonment Unilateral termination of care by the EMT without the patient's consent and without making provisions for transferring care to another medical professional with the skills and training necessary to meet the needs of the patient.

abdomen The body cavity that contains the major organs of digestion and excretion. It is located below the diaphragm and above the pelvis.

abdominal aortic aneurysm (AAA) A condition in which the walls of the aorta in the abdomen weaken and blood leaks into the layers of the vessel, causing it to bulge.

abdominal-thrust maneuver The preferred method to dislodge a severe airway obstruction in adults and children; also called the Heimlich maneuver.

abduction Motion of a limb away from the midline.

abrasion Loss or damage of the superficial layer of skin as a result of a body part rubbing or scraping across a rough or hard surface.

abruptio placenta A premature separation of the placenta from the wall of the uterus.

absorption The process by which medications travel through body tissues until they reach the bloodstream.

access Gaining entry to an enclosed area and reaching a patient.

access port A sealed hub on an administration set designed for sterile access to the intravenous fluid.

accessory muscles The secondary muscles of respiration. They include the neck muscles (sternocleidomastoids), the chest pectoralis major muscles, and the abdominal muscles.

acetabulum The depression on the lateral pelvis where its three component bones join, in which the femoral head fits snugly.

acidosis A pathologic condition that results from the accumulation of acids in the body.

acromioclavicular (AC) joint A simple joint where the bony projections of the scapula and the clavicle meet at the top of the shoulder.

action The therapeutic effect of a medication on the body.

activated charcoal An oral medication that binds and adsorbs ingested toxins in the gastrointestinal tract for treatment of some poisonings and medication overdoses. Charcoal is ground into a very fine powder that provides the greatest possible surface area for binding medications that have been taken by mouth; it is carried on the EMS unit.

activities of daily living The basic activities a person usually accomplishes during a normal day, such as eating, dressing, and bathing.

acute abdomen A condition of sudden onset of pain within the abdomen, usually indicating peritonitis; immediate medical or surgical treatment is necessary.

acute coronary syndrome A term used to describe a group of symptoms caused by myocardial ischemia; includes angina and myocardial infarction.

acute myocardial infarction (AMI) A heart attack; death of heart muscle following obstruction of blood flow to it. Acute in this context means "new" or "happening right now."

acute stress reactions Reaction to stress that occurs during a stressful situation.

Adam's apple The firm prominence in the upper part of the larynx formed by the thyroid cartilage. It is more prominent in men than in women.

addiction A state of overwhelming obsession or physical need to continue the use of a drug or agent.

adduction Motion of a limb toward the midline.

adenosine triphosphate (ATP) The nucleotide involved in energy metabolism; used to store energy.

adolescents Persons who are 12 to 18 years of age.

adrenal glands Endocrine glands located on top of the kidneys that release adrenaline when stimulated by the sympathetic nervous system.

adrenergic Pertaining to nerves that release the neurotransmitter norepinephrine, or noradrenaline (such as adrenergic nerves, adrenergic response). The term also pertains to the receptors acted on by norepinephrine, that is, the adrenergic receptors.

adsorption The process of binding or sticking to a surface.

advance directive Written documentation that specifies medical treatment for a competent patient should the patient become unable to make decisions; also called a living will or health care directive.

advanced EMT (AEMT) An individual who has training in specific aspects of advanced life support, such as intravenous therapy, and the administration of certain emergency medications.

advanced life support (ALS) Advanced lifesaving procedures, some of which are now being provided by the EMT.

adventitious breath sounds Abnormal breath sounds such as wheezes, rhonchi, and rales.

aerobic metabolism Metabolism that can proceed only in the presence of oxygen.

afterload The force or resistance against which the heart pumps.

aging The process by which the temporary bond between the organophosphate and acetylcholinesterase undergoes hydrolysis, resulting in a permanent covalent bond.

agitated delirium A condition of disorientation, confusion, and possible hallucinations coupled with purposeless, restless physical activity.

agonal gasps Slow, shallow, irregular breaths or occasional gasping breaths; sometimes seen in dying patients.

agonist A medication that causes stimulation of receptors.

air ambulances Fixed-wing aircraft and helicopters that have been modified for medical care; used to evacuate and transport patients with life-threatening injuries to treatment facilities.

air embolism The presence of air in the veins, which can lead to cardiac arrest if it enters the heart.

airborne transmission The spread of an organism in aerosol form.

airway The upper airway tract or the passage above the larynx, which includes the nose, mouth, and throat.

alkalosis The buildup of excess base (lack of acids) in the body fluids.

allergen A substance that causes an allergic reaction.

allergic reaction The body's exaggerated immune response to an internal or surface agent.

alpha A type of energy that is emitted from a strong radiologic source; it is the least harmful penetrating type of radiation and cannot travel fast or through most objects.

alpha-adrenergic receptors Portions of the nervous system that, when stimulated, can cause constriction of blood vessels.

altered mental status A change in the way a person thinks and behaves that may signal disease in the central nervous system or elsewhere in the body.

alveolar ventilation The volume of air that reaches the alveoli. It is determined by subtracting the amount of dead space air from the tidal volume.

alveoli The air sacs of the lungs in which the exchange of oxygen and carbon dioxide takes place.

ambient temperature The temperature of the surrounding environment.

ambulance A specialized vehicle for treating and transporting sick and injured patients.

American Standard System A safety system for large oxygen cylinders, designed to prevent the accidental attachment of a regulator to a cylinder containing the wrong type of gas.

Americans With Disabilities Act (ADA) Comprehensive legislation that is designed to protect individuals with disabilities against discrimination.

amniotic sac The fluid-filled, baglike membrane in which the fetus develops.

amputation An injury in which part of the body is completely severed.

anaerobic metabolism The metabolism that takes place in the absence of oxygen; the principle product is lactic acid.

anaphylaxis (anaphylactic shock) An extreme, life-threatening systemic allergic reaction that may include shock and respiratory failure.

anatomic position The position of reference in which the patient stands facing you, arms at the side, with the palms of the hands forward.

aneurysm A swelling or enlargement of part of a blood vessel, resulting from weakening of the vessel wall.

angina pectoris Transient (short-lived) chest discomfort caused by partial or temporary blockage of blood flow to the heart muscle.

anisocoria Naturally occurring uneven pupil size.

antagonist A medication that binds to a receptor and blocks other medications.

anterior The front surface of the body; the side facing you in the standard anatomic position.

anterograde (posttraumatic) amnesia Inability to remember events after an injury.

anthrax A disease caused by deadly bacteria (Bacillus anthracis) that lay dormant in a spore (protective shell); the germ is released from the spore when exposed to the optimal temperature and moisture. The routes of entry are inhalation, cutaneous, and gastrointestinal (from consuming food that contains spores).

antidote A substance that is used to neutralize or counteract a poison.

antivenin A serum that counteracts the effect of venom from an animal or insect.

anxious-avoidant attachment A bond between an infant and his or her parent or caregiver in which the infant is repeatedly rejected and develops an isolated lifestyle that does not depend on the support and care of others.

aorta The main artery that receives blood from the left ventricle and delivers it to all the other arteries that carry blood to the tissues of the body.

aortic aneurysm A weakness in the wall of the aorta that makes it susceptible to rupture.

aortic valve The one-way valve that lies between the left ventricle and the aorta and keeps blood from flowing back into the left ventricle after the left ventricle ejects its blood into the aorta; one of four heart valves.

apex (plural apices) The pointed extremity of a conical structure.

Apgar score A scoring system for assessing the status of a newborn that assigns a number value to each of five areas of assessment.

aphasia The inability to understand and/or produce speech.

aplastic crisis A condition in which the body stops producing red blood cells; typically caused by infection.

apnea Absence of spontaneous breathing.

apneustic center Portion of the pons that increases the length of inspiration and decreases the respiratory rate.

apparent life-threatening event (ALTE) An event that causes unresponsiveness, cyanosis, and apnea in an infant, who then resumes breathing with stimulation.

appendicitis Inflammation of the appendix.

appendicular skeleton The portion of the skeletal system that comprises the arms, legs, pelvis, and shoulder girdle.

appendix A small tubular structure that is attached to the lower border of the cecum in the lower right quadrant of the abdomen.

applied ethics The manner in which principles of ethics are incorporated into professional conduct.

arrhythmia An irregular or abnormal heart rhythm.

arterial air embolism Air bubbles in the arterial blood vessels.

arterioles The smallest branches of arteries leading to the vast network of capillaries.

artery A blood vessel, consisting of three layers of tissue and smooth muscle that carries blood away from the heart.

articular cartilage A pearly layer of specialized cartilage covering the articular surfaces (contact surfaces on the ends) of bones in synovial joints.

ascites Fluid in the abdomen.

aspiration In the context of airway, the introduction of vomitus or other foreign material into the lungs.

aspirin (acetylsalicylic acid or ASA) A medication that is an antipyretic (reduces fever), analgesic (reduces pain), anti-inflammatory (reduces inflammation), and potent inhibitor of platelet aggregation (clumping).

assault Unlawfully placing a patient in fear of bodily harm.

asthma An acute spasm of the smaller air passages, called bronchioles, associated with excessive mucus production and with swelling of the mucous lining of the respiratory passages.

asystole The complete absence of any electrical cardiac activity, appearing as a straight or almost straight line on an ECG strip.

ataxic respirations Irregular, ineffective respirations that may or may not have an identifiable pattern.

atelectasis Collapse of the alveolar air spaces of the lungs.

atherosclerosis A disorder in which cholesterol and calcium build up inside the walls of the blood vessels, forming plaque, which eventually leads to partial or complete blockage of blood flow.

atrium One of two (right and left) upper chambers of the heart. The right atrium receives blood from the vena cava and delivers it to the right ventricle. The left atrium receives blood from pulmonary veins and delivers it to the left ventricle.

aura A sensation experienced prior to a seizure; serves as a warning sign that a seizure is about to occur.

auscultate To listen to sounds within an organ with a stethoscope.

automated external defibrillator (AED) A device that detects treatable life-threatening cardiac arrhythmias (ventricular fibrillation and ventricular tachycardia) and delivers the appropriate electrical shock to the patient.

automatic transport ventilator (ATV) A ventilation device attached to a control box that allows the variables of ventilation to be set. It frees the EMT to perform other tasks while the patient is being ventilated.

automaticity The ability of cardiac muscle cells to contract without stimulation from the nervous system.

autonomic nervous system The part of the nervous system that regulates involuntary functions, such as heart rate, blood pressure, digestion, and sweating.

AVPU scale A method of assessing the level of consciousness by determining whether the patient is awake and alert, responsive to verbal stimuli or pain, or unresponsive; used principally early in the assessment process.

avulsion An injury in which soft tissue is torn completely loose or is hanging as a flap.

axial skeleton The part of the skeleton comprising the skull, spinal column, and rib cage.

backboard A device that is used to provide support to a patient who is suspected of having a hip, pelvic, spinal, or lower extremity injury. Also called a spine board, trauma board, and longboard.

bacteria Microorganisms that reproduce by binary fission. These single-cell creatures reproduce rapidly. Some can form spores (encysted variants) when environmental conditions are harsh.

bacterial vaginosis An overgrowth of bacteria in the vagina; characterized by itching, burning, or pain, and possibly a "fishy" smelling discharge.

bag-mask device A device with a one-way valve and a face mask attached to a ventilation bag; when attached to a reservoir and connected to oxygen, it delivers more than 90% supplemental oxygen.

ball-and-socket joint A joint that allows internal and external rotation, as well as bending.

bariatrics A branch of medicine concerned with the management (prevention or control) of obesity and allied diseases.

barotrauma Injury resulting from pressure disequilibrium across body surfaces; for example, from too much pressure in the lungs.

barrier device A protective item, such as a pocket mask with a valve, that limits exposure to a patient's body fluids.

base station Any radio hardware containing a transmitter and receiver that is located in a fixed place.

basic life support (BLS) Noninvasive emergency lifesaving care that is used to treat medical conditions, including airway obstruction, respiratory arrest, and cardiac arrest.

basilar skull fractures Usually occur following diffuse impact to the head (such as falls, motor vehicle crashes); generally result from extension of a linear fracture to the base of the skull and can be difficult to diagnose with a radiograph (x-ray).

basket stretcher A rigid stretcher commonly used in technical and water rescues that surrounds and supports the patient yet allows water to drain through holes in the bottom. Also called a Stokes litter.

battery Touching a patient or providing emergency care without consent.

Battle's sign Bruising behind an ear over the mastoid process that may indicate a skull fracture.

behavior How a person functions or acts in response to his or her environment.

behavioral crisis The point at which a person's reactions to events interfere with activities of daily living; this becomes a psychiatric emergency when it causes a major life interruption, such as attempted suicide.

bends Common name for decompression sickness.

beta A type of energy that is emitted from a strong radiologic source; is slightly more penetrating than alpha and requires a layer of clothing to stop it.

beta-adrenergic receptors Portions of the nervous system that, when stimulated, can cause an increase in the force of contraction of the heart, an increased heart rate, and bronchial dilation.

biceps The large muscle that covers the front of the humerus.

bilateral A body part or condition that appears on both sides of the midline.

bile ducts The ducts that convey bile between the liver and the intestine.

bills of lading The shipping papers used for transport of chemicals over roads and highways. Also referred to as freight bills.

bioethics The study of ethics related to issues that arise in health care.

birth canal The vagina and cervix.

blanching Turning white.

blind spots Areas of the road that are blocked from your sight by your own vehicle or mirrors.

blood pressure The pressure of circulating blood against the walls of the arteries.

bloodborne pathogens Pathogenic microorganisms that are present in human blood and can cause disease in humans. These pathogens include, but are not limited to, hepatitis B virus and human immunodeficiency virus (HIV).

bloody show A small amount of blood at the vagina that appears at the beginning of labor and may include a plug of pink-tinged mucus that is discharged when the cervix begins to dilate.

blowout fracture A fracture of the orbit or of the bones that support the floor of the orbit.

blunt trauma An impact on the body by objects that cause injury without penetrating soft tissues or internal organs and cavities.

B-NICE A memory device to recall the types of weapons of mass destruction: biologic, nuclear, incendiary, chemical, and explosive.

bonding The formation of a close, personal relationship.

botulinum Produced by bacteria, this is a very potent neurotoxin. When introduced into the body, this neurotoxin affects the nervous system's ability to function and causes botulism.

brachial artery The major vessel in the upper extremity that supplies blood to the arm.

bradycardia A slow heart rate, less than 60 beats/min.

bradypnea Slow respiratory rate; ominous sign in a child that indicates impending respiratory arrest.

brain The controlling organ of the body and center of consciousness; functions include perception, control of reactions to the environment, emotional responses, and judgment.

brain stem The area of the brain between the spinal cord and cerebrum, surrounded by the cerebellum; controls functions that are necessary for life, such as respiration.

breach of confidentiality Disclosure of information without proper authorization.

breath-holding syncope Loss of consciousness caused by a decreased breathing stimulus.

breath sounds An indication of air movement in the lungs, usually assessed with a stethoscope.

breech presentation A delivery in which the buttocks come out first.

bronchial breath sounds Normal breath sounds made by air moving through the bronchi.

bronchioles Subdivision of the smaller bronchi in the lungs; made of smooth muscle and dilate or constrict in response to various stimuli.

bronchiolitis Inflammation of the bronchioles that usually occurs in children younger than 2 years and is often caused by the respiratory syncytial virus.

bronchitis An acute or chronic inflammation of the lung that may damage lung tissue; usually associated with cough and production of sputum and, depending on its cause, sometimes fever.

buboes Enlarged lymph nodes (up to the size of a tennis ball) that were characteristic in people infected with the bubonic plague.

bubonic plague An epidemic that spread throughout Europe in the Middle Ages, causing more than 25 million deaths, also called the Black Death; transmitted by infected fleas and characterized by acute malaise, fever, and the formation of tender, enlarged, inflamed lymph nodes that appear as lesions, called buboes.

burns Injuries in which soft-tissue damage occurs as a result from thermal heat, frictional heat, toxic chemicals, electricity, or nuclear radiation.

calcaneus The heel bone.

capillaries The small blood vessels that connect arterioles and venules; various substances pass through capillary walls, into and out of the interstitial fluid, and then on to the cells.

capillary refill A test that evaluates distal circulatory system function by squeezing (blanching) blood from an area such as a nail bed and watching the speed of its return after releasing the pressure.

capillary vessels The tiny blood vessels between the arterioles and venules that permit transfer of oxygen, carbon dioxide, nutrients, and waste between body tissues and the blood.

capnography A noninvasive method that can quickly and efficiently provide information on a patient's ventilatory status, circulation, and metabolism.

capnometry The use of a capnometer, a device that measures the amount of expired carbon dioxide.

carbon dioxide A component of air that typically makes up 0.3% of air at sea level; also a waste product exhaled during expiration by the respiratory system.

carbon dioxide retention A condition characterized by a chronically high blood level of carbon dioxide in which the respiratory center no longer responds to high blood levels of carbon dioxide.

carbon monoxide An odorless, highly poisonous gas that results from incomplete oxidation of carbon in combustion.

carboys Glass, plastic, or steel containers, ranging in volume from 5 to 15 gallons.

cardiac arrest A state in which the heart fails to generate effective and detectable blood flow; pulses are not palpable in cardiac arrest, even if muscular and electrical activity continues in the heart.

cardiac monitoring The act of viewing the electrical activity of the heart through the use of an ECG machine or cardiac monitor.

cardiac muscle The heart muscle.

cardiac output A measure of the volume of blood circulated by the heart in 1 minute, calculated by multiplying the stroke volume by the heart rate.

cardiac tamponade (pericardial tamponade) Compression of the heart as the result of buildup of blood or other fluid in the pericardial sac, leading to decreased cardiac output.

cardiogenic shock A state in which not enough oxygen is delivered to the tissues of the body, caused by low output of blood from the heart. It can be a severe complication of a large acute myocardial infarction, as well as other conditions.

cardiopulmonary resuscitation (CPR) The combination of rescue breathing and chest compressions used to establish adequate ventilation and circulation in a patient who is not breathing and has no pulse.

carina Point at which the trachea bifurcates (divides) into the left and right mainstem bronchi.

carotid artery The major artery that supplies blood to the head and brain.

cartilage The support structure of the skeletal system that provides cushioning between bones; also forms the nasal septum and portions of the outer ear.

casualty collection area An area set up by physicians, nurses, and other hospital staff near a major disaster scene where patients can receive further triage and medical care.

cataracts Clouding of the lens of the eye or its surrounding transparent membranes.

catheter A flexible, hollow structure that drains or delivers fluids.

catheter shear The cutting of the catheter by the needle during improper rethreading of the catheter with the needle; the severed piece can then enter the circulatory system.

cavitation A phenomenon in which speed causes a bullet to generate pressure waves, which cause damage distant from the bullet's path.

cecum The first part of the large intestine, into which the ileum opens.

cellular telephone A low-power portable radio that communicates through an interconnected series of repeater stations called "cells."

Centers for Disease Control and Prevention (CDC) The primary federal agency that conducts and supports public health activities in the United States. The CDC is part of the US Department of Health and Human Services.

central nervous system (CNS) The brain and spinal cord.

central pulses Pulses that are closest to the core (central) part of the body where the vital organs are located; include the carotid, femoral, and apical pulses.

cerebellum One of the three major subdivisions of the brain, sometimes called the "little brain"; coordinates the various activities of the brain, particularly fine body movements.

cerebral edema Swelling of the brain.

cerebral palsy A term for a group of disorders characterized by poorly controlled body movement.

cerebrospinal fluid (CSF) Fluid produced in the ventricles of the brain that flows in the subarachnoid space and bathes the meninges.

cerebrovascular accident (CVA) An interruption of blood flow to the brain that results in the loss of brain function. Also called a stroke.

cerebrum The largest part of the three subdivisions of the brain, sometimes called the "gray matter"; made up of several lobes that control movement, hearing, balance, speech, visual perception, emotions, and personality.

certification A process in which a person, an institution, or a program is evaluated and recognized as meeting certain predetermined standards to provide safe and ethical care.

cervical spine The portion of the spinal column consisting of the first seven vertebrae that lie in the neck.

cervix Narrowest portion of the uterus that opens into the vagina.

channel An assigned frequency or frequencies that are used to carry voice and/or data communications.

Chemical Transportation Emergency Center (CHEMTREC) An agency that assists emergency personnel in identifying and handling hazardous materials transport incidents.

chemoreceptors Monitor the levels of O_2, CO_2, and the pH of the cerebrospinal fluid and then provide feedback to the respiratory centers to modify the rate and depth of breathing based on the body's needs at any given time.

chest leads The leads that are used only with a 12-lead ECG and must be placed exactly; includes leads V_1, V_2, V_3, V_4, V_5, and V_6.

chief complaint The reason a patient called for help; also, the patient's response to questions such as "What's wrong?" or "What happened?"

child abuse A general term applying to all forms of child abuse and neglect.

chlamydia A sexually transmitted disease caused by the bacterium *Chlamydia trachomatis*.

chlorine (CL) The first chemical agent ever used in warfare. It has a distinct odor of bleach and creates a green haze when released as a gas. Initially it produces upper airway irritation and a choking sensation.

cholecystitis Inflammation of the gallbladder.

chordae tendineae Thin bands of fibrous tissue that attach to the valves in the heart and prevent them from inverting.

chronic bronchitis Irritation of the major lung passageways from infectious disease or irritants such as smoke.

chronic obstructive pulmonary disease (COPD) A slow process of dilation and disruption of the airways and alveoli caused by chronic bronchial obstruction.

chyme The name of the substance that leaves the stomach. It is a combination of all of the eaten foods with added stomach acids.

circulatory system The complex arrangement of connected tubes, including the arteries, arterioles, capillaries, venules, and veins, that moves blood, oxygen, nutrients, carbon dioxide, and cellular waste throughout the body.

clavicle The collarbone; it is lateral to the sternum and anterior to the scapula.

cleaning The process of removing dirt, dust, blood, or other visible contaminants from a surface.

closed abdominal injury An injury in which there is soft-tissue damage inside the body but the skin remains intact.

closed chest injury An injury to the chest in which the skin is not broken, usually caused by blunt trauma.

close-ended questions Questions that can be answered in short or single word responses.

closed fracture A fracture in which the skin is not broken.

closed head injury Injury in which the brain has been injured but the skin has not been broken and there is no obvious bleeding.

closed injuries Injuries in which damage occurs beneath the skin or mucous membrane but the surface remains intact.

coagulate To form a clot to plug an opening in an injured blood vessel and stop bleeding.

coccyx The last three or four vertebrae of the spine; the tailbone.

cold zone A safe area at a hazardous materials incident for the agencies involved in the operations. The incident commander, the command post, EMS providers, and other support functions necessary to control the incident should be located in the cold zone. Also referred to as the clean zone or the support zone.

colorimetric devices Capnometer or end-tidal carbon dioxide detectors are devices that use a chemical reaction to detect the amount of carbon dioxide present in expired gases by changing colors (qualitative measurement rather than quantitative).

colostomy A surgical procedure to establish an opening between the colon and the surface of the body.

coma A state of profound unconsciousness from which one cannot be roused.

Combitube A multilumen airway device that consists of a single tube with two lumens, two balloons, and two ventilation ports; an alternative airway device if endotracheal intubation is not possible or has failed.

command In incident command, the position that oversees the incident, establishes the objectives and priorities, and from there develops a response plan.

command post The location of the incident commander at the scene of an emergency and where command, coordination, control, and communication are centralized.

common cold A viral infection usually associated with swollen nasal mucous membranes and the production of fluid from the sinuses and nose.

commotio cordis A blunt chest injury caused by a sudden, direct blow to the chest that occurs only during the critical portion of a person's heartbeat.

communicable disease A disease that can be spread from one person or species to another.

communication The transmission of information to another person—verbally or through body language.

compartment syndrome Swelling in a confined space that produces dangerous pressure; may cut off blood flow or damage sensitive tissue; frequently seen in fractures below the elbow or knee in children.

compensated shock The early stage of shock, in which the body can still compensate for blood loss.

compensatory damages Damages awarded in a civil suit that are intended to restore the plaintiff to the same condition that he or she was in prior to the incident complained about in the lawsuit.

competent Able to make rational decisions about personal well-being.

complex access Complicated entry that requires special tools and training and includes breaking windows or using other force.

compliance The ability of the alveoli to expand when air is drawn in during inhalation.

concussion A temporary loss or alteration of part or all of the brain's abilities to function without actual physical damage to the brain.

conduction The loss of heat by direct contact (eg, when a body part comes into contact with a colder object).

congestive heart failure (CHF) A disorder in which the heart loses part of its ability to effectively pump blood, usually as a result of damage to the heart muscle and usually resulting in a backup of fluid into the lungs.

conjunctiva The delicate membrane that lines the eyelids and covers the exposed surface of the eye.

conjunctivitis Inflammation of the conjunctiva.

connecting nerves Nerves in the spinal cord that connect the motor and sensory nerves.

consent Permission to render care.

contact burn A burn caused by direct contact with a hot object.

contact hazard A hazardous agent that gives off very little or no vapors; the skin is the primary route for this type of chemical to enter the body; also called a skin hazard.

contagious An infectious disease that can be transmitted to another; communicable. A person who has a contagious disease and can transmit it to another person might be considered "contagious."

container Any vessel or receptacle that holds material, including storage vessels, pipelines, and packaging.

contamination The presence of infectious organisms or foreign bodies on or in objects such as dressings, water, food, needles, wounds, or a patient's body.

continuous positive airway pressure (CPAP) A method of ventilation used primarily in the treatment of critically ill patients with respiratory distress; can prevent the need for endotracheal intubation.

continuous quality improvement (CQI) A system of internal and external reviews and audits of all aspects of an EMS system.

contraindications Conditions that make a particular medication or treatment inappropriate; for example, a condition in which a medication should not be given because it would not help or may actually harm a patient.

contributary negligence A legal defense that may be raised when the defendant feels that the conduct of the plaintiff somehow contributed to any injuries or damages that were sustained by the plaintiff.

control zones Areas at a hazardous materials incident that are designated as hot, warm, or cold, based on safety issues and the degree of hazard found there.

contusion A bruise from an injury that causes bleeding beneath the skin without breaking the skin.

convection The loss of body heat caused by air movement (eg, breeze blowing across the body).

conventional reasoning A type of reasoning in which a child looks for approval from peers and society.

core temperature The temperature of the central part of the body (eg, the heart, lungs, and vital organs).

cornea The transparent tissue layer in front of the pupil and iris of the eye.

coronal plane An imaginary plane where the body is cut into front and back parts.

coronary arteries The blood vessels that carry blood and nutrients to the heart muscle.

coup-contrecoup injury Dual impacting of the brain into the skull; coup injury occurs at the point of impact; contrecoup injury occurs on the opposite side of impact, as the brain rebounds.

cover and concealment The tactical use of an impenetrable barrier for protection.

covert An act in which the public safety community generally has no prior knowledge of the time, location, or nature of the attack.

CPR board A device that provides a firm surface under the patient's torso.

cranium The area of the head above the ears and eyes; the skull. The cranium contains the brain.

crepitus A grating or grinding sensation caused by fractured bone ends or joints rubbing together; also air bubbles under the skin that produce a crackling sound or crinkly feeling.

cricoid cartilage A firm ridge of cartilage that forms the lower part of the larynx.

cricothyroid membrane A thin sheet of fascia that connects the thyroid and cricoid cartilages that make up the larynx.

critical incident stress management (CISM) A process that confronts the responses to critical incidents and defuses them, directing the emergency services personnel toward physical and emotional equilibrium.

cross-contamination Occurs when a person is contaminated by an agent as a result of coming into contact with another contaminated person.

croup An inflammatory disease of the upper respiratory system that may cause a partial airway obstruction and is characterized by a barking cough; usually seen in children.

crowning The appearance of the infant's head at the vaginal opening during labor.

crush syndrome Significant metabolic derangement that develops when crushed extremities or body parts remain trapped for prolonged periods. This can lead to renal failure and death.

crushing injury An injury that occurs when a great amount of force is applied to the body.

cultural imposition When one person imposes his or her beliefs, values, and practices on another because he or she believe his or her ideals are superior.

cumulative stress reactions Prolonged or excessive stress.

cushion of safety Keeping a safe distance between your vehicle and other vehicles on any side of you.

cyanide An agent that affects the body's ability to use oxygen. It is a colorless gas that has an odor similar to almonds. The effects begin on the cellular level and are very rapidly seen at the organ and system levels.

cyanosis A bluish gray skin color that is caused by a reduced level of oxygen in the blood.

cylinders Portable, compressed gas containers used to hold liquids and gases. Uninsulated compressed gas cylinders are used to store substances such as nitrogen, argon, helium, and oxygen. They have a range of sizes and internal pressures.

cystitis Inflammation of the bladder.

danger zone (hot zone) An area where individuals can be exposed to hazards such as sharp metal edges, broken glass, toxic substances, lethal rays, or ignition or explosion of hazardous materials.

DCAP-BTLS A mnemonic for assessment in which each area of the body is evaluated for Deformities, Contusions, Abrasions, Punctures/penetrations, Burns, Tenderness, Lacerations, and Swelling.

dead space The portion of the tidal volume that does not reach the alveoli and thus does not participate in gas exchange.

decay A natural process in which a material that is unstable attempts to stabilize itself by changing its structure.

deceleration The slowing of an object.

decision-making capacity Ability to understand and process information and make a choice regarding appropriate medical care.

decompensated shock The late stage of shock when blood pressure is falling.

decompression sickness A painful condition seen in divers who ascend too quickly, in which gas, especially nitrogen, forms bubbles in blood vessels and other tissues; also called "the bends."

decontamination The process of removing or neutralizing and properly disposing of hazardous materials from equipment, patients, and rescue personnel.

decontamination area The designated area in a hazardous materials incident where all patients and rescuers must be decontaminated before going to another area.

decubitus ulcers Also known as bedsores, they are caused by the pressure of skin against a surface for long periods. These sores can range from a pink discoloration of the skin to a deep wound that may invade into bone or organs.

dedicated line A special telephone line that is used for specific point-to-point communications; also known as a "hotline."

deep Further inside the body and away from the skin.

deep venous thrombosis The formation of a blood clot within the larger veins of an extremity, typically following a period of prolonged immobilization.

defamation The communication of false information about a person that is damaging to that person's reputation or standing in the community.

defibrillate To shock a fibrillating (chaotically beating) heart with specialized electrical current in an attempt to restore a normal, rhythmic beat.

dehydration Loss of water from the tissues of the body.

delayed stress reaction Reaction to stress that occurs after a stressful situation.

delirium A more or less sudden change in mental status marked by the inability to focus, think logically, and maintain attention.

delirium tremens (DTs) A severe withdrawal syndrome seen in alcoholics who are deprived of ethyl alcohol; characterized by restlessness, fever, sweating, disorientation, agitation, and seizures; can be fatal if untreated.

dementia The slow onset of progressive disorientation, shortened attention span, and loss of cognitive function.

demobilization The process of directing responders to return to their facilities when work at a disaster or mass-casualty incident has finished, at least for those particular responders.

dependent edema Swelling in the part of the body closest to the ground, caused by collection of fluid in the tissues; a possible sign of congestive heart failure.

dependent lividity Blood settling to the lowest point of the body, causing discoloration of the skin.

depositions Oral questions asked of parties and witnesses under oath.

depression A persistent mood of sadness, despair, and discouragement; may be a symptom of many different mental and physical disorders, or it may be a disorder on its own.

dermis The inner layer of the skin, containing hair follicles, sweat glands, nerve endings, and blood vessels.

designated officer The individual in the department who is charged with the responsibility of managing exposures and infection control issues.

developmental disability Insufficient development of the brain, resulting in some level of dysfunction or impairment.

diabetes mellitus A metabolic disorder in which the ability to metabolize carbohydrates (sugars) is impaired, usually because of a lack of insulin.

diabetic ketoacidosis (DKA) A form of hyperglycemia in uncontrolled diabetes in which certain acids accumulate when insulin is not available.

diamond carry A carrying technique in which one EMT is located at the head end of the stretcher or backboard, one at the foot end, and one at each side of the patient; each of the two EMTs at the sides uses one hand to support the stretcher/backboard so that all are able to face forward as they walk.

diaphoretic Characterized by profuse sweating.

diaphragm A muscular dome that forms the undersurface of the thorax, separating the chest from the abdominal cavity. Contraction of the diaphragm (and the chest wall muscles) brings air into the lungs. Relaxation allows air to be expelled from the lungs.

diastole The relaxation, or period of relaxation, of the heart, especially of the ventricles.

diastolic pressure The pressure that remains in the arteries during the relaxing phase of the heart's cycle (diastole) when the left ventricle is at rest.

diffusion A process in which molecules move from an area of higher concentration to an area of lower concentration.

digestion The processing of food that nourishes the individual cells of the body.

dilation Widening of a tubular structure such as a coronary artery.

diphtheria An infectious disease in which a membrane forms, lining the pharynx; this lining can severely obstruct the passage of air into the larynx.

direct contact Exposure or transmission of a communicable disease from one person to another by physical contact.

direct ground lift A lifting technique that is used for patients who are found lying supine on the ground with no suspected spinal injury.

dirty bomb Name given to a bomb that is used as a radiologic dispersal device.

disaster A widespread event that disrupts community resources and functions, in turn threatening public safety, citizens' lives, and property.

discovery The phase of a civil suit where the plaintiff and defense obtain information from each other that will enable the attorneys to have a better understanding of the case, which will assist them in negotiating a possible settlement or in preparing for trial. Discovery includes depositions, interrogatories, and demands for production of records.

disease vector An animal that spreads a disease, once infected, to another animal.

disinfection The killing of pathogenic agents by direct application of chemicals.

dislocation Disruption of a joint in which ligaments are damaged and the bone ends are completely displaced.

displaced fracture A fracture in which bone fragments are separated from one another and not in anatomic alignment.

dissecting aneurysm A condition in which the inner layers of an artery, such as the aorta, become separated, allowing blood (at high pressures) to flow between the layers.

dissemination The means by which a terrorist will spread a disease, for example, by poisoning the water supply or aerosolizing the agent into the air or ventilation system of a building.

distal Farther from the trunk or nearer to the free end of the extremity.

distraction The action of pulling the spine along its length.

distributive shock A condition that occurs when there is widespread dilation of the small arterioles, small venules, or both.

diverticulitis Bulging out of intestinal rings in small pockets at weak areas in the muscle walls, creating abdominal discomfort.

diving reflex Slowing of the heart rate caused by submersion in cold water.

do not resuscitate (DNR) orders Written documentation by a physician giving permission to medical personnel to not attempt resuscitation in the event of cardiac arrest.

documentation The written portion of the EMT's patient interaction. This becomes part of the patient's permanent medical record.

dorsal The posterior surface of the body, including the back of the hand.

dorsal respiratory group (DRG) A portion of the medulla oblongata where the primary respiratory pacemaker is found.

dorsalis pedis artery The artery on the anterior surface of the foot between the first and second metatarsals.

dose The amount of medication given on the basis of the patient's size and age.

Down syndrome A genetic chromosomal defect that can occur during fetal development and that results in mental retardation as well as certain physical characteristics, such as a round head with a flat occiput and slanted, wide-set eyes.

drag Resistance that slows a projectile, such as air.

drip chamber The area of the administration set where fluid accumulates so that the tubing remains filled with fluid.

drip sets Another name for administration sets.

drowning The process of experiencing respiratory impairment from submersion or immersion in liquid.

drums Barrel-like containers used to store a wide variety of substances, including food-grade materials, corrosives, flammable liquids, and grease. Drums may be constructed of low-carbon steel, polyethylene, cardboard, stainless steel, nickel, or other materials.

DuoDote auto-injector A nerve agent antidote kit containing atropine and pralidoxime chloride; delivered as a single dose through one needle.

duplex The ability to transmit and receive simultaneously.

durable power of attorney for health care A type of advance directive executed by a competent adult that appoints another individual to make medical treatment decisions on his or her behalf in the event that the person making the appointment loses decision-making capacity.

duty to act A medicolegal term relating to certain personnel who either by statute or by function have a responsibility to provide care.

dysarthria Slurred speech.

dysbarism injuries Any signs and symptoms caused by the difference between the surrounding atmospheric pressure and the total gas pressure in various tissues, fluids, and cavities of the body.

dyspnea Shortness of breath or difficulty breathing.

early adults Persons who are 19 to 40 years of age.

ecchymosis Bruising or discoloration associated with bleeding within or under the skin.

ECG Electrocardiogram; an electronic tracing of the heart's electrical activity through leads, which originate in the electrocardiograph machine and contain electrodes that attach to the patient's chest and/or limbs.

eclampsia Seizures (convulsions) resulting from severe hypertension in a pregnant woman.

ectopic pregnancy A pregnancy that develops outside the uterus, typically in a fallopian tube.

edema The presence of abnormally large amounts of fluid between cells in body tissues, causing swelling of the affected area.

elder abuse Any action on the part of an older person's family member, caregiver, or other associated person that takes advantage of the older person, his or her property, or emotional state; also called granny beating and parent battering.

electrical conduction system A network of special cells in the heart through which an electrical current flows, causing contractions of the heart that produce pumping of blood.

emancipated minors A person who is under the legal age in a given state but, because of other circumstances, is legally considered an adult.

embolus A blood clot or other substance in the circulatory system that travels to a blood vessel where it causes a blockage.

embryo The fertilized egg that is the early stages of a fetus.

emergency A serious situation, such as injury or illness, that threatens the life or welfare of a person or group of people and requires immediate intervention.

emergency doctrine The principle of law that permits a health care provider to treat a patient in an emergency situation when the patient is incapable of granting consent because of an altered level of consciousness, disability, the effects of drugs or alcohol, or the patient's age.

emergency medical care Immediate care or treatment.

emergency medical dispatch (EMD) A system that assists dispatchers in selecting appropriate units to respond to a particular call for assistance and in providing callers with vital instructions until the arrival of EMS crews.

emergency medical responder (EMR) The first trained individual, such as a police officer, fire fighter, lifeguard, or other rescuer, to arrive at the scene of an emergency to provide initial medical assistance.

emergency medical services (EMS) A multidisciplinary system that represents the combined efforts of several professionals and agencies to provide prehospital emergency care to the sick and injured.

emergency medical technician (EMT) An individual who has training in basic life support, including automated external defibrillation, use of a definitive airway adjunct, and assisting patients with certain medications.

emergency move A move in which the patient is dragged or pulled from a dangerous scene before assessment and care are provided.

Emergency Response Guidebook A preliminary action guide for first responders operating at a hazardous materials incident in coordination with the US Department of Transportation's labels and placards marking system. The ERG was jointly developed by the DOT, the Secretariat of Communications and Transportation of Mexico, and Transport Canada.

emesis Vomiting.

emphysema A disease of the lungs in which there is extreme dilation and eventual destruction of the pulmonary alveoli with poor exchange of oxygen and carbon dioxide; it is one form of chronic obstructive pulmonary disease.

EMT-administered medication When the EMT directly administers the medication to the patient.

endocrine glands Glands that secrete or release chemicals that are used inside the body.

endocrine system The complex message and control system that integrates many body functions, including the release of hormones.

endometrium The lining of the inside of the uterus.

endotracheal intubation Insertion of an endotracheal tube directly through the larynx between the vocal cords and into the trachea to maintain and protect an airway.

end-tidal carbon dioxide detectors Plastic, disposable indicators that signal by color change when an endotracheal tube is in the proper place.

end-tidal CO_2 The amount of carbon dioxide present in exhaled breath.

enteral medications Medications that enter the body through the digestive system.

entrapment To be caught (trapped) within a vehicle, room, or container with no way out or to have a limb or other body part trapped.

envenomation The act of injecting venom.

enzymes Catalysts designed to speed up the rate of specific biochemical reactions.

epidemic Occurs when new cases of a disease occur in a human population and substantially exceed what is "expected," based on recent experience.

epidermis The outer layer of skin that acts as a watertight protective covering.

epidural hematoma An accumulation of blood between the skull and the dura mater.

epiglottis A thin, leaf-shaped valve that allows air to pass into the trachea but prevents food and liquid from entering.

epiglottitis A disease in which the epiglottis becomes inflamed and enlarged and may cause an upper airway obstruction.

epinephrine A substance produced by the body (commonly called adrenaline), and a drug produced by pharmaceutical companies that increases pulse rate and blood pressure; the drug of choice for an anaphylactic reaction.

epistaxis A nosebleed.

esophagus A collapsible tube that extends from the pharynx to the stomach; contractions of the muscle in the wall of the esophagus propel food and liquids through it to the stomach.

ethics The philosophy of right and wrong, of moral duties, and of ideal professional behavior.

ethnocentrism When a person considers his or her own cultural values as more important when interacting with people of a different culture.

eustachian tube A branch of the internal auditory canal that connects the middle ear to the oropharynx.

evaporation Conversion of water or another fluid from a liquid to a gas.

evisceration The displacement of organs outside of the body.

exhalation The passive part of the breathing process in which the diaphragm and the intercostal muscles relax, forcing air out of the lungs.

expiratory reserve volume The amount of air that can be exhaled following a normal exhalation; average volume is about 1,200 mL.

exposure A situation in which a person has had contact with blood, body fluids, tissues, or airborne particles in a manner that suggests disease transmission may occur.

expressed consent A type of consent in which a patient gives express authorization for provision of care or transport.

extend To straighten.

extension The straightening of a joint.

external auditory canal The ear canal; leads to the tympanic membrane.

external jugular IV IV access established in the external jugular vein of the neck.

external respiration The exchange of gases between the lungs and the blood cells in the pulmonary capillaries; also called pulmonary respiration.

extremity lift A lifting technique that is used for patients who are supine or in a sitting position with no suspected extremity or spinal injuries.

extrication Removal of a patient from entrapment or a dangerous situation or position, such as removal from a wrecked vehicle, industrial accident, or building collapse.

extrication supervisor In incident command, the person appointed to determine the type of equipment and resources needed for a situation involving extrication or special rescue; also called the rescue officer.

extubation Removal of a tube after it has been placed.

eyes-forward position A head position in which the patient's eyes are looking straight ahead and the head and torso are in line.

fallopian tubes The tubes that connect each ovary with the uterus and are the primary location for fertilization of the ovum.

false imprisonment The confinement of a person without legal authority or the person's consent.

fascia The fiberlike connective tissue that covers arteries, veins, tendons, and ligaments.

febrile seizures Seizures that result from sudden high fevers, particularly in children.

Federal Communications Commission (FCC) The federal agency that has jurisdiction over interstate and international telephone and telegraph services and satellite communications, all of which may involve EMS activity.

femoral artery The principal artery of the thigh, a continuation of the external iliac artery. It supplies blood to the lower abdominal wall, external genitalia, and legs. It can be palpated in the groin area.

femoral head The proximal end of the femur, articulating with the acetabulum to form the hip joint.

femur The thighbone; the longest and one of the strongest bones in the body.

fetal alcohol syndrome A condition of infants who are born to women who consume alcohol during pregnancy; characterized by growth and physical problems, mental retardation, and a variety of congenital abnormalities.

fetus The developing, unborn infant inside the uterus.

fibula The outer and smaller bone of the two bones of the lower leg.

finance In incident command, the position in an incident responsible for accounting of all expenditures.

first-responder vehicles Specialized vehicles used to transport EMS equipment and personnel to the scenes of medical emergencies.

flail chest A condition in which two or more ribs are fractured in two or more places or in association with a fracture of the sternum so that a segment of the chest wall is effectively detached from the rest of the thoracic cage.

flame burn A burn caused by an open flame.

flank The posterior region below the margin of the lower rib cage.

flash burn A burn caused by exposure to very intense heat, such as in an explosion.

flex To bend.

flexible stretcher A stretcher that is a rigid carrying device when secured around a patient but can be folded or rolled when not in use.

flexion The bending of a joint.

flutter valve A one-way valve that allows air to leave the chest cavity but not return; formed by taping three sides of an occlusive dressing to the chest wall, leaving the fourth side open as a valve.

focused assessment A type of physical assessment that is typically performed on patients who have sustained nonsignificant mechanisms of injury or on responsive medical patients. This type of examination is based on the chief complaint and focuses on one body system or part.

fontanelles Areas where the infant's skull has not fused together; usually disappear at approximately 18 months of age.

foodborne transmission The contamination of food or water with an organism than can cause disease.

foramen magnum A large opening at the base of the skull through which the brain connects to the spinal cord.

forcible restraint The act of physically preventing an individual from initiating any physical action.

four-person log roll The recommended procedure for moving a patient with a suspected spinal injury from the ground to a long backboard.

fracture A break in the continuity of a bone.

freelancing When individual units or different organizations make independent and often inefficient decisions about the next appropriate action.

freight bills The shipping papers used for transport of chemicals along roads and highways. Also referred to as bills of lading.

frostbite Damage to tissues as the result of exposure to cold; frozen or partially frozen body parts are frostbitten.

full-body scan A systematic head-to-toe examination that is performed during the secondary assessment on a patient who has sustained a significant mechanism of injury, is unconscious, or is in critical condition.

full-thickness (third-degree) burns Burns that affects all skin layers and may affect the subcutaneous layers, muscle, bone, and internal organs, leaving the area dry, leathery, and white, dark brown, or charred.

functional disorder A disorder in which there is no known physiologic reason for the abnormal functioning of an organ or organ system.

fundus The dome-shaped top of the uterus.

G agents Early nerve agents that were developed by German scientists in the period after World War I and into World War II. There are three such agents: sarin, soman, and tabun.

gag reflex A normal reflex mechanism that causes retching; activated by touching the soft palate or the back of the throat.

gallbladder A sac on the undersurface of the liver that collects bile from the liver and discharges it into the duodenum through the common bile duct.

gamma (x-ray) A type of energy that is emitted from a strong radiologic source that is far faster and stronger than alpha and beta rays. These rays easily penetrate through the human body and require lead or several inches of concrete to prevent penetration.

gastric distention A condition in which air fills the stomach, often as a result of high volume and pressure during artificial ventilation.

gastric tube An advanced airway adjunct that provides a channel directly into a patient's stomach, allowing for removal of gas, blood, and toxins and for instilling medications and nutrition.

gauge A measure of the interior diameter of the catheter. It is inversely proportional to the true diameter of the catheter.

gel A semiliquid substance that is administered orally in capsule form or through plastic tubes.

general adaptation syndrome The body's response to stress that begins with an alarm response, followed by a stage of reaction and resistance, and then recovery or, if the stress is prolonged, exhaustion.

general impression The overall initial impression that determines the priority for patient care; based on the patient's surroundings, the mechanism of injury, signs and symptoms, and the chief complaint.

generalized seizure A seizure characterized by severe twitching of all of the body's muscles that may last several minutes or more; formerly known as a grand mal seizure.

generalized tonic-clonic seizure A seizure that features rhythmic back-and-forth motion of an extremity and body stiffness.

generic name The original chemical name of a medication (in contrast with one of its "trade names"); the name is not capitalized.

genital system The reproductive system in males and females.

geriatrics The assessment and treatment of disease in someone who is 65 years or older.

germinal layer The deepest layer of the epidermis where new skin cells are formed.

gestational diabetes Diabetes that develops during pregnancy in women who did not have diabetes before pregnancy.

Glasgow Coma Scale (GCS) score An evaluation tool used to determine level of consciousness, which evaluates and assigns point values (scores) for eye opening, verbal response, and motor response, which are then totaled; effective in helping predict patient outcomes.

glenoid fossa The part of the scapula that joins with the humeral head to form the glenohumeral joint.

globe The eyeball.

glottis The space in between the vocal cords that is the narrowest portion of the adult's airway; also called the glottic opening.

glucose One of the basic sugars; it is the primary fuel, in conjunction with oxygen, for cellular metabolism.

Golden Period The time from injury to definitive care, during which treatment of shock and traumatic injuries should occur because survival potential is best.

gonorrhea A sexually transmitted disease caused by *Neisseria gonorrhoeae*.

good air exchange A term used to distinguish the degree of distress in a patient with a mild airway obstruction. With good air exchange, the patient is still conscious and able to cough forcefully, although wheezing may be heard.

Good Samaritan laws Statutory provisions enacted by many states to protect citizens from liability for errors and omissions in giving good faith emergency medical care, unless there is wanton, gross, or willful negligence.

governmental immunity If your service is covered by immunity, it may mean that you cannot be sued or it may limit the amount of the monetary judgment that the plaintiff may recover; generally applies only to EMS services that are operated by municipalities or other governmental entities.

greater trochanter A bony prominence on the proximal lateral side of the thigh, just below the hip joint.

gross negligence Conduct that constitutes a willful or reckless disregard for a duty or standard of care.

grunting An "uh" sound heard during exhalation; reflects the child's attempt to keep the alveoli open; a sign of increased work of breathing.

guarding Involuntary muscle contractions (spasms) of the abdominal wall in an effort to protect an inflamed abdomen; a sign of peritonitis.

hair follicles The small organs that produce hair.

hallucinogens Agents that produce false perceptions in any one of the five senses.

hay fever An allergic response usually to outdoor airborne allergens such as pollen or sometimes indoor allergens such as dust mites or pet dander; also called allergic rhinitis.

hazardous material Any substance that is toxic, poisonous, radioactive, flammable, or explosive and causes injury or death with exposure.

hazardous materials (HazMat) incident An incident in which a hazardous material is no longer properly contained and isolated.

head tilt–chin lift maneuver A combination of two movements to open the airway by tilting the forehead back and lifting the chin; not used for trauma patients.

health care directive A written document that specifies medical treatment for a competent patient, should he or she become unable to make decisions. Also known as an advance directive or a living will.

health care proxies A type of advance directive executed by a competent adult that appoints another individual to make medical treatment decisions on his or her behalf in the event that the person making the appointment loses decision making capacity. Also known as a durable power of attorney for health care.

Health Insurance Portability and Accountability Act (HIPAA) Federal legislation passed in 1996. Its main effect in EMS is in limiting availability of patients' health care information and penalizing violations of patient privacy.

heart A hollow muscular organ that pumps blood throughout the body.

heart rate The number of heartbeats during a specific time.

heat cramps Painful muscle spasms usually associated with vigorous activity in a hot environment.

heat exhaustion A form of heat injury in which the body loses significant amounts of fluid and electrolytes because of heavy sweating; also called heat prostration or heat collapse.

heatstroke A life-threatening condition of severe hyperthermia caused by exposure to excessive natural or artificial heat, marked by warm, dry skin; severely altered mental status; and often irreversible coma.

hematemesis Vomited blood.

hematology The study and prevention of blood-related disorders.

hematoma A mass of blood in the soft tissues beneath the skin.

hematuria Blood in the urine.

hemiparesis Weakness on one side of the body.

hemolytic crisis A rapid destruction of red blood cells that occurs faster than the body's ability to create new cells.

hemophilia A congenital abnormality in which the body is unable to produce clots, which results in uncontrollable bleeding.

hemopneumothorax The accumulation of blood and air in the pleural space of the chest.

hemoptysis Coughing up blood.

hemorrhage Bleeding.

hemorrhagic stroke One of the two main types of stroke; occurs as a result of bleeding inside the brain.

hemothorax A collection of blood in the pleural cavity.

hepatitis Inflammation of the liver, usually caused by a viral infection, that causes fever, loss of appetite, jaundice, fatigue, and altered liver function.

Hering-Breuer reflex A protective mechanism that terminates inhalation, thus preventing overexpansion of the lungs.

hernia The protrusion of a loop of an organ or tissue through an abnormal body opening.

herpes simplex Virus caused by human herpesviruses 1 and 2, characterized by small blisters whose location depends on the type of virus. Type 2 results in blisters on the genital area, while type 1 results in blisters in nongenital areas.

high-level disinfection The killing of pathogenic agents by using potent means of disinfection.

hinge joints Joints that can bend and straighten but cannot rotate; they restrict motion to one plane.

histamines Substances released by the immune system in allergic reactions that are responsible for many of the symptoms of anaphylaxis, such as vasodilation.

history taking A step within the patient assessment process that provides detail about the patient's chief complaint and an account of the patient's signs and symptoms.

hollow organs Structures through which materials pass, such as the stomach, small intestines, large intestines, ureters, and bladder.

homeostasis A balance of all systems of the body.

hormones Substances formed in specialized organs or glands and carried to another organ or group of cells in the same organism. Hormones regulate many body functions, including metabolism, growth, and body temperature.

host The organism or individual that is attacked by the infecting agent.

hot zone The area immediately surrounding a hazardous materials spill/incident site that is directly dangerous to life and health. All personnel working in the hot zone must wear complete, appropriate protective clothing and equipment. Entry requires approval by the incident commander or other designated officer.

human immunodeficiency virus (HIV) Acquired immunodeficiency syndrome (AIDS) is caused by HIV, which damages the cells in the body's immune system so that the body is unable to fight infection or certain cancers.

humerus The supporting bone of the upper arm.

hydroplaning A condition in which the tires of a vehicle may be lifted off the road surface as water "piles up" under them, making the vehicle feel as though it is floating.

hydrostatic pressure The pressure of water against the walls of its container.

hymenoptera A family of insects that includes bees, wasps, ants, and yellow jackets.

hypercarbia Increased carbon dioxide level in the bloodstream.

hyperglycemia An abnormally high glucose level in the blood.

hyperglycemic crisis A state of unconsciousness resulting from several problems, including ketoacidosis, dehydration because of excessive urination, and hyperglycemia.

hypertension Blood pressure that is higher than the normal range.

hypertensive emergency An emergency situation created by excessively high blood pressure, which can lead to serious complications such as stroke or aneurysm.

hyperthermia A condition in which the body core temperature rises to 101°F (38.3°C) or more.

hyperventilation Rapid or deep breathing that lowers the blood carbon dioxide level below normal.

hyperventilation syndrome (panic attack) This syndrome occurs in the absence of other physical problems. The respirations of a person who is experiencing hyperventilation syndrome may be as high as 40 shallow breaths/min or as low as only 20 very deep breaths/min.

hypnotic A sleep-inducing effect or agent.

hypoglycemia A condition characterized by a low blood glucose level.

hypoglycemic crisis Severe hypoglycemia resulting in changes in mental status.

hypoperfusion A condition that occurs when the level of tissue perfusion decreases below that needed to maintain normal cellular functions; also called shock.

hypotension Blood pressure that is lower than the normal range.

hypothermia A condition in which the internal body temperature falls below 95°F (35°C), usually as a result of prolonged exposure to cool or freezing temperatures.

hypovolemic shock A condition in which low blood volume, due to massive internal or external bleeding or extensive loss of body water, results in inadequate perfusion.

hypoxia A dangerous condition in which the body tissues and cells do not have enough oxygen.

hypoxic drive A "backup system" to control respiration; senses drops in the oxygen level in the blood.

ileostomy A surgical procedure to create an opening between the small intestine and the surface of the body.

ileus Paralysis of the bowel, arising from any one of several causes; stops contractions that move material through the intestine.

ilium One of three bones that fuse to form the pelvic ring.

immune The body's ability to protect itself from acquiring a disease.

immune response The body's response to substance perceived by the body as foreign.

immune system The body system that includes all of the structures and processes designed to mount a defense against foreign substances and disease-causing agents.

immunology The study of the body's immune system.

impedance threshold device (ITD) A valve device placed between the endotracheal tube and a bag-mask device that limits the amount of air entering the lungs during the recoil phase between chest compressions.

implied consent Type of consent in which a patient who is unable to give consent is given treatment under the legal assumption that he or she would want treatment.

in loco parentis Refers to the legal responsibility of a person or organization to take on some of the functions and responsibilities of a parent.

incident action plan An oral or written plan stating general objectives reflecting the overall strategy for managing an incident.

incident command system (ICS) A system implemented to manage disasters and mass-casualty incidents in which section chiefs, including finance, logistics, operations, and planning, report to the incident commander.

incident commander (IC) The overall leader of the incident command system to whom commanders or leaders of incident command system divisions report.

incision A sharp, smooth cut.

incontinence Loss of bowel and/or bladder control; may be the result of a generalized seizure.

incubation The period of time from a person being exposed to a disease to the time when symptoms begin.

index of suspicion Awareness that unseen life-threatening injuries may exist when determining the mechanism of injury.

indications The therapeutic uses for a specific medication.

indirect contact Exposure or transmission of disease from one person to another by contact with a contaminated object.

infancy The first year of life.

infants Persons who are from 1 month to 1 year of age.

infarction Death of a body tissue, usually caused by interruption of its blood supply.

infection The abnormal invasion of a host or host tissues by organisms such as bacteria, viruses, or parasites, with or without signs or symptoms of disease.

infection control Procedures to reduce transmission of infection among patients and health care personnel.

infectious disease A medical condition caused by the growth and spread of small, harmful organisms within the body.

inferior The part of the body or any body part nearer to the feet.

inferior vena cava One of the two largest veins in the body; carries blood from the lower extremities and the pelvic and the abdominal organs to the heart.

infiltration The escape of fluid into the surrounding tissue when the IV catheter is not in the vein.

influenza type A Virus that has crossed the animal/human barrier and has infected humans, recently reaching a pandemic level with the H1N1 strain.

informed consent Permission for treatment given by a competent patient after the potential risks, benefits, and alternatives to treatment have been explained.

ingestion Swallowing; taking a substance by mouth.

inhalation Breathing into the lungs; a medication delivery route.

inspiratory reserve volume The amount of air that can be inhaled after a normal inhalation; the amount of air that can be inhaled in addition to the normal tidal volume.

insulin A hormone produced by the islets of Langerhans (endocrine gland located throughout the pancreas) that enables glucose in the blood to enter cells; used in synthetic form to treat and control diabetes mellitus.

intended effect The effect that a medication is expected to have on the body.

intermodal tanks Shipping and storage vessels that can be either pressurized or nonpressurized.

internal respiration The exchange of gases between the blood cells and the tissues.

international terrorism Terrorism that is carried out by people in a country other than their own; also known as cross-border terrorism.

interrogatories Written questions that the defense and plaintiff send to one other.

interstitial space The space in between the cells.

intervertebral disk The cushion that lies between two vertebrae.

intracerebral hematoma Bleeding within the brain tissue (parenchyma) itself; also referred to as an intraparenchymal hematoma.

intracranial pressure (ICP) The pressure within the cranial vault.

intramuscular (IM) injection An injection into a muscle; a medication delivery route.

intranasal (IN) A delivery route in which a medication is pushed through a specialized atomizer device called a mucosal atomizer device (MAD) into the naris.

intraosseous (IO) Into the bone; a medication delivery route.

intraosseous (IO) needles Rigid, boring catheters placed into a bone to provide intravenous fluids.

intrapulmonary shunting Bypassing of oxygen-poor blood past nonfunctional alveoli to the left side of the heart.

intravenous (IV) injection An injection directly into a vein; a medication delivery route.

intravenous (IV) therapy The delivery of medication directly into a vein.

involuntary activities Actions of the body that are not under a person's conscious control.

involuntary muscle The muscle over which a person has no conscious control. It is found in many automatic regulating systems of the body.

ionizing radiation Energy that is emitted in the form of rays, or particles.

iris The muscle and surrounding tissue behind the cornea that dilate and constrict the pupil, regulating the amount of light that enters the eye; pigment in this tissue gives the eye its color.

irreversible shock The final stage of shock, resulting in death.

ischemia A lack of oxygen that deprives tissues of necessary nutrients, resulting from partial or complete blockage of blood flow; potentially reversible because permanent injury has not yet occurred.

ischemic stroke One of the two main types of stroke; occurs when blood flow to a particular part of the brain is cut off by a blockage (eg, a clot) inside a blood vessel.

ischium One of three bones that fuse to form the pelvic ring.

isotonic crystalloids Intravenous solutions that do not cause a fluid shift into or out of the cell; examples include normal saline and lactated Ringer's solutions.

jaundice Yellow skin or sclera that is caused by liver disease or dysfunction.

jaw-thrust maneuver Technique to open the airway by placing the fingers behind the angle of the jaw and bringing the jaw forward; used for patients who may have a cervical spine injury.

joint The place where two bones come into contact.

joint capsule The fibrous sac that encloses a joint.

joint Information center An area designated by the incident commander, or a designee, in which public information officers from multiple agencies disseminate information about the incident.

jugular vein distention A visual bulging of the jugular veins in the neck that can be caused by fluid overload, pressure in the chest, cardiac tamponade, or tension pneumothorax.

jump kit A portable kit containing items that are used in the initial care of the patient.

JumpSTART triage A sorting system for pediatric patients younger than 8 years or weighing less than 100 lb. There is a minor adaptation for infants since they cannot ambulate on their own.

Kehr sign Left shoulder pain caused by blood in the peritoneal cavity.

kidnapping The seizing, confining, abducting, or carrying away of a person by force, including transporting a competent adult for medical treatment without his or her consent.

kidney stones Solid crystalline masses formed in the kidney, resulting from an excess of insoluble salts or uric acid crystallizing in the urine; may become trapped anywhere along the urinary tract.

kidneys Two retroperitoneal organs that excrete the end products of metabolism as urine and regulate the body's salt and water content.

kinetic energy The energy of a moving object.

King LT A disposal supraglottic airway used as an alternative to tracheal or mask ventilation.

Kussmaul respirations Deep, rapid breathing; usually the result of an accumulation of certain acids when insulin is not available in the body.

kyphosis A forward curling of the back caused by an abnormal increase in the curvature of the spine.

labia majora Outer fleshy "lips" covered with pubic hair that protect the vagina.

labia minora Inner fleshy "lips" devoid of pubic hair that protect the vagina.

labored breathing Breathing that requires greater than normal effort; may be slower or faster than normal and usually requires the use of accessory muscles.

laceration A jagged, open wound.

lacrimal glands The glands that produce fluids to keep the eye moist; also called tear glands.

lactic acid A metabolic end product of the breakdown of glucose that accumulates when metabolism proceeds in the absence of oxygen.

large intestine The portion of the digestive tube that encircles the abdomen around the small bowel, consisting of the cecum, the colon, and the rectum. It helps regulate water balance and eliminate solid waste.

laryngeal mask airway An advanced airway device that is blindly inserted into the mouth to isolate the larynx for direct ventilation; consists of a tube and a mask or cuff that inflates to seal around the laryngeal opening.

laryngoscope An instrument used to give a direct view of the patient's vocal cords during endotracheal intubation.

larynx A complex structure formed by many independent cartilaginous structures that all work together; where the upper airway ends and the lower airway begins; also called the voice box.

late adults Persons who are 61 years old or older.

lateral In anatomy, parts of the body that lie farther from the midline. Also called outer structures.

LD_{50} The amount of an agent or substance that will kill 50% of people who are exposed to this level.

lens The transparent part of the eye through which images are focused on the retina.

lesser trochanter The projection on the medial/superior portion of the femur.

leukotrienes Chemical substances that contribute to anaphylaxis; released by the immune system in allergic reactions.

Lewisite (L) A blistering agent that has a rapid onset of symptoms and produces immediate, intense pain and discomfort on contact.

liaison officer In incident command, the person who relays information, concerns, and requests among responding agencies.

libel False and damaging information about a person that is communicated in writing.

licensure The process whereby a competent authority, usually the state, allows individuals to perform a regulated act.

life expectancy The average amount of years a person can be expected to live.

ligament A band of fibrous tissue that connects bones to bones. It supports and strengthens a joint.

lightening A sensation felt by a pregnant patient when the fetus positions itself for delivery.

limb leads The four leads used with a 4-lead ECG; placed on or close to the right arm, left arm, right leg, and left leg.

limb presentation A delivery in which the presenting part is a single arm, leg, or foot.

linear skull fractures Account for 80% of skull fractures; also referred to as nondisplaced skull fractures; commonly occur in the temporal-parietal region of the skull; not associated with deformities to the skull.

liver A large solid organ that lies in the right upper quadrant immediately below the diaphragm; it produces bile, stores glucose for immediate use by the body, and produces many substances that help regulate immune responses.

load-distributing band (LDB) A circumferential chest compression device composed of a constricting band and backboard that is either electrically or pneumatically driven to compress the heart by putting inward pressure on the thorax.

logistics In incident command, the position that helps procure and stockpile equipment and supplies during an incident.

lumbar spine The lower part of the back, formed by the lowest five nonfused vertebrae; also called the dorsal spine.

lumen The inside diameter of an artery or other hollow structure.

lymph nodes The area of the lymphatic system where infection-fighting cells are housed.

lymphatic system A passive circulatory system that transports a plasmalike liquid called lymph, a thin fluid that bathes the tissues of the body.

macrodrip set An administration set named for the large orifice between the piercing spike and the drip chamber; allows for rapid fluid flow into the vascular system.

mandible The bone of the lower jaw.

manually triggered ventilation device A fixed flow/rate ventilation device that delivers a breath every time its button is pushed; also referred to as a flow-restricted, oxygen-powered ventilation device.

manubrium The upper quarter of the sternum.

Mark 1 Nerve Agent Antidote Kit (NAAK) A nerve agent antidote kit containing two auto-injectors containing atropine and pralidoxime chloride.

mass-casualty incident (MCI) An emergency situation involving three or more patients that can place great demand on the equipment or personnel of the EMS system or has the potential to produce multiple casualties.

mastoid process The prominent bony mass at the base of the skull about 1" posterior to the external opening of the ear.

material safety data sheet (MSDS) A form, provided by manufacturers and compounders (blenders) of chemicals, containing information about chemical composition, physical and chemical properties, health and safety hazards, emergency response, and waste disposal of a specific material.

maxillae The upper jawbones that assist in the formation of the orbit, the nasal cavity, and the palate and hold the upper teeth.

mechanical piston device A device that depresses the sternum via a compressed gas-powered plunger mounted on a backboard.

mechanism of injury (MOI) The way in which traumatic injuries occur; the forces that act on the body to cause damage.

meconium A dark green material in the amniotic fluid that can indicate distress or disease in the newborn; the meconium can be aspirated into the infant's lungs during delivery; the infant's first bowel movement.

MED channels VHF and UHF channels that the Federal Communications Commission has designated exclusively for EMS use.

medial Parts of the body that lie closer to the midline; also called inner structures.

mediastinum Space within the chest that contains the heart, major blood vessels, vagus nerve, trachea, major bronchi, and esophagus; located between the two lungs.

medical control Physician instructions that are given directly by radio or cell phone (online/direct) or indirectly by protocol/guidelines (off-line/indirect), as authorized by the medical director of the service program.

medical director The physician who authorizes or delegates to the EMT the authority to provide medical care in the field.

medical emergencies Emergencies that require EMS attention because of illnesses or conditions not caused by an outside force.

medical incident command A branch of operations in a unified command system, whose three designated sector positions are triage, treatment, and transport.

medication A chemical substance that is used to treat or prevent disease or relieve pain.

medicolegal A term relating to medical jurisprudence (law) or forensic medicine.

medivac Medical evacuation of a patient by helicopter.

medulla oblongata Nerve tissue that is continuous inferiorly with the spinal cord; serves as a conduction pathway for ascending and descending nerve tracts; coordinates heart rate, blood vessel diameter, breathing, swallowing, vomiting, coughing, and sneezing.

melena Black, foul-smelling, tarry stool containing digested blood.

meninges Three distinct layers of tissue that surround and protect the brain and the spinal cord within the skull and the spinal canal.

meningitis An inflammation of the meningeal coverings of the brain and spinal cord; it is usually caused by a virus or a bacterium.

meningococcal meningitis An inflammation of the meningeal coverings of the brain and spinal cord; can be highly contagious.

metabolism (cellular respiration) The biochemical processes that result in production of energy from nutrients within the cells.

metered-dose inhaler (MDI) A miniature spray canister used to direct medications through the mouth and into the lungs.

methicillin-resistant *Staphylococcus aureus* (MRSA) A bacterium that causes infections in different parts of the body and is often resistant to commonly used antibiotics; can be found on the skin, in surgical wounds, in the bloodstream, lungs, and urinary tract.

microdrip set An administration set named for the small orifice between the piercing spike and the drip chamber; allows for carefully controlled fluid flow and is ideally suited for medication administration.

midbrain The part of the brain that is responsible for helping to regulate the level of consciousness.

middle adults Persons who are 41 to 60 years of age.

midsagittal plane (midline) An imaginary vertical line drawn from the middle of the forehead through the nose and the umbilicus (navel) to the floor.

mild airway obstruction Occurs when a foreign body partially obstructs the patient's airway. The patient is able to move adequate amounts of air, but also experiences some degree of respiratory distress.

minute ventilation The volume of air moved through the lungs in 1 minute minus the dead space; calculated by multiplying tidal volume (minus dead space) and respiratory rate; also referred to as minute volume.

minute volume The amount of air that moves in and out of the lungs per minute minus the dead space. Also called minute ventilation.

miosis Excessively constricted pupil; often bilateral after exposure to nerve agents.

miscarriage The passage of the fetus and placenta before 20 weeks; spontaneous abortion.

mobile data terminals (MDT) Small computer terminals inside ambulances that directly receive data from the dispatch center.

morality A code of conduct that can be defined by society, religion, or a person, affecting character, conduct, and conscience.

morgue supervisor In incident command, the person who works with area medical examiners, coroners, and law enforcement agencies to coordinate the disposition of dead victims.

moro reflex An infant reflex in which, when an infant is caught off guard, the infant opens his or her arms wide, spreads the fingers, and seems to grab at things.

motor nerves Nerves that carry information from the central nervous system to the muscles of the body.

mucosal atomizer device (MAD) A device that is used to change a liquid medication into a spray and push it into a nostril.

mucous membranes The lining of body cavities and passages that communicate directly or indirectly with the environment outside the body.

mucus The opaque, sticky secretion of the mucous membranes that lubricates the body openings.

multigravida A woman who has had previous pregnancies.

multilumen airways Advanced airway devices, such as the esophageal tracheal Combitube and the pharyngeotracheal lumen airway, that have multiple tubes to aid in ventilation and will work whether placed in the trachea or esophagus.

multisystem trauma Trauma that affects more than one body system.

musculoskeletal system The bones and voluntary muscles of the body.

mutagen A substance that mutates, damages, and changes the structures of DNA in the body's cells.

mutual aid response An agreement between neighboring EMS systems to respond to mass-casualty incidents or disasters in each other's region when local resources are insufficient to handle the response.

myocardial contractility The ability of the heart muscle to contract.

myocardial contusion A bruise of the heart muscle.

myocardium The heart muscle.

nares The external openings of the nostrils. A single nostril opening is called a naris.

nasal cannula An oxygen-delivery device in which oxygen flows through two small, tubelike prongs that fit into the patient's nostrils; delivers 24% to 44% supplemental oxygen, depending on the flow rate.

nasal flaring Flaring out of the nostrils, indicating that there is an airway obstruction.

nasopharyngeal (nasal) airway Airway adjunct inserted into the nostril of an unresponsive patient, or a patient with an altered level of consciousness who is unable to maintain airway patency independently.

nasopharynx The nasal cavity; formed by the union of facial bones and protects the respiratory tract from contaminants.

National EMS Scope of Practice Model A document created by the National Highway Traffic Safety Administration (NHTSA) that outlines the skills performed by various EMS providers.

National Incident Management System (NIMS) A Department of Homeland Security system designed to enable federal, state, and local governments and private-sector and nongovernmental organizations to effectively and efficiently prepare for, prevent, respond to, and recover from domestic incidents, regardless of cause, size, or complexity, including acts of catastrophic terrorism.

nature of illness (NOI) The general type of illness a patient is experiencing.

neglect Refusal or failure on the part of the caregiver to provide life necessities.

negligence per se A theory that may be used when the conduct of the person being sued is alleged to have occurred in clear violation of a statute.

negligence Failure to provide the same care that a person with similar training would provide.

Neisseria meningitides A form of bacterial meningitis characterized by rapid onset of symptoms, often leading to shock and death.

neonate Persons who are birth to 1 month of age.

nephrons The basic filtering units in the kidneys.

nerve agents A class of chemical called organophosphates; they function by blocking an essential enzyme in the nervous system, which causes the body's organs to become overstimulated and burn out.

nervous system The system that controls virtually all activities of the body, both voluntary and involuntary.

neurogenic shock Circulatory failure caused by paralysis of the nerves that control the size of the blood vessels, leading to widespread dilation; seen in patients with spinal cord injuries.

neuropathy A group of conditions in which the nerves leaving the spinal cord are damaged, resulting in distortion of signals to or from the brain.

neurotoxins Biologic agents that are the most deadly substances known to humans; they include botulinum toxin and ricin.

neutron radiation The type of energy that is emitted from a strong radiologic source; neutron particles are among the most powerful forms of radiation. Neutrons easily penetrate through lead and require several feet of concrete to stop them.

nitroglycerin A medication that increases cardiac perfusion by causing arteries to dilate; you may be allowed to help the patient self-administer the medication.

noise Anything that dampens or obscures the true meaning of a message.

nonbulk storage vessels Any container other than bulk storage containers such as drums, bags, compressed gas cylinders, and cryogenic containers. Nonbulk storage vessels hold commonly used commercial and industrial chemicals such as solvents, industrial cleaners, and compounds.

nondisplaced fracture A simple crack in the bone that has not caused the bone to move from its normal anatomic position; also called a hairline fracture.

nonrebreathing mask A combination mask and reservoir bag system that is the preferred way to give oxygen in the prehospital setting; delivers up to 90% inspired oxygen and prevents inhaling the exhaled gases (carbon dioxide).

norepinephrine A neurotransmitter and drug sometimes used in the treatment of shock; produces vasoconstriction through its alpha-stimulator properties.

normal sinus rhythm A rhythm that has consistent P waves, consistent P-R intervals, and a regular heart rate of between 60 and 100 beats/min.

nuchal cord An umbilical cord that is wrapped around the infant's neck.

obesity A condition in which a person has an excessive amount of body fat.

obstructive shock Shock that occurs when there is a block to blood flow in the heart or great vessels, causing an insufficient blood supply to the body's tissues.

occiput The most posterior portion of the cranium.

occlusion A blockage, usually of a tubular structure such as a blood vessel.

occlusive dressing A dressing made of Vaseline-impregnated gauze, aluminum foil, or plastic that protects a wound from air and bacteria.

Occupational Safety and Health Administration (OSHA) The federal regulatory compliance agency that develops, publishes, and enforces guidelines concerning safety in the workplace.

off-gassing The emitting of an agent after exposure, for example from a person's clothes that have been exposed to the agent.

oncotic pressure The pressure of water to move, typically into the capillary, as the result of the presence of plasma proteins.

open abdominal injury An injury in which there is a break in the surface of the skin or mucous membrane, exposing deeper tissue to potential contamination.

open chest injury An injury to the chest in which the chest wall itself is penetrated by a fractured rib or, more frequently, by an external object such as a bullet or knife.

open-ended questions Questions for which the patient must provide detail to give an answer.

open fracture Any break in a bone in which the overlying skin has been damaged.

open head injury Injury to the head often caused by a penetrating object in which there may be bleeding and exposed brain tissue.

open injuries Injuries in which there is a break in the surface of the skin or the mucous membrane, exposing deeper tissue to potential contamination.

open pneumothorax An open or penetrating chest wall wound through which air passes during inspiration and expiration, creating a sucking sound; also referred to as a sucking chest wound.

operations In incident command, the position that carries out the orders of the commander to help resolve the incident.

opioids Any drug or agent with actions similar to morphine.

OPQRST An abbreviation for key terms used in evaluating a patient's pain: Onset, Provocation or Palliation, Quality, Region/radiation, Severity, and Timing of pain.

optic nerve A cranial nerve that transmits visual information to the brain.

oral By mouth; a medication delivery route.

oral glucose A simple sugar that is readily absorbed by the bloodstream; it is carried on the EMS unit.

orbit The eye socket, made up of the maxilla and zygoma.

organic brain syndrome Temporary or permanent dysfunction of the brain, caused by a disturbance in the physical or physiologic functioning of brain tissue.

orientation The mental status of a patient as measured by memory of person (name), place (current location), time (current year, month, and approximate date), and event (what happened).

oropharyngeal (oral) airway Airway adjunct inserted into the mouth of an unresponsive patient to keep the tongue from blocking the upper airway and to facilitate suctioning the airway, if necessary.

oropharynx Forms the posterior portion of the oral cavity, which is bordered superiorly by the hard and soft palates, laterally by the cheeks, and inferiorly by the tongue.

orotracheal intubation Endotracheal intubation through the mouth.

osteoporosis A generalized bone disease, commonly associated with postmenopausal women, in which there is a reduction in the amount of bone mass leading to fractures after minimal trauma in either sex.

ovaries The primary female reproductive organs that produce an ovum, or egg, that, if fertilized, will develop into a fetus.

overdose An excessive quantity of a drug which, when taken or administered, can have toxic or lethal consequences.

over-the-counter (OTC) medications Medications that may be purchased directly by a patient without a prescription.

ovulation The process in which an ovum is released by a follicle.

oxygen A gas that all cells need for metabolism; the heart and brain, especially, cannot function without oxygen.

oxygenation The process of delivering oxygen to the blood by diffusion from the alveoli following inhalation into the lungs.

paging The use of a radio signal and a voice or digital message that is transmitted to pagers ("beepers") or desktop monitor radios.

palmar The forward facing part of the hand in the anatomic position.

palmar grasp An infant reflex that occurs when something is placed in the infant's palm; the infant grasps the object.

palpate To examine by touch.

pancreas A flat, solid organ that lies below the liver and the stomach; it is a major source of digestive enzymes and produces the hormone insulin.

pancreatitis Inflammation of the pancreas.

pandemic An outbreak that occurs on a global scale.

paradoxical motion The motion of the portion of the chest wall that is detached in a flail chest; the motion—in during inhalation, out during exhalation—is exactly the opposite of normal chest wall motion during breathing.

paramedic An individual who has extensive training in advanced life support, including endotracheal intubation, emergency pharmacology, cardiac monitoring, and other advanced assessment and treatment skills.

parasympathetic nervous system A subdivision of the autonomic nervous system, involved in control of involuntary, vegetative functions, mediated largely by the vagus nerve through the chemical acetylcholine.

parenteral medications Medications that enter the body by a route other than the digestive tract, skin, or mucous membranes.

parietal pleura Thin membrane that lines the chest cavity.

parietal regions The areas between the temporal and occipital regions of the cranium.

partial pressure The term used to describe the amount of gas in air or dissolved in fluid, such as blood.

partial seizure A seizure affecting a limited portion of the brain.

partial-thickness (second-degree) burns Burns affecting the epidermis and some portion of the dermis but not the subcutaneous tissue; characterized by blisters and skin that is white to red, moist, and mottled.

patella The kneecap; a specialized bone that lies within the tendon of the quadriceps muscle.

patent Open, clear of obstruction.

pathogen A microorganism that is capable of causing disease in a susceptible host.

pathophysiology The study of how normal physiologic processes are affected by disease.

patient autonomy The right of a patient to make informed choices regading his or her health care.

patient care report (PCR) The legal document used to record all patient care activities. This report has direct patient care functions but also administrative and quality control functions. PCRs are also known as prehospital care reports.

patient-assisted medication When the EMT assists the patient with the administration of his or her own medication.

pediatric assessment triangle (PAT) A structured assessment tool that allows you to rapidly form a general impression of the infant or child without touching him or her; consists of assessing appearance, work of breathing, and circulation to the skin.

pediatric resuscitation tape measure A tape used to estimate an infant or child's weight on the basis of length; appropriate drug doses and equipment sizes are listed on the tape.

pediatrics A specialized medical practice devoted to the care of the young.

peer-assisted medication When the EMT adminsters medication to him or herself or to a partner.

pelvic binders Used to splint the bony pelvis to reduce hemorrhage from bone ends, venous disruption, and pain.

pelvic inflammatory disease (PID) An infection of the fallopian tubes and the surrounding tissues of the pelvis.

penetrating trauma Injury caused by objects, such as knives and bullets, that pierce the surface of the body and damage internal tissues and organs.

penetrating wound An injury resulting from a sharp, pointed object.

peptic ulcer disease An abrasion of the stomach or small intestine.

per os (PO) Through the mouth; a medication delivery route; same as oral.

per rectum (PR) Through the rectum; a medication delivery route.

perfusion Circulation of blood within an organ or tissue in adequate amounts to meet the current needs of the cells.

pericardium The fibrous sac that surrounds the heart.

perineum The area of skin between the vagina and the anus.

peripheral nervous system The part of the nervous system that consists of 31 pairs of spinal nerves and 12 pairs of cranial nerves. These peripheral nerves may be sensory nerves, motor nerves, or connecting nerves.

peristalsis The wavelike contraction of smooth muscle by which the ureters or other tubular organs propel their contents.

peritoneal cavity The abdominal cavity.

peritoneum The membrane lining the abdominal cavity (parietal peritoneum) and covering the abdominal organs (visceral peritoneum).

peritonitis Inflammation of the peritoneum.

persistency Term used to describe how long a chemical agent will stay on a surface before it evaporates.

personal protective equipment (PPE) Clothing or specialized equipment that provides protection to the wearer.

personal protective equipment (PPE) levels Measures of the amount and type of protective equipment that an individual needs to avoid injury during contact with a hazardous material.

pertinent negatives Negative findings that warrant no care or intervention.

pertussis (whooping cough) An airborne bacterial infection that affects mostly children younger than 6 years. Patients will be feverish and exhibit a "whoop" sound on inspiration after a coughing attack; highly contagious through droplet infection.

pharmacodynamics The process by which a medication works on the body.

pharmacology The study of the properties and effects of medications.

pharyngeotracheal lumen airway A multilumen airway that consists of two tubes, two masks, and a bite block.

phlebitis Inflammation of a vein; often associated with a clot in the vein.

phosgene A pulmonary agent that is a product of combustion, such as might be produced in a fire at a textile factory or house or from metalwork or burning Freon. Phosgene is a very potent agent that has a delayed onset of symptoms, usually hours.

phosgene oxime (CX) A blistering agent that has a rapid onset of symptoms and produces immediate, intense pain and discomfort on contact.

phrenic nerve Nerve that innervates the diaphragm; necessary for adequate breathing to occur.

piercing spike The hard, sharpened plastic spike on the end of the administration set designed to pierce the sterile membrane of the intravenous bag.

pin-indexing system A system established for portable cylinders to ensure that a regulator is not connected to a cylinder containing the wrong type of gas.

pinna The external, visible part of the ear.

placards Signage required to be placed on all four sides of highway transport vehicles, railroad tank cars, and other forms of hazardous materials transportation; the sign identifies the hazardous contents of the vehicle, using a standardization system with 10¾-inch diamond-shaped indicators.

placenta The tissue attached to the uterine wall that nourishes the fetus through the umbilical cord.

placenta previa A condition in which the placenta develops over and covers the cervix.

planning In incident command, the position that ultimately produces a plan to resolve any incident.

plantar The bottom surface of the foot.

plasma A sticky, yellow fluid that carries the blood cells and nutrients and transports cellular waste material to the organs of excretion.

platelets Tiny, disk-shaped elements that are much smaller than the cells; they are essential in the initial formation of a blood clot, the mechanism that stops bleeding.

pleura The serous membranes covering the lungs and lining the thoracic cavity, completely enclosing a potential space known as the pleural space.

pleural effusion A collection of fluid between the lung and chest wall that may compress the lung.

pleural space The potential space between the parietal pleura and the visceral pleura. It is described as "potential" because under normal conditions, the space does not exist.

pleuritic chest pain Sharp, stabbing pain in the chest that is worsened by a deep breath or other chest wall movement; often caused by inflammation or irritation of the pleura.

pneumonia An inflammation/infection of the lung from a bacterial, viral, or fungal cause.

pneumonic plague A lung infection, also known as plague pneumonia, that is the result of inhalation of plague-causing bacteria.

pneumonitis Inflammation of the lung.

pneumotaxic (pontine) center A portion of the pons that assists in creating shorter, faster respirations.

pneumothorax A partial or complete accumulation of air in the pleural space.

point tenderness Tenderness that is sharply localized at the site of the injury, found by gently palpating along the bone with the tip of one finger.

points of distribution (PODs) Existing facilities that are established in a time of need for the mass distribution of antibiotics, antidotes, vaccinations, and other medications and supplies.

poison A substance whose chemical action could damage structures or impair function when introduced into the body.

polydipsia Excessive thirst that persists for long periods, despite reasonable fluid intake; often the result of excessive urination.

polyphagia Excessive eating; in diabetes, the inability to use glucose properly can cause a sense of hunger.

polypharmacy The simultaneous use of multiple medications as typically seen in elderly people.

polyuria The passage of an unusually large volume of urine in a given period; in diabetes, this can result from the wasting of glucose in the urine.

pons An organ that lies below the midbrain and above the medulla and contains numerous important nerve fibers, including those for sleep, respiration, and the medullary respiratory center.

poor air exchange A term used to describe the degree of distress in a patient with a mild airway obstruction. With poor air exchange, the patient often has a weak, ineffective cough, increased difficulty breathing, or possible cyanosis and may produce a high-pitched noise during inhalation (stridor).

portable stretcher A stretcher with a strong rectangular tubular metal frame and rigid fabric stretched across it.

position of function A hand position in which the wrist is slightly dorsiflexed and all finger joints are moderately flexed.

positive end-expiratory pressure (PEEP) Mechanical maintenance of pressure in the airway at the end of expiration to increase the volume of gas remaining in the lungs.

postconventional reasoning A type of reasoning in which a child bases decisions on his or her conscience.

posterior In anatomy, the back surface of the body; the side away from you in the standard anatomic position.

posterior tibial artery The artery just behind the medial malleolus; supplies blood to the foot.

postictal state A period following a seizure that lasts between 5 and 30 minutes; characterized by labored respirations and some degree of altered mental status.

posttraumatic stress disorder (PTSD) A delayed stress reaction to a prior incident. This delayed reaction is often the result of one or more unresolved issues concerning the incident.

potential energy The product of mass, gravity, and height, which is converted into kinetic energy and results in injury, such as from a fall.

power grip A technique in which the litter or backboard is gripped by inserting each hand under the handle with the palm facing up and the thumb extended, fully supporting the underside of the handle on the curved palm with the fingers and thumb.

power lift A lifting technique in which the EMT's back is held upright, with legs bent, and the patient is lifted when the EMT straightens the legs to raise the upper body and arms.

precedence Basing current action on lessons, rules, or guidelines derived from previous similar experiences.

preconventional reasoning A type of reasoning in which a child acts almost purely to avoid punishment to get what he or she wants.

preeclampsia A condition of late pregnancy that involves headache, visual changes, and swelling of the hands and feet; also called pregnancy-induced hypertension.

pregnancy-induced hypertension A condition of late pregnancy that involves headache, visual changes, and swelling of the hands and feet; also called preeclampsia.

preload The precontraction pressure in the heart as the volume of blood builds up.

presbycusis An age-related condition of the ear that produces progressive bilateral hearing loss that is most noted at higher frequencies.

preschoolers Persons who are 3 to 6 years of age.

prescription medications Medications that are distributed to patients only by pharmacists according to a physician's order.

presentation The position in which an infant is born; the part of the infant that appears first.

pressure point A point where a blood vessel lies near a bone.

primary (direct) injury An injury to the brain and its associated structures that is a direct result of impact to the head.

primary assessment A step within the patient assessment process that identifies and initiates treatment of immediate and potential life threats.

primary blast injury Injuries caused by an explosive pressure wave on the hollow organs of the body.

primary prevention Efforts to prevent an injury or illness from ever occurring.

primary service area (PSA) The designated area in which the EMS service is responsible for the provision of prehospital emergency care and transportation to the hospital.

primary triage A type of patient sorting used to rapidly categorize patients; the focus is on speed in locating all patients and determining an initial priority as their conditions warrant.

primigravida A woman who is experiencing her first pregnancy.

projectile Any object propelled by force, such as a bullet by a weapon.

prolapse of the umbilical cord A situation in which the umbilical cord comes out of the vagina before the infant.

prostate gland A small gland that surrounds the male urethra where it emerges from the urinary bladder; it secretes a fluid that is part of the ejaculatory fluid.

protected health information (PHI) Any information about health status, provision of health care, or payment for health care that can be linked to an individual. This is interpreted rather broadly and includes any part of a patient's medical record or payment history.

proxemics The study of space between people and its effects on communication.

proximal Closer to the trunk.

proximal tibia Anatomic location for intraosseous catheter insertion; the wide portion of the tibia located directly below the knee.

proximate causation When a person who has a duty abuses it, and causes harm to another individual; the EMT, the agency, and/or the medical director may be sued for negligence.

psychiatric disorder An illness with psychological or behavioral symptoms and/or impairment in functioning caused by a social, psychological, genetic, physical, chemical, or biologic disturbance.

psychiatric emergency An emergency in which abnormal behavior threatens a person's own health and safety or the health and safety of another person, for example when a person becomes suicidal, homicidal, or has a psychotic episode.

psychogenic shock Shock caused by a sudden, temporary reduction in blood supply to the brain that causes fainting (syncope).

psychosis A mental disorder characterized by the loss of contact with reality.

pubic symphysis A hard bony prominence that is found in the midline in the lowermost portion of the abdomen.

pubis One of three bones that fuse to form the pelvic ring.

public health Focused on examining the health needs of entire populations with the goal of preventing health problems.

public information officer (PIO) In incident command, the person who keeps the public informed and relates any information to the press.

public safety access point A call center, staffed by trained personnel who are responsible for managing requests for police, firefighting, and ambulance services.

pulmonary artery The major artery leading from the right ventricle of the heart to the lungs; it carries oxygen-poor blood.

pulmonary blast injuries Pulmonary trauma resulting from short-range exposure to the detonation of explosives.

pulmonary circulation The flow of blood from the right ventricle through the pulmonary arteries and all of their branches and capillaries in the lungs and back to the left atrium through the venules and pulmonary veins; also called the lesser circulation.

pulmonary contusion Injury or bruising of lung tissue that results in hemorrhage.

pulmonary edema A buildup of fluid in the lungs, usually as a result of congestive heart failure.

pulmonary embolism A blood clot that breaks off from a large vein and travels to the blood vessels of the lung, causing obstruction of blood flow.

pulmonary veins The four veins that return oxygenated blood from the lungs to the left atrium of the heart.

pulse The pressure wave that occurs as each heartbeat causes a surge in the blood circulating through the arteries.

pulse oximetry An assessment tool that measures oxygen saturation of hemoglobin in the capillary beds.

punitive damages Damages that are sometimes awarded in a civil suit when the conduct of the defendant was intentional or constituted a reckless disregard for the safety of the public.

pupil The circular opening in the middle of the iris that admits light to the back of the eye.

putrefaction Decomposition of body tissues.

quadrants The way to describe the sections of the abdominal cavity. Imagine two lines intersecting at the umbilicus dividing the abdomen into four equal areas.

quality control The responsibility of the medical director to ensure that the appropriate medical care standards are met by EMTs on each call.

rabid Describes an animal that is infected with rabies.

raccoon eyes Bruising under the eyes that may indicate a skull fracture.

radial artery The major artery in the forearm; it is palpable at the wrist on the thumb side.

radiation The transfer of heat to colder objects in the environment by radiant energy, for example heat gain from a fire.

radioactive material Any material that emits radiation.

radiologic dispersal device (RDD) Any container that is designed to disperse radioactive material.

radius The bone on the thumb side of the forearm.

rales A crackling, rattling breath sound that signals fluid in the air spaces of the lungs; also called crackles.

rape Sexual intercourse inflicted forcibly on another person, against that person's will.

rapid extrication technique A technique to move a patient from a sitting position inside a vehicle to supine on a backboard in less than 1 minute when conditions do not allow for standard immobilization.

rapport A trusting relationship that you build with your patient.

reassessment A step within the patient assessment process that is performed at regular intervals to identify and treat changes in a patient's condition. A patient in unstable condition should be reassessed every 5 minutes, whereas a patient in stable condition should be reassessed every 15 minutes.

recovery position A side-lying position used to maintain a clear airway in unconscious patients without injuries who are breathing adequately.

rectum The lowermost end of the colon.

red blood cells Cells that carry oxygen to the body's tissues; also called erythrocytes.

reduce Return a dislocated joint or fractured bone to its normal position; set.

referred pain Pain felt in an area of the body other than the area where the cause of pain is located.

rehabilitation area The area that provides protection and treatment to fire fighters and other personnel working at an emergency. Here, work-

ers are medically monitored and receive any needed care as they enter and leave the scene.

rehabilitation supervisor In incident command, the person who establishes an area that provides protection for responders from the elements and the situation.

renal pelvis A cone-shaped collecting area that connects the ureter and the kidney.

repeater A special base station radio that receives messages and signals on one frequency and then automatically retransmits them on a second frequency.

res ipsa loquitor When the EMT or an EMS service is held liable even when the plaintiff is unable to clearly demonstrate how an injury occurred.

rescue supervisor In incident command, the person appointed to determine the type of equipment and resources needed for a situation involving extrication or special rescue; also called the extrication officer.

residual volume The air that remains in the lungs after maximal expiration.

respiration The process of exchanging oxygen and carbon dioxide.

respiratory syncytial virus (RSV) A virus that causes an infection of the lungs and breathing passages; can lead to other serious illnesses that affect the lungs or heart, such as bronchiolitis and pneumonia. RSV is highly contagious and spread through droplets.

respiratory system All the structures of the body that contribute to the process of breathing, consisting of the upper and lower airways and their component parts.

responsiveness The way in which a patient responds to external stimuli, including verbal stimuli (sound), tactile stimuli (touch), and painful stimuli.

reticular activating system Located in the upper brain stem; responsible for maintenance of consciousness, specifically one's level of arousal.

retina The light-sensitive area of the eye where images are projected; a layer of cells at the back of the eye that changes the light image into electrical impulses, which are carried by the optic nerve to the brain.

retinal detachment Separation of the retina from its attachments at the back of the eye.

retractions Movements in which the skin pulls in around the ribs during inspiration.

retrograde amnesia The inability to remember events leading up to a head injury.

retroperitoneal Behind the abdominal cavity.

retroperitoneal space The space between the abdominal cavity and the posterior abdominal wall, containing the kidneys, certain large vessels, and parts of the gastrointestinal tract.

reverse triage A triage process in which efforts are focused on those who are in respiratory and cardiac arrest, and different from conventional triage where such patients would be classified as deceased. Used in triaging multiple victims of a lightning strike.

Revised Trauma Score (RTS) A scoring system used for patients with head trauma.

rhonchi Coarse, low-pitched breath sounds heard in patients with chronic mucus in the upper airways.

ricin A neurotoxin derived from mash that is left from the castor bean; causes pulmonary edema and respiratory and circulatory failure leading to death.

rigor mortis Stiffening of the body; a definitive sign of death.

rooting reflex An infant reflex that occurs when something touches an infant's cheek, and the infant instinctively turns his or her head toward the touch.

route of exposure The manner by which a toxic substance enters the body.

rule of nines A system that assigns percentages to sections of the body, allowing calculation of the amount of skin surface involved in the burn area.

sacroiliac joint The connection point between the pelvis and the vertebral column.

sacrum One of three bones (sacrum and two pelvic bones) that make up the pelvic ring; consists of five fused sacral vertebrae.

safe zone An area of protection providing safety from the danger zone (hot zone).

safety officer In incident command, the person who gives the "go ahead" to a plan or who may stop an operation when rescuer safety is an issue.

sagittal (lateral) plane An imaginary line where the body is cut into left and right parts.

saline locks (buff caps) Special types of intravenous apparatus, also called heparin caps and heparin locks.

salivary glands The glands that produce saliva to keep the mouth and pharynx moist.

SAMPLE history A brief history of a patient's condition to determine signs and symptoms, allergies, medications, pertinent past history, last oral intake, and events leading to the injury or illness.

sarin (GB) A nerve agent that is one of the G agents; a highly volatile colorless and odorless liquid that turns from liquid to gas within seconds to minutes at room temperature.

scald burn A burn caused by hot liquids.

scalp The thick skin covering the cranium, which usually bears hair.

scanner A radio receiver that searches or "scans" across several frequencies until the message is completed; the process is then repeated.

scapula The shoulder blade.

scene size-up A step within the patient assessment process that involves a quick assessment of the scene and the surroundings to provide information about scene safety and the mechanism of injury or nature of illness before you enter and begin patient care.

school age A person who is 6 to 12 years of age.

sciatic nerve The major nerve to the lower extremities; controls much of muscle function in the leg and sensation in most of the leg and foot.

sclera The tough, fibrous, white portion of the eye that protects the more delicate inner structures.

scoop stretcher A stretcher that is designed to be split into two or four sections that can be fitted around a patient who is lying on the ground or other relatively flat surface; also called an orthopedic stretcher.

scope of practice Most commonly defined by state law; outlines the care you are able to provide for the patient.

SCUBA A system that delivers air to the mouth and lungs at various atmospheric pressures, increasing with the depth of the dive; stands for self-contained underwater breathing apparatus.

sebaceous glands Glands that produce an oily substance called sebum, which discharges along the shafts of the hairs.

secondary assessment A step within the patient assessment process in which a systematic physical examination of the patient is performed. The examination may be a systematic full-body scan or a systematic assessment that focuses on a certain area or region of the body, often determined through the chief complaint.

secondary blast injury A penetrating or nonpenetrating injury caused by ordnance projectiles or secondary missiles.

secondary containment An engineered method to control spilled or released product if the main containment vessel fails.

secondary device An additional explosive used by terrorists, set to explode after the initial bomb.

secondary (indirect) injury The "after effects" of the primary injury; includes abnormal processes such as cerebral edema, increased intracranial pressure, cerebral ischemia and hypoxia, and infection; onset is often delayed following the primary brain injury.

secondary prevention Efforts to limit the effects of an injury or illness that you cannot completely prevent.

secondary triage A type of patient sorting used in the treatment sector that involves retriage of patients.

secure attachment A bond between an infant and his or her parent or caregiver, in which the infant understands that his or her parents or caregivers will be responsive to his or her needs and take care of him or her when he or she needs help.

sedative A substance that decreases activity and excitement.

seizure Generalized, uncoordinated muscular activity associated with loss of consciousness; a convulsion.

self-contained breathing apparatus (SCBA) Respirator with independent air supply used by fire fighters to enter toxic and otherwise dangerous atmospheres.

semen Seminal fluid ejaculated from the penis and containing sperm.

seminal vesicles Storage sacs for sperm and seminal fluid, which empty into the urethra at the prostate.

sensitization Developing a sensitivity to a substance that initially caused no allergic reaction.

sensorineural deafness A permanent lack of hearing caused by a lesion or damage of the inner ear.

sensory nerves The nerves that carry sensations of touch, taste, heat, cold, pain, and other modalities from the body to the central nervous system.

septic shock Shock caused by severe infection, usually a bacterial infection.

severe acute respiratory syndrome (SARS) Potentially life-threatening viral infection that usually starts with flulike symptoms.

severe airway obstruction Occurs when a foreign body completely obstructs the patient's airway. Patients cannot breathe, talk, or cough.

sexual assault An attack against a person that is sexual in nature, the most common of which is rape.

shaken baby syndrome A syndrome seen in abused infants and children; the patient has been subjected to violent, whiplash-type shaking injuries inflicted by the abusing individual that may cause coma, seizures, and increased intracranial pressure due to tearing of the cerebral veins with consequent bleeding into the brain.

shallow respirations Respirations that are charcterized by little movement of the chest wall (reduced tidal volume) or poor chest excursion.

shock A condition in which the circulatory system fails to provide sufficient circulation to enable every body part to perform its function; also called hypoperfusion.

shoulder girdle The proximal portion of the upper extremity, made up of the clavicle, the scapula, and the humerus.

shunts Tubes that drain fluid from the brain to another part of the body outside of the brain, such as the abdomen; lowers pressure in the brain.

sickle cell disease A hereditary disease that causes normal, round red blood cells to become oblong, or sickle shaped.

side effects Any effects of a medication other than the desired ones.

sign Objective findings that can be seen, heard, felt, smelled, or measured.

simple access Access that is easily achieved without the use of tools or force.

simple pneumothorax Any pneumothorax that is free from significant physiologic changes and does not cause drastic changes in the vital signs of the patient.

simplex Single-frequency radio; transmissions can occur in either direction but not simultaneously in both; when one party transmits, the other can only receive, and the party that is transmitting is unable to receive.

single command system A command system in which one person is in charge; generally used with small incidents that involve only one responding agency or one jurisdiction.

sinus bradycardia A rhythm that has consistent P waves, consistent P-R intervals, and a regular heart rate that is less than 60 beats/min.

sinus rhythm A rhythm in which the sinoatrial node acts as the pacemaker.

sinus tachycardia A rhythm that has consistent P waves, consistent P-R intervals, and a regular heart rate that is more than 100 beats/min.

size-up The ongoing process of information gathering and scene evaluation to determine appropriate strategies and tactics to manage an emergency.

skeletal muscle Muscle that is attached to bones and usually crosses at least one joint; striated, or voluntary, muscle.

skeleton The framework that gives the body its recognizable form; also designed to allow motion of the body and protection of vital organs.

slander False and damaging information about a person that is communicated by the spoken word.

sling A bandage or material that helps to support the weight of an injured upper extremity.

small intestine The portion of the digestive tube between the stomach and the cecum, consisting of the duodenum, jejunum, and ileum.

smallpox A highly contagious disease; it is most contagious when blisters begin to form.

small-volume nebulizer A respiratory device that holds liquid medicine that is turned into a fine mist. The patient inhales the medication into the airways and lungs as a treatment for conditions like asthma.

smooth muscle Involuntary muscle; it constitutes the bulk of the gastrointestinal tract and is present in nearly every organ to regulate automatic activity.

sniffing position An upright position in which the patient's head and chin are thrust slightly forward to keep the airway open.

solid organs Solid masses of tissue where much of the chemical work of the body takes place (eg, the liver, spleen, pancreas, and kidneys).

solution A liquid mixture that cannot be separated by filtering or allowing the mixture to stand.

soman (GD) A nerve agent that is one of the G agents; twice as persistent as sarin and five times as lethal; it has a fruity odor, as a result of the type of alcohol used in the agent, and is a contact and an inhalation hazard that can enter the body through skin absorption and through the respiratory tract.

somatic nervous system The part of the nervous system that regulates activities over which there is voluntary control.

span of control In incident command, the subordinate positions under the commander's direction to which the workload is distributed; the supervisor/worker ratio.

Special Atomic Demolition Munitions (SADM) Small suitcase-sized nuclear weapons that were designed to destroy individual targets, such as important buildings, bridges, tunnels, and large ships.

special weapons and tactics team (SWAT) A specialized law enforcement tactical unit.

sphincters Muscles arranged in circles that are able to decrease the diameter of tubes. Examples are found within the rectum, bladder, and blood vessels.

sphygmomanometer A device used to measure blood pressure.

spina bifida A development defect in which a portion of the spinal cord or meninges may protrude outside of the vertebrae and possibly even outside of the body, usually at the lower third of the spine in the lumbar area.

spinal cord An extension of the brain, composed of virtually all the nerves carrying messages between the brain and the rest of the body. It lies inside of and is protected by the spinal canal.

splenic sequestration crisis An acute, painful enlargement of the spleen caused by sickle cell disease.

splint A flexible or rigid appliance used to protect and maintain the position of an injured extremity.

spontaneous pneumothorax A pneumothorax that occurs when a weak area on the lung ruptures in the absence of major injury, allowing air to leak into the pleural space.

spontaneous respirations Breathing that occurs with no assistance.

spotter A person who assists a driver in backing up an ambulance to compensate for blind spots at the back of the vehicle.

sprain A joint injury involving damage to supporting ligaments, and sometimes partial or temporary dislocation of bone ends.

staging supervisor In incident command, the person who locates an area to stage equipment and personnel and tracks unit arrival and deployment from the staging area.

stair chair A lightweight folding device that is used to carry a conscious, seated patient up or down stairs.

standard of care Written, accepted levels of emergency care expected by reason of training and profession; written by legal or professional organizations so that patients are not exposed to unreasonable risk or harm.

standard precautions Protective measures that have traditionally been developed by the Centers for Disease Control and Prevention for use in dealing with objects, blood, body fluids, and other potential exposure risks of communicable disease.

standing orders Written documents, signed by the EMS system's medical director, that outline specific directions, permissions, and sometimes prohibitions regarding patient care; also called protocols.

Star of Life® The six-pointed star that identifies vehicles that meet federal specifications as licensed or certified ambulances.

START triage A patient sorting process that stands for Simple Triage And Rapid Treatment and uses a limited assessment of the patient's ability to walk, respiratory status, hemodynamic status, and neurologic status.

state-sponsored terrorism Terrorism that is funded and/or supported by nations that hold close ties with terrorist groups.

status epilepticus A condition in which seizures recur every few minutes or last more than 30 minutes.

statute of limitations The time within which a case must be commenced.

steam burn A burn caused by exposure to hot steam.

STEMI (ST-segment elevation myocardial infarction) Elevation of the ST segment of the 12-lead ECG that is likely evidence that the patient is having a heart attack.

sterilization A process, such as heating, that removes microbial contamination.

sternocleidomastoid muscles The muscles on either side of the neck that allow movement of the head.

sternum The breastbone.

stimulant An agent that produces an excited state.

stoma An opening through the skin and into an organ or other structure; a stoma in the neck connects the trachea directly to the skin.

strain Stretching or tearing of a muscle; also called a muscle pull.

strangulation Complete obstruction of blood circulation in a given organ as a result of compression or entrapment; an emergency situation causing death of tissue.

stratum corneal layer The outermost or dead layer of the skin.

stridor A high-pitched noise heard primarily on inspiration.

stroke An interruption of blood flow to the brain that results in the loss of brain function; also called a cerebrovascular accident (CVA).

stroke volume (SV) The volume of blood pumped forward with each ventricular contraction.

structure fire A fire in a house, apartment building, office, school, plant, warehouse, or other building.

stylet A plastic-coated wire that gives added rigidity and shape to the endotracheal tube.

subarachnoid hemorrhage Bleeding into the subarachnoid space, where the cerebrospinal fluid circulates.

subcutaneous emphysema A characteristic crackling sensation felt on palpation of the skin, caused by the presence of air in soft tissues.

subcutaneous (SC) injection Injection into the tissue between the skin and muscle; a medication delivery route.

subcutaneous tissue Tissue, largely fat, that lies directly under the dermis and serves as an insulator of the body.

subdural hematoma An accumulation of blood beneath the dura mater but outside the brain.

sublingual (SL) Under the tongue; a medication delivery route.

subluxation A partial or incomplete dislocation.

substance abuse The misuse of any substance to produce some desired effect.

sucking chest wound An open or penetrating chest wall wound through which air passes during inspiration and expiration, creating a sucking sound. See also open pneumothorax.

sucking reflex An infant reflex in which the infant starts sucking when his or her lips are stroked.

suction catheter A hollow, cylindrical device used to remove fluid from the patient's airway.

sudden infant death syndrome (SIDS) Death of an infant or young child that remains unexplained after a complete autopsy.

sulfur mustard (H) A vesicant; it is a brownish, yellowish oily substance that is generally considered very persistent; has the distinct smell of garlic or mustard and, when released, is quickly absorbed into the skin and/or mucous membranes and begins an irreversible process of damaging the cells.

superficial Closer to or on the skin.

superficial (first-degree) burns Burns affecting only the epidermis; characterized by skin that is red but not blistered or actually burned through.

superior The part of the body or any body part nearer to the head.

superior vena cava One of the two largest veins in the body; carries blood from the upper extremities, head, neck, and chest into the heart.

supine hypotensive syndrome Low blood pressure resulting from compression of the inferior vena cava by the weight of the pregnant uterus when the mother is supine.

surfactant A liquid protein substance that coats the alveoli in the lungs, decreases alveolar surface tension, and keeps the alveoli expanded; a low level in a premature infant contributes to respiratory distress syndrome.

suspension A mixture of ground particles that are distributed evenly throughout a liquid but do not dissolve.

swathe A bandage that passes around the chest to secure an injured arm to the chest.

sweat glands The glands that secrete sweat, located in the dermal layer of the skin.

sympathetic nervous system The part of the autonomic nervous system that controls active functions such as responding to fear (also known as the "fight-or-flight" system).

symphysis A type of joint that has grown together forming a very stable connection.

symptom Subjective findings that the patient feels but that can be identified only by the patient.

syncope A fainting spell or transient loss of consciousness, often caused by an interruption of blood flow to the brain.

syndromic surveillance The monitoring, usually by local or state health departments, of patients presenting to emergency departments and alternative care facilities, the recording of EMS call volume, and the use of over-the-counter medications.

synovial fluid The small amount of liquid within a joint used as lubrication.

synovial membrane The lining of a joint that secretes synovial fluid into the joint space.

systemic circulation The portion of the circulatory system outside of the heart and lungs.

systemic complication A moderate to severe complication affecting the systems of the body; after administration of medications, the reaction might be systemic.

systemic vascular resistance (SVR) The resistance that blood must overcome to be able to move within the blood vessels. SVR is related to the amount of dilation or constriction in the blood vessel.

systole The contraction, or period of contraction, of the heart, especially that of the ventricles.

systolic pressure The increased pressure in an artery with each contraction of the ventricles (systole).

tabun (GA) A nerve agent that is one of the G agents; is 36 times more persistent than sarin and approximately half as lethal; has a fruity smell and is unique because the components used to manufacture the agent are easy to acquire and the agent is easy to manufacture.

tachycardia A rapid heart rate, more than 100 beats/min.

tachypnea Increased respiratory rate.

tactical situation A hostage, robbery, or other situation in which armed conflict is threatened or shots have been fired and the threat of violence remains.

technical rescue group A team of individuals from one or more departments in a region who are trained and on call for certain types of technical rescue.

technical rescue situation A rescue that requires special technical skills and equipment in one of many specialized rescue areas, such as technical rope rescue, cave rescue, and dive rescue.

telemetry A process in which electronic signals are converted into coded, audible signals; these signals can then be transmitted by radio or telephone to a receiver with a decoder at the hospital.

temporal regions The lateral portions on each side of the cranium.

temporomandibular joint (TMJ) The joint formed where the mandible and cranium meet, just in front of the ear.

tendons The fibrous connective tissue that attaches muscle to bone.

tension pneumothorax A life-threatening collection of air within the pleural space; the volume and pressure have both collasped the involved lung and caused a shift of the mediastinal structures to the opposite side.

terminal drop hypothesis The theory that a person's mental function declines in the last 5 years of life.

termination of command The end of the incident command structure when an incident draws to a close.

tertiary blast injury An injury from whole body displacement and subsequent traumatic impact with environmental objects.

testicle A male genital gland that contains specialized cells that produce hormones and sperm.

therapeutic communication Verbal and nonverbal communication techniques that encourage patients to express their feelings and to achieve a positive relationship.

thermal burns Burns caused by heat.

thoracic cage The chest or rib cage.

thoracic cavity The chest cavity that contains the heart, lungs, esophagus, and great vessels.

thoracic spine The 12 vertebrae that lie between the cervical vertebrae and the lumbar vertebrae. One pair of ribs is attached to each of the thoracic vertebrae.

thorax The chest cavity that contains the heart, lungs, esophagus, and great vessels.

thromboembolism A blood clot that has formed within a blood vessel and is floating within the bloodstream.

thrombophilia A tendency toward the development of blood clots as a result of an abnormality of the system of coagulation.

thrombosis A blood clot, either in the arterial or venous system.

thyroid cartilage A firm prominence of cartilage that forms the upper part of the larynx; the Adam's apple.

tibia The shin bone, the larger of the two bones of the lower leg.

tidal volume The amount of air (in milliliters) that is moved in or out of the lungs during one breath.

toddlers Persons who are 1 to 3 years of age.

tolerance The need for increasing amounts of a drug to obtain the same effect.

tonic-clonic seizure A type of seizure that features rhythmic back-and-forth motion of an extremity and body stiffness.

tonsil tips Large, semirigid suction tips recommended for suctioning the pharynx; also called Yankauer tips.

topical medications Lotions, creams, and ointments that are applied to the surface of the skin and affect only that area; a medication delivery route.

topographic anatomy The superficial landmarks of the body that serve as guides to the structures that lie beneath them.

torso The trunk without the head and limbs.

tort A wrongful act that gives rise to a civil suit.

tourniquet The bleeding control method used when a wound continues to bleed despite the use of direct pressure and elevation; useful if a patient is bleeding severely from a partial or complete amputation.

toxicity levels Measures of the risk that a hazardous material poses to the health of an individual who comes into contact with it.

toxicology The study of toxic or poisonous substances.

toxin A poison or harmful substance produced by bacteria, animals, or plants.

trachea The windpipe; the main trunk for air passing to and from the lungs.

tracheostomy Surgical opening into the trachea.

tracheostomy tube Plastic tube placed within the tracheostomy site (stoma).

traction Longitudinal force applied to a structure.

trade name The brand name that a manufacturer gives a medication; the name is capitalized.

tragus The small, rounded, fleshy bulge that lies immediately anterior to the ear canal.

trajectory The path a projectile takes once it is propelled.

transcutaneous (transdermal) Through the skin; a medication delivery route.

transient ischemic attack (TIA) A disorder of the brain in which brain cells temporarily stop working because of insufficient oxygen, causing strokelike symptoms that resolve completely within 24 hours of onset.

transmission The way in which an infectious disease is spread: contact, airborne, by vehicles, or by vectors.

transportation area The area in a mass-casualty incident where ambulances and crews are organized to transport patients from the treatment area to receiving hospitals.

transportation supervisor The individual in charge of the transportation sector in a mass-casualty incident who assigns patients from the treatment area to awaiting ambulances in the transportation area.

transverse (axial) plane An imaginary line where the body is cut into top and bottom parts.

trauma emergencies Emergencies that are the result of physical forces applied to a patient's body.

trauma score A score that relates to the likelihood of patient survival with the exception of a severe head injury. It calculates a number from 1 to 16, with 16 being the best possible score. It takes into account the Glasgow Coma Scale (GCS) score, respiratory rate, respiratory expansion, systolic blood pressure, and capillary refill.

traumatic asphyxia A pattern of injuries seen after a severe force is applied to the chest, forcing blood from the great vessels back into the head and neck.

traumatic brain injury (TBI) A traumatic insult to the brain capable of producing physical, intellectual, emotional, social, and vocational changes.

treatment area The location in a mass-casualty incident where patients are brought after being triaged and assigned a priority, where they are reassessed, treated, and monitored until transport to the hospital.

treatment supervisor The individual, usually a physician, who is in charge of and directs EMS personnel at the treatment area in a mass-casualty incident.

triage The process of sorting patients based on the severity of injury and medical need to establish treatment and transportation priorities.

triage supervisor The individual in charge of the incident command triage sector who directs the sorting of patients into triage categories in a mass-casualty incident.

triceps The muscle in the back of the upper arm.

tripod position An upright position in which the patient leans forward onto two arms stretched forward and thrusts the head and chin forward.

trunking Telecommunication systems that allow a computer to maximize utilization of a group of frequencies.

trust and mistrust A phrase that refers to a stage of development from birth to approximately 18 months of age, during which infants gain trust of their parents or caregivers if their world is planned, organized, and routine.

tuberculosis (TB) A chronic bacterial disease, caused by *Mycobacterium tuberculosis*, that usually affects the lungs but can also affect other organs such as the brain and kidneys.

tunica media The middle and thickest layer of tissue of a blood vessel wall, composed of elastic tissue and smooth muscle cells that allow the vessel to expand or contract in response to changes in blood pressure and tissue demand.

turbinates Layers of bone within the nasal cavity.

turgor The ability of the skin to resist deformation; tested by gently pinching skin on the forehead or back of the hand.

two- to three-word dyspnea A severe breathing problem in which a patient can speak only two to three words at a time without pausing to take a breath.

tympanic membrane The eardrum; a thin, semitransparent membrane in the middle ear that transmits sound vibrations to the internal ear by means of auditory ossicles.

type 1 diabetes The type of diabetic disease that typically develops in childhood and requires synthetic insulin for proper treatment and control.

type 2 diabetes The type of diabetic disease that typically develops in later life and often can be controlled through diet and oral medications.

UHF (ultra-high frequency) Radio frequencies between 300 and 3,000 MHz.

ulna The inner bone of the forearm, on the side opposite the thumb.

umbilical cord The conduit connecting mother to infant via the placenta; contains two arteries and one vein.

unified command system A command system used in larger incidents in which there is a multiagency response or multiple jurisdictions are involved.

unilateral pedal edema Pedal edema is a swelling of the foot and ankle caused by fluid overload; unilateral would present in only one extremity.

unintended effect Actions that are undesirable but pose little risk to the patient.

untoward effects Actions that can be harmful to the patient.

uremia Severe kidney failure resulting in the buildup of waste products within the blood. Eventually brain functions will be impaired.

ureter A small, hollow tube that carries urine from the kidneys to the bladder.

urethra The canal that conveys urine from the bladder to outside the body.

urinary bladder A sac behind the pubic symphysis made of smooth muscle that collects and stores urine.

urinary system The organs that control the discharge of certain waste materials filtered from the blood and excreted as urine.

urinary tract infection (UTI) Infections, usually of the lower urinary tract (urethra and bladder), which occur when normal flora bacteria enter the urethra and grow.

urticaria Small spots of generalized itching and/or burning that appear as multiple raised areas on the skin; hives.

uterus The muscular organ where the fetus grows, also called the womb; responsible for contractions during labor.

V agent (VX) One of the G agents; it is a clear, oily agent that has no odor and looks like baby oil; more than 100 times more lethal than sarin and is extremely persistent.

vagina The outermost cavity of a woman's reproductive system; the lower part of the birth canal.

vapor hazard An agent that enters the body through the respiratory tract.

vasa deferentia The spermatic duct of the testicles; also called vas deferens.

vasoconstriction Narrowing of a blood vessel, such as with hypoperfusion or cold extremities.

vaso-occlusive crisis Ischemia and pain caused by sickle-shaped red blood cells that obstruct blood flow to a portion of the body.

vector-borne transmission The use of an animal to spread an organism from one person or place to another.

veins The blood vessels that carry blood from the tissues to the heart.

ventilation Exchange of air between the lungs and the environment, spontaneously by the patient or with assistance from another person, such as an EMT.

ventral The anterior surface of the body.

ventral respiratory group (VRG) A portion of the medulla oblongata that is responsible for modulating breathing during speech.

ventricle One of two (right and left) lower chambers of the heart. The left ventricle receives blood from the left atrium (upper chamber) and delivers blood to the aorta. The right ventricle receives blood from the right atrium and pumps it into the pulmonary artery.

ventricular fibrillation Disorganized, ineffective twitching of the ventricles, resulting in no blood flow and a state of cardiac arrest.

ventricular tachycardia A rapid heart rhythm in which the electrical impulse begins in the ventricle (instead of the atrium), which may result in inadequate blood flow and eventually deteriorate into cardiac arrest.

venules Very small, thin-walled vessels.

vertebrae The 33 bones that make up the spinal column.

vertex presentation A delivery in which the head comes out first.

vesicants Blister agents; the primary route of entry for vesicants is through the skin.

vesicular breath sounds Normal breath sounds made by air moving in and out of the alveoli.

VHF (very high frequency) Radio frequencies between 30 and 300 MHz; the VHF spectrum is further divided into "high" and "low" bands.

viral hemorrhagic fevers (VHF) A group of diseases caused by viruses that include the Ebola, Rift Valley, and yellow fevers, among others. This group of viruses causes the blood in the body to seep out from the tissues and blood vessels.

virulence The strength or ability of a pathogen to produce disease.

viruses Germs that require a living host to multiply and survive.

visceral pleura Thin membrane that covers the lungs.

vital capacity The amount of air that can be forcibly expelled from the lungs after breathing in as deeply as possible.

vital signs The key signs that are used to evaluate the patient's overall condition, including respirations, pulse, blood pressure, level of consciousness, and skin characteristics.

vocal cords Thin white bands of tough muscular tissue that are lateral borders of the glottis and serve as the primary center for speech production.

volatility A term used to describe how long a chemical agent will stay on a surface before it evaporates.

voluntary activities Actions that we consciously perform, in which sensory input or conscious thought determines a specific muscular activity.

voluntary muscle Muscle that is under direct voluntary control of the brain and can be contracted or relaxed at will; skeletal, or striated, muscle.

vomitus Vomited material.

V/Q ratio A measurement that examines how much gas is being moved effectively and how much blood is gaining access to the alveoli.

warm zone The area located between the hot zone and the cold zone at a hazardous materials incident. The decontamination corridor is located in the warm zone.

weapon of mass casualty (WMC) Any agent designed to bring about mass death, casualties, and/or massive damage to property and infrastructure (bridges, tunnels, airports, and seaports); also known as a weapon of mass destruction (WMD).

weapon of mass destruction (WMD) Any agent designed to bring about mass death, casualties, and/or massive damage to property and infrastructure (bridges, tunnels, airports, and seaports); also known as a weapon of mass casualty (WMC).

weaponization The creation of a weapon from a biologic agent generally found in nature and that causes disease; the agent is cultivated, synthesized, and/or mutated to maximize the target population's exposure to the germ.

wheal A raised, swollen, well-defined area on the skin resulting from an insect bite or allergic reaction.

wheeled ambulance stretcher A specially designed stretcher that can be rolled along the ground. A collapsible undercarriage allows it to be loaded into the ambulance. Also called the stretcher or an ambulance stretcher.

wheezing The production of whistling sounds during expiration such as occurs in asthma and bronchiolitis.

white blood cells Blood cells that have a role in the body's immune defense mechanisms against infection; also called leukocytes.

work The product of force times distance.

work of breathing An indicator of oxygenation and ventilation. Work of breathing reflects the child's attempt to compensate for hypoxia.

xiphoid process The narrow, cartilaginous lower tip of the sternum.

zone of injury The area of potentially damaged soft tissue, adjacent nerves, and blood vessels surrounding an injury to a bone or a joint.

zygomas The quadrangular bones of the cheek, articulating with the frontal bone, the maxillae, the zygomatic processes of the temporal bone, and the great wings of the sphenoid bone.

Stock, Inc; **24-17C** © E.M. Singletary, MD. Used with permission; **24-24** © Chuck Stewart, MD

Chapter 25
Opener ©.E.M. Singletary, MD. Used with permission; **25-7** Courtesy of Rhonda Beck; **25-10** © E.M. Singletary, MD. Used with permission; **25-14** Courtesy of John T. Halgren, MD, University of Nebraska Medical Center

Chapter 26
26-16 © Joe Gough/ShutterStock, Inc.

Chapter 27
Opener © PHT/Photo Researchers, Inc; **27-5** Courtesy of ED, Royal North Shore Hospital/NSW Institute of Trauma & Injury; **27-14** © Chuck Stewart, MD.

Chapter 28
Opener © Chuck Stewart, MD; **28-6** © M. English, MD/Custom Medical Stock Photo; **28-7** © Dr. P. Marazzi/Photo Researchers, Inc; **28-8, 28-9** © Wellcome Photo Library/Custom Medical Stock Photo

Chapter 29
Opener © Chuck Stewart, MD; **29-17** © Chuck Stewart, MD; **29-21** © Chuck Stewart, MD; **29-22** © Dr. P. Marazzi/Photo Researchers, Inc; **29-23** © Sean Gladwell/Dreamstime.com; **29-27** Courtesy of Reel Research and Development, Inc. (www.reel.com); **29-28** © Sam Medical Products®; **29-29** © Shea, MD/Custom Medical Stock Photo

Chapter 30
Opener Courtesy of BM1 Kevin Erwin/US Coast Guard; **30-3C** © Chuck Stewart, MD; **30-4** Courtesy of Dr. Jack Poland/CDC; **30-5** Courtesy of Neil Malcom Winkelmann; **30-10** © Jones and Bartlett Publishers. Courtesy of Ellis & Associates; **30-11** Courtesy of Perry Baromedical Corporation; **30-12** © Crystal Kirk/ShutterStock, Inc; **30-13** Courtesy of Kenneth Cramer, Monmouth College; **30-14** Courtesy of Department of Entomology, University of Nebraska; **30-15A** Courtesy of Ray Rauch/US Fish and Wildlife Service; **30-15B** Courtesy of Luther C. Goldman/US Fish and Wildlife Service; **30-15C** © Amee Cross/ShutterStock, Inc; **30-15D** © SuperStock/Alamy Images; **30-18** © Visual&Written SL/Alamy Images; **30-19** © Joao Estevao A. Freitas (jefras)/ShutterStock, Inc; **30-20** © E.M. Singletary, MD. Used with permission; **30-21A** © Creatas/Alamy Images; **30-21B** Courtesy of NOAA; **30-21C** © Photos.com

Chapter 31
31-12 Courtesy of David J. Burchfield, MD

Chapter 32
32-11 © Photos.com; **32-12, 32-13, 32-14** Courtesy Health Resources and Services Administration, Maternal and Child Health Bureau, Emergency Medical Services for Children Program; **32-33** © Mediscan/Visuals Unlimited; **32-35** Courtesy of Ronald Dieckmann, MD; **32-38** © Chuck Stewart, MD; **32-39** © Lou Romig MD, 2002; **32-40** Courtesy of Ronald Dieckmann, MD

Chapter 33
33-1 Courtesy of the National Cancer Institute; **33-3** © Photodisc; **33-10** © Jeff Greenberg/PhotoEdit, Inc.

Chapter 34
Opener © Richard Levine/Alamy Images; **34-1** © PhotoCreate/ShutterStock, Inc; **34-2** Courtesy of the Guide Dog Foundation for the Blind. Photographed by Christopher Appoldt; **34-5** © Sally and Richard Greenhill/Alamy Images; **34-6** © Biophoto Associates/Photo Researchers, Inc; **34-7** Portex® Blue Line® Ultra Tracheostomy courtesy of Smiths Medical; **34-8** © ResMed 2010. Used with permission; **34-11** © DELOCHE/age fotostock

Chapter 35
35-20 © Dr. P. Marazzi/Photo Researchers, Inc; **35-22, 35-23** Courtesy of Stryker Medical

Chapter 36
36-1 © National Library of Medicine; **36-3B** Courtesy of Captain David Jackson, Saginaw Township Fire Department; **36-3C** © Kevin Norris/ShutterStock, Inc; **36-10A** Courtesy of Ferno Washington, Inc; **36-13** LIFEPAK® 1000 Defibrillator (AED) courtesy of Physio-Control. Used with Permission of Physio-Control, Inc, and according to the Material

Release Form provided by Physio-Control; **36-19** © Mark Terrill/AP Photos; **36-20** © John Sartin/ShutterStock, Inc; **36-22** © Andrew Poertner, *Roswell Daily Record*/AP Photos; **36-26** © Gary Lloyd, *The Decatur Daily*/AP Photos; **36-28** Courtesy of Bryan Dahlberg/FEMA; **36-29A** © Ralph Duenas/www.jetwashimages.com; **36-29B** Courtesy of Ed Edahl/FEMA; **36-30** © Mark C. Ide

Chapter 37
Opener © Glen E. Ellman; **37-2** © Tony Freeman/PhotoEdit, Inc; **37-3** © Mark C. Ide; **37-5** © Keith D. Cullom; **37-6** © Mark C. Ide; **37-7, 37-9, 37-10** © Keith D. Cullom; **37-12** © Kathy Easthagen, *The Minnesota Daily*/AP Photos

Chapter 38
38-2 Courtesy of Captain David Jackson, Saginaw Township Fire Department; **38-5** © Edward Keating, POOL/AP Photos; **38-6** Courtesy of Michael Rieger/FEMA; **38-8** © Suzanne Kreiter, *The Boston Globe*/Landov; **38-9** Courtesy of Journalist 1st Class Mark D. Faram/US Navy; **38-11A** Courtesy of Rob L. Jackson/US Marines; **38-11B** Courtesy of George Roarty/Virginia Department of Emergency Management; **38-12** Courtesy of EMD Chemicals, Inc; **38-13** © Ulrich Mueller/ShutterStock, Inc; **38-14** Courtesy of Tank Service, Inc; **38-15** Courtesy of UBH International Ltd; **38-16** Courtesy of USDA; **38-17** Courtesy of EMD Chemicals, Inc; **38-18** Courtesy of the US Department of Transportation; **38-19** © Mark Winfrey/ShutterStock, Inc; **38-21** Courtesy of the US Department of Transportation; **38-22** Courtesy of Tanner Industries, Inc, Southampton, PA; **38-23** Courtesy of RSI Logistics; **38-25** © *South Florida Sun-Sentinel*, MCT/Landov; **38-26** Courtesy of Airman 1st Class Scherrie Gates/US Air Force; **38-27C** Courtesy of The DuPont Company

Chapter 39
Opener Courtesy of Photographer's Mate 2nd Class Bob Houlihan, US Navy; **39-1** © Reuters, STR/Landov; **39-2** © Rick Bowmer/AP Photos; **39-3** © Todd Hollis/AP Photos; **39-5** © Gary Stelzer, *Middletown Journal*/AP Photos; **39-6** © Dennis MacDonald/Alamy Images; **39-7** © pbpgalleries/Alamy Images; **39-8** Courtesy of Dr. Saeed Keshavarz/RCCI, Research Center of Chemical Injuries/IRAN; **39-10** © Chiaki Tsukumo/AP Photos; **39-11** Courtesy of CDC; **39-12** Courtesy of Professor Robert Swanepoel/National Institute for Communicable Disease, South Africa; **39-13** Courtesy of James H. Steele/CDC; **39-14** Courtesy of CDC; **39-15** Courtesy of Brian Prechtel/USDA; **39-16** Courtesy of the Strategic National Stockpile/CDC

Chapter 40
40-4B Courtesy of James P. Thomas, MD (voicedoctor.net); **40-15** The LIFEPAK® 15 Defibrillator monitor courtesy of Physio-Control. Used with permission of Physio-Control, Inc, and according to the Material Release Form provided by Physio-Control; **40-18** Courtesy of King Systems; **40-21** Courtesy of Respironics, Inc, Murraysville, PA. All rights reserved; **40-32** Courtesy of Rhonda Beck; **40-37** Courtesy of Henry Geiter, Jr, RN, CCRN [www.Nurse411.com]. Copyright 2009; **40-40, 40-41, 40-42, 40-43, 40-44** From *Arrhythmia Recognition: The Art of Interpretation*, courtesy of Tomas B. Garcia, MD.

Unless otherwise indicated, all photographs and illustrations are under copyright of Jones and Bartlett Publishers, LLC, courtesy of Maryland Institute for Emergency Medical Services Systems, or have been provided by the American Academy of Orthopaedic Surgeons.